Comprehensive Organometallic Chemistry II

A Review of the Literature 1982–1994

Comprehensive Organometallic Chemistry II
A Review of the Literature 1982–1994

Editors-in-Chief

Edward W. Abel
University of Exeter, UK

F. Gordon A. Stone
Baylor University, Waco, TX, USA

Geoffrey Wilkinson
Imperial College of Science, Technology and Medicine, London, UK

Volume 11
MAIN-GROUP METAL ORGANOMETALLICS
IN ORGANIC SYNTHESIS

Volume Editor

Alexander McKillop
University of East Anglia, Norwich, UK

PERGAMON

UK	Elsevier Science Ltd., The Boulevard, Langford Lane, Kidlington, Oxford OX5 1GB, UK
USA	Elsevier Science Inc., 660 White Plains Road, Tarrytown, New York 10591-5153, USA
JAPAN	Elsevier Science Japan, Tsunashima Building Annex, 3-20-12 Yushima, Bunkyo-ku, Tokyo 113, Japan

Copyright © 1995 Elsevier Science Ltd.

All rights reserved. No part of this publication may be reproduced, stored in any retrieval system or transmitted in any form or by any means: electronic, electrostatic, magnetic tape, mechanical, photocopying, recording or otherwise, without permission in writing from the publishers.

First edition 1995

Library of Congress Cataloging in Publication Data
Comprehensive organometallic chemistry II : a review of the literature 1982–1994 / editors-in-chief, Edward W. Abel, F. Gordon A. Stone, Geoffrey Wilkinson
    p. cm.
    Includes indexes.
    1. Organometallic chemistry.   I. Abel, Edward W.   II. Stone, F. Gordon A.   III. Wilkinson, Geoffrey.
    QD411.C652   1995
    547'.05—dc20   95–7030

British Library Cataloguing in Publication Data
A catalogue record for this book is available from the British Library.

ISBN 0-08-040608-4 (set : alk. paper)
ISBN 0-08-042318-3 (Volume 11)

Important note
For safety reasons, readers should always consult the list of abbreviations on p. xi before making use of the experimental details provided.

∞™ The paper used in this publication meets the minimum requirements of the American National Standard for Information Sciences—Permanence of Paper for Printed Library Materials, ANSI Z39.48–1984.

Chemical structures drawn by Synopsys Scientific Systems Ltd., Leeds, UK.

Printed and bound in Great Britain by BPC Wheatons Ltd., Exeter, UK.

Contents

Preface		vii
Preface to 'Comprehensive Organometallic Chemistry'		vii
Contributors to Volume 11		ix
Abbreviations		xi
Contents of All Volumes		xv
1	Lithium V. SNIECKUS, M. GRAY and M. TINKL, *University of Waterloo, ON, Canada*	1
2	Sodium and Potassium A. MORDINI, *Università di Firenze, Italy*	93
3	Magnesium J. M. BROWN and S. K. ARMSTRONG, *University of Oxford, UK*	129
4	Zinc and Cadmium P. KNOCHEL, *Philipps-Universität Marburg, Germany*	159
5	Boron M. VAULTIER and B. CARBONI, *Université de Rennes I, France*	191
6	Aluminum J. J. EISCH, *State University of New York at Binghamton, NY, USA*	277
7	Silicon E. W. COLVIN, *University of Glasgow, UK*	313
8	Tin T. SATO, *Waseda University, Tokyo, Japan*	355
9	Mercury R. C. LAROCK, *Iowa State University, Ames, IA, USA*	389
10	Thallium I. E. MARKO and C. W. LEUNG, *Université Catholique de Louvain, Louvain-la-Neuve, Belgium*	437
11	Lead J. T. PINHEY, *University of Sydney, NSW, Australia*	461
12	Antimony and Bismuth Y.-Z. HUANG and Z.-L. ZHOU, *Shanghai Institute of Organic Chemistry, People's Republic of China*	487
13	Selenium A. KRIEF, *Facultés Universitaires Notre-Dame de La Paix, Namur, Belgium*	515
14	Tellurium N. PETRAGNANI, *Universidade de São Paulo, Brazil*	571
Author Index		603
Subject Index		653

Preface

'Comprehensive Organometallic Chemistry', published in 1982, was well received and remains very highly cited in the primary journal literature. Since its publication, studies on the chemistry of molecules with carbon–metal bonds have continued to expand rapidly. This is due to many factors, ranging from the sheer intellectual challenge and excitement provided by the continuing production of novel results, which demand new ideas, through to the successful application of organometallic species in organic syntheses, the generation of living catalysts for polymerization, and the synthesis of precursors for materials employed in the electronic and ceramic industries. For many reasons, therefore, we judged it timely to update 'Comprehensive Organometallic Chemistry' with a new work.

Due to the scope and depth of this area of chemistry, to have merely revised each of the original nine volumes did not seem the most user-friendly or cost-effective procedure to follow. As a consequence of the sheer bulk of the literature of the subject, a revised edition would necessarily require either the elimination of much chemistry of archival value but which is still important, or the production of a set of volumes significantly larger in number than the original nine. Accordingly, we decided it would be best to use the original work as a basis for new volumes focusing on organometallic chemistry reported since 1982, with reference back to the original work when necessary. For ease of use the new volumes maintain the same general structure as employed previously but reflect the changes in substance and direction the field has undergone in the last ten years. Thus it is not surprising that the largest volume in the new work concerns the role of the transition elements in metal-mediated organic syntheses.

The expansion of organometallic chemistry since the early 1980s also led us to decide that an updating of 'Comprehensive Organometallic Chemistry' would be more effectively accomplished if each volume had one or two editors who would be responsible both for recruiting experts for the Herculean task of writing the many chapters of each volume and for overseeing the content. We are deeply indebted to the volume editors and their authors for the time and effort they have given to the project.

As with the original 'Comprehensive Organometallic Chemistry', published some thirteen years ago, we hope this new version will serve as a pivotal reference point for new work and will function to generate new ideas and perceptions for the continued advance of what will surely continue as a vibrant area of chemistry.

Edward W. Abel
Exeter, UK

F. Gordon A. Stone
Waco, Texas, USA

Geoffrey Wilkinson
London, UK

Preface to 'Comprehensive Organometallic Chemistry'

Although the discovery of the platinum complex that we now know to be the first π-alkene complex, $K[PtCl_3(C_2H_4)]$, by Zeise in 1827 preceded Frankland's discovery (1849) of diethylzinc, it was the latter that initiated the rapidly developing interest during the latter half of the nineteenth century in compounds with organic groups bound to the elements. This era may be considered to have reached its apex in the discovery by Grignard of the magnesium reagents which occupy a special place because of their ease of synthesis and reactivity. With the exception of trimethylplatinum chloride discovered by Pope, Peachy and Gibson in 1907 by use of the Grignard reagent, attempts to make stable transition metal alkyls and aryls corresponding to those of main group elements met with little success, although it is worth recalling that even in 1919 Hein and his co-workers were describing the 'polyphenyl-chromium' compounds now known to be arene complexes.

The other major area of organometallic compounds, namely metal compounds of carbon monoxide, originated in the work starting in 1868 of Schützenberger and later of Mond and his co-workers and was subsequently developed especially by Hieber and his students. During the first half of this century, aided by the use of magnesium and, later, lithium reagents the development of main group organo chemistry was quite rapid, while from about 1920 metal carbonyl chemistry and catalytic reactions of carbon monoxide began to assume importance.

In 1937 Krause and von Grosse published their classic book 'Die Chemie der Metallorganischen Verbindungen'. Almost 1000 pages in length, it listed scores of compounds, mostly involving metals of the main groups of the periodic table. Compounds of the transition elements could be dismissed in 40 pages. Indeed, even in 1956 the stimulating 197-page monograph 'Organometallic Compounds' by Coates adequately reviewed organo transition metal complexes within 27 pages.

Although exceedingly important industrial processes in which transition metals were used for catalysis of organic reactions were developed in the 1930s, mainly in Germany by Reppe, Koch, Roelen, Fischer and Tropsch and others, the most dramatic growth in our knowledge of organometallic chemistry, particularly of transition metals, has stemmed from discoveries made in the middle years of this century. The introduction in the same period of physical methods of structure determination (infrared, nuclear magnetic resonance, and especially single-crystal X-ray diffraction) as routine techniques to be used by preparative chemists allowed increasingly sophisticated exploitation of discoveries. Following the recognition of the structure of ferrocene, other major advances quickly followed, including the isolation of a host of related π-complexes, the synthesis of a plethora of organometallic compounds containing metal–metal bonds, the characterization of low-valent metal species in which hydrocarbons are the only ligands, and the recognition from dynamic NMR spectra that ligand site exchange and tautomerism were common features in organometallic and metal carbonyl chemistry. The discovery of alkene polymerization using aluminium alkyl–titanium chloride systems by Ziegler and Natta and of the Wacker palladium–copper catalysed ethylene oxidation led to enormous developments in these areas.

In the last two decades, organometallic chemistry has grown more rapidly in scope than have the classical divisions of chemistry, leading to publications in journals of all national chemical societies, the appearance of primary journals specifically concerned with the topic, and the growth of annual review volumes designed to assist researchers to keep abreast of accelerating developments.

Organometallic chemistry has become a mature area of science which will obviously continue to grow. We believe that this is an appropriate time to produce a comprehensive review of the subject, treating organo derivatives in the widest sense of both main group and transition elements. Although advances in transition metal chemistry have appeared to dominate progress in recent years, spectacular progress has, nevertheless, also been made in our knowledge of organo compounds of main group elements such as aluminium, boron, lithium and silicon.

In these Volumes we have assembled a compendium of knowledge covering contemporary organometallic and carbon monoxide chemistry. In addition to reviewing the chemistry of the elements individually, two Volumes survey the use of organometallic species in organic synthesis and in catalysis, especially of industrial utility. Within the other Volumes are sections devoted to such diverse topics as the nature of carbon–metal bonds, the dynamic behaviour of organometallic compounds in solution, heteronuclear metal–metal bonded compounds, and the impact of organometallic compounds on the environment. The Volumes provide a unique record, especially of the intensive studies conducted during the past 25 years. The last Volume of indexes of various kinds will assist readers seeking information on the properties and synthesis of compounds and on earlier reviews.

As Editors, we are deeply indebted to all those who have given their time and effort to this project. Our Contributors are among the most active research workers in those areas of the subject that they have reviewed and they have well justified international reputations for their scholarship. We thank them sincerely for their cooperation.

Finally, we believe that 'Comprehensive Organometallic Chemistry', as well as providing a lasting source of information, will provide the stimulus for many new discoveries since we do not believe it possible to read any of the articles without generating ideas for further research.

E. W. ABEL
Exeter

G. WILKINSON
London

F. G. A. STONE
Bristol

Contributors to Volume 11

Dr. S. K. Armstrong
The Dyson Perrins Laboratory, University of Oxford, South Parks Road, Oxford, OX1 3QY, UK

Dr. J. M. Brown
The Dyson Perrins Laboratory, University of Oxford, South Parks Road, Oxford, OX1 3QY, UK

Dr. B. Carboni
Groupe de Recherches de Physiochimie Structurale, Campus de Beaulieu, Université de Rennes I, Avenue du Général Ledere, F-35042 Rennes Cédex, France

Dr. E. W. Colvin
Department of Chemistry, University of Glasgow, Glasgow, G12 8QQ, UK

Professor J. J. Eisch
Department of Chemistry, State University of New York at Binghamton, PO Box 6000, Binghamton, NY 13902-6000, USA

Dr. M. Gray
The Guelph–Waterloo Center for Graduate Work in Chemistry, Waterloo Campus, Department of Chemistry, University of Waterloo, Waterloo, Ontario N2L 3G1, Canada

Professor Y.-Z. Huang
Organometallic Chemistry Laboratory, Shanghai Institute of Organic Chemistry, Academia Sinica, 345 Lingling Lu, Shanghai 200032, People's Republic of China

Professor P. Knochel
Fachbereich Chemie, Philipps-Universität Marburg, Hans-Meerwein-Strasse, D-3550 Marburg, Germany

Professor A. Krief
Laboratoire de Chimie Organique, Facultés Universitaires Notre-Dame de la Paix, 61 rue de Bruxelles, B-5000 Namur, Belgium

Professor R. C. Larock
Department of Chemistry, 3831 Gilman Hall, Iowa State University, Ames, IA 50011-3111, USA

Dr. C. W. Leung
Laboratoire de Chimie Organique, Bâtiment Lavoisier, Université Catholique de Louvain, Place Louis Pasteur 1, B-1348 Louvain-la-Neuve, Belgium

Dr. I. E. Markó
Laboratoire de Chimie Organique, Bâtiment Lavoisier, Université Catholique de Louvain, Place Louis Pasteur 1, B-1348 Louvain-la-Neuve, Belgium

Dr. A. Mordini
Dipartimento de Chimica Organica dell'Università di Firenze, via Gino Capponi 9, I-50121 Firenze, Italy

Professor N. Petragnani
Instituto de Quimica, Universidade de São Paulo, Cx. P. 20780-CEP, 01498 São Paulo, Brazil

Professor J. T. Pinhey
Department of Organic Chemistry, University of Sydney, Sydney, NSW 2006, Australia

Professor T. Sato
Department of Applied Chemistry, School of Science and Engineering, Waseda University, Ookubo 3, Shinjuku-ku, Tokyo 169, Japan

Professor V. Snieckus
The Guelph–Waterloo Center for Graduate Work in Chemistry, Waterloo Campus, Department of Chemistry, University of Waterloo, Waterloo, Ontario N2L 3G1, Canada

Dr. M. Tinkl
The Guelph–Waterloo Center for Graduate Work in Chemistry, Waterloo Campus, Department of Chemistry, University of Waterloo, Waterloo, Ontario N2L 3G1, Canada

Dr. M. Vaultier
Groupe de Recherches de Physiochimie Structurale, Campus de Beaulieu, Université de Rennes I, Avenue du Général Ledere, F-35042 Rennes Cédex, France

Dr. Z.-L. Zhou
Organometallic Chemistry Laboratory, Shanghai Institute of Organic Chemistry, Academia Sinica, 345 Lingling Lu, Shanghai 200032, People's Republic of China

Abbreviations

The abbreviations used throughout 'Comprehensive Organometallic Chemistry II' are consistent with those used in 'Comprehensive Organometallic Chemistry' and with other standard texts in this area. The abbreviations in some instances may differ from those commonly used in other branches of chemistry.

Ac	acetyl
acac	acetylacetonate
AIBN	2,2'-azobisisobutyronitrile
Ar	aryl
arphos	1-(diphenylphosphino)-2-(diphenylarsino)ethane
Azb	azobenzene
9-BBN	9-borabicyclo[3.3.1]nonyl
9-BBN-H	9-borabicyclo[3.3.1]nonane
BHT	2,6-di-t-butyl-4-methylphenol (butylated hydroxytoluene)
bipy	2,2'-bipyridyl
t-BOC	t-butoxycarbonyl
bsa	N,O-bis(trimethylsilyl)acetamide
bstfa	N,O-bis(trimethylsilyl)trifluoroacetamide
btaf	benzyltrimethylammonium fluoride
Bz	benzyl
can	ceric ammonium nitrate
cbd	cyclobutadiene
1,5,9-cdt	cyclododeca-1,5,9-triene
chd	cyclohexadiene
chpt	cycloheptatriene
[Co]	cobalamin
(Co)	cobaloxime [Co(DMG)$_2$] derivative
cod	1,5-cyclooctadiene
cot	cyclooctatetraene
Cp	η^5-cyclopentadienyl
Cp*	pentamethylcyclopentadienyl
18-crown-6	1,4,7,10,13,16-hexaoxacyclooctadecane
CSA	camphorsulfonic acid
csi	chlorosulfonyl isocyanate
Cy	cyclohexyl
dabco	1,4-diazabicyclo[2.2.2]octane
dba	dibenzylideneacetone
dbn	1,5-diazabicyclo[4.3.0]non-5-ene
dbu	1,8-diazabicyclo[5.4.0]undec-7-ene
dcc	dicyclohexylcarbodiimide
dcpe	1,2-bis(dicyclohexylphosphino)ethane
ddq	2,3-dichloro-5,6-dicyano-1,4-benzoquinone
deac	diethylaluminum chloride
dead	diethyl azodicarboxylate
depe	1,2-bis(diethylphosphino)ethane
depm	1,2-bis(diethylphosphino)methane
det	diethyl tartrate (+ or −)

DHP	dihydropyran
diars	1,2-bis(dimethylarsino)benzene
dibal-H	diisobutylaluminum hydride
dien	diethylenetriamine
DIGLYME	bis(2-methoxyethyl)ether
diop	2,3-*O*-isopropylidene-2,3-dihydroxy-1,4-bis(diphenylphosphino)butane
dipt	diisopropyl tartrate (+ or −)
dma	dimethylacetamide
dmac	dimethylaluminum chloride
DMAD	dimethyl acetylenedicarboxylate
dmap	4-dimethylaminopyridine
DME	dimethoxyethane
DMF	*N*,*N*'-dimethylformamide
DMG	dimethylglyoximate
DMI	*N*,*N*'-dimethylimidazalone
dmpe	1,2-bis(dimethylphosphino)ethane
dmpm	bis(dimethylphosphino)methane
DMSO	dimethyl sulfoxide
dmtsf	dimethyl(methylthio)sulfonium fluoroborate
dpam	bis(diphenylarsino)methane
dppb	1,4-bis(diphenylphosphino)butane
dppe	1,2-bis(diphenylphosphino)ethane
dppf	1,1'-bis(diphenylphosphino)ferrocene
dpph	1,6-bis(diphenylphosphino)hexane
dppm	bis(diphenylphosphino)methane
dppp	1,3-bis(diphenylphosphino)propane
eadc	ethylaluminum dichloride
edta	ethylenediaminetetraacetate
eedq	*N*-ethoxycarbonyl-2-ethoxy-1,2-dihydroquinoline
en	ethylene-1,2-diamine
Et_2O	diethyl ether
F_6acac	hexafluoroacetylacetonate
Fc	ferrocenyl
Fp	$Fe(CO)_2Cp$
HFA	hexafluoroacetone
hfacac	hexafluoroacetylacetonate
hfb	hexafluorobut-2-yne
HMPA	hexamethylphosphoramide
hobt	hydroxybenzotriazole
$IpcBH_2$	isopinocampheylborane
Ipc_2BH	diisopinocampheylborane
kapa	potassium 3-aminopropylamide
K-selectride	potassium tri-*s*-butylborohydride
LAH	lithium aluminum hydride
LDA	lithium diisopropylamide
LICA	lithium isopropylcyclohexylamide
LITMP	lithium tetramethylpiperidide
L-selectride	lithium tri-*s*-butylborohydride
LTA	lead tetraacetate
mcpba	*m*-chloroperbenzoic acid
MeCN	acetonitrile
MEM	methoxyethoxymethyl
MEM-Cl	β-methoxyethoxymethyl chloride

Mes	mesityl
mma	methyl methacrylate
mmc	methylmagnesium carbonate
MOM	methoxymethyl
Ms	methanesulfonyl
MSA	methanesulfonic acid
MsCl	methanesulfonyl chloride
nap	1-naphthyl
nbd	norbornadiene
NBS	*N*-bromosuccinimide
NCS	*N*-chlorosuccinimide
nmo	*N*-methylmorpholine *N*-oxide
NMP	*N*-methyl-2-pyrrolidone
Nu$^-$	nucleophile
ox	oxalate
pcc	pyridinium chlorochromate
pdc	pyridinium dichromate
phen	1,10-phenanthroline
phth	phthaloyl
ppa	polyphosphoric acid
ppe	polyphosphate ester
[PPN]$^+$	[(Ph$_3$P)$_2$N]$^+$
ppts	pyridinium *p*-toluenesulfonate
py	pyridine
pz	pyrazolyl
Red-Al	sodium bis(2-methoxyethoxy)aluminum dihydride
sal	salicylaldehyde
salen	*N,N'*-bis(salicylaldehydo)ethylenediamine
SEM	β-trimethylsilylethoxymethyl
tas	tris(diethylamino)sulfonium
tasf	tris(diethylamino)sulfonium difluorotrimethylsilicate
tbaf	tetra-*n*-butylammonium fluoride
TBDMS	*t*-butyldimethylsilyl
TBDMS-Cl	*t*-butyldimethylsilyl chloride
TBDPS	*t*-butyldiphenylsilyl
tbhp	*t*-butyl hydroperoxide
TCE	2,2,2-trichloroethanol
TCNE	tetracyanoethene
TCNQ	7,7,8,8-tetracyanoquinodimethane
terpy	2,2':6',2''-terpyridyl
tes	triethylsilyl
Tf	triflyl (trifluoromethanesulfonyl)
TFA	trifluoracetic acid
TFAA	trifluoroacetic anhydride
tfacac	trifluoroacetylacetonate
THF	tetrahydrofuran
THP	tetrahydropyranyl
tipbs-Cl	2,4,6-triisopropylbenzenesulfonyl chloride
tips-Cl	1,3-dichloro-1,1,3,3-tetraisopropyldisiloxane
TMEDA	tetramethylethylenediamine [1,2-bis(dimethylamino)ethane]
TMS	trimethylsilyl
TMS-Cl	trimethylsilyl chloride
TMS-CN	trimethylsilyl cyanide
Tol	tolyl
tpp	*meso*-tetraphenylporphyrin

Tr	trityl (triphenylmethyl)
tren	2,2',2''-triaminotriethylamine
trien	triethylenetetraamine
triphos	1,1,1-tris(diphenylphosphinomethyl)ethane
Ts	tosyl
TsMIC	tosylmethyl isocyanide
ttfa	thallium trifluoroacetate

Contents of All Volumes

Volume 1   Lithium, Beryllium, and Boron Groups
1   Alkali Metals
2   Beryllium
3   Magnesium, Calcium, Strontium and Barium
4   Compounds with Three- or Four-coordinate Boron, Emphasizing Cyclic Systems
5   Boron Rings Ligated to Metals
6   Polyhedral Carbaboranes
7   Main-group Heteroboranes
8   Metallaboranes
9   Transition Metal Metallacarbaboranes
10 Aluminum
11 Gallium, Indium and Thallium, Excluding Transition Metal Derivatives
12 Transition Metal Complexes of Aluminum, Gallium, Indium and Thallium
Author Index
Subject Index

Volume 2   Silicon Group, Arsenic, Antimony, and Bismuth
1   Organosilanes
2   Carbacyclic Silanes
3   Organopolysilanes
4   Silicones
5   Germanium
6   Tin
7   Lead
8   Arsenic, Antimony and Bismuth
Author Index
Subject Index

Volume 3   Copper and Zinc Groups
1   Gold
2   Copper and Silver
3   Mercury
4   Cadmium and Zinc
Author Index
Subject Index

Volume 4   Scandium, Yttrium, Lanthanides and Actinides, and Titanium Group
1   Zero Oxidation State Complexes of Scandium, Yttrium and the Lanthanide Elements
2   Scandium, Yttrium, and the Lanthanide and Actinide Elements, Excluding their Zero Oxidation State Complexes
3   Titanium Complexes in Oxidation States Zero and Below
4   Titanium Complexes in Oxidation States +2 and +3
5   Titanium Complexes in Oxidation State +4
6   Zirconium and Hafnium Complexes in Oxidation States Zero and Below
7   Metallocene(II) Complexes of Zirconium and Hafnium
8   Zirconium and Hafnium Compounds in Oxidation State +3
9   Bis(cyclopentadienyl)zirconium and -hafnium Halide Complexes in Oxidation State +4
10 Bis(cyclopentadienyl) Metal(IV) Compounds with Si, Ge, Sn, N, P, As, Sb, O, S, Se, Te or Transition Metal-centred Ligands
11 Zirconium and Hafnium Complexes in Oxidation State +4

12 Cationic Organozirconium and Organohafnium Complexes
13 Cyclooctatetraene Complexes of Zirconium and Hafnium
Author Index
Subject Index

Volume 5 Vanadium and Chromium Groups
1 Vanadium
2 Niobium and Tantalum
3 Hexacarbonyls and Carbonyl Complexes of Carbon σ-Bonded Ligands of Chromium, Molybdenum and Tungsten
4 Carbonyl Complexes of Noncarbon σ-Bonded Ligands of Chromium, Molybdenum and Tungsten
5 Organometallic Complexes of Chromium, Molybdenum and Tungsten without Carbonyl Ligands
6 π-Complexes of Chromium, Molybdenum and Tungsten, Excluding those of Cyclopentadienyls and Arenes
7 Cyclopentadienyl Complexes of Chromium, Molybdenum and Tungsten
8 Arene and Heteroarene Complexes of Chromium, Molybdenum and Tungsten
Author Index
Subject Index

Volume 6 Manganese Group
1 Manganese Carbonyls and Manganese Carbonyl Halides
2 Manganese Alkyls and Hydrides
3 Manganese Complexes Containing Nonmetallic Elements
4 Manganese Hydrocarbon Complexes Excluding Cyclopentadienyl
5 Cyclopentadienyl Manganese Complexes
6 Manganese Nitrosyl and Isonitrile Complexes
7 High-valent Organomanganese Compounds
8 Technetium
9 Low-valent Organorhenium Compounds
10 High-valent Organorhenium Compounds
Author Index
Subject Index

Volume 7 Iron, Ruthenium, and Osmium
1 Iron Compounds without Hydrocarbon Ligands
2 Mononuclear Iron Compounds with η^1–η^6 Hydrocarbon Ligands
3 Dinuclear Iron Compounds with Hydrocarbon Ligands
4 Polynuclear Iron Compounds with Hydrocarbon Ligands
5 Introduction to Organoruthenium and Organoosmium Chemistry
6 Mononuclear Complexes of Ruthenium and Osmium Containing η^1 Carbon Ligands
7 Complexes of Ruthenium and Osmium Containing η^2–η^6 Hydrocarbon Ligands: (i) Complexes not Containing Cyclobutadiene, Cyclopentadienyl or η-Arene Coligands
8 Complexes of Ruthenium and Osmium Containing η^2–η^6 Hydrocarbon Ligands: (ii) Complexes Containing Four- and Five-membered Rings (Including MCp(arene) Complexes)
9 Complexes of Ruthenium and Osmium Containing η^2–η^6 Hydrocarbon Ligands: (iii) Complexes Containing Six-, Seven- and Eight-membered Rings
10 Ruthenocenes and Osmocenes
11 Binuclear Complexes of Ruthenium and Osmium Containing Metal–Metal Bonds
12 Trinuclear Clusters of Ruthenium and Osmium: (i) Introduction and Simple Neutral, Anionic and Hydrido Clusters
13 Trinuclear Clusters of Ruthenium and Osmium: (ii) Hydrocarbon Ligands on Metal Clusters
14 Trinuclear Clusters of Ruthenium and Osmium: (iii) Clusters with Metal–Carbon Bonds to Heteroatom Ligands
15 Tetranuclear Clusters of Ruthenium and Osmium
16 Medium- and High-nuclearity Clusters of Ruthenium and Osmium
Author Index
Subject Index

Volume 8 Cobalt, Rhodium, and Iridium
1   Cobalt
2   Rhodium
3   Iridium
4   Cluster Complexes of Cobalt, Rhodium, and Iridium
Author Index
Subject Index

Volume 9 Nickel, Palladium, and Platinum
1   Nickel Complexes with Carbonyl, Isocyanide, and Carbene Ligands
2   Nickel–Carbon σ-Bonded Complexes
3   Nickel–Carbon π-Bonded Complexes
4   Palladium Complexes with Carbonyl, Isocyanide and Carbene Ligands
5   Palladium–Carbon σ-Bonded Complexes
6   Palladium–Carbon π-Bonded Complexes
7   Platinum Complexes with Carbonyl, Isocyanide and Carbene Ligands
8   Platinum–Carbon σ-Bonded Complexes
9   Platinum–Carbon π-Bonded Complexes
Author Index
Subject Index

Volume 10 Heteronuclear Metal–Metal Bonds
1   Synthesis of Compounds Containing Heteronuclear Metal–Metal Bonds
2   Heterodinuclear Compounds
3   Heteronuclear Clusters Containing C_1, C_2, C_3, ..., C_n Acyclic Hydrocarbyl Ligands
4   Binary Carbonyls, Carbonyls plus Hydrides, Carbonyls plus Phosphines, Cyclic Hydrocarbyls and Main-group Ligands without Acyclic Hydrocarbyls
5   Cluster Complexes with Bonds Between Transition Elements and Copper, Silver and Gold
6   Cluster Complexes with Bonds Between Transition Elements and Zinc, Cadmium, and Mercury
7   Catalysis and Related Reactions with Compounds Containing Heteronuclear Metal–Metal Bonds
Author Index
Subject Index

Volume 11 Main-group Metal Organometallics in Organic Synthesis
1   Lithium
2   Sodium and Potassium
3   Magnesium
4   Zinc and Cadmium
5   Boron
6   Aluminum
7   Silicon
8   Tin
9   Mercury
10 Thallium
11 Lead
12 Antimony and Bismuth
13 Selenium
14 Tellurium
Author Index
Subject Index

Volume 12 Transition Metal Organometallics in Organic Synthesis
1   Introduction and Fundamentals
2   Transition Metal Hydrides: Hydrocarboxylation, Hydroformylation, and Asymmetric Hydrogenation
3.1   Transition Metal Alkyl Complexes from Hydrometallation
3.2   Transition Metal Alkyl Complexes from RLi and CuX
3.3   Transition Metal Alkyl Complexes: Main-group Transmetallation and Insertion Chemistry
3.4   Transition Metal Alkyl Complexes: Oxidative Addition and Transmetallation

3.5 Transition Metal Alkyl Complexes: Oxidative Addition and Insertion
3.6 Transition Metal Alkyl Complexes: Multiple Insertion Cascades
3.7 Transition Metal Alkyl Complexes: Reductive Dimerization of Alkenes and Alkynes
4 Transition Metal Carbonyl Complexes
5.1 Transition Metal Carbene Complexes: Cyclopropanation
5.2 Transition Metal Carbene Complexes: Diazodecomposition, Ylide, and Insertion
5.3 Transition Metal Carbene Complexes: Alkyne and Vinyl Ketene Chemistry
5.4 Transition Metal Carbene Complexes: Photochemical Reactions of Carbene Complexes
5.5 Transition Metal Carbene Complexes: Tebbe's Reagent and Related Nucleophilic Alkylidenes
6.1 Transition Metal Alkene, Diene, and Dienyl Complexes: Nucleophilic Attack on Alkene Complexes
6.2 Transition Metal Alkene, Diene, and Dienyl Complexes: Complexation of Dienes for Protection
6.3 Transition Metal Alkene, Diene, and Dienyl Complexes: Nucleophilic Attack on Diene and Dienyl Complexes
7.1 Transition Metal Alkyne Complexes: Transition Metal-stabilized Propargyl Systems
7.2 Transition Metal Alkyne Complexes: Pauson–Khand Reaction
7.3 Transition Metal Alkyne Complexes: Transition Metal-catalyzed Cyclotrimerization
7.4 Transition Metal Alkyne Complexes: Zirconium–Benzyne Complexes
8.1 Transition Metal Allyl Complexes: Telomerization of Dienes
8.2 Transition Metal Allyl Complexes: Pd, W, Mo-assisted Nucleophilic Attack
8.3 Transition Metal Allyl Complexes: Intramolecular Alkene and Alkyne Insertions
8.4 Transition Metal Allyl Complexes: Trimethylene Methane Complexes
8.5 Transition Metal Allyl Complexes: π-Allylnickel Halides and Other π-Allyl Complexes Excluding Palladium
9.1 Transition Metal Arene Complexes: Nucleophilic Addition
9.2 Transition Metal Arene Complexes: Ring Lithiation
9.3 Transition Metal Arene Complexes: Side-chain Activation and Control of Stereochemistry
10 Synthetically Useful Coupling Reactions Promoted by Ti, V, Nb, W, Mo Reagents
11.1 Transition Metal-catalyzed Oxidations: Asymmetric Epoxidation
11.2 Transition Metal-catalyzed Oxidations: Asymmetric Hydroxylation
11.3 Transition Metal-catalyzed Oxidations: Other Oxidations
12.1 Transition Metals in Polymer Synthesis: Ziegler–Natta Reaction
12.2 Transition Metals in Polymer Synthesis: Ring-opening Metathesis Polymerization and Other Transition Metal Polymerization Techniques
Author Index
Subject Index

Volume 13 Structure Index
Structures of Organometallic Compounds Determined by Diffraction Methods

Volume 14 Cumulative Indexes
Cumulative Formula Index
Cumulative Subject Index

1
Lithium

MATTHEW GRAY, MICHAEL TINKL and VICTOR SNIECKUS
University of Waterloo, ON, Canada

1.1 INTRODUCTION	2
1.2 LITHIUM REAGENTS	3
1.2.1 Organolithiums ($pK_a \cong 35–50$) and Additives	3
1.2.1.1 Titration methods	3
1.2.2 Lithium Amides	4
1.2.2.1 Lithium diisopropylamide (LDA; $pK_a = 35.7$)	4
1.2.2.2 Lithium diethylamide ($pK_a = 31.7$)	5
1.2.2.3 Lithium pyrrolidide	5
1.2.2.4 Lithium piperidide ($pK_a = 30.7$)	5
1.2.2.5 Lithium 2,2,6,6-tetramethylpiperidide (LiTMP; $pK_a = 37.3$)	5
1.2.2.6 Lithium bis(trimethylsilyl)amide or lithium hexamethyldisilazide (LiHMDS; $pK_a = 29.5$)	5
1.2.2.7 Miscellaneous lithium amides	5
1.3 sp^3 CARBANIONS	6
1.3.1 Unstabilized sp^3 Carbanions	6
1.3.1.1 Lithium–proton exchange	6
1.3.1.2 Lithium–halogen exchange	6
1.3.1.3 Reductive cleavage of heterocyclic alkanes	12
1.3.1.4 Miscellaneous exchange reactions	14
1.3.2 α-Nitrogen sp^3 Carbanions	15
1.3.3 α-Oxygen sp^3 Carbanions	21
1.3.3.1 Configurational stability and enantioinduction	22
1.3.3.2 α-Oxylithio rearrangements	24
1.3.4 α-Silicon sp^3 Carbanions	25
1.3.4.1 The Peterson reaction	26
1.3.5 α-Sulfur- and α-Selenium-stabilized Carbanions	28
1.3.5.1 Generation	28
1.3.5.2 Subsequent transformations	29
1.3.5.3 α-Sulfenyl and selenyl carbanions	29
1.3.5.4 α-Sulfenyl and selenyl allylic carbanions	30
1.3.5.5 α-Sulfinyl carbanions	32
1.3.5.6 α-Sulfinyl allylic carbanions	35
1.3.5.7 α-Sulfonyl carbanions	36
1.3.5.8 α-Sulfonyl allylic carbanions	39
1.3.5.9 Acyl anion equivalents	39
1.3.5.10 Sulfoximines	41
1.4 ARYLLITHIUM AND HETEROARYLLITHIUM ANIONS	42
1.4.1 Introduction	42
1.4.2 Directed ortho-Metallation Based Methods	42
1.4.3 Arene Metallation by Lithium–Proton Exchange	42
1.4.3.1 Carbon-based directed metallation groups	42
1.4.3.2 Heteroatom-based DMGs	48
1.4.3.3 Benzyllithiums	55
1.4.3.4 Lithiated arene chromium tricarbonyl complexes	63
1.4.3.5 Lithiated ferrocenes	64
1.4.4 Aryllithiums by Metal–Halogen Exchange	66
1.4.5 Directed ortho-Metallation of Heteroaromatics	70

1.4.5.1 Introduction and overview	70
1.4.5.2 Metallation of five-membered ring heteroaromatics	71
1.4.5.3 Metallation of six-membered ring heteroaromatics	75
1.5 REFERENCES	81

1.1 INTRODUCTION

Organolithiums and carbanions are words in association which, in turn, translate into the most important modern synthetic methods of carbon–carbon bond formation. In 1982, Wakefield, in his excellent review for *COMC-I*,[1] treated organolithiums as part of a review on alkali and alkaline earth metals, citing 605 references in total. In 1994, a CAS Online search for the last decade using a lithium and enolate combination of key words generated 903 references. Hence, the impossibility of the task at hand in the area of organolithium chemistry to fulfil the aim of the book title. As reinforcement for this point, the distinguished Houben–Weyl series states apologetically that "*Eine vollständige Behandlung des umfangreichen Themas 'Carbanionen' ist auch im Rahmen des Houben-Weyl nicht möglich*" ("The complete treatment of the large area of 'Carbanions' is also impossible within the framework of Houben–Weyl").[2]

Fortunately, carbanions have not suffered neglect in this age of extensive and overlapping reviews and the 1980s and early 1990s has witnessed a number of noteworthy treatises[2–10] in diverse aspects of structure, reactivity, and synthesis of organolithiums. Thus, with some justification but also with regret, ruthless limits have been set in the preparation of this chapter: structure[2,11–13] and mechanistic[14] studies, attempting to give insight to collections of empirical observations,[15] have been completely omitted; reactivity investigations,[16] which are clearly of more than peripheral value to synthesis, have been excluded; and the important topics of enolates,[17,18] sp^2 organolithium anions,[19] sp organolithium anions[20] and carbanionic rearrangements,[21] which have seen imposing advances since the 1980s are not covered.

The reader may then ask, with justification, what is covered? Mindful of the accessible and recent sources of organolithium chemistry, both comprehensive and specialized, this chapter presents topics, hopefully viewed as objectively selected, in which significant and exciting synthetic advances have occurred since the late 1970s. Following a short section on lithium reagents (Section 1.2)[22] (perhaps appropriately placed near the bench), unstabilized sp^3 carbanions (Section 1.3.1), obtained by standard but also new methods,[23] are treated, illustrating a multitude of new useful synthetic pathways. α-Heteroatom sp^3 carbanions (Sections 1.3.2–1.3.5) follow. An especially strong impact in synthesis is due to α-nitrogen systems[24,25] which are readily generated, provide new methodology for amine synthesis, may be applied in natural product construction, and, most recently, participate in chiral base-induced asymmetric deprotonation. Advances in α-oxygen carbanions are presented next (Section 1.3.3). These species, usually indirectly derived by metal–heteroatom exchange and exhibiting configurational stability,[26] have also enjoyed considerable application in context of the [2,3]-Wittig rearrangement in natural product synthesis. As evidenced from Section 1.3.4, an enormous body of literature has accumulated in the use of α-silicon carbanions in synthesis,[27] with the major advances noted in new variants of the Peterson alkenation[28] and in allyl silane chemistry.[29] A combined section on α-sulfur- and α-selenium-stabilized carbanions (Section 1.3.5) emphasizes sulfur and, in particular, sulfoxide- and sulfone-mediated transformations where key advances in their use in enantioselective[30,31] and Julia reactions,[32] in addition to continuing use as acyl anion equivalents, have been made.

Being nurtured in the late 1970s,[33] the area of aryllithium chemistry (Section 1.4) has found an entrenched niche in the directed *ortho*-metallation (DoM) reaction[34,35] to the immense benefit of regioselective, small-scale and bulk synthesis of polysubstituted aromatics. Subdivided into carbon- (Section 1.4.3.1) and heteroatom- (Section 1.4.3.2) based directed metallation groups (DMGs), this section displays, by selected examples, the diversity of solutions that DoM chemistry has brought to methodological and total synthesis problems in the aromatic arena. Benzyl carbanions bearing *ortho*-DMGs[36] are given a separate section (Section 1.4.3.3) to emphasize their value in chain extension, ring annelation, and complex molecule construction. As described in short sections, lithiated chromium tricarbonyl complexes (Section 1.4.3.4) and ferrocenes (Section 1.4.3.5) show, respectively, surprising reactivity patterns and promising potential for the design of chiral catalysts. Section 1.4.4, devoted to the traditional lithium–halogen exchange, attests to the continuing dependence of synthetic chemists on this process for assured generation of aryllithiums. Alongside the growth of DoM chemistry, the field of HETDoM (five-membered (Section 1.4.5.2); six-membered (Section 1.4.5.3))[37,38] has experienced accelerated activity, particularly in natural product construction and in hitherto unmetallated heterocyclic substrates.

In view of the limited coverage, most recent reviews are cited wherever possible as leads to specific topics but also in attempts to predict the mental connections chemists make in developing synthetic strategies. Criteria of broad synthetic value, meritous application, and innovation were used for inclusion of schemes. The critical reader will tend to conclude that, although new methods abound and ingenious applications have emerged, the experimental conditions in organolithium chemistry are highly variable for seemingly similar reactions and minor changes are crucial for optimum results. For this kind of information as well as practical tips, we have fortunately been guided since the 1980s by Brandsma[7,8,34] who has checked more than 2000 organolithium reactions.

1.2 LITHIUM REAGENTS[22,39]

1.2.1 Organolithiums ($pK_a \cong 35$–50) and Additives

1.2.1.1 Titration methods

A myriad of titration methods are known,[40] varying in reproducibility, ease of operation, availability of reagents, and clarity of end-point.[6] The choice appears to depend on the lithium base and on personal preference.[41] New techniques are frequently published, often claiming certain advantages over existing systems.[42–7]

(i) n-Butyllithium

n-Butyllithium[7,48–50] is arguably the most commonly used alkyllithium reagent in synthesis, and is commercially available in hydrocarbon solvents. *In situ* generation involves the use of n-butyl chloride and lithium wire in THF,[51] and is also applicable to s- and t-butyllithium. Halogen-free primary and secondary alkyllithiums may be prepared by treatment of aliphatic dialkyl sulfates with excess lithium powder in the presence of catalytic naphthalene.[52] New reagents such as n-hexyllithium may become increasingly attractive as alternatives to n-butyllithium, especially in industry, due to the absence of gaseous by-products.

(ii) s-Butyllithium

Commercially available in hydrocarbons, s-butyllithium is a more hindered and stronger base than n-butyllithium. A perhaps more environmentally advantageous substitute for large-scale applications is 2-ethylhexyllithium.

(iii) t-Butyllithium

t-Butyllithium,[6,53] a very strong and hindered base, can also be prepared as stated above. Commercial t-butyllithium is sold in hydrocarbons (normally pentane). It is now available in heptane, which has advantages of safer handling and smaller concentration fluctuations during storage.

(iv) Methyllithium

Methyllithium is available in ether (with or without lithium halide) or in THF/cumene (nonpyrophoric). Halide-free methyllithium can be prepared by the reaction of n-butyllithium with iodomethane in hexane,[54] or from lithium metal and methyl chloride in ether.[55]

(v) Phenyllithium

Commercial phenyllithium[6,56] is available in cyclohexane/ether solution. A convenient means of preparation involves treatment of bromobenzene with n-butyllithium (1 equiv.) or t-butyllithium (2 equiv.) at low temperature in THF.

(vi) Lithium naphthalide, LiDBN and LiDBB

Radical anions, such as lithium naphthalide, LiDBN (lithium di-*t*-butylnaphalide) and LiDBB (lithium di-*t*-butylbiphenyl) are prepared by treatment of lithium powder with an arene and used for the generation of carbanions from alkyl halides, phenyl sulfides, alkyl sulfates, nitriles and heterocyclic alkanes as described in Section 1.3.1.

(vii) Mixed bases

The "superbases" developed by Lochmann and Schlosser (LICKOR (butyllithium/potassium *t*-butoxide), LIDAKOR (lithium diisopropylamide/potassium *t*-butoxide)) are extremely strong mixed-base systems exhibiting different reactivities to simple organolithiums or -sodiums.[57-61] It is possible that the alkoxide enhances the reactivity of the alkyllithium by the initial formation of mixed clusters.

(viii) Additives

(a) TMEDA. Metallation rate enhancement in the presence of TMEDA (even catalytic) is a common experimental observation,[35,62] which is generally rationalized in terms of deaggregation effects.[63,64] Evidence for Li–TMEDA chelates (x ray,[12,65] NMR[66]) and TMEDA-stabilized intermediates[67] has been presented. Significant regiochemical effects of TMEDA in directed *ortho*-metallation have been noted.[35,68] Collum's informative account includes general guidelines on the use of TMEDA.[63]

(b) Other additives. Several other additives are used to modify the reactivity of alkyllithium reagents. Hexamethylphosphoramide (HMPA) is used as a cosolvent to break up aggregates and increase ionization and suppress radical processes (a less toxic alternative is DMPU (*N,N'*-dimethylpropyleneurea)[69]). As with TMEDA, some uncertainty remains as regards general conclusions about solution structures.[70] 1,4-Diazabicyclo[2.2.2]octane (DABCO)[71,72] has seen limited use, whereas PMDTA (N,N,N',N'',N'''-pentamethyldiethylenetriamine) has found utility in regiocontrol of metallations,[73] and in the structural analysis of organolithiums. The tendency of alkyllithiums to cleave THF at ambient temperatures (0–25 °C) is greatly diminished in the presence of magnesium 2-ethoxyethoxide (prepared from ethylene glycol ethyl ether (Cellosolve) and magnesium powder[74]). Although direct metallating ability is reduced, preparation of alkyl- and aryllithiums by metal halogen exchange (with lithium metal or *n*-butyllithium) in the presence of Mg(OCH$_2$CH$_2$OEt)$_2$ suppresses reaction with THF and (for aryl substrates) side reactions such as *ortho*-metallation or Wittig rearrangement.[75] The magnesium alkoxide, present in less than stoichiometric quantities (0.33–0.50 equiv.), is believed to form a complex with the organolithium, rather than initiating simple Grignard generation. The addition of lithium salts LiX (X = Br, Cl, ClO$_4$, OSO$_2$R, etc.) can alter the reactivity of alkyllithiums by the formation of mixed aggregates and an increase in solubility.[76]

1.2.2 Lithium Amides

Organolithium amides used for carbanion generation are commercially available or are routinely generated *in situ* immediately prior to use by treatment of the corresponding amine with alkyllithium or, less frequently, lithium metal. A complicating feature is the ability of lithium dialkylamides[77] to reduce alkyl and aryl halides, triflates, ketones, and aldehydes, among other groups by either hydride transfer or single-electron transfer (SET) mechanisms. A rarer complication is the *ipso* substitution of aromatic ethers by lithium amides.[78]

1.2.2.1 Lithium diisopropylamide (LDA; pK_a = 35.7)

LDA is the household nonnucleophilic base[79] for the generation of kinetic enolates,[80] α-heteroatom allylic,[81,82] and aromatic and heteroaromatic carbanions,[35,36,38] and, more obscurely, for dehydrohalogenation of a β-bromo carboxylic acid.[83] LDA, commercially available as a solid or stabilized solutions of various concentrations, is unstable in most ethereal solvents and HMPA above 0 °C, but may be stored for prolonged periods at room temperature in hydrocarbon solvents.[84] LDA is prepared *in situ* from *n*-butyllithium and anhydrous diisopropylamine,[85] or from lithium metal and diisopropylamine in the presence of styrene.[86,87] An ultrasound-promoted preparation of LDA uses *n*-butyl chloride, lithium metal, and diisopropylamine.[51]

1.2.2.2 Lithium diethylamide ($pK_a = 31.7$)

Lithium diethylamide is a strong, nonnucleophilic base[88] which can be superior to LDA for enolate formation.[89] It is commercially available and may be prepared by treatment of diethylamine with phenyllithium or butyllithium or, alternatively, by reaction of diethylamine with lithium metal in the presence of HMPA[90] or an electron acceptor.[86]

1.2.2.3 Lithium pyrrolidide

Lithium pyrrolidide is a nucleophilic base, mainly used for the reduction of nonenolizable ketones,[91] carbonylation,[92] nucleophilic additions,[93,94] and nucleophilic aromatic *ipso* substitutions.[78] It is prepared by reaction of pyrrolidine with methyllithium or *n*-butyllithium.[78,91]

1.2.2.4 Lithium piperidide ($pK_a = 30.7$)

Lithium piperidide is a strong, nucleophilic base[88] used for benzyne generation[78] and for dehydrohalogenation.[95] It is prepared by treatment of piperidine with phenyl- or *n*-butyllithium.[96]

1.2.2.5 Lithium 2,2,6,6-tetramethylpiperidide (LiTMP; $pK_a = 37.3$)

Owing to its highly hindered, nonnucleophilic nature, LiTMP[79] is capable of inducing selective deprotonations in the presence of electrophilic groups and is compatible with select electrophiles (e.g., TMS-Cl) at low temperatures, thereby allowing its use for *in situ* quenching of thermodynamically derived carbanions.[97–101] It has been used for the cyclopropanation of alkenes with allylic halides,[102] for anthracene synthesis,[103] and for the dehydrohalogenation of alkanes[104] with excellent Hofmann selectivity. LiTMP·LiBr has been employed for the highly selective formation of (*E*)-lithium enolates.[105,106] LiTMP/TMEDA is effective for the diastereoselective Carroll rearrangement.[107] LiTMP is prepared as an ethereal solution by reaction of tetramethylpiperidine with *n*-butyllithium,[108] or with *n*-butyl chloride and lithium metal under ultrasound conditions.[51]

1.2.2.6 Lithium bis(trimethylsilyl)amide or lithium hexamethyldisilazide (LiHMDS; $pK_a = 29.5$)

LiHMDS is a strong nonnucleophilic base[79] showing higher stability than LDA. It has been advantageously used for the regioselective deprotonation of aromatic methyl ketones,[109] for the generation of kinetic enolates,[110] and for the stereospecific formation of ester enolates.[111] It has been applied in the Darzens condensation,[112] ester enolate Claisen rearrangement,[113] and intramolecular double Michael addition.[114] LiHMDS undergoes reaction with allylic chlorides,[115] propargyl bromide,[116] and aldehydes[117,118] to give synthetically useful products. Its reaction with β-bromo ketones provides access to β-ketosilanes.[119] Commercially available as a solid or as solutions in hexanes, cyclohexane, or THF, LiHMDS is also conveniently prepared *in situ* by treating hexamethyldisilazane with *n*-butyllithium.[120]

1.2.2.7 Miscellaneous lithium amides

Lithium cyclohexylisopropylamide (LCI) is superior to LDA for the directed *ortho*-metallation reaction of thiosalicylamides and for benzyne generation from halobenzenes.[121] LiN(Bu^t)SiMePh$_2$ exhibits greater selectivity compared with LDA and LiHMDS for kinetic enolate generation.[122] Lithium *t*-octyl-*t*-butylamide has been used for the selective formation of kinetic enolates,[123] for example, in the total synthesis of aphidicolin.[124] Lithium morpholide and *N*-lithio-*N*,*N'*,*N'*-trimethylethylenediamine have been used for the latent formation of carbinol amine oxides for directed *ortho*-metallation chemistry.[125] Lithium *N*-benzyltrimethylsilylamide (LSA) adds in 1,4 fashion to α,β-unsaturated esters and has been used for the synthesis of amino acid analogues and lactones.[126,127] *N*-Lithioethylenediamine (LEDA) promotes the isomerization of primary allylic alcohols to aldehydes.[128] Chiral lithium L-α,α'-dimethylbenzylamide has been shown to give moderate enantioselectivity in the deprotonation of cyclohexene oxide.[129] A number of homochiral lithium amides as either bases in asymmetric deprotonations or as noncovalent bound chiral auxiliaries in aldol reactions have been developed.[130-3] Lithium 1-(dimethylamino)-naphthalide (LDMAN) has been used for Peterson alkenations[134] and selective reductive lithiations of thioethers.[135]

1.3 sp^3 CARBANIONS

1.3.1 Unstabilized sp^3 Carbanions

1.3.1.1 Lithium–proton exchange

Metal–proton exchange of normal, unactivated alkanes by strong bases usually requires drastic conditions. In contrast, strained carbocyclic systems bear C–H bonds of increased s-character and therefore exhibit enhanced acidity, especially in the proximity of base-coordinating groups, resulting in deprotonation by a process similar to the DoM reaction (see Section 1.4). For example, metallation of cyclopropane requires the use of superbase,[136] whereas an amide-substituted derivative undergoes smooth lithiation with LiTMP.[137] Similarly, the alkoxy group promotes lithiation of the cyclopropane derivative (**1**) to form the monoalkylated compound as the major product (Equation (1)).[138]

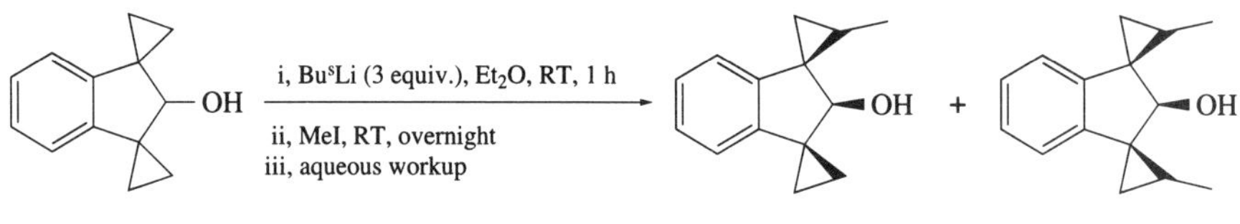

The same principle applies for the lithiation of highly strained cubanes. The cubane amide (**2**), undergoes metallation upon treatment with LiTMP, thus allowing entry to otherwise inaccessible substituted cubane derivatives (Equation (2)).[139]

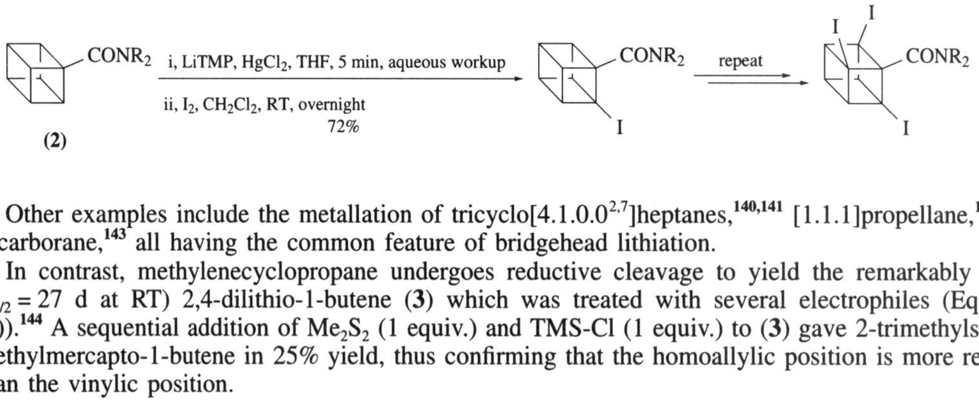

Other examples include the metallation of tricyclo[4.1.0.0^{2,7}]heptanes,[140,141] [1.1.1]propellane,[142] and o-carborane,[143] all having the common feature of bridgehead lithiation.

In contrast, methylenecyclopropane undergoes reductive cleavage to yield the remarkably stable ($t_{1/2}$ = 27 d at RT) 2,4-dilithio-1-butene (**3**) which was treated with several electrophiles (Equation (3)).[144] A sequential addition of Me$_2$S$_2$ (1 equiv.) and TMS-Cl (1 equiv.) to (**3**) gave 2-trimethylsilyl-4-methylmercapto-1-butene in 25% yield, thus confirming that the homoallylic position is more reactive than the vinylic position.

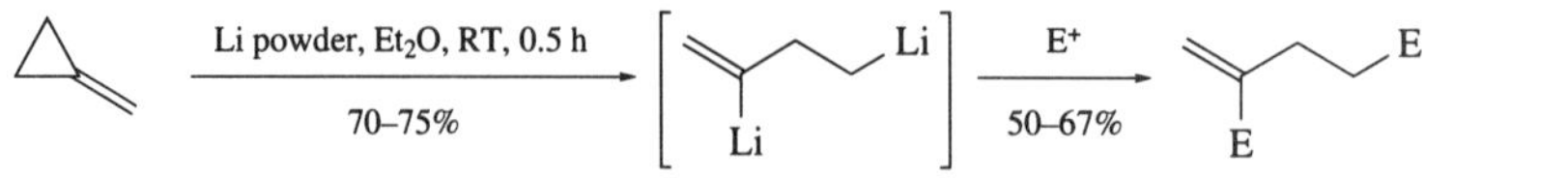

1.3.1.2 Lithium–halogen exchange

(i) Lithium metal

Chloroalkanes are suitable for lithium–halogen exchange using lithium metal. For example, the preparation of halide-free methyllithium[55] and 2-bornyllithium[145] proceeds in good yield by treatment of the corresponding alkyl chloride with lithium metal. Reaction conditions at room temperature or above cause decomposition of ethereal solvents and appear to limit this methodology to the use of hydrocarbon solvents or low temperature. However, the addition of a magnesium alkoxide prevents the reaction of organolithiums with ethereal solvents at room temperature.[75] With cyclohexyl chloride, in the absence of magnesium alkoxide, 3-cyclohexylpropanoic acid is formed in 60% yield, presumably by reaction of ethene, a decomposition product of THF, with cyclohexyllithium and subsequent CO_2 quench (Scheme 1).[75]

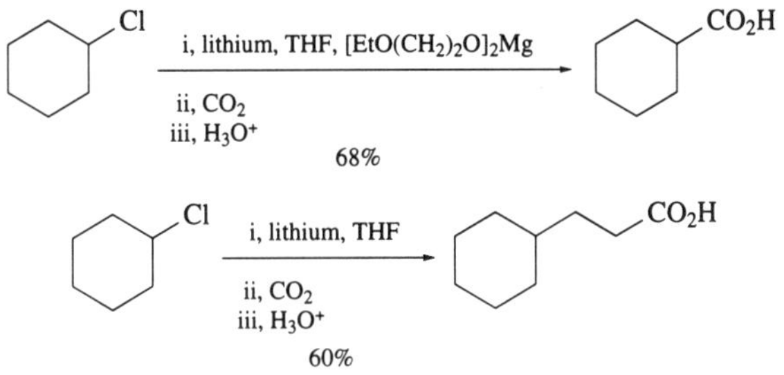

Scheme 1

Freeman and Hutchinson[146] reported the facile and high yielding metallation of several primary, secondary, and tertiary alkyl halides using an excess of lithium radical anion, such as LiDBB[147] or LiDBN, which were found to be superior to lithium naphthalide. Rawson and Meyers made the additional observation that LiDBB-derived alkyllithium species react much faster with electrophilic naphthyloxazolines when compared with commercially available butyllithium or a lithium naphthalide-derived alkyllithium.[148] They proposed that the aggregation properties are altered when LiDBB is employed, which increases the reactivity of organolithium species.

Alcohol or acetal functionalized alkyl chlorides may be lithiated either with lithium metal, accompanied by a catalytic[149] or stoichiometric[150-2] amount of naphthalene, or LiDBB, thus providing high yield access to various functionalized alcohols, aldehydes, hemiacetals, and δ-lactones (Table 1). In the same manner, 3-bromopropanoic acid may be dilithiated.[153] The utility of this method is demonstrated by the synthesis of the brevicomin and frontalin skeleton (**5**) via the naked masked lithium bishomoenolate intermediate (**4**)[154] (Equation (4)).

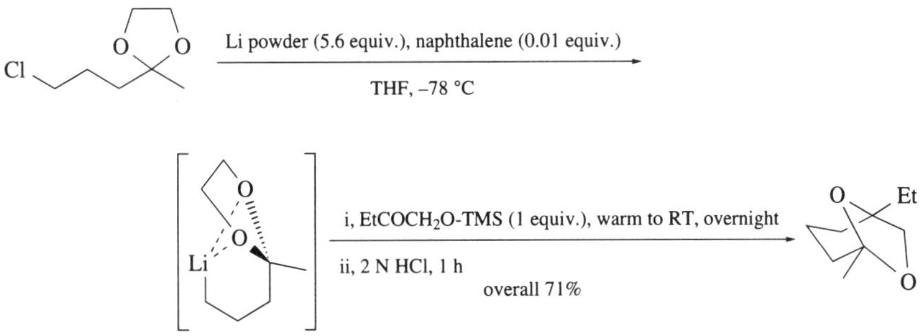

Lithium naphthalide was found to be a very effective reagent for the formation of β- and γ-nitrogen-functionalized and γ- and ε-oxygen-functionalized organolithium compounds (**6**) and (**7**) (Scheme 2), which were coupled with bromides and iodides.[155] Selected examples are presented in Table 2. The mechanism was shown to involve a halogen transfer from the organohalide to the β, γ, or ε position of (**6**) and (**7**).

As described in Section 1.2, sonication[156] promotes the lithiation of alkyl halides[51,157,158] with lithium metal. Particularly effective is the one-pot modified Bouveault reaction (Table 3).

(ii) Organolithium reagents

Until recently, the formation of unstabilized sp^3 alkyllithiums using halogen–lithium exchange with organolithium bases has met with limited success, whereas lithium–halogen exchange has been widely used for the generation of alkenyl- and aryllithiums.[159,160] However, the preparation of alkyllithiums by

Table 1 Naphthalene-promoted lithiations of substituted alkyl chlorides.

Alkyl chloride	Electrophile	Product	Method[a]	Yield (%)	Ref.
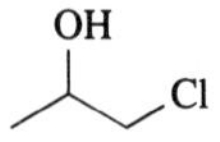 OH-CH(Me)-CH$_2$Cl	Pr^{i}CHO	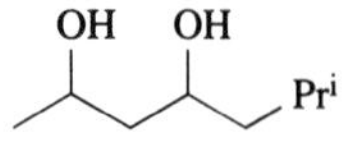 OH-CH(Me)-CH$_2$-CH(OH)-Pr^i	A	68	150
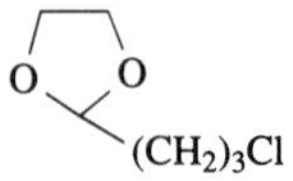 dioxolane-(CH$_2$)$_3$Cl	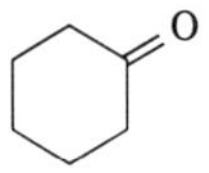 cyclohexanone	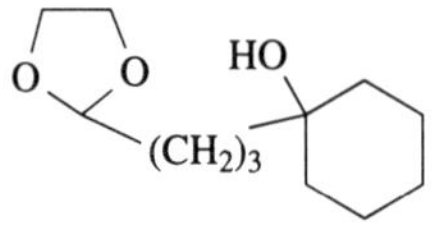 dioxolane-(CH$_2$)$_3$-C(OH)(cyclohexyl)	B	75	150
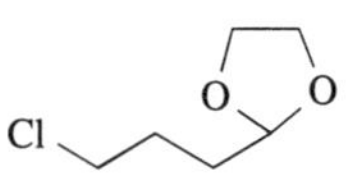 Cl-CH$_2$-dioxolane	R^{1}R^{2}CO	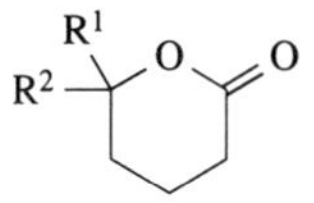 δ-lactone with R^{1}R^2	C	30–62	151
BrCH$_2$CH$_2$CO$_2$H	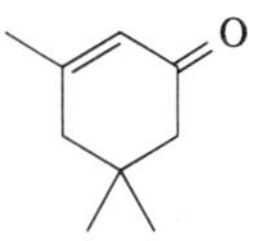 isophorone	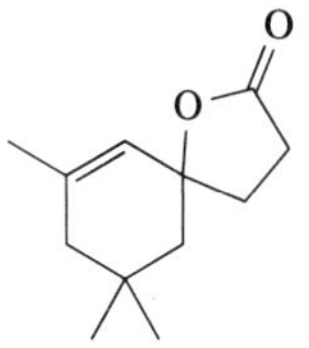 spiro-γ-butyrolactone	D	38	154

[a] A: i, RCl, Bu^{n}Li, THF, –78 °C, 15 min; ii, Li powder (10 equiv.), naphthalene (0.01 equiv.), THF, –78 °C, 1–3 h; iii, E$^+$ (1.2 equiv.), warm to RT, 3 h; iv, hydrolysis. B: i, RCl, Li powder (10 equiv.), naphthalene (0.01 equiv.), THF, –78 °C, 1–3 h; ii, E$^+$ (1.2 equiv.), warm to RT, 3 h; iii, HCl(aq.); hydrolysis. C: i, RCl, Li naphthalide (2.4 equiv.), THF, –78 °C, 7 h; ii, E$^+$ (1.0 equiv), warm to RT, overnight, iii, H$_2$O; iv, 2 N HCl/THF; v, CrO$_3$, H$_2$SO$_4$, Me$_2$CO. D: i, RBr, Bu^{n}Li (1 equiv.), THF, –78 °C, 30 min; ii, Li naphthalide (1.67 equiv.), –78 °C, 1 h; iii, E$^+$, –78 °C, 1 h, then RT, hydrolysis.

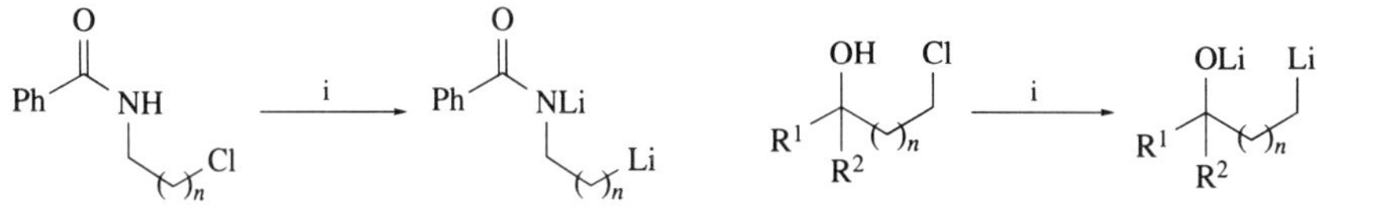

PhC(O)NH-(CH$_2$)$_n$-Cl →^{i} PhC(O)NLi-(CH$_2$)$_n$-Li

(**6**) $n = 1, 2$

R^{1}R^{2}C(OH)-(CH$_2$)$_n$-Cl →^{i} R^{1}R^{2}C(OLi)-(CH$_2$)$_n$-Li

(**7**)
(a) R^1 = Ph, R^2 = H, $n = 1$
(b) R^1 = R^2 = allyl, $n = 1, 3$

i, Bu^{n}Li, THF, –78 °C, 30 min then Li naphthalide (2 equiv.), –78 °C, 6–8 h

Scheme 2

this method is limited by the reversibility of the reaction, resulting in a mixture of two organometallic species.[161] Furthermore, alkyl halides are susceptible to coupling, β-elimination, and α-metallation when treated with an organolithium base.[6,49,162-4]

Two pathways have been suggested for the mechanism of this reaction.[165] One proposed mechanism involves SET from an alkyllithium to an alkyl halide, generating radical intermediates,[166] which the other includes nucleophilic attack of the organolithium on the halide to form an ate complex as either an intermediate or a transition state.[167,168] The two mechanisms may be distinguished by stereochemical[165] and regiochemical[169] outcomes of cyclization reactions. The overall evidence suggests that primary alkyl iodides follow the latter mechanism in contrast to alkyl bromides, which predominately follow the radical-mediated SET mechanism. Generally, alkyllithium bases are not suitable for the metal–halogen exchange reaction with primary alkyl chlorides due to predominant elimination and α-lithiation pathways.[6,49,162-4]

A comparative study of metal–halogen exchange of 1-halobicyclo[1.1.1]pentane with t-butyllithium demonstrates that the chloro derivative defies metal–halogen exchange, whereas the bromo analogue undergoes elimination to form [1.1.1]propellane.[170] In contrast, 1-iodobicyclo[1.1.1]pentane is metallated with t-butyllithium in high yield.[171] Thus, the primary alkyl bromide $ortho$-ester (**8**) was found to metallate smoothly, giving access to 4-hydroxy-substituted esters[172] (Equation (5)), presumably via the stabilized intermediate (**9**), which resembles the acetal-stabilized alkyllithium (**4**).

Table 2 Coupling of organolithiums (**6**) and (**7**) with aryl and vinyl halides.

Organolithium	n	Halide	Product	Yield[a] (%)
(**6**)	1	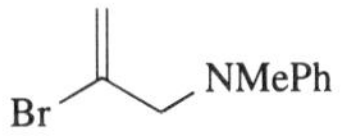 CH₂=C(Br)CH₂NMePh	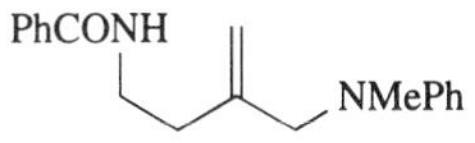 PhCONH(CH₂)₂C(=CH₂)CH₂NMePh	95
(**6**)	1	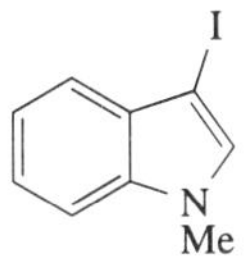 3-iodo-1-methylindole	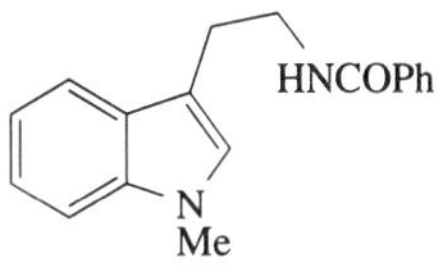 3-(CH₂CH₂NHCOPh)-1-methylindole	85
(**6**)	2	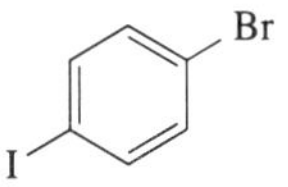 1-bromo-4-iodobenzene	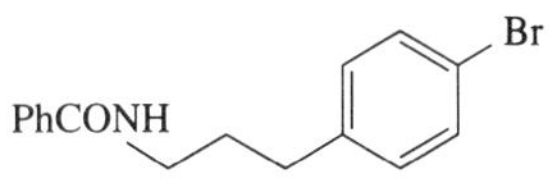 PhCONH(CH₂)₃-C₆H₄-Br	68
(**7b**)	1	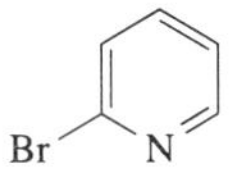 2-bromopyridine	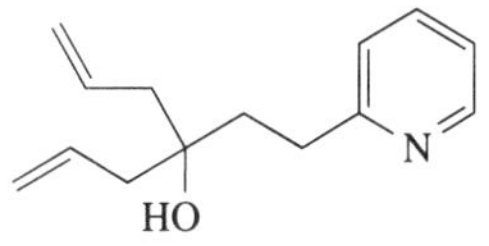 4-allyl-4-hydroxy-1-(2-pyridyl)hept-6-ene (tertiary alcohol with two allyl, one CH₂CH₂-2-pyridyl, OH)	92
(**6**)[b] (**7a**)[b]	2 1	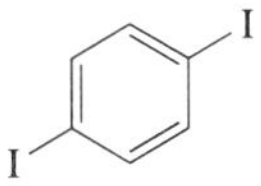 1,4-diiodobenzene	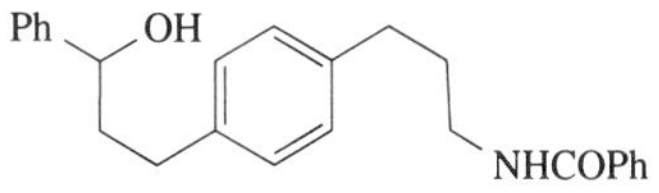 Ph-CH(OH)-CH₂CH₂-C₆H₄-(CH₂)₃NHCOPh	65

[a] Yields based on starting material. [b] 1 equiv. of each (**6**) and (**7a**) was subsequently added to 1,4-diiodobenzene.

Table 3 Ultrasound-mediated lithium–halogen exchange/*in situ* electrophile quench.

Alkyl halide[a]	Electrophile	Product	Yield (%)	Ref.
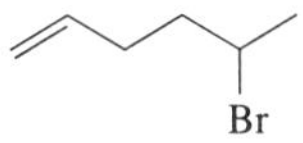 5-bromohex-1-ene	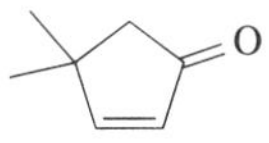 4,4-dimethylcyclopent-2-enone	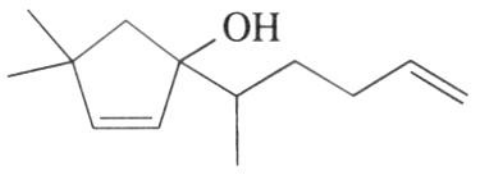 1-(hex-5-en-2-yl)-4,4-dimethylcyclopent-2-en-1-ol	70	158
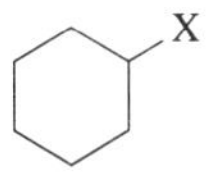 cyclohexyl-X	DMF	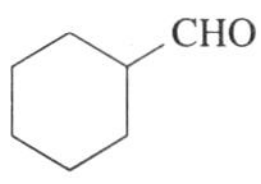 cyclohexanecarbaldehyde	70 (X = Cl); 76 (X = Br)	159
Br(CH₂)ₙBr	DMF	OHC(CH₂)ₙCHO	64–84	159

[a] General method: alkyl halide (1 equiv.), DMF (1 equiv.), Li sand (2.1 equiv. per halogen); 10–20 °C, THF,))) (40 kHz), 5–15 min then hydrolysis.

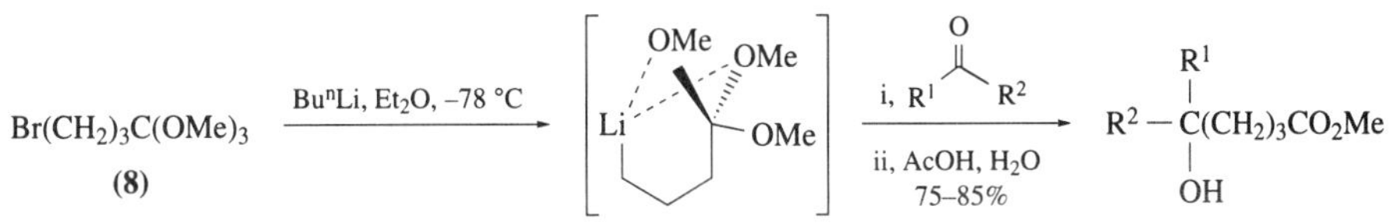

$$Br(CH_2)_3C(OMe)_3 \xrightarrow{Bu^nLi, Et_2O, -78\ °C} [\text{Li-}(9)\text{-C(OMe)}_3] \xrightarrow[\text{ii, AcOH, H}_2\text{O}]{\text{i, R}^1\text{COR}^2} R^2-\underset{OH}{\overset{R^1}{C}}(CH_2)_3CO_2Me \quad (5)$$

$$75\text{–}85\%$$

(**8**) → (**9**)

A similar stabilization effect applies to the lithiation of the racemic 1,1-dibromo-3-(trimethylsilyloxy)alkane (**10**). High diastereoselectivity was attained when the lithium intermediate was trapped with a 2-phenyl-1,3,2-dioxaborolane to form the boronate (**11**). The diol (**12**) was liberated using a standard oxidation–deprotection protocol (Scheme 3).[173]

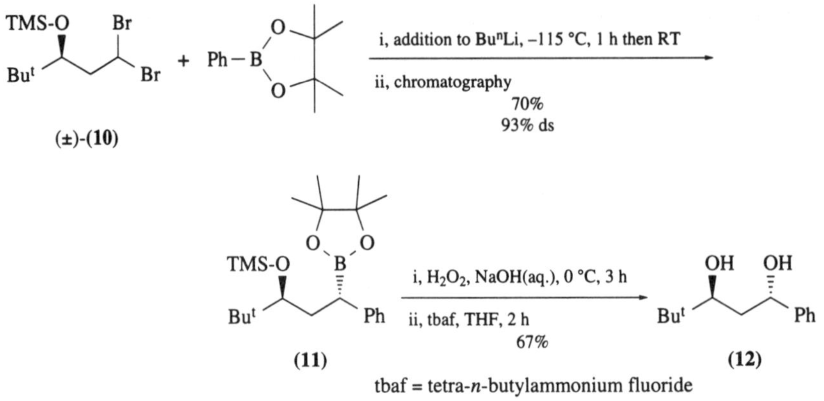

Scheme 3

Synthetic application of the metal–halogen exchange process for primary alkyl iodides requires that the equilibrium is irreversibly shifted towards the desired alkyllithium. For example, the exchange can be controlled by the choice of a solvent from which the alkyllithium product will precipitate, as shown in Equation (6) for the industrial preparation of methyllithium.[54]

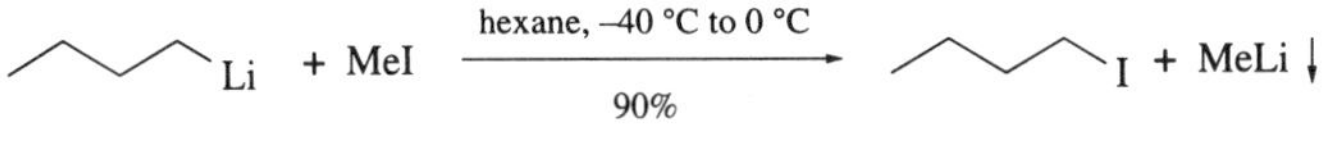

$$\text{BuLi} + \text{MeI} \xrightarrow[90\%]{\text{hexane}, -40\,°\text{C to } 0\,°\text{C}} \text{BuI} + \text{MeLi}\downarrow \qquad (6)$$

Primary alkyllithiums may be generally and conveniently prepared in high yield by reaction of the corresponding primary alkyl iodides with ca. 2 equiv. of *t*-butyllithium, as illustrated in Table 4.[174,175] This reaction is performed in a hydrocarbon/diethyl ether solvent mixture at −78 °C. On warming to room temperature, any residual *t*-butyllithium is removed by the formation of ethene and lithium ethoxide.[176] The lithium–iodine exchange is so rapid that it can occasionally be performed in the presence of reactive electrophilic functional groups such as ketones[177-9] and Michael acceptors.[180-2] When α,ω-diiodoalkanes are employed for the exchange reaction, the outcome depends on the stoichiometric quantity of *t*-butyllithium used. Bailey[183] showed that the reaction of primary diiodoalkanes with 2 equiv. of *t*-butyllithium forms three-, four-, and five-membered carbocycles, provided that the lithium and iodine atoms are separated by less than six carbons. The intermediate α-lithio-ω-iodoalkane normally undergoes smooth intramolecular coupling above −23 °C. Performing this reaction with 4 equiv. of *t*-butyllithium leads to a clean exchange of both iodine atoms. Selected examples for the preparation of alkyllithiums and products derived from intramolecular cyclizations are given in Table 4.

Lithiation of unsaturated iodoalkanes allows access to functionalized carbocycles by intramolecular cyclization. 5-Hexenyllithium derivatives undergo a 5-*exo-trig* cyclization upon warming to room temperature to give cyclopentylmethyl-containing products in a regio- and stereospecific manner.[185] Moreover, the addition of lithiophilic additives such as TMEDA, PMDTA, or THF facilitates the cyclization process, especially when quaternary centers or strained systems are formed.[186] The synthetic value of this methodology is demonstrated by the preparation of (±)-cuparene (**13**), a sterically demanding sesquiterpene bearing two neighboring quaternary centers (Scheme 4).[187]

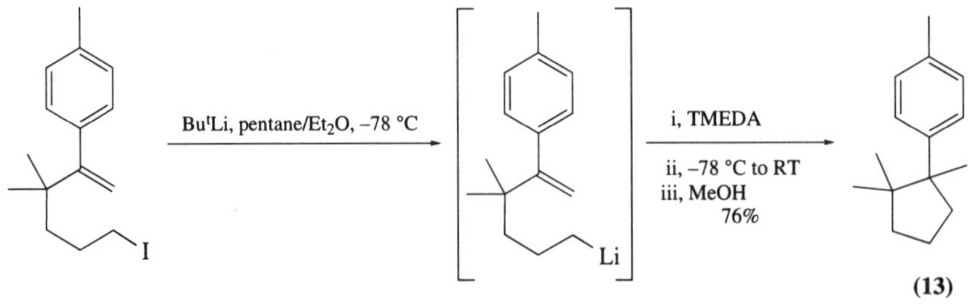

Scheme 4

In contrast, 6-alkyl-substituted 5-hexenyllithiums do not cyclize to form five-membered rings. However, a 6-chloro-substituted 5-hexenyllithium derivative was shown to undergo cyclization via a

Table 4 Preparation of primary alkyllithiums by lithium–iodine exchange using Bu^{t}Li.

Alkyl iodide[a]	Electrophile	Product	Yield (%)	Ref.
Ph(CH$_2$)$_2$I	PhCHO	Ph(CH$_2$)$_2$CH(OH)Ph	91	175
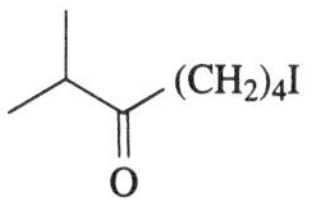 (CH$_3$)$_2$CHC(O)(CH$_2$)$_4$I		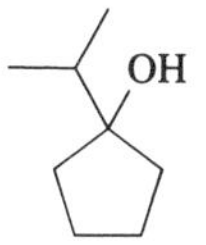 1-(1-hydroxy-2-methylpropyl)cyclopentanol structure (cyclopentane with C(OH)CHMe$_2$)	82	178
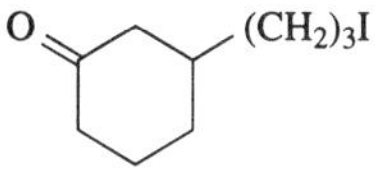 3-(3-iodopropyl)cyclohexanone	H$_2$O	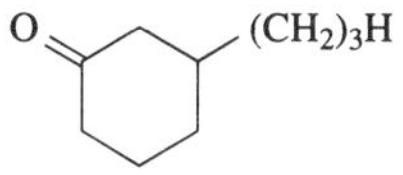 3-propylcyclohexanone	41	179
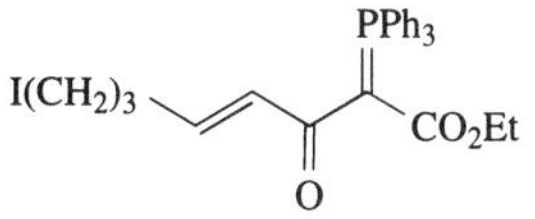 I(CH$_2$)$_3$CH=CHC(O)C(=PPh$_3$)CO$_2$Et	MeI	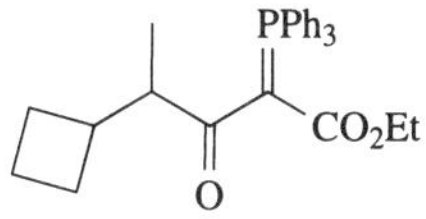 cyclobutyl-CH(Me)C(O)C(=PPh$_3$)CO$_2$Et	73	181
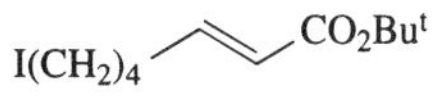 I(CH$_2$)$_4$CH=CHCO$_2$Bu^t		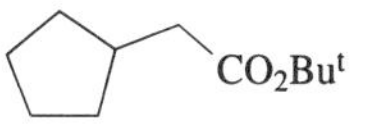 cyclopentyl-CH$_2$CO$_2$Bu^t	82	182
I(CH$_2$)$_5$I	c	Li(CH$_2$)$_5$Li	80[b]	176
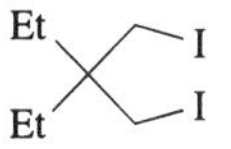 Et$_2$C(CH$_2$I)$_2$		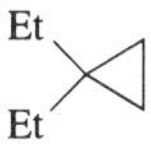 1,1-diethylcyclopropane	93	184
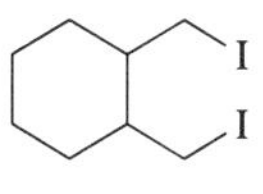 cis-1,2-bis(iodomethyl)cyclohexane		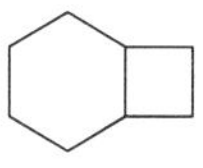 bicyclo[4.2.0]octane	97	184
I(CH$_2$)$_5$I		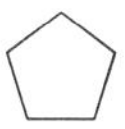 cyclopentane	99	184

[a] The primary alkyl iodides are readily prepared from chlorides, bromides, tosylates, or mesylates using the Finkelstein reaction.[185] [b] GLC yield of bis(TMS) adduct. [c] 4 equiv. of Bu^{t}Li were used.

carbenoid intermediate to afford a mixture of methylenecyclopentane and bicyclo[3.1.0]hexane.[188] 6-TMS- or 6-phenyl-substituted 5-hexenyllithiums undergo smooth cyclization and the resulting stabilized anion may be trapped with electrophiles to give cyclopentanes in high yields (Equation (7)).[185]

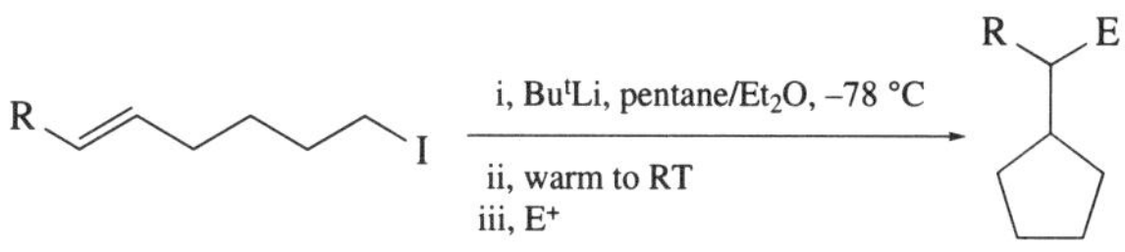

$$R-CH=CH-(CH_2)_4-I \xrightarrow[\text{ii, warm to RT}]{\text{i, Bu}^t\text{Li, pentane/Et}_2\text{O, }-78\,°C} \text{cyclopentyl-CH(R)(E)} \quad (7)$$
iii, E$^+$

R = Ph 72–95%, (E:Z) = 75:25
R = TMS 72–84%, (E) = 100
E = H, D, CO$_2$H, CHO, (Me)$_2$C(OH), MeCH(OH), TMS

The ease of this reaction allows rational design of tandem cyclization routes to polycarbocyclic systems such as [4.3.3]propellanes (**14**),[189] stereoisomerically pure *endo*-2-substituted bicyclo[2.2.1]heptanes (**15**) and *trans*-fused bicyclo[3.3.0]octanes (**16**),[190] (Scheme 5).

The employment of LDA for the lithium–iodine exchange of a 6-iodo-1-hexene derivative is unsatisfactory since radical, carbanion, and carbene pathways occur simultaneously to yield a mixture of five products.[169]

Cyclization of 3,4-pentadien-1-yllithium anions derived from allenes may also be effected. The acyclic anion may be trapped at −78 °C with electrophiles; however, on warming to room temperature, either cyclization or 1,3 shift of the lithium species occurs, and the ratio of the cyclized and rearranged products depends on the substrate, electrophile, and additive used.[191] Similarly, the corresponding anion

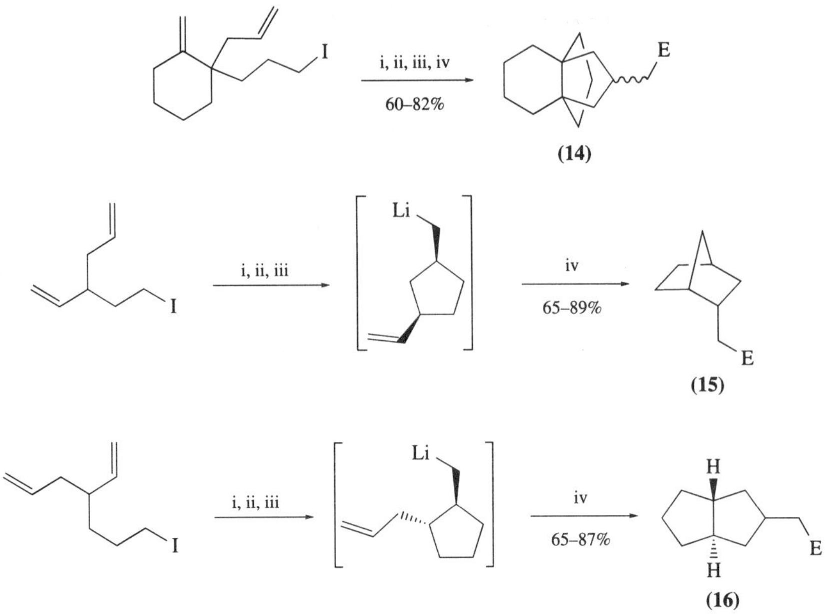

Scheme 5

i, Bu^{t}Li, hexane/Et$_2$O, −78 °C; ii, TMEDA, −78 °C; iii, warm to RT;
iv, E$^+$, (E = H, D, CO$_2$H, CH$_2$OH, CHO, Bu^{t}CH(OH), TMS, SnBu$^n_3$, I)

of 5,6-heptadien-1-yl iodides may also be trapped at −100 °C with electrophiles. On warming to room temperature and subsequent electrophile quench, a mixture of products originating from anionic shifts and addition reactions was obtained.[192]

Alkynic alkyllithiums also undergo highly stereoselective cyclizations to form four-, five-, or six-membered rings bearing an exocyclic lithiomethylidene moiety,[193,194] for example, the synthesis of a cyclopentylidene derivative (Scheme 6).[184] When treated with *t*-butyllithium, 4-oxa-5-hexenyliodines also undergo the familiar 5-*exo*-trig cyclization to form tetrahydrofuran derivatives, which, however, then spontaneously follow a [1,4]-Wittig rearrangement pathway to form 4-alken-1-ols (Equation (8)).[195]

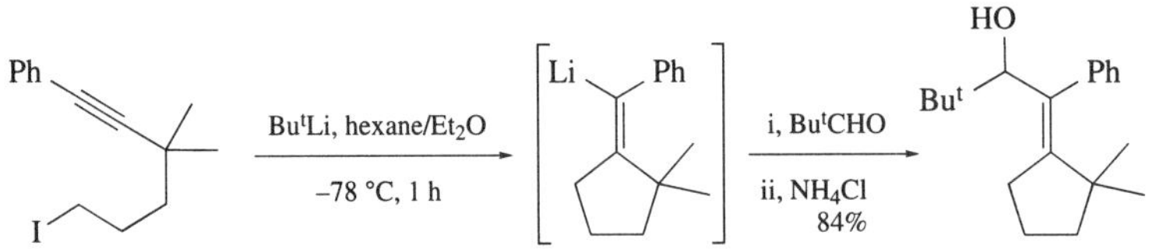

Scheme 6

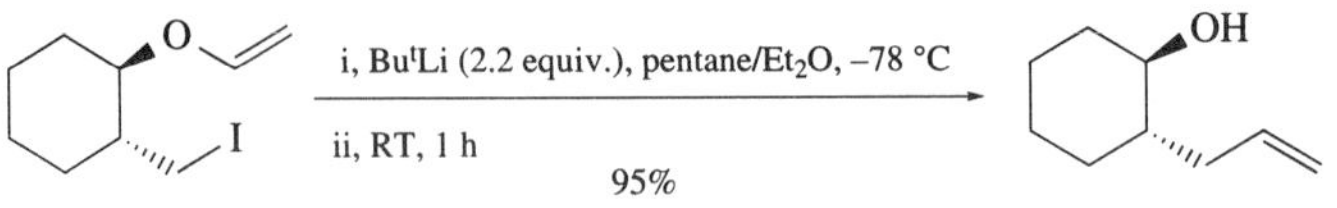

1.3.1.3 Reductive cleavage of heterocyclic alkanes

Epoxides are reductively cleaved by means of LiDBB and other radical anion salts of aromatic hydrocarbons[147,196] to form synthetically useful β-lithioalkoxides, which were previously available only via organomercury compounds or β-chlorohydrins.[197] Reaction with ketones and aldehydes gives 1,3-diols which undergo cyclization upon treatment with methanesulfonyl chloride (MsCl) or acid to give spirocycles (Scheme 15).[198]

Similarly, oxetanes are readily cleaved to form γ-lithioalkoxides (Scheme 8). Subsequent treatment with electrophiles gives rise to substituted 1,4-diols, lactones, and tetrahydrofurans.[199] Reactions of the dianion (**17**) with lactones provides a short and convenient synthesis of pheromones (**19**) and (**20**) Table

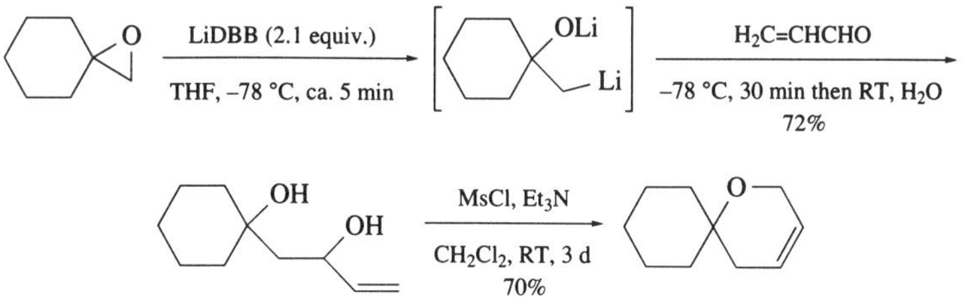

Scheme 7

5, most efficiently, however, with the cerium derivative (**18**).[200] Treatment of dianions (**17**) with $Cr(CO)_6$ results in the formation of pentacarbonyl(2-oxacyclopentylidene)chromium complexes.[201] The synthetic utility of this method is illustrated in Table 5.

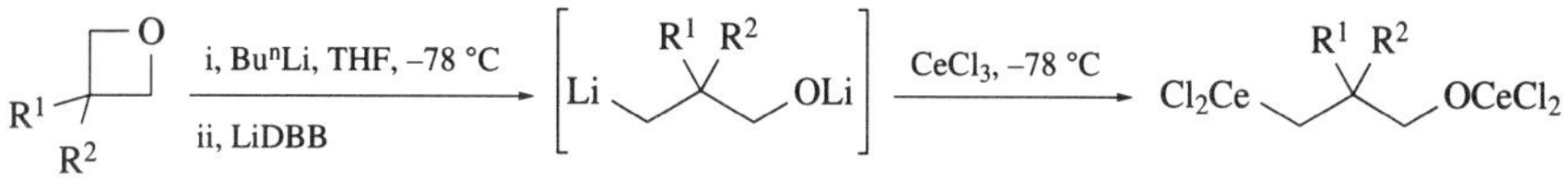

Scheme 8

Table 5 Reaction of γ-metal alkoxides (**17**) and (**18**) with electrophiles.

Metal alkoxide	Electrophile[a]	Product	Yield (%)	Ref.
(**17**) $R^1 = R^2 = H$	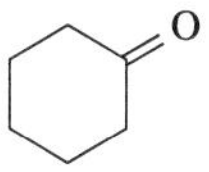 cyclohexanone	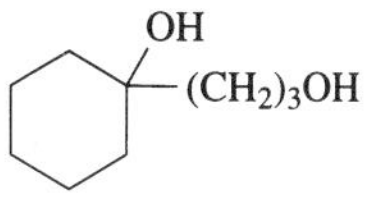 1-hydroxy-1-(3-hydroxypropyl)cyclohexane	56	200
(**17**) $R^1 = R^2 = Me$	CO_2	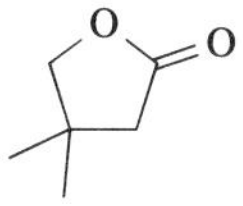 β,β-dimethyl-γ-butyrolactone	53	200
(**17**) $R^1 = R^2 = Me$	$MeCOCO_2Et$	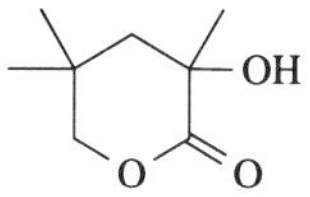 hydroxy-dimethyl-δ-valerolactone	42	200
(**17**) $R^1 = R^2 = Me$	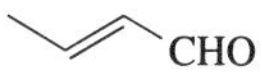 CH₃CH=CHCHO	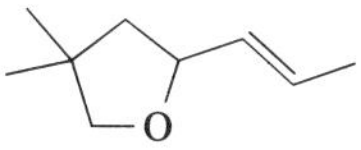 2-propenyl-dimethyl-tetrahydrofuran	70	200
(**17**) $R^1 = R^2 = H$ (**18**) $R^1 = R^2 = H$	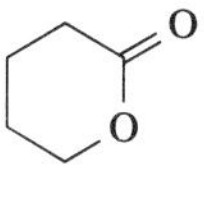 δ-valerolactone	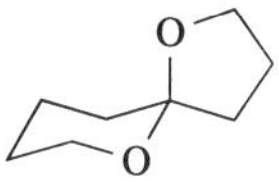 (**19**)	22 69	201
(**18**) $R^1 = R^2 = H$	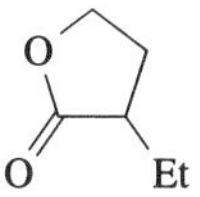 α-ethyl-γ-butyrolactone	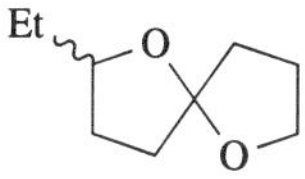 (**20**)	60	201
(**17**) $R^1 = R^2 = Me$	i, $Cr(CO)_6$; ii, $Me_3O^+ BF_4^-$	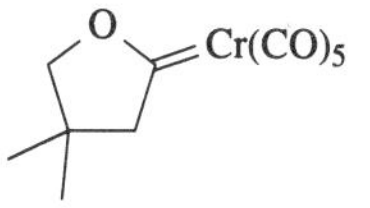 dimethyl-tetrahydrofuranylidene-Cr(CO)₅	53	202

[a] Followed by acid hydrolysis.

Tetrahydrofurans are also cleaved in the presence of $BF_3 \cdot Et_2O$ with equimolar amounts of LiDBB[202] or lithium powder in the presence of catalytic amounts of naphthalene[203] to form δ-lithiobutoxides. The resulting dianionic species, when treated with ketones and aldehydes, give 1,5-diols and, with benzaldehydes and cyclohexenones, yield tetrahydropyrans. For example, treatment of this dianion with lactones offers a very simple entry into the major (**21**) and minor (**22**) components of the olive fruit fly pheromone (Scheme 9).[202]

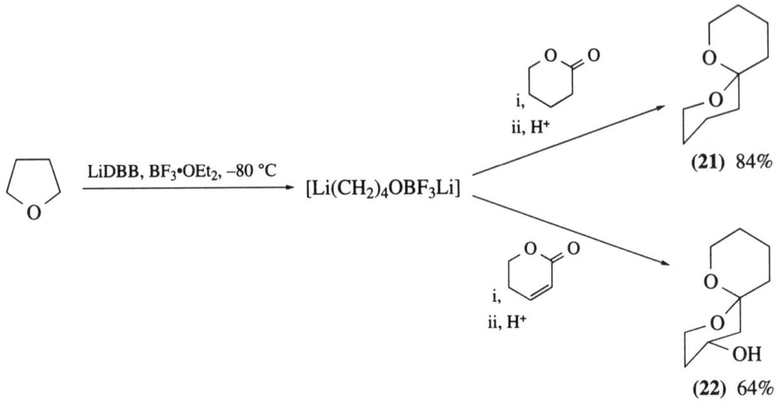

Scheme 9

The reductive opening of aziridines with excess lithium powder in the presence of catalytic naphthalene leads to the corresponding *N*-lithio-2-lithioethylamine (**23**), which, upon treatment with electrophiles, furnishes β-substituted secondary amines (**24**).[204] The reaction may be applied to the chiral aziridine (**25**), yielding the corresponding chiral amine (**26**) as a single diastereoisomer in good yield (Scheme 10).[204]

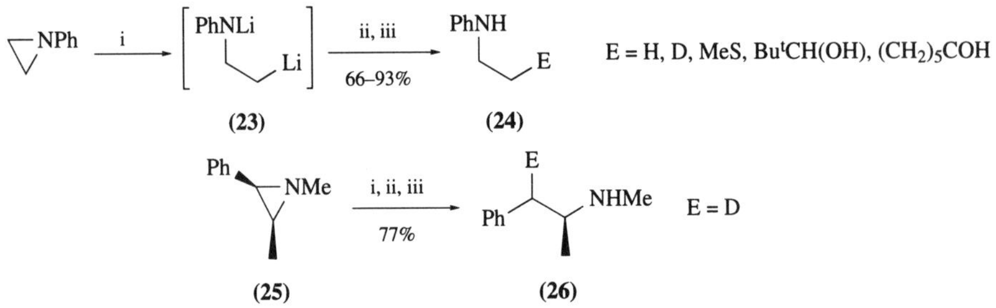

i, Li powder (5 equiv.), naphthalene (0.05 equiv.), THF, −78 °C, 6 h;
ii, E$^+$ (1.1 equiv.), THF, −78 °C to RT, 3 h; iii, H$_2$O

Scheme 10

The same principle applies for the LiDBB-mediated cleavage of azetidines to give *N*-lithio-3-lithiopropylamines, affording γ-substituted secondary amines after treatment with electrophiles.[205] In both aziridine and azetidine examples, a phenyl substituent is required either on the nitrogen or on the α-position, presumably for the stabilization of radical intermediates.

1.3.1.4 Miscellaneous exchange reactions

The lithium–tin exchange reaction has also been used for the formation of sp^3-hybridized lithio carbanions. For example, Broka *et al.* developed an elegant heteroalkenyl version of this technique for the stereospecific synthesis of substituted tetrahydrofurans (Equation (9)).[206] The facile loss of alkoxide promotes the ring formation process via intramolecular $S_N{}'$ cyclization of the generated α-oxa-5-hexenyllithium. The corresponding alkyl-substituted alkene precursor resisted cyclization, presumably due to the formation of a relatively unstable secondary carbanion.[185,206]

The metal–metal exchange tactic was used for the selective formation of stabilized β-lithio enolates such as (**27**). This *in situ* ketone protection method allows access to β-substituted ketones and enol ethers (Scheme 11).[207]

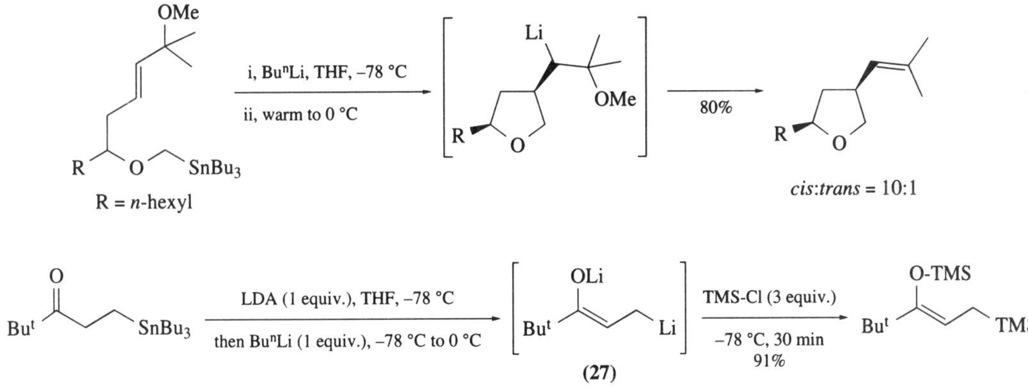

Scheme 11

Lithio sp^3 carbanions may also be formed by a lithium–selenium exchange process. Scheme 12 illustrates a cyclization–methanol quench sequence leading to a stereoisomeric mixture of cyclopentane derivatives.[208]

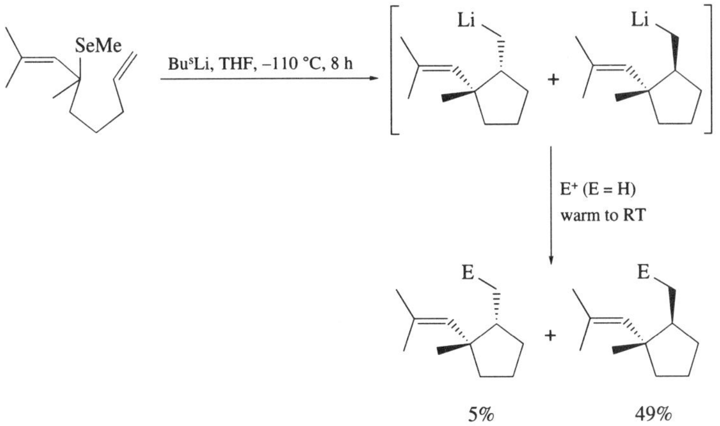

Scheme 12

Other entries to organolithiums include lithium–tellurium exchange[209] and the reaction of phenyl sulfides with excess lithium powder in the presence of a catalytic amount of naphthalene,[149] LiDBB,[210–13] or LDMAN.[214] Alkyllithium reagents may also be generated by treatment of dialkyl sulfates[52] with lithium powder and a catalytic amount of naphthalene and by reductive decyanation of nitriles[215] with lithium and a catalytic amount of 4,4'-di-t-butylbiphenyl. These methods are summarized in Table 6.

1.3.2 α-Nitrogen sp^3 Carbanions

Since the 1980s, considerable work has been devoted to the elaboration of amines via α-lithioamine equivalents (Scheme 13) with increasing emphasis on asymmetric synthetic applications, for example, α-amino acids, β-amino alcohols, and alkaloids.[24,25,216,217] α-Metallation of unactivated tertiary amines may be achieved with a s-butyllithium/potassium t-butoxide mixture (Equation (10)).[218] Attempts to use other alkyllithium components were unsuccessful. Similarly, TMEDA can be metallated at the methyl (Bu^{t}Li, −78 °C to RT, pentane) or the methylene (LICKOR, −78 °C to RT, pentane) sites.[219] The internal coordinating ability of the extra dimethylamino moiety is likely to assist in the metallation process. Indirect methods for α-amino carbanion generation via thio[220] or stannyl precursors[221] are available.

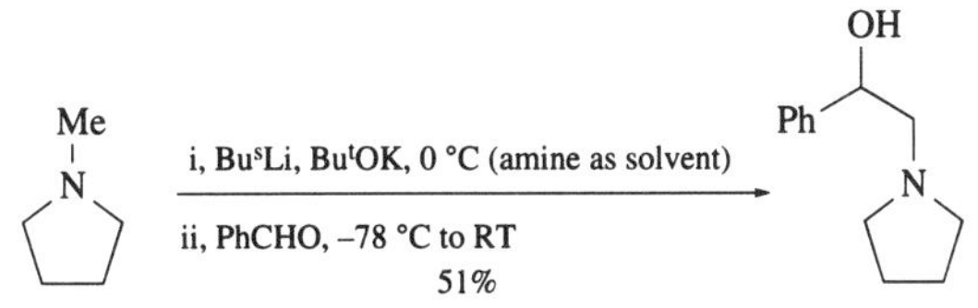

Table 6 Organolithiums by heteroatom exchange and their reactions with electrophiles.

Substrate	Lithium source, E^+	Product	Yield (%)	Ref.
Bu^n_2Te	2 Bu^sLi, PhCHO	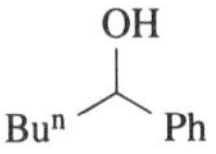 Bu^n-CH(OH)-Ph	86	210
MeSPh	Li, naphthalene[a], 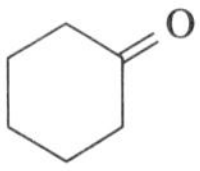 cyclohexanone	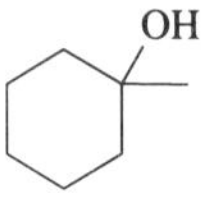 1-methylcyclohexanol	92	150
 bicyclic-SPh	Li, LiDBB, PhCHO	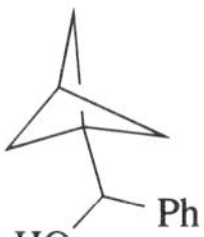 bicyclic-CH(OH)Ph	57	211
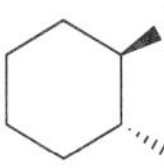 cyclohexyl(OMe)(CH$_2$SPh)	LiDBB, CO_2	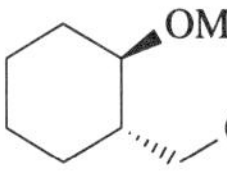 cyclohexyl(OMe)(CH$_2$CO$_2$H)	81	214
$(MeO)_2SO_2$	Li, naphthalene[a], Ph_2CO	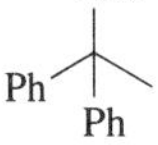 $Ph_2C(OH)Me$	99	53
PhCN	Li, DBB[a], TMS-Cl	PhTMS	63	216

[a] Catalytic amount used.

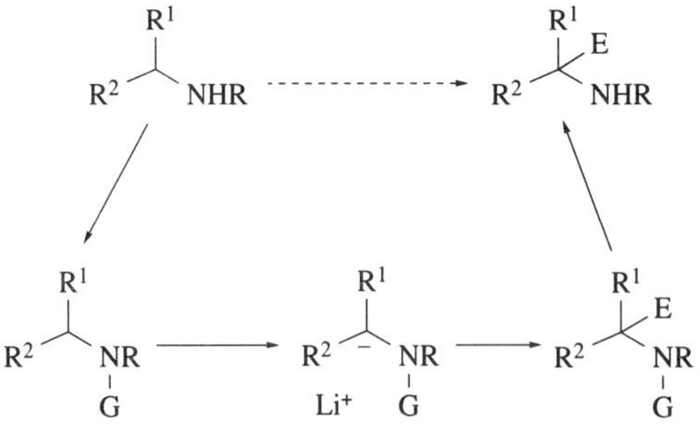

Scheme 13

G = activating group

The use of temporary activating groups on nitrogen in order to facilitate α-nitrogen functionalization has been advantageously exploited in synthetic problems with notable contributions from the laboratories of Beak (amides, urethanes) and Meyers (formamidines).

Unlike nitrosoamine anions, which are resonance stabilized, α-amide, urethane, and formamidine anions are "dipole-stabilized" carbanions.[216,222] Here, the formally sp^3 carbanion is orthogonal to the π-bond of the activating group, permitting a stabilizing interaction between the α-lithium and the carbonyl or imino moiety.[223] The result is differing stereochemistry in subsequent alkylation (Equation (11)).

Investigation of the metallation mechanism in these systems using stopped-flow IR techniques provided evidence for premetallation coordination complexes.[224] In the generation of α-lithiated cyclic formamidines, SET processes sometimes compete; this may be overcome by the addition of HMPA or transmetallation to the cuprate.[225] Substrates which possess allyl or benzyl groups β-positioned to nitrogen do not suffer these complications.

Although N-nitrosamines have been comprehensively studied by Seebach and Saavedra,[226] their synthetic utility is compromised by carcinogenic and mutagenic properties. Equation (12) illustrates the use of N-nitroso-2,2-dimethyloxazolidine as a masked 2-lithio-2-aminoethanol.[226]

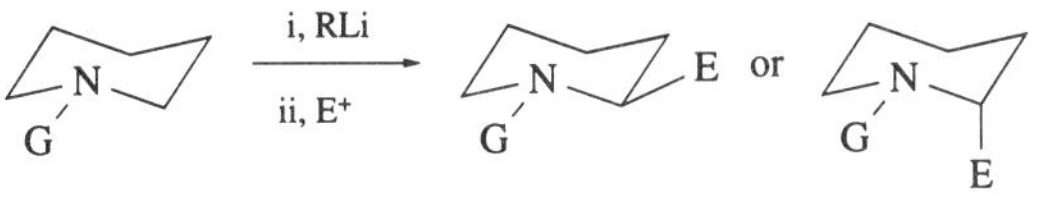

i, RLi
ii, E+
N
G
E
or

(11)

(28) **(29)**

N-G = amide (e.g., pivaloyl), formamidine (e.g. N-CH=NBu^t): **(28)**
N-G = nitrosamine: **(29)**

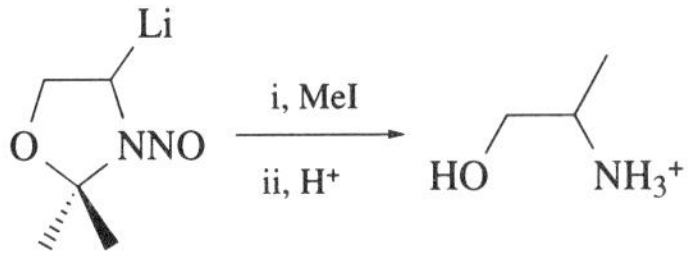

Li
O
NNO
i, MeI
ii, H+
HO
NH3+

(12)

Due to lack of reliable and mild methods for subsequent cleavage, acyl groups such as pivaloyl[227,228] (Equation (13)) and 2,2-diethylbutanoyl[229] (Equation (14)) have been superseded by urethane and formamidine activators.

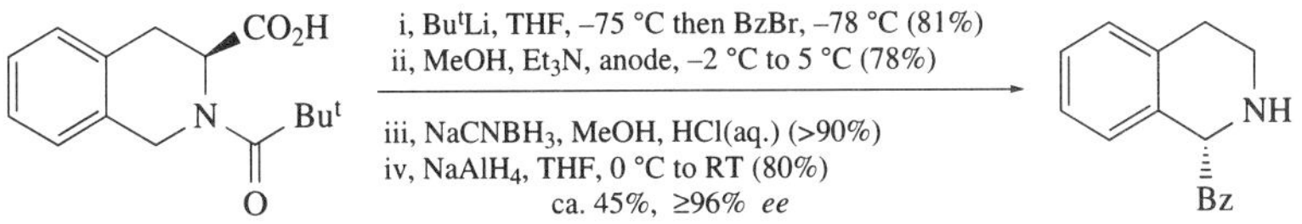

CO2H
N
But
O
i, ButLi, THF, –75 °C then BzBr, –78 °C (81%)
ii, MeOH, Et3N, anode, –2 °C to 5 °C (78%)
iii, NaCNBH3, MeOH, HCl(aq.) (>90%)
iv, NaAlH4, THF, 0 °C to RT (80%)
ca. 45%, ≥96% ee
NH
Bz

(13)

N
Et
Et
Et
O
i, BusLi, TMEDA, Et2O, –78 °C to RT
then EtCHO, –78 °C to RT (65%, 1:1 mixture)
ii, conc. HCl, MeOH, reflux (99%)
iii, KOBut, diglyme, H2O, reflux (95%)
61%
N
H H
Et
OH
threo isomer

(14)

α,α'-Dialkylation of lithiated formamidine and *t*-BOC derivatives of piperidines (Equation (15))[230] lead to *cis* and *trans* products, respectively. The steric bulk of the *t*-BOC function precludes an equatorial lithio species due to interaction with the existing α-substituent, introduced by the first alkylation step.

H
N
Li
N
But
But
O
Li
N
ButO
But

(15)

Selected examples of α,α'-dialkylated piperidines, including natural products, obtained via α-lithiation are depicted in Equation (16) and Scheme 14.[223,225]

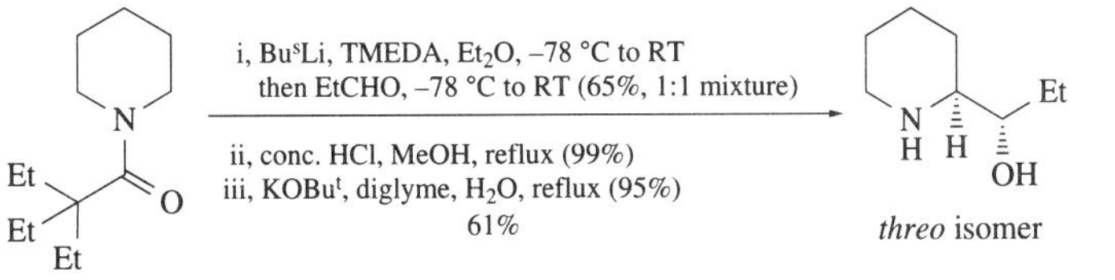

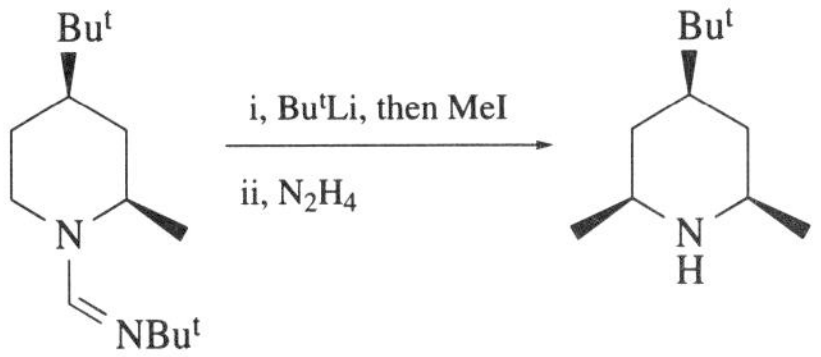

But
N
NBut
i, ButLi, then MeI
ii, N2H4
But
N
H

(16)

In contrast with lithio *N*-formamidinopiperidines, which require a bulky 4-substituent to promote conformational stability, and hence, diastereoselectivity, the corresponding α-lithiopyrrolidines appear conformationally rigid in that they lead to high *threo* selectivity in reactions with aromatic aldehydes (Equation (17).[231]

As part of a continuing exploration of the synthetic utility of *N*-*t*-BOC heterocycles, Beak and Yum reported the use of *N*-*t*-BOC-2-methyltetrahydro-1,3-oxazine as a 1-lithio-3-hydroxy-1-propylamine equivalent (Equation (18)).[232]

Chiral base-induced asymmetric deprotonation is a recent rapidly growing field.[233] Thus, in the presence of sparteine, a naturally occurring and commercially available chiral diamine ligand, lithiated

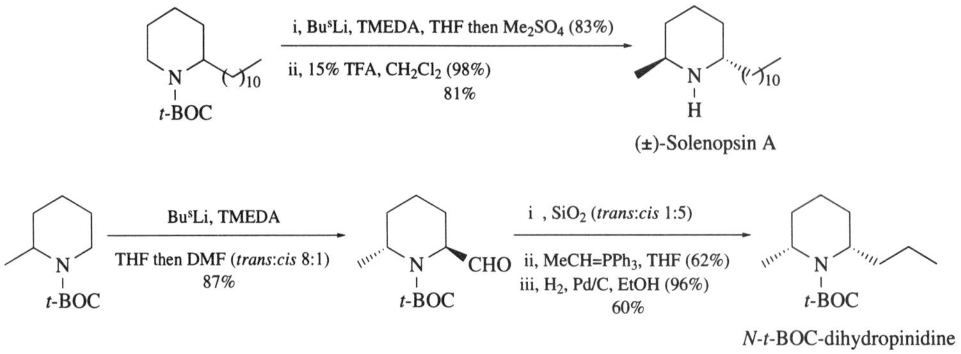

Scheme 14

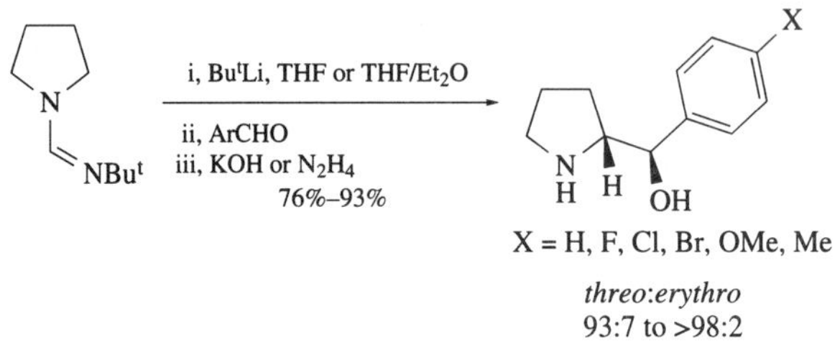

N-t-BOC-pyrrolidine has been shown to undergo alkylation with a variety of electrophiles with greater than 90% ee (Equation (19)).[234] Diastereoselective amplification in a sequential lithiation–substitution sequence was also demonstrated.

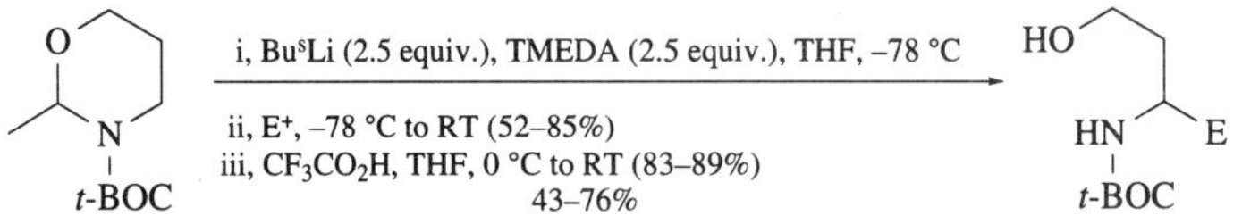

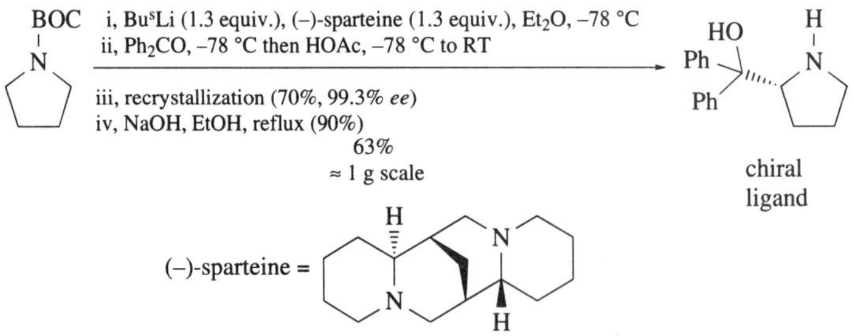

Mainly owing to the work of Meyers, chiral formamidines[235] are emerging as highly useful auxiliaries, especially in the synthesis of alkaloids (Equation (20) and Scheme 15).[236,237] Thus, asymmetric alkylation of (**30**) followed by N-formamidine to methyl interconversion and oxidative coupling gave the isopavine (**31**) in 46% overall yield.

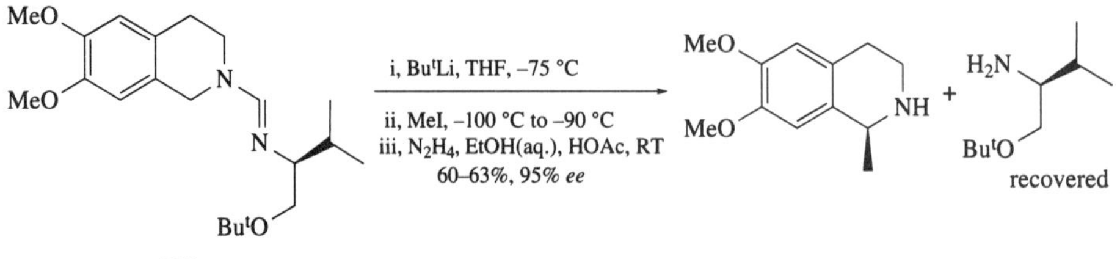

Chiral formamidines have been used in the context of an intramolecular benzyne trapping reaction.[238]

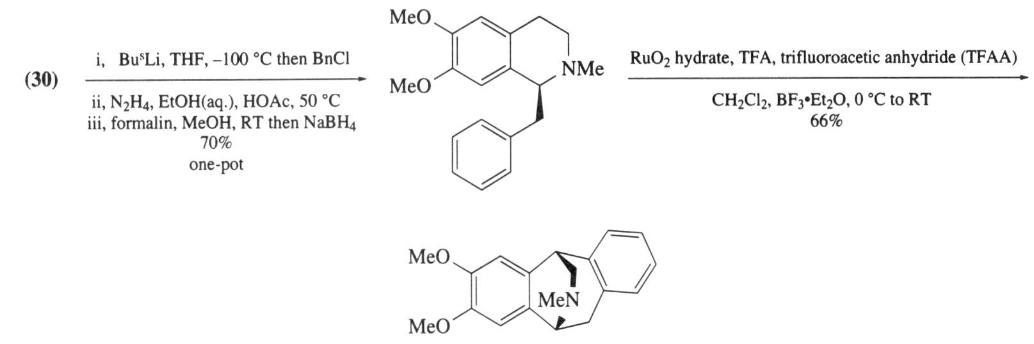

Scheme 15

The less rigid azepine appears to have no negative effect on the degree of asymmetric induction, allowing access to berberine-like systems (Equation (21)).[239]

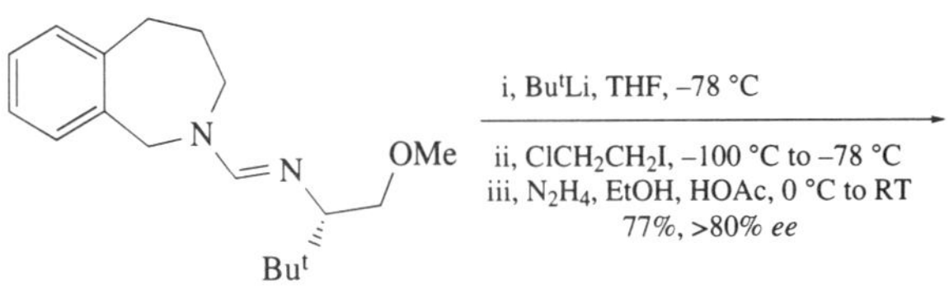

(−)-Alloyohimbane was constructed in seven steps from the β-carboline formamidine (**32**) in 31% overall yield (Scheme 16).[240] The α-stereocenter generated by asymmetric alkylation (92% ee) was used to control the stereochemical outcome of a subsequent Diels–Alder reaction.

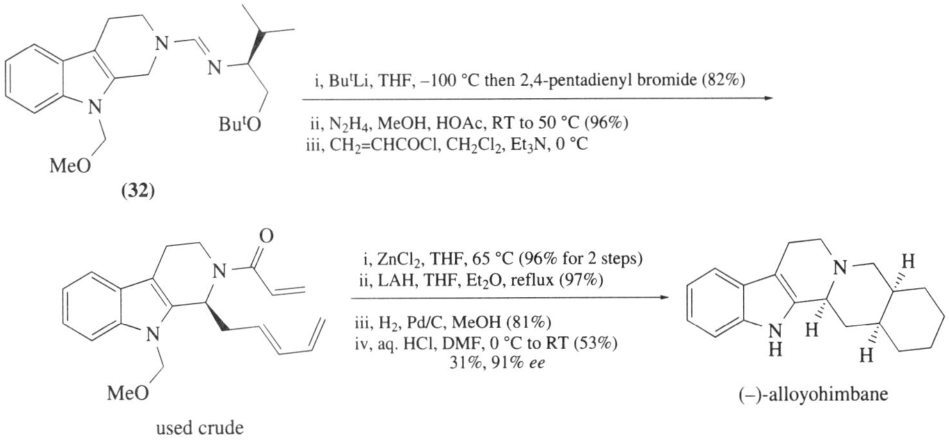

Scheme 16

Chiral quaternary-substituted tetrahydroisoquinolines are also available by a double alkylation strategy (Equation (22)).[241] The less sterically demanding methoxy-containing formamidine auxiliary is required to allow deprotonation at the tertiary center; the valine *t*-butyl ether derived auxiliary (Equation (20) and Scheme 16) did not allow this deprotonation. Generally, the second alkylation is somewhat less efficient than the first. Formamidines derived from *o*-methoxymethylaniline[242] also allow the removal of tertiary protons of 1-substituted tetrahydroisoquinolines, owing to increased α-acidity of the extended π-system. Thus, in general, a combination of ease of introduction, removal, and availability of chiral variants sets formamidines apart from other stabilized α-*N* anions.

Although achiral, an operationally simple procedure for carbon–carbon bond formation at the α-position of secondary amines involves the use of carbon dioxide as the nitrogen protecting–directing group (Equation (23)).[243]

Although not widely tested, *N*-phosphorylamides provide reasonably nucleophilic α-lithio carbanions and may be hydrolysed under relatively mild conditions (Equation (24)).[227]

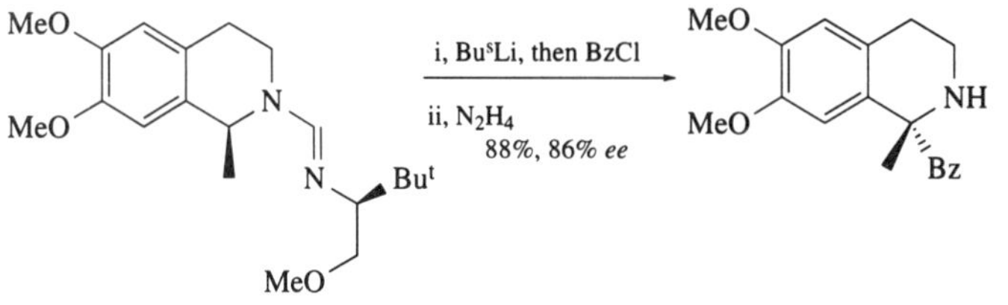

N-*t*-BOC-activated chiral imidazoline (**33**) may be alkylated diastereoselectively at the activated α-position to give *trans*-substituted products with good to excellent stereoselectivity (Scheme 17). A second alkylation followed by hydrolysis provides an efficient route to α,β-diamino acids.[244]

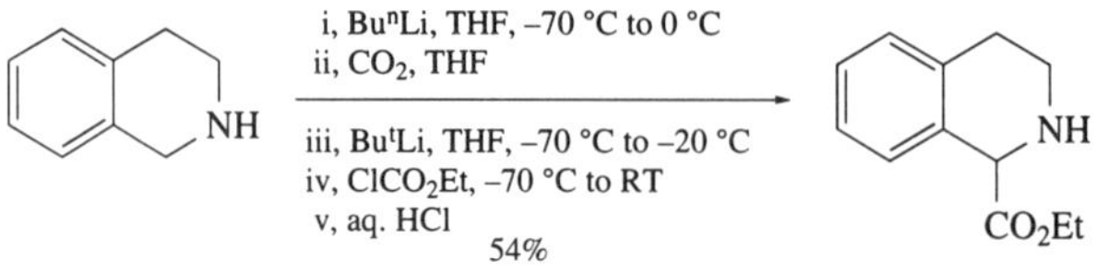

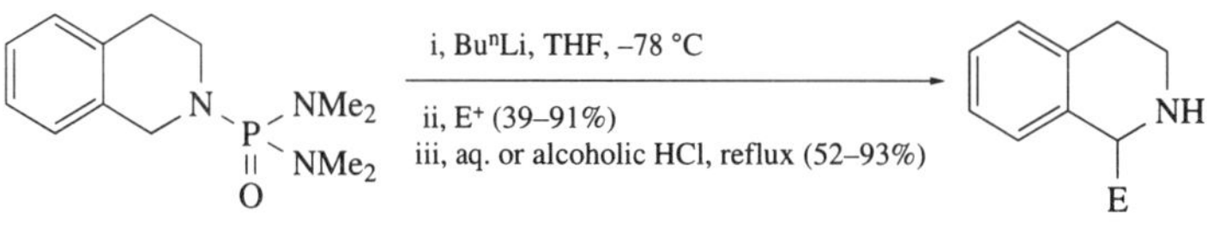

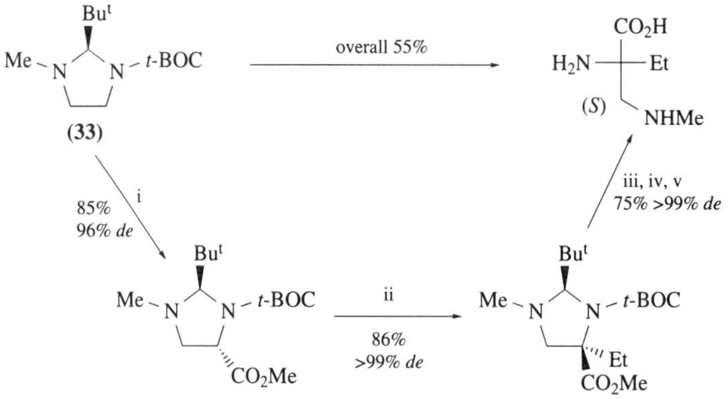

i, Bu^{t}Li, TMEDA, Et$_2$O, −75 °C then CO$_2$/H$_2$O/CH$_2$N$_2$ (85%, >98:2);

ii, Bu^{t}Li, TMEDA, Et$_2$O, −75 °C then EtI (86%, >99% *de*); iii, TFA, CH$_2$Cl$_2$, 4 °C to RT then H$_2$O;

iv, NaOH(aq.), RT; v, cation exchange chromatography then recrystallization (75%, >99% *de*)

Scheme 17

Gawley and co-workers have shown that chiral *N*-oxazolidinoyl and *N*-oxazolinoyl activators permit stereoselective α-*N* alkylation (Scheme 18 and Equation (25)).[245,246] A limited study of *N*-oxazolidinoyl nonbenzylic systems shows 100% stereoselectivity; however, depending on the choice of electrophile, competing radical pathways afford virtually racemic and oxidation products.[247] These results may be contrasted with valinol-derived chiral formamidines lacking additional benzylic or allylic activation, where bidentate chelation complexes the metallating agent and precludes α-metallation.[248]

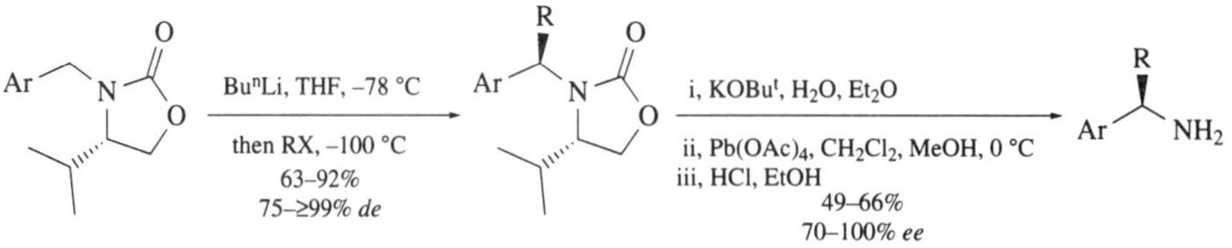

Scheme 18

α-*N*-Organolithiums generated via tin–lithium exchange show reasonable configurational stability and undergo electrophile quench with retention of configuration.[249,250] A stable chelated lithio

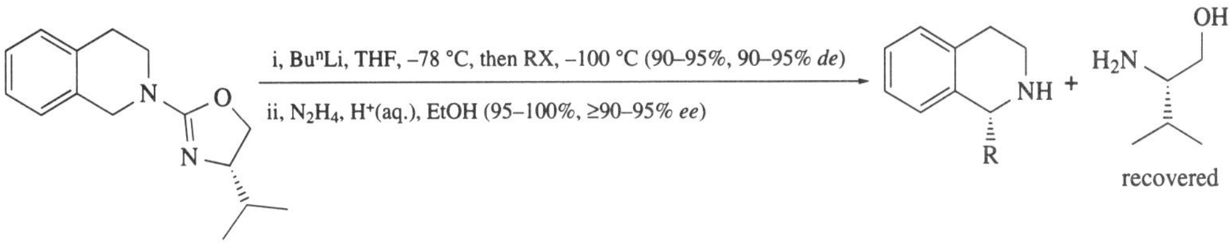

[Equation (25) scheme]

diastereomeric species is formed from the stannyl imidazolidinone (**34**) with retention and quenched with retention (Equation (26)).[249] The epimeric α-stannane (R = Me) gives a 2.4:1 mixture of α and β products after Bu₃SnCl quench, indicating considerable epimerization.

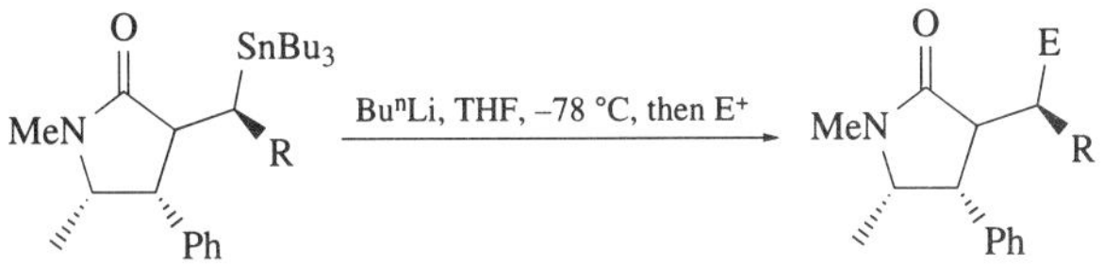

[Equation (26) scheme]

R = Et, E = CO$_2$Et (83%, α:β = 8:92)
R = Me, E = D (63%, α:β = 0:100)

In an observation of considerable mechanistic and synthetic interest, Gawley has demonstrated that unchelated α-amino organolithiums derived from chiral tin precursors are configurationally stable at temperatures as high as –40 °C for up to 45 min in the presence of TMEDA (Equation (27)).[250]

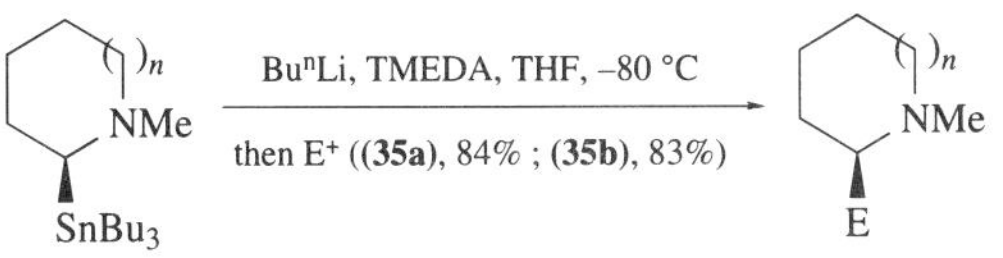

[Equation (27) scheme]

(**35**)
(a) 100% ee, n = 2
(b) 94% ee, n = 1

99% ee, n = 2, E = CO$_2$Et
94% ee, n = 1, E = CO$_2$H

An effective procedure for *in situ* activation of tertiary amines towards α-metallation involving boron trifluoride coordination of nitrogen[251] has been applied to the synthesis of the spirobenzylisoquinoline alkaloid yenhusomidine (Equation (28)). Nonbenzylic and nonallylic α-amino carbanions may also be generated by this method.

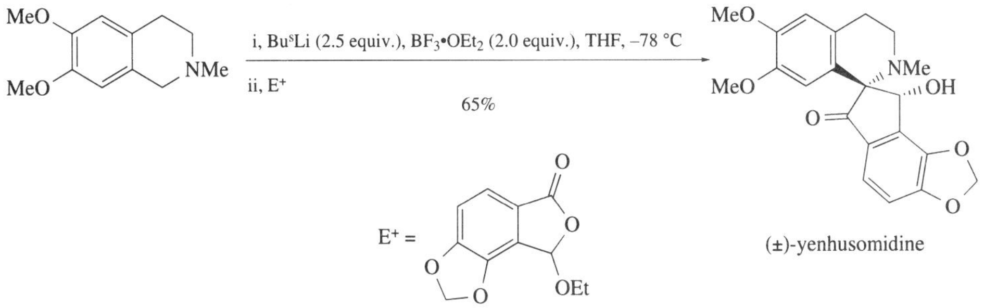

[Equation (28) scheme]

(±)-yenhusomidine

1.3.3 α-Oxygen sp^3 Carbanions

Inductive and base-coordinating vs. polarizability and strong mesomeric properties of oxygen play competing and contradictory roles for the synthetically useful generation of α-oxygen-stabilized carbanions. The carbenoid-like nature of α-lithiated ethers is an additional factor.[252] The common routes for α-deprotonation of nonbenzylic ethers include metal–heteroatom exchange of thio[135,214] and chloro[253] precursors using lithium naphthalide, LDMAN, LiDBB and, most significantly, tin–lithium exchange procedures.[254] For example, a practical route to cyclobutanones employs a *t*-butyllithium-mediated bromo–lithio exchange (Equation (29)).[255]

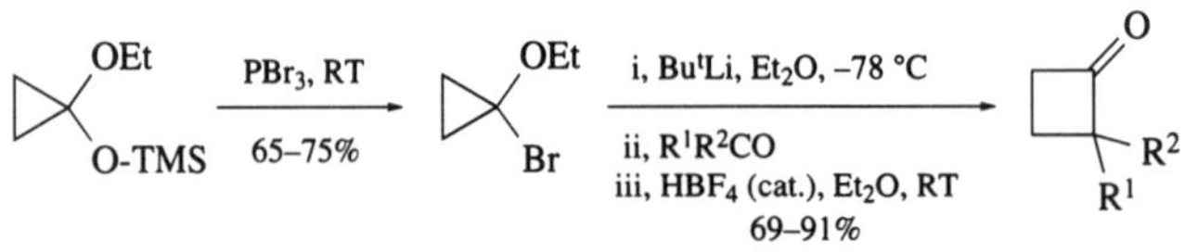

$$\text{(29)}$$

α-Functionalization of bare benzyl alcohols is feasible by *in situ* CO_2 activation, conditions which avoid normal *ortho*-lithiation and [1,2]-Wittig rearrangement.[256] The latter may also be suppressed by chromium tricarbonyl complexation.[257]

The $HOCH_2^-$ equivalent may be generated by direct deprotonation or, less practically, from the corresponding stannyl precursor (advantage: inert by-products; detriment: requirement of excess base) (Scheme 19).[258] With *gem*-dialkoxy derivatives, this tactic provides an acyl anion equivalent.[259] The synthetic use of the lithium–tin exchange reaction has been comprehensively studied.[254] Intramolecular trapping of an α-alkoxy lithio species by an allyl ether proceeds with high stereoselectivity (Equation (30)).[206]

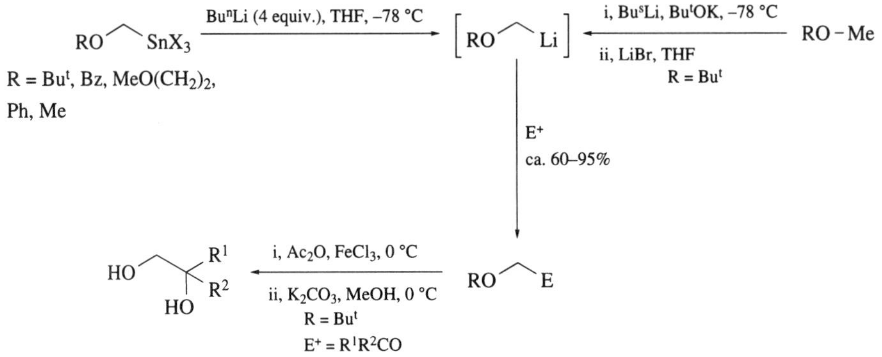

Scheme 19

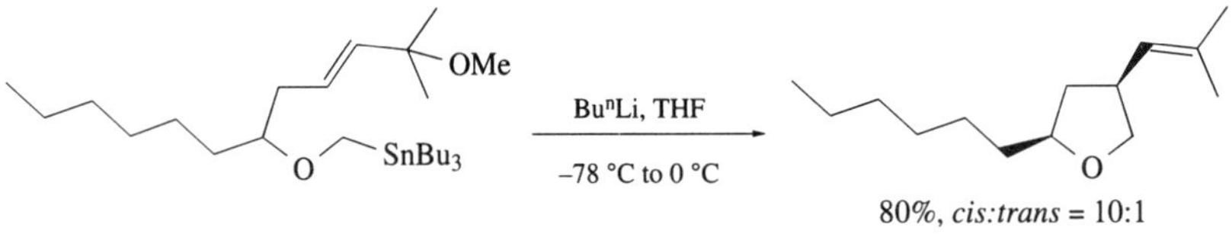

$$\text{(30)}$$

The reductive lithiation of 4-thio-substituted 1,3-dioxanes constitutes a versatile methodology for the preparation of optically active and functionalized 1,3-diols (Equation (31)).[260] Phenylthio variants are equally effective.

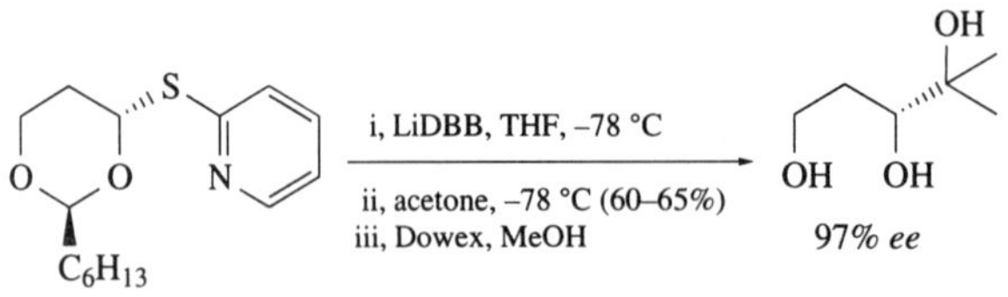

$$\text{(31)}$$

New formyl and alkoxycarbonyl anion equivalents are realized via mono- and disilyl-substituted methoxymethane. Mild electrochemical removal of silicon and the tolerance of the α-silyl ether moiety toward diverse reactions are attractive aspects (Scheme 20).[261]

1.3.3.1 Configurational stability and enantioinduction

The configurational stability of α-alkoxyorganolithiums, first demonstrated by Still and Sreekumar,[262] has spawned much synthetic interest.[26,263]

Primary and secondary benzyl *N,N*-diisopropylcarbamates are configurationally stable at low temperature,[264,265] but their reactions are capricious (Equation (32) and Table 7).[264] Solvent[265] and nitrogen substituent[266] influence has been noted.

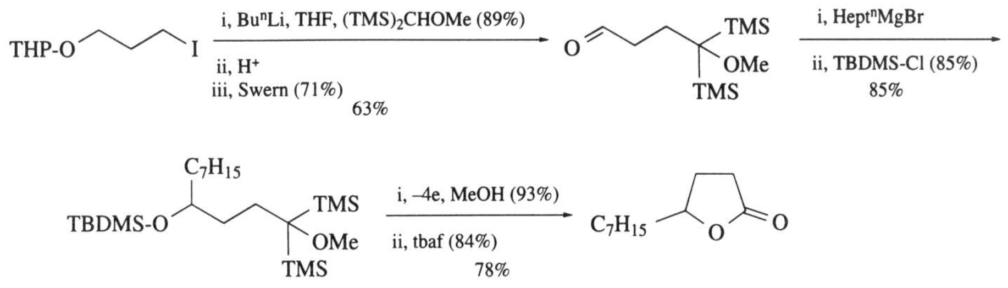

Scheme 20

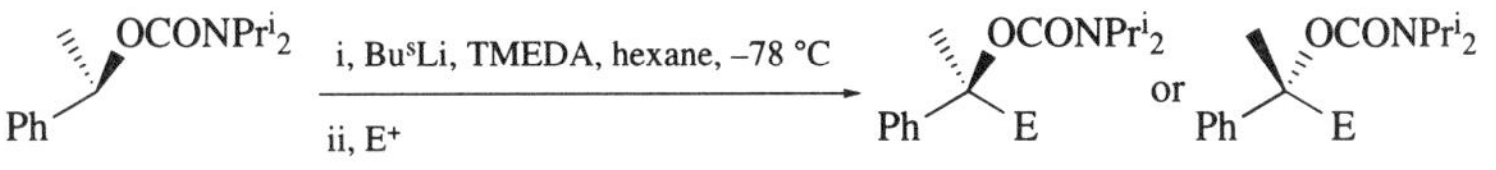

(32)

Table 7 Reaction of chiral α-oxybenzyllithium of (**36**) with electrophiles.

E^+	Product	Yield (%)	ee (%)
MeOH	(**37**)	80	80
AcOH	(**38**)	75	80
Pr^{n}Br	(**37**)	77	85
(MeO)$_2$CO	(**37**)	85	90
ClCO$_2$Me	(**38**)	90	85
CO$_2$, CH$_2$N$_2$	(**38**)	84	84

Enantiotopic discrimination of hindered aliphatic carbamates under the influence of (−)-sparteine is a simple and efficient process for chiral alcohol synthesis (Scheme 21).[267,268] The spirocyclic oxazolidino carbamate is removed via a mild two-step, one-pot procedure.

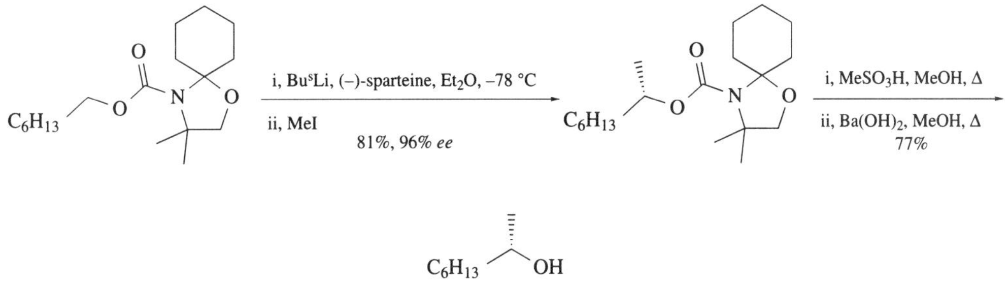

Scheme 21

Where a β-prostereogenic center is present, (−)-sparteine-assisted lithiation results in kinetic resolution at the α-position and high diastereoselectivity in a subsequent electrophilic substitution (Equation (33)).[269]

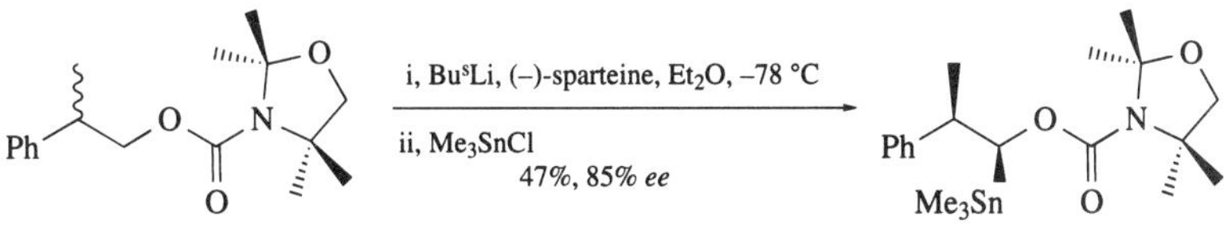

(33)

A variety of functionalized aliphatic carbamates have been investigated, including 2- and 3-amino[270,271] and 3- and 4-alkoxy[272] substrates.

Enantioselective deprotonation of allyl carbamates may also be effected via lithiation in the presence of (−)-sparteine (Equation (34)). Use of a full equivalent of alkyllithium–chiral amine gave the same result, indicating a substantial rate difference for the two enantiomeric reactions.[273,274]

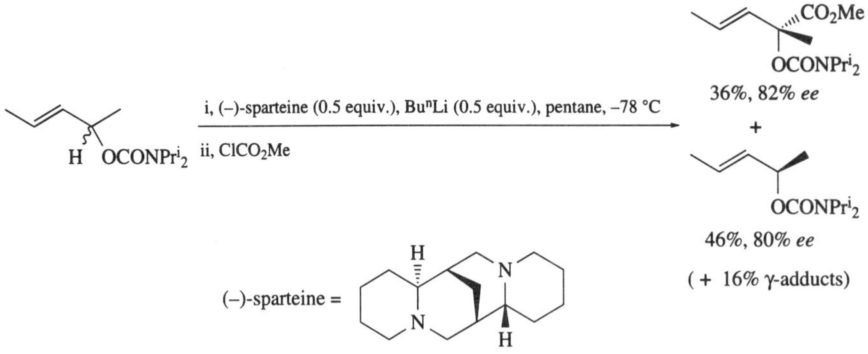

$$\text{(34)}$$

The configurational stability of α-alkoxyalkyllithiums[262,264,275] allows the use of enantiopure stannanes as starting materials for diverse synthetic operations (Equations (35);[276,277] and (38)[278]).

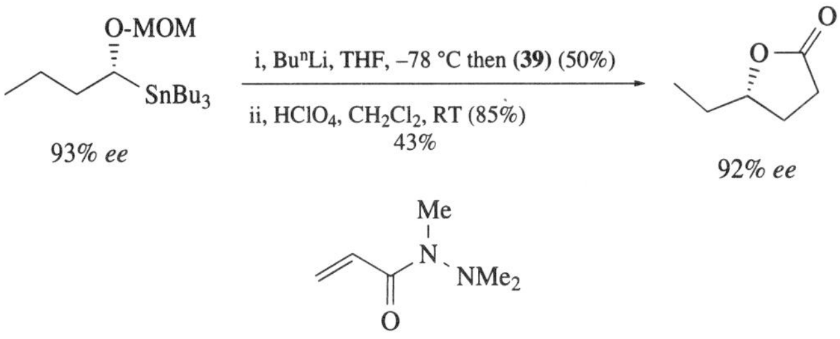

$$\text{(35)}$$

1.3.3.2 α-Oxylithio rearrangements

[2,3]-Wittig rearrangements, for example, Equation (36),[279] have found widespread application in synthetic practice.[280]

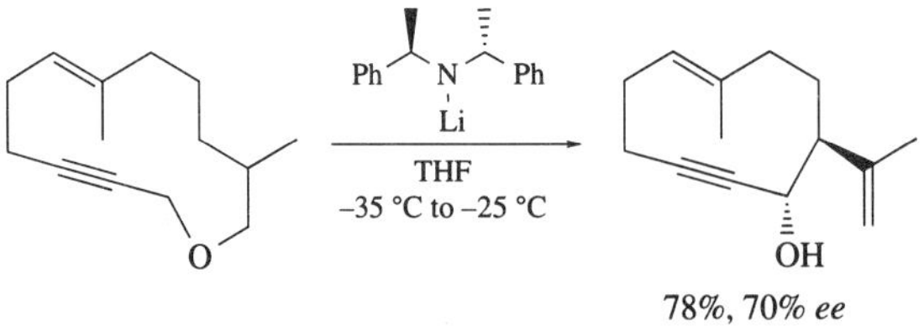

$$\text{(36)}$$

The Brook and reverse-Brook (sila-Wittig) rearrangement have enjoyed substantial synthetic utility.[278] In this equilibrium process involving transient α-oxy lithium species (Equation (37)), anion (41) is favored over (40) by virtue of negative charge preference on oxygen. Thus, excess base promotes O–Si cleavage, whereas catalytic base leads to C → O silyl transfer. R group stability, for example, aryl, favors anion (40).

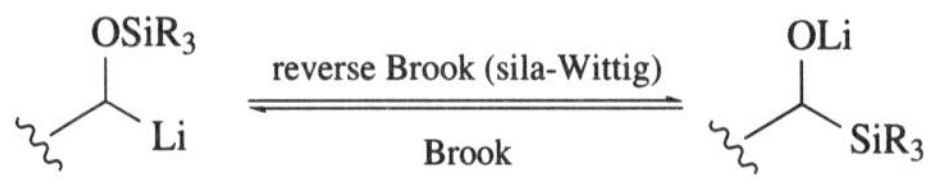

$$\text{(37)}$$

Among the significant mechanistic studies,[278,281] transmetallation tactics for triggering the Brook process on aliphatic substrates have shown that the rearrangement is intramolecular, nonradical, and occurs with retention of configuration at carbon (Equation (38)).[278] In contrast, when using aryl substrates, the reaction proceeds with complete inversion of stereochemistry. [1,n]-Silyl C → O migrations have been studied.[282]

The synthetically useful α,β-unsaturated acyl silanes[283,284] have been prepared via sila-Wittig rearrangement of α-silyloxy allyllithiums (Equation (39)).[285,286]

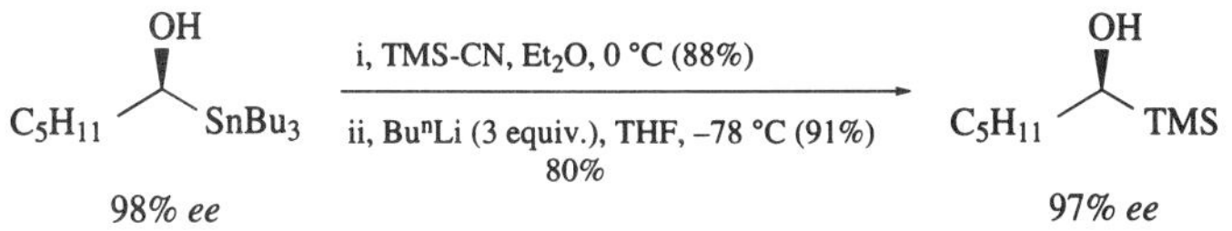

$$C_5H_{11}\underset{98\%\ ee}{\overset{OH}{\rightarrowtail}} SnBu_3 \xrightarrow[\text{ii, Bu}^n\text{Li (3 equiv.), THF, }-78\ °C\ (91\%)]{\text{i, TMS-CN, Et}_2\text{O, 0 °C (88\%)}} C_5H_{11}\underset{97\%\ ee}{\overset{OH}{\rightarrowtail}} TMS \quad (38)$$

$$R^2\underset{R^1}{\diagdown}\!=\!\diagup OH \xrightarrow[\substack{\text{iii, Bu}^t\text{Li (3 equiv.), }-78\ °C\ \text{or }-78\ °C\ \text{to }-30\ °C \\ \text{iv, NH}_4\text{Cl} \\ \text{v, Swern or Corey-Kim oxidation} \\ 64-79\%}]{\substack{\text{i, Bu}^n\text{Li, THF, }-78\ °C \\ \text{ii, TMS-Cl}}} R^2\underset{R^1}{\diagdown}\!=\!\diagup\overset{O}{\underset{}{\text{C}}}\text{—TMS} \quad (39)$$

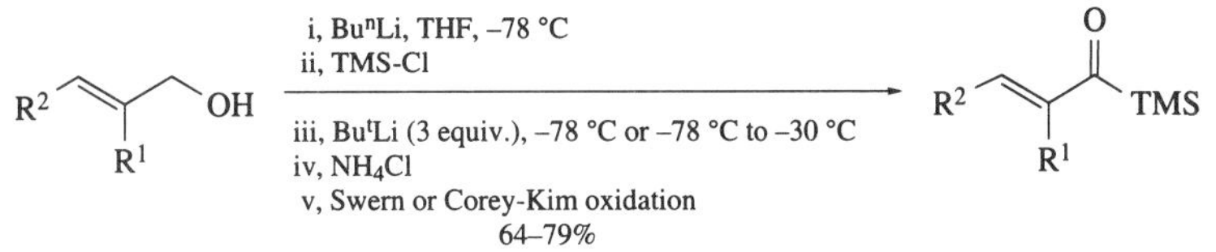

Silyloxy enynes may be generated stereoselectively by a sequence of carbanion addition to an acyl silane and Brook rearrangement–elimination process (Equation (40)).[287] A related stereocontrolled method for allylic alcohol synthesis involves addition of α-sulfur-stabilized carbanions to trimethylsilylethylene oxide.[288]

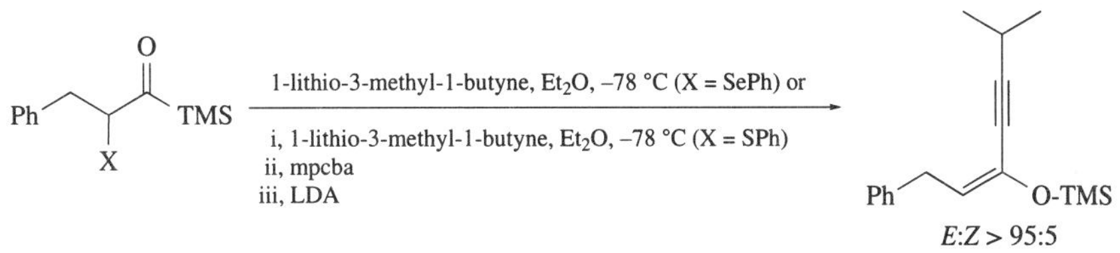

$$\text{Ph}\diagdown\!\!\underset{X}{\overset{O}{\text{C}}}\!\!\diagup\text{TMS} \xrightarrow[\substack{\text{i, 1-lithio-3-methyl-1-butyne, Et}_2\text{O, }-78\ °C\ (X=\text{SPh}) \\ \text{ii, mpcba} \\ \text{iii, LDA}}]{\text{1-lithio-3-methyl-1-butyne, Et}_2\text{O, }-78\ °C\ (X=\text{SePh})\ \text{or}} \underset{E:Z > 95:5}{\text{Ph}\diagdown\!=\!\diagup\text{O-TMS}\cdots} \quad (40)$$

1.3.4 α-Silicon sp^3 Carbanions

α-Lithiated silanes, stabilized by Si–C σ*-orbital–α-carbanion p-orbital hyperconjugation, are versatile synthetic species for the preparation of alkenes, functionalized alkenes, carbonyl compounds, and epoxides.[27,118] They can be prepared (Scheme 22) by: (i) direct deprotonation, (ii) metal–halogen exchange, (iii) transmetallation and other exchange reactions, and (iv) addition of an organometallic species to a vinylsilane. Thus, in method (i), due to the inherent stabilizing ability of silicon, even alkylsilanes can be deprotonated with alkyllithiums (Equation (41)).[289] Method (ii) is a useful route to the desired organometallics, but relies on the availability of the requisite α-halosilane.[290] Among heteroatoms which undergo exchange reactions (iii), most notable are sulfur (using reagents such as lithium naphthalide[291,292] (Equation (42), LiDBB,[213] and LDMAN[293]), selenium and tin (where alkyllithiums suffice),[294,295] and silicon (requiring merely alkali metal alkoxides).[296] As expected from the α-anion-stabilizing effect of silicon, the conjugate addition of alkyllithiums to vinylsilanes (method (iv); Scheme 22) proceeds in a regioselective manner.[297]

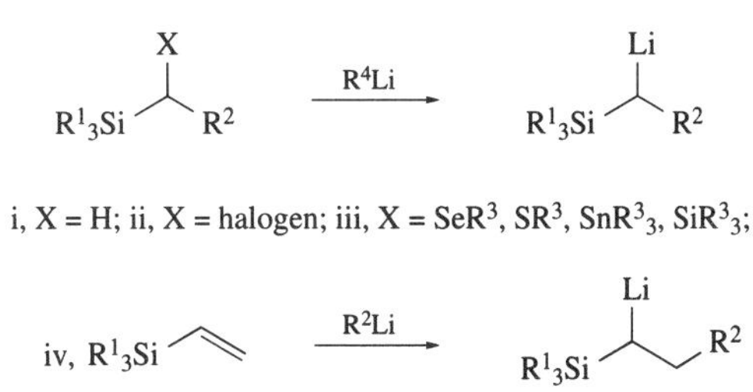

$$R^1{}_3Si\underset{R^2}{\diagdown}\!\!\diagup X \xrightarrow{R^4Li} R^1{}_3Si\underset{R^2}{\diagdown}\!\!\diagup Li$$

i, X = H; ii, X = halogen; iii, X = SeR^3, SR^3, SnR$^3{}_3$, SiR$^3{}_3$;

$$\text{iv, } R^1{}_3Si\diagup\!\!\!\diagdown \xrightarrow{R^2Li} R^1{}_3Si\diagdown\!\!\!\underset{}{\overset{Li}{\text{C}}}\!\!\diagup R^2$$

Scheme 22

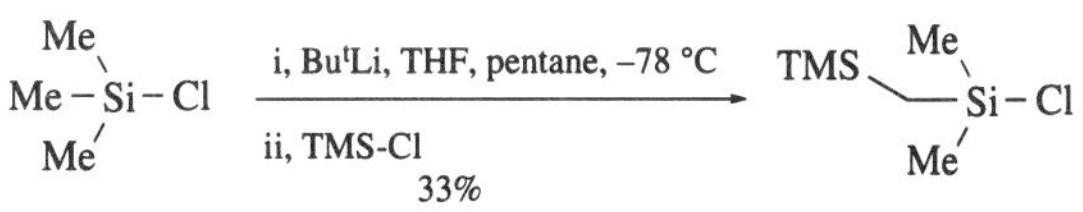

$$\underset{\text{Me}}{\overset{\text{Me}}{\text{Me}}}\!\!-\!\text{Si}\!-\!\text{Cl} \xrightarrow[\substack{\text{ii, TMS-Cl} \\ 33\%}]{\text{i, Bu}^t\text{Li, THF, pentane, }-78\ °C} \text{TMS}\underset{\text{Me}}{\overset{\text{Me}}{\diagdown}}\!\!\text{Si}\!-\!\text{Cl} \quad (41)$$

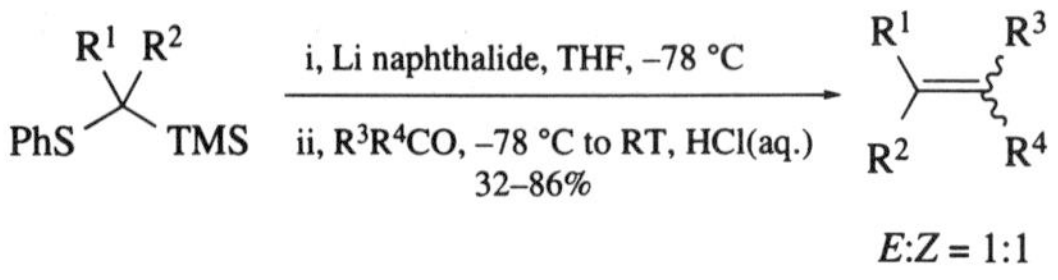

R1 R2
PhS TMS
i, Li naphthalide, THF, –78 °C
ii, R3R4CO, –78 °C to RT, HCl(aq.)
32–86%
R1 R3
R2 R4
E:Z = 1:1

(42)

1.3.4.1 The Peterson reaction

The Peterson alkenation represents the most expansive use of α-silyl carbanions.[28,298] In the first step, a mixture of diastereomeric β-alkoxy silanes is generated which may undergo *in situ* elimination to afford alkenes (Scheme 23). The *in situ* elimination, the concerted nature of which is under debate, requires the presence of an electron-withdrawing group and/or ionic alkoxide and is facilitated by aryl substitution on silicon.[299] Isolation and separation of the intermediate β-hydroxysilanes is possible, and allows for stereochemical control of a subsequent elimination by the choice of reaction conditions: alkali metal hydride gives *syn* elimination, whereas acid treatment promotes an *anti* pathway (Scheme 23). Thionyl chloride or fluoride anion (after pretreatment of the hydroxy function with acetyl chloride) can also be used to effect alkene formation.[300]

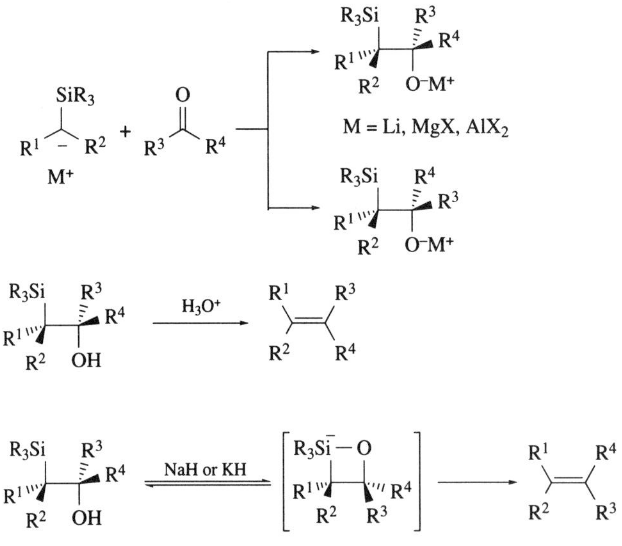

M = Li, MgX, AlX2
H3O+
NaH or KH

Scheme 23

In addition to simple Peterson reactions, there exists a myriad of variants incorporating extra functionality in the α-silyl carbanion,[28] including unsaturation (see below), carbonyl,[301,302] imine,[303] nitrogen,[304] sulfur,[305,306] selenium,[307] silicon,[308] tin,[309] phosphorus,[310,311] halogen (where α,β-epoxytrimethylsilanes are produced),[312] oxygen,[306,313,314] and boron,[315] as well as two or more functional groups. Selected synthetic examples are depicted in Scheme 24 and Equation (43).[305,306]

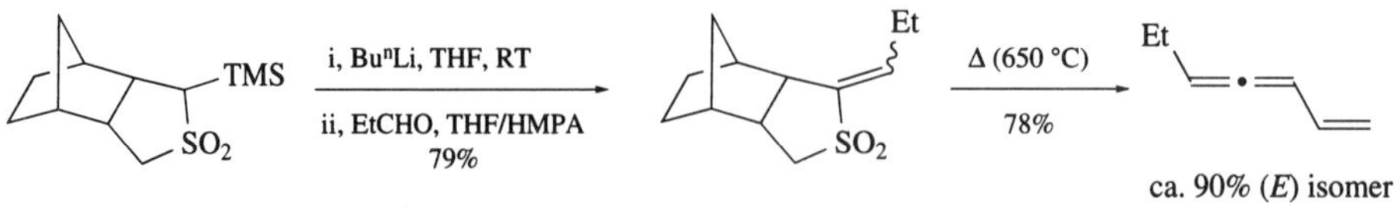

TMS
SO2
i, BunLi, THF, RT
ii, EtCHO, THF/HMPA
79%
Et
Δ (650 °C)
78%
ca. 90% (E) isomer

Scheme 24

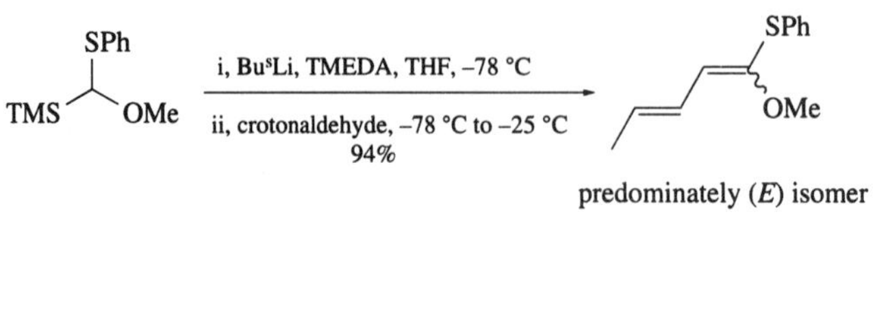

SPh
TMS OMe
i, BusLi, TMEDA, THF, –78 °C
ii, crotonaldehyde, –78 °C to –25 °C
94%
predominately (E) isomer

(43)

Reaction with nitrogen-based electrophiles such as imines, oximes, nitrones, and nitriles leads to Peterson products often with high stereoselectivity (Equation (44)).[316,317]

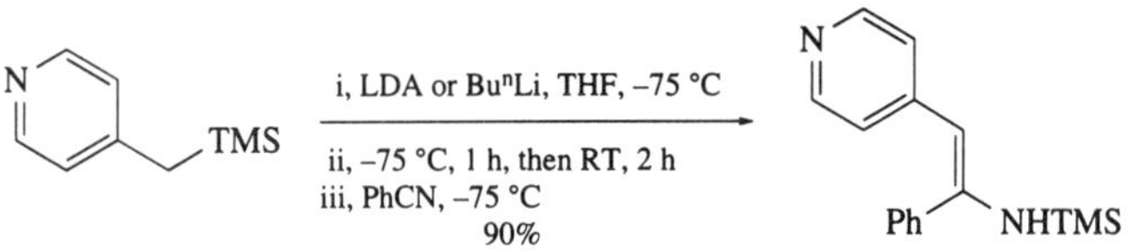

$$\text{4-Py-CH}_2\text{-TMS} \xrightarrow[\substack{\text{ii, } -75°C, 1 \text{ h, then RT, 2 h} \\ \text{iii, PhCN, } -75°C \\ 90\%}]{\text{i, LDA or Bu}^n\text{Li, THF, } -75°C} \text{4-Py-CH=C(Ph)(NHTMS)} \quad (44)$$

Intramolecular Peterson alkenation forms the basis of the Bartoli indole synthesis, a general and mild method for 2- and 2,3-substituted indole construction (Equation (45)).[318]

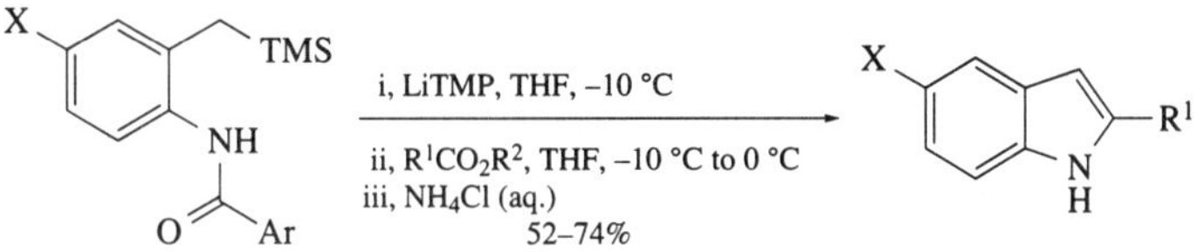

$$\quad (45)$$

with conditions: i, LiTMP, THF, −10 °C; ii, R^{1}CO$_2$R^2, THF, −10 °C to 0 °C; iii, NH$_4$Cl (aq.); 52–74%

The *in situ* activation of TMS–methanol as its lithium carbonate leads to a methanol dianion synthetic equivalent and thereby provides a route to α-hydroxy ketones and homologated aldehydes (Equation (46)).[314]

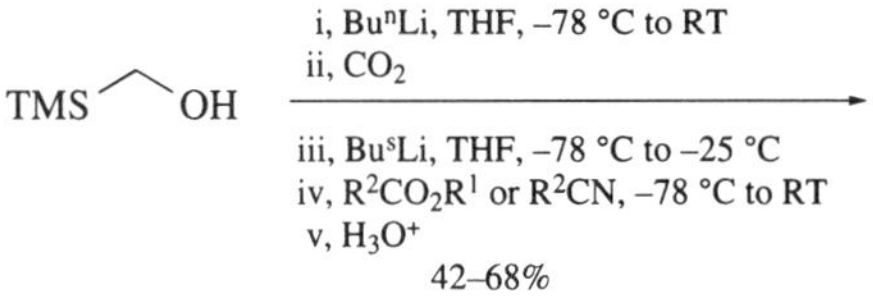

$$\text{TMS-CH}_2\text{-OH} \xrightarrow[\substack{\text{iii, Bu}^s\text{Li, THF, } -78°C \text{ to } -25°C \\ \text{iv, R}^2\text{CO}_2\text{R}^1 \text{ or R}^2\text{CN, } -78°C \text{ to RT} \\ \text{v, H}_3\text{O}^+ \\ 42-68\%}]{\substack{\text{i, Bu}^n\text{Li, THF, } -78°C \text{ to RT} \\ \text{ii, CO}_2}} \text{R}^2\text{C(O)CH(OH)} \quad (46)$$

The reaction of α-silyl organolithiums with epoxy tosylates constitutes a method for the construction of cyclopentanols (including chiral systems) which involves a cascade process including a 1,4-Brook rearrangement (Equation (47)).[319]

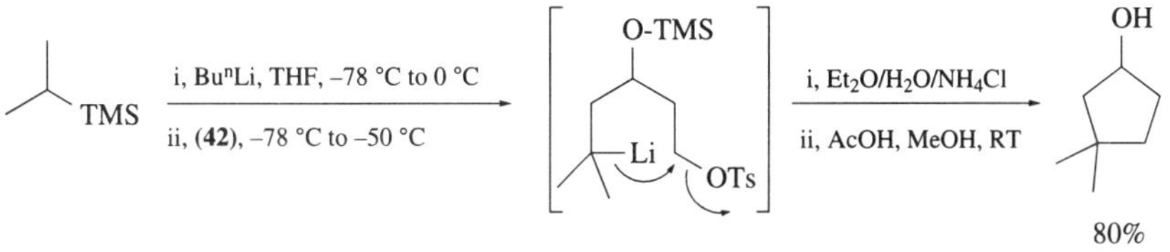

$$\quad (47)$$

with conditions: i, Bu^{n}Li, THF, −78 °C to 0 °C; ii, (42), −78 °C to −50 °C; then i, Et$_2$O/H$_2$O/NH$_4$Cl; ii, AcOH, MeOH, RT; 80%

(42) = epoxide-CH$_2$CH$_2$-OTs

A problem associated with ambident anions, such as α-silylallyl examples, is control of regioselectivity. Chan and co-workers have demonstrated the use of a chelating substituent on silicon to effect preferential regioselective α-alkylation (Equation (48)).[29,320–2] Dimethylphenylsilylallyl substrates have also been used.[323]

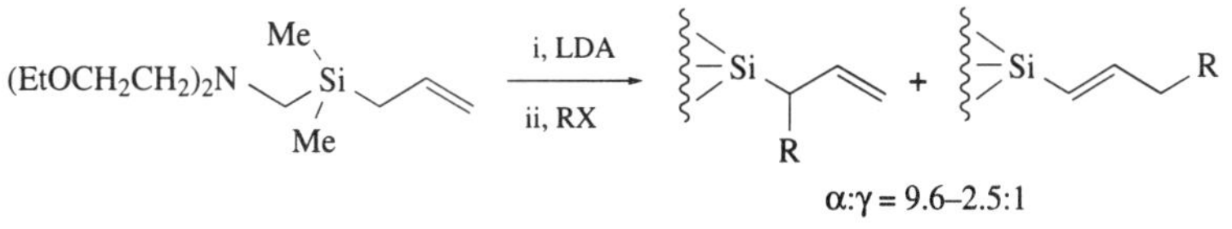

$$(\text{EtOCH}_2\text{CH}_2)_2\text{N-SiMe}_2\text{-CH}_2\text{-CH=CH}_2 \xrightarrow[\text{ii, RX}]{\text{i, LDA}} \text{Si-CH(R)-CH=CH}_2 + \text{Si-CH}_2\text{-CH=CH-R} \quad (48)$$

α:γ = 9.6–2.5:1

R = Me, *n*-C$_6$H$_{13}$, allyl, γ,γ-dimethylallyl

Silylcinnamyl carbanions with chiral bidentate auxiliaries (**43**) can be α-alkylated in good yield with high enantioselectivity (>90% *ee*) and good regioselectivity (ca. 10:1 α:γ) using Bu^{s}Li in nonpolar solvents (Et$_2$O, toluene). The process is most efficient with small electrophiles, although loss of regioselectivity with more sterically demanding electrophiles can be offset somewhat by incorporation of electron-donating aryl substituents.[29]

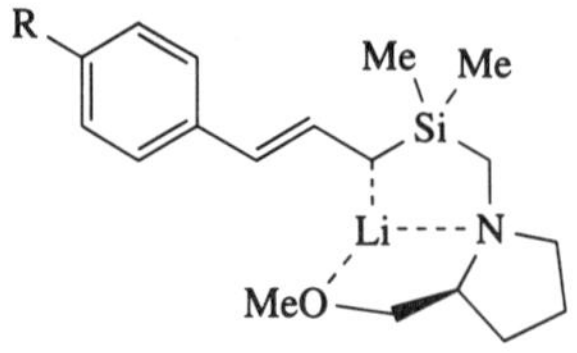

R = H, MeO

(43)

This methodology has been also applied to chiral auxiliaries bearing benzyl[324] (Equation (49)) and propargyl[322] substrates.

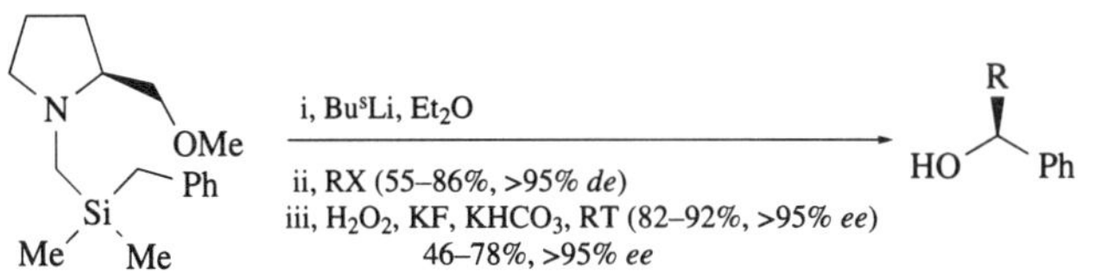

i, Bu^{s}Li, Et$_2$O
ii, RX (55–86%, >95% de)
iii, H$_2$O$_2$, KF, KHCO$_3$, RT (82–92%, >95% ee)
46–78%, >95% ee

(49)

Michael addition of organolithiums to α-thio vinylsilanes and alkylation of α-thio silylalkanes leads to intermediates which undergo sila-Pummerer reaction to give ketones (Scheme 25).[292]

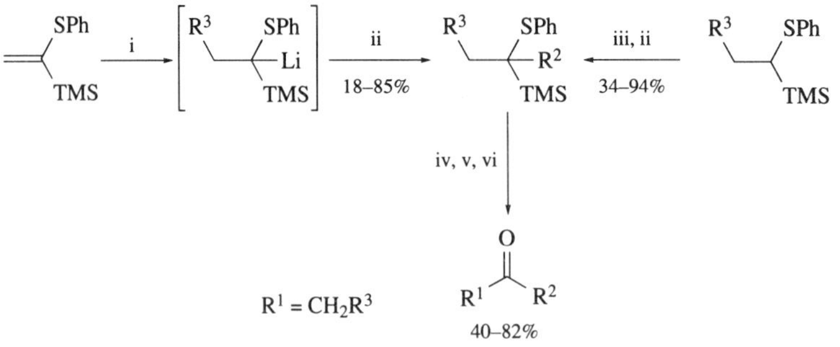

$R^1 = CH_2R^3$

18–85%   34–94%   40–82%

i, R^{3}Li, Et$_2$O, TMEDA, 0 °C; ii, R^{2}X; iii, Bu^{n}Li, hexane, TMEDA, 0 °C;
iv, mcpba, CH$_2$Cl$_2$, −23 °C to RT; v, Δ; vi, HCl (aq.) or Hg(OAc)$_2$(aq.), THF

Scheme 25

1.3.5 α-Sulfur- and α-Selenium-stabilized Carbanions

1.3.5.1 Generation

The chemistry of sulfur- and selenium-stabilized carbanions[325–34] is inextricably linked. α-Sulfur or selenium (at various oxidation levels) carbanions are similarly stabilized as a consequence of polarizability and hyperconjugation effects (PhCH$_2$SPh, pK_a ~32; PhCH$_2$SePh, pK_a ~35[335]) and their acidity is strongly affected by other heteroatoms.[331] Sulfides are readily lithiated by n-butyllithium in the presence of TMEDA or dabco. Heteroatom–lithio exchange reactions (e.g., Br, Sn, Se, or S (from dithianes)) provide alternative approaches to α-sulfenylcarbanions. In particular, lithium–tin exchange is more efficient than a Li/H process for 2-lithio dithiane generation.

In general, direct lithiation of selenides with alkyllithiums results in preferential lithium–selenium exchange rather than the desired Li/H process. Where pK_a requirements permit, this problem may be circumvented by the employment of less selenophilic bases (LDA, LiHMDS, or LiTMP).

Heteroatom–lithium exchange processes constitute complementary methods to direct deprotonation for α-sulfur and α-selenium carbanion generation. The approximate reactivity order Sn > Br > Se (relative rates ca. 10^3, 1, 10^{-1}) holds. Due to the ready preparation of selenoacetals and ketals, lithium–selenium exchange is a widely used method.

As well appreciated, α-sulfoxide and α-selenoxide carbanions are easier to generate than their lower oxidation counterparts. Sulfones afford stable anions with broad synthetic application, whereas selenones require *in situ* generation and are of limited use owing to thermal instability and moisture sensitivity. α,α-(Geminal)dianions derived from sulfones offer interesting reactivity and chemoselectivity characteristics.[336]

1.3.5.2 *Subsequent transformations*

Reductive removal of both sulfide and selenide activators may be performed with Raney nickel, Li/Et$_2$NH, lithium naphthalide, and R$_3$SnH among other reagents. Sulfoxide (Raney nickel or Li/Et$_2$NH) and sulfone (Na/Hg amalgam) reductive removal is well known. Oxidative elimination (NaIO$_4$, mcpba, H$_2$O$_2$) of sulfides and selenides, occurring via sulfoxide and selenoxide intermediates, is a well-traveled route to alkenes, especially for selenoxides, for which thermal elimination rates are very fast. Alkenes are also obtained from the addition of iodomethylstannanes or silanes to lithiated sulfones, followed by treatment with tbaf; the stannyl variant is particularly useful for hindered sulfones.[337] [2,3]-Sigmatropic rearrangement of allyl substrates, promoted by a thiophile and selenophile such as an amine or a phosphite, constitutes a stereospecific synthesis of allylic alcohols.[82] Base-induced double elimination of β-acetoxy-δ-amido sulfones[338] and reductive elimination of β-mesyl sulfones[339] constitute other routes to alkenes. SO$_2$ extrusion via thermal and photochemical Ramberg–Bäcklund provide interesting C–C bond connectivity and enhance the utility of α-sulfonyl carbanions.

Regeneration of carbonyls from thioacetals and ketals, not always a trivial matter, has been carried out traditionally by metal-ion-promoted (Hg^{2+}, Cu^{2+}, or Ag$^+$) or oxidative (NCS or Tl(NO$_3$)$_3$) hydrolysis. A method which is mild, tolerates many functional groups, and gives high yields employs bis(trifluoroacetoxy)iodobenzene in aqueous medium; when an alcoholic solvent is used, acetals are generated directly from thioacetals.[340] Selenoketals tend to be hydrolyzed somewhat more slowly than their sulfur counterparts. Oxidation of α-silyl sulfides and selenides leads, via sila- and sila-seleno-Pummerer rearrangement, respectively, to carbonyl compounds.[341]

For sulfone to carbonyl conversion, α-lithiation followed by reaction with bis(trimethylsilyl)peroxide has been established. Yields are good to excellent, and ^{18}O incorporation is an option.[342]

1.3.5.3 *α-Sulfenyl and selenyl carbanions*

α-Lithiated sulfide and sulfoxide carbanions play a prominent role in an overall sulfide to thiolactone conversion which is part of the synthesis of methynolide (Scheme 26).[343]

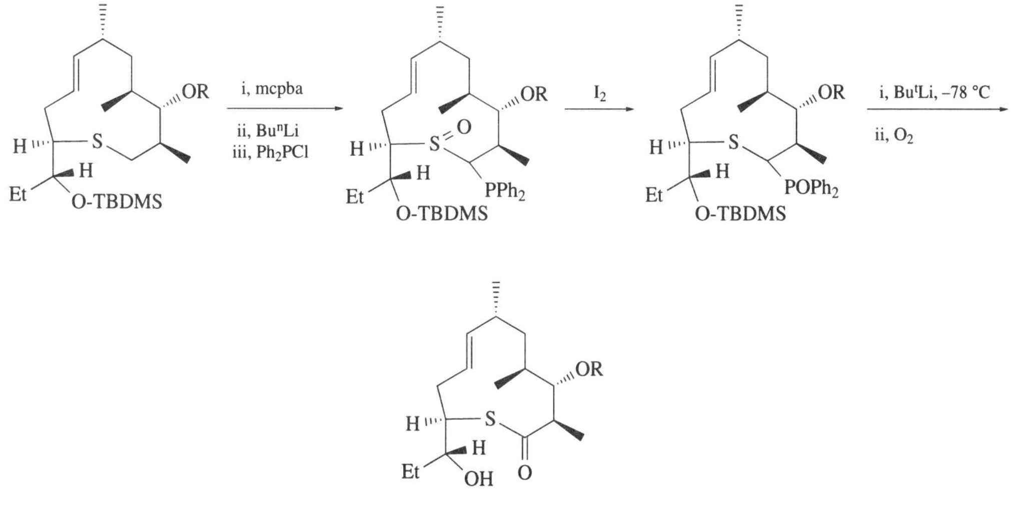

Scheme 26

Addition of an α-lithio selenide to a cyclobutanone, followed by methyl fluorosulfonate-induced ring expansion provides a synthesis of α-cuparenone, an aromatic monoterpene (Equation (50)).[344] A related ring expansion of spirocyclobutanones at the selenoxide level has been reported.[345]

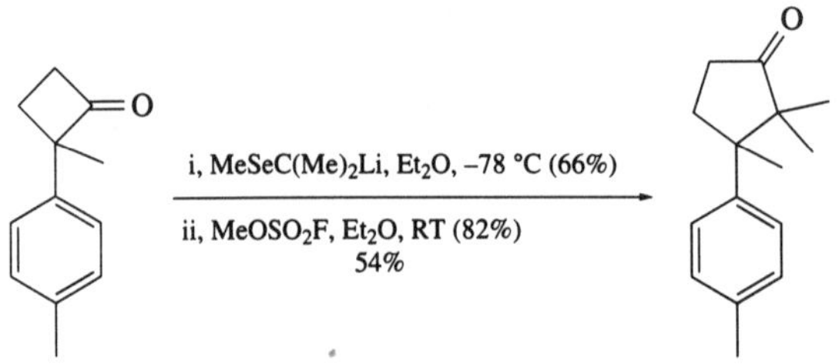

Alkylation of an α-selenenyl carbanion, generated by lithium–selenium exchange, constitutes a key step in a vitamin D synthesis (Equation (51)).[346] Reductive cleavage of the selenide moiety involved temporary masking of the triene by SO_2 cycloaddition.

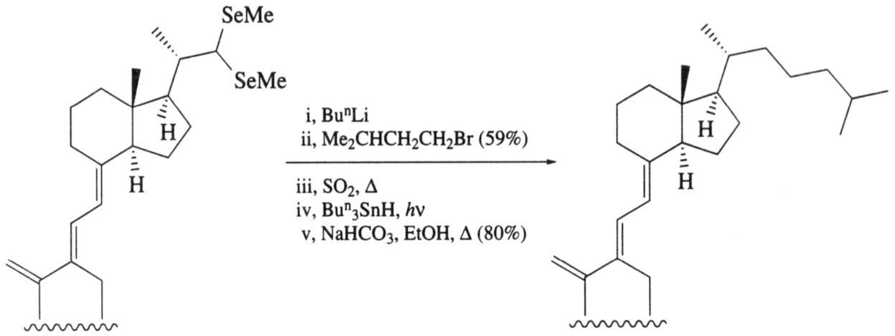

The exocyclic methylene of the insect defense component (±)-pederin was incorporated by Michael addition of phenylselenomethyllithium followed by eventual oxidative ($NaIO_4$) elimination (Scheme 27).[347]

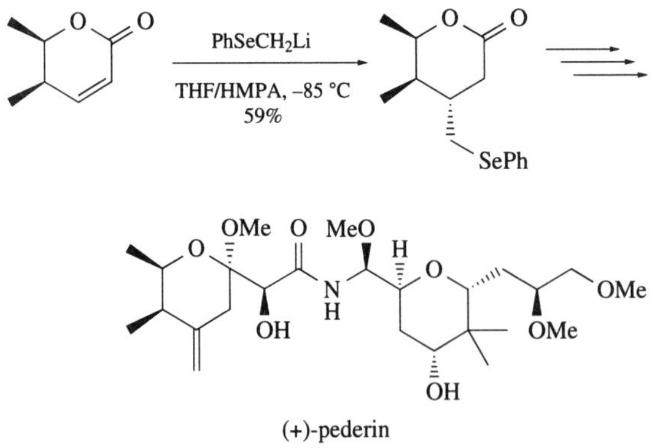

Scheme 27

1.3.5.4 α-Sulfenyl and selenyl allylic carbanions

Originating with key observations by Evans and Andrews,[82] alkylation of allylic sulfides in combination with [2,3]-sigmatropic rearrangement of derived substituted allylic sulfoxides has enjoyed extensive synthetic application, for example, a convenient procedure for the preparation of rosefuran (Scheme 28).[348]

Regioselective allyl sulfide alkylation is a key step in the synthesis of the insect attractant (±)-eldanolide (Scheme 29).[349]

The intramolecular α-lithio allyl sulfide–electrophile reaction is an effective method for the formation of macrocycles, for example, in a convergent synthesis of the marine diterpene (±)-obscuranatin (Scheme 30),[350] and a route to the nucleus of crassin acetate.[351]

Successive regiospecific alkylation and thioallylic rearrangement of α-methoxyallyl sulfides is a key sequence involved in the preparation of substituted α,β-unsaturated carbonyl compounds (Scheme 31).[352] The starting material thus behaves as a homoenolate dianion equivalent.

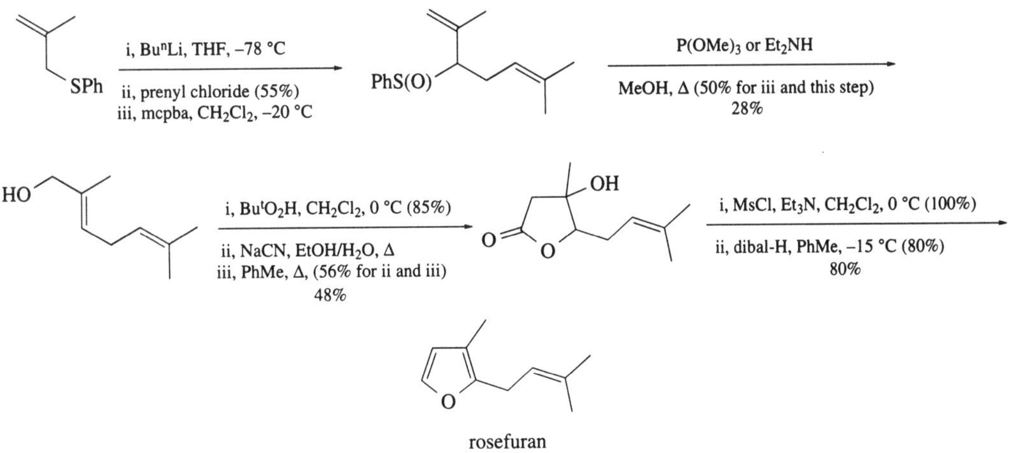

Scheme 28

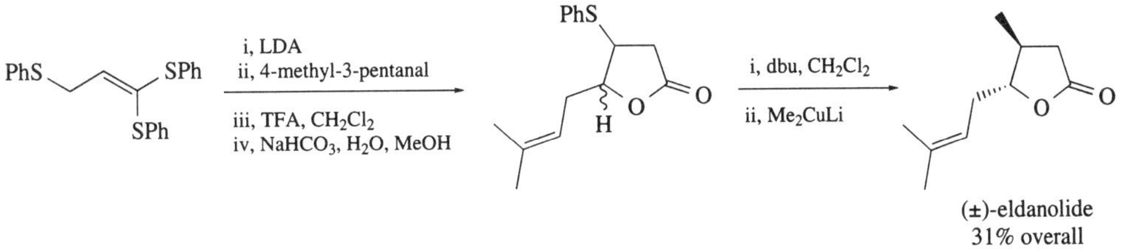

Scheme 29

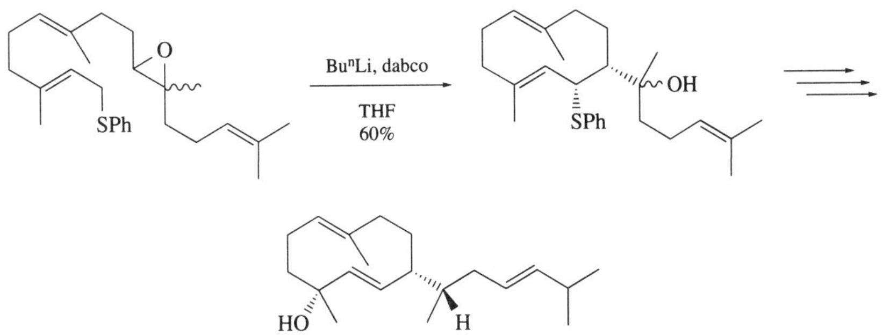

Scheme 30

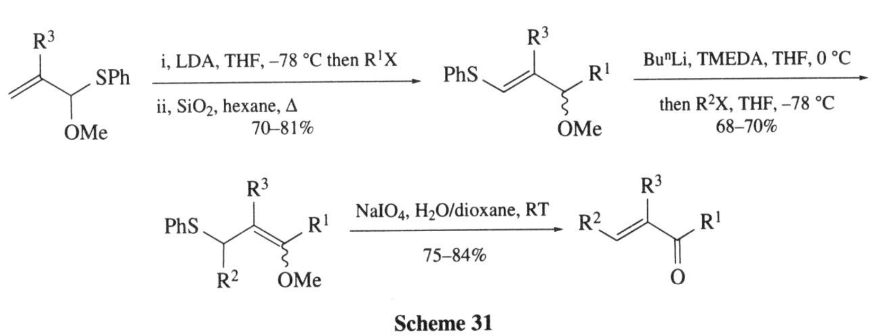

Scheme 31

Repetitive allyl dithiocarbamate alkylations and [3,3]-sigmatropic rearrangements are integral to the strategy of a highly convergent route to the tetraenoic acid, β-parinaric acid. The route is initiated by allyl dithiocarbamate α-metallation and involves a mild thiocarbamate elimination (Scheme 32).[353]

α-Carbanions of allyl selenides have seen considerably less use. En route to the antitumor cembranoid asperdiol, two highly functionalized fragments were coupled via a lithiated allyl selenide (Equation (52)).[354] Reductive removal of selenium, functional group interconversion, and macrocyclization completed the synthesis.

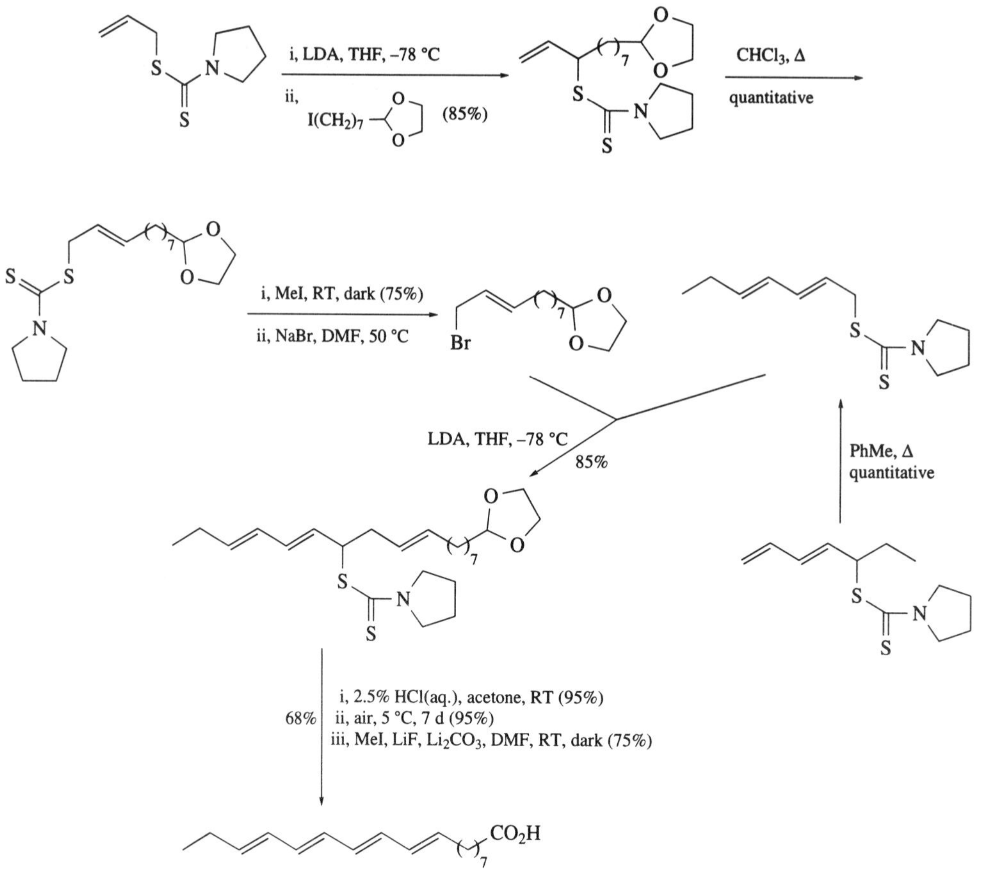

β-parinaric acid

Scheme 32

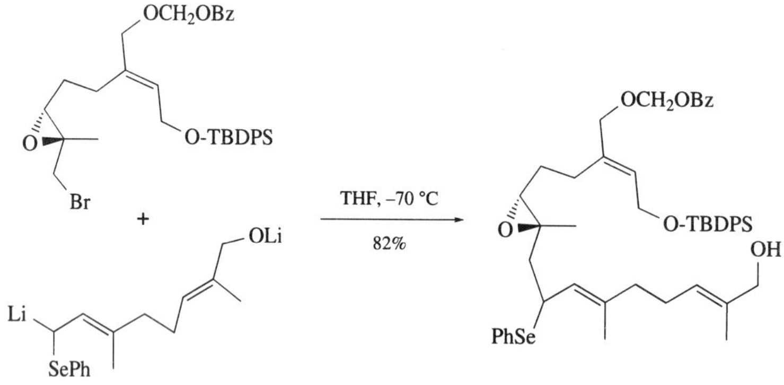

(52)

γ-Seleno allyl selenides, readily prepared from 1,3-dichloropropene, are useful precursors for substituted α,β-unsaturated systems (Equation (53)).[355]

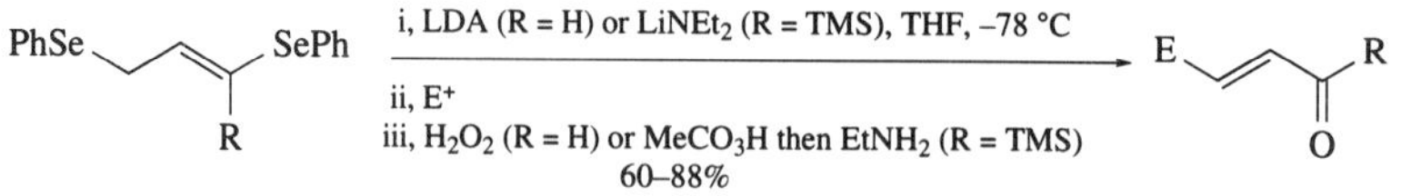

(53)

1.3.5.5 α-Sulfinyl carbanions

Functionalized dienes are obtained from α,β-unsaturated acyl silanes by treatment with phenysulfonylmethyllithium, a reaction which involves a Brook rearrangement and a sulfoxide elimination (Equation (54)).[356]

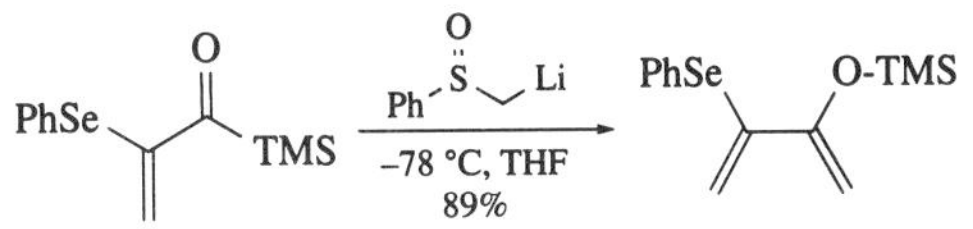

(54)

Intramolecular acylation of a lithio sulfoxide constitutes a key step in the synthesis of the potent hypocholesterolemic mevinolin fungal metabolites (Scheme 33).[357]

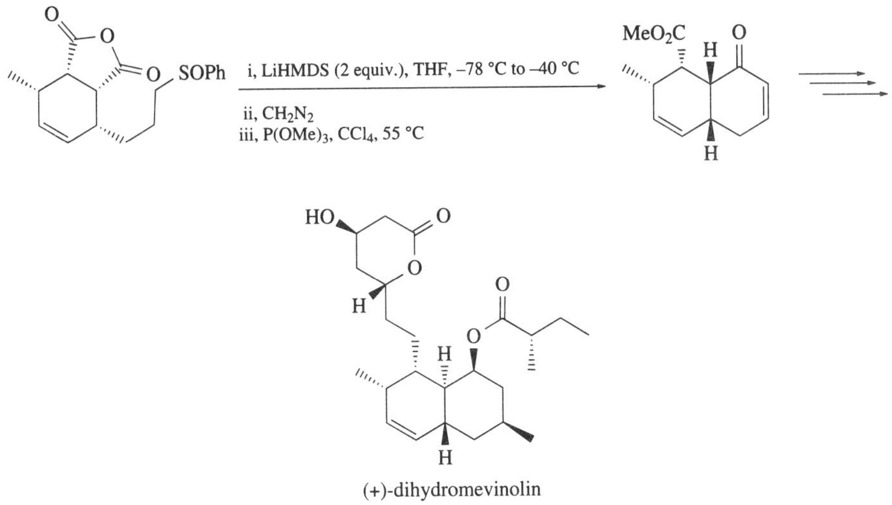

(+)-dihydromevinolin

Scheme 33

α-Lithio halomethyl sulfoxides serve as dual-reactivity species in reaction with imines leading to aziridines and subsequently pyrrole derivatives (Scheme 34).[358]

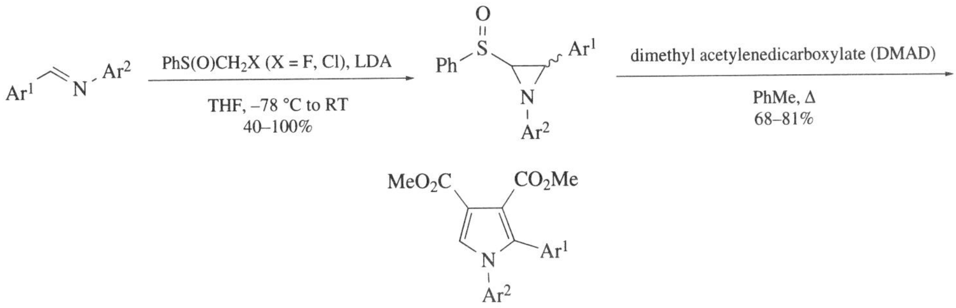

Scheme 34

The value of optically active sulfoxides in enantioselective synthesis has been extensively demonstrated.[30,31,359–61] A front runner in chiral sulfoxide chemistry, Solladié, has used the addition of chiral α-sulfinyl species to esters to provide a chiral auxiliary for subsequent enantioselective ketone reduction. Syntheses of the macrolide lasiodiplodin and of an eicosatetraenoic acid fragment are illustrative (Scheme 35).[362]

A chiral α-sulfinyl imine is the starting point for a short synthesis of indolizidine alkaloids (Scheme 36). A one-pot alkylation–annulation followed by sodium borohydride reduction (4:1 (8eS):(8eR)) led to an epimeric mixture of sulfoxides (**44**). A second alkylation followed by elimination afforded (−)-elaeokanine B.[363]

Alkylation of a chiral lithio α-chloromethyl sulfoxide occurs with poor 1,2-asymmetric induction; however, in combination with a Darzens reaction, the two diastereomers led to *cis*- and *trans*-disparlure, the former being an insect pheromone (Scheme 37).[364]

Alkylation of a lithiated chiral cyclic *trans*-sulfoxide (**45**), derived by the agency of a homochiral lithium amide base, gave moderate enantioselective induction (Scheme 38).[132] Enantiomeric enrichment of the carbinol product (**46**) by repeated recrystallization afforded optically pure material which was stereopecifically transformed into the functionalized epoxide (**47**).

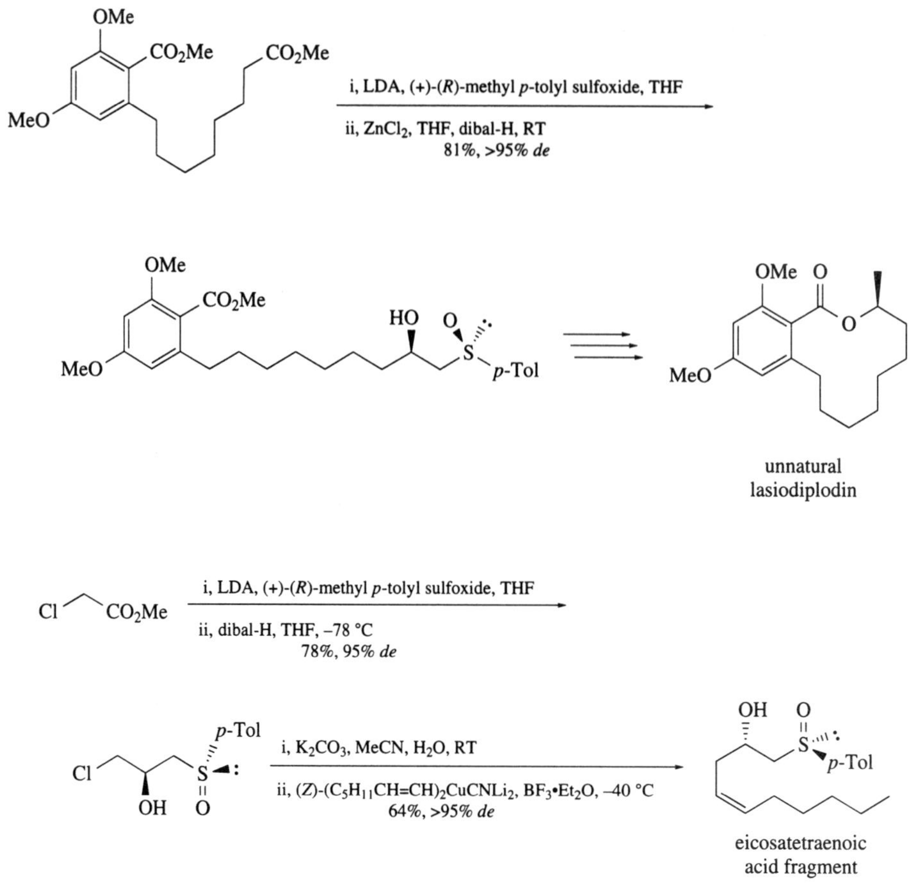

Scheme 35

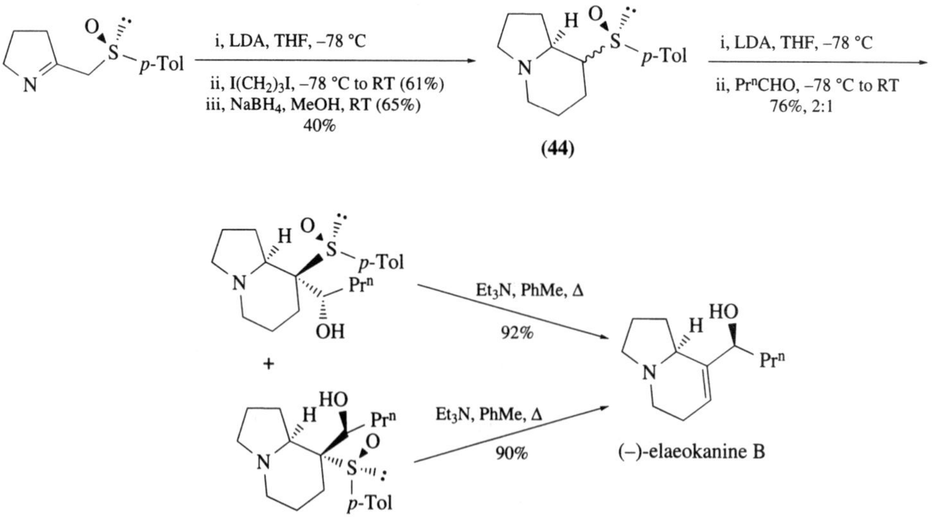

Scheme 36

Addition of chiral α-lithiomethyl *p*-tolyl sulfoxide to a dihydroisoquinoline occurred in high yield and excellent diastereoselectivity, thereby providing an efficient route to the alkaloid (*R*)-(+)-tetrahydropalmatine (Scheme 39).[365] The observed diastereoselection is believed to arise via proton transfer within the initial adduct, followed by a retro-Michael reaction. Reductive *N*-alkylation, intramolecular Pummerer rearrangement, and reductive desulfurization concluded the synthesis. Asymmetric addition of chiral α-lithio sulfoxides to dihydroisoquinoline *N*-oxides in the presence of a chiral alkoxide has been reported.[366]

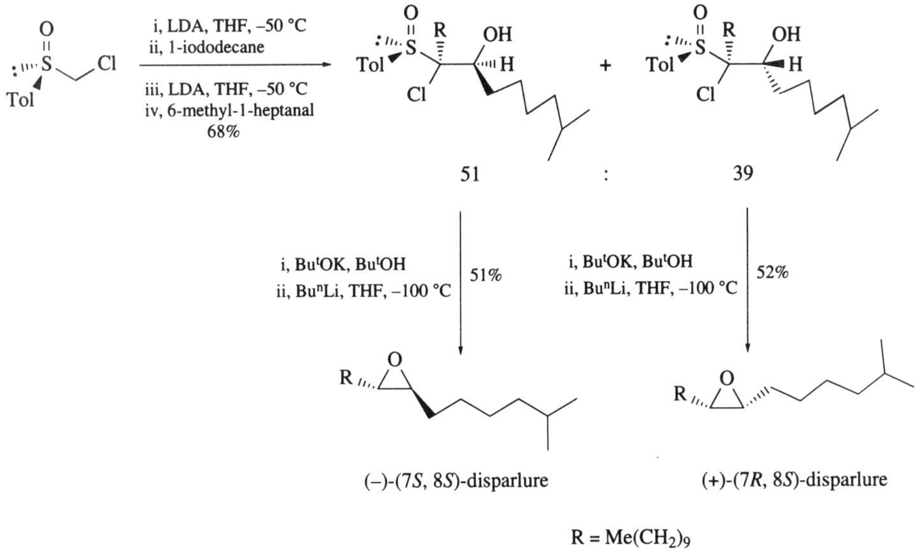

Scheme 37

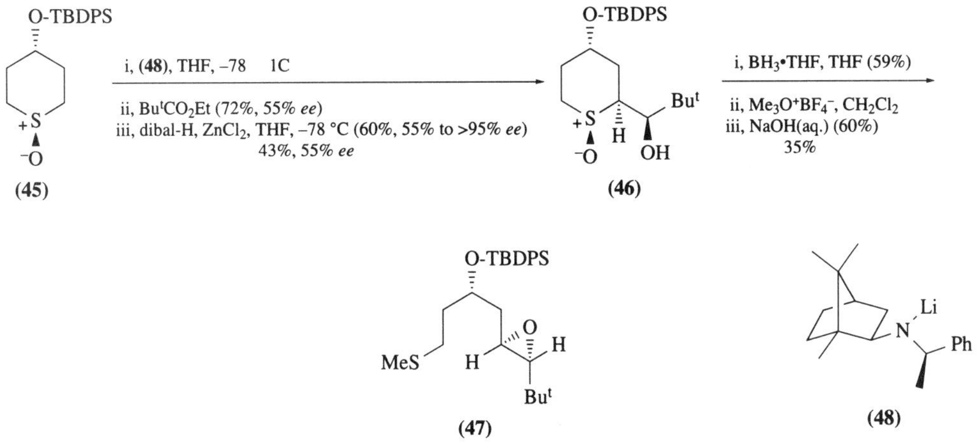

Scheme 38

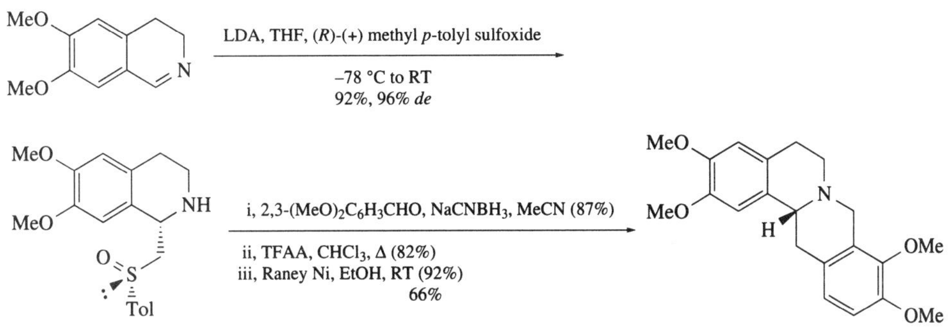

(*R*)-(+)-tetrahydropalmatine

Scheme 39

1.3.5.6 α-Sulfinyl allylic carbanions

The regioselectivity for addition of electrophiles to the ambident lithio species derived from allyl sulfoxides is dependent on the structure of both the substrate and the electrophile. α-Alkylation is often preponderant, especially when the γ-position is mono- or disubstituted. The regioselectivity is lower and less predictable for carbonyl addition processes than the corresponding alkylation reactions with alkyl halides.[30,82] Better regiocontrol, in combination with facile sulfide oxidation, makes the α-alkylation of allylic sulfides the preferred route to α-substituted allylic sulfoxides (e.g., Scheme 28). Good regiocontrol and variable stereocontrol (2:1 to 28:1 *ds*) in addition to carbonyl compounds can be achieved by adding substantial amounts of HMPA to the lithio species (Scheme 40).[367] Incorporation of a coordinating ligand at sulfur (i.e., pyridyl) is an alternative tactic.[368]

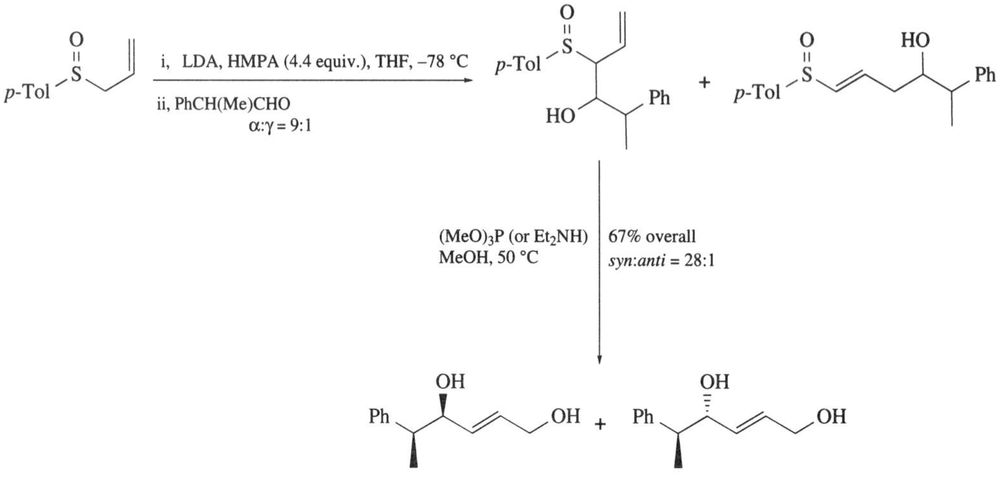

Scheme 40

1.3.5.7 α-Sulfonyl carbanions

In contrast to the corresponding chloro and bromo species which form epoxides, α-lithiofluoromethyl phenyl sulfone undergoes reaction with carbonyl systems to afford stable β-fluoro alcohols which can be dehydrated (overall Wittig process) and desulfonylated to give fluoroalkenes (Scheme 41).[369] Reaction of α-sulfonyl carbanions with halogenocarbenoids furnishes alkenes or haloalkenes, depending on the carbenoid used.[370,371]

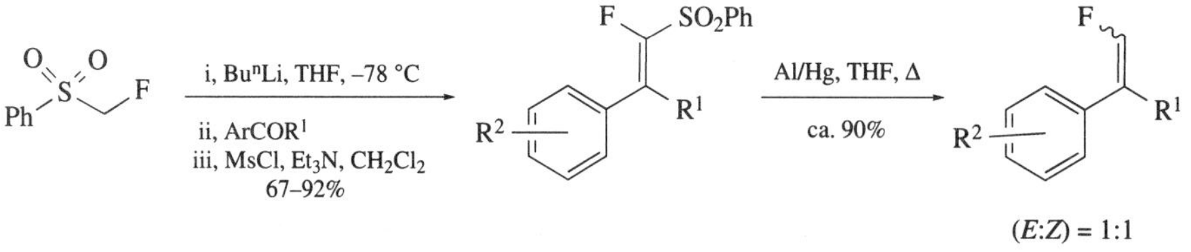

Scheme 41

Successive triple alkylation of methyl α-triflylmethyl sulfone followed by oxidation and Ramberg–Bäcklund SO_2 extrusion provides yet another route to dihydrojasmone (Scheme 42).[372,373]

Addition of chiral electrophiles to steroidal α-lithio sulfones affords enantiomeric alcohols which, upon sulfone excision, furnishes cerebrosterol (**50**) and its 24-epimer (**49**), compounds of physiological interest (Scheme 43).[374]

α-Lithio sulfone chemistry in tandem with reduction of the sulfonyl group has found much application in the synthesis of naturally occurring alkenic compounds, for example, vitamins[375] and pheromones (Scheme 44).[376]

Using mannitol as a starting point, α-lithio sulfone methodology served well in the attachment of an optically active chain en route to the biologically significant (+)-brefeldin A (Scheme 45).[377]

The homoenolate equivalent, α-lithio 3-phenylsulfonylorthopropionate was treated with a homochiral epoxide in a convergent route to (−)-argentilactone (Equation (55)).[378]

In a formal synthesis of the immunosuppressant FK-506, Danishefsky used the coupling of an α-lithio sulfone with a highly functionalized aldehyde (Equation (56)).[379] A similar use of α-lithio sulfone chemistry for the coupling of highly functionalized fragments is involved in the Isobe synthesis of the naturally occurring polyether, okadaic acid (Scheme 46).[380]

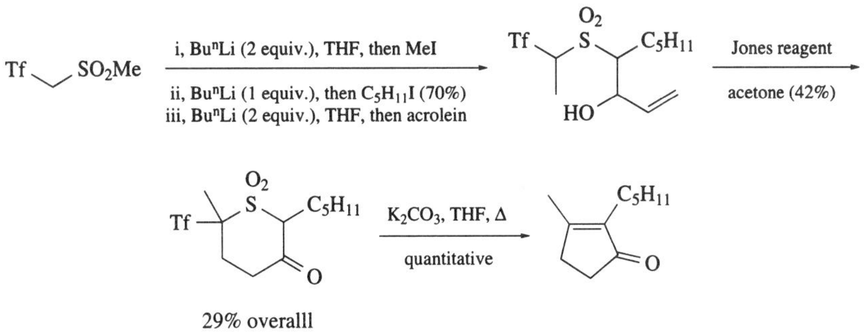

Scheme 42

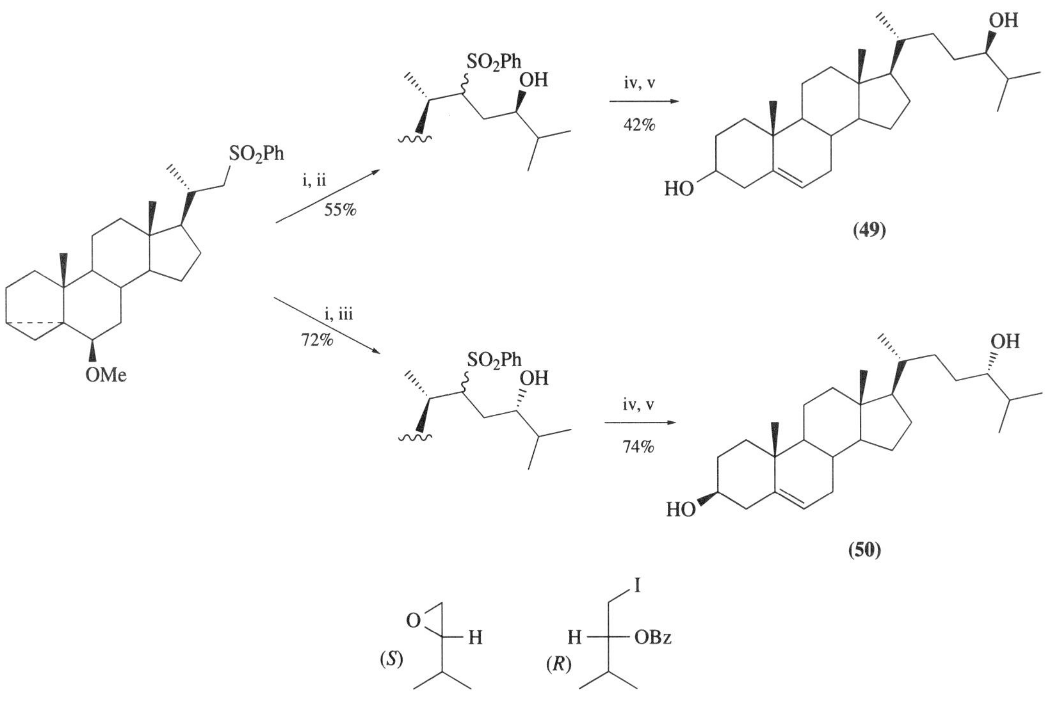

Scheme 43

i, LDA, THF, −60 °C; ii, (**51**) (prepared *in situ*), −60 °C to RT (55%); iii, (**52**), −60 °C to RT (72%); iv, Li, Et$_2$NH, 0 °C; v, *p*-TsOH, dioxan/H$_2$O, Δ

Scheme 44

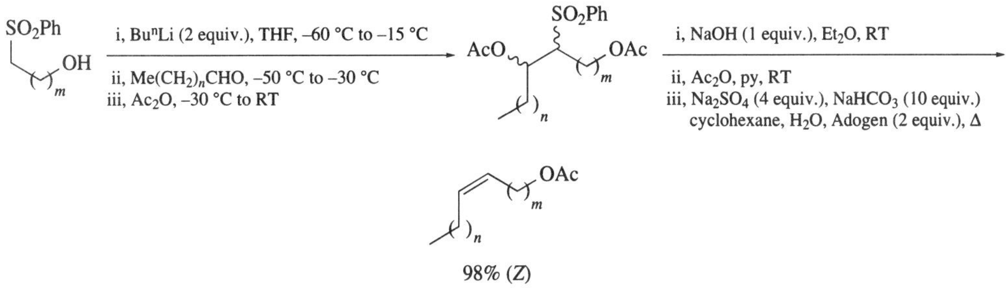

$n = 2, m = 7; n = 1, m = 8; n = 3, m = 8$

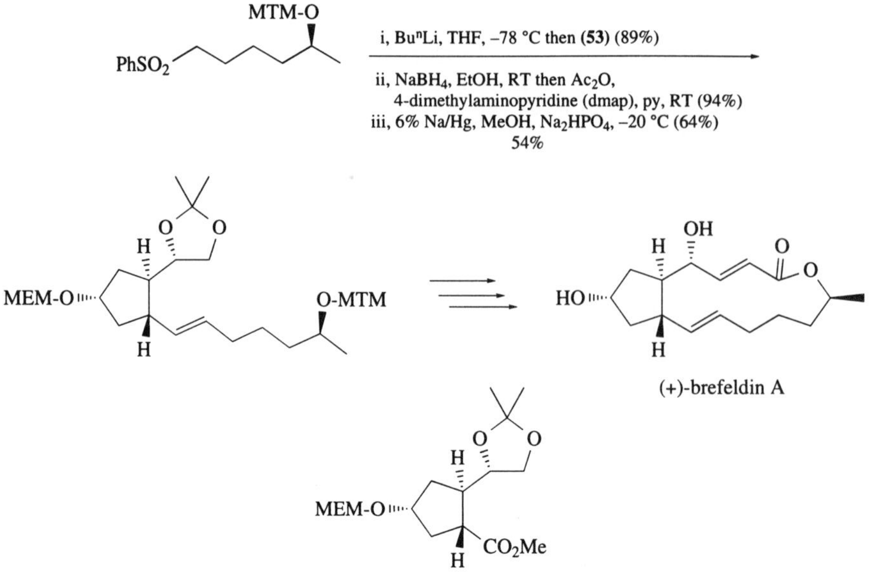

(53)

Scheme 45

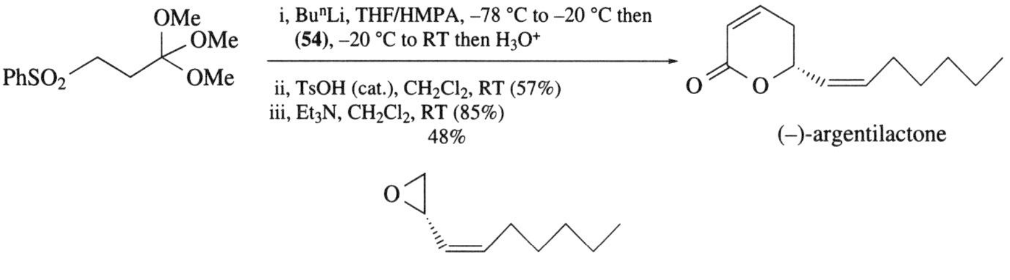

(54)

(55)

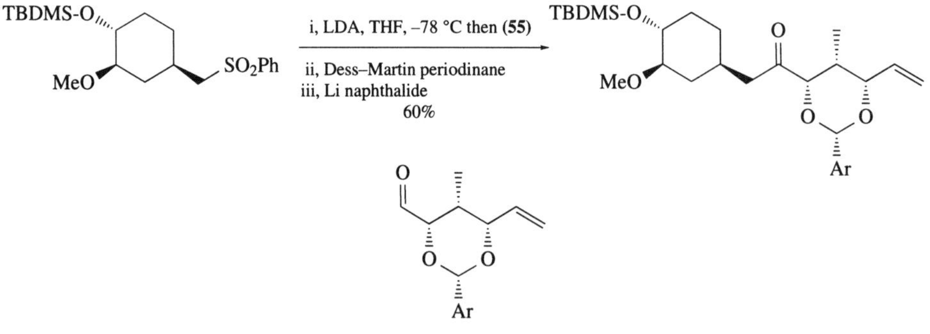

(55)

(56)

A concise, stereocontrolled, and high-yield synthesis of heliannuol A hinges on an intramolecular sulfone cyclization conducted under high dilution (Equation (57)).[32] Desulfonation and methyl ether cleavage afforded the natural product.

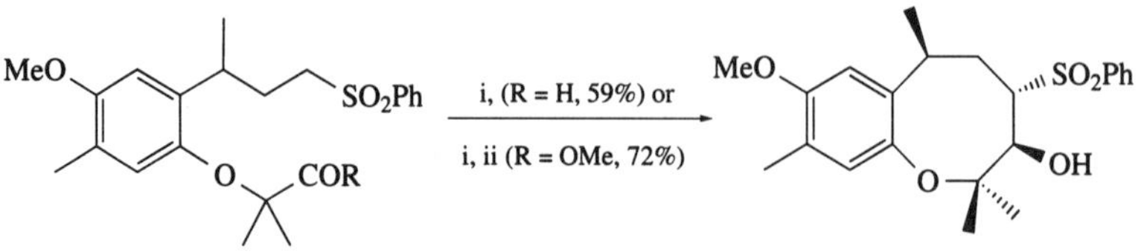

(57)

i, LiHMDS (2 equiv.), THF, 0 °C, substrate 0.005 M in THF; ii, $NaBH_4$, MeOH, 0 °C

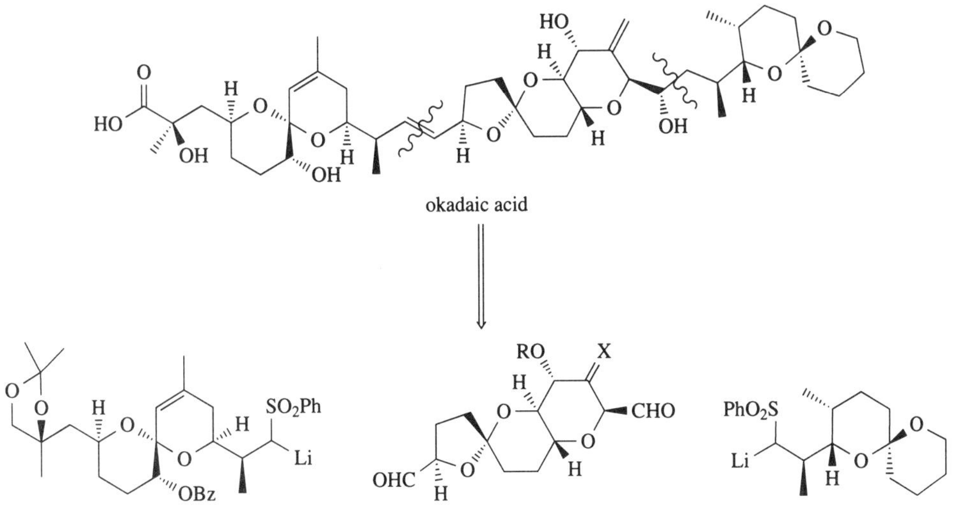

Scheme 46

1.3.5.8 α-Sulfonyl allylic carbanions

Regiospecific alkylation of α-lithio 3-methyl-3-sulfolene followed by cheletropic SO_2 extrusion constitutes an efficient route to monoterpenes (Equation (58)).[381,382]

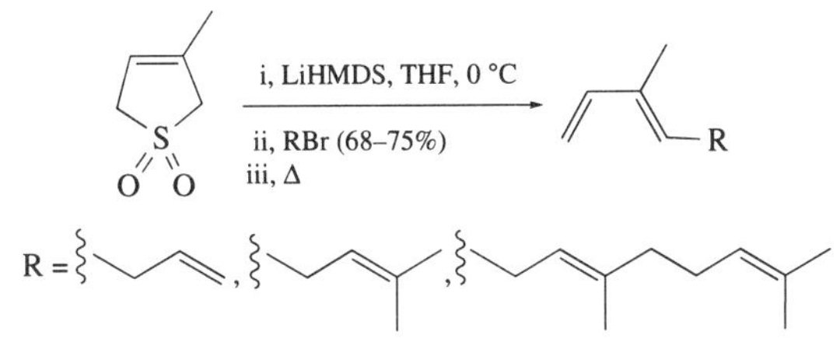

$$\text{(58)}$$

Thermodynamic conditions allow the conversion of the initial 1,o-dilithio species of allyl phenyl sulfone into the *gem*-dianion. Subsequent cycloalkylation with biselectrophiles proceeded in high yield and with good diastereoselectivity (Equation (59)).[383]

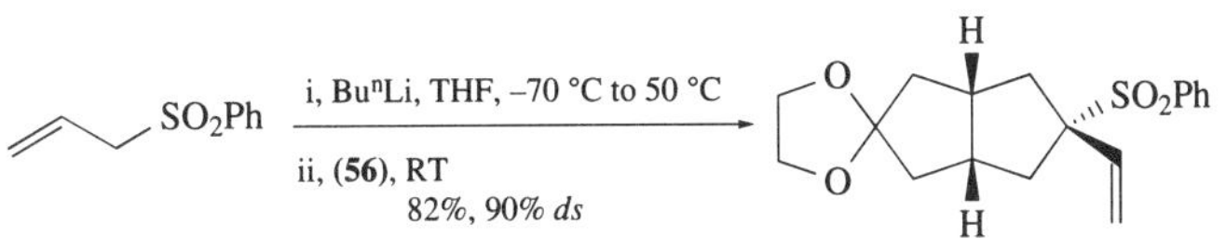

$$\text{(59)}$$

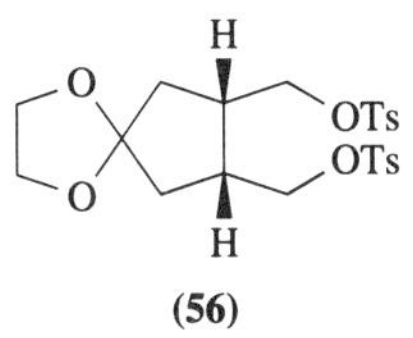

1.3.5.9 Acyl anion equivalents

A convergent synthesis of the antibiotic calcimycin (A23187) involved α-lithio dithiane chemistry to prepare an open-chain precursor to the dioxaspiro segment (Equation (60)).[384] Among numerous applications of lithiodithianes is the outstanding contribution by Kishi in the construction of debromoaplysiatoxin and aplysiatoxin.[385]

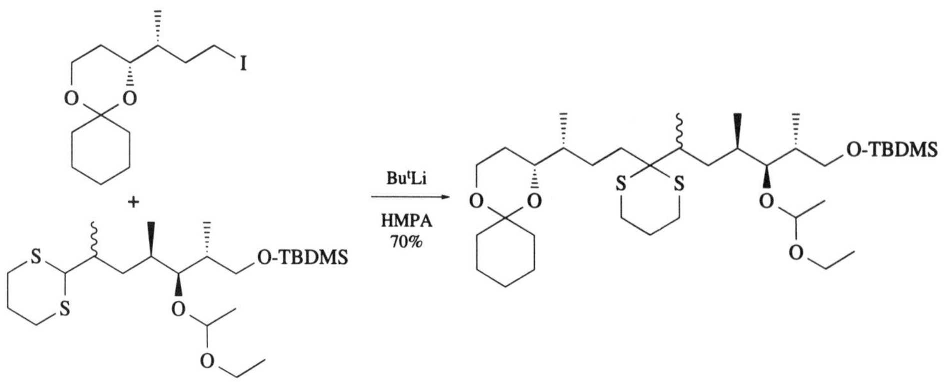

Modest to good 1,6-asymmetric induction may be achieved in the reaction of a variety of lithiated chiral α-(1,3-dithian-2-yl) acetals with aldehydes (Equation (61)).[386]

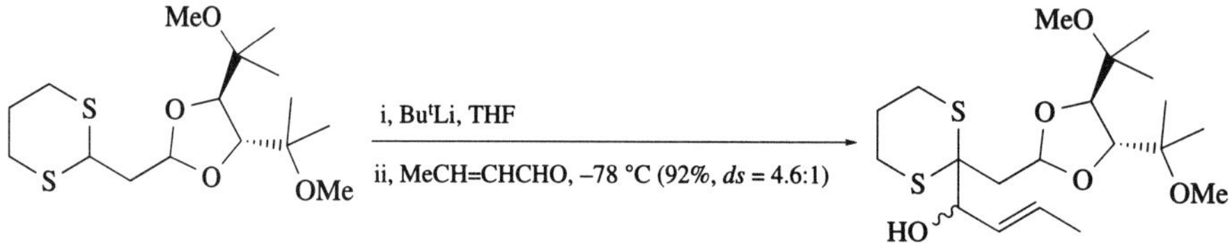

Owing to a relatively high acidity (pK_a ~23 (THF)), a range of common lithio bases may be used for deprotonation of ethynyldithianes, the γ-silyl derivatives being the most effective substrates.[387] This acyl anion equivalent was used in a biomimetic approach to the northern portion of the corticoids (Scheme 47).[388]

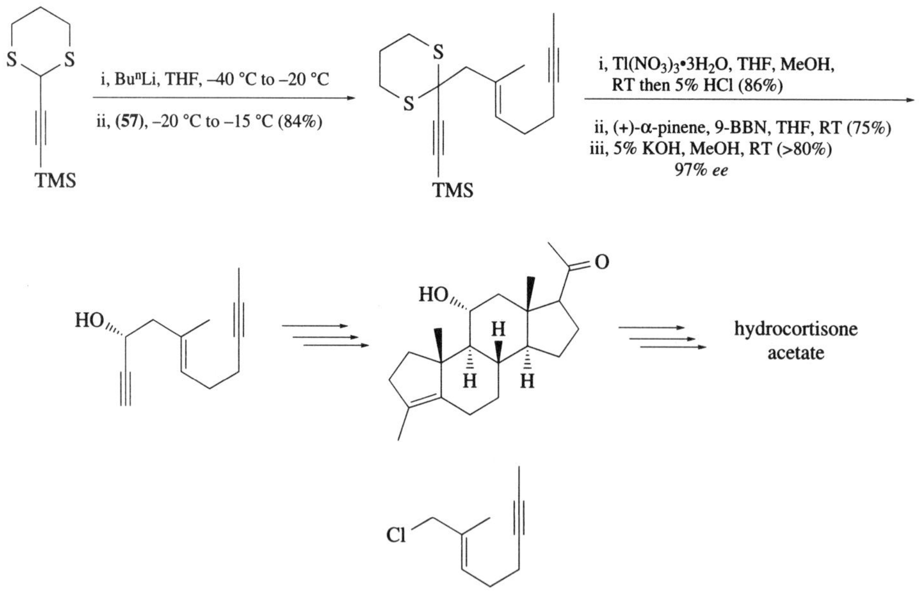

Scheme 47

In a synthesis of the potent prostacyclin analogue (+)-isocarbacyclin, a thioacetal S-oxide served as a useful acyl anion equivalent which has the advantage of liberating the carbonyl functionality under mild conditions (Scheme 48).[389]

Selenoacetals may be used in diastereoselective reactions. The presence of a chiral centre in the β-position induces diastereoselection in a lithium–selenium exchange which may be preserved (under strict conditions) in reaction with carbonyl compounds (Equation (62), R^1 = Ph).[390] The methylthio derivative undergoes rapid epimerization leading to the opposite stereochemical result.

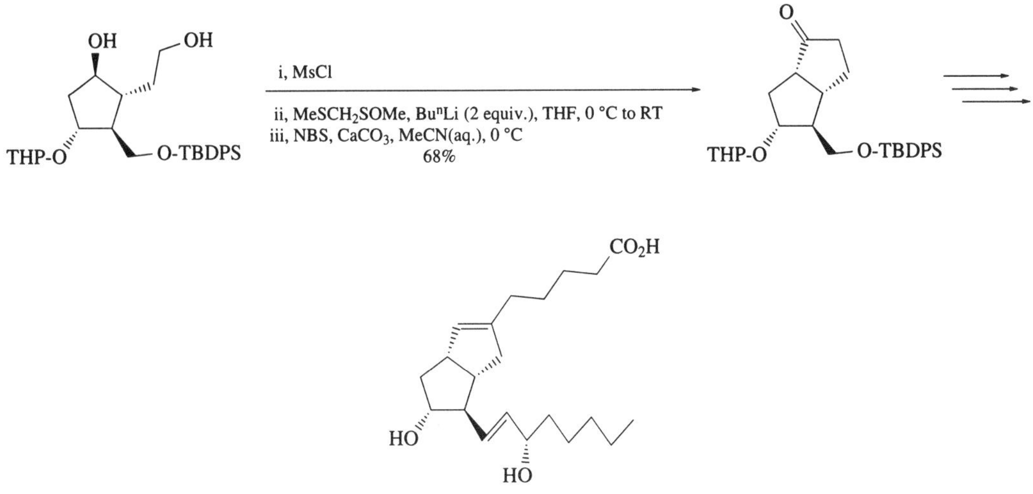

Scheme 48

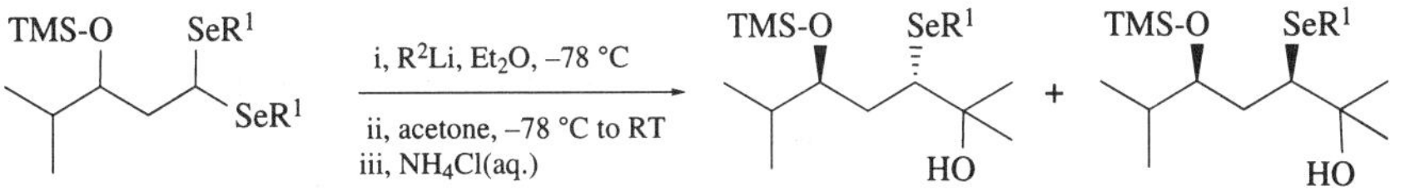

$$R^2Li = Bu^nLi, R^1 = Ph \quad 9:91 \ (92\%)$$
$$R^2Li = Bu^sLi, R^1 = Me \quad 84:16 \ (79\%)$$

(62)

Bisalkylation of a lithiated α-silyldithiane with chiral epoxides affords enantiopure 1,5-diols, suitable for further functionalization. The process presumably involves a 1,4-Brook rearrangement (Equation (63)).[391]

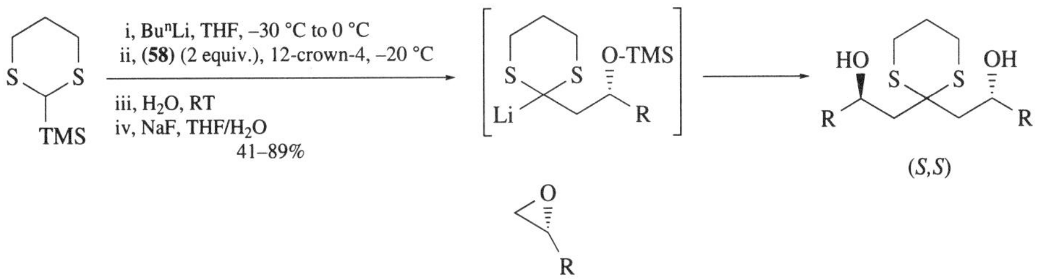

(63)

1.3.5.10 Sulfoximines

Chiral sulfonimidoyl-stabilized carbanions, pioneered by Johnson,[392] have found wide application for the generation–resolution of a variety of systems, for example, alcohols, ketones, and alkenes. This protocol involves the addition of an organolithium to a prochiral compound, resolution by chromatographic separation of the derived diastereomeric mixture, and excision of the chiral sulfur moiety by reduction or elimination. Illustrative of application is the efficient construction of the carbacyclin skeleton (Scheme 49).[393]

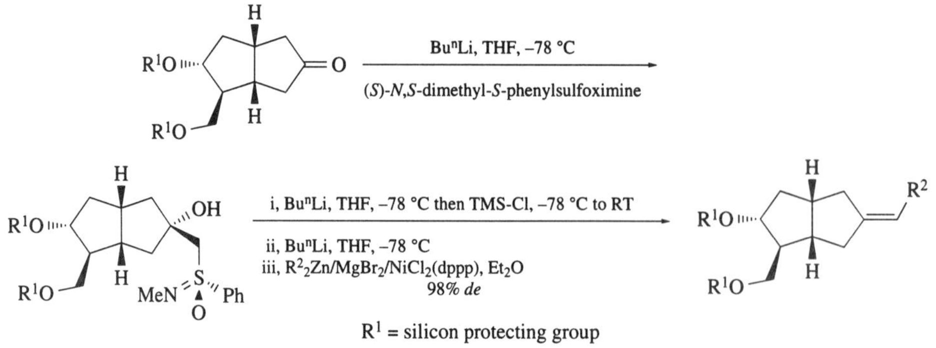

Scheme 49

1.4 ARYLLITHIUM AND HETEROARYLLITHIUM ANIONS

1.4.1 Introduction

The classical metal–halogen exchange reaction for the generation of aryl- and heteroaryllithiums has been superseded by the emergence of the DoM reaction which allows, by the selective use of a variety of DMGs, to rationally attack problems of regiospecific construction of polysubstituted aromatics and heteroaromatics.[1,9,33-5,37,38,394-7] In spite of mechanistic uncertainty,[398] efficient and abbreviated sequences utilizing readily accessible DMGs, suitable for further manipulation, have become the norm. The increasing number of DoM reactions on a commercial scale[399] attests to the increasing general utility of this methodology.

1.4.2 Directed *ortho*-Metallation Based Methods

Compared with benzene, the kinetic deprotonations of DMG-bearing aromatics are enhanced by coordination (e.g., OMe or NR_2), inductive (e.g., F or CF_3), or combined (e.g., $CONR_2$, SO_2NR_2) effects of the heteroatom substituents. In a qualitatively predictive sense, coordination dominates inductive effects for DMGs on the same aromatic nucleus. The use of a polar solvent or coordinating additive (e.g., TMEDA) can shift the selectivity to favor deprotonation *ortho* to inductive DMGs. The deprotonation event follows with a sizable kinetic isotope effect, leading to a coordinatively stabilized *ortho*-DMG lithiated aromatic undoubtedly as an aggregated, oligomeric species. The continuing accumulation of carbon- and heteroatom-based DMGs indicates present and future synthetic potential; inter- and intramolecular competition studies have provided a qualitative hierarchy of DMG potency for evaluation in synthetic problems.[35,400-10]

1.4.3 Arene Metallation by Lithium–Proton Exchange

1.4.3.1 Carbon-based directed metallation groups

Amides (secondary and tertiary)[35] and oxazolines[411] constitute the most widely used carbon-based DMGs, and have served decisively in the construction of complex aromatics, including natural products and pharmacologically active molecules (Schemes 50–55, Equation (64)).[412-20] Conditions have (finally) been reported (Bu^{s}Li/TMEDA, THF, −90 °C) for the effective *ortho*-lithiation of benzoic acids, a disclosure of obvious synthetic potential, especially for industrial applications.[410] *ortho*-Lithiation of benzoic esters can suffer from competing self-condensation; however, under Martin's *in situ* trap conditions,[97,123] hindered esters (isopropyl,[97] *t*-butyl[421]) are efficiently alkylated (Equation (65)). Alternatively, benzoic esters (even methyl esters) can be *ortho*-metallated in high yield, at ambient temperature, using Eaton's $(TMP)_2Mg$ reagent.[422]

The introduction of the remote metallation concept, based on the complex induced proximity effect (CIPE),[423] has led to the development of a general method for biaryl amide to fluorenone conversion which complements the classical Friedel–Crafts regimen and allows the preparation of naturally occurring systems (Equation (64)).[420]

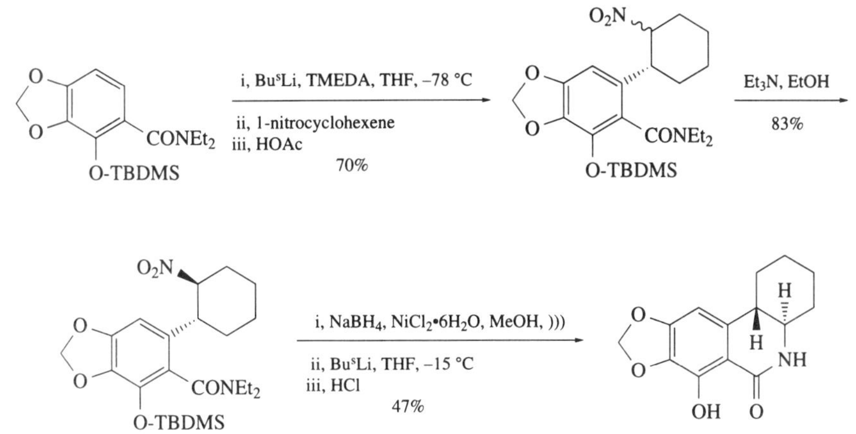

pancratistatin model

Scheme 50

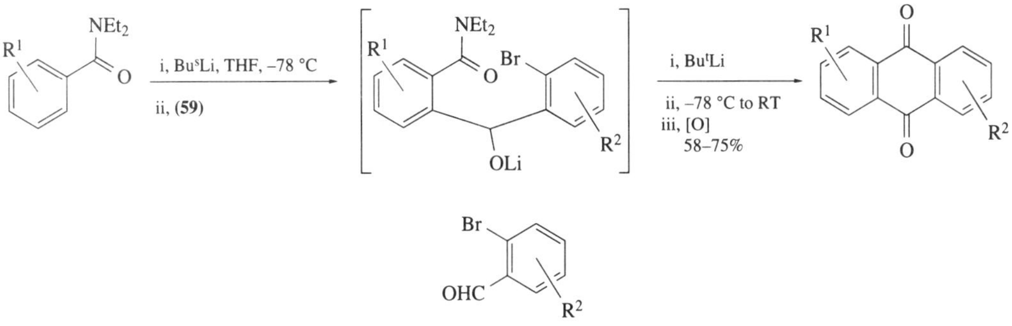

(59)

Scheme 51

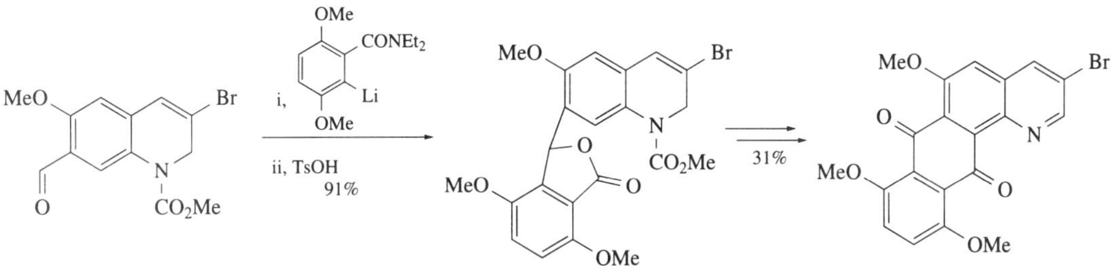

subunit of dynemicin A

Scheme 52

To heighten synthetic utility, several regimens for the conversion of oxazolines[411,424] and the recalcitrant amides[35] to other, synthetically useful functional groups (for oxazolines: carboxylic acids, aldehydes, nitriles, *N*-vinylamides, β-phenethylamines, amino alcohols; for amides: carboxylic acids, aldehydes, alcohols) have been developed. This problem is minor if the electrophiles introduced contain heteroatoms to facilitate hydrolysis by anchimeric assistance, for example, Schemes 52, 54, and 55. Solutions to this problem have been provided in several amide variants (Schemes 56,[425] 57,[396] and 58[426]). Scheme 57 illustrates the use of α,α-disilylation in order to create sufficient steric hindrance to prevent RLi attack on the normally vulnerable dimethylamide which, once unmasked, behaves well in hydrolytic and reductive conversions. The bis(silyl) amide DMG exhibits additional synthetic utility for heteroring annelation and acyl enamine formation via a Peterson-like processes. Secondary amides are strong DMGs which are more susceptible to hydrolysis. Dianionic species, generated from *N*-allyl- and *N*-vinylamides, overcome solubility problems observed with other secondary amide DMGs and provide a route to *ortho*-substituted primary amides whose conversion to other functional groups by

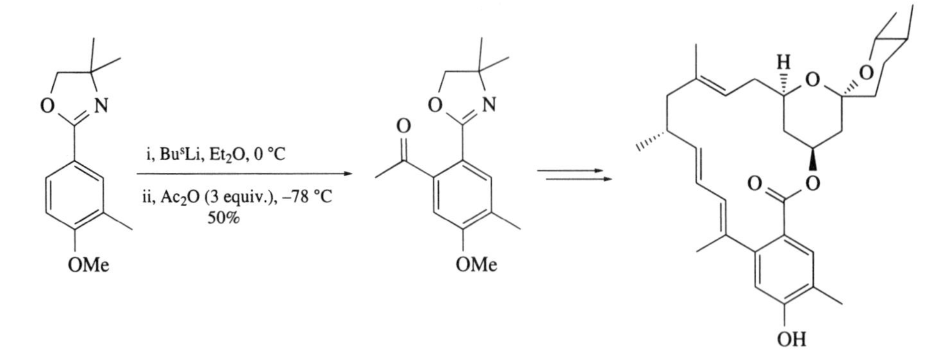

milbemycin β_3

Scheme 53

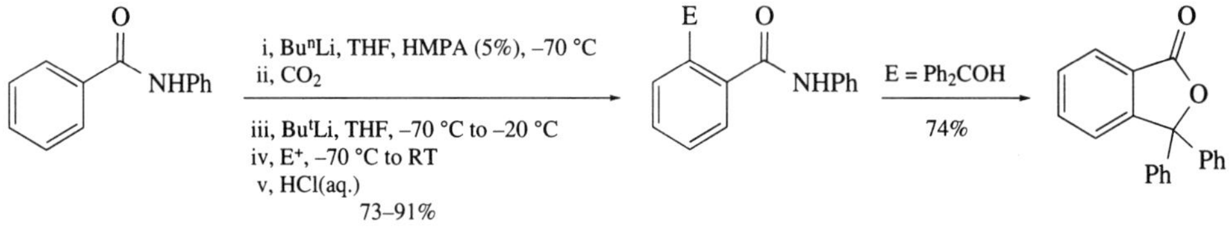

Scheme 54

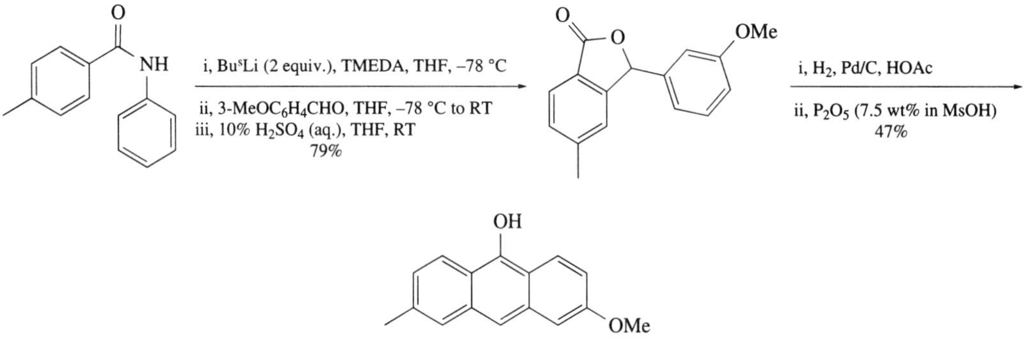

precursor to peptide spacer

Scheme 55

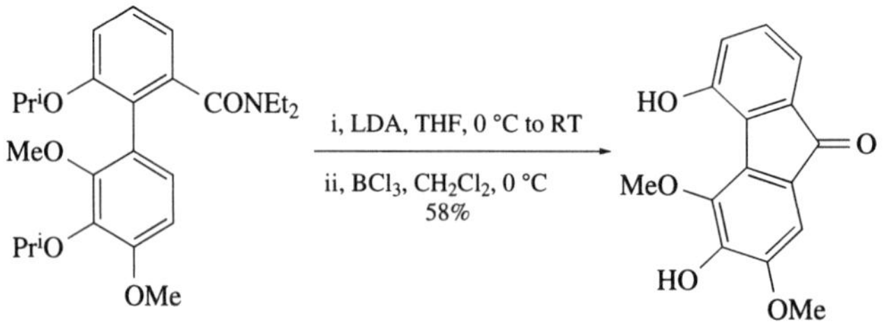

dengibsinin

(64)

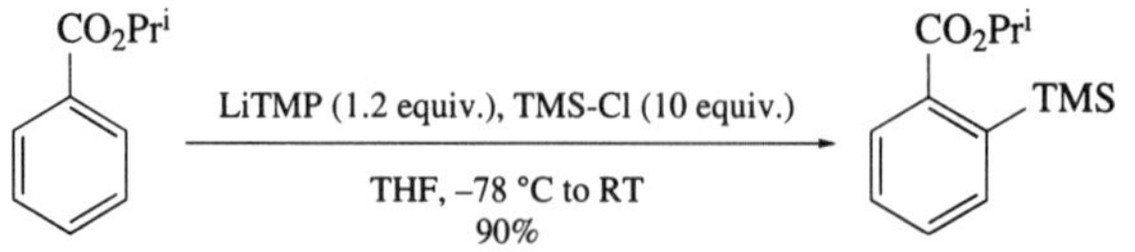

(65)

conventional means is less problematical (Scheme 58). The future development of the CO_2-DMG may render the hydrolytic difficulties of amides (and oxazolines) a nonissue.[410]

Tetrazole (Equation 66)[427] and imidazoline[428] DMGs have attracted recent attention as means of accessing biologically interesting aromatics.

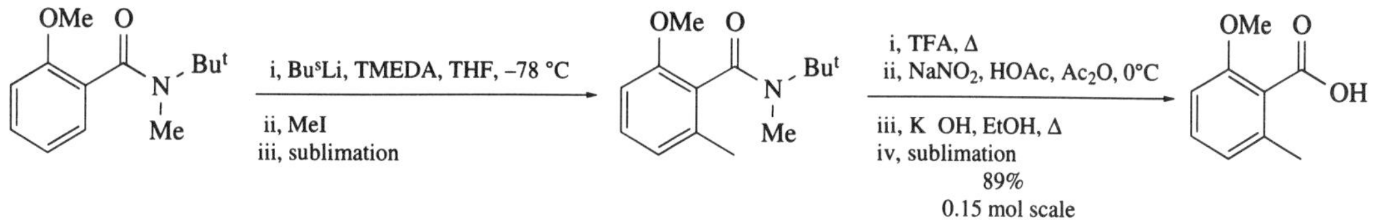

Scheme 56

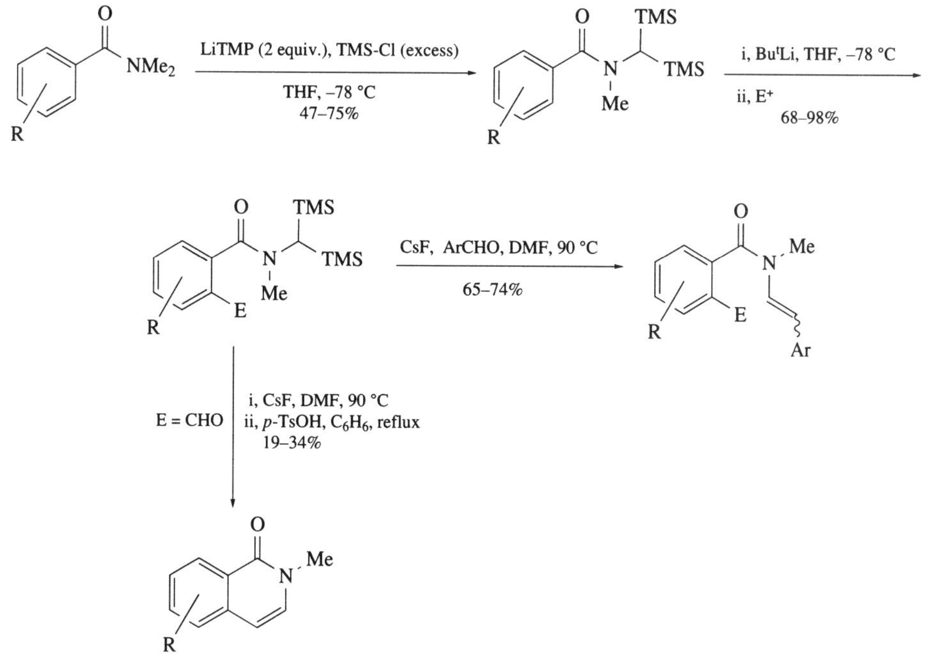

Scheme 57

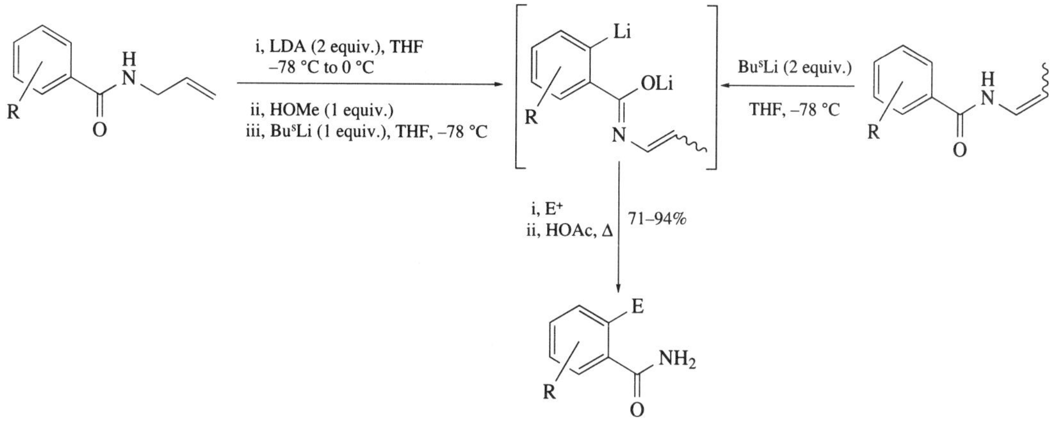

Scheme 58

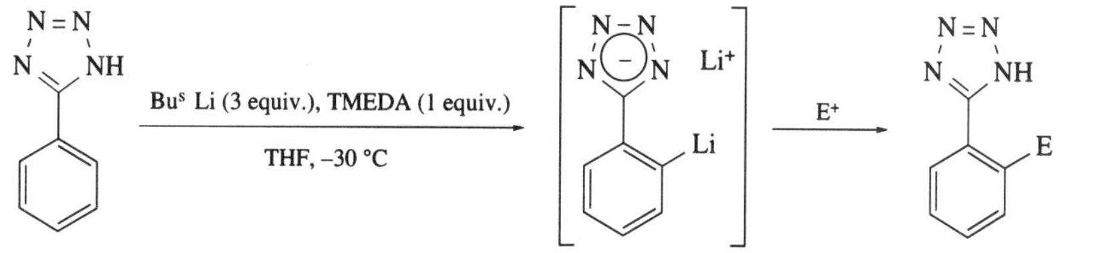

$$E = Me\ (78\%),\ MeCH(OH)\ (95\%),\ CH_2=CHCH(OH)\ (51\%) \tag{66}$$

An interesting result, achieved by Brandsma et al., is the double deprotonation of phenylacetylene under superbase conditions, providing a route to fused heterocycles (Scheme 59).[429]

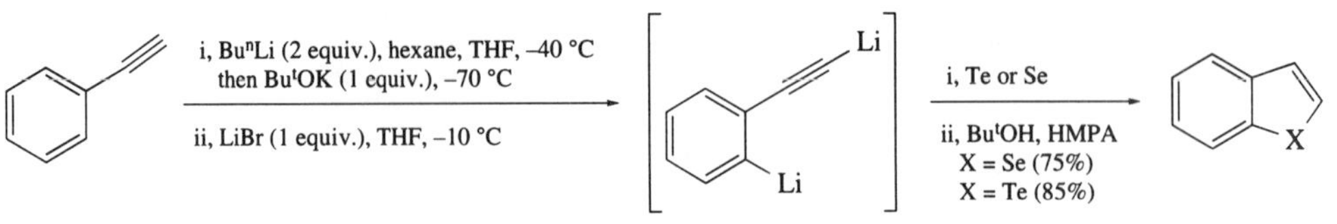

Scheme 59

Although advances in the use of imine DMGs have occurred,[430] the α-amino alkoxide, generated *in situ*, has emerged, especially through the contributions of D. L. Comins, as the most widely used masked aldehyde DMG (Schemes 60 and 61).[125,431,432]

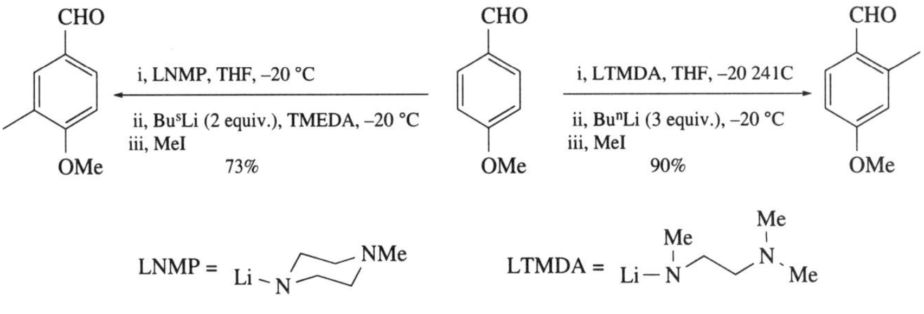

Scheme 60

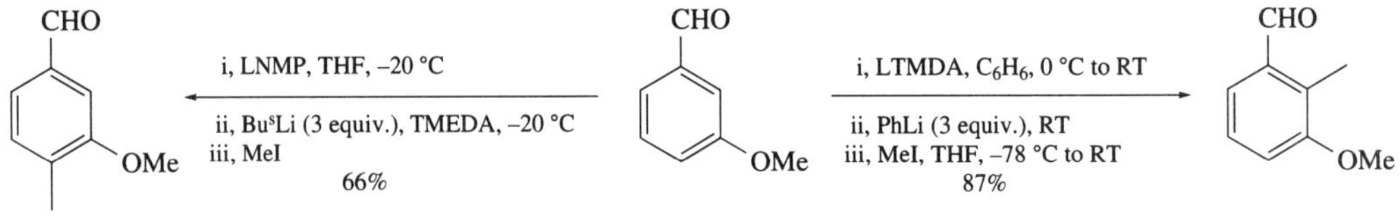

Scheme 61

Although not widely exploited, hindered acetal DMGs provide an alternative route to *ortho*-substituted benzaldehydes (Equation (67)).[433]

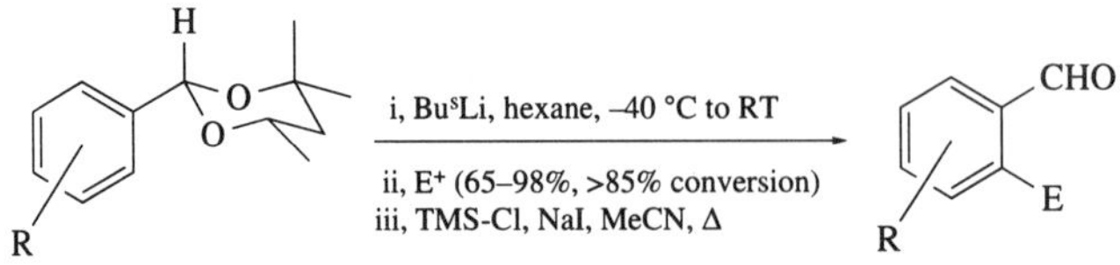

In spite of being low on the scale of potency,[35] the vintage CH_2NR_2 and CH_2N^-R DMGs continue to see synthetic application.[434] N-Silyl variants have provided a means of indirectly metallating primary benzylamines (Equations (68) and (69) and Scheme 62).[435-7] N-Pivaloylbenzylamines require a substituent at the α-position to prevent competing α-metallation.[438]

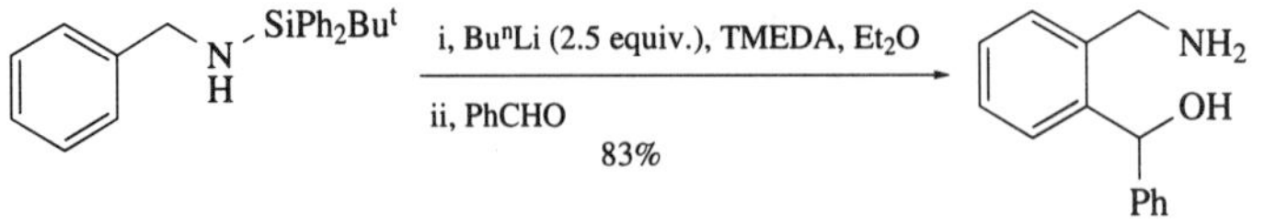

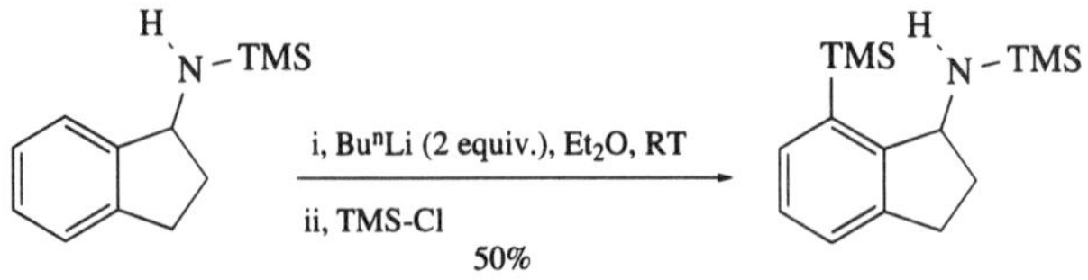

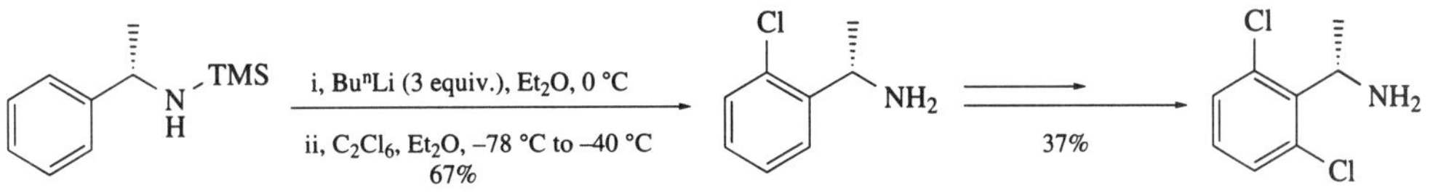

Scheme 62

The CH$_2$CH$_2$NR$_2$ DMG, a weaker homologue, is still useful in the context of regioselective, condition-dependent, synthesis of dopamine derivatives (Scheme 63).[439]

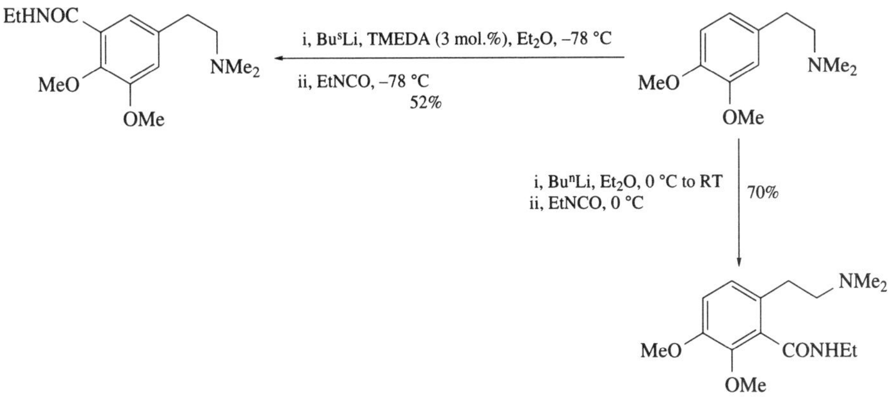

Scheme 63

In the 2-m-anisyl-N-pivaloylethylamine (**60**), regiocontrol of lithiation is dependent on the choice of base (Scheme 64).[440] Under n-butyllithium conditions, dihydroisoquinolines, which are difficult to prepare by classical Bischler–Napieralski and related methods, are obtained.

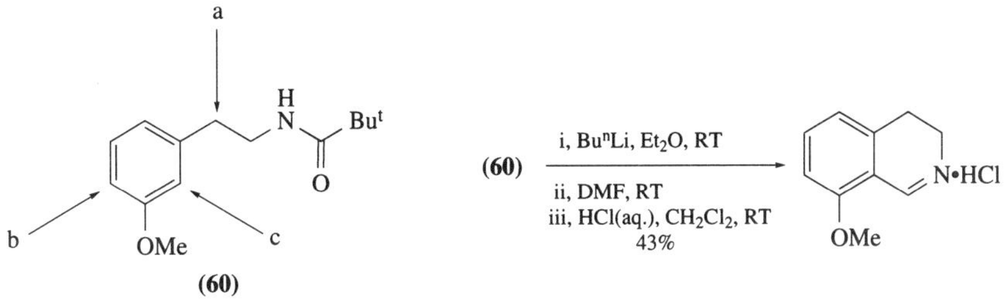

a: i, Bu^{t}Li, THF, –50 °C; ii, CO$_2$; iii, H$^+$ (77%)
b: i, LICKOR, THF, –95 °C; ii, CO$_2$; iii, H$^+$ (58%)
c: i, Bu^{n}Li, Et$_2$O, RT; ii, CO$_2$; iii, H$^+$ (61%)

Scheme 64

Although widely used, the CH$_2$O$^-$ DMG suffers from incomplete *ortho*-deprotonation.[441] O-Protected systems show advantage, especially in cooperation with *m*-alkoxy substituents (Equation (70)).[442]

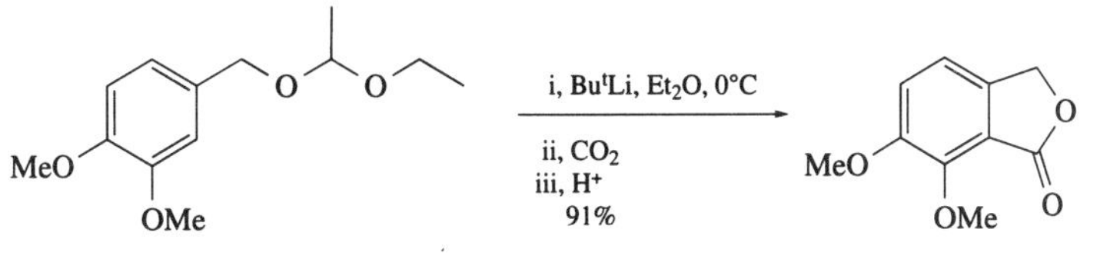

(70)

Benzylic O-carbamates show dual metallation characteristics depending on conditions and substrates, that is, *ortho*-metallation followed by Snieckus[266] rearrangement or benzylic metallation followed (in absence of electrophile quench) by 1,2-Wittig rearrangement (Scheme 65).

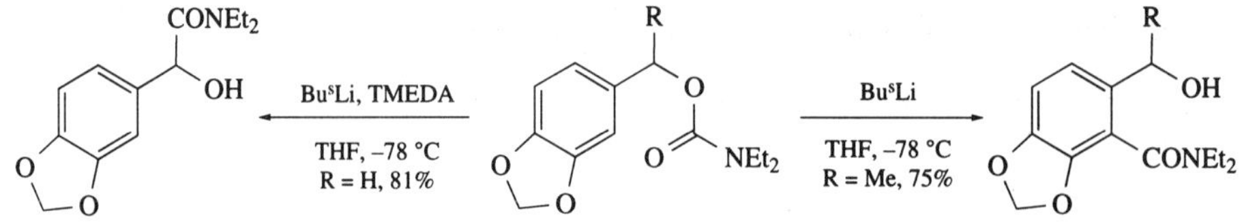

1.4.3.2 Heteroatom-based DMGs

(i) Aryl ethers (DMG = OR)

The formal formation of *ortho*-lithiated lithio phenoxide may be achieved directly[443–6] or by metal–halogen exchange[447] using excess alkyllithiums and leads, after electrophilic quench to several *ortho*-substituted phenols in moderate to good yields. Solubility problems of the dilithio species often have a limiting effect and this technique has not been widely exploited.

The modestly reactive OMe DMG, with which DoM chemistry was born,[33] has been extensively used.[448–52] As an alternative method to Bu^{n}Li for the metallation of anisole, Luche devised an ultrasound-aided *in situ* formation of *s*-butyllithium from the corresponding chloride (Equation (71)).[51]

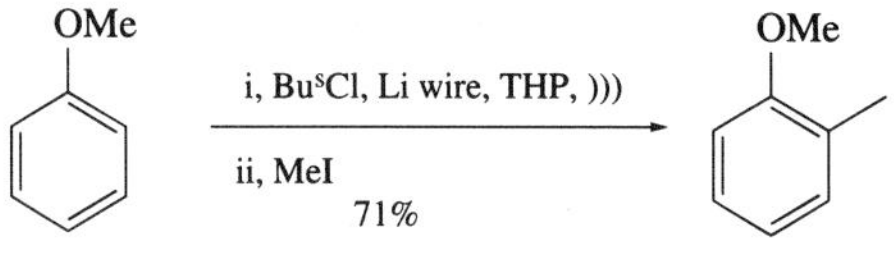

OMe has found extensive synthetic use in conjunction with other carbon- and heteroatom-based DMGs, especially when in a 1,3 relationship.[35] Scheme 66 illustrates the impelling forces of coordinating (OMe) and inductive (F) effects dictated by choice of base or additive.[453] Extension to dimetallation of *o*- and *p*-dimethoxybenzenes with *n*-butyllithium/TMEDA has as yet unexplored synthetic potential (Scheme 67).[454]

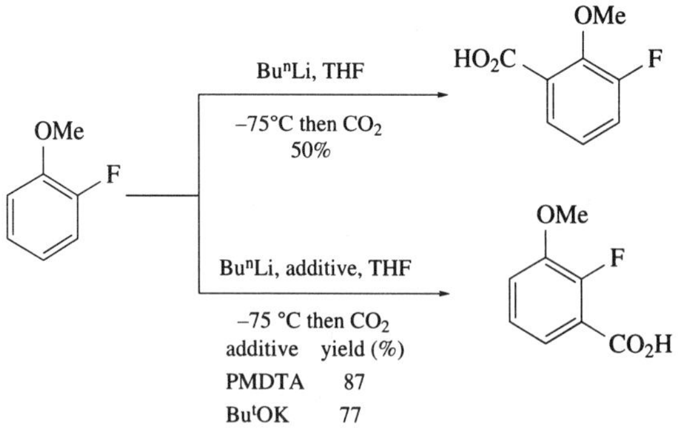

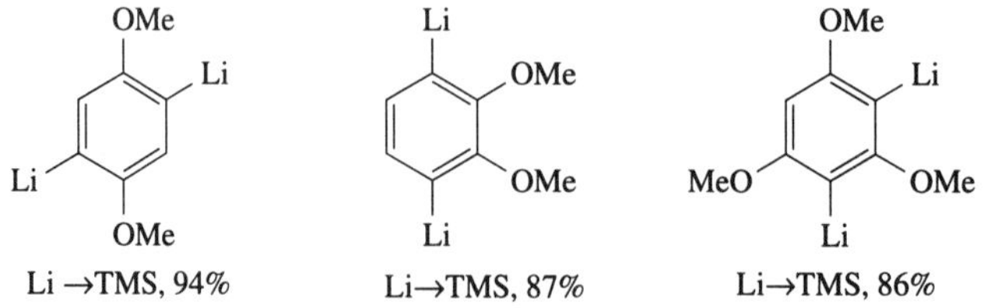

The methoxymethoxy (O-MOM) ether is a relatively powerful DMG and may be derived from phenols either by the agency of ClCH$_2$OMe (regulated carcinogen) or, less apprehensively, an acid-catalyzed reaction with dimethoxymethane.[455,456] Comparative experiments[457,458] suggest that coordination effects play a considerable role in facilitating selective *ortho*-deprotonation of O-MOM compared with OMe, for example, 1-methoxymethoxy-5-methoxynaphthalene is selectively lithiated at C-2 (Equation (72)).[458]

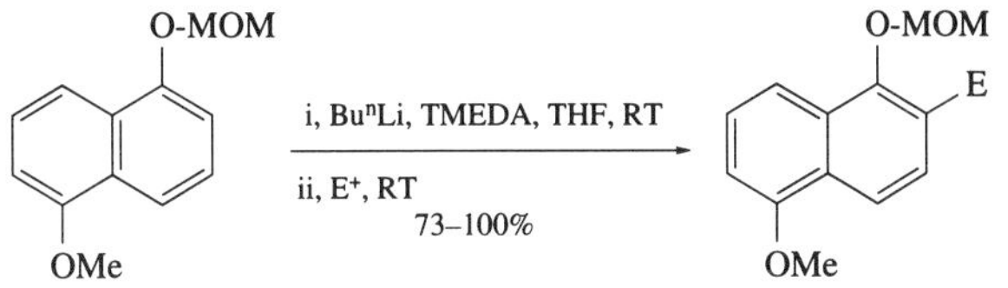

(72)

E = n-C_9H_{19}CH(OH), Bu^{n}CH(OH), CHO, Me

Among the multitude of applications of the O-MOM DMG,[456,459–61] the synthesis of pharmaceutically relevant coumarins (Scheme 68)[462] and intermediates for fridamycin E (Scheme 69) are illustrative.[463] In the latter case, *ipso* substitution or addition at position C-9 and C-10 of the central anthracene ring posed a general problem which was circumvented by the use of LICKOR base.[464]

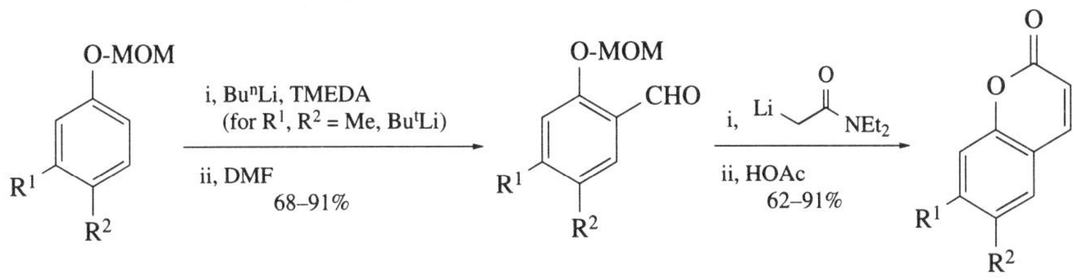

R^1 = Me, R^2 = H
R^1 = H, R^2 = H, Cl, Me, Ph

Scheme 68

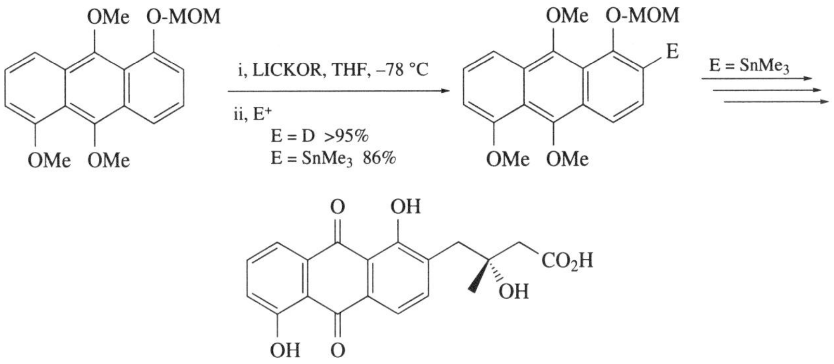

fridamycin A

Scheme 69

Although costly to prepare, O-SEM represents an excellent phenolic DMG (Scheme 70) (SEM = 2-(trimethylsilyl)ethoxymethoxy).[465]

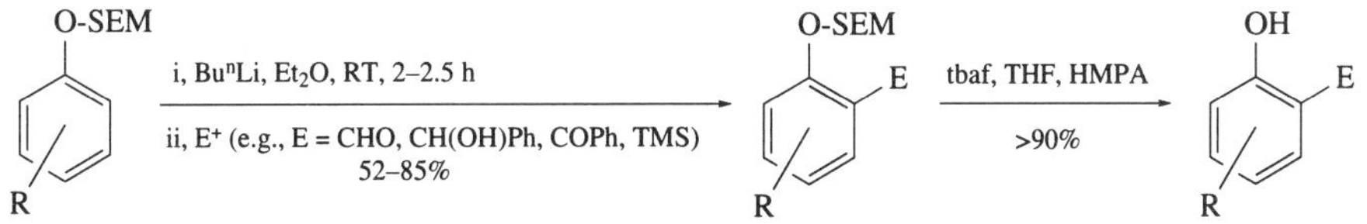

Scheme 70

In a qualitative hierarchical evaluation,[35] the *O*-carbamate ($OCONR_2$) has emerged as the most powerful DMG. The additional feature of the anionic *ortho*-Fries rearrangement encourages its use for around-the-ring DoM and allows diversity and flexibility in the construction of polysubstituted aromatics (Scheme 71).[466]

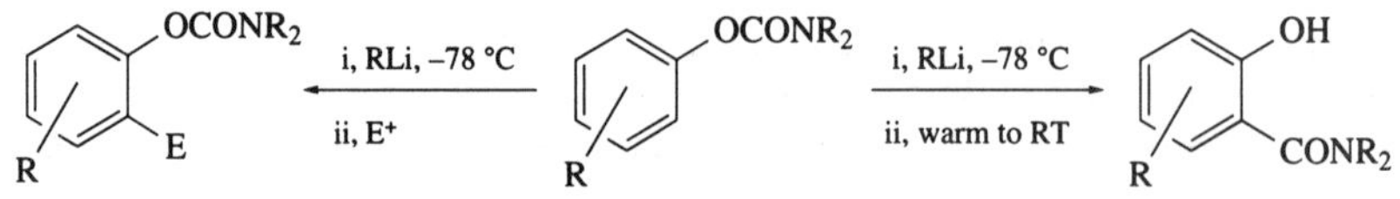

Scheme 71

The anionic *ortho*-Fries rearrangement has been demonstrated for other groups (OC(O)R,[467] OP(O)(NR$_2$)$_2$,[407,468] OP(O)(OR)$_2$,[469] OP(O)(OR)(OAr)[470]), some through the expediency of metal–halogen exchange.[471] Key use of DoM and anionic migration characteristics of the OCONEt$_2$ group is illustrated in the synthesis of ochratoxin B (Scheme 72),[472] and pancratistatin (Scheme 73).[473]

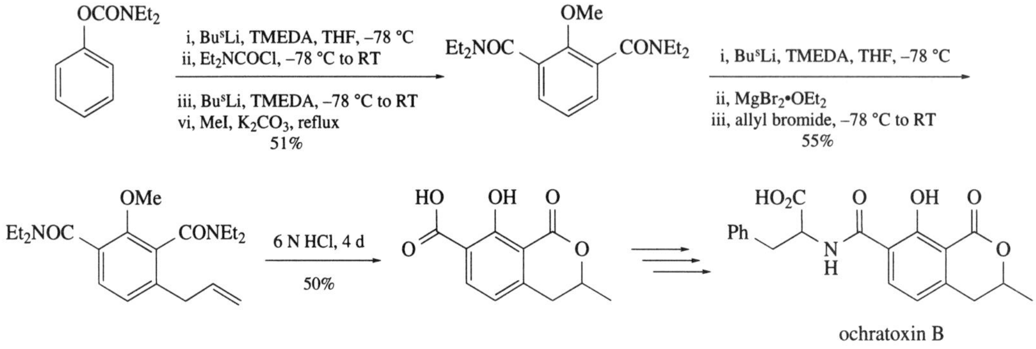

Scheme 72

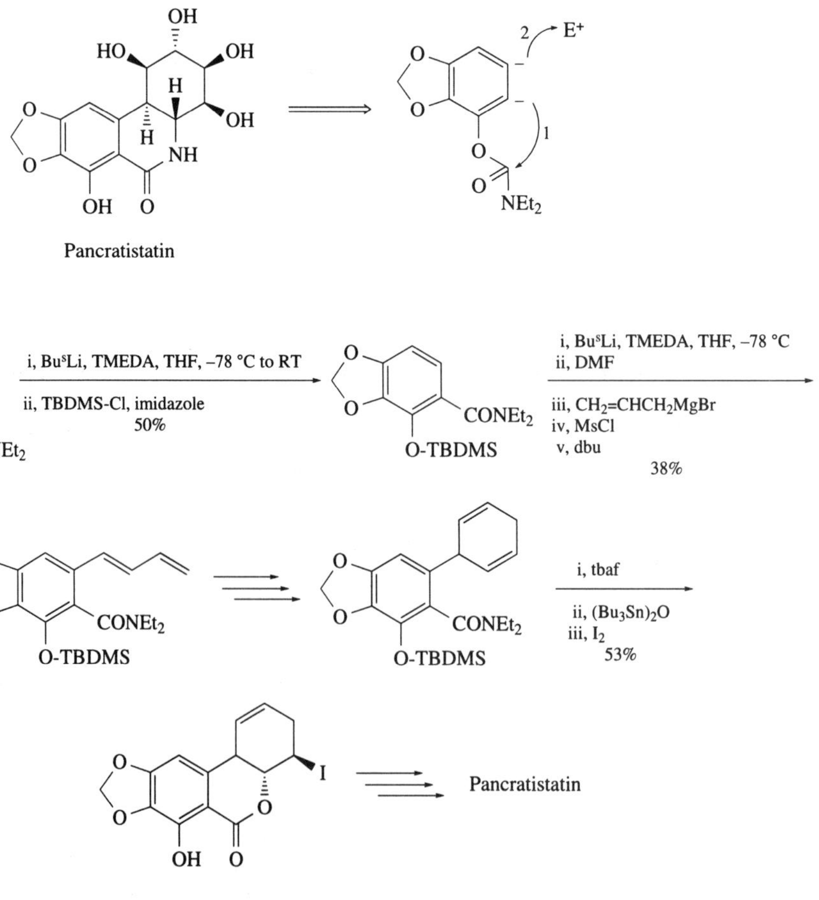

Scheme 73

The analogous *O*-thiocarbamate (OCSNR$_2$) serves as a DMG and allows the regiospecific preparation of *ortho*-substituted thiophenols via an intermediate Newman–Kwart (*O*-thiocarbamate to *S*-thiocarbamate) rearrangement step (Scheme 74).[474] In a similar manner to the oxygen analogue, the *O*-thiocarbamate DMG also undergoes anionic *ortho*-Fries migration.[475]

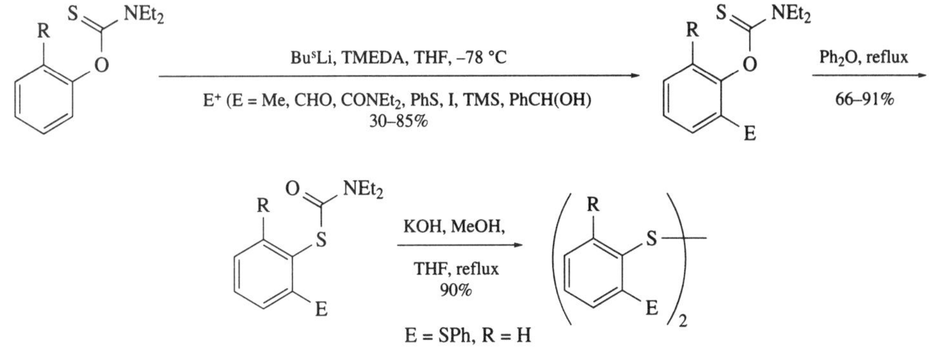

Scheme 74

The phosphoramidate [OP(O)[NMe$_2$]$_2$ DMG requires very low temperatures for clean lithiation–electrophile introduction (Scheme 75).[407] For the observation of smooth anionic *ortho*-Fries reaction, the choice of phosphoramidate substrate is crucial.[468]

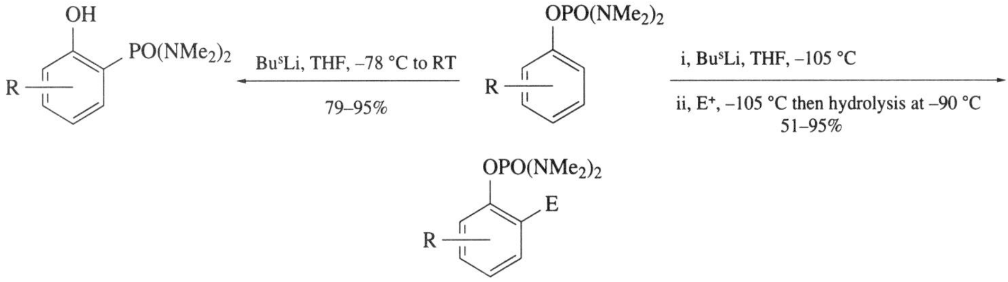

E = TMS, PhS, CH(OH)Me, CH(OH)C$_6$H$_4$NO$_2$-4, CH(OH)C$_6$H$_4$OMe-4, Ph$_2$C(OH)

Scheme 75

Biaryl *O*-carbamates without 3-substitution undergo rapid and high-yield anionic Fries rearrangement (Scheme 76).[476] However, silylative 3-protection allows remote lithiation, ring-to-ring carbamoyl transfer to afford highly substituted biarylamides. Their further transformation into dibenzo[*b,d*]pyranones and fluorenones, including natural products, has been achieved (see also Equation (64)).

(ii) Anilines (DMG = N-TMS, N-CO$_2$Bu^t, or N-COBu^t)

Since their discovery by Gschwend[477] and Muchowski,[478] the *N*-pivaloyl[479–83] and *N-t*-BOC[481–5] DMGs have been widely adapted. For example, a high-yielding modified Friedländer synthesis of quinolines has been devised (Equation (73)).[486] An interesting anionic *ortho*-Fries rearrangement of an *N*-tolyl-*N*-phenylpivalamide with excess alkyllithium has been reported (Equation (74)).[471] Isatins are available using both *N*-pivaloyl and *N-t*-BOC DMGs (Scheme 77).[487]

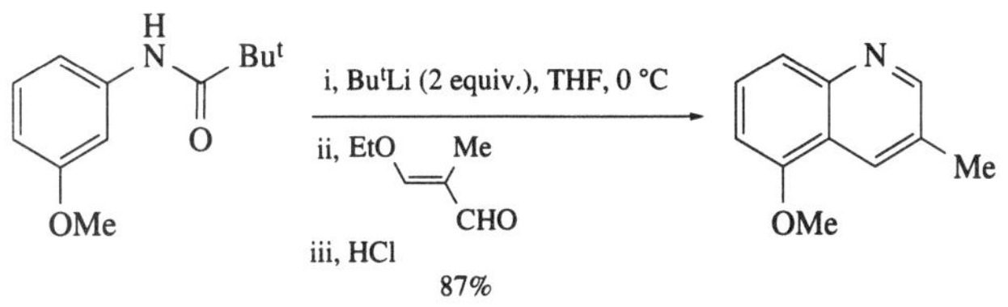

(73)

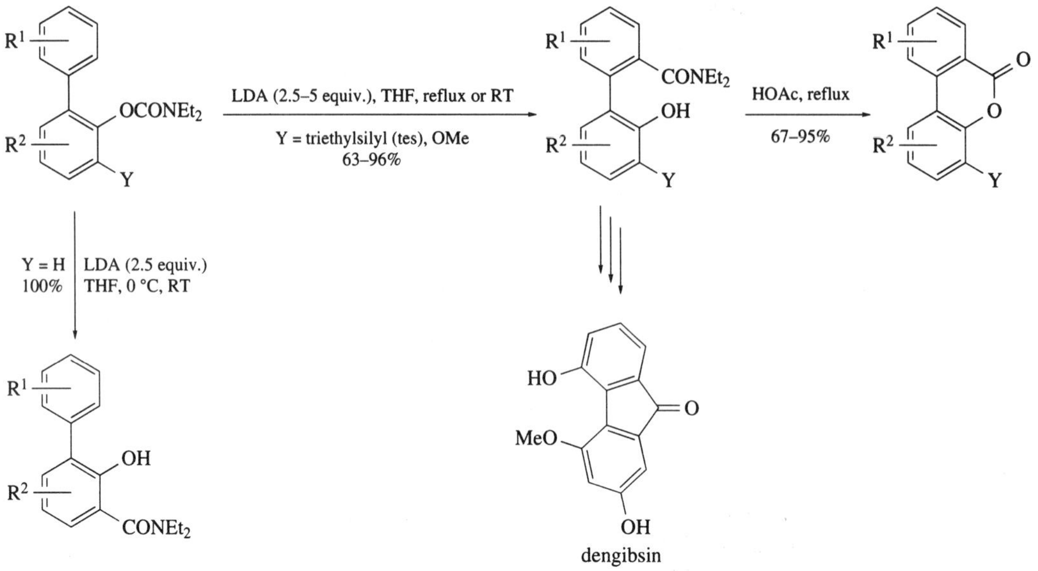

Scheme 76

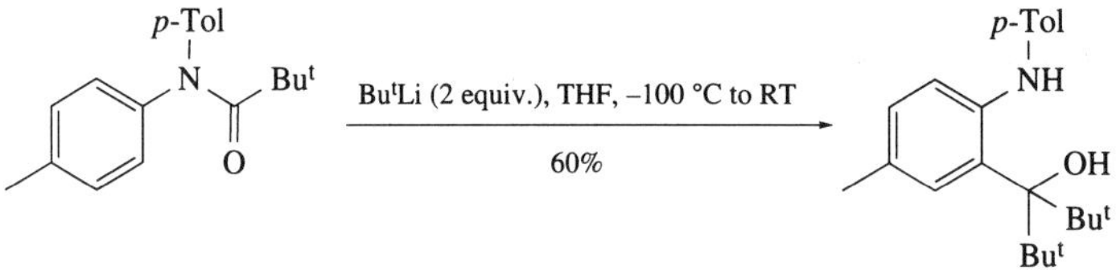

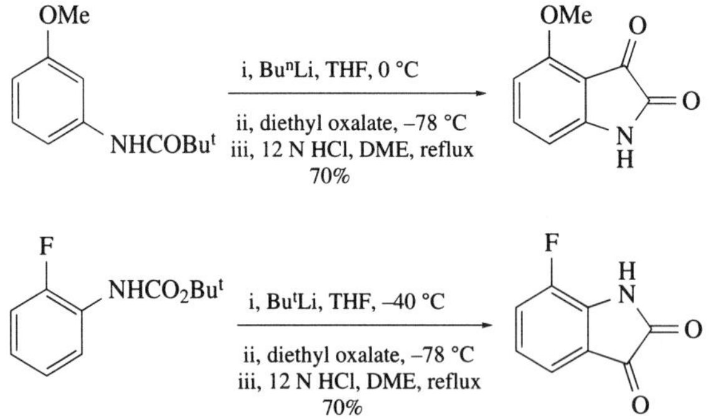

Scheme 77

The efficacious use of the *N*-TMS group in the one-pot synthesis of a tetrahydrocannabinol analogue (Equation (75))[488] invites broader application of this potential DMG.

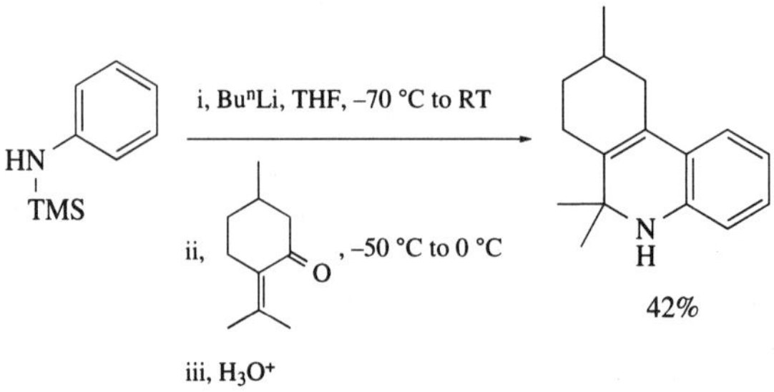

ortho-Substituted anilines and naphthylamines are available by the Katritzky CO_2 activation method (Equation (76)).[489]

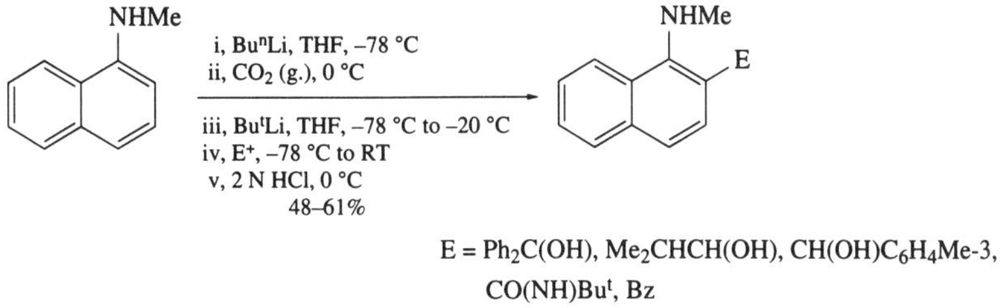

[Scheme/Reaction 76]

1-naphthyl-NHMe → 2-E-1-naphthyl-NHMe

i, Bu^{n}Li, THF, −78 °C
ii, CO$_2$ (g.), 0 °C
iii, Bu^{t}Li, THF, −78 °C to −20 °C
iv, E$^+$, −78 °C to RT
v, 2 N HCl, 0 °C
48–61%

E = Ph$_2$C(OH), Me$_2$CHCH(OH), CH(OH)C$_6$H$_4$Me-3, CO(NH)Bu^t, Bz

(76)

In order to evaluate the synthetic potential of the isocyanide DMG, further investigation is needed. In one reported example, treatment of phenyl isocyanide with *t*-butyllithium/TMEDA (2 equiv.) results in α-addition plus *ortho*-lithiation and gives, upon electrophile quench, *ortho*-substituted anilines and 3-metaloindolines.[490]

(iii) Sulfur derivatives

Perhaps owing to the strong coordination affinity of sulfur, thiophenol undergoes, in contrast with aniline, smooth *ortho*-metallation[446,491] leading to synthetically useful yields of substituted products including 2-phosphino- and 2-phosphinylbenezenethiol ligands (Scheme 78).[491]

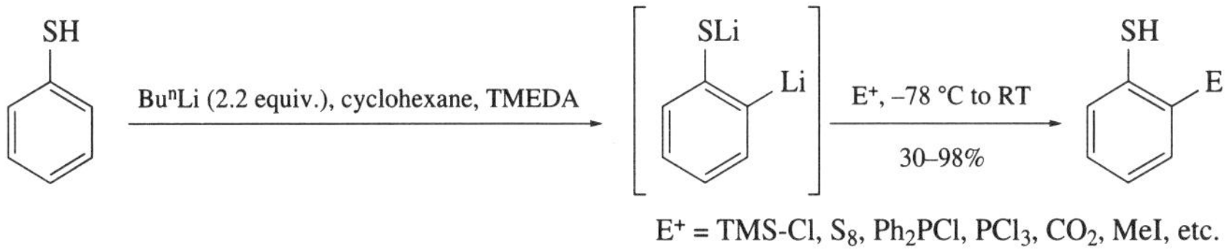

[Scheme 78: PhSH → (o-Li-C$_6$H$_4$-SLi) → 2-E-C$_6$H$_4$-SH]

Bu^{n}Li (2.2 equiv.), cyclohexane, TMEDA; E$^+$, −78 °C to RT; 30–98%

E$^+$ = TMS-Cl, S$_8$, Ph$_2$PCl, PCl$_3$, CO$_2$, MeI, etc.

Scheme 78

t-Butyl sulfones[406] and sulfoxides[408] (Scheme 79) serve as powerful DMGs (e.g., SO$_2$Bu^t > CONEt$_2$).

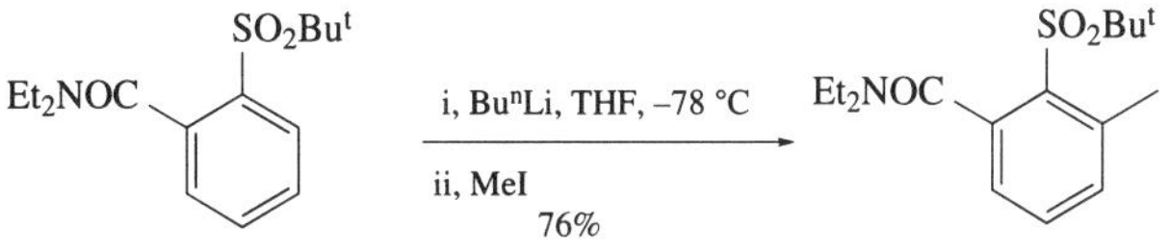

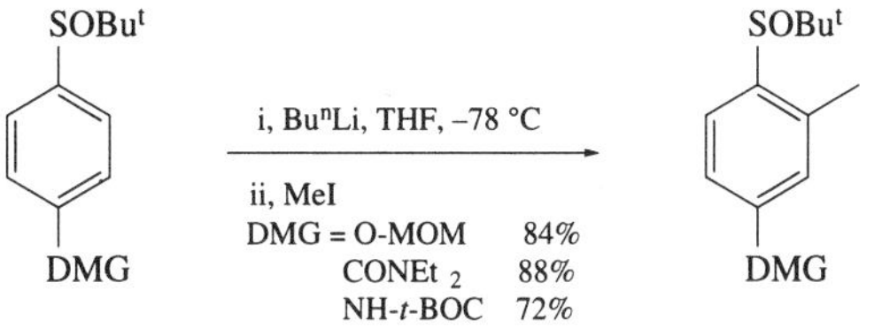

[Scheme 79 top: 2-(SO$_2$Bu^t)-C$_6$H$_4$-CONEt$_2$ → 3-Me product]

i, Bu^{n}Li, THF, −78 °C
ii, MeI
76%

[Scheme 79 bottom: 4-DMG-C$_6$H$_4$-SOBu^t → 3-Me product]

i, Bu^{n}Li, THF, −78 °C
ii, MeI

DMG = O-MOM 84%
CONEt$_2$ 88%
NH-*t*-BOC 72%

Scheme 79

Chiral sulfoxide DMGs have seen use for asymmetric inductions and may allow access to both enantiomers (Scheme 80).[492] The use of butyllithium as a base resulted in either metal–halogen exchange or addition to the sulfoxide moiety. Coupling of the chiral sulfoxide DoM chemistry with its nucleophilic displacement by Grignard reagents provides a route to axially chiral binaphthyls in excellent optical purity (Scheme 81).[493]

The known facile *ortho*-deprotonation of diaryl sulfones[97,329,494] has been applied to the preparation of boron compounds for use as toner composites in photocopiers.[495] Sulfonates (SO$_3$R) appear to be DMGs worthy of further exploration.[496]

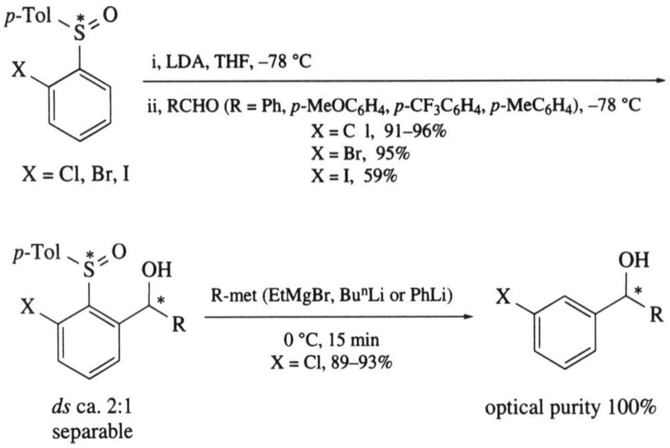

Scheme 80

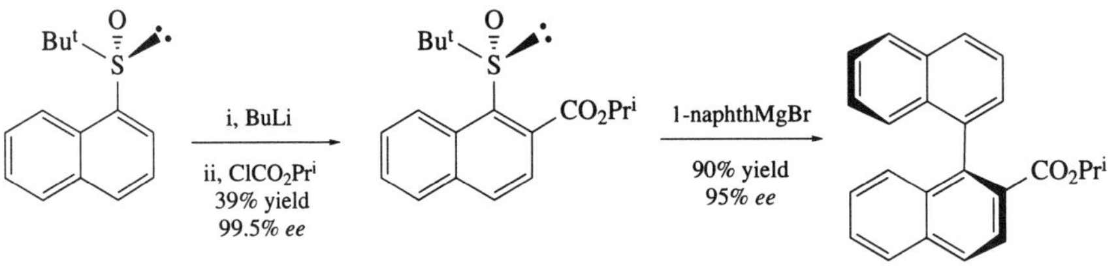

Scheme 81

(iv) Aryl halides (DMG = F)

As is well known, iodo- and bromobenzenes undergo diffusion-controlled alkyllithium-induced metal–halogen exchange reactions (see Section 1.4.4). The chloro group is a weak DMG and requires very low-temperature metallation conditions in order to prevent benzyne formation, for example, *ortho*-lithiated chlorobenzenes formed at −105 °C may be used in Michael reactions.[497] In contrast, the fluoro substitutent may be considered to be a strong DMG.[498,499] Intramolecular competition studies on *p*-fluorophenyl-*O*-carbamate showed that, with an appropriate choice of base, preferential deprotonation *ortho* to fluoro can be achieved.[409] Furthermore, 2-fluoroanisole undergoes exclusive 3-metallation using LICKOR or a strong complexing additive.[453] In a similar comparison between fluoro and amino DMGs, metallation occurs selectively *ortho* to fluoro.[73] An interesting application involves the smooth in-between deprotonation of a 1-chloro-3-fluoro-6-iodobenzene as the initial step of a route leading to benzene ring-substituted benzo[*b*]thiophene-2-carboxylates (Scheme 82).[500] A similar metallation was applied for the formation of biologically interesting benzo[*b*]thiophene derivatives.[501]

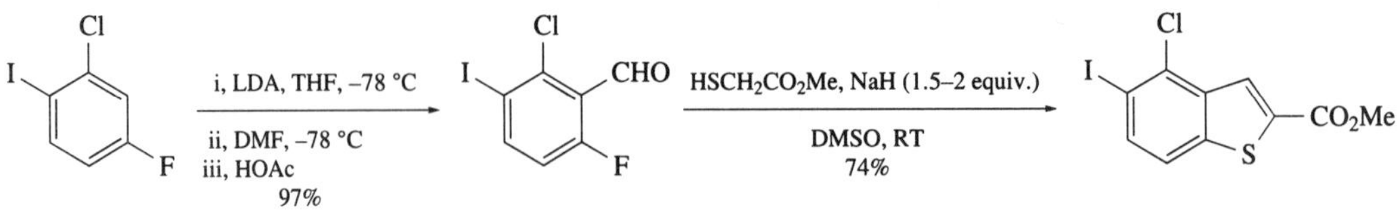

Scheme 82

As an indication of further scope, *ortho*-fluoro and tolyl anions may be formed as a function of type of base (Scheme 83).[502]

(v) Miscellaneous heteroatom-based DMGs

ortho-Lithiated heteroatom-tethered alkynes, obtained in a complex reaction, provide a one-pot route to 2,3-disubstituted benzofurans, indoles, and benzothiophenes (Scheme 84).[503,504]

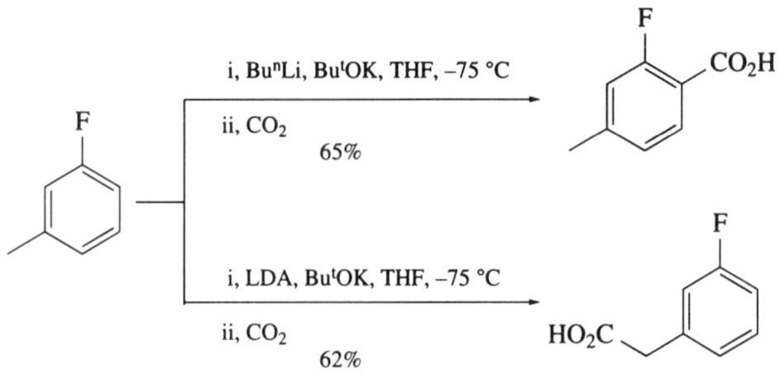

Scheme 83

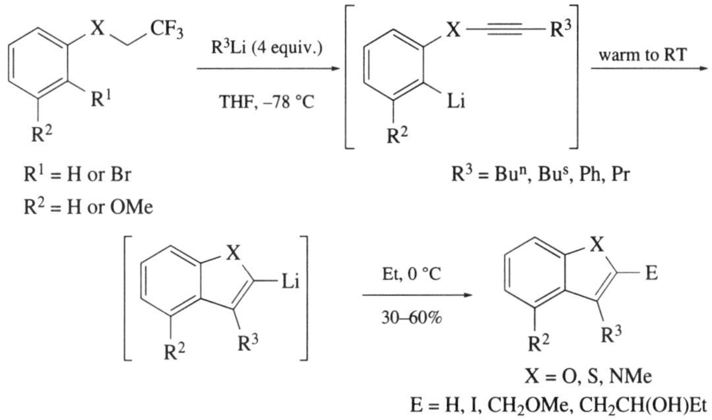

Scheme 84

Trimethylethylenediamino-substituted diphenylsilanes undergo DoM and were transformed *in situ* to the more stable isopropoxysilane derivatives leading, after oxidation, to substituted phenols (Scheme 85).[505]

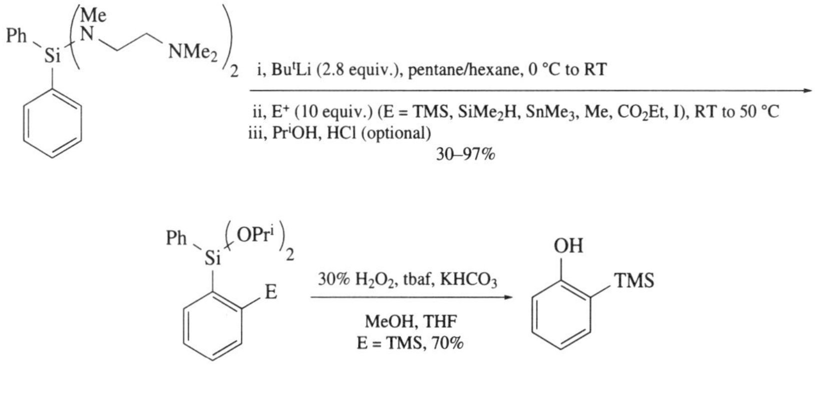

Scheme 85

Although in as yet an evolutionary stage, phosphorus-based DMGs have been used for the preparation of ligands for asymmetric catalysis.[506]

1.4.3.3 Benzyllithiums

One of the pioneers in superbase metallation, Lochmann, demonstrated, only in 1991, the effective metallation of toluene (Equation (77)).[507] Among the numerous and comprehensive contributions by Schlosser[3] to superbase chemistry is the innovative synthesis of the analgesic ibuprofen, involving three consecutive metallation steps (Equation (78)).[508]

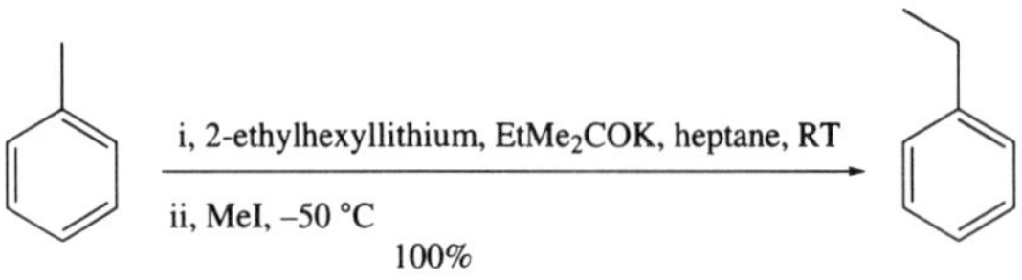

Preliminary observations by Beak *et al.* indicate that benzylic deprotonation in phenyl propionamides, presumably driven by CIPE, allows diastereoselective alkylation (Equation (79)).[509] Sparteine-mediated enantioselective substitutions of these systems have been reported.[510]

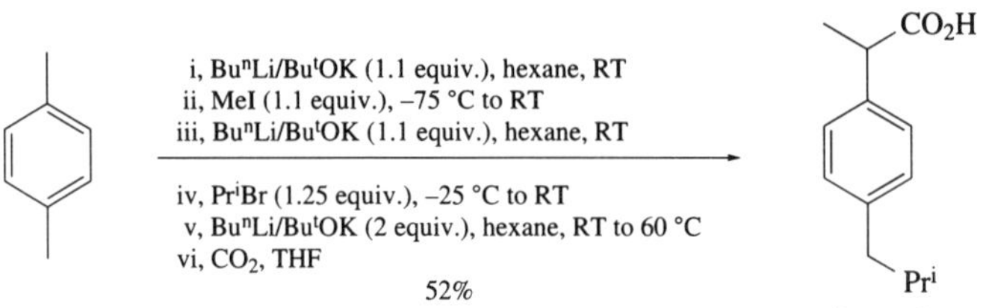

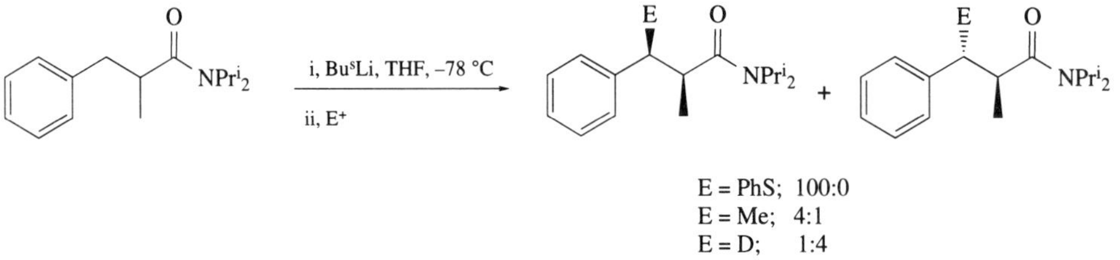

E = PhS; 100:0
E = Me;   4:1
E = D;   1:4

Racemic *o*-alkylbenzamides undergo enantioselective alkylations in the presence of (−)-sparteine with good to high *ee*s. The production of opposite enantiomers with the same chiral ligand by simply changing the leaving group in the electrophile was observed; an apparently configurationally labile carbanion has been shown to undergo enantioselective alkylation via competing (−)-sparteine-complexed transition states (Scheme 86).[511]

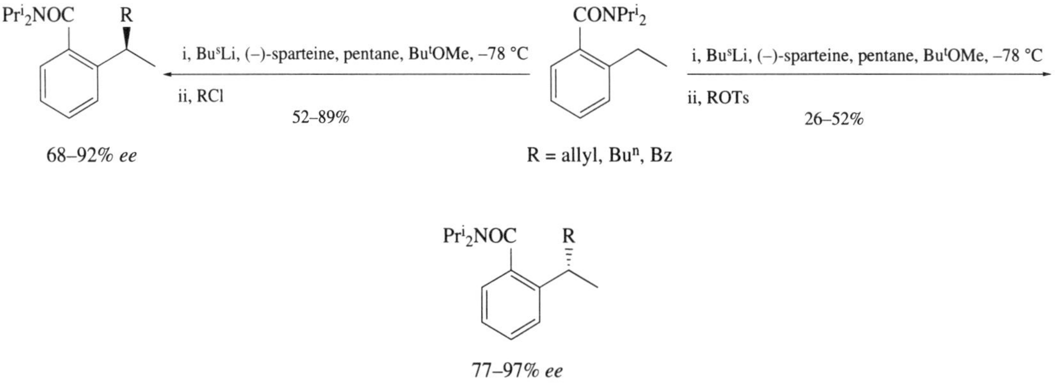

R = allyl, Bu^n, Bz

Scheme 86

Intramolecular cyclization of a tolyl anion derived by selenium–lithium exchange leads to stereoselective cyclopentane formation with a strong solvent dependence (Equation (80)).[512] LiDBB-mediated sulfur–lithium exchange triggers a stereoselective 1,5-silicon migration (Equation (81)).[513]

Directed metallation of *ortho*-DMG containing toluenes is a blossoming area of organolithium chemistry, especially for natural product assembly and asymmetric carbon–carbon bond formation.[36] Simplified, regiospecific construction of isoquinolones (Equation (82)),[514] naphthols and naphthyl amines (Scheme 87),[515] naphthols and indanones (Scheme 88),[516] as well as applications to natural product synthesis (Equation (83)),[517,518] Scheme 89,[519] and Scheme 90[520]) have been documented.

α,α-Disilylation of *ortho*-toluamides allows Ar–methyl protection during DoM chemistry, eventually leading to *peri*-methyl-substituted anthraquinones, including natural products (Scheme 91).[521]

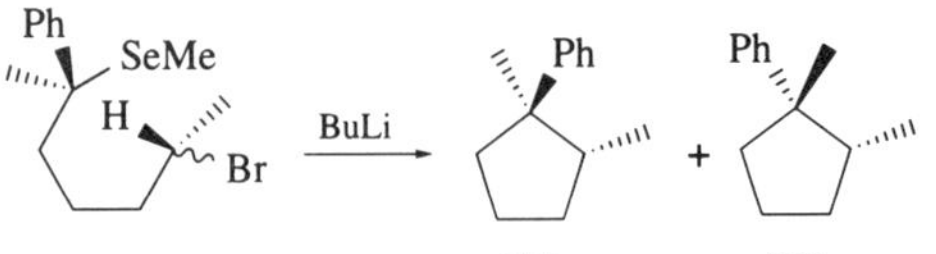

Bu^{s}Li, Et$_2$O or Et$_2$O/HMPA: 80%, (61):(62) = 80:20
Bu^{s}Li, pentane: 73%, (61):(62) = 19:81

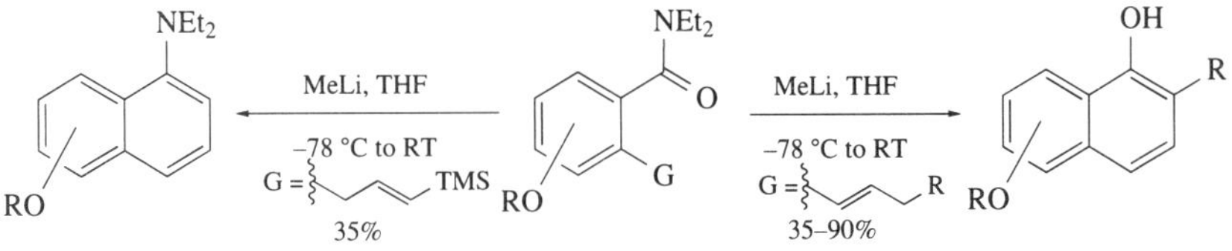

Scheme 87

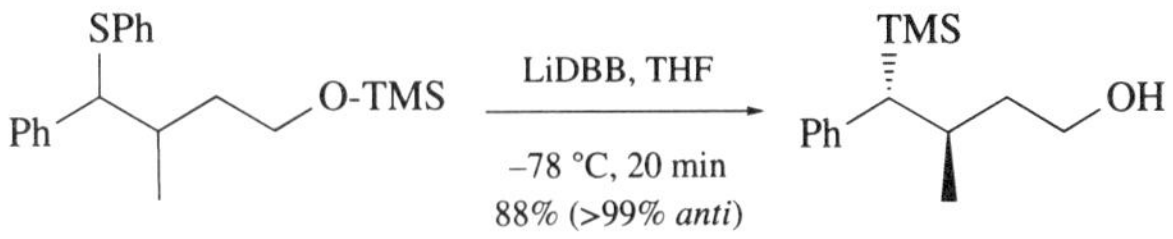

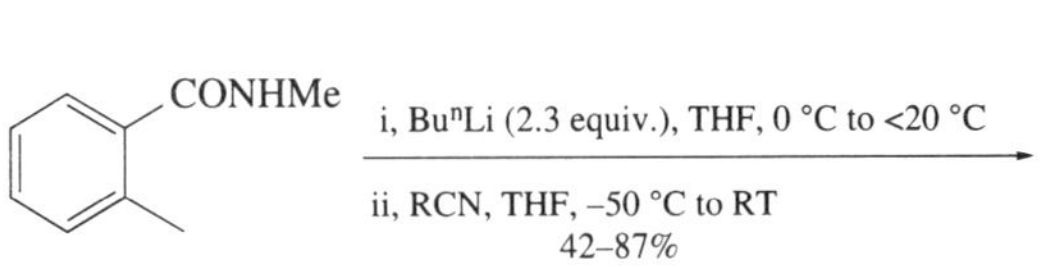

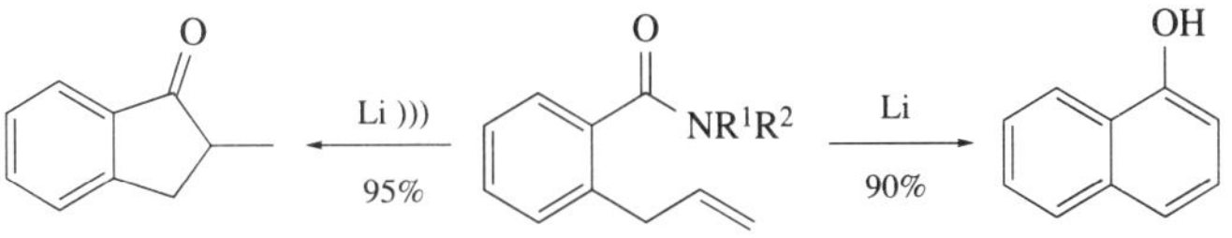

Scheme 88

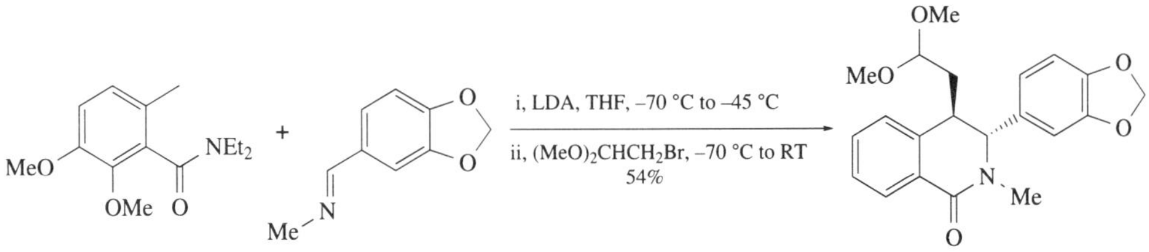

Scheme 89

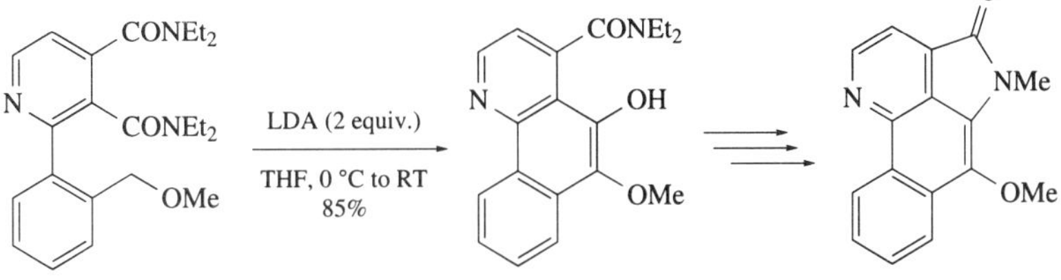

The original work of Creger[522] demonstrating the synthetic viability of dilithiated *ortho*-toluic acid triggered multifold applications,[523] among which the construction of tetralones (Equation (84))[524] and chiral isochromanone natural products (Equation (85))[525] from toluates, are illustrative.

o-Toluonitrile anions, formed via an interesting sequence involving benzyne addition and cyano migration, provide a facile assembly of chromanones (Scheme 92).[526]

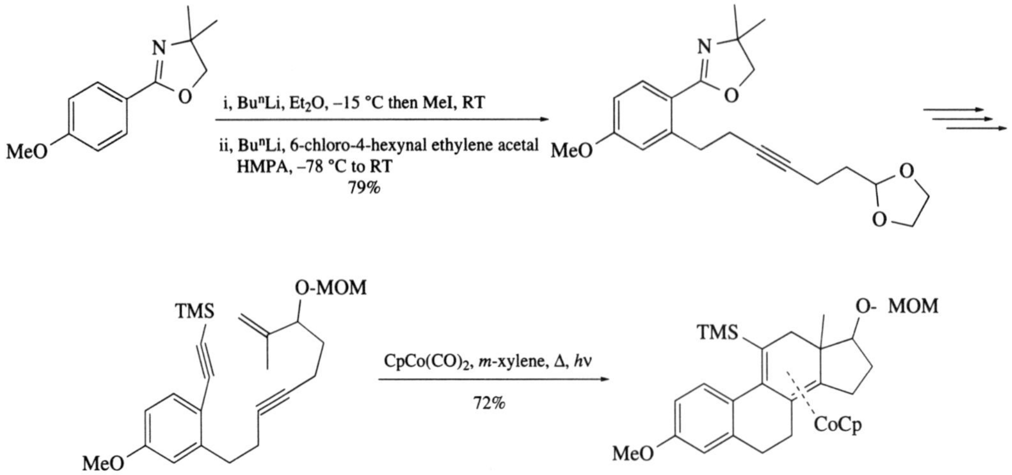

Scheme 90

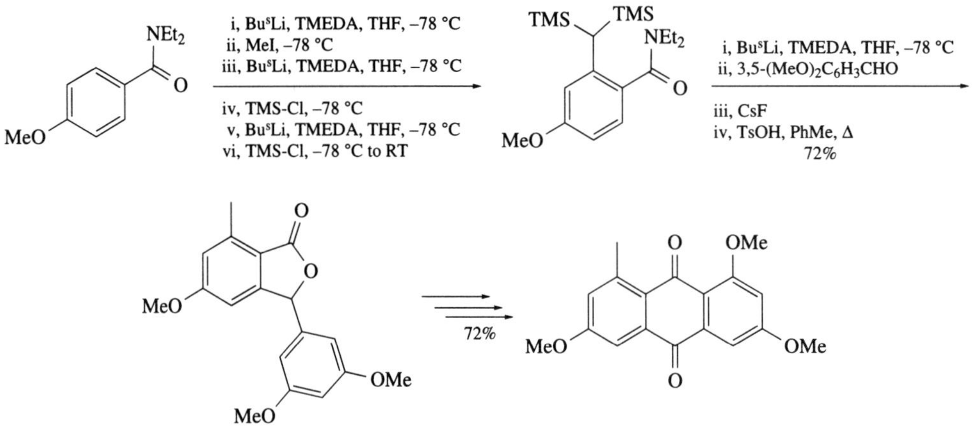

Scheme 91

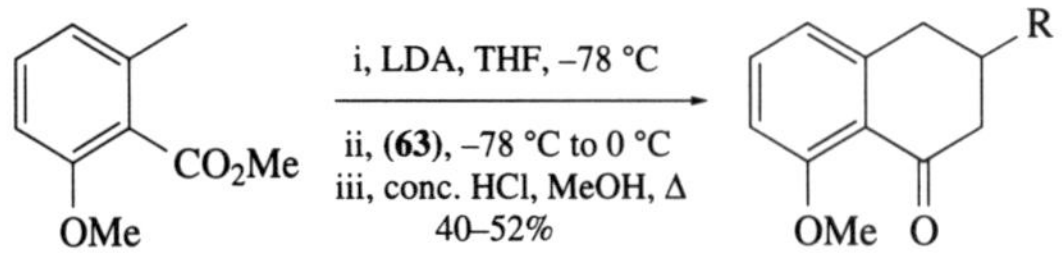

(84)

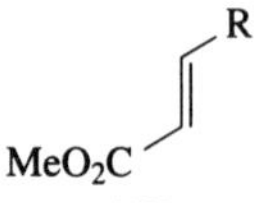

(63)

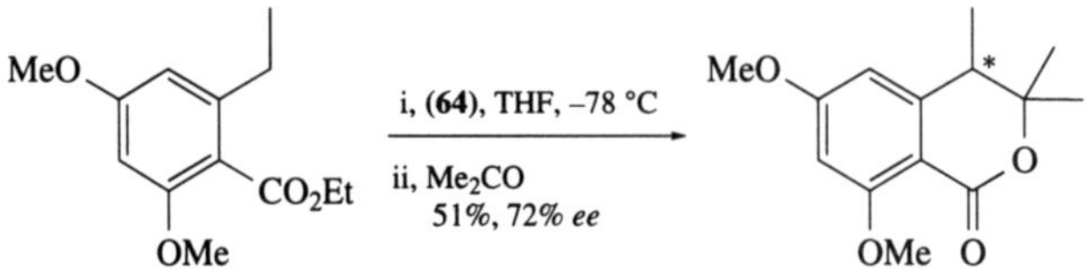

(85)

citrinin model

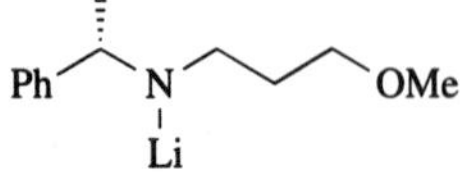

(64)

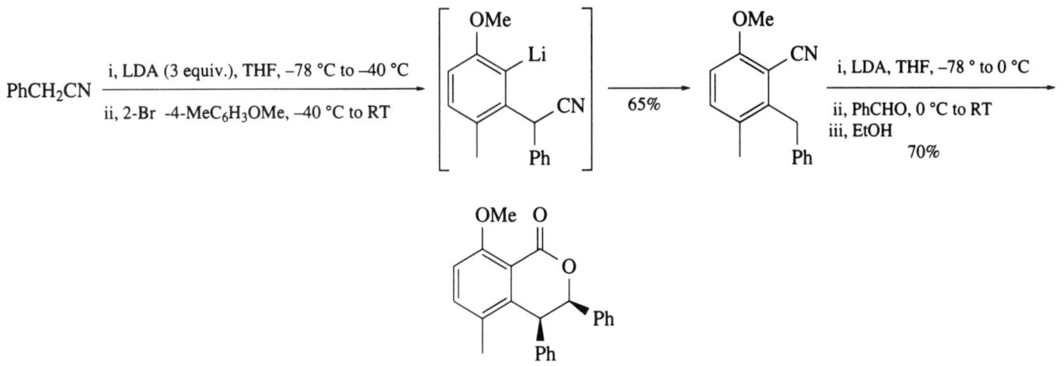

Scheme 92

o-Toluidines with appropriate activation have been used for chain extension (Scheme 93),[489] 3-substituted (Scheme 94)[527,528] and 2-substituted (Scheme 95)[529] indole construction, and for a modified Madelung indole synthesis (Scheme 96).[530]

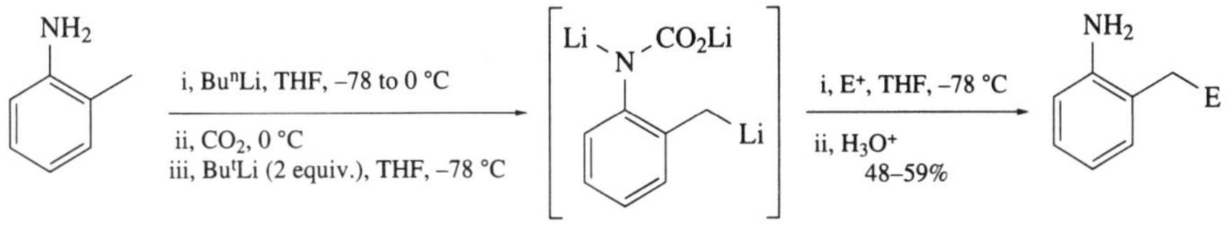

Scheme 93

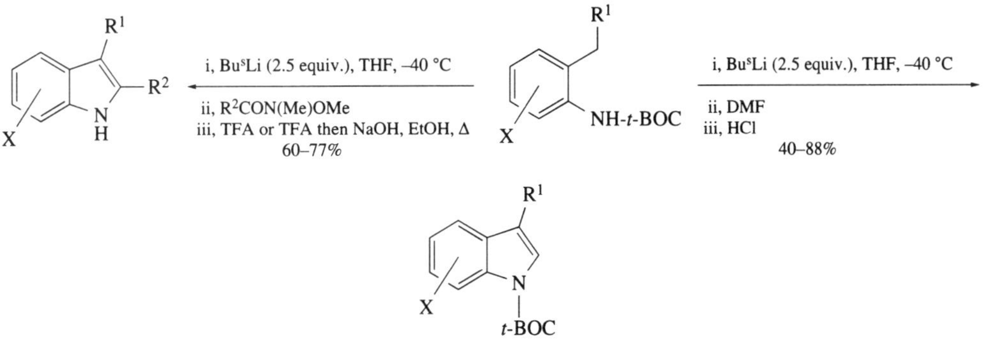

Scheme 94

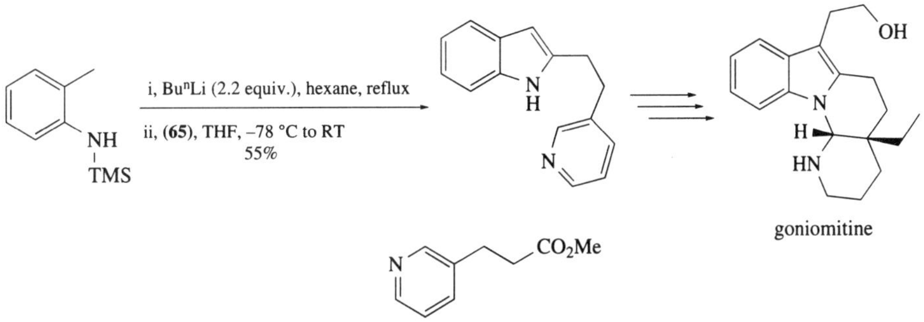

Scheme 95

A versatile method for the synthesis of substituted indoles, including tryptophols, utilizes intramolecular cyclization of an *o*-isocyanobenzyllithium (Scheme 97).[531]

Implementation of superbase conditions permits lateral functionalization of cresol (Scheme 98).[446]

Among the use of DMG-activated *o*-cresols, the simple *o*-methyl carbamate metallation (Scheme 99)[405]

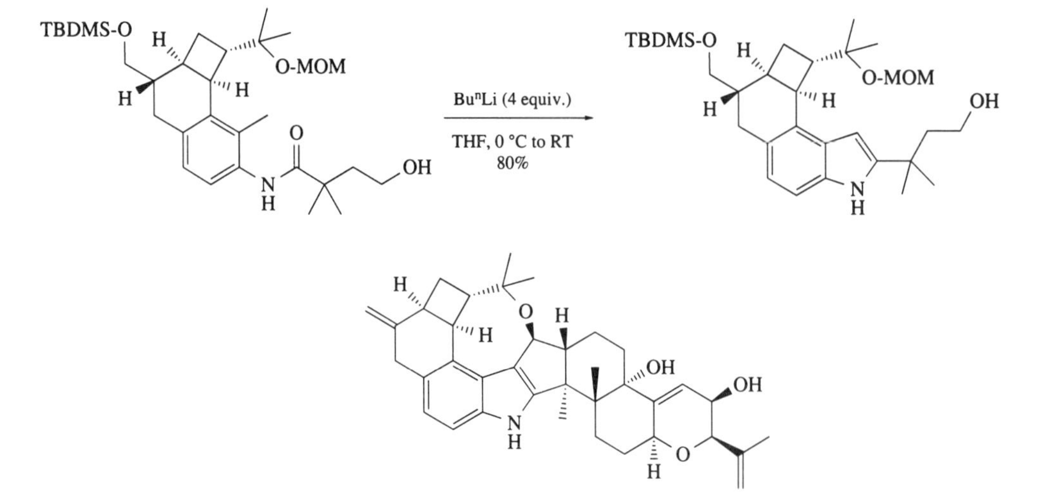

Scheme 96

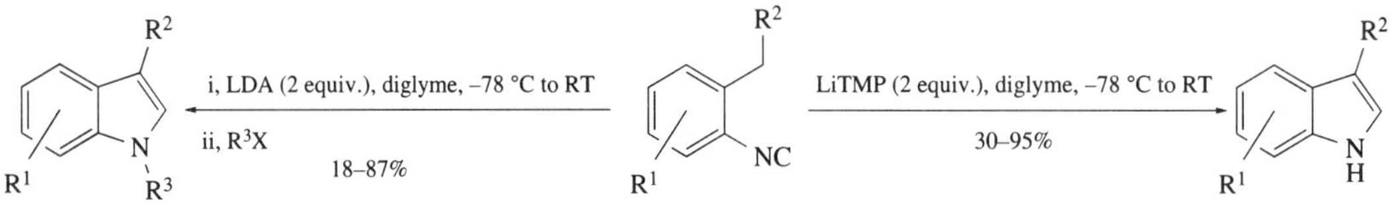

Scheme 97

and the complex "*peri*"-O-MOM indene deprotonation en route to fredericamycin (Scheme 100)[532] are illustrative of the two extremes.

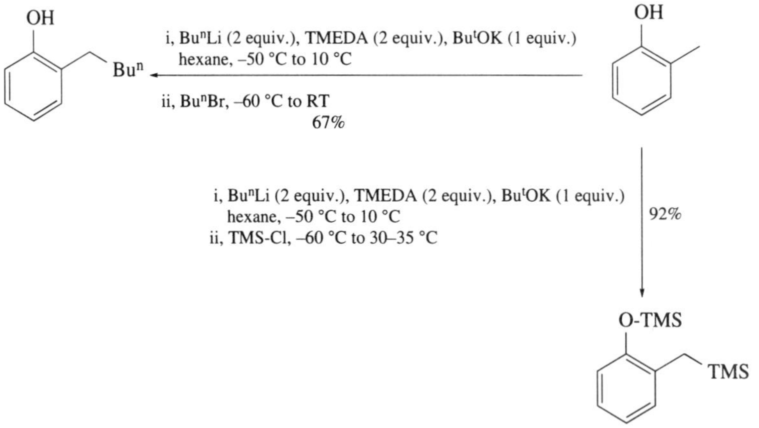

Scheme 98

o-Carboxamide-activated benzylic lithiation is central to the two-pot synthesis of the isocoumarin natural product hydrangenol (Scheme 101; see also Equation (82)).[533]

In a more elaborate application, remote metallation–carbamoyl migration followed by vinylogous benzylic deprotonation–cyclization leads to a phenanthrol derivative in a one-pot process; this constitutes a model study of the successful route to gymnopusin, a naturally occurring phenanthrene (Equation (86); see also Scheme 89).[534]

Although imine DMGs suffer from complications due to competitive RLi addition, they serve admirably in the context of *ortho*-methyl deprotonation[430,535] as demonstrated in industrial scale-up examples which illustrate the use of catalytic amounts of the expensive TMP (Scheme 102).[430]

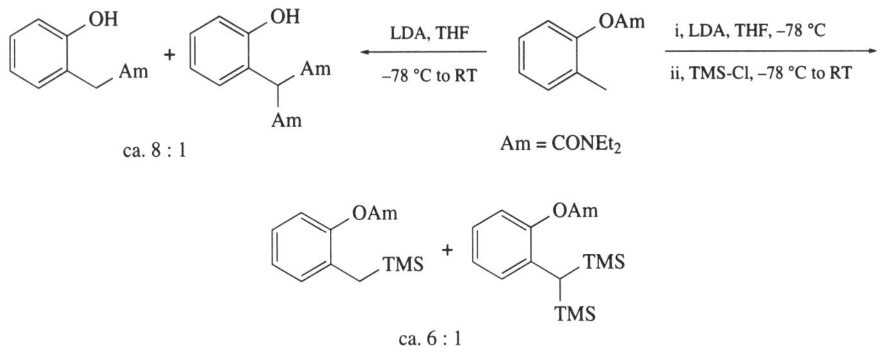

Scheme 99

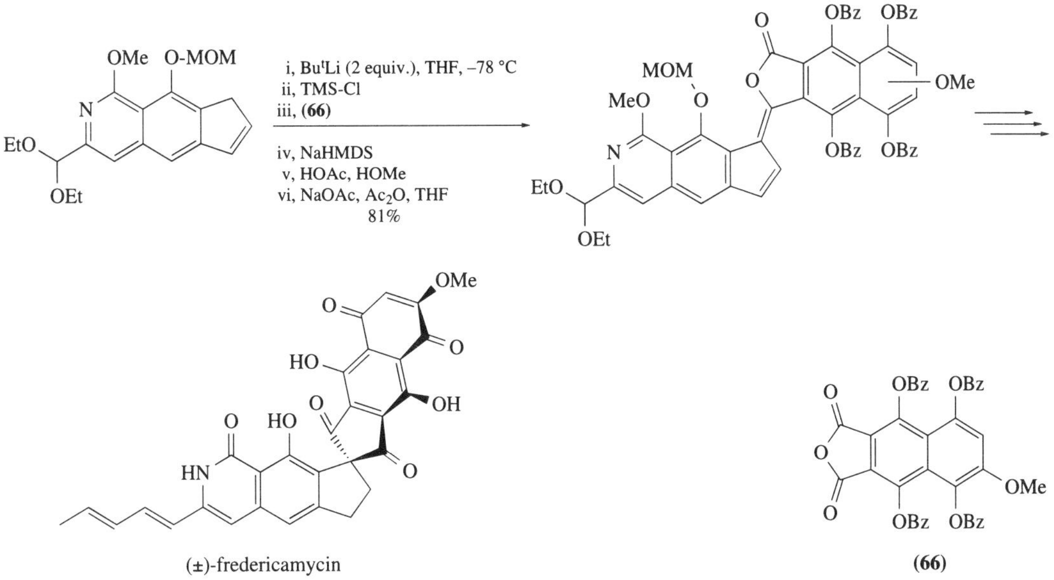

Scheme 100

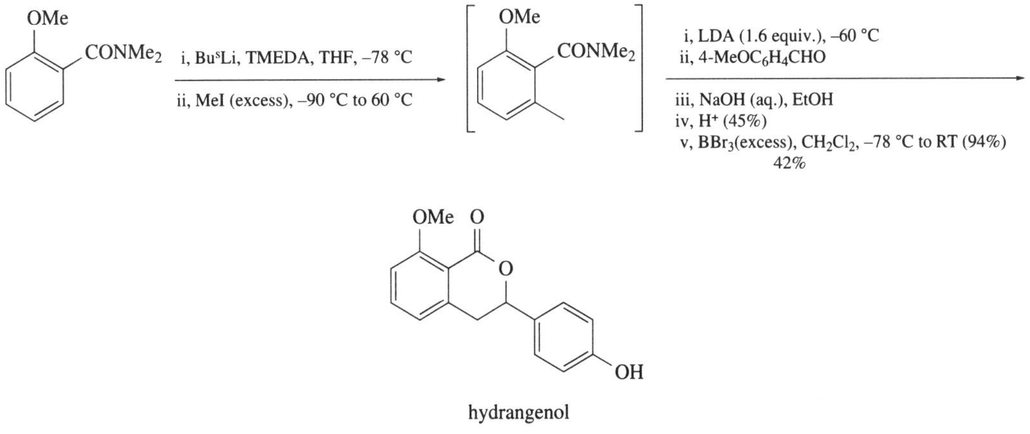

(±)-fredericamycin

(66)

Scheme 101

hydrangenol

By analogy with the *o*-toluidine chemistry illustrated earlier (Scheme 94), isoquinolines can be accessed by lateral lithiation of *t*-BOC-activated *o*-methylbenzylamines (Scheme 103).[536]

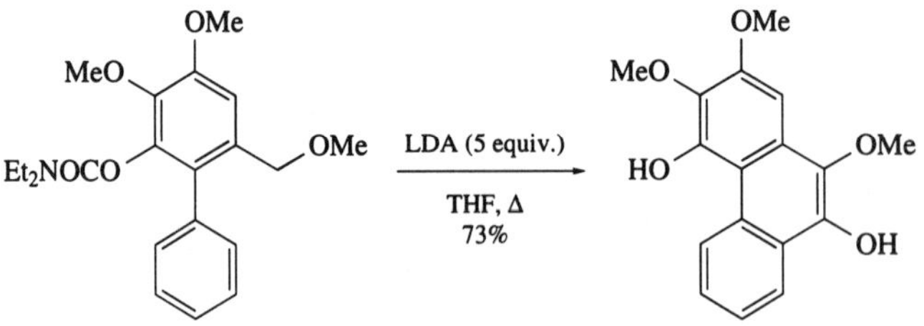

(86)

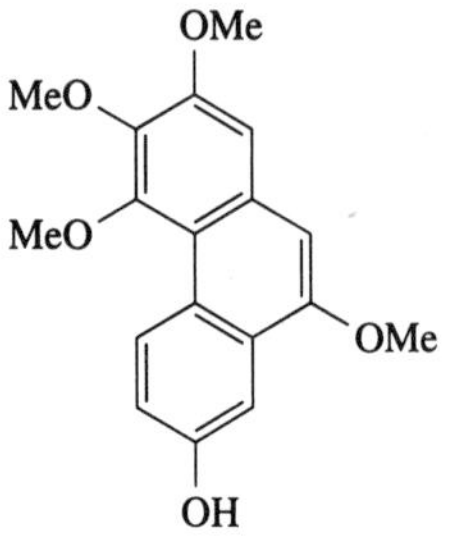

gymnopusin

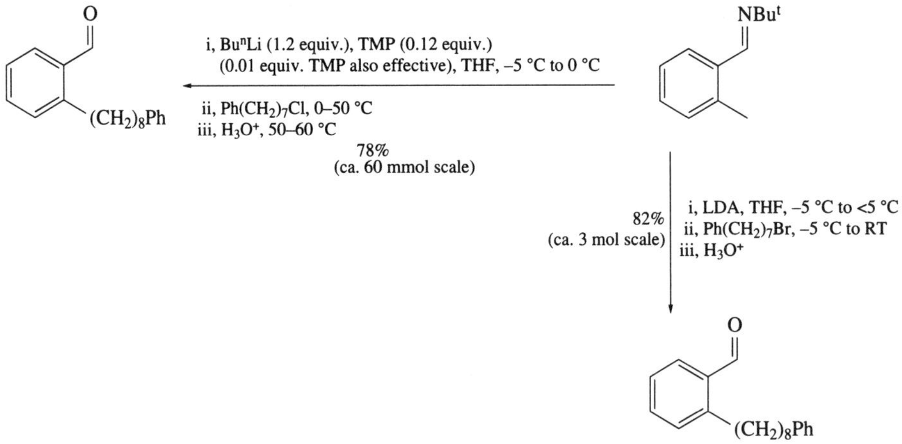

Scheme 102

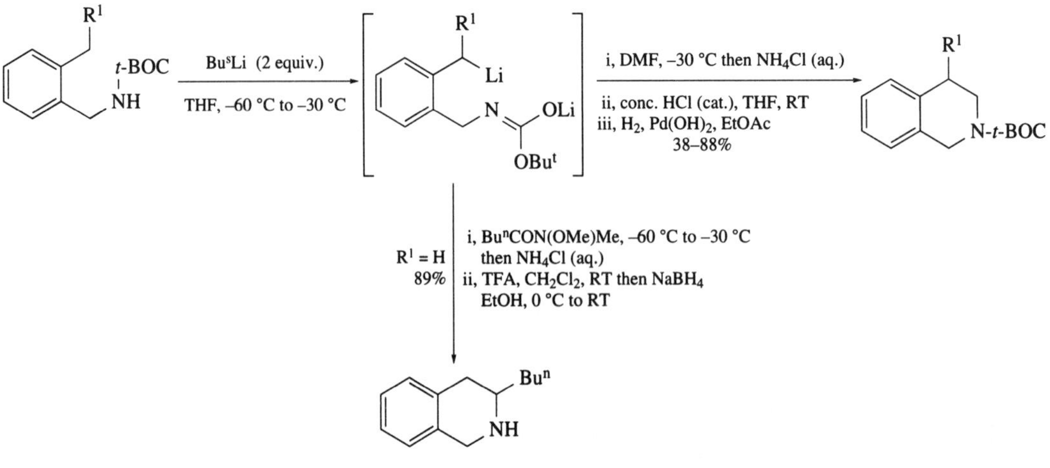

Scheme 103

1.4.3.4 Lithiated arene chromium tricarbonyl complexes

The reactivity of benzenes is considerably altered by $Cr(CO)_3$ complexation, allowing interesting and sometimes highly useful synthetic operations.[537–41] Aside from properties of enhanced reactivity towards nucleophiles, facial selectivity,[539] and benzylic carbocation stabilization[542] that the $Cr(CO)_3$ moiety confers, its anion-stabilizing property, favoring ring[543–8] and benzylic[549,550] deprotonation, are relevant here. Lithiated arene–$Cr(CO)_3$ complexes bearing OMe, F, and Cl DMGs afford, upon electrophile quench, 1,2-disubstituted products.[543–8] When DMG = F, advantage may be taken of the known nucleophilic displacement of the fluoro group to provide routes to certain heterocycles (Scheme 104).[551,552] In a competition experiment, 4- and 2-fluoroanisole complexes undergo selective deprotonation *ortho* to the fluoro substituent.[553,554] Competition experiments on *p*-disubstituted arene–$Cr(CO)_3$ complexes have established[545] the DMG hierarchy F > CONHR > NHCOR > CH_2NR_2 ≈ -OMe ≫ -CH_2OMe. Confirmatory of this order is the preparation of quinazolinedione complexes (Equation (87)).[554] $Cr(CO)_3$ complexes of benzyl alkyl ethers and sulfides undergo benzylic lithiation without Wittig and related rearrangements.[555] Dilithiated arene–$Cr(CO)_3$ complexes have been obtained by lithium–tin exchange.[556]

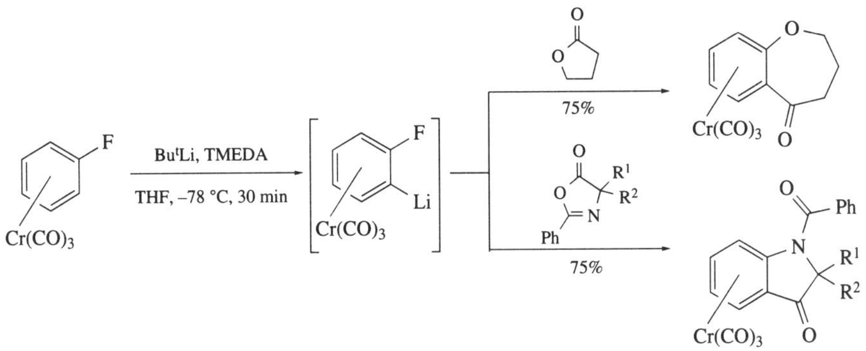

Scheme 104

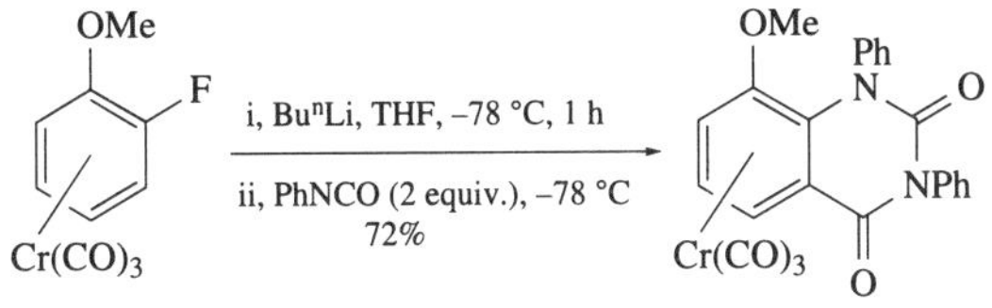

(87)

As an indication of the acidifying power of the $Cr(CO)_3$ moiety, the metallation of uncomplexed and complexed benzo-1,4-dioxin may be compared (Scheme 105).[557]

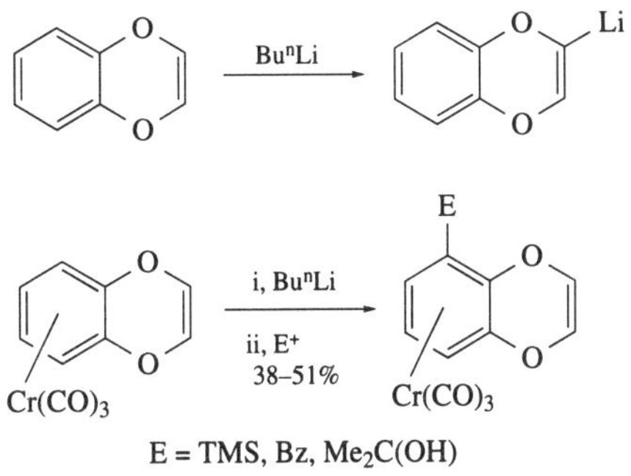

E = TMS, Bz, $Me_2C(OH)$

Scheme 105

The $Cr(CO)_3$ complex of the alkaloid dihydrocryptopine has been shown to undergo regioselective metallation–methylation (Equation (88)).[558]

1,2- and 1,3-disubstituted arene–$Cr(CO)_3$ complexes with different substituents exist in two enantiomeric forms due to planar chirality. This property, combined with facial selectivity, allows their use as intermediates[559–62] and catalysts[563–5] in asymmetric synthesis. For example, lithiation and subsequent electrophile quench of a chiral benzaldehyde acetal complex proceeds with excellent diastereoselectivity (Scheme 106).[562] Complexes bearing chiral benzylamino and benzyl ether groups behave analogously.[566] Thus, the readily accessible chiral phenethylamine complex undergoes sequential reaction with Bu^{t}Li and diethyl ketone to give exclusively one diastereomeric complex (Scheme 107);[564] this material was used as a catalyst for the enantioselective ethylation of benzaldehyde with diethylzinc.

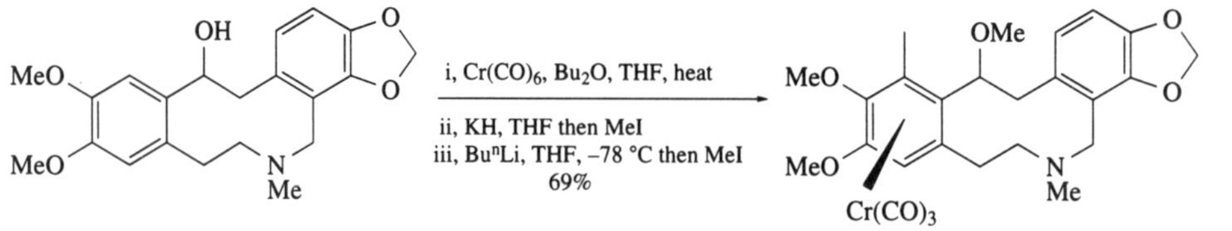

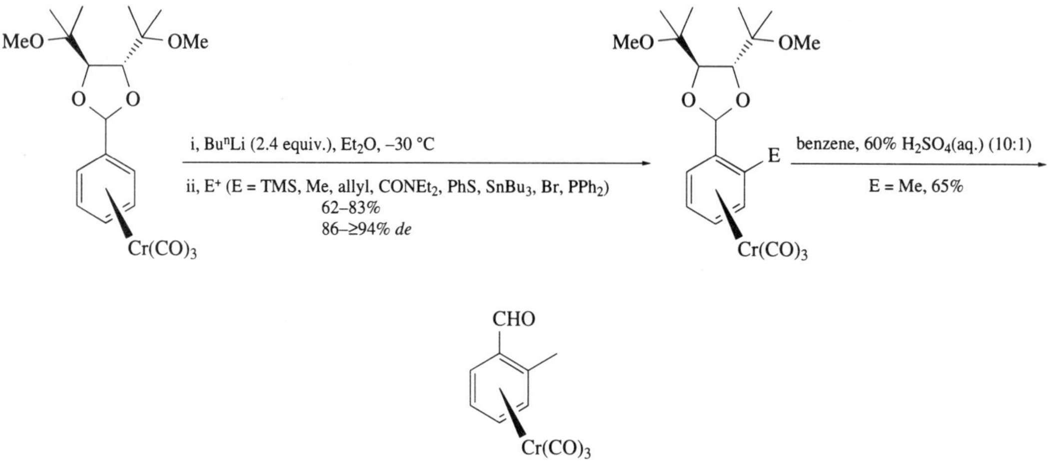

Scheme 106

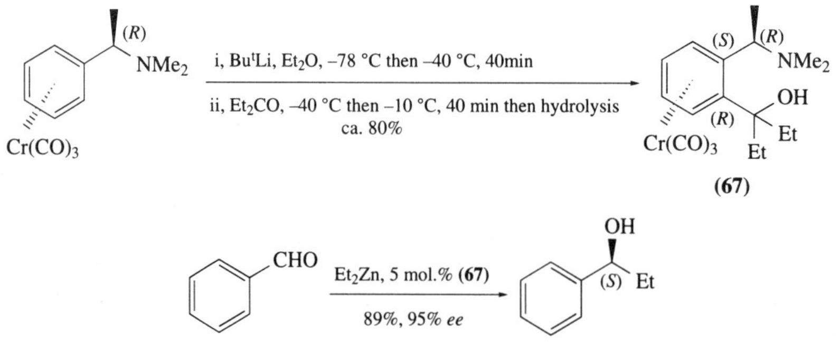

Scheme 107

A chiral tetralol complex, available via enantioselective reduction of the noncomplexed racemic ketone precursor, was used in a sequence of two successive stereo- and regioselective benzylic deprotonations–alkylations en route to a total synthesis of (1S,4S)-7,8-dihydroxycalamene (Scheme 108).[567] This conversion proceeded without loss of enantiomeric purity in an overall yield of 24% over 11 steps and the strategy has been extended[568] to an unsaturated calamenene derivative.

Independent findings that chiral arene–Cr(CO)$_3$ complexes are directly accessible by enantioselective lithiation of achiral educts with either a chiral lithium amide base (Equation (89))[569] or with Bu^{n}Li in the presence of a chiral diamine (Equation (90))[570] bodes well for future activity in this area.

1.4.3.5 Lithiated ferrocenes

Monolithiated ferrocene[571] may be generated by direct lithiation,[572] indirectly by transmetallation of chloromercuriferrocene,[573,574] and by metal–halogen exchange of bromoferrocene.[575] 1,1'-Dilithiofer-

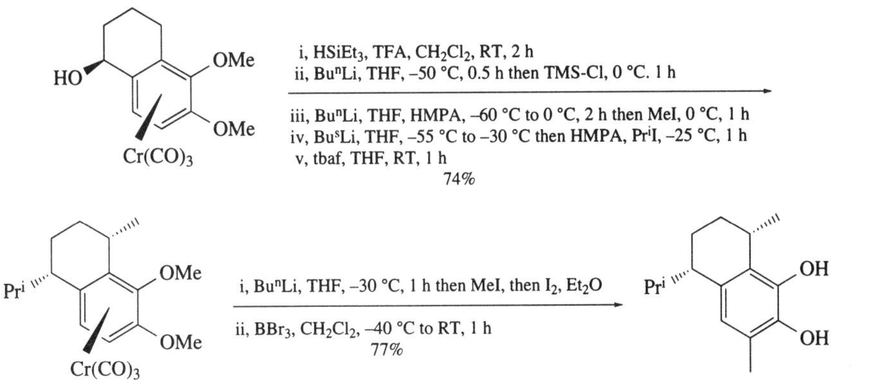

(1*S*, 4*S*)-7,8-dihydroxycalamene

Scheme 108

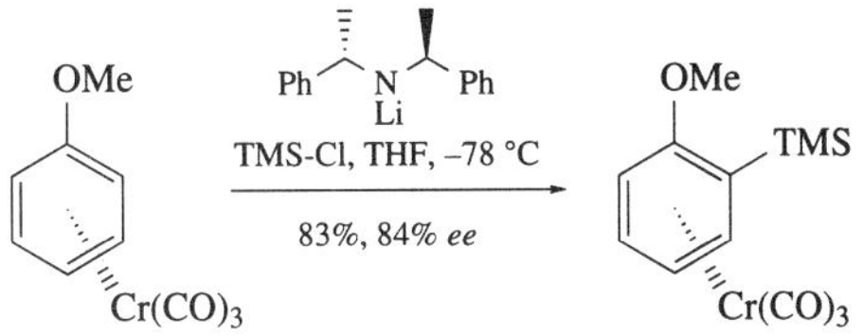

(89)

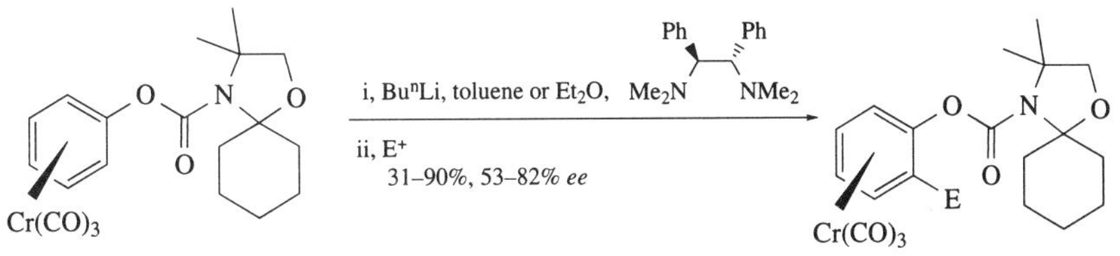

(90)

E = DMF, ClCOMe, PPh_2Cl, $ClCO_2Me$, $BrCF_2CF_2Br$

rocene may be prepared either by reaction with 2.5 equiv. Bu^{n}Li/TMEDA[576] or with Bu^{t}Li.[577] The current focus on the use of chiral ferrocenes as catalysts in asymmetric synthesis[578,579] owes its origin in the ground-breaking work of Ugi and co-workers,[580–2] who found that chiral (1-ferrocenylethyl)-dimethylamine, conveniently available as both antipodes by chiral resolution,[583] undergoes selective *ortho*-lithiation. This has led to ready access to chiral ferrocenyl amino alcohols (Scheme 109)[584] which have been used as efficient catalysts in the enantioselective addition of alkylzinc reagents to aldehydes (Equation (91)).[584–7] Secondary alcohols are obtained in high optical purity but the best results are achieved for benzaldehyde. The (*S,R*) antipodal catalyst gives rise to secondary alcohols of (*R*) configuration.

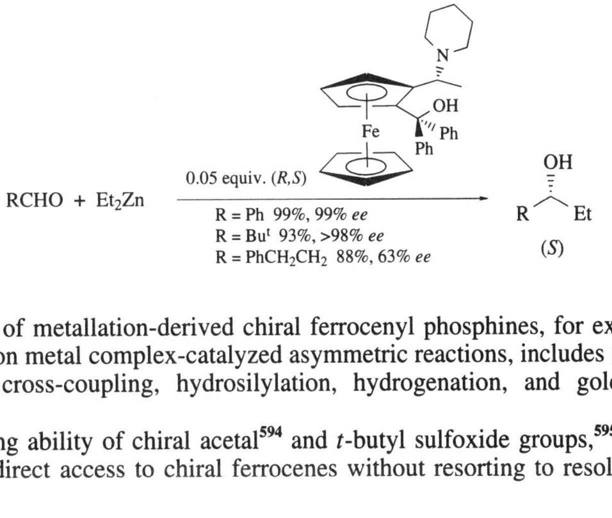

(91)

The efficient use of metallation-derived chiral ferrocenyl phosphines, for example, (Scheme 110)[588] as ligands in transition metal complex-catalyzed asymmetric reactions, includes asymmetric allylation of enolates, Grignard cross-coupling, hydrosilylation, hydrogenation, and gold-catalyzed aldol reactions.[579,588–93]

Using the directing ability of chiral acetal[594] and *t*-butyl sulfoxide groups,[595] Kagen and co-workers have demonstrated direct access to chiral ferrocenes without resorting to resolution (Scheme 111).

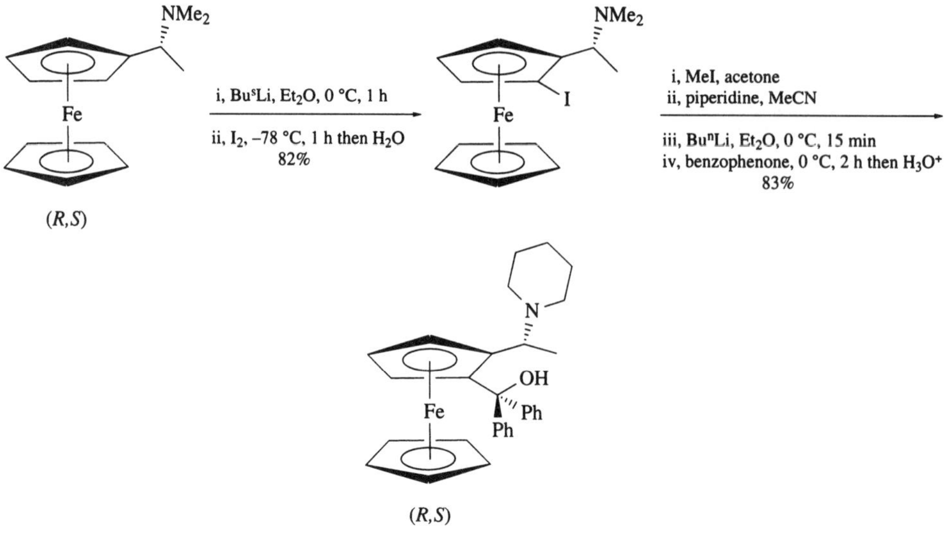

Scheme 109

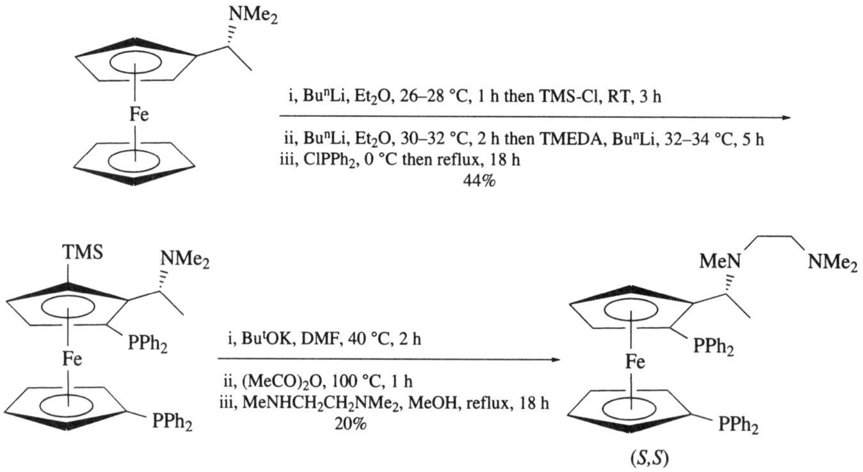

Scheme 110

1.4.4 Aryllithiums by Metal–Halogen Exchange

Lithium–iodo or –bromo (but not –chloro) exchange is a traditional, still mechanistically controversial, route to aryllithiums which is also valuable for the formation of *ortho*-DMG lithio species which cannot be efficiently generated directly. Furthermore, this reaction, in view of its diffusion controlled rates, is normally not compromised by the presence of certain electrophilic (e.g., NO_2, CO_2R) and highly acidic (conjugatively stabilized Me) groups. For the synthesis of 1,2-disubstituted systems and, therefore, for annelation to aromatic rings, the single disadvantage is the requirement of *ortho*-halo intermediates.

The use of lithium metal under the influence of ultrasound provides an alternate technique for the *in situ* generation of aryllithiums (Scheme 112 and Equation (92)).[596,597] Luche and Einhorn have applied this chemistry to *ortho*-metallation of the Bouveault intermediate, supplying a complementary protocol to that of Comins (Scheme 112).[125] Optimum conditions were achieved using tetrahydropyran as solvent.

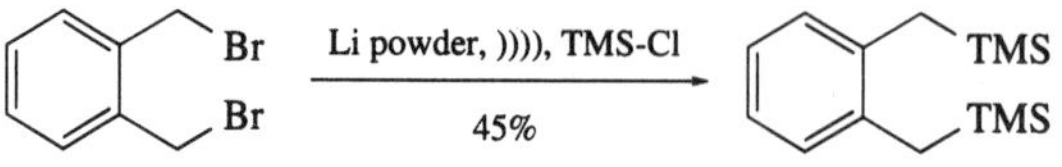

(92)

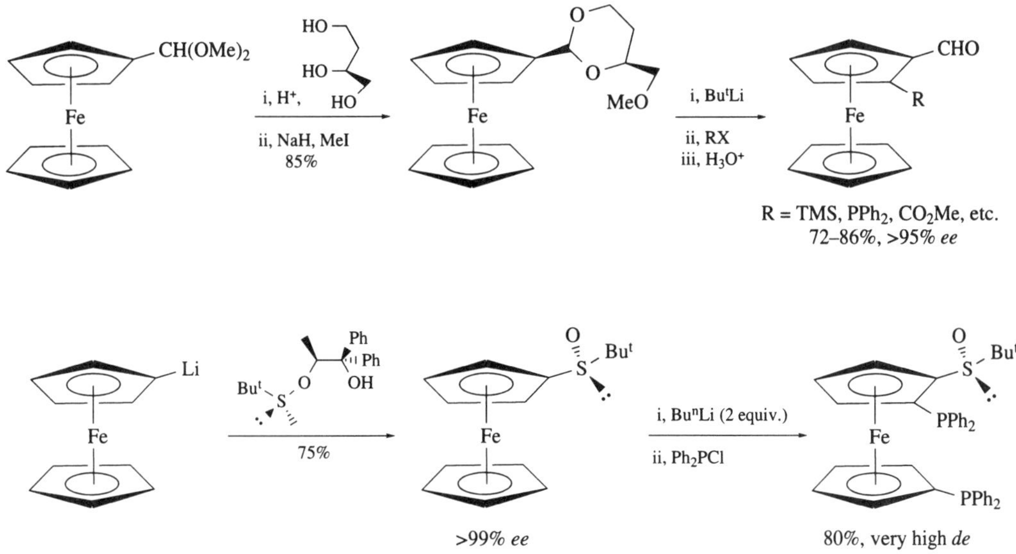

Scheme 111

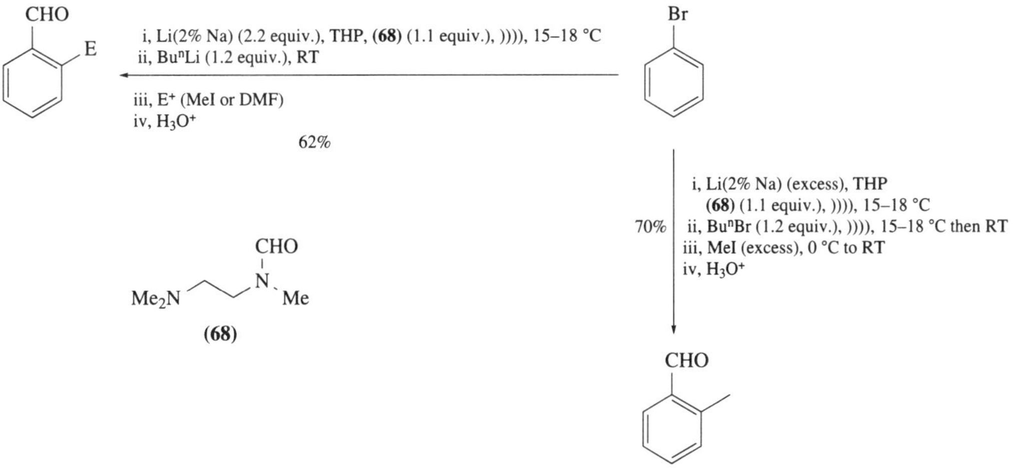

Scheme 112

An extensive literature attests to the value of the metal–halogen exchange tactic for carbocyclic (Equation (93))[194] and heterocyclic (Equation (94),[598,599] Scheme 113,[600] Scheme 114,[601] and Equation (95)[602]) ring annelation, strained ring formation (Scheme 115),[603,604] and construction of natural products (Scheme 116[605,606] and Equation (96)[607]).

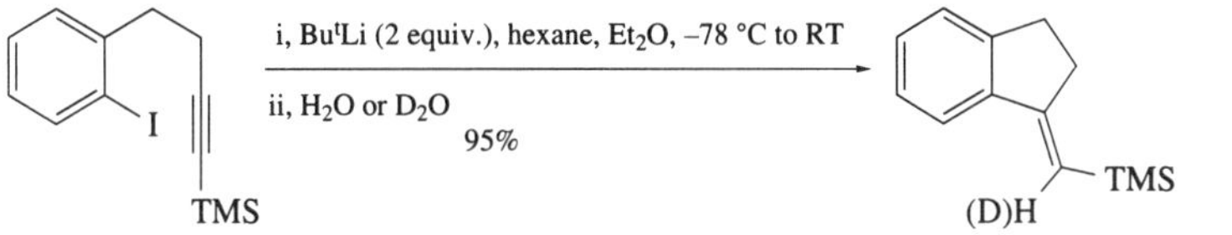

$$\text{(93)}$$

Nitrogen-protected (pivaloyl, t-BOC, trifluoroacetyl) 2-bromoaniline serves as a starting material for tetrahydrocarbazole (Equation (97)).[608,609]

Use of readily available chiral tethers provides a powerful strategy for the construction of axially chiral biaryls by metal–halogen exchange, lithium–copper transmetallation, and intramolecular coupling (Equation (98)).[610] The chiral tether can be removed by hydrogenation. The product (**69**) constitutes a model for the natural product, tellimagradin II.

The Parham cycliacylation[611] is a venerable methodology for carbanionic ring annelation (Equation (99))[612] (see also Equation (94) and Scheme 115). Parham-type anionic cyclization of o-bromobenzyl carbamates affords phthalides in high yield (Scheme 117), a protocol which has been applied to the synthesis of aristocularine alkaloids.[613]

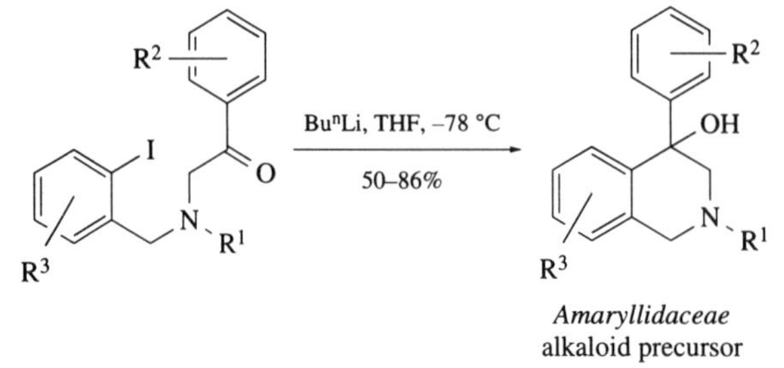

Scheme 113

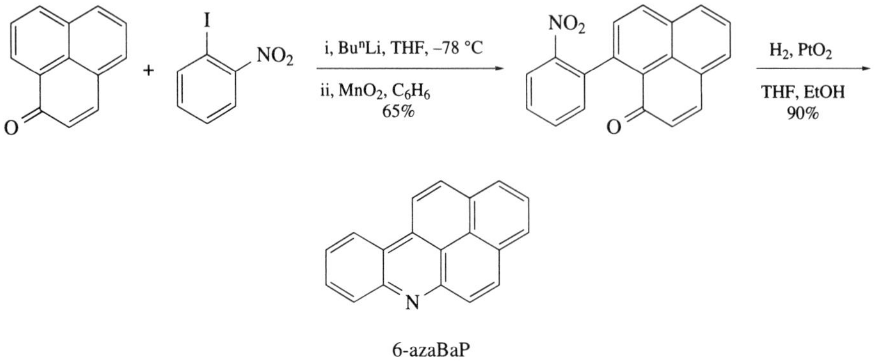

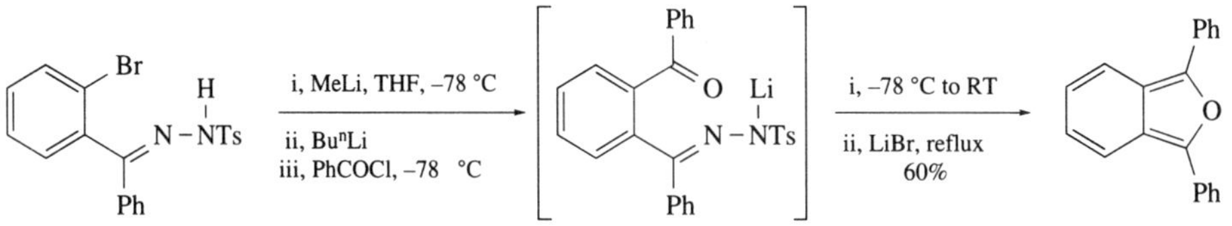

Scheme 114

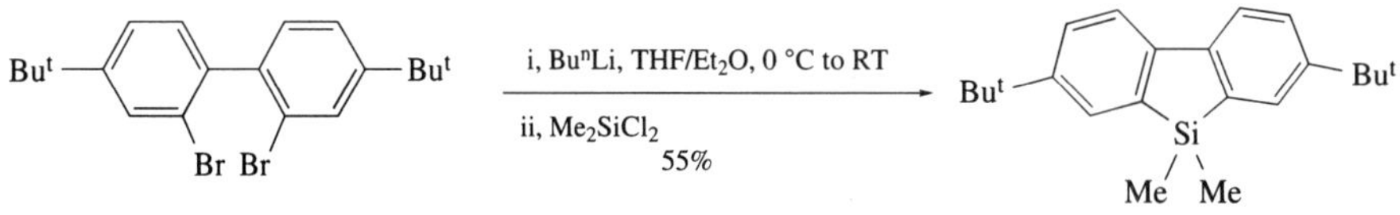

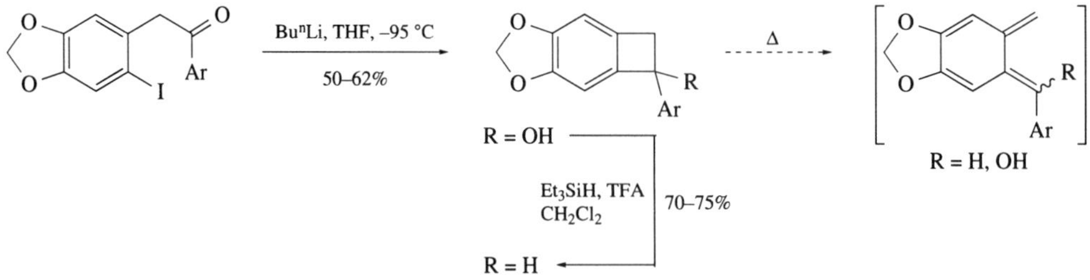

Scheme 115

A high-yielding route to the benzazaphospholine framework involves preformation of an α-metaloamine in order to preclude internal proton transfer on subsequent treatment with *t*-butyllithium (Equation (100)).[614]

o-Bromoanilines containing electrophilic nitrogen substituents undergo metal–halogen induced rearrangements to afford *ortho*-substituted anilines (Equation (101)).[471] This process may, in certain cases, be accomplished by a DoM reaction (Equation (74)). An analogous reaction has been demonstrated on aryl dialkyl phosphates (Equation (102)).[469]

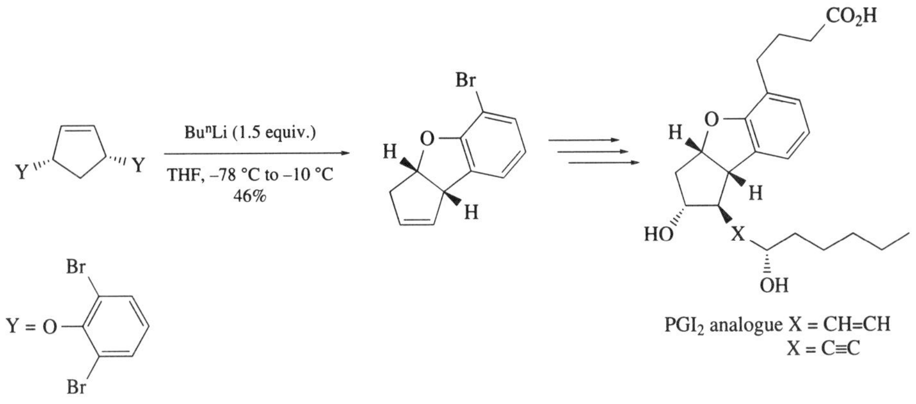

Scheme 116

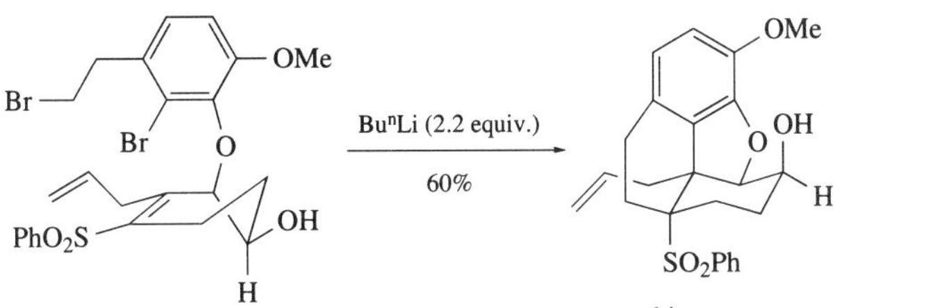

morphine precursor

(96)

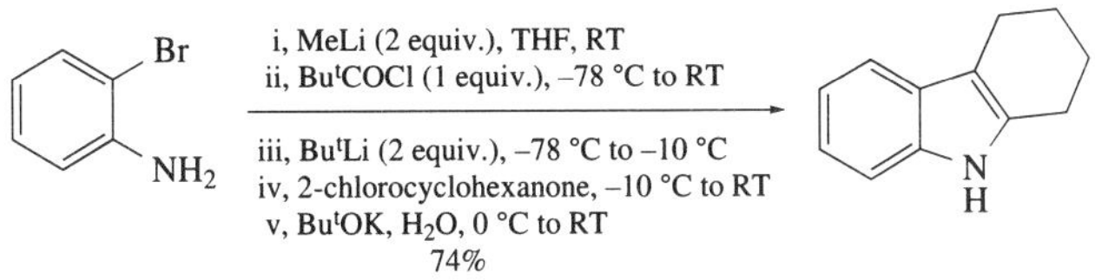

(97)

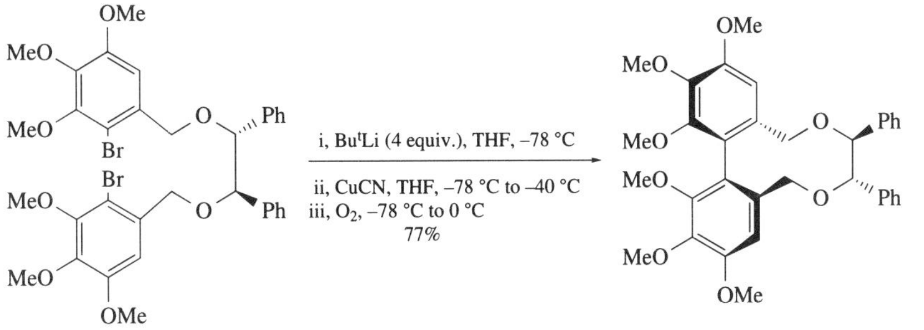

(69) one isomer

(98)

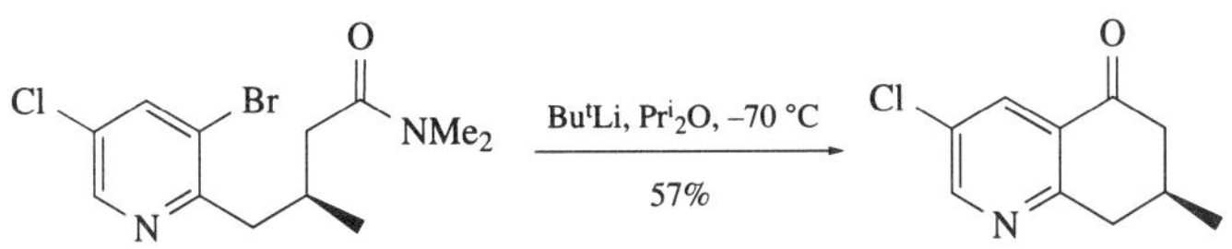

(99)

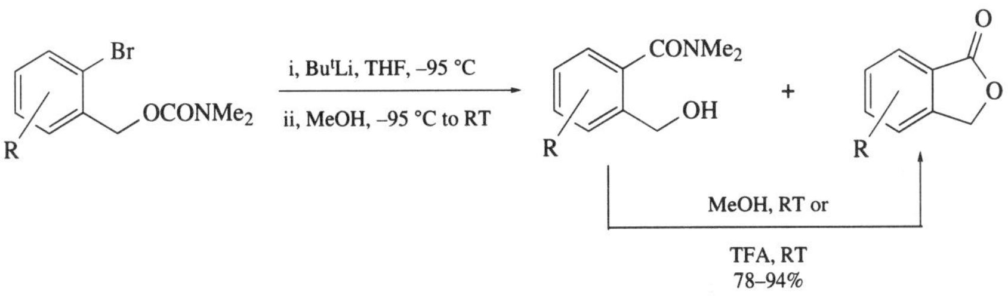

Scheme 117

(100)

(101)

R^1 = COPh or COBu^t, (Bu^{t}Li), 81–85% } R^2 = H
R^1 = P(O)Ph_2, (Bu^{t}Li), 48%
R^1 = TMS, (MeLi), 52%, R^2 = Me (intermediate trapped with MeI)

(102)

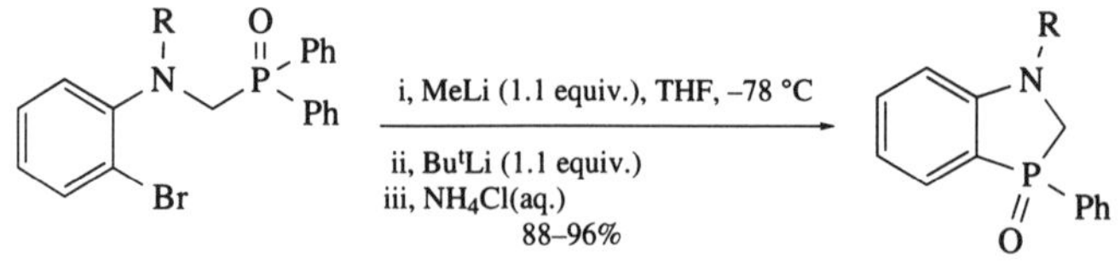

Presumably as a consequence of a CIPE, selective metal–bromo exchange *ortho* to DMGs in aryl dibromides has been demonstrated and begs exploitation (Equation (103)).[615]

(103)

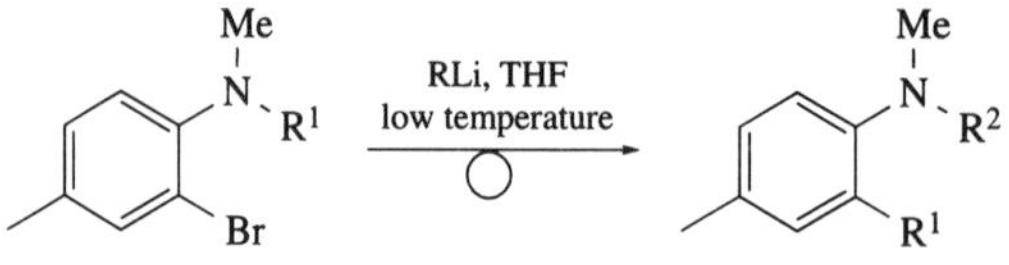

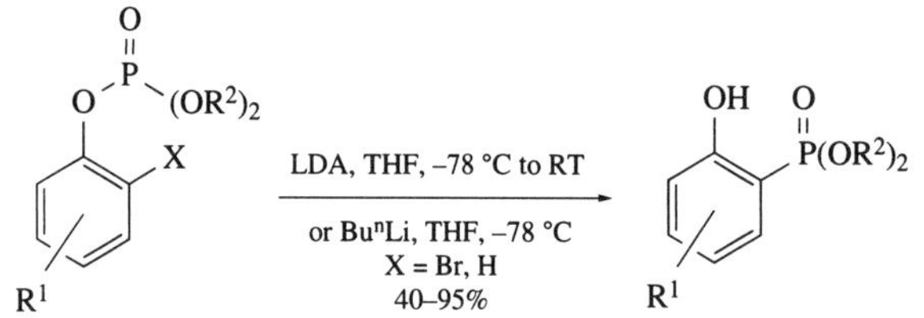

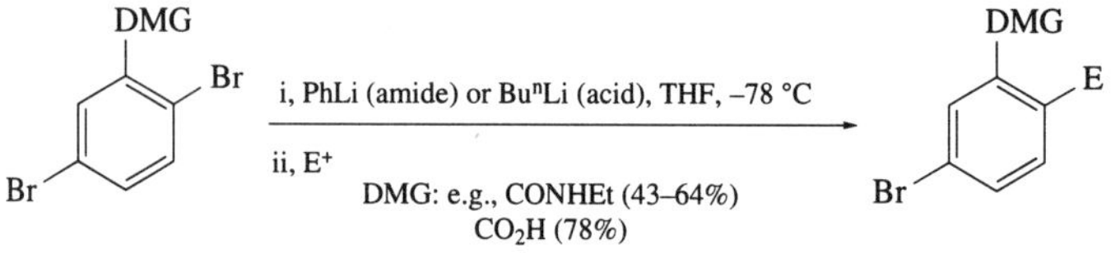

1.4.5 Directed *ortho*-Metallation of Heteroaromatics

1.4.5.1 Introduction and overview

Heteroaromatic-directed *ortho*-metallation (HETDoM) reactions are of increasing interest, fueled especially in pharmaceutical and agrochemical industry by the need for regioselective synthesis of specific substitution patterns which are difficult to achieve by *de novo* or classical (e.g., electrophilic substitution) and indirect (e.g., functional group interconversion) routes. The pioneering investigations by Gronowitz[616] constituted early demonstration of the synthetic potential of metallation chemistry in simple five-membered heterocycles (furans, thiophenes). In contrast, the knowledge that RLi and RMgX reagents readily add to pyridines delayed development of analogous reactivity studies in six-membered heterocyclic ring systems until the 1960s when Abramovitch *et al.*[617] and Chambers and co-workers[618,619] independently showed that metallation of pyridine *N*-oxides and polyfluoropyridines may be achieved. Not surprisingly, HETDoM chemistry followed on the heels of the flurry of activity in the aromatic area and, in the period 1970–1980, contributions from many laboratories provided increasing confidence that synthetically useful reactions could be achieved, especially on pyridines.[38] Nevertheless, the field in the early 1990s is at an early stage of development and many heterocycles, for example, diazines and condensed systems, which possess rather individual reactivities, have not been HETDoM tested. While our knowledge of HETDoM mechanisms is even more primitive than DoM, results during the 1980s in the HETDoM field provide promise for synthesis, especially when it is linked to other modern synthetic strategies, for example, transition metal-catalyzed cross-coupling, $S_{RN}1$ reactions, and nucleophilic heteroaromatic substitution, among others.

1.4.5.2 Metallation of five-membered ring heteroaromatics

(i) Non-DMG-containing furans and thiophenes

Kinetic 2-deprotonation (*n*-butyllithium/hexane) shows the order furan > thiophene ≫ *N*-methylpyrrole, while thermodynamic conditions (*n*-butyllithium/TMEDA) follow the order thiophene > furan ≫ *N*-methylpyrrole.[620] 2,5-Dilithiated species of furan and thiophene,[621-3] and 2,4- or 2,5-dilithio derivatives of *N*-methylpyrrole may be produced, generally with good efficacy.[621] Although pyrroles generally show poor 2-deprotonation facility,[37] *N*-methylpyrrole may be smoothly lithiated with *n*-butyllithium/TMEDA,[624] *t*-butyllithium/THF,[624] and *n*-butyllithium/THF/hexane.[625]

2-Lithiofuran (furyllithium), generated with *n*-butyllithium or under ultrasound,[51] is a widely useful synthetic intermediate[626] which leads to diversity of application, for example, 2-furanalcohols,[627] highly substituted pyranones,[628-31] 5-substituted-2-pyrrolidinones,[632] and sympathomimetic amines[633] (Scheme 118).

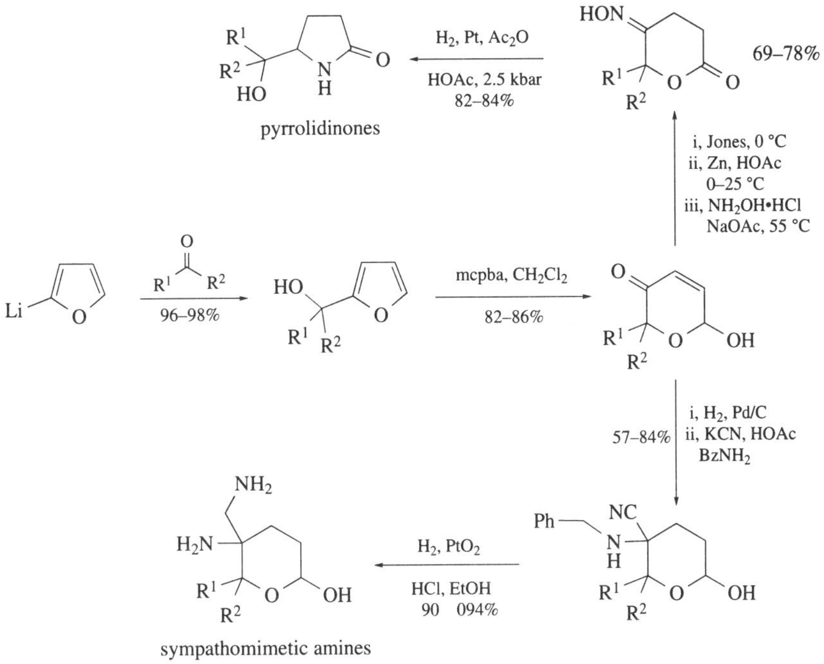

Scheme 118

2-Furyllithium is also the starting point for the syntheses of pharmaceutically significant furochromone and furocoumarin natural products, for example, khellin (**70**) and pimpinellin (**71**),[634] some involving intermediate chromium carbene complexes.[634,635]

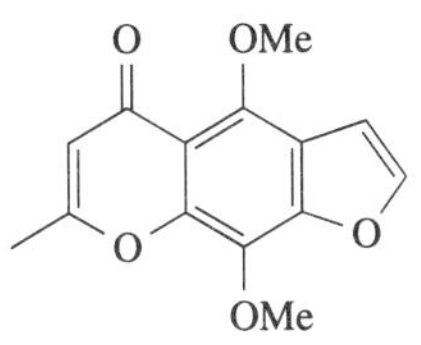

(**70**) 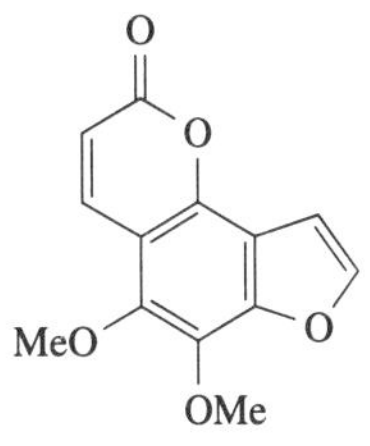   (**71**)

As a further selected illustration of lithiofuran utility, the synthesis of a trioxa-tricyclic subunit of saponaceolides was accomplished using tandem lithiation–electrophile quench followed by oxidative rearrangement (Scheme 119).[636] 2-Lithiothiophene has equally significant utility, the industrial scale preparation of a thiophene derivative with antiinflammatory properties constituting a specific example.[637]

Of potentially general value is the lithiation–oxidation sequence of furans and thiophenes leading to Δ^3-butenolides,[638] Δ^2-butenolides,[639,640] and thiobutenolides[641-4] (Scheme 120).

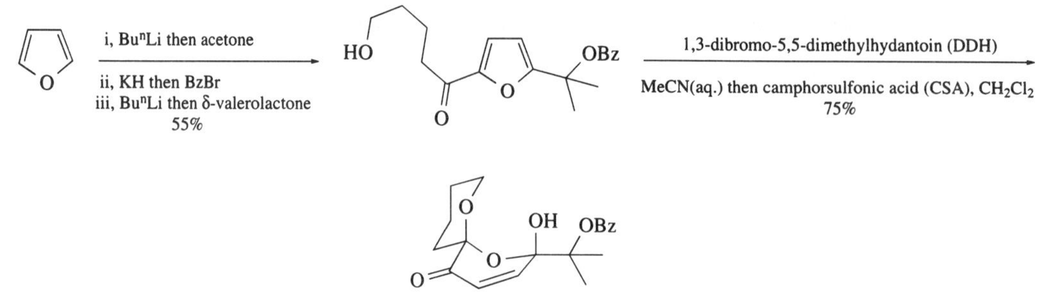

Scheme 119

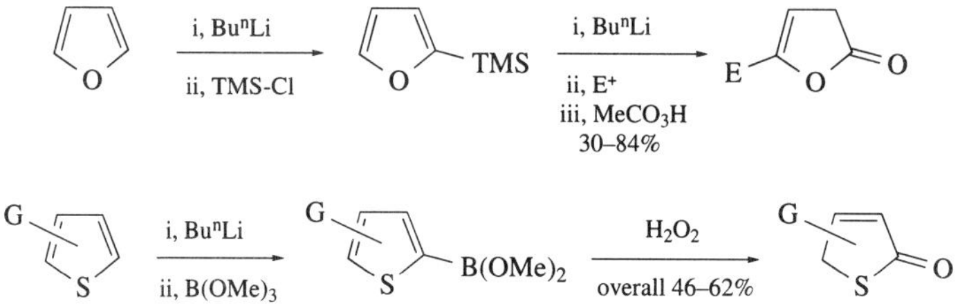

Scheme 120

(ii) Directed ortho-metallation reactions

(a) Furans and thiophenes. A large number of C-2-substituted thiophenes and furans[33,616,645-7], in spite of possessing potential DMGs, show deprotonation at the inherently acidic C-5 site with butyllithiums or LDA (Table 8). However, as is evident from activity during the 1980s, the HETDoM chemistry of an equally large number of other DMG-containing systems has provided rich rewards in synthetic use (Tables 9 and 10). Representative cases show that deprotonation is determined by the nature of the DMG and may be dictated by the choice of base and reaction conditions. A case in point is the DoM chemistry of 2-(2-imidazolino)thiophene and -furan where a slight alteration of conditions leads to a dramatic change in regioselectivity.[648] 5-Methyl-2-(2-imidazolidino)furan, however, did not undergo the expected C-3 metallation.[649] The 3-(α-amino alkoxide) DMG offers selective C-2 or C-5 lithiation of furan depending on the nature of the derived lithio amide.[125,650] Of special note is the extensive utility of the CO_2^- DMG in HETDoM chemistry, in contrast with the corresponding aromatic area in which this DMG emerged only recently.[410]

HETDoM methodology is beginning to enjoy widespread application in the preparation of highly functionalized heterocyclic systems. For example, a formal synthesis of (−)-canadensolide (Scheme 121) involves the use of the CH_2O^- DMG for the introduction of a secondary alcohol, which was then kinetically resolved using Sharpless reagents.[682] The construction of the antiinflammatory diterpene kalloide (Scheme 122) is rationally based on DMG incorporation and further modification.[664]

Other examples of HETDoM application in target-oriented synthesis include the pharmacologically active sesquiterpenes confertin,[684] gnididione,[685] furochromone khellin,[681] and manoalide.[686]

(b) Pyrroles. A body of literature has accumulated on latent N-DMG pyrroles for C-2 lithiation: SO_2Ph,[624] CO_2Bu^t,[624] NMe_2,[687] SEM,[688,689] SiR_3,[690] CO_2^- (Scheme 123),[691] CON^-Bu^t (Equation (104))[692] and, arguably, valuable are the CO_2Bu^t, CON^-Bu^t, CO_2^-, and SEM groups for which N-functionalization is facile and deprotection occurs under mild conditions. A recent application concerns the synthesis of routiennocin, an ionophore antibiotic (Scheme 124).[689]

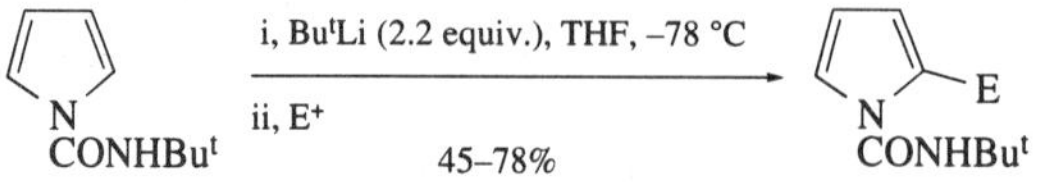

$E = CO_2H, PhCH(OH), TMS, D, Me_2CH(OH)$

(104)

Table 8 Selective C-5 lithiation of C-2 substituted thiophenes and furans.

Substrate	R	Metallation conditions	Ref.
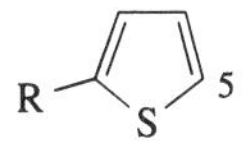 R─[thiophene]─5	CO_2^-	LDA (2 equiv.), THF, −78 °C	337, 652
	Me	Bu^nLi, THF, 0 °C	653
	$CONEt_2$	LDA or Bu^sLi, THF, −78 °C	654
	C(OMe)NR	Bu^nLi, Bu^sLi, LDA in THF or DME, −78 °C	655
	CHO^-NR_2	Bu^nLi, TMEDA, THF, −23 °C	126, 656
	$SO_2N^-Bu^t$	Bu^nLi (1.95 equiv.), THF, ≤−20 °C or LDA (2.1 equiv.), THF, −78 °C	657
	2-pyridyl	Bu^nLi, THF, −20 °C	658
	4-pyrimidyl	LDA, THF, −78 °C	658
	I	LDA, THF, −40 to 0 °C	659
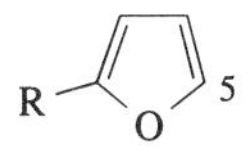 R─[furan]─5	CO_2^-	LDA, THF, −78 °C	337, 652, 660, 661
	Me	Bu^nLi, TMEDA, 0 °C	662
	2-oxazolino	LDA (2 equiv.), TMEDA, 0 °C	663
	C(OMe)NR	Bu^nLi, Bu^sLi or LDA in THF, −78 °C	655
	CHO^-NR_2	Bu^nLi, TMEDA, THF, −23 °C	126, 656
	CH=NOH	Bu^nLi, TMEDA, THF, −78 °C	664
	CHO-TBDMS	Bu^nLi, THF	665

Table 9 DMG-assisted selective metallation of thiophenes.

Substrate	DMG	Metallation conditions	Position	Ref.
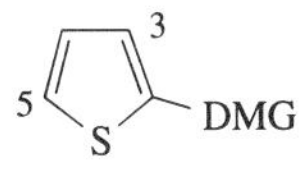 2-DMG thiophene	$CONEt_2$	Bu^sLi (1 equiv.), TMEDA, THF, −78 °C	C-5	666
		Bu^sLi (2 equiv.), TMEDA, THF, −78 °C	C-3, C-5	
	CO^-NBu^t	Bu^nLi or Bu^sLi, (2 equiv.), THF or DME −78 to 0 °C	C-3, C-5	654
		LDA, THF, 0 °C	C-5	
	CO_2^-	LDA (2 equiv.), THF, −78 °C	C-5	337, 652
		Bu^nLi (2 equiv.), THF, −78 °C	C-3	661
	2-oxazolino	LDA or Bu^nLi, hexane or Et_2O, −78 °C	C-3	667–9
		LDA, THF, −78 to 0 °C	C-5	667
		Bu^sLi, (3.3 equiv.), THF, −20 °C	C-3, C-5	670
	2-imadazolino	Bu^nLi (2.5 equiv.), THF or DME, −78 °C	C-3	649
		Bu^nLi (2.5 equiv.), THF, −20 °C	C-5	
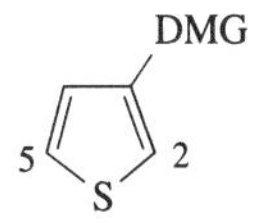 3-DMG thiophene	$CONEt_2$	Bu^sLi, TMEDA, Et_2O, −78 °C	C-2	668, 671, 672
	CHO^-NR_2	LDA, Bu^nLi, THF, −20 °C	C-2	126, 656
	CO_2SiR_3	LDA, THF, HMPA, −78 °C	SiR_3 to C-2	673
	CO_2^-	LDA (2 equiv.), THF, −78 °C	C-2	337, 652
	$CH=CHArCO_2^-$	LDA, Et_2O, RT	C-2	674
	C≡CH	Bu^nLi, Bu^tOK, THF	C-2	675
	CH_2O-TBDMS	Bu^nLi, THF, HMPA, −20 °C	SiR_3 to C-2	676
	$SO_2N^-Bu^t$	Bu^nLi (2 equiv.)	C-2	677
	N^-COR	Bu^nLi (2 equiv.), THF, −78 °C	C-2	678
	OAr	PhLi (2 equiv.), Et_2O, 0 °C	C-2	679

In general, pyrroles bearing latent nitrogen substituents and 2-DMGs cannot be cleanly and regiospecifically deprotonated. However, N-methylpyrrole is metallated at C-3 in moderate yield in the presence of 2-DMGs such as CON^-Bu^t[653] or 2-oxazolino.[679] Alternatively, C-3 selective lithiation may be achieved in excellent yield by metal–halogen exchange on nitrogen-protected 3-halopyrroles.[37] N-Methylpyrroles bearing 3-DMGs have been sparsely investigated.[37]

(c) Indoles. The availability of a broad range of N-DMGs for smooth C-2-metallation of indoles has allowed extensive synthetic application.[37] An illustrative case involves C-2 deprotonation of an *in situ* generated N-CO_2^- DMG as a starting point in the synthesis of yuehchukene (Scheme 125).[693]

Table 10 DMG-assisted selective metallation of furans.

Substrate	DMG	Metallation conditions	Position	Ref.
2-DMG furan	CONEt$_2$	Bu^{s}Li, TMEDA, THF, −78 °C	ring opening	666
	CON$^-$Bu^t	Bu^{s}Li, DME, −78 °C	C-3	654
		Bu^{n}Li, DME, −10 °C	C-5	
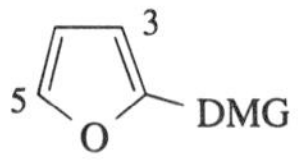	2-oxazolino	Bu^{n}Li, DME, −78 °C	C-3 >> C-5	680
		Bu^{s}Li (3.3 equiv.), TMEDA, THF, −78 °C	C-3, C-5	670
	2-imadazolino	Bu^{n}Li (2.5 equiv.), THF or DME, −78 °C	C-3	649
		Bu^{n}Li (2.5 equiv), THF, −20 °C	C-5	
	OPONMe$_2$	Bu^{n}Li, THF, −75 °C	C-3	681
3-DMG furan	CONEt$_2$	Bu^{s}Li, TMEDA, Et$_2$O, −78 °C	C-2	671, 672
	CHO$^-$NR$_2$	Bu^{n}Li, THF, −23 °C (NR$_2$ = TMDA)^a	C-2	126, 656
		Bu^{n}Li, THF, −78 °C (NR$_2$ = morph)	C-5	651
	CO$_2$SiR$_3$	LDA, THF, HMPA, −78 °C	SiR$_3$ to C-2	673
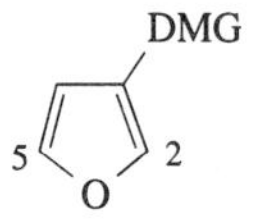	CO$_2^-$	LDA (2 equiv.), THF, −78 °C	C-2	337, 660, 682
	CH$_2$O-TBDMS	Bu^{n}Li, THF, HMPA, −20 °C	SiR$_3$ to C-2	676
	CH$_2$O$^-$	Bu^{n}Li (2 equiv.), THF, −78 °C	C-2	640, 683
	SO$_2$N$^-$Bu^t	Bu^{n}Li (2 equiv.)	C-2	677
	Ts	Bu^{n}Li, THF, −70 °C	C-2	684
	OMe	Bu^{n}Li (3 equiv.), TMEDA, hexane, reflux	C-2, C-5	662

^{a} TMDA = N,N,N'-trimethylethylenediamine.

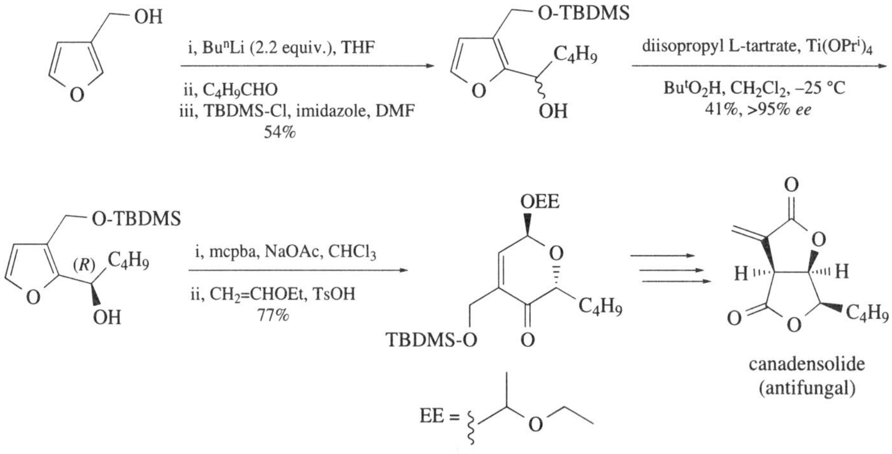

Scheme 121

Indole metallation chemistry is dramatically altered by Cr(CO)$_3$ complexation. The electron-withdrawing effect of the chromium moiety makes 2-, 4-, 6-, and 7-substituted indoles selectively available, the regiochemical outcome depending on the indole substitution pattern (Scheme 126).[694–8] These observations allowed the achievement of an elegant synthesis of (±)-chuangxinmycin methyl ester.[699]

In an important development, Iwao et al. have demonstrated that C-7-substituted indoles are indirectly available via metallation of N-t-BOC-indoline and have applied this protocol in a short and convergent synthesis of several *Amaryllidaceae* alkaloids (Scheme 12).[700]

As expected, nitrogen-protected 3-DMG indoles undergo efficient 2-deprotonation especially if N-DMGs are incorporated (Table 11).[701] Such combinations are synthetically useful, for example, in a regiospecific synthesis of ellipticine alkaloids.[702]

Using bulky nitrogen protection in gramine prevents 2-lithiation by the sterically demanding t-butyllithium and allows the establishment of a general protocol for 4-substitution, an application of which has been demonstrated (Scheme 128).[703]

An as yet rare case of DoM chemistry in the benzene ring of indole involves the 5-O-indole carbamate (Scheme 129).[704] Depending on the reaction conditions, the combination of a bulky nitrogen-protecting group and a strong DMG promotes selective C-4-metallation as evidenced by the

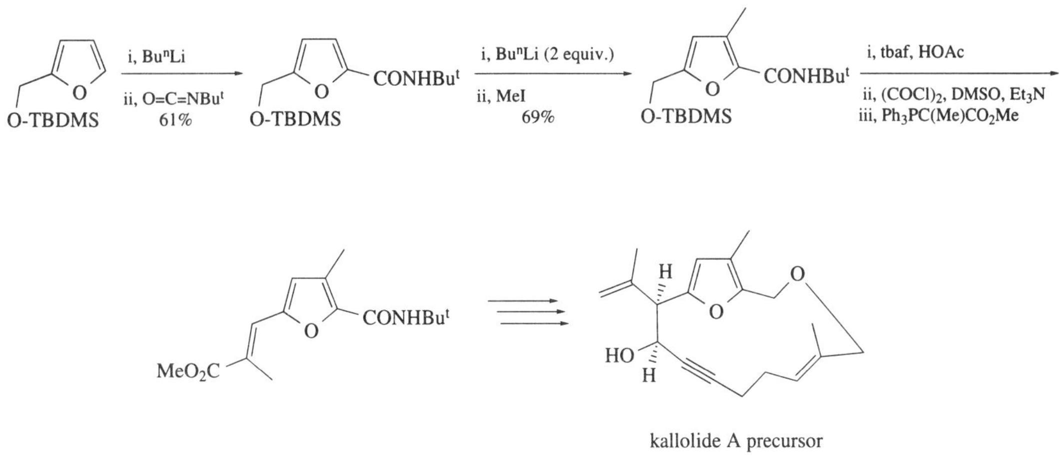

Scheme 122

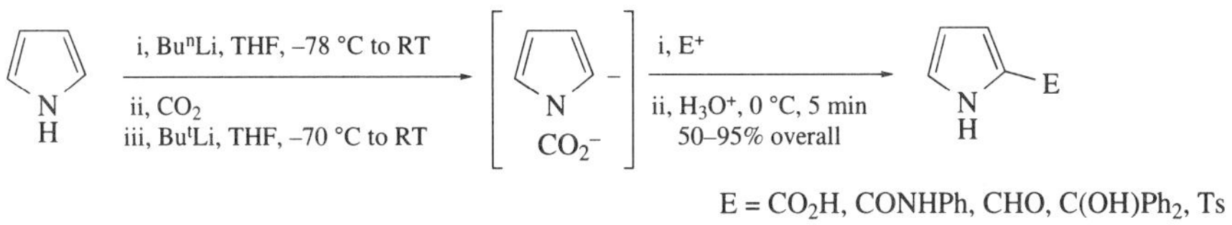

Scheme 123

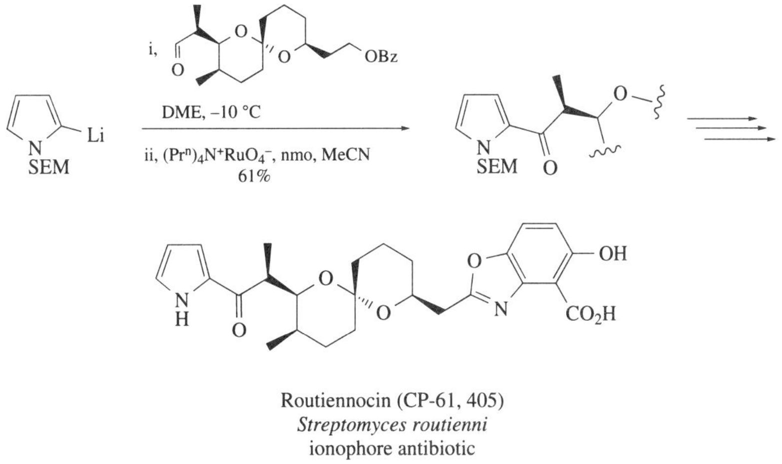

Scheme 124

incorporation of a variety of electrophiles. Further, 4-TMS protection allows 6-lithiation–electrophile introduction as well as anionic *ortho*-Fries rearrangement, thus providing access to new combinations of 4,5,6-substituted indoles.

(d) Miscellaneous. A comprehensive and recent review[37] deals with the metallation of pyrazole, indazole, triazoles, azaindolizine, isoxazole, isothiazole, oxazole, thiazole, and diazole heterocycles.

1.4.5.3 Metallation of six-membered ring heteroaromatics

A steadily increasing stream of literature attests to the utility of DoM chemistry in diazine, quinoline, but especially pyridine heterocycles.[37,38]

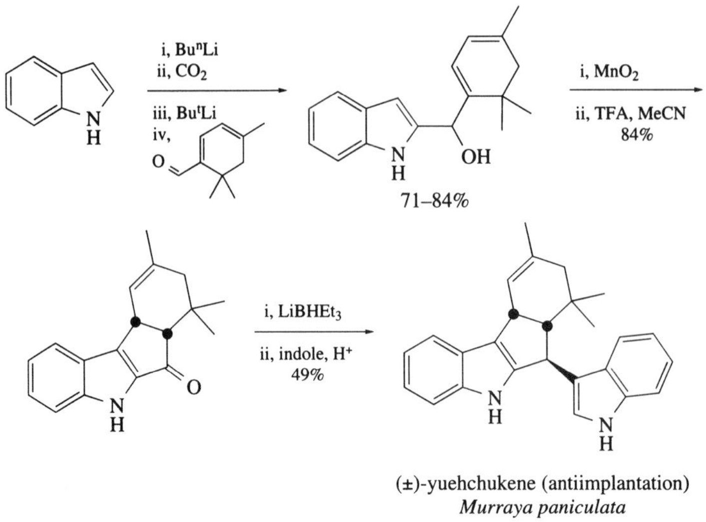

Scheme 125

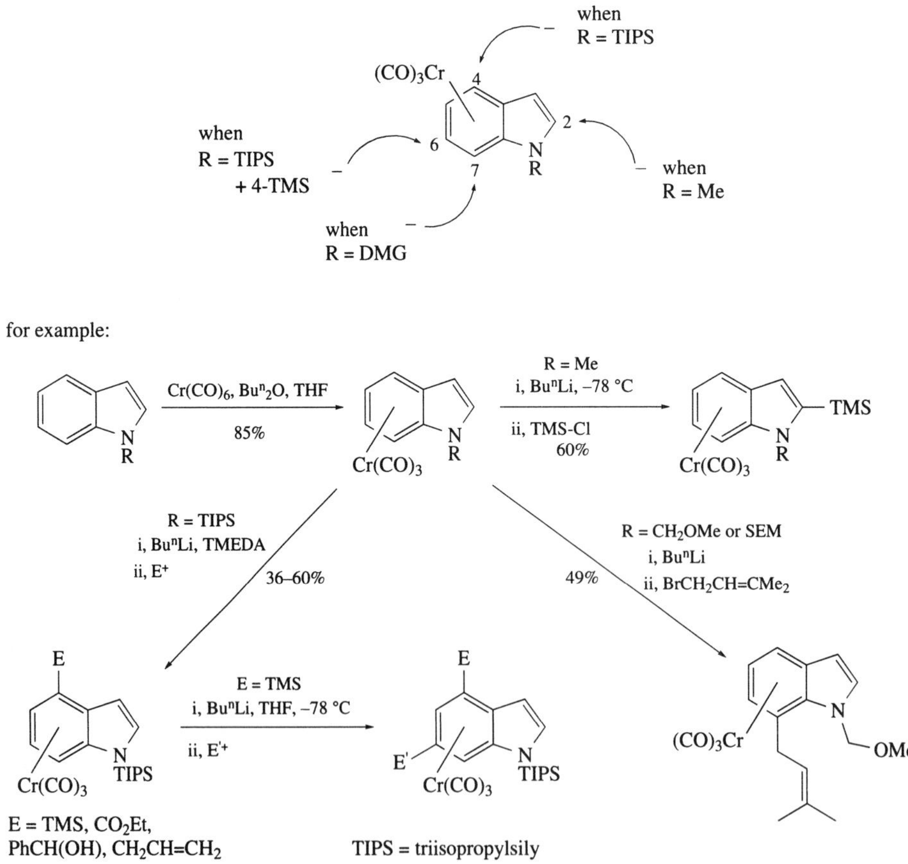

Scheme 126

(i) Pyridines

Among the rapidly accumulating DoM studies on pyridines,[37,38] halogen DMGs have provided most significant and diverse application. Illustrative are LDA metallation of iodopyridines and subsequent migration ("halogen tango") to give stabilized iodolithiopyridines which can be quenched with

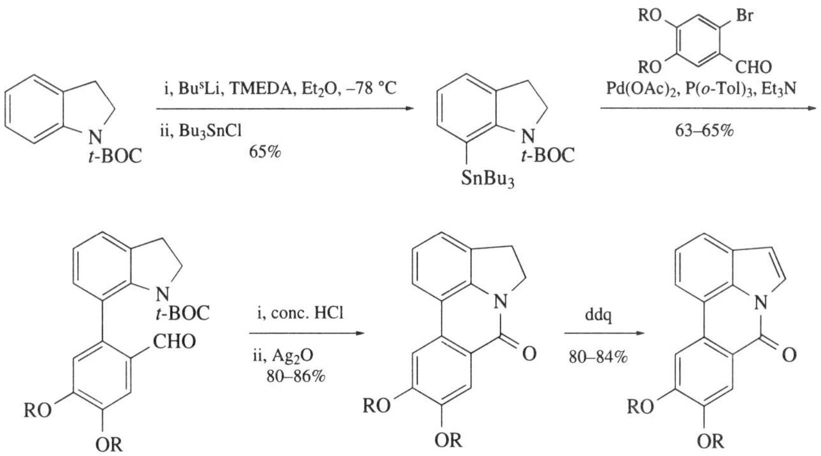

Scheme 127

R, R = -CH$_2$- anhydrolycorin-7-one
R = Me oxoassoanine

R, R = -CH$_2$- hippadine
R = Me pratosine

Table 11 3-DMG-assisted 2-lithiation of indoles.

Substrate	R	Conditions	E$^+$	Yield (%)
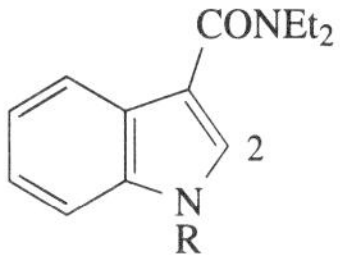 3-CONEt$_2$ indole (2-position)	CH$_2$OMe	Bu^{n}Li, THF, –78 °C	TMS-Cl, MeI, EtI, PhCHO, Pr^{n}CHO, PhCOMe	26–95
	SO$_2$Ph	Bu^{n}Li, THF, –78 °C	MeI, EtI, PhCHO, Pr^{n}CHO	70–88
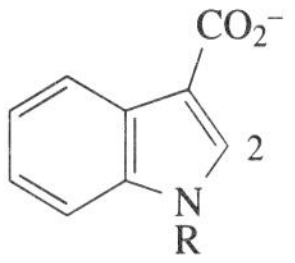 3-CO$_2^-$ indole (2-position)	Me	LDA, THF, –78 °C	TMS-Cl, MeI, PhCHO, PrCHO, Ph$_2$CO, Me$_2$CO, PhCOMe	40–90
	CH$_2$OMe	Bu^{n}Li, THF, –78 °C	MeI, PhCHO, Pr^{n}CHO, Ph$_2$CO, EtCOMe	18–70

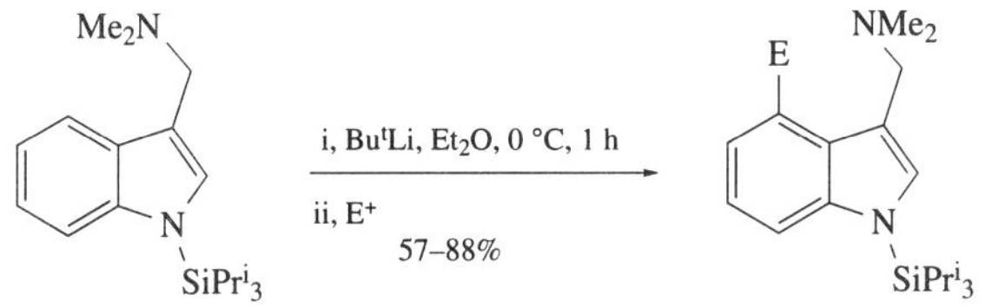

for example E = Cl:

E = CHO, TMS, Bu$_3$Sn, PhS, I, Br, Cl, NH$_2$(N$_3$CH$_2$TMS)

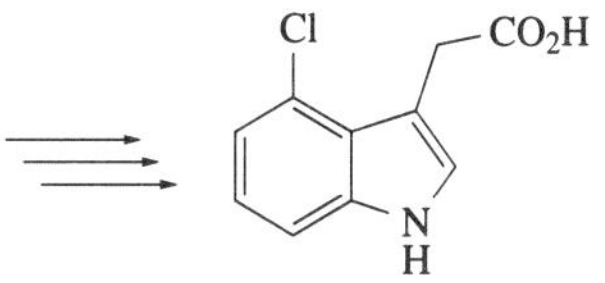

4-chloroindole-3-acetic acid
(plant growth hormone)

Scheme 128

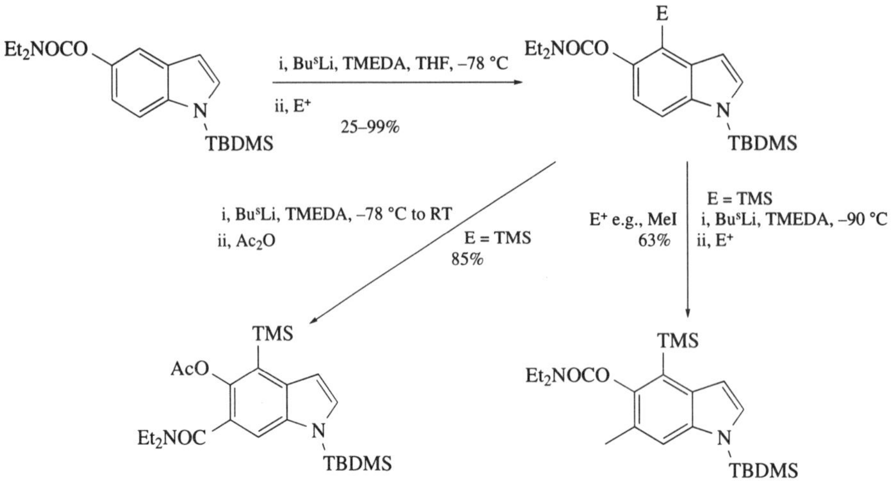

Scheme 129

electrophiles to generate polysubstituted systems (Scheme 130),[705] an improved preparation of 2-bromonicotinaldehyde and acid via 2-bromopyridine (Scheme 131),[706] and the metallation of 4-chloropyridine, a cross-coupling partner in the synthesis of benzonaphthyridinones (Scheme 132).[707]

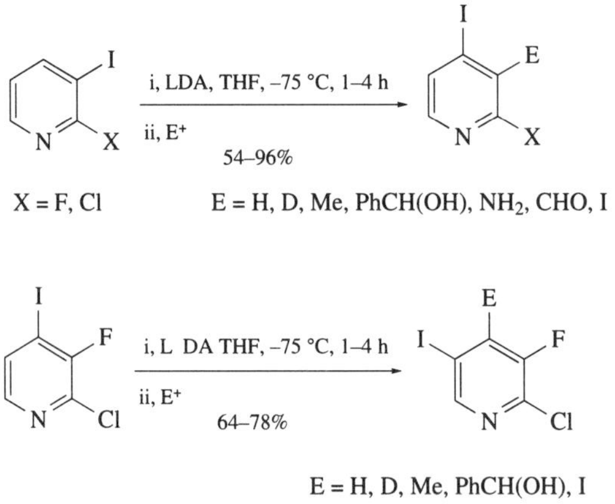

Scheme 130

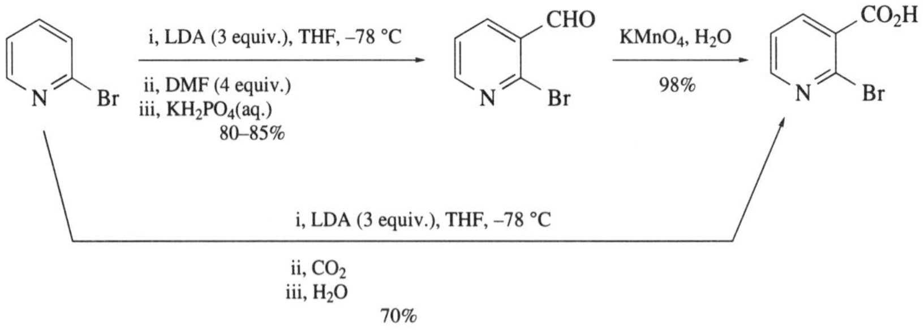

Scheme 131

2-Fluoroquinoline metallation as the first step in the enantioselective construction of a carbon nucleoside serves as an example of the less widely explored DoM of haloquinolines (Scheme 133).[708]

Lithiated species of all isomeric pyridyl *O*-carbamates are useful synthetic intermediates.[709] The 3-*O*-carbamate is thereby efficiently converted into the 4-silylated derivative which serves for 2-deprotonation and provides, after silyl deprotection, a route to correspondingly substituted 3-pyridinols (Scheme 134).[710] Some derived intermediates are useful precursors to 3,4-pyridynes.[711]

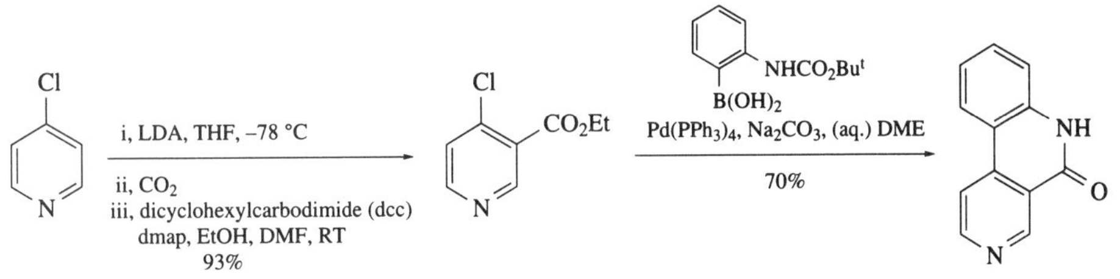

Scheme 132

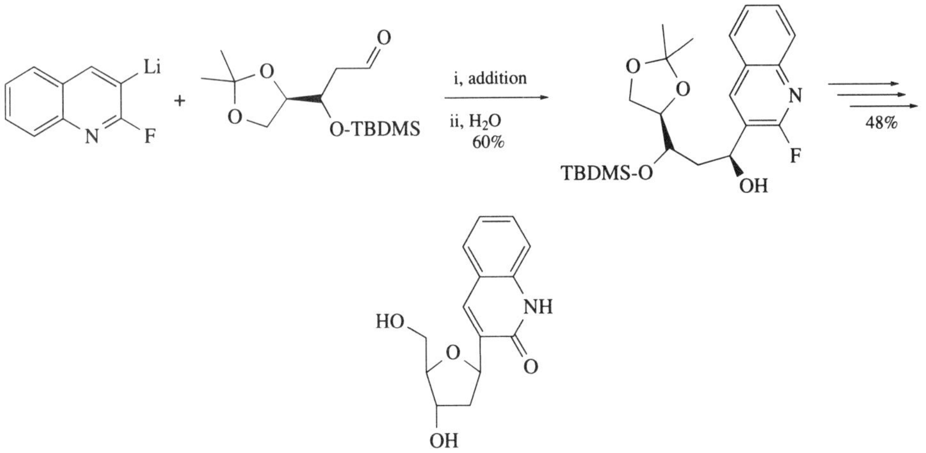

Scheme 133

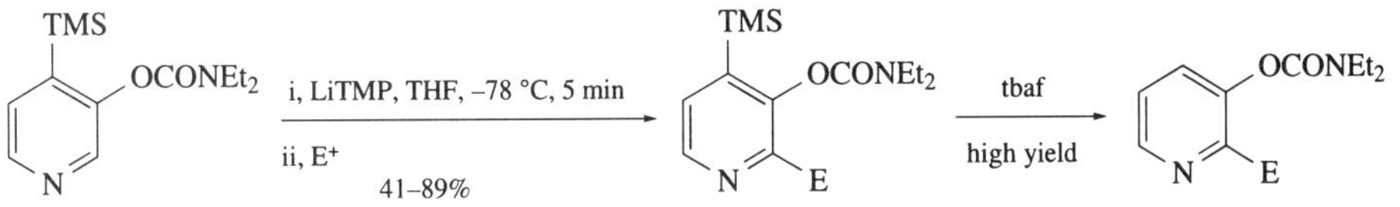

Scheme 134

E = D, Pr^{i}CH(OH), PhCH(OH), Ph$_2$C(OH), PhCO, Cl, SPh, TMS, SePh

Following systematic investigations of isomeric *N*-pivaloylaminopyridines,[712] Quéguiner *et al.* developed an application of this DMG in the synthesis of a streptonigrin alkaloid model (Scheme 135).[713]

Carboxamide DMGs have been widely used for the selective metallation of pyridines,[37,38] as illustrated in the synthesis of sesbanine (Scheme 136),[714] and a formal synthesis of the marine alkaloid amphimedine.[707]

DoM reactions of quinolines constitute key steps in the construction of dynemicin A.[415]

(ii) Diazines

Diazines are generally more susceptible to alkyllithium addition compared to pyridines, but methoxy substituted diazines *ortho*-lithiate cleanly and incorporate electrophiles in high yield.[715] Among the initial studies by Quéguiner and co-workers[37,38,715] are metallations of the pyridazine nucleus,[716] which showed that in some cases significant excess of lithium amide base is needed for the deprotonation and that the position of lithiation is strongly base dependent (Scheme 137).[717] 3-O-MOM pyridazines were found to be reluctant metallation substrates.

LiTMP-induced deprotonation of a methoxypyridazine played a pivotal role in an efficient synthesis of the antidepressant minaprine (Scheme 138).[718]

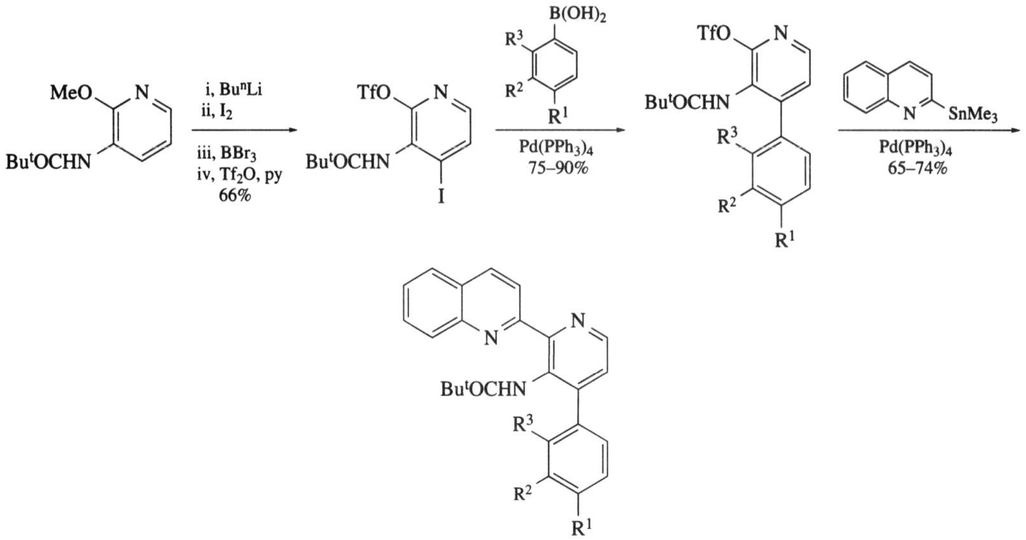

Scheme 135

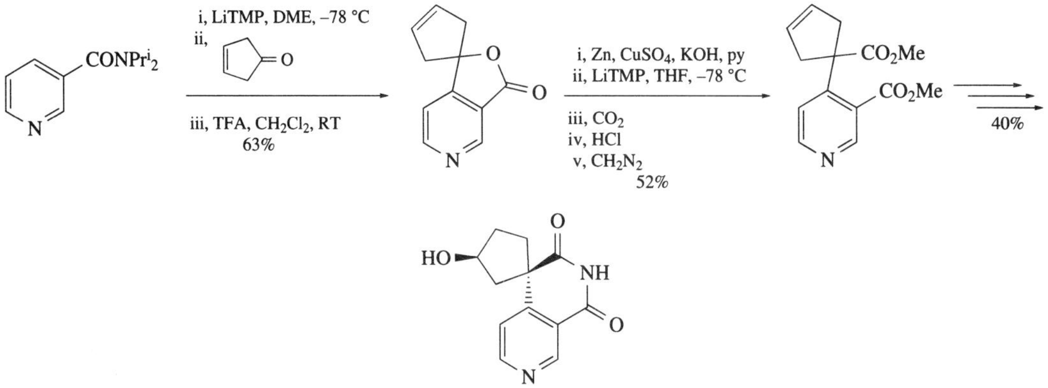

(±)-sesbanine

Scheme 136

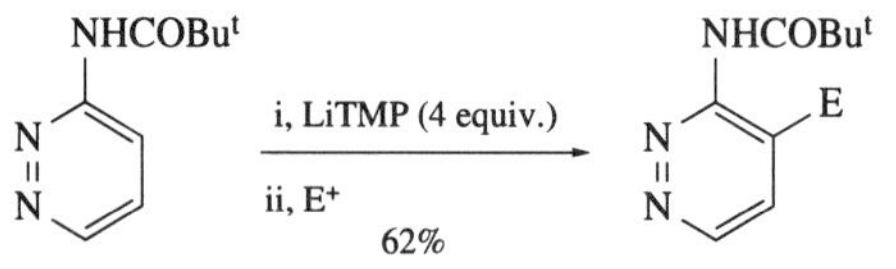

E = PhCH(OH)

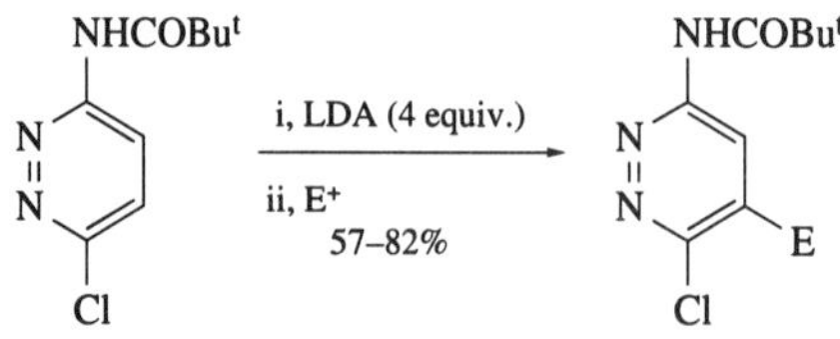

E = MeCH(OH), PhCH(OH)

Scheme 137

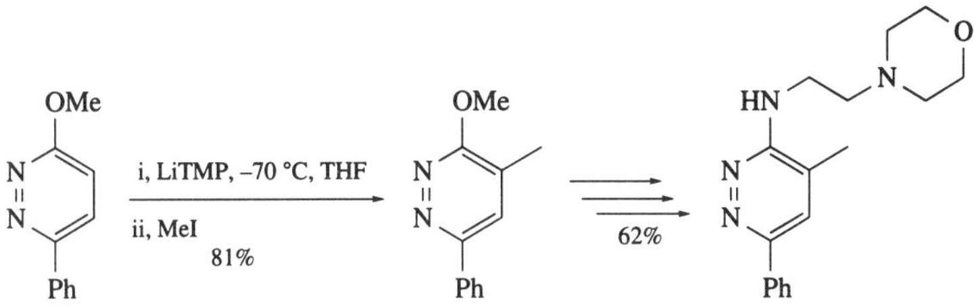

Scheme 138

Pyrimidine DoM chemistry is of older vintage compared with that of pyridazine derivatives.[37,38] The problematic selective C-2 or C-4 lithiation may be conveniently achieved by tin–lithium or iodine–lithium exchange and stannylated precursors are accessible via reaction of tributylstannyllithium with halopyrimidines (Scheme 139).[719] Notably, the 4-stannylated derivative is available either by lithium–halogen exchange or nucleophilic substitution.

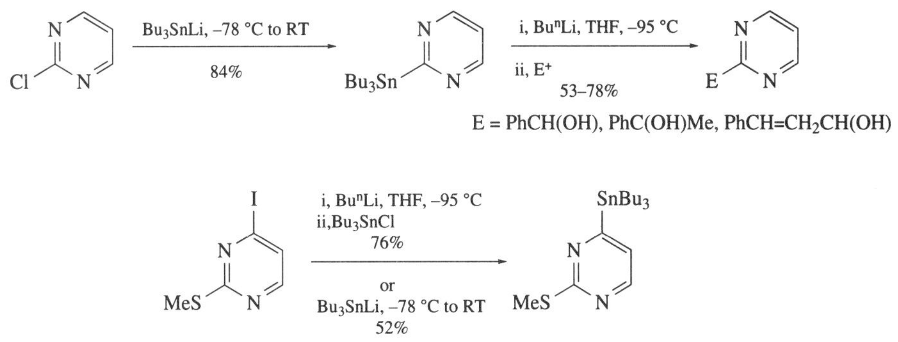

Scheme 139

The metallation of 4-chloro- and 4-methoxypyrimidine derivatives constitutes a key initial event in the preparation of analogues of biomolecules bacimethrin (**72**) and trimethoprim (**73**).[720,721]

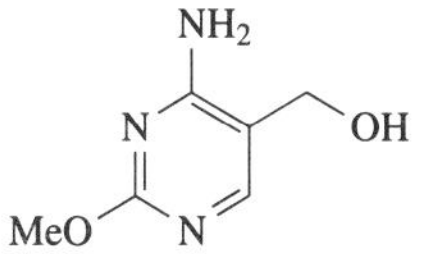

Preliminary results on the DoM chemistry of pyrazines involves the OMe, *N*-pivaloyl, *t*-butylamide, and chlorine DMGs.[101] The 2-*t*-butylamide DMG, however, led to deprotonation at the unexpected C-5 position.[101]

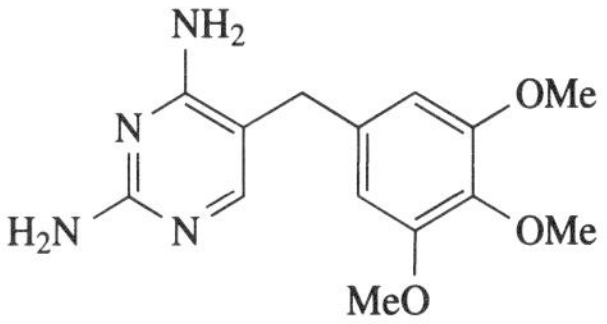

1.5 REFERENCES

1. B. J. Wakefield, in 'COMC-I', vol. 7, p. 1.
2. M. Hanack (ed), 'Houben-Weyl Methoden der Organischen Chemie', 4th edn., Thieme, Stuttgart, 1993, vol. E19d.
3. M. Schlosser (ed.), 'Organometallics in Synthesis. A Manual', Wiley, Chichester, 1994.
4. V. Snieckus (ed.), 'Advances in Carbanion Chemistry', Jai Press, Greenwich, CT, 1992, vol. 1.
5. B. M. Trost (ed.), 'Comprehensive Organic Synthesis', 1st edn., Pergamon, Oxford, 1991.
6. B. J. Wakefield, 'Organolithium Methods', 1st edn., Academic Press, London, 1988.
7. L. Brandsma and H. D. Verkruijsse, 'Preparative Polar Organometallic Chemistry 1', 1st edn., Springer, Berlin, 1987.
8. L. Brandsma, 'Preparative Polar Organometallic Chemistry 2', 1st edn., Springer, Berlin, 1990.

9. J. L. Wardell, in 'The Chemistry of the Metal–Carbon Bond. The Use of Organometallic Compounds in Organic Synthesis', 1st edn., ed. F. R. Hartley, Wiley, Chichester, 1987, vol. 4, p. 1.
10. LITHIUMLINK, a pamphlet available from: FMC, Lithium Division, 449 North Cox Road, Gastonia, NC 28 054, USA. Europe: Lithium Division, Commercial Road, Bromborough, Merseyside, L62 3NL, UK.
11. W. Bauer and P. von Ragué Scheyler, in 'Advances in Carbanion Chemistry', ed. V. Snieckus, Jai Press, Greenwich, CT, 1992, vol. 1, p. 89.
12. P. G. Williard, in 'Comprehensive Organic Synthesis', eds. B. M. Trost and I. Fleming, Pergamon, Oxford, 1991, vol. 1, p. 1.
13. G. Fraenkel, H. Hsu and B. M. Su, in 'Lithium: Current Applications in Science, Medicine and Technology', ed. R. O. Bach, Wiley, New York, 1985, p. 273.
14. M. Schlosser (ed.), 'Mechanistic Aspects of Polar Organometallic Chemistry', Tetrahedron Symposia-in-Print Number 55, *Tetrahedron*, 1994, **50**, 5845.
15. G. Boche, *Angew. Chem., Int. Ed. Engl.*, 1989, **28**, 277.
16. M. Schlosser, in 'Organometallics in Synthesis. A Manual', ed. M. Schlosser, Wiley, Chichester, 1994, p. 18.
17. M. Braun, in 'Advances in Carbanion Chemistry', ed. V. Snieckus, Jai Press, Greenwich, CT, 1992, vol. 1, p. 177.
18. C. H. Heathcock, in 'Comprehensive Organic Synthesis', eds. B. M. Trost and I. Fleming, Pergamon, Oxford, 1991, vol. 2, p. 181.
19. M. Braun, in 'Houben-Weyl Methoden der Organischen Chemie', 4th edn., ed. M. Hanack, Georg Thieme Verlag, Stuttgart, 1993, vol. E19d, p. 171.
20. J. Backes, in 'Houben-Weyl Methoden der Organischen Chemie', 4th edn., ed. M. Hanack, Georg Thieme Verlag, Stuttgart, 1993, vol. E19d, p. 114.
21. J. R. Hanson in 'Comprehensive Organic Synthesis', eds. B. M. Trost and I. Fleming, Pergamon, 1991, vol. 3, p. 705–1014.
22. For a detailed description of lithium reagents, see: 'Encyclopedia of Reagents for Organic Synthesis' (EROS), ed. L. Paquette, Wiley, Chichester, 1995.
23. W. F. Bailey, LITHIUMLINK, Spring 1994, FMC. North America: Lithium Division, 449 North Cox Road, Gastonia, NC 28054, USA. Europe: Lithium Division, Commercial Road, Bromborough, Merseyside, L62 3NL, UK.
24. R. E. Gawley and K. Rein, in 'Comprehensive Organic Synthesis', 1st edn., eds. B. M. Trost and I. Fleming, Pergamon, Oxford, 1991, vol. 1, p. 459.
25. R. E. Gawley and K. Rein, in 'Comprehensive Organic Synthesis', 1st edn., eds. B. M. Trost and I. Fleming, Pergamon, Oxford, 1991, vol. 3, p. 65.
26. P. Knochel, *Angew. Chem., Int. Ed. Engl.*, 1992, **31**, 1459.
27. J. S. Panek, in 'Comprehensive Organic Synthesis', eds. B. M. Trost and I. Fleming, Pergamon, Oxford, 1991, vol. 1, p. 579.
28. D. J. Ager, *Org. React.*, 1990, **38**, 1.
29. S. Lamothe, K. L. Cook and T. H. Chan, *Can. J. Chem.*, 1992, **70**, 1733.
30. D. H. Hua, in 'Advances in Carbanion Chemistry', 1st edn., ed. V. Snieckus, Jai Press, Greenwich, CT, 1992, vol. 1, p. 249.
31. G. Solladié, in 'Perspectives in the Organic Chemistry of Sulfur', eds. B. Zwanenburg and A. J. H. Klunder, Elsevier, Amsterdam, 1987, p. 293.
32. E. L. Grimm, S. Levac and L. A. Trimble, *Tetrahedron Lett.*, 1994, **35**, 6847 and references cited therein.
33. H. W. Gschwend and H. R. Rodriguez, *Org. React.*, 1979, **26**, 1.
34. L. Brandsma, in 'Houben-Weyl Methoden der Organischen Chemie', 4th edn., ed. M. Hanack, Georg Theime Verlag, Stuttgart, 1993, vol. E19d, p. 369.
35. V. Snieckus, *Chem. Rev.*, 1990, **90**, 879.
36. R. D. Clark and Jahangir, *Org. React.*, 1995, **47**, 1.
37. G. W. Rewcastle and A. R. Katritzky, *Adv. Heterocyclic Chem.*, 1993, **56**, 155.
38. G. Quéguiner, F. Marsais, V. Snieckus and J. Epsztajn, *Adv. Heterocyclic Chem.*, 1991, **52**, 187.
39. Selected list of suppliers: Aldrich, FMC, Janssen, Lancaster, Fluka, Chemetall GmbH.
40. A review is available from FMC. North America: Lithium Division, 449 North Cox Road, Gastonia, NC 28 054, USA. Europe: Lithium Division, Commercial Road, Bromborough, Merseyside, L62 3NL, UK.
41. For example, the Snieckus group uses the method of S. C. Watson and J. F. Eastham, *J. Organomet. Chem.*, 1967, **9**, 165, for butyllithiums, whereas the Comins group uses that of W. G. Kofron and L. M. Baclawski, *J. Org. Chem.*, 1976, **41**, 1879.
42. J. Suffert, *J. Org. Chem.*, 1989, **54**, 509.
43. Y. Aso, H. Yamashita, T. Otsubo and F. Ogura, *J. Org. Chem.*, 1989, **54**, 5627.
44. R. Miranda, A. Hernández, E. Angeles, A. Cabrera, M. Salmón and P. Joseph-Nathan, *Analyst (London)*, 1990, **115**, 1483.
45. H. Kiljunen and T. A. Hase, *J. Org. Chem.*, 1991, **56**, 6950.
46. L. Duhamel and J.-C. Plaquevent, *J. Organomet. Chem.*, 1993, **448**, 1.
47. S. W. McCombie, '2-[(*t*-Butylamino)carbonyl]-3-methylpyridine, a new RLi indicator.' Schering-Plough Research Institute, 2015 Galloping Hill Road, Kenilworth, NJ 07033-0539, USA, personal communication. See also D. P. Schumacher, B. L. Murphy, J. E. Clark, P. Tahbaz and T. A. Mann, *J. Org. Chem.*, 1989, **54**, 2242.
48. H. Gilman, W. Langham and F. W. Moore, *J. Am. Chem. Soc.*, 1940, **62**, 2327.
49. R. G. Jones and H. Gilman, *Org. React.*, 1951, **6**, 339.
50. M. Casey, J. Leonard, J. L. Lygo and G. Procter, 'Advanced Practical Organic Chemistry', 1st edn., Blackie, Glasgow, 1990, p. 229.
51. J. Einhorn and J. L. Luche, *J. Org. Chem.*, 1987, **52**, 4124.
52. D. Guijarro, G. Guillena, B. Mancheño and M. Yus, *Tetrahedron*, 1994, **50**, 3427 and references cited therein.
53. L. F. Fieser and M. Fieser, 'Reagents for Organic Synthesis Volume 1', 1st edn., Wiley, New York, 1967, p. 96.
54. R. C. Morrison and T. L. Rathman (FMC Lithium Corp.) *US Pat.* 4 976 886 (1990).
55. M. J. Lusch, W. V. Phillips, R. F. Sieloff, G. S. Nomura and H. O. House, *Org. Synth.*, 1984, **62**, 101.
56. H. Gilman and J. W. Morton, Jr., *Org. React.*, 1954, **8**, 258.
57. M. Schlosser and S. Strunk, *Tetrahedron Lett.*, 1984, **25**, 741.
58. L. Lochmann and J. Trekoval, *Collect. Czech Chem. Commun.*, 1986, **51**, 1439.

59. A. Mordini, in 'Advances in Carbanion Chemistry', 1st edn., ed. V. Snieckus, Jai Press, Greenwich, CT, 1992, vol. 1, p. 1.
60. P. Caubère, *Chem. Rev.*, 1993, **93**, 2317.
61. S. Cabiddu, C. Fattuoni, C. Floris, S. Melis and A. Serci, *Tetrahedron*, 1994, **50**, 6037.
62. D. W. Slocum, R. Moon, J. Thompson, D. S. Coffey, J. D. Li, M. G. Slocum, A. Siegeland and R. Gayton-Garcia, *Tetrahedron Lett.*, 1994, **35**, 385.
63. D. B. Collum, *Acc. Chem. Res.*, 1992, **25**, 448.
64. M. Schlosser, in 'Organometallics in Synthesis. A Manual', ed. M. Schlosser, Wiley, Chichester, 1994, p. 71.
65. M. Schlosser, in 'Organometallics in Synthesis. A Manual', ed. M. Schlosser, Wiley, Chichester, 1994, p. 13.
66. C. Lambert and P. von Ragué Schleyer, in 'Houben-Weyl Methoden der organischen Chemie', 4th edn., ed. M. Hanack, Georg Thieme Verlag, Stuttgart, 1993, vol. E19d, p. 1.
67. N. A. A. Al-Jabar and A. G. Massey, *J. Organomet. Chem.*, 1985, **288**, 145.
68. M. Khaldi, F. Chrétien and Y. Chapleur, *Tetrahedron Lett.*, 1994, **35**, 401.
69. T. Mukhopadhyay and D. Seebach, *Helv. Chim. Acta*, 1982, **65**, 385.
70. F. E. Romesberg, J. H. Gilchrist, A. T. Harrison, D. J. Fuller and D. B. Collum, *J. Am. Chem. Soc.*, 1991, **113**, 5751.
71. E. M. Kaiser and H. H. Yun, *J. Org. Chem.*, 1970, **35**, 1348.
72. R. R. Fraser and T. S. Mansour, *Tetrahedron Lett.*, 1986, **27**, 331.
73. S. Takagishi, G. Katsoulos and M. Schlosser, *Synlett*, 1992, 360 and references cited therein.
74. C. G. Screttas and M. Micha-Screttas, *Organometallics*, 1984, **3**, 904.
75. C. G. Screttas and B. R. Steele, *J. Org. Chem.*, 1989, **54**, 1013.
76. D. Seebach, *Angew. Chem., Int. Ed. Engl.*, 1988, **27**, 1624.
77. For a review see: M. Majewski and D. M. Gleave, *J. Organomet. Chem.*, 1994, **470**, 1.
78. W. ten Hoeve, C. G. Kruse, J. M. Luteyn, J. R. G. Thiecke and H. Wynberg, *J. Org. Chem.*, 1993, **58**, 5101.
79. R. R. Fraser and T. S. Mansour, *J. Org. Chem.*, 1984, **49**, 3442.
80. D. Caine, in 'Comprehensive Organic Synthesis', eds. B. M. Trost and I. Fleming, Pergamon, Oxford, 1991, vol. 3, p. 1.
81. Y. Yamamoto, in 'Comprehensive Organic Synthesis', ed. C. H. Heathcock, Pergamon, Oxford, 1991, vol. 2, p. 55.
82. D. A. Evans and G. C. Andrews, *Acc. Chem. Res.*, 1974, **7**, 147.
83. L. Maciejewski, M. Martin, G. Ricart and J. Brocard, *Synth. Commun.*, 1988, **18**, 1757.
84. H. O. House, W. V. Phillips, T. S. B. Sayer and C.-C. Yau, *J. Org. Chem.*, 1978, **43**, 700.
85. L. Brandsma and H. D. Verkruijsse, 'Preparative Polar Organometallic Chemistry 1', Springer, Berlin, 1987, p. 19.
86. F. Gaudemar-Bardone and M. Gaudemar, *Synthesis*, 1979, 463.
87. M. T. Reetz and W. F. Maier, *Liebigs Ann. Chem.*, 1980, 1471.
88. H. Ahlbrecht and G. Schneider, *Tetrahedron*, 1986, **42**, 4729.
89. D. Seebach and D. Wasmuth, *Angew. Chem., Int. Ed. Engl.*, 1981, **20**, 971.
90. H. Normant, T. Cuvigny and D. Reisdorf, *C. R. Acad. Sci. Ser. C*, 1969, **268**, 521.
91. G. Wittig and A. Hesse, *Liebigs Ann. Chem.*, 1971, **746**, 149.
92. C. M. Lindsay and D. A. Widdowson, *J. Chem. Soc., Perkin Trans. 1*, 1988, 569.
93. D. N. Reinhoudt and C. G. Kouwenhoven, *Recl. Trav. Chim. Pays-Bas*, 1976, **95**, 67.
94. B.-A. Feit, S. Dickerman, D. Masrawe and A. Fishman, *J. Chem. Soc., Perkin Trans. 1*, 1988, 927.
95. J. Villieras, C. Baquet and J. F. Normant, *J. Organomet. Chem.*, 1975, **97**, 355.
96. G. E. Gream, D. K. C. Hodgeman and R. H. Prager, *Aust. J. Chem.*, 1972, **25**, 569.
97. T. D. Krizan and J. C. Martin, *J. Am. Chem. Soc.*, 1983, **105**, 6155.
98. R. R. Fraser and S. Savard, *Can. J. Chem.*, 1986, **64**, 621.
99. P. Beak and B. Lee, *J. Org. Chem.*, 1989, **54**, 458.
100. P. E. Eaton and R. M. Martin, *J. Org. Chem.*, 1988, **53**, 2728.
101. A. Turck, N. Plé, D. Trohay, B. Ndzi and G. Quéguiner, *J. Heterocycl. Chem.*, 1992, **29**, 699 and references cited therein.
102. I. Jefferies, M. Julia, J.-N. Verpaux and T. Zahneisen, *Synlett*, 1991, 647.
103. J. J. Fitzgerald, N. E. Drysdale and R. A. Olofson, *J. Org. Chem.*, 1992, **57**, 7122.
104. I. E. Kopka, M. A. Nowak and M. W. Rathke, *Synth. Commun.*, 1986, **16**, 27.
105. P. L. Hall, J. H. Gilchrist and D. B. Collum, *J. Am. Chem. Soc.*, 1991, **113**, 9571.
106. P. L. Hall, J. H. Gilchrist, A. T. Harrison, D. J. Fuller and D. B. Collum, *J. Am. Chem. Soc.*, 1991, **113**, 9575.
107. J. C. Gilbert and T. A. Kelly, *Tetrahedron*, 1988, **44**, 7587.
108. L. A. Flippin, J. M. Muchowski and D. S. Carter, *J. Org. Chem.*, 1993, **58**, 2463.
109. W. Murray, M. Wachter, D. Barton and Y. Forero-Kelly, *Synthesis*, 1991, 18.
110. M. Tanabe and D. F. Crowe, *J. Chem. Soc., Chem. Commun.*, 1973, 564.
111. M. Es-Sayed, C. Gratkowski, N. Krass, A. I. Meyers and A. de Meijere, *Synlett*, 1992, 962.
112. R. F. Borch, *Tetrahedron Lett.*, 1972, 3761.
113. R. E. Ireland, P. Wipf and J. D. Amstrong, III, *J. Org. Chem.*, 1991, **56**, 650.
114. M. Ihara, M. Suzuki, K. Fukumoto and C. Kabuto, *J. Am. Chem. Soc.*, 1990, **112**, 1164.
115. T. Murai, M. Yamamoto, S. Kondo and S. Kato, *J. Org. Chem.*, 1993, **58**, 7440.
116. R. J. P. Corriu, V. Huynh, J. Iqbal, J. J. E. Moreau and C. Vernhet, *Tetrahedron*, 1992, **48**, 6231.
117. P. Andreoli *et al.*, *Tetrahedron*, 1991, **47**, 9061.
118. E. W. Colvin, 'Silicon Reagents in Organic Synthesis', 1st edn., Academic Press, London, 1988.
119. P. Sampson, G. B. Hammond and D. F. Wiemer, *J. Org. Chem.*, 1986, **51**, 4342.
120. M. W. Rathke, *Org. Synth.*, 1988, **VI**, 598.
121. M. Watanabe, M. Date, M. Tsukazaki and S. Furukawa, *Chem. Pharm. Bull.*, 1989, **37**, 36.
122. J. A. Prieto, J. Suarez and G. L. Larson, *Synth. Commun.*, 1988, **18**, 253.
123. E. J. Corey and A. W. Gross, *Tetrahedron Lett.*, 1984, **25**, 495.
124. E. J. Corey and A. W. Gross, *Tetrahedron Lett.*, 1984, **25**, 491.
125. D. L. Comins, *Synlett*, 1992, 615.
126. T. Uyehara, N. Shida and Y. Yamamoto, *J. Chem. Soc., Chem. Commun.*, 1989, 113.
127. T. Uyehara, N. Asao and Y. Yamamoto, *J. Chem. Soc., Chem. Commun.*, 1989, 753.
128. H. M. R. Hoffmann, A. Köver and D. Pauluth, *J. Chem. Soc., Chem. Commun.*, 1985, 812.

129. J. K. Whitesell and S. W. Felman, *J. Org. Chem.*, 1980, **45**, 755.
130. Y. Landais and P. Ogay, *Tetrahedron: Asymmetry*, 1994, **5**, 541.
131. Y. Hasegawa, H. Kawasaki and K. Koga, *Tetrahedron Lett.*, 1993, **34**, 1963.
132. R. Armer, M. J. Begley, P. J. Cox, A. Persad and N. S. Simpkins, *J. Chem. Soc., Perkin Trans. 1*, 1993, 3099.
133. B. J. Bunn, N. S. Simpkins, Z. Spavold and M. J. Crimmin, *J. Chem. Soc., Perkin Trans. 1*, 1993, 3113.
134. T. Cohen, J. P. Sherbine, J. R. Matz, R. R. Hutchins, B. M. McHenry and P. R. Willey, *J. Am. Chem. Soc.*, 1984, **106**, 3245.
135. T. Cohen and M.-T. Lin, *J. Am. Chem. Soc.*, 1984, **106**, 1130.
136. M. Schlosser, J. Hartmann, M. Stähle, J. Kramar, A. Walde and A. Mordini, *Chimia*, 1986, **40**, 306.
137. P. E. Eaton, R. G. Daniels, D. Casucci, G. T. Cunkle and P. Engel, *J. Org. Chem.*, 1987, **52**, 2100.
138. D. J. Gallagher, C. G. Garrett, R. P. Lemieux and P. Beak, *J. Org. Chem.*, 1991, **56**, 853.
139. P. E. Eaton and G. Castaldi, *J. Am. Chem. Soc.*, 1985, **107**, 724.
140. G. Szeimies, F. Philipp, O. Baumgärtel and J. Harnisch, *Tetrahedron Lett.*, 1977, 2135.
141. Y. Igarashi, Y. Kabe, T. Hagiwara and W. Ando, *Tetrahedron*, 1992, **48**, 89.
142. A.-D. Schlüter, *Angew. Chem., Int. Ed. Engl.*, 1988, **27**, 296.
143. T. Ghosh, H. L. Gingrich, C. K. Kam, E. C. Mobraaten and M. Jones, Jr., *J. Am. Chem. Soc.*, 1991, **113**, 1313.
144. A. Maercker and K.-D. Klein, *Angew. Chem., Int. Ed. Engl.*, 1989, **28**, 83.
145. G. W. Erickson and J. L. Fry, *J. Org. Chem.*, 1987, **52**, 462.
146. P. K. Freeman and L. L. Hutchinson, *J. Org. Chem.*, 1980, **45**, 1924.
147. B. Mudryk and T. Cohen, *Org. Synth.*, 1993, **72**, 173.
148. D. J. Rawson and A. I. Meyers, *Tetrahedron Lett.*, 1991, **32**, 2095.
149. D. J. Ramón, and M. Yus, *J. Chem. Soc., Chem. Commun.*, 1991, 398.
150. D. J. Ramón and M. Yus, *Tetrahedron Lett.*, 1990, **31**, 3767.
151. D. J. Ramón and M. Yus, *Tetrahedron Lett.*, 1990, **31**, 3763.
152. D. J. Ramón and M. Yus, *J. Org. Chem.*, 1991, **56**, 3825.
153. D. Caine and C.-R. Lin, *Synth. Commun.*, 1994, **24**, 2473.
154. M. Yus and D. J. Ramón, *J. Org. Chem.*, 1992, **57**, 750.
155. J. Barluenga, J. M. Montserrat and J. Flórez, *J. Org. Chem.*, 1993, **58**, 5976.
156. S. V. Ley and C. M. R. Low, 'Ultrasound in Synthesis', Springer, Berlin, 1989.
157. M. Ihara, M. Katogi, K. Fukumoto and T. Kametani, *J. Chem. Soc., Chem. Commun.*, 1987, 721.
158. C. Pétrier, A. L. Gemal and J.-L. Luche, *Tetrahedron Lett.*, 1982, **23**, 3361.
159. G. Wittig, U. Pockels and H. Droge, *Chem. Ber.*, 1938, **71**, 1903.
160. H. Gilman, W. Langham and A. L. Jacoby, *J. Am. Chem. Soc.*, 1939, **61**, 106.
161. D. E. Applequist and D. F. O'Brien, *J. Am. Chem. Soc.*, 1963, **85**, 743.
162. R. G. Jones and H. Gilman, *Chem. Rev.*, 1954, 835.
163. B. J. Wakefield, in 'Comprehensive Organic Chemistry', eds. D. Barton and W. D. Ollis, Pergamon, Oxford, 1979, vol. 3, p. 943.
164. J. L. Wardell, in 'COMC-I' vol. 1, p. 43.
165. W. F. Bailey and J. J. Patricia, *J. Organomet. Chem.*, 1988, **352**, 1.
166. D. R. Myers and M. Jones, Jr., *Tetrahedron Lett.*, 1991, **32**, 2203.
167. H. J. Reich, D. P. Green and N. H. Phillips, *J. Am. Chem. Soc.*, 1991, **113**, 1414.
168. P. Beak and D. J. Allen, *J. Am. Chem. Soc.*, 1992, **114**, 3420.
169. E. C. Ashby, B. Park, G. S. Patil, K. Gadru and R. Gurumurthy, *J. Org. Chem.*, 1993, **58**, 424.
170. E. W. Della, D. K. Taylor and J. Tsanaktsidis, *Tetrahedron Lett.*, 1990, **31**, 5219.
171. E. W. Della and D. K. Taylor, *Aust. J. Chem.*, 1991, **44**, 881.
172. B. C. Borer and R. J. K. Taylor, *Synlett*, 1990, 601.
173. R. W. Hoffmann, M. Bewersdorf, M. Krüger, W. Mikolaiski and R. Stürmer, *Chem. Ber.*, 1991, **124**, 1243.
174. W. F. Bailey and E. R. Punzalan, *J. Org. Chem.*, 1990, **55**, 5404.
175. E. Negishi, D. R. Swanson and C. J. Rousset, *J. Org. Chem.*, 1990, **55**, 5406.
176. T. F. Bates, M. T. Clarke and R. D. Thomas, *J. Am. Chem. Soc.*, 1988, **110**, 5109.
177. M. P. Cooke, Jr. and I. N. Houpis, *Tetrahedron Lett.*, 1985, **26**, 4987.
178. G. A. Molander and J. A. McKie, *J. Org. Chem.*, 1991, **56**, 4112.
179. P. G. Gassman and N. J. O'Reilly, *J. Org. Chem.*, 1987, **52**, 2481.
180. M. P. Cooke, Jr. and R. K. Widener, *J. Org. Chem.*, 1987, **52**, 1381.
181. M. P. Cooke, Jr., *J. Org. Chem.*, 1984, **49**, 1144.
182. M. P. Cooke, Jr., *J. Org. Chem.*, 1992, **57**, 1495.
183. W. F. Bailey, R. P. Gagnier and J. J. Patricia, *J. Org. Chem.*, 1984, **49**, 2098.
184. W. F. Bailey and T. V. Ovaska, *J. Am. Chem. Soc.*, 1993, **115**, 3080.
185. W. F. Bailey and K. V. Gavaskar, *Tetrahedron*, 1994, **50**, 5957.
186. W. F. Bailey *et al.*, *J. Am. Chem. Soc.*, 1991, **113**, 5720.
187. W. F. Bailey and A. D. Khanolkar, *Tetrahedron*, 1991, **47**, 7727.
188. W. R. Dolbier, Jr. and Y. Chen, *J. Org. Chem.*, 1992, **57**, 1947.
189. W. F. Bailey and K. Rossi, *J. Am. Chem. Soc.*, 1989, **111**, 765.
190. W. F. Bailey, A. D. Khanolkar and K. V. Gavaskar, *J. Am. Chem. Soc.*, 1992, **114**, 8053.
191. J. K. Crandall and T. A. Ayers, *J. Org. Chem.*, 1992, **57**, 2993.
192. J. K. Crandall and T. A. Ayers, *Tetrahedron Lett.*, 1992, **33**, 5311.
193. W. F. Bailey and T. V. Ovaska, *Tetrahedron Lett*, 1990, **31**, 627.
194. G. Wu, F. E. Cederbaum and E. Negishi, *Tetrahedron Lett.*, 1990, **31**, 493.
195. W. F. Bailey and L. M. J. Zarcone, *Tetrahedron Lett.*, 1991, **32**, 4425.
196. E. Bartmann, *Angew. Chem., Int. Ed. Engl.*, 1986, **25**, 653.
197. C. Nájera, M. Yus and D. Seebach, *Helv. Chim. Acta*, 1984, **67**, 289.
198. T. Cohen, I.-H. Jeong, B. Mudryk, M. Bhupathy and M. M. A. Awad, *J. Org. Chem.*, 1990, **55**, 1528.
199. B. Mudryk and T. Cohen, *J. Org. Chem.*, 1989, **54**, 5657.
200. B. Mudryk, C. A. Shook and T. Cohen, *J. Am. Chem. Soc.*, 1990, **112**, 6389.

201. E. Licandro, S. Maiorana, A. Papagni and A. Zanotti-Gerosa, *J. Chem. Soc., Chem. Commun.*, 1992, 1623.
202. B. Mudryk and T. Cohen, *J. Am. Chem. Soc.*, 1991, **113**, 1866.
203. D. J. Ramón and M. Yus, *Tetrahedron*, 1992, **48**, 3585.
204. J. Almena, F. Foubelo and M. Yus, *Tetrahedron Lett.*, 1993, **34**, 1649.
205. J. Almena, F. Foubelo and M. Yus, *Tetrahedron*, 1994, **50**, 5775.
206. C. A. Broka, W. J. Lee and T. Shen, *J. Org. Chem.*, 1988, **53**, 1336.
207. H. Nakahira, I. Ryu, M. Ikebe, N. Kambe and N. Sonoda, *Angew. Chem., Int. Ed. Engl.*, 1991, **30**, 177.
208. A. Krief, D. Derouane and W. Dumont, *Synlett*, 1992, 907.
209. T. Hiiro, N. Kambe, A. Ogawa, N. Miyoshi, S. Murai and N. Sonoda, *Angew. Chem., Int. Ed. Engl.*, 1987, **26**, 1187.
210. K. B. Wiberg and S. T. Waddell, *J. Am. Chem. Soc.*, 1990, **112**, 2194.
211. L. A. Paquette, D. R. Lagerwall, J. L. King and S. Niwayama, *Tetrahedron Lett.*, 1991, **32**, 6529.
212. G. Stork and S. D. Rychnovsky, *J. Am. Chem. Soc.*, 1987, **109**, 1565.
213. C. Rücker, *J. Organomet. Chem.*, 1986, **310**, 135.
214. T. Cohen and M. Bhupathy, *Acc. Chem. Res.*, 1989, **22**, 152.
215. D. Guijarro and M. Yus, *Tetrahedron*, 1994, **50**, 3447.
216. P. Beak, W. J. Zajdel and D. B. Reitz, *Chem. Rev.*, 1984, **84**, 471.
217. V. K. Aggarwal, *Angew. Chem., Int. Ed. Engl.*, 1994, **33**, 175.
218. H. Ahlbrecht and H. Dollinger, *Tetrahedron Lett.*, 1984, **25**, 1353.
219. F. H. Köhler, N. Hertkorn and J. Blümel, *Chem. Ber.*, 1987, **120**, 2081.
220. T. Tsunoda, K. Fujiwara, Y.-I. Yamamoto and S. Itô, *Tetrahedron Lett.*, 1991, **32**, 1975.
221. H. Ahlbrecht, J. Harbach, T. Hauck and H.-O. Kalinowski, *Chem. Ber.*, 1992, **125**, 1753.
222. L. J. Bartolotti and R. E. Gawley, *J. Org. Chem.*, 1989, **54**, 2980.
223. P. Beak and W. K. Lee, *J. Org. Chem.*, 1993, **58**, 1109 and references cited therein.
224. D. R. Hay, Z. Song, S. G. Smith and P. Beak, *J. Am. Chem. Soc.*, 1988, **110**, 8145 and references cited therein.
225. A. I. Meyers, P. D. Edwards, W. F. Rieker and T. R. Bailey, *J. Am. Chem. Soc.*, 1984, **106**, 3270.
226. For a review see: J. E. Saavedra, *Org. Prep. Proceed. Int.*, 1987, **19**, 83.
227. D. Seebach, J.-J. Lohmann, M. A. Syfrig and M. Yoshifuji, *Tetrahedron*, 1983, **39**, 1963.
228. I. M. P. Huber and D. Seebach, *Helv. Chim. Acta*, 1987, **70**, 1944.
229. P. Beak and W. J. Zajdel, *J. Am. Chem. Soc.*, 1984, **106**, 1010.
230. T. T. Shawe and A. I. Meyers, *J. Org. Chem.*, 1991, **56**, 2751.
231. M. A. Sanner, *Tetrahedron Lett.*, 1989, **30**, 1909.
232. P. Beak and E. K. Yum, *J. Org. Chem.*, 1993, **58**, 823.
233. P. J. Cox and N. S. Simpkins, *Tetrahedron: Asymmetry*, 1991, **2**, 1.
234. P. Beak, S. T. Kerrick, S. Wu and J. Chu, *J. Am. Chem. Soc.*, 1994, **116**, 3231.
235. A. I. Meyers, *Tetrahedron*, 1992, **48**, 2589.
236. A. I. Meyers, M. Boes and D. A. Dickman, *Org. Synth.*, 1988, **67**, 60.
237. L. Gottlieb and A. I. Meyers, *J. Org. Chem.*, 1990, **55**, 5659.
238. T. H. Sielecki and A. I. Meyers, *J. Org. Chem.*, 1992, **57**, 3673.
239. A. I. Meyers and R. H. Hutchings, *Tetrahedron*, 1993, **49**, 1807.
240. A. I. Meyers, T. K. Highsmith and P. T. Buonora, *J. Org. Chem.*, 1991, **56**, 2960.
241. A. I. Meyers, M. A. Gonzalez, V. Struzka, A. Akahane, J. Guiles and J. S. Warmus, *Tetrahedron Lett.*, 1991, **32**, 5501.
242. M. A. Gonzalez and A. I. Meyers, *Tetrahedron Lett.*, 1989, **30**, 47.
243. A. R. Katritzky and K. Akutagawa, *Tetrahedron*, 1986, **42**, 2571.
244. E. Pfammatter and D. Seebach, *Liebigs Ann. Chem.*, 1991, 1323.
245. R. E. Gawley, K. Rein and S. Chemburkar, *J. Org. Chem.*, 1989, **54**, 3002.
246. K. Rein, M. Goicoechea-Pappas, T. V. Anklekar, G. C. Hart, G. A. Smith and R. E. Gawley, *J. Am. Chem. Soc.*, 1989, **111**, 2211.
247. R. E. Gawley, G. C. Hart and L. J. Bartolotti, *J. Org. Chem.*, 1989, **54**, 175.
248. A. I. Meyers, D. A. Dickman and T. R. Bailey, *J. Am. Chem. Soc.*, 1985, **107**, 7974.
249. W. H. Pearson, A. C. Lindbeck and J. W. Kampf, *J. Am. Chem. Soc.*, 1993, **115**, 2622 and references cited therein.
250. R. E. Gawley and Q. Zhang, *J. Am. Chem. Soc.*, 1993, **115**, 7515 and references cited therein.
251. S. V. Kessar, R. Vohra, N. P. Kaur, K. N. Singh and P. Singh, *J. Chem. Soc., Chem. Commun.*, 1994, 1327 and references cited therein.
252. G. Boche, F. Bosold, J. C. W. Lohrenz, A. Opel and P. Zulauf, *Chem. Ber.*, 1993, **126**, 1873 and references cited therein.
253. J.-M. Lancelin, L. Morin-Allory and P. Sinaÿ, *J. Chem. Soc., Chem. Commun.*, 1984, 355.
254. J. S. Sawyer, A. Kucerovy, T. L. Macdonald and G. J. McGarvey, *J. Am. Chem. Soc.*, 1988, **110**, 842.
255. R. C. Gadwood, M. R. Rubino, S. C. Nagarajan and S. T. Michel, *J. Org. Chem.*, 1985, **50**, 3255.
256. A. R. Katritzky, W.-Q. Fan and K. Akutagawa, *Synthesis*, 1987, 415.
257. J. Blagg, S. G. Davies, C. L. Goodfellow and K. H. Sutton, *J. Chem. Soc., Perkin Trans. 1*, 1987, 1805.
258. E. J. Corey and T. M. Eckrich, *Tetrahedron Lett.*, 1983, **24**, 3165.
259. J.-P. Quintard, B. Elissondo and M. Pereyre, *J. Organomet. Chem.*, 1981, **212**, C31.
260. S. D. Rychnovsky, K. Plzak and D. Pickering, *Tetrahedron Lett.*, 1994, **35**, 6799 and references cited therein.
261. J. I. Yoshida, S. I. Matsunaga and S. Isoe, *Tetrahedron Lett.*, 1989, **30**, 219.
262. W. C. Still and C. Sreekumar, *J. Am. Chem. Soc.*, 1980, **102**, 1201.
263. P. Lesimple, J.-M. Beau and P. Sinaÿ, *J. Chem. Soc., Chem. Commun.*, 1985, 894.
264. D. Hoppe, A. Carstens and T. Krämer, *Angew. Chem., Int. Ed. Engl.*, 1990, **29**, 1424.
265. R. W. Hoffmann, T. Rühl and J. Harbach, *Liebigs Ann. Chem.*, 1992, 725.
266. P. Zhang and R. E. Gawley, *J. Org. Chem.*, 1993, **58**, 3223 and references cited therein.
267. D. Hoppe, F. Hintze and P. Tebben, *Angew. Chem., Int. Ed. Engl.*, 1990, **29**, 1422.
268. F. Hintze and D. Hoppe, *Synthesis*, 1992, 1216.
269. J. Haller, T. Hense and D. Hoppe, *Synlett*, 1993, 726.
270. P. Sommerfeld and D. Hoppe, *Synlett*, 1992, 764.
271. J. Schwerdtfeger and D. Hoppe, *Angew. Chem., Int. Ed. Engl.*, 1992, **31**, 1505.

272. H. Ahrens, M. Paetow and D. Hoppe, *Tetrahedron Lett.*, 1992, **33**, 5327 and Reference 3 cited therein.
273. O. Zschage and D. Hoppe, *Tetrahedron*, 1992, **48**, 8389 and References 3 cited therein.
274. For a thorough review covering early work on metallated allyl carbamates see: D. Hoppe, *Angew. Chem., Int. Ed. Engl.*, 1984, **23**, 932.
275. D. S. Matteson, P. B. Tripathy, A. Sarkar and K. M. Sadhu, *J. Am. Chem. Soc.*, 1989, **111**, 4399.
276. J. M. Chong and E. K. Mar, *Tetrahedron Lett.*, 1990, **31**, 1981.
277. P. C.-M. Chan and J. M. Chong, *Tetrahedron Lett.*, 1990, **31**, 1985.
278. R. J. Lindermann and A. Ghannam, *J. Am. Chem. Soc.*, 1990, **112**, 2392 and references cited therein.
279. K. Mikami and T. Nakai, *Synthesis*, 1991, 594 and references cited therein.
280. J. A. Marshall, in 'Comprehensive Organic Synthesis', eds. B. M. Trost and I. Fleming, Pergamon, Oxford, 1991, vol. 3, p. 975.
281. G. Boche *et al.*, *Chem. Ber.*, 1992, **125**, 2265.
282. R. Hoffmann and R. Brückner, *Chem. Ber.*, 1992, **125**, 2731.
283. A. Ricci and A. Degl'Innocenti, *Synthesis*, 1989, 647.
284. P. C. Bulman Page, S. S. Klair and S. Rosenthal, *Chem. Soc. Rev.*, 1990, **19**, 147.
285. R. L. Danheiser, D. M. Fink, K. Okano, Y.-M. Tsai and S. W. Szczepanski, *J. Org. Chem.*, 1985, **50**, 5393.
286. M. E. Scheller and B. Frei, *Helv. Chim. Acta*, 1986, **69**, 44.
287. H. J. Reich, R. C. Holtan and C. Bolm, *J. Am. Chem. Soc.*, 1990, **112**, 5609 and references cited therein.
288. P. Jankowski, S. Marczak, M. Masnyk and J. Wicha, *J. Organomet. Chem.*, 1991, **403**, 49.
289. G. A. Gornowicz and R. West, *J. Am. Chem. Soc.*, 1968, **90**, 4478.
290. R. Anderson, *Synthesis*, 1985, 717.
291. D. J. Ager, *J. Chem. Soc., Perkin Trans. 1*, 1986, 183.
292. D. J. Ager, *J. Chem. Soc., Perkin Trans. 1*, 1986, 195 and references cited therein.
293. P. A. Brown, R. V. Bonnert, P. R. Jenkins and M. R. Selim, *Tetrahedron Lett.*, 1987, **28**, 693 and references cited therein.
294. W. Dumont and A. Krief, *Angew. Chem., Int. Ed. Engl.*, 1976, **15**, 161.
295. D. E. Seitz and A. Zapata, *Synthesis*, 1981, 557.
296. A. R. Bassindale, R. J. Ellis and P. G. Taylor, *Tetrahedron Lett.*, 1984, **25**, 2705.
297. K. Tamao, R. Kanatani and M. Kumada, *Tetrahedron Lett.*, 1984, **25**, 1905 and references cited therein.
298. D. J. Ager, *Synthesis*, 1984, 384.
299. T. H. Chan and E. Chang, *J. Org. Chem.*, 1974, **39**, 3264.
300. P. W. K. Lau and T. H. Chan, *Tetrahedron Lett.*, 1978, 2383.
301. R. S. Budhram, V. A. Palaniswamy and E. J. Eisenbraun, *J. Org. Chem.*, 1986, **51**, 1402.
302. H.-J. Bergmann, R. Mayrhofer and H.-H. Otto, *Arch. Pharm.*, 1986, **319**, 203.
303. R. H. Schlessinger, M. A. Poss, S. Richardson and P. Lin, *Tetrahedron Lett.*, 1985, **26**, 2391.
304. J.-C. Cuevas, P. Patil and V. Snieckus, *Tetrahedron Lett.*, 1989, **30**, 5841.
305. R. Bloch, D. Hassan and X. Mandard, *Tetrahedron Lett.*, 1983, **24**, 4691.
306. S. Hackett and T. Livinghouse, *J. Org. Chem.*, 1986, **51**, 879.
307. J. V. Comasseto, *J. Organomet. Chem.*, 1983, **253**, 131.
308. A. R. Bassindale, R. J. Ellis, J. C.-Y. Lau and P. G. Taylor, *J. Chem. Soc., Perkin Trans. 2*, 1986, 593.
309. D. A. Ager, G. E. Cooke, M. B. East, S. J. Mole, A. Rampersaud and V. J. Webb, *Organometallics*, 1986, **5**, 1906.
310. J. Binder and E. Zbiral, *Tetrahedron Lett.*, 1986, **27**, 5829.
311. E. E. Aboujaoude, S. Liétjé, N. Collignon, M. P. Teulade and P. Savignac, *Synthesis*, 1986, 934.
312. C. Burford, F. Cooke, G. Roy and P. Magnus, *Tetrahedron*, 1983, **39**, 867.
313. S. Kanemasa, J. Tanaka, H. Nagahama and O. Tsuge, *Bull. Chem. Soc. Jpn.*, 1985, **58**, 3385.
314. A. R. Katritzky and S. Sengupta, *Tetrahedron Lett.*, 1987, **28**, 1847.
315. J. W. Wilson, A. Pelter, M. V. Garad and R. Pardasani, *Tetrahedron*, 1993, **49**, 2979.
316. T. Konakahara and Y. Takagi, *Tetrahedron Lett.*, 1980, **21**, 2073.
317. T. Konakahara and Y. Kurosaki, *J. Chem. Res. (S)*, 1989, 130 and references cited therein.
318. G. Bartoli, G. Palmieri, M. Petrini, M. Bosco and R. Dalpozzo, *Tetrahedron*, 1990, **46**, 1379.
319. M.-R. Fischer, A. Kirschning, T. Michel and E. Schaumann, *Angew. Chem., Int. Ed. Engl.*, 1994, **33**, 217.
320. T. H. Chan *et al.*, 'Silicon Chemistry', 1st edn., eds. J. Y. Corey, E. R. Corey and P. P. Gaspar, Ellis Horwood, Chichester, 1988, p. 49.
321. R. F. Horvath and T. H. Chan, *J. Org. Chem.*, 1989, **54**, 317.
322. R. C. Hartley, S. Lamothe and T. H. Chan, *Tetrahedron Lett.*, 1993, **34**, 1449.
323. H. Uno, *Bull. Chem. Soc. Jpn.*, 1986, **59**, 2471.
324. T. H. Chan and P. Pellon, *J. Am. Chem. Soc.*, 1989, **111**, 8737.
325. K. Ogura, in 'Comprehensive Organic Synthesis', 1st edn., eds. B. M. Trost and I. Fleming, Pergamon, Oxford, 1991, vol. 1, p. 505.
326. A. Krief, in 'Comprehensive Organic Synthesis', 1st edn., eds. B. M. Trost and I. Fleming, Pergamon, Oxford, 1991, vol. 3, p. 85.
327. A. Krief, in 'Comprehensive Organic Synthesis', 1st edn., eds. B. M. Trost and I. Fleming, Pergamon, Oxford, 1991, vol. 1, p. 629.
328. F. M. Stoyanovich, in 'Chemistry of Organosulfur Compounds. General Problems', 1st edn., ed. L. I. Belen'kii, Ellis Horwood, Chichester, 1990, p. 98.
329. K. Tanaka and A. Kaji, in 'The Chemistry of Sulphones and Sulphoxides', eds. S. Patai, Z. Rappoport and C. Stirling, Wiley, Chichester, 1988, p. 759.
330. G. H. Posner, in 'The Chemistry of Sulphones and Sulphoxides', eds. S. Patai, Z. Rappoport and C. Stirling, Wiley, Chichester, 1988, p. 823.
331. H. J. Reich, in 'Organoselenium Chemistry', 1st edn., ed. D. Liotta, Wiley, New York, 1987, p. 243.
332. T. G. Back, in 'The Chemistry of Selenium and Tellurium Compounds', 1st edn., ed. S. Patai, Wiley, Chichester, 1987, vol 2 p. 91.
333. A. Krief, in 'The Chemistry of Organic Selenium and Tellurium Compounds', ed. S. Patai, Wiley, Chichester, 1987, vol. 2 p. 675.

334. C. Paulmier, 'Selenium Reagents and Intermediates in Organic Synthesis', 1st edn., Pergamon, Oxford, 1986.
335. D. Seebach and N. Peleties, *Chem. Ber.*, 1972, **105**, 511.
336. C. M. Thompson and D. L. C. Green, *Tetrahedron*, 1991, **47**, 4223.
337. B. A. Pearlman, S. R. Putt and J. A. Fleming, *J. Org. Chem.*, 1985, **50**, 3622.
338. T. Mandai, T. Moriyama, K. Tsujimoto, M. Kawada and J. Otera, *Tetrahedron Lett.*, 1986, **27**, 603.
339. C.-N. Hsiao and H. Shechter, *Tetrahedron Lett.*, 1982, **23**, 1963.
340. G. Stork and K. Zhao, *Tetrahedron Lett.*, 1989, **30**, 287.
341. O. De Lucchi, U. Miotti and G. Modena, *Org. React.*, 1991, **40**, 157.
342. J. R. Hwu, *J. Org. Chem.*, 1983, **48**, 4432.
343. E. Vedejs, in 'Perspectives in the Organic Chemistry of Sulfur', eds. B. Zwanenburg and A. J. H. Klunder, Elsevier, Amsterdam, 1987, p. 75.
344. S. Halazy, F. Zutterman and A. Krief, *Tetrahedron Lett.*, 1982, **23**, 4385.
345. R. C. Gadwood, *J. Org. Chem.*, 1983, **48**, 2098.
346. M. J. Calverley, *Tetrahedron Lett.*, 1987, **28**, 1337.
347. T. Willson, P. Kocienski, A. Faller and S. Campbell, *J. Chem. Soc., Chem. Commun.*, 1987, 106.
348. S. Takano, M. Morimoto, S. Satoh and K. Ogasawara, *Chem. Lett.*, 1984, 1261.
349. E. Dziadulewicz and T. Gallagher, *Tetrahedron Lett.*, 1985, **26**, 4547.
350. M. Kodama, K. Okumura, K. Kobayashi, T. Tsunoda and S. Itô, *Tetrahedron Lett.*, 1984, **25**, 5781.
351. W. G. Dauben, R. K. Saugier and I. Fleischhauer, *J. Org. Chem.*, 1985, **50**, 3767.
352. T. Mandai, T. Moriyama, Y. Nakayama, K. Sugino, M. Kawada and J. Otera, *Tetrahedron Lett.*, 1984, **25**, 5913.
353. T. Hayashi and T. Oishi, *Chem. Lett.*, 1985, 413.
354. W. C. Still and D. Mobilio, *J. Org. Chem.*, 1983, **48**, 4785.
355. H. J. Reich, M. C. Clark and W. W. Willis, Jr., *J. Org. Chem.*, 1982, **47**, 1618.
356. H. J. Reich, M. J. Kelly, R. E. Olson and R. C. Holtan, *Tetrahedron*, 1983, **39**, 949.
357. J. R. Falck and Y.-L. Yang, *Tetrahedron Lett.*, 1984, **25**, 3563.
358. C. Mahidol, V. Reutrakul, V. Prapansiri and C. Panyachotipun, *Chem. Lett.*, 1984, 969.
359. G. Demailly, C. Greck and G. Solladié, *Tetrahedron Lett.*, 1984, **25**, 4113.
360. G. Solladié, *Chimia*, 1984, **38**, 233.
361. G. Solladié, *Pure Appl. Chem.*, 1988, **60**, 1699.
362. G. Solladié, C. Hamdouchi and C. Ziani-Chérif, *Tetrahedron: Asymmetry*, 1991, **2**, 457 and references cited therein.
363. D. H. Hua, S. N. Bharathi, P. D. Robinson and A. Tsujimoto, *J. Org. Chem.*, 1990, **55**, 2128.
364. T. Satoh, T. Oohara, Y. Ueda and K. Yamakawa, *Tetrahedron Lett.*, 1988, **29**, 313.
365. S. G. Pyne and B. Dikic, *J. Org. Chem.*, 1990, **55**, 1932.
366. S.-I. Murahashi, J. Sun and T. Tsuda, *Tetrahedron Lett.*, 1993, **34**, 2645.
367. R. Annunziata, M. Cinquini, F. Cozzi, L. Raimondi and S. Stefanelli, *Tetrahedron*, 1986, **42**, 5443.
368. R. Annunziata, M. Cinquini, F. Cozzi, L. Raimondi and S. Stefanelli, *Tetrahedron*, 1986, **42**, 5451.
369. M. Inbasekaran, N. P. Peet, J. R. McCarthy and M. E. LeTourneau, *J. Chem. Soc., Chem. Commun.*, 1985, 678.
370. P. Charreau, M. Julia and J.-N. Verpeaux, *J. Organomet. Chem.*, 1989, **379**, 201.
371. C. De Lima, M. Julia and J.-N. Verpeaux, *Synlett*, 1992, 133.
372. J. B. Hendrickson and P. S. Palumbo, *J. Org. Chem.*, 1985, **50**, 2110.
373. J. B. Hendrickson, G. J. Boudreaux and P. S. Palumbo, *J. Am. Chem. Soc.*, 1986, **108**, 2358.
374. P. Koch, Y. Nakatani, B. Luu and G. Ourisson, *Bull. Soc. Chim. Fr.*, 1983, **II**, 189.
375. A. Krief, in 'Comprehensive Organic Synthesis', 1st edn., eds. B. M. Trost and I. Fleming, Pergamon, Oxford, 1991, vol. 3, p.169.
376. M. Julia and J.-P. Stacino, *Tetrahedron*, 1986, **42**, 2469.
377. B. M. Trost, J. Lynch, P. Renaut and D. H. Steinman, *J. Am. Chem. Soc.*, 1986, **108**, 284.
378. J. C. Carretero and L. Ghosez, *Tetrahedron Lett.*, 1988, **29**, 2059.
379. A. B. Jones, A. Villalobos, R. G. Linde, II and S. J. Danishefsky, *J. Org. Chem.*, 1990, **55**, 2786.
380. M. Isobe, Y. Ichikawa and T. Goto, *Tetrahedron Lett.*, 1986, **27**, 963.
381. T.-S. Chou, H.-H. Tso and L.-J. Chang, *J. Chem. Soc., Chem. Commun.*, 1984, 1323.
382. For earlier work, see: R. Bloch and J. Abecassis, *Tetrahedron Lett.*, 1982, **23**, 3277.
383. J. Vollhardt, H.-J. Gais and K. L. Lukas, *Angew. Chem., Int. Ed. Engl.*, 1985, **24**, 610.
384. Y. Nakahara, A. Fujita, K. Beppu and T. Ogawa, *Tetrahedron*, 1986, **42**, 6465.
385. P.-u. Park, C. A. Broka, B. F. Johnson and Y. Kishi, *J. Am. Chem. Soc.*, 1987, **109**, 6205.
386. H. Chikashita, T. Yuasa and K. Itoh, *Chem. Lett.*, 1992, 1457.
387. N. H. Andersen, A. D. Denniston and D. A. McCrae, *J. Org. Chem.*, 1982, **47**, 1145.
388. W. S. Johnson, B. Frei and A. S. Gopalan, *J. Org. Chem.*, 1981, **46**, 1512.
389. Y. Torisawa, H. Okabe and S. Ikegami, *J. Chem. Soc., Chem. Commun.*, 1984, 1602.
390. R. W. Hoffmann and M. Bewersdorf, *Liebigs Ann. Chem.*, 1992, 643 and References 7 and 8 cited therein.
391. L. F. Tietze, H. Geissler, J. A. Gewert and U. Jakobi, *Synlett*, 1994, 511.
392. C. R. Johnson and T. D. Penning, *J. Am. Chem. Soc.*, 1988, **110**, 4726.
393. I. Erdelmeier and H.-J. Gais, *J. Am. Chem. Soc.*, 1989, **111**, 1125 and Reference 10 cited therein.
394. P. Beak and V. Snieckus, *Acc. Chem. Res.*, 1982, **15**, 306.
395. N. S. Narasimhan and R. S. Mali, *Synthesis*, 1983, 957.
396. V. Snieckus, *Pure and Appl. Chem.*, 1990, **62**, 671.
397. V. Snieckus, *Pure and Appl. Chem.*, 1990, **62**, 2047.
398. W. Bauer and P. von Ragué Schleyer, *J. Am. Chem. Soc.*, 1989, **111**, 7191.
399. F. Totter and P. Rittmeyer, in 'Organometallics in Synthesis. A Manual', ed. M. Schlosser, Wiley, Chichester, 1994, p. 167.
400. D. W. Slocum and C. A. Jennings, *J. Org. Chem.*, 1976, **41**, 3653.
401. P. Beak and R. A. Brown, *J. Org. Chem.*, 1979, **44**, 4463.
402. A. I. Meyers and K. Lutomski, *J. Org. Chem.*, 1979, **44**, 4464.
403. P. Beak and R. A. Brown, *J. Org. Chem.*, 1982, **47**, 34.
404. P. Beak, A. Tse, J. Hawkins, C.-W. Chen and S. Mills, *Tetrahedron*, 1983, **39**, 1983.

405. M. A. J. Miah, Ph.D. Thesis, University of Waterloo, 1985.
406. M. Iwao, T. Iihama, K. K. Mahalanabis, H. Perrier and V. Snieckus, *J. Org. Chem.*, 1989, **54**, 24.
407. M. Watanabe, M. Date, K. Kawanishi, T. Hori and S. Furukawa, *Chem. Pharm. Bull.*, 1990, **38**, 2637.
408. C. Quesnelle, T. Iihama, T. Aubert, H. Perrier and V. Snieckus, *Tetrahedron Lett.*, 1992, **33**, 2625.
409. A. J. Bridges, A. Lee, E. C. Maduakor and C. E. Schwartz, *Tetrahedron Lett.*, 1992, **33**, 7495.
410. J. Mortier, J. Moyroud, B. Bennetau and P. A. Cain, *J. Org. Chem.*, 1994, **59**, 4042.
411. T. G. Gant and A. I. Meyers, *Tetrahedron*, 1994, **50**, 2297.
412. R. S. C. Lopes, C. C. Lopes and C. H. Heathcock, *Tetrahedron Lett.*, 1992, **33**, 6775.
413. X. Wang and V. Snieckus, *Synlett*, 1990, 313.
414. H. Chikashita, J. A. Porco, Jr., T. J. Stout, J. Clardy and S. L. Schreiber, *J. Org. Chem.*, 1991, **56**, 1692.
415. For an approach to dynemicin A involving benzamide and quinoline metallation see: K. C. Nicolaou, J. L. Gross, M. A. Kerr, R. H. Lemus, K. Ikeda and K. Ohe, *Angew. Chem., Int. Ed. Engl.*, 1994, **33**, 781.
416. A. B. Smith, III, S. R. Schow, J. D. Bloom, A. S. Thompson and K. N. Winzenberg, *J. Am. Chem. Soc.*, 1982, **104**, 4015.
417. A. R. Katritzky and W.-Q. Fan, *Org. Prep. Proceed. Int.*, 1987, **19**, 263.
418. J. W. Lyga and G. A. Meier, *Synth. Commun.*, 1994, **24**, 2491.
419. L. Engman and A. Hallberg, *J. Org. Chem.*, 1989, **54**, 2964.
420. J.-M. Fu, B.-P. Zhao, M. J. Sharp and V. Snieckus, *J. Org. Chem.*, 1991, **56**, 1683.
421. Application of Martin's conditions to the *ortho*-metallation of *t*-butyl benzoate will be detailed in the forthcoming book: V. Snieckus et al, '*ortho*-Directed Metallation. A Practical Approach', Oxford University Press, Oxford, 1996.
422. P. E. Eaton, C.-H. Lee and Y. Xiong, *J. Am. Chem. Soc.*, 1989, **111**, 8016.
423. P. Beak and A. I. Meyers, *Acc. Chem. Res.*, 1986, **19**, 356.
424. M. Reuman and A. I. Meyers, *Tetrahedron*, 1985, **41**, 837.
425. D. B. Reitz and S. M. Massey, *J. Org. Chem.*, 1990, **55**, 1375.
426. L. E. Fisher, J. M. Muchowski and R. D. Clark, *J. Org. Chem.*, 1992, **57**, 2700.
427. L. A. Flippin, *Tetrahedron Lett.*, 1991, **32**, 6857.
428. W. J. Houlihan and V. A. Parrino, *J. Org. Chem.*, 1982, **47**, 5177.
429. P. A. A. Klusener, J. C. Hanekamp, L. Brandsma and P. von Ragué Schleyer, *J. Org. Chem.*, 1990, **55**, 1311.
430. M. A. Forth, M. B. Mitchell, S. A. C. Smith, K. Gombatz and L. Snyder, *J. Org. Chem.*, 1994, **59**, 2616 and references cited therein.
431. D. L. Comins and J. D. Brown, *J. Org. Chem.*, 1984, **49**, 1078.
432. D. L. Comins and J. D. Brown, *J. Org. Chem.*, 1989, **54**, 3730.
433. A. L. Campbell and I. K. Khanna, *Tetrahedron Lett.*, 1986, **27**, 3963.
434. R. S. Mali, S. D. Patil and S. L. Patil, *Tetrahedron*, 1986, **42**, 2075.
435. R. P. Robinson, K. M. Donahue and N. A. Saccomano, *Tetrahedron Lett.*, 1989, **30**, 5203.
436. S. A. Burns, R. J. P. Corriu, V. Huynh and J. J. E. Moreau, *J. Organomet. Chem.*, 1987, **333**, 281.
437. R. P. Polniaszek and C. R. Kaufman, *J. Am. Chem. Soc.*, 1989, **111**, 4859.
438. W. Oppolzer, M. Wills, M. J. Kelly, M. Signer and J. Blagg, *Tetrahedron Lett.*, 1990, **31**, 5015 and Reference 9 cited therein.
439. C. D. Liang, *Tetrahedron Lett.*, 1986, **27**, 1971 and Reference 8 cited therein.
440. M. Schlosser and G. Simig, *Tetrahedron Lett.*, 1991, **32**, 1965 and Reference 3 cited therein.
441. B. M. Trost, G. T. Rivers and J. M. Gold, *J. Org. Chem.*, 1980, **45**, 1835.
442. E. Napolitano, E. Giannone, R. Fiaschi and A. Marsili, *J. Org. Chem.*, 1983, **48**, 3653.
443. G. H. Posner and K. A. Canella, *J. Am. Chem. Soc.*, 1985, **107**, 2571.
444. G. Coll, J. Morey, A. Costa and J. M. Saá, *J. Org. Chem.*, 1988, **53**, 5345.
445. G. A. Suñer, P. M. Deyá and J. M. Saá, *J. Am. Chem. Soc.*, 1990, **112**, 1467.
446. H. Andringa, H. D. Verkruijsse, L. Brandsma and L. Lochmann, *J. Organomet. Chem.*, 1990, **393**, 307 and references cited therein.
447. J. J. Talley and I. A. Evans, *J. Org. Chem.*, 1984, **49**, 5267.
448. M. Lofthagen, R. Vernon-Clark, K. K. Baldridge and J. S. Siegel, *J. Org. Chem.*, 1992, **57**, 61.
449. J. Morey, A. Costa, P. M. Deyá, G. Suñer and J. M. Saá, *J. Org. Chem.*, 1990, **55**, 3902.
450. R. S. Michalak, D. R. Myers, J. L. Parsons, P. A. Risbood, R. D. Haugwitz and V. L. Narayanan, *Tetrahedron Lett.*, 1989, **30**, 4783.
451. D. J. Cram *et al.*, *J. Am. Chem. Soc.*, 1988, **110**, 2554.
452. A. Costa and J. M. Saá, *Tetrahedron Lett.*, 1987, **28**, 5551.
453. G. Katsoulos, S. Takagishi and M. Schlosser, *Synlett*, 1991, 731.
454. G. P. Crowther, R. J. Sundberg and A. M. Sarpeshkar, *J. Org. Chem.*, 1984, **49**, 4657.
455. J. P. Yardley and H. Fletcher, III, *Synthesis*, 1976, 244.
456. A. R. Katritzky, H. Lang and X. Lan, *Synth. Commun.*, 1993, **23**, 1175.
457. R. C. Ronald and M. R. Winkle, *Tetrahedron*, 1983, **39**, 2031.
458. T. Kamikawa and I. Kubo, *Synthesis*, 1986, 431.
459. S. Jeganathan, M. Tsukamoto and M. Schlosser, *Synthesis*, 1990, 109.
460. G. A. Kraus, P. J. Thomas and M. D. Schwinden, *Tetrahedron Lett.*, 1990, **31**, 1819.
461. I. Kubo, T. Kamikawa and I. Miura, *Tetrahedron Lett.*, 1983, **24**, 3825.
462. R. G. Harvey, C. Cortez, T. P. Ananthanarayan and S. Schmolka, *J. Org. Chem.*, 1988, **53**, 3936.
463. T. Matsumoto, H. Jona, M. Katsuki and K. Suzuki, *Tetrahedron Lett.*, 1991, **32**, 5103.
464. T. Matsumoto, H. Kakigi and K. Suzuki, *Tetrahedron Lett.*, 1991, **32**, 4337.
465. S. Sengupta and V. Snieckus, *Tetrahedron Lett.*, 1990, **31**, 4267.
466. M. P. Sibi and V. Snieckus, *J. Org. Chem.*, 1983, **48**, 1935.
467. J. A. Miller, *J. Org. Chem.*, 1987, **52**, 322.
468. A. M. Jardine, S. M. Vather and T. A. Modro, *J. Org. Chem.*, 1988, **53**, 3983.
469. D. A. Casteel and S. P. Peri, *Synthesis*, 1991, 691 and references cited therein.
470. B. Dhawan and D. Redmore, *J. Org. Chem.*, 1991, **56**, 833.
471. D. Hellwinkel, F. Lämmerzahl and G. Hofmann, *Chem. Ber.*, 1983, **116**, 3375.

472. M. P. Sibi, S. Chattopadhyay, J. W. Dankwardt and V. Snieckus, *J. Am. Chem. Soc.*, 1985, **107**, 6312.
473. S. Danishefsky and J. Y. Lee, *J. Am. Chem. Soc.*, 1989, **111**, 4829.
474. F. Beaulieu and V. Snieckus, *Synthesis*, 1992, 112.
475. F. Beaulieu, Ph.D. Thesis, University of Waterloo, 1994.
476. W. Wang and V. Snieckus, *J. Org. Chem.*, 1992, **57**, 424.
477. W. Fuhrer and H. W. Gschwend, *J. Org. Chem.*, 1979, **44**, 1133.
478. J. M. Muchowski and M. C. Venuti, *J. Org. Chem.*, 1980, **45**, 4798.
479. L. R. Hillis and S. J. Gould, *J. Org. Chem.*, 1985, **50**, 718.
480. R. M. Soll, C. Guinosso and A. Asselin, *J. Org. Chem.*, 1988, **53**, 2844.
481. T. J. Thornton and M. Jarman, *Synthesis*, 1990, 295.
482. S. J. Gould and R. L. Eisenberg, *J. Org. Chem.*, 1991, **56**, 6666.
483. M. Iwao, *Heterocycles*, 1994, **38**, 45.
484. J. E. Macdonald and G. S. Poindexter, *Tetrahedron Lett.*, 1987, **28**, 1851.
485. S. Bengtsson and T. Högberg, *J. Org. Chem.*, 1989, **54**, 4549.
486. I.-S. Cho, L. Gong and J. M. Muchowski, *J. Org. Chem.*, 1991, **56**, 7288.
487. P. Hewawasam and N. A. Meanwell, *Tetrahedron Lett.*, 1994, **35**, 7303.
488. P. Pedaja, C. Westerlund and A. Hallberg, *J. Heterocycl. Chem.*, 1986, **23**, 1353.
489. A. R. Katritzky, M. Black and W.-Q. Fan, *J. Org. Chem.*, 1991, **56**, 5045.
490. H. M. Walborsky and P. Ronman, *J. Org. Chem.*, 1978, **43**, 731.
491. E. Block, G. Ofori-Okai and J. Zubieta, *J. Am. Chem. Soc.*, 1989, **111**, 2327 and Reference 2 cited therein.
492. S. Ogawa and N. Furukawa, *J. Org. Chem.*, 1991, **56**, 5723.
493. R. W. Baker, G. R. Pocock, M. V. Sargent and E. Twiss, *Tetrahedron: Asymmetry*, 1993, **4**, 2423.
494. W. E. Truce and M. F. Amos, *J. Am. Chem. Soc.*, 1951, **73**, 3013.
495. B. R. Hsieh, R. J. Gruber, J. L. Haack, (Xerox Corp. USA) *US Pat.* 4 898 802 (1990).
496. J. N. Bonfiglio, *J. Org. Chem.*, 1986, **51**, 2833.
497. M. Iwao, *J. Org. Chem.*, 1990, **55**, 3622.
498. A. K. Sinhababu, M. Kawase and R. T. Borchardt, *Tetrahedron Lett.*, 1987, **28**, 4139.
499. D. C. Furlano, S. N. Calderon, G. Chen and K. L. Kirk, *J. Org. Chem.*, 1988, **53**, 3145.
500. A. J. Bridges, A. Lee, E. C. Maduakor and C. E. Schwartz, *Tetrahedron Lett.*, 1992, **33**, 7499.
501. A. J. Bridges, A. Lee, C. E. Schwartz, M. J. Towle and B. A. Littlefield, *Biorg. Med. Chem.*, 1993, **1**, 403.
502. S. Takagishi and M. Schlosser, *Synlett*, 1991, 119.
503. R. Subramanian and F. Johnson, *J. Org. Chem.*, 1985, **50**, 5430.
504. F. Johnson and R. Subramanian, *J. Org. Chem.*, 1986, **51**, 5040.
505. K. Tamao, H. Yao, Y. Tsutsumi, H. Abe, T. Hayashi and Y. Ito, *Tetrahedron Lett.*, 1990, **31**, 2925.
506. J. M. Brown and S. Woodward, *J. Org. Chem.*, 1991, **56**, 6803 and references cited therein.
507. L. Lochmann and J. Petránek, *Tetrahedron Lett.*, 1991, **32**, 1483.
508. F. Faigl and M. Schlosser, *Tetrahedron Lett.*, 1991, **32**, 3369.
509. G. P. Lutz, A. P. Wallin, S. T. Kerrick and P. Beak, *J. Org. Chem.*, 1991, **56**, 4938.
510. P. Beak and H. Du, *J. Am. Chem. Soc.*, 1993, **115**, 2516.
511. S. Thayumanavan, S. Lee, C. Liu and P. Beak, *J. Am. Chem. Soc.*, 1994, **116**, 9755.
512. A. Krief, M. Hobe, W. Dumont, E. Badaoui, E. Guittet and G. Evrard, *Tetrahedron Lett.*, 1992, **33**, 3381.
513. S. Marumoto and I. Kuwajima, *J. Am. Chem. Soc.*, 1993, **115**, 9021.
514. G. S. Poindexter, *J. Org. Chem.*, 1982, **47**, 3787.
515. M. P. Sibi, J. W. Dankwardt and V. Snieckus, *J. Org. Chem.*, 1986, **51**, 271.
516. J. L. Luche, C. Einhorn, J. Einhorn and J. V. Sinisterra-Gago, *Tetrahedron Lett.*, 1990, **31**, 4125.
517. R. D. Clark and Jahangir, *J. Org. Chem.*, 1988, **53**, 2378.
518. R. D. Clark and Jahangir, *J. Org. Chem.*, 1989, **54**, 1174.
519. X. Wang and V. Snieckus, *Tetrahedron Lett.*, 1991, **32**, 4883.
520. J.-C. Clinet, E. Duñach and K. P. C. Vollhardt, *J. Am. Chem. Soc.*, 1983, **105**, 6710.
521. R. J. Mills and V. Snieckus, *J. Org. Chem.*, 1989, **54**, 4386.
522. P. L. Creger, *J. Am. Chem. Soc.*, 1970, **92**, 1396.
523. L. Kopanski, M. Klaar and W. Steglich, *Liebigs Ann. Chem.*, 1982, 1280.
524. B. Tarnchompoo, C. Thebtaranonth and Y. Thebtaranonth, *Synthesis*, 1986, 785.
525. A. C. Regan and J. Staunton, *J. Chem. Soc., Chem. Commun.*, 1987, 520.
526. L. Crenshaw, S. P. Khanapure, U. Siriwardane and E. R. Biehl, *Tetrahedron Lett.*, 1988, **29**, 3777.
527. R. D. Clark, J. M. Muchowski, M. Souchet and D. B. Repke, *Synlett*, 1990, 207.
528. R. D. Clark, J. M. Muchowski, L. E. Fisher, L. A. Flippin, D. B. Repke and M. Souchet, *Synthesis*, 1991, 871.
529. C. Hashimoto and H.-P. Husson, *Tetrahedron Lett.*, 1988, **29**, 4563 and Reference 6 cited therein.
530. J. Haseltine, M. Visnick and A. B. Smith, III, *J. Org. Chem.*, 1988, **53**, 6160.
531. Y. Ito, K. Kobayashi, N. Seko and T. Saegusa, *Bull. Chem. Soc. Jpn.*, 1984, **57**, 73.
532. T. R. Kelly, S. H. Bell, N. Ohashi and R. J. Armstrong-Chong, *J. Am. Chem. Soc.*, 1988, **110**, 6471.
533. M. Watanabe, M. Sahara, S. Furukawa, R. J. Billedeau and V. Snieckus, *J. Org. Chem.*, 1984, **49**, 742.
534. X. Wang and V. Snieckus, *Tetrahedron Lett.*, 1991, **32**, 4879.
535. L. A. Flippin and J. M. Muchowski, *J. Org. Chem.*, 1993, **58**, 2631.
536. R. D. Clark, Jahangir and J. A. Langston, *Can. J. Chem.*, 1994, **72**, 23.
537. S. G. Davies, 'Organotransition Metal Chemistry: Applications to Organic Synthesis', Pergamon, Oxford, 1982.
538. P. J. Harrington, 'Transition Metals in Total Synthesis', Wiley, New York, 1990, p. 317.
539. M. F. Semmelhack, in 'Comprehensive Organic Synthesis', eds. B. M. Trost and I. Fleming, Pergamon, Oxford, 1991, vol. 4, p. 517.
540. L. S. Hegedus, 'Transition Metals in the Synthesis of Complex Organic Molecules', University Science Books, Mill Valley, CA, 1994, p. 307.
541. F. J. McQuillin, D. G. Parker and G. R. Stephenson, 'Transition Metal Organometallics for Organic Synthesis', 1st edn., Cambridge University Press, Cambridge, 1991.

542. S. G. Davies and T. J. Donohoe, *Synlett*, 1993, 323.
543. M. F. Semmelhack, J. Bisaha and M. Czarny, *J. Am. Chem. Soc.*, 1979, **101**, 768.
544. R. J. Card and W. S. Trahanovsky, *J. Org. Chem.*, 1980, **45**, 2560.
545. J. P. Gilday, J. T. Negri and D. A. Widdowson, *Tetrahedron*, 1989, **45**, 4605.
546. P. J. Dickens, J. P. Gilday, J. T. Negri and D. A. Widdowson, *Pure Appl. Chem.*, 1990, **62**, 575.
547. E. P. Kündig, C. Perret and B. Rudolph, *Helv. Chim. Acta*, 1990, **73**, 1970.
548. M. F. Semmelhack and A. Zask, *J. Am. Chem. Soc.*, 1983, **105**, 2034.
549. S. G. Davies et al., *Philos. Trans. R. Soc. London A*, 1988, **326**, 619.
550. S. G. Davies, *Chem. Ind.*, 1986, 506.
551. M. Ghavshou and D. A. Widdowson, *J. Chem. Soc., Perkin Trans. 1*, 1983, 3065.
552. P. J. Dickens, A. M. Z. Slawin, D. A. Widdowson and D. J. Williams, *Tetrahedron Lett.*, 1988, **29**, 103.
553. J. P. Gilday and D. A. Widdowson, *Tetrahedron Lett.*, 1986, **27**, 5525.
554. J. P. Gilday and D. A. Widdowson, *J. Chem. Soc., Chem. Commun.*, 1986, 1235.
555. J. Blagg, S. G. Davies, N. J. Holman, C. A. Laughton and B. E. Mobbs, *J. Chem. Soc., Perkin Trans. 1*, 1986, 1581.
556. M. E. Wright, *Organometallics*, 1989, **8**, 407.
557. T. V. Lee, A. J. Leigh and C. B. Chapleo, *Tetrahedron Lett.*, 1989, **30**, 5519.
558. S. G. Davies, C. L. Goodfellow, J. M. Peach and A. Waller, *J. Chem. Soc., Perkin Trans. 1*, 1991, 1019.
559. A. Solladié-Cavallo, in 'Advances in Metal-Organic Chemistry', ed. L. S. Liebeskind, Jai Press, Greenwich, CT, 1989, vol. 1, p. 99.
560. S. G. Davies, S. J. Coote and C. L. Goodfellow, in 'Advances in Metal-Organic Chemistry', ed. L. S. Liebeskind, Jai Press, Greenwich, CT, 1991, vol. 2, p. 1.
561. M. Uemura in 'Advances in Metal-Organic Chemistry', ed. L. S. Liebeskind, Jai Press, Greenwich, CT, 1991, vol. 2, p. 195.
562. Y. Kondo, J. R. Green and J. Ho, *J. Org. Chem.*, 1993, **58**, 6182.
563. M. Sodeoka and M. Shibasaki, *Synthesis*, 1993, 643.
564. M. Uemura, R. Miyake, K. Nakayama, M. Shiro and Y. Hayashi, *J. Org. Chem.*, 1993, **58**, 1238.
565. G. B. Jones and S. B. Heaton, *Tetrahedron: Asymmetry*, 1993, **4**, 261.
566. J. A. Heppert, J. Aubé, M. E. Thomas-Miller, M. L. Miligan and F. Takusagawa, *Organometallics*, 1990, **9**, 727.
567. H.-G. Schmalz, J. Hollander, M. Arnold and G. Dürner, *Tetrahedron Lett.*, 1993, **34**, 6259.
568. H.-G. Schmalz, M. Arnold, J. Hollander and J. W. Bats, *Angew. Chem., Int. Ed. Engl.*, 1994, **33**, 109.
569. D. A. Price, N. S. Simpkins, A. M. MacLeod and A. P. Watt, *J. Org. Chem.*, 1994, **59**, 1961.
570. M. Uemura, Y. Hayashi and Y. Hayashi, *Tetrahedron: Asymmetry*, 1994, **5**, 1427.
571. D. W. Slocum et al., *J. Chem. Educ.*, 1969, **46**, 144.
572. F. Rebière, O. Samuel and H. B. Kagan, *Tetrahedron Lett.*, 1990, **31**, 3121.
573. D. Seyferth, H. P. Hofmann, R. Burton and J. F. Helling, *Inorg. Chem.*, 1962, **1**, 227.
574. W. Reeve and E. F. Group, Jr., *J. Org. Chem.*, 1967, **32**, 122.
575. A. N. Nesmeyanov, N. N. Sedova, V. A. Sazonova and S. K. Moiseev, *J. Organomet. Chem.*, 1980, **185**, C6.
576. J. J. Bishop, A. Davidson, M. L. Katcher, D. W. Lichtenberg, R. E. Merrill and J. C. Smart, *J. Organomet. Chem.*, 1971, **27**, 241.
577. H. R. Allcock, K. D. Lavin, G. H. Riding, P. R. Suszko and R. R. Whittle, *J. Am. Chem. Soc.*, 1984, **106**, 2337.
578. T. Hayashi and M. Kumada, *Acc. Chem. Res.*, 1982, **15**, 395.
579. R. Noyori, 'Asymmetric Catalysis in Organic Synthesis', Wiley, New York, 1994.
580. D. Marquarding, H. Klusacek, G. Gokel, P. Hoffmann and I. Ugi, *J. Am. Chem. Soc.*, 1970, **92**, 5389.
581. G. Gokel, P. Hoffmann, H. Klusacek, D. Marquarding, E. Ruch and I. Ugi, *Angew. Chem., Int. Ed. Engl.*, 1970, **9**, 64.
582. G. W. Gokel, D. Marquarding and I. K. Ugi, *J. Org. Chem.*, 1972, **37**, 3052.
583. Alternatively, racemic 1,2-substituted ferrocenyl derivates may be resolved by an enzyme-mediated process: G. Nicolosi, A. Patti, R. Morrone and M. Piattelli, *Tetrahedron: Asymmetry*, 1994, **5**, 1275.
584. M. Watanabe, S. Araki, Y. Butsugan and M. Uemura, *J. Org. Chem.*, 1991, **56**, 2218.
585. G. Nicolosi, A. Patti, R. Morrone and M. Piattelli, *Tetrahedron: Asymmetry*, 1994, **5**, 1639.
586. H. Wally, M. Widhalm, W. Weissensteiner and K. Schlögl, *Tetrahedron: Asymmetry*, 1993, **4**, 285.
587. M. Watanabe, N. Hashimoto, S. Araki and Y. Butsugan, *J. Org. Chem.*, 1992, **57**, 742.
588. A. Togni and S. D. Pastor, *J. Org. Chem.*, 1990, **55**, 1649.
589. M. O. Okoroafor, D. L. Ward and C. H. Brubaker, Jr., *Organometallics*, 1988, **7**, 1504.
590. T. Hayashi, K. Kanehira, T. Hagihara and M. Kumada, *J. Org. Chem.*, 1988, **53**, 113.
591. S. D. Pastor and A. Togni, *J. Am. Chem. Soc.*, 1989, **111**, 2333.
592. A. Togni and R. Häusel, *Synlett*, 1990, 633.
593. M. Sawamura, R. Kuwano and Y. Ito, *Angew. Chem., Int. Ed. Engl.*, 1994, **33**, 111.
594. O. Riant, O. Samuel and H. B. Kagan, *J. Am. Chem. Soc.*, 1993, **115**, 5835.
595. F. Rebière, O. Riant, L. Ricard and H. B. Kagan, *Angew. Chem. Int. Ed. Engl.*, 1993, **32**, 568.
596. J. Einhorn and J. L. Luche, *Tetrahedron Lett.*, 1986, **27**, 1793.
597. P. Boudjouk, R. Sooriyakumaran and B.-H. Han, *J. Org. Chem.*, 1986, **51**, 2818.
598. M. Kihara, M. Kashimoto, Y. Kobayashi and S. Kobayashi, *Tetrahedron Lett.*, 1990, **31**, 5347.
599. M. Kihara, M. Ikeuchi, K. Jinno, M. Kashimoto, Y. Kobayashi and Y. Nagao, *Tetrahedron*, 1993, **49**, 1017.
600. K. Fukuhara, N. Miyata and S. Kamiya, *Tetrahedron Lett.*, 1990, **31**, 3743.
601. J. T. Sharp and C. E. D. Skinner, *Tetrahedron Lett.*, 1986, **27**, 869.
602. J. Y. Corey and L. S. Chang, *J. Organomet. Chem.*, 1986, **307**, 7.
603. I. S. Aidhen and N. S. Narasimhan, *Tetrahedron Lett.*, 1991, **32**, 2171.
604. I. S. Aidhen and J. R. Ahuja, *Tetrahedron Lett.*, 1992, **33**, 5431.
605. K. Ohno, H. Nishiyama, H. Nagase, K. Matsumoto and M. Ishikawa, *Tetrahedron Lett.*, 1990, **31**, 4489.
606. H. Nishiyama et al., *J. Org. Chem.*, 1992, **57**, 407.
607. J. E. Toth and P. L. Fuchs, *J. Org. Chem.*, 1987, **52**, 473.
608. P. A. Wender and A. W. White, *Tetrahedron*, 1983, **39**, 3767.
609. J. Graham, A. Ninan, K. Reza, M. Sainsbury and H. G. Shertzer, *Tetrahedron*, 1992, **48**, 167.

610. B. H. Lipshutz, F. Kayser and Z.-P. Liu, *Angew. Chem., Int. Ed. Engl.*, 1994, **33**, 1842.
611. W. E. Parham and C. K. Bradsher, *Acc. Chem. Res.*, 1982, **15**, 300.
612. G. J. Quallich, D. E. Fox, R. C. Friedmann and C. W. Murtiashaw, *J. Org. Chem.*, 1992, **57**, 761 and Reference 12 cited therein.
613. M. R. Paleo, C. Lamas, L. Castedo and D. Domínguez, *J. Org. Chem.*, 1992, **57**, 2029.
614. A. Couture, E. Deniau and P. Grandclaudon, *J. Chem. Soc., Chem. Commun.*, 1994, 1329.
615. P. Beak, T. J. Musick, C. Liu, T. Copper and D. J. Gallagher, *J. Org. Chem.*, 1993, **58**, 7330.
616. S. Gronowitz, in 'The Chemistry of Hetreocyclic Compounds', eds. A. R. Weisberger and E. C. Taylor, Wiley, New York, 1985, vol. 44.
617. R. A. Abramovitch, M. Saha, E. M. Smith and R. T. Coutts, *J. Am. Chem. Soc.*, 1967, **89**, 1537.
618. R. D. Chambers, F. G. Drakesmith and W. K. R. Musgrave, *J. Chem. Soc.*, 1965, 5045.
619. R. D. Chambers, C. A. Heaton, W. K. R. Musgrave and L. Chadwick, *J. Chem. Soc. C*, 1969, 1700.
620. C. W. Bird and G. W. H. Cheeseman, in 'Comprehensive Heterocyclic Chemistry', eds. C. W. Bird and G. W. H. Cheeseman, Pergamon, Oxford, 1984, vol. 4, p. 59.
621. D. J. Chadwick and C. Willbe, *J. Chem. Soc., Perkin Trans. 1*, 1977, 887.
622. B. L. Feringa, R. Hulst, R. Rikers and L. Brandsma, *Synthesis*, 1988, 316.
623. J. M. Brown and L. R. Canning, *J. Chem. Soc., Chem. Commun.*, 1983, 460.
624. I. Hasan, E. R. Marinelli, L.-C. C. Lin, F. W. Fowler and A. B. Levy, *J. Org. Chem.*, 1981, **46**, 157.
625. J. M. Brittain, R. A. Jones, J. Sepulveda Arques and T. Azhar Saliente, *Synth. Commun.*, 1982, **12**, 231.
626. M. V. Sargent and F. M. Dean, in 'Comprehensive Heterocyclic Chemistry', eds. C. W. Bird and G. W. H. Cheeseman, Pergamon, Oxford, 1984, vol. 4, p. 651.
627. A. v. Oeveren, W. Menge and B. L. Feringa, *Tetrahedron Lett.*, 1989, **30**, 6427.
628. M. P. Georgiadis, E. A. Couladouros, M. G. Polissiou, S. E. Filippakis, D. Mentzafos and A. Terzis, *J. Org. Chem.*, 1982, **47**, 3054.
629. F. Perron and K. F. Albizati, *J. Org. Chem.*, 1989, **54**, 2044.
630. J. Raczko, A. Golebiowski, J. W. Krajewski, P. Gluzinski and J. Jurczak, *Tetrahedron Lett.*, 1990, **31**, 3797.
631. T. Honda, M. Imai, K. Keino and M. Tsubuki, *J. Chem. Soc., Perkin Trans. 1*, 1990, 2677.
632. M. P. Georgiadis, S. A. Haroutounian and C. D. Apostolopoulos, *Synthesis*, 1991, 379.
633. M. P. Georgiadis, S. A. Haroutounian and J. C. Bailar, Jr., *J. Heterocycl. Chem.*, 1988, **25**, 995.
634. M. W. Reed and H. W. Moore, *J. Org. Chem.*, 1988, **53**, 4166.
635. W. D. Wulff, J. S. McCallum and F.-A. Kunng, *J. Am. Chem. Soc.*, 1988, **110**, 7419.
636. R. A. De Haan, M. J. Heeg and K. F. Albizati, *J. Org. Chem.*, 1993, **58**, 291.
637. E. Braye, (Parcor, Paris, France) *US Pat.* 4 127 580 (1978).
638. I. Kuwajima and H. Urabe, *Tetrahedron Lett.*, 1981, **22**, 5191.
639. D. Goldsmith, D. Liotta, M. Saindane, L. Waykole and P. Bowen, *Tetrahedron Lett.*, 1983, **24**, 5835.
640. G. C. M. Lee, E. T. Syage, D. A. Harcourt, J. M. Holmes and M. E. Garst, *J. Org. Chem.*, 1991, **56**, 7007.
641. A.-B. Hörnfeldt, *Acta Chem. Scand.*, 1967, **21**, 1952.
642. A.-B. Hörnfeldt and P.-O. Sundberg, *Acta Chem. Scand.*, 1972, **26**, 31.
643. R. Kiesewetter and P. Margaretha, *Helv. Chim. Acta*, 1985, **68**, 2350.
644. S. Gronowitz and A.-B. Hörnfeldt, in 'Thiophene and its Derivatives', ed. S. Gronowitz, Wiley, New York, 1986, vol. 44, p. 1.
645. S. Gronowitz, in 'Organic Sulfur Chemistry: Structure, Mechanism and Synthesis', ed. C. J. M. Stirling, Butterworth, London, 1975, p. 203.
646. F. M. Dean, in 'Advances in Heterocyclic Chemistry', ed. A. R. Katritzky, Academic Press, New York, 1981, vol. 30, p. 167.
647. F. M. Dean, in 'Advances in Heterocyclic Chemistry', ed. A. R. Katritzky, Academic Press, New York, 1982, vol. 31, p. 237.
648. D. J. Chadwick and D. S. Ennis, *Tetrahedron*, 1991, **47**, 9901.
649. C. Domínguez, A. G. Csáky and J. Plumet, *Tetrahedron*, 1992, **48**, 149.
650. G. C. M. Lee, J. M. Holmes, D. A. Harcourt and M. E. Garst, *J. Org. Chem.*, 1992, **57**, 3126.
651. D. W. Knight and A. P. Nott, *J. Chem. Soc., Perkin Trans. 1*, 1983, 791.
652. W. L. Whipple and H. J. Reich, *J. Org. Chem.*, 1991, **56**, 2911.
653. A. J. Carpenter and D. J. Chadwick, *J. Org. Chem.*, 1985, **50**, 4362.
654. R. A. Barcock, D. J. Chadwick, R. C. Storr, L. S. Fuller and J. H. Young, *Tetrahedron*, 1994, **50**, 4149.
655. D. L. Comins and M. O. Killpack, *J. Org. Chem.*, 1987, **52**, 104.
656. S. L. Graham and T. H. Scholz, *J. Org. Chem.*, 1991, **56**, 4260.
657. F. Lucchesini, *Tetrahedron*, 1992, **48**, 9951.
658. P. T. De Sousa, Jr. and R. J. K. Taylor, *Synlett*, 1990, 755.
659. D. W. Knight and A. P. Nott, *J. Chem. Soc., Perkin Trans. 1*, 1981, 1125.
660. A. J. Carpenter and D. J. Chadwick, *Tetrahedron Lett.*, 1985, **26**, 1777.
661. C. H. Eugster, M. Balmer, R. Prewo and J. H. Bieri, *Helv. Chim. Acta*, 1981, **64**, 2636.
662. D. S. Ennis and T. L. Gilchrist, *Tetrahedron*, 1990, **46**, 2623.
663. D. J. Ager, *Tetrahedron Lett.*, 1983, **24**, 5441.
664. J. A. Marshall and D. J. Nelson, *Tetrahedron Lett.*, 1988, **29**, 741.
665. E. G. Doadt and V. Snieckus, *Tetrahedron Lett.*, 1985, **26**, 1149.
666. A. J. Carpenter and D. J. Chadwick, *J. Chem. Soc., Perkin Trans. 1*, 1985, 173.
667. M. G. Reinecke and L. J. Chen, *Acta Chem. Scand.*, 1993, **47**, 318.
668. P. Ribéreau and G. Quéguiner, *Tetrahedron*, 1984, **40**, 2107.
669. A. J. Carpenter and D. J. Chadwick, *Tetrahedron Lett.*, 1985, **26**, 5335.
670. M. Watanabe and V. Snieckus, *J. Am. Chem. Soc.*, 1980, **102**, 1457.
671. M. A. F. Brandão, A. B. de Oliveira and V. Snieckus, *Tetrahedron Lett.*, 1993, **34**, 2437.
672. G. Beese and B. A. Keay, *Synlett*, 1991, 33.
673. A. Hallberg and P. Pedaja, *Tetrahedron*, 1983, **39**, 819.

674. D. Solooki, V. O. Kennedy, C. A. Tessier and W. J. Youngs, *Synlett*, 1990, 427.
675. E. J. Bures and B. A. Keay, *Tetrahedron Lett.*, 1987, **28**, 5965.
676. J. Cuomo, S. K. Gee and S. C. Hartzell, in 'Synthesis and Chemistry of Agrochemicals II', eds. D. R. Baker, J. G. Fenyes and W. K. Moberg, ACS Symposium Series No. 443, American Chemical Society, Washington, DC, 1990.
677. M. Prats, C. Gálvez, Y. Gasanz and A. Rodriguez, *J. Org. Chem.*, 1992, **57**, 2184.
678. J. W. H. Watthey and M. Desai, *J. Org. Chem.*, 1982, **47**, 1755.
679. D. J. Chadwick, M. V. McKnight and R. Ngochindo, *J. Chem. Soc., Perkin Trans. 1*, 1982, 1343.
680. J. H. Näsman, N. Kopola and G. Pensar, *Tetrahedron Lett.*, 1986, **27**, 1391.
681. R. B. Gammill and B. R. Hyde, *J. Org. Chem.*, 1983, **48**, 3863.
682. T. Honda, Y. Kobayashi and M. Tsubuki, *Tetrahedron Lett.*, 1990, **31**, 4891.
683. S. W. McCombie, B. B. Shankar and A. K. Ganguly, *Tetrahedron Lett.*, 1987, **28**, 4123.
684. A. G. Schultz and L. A. Motyka, *J. Am. Chem. Soc.*, 1982, **104**, 5800.
685. C. P. Dell and D. W. Knight, *J. Chem. Soc., Chem. Commun.*, 1987, 349.
686. G. C. M. Lee (Allergan Inc., USA) *US Pat.*, 4 935 530 (1990).
687. G. R. Martinez, P. A. Grieco and C. V. Srinivasan, *J. Org. Chem.*, 1981, **46**, 3760.
688. M. P. Edwards, A. M. Doherty, S. V. Ley and H. M. Organ, *Tetrahedron*, 1986, **42**, 3723.
689. N. R. Kotecha, S. V. Ley and S. Mantegani, *Synlett*, 1992, 395.
690. D. J. Chadwick and S. T. Hodgson, *J. Chem. Soc., Perkin Trans. 1*, 1982, 1833.
691. A. R. Katritzky and K. Akutagawa, *Org. Prep. Proceed. Int.*, 1988, **20**, 585.
692. M. Gharpure, A. Stoller, F. Bellamy, G. Firnau and V. Snieckus, *Synthesis*, 1991, 1079.
693. J. Bergman and L. Venemalm, *Tetrahedron*, 1992, **48**, 759.
694. G. Nechvatal, D. A. Widdowson and D. J. Williams, *J. Chem. Soc., Chem. Commun.*, 1981, 1260.
695. G. Nechvatal and D. A. Widdowson, *J. Chem. Soc., Chem. Commun.*, 1982, 467.
696. N. F. Masters, N. Mathews, G. Nechvatal and D. A. Widdowson, *Tetrahedron*, 1989, **45**, 5955.
697. D. A. Widdowson, *Philos. Trans. R. Soc. London, A*, 1988, **326**, 595.
698. P. J. Beswick, C. S. Greenwood, T. J. Mowlem, G. Nechvatal and D. A. Widdowson, *Tetrahedron*, 1988, **44**, 7325.
699. M. J. Dickens, T. J. Mowlem, D. A. Widdowson, A. M. Z. Slawin and D. J. Williams, *J. Chem. Soc., Perkin Trans. 1*, 1992, 323.
700. M. Iwao, H. Takehara, S. Obata and M. Watanabe, *Heterocycles*, 1994, **38**, 1717.
701. C. D. Buttery, R. G. Jones and D. W. Knight, *Synlett*, 1991, 315.
702. G. W. Gribble *et al.*, *J. Org. Chem.*, 1992, **57**, 5878.
703. M. Iwao, *Heterocycles*, 1993, **36**, 29.
704. E. J. Griffen, D. G. Roe and V. Snieckus, *J. Org. Chem.*, 1995, **60**, 1484.
705. P. Rocca *et al.*, *J. Org. Chem.*, 1993, **58**, 7832.
706. P. Melnyk, J. Gasche and C. Thal, *Synth. Commun.*, 1993, **23**, 2727.
707. F. Guillier *et al.*, *J. Org. Chem.*, 1995, **60**, 292.
708. M. S. Solomon and P. B. Hopkins, *Tetrahedron Lett.*, 1991, **32**, 3297.
709. M. A. J. Miah and V. Snieckus, *J. Org. Chem.*, 1985, **50**, 5436.
710. M. Tsukazaki and V. Snieckus, *Heterocycles*, 1993, **35**, 689.
711. M. Tsukazaki and V. Snieckus, *Heterocycles*, 1992, **33**, 533.
712. L. Estel, F. Linard, F. Marsais, A. Godard and G. Quéguiner, *J. Heterocycl. Chem.*, 1989, **26**, 105 and references cited therein.
713. A. Godard, J.-C. Rovera, F. Marsais, N. Plé and G. Quéguiner, *Tetrahedron*, 1992, **48**, 4123.
714. M. Iwao and T. Kuraishi, *Tetrahedron Lett.*, 1983, **24**, 2649.
715. R. J. Mattson and C. P. Sloan, *J. Org. Chem.*, 1990, **55**, 3410.
716. A. Turck, N. Plé, L. Mojovic and G. Quéguiner, *J. Heterocyclic Chem.*, 1990, **27**, 1377.
717. A. Turck *et al.*, *Tetrahedron*, 1993, **49**, 599.
718. A. Turck, N. Plé, L. Mojovic and G. Quéguiner, *Bull. Soc. Chim. Fr.*, 1993, **130**, 488.
719. J. Sandosham and K. Undheim, *Tetrahedron*, 1994, **50**, 275.
720. N. Plé, A. Turck, F. Bardin and G. Quéguiner, *J. Heterocyclic Chem.*, 1992, **29**, 467.
721. N. Plé, A. Turck, P. Martin, S. Barbey and G. Quéguiner, *Tetrahedron Lett.*, 1993, **34**, 1605.

2
Sodium and Potassium

ALESSANDRO MORDINI
Università di Firenze, Italy

2.1 INTRODUCTION	93
2.1.1 Literature	93
2.1.2 Background	94
2.2 SODIUM- AND POTASSIUM-CONTAINING MIXED-METAL REAGENTS	94
2.2.1 Generalities	94
2.2.2 Superbases and their Constitution	95
2.2.3 Nonorganometallic Mixed-metal Reagents	96
2.3 METALLATION OF HYDROCARBONS	96
2.3.1 Metallation of Resonance-active Positions	97
2.3.1.1 Alkenes	97
2.3.1.2 The chemical behaviour of allylsodium and allylpotassium compounds	97
2.3.1.3 Conjugated and homoconjugated dienes and polyenes	103
2.3.1.4 Nonconjugated and nonhomoconjugated dienes	105
2.3.1.5 Allylarenes	108
2.3.1.6 Alkylarenes	108
2.3.1.7 Allenes and alkynes	110
2.3.2 Metallation of Resonance-inactive Positions	110
2.3.2.1 Alkenes and dienes	110
2.3.2.2 Arenes	112
2.3.2.3 Cyclopropane derivatives	113
2.4 METALLATION OF HETEROSUBSTITUTED ALKENES	114
2.4.1 Heterosubstituted Allylmetallic Compounds	114
2.4.2 Nitrogen-containing Alkenes	115
2.4.3 Oxygen- and Sulfur-containing Alkenes	116
2.4.4 Silicon-containing Alkenes	117
2.4.5 Heterocyclic Compounds	119
2.4.6 Heterosubstituted Arenes	121
2.5 REFERENCES	124

2.1 INTRODUCTION

2.1.1 Literature

There are a few pertinent reviews and books on aspects of the chemistry of organoalkali metal compounds. Most of them, such as a volume in the Houben-Weyl series[1] by Schöllkopf, the book *Struktur and Reaktivität polarer Organometalle*[2] by Schlosser, the two books *Preparative Polar Organometallic Chemistry 1* by Brandsma and Verkruijsse and *Preparative Polar Organometallic Chemistry 2*[3] by Brandsma, Chapter 1 in the book *Organometallics in Synthesis. A Manual* edited by Schlosser[4] and the chapters in *COMC-I* by Wardell[5] and Wakefield[6] contain general considerations concerning organic derivatives of lithium, sodium and potassium. Only a very few general reports are concerned with the chemistry of organosodium and organopotassium compounds and none of them has been published since the mid-1960s. In detail, two reviews by Schlosser published in 1964[7] are the most recent, following the articles by Morton[8] and Benkeser[9] in 1944 and 1957, respectively. Most of the reports in this field have been devoted to the structure and reactivity of the superbases which are highly reactive organometallic species containing lithium and sodium or potassium. Superbase chemistry was reviewed by Schlosser in 1988,[10] in 1992,[11] in 1993[12] and in 1994,[13] and by Mordini in 1992.[14] A report dealing with 'unimetal superbases' has been published by Caubère.[15]

2.1.2 Background

A general comment on the chemistry of organosodium and organopotassium compounds is that they have found a limited number of applications in organic synthesis,[6] due to the difficulties in preparing and handling these organometallic reagents. The synthetic utility of real organosodium and organopotassium compounds is confined mainly to the work of Morton[8] and Benkeser[9] in the 1940s and 1950s. At that time, the metallation by organosodium and organopotassium reagents of hydrocarbons at allyl,[16,17] benzyl,[18–22] aryl,[23,24] vinyl[25] and cyclopropyl[26] positions as well as Alfin-type butadiene polymerization catalysed by the allylsodium/sodium isopropoxide mixture[27] were studied. Later, the finding that addition of a sodium or a potassium alcoholate to an organosodium reagent improved its metallating ability[16,28] was applied to the metallation of benzene[29,30] and other alkylaryl hydrocarbons,[18,31] and in Wurtz-type coupling reactions between organosodium compounds and alkyl halides.[32] The need for strong metallating agents capable of deprotonating weakly acidic hydrocarbons was the driving force for these efforts. However, as revealed by a closer examination of the experimental procedures, the use of organosodium and organopotassium reagents presents some serious drawbacks: the preparation of organosodium and organopotassium compounds usually requires high-speed stirring and special equipment; despite the large excess of substrates and the very long reaction times used, unsatisfactory yields are often obtained. Thus, pure organosodium and organopotassium reagents were partially abandoned when new strong organometallic bases were discovered. The main reason for the resurgence of interest in organosodium and potassium chemistry was the discovery by Schlosser in 1967[33] that an equimolar mixture of butyllithium and potassium t-butoxide constituted a reagent with exceptionally high metallating power. The subject of this chapter, organosodium and organopotassium reagents in organic synthesis, consists mainly of the recent developments in superbase chemistry.

The first questions to answer now are: what have superbases in common with organosodium and organopotassium compounds, and how are they related to each other? The answers are not trivial and many investigations have been carried out in order to understand if the use of a superbasic reagent is equivalent to the use of a neat organosodium or organopotassium reagent in terms of the organometallic intermediate produced after the metallation reaction. This has been the object of an NMR investigation[34] which has clearly shown that metallation with the superbasic reagents butyllithium/potassium t-butoxide and butyllithium/sodium t-butoxide affords the corresponding organopotassium and organosodium derivatives, respectively. This result confirms previous findings by Lím and co-workers that reaction of triphenylmethane with butyllithium/sodium t-butoxide leads to triphenylmethylsodium[35] and that butyllithium/potassium t-butoxide reacts with toluene to give benzylpotassium[35,36] and with benzene to give phenylpotassium,[37] and by Schlosser and co-workers, who prepared benzyl,[33,38] allyl[39,40] and pentadienyl[41] potassium compounds by metallation with butyllithium/potassium t-butoxide.

2.2 SODIUM- AND POTASSIUM-CONTAINING MIXED-METAL REAGENTS

2.2.1 Generalities

This chapter deals mainly with the use of organosodium and organopotassium reagents in organic synthesis; therefore, we shall focus on the most useful synthetic processes involving these organometallic species. The prime objective of an organic chemist using polar organometallic reagents is to selectively replace a hydrogen atom by a metal in a hydrocarbon substrate and, subsequently, by a functional group. Most of the reports on organoalkali metal reagents are concerned with this concept: a metallation reaction by means of a suitable organometallic base and the subsequent interception of the new organometallic species with an electrophilic reagent (Equation (1)). We shall devote most of our attention to the practical advantages that superbasic reagents can offer in such reaction sequences.

$$-\overset{\diagdown}{\underset{\diagup}{C}}-H \xrightarrow{MR} -\overset{\diagdown}{\underset{\diagup}{C}}-M \xrightarrow{EX} -\overset{\diagdown}{\underset{\diagup}{C}}-E \qquad (1)$$
$$ RH MX$$

Many sodium- and potassium-containing mixed-metal reagents have been reported. Most contain an equimolar amount of an organolithium reagent and a sodium or a potassium alkoxide. Other mixed-metal reagents are made by mixing an organosodium reagent with a sodium or a potassium alcoholate. It is clear that the greater spread of R^1Li/MOR^2 mixtures ('LICNAOR' or 'LICKOR') is due to their ease of preparation. Organolithium compounds are easily accessible if not commercially available; they can be easily handled, and are far more stable than the corresponding sodium analogues. Thus, R^1Li/MOR^2 mixtures are the reagents of choice unless an exceptional situation warrants a different approach.

2.2.2 Superbases and their Constitution

According to Caubère,[15] the term superbase should refer only to an organometallic base resulting from a mixing of two (or more) bases leading to new basic species. Thus, the term superbase does not necessarily mean that the reagent is thermodynamically or kinetically stronger than a normal base, but refers to a newly created organometallic entity having modified properties in terms of its reactivity potential as well as its regio-, stereo- and typoselectivity.

Since the original work of Morton and co-workers,[16,28] the precise nature of the alcoholate/organometallic complexes has repeatedly been the subject of considerable speculation. However, little is known for sure. A long and still open debate concerns the constitution of the superbasic reagents. We do not need to describe this matter in detail because this has been done previously.[11-14] It suffices to point out that the real structural identity of superbases remains obscure even now. The superbasic reagent butyllithium/potassium t-butoxide was conceived by Schlosser[33] as a mixed-metal reagent. He postulated a 'mixed aggregate' type interaction[2,42] which strongly polarizes the organometallic bonds, leading to an increase in reactivity compared with the neat organolithium reagent. At this stage it is not useful to hypothesize on the real constitution of these aggregates, which may also exist as equilibria of several kinds of mixed aggregates having different stoichiometry and undergoing a morphological evolution in the course of the reaction (Equation (2)).[11]

$$(R^1Li)_m + (KOR^2)_n \longrightarrow (R^1)_m(Li+K)_{m+n}(OR^2)_n \qquad (2)$$

By mixing an organolithium reagent with a sodium[43] or a potassium[44] alcoholate, an organometallic precipitate can be produced in some instances which mainly consists of organosodium or organopotassium compounds. This result induced the authors to conclude that in a superbasic mixture a metathetical exchange occurs, producing a new reactive organometallic species. The existence of discrete species rather than mixed aggregates has been recently supported by NMR investigations[45] coupled with MNDO (modified neglect of diatomic overlap) calculations on the triphenylmethyllithium/caesium 3-ethylheptoxide mixture. The authors found that this particular mixture in THF does not form any mixed aggregate, but rather individual species (**1**) and (**2**) in which a metal exchange has taken place (Equation (3)).

$$Ph_3C-Li + CsO-CH(Et)(C_5H_{11}) \longrightarrow Ph_3C-Cs + LiO-CH(Et)(C_5H_{11}) \qquad (3)$$
$$(\mathbf{1}) (\mathbf{2})$$

On the other hand, the existence of mixed organolithium/metal alkoxide aggregates is established beyond any doubt. For example, butyllithium was found to form tetrameric aggregates with lithium t-butoxide in THF solution.[46] A variety of additional evidence[15] supports the idea that the formation of mixed aggregates is a general phenomenon occurring whenever metal species are mixed.

The simple metathetical exchange between lithium and a second alkali metal has also been ruled out by comparative experiments. Butylpotassium, well-studied since 1937,[47] is known to abstract (like butylsodium) concomitantly allylic and vinylic protons from alkenes,[48] whereas the butyllithium/potassium t-butoxide mixture offers a greater selectivity and higher yields.[49,50] The butyllithium/potassium t-butoxide mixture and the butylpotassium reagent also exhibit striking differences in their stability towards ethereal solvents.[51] The mixture survives for hours in THF at −50 °C, whereas neat butylpotassium reacts with THF instantaneously, even below −100 °C, to give 2-tetrahydrofuryl potassium as the main product, which then decomposes at around −50 °C to ethene and the potassium enolate of acetaldehyde (Equation (4)).[51]

$$\text{(THF)} \longrightarrow \text{(2-tetrahydrofuryl-K)} \longrightarrow KO\diagup\!\!\!\diagdown + H_2C=CH_2 \qquad (4)$$

We have summarized the few safe conclusions which can currently be drawn concerning the nature of superbasic reagents:
  (i) when an alcoholate and an organolithium reagent are combined, a mixed aggregate is probably formed which contains strongly polarized organometallic bonds endowed with high reactivity;
  (ii) the metallation product after reaction with a superbasic reagent contains mainly, although not exclusively, the heavier alkali metal;
  (iii) the butyllithium/alkali metal alkoxide mixtures show different chemical behaviour compared with simple butylsodium or butylpotassium and offer major practical advantages over the latter reagents.

The potentialities of superbases in synthesis have been exploited by Schlosser, who first envisaged the butyllithium/potassium t-butoxide mixture as a unique organometallic reagent offering a very high reactivity and an exceptional selectivity at the same time. For this reason the equimolar mixture of butyllithium and potassium t-butoxide is often known as Schlosser's base.

In the wake of the successful employment of Schlosser's base, a variety of superbasic reagents have been studied, showing sometimes a modulated behaviour. It must be pointed out, however, that the simple substitution of one component with a similar reagent (e.g., potassium t-butoxide with another potassium alcoholate) affects the overall behaviour of the base only to a small extent.

2.2.3 Nonorganometallic Mixed-metal Reagents

Activation due to the formation of aggregates between ionic species is not limited to the mixture constituted by organometallic reagents and metal alcoholates. The so-called 'unimetal superbases',[15] in which complexation occurs between a sodium alcoholate and sodium hydride or sodium amide, have been discovered and widely studied by Caubère. The mixing of two metal-containing species has been shown to result in an enhancement of both nucleophilic and basic properties, which has led to their employment in a variety of applications, among which proton abstraction, anionic polymerization and, to a larger extent, elimination and reduction[52] have been the most studied.

The basic and scarcely nucleophilic properties of hindered lithium amides can also be significantly enhanced by the addition of potassium t-butoxide. The mixed-metal reagent obtained by mixing lithium diisopropylamide and potassium t-butoxide (LIDAKOR),[53] for instance, has found a wide range of synthetically useful applications as a metallating agent as well as a promoter of alkene, epoxide and unsaturated alcohol isomerizations.

2.3 METALLATION OF HYDROCARBONS

Superbases have mainly been employed as metallating agents of weakly acidic hydrocarbons and are currently the only reagents that allow the efficacious metallation of alkenes and cyclopropanes. When considering the various substrates of metallation, we distinguish between aliphatic and aromatic hydrocarbons and, for both categories, between resonance and geometry activated substrates.[54] Resonance-activated positions are those neighbouring double and triple bonds or aromatic rings thus giving rise, after hydrogen–metal exchange, to allyl-, propargyl- or benzyl-type organometallic species. In the class of geometry-activated substrates are cyclopropanes, and alkenes when metallated at an alkenic position.

2.3.1 Metallation of Resonance-active Positions

2.3.1.1 Alkenes

The stoichiometric mixture of butyllithium and potassium *t*-butoxide has proved to be particularly suited to the selective metallation of resonance-active hydrocarbons. The order of reactivity of allylic alkyl groups is methyl > methylene > methine. Schlosser's base is in general able to exchange protons of the first two classes whereas, in order to deprotonate methine centres, trimethylsilylmethylpotassium was found to be more effective.[55] Intensive and extensive metallation studies have been carried out showing that the allylpotassium compounds thus obtained may find valuable applications in organic synthesis. Table 1 summarizes typical examples of alkene metallations.

Table 1 Representative examples of metallations by sodium- and potassium-containing bases (reference citations appear below each diagram).

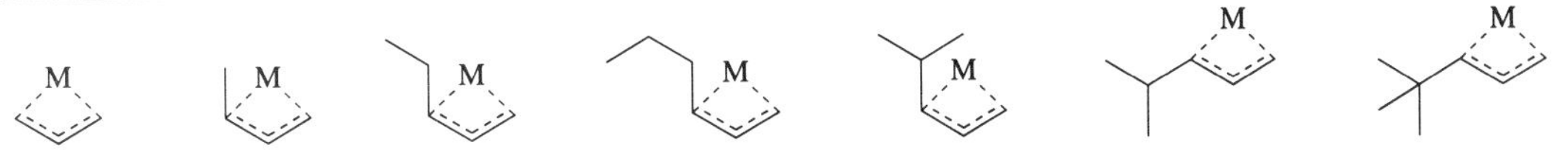

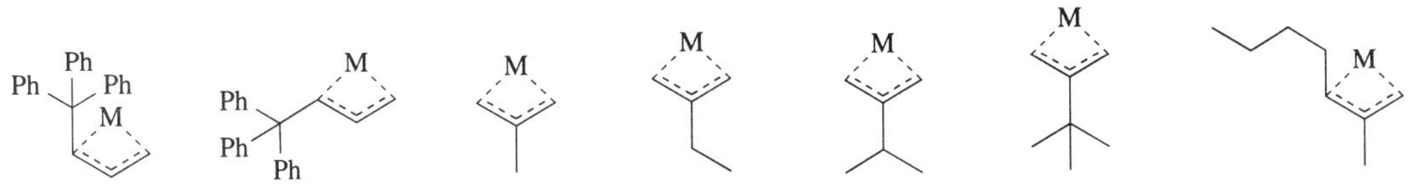

2.3.1.2 The chemical behaviour of allylsodium and allylpotassium compounds

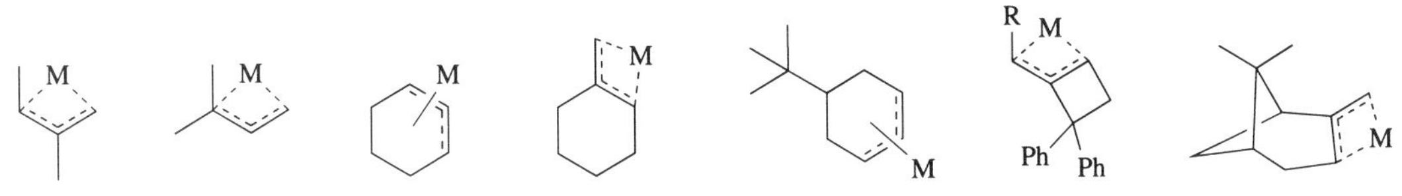

Allylsodium and allylpotassium compounds, as generated by the metallation reaction of alkenes with superbases, have a chemical behaviour which differs considerably from that of equally polar allyl-type lithium, magnesium and zinc species. The latter, for instance, react mainly to give the branched product, resulting from electrophilic attack at the inner, more substituted position of the allyl moiety (Scheme 1, route a), while allylsodium and potassium compounds give preferentially, if not exclusively, the derivative carrying the substituent at the terminal position (route b).

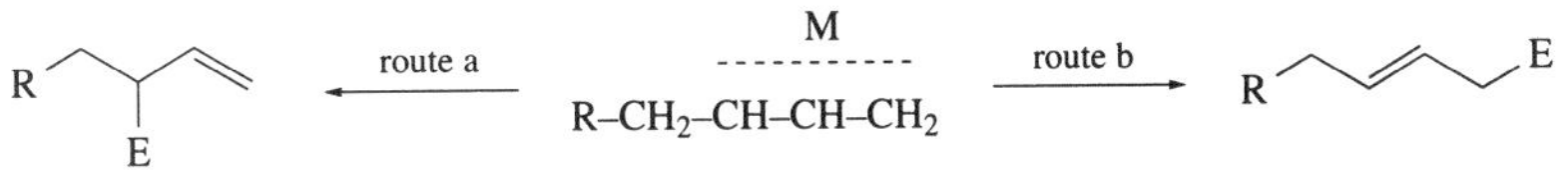

Scheme 1

The explanation for this dichotomy is not yet clear. The different behaviour may reflect to some extent a difference in the structure.[11] Careful NMR studies[60,69,70] based upon the Saunders' isotopic perturbation technique have revealed a π-bonded trihapto (η^3) structure of the 2-alkenyllithium and 2-alkenylpotassium compounds.[71] In contrast, the 2-alkenylmagnesium compounds have a monohapto (η^1) σ-structure with the metal tightly bonded to the primary carbon (Figure 1).[11]

The primary effect of this structural situation is that the metal–carbon bond does not have enough charge density to be attacked by the electrophile, which therefore must approach the sterically more hindered position of the allyl moiety. A completely different situation pertains for 2-alkenylpotassium compounds. They have a rather symmetrical trihapto (η^3) π-structure in which the charge density is almost equivalent at both termini of the allyl unit. Thus the electrophile will preferentially approach the

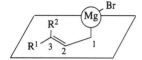 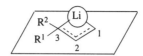 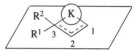

Figure 1 Monohapto structure of 2-alkenylmagnesium compounds compared with the trihapto structures of 2-alkenyllithium and 2-alkenylpotassium analogues.

less hindered side, giving the product with regiochemistry opposite to that obtained with the 2-alkenylmagnesium compound. The sodium derivatives, although not extensively studied, behave similarly to the 2-alkenylpotassium species. Some examples of the regiochemical behaviour of 2-alkenylmagnesium and potassium reagents with electrophiles are shown in Table 2.

Table 2 Reaction of 2-alkenyl- and 3-methyl-2-alkenylmagnesium bromide or the corresponding organopotassium compounds with electrophiles in THF at −75 °C: ratios of attack at the allylic-3 vs. allylic-1 positions.

EX	R¹⌒MgBr	R¹⌒K	R¹/R²⌒MgBr	R¹/R²⌒K
TMS-Cl		<5:95		0:100
FB(OMe)₂ᵃ	90:10	<5:95	80:20	<5:95
MeI	65:35	15:85	90:10	5:95
H₂C=CHCH₂Br	65:35		70:30	20:80
(CH₂)₂Oᵇ	<95:5	35:65	<95:5	20:80
CH₂O	<95:5	35:65		30:70
CO₂	<95:5	30:70	95:5	<5:95

ᵃ The resulting 2-alkeneboronates were converted into 2-alken-1-ols with alkaline hydrogen peroxide. ᵇ Oxirane.

The regiochemical outcome of the reactions of allylmetallic compounds with electrophiles depends on many factors other than the metal. The solvent, the temperature, the substituents on the allylic moiety and the electrophile all play an important role in orienting the course of reaction. Therefore, the above concepts and results are only partially valid.

The high degree of regioselection often found with allylpotassium reagents has been widely used for the synthesis of natural products. Potassium and sodium can also easily be replaced by other alkali metals,[3,38] transition metals or metalloids,[72] offering a valuable option for tuning reactivity as well as regioselectivity.

A further important difference between the allylic derivatives of magnesium, lithium, sodium and potassium relies on their configuration. Once the allylmetallic derivative (3) is produced, it may adopt either an *endo* (Z) or an *exo* (E) configuration if the alkyl group R occupies the inward or the outward position at the allylic termini (Scheme 2). The two stereoisomers are interconverted through a torsional isomerization process. For synthetic purposes one has to know both the *endo:exo* distribution for the different allylmetallic species and the time required to establish the equilibrium. Allylmagnesium and allyllithium compounds undergo rapid torsional isomerization, the energy of activation falling in the range 41.8–62.8 kJ mol⁻¹.[73] Sodium, potassium and caesium analogues have rotational barriers of 62.8–104.6 kJ mol⁻¹ and so it requires several hours, without catalysis, to reach the *endo:exo* equilibrium. Moreover, the two classes differ remarkably in the *endo:exo* equilibrium distribution. When the metal is magnesium or lithium there is little discrimination between the two stereoisomers, whereas sodium, potassium and caesium clearly prefer the *endo* form despite the larger steric hindrance. The *endo:exo* ratios for many allylmetallic compounds, as a function of the metal atom, the organic residue and the solvent, have been determined by intercepting the organometallic intermediate with a suitable electrophile and analysing the reaction mixture by gas chromatography.[40] In order to obtain meaningful results, the electrophile must react with the organometallic intermediate faster than the latter undergoes torsional isomerization, and it must preferentially attack the unsubstituted carbon atom of the allyl moiety, thus avoiding the loss of stereochemical information. The results of such investigations are shown in Table 3, in which *endo:exo* ratios obtained in hexane suspension are compiled. In THF solution the data are quantitatively similar, although the *endo* form is frequently favoured to a somewhat larger extent. The *endo* preference is more pronounced for the heavier alkali metals and, predictably, diminishes on increasing the steric hindrance of the alkyl group R. If the substituent is a tertiary group, the *exo* species becomes the thermodynamically favoured isomer even if, as revealed by a comparison with the corresponding hydrocarbon 4,4-dimethyl-2-pentene, the 2-alkenylpotassium compound still takes advantage of some electronic effect.

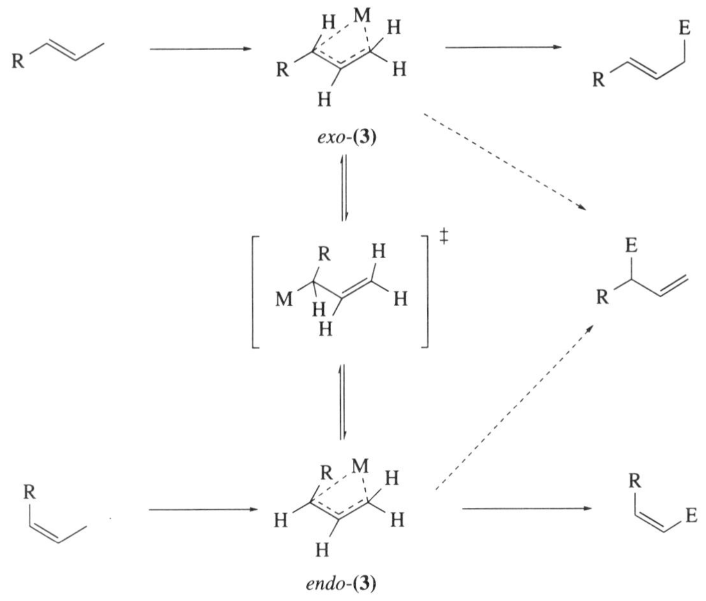

Scheme 2

Table 3 *Endo:exo* ratios of 2-alkenylmetal compounds (3) (M = Li, Na, K, Cs) after torsional equilibration, and *cis:trans* equilibrium ratios of the corresponding 2-alkenes (M = H).

R	M = H	M = Li	M = Na	M = K	M = Cs
Me	1:5	2:1	10:1	100:1	500:1
Et	1:5			15:1	
Pr^i	1:5			5:1	
Bu^t	1:10 000			1:10	

This surprising *endo* preference has been attributed[40] for a long time to a hydrogen-bond-like interaction between homoallylic C–H bonds and the electron-rich terminus of the allyl moiety. More recently, a new explanation[12] has been advanced on the basis of first-order electrostatic principles. The deprotonation of a symmetrically substituted alkene leads to an increase of the charge density mainly at the allylic termini, and this induces an inevitable elongation of all the bonds connecting alkyl groups to the allyl moiety. The *endo* bond is less affected by this elongation than the *exo* bond.

Alkyl substitution plays a large role in determining the *endo:exo* ratio of 2-alkenylmetallic compounds. Thus, a methyl (or any other alkyl) group in the central position of the allylic moiety raises the thermodynamic *endo:exo* ratio to levels of 50:1 to 500:1, due to steric repulsion suffered by the *exo* stereoisomer.[12] Finally, if there is both a methyl and an alkyl group at one of the allylic termini, they compete for the more favourable *endo* position and the smaller substituent is favoured, giving an *endo:exo* ratio of the order of 1:25 (Scheme 3).[12]

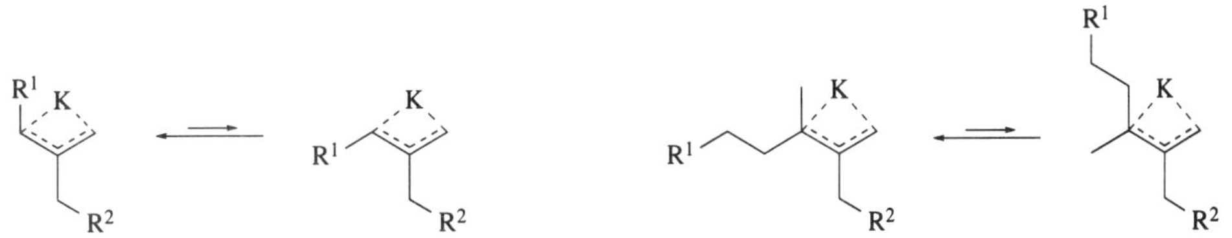

Scheme 3

The more obvious consequence of the *endo* preference and the slow torsional equilibration of 2-alkenylsodium and 2-alkenylpotassium compounds is the possibility of achieving a very high degree of stereocontrol in their reactions with electrophilic reagents. The metallation of a stereochemically pure alkene with Schlosser's base, followed by interception with an electrophile before onset of torsional isomerization, leads through a 'stereoconservative' process[74] to a pure product, having the same configuration as the starting material. If, on the other hand, the allylpotassium reagent is allowed to equilibrate, the more stable *endo* intermediate is formed and, after reaction with an electrophile, the pure *cis* product is obtained through a 'stereoselective' process[39,74] regardless of the stereochemistry of the precursor.

Many stereocontrolled carbon–carbon linking reactions have been described illustrating these principles. In particular, the synthesis of natural insect pheromones has attracted considerable effort.[10–12] *cis*-2-Heptene was metallated with Schlosser's base and then treated with 3-(2-tetrahydropyranyl)propyl iodide (Equation (5)). The resulting acetal was hydrolysed and esterified with isovaleroyl chloride to afford (**4**),[12] the sex attractant (31%) of female *Nudaurelia cytherea cytherea*,[75] the pine emperor moth.

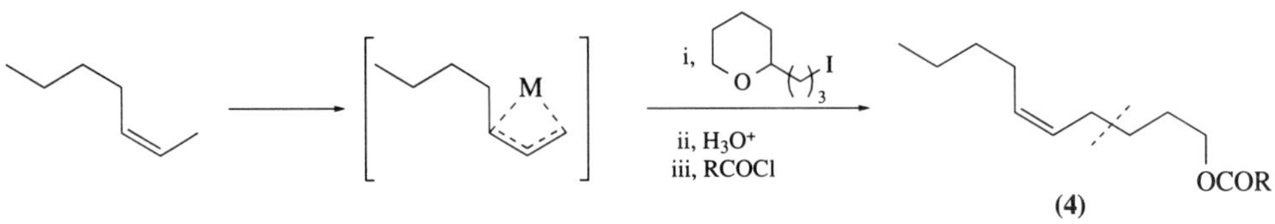

$$R = CH_2Pr^i \tag{5}$$

A pheromone component of the flat grain beetle *Cryptolestes pusillus* (**5**)[76] and the sex attractant of the female grape berry moth *Paralobesia viteana* (**6**)[77] were both synthesized from 10-undecen-1-ol (**7**) (Scheme 4).[12] Metallation of the alcoholate or the *O*-methoxymethyl-protected alcohol gave an allylpotassium intermediate (**8**) which was then treated either with carbon dioxide and cyclized to (*Z*)-3-dodecen-12-olide (**5**) (40%) or with methyl iodide, to afford the alkene (**6**) (69%), the acetate of which (R^2 = Ac) is the expected pheromone.

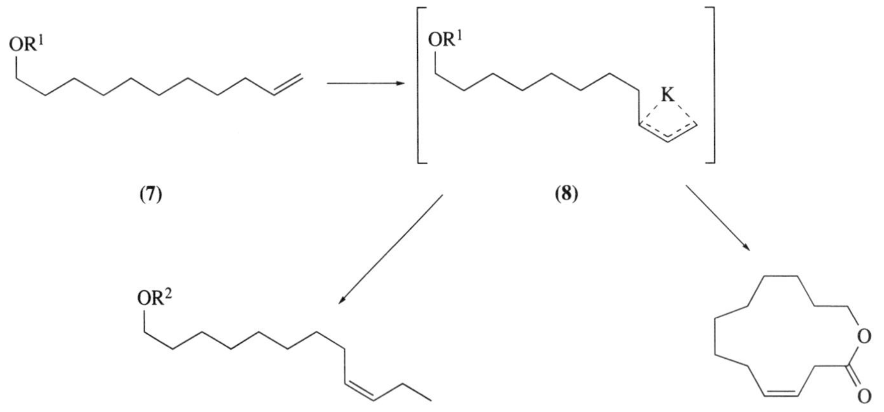

R^1 = H, CH$_2$OMe

Scheme 4

Metallation of 1-nonene with Schlosser's base gives the 2-alkenylpotassium compound (**9**). When this was allowed to undergo torsional equilibration and then treated with carbon dioxide, (*Z*)-3-decenoic acid (**10**) (38%), a pheromone of the virgin furniture carpet beetle *Anthrenus flavipes*,[78] was obtained (Equation (6)).[12]

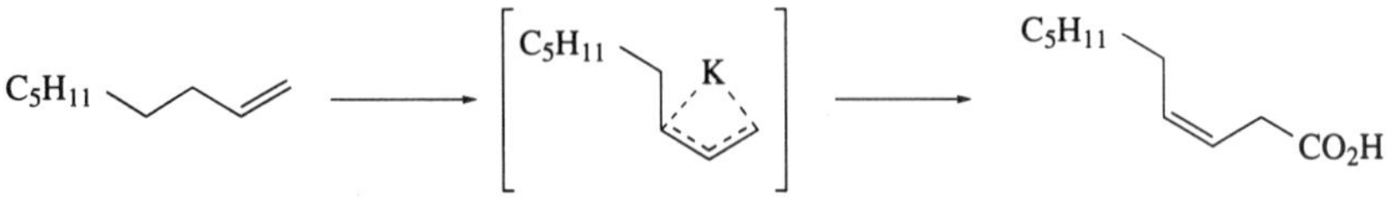

$$\tag{6}$$

A similar reaction sequence was applied to 1-undecene. The use of the electrophile fluorodimethoxyboron followed by alkaline hydrogen peroxide oxidation[62] introduced a hydroxyl group, affording (Z)-2-undecen-1-ol (**11**) (Equation (7)). The latter was then converted to the acetate and this, via copper-catalysed coupling[79] with dodecylmagnesium bromide, gave (Z)-9-tricosene[80] ('muscalure' (**12**), 77%), the sex attractant of the common house fly *Musca domestica*.

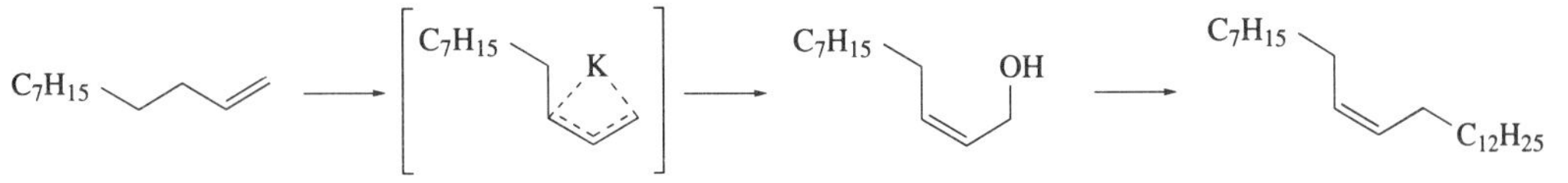

A remarkable application of superbase-metallated alkenes was reported by Djerassi and co-workers,[64] who prepared a wide variety of Δ^{24} and $\Delta^{24(28)}$ sterols such as 'desmosterol' (**13**) via the coupling of prenylpotassium (3-methyl-2-butenylpotassium (**14**)) with steroidal iodides (Equation (8)). Perfect tail-to-tail regioselectivity was achieved by adding a cuprate complex.

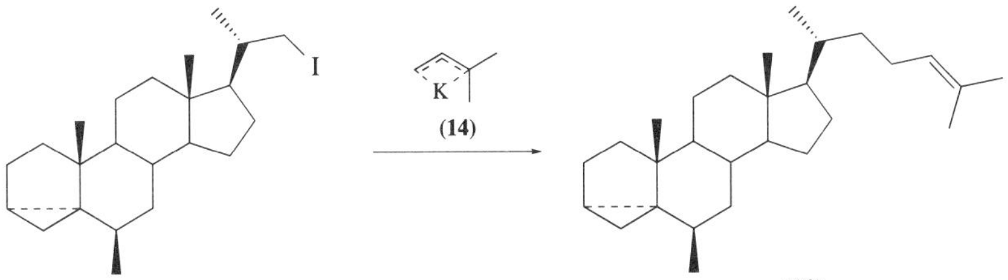

When prenylpotassium (**14**) was allowed to react with 8-bromonortricyclene (**15**), α-santalene (**16**) was produced which was submitted again to metallation with Schlosser's base followed by borylation and oxidation.[63] Allowance for torsional equilibration produced the (Z)-α-santalol (**17**), one of the main constituents of the highly prized Sandalwood oil,[81] with no trace of the (E) isomer (Equation (9)). The overall sequence could be contracted to a one-flask protocol, still in quite an acceptable yield.[63]

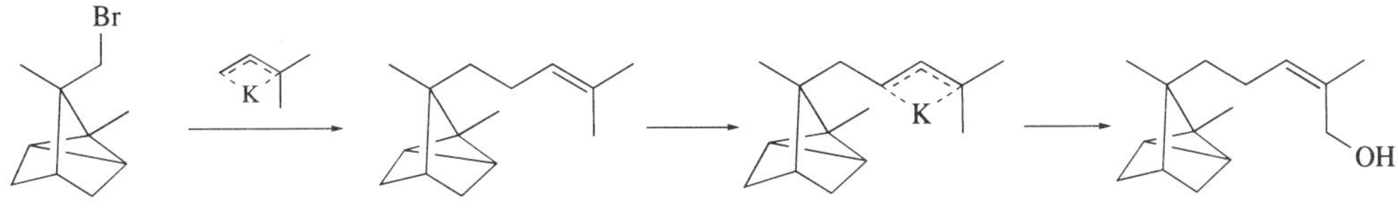

Taking advantage of the regio- and stereochemical features of alkenylpotassium compounds, it is easy to produce stereochemically pure 2-alkene-boronates or -boranes via reaction with a suitable electrophile. Thus, 2-alkene-boronates were prepared using fluorodimethoxyboron[62] as electrophile and used in stereoselective reactions with carbonyl compounds, *erythro* adducts being produced from (Z)-alkenylboronates and *threo* adducts from the (E) isomers.[57] This behaviour was successfully applied to the synthesis of the (S,S)-enantiomer of 4-methyl-3-heptanol (**18**),[72] the major constituent of the aggregation pheromone of the elm bark beetle *Scolytus multistriatus* Marsham[82] and *cis*-5-butyl-4-methyltetrahydro-2-furanone (**19**),[83] known as 'quercus lactone' or 'oak lactone' (Scheme 5). Both were prepared starting from *cis*-2-butene via addition of (Z)-2-butenedimethoxyborane (**20**) to propanal, followed by methylation to the alkenol (**21**) and hydrogenation for the former, and to pentanal followed by hydroboration to the diol (**22**) and cyclization for the latter.

The metallation of 2-butene with butyllithium/potassium *t*-butoxide followed by reaction with a suitable boron electrophile has been reported[84] to be the most reliable method so far for the preparation of optically active crotylboranes such as (**23**)–(**25**) with a high degree of geometrical purity. Chiral substituents on boron induce very high enantioselectivities in addition to the already mentioned regio- and diastereoselectivities in reaction with carbonyl compounds.

Dialkylallylboranes have also been prepared by Zaidlewicz[85] using trimethylsilylmethylpotassium as metallating agent and dialkylchloroboranes as trapping reagents. The organoboron derivatives thus obtained have been employed for the synthesis of allylic alcohols via oxidation,[85] for double bond

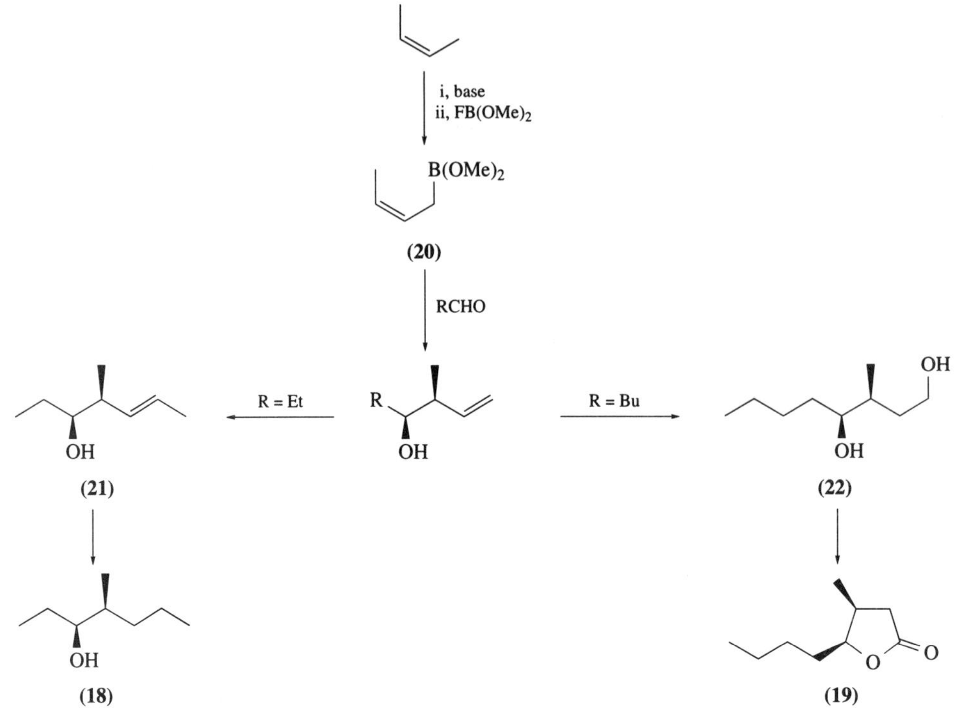

Scheme 5

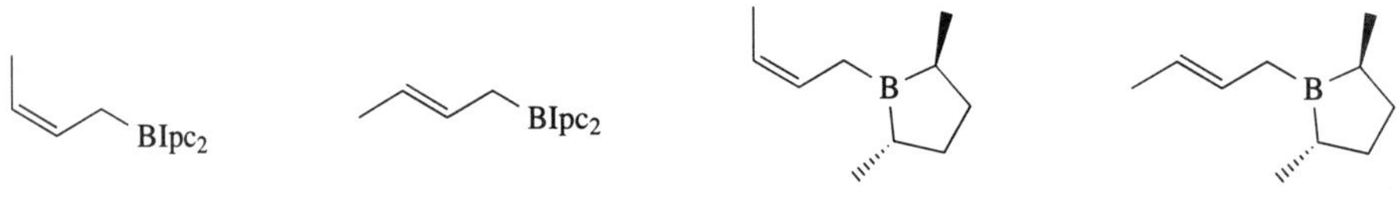

(*Z*)-(**23**) (*E*)-(**23**) (*Z*)-(**24**) (*E*)-(**24**)

Ipc = isopinocampheyl

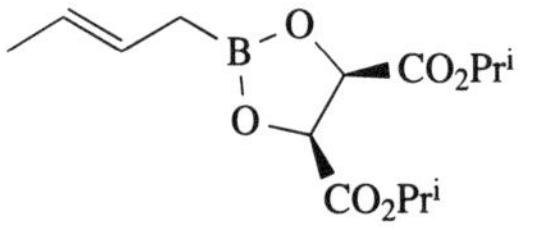

(25)

isomerization via hydrolysis,[85] for deuterium incorporation via reaction with deuterium oxide,[86] and for the synthesis of homoallylic alcohols and α,β-unsaturated carbonyl compounds via reaction with aldehydes.[87] As an application, α- and γ-damascone (**26**) and (**27**) have been prepared by this route (Scheme 6).[85]

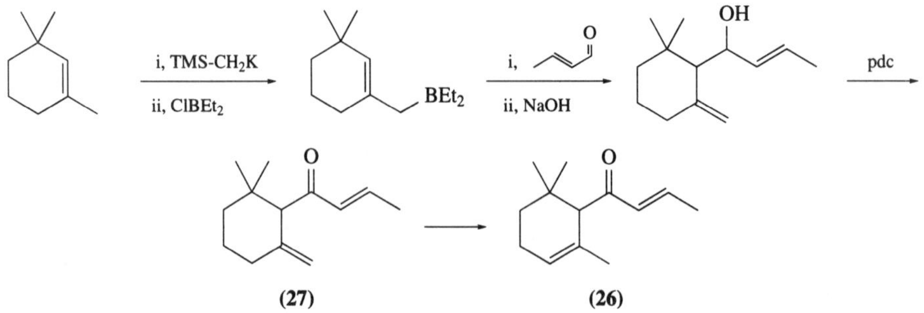

(27) **(26)**

pdc = pyridinium dichromate

Scheme 6

An easy and efficient procedure for the conversion of α-pinene to the more valuable β-pinene has been reported.[68] The α-pinene was metallated with Schlosser's base and the resulting allylpotassium species (**28**) transformed to the stable 'ate' complex (**29**) which, upon hydrolysis, afforded β-pinene (Equation (10)).

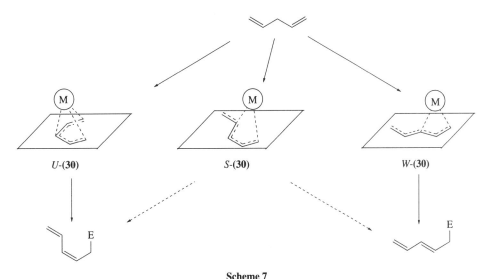

2.3.1.3 Conjugated and homoconjugated dienes and polyenes

Conjugated[88-91] and homoconjugated[41,56,76,92] dienes can be easily metallated at the activated allylic position, giving rise to organometallics of the pentadienyl type with all five carbon atoms placed in the same plane in order to gain maximum resonance energy. The conformation adopted by a pentadienyl-type anion has been the object of careful NMR and synthetic studies. In principle, a vinyl moiety attached to the terminal position of an allyl system may occupy an *endo* or an *exo* position and, in both cases, it may be twisted inward or outward. In this way three coplanar species result: a horseshoe-like *U* shape, a sickle-like *S* shape and a zigzag-like *W* shape (Scheme 7).

Scheme 7

As long as we deal with counterion-free carbanions, the *W* form should be favoured for obvious reasons: absence of steric strain and optimum charge separation. However, if we consider the pentadienyl system as a tight contact species, as they are in the condensed phase, the importance of the metallic counterion in designing the conformation must be taken into account. Thus, as NMR studies[93] have revealed, the pentadienyllithium (**30**) (M = Li) adopts the *W* form. This conclusion has been confirmed by chemical transformation to derivatives having the *trans* configuration.[41] Larger metals such as potassium or caesium have a size that allows them to establish binding interactions with all three electron-rich centres of the pentadienyl backbone if the latter is coiled up to adopt the *U* conformation. Pentadienylpotassium (**30**) (M = K) in THF has been found to exist in the pentahapto (η^5) *U* shape, as established by NMR spectroscopic data[94] and by the formation of (Z)-2,4-pentadienyl derivatives after treatment with electrophiles.[41,95] Alkyl substitution may affect the conformational equilibria. Thus, a methyl group in the central position forces both the lithium and the potassium species to adopt a *W* shape (**31**),[41] this being the only conformation that allows for a double *endo* position of the methyl group. On the other hand, 2,4-dimethyl-2,4-pentadienyllithium (**32**) (M = Li) and 2,4-dimethyl-2,4-pentadienylpotassium (**32**) (M = K) both prefer the *U* shape due to a pronounced steric strain generated by the two methyl groups in the *W* form.[41]

If the alkyl substitution does not produce a symmetrical pentadienyl system, then the number of possible conformations increases, even if only a few of them are actually adopted. Thus, 2-methyl-2,4-

W-(**31**)

U-(**32**)

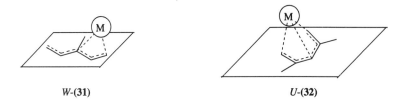

pentadienyllithium (**33**) (M = Li) and 2-methyl-2,4-pentadienylpotassium (**33**) (M = K) adopt the *W* and *U* shapes, respectively, out of the four possible conformations (Equation (11)).

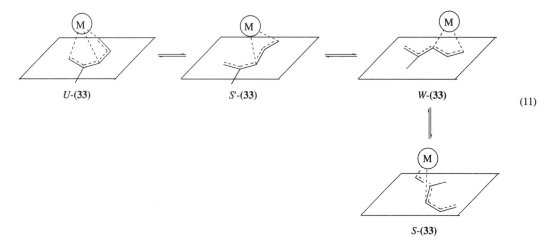

U-(**33**) *S'*-(**33**) *W*-(**33**) *S*-(**33**) (11)

When the alkyl substituent is located at the terminal position as in (**34**), eight possible conformations can be drawn. It has been found, however, that only three of them are adopted. The organolithium compound exists exclusively in the *endo-W* form if generated from lithiation of a *cis*-1,4-alkadiene and mainly in the *exo-W* shape if derived from a *trans*-1,4-alkadiene. The pentadienylpotassium adopts the *exo-U* shape when produced by treatment of a *trans*-1,4-diene with an organopotassium reagent (Equation (12)).[74]

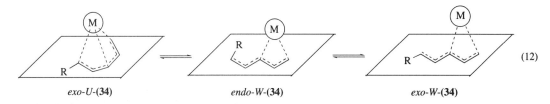

exo-U-(**34**) *endo-W*-(**34**) *exo-W*-(**34**) (12)

For a heptatrienyl system similar considerations may be applied even if the number of possible resonance-stabilized structures increases considerably. The unsubstituted heptatrienylpotassium species (**35**) (M = K) exists as an 85:15 mixture of two conformations (*W-W*-(**35**) and *U-W*-(**35**)) out of the 10 different possibilities (Equation (13)).[12] Primary alkyl substituents at the antinodal points favour the *W-W* conformation by analogy with the pentadienyl series.

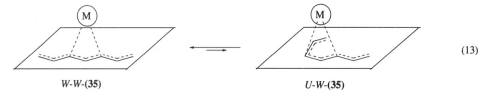

W-W-(**35**) *U-W*-(**35**) (13)

Understandably, stereoselective reactions with electrophiles become less realistic under circumstances in which a large number of conformations are possible. Nevertheless, some applications have been reported concerning the use of pentadienyl and heptatrienylmetallic species. (*E*)-2-(2-Butenyl)-1,3,3-trimethylcyclohexene (**36**), for example, was metallated by trimethylsilylmethylpotassium and, after borylation and oxidation, gave rise to β-ionol (**37**) in 30% yield, which derives from a *W*-shaped pentadienylpotassium compound (**38**) (Equation (14)).[12]

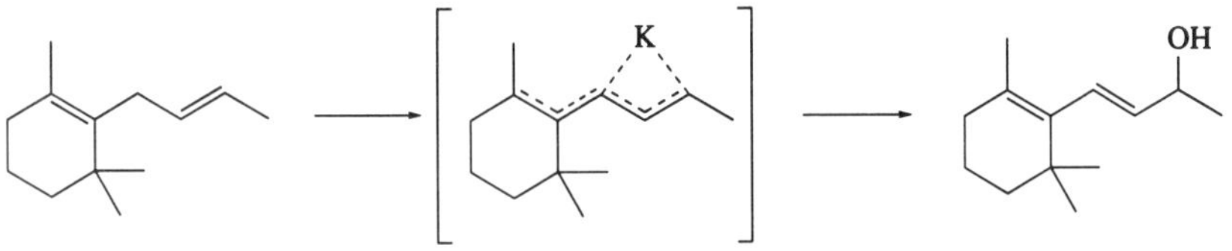

Metallation of β-ocimene (**39**) (3,7-dimethyl-1,3,6-octatriene) with both an organolithium and organopotassium reagent generated the heptatrienylmetallic species (**40**), which upon hydroxylation led to (2E,4E)-4,5-didehydrogeraniol (**41**) in 68% yield, practically stereochemically pure (Equation (15)).[12]

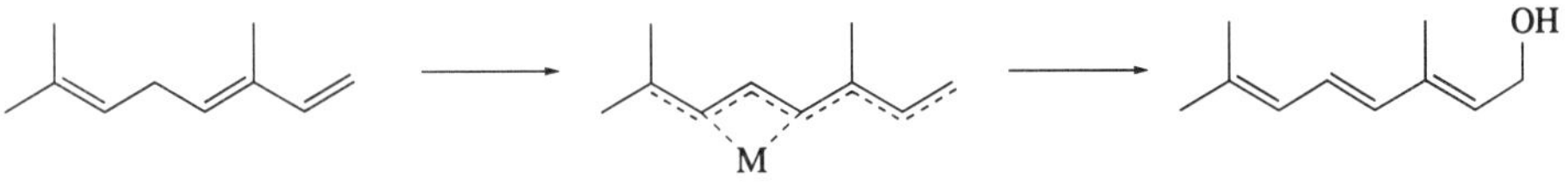

Pentadienylmetallic species generated in cyclic compounds may also adopt different conformations, but some of them are often disfavoured for steric constraints. 7,8-Dehydrocholesterol (**42**) (provitamin D$_3$) can be metallated by Schlosser's base or by trimethylsilylmethylpotassium at the three allylic positions. After reaction with electrophiles only the 7-substituted regioisomer (**43**) was obtained, which corresponds to deprotonation at position 14 to afford an S-shaped pentadienylpotassium intermediate (**44**) (Equation (16)). Deprotonation at positions 4 and 9 seems disfavoured for steric reasons, even if the latter would produce a U-shaped pentadienyl system.[91]

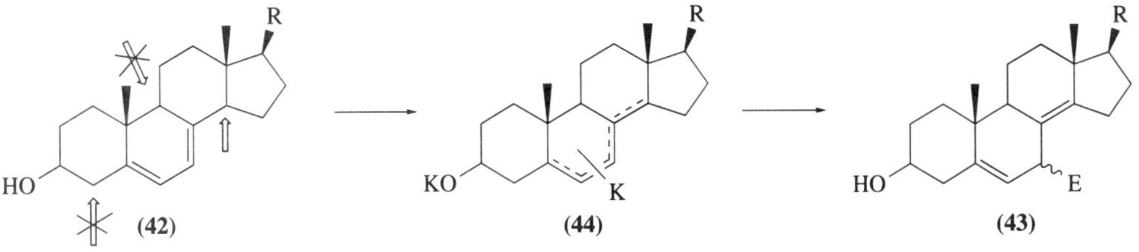

Metallation of α-terpinene (**45**) with Schlosser's base and subsequent borylation–oxidation afforded the monoterpene alcohol anthemol (**46**) as the unique polar product, derived from proton abstraction from the methyl group (Scheme 8). Actually, the metallation, followed by reaction with fluorodimethoxyboron, gave all the possible regioisomers (**47**)–(**49**) but, after oxidation, only one alcohol was formed because the other two boronates produced 1-isopropyl-4-methylbenzene (p-cymene) via elimination.[90]

2.3.1.4 Nonconjugated and nonhomoconjugated dienes

When the two double bonds of a diene are separated by more than one methylene or methine group, only simple allylmetal rather than extensively delocalized pentadienylmetal species can be formed upon metallation. Some efforts[10] have been made to understand under which conditions a diene can be either mono- or dimetallated. The crucial variable is the solvent. In a polar medium (e.g., THF) the second deprotonation is retarded compared with the first, thus favouring the formation of monometallated species. In a nonpolar medium (hydrocarbon solvent) the second deprotonation is accelerated, probably because the metal is complexed intramolecularly by the available double bond.[96] Hence, 1,5-hexadiene, 1,7-octadiene, 1,8-nonadiene and 1,9-decadiene were found to be easily monometallated by Schlosser's base (LICKOR) in THF, no matter whether the base was used in stoichiometric amounts or in excess. The dimetallated species (with *endo,endo* configuration) were invariably obtained by working in hexane suspension, except for the short 1,5-hexadiene (Scheme 9). Apparently the nearby allylmetallic moiety in THF solution protects the second alkenic bond against metallation.[96] This effect should diminish with distance and, while 1,10-undecadiene was still exclusively monometallated, 1,13-tetradecadiene underwent extensive dimetallation also in THF solution. The former result was used[96] for the synthesis

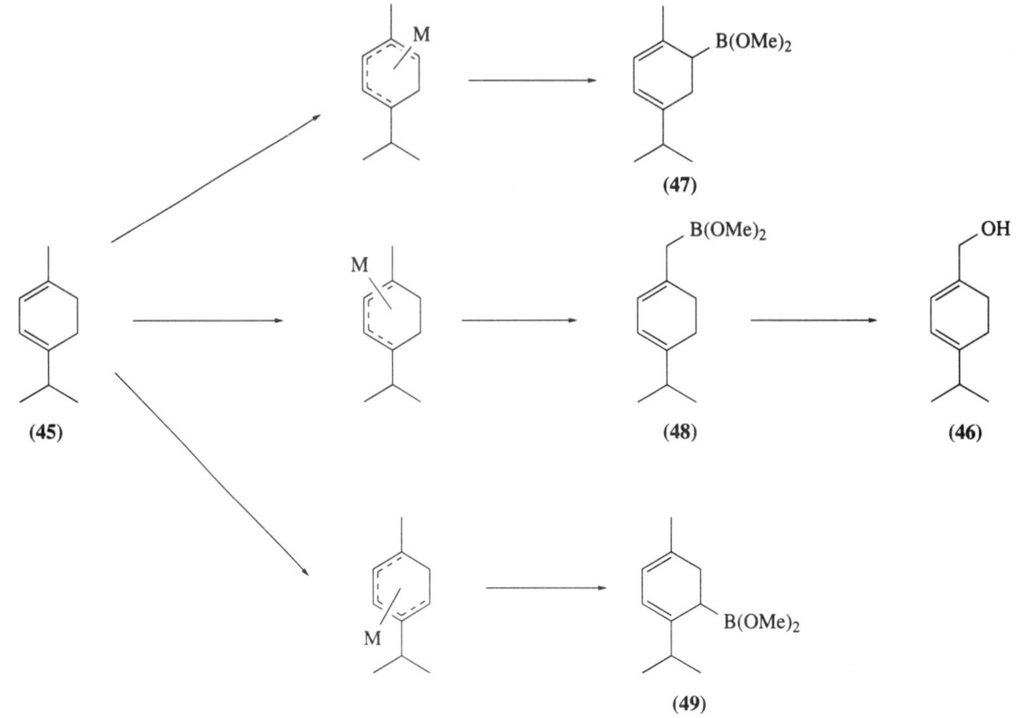

Scheme 8

of (Z)-5,13-tetradecadien-1-ol (**50**) in 17% yield (Equation (17)), the acetate of which is the major constituent of the mandibular secretion of carpenter moth (*Cossus cossus*) larvae.[97]

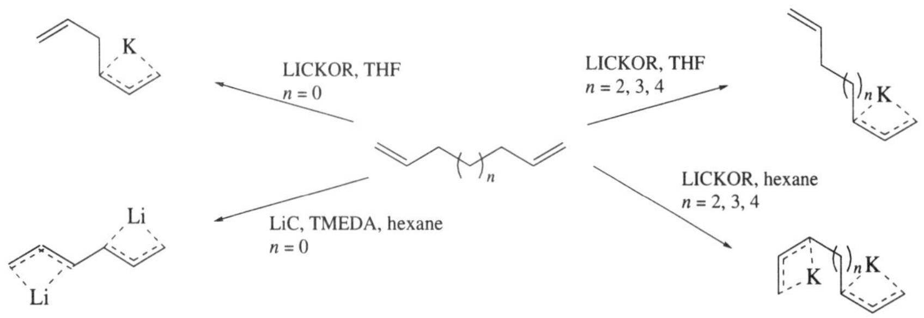

Scheme 9

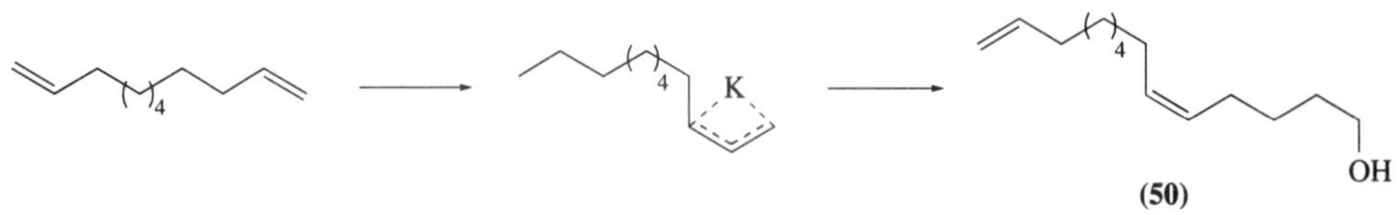

Monometallation of (Z)-1,12-heptadecadiene (**51**) with Schlosser's base in THF followed by treatment with paraformaldehyde gave[96] a 2:3 mixture of regioisomeric alcohols, the major (22%) being (3Z,13Z)-3,13-octadecadien-1-ol (**52**) (Equation (18)), the acetate of which is the sex attractant of the peach tree borer moth *Sanninoidea exitiosa*.[98]

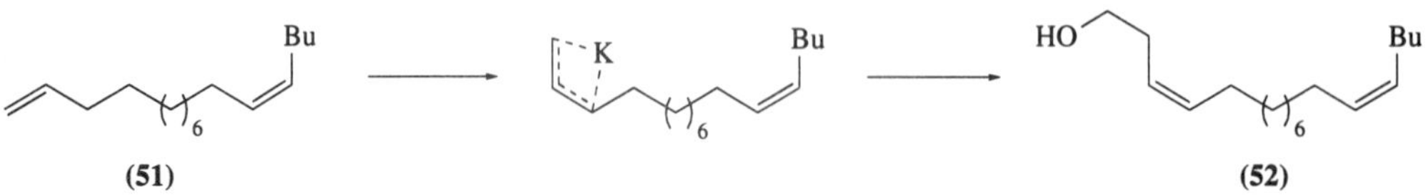

Bicyclo[3.2.1]octa-2,6-diene (**53**) has been selectively deprotonated[99] at the allylic position by Schlosser's base in THF at −60 °C giving, after reaction with trimethylchlorostannane, a 14:86 mixture of *endo*- and *exo*-4-trimethylstannylbicyclo[3.2.1]octa-2,6-dienes (*endo*- and *exo*-(**54**)) in an 85% global yield (Scheme 10). If the same substrate was treated with *t*-butyllithium/TMEDA, metallation occurred at the vinylic positions.

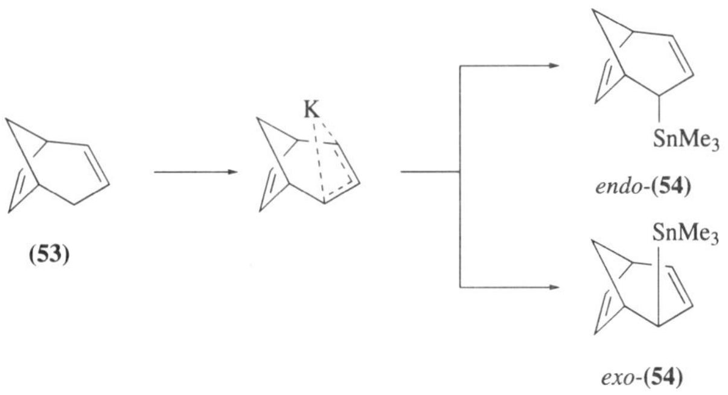

Scheme 10

When trying to replace a hydrogen atom from a 1,3-diene by an alkali metal an undesired addition onto the double bond may occur. Thus, while 2,3-dimethyl-1,3-butadiene can be mono- and dimetallated[88,100] using Schlosser's base, isoprene, under the same reaction conditions, undergoes addition to produce dimers. As isoprenylation is an important step often encountered in the synthesis of natural products, many efforts have been devoted to generate an isoprenylmetallic reagent. As Brandsma and co-workers[101,102] found, the combination of lithium 2,2,6,6-tetramethylpiperidide and potassium *t*-butoxide is able to produce the desired isoprenylpotassium intermediate (**55**), which was trapped with oxiranes and alkyl halides in acceptable yields (Equation (19)).

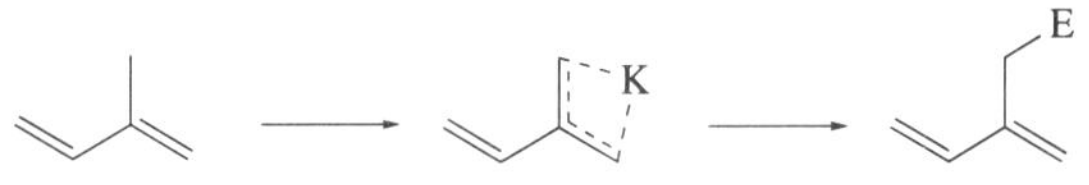

(19)

This intermediate was employed by Brown and Randad[103] to prepare *B*-2'-isoprenyl-9-borabicyclo[3.3.1]nonane and *B*-2'-isoprenyldiisopinocampheylborane (**56**). The latter reagent can be used for highly enantioselective addition reactions to aldehydes, as demonstrated by the synthesis of both antipodes of ipsenol (**57**) and ipsdienol (**58**) (Scheme 11), two aggregation pheromones of the bark beetle *Ips paraconfusus* Lanier.[104]

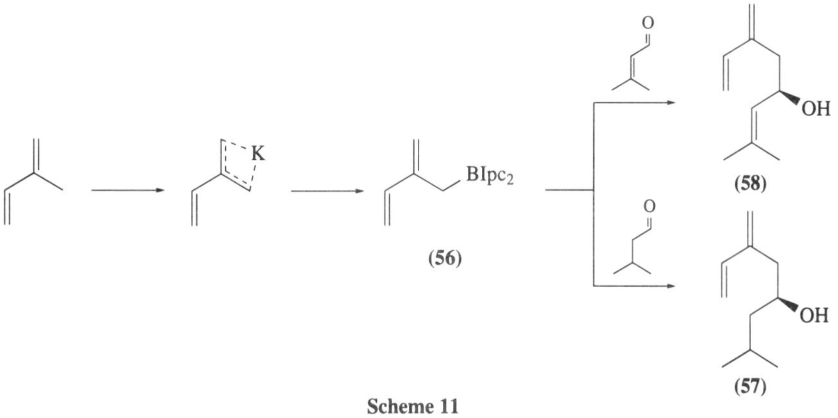

Scheme 11

2.3.1.5 Allylarenes

When a phenyl ring is located at the terminal position of an allylmetallic system, the area of delocalization increases provided coplanarity is possible. This is the case if the phenyl group occupies an *exo* position, while in the *endo* form the steric repulsion between the *ortho* and ω hydrogen atoms forces the ring to rotate out of the plane. Thus, phenylallylpotassium (**59**) (M = K) has been found[12] to exist mainly as the *exo* conformer, the *cis:trans* ratios of electrophilically substituted products varying from 14:86 in THF to 5:95 in hexane (Equation (20)).[12,33] Similar results have been obtained with phenylallyllithium (**59**) (M = Li).[12]

$$\text{exo-}(\mathbf{59}) \rightleftharpoons \text{endo-}(\mathbf{59}) \qquad (20)$$

M = K, Li

Taking advantage of the *exo* preference of allylarenes, the three lignin building blocks coumaryl, coniferyl and sinapyl alcohol (**60**)–(**62**) have been prepared in 76–81% yield as pure (E) isomers.[105] The one-pot reaction sequence consisted of the metallation of the allylarenes chavicol, eugenol and 5-allylpyrogallol-1,3-dimethyl ether with Schlosser's base in THF a −95 °C followed by borylation and oxidation (Equation (21)).

$$\qquad (21)$$

(**60**) $X^1 = X^2 = H$
(**61**) $X^1 = OMe, X^2 = H$
(**62**) $X^1 = X^2 = OMe$

2.3.1.6 Alkylarenes

When alkylbenzene derivatives are submitted to metallation, the metal may displace either a benzylic (α) or a ring hydrogen atom (preferably at the sterically unhindered *meta* or *para* positions). Studies with isopropylbenzene (cumene) have shown that the use of pentylsodium results mainly in ring metallation (*meta:para* = 1:1), while pentylpotassium after 20 h in heptane leads to the α-metallated product exclusively (Figure 2). The regioisomeric distribution found with pentylsodium was explained by Benkeser *et al.*,[21] who suggested that metallation occurred first at the ring and then a slow transmetallation produced the thermodynamically more stable benzylic isomer. With the pentylpotassium reagent, formation of the resonance-stabilized species is favoured. However, at a short reaction time (3 h) a 42:58 benzylic vs. ring metallation was observed, showing again that in the first stage all the regioisomers are produced followed by rearrangement to the resonance-stabilized species. The different behaviour of alkylsodium and alkylpotassium compounds has been attributed[20,21] to the difference in their ionic character.

Figure 2 Possible products from the metallations of alkylbenzenes.

Metallation with superbases has revealed a marked reagent dependence.[49,106] Schlosser's base gives mainly ring-metallated products regardless of whether a paraffinic or ethereal solvent is employed.

Trimethylsilylmethylpotassium in THF affords exclusively the α-metallated derivatives in good yields. Mainly α-metallation (95:5) has again been achieved by using TMEDA-activated pentylsodium.[107] The same reagent deprotonates ethylbenzene[108] at the benzylic α-position giving, after carboxylation and esterification, methyl 2-phenylpropanoate (**63**) in nearly quantitative yield (Equation (22)).

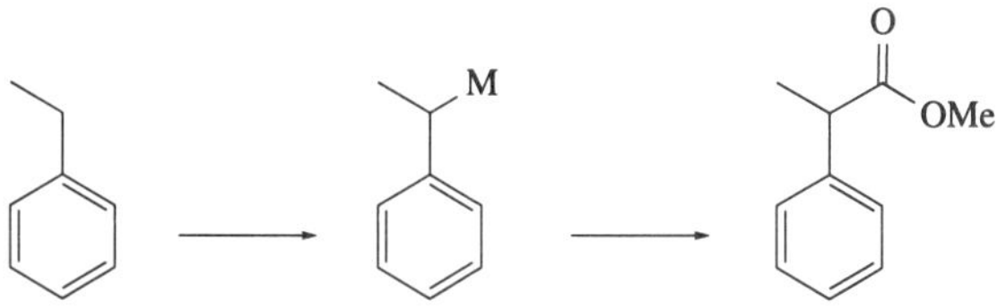

$$\text{(22)}$$

Toluene can be cleanly metallated at the benzylic position by Schlosser's base[33] or similar superbasic mixtures[36,109] as well as alkylpotassium reagents.[20] Contamination by ring-metallated products becomes important when alkylsodium reagents are used.[20] The utility of a superbasic approach is demonstrated by the fact that TMEDA-activated butyllithium gives rise to some contamination by ring-metallated products.[110]

The relative reactivity of different alkyl groups on a phenyl ring parallels that encountered in alkenes: methyl > methylene > methine. This discrimination between benzylic positions has been elegantly used to design a one-pot synthesis of ibuprofen (**64**),[111] an analgesic with both antiinflammatory and antirheumatic properties. The synthesis has been accomplished[112] by sequential deprotonation with Schlosser's base and alkylation, starting with *p*-xylene (Scheme 12).

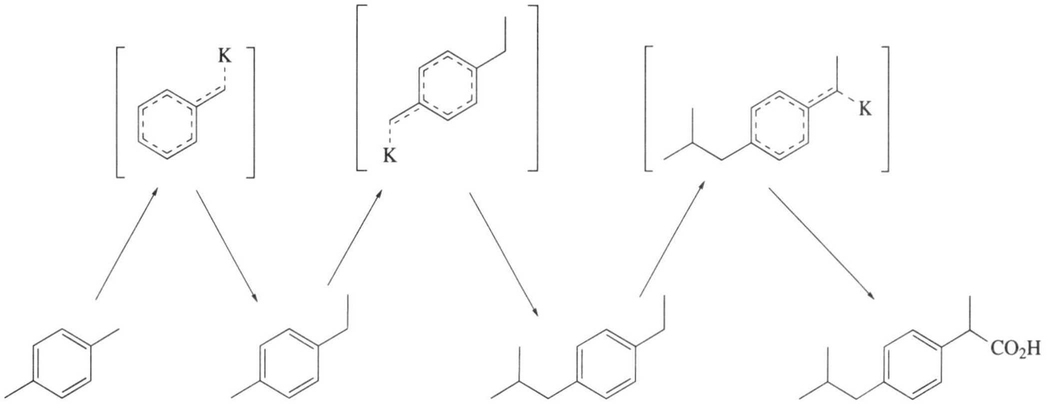

Scheme 12

Direct dimetallation of xylenes,[113] cresols[114] and symmetrical dimethylbiphenyls,[115] trimetallation of mesitylene,[116] isomeric trimethylbenzenes[116] and dimethylphenols,[117] and tetrametallation of 3,3',5,5'-tetramethylbiphenyl[118] have been accomplished by means of superbases.[119]

2,6-Dimethylphenyl alkyl ethers (**65**), when treated with 2 equiv. of the butyllithium/potassium *t*-butoxide mixture undergo a Wittig 1,4 migration of the oxygen-attached alkyl group to one of the benzylic carbon positions, ultimately leading to a species of type (**66**).[120] When the alkyl R group is methyl, the rearrangement can be largely avoided by working at room temperature. Interception with electrophiles then affords 2,6-disubstituted anisoles, such as *meta*-cyclophanes (**67**) (*n* = 8–15) and analogous macrocycles (Scheme 13).[121]

As mentioned previously, the exchange between two alkali metals may play a crucial role in the regiochemical outcome of the reaction between an electrophile and an organometallic species. For instance, benzylpotassium compounds (**68**), easily obtained by superbase metallation of primary alkylbenzenes (**69**), react with formaldehyde with a regioselectivity opposite to the corresponding magnesium species (**70**), which can be produced by treatment of the benzyl-type organometallic with an equimolar amount of magnesium dibromide.[122] The organopotassium compounds react with formaldehyde exclusively at the benzylic α-position, affording 2-arylethyl alcohols (**71**), while the magnesium derivatives bind the aldehyde preferentially at the *ortho* position, leading to benzylic alcohols (**72**) (Scheme 14).

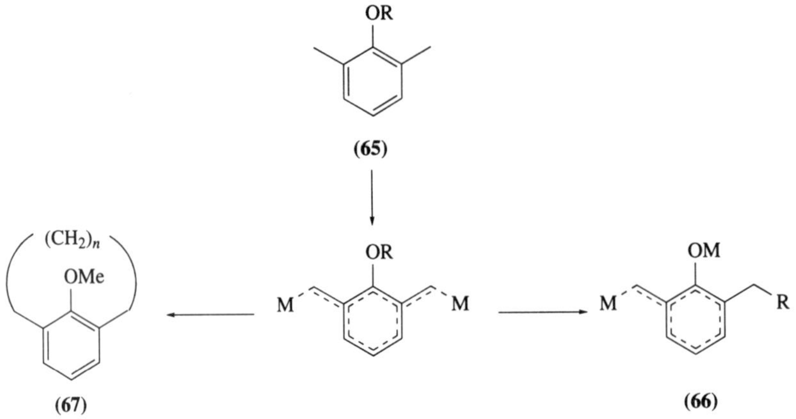

110   *Sodium and Potassium*

Scheme 13

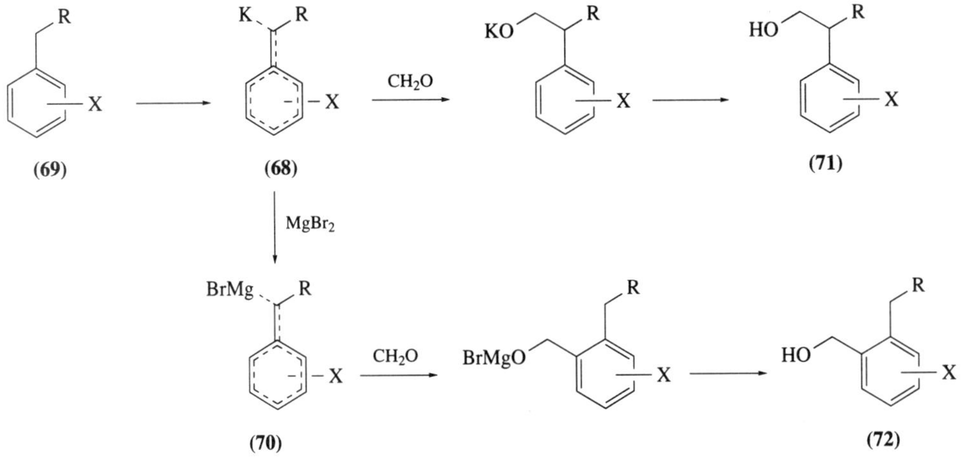

Scheme 14

R = H, alkyl; X = H, OMe, alkyl

2.3.1.7 Allenes and alkynes

Alkynic and allenic substrates have so far received only limited attention as substrates for metallation studies. 2-Alkynes undergo deprotonation with Schlosser's base at the terminal methyl group[123] leading, after reaction with alkyl halides, mainly to the allenic derivative (**73**) together with minor amounts of the alkynyl compound (**74**), and with carbon disulfide to a thiophene derivative (**75**) (Scheme 15). The same product results from a substituted allene through the same sequence.[123]

If the substrate contains an additional conjugated triple bond, as in a 2,4-diyne system,[124] the metallation followed by reaction with carbon disulfide leads to condensed bicyclic thiophene derivatives. α,β-Unsaturated alkynes can be deprotonated with 2 equiv. of Schlosser's base and, after potassium–lithium exchange with lithium bromide and consecutive addition of an electrophile, products (**76**) are obtained, deriving from attack at the more reactive allylic position,[125,126] or (**77**) if only vinylic hydrogen atoms are present (Scheme 16).

2.3.2 Metallation of Resonance-inactive Positions

2.3.2.1 Alkenes and dienes

Alkenes lacking hydrogen atoms at allylic positions are much less acidic than ordinary unsaturated hydrocarbons. The same is true for bridgehead positions in bicyclic structures where orthogonality impedes resonance between a deprotonated centre and an adjacent double bond. Superbases

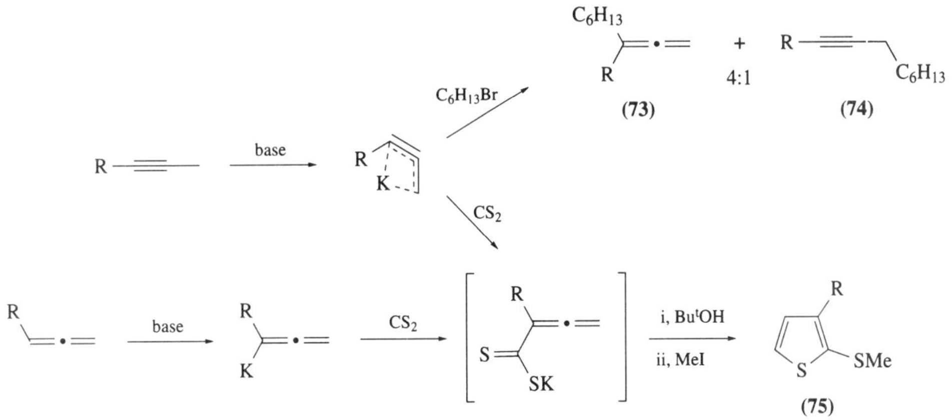

Scheme 15

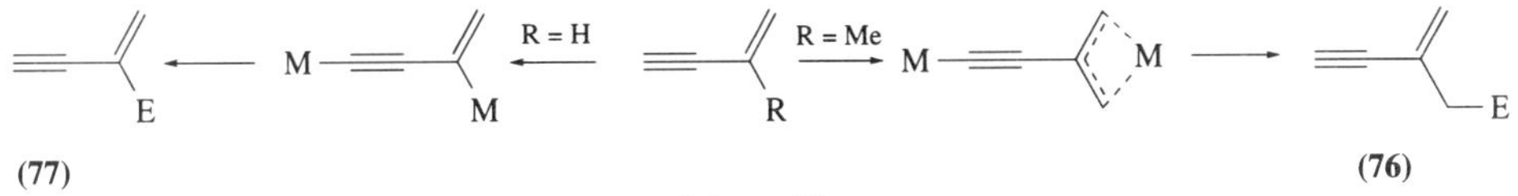

Scheme 16

regioselectively exchange allylic protons in alkenes whenever there is a choice. A few examples of metallation of alkenic C–H bonds are known. The superbasic mixture butyllithium/sodium *t*-butoxide has been found to outperform Schlosser's base in the abstraction of vinylic hydrogen atoms. It has, for instance, been successfully employed in the metallation of bicyclo[2.2.1]hept-2-ene (norbornene (**78**))[127] and bicyclo[2.2.1]hepta-2,5-diene (norbornadiene (**79**)).[127] Previous attempts to accomplish this latter type of reaction with pentylsodium had failed due to decomposition of the organometallic intermediate (**80**) to sodium cyclopentadienide (**81**) and sodium acetylide (Scheme 17).[128]

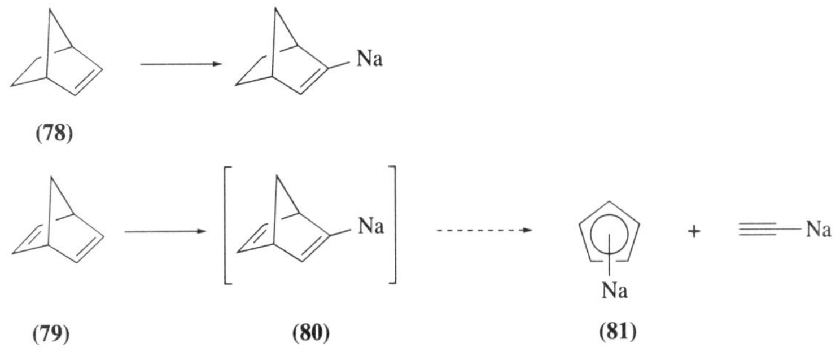

Scheme 17

2-Bornene (**82**) can be metallated with Schlosser's base[129] to give an 87:13 mixture of dimethylphenylsilyl derivates (**83**) and (**84**) in 80% overall yield after reaction with dimethylphenylchlorosilane (Equation (23)). By exchanging the phenyl group for a chlorine atom[130] the method has been extended to the preparation of optically active organochlorosilanes which can be employed in the synthesis of enantiomerically pure compounds.[129]

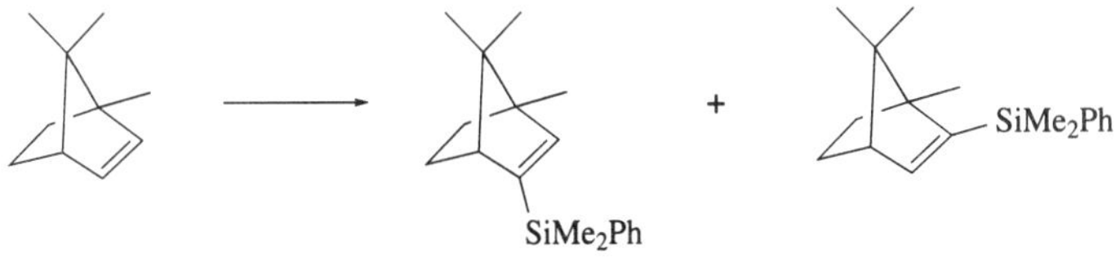

As competition experiments[127] have revealed, norbornadiene is roughly 30 times more reactive than norbornene. An enhancement of kinetic acidity has also been shown in the metallation of bicyclo[3.2.0]hepta-2,6-diene, which is roughly five times more reactive than bicyclo[3.2.0]hept-6-

ene.[127] This behaviour has been attributed to a polarization of π electrons due to the proximity of the homoconjugated double bond.

The selectivity observed with 1,3,5-cycloheptatriene is particularly noteworthy. Consecutive treatment with the butyllithium/sodium *t*-butoxide reagent and trimethylchlorosilane gave 2-trimethylsilyl-1,3,5-cycloheptatriene (**85**) as the unique product, without any trace of the 1-, 3- or 7-substituted isomers (Equation (24)).[127]

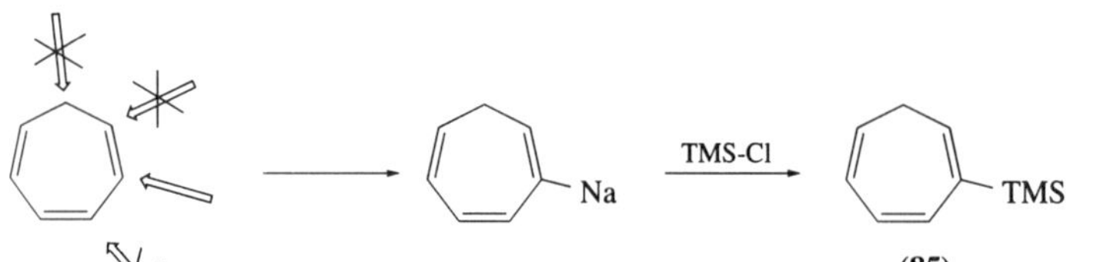

Even more strongly basic mixtures are required to abstract protons from alkenic positions which are not activated by either ring strain or conjugation. Pentylsodium activated by disodium pinacolate or potassium *t*-butoxide was found to selectively convert 3,3-dimethyl-1-butene,[50] camphene,[49,50] bicyclo[2.2.2]oct-2-ene[50] and spiro[4.4]nona-1,3-diene[50] into sodium derivatives (**86**)–(**89**) bearing the metal at an alkenic carbon atom.

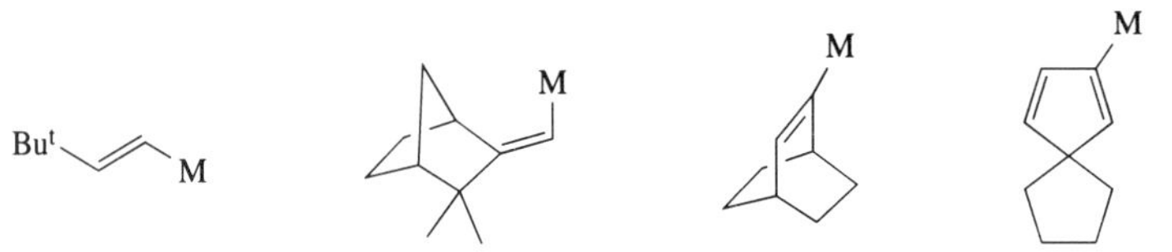

Brandsma *et al.* have studied the simultaneous activation of butyllithium by potassium *t*-butoxide and TMEDA,[131] and employed the resulting superbasic mixture for the metallation of ethene. The hydrocarbon-soluble mixtures butylsodium/TMEDA[43] and butylpotassium/TMEDA,[44] obtained by metal–metal exchange between butyllithium and sodium or potassium *t*-butoxide and solubilization of the resulting butylsodium and butylpotassium in hexane/TMEDA, have also been studied.

2.3.2.2 Arenes

Benzene and many other aromatic hydrocarbons are readily metallated by superbasic reagents. Following the first demonstration of benzene metallation by Schorigin,[132] using ethylsodium, a large number of experiments have been reported in which pentylsodium,[23,133] pentylsodium/metal alkoxides mixtures[23,134] and alkylpotassium reagents[24] have been used. Monometallation of benzene with butyllithium/potassium *t*-butoxide using the substrate as a cosolvent was the first revelation of the enormous reactivity of Schlosser's base.[33] According to a more recent study,[135] clean monometallation can be accomplished with stoichiometric amounts of benzene only if the reaction is carried out in THF solution. Dimetallation appears to be inevitable in hexane suspension and it may even become dominant if the superbasic mixture is used in excess.[136] *Ortho* isomers have only been identified in trace amounts and the *meta:para* ratios are always somewhat smaller than unity.[37,135] *t*-Butylbenzene[135] and (1-methylcyclopropyl)benzene[135] are converted by Schlosser's base into monometallated derivatives (**90**)–(**93**) with *meta:para* ratios of approximately 1:1. Under the same reaction conditions, 1,3-di-*t*-butylbenzene gives the *meta*-substituted compound (**94**). Several previous attempts to carry out the same reaction with alkylsodium[30] or alkylpotassium[22] gave only very poor yields.

Alkynyl substituents enhance the reactivity of arenes.[137] *Ortho*-substituted phenylacetylenes (**95**) are obtained by metallation of phenylacetylene with Schlosser's base and subsequent interception of the dipotassium intermediate (**96**) with electrophiles (Equation (25)).[137,138]

On the other hand, (3,3-dimethyl-1-butenyl)benzene (**97**), (3-hydroxy-3-methyl-1-butynyl)benzene (**98**), (3-hydroxy-3,4,4-trimethyl-1-pentynyl)benzene (**99**) and (1,3-butadyn-1-yl)benzene (**100**) form mixtures of *ortho*, *meta* and *para* isomers in the ratios 9:78:13, 69:25:6, 77:18:5 and 45:44:11, respectively,[137] showing that the *ortho* specificity is unique for an acetylide function directly attached to the aromatic ring.

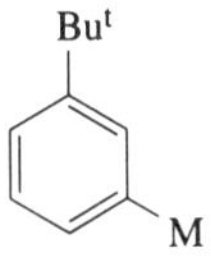

Sodium and Potassium

(90)  (91) 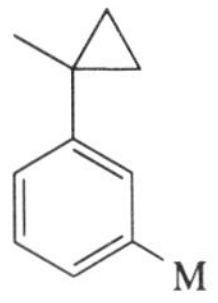 (92) 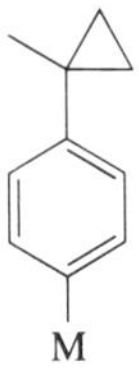 (93) 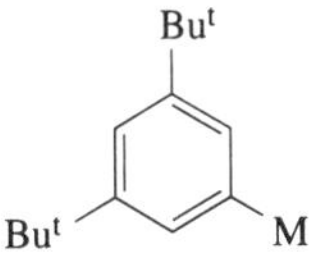 (94)

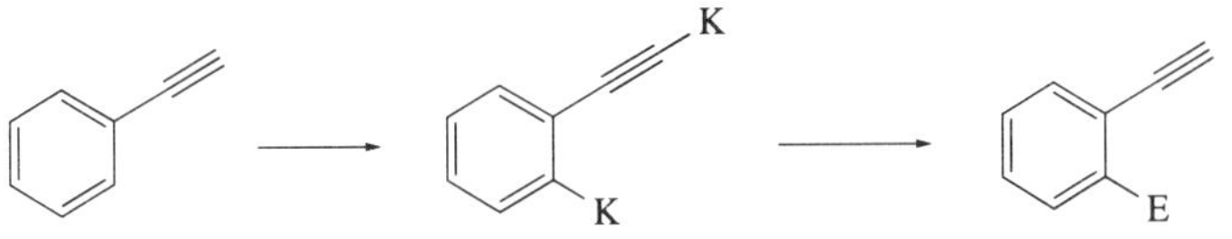

(25)

(96) (95)

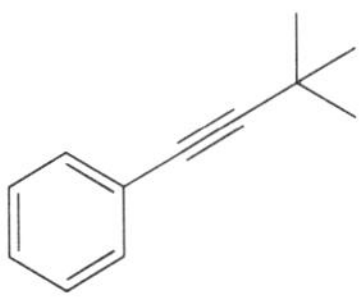

(97) 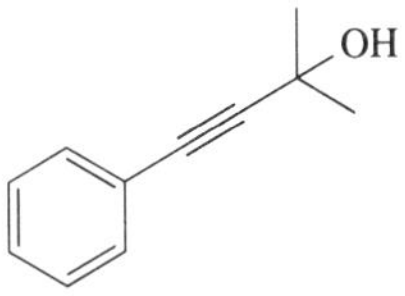 (98) 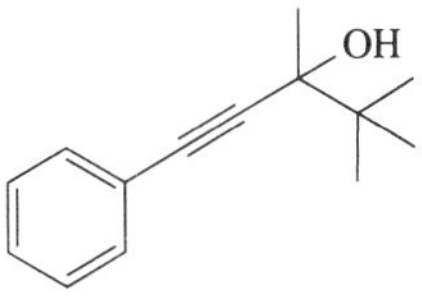 (99) 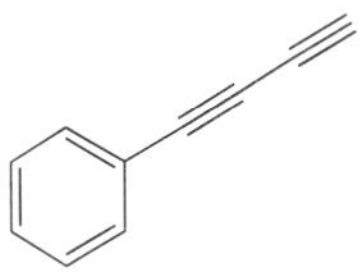 (100)

2.3.2.3 Cyclopropane derivatives

Cyclopropanes are very weak acids. Their metallation being slow and leading to poor yields of products, they constitute a bench mark to probe the effectiveness of a superbasic reagent. Cyclopropane, when used as a cosolvent, afforded not more than 20% of cyclopropanecarboxylic acid (**101**) when consecutively treated with pentylsodium/sodium isopropoxide (6.5 h) and carbon dioxide.[26] The presence of a phenyl group increases the acidity of cyclopropane and thus renders its metallation easier. Phenylcyclopropane undergoes deprotonation[54] on the cyclopropyl ring by Schlosser's base (LICKOR) or trimethylsilylmethylpotassium in THF to give (1-methylcyclopropyl)benzene (**102**) in 23% and 33% yield, respectively (Scheme 18). The use of pentylsodium in pentane generates organometallic species resulting from the deprotonation of the aromatic ring.[54]

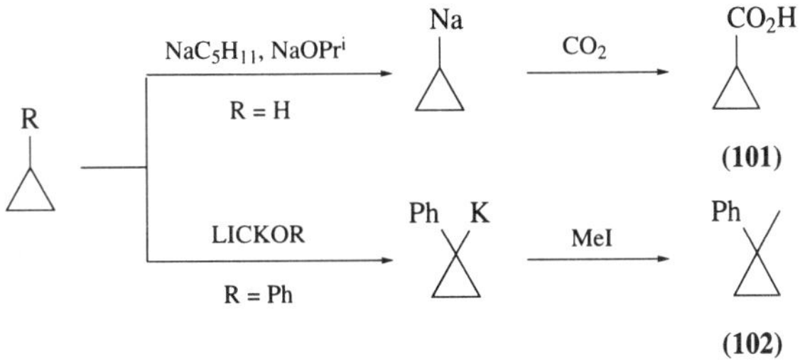

Scheme 18

Polycyclic compounds containing a cyclopropane ring undergo metallation with relative ease.[50] The use of alkylsodium reagents always involves very long reaction times and poor yields are obtained despite the large excess of hydrocarbon substrate often employed. Activation by sodium or potassium alcoholates gives a substantial improvement. Tricyclo[2.2.1.0^{2,6}]heptane (nortricyclane (**103**) (M = H)) gives only a 5% yield of tricyclo[2.2.1.0^{2,6}]heptanoic acid (**103**) (M = CO$_2$H) after treatment with pentylsodium over 500 h and reaction with carbon dioxide,[25] whereas consecutive treatment with pentylsodium/potassium *t*-butoxide (over 24 h) and trimethylchlorosilane allows isolation of 1-trimethylsilyltricyclo[2.2.1.0^{2,6}]heptane (**103**) (M = TMS) in 60% yield.[49,50] Bicyclo[4.1.0]heptane (norcarane (**104**) (M = H)) requires 100 h of reaction at 25 °C with pentylsodium to give the sodium derivative (**104**) (M = Na) which, upon carboxylation, affords 3% of bicyclo[4.1.0]-7-heptanoic acid (**104**) (M = CO$_2$H). The reaction time can be reduced to 50 h and the yield is increased to 30% by adding potassium *t*-butoxide.[50] Spiro[2.2]pentane (spiropentane (**105**) (M = H)) is metallated in 36 h at 25 °C by the pentylsodium/potassium *t*-butoxide reagent in pentane, affording a mixture of mono- and dimetallated products ((**105**) and (**106**), M = K) in a 27% global yield[139] (mono- and dimetallation ratio 50:50, after carboxylation). The use of butylpotassium alone affords only 2% of monometallated product.

(**103**) 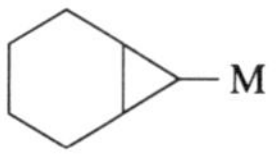   (**104**)    (**105**) 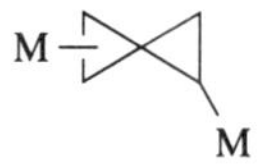   (**106**)

Norcarane and spiropentane are probably the least acidic hydrocarbons to have been metallated in a stoichiometric reaction.

2.4 METALLATION OF HETEROSUBSTITUTED ALKENES

2.4.1 Heterosubstituted Allylmetallic Compounds

Two important questions arise when functional groups are attached to the allyl moiety. Are they compatible with superbasic metallation and how do they affect the regio- and stereochemical outcome of the reaction? The first problem can be often solved by an appropriate choice of the reaction conditions as will be shown in several of the following examples. The regiochemistry of electrophilic attack of allylmetal species is a more complex issue (Scheme 19).[140-3]

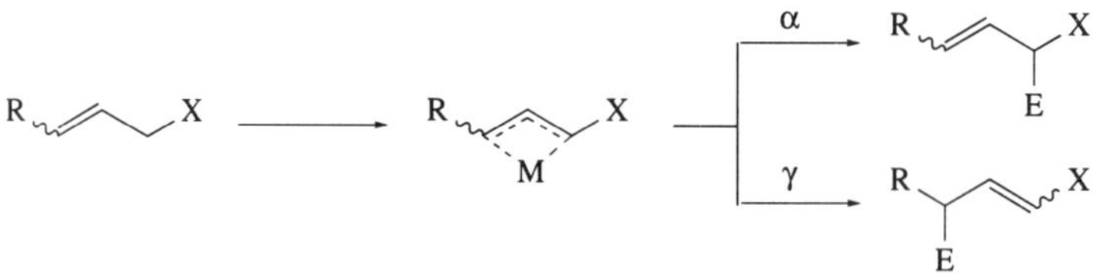

Scheme 19

The α:γ selectivity is related to many factors such as the nature of the heteroatom, the organometallic base, the substituent bonded to the heteroatom, the electrophile, the solvent, and the reaction conditions. Hence it is quite difficult to rationalize the behaviour of heterosubstituted allylmetallic derivatives in their reactions with electrophiles. It has been proposed as a rule of thumb[140] that, for a given counterion, say lithium, anion-destabilizing substituents (e.g., OR, NR$_2$) favour the attack of alkyl halides and protons at the γ-position, the site of higher electron density, while carbonyl compounds react preferentially at the α-position. The complementary regioselectivity is encountered when anion-stabilizing substituents (e.g., SR, BR$_2$) are bonded to the allyl moiety.

Butyllithium, *s*-butyllithium and *t*-butyllithium are powerful enough to generate most of the heterosubstituted allylmetals. Sometimes activators (e.g., TMEDA) or a cosolvent (e.g., HMPA) are required. Nevertheless, the use of superbasic reagents may still be advisable since a change of the counterion may considerably influence the regio- and stereoselectivity of the reaction with electrophiles.

2.4.2 Nitrogen-containing Alkenes

The relatively low acidity of allylamines and enamines requires the use of activated bases in order to successfully produce the allylmetallic species. This subject has been extensively studied by Ahlbrecht and co-workers, who reported successful metallations of unsubstituted,[144] alkyl-[145] or silyl-substituted[145] allylamines (**107**) or enamines (**108**) with the superbasic mixture butyllithium/potassium *t*-butoxide or *t*-butyllithium/potassium *t*-butoxide, and the deprotonation of carbomethoxy-substituted[146] substrates with lithium diisopropylamide/potassium *t*-butoxide. The allylpotassium species (**109**) thus obtained react with electrophilic reagents either exclusively or predominantly at the γ position (Scheme 20). Hence, alkyl halides, deuterium oxide, trimethylchlorosilane and epoxides give the enamine derivatives (**110**), which can be hydrolysed by dilute hydrochloric acid to afford three-carbon homologated carbonyl compounds (**111**). Aldehydes and ketones react less selectively, giving an approximate 1:1 mixture of α and γ regioisomers, the latter (**112**) being easily cyclized upon heating to the dihydrofuran derivative (**113**). The substituents on nitrogen also seem to influence the outcome of the reaction. When *N,N*-bistrimethylsilylallylamine (**114**), for example, is submitted to the action of the superbasic reagent lithium diisopropylamide/potassium *t*-butoxide followed by treatment with a carbonyl compound, the 2-aza-1,3-diene species (**115**) carrying a trimethylsilyl group on the allylic position is formed (Scheme 21).[147] The formation of (**115**) has been attributed to a novel 1,4-migration of a trimethylsilyl group from nitrogen to carbon occurring after the metallation step, followed by attack of the carbonyl compound at nitrogen and elimination of trimethylsilyl salt. 2-Aza-1,3-dienes are formed with complete regio- and stereocontrol: only the products with *cis* configuration at the C=C bond and *trans* configuration at the C=N bond are obtained. Interestingly, on changing the metal atom a significant variation of the stereochemical outcome occurs. Lithium and, to a greater extent, magnesium show a lower degree of stereoselection with respect to the C=C bond. The former gives a mixture of (*Z*) and (*E*) isomers in an 80:20 ratio, which is reversed (33:67) with magnesium.

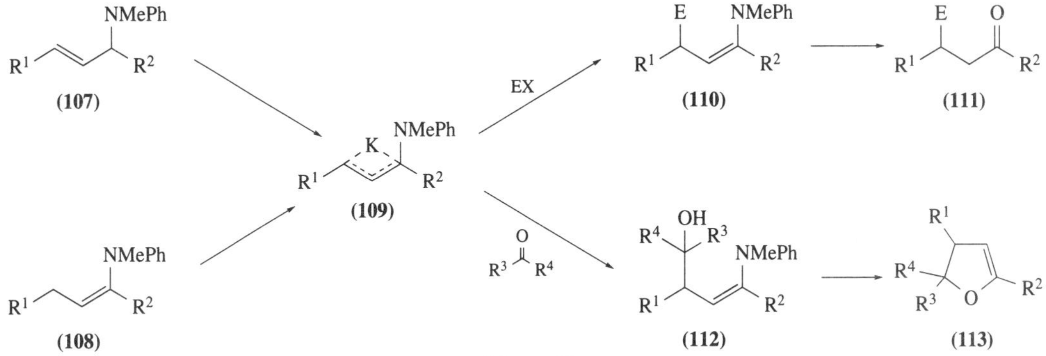

Scheme 20

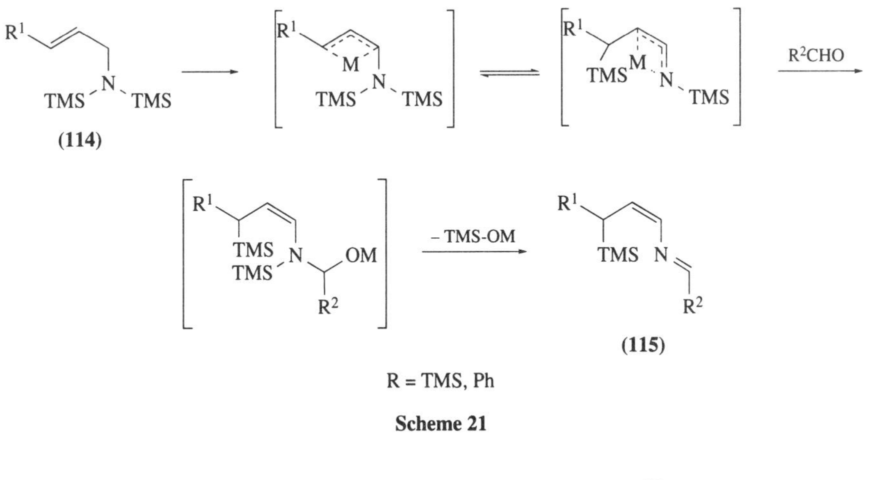

Scheme 21

R = TMS, Ph

Secondary allyl amines can also be deprotonated with lithium bases,[148] provided 2 equiv. of the base are used. When remote allylic protons are present, the initially formed negative charge on nitrogen disfavours deprotonation at the position close to it and directs metallation at the remote site. This has been found to be the case in the lithiation of methallyl amides,[149] and has been recently applied to the elaboration of allyl amines derived from α-aminoacids through metallation with superbasic reagents.[150] Amines (**116**), easily obtained from *t*-butoxycarbonyl-protected (*t*-BOC-protected) amino acids via reduction[151] to the aminoaldehydes and *cis*-selective Wittig alkenation,[152] have been submitted to metallation with Schlosser's base followed by reaction with trimethylchlorosilane,[150] alkyl iodides,[153] deuterium oxide[153] and allyl bromide.[153] Attack of the electrophile always occurs at the γ position to give the (Z)-configured allyl amines (**117**) (Equation (26)).

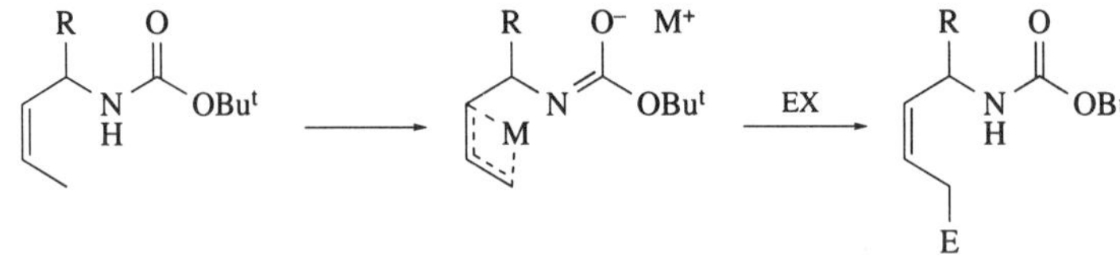

$$\text{(116)} \longrightarrow \text{[M intermediate]} \xrightarrow{EX} \text{(117)} \quad (26)$$

R = Me, Pr^i, Bu^i, Bu^s; EX = TMS-Cl, BuI, D$_2$O, H$_2$C=CHCH$_2$Br

2.4.3 Oxygen- and Sulfur-containing Alkenes

Enethers and enethioethers without allylic hydrogen atoms are reported to be metallated with Schlosser's base in the vinylic α-position,[3,154,155] activated by the heteroatom (Figure 3). When R is a phenyl group, the thioether is still deprotonated at the double bond (**118**), while the oxygen analogue undergoes initial ring metallation followed by abstraction of the vinylic proton (**119**).[156,157]

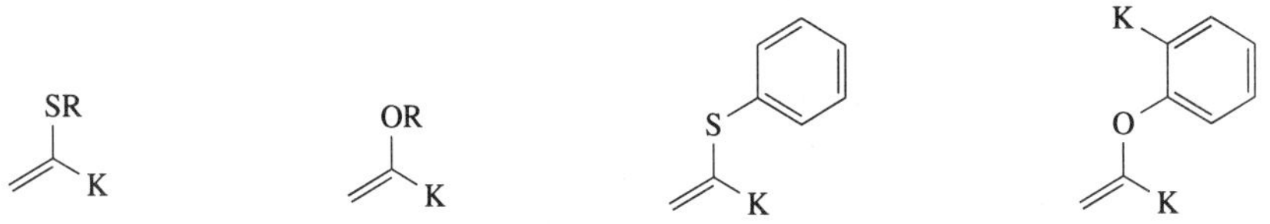

Figure 3 Behaviour of enethers and enethioethers towards metallation with Schlosser's base.

When an alkyl group is present on the double bond, proton abstraction may occur at both the allylic and the vinylic positions, giving organometallic intermediates (**120**) and (**121**). This leads, after reaction with electrophiles, to the formation of three regioisomeric products derived by attack at the two termini of the allyl moiety and at the vinylic position. It has been found[154] that enethioethers (Y = S) undergo metallation at both sites, the proportions depending on the substitution at the terminal position of the allyl moiety. Thus, when R^2 = R^3 = H a mixture of (**120**) and (**121**) in a 29:71 ratio is obtained while, when the steric hindrance increases (R^2 = Me, R^3 = H; R^2 = Me, R^3 = Me), only the vinylmetallic compound (**121**) is formed. 1,3-Dienylethers and thioethers can also be regiospecifically deprotonated at the α-position[158] to the dienylmetallic species.

(**120**) Y = O, S   (**121**) Y = O, S

Allyl ethers and allyl sulfides are acidic enough[159] to be deprotonated by simple organolithium reagents. However, the need for good regio- and stereocontrol leads to the choice of superbasic reagents which, through the formation of allylpotassium species, may offer good alternative solutions for a fine tuning of regio- and stereoselection. It has been found,[56] for instance, that allyl phenyl sulfide (**122**) (Y = S), when treated sequentially with Schlosser's base and methyl iodide, gives a mixture of α and γ regioisomers (**123**) and (**124**) (Y = S), which is oriented towards the α isomer when THF is used as solvent and, to a lesser extent, towards the γ isomer when petroleum ether is used (Scheme 22). At the same time, the use of the superbasic reagents allows control of the stereochemistry of the γ product: the (Z)-enethioether (Z)-(**124**) (Y = S) is obtained as the major isomer with (Z):(E) ratios ranging from 1.8 to 3.7 when the reaction is performed in hexane. Allyl phenyl ether (**122**) (Y = O) is also easily metallated by Schlosser's base[56] and affords a mixture of α and γ regioisomers (**123**) and (**124**) (Y = O) in a 25:75 mixture (63% global yield) after reaction with methyl iodide.

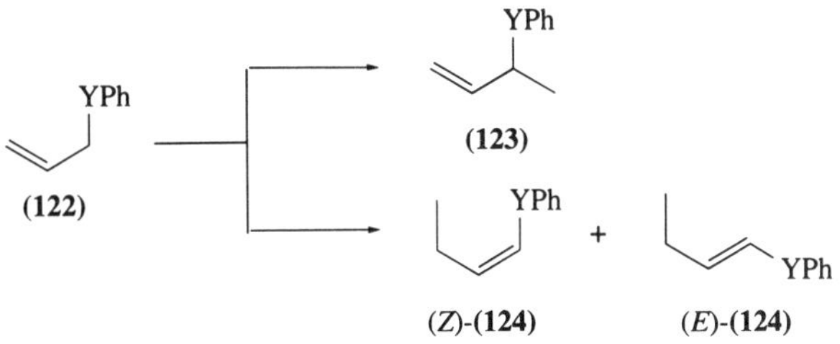

Scheme 22

It must be pointed out that 1-oxyallyl anions (**125**) are prone to undergo Wittig rearrangement,[160] giving α-vinyl alcoholates (**126**) (by 1,2 migration) and enolates (**127**) (by 1,4 migration) (Equation (27)). In order to avoid the rearrangement, the mixture must be kept at very low temperature (below −50 °C) before addition of the electrophile. On the other hand, if migration is desired, it suffices to warm it to −20 °C. The Wittig rearrangement has been mainly studied for organolithium species but it has been found[161] that the allylpotassium species, generated by treatment of the allyl ether with a superbasic reagent, predominantly undergo 1,4 migration with good selectivity.

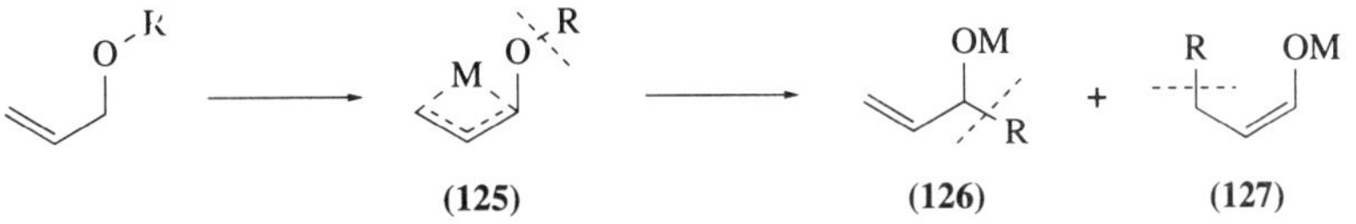

The metallation of allyl ethers and sulfides with superbases has been used for the synthesis of 4'-oxa-α-santalol (**128**) (Y = O)[162] and 4'-thia-α-santalol (**128**) (Y = S),[163] two isosteres[164] of the naturally occurring α-santalol (Equation (28)). The metallated allylmetallic species (**129**) were submitted to reaction with fluorodimethoxyboron and subsequent oxidation with alkaline hydrogen peroxide.

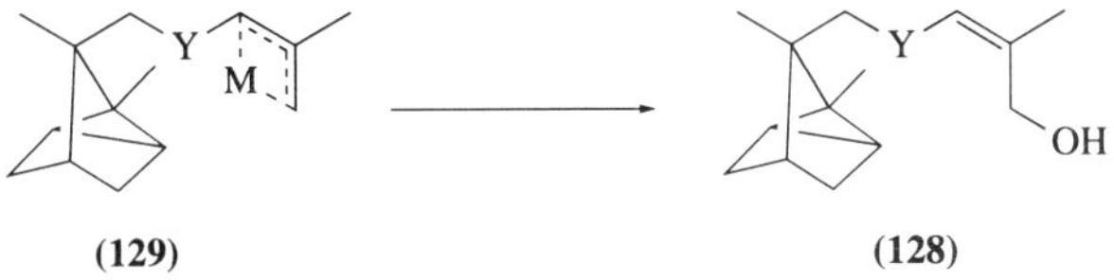

2.4.4 Silicon-containing Alkenes

Allyltrimethylsilane undergoes easy α-metallation by Schlosser's base,[165] trimethylsilylmethylpotassium[40] or butyllithium/TMEDA.[166] Alkylation of the organometallic species (**130**) occurs preferentially at the γ-position (**132**), the α:γ ratio (**131**):(**132**) varying from 40:60 with butyllithium/TMEDA to 15:85 when Schlosser's base is used; with perfect stereocontrol, the (E)-vinylsilane (**132**) is formed as the exclusive product (Scheme 23).

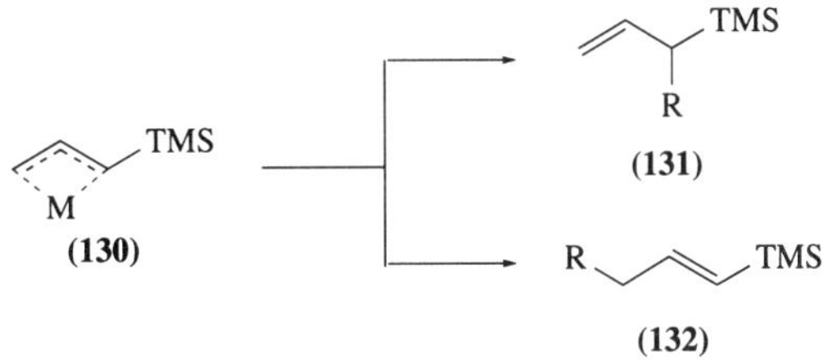

Scheme 23

The regioselective outcome may change significantly if the allylsilane contains substituents on the allylic moiety or on the silicon atom which are able to exert either a steric or a chelating effect. Thus, 2-alkenyltrimethylsilane (**133**) undergoes deprotonation[167] at the methylene group close to silicon and the α:γ ratio of alkylated products (**134**) and (**135**) varies depending on the size of both the alkyl substituents on the double bond and the alkyl residue of the alkylating agent (Scheme 24). A methyl group in position 3 is enough to direct attack of methyl iodide at the α-position, but if butyl iodide is employed a 50:50 mixture of the two regioisomers is formed. When a bulkier substituent is present on the double bond the α:γ ratio is even more pronounced towards the α species. The stereocontrol is also very good: vinylsilanes (**135**) are always obtained in the (*E*) configuration due to steric hindrance of the trimethylsilyl group, whereas the allylsilanes (**134**) can be prepared in the (*Z*) or (*E*) configuration depending on the reaction conditions. Allowance for equilibration at −50 °C in THF produces the (*Z*) isomer regardless of the configuration of the starting 2-alkenyltrimethylsilane, while in order to obtain the (*E*) isomer the reaction has to be performed in hexane at 25 °C on the starting silane having the same configuration.

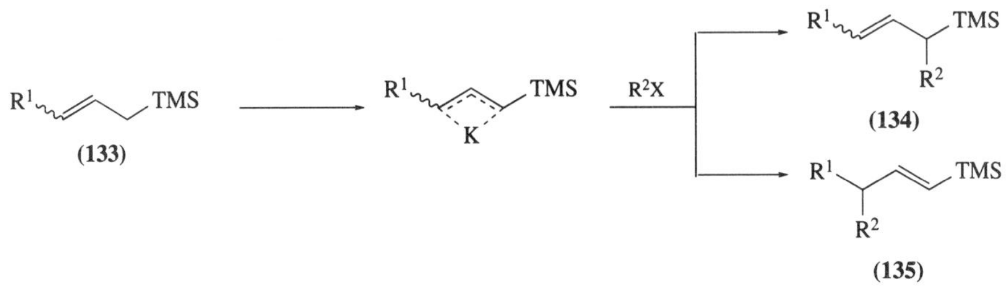

R^1	R^2	*(134)*	*(135)*
Me	Me	98	2
Me	Bu	50	50
Pr	Me	98	2
Pr	Bu	67	33
Et(Me)CH	Me	98	2
Et(Me)CH	Bu	80	20

Scheme 24

A methyl group in the central position of the allylic moiety does not appreciably affect the regioselectivity results,[168] while a wide range of α:γ ratios has been found by using alkoxy or amino substituents on silicon.[169]

If the trimethylsilyl group is not directly bonded to the allyl moiety but is located one carbon atom distant as in 4-trimethylsilyl-1-butene (**136**), metallation leads either predominantly or exclusively to an *endo* allylmetallic species (**137**) which, after trapping with electrophiles, is converted into *cis*-substituted allylsilanes (**138**) (Equation (29)).[170]

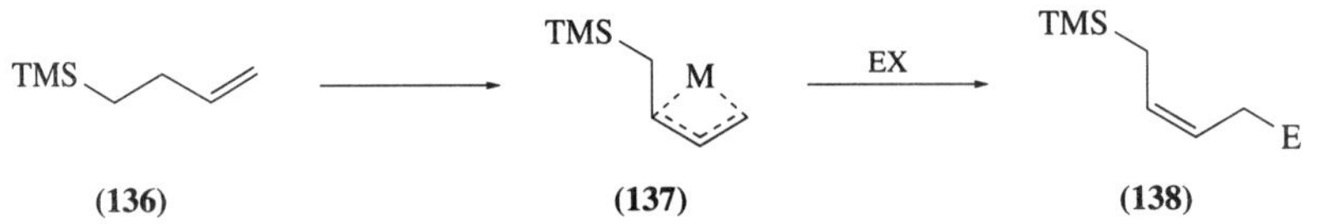

EX = TMS-Cl, BuI, BzBr, PhCHO, PhSSPh

Superbase metallation of arylic and vinylic silanes has also been studied. Phenyltrimethylsilane (**139**) is metallated by the butyllithium/sodium *t*-butoxide reagent to a mixture of regioisomeric sodium derivatives, the alkylmetal derivative being the major one (14:14:72).[171] In order to achieve exclusive deprotonation of the phenyl ring, triisopropylsilylbenzene must be used as the substrate.[135] Similar results have also been found in the metallation of 2-trimethylsilylnorbornadiene (**140**)[172] and 2-trimethylsilylcycloheptatriene (**141**).[171] Both substrates undergo deprotonation mainly at the symmetrical position but variable amounts of products derived from the hydrogen–metal exchange on the methyl bonded to silicon are found. In a competitive experiment, phenyltrimethylsilane has been found to react 15 times faster than tetramethylsilane, thus showing that some enhancement of metallation rates occurs when the trimethylsilyl group is attached to alkenic or aromatic carbon atoms.

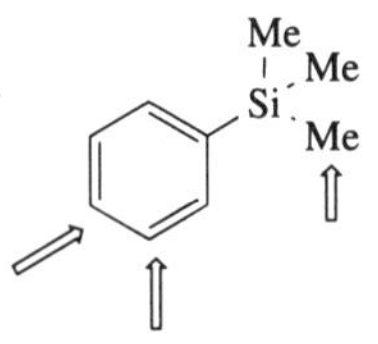

2.4.5 Heterocyclic Compounds

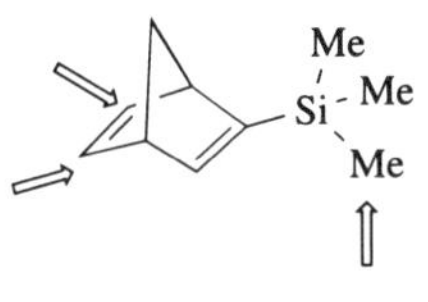

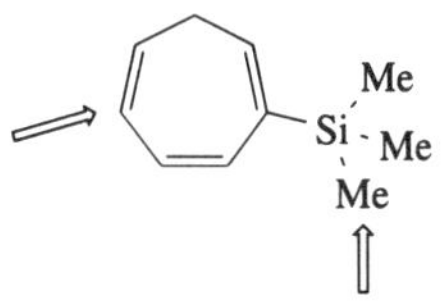

Furan and thiophene are more acidic than benzene and hence they can be easily α-metallated by butyllithium in ether or THF[173,174] and even dimetallated in the 2,5 positions with TMEDA-activated butyllithium in refluxing hexane[173] or the butyllithium/potassium *t*-butoxide/TMEDA reagent in pentane at low temperature.[175] 1-Methylpyrrole is poorly metallated compared with the oxygen and sulfur analogues; monometallation occurs using Schlosser's base in THF/hexane at −80 °C.[3] *N*-Phenylpyrrole is dimetallated (**142**)[176] in the *ortho* and α-positions when treated with butyllithium/TMEDA in ether (at 0 °C) or hexane (at 65 °C). The α-monometallated derivative (**143**) can be obtained by using Schlosser's base or by prolonging reaction times, the dimetallated species undergoing a slow transmetallation with unconsumed substrate (Figure 4).

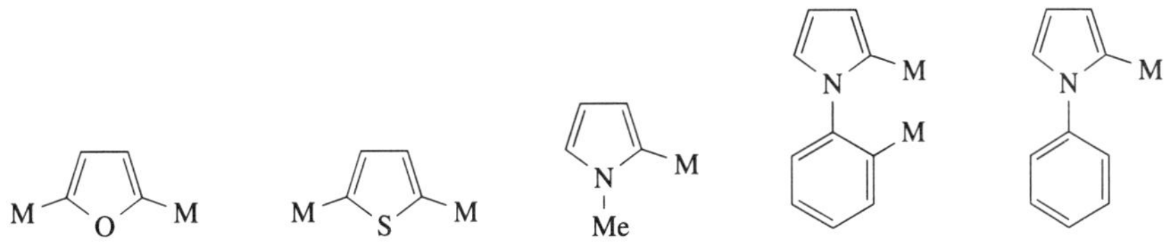

Figure 4 Metallation of some five-membered heterocyclic compounds.

Pyridine is even more difficult to metallate. Use of lithium diisopropylamide (LIDA) in HMPA leads only to 2,2'-bipyridine,[177] while monometallation is achieved using Schlosser's base in THF at −100 °C,[178] the regioselection depending on the reaction conditions. Polar conditions favour the 4-metallated compound (**144**) while in apolar media the 2-substituted product (**145**) is formed preferentially. 4*H*-Pyran, 4-methyl-4*H*-pyran and 1,4-dimethyldihydropyridine are regiospecifically metallated at the position next to the heteroatom, (**146**), (**147**) and (**148**), by the superbasic reagent or by trimethylsilylmethylpotassium.[179] The metallation of cyclic vinyl ethers with Schlosser's base has been shown to be a very useful method to access *C*-glycosides (**149**).[180-2]

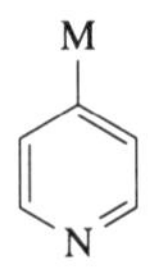

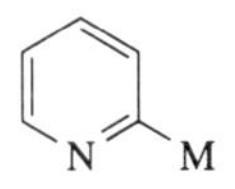

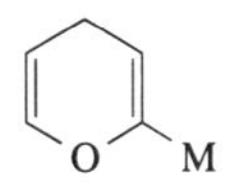

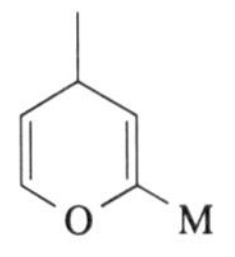

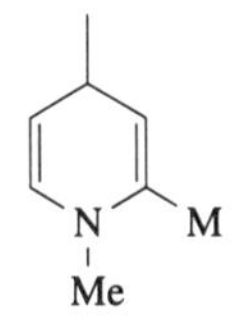

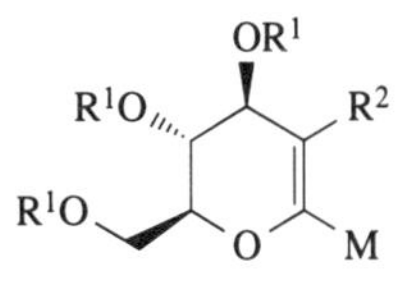

1-Methyl-1,4-dihydropyridine (**150**) gives different products depending on the metallating agent.[179] Schlosser's base (LICKOR) attacks again the 'geometrically acidified'[7] 2 position, while trimethylsilylmethylpotassium (KQ) prefers the 'electronically acidified'[7] 4 position, thus giving an organometallic 8π system. A comparison between 1,4-dimethyl-1,2,3,6-tetrahydropyridine (**151**) and 4-methyl-3,6-dihydro-2*H*-pyran (**152**) reveals an interesting difference in behaviour towards metallation. The former, when treated with trimethylsilylmethylpotassium, undergoes hydrogen–potassium exchange at the allylic position 3 while the pyran is metallated by butylpotassium at the oxygen-adjacent allylic position.[65] The structurally related 1-methylcyclohexene (**153**) undergoes metallation only at the methyl group (Figure 5). The different behaviour of the heterocyclic analogues may be explained in terms of anchimeric assistance of the heteroatom at the α or β position depending on the most advantageous transition state.[65]

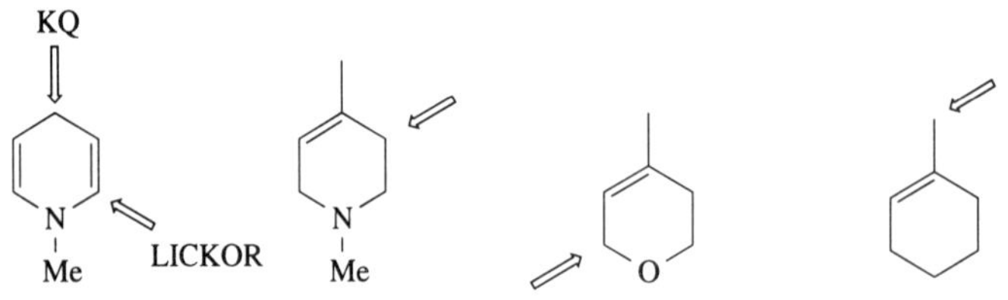

Figure 5 Sites of LICKOR attack.

When the dihydropyran (**152**) is submitted to the action of LIDA in the presence of potassium *t*-butoxide (10%), another pathway is followed, leading, through ring opening, to 2,4-dienolate (**154**) (Equation (30)).[53] The reaction has been applied to several dihydropyrans giving the corresponding alkadienols in nearly quantitative yield and unchanged initial (Z) configuration.[183]

$$(30)$$

Even saturated heterocyclic compounds can be deprotonated by superbasic reagents. As already mentioned, Schlosser's base survives in THF for hours if kept at −50 °C, but butylpotassium quickly generates 2-tetrahydrofurylpotassium (**155**) (M = K) when dissolved in THF, even at −75 °C.[51] Tetrahydropyran is also attacked by butylpotassium at the α position (**156**), giving 68% of the trimethylsilyl derivative after trapping with trimethylchlorosilane.[51] Cyclic *N*-methylamines have been deprotonated by *s*-butyllithium/potassium *t*-butoxide to give aminomethyl potassium species (**157**) and (**158**).[184] Substituted dithianes have been successfully deprotonated by butyllithium/sodium *t*-butoxide[185] to the sodium species (**159**). Use of the superbasic mixture has been found to be much more effective than TMEDA-activated butyllithium and gives high yields of functionalized products.

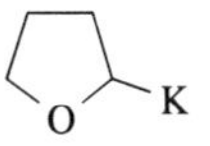

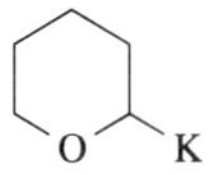

The LIDA/potassium *t*-butoxide base has been successfully employed for the isomerization of oxiranes[186] to allyl alcohols, revealing some advantages concerning the regio- and stereoselective outcome compared with previously reported experiments with nonactivated lithium amides,[187] in which the formation of several by-products was observed. The superbasic reagent promotes clean conversion to allyl alcohols with good stereocontrol, which has been rationalized[186] by assuming a *syn*-periplanar mechanism (Equation (31)).

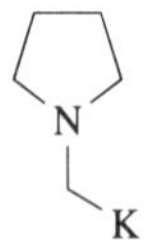

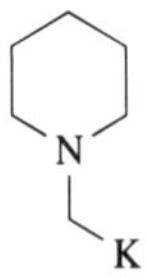

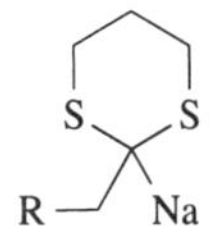

The same isomerization reaction has also been performed on alkoxy-substituted oxiranes[188,189] with perfect regiocontrol, only the methylenic protons adjacent to the oxygen being abstracted. Stereocontrol has been shown to be due to alkyl substitution on the oxirane ring and to the starting configuration. Thus, *cis*-oxiranes always give better selectivity than the *trans* isomers, and 2-alkyl substituted substrates always afford the (*E*)-enether as the unique product.

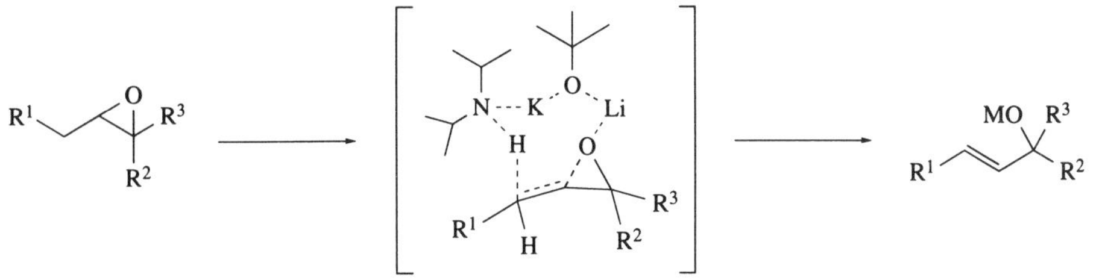

2.4.6 Heterosubstituted Arenes

Since the simultaneous discovery by Gilman[190] and Wittig[191] of anisole *ortho* lithiation, '*ortho*-directed metallations' have been systematically explored, opening new pathways to a variety of natural and synthetic products.[142,192] The activation provided by heterosubstituents is usually strong enough to allow deprotonation of the aryl moiety with organolithium reagents, but the use of superbases often results in different regiochemical behaviour. In particular, the appropriate choice of metallating agent may allow the optional exploitation of different site selectivities to be achieved. We shall restrict our summary to some typical examples of heterosubstituted arene metallation by superbases. Fluorobenzene can be *ortho*-lithiated by butyllithium in THF[193] (**160**) (M = Li) to give the carboxylation product in 60% yield. The same organometallic intermediate (**160**) (M = K) is quantitatively formed if Schlosser's base is employed.[194,195] Butyllithium/sodium *t*-butoxide is also able to introduce a metal atom into the *ortho* position of fluorobenzene, but the resulting intermediate (**160**) (M = Na) decomposes to 1,2-dehydrobenzene (benzyne) under the reaction conditions, producing triphenylene (**161**) in 66% yield (Equation (32)).[196]

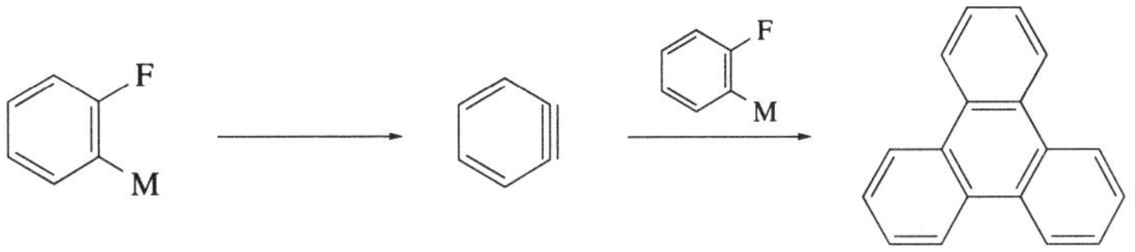

Both 1,2- and 1,4-difluorobenzene react with Schlosser's base to give, after carboxylation, 83% of 2,3- and 65% of 2,5-difluorobenzoic acid, (**162**) and (**163**), respectively (Scheme 25).[194]

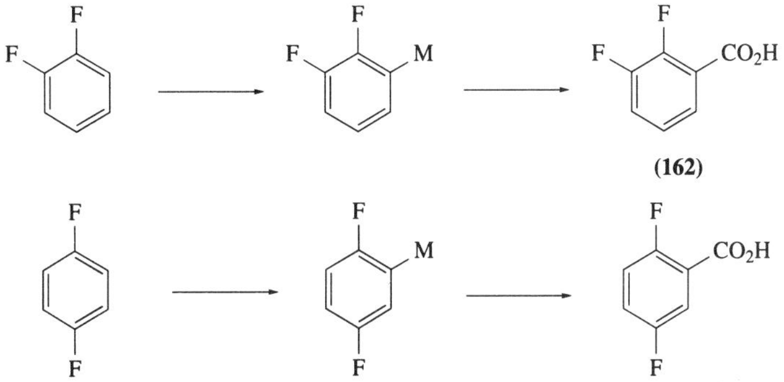

Scheme 25

Fluorotoluenes can be deprotonated either at the benzylic or the aromatic position adjacent to the halogen, these two positions being the most thermodynamically and kinetically acidic. The superbase invariably attacks the aromatic positions[197] giving, after carboxylation, 2-fluoro-3-, 2-fluoro-4- and

2-fluoro-5-methylbenzoic acid, (164), (165) and (166) (M = CO$_2$H), in 60%, 65% and 60% yields, respectively (Figure 6), thus considerably improving previous results with butyllithium.[193]

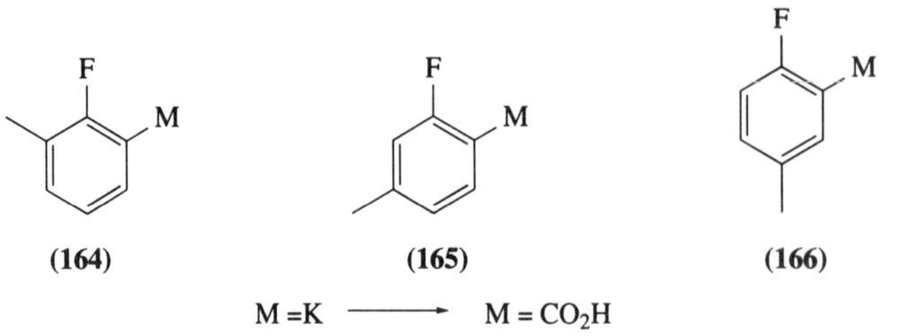

(164)   (165)   (166)

M = K ⟶ M = CO$_2$H

Figure 6 Deprotonation of metallated fluorotoluenes and subsequent carboxylation to yield fluorobenzoic acids.

The benzylic position can instead be selectively deprotonated by using LIDA/potassium t-butoxide under similar reaction conditions.[197] The isomeric 2-, 3- and 4-fluorophenylacetic acids (167) were obtained (Equation (33)) after reaction with carbon dioxide, in 37%, 62% and 7% yields, the *ortho* and *meta* isomers isomerically pure and the *para* isomer as a contaminant to the main product, 2-fluoro-5-methylbenzoic acid (166) (M = CO$_2$H, 53%) resulting from the metallation of an aromatic position.[197] This significantly different behaviour has been rationalized by considering that the LIDA/potassium t-butoxide reagent is much less basic than Schlosser's base and hence more discriminating between the two possible deprotonation pathways leading ultimately to the thermodynamically favoured organometallic species.[197]

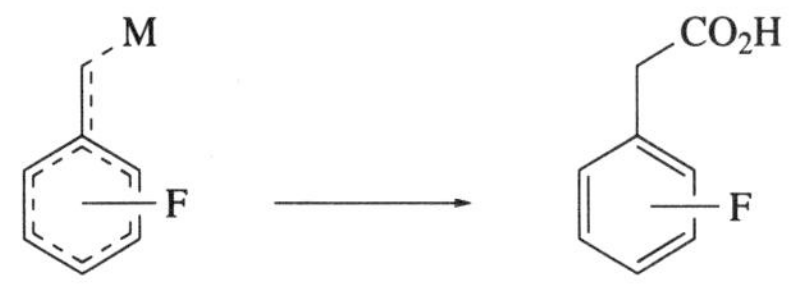

(167)   (33)

The trifluoromethyl group is also an *ortho*-directing group. Trifluoromethylbenzene is deprotonated by butyllithium to give a 71:28:1 mixture of *ortho*-, *meta*- and *para*-lithiated species.[198] In contrast, Schlosser's base gives the *ortho*-metallated intermediate (168) in quantitative yield and isomerically pure.[194] Metallation of the three isomers of fluoro(trifluoromethyl)benzene has clearly shown[194] that a fluorine atom directly attached to the aromatic ring outperforms three fluorine atoms, one carbon atom further away, in directing deprotonation. The three intermediates (169), (170) and (171) are obtained after metallation with Schlosser's base, and subsequent carboxylation gives the 2-fluoro-3-, 2-fluoro-6- and 2-fluoro-5-trifluoromethylbenzoic acids in 55%, 46% and 55% yields, respectively (Figure 7).[194]

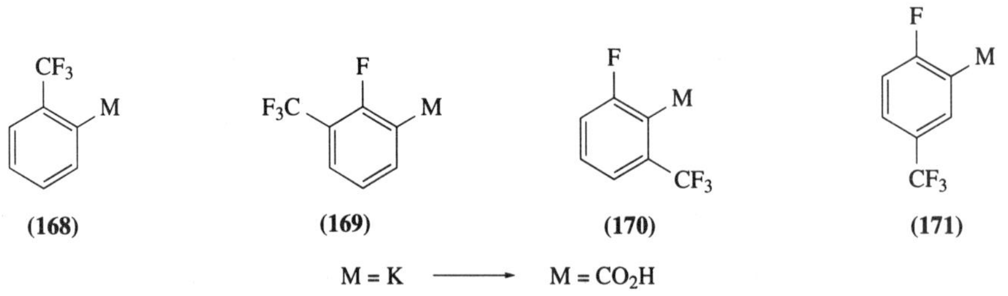

(168)   (169)   (170)   (171)

M = K ⟶ M = CO$_2$H

Figure 7 Metallated intermediates arising from the *ortho*-directing nature of the trifluoromethyl group.

When two trifluoromethyl groups are present on a phenyl ring, an effect due to steric hindrance may be observed. 1,3-Bis(trifluoromethyl)benzene is attacked only partially by butyllithium at the 2 position (172) (M = Li) and mainly at the 4 position (173) (M = Li) flanked by only one trifluoromethyl group,[199,200] while TMEDA-activated butyllithium produces only the 2-metallated intermediate (172) (M = Li) in low yield (30%).[201] Metallation with Schlosser's base affords a 78% yield of 2,6-bis(trifluoromethyl)benzoic acid (174) (M = CO$_2$H) after reaction with carbon dioxide.[194]

The metallation of fluoroanisoles is a very clear example of optional regioselectivity due to the organometallic reagent.[202] Both *o*- and *p*-fluoroanisole, (175) and (176), undergo deprotonation by butyllithium in ethereal solvent at the position next to the methoxy group, both in 50% yield after carboxylation.[202] The use of Schlosser's base (LICKOR) or butyllithium activated by N,N,N',N'',N''-pentamethyldiethylenetriamine (PMDTA) involves a reversal of regioselectivity and gives the organometallic intermediates with the metal close to the fluorine atom (Figure 8). After reaction with carbon dioxide they are converted into 2-fluoro-3-methoxybenzoic acid and the 2-fluoro-5-methoxybenzoic acid in 87% and 85% yields, respectively, using butyllithium/PMDTA, and 77% and 51% yields, respectively, using the superbasic reagent.[202]

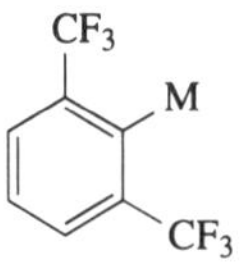

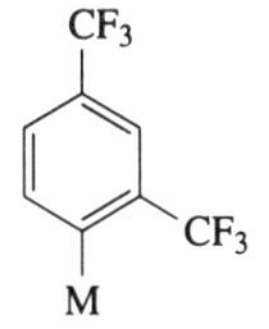

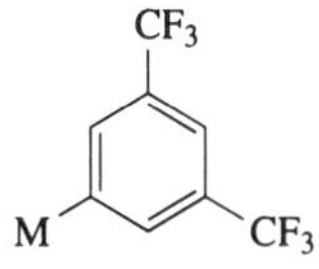

Figure 8 Metallation of fluoroanisoles showing regioselectivity due to the organometallic reagent.

A similar optional site selectivity based on the reagent has been found in the metallation of fluorine- and trifluoromethyl-substituted anilines.[203] The N-BOC-protected substrates have been submitted to metallation with t-butyllithium in THF, which showed that the BOC–amido moiety is a stronger directing group than the fluorine and trifluoromethyl substituents. However, when the superbasic mixture t-butyllithium/potassium t-butoxide is used in the metallation of N-BOC-4-fluoroaniline (**177**), the metal exchanges a hydrogen atom *ortho* to the fluorine.[203] 2-*m*-Anisyl-*N*-pivaloylethylamine (**178**) offers three possible deprotonation pathways given the appropriate reagents (Figure 9). Thus, butyllithium in ether and t-butyllithium in THF attack the aromatic position between the two substituents and the benzylic position, respectively. With Schlosser's base (LICKOR) in THF at −100 °C, on the other hand, hydrogen–metal exchange occurs at the aromatic position which is adjacent to the methoxy group but distant from the alkyl side chain (58% of the acid after carboxylation).[204]

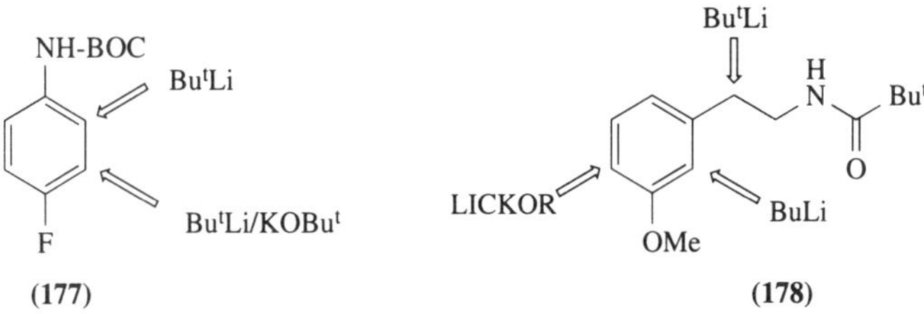

Figure 9 Site selectivity of metallation reagents.

The use of Schlosser's base is necessary when the ring functionalization of 2,3- and 2,5-difluoroaniline is desired.[206] Thus, the N-BOC derivatives have been deprotonated at the position near to one fluorine atom giving, after carboxylation and esterification, methyl 4-(t-butoxycarbonylamino)-2,3-difluorobenzoate (**179**) and -2,5-difluorobenzoate (**180**) in 42% and 11% yields, respectively, the latter being contaminated by 3% of the isomeric methyl 2-(t-butoxycarbonylamino)-3,6-difluorobenzoate (**181**).

All the examples of site selectivity reported require explanation. A schematic rationalization has been proposed based on the character of the heterosubstituent, which can be electron-withdrawing because of the inductive polarization of the σ-bond and electron-donating because of the lone pair on the heteroatom. Thus, if a substituent such as an alkoxy or an amino group is capable of complexing the organometallic reagent, the favoured metallation site is the *ortho* position and hydrogen–metal exchange occurs under quasiintramolecular conditions. If the organometallic reagent is used in conjunction with activating compounds such as potassium t-butoxide, TMEDA or PMDTA, it has no tendency to combine with a heteroatom, being already coordinatively saturated. Thus, the only activation which remains

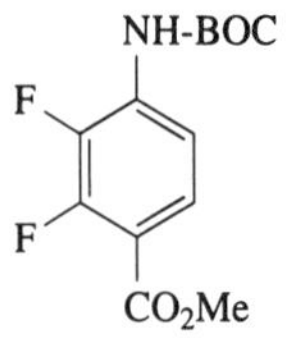

(179) 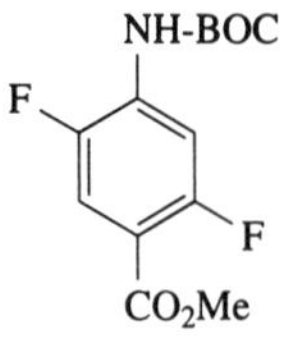 (180) 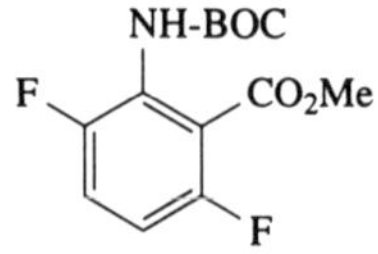 (181)

operative is due to the inductive effect, which is more efficient for the highly electronegative halides than for oxygen or nitrogen.

Phenol can be dimetallated[206] to the dipotassium species (**182**) using the butyllithium/potassium *t*-butoxide/TMEDA mixture (LICKOR/TMEDA) in hexane. Similar results have been obtained with a superbase made from 2-ethylhexyllithium and potassium 3-methyl-3-pentoxide.[206] The dilithio derivative (**183**) of phenol is, however, easily obtained in 67% yield after reaction with trimethylchlorosilane, by simple metallation with 2.8 equiv. of *t*-butyllithium in tetrahydropyran (Equation (34)).[208]

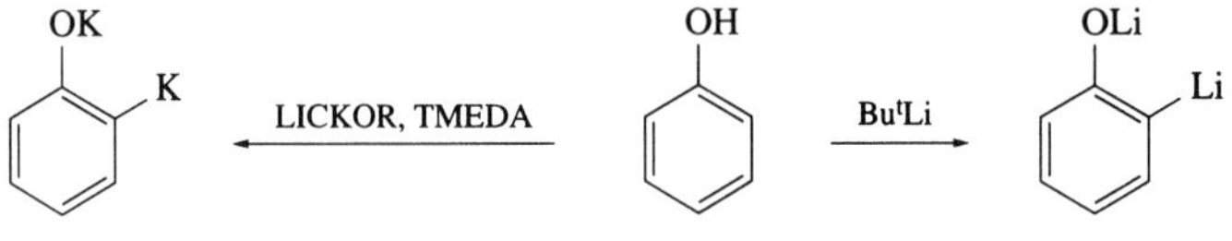

$$\text{(182)} \xleftarrow{\text{LICKOR, TMEDA}} \text{PhOH} \xrightarrow{\text{Bu}^t\text{Li}} \text{(183)} \qquad (34)$$

By contrast, thiophenol is not dimetallated by Schlosser's base or butyllithium/potassium *t*-butoxide/TMEDA,[206] whereas it can be doubly deprotonated to the dilithio species (**184**) by butyllithium/TMEDA in cyclohexane.[208] When an alkyl group is bonded to sulfur the use of Schlosser's base is recommended if one wishes to obtain *ortho*-α-dimetallated products (**185**). Butyllithium alone is just able to exchange the *ortho* proton in alkylthiobenzenes and both the *ortho* and α-protons in methylthiobenzene (Scheme 26, (**186**)).[209]

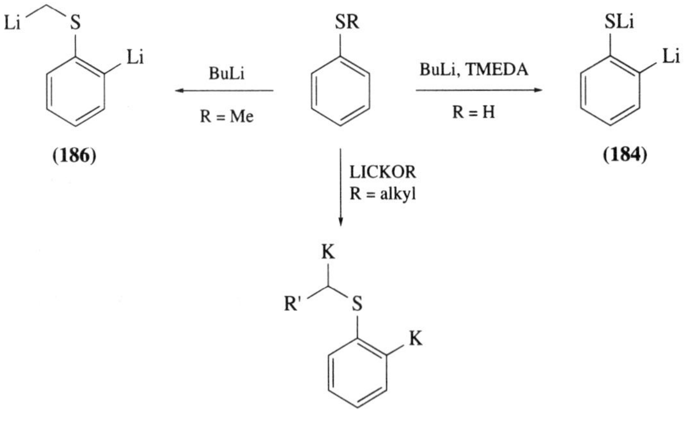

Scheme 26

2.5 REFERENCES

1. U. Schöllkopf, in 'Houben-Weyl, Methoden der organischen Chemie', ed. E. Müller, Thieme, Stuttgart, 1970, vol. 13/1, p. 116.
2. M. Schlosser, 'Struktur and Reaktivität polarer Organometalle', Springer, Berlin, 1973.
3. L. Brandsma and H. D. Verkruijsse, 'Preparative Polar Organometallic Chemistry 1', Springer, Berlin, 1987; L. Brandsma, 'Preparative Polar Organometallic Chemistry 2', Springer, Berlin, 1990.
4. M. Schlosser (ed.), 'Organometallics in Synthesis. A Manual', Wiley, London, 1994.
5. J. L. Wardell, in 'COMC-I', vol. 1, p. 43.
6. B. J. Wakefield, in 'COMC-I' vol. 7, p. 1.
7. M. Schlosser, *Angew. Chem.*, 1964, **76**, 124 and 258; *Angew. Chem., Int. Ed. Engl.*, 1964, **3**, 287 and 362.
8. A. A. Morton, *Chem. Rev.*, 1944, **35**, 1.

9. R. A. Benkeser, D. J. Foster, D. M. Sauve and J. F. Nobis, *Chem. Rev.*, 1957, **57**, 867.
10. M. Schlosser, *Pure Appl. Chem.*, 1988, **60**, 1627.
11. M. Schlosser, in 'Modern Synthetic Methods', ed. R. Scheffold, Helvetica Chimica Acta, Basel, 1992, vol. 6, p. 227.
12. M. Schlosser, O. Desponds, R. Lehmann, E. Moret and G. Rauchschwalbe, *Tetrahedron*, 1993, **49**, 10 175.
13. M. Schlosser, F. Faigl, L. Franzini, H. Geneste, G. Katsoulos and G.-F. Zhong, *Pure. Appl. Chem.*, 1994, **66**, 1423.
14. A. Mordini, in 'Advances in Carbanion Chemistry', ed. V. Snieckus, Jai Press, Greenwich, CT, 1992, vol. 1, p. 1.
15. P. Caubère, *Chem Rev.*, 1993, **93**, 2317.
16. A. A. Morton and M. E. T. Holden, *J. Am. Chem. Soc.*, 1947, **69**, 1675.
17. A. A. Morton, M. L. Brown, M. E. T. Holden, R. L. Letsinger and E. E. Magat, *J. Am. Chem. Soc.*, 1945, **67**, 2224.
18. A. A. Morton and J. L. Eisenmann, *J. Org. Chem.*, 1958, **23**, 1469.
19. A. A. Morton and I. Hechenbleikner, *J. Am. Chem. Soc.*, 1936, **58**, 2599; R. A. Benkeser, A. E. Trevillyan and J. Hooz, *J. Am. Chem. Soc.*, 1962, **84**, 4971; A. A. Morton, J. T. Massengale and M. L. Brown, *J. Am. Chem. Soc.*, 1945, **67**, 1620; A. A. Morton and E. J. Lanpher, *J. Org. Chem.*, 1958, **23**, 1636.
20. C. D. Broaddus, *J. Am. Chem. Soc.*, 1966, **88**, 4174.
21. R. A. Benkeser and T. V. Liston, *J. Am. Chem. Soc.*, 1960, **82**, 3221; R. A. Benkeser, J. Hooz, T. V. Liston and A. E. Trevillyan, *J. Am. Chem. Soc.*, 1963, **85**, 3984.
22. D. Bryce-Smith, *J. Chem. Soc.*, 1954, 1079; D. Bryce-Smith, *J. Chem. Soc.*, 1963, 5983.
23. A. A. Morton, E. L. Little and W. O. Strong, *J. Am. Chem. Soc.*, 1943, **65**, 1339.
24. D. Bryce-Smith and E. E. Turner, *J. Chem. Soc.*, 1953, 861.
25. R. A. Finnegan and R. S. McNees, *Chem. Ind.*, 1961, 1450; R. A. Finnegan and R. S. McNees, *J. Org. Chem.*, 1964, **29**, 3234; 3241.
26. E. J. Lanpher, L. M. Redman and A. A. Morton, *J. Org. Chem.*, 1958, **23**, 1370.
27. A. A. Morton, E. E. Magat and R. L. Letsinger, *J. Am. Chem. Soc.*, 1947, **69**, 950.
28. A. A. Morton *et al.*, *J. Am. Chem. Soc.*, 1950, **72**, 3785.
29. A. A. Morton and C. E. Claff, *J. Am. Chem. Soc.*, 1954, **76**, 4935.
30. A. A. Morton, C. E. Claff and F. W. Collins, *J. Org. Chem.*, 1955, **20**, 428.
31. R. A. Benkeser, T. F. Crimmins and W. H. Tong, *J. Am. Chem. Soc.*, 1968, **90**, 4366.
32. A. A. Morton and A. E. Brachman, *J. Am. Chem. Soc.*, 1951, **73**, 4363.
33. M. Schlosser, *J. Organomet. Chem.*, 1967, **8**, 9.
34. G. Boche and H. Etzrodt, *Tetrahedron Lett.*, 1983, **24**, 5477.
35. L. Lochmann, J. Pospíšil and D. Lím, *Tetrahedron Lett.*, 1966, **7**, 257.
36. L. Lochmann and J. Trekoval, *J. Organomet. Chem.*, 1987, **326**, 1.
37. L. Lochmann, *Collect. Czech. Chem. Commun.*, 1987, **52**, 2710.
38. M. Schlosser and J. Hartmann, *Angew. Chem.*, 1973, **85**, 544; *Angew. Chem., Int. Ed. Engl.*, 1973, **12**, 508.
39. M. Schlosser, J. Hartmann and V. David, *Helv. Chim. Acta*, 1974, **57**, 1567.
40. M. Schlosser and J. Hartmann, *J. Am. Chem. Soc.*, 1976, **98**, 4674.
41. M. Schlosser and G. Rauchschwalbe, *J. Am. Chem. Soc.*, 1978, **100**, 3258.
42. Review on the concept of aggregative activation: P. Caubère, *Rev. Heteroatom Chem.*, 1991, **4**, 78.
43. C. Schade, W. Bauer and P. v. R. Schleyer, *J. Organomet. Chem.*, 1985, **295**, C25.
44. R. Pi, W. Bauer, B. Brix, C. Schade and P. v. R. Schleyer, *J. Organomet. Chem.*, 1986, **306**, C1.
45. W. Bauer and L. Lochmann, *J. Am. Chem. Soc.*, 1992, **114**, 7482.
46. J. F. McGarrity and C. A. Ogle, *J. Am. Chem. Soc.*, 1985, **107**, 1805; J. F. McGarrity, C. A. Ogle, Z. Brich and H. R. Loosli, *J. Am. Chem. Soc.*, 1985, **107**, 1810.
47. H. Gilman and I. C. Bailie, *J. Org. Chem.*, 1937, **2**, 84.
48. C. D. Broaddus, T. J. Logan and T. J. Flautt, *J. Org. Chem.*, 1963, **28**, 1174; C. D. Broaddus, *J. Org. Chem.*, 1964, **29**, 2689.
49. J. Hartmann and M. Schlosser, *Helv. Chim. Acta*, 1976, **59**, 453.
50. M. Schlosser, J. Hartmann, M. Stähle, J. Kramař, A. Walde and A. Mordini, *Chimia*, 1986, **40**, 306.
51. R. Lehmann and M. Schlosser, *Tetrahedron*, 1984, **25**, 745.
52. Reviews: P. Caubère, *Acc. Chem. Res.*, 1974, **7**, 301; *Topics Curr. Chem.*, 1978, **33**, 50; *Angew. Chem.*, 1983, **95**, 597; *Angew. Chem., Int. Ed. Engl.*, 1983, **22**, 599.
53. C. Margot and M. Schlosser, *Tetrahedron Lett.*, 1985, **26**, 1035.
54. M. Schlosser and P. Schneider, *Helv. Chim. Acta*, 1980, **63**, 2404.
55. J. Hartmann and M. Schlosser, *Synthesis*, 1975, 328.
56. J. Hartmann, R. Muthukrishnan and M. Schlosser, *Helv. Chim. Acta*, 1974, **57**, 2261.
57. K. Fujita and M. Schlosser, *Helv. Chim. Acta*, 1982, **65**, 1258.
58. M. Stähle, J. Hartmann and M. Schlosser, *Helv. Chim. Acta*, 1977, **60**, 1730.
59. E. Moret, J. Fürrer and M. Schlosser, *Tetrahedron*, 1988, **44**, 3539.
60. M. Schlosser, R. Dahan and S. Cottens, *Helv. Chim. Acta*, 1984, **67**, 284.
61. M. Schlosser and M. Stähle, *Angew. Chem.*, 1980, **92**, 497; *Angew. Chem., Int. Ed. Engl.*, 1980, **19**, 487.
62. G. Rauchschwalbe and M. Schlosser, *Helv. Chim. Acta*, 1975, **58**, 1094.
63. M. Schlosser and G.-f. Zhong, *Tetrahedron Lett.*, 1993, **34**, 5441.
64. J. L. Giner, C. Margot and C. Djerassi, *J. Org. Chem.*, 1989, **54**, 2117.
65. E. Moret, P. Schneider, C. Margot, M. Stähle and M. Schlosser, *Chimia*, 1985, **39**, 231.
66. M. Schlosser, R. Lehmann and T. Jenny, *J. Organomet. Chem.*, 1990, **389**, 149.
67. A. Maercker and W. Berkulin, *Chem. Ber.*, 1990, **123**, 185.
68. H. C. Brown, M. Zaidlewicz and K. S. Bhat, *J. Org. Chem.*, 1989 **54**, 1764.
69. M. Stähle and M. Schlosser, *J. Organomet. Chem.*, 1981, **220**, 277.
70. See also: U. Schümann, E. Weiss, H. Dietrich and W. Mahdi, *J. Organomet. Chem.*, 1987, **322**, 299; S. Brownstein, S. Bywater and D. J. Worsfold, *J. Organomet. Chem.*, 1980, **199**, 1.
71. G. M. Whitesides, J. E. Nordlander and J. D. Roberts, *J. Am. Chem. Soc.*, 1962, **84**, 2010 and references cited therein.
72. M. Schlosser and K. Fujita, *Angew. Chem.*, 1982, **94**, 320; *Angew. Chem., Int. Ed. Engl.*, 1982, **21**, 309.

73. J. E. Nordlander and J. D. Roberts, *J. Am. Chem. Soc.*, 1959, **81**, 1769; P. West, J. I. Purmort and S. V. McKinley, *J. Am. Chem. Soc.*, 1968, **90**, 797; H. E. Zieger and J. D. Roberts, *J. Org. Chem.*, 1969, **34**, 1976.
74. H. Bosshardt and M. Schlosser, *Helv. Chim. Acta*, 1980, **63**, 2393.
75. H. E. Henderson *et al.*, *J. Chem. Soc., Chem. Commun.*, 1972, 686.
76. J. G. Millar, H. D. Pierce, A. M. Pierce, A. C. Oehlschlager, J. H. Borden and A. V. Barak, *J. Chem. Ecol.*, 1985, **11**, 1053.
77. W. L. Roelofs, J. P. Tette, E. F. Taschenberg and A. Cameau, *J. Insect Physiol.*, 1971, **17**, 2235.
78. H. Fukui, F. Matsumura, M. C. Ma and W. G. Burkholder, *Tetrahedron Lett.*, 1974, **15**, 3563.
79. G. Fouquet and M. Schlosser, *Angew. Chem.*, 1974, **86**, 50; *Angew. Chem., Int. Ed. Engl.*, 1974, **13**, 82; M. Schlosser and H. Bossert, *Tetrahedron*, 1991, **47**, 6287.
80. A. M. Moiseenkov, B. Schaub, C. Margot and M. Schlosser, *Tetrahedron Lett.*, 1985, **26**, 305.
81. J. L. Simonsen and D. H. R. Barton, 'The Terpenes', 2nd edn., Cambridge University Press, Cambridge, 1951, vol. 3, p. 178; E. Demole, C. Demole and P. Enggist, *Helv. Chim. Acta*, 1976, **59**, 737.
82. K. Mori, *Tetrahedron*, 1977, **33**, 289.
83. E. Moret and M. Schlosser, *Tetrahedron Lett.*, 1984, **25**, 4491.
84. H. C. Brown and K. S. Bhat, *J. Am. Chem. Soc.*, 1986, **108**, 293; J. Garcia, B. M. Kim and S. Masamune, *J. Org. Chem.*, 1987, **52**, 4831; W. R. Roush and R. L. Halterman, *J. Am. Chem. Soc.*, 1986, **108**, 294.
85. M. Zaidlewicz, *J. Organomet. Chem.*, 1985, **293**, 139.
86. M. Zaidlewicz and C. S. Panda, *Synthesis*, 1987, 645.
87. M. Zaidlewicz, *Tetrahedron Lett.*, 1986, **27**, 5135; M. Zaidlewicz, *Synthesis*, 1988, 701; M. Zaidlewicz, *J. Organomet. Chem.*, 1991, **409**, 103.
88. J. J. Bahl, R. B. Bates and B. Gordon, *J. Org. Chem.*, 1979, **44**, 2290.
89. A. Rusinko, N. S. Mills and P. Morse, *J. Org. Chem.*, 1982, **47**, 5198; W. E. Paget, K. Smith, M. G. Hutchings and G. E. Martin, *J. Chem. Res. (S)*, 1983, 30; *J. Chem. Res. (M)*, 1983, 0327.
90. M. Schlosser, H. Bosshardt, A. Walde and M. Stähle, *Angew. Chem.*, 1980, **92**, 302; *Angew. Chem., Int. Ed. Engl.*, 1980, **19**, 303.
91. E. Moret and M. Schlosser, *Tetrahedron Lett.*, 1984, **25**, 1449.
92. M. G. Hutchings, W. E. Paget and K. Smith, *J. Chem. Res. (S)*, 1983, 31; *J. Chem. Res. (M)*, 1983, 0342.
93. R. B. Bates, D. W. Gosselink and J. A. Kaczynski, *Tetrahedron Lett.*, 1967, **8**, 205; W. T. Ford and M. Newcomb, *J. Am. Chem. Soc.*, 1974, **96**, 309.
94. H. Yasuda, M. Yamauchi, Y. Ohnuma and A. Nakamura, *Bull. Chem. Soc. Jpn.*, 1981, **54**, 1481.
95. H. Yasuda, Y. Ohnuma, M. Yamauchi, H. Tani and A. Nakamura, *Bull. Chem. Soc. Jpn.*, 1979, **52**, 2036.
96. E. Moret, O. Desponds and M. Schlosser, *J. Organomet. Chem.*, 1991, **409**, 83.
97. L. Garanti, A. Marchesini, U. M. Pagnoni and R. Trave, *Gazz. Chim. Ital.*, 1976, **106**, 187.
98. J. H. Tumlinson, C. E. Yonce, R. E. Doolittle, R. R. Heath, C. R. Gentry and E. R. Mitchell, *Science*, 1974, **185**, 614.
99. N. Hertkorn and F. H. Köhler, *J. Organomet. Chem.*, 1988, **355**, 19.
100. B. Gordon, M. Blumenthal, A. E. Mera and R. J. Kumpf, *J. Org. Chem.*, 1985, **50**, 1540.
101. P. A. A. Klusener, H. H. Hommes, H. D. Verkruijsse and L. Brandsma, *J. Chem. Soc., Chem. Commun.*, 1985, 1677.
102. P. A. A. Klusener, L. Tip and L. Brandsma, *Tetrahedron*, 1991, **47**, 2041.
103. H. C. Brown and R. S. Randad, *Tetrahedron*, 1990, **46**, 4463.
104. R. M. Silverstein, J. O. Rodin and D. L. Wood, *Science*, 1966, **154**, 509.
105. L. Rothen and M. Schlosser, *Tetrahedron Lett.*, 1991, **32**, 2475.
106. M. Schlosser and S. Strunk, *Tetrahedron Lett.*, 1984, **25**, 741.
107. T. F. Crimmins and C. M. Chan, *J. Org. Chem.*, 1976, **41**, 1870.
108. T. F. Crimmins and E. M. Rather, *J. Org. Chem.*, 1978, **43**, 2170.
109. L. Lochmann and J. Petránek, *Tetrahedron Lett.*, 1991, **32**, 1483.
110. A. J. Chalk and T. J. Hoogeboom, *J. Organomet. Chem.*, 1968, **11**, 615; C. D. Broaddus, *J. Org. Chem.*, 1970, **35**, 10.
111. T. Amano *et al.*, *Chem. Pharm. Bull.*, 1986, **34**, 4653.
112. F. Faigl and M. Schlosser, *Tetrahedron Lett.*, 1991, **32**, 3369.
113. R. B. Bates and C. A. Ogle, *J. Org. Chem.*, 1982, **47**, 3949; R. D. Bach and R. C. Klix, *Tetrahedron Lett.*, 1986, **27**, 1983.
114. R. B. Bates and T. J. Siahaan, *J. Org. Chem.*, 1986, **51**, 1432.
115. R. B. Bates, F. A. Camou, V. V. Kane, P. K. Mishra, K. Suvannachut and J. J. White, *J. Org. Chem.*, 1989, **54**, 311.
116. R. B. Bates, B. A. Hess, C. A. Ogle and L. J. Schaad, *J. Am. Chem. Soc.*, 1981, **103**, 5052.
117. R. B. Bates, T. J. Siahaan, K. Suvannachut, S. K. Vasey and K. M. Yager, *J. Org. Chem.*, 1987, **52**, 4605.
118. B. Gordon, III and J. E. Loftus, *J. Org. Chem.*, 1986, **51**, 1618.
119. C. E. Barry, III, *et al.*, *Synlett*, 1991, 207.
120. R. B. Bates, T. J. Siahaan and K. Suvannachut, *J. Org. Chem.*, 1990, **55**, 1328.
121. R. B. Bates, S. Gangwar, V. V. Kane, K. Suvannachut and S. R. Taylor, *J. Org. Chem.*, 1991, **56**, 1696.
122. Y. Guggisberg, F. Faigl and M. Schlosser, *J. Organomet. Chem.*, 1991, **415**, 1.
123. R. L. P. de Jong and L. Brandsma, *J. Organomet. Chem.*, 1982, **238**, C17.
124. R. L. P. de Jong and L. Brandsma, *J. Chem. Soc., Chem. Commun.*, 1983, 1056.
125. P. A. A. Klusener, W. Kulik and L. Brandsma, *J. Org. Chem.*, 1987, **52**, 5261; W. Kulik, H. D. Verkruijsse, R. L. P. de Jong, H. Hommes and L. Brandsma, *Tetrahedron Lett.*, 1983, **24**, 2203.
126. L. Brandsma *et al.*, *Recl. Trav. Chim. Pays-Bas*, 1988, **107**, 286.
127. M. Stähle, R. Lehmann, J. Kramař and M. Schlosser, *Chimia*, 1985, **39**, 229; H. D. Verkruijsse and L. Brandsma, *Recl. Trav. Chim. Pays-Bas*, 1986, **105**, 66.
128. R. A. Finnegan and R. S. McNees, *Tetrahedron Lett.*, 1962, **3**, 755; R. A. Finnegan and R. S. McNees, *J. Org. Chem.*, 1964, **29**, 3234.
129. C. Nativi, N. Ravidà, A. Ricci, G. Seconi and M. Taddei, *J. Org. Chem.*, 1991, **56**, 1951.
130. C. Eaborn, *J. Organomet. Chem.*, 1975, **100**, 43.
131. L. Brandsma, H. D. Verkruijsse, C. Schade and P. v. R. Schleyer, *J. Chem. Soc., Chem. Commun.*, 1986, 260.
132. P. Schorigin, *Ber. Dtsch. Chem. Ges* 1908, **41**, 2711; 2723 P. Schorigin, *Ber. Dtsch. Chem. Ges* 1910, **43**, 1938.
133. A. A. Morton and I. Hechenbleikner, *J. Am. Chem. Soc.*, 1936, **58**, 1024; A. A. Morton, F. Fallwell, Jr. and L. Palmer, *J. Am. Chem. Soc.*, 1938, **60**, 1426; A. A. Morton and F. Fallwell, Jr., *J. Am. Chem. Soc.*, 1938, **60**, 1429; 1924.

134. A. A. Morton, G. M. Richardson and A. T. Hallowell, *J. Am. Chem. Soc.*, 1941, **63**, 327.
135. M. Schlosser, J. H. Choi and S. Takagishi, *Tetrahedron*, 1990, **46**, 5633.
136. See also: L. Lochmann, M. Fossatelli and L. Brandsma, *Recl. Trav. Chim. Pays-Bas*, 1990, **109**, 529.
137. P. A. A. Klusener, J. C. Hanekamp, L. Brandsma and P. v. R. Schleyer, *J. Org. Chem.*, 1990, **55**, 1311.
138. H. Hommes, H. D. Verkruijsse and L. Brandsma, *Tetrahedron Lett.*, 1981, **22**, 2495; H. Hommes, H. D. Verkruijsse and L. Brandsma, *J. Chem. Soc., Chem. Commun.*, 1981, 366; L. Brandsma, H. Hommes, H. D. Verkruijsse and R. L. P. de Jong, *Recl. Trav. Chim. Pays-Bas*, 1985, **104**, 226.
139. A. Mordini and M. Schlosser, *J. Organomet. Chem.*, in press.
140. Y. Yamamoto, in 'Comprehensive Organic Synthesis', ed. C. H. Heathcock, Pergamon, Oxford, 1991, vol. 2, p. 55.
141. Y. Yamamoto and N. Asao, *Chem. Rev.*, 1993, **93**, 2207.
142. H. W. Gschwend and H. R. Rodriguez, *Org. React.*, 1979, **26**, 1.
143. D. Hoppe, *Angew. Chem.*, 1984, **96**, 930; *Angew. Chem., Int. Ed. Engl.*, 1984, **23**, 932; D. Hoppe, T. Krämer, J.-R. Schwark and O. Zschage, *Pure Appl. Chem.*, 1990, **62**, 1999; P. Beak and A. I. Meyers, *Acc. Chem. Res.*, 1986, **19**, 356.
144. H. Ahlbrecht and J. Eichler, *Synthesis*, 1974, 672.
145. H. Ahlbrecht and C. S. Sudheendranath, *Synthesis*, 1982, 717.
146. H. Ahlbrecht and H. Simon, *Synthesis*, 1983, 58; H. Ahlbrecht and H. Simon, *Synthesis*, 1983, 61.
147. A. Degl'Innocenti, A. Mordini, D. Pinzani, G. Reginato and A. Ricci, *Synlett*, 1991, 712.
148. J. E. Resek and P. Beak, *Tetrahedron Lett.*, 1993, **34**, 3043.
149. D. J. Kempf, *Tetrahedron Lett.*, 1989, **30**, 2029.
150. M. Franciotti, A. Mordini and M. Taddei, *Synlett*, 1992, 137; M. Franciotti, A. Mann, A. Mordini and M. Taddei, *Tetrahedron Lett.*, 1993, **34**, 1355.
151. J.-A. Fehrentz and B. Castro, *Synthesis*, 1983, 676.
152. M. Schlosser, G. Müller and K. F. Christmann, *Angew. Chem.*, 1966, **78**, 677; *Angew. Chem., Int. Ed. Engl.*, 1966, **5**, 667; M. Schlosser, *Top. Stereochem.*, 1970, **5**, 1.
153. C. Canali, A. Mordini, S. Pecchi, G. Reginato and A. Ricci, unpublished results presented at the 7th IUPAC Symposium on Organo-Metallic Chemistry directed towards Organic Synthesis, 19–23 September, 1993, Kobe, Japan.
154. J. Hartmann, M. Stähle and M. Schlosser, *Synthesis*, 1974, 888.
155. H. D. Verkruijsse, L. Brandsma and P. v. R. Schleyer, *J. Organomet. Chem.*, 1987, **332**, 99.
156. R. Muthukrishnan and M. Schlosser, *Helv. Chim. Acta*, 1976, **59**, 13.
157. See also: K. Oshima, K. Shimoji, H. Takahashi, H. Yamamoto and H. Nozaki, *J. Am. Chem. Soc.*, 1973, **95**, 2694; U. Schöllkopf and P. Hänssle, *Liebigs Ann. Chem.*, 1972, **763**, 208.
158. R. H. Everhardus, R. Gräfing and L. Brandsma, *Recl. Trav. Chim. Pays-Bas*, 1978, **97**, 69.
159. Replacement of a hydrogen atom with a thiophenyl group increases the acidity by 9.1 pK units: F. G. Bordwell, W. S. Matthews and N. R. Vanier, *J. Am. Chem. Soc.* 1975, **97**, 442.
160. G. Wittig, *Angew. Chem.*, 1954, **66**, 10; U. Schöllkopf, *Angew. Chem.*, 1970, **82**, 795; *Angew. Chem., Int. Ed. Engl.*, 1970, **9**, 763.
161. M. Schlosser and S. Strunk, *Tetrahedron*, 1989, **45**, 2649.
162. G.-f. Zhong and M. Schlosser, *Tetrahedron Lett.*, 1993, **34**, 6265.
163. A. Mordini, S. Pecchi and G. Capozzi, *Tetrahedron*, 1994, **50**, 6029.
164. T. Nogrady, 'Medicinal Chemistry. A Biochemical Approach', Oxford University Press, New York, 1985, p. 384–385.
165. K. Koumaglo and T. H. Chan, *Tetrahedron Lett.*, 1984, **25**, 717.
166. See for instance: R. J. P. Corriu and J. Masse, *J. Organomet. Chem.*, 1973, **57**, C5; R. J. P. Corriu, J. Masse and D. Samate, *J. Organomet. Chem.*, 1975, **93**, 71.
167. A. Mordini, G. Palio, A. Ricci and M. Taddei, *Tetrahedron Lett.*, 1988, **29**, 4991.
168. L.-H. Li, D. Wang and T. H. Chan, *Tetrahedron Lett.*, 1991, **32**, 2879.
169. K. Tamao, E. Nakajo and Y. Ito, *Synth. Commun.*, 1987, **17**, 1637; *J. Org. Chem.*, 1987, **52**, 4412.
170. A. Degl'Innocenti, A. Mordini, L. Pagliai and A. Ricci, *Synlett*, 1991, 155.
171. A. Mordini, Ph.D. Thesis, University of Florence, 1987.
172. A. Mordini and M. Schlosser, *Chimia*, 1986, **40**, 309.
173. D. J. Chadwick and C. Willbe, *J. Chem. Soc., Perkin Trans. 1*, 1977, 887.
174. V. Ramanathan and R. Levine, *J. Org. Chem.*, 1962, **27**, 1216.
175. B. L. Feringa, R. Hulst, R. Rikers and L. Brandsma, *Synthesis*, 1988, 316.
176. F. Faigl and M. Schlosser, *Tetrahedron*, 1993, **49**, 10 271.
177. A. J. Clarke, S. McNamara and O. Meth-Cohn, *Tetrahedron Lett.*, 1974, **15**, 2373.
178. J. Verbeek, A. V. E. George, R. L. P. de Jong and L. Brandsma, *J. Chem. Soc., Chem. Commun.*, 1984, 257; J. Verbeek and L. Brandsma, *J. Org. Chem.*, 1984, **49**, 3857.
179. M. Schlosser and P. Schneider, *Angew. Chem.*, 1979, **91**, 515; *Angew. Chem., Int. Ed. Engl.*, 1979, **18**, 489.
180. S. Hanessian, M. Martin and R. C. Desai, *J. Chem. Soc., Chem. Commun.*, 1986, 926.
181. R. R. Schmidt, R. Preuss and R. Betz, *Tetrahedron Lett.*, 1987, **28**, 6591.
182. See also: R. K. Boeckman, Jr., A. B. Charette, T. Asberom and B. H. Johnston, *J. Am. Chem. Soc.*, 1991, **113**, 5337.
183. C. Margot, M. Rizzolio and M. Schlosser, *Tetrahedron*, 1990, **46**, 2411.
184. H. Ahlbrecht and H. Dollinger, *Tetrahedron Lett.*, 1984, **25**, 1353.
185. B. H. Lipshutz and E. Garcia, *Tetrahedron Lett.*, 1990, **31**, 7261.
186. A. Mordini, E. Ben Rayana, C. Margot and M. Schlosser, *Tetrahedron*, 1990, **46**, 2401.
187. J. K. Crandall and M. Apparu, *Org. React.*, 1983, **29**, 345.
188. A. Degl'Innocenti, A. Mordini, S. Pecchi, D. Pinzani, G. Reginato and A. Ricci, *Synlett*, 1992, 753, 803.
189. A. Mordini *et al.*, *J. Org. Chem.*, 1994, **59**, 4784.
190. H. Gilman and R. L. Bebb, *J. Am. Chem. Soc.*, 1939, **61**, 109.
191. G. Wittig and G. Fuhrmann, *Chem. Ber.*, 1940, **73**, 1197.
192. N. S. Narasimhan and R. S. Mali, *Synthesis*, 1983, 957; *Topics Curr. Chem.*, 1987, **138**, 63; V. Snieckus, *Bull. Soc. Chim. Fr.*, 1988, 67; *Chem. Rev.*, 1990, **90**, 879.
193. H. Gilman and T. S. Soddy, *J. Org. Chem.*, 1957, **22**, 1715.
194. M. Schlosser, G. Katsoulos and S. Takagishi, *Synlett*, 1990, 747.

195. M. Fossatelli, H. D. Verkruijsse and L. Brandsma, *Synth. Commun.*, 1990, **20**, 1701.
196. M. Fossatelli and L. Brandsma, *Synthesis*, 1992, 756.
197. S. Takagishi and M. Schlosser, *Synlett*, 1991, 119.
198. J. D. Roberts and D. Y. Curtin, *J. Am. Chem. Soc.*, 1946, **68**, 1658; D. A. Shirley, J. R. Johnson, Jr. and J. P. Hendrix, *J. Organomet. Chem.*, 1968, **11**, 209.
199. P. Aeberli and W. J. Houlihan, *J. Organomet. Chem.*, 1974, **67**, 321.
200. K. D. Bartle, G. Hallas and J. D. Hepworth, *Org. Magn. Res.*, 1973, **5**, 479.
201. D. E. Grocock, T. K. Jones, G. Hallas and J. D. Hepworth, *J. Chem. Soc. C*, 1971, 3305.
202. G. Katsoulos, S. Takagishi and M. Schlosser, *Synlett*, 1991, 731.
203. S. Takagishi, G. Katsoulos and M. Schlosser, *Synlett*, 1992, 360.
204. M. Schlosser and G. Simig, *Tetrahedron Lett.*, 1991, **32**, 1965.
205. T. J. Thornton and M. Jarman, *Synthesis*, 1990, 295.
206. H. Andringa, H. D. Verkruijsse, L. Brandsma and L. Lochmann, *J. Organomet. Chem.*, 1990, **393**, 307.
207. G. H. Posner and K. A. Canella, *J. Am. Chem. Soc.*, 1985, **107**, 2571.
208. G. D. Figuly, C. K. Loop and J. C. Martin, *J. Am. Chem. Soc.*, 1989, **111**, 654; E. Block *et al.*, *J. Am. Chem. Soc.*, 1989, **111**, 658; K. Smith, C. M. Lindsay and G. J. Pritchard, *J. Am. Chem. Soc.*, 1989, **111**, 665.
209. S. Cabiddu, C. Fattuoni, C. Floris, G. Gelli and S. Melis, *J. Organomet. Chem.*, 1992, **441**, 197; S. Cabiddu, C. Fattuoni, C. Floris, S. Melis and A. Serci, *Tetrahedron*, 1994, **50**, 6037.

3
Magnesium

JOHN M. BROWN and SUSAN K. ARMSTRONG
University of Oxford, UK

3.1 GENERAL CONSIDERATIONS	129
3.1.1 Scope	129
3.1.2 Modification of Reactivity; Organocerium Reagents	130
3.1.3 Overview of Asymmetric Synthesis	130
3.2 ADDITION TO CARBON–CARBON MULTIPLE BONDS	131
3.2.1 Carbomagnesiation of Alkenes	131
3.2.2 Magnesium 'Ene' Reactions	131
3.2.3 Conjugate Additions	132
3.2.4 Biaryl and Related Couplings	133
3.3 ADDITION TO CARBON-NITROGEN MULTIPLE BONDS	134
3.3.1 Imines	134
3.3.2 Diastereoselective Attack on C=N Double Bonds (Excluding Pyridines)	136
3.3.2.1 N-chiral imines and heteroimines	136
3.3.2.2 C-chiral imines and N-heteroimines	137
3.3.2.3 Other possibilities	139
3.3.3 Addition to Nitriles and Isothiocyanates	139
3.3.4 Pyridines and Related Reactants	141
3.3.4.1 Nonstereoselective addition	141
3.3.4.2 Stereoselective attack on pyridine derivatives	144
3.4 ADDITION TO AND SUBSTITUTION OF CARBON–OXYGEN BONDS	145
3.4.1 Diastereoselective Additions to Aldehydes and Ketones	145
3.4.1.1 Reaction between two prochiral compounds	145
3.4.1.2 Addition to α-chiral aldehydes and ketones	146
3.4.1.3 Addition to aldehydes and ketones with chiral auxiliaries	148
3.4.1.4 Chiral Grignard reagents	150
3.4.2 Enantioselective Additions to Aldehydes and Ketones	150
3.4.3 Additions to Carboxylic Acid Derivatives	151
3.4.4 Acetals	153
3.5 REFERENCES	154

3.1 GENERAL CONSIDERATIONS

3.1.1 Scope

The application of Grignard reagents is manifestly widespread in organic synthesis. Increasingly, the chemistry involves a transmetallation (e.g., to copper or palladium species present in catalytic amounts) so that the true synthetic reagent is an organocuprate or organopalladate. Such reactions are treated in the context of the metal providing the catalytic species, and some important topics not covered in depth are summarized here. The quality of magnesium used in synthesis of the Grignard reagent can be critical, especially at the extremes of alkyl halide reactivity, and many methods are available for preparing activated magnesium[1-7] either by mechanical degradation of magnesium turnings, by distillation of the metal, by solubilization with anthracene or by ultrasound. The detailed pathway of Grignard reagent formation, whether through radicals freely diffusing in solution or constrained at the magnesium surface, has been studied in depth and debated,[8,9] and several groups have pursued the mechanism of reaction with electrophiles,[10-14] with much emphasis on the role of electron transfer.[12,15-21] The challenges associated with 'difficult' Grignard reagents such as short-chain α,ω-diorganomagnesium compounds have provided a stimulus for their successful solution.[22,23]

3.1.2 Modification of Reactivity; Organocerium Reagents

Organomagnesium compounds prepared in ether solvents are strongly basic, leading to well-recognized side reactions in synthesis, particularly with readily enolized aldehydes and ketones. In some cases this can lead to low yields or even complete absence of the desired product. A common recourse, following Imamoto,[24] is to carry out the reaction in the presence of cerium chloride, converting R–Mg into the less basic [R–CeCl$_3$]MgX ate-complex. The two components are normally mixed and stirred at low temperature before the addition of the electrophile. Some examples[22,23,25-7] where this procedure has been used successfully are given in Equations (1),[24] (2),[25] (3)[26] and (4)[27]. The effects are quite dramatic in many cases, diverting a completely ineffective reaction into a high-yielding one. In the alkylation of hydrazones, RLi/CeCl$_3$ is effective in cases where the Grignard equivalent is not.[28]

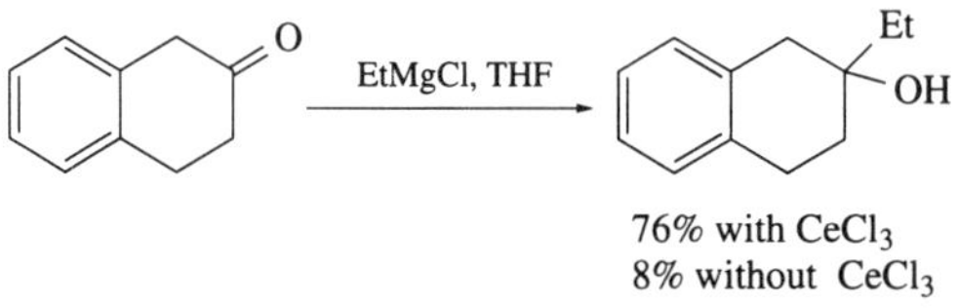

$$\text{α-tetralone} \xrightarrow{\text{EtMgCl, THF}} \text{2-ethyl-2-hydroxytetralin} \quad (1)$$

76% with CeCl$_3$
8% without CeCl$_3$

$$\text{MeO-CH(OMe)-(CH}_2\text{)}_n\text{-CO}_2\text{Me} \xrightarrow[\text{CeCl}_3]{\text{ClMgCH}_2\text{TMS}} \text{MeO-CH(OMe)-(CH}_2\text{)}_n\text{-C(=CH}_2\text{)-CH}_2\text{TMS} \quad (2)$$

$n = 4, 78\%; n = 3, 42\%; n = 1, 48\%$

$$\text{PhCOCH}_2\text{COR'} + \text{RMgX} \xrightarrow{\text{CeCl}_3, \text{THF}, 0\,°\text{C}} \text{PhCOCH}_2\text{C(OH)(R)} \quad (3)$$

70–88%

$$\text{C}_6\text{H}_{11}\text{NO}_2 + \text{PrMgCl (excess)} \xrightarrow[\text{ii, 10\% AcOH}]{\text{i, CeCl}_3, \text{THF}, -40\,°\text{C}} \text{C}_6\text{H}_{11}\text{N(OH)Pr} \quad (4)$$

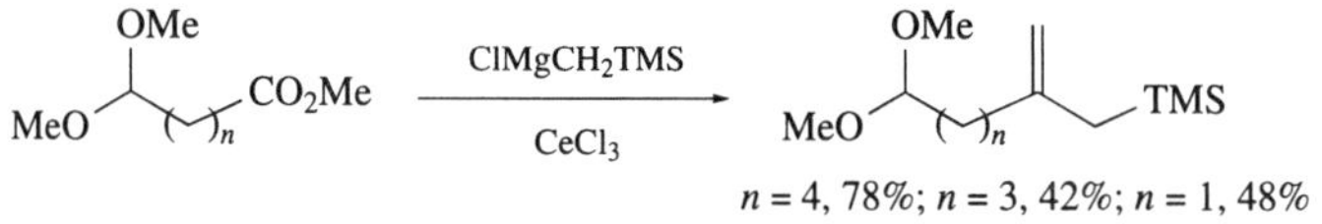

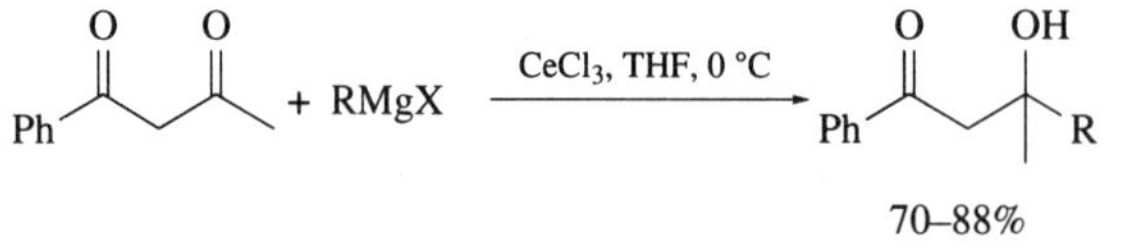

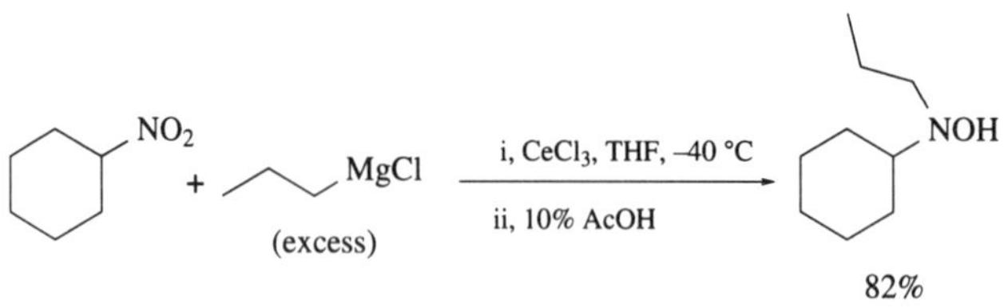

82%

3.1.3 Overview of Asymmetric Synthesis

The addition of Grignard reagents to aldehydes, prochiral ketones and electrophilic alkenes is an asymmetric synthesis whose potential importance has never been adequately realized. Examples are scattered through the text and will be discussed in context, but there are relatively few cases where the Grignard reaction is the best general method of asymmetric synthesis; a notable exception is in Section 3.2.1.

3.2 ADDITION TO CARBON–CARBON MULTIPLE BONDS

3.2.1 Carbomagnesiation of Alkenes

The direct addition of RMgX to alkenes is catalysed by Cp_2ZrR_2 via a transmetallation involving a metallocyclopentane, and is rendered synthetically useful when there is an internal basic site in the alkene.[29] The reaction is highly diastereoselective for allylic alcohols and their ethers, giving the Grignard adduct as the formal product which may be converted into an alcohol by O_2 oxidation or a carboxylic acid by carboxylation, or it may simply be reduced by protonolysis. Examples are given in Equations (5)–(8).[30,31] With the enantiomerically pure catalyst shown, there is a significant difference in the diastereoselectivity of reaction, depending on whether the (R)- or (S)-enantiomer of allylic alcohol is employed.[32] For 3,4-dihydropyran or 3,4-dihydrofuran, the same catalyst effects the addition of ethyl- or propylmagnesium chloride in 95–97% ee.[33] If the dihydropyran possesses a substituent at the 2-position, then efficient kinetic resolution occurs,[34] with an (S)-factor greater than 20. As the reagent is catalytic, and complex products are formed from simple starting materials with a high degree of stereocontrol, this represents a significant complement to existing asymmetric syntheses.

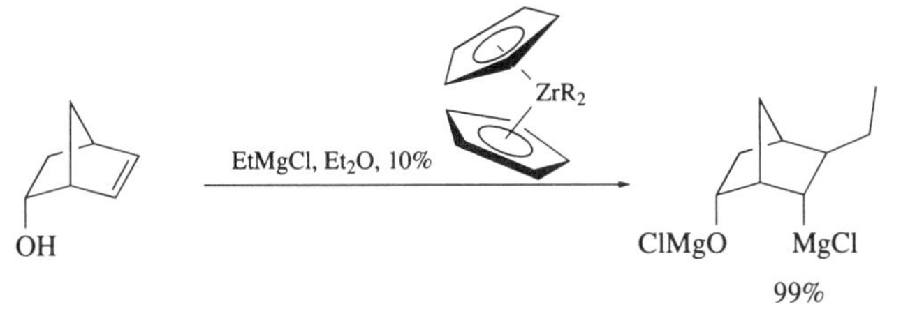

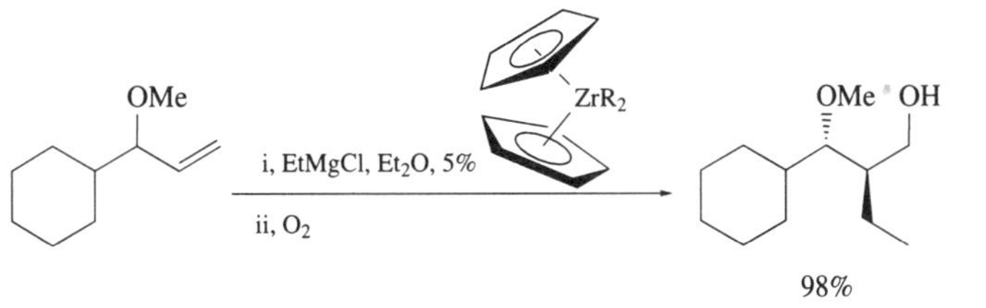

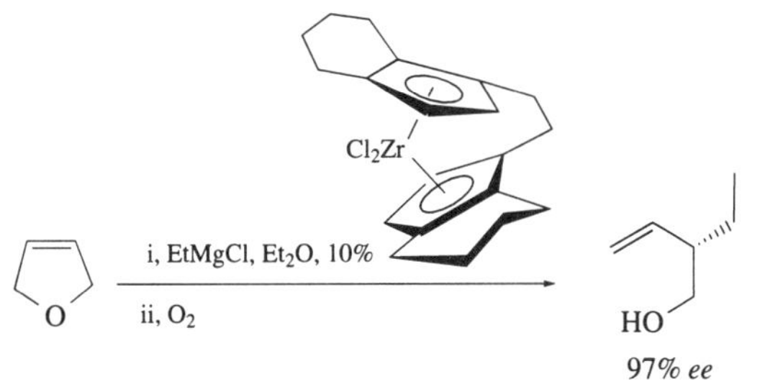

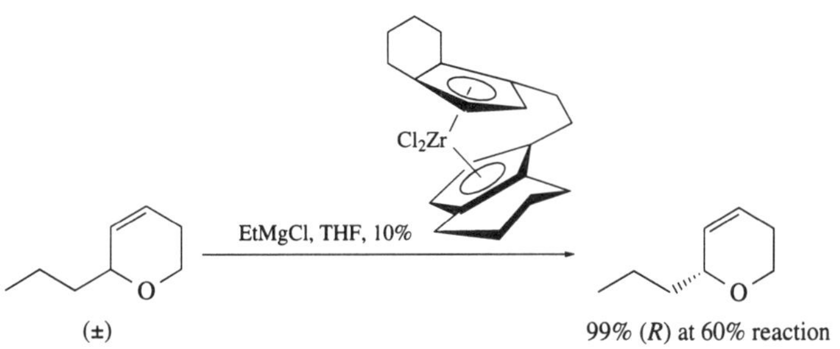

3.2.2 Magnesium 'Ene' Reactions

Allylic Grignard reagents are capable of adding to alkenes without external reagents by a metallo-ene mechanism.[35] The synthetic applications of this observation have come mainly from Oppolzer's laboratory, and lead to the distinction between two possible pathways designated types I and II, depending on whether the magnesium becomes bonded to the terminal or to the internal carbon of the alkene. Fairly vigorous conditions are required to promote the reaction, but high levels of regio- and stereoselectivity can be achieved (Equations (9) and (10)).[36]

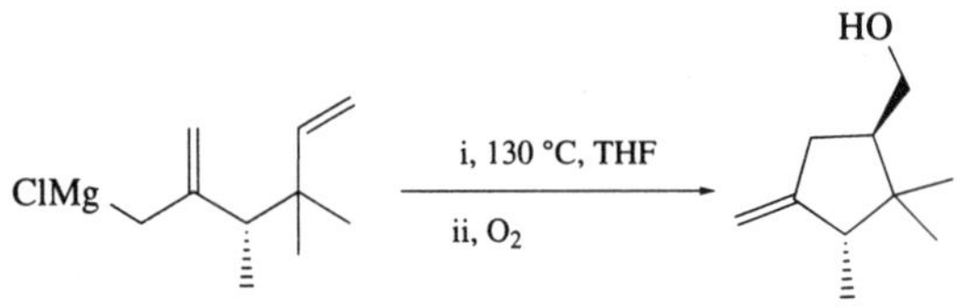

$$\text{(9)}$$

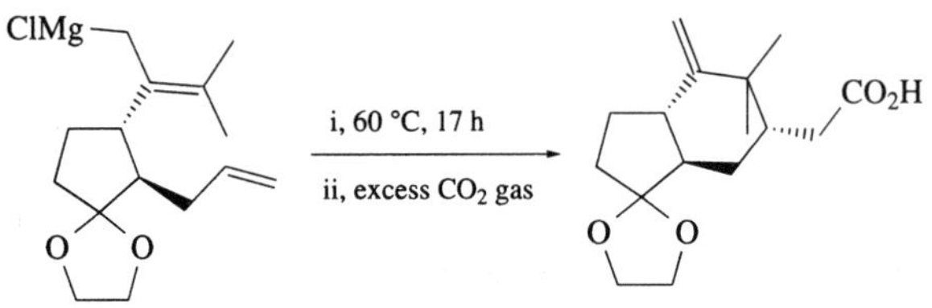

$$\text{(10)}$$

3.2.3 Conjugate Additions

Characteristically, Grignard reactions of conjugated carbonyl compounds do not give high yields of the 1,4-addition product unless the reaction is catalysed by cuprates or the carbonyl site is sterically blocked; a change in regiochemistry towards conjugate addition occurs in the presence of complexed zinc salts in catalytic[37] or stoichiometric[38] quantity, where the reactive species could be a zincate, R_3ZnMgX. The most successful method of asymmetric synthesis in this field is due to Mukaiyama,[39] and is long-standing; it consists simply of reacting a large excess of the Grignard reagent with the ephedrine amide of the α,β-unsaturated acid (Equation (11)), which leads to high diastereomer excess under carefully controlled conditions. The experiment has been refined recently by replacing ephedrine with the readily available chiral auxiliary 2(o-fluorobenzylamino)-1-butanol, when comparable selectivity is obtained under simpler conditions[40] (Equation (12)). The cyclic ephedrine derivative of benzylidenemalonates (for which the (Z)-isomer is preferred in synthesis by Knoevenagel condensation) reacts with Grignard reagents in a stereospecific manner[41] (Equation (13)) which can be rationalized by x-ray analysis of the structure of the reactant. There is a comprehensive review on asymmetric conjugate addition, which places Grignard reagents in a fuller context[42] (Schemes 1 and 2 exemplify Posner's work as reviewed in Ref. 42).

$$\text{(11)}$$

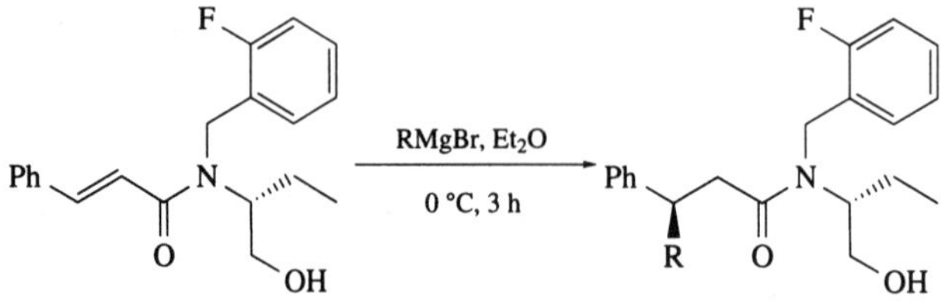

up to 99% de

$$\text{(12)}$$

>92% de by ^{19}F NMR

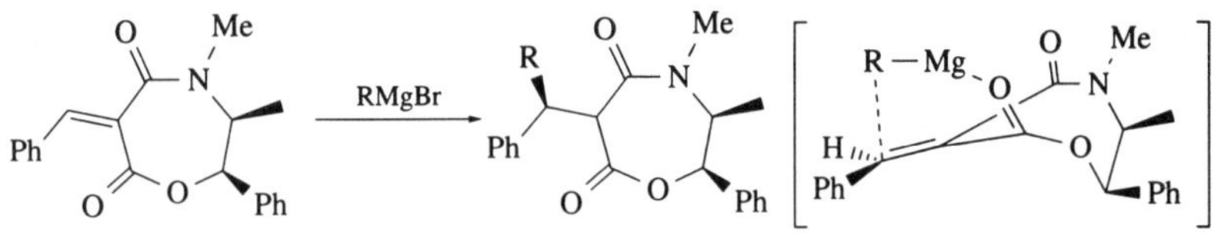

3.2.4 Biaryl and Related Couplings

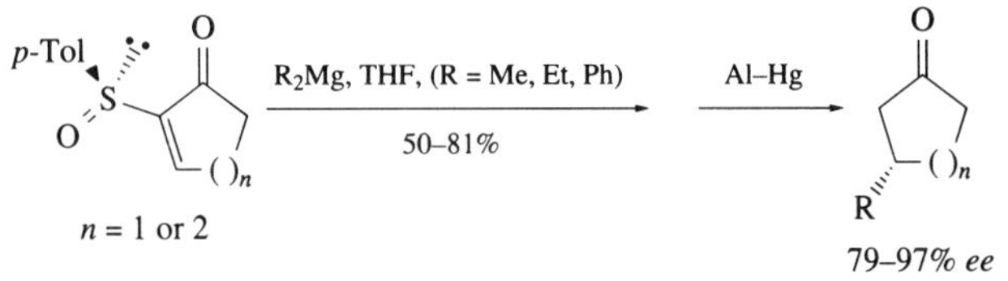

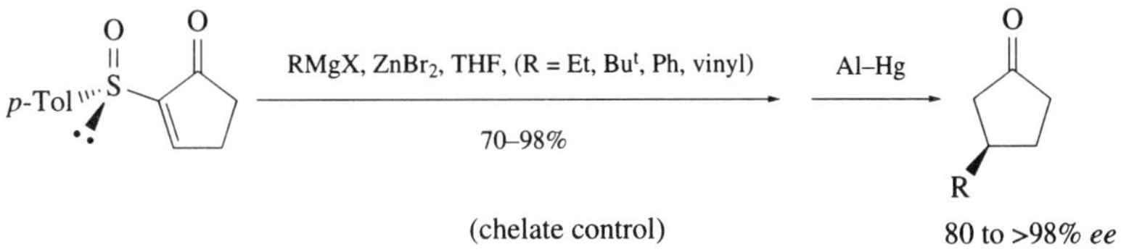

A simple and widely useful method for biaryl coupling was originally due to Meyers and co-workers.[43-5] This consists of reacting an aromatic Grignard reagent with an *o*-methoxyaryloxazoline, and the magnesium-assisted bond-forming process is shown in Equation (14); there are many applications in natural product synthesis. When the second *ortho*-positions of the coupling partners are both blocked as shown, reaction of an enantiomerically pure oxazoline leads to the preponderance of a single diastereomer of the biaryl. Interestingly, the *t*-butyldimethylsilyl (TBDMS) ether (**1**) gives opposite diastereoselectivity to that given by the free benzylic alcohol. A different mechanism is involved in the biaryl synthesis from arylmagnesium halides and polyhalobenzenes (Equation (15)), which probably involves the formation of an aryne by transmetallation and halide elimination; the aryne is trapped by more Grignard reagent.[46-8] A further efficient method involves the reaction of aryl sulfoxides with Grignard reagents; this requires first the addition of reagent to sulfur to give a formally tetravalent sulfurane, and then a fragmentation of axial and equatorial substituents to form the new C–C bond. In this way a variety of bipyridyls and arylpyridyls can be synthesized.[49-52] Further, the same procedure can be utilized to prepare atropisomerically chiral biaryls (naphthalenes, isoquinolines) with effective asymmetric induction from the sulfoxide to the inter-ring axis (Equation (16)).[53-5]

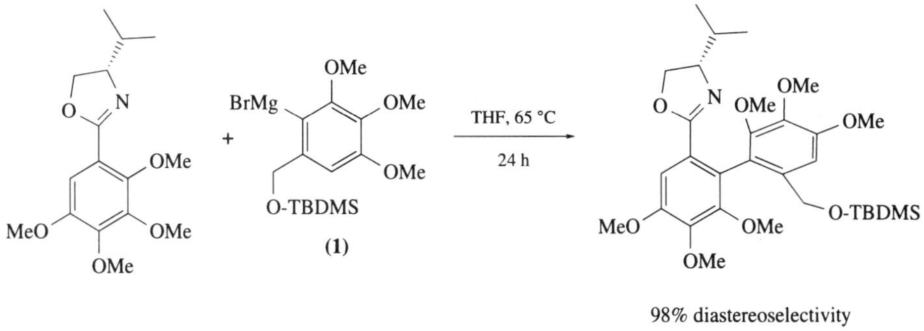

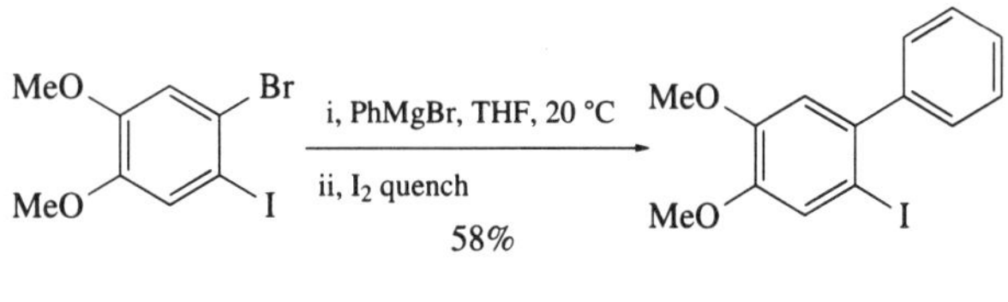

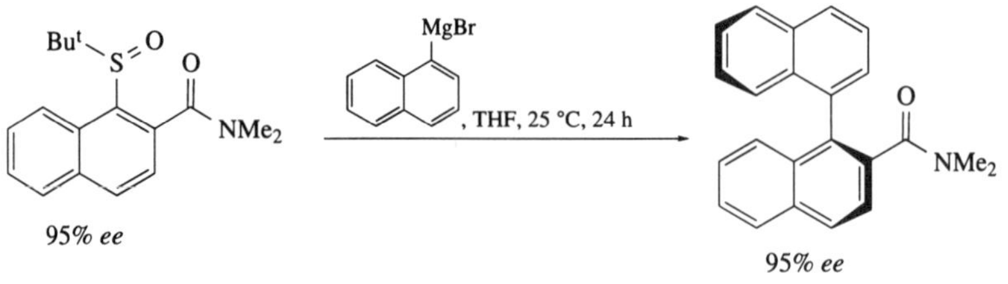

3.3 ADDITION TO CARBON-NITROGEN MULTIPLE BONDS

3.3.1 Imines

In the vast majority of cases, regardless of the substituents on carbon or nitrogen, the primary reaction of Grignards with C=N double bonds is simple addition. If the double bond is conjugated, 1,4-attack may be observed instead,[56] sometimes with further reaction.[57] Florio and co-workers have shown that the organomagnesium may act as a base in some circumstances (see Scheme 5 for an example).[58–60] The only other exceptions involve aromatic polyazaheterocycles, which will be discussed later. Frequently—and necessarily in the case of simple alkyl- or aryl-substituted imines—no further reaction occurs, and the isolated product is then the saturated amine.[61–3] A few of the many published examples are shown in Schemes 3 and 4 and Equation (17).

If the carbon of the original C=N double bond bears a heteroatom-based leaving group, such as Cl, OR, or SR, this may be displaced to form a new imine, which may be isolated,[58] or may undergo further attack,[56] or occasionally reduction,[64] by the Grignard reagent. In the case of N,S-heterocycles, this may lead to S,S-coupled dimeric products.[65,66] If the nitrogen of the original C=N double bond bears a leaving group, rearrangement of the carbon skeleton may occur.[67,68] Examples of these manifold pathways are collected in Schemes 5–7 and Equations (18)–(20).

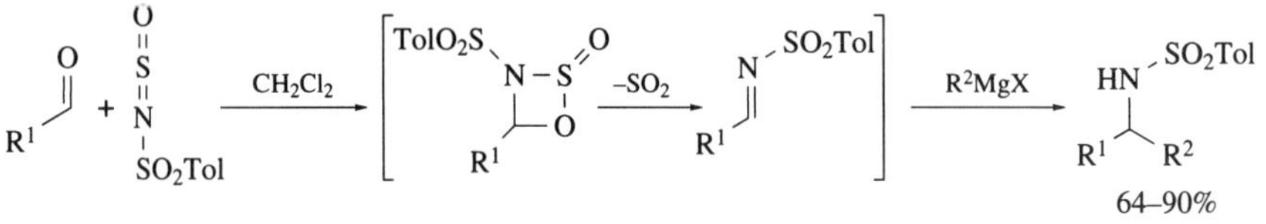

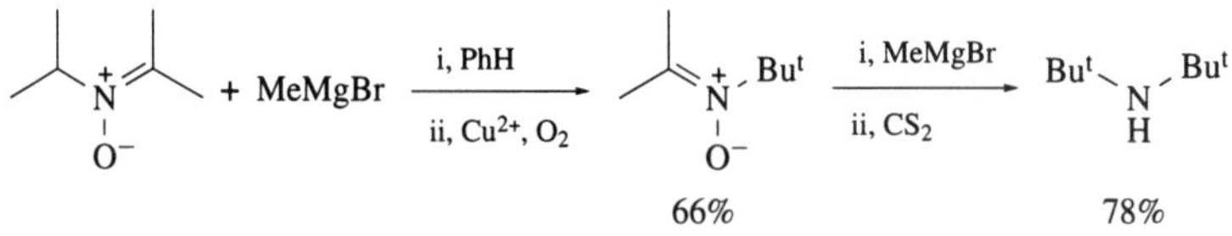

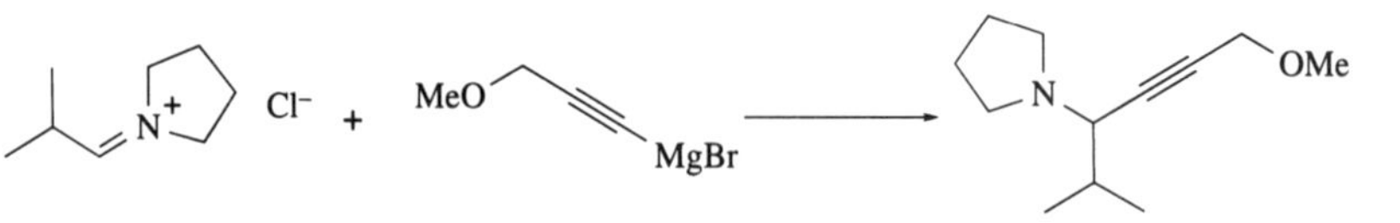

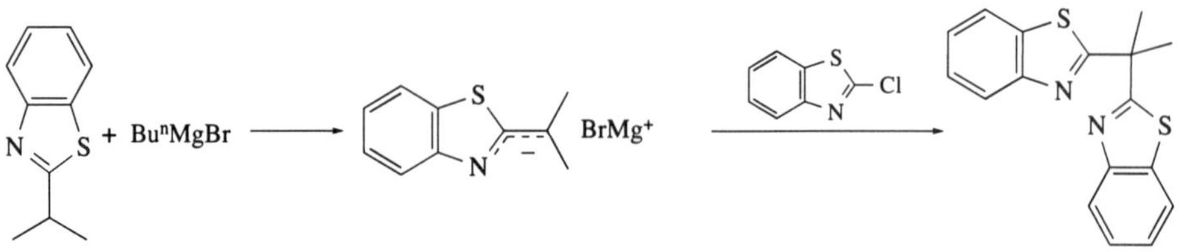

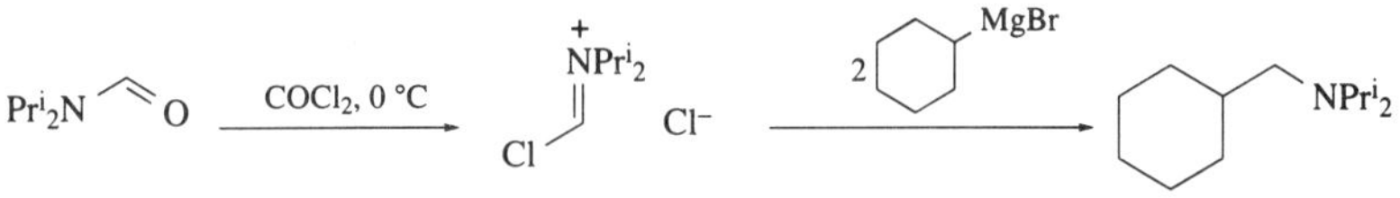

Scheme 6

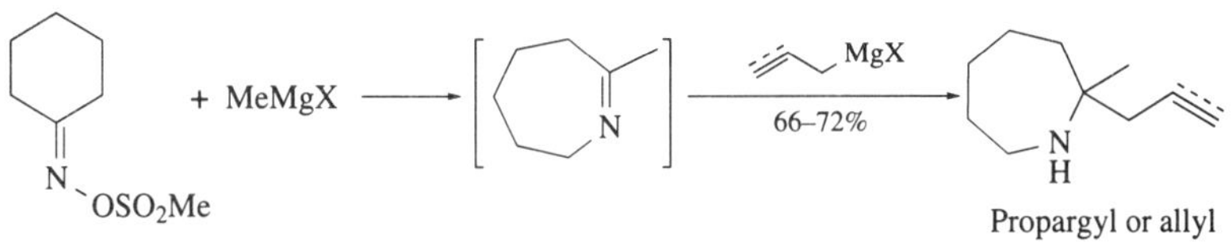

Scheme 7

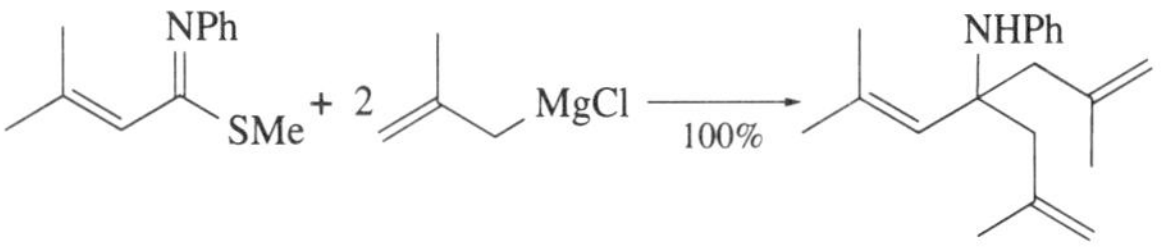

$$\text{(18)}$$

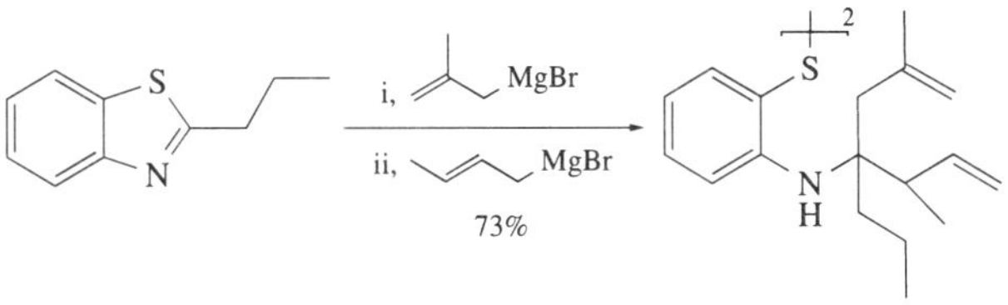

$$\text{(19)}$$

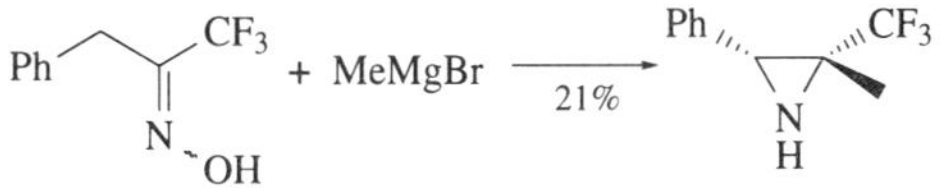

$$\text{(20)}$$

The mode of reaction of organomagnesium compounds with aromatic heterocycles containing two or more nitrogen atoms depends largely on steric factors, although electronic factors may also influence the regiochemistry of addition (Equations (21)–(23)). Thus quinoxaline (**2**), studied by Florio and co-workers, may be attacked by one or two equivalents of Grignard reagent, depending on the degree of steric hindrance (Equations (21) and (22)).[69] The total regioselectivity of attack of Grignard reagents on 5-cyano-2-methylthiopyrimidine (**3**), exemplified in Equation (23), may be the result of electronic as well as steric considerations.[70]

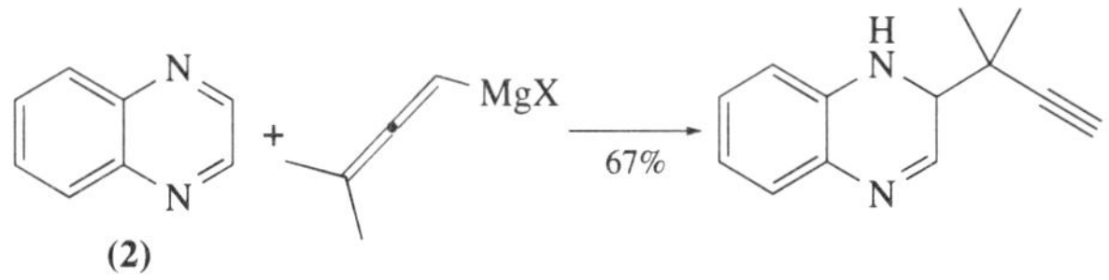

$$\text{(21)}$$

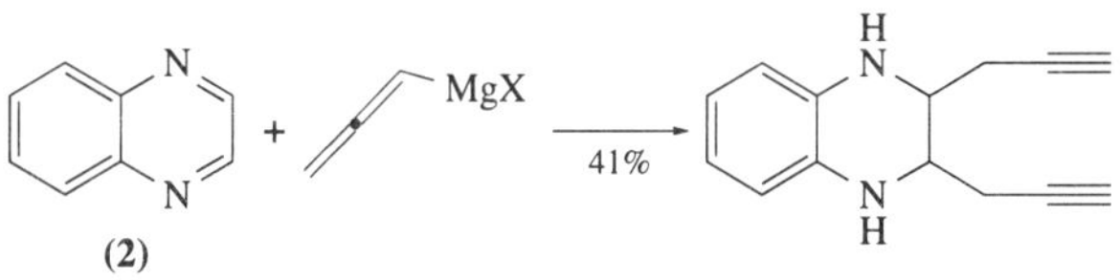

$$\text{(22)}$$

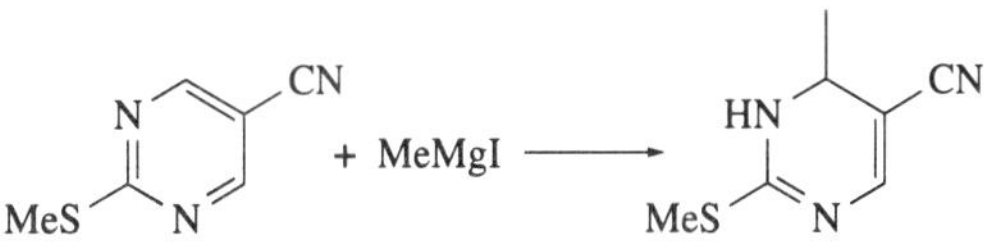

$$\text{(23)}$$

3.3.2 Diastereoselective Attack on C=N Double Bonds (Excluding Pyridines)

3.3.2.1 N-chiral imines and heteroimines

Stereoselective organomagnesium attack on imines and N-heteroimines (hydrazones, etc.) may be mediated by a chiral group on either carbon or nitrogen of the imine. Attempts to achieve stereoselectivity without a chiral starting material, by the addition of crotyl Grignards to imines, gave only low selectivities, better results being obtained with organoboron reagents.[71] N-chiral hydrazines have the disadvantage that the chiral centre is at some distance from the reaction site. Perhaps because of this, an oxygen substituent is generally included in the chiral group. Chelation of a metal to this oxygen and to a hydrazine nitrogen controls the substrate geometry in the transition state. Cerium chloride may be added to the reaction mixture for this reason,[72,73] although in its absence the magnesium itself presumably forms a cyclic chelate.[74] Very high diastereoselectivity has been obtained in favourable cases, as shown in Equations (24)–(26).

R = Me, Et, one isomer only
R = $ArCH_2$, 8–16% *de*

(24)

97% *de*

(25)

86 to >96% *de*

(26)

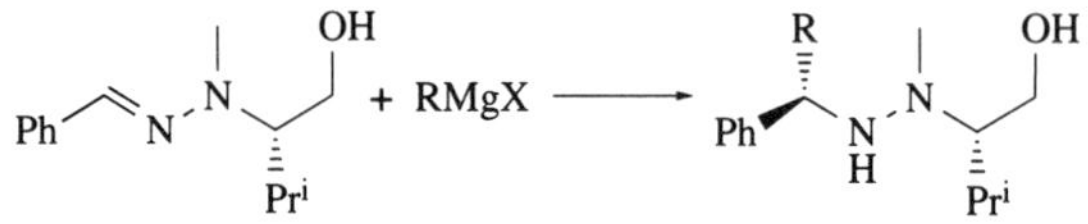

N-chiral imines and nitrones generally have a chiral centre attached directly to the imine (or nitrone) nitrogen, thus bringing it closer to the reaction site. Potentially chelating oxygen groups are still present in the majority of chiral groups used, and support for a chelated transition state, at least in some cases, comes from Coates' observation that the degree and even the direction of selectivity may depend on whether a methyl ether or a bulky silyl ether is used for this purpose.[75] As this observation suggests, selectivity in these reactions is very variable (Equations (27)–(33)). Takahashi obtained total selectivity with a relatively simple chiral auxiliary, similar to that used later by Coates, which has the advantage that both enantiomers are available.[76] A sugar-derived nitrone, alternatively, gave low selectivity,[77] as did an apparently strongly chelating and very sterically demanding auxiliary, used to try to control 1,4-addition to an α,β-unsaturated imine.[78] Excellent selectivity has also been achieved using an *N*-sulfinylimine.[79]

95 : 5

(27)

64 : 36

(28)

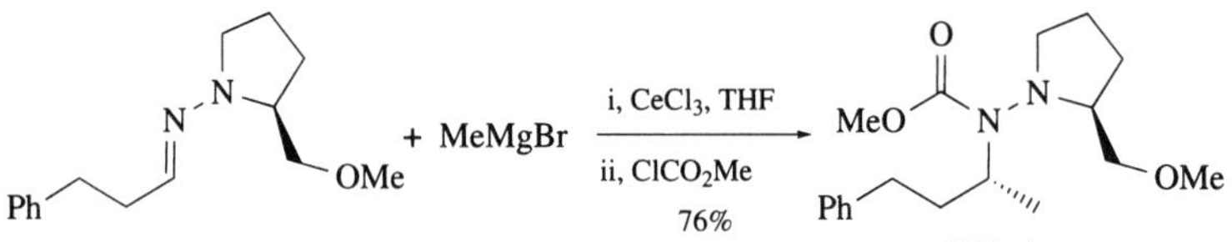

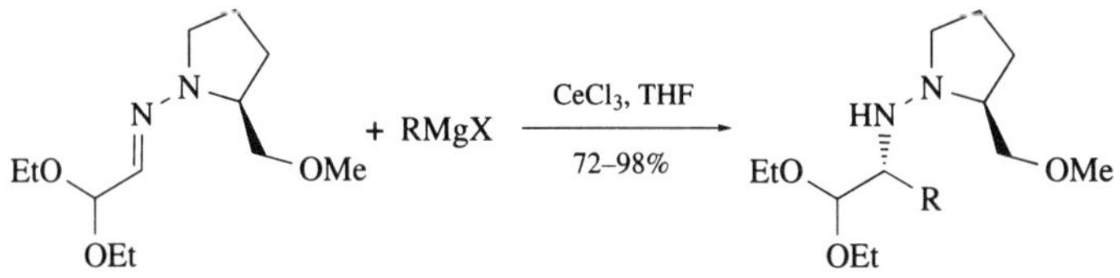

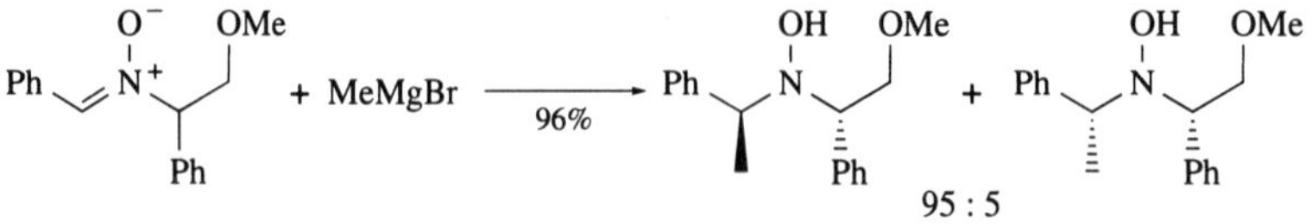

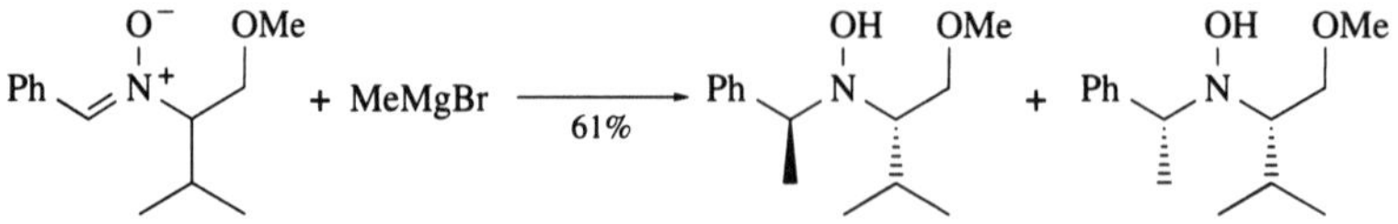

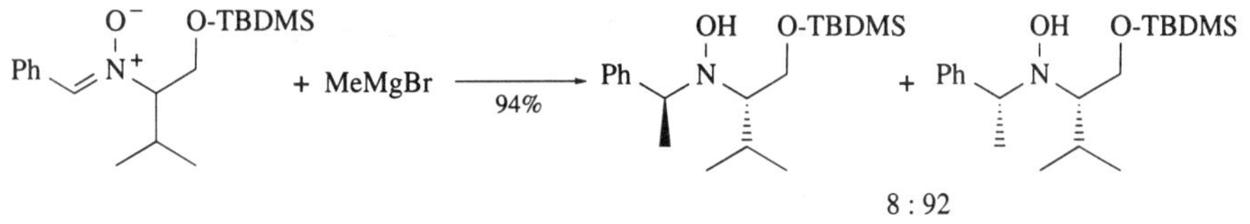

3.3.2.2 C-chiral imines and N-heteroimines

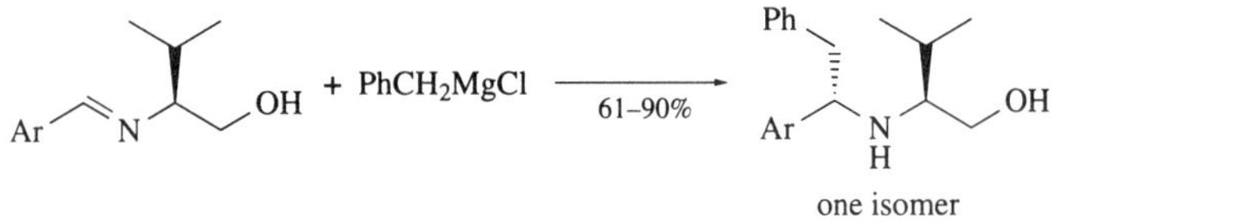

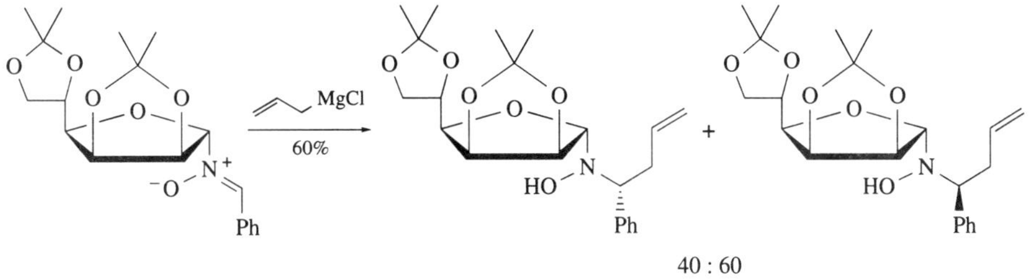

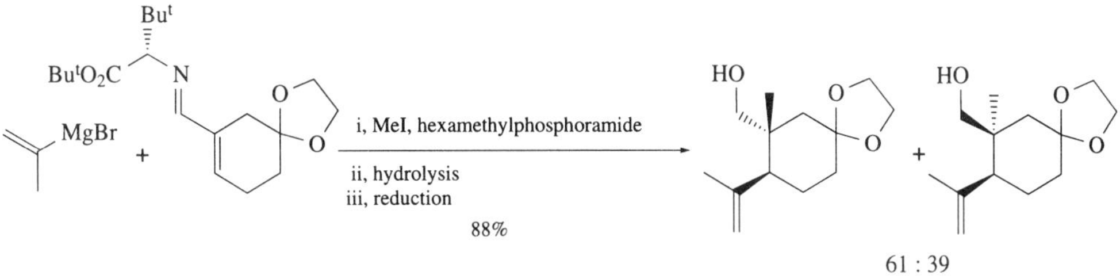

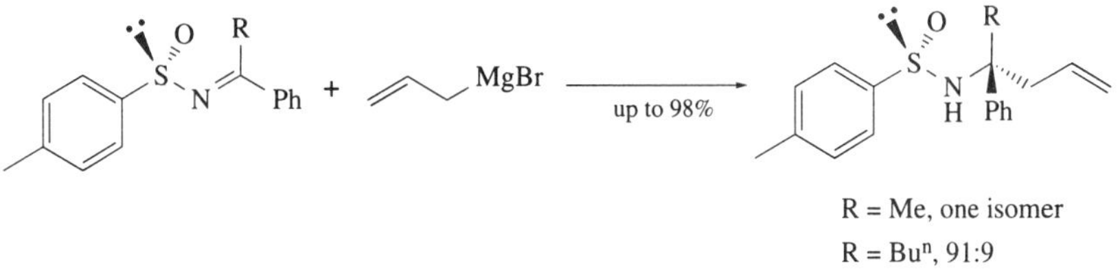

C-chiral imines (and their N-hetero analogues) can have the chiral centre even closer to the site of attack of the nucleophile than their N-chiral counterparts, and the products of organomagnesium addition can then have two (or more) adjacent stereogenic centres. The chiral moiety must remain in the product, unlike the N-chiral cases above, in which an auxiliary could be removed from the nitrogen atom after the reaction. This allows rather less choice in the structure of the chiral moiety, but again potentially chelating oxygen groups have been used in some cases (Equations (34)–(36)). CeCl$_3$ has been used as an additive, sometimes to promote reactions which do not otherwise occur.[80,81] Addition of ZnBr$_2$ to the reaction mixture has been shown to enhance the yields dramatically, but the diastereoselectivity very little; curiously, the direction of selectivity is critically dependent on the organic portion of the Grignard reagent in these cases.[82] In another case, diastereoselectivities were improved by mixing AlCl$_3$ with the organomagnesium reagent before the reaction, but the nature of the reagent (RMg or RAl) is in doubt here.[83] Reetz et al. have shown that α-amino imines can also be used successfully (Equation (35)).[84] Interestingly, lower selectivities are generally obtained when the chiral portion of the molecule does *not* bear a potentially chelating heteroatom (Equations (37) and (38)).[71,85]

In several cases, more remote chiral groups have been used, not in the carbon skeleton of the imine but attached to the carbon atom via a heteroatom, as an auxiliary (Aux). Although the chiral moiety is sometimes many bonds removed from the reactive centre, chelation is always a possibility, and very high selectivites have been observed (Equations (39) and (40) and Scheme 8).[86–8]

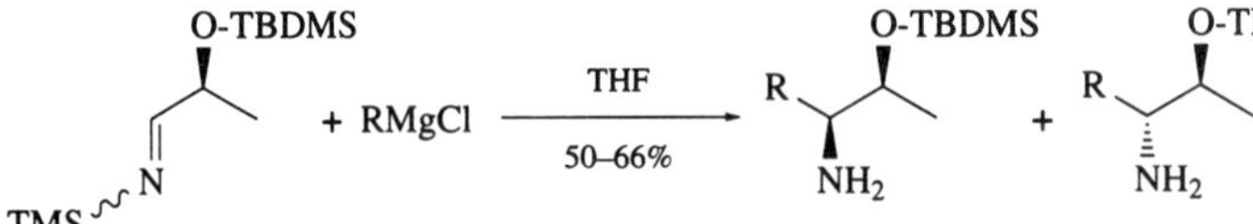

(34)

R = Bz, 100:0
R = allyl, 18:82
R = alkyl, in the presence of $ZnBr_2$, 96%, 9:91

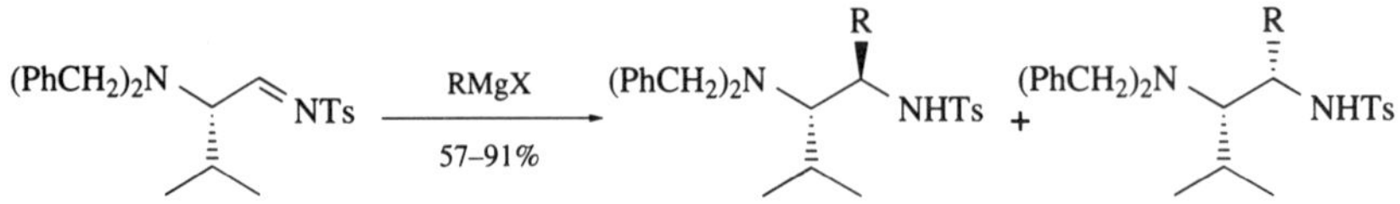

(35)

≥92:8
R = alkyl, Ph, vinyl, allyl

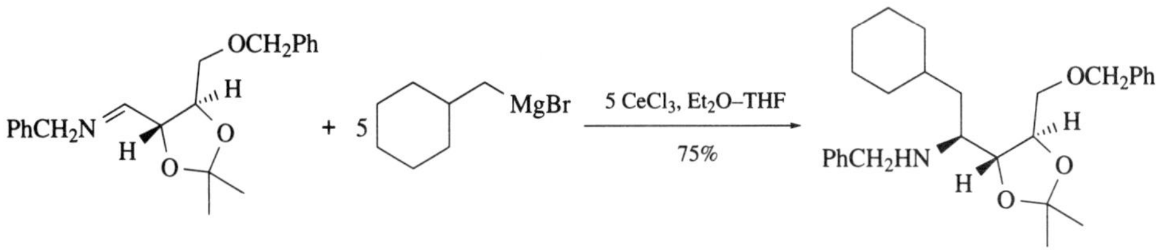

(36)

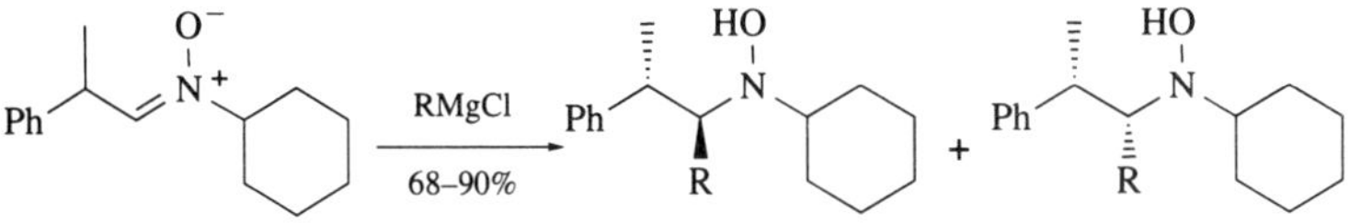

(37)

R = alkyl
67:33 to 83:17

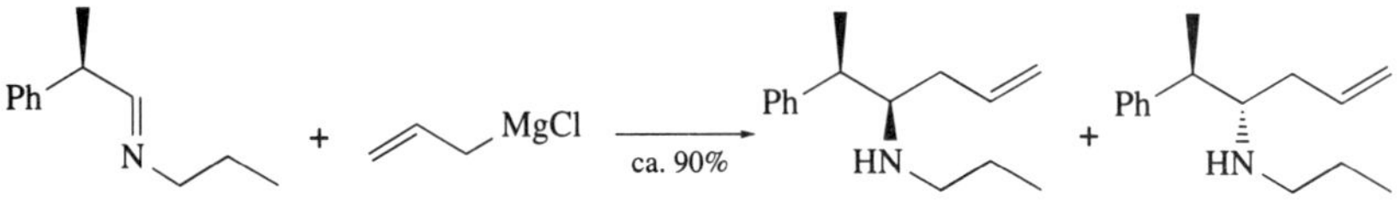

(38)

84 : 16

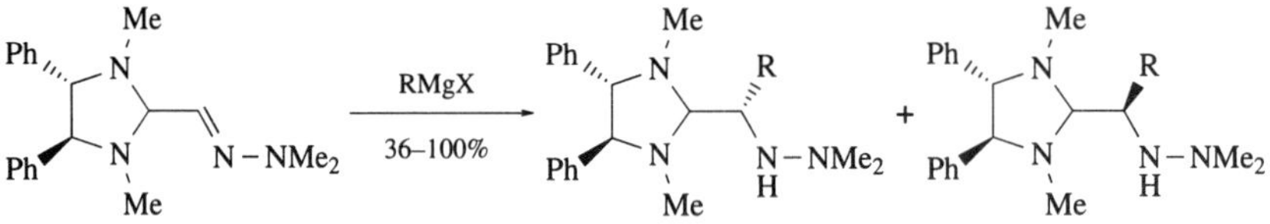

(39)

82:18 to 100:0

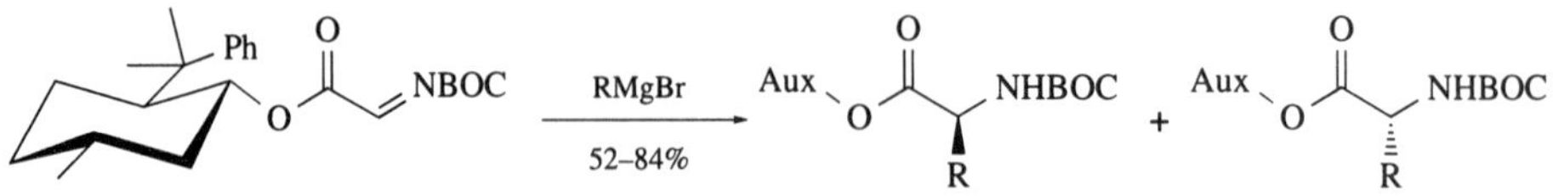

(40)

BOC = *t*-butoxycarbonyl

51 to >98% *de*

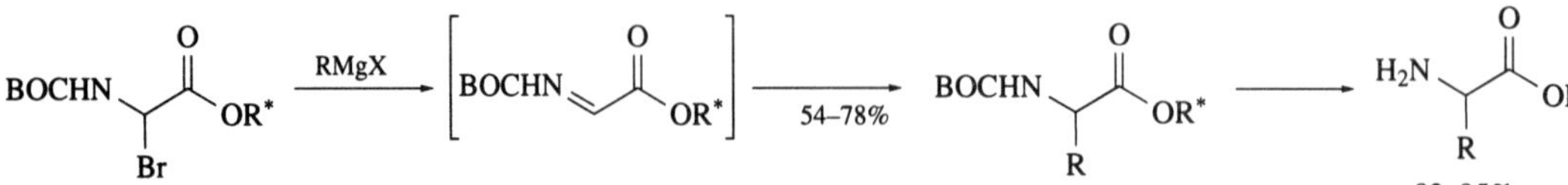

82–95% *ee*

Scheme 8

3.3.2.3 Other possibilities

Placing chiral groups on both carbon and nitrogen of the imine has been shown to give double diastereomeric induction (Scheme 9 and Equations (41) and (42)).[83] In several cases chiral cyclic imines have been used; whether or not these benefit from double diastereomeric induction would be harder to determine.[83,89–91] The observed selectivities vary, but can be excellent. There has also been one report of the addition of a chiral Grignard reagent to imines (Scheme 10).[92] This complements the analogous addition to aldehydes, since it gave opposite selectivity.[92]

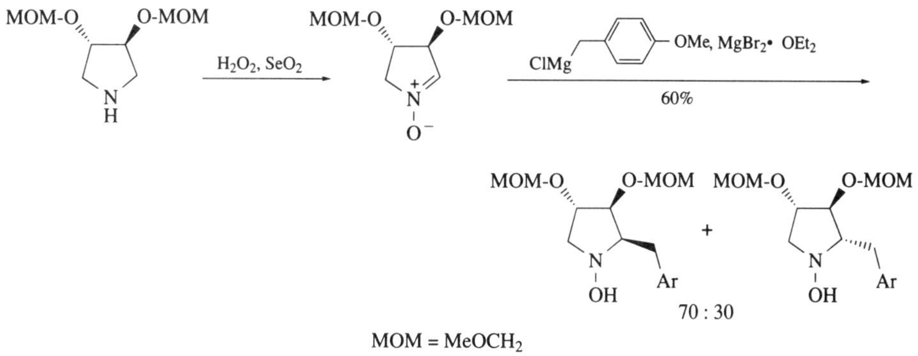

Scheme 9

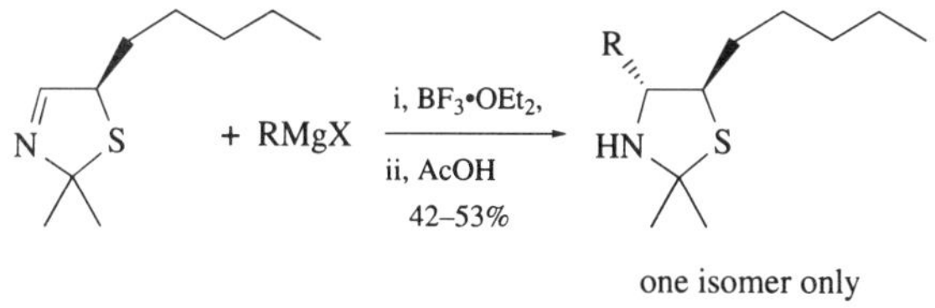

MOM = MeOCH$_2$

(41) one isomer only

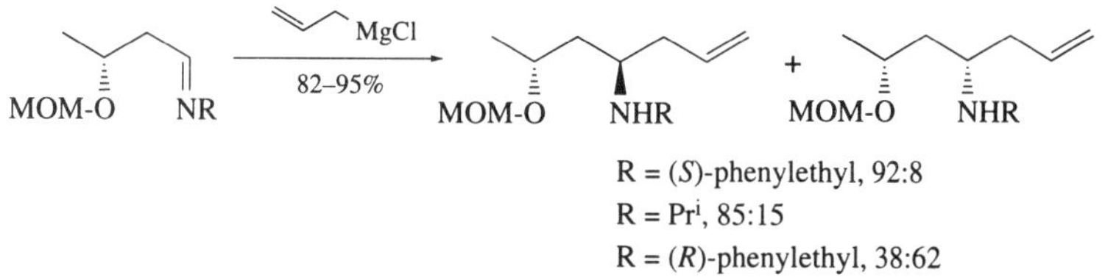

(42)
R = (S)-phenylethyl, 92:8
R = Pr^i, 85:15
R = (R)-phenylethyl, 38:62

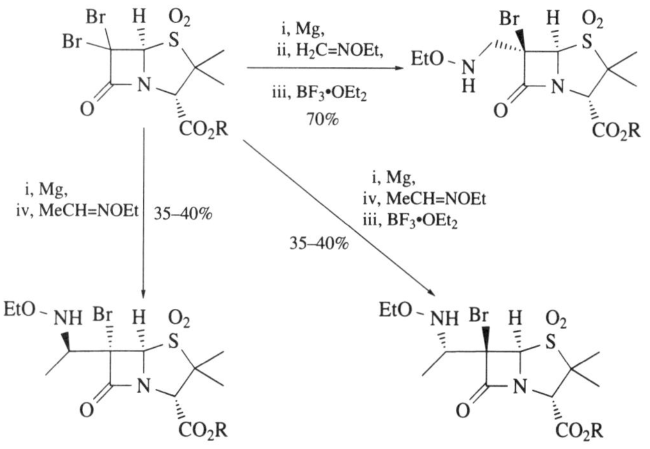

Scheme 10

3.3.3 Addition to Nitriles and Isothiocyanates

The initial product of organomagnesium attack on a nitrile is an *N*-magnesioimine. This may suffer a variety of fates. It may be protonated, and the imine isolated;[93] or hydrolysed, and the ketone isolated;[94,95] or it may undergo further chemistry. The *N*-magnesioimine may be attacked by a nucleophile[96]—which can be an excess of the organomagnesium reagent[97]—or may itself attack an electrophile, either intra-[93] or intermolecularly.[98] Reductive decyanation can be a significant side reaction with hindered nitriles.[99] Use of an unsaturated organomagnesium may give rise to an azadiene which can dimerize by cycloaddition.[97] Representative examples are collected in Equations (43)–(48) and Scheme 11. Grignard attack on an isocyanate usually leads to an amide;[100] the reaction may be improved by titanium catalysis.[56,101] In the analogous attack on isothiocyanates, Masson *et al.* have shown how the thioamide may be alkylated α to the unsaturated carbon in a one-pot reaction. Some of these reactions are shown in Schemes 12 and 13 and Equation (49).[102]

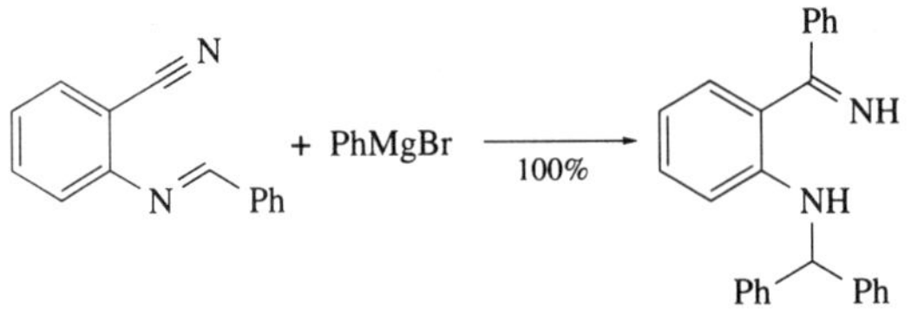

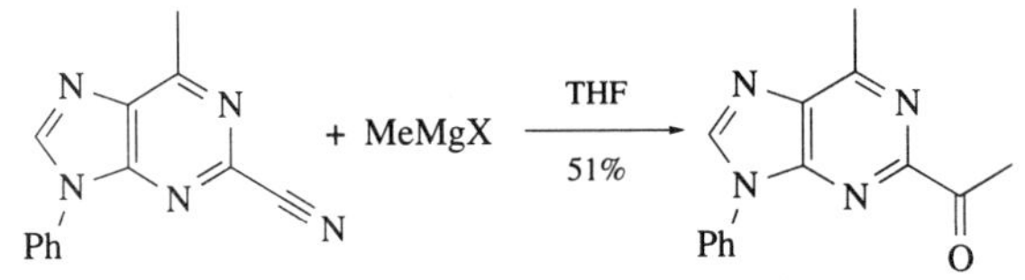

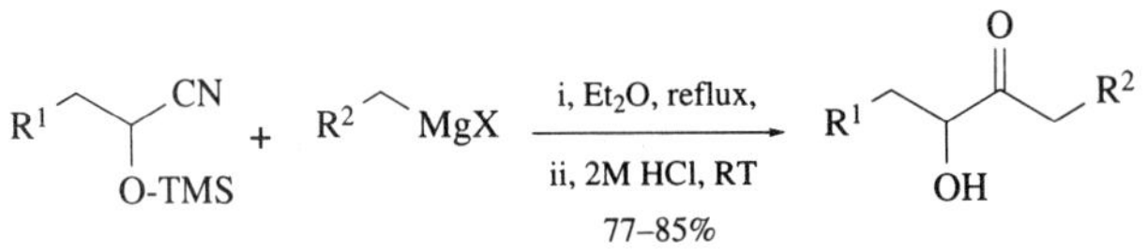

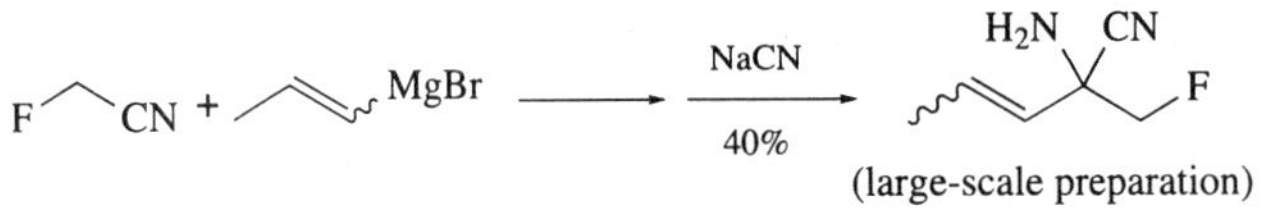

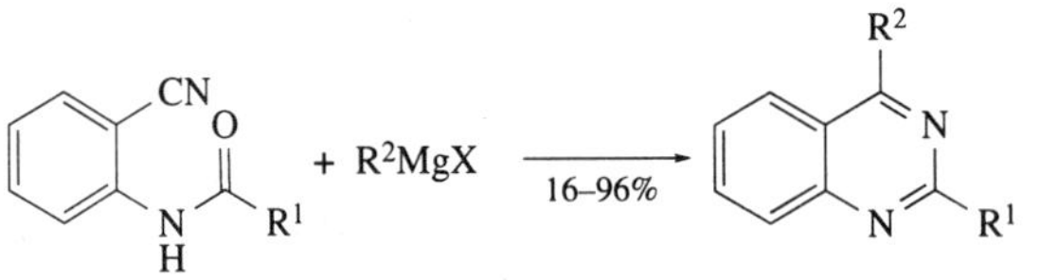

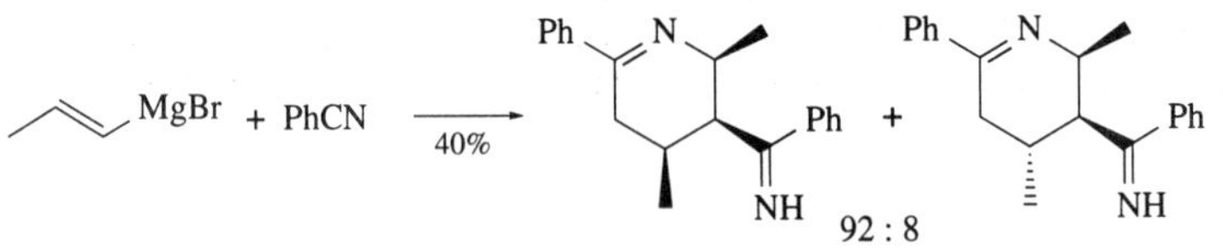

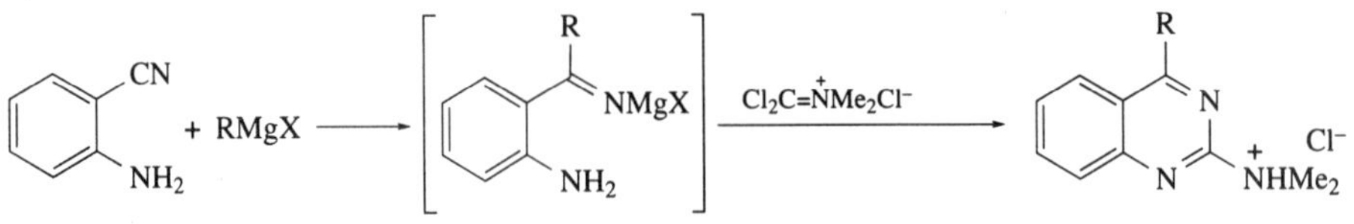

Scheme 11

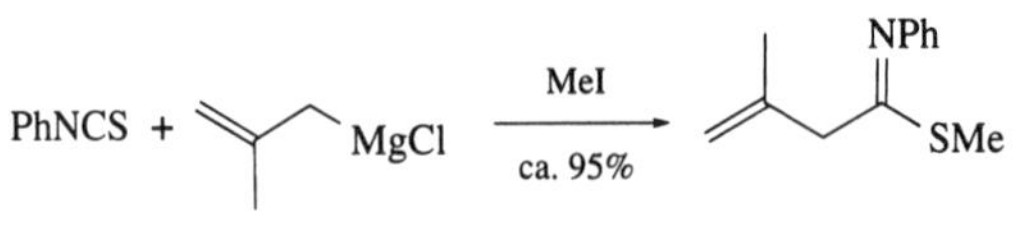

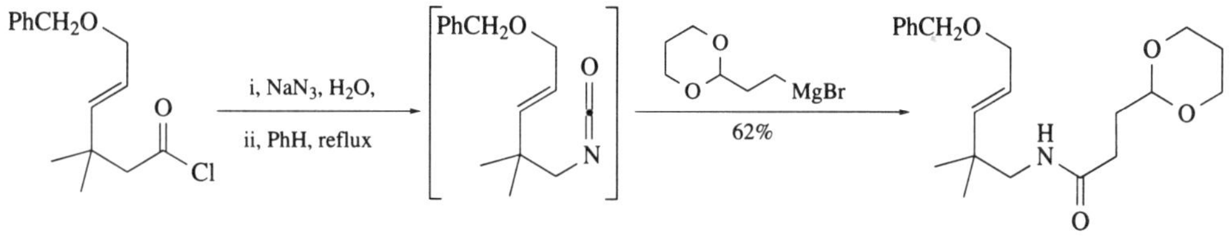

Magnesium 141

Scheme 12

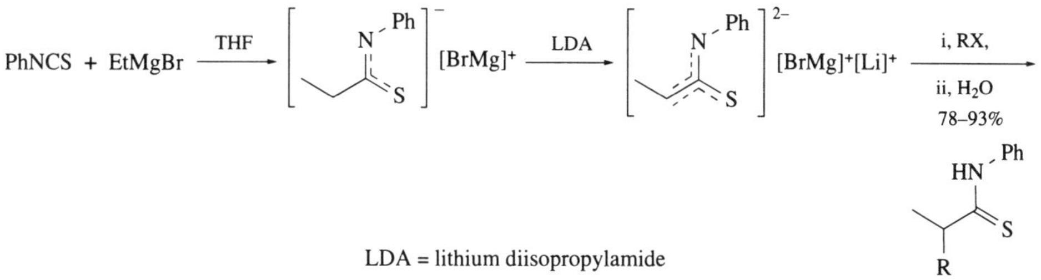

Scheme 13

LDA = lithium diisopropylamide

3.3.4 Pyridines and Related Reactants

3.3.4.1 Nonstereoselective addition

Pyridines are generally inactive towards organomagnesium compounds, unless they bear a suitably placed leaving group such as bromine[103] (Equation (50)); chlorine and methoxy groups are not sufficiently good leaving groups for this reaction (see below). They must normally be activated by placing an electron-withdrawing substituent on nitrogen. This activates both 2- and 4-positions, so that the question of regioselectivity becomes important. With N-formylpyridinium salts, the regioselectivity is highly dependent on the attacking organic fragment (Scheme 14 and Equations (51) and (52)). Alkyl-, allyl- and phenylmagnesium halides tend to give mixtures of 2- and 4-substituted dihydropyridines in which the former predominate unless the alkyl group is very bulky (e.g., Pr^i).[104–6] On the other hand, preferential or even exclusive 4-substitution has been reported for electron-rich benzylic Grignard reagents.[107] Comins and Abdullah, somewhat surprisingly, have obtained exclusively 1,2-addition using 4-functionalized alkyl Grignards.[108] Yamaguchi et al. have shown that vinyl- and alkynylmagnesium halides also give exclusive 1,2-addition, perhaps because of their smaller size at the reactive centre. They have used this observation in the regioselective formation of 2,6-disubstituted dihydropyridine derivatives.[104] Formates are of course not the only activated pyridine derivatives available. Although acyl- and benzoyl-activated pyridines behave in almost the same way as N-formylpyridines, Webb has shown that Grignard attack on N-carbonate pyridinium chloride occurs exclusively at the 2-position for arylmagnesium halides as well as vinyl and alkynyl ones.[109] By contrast, N-silylpyridinium salts react only in the 4-position, presumably because of the bulk of the silyl group (Scheme 15).[110]

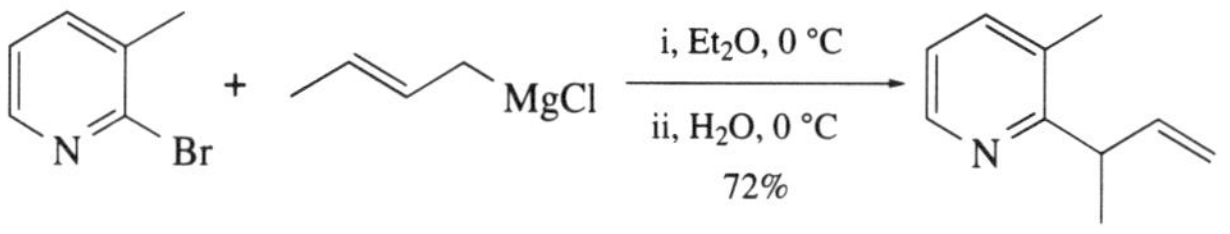

(50)

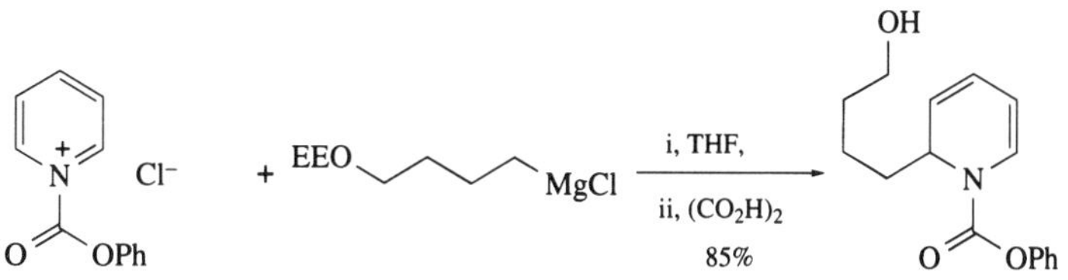

(51)

EEO = EtOCH$_2$CH$_2$O

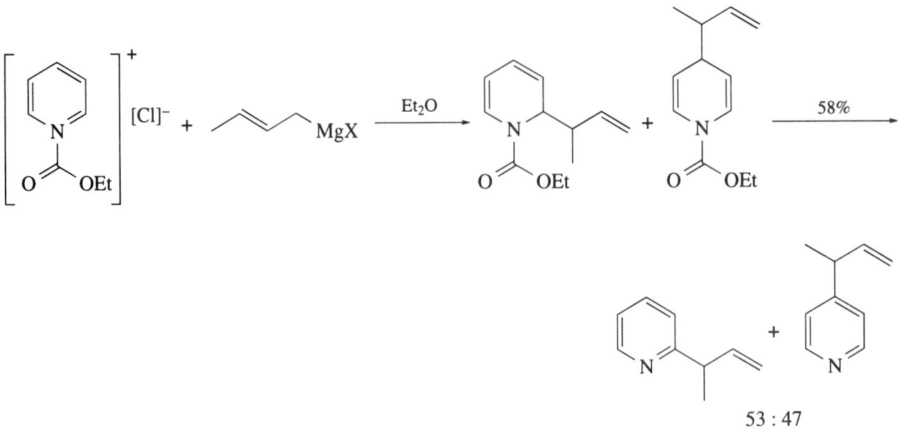

Scheme 14

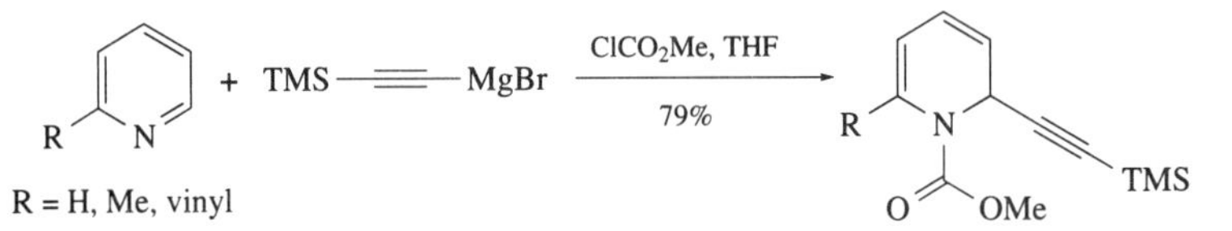

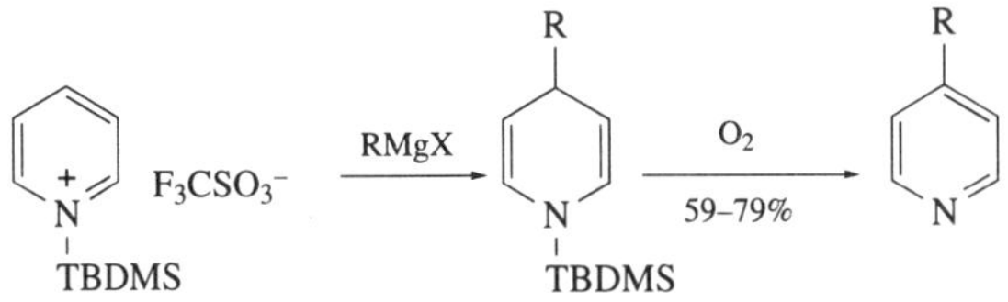

Scheme 15

Perhaps the most obvious way to control the regiochemistry of the addition is to place a blocking group on the 4-position of the pyridine ring. Comins and co-workers have used this strategy extensively in a number of natural product syntheses. The blocking group may be a halide or a methyl group, which is unchanged by the reaction (Equation (53)),[111] but frequently a methoxy group has been used. This becomes an enol ether, which is hydrolysed in workup so that a dihydropyridone is the isolated product (Equation (54)).[112] A further possibility is to use a group such as SnMe$_3$, which can be removed under mild conditions after the reaction to give a simple 2-substituted dihydropyridine (Scheme 16).[113]

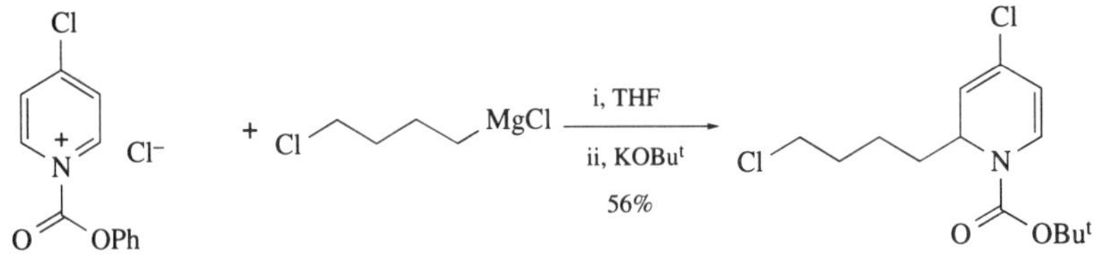

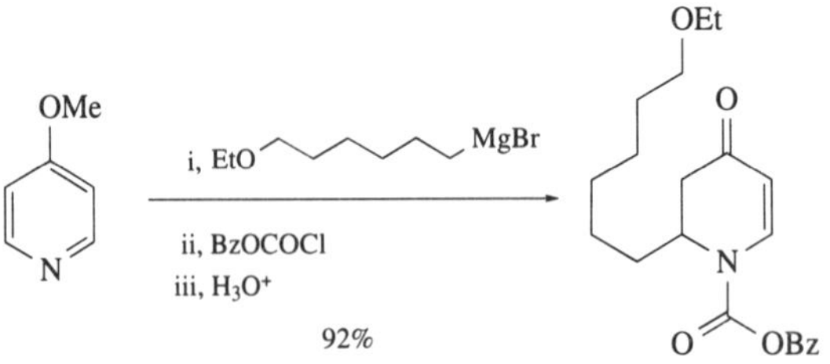

An alternative strategy which Comins and co-workers have also used successfully is to place a large blocking group on the 5-position of the pyridine ring, thereby sterically hindering attack at the 4-position. Two groups have proved large enough to be useful: SnCy$_3$ (Equation (55))[114] and SiPr$^i_3$ (Equation (56));[115] the second of these gives better control. Smaller groups such as CO$_2$Bu^t confer less

Scheme 16

useful levels of regioselectivity (Equation (57)).[116] After the reaction the blocking group may be removed altogether,[117] or used to introduce another group at the 5-position.[118] Used in conjunction with a 4-methoxy group, the latter method gives a regiocontrolled synthesis of 1,2,5-trisubstituted-2,3-dihydro-4-pyridones, and hence a variety of fused-ring polycyclic compounds (Scheme 17).

(55)

(56)

(57)

Scheme 17

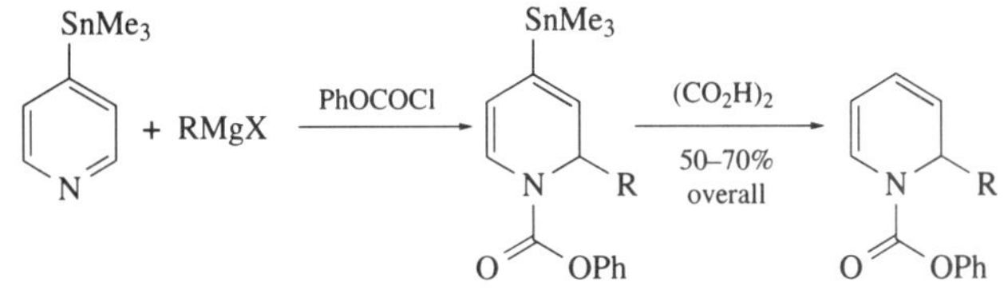

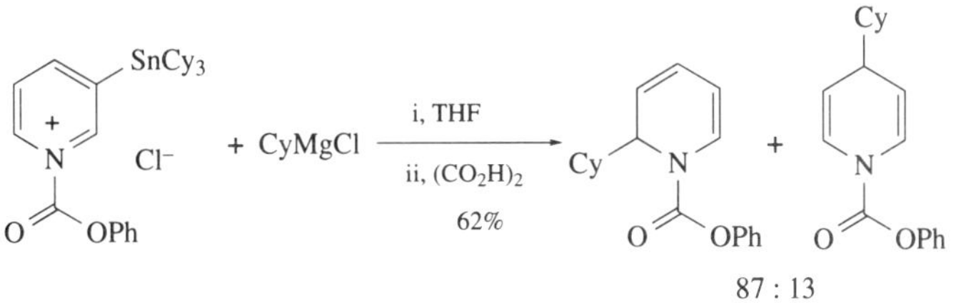

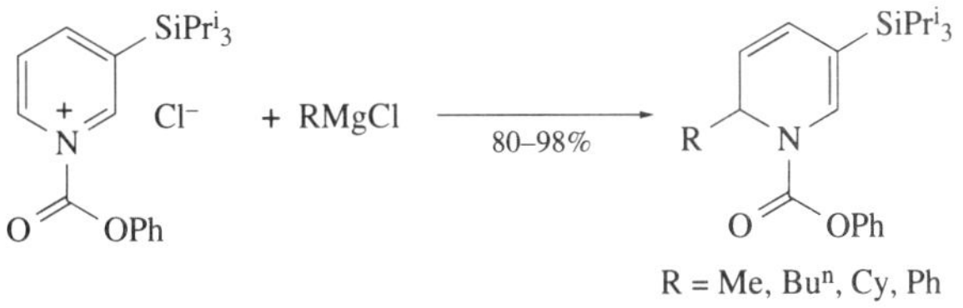

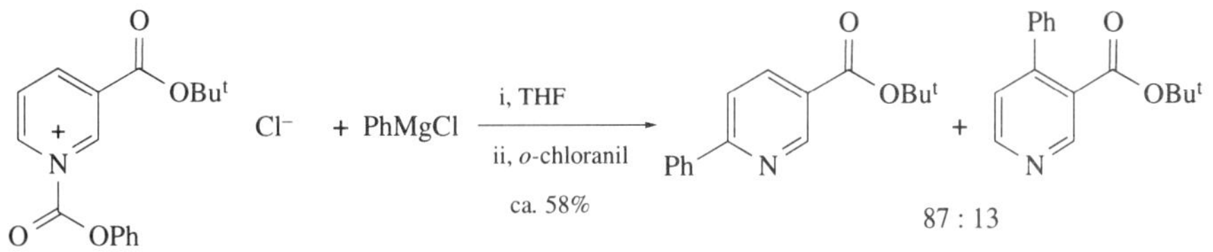

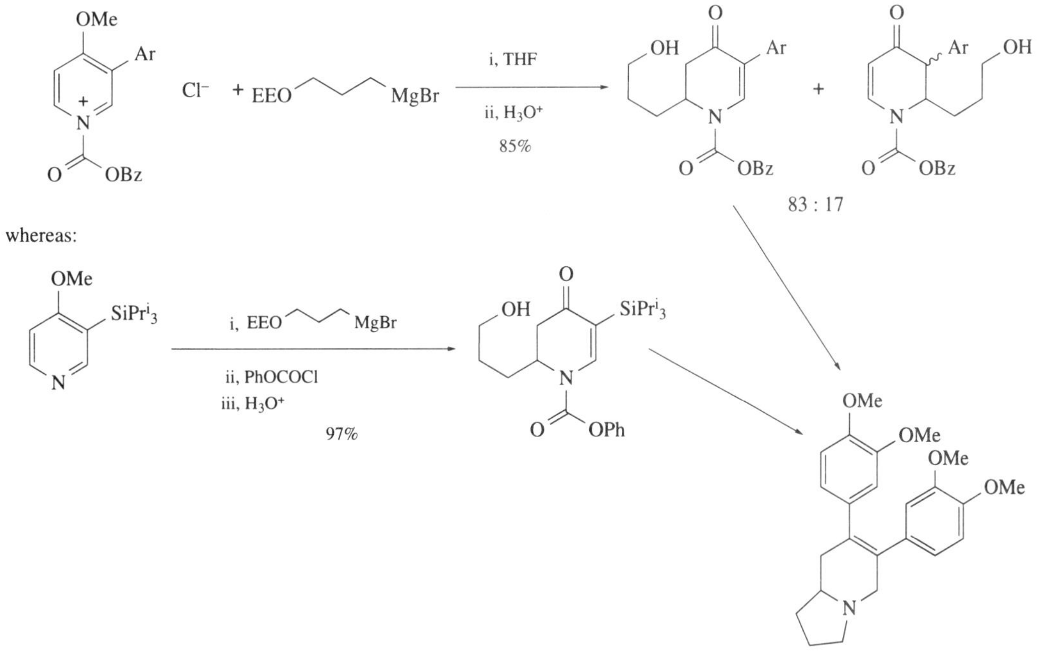

3.3.4.2 Stereoselective attack on pyridine derivatives

Following his investigation of the regiochemical problem (see Section 3.3.4.1), Comins has shown that considerable stereoselectivity can be induced in organomagnesium addition to pyridine derivatives by using a chiral activating group on nitrogen. Although menthyl formate gave only moderate selectivity, good to excellent selectivity has been achieved using 8-phenyl- or 8-(4-phenoxyphenyl)-menthyl formate. This has been combined with the regiodirecting strategies discussed above in several natural product syntheses. Thus, (−)-N-methylconiine has been synthesized from 3-triisopropylstannylpyridine (Scheme 18),[119] and (−)-coniine hydrochloride from 4-methoxy-3-triisopropylsilylpyridine (Scheme 19),[120] while a synthesis of elaeokanine C from the latter starting material gave the unnatural enantiomer (Scheme 20).[121] (The last step in this latter synthesis also involves an organomagnesium reagent, in a cerium-mediated Grignard addition to an amide.) An enantioselective synthesis of solenopsin-a relies on a similar strategy.[122] In the absence of a bulky group on the 3- (or 5-) position of the pyridine ring, high stereoselectivity can be achieved only by using an exceptionally bulky nucleophile such as triphenylsilylmagnesium bromide.[123a] Rather lower selectivity may be obtained using a variety of more conventional organomagnesium reagents (Equation (58)).[123b]

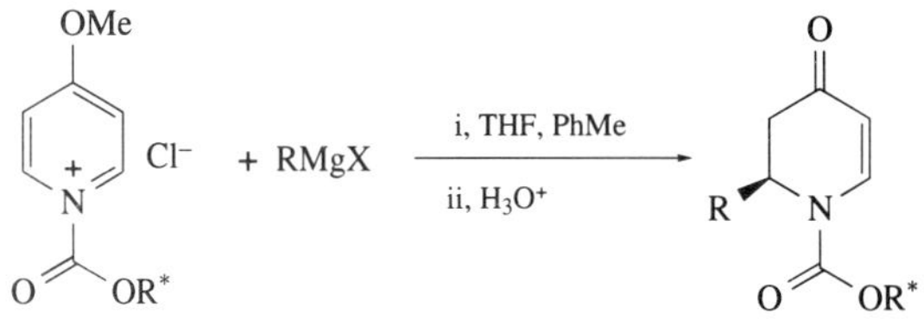

$$R = Ph, 30\% \ de, 83\% \ \text{yield}$$
$$R = Ph_3Si, 96\% \ de, 88\% \ \text{yield}$$

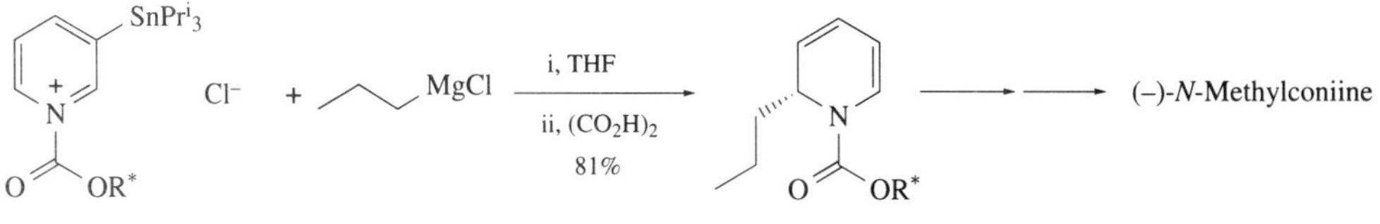

(58)

$R^* = (−)$-8-phenylmenthyl

Scheme 18

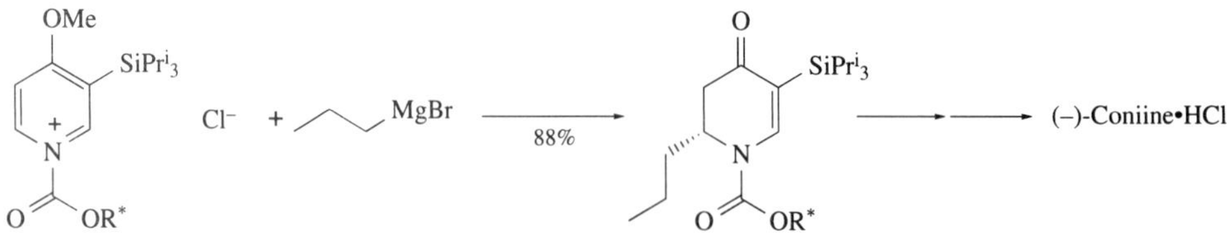

Scheme 19

Two other approaches to stereoselective attack on a pyridine derivative have been reported (Equations (59) and (60)). Génisson et al. have obtained 1,2,3,4-tetrasubstituted 1,2,5,6-tetrahydropyridine derivatives in up to 86% de using N-1-phenylethylpyridinium salts (Equation (60)),[124] while Schultz et al. used a rather different strategy to obtain regio- and stereoselectivity.[125]

Elaeokanine C

Scheme 20

(59)

R^1 = Me, CH_2Ph; R^2 = Me, Bu^n, Bu^i, Ph, vinyl
but when R^2 = allyl, regio- and stereoselectivity are much lower

(60)

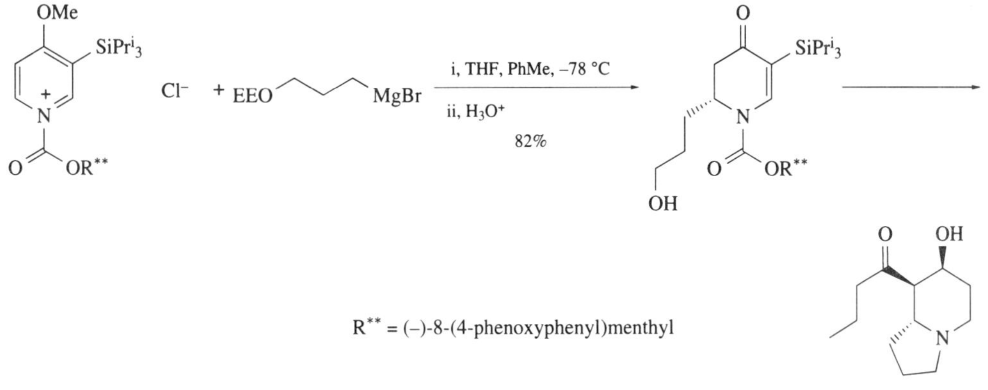

3.4 ADDITION TO AND SUBSTITUTION OF CARBON–OXYGEN BONDS

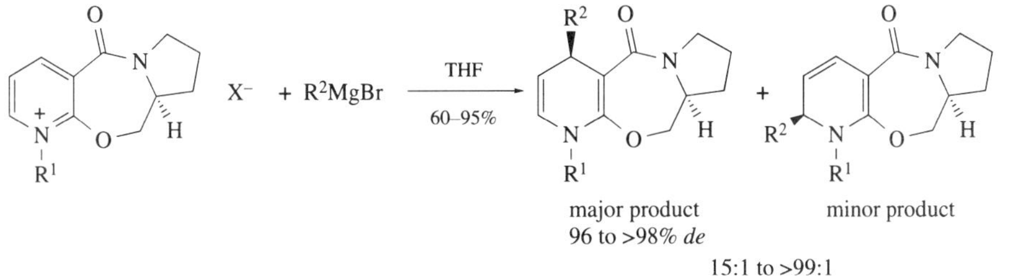

3.4.1 Diastereoselective Additions to Aldehydes and Ketones

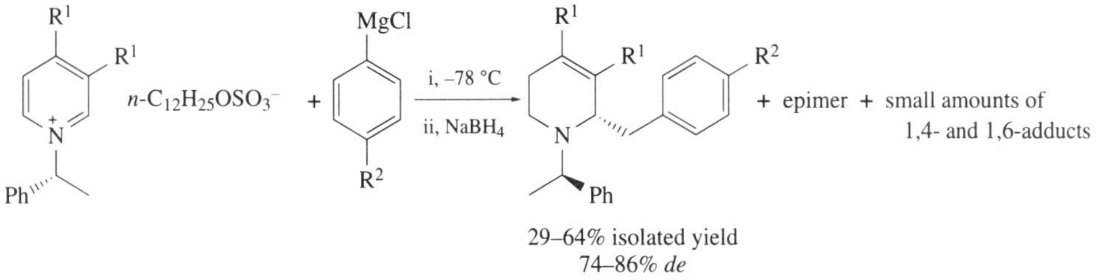

3.4.1.1 Reaction between two prochiral compounds

Crotylmagnesium halides (where the magnesium is bound to the primary carbon) react with aldehydes and prochiral ketones to give *syn* and *anti* diastereoisomeric alcohols. The selectivity of this reaction varies from negligible to complete, depending on the structure of the electrophile (Equations (61) and (62)).[126] Stereoselectivity is poor in the reaction of a cyclopentenylmethylmagnesium halide with aldehydes even at −95 °C, but the boronate complex formed by addition of 9-methoxy-9-borabicyclo[3.3.1]nonane reacts with satisfactory selectivity (Equation (63)).[127] Seebach and co-workers found good diastereoselectivity in the attack of the benzylic Grignard reagent shown (here classified as a formally prochiral benzyl anion, although the racemization process may be relatively slow) on a variety of prochiral aldehydes and also on benzophenone (Equation (64)).[128]

3.4.1.2 Addition to α-chiral aldehydes and ketones

One of the long-standing problems in asymmetric synthesis is the control of stereochemistry in addition of organometallic reagents to ketones and aldehydes with an adjacent stereocentre. The rationalization of this reaction has largely been due to Felkin and Anh, and their refinement of Cram's rule is used as the standard model for rationalization, as shown in Figure 1. The literature is conveniently subdivided into simple, polarity-controlled and chelate-controlled additions, and clear guidelines are provided in a review by Reetz.[129] Two significant methods for improving (or reversing!) the stereoselectivity are reported, exemplified by Equations (65) and (66). The first involves the addition of bulky Lewis acids to the reaction mixture,[130] which effectively changes the stereochemical preference. To reinforce the natural (Cram) selectivity, a simple strategy of changing the spectator substituent on magnesium to an oxygen function[131] works well. Even so, the prediction of stereochemistry in a particular case is fraught with difficulty, because the reaction can be sensitive to solvent, temperature and reagent structure; the steroidal and alicyclic examples[132,133] in Equations (67) and (68) serve to illustrate this.

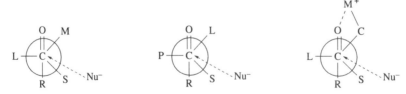

Figure 1 Standard model for rationalization of addition to α-chiral aldehydes and ketones. Groups in reactive conformation are designated: L, large; M, medium-sized; S, small; and P, polar. C represents a chelating group.

Insofar as chelate-controlled additions are concerned, Still and McDonald[134] have provided a model for Grignard addition steps in many natural product syntheses. The strategy is a simple one, in that methoxyethoxymethyl (MEM) and other α-alkoxyketones undergo quite specific *syn*-addition of Grignard reagents at low temperature in THF[134] (Cram's original work was in Et$_2$O). Notwithstanding the high diastereoselectivity, this method is less successful for α-alkoxyaldehydes but in such cases

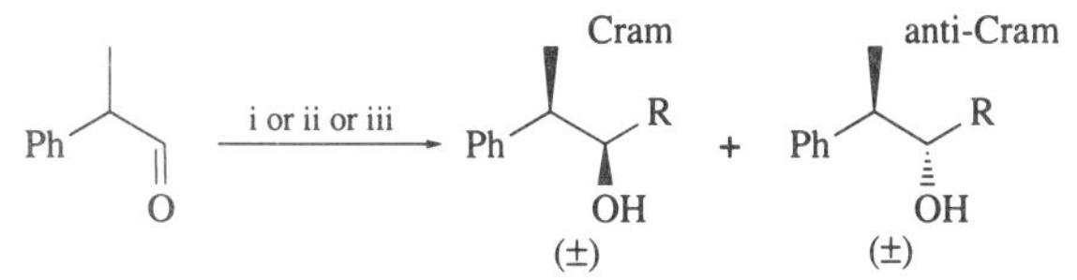

Equation (65)

i, MeMgOSO$_2$CF$_3$, THF, –78 °C   90 : 10
ii, EtMgBr, –78 °C   84 : 16
iii, EtMgBr, AlR$_3$, toluene/Et$_2$O, –78 °C   13 : 87

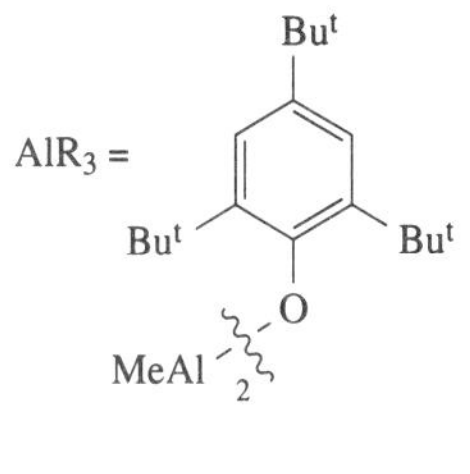

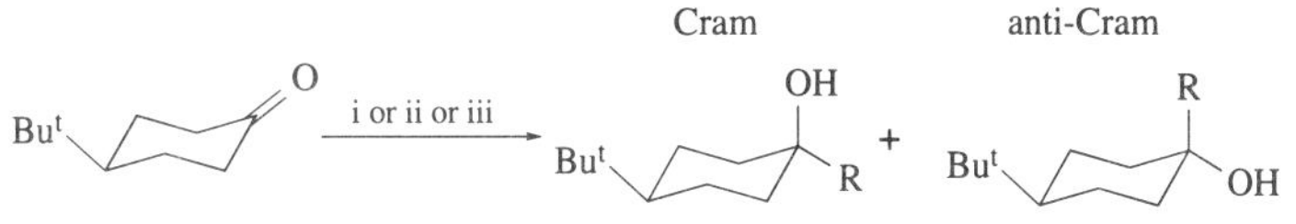

AlR$_3$ = (2,6-di-Bu^t-4-Bu^t-phenoxy)(MeAl)$_2$

Equation (66)

i, MeMgOSO$_2$C$_6$H$_2$Me$_3$, THF, –78 °C   90 : 10
ii, EtMgBr, –78 °C   48 : 52
iii, EtMgBr, AlR$_3$, toluene/Et$_2$O, –78 °C   0 : 100

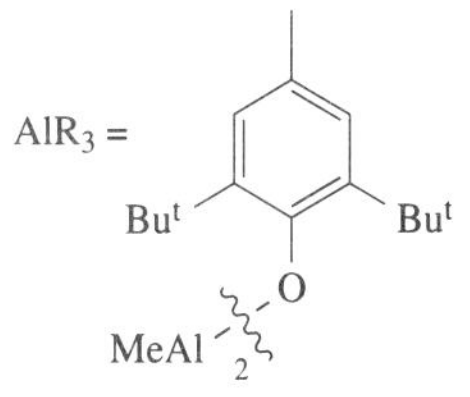

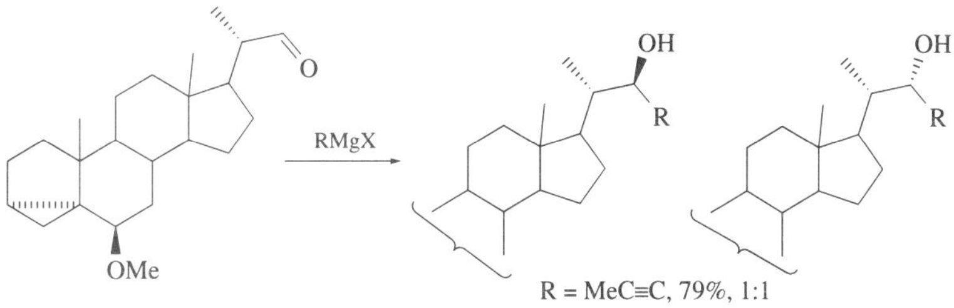

AlR$_3$ = (2,6-di-Bu^t-4-Me-phenoxy)(MeAl)$_2$

Equation (67)

R = MeC≡C, 79%, 1:1
R = Pr^n, 100%, 100:0

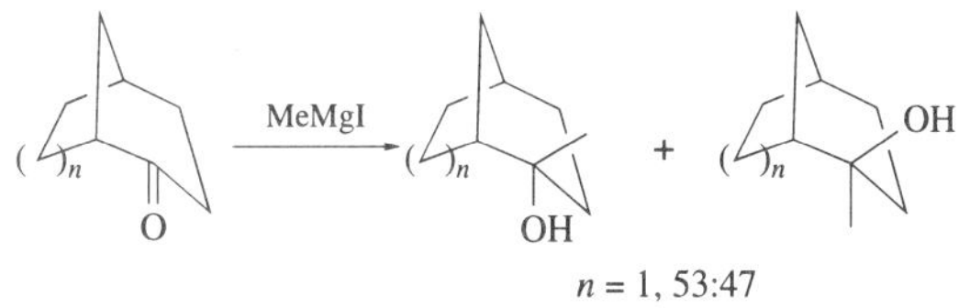

Equation (68)

n = 1, 53:47
n = 2, 95:5, 90%

carrying out the reaction in Et$_2$O in the presence of ZnBr$_2$ gives >97% *syn*-selectivity.[135] Many examples of chelate-controlled additions have been reported, including ones involving sugar-based aldehydes and ketones,[136,137] glyceraldehyde derivatives[138] and epoxyketones,[139–41] among others. The stereochemical course of Grignard additions to glyceraldehyde acetals is difficult to control, and dependent on the reagent. Examples are given in Equations (69)–(73). Epoxy ketones and aldehydes normally give rather poor stereochemical control, but a silyl group at the α- or β-carbon provides the bulk required for a stereoselective process believed to occur under chelate control.

Of course, these addition reactions should be set in context. Many other organometallics have been utilized in diastereoselective additions to chiral aldehydes and ketones; often the best results are obtained with more oxophilic compounds, for example, of boron or titanium.

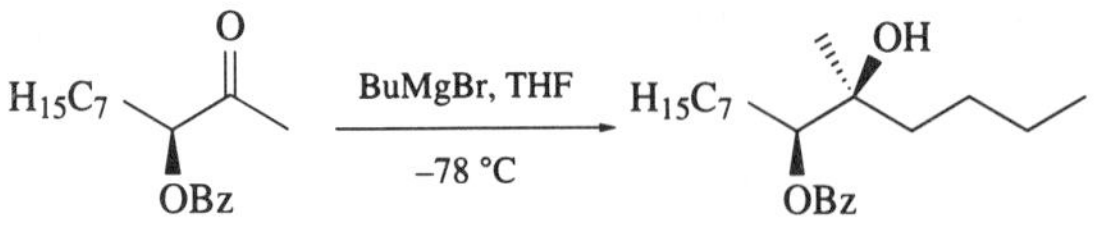

(69)

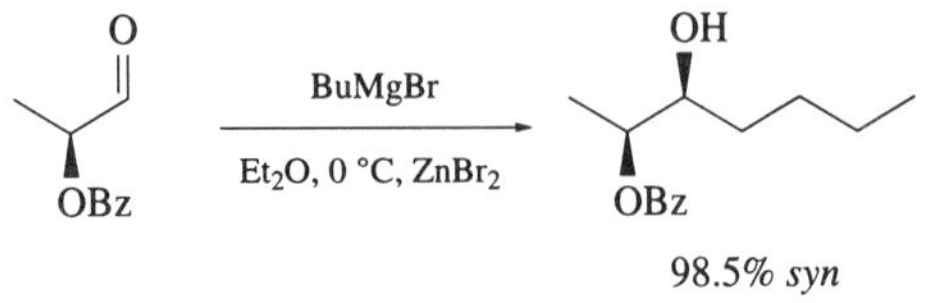

(70)
98.5% syn

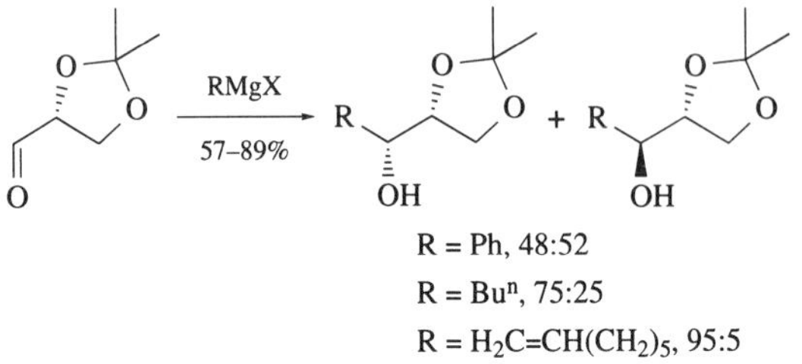

(71)
R = Ph, 48:52
R = Bu^n, 75:25
R = H$_2$C=CH(CH$_2$)$_5$, 95:5

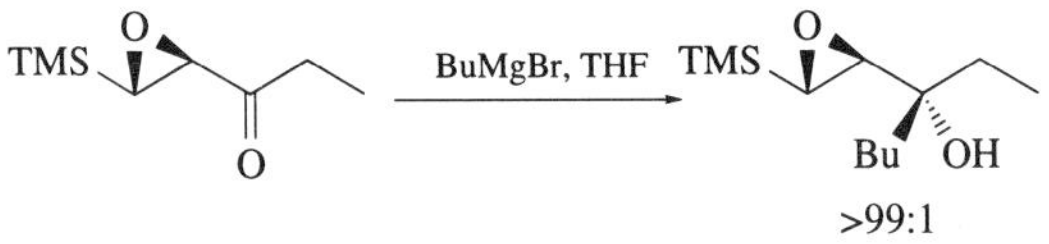

(72)
>99:1

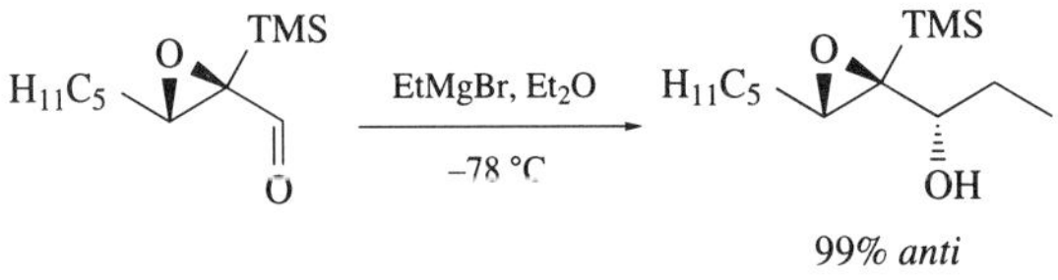

(73)
99% anti

3.4.1.3 Addition to aldehydes and ketones with chiral auxiliaries

A frequent application of chiral auxiliaries in Grignard additions to ketones is when the auxiliary protects one C=O of an α-dicarbonyl compound. The prolinol derivative shown[142,143] (Scheme 21) can be used to control the first step in an ingenious synthesis of *t*-allylic alcohols; the reverse selectivity may be obtained in this first step when cyclohexyllithium is employed in the presence of 1,8-diazabicyclo[5.4.0]undec-7-ene (dbu). This stereoselective RMgX addition works presumably because approach to one face of the carbonyl group is disfavoured sterically. For the tartrate-derived acetals developed by Annoura and co-workers (Equation (74)),[144] it is more probable that a pendant –OMe group coordinates to the incoming R–Mg, but then only one approach is feasible. Tartaric acid forms both the chiral auxiliary and the template for an asymmetric synthesis of butenolides which involves two sequential stereoselective Grignard additions, both with a high degree of stereocontrol[145] (Scheme 22).

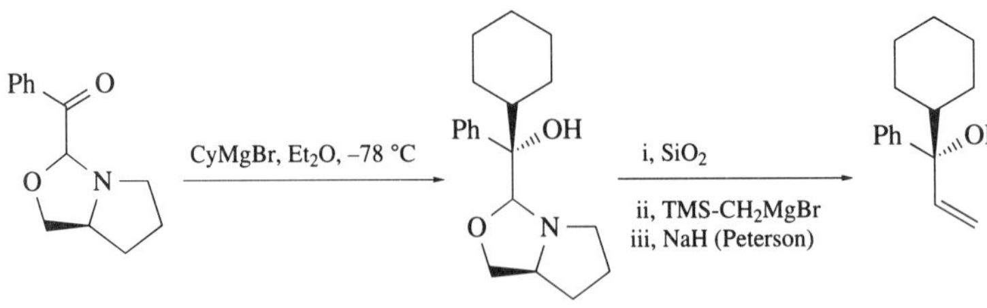

Scheme 21

The ubiquitous oxazolidines derived from aldehydes and chiral β-aminoethanol derivatives can be employed in two distinct ways. First the formal monoderivative of *o*-phthalaldehyde and *N*-isopropylvalinol adds Grignard reagents under carefully defined conditions to give an intermediate which can readily be converted into the corresponding phthalide (Scheme 23).[146,147] The oxazolidine

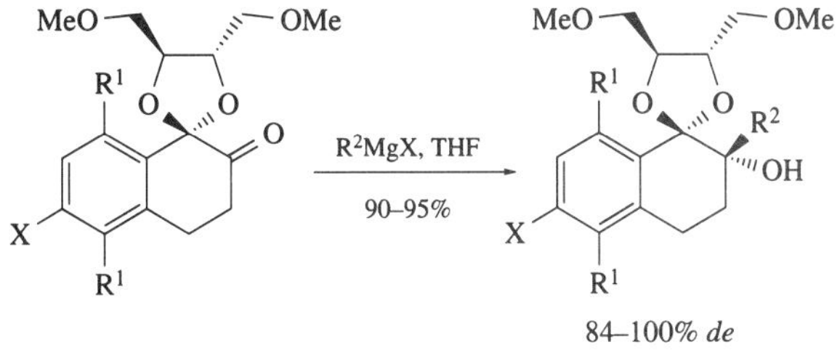

(74)

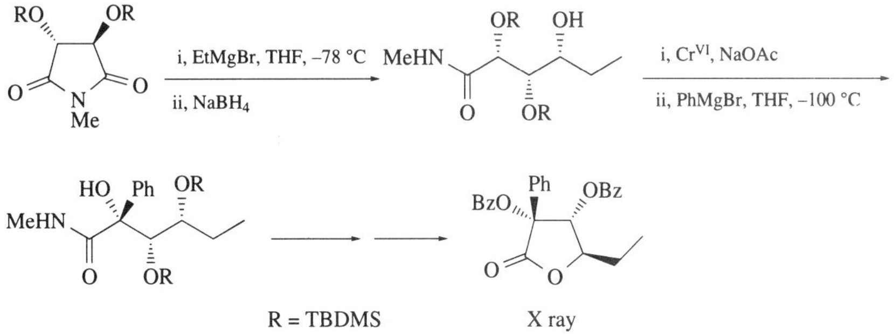

Scheme 22

from benzaldehyde undergoes directed BuLi *ortho*-metallation and transmetallation with $MgBr_2$ to give a reagent capable of highly stereo- and enantioselective addition to an anhydride (Equation (75)).[148] In order to 'reach' from the auxiliary to the carbonyl group and provide recognition, it was proposed that two molecules of the Grignard reagent were involved at the transition state. Other chiral auxiliaries which have been used include a BINOL-derived system[149] where one arm functions as a Lewis base towards the magnesium atom and the other carries the carbonyl group (Equation (76), BINOL = 2,2'-dihydroxy-1,1'-binaphthyl).

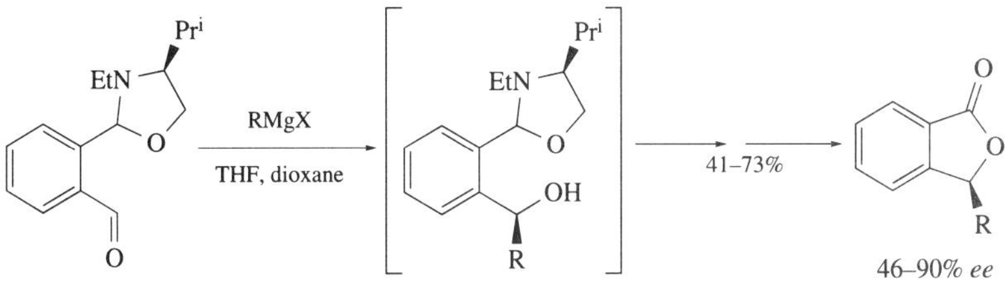

Scheme 23

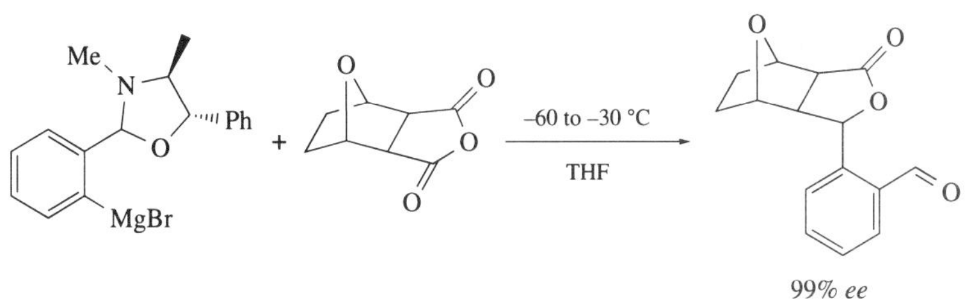

(75)

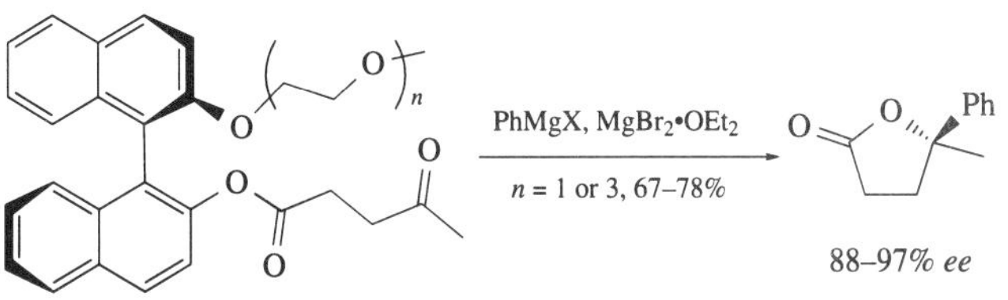

(76)

3.4.1.4 Chiral Grignard reagents

α-Chiral Grignard reagents are configurationally unstable, but the rate of racemization depends very much on the substituents at the stereogenic centre; there is a lack of systematic work on the diastereoselectivity of their reactions with prochiral carbonyl compounds. For configurational stability, the stereogenic centre must be at least one carbon distant from the metal. Miles *et al.* have achieved remarkably good stereoselectivity (up to 10:1) with a γ-chiral Grignard reagent which probably involves MeO coordination to magnesium (Equation (77)),[150] while a menthol-derived Grignard reagent was less selective (21–66% *de*).[151] Volkmann and co-workers have reported several penicillin-derived Grignard reagents, one of which gave only one product with acetaldehyde (Equation (78)).[152,153] Nitrogen heterocycles[154] and chiral acetals (Scheme 24)[155] have been used as auxiliaries in Grignard reagents, with moderate success (30–88% *de*). Using a γ-chiral Grignard reagent and his menthol-derived chiral auxiliary, Eliel has demonstrated a degree of double asymmetric induction (Equations (79) and (80)).[156]

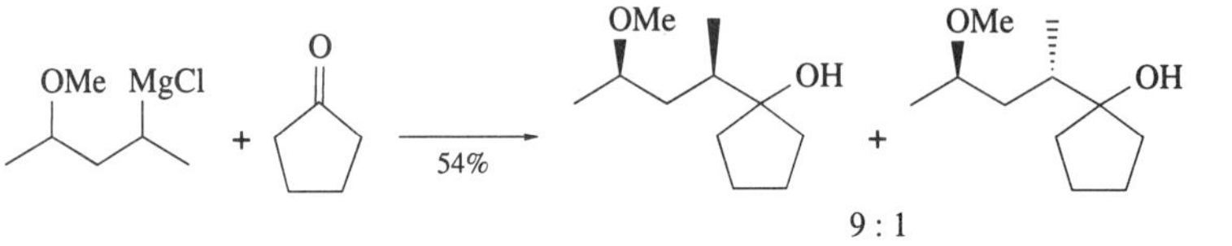

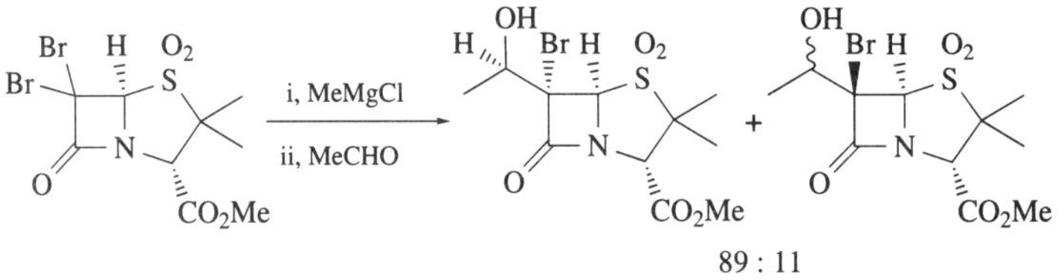

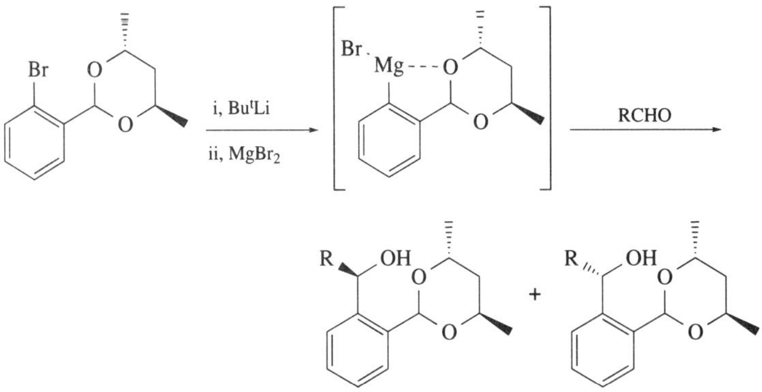

Scheme 24

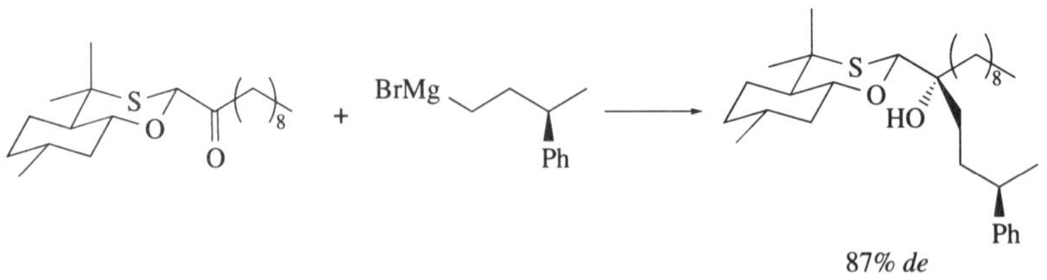

3.4.2 Enantioselective Additions to Aldehydes and Ketones

There are rather few successful experiments, despite much effort, employing a chiral additive to make the addition of an organomagnesium compound to a carbonyl group enantioselective (Equations (81) and (82)). The species involved in the addition may be complex aggregates, making predictive

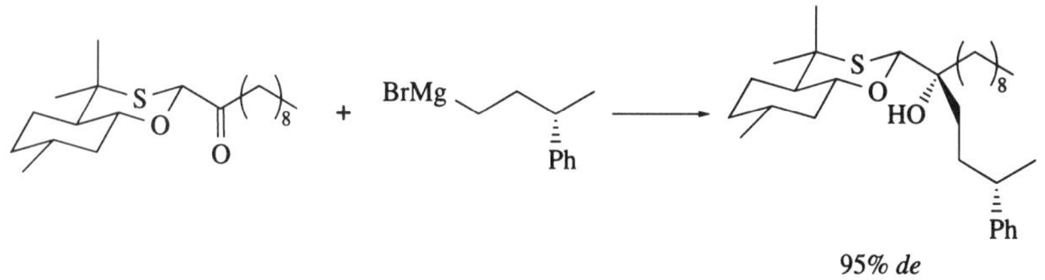

(80) 95% de

analysis difficult. Thus, Seebach and co-workers use three equivalents of Grignard reagent to one equivalent of their tartaric acid based auxiliary TADDOL (α,α,α',α'-tetraaryl-2,2-dimethyl-1,3-dioxolane-4,5-dimethanol) to generate enantiomerically enriched alcohols from conjugated or aromatic ketones.[157] Noyori has utilized a complex of the lithium salt of BINOL, admixed with the Grignard reagent, and carried out the addition at very low temperatures. Enantioselectivity is higher with aldehydes than with ketones, and although the enantiomeric excesses are respectable, the operational simplicity of catalysed organozinc additions to aromatic aldehydes makes that the preferred method where appropriate (see Chapter 4, this volume).[158] Asymmetric catalysts have also been used, especially diamine complexes of aluminum[159] or zinc; the latter give rather low enantiomeric excesses at present (maximum 26%).[42,160] Better ones (68–82%) have been obtained by mixing the organomagnesium and $ZnCl_2$ before adding ligand and substrate, although the exact roles of the two metals in the reactive complex remains in doubt.[161]

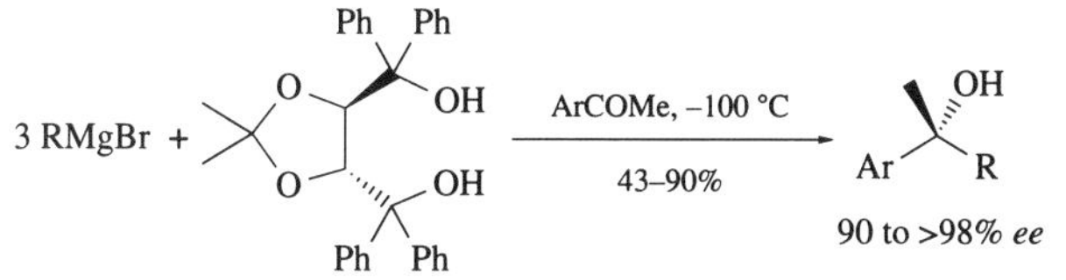

(81) 43–90%, 90 to >98% ee

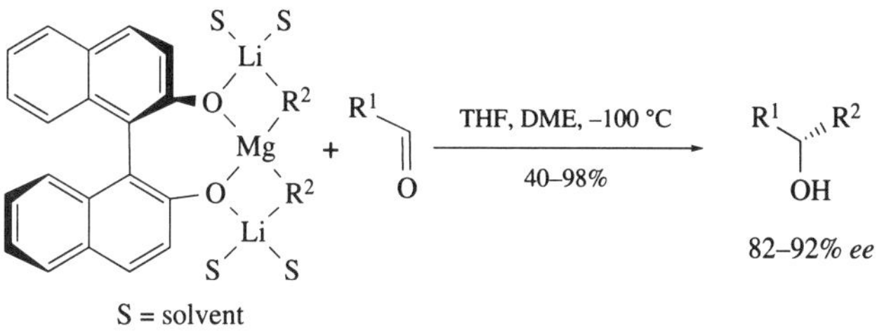

(82) 40–98%, 82–92% ee

S = solvent

3.4.3 Additions to Carboxylic Acid Derivatives

Allylmagnesium halides are often more convenient and easy to prepare than are vinylmagnesium halides. By carrying out the Grignard addition of allyls to esters, and isomerizing the initially formed β,γ-unsaturated ketones *in situ*, Fehr and co-workers have provided a successful route to a number of conjugated ketones useful as perfumery intermediates (Equation (83) and Scheme 25).[162] By far the most common reaction of RMgX with amides is monoaddition, to give ketones on workup. This has also been intensively studied with formamides in order to optimize the synthesis of aldehydes (Scheme 26)[163,164] or oxalyl diamides (Equation (84)),[165–7] and can be carried out in the presence of other groups such as esters (Equation (85))[168] or silyl ethers (Equation (86))[169] which are sensitive to Grignard attack. Starting from a urea, two different Grignard reagents may be added sequentially to give an unsymmetrical ketone via an amide (Scheme 27);[170] similarly, Comins has shown how to control Grignard addition to amides to give either ketones or unsymmetrical tertiary alcohols (Scheme 28).[171]

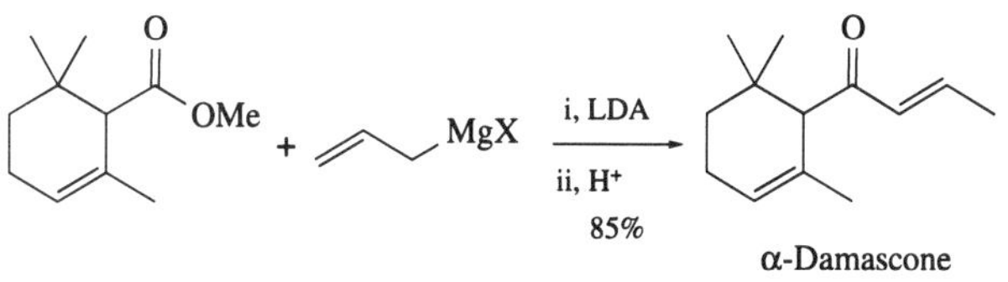

(83) 85%, α-Damascone

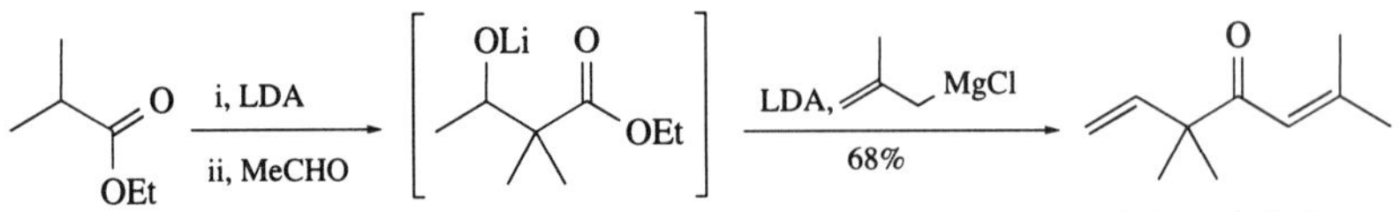

Scheme 25

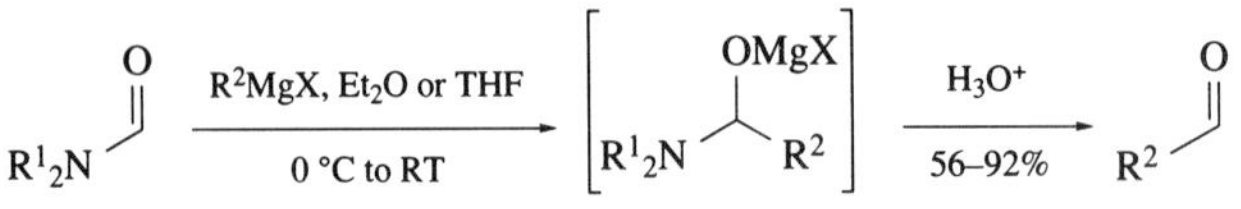

Scheme 26

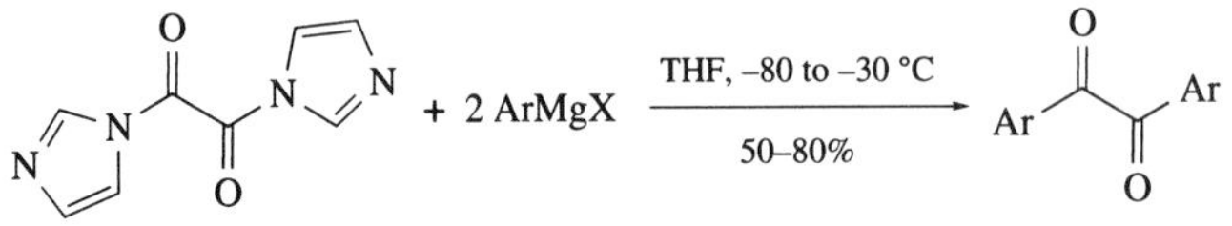

$$R^1{}_2N-CHO \xrightarrow[0\ °C\ to\ RT]{R^2MgX,\ Et_2O\ or\ THF} \left[R^1{}_2N-\underset{R^2}{\overset{OMgX}{C}}H \right] \xrightarrow{H_3O^+} R^2-CHO \quad 56–92\%$$

$$\text{(imidazole diamide)} + 2\ ArMgX \xrightarrow[50–80\%]{THF,\ -80\ to\ -30\ °C} Ar-CO-CO-Ar \quad (84)$$

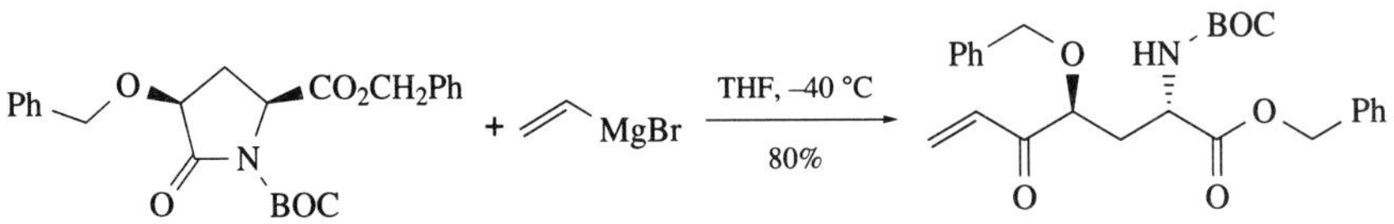

(85)

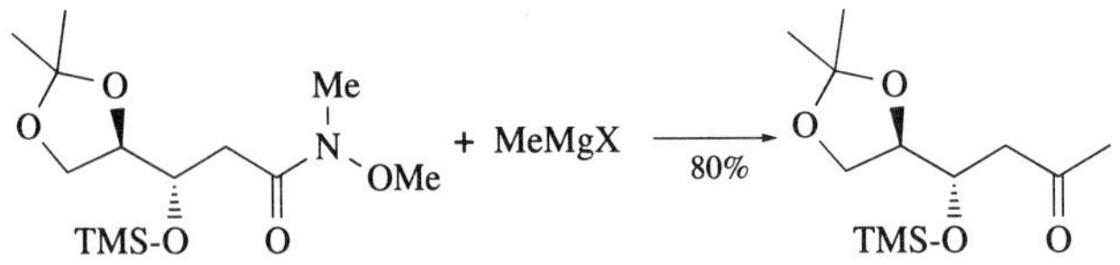

(86)

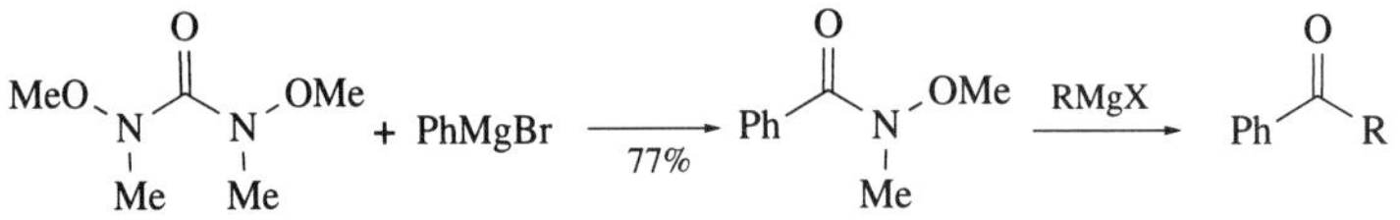

Scheme 27

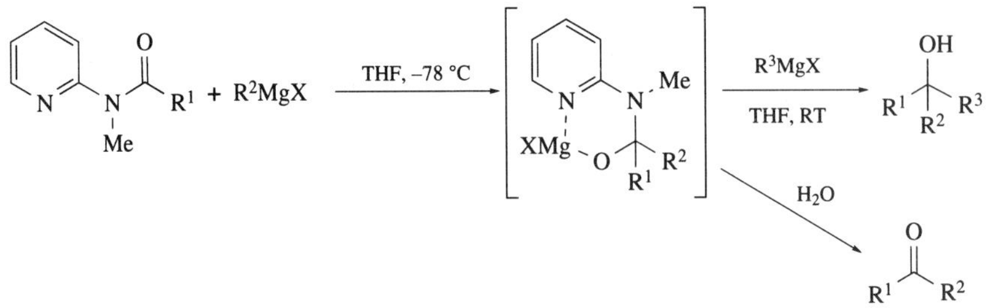

Scheme 28

In a similar manner (Equations (87)–(90)), cyclic imides may undergo single attack by RMgX to give an amidoketone.[172,173] The intermediate N,O-acetal may be reduced *in situ* to give an amido alcohol,[174] it may be oxidized back to an amide[175] or it may undergo further intramolecular reaction.[176] Again, this reaction may be carried out in the presence of esters and silyl ethers. Other reactions of amides with organomagnesium compounds include the formation of imines which are reduced *in situ* to amines.[177]

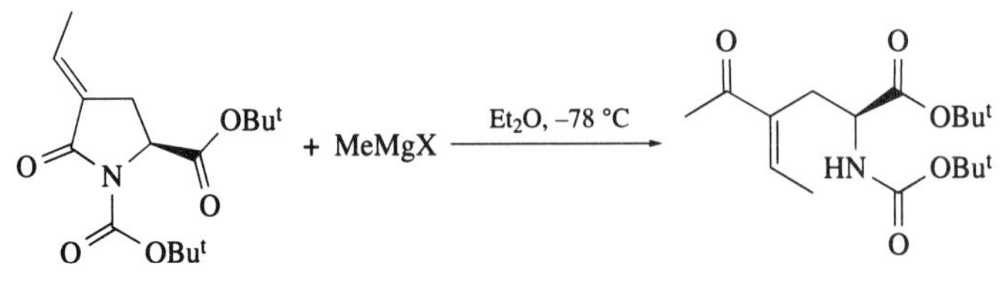

(87)

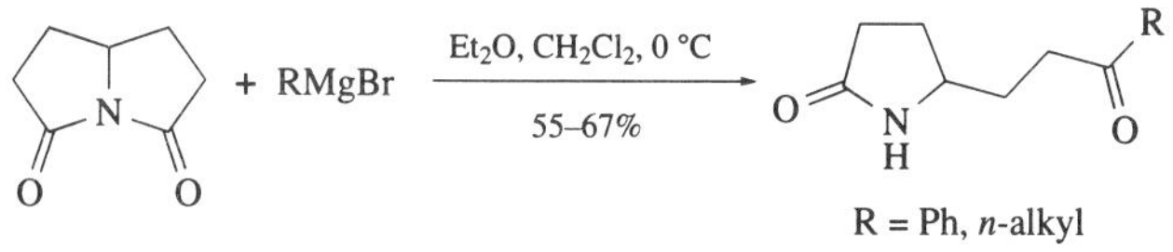

3.4.4 Acetals

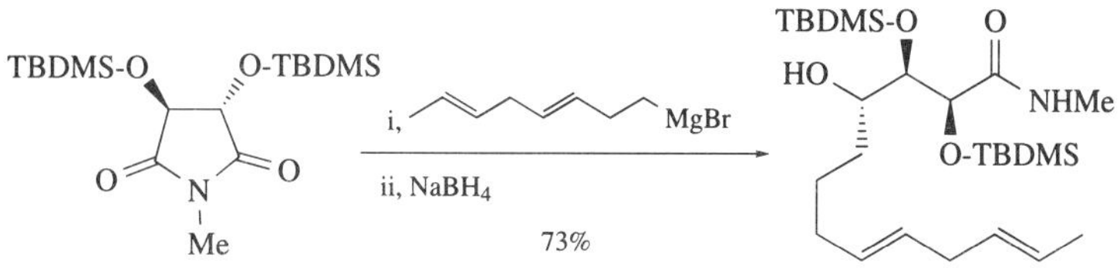

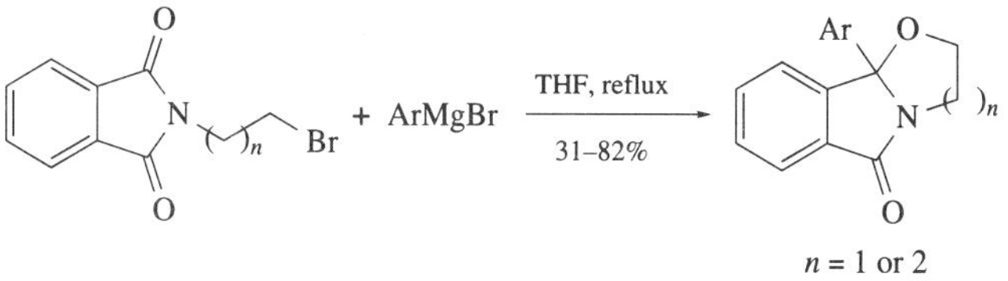

The reaction of acetals derived from a variety of ketones and aldehydes with Grignard reagents is promoted by TiCl$_4$, and leads to clean displacement of one oxygen by the alkyl or aryl group of the nucleophile (Equations (91) and (92)).[178] For mixed aryl alkyl acetals the aryloxy group is displaced selectively and *p*-chloro or 2,4-dichlorophenoxy groups work best. The cyclic hemiacetal of Equation (93) reacts with allyl or alkylmagnesium halides with good diastereoselectivity, but only the allylic reagents work in high yield.[179] Related strategies for obtaining stereocontrol in the reaction of Grignard reagents with oxazolidines are recorded; removal of the resulting nitrogen side chain of the ring-opened product can cause difficulty and limit the applicability even though the diastereoselectivity is good, and opposite for magnesium and lithium reagents.[180,181] In the synthesis of monomorine shown in Scheme 29, the side chain is released during cyclization.[182] Chelation of magnesium facilitates the ring opening of a protected ribose, and trapping of the iminium ion (Equation (94)), since the corresponding reaction with the 5-deoxy derivative is much slower.[183] For reaction of other C–O single bonds with Grignard reagents, there needs to be an alternative source of activation. Thus, allylic phosphates react with allylic Grignard reagents with high S_N2/S_E2' regiochemistry, as exemplified by Equation (95); the reaction is unique to the phosphate leaving group and implies possible Mg···O=P coordination.[184]

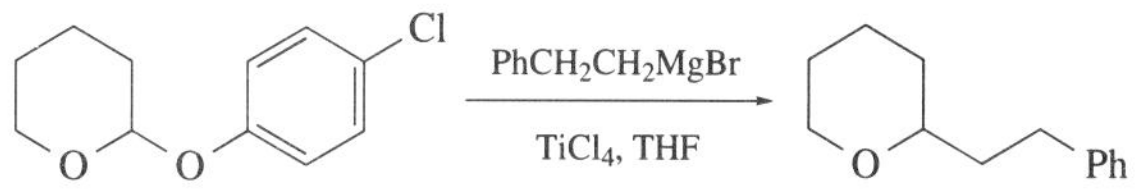

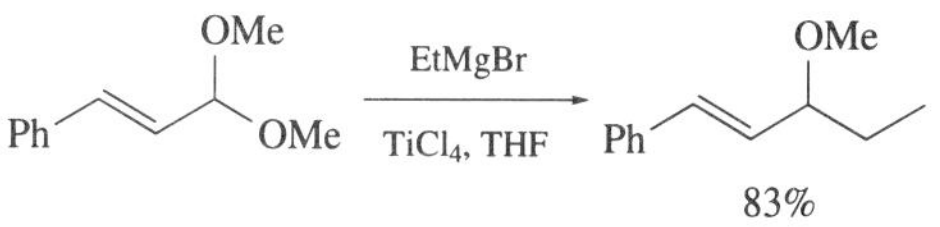

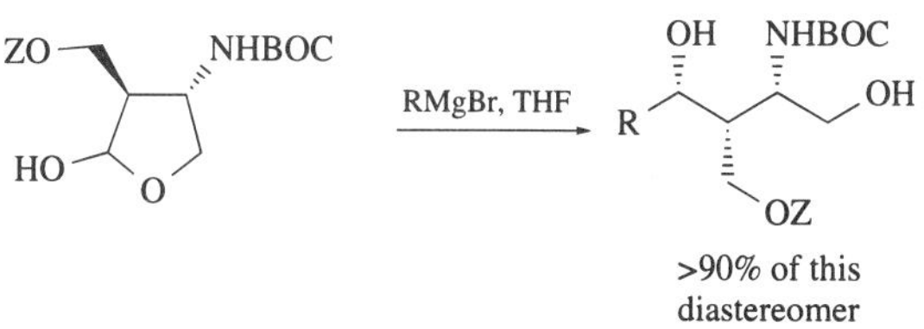

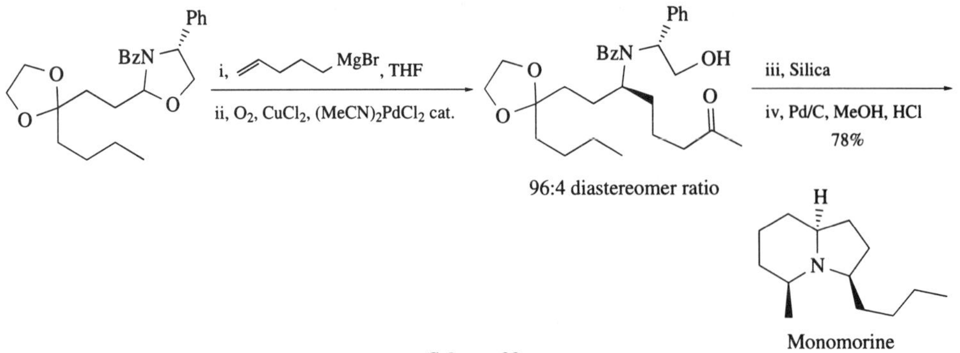

Scheme 29

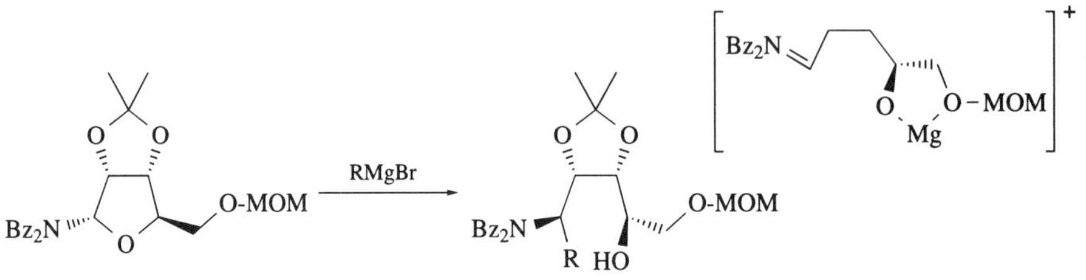

(94)

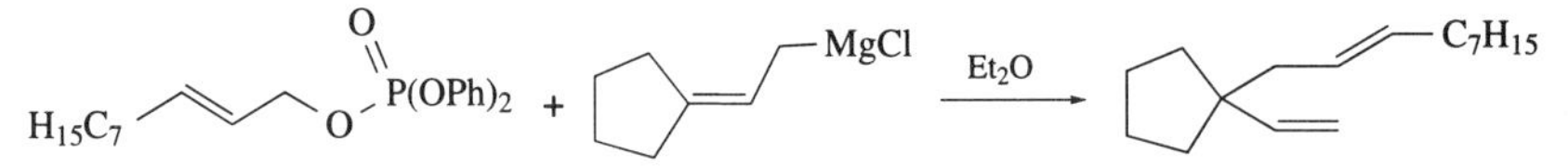

(95)

3.5 REFERENCES

1. A. Fürstner, *Angew. Chem., Int. Ed. Engl.*, 1993, **32**, 164.
2. B. Bogdanović, P. Bons, S. Konstantinović, M. Schwickardi and U. Westeppe, *Chem. Ber.*, 1993, **126**, 1371.
3. K. V. Baker, J. M. Brown, N. Hughes, A. J. Skarnulis and A. Sexton, *J. Org. Chem.*, 1991, **56**, 698.
4. M. Orchin, *J. Chem. Ed.*, 1989, **66**, 586.
5. T. P. Burns and R. D. Rieke, *J. Org. Chem.*, 1987, **52**, 3674.
6. R. D. Rieke, S. E. Bales, P. M. Hundnall, T. P. Burns and G. S. Poindexter, *Org. Synth.*, 1988, 845.
7. A. Boudin, G. Cerveau, C. Chuit, R. J. P. Corriu and C. Reye, *Tetrahedron*, 1989, **45**, 171.
8. J. F. Garst, *Acc. Chem. Res.*, 1991, **24**, 95.
9. H. M. Walborsky, *Acc. Chem. Res.*, 1990, **23**, 286.
10. T. Holm, *Acta Chem. Scand.*, 1991, **45**, 925.
11. M. Chanon, *Acta Chem. Scand.*, 1992, **46**, 695.
12. T. Holm, *Acta Chem. Scand. B*, 1988, **42**, 685.
13. J. Toullec, M. Mladenova, F. Gaudemar-Bardone and B. Blagoev, *J. Org. Chem.*, 1985, **50**, 2563.
14. C. Bernardon, *J. Organomet. Chem.*, 1989, **367**, 11.
15. T. Holm, *Acta Chem. Scand. B*, 1983, **37**, 567.
16. Y. C. Liu and H. S. Dang, *Acta Chim. Sinica*, 1985, **43**, 1079.
17. D. Nasipuri and A. Saha, *Indian. J. Chem., Sect. B*, 1990, **29**, 474.
18. S. K. Taylor, S. G. Bennett, K. J. Heinz and L. K. Lashley, *J. Org. Chem.*, 1981, **46**, 2194.
19. K. Muraoka, M. Nojima, S. Kusabayashi and S. Nagase, *J. Chem. Soc., Perkin Trans. 2*, 1986, 761.
20. W. F. Bailey and A. A. Croteau, *Tetrahedron Lett.*, 1981, **22**, 545.
21. J. W. Cheng and F.-T. Luo, *Tetrahedron Lett.*, 1988, **29**, 1293.
22. F. Bickelhaupt, *Angew. Chem.*, 1987, **99**, 1020.
23. For general preparative descriptions, see (a) L. Brandsma and H. Verkruijsse, 'Preparative Polar Organometallic Chemistry', Springer, Berlin, 1987, vol. 1; (b) L. Brandsma, 'Preparative Polar Organometallic Chemistry', Springer, Berlin, 1990, vol. 2.
24. T. Imamoto, N. Takiyama, K. Nakamura, T. Hatajima and Y. Kamiya, *J. Am. Chem. Soc.*, 1989, **111**, 4392.
25. T. V. Lee, J. A. Channon, C. Cregg, J. R. Porter, F. S. Roden and T.-L. Yeoh, *Tetrahedron*, 1989, **45**, 5877.
26. G. Bartoli, E. Marcantoni and M. Petrini, *Angew. Chem., Int. Ed. Engl.*, 1993, **32**, 1061.
27. G. Bartoli, E. Marcantoni and M. Petrini, *J. Chem. Soc., Chem. Comm.*, 1993, 1373.
28. S. E. Denmark, J. P. Edwards and O. Nicaise, *J. Org. Chem.*, 1993, **58**, 569.
29. A. H. Hoveyda, J. P. Morken, A. F. Houri and Z. Xu, *J. Am. Chem. Soc.*, 1992, **114**, 6692.
30. A. F. Houri, M. T. Didiuk, Z. Xu, N. R. Horan and A. H. Hoveyda, *J. Am. Chem. Soc.*, 1993, **115**, 6614.
31. A. H. Hoveyda, Z. Xu, J. P. Morken and A. F. Houri, *J. Am. Chem. Soc.*, 1991, **113**, 8950.
32. A. H. Hoveyda and J. P. Morken, *J. Org. Chem.*, 1993, **58**, 4237.
33. J. P. Morken, M. T. Didiuk and A. H. Hoveyda, *J. Am. Chem. Soc.*, 1993, **115**, 6997.
34. J. P. Morken, M. T. Didiuk, M. S. Visser and A. H. Hoveyda, *J. Am. Chem. Soc.*, 1994, **116**, 3123.
35. H. Lehmkuhl, *Bull. Soc. Chim. Fr., Part 2*, 1981, 88.

36. (a) W. Oppolzer and R. Pitteloud, *J. Am. Chem. Soc.*, 1982, **104**, 6478; (b) W. Oppolzer and P. Schneider, *Helv. Chim. Acta*, 1986, **69**, 1817.
37. J. F. G. A. Jansen and B. L. Feringa, *J. Chem. Soc., Chem. Commun.*, 1989, 741.
38. R. A. Kjonaas and E. J. Vawter, *J. Org. Chem.*, 1986, **51**, 3993.
39. T. Mukaiyama and N. Iwasawa, *Chem. Lett.*, 1981, 913.
40. J. Touet, C. Le Grumelec, F. Huet and E. Brown, *Tetrahedron: Asymmetry*, 1993, **4**, 1469.
41. L. F. Tietze, S. Brand, T. Pfeiffer, J. Antel, K. Harms and G. M. Sheldrick, *J. Am. Chem. Soc.*, 1987, **109**, 921.
42. B. E. Rossiter and N. M. Swingle, *Chem. Rev.*, 1992, **92**, 771.
43. T. G. Gant and A. I. Meyers, *Tetrahedron*, 1994, **50**, 2297.
44. A. I. Meyers, A. Meier and D. J. Rawson, *Tetrahedron Lett.*, 1992, **33**, 853.
45. M. A. Rizzacasa and M. V. Sargent, *J. Chem. Soc. Perkin Trans. 1*, 1991, 841.
46. H. Hart, K. Harada and C.-J. F. Du, *J. Org. Chem.*, 1985, **50**, 3104.
47. C.-J. F. Du and H. Hart, *J. Org. Chem.*, 1987, **52**, 4311.
48. T. Ghosh and H. Hart, *J. Org. Chem.*, 1988, **53**, 3555.
49. T. Kawai, N. Furukawa and S. Oae, *Tetrahedron Lett.*, 1984, **25**, 2549.
50. N. Furukawa, T. Shibutani and H. Fujihara, *Tetrahedron Lett.*, 1987, **28**, 5845.
51. N. Furukawa, T. Shibutani and H. Fujihara, *Tetrahedron Lett.*, 1989, **30**, 7091.
52. T. Shibutani, H. Fujihara and N. Furukawa, *Tetrahedron Lett.*, 1991, **32**, 2943.
53. R. W. Baker, G. R. Pocock and M. V. Sargent, *J. Chem. Soc., Chem. Commun.*, 1993, 1489.
54. R. W. Baker, G. R. Pocock, M. V. Sargent and E. Twiss, *Tetrahedron: Asymmetry*, 1993, **4**, 2423.
55. R. W. Baker, S. O. Rea, M. V. Sargent, E. M. C. Schenkelaars, B. W. Skelton and A. H. White, *Tetrahedron: Asymmetry*, 1994, **5**, 45.
56. M. El-Jazouli, S. Masson and A. Thuillier, *Bull. Soc. Chim. Fr. Part 2*, 1988, **5**, 875.
57. S. Inoue, O. Suzuki and K. Sato, *Bull. Chem. Soc. Jpn.*, 1989, **62**, 1601 and references therein.
58. F. Babudri, S. Florio, G. Ingrosso and A. M. Turco, *Heterocycles*, 1986, **24**, 2215 and references therein.
59. S. Florio, G. Ingrosso and R. Sgarra, *Tetrahedron*, 1985, **41**, 3091.
60. F. Babudri, S. Florio and L. Ronzini, *Tetrahedron*, 1986, **42**, 3905.
61. J. Sisko and S. M. Weinreb, *J. Org. Chem.*, 1990, **55**, 393.
62. M. A. Schwartz and X. Hu, *Tetrahedron Lett.*, 1992, **33**, 1689.
63. G. Courtois and P. Miginiac, *Bull. Soc. Chim. Fr., Part 2*, 1983, 21.
64. G. Wieland and G. Simchen, *Liebigs Ann. Chem.*, 1985, 2178.
65. F. Babudri, S. Florio and G. Inguscio, *Synthesis*, 1985, 522 and references therein.
66. S. Florio, E. Epifani and G. Ingrosso, *Tetrahedron*, 1984, **40**, 4527.
67. C. Felix, A. Laurent, S. Lesniak and P. Mison, *J. Chem. Res. (S)*, 1991, 32; *(M)*, 1991, 0301.
68. K. Hattori, K. Maruoka and H. Yamamoto, *Tetrahedron Lett.*, 1982, **23**, 3395.
69. E. Epifani, S. Florio, G. Ingrosso, R. Sgarra and F. Stasi, *Tetrahedron*, 1987, **43**, 2769.
70. F. Rise, L. Ongstad, M. Gacek and K. Undheim, *Acta Chem. Scand. Ser. B*, 1983, **37**, 613.
71. Y. Yamamoto, S. Nishii, K. Maruyama, T. Komatsu and W. Ito, *J. Am. Chem. Soc.*, 1986, **108**, 7778 and references therein.
72. S. E. Denmark, T. Weber and D. W. Piotrowski, *J. Am. Chem. Soc.*, 1987, **109**, 2224.
73. D. Enders, R. Funk, M. Klatt, G. Raabe and E. R. Hovestreydt, *Angew. Chem., Int. Edn. Engl.*, 1993, **32**, 418 and references therein.
74. H. Takahashi and Y. Suzuki, *Chem. Pharm. Bull.*, 1983, **31**, 4295 and references therein.
75. Z.-Y. Chang and R. M. Coates, *J. Org. Chem.*, 1990, **55**, 3464 and 3475.
76. H. Takahashi, Y. Chida, T. Suzuki, S. Yanaura, Y. Suzuki and C. Masuda, *Chem. Pharm. Bull.*, 1983, **31**, 1659 and references therein.
77. F. Mancini, M. G. Piazza and C. Trombini, *J. Org. Chem.*, 1991, **56**, 4246.
78. K. Tomioka, F. Masumi, T. Yamashita and K. Koga, *Tetrahedron Lett.*, 1984, **25**, 333.
79. D. H. Hua, S. W. Miao, J. S. Chen and S. Iguchi, *J. Org. Chem.*, 1991, **56**, 4.
80. Y. Kobayashi, K. Nakatani, Y. Ito and S. Terashima, *Chem. Lett.*, 1990, 1709.
81. T. Matsumoto, *et al.*, *Tetrahedron Lett.*, 1990, **31**, 4175.
82. G. Cainelli, D. Giacomini, E. Mezzina, M. Panunzio and P. Zarantonello, *Tetrahedron Lett.*, 1991, **32**, 2967.
83. Y. Yamamoto, T. Komatsu and K. Maruyama, *J. Chem. Soc., Chem. Commun.*, 1985, 814.
84. M. T. Reetz, R. Jaeger, R. Drewlies and M. Hübel, *Angew. Chem., Int. Ed. Engl.*, 1991, **30**, 103.
85. M. P. Cowling, P. R. Jenkins and K. Cooper, *J. Chem. Soc., Chem. Commun.*, 1988, 1503.
86. A. Alexakis, N. Lensen, J.-P. Trancher and P. Mangeney, *J. Org. Chem.*, 1992, **57**, 4563.
87. D. P. G. Hamon, R. A. Massy-Westropp and P. Razzino, *Tetrahedron*, 1992, **48**, 5163.
88. P. Ermert, J. Meyer, C. Stucki, J. Schneebeli and J.-P. Obrecht, *Tetrahedron Lett.*, 1988, **29**, 1265.
89. R. Ballini, E. Marcantoni and M. Petrini, *J. Org. Chem.*, 1992, **57**, 1316.
90. C. N. Meltz and R. A. Volkmann, *Tetrahedron Lett.*, 1983, **24**, 4503.
91. N. Bischofberger, *Tetrahedron Lett.*, 1989, **30**, 1621.
92. D. K. Pirie, W. M. Welch, P. D. Weeks and R. A. Volkmann, *Tetrahedron Lett.*, 1986, **27**, 1549.
93. L. Strekowski, M. T. Cegla, D. B. Harden, J. L. Mokrosz and M. J. Mokrosz, *Tetrahedron Lett.*, 1988, **29**, 4265.
94. K.-I. Tanji and T. Higashino, *Heterocycles*, 1990, **30**, 435.
95. M. Gill, M. J. Kiefel and D. A. Lally, *Tetrahedron Lett.*, 1986, **27**, 1933 and references therein.
96. F. Gerhart, J.-P. François, M. Kolb, M. Laskovics and J.-F. LeBorgne, *J. Fluorine Chem.*, 1990, **50**, 243.
97. M. A. Grassberger, A. Horvath and G. Schulz, *Tetrahedron Lett.*, 1991, **32**, 7393.
98. J. Bergman, A. Brynolf, B. Elman and E. Vuorinen, *Tetrahedron*, 1986, **42**, 3697.
99. S.S. Kulp and A. Romanelli, *Org. Prep. Proc. Int.*, 1992, **24**, 7.
100. See for example D. J. Hart and T.-K. Yang, *Tetrahedron Lett.*, 1982, **23**, 2761.
101. (a) Y. Zhang, J. Jiang and Y. Chen, *Tetrahedron Lett.*, 1987, **28**, 3815; (b) Y. Zhang, J. Jiang and Z. Zhang, *Tetrahedron Lett.*, 1988, **29**, 651.
102. S. Masson, V. Mothies and A. Thuillier, *Tetrahedron*, 1984, **40**, 1573.
103. D. Pini, R. Lazzaroni, S. Bertozzi and P. Salvadori, *Gaz. Chim. Ital.*, 1983, **113**, 227.

104. R. Yamaguchi, Y. Nakazono, T. Matsuki, E.-I. Hata and M. Kawanisi, *Bull. Chem. Soc., Jpn.*, 1987, **60**, 215 and references therein.
105. G. Courtois, A. Al-Arnaout and L. Miginiac, *Tetrahedron Lett.*, 1985, **26**, 1027.
106. D. L. Comins and A. H. Abdullah, *J. Org. Chem.*, 1982, **47**, 4315.
107. M.-J. Shiao and W.-L. Chia, *Synth. Commun.*, 1991, **21**, 401.
108. D. L. Comins and A. H. Abdullah, *Tetrahedron Lett.*, 1985, **26**, 43 and references therein.
109. T. R. Webb, *Tetrahedron Lett.*, 1985, **26**, 3191.
110. K.-Y. Akiba, Y. Iseki and M. Wada, *Tetrahedron Lett.*, 1982, **23**, 3935.
111. D. L. Comins, M. A. Weglarz and S. O'Connor, *Tetrahedron Lett.*, 1988, **29**, 1751 and references therein.
112. D. L. Comins and R. S. Al-awar, *J. Org. Chem.*, 1992, **57**, 4098 and references therein.
113. D. L. Comins, A. H. Abdullah and N. B. Mantlo, *Tetrahedron Lett.*, 1984, **25**, 4867.
114. D. L. Comins and N. B. Mantlo, *Tetrahedron Lett.*, 1987, **28**, 759.
115. D. L. Comins and Y. C. Myoung, *J. Org. Chem.*, 1990, **55**, 292.
116. D. L. Comins, E. D. Stroud and J. J. Herrick, *Heterocycles*, 1984, **22**, 151.
117. D. L. Comins and A. Dehghani, *Tetrahedron Lett.*, 1991, **32**, 5697.
118. D. L. Comins and L. A. Morgan, *Tetrahedron Lett.*, 1991, **32**, 5919.
119. D. L. Comins, H. Hong and J. M. Salvador, *J. Org. Chem.*, 1991, **56**, 7197.
120. R. S. Al-awar, S. P. Joseph and D. L. Comins, *Tetrahedron Lett.*, 1992, **33**, 7635.
121. D. L. Comins and H. Hong, *J. Am. Chem. Soc.*, 1991, **113**, 6672.
122. D. L. Comins and N. R. Benjelloun, *Tetrahedron Lett.*, 1994, **35**, 829.
123. (a) D. L. Comins and M. O. Killpack, *J. Am. Chem. Soc.*, 1992, **114**, 10 972; (b) D. L. Comins, R. R. Goehring, S. P. Joseph and S. O'Connor, *J. Org. Chem.*, 1990, **55**, 2574.
124. Y. Génisson, C. Marazano and B. C. Das, *J. Org. Chem.*, 1993, **58**, 2052.
125. A. G. Schultz, L. Flood and J. P. Springer, *J. Org. Chem.*, 1986, **51**, 838.
126. T. Zair, C. Santelli-Rouvier and M. Santelli, *J. Org. Chem.*, 1993, **58**, 2686.
127. S. Habaue, K. Yasue, A. Yanagisawa and H. Yamamoto, *Synlett*, 1993, 788.
128. D. Seebach and M. A. Syfrig, *Angew. Chem., Int. Ed. Engl.*, 1984, **23**, 248.
129. M. T. Reetz, *Angew. Chem., Int. Ed. Engl.*, 1984, **23**, 556.
130. K. Maruoka, T. Itoh, M. Sakurai, K. Nonoshita and H. Yamamoto, *J. Am. Chem. Soc.*, 1988, **110**, 3588.
131. M. T. Reetz, N. Harmat and R. Mahrwald, *Angew. Chem., Int. Ed. Engl.*, 1992, **31**, 342.
132. Y. Hirano and C. Djerassi, *J. Org. Chem.*, 1982, **47**, 2420.
133. A. Guerriero, F. Pietra, M. Cavazza and F. Del Cima, *J. Chem. Soc., Perkin Trans. 1*, 1982, 979.
134. W. C. Still and J. H. McDonald III, *Tetrahedron Lett.*, 1980, **21**, 1031.
135. M. Asami and R. Kimura, *Chem. Lett.*, 1985, 1221.
136. W. Schmid and G. M. Whitesides, *J. Am. Chem. Soc.*, 1990, **112**, 9670.
137. B. Reitstøen, L. Kilaas and T. Anthonsen, *Acta Chem. Scand., Ser. B*, 1986, **40**, 440.
138. J. Jurczak, S. Pikul and T. Bauer, *Tetrahedron*, 1986, **42**, 447.
139. H. Urabe and F. Sato, *J. Synth. Org. Chem. Jpn.*, 1993, **51**, 14.
140. S. Okamoto, H. Tsujiyama, T. Yoshino and F. Sato, *Tetrahedron Lett.*, 1991, **32**, 5789.
141. Y. Takeda, T. Matsumoto and F. Sato, *J. Org. Chem.*, 1986, **51**, 4728.
142. Y. Ukaji, K. Yamamoto, M. Fukui and T. Fujisawa, *Tetrahedron Lett.*, 1991, **32**, 2919.
143. M. Asami and T. Mukaiyama, *Chem. Lett.*, 1983, 93.
144. H. Fujioka *et al.*, *Chem. Pharm. Bull.*, 1989, **37**, 1488.
145. H. Yoda, H. Kitayama, K. Takabe and A. Kakehi, *Tetrahedron: Asymmetry*, 1993, **4**, 1759.
146. H. Takahashi, T. Tsubuki and K. Higashiyama, *Chem. Pharm. Bull.*, 1991, **39**, 3136.
147. A. I. Meyers, M. A. Hanagan, L. M. Trefonas and R. J. Baker, *Tetrahedron*, 1983, **39**, 1991.
148. S. D. Real, D. R. Kronenthal and H. Y. Wu, *Tetrahedron Lett.*, 1993, **34**, 8063.
149. Y. Tamai, M. Akiyama, A. Okamura and S. Miyano, *J. Chem. Soc., Chem. Commun.*, 1992, 687.
150. W. H. Miles, S. L. Rivera and J. D. del Rosario, *Tetrahedron Lett.*, 1992, **33**, 305.
151. M. Falorni, L. Lardicci, G. Uccello-Barretta and G. Giacomelli, *Gazz. Chim. Ital.*, 1988, **118**, 495.
152. B. B. Brown and R. A. Volkmann, *Tetrahedron Lett.*, 1986, **27**, 1545.
153. L. A. Reed, III, D. A. Charleson and R. A. Volkmann, *Tetrahedron Lett.*, 1987, **28**, 3431.
154. P. Zhang and R. E. Gawley, *Tetrahedron Lett.*, 1992, **33**, 2945.
155. M. Kaino, K. Ishihara and H. Yamamoto, *Bull. Chem. Soc. Jpn.*, 1989, **62**, 3736.
156. T. Kogure and E. L. Eliel, *J. Org. Chem.*, 1984, **49**, 576.
157. B. Weber and D. Seebach, *Angew. Chem., Int. Ed. Engl.*, 1992, **31**, 84.
158. R. Noyori and M. Kitamura, *Angew. Chem., Int. Ed. Engl.*, 1991, **30**, 49.
159. K. Tomioka, M. Nakajima and K. Koga, *Tetrahedron Lett.*, 1987, **28**, 1291.
160. J. F. G. A. Jansen and B. L. Feringa, *J. Org. Chem.*, 1990, **55**, 4168 and references therein.
161. K. Soai, Y. Kawase and A. Oshio, *J. Chem. Soc., Perkin Trans. 1*, 1991, 1613.
162. C. Fehr, *Chimia*, 1991, **45**, 253 and references therein.
163. G. A. Olah, L. Ohannesian and M. Arvanaghi, *J. Org. Chem.*, 1984, **49**, 3856 and references therein.
164. W. Amaratunga and J. M. J. Fréchet, *Tetrahedron Lett.*, 1983, **24**, 1143.
165. R. H. Mitchell and V. S. Iyer, *Tetrahedron Lett.*, 1993, **34**, 3683.
166. U. T. Mueller-Westerhoff and M. Zhou, *Tetrahedron Lett.*, 1993, **34**, 571.
167. M. P. Sibi, R. Sharma and K. L. Paulson, *Tetrahedron Lett.*, 1992, **33**, 1941.
168. T. Ohta, A. Hosoi and S. Nozoe, *Tetrahedron Lett.*, 1988, **29**, 329.
169. C. Palomo, J. M. Aizpurua, N. Aurrekoetxea and M. C. López, *Tetrahedron Lett.*, 1991, **32**, 2525.
170. W. L. Whipple and H. J. Reich, *J. Org. Chem.*, 1991, **56**, 2911 and references therein.
171. D. L. Comins, *Synlett*, 1992, 615.
172. C. M. Moody and D. W. Young, *Tetrahedron Lett.*, 1993, **34**, 4667 and references therein.
173. K. Takahashi and A. Brossi, *Heterocycles*, 1988, **27**, 2475 and references therein.
174. H. Yoda, T. Katagiri and K. Takabe, *Tetrahedron Lett.*, 1991, **32**, 6771 and references therein.

175. C. G. Screttas and B. R. Steele, *J. Org. Chem.*, 1988, **53**, 5151.
176. L. L. Braun, J. Xia, M. K. Ahmed, E. I. Isaacson and M. J. Cooney, *J. Heterocycl. Chem.*, 1989, **26**, 1441.
177. B. L. Feringa and J. F. G. A. Jansen, *Synthesis*, 1988, 184.
178. H. Ishikawa, T. Mukaiyama and S. Ikeda, *Bull. Chem. Soc. Jpn.*, 1981, **54**, 776.
179. H. Yoda, Y. Nakagami and K. Takabe, *Tetrahedron: Asymmetry*, 1994, **5**, 169.
180. K. Higashiyama, H. Inoue and H. Takahashi, *Tetrahedron*, 1994, **50**, 1083.
181. K. Higashiyama, H. Inoue and H. Takahashi, *Tetrahedron Lett.*, 1992, **33**, 235.
182. K. Higashiyama, K. Nakahata and H. Takahashi, *J. Chem. Soc., Perkin Trans. 1*, 1994, 351.
183. M. Nagai, J. J. Gaudino and C. S. Wilcox, *Synthesis*, 1992, 163.
184. A. Yanagisawa, H. Hibino, N. Nomura and H. Yamamoto, *J. Am. Chem. Soc.*, 1993, **115**, 5879.

4
Zinc and Cadmium

PAUL KNOCHEL

Philipps-Universität Marburg, Germany

4.1 ZINC REAGENTS IN ORGANIC SYNTHESIS	159
4.1.1 Introduction	159
4.1.2 Preparation of Organozinc Compounds	160
4.1.2.1 The direct insertion of zinc metal	160
4.1.2.2 The iodine–zinc exchange reaction	162
4.1.2.3 Preparation of organozinc compounds via transmetallations	163
4.1.3 Reactivity of Organozinc Compounds	165
4.1.3.1 Substitution reactions	165
4.1.3.2 Addition reactions	169
4.1.3.3 Barbier reactions	178
4.1.4 Reactivity of Zinc Carbenoids	178
4.1.4.1 Cyclopropanations	178
4.1.4.2 Methylene homologation and related reactions	179
4.1.5 Preparation and Reactions of Zincates	182
4.1.6 Preparation and Reactivity of Bimetallic Reagents of Zinc	182
4.2 CADMIUM REAGENTS IN ORGANIC SYNTHESIS	183
4.3 REFERENCES	185

4.1 ZINC REAGENTS IN ORGANIC SYNTHESIS

4.1.1 Introduction

Since 1982, the applications of organozinc reagents in organic synthesis have increased in an exponential manner.[1] Besides the Reformatsky reaction[2] and the Simmons–Smith reaction,[3] a number of useful carbon–carbon bond forming reactions using zinc organometallics have been reported.[4] This review will emphasize these developments, especially those allowing stereoselective transformations. For a long time, synthetic applications of organozinc derivatives[5] have been hampered by their low reactivity and only a limited number of electrophiles, such as oxygen,[6] halogens,[6] or pseudohalogens[7] react directly with alkylzinc derivatives. The low polarity of the carbon–zinc bond (comparable to the carbon–tin bond) and the relatively moderate Lewis acidity of the zinc(II) center are responsible for this inertness. However, the presence of empty low-lying orbitals at zinc allows easy transmetallations of zinc organometallics. Numerous transition metal salts (of Ti, Mo, Ta, Nb, V, Pd, Ni, Pt, Cu ...) undergo fast transmetallations with zinc species to give transition metal organometallics which show excellent reactivity toward many classes of organic electrophiles (Section 4.1.3). Transmetallations with catalytic amounts of palladium salts[8] or stoichiometric amounts of CuCN·2LiCl[9] have proved especially useful synthetically. The low reactivity of zinc organometallics in conjunction with their high transmetallation ability constitutes a unique advantage. Their high functional group tolerance allows the preparation of a wide range of polyfunctional zinc reagents which can be activated toward reaction with electrophiles by the addition of the appropriate transition metal salt.[4] Organozinc compounds are therefore ideal intermediates for the synthesis of complex polyfunctional molecules. Dialkylzincs have also proved to be excellent reagents for enantioselective synthesis, and their addition to aldehydes in the presence of chiral additives furnishes secondary alcohols with excellent enantioselectivity[10] (Section 4.1.3.2(i)). Zinc carbenoids such as ICH_2ZnI have recently found applications as methylene homologation reagents in addition to their classical use as cyclopropanation agents (Section 4.1.4). The moderate reactivity of dialkylzincs (R_2Zn) and alkylzinc halides (RZnX) can be enhanced by the formation of zincates (R_3ZnM), which readily undergo 1,4-additions (Section 4.1.5). A range of bimetallic zinc reagents have also been prepared and used in organic synthesis as alkenation reagents and multicoupling reagents (Section 4.1.6).

4.1.2 Preparation of Organozinc Compounds

4.1.2.1 The direct insertion of zinc metal

The direct insertion of zinc into organic halides is the most general preparation of organozinc halides. The insertion rate depends on the structure of the organic halide, the reaction conditions (solvent, temperature, concentration) and the zinc activation. The procedure using zinc dust (2–3 equiv.) in THF reported originally by Gaudemar[11] has proved to be very general. The addition of primary alkyl iodides, added as a 2.5–3 M solution, to zinc dust successively activated with 1,2-dibromoethane (4–5 mol.%) and chlorotrimethylsilane (1 mol.%)[9,12] leads to complete insertion within 2–3 h between 35°C and 40 °C, whereas secondary alkyl iodides react even faster (0.5–1 h at 25 °C).[9] Remarkably, a wide range of functional groups can be present in the alkyl iodide FG–RI (FG = ester, ketone, cyanide, halide, amine, amide, sulfoxide, sulfide, sulfone, thioester, boronic ester, enone, and phosphate, see (1)–(4)); however, the presence of a hydroxy, nitro, or azido function interferes with the zinc insertion (Equation (1)).[4]

$$FG-RX + Zn \xrightarrow[>85\%]{THF, 5-45\,°C} FG-RZnX \quad (1)$$

X = I, Br; FG = CO_2R, enoate, CN, enone, halide, $(RCO)_2N$, $(TMS)_2N$, RNH, NH_2, RCONH, $(RO)_3Si$, $(RO)_2P(O)$, RS, RSO, RSO_2, PhCOS; R = alkyl, aryl, benzyl, allyl

Alkyl chlorides or bromides do not usually react under these conditions and only the reactive benzylic and allylic bromides or chlorides react directly with zinc dust.[11,13] However, polar solvents such as N,N-dimethylacetamide (dma) or N,N'-dimethylpropyleneurea (dmpu) in the presence of a catalytic amount of LiI allows the use of primary alkyl chlorides, bromides, sulfonates, and phosphates as precursors for the formation of alkylzinc derivatives (Equation (2)).[14] Alkenyl and aryl iodides do not insert zinc in THF solution and react only in polar solvents at elevated temperatures.[15] However, if the alkenyl iodide is conjugated with a carbonyl group, as in the case of 3-iodo-2-cyclohexen-1-one, then the zinc insertion proceeds readily, leading to the zinc derivative (5) under relatively mild conditions (THF, 25–50 °C, 0.5 h, >85% yield).[4] Zinc homoenolates (6) and other alkylzinc iodides have been prepared in mixtures of benzene and dma[16–18] or hexamethylphosphoric triamide (HMPT).[19] Reagents of type (6) can also be obtained by the ring opening of silyloxy cyclopropanes (7), and constitute an important class of three-carbon nucleophiles with general synthetic utility (Equation (3) and Scheme 1).[20] The use of highly activated zinc powders, obtained from the *in situ* reduction of zinc chloride with lithium naphthalenide in THF, allows the conversion of alkyl bromides to the corresponding zinc reagents at 25 °C within a few hours; even aromatic bromides lead, under these conditions, to arylzinc bromides in excellent yields (Equation (4)).[21,22] Similarly, reactive zinc is obtained by the reduction of zinc chloride with potassium–graphite[23–5] (Equation (5)).[25]

$$RX \xrightarrow[\text{dma or dmpu, 40-80 °C, 2-12 h}]{\text{Zn, MI(0.2 equiv.), MBr (1.0 equiv.)}} RZnX \quad (2)$$

X = Cl, Br, OMs, OTs, OP(O)(OPh)$_2$; M = Li, Na, Cs

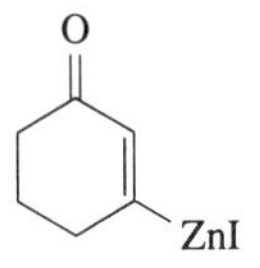

(5)

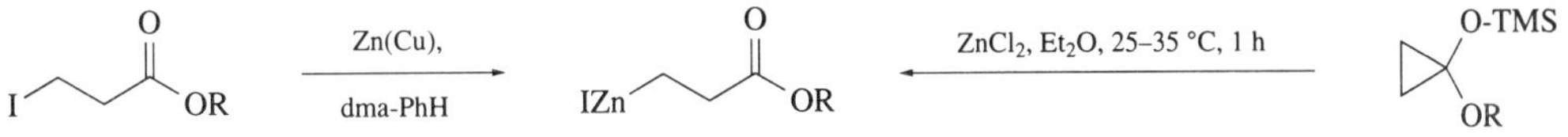

(3)

$$ZnCl_2 \xrightarrow{\text{Li·naphthalene}} Zn^* \xrightarrow{\text{FG-RX (X = Br, I), THF, 25-60 °C}} \text{FG-RZnX} \quad (4)$$

$$\text{K-graphite} \xrightarrow{ZnCl_2, AgOAc} \text{Zn(Ag)-graphite} \xrightarrow{\text{FG-ArI, THF, 20-40 °C}} \text{FG-ArZnI} \quad (5)$$

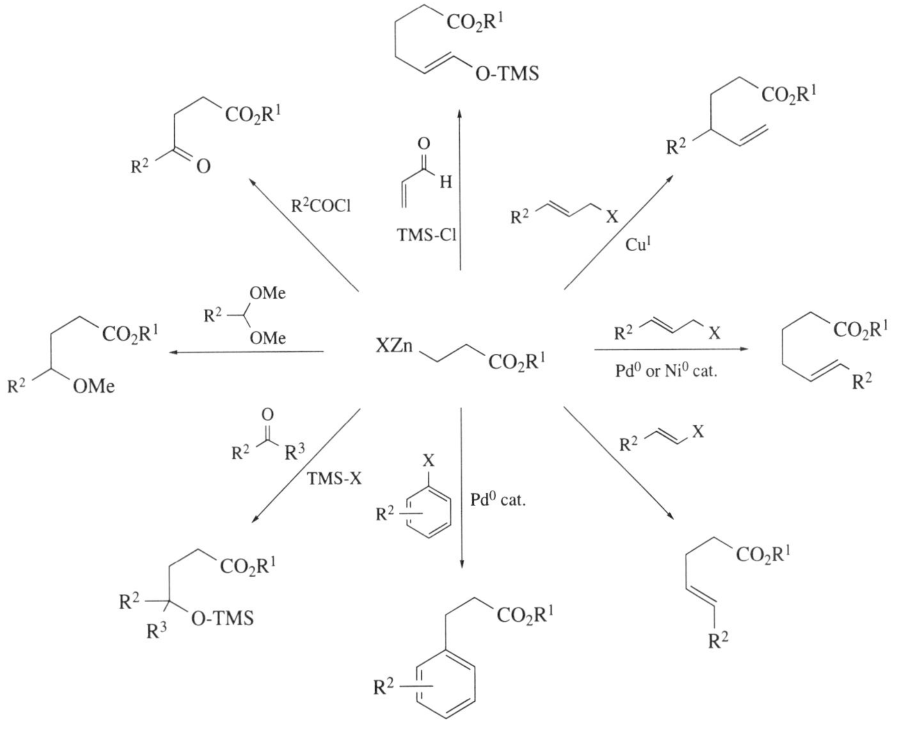

Scheme 1

The direct reaction of alkyl, alkenyl, and aromatic bromides with lithium metal and zinc chloride under ultrasonic irradiation affords the corresponding zinc reagents under mild conditions and in excellent yields (Equation (6)).[26] The sonication of zinc powder[27] increases its reactivity and facilitates various reactions such as the Reformatsky reaction,[28] the carboxylation of perfluoroalkyl iodides[29] and cyclopropanation.[30] Fluorinated organozinc reagents can be prepared by direct insertion of zinc using polar solvents. This class of zinc compounds has found numerous applications in synthesis.[31] Thus, polyfluoroalkyl iodides cleanly give the corresponding perfluorozinc iodides under mild conditions (Equation (7)).[32] Difluorodihalomethanes react with zinc metal in a complex manner and produce a mixture of bis(trifluoromethyl)zinc and trifluoromethylzinc halides in high yields (Equation (8)).[33] Whereas alkenyl bromides do not readily undergo zinc insertion, the presence of a trifluoromethyl group in the α-position considerably facilitates the reaction,[34] and bromotrifluoropropene (**8**) is converted to the alkenylzinc bromide (**9**) in 93% yield (Equation (9)).[35] Electrochemical activation of zinc can be achieved by the electrochemical oxidation of zinc in the presence of various organic halides, and the resulting organozinc species have been successfully used for synthetic applications.[36] Metal vapors of zinc can be used to generate zinc reagents under very mild conditions.[37] Finally, it should be mentioned that various allylic zinc reagents such as (**10**)–(**12**) can be prepared in excellent yields from the corresponding allylic bromides and zinc in THF.[11] The tendency to form Wurtz-coupling products increases with the presence of electron-donating substituents. Whereas (**10**)[38] can be prepared in high yields at 25 °C, the preparation of (**11**) has to be performed below 10 °C[11], and (**12**) can only be obtained in good yields if prepared below −5 °C. Cinnamylzinc bromide has to be prepared below −15 °C.[11] A similar pattern of behavior is observed with electron-rich benzylic halides. Thus, under typical conditions[13] (THF, 5 °C, 2 h), the benzylic bromide (**13**) reacts with zinc to give only the Wurtz-coupling product. However, reaction of the readily available corresponding phosphate (**14**) with zinc in dmpu in the presence of LiI (0.2 equiv.) gives, after 12 h at 50 °C, the desired zinc reagent (**15**) without the formation of any self-coupling product (Equation (10)).[14]

$$p\text{-TolBr} \xrightarrow{\text{Li, ZnBr}_2\text{, Et}_2\text{O, sonication, 25 °C, 0.5 h}} (p\text{-Tol})_2\text{Zn} \quad (6)$$

$$\text{C}_4\text{F}_9\text{I} \xrightarrow[70\%]{\text{Zn/Cu couple, dioxane, 25 °C, 0.5 h}} \text{C}_4\text{F}_9\text{ZnI} \quad (7)$$

$$\text{Zn} + \text{CF}_3\text{X}_2 \xrightarrow[80-95\%]{\text{DMF, 25 °C}} \text{CF}_3\text{ZnX} + (\text{CF}_3)_2\text{Zn} \quad (8)$$
$$X = \text{Cl, Br}$$

$$\underset{(\mathbf{8})}{\text{CF}_3\text{-C(=CH}_2)\text{-Br}} + \text{Zn(Ag)} \xrightarrow[93\%]{\text{TMEDA, THF, 60 °C, 9 h}} \underset{(\mathbf{9})}{\text{CF}_3\text{-C(=CH}_2)\text{-ZnBr·TMEDA}} \quad (9)$$

(**10**) CH2=C(CO2Et)CH2ZnBr   (**11**) CH2=CH-CH2ZnBr   (**12**) CH2=C(CH(OMe)O-)CH2ZnBr

$$\left(\text{piperonyl}\right)_2 \xleftarrow[100\%]{\text{Zn, THF, 0 °C, 2 h}} \underset{\substack{(\mathbf{13})\ X=\text{Br}\\(\mathbf{14})\ X=\text{OP(O)(OEt)}_2}}{\text{piperonyl-CH}_2X} \xrightarrow[100\%]{\text{Zn, LiI cat., dmpu, 50 °C, 12 h}} \underset{(\mathbf{15})}{\text{piperonyl-CH}_2\text{Zn-OP(O)(OEt)}_2} \quad (10)$$

4.1.2.2 The iodine–zinc exchange reaction

The preparation of diorganozincs was for a long time limited to the preparation of nonfunctionalized reagents, since they were obtained by transmetallation using organolithiums or Grignard reagents.[6] These zinc derivatives are of special importance because of their application in catalytic asymmetric additions to aldehydes.[10] Furukawa has reported that diiodomethane reacts readily with diethylzinc to give ethyl(iodomethyl)zinc (**16**) and ethyl iodide via an iodine–zinc exchange (Equation (11)).[39] The reaction has been extended to a variety of functionalized primary iodides, allowing unique access to functionalized dialkylzincs (Equation (12)).[40] The exchange reaction is catalyzed by the addition of transition metals; 0.3 mol.% of CuI or CuCN reduces the reaction times by half and avoids the use of a large excess of diethylzinc.[41] Interestingly, catalysis with palladium or nickel salts is also effective, providing, however, not dialkylzincs, but alkylzinc iodides (Equation (13)).[42] The mechanism of the reaction seems to be radical in nature and the process has been used to perform highly diastereoselective radical cyclization reactions leading to zinc organometallics which can be subsequently trapped by an electrophile (Equation (14)).[43] Polyfluorinated iodides such as CF_3I or $\text{C}_6\text{F}_5\text{I}$ react with dialkylzincs to give $(\text{CF}_3)_2\text{Zn}$ and $(\text{C}_6\text{F}_5)_2\text{Zn}$ respectively. Reactions of other polyfluorinated iodides do not proceed cleanly.[44]

$$\text{ICH}_2\text{I} + \text{Et}_2\text{Zn} \xrightarrow[>90\%]{\text{THF}, -40\,°\text{C}} \underset{(\mathbf{16})}{\text{ICH}_2\text{ZnEt}} + \text{EtI} \quad (11)$$

$$\text{FG-RCH}_2\text{I} + \text{Et}_2\text{Zn} \xrightarrow[\substack{\text{ii, } 50\,°\text{C, } 0.1\text{ mm Hg, 1 h} \\ \text{ca. } 80\%}]{\text{i, neat, } 25–50\,°\text{C}, 2–20\text{ h, CuI (0.3 mol.\%)}} (\text{FG-RCH}_2)_2\text{Zn} \quad (12)$$

$$\text{Oct}_2\text{Zn} \xleftarrow[80\%]{\text{Et}_2\text{Zn, neat, } 50\,°\text{C, 12 h, CuI (0.3 mol.\%)}} \text{OctI} \xrightarrow[75–80\%]{\text{Et}_2\text{Zn, THF, } 25\,°\text{C, 2 h, PdCl}_2(\text{dppf})\ (1.5\text{ mol.\%})} \text{OctZnI} \quad (13)$$

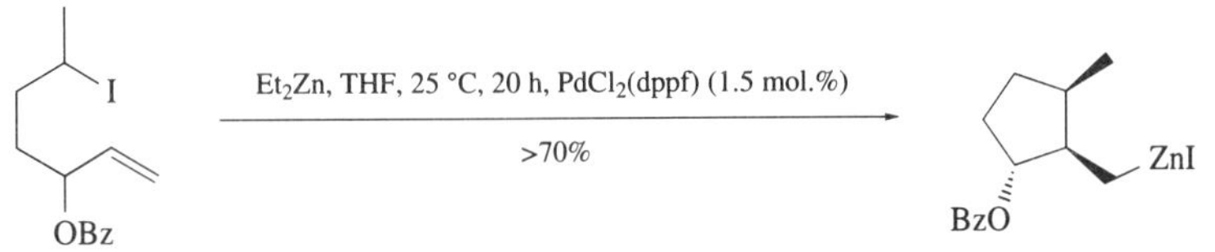

(14)

4.1.2.3 Preparation of organozinc compounds via transmetallations

(i) Lithium–zinc exchange

Main-group organometallics of electropositive metals (Li, Mg, Al) are readily transmetallated with zinc salts to give zinc organometallics[6] (Equations (15) and (16)[45,46] and Scheme 2[47]). This procedure allows the stabilization of highly reactive lithium organometallics. Thus, whereas 1,2,2-trifluoroethenyllithium (**17**) is a very unstable carbenoid (stable only below −100 °C), the addition of ZnX_2 furnishes a very stable zinc reagent (**18**) which has considerable synthetic utility (Equation (17)).[48] Similarly, 2-hydropentafluoropropene has been converted to the corresponding zinc reagent (**19**) via a highly reactive lithium intermediate (Equation (18)).[34] It is possible to extend these transmetallations to the preparation of functionalized alkenyl- and arylzinc halides.[49] Alkenyl- and aryllithiums are less reactive than their alkyl counterparts, and tolerate at low temperature the presence of several electrophilic functions (ester, azide, nitrile, nitro, halide). Thus, the low-temperature iodine–zinc exchange of functionalized iodides of type (**20**) provides the corresponding functionalized lithium derivatives (**21**) which, after the addition of ZnI_2, lead to the polyfunctional zinc reagents (**22**) (Equation (19)). The transmetallation of alkylmagnesium halides with zinc chloride in ether followed by the addition of 1,4-dioxane constitutes a convenient method for the preparation of salt-free dialkylzincs (Equation (20)).[50]

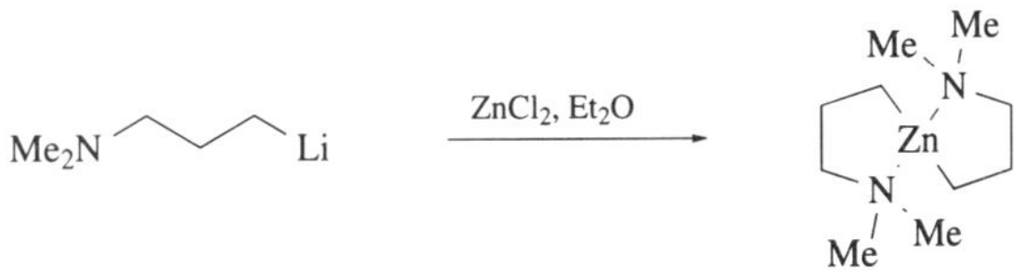

(15)

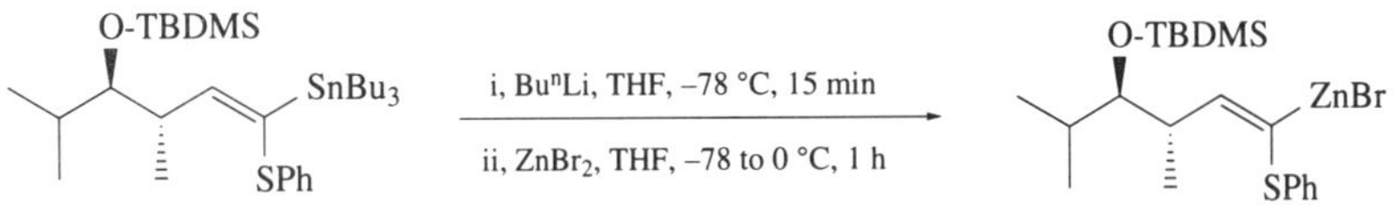

(16)

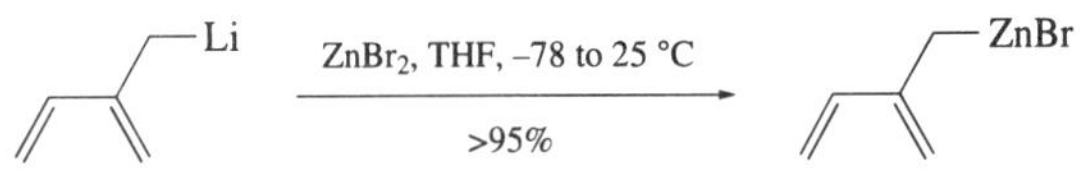

$$2\ \text{LiCHCl}_2 \xrightarrow{\text{ZnCl}_2\ (0.5\text{ equiv.}), \text{THF}, -78\text{ to }25\,°\text{C}} \text{Zn}(\text{CHCl}_2)_2$$

Scheme 2

$$F_2C=CFCl \xrightarrow{\text{BuLi, THF-Et}_2\text{O (3:2), -100 °C}} [F_2C=CFLi] \xrightarrow{\text{ZnX}_2, -100\text{ °C}} F_2C=CFZnX \quad (17)$$
(17)   (18) stable at 25 °C

$$\underset{H}{\overset{F_3C}{>}}=CF_2 \xrightarrow{\text{LDA, -78 °C}} \left[\underset{Li}{\overset{F_3C}{>}}=CF_2\right] \xrightarrow{\text{ZnI}_2, -78\text{ °C}} \underset{IZn}{\overset{F_3C}{>}}=CF_2 \quad (18)$$
(19)

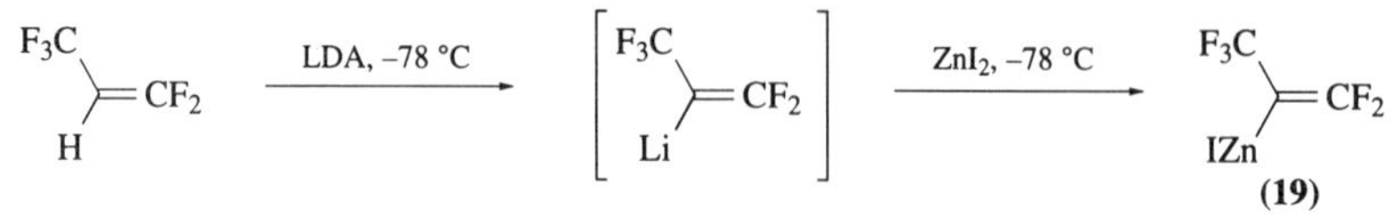

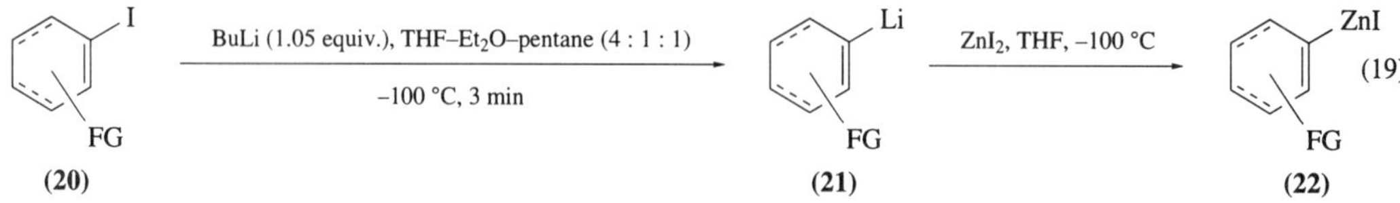

Ar–I (20) with FG →[BuLi (1.05 equiv.), THF–Et$_2$O–pentane (4:1:1), –100 °C, 3 min]→ Ar–Li (21) →[ZnI$_2$, THF, –100 °C]→ Ar–ZnI (22) (19)

FG = CO$_2$R, CN, Cl, N$_3$, NO$_2$

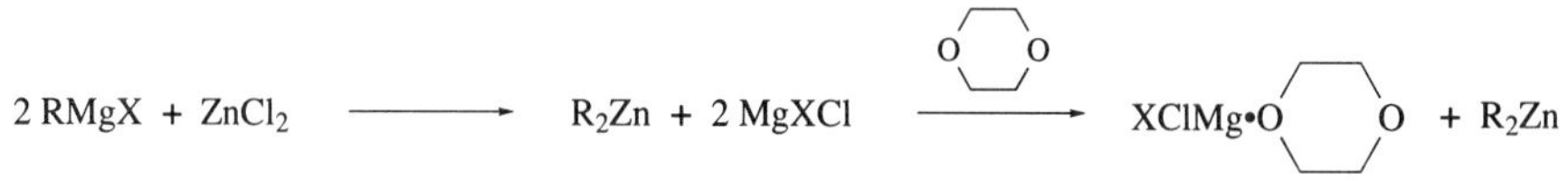

$$2\ RMgX + ZnCl_2 \longrightarrow R_2Zn + 2\ MgXCl \xrightarrow{\text{dioxane}} XClMg \cdot O\text{-}O + R_2Zn \quad (20)$$

(ii) Boron–zinc exchange

Organoboranes readily undergo transmetallation reactions with diorganozinc compounds. The reaction is reported to be an equilibrium[6] and finds, under appropriate reaction conditions, interesting synthetic applications (Equation (21)).[6] The reaction of Me$_2$Zn with triallyl- or tribenzylborane is an excellent method for preparing diallyl- and dibenzylzinc. The driving force of the reaction is the formation of volatile BMc$_3$ (b.p. = –22 °C; Equation (22)).[51] The reaction is well suited for the preparation of di(alkenyl)zincs, and alkenylboranes react with diethyl- or dimethylzinc within a few minutes at 0 °C (Equation (23)).[52] A related preparation of dialkylzincs can also be performed (Equation (24)).[53]

$$3\ R^1_2Zn + 2\ R^2_3B \rightleftharpoons 3\ R^2_2Zn + 2\ R^1_3B \quad (21)$$

$$3\ Me_2Zn + 2\ (\text{allyl})_3B \longrightarrow 3\ (\text{allyl})_2Zn + 2\ Me_3B \quad (22)$$

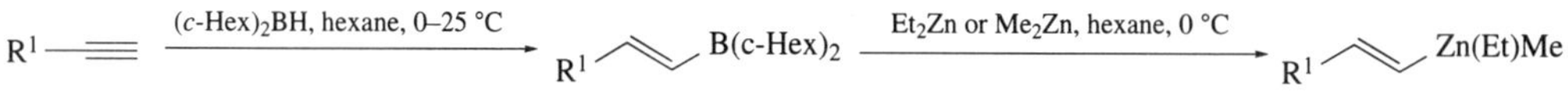

$$R^1{-}{\equiv}{-} \xrightarrow{\text{(c-Hex)}_2\text{BH, hexane, 0-25 °C}} R^1{-}CH{=}CH{-}B(c\text{-Hex})_2 \xrightarrow{\text{Et}_2\text{Zn or Me}_2\text{Zn, hexane, 0 °C}} R^1{-}CH{=}CH{-}Zn(Et)Me \quad (23)$$

$$\text{(β-pinene methylene)} \xrightarrow[\text{Et}_2\text{B}]{\text{Et}_2\text{BBr, NaH, THF, -78 to 25 °C, 7 h}} \text{(pinanyl-CH}_2\text{-BEt}_2\text{)} \xrightarrow[\text{ii, 0.1 mm Hg, 50 °C, 1 h}]{\text{i, Et}_2\text{Zn, hexanes, 0-25 °C}} (\text{pinanyl-CH}_2)_2\text{Zn} \quad (24)$$

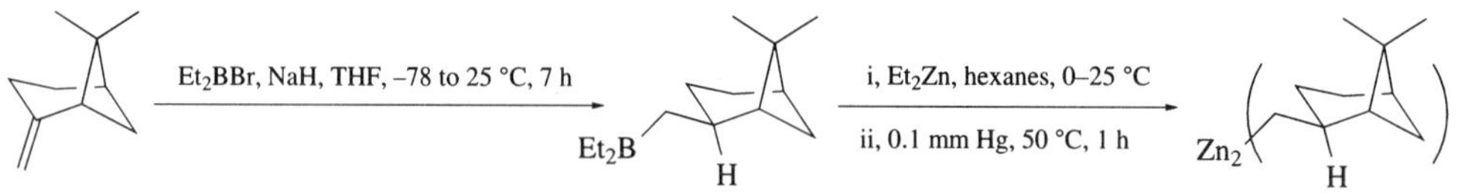

(iii) Mercury–zinc exchange

Diorganomercurials react efficiently with zinc metal to give diorganozincs,[6,54] and the reaction gives access to various functionalized dialkylzincs. The transmetallation is usually complete within a few hours at reflux in toluene; however, the addition of zinc halides catalyzes the reaction and a complete conversion can be achieved under these conditions at 60 °C after 2 h (Equation (25)).[55] It is interesting to note that the reverse transmetallation can be achieved by treating an alkylzinc halide with calomel (Hg$_2$Cl$_2$) in THF (Equation (26)).[55] A related transmetallation of bis(trifluoromethyl)mercury with dimethylzinc in pyridine produces a complex of bis(trifluoromethyl)zinc and pyridine in 82% yield.[56]

$$(FG\text{-}R)_2Hg \xrightarrow{\text{Zn, PhMe, 110 °C, 3-5 h}} (FG\text{-}R)_2Zn + Hg^0 \quad (25)$$

$$\text{FG-RZn} + \text{Hg}_2\text{Cl}_2 \xrightarrow[61-89\%]{\text{THF}, -50 \text{ to } 20\,°\text{C}, 2\,\text{h}} (\text{FG-R})_2\text{Hg} + 2\,\text{ZnX}_2 + \text{Hg}^0 \quad (26)$$

4.1.3 Reactivity of Organozinc Compounds

4.1.3.1 Substitution reactions

(i) Coupling reactions with organic halides and related electrophiles

Alkylzinc halides and dialkylzincs are relatively unreactive organometallics[6,57] and transmetallations to more reactive species are required for most reactions with electrophiles. Exceptions are reactions with oxidation reagents, halogens, and pseudohalogens.[6] Thus, tosyl cyanide reacts readily with various classes of organozinc halides.[7] Interestingly, whereas a benzylic zinc derivative affords the *ortho*-substitution product, transmetallation of the benzylic zinc bromide with CuCN·2LiCl affords the substitution product at the benzylic position (Equation (27)).[7] In contrast, allylic zinc reagents show far higher reactivity due to the more ionic character of the carbon–zinc bond. A recent review is available.[58] Metallic halides (MX_n) react differently with organozinc compounds. Thus, TMS-Cl is usually not reactive enough, but with the heterocyclic zinc reagent (**23**), a silylated heterocycle (**24**) is obtained (Equation (28)). On the other hand, trialkylchlorostannanes R_3SnCl react more readily and, for example, stannylated phosphabenzenes have been prepared in this way (Equation (29)).[59]

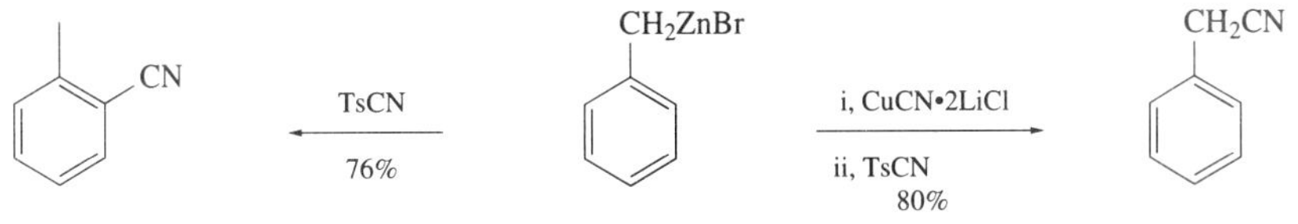

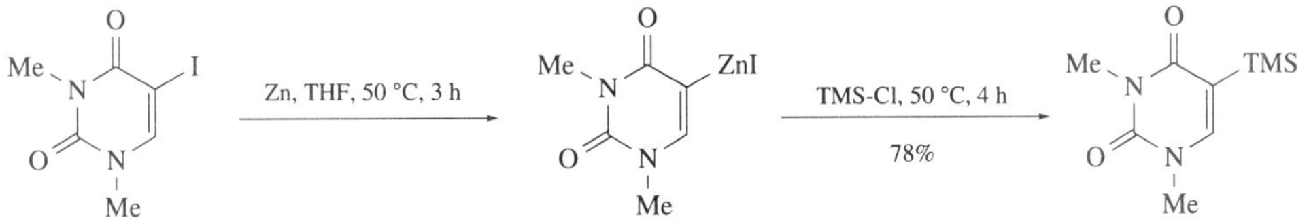

For many substitution reactions a transmetallation of a zinc compound to the corresponding copper reagent using the THF-soluble copper salt CuCN·2LiCl is required in order to obtain an organometallic species of sufficient reactivity.[9] Organozinc halides or diorganozincs can be allylated with allylic halides in the presence of catalytic or stoichiometric amounts of copper(I) salts. These reactions proceed with very high S_N2' selectivity[60-2] (usually $S_N2'/S_N2 = 95-80/5-20$), whereas, in the presence of nickel or palladium salts,[20] the reaction with allylic halides preferentially produces the S_N2 substitution product (Equations (30) and (31)).[63,64] γ-Mesyloxy-α,β-unsaturated esters[65] and 3-carbomethoxy-2-propenyl bromide[61] react with zinc reagents with complete S_N2' selectivity (Equation (32)).[61] 1,3-Dichloropropenes of type (**25**) (Y = SePh, SPh) also react with the zinc–copper reagents RCu(CN)ZnI (prepared by the addition of CuCN·2LiCl to RZnI)[9,66] and afford, after two successive S_N2' substitutions, polyfunctionalized 1,3-disubstituted products of type (**26**) in excellent yields (Equation (33)).[67] Allylic phosphates react readily with zinc–copper reagents, and this procedure has been applied to a short preparation of isocarbacyclins (Equation (34)).[68]

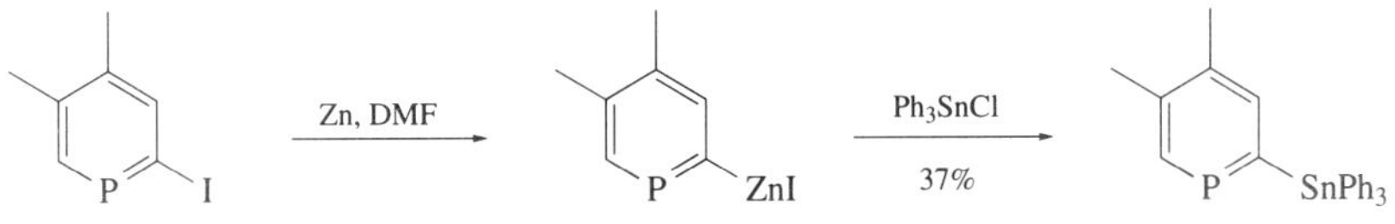

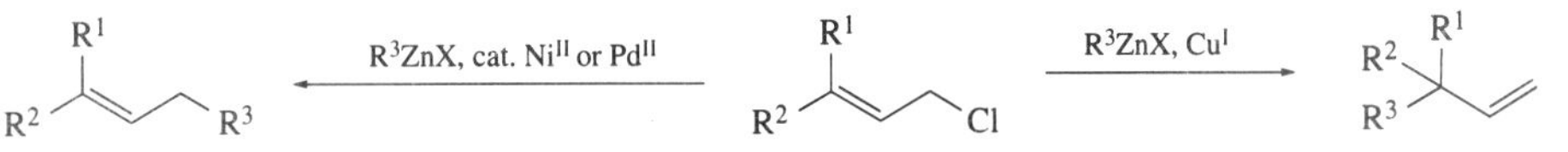

Allylic substitutions of propargylic halides or tosylates furnish allenes[61,69] or dienes[70] (Equation (35)).[69] Cationic pentadienyliron or molybdenum complexes can be alkylated by the zinc–copper reagents FG-RCu(CN)ZnX, and this process provides unique access to polyfunctional transition metal diene complexes[71,72] which can be readily cyclized under basic conditions in the presence of CO (Equation (36)).[71] Benzylic halides do not react directly with zinc reagents, but the zinc–copper species obtained by the reaction of functionalized dialkylzincs with $Me_2Cu(CN)(MgCl)_2$ in dmpu provides the substitution products in excellent yields (Equation (37)).[73] Activated alkenyl halides such as 3-iodo-2-cyclohexenone or related reagents such as phenylsulfonylmethylidenemalonate undergo a smooth addition–elimination with the zinc–copper reagents RCu(CN)ZnI[74] (Equation (38)).[75] A selective double addition–elimination on 3,4-dichlorocyclobutane-1,2-dione (**27**) produces a range of functionalized 3,4-disubstituted cyclobutene-1,2-diones (Equation (39)).[76] The addition of readily available α-acetoxyalkyl-zinc–copper reagents[77] such as (**28**) to (E)-2-ethylsulfonyl-1-nitroethylene (**29**) constitutes a unique preparation of γ-acetoxy nitroalkenes (**30**) (Equation (40)).[78] Palladium catalysis facilitates a variety of coupling reactions between organozinc derivatives and organic halides. Pioneered by Negishi,[8] this coupling method is one of the most powerful tools for forming carbon–carbon bonds. It possesses an extraordinary functionality tolerance and can be applied to a wide range of different classes of organic halides. Recent reviews on this topic are available.[79] The palladium catalyzed coupling of alkenyl bromides with allylic zinc reagents produces skipped dienes (**31**) (Equation (41)).[80] The coupling reaction with substituted allylic zinc reagents affords a mixture of substitution products resulting from attack at both ends of the allylic system.[63] The coupling of alkenyl or aryl iodides with various zinc reagents has an exceptional synthetic potential (Equations (42),[81] (43),[82] and (44)[83]).

(37)

(38)

(39)

(27)

(40)

(28) **(29)** **(30)**

(41)

(31)

$$\mathrm{TMS{-}C{\equiv}C(CH_2)_2ZnCl} \xrightarrow[83\%]{\mathrm{PhI,\ THF,\ 5\%\ Pd(PPh_3)_4}} \mathrm{TMS{-}C{\equiv}C(CH_2)_2Ph} \qquad (42)$$

(43)

(44)

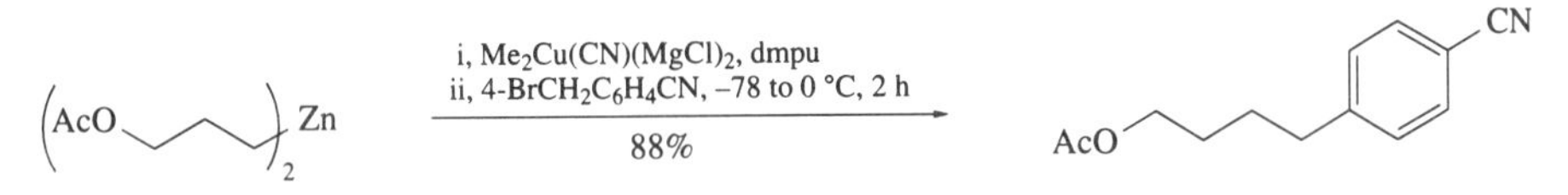

Secondary benzylic zinc halides react enantioselectively with alkenyl bromides in the presence of a chiral palladium catalyst (**32**) to give coupling products of up to 86% *ee* (Equation (45)).[84] Several extensions to the preparation of heterocycles have been reported (Equations (46)[85] and (47)[86]).[79] Alkynyl zinc halides, readily obtained from the corresponding alkyne, can be coupled in the presence of a palladium catalyst to alkenyl iodides (Equation (48)).[87] (*E*)-Alkenyl halides undergo oxidative addition with palladium(0) complexes more readily than (*Z*)-alkenyl halides. This has been exploited for the preparation of stereochemically enriched or pure (*E*)-enynes by reacting (*E*)/(*Z*) mixtures of alkenyl halides with an appropriate amount of an alkynylzinc reagent.[88] Highly functionalized unsaturated molecules can be obtained by palladium catalyzed coupling reactions (Equation (49)).[62] Alkenyl triflates can also be used instead of alkenyl halides.[19] Zinc organometallic reagents treated with CuCN·2LiCl react well with 1-iodo- or 1-bromoalkynes and provide functionalized alkylated alkynes in excellent yields (Equation (50)).[89] Direct coupling between C_{sp^3} centers is difficult using palladium catalysis. However, treatment of dialkylzincs with one equivalent of $Me_2Cu(CN)(MgCl)_2$ produces a reagent which couples readily with various primary iodides in dmpu (Equation (51)).[73]

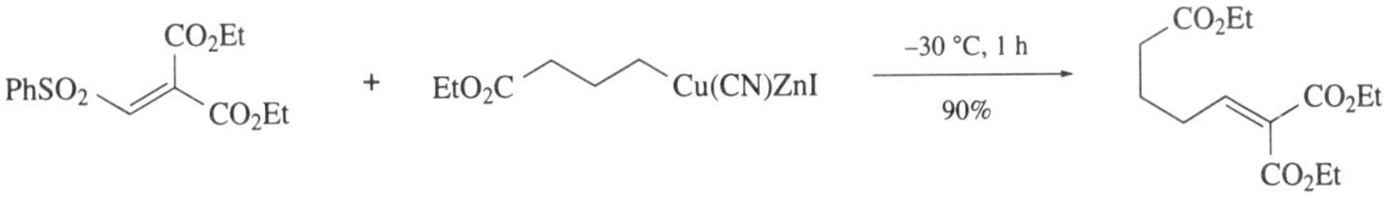

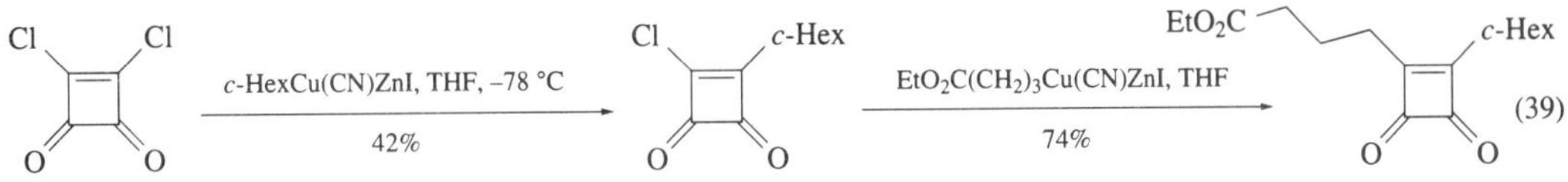

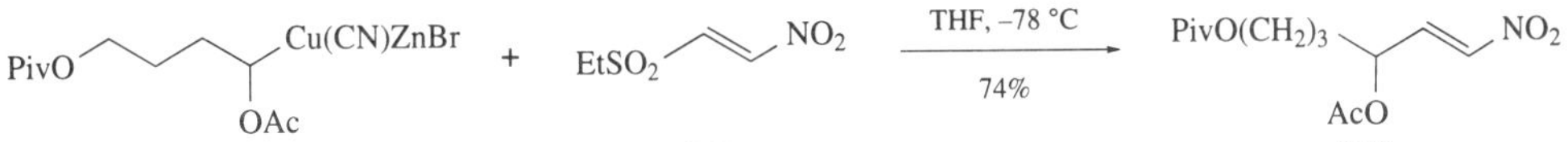

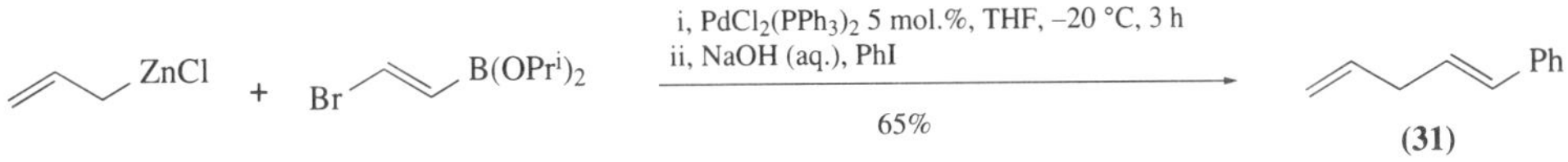

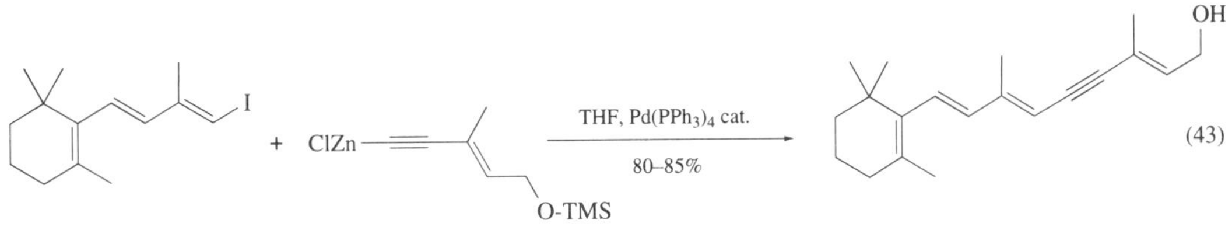

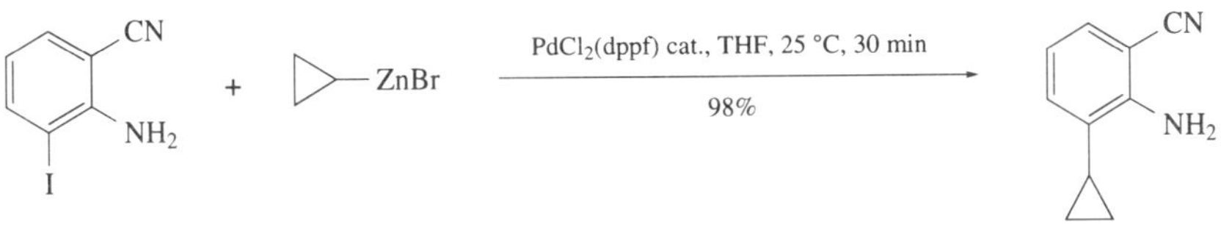

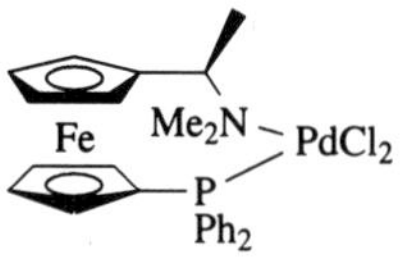

PhCH(Me)ZnX + CH₂=CHCH₂Br →[(32) 5 mol.%, THF, 0 °C, 21 h; 95%, 86% ee] Ph-CH(Me)-CH=CH₂ (45)

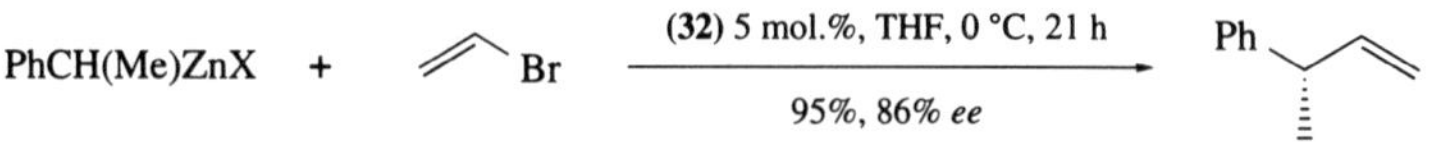

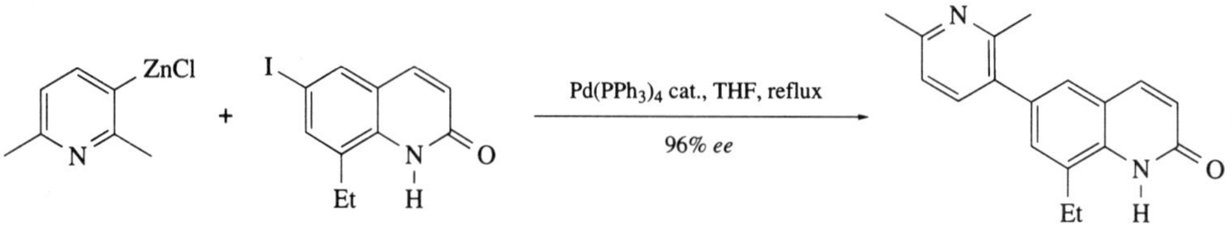

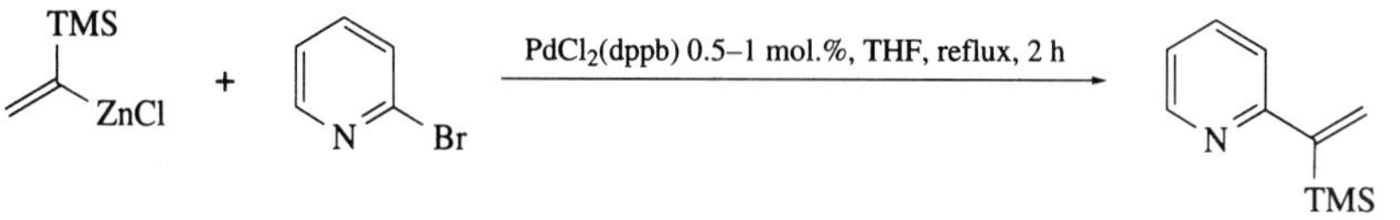

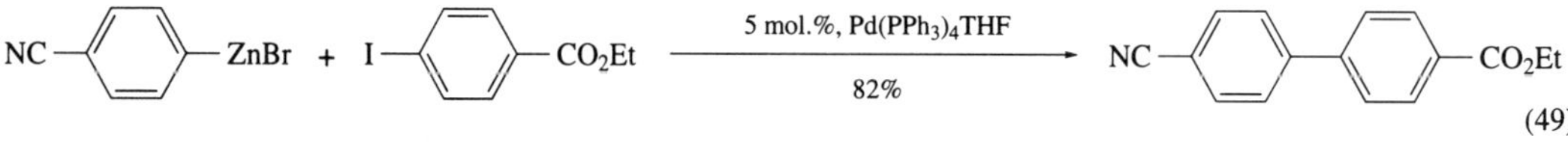

$$\text{ArZnCl} + \text{ArI} \xrightarrow{\text{Pd(PPh}_3)_4 \text{ cat., THF, reflux}, 96\% ee} \text{Ar-Ar}' \quad (46)$$

$$\text{CH}_2\text{=C(TMS)ZnCl} + \text{2-BrPy} \xrightarrow{\text{PdCl}_2(\text{dppb}) \ 0.5-1 \ \text{mol.\%, THF, reflux, 2 h}} \text{2-Py-C(TMS)=CH}_2 \quad (47)$$

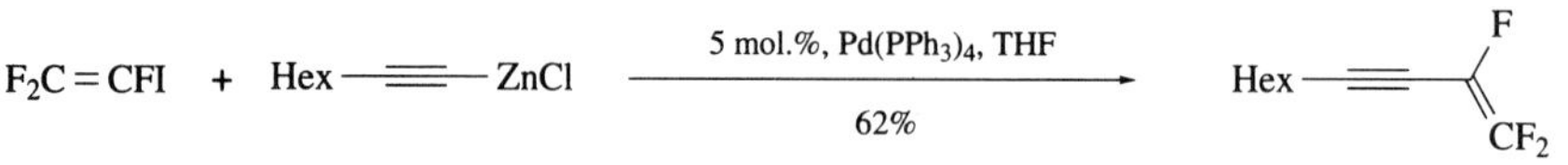

$$F_2C=CFI + \text{Hex}-\!\!\equiv\!\!-\text{ZnCl} \xrightarrow{5 \ \text{mol.\%, Pd(PPh}_3)_4, \text{THF}, 62\%} \text{Hex}-\!\!\equiv\!\!-CF=CF_2 \quad (48)$$

$$\text{NC-C}_6\text{H}_4\text{-ZnBr} + \text{I-C}_6\text{H}_4\text{-CO}_2\text{Et} \xrightarrow{5 \ \text{mol.\%, Pd(PPh}_3)_4 \text{THF}, 82\%} \text{NC-C}_6\text{H}_4\text{-C}_6\text{H}_4\text{-CO}_2\text{Et} \quad (49)$$

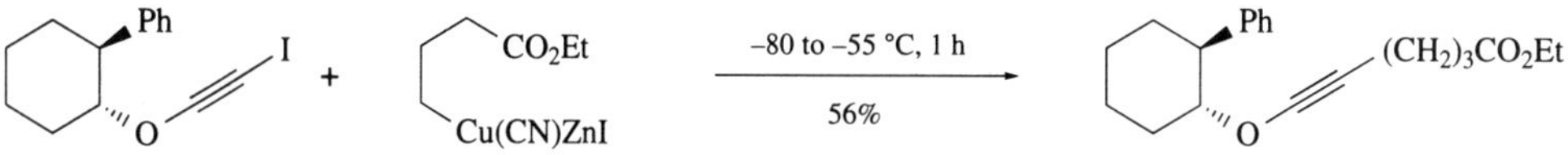

(50) trans-2-phenylcyclohexyl propargyl ether coupling with ICu(CN)ZnI-(CH₂)₃CO₂Et, −80 to −55 °C, 1 h, 56%

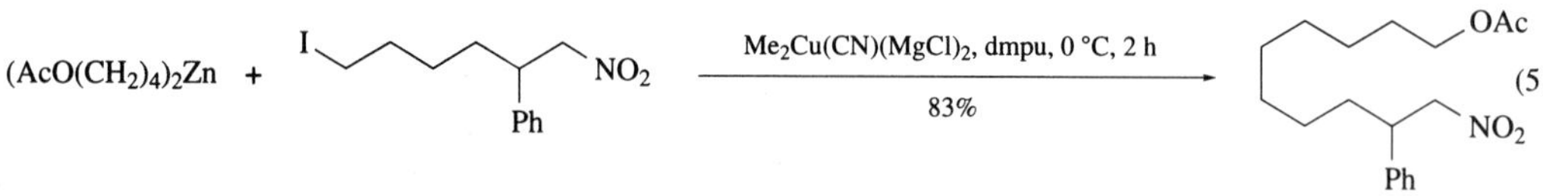

(AcO(CH₂)₄)₂Zn + I-CH(Ph)-(CH₂)₃-NO₂ →[Me₂Cu(CN)(MgCl)₂, dmpu, 0 °C, 2 h; 83%] AcO(CH₂)₆-CH(Ph)-CH₂NO₂ (51)

(ii) Coupling reactions with acid chlorides

The uncatalyzed reaction of alkylzinc halides or dialkylzincs with acid chlorides proceeds sluggishly and results in moderate yields.[6] In THF, the reaction is complicated by a zinc(II)-catalyzed ring opening by the acid chloride. In contrast, polyfunctional ketones can be prepared by performing the acylation in the presence of copper salts (stoichiometric or catalytic)[9] or palladium(0) complexes. Alkyl, aryl, or benzylic zinc reagents can be used with equal success (Equations (52),[9] (53),[62] (54),[90] and (55)[91]). The direct acylation of organozinc halides with carbon monoxide and an allylic benzoate in the presence of Pd(PPh₃)₄ (5 mol.%) provides δ-ketoesters in high yields (Equation (56)).[92] Finally, whereas the direct addition of allylic zinc reagents with acid chlorides or anhydrides is complicated by the further reaction of the allylic zinc reagents[58] with the allylic ketone produced, an equivalent reaction can be realized by using a nitrile as acylation agent (Equation (57)).[93]

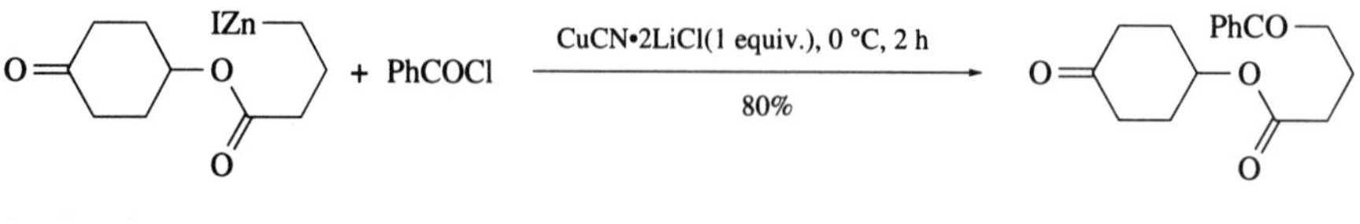

(52) 4-oxocyclohexyl ester with CH₂ZnI + PhCOCl →[CuCN·2LiCl (1 equiv.), 0 °C, 2 h; 80%] PhCO-substituted product

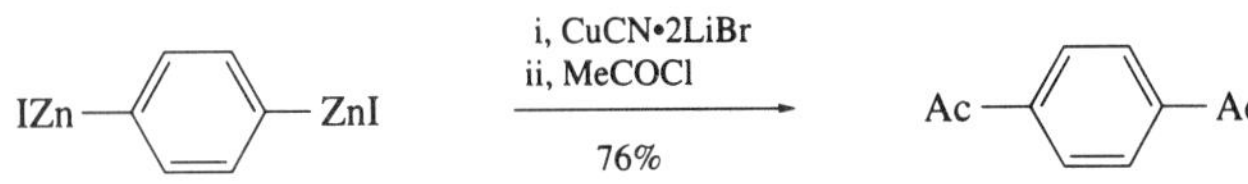

4.1.3.2 Addition reactions

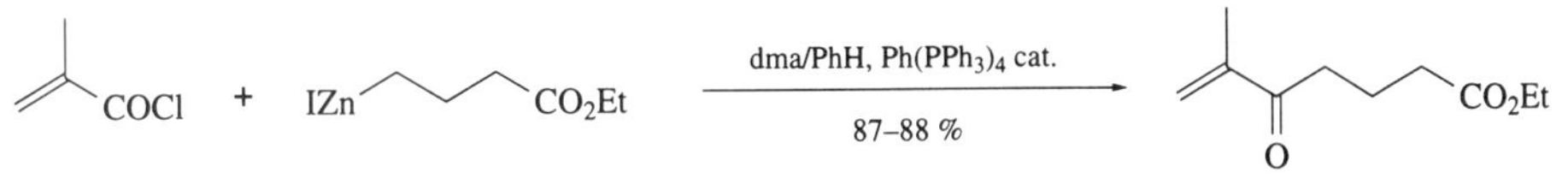

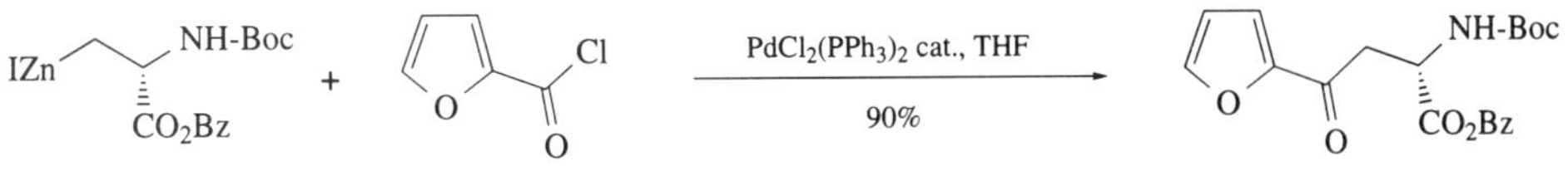

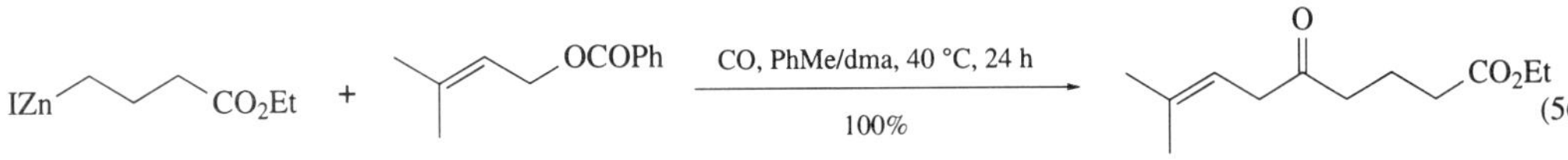

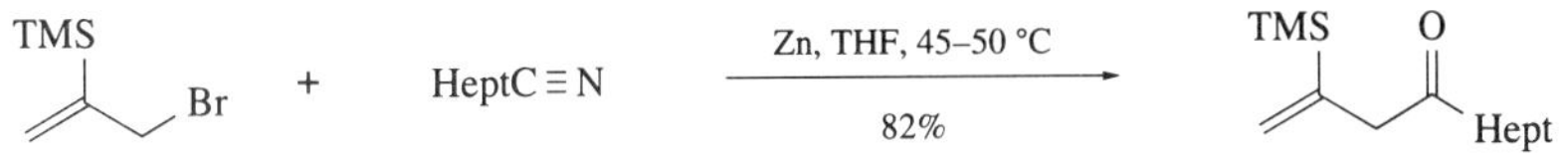

(i) Addition to carbonyl derivatives and related electrophiles

With the exception of the more reactive allylic and propargylic zinc compounds, organozinc derivatives do not add readily to aldehydes or ketones and the addition of catalysts is required.[94] The addition of allylic zinc halides to aldehydes or ketones produces homoallylic alcohols. If unsymmetrically substituted allylic zinc reagents are used, the formation of the new carbon–carbon bond occurs preferentially from the most substituted end of the allylic system (Equation (58)).[6,11,58,94,95] The addition of functionalized allylic zinc halides to aldehydes or ketones provides useful access to five-membered carbo- and heterocycles (Equations (59),[96] (60),[97] and (61)[38,98]). The insertion of zinc into propargylic bromides provides allenic and propargylic zinc reagents. Their addition to aldehydes and ketones generally affords a mixture of homopropargylic and allenic alcohols (Equation (62)).[6,58,94,99] Excellent diastereoselectivities are obtained for the addition of trimethylsilyl-substituted allenic zinc halides to aldehydes, providing *anti*-homopropargylic alcohols (Equation (63)).[100] Various types of zinc reagents add, with good to excellent stereoselectivity, to chiral aldehydes (Equations (64)[101] and (65)[102]). In some cases, an excellent chelation-controlled diastereoselectivity is observed (Equations (66)[103] and (67)[104]). Perfluoroalkylzincs have been added to chromium tricarbonyl complexes of aromatic aldehydes with fair diastereoselectivity (44–66% *de*).[105] Low reactivity of alkylzinc derivatives towards ketones has been observed. The reaction is often complicated by a β-hydride elimination of the zinc reagent leading to reduced products.[6,58] Reactive ketones, such as menthyl phenylglyoxylate, give α-substituted menthyl mandelates with good stereoselectivity (71–88% *de*).[106]

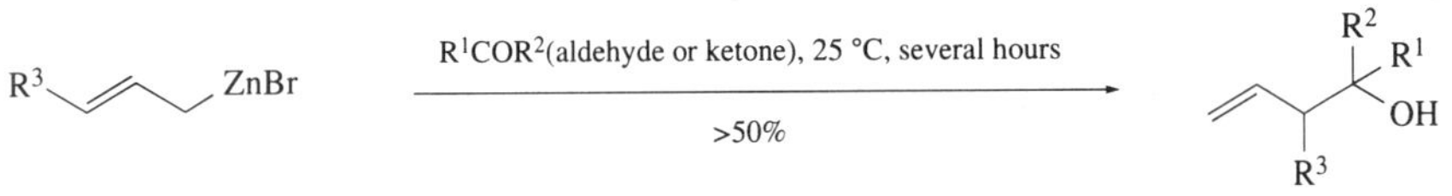

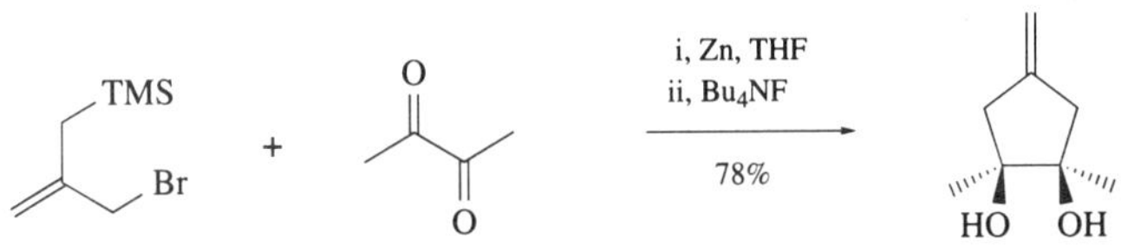

The addition of Lewis acids, such as $Ti(OPr^i)_nCl_{4-n}$,[107] TMS-Cl, or $BF_3 \cdot OEt_2$,[108] catalyzes the reaction of zinc reagents with aldehydes. Excellent chemoselectivity is observed. Thus, the addition of a copper–zinc reagent FG-RCu(CN)ZnI to cinnamaldehyde in the presence of $BF_3 \cdot OEt_2$ produces only the

(60)

(61)

(62)

(63)

(64)

anti : *syn* = 92 : 8

(65)

anti : *syn* = 2 : 98

(66)

syn : *anti* = 86 : 14

(67)

syn : *anti* = 85 : 15

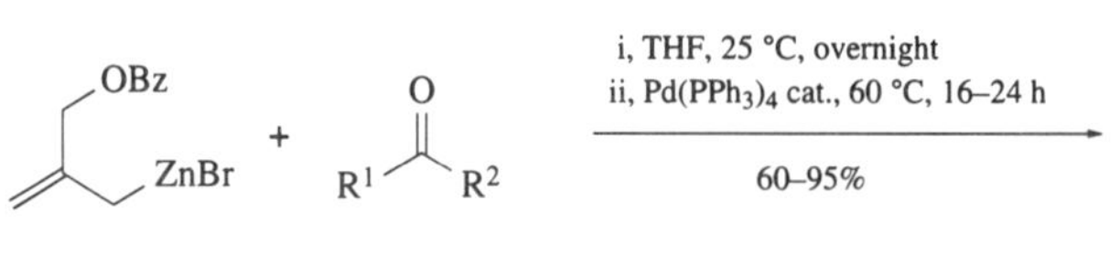

1,2-adduct, whereas in the presence of TMS-Cl, 1,4-addition is observed.[108] The addition of FG-RCu(CN)ZnI to α-chiral aldehydes occurs with stereoselectivity comparable to organotitanium reagents[109] (Equation (68)).[108] Various reactions derived from the classical Reformatsky reaction have been reported,[2] and optically active β-hydroxy nitriles have been obtained in up to 93% *ee* by the enantioselective addition of cyanomethylzinc bromide to aldehydes and ketones in the presence of a chiral catalyst.[110] Functionalized (bromomethyl)oxazoles of type (**33**) can be readily converted to the corresponding zinc reagents (**34**) which add rapidly to aldehydes and ketones (Equation (69)).[111] Nonactivated imines react only with the highly reactive allylic and propargylic zinc compounds.[58]

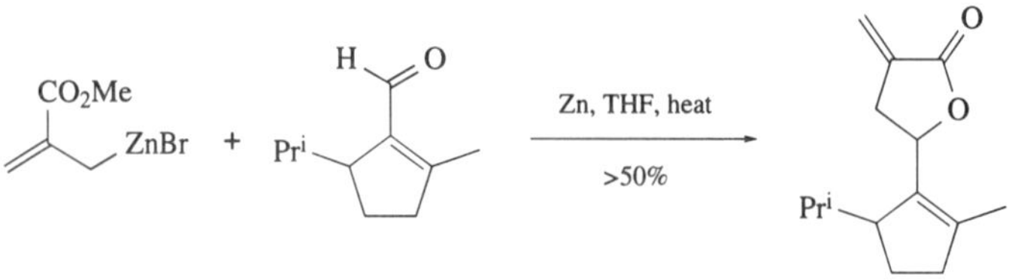

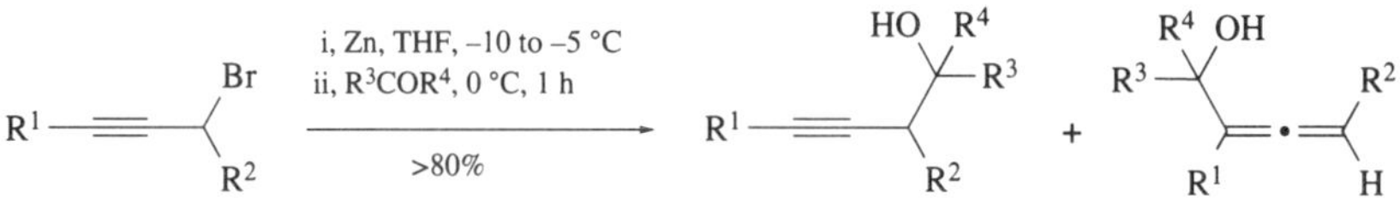

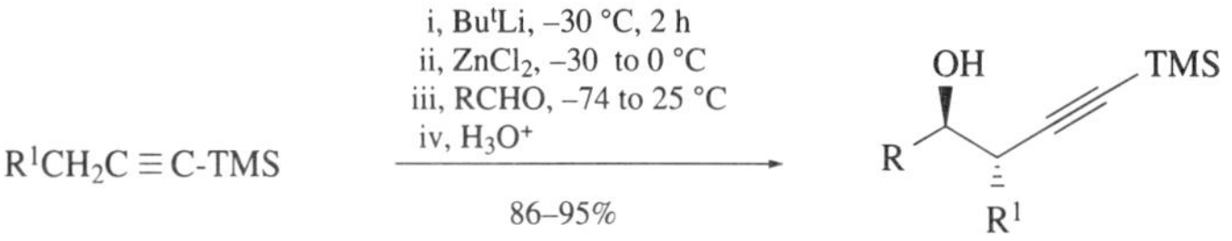

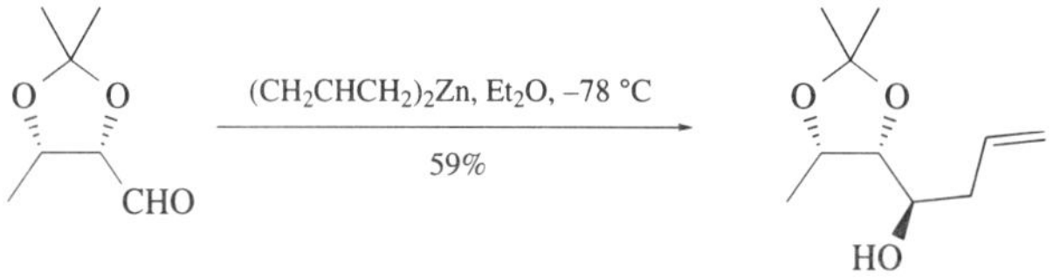

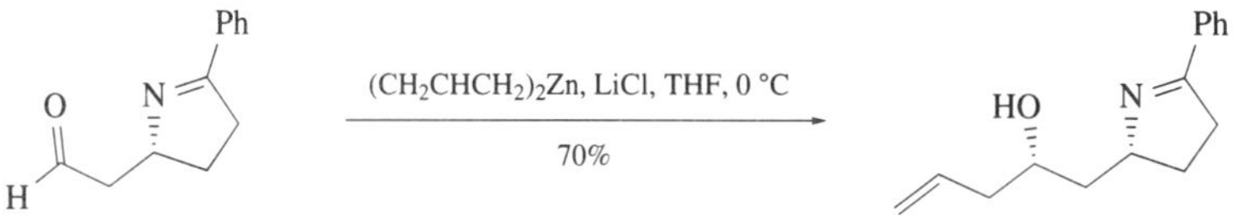

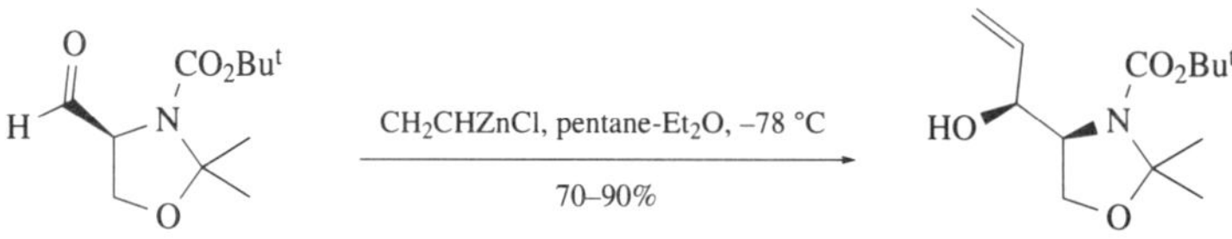

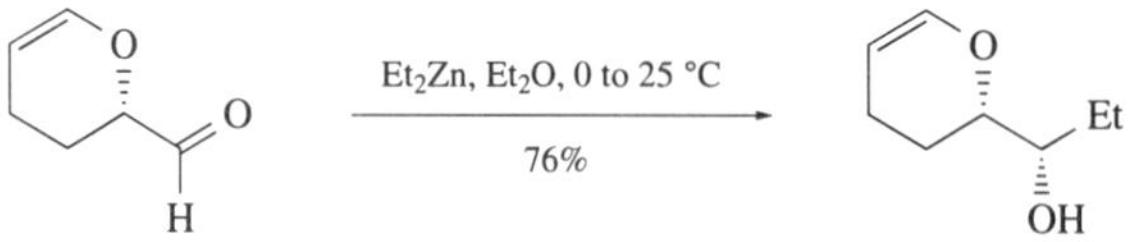

Chelation control is observed for the addition of allylzinc bromide to α- and β-alkoxyimines.[112] Dialkylzincs add to activated imines in the presence of TiCl$_4$ in satisfactory yields.[113] Imines generated *in situ* from α-amino thioethers provide α-alkylated amines (Equation (70)).[114] Dialkylzincs add quantitatively to α-diimines or iminium ions, the organic group being transferred to nitrogen allowing a new synthesis of condensed heterocycles. (Equation (71)).[115] The addition of dialkylzincs to N-diphenylphosphinoylimines (**35**) in the presence of a chiral catalyst furnishes chiral amines (75–98% *ee*) after acidic hydrolysis (Equation (72)).[116]

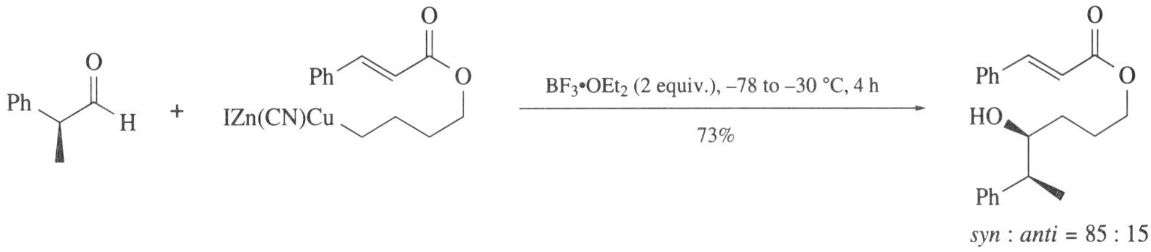

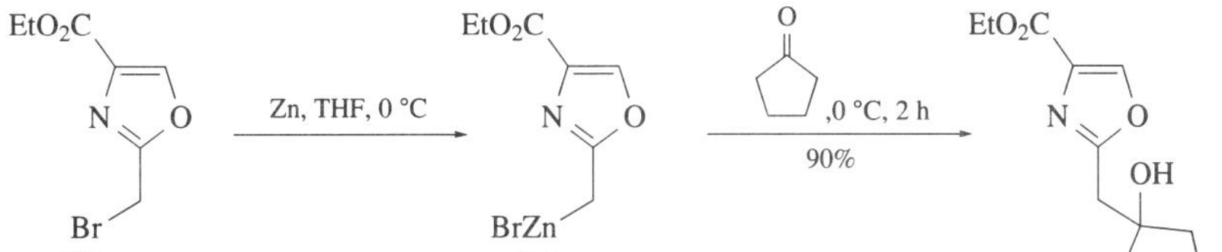

Pyridinium salts (Equations (73)[117] and (74)[118]) or γ-alkoxy-α,β-unsaturated piperidines (Equation (75))[119] react with various alkylzinc halides or the corresponding zinc–copper reagents and provide an expedient route to various polyfunctional nitrogen heterocycles. As mentioned previously, the low reactivity of organozinc compounds toward aldehydes necessitates the addition of catalysts to achieve these addition reactions. This condition makes diorganozincs ideal reagents for enantioselective additions using a chiral catalyst.[10,94,120] Oguni found that (S)-leucinol (**36**) catalyzes the addition of diethylzinc to benzaldehyde with 48% *ee*.[121] Since this initial discovery, a wide range of chiral amino alcohols and related reagents (**37**),[122] (**38**),[123] (**39**),[124] (**40**),[125] (**41**),[126] (**42**),[127] (**43**),[50,128] have been found to catalyze the addition of diethylzinc to aldehydes with high enantioselectivity (up to 99% *ee*). The catalyst (**37**), which is especially effective for aromatic aldehydes, has a particularly high catalytic activity and addition of diethylzinc to benzaldehyde proceeds within 6 h at 0 °C (98% *ee*).[122,129] A range of readily available pyrrolidinylmethanols, such as (**38**)[123] derived from proline, give excellent enantioselectivities with both aliphatic and aromatic aldehydes (Equation (76)).[123,130]

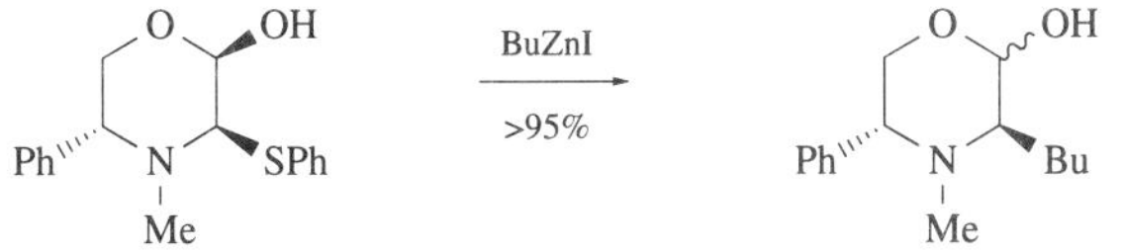

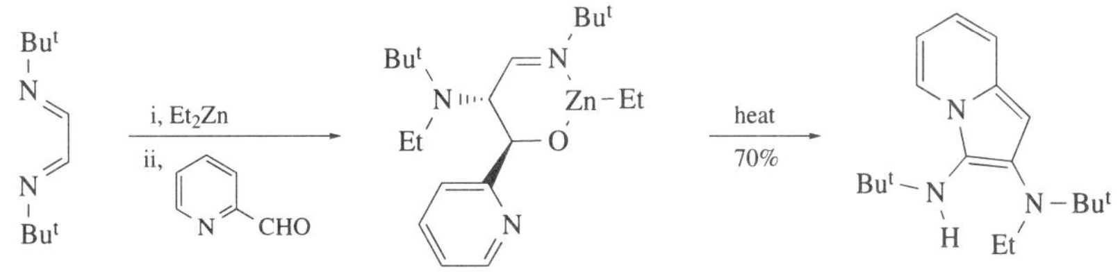

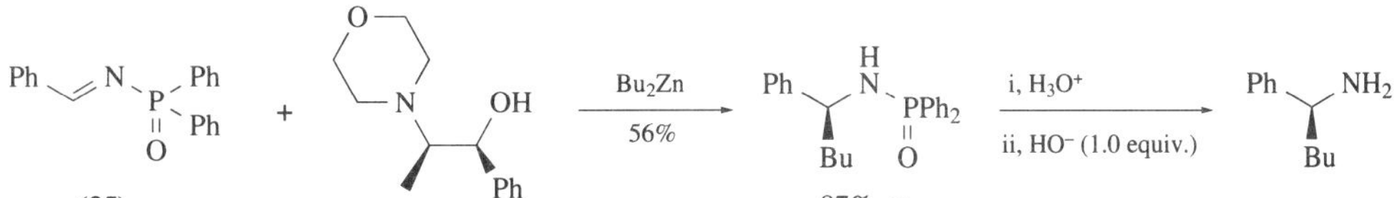

Catalysis in the solid state using the chiral quaternary ammonium salt (**39**) is also possible.[124] Chiral metal complexes such as oxazaborolidines transfer an ethyl group to aldehydes in 52–95% *ee*.[131] Very active catalysts such as (**42**) and (**43**) have been developed.[50,127,128] Their very high catalytic activity allows extension of the enantioselective addition to higher dialkylzincs (Equation (77))[50,128] and to functionalized dialkylzincs or polyfunctional aldehydes (Equations (78)–(80)).[40,41,132] A new enantioselective synthesis of lactones is based on the enantioselective addition of diethylzinc to 3- or 4-formyl esters (Equation (81)).[133] Ketoaldehydes react chemo- and enantioselectively with diethylzinc in the presence of catalysts of type (**40**) (Equation (82)).[134] Functionalized alkenylzincs can be readily prepared from the corresponding alkenylboranes and added with excellent enantioselectivity to aldehydes.[52] An intramolecular version of the reaction (Equation (83)) facilitated an elegant access to a key intermediate in a synthesis of (R)-(−)-muscone.[135] The mechanism of the enantioselective addition has been studied in detail[10,94,120,136] and reactive bimetallic intermediates of type (**44**) or (**45**) have been proposed.[10] One zinc atom provides an electrophilic activation, the other a nucleophilic activation. With several catalytic systems, an asymmetric amplification is observed.[121,122] This phenomenon is explained by diastereomeric interactions between the two enantiomeric catalytic entities.[10,120,122]

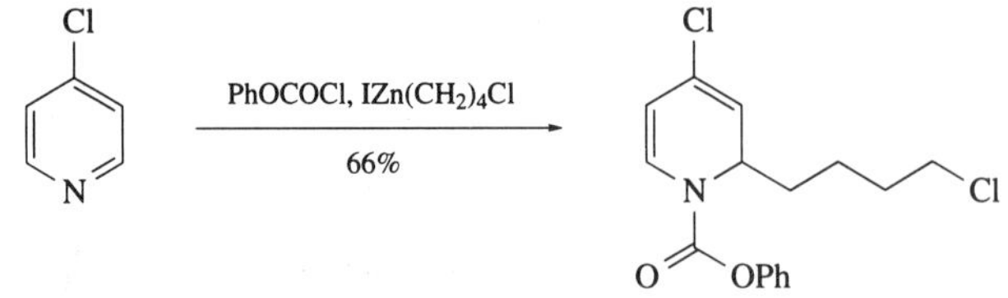

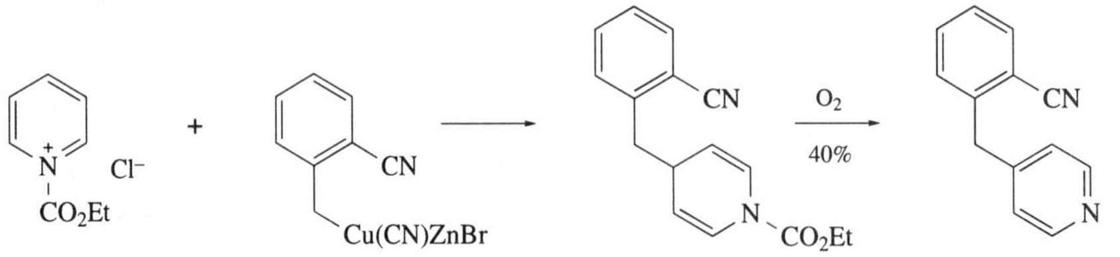

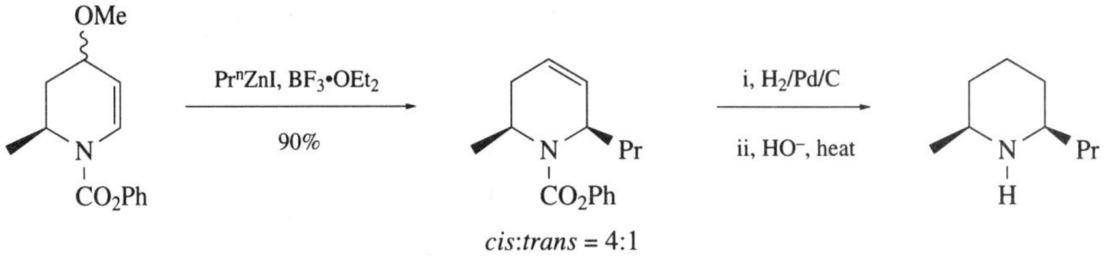

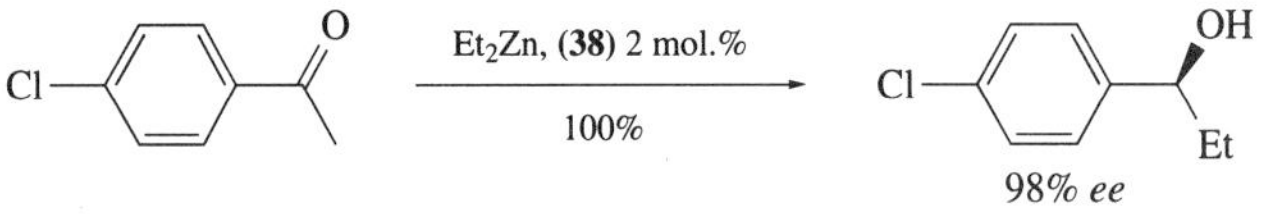

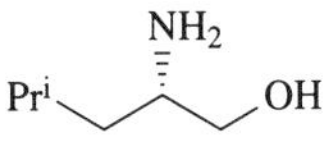

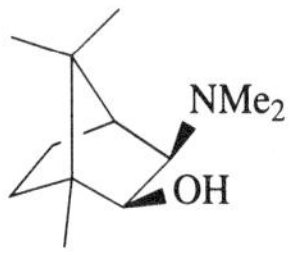

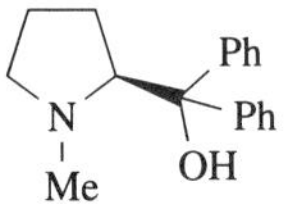

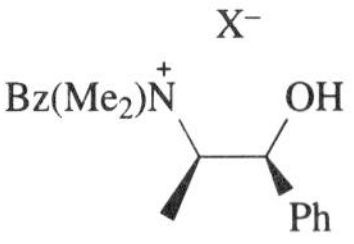

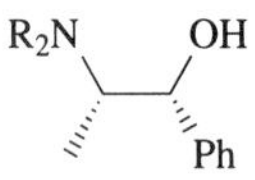

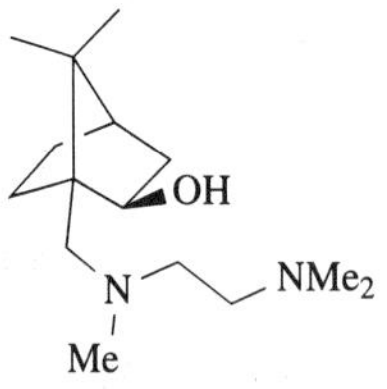

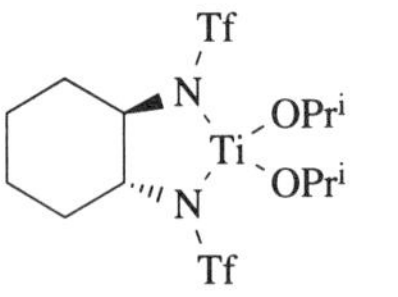

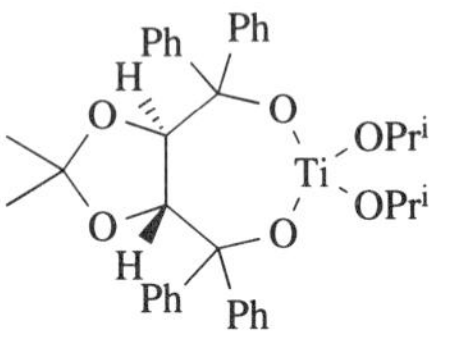

(77)

(78)

(79)

(80)

(81)

(82)

(83)

(44) **(45)**

L:large; S:small

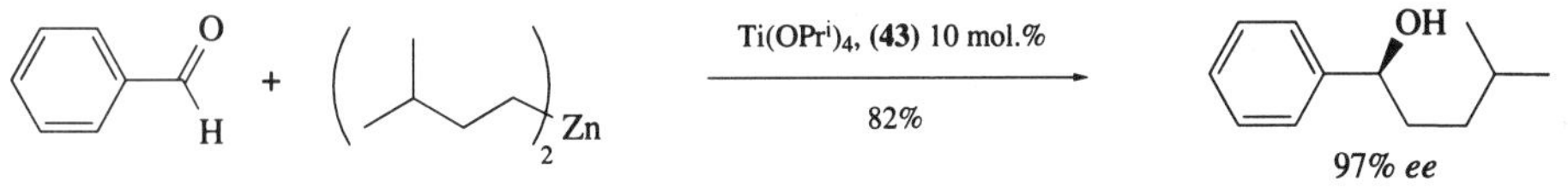

(ii) Michael additions

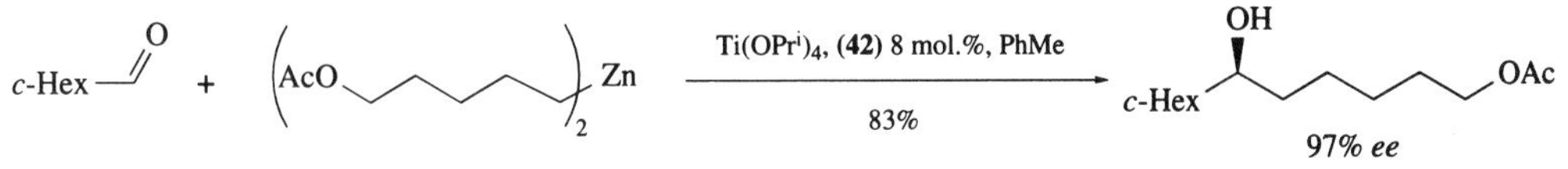

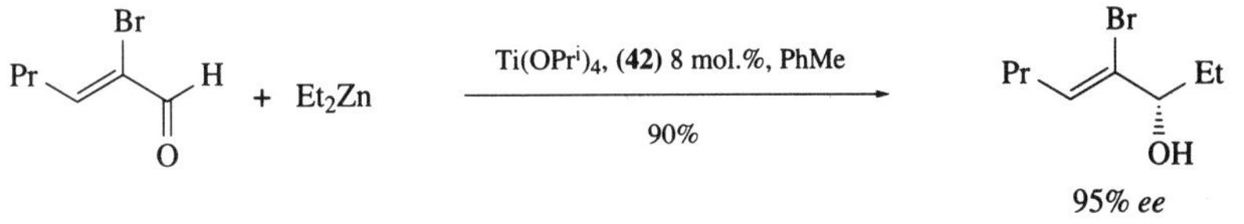

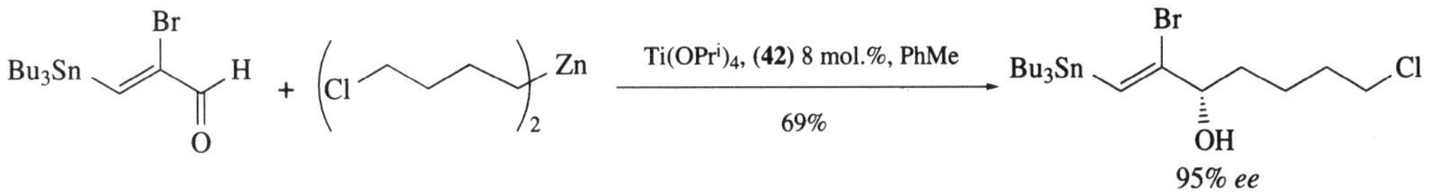

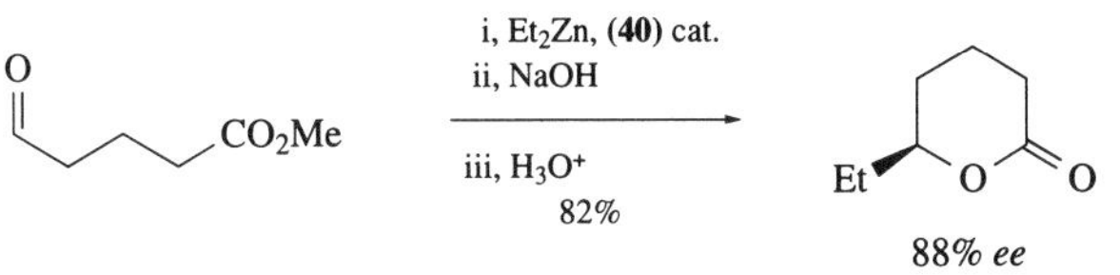

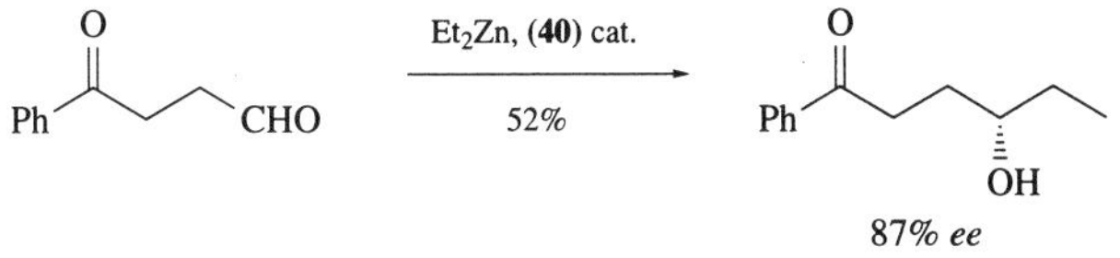

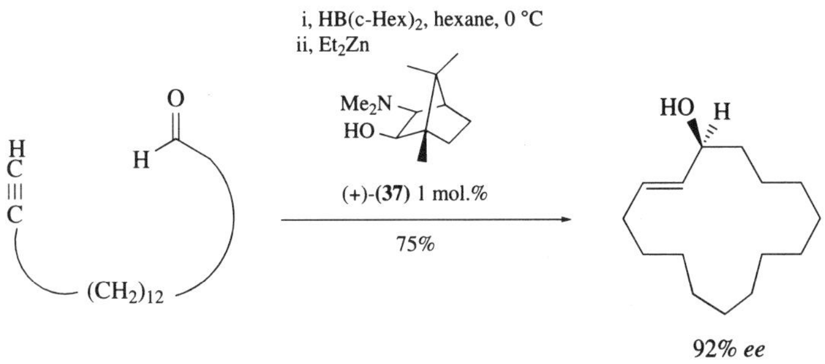

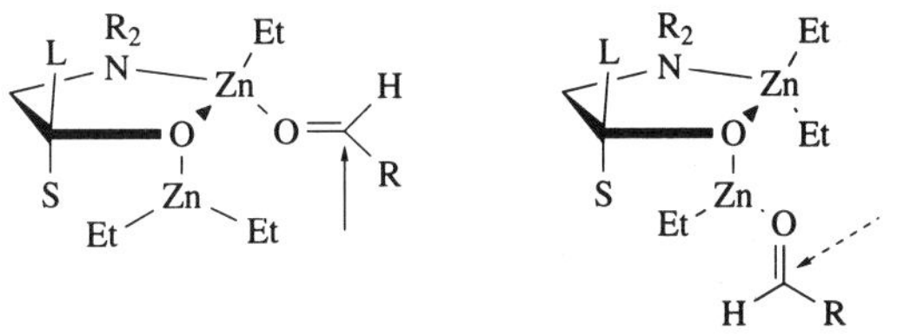

Dialkylzincs only undergo Michael additions to a limited range of α,β-unsaturated carbonyl compounds. Diphenylzinc adds cleanly in a 1,4-fashion to benzylideneacetophenone, but the reaction of dibutylzinc with benzylideneacetone produces only the corresponding zinc enolate.[6] Carrying out these reactions in the presence of stoichiometric amounts of copper(I) or catalytic amounts of nickel(II) salts considerably improves the scope of the reaction. Thus, the formation of the zinc–copper reagent FG-RCu(CN)ZnI allows the addition to various β-monosubstituted enones (Equations (84) and (85)).[9,137] However, β-disubstituted enones or unsaturated esters do not react under these conditions. By performing the addition reactions in HMPA, the addition to β-disubstituted enones as well as to methyl acrylate proceeds well (Equations (86) and (87)).[61] Another alternative consists of treating β-disubstituted enones with FG-RCu(CN)ZnI in the presence of $BF_3 \cdot OEt_2$ (Equation (88)). Interesting cyclizations are observed if a 3-cyanoalkyl chain is attached at position 3 leading to an intermediate difluoroboron enolate (**46**) (Equation (89)).[138] Arylzinc compounds generated by electroreduction from the corresponding chloride or bromide using a sacrificial zinc electrode readily undergo Michael addition reactions with enones after their transmetallation with $CuCN \cdot 2LiCl$.[36] Dialkylzincs have also been efficiently added to enones after transmetallation to the corresponding copper species.[40] The Michael addition of functionalized zinc–copper organometallics has been applied to the synthesis of prostacyclins, prostaglandins, and related molecules (Equation (90)).[139] Nickel catalysis was first reported by Luche.[140] The use of catalytic amounts of a chiral ligand leads to the 1,4-adducts with good enantioselectivity.[141] The scope of the reaction seems to be limited to unsaturated aromatic ketones.[141,142]

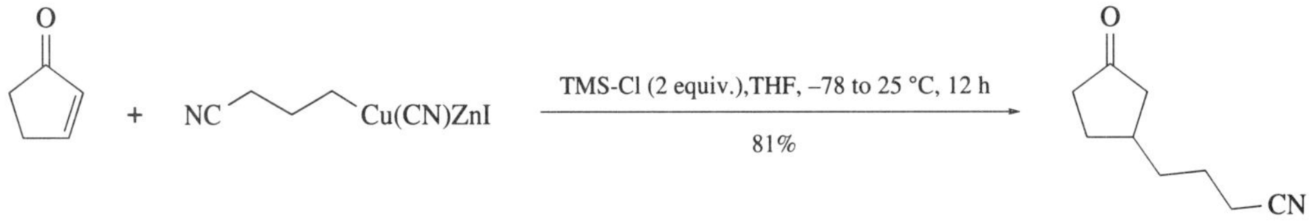

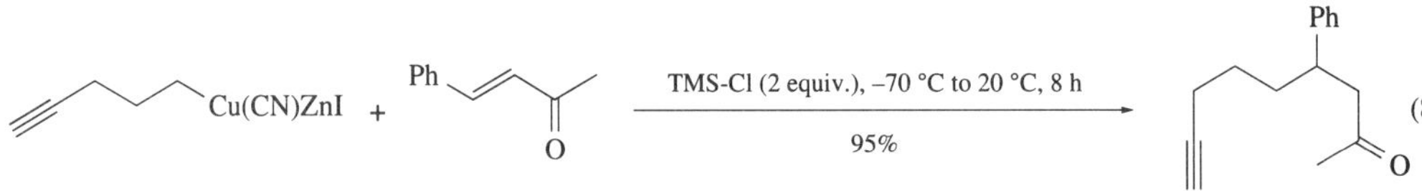

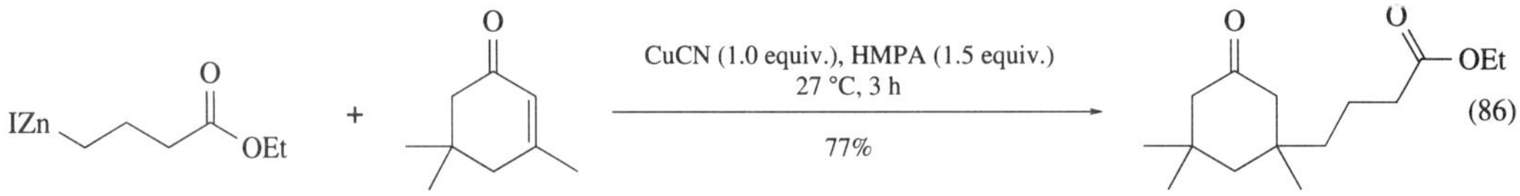

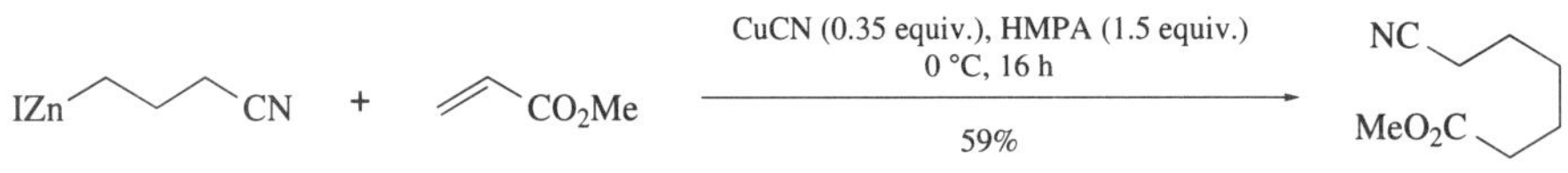

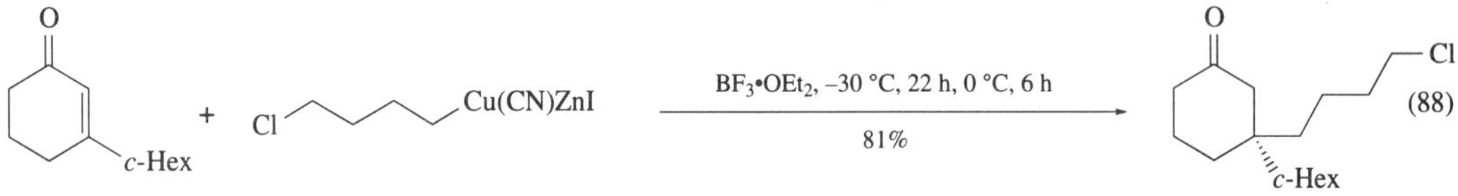

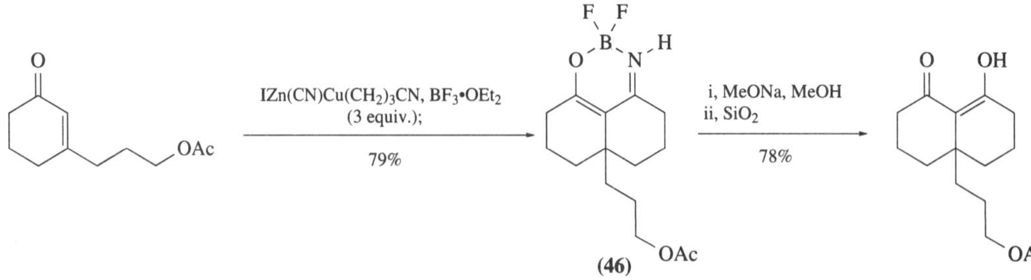

The addition of a chiral ligand without nickel salts also furnishes the desired β-substituted ketones with up to 94% *ee* (Equation (91)).[142] Nitroalkenes constitute an excellent class of Michael acceptors affording, after addition, functionalized nitroalkanes, which are important intermediates in synthesis. The addition of zinc–copper reagents proceeds very smoothly and provides a route to a wide range of polyfunctional nitroalkanes (Equation (92)).[78,143] An addition–elimination of functionalized zinc–copper reagents to nitroalkenes bearing a leaving group in the β-position provides a synthesis of pure (E)-nitroalkenes (Equation (40)).[78] The preparation of 2,2-disubstituted 1-nitroalkenes is possible by trapping the Michael adducts with phenylselenyl bromide and subsequent oxidative elimination (Equation (93)).[144] Intramolecular 1,4-additions of functionalized zinc reagents have been reported. A mechanistic study showed that only a small part of the cyclization product is produced through a radical mechanism and that most of the reaction proceeds via an organometallic pathway (Equation (94)).[145]

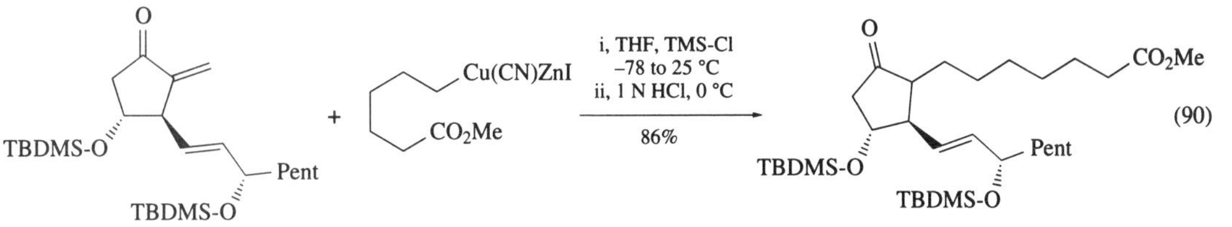

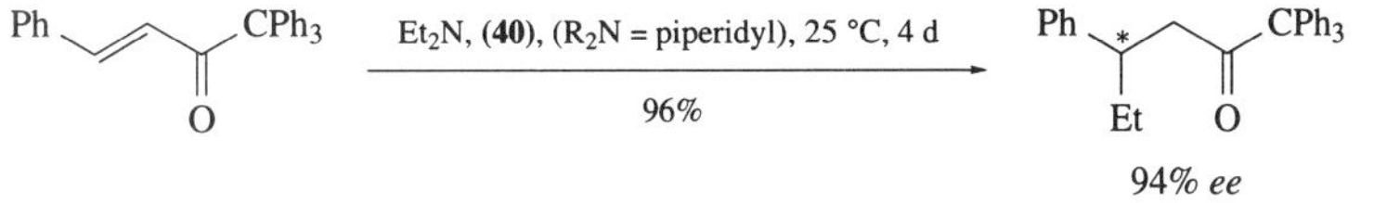

(iii) Carbozincation of alkynes and alkenes

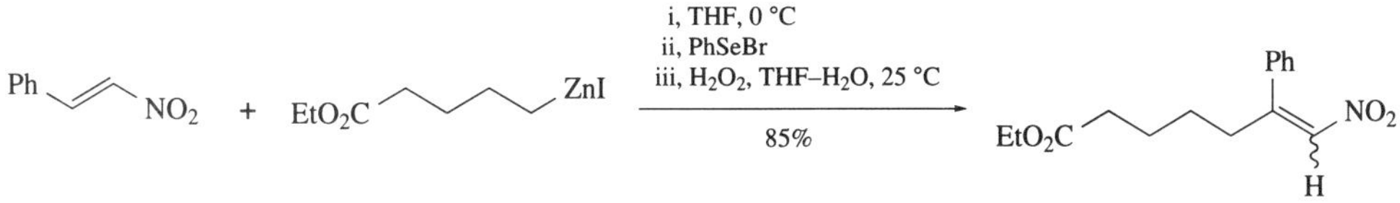

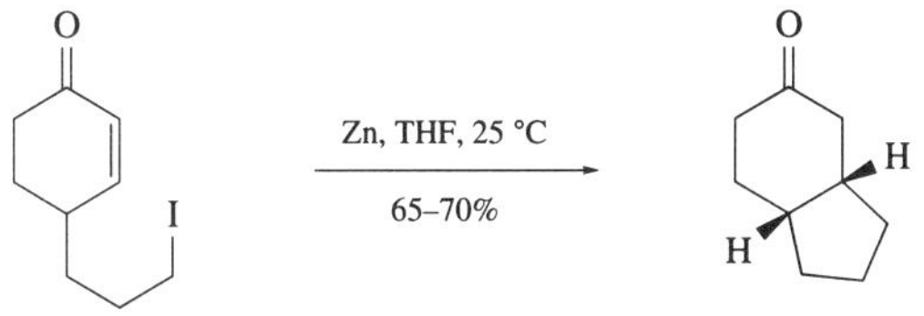

Diallyl- and di-t-butylzinc add to terminal alkenes under relatively harsh conditions (20–90 °C).[146] The carbozincation of ethene with these reagents takes place under somewhat milder conditions (Equation (95)).[146] The addition of diallylzincs to strained alkenes such as cyclopropene derivatives proceeds in quantitative yields and under very mild conditions.[146] Diallylzinc and allylzinc bromide have moderate thermal stability and undergo a carbozincation of the allylic double bond providing a 1,3-diorganometallic species (Equation (96)).[147] This propensity of allylic zinc reagents to add to double bonds is shown by their smooth addition to alkenyl organometallic species. Thus, Gaudemar found that allylic zinc reagents readily add to alkenylmagnesium bromides to give 1,1-dimetallic compounds of magnesium and zinc (Equation (97)).[148–50] These dimetallic reagents have proved to have synthetic applications. Thus, allylzincation using the functionalized allylic zinc reagent (**47**) diastereoselectively provides a 1,1-dimetallic species which, after a direct oxidation using air, furnishes *anti*-aldol products (Equation (98)).[151] Moreover, these 1,1-dimetallic reagents have been found to add readily to alkylidenemalonates to give (Z)-alkenes with excellent stereoselectivity,[152] and the reaction has been applied to the synthesis of a pheromone (Equation (99)).[152] The intramolecular carbozincation of 5-hexenyl iodides has been described.[42,43,153] It provides an excellent method for the diastereoselective preparation of trisubstituted cyclopentylzinc compounds (Equation (14))[43] as well as polycyclic zinc reagents (Equation (100)).[43]

Activated alkenes such as α,β-unsaturated acetals react readily with allylic zinc reagents in the presence of a nickel catalyst (Equation (101)).[154] The addition of the reactive allylic zinc compounds to

(95)

(96)

(97)

(98)

(47) E:Z = 12 : 88 E:Z = 12 : 88

(99)

(100)

alkynes has been extensively studied.[58,155] Some other less reactive organozinc derivatives add to alkynes, but the scope of these additions is limited. Di-*t*-butylzinc only adds to phenylacetylene to give the *anti* addition product, (*Z*)-3,3-dimethyl-1-phenylbutene.[156] Interestingly, zinc malonates such as (**48**) are sufficiently reactive to add to alkynes in fair yields (Equation (102)).[157] The addition of propargylic zinc bromides to terminal alkynes affords 1,4-enynes.[158] Transition metal catalysis considerably improves the scope of the carbozincation. Dialkylzincs add in the presence of Cp_2ZrI_2 to terminal and internal alkynes to give the *syn* adducts with good regioselectivity (70–90%) and very high stereoselectivity (>98%) (Equation (103)).[159] Alkynylzinc derivatives undergo the addition of diethylzinc under these conditions to give 1,1-dimetallic alkenes which, after treatment with iodine, undergo cyclization (Equation (104)).[160] Functionalized allylic zinc compounds add to unactivated alkynes intramolecularly. These reactions have been elegantly applied to the synthesis of 1,5-annelated-4-methylenecyclopentenes (Equation (105)).[161]

(101)

(102)

(**48**) E = CO_2Et or CN

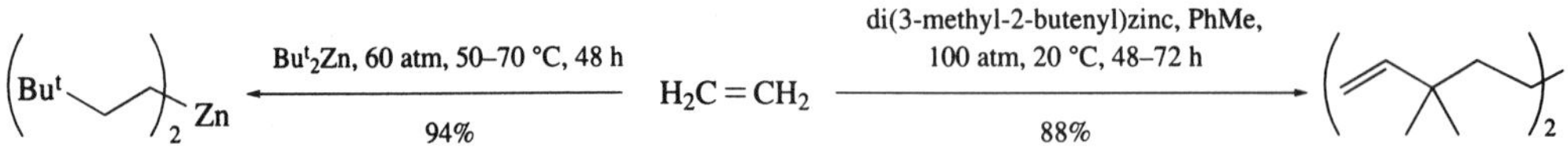

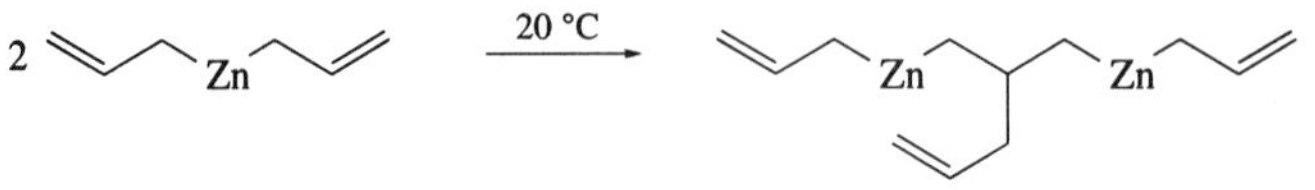

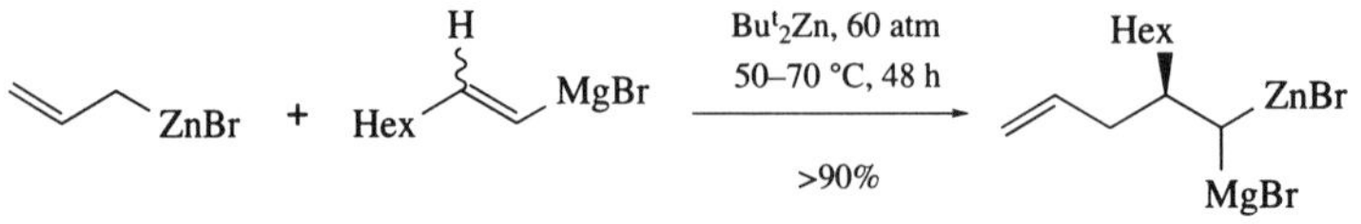

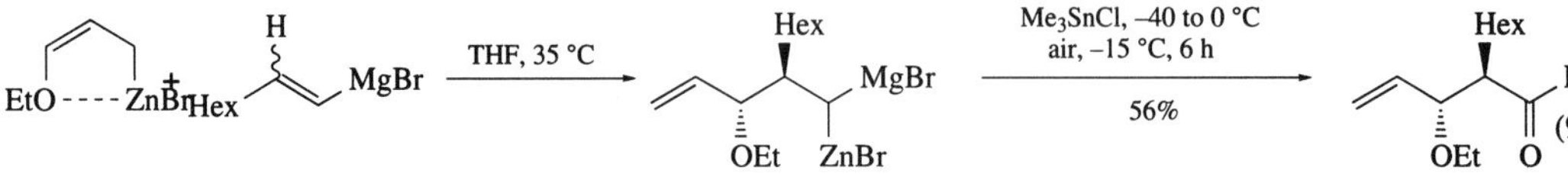

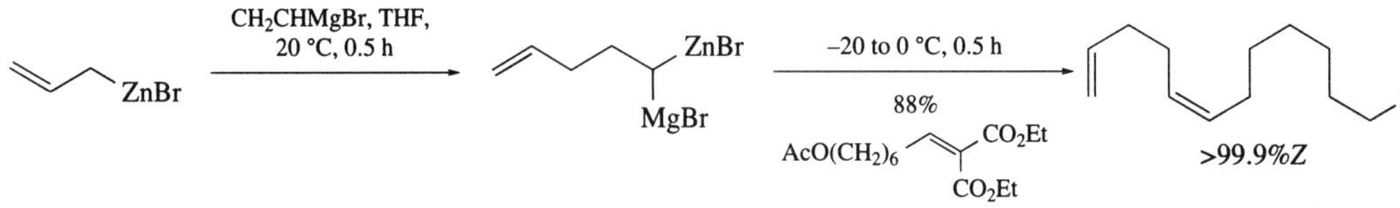

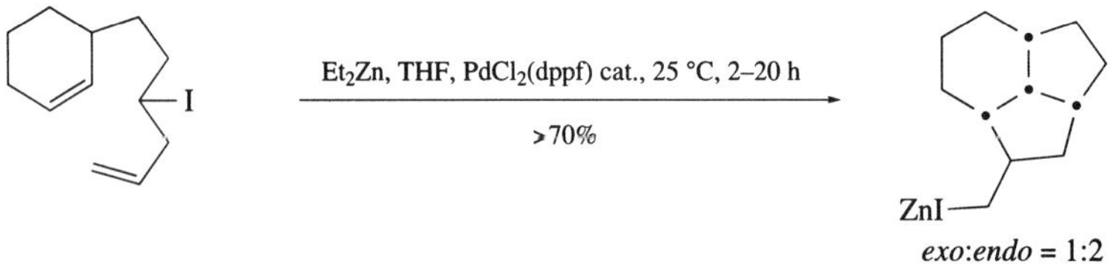

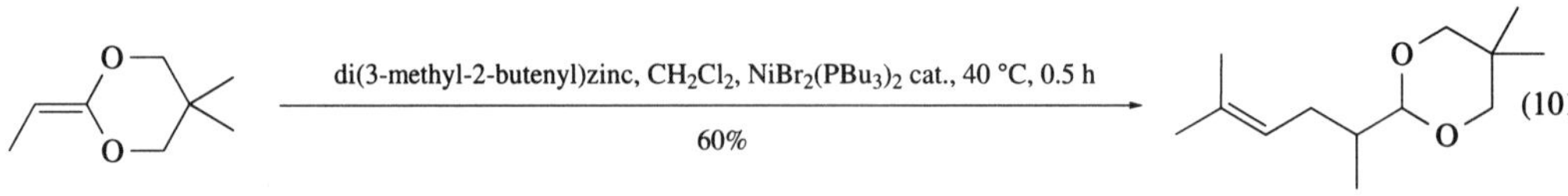

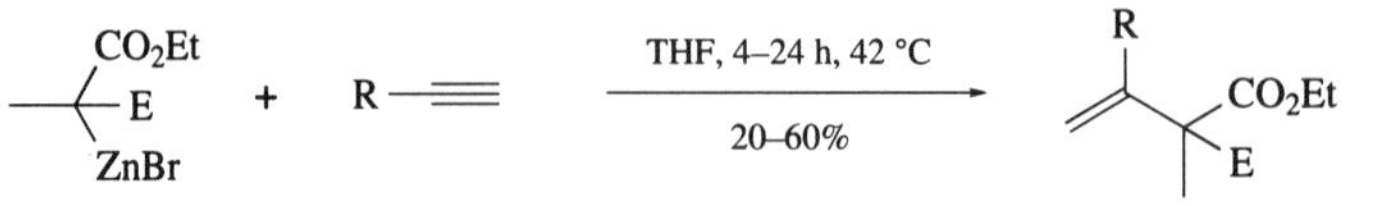

Activated alkynes such as ethyl propiolate react very easily with the mixed zinc–copper species FG-RCu(CN)ZnI to give the *syn* addition products. Performing the reaction at higher temperature and in the presence of an excess of TMS-Cl leads to an (*E*)-α-silylated acrylic ester (Equation (106)).[89] The

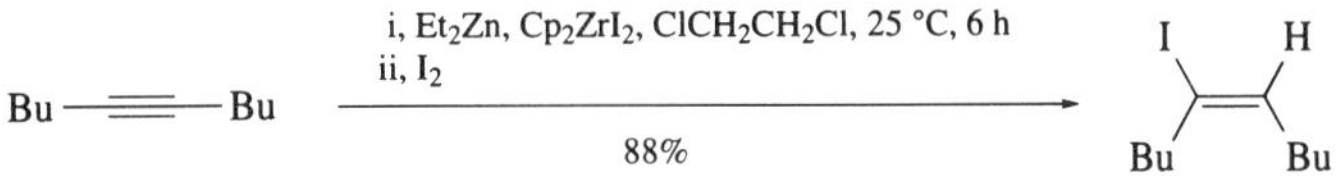

Zinc and Cadmium 177

$$Bu-\equiv-Bu \xrightarrow[88\%]{\text{i, } Et_2Zn, Cp_2ZrI_2, ClCH_2CH_2Cl, 25\,°C, 6\,h \atop \text{ii, } I_2} \underset{Bu}{\overset{I}{\diagdown}}C=C\underset{Bu}{\overset{H}{\diagup}} \quad (103)$$

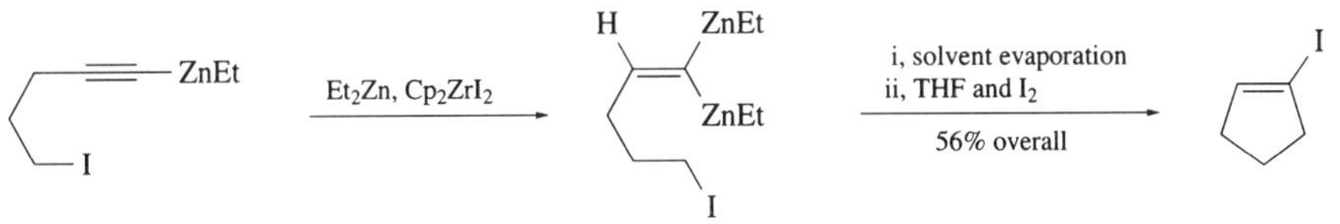

(104) (structure diagram)

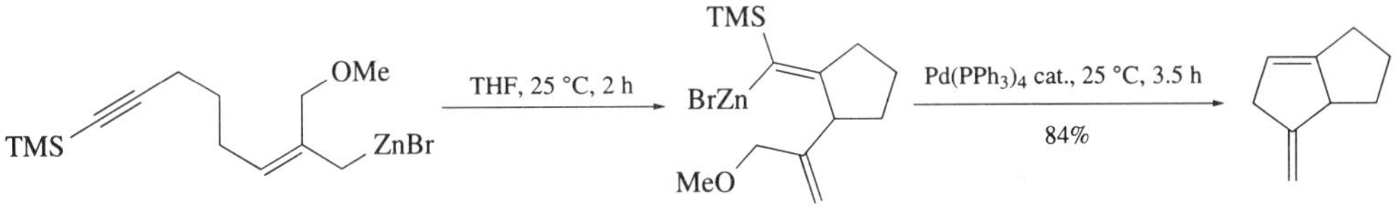

(105) (structure diagram)

addition to propiolamide proceeds in excellent yield despite the presence of the relatively acidic protons of the NH$_2$ group (Equation (107)).[137] The addition of a zinc enolate to an acetylenic ester in HMPA produces highly functionalized cyclopentenones and the method has found applications in natural product synthesis (Equation (108)).[162] Transmetallation of zinc reagents to organocopper compounds provides species which undergo *syn*-addition to several classes of alkynes. Thus, whereas zinc–copper mixed compounds FG-RCu(CN)ZnI do not add readily to unactivated alkynes, treatment of an alkylzinc halide with Me$_2$Cu(CN)Li$_2$ produces a copper–zinc species which adds smoothly to alkynyl thioethers leading stereospecifically, after trapping of the intermediate alkenylcopper with an electrophile, to tri- or tetrasubstituted alkenes (Equation (109)).[163] Only secondary alkyl organometallics add to ethyne in satisfactory yields (Equation (110)).[163] The intramolecular carbometallation of alkynes allows the preparation of highly functionalized *exo*-alkylidenecyclopentanone derivatives (Equation (111)).[163]

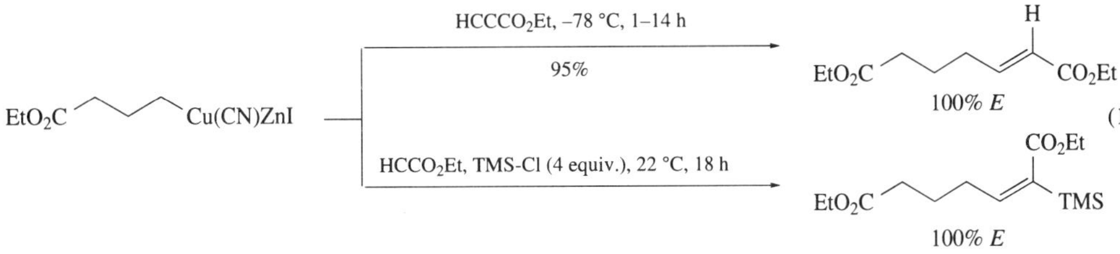

(106) Reaction of EtO$_2$C(CH$_2$)$_3$Cu(CN)ZnI with HCCCO$_2$Et, −78 °C, 1–14 h, 95%, gives (E)-alkene product (100% E); with HCCCO$_2$Et, TMS-Cl (4 equiv.), 22 °C, 18 h, gives TMS-substituted product (100% E).

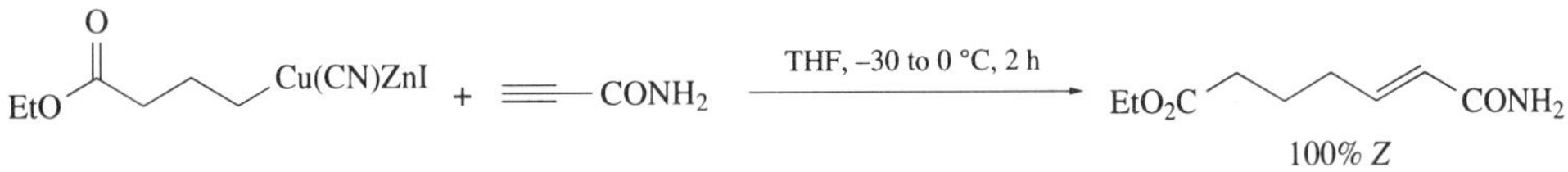

(107) EtO$_2$C(CH$_2$)$_3$Cu(CN)ZnI + HC≡C-CONH$_2$ → THF, −30 to 0 °C, 2 h → EtO$_2$C(CH$_2$)$_3$CH=CH-CONH$_2$ (100% Z)

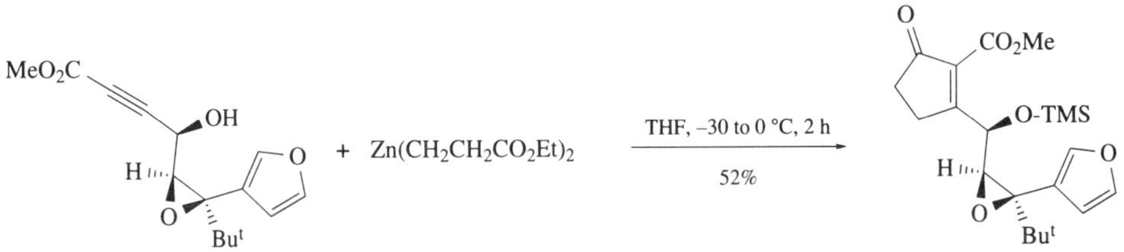

(108) Reaction with Zn(CH$_2$CH$_2$CO$_2$Et)$_2$, THF, −30 to 0 °C, 2 h, 52%.

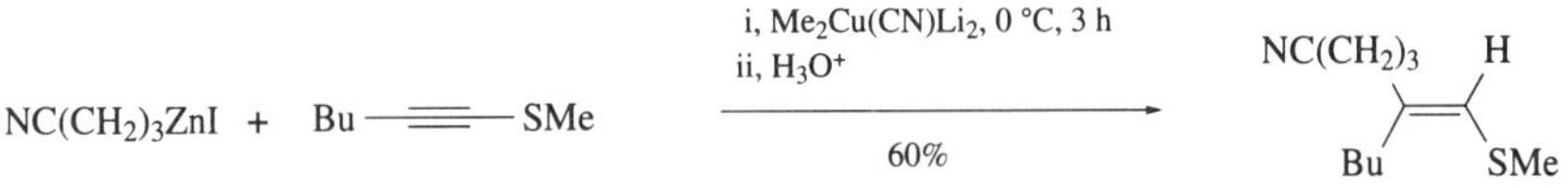

(109) NC(CH$_2$)$_3$ZnI + Bu–≡–SMe → i, Me$_2$Cu(CN)Li$_2$, 0 °C, 3 h; ii, H$_3$O$^+$, 60%.

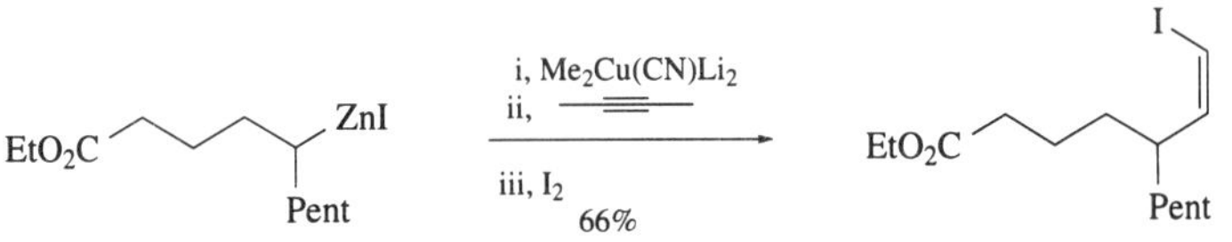

(110) EtO$_2$C-(CH$_2$)$_3$-CH(Pent)-ZnI → i, Me$_2$Cu(CN)Li$_2$; ii, HC≡CH; iii, I$_2$, 66%.

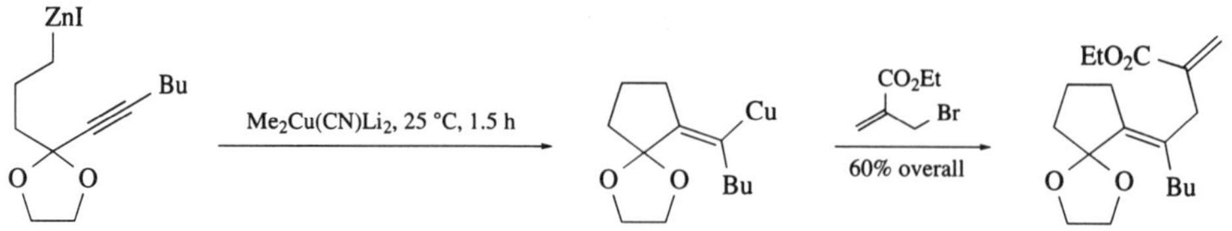

4.1.3.3 Barbier reactions

Barbier reactions,[164] which involve the mixing of zinc dust, the organic halide, and the electrophile, have found increasing applications in organic synthesis over the years. The nature of the zinc organometallic intermediate formed during these reactions is not clear. It cannot be an alkylzinc halide since many of these reactions are performed in the presence of water.[165] Furthermore, it seems not to be a free radical since 5-hexenyl halides do not undergo ring closure under these reaction conditions (Equation (112)).[165] This experimental procedure constitutes an excellent method for the preparation of homoallylic alcohols.[166,167] Especially noteworthy is the tolerance of the presence of water, and thus the potential for applications to hydroxy aldehydes (Equation (113)).[166] Conjugate additions can also be performed under Barbier conditions to unsaturated aldehydes or ketones.[26,168] An impressive four-component reaction has been reported, and the heating of an unsaturated nitrile, a ketone, an alkyl iodide, and zinc in acetonitrile affords β-hydroxy nitriles in excellent yields (Equation (114)).[169] This method allows the preparation of γ-butyrolactones by treatment of dimethyl maleate with a carbonyl compound and zinc.[170] Stereoinduction for the zinc-mediated addition of alkyl iodides to chiral alkylidenemalonitriles has been observed.[171] Zinc–Barbier reactions have also been used to open epoxides[172] and to prepare allylstannanes.[173]

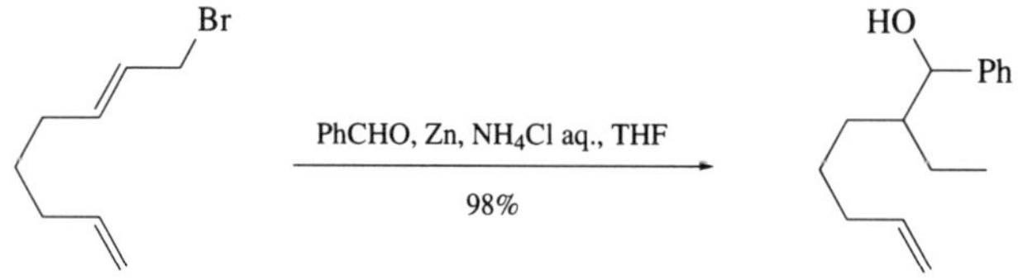

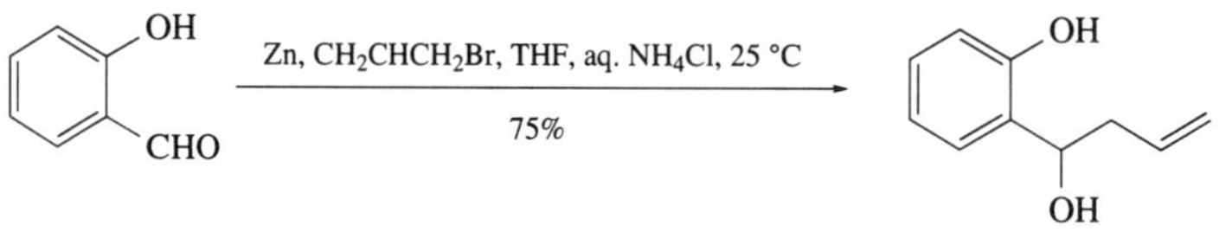

4.1.4 Reactivity of Zinc Carbenoids

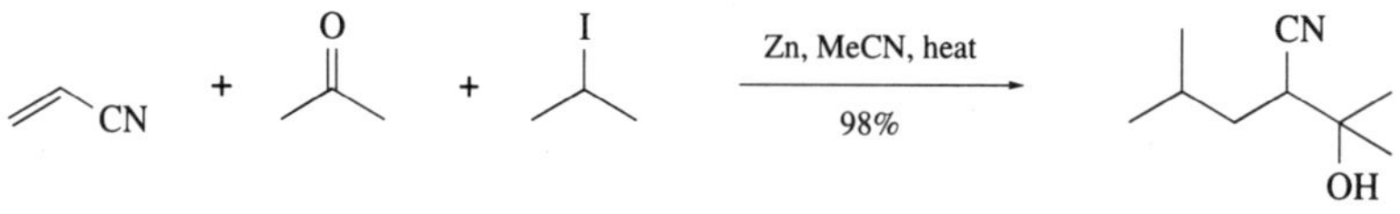

4.1.4.1 Cyclopropanations

The procedure of Simmons and Smith has been improved and adapted to the different classes of alkenes.[1,3] Iodomethylzinc derivatives were originally generated by the reaction of diiodomethane with a zinc–copper couple (Simmons–Smith reaction).[3] A more reactive zinc–silver couple has been introduced by Conia[174] and has proved to be useful for the preparation of strained molecules and for the cyclopropanation of electron-poor molecules (Equation (115)).[174] Treatment of diiodomethane with diethylzinc (Furukawa reaction)[39] or with ethylzinc iodide (Sawada reaction)[175] produces iodomethylzinc derivatives which are excellent cyclopropanation agents (Equation (116)).[176] The presence of a hydroxy or amino group in proximity to the double bond facilitates and directs the cyclopropanation. This chelation has been exploited for the preparation of optically active cyclopropanes (Equations (117)[177] and (118)[178]). The exact nature of the cyclopropanating agent is not known, but solution- and solid-state structural studies of various (halomethyl)zinc reagents have been performed.[179] The reaction of diethylzinc and bromoform with alkenes in the presence of oxygen produces monobromocyclopropanes in good yields (Equation (119)).[180] A novel type of cyclopropanation involving the reaction of bromoform and diethylzinc with ketone acetals gives spiro- and bicyclic esters (Equation (120)).[181]

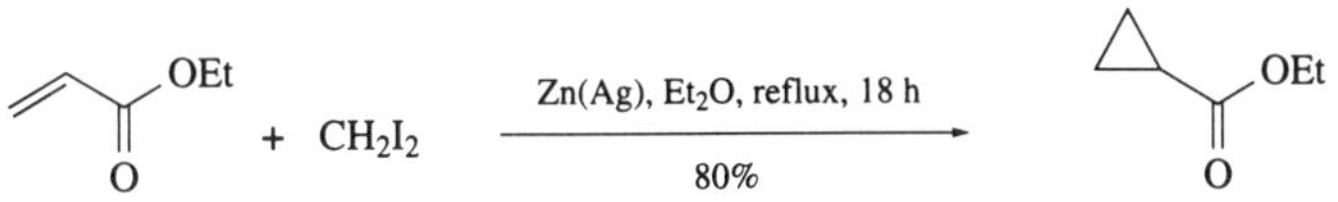

$$\text{CH}_2=\text{CHCO}_2\text{Et} + \text{CH}_2\text{I}_2 \xrightarrow[80\%]{\text{Zn(Ag), Et}_2\text{O, reflux, 18 h}} \text{cyclopropyl-CO}_2\text{Et} \quad (115)$$

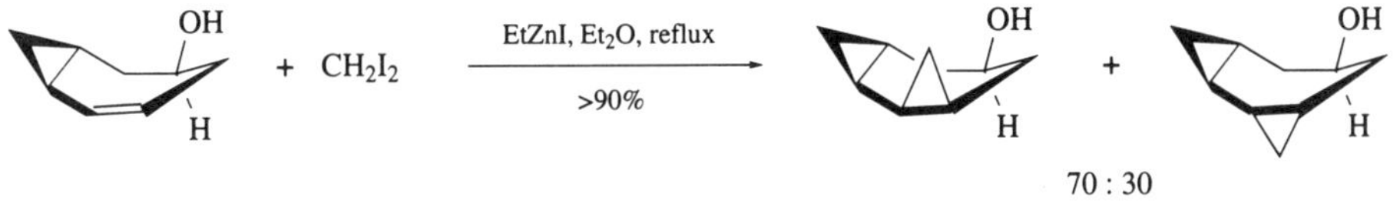

(116) allylic alcohol + CH$_2$I$_2$ → cyclopropanated products (70:30), EtZnI, Et$_2$O, reflux, >90%

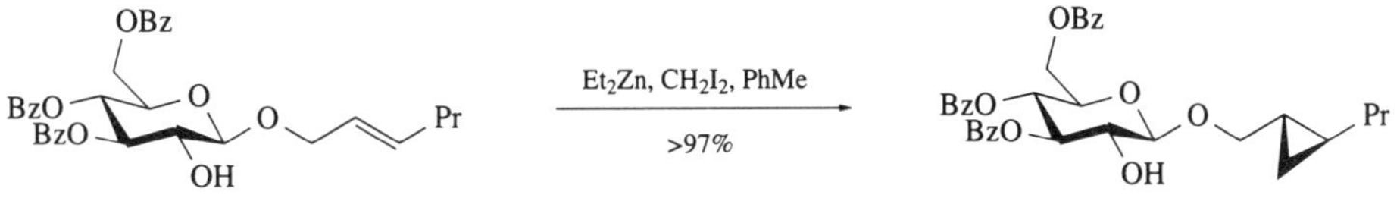

(117) sugar allyl ether + Et$_2$Zn, CH$_2$I$_2$, PhMe, >97% → cyclopropanated sugar

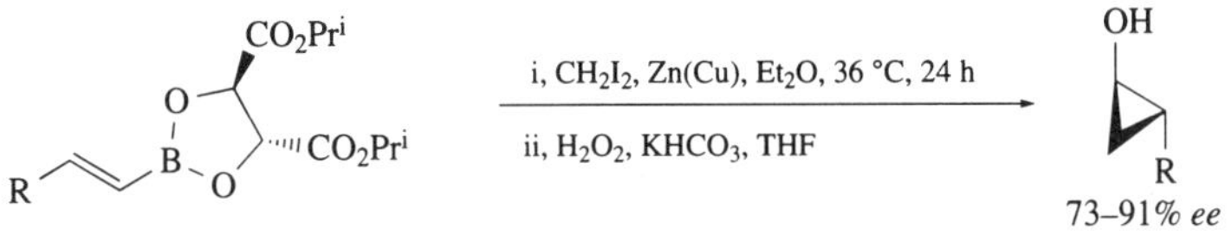

(118) vinyl boronate with chiral diol + i, CH$_2$I$_2$, Zn(Cu), Et$_2$O, 36 °C, 24 h; ii, H$_2$O$_2$, KHCO$_3$, THF → cyclopropyl carbinol R, 73–91% ee

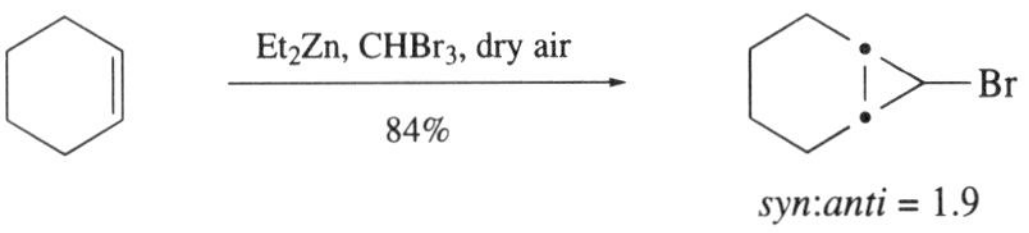

(119) cyclohexene + Et$_2$Zn, CHBr$_3$, dry air, 84% → bromocyclopropane fused to cyclohexane, syn:anti = 1.9

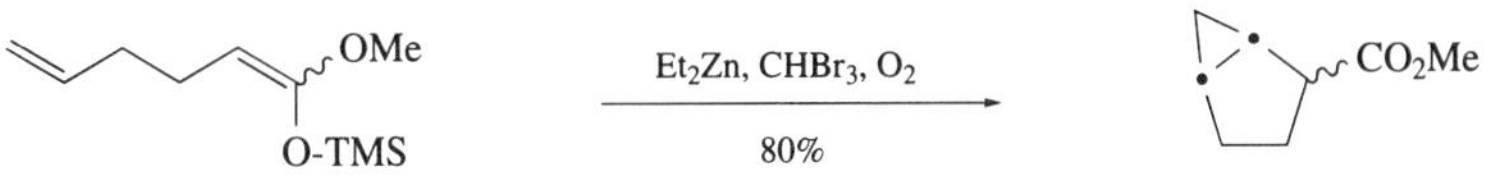

(120) silyl ketene acetal + Et$_2$Zn, CHBr$_3$, O$_2$, 80% → bicyclic CO$_2$Me product

4.1.4.2 Methylene homologation and related reactions

Iodomethylzinc derivatives can be used as efficient reagents for the methylene homologation of organometallics, especially organocopper species (Equation (121)).[182] The reaction proceeds via a 1,2-migration (see (49)). Similarly, the lithium carbenoid (50) undergoes an insertion reaction with dibutylzinc (Equation (122)).[183] These homologations can be used to convert alkenyl- and alkynylcopper–zinc reagents, respectively, into allylic and propargylic organometallics which can be trapped by an aldehyde, ketone, imine, or formate leading to homoallylic and homopropargylic alcohols (Equation (123)).[184]

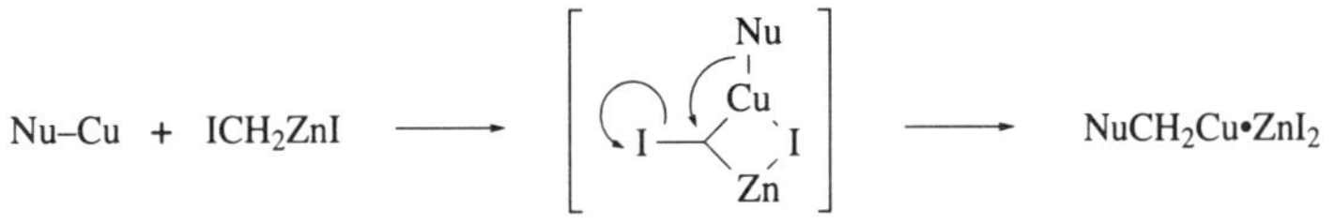

$$\text{Nu–Cu} + \text{ICH}_2\text{ZnI} \longrightarrow \begin{bmatrix} \text{Nu–Cu intermediate with I, Zn bridging} \\ (49) \end{bmatrix} \longrightarrow \text{NuCH}_2\text{Cu}\cdot\text{ZnI}_2 \quad (121)$$

Nu: CN, CH(R)CN, NR$_2$, S-alkyl, Ar, 2-thienyl, alkynyl, alkenyl

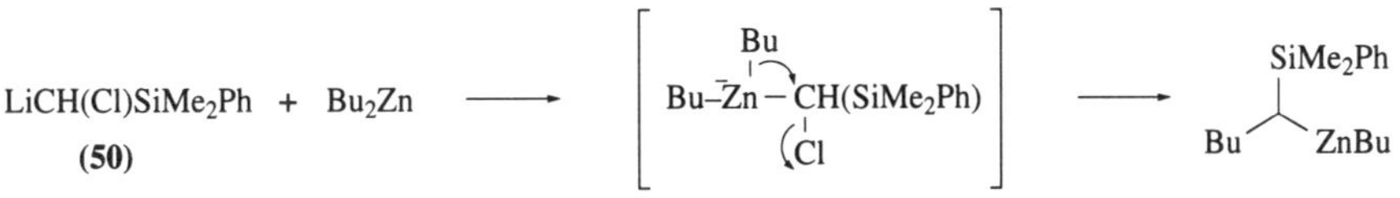

$$\text{LiCH(Cl)SiMe}_2\text{Ph} + \text{Bu}_2\text{Zn} \longrightarrow \begin{bmatrix} \text{Bu–Zn–CH(SiMe}_2\text{Ph)} \\ \text{Cl} \end{bmatrix} \longrightarrow \text{Bu–CH(SiMe}_2\text{Ph)–ZnBu} \quad (122)$$
(50)

The homologation reaction can be applied to perform new cyclization reactions (Equation (124)).[66] The carbocupration of acetylenic esters with functionalized zinc–copper reagents, FG-RCu(CN)ZnI, provides functionalized alkenylcoppers (**51**) which readily insert ICH$_2$ZnI. By performing the reaction in the presence of an aldehyde or ketone, the resulting allylic zinc–copper reagent (**52**) reacts stereoselectively with these electrophiles to give cis-α-methylene-γ-butyrolactones (**53**) (Scheme 3).[185]

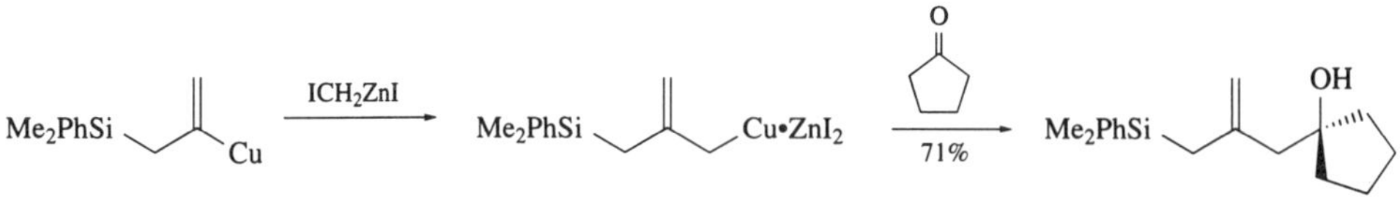

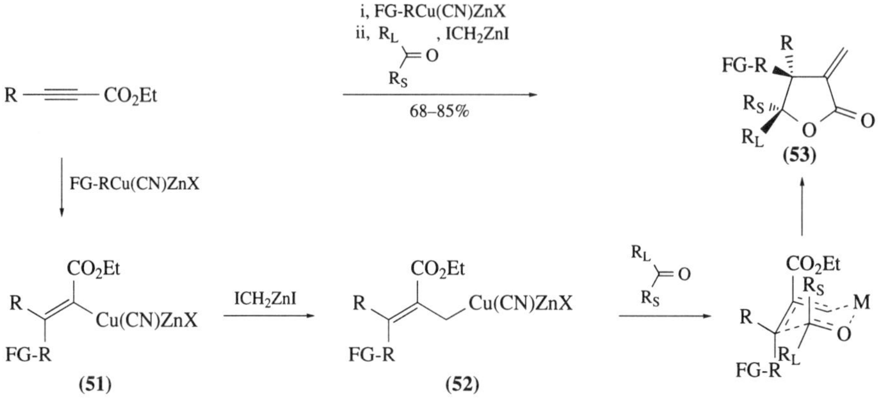

Scheme 3

Geminal dibromocyclopropanes (**54**) react with lithium zincates (R$_3$ZnLi) to give cyclopropylzinc carbenoids (**55**) which, after a 1,2-migration, give mixed dialkylzincs (Equation (125)).[186] This reaction pattern is quite general and both 1,1-dibromoalkenes[187] and 1,1-dibromoalkanes[188] undergo bromine–zinc exchange with the zincates R$_3$ZnLi to give the corresponding carbenoids (Equations (126)[187] and (127)[188]) which, after migration and trapping with an electrophile, furnish products of type (**56**) and (**57**).

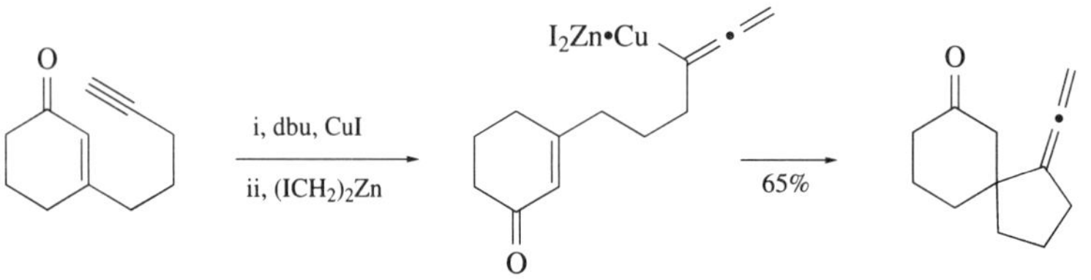

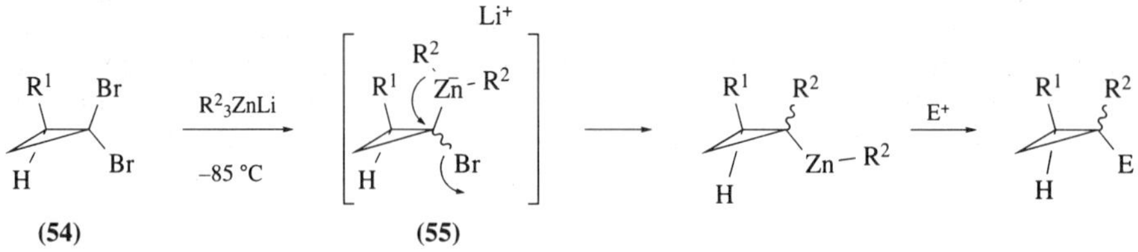

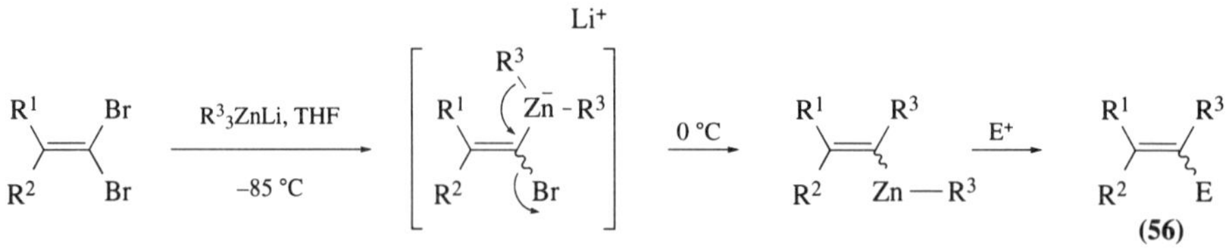

Selective polymethylene insertion results from the treatment of an alkynylcopper with an excess of iodomethylzinc iodide (Scheme 4).[189] The first methylene insertion affords a propargylic organometallic (**58**) which is in equilibrium with an allenyl organometallic (**59**). The insertion of a second methylene unit leads to an allylic reagent (**60**) which is in equilibrium with the dienic copper compound (**61**). A third methylene insertion furnishes a highly reactive allylzinc–copper species (**62**), which immediately

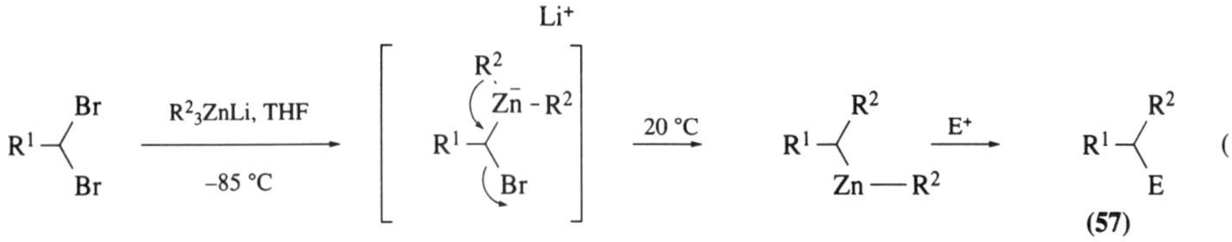

(57)

(127)

R^1 = Oct, $PhCH_2CH_2$, THP-$O(CH_2)_4$; E^+ = R^3COCl, ArBr, alkenyl-Br

undergoes a fourth insertion reaction to give a relatively unreactive alkylcopper (**63**) which can be further trapped with an electrophile.[189]

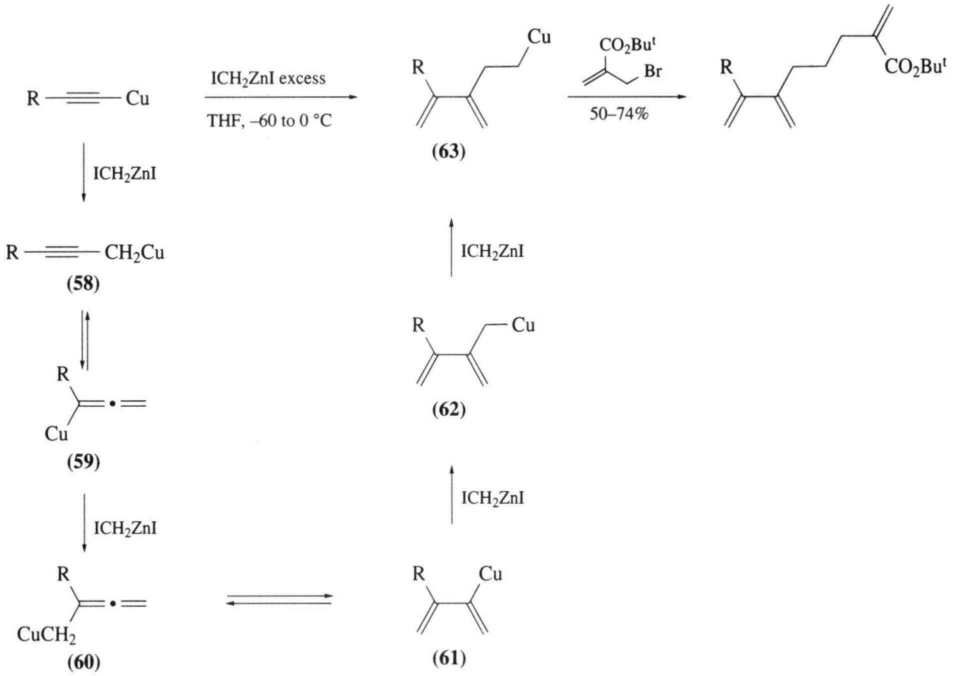

Scheme 4

Metallated propargylic ethers show a reactivity pattern dependent on the nature of the electrophile. In the presence of a reactive electrophile, such as benzaldehyde, the propargylic organometallic formed after the first methylene homologation adds to the aldehyde to give a homopropargylic alcohol (**64**). With a less reactive electrophile, such as cyclopentanone, the propargylic organometallic is not reactive enough to react with the ketone and is further homologated by two methylene units to produce a reactive allylic species which adds to cyclopentanone to give a hydroxydiene (**65**) (Equation (128)).[189] Polyfluorinated zinc carbenoids can be used for performing multiple difluoromethylene homologation reactions.[190]

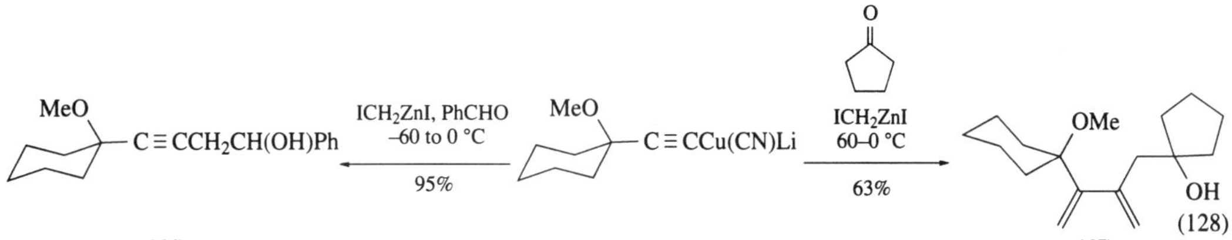

(64)

(65)

(128)

4.1.5 Preparation and Reactions of Zincates

Lithium triorganozincates (R_3ZnLi) are prepared by the reaction of an alkyllithium (3 equiv.) with zinc chloride or by the treatment of a dialkylzinc with an alkyllithium (1 equiv.) in ether or THF.[191] Lithium or magnesium trialkylzincates readily undergo 1,4-addition reactions to α,β-unsaturated ketones. The nature of the lithium salts present strongly influences the yield of the reaction.[192] Lithium trimethylzincate is one of the least reactive triorganozincates and the methyl group is usually transferred with moderate yield.[191] Methylation with Me_3ZnLi can be catalyzed by cobalt complexes (Equation (129)).[191] The low transfer ability of a methyl group has been exploited, and mixed alkyldimethylzincates selectively transfer the alkyl group in Michael additions (Equation (130)).[191,193] Related mixed heterozincates $R_2(OBu^t)ZnMgBr \cdot TMEDA$ prepared from RMgX (2 equiv.), $ZnCl_2 \cdot TMEDA$ and Bu^tOK react readily with enones to give the 1,4-adducts in satisfactory yields. By using chiral nitrogen chelating ligands, an asymmetric 1,4-addition can be achieved with moderate enantioselectivity.[194] The addition of Bu_3ZnLi to nitrostyrene in a chiral solvent mixture pentane-(S,S)-1,4-dimethylamino-2,3-dimethoxybutane (DDB) affords an optically active nitroalkane (Equation (131)).[195] Triorganozincates undergo halogen–zinc exchange reactions with a range of 1,1-dibromo-substituted alkenes or alkanes.[186–8]

$$\text{cyclohexenone} + Me_3ZnLi \xrightarrow[100\%]{Co(acac)_3,\ 2\ mol.\%,\ THF,\ -78\ °C,\ 1\ h} \text{3-methylcyclohexanone} \quad (129)$$

$$\text{cyclohexenone} + BuMe_2ZnLi \xrightarrow[95\%]{THF,\ -78\ °C} \text{3-methylcyclohexanone} + \text{3-butylcyclohexanone} \quad (130)$$
$$3 : 92$$

$$Ph\text{—}CH=CH\text{—}NO_2 \xrightarrow[60\%]{Bu_3ZnLi,\ pentane–DDB,\ -78\ °C} Ph\text{—}CH(Bu)\text{—}CH_2NO_2 \quad (131)$$

4.1.6 Preparation and Reactivity of Bimetallic Reagents of Zinc

A range of mixed 1,1-bimetallic reagents of zinc have been prepared by the direct insertion of zinc metal[196,197] (Equation (132)),[197] the allylzincation of alkenyl-magnesium, -lithium, or -aluminum compounds (see Section 4.1.3.2(iii)), or the hydrozirconation of alkenylzinc derivatives.[198] The mixed zinc–magnesium bimetallic species have been used to form carbon–carbon bonds in a selective manner with a range of electrophiles (Scheme 5).[148–52] Alkenation reactions can be performed with a number of these reagents. Aldehydes are converted into (E)-alkenes, with mixed zinc–magnesium bimetallics in the presence of $BF_3 \cdot OEt_2$ or mixed zinc–zirconium reagents,[198] whereas the reaction of alkylidenemalonates with zinc–magnesium bimetallics affords (Z)-alkenes with very high stereoselectivity (Section 4.1.3.2(iii)), Equation (99)).[152] The reaction of diiodomethane with zinc dust in THF produces $CH_2(ZnI)_2$.[196] Its transmetallation with dichlorotitanocene gives a deep red solution of titanocene methylene–zinc halide complex. This reagent methylenates ketones and converts nitriles into methyl ketones (Equation (133)).[196] Similarly, the Nysted reagent (66) directly converts carbonyl groups of steroids into terminal alkenes (Equation (134)).[196] Procedures in which the bimetallic reagents have been generated *in situ* are of great preparative importance. Thus, Takai showed that various types of carbonyl derivatives can be methylenated by a mixture of zinc, a dibromoalkane, and titanium tetrachloride (Equation (135)).[199] A modification developed by Lombardo provides a mild, nonbasic method for the methylenation of ketones. The Lombardo reagent is compatible with the presence of various functionalities (Equation (136)).[200] Whereas 1,2- and 1,3-bimetallics of zinc cannot be prepared,[201] higher homologues are readily available. Their reaction with $CuCN \cdot 2LiCl$ provides mixed 1,n-bimetallics of zinc and copper which react selectively with two different electrophiles (Equation (137)).[202]

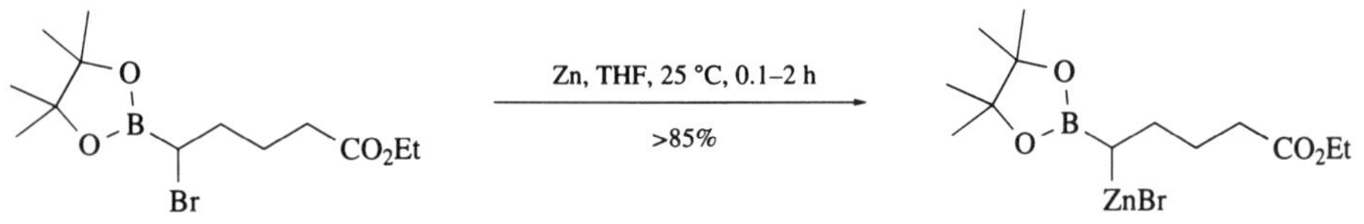

4.2 CADMIUM REAGENTS IN ORGANIC SYNTHESIS

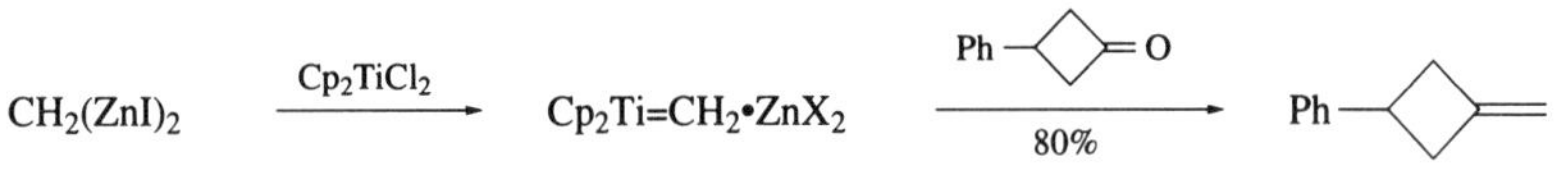

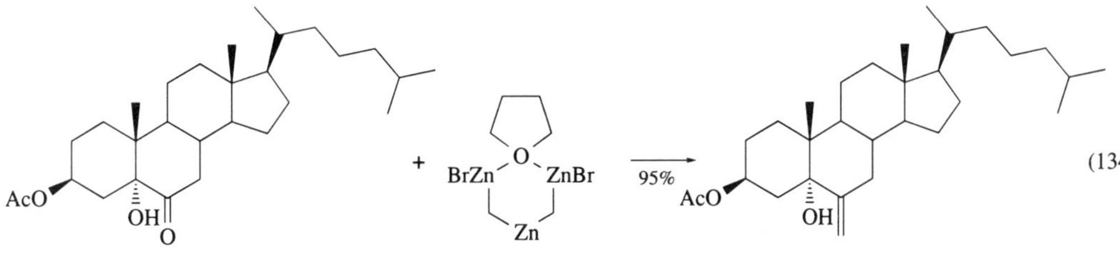

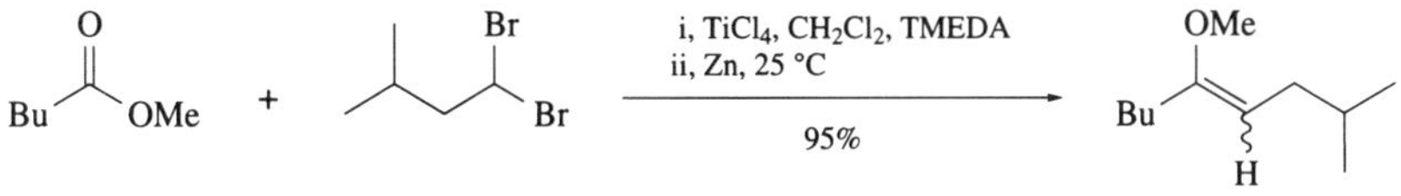

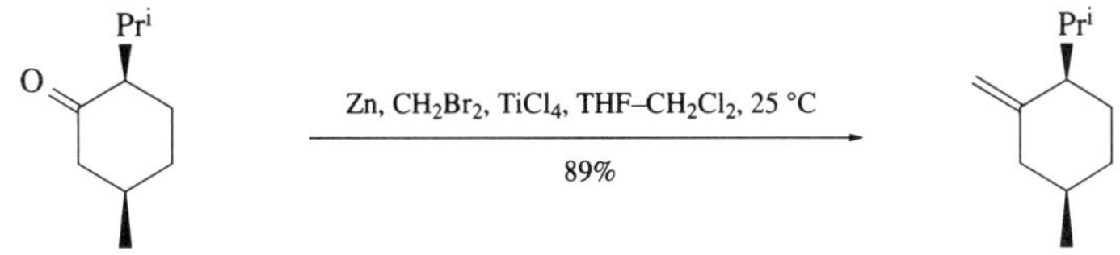

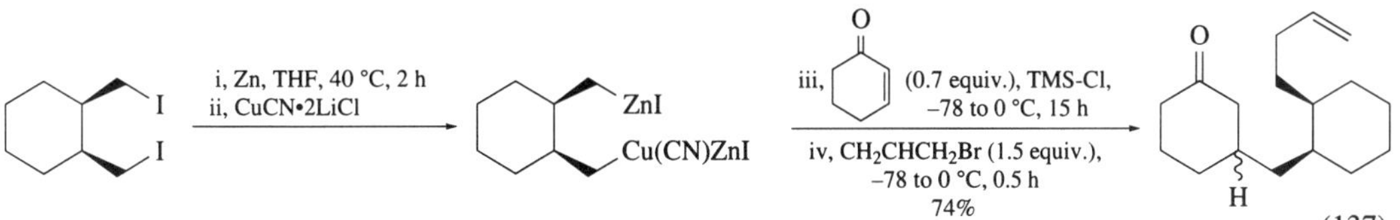

Organocadmium compounds have found far less synthetic application than zinc organometallics.[94,203,204] This may be due to their somewhat reduced reactivity and their thermal and photochemical instability, as well as to their toxicity.[203] Since 1982, only a few synthetic applications have been reported.[1] Alkylcadmium derivatives are prepared by similar methods to organozincs.[204] The reaction of lithium naphthalenide with cadmium chloride in THF produces a highly reactive form of cadmium which reacts within 3 h at 25 °C with benzylic bromides and within 18 h in refluxing THF with iodobenzene. These organocadmium reagents add to allyl bromide to give the expected allylated products and couple with acid chlorides furnishing the corresponding ketones in fair yields (Equation (138)).[205] Interestingly, the cadmium reagent (**67**) prepared by the direct insertion of cadmium powder into iodomethyl pivalate reacts, in the presence of CuCN·2LiCl, with acid chlorides in better yields than the corresponding zinc reagent (Equation (139)).[77] A Barbier procedure using a cadmium powder suspension, an allylic or propargylic bromide, and a carbonyl compound allows the addition of allylic and propargylic groups to aldehydes, ketones, and imines in the presence of tetrabutylammonium bromide (Equation 140).[206] Cyclohexadienyliron tricarbonyl tetrafluoroborate reacts with organocadmium reagents to give, after oxidative demetallation, substituted cyclohexadienes.[207]

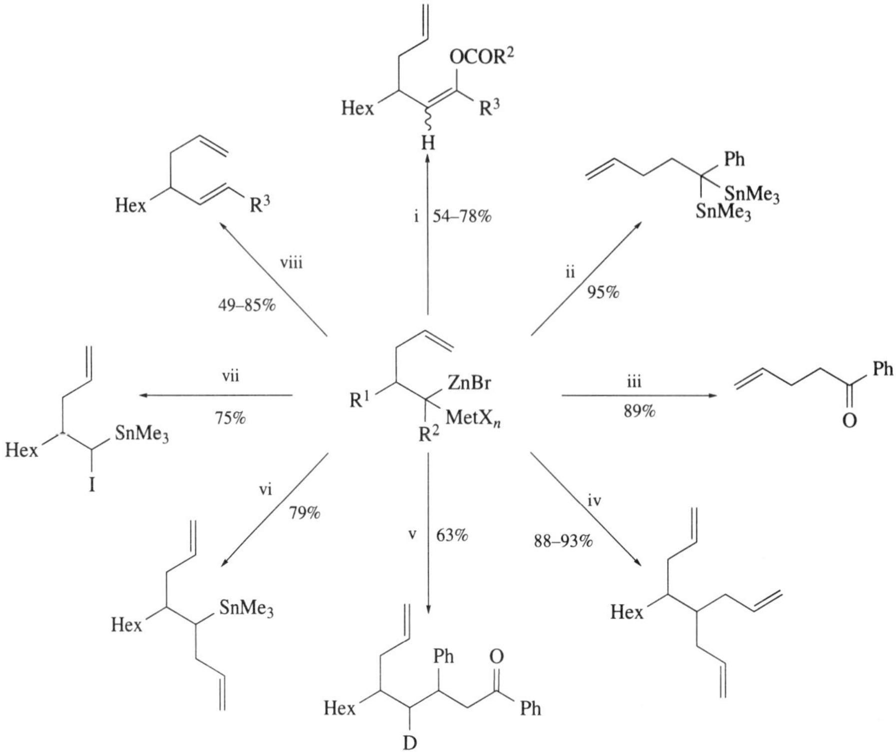

i, R^1 = Hex, R^2 = H, MetX$_n$ = MgBr; CuCN (1 equiv.), −35 °C, 0.25 h, RCOCl (3 equiv.), −10 °C, 2 h; ii, R^1 = H, R^2 = Ph, MetX$_n$ = MgBr; CuCN (1 equiv.), −20 °C, 0.5 h, then Me$_3$SnCl (2 equiv.), iii, R^1 = H, R^2 = Ph, MetX$_n$ = MgBr; Me$_3$SnCl (1 equiv.), −40 to 0 °C, 0.3 h, then dry air, −10 to 0 °C, 0.5 h; iv, R^1 = Hex, R^2 = H, MetX$_n$ = MgBr or Li; CuCN (1 equiv.), −35 °C, 0.25 h then CH$_2$CHCH$_2$Br (excess), −40 to 0 °C, 0.5 h; v, R^1 = Hex, R^2 = H, MetX$_n$ = Li; benzylideneacetophenone (1.0 equiv.), −78 to 0 °C, 1 h, D$_3$O$^+$; vi, R^1 = Hex, R^2 = H, MetX$_n$ = Li, Me$_3$SnCl (1 equiv.), −40 to 0 °C, 0.5 h, CuCN (1 equiv.), −30 °C, 0.5 h, CH$_2$CHCH$_2$Br (3 equiv.), −30 to 10 °C, 0.5 h; vii, R^1 = Hex, R^2 = H, MetX$_n$ = MgBr ; Me$_3$SnCl (1 equiv.), −25 to 0 °C, 0.5 h, I$_2$ (1 equiv.), −78 °C, 0.5 h; viii, R^1 = Hex, R^2 = H, MetX$_n$ = MgBr, BF$_3$·OEt$_2$ (1 equiv.), −90 °C, RCHO, −90 to −50 °C, 0.25 h (formation of the (E)-isomer in over 90% selectivity)

Scheme 5

Bis(trifluoromethyl)cadmium, which was prepared by transmetallation of (CF$_3$)$_2$Hg with Me$_2$Cd in dimethoxyethane, reacts with acyl halides to give acyl fluorides in excellent yields, together with difluorocarbene, which can be trapped by the addition of tetramethylethylene affording the expected difluoroethylene derivative in 53% yield.[208] Organocadmiums add cleanly to the cyano group of ethyl cyanoformate in the presence of ZnCl$_2$ to give α-keto esters (Equation (141)).[209]

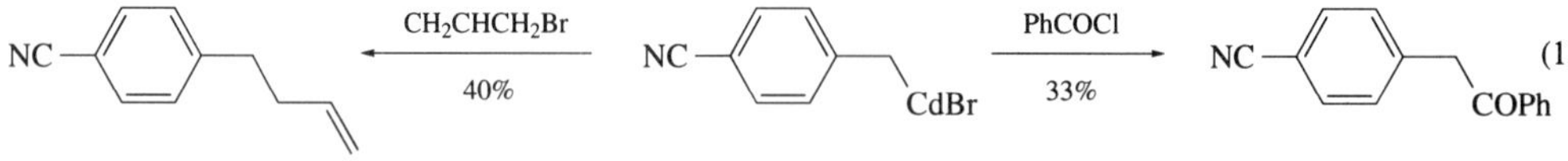

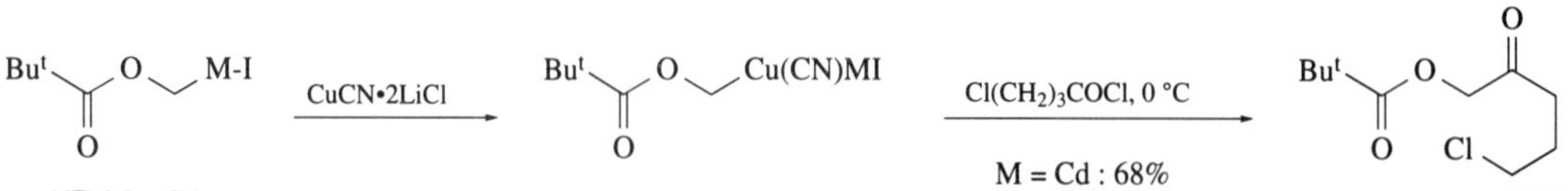

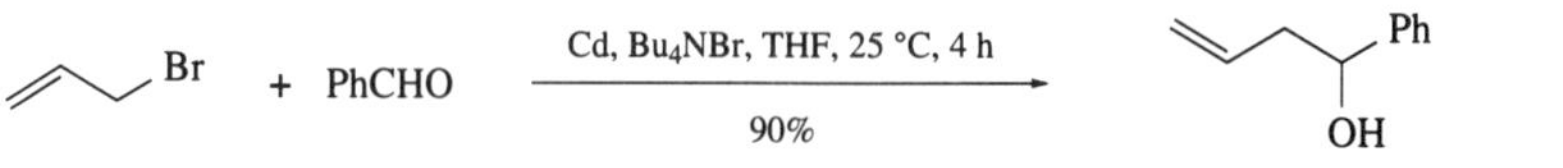

$$\text{Cyclohexyl-CdBr} + \text{N} \equiv \text{CCO}_2\text{Et} \xrightarrow[66\%]{\text{i, ZnCl}_2 \\ \text{ii, aq. HCl}} \text{Cyclohexyl-C(=O)-C(=O)OEt} \quad (141)$$

4.3 REFERENCES

1. W. Carruthers, in 'COMC-I', vol. 7, p. 661.
2. (a) M. W. Rathke, *Org. React.*, 1975, **22**, 423; (b) A. Fürstner, *Synthesis*, 1989, 571.
3. H. E. Simmons, T. L. Cairns, S. A. Vladuchick and C. M. Hoiness, *Org. React.*, 1973, **20**, 1; K.-P. Zeller and H. Gugel in Houben-Weyl, 'Methoden der Organischen Chemie', ed. M. Regitz, Georg Thieme Verlag, Stuttgart, 1989, vol. E 19b, p. 195.
4. P. Knochel and R. D. Singer, *Chem. Rev.*, 1993, **93**, 2117.
5. I. B. Gorrell, A. Looney and G. Parkin, *J. Chem. Soc., Chem. Commun.*, 1990, 220.
6. K. Nützel in Houben-Weyl, 'Methoden der organischen Chemie', ed. E. Müller, Georg Thieme Verlag, Stuttgart, 1973, vol. 13/2a, p. 552.
7. I. Klement, K. Lennick, C. E. Tucker and P. Knochel, *Tetrahedron Lett.*, 1993, **34**, 4623.
8. (a) E. Negishi, L. F. Valente and M. Kobayashi, *J. Am. Chem. Soc.*, 1980, **102**, 3298; (b) M. Kobayashi and E. Negishi *J. Org. Chem.* 1980, **45**, 5223.; (c) E. Negishi, *Acc. Chem. Res.*, 1982, **15**, 340; (d) E. Negishi, V. Bagheri, S. Chatterjee, F.-T. Luo, J. A. Miller and A. T. Stoll, *Tetrahedron Lett.*, 1983, **24**, 5181; (e) R. A. Grey, *J. Org. Chem.* 1984, **49**, 2288; (f) E. Nakamura and I. Kuwajima, *Tetrahedron Lett.*, 1986, **27**, 83.
9. P. Knochel, M. C. P. Yeh, S. C. Berk and J. Talbert, *J. Org. Chem.*, 1988, **53**, 2390.
10. (a) K. Soai and S. Niwa, *Chem. Rev.*, 1992, **92**, 833; (b) R. O. Duthaler and A. Hafner, *Chem. Rev.*, 1992, **92**, 807.
11. M. Gaudemar, *Bull. Soc. Chim. Fr.*, 1962, 974.
12. E. Erdik, *Tetrahedron*, 1987, **43**, 2203.
13. (a) S. C. Berk, M. C. P. Yeh, N. Jeong and P. Knochel, *Organometallics*, 1990, **9**, 3053; (b) H. G. Chen, C. Hoechstetter and P. Knochel, *Tetrahedron Lett.*, 1989, **30**, 4795.
14. C. Jubert and P. Knochel, *J. Org. Chem.*, 1992, **57**, 5425.
15. (a) T. N. Majid and P. Knochel, *Tetrahedron Lett.* 1990, **31**, 4413; (b) J. Grondin, M. Sebban, P. Vottero, H. Blancou and A. Commeyras, *J. Organomet. Chem.*, 1989, **362**, 237; (c) C. Janakiram Rao and P. Knochel, *J. Org. Chem.*, 1991, **56**, 4593; (d) P. Knochel, and C. Janakiram Rao, *Tetrahedron*, 1993, **49**, 29.
16. (a) Y. Tamaru, H. Ochiai, T. Nakamura, K. Tsubaki, and Z. Yoshida, *Tetrahedron Lett.*, 1985, **26**, 5559; (b) Y. Tamaru, H. Ochiai, F. Sanda and Z. Yoshida, *Tetrahedron Lett.*, 1985, **26**, 5529.
17. Y. Tamaru, H. Ochiai, T. Nakamura and Z. Yoshida, *Tetrahedron Lett.*, 1986, **27**, 955.
18. Y. Tamaru in 'New Trends in Organometallic Chemistry', ed. H. Sakurai, Tohoku University, Sendai, 1990, 304.
19. Y. Tamaru, H. Ochiai, T. Nakamura and Z. Yoshida, *Angew. Chem.*, 1987, **99**, 1193.
20. (a) E. Nakamura, S. Aoki, K. Sekiya, H. Oshino and I. Kuwajima, *J. Am. Chem. Soc.*, 1987, **109**, 8056; E. Nakamura and I. Kuwajima, *J. Am. Chem. Soc.*, 1984, **106**, 3368; (b) H. Oshino, E. Nakamura and I. Kuwajima, *J. Org. Chem.*, 1985, **50**, 2802; (c) E. Nakamura and I. Kuwajima, *Tetrahedron Lett.*, 1986, **27**, 83; (d) E. Nakamura, K. Sekiya and I. Kuwajima, *Tetrahedron Lett.*, 1987, **28**, 337.
21. (a) R. D. Rieke, S. J. Uhm and P. M. Hudnall, *J. Chem. Soc., Chem. Commun.*, 1973, 269; (b) R. D. Rieke, P. T.-J. Li, T. P. Burns and S. J. Uhm, *J. Org. Chem.*, 1981, **46**, 4323; (c) R. D. Rieke and S. J. Uhm, *Synthesis*, 1975, 452; (d) R. T. Arnold and S. T. Kulenovic, *Synth. Commun*, 1977, **7**, 223; (e) H. Bönnemann, W. Brijoux and T. Joussen, *Angew. Chem., Int. Ed. Engl*, 1990, **29**, 273; (f) F. J. M. Freijee, J. W. F. L. Seetz, O. S. Akkerman and F. Bickelhaupt, *J. Organomet. Chem.*, 1982, **224**, 217.
22. R. D. Rieke, *Science*, 1989, **246**, 1260.
23. (a) D. Braga, A. Ripamonti, D. Savoia, C. Trombini and A. Umani-Ronchi, *J. Chem. Soc., Chem. Commun.*, 1978, 927; (b) G. P. Boldrini, D. Savoia, E. Tagliavini, C. Trombini and A. Umani-Ronchi, *J. Org. Chem.*, 1983, **48**, 4108.
24. (a) R. Csuk, B. I. Glänzer and A. Fürstner, *Adv. Organomet. Chem.*, 1988, **28**, 85; (b) A. Fürstner, *Angew. Chem. Int. Ed. Engl.*, 1993, **32**, 164.
25. A. Fürstner, R. Singer and P. Knochel, *Tetrahedron Lett.*, 1994, **35**, 1047.
26. (a) C. Einhorn, J. Einhorn and J.-L. Luche, *Synthesis*, 1989, 787; (b) J. L. Luche, C. Pétrier and C. Dupuy, *Tetrahedron Lett.*, 1984, **25**, 3463; (c) C. Pétrier, J. C. de Souza Barbosa, C. Dupuy and J.-L. Luche, *J. Org. Chem.*, 1985, **50**, 5761; (d) E. J. Corey, J. O. Link and Y. Shao, *Tetrahedron Lett.*, 1992, **33**, 3435; (e) P. Knochel and J. F. Normant, *Tetrahedron Lett.*, 1984, **25**, 1475.
27. (a) K. S. Suslick and S. J. Doktycz, *J. Am. Chem. Soc.*, 1989, **111**, 2342; (b) J. L. Luche, C. Einhorn, J. Einhorn and J. V. Sinisterra-Gago, *Tetrahedron Lett.*, 1990, **31**, 4125.
28. (a) B.-H. Han, and P. Boudjouk, *J. Org. Chem.*, 1982, **47**, 5030; (b) B.-H. Han and P. Boudjouk, *J. Org. Chem.*, 1982, **47**, 751; (c) P. Boudjouk, D. P. Thompson; W. H. Ohrbom and B.-H. Han, *Organometallics*, 1986, **5**, 1257.
29. (a) T. Kitazume and N. Ishikawa, *Chem. Lett.*, 1981, 1679; (b) T. Kutazume and N. Ishikawa, *Chem. Lett.*, 1982, 137; (c) T. Kitazume and N. Ishikawa, *Chem. Lett.*, 1982, 1453.
30. O. Repic and S. Vogt, *Tetrahedron Lett.*, 1982, **23**, 2729.
31. D. J. Burton and Z.-Y. Yang, *Tetrahedron*, 1992, **48**, 189.
32. (a) H. Blancou and A. Commeyras, *J. Fluorine Chem.*, 1982, **20**, 255; (b) S. Benefice, H. Blancou and A. Commeyras, *J. Fluorine Chem.*, 1983, **23**, 47; (c) S. Benefice-Malouet, H. Blancou and A. Commeyras, *J. Fluorine Chem.*, 1985, **30**, 171; (d) H. Blancou, P. Moreau and A. Commeyras, *Tetrahedron*, 1977, **33**, 2061; (e) A. Sekiya and N. Ishikawa, *Chem. Lett.*, 1977, 81.
33. (a) D. J. Burton and D. M. Wiemers, *J. Am. Chem. Soc.*, 1985, **107**, 5014; (b) D. M. Wiemers and D. J. Burton, *J. Am. Chem. Soc.*, 1986, **108**, 832.

34. (a) P. A. Morken, H. Lu, A. Nakamura and D. J. Burton, *Tetrahedron Lett.*, 1991, **32**, 4271; (b) S. W. Hansen, T. D. Spawn and D. J. Burton, *J. Fluorine Chem.*, 1987, **35**, 415; (c) P. A. Morken and D. J. Burton, *J. Org. Chem.*, 1993, **58**, 1167; (d) J. P. Gillet, R. Sauvêtre and J. F. Normant, *Synthesis*, 1986, 538.
35. B. Jiang and Y. Xu, *J. Org. Chem.*, 1991, **56**, 7336.
36. (a) J. J. Habeeb, A. Osman and D. G. Tuck, *J. Organomet. Chem.*, 1980, **185**, 117; (b) F. F. Said and D. G. Tuck, *J. Organomet. Chem.*, 1982, **224**, 121; (c) A. Conan, S. Sibille and J. Périchon, *J. Org. Chem.*, 1991, **56**, 2018; (d) S. Sibille, V. Ratovelomanana and J. Périchon, *J. Chem. Soc., Chem. Commun.*, 1992, 283; (e) J. Chaussard, J.-C. Folest, J.-Y. Nédélec, J. Périchon, S. Sibille and M. Troupel, *Synthesis*, 1990, 369; (f) S. Sibille, V. Ratovelomanana, J. Y. Nédélec and J. Périchon, *Synlett*, 1993, 425.
37. (a) K. J. Klabunde and R. Campostrini, *J. Fluorine Chem.*, 1989, **42**, 93; (b) M. A. Guerra, T. R. Bierschenk and R. J. Lagow, *J. Am. Chem. Soc.*, 1986, **108**, 4103; (c) K. J. Klabunde, M. S. Key and J. Y. F. Low, *J. Am. Chem. Soc.*, 1972, **94**, 999; (d) K. J. Klabunde, *Angew. Chem., Int. Ed. Engl.*, 1975, **14**, 287.
38. (a) N. El Alami, C. Belaud and J. Villiéras, *J. Organomet. Chem.*, 1987, **319**, 303; (b) N. El Alami, C. Belaud and J. Villiéras, *J. Organomet. Chem.*, 1988, **348**, 1; (c) C. Belaud, C. Roussakis, Y. Letourneux, N. El Alami and J. Villiéras, *Synth. Commun.*, 1985, **15**, 1233; (d) N. E. Alami, C. Belaud and J. Villiéras, *Tetrahedron Lett.*, 1987, **28**, 59.
39. J. Furukawa and N. Kawabata, *Adv. Organomet. Chem.*, 1974, **12**, 83.
40. M. J. Rozema, S. AchyuthaRao, and P. Knochel, *J. Org. Chem.*, 1992, **57**, 1956.
41. M. J. Rozema, C. Eisenberg, H. Lütjens, R. Ostwald, K. Belyk and P. Knochel, *Tetrahedron Lett.*, 1993, **34**, 3115.
42. H. Stadtmüller, R. Lentz, C. E. Tucker, T. Stüdemann, W. Dörner and P. Knochel, *J. Am. Chem. Soc.*, 1993, **115**, 7027.
43. H. Stadtmüller, C. E. Tucker, A. Vaupel and P. Knochel, *Tetrahedron Lett.*, 1993, **34**, 7911.
44. H. Lange and D. Naumann, *J. Fluorine Chem.*, 1984, **26**, 435.
45. (a) K. H. Thiele, E. Langguth and G. E. Müller, *Z. Anorg. Allg. Chem.*, 1980, **462**, 152; (b) H.-O. Fröhlich, B. Kosan, B. Müller and W. Hiller, *J. Organomet. Chem.*, 1989, **441**, 177.
46. A. Pimm, P. Kociénski and S. D. A. Street, *Synlett*, 1992, 886.
47. (a) M. F. Semmelhack and E. J. Fewkes, *Tetrahedron Lett.*, 1987, **28**, 1497; (b) G. Köbrich and H. R. Merkle, *Chem. Ber.*, 1966, **99**, 1782.
48. (a) J.-P. Gillet, R. Sauvêtre and J. F. Normant, *Tetrahedron Lett.*, 1985, **26**, 3999; (b) F. Tellier, R. Sauvêtre and J. F. Normant, *J. Organomet. Chem.*, 1986, **303**, 309; (c) F. Tellier, R. Sauvêtre, J. F. Normant, Y. Dromzee and Y. Jeannin, *J. Organomet. Chem.*, 1987, **331**, 281; (d) P. Martinet, R. Sauvêtre and J. F. Normant, *J. Organomet. Chem.*, 1989, **367**, 1.
49. (a) C. E. Tucker, T. N. Majid and P. Knochel, *J. Am. Chem. Soc.*, 1992, **114**, 3983; (b) G. Cahiez, P. Venegas, C. E. Tucker, T. N. Majid and P. Knochel, *J. Chem. Soc., Chem. Commun.*, 1992, 1406.
50. (a) J. L. von dem Bussche-Hünnefeld and D. Seebach, *Tetrahedron*, 1992, **48**, 5719; (b) B. Schmidt and D. Seebach, *Angew. Chem. Int. Ed. Engl.*, 1991, **30**, 99; (c) B. Schmidt and D. Seebach, *Angew. Chem., Int. Ed. Engl.*, 1991, **30**, 1321.
51. (a) K.-H. Thiele and P. Zdunneck, *J. Organomet. Chem.*, 1965, **4**, 10; (b) K.-H. Thiele, G. Engelhardt, J. Köhler and M. Arnstedt, *J. Organomet. Chem.*, 1967, **9**, 385.
52. (a) W. Oppolzer and R. N. Radinov, *Helv. Chim. Acta*, 1992, **75**, 170; (b) M. Srebnik, *Tetrahedron Lett.*, 1991, **32**, 2449.
53. F. Langer, J. Waas and P. Knochel, *Tetrahedron Lett.*, 1993, **34**, 5261.
54. E. Frankland and D. F. Duppa, *Liebigs Ann Chem.*, 1864, **130**, 117.
55. M. J. Rozema, D. Rajagopal, C. E. Tucker and P. Knochel, *J. Organomet. Chem.*, 1992, **438**, 11.
56. (a) E. K. S. Liu, *Inorg. Chem.*, 1980, **19**, 266; (b) E. K. S. Liu and L. B. Asprey, *J. Organomet. Chem.*, 1979, **169**, 249.
57. (a) C. Eaborn, N. Retta and J. D. Smith, *J. Organomet. Chem.*, 1980, **190**, 101; (b) F. Aigbirhio, C. Eaborn, A. Habtemariam and J. D. Smith, *J. Chem. Soc., Chem. Commun.*, 1990, 1471.
58. L. Miginiac, 'The Chemistry of the Metal–Carbon Bond', eds. F. R. Harley and S. Patai, Wiley, New York, 1985, vol. 3, p. 99.
59. H. T. Teunissen and F. Bickelhaupt, *Tetrahedron Lett.*, 1992, **33**, 3537.
60. P. Knochel and J. F. Normant, *Tetrahedron Lett.*, 1986, **27**, 4431.
61. (a) H. Ochiai, Y. Tamaru, K. Tsubaki and Z. Yoshida, *J. Org. Chem.*, 1987, **52**, 4418; (b) Y. Tamaru, H. Tanigawa, T. Yamamoto and Z. Yoshida, *Angew. Chem., Int. Ed. Engl.*, 1989, **28**, 351.
62. L. Zhu, R. M. Wehmeyer and R. D. Rieke, *J. Org. Chem.*, 1991, **56**, 1445.
63. K. Sekiya and E. Nakamura, *Tetrahedron Lett.*, 1988, **29**, 5155.
64. M. C. P. Yeh and P. Knochel, *Tetrahedron Lett.*, 1988, **29**, 2395.
65. (a) T. Ibuka et al., *Tetrahedron Lett.*, 1992, **33**, 3783; (b) T. Ibuka et al., *J. Org. Chem.*, 1993, **58**, 1207; (c) Y. Yamamoto, Y. Chounan, M. Tanaka and T. Ibuka, *J. Org. Chem.*, 1992, **57**, 1024.
66. P. Knochel, M. J. Rozema, C. E. Tucker, C. Retherford, M. Furlong and S. AchyuthaRao, *Pure Appl. Chem.*, 1992, **64**, 361.
67. H. G. Chen, J. L. Gage, S. D. Barrett and P. Knochel, *Tetrahedron Lett.*, 1990, **31**, 1829.
68. T. Tanaka, K. Bannai, A. Hazato, M. Koga, S. Kurozumi and Y. Kato, *Tetrahedron*, 1991, **47**, 1861.
69. (a) M. J. Dunn and R. F. W. Jackson, *J. Chem. Soc., Chem. Commun.*, 1992, 319; (b) R. F. W. Jackson, N. Wishart, A. Wood, K. James and M. J. Wythes, *J. Org. Chem.*, 1992, **57**, 3397.
70. L. Zhu and R. D. Rieke, *Tetrahedron Lett.*, 1991, **32**, 2865.
71. (a) M.-C. P. Yeh, M.-L. Sun and S.-K. Lin, *Tetrahedron Lett.*, 1991, **32**, 113; (b) M.-C. P. Yeh and S.-I. Tau, *J. Chem. Soc., Chem. Commun.*, 1992, 13; (c) M.-C. P. Yeh, C.-J. Tsou, C.-N. Chuang and H.-C. Lin, *J. Chem. Soc., Chem. Commun.*, 1992, 890; (d) M.-C. P. Yeh, B.-A. Sheu, H.-W. Fu, S.-I. Tau and L.-W. Chuang, *J. Am. Chem. Soc.*, 1993, **115**, 5941.
72. M. J. Dunn, R. F. W. Jackson and G. R. Stephenson, *Synlett*, 1992, 905.
73. C. E. Tucker and P. Knochel, *J. Org. Chem.*, 1993, **58**, 4781.
74. (a) C. Retherford, T.-S. Chou, R. M. Schelkun and P. Knochel, *Tetrahedron Lett.*, 1990, **31**, 1833; (b) T. N. Majid, M. C. P. Yeh and P. Knochel, *Tetrahedron Lett.*, 1989, **30**, 5069.
75. C. E. Tucker, S. AchyuthaRao and P. Knochel, *J. Org. Chem.*, 1990, **55**, 5446.
76. A. Sidduri, N. Budries, R. M. Laine and P. Knochel, *Tetrahedron Lett.*, 1992, **33**, 7515.
77. (a) P. Knochel, T.-S. Chou, H. G. Chen, M. C. P. Yeh and M. J. Rozema, *J. Org. Chem.*, 1989, **54**, 5202; (b) T.-S. Chou and P. Knochel, *J. Org. Chem.*, 1990, **55**, 4791; (c) P. Knochel, T.-S. Chou, C. Jubert and D. Rajagopal, *J. Org. Chem.*, 1993, **58**, 588.
78. C. Jubert and P. Knochel, *J. Org. Chem.*, 1992, **57**, 5431.

79. (a) D. C. Billington in 'Comprehensive Organic Synthesis' eds. B. M. Trost, I. Fleming and G. Pattenden, Pergamon, Oxford, vol. 3, p. 413; (b) K. Tamao *ibid.* p. 435; (c) D. W. Knight *ibid.* p. 481; (d) K. Sonogashira *ibid.* p. 521 and p. 551; (e) E. Erdik, *Tetrahedron*, 1992, **48**, 9577.
80. S. Hyuga, N. Yamashina, S. Hara and A. Suzuki, *Chem. Lett.*, 1988, 809.
81. E. Negishi, H. Matsushita, M. Kobayashi and C. L. Rand, *Tetrahedron Lett.*, 1983, **24**, 3823.
82. E. Negishi and Z. Owczarczyk, *Tetrahedron Lett.*, 1991, **32**, 6683.
83. J. B. Campbell, Jr., J. W. Firor and T. W. Davenport, *Synth. Commun.*, 1989, **19**, 2265.
84. T. Hayashi, T. Hagihara, Y. Katsuro and M. Kumada, *Bull. Chem. Soc. Jpn.*, 1983, **56**, 363.
85. A. S. Bell, D. A. Roberts and K. S. Ruddock, *Synthesis*, 1987, 843.
86. A. Minato, K. Suzuki, K. Tamao and M. Kumada, *Tetrahedron Lett.*, 1984, **25**, 83.
87. F. Tellier, R. Sauvêtre and J. F. Normant, *J. Organomet. Chem.*, 1987, **328**, 1.
88. (a) A. Carpita and R. Rossi, *Tetrahedron Lett.*, 1986, **27**, 4351; (b) B. P. Andreini, A. Carpita and R. Rossi, *Tetrahedron Lett.*, 1986, **27**, 5533; (c) B. P. Andreini, A. Carpita and R. Rossi, *Tetrahedron Lett.*, 1988, **29**, 2239.
89. (a) M. C. P. Yeh and P. Knochel, *Tetrahedron Lett.*, 1989, **30**, 4799; (b) H. Sörensen and A. E. Greene, *Tetrahedron Lett.*, 1990, **31**, 7597.
90. Y. Tamaru, H. Ochiai, T. Nakamura and Z. Yoshida, *Org. Synth.*, 1988, **67**, 98.
91. (a) R. F. W. Jackson, K. James, M. J. Wythes and A. Wood, *J. Chem. Soc., Chem. Commun.*, 1989, 644; (b) R. F. W. Jackson, M. J. Wythes and A. Wood, *Tetrahedron Lett.*, 1989, **30**, 5941; (c) R. F. W. Jackson, A. Wood and M. J. Wythes, *Synlett*, 1990, 735; (d) R. F. W. Jackson, N. Wishart and M. J. Wythes, *J. Chem. Soc., Chem. Commun.*, 1992, 1587; (e) R. F. W. Jackson; N. Wishart and M. J. Wythes, *Synlett*, 1993, 219.
92. Y. Tamaru, K. Yasui, H. Takanabe, S. Tanaka and K. Fugami, *Angew. Chem.*, 1992, **104**, 662.
93. (a) P. Knochel and J. F. Normant, *Tetrahedron Lett.*, 1984, **25**, 4383; (b) G. Rousseau and J. M. Conia, *Tetrahedron Lett.*, 1981, **22**, 649; (c) G. Rousseau and J. Drouin, *Tetrahedron*, 1983, **39**, 2307.
94. P. Knochel in 'Comprehensive Organic Synthesis', eds. B. M. Trost, I. Fleming and S. L. Schreiber, Pergamon, Oxford, vol. 1, p. 211.
95. (a) M. Gaudemar, *Bull. Soc. Chim. Fr.*, 1963, 1475; (b) B. Maurer and A. Hauser, *Helv. Chim. Acta*, 1982, **65**, 462; (c) L. Miginiac-Groizeleau, P. Miginiac and C. Prévost, *Bull. Soc. Chim. Fr.*, 1965, 3560; (d) M. Taniguchi, K. Oshima and K. Utimoto, *Chem. Lett.*, 1992, 2135.
96. G. A. Molander and D. C. Shubert, *J. Am. Chem. Soc.*, 1986, **108**, 4683.
97. (a) J. van der Louw, J. L. van der Baan, H. Stichter, G. J. J. Out, F. Bickelhaupt and G. W. Klumpp, *Tetrahedron Lett.*, 1988, **29**, 3579; (b) J. van der Louw *et al.*, *Tetrahedron*, 1992, **48**, 9877.
98. (a) J. F. Ruppert, M. A. Avery and J. D. White, *J. Chem. Soc., Chem. Commun.*, 1976, 978; (b) E. Öhler, K. Reininger and U. Schmidt, *Angew. Chem.*, 1970, **82**, 480; (c) H. Mattes and C. Benezra, *Tetrahedron Lett.*, 1985, **26**, 5697; (d) P. Knochel and J. F. Normant, *J. Organomet. Chem.*, 1986, **309**, 1; (e) P. Auvray, P. Knochel and J. F. Normant, *Tetrahedron*, 1988, **44**, 4495, 4509.
99. (a) J.-L. Moreau in 'The Chemistry of Ketenes, Allenes and Related Compounds', ed. S. Patai, Wiley, New York, 1980, p. 363; (b) J.-L. Moreau and M. Gaudemar, *Bull. Soc. Chim. Fr.*, 1970, 2171, 2175; (c) J.-L. Moreau, *Bull. Soc. Chim. Fr.*, 1975, 1248; (d) M. Suzuki, Y. Morita and R. Noyori, *J. Org. Chem.*, 1990, **55**, 441.
100. G. Zweifel and G. Hahn, *J. Org. Chem.*, 1984, **49**, 4565.
101. (a) G. Fronza, C. Fuganti and P. Grasselli, *J. Chem. Soc., Chem. Commun.*, 1980, 442; (b) C. Fuganti, P. Grasselli, S. Servi, F. Spreafico and C. Zirotti, *J. Org. Chem.*, 1984, **49**, 4087; (c) J. Mulzer, M. Kappert, G. Huttner and I. Jibril, *Angew. Chem., Int. Ed. Engl.*, 1984, **23**, 704; (d) T. Fujisawa, E. Kojima, T. Itoh and T. Sato, *Tetrahedron Lett.*, 1985, **26** 6089; (e) T. Fujisawa, I. Takemura and Y. Ukaji, *Tetrahedron Lett.*, 1990, **31**, 5479.
102. K. R. Overly, J. M. Williams and G. J. McGarvey, *Tetrahedron Lett.*, 1990, **31**, 4573.
103. (a) R. S. Coleman and A. J. Carpenter, *Tetrahedron Lett.*, 1992, **33**, 1697; (b) W. J. Thompson, T. J. Tucker, J. E. Schwering and J. L. Barnes, *Tetrahedron Lett.*, 1990, **31**, 6819.
104. M. Bhupathy and T. Cohen, *Tetrahedron Lett.*, 1985, **26**, 2619.
105. A. Solladie-Cavallo, D. Farkhani, S. Fritz, T. Lazrak and J. Suffert, *Tetrahedron Lett.*, 1984, **25**, 4117.
106. G. Boireau, A. Deberly and D. Abenhaïm, *Tetrahedron Lett.*, 1988, **29**, 2175.
107. (a) H. Ochiai, T. Nishihara, Y. Tamaru and Z. Yoshida, *J. Org. Chem.*, 1988, **53**, 1343; (b) Y. Tamaru, T. Nakamura, M. Sakaguchi, H. Ochiai and Z. Yoshida, *J. Chem. Soc., Chem. Commun.*, 1988, 610; (c) A. E. DeCamp; A. T. Kawaguchi, R. P. Volante and I. Shinkai, *Tetrahedron Lett.*, 1991, **32**, 1867.
108. (a) M. C. P. Yeh, P. Knochel and L. E. Santa, *Tetrahedron Lett.*, 1988, **29**, 3887; (b) M. C. P. Yeh, H. G. Chen and P. Knochel, *Org. Synth.*, 1991, **70**, 195; (c) P. Quinton and T. Le Gall, *Tetrahedron Lett.*, 1991, **32**, 4909.
109. (a) M. T. Reetz, 'Organotitanium Reagents in Organic Synthesis', Springer, Berlin, 1986; (b) B. Weidmann and D. Seebach, *Angew. Chem., Int. Ed. Engl.*, 1983, **22**, 31.
110. K. Soai, Y. Hirose and S. Sakata, *Tetrahedron: Asymmetry*, 1992, **3**, 677.
111. A. R. Gangloff, B. Akermark and P. Helquist, *J. Org. Chem.*, 1992, **57**, 4797.
112. (a) Y. Yamamoto, T. Komatsu and K. Maruyama, *J. Chem. Soc., Chem. Commun.*, 1985, 814; (b) J. Pornet and L. Miginiac, *Bull. Soc. Chim. Fr.*, 1975, 841.
113. J. Yamada, H. Satô and Y. Yamamoto, *Tetrahedron Lett.*, 1989, **30**, 5611.
114. (a) C. Agami, F. Couty, J.-C. Daran, B. Prince and C. Puchot, *Tetrahedron Lett.*, 1990, **31**, 2889; (b) C. Agami, F. Couty, M. Poursoulis and J. Vaissermann, *Tetrahedron*, 1992, **48**, 431; (c) M. Eguchi, Q. Zeng, A. Korda and I. Ojima, *Tetrahedron Lett.*, 1993, **34**, 915; (d) C. Andrés *et al.*, *Tetrahedron Lett.*, 1992, **33**, 4743.
115. (a) E. Wissing, R. W. A. Havenith, J. Boersma and G. van Koten, *Tetrahedron Lett.*, 1992, **33**, 7933; (b) F. H. van der Steen, H. Kleijn, J. T. B. H. Jastrzebski and G. van Koten, *J. Org. Chem.*, 1991, **56**, 5147.
116. (a) K. Soai, T. Hatanaka and T. Miyazawa, *J. Chem. Soc., Chem. Commun.*, 1992, 1097; (b) A. R. Katritzky and P. A. Harris, *Tetrahedron: Asymmetry*, 1992, **3**, 437; (c) Y. A. Dembélé, C. Belaud, P. Hitchcock and J. Villiéras, *Tetrahedron: Asymmetry*, 1992, **3**, 351.
117. D. L. Comins and S. O'Connor, *Tetrahedron Lett.*, 1987, **28**, 1843.
118. (a) W.-L. Chia and M.-J. Shiao, *Tetrahedron Lett.*, 1991, **32**, 2033; (b) T.-L. Shing, W.-L. Chia, M.-J. Shiao and T.-Y. Chau, *Synthesis*, 1991, 849; (c) M.-J. Shiao, K.-H. Liu and L.-G. Lin, *Synlett*, 1992, 655.

119. (a) D. L. Comins and M. A. Foley, *Tetrahedron Lett.*, 1988, **29**, 6711; (b) J.-L. Bettiol and R. J. Sundberg, *J. Org. Chem.*, 1993, **58**, 814.
120. (a) D. A. Evans, *Science*, 1988, **240**, 420; (b) R. Noyori and M. Kitamura, *Angew. Chem., Int. Ed. Engl.*, 1991, **30**, 49.
121. (a) N. Oguni and T. Omi, *Tetrahedron Lett.*, 1984, **25**, 2823; (b) N. Oguni, Y. Matsuda and T. Kaneko, *J. Am. Chem. Soc.*, 1988, **110**, 7877.
122. (a) M. Kitamura, S. Okada, S. Suga and R. Noyori, *J. Am. Chem. Soc.*, 1989, **111**, 4028; (b) S. Itsuno and J. M. J. Fréchet, *J. Org. Chem.*, 1987, **55**, 4140; (c) S. Itsuno, Y. Sakurai, K. Ito, T. Maruyama, S. Nakahama and J. M. J. Fréchet, *J. Org. Chem.*, 1990, **55**, 304.
123. (a) K. Soai, A. Ookawa, K. Ogawa and T. Kaba, *J. Chem. Soc., Chem. Commun.*, 1987, 467; (b) K. Soai, A. Ookawa, T. Kaba and K. Ogawa, *J. Am. Chem. Soc.*, 1987, **109**, 7111.
124. K. Soai and M. Watanabe, *J. Chem. Soc., Chem. Commun.*, 1990, 43.
125. K. Soai, S. Yokoyama, K. Ebihara and T. Hayasaka, *J. Chem. Soc., Chem. Commun.*, 1987, 1690.
126. (a) W. Oppolzer and R. N. Radinov, *Tetrahedron Lett.*, 1988, **29**, 5645; (b) W. Oppolzer and R. N. Radinov, *Tetrahedron Lett.*, 1991, **32**, 5777.
127. (a) M. Yoshioka, T. Kawakita and M. Ohno, *Tetrahedron Lett.*, 1989, **30**, 1657; (b) H. Takahashi, T. Kawakita, M. Yoshioka, S. Kobayashi and M. Ohno, *Tetrahedron Lett.*, 1989, **30**, 7095; (c) H. Takahashi, T. Kawakita, M. Ohno, M. Yoshioka and S. Kobayashi, *Tetrahedron*, 1992, **48**, 5691.
128. (a) A. K. Beck *et al.*, *Chimia*, 1991, **45**, 238; (b) D. Seebach, L. Behrendt and D. Felix, *Angew. Chem., Int. Ed. Engl.*, 1991, **30**, 1008; (c) D. Seebach, D. A. Plattner, A. K. Beck; Y. M. Wang, D. Hunziker and W. Petter, *Helv. Chim. Acta*, 1992, **75**, 2171.
129. R. Noyori, S. Suga, K. Kawai, S. Okada, M. Kitamura, N. Oguni, M. Hayashi, T. Kaneko and Y. Matsuda, *J. Organomet. Chem.*, 1990, **382**, 19.
130. (a) K. Soai and A. Ookawa, *J. Chem. Soc., Chem. Commun.*, 1986, 412; (b) A. Ookawa and K. Soai, *J. Chem. Soc. Perkin 1*, 1987, 1465; (c) M. Asami and S. Inoue, *Chem. Lett.*, 1991, 685.
131. N. N. Joshi, M. Srebnik and H. C. Brown, *Tetrahedron Lett.*, 1989, **30**, 5551.
132. (a) W. Brieden, R. Ostwald and P. Knochel, *Angew. Chem., Int. Ed. Engl.*, 1993, **32**, 582; (b) P. Knochel, W. Brieden, C. Eisenberg and M. J. Rozema, *Tetrahedron Lett.*, 1993, **34**, in press.
133. K. Soai, S. Yokoyama, T. Hayasaka and K. Ebihara, *Chem. Lett.*, 1988, 843.
134. (a) K. Soai, M. Watanabe and M. Koyano, *J. Chem. Soc., Chem. Commun.*, 1989, 534; (b) K. Soai, H. Hori and M. Kawahara, *Tetrahedron: Asymmetry*, 1990, **1**, 769.
135. W. Oppolzer and R. N. Radinov, *J. Am. Chem. Soc.*, 1993, **115**, 1593.
136. E. J. Corey, P.-W. Yuen, F. J. Hannon and D. A. Wierda, *J. Org. Chem.*, 1990, **55**, 784.
137. (a) H. P. Knoess, M. T. Furlong, M. J. Rozema and P. Knochel, *J. Org. Chem.*, 1991, **56**, 5974; (b) T. N. Majid and P. Knochel, *Tetrahedron Lett.*, 1990, **31**, 4413; (c) S. AchyuthaRao, C. E. Tucker and P. Knochel, *Tetrahedron Lett.*, 1990, **31**, 7575; (d) S. AchyuthaRao, T.-S. Chou, I. Schipor and P. Knochel, *Tetrahedron*, 1992, **48**, 2025; (e) S. Kim and J. M. Lee, *Tetrahedron Lett.*, 1990, **31**, 7627.
138. M. C. P. Yeh, P. Knochel, W. M. Butler and S. C. Berk, *Tetrahedron Lett.*, 1988, **29**, 6693.
139. (a) H. Tsujiyama, N. Ono, T. Yoshino, S. Okamoto and F. Sato, *Tetrahedron Lett.*, 1990, **31**, 4481; (b) T. Yoshino, S. Okamoto and F. Sato, *J. Org. Chem.*, 1991, **56**, 3205.
140. A. E. Greene, J.-P. Lansard, J.-L. Luche and C. Pétrier, *J. Org. Chem.*, 1984, **49**, 931.
141. (a) K. Soai, T. Hayasaka, S. Ugajin and S. Yokoyama, *Chem. Lett.*, 1988, 1571; (b) K. Soai, S. Yokoyama, T. Hayasaka and K. Ebihara, *J. Org. Chem.*, 1988, **53**, 4149; (c) K. Soai, T. Hayasaka and S. Ugajin, *J. Chem. Soc., Chem. Commun.*, 1989, 516; (d) C. Bolm and M. Ewald, *Tetrahedron Lett.*, 1990, **31**, 5011; (e) C. Bolm, *Tetrahedron: Asymmetry*, 1991, **2**, 701; (f) C. Bolm, M. Felder and J. Müller, *Synlett*, 1992, 439; (g) J. F. G. A. Jansen and B. L. Feringa, *Tetrahedron: Asymmetry*, 1992, **3**, 581; (h) M. Uemura, R. Miyake, K. Nakayama and Y. Hayashi, *Tetrahedron: Asymmetry*, 1992, **3**, 713.
142. K. Soai, M. Okudo and M. Okamoto, *Tetrahedron Lett.*, 1991, **32**, 95.
143. (a) C. Retherford, M. C. P. Yeh, I. Schipor, H. G. Chen and P. Knochel, *J. Org. Chem.*, 1989, **54**, 5200; (b) C. Retherford and P. Knochel, *Tetrahedron Lett.*, 1991, **32**, 441.
144. S. E. Denmark and L. R. Marcin, *J. Org. Chem.*, 1993, **58**, 3850.
145. B. S. Bronk, S. J. Lippard and R. L. Danheiser, *Organometallics*, 1993, **12**, 3340.
146. (a) H. Lehmkuhl and O. Olbrysch, *Liebigs Ann. Chem.*, 1975, 1162; (b) H. Lehmkuhl, I. Döring and H. Nehl, *J. Organomet. Chem.*, 1981, **221**, 123; (c) H. Lehmkuhl and H. Nehl, *J. Organomet. Chem.*, 1981, **221**, 131.
147. (a) G. Courtois and L. Miginiac, *J. Organomet. Chem.*, 1973, **52**, 241; (b) H. Lehmkuhl, I. Döring, R. McLane and H. Nehl, *J. Organomet. Chem.*, 1981, **221**, 1.
148. (a) M. Gaudemar, *C. R. Hebd. Seances Acad. Sci., Ser. C.*, 1971, **273**, 1669; (b) Y. Frangin and M. Gaudemar, *C. R. Hebd. Seances Acad. Sci., Ser. C.*, 1974, **278**, 885; (c) M. Bellassoued, Y. Frangin and M. Gaudemar, *Synthesis*, 1977, 205.
149. (a) P. Knochel and J. F. Normant, *Tetrahedron Lett.*, 1986, **27**, 1039, 1043, 4427, 5727; (b) P. Knochel, M. C. P. Yeh and C. Xiao, *Organometallics*, 1989, **8**, 2831; (c) D. Beruben, I. Marek, L. Labaudinière and J.-F. Normant, *Tetrahedron Lett.*, 1993, **34**, 2303; (d) J.-F. Normant, J.-C. Quirion, Y. Masuda and A. Alexakis, *Tetrahedron Lett.*, 1990, **31**, 2879; (e) J.-F. Normant, J.-C. Quirion, A. Alexakis and Y. Masuda, *Tetrahedron Lett.*, 1989, **30**, 3955; (f) J.-F. Normant and J.-C. Quirion, *Tetrahedron Lett.*, 1989, **30**, 3959; (g) I. Marek, J.-M. Lefrançois and J.-F. Normant, *Synlett*, 1992, 633; (h) I. Marek, J.-M. Lefrançois and J.-F. Normant, *Tetrahedron Lett.*, 1992, **33**, 1747; (i) I. Marek, J.-M. Lefrançois, and J.-F. Normant, *Tetrahedron Lett.*, 1991, **32**, 5969; (j) I. Marek and J.-F. Normant, *Tetrahedron Lett.*, 1991, **32**, 5973.
150. P. Knochel, in 'Comprehensive Organic Synthesis', eds. B. M. Trost, I. Fleming and M. F. Semmelhack, Pergamon, Oxford, 1991, vol. 4, p. 865.
151. P. Knochel, C. Xiao and M. C. P. Yeh, *Tetrahedron Lett.*, 1988, **29**, 6697.
152. C. E. Tucker and P. Knochel, *Synthesis*, 1993, 530.
153. (a) C. Meyer, I. Marek, G. Courtemanche and J.-F. Normant, *Synlett*, 1993, 266; (b) G. Courtois, A. Masson and L. Miginiac, *C. R. Hebd. Seances Acad. Sci., Ser., C.*, 1978, **286**, 265.
154. A. Yanagisawa, S. Habaue and H. Yamamoto, *J. Am. Chem. Soc.*, 1989, **111**, 366.
155. (a) G. Courtois and L. Miginiac, *J. Organomet. Chem.*, 1974, **69**, 1; (b) F. Bernardou and L. Miginiac, *Tetrahedron Lett.*, 1976, 3083; (c) E. Negishi and J. A. Miller, *J. Am. Chem. Soc.*, 1983, **105**, 6761; (d) G. A. Molander, *J. Org. Chem.*, 1983, **48**, 5409.

156. J. Auger, G. Courtois and L. Miginiac, *J. Organomet. Chem.*, 1977, **133**, 285.
157. (a) K. E. Schulte, G. Rücker and J. Feldkamp, *Chem. Ber.*, 1972, **105**, 24; (b) M. T. Bertrand, G. Courtois and L. Miginiac, *Tetrahedron Lett.*, 1974, 1945 and 1975, 3147.
158. M. Bellassoued, Y. Frangin and M. Gaudemar, *J. Organomet. Chem.*, 1979, **166**, 1.
159. E. Negishi, D. E. Van Horn, T. Yoshida and C. L. Rand, *Organometallics*, 1983, **2**, 563.
160. (a) E. Negishi, H. Sawada, J. M. Tour and Y. Wei, *J. Org. Chem.*, 1988, **53**, 913; (b) E. Negishi, *Acc. Chem. Res.*, 1987, **20**, 65.
161. (a) J. van der Louw *et al.*, *Tetrahedron Lett.*, 1989, **30**, 4453; (b) J. van der Louw, J. L. van der Baan, F. Bickelhaupt and G. W. Klumpp, *Tetrahedron Lett.*, 1987, **28**, 2889; (c) J. van der Louw, J. L. van der Baan, F. J. J. de Kanter, F. Bickelhaupt and G. W. Klumpp, *Tetrahedron*, 1992, **48**, 6087; (d) J. van der Louw *et al.*, *Tetrahedron*, 1992, **48**, 6105.
162. (a) M. T. Crimmins and P. G. Nantermet, *J. Org. Chem.*, 1990, **55**, 4235; (b) M. T. Crimmins *et al.*, *J. Org. Chem.*, 1993, **58**, 1038; (c) M. T. Crimmins, D. K. Jung and J. L. Gray, *J. Am. Chem. Soc.*, 1993, **115**, 3146.
163. S. AchyuthaRao and P. Knochel, *J. Am. Chem. Soc.*, 1991, **113**, 5735.
164. C. Blomberg and F. A. Hartog, *Synthesis*, 1977, 18.
165. S. R. Wilson and M. E. Guazzaroni, *J. Org. Chem.*, 1989, **54**, 3087.
166. (a) C. Pétrier and J. L. Luche, *J. Org. Chem.*, 1985, **50**, 910; (b) C. Pétrier, J. Einhorn and J. L. Luche, *Tetrahedron Lett.*, 1985, **26**, 1449; (c) C. Einhorn and J. L. Luche, *J. Organomet. Chem.*, 1987, **322**, 177.
167. (a) T. Shono, M. Ishifune and S. Kashimura, *Chem. Lett.*, 1990, 449; (b) H. Waldmann, *Synlett*, 1990, 627; (c) Y. Oda, S. Matsuo and K. Saito, *Tetrahedron Lett.*, 1992, **33**, 97.
168. (a) J. C. de Souza Barbosa, C. Pétrier and J. L. Luche, *Tetrahedron Lett.*, 1985, **26**, 829; (b) C. Pétrier, C. Dupuy and J. L. Luche, *Tetrahedron Lett.*, 1986, **27**, 3149.
169. T. Shono, I. Nishiguchi and M. Sasaki, *J. Am. Chem. Soc.*, 1978, **100**, 4314.
170. T. Shono *et al.*, *Chem. Lett.*, 1981, 1217.
171. (a) B. Giese, W. Damm, M. Roth and M. Zehnder, *Synlett*, 1992, 441; (b) P. Erdmann, J. Schäfer, R. Springer, H.-G. Zeitz and B. Giese, *Helv. Chim. Acta*, 1992, **75**, 638.
172. L. A. Sarandeses, A. Mouriño and J.-L. Luche, *J. Chem. Soc., Chem. Commun.*, 1992, 798.
173. T. Carofiglio, D. Marton and G. Tagliavini, *Organometallics*, 1992, **11**, 2961.
174. (a) J. M. Denis, C. Girard and J. M. Conia, *Synthesis*, 1972, 549; (b) C. Girard and J. M. Conia, *Tetrahedron Lett.*, 1974, 3327; (c) C. Girard, P. Amice, J. P. Barnier and J. M. Conia, *Tetrahedron Lett.*, 1974, 3329; (d) J. M. Conia, *Pure and Appl. Chem.*, 1975, **43**, 317; (e) L. Fitjer *et al.*, *Tetrahedron*, 1984, **40**, 4337.
175. S. Sawada and Y. Inouye, *Bull. Soc. Chim. Jpn.*, 1969, **42**, 2669.
176. M. R. Detty and L. A. Paquette, *J. Am. Chem. Soc.*, 1977, **99**, 821.
177. (a) A. B. Charette, B. Côté and J.-F. Marcoux, *J. Am. Chem. Soc.*, 1991, **113**, 8166; (b) I. Arai, A. Mori and H. Yamamoto, *J. Am. Chem. Soc.*, 1985, **107**, 8254; (c) E. A. Mash and K. A. Nelson, *J. Am. Chem. Soc.*, 1985, **107**, 8256; (d) E. A. Mash and D. S. Torok, *J. Org. Chem.*, 1989, **54**, 250; (e) E. A. Mash, K. A. Nelson, S. Van Deusen and S. B. Hemperly, *Org. Synth.*, 1989, **68**, 92; (f) E. A. Mash and S. B. Hemperly, *J. Org. Chem.*, 1990, **55**, 2055; (g) T. Sugimura, T. Futagawa, M. Yoshikawa and A. Tai, *Tetrahedron Lett.*, 1989, **30**, 3807.
178. T. Imai, H. Mineta and S. Nishida, *J. Org. Chem.*, 1990, **55**, 4986.
179. (a) S. E. Denmark, J. P. Edwards and S. R. Wilson, *J. Am. Chem. Soc.*, 1991, **113**, 723; (b) S. E. Denmark and J. P. Edwards, *J. Org. Chem.*, 1991, **56**, 6974; (c) S. E. Denmark; J. P. Edwards and S. R. Wilson, *J. Am. Chem. Soc.*, 1992, **114**, 2592.
180. S. Miyano, Y. Matsumoto and H. Hashimoto, *J. Chem. Soc., Chem. Commun.*, 1975, 364.
181. G. Rousseau and N. Slougui, *J. Am. Chem. Soc.*, 1984, **106**, 7283.
182. (a) A. Sidduri, M. J. Rozema and P. Knochel, *J. Org. Chem.*, 1993, **58**, 2694; (b) P. Knochel, N. Jeong, M. J. Rozema and M. C. P. Yeh, *J. Am. Chem. Soc.*, 1989, **111**, 6474.
183. (a) E. Negishi and K. Akiyoshi, *J. Am. Chem. Soc.*, 1988, **110**, 646; (b) E. Negishi, K. Akiyoshi, B. O'Connor, K. Takagi and G. Wu, *J. Am. Chem. Soc.*, 1989, **111**, 3089.
184. (a) P. Knochel and S. AchyuthaRao, *J. Am. Chem. Soc.*, 1990, **112**, 6146; (b) M. R. Burns and J. K. Coward, *J. Org. Chem.*, 1993, **58**, 528.
185. S. AchyuthaRao and P. Knochel, *J. Am. Chem. Soc.*, 1992, **114**, 7579.
186. T. Harada, K. Hattori, T. Katsuhira and A. Oku, *Tetrahedron Lett.*, 1989, **30**, 6035, 6039.
187. (a) T. Harada, D. Hara, K. Hattori and A. Oku, *Tetrahedron Lett.*, 1988, **29**, 3821; (b) T. Harada *et al.*, *J. Org. Chem.*, 1993, **58**, 4897.
188. (a) T. Harada, Y. Kotani, T. Katsuhira and A. Oku, *Tetrahedron Lett.*, 1991, **32**, 1573; (b) T. Harada, T. Katsuhira and A. Oku, *J. Org. Chem.*, 1992, **57**, 5805.
189. M. J. Rozema and P. Knochel, *Tetrahedron Lett.*, 1991, **32**, 1855.
190. Z.-Y. Yang, D. M. Wiemers and D. J. Burton, *J. Am. Chem. Soc.*, 1992, **114**, 4402.
191. (a) W. Tückmantel, K. Oshima and H. Nozaki, *Chem. Ber.*, 1986, **119**, 1581; (b) R. M. Fabicon, M. Parvez and H. G. Richey, Jr., *J. Am. Chem. Soc.*, 1991, **113**, 1412; (c) R. M. Fabicon, A. D. Pajerski and H. G. Richey, Jr., *J. Am. Chem. Soc.*, 1991, **113**, 6680; (d) A. P. Purdy and C. F. George, *Organometallics*, 1992, **11**, 1955.
192. M. Isobe, S. Kondo, N. Nagasawa and T. Goto, *Chem. Lett.*, 1977, 679.
193. (a) R. A. Kjonaas and R. K. Hoffer, *J. Org. Chem.*, 1988, **53**, 4133; (b) R. A. Kjonaas and E. J. Vawter, *J. Org. Chem.*, 1986, **51**, 3993.
194. (a) J. F. G. A. Jansen and B. L. Feringa, *Tetrahedron Lett.*, 1988, **29**, 3593; (b) J. F. G. A. Jansen and B. L. Feringa, *J. Chem. Soc., Chem. Commun.*, 1989, 741; (c) J. F. G. A. Jansen and B. L. Feringa, *J. Org. Chem.*, 1990, **55**, 4168.
195. D. Seebach and W. Langer, *Helv. Chim. Acta*, 1979, **62**, 1701 and 1710.
196. (a) J. J. Eisch and A. Piotrowski, *Tetrahedron Lett.*, 1983, **24**, 2043; (b) L. N. Nysted, US Pat. 3 865 848 (1975) (*Chem. Abstr.*, 1975, **83**, 10 406q); (c) L. N. Nysted, US Pat. 3 960 904 (1976) (*Chem. Abstr.*, 1976, **85**, 94 618 n).
197. (a) P. Knochel, *J. Am. Chem. Soc.*, 1990, **112**, 7431; (b) J. R. Waas, A. Sidduri and P. Knochel, *Tetrahedron Lett.*, 1992, **33**, 3717.
198. C. E. Tucker and P. Knochel, *J. Am. Chem. Soc.*, 1991, **113**, 9888.

199. (a) K. Takai, Y. Hotta, K. Oshima and H. Nozaki, *Tetrahedron Lett.*, 1978, 2417; (b) T. Okazoe, K. Takai, K. Oshima and K. Utimoto, *J. Org. Chem.*, 1987, **52**, 4410; (c) K. Takai, O. Fujimura, Y. Kataoka and K. Utimoto, *Tetrahedron Lett.*, 1989, **30**, 211; (d) K. Takai, Y. Kataoka, T. Okazoe and K. Utimoto, *Tetrahedron Lett.*, 1988, **29**, 1065; (e) K. Takai, M. Tezuka, Y. Kataoka and K. Utimoto, *Synlett*, 1989, 27.
200. L. Lombardo, *Org. Synth.*, 1987, **65**, 81.
201. (a) F. Bickelhaupt, *Pure Appl. Chem.*, 1986, **58**, 537; (b) F. Bickelhaupt, *Angew. Chem., Int. Ed. Engl.*, 1987, **26**, 990; (c) S. AchyuthaRao and M. Periasamy, *Tetrahedron Lett.*, 1988, **29**, 1583.
202. S. AchyuthaRao and P. Knochel, *J. Org. Chem.*, 1991, **56**, 4591.
203. K. Nützel, in Houben-Weyl, 'Methoden der organischen Chemie', ed. E. Müller, Georg Thieme Verlag, Stuttgart, 1973, vol. 13/2a, p. 859.
204. P. R. Jones and P. J. Desio, *Chem. Rev.*, 1978, **78**, 491.
205. E. R. Burkhardt and R. D. Rieke, *J. Org. Chem.*, 1985, **50**, 416.
206. B. Sain, D. Prajapati and J. S. Sandhu, *Tetrahedron Lett.*, 1992, **33**, 4795.
207. (a) A. J. Pearson and J. Yoon, *Tetrahedron Lett.*, 1985, **26**, 2399; (b) I. Wölfle, S. Chan and G. B. Schuster, *J. Org. Chem.*, 1991, **56**, 7313.
208. L. J. Krause and J. A. Morrison, *J. Chem. Soc., Chem. Commun.*, 1980, 671.
209. Y. Akiyama, T. Kawasaki and M. Sakamoto, *Chem. Lett.*, 1983, 1231.

5
Boron

MICHEL VAULTIER and BERTRAND CARBONI
Université de Rennes I, France

5.1 INTRODUCTION	192
5.1.1 Scope	192
5.1.2 Safety Considerations	192
5.1.3 General Considerations	192
5.2 SYNTHESIS OF ORGANOBORON COMPOUNDS	193
5.2.1 By Creation of a Carbon–Boron Bond	193
5.2.1.1 Hydroboration	193
5.2.1.2 Haloboration	201
5.2.1.3 Transmetallation	202
5.2.2 From One Organoboron Compound to Another	204
5.2.2.1 Modification of substitution at boron	204
5.2.2.2 Insertion of a –CXY group into a carbon–boron bond	207
5.2.2.3 Reactions of α-haloboranes with nucleophilic reagents	207
5.2.2.4 Reactions of boron-stabilized carbanions	208
5.2.2.5 Addition to α,β-unsaturated organoboron compounds	211
5.3 SYNTHETIC APPLICATIONS OF ORGANOBORON COMPOUNDS	213
5.3.1 Replacement of a Carbon–Boron Bond by a Carbon–Hydrogen or Carbon–Heteroatom Bond	213
5.3.1.1 Protonolysis of organoboranes	213
5.3.1.2 Oxidation	214
5.3.1.3 Replacement of a carbon–boron bond by a carbon–nitrogen bond	216
5.3.1.4 Halogenolysis	218
5.3.1.5 Transmetallation	220
5.3.2 Carbon–Carbon Bond Formation via a Transient Four-coordinate Boron	220
5.3.2.1 Rearrangements of α-haloorganoborates	221
5.3.2.2 Carbonylation	225
5.3.2.3 Reactions of organoborates with electrophiles	227
5.3.2.4 Cross-coupling reactions of organoboranes	231
5.3.2.5 Boron-stabilized carbanions in organic synthesis	234
5.3.2.6 Reactions of allylboranes and related compounds	235
5.3.2.7 Boron-mediated aldol methodology	240
5.3.3 Radical Reactions of Organoboranes	241
5.3.4 Miscellaneous Reactions of Boron Derivatives	244
5.4 ASYMMETRIC SYNTHESIS	247
5.4.1 Hydroboration	247
5.4.2 Homologation Reactions	249
5.4.3 Asymmetric Allylborations	251
5.4.4 Asymmetric Aldol Reactions using Chiral Reagents	256
5.4.4.1 Syn-selective aldol reactions	256
5.4.4.2 Anti-selective aldol reactions	257
5.4.5 Enantioselective Reduction	258
5.4.5.1 Oxazaborolidines in enantioselective reduction of ketones	259
5.4.5.2 Chiral organoboranes based on α-pinene	260
5.4.5.3 Other enantioselective boron reducing agents	261
5.4.6 Other Asymmetric Boron-mediated Reactions	262
5.5 REFERENCES	264

5.1 INTRODUCTION

5.1.1 Scope

This chapter aims to update the applications of boron chemistry in organic synthesis covered in *COMC-I*.[1] The literature was comprehensively searched during the early part of 1993. Since 1982 boron chemistry has been the subject of tremendous efforts all over the world, thus producing a wealth of information. Certain areas are expanding very rapidly, such as the catalytic or stoichiometric use of organoboranes in asymmetric synthesis. New important fields with no immediate applications to organic synthesis in terms of methodology are not mentioned in this review. These include classical and nonclassical methyleneboranes,[2] carboranes,[3-10] scorpionates,[11] multilayered sandwich complexes,[12] polyhedral boron halides,[13] new materials based on boron,[14] biological applications,[3,15-21] boron-centered cations,[22] and ^{11}B NMR applications.[23,24]

This chapter has been organized in three main parts. The first is devoted to the synthesis of organoboranes since we felt it would be useful to select methods and related papers where the isolation and characterization of boranes are described. The second part describes the synthetic applications of organoboron compounds with the exception of work dealing with asymmetric synthesis, which is the topic of third part of this review. Some background is given where necessary. Since 1981 several monographs on this subject have been published, including a three-volume extensive review by Köster (*Methoden der Organische Chemie*, 1984)[25] which is a major source of information, as well as books,[3,26,27] annual surveys,[28] and chapters within more general books.[29,30]

5.1.2 Safety Considerations

Provided that normal precautions are taken, borane–THF and borane–dimethyl sulfide complexes, and alkyl-, dialkyl-, and trialkylboranes present no more hazard to the chemist than typical Grignard or lithium reagents. Spontaneous ignition in air is observed with low molecular weight boranes such as Me$_3$B,[31] but the hazard decreases as the molecular weight increases, and tributylborane does not normally ignite spontaneously. Fire risks with boronic esters R^{1}B(OR^2)$_2$ are no different from common organic compounds and boronic acids are stable in air, although some may autoxidize if dry[32] or even when moist.[33] The toxicities of water-soluble boronic acids and esters are low. Some fat-soluble boronic acids show some significant toxicities although they are not highly hazardous.[34,35] Diborane is much more toxic, with the maximum permissible concentration for workers set at 0.1 ppm.[36] There are apparently no major environmental and disposal problems with organoboron compounds since they are readily oxidized to alcohols and boric acid, and are unlikely to persist very long in the presence of oxygen.

5.1.3 General Considerations

The chemistry of organic boron compounds shows peculiar characteristics as a result of the electronic structure of the boron atom (three valence electrons and four orbitals) which has a trigonal valence state and sp^2 hybridization in BX$_3$ derivatives. The most important feature of these compounds is their ability to form coordinate entities containing a tetrahedral boron atom, including neutral complexes and ionic compounds (Figure 1). This accounts for the unique chemistry of this element. An organoborate anion with a leaving group (LG) in a position α to boron undergoes spontaneous 1,2-migration while retaining the configuration of the migrating group (Equation (1)).

$$D \cdot BX_3 \qquad [X_3B-Y]^- M^+$$

D = donor

Figure 1 Neutral complexes and ionic boron compounds.

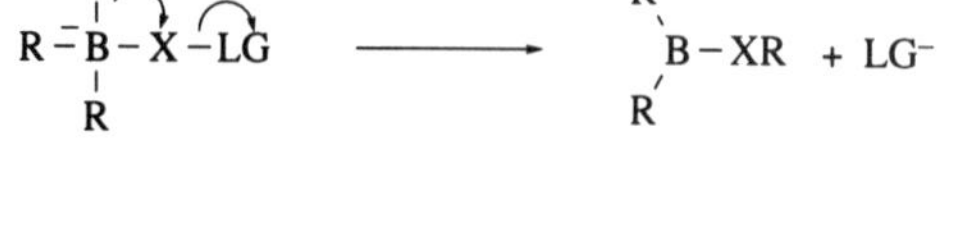

$$R-\underset{R}{\underset{|}{B}}-X-LG \longrightarrow \underset{R}{\overset{R}{B}}-XR + LG^- \qquad (1)$$

The formation of carbon–carbon bonds by such a transfer from boron to an α-carbon is an important process and most synthetically useful ionic reactions of organoboron compounds rely on this principle. For example, homologation is among the most well-known reactions of this type, as illustrated by the work of Matteson *et al.* It is now a very efficient method for the asymmetric synthesis of a variety of optically active compounds with very high enantiomeric excesses. The reaction of organoborates with electrophiles in the presence of a palladium catalyst, that is, the so-called Suzuki coupling reaction, has also received many synthetic applications. The hydroboration reaction is still under active investigation, in particular because of the discovery by Nöth *et al.* that this reaction may be catalyzed by some transition metal complexes. Many important results have been recorded in the pericyclic reactions of β,γ-unsaturated boron compounds with aldehydes and ketones. These include the synthesis of homoallylic alcohols and aldols from optically active allylboranes and enol borinates with asymmetric induction levels ranging up to 100%. Vinylboranes are attractive partners in Diels–Alder reactions and show interesting reactivity when properly substituted. Dienylborane chemistry is also burgeoning. These compounds may find synthetic applications in tandem sequences, including cycloaddition followed by oxidation or amination with azides, for example. Radical reactions of organoboranes and supramolecular chemistry are emerging areas which should provide important developments in the future. Especially noteworthy is the enormous development of highly efficient asymmetric syntheses based on boron catalysts. Remarkable results have been obtained by Corey and others with the enzyme-like reduction of prochiral ketones by borane in the presence of optically active oxazaborolidines, as well as in the Diels–Alder reaction, the Mukaiyama aldol process, and various miscellaneous reactions. No doubt these new results are only the beginning of the exploration of the new continent discovered by Brown.

5.2 SYNTHESIS OF ORGANOBORON COMPOUNDS

5.2.1 By Creation of a Carbon–Boron Bond

5.2.1.1 Hydroboration

The addition of borane R^1R^2BH to carbon–carbon multiple bonds, namely, the hydroboration reaction,[37] has proven to be one of the most versatile routes to organoborane compounds (Equations (2) and (3)). Excellent reviews of this area have already appeared;[1,30,38,39] this section will focus only on some representative examples and on the most recent developments in this field. Asymmetric hydroboration via chiral boranes is discussed in Section 5.4.1.

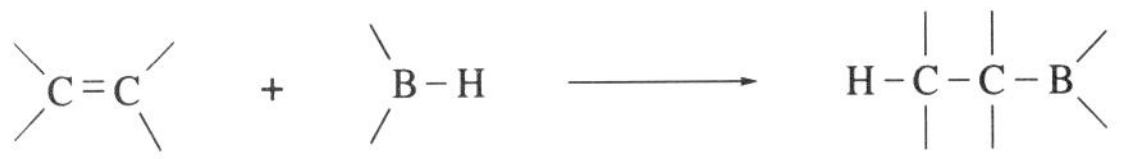

$$\text{C=C} \quad + \quad \text{B–H} \quad \longrightarrow \quad \text{H–C–C–B} \qquad (2)$$

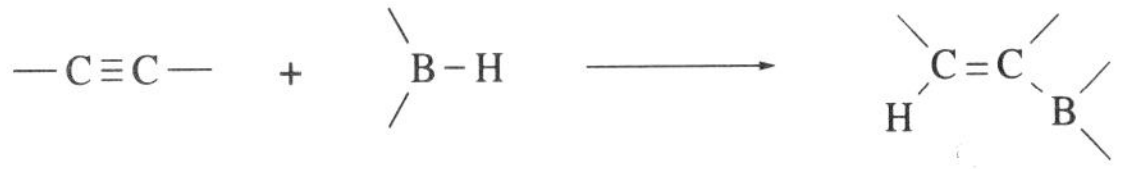

$$\text{–C≡C–} \quad + \quad \text{B–H} \quad \longrightarrow \quad \text{C=C(H)(B)} \qquad (3)$$

Hydroboration exhibits the following general characteristics: (i) addition occurs in a *cis*-fashion; and (ii) the boron atom preferentially goes to the less sterically crowded site. For example, a single enantiomer is produced from (α)-pinene[40] (Equation (4)). Such complete selectivity is, however, unusual and the stereochemical outcome of these reactions will be generally governed by steric and electronic influences.

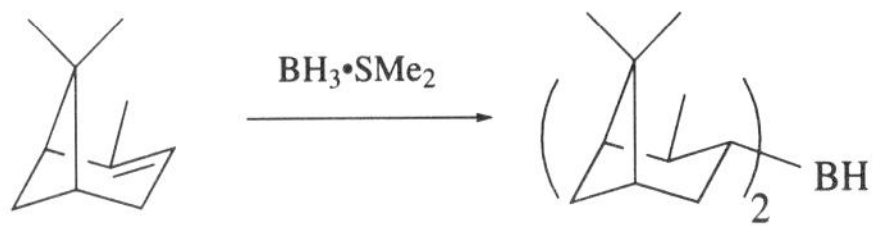

$$\text{(α-pinene)} \xrightarrow{BH_3 \cdot SMe_2} \text{(pinene)}_2 BH \qquad (4)$$

(i) Hydroboration with diborane and borane complexes

Earlier reactions were generally performed with *in situ* generated diborane. Commercially available borane–Lewis base complexes are now preferred (Figure 2). The hydroboration characteristics of all these reagents are broadly comparable. Alkenes give first monoalkylboranes, then dialkyl- and finally trialkylboranes. The steric hindrance around the double bond reduces the rates of the second and third steps and allows, in some cases, the clean formation of di- or monoalkylboranes (Equations (5)–(7)). Intramolecular complexation of boron by an heteroatom induces monohydroboration (Equations (8) and (9)).[41,42] For simple acyclic and cyclic alkenes, regioselectivity, which is evaluated in most cases after oxidation (see Section 5.3.1.2), is usually dominated by steric effects (Figure 3).[43] It should be noted that some of these values are the average of successive hydroboration steps, the regioselectivity increasing from the first to the last stage. Proximate functional groups can also exert powerful influences on regio- and stereochemistry. For example, hydroboration of various alkenylsilanes revealed important directive effects of silicon. An increased electron-withdrawing ability of the silyl group enhances placement of the boron atom in the β-position to silicon (Figure 4).[44,45] The hydroboration of a variety of allylsilanes has also been reported.[46] Similar effects are observed with tin.[47]

$Me_2S·BH_3$ $THF·BH_3$ $R_3N·BH_3$ $R_3P·BH_3$

Figure 2 Some common sources of borane.

Figure 3 Regioselectivity in the addition of $BH_3·THF$.

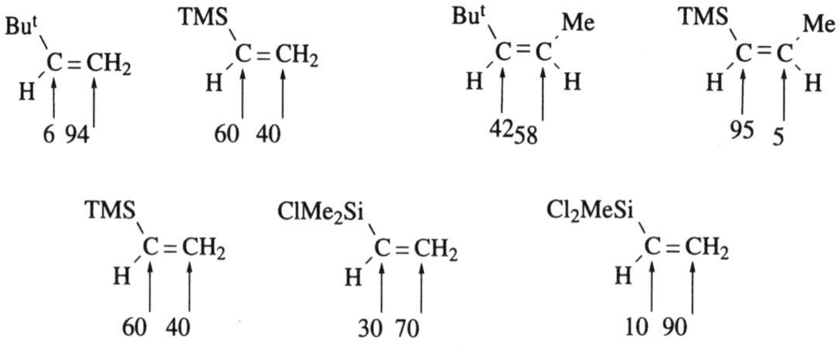

Figure 4 Percentage boron addition with $BH_3·THF$.

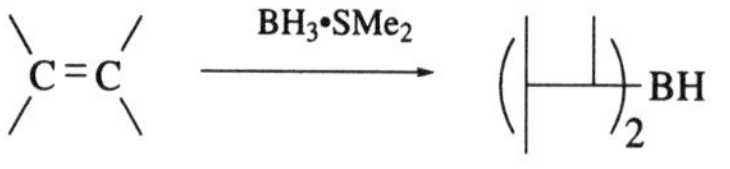

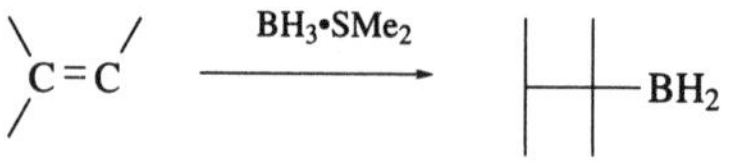

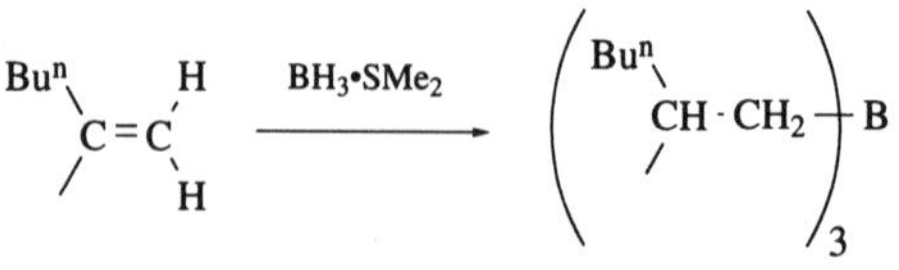

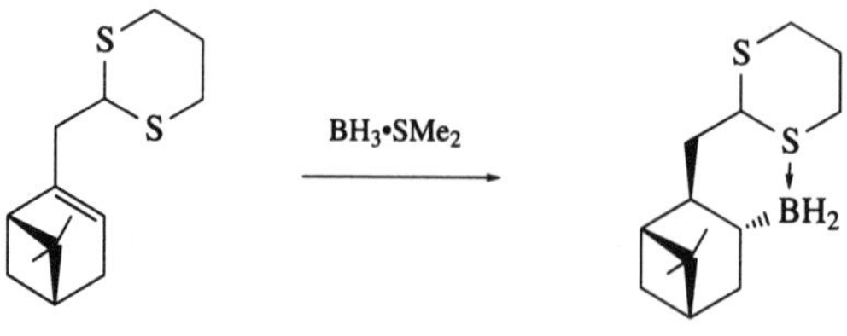

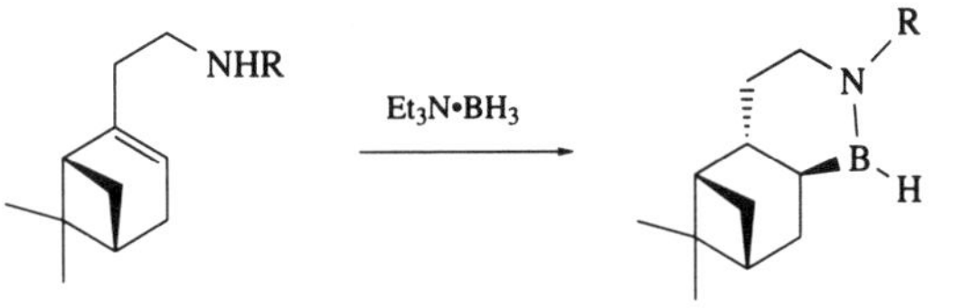

Substrate-controlled hydroboration of allyl systems has been explored.[48] Acyclic secondary allylic alcohols yield, after oxidation, predominantly *anti*-1,3-diols (Equation (10)).[49] The outcome depends on the degree of substitution, and the most sterically demanding boranes afford better stereoselectivity. It can be reversed by the use of a catalyst[50] or a vinyl ether.[51] Hydroboration of allylamines and derivatives parallels the reaction of allyl alcohols and ethers.[52–6] Internal alkenes with an allylic nitrogen-bearing stereocenter undergo regio- and steroselective hydroborations. A 1,3-strain has been proposed to be a dominant stereocontrol element (Equation (11)).[57] α-Alkoxy-β,γ-unsaturated esters undergo an alkoxy-directed hydroboration with useful levels of diastereoselectivity favoring the *anti*-diastereoisomer (Scheme 1).[58]

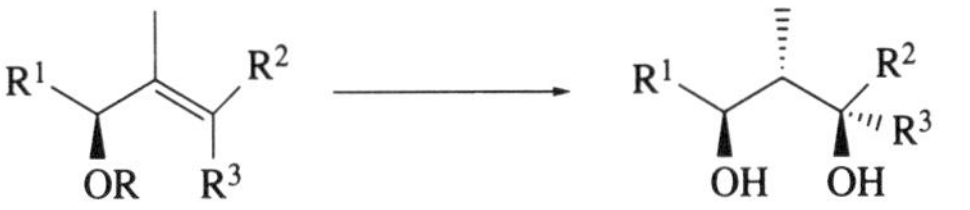

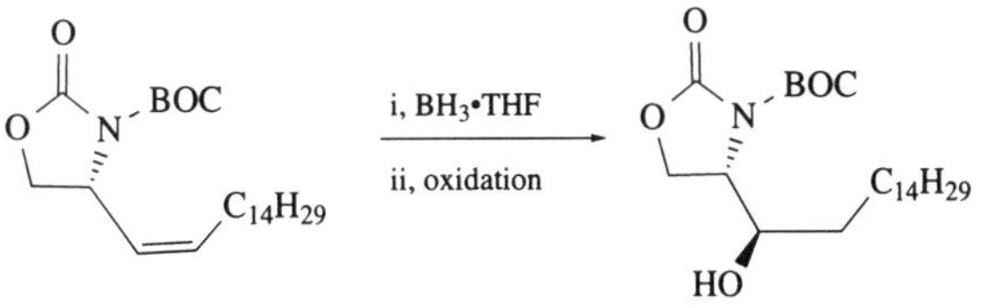

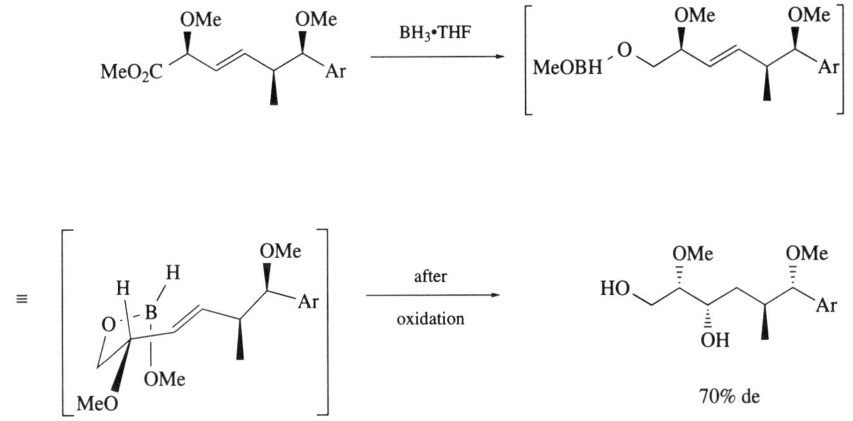

Scheme 1

When hydroboration places the boron atom in a position β to a potential leaving group (halogen, OR, NR$_2$), elimination occurs to afford the corresponding alkenes (Equation (12)).[59–61] An unusual rearrangement was observed in the hydroboration of β,β-disubstituted enamines (Equation (13)).[62]

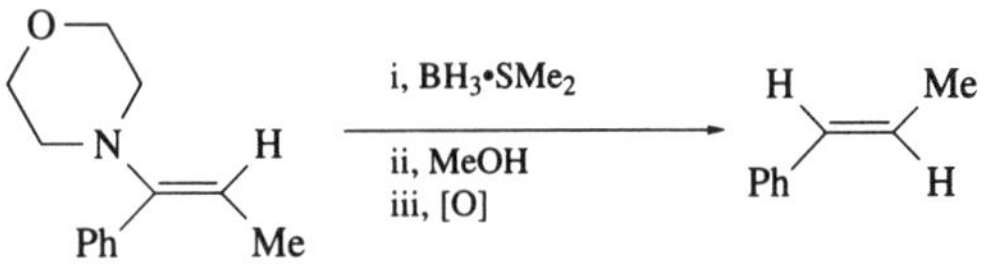

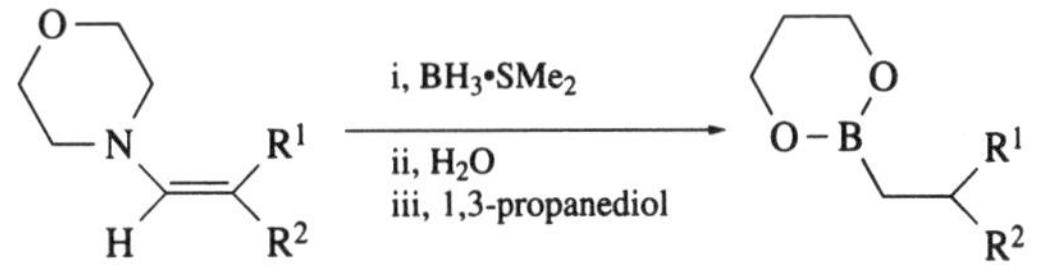

Hydroboration of functional alkenes is rarely restricted to the use of BH_3. The same trends are observed with other boranes, which are mostly employed to improve the regio- and stereochemical outcomes of the reactions. Only special cases of behavior will be reported in the following sections. More extensive accounts of functional alkene hydroborations are reported elsewhere.[1,26]

Alkynes usually give complex reaction mixtures with BH_3. Predominant formation of (E)-trialkenylboranes is often accompanied by the formation of 1,1-diboryl derivatives (Equation (14)).[63]

$$R-C\equiv C-H \xrightarrow{BH_3} R-CH_2-CH\begin{matrix}B\diagup\\ \diagdown B\end{matrix} \qquad (14)$$

The hydroboration of 1,n-dienes with $BH_3 \cdot THF$ or $BH_3 \cdot SMe_2$ can also lead to complex mixtures of regioisomers, cyclic and open-chain products, and oligomeric species. However, under controlled conditions this approach has been used successfully to prepare important dialkylboranes such as 9-BBN–H[64] and 3,5-dimethylborinane (Equations (15) and (16)).[65]

$$\text{(cyclooctadiene)} \xrightarrow[\text{ii, heat}]{\text{i, } BH_3 \cdot THF} \text{9-BBN}-B-H \qquad (15)$$

$$\text{(2,4-dimethyl-1,4-pentadiene)} \xrightarrow[\text{ii, heat}]{\text{i, } BH_3 \cdot THF} \text{3,5-dimethylborinane-BH} \qquad (16)$$

Other sources of BH_3 have been also used for hydroborations including lithium borohydride/ethyl acetate,[66] sodium acetoxyborohydride,[67] sodium borohydride in combination with transition metal halides,[68,69] or benzyltriethylammonium borohydride in the presence of chlorotrimethylsilane[70] or titanium tetrachloride.[71] Alcohols are obtained regio- and stereoselectively in good yields by electrolysis of sodium borohydride in the presence of alkenes followed by treatment with alkaline hydrogen peroxide.[72]

(ii) Hydroboration with monosubstituted boranes

Monosubstituted boranes are mainly prepared by:
(a) hydroboration of hindered alkenes with borane complexes ($ThexBH_2$[73]);
(b) successive treatment of boronic esters with LAH and hydrogen chloride or trimethylsilyl chloride (RBH_2; where R = Me,[74] 2,4,6-triisopropylphenyl[75] ($TripBH_2$)); or
(c) displacement of 2,3-dimethyl-2-butene after addition of TMEDA or NEt_3 ($IpcBH_2$[76]).

Monohaloborane–dimethyl sulfide complexes are obtained via redistribution reactions from $BH_3 \cdot SMe_2$ and BX_3 (Figure 5).[77]

Except for $H_2BX \cdot SMe_2$, these boranes exist primarily as dimers in THF in equilibrium with variable amounts of monomers. They are not stable for long periods and tend to redistribute to give mixtures of R_2BH, R_3B and BH_3. The presence of two boron–hydrogen bonds in monosubstituted boranes usually induces the formation of trialkylboranes by consumption of two equivalents of alkenes. However, alkenes with medium steric requirements can be converted to the corresponding monoalkylthexylborane by reaction with thexylborane, whereas unhindered alkenes give thexyldialkylboranes (Scheme 2).[73,78]

$MeBH_2$ and $TripBH_2$ are the only monoalkylboranes capable of clean monoaddition to 1-alkenes. This allows the preparation of totally mixed trialkylboranes (Equation (17)).[74,75,79] $H_2BX \cdot SMe_2$ (X = Cl, Br, I) will hydroborate alkenes rapidly and quantitatively to give the corresponding dialkylhaloboranes (Equation (18)).[80,81] With monosubstituted boranes, regioselectivities are usually higher than those observed in BH_3 reactions, with a quite surprising result in the case of unhindered methylborane (Figure 6).[73,74,80] Monosubstituted boranes have definite advantages over BH_3 in the hydroboration of dienes since they give rise to much simpler products and in several cases, after isomerization of the initial products, afford single cyclic compounds (Equation (19)).[82,83]

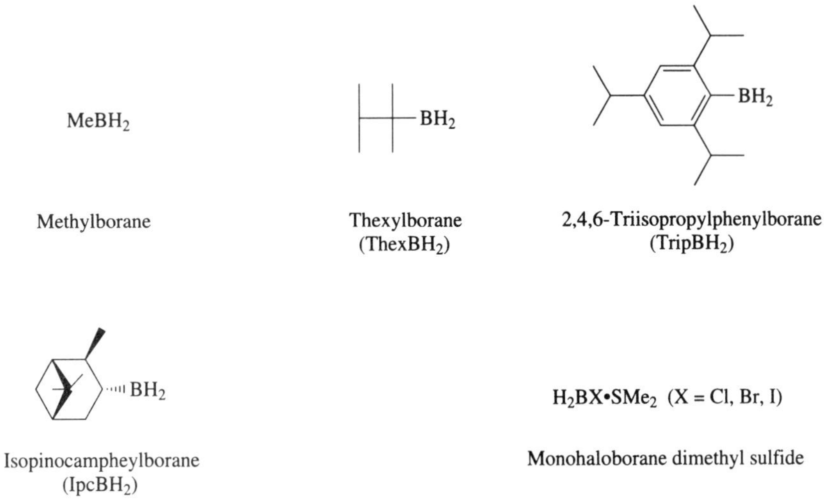

MeBH₂

Methylborane

⊢BH₂

Thexylborane
(ThexBH₂)

2,4,6-Triisopropylphenylborane
(TripBH₂)

Isopinocampheylborane
(IpcBH₂)

$H_2BX \cdot SMe_2$ (X = Cl, Br, I)

Monohaloborane dimethyl sulfide

Figure 5 Main monosubstituted boranes.

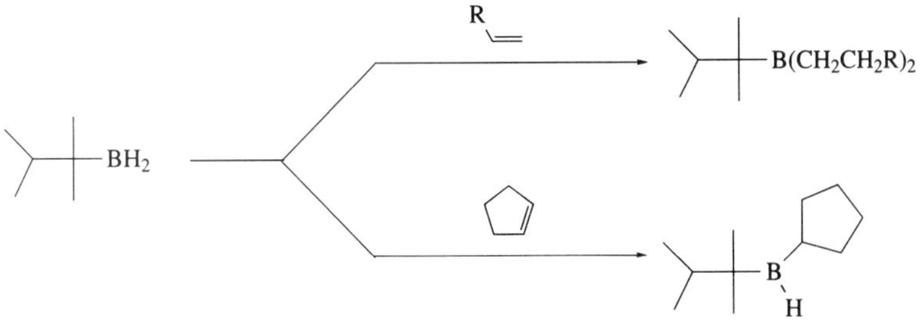

Scheme 2

$$MeBH_2 \xrightarrow{\text{i, alkene A}}_{\text{ii, alkene B}} MeBR^AR^B \qquad (17)$$

$$2 \text{ alkene} + H_2BX \cdot SMe_2 \longrightarrow R_2BX \cdot SMe_2 \qquad (18)$$

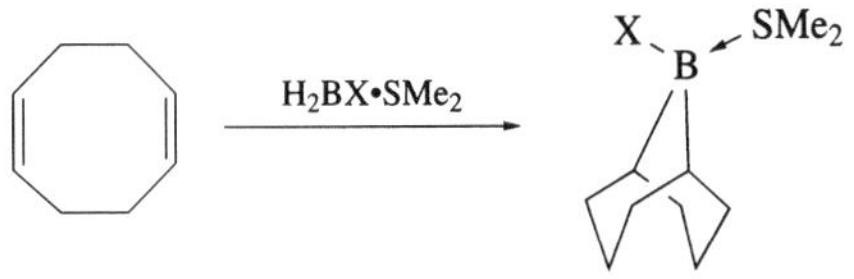

$$\text{(cyclooctadiene)} \xrightarrow{H_2BX \cdot SMe_2} \text{(B-bicyclic with X, SMe}_2\text{)} \qquad (19)$$

	ThexBH₂	MeBH₂	TripBH₂	H₂BX
Bu^n ←	<6	1.5	0.6	<0.8
←	>94	98.5	99.4	>99.2

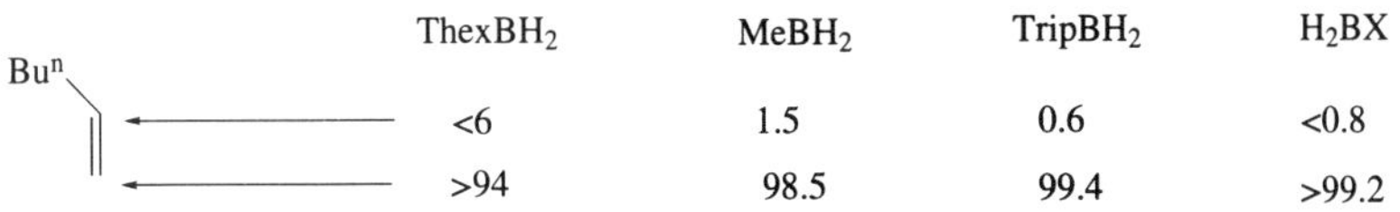

Figure 6 Percentage boron atom addition.

Dialkenylchloroboranes are cleanly produced in ether from internal alkynes whereas monohydroboration was observed in THF (Equations (20) and (21)).[81]

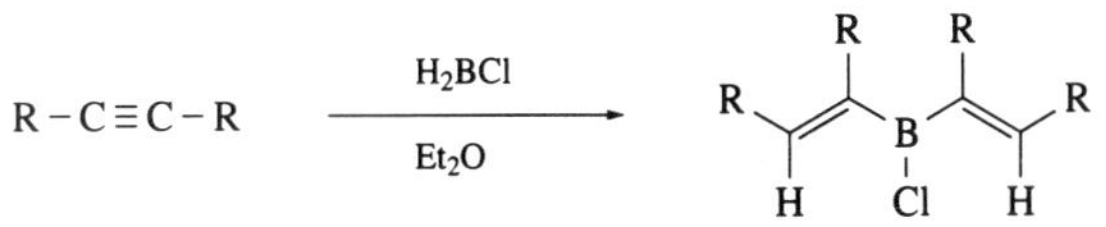

$$R-C \equiv C-R \xrightarrow[Et_2O]{H_2BCl} \underset{H \quad Cl \quad H}{\overset{R \quad R}{\text{(divinylchloroborane)}}} \qquad (20)$$

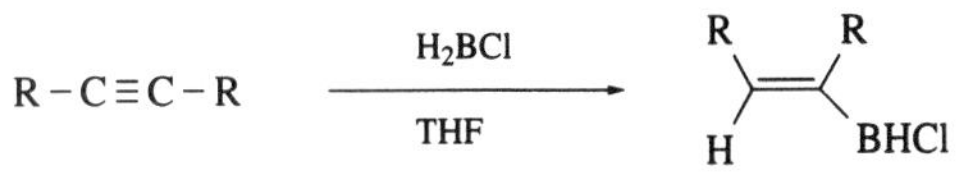

$$R-C \equiv C-R \xrightarrow[THF]{H_2BCl} \underset{H \quad BHCl}{\overset{R \quad R}{\text{(vinylborane)}}} \qquad (21)$$

(iii) Hydroboration with disubstituted boranes

A wide variety of disubstituted boranes have been used as hydroborating agents. They are prepared *via* different routes:
(a) hydroboration of alkenes or dienes (Sia$_2$BH;[27] dicyclohexylborane (Chx$_2$BH);[27] Ipc$_2$BH;[40] 9-BBN–H;[64] thexylchloroborane[84,85]);
(b) redistribution reactions (HBX$_2$·SMe$_2$, where X = Cl, Br, I[77]);
(c) treatment of a diol with BH$_3$·SMe$_2$ (catecholborane,[27] pinacolborane[86]);
(d) reduction of a haloborane (uncomplexed HBCl$_2$;[87] R$_2$BH, dimesitylborane[27,88]) (Figure 7).

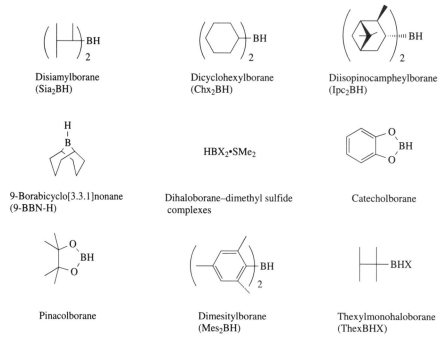

Figure 7 Main disubstituted boranes.

9-BBN-H is a useful hydroborating agent which exhibits high regio- and stereoselectivity and remarkable sensitivity to the structural features of alkenes. Disiamyl- and dicyclohexylborane show very similar characteristics (Figure 8).[27]

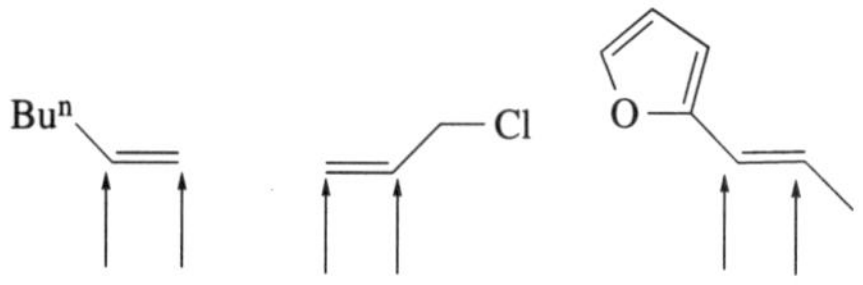

	Bu^n		Cl		furan	
BH$_3$•THF	6	94	60	40	91	9
Sia$_2$BH	1	99	95	5	95	5
9-BBNH	0.1	99.9	98.9	1.1	96	4

Figure 8 Percentage boron atom addition.

Stereoselective 1,4-addition of dialkylboranes to α,β-unsaturated ketones represents a simple approach to enol borinates (Equation (22)).[89,90]

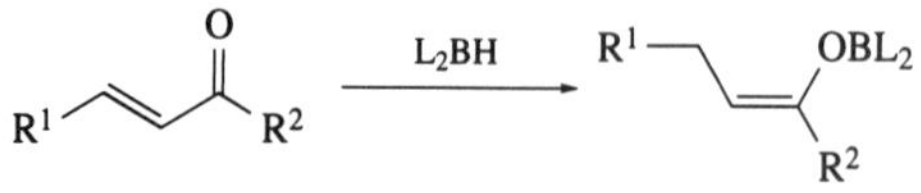

$$R^1\text{-CH=CH-C(O)-}R^2 \xrightarrow{L_2BH} R^1\text{-CH}_2\text{-CH=C(OBL}_2\text{)-}R^2 \tag{22}$$

Simple dialkoxyboranes (RO)$_2$BH readily undergo disproportionation, except for pinacolborane which is more stable and hydroborates alkenes in high yield.[86] Catecholborane, which can be distilled without noticeable decomposition, has been more deeply investigated. It provides a useful route to functionalized substituted boronic esters with regio- and stereoselectivities closely related to those observed with disiamylborane.[27] Temperatures as high as 70–100 °C are sometimes necessary to hydroborate alkenes. Fortunately, Männig and Nöth demonstrated in 1985 that catalytic amounts of Wilkinson's catalyst, RhCl(PPh$_3$)$_3$ accelerated this reaction.[91] Advances in several other groups have now greatly expanded the scope of this new approach. In addition to rate enhancement, transition metal-promoted hydroboration shows complementary chemo-, regio- and stereoselectivities (Scheme 3). A review of this promising area appeared in 1991.[92] For all of these reactions, the exact catalyst composition is crucial and recent reports underline the complexity of the Wilkinson's catalyst–catechol hydroboration system.[93–100] More fundamental studies of transition metal–boryl complexes will be necessary for a better understanding of the exact mechanisms of these processes. The enantioselective version with optically active rhodium catalysts is discussed in Section 5.4.1.

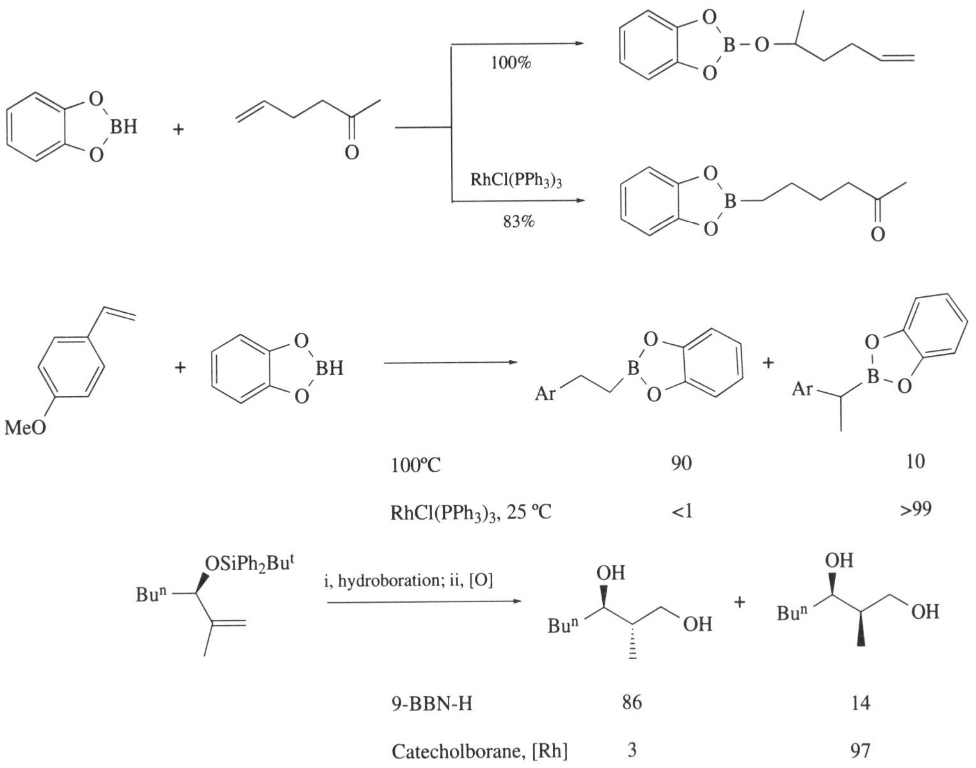

Scheme 3

Reactions of dihaloboranes (HBX$_2$, where X = Cl, Br, I) with alkenes have been investigated in detail.[101] HBBr$_2$·SMe$_2$ is a synthetically important reagent used for the regioselective preparation of dibromoalkylboranes, which are precursors of several other interesting boranes (Scheme 4).[102,103]

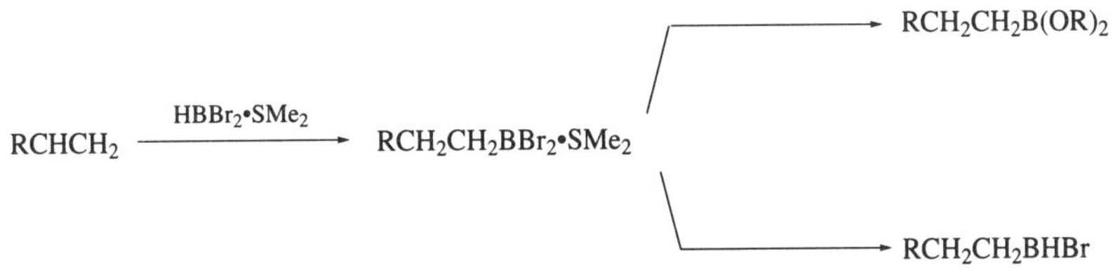

Scheme 4

Although several monoalkylhaloboranes have been prepared, thexylmonochloroborane has been almost the only reagent to receive detailed investigation. It shows great sensitivity to steric factors and regioselectivities are comparable to those of 9-BBN–H.[84,85] Synthesis of mixed triorganylboranes from monoalkylhaloboranes have been reported (Scheme 5).[104,105]

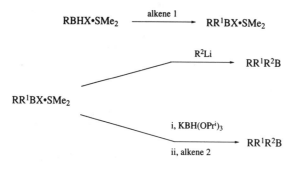

Scheme 5

Directive effects in the hydroboration of heterocyclic alkenes have been reviewed recently,[106] while hydroboration of allenes with disubstituted boranes has found useful application in the preparation of some allylboranes (Equation (23)).[107–10]

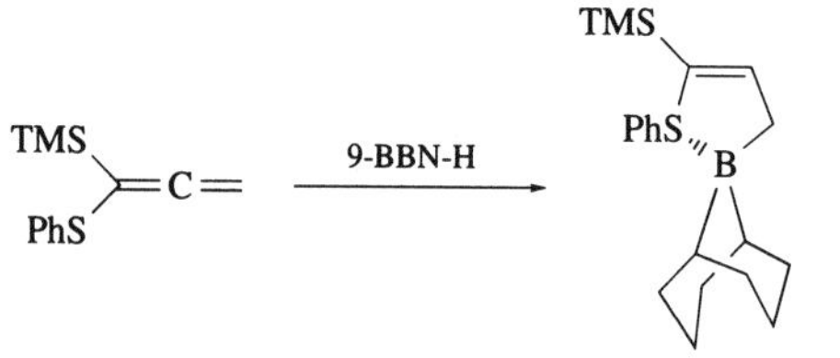

$$\text{TMS}\diagup\!\!\!\!\diagdown\text{PhS}\!=\!C\!=\quad\xrightarrow{\text{9-BBN-H}}\quad \text{(allylborane product)} \qquad (23)$$

Alkynes are readily converted to the corresponding vinylboranes under appropriate conditions which prevent possible dihydroboration reactions (Scheme 6).[86,87,111–26] Monohydroboration of 1-haloalkynes with dialkylboranes,[127] alkylmonohaloborane,[128,129] and dibromoborane–dimethyl sulfide[130] occurs regio- and stereoselectively to give the corresponding alkenylboranes, which are useful synthetic intermediates, as illustrated in Scheme 7. Hydroboration of thioacetylenes with catecholborane in the presence of a nickel catalyst allows a complete reversal of the regioselectivity (Equations (24) and (25)).[119]

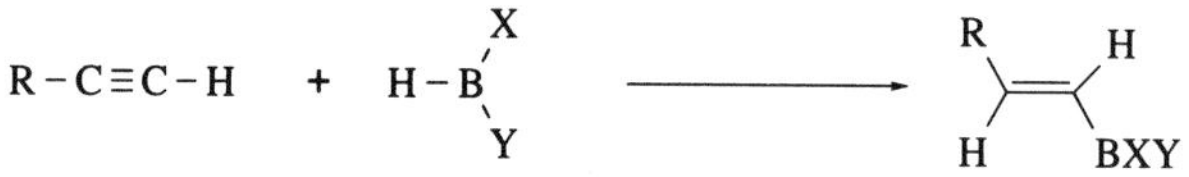

$$R-C\equiv C-H \;+\; H-B{\diagup X \atop \diagdown Y} \longrightarrow \begin{array}{c} R \quad H \\ \diagup\!\!\!=\!\!\!\diagdown \\ H \quad BXY \end{array}$$

HBXY = 9-BBN-H, dicyclohexylborane, catecholborane, dialkylborane, dihalogenoborane, pinacolborane, Ipc$_2$BH, thexylmonohalogenoborane

Scheme 6

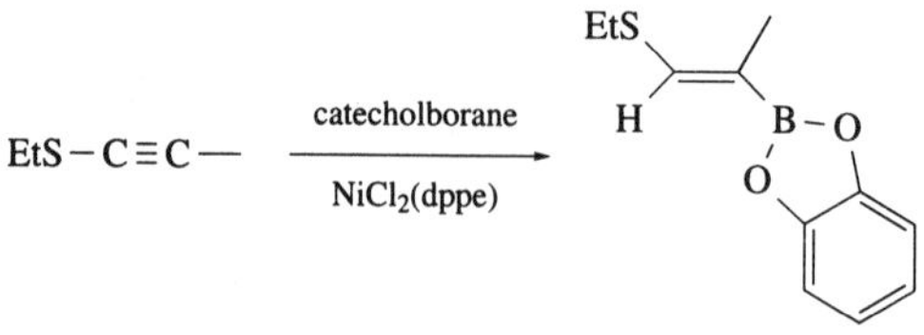

$$\text{EtS}-C\equiv C- \quad\xrightarrow[\text{NiCl}_2(\text{dppe})]{\text{catecholborane}}\quad \text{(product)} \qquad (24)$$

>99:1

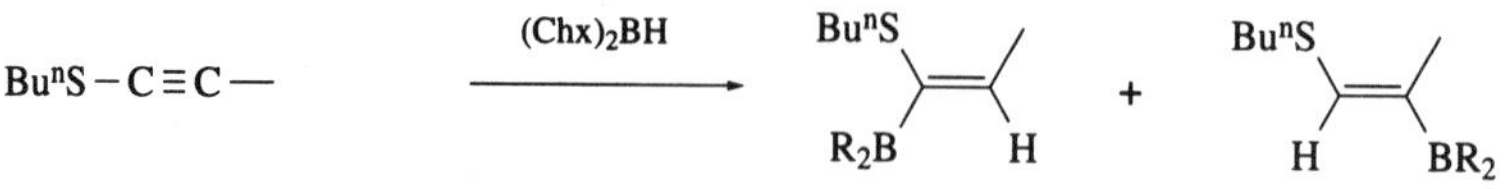

$$\text{Bu}^n\text{S}-C\equiv C- \quad\xrightarrow{(\text{Chx})_2\text{BH}}\quad \text{(product A)} \;+\; \text{(product B)} \qquad (25)$$

83 : 17

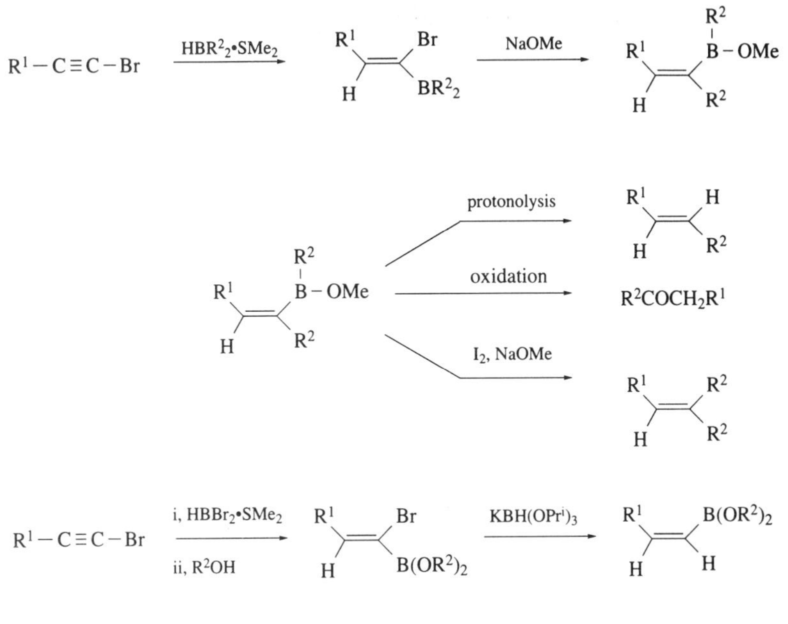

Scheme 7

Monohydroboration of 1,3-enynes with dialkylboranes or catecholborane furnished the corresponding 1,3-dienylboranes, which can be converted to various valuable synthetic intermediates.[131-7]

(iv) Mechanism

The mechanistic details of hydroboration reactions have intrigued experimental and theoretical chemists for many years and a number of reviews have appeared towards the end of the 1980s and in the early 1990s.[27,30,39] The hydroboration of alkenes with $BH_3 \cdot THF$ involves successive steps and is too complex to furnish useful kinetic data. By contrast, for 9-BBN–H and $HBBr_2$–SMe_2, the mechanisms are now well established. Prior dissociation of the 9-BBN–H dimer, with the assistance of a donor molecule or the dissociation of the haloborane–dimethyl sulfide complex, is followed by reaction of the free monomeric borane with the alkene. Molecular orbital calculations have been performed mainly on the reaction of BH_3 with ethene.[138] Simple substituted alkenes and boranes have also been studied.[139,140] It is generally accepted that a three-centre π-complex precedes the formation of a four-centre transition state with no or little activation energy. The exact geometry of transition structures is still open to discussion.

5.2.1.2 Haloboration

The haloboration reaction is the addition of a boron–halogen bond across carbon–carbon multiple bonds.[141] Although haloboranes react with alkenes, these reactions give rise to mixtures of compounds and have not yet been brought to the level of a useful synthesis of saturated dihaloboranes.[142] *B*-bromo or *B*-iodo-9-borabicyclo[3.3.1]nonane (*B*-X-9-BBN), tribromoborane, and some other haloboranes react readily with 1-alkynes through a Markovnikov *cis*-addition of the X–B moiety to C≡C bonds, leading to (*Z*)-2-halo-1-alkenylboranes (Equation (26)).[143] The reaction occurs stereo-, regio-, and chemoselectively at terminal triple bonds but not at internal triple bonds, or terminal or internal double bonds. The addition of *B*-Br-9-BBN to an equimolar mixture of 1-octyne and 2-octyne gives only 2-bromo-1-octene after protonolysis, whereas 2-octyne remains unchanged. Both 1- and 2-octenes do not react with *B*-Br-9-BBN under the same conditions.[144] Functional groups such as ester or halogen tolerate the reaction conditions. *B*-I-9-BBN may also be used for the reaction, but not *B*-Cl-9-BBN, which is inert as a haloborating agent. Tribromoborane is also known to be a selective haloborating agent of 1-alkynes when used at low temperatures (−78 °C).[145,146] That (*Z*)-2-bromo- or 2-iodo-1-alkenylboranes are formed quantitatively, stereo-, and regioselectively, is proven by protonolysis and deuterolysis (Scheme 8).

$$RC{\equiv}CH + XBL_2 \longrightarrow RCX{=}CHBL_2 \qquad (26)$$

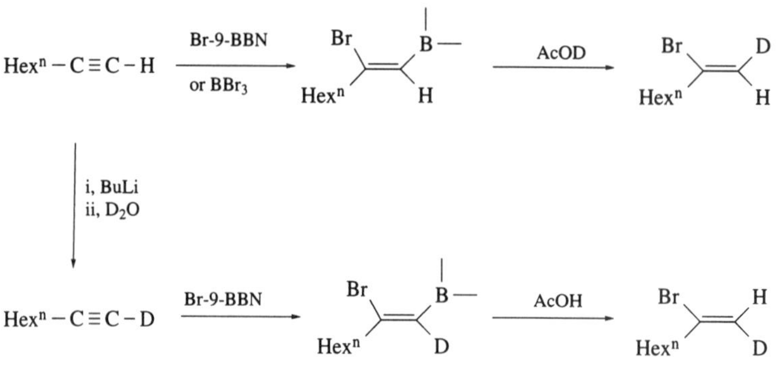

Scheme 8

When less reactive chloroboranes are used, mixtures of stereoisomers and sometimes of regioisomers are obtained.[147] The same problems are observed with 1-haloalkynes and B-Br-9-BBN.[148] Bromoboration with BBr_3 is kinetically controlled at $-80\,°C$, giving the *cis*-addition products, whereas thermodynamic stability favors the *trans* isomer.[145] The 2-halo-1-alkenylboranes formed in these reactions may be either isolated or engaged in further reactions as will be shown later.[149,150]

A simple synthesis of (2-bromoallyl)diphenoxyborane by the bromoboration of allene has been described recently (Equation (27)).[151] The primary adduct leads to the diphenoxyborane by treatment at low temperature with a slight excess of anisole, a method which is essentially neutral.

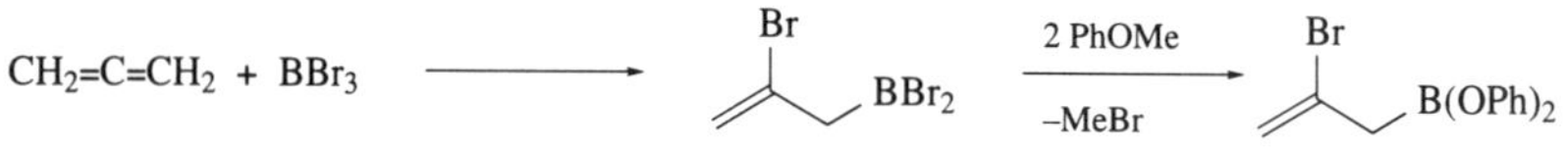

$$CH_2{=}C{=}CH_2 + BBr_3 \longrightarrow \text{(Br)CH}_2{=}C\text{-CH}_2BBr_2 \xrightarrow[-MeBr]{2\,PhOMe} \text{(Br)CH}_2{=}C\text{-CH}_2B(OPh)_2 \qquad (27)$$

5.2.1.3 Transmetallation

Hydroboration and related reactions are not the only methods available for the preparation of organoboranes. A number of alternative routes to a variety of organoboranes have been examined. Among these are the transmetallation reactions, which offer a broadly applicable route to organoboranes (Equation (28)). An excellent general insight into this area has been given by Pelter *et al*.[27] It appears that making well-founded and logical decisions about the optimum organometallic reagents and boron substrates for a given preparation is somehow difficult owing to the lack of comparative data available.

$$3\,R_nM + nBX_3 \longrightarrow nBR_3 + 3\,MX_n \qquad (28)$$

(i) From organoalkali and organomagnesium compounds

The replacement of an electronegative group bound to boron by an organic residue through the reaction with an organometallic reagent occurs as a two-stage process (Equation (29)). For this process to be useful, step (a) must be carried out efficiently and the product liberated (step (b)) subsequently. Otherwise, if the first intermediate is unstable and collapses to give a new borane, this compound may be reactive towards the organometallic reagent and further undesired reaction is likely. Much progress has been made in this area and good methods are now available to control steps (a) and (b), and therefore to prepare boranes from organolithium and organomagnesium compounds. Brown *et al*. reported that boronic esters can be prepared in excellent yield from the reaction of organolithium reagents with triisopropoxyborane. The boronic esters were liberated from the initial product by the addition of anhydrous HCl,[152-4] or acid chlorides,[155] or by pyrolysis,[155] and were isolated in good yields (Equation (30)). A variety of 1-alkynyldiisopropoxyboranes has been prepared in high yields from 1-lithio or 1-magnesio-1-alkynes.[156,157] This procedure has been extended to the preparation of borinic esters from Grignard reagents and triisopropoxyborane.[158] The use of other boron derivatives, including BF_3, BCl_3, $B(OMe)_3$, $FB(OMe)_2$, $ClB(OMe)_2$, $(EtO)_3B$, and $(Bu^iO)_3B$, led to mixtures of boronic and borinic esters, trialkylboranes and lithium tetraalkylborates, as did the use of Grignard reagents.[152,153] Chlorobis(diisopropylamino)borane seems to be the reagent of choice for the borylation of organomagnesium and organolithium reagents, giving boronic derivatives uniquely (Equation (31)).[159-62]

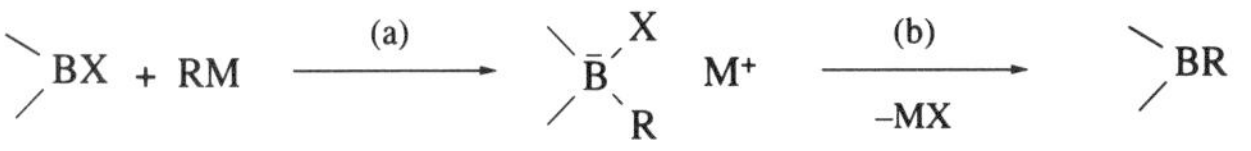

$$\ce{>BX + RM ->[(a)] >B^{-}(X)(R) M^+ ->[(b)][-MX] >BR} \quad (29)$$

$$\ce{(Pr^iO)3B + RLi ->[-78^\circ C][Et2O] RB(OPr^i)3Li ->[HCl] RB(OPr^i)2 + Pr^iOH + LiCl} \quad (30)$$

$$\ce{R^1MgX\ or\ R^1Li + ClB(NPr^i_2)2 -> R^1B(NPr^i_2)2} \quad (31)$$

R^1 = alkyl, aryl, alkynyl, N_2CH, $(R^2_2N)_2BCN_2$

Interestingly, a chiral borate can differentiate one isomer with high kinetic diastereoselectivity from a racemic mixture of the Grignard reagents, leading to one allylboronate with a 94% diastereomeric excess (Equation (32)).[163]

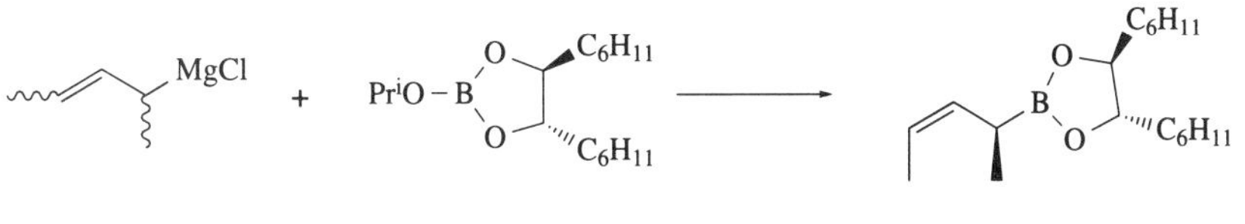

$$\text{(Equation 32)} \quad 94\%\,de \quad (32)$$

Arylboronic acids can also be prepared from BH_3 and arylmagnesium halides,[164,165] or by borylation with trimethoxyborane of functionalized aryllithium derivatives, which are directly accessible by aromatic-directed metallation,[166-8] or by lithium–iodine[169] or magnesium–bromine exchange.[170] Systematic studies of the synthesis of borinic esters from boronic esters and alkynyllithium,[171] alkenyl- and alkyllithium[172-4] and -magnesium[174] reagents have been reported.

Triorganylboranes are now easily prepared within a few hours in essentially quantitative yields in a rapid and efficient manner by the direct reaction of magnesium with an organic halide and boron trifluoride etherate in diethyl ether.[175,176] This procedure can be readily applied to the synthesis of large quantities of organoboranes. Ultrasound dramatically accelerates the slow reactions. The same reaction leads to tetraorganylborates in THF.[177] Triorganylboranes have also been prepared by the reaction of lithium (or magnesium) dialkylcuprates with B-Cl-9-BBN[178-80] or chlorothexylborane,[181] or of allylic organopotassium compounds with B-chloro reagents.[182] Aryldimesitylboranes are accessible by the reaction of dimesitylfluoroborane with aromatic Grignards.[183] Ethano-bridged triarylboranes were synthesized from the corresponding tribromoarenes and could be obtained in optically pure form (Equation (33)). This is the first example of optical resolution of triarylboranes.[184]

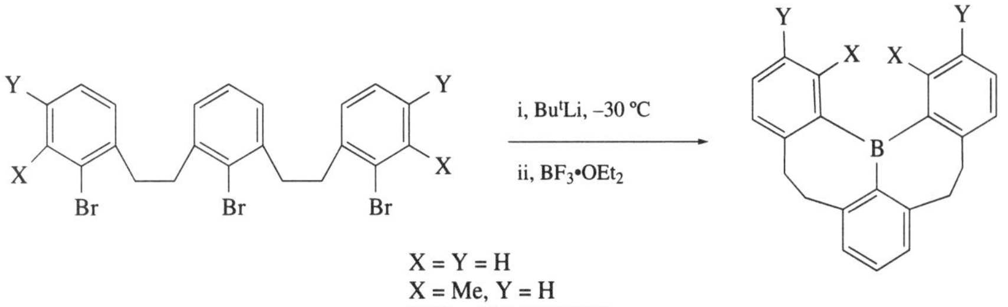

$$\text{(Equation 33)} \quad (33)$$

X = Y = H
X = Me, Y = H
X,Y = CHCHCHCH

Borylation of dicarbanions provides an efficient method for the synthesis of unsaturated boron–carbon ring systems. 2,5-Dihydro-1H-boroles can easily be obtained from magnesium butadiene $MgC_4H_6\,(THF)_2$ (Equation (34)).[185] These cyclic allylboranes easily react with aldehydes, ketones and ketenes to give new cyclic borinates.[186] A wider range of applications becomes possible when the dicarbanions used are generated from alkenes by metallation.[187-9]

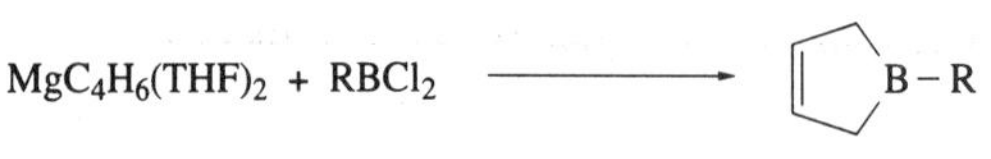

$$\text{MgC}_4\text{H}_6(\text{THF})_2 + \text{RBCl}_2 \longrightarrow \text{[cyclopentenyl]B-R} \qquad (34)$$

(ii) From other organometallic compounds

When, in Equation (28), M is an electropositive metal and X is a halide, reaction proceeds towards the formation of the borane. Advances have been made in this area, especially when M is silicon, tin and zirconium. Haubold *et al.* used exchange reactions between Ar-TMS and BX_3 (or R^1BX_2, R^1R^2BX or R^1_2BX, with X = Cl or Br) to prepare aryl and diarylhaloboranes.[190] This *ipso* borodesilylation has also been used by Snieckus and co-workers to prepare functionalized arylboronic acids.[167] The first asymmetric synthesis of a β-binaphthylborane was realized using this methodology;[191] dibromo and dichlorovinylboranes were prepared similarly (Equations (35) and (36)).[192]

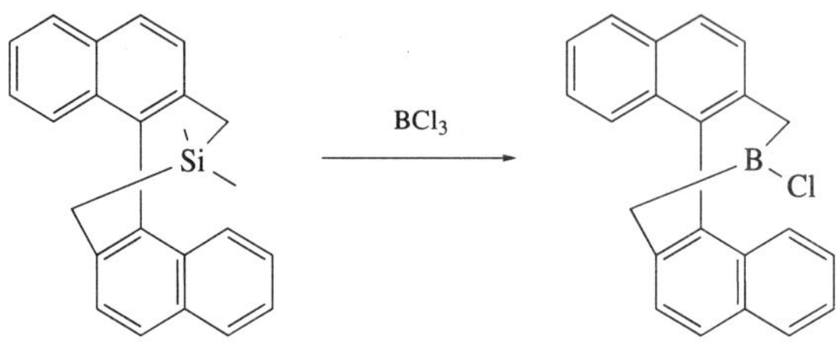

$$\text{[binaphthyl-Si(Me)}_2\text{]} \xrightarrow{BCl_3} \text{[binaphthyl-B-Cl]} \qquad (35)$$

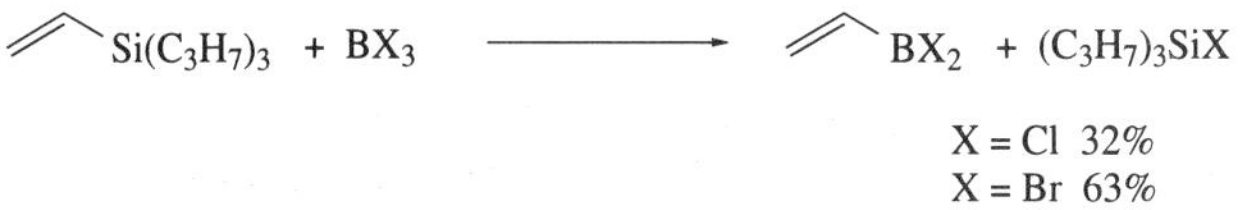

$$\text{CH}_2=\text{CH-Si}(C_3H_7)_3 + BX_3 \longrightarrow \text{CH}_2=\text{CH-BX}_2 + (C_3H_7)_3\text{SiX} \qquad (36)$$

X = Cl 32%
X = Br 63%

Vinyl-9-BBN and other dialkylvinylboranes are easily prepared by reaction of tributylvinyltin with a variety of dialkylbromoboranes,[193,194] and the first *C*-unsubstituted borepin was prepared from 1,1-dibutylstannepin and $MeBBr_2$.[195] The corresponding benzo-fused borepin was also obtained the same way.[196] Chiral allyl-, allenyl-, and propargylboranes were obtained from allyl-, allenyl-, and propargyltin derivatives respectively and the corresponding cyclic bromoborane (Equation (37)).[197,198]

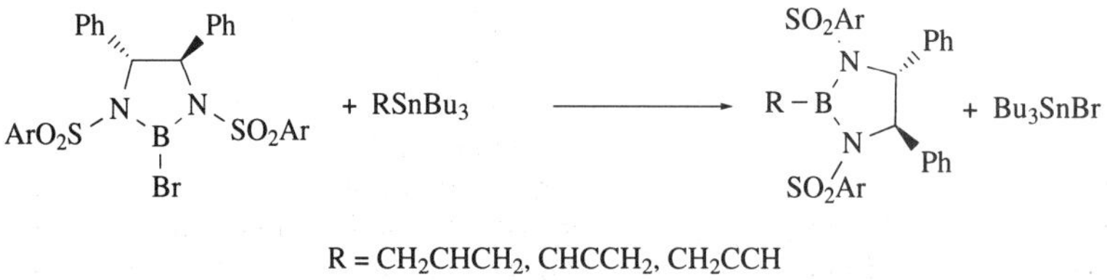

$$\text{[chiral diazaborolidine-Br]} + \text{RSnBu}_3 \longrightarrow \text{[chiral diazaborolidine-R]} + \text{Bu}_3\text{SnBr} \qquad (37)$$

R = CH_2CHCH_2, $CHCCH_2$, CH_2CCH

A dienyl or an alkenyl unit can be transferred from zirconium to boron. This process usually occurs in very high yield with retention of configuration. The combination of the facile hydrozirconation and unique isomerization reactions of organozirconium reagents with the well-established chemistry of boron expands the function of both metals for use in organic synthesis. Thus, dienyl-[199,200] and vinylboranes[201,202] are easily prepared. Cyclic boranes are accessible from zirconium metallacycles[203] and alkyl groups can also be transferred.[204,205]

5.2.2 From One Organoboron Compound to Another

5.2.2.1 Modification of substitution at boron

(i) Redistribution reaction

Pure unsymmetrical trialkylboranes are generally quite stable to redistribution below ca. 100 °C.[206] However, it is important to keep in mind that these reactions are catalyzed by B–H species and by electron-deficient organometallic compounds such as trialkylalanes. Redistribution between different kinds of boron compounds can have practical value if the equilibrium constant strongly favors the desired compound or if the addition of an excess of one reagent forces the reaction to completion (Equation (38)).

$$R_3B + BX_3 \rightleftharpoons R_2BX + RBX_2 \qquad (38)$$

Cyclic boronates or borinates can be cleanly obtained under stoichiometric conditions (Equations (39) and (40)),[207,208] and alkyldibromoboranes and dialkylbromoboranes can be synthesized in the same way (Equations (41) and (42)).[209]

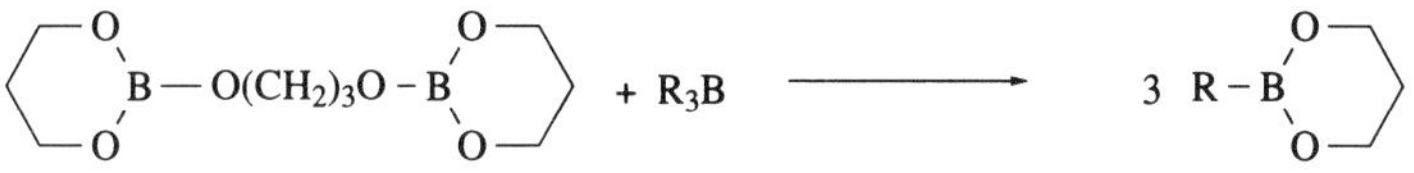

$$\text{(cyclic)}B-O(CH_2)_3O-B\text{(cyclic)} + R_3B \longrightarrow 3\ R-B\text{(cyclic)} \qquad (39)$$

$$2\ R_3B + B(OAr)_3 \longrightarrow 3\ R_2BOAr \qquad (40)$$

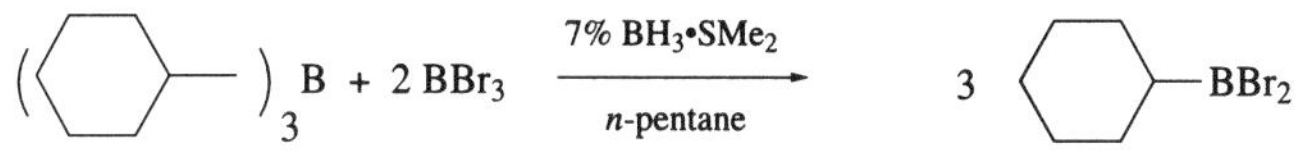

$$(C_6H_{11}-)_3B + 2\ BBr_3 \xrightarrow[n\text{-pentane}]{7\%\ BH_3\cdot SMe_2} 3\ C_6H_{11}-BBr_2 \qquad (41)$$

$$(C_6H_{11}-)_3B + BBr_3 \xrightarrow[n\text{-pentane}]{7\%\ BH_3\cdot SMe_2} 3\ (C_6H_{11}-)_2BBr \qquad (42)$$

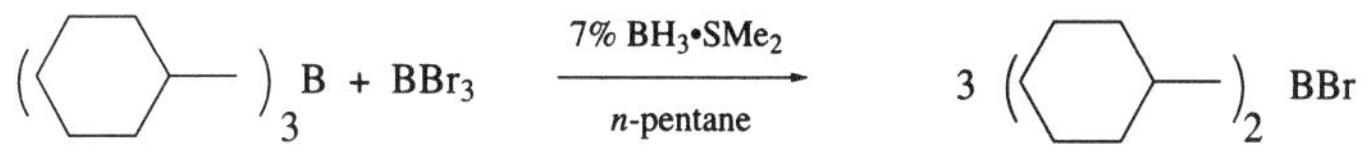

Enantiomerically pure diethyl alkylboronates, R*B(OEt)$_2$ are effectively converted into the corresponding chiral alkyldichloroboranes R*BCl$_2$ by treatment with boron trichloride in the presence of a catalytic amount of anhydrous FeCl$_3$.[210] A simple conversion of anhydrides of boronic and borinic acids to the corresponding organodihaloboranes and diorganohaloboranes by reaction with trihaloboranes in hydrocarbons has also been described.[211]

(ii) Alkyl- and alkenylboronic esters from refunctionalization of hydroboration products

Isopinocampheylboranes and diisopinocampheylboranes react with aldehydes and ketones to liberate α-pinene (Equation (43)).[212] Borinates, boronates and the corresponding acids are therefore easily accessible. Treatment with acetaldehyde in excess is commonly used because of its very easy removal.[212] Nevertheless, this reaction is sluggish and trichloroacetaldehyde or benzaldehyde in the presence of 5 mol.% BF$_3$·Et$_2$O seems to be more efficient.[213] Enantiomerically pure alkylborinates and boronates have been prepared along these lines, that is, asymmetric hydroboration (see Section 5.4.1) followed by refunctionalization.[76,214-17] Controlled treatment of B-alkyldiisopinocampheylboranes (Ipc$_2$BR*), obtained by asymmetric hydroboration of appropriate alkenes with aldehydes, produces kinetically resolved borinic esters. This has been developed into a simple and efficient *in situ* procedure for converting the initial hydroboration products of 81–96% ee to the corresponding boronic esters of ≥99% ee.[213] A new class of electron-deficient alkenes, that is, vinylboronates β-substituted by an electron-withdrawing group, has been obtained from the corresponding B-alkenyldiisopinocampheylboranes (Equation (44)).[123,218]

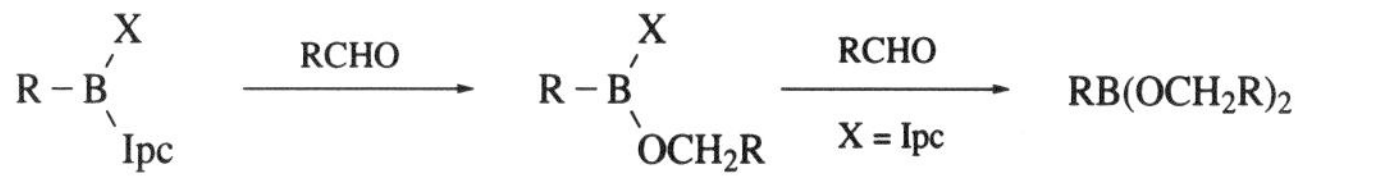

$$R-B(X)(Ipc) \xrightarrow{RCHO} R-B(X)(OCH_2R) \xrightarrow[X=Ipc]{RCHO} RB(OCH_2R)_2 \qquad (43)$$

Alkyl- and alkenyldibromoborane–dimethyl sulfide complexes, readily obtained by the hydroboration of alkenes and alkynes with HBBr$_2$·SMe$_2$, react with water, alcohols, or glycols to give the corresponding boronic acids or esters, respectively.[219]

(iii) Addition compounds of alkali-metal hydrides

Boronic and borinic esters readily react with LAH or monoethoxyaluminohydride at 0 °C to give the corresponding lithium monoorganylborohydrides[220] or lithium dialkylborohydrides,[221] respectively, in

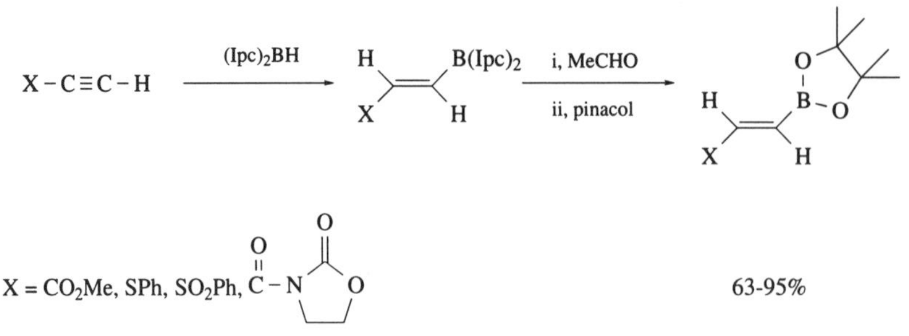

$$X-C\equiv C-H \xrightarrow{(Ipc)_2BH} \underset{X}{\overset{H}{>}}=\underset{H}{\overset{B(Ipc)_2}{<}} \xrightarrow[\text{ii, pinacol}]{\text{i, MeCHO}} \underset{X}{\overset{H}{>}}=\underset{H}{\overset{Bpin}{<}} \quad (44)$$

$$X = CO_2Me, SPh, SO_2Ph, \overset{O}{\underset{\parallel}{C}}-N\overset{O}{\underset{O}{\diagup}} \qquad 63\text{-}95\%$$

quantitative yields (Equations (45) and (46)). Under these reaction conditions, the dialkoxyalane precipitates quantitatively, leaving after filtration an aluminum-free solution of pure lithium borohydrides. These reactions are broadly applicable to the synthesis of a representative variety of very stable lithium monoalkyl- and dialkylborohydrides, eventually of 100% optical purity.[214] Lithium dialkylborohydrides are also quantitatively accessible from trialkylboranes and LAH.[222] The generation of active borane derivatives from the corresponding borohydrides can be realized with a variety of reagents including methyl iodide, trimethylsilyl chloride, trimethylsilyl methanesulfonate, and ethereal hydrogen chloride in solvents such as pentane, ether, and THF.[223] For example, methylborane and dimethylborane are cleanly liberated from the corresponding lithium borohydrides by protonation with hydrogen chloride in ether (Equations (47) and (48)).[224]

$$R^1B(OR^2)_2 + LAH \xrightarrow[0.25\text{ h}]{Et_2O\text{-pentane, }0\,°C} R^1BH_3Li + (R^2O)_2AlH \qquad (45)$$

$$R^1{}_2BOR^2 + LiEtOAlH_3 \xrightarrow[0.25\text{ h}]{Et_2O\text{-pentane, }0\,°C} R^1{}_2BH_2Li + (R^2O)EtOAlH \qquad (46)$$

$$2\text{ LiMeBH}_3 + 2\text{ HCl} \longrightarrow (MeBH_2)_2 + 2\text{ H}_2 + 2\text{ LiCl} \qquad (47)$$

$$2\text{ LiMe}_2BH_2 + 2\text{ HCl} \longrightarrow (Me_2BH)_2 + 2\text{ H}_2 + 2\text{ LiCl} \qquad (48)$$

A controlled and sequential hydroboration of alkenes with monoorganylboranes in THF and synthesis of mixed borinic esters and trialkylboranes can be achieved.[79] These methods are very convenient for the synthesis of chiral boranes.[216,217] The monalkylborohydrides, on treatment with 1, 2, or 3 equivalents of hydrogen chloride in ether, give monoalkylboranes,[223] monoalkylchloroboranes, or monoalkyldichloroboranes, respectively.[225]

(iv) Miscellaneous

A large number of possibilities exist for the transformation of organyldiaminoboranes and diorganylaminoboranes into other boranes by refunctionalization. It is beyond the scope of this article to discuss these transformations, which have been thoroughly reviewed elsewhere.[25] The coordination chemistry of boranes has gained importance in certain areas. Thus, monoethanolamine provides a simple means for precipitating various borinic acid-derived chelates from borinate solutions.[226] Boronates are also known to form chelate compounds with diethanolamine.[227,228] Since these crystalline chelates are stable in air, they have also been used to store borinates and boronates. More importantly, they have been used for upgrading borinate and boronate optical purities to values approaching 100% ee.[229] Various other chelates have been reported, including bicyclic organylboronic esters derived from iminodiacetic acids[230] and α-aminodiacids.[231] Strong intramolecular N–B coordination, which confers rigidity to these bicyclic structures, has been shown by ^{11}B and ^{13}C NMR spectroscopy.[230] A number of x-ray structure determinations have been performed on polycyclic structures, providing a wealth of structural information.[232–4]

5.2.2.2 Insertion of a –CXY group into a carbon–boron bond

In view of their synthetic interest, the reactions of organoboranes with carbenoid reagents bearing a potential leaving group have been particularly well documented since the early 1980s. These homologation processes have common initial steps in which an intermediate organoborate rearranges by a 1,2-migration (Equation (49)).

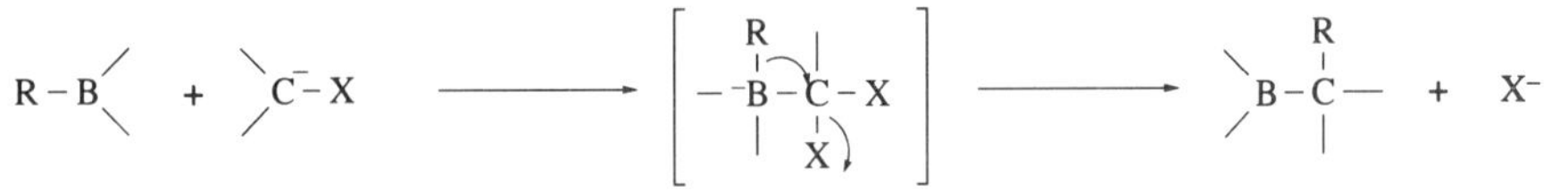

$$R-B\diagup + \diagup C-X \longrightarrow \left[-B-C-X \atop X \right] \longrightarrow B-C- + X^- \qquad (49)$$

The difficulties encountered in the generation of unstable carbenoids such as LiCCl$_3$, LiCHCl$_2$, or LiCH$_2$Cl, have been solved by using *in situ* procedures. In many instances, the initial rearrangement product was not isolated but directly converted to alcohols, ketones, or other functional derivatives. Boronic esters, whose stability generally allows rigorous purification, give homologated products which can be isolated and well characterized.[235,236] Their reactivity is only illustrated in this section and supplementary examples of –CXY insertions in the carbon–boron bond of other organoboranes can be found in Section 5.3.2.1(ii). Chiral applications, which constitute one of the most promising aspects of this chemistry, are treated separately in Section 5.4.2.

As previously mentioned, the addition of boronic esters to LiCHXY affords ate complexes which undergo clean transfer reactions to provide the desired α-organyl-substituted boronic esters (Scheme 9).[237–42] The preparation of allylboronic esters from (α-chloroallyl)lithium[243] (Equation (50)) and the successful ring enlargement of *B*-methoxyboracyclanes[244] (Equation (51)) are also noteworthy.

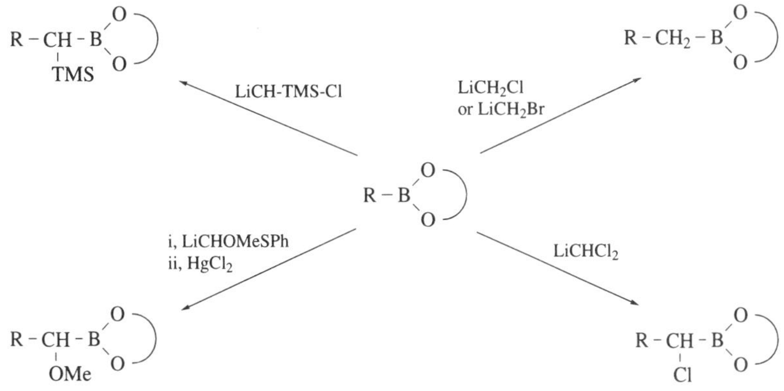

Scheme 9

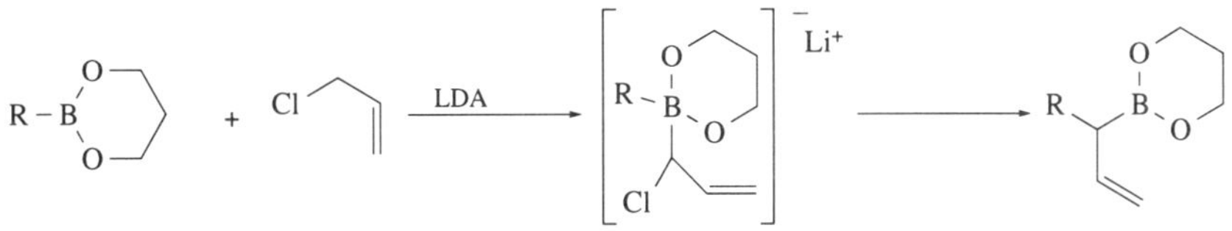

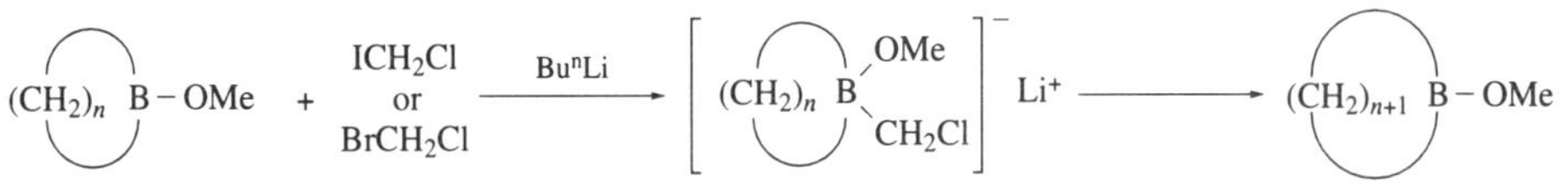

5.2.2.3 Reactions of α-haloboranes with nucleophilic reagents

As in the preceding section, only the reactions of α-haloboronic esters which afford stable isolable boron compounds are developed here. Other examples of such reactions with other organoboranes can be found in Section 5.3.2.1 and the asymmetric version with chiral boronic esters in Section 5.4.2.

One of the most convenient routes to α-haloboronic esters is the homologation reaction, but other approaches can also be used, in particular, the addition of lithium or Grignard reagents to dichloromethylboronic esters, Cl$_2$CHB(OR)$_2$ (*vide infra*). The mechanism for the reaction of an α-haloboronic ester with a nucleophile is very similar to the –CXY insertion described in Section 5.2.2.2. The formation of an ate complex is followed by a 1,2-migration from boron to the α-carbon center by displacement of a leaving group. This offers a convenient way to create a new C–H, C–C or C–heteroatom bond with, as has been proven in chiral series, very high stereoselectivity since the halogen atom is displaced with clean inversion (Equation (52)). A wide range of nucleophiles have been explored.[235,236] For example, amines,[245,246] ester enolates,[247,248] malonic ester anion,[249] and hydride[250] displace halogen from α-haloalkylboronic esters (Scheme 10), and stereodefined allylic boronic esters can be prepared from alkenyllithiums or alkenylmagnesium halides (Equations (53) and (54)).[251,252] This methodology is not restricted to the use of α-halomethylboronic esters. α-Hetero-substituted allylboronates,[253] 1-alkenylboronic esters,[130,254] 1,3-dienyl-2-boronic esters,[255] and boron analogues of carboxylic α-amino acids[3,19,20,235,236,256] are readily prepared via similar approaches (Scheme 11).

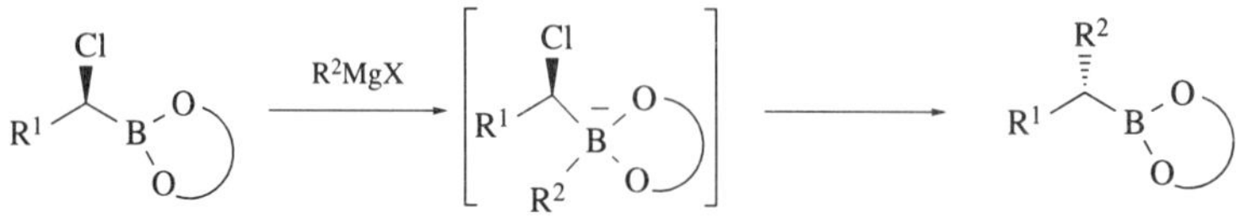

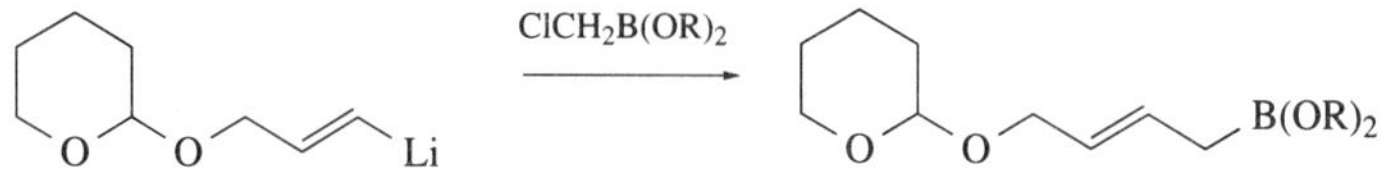

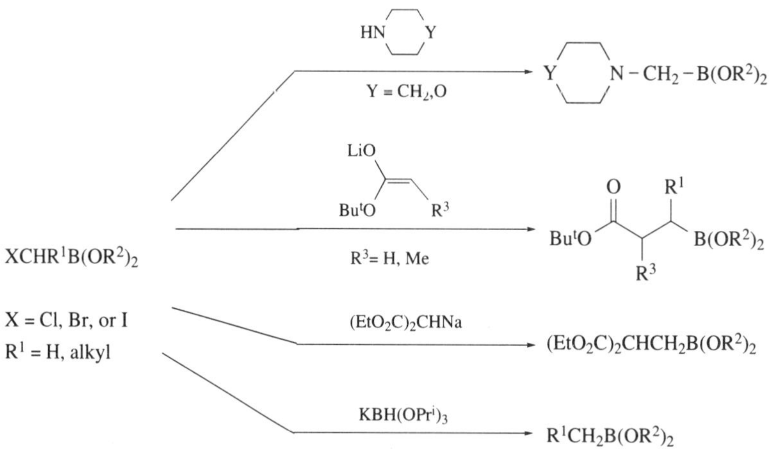

Scheme 10

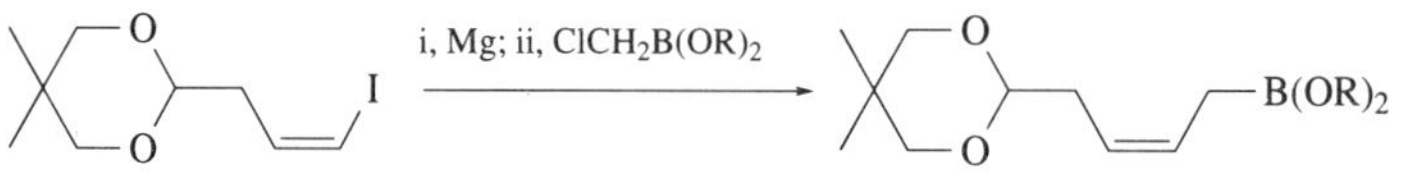

5.2.2.4 Reactions of boron-stabilized carbanions

This area has been reviewed several times.[3,27,257–61] Three methods are available for the production of boron-stabilized carbanions. The first is the selective cleavage of a carbon–heteroatom bond from an α-substituted organoborane such as 1,1-diborylalkanes, borylstannylmethanes, or boryltrimethylsilyl-methanes (Equations (55)–(58)). This first process is a mild and general one for the production of boron-stabilized carbanions. It suffers from a wasteful use of base and boron,[262–5] and from difficult access to *gem*-diboryl compounds within a chain and substituted *gem*-diboronates.[265] The addition of bases to borylstannylalkanes generally leads to cleavage of the tin–carbon bond, lithium thiophenoxide being the most efficient.[266] Desilylation of dimesitylboryl(trimethylsilyl)alkanes also gives α-borylcarbanions.[267] Boron-stabilized alkenyl carbanions are obtained in this way.[260,268]

The second method involves the deprotonation of simple alkylboranes, as shown in Equation (59). Rathke and Kow succeeded in the deprotonation of *B*-methyl-9-BBN with LiTMP.[269,270] This deprotonation process could compete with the essentially irreversible attack by base on the boron atom to form an "ate" complex, and the necessity of introducing a major element of steric hindrance into

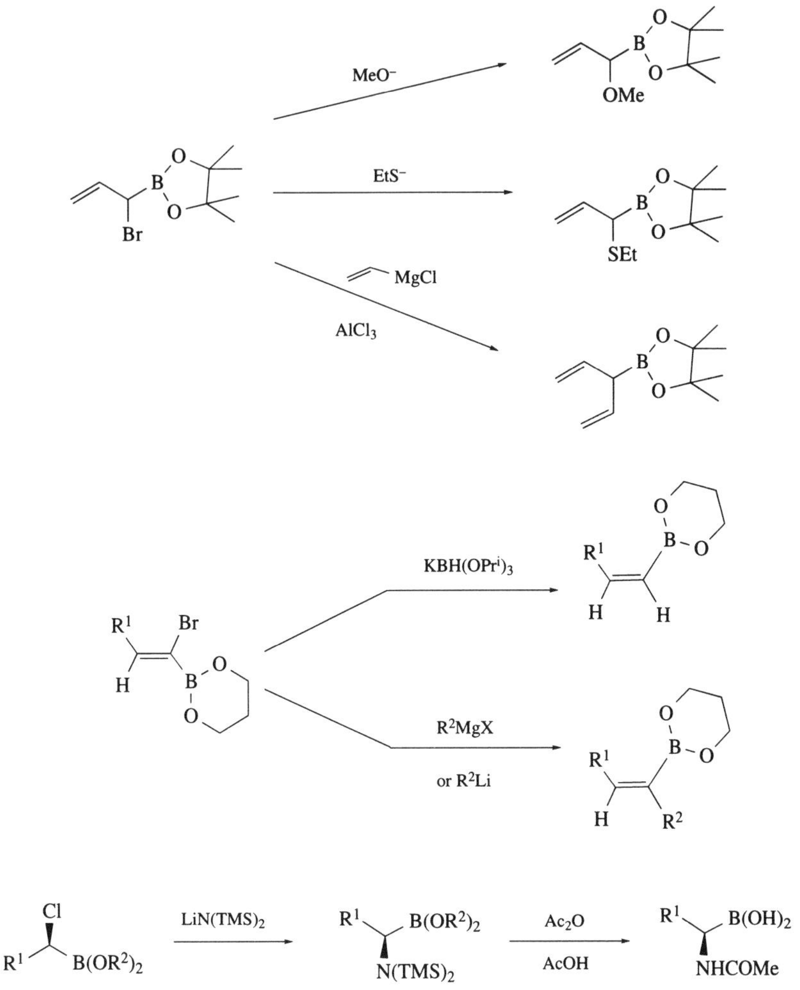

Scheme 11

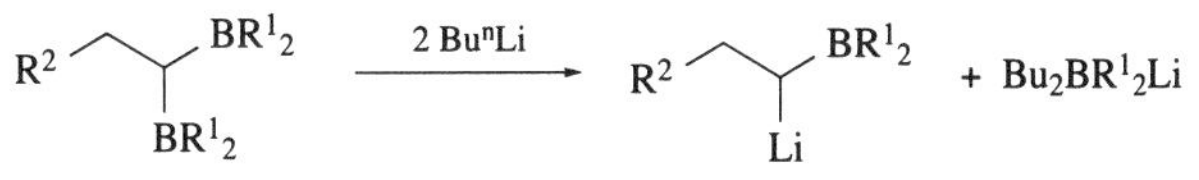

$$R^2\text{-CH}(BR^1_2)_2 \xrightarrow{2\,Bu^nLi} R^2\text{-C}(BR^1_2)(Li) + Bu_2BR^1_2Li \quad (55)$$

BR^1_2 = B(cyclohexyl)$_2$, B(Sia)$_2$, B(OR)$_2$, 9-BBN

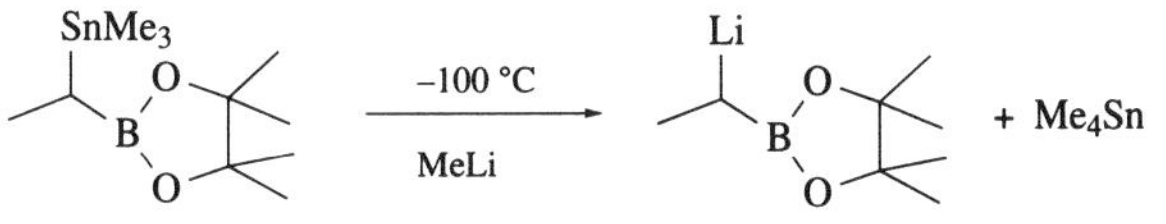

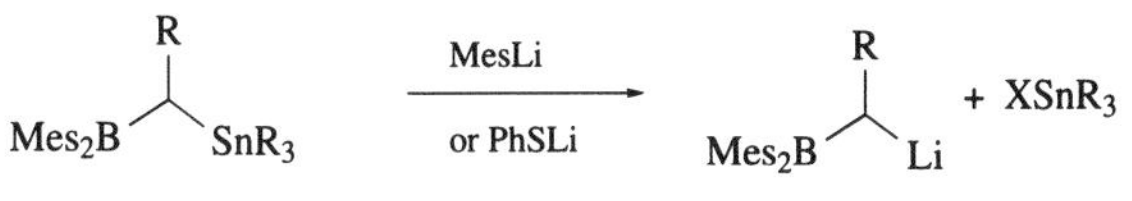

$$\text{(pinB)CH(SnMe}_3)\text{CH}_3 \xrightarrow[\text{MeLi}]{-100\,°C} \text{(pinB)CH(Li)CH}_3 + Me_4Sn \quad (56)$$

$$Mes_2B\text{-CR(SnR}_3) \xrightarrow{\text{MesLi or PhSLi}} Mes_2B\text{-CR(Li)} + XSnR_3 \quad (57)$$

$$Mes_2B\text{-CR(TMS)} \xrightarrow{\text{LiF}} Mes_2B\text{-CR(Li)} + \text{TMS-F} \quad (58)$$

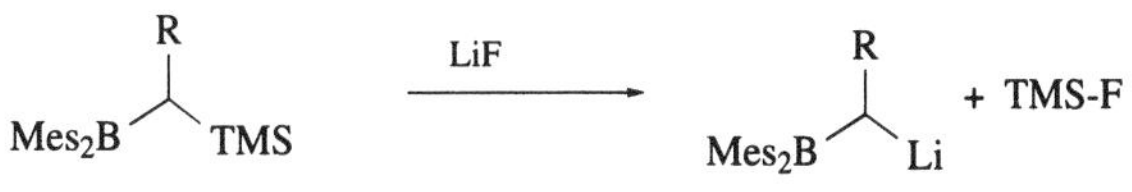

either the base, the borane or both was immediately recognized. Deprotonation with LiTMP is successful for boronates with X = Ph,[271] SPh,[272,273] TMS,[237,274] Ph$_3$P$^+$,[237] CH=CH$_2$,[237] and B(OR)$_2$[271] but failed with X = H[271] (Equation (60)), thus defining the structural limits for deprotonation of boronates. The dimesitylboryl group has been shown to be an excellent sterically hindered anion stabilizing group, as demonstrated first by Wilson.[275] Further studies have outlined the influence of the nature of the base used,[276] and shown that a variety of substituted anions Mes$_2$BCHLiX, in which X = SPh, TMS, BMes$_2$, are available by deprotonation of the corresponding boranes.[277] Allyl carbanions are also accessible by deprotonation of allylboranes.[278]

$$R^2\!\!-\!\!CH(BR^1{}_2)\!-\!\!R^2 + LiY \longrightarrow R^2\!\!-\!\!CLi(BR^1{}_2)\!-\!\!R^2 + YH \tag{59}$$

$$X\!-\!CH_2\!-\!B(OCH_2CH_2CH_2O) \xrightarrow[-100\,°C]{LiTMP} X\!-\!CHLi\!-\!B(OCH_2CH_2CH_2O) \tag{60}$$

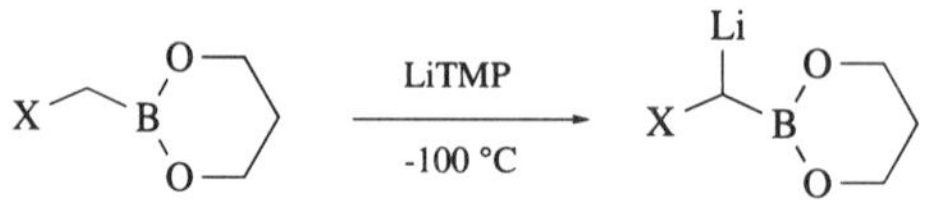

The third method consists of a conjugate addition to alkenylboranes (Equation (61)). A wide variety of organometallics have been added to 1-dimesitylboryl-1-trimethylsilylethylene.[279] Adducts were not obtained either with the lithium enolate of pinacolone or with lithium acetylide. β-Substitution is tolerated in the absence of allylic hydrogens.

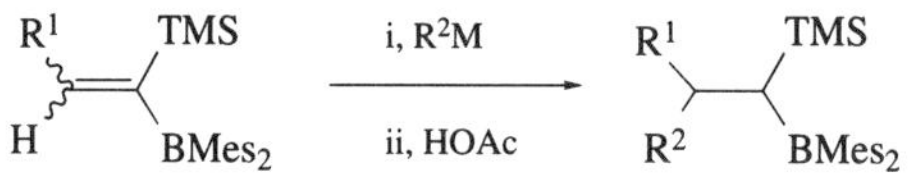

$$\underset{H}{\overset{R^1}{\diagdown}}\!C\!=\!C\underset{BMes_2}{\overset{TMS}{\diagup}} \xrightarrow[ii,\ HOAc]{i,\ R^2M} \underset{R^2}{\overset{R^1}{\diagdown}}\!CH\!-\!C\underset{BMes_2}{\overset{TMS}{\diagup}} \tag{61}$$

R^{2}M = Bu^{n}Li, Bu^{n}MgCl, Bu$^n{}_2$(CN)CuLi$_2$, PhLi, Bu^{t}Li, LiCH$_2$CO$_2$Bu^t

A general preparation of α-(dialkoxyboryl)alkylzinc and -copper organometallics from the α-bromo or -iodo derivatives has been reported (Equation (62)).[280] It is noteworthy that these organometallics are able to accommodate functional groups such as esters.

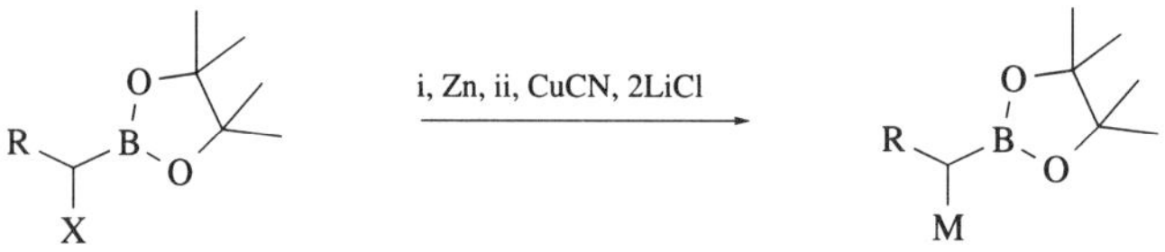

$$R\!-\!CHX\!-\!Bpin \xrightarrow{i,\ Zn,\ ii,\ CuCN,\ 2LiCl} R\!-\!CHM\!-\!Bpin \tag{62}$$

X = Br, I   M = ZnX, Cu(CN)ZnX

The reaction of α-boron-stabilized organometallics with electrophiles leads to new boranes (Equation (63)) which may be isolated, depending on the nature of the electrophilic reagent. Numerous carbanions derived from α-(phenylthio)boronic esters and efficiently generated in THF by deprotonation with lithium diisopropylamide (LDA) react with primary alkyl halides to give pinacol 1-(phenylthio)alkane-1-boronates which, in turn, can also be deprotonated with LDA and alkylated with alkyl halides to introduce a second alkyl group (Equation (64)).[272,273] Yields are generally good. Alkylations of boron-stabilized carbanions are chemoselective and have been carried out with primary alkyl halides containing acetal,[257,281] alkene,[279] alkyne,[257] halide,[281] cyano,[281] ester,[281] and tosylate[281] groups, though a keto group was not tolerated.[257] Poor results were obtained with secondary halides.[272,273] The alkylation of dimesitylboryl compounds also proceeds very well and the alkylation products have been directly oxidized to the corresponding alcohols (see Section 5.3.2.5).

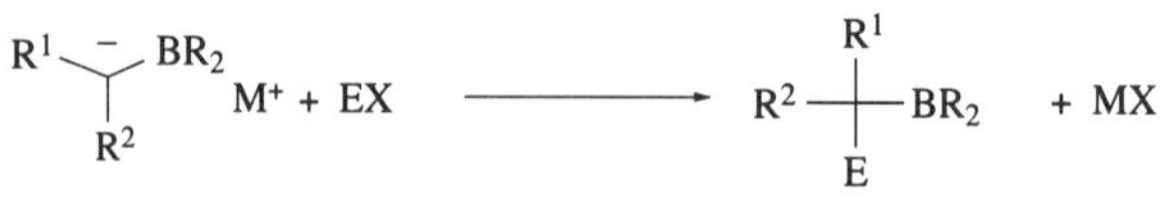

$$\underset{R^2}{\overset{R^1}{\diagdown}}\!C^-\!BR_2\ M^+ + EX \longrightarrow \underset{E}{\overset{R^1}{|}}\!CR^2\!-\!BR_2 + MX \tag{63}$$

$$PhSCH_2\!-\!Bpin \xrightarrow{i,\ LDA;\ ii,\ R^1X} PhSCHR^1\!-\!Bpin \xrightarrow{iii,\ LDA;\ iv,\ R^2X} PhSCR^1R^2\!-\!Bpin \tag{64}$$

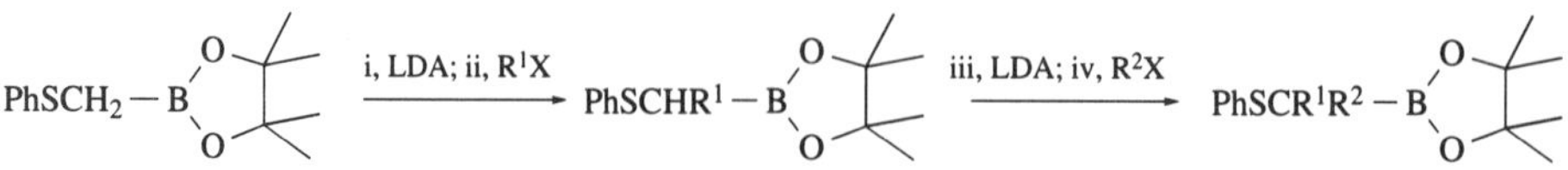

Metal halides R$_n$MX react readily with boron-stabilized carbanions for M = Si, Ge, Sn, or Pb. Further reaction with Bu^{n}Li followed by the addition of a new metal halide led to mixed compounds in good yields (Scheme 12).[257] It is also possible to obtain organometallics containing α-protons.[282] The same kind of reaction has been used to make a set of dimesitylboryl compounds (Equation (65)) in isolated yields of ca. 60%.[277]

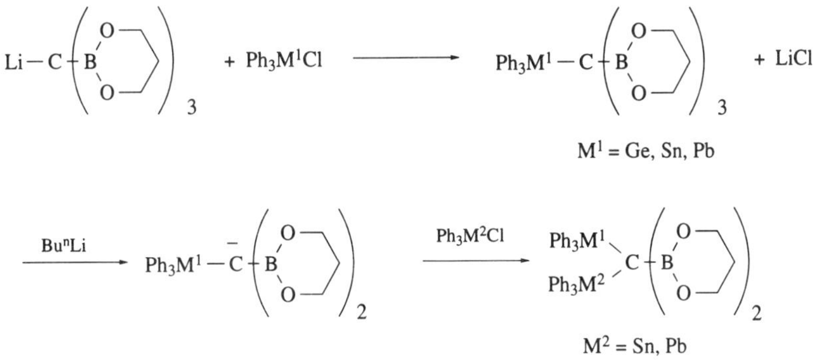

$$\text{Li}-\text{C}\left(\text{B}\begin{pmatrix}\text{O}\\\text{O}\end{pmatrix}\right)_3 + \text{Ph}_3\text{M}^1\text{Cl} \longrightarrow \text{Ph}_3\text{M}^1-\text{C}\left(\text{B}\begin{pmatrix}\text{O}\\\text{O}\end{pmatrix}\right)_3 + \text{LiCl}$$

$$M^1 = \text{Ge, Sn, Pb}$$

$$\xrightarrow{\text{Bu}^n\text{Li}} \text{Ph}_3\text{M}^1-\overset{-}{\text{C}}\left(\text{B}\begin{pmatrix}\text{O}\\\text{O}\end{pmatrix}\right)_2 \xrightarrow{\text{Ph}_3\text{M}^2\text{Cl}} \begin{matrix}\text{Ph}_3\text{M}^1\\\text{Ph}_3\text{M}^2\end{matrix}\text{C}\left(\text{B}\begin{pmatrix}\text{O}\\\text{O}\end{pmatrix}\right)_2$$

$$M^2 = \text{Sn, Pb}$$

Scheme 12

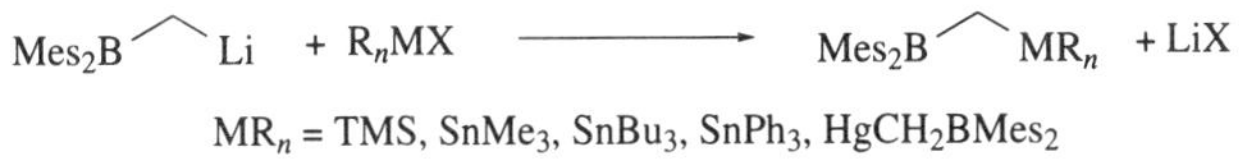

$$\text{Mes}_2\text{B}\frown\text{Li} + R_n\text{MX} \longrightarrow \text{Mes}_2\text{B}\frown\text{MR}_n + \text{LiX} \quad (65)$$

$$MR_n = \text{TMS, SnMe}_3, \text{SnBu}_3, \text{SnPh}_3, \text{HgCH}_2\text{BMes}_2$$

The reaction of α-(dialkoxyboryl)alkylzinc and -copper organometallics with electrophiles described by Knochel and Singer constitutes a powerful method for the preparation of functionalized boronates in very high yields (Equation (66)).[280] These organometallics are able to accommodate functional groups and display excellent reactivity toward a broad range of electrophiles such as allylic halides, aldehydes, acyl chlorides, and various types of Michael acceptors.

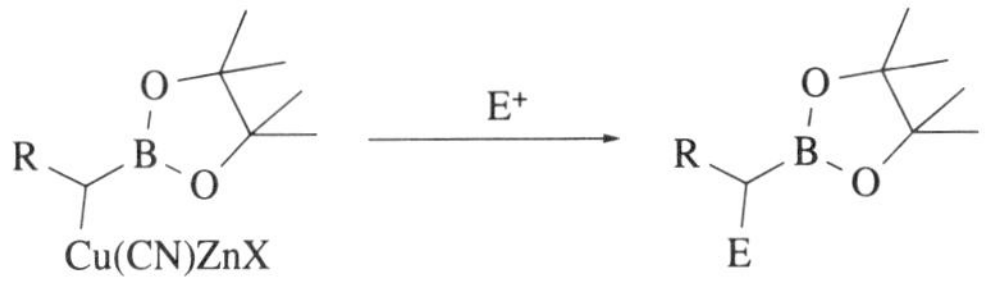

$$\underset{\text{Cu(CN)ZnX}}{R\diagdown\text{B}\diagup\overset{\text{O}}{\underset{\text{O}}{\diagup}}} \xrightarrow{E^+} R\diagdown\underset{E}{\text{B}}\diagup\overset{\text{O}}{\underset{\text{O}}{\diagup}} \quad (66)$$

5.2.2.5 Addition to α,β-unsaturated organoboron compounds

The idea that a trivalent boron behaves like a conventional electron-withdrawing group led to the use of vinylboranes as dienophiles. The Diels–Alder cycloaddition and other pericyclic reactions such as 1,3-dipolar cycloadditions or [2 + 1] cyclopropanation, for example, are interesting because they should produce new boranes which are not otherwise easily accessible.

(i) Diels–Alder cycloaddition

Diels–Alder cycloadditions involving vinylboranes are rare[283-5] and vinylboronic esters were shown to be not very reactive towards 1,3-dienes.[286,287] However, this area was recently reactivated with the discovery of much more reactive vinylboranes such as vinyl-9-BBN,[193] vinylboronates β-substituted by an electron-withdrawing group,[123,218,288] and vinyldichloroboranes (Figure 9).[289] These dienophiles react smoothly with typical dienes to give new cyclohexenylboranes in good yields (Equation (67)).[123,193,218,285,288–90] Typically, vinyl-9-BBN reacts in a few hours at room temperature with common dienes. An excellent regioselectivity was observed with unsymmetrical dienes.[6] Vinylboronates β-substituted with an electron-withdrawing group also show good reactivity. For example, the dienophile with R^1 = COCl and R^2 = alkoxy reacts with cyclopentadiene at −10 °C in 5 h, giving a 95:5 mixture of endo and exo isomers.[290] Vinyldichloroborane also appears to be a reactive dienophile and gives adducts with cyclopentadiene below room temperature.[289] Optically active cyclohexenylboranes were obtained when 1,3-oxazolidin-2-one derivatives of 3-borylpropenoic acids were used in the presence of a chiral titanium reagent at low temperature.[218]

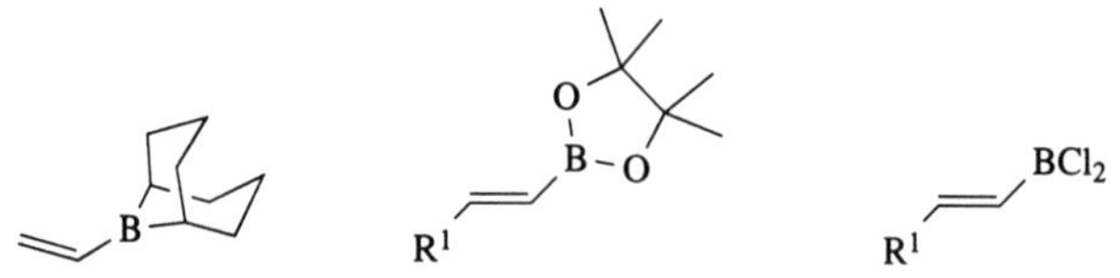

R^1 = COX, SO_2Ar, etc. R^1 = H, alkyl

Figure 9 New boron dienophiles.

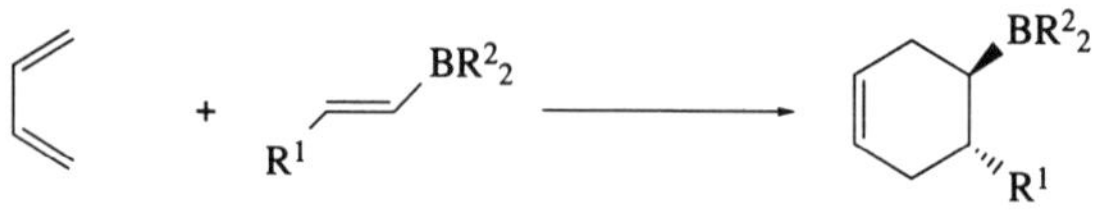

(67)

(ii) Miscellaneous additions

Very few examples of 1,3-dipolar cycloadditions to vinylboranes have been reported. Dibutyl vinylboronate[291] and pinacol vinylboronate[292,293] react with nitrile oxides to give the corresponding isoxazolines in good yields (Equation (68)). New boronates were obtained from these isoxazolines by Matteson's homologation.[293] Cycloaddition of diazo compounds to alkenyl boronic esters yielded 2-pyrazolines after a 1,3-migration of the borylated group.[292,294,295]

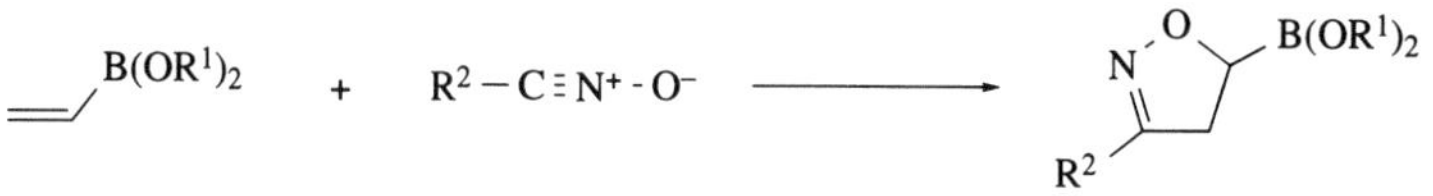

(68)

1-Alkenylboronates are good precursors for cyclopropylboronates either by methylene transfer from diazo derivatives in the presence of catalytic amounts of $Pd(OAc)_2$, or from the carbenoid generated from diiodomethane/diethylzinc (Equation (69)).[296,297] Enantiomerically pure cyclopropylboronates could be obtained by use of a chiral catalyst in this last reaction.[298]

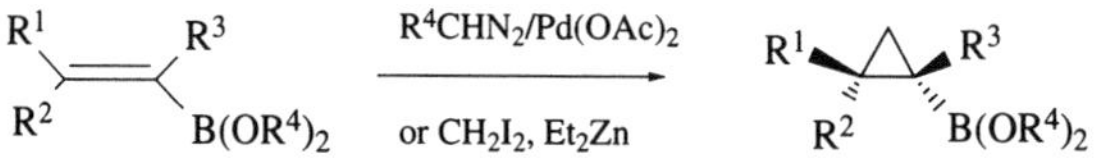

(69)

A wide variety of functionalized boronic esters are available via radical addition to α,β-unsaturated boronates (Scheme 13).[26,235,289–302]

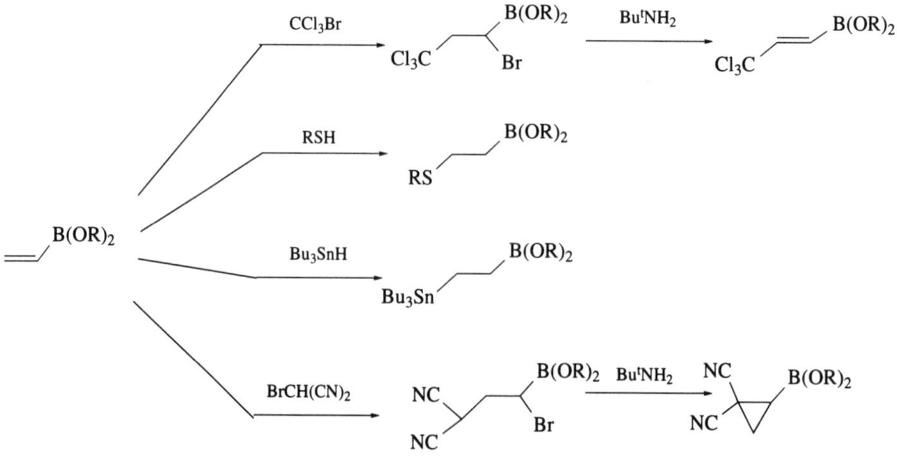

Scheme 13

Recent developments in iodosulfonylation[303,304] and inter-[305] and intramolecular[305,306] alkyl radical addition have been reported.

Strongly electrophilic reagents (Br$_2$, HBr, HI) also add to the double bond of alkenylboronic esters (Scheme 14).[26,295,307]

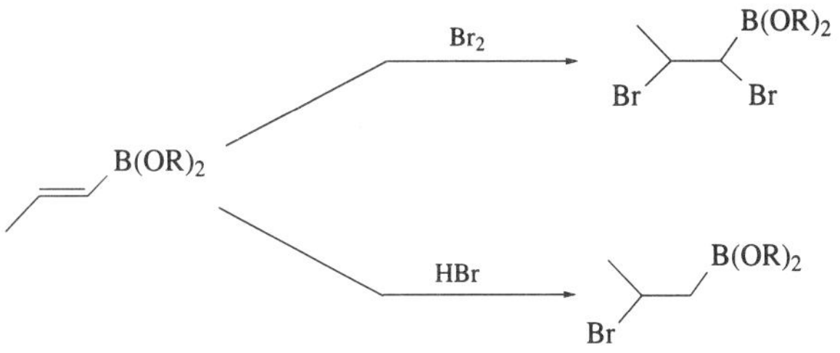

Scheme 14

The reactivity of substituted alkenyl-, allenyl-, and alkynyl boronic esters towards radicals and electrophilic reagents has also been investigated.[26,295] The stereospecific synthesis of alkenes from alkenylboranes is discussed in Section 5.3.1.4. They effectively involve electrophilic reactions of iodine or cyanogen bromide. The primary adducts are not isolated, but directly heated with a base.

5.3 SYNTHETIC APPLICATIONS OF ORGANOBORON COMPOUNDS

5.3.1 Replacement of a Carbon–Boron Bond by a Carbon–Hydrogen or Carbon–Heteroatom Bond

5.3.1.1 Protonolysis of organoboranes

The protonolysis of boranes is an old, but effective reaction. The overall process consists of reducing an alkene to an alkane, or an alkyne to either an alkene of defined stereochemistry or an alkane via a hydroboration reaction.[39] Trialkylboranes are inert toward water[308] and strong mineral acids with the exception of anhydrous HF,[309] but react readily with carboxylic acids to liberate the corresponding alkanes.[310] This reactivity may be ascribed to coordination of the carbonyl oxygen atom of the carboxylic acid to the boron atom of the organoborane, followed by intramolecular proton transfer (Equation (70)).[310] Protonolysis of the first alkyl group is usually straightforward but removal of second and third alkyl groups becomes more difficult. Elevated temperatures are required for protonolysis of the third alkyl group.[310] Replacement of alkyl groups by acyloxy substituents renders the boron atom less acidic and makes the corresponding intermediates less reactive towards protonolysis.

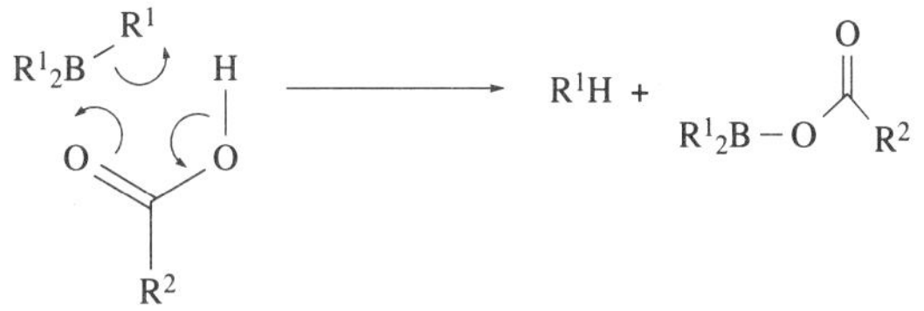

$$R^1{}_2B\text{-}R^1\cdots H\text{-}O\text{-}C(=O)R^2 \longrightarrow R^1H + R^1{}_2B\text{-}O\text{-}C(=O)R^2 \tag{70}$$

Steric requirements parallel the reactivity where primary alkyl groups undergo protonolysis under mild conditions, whereas secondary, tertiary and hindered alkyl groups require harsher conditions. The protonolysis reaction proceeds while retaining the configuration at the carbon atom originally attached to boron.[310,311] A set of experiments involving hydroboration or deuteroboration of norbornene, followed by protonolysis or deuterolysis, provides the corresponding norbornanes (Scheme 15).[310] Products sensitive to the action of hot carboxylic acids may present difficulties such as racemization, as has been reported for the preparation of L-menthene from D-limonene.[310]

Hydroboration with a dialkylborane followed by protonolysis with a carboxylic acid is of general use for the conversion of alkynes into alkenes (Equation (71)).[312–14] Methanol provides milder conditions in certain cases[312] and has been used for the protonolysis of triphenylborane.[315] This process provides a general, stereospecific and versatile route to enynes and dienes (Equations (72) and (73)).[316]

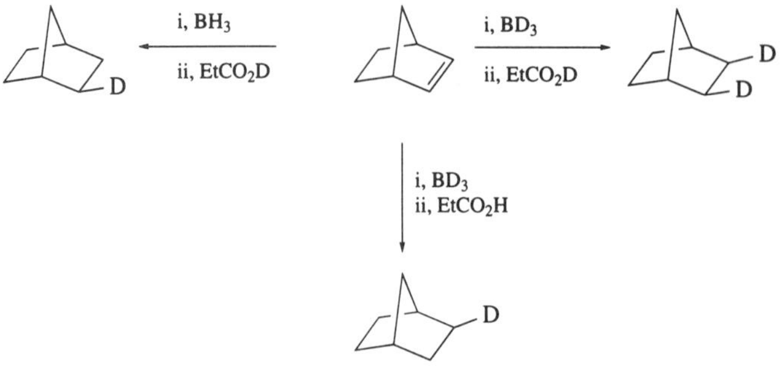

Scheme 15

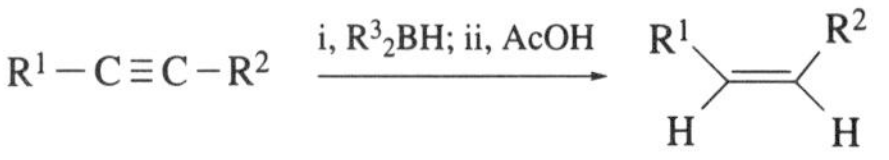

$$R^1-C\equiv C-R^2 \xrightarrow{\text{i, } R^3{}_2BH; \text{ ii, AcOH}} \begin{array}{c} R^1 \\ \diagup \\ H \end{array} = \begin{array}{c} R^2 \\ \diagdown \\ H \end{array} \quad (71)$$

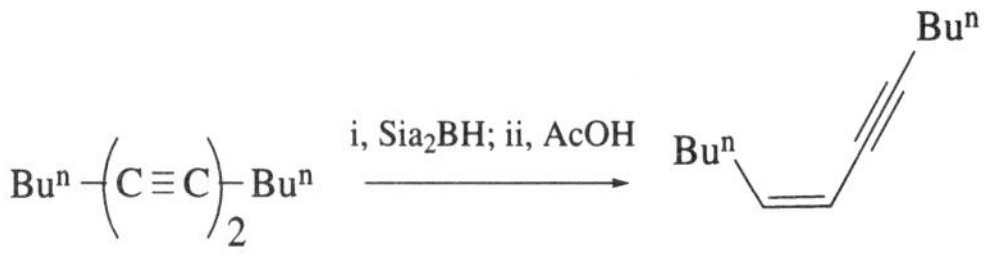

$$Bu^n-(C\equiv C)_2-Bu^n \xrightarrow{\text{i, Sia}_2BH; \text{ ii, AcOH}} Bu^n\diagdown\!\!\!\diagup\equiv\!\!\!\diagdown Bu^n \quad (72)$$

$$Bu^n-(C\equiv C)_2-Bu^n \xrightarrow{\text{i, 2 Cy}_2BH \text{ ii, AcOH}} Bu^n\diagdown\!\!\!=\!\!\!\diagup\!\!=\!\!\!\diagdown Bu^n \quad (73)$$

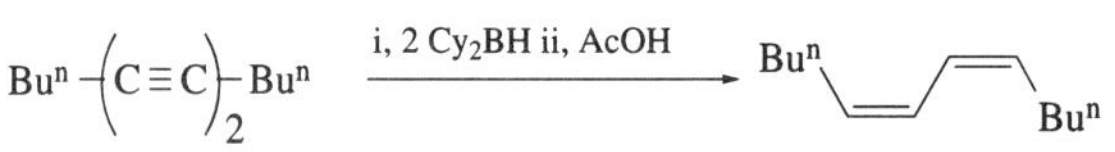

Various insect pheromones with straight-chain (Z)-monoalkene structures have been prepared in high yield and stereochemical purity.[317] The incorporation of functional groups is also easily achieved. The process can also be used for stereospecific synthesis of (E)-deuterioalkenes.[318] Bromoboration of alkynes followed by protonolysis leads efficiently to 2-bromo-1-alkenes. Deuterolysis produces (Z)-1-deuterio-2-bromo-1-alkenes (Equation (74)).[144]

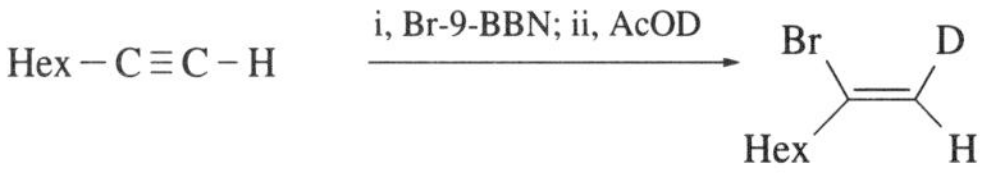

$$\text{Hex}-C\equiv C-H \xrightarrow{\text{i, Br-9-BBN; ii, AcOD}} \begin{array}{c} Br \\ \diagup \\ Hex \end{array} = \begin{array}{c} D \\ \diagdown \\ H \end{array} \quad (74)$$

5.3.1.2 *Oxidation*

Organoboranes are usually handled under an inert atmosphere because they are easily oxidized with oxygen. This autoxidation reaction is used for the preparation of alcohols and alkyl hydroperoxides. Radical processes of the addition of organoboranes to unsaturated compounds are initiated by oxygen (see Section 5.3.3). Steric hindrance and substitution at boron with groups having lone pairs of electrons on the boron-bound atom increase the stability towards oxidation. For example, trimesitylborane and boronic esters are stable towards autoxidation for long periods, as are Lewis base complexes. Nevertheless, almost every organoboron compound can be oxidized under appropriate conditions and this has been reviewed extensively as subsections of general reviews on boron chemistry.[1,26,27,319,320] Only the essential features are reported here. Generally speaking, alkylboranes are converted into alcohols, hydroperoxides, or carbonyl compounds, depending on the reagent used. Arylboranes lead to phenols and alkenylboranes are precursors for aldehydes or ketones.[27]

(i) Autoxidation

The reaction of an organoborane with oxygen can be controlled to give a quantitative conversion of all three alkyl groups on boron to the corresponding alcohol (Equation (75)).[321] The relative rates of oxidation of alkyl groups are in the order: tertiary > secondary > primary, consistent with a free radical mechanism. However, the selective removal of one alkyl group in the presence of another is not possible. In the case of alkenyldialkyl- or alkyldialkenylboranes, the selective oxidation of alkyl groups provides the corresponding alkenylboronic or dialkenylborinic acids respectively in good yields.[321] In contrast to oxidation by alkaline hydrogen peroxide, only a portion of the autoxidation reaction proceeds through alkyl radicals, resulting in loss of stereoselectivity (Equations (76) and (77)).[27,321]

$$R_3B \xrightarrow{\text{i, 1.5 O}_2\text{, THF, 0 °C; ii, NaOH}} 3 \text{ ROH} \qquad (75)$$

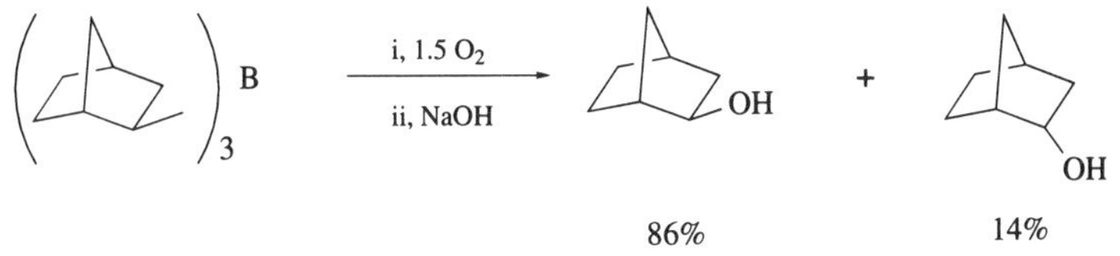

$$(\text{norbornyl})_3B \xrightarrow[\text{ii, NaOH}]{\text{i, 1.5 O}_2} \text{exo-norbornanol} + \text{endo-norbornanol} \qquad (76)$$
$$\qquad\qquad\qquad\qquad\qquad\qquad 86\% \qquad\qquad 14\%$$

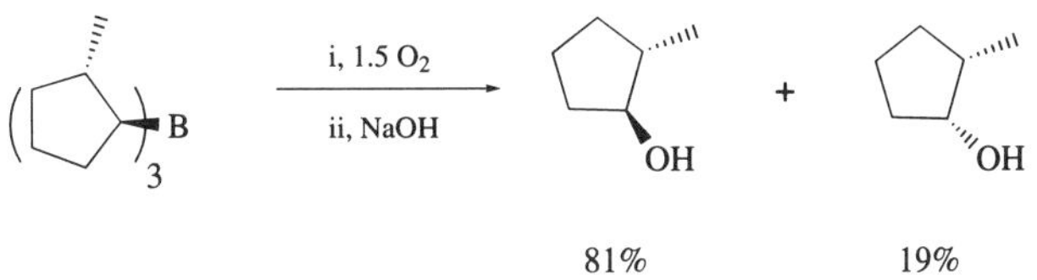

$$(\text{trans-2-methylcyclopentyl})_3B \xrightarrow[\text{ii, NaOH}]{\text{i, 1.5 O}_2} \text{cis-alcohol} + \text{trans-alcohol} \qquad (77)$$
$$\qquad\qquad\qquad\qquad\qquad\qquad 81\% \qquad\qquad 19\%$$

The low temperature autoxidation of organoboranes in THF gives the corresponding alkyl hydroperoxides in excellent yields upon treatment with hydrogen peroxide (Equation (78)). However, only two of the three alkyl groups on boron are used for the formation of hydroperoxides.[322] This method appears to be most promising for isotope incorporation studies since oxygen isotopes are readily available as oxygen gas. A number of ^{17}O-labeled alcohols have been prepared in good yields.[323] ^{15}O-Labeled butanol, the first ^{15}O-labeled compound (half-life 2.0 min !), was prepared by the direct reaction of ^{15}O-labeled oxygen gas (Equation (79)).[324,325] A procedure using an alkylborane polymer has been developed.[326] Ether solutions of alkyldichloroboranes offer the method of choice for the synthesis of hydroperoxides when maximum use of an alkyl group is desired (Equation (80)).[322]

$$R_3B \xrightarrow{\text{i, 2 O}_2\text{, –78 °C, THF}} RB(O_2R)_2 \xrightarrow{\text{ii, 0 °C; iii, 30\%H}_2\text{O}_2} 2 \text{ RO}_2H + \text{ROH} + B(OH)_3 \qquad (78)$$

$$(Bu^n)_3B \xrightarrow[\text{H}_2\text{O}]{^{15}O_2} Bu^{n15}OH \qquad (79)$$

$$RBCl_2 \xrightarrow[\text{–78 °C}]{O_2\text{, Et}_2O} ROOBCl_2 \xrightarrow{H_2O} RO_2H \qquad (80)$$

(ii) Oxidation of boranes to alcohols

The reaction of organoboranes with alkaline hydrogen peroxide (Equation (81)), although one of the oldest, remains the most widely used method for the production of alcohols from boranes. The reaction proceeds with retention of configuration for the three alkyl groups and therefore provides a general route to alcohols of high enantiomeric purity.[27,320] Under the usual conditions, this reaction will tolerate a number of functional groups.[327,328] It may be advantageous in certain cases to use acetate or phosphate buffers at pH ~8.[329] The reactivity trend $R_3B > R_2BX > RBX_2$ (X = OH, OR) is consistent with the reduced Lewis acidity at the boron atom. Anhydrous trimethylamine N-oxide[330,331] or the commercially available dihydrate[332] are very mild oxidizing agents for a wide variety of boron derivatives (Equation (82)). The first reagent is readily prepared[333] and must be used for the oxidation of alkenylboranes in the absence of a proton source.[334] Sequential oxidation[335] is possible with the following order: tertiary alkyl > secondary cycloalkyl > secondary alkyl > primary alkyl > branched primary alkyl > vinyl. Useful differentiations have been found (Equations (83) and (84)).[115,336] As in the case with alkaline hydrogen peroxide, the reaction proceeds with retention of configuration[337] and accommodates many functional groups including alkyl sulfides[338] and cyclopropylboranes.[339] Acylsilanes are accessible according to Equation (85).[122]

$$R_3B \xrightarrow[3\ H_2O_2]{NaOH} 3\ ROH + NaB(OH)_4 \quad (81)$$

$$R_3B + 3\ Me_3NO \longrightarrow 3\ Me_3N + (RO)_3B \xrightarrow{H_2O} 3\ ROH \quad (82)$$

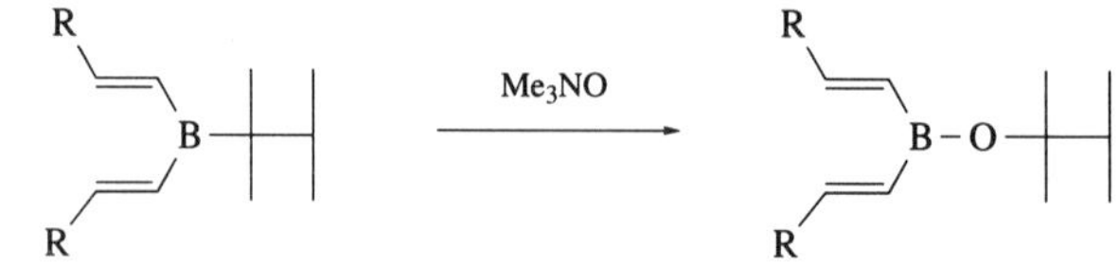

(83)

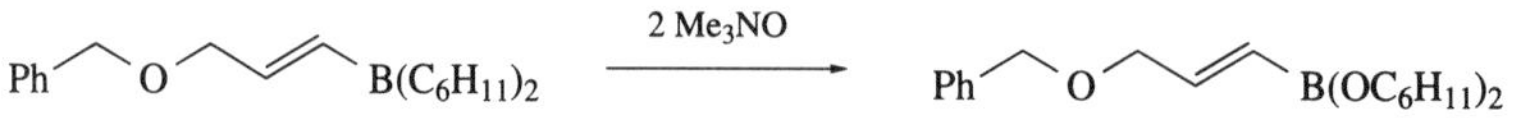

(84)

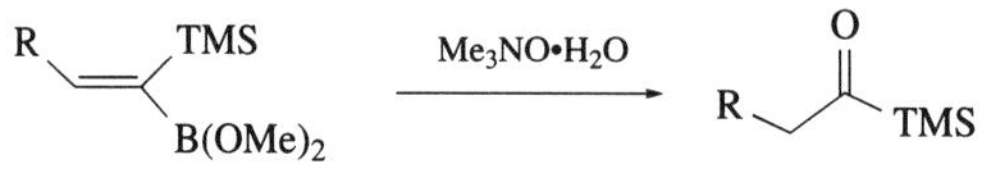

(85)

Some convenient, readily available and inexpensive reagents including sodium perborate[340] and sodium percarbonate[341,342] have been shown to efficiently oxidize organoboranes under very mild conditions. Sodium perborate has been used for the selective oxidation of cyclopropylboronates to cyclopropanols without overoxidation.[297]

(iii) Oxidation of boranes to carbonyl compounds

Hydroboration of alkenes followed by chromic acid oxidation at pH < 3, provides a good synthesis of ketones from alkenes.[27,320] Secondary alcohols may be obtained at pH 3–7.[343] Representative cycloalkanones and α-methylcycloalkanones have been prepared from the corresponding alkenes via hydroboration, followed by chromic acid oxidation (Equation (86)).[344]

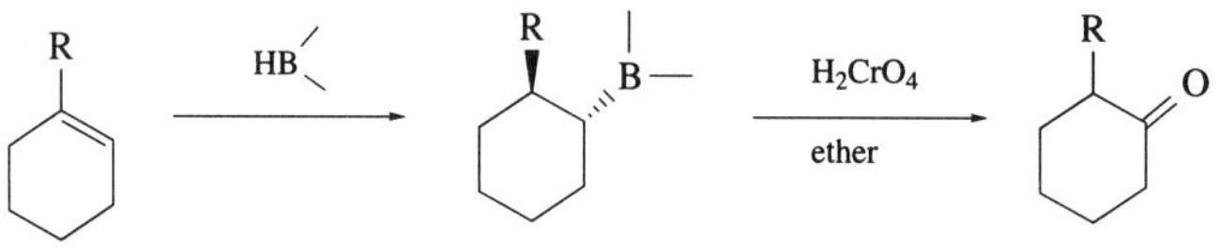

(86)

R = alkyl, aryl

Pyridinium chlorochromate oxidizes secondary alkylboranes to ketones and primary alkylboranes to aldehydes under mild alkaline and anhydrous conditions, in good yields. Dialkylchloroboranes and thexylalkylchloroboranes are also good partners in these reactions.[345] The direct oxidation of organoboranes from terminal alkenes to carboxylic acids may be achieved in a convenient manner with pyridinium dichromate, sodium dichromate in aqueous sulfuric acid, and chromium trioxide in 90% aqueous acetic acid in good yields, with retention of the structure of the organic group attached to boron (Equation (87)).[346,347]

$$RCH_2-B{\diagdown \atop \diagup} \longrightarrow RCOOH \quad (87)$$

5.3.1.3 Replacement of a carbon–boron bond by a carbon–nitrogen bond

(i) Synthesis of primary amines

The reaction of amines bearing a leaving group on nitrogen such as hydroxylamine-O-sulfonic acid, chloramine, or mesitylenesulfonylhydroxylamine, with organoboranes leads to primary alkylamines (Scheme 16).[320] This reaction proceeds with strict retention of configuration at carbon.[348] A convenient procedure using an *in situ* generation of chloramine, which is unstable and cannot be stored, from sodium hypochlorite and ammonium hydroxide gives primary amines in good yields (72–96%).[349] Nitrogen-15-labeled primary amines are easily accessible from $^{15}NH_4OH$ using this method.[350]

$$R_3B + NH_2X \longrightarrow [R_3\overset{-}{B}-\overset{+}{N}H_2X] \longrightarrow R_2B\overset{+}{N}H_2R, X^- \xrightarrow{NaOH} RNH_2 + R_2BOH + NaX$$

Scheme 16

Triorganylboranes, R_3B, and diorganylborinic esters, $R^1{}_2BOR^2$, react with chloramine or hydroxylamine-O-sulfonic acid derivatives to give primary amines in yields limited to 67% for R_3B and 50% for $R^1{}_2BOR^2$. This problem can be overcome by using monoorganyldimethylboranes, which are readily accessible from alkenes and dimethylborane. These boranes provide the desired primary amines in isolated yields from 73% to 95%.[351,352] Nitrogen-13-labeled γ-aminobutyric acid and putrescine have been prepared from the corresponding organyldimethylboranes and ^{13}N-labeled chloramine.[353]

Hydrazoic acid, HN_3, either generated from sodium azide and an aqueous acid[354] or from trimethylsilyl azide and methanol,[355] reacts with trialkylboranes to give primary amines in good to excellent yields (43–87%).

Primary amines can be obtained regio-, stereo-, and enantioselectively from enantiomerically pure boronic esters prepared by the asymmetric hydroboration of prochiral alkenes (Equation (88)).[356]

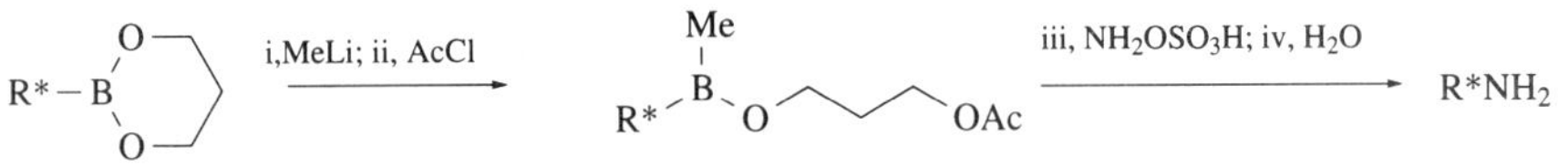

$$R^*-B\overset{O}{\underset{O}{\diagdown}}\bigg) \xrightarrow{i, MeLi; ii, AcCl} R^*-\overset{Me}{\underset{O}{B}}{\diagdown}OAc \xrightarrow{iii, NH_2OSO_3H; iv, H_2O} R^*NH_2 \quad (88)$$

Alkylboranes react rapidly at low temperature with lithium or potassium t-butyl-N-tosyloxycarbamate to give the corresponding N-BOC-protected primary amines in yields from 34% to 81% (Equation (89)).[357] Interestingly, N-(p-toluenesulfonyl)-protected primary amines have been obtained in good to excellent yields by reaction of [N-(p-toluenesulfonyl)imino]phenyliodinane, a hypervalent iodonium ylide, with trialkylboranes (Equation (90)).[358]

$$TsONH-BOC \longrightarrow TsON(M)-BOC \xrightarrow{R_3B} RNH-BOC \quad (89)$$

$$M = Li, K$$

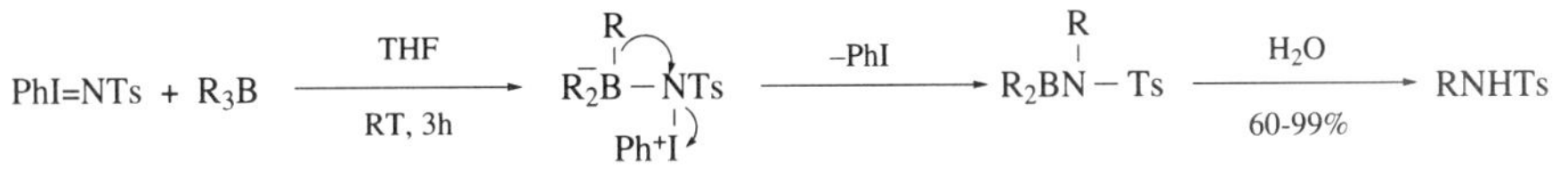

$$PhI=NTs + R_3B \xrightarrow[RT, 3h]{THF} \underset{Ph^+I}{R_2\overset{R}{\overset{|}{B}}-\overset{-}{N}Ts} \xrightarrow{-PhI} R_2\overset{R}{\underset{|}{B}N}-Ts \xrightarrow[60-99\%]{H_2O} RNHTs \quad (90)$$

(ii) Synthesis of secondary amines

Organoboranes react with azides in boiling xylene to evolve nitrogen and form dialkyl-(dialkylamino)boranes, which give secondary amines on hydrolysis or methanolysis and a borinic acid derivative.[359] Only one alkyl group is used. The use of organyldichloroboranes[360] is better, as they require milder conditions and no alkyl group is lost (Equation (91)). These reactions proceed with retention of configuration,[361] and this secondary amine synthesis is highly efficient in terms of chemoselectivity, yields, and wide applicability.[362–4] The scope of this reaction has been widened, since dichloroboranes can be generated *in situ* from organyl(bis-dialkylamino)boranes in the presence of azides by addition of an excess of dry HCl. Thus, secondary amine hydrochlorides are produced according to Equation (92).[365]

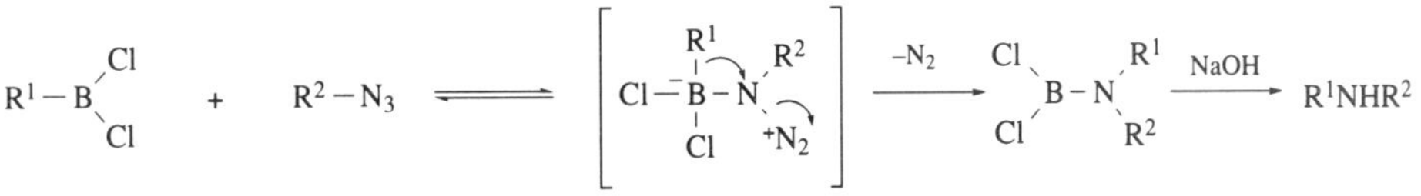

$$R^1-B\overset{Cl}{\underset{Cl}{\diagdown}} + R^2-N_3 \rightleftharpoons \left[Cl-\overset{R^1}{\underset{Cl}{\overset{|}{B}}}-\overset{R^2}{\underset{+N_2}{\overset{|}{N}}}\right] \xrightarrow{-N_2} \overset{Cl}{\underset{Cl}{\diagdown}}B-\overset{R^1}{\underset{R^2}{\diagdown}}N \xrightarrow{NaOH} R^1NHR^2 \quad (91)$$

$$R^1B(NPr^i{}_2)_2 + R^2N_3 \xrightarrow{dry\ HCl} R^1R^2NH, HCl \quad (92)$$

The reaction of α-chiral organyldichloroboranes generated from the corresponding boronates with organyl azides provides a synthesis of secondary amines with high enantiomeric purities (Equation (93)).[225] Several applications of this secondary amine synthesis have appeared in the literature. Aziridines are accessible from 2-iodoalkyl azides.[366] A convenient general new route to open-chain polyamines has been developed using the reductive alkylation of aliphatic amino azides by (ω-halogenoalkyl)dichloroboranes (Scheme 17).[367,368] A convergent synthesis of symmetrical diamines and polyamines from α,ω-diazidoalkanes has been developed using this methodology,[369] and the intramolecular version of this secondary amine synthesis is a general stereoselective route to pyrrolidines and piperidines. Precursors can be ω-azidoalkylboronates,[370,371] ω-azidoborinates,[372] or ω-azidoalkenes (Scheme 18).[371,373]

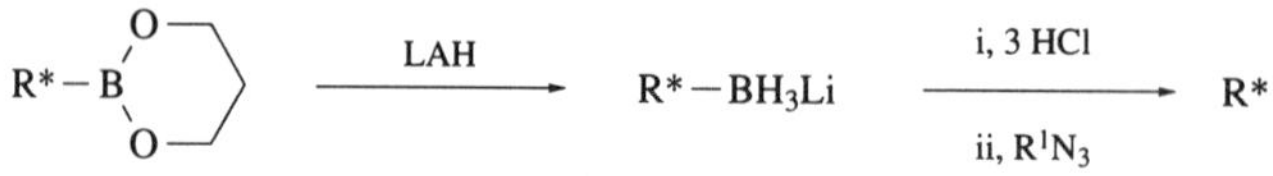

$$R^*-B\overset{O}{\underset{O}{\diagdown}}\xrightarrow{LAH} R^*-BH_3Li \xrightarrow[ii, R^1N_3]{i, 3 HCl} R^*-NHR^1 \quad (93)$$

$$R^1R^2N-[\ ]-N_3, HCl \xrightarrow{i, Br-[\ ]-BCl_2; ii, MeOH} R^1R^2N-[\ ]-NH-[\ ]-Br, 2 HCl \xrightarrow{iii, NaN_3; iv, NaOH}$$

$$R^1R^2N-[\ ]-NH-[\ ]-N_3 \xrightarrow{v, 2 HCl; vi, RBCl_2} R^1R^2N-[\ ]-NH-[\ ]-NHR, 2 HCl$$

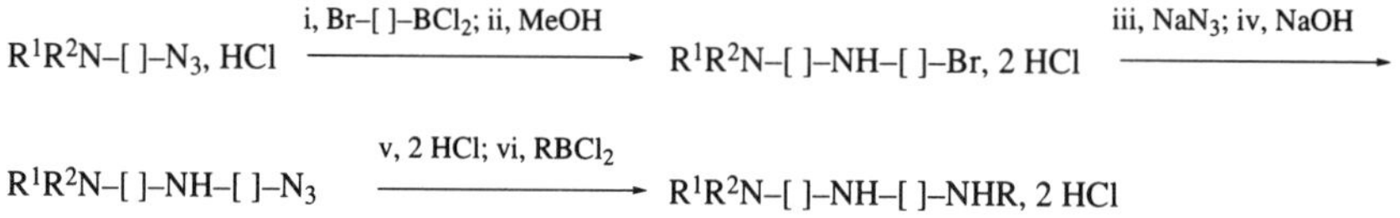

Scheme 17

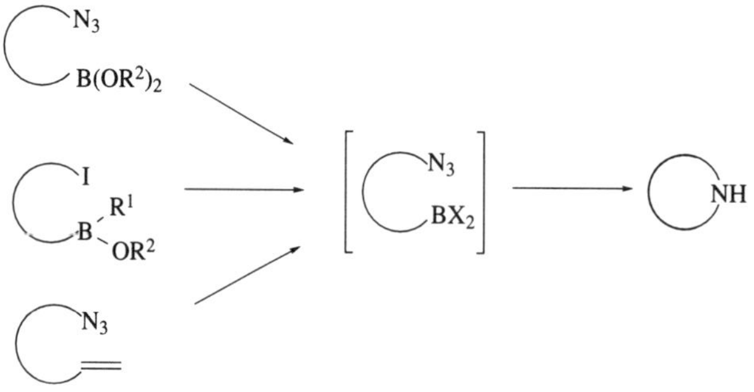

Scheme 18

Routes to bicyclic nitrogen heterocycles like perhydroindoles,[371] pyrrolizidines,[371] *trans*-decahydroquinolines and *trans*-octahydroindoles[374,375] have been reported and further developments in this area related to alkaloid synthesis are highly predictable.

5.3.1.4 *Halogenolysis*

Halogenolysis of organoboranes has been reviewed in 1991.[320] The *ipso* fluorination of arylboronic acids and some derivatives with caesium fluoroxysulfate (CsSO$_4$F) has been developed as a new synthesis of aryl fluorides. The reaction proceeds with a wide range of functionalized aromatics and is strongly solvent dependent (Equation (94)).[376]

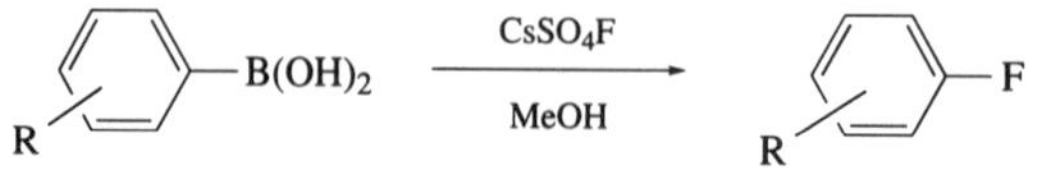

$$R-C_6H_4-B(OH)_2 \xrightarrow[MeOH]{CsSO_4F} R-C_6H_4-F \quad (94)$$

Chlorination of trialkylboranes has not been well studied. Nevertheless, two efficient methods have appeared. The free radical reaction of nitrogen trichloride (NCl$_3$) with boranes constitutes a valuable method for the anti-Markovnikov hydrochlorination of alkenes in 56–94% yield (Equation (95)).[377] A disadvantage of this procedure is that only two groups of the trialkylboranes are readily accommodated in the reaction, but this may be circumvented by the use of mixed alkylborane derivatives. Stereochemically pure (*E*)-1-chloro-(or bromo)alk-1-enes are obtained in good yields from alk-1-ynes via hydroboration with dialkylboranes followed by reactions with copper(II) halides in the presence of a small amount of water, in polar aprotic solvents (Equation (96)).[378]

$$R_3B + 3\,NCl_3 \xrightarrow[0\,°C]{CH_2Cl_2} 3RCl + B(NCl_2)_3 \qquad (95)$$

$$R^1\!\!\equiv\!\!-H \xrightarrow[THF]{R^2{}_2BH} \underset{H\quad BR^2{}_2}{\overset{R^1\quad H}{\diagup\!\!=\!\!\diagdown}} \xrightarrow[ii,\,H_2O]{i,\,HMPA,\,CuX_2} \underset{H\quad X}{\overset{R^1\quad H}{\diagup\!\!=\!\!\diagdown}} \qquad (96)$$

R^1 = alkyl, aryl R^2 = Sia   X = Cl, Br   72–89%

Hydroboration, followed by sodium methoxide-induced bromination, provides a convenient procedure for the conversion of alkenes to alkyl bromides (Equation (97)).[379] Bromination of tri-*exo*-norbornylborane shows that the 2-bromonorbornane obtained is predominantly *endo*, indicative of an inversion of configuration. The use of sodium bromide in the presence of chloramine-T allows the incorporation of bromine.[380] The dark reaction of bromine with trialkylboranes in methylene chloride provides a convenient procedure for the anti-Markovnikov hydrobromination of internal alkenes.[381] (Z)-1,2-Dihalo-1-alkenes can be prepared stereo- and regioselectively in good yields from 1-alkynes by bromoboration with tribromoborane, followed by treatment with iodine chloride or bromine chloride in the presence of sodium acetate (Equation (98)).[382]

$$RCH{=}CH_2 \xrightarrow[ii,\,Br_2,\,NaOMe]{i,\,BH_3} RCH_2CH_2Br \qquad (97)$$

$$R\!\!\equiv\!\!-H + BBr_3 \longrightarrow \underset{Br\quad BBr_2}{\overset{R\quad H}{\diagup\!\!=\!\!\diagdown}} \xrightarrow[AcONa]{XCl} \underset{Br\quad X}{\overset{R\quad H}{\diagup\!\!=\!\!\diagdown}} \qquad (98)$$

R = alkyl, aryl   X = Br, I   61–84% yield

Iodinolysis of C–B bonds parallels brominolysis in the presence of sodium hydroxide or sodium methoxide and provides a convenient route for the anti-Markovnikov hydroiodination of alkenes under mild conditions.[383] The *in situ* oxidation of sodium iodide by mild oxidants such as chloramine-T in the presence of an organoborane provides an efficient methodology for the synthesis of iodine-labeled materials.[384] The reactions of (E)-1-alkenylboronic acids and esters and (E)-1-alkenyl-dibromoborane–dimethyl sulfide complexes with iodine in the presence of base give (E)-1-iodo-1-alkenes in excellent yields (51–85%) and high stereochemical purities (Equation (99)).[385]

$$\underset{H\quad B(OH)_2}{\overset{R\quad H}{\diagup\!\!=\!\!\diagdown}} \xrightarrow[NaOH]{I_2} \underset{H\quad I}{\overset{R\quad H}{\diagup\!\!=\!\!\diagdown}} \qquad (99)$$

R = alkyl, aryl

Addition of bromine at −25 °C to (Z)-1-alkenylboronic esters followed by treatment with sodium methoxide results in inversion of configuration and formation of the corresponding (E)-1-bromo-1-alkenes (79–81% yield) (Equation (100)).[385] Similarly, (Z)-1-alkenylboronic acids afford (E)-1-chloro-1-alkenes. (E)-Alkenylboronic acids add bromine readily at low temperatures to produce intermediates which are converted by a base into (Z)-1-bromo-1-alkenes in 99% stereochemical purities (Equation (101)).[386]

$$\underset{H\quad H}{\overset{R^1\quad B(OR^2)_2}{\diagup\!\!=\!\!\diagdown}} \xrightarrow{i,\,Br_2;\,ii,\,NaOMe} \underset{H\quad Br}{\overset{R^1\quad H}{\diagup\!\!=\!\!\diagdown}} \qquad (100)$$

$$\underset{H\quad B(OH)_2}{\overset{R\quad H}{\diagup\!\!=\!\!\diagdown}} \xrightarrow{i,\,Br_2;\,ii,\,NaOMe/MeOH} \underset{H\quad H}{\overset{R\quad Br}{\diagup\!\!=\!\!\diagdown}} \qquad (101)$$

A synthesis of radioiodinated vinyl iodide involves the *in situ* oxidation of a vinylboronic acid in the presence of a sodium iodide.[387] (E)-1-Alkenylboronic acids react with iodine on γ-alumina to give a mixture of (E)- and (Z)-1-alkenyl iodides, whereas internal vinylboronic acids do not yield vinyl iodides.[388]

5.3.1.5 *Transmetallation*

A wide variety of organometallic compounds may be produced by exchange reactions between organoboranes and metal derivatives. This occurs most readily for the less electropositive elements such as mercury and has been described for lead, thallium, tin, aluminum, zinc, magnesium, and beryllium.[27] The most significant results reported recently in this area concern boron–zinc and, to a lesser extent, boron–copper exchange. This transmetallation is very useful for the stereospecific preparation of bis-(alkenyl)zincs.[389] These organometallics react with aldehydes to give secondary allylic alcohols, the alkenyl group transfer occurring with retention of configuration of the double bond. Interestingly, the reactivity of alkenylzinc species is enhanced in the presence of 10% of an *N*-methylpiperidine.[389] This transmetallation is very fast in hexane, leading within a few minutes at 0 °C to alkenylzincs which react with aldehydes in the presence of a catalytic amount of (−)-3-*exo*-(dimethylamino)isoborneol (DAIB) to provide secondary (*E*)-allyl alcohols, usually in 70–95% yield with 79–98% enantiomeric excess (Equation (102)).[390] An intramolecular version of this sequence led to a very elegant synthesis of (*R*)-(−)-muscone.[391] A related transmetallation using 2-alkyl-1,3-dithia-2-borolanes or alkyldiethylboranes and diethyl- or dimethylzinc provides dialkylzincs.[392] The activation of vinylboranes as vinyl anions via transmetallation from boron to copper, is possible by using a combination of copper(I) salts and nucleophilic species for boron ate complex formation. Methylcopper,[393] cuprous halide alkyllithium,[394] and cuprous halide–sodium methoxide[395] have been used successfully. Fluoride anion activates vinyldialkylboranes, presumably via ate complex formation, which in turn react with acid chlorides or chloroformates in the presence of cuprous iodide to afford 2,2-difluorovinyl carbonyl compounds in 63–84% yields (Equation (103)).[396]

$$R^1\text{———}H \xrightarrow{\text{i, HB(Chx)}_2;\ \text{ii, Et}_2\text{Zn}} \underset{H}{\overset{R^1}{\diagup}}=\underset{\text{ZnEt}}{\overset{H}{\diagdown}} \xrightarrow{\text{iii, R}^2\text{CHO, 1% (−)DAIB}} \underset{H}{\overset{R^1}{\diagup}}=\overset{H}{\diagdown}\text{—CH(OH)R}^2 \qquad (102)$$

$$\underset{F}{\overset{F}{\diagdown}}\text{C}=\underset{R^1}{\overset{BR^1_2}{\diagup}} \xrightarrow{\text{i, LiF or CsF; ii, CuI; iii, R}^2\text{COCl}} \underset{F}{\overset{F}{\diagdown}}\text{C}=\underset{R^1}{\overset{COR^2}{\diagup}} \qquad (103)$$

5.3.2 Carbon–Carbon Bond Formation via a Transient Four-coordinate Boron

Organoboranes are more or less strong Lewis acids because of the presence of an empty *p* orbital on the boron atom. They react easily with nucleophiles to form organoborates which, in most cases, do not undergo any further spontaneous reaction. However, the presence in an α-position of a leaving group or the creation of an electron-deficient center induces an instantaneous 1,2-transfer from boron to the adjacent carbon atom. As can be seen below and in the following sections, this rearrangement dominates much of the ionic reactivity of organoboranes. Reviews and books have highlighted many uses of such reactions in organic synthesis.[27,30,397–401]

Such a favorable situation can be encountered when:

(i) an organoborane bearing an α-leaving group reacts with a nucleophile (Equation (104));
(ii) an organoborane is treated with a nucleophilic reagent possessing a leaving group (Equation (105));
(iii) a stable α,β-unsaturated organoborate reacts with an electrophile (Equation (106)).

For most of these rearrangements, and it has been unambiguously demonstrated for a number of reactions, there is complete retention of the configuration of the migrating group and clean inversion at the nucleophuge-bearing carbon center.

$$\underset{R'}{\overset{R}{\diagdown}}B\text{—}\underset{X}{\overset{}{C}} \xrightarrow{\text{Nu}^-} \left[\underset{R'\ X}{\overset{R}{\text{Nu—B—C}}}\right] \longrightarrow \underset{R'}{\overset{\text{Nu}}{\diagdown}}B\text{—}\overset{R}{\underset{}{C}}\text{—} \qquad (104)$$

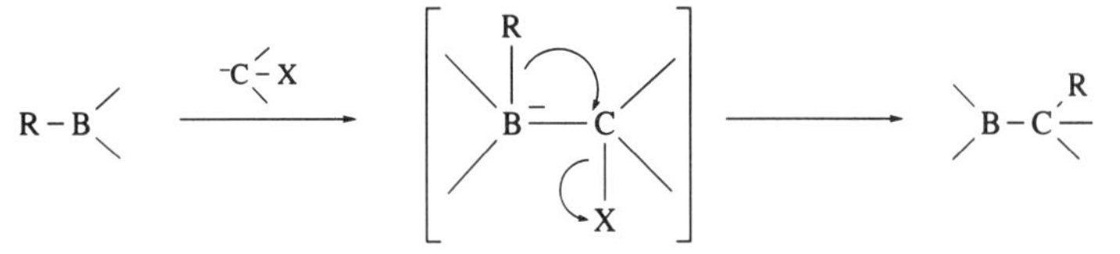

$$R-B\diagup \xrightarrow{-\underset{|}{C}-X} \left[\diagup\hspace{-0.5em}\underset{\diagdown}{B}\!\!-\!\!C\diagdown \right] \longrightarrow \diagup B-C\underset{\diagdown}{\diagup}\overset{R}{-} \quad (105)$$

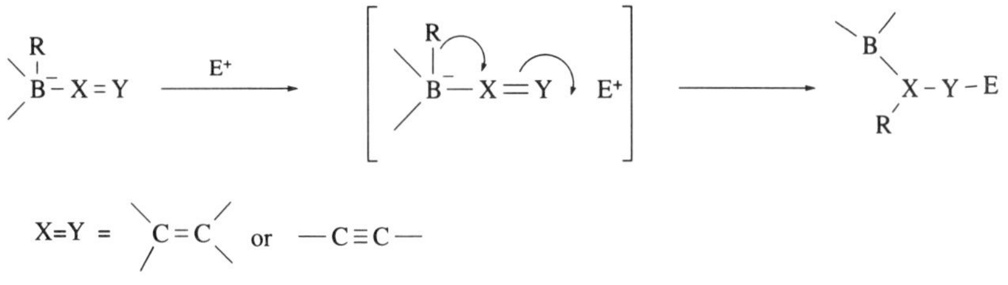

$$\underset{\diagup}{\overset{R}{\diagdown}}\!\!\underset{|}{B}\!-\!X\!=\!Y \xrightarrow{E^+} \left[\underset{\diagup}{\overset{R}{\diagdown}}\!\underset{|}{B}\!-\!X\!=\!Y \ E^+ \right] \longrightarrow \underset{R'}{\overset{\diagdown B \diagup}{}} \underset{}{X-Y-E} \quad (106)$$

$$X=Y = \diagup C=C \diagdown \quad \text{or} \quad -C\equiv C-$$

5.3.2.1 Rearrangements of α-haloorganoborates

(i) From α-haloorganoboranes and nucleophilic reagents and related process

Trialkylboranes are easily α-brominated by bromine to produce hydrogen bromide and α-bromoalkyldialkylboranes which can be isolated after rapid removal of hydrogen bromide (Equation (107)).[402] These highly reactive species undergo easy rearrangements in the presence of nucleophiles and are rarely isolated.[403] They are converted to more stable derivatives as illustrated by the conversion of 9-alkyl-9-BBN to bicyclo[3.3.0]octanol (Equation (108)).[404] α-Haloalkylboranes rearrange with clean inversion at the carbon bearing the halide (Scheme 19).[405]

$$R^1{}_2B\text{–}CH_2R^2 + Br_2 \longrightarrow HBr + R^1{}_2B\text{–}\underset{\underset{Br}{|}}{C}HR^2 \longrightarrow R^1{}_2BBr + R^2CH_2Br \quad (107)$$

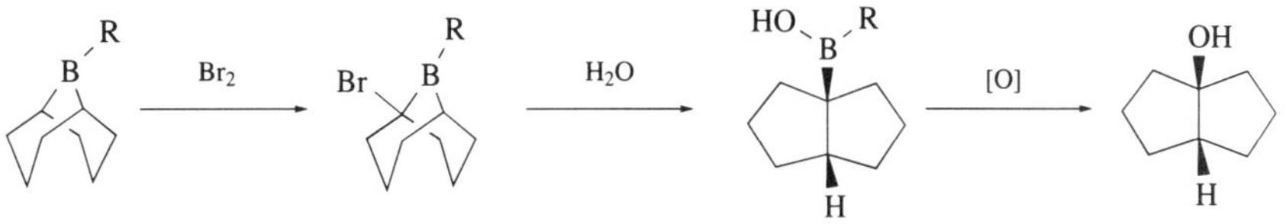

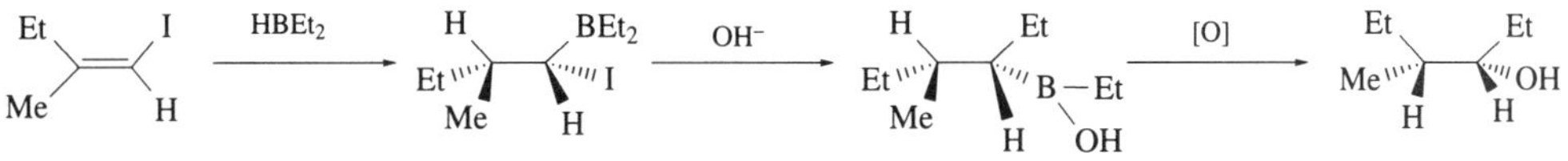

Scheme 19

α-Haloalkylboronic esters can also be prepared by bromination of boronic esters[406] although other routes have been successfully employed (see Section 5.2.2.2). They are much less reactive than their dialkyl analogues and can be isolated without rearrangement. Nevertheless, they exhibit very similar reactivity and can be involved in numerous useful synthetic transformations. Their chemistry is discussed in Section 5.2.2.3.

1-Halo-1-alkenylboranes are regio- and stereospecifically prepared by hydroboration of 1-halo-1-alkynes. Treatment with nucleophiles results in the displacement of halide with inversion of the configuration, giving a new borane which can be further transformed into a large variety of compounds (Scheme 20).[127,407-9]

Hydroboration of 3-chloro-1-iodo-1-propyne with dialkylboranes gives (Z)-(3-chloro-1-iodo-1-propenyl)dialkylboranes exclusively. The reaction of such alkenyldialkylboranes with alkyllithiums results in the intramolecular displacement of the halogens by the alkyl group to provide modified alkenylboranes with a new carbon–boron bond (Scheme 21).[410,411]

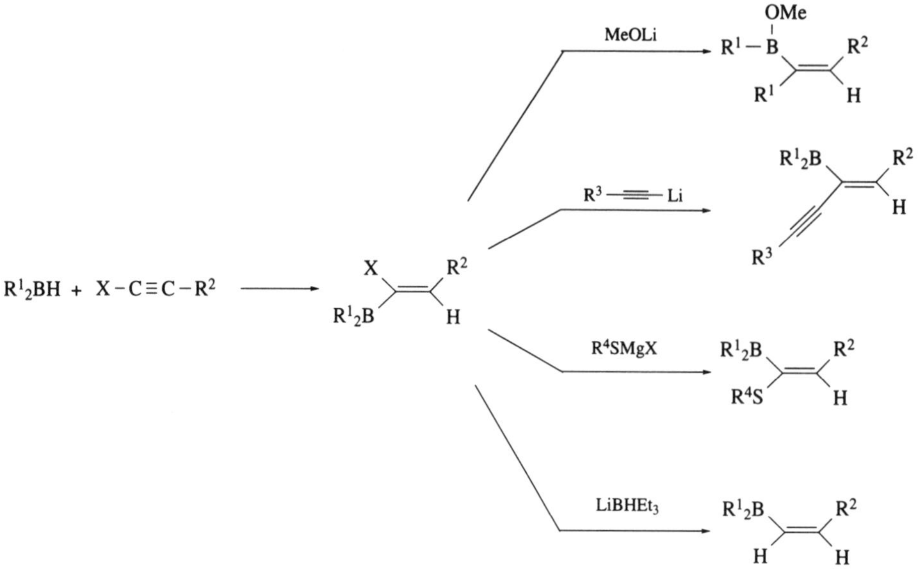

Scheme 20

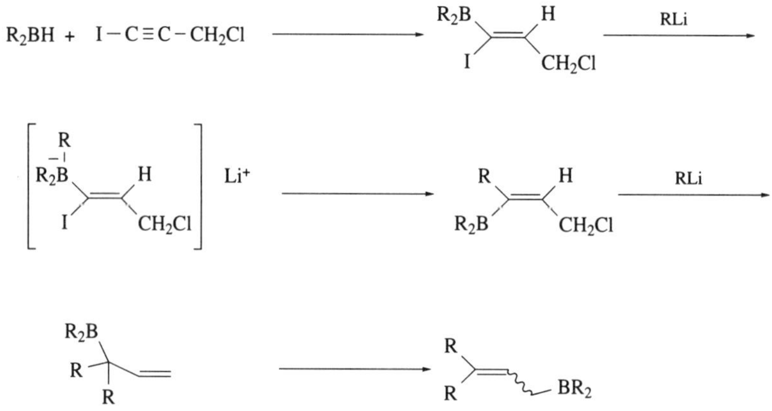

Scheme 21

(ii) From organoboranes and nucleophiles bearing an α-leaving group

Carbanionic reagents bearing potential leaving group(s) homologate organoboranes. As described previously, the reaction proceeds by formation of an ate complex followed by a 1,2-migration of the organyl group(s) from boron to the α-carbon by displacement of the leaving group. A new carbon–carbon bond has thus been created. A large number of reagents have been successfully applied to such transformations. In most cases, only one of the three alkyl groups of the trialkylborane is used. The use of 9-BBN or thexyl derivatives can circumvent this difficulty. Boronic esters which possess only a single alkyl or aryl group are also good partners in these homologation reactions. Nevertheless, several reagents used for the one carbon homologation fail to react with boronic esters. Their chemistry has been discussed in Section 5.2.2.2 and the use of chiral boronic esters is developed in Section 5.4.2.

(a) Anions of α-haloesters, ketones, nitriles, and sulfones. Trialkylboranes react with ethyl bromoacetate in the presence of potassium *t*-butoxide in *t*-butanol to provide the corresponding ester in good yields. The α-borylated ester, presumably formed initially in the rearrangement, is not stable, but rearranges to the boron enolate which is rapidly protonolyzed to give the α-alkylated carbonyl compound (Scheme 22). Similar reactions can be achieved with α-haloketones, α-halonitriles, and α-halosulfones.[27,30]

(b) Diazo compounds. The reaction of organoboranes with α-diazocarbonyl compounds also produces enol borinates[412] (mainly as (*E*)-isomers), which are useful intermediates for aldol and related reactions

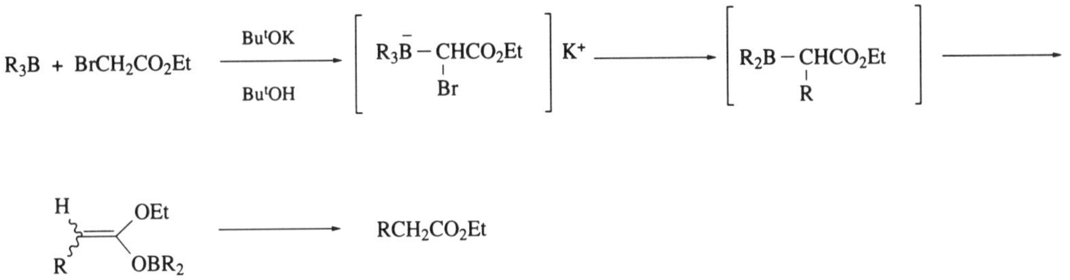

$$R_3B + BrCH_2CO_2Et \xrightarrow[Bu^tOH]{Bu^tOK} \left[R_3\bar{B}-\underset{Br}{CH}CO_2Et \right] K^+ \longrightarrow \left[R_2B-\underset{R}{CH}CO_2Et \right] \longrightarrow$$

$$\underset{R}{\overset{H}{>}}C=C\underset{OBR_2}{\overset{OEt}{<}} \longrightarrow RCH_2CO_2Et$$

Scheme 22

(see Section 5.3.2.7) (Scheme 23). Alkyl-,[413] alkenyl-,[414] and aryldichloroboranes[413] also react readily with ethyl diazoacetate. Competitive migration of chlorine from boron to carbon is observed (Equation (109)). (Dimethylamino)-bis(trifluoromethyl)borane, in contrast, reacts in an entirely different way and the stable borylated ethyl diazoacetate is formed in almost quantitative yield (Equation (110)).[415]

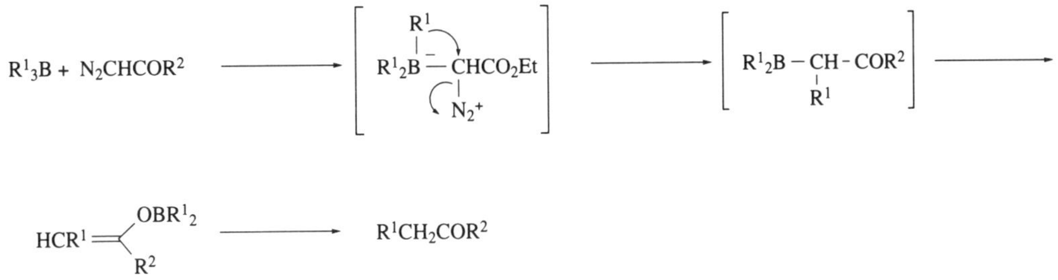

$$R^1{}_3B + N_2CHCOR^2 \longrightarrow \left[R^1{}_2\bar{B}-\underset{N_2^+}{CHCO_2Et}\overset{R^1}{|} \right] \longrightarrow \left[R^1{}_2B-\underset{R^1}{CH}-COR^2 \right] \longrightarrow$$

$$HCR^1=C\underset{R^2}{\overset{OBR^1{}_2}{<}} \longrightarrow R^1CH_2COR^2$$

Scheme 23

$$RBCl_2 + N_2CHCO_2Et \longrightarrow RCH_2CO_2Et + ClCH_2CO_2Et \quad (109)$$

R = alkyl, (E)- or (Z)-alkenyl, aryl.

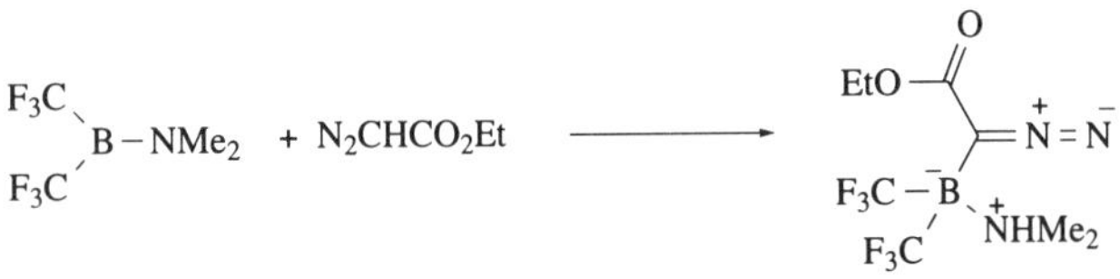

$$\underset{F_3C}{\overset{F_3C}{>}}B-NMe_2 + N_2CHCO_2Et \longrightarrow \underset{F_3C}{\overset{F_3C}{>}}\underset{\overset{+}{N}HMe_2}{\overset{EtO-\overset{O}{\overset{\|}{C}}-\overset{+}{C}=\overset{-}{N}=\overset{-}{N}}{B}} \quad (110)$$

(c) Haloform, dihalomethane anions, and related species. No reaction of trialkylboranes with a monohalomethyllithium, LiCH$_2$X, has been hithertho reported, but other reagents bearing a single leaving group have been studied including sulfonium, ammonium, arsonium, phosphonium, and sulfoxonium ylides (Equation (111)). Polyhomologation may occur and therefore limits the synthetic applicability of these approaches.[27] Use of the α-lithio derivative of a methylorganyl thioether followed by treatment with an alkylating agent is a useful alternative procedure (Scheme 24),[416] and the reactions of trialkylboranes with α-lithio thioacetals[417] or 1,3-benzodithiols[418] are used to prepare aldehydes or ketones in good yields (Scheme 25).

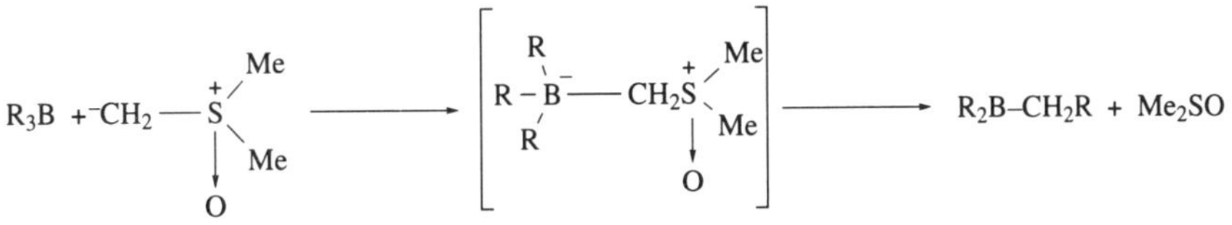

$$R_3B + {}^-CH_2-\overset{\overset{+}{S}(Me)(Me)}{\underset{O}{|}} \longrightarrow \left[R-\underset{R}{\overset{R}{B}}-CH_2\overset{+}{S}\underset{\underset{O}{|}}{\overset{Me}{<}}_{Me} \right] \longrightarrow R_2B-CH_2R + Me_2SO \quad (111)$$

Direct synthesis of carboxylic acids via a two-carbon homologation has been achieved by the reaction with the dianion of phenoxyacetic acid (Equation (112)).[419]

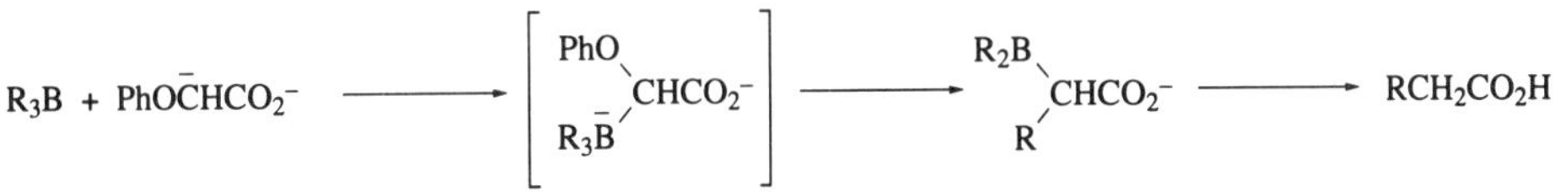

$$R_3B + PhO\bar{C}HCO_2^- \longrightarrow \left[\underset{R_3\bar{B}}{\overset{PhO}{>}}CHCO_2^- \right] \longrightarrow \underset{R}{\overset{R_2B}{>}}CHCO_2^- \longrightarrow RCH_2CO_2H \quad (112)$$

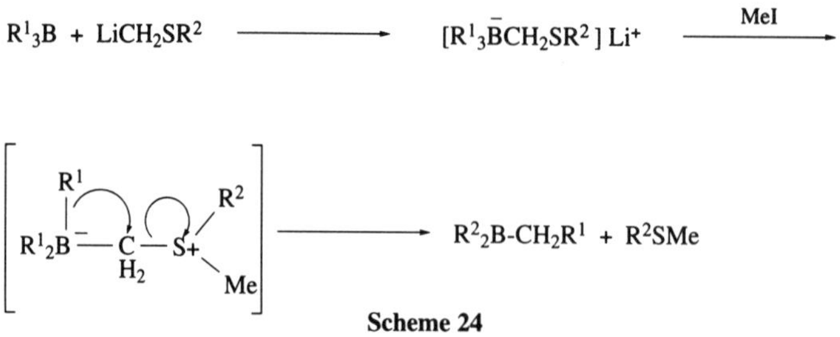

$R^1_3B + LiCH_2SR^2 \longrightarrow [R^1_3\bar{B}CH_2SR^2]\,Li^+ \xrightarrow{MeI}$

$$\left[R^1_2B\underset{H_2}{-}\overset{R^1}{\underset{|}{C}}-\overset{R^2}{\underset{Me}{S+}} \right] \longrightarrow R^2_2B\text{-}CH_2R^1 + R^2SMe$$

Scheme 24

$R^1_3B + R^2\bar{C}(SPh)_2\,Li^+ \longrightarrow R^1_3\bar{B}\text{-}CR^2(SPh)_2\,Li^+ \longrightarrow R^1B\text{-}CHR^2SPh \xrightarrow{[O]} R^2R^1C=O$

$R^2 = H,\ alkyl$

Scheme 25

Reagents possessing two leaving groups are acyl carbanion equivalents. Their double migration reactions are illustrated by the dichloromethyllithium additions to trialkylboranes which give, after oxidation, the corresponding secondary alcohols. These reactions are highly sensitive to the steric requirements of the starting borane and best results are obtained from dialkylborinic esters (Scheme 26).[420]

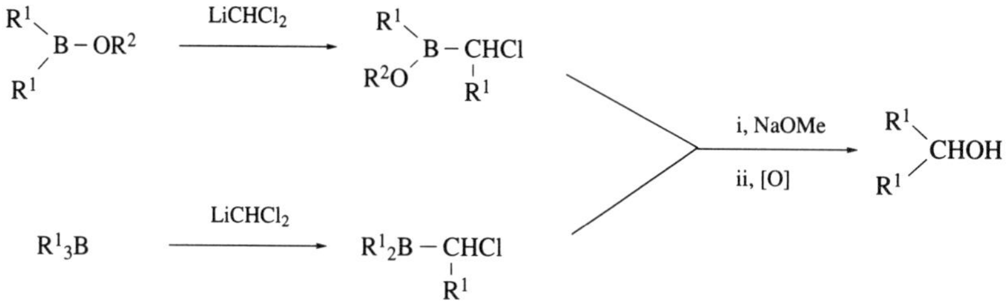

Scheme 26

Addition of mercuric chloride to α-arylthioorganoboranes generated from trialkylboranes and lithiated thioacetals is necessary to induce a second rearrangement (Equation (113)).[421,422]

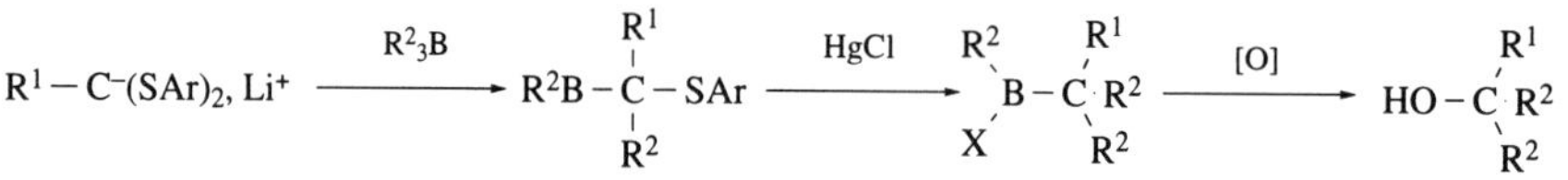

$$R^1-C^-(SAr)_2,\,Li^+ \xrightarrow{R^2_3B} R^2B-\underset{R^2}{\overset{R^1}{\underset{|}{C}}}-SAr \xrightarrow{HgCl} \underset{X}{\overset{R^2}{\diagup}}B-\underset{R^2}{\overset{R^1}{\underset{|}{C}}}R^2 \xrightarrow{[O]} HO-\underset{R^2}{\overset{R^1}{\underset{|}{C}}}R^2 \qquad (113)$$

Trimethylsilylhalomethyllithium successfully homologates trialkylboranes to α-trimethylsilylalkylboranes[423] and has also found interesting synthetic applications in the field of boronic esters (Equation (114)).[237] Closely related reactions are those of α-methoxyalkenyllithium,[424–8] 2-lithiofuran,[429] 2-bromo-6-lithiopyridine,[430] and lithium chloro- or acetoxypropargylide (Equation (115), Schemes 27 and 28).[431–4]

$$TMS\text{-}\bar{C}HClLi^+ \xrightarrow{R_3B} R_2B\text{-}CHR\text{-}TMS + LiCl \qquad (114)$$

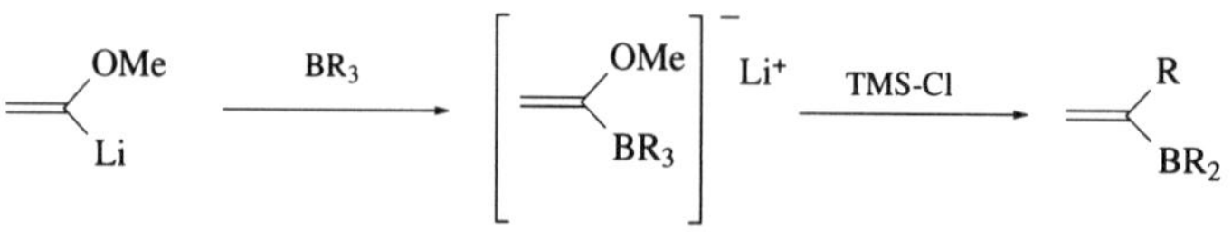

(115)

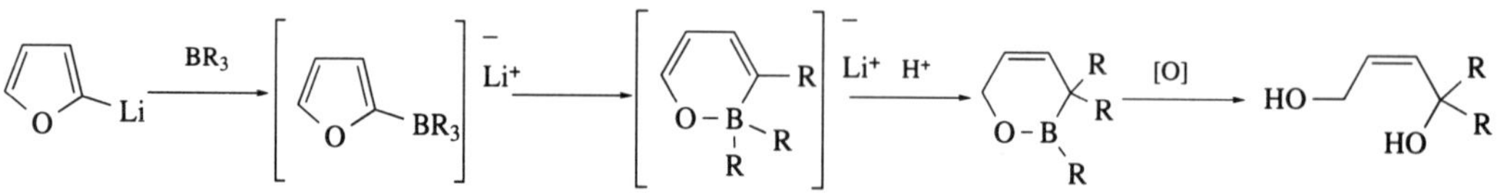

Scheme 27

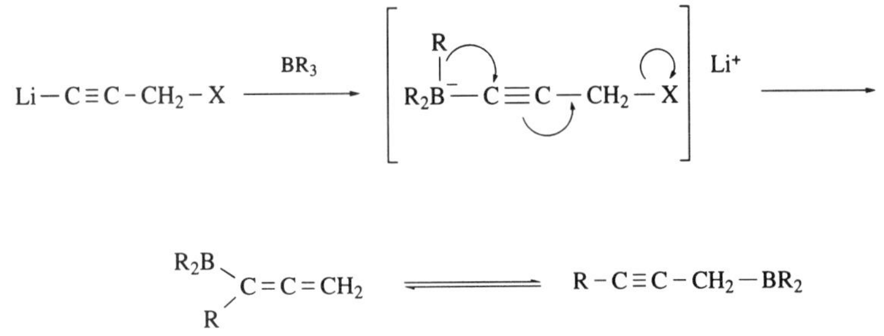

$$\text{Li}-\text{C}\equiv\text{C}-\text{CH}_2-\text{X} \xrightarrow{\text{BR}_3} \left[\begin{array}{c} \text{R} \\ \text{R}_2\text{B}-\text{C}\equiv\text{C}-\text{CH}_2-\text{X} \end{array} \right] \text{Li}^+ \longrightarrow$$

$$\begin{array}{c} \text{R}_2\text{B} \\ \phantom{\text{R}_2\text{B}}\diagdown \\ \text{R}\diagup \end{array} \text{C}=\text{C}=\text{CH}_2 \rightleftharpoons \text{R}-\text{C}\equiv\text{C}-\text{CH}_2-\text{BR}_2$$

X = OAc, Cl

Scheme 28

Anions generated from chlorodifluoromethane and sterically hindered bases react with trialkylboranes to give tertiary alcohols after oxidation (Equation (116)).[435] The reaction with dichloromethyl methyl ether (DCME) can also be used to prepare hindered trialkylcarbinols and provides a useful alternative to the carbonylation reaction or the reaction of cyanoborates with electrophiles (Equation (117)).[30,436]

$$\text{R}_3\text{B} \xrightarrow{\text{i, HCCl}_2\text{F, Et}_3\text{COLi; ii, [O]}} \text{R}_3\text{COH} \qquad (116)$$

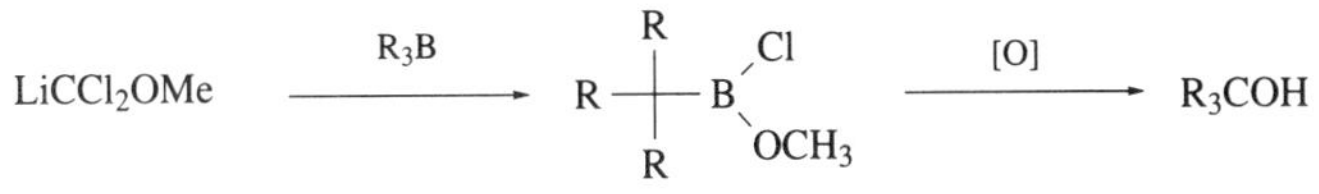

$$\text{LiCCl}_2\text{OMe} \xrightarrow{\text{R}_3\text{B}} \text{R}-\underset{\text{R}}{\overset{\text{R}}{\text{C}}}-\text{B}\underset{\text{OCH}_3}{\overset{\text{Cl}}{\diagup}} \xrightarrow{\text{[O]}} \text{R}_3\text{COH} \qquad (117)$$

Organoboranes react with tris(phenylthio)methyllithium. In the presence of HgCl$_2$, tertiary alcohols are produced after oxidation. Ketones are obtained when mercuric chloride is omitted (Scheme 29).[437]

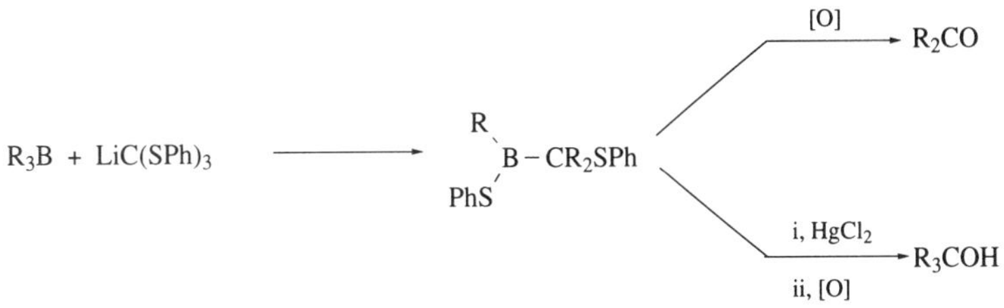

$$\text{R}_3\text{B} + \text{LiC(SPh)}_3 \longrightarrow \underset{\text{PhS}}{\overset{\text{R}}{\diagdown}}\text{B}-\text{CR}_2\text{SPh} \begin{array}{c} \xrightarrow{\text{[O]}} \text{R}_2\text{CO} \\ \\ \xrightarrow[\text{ii, [O]}]{\text{i, HgCl}_2} \text{R}_3\text{COH} \end{array}$$

Scheme 29

5.3.2.2 Carbonylation

The reaction of organoboranes with carbon monoxide results in the transfer of one to three alkyl groups from boron to the adjacent carbon depending on the experimental conditions (Scheme 30). This chemistry has been reviewed recently[27,30] and only a few synthetic applications are reported here.

Carbonylation in the presence of hydrides is best run with lithium trimethoxyaluminum hydride (LTMA) or potassium triisopropylborohydride (KIPBH). 9-Alkyl-9-BBN undergoes selective migration of the alkyl group with the usual retention of configuration, thus providing the corresponding alcohol or aldehyde (Scheme 31).[438] In the absence of a hydride source, two successive migrations occur. The use of a thexyl group, which shows low migratory aptitude, allows the preparation of unsymmetrical ketones (Equations (118) and (119)).[439,440]

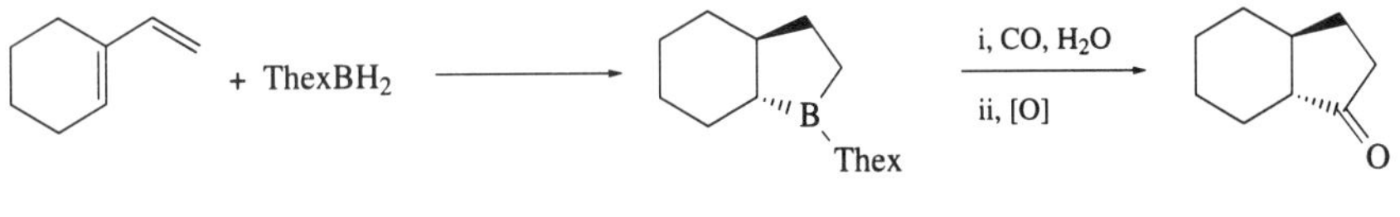

$$\text{cyclohexenyl} + \text{ThexBH}_2 \longrightarrow \text{cyclohexyl-B(Thex)} \xrightarrow[\text{ii, [O]}]{\text{i, CO, H}_2\text{O}} \text{bicyclic ketone} \qquad (118)$$

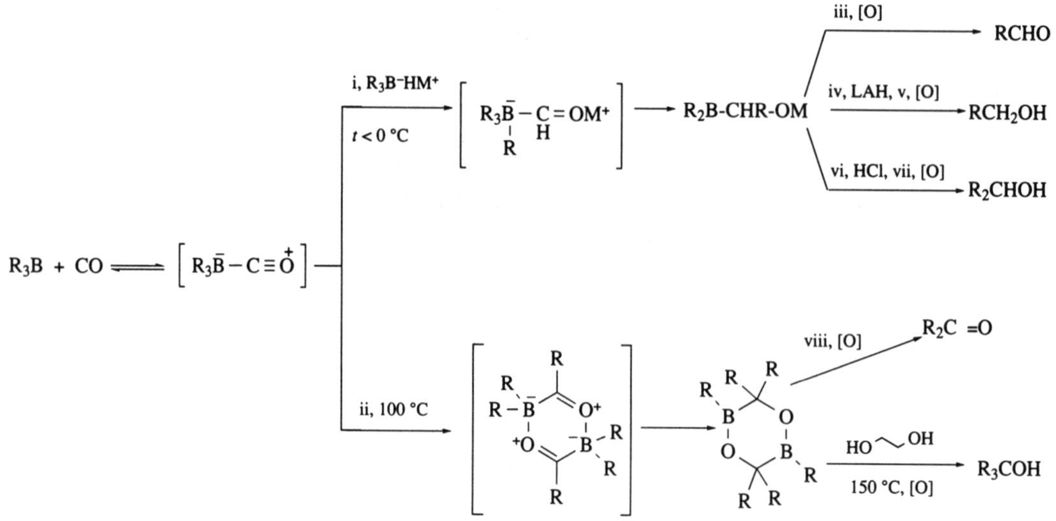

Scheme 30

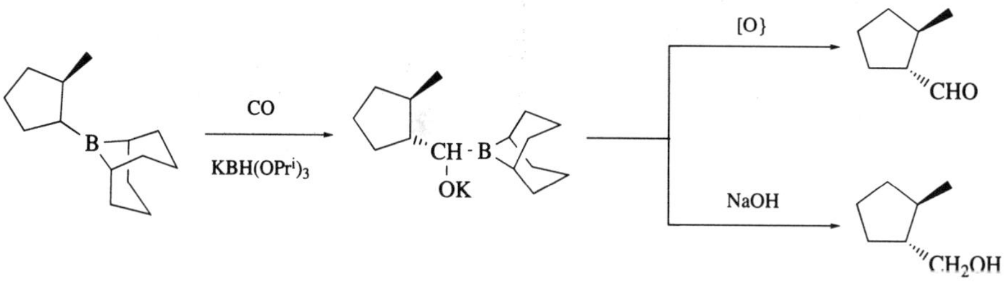

Scheme 31

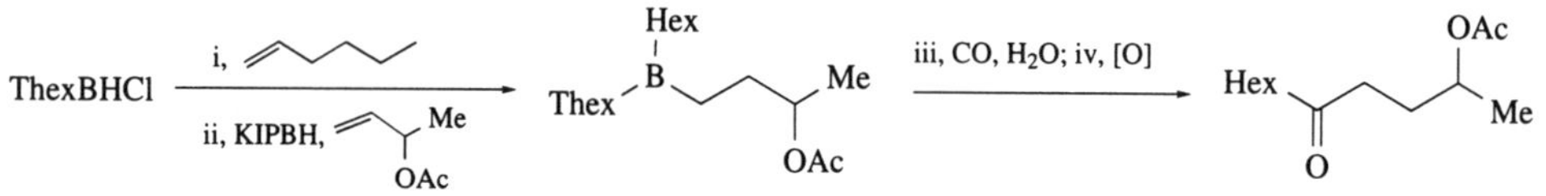

Reaction of NaCo(CO)$_4$ with R$_2$BI under carbon monoxide at atmospheric pressure and room temperature readily gives dialkyl ketones in good yields after oxidation (Equation (120)).[441] At higher temperature and in the presence of ethylene glycol, a third migration is observed. This reaction provides access to tertiary alcohols and can be combined with hydroboration to produce complex structures from polyenes. This "stitching" and "riveting" sequence has been applied to the preparation of polycyclic alcohols (Equation (121)).[442]

$$R_2BI + NaCo(CO)_4 \xrightarrow{\text{i, CO; ii, H}_2\text{O}_2/\text{OH}^-} R_2C=O \quad (120)$$

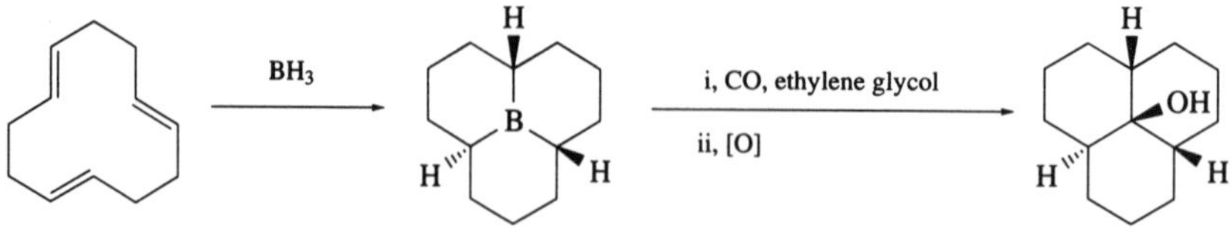

Although the carbonylation reaction has opened new routes to various new structures and has found important applications in the synthesis of radiolabeled compounds,[384] the more recent DCME (see Section 5.3.2.1(ii)(c)) and cyanoborate (see Section 5.3.2.3(iv)) approaches, which do not require special apparatus, are usually more convenient for laboratory operations.

5.3.2.3 Reactions of organoborates with electrophiles

Simple tetraorganylborates are quite unreactive towards electrophiles. Nevertheless, an alkyl group transfer can occur with some selectivity in the presence of acyl halides or powerful alkylating agents (Equation (122)).[443] In addition to these reactions, a hydride transfer from lithium "ate" complexes, mainly from B-alkyl-9-BBN–H derivatives, has been observed. Reduction of ketones occurs with a fair degree of stereocontrol.[444] A similar process has been applied to the preparation of some representative 1-substituted cis-bicyclo[3.3.0]octanes (Equation (123)).[445]

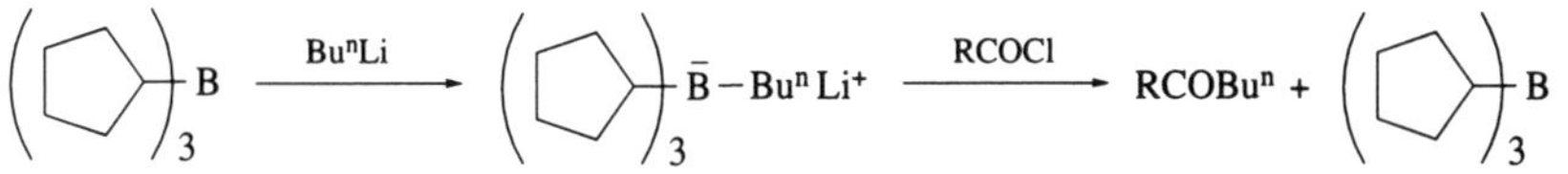

$$\left(\text{cyclopentyl}\right)_3 B \xrightarrow{\text{Bu}^n\text{Li}} \left(\text{cyclopentyl}\right)_3 \bar{B}-\text{Bu}^n \text{Li}^+ \xrightarrow{\text{RCOCl}} \text{RCOBu}^n + \left(\text{cyclopentyl}\right)_3 B \quad (122)$$

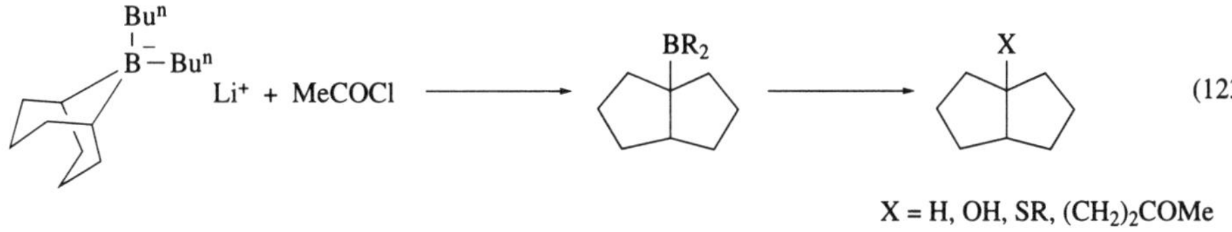

$$\text{9-BBN-Bu}^n \text{Li}^+ + \text{MeCOCl} \longrightarrow \text{bicyclic-BR}_2 \longrightarrow \text{bicyclic-X} \quad (123)$$

X = H, OH, SR, (CH$_2$)$_2$COMe

Copper(I) methyltrialkylborates behave differently. For example, reactions with (E)- and (Z)-β-bromoacrylates occur smoothly to afford the corresponding α,β-unsaturated esters with conservation of the geometry of the double bond (Scheme 32).[446] But most of the useful synthetic applications of organoborates are in the intramolecular transfer reactions of alkenyl- and alkynylborates. As shown in Scheme 33, the addition of an electrophile EX induces the formation of a new carbon–carbon bond via a 1,2-migration. These reactions are quite general and a large variety of α,β-unsaturated organoborates and electrophiles have been used to produce the corresponding alkyl- or alkenylboranes.[27,401] Except in a few cases, these products were not isolated, but directly oxidized or protonolyzed.

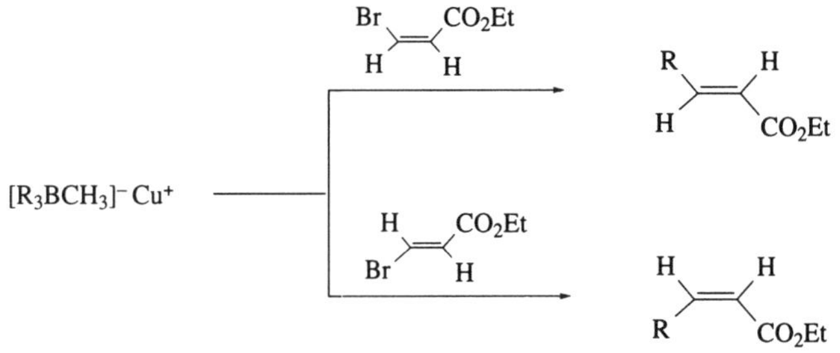

Scheme 32

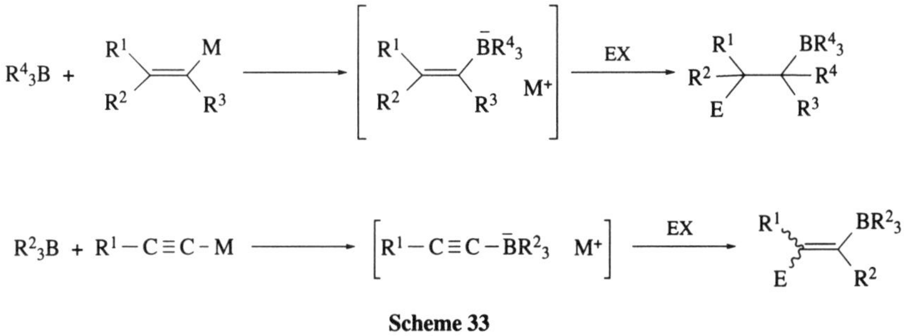

Scheme 33

(i) Alkynylborate chemistry

Protonation of lithium ethynyltrialkylborates with hydrochloric acid may be controlled to produce "Markovnikov" alkenylboranes (Scheme 34).[447,448] An excess of acid causes a facile second migration. A mixture of (E)- and (Z)-isomers is obtained from alkynyltrialkylborates, both of which give the same ketone after oxidation (Equation (124)).[449] Other electrophilic reagents have been used in similar reactions (Scheme 35) and the stereochemical outcome depends on several factors, including the nature of the electrophile, the organoborate, and solvent.[27]

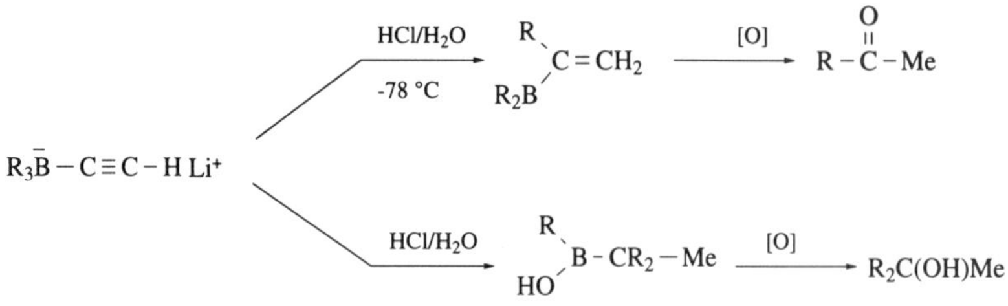

Scheme 34

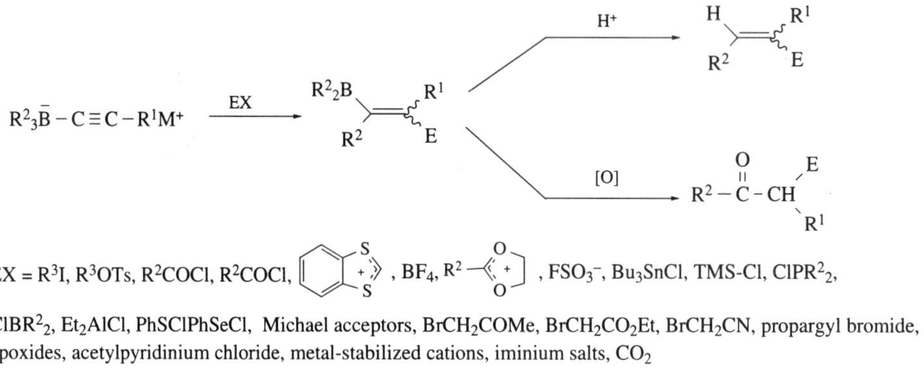

Scheme 35

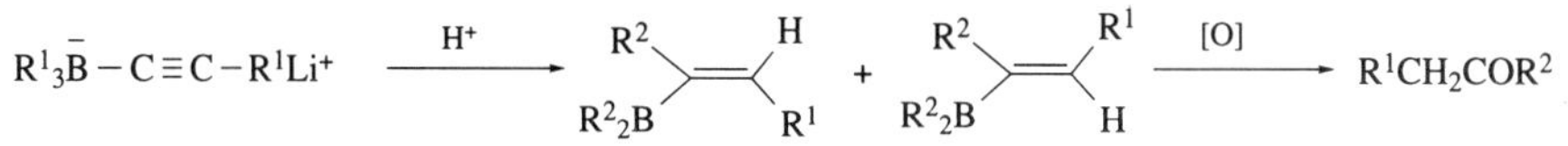

EX = R^3I, R^3OTs, R^2COCl, R^2COCl, [benzodithiolylium], BF_4, R^2-[dioxolanylium], FSO_3^-, Bu_3SnCl, TMS-Cl, $ClPR^2{}_2$, $ClBR^2{}_2$, Et_2AlCl, PhSCl PhSeCl, Michael acceptors, $BrCH_2COMe$, $BrCH_2CO_2Et$, $BrCH_2CN$, propargyl bromide, epoxides, acetylpyridinium chloride, metal-stabilized cations, iminium salts, CO_2

$$R^1{}_3\bar{B}-C\equiv C-R^1Li^+ \xrightarrow{H^+} \begin{array}{c}R^2\\ \diagup\\ R^2{}_2B\end{array}\!\!=\!\!\begin{array}{c}H\\ \diagdown\\ R^1\end{array} + \begin{array}{c}R^2\\ \diagup\\ R^2{}_2B\end{array}\!\!=\!\!\begin{array}{c}R^1\\ \diagdown\\ H\end{array} \xrightarrow{[O]} R^1CH_2COR^2 \quad (124)$$

Good stereoselectivity giving the migrating alkyl group *trans* to the electrophile has been observed with tributyltin chloride,[394,450] dialkylchloroboranes,[451] dialkylchlorophosphines,[452] and trimethylsilyl chloride.[453] Inverse stereochemistry is observed with α-haloketones, α-halonitriles, α-haloesters, propargyl bromide,[454] and 1,3-dioxolan-2-ylium salts.[455] Alkylating agents possessing two leaving groups such as dihalomethanes,[456] and ortho esters in the presence of titanium tetrachloride,[457] induce a double migration (Equation (125)). Intramolecular versions of these rearrangements have found valuable synthetic applications in the preparation of (Z)-γ-bisabolene[458] and exocyclic alkenes[459] (Equation (126) and Scheme 36).

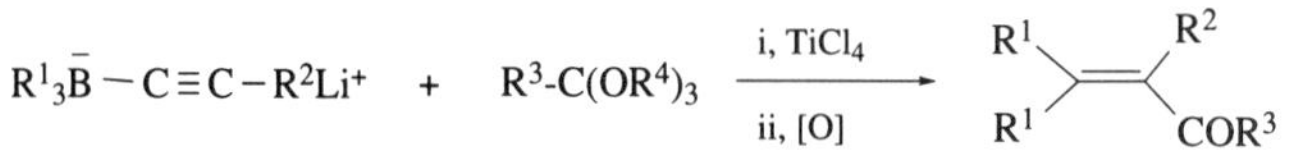

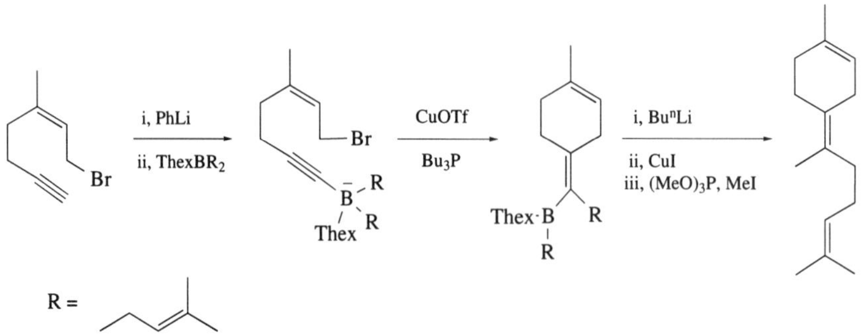

$$R^1{}_3\bar{B}-C\equiv C-R^2Li^+ + R^3-C(OR^4)_3 \xrightarrow[\text{ii, [O]}]{\text{i, TiCl}_4} \begin{array}{c}R^1\\ \diagup\\ R^1\end{array}\!\!=\!\!\begin{array}{c}R^2\\ \diagdown\\ COR^3\end{array} \quad (125)$$

Lithium 1-alkynyltrialkylborates also react rapidly with iodine. This reaction appears to involve electrophilic attack of the triple bond by iodine and subsequent 1,2-migration of an alkyl group from boron to carbon. The β-iodovinylboranes spontaneously undergo β-elimination to give alkynes in high yields (Equation (127)).[460-2] The relative migratory aptitudes of alkyl groups in a mixed borate roughly follow the order secondary alkyl > primary alkyl > tertiary alkyl.[463-5] Best results have been obtained with the combined use of the thexyl and methoxy "blocking" groups. This reaction can be extended to the synthesis of enynes, dienes, diynes, and butatrienes.[466]

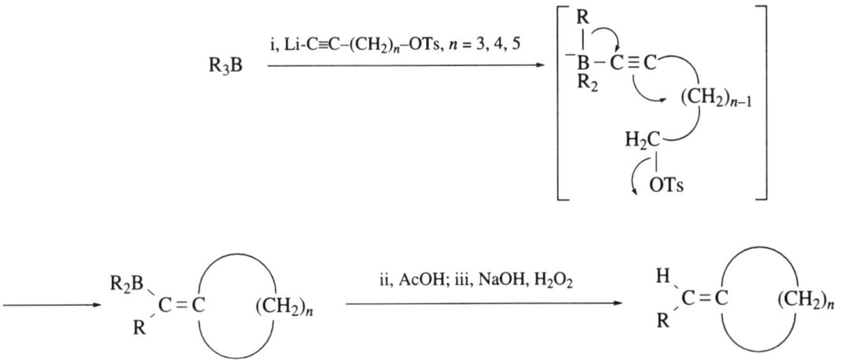

$$R^1{}_3\bar{B}-C\equiv C-R^2, Li^+ \xrightarrow{I_2} R^1{}_2BCR^1=CR^2I \xrightarrow{-R^1{}_2BI} R^1-C\equiv C-R^2 \quad (127)$$

(ii) Alkenylborate chemistry

Although they have been less well studied, most reactions of alkenylborates are comparable to those described for alkynylborates in Section 5.3.2.3(i). Protonation[467] and alkylation with alkyl halides,[427] aldehydes,[468] or epoxides[469] yield after oxidation, the corresponding alcohols or diols. The use of iodine provides convenient procedures for the synthesis of alkenes (Equations (128)–(130)).[470-3]

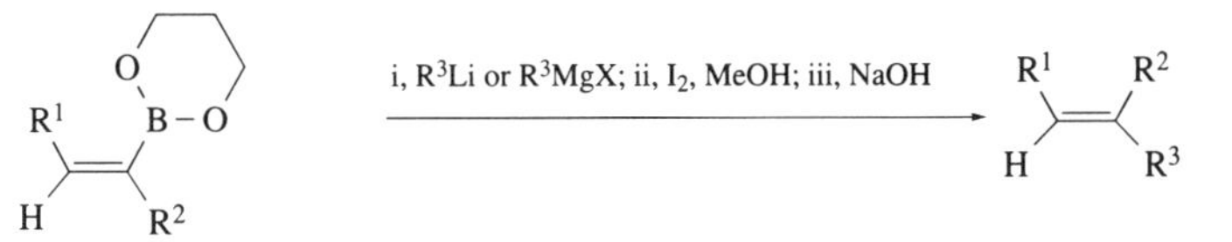

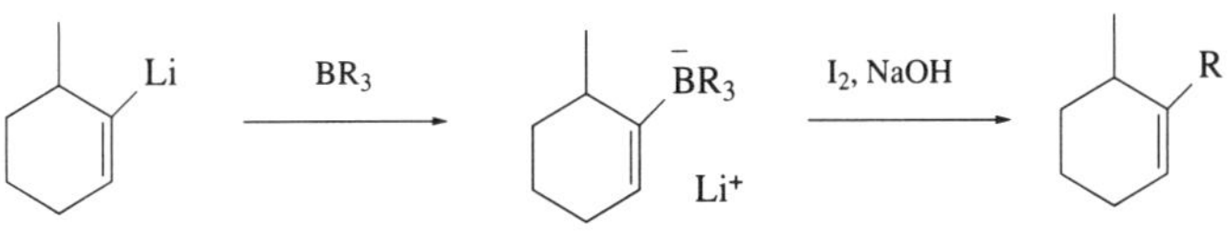

(iii) Aryl and heteroarylborate chemistry

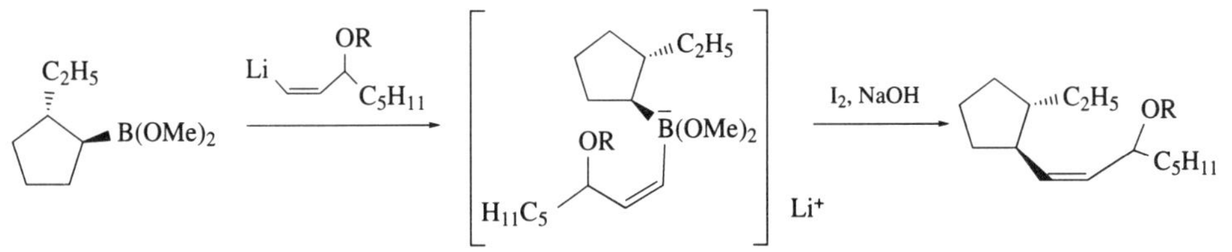

Aromatic organoborates behave to some extent like alkenylorganoborates. Reactions of indole derivatives are particularly noteworthy (Scheme 37),[474-6] and other heteroaryl systems such as 2-furyl, 2-pyrrolyl, and 2-thienyl have also been used in similar reactions.[27]

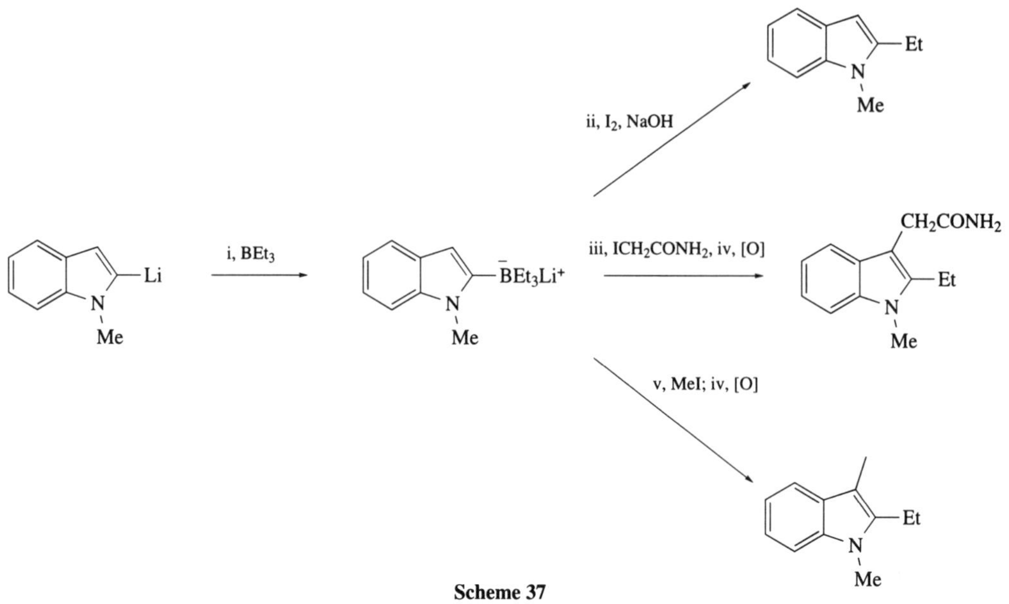

Scheme 37

(iv) The cyanoborate process

Although alkali-metal cyanotriorganylborates are stable salts, sequential 1,2-shifts from boron to carbon can be induced upon treatment with electrophiles. Acylating agents are the most effective and the cyanoborate process is a convenient alternative to the carbonylation and DCME reactions.[27,30] The proposed mechanism is illustrated in Scheme 38. Many types of structure can be readily produced in good yields. The use of thexyldialkylcyanoborates avoids wastage of the alkyl residues and cyclic ketones are obtained from 1,n-dienes by a hydroboration–cyanidation sequence (Equation (131)).[477] If the reaction is carried out with an excess of trifluoroacetic anhydride at higher temperature, a third migration occurs and tertiary alcohols are produced (Equation (132)).[478]

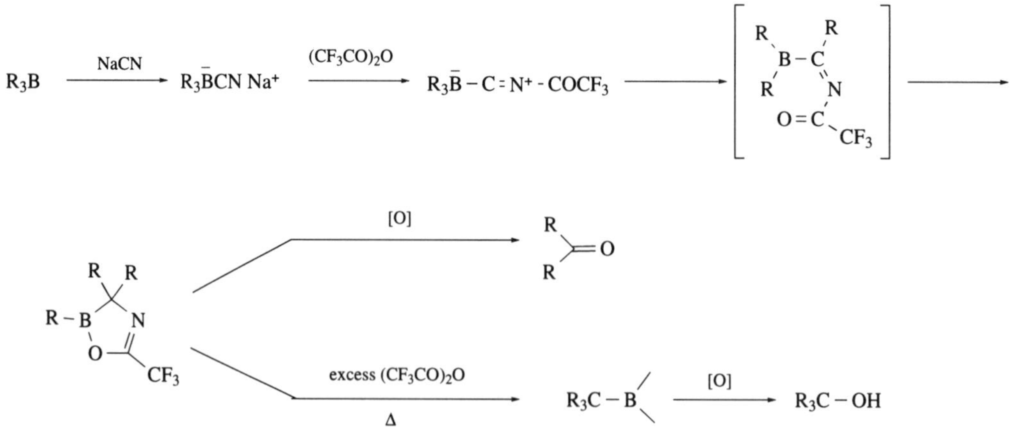

Scheme 38

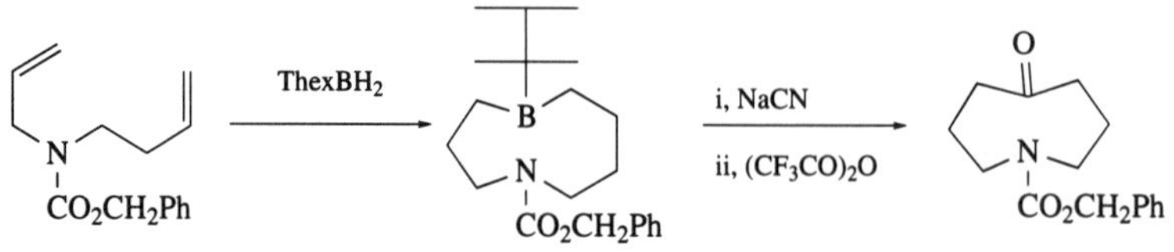

(131)

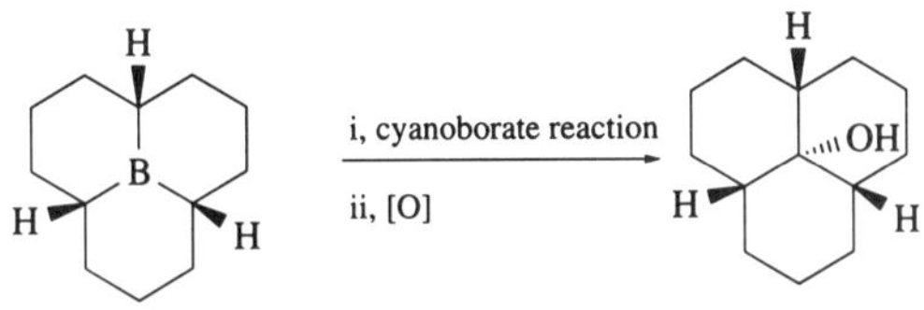

5.3.2.4 Cross-coupling reactions of organoboranes

Palladium-catalyzed cross-coupling of alkenylboranes with vinylic halides was discovered by Suzuki and co-workers and can be effected in the presence of an alkoxide or hydroxide base (Equation (133)).[479] The scope of these reactions has been widely extended over the past decade and progress reviewed several times.[38,143,401,480-3]

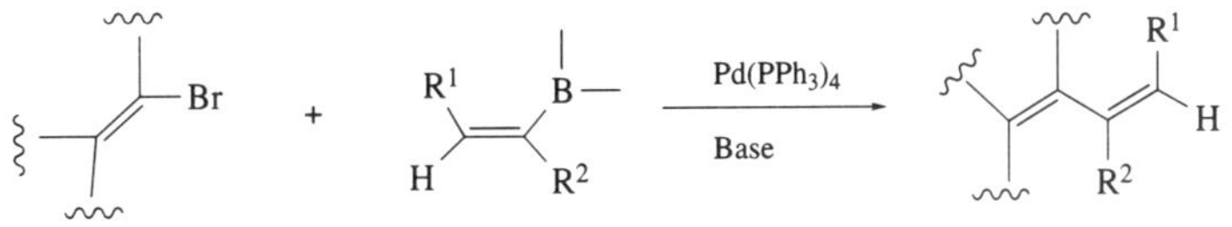

(i) Mechanism

The principal features of this cross-coupling reaction are as follows.[484] Only catalytic amounts of palladium complexes (1–3 mol.%) are required. The original configurations of both the starting alkenylboranes and the haloalkenes are retained in this regio- and stereospecific process. As shown in Scheme 39, 2 equiv. of base are required in the catalytic cycle. One equivalent is used to quaternize the boron derivative. The second is consumed in the metathetical displacement to form organopalladium alkoxide (R^1–Pd–OR^2) or organopalladium hydroxide (R–Pd–OH), which are believed to be more reactive than the organopalladium halide. This mechanism clearly shows the role of the base and therefore explains the failure of the coupling reaction in the absence of base.[485]

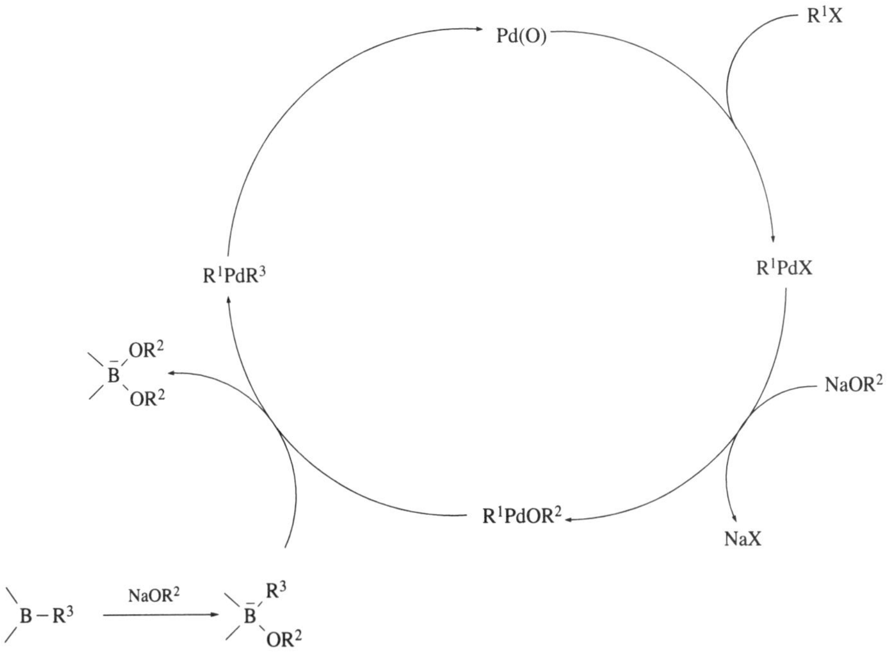

Scheme 39

(ii) Cross-coupling reactions of vinylboron derivatives

The palladium-catalyzed cross-coupling reaction of (*E*)- or (*Z*)-1-alkenylboranes, boronates or boronic acids with either (*E*)- or (*Z*)-1-alkenyl halides is a general method for the synthesis of (*E,E*)-, (*E,Z*)-, (*Z,E*)- or (*Z,Z*)-conjugated alkadienes in a stereo- and regiospecific manner.[38,401,481,484,486] The versatility of this method has been demonstrated by the stereospecific synthesis of natural products bearing conjugated alkadiene structures[38] such as bombykol[486,487] and its geometrical isomers (Equation (134)).

$$\text{HO}\text{-(CH}_2\text{)}_6\text{-CH=CH-B(OR)}_2 + \text{CH}_2\text{=CHBr} \xrightarrow[82\%]{\text{PdL}_4, \text{KOH}} \text{HO-(CH}_2\text{)}_6\text{-CH=CH-CH=CH-CH}_2\text{CH}_2\text{CH}_3 \quad (134)$$

A dramatic rate enhancement realized by use of thallium hydroxide as base was reported by Kishi and co-workers.[488] Under these conditions, coupling can be achieved at 0 °C almost instantaneously, thus allowing its application to substrates with sensitive functional groups as well as with large molecular weights. These efficient coupling conditions have been applied to the palytoxin synthesis and other biologically interesting and complex diene compounds.[489–92] 3-Bromo-2-alkenoates and 3-halo-2-alken-1-ones have been used as partners in this type of cross-coupling reaction with 1-alkenylboronates, leading to 2,4-alkadienoates[493] and -dienones[494,495] respectively.

The alkenylboron derivatives react not only with 1-alkenyl halides but also with a variety of organic halides, including 1-bromo-1-alkynes,[484] aryl halides,[487,496] and allylic or benzylic halides.[481,497] Pd(PPh$_3$)$_4$ and PdCl$_2$(PPh$_3$)$_2$ are good catalysts.[498] Two direct and stereocontrolled syntheses of trisubstituted alkenes by stepwise processes have been described. α-Bromovinylboronates react with organolithiums to give α-substituted (*Z*)-vinylboronates which undergo Suzuki cross-coupling to give trisubstituted alkenes (Equation (135)).[486] The palladium stepwise cross-coupling reaction of (*E*)-(2-bromo-1-alkenyl) dibromo- or diisopropoxyboranes, first with organozinc chloride derivatives and then with organic halides, gives rise to a wealth of unsaturated compounds (Scheme 40).[499–501] This widely applicable methodology has been used in a synthesis of prostaglandin B$_1$ methyl ester.[502]

$$\begin{array}{c}R^1\\ \diagdown \\ H \end{array}\!\!\!=\!\!\!\begin{array}{c}Br\\ \diagup \\ B(OPr^i)_2\end{array} \xrightarrow{i, R^2Li;\ ii, R^3X, Pd(PPh_3)_4, KOH} \begin{array}{c}R^1\\ \diagdown \\ H\end{array}\!\!\!=\!\!\!\begin{array}{c}R^3\\ \diagup \\ R^2\end{array} \quad (135)$$

(iii) Cross-coupling reactions of arylboronic acids

Benzeneboronic acid reacts with aryl halides in the presence of tetrakis(triphenylphosphine)-palladium(0) and 2 equiv. of aqueous sodium carbonate to give the corresponding biaryl derivatives (Equation (136)).[483,485] Gronowitz and co-workers have extended this coupling reaction to the synthesis of biaryls, heterobiaryls, tetraaryls, and condensed ring systems, and have shown that the use of 1,2-dimethoxyethane as solvent can suppress the deboronations sometimes observed under Suzuki coupling conditions.[503–5] A variety of palladium catalysts have been employed with success including Pd(dppb)Cl$_2$,[506] Pd(OAc)$_2$,[507] PdCl$_2$,[508] and Pd(dppf)(OAc)$_2$.[509] Pd[PPh$_2$(*m*-C$_6$H$_4$SO$_3$M)]$_3$, M = Na, K, is water soluble and has been shown to catalyze the cross-coupling process in water as solvent.[510] The best results are achieved with the use of relatively weak bases, usually sodium carbonate,[504,511] Et$_3$N,[509,512] TlOH,[488,489] Ba(OH)$_2$,[513] or K$_3$PO$_4$.[513] A very interesting feature of these coupling reactions is their chemoselectivity. They can tolerate almost all kinds of functionality on either coupling partner. Accordingly, this allows the synthesis of biaryls suitably substituted for further cyclization to polycyclic systems. This synthesis of condensed ring systems, as well as multiple aryl couplings leading to polyaryls or even polymers, has been thoroughly reviewed by Martin and Yang.[483] Directed *ortho*-metallation of heterocycles such as pyridines and diazines in conjunction with palladium-catalyzed cross-coupling reactions, constitutes a new methodology for the functionalization of π-deficient heterocycles.[514,515] Several total syntheses of alkaloids have been completed using these methodologies, including streptonigrin,[516] lavendamycin analogues,[517,518] and fascaplysin.[519] Convenient access to 4-aryl-1,2,3,6-tetrahydropyridine-1-carboxylates[520] and 3-indolyl-1,2,5,6-tetrahydropyridines[521] have been designed. Recently, aryl and vinyl triflates have been used as partners in palladium-catalyzed cross-coupling reactions with aryl- and 1-heterarylboron compounds.[522] In the presence of potassium phosphate, the reaction takes place readily in high yield in THF or dioxane.[523,524]

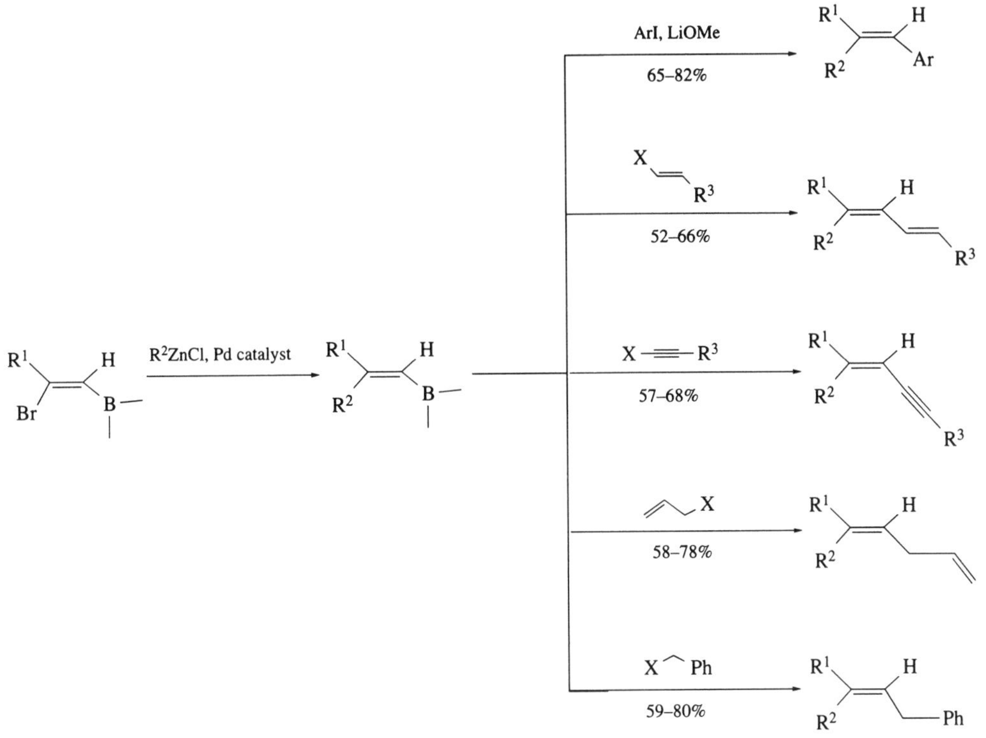

Scheme 40

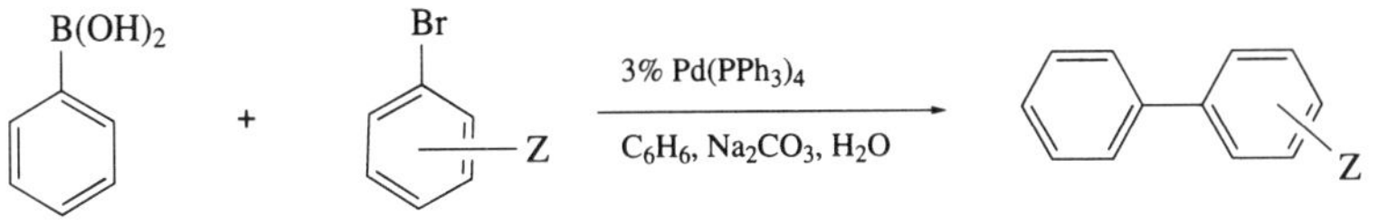

The palladium-catalyzed cross-coupling reaction of arylboronic acids with tyrosine triflate leads to a simple and efficient asymmetric synthesis of 4-arylphenylalanines.[525]

Sodium tetraarylborates have also been found to couple efficiently with vinyl and aryl triflates in DMF in the presence of Pd(PPh$_3$)$_4$ to afford arylalkenes and biaryls in good yields under mild conditions.[526,527]

(iv) Cross-coupling reactions of B-alkyl-9-BBN derivatives

Coupling reactions of organometallic reagents with alkyl groups having sp^3 carbons containing β-hydrogens have been severely limited due to competitive side reactions.[528] A solution to this problem may be the use of B-alkyl-9-BBN. These boranes, readily obtainable from alkenes by hydroboration with 9-BBN–H, undergo cross-coupling reaction with 1-halo-1-alkenes or haloarenes in the presence of a catalytic amount of dichloro-[1,1'-bis(diphenylphosphino)ferrocene]palladium(II) [PdCl$_2$(dppf)] and bases such as sodium hydroxide or potassium carbonate and phosphate, to give the corresponding alkenes and arenes chemo- and stereoselectively in good yields (Equations (137) and (138)).[528–30]

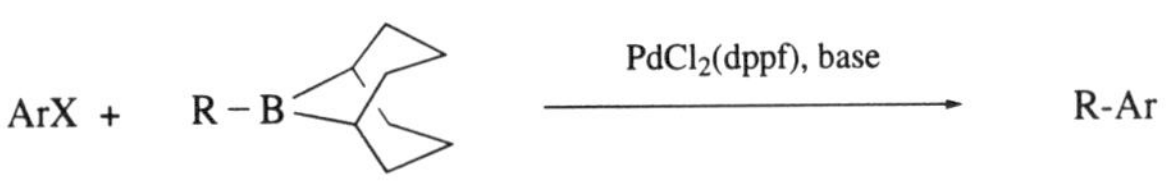

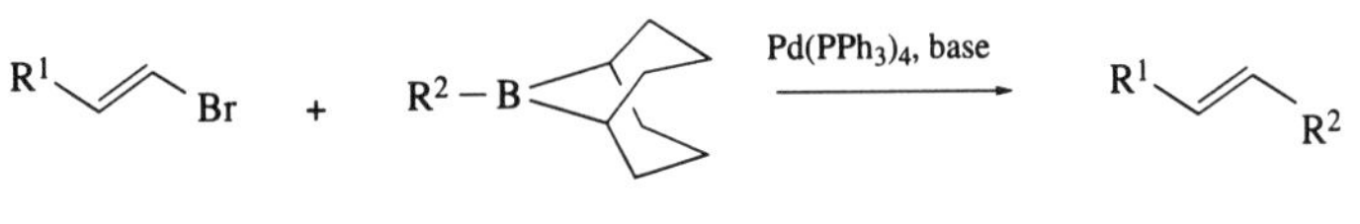

9-Alkyl-9-BBN derivatives undergo cross-coupling reaction with iodoalkanes possessing β-hydrogens, in the presence of a catalytic amount of $Pd(PPh_3)_4$ and K_3PO_4, to give the coupling products in good yields.[531] The hydroboration of haloalkadienes, followed by intramolecular cross-coupling, constitutes a short and efficient procedure for the synthesis of cycloalkenes, benzo-fused cycloalkenes, and exocyclic alkenes (Equations (139) and (140)).[524,528]

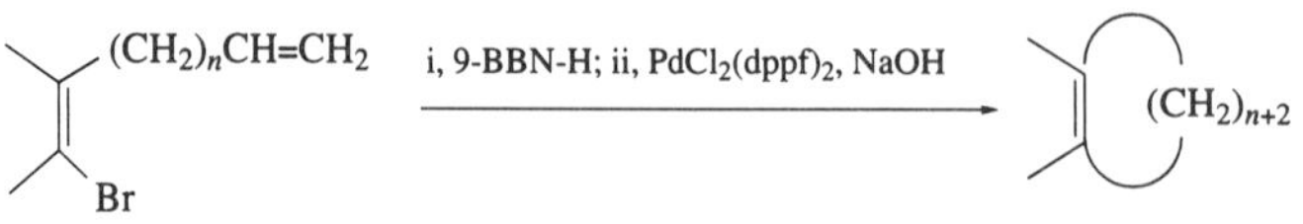

$$\text{(139)}$$

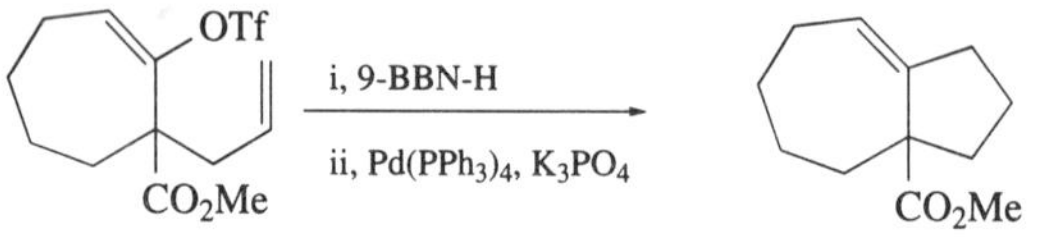

$$\text{(140)}$$

Efficient palladium-catalyzed cross-coupling of vinyl, alkynyl, and aryl bromides with trialkylboranes in the presence of $PdCl_2(dppf)$[532] or with Markovnikov vinylboranes[533] have been reported. Methylation[534] and trimethylsilylation[535] leading to allyl-, benzyl-, and propargylsilanes proceed efficiently.

Cross-coupling reactions of 9-alkyl-9-BBN with 1-haloalkenes or 1-iodoalkenes under a carbon monoxide atmosphere constitutes a direct synthesis of α,β-unsaturated ketones[536] or unsymmetrical ketones.[537] t-Butyl isocyanide can be used as the carbonylating agent in this reaction.[538]

5.3.2.5 Boron-stabilized carbanions in organic synthesis

The reaction of boron-stabilized carbanions with electrophiles may be useful for the preparation of new boranes[261] (see also Section 5.2.2.4). When coupled with oxidation, these reactions lead to alcohols. Hence, alkylations of anions derived from B-alkyldimesitylboranes are efficient reactions which offer general syntheses of primary and secondary alcohols (Equations (141)–(143)).[539]

$$Mes_2BCH_3 \xrightarrow{MesLi} Mes_2BCH_2Li \xrightarrow{R^1X} Mes_2BCH_2R^1 \xrightarrow{[O]} R^1CH_2OH \quad (141)$$

$$Mes_2BCH_2R^1 \xrightarrow[\text{ii, }R^2X]{\text{i, MesLi}} Mes_2BCHR^1R^2 \xrightarrow{[O]} R^1R^2CHOH \quad (142)$$

$$Mes_2BCHR^1R^2 \xrightarrow[\text{ii, }R^3X]{\text{i, MesLi}} Mes_2BCR^1R^2R^3 \xrightarrow{\text{iii, [O]}} R^1R^2R^3COH \quad (143)$$

Although t-alkylboranes can be prepared efficiently, they cannot be readily oxidized to the corresponding t-alcohols. α-Boryl anions are also versatile reagents for the introduction of the hydroxymethyl or hydroxy ether group into a wide range of mono-, di-, and trisubstituted oxiranes (Scheme 41).[540] These reactions are generally regiospecific and proceed under steric control with inversion of configuration. Boron-stabilized alkenyl carbanions can be produced by specific displacement of appropriate gem-dimetalloalkenes, alkylated, and the resulting borane oxidized to produce ketones in good yields (Equation (144)).[268]

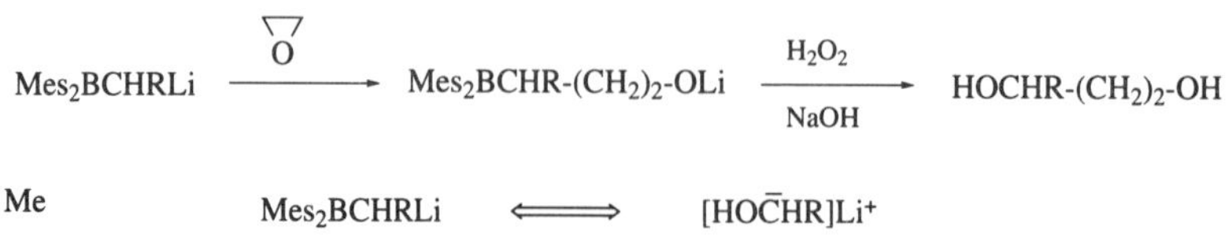

$$Mes_2BCHRLi \xrightarrow{\triangle\text{O}} Mes_2BCHR-(CH_2)_2-OLi \xrightarrow[\text{NaOH}]{H_2O_2} HOCHR-(CH_2)_2-OH$$

R = H, Me $Mes_2BCHRLi \rightleftharpoons [HO\overline{C}HR]Li^+$

Scheme 41

$$R^1CH=\bar{C}BR^2{}_2 \; Li^+ \xrightarrow{\text{i, } R^3X; \text{ ii, NaOAc/H}_2\text{O}_2} R^1CH_2COR^3 \quad (144)$$

α-Iodo and α-bromoalkylboronic esters readily insert zinc dust in THF to give 1,1-bimetallics of boron and zinc (see Section 5.2.2.4). α-Iodoalkenylboronic esters also react with zinc dust in dry N,N-dimethylacetamide and furnish 1,1-boron, zinc alkenylbimetallics (Equation (145)).[541] After transmetallation, the 1,1-boron, copper bimetallics react with a wide range of electrophiles to give polyfunctional boronic esters. Polyfunctional ketones are produced in good to excellent yields by oxidation.[541]

$$RCH=C\begin{smallmatrix}B(O)O\\X\end{smallmatrix} \xrightarrow{\text{i, E}^+; \text{ ii, [O]}} RCH_2-C\begin{smallmatrix}O\\E\end{smallmatrix} \quad (145)$$

X = ZnI
X = Cu(CN)ZnI

Boron-stabilized carbanions have long been known to react with benzaldehyde and ketones to yield alkenes.[269,542] The use of dimesitylboryl-stabilized carbanions in boron-Wittig reactions has been recently reviewed.[261] The general trends are as follows.
(i) Aromatic ketones give the corresponding alkenes directly in good yields (70–90%)[543] whereas benzaldehyde gives complex mixtures. If the condensation reactions followed by oxidation are carried out at low temperature, *erythro*-1,2-diols are obtained in good yields.[544]
(ii) (*E*) or (*Z*)-alkenes may be obtained at will according to the type of quench used. The addition of trimethylsilyl chloride at −110 °C and treatment of the resulting trimethylsilyl ether with HF/MeCN gives (*E*)-alkenes in good yields.[545] By contrast, the use of trifluoroacetic anhydride, again at low temperature, followed by decomposition of the intermediate at 25 °C gives (*Z*)-alkenes.[545] The reactions with aliphatic aldehydes, depending on reaction conditions and substitution at the carbanionic center, give different results including an unexpected redox process yielding ketones.[546]

The versatility of the boron-Wittig reaction is also illustrated by a new allene synthesis involving the reaction of boron-stabilized alkenyl carbanions with aldehydes (22–65% yields) (Equation (146)).[547]

$$R^1CH=\bar{C}BR^2{}_2 \; Li^+ \xrightarrow{\text{i, } R^3CHO; \text{ ii, CF}_3\text{CO}_2\text{H}} R^1CH=C=CHR^3 \quad (146)$$

5.3.2.6 Reactions of allylboranes and related compounds

Allylic boron compounds show specific behavior with respect to different classes of organic compounds. Allylation of aldehydes and ketones is probably one of the most widely studied aspects of their reactivity, but other reactions of allylboranes should also be expected to find useful synthetic applications. The preparation and the properties of these reagents have been the subjects of a number of reviews.[26,27,30,548–51] A characteristic feature of these boron derivatives is their ability to undergo metallotropic rearrangement at a rate depending on substituents and temperature. For example, attempts to prepare the (*E*)- or (*Z*)-isomers of tricrotylborane give the equilibrium mixture rather than a pure product (Equation (147)). However, dialkylallylboranes can be used in diastereoselective reactions either at low temperature, which slows down the equilibration rate, or if the predominant isomer is the desired one. By contrast, allylboronic esters are relatively stable towards isomerization. This explains their wide use in the synthesis of homoallylic alcohols. With two amino substituents, allylboranes are particularly resistant to boratropy.[552,553]

$$R^2\!\!\diagdown\!\!\diagup\!\!\diagdown\!BR^1{}_2 \rightleftharpoons \underset{BR^1{}_2}{R^2\!\!\diagdown\!\!\diagup\!\!\diagdown} \rightleftharpoons \underset{R^2}{\diagup\!\!\diagdown\!\!\diagup\!BR^1{}_2} \quad (147)$$

Allylboranes are accessible by different routes as illustrated in Equations (148)–(153):
(i) transmetallation of allylmetals;[197,554–7]
(ii) coupling reactions of 1-halo-1-alkenes with zinc derivatives;[558,559]

(iii) reaction of an α-haloalkylboronate with an alkenyl organometallic reagent;[560–2]
(iv) hydroboration of 1,2- or 1,3-dienes;[108,563] and
(v) homologation of an alkenylboronate.[238,564,565]

Many types of allylboranes are available by these approaches. The recent syntheses of a series of new boranes are noteworthy (Figure 10).[549,550,566]

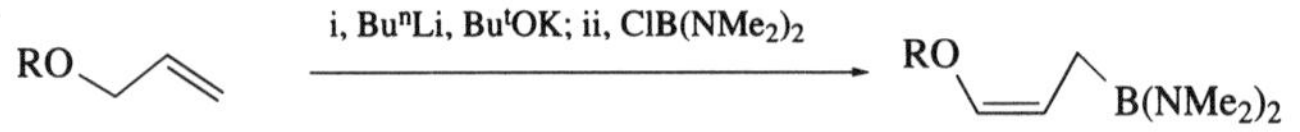
RO
i, BuⁿLi, BuᵗOK; ii, ClB(NMe2)2
RO
B(NMe2)2
(148)

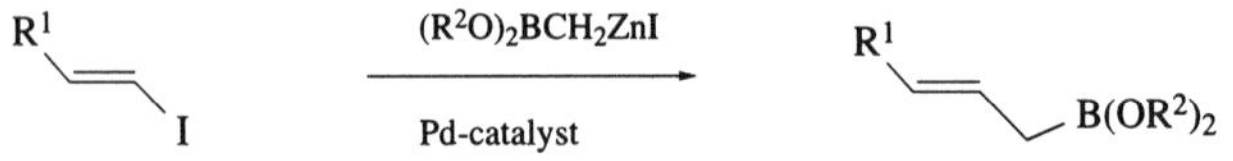
R1
I
(R2O)2BCH2ZnI
Pd-catalyst
R1
B(OR2)2
(149)

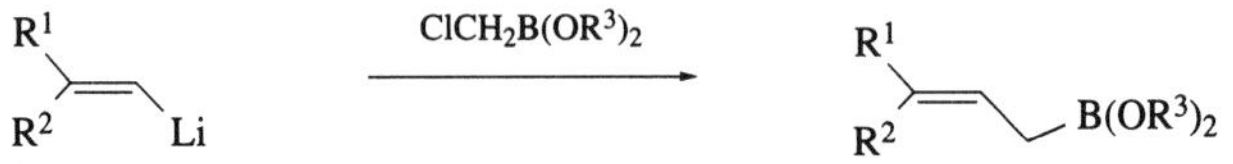
R1
R2
Li
ClCH2B(OR3)2
R1
R2
B(OR3)2
(150)

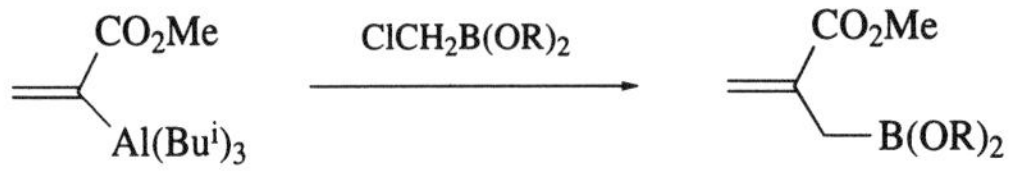
CO2Me
Al(Buⁱ)3
ClCH2B(OR)2
CO2Me
B(OR)2
(151)

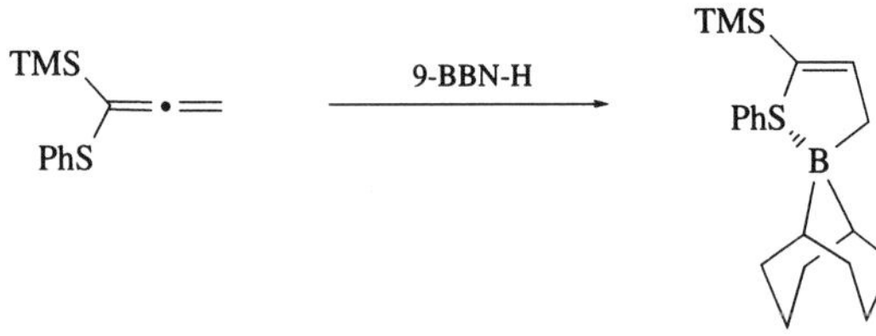
TMS
PhS
9-BBN-H
TMS
PhS
B
(152)

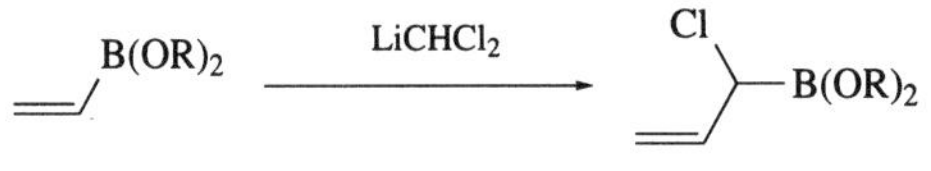
B(OR)2
LiCHCl2
Cl
B(OR)2
(153)

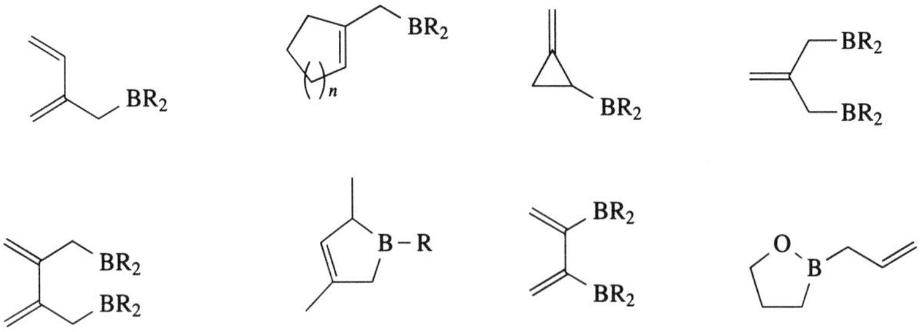
BR2
BR2
n
BR2
BR2
BR2
BR2
BR2
B–R
BR2
BR2
O
B

Figure 10 Some new allylboranes.

(i) Addition of allylboranes to aldehydes, ketones, and related species

Allylboranes react with carbonyl compounds to form a new carbon–carbon bond with allylic rearrangement (see also Section 5.4.3) (Scheme 42). A systematic study has revealed the importance of the solvent, the structure of the aldehyde, the temperature, and the nature of the substitution at boron on the rate of allylboration.[567] In particular, allyldialkylboranes have proven to be among the most reactive, but also the least configurationally stable, of the allylboron reagents.

Addition of (*E*)- and (*Z*)-crotylboranes to aldehydes proceeds with high simple diastereoselection (Scheme 43). These results have been rationalized in terms of a chairlike cyclic six-membered transition state. However, the energy difference is not so large compared with a boatlike transition state and both possibilities must be considered in a discussion of the diastereoselection. (*E*)-Allylboranes give rise mainly to the *anti* diastereoisomer whereas (*Z*)-allylboranes give rise to the *syn* diastereoisomer.[548,551]

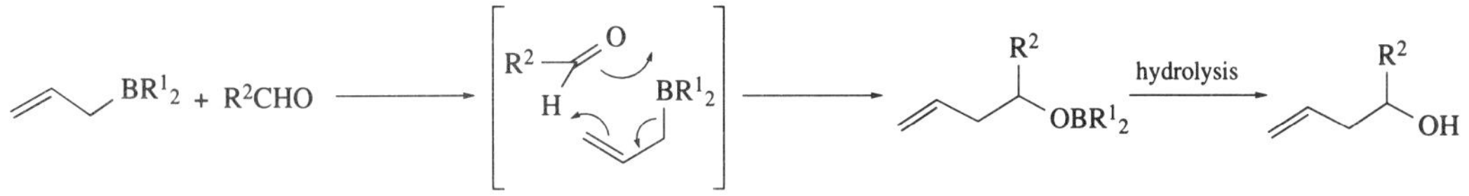

Scheme 42

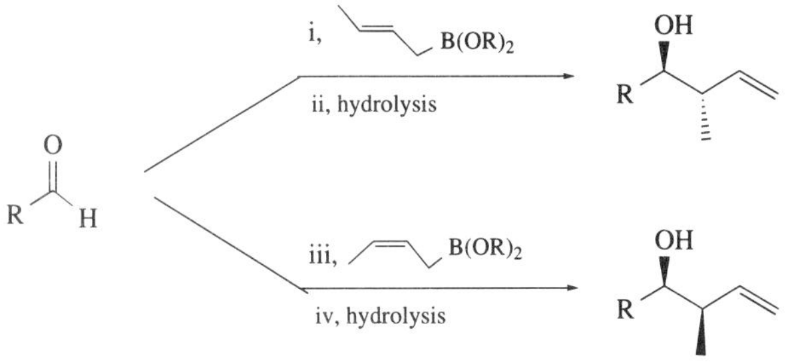

Scheme 43

γ-Trimethylsilyl-substituted allyldialkylboranes smoothly condense with aldehydes and ketones to give (*E*)- or (*Z*)-1,3-butadienes depending on the acidic or basic work-up (Scheme 44).[108,109]

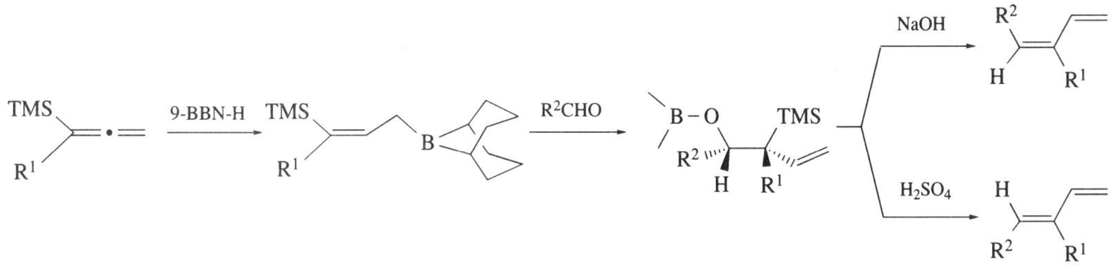

Scheme 44

Addition of α-substituted allylboronates to aldehydes yields preferentially the (*Z*)-isomers. The origin of this stereoselectivity has been discussed recently (Equation (154)).[568,569]

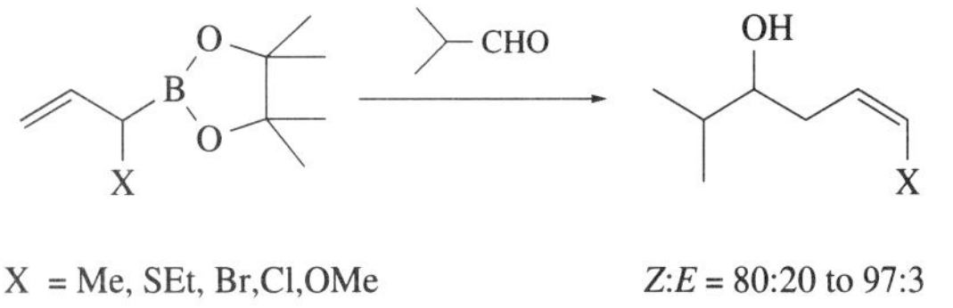

$$X = Me, SEt, Br, Cl, OMe \qquad Z:E = 80:20 \text{ to } 97:3 \tag{154}$$

Diastereomerically pure *syn*-homoallylic alcohols are obtained from (*Z*)-pentenylboronates and aldehydes. In contrast, addition reactions of the corresponding (*E*)-derivatives give mixture of *anti* isomers (Scheme 45).[570] (For addition of (*E*)- and (*Z*)-pentenylboronates to ketones, see Hoffmann and Sander.[571])

The introduction of a chiral center in the α-position of the aldehyde causes a 1,2-asymmetric induction. As a result, a preferred attack on the *re*- or the *si*-face may occur and different diastereoisomeric adducts will be formed. Diastereofacial selectivity is then determined both by the substitution of the borane and by the electronic structure of the aldehyde. For example, the reactions of (*Z*)-crotylboronates with glyceraldehyde acetonide are highly stereoselective. In contrast, (*E*)-crotylboronates are clearly less efficient. The selectivity is intermediate for the parent derivative (Scheme 46).[572]

Intramolecular allylboration reactions of (8-oxo-2-octenyl)boronates proceeds with excellent diastereoselectivity. The (*E*)-isomer yields the *trans*-2-vinylcyclohexanol with 99.5% diastereoselectivity and the corresponding (*Z*)-isomer leads to the *cis* isomer with diastereoselectivity >99.8% (Equation (155)).[559] Other related species including bromoacetaldehyde,[573] α-hydroxyketones,[574] α-carboxylic ketones,[575] imines,[576–9] α-iminoesters,[580] ethylidenemalonates,[581] oximes,[582] and sulfenimides[583] also react similarly, sometimes with high stereocontrol. In addition to allylboranes, allenyl-

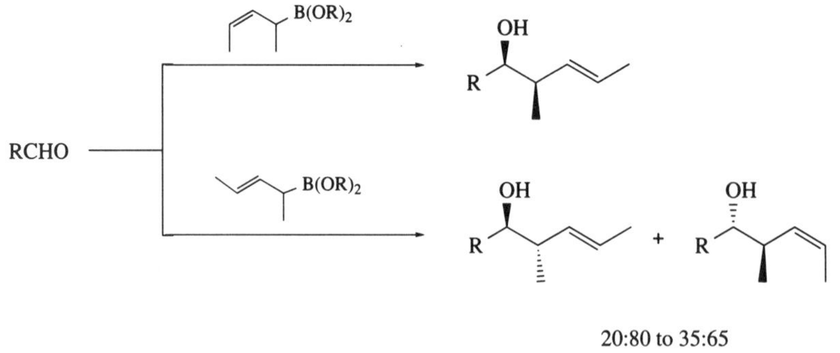

Scheme 45

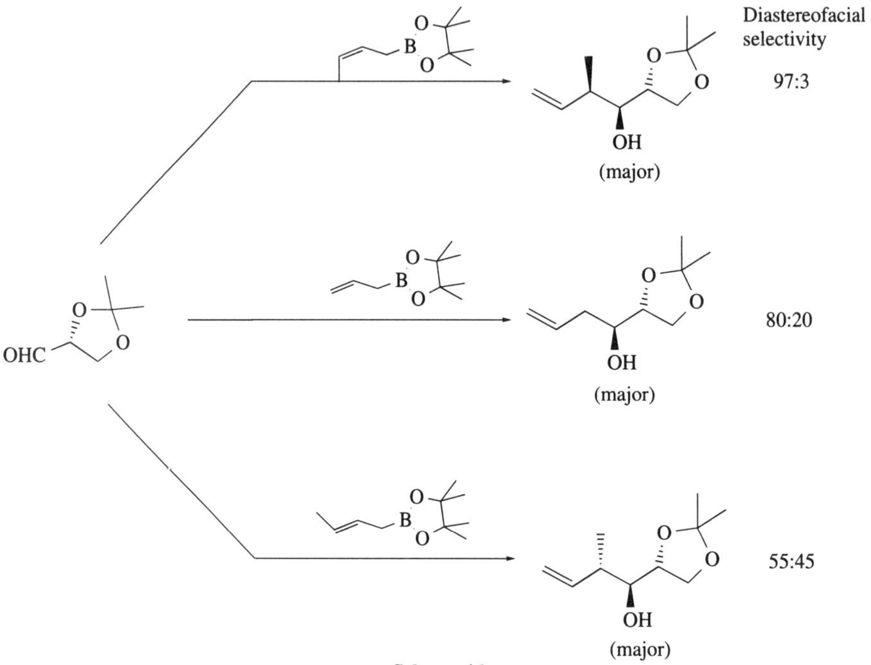

20:80 to 35:65

Scheme 46

Diastereofacial selectivity
97:3
80:20
55:45

and propargylboronic derivatives display similar physical and chemical properties. On warming, rearrangement of allenylboranes yields propargylboranes and these two species react with carbonyl derivatives in a closely related manner to that described previously.[584] They have generally been prepared by transmetallation[197,585–7] or by rearrangement of an α-halo- or α-acetoxyalkynylborate (Scheme 47).[588] Allenic boranes react with aldehydes to give homopropargylic alcohols while propargylic boranes yield allenic alcohols (Scheme 48).[586–90]

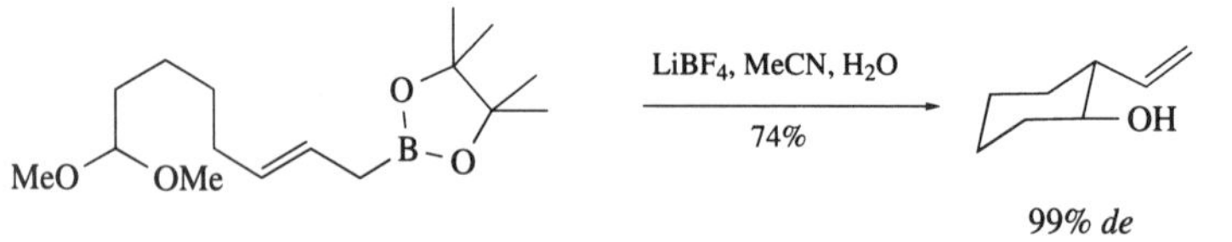

$$\text{(155)}$$

74%
99% de

(ii) Other aspects of the reactivity of allylboranes

The presence of a double bond is responsible for the particular reactivity of allylboranes. For example, their complexing ability is higher than that of their alkyl analogues. They are also readily protonolyzed. Most of the reactions of allylboranes proceed with the participation of the boron–allyl system involving allylic rearrangement. However, oxidation with H_2O_2/OH^-, reaction with ethyl diazoacetate and homologation with Cl_2CHLi are known to occur by direct rupture of the boron–carbon bond. Numerous other aspects of the particular reactivity of allylboranes have been extensively

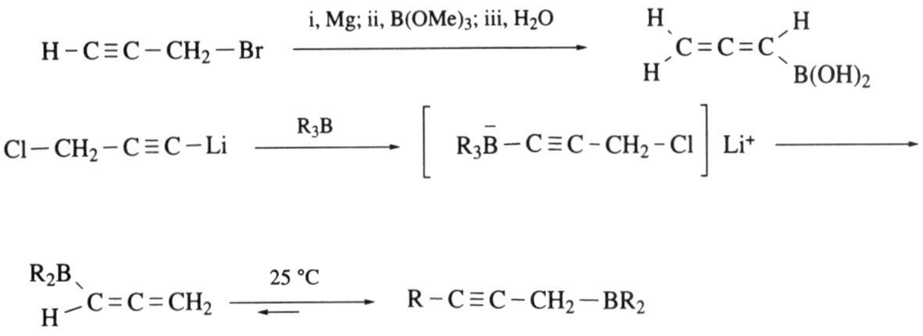

$$H-C\equiv C-CH_2-Br \xrightarrow{\text{i, Mg; ii, B(OMe)}_3\text{; iii, H}_2\text{O}} \underset{H}{\overset{H}{>}}C=C=C\underset{B(OH)_2}{\overset{H}{<}}$$

$$Cl-CH_2-C\equiv C-Li \xrightarrow{R_3B} \left[R_3\bar{B}-C\equiv C-CH_2-Cl \right] Li^+ \longrightarrow$$

$$\underset{H}{\overset{R_2B}{>}}C=C=CH_2 \xrightleftharpoons{25\,^\circ C} R-C\equiv C-CH_2-BR_2$$

Scheme 47

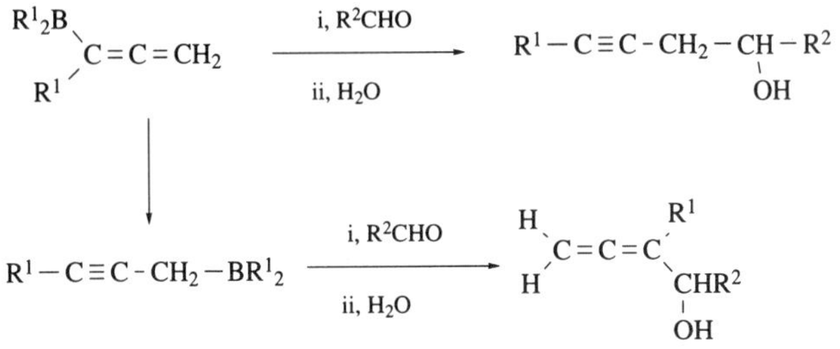

$$\underset{R^1}{\overset{R^1{}_2B}{>}}C=C=CH_2 \xrightarrow[\text{ii, H}_2\text{O}]{\text{i, R}^2\text{CHO}} R^1-C\equiv C-CH_2-\underset{OH}{\overset{}{CH}}-R^2$$

$$R^1-C\equiv C-CH_2-BR^1{}_2 \xrightarrow[\text{ii, H}_2\text{O}]{\text{i, R}^2\text{CHO}} \underset{H}{\overset{H}{>}}C=C=C\underset{\underset{OH}{CHR^2}}{\overset{R^1}{<}}$$

Scheme 48

studied.[26,27,30,549–51] The reactions of triallylboranes with alkynes are of particular interest. 3-Substituted-1,5-diallyl-1-boracyclohex-2-enes or 7-substituted derivatives of 3-allyl-3-borabicyclo[3.3.1]non-6-ene are obtained depending on the reaction temperature (Scheme 49).[591] Further transformations give access to a large variety of new organoboranes (Equations (156) and (157)).[549]

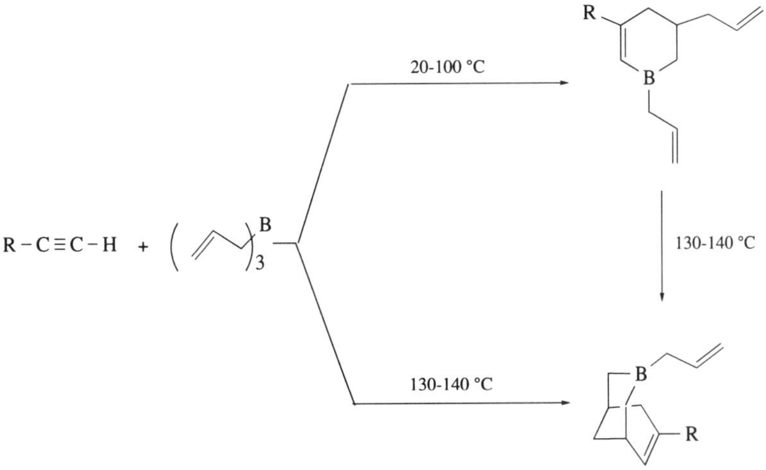

Scheme 49

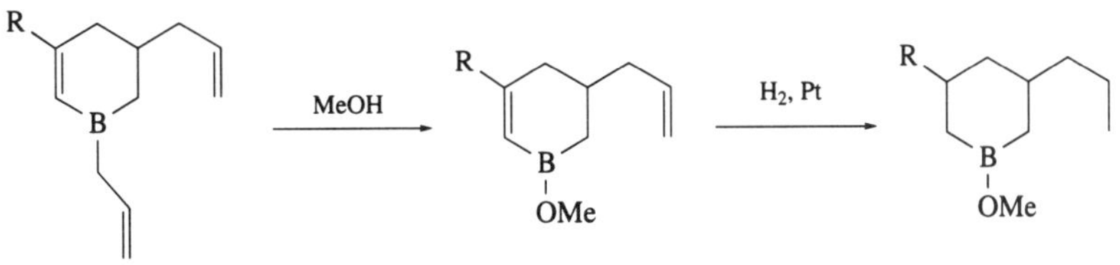

(156)

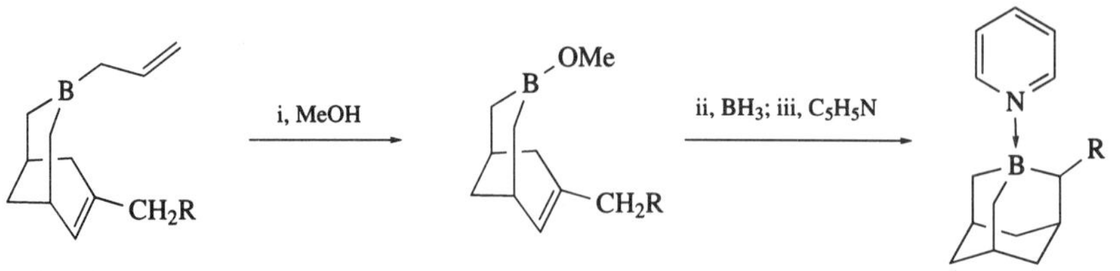

(157)

5.3.2.7 Boron-mediated aldol methodology

Alkenyloxyboranes were soon recognized as interesting enolate reagents for the directed aldol reaction,[592,593] and this methodology has become one of the most important procedures for diastereoselective and enantioselective C–C bond formation (see also Section 5.4.4).[594–600]

(i) Preparation of boron enolates

A variety of protocols are available to prepare boron enolates.[601] Most of the methods initially developed did not use direct enolization procedures; these include the oxidation of vinylboronates[602–4] and reactions of α-diazocarbonyl compounds,[412,414,601,605] sulfur ylides,[606–8] and halogen-substituted enolates[609] with trialkylboranes. Organoboranes react with α,β-unsaturated aldehydes and ketones to give vinyloxyboranes (Equation (158)).[610–17] The available data indicate that the reaction proceeds by a radical chain mechanism (see also Section 5.3.3).[616,617] Acylation of boron-stabilized carbanions with acid chlorides, acid anhydrides, or carboxylic esters produces vinyloxyboranes.[237,273,618] Et_3B has been shown to react with α-iodo ketones at low temperature to give the corresponding alkenyloxyboranes.[619]

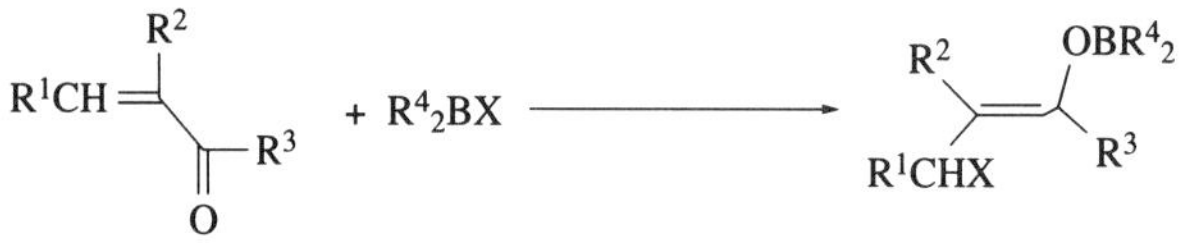

$$R^1CH=\underset{O}{\overset{R^2}{C}}-R^3 + R^4{}_2BX \longrightarrow \underset{R^1CHX}{\overset{R^2}{C}}=\underset{R^3}{\overset{OBR^4{}_2}{C}} \quad (158)$$

X = H, alkyl, 1-alkenyl, 1-alkynyl, SR, Br

The most widely used method for the production of alkenyloxyboranes, however, is the enolization of ketones. Since it was soon recognized that the diastereoselectivities of aldol reactions with aldehydes correspond to the Z/E ratios of alkenyloxyboranes,[593,620] general methods for the highly stereoselective enolization of ketones have been designed. This can be carried out under mild conditions by reaction of ketones with Lewis acids of the type L_2BX which have two alkyl ligands and one electronegative group X (Cl, OTf, etc.) in the presence of a tertiary amine ($Pr^i{}_2NEt$ or Et_3N).[595,600] Systematic studies by Evans et al.[621,622] and Masamune and co-workers[600] revealed that the structural features of the ketone, tertiary amine and Lewis acid L_2BX individually contribute to the observed enolate stereoselection. Dialkylboron chloride and triflate reagents with bulky ligands (e.g., L = Chx or Ipc) give different regio- and stereoselectivities for butanone and diethyl ketone (Scheme 50).[623–9] A tentative rationale for these observations has been given by Goodman and Paterson[630] by consideration of the conformations and electronic properties of the intermediate ketone–L_2BX complexes. An alternative explanation of these stereoselectivities as a function of X in L_2BX has been proposed by Corey and Kim,[631] in which enolization stereoselectivity would depend on triflate being a better leaving group than chloride. Triflate reagents would form (Z)-enol borinates through an E1-like mechanism, whereas chloride reagents would form (E)-enol borinates by an E2-like mechanism. An examination of the effect of the leaving group X in L_2BX has been done by Brown et al.[632] They found that L_2BX with good leaving groups such as triflate, mesylate, and iodide favor the formation of (Z)-enol borinates. A large number of Lewis acids L_2BX are available for enolization of ketones under kinetic conditions some of which are given in Scheme 51.[618,621–5,627,628,632–8]

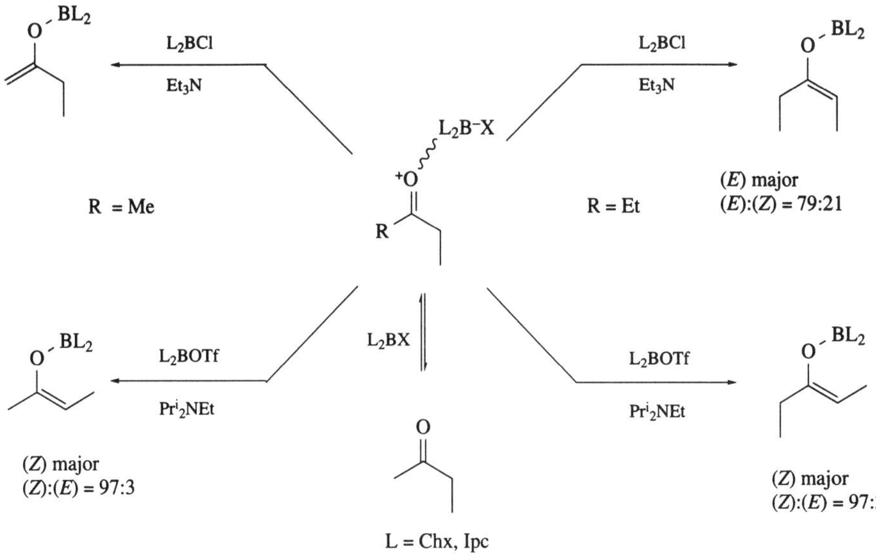

Scheme 50

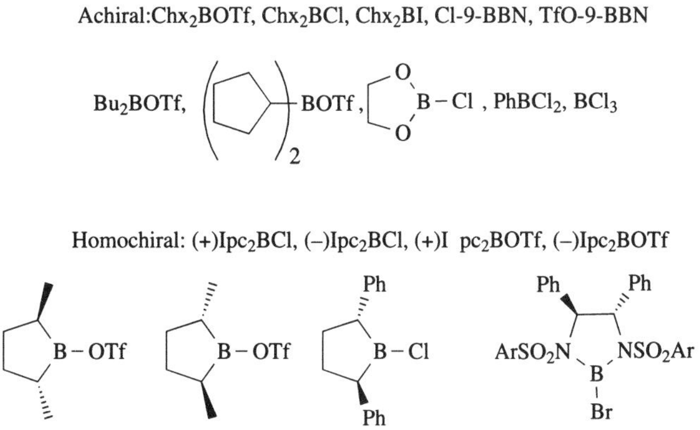

Achiral: Chx$_2$BOTf, Chx$_2$BCl, Chx$_2$BI, Cl-9-BBN, TfO-9-BBN

Bu$_2$BOTf, (cyclopentyl)$_2$BOTf, (dioxaborolane)B–Cl, PhBCl$_2$, BCl$_3$

Homochiral: (+)Ipc$_2$BCl, (−)Ipc$_2$BCl, (+)Ipc$_2$BOTf, (−)Ipc$_2$BOTf

Scheme 51

(ii) Diastereoselectivity of boron-mediated aldolization

The following general trend is observed: the ratio *syn/anti* of aldols depends on the enolate geometry. (Z)-Enolborinates are stereodivergent, the (Z)-enolates giving the *syn*-aldols and the (E)-enolates the *anti*-aldols.[595–600] This is illustrated in Scheme 52.[639] Conversely, enolboronates are stereoconvergent, both (E)- and (Z)-enolates leading to the same *syn*-aldol.[640,641] Explanations to account for these results are based on chairlike transition states originally proposed by Zimmerman and Traxler.[642] The existence of boatlike structures has been proposed by Evans *et al.*[595] to explain the behavior of certain boron enolates and by Hoffmann *et al.*[603] and Gennari *et al.*[643] to rationalize the *syn*-selectivity of *E*-dialkoxyvinylboranes.[604] More sophisticated calculations by Houk and co-workers,[644,645] Gennari and co-workers,[646] and Bernardi *et al.*[647] suggest three possible transition structures corresponding to highly asynchronous "pericyclic" transition states where the metal–oxygen bond is almost completely formed and the C–C bond still very long. At the MC-SCF level the twist-boat structure is more stable than the chair structure by 7.1 kJ mol^{-1} and more stable than the half-chair structure by 10.5 kJ mol^{-1}.

5.3.3 Radical Reactions of Organoboranes

Two types of process predominate:[27] (i) bimolecular homolytic substitution at boron, and (ii) abstraction of an α-hydrogen atom (Scheme 53). The former type of reactivity has been demonstrated

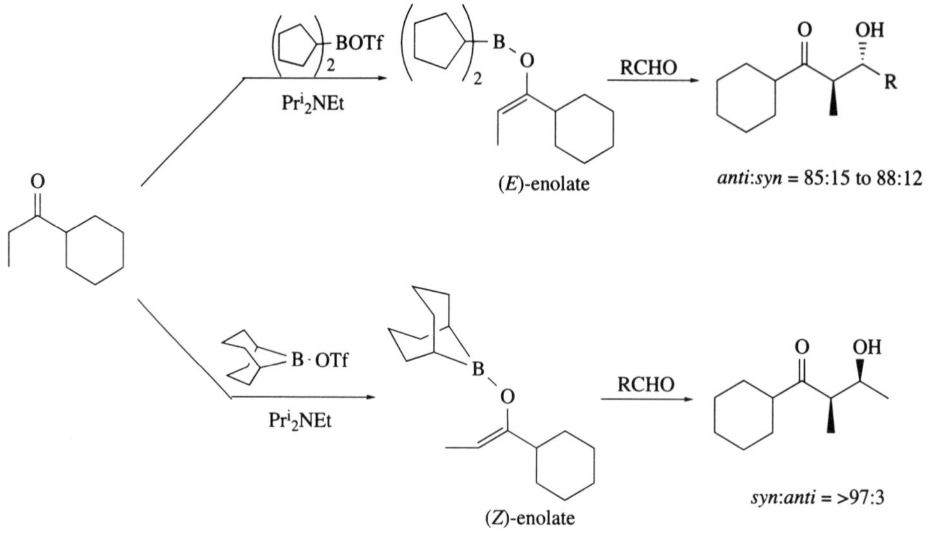

Scheme 52

in the autoxidation of alkylboranes (see Section 5.3.1.2(i)). Alkyl peroxides or alcohols are obtained depending on the reaction conditions. Similarly, dialkylthioboronic esters are prepared from trialkylboranes and thiols.[648] The formation of α-bromoalkylboranes represents the main example of α-hydrogen abstraction (see Section 5.3.2.1(i)).

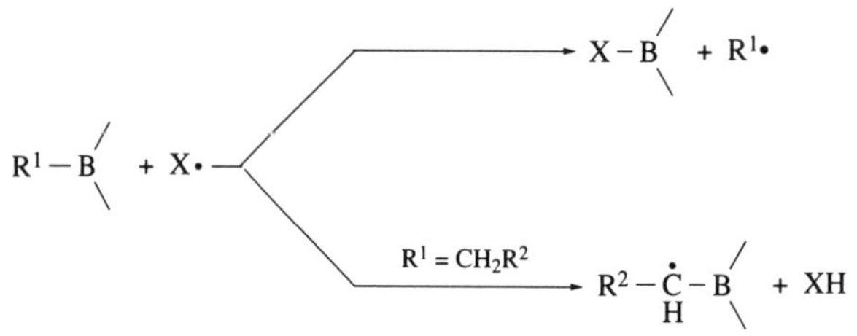

Scheme 53

1,4-Addition reaction of organoboranes to α,β-unsaturated carbonyl compounds has found valuable synthetic applications. The mechanism involves the generation of a radical under the influence of oxygen and its addition to the carbon–carbon double bond. The intermediate is trapped by the trialkylborane to form the enolborinate which can be hydrolyzed (Scheme 54). Most of this chemistry has been reviewed.[127,649] The reaction is general for all types of trialkylboranes and numerous derivatives of α,β-unsaturated ketones, esters, aldehydes, nitriles, and epoxides (Equations (159)–(161)).

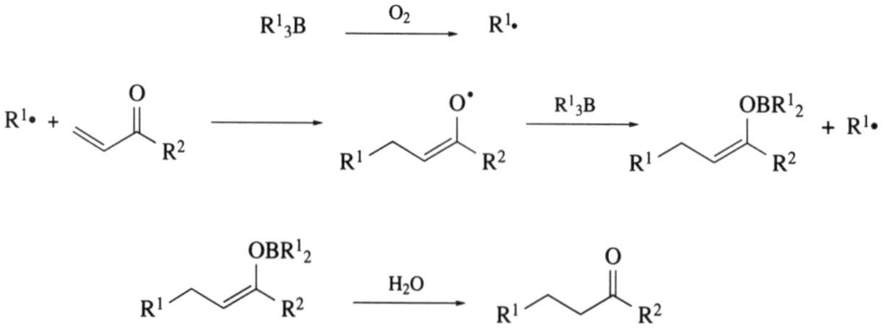

Scheme 54

Triethylborane has also been used as an initiator and terminator of free radical additions. The resulting boron enolates were efficiently trapped by carbonyl compounds or methanol (Scheme 55).[650]

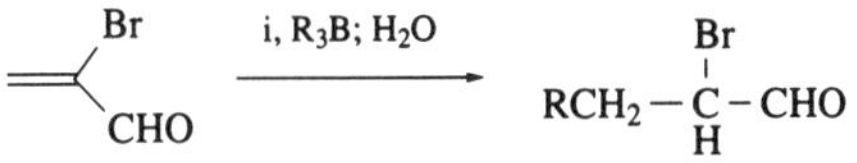

$$\text{CH}_2=\text{C(Br)CHO} \xrightarrow{\text{i, R}_3\text{B; H}_2\text{O}} \text{RCH}_2-\underset{\text{H}}{\overset{\text{Br}}{\text{C}}}-\text{CHO} \quad (159)$$

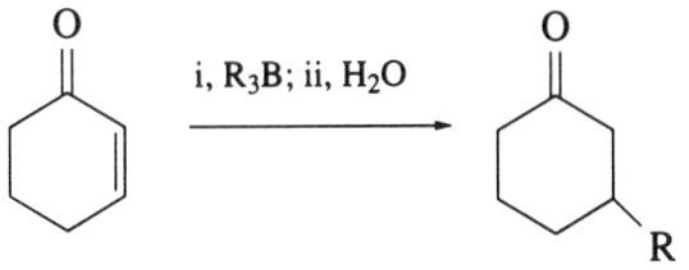

$$\text{cyclohexenone} \xrightarrow{\text{i, R}_3\text{B; ii, H}_2\text{O}} \text{3-R-cyclohexanone} \quad (160)$$

$$\text{vinyloxirane} \xrightarrow{\text{i, R}_3\text{B; ii, H}_2\text{O}} \text{R-CH=CH-CH}_2\text{OH} \quad (161)$$

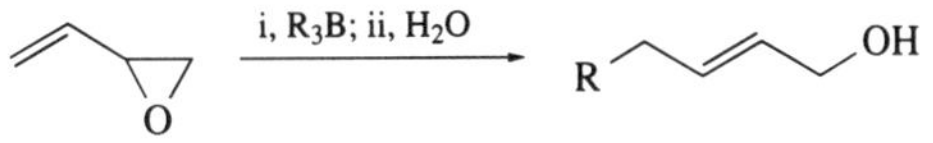

Treatment of terminal alkynes with secondary or tertiary alkyl iodides in the presence of triethylborane provides alkenyl iodides in good yields (Equation (162)).[651] Et_3B induces a facile radical addition of R_3SnH to alkynic compounds to give vinylstannanes (Equation (163)).[652] Stereoselective addition of organosilanes has been also studied.[653] Radical ring closure onto an aldehyde, starting from a selenide, is realized by the stannane–triethylborane–air method (Equation (164)).[654] Thiocarbonyl derivatives of alcohols are readily reduced by diphenylsilane or tributyltin hydride as hydrogen donor and triethylborane–air as initiator (Equation (165)).[655,656] Numerous other applications of triethylborane-induced radical reactions have been reported in the early 1990s.[657–61]

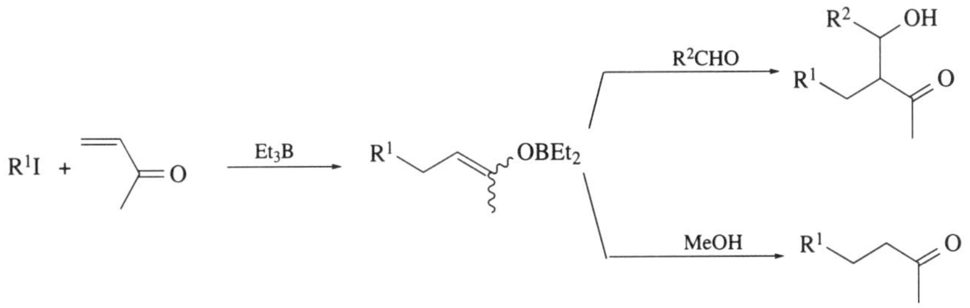

Scheme 55

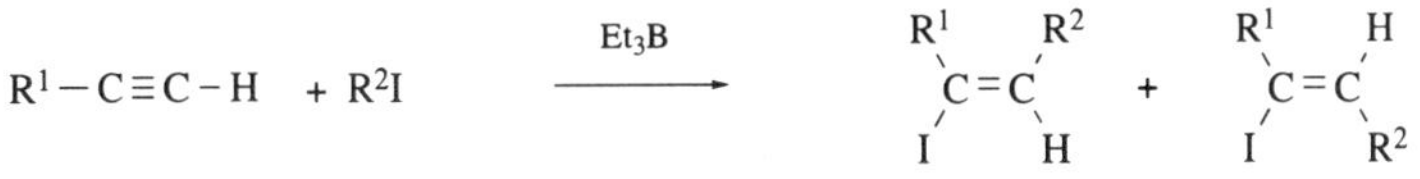

$$R^1-C\equiv C-H + R^2I \xrightarrow{Et_3B} \underset{I\quad H}{\overset{R^1\quad R^2}{C=C}} + \underset{I\quad R^2}{\overset{R^1\quad H}{C=C}} \quad (162)$$

$$R^1-C\equiv C-H + R_3SnH \xrightarrow{Et_3B} \underset{H\quad SnR_3}{\overset{R^1\quad H}{C=C}} + \underset{H\quad H}{\overset{R^1\quad SnR_3}{C=C}} \quad (163)$$

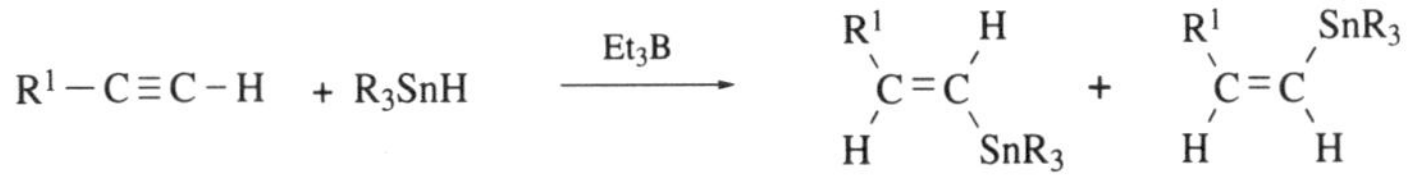

$$\text{(selenide aldehyde)} \xrightarrow[\text{Et}_3\text{B, air}]{\text{Ph}_3\text{SnH}} \text{(bicyclic alcohol)} \quad (164)$$

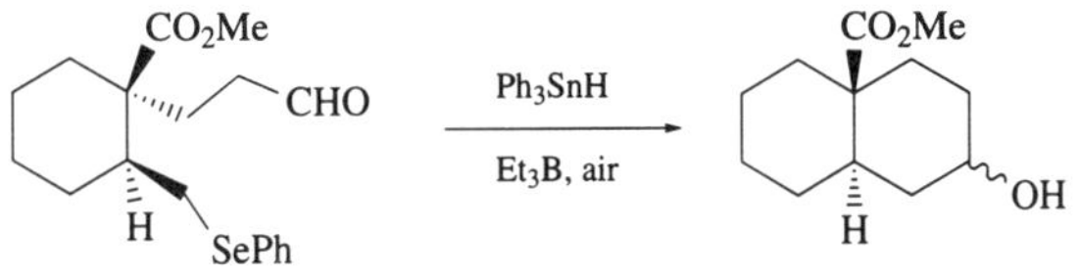

$$R^1O-\underset{S}{\overset{\|}{C}}-OR^2 \xrightarrow[\text{Et}_3\text{B, air}]{\text{Ph}_2\text{SiH}_2} R^1H \quad (165)$$

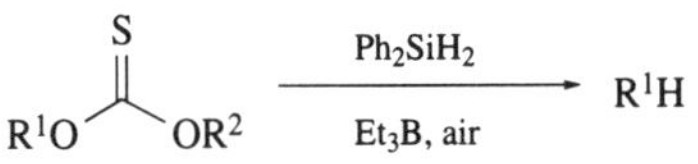

Radical additions to α,β-unsaturated organoboranes are discussed in Section 5.2.2.5.

Amine–borane complexes act as donor polarity reversal catalysts and promote hydrogen abstractions from an electron-deficient α-C–H group in an ester, ketone, or nitrile. The new radical formed in this reaction adds to vinyloxirane and allyl t-butyl peroxide (Scheme 56).[662]

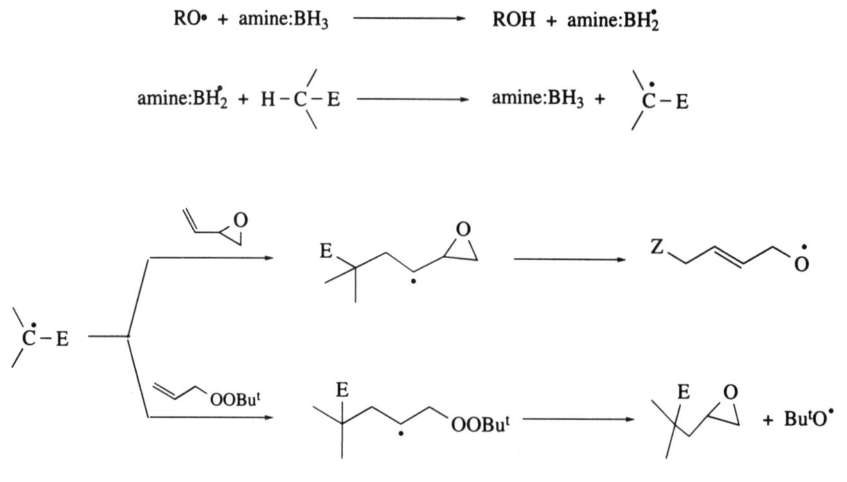

Scheme 56

5.3.4 Miscellaneous Reactions of Boron Derivatives

Boranes and borohydride reagents are used in thousands of reductions each year. Detailed surveys of these reactions have been published[27,663,664] and only some new synthetic developments are reported here. Asymmetric reduction with chirally modified boron reagents is treated in Section 5.4.5.

Lithium aminoborohydrides obtained by the reaction of n-butyllithium with amine boranes constitute a new class of powerful reducing agents, as illustrated in the regiospecific reductions of α,β-unsaturated carbonyl compounds (Scheme 57).[665–7] (For reductions with sodium dimethylaminoborohydride, see Hutchins et al.[668])

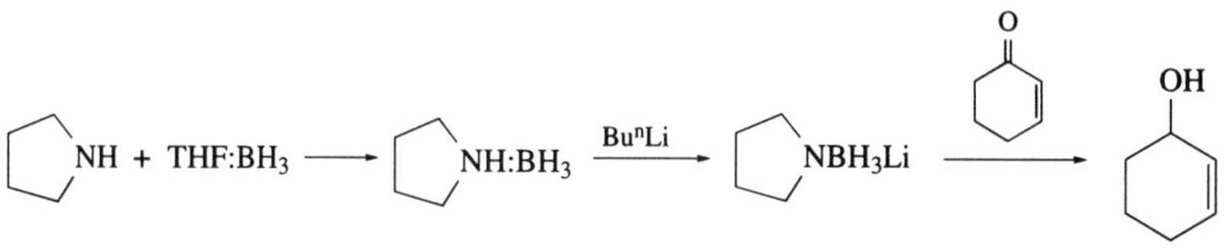

Scheme 57

Convenient procedures for the reduction of amides, nitriles, carboxylic acids, and esters using $NaBH_4/I_2$,[669,670] $NaBH_4$/catechol, and $NaBH_4/CF_3CO_2H$[671] have been reported in the early 1990s. The diiodoborane- and triiodoborane-N,N-diethylaniline complexes also show some interesting selectivities in the iodination of alcohols, the reductive iodination of carbonyl compounds,[672] the deoxygenation of sulfoxides,[673] and the cleavage of ethers, acetals, and lactones.[674,675] Thexylmonohaloboranes have been shown to be attractive selective reducing reagents, especially for the conversion of carboxylic acids to the corresponding aldehydes.[676,677] Lithium di-(2,4,6-triisopropylphenyl)ethylborane reduces substituted cyclohexanones with very high diastereoselectivities.[678]

The reactivity of disubstituted boron bromide reagents towards a variety of functional groups has been extensively studied.[679] Cleavage of a carbon–oxygen bond can be realized in a mild and regioselective fashion as illustrated by the representative examples in Equations (166) and (167).[680,681]

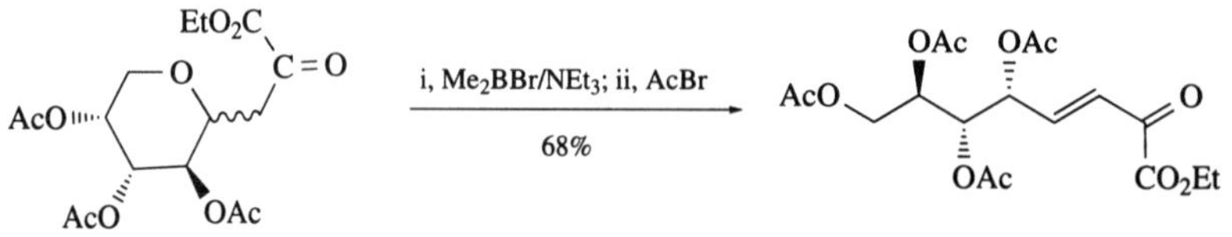

(166)

Highly selective 1,3-asymmetric inductions via boron chelates have been observed (Equation (168)).[682–5] A similar boron-complexed intermediate has been postulated to explain the observed diastereoselectivities in the reduction of a ketoboronate (Equation (169)).[686] High levels of asymmetric induction are obtained in a tandem hydroboration/intramolecular carbonyl reduction of α,ω-unsaturated

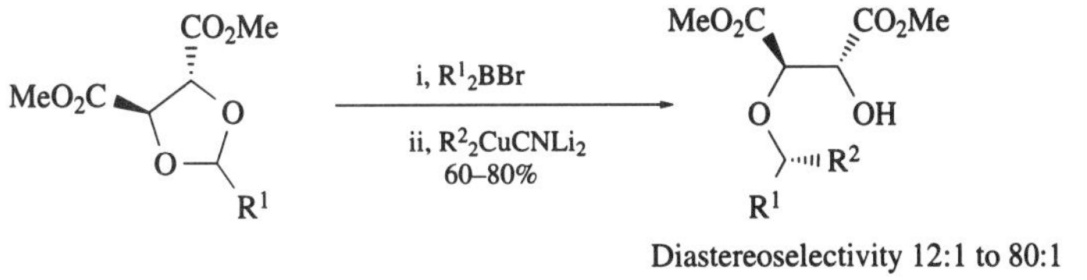

(167) Diastereoselectivity 12:1 to 80:1

ketones, as illustrated in Equation (170),[687,688] and the formation of cyclic intermediates has also been proposed to account for the highly chemoselective reductions of a substituted malonic monoester[689] and (*S*)-malic diethyl ester (Equations (171) and (172)).[690]

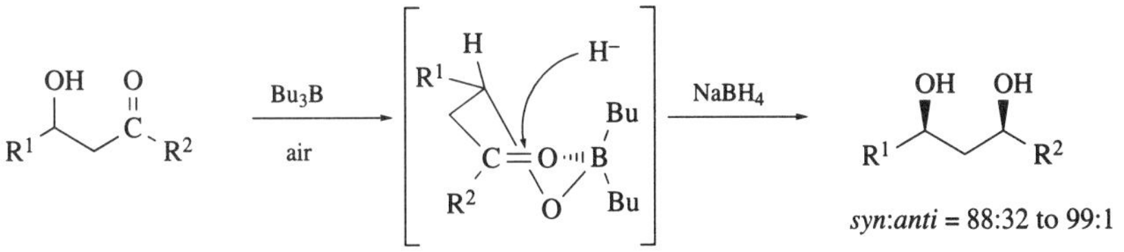

(168) syn:anti = 88:32 to 99:1

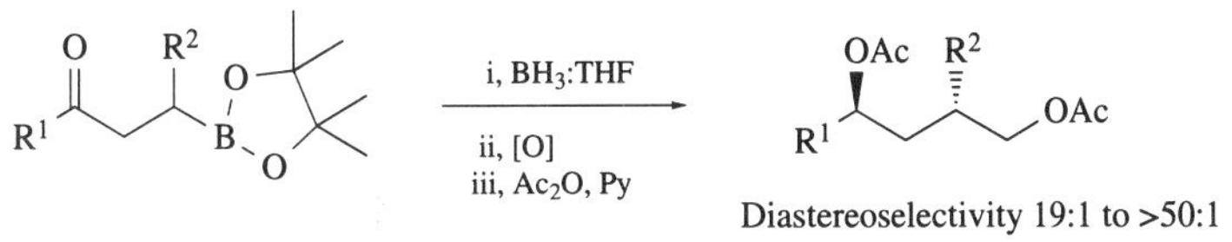

(169) Diastereoselectivity 19:1 to >50:1

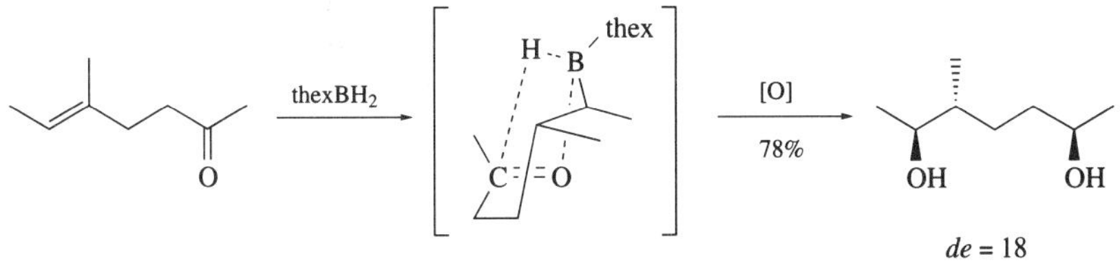

(170) de = 18

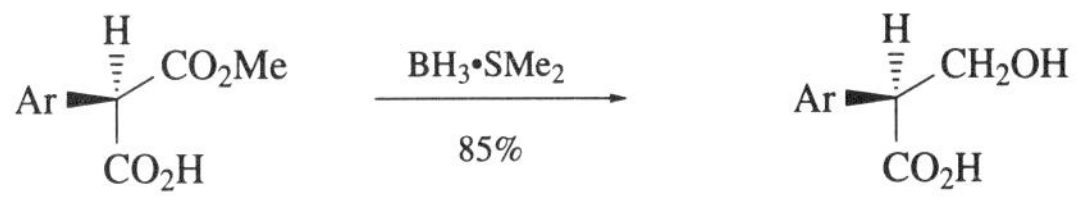

(171)

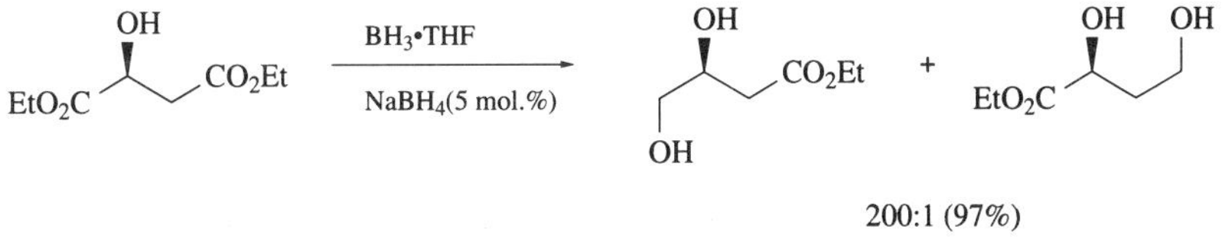

(172) 200:1 (97%)

The use of boron-templated cyclization has been explored in the synthesis of spermine alkaloids and macrocyclic hosts containing convergent hydroxyl groups (Equations (173) and (174)).[691,692] The selective synthesis of mono-*N*-substituted derivatives of tetraazamacrocycles can be achieved using a boron protection (Equation (175)).[693]

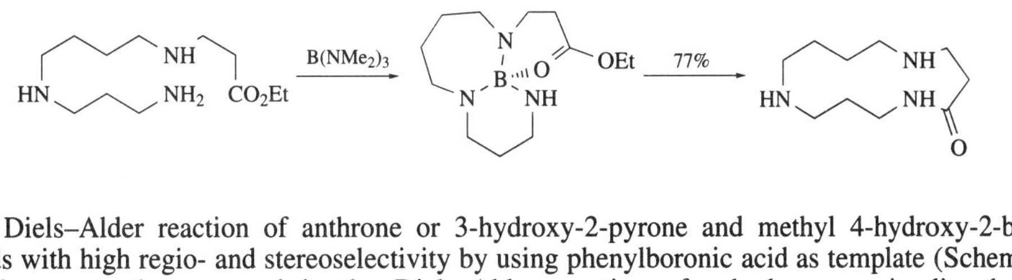

(173)

The Diels–Alder reaction of anthrone or 3-hydroxy-2-pyrone and methyl 4-hydroxy-2-butenoate proceeds with high regio- and stereoselectivity by using phenylboronic acid as template (Scheme 58).[694] A similar approach was used in the Diels–Alder reaction of α-hydroxy-*o*-quinodimethane with 4-methyl-2-hydroxybutenoate.[695]

New boron-containing crown ethers have been synthesized. They show interesting binding properties and, for example, recognized simultaneously and selectively alcohol and amine (Equation (176)).[170,696,697]

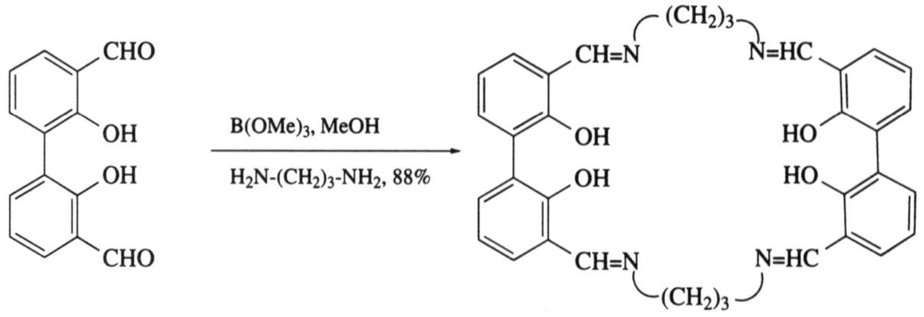

(174)

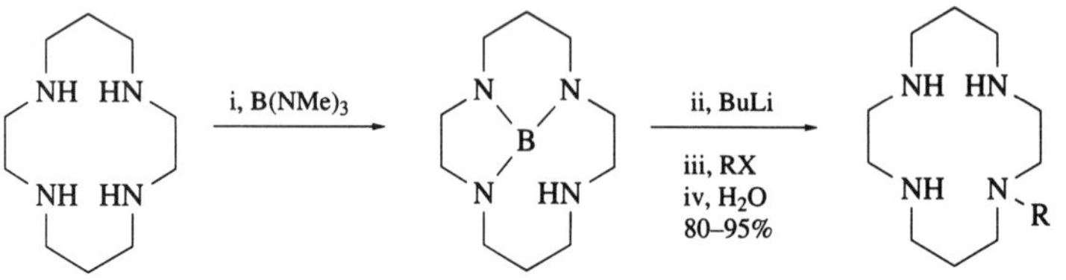

(175)

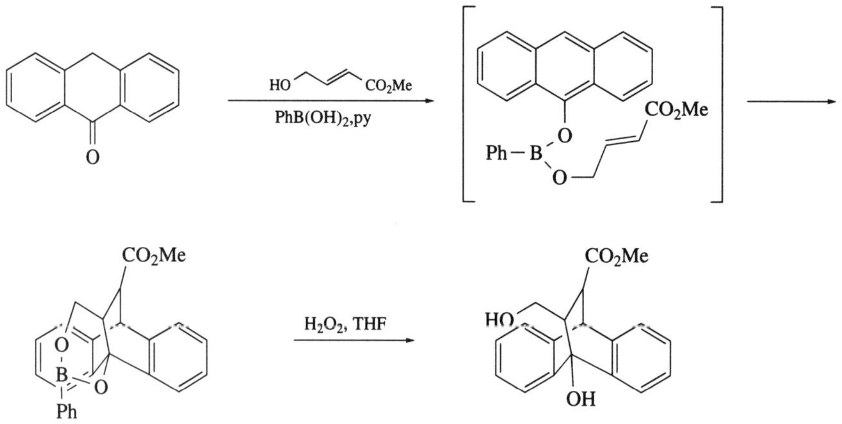

Scheme 58

Another type of macrocyclic host has been prepared and converted upon complexation with boron into a chiral coronand (Equation (177)).[698]

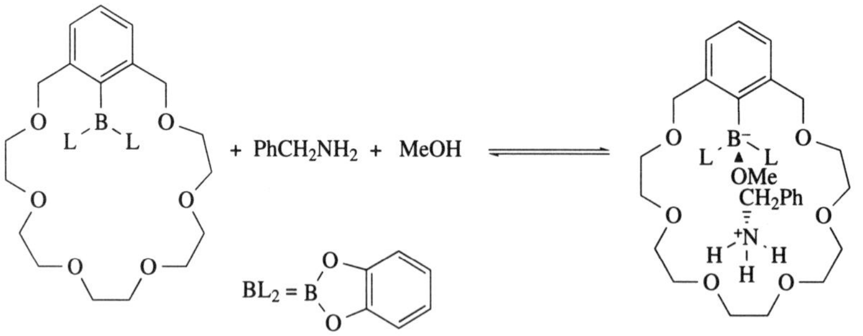

(176)

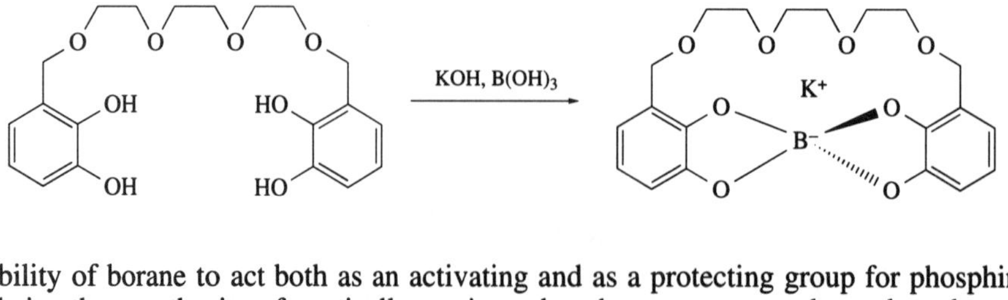

(177)

The ability of borane to act both as an activating and as a protecting group for phosphines has been exploited in the synthesis of optically active phosphorus compounds and a large variety of functionalized phosphines (Schemes 59 and 60, Equation (178)).[699–705]

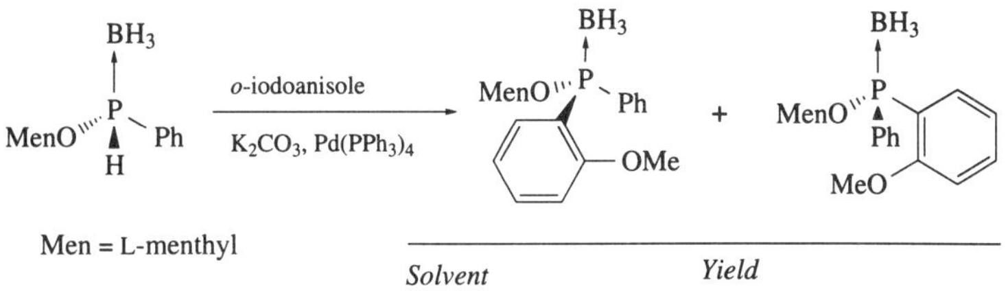

Scheme 59

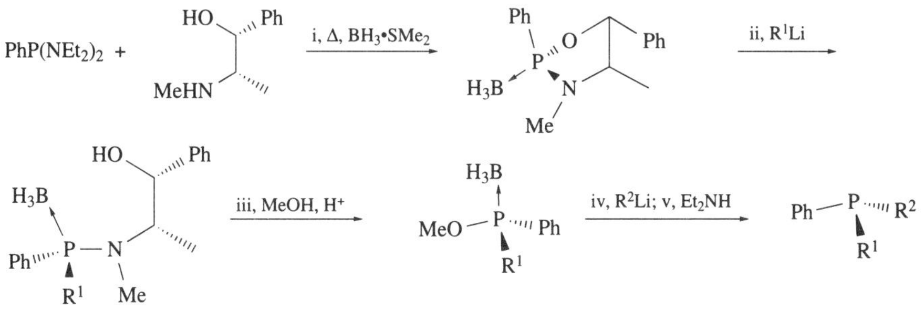

Scheme 60

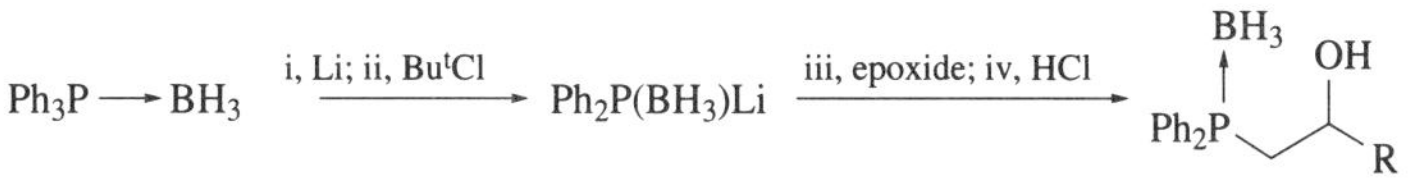

$$Ph_3P \longrightarrow BH_3 \xrightarrow{\text{i, Li; ii, Bu}^t\text{Cl}} Ph_2P(BH_3)Li \xrightarrow{\text{iii, epoxide; iv, HCl}} Ph_2P\overset{BH_3}{\underset{}{\diagup}}\hspace{-0.5em}\diagdown\hspace{-0.5em}\overset{OH}{\underset{R}{\diagup}} \quad (178)$$

5.4 ASYMMETRIC SYNTHESIS

5.4.1 Hydroboration

Asymmetric hydroboration has proved to be a highly efficient reaction for the synthesis of chiral organoboranes. Several reviews describe recent advances in this area.[27,30,39,706-8] Most of the hydroborating agents of high enantiomeric purity have been readily prepared from available low-cost terpenes.[40,41,76,709-13] The synthesis of (R,R)- and (S,S)-2,5-dimethylborolane, has been achieved from (diethylamino)dichloroborane and the Grignard reagent prepared from 2,5-dibromohexane, followed by a resolution step (Figure 11).[714]

No chiral borane gives high asymmetric induction for all types of alkenes. Low enantiomeric excesses are obtained in the hydroboration of 1,1-disubstituted alkenes. Optimum results are achieved with diisopinocampheylborane. For other alkenes, 2,5-dimethylborolane appears to be the best chiral hydroborating agent, but it is also probably the least convenient to prepare. Enantioselectivities in the hydroboration of cis-alkenes with di-2-isocaranylborane and diisopinocampheylborane are comparable to that achieved with 2,5-dimethylborolane. Mono-(2-ethylapoisopinocampheyl)borane, monoisopinocampheylborane and dilongifolylborane give best results for trans-disubstituted, trisubstituted cyclic or acyclic alkenes.

In addition to their easy accessibility in very high enantiomeric purity, mono- and diisopinocampheylborane offer supplementary advantages. The optical purity of a monoisopinocampheylalkylborane can be upgraded by suspending the solid borane in THF and allowing the reaction mixture to age for 12 h at 0 °C. The crystalline product then obtained was >99% optically pure (Scheme 61).[40] However, this purification step depends on the chosen models and may fail when more complex substrates are used. Isopinocampheyl derivatives can be converted under mild conditions to the corresponding alkylboronic esters and acids of high optical purity and these are valuable synthetic intermediates (see Section 5.2.2.1(ii)).

The hydroboration of cycloalka-1,3-dienes with diisopinocampheylborane provides the highly enantiomerically pure corresponding allylboranes without racemization together with a minor homoallylic derivative which does not interfere in further allylboration reactions (Equation (179)).[715]

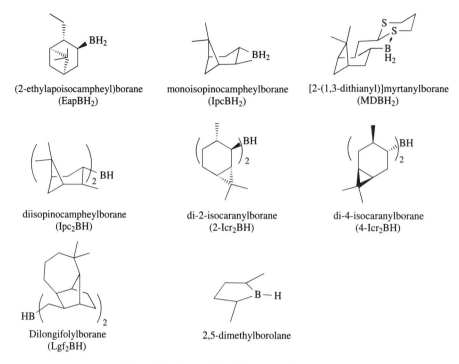

Figure 11 Some chiral hydroborating agents.

Asymmetric hydroboration of representative heterocyclic alkenes has been also investigated. After treatment with acetaldehyde, the corresponding boronic esters are obtained in a very high enantiomeric purity (Scheme 62).[106]

A number of useful synthetic intermediates have been prepared using asymmetric hydroboration as a key step; examples are given in Equations (180)–(182).[716-19]

Enantioselective hydroborations using optically active rhodium catalysts have been reported in the early 1990s.[92] Chiral induction depends on substrate structure, ligands, and, to a lesser extent, catalyst type. The best results are obtained for the hydroboration of styrene derivatives with a regioselectivity

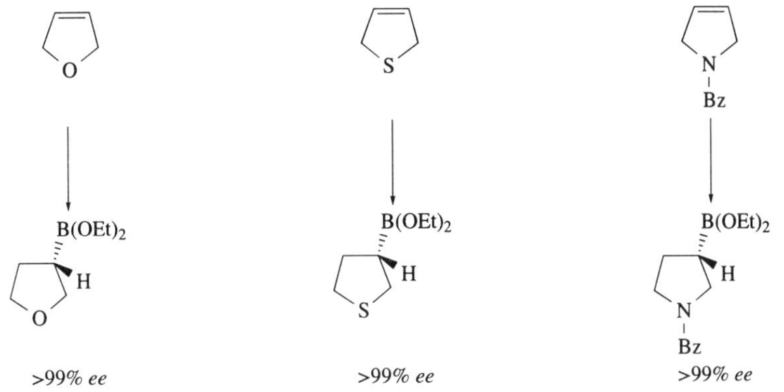

O
S
N
Bz
B(OEt)2
H
O
B(OEt)2
H
S
B(OEt)2
H
N
Bz
>99% ee
>99% ee
>99% ee

Scheme 62

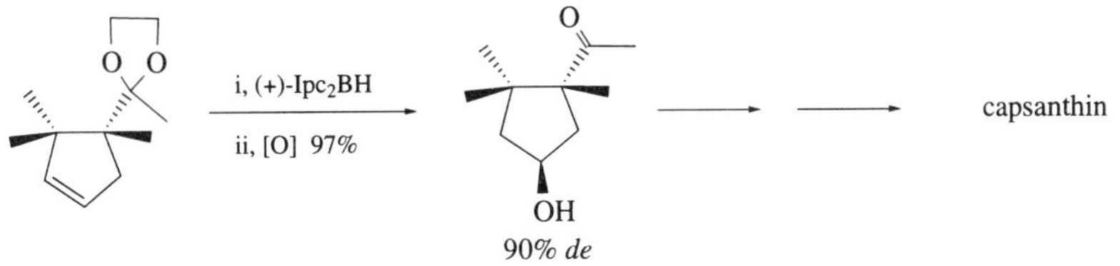

O O
i, (+)-Ipc2BH
ii, [O] 97%
O
OH
90% de
capsanthin

(181)

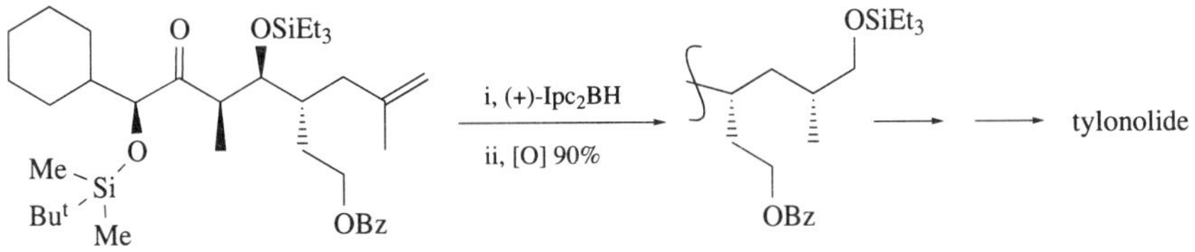

O
OSiEt3
Me
Si
But
Me
O
OBz
i, (+)-Ipc2BH
ii, [O] 90%
OSiEt3
OBz
tylonolide

(182)

opposite to that observed in uncatalyzed hydroboration (Equation (183)).[719,720] Although other models have been studied with less success,[721–4] these preliminary studies are encouraging and more work is required to understand and, later, to control the enantioselectivity in catalyzed hydroborations.

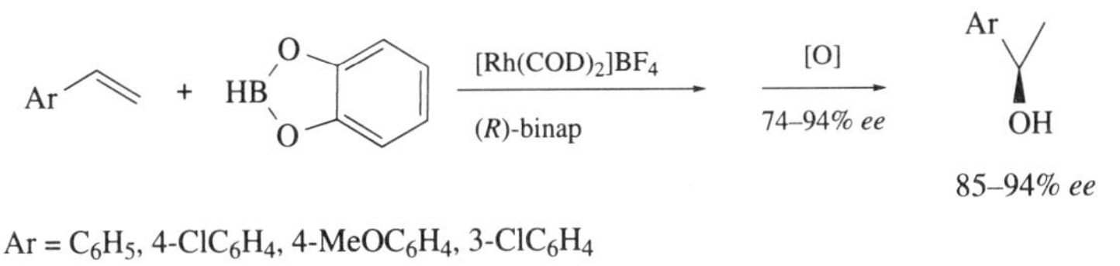

Ar
+
HB
O
O
[Rh(COD)2]BF4
(R)-binap
[O]
74–94% ee
Ar
OH
85–94% ee
Ar = C6H5, 4-ClC6H4, 4-MeOC6H4, 3-ClC6H4

(183)

5.4.2 Homologation Reactions

The synthetic potential of the reaction of dichloromethyllithium with boronic esters to form homologated 1-chloroalkylboronic esters has been amply demonstrated and a large variety of structures are now available via this elegant process (Equation (184)) (see Sections 5.2.2.2 and 5.2.2.3). The possibility of controlling the absolute configuration of the newly created asymmetric center is of major interest.[235,236,725] (1*S*,2*S*,3*R*,5*S*)-2,3-Pinanediol, designated as (*s*)-pinanediol because it directs the formation of (1*S*)-1-chloroalkylboronic esters, was reported to be the first really efficient chiral directing group. Homologation of (*s*)-pinanediol *n*-butylboronate with dichloromethyllithium was highly diastereoselective. After treatment with methylmagnesium bromide and oxidation, which respectively occurs with complete inversion and complete retention of configuration, (*S*)-2-hexanol was obtained in 80% enantiomeric excess (Scheme 63).[227] Unfortunately, yields and diastereoselection decrease dramatically with isobutyl- or benzyloxymethyl boronic esters. Zinc chloride catalysis was discovered in response to these unexpected difficulties, which appeared to be caused by slow and inefficient rearrangement of the intermediate borate. Furthermore, kinetics studies revealed that zinc chloride greatly retarded epimerization of the α-chloroboronic ester (Equation (185)).[726,727] Very high diastereoselection (typically > 99%) and high yields can now be obtained. This process can be repeated to introduce a second chiral center, as illustrated by the enantioselective synthesis of α-aminoacids and asymmetrically deuterated glycerol (Schemes 64 and 65).[728–30] Reactions of (*s*)-pinanediol boronic esters with 1,1-dichloroethyllithium have also been described. The chiral direction of insertion of the CMeCl group in the carbon–boron bond and the diastereoselectivity are clearly dependent on the nature of the migrating group.[731]

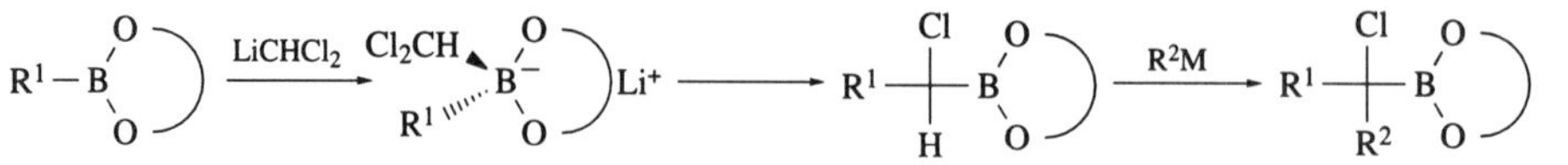

(184)

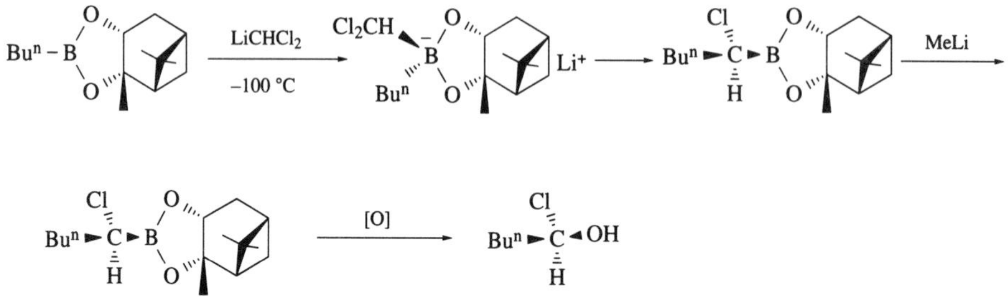

>80% *ee* (60%)

Scheme 63

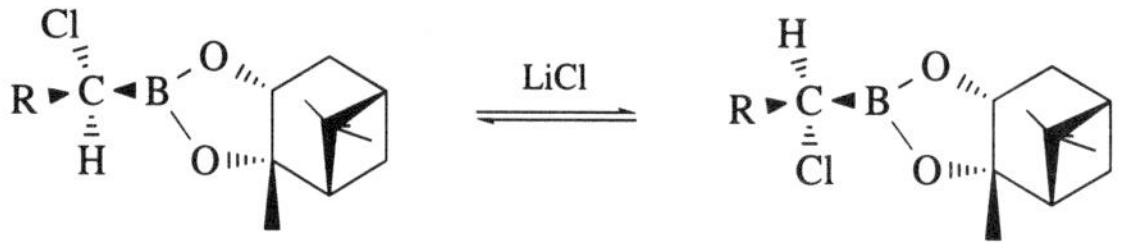

(185)

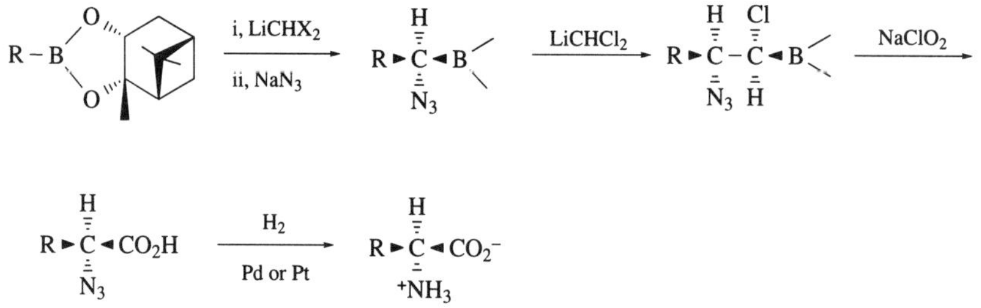

92–98% *ee*
32–63%

Scheme 64

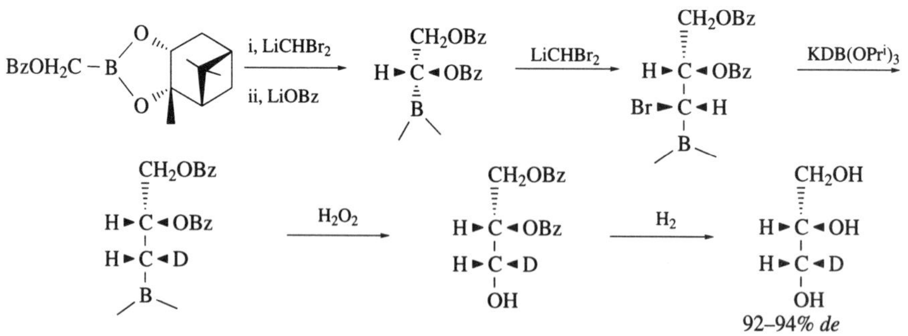

92–94% *de*
overall yield 29%

Scheme 65

Although pinanediol esters have good chiral directing properties, they do have some disadvantages. They are exceptionally stable to hydrolysis or transesterification, but more fundamentally, the two faces of the boron atom are not equivalent. Nucleophiles attack on the less hindered face. Thus, addition of dichloromethyllithium to (*s*)-pinanediol *n*-butylboronate gave (α*S*)-α-chloroboronic esters (80% *de* without $ZnCl_2$). Addition of *n*-butyllithium to (*s*)-pinanediol dichloromethylboronate gave a diastereoisomeric borate intermediate which yielded (αR)-α-chloroboronic ester in only 31% *de* (Scheme 66).[227,732] Both processes carried out with C-2 symmetric chiral directing groups yield the same intermediate and, therefore, the same results are obtained from the reactions of boronic esters with dichloromethyllithium and from dichloromethylboronate and alkyllithium.

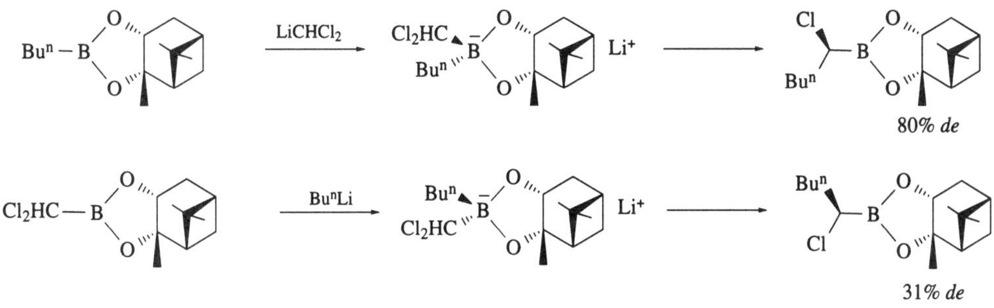

Scheme 66

(*R,R*)-2,3-Butanediol butylboronate was homologated with lithiodichloromethane/zinc chloride in 95% diastereoisomeric purity.[733] The butanediol esters are easily hydrolyzed and recrystallization gave boronic acid of *ee* ≥ 99%. However, this hydrolysis may fail in a few circumstances. Secondary products and racemization have also been observed.[734]

(*S,S*)-Diisopropylethanediol and (*S,S*)-dicyclohexylethanediol are more stable towards water and possess high chiral directing power (Equations (186) and (187)).[735-7]

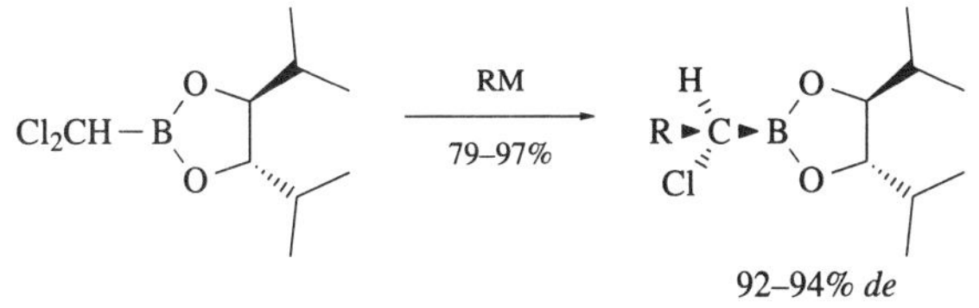

5.4.3 Asymmetric Allylborations

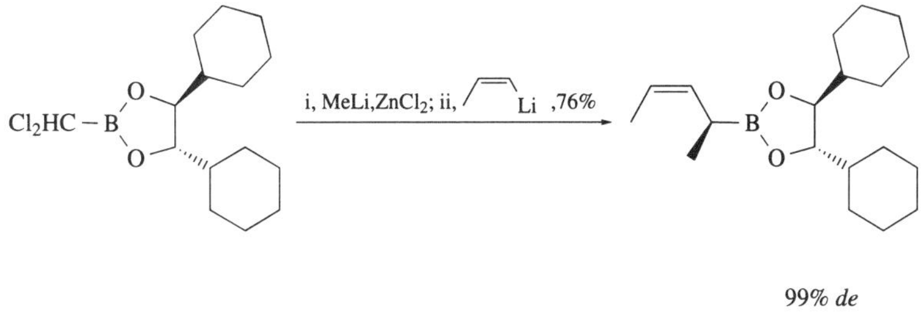

Allylboranes react with aldehydes under mild conditions to yield homoallylic alcohols (see Section 5.3.2.6). The use of chiral allylboron reagents which can give high enantio- and diastereoselectivities has rapidly become particularly attractive. Excellent reviews cover results from the early 1990s in this area,[548,551] and we focus only on some representative and some new developments. Catalytic asymmetric allylation is reported in Section 5.4.6. A number of efficient chiral allylboranes have been reported (Figure 12).[197,738-47]

In the absence of magnesium salts, chiral *B*-allyldi-(2-isocaranyl)borane and *B*-allyldiisopinocampheylborane react almost instantaneously with aldehydes at −100 °C giving, after oxidation, allylic alcohols with optical purities approaching 100% *ee*.[747] They seem to be the most highly enantioselective allylboranes yet described, but the tartrate ester-modified allylboronates are attractive alternatives owing to their ease of preparation and their stability.[573,741,748-51]

B-Allyl-2-(trimethylsilyl)borolane also gives very high enantiomeric excess with aldehydes, but this reagent requires a multistep preparation of the racemic auxiliaries followed by resolution.[744] In the reactions with achiral aldehydes, the introduction of a substituent in a γ-position results in the creation of two chiral centers. A large variety of such reagents with the same chiral boron ligands as previously

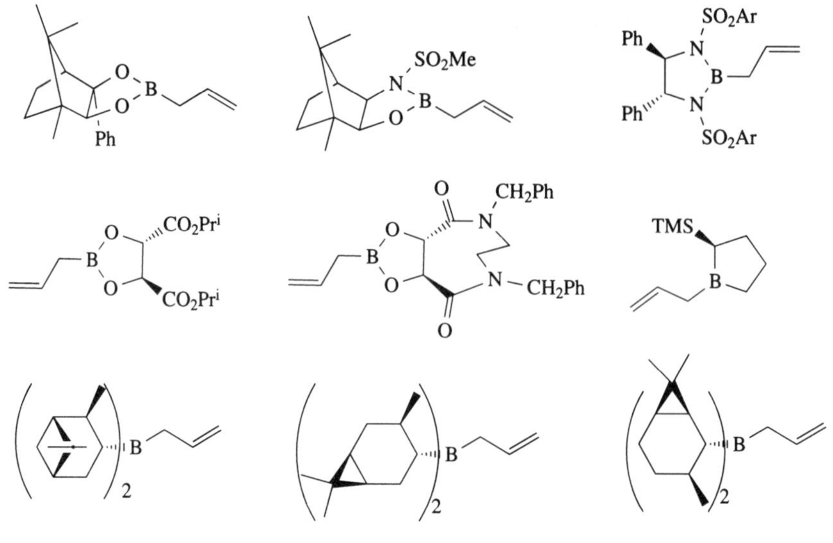

Figure 12 Some chiral allylboranes.

used for the allyl derivatives have been developed.[548,551] They exhibit good to excellent enantioselectivity and very high simple diastereoselectivity. Representative examples are illustrated in Equations (188)–(192).[752–7] Other chiral allyl and allenylboranes have also been reported (Equations (193)–(195)).[198,715,758–61]

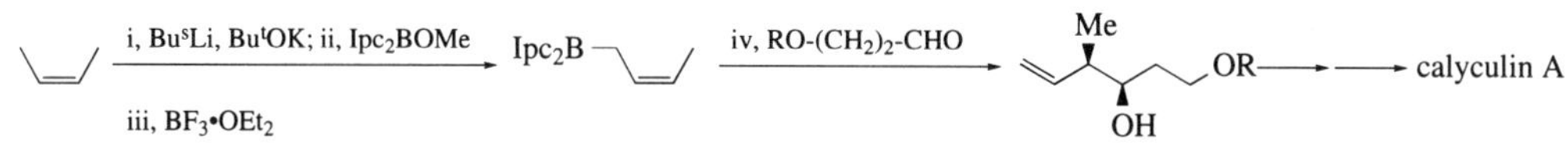

(188)

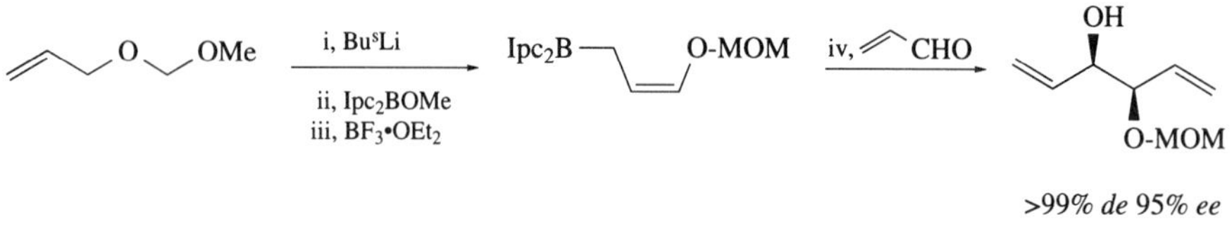

(189)

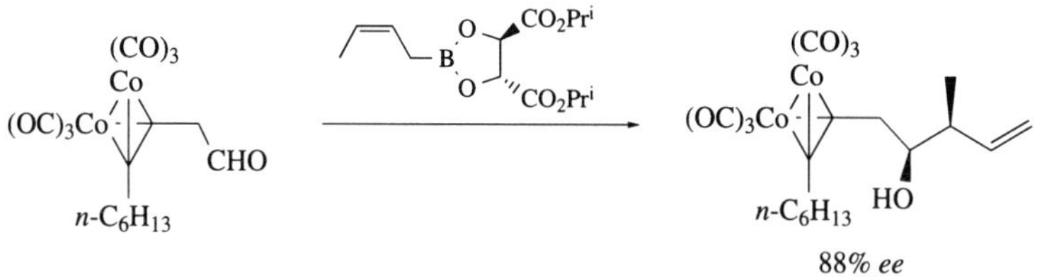

(190)

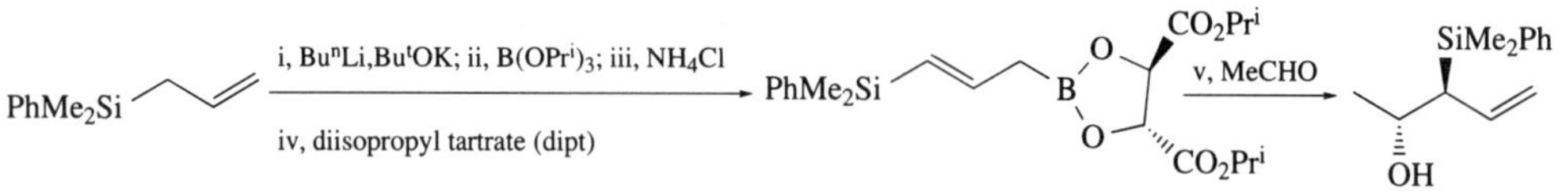

(191)

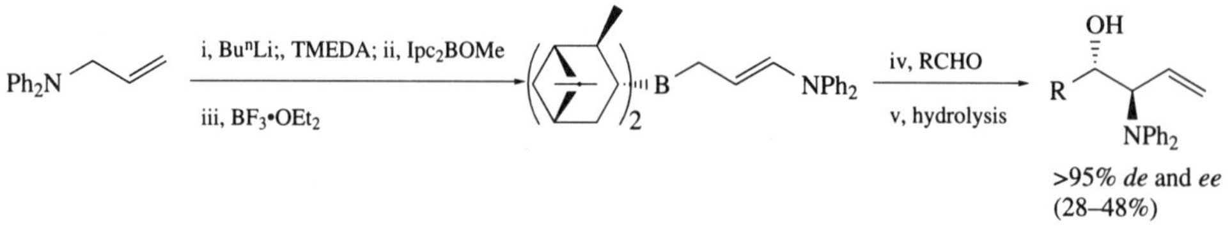

(192)

Another approach to stereocontrolled additions to carbonyl compounds has been the use of allylboranes with stereocenters at the α-position of the allyl or crotyl unit.[762] A variety of α-hetero-substituted allylboronates have been prepared. Halogens and other polar substituents show a strong preference to adopt an axial position in the transition state which leads to homoallylic alcohols with a (Z)-double bond (Equation (196)).[763-5] Similarly, reactions of optically active 1-chlorocrotylboronic esters with prochiral aldehydes proceeds with a high level of chirality transfer (Equation (197)).[765]

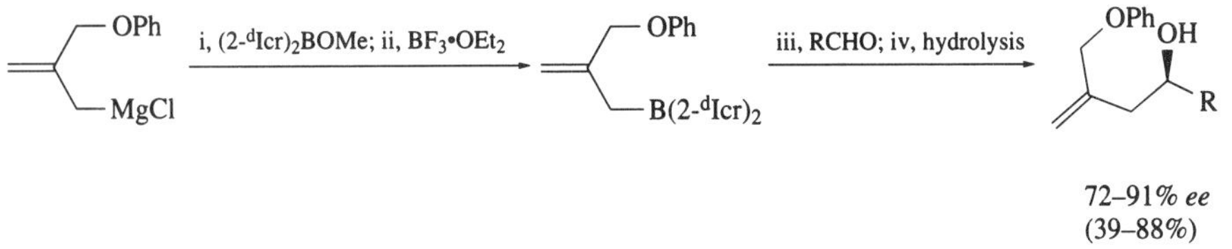

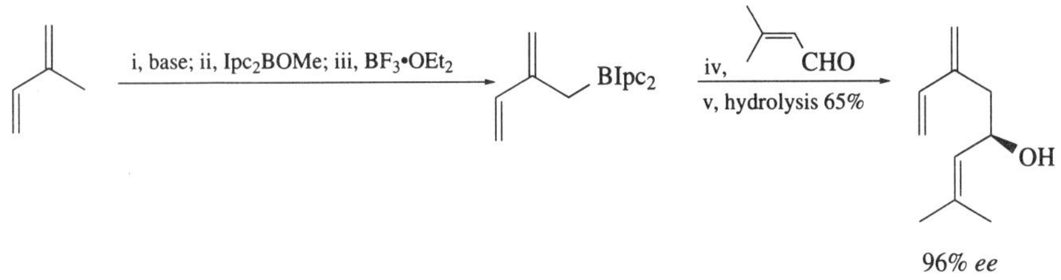

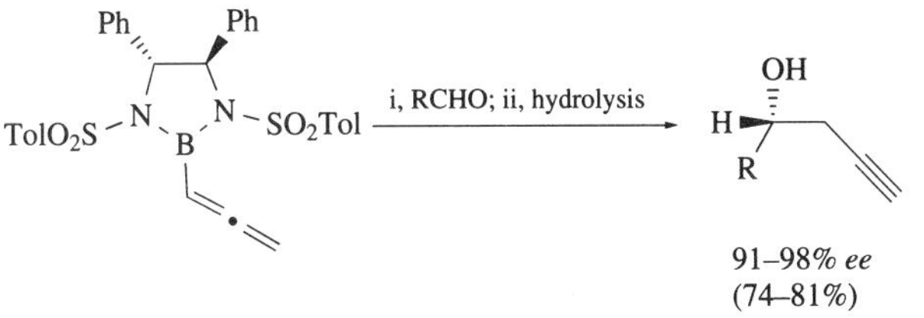

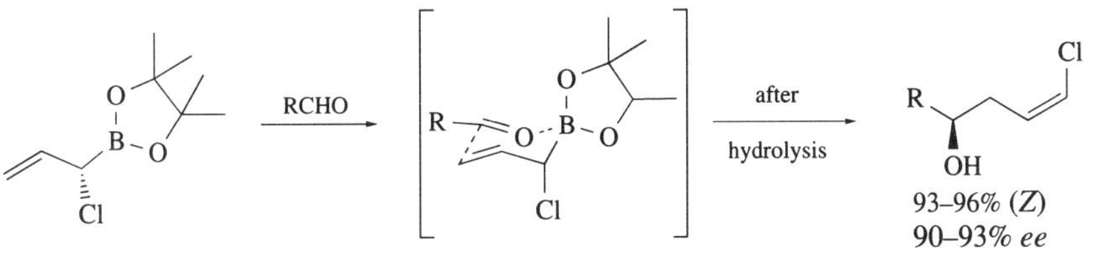

The enantiomerically pure (Z)-pentenylboronate prepared by using 1,2-dicyclohexyl-1,2-ethanediol as chiral auxilliary reacts with benzaldehyde to give the *syn*-(*E*)-homoallylic alcohol (Scheme 67). The α-methyl group prefers an equatorial position in order to avoid 1,3-interaction with the other Me group.[163,737,766]

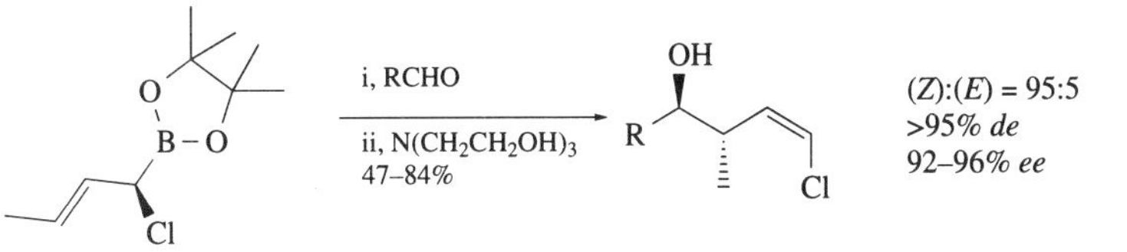

A force-field modeling of the transition state has been developed to gain some insight into the origins of stereoselectivity in the addition of chiral allyl- and crotylboranes to aldehydes.[767]

Significant improvements in diastereoselectivity have been achieved by using double asymmetric induction. High levels of diastereoselection are relatively easily obtained in matched double asymmetric reactions since intrinsic diastereofacial preferences of the C=O electrophile and the chiral allylborane are cooperative. It is more difficult to achieve high diastereoselection in mismatched situations (Scheme 68).[741,742,768] Similarly, (*R*)-α-chloroallylpinacolboronate and D-glyceraldehyde acetonide are a matched pair and react with very good selectivity, but mismatched diastereoselection using the (*S*)-boronate is clearly lower (Scheme 69).[764]

In addition to these results, valuable applications of double asymmetric allylboration reactions have been reported. A stereochemically general strategy for the synthesis of 2-deoxyhexoses from epoxyaldehydes has been developed as illustrated in Scheme 70. It is worth noting that the enantiomeric purity of the major diastereoisomer is greater than that of the epoxyaldehyde precursor.[769]

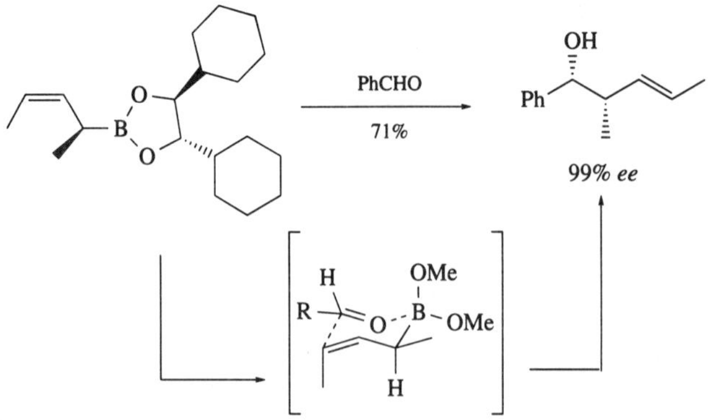

Scheme 67

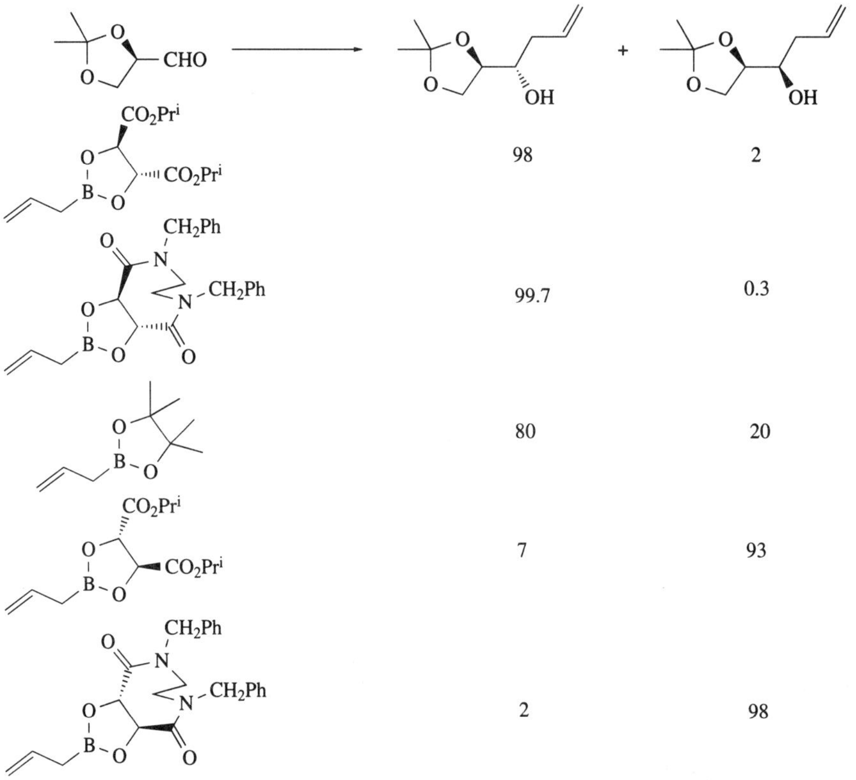

Scheme 68

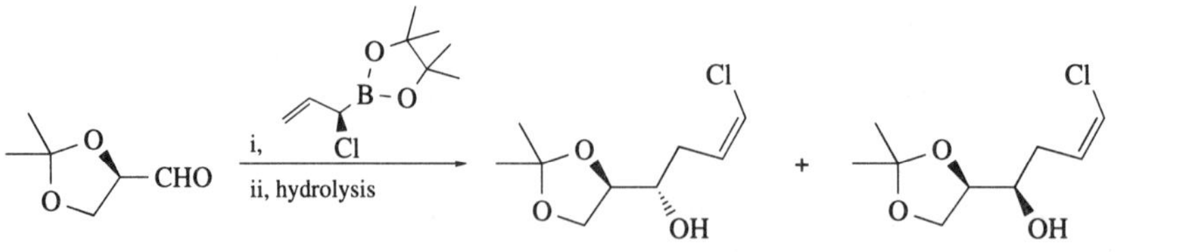

(*R*)	94	4 (+ (*E*)-alkene)
(*S*)	11	77 (+ (*E*)-alkene)

Scheme 69

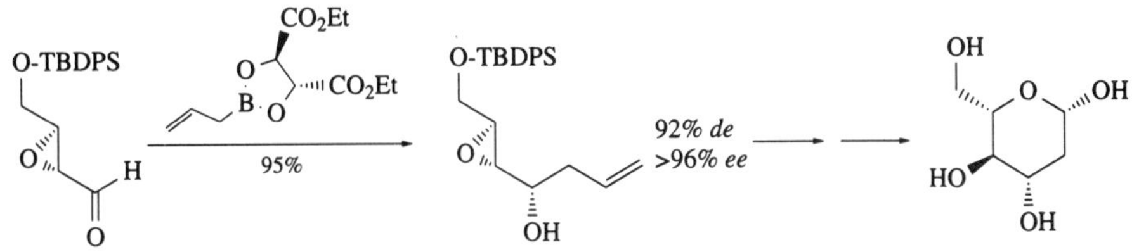

Scheme 70

The (Z)-(S)-1-methyl-2-butenylboronic ester of (S,S)-1,2-dicyclohexyl-1,2-ethanediol was employed as asymmetric allyboration agent in the synthesis of (9S)-dihydroerythronolide A (Scheme 71) and benzoyl-pedamide.[770,771]

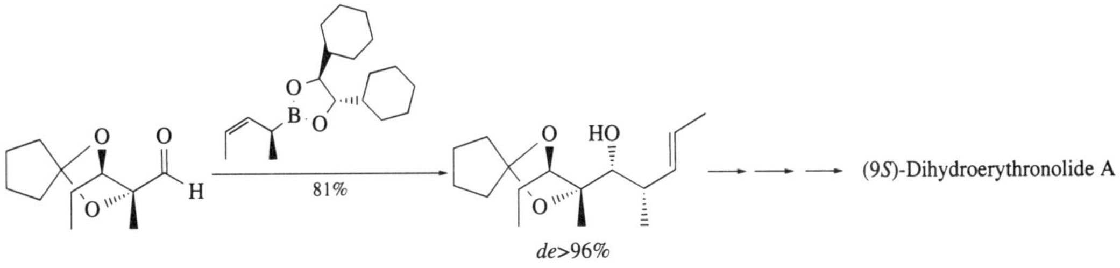

Scheme 71

The enantioselectivity of the asymmetric allylboration of diene aldehydes is significantly improved by using the iron tricarbonyl complexes as substrate surrogates (Equation (198)).[772]

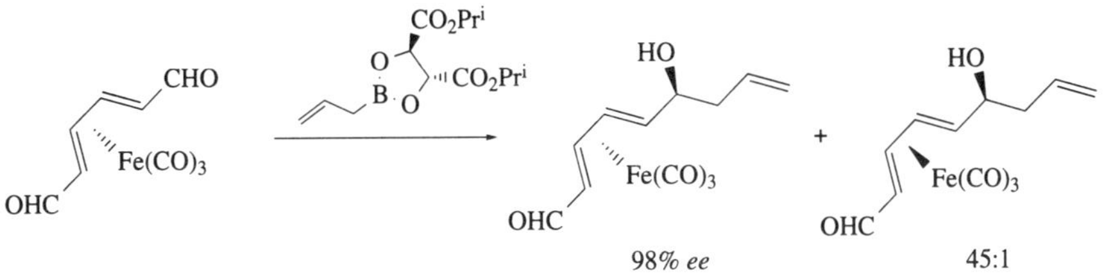

Asymmetric allylboration with [(Z)-γ-(methoxymethoxy)allyl]diisopinocampheylborane was used as a key step in the synthesis of castanospermine analogues (Scheme 72).[773,774]

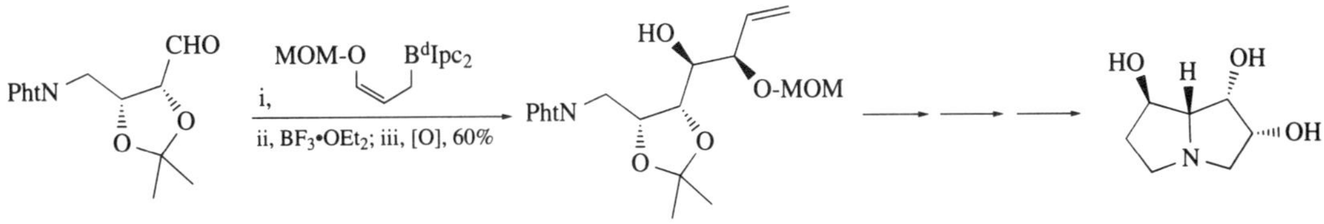

Scheme 72

A single isomer was detected in the reaction of (E)-B-[3-((diisopropylamino)dimethylsilyl)-allyl]diisopinocampheylborane with a serinal derivative (Equation (199)).[775]

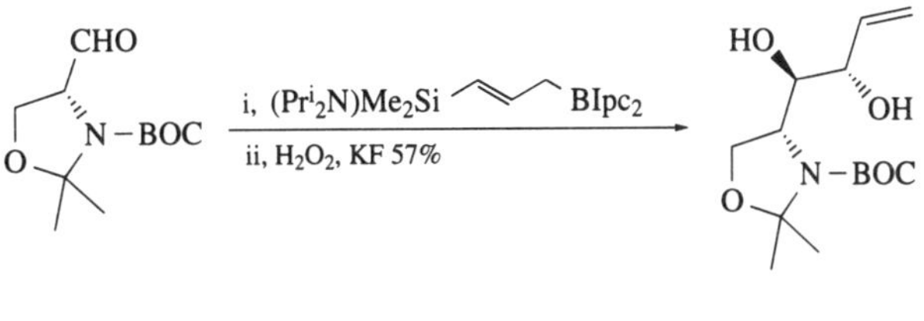

All stereocenters in the trioxadecalin ring of a precursor of mycalamides A, B and onnamide A are controlled by using asymmetric allylboration reactions (Scheme 73).[756]

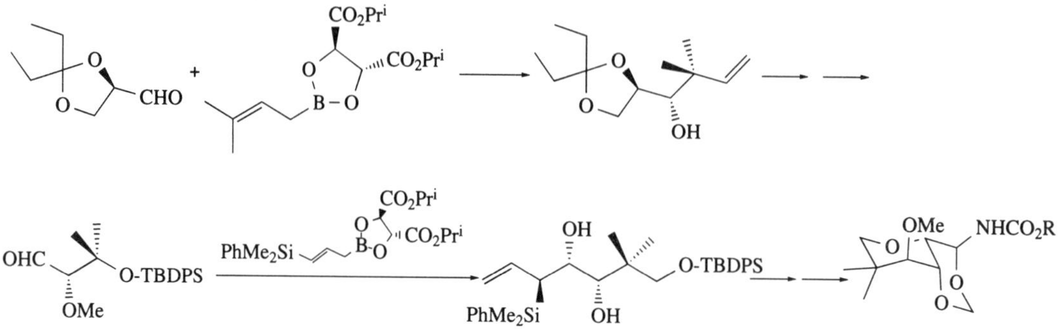

Scheme 73

5.4.4 Asymmetric Aldol Reactions using Chiral Reagents

Formation of carbon–carbon bonds in a diastereo- and enantiocontrolled manner is an increasingly important prerequisite for the efficient chemical synthesis of chiral compounds. Among the many different methods presently available for enantioselective synthesis, the aldol reaction of boron enolates is especially useful for attaining absolute stereocontrol in the synthesis of β-hydroxycarbonyl compounds. It is beyond the scope of the present review of boron chemistry to present a complete account on this important topic. Many pertinent reviews have appeared since 1983[599,600,776–80] and progress appears in the literature every week. Only general trends will be given here.

In the reaction of a chiral aldehyde with an achiral enolate, the facial selectivity inherent to the aldehyde is often referred to as the Cram/anti-Cram ratio.[596,597] Control of the apparent facial selectivity of a substrate in any reaction, including of course the aldol reaction, has been a major challenge in the organic synthesis of acyclic systems and has led to the emergence of a new strategy based on the rule of double asymmetric synthesis.[597]

5.4.4.1 Syn-selective aldol reactions

Substrate-controlled aldol reactions using chiral boron enolates that exhibit facial selectivity of >100:1 have been designed.[781,782] For example, reaction of either enolate (**1**) or (**2**) with a set of aldehydes provides 2,3-*syn*-diastereomeric aldol products with stereoselection of >500:1 (Equations (200) and (201)).[782] A typical example of double asymmetric synthesis is shown in Equation (202).[783] It is interesting to note that the facial selectivity of this α-chiral aldehyde is only 1.75:1 (Equation (203)). These experiments among others confirm the validity of the rule of double asymmetric synthesis. Meyers and Yamamoto reported the use of boron azaenolates derived from enantiomerically homogeneous oxazolines[784] and Oppolzer et al. the use of homogeneous bornane sultam-derived boron enolates to give *syn*-aldol products with extremely high diastereoselectivities.[785]

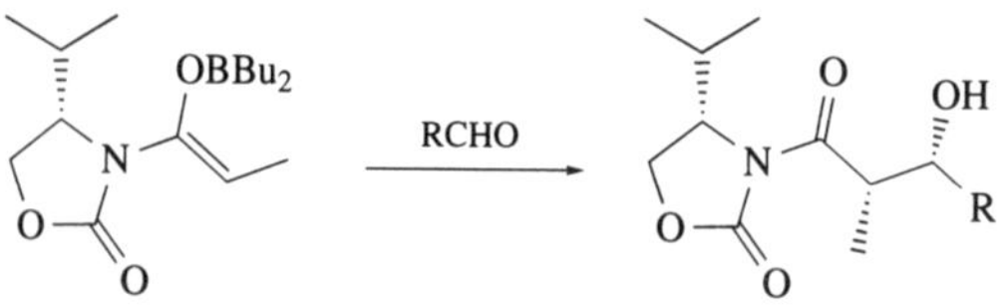

$$R = Pr^i \quad 497:1$$
$$R = Ph \quad >500:1$$
(200)

Analogous *syn*-aldol reactions of the enol borinate (*S*)-(**3**) with achiral boron reagents are nonselective. However, reagent control from (−)- or (+)-Ipc$_2$BOTf leads to the *syn–anti* or *syn–syn* aldol with >76% *de* (Equations (204) and (205)).[786]

(201)

(2)

R = Pr^i 1:500
R = Ph <1:500

(202)

1 : 400

(203)

1.75 : 1

(204)

(*S*)(**3**)

78–84% *de*

(205)

(*S*)(**4**)

76–80% *de*

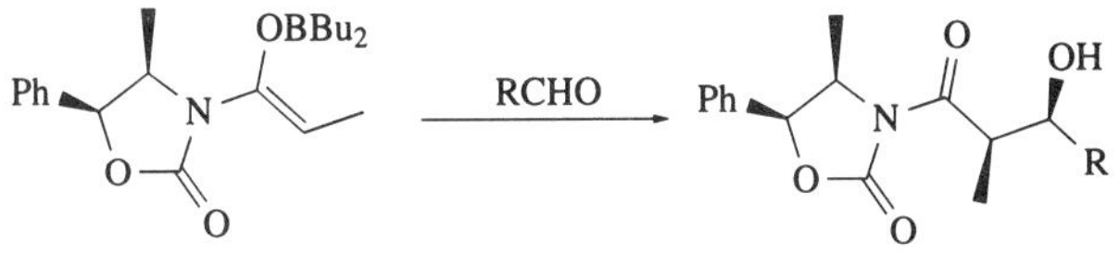

Highly diastereoselective substrate-controlled aldol condensations are reported with α,β-chiral ethyl ketones giving (*Z*)-enolates by enolization with 9-BBN–OTf or Bu_2BOTf and Pr^i_2NEt. The *syn–syn* adducts are typically obtained with around 90% *de*.[787] This process can be combined with kinetic resolution by using a racemic starting ketone.[788]

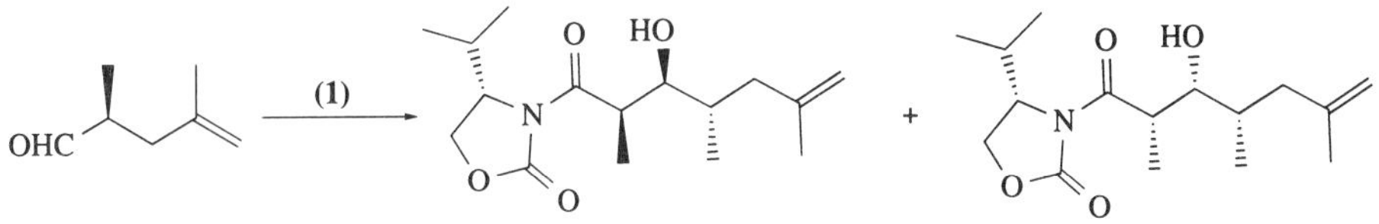

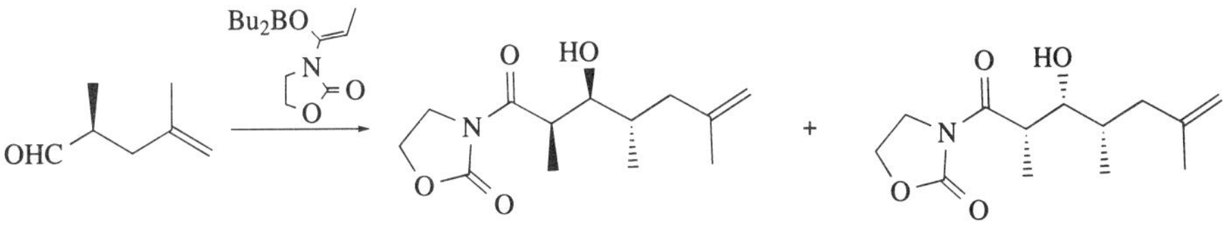

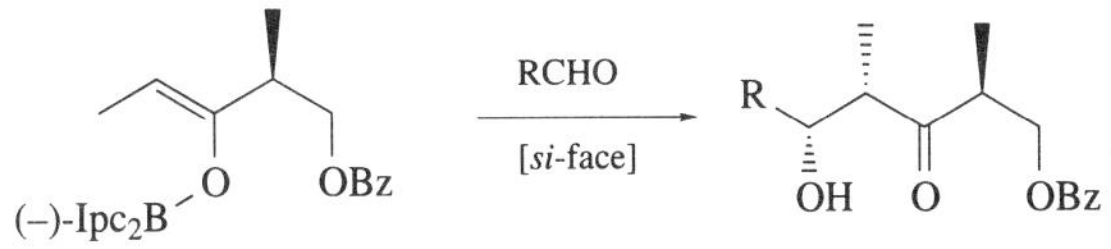

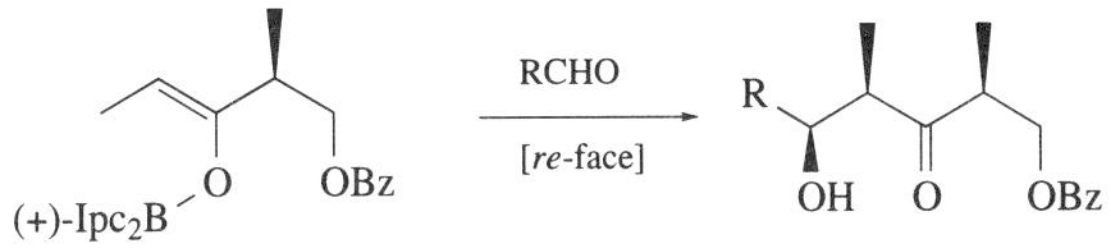

Bis(sulfonamides) derived from (*R*,*R*)- or (*S*,*S*)-1,2-diamino-1,2-diphenylethane have been shown by Corey *et al.* to be excellent controller systems in enantioselective aldol reactions. *Syn*-aldol adducts are obtained from diethyl ketone,[789] thioesters,[790] and various aldehydes in very high enantiomeric excess.

5.4.4.2 Anti-*selective aldol reactions*

Effective asymmetric synthesis via direct reagent-controlled aldol condensation with (*E*)-enol borinates to give *anti*-aldol adducts has been reported.[784] Masamune *et al.* designed a set of chiral *trans*-2,5-dimethylborolane-based reagents that give *anti*:*syn* > 30:1 with *ee* > 97% for the *anti* aldols (Equation (206)).[636] *Anti*-aldol products were also obtained from *t*-butyl propionate with excellent enantioselectivity by Corey and co-workers according to Equation (207).[631,638]

Ipc ligands are ineffective for asymmetric *anti*-aldol reactions of ethyl ketones. Gennari *et al.* have proposed a new boron reagent, designed with the help of a computer, that gives good levels of enantioselectivity (56–88% *ee*).[629] An efficient substrate control is possible in the *anti*-aldol reaction of the (*E*)-enol dicyclohexylborinate(s) which gives the *anti–anti* adduct in ≥90% *de* by attack of the *re* face of the aldehyde (Equation (208)).[791]

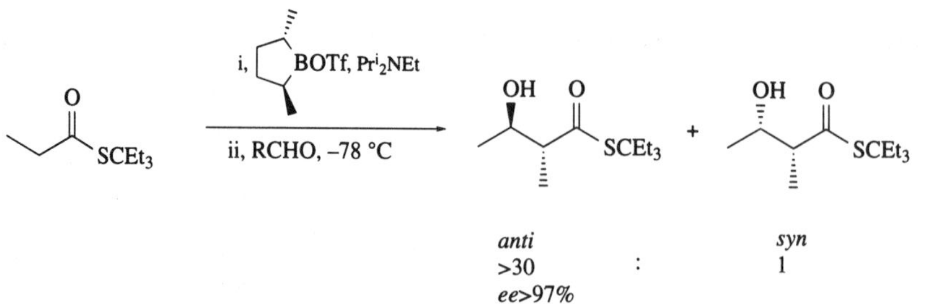

Enolization of α,β-chiral ethyl ketones by dicyclohexylboron chloride and Et_3N gives the (E)-enolates which attack the si-face of an aldehyde such as methacrolein to give mainly the anti-syn adducts.[792,793]

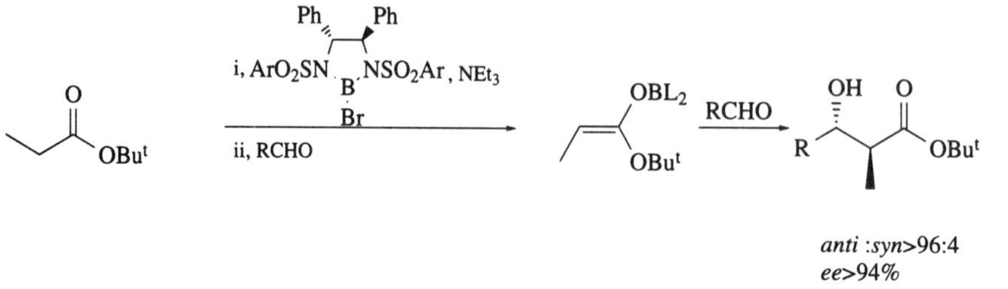

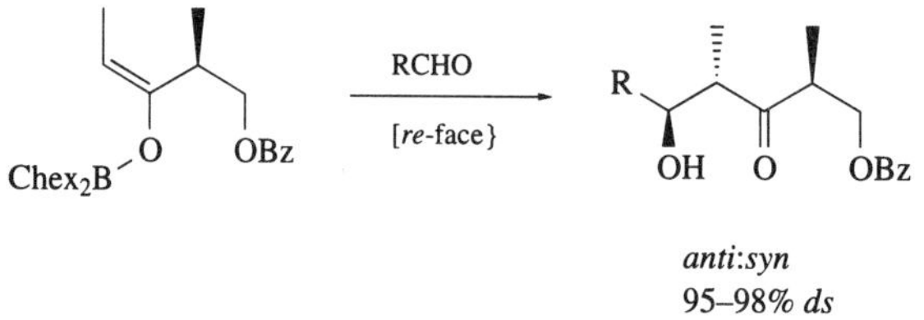

A novel route to optically active anti-aldols of certain aldehydes has been discovered by Heathcock and co-workers.[794] The boron enolates derived from Evans' propionimides react with various aldehydes to give anti-aldols when an excess of dibutylboron triflate is used in the enolization. The excess of Lewis acid activates the carbonyl of the aldehyde by complexation and the reaction can occur through an open transition state (Equation (209)). The aldehyde can also be precomplexed with an external Lewis acid. The observed anti–syn ratios are different for various Lewis acids and also stoichiometry. This allows optimization of the reaction conditions for a given aldehyde so as to obtain the desired stereoisomer.[779]

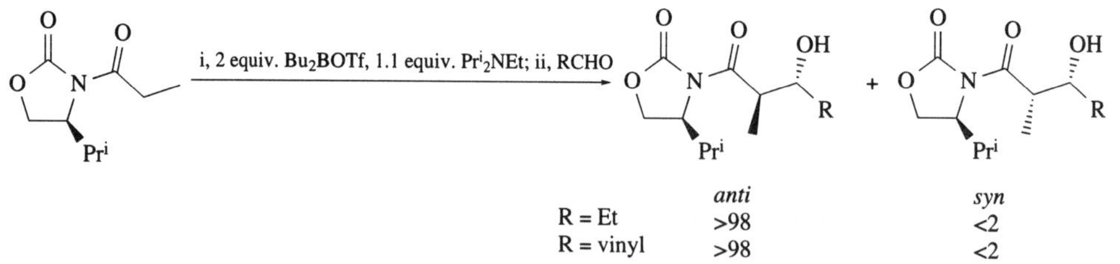

These aldol methodologies have been applied to the total synthesis of numerous natural products of the polypropionate family.[599,776–80] It is also interesting to mention the numerous investigations on transition-state geometry and the origins of stereoselectivity in chiral boron enolate aldol reactions.[630,645,646,795–800]

5.4.5 Enantioselective Reduction

Enantioselective reduction of prochiral ketones to optically active alcohols occupies a position of prime importance in organic synthesis. Since 1984, oxazaborolidine chemistry and asymmetric reduction with chiral α-pinene-based reagents have been intensively developed.[708,801-5] A few other different types of boron chiral reagents have also been reported as efficient enantioselective reducing agents.[42,806–14]

5.4.5.1 Oxazaborolidines in enantioselective reduction of ketones

Earlier examples of oxazaborolidines for stoichiometric enantioselective reduction of ketones were reported by Itsuno and co-workers (Equation (210)).[815] Further studies led to improvements in the catalyst and modification of the reaction conditions. However, a crucial advance was achieved simultaneously by Itsuno et al.[816] and by Corey et al.,[817] who developed a very efficient process with borane–tetrahydrofuran as stoichiometric reductant and an oxazaborolidine as catalyst (Schemes 74 and 75). A reaction mechanism has been suggested (Scheme 76).[817] A Lewis acid–base adduct is first formed from the oxazaborolidine and BH_3. The prochiral ketone then coordinates with the boron of the oxazoborolidine cis to the BH_3. After intramolecular hydrogen transfer via a six-membered transition state on the re face of the carbonyl compound, the alkoxyborane moiety is eliminated and thus regenerates the oxazaborolidine.[801,804,805] The observed enantioselectivity is thought to be controlled by steric factors which force the larger group to occupy the less hindered exo face of the ring system (Figure 13). Due to its enzyme-like behavior, the oxazaborolidine which brings together the reductant and the ketone has been described as a "chemzyme."

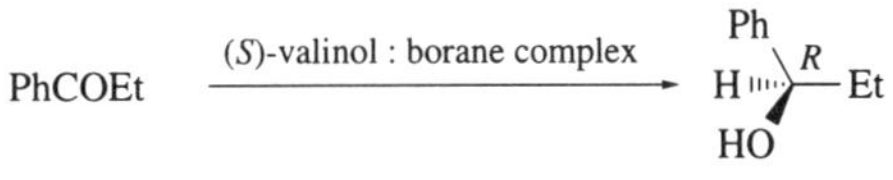

$$\text{PhCOEt} \xrightarrow{\text{(S)-valinol : borane complex}} \begin{array}{c} \text{Ph} \\ \text{H} \overset{R}{\diagdown} \text{Et} \\ \text{HO} \end{array} \quad (210)$$

60% ee

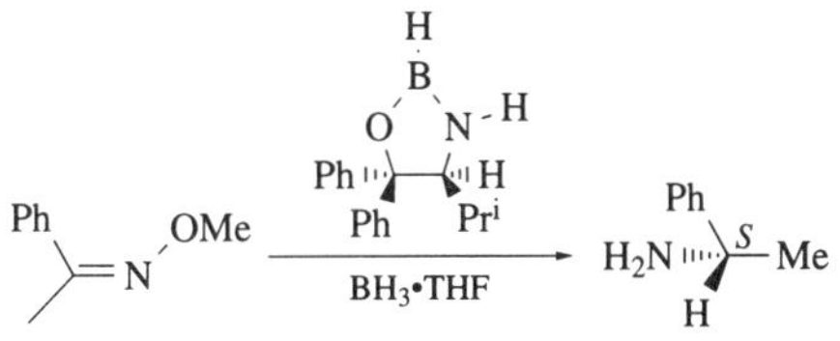

Oxime ether	BH_3	Oxazaborolidine	%ee
1	4	1	95
4	4	1	90
10	10	1	52

Scheme 74

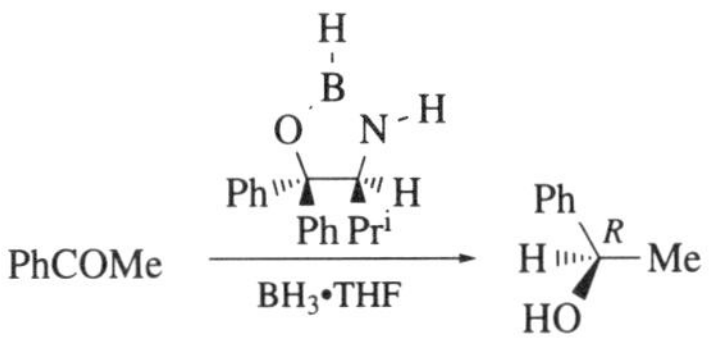

Ketone	BH_3	Oxazaborolidine	% ee
1	2	1	97
1	1.2	0.025	95
1	1.2	0.005	80

Scheme 75

Mechanistic details of the catalysis were later investigated using ab initio molecular orbital methods.[818,819] Since the pioneering reports, numerous modifications and improvements have been reported.[820–8] The preparation and many of the properties of these highly efficient reagents have been reviewed.[801,803–5] Direct comparison between these catalysts is not always easy, but the proline-derived oxazaborolidine seems to be the most efficient reagent. A practical, large-scale procedure for the preparation of (S)-(−)-2-(diphenylhydroxymethyl)pyrrolidine and the corresponding oxazaborolidine–borane complex is worth noting.[829]

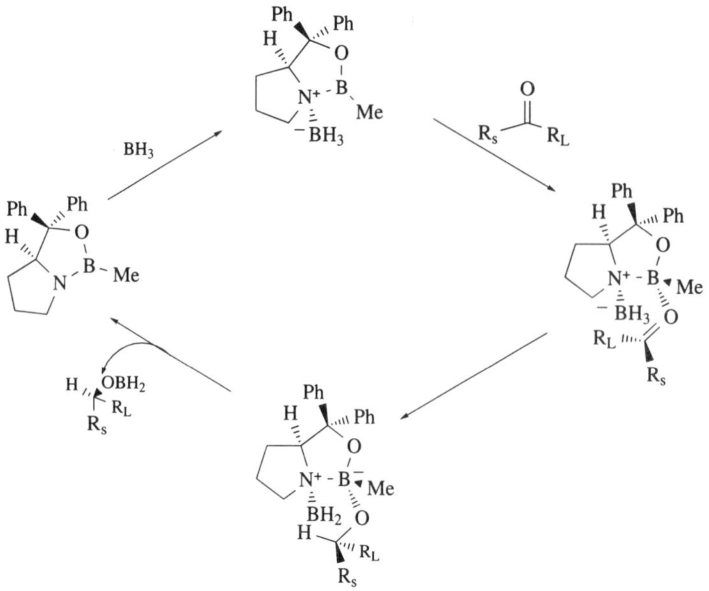

Scheme 76

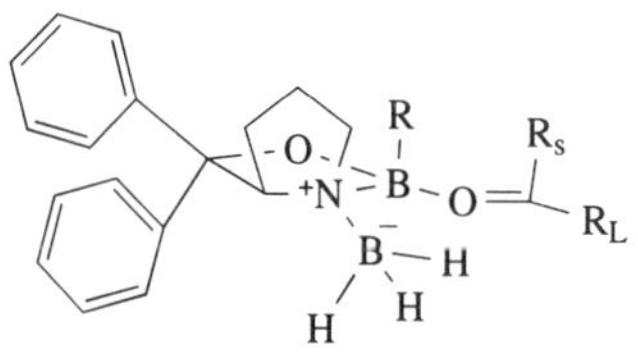

Figure 13 Six-membered transition state.

Less attention has been devoted to the asymmetric reduction of C=N double bonds. For ketoxime ethers, best optical yields are usually obtained with a reducing agent based on a chiral aminoalcohol and borane.[830,831] However, the combination of sodium borohydride, zirconium tetrachloride and a chiral aminoalcohol should also lead to promising results.[832]

Reductions of imines were carried out under similar conditions to those applied to ketones. *N*-Phenyl imines derived from aromatic ketones were converted into *N*-phenyl secondary amines with high enantiomeric excess. In the case of *N*-alkyl ketimine derivatives, the reduction afforded lower optical inductions.[833] Recently, chiral dialkoxyborane reagents have also been used in the presence of MgBr$_2$–OEt$_2$. The characteristic feature of this reagent is the ability to reduce 3,4-dihydroisoquinoline, which was not reduced by Itsuno's reagent, but with a low optical yield. Moderate enantioselectivities are obtained for dihydro-β-carboline.[834]

5.4.5.2 Chiral organoboranes based on α-pinene

Although the use of a catalytic reducing agent is *a priori* very attractive, stoichiometric reagents may also be competitive or even preferable. The excellent results that have been obtained in asymmetric hydroboration with mono- and diisopinocampheylboranes led to the development of α-pinene-based asymmetric reducing agents which are good alternatives to the oxazaborolidine methodology. The synthesis and the reactivity of these readily available chiral organoboranes and borohydrides has been reviewed (Figure 14).[708,802,803] A comparative study of the asymmetric reduction of representative ketones showed that, with only minor exceptions, excellent results can be obtained with alpine borane, diisopinocampheylborane or *B*-chlorodiiso-2-ethylapopinocampheylborane.[802] A number of successful applications of this powerful methodology have been reported.[835–8]

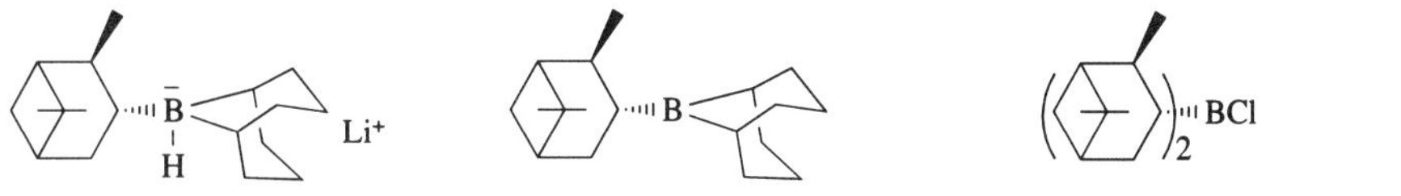

alpine hydride   alpine borane   *B*-chlorodiisopinocampheylborane

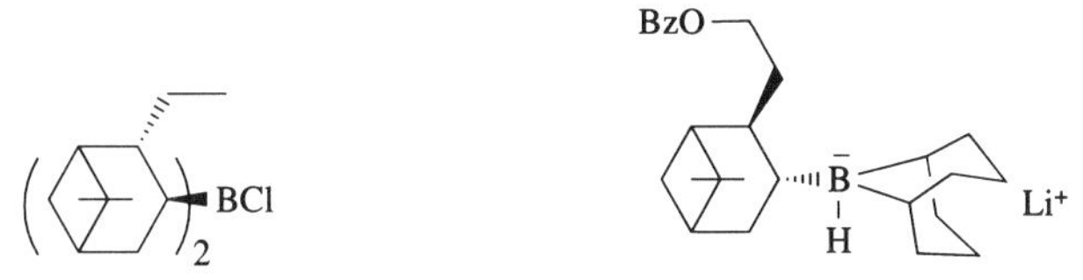

B-chlorodiiso-2-ethylapopinocampheylborane   N*B*-enantride

Figure 14 Main chiral α-pinene reducing agents.

5.4.5.3 Other enantioselective boron reducing agents

In addition to oxazaborolidines, oxazaphospholidines have been reported to give >99% *ee* in the stoichiometric reduction of acetophenone, ethyl acetoacetate, and isopropyl methyl ketone (Figure 15).[808,809] Chiral terpenic 1,2-azaboracyclohexane–borane complexes give also promising results in stoichiometric reduction of acyclic ketones.[42]

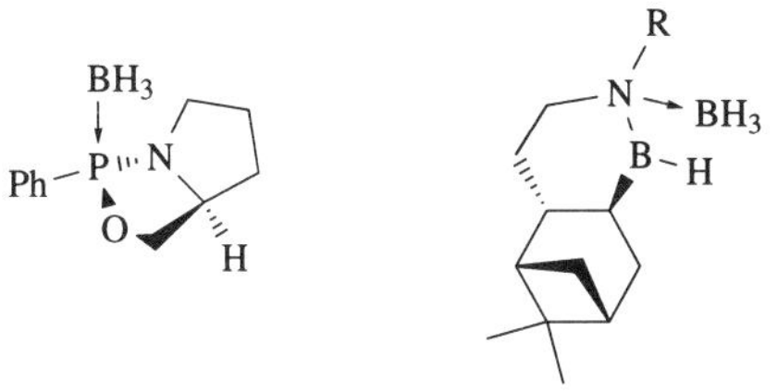

Figure 15 Some enantioselective boron reducing agents.

High enantioselectivity has been observed in the reduction of 2-butanone with a mixture of 2,5-dimethylborolanyl mesylate (0.2 equiv.) and 2,5-dimethylborolane (1.0 equiv.). With pure 2,5-dimethylborolane, the *ee* is only 4% (Scheme 77).[806,807]

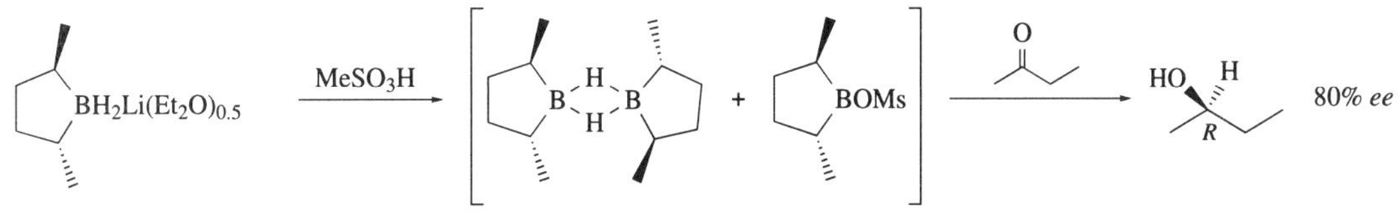

(*R*,*R*)

Scheme 77

A new class of bimetallic complexes were synthesized in 1993 as possible templates for asymmetric catalytic reactions that involve substrate precoordination to a Lewis site remote from the transition metal site (Figure 16).[810,811]

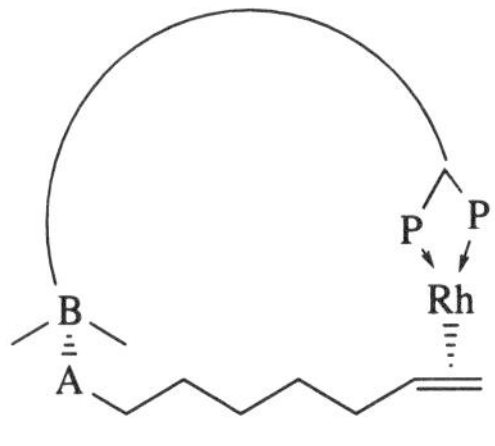

Figure 16 Proposed interactions of substrate with bimetallic complexes.

5.4.6 Other Asymmetric Boron-mediated Reactions

Increasing interest has been devoted in recent years to the development of chiral Lewis acids that mediate catalytic asymmetric reactions. Chiral oxazaborolidines which catalyze the reduction of prochiral ketones (see Section 5.4.5) have also been used in several other asymmetric reactions. The synthetic and mechanistic aspects of asymmetric boron-catalyzed reactions have been reviewed in the early 1990s,[801,805,839,840] and only some representative examples and new applications are reported here.

Optically active diborates catalyze the Diels–Alder reaction of methacrolein with cyclopentadiene with high stereo- and enantioselectivity (Scheme 78).[841]

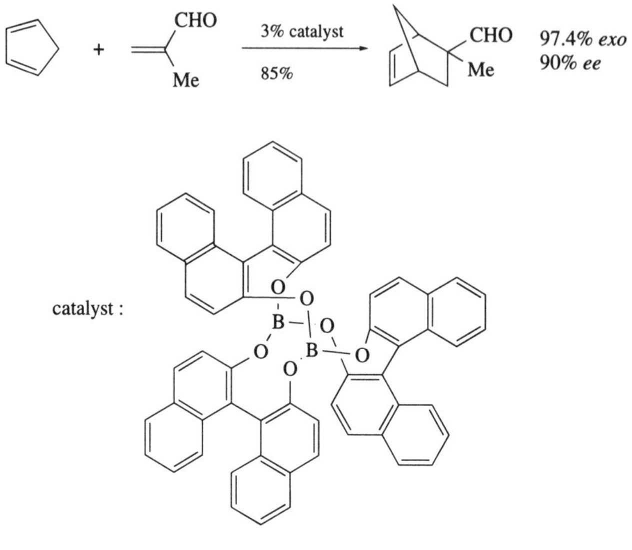

Scheme 78

Takasu and Yamamoto et al.[842] and Helmchen and co-workers[843,844] independently prepared oxazaborolidines derived from the 2,4,6-triisopropylbenzenesulfonamide of α-amino acids and borane, and applied them to enantioselective Diels–Alder reactions (Equations (211) and (212)). Another approach has been suggested by Yamamoto and co-workers,[845–7] who obtained very good results for the reaction of acrylic acid and cyclopentadiene in the presence of a chiral acyloxyborane (Scheme 79). The hetero Diels–Alder reaction has also been investigated.[848] Asymmetric aza-Diels–Alder reactions of imines promoted by an *in situ* generated binaphthol boron complex afford products of high enantiomeric purity (Equation (213)).[849,850]

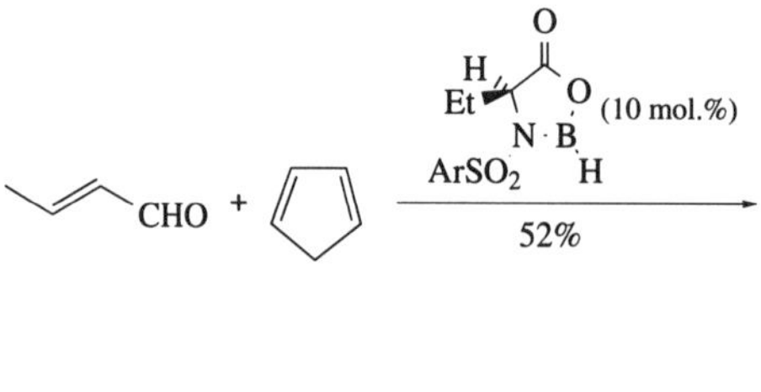

(211)

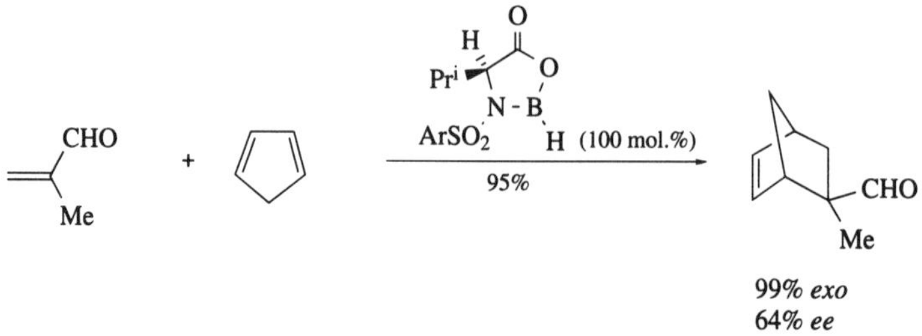

93% endo
54% ee

(212)

99% exo
64% ee

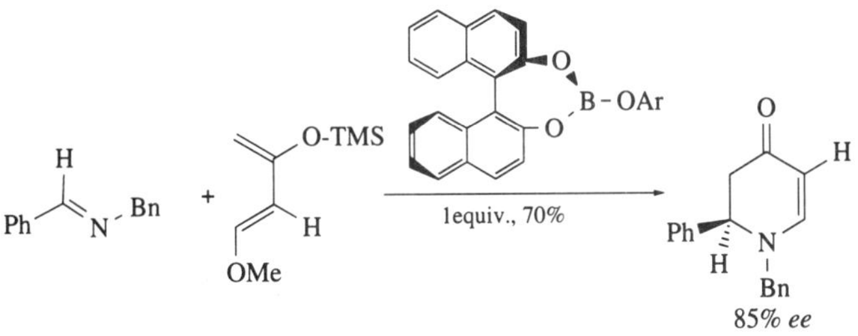

(213)

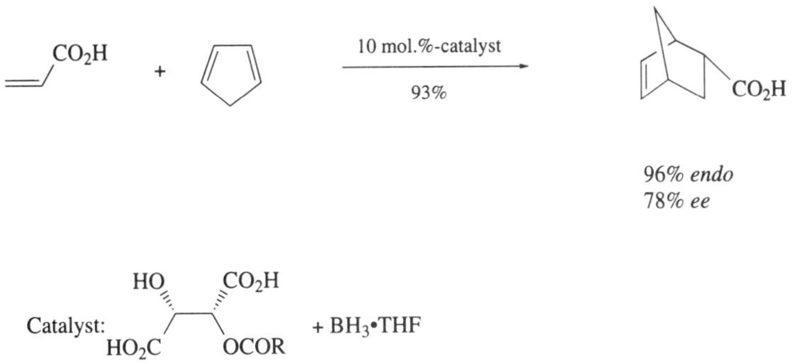

Scheme 79

The reaction of juglone with different dienes catalyzed by a combination of trimethyl borate and (R,R)-tartaric acid diarylamide gives the chiral adduct with high enantioselectivity (Equation (214)).[851,852]

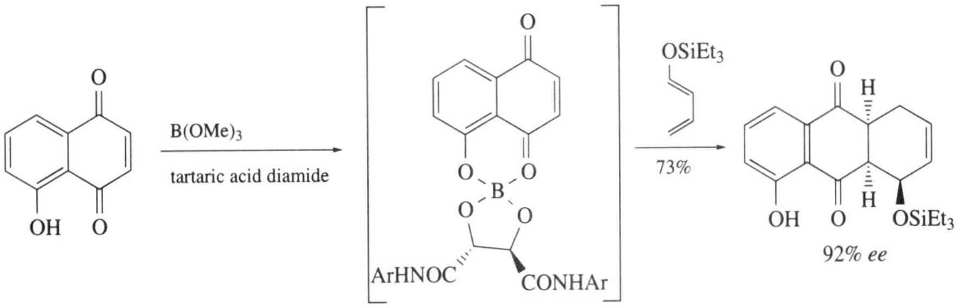

(214)

Corey and co-workers have reported Diels–Alder reactions catalyzed by oxazaborolidines derived from N-tosyl-L-α-tryptophan. For example, cyclopentadiene and 2-bromoacrolein give the corresponding adduct with >200:1 enantioselectivity (*exo:endo* = 96:4) (Scheme 80).[853–6] Promising results have also been reported with chiral alkyldichloroboranes[857,858] and adducts of a chiral prolinol derivative and boron tribromide.[859]

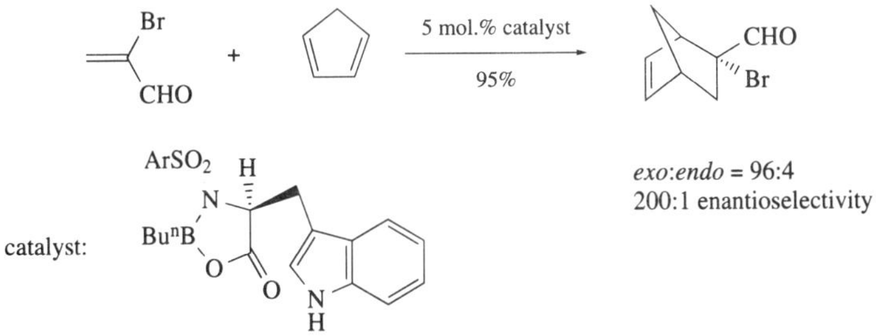

Scheme 80

Mukaiyama-aldol reactions catalyzed by chiral Lewis acids such as (acyloxy)borane complexes,[860] (R)- or (S)-binaphthol-derived borates,[861] tryptophan,[862] or valine-derived[863,864] oxazaborolidines give very high levels of asymmetric induction and this is becoming an important and expanding area of research.

Further developments of other important reactions such as allylation[865,866] or hydrocyanation[640] of aldehydes, enantioselective hydrogen-atom abstraction,[867,868] diethylzinc addition to aldehydes,[869,870] and atropo-enantioselective ring opening of lactones[871] are expected.

5.5 REFERENCES

1. H. C. Brown, M. Zaidlewicz and E. Negishi, in 'COMC-I', vol. 7, p. 111; 161; 229; 270; 305; 326; 354; 459.
2. A. Berndt, *Angew. Chem., Int. Ed. Engl.*, 1993, **32**, 985.
3. J. F. Liebman, A. Greenberg and R. E. Williams (eds.), 'Advances in Boron and the Boranes', VCH, New York, 1988 p. 350; 353.
4. R. E. Williams, *Chem. Rev.*, 1992, **92**, 177.
5. V. I. Bregadze, *Chem. Rev.*, 1992, **92**, 209.
6. B. Štíbr, *Chem. Rev.*, 1992, **92**, 225.
7. R. N. Grimes, *Chem. Rev.*, 1992, **92**, 251.
8. J. Plešek, *Chem. Rev.*, 1992, **92**, 269.
9. L. A. Leites, *Chem. Rev.*, 1992, **92**, 279.
10. A. K. Saxena and N. S. Hosmane, *Chem. Rev.*, 1993, **93**, 1081.
11. S. Trofimenko, *Chem. Rev.*, 1993, **93**, 943.
12. W. Siebert, *Angew. Chem., Int. Ed. Engl.*, 1985, **24**, 943.
13. J. A. Morrison, *Chem. Rev.*, 1991, **91**, 35.
14. R. T. Paine and C. K. Narula, *Chem. Rev.*, 1990, **90**, 73.
15. M. F. Hawthorne, *Angew. Chem., Int. Ed. Engl.*, 1993, **32**, 950.
16. R. F. Barth, A. H. Soloway, R. G. Fairchild and R. M. Brugger, *Cancer*, 1992, **70**, 2995.
17. I. H. Hall, E. S. Hall, M. C. Miller, III, A. Sood and B. F. Spielvogel, *Amino Acids*, 1993, **4**, 287.
18. I. H. Hall *et al.*, *Biomed. Pharmacother.*, 1993, **47**, 79.
19. M. S. L. Lim, E. R. Johnston and C. A. Kettner, *J. Med. Chem.*, 1993, **36**, 1831.
20. C. Tapparelli *et al.*, *J. Biol. Chem.*, 1993, **268**, 4734.
21. W. G. Gutheil and W. W. Bachovchin, *Biochemistry*, 1993, **32**, 8723.
22. P. Kölle and H. Nöth, *Chem. Rev.*, 1985, **85**, 399.
23. H. Nöth and B. Wrackmeyer, 'Nuclear Magnetic Resonance Spectroscopy of Boron Compounds', Springer, Berlin, 1978.
24. S. Heřmánek, *Chem. Rev.*, 1992, **92**, 325.
25. R. Köster, 'Methoden der Organischen Chemie', Thieme, Stuttgart, 1984, vol. XIII, 3a, 3b, 3c.
26. B. M. Mikhaïlov and Y. N. Bubnov, in 'Organoboron Compounds in Organic Synthesis', Harwood Academic, Glasgow, 1984, p. 29; 117; 255; 316; 530; 571; 682.
27. A. Pelter, K. Smith and H. C. Brown, in 'Borane Reagents', Academic Press, New York, 1988, p. 103; 125; 165; 187; 192; 194; 201; 210; 240; 242; 254; 256; 262; 265; 274; 280; 283; 291; 301; 309; 336; 426; 428; 429; 458.
28. G. W. Kabalka and R. C. Marks, *J. Organomet. Chem.*, 1993, **457**, 25 and previous annual surveys.
29. E. Negishi and M. J. Idacavage, *Org. React.*, 1985, **33**, 1.
30. D. S. Matteson, in 'The Chemistry of the Metal–Carbon Bond', ed. F. R. Hartley, Wiley, New York, 1987, vol. 4, p. 307; 330; 346; 350; 351; 353; 357; 368; 387.
31. C. H. Bamford and D. M. Newitt, *J. Chem. Soc.*, 1946, 695.
32. D. S. Matteson, *J. Am. Chem. Soc.*, 1960, **82**, 4228.
33. E. Krause and R. Nitsche, *Chem. Ber.*, 1921, **54**, 2784.
34. A. H. Soloway, B. Whitman and J. R. Messer, *J. Med. Pharm. Chem.*, 1962, **5**, 191.
35. D. S. Matteson, A. H. Soloway, D. W. Tomlinson, J. D. Campbell and G. A. Nixon, *J. Med. Chem.*, 1964, **7**, 640.
36. R. L. Hughes, I. C. Smith and E. W. Lawless, in 'Production of the Boranes and Related Research', ed. R. T. Holtzmann, Academic Press, New York, 1967, p. 291.
37. (a) H. C. Brown and B. C. Subba Rao, *J. Am. Chem. Soc.*, 1956, **78**, 5694; (b) H. C. Brown and B. C. Subba Rao *J. Org. Chem.*, 1957, **22**, 1136.
38. A. Suzuki and R. S. Dhillon, *Top. Curr. Chem.*, 1986, **130**, 23.
39. K. Smith and A. Pelter, in 'Comprehensive Organic Synthesis', eds. B. M. Trost and I. Fleming, Pergamon, Oxford, 1991, vol. 8, p. 703; 720; 724.
40. H. C. Brown, M. C. Desai and P. K. Jadhav, *J. Org. Chem.*, 1982, **47**, 5065.
41. R. Kiesgen de Richter, M. Bonato, M. Follet and J.-M. Kamenka, *J. Org. Chem.*, 1990, **55**, 2855.
42. M. M. Midland and A. Kazubski, *J. Org. Chem.*, 1992, **57**, 2953.
43. H. C. Brown and G. Zweifel, *J. Am. Chem. Soc.*, 1960, **82**, 4708.
44. H. C. Brown and M. V. Rangaishenvi, in 'Chemistry and Technology of Silicon and Tin, Proceedings of the Asian Network in Analytical Inorganic Chemistry International Chemistry Conference on Silicon and Tin', eds. V. G. K. Das, S. W. Ng and M. Gielen, Oxford University Press, Oxford, 1992, p. 3.
45. J. A. Soderquist and H. C. Brown, *J. Org. Chem.*, 1980, **45**, 3571.
46. I. Fleming and N. J. Lawrence, *J. Chem. Soc., Perkin Trans. 1*, 1992, 3309.
47. J. A. Akers and T. A. Bryson, *Synth. Commun.*, 1990, **20**, 3453.
48. D. J. Ager and M. B. East, *Tetrahedron*, 1992, **48**, 2803.
49. W. C. Still and J. C. Barrish, *J. Am. Chem. Soc.*, 1983, **105**, 2487.
50. K. Burgess, J. Cassidy and M. J. Ohlmeyer, *J. Org. Chem.*, 1991, **56**, 1020.
51. T. Harada, Y. Matsuda, J. Uchimura and A. Oku, *J. Chem. Soc., Chem. Commun.*, 1989, 1429.
52. Z. Polívka, V. Kubelka, N. Holubová and M. Ferles, *Collect. Czech. Chem. Commun.*, 1970, **35**, 1131.
53. M. Baboulène, J.-L. Torregrosa, V. Spéziale and A. Lattes, *Bull. Soc. Chim. Fr.*, 1980, 565.
54. Z. Benmaarouf-Khallaayoun, M. Baboulène, V. Spéziale and A. Lattes, *J. Organomet. Chem.*, 1985, **289**, 309.
55. A. Dicko, M. Montury and M. Baboulène, *Tetrahedron Lett.*, 1987, **28**, 6041.

56. K. Burgess and M. J. Ohlmeyer, *J. Org. Chem.*, 1991, **56**, 1027.
57. M. P. Sibi and B. Li, *Tetrahedron Lett.*, 1992, **33**, 4115.
58. J. S. Panek and F. Xu, *J. Org. Chem.*, 1992, **57**, 5288.
59. C. T. Goralski, B. Singaram and H. C. Brown, *J. Org. Chem.*, 1987, **52**, 4014.
60. B. Singaram, M. V. Rangaishenvi, H. C. Brown, C. T. Goyalski and D. L. Hasha, *J. Org. Chem.*, 1991, **56**, 1543.
61. B. Singaram, C. T. Goralski and G. B. Fisher, *J. Org. Chem.*, 1991, **56**, 5691.
62. G. B. Fisher, J. J. Juarez-Brambila, C. T. Goralski, W. T. Wipke and B. Singaram, *J. Am. Chem. Soc.*, 1993, **115**, 440.
63. G. Zweifel and H. Arzoumanian, *J. Am. Chem. Soc.*, 1967, **89**, 291.
64. J. A. Soderquist and H. C. Brown, *J. Org. Chem.*, 1981, **46**, 4599.
65. E. Negishi and H. C. Brown, *J. Am. Chem. Soc.*, 1973, **95**, 6757.
66. H. C. Brown, V. Somayaji and S. Narasimhan, *J. Org. Chem.*, 1984, **49**, 4822.
67. R. S. Dhillion, K. Nayyar and J. Singh, *Tetrahedron Lett.*, 1992, **33**, 6015.
68. N. Satyanarayana and M. Periasamy, *Tetrahedron Lett.*, 1984, **25**, 2501.
69. S. Kano, Y. Tanaka and S. Hibino, *J. Chem. Soc., Chem. Commun.*, 1980, 414.
70. S. Baskaran, V. Gupta, N. Chidambaram and S. Chandrasekaran, *J. Chem. Soc., Chem. Commun.*, 1989, 903.
71. K. S. Ravi Kumar, S. Baskaran and S. Chandrasekaran, *Tetrahedron Lett.*, 1993, **34**, 171.
72. R. Shundo, Y. Matsubara, I. Nishiguchi and T. Hirashima, *Bull. Chem. Soc. Jpn.*, 1992, **65**, 530.
73. E. Negishi and H. C. Brown, *Synthesis*, 1974, 77.
74. M. Srebnik, T. E. Cole and H. C. Brown, *J. Org. Chem.*, 1990, **55**, 5051.
75. A. Pelter, K. Smith, D. Buss and Z. Jin, *Heteroatom Chem.*, 1992, **3**, 275.
76. H. C. Brown and B. Singaram, *J. Am. Chem. Soc.*, 1984, **106**, 1797.
77. K. Kinberger and W. Siebert, *Z. Naturforsch., B: Chem. Sci.* 1975, **30**, 55.
78. H. C. Brown, E. Negishi and J. J. Katz, *J. Am. Chem. Soc.*, 1975, **97**, 2791.
79. M. Srebnik, T. E. Cole, P. V. Ramachandran and H. C. Brown, *J. Org. Chem.*, 1989, **54**, 6085.
80. H. C. Brown, N. Ravindran and S. U. Kulkarni, *J. Org. Chem.*, 1979, **44**, 2417.
81. H. C. Brown and M. Zaidlewicz, *Chem. Stosow.*, 1982, **26**, 155.
82. H. C. Brown and E. Negishi, *Tetrahedron*, 1977, **33**, 2331.
83. H. C. Brown and S. U. Kulkarni, *J. Org. Chem.*, 1979, **44**, 2422.
84. H. C. Brown, J. A. Sikorski, S. U. Kulkarni and H. D. Lee, *J. Org. Chem.*, 1982, **47**, 863.
85. H. C. Brown and J. A. Sikorski, *Organometallics*, 1982, **1**, 28.
86. C. E. Tucker, J. Davidson and P. Knochel, *J. Org. Chem.*, 1992, **57**, 3482.
87. R. Soundararajan and D. S. Matteson, *J. Org. Chem.*, 1990, **55**, 2274.
88. A. Pelter, S. Singaram and H. C. Brown, *Tetrahedron Lett.*, 1983, **24**, 1433.
89. G. P. Boldrini, M. Bortolotti, F. Mancini, E. Tagliavini, C. Trombini and A. Umani-Ronchi, *J. Org. Chem.*, 1991, **56**, 5820.
90. Y. Matsumoto and T. Hayashi, *Synlett*, 1991, 349.
91. D. Männig and H. Nöth, *Angew. Chem., Int. Ed. Engl.*, 1985, **24**, 878.
92. K. Burgess and M. J. Ohlmeyer, *Chem. Rev.*, 1991, **91**, 1179.
93. K. Burgess, W. A. van der Donk, S. A. Westcott, T. B. Marder, R. T. Baker and J. C. Calabrese, *J. Am. Chem. Soc.*, 1992, **114**, 9350.
94. D. A. Evans, G. C. Fu and A. H. Hoveyda, *J. Am. Chem. Soc.*, 1992, **114**, 6671.
95. D. A. Evans, G. C. Fu and B. A. Anderson, *J. Am. Chem. Soc.*, 1992, **114**, 6679.
96. S. A. Westcott, H. P. Blom, T. B. Marder and R. T. Baker, *J. Am. Chem. Soc.*, 1992, **114**, 8863.
97. R. T. Baker, J. C. Calabrese, S. A. Westcott, P. Nguyen and T. B. Marder, *J. Am. Chem. Soc.*, 1993, **115**, 4367.
98. J. F. Hartwig and S. Huber, *J. Am. Chem. Soc.*, 1993, **115**, 4908.
99. S. A. Westcott, H. P. Blom, T. B. Marder, R. T. Baker and J. C. Calabrese, *Inorg. Chem.*, 1993, **32**, 2175.
100. S. A. Westcott, T. B. Marder and R. T. Baker, *Organometallics*, 1993, **12**, 975.
101. H. C. Brown and U. S. Racherla, *J. Org. Chem.*, 1986, **51**, 895.
102. H. C. Brown, N. G. Bhat and V. Somayaji, *Organometallics*, 1983, **2**, 1311.
103. H. C. Brown and D. Basavaiah, *Synthesis*, 1983, 283.
104. H. C. Brown, U. S. Racherla and S. M. Singh, *Synthesis*, 1984, 923.
105. G. Zweifel and N. R. Pearson, *J. Am. Chem. Soc.*, 1980, **102**, 5919.
106. H. C. Brown and M. V. Rangaishenvi, *J. Heterocycl. Chem.*, 1990, **27**, 13.
107. W. H. Pearson, K.-C. Lin and Y.-F. Poon, *J. Org. Chem.*, 1989, **54**, 5814.
108. K. K. Wang, C. Liu, Y. G. Gu, F. N. Burnett and P. D. Sattsangi, *J. Org. Chem.*, 1991, **56**, 1914.
109. Y. G. Gu and K. K. Wang, *Tetrahedron Lett.*, 1991, **32**, 3029.
110. P. D. Sattsangi and K. K. Wang, *Tetrahedron Lett.*, 1992, **33**, 5025.
111. H. C. Brown, C. G. Scouten and R. Liotta, *J. Am. Chem. Soc.*, 1979, **101**, 96.
112. D. A. Singleton and J. P. Martinez, *Tetrahedron Lett.*, 1991, **32**, 7365.
113. J. A. Soderquist, J. C. Colberg and L. Del Valle, *J. Am. Chem. Soc.*, 1989, **111**, 4873.
114. J. A. Soderquist and B. Santiago, *Tetrahedron Lett.*, 1990, **31**, 5113.
115. R. W. Hoffmann and S. Dresely, *Synthesis*, 1988, 103.
116. M. Hoshi, Y. Masuda and A. Arase, *Bull. Chem. Soc. Jpn.*, 1990, **63**, 447.
117. C. F. Lane and G. W. Kabalka, *Tetrahedron*, 1976, **32**, 981.
118. Y. Suseela, A. S. B. Prasad and M. Periasamy, *J. Chem. Soc., Chem. Commun.*, 1990, 446.
119. I. D. Gridnev, N. Miyaura and A. Suzuki, *Organometallics*, 1993, **12**, 589.
120. H. C. Brown, D. Basavaiah and S. U. Kulkarni, *J. Organomet. Chem.*, 1982, **225**, 63.
121. H. C. Brown and J. B. Campbell, Jr., *J. Org. Chem.*, 1980, **45**, 389.
122. A. Hassner and J. A. Soderquist, *J. Organomet. Chem.*, 1977, **131**, C1.
123. C. Rasset-Deloge, P. Martinez-Fresneda and M. Vaultier, *Bull. Soc. Chim. Fr.*, 1992, **129**, 285.
124. A. Kamabuchi, N. Miyaura and A. Suzuki, *Tetrahedron Lett.*, 1993, **34**, 4827.
125. A. Kamabuchi, T. Moriya, N. Miyaura and A. Suzuki, *Synth. Commun.*, 1993, **23**, 2851.
126. J. S. Cha, S. J. Min, J. M. Kim and O. O. Kwon, *Tetrahedron Lett.*, 1993, **34**, 5113.
127. H. C. Brown and D. Basavaiah, *J. Org. Chem.*, 1982, **47**, 754.

128. H. C. Brown, D. Basavaiah and S. U. Kulkarni, *J. Org. Chem.*, 1982, **47**, 3808.
129. H. C. Brown, N. G. Bhat and D. Basavaiah, *Synthesis*, 1983, 885.
130. H. C. Brown and T. Imai, *Organometallics*, 1984, **3**, 1392.
131. G. Zweifel, G. R. Hahn and T. M. Shoup, *J. Org. Chem.*, 1987, **52**, 5484.
132. G. Zweifel and T. M. Shoup, *Synthesis*, 1988, 130.
133. M. Vaultier, F. Truchet, B. Carboni, R. W. Hoffmann and I. Denne, *Tetrahedron Lett.*, 1987, **28**, 4169.
134. G. Zweifel, M. Ramin Najafi and S. Rajagopalan, *Tetrahedron Lett.*, 1988, **29**, 1895.
135. T. M. Shoup and G. Zweifel, *Synth. Commun.*, 1993, **23**, 2937.
136. Y. Matsumoto, M. Naito and T. Hayashi, *Organometallics*, 1992, **11**, 2732.
137. Y. S. Lee, W. Leong and G. Zweifel, *Heteroatom Chem.*, 1992, **3**, 227.
138. K. N. Houk, N. G. Rondan, Y.-D. Wu, J. T. Metz and M. N. Paddon-Row, *Tetrahedron*, 1984, **40**, 2257.
139. N. J. R. Van Eikema Hommes and P. Von Ragué Schleyer, *J. Org. Chem.*, 1991, **56**, 4074 and references therein.
140. X. Wang, Y. Li, Y.-D. Wu, M. N. Paddon-Row, N. G. Rondan and K. N. Houk, *J. Org. Chem.*, 1990, **55**, 2601 and references therein.
141. M. F. Lappert and B. Prokai, *J. Organomet. Chem.*, 1964, **1**, 384.
142. F. Joy, M. F. Lappert and B. Prokai, *J. Organomet. Chem.*, 1966, **5**, 506.
143. A. Suzuki, *Pure Appl. Chem.*, 1986, **58**, 629.
144. S. Hara, H. Dojo, S. Takinami and A. Suzuki, *Tetrahedron Lett.*, 1983, **24**, 731.
145. J. R. Blackborow, *J. Organomet. Chem.*, 1977, **128**, 161.
146. Y. Satoh, T. Tayano, H. Koshino, S. Hara and A. Suzuki, *Synthesis*, 1985, 406.
147. J. R. Blackborow, *J. Chem. Soc., Perkin Trans. 2*, 1973, 1989.
148. S. Hara, T. Kato and A. Suzuki, *Synthesis*, 1983, 1005.
149. R.-J. Binnewirtz, H. Klingenberger, R. Welte and P. Paetzold, *Chem. Ber.*, 1983, **116**, 1271.
150. W. Siebert, R. Full, J. Edwin and K. Kinberger, *Chem. Ber.*, 1978, **111**, 823.
151. S. Hara and A. Suzuki, *Tetrahedron Lett.*, 1991, **32**, 6749.
152. H. C. Brown and T. E. Cole, *Organometallics*, 1983, **2**, 1316.
153. H. C. Brown and T. E. Cole, *Organometallics*, 1985, **4**, 816.
154. D. S. Matteson and G. D. Hurst, *Organometallics*, 1986, **5**, 1465.
155. H. C. Brown, M. Srebnik and T. E. Cole, *Organometallics*, 1986, **5**, 2300.
156. H. C. Brown, N. G. Bhat and M. Srebnik, *Tetrahedron Lett.*, 1988, **29**, 2631.
157. D. S. Matteson and K. Peacock, *J. Org. Chem.*, 1963, **28**, 3.
158. T. E. Cole and B. D. Haly, *Organometallics*, 1992, **11**, 652.
159. P.-Y. Chavant and M. Vaultier, *J. Organomet. Chem.*, 1993, **455**, 37.
160. C. Blanchard, G. Mignani, F. Leising and M. Vaultier, Rhône Poulenc, FR-2 699 926 (1993/1994).
161. M.-P. Arthur, A. Baceiredo and G. Bertrand, *J. Am. Chem. Soc.*, 1991, **113**, 5856.
162. M.-P. Arthur, H. P. Goodwin, A. Baceiredo, K. B. Dillon and G. Bertrand, *Organometallics*, 1991, **10**, 3205.
163. R. Stürmer, *Angew. Chem., Int. Ed. Engl.*, 1990, **29**, 59.
164. G. W. Kabalka, U. Sastry, K. A. R. Sastry, F. F. Knapp and P. C. Srivastava *J. Organomet. Chem.*, 1983, **259**, 269.
165. G. W. Kabalka, R. S. Varma, Y.-Z. Gai and R. M. Baldwin, *Tetrahedron Lett.*, 1986, **27**, 3843.
166. M. J. Sharp and V. Snieckus, *Tetrahedron Lett.*, 1985, **26**, 5997.
167. M. J. Sharp, W. Cheng and V. Snieckus, *Tetrahedron Lett.*, 1987, **28**, 5093.
168. J. Altman, H. Boehnke, A. Steigel and G. Wulff, *J. Organomet. Chem.*, 1986, **309**, 241.
169. H. E. Katz, *Organometallics*, 1986, **5**, 2308.
170. S. M. Tuladhar and C. D'Silva, *Tetrahedron Lett.*, 1992, **33**, 265.
171. H. C. Brown and M. Srebnik, *Organometallics*, 1987, **6**, 629.
172. H. C. Brown, T. E. Cole and M. Srebnik, *Organometallics*, 1985, **4**, 1788.
173. H. C. Brown and V. K. Mahindroo, *Synlett*, 1992, 626.
174. H. C. Brown, N. Vasumathi and N. N. Joshi, *Organometallics*, 1993, **12**, 1058.
175. H. C. Brown and U. S. Racherla, *J. Org. Chem.*, 1986, **51**, 427.
176. V. Dimitrov, K.-H. Thiele and A. Zschunke, *Z. Anorg. Allg. Chem.*, 1982, **494**, 144.
177. H. C. Brown and U. S. Racherla, *Organometallics*, 1986, **5**, 391.
178. C. G. Whiteley, *J. Chem. Soc., Chem. Commun.*, 1981, 5.
179. C. G. Whiteley, *S. Afr. J. Chem.*, 1982, **35**, 9.
180. C. G. Whiteley and I. Zwane, *J. Org. Chem.*, 1985, **50**, 1969.
181. C. G. Whiteley, *Tetrahedron Lett.*, 1984, **25**, 5563.
182. M. Zaidlewicz, *J. Organomet. Chem.*, 1985, **293**, 139.
183. K. Okada, T. Sugawa and M. Oda, *J. Chem. Soc., Chem. Commun.*, 1992, 74.
184. K. Okada, H. Inokawa and M. Oda, *Tetrahedron Lett.*, 1991, **32**, 6363.
185. G. E. Herberich *et al.*, *Chem. Ber.*, 1986, **119**, 420.
186. G. E. Herberich, U. Englert and S. Wang, *Chem. Ber.*, 1993, **126**, 297.
187. G. E. Herberich, B. Hessner and M. Hostalek, *Angew. Chem., Int. Ed. Engl.*, 1986, **25**, 642.
188. G. E. Herberich, U. Eigendorf and C. Ganter, *J. Organomet. Chem.*, 1991, **402**, C17.
189. G. E. Herberich, U. Eigendorf and U. Englert, *Chem. Ber.*, 1993, **126**, 1397.
190. W. Haubold, J. Herdtle, W. Gollinger and W. Einholz, *J. Organomet. Chem.*, 1986, **315**, 1.
191. U. M. Gross, M. Bartels and D. Kaufmann, *J. Organomet. Chem.*, 1988, **344**, 277.
192. I. Mikhail and D. Kaufmann, *J. Organomet. Chem.*, 1990, **398**, 53.
193. D. A. Singleton and J. P. Martinez, *J. Am. Chem. Soc.*, 1990, **112**, 7423.
194. D. A. Singleton, J. P. Martinez, J. V. Watson and G. M. Ndip, *Tetrahedron*, 1992, **48**, 5831.
195. Y. Nakadaira, R. Sato and H. Sakurai, *Chem. Lett.*, 1987, 1451.
196. A. J. Ashe, III, J. W. Kampf, C. M. Kausch, H. Konishi, M. O. Kristen and J. Kroker, *Organometallics*, 1990, **9**, 2944.
197. E. J. Corey, C.-M. Yu and S. S. Kim, *J. Am. Chem. Soc.*, 1989, **111**, 5495.
198. E. J. Corey, C.-M. Yu and D.-H. Lee, *J. Am. Chem. Soc.*, 1990, **112**, 878.
199. M. D. Fryzuk, G. S. Bates and C. Stone, *J. Org. Chem.*, 1988, **53**, 4425.

200. M. D. Fryzuk, C. Stone, R. F. Alex and R. K. Chadha, *Tetrahedron Lett.*, 1988, **29**, 3915.
201. T. E. Cole, R. Quintanilla and S. Rodewald, *Organometallics*, 1991, **10**, 3777.
202. T. E. Cole and R. Quintanilla, *J. Org. Chem.*, 1992, **57**, 7366.
203. P. J. Fagan, E. G. Burns and J. C. Calabrese, *J. Am. Chem. Soc.*, 1988, **110**, 2979.
204. J. A. Marsella and K. G. Caulton, *J. Am. Chem. Soc.*, 1982, **104**, 2361.
205. T. E. Cole, S. Rodewald and C. L. Watson, *Tetrahedron Lett.*, 1992, **33**, 5295.
206. R. Köster and G. Bruno, *Ann. Chem.*, 1960, **629**, 89.
207. H. C. Brown and S. K. Gupta, *J. Am. Chem. Soc.*, 1970, **92**, 6983.
208. H. C. Brown and S. K. Gupta, *J. Am. Chem. Soc.*, 1971, **93**, 2802.
209. H. C. Brown, D. Basavaiah and N. G. Bhat, *Organometallics*, 1983, **2**, 1309.
210. H. C. Brown, A. M. Salunkhe and A. B. Argade, *Organometallics*, 1992, **11**, 3094.
211. T. E. Cole, R. Quintanilla, B. M. Smith and D. Hurst, *Tetrahedron Lett.*, 1992, **33**, 2761.
212. H. C. Brown, P. K. Jadhav and M. C. Desai, *J. Am. Chem. Soc.*, 1982, **104**, 4303.
213. N. N. Joshi, C. Pyun, V. K. Mahindroo, B. Singaram and H. C. Brown, *J. Org. Chem.*, 1992, **57**, 504.
214. H. C. Brown, B. Singaram and T. E. Cole, *J. Am. Chem. Soc.*, 1985, **107**, 460.
215. H. C. Brown, J. V. N. Vara Prasad, A. K. Gupta and R. K. Bakshi, *J. Org. Chem.*, 1987, **52**, 310.
216. H. C. Brown, R. K. Bakshi and B. Singaram, *J. Am. Chem. Soc.*, 1988, **110**, 1529.
217. H. C. Brown, N. N. Joshi, C. Pyun and B. Singaram, *J. Am. Chem. Soc.*, 1989, **111**, 1754.
218. K. Narasaka and I. Yamamoto, *Tetrahedron*, 1992, **48**, 5743.
219. H. C. Brown, N. G. Bhat and V. Somayaji, *Organometallics*, 1983, **2**, 1311.
220. B. Singaram, T. E. Cole and H. C. Brown, *Organometallics*, 1984, **3**, 774.
221. B. Singaram, T. E. Cole and H. C. Brown, *Organometallics*, 1984, **3**, 1520.
222. H. C. Brown, M. V. Rangaishenvi and U. S. Racherla, *J. Org. Chem.*, 1987, **52**, 728.
223. T. E. Cole, R. K. Bakshi, M. Srebnik, B. Singaram and H. C. Brown, *Organometallics*, 1986, **5**, 2303.
224. H. C. Brown, T. E. Cole, M. Srebnik and K.-W. Kim, *J. Org. Chem.*, 1986, **51**, 4925.
225. H. C. Brown, A. M. Salunkhe and B. Singaram, *J. Org. Chem.*, 1991, **56**, 1170.
226. H. C. Brown, P. K. Jadhav and K. S. Bhat, *J. Am. Chem. Soc.*, 1985, **107**, 2564.
227. D. S. Matteson, R. Ray, R. R. Rocks and D. J. S. Tsai, *Organometallics*, 1983, **2**, 1536.
228. H. C. Brown and J. V. N. Vara Prasad, *J. Am. Chem. Soc.*, 1986, **108**, 2049.
229. H. C. Brown and J. V. N. Vara Prasad, *J. Org. Chem.*, 1986, **51**, 4526.
230. T. Mancilla, R. Contreras and B. Wrackmeyer, *J. Organomet. Chem.*, 1986, **307**, 1.
231. B. Garrigues, M. Mulliez and A. Rahariniriana, *J. Organomet. Chem.*, 1986, **302**, 153.
232. H. Amt, W. Kliegel, S. J. Rettig and J. Trotter, *Can. J. Chem.*, 1988, **66**, 1117.
233. N. Farfan, P. Joseph-Nathan, L. M. Chiquete and R. Contreras, *J. Organomet. Chem.*, 1988, **348**, 149.
234. S. Toyota and M. ōki, *Bull. Chem. Soc. Jpn.*, 1992, **65**, 1832.
235. D. S. Matteson, *Chem. Rev.*, 1989, **89**, 1535.
236. D. S. Matteson, *Tetrahedron*, 1989, **45**, 1859.
237. D. S. Matteson and D. Majumdar, *Organometallics*, 1983, **2**, 230.
238. D. S. Matteson and D. Majumdar, *Organometallics*, 1983, **2**, 1529.
239. K. M. Sadhu and D. S. Matteson, *Organometallics*, 1985, **4**, 1687.
240. H. C. Brown, S. M. Singh and M. V. Rangaishenvi, *J. Org. Chem.*, 1986, **51**, 3150.
241. T. J. Michnick and D. S. Matteson, *Synlett*, 1991, 631.
242. H. C. Brown and T. Imai, *J. Am. Chem. Soc.*, 1983, **105**, 6285.
243. H. C. Brown, M. V. Rangaishenvi and S. Jayaraman, *Organometallics*, 1992, **11**, 1948.
244. H. C. Brown, A. S. Phadke and M. V. Rangaishenvi, *Heteroatom Chem.*, 1990, **1**, 83.
245. D. S. Matteson and T.-C. Cheng, *J. Org. Chem.*, 1968, **33**, 3055.
246. D. S. Matteson and D. Majumdar, *J. Organomet. Chem.*, 1979, **170**, 259.
247. R. J. Mears and A. Whiting, *Tetrahedron*, 1993, **49**, 177.
248. D. S. Matteson and T. J. Michnick, *Organometallics*, 1990, **9**, 3171.
249. D. H. Kinder and M. M. Ames, *J. Org. Chem.*, 1987, **52**, 2452.
250. H. C. Brown and S. M. Singh, *Organometallics*, 1986, **5**, 994.
251. P. G. M. Wuts, P. A. Thomson and G. R. Callen, *J. Org. Chem.*, 1983, **48**, 5398.
252. P. G. M. Wuts and S. S. Bigelow, *J. Org. Chem.*, 1988, **53**, 5023.
253. R. W. Hoffmann and B. Landmann, *Chem. Ber.*, 1986, **119**, 1039.
254. H. C. Brown, T. Imai and N. G. Bhat, *J. Org. Chem.*, 1986, **51**, 5277.
255. H. C. Brown, N. G. Bhat and R. R. Iyer, *Tetrahedron Lett.*, 1991, **32**, 3655.
256. D. S. Matteson and K. M. Sadhu, *Organometallics*, 1984, **3**, 614.
257. D. S. Matteson, *Synthesis*, 1975, 147.
258. A. Pelter and K. Smith, in 'Comprehensive Organic Chemistry', eds. D. H. R. Barton and W. D. Ollis, Pergamon, Oxford, 1979, vol. 3, p. 689.
259. A. Pelter, *Chem. Soc. Rev.*, 1982, **11**, 191.
260. A. Pelter, K. Smith and K. Jones, *Pure Appl. Chem.*, 1991, **63**, 403.
261. A. Pelter and K. Smith, in 'Comprehensive Organic Synthesis', eds. B. M. Trost and I. Fleming, Pergamon, Oxford, 1991, vol. 1, p. 487.
262. G. Cainelli, G. Dal Bello and G. Zubiani, *Tetrahedron Lett.*, 1965, 3429.
263. T. Mukaiyama, M. Murakami, T. Oriyama and M. Yamaguchi, *Chem. Lett.*, 1981, 1193.
264. H. C. Brown and G. Zweifel, *J. Am. Chem. Soc.*, 1961, **83**, 3834.
265. D. S. Matteson, R. J. Moody and P. K. Jesthi, *J. Am. Chem. Soc.*, 1975, **97**, 5608.
266. D. S. Matteson and J. W. Wilson, *Organometallics*, 1985, **4**, 1690.
267. D. J. S. Tsai and D. S. Matteson, *Organometallics*, 1983, **2**, 236.
268. A. Pelter, K. Smith, D. E. Parry and K. D. Jones, *Aust. J. Chem.*, 1992, **45**, 57.
269. M. W. Rathke and R. Kow, *J. Am. Chem. Soc.*, 1972, **94**, 6854.
270. R. Kow and M. W. Rathke, *J. Am. Chem. Soc.*, 1973, **95**, 2715.

271. D. S. Matteson and R. J. Moody, *Organometallics*, 1982, **1**, 20.
272. D. S. Matteson and K. Arne, *J. Am. Chem. Soc.*, 1978, **100**, 1325.
273. D. S. Matteson and K. Arne, *Organometallics*, 1982, **1**, 280.
274. D. S. Matteson and D. Majumdar, *J. Chem. Soc., Chem. Commun.*, 1980, 39.
275. J. W. Wilson, *J. Organomet. Chem.*, 1980, **186**, 297.
276. A. Pelter, B. Singaram, L. Williams and J. W. Wilson, *Tetrahedron Lett.*, 1983, **24**, 623.
277. M. V. Garad, A. Pelter, B. Singaram and J. W. Wilson, *Tetrahedron Lett.*, 1983, **24**, 637.
278. A. Pelter, B. Singaram and J. W. Wilson, *Tetrahedron Lett.*, 1983, **24**, 631.
279. M. P. Cooke, Jr. and R. K. Widener, *J. Am. Chem. Soc.*, 1987, **109**, 931.
280. P. Knochel and R. D. Singer, *Chem. Rev.*, 1993, **93**, 2117.
281. D. S. Matteson and R. J. Moody, *J. Am. Chem. Soc.*, 1977, **99**, 3196.
282. D. S. Matteson and P. K. Jesthi, *J. Organomet. Chem.*, 1976, **110**, 25.
283. D. S. Matteson and J. O. Waldbillig, *J. Org. Chem.*, 1963, **28**, 366.
284. D. S. Matteson and M. L. Talbot, *J. Am. Chem. Soc.*, 1967, **89**, 1123.
285. G. Coindard and J. Braun, *Bull. Soc. Chim. Fr.*, 1972, 817.
286. D. A. Evans, W. L. Scott and L. K. Truesdale, *Tetrahedron Lett.*, 1972, 121.
287. D. A. Evans, A. M. Golob, N. S. Mandel and G. S. Mandel, *J. Am. Chem. Soc.*, 1978, **100**, 8170.
288. P. Martinez-Fresneda and M. Vaultier, *Tetrahedron Lett.*, 1989, **30**, 2929.
289. N. Noiret, A. Youssofi, B. Carboni and M. Vaultier, *J. Chem. Soc. Chem. Commun.*, 1992, 1105.
290. C. Rasset-Deloge and M. Vaultier, *Tetrahedron*, 1994, **50**, 3397.
291. G. Bianchi, A. Cogoli and P. Gruenanger, *Ricerca Sci.*, 1966, **36**, 132.
292. M. Jazouli, Ph.D. Thesis, University of Rennes, 1991.
293. R. H. Wallace and K. K. Zong, *Tetrahedron Lett.*, 1992, **33**, 6941.
294. D. S. Matteson, *J. Org. Chem.*, 1962, **27**, 4293.
295. D. S. Matteson, *Organomet. Chem. Rev.*, 1966, **1**, 1.
296. P. Fontani, B. Carboni, M. Vaultier and R. Carrié, *Tetrahedron Lett.*, 1989, **30**, 4815.
297. P. Fontani, B. Carboni, M. Vaultier and G. Maas, *Synthesis*, 1991, 605.
298. T. Imai, H. Mineta and S. Nishida, *J. Org. Chem.*, 1990, **55**, 4986.
299. W. G. Woods and I. S. Bengelsdorf, *J. Org. Chem.*, 1966, **31**, 2769.
300. R. H. Fish, *J. Organomet. Chem.*, 1972, **42**, 345.
301. D. J. Pasto, J. Chow and S. K. Arora, *Tetrahedron*, 1969, **25**, 1557.
302. D. S. Matteson and G. D. Schaumberg, *J. Org. Chem.*, 1966, **31**, 726.
303. N. Guennouni, C. Rasset-Deloge, B. Carboni and M. Vaultier, *Synlett*, 1992, 581.
304. M. Vaultier, A. El Louzi, S. L. Titouani and M. Soufiaoui, *Synlett*, 1991, 267.
305. N. Guennouni, Ph.D. Thesis, University of Rennes, 1994.
306. M. P. Cooke, Jr., *J. Org. Chem.*, 1992, **57**, 1495.
307. G. Goindord, J. Blaum and P. Cadiot, *Bull. Soc. Chim. Fr.*, 1972, 811.
308. D. Ulmschneider and J. Goubeau, *Chem. Ber.*, 1957, **90**, 2733.
309. G. A. Olah, P. W. Westerman, Y. K. Mo and G. Klopman, *J. Am. Chem. Soc.*, 1972, **94**, 7859.
310. H. C. Brown and K. J. Murray, *Tetrahedron*, 1986, **42**, 5497.
311. G. W. Kabalka, R. J. Newton, Jr. and J. Jacobus, *J. Org. Chem.*, 1979, **44**, 4185.
312. H. C. Brown and G. A. Molander, *J. Org. Chem.*, 1986, **51**, 4512.
313. C. D. Blue and D. J. Nelson, *J. Org. Chem.*, 1983, **48**, 4538.
314. S. Rajagopalan and G. Zweifel, *Synthesis*, 1984, 113.
315. P. J. Domaille, J. D. Druliner, L. W. Gosser, J. M. Read, Jr., E. R. Schmelzer and W. R. Stevens, *J. Org. Chem.*, 1985, **50**, 189.
316. G. Zweifel and N. L. Polston, *J. Am. Chem. Soc.*, 1970, **92**, 4068.
317. H. C. Brown and K. K. Wang, *J. Org. Chem.*, 1986, **51**, 4514.
318. R. W. Murray and G. J. Williams, *J. Org. Chem.*, 1969, **34**, 1896.
319. T. Onak, 'Organoborane Chemistry', Academic Press, New York, 1975.
320. A. Pelter and K. Smith, in 'Comprehensive Organic Synthesis', eds. B. M. Trost and I. Fleming, Pergamon, Oxford, 1991, vol. 7, p. 593; 604; 606.
321. H. C. Brown, M. M. Midland and G. W. Kabalka, *Tetrahedron*, 1986, **42**, 5523.
322. H. C. Brown and M. M. Midland, *Tetrahedron*, 1987, **43**, 4059.
323. G. W. Kabalka, T. J. Reed and S. A. Kunda, *Synth. Commun.*, 1983, **13**, 737.
324. G. W. Kabalka *et al.*, *Appl. Radiat. Isot.*, 1985, **36**, 853.
325. M. S. Berridge, M. P. Franceschini, T. J. Tewson and K. L. Gould, *J. Nucl. Med.*, 1986, **27**, 834.
326. G. W. Kabalka, *3rd Int. Symp. Synth. Appl. Isotopically Labelled Compds., Innsbruck, Austria*, 1988, 537.
327. G. Zweifel and H. C. Brown, *Org. React.*, 1963, **13**, 1.
328. H. C. Brown, C. Snyder, B. C. Subba Rao and G. Zweifel, *Tetrahedron*, 1986, **42**, 5505.
329. A. Pelter, L. Hughes and J. M. Rao, *J. Chem. Soc., Perkin Trans. 1*, 1982, 719.
330. R. Köster and Y. Morita, *Justus Liebigs Ann. Chem.*, 1967, **704**, 70.
331. R. Köster, S. Awra and P. Bingu, *Angew. Chem., Int. Ed. Engl.*, 1969, **8**, 205.
332. G. W. Kabalka and H. C. Hedgecock, Jr., *J. Org. Chem.*, 1975, **40**, 1776.
333. J. A. Soderquist and C. L. Anderson, *Tetrahedron Lett.*, 1986, **27**, 3961.
334. G. P. Boldorini, L. Lodi, E. Tagliavini, C. Hombini and A. Umani-Ronchi, *J. Organomet. Chem.*, 1987, **336**, 23.
335. J. A. Soderquist and M. Ramin Najafi, *J. Org. Chem.*, 1986, **51**, 1330.
336. G. Zweifel, N. L. Polston and C. C. Whitney, *J. Am. Chem. Soc.*, 1968, **90**, 6243.
337. A. G. Davies and B. P. Roberts, *J. Chem. Soc. C*, 1968, 1474.
338. A. Pelter, P. Rupani and P. Stewart, *J. Chem. Soc., Chem. Commun.*, 1981, 164.
339. R. L. Danheiser and A. C. Savoca, *J. Org. Chem.*, 1985, **50**, 2401.
340. G. W. Kabalka, T. M. Shoup and N. M. Goudgaon, *Tetrahedron Lett.*, 1989, **30**, 1483.
341. G. W. Kabalka, P. P. Wadgaonkar and T. M. Shoup, *Tetrahedron Lett.*, 1989, **30**, 5103.

342. G. W. Kabalka, P. P. Wadgaonkar and T. M. Shoup, *Organometallics*, 1990, **9**, 1316.
343. J. C. Ware and T. G. Traylor, *J. Am. Chem. Soc.*, 1963, **85**, 3026.
344. H. C. Brown and C. P. Garg, *Tetrahedron*, 1986, **42**, 5511.
345. H. C. Brown, S. U. Kulkarni, C. G. Rao and V. D. Patil, *Tetrahedron*, 1986, **42**, 5515.
346. U. S. Racherla, V. V. Khanna and H. C. Brown, *Tetrahedron Lett.*, 1992, **33**, 1037.
347. H. C. Brown, S. U. Kulkarni, V. V. Khanna, V. D. Patil and U. S. Racherla, *J. Org. Chem.*, 1992, **57**, 6173.
348. L. Verbit and P. J. Heffron, *J. Org. Chem.*, 1967, **32**, 3199.
349. G. W. Kabalka, K. A. R. Sastry, G. W. McCollum and H. Yoshioka, *J. Org. Chem.*, 1981, **46**, 4296.
350. G. W. Kabalka, K. A. R. Sastry, G. W. McCollum and C. A. Lane, *J. Chem. Soc., Chem. Commun.*, 1982, 62.
351. H. C. Brown, K. W. Kim, M. Srebnik and B. Singaram, *Tetrahedron*, 1987, **43**, 4071.
352. G. W. Kabalka, Z. Wang and N. M. Goudgaon, *Synth. Commun.*, 1989, **19**, 2409.
353. G. W. Kabalka, Z. Wang, J. F. Green and M. M. Goodman, *Appl. Radiat. Isot.*, 1992, **43**, 389.
354. G. W. Kabalka, D. A. Henderson and R. S. Varma, *Organometallics*, 1987, **6**, 1369.
355. G. W. Kabalka, N. M. Goudgaon and Y. Liang, *Synth. Commun.*, 1988, **18**, 1363.
356. H. C. Brown, K.-W. Kim, T. E. Cole and B. Singaram, *J. Am. Chem. Soc.*, 1986, **108**, 6761.
357. J.-P. Genêt, J. Hajicek, L. Bischoff and C. Greck, *Tetrahedron Lett.*, 1992, **33**, 2677.
358. R. Y. Yang and L. X. Dai, *Synthesis*, 1993, 481.
359. A. Suzuki, S. Sono, M. Itoh, H. C. Brown and M. M. Midland, *J. Am. Chem. Soc.*, 1971, **93**, 4329.
360. H. C. Brown, M. M. Midland and A. B. Levy, *J. Am. Chem. Soc.*, 1972, **94**, 3662.
361. H. C. Brown, M. M. Midland and A. B. Levy, *J. Am. Chem. Soc.*, 1973, **95**, 2394.
362. B. Carboni, M. Vaultier and R. Carrié, *Tetrahedron*, 1987, **43**, 1799.
363. B. Carboni, M. Vaultier, T. Courgeon and R. Carrié, *Bull. Soc. Chim. Fr.*, 1989, 844.
364. H. C. Brown, M. M. Midland, A. B. Levy, A. Suzuki, S. Sono and M. Itoh, *Tetrahedron*, 1987, **43**, 4079.
365. P.-Y. Chavant, F. Lhermitte and M. Vaultier, *Synlett*, 1993, 519.
366. A. B. Levy and H. C. Brown, *J. Am. Chem. Soc.*, 1973, **95**, 4067.
367. A. Benalil, B. Carboni and M. Vaultier, *Tetrahedron*, 1991, **47**, 8177.
368. B. Carboni, A. Benalil and M. Vaultier, *J. Org. Chem.*, 1993, **58**, 3736.
369. M. Vaultier, B. Carboni and P. M. Fresneda, *Synth. Commun.*, 1992, **22**, 665.
370. J.-M. Jego, B. Carboni, M. Vaultier and R. Carrié, *J. Chem. Soc., Chem. Commun.*, 1989, 142.
371. J.-M. Jego, B. Carboni and M. Vaultier, *Bull. Soc. Chim. Fr.*, 1992, **129**, 554.
372. Y. N. Bubnov, M. E. Gursky and D. G. Pershin, *J. Organomet. Chem.*, 1991, **412**, 1.
373. D. A. Evans and A. E. Weber, *J. Am. Chem. Soc.*, 1987, **109**, 7151.
374. J.-M. Jego, B. Carboni, A. Youssofi and M. Vaultier, *Synlett*, 1993, 595.
375. H. C. Brown and A. M. Salunkhe, *Tetrahedron Lett.*, 1993, **34**, 1265.
376. L. J. Diorazio, D. A. Widdowson and J. M. Clough, *Tetrahedron*, 1992, **48**, 8073.
377. H. C. Brown and N. R. De Lue, *Tetrahedron*, 1988, **44**, 2785.
378. Y. Masuda, M. Hoshi and A. Arase, *J. Chem. Soc., Perkin Trans. 1*, 1992, 2725.
379. H. C. Brown and C. F. Lane, *Tetrahedron*, 1988, **44**, 2763.
380. G. W. Kabalka, K. A. R. Sastry, H. C. Hsu and M. D. Hylarides, *J. Org. Chem.*, 1981, **46**, 3113.
381. H. C. Brown, C. F. Lane and N. R. De Lue, *Tetrahedron*, 1988, **44**, 2773.
382. S. Hara, T. Kato, H. Shimizu and A. Suzuki, *Tetrahedron Lett.*, 1985, **26**, 1065.
383. H. C. Brown, M. W. Rathke, M. M. Rogić and N. R. De Lue, *Tetrahedron*, 1988, **44**, 2751.
384. G. W. Kabalka and R. S. Varma, *Tetrahedron*, 1989, **45**, 6601.
385. H. C. Brown, T. Hamaoka, N. Ravindran, C. Subrahmanyam, V. Somayaji and N. G. Bhat, *J. Org. Chem.*, 1989, **54**, 6075.
386. H. C. Brown *et al.*, *J. Org. Chem.*, 1989, **54**, 6068.
387. G. W. Kabalka, K. A. R. Sastry and V. Somayaji, *Heterocycles*, 1982, **18**, 157.
388. W. R. Sponholtz, III, R. M. Pagni, G. W. Kabalka, J. F. Green and L. C. Tan, *J. Org. Chem.*, 1991, **56**, 5700.
389. M. Srebnik, *Tetrahedron Lett.*, 1991, **32**, 2449.
390. W. Oppolzer and R. N. Radinov, *Helv. Chim. Acta*, 1992, **75**, 170.
391. W. Oppolzer and R. N. Radinov, *J. Am. Chem. Soc.*, 1993, **115**, 1593.
392. F. Langer, J. Waas and P. Knochel, *Tetrahedron Lett.*, 1993, **34**, 5261.
393. H. Yatagai, *J. Org. Chem.*, 1980, **45**, 1640.
394. K.-H. Chu and K. K. Wang, *J. Org. Chem.*, 1986, **51**, 767.
395. H. C. Brown and G. A. Molander, *J. Org. Chem.*, 1981, **46**, 645.
396. J. Ichikawa, S. Hamada, T. Sonoda and H. Kobayashi, *Tetrahedron Lett.*, 1992, **33**, 337.
397. E. Negishi, *J. Organomet. Chem.*, 1976, **108**, 281.
398. E. Negishi, *J. Chem. Ed.*, 1975, **52**, 159.
399. G. M. L. Cragg and K. R. Koch, *Chem. Soc. Rev.*, 1977, **6**, 393.
400. A. Suzuki, *Acc. Chem. Res.*, 1982, **15**, 178.
401. A. Suzuki, *Top. Curr. Chem.*, 1983, **112**, 67.
402. H. C. Brown and Y. Yamamoto, *J. Am. Chem. Soc.*, 1971, **93**, 2796.
403. J. Rathke and R. Schaeffer, *Inorg. Chem.*, 1972, **11**, 1150.
404. Y. Yamamoto and H. C. Brown, *J. Org. Chem.*, 1974, **39**, 861.
405. M. M. Midland, A. R. Zolopa and R. L. Halterman, *J. Am. Chem. Soc.*, 1979, **101**, 248.
406. H. C. Brown, N. R. De Lue, Y. Yamamoto and K. Maruyama, *J. Org. Chem.*, 1977, **42**, 3252.
407. A. Arase, M. Hoshi and Y. Masuda, *Chem. Lett.*, 1984, 2093.
408. M. Hoshi, Y. Masuda and A. Arase, *J. Chem. Soc., Chem. Commun.*, 1985, 1068.
409. E. Negishi, R. M. Williams, G. Lew and T. Yoshida, *J. Organomet. Chem.*, 1975, **92**, C4.
410. Y. Masuda, M. Hoshi and A. Arase, *Bull. Chem. Soc. Jpn.*, 1992, **65**, 3294.
411. M. Hoshi, Y. Masuda and A. Arase, *Bull. Chem. Soc. Jpn.*, 1992, **65**, 685.
412. H. Kono and J. Hooz, in 'Organic Syntheses', Wiley, New York, 1988, vol. 6, p. 919 and references therein.
413. J. Hooz, J. N. Bridson, J. G. Calzada, H. C. Brown, M. M. Midland and A. B. Levy, *J. Org. Chem.*, 1973, **38**, 2574.
414. H. C. Brown and A. M. Salunkhe, *Synlett*, 1991, 684.

415. A. Ansorge, D. J. Brauer, H. Bürger, T. Hagen and G. Pawelke, *Angew. Chem., Int. Ed. Engl.*, 1993, **32**, 384.
416. E. Negishi, T. Yoshida, A. Silveira, Jr. and B. L. Chiou, *J. Org. Chem.*, 1975, **40**, 814.
417. S. Yamamoto, M. Shiono and T. Mukaiyama, *Chem. Lett.*, 1973, 961.
418. S. Ncube, A. Pelter and K. Smith, *Tetrahedron Lett.*, 1979, 1893.
419. S. Hara, K. Kishimura, A. Suzuki and R. S. Dhillon, *J. Org. Chem.*, 1990, **55**, 6356.
420. H. C. Brown, T. Imai, P. T. Perumal and B. Singaram, *J. Org. Chem.*, 1985, **50**, 4032.
421. R. J. Hughes, S. Ncube, A. Pelter, K. Smith, E. Negishi and T. Yoshida, *J. Chem. Soc., Perkin Trans. 1*, 1977, 1172.
422. S. Ncube, A. Pelter and K. Smith, *Tetrahedron Lett.*, 1979, 1895.
423. G. L. Larson, R. Arguelles, O. Rosario and S. Sandoval, *J. Organomet. Chem.*, 1980, **198**, 15.
424. J. A. Soderquist and I. Rivera, *Tetrahedron Lett.*, 1989, **30**, 3919.
425. A. B. Levy and S. J. Schwartz, *Tetrahedron Lett.*, 1976, 2201.
426. T. Yogo and A. Suzuki, *Chem. Lett.*, 1980, 591.
427. S. Birkinshaw and P. Kocieński, *Tetrahedron Lett.*, 1991, **32**, 6961.
428. J. Koshino, T. Sugawara and A. Suzuki, *Heterocycles*, 1984, **22**, 489.
429. A. Suzuki, N. Miyaura and M. Itoh, *Tetrahedron*, 1971, **27**, 2775.
430. K. Utimoto, N. Sakai, M. Obayashi and H. Nozaki, *Tetrahedron*, 1976, **32**, 769.
431. M. M. Midland, *J. Org. Chem.*, 1977, **42**, 2650.
432. M. M. Midland and D. C. McDowell, *J. Organomet. Chem.*, 1978, **156**, C5.
433. M. M. Midland and S. B. Preston, *J. Org. Chem.*, 1980, **45**, 747.
434. G. Zweifel and N. R. Pearson, *J. Org. Chem.*, 1981, **46**, 829.
435. H. C. Brown, B. A. Carlson and R. H. Prager, *J. Am. Chem. Soc.*, 1971, **93**, 2070.
436. B. A. Carlson and H. C. Brown, in 'Organic Syntheses', John Wiley, New York, 1988, vol. 6, p. 137 and references therein.
437. A. Pelter and J. M. Rao, *J. Organomet. Chem.*, 1985, **285**, 65.
438. H. C. Brown, J. L. Hubbard and K. Smith, *Synthesis*, 1979, 701.
439. H. C. Brown and E. Negishi, *J. Chem. Soc., Chem. Commun.*, 1968, 594.
440. H. C. Brown, D. Basavaiah and U. S. Racherla, *Synthesis*, 1983, 886.
441. A. Devasagayaraj, M. L. Narayama Rao and M. Periasamy, *J. Organomet. Chem.*, 1991, **421**, 147.
442. H. C. Brown and W. C. Dickason, *J. Am. Chem. Soc.*, 1969, **91**, 1226.
443. E. Negishi, K. W. Chiu and T. Yosida, *J. Org. Chem.*, 1975, **40**, 1676.
444. Y. Yamamoto, H. Toi, A. Sonoda and S. I. Murahashi, *J. Am. Chem. Soc.*, 1976, **98**, 1965.
445. G. W. Kramer and H. C. Brown, *J. Org. Chem.*, 1977, **42**, 2832.
446. N. Sasaki, N. Miyaura, M. Itoh and A. Suzuki, *Tetrahedron Lett.*, 1977, 3369.
447. H. C. Brown, A. B. Levy and M. M. Midland, *J. Am. Chem. Soc.*, 1975, **97**, 5017.
448. M. M. Midland and H. C. Brown, *J. Org. Chem.*, 1975, **40**, 2845.
449. A. Pelter, C. R. Harrison, C. Subrahmanyam and D. Kirkpatrick, *J. Chem. Soc., Perkin Trans. 1*, 1976, 2435.
450. K. K. Wang and K. H. Chu, *J. Org. Chem.*, 1984, **49**, 5175.
451. P. Binger and R. Köster, *Tetrahedron Lett.*, 1965, 1901.
452. R. Köster and L. A. Hagelee, *Synthesis*, 1976, 118.
453. P. Binger and R. Köster, *Synthesis*, 1973, 309.
454. A. Pelter, K. J. Gould and C. R. Harrison, *J. Chem. Soc., Perkin Trans. 1*, 1976, 2428.
455. A. Pelter and M. E. Colclough, *Tetrahedron Lett.*, 1986, **27**, 1935.
456. A. Pelter and C. R. Harrison, *J. Chem. Soc., Chem. Commun.*, 1974, 828.
457. S. Hara, H. Dojo and A. Suzuki, *Chem. Lett.*, 1983, 285.
458. E. J. Corey and W. L. Seibel, *Tetrahedron Lett.*, 1986, **27**, 909.
459. E. I. Negishi, T. Nguyen, L. D. Boardman, H. Sawada and J. A. Morrison, *Heteroatom Chem.*, 1992, **3**, 293.
460. A. Suzuki et al., *J. Org. Chem.*, 1986, **51**, 4507.
461. H. C. Brown, D. Basavaiah and N. G. Bhat, *J. Org. Chem.*, 1986, **51**, 4518.
462. J. A. Sikorski, N. G. Bhat, T. E. Cole, K. K. Wang and H. C. Brown, *J. Org. Chem.*, 1986, **51**, 4521.
463. S. W. Slayden, *J. Org. Chem.*, 1981, **46**, 2311.
464. S. W. Slayden, *J. Org. Chem.*, 1982, **47**, 2753.
465. A. Pelter and R. A. Drake, *Tetrahedron Lett.*, 1988, **29**, 4181.
466. E. I. Negishi, T. Yoshida, A. Abramovitch, G. Lew and R. W. Williams, *Tetrahedron*, 1991, **47**, 343.
467. G. Zweifel and R. P. Fisher, *Synthesis*, 1974, 339.
468. K. Utimoto, K. Uchida and H. Nozaki, *Chem. Lett.*, 1974, 1493.
469. K. Utimoto, K. Uchida and H. Nozaki, *Tetrahedron Lett.*, 1973, 4527.
470. H. C. Brown, D. Basavaiah, S. U. Kulkarni, N. G. Bhat and J. V. N. Vara Prasad, *J. Org. Chem.*, 1988, **53**, 239.
471. H. C. Brown and N. G. Bhat, *J. Org. Chem.*, 1988, **53**, 6009.
472. T. Baba, K. Avasthi and A. Suzuki, *Bull. Soc. Chem. Jpn.*, 1983, **56**, 1571.
473. D. A. Evans, T. C. Crawford, R. C. Thomas and J. A. Walker, *J. Org. Chem.*, 1976, **41**, 3947.
474. A. B. Levy, *J. Org. Chem.*, 1978, **43**, 4684.
475. A. B. Levy, *Tetrahedron Lett.*, 1979, 4021.
476. M. Ishikura, M. Terashima, K. Okamura and T. Date, *J. Chem. Soc., Chem. Commun.*, 1991, 1219.
477. M. E. Garst and J. N. Bonfiglio, *Tetrahedron Lett.*, 1981, **22**, 2075.
478. A. Pelter, P. J. Maddocks and K. Smith, *J. Chem. Soc., Chem. Commun.*, 1978, 805.
479. N. Miyaura, K. Yamada and A. Suzuki, *Tetrahedron Lett.*, 1979, **19**, 3437.
480. R. F. Heck, in 'Palladium Reagents in Organic Syntheses', eds. A. R. Katritzky, O. Meth-Cohn and C. W. Rees, Academic Press, New York, 1985, p. 179.
481. A. Suzuki, *Pure Appl. Chem.*, 1985, **57**, 1749.
482. A. Suzuki, *Pure Appl. Chem.*, 1991, **63**, 419.
483. A. R. Martin and Y. Yang, *Acta Chem. Scand.*, 1993, **47**, 221.
484. N. Miyaura, K. Yamada, H. Suginome and A. Suzuki, *J. Am. Chem. Soc.*, 1985, **107**, 972.
485. N. Miyaura, T. Yanagi and A. Suzuki, *Synth. Commun.*, 1981, **11**, 513.
486. N. Miyaura, H. Suginome and A. Suzuki, *Tetrahedron*, 1983, **39**, 3271.

487. N. Miyaura, M. Satoh and A. Suzuki, *Tetrahedron Lett.*, 1986, **27**, 3745.
488. J. Uenishi, J. M. Beau, R. W. Armstrong and Y. Kishi, *J. Am. Chem. Soc.*, 1987, **109**, 4756.
489. W. R. Roush, B. B. Brown and S. E. Drozda, *Tetrahedron Lett.*, 1988, **29**, 3541.
490. W. R. Roush, K. J. Moriarty and B. B. Brown, *Tetrahedron Lett.*, 1990, **31**, 6509.
491. K. C. Nicolaou, J. Y. Ramphal, J. M. Palazon and R. A. Spanevello, *Angew. Chem., Int. Ed. Engl.*, 1989, **28**, 587.
492. A. R. de Lera, A. Torrado, B. Iglesias and S. López, *Tetrahedron Lett.*, 1992, **33**, 6205.
493. T. Yanagi, T. Oh-e, N. Miyaura and A. Suzuki, *Bull. Chem. Soc. Jpn.*, 1989, **62**, 3892.
494. M. V. Mavrov, N. A. Urdaneta, N. K. Hao and E. P. Serebkyakov, *Bull. Acad. Sci. USSR, Div. Chem. Sci.*, 1987, **36**, 2447.
495. N. Satoh, T. Ishiyama, N. Miyaura and A. Suzuki, *Bull. Chem. Soc. Jpn.*, 1987, **60**, 3471.
496. M. Satoh, N. Miyaura and A. Suzuki, *Chem. Lett.*, 1986, 1329.
497. N. Miyaura, T. Yano and A. Suzuki, *Tetrahedron Lett.*, 1980, **21**, 2865.
498. N. Miyaura and A. Suzuki, *Org. Synth.*, 1989, **68**, 130.
499. S. Hyuga, Y. Chiba, N. Yamashina, S. Hara and A. Suzuki, *Chem. Lett.*, 1987, 1757; 1988, 809.
500. Y. Satoh, H. Serizawa, N. Miyaura, S. Hara and A. Suzuki, *Tetrahedron Lett.*, 1988, **29**, 1811.
501. N. Yamashina, S. Hyuga, S. Hara and A. Suzuki, *Tetrahedron Lett.*, 1989, **30**, 6555.
502. S. Hyuga, S. Hara and A. Suzuki, *Bull. Chem. Soc. Jpn.*, 1992, **65**, 2303.
503. S. Gronowitz and K. Lawitz, *Chem. Scr.*, 1983, **22**, 265.
504. S. Gronowitz, V. Bobosik and K. Lawitz, *Chem. Scr.*, 1984, **23**, 120.
505. S. Gronowitz and K. Lawitz, *Chem. Scr.*, 1984, **24**, 5.
506. M. B. Mitchell and P. J. Wallbank, *Tetrahedron Lett.*, 1991, **32**, 2273.
507. N. A. Bumagin, V. V. Bykov and I. P. Beletskaya, *Izv. Akad. Nauk SSSR, Ser. Khim.*, 1989, **10**, 2394 (*Engl. Transl.*, 1992, 2206).
508. N. A. Bumagin, V. V. Bykov and I. P. Beletskaya, *Dokl. Akad. Nauk SSSR*, 1990, **315**, 1133 (*Engl. Transl.*, 1991, 357).
509. W. J. Thompson, J. H. Jones, P. A. Lyle and J. E. Thies, *J. Org. Chem.*, 1988, **53**, 2052.
510. A. L. Casalmuovo and J. C. Calabrese, *J. Am. Chem. Soc.*, 1990, **112**, 4324.
511. Y. Yang, A. B. Hörnfeldt and S. Gronowitz, *J. Heterocycl. Chem.*, 1989, **26**, 865.
512. Y. Yang, *Synth. Commun.*, 1989, **19**, 1001.
513. T. Watanabe, N. Miyaura and A. Suzuki, *Synlett*, 1992, 207.
514. A. Godard, A. Turck, N. Plé, F. Marsais and G. Quéguiner, *Trends Heterocycl. Chem.*, 1993, **3**, 19.
515. P. Rocca, F. Marsais, A. Godard and G. Quéguiner, *Tetrahedron*, 1993, **49**, 3325.
516. A. Godard, J. C. Rovera, F. Marsais, N. Plé and G. Quéguiner, *Tetrahedron*, 1992, **48**, 4123.
517. A. Godard, P. Rocca, J. M. Fourquez, J. C. Rovera, F. Marsais and G. Quéguiner, *Tetrahedron Lett.*, 1993, **34**, 7919.
518. P. Rocca, F. Marsais, A. Godard and G. Quéguiner, *Tetrahedron Lett.*, 1993, **34**, 2937.
519. P. Rocca, F. Marsais, A. Godard and G. Quéguiner, *Tetrahedron Lett.*, 1993, **34**, 7917.
520. D. J. Wustrow and L. D. Wise, *Synthesis*, 1991, 993.
521. Q. Zheng, Y. Yang and A. R. Martin, *Tetrahedron Lett.*, 1993, **34**, 2235.
522. A. Huth, I. Beetz and I. Schumann, *Tetrahedron*, 1989, **45**, 6679.
523. T. Oh-e, N. Miyaura and A. Suzuki, *Synlett*, 1990, 221.
524. T. Oh-e, N. Miyaura and A. Suzuki, *J. Org. Chem.*, 1993, **58**, 2201.
525. W. C. Shieh and J. A. Carlson, *J. Org. Chem.*, 1992, **57**, 379.
526. P. G. Ciattini, E. Morera and G. Ortar, *Tetrahedron Lett.*, 1992, **33**, 4815.
527. C. S. Cho, K. Itotani and S. Uemura, *J. Organomet. Chem.*, 1993, **443**, 253.
528. N. Miyaura, T. Ishiyama, H. Sasaki, M. Ishikawa, M. Satoh and A. Suzuki, *J. Am. Chem. Soc.*, 1989, **111**, 314.
529. S. Abe, N. Miyaura and A. Suzuki, *Bull. Chem. Soc. Jpn.*, 1992, **65**, 2863.
530. I. Rivera, J. C. Colberg and J. A. Soderquist, *Tetrahedron Lett.*, 1992, **33**, 6919.
531. T. Ishiyama, S. Abe, N. Miyaura and A. Suzuki, *Chem. Lett.*, 1992, 691.
532. N. Miyaura, T. Ishiyama, M. Ishikawa and A. Suzuki, *Tetrahedron Lett.*, 1986, **27**, 6369.
533. I. Rivera and J. A. Soderquist, *Tetrahedron Lett.*, 1991, **32**, 2311.
534. J. A. Soderquist and B. Santiago, *Tetrahedron Lett.*, 1990, **31**, 5541.
535. J. A. Soderquist, B. Santiago and I. Rivera, *Tetrahedron Lett.*, 1990, **31**, 4981.
536. T. Ishiyama, N. Miyaura and A. Suzuki, *Bull. Chem. Soc. Jpn.*, 1991, **64**, 1999.
537. T. Ishiyama, N. Miyaura and A. Suzuki, *Tetrahedron Lett.*, 1991, **32**, 6923.
538. T. Ishiyama, T. Oh-e, N. Miyaura and A. Suzuki, *Tetrahedron Lett.*, 1992, **33**, 4465.
539. A. Pelter, L. Warren and J. W. Wilson, *Tetrahedron*, 1993, **49**, 2988.
540. A. Pelter, G. F. Vaughan-Williams and R. M. Rosser, *Tetrahedron*, 1993, **49**, 3007.
541. J. R. Waas, A. Sidduri and P. Knochel, *Tetrahedron Lett.*, 1992, **33**, 3717.
542. G. Cainelli, G. Dal Bello and G. Zubiani, *Tetrahedron Lett.*, 1966, 4315.
543. A. Pelter, B. Singaram and J. W. Wilson, *Tetrahedron Lett.*, 1983, **24**, 635.
544. A. Pelter, D. Buss and A. Pitchford, *Tetrahedron Lett.*, 1985, **26**, 5093.
545. A. Pelter, D. Buss and M. E. Colclough, *J. Chem. Soc., Chem. Commun.*, 1987, 297.
546. A. Pelter, K. Smith, S. Elgendy and M. Rowlands, *Tetrahedron Lett.*, 1989, **30**, 5643; 5647.
547. A. Pelter, K. Smith and K. D. Jones, *J. Chem. Soc., Perkin Trans. 1*, 1992, 747.
548. W. R. Roush, in 'Comprehensive Organic Synthesis', eds. B. M. Trost and I. Fleming, Pergamon, Oxford, 1991, vol. 2, p. 1.
549. Y. N. Bubnov, *Pure Appl. Chem.*, 1987, **59**, 895.
550. Y. N. Bubnov, *Pure Appl. Chem.*, 1991, **63**, 361.
551. Y. Yamamoto and N. Asao, *Chem. Rev.*, 1993, **93**, 2207.
552. R. W. Hoffmann, *Angew. Chem., Int. Ed. Engl.*, 1982, **21**, 555.
553. Y. N. Bubnov, M. E. Gurskii, I. D. Gridnev, A. V. Ignatenko, Y. A. Ustynyuk and V. I. Mstislavsky, *J. Organomet. Chem.*, 1992, **424**, 127.
554. P. G. M. Wuts and S. S. Bigelow, *J. Org. Chem.*, 1982, **47**, 2498.
555. P. G. M. Wuts and S. S. Bigelow, *Synth. Commun.*, 1982, **12**, 779.
556. R. W. Hoffmann and B. Kemper, *Tetrahedron Lett.*, 1982, **23**, 845.

557. R. W. Hoffmann and R. Metternich, *Liebigs Ann. Chem.*, 1985, 2390.
558. T. Watanabe, N. Miyaura and A. Suzuki, *J. Organomet. Chem.*, 1993, **444**, C1.
559. R. W. Hoffmann, T. Sander and A. Hense, *Liebigs Ann. Chem.*, 1993, 771.
560. R. W. Hoffmann and A. Schlapbach, *Liebigs Ann. Chem.*, 1990, 1243.
561. V. Nyzam, C. Belaud and J. Villieras, *Tetrahedron Lett.*, 1993, **34**, 6899.
562. R. W. Hoffmann and G. Niel, *Liebigs Ann. Chem.*, 1991, 1195.
563. M. Satoh, Y. Nomoto, N. Miyaura and A. Suzuki, *Tetrahedron Lett.*, 1989, **30**, 3789.
564. D. S. Matteson and D. Majumdar, *J. Am. Chem. Soc.*, 1980, **102**, 7588.
565. M. Sato, Y. Yamamoto, S. Hara and A. Suzuki, *Tetrahedron Lett.*, 1993, **34**, 7071.
566. W. Zhou, S. Liang, S. Yu and W. Luo, *J. Organomet. Chem.*, 1993, **452**, 13 and references therein.
567. H. C. Brown, U. S. Racherla and P. J. Pellechia, *J. Org. Chem.*, 1990, **55**, 1868.
568. R. W. Hoffmann and B. Landmann, *Tetrahedron Lett.*, 1983, **24**, 3209.
569. R. W. Hoffmann and J. J. Wolff, *Chem. Ber.*, 1991, **124**, 563.
570. M. W. Andersen, B. Hildebrandt, G. Köster and R. W. Hoffmann, *Chem. Ber.*, 1989, **122**, 1777.
571. R. W. Hoffmann and T. Sander, *Chem. Ber.*, 1990, **123**, 145.
572. W. R. Roush, M. A. Adam, A. E. Walts and D. J. Harris, *J. Am. Chem. Soc.*, 1986, **108**, 3422.
573. Y. N. Bubnov, L. I. Lavrinovich, A. Y. Zykov and A. V. Ignatenko, *Mendeleev Commun.*, 1992, 86.
574. Z. Wang, X. J. Meng and G. W. Kabalka, *Tetrahedron Lett.*, 1991, **32**, 1945.
575. Z. Wang, X. J. Meng and G. W. Kabalka, *Tetrahedron Lett.*, 1991, **32**, 5677.
576. Y. Yamamoto, T. Komatsu and K. Maruyama, *J. Am. Chem. Soc.*, 1984, **106**, 5031.
577. Y. Yamamoto, T. Komatsu and K. Maruyama, *J. Org. Chem.*, 1985, **50**, 3115.
578. B. Guyot, J. Pornet and L. Miginiac, *Synth. Commun.*, 1990, **20**, 2409.
579. P. G. M. Wuts and Y. W. Jung, *J. Org. Chem.*, 1991, **56**, 365.
580. Y. Yamamoto, W. Ito and K. Maruyama, *J. Chem. Soc., Chem. Commun.*, 1985, 1131.
581. Y. Yamamoto and S. Nishii, *J. Org. Chem.*, 1988, **53**, 3597.
582. R. W. Hoffmann and A. Endesfelder, *Liebigs Ann. Chem.*, 1987, 215.
583. P. G. M. Wuts and Y. W. Jung, *Tetrahedron Lett.*, 1986, **27**, 2079.
584. Y. Yamamoto, in 'Comprehensive Organic Synthesis', eds. B. M. Trost and I. Fleming, Pergamon, Oxford, 1991, vol. 2, p. 81.
585. E. Favre and M. Gaudemar, *Bull. Soc. Chim. Fr.*, 1968, 3724.
586. K. K. Wang, S. S. Nikam and C. D. Ho, *J. Org. Chem.*, 1983, **48**, 5376.
587. H. C. Brown, U. R. Khise and U. S. Racherla, *Tetrahedron Lett.*, 1993, **34**, 15 and references therein.
588. G. Zweifel, S. J. Backlund and T. Leung, *J. Am. Chem. Soc.*, 1978, **100**, 5561.
589. K. K. Wang and C. Liu, *J. Org. Chem.*, 1985, **50**, 2578.
590. E. Favre and M. Gaudemar, *J. Organomet. Chem.*, 1974, **76**, 297; 305.
591. Y. N. Bubnov, T. V. Potapova and M. E. Gurskii, *J. Organomet. Chem.*, 1991, **412**, 311 and references therein.
592. M. Muraki, K. Inomata and T. Mukaiyama, *Bull. Chem. Soc. Jpn.*, 1975, **48**, 3200.
593. W. Fenzl and R. Köster, *Liebigs Ann. Chem.*, 1975, 1322.
594. T. Mukaiyama, *Org. React.*, 1982, **28**, 203.
595. D. A. Evans, J. V. Nelson and T. R. Taber, in 'Topics in Stereochemistry', Wiley, New York, 1982, vol. 13, p. 1.
596. C. H. Heathcock, in 'Asymmetric Synthesis', ed. J. D. Morrison, Academic Press, New York, 1984, vol. 3, p. iii; 154.
597. S. Masamune, W. Choy, J. S. Petersen and L. R. Sita, *Angew. Chem., Int. Ed. Engl.*, 1985, **24**, 1.
598. M. Braun, *Angew. Chem., Int. Ed. Engl.*, 1987, **26**, 24.
599. R. W. Hoffmann, *Angew. Chem., Int. Ed. Engl.*, 1987, **26**, 489.
600. B. M. Kim, S. F. Williams and S. Masamune, in 'Comprehensive Organic Synthesis', eds. B. M. Trost and I. Fleming, Pergamon, Oxford, 1991, vol. 2, p. 239.
601. A. Pelter, in 'Methoden der Organische Chemie', eds. H. Kropf and E. Schaumann, Thieme, Stuttgart, 1993, vol. E15 (1), p. 373.
602. R. W. Hoffmann, K. Ditrich and S. Fröch, *Liebigs Ann. Chem.*, 1987, 977.
603. R. W. Hoffmann, K. Ditrich, S. Fröch and D. Cremer, *Tetrahedron*, 1985, **41**, 5517.
604. T. Basile, S. Biondi, G. P. Boldrini, E. Tagliavini, C. Trombini and A. Umani-Ronchi, *J. Chem. Soc., Perkin Trans. 1*, 1989, 1025.
605. J. Hooz, J. Oudenes, J. L. Roberts and A. Benderly, *J. Org. Chem.*, 1987, **52**, 1347.
606. J. J. Tufariello, L. T. C. Lee and P. W. Wojtkowski, *J. Am. Chem. Soc.*, 1967, **89**, 6804.
607. D. J. Pasto and P. W. Wojtkowski, *Tetrahedron Lett.*, 1970, 215.
608. M. Z. Deng, N. S. Li and Y. Z. Huang, *J. Chem. Soc., Chem. Commun.*, 1993, 65.
609. H. C. Brown, M. M. Rogić and M. W. Rathke, *J. Am. Chem. Soc.*, 1968, **90**, 6218.
610. G. P. Boldrini, F. Mancini, E. Tagliavini, C. Trombini and A. Umani-Ronchi, *J. Chem. Soc., Chem. Commun.*, 1990, 1680.
611. G. P. Boldrini, M. Bortolotti, E. Tagliavini, C. Trombini and A. Umani-Ronchi, *Tetrahedron Lett.*, 1991, **32**, 1229.
612. Y. Satoh, H. Serizawa, S. Hara and A. Suzuki, *J. Am. Chem. Soc.*, 1985, **107**, 5225.
613. G. A. Molander and H. C. Brown, *J. Org. Chem.*, 1977, **42**, 3106.
614. K. Inomata and T. Mukaiyama, *Bull. Chem. Soc. Jpn.*, 1971, **44**, 3215.
615. H. Shimizu, S. Hara and A. Suzuki, *Synth. Commun.*, 1990, **20**, 549.
616. A. G. Davies, *Pure Appl. Chem.*, 1974, **39**, 497.
617. A. G. Davies and B. P. Roberts, *Acc. Chem. Res.*, 1972, **5**, 387.
618. T. Mukaiyama, M. Murakami, T. Oriyama and M. Yamaguchi, *Chem. Lett.*, 1981, 1193.
619. K. Maruoka, N. Hirayama and H. Yamamoto, *Polyhedron*, 1990, **9**, 223.
620. S. Masamune, S. Mori, D. E. Van Horn and D. W. Brooks, *Tetrahedron Lett.*, 1979, 1665.
621. D. A. Evans, E. Vogel and J. V. Nelson, *J. Am. Chem. Soc.*, 1979, **101**, 6120.
622. D. A. Evans, J. V. Nelson, E. Vogel and T. R. Taber, *J. Am. Chem. Soc.*, 1981, **103**, 3099.
623. H. C. Brown, R. K. Dhar, R. K. Bakshi, P. K. Pandiarajan and B. Singaram, *J. Am. Chem. Soc.*, 1989, **111**, 3441.
624. H. C. Brown, R. K. Dhar, K. Ganesan and B. Singaram, *J. Org. Chem.*, 1992, **57**, 499.
625. H. C. Brown, R. K. Dhar, K. Ganesan and B. Singaram, *J. Org. Chem.*, 1992, **57**, 2716.

626. I. Paterson, M. A. Lister and C. K. McClure, *Tetrahedron Lett.*, 1986, **27**, 4787.
627. I. Paterson and J. M. Goodman, *Tetrahedron Lett.*, 1989, **30**, 997.
628. I. Paterson, J. M. Goodman, M. A. Lister, R. C. Schumann, C. K. McClure and R. D. Norcross, *Tetrahedron*, 1990, **46**, 4663.
629. C. Gennari *et al.*, *J. Org. Chem.*, 1992, **57**, 5173.
630. J. M. Goodman and I. Paterson, *Tetrahedron Lett.*, 1992, **33**, 7223.
631. E. J. Corey and S. S. Kim, *J. Am. Chem. Soc.*, 1990, **112**, 4976.
632. H. C. Brown, K. Ganesan and R. K. Dhar, *J. Org. Chem.*, 1993, **58**, 147.
633. C. Gennari, L. Colombo and G. Poli, *Tetrahedron Lett.*, 1984, **25**, 2279.
634. H. Hamana, K. Sasakura and T. Sugasawa, *Chem. Lett.*, 1984, 1729.
635. H. F. Chow and D. Seebach, *Helv. Chim. Acta*, 1986, **69**, 604.
636. S. Masamune, T. Sato, B. M. Kim and T. A. Wollmann, *J. Am. Chem. Soc.*, 1986, **108**, 8279.
637. M. T. Reetz, F. Kunisch and P. Heitmann, *Tetrahedron Lett.*, 1986, **27**, 4721.
638. E. J. Corey and D. H. Lee, *Tetrahedron Lett.*, 1993, **34**, 1737 and references therein.
639. D. E. Van Horn and S. Masamune, *Tetrahedron Lett.*, 1979, 2229.
640. C. Gennari, S. Cardani, L. Colombo and C. Scolastico, *Tetrahedron Lett.*, 1984, **25**, 2283.
641. R. W. Hoffmann and K. Ditrich, *Tetrahedron Lett.*, 1984, **25**, 1781.
642. H. E. Zimmerman and M. D. Traxler, *J. Am. Chem. Soc.*, 1957, **79**, 1920.
643. C. Gennari, L. Colombo, C. Scolastico and R. Todeschini, *Tetrahedron*, 1984, **40**, 4051.
644. Y. Li, M. N. Paddon-Row and K. N. Houk, *J. Am. Chem. Soc.*, 1988, **110**, 3684.
645. Y. Li, M. N. Paddon-Row and K. N. Houk, *J. Org. Chem.*, 1990, **55**, 481.
646. A. Bernardi, A. M. Capelli, C. Gennari, J. M. Goodman and I. Paterson, *J. Org. Chem.*, 1990, **55**, 3576.
647. F. Bernardi, M. A. Robb, G. Suzzi-Valli, E. Tagliavini, C. Trombini and A. Umani-Ronchi, *J. Org. Chem.*, 1991, **56**, 6472.
648. H. C. Brown and M. M. Midland, *J. Am. Chem. Soc.*, 1971, **93**, 3291.
649. G. W. Kabalka, *Intra-Sci. Chem. Rep.*, 1973, **7**, 57.
650. K. Nozaki, K. Oshima and K. Utimoto, *Bull. Chem. Soc. Jpn.*, 1991, **64**, 403.
651. Y. Ichinose, S. I. Matsunaga, K. Fugami, K. Oshima and K. Utimoto, *Tetrahedron Lett.*, 1989, **30**, 3155.
652. K. Nozaki, K. Oshima and K. Utimoto, *Tetrahedron*, 1989, **45**, 923.
653. K. Miura, K. Oshima and K. Utimoto, *Bull. Chem. Soc. Jpn.*, 1993, **66**, 2356.
654. D. L. J. Clive and M. H. D. Postema, *J. Chem. Soc., Chem. Commun.*, 1993, 429; 1240.
655. D. H. R. Barton, D. O. Jang and J. C. Jaszberenyi, *Tetrahedron Lett.*, 1990, **31**, 4681.
656. K. Nozaki, K. Oshima and K. Utimoto, *Tetrahedron Lett.*, 1988, **29**, 6125; 6127.
657. K. Nozaki, K. Oshima and K. Utimoto, *Bull. Chem. Soc. Jpn.*, 1991, **64**, 2585.
658. M. Taniguchi, K. Nozaki, K. Miura, K. Oshima and K. Utimoto, *Bull. Chem. Soc. Jpn.*, 1992, **65**, 349.
659. K. Miura, J. Sugimoto, K. Oshima and K. Utimoto, *Bull. Chem. Soc. Jpn.*, 1992, **65**, 1513.
660. K. Matsumoto, K. Miura, K. Oshima and K. Utimoto, *Tetrahedron Lett.*, 1992, **33**, 7031.
661. E. Kawashima *et al.*, *Tetrahedron Lett.*, 1993, **34**, 1317.
662. H. S. Dang and B. P. Roberts, *J. Chem. Soc., Perkin Trans. 1*, 1993, 891 and references therein.
663. B. M. Trost and I. Fleming (eds.), 'Comprehensive Organic Synthesis', Pergamon, Oxford, 1991, vol. 8.
664. J. Seyden-Penne, 'Réduction par les alumino- et borohydrures en synthèse organique', Lavoisier, Paris, 1988.
665. G. B. Fisher, J. Harrison, J. C. Fuller, C. T. Goralski and B. Singaram, *Tetrahedron Lett.*, 1992, **33**, 4533.
666. J. C. Fuller, E. L. Stangeland, C. T. Goralski and B. Singaram, *Tetrahedron Lett.*, 1993, **34**, 257.
667. G. B. Fisher, J. C. Fuller, J. Harrison, C. T. Goralski and B. Singaram, *Tetrahedron Lett.*, 1993, **34**, 1091.
668. R. O. Hutchins, K. Learn, F. El-Telbany and Y. P. Stercho, *J. Org. Chem.*, 1984, **49**, 2438.
669. J. V. Bhaskar Kanth and M. Periasamy, *J. Org. Chem.*, 1991, **56**, 5964.
670. A. S. Bhanu Prasad, J. V. Bhaskar Kanth and M. Periasamy, *Tetrahedron*, 1992, **48**, 4623.
671. Y. Suseela and M. Periasamy, *Tetrahedron*, 1992, **48**, 371.
672. C. Khishan Reddy and M. Periasamy, *Tetrahedron*, 1992, **48**, 8329.
673. C. Narayana, S. Padmanabhan and G. W. Kabalka, *Synlett*, 1991, 125.
674. C. Narayana, S. Padmanabhan and G. W. Kabalka, *Tetrahedron Lett.*, 1990, **31**, 6977.
675. C. Narayana, N. K. Reddy and G. W. Kabalka, *Tetrahedron Lett.*, 1991, **32**, 6855.
676. H. C. Brown, B. Nazer, J. S. Cha and J. A. Sikorski, *J. Org. Chem.*, 1986, **51**, 5264.
677. J. S. Cha, S. J. Min, J. M. Kim, O. O. Kwon and M. K. Jeoung, *Org. Prep. Proced. Int.*, 1993, **25**, 466.
678. K. Smith, A. Pelter and A. Norbury, *Tetrahedron Lett.*, 1991, **32**, 6243.
679. Y. Guindon, P. C. Anderson, C. Yoakim, Y. Girard, S. Berthiaume and H. E. Morton, *Pure Appl. Chem.*, 1988, **60**, 1705.
680. S. Abel, T. Linker and B. Giese, *Synlett*, 1991, 171.
681. Y. Guindon, B. Simoneau, C. Yoakim, V. Gorys, R. Lemieux and W. Ogilvie, *Tetrahedron Lett.*, 1991, **32**, 5453.
682. K. Narasaka and F. C. Pai, *Tetrahedron*, 1984, **40**, 2233.
683. F. G. Kathawala *et al.*, *Helv. Chim. Acta*, 1986, **69**, 803.
684. K. M. Chen, K. G. Gunderson, G. E. Hardtmann, K. Prasad, O. Repić and M. J. Shapiro, *Chem. Lett.*, 1987, 1923.
685. K. M. Chen, G. E. Hardtmann, K. Prasad, O. Repić and M. J. Shapiro, *Tetrahedron Lett.*, 1987, **28**, 155.
686. G. A. Molander and K. L. Bobbitt *J. Am. Chem. Soc.*, 1993, **115**, 7517.
687. T. Harada, Y. Matsuda, S. Imanaka and A. Oku, *J. Chem. Soc., Chem. Commun.*, 1990, 1641.
688. T. Harada, S. Imanaka, Y. Ohyama, Y. Matsuda and A. Oku, *Tetrahedron Lett.*, 1992, **33**, 5807.
689. A. Fadel, J. L. Canet and J. Salaun, *Tetrahedron Lett.*, 1989, **30**, 6687.
690. S. Saito, T. Ishikawa, A. Kuroda, K. Koga and T. Morikawe, *Tetrahedron*, 1992, **48**, 4067.
691. H. Yamamoto and K. Maruoka, *J. Am. Chem. Soc.*, 1981, **103**, 6133.
692. W. Moneta, P. Baret and J. L. Pierre, *Bull. Soc. Chim. Fr.*, 1988, 995.
693. H. Bernard, J. J. Yaouang, J. C. Clément, H. des Abbayes and H. Handel, *Tetrahedron Lett.*, 1991, **32**, 639.
694. K. Narasaka, S. Shimada, K. Osoda and N. Iwasawa, *Synthesis*, 1991, 1171.
695. S. Shimada, K. Osada and K. Narasaka, *Bull. Chem. Soc. Jpn.*, 1993, **66**, 1254.
696. M. T. Reetz, C. M. Niemeyer, M. Hermes and R. Goddard, *Angew. Chem., Int. Ed. Engl.*, 1992, **31**, 1017.
697. M. T. Reetz, C. M. Niemeyer and K. Harms, *Angew. Chem., Int. Ed. Engl.*, 1991, **30**, 1472; 1474.

698. Y. Kobuke, Y. Sumida, M. Hayashi and H. Ogoshi, *Angew. Chem., Int. Ed. Engl.*, 1991, **30**, 1496.
699. T. Imamoto, *Pure Appl. Chem.*, 1993, **65**, 655 and references therein.
700. P. Pellon, *Tetrahedron Lett.*, 1992, **33**, 4451.
701. Y. Gourdel, A. Ghanimi, P. Pellon and M. Le Corre, *Tetrahedron Lett.*, 1993, **34**, 1011.
702. H. Brisset, Y. Gourdel, P. Pellon and M. Le Corre, *Tetrahedron Lett.*, 1993, **34**, 4523.
703. S. Juge, M. Stephan, S. Achi and J.-P. Genêt, *Phosphorus Sulfur Silicon*, 1990, **49/50**, 267.
704. S. Juge, M. Stephan, J. A. Laffitte and J.-P. Genêt, *Tetrahedron Lett.*, 1990, **31**, 6357.
705. S. Juge, M. Stephan, R. Merdes, J.-P. Genêt and S. Halut-Desportes, *J. Chem. Soc., Chem. Commun.*, 1993, 531.
706. H. C. Brown and B. Singaram, *Acc. Chem. Res.*, 1988, **21**, 287.
707. A. U. Rahman, *Stud. Nat. Prod. Chem.*, 1991, **8**, 463.
708. H. C. Brown and P. V. Ramachandran, *Pure Appl. Chem.*, 1991, **63**, 307.
709. H. C. Brown *et al.*, *J. Org. Chem.*, 1988, **53**, 5513.
710. H. C. Brown and N. N. Joshi, *J. Org. Chem.*, 1988, **53**, 4059.
711. P. E. Peterson and M. Stepanian, *J. Org. Chem.*, 1988, **53**, 1903.
712. H. C. Brown, J. V. N. Vara Prasad and M. Zaidlewicz, *J. Org. Chem.*, 1988, **53**, 2911.
713. P. K. Jadhav and H. C. Brown, *J. Org. Chem.*, 1981, **46**, 2988.
714. S. Masamune, B. M. Kim, J. S. Petersen, T. Sato and S. J. Veenstra, *J. Am. Chem. Soc.*, 1985, **107**, 4549.
715. H. C. Brown, K. S. Bhat and P. K. Jadhav, *J. Chem. Soc., Perkin Trans. 1*, 1991, 2633.
716. A. Rüttimann and H. Mayer, *Helv. Chim. Acta*, 1980, **63**, 1456.
717. A. Rüttimann, G. Englert, H. Mayer, G. P. Moss and B. C. L. Weedon, *Helv. Chim. Acta*, 1983, **66**, 1939.
718. S. Masamune, L. D. L. Lu, W. P. Jackson, T. Kaiho and T. Toyoda, *J. Am. Chem. Soc.*, 1982, **104**, 5523.
719. T. Hayashi, Y. Matsumoto and Y. Ito, *Tetrahedron: Asymmetry*, 1991, **2**, 601.
720. J. Zhang, B. Lou, G. Guo and L. Dai, *J. Org. Chem.*, 1991, **56**, 1670.
721. K. Burgess, W. A. van der Donk and M. J. Ohlmeyer, *Tetrahedron: Asymmetry*, 1991, **2**, 613.
722. Y. Matsumoto and T. Hayashi, *Tetrahedron Lett.*, 1991, **32**, 3387.
723. Y. Matsumoto, M. Naito and T. Hayashi, *Organometallics*, 1992, **11**, 2732.
724. Y. Matsumoto, M. Naito, Y. Uozumi and T. Hayashi, *J. Chem. Soc., Chem. Commun.*, 1993, 1468.
725. D. S. Matteson, *Pure Appl. Chem.*, 1991, **63**, 339.
726. D. S. Matteson and E. Erdik, *Organometallics*, 1983, **2**, 1083.
727. D. S. Matteson, K. M. Sadhu and M. L. Peterson, *J. Am. Chem. Soc.*, 1986, **108**, 810.
728. D. S. Matteson and E. C. Beedle, *Tetrahedron Lett.*, 1987, **28**, 4499.
729. D. S. Matteson, A. A. Kandil and R. Soundararajan, *J. Am. Chem. Soc.*, 1990, **112**, 3964.
730. D. S. Matteson and M. L. Peterson, *J. Org. Chem.*, 1987, **52**, 5116.
731. D. S. Matteson and G. D. Hurst, *Heteroatom Chem.*, 1990, **1**, 65.
732. D. J. S. Tsai, P. K. Jesthi and D. S. Matteson, *Organometallics*, 1983, **3**, 1543.
733. K. M. Sadhu, D. S. Matteson, G. D. Hurst and J. M. Kurosky, *Organometallics*, 1984, **3**, 804.
734. D. S. Matteson and J. D. Campbell, *Heteroatom Chem.*, 1990, **1**, 109.
735. D. S. Matteson and A. A. Kandil, *Tetrahedron Lett.*, 1986, **27**, 3831.
736. D. S. Matteson, P. B. Tripathy, A. Sarkar and K. M. Sadhu, *J. Am. Chem. Soc.*, 1989, **111**, 4399.
737. K. Ditrich, T. Bube, R. Stürmer and R. W. Hoffmann, *Angew. Chem., Int. Ed. Engl.*, 1986, **25**, 1028.
738. T. Herold, U. Schrott and R. W. Hoffmann, *Chem. Ber.*, 1981, **114**, 359.
739. R. W. Hoffmann and T. Herold, *Chem. Ber.*, 1981, **114**, 375.
740. M. T. Reetz and T. Zierke, *Chem. Ind.*, 1988, 663.
741. W. R. Roush, L. K. Hoong, M. A. J. Palmer and J. C. Park, *J. Org. Chem.*, 1990, **55**, 4109.
742. W. R. Roush and L. Banfi, *J. Am. Chem. Soc.*, 1988, **110**, 3979.
743. W. R. Roush, A. M. Ratz and J. A. Jablonowski, *J. Org. Chem.*, 1992, **57**, 2047.
744. R. P. Short and S. Masamune, *J. Am. Chem. Soc.*, 1989, **111**, 1892.
745. P. K. Jadhav, K. S. Bhat, P. T. Perumal and H. C. Brown, *J. Org. Chem.*, 1986, **51**, 432.
746. H. C. Brown, R. S. Randad, K. S. Bhat, M. Zaidlewicz and U. S. Racherla, *J. Am. Chem. Soc.*, 1990, **112**, 2389.
747. U. S. Racherla and H. C. Brown, *J. Org. Chem.*, 1991, **56**, 401.
748. W. R. Roush, X. Lin and J. A. Straub, *J. Org. Chem.*, 1991, **56**, 1649.
749. T. Nguyen, D. Sherman, D. Ball, M. Solow and B. Singaram, *Tetrahedron: Asymmetry*, 1993, **4**, 189.
750. Z. Wang and D. Deschênes, *J. Am. Chem. Soc.*, 1992, **114**, 1090.
751. U. S. Racherla, Y. Liao and H. C. Brown, *J. Org. Chem.*, 1992, **57**, 6614.
752. A. G. M. Barrett, J. J. Edmunds, K. Horita and C. J. Parkinson, *J. Chem. Soc., Chem. Commun.*, 1992, 1236.
753. H. C. Brown, P. K. Jadhav and K. S. Bhat, *J. Am. Chem. Soc.*, 1988, **110**, 1535.
754. W. R. Roush and J. C. Park, *Tetrahedron Lett.*, 1991, **32**, 6285.
755. W. R. Roush and P. T. Grover, *Tetrahedron*, 1992, **48**, 1981.
756. W. R. Roush and T. G. Marron, *Tetrahedron Lett.*, 1993, **34**, 5421.
757. A. G. M. Barrett and M. A. Seefeld, *Tetrahedron*, 1993, **49**, 7857.
758. T. A. J. Van der Heide, J. L. Van der Baan, E. A. Bijpost, F. J. J. de Kanter, F. Bickelhaupt and G. Klumpp, *Tetrahedron Lett.*, 1993, **34**, 4655.
759. H. C. Brown and R. S. Randad, *Tetrahedron*, 1990, **46**, 4463.
760. N. Ikeda, I. Arai and H. Yamamoto, *J. Am. Chem. Soc.*, 1986, **108**, 483.
761. Y. N. Bubnov and M. Y. Etinger, *Izv. Akad. Nauk SSSR, Ser. Khim.*, 1988, **7**, 1687. (*Engl. Transl.*, 1989, 1502).
762. R. W. Hoffmann, *Pure Appl. Chem.*, 1988, **60**, 123.
763. R. W. Hoffmann and B. Landmann, *Angew. Chem., Int. Ed. Engl.*, 1984, **23**, 437.
764. R. W. Hoffmann and B. Landmann, *Chem. Ber.*, 1986, **119**, 2013.
765. R. W. Hoffmann and S. Dresely, *Angew. Chem., Int. Ed. Engl.*, 1986, **25**, 189.
766. R. W. Hoffmann, K. Ditrich, G. Köster and R. Stürmer, *Chem. Ber.*, 1989, **122**, 1783.
767. A. Vulpetti, M. Gardner, C. Gennari, A. Bernardi, J. M. Goodman and I. Paterson, *J. Org. Chem.*, 1993, **58**, 1711.
768. W. R. Roush, A. E. Walts and L. K. Hoong, *J. Am. Chem. Soc.*, 1985, **107**, 8186.
769. W. R. Roush, J. A. Straub and M. S. Van Nieuwenhze, *J. Org. Chem.*, 1991, **56**, 1636.

770. R. Stürmer, K. Ritter and R. W. Hoffmann, *Angew. Chem., Int. Ed. Engl.*, 1993, **32**, 101.
771. R. W. Hoffmann and A. Schlapbach, *Tetrahedron*, 1992, **48**, 1959.
772. W. R. Roush and J. C. Park, *Tetrahedron Lett.*, 1990, **31**, 4707.
773. K. Burgess and I. Henderson, *Tetrahedron Lett.*, 1990, **31**, 6949.
774. K. Burgess, D. A. Chaplin, I. Henderson, Y. T. Pan and A. D. Elbein, *J. Org. Chem.*, 1992, **57**, 1103.
775. A. G. M. Barrett and J. W. Malecha, *J. Org. Chem.*, 1991, **56**, 5243.
776. I. Paterson and M. M. Mansuri, *Tetrahedron*, 1985, **41**, 3569.
777. S. Masamune and P. A. McCarthy, in 'Macrolide Antibiotics Chemistry, Biology and Practice', ed. S. Omura, Academic Press, Orlando, 1984, chap. 4.
778. M. T. Reetz, *Pure Appl. Chem.*, 1988, **60**, 1607.
779. C. H. Heathcock, *Aldrichimica Acta*, 1990, **23**, 99.
780. I. Paterson, *Pure Appl. Chem.*, 1992, **64**, 1821.
781. S. Masamune, W. Choy, F. A. J. Kerdesky and B. Imperiali, *J. Am. Chem. Soc.*, 1981, **103**, 1566.
782. D. A. Evans, J. Bartroli and T. L. Shih, *J. Am. Chem. Soc.*, 1981, **103**, 2127.
783. D. A. Evans and J. Bartroli, *Tetrahedron Lett.*, 1982, **23**, 807.
784. A. I. Meyers and Y. Yamamoto, *Tetrahedron*, 1984, **40**, 2309.
785. W. Oppolzer, J. Blagg, I. Rodriguez and E. Walther, *J. Am. Chem. Soc.*, 1990, **112**, 2767.
786. I. Paterson and M. A. Lister, *Tetrahedron Lett.*, 1988, **29**, 585.
787. I. Paterson and C. K. McClure, *Tetrahedron Lett.*, 1987, **28**, 1229.
788. I. Paterson, C. K. McClure and R. C. Schumann, *Tetrahedron Lett.*, 1989, **30**, 1293.
789. E. J. Corey, R. Imwinkelried, S. Pikul and Y. B. Xiang, *J. Am. Chem. Soc.*, 1989, **111**, 5493.
790. E. J. Corey and H. C. Huang, *Tetrahedron Lett.*, 1989, **30**, 5235.
791. I. Paterson, J. M. Goodman and M. Isaka, *Tetrahedron Lett.*, 1989, **30**, 7121.
792. I. Paterson, A. N. Hulme and D. J. Wallace, *Tetrahedron Lett.*, 1991, **32**, 7601.
793. D. A. Evans, H. P. Ng, J. S. Clark and D. L. Rieger, *Tetrahedron*, 1992, **48**, 2127.
794. H. Danda, M. M. Hansen and C. H. Heathcock, *J. Org. Chem.*, 1990, **55**, 173.
795. Y. Li, M. N. Paddon-Row and K. N. Houk, *J. Am. Chem. Soc.*, 1988, **110**, 3684; 7260.
796. J. M. Goodman, S. D. Kahn and I. Paterson, *J. Org. Chem.*, 1990, **55**, 3295.
797. S. E. Denmark and B. R. Henke, *J. Am. Chem. Soc.*, 1991, **113**, 2177.
798. A. Bernardi *et al.*, *Tetrahedron*, 1991, **47**, 3471.
799. A. Bernardi *et al.*, *Tetrahedron*, 1992, **48**, 4183.
800. A. Vulpetti, A. Bernardi, C. Gennari, J. M. Goodman and I. Paterson, *Tetrahedron*, 1993, **49**, 685.
801. L. Deloux and M. Srebnik, *Chem. Rev.*, 1993, **93**, 763.
802. H. C. Brown and P. V. Ramachandran, *Acc. Chem. Res.*, 1992, **25**, 16.
803. V. K. Singh, *Synthesis*, 1992, 605.
804. B. B. Lohray and V. Bhushan, *Angew. Chem., Int. Ed. Engl.*, 1992, **31**, 729.
805. S. Wallbaum and J. Martens, *Tetrahedron: Asymmetry*, 1992, **3**, 1475.
806. T. Imai, T. Tamura A. Yamamuro, T. Sato, T. A. Wollmann, R. M. Kennedy and S. Masamune, *J. Am. Chem. Soc.*, 1986, **108**, 7402.
807. S. Masamune, R. M. Kennedy, J. S. Petersen, K. N. Houk and Y. D. Wu, *J. Am. Chem. Soc.*, 1986, **108**, 7404.
808. G. Buono, J. M. Brunel, B. Faure and O. Pardigon, *Phosphorus Sulfur Silicon*, 1993, **75**, 43.
809. J. M. Brunel, O. Pardigon, B. Faure and G. Buono, *J. Chem. Soc., Chem. Commun.*, 1992, 287.
810. A. Börner, J. Ward, K. Kortus and H. B. Kagan, *Tetrahedron: Asymmetry*, 1993, **4**, 2219.
811. L. B. Fields and E. N. Jacobsen, *Tetrahedron: Asymmetry*, 1993, **4**, 2229.
812. B. Burns, E. Merifield, M. F. Mahon, K. C. Molloy and M. Wills, *J. Chem. Soc., Perkin Trans. 1*, 1993, 2243.
813. H. Sakuraba, N. Inomata and Y. Tanaka, *J. Org. Chem.*, 1989, **54**, 3482.
814. J. V. Bhaskar Kanth and M. Periasamy, *J. Chem. Soc., Chem. Commun.*, 1990, 1145.
815. A. Hirao, S. Itsuno, S. Nakahama and N. Yamazaki, *J. Chem. Soc., Chem. Commun.*, 1981, 315.
816. S. Itsuno, Y. Sakurai, K. Ito, A. Hirao and S. Nakahama, *Bull. Chem. Soc. Jpn.*, 1987, **60**, 395.
817. E. J. Corey, R. K. Bakshi and S. Shibata, *J. Am. Chem. Soc.*, 1987, **109**, 5551.
818. V. Nevalainen, *Tetrahedron: Asymmetry*, 1993, **4**, 1597 and references therein.
819. D. K. Jones, D. C. Liotta, I. Shinkai and D. J. Mathre, *J. Org. Chem.*, 1993, **58**, 799.
820. H. Tlahuext and R. Contreras, *Tetrahedron: Asymmetry*, 1992, **3**, 1145.
821. E. J. Corey, J. O. Link and R. K. Bakshi, *Tetrahedron Lett.*, 1992, **33**, 7107.
822. B. T. Cho and Y. S. Chun, *Tetrahedron: Asymmetry*, 1992, **3**, 1539.
823. G. J. Quallich and T. M. Woodall, *Tetrahedron Lett.*, 1993, **34**, 785.
824. R. Berenguer, J. Garcia, M. Gonzalez and J. Vilarrasa, *Tetrahedron: Asymmetry*, 1993, **4**, 13.
825. G. J. Quallich and T. M. Woodall, *Tetrahedron Lett.*, 1993, **34**, 4145.
826. C. Dauelsberg and J. Martens, *Synth. Commun.*, 1993, **23**, 2091.
827. J. M. Brunel, M. Maffei and G. Buono, *Tetrahedron: Asymmetry*, 1993, **4**, 2255.
828. E. J. Corey and J. Lee, *J. Am. Chem. Soc.*, 1993, **115**, 8873.
829. D. J. Mathre *et al.*, *J. Org. Chem.*, 1993, **58**, 2880.
830. S. Itsuno *et al.*, *J. Chem. Soc., Perkin Trans. 1*, 1985, 2039.
831. S. Itsuno *et al.*, *J. Chem. Soc., Perkin Trans. 1*, 1985, 2615.
832. S. Itsuno, Y. Sakurai, K. Shimizu and K. Ito, *J. Chem. Soc., Perkin Trans. 1*, 1990, 1859.
833. B. T. Cho and Y. S. Chun, *Tetrahedron: Asymmetry*, 1992, **3**, 1583.
834. M. Nakagawa, T. Kawate, T. Kakikawa, H. Yamada, T. Matsui and T. Hino, *Tetrahedron*, 1993, **49**, 1739.
835. P. V. Ramachandran, A. V. Teodorović, M. V. Rangaishenvi and H. C. Brown, *J. Org. Chem.*, 1992, **57**, 2379.
836. C. Bolm, M. Ewald, M. Felder and G. Schlingloff, *Chem. Ber.*, 1992, **125**, 1169.
837. P. V. Ramachandran, A. Teodorović and H. C. Brown, *Tetrahedron*, 1993, **49**, 1725.
838. T. Toya, H. Nagase and T. Honda, *Tetrahedron: Asymmetry.*, 1993, **4**, 1537.
839. H. B. Kagan and O. Riant, *Chem. Rev.*, 1992, **92**, 1007.
840. U. Pindur, G. Lutz and C. Otto, *Chem. Rev.*, 1993, **93**, 741.

841. D. Kaufmann and R. Boese, *Angew. Chem., Int. Ed. Engl.*, 1990, **29**, 545.
842. M. Takasu and H. Yamamoto, *Synlett*, 1990, 194.
843. D. Sartor, J. Saffrich and G. Helmchen, *Synlett*, 1990, 197.
844. D. Sartor, J. Saffrich, G. Helmchen, C. J. Richards and H. Lambert, *Tetrahedron: Asymmetry*, 1991, **2**, 639.
845. K. Furuta, Y. Miwa, K. Iwanaga and H. Yamamoto, *J. Am. Chem. Soc.*, 1988, **110**, 6254.
846. K. Furuta, S. Shimizu, Y. Miwa and H. Yamamoto, *J. Org. Chem.*, 1989, **54**, 1481.
847. K. Furuta, A. Kanematsu, H. Yamamoto and S. Takaoka, *Tetrahedron Lett.*, 1989, **30**, 7231.
848. Q. Gao, T. Maruyama, M. Mouri and H. Yamamoto, *J. Org. Chem.*, 1992, **57**, 1951.
849. K. Hattori and H. Yamamoto, *J. Org. Chem.*, 1992, **57**, 3264.
850. K. Hattori and H. Yamamoto, *Synlett*, 1993, 129.
851. K. Maruoka, M. Sakurai, J. Fujiwara and H. Yamamoto, *Tetrahedron Lett.*, 1986, **27**, 4895.
852. T. R. Kelly, A. Whiting and N. S. Chandrakumar, *J. Am. Chem. Soc.*, 1986, **108**, 3510.
853. E. J. Corey and T. P. Loh, *J. Am. Chem. Soc.*, 1991, **113**, 8966.
854. E. J. Corey and S. Sarshar, *J. Am. Chem. Soc.*, 1992, **114**, 7938.
855. E. J. Corey, T. P. Loh, T. D. Roper, M. D. Azimioara and M. C. Noe, *J. Am. Chem. Soc.*, 1992, **114**, 8290.
856. J. P. G. Seerden and H. W. Scheeren, *Tetrahedron Lett.*, 1993, **34**, 2669.
857. J. M. Hawkins and S. Loren, *J. Am. Chem. Soc.*, 1991, **113**, 7794.
858. G. Bir and D. Kaufmann, *Tetrahedron Lett.*, 1987, **28**, 777.
859. S. Kobayashi, M. Murakami, T. Harada and T. Mukaiyama, *Chem. Lett.*, 1991, 1341.
860. K. Furuta, T. Maruyama and H. Yamamoto, *J. Am. Chem. Soc.*, 1991, **113**, 1041.
861. K. Hattori, M. Miyata and H. Yamamoto, *J. Am. Chem. Soc.*, 1993, **115**, 1151.
862. E. J. Corey, C. L. Cywin and T. D. Roper, *Tetrahedron Lett.*, 1992, **33**, 6907.
863. S. Kiyooka, Y. Kaneko and K. Kume, *Tetrahedron Lett.*, 1992, **33**, 4927.
864. S. Kiyooka, Y. Kaneko, M. Komura, H. Matsuo and M. Nakano, *J. Org. Chem.*, 1991, **56**, 2276.
865. K. Furuta, M. Mouri and H. Yamamoto, *Synlett*, 1991, 561.
866. J. A. Marshall and Y. Tang, *Synlett*, 1992, 653.
867. P. L. H. Mok and B. P. Roberts, *J. Chem. Soc., Chem. Commun.*, 1991, 150.
868. P. L. H. Mok and B. P. Roberts, *Tetrahedron Lett.*, 1992, **33**, 7249.
869. N. Oguni and T. Omi, *Tetrahedron Lett.*, 1984, **25**, 2823.
870. N. N. Joshi, M. Srebnik and H. C. Brown, *Tetrahedron Lett.*, 1989, **30**, 5551.
871. G. Bringmann and T. Hartung, *Angew. Chem., Int. Ed. Engl.*, 1992, **31**, 761.

6
Aluminum

JOHN J. EISCH

State University of New York at Binghamton, NY, USA

6.1 STATUS OF ORGANOALUMINUM REAGENTS IN ORGANIC SYNTHESIS	278
6.1.1 Historical Development	278
6.1.1.1 Organometallics in synthesis	278
6.1.1.2 Discovery of carbalumination and hydroalumination	279
6.1.1.3 Pivotal role of Lewis bases	280
6.1.1.4 Unexpected effects of transition metal salts	281
6.1.1.5 Impact on organometallics of other metals	282
6.1.1.6 Focus of the present treatment relative to recent reviews	283
6.1.2 Scope and Limitations of Organoaluminum Reactions	283
6.1.2.1 Addition reactions of aluminum–carbon bonds	283
6.1.2.2 Addition reactions of aluminum–hydrogen bonds	284
6.1.2.3 Metathesis reactions of Al–C and Al–H bonds with nonmetallic compounds	285
6.1.2.4 Reactions of aluminum–carbon bonds with metal salts	286
6.1.2.5 Reactions catalyzed by organoaluminum Lewis acids	287
6.1.2.6 Thermal transformations of aluminum–carbon bonds	288
6.2 REACTIVITY, STEREOCHEMISTRY, AND MECHANISMS OF REACTION FOR ALUMINUM REAGENTS	289
6.2.1 Gradations in Reactivity of Aluminum–Hydrogen and Aluminum–Carbon Bonds	289
6.2.1.1 General considerations	289
6.2.1.2 Polar factors	289
6.2.1.3 Steric factors	290
6.2.1.4 Role of ligands around aluminum	290
6.2.2 Stereochemistry	291
6.2.2.1 Locoselectivity	291
6.2.2.2 Regioselectivity	291
6.2.2.3 Stereoselectivity	292
6.2.3 Mechanisms of Organoaluminum Reactions	292
6.2.3.1 Electrophilic attack by aluminum	292
6.2.3.2 Nucleophilic attack by an aluminate complex	293
6.2.3.3 Homolytic pathways	293
6.2.3.4 Intermediates	293
6.3 AVAILABILITY OF ALUMINUM REAGENTS	294
6.3.1 Commercially Available Compounds	294
6.3.1.1 Aluminum alkyls	294
6.3.1.2 Alkylaluminum hydrides	294
6.3.1.3 Alkylaluminum halides	295
6.3.1.4 Alkylaluminum alkoxides and aluminoxanes	295
6.3.2 Accessible, Specialized Reagents	295
6.3.2.1 Aluminum aryls	295
6.3.2.2 Unsymmetrical vinylaluminums	296
6.3.2.3 Unsymmetrical aluminum acetylides or cyanides	296
6.3.2.4 Aluminum alkoxides, aryloxides, oxides, and related sulfur, selenium, and nitrogen derivatives	296
6.3.3 Specialized Organoaluminum Reagents for Carbon–Carbon Bond Formation	296
6.3.3.1 Stoichiometric reagents	296
6.3.3.2 Catalysts	297
6.4 CARBON–HYDROGEN BONDS MADE BY ORGANOALUMINUM HYDRIDES	297
6.4.1 Alkanes from Alkenes	297

6.4.2 Alkenes from Alkynes	298
6.4.3 Alcohols or Aldehydes from Carbonyl Compounds or Epoxides	298
6.4.4 Methylene Derivatives from Alcohols or Carbonyls	299
6.4.5 Ethers from Acetals, Ketals, or Orthoformates	300
6.4.6 Amines or Imines from Nitriles or Enamines	300
6.4.7 Hydrocarbons from Halides	301
6.5 CARBON–HALOGEN BONDS	302
6.6 CARBON–OXYGEN BONDS MADE BY OXIDATION OF ALUMINUM ORGANYLS	302
6.6.1 Alcohols from Aluminum Alkyls	302
6.6.2 Ethers from Aluminum Vinyls	303
6.7 CARBON–SULFUR AND CARBON–HETEROATOM BONDS	303
6.8 CARBON–NITROGEN BONDS	305
6.9 CARBON–METAL BONDS	306
6.10 CARBON–CARBON BONDS	306
6.10.1 Carbaluminations	306
6.10.2 Carbodealuminations	307
6.11 TRANSITION METAL-PROMOTED REACTIONS OF ALUMINUM REAGENTS	308
6.12 REFERENCES	309

6.1 STATUS OF ORGANOALUMINUM REAGENTS IN ORGANIC SYNTHESIS

6.1.1 Historical Development

6.1.1.1 Organometallics in synthesis

Shortly after Frankland first prepared the simple alkyl derivatives of zinc in 1849, chemists began to explore what possible synthetic applications zinc alkyls might have in organic chemistry. Through the composite researches of Frankland, Butlerow, Zaitsev, Reformatsky, Freund, and many others,[1] the basic chemical responses of the carbon–zinc bond to Brønsted acids, to oxidants, to organic halides, and to carbonyl derivatives were thoroughly studied and accurately described. As a result, even though organozinc reagents for general organic synthesis were eventually replaced by more reactive alkyls of other metals, the pioneering work with zinc alkyls had already identified many aspects of organic synthesis where the new metal alkyl could offer a distinct advantage. Thus, it was recognized that R_2Zn could add to the C=O bond in aldehydes or in ketones, but not to that in CO_2. (In fact, the preparation and reactions of zinc alkyls could be conducted under a protective atmosphere of CO_2.) Hence, after Grignard first prepared his class of RMgX reagents in 1900,[2] he realized the significance of the observation that they readily reacted with CO_2 to yield ultimately carboxylic acids, a gain for organic synthesis over the use of R_2Zn.[3] Likewise, when Ziegler uncovered a more feasible route to organolithium compounds from organic halides in 1930, he wished to find an inert solvent in which to examine possible electrical conductivity. His choice of pyridine arose partly from the chemical inertness of pyridine toward RMgX and Ph_3CNa. When he mixed RLi with pyridine and observed addition of RLi to the C=N bond, he could immediately recognize at least one clear advantage of RLi over RMgX for organic synthesis; many other merits of RLi were to be subsequently recognized by Ziegler,[4] by Gilman,[5,6] and by Wittig.[7]

Finally, the introduction of unsolvated aluminum alkyls by Ziegler and co-workers in 1952[8] is noteworthy, because such alkyls complement the synthetic capabilities of zinc, magnesium, and lithium alkyls in a most valuable manner: they allow the smooth and often stereoselective addition of C–Al and H–Al bonds to unconjugated carbon–carbon unsaturation.

Through the use of these four types of organometallics, moreover, the preparation of the organometallics of almost all the metals of groups 1–15 can be achieved by metal–metal exchange (Equations (1) and (2)), metal displacement (Equation (3)) and metal addition (Equation (4)). Furthermore, the organyl substituent on a given metal can be readily altered by well-known exchange processes (Li–H, Li–Br, Li–M; Equations (5)–(7)) to produce the desired RLi. Then the R group can be finally transferred to the desired metal by lithium–metal exchange (cf., e.g., Equation (8)).

$$2\ PhMgBr + HgCl_2 \longrightarrow Ph_2Hg + 2\ MgBrCl \qquad (1)$$

$$Cp_2TiCl_2 + 2\ MeLi \longrightarrow Cp_2TiMe_2 + 2\ LiCl \qquad (2)$$

$$Ph_2Hg + 2\ Na \longrightarrow 2\ PhNa + Hg \qquad (3)$$

$$\text{naphthalene} + 2\ Li \xrightarrow{\text{donor solvent}} \text{naphthalene}^{2-}\ 2\ Li^+ \qquad (4)$$

$$Ph\text{—}{\equiv}\text{—}H \xrightarrow[-Bu^nH]{Bu^nLi} Ph\text{—}{\equiv}\text{—}Li \qquad (5)$$

$$\text{1,4-}Br_2C_6H_4 \xrightarrow[-Bu^nBr]{Bu^nLi} \text{4-}Br\text{-}C_6H_4\text{-}Li \qquad (6)$$

$$PhOCH_2SnBu^n_3 \xrightarrow[-Bu^n_4Sn]{Bu^nLi} PhOCH_2Li \qquad (7)$$

$$Ph\text{—}{\equiv}\text{—}Li \xrightarrow[-LiCl]{Et_2AlCl} Ph\text{—}{\equiv}\text{—}AlEt_2 \qquad (8)$$

The transformations typified by Equations (1)–(8) currently place a wide array of organometallics, comprising most of the metals and many organyl groups, at the disposal of the synthetic chemist, who has gained a new appreciation for organometallics previously undervalued or dismissed as inferior to those of magnesium, lithium, or aluminum. A case in point is the initial importance, and the gradual decline in the value, of zinc alkyls for synthesis in the nineteenth century. In the period 1930–1970 zinc reagents began to be recognized again for their special strengths in synthesis, such as the formation of quaternary carbons from *t*-alkyl halides,[9] the generation of carbenoids for cyclopropanation,[10,11] and the preparation of powerful, Wittig-like methylenating agents.[12] This rejuvenation in organozinc chemistry should caution chemists never to consign any class of organometallics to permanent desuetude. The fine gradations in the reactivity of various C–M bonds are often the necessary prerequisite for great selectivity in C–C and C–H bond formation.

6.1.1.2 Discovery of carbalumination and hydroalumination

Isolated instances of the carbalumination of the C=C bond in α,β-unsaturated ketones (Equation (9))[13] and the hydroalumination of highly polar aldehydes (Equation (10))[14] had been observed prior to World War II. However, it was the fundamental investigations of Ziegler and co-workers in the immediate postwar years on metal hydride additions to ethylene that uncovered the generality and feasibility of both carbalumination and hydroalumination for both unconjugated alkenes and alkynes,[15–17] as well as for a wide variety of carbonyl and related compounds.[18] Beginning with the abortive attempt to add LiH to ethene, Ziegler extended such efforts to LAH and then to AlH_3 and achieved resounding success in hydroaluminating an ordinary C=C bond for the first time (Scheme 1). Moreover, in the absence of Lewis bases such as ethers or amines, the generated triethylaluminum (**1**) underwent insertion of further ethene units to yield higher aluminum alkyls ((**2**): *a, b, c* = 1–12). With this finding, Ziegler had at the same time discovered the carbalumination of ordinary C=C bonds (Equation (11)). Both carbalumination and hydroalumination were subsequently shown to be applicable to α-alkenes as well, leading to alkene dimerizations via carbalumination (Scheme 2)[16,19] or to hydroalumination via aluminum hydride transfer (Equation (12)).[20]

$$\underset{\text{(PhHC=C(Ph)CHO)}}{\text{PhCH=C(Ph)CHO}} \xrightarrow{Ph_3Al} \underset{\text{Ph}_2AlO\text{-}C(Ph)=CH\text{-}CHPh\text{-}Ph}{\text{Ph}_2AlO\text{-}C(Ph)=CH\text{-}CHPh\text{-}Ph} \qquad (9)$$

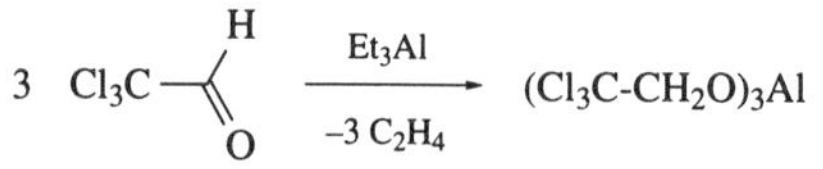

$$3\ Cl_3C-CH(H)=O \xrightarrow[-3\ C_2H_4]{Et_3Al} (Cl_3C-CH_2O)_3Al \quad (10)$$

$$LiAlEt_4 \xleftarrow[n=4]{LAH} n\ H_2C=CH_2 \xrightarrow[n=3]{AlH_3} Et_3Al\ (1)$$

Scheme 1

$$Et_3Al + n\ H_2C=CH_2 \xrightarrow{exothermic} \begin{array}{c} Et-(CH_2CH_2)_a \\ Et-(CH_2CH_2)_b-Al \\ Et-(CH_2CH_2)_c \end{array} \quad (11)$$

(1) \hspace{4cm} (2)

$$Pr^n{}_3Al \xrightarrow{H_2C=CHMe} Pr^n{}_2Al-\overset{Pr^n}{\diagup} \xrightarrow{-Pr^n{}_2AlH} \overset{}{\diagup\!\!\diagdown}_{Pr^n}$$

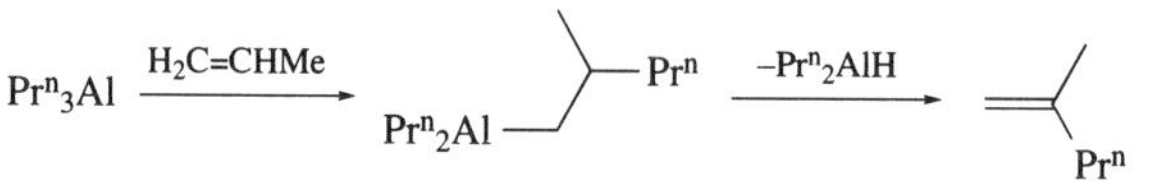

Scheme 2

$$3\ H_2C=CHR \xrightarrow[-C_4H_8^i]{Bu^i{}_3Al} (RCH_2CH_2)_3Al \quad (12)$$

The comparable feasibility of both carbalumination and hydroalumination for alkynes was established by Wilke and co-workers,[21,22] who further showed that such additions can occur in a kinetically controlled *syn* fashion (Equations (13) and (14)). Extensive studies of the hydroalumination of unsymmetrically substituted alkynes by the J. J. Eisch group have shown how markedly both the stereochemistry and the regiochemistry of such additions depend upon the steric and polar nature of the substituents.[23,24]

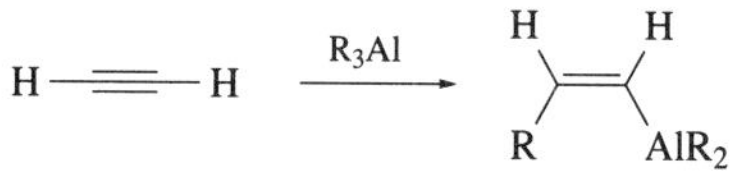

$$H-\!\!\equiv\!\!-H \xrightarrow{R_3Al} \begin{array}{c} H \quad\quad H \\ \diagup\!\!=\!\!\diagdown \\ R \quad\quad AlR_2 \end{array} \quad (13)$$

$$R^1-\!\!\equiv\!\!-R^1 \xrightarrow{R^2{}_2AlH} \begin{array}{c} R^1 \quad\quad R^1 \\ \diagup\!\!=\!\!\diagdown \\ H \quad\quad AlR^2{}_2 \end{array} \quad (14)$$

6.1.1.3 Pivotal role of Lewis bases

The discovery of the carbalumination and the hydroalumination of ordinary C=C and C≡C bonds with organoaluminum compounds by Ziegler and co-workers can most likely be attributed to their fortunate choice of hydrocarbon media for their reactions.[25] Detailed mechanistic studies have later shown that Lewis bases can drastically retard or prevent altogether the addition of C–Al and H–Al bonds to carbon–carbon unsaturation.[26–8] The kinetic reason for this is that the rate-determining step of such additions is the electrophilic attack (**6**) → (**8**) of the tricoordinate monomer (**4**) on the unsaturated substrate (**5**). Crucial intermediate (**4**), in hydrocarbon media, must be generated by the dissociation of oligomer (**3**); in donor media (D = R_2O, R_3N) (**3**) will complex with D (**7**), and again free (**3**) will have to be generated by dissociation (Scheme 3).

Not only the rate but also the stereochemistry of hydroalumination is strongly influenced by the presence of Lewis bases. An impressive illustration is the behavior of phenyl(trimethylsilyl)acetylene ((**9**), Scheme 4). Treatment of (**9**) with $Bu^i{}_2AlH$ in the presence of one or more equivalents of Lewis base (D = THF, R_3N, etc.) relatively slowly yields adduct (**11**) (>95%), presumably by way of the kinetically determined formation of (**10**). On the other hand, treatment of (**9**) with $Bu^i{}_2AlH$ in pure hydrocarbon media forms (**12**) (>95%). That (**12**) is formed via isomerization of first-formed (**10**) is made most probable through the observation that treatment of (**11**) with $Bu^i{}_2AlCl$ immediately forms (**12**). From this result it is clear that (**10**), but not (**11**), is able to undergo relatively facile rotation about the C=C bond at 25 °C and thus isomerize to (**12**) (possibly via (**13**), Scheme 4).[29,30]

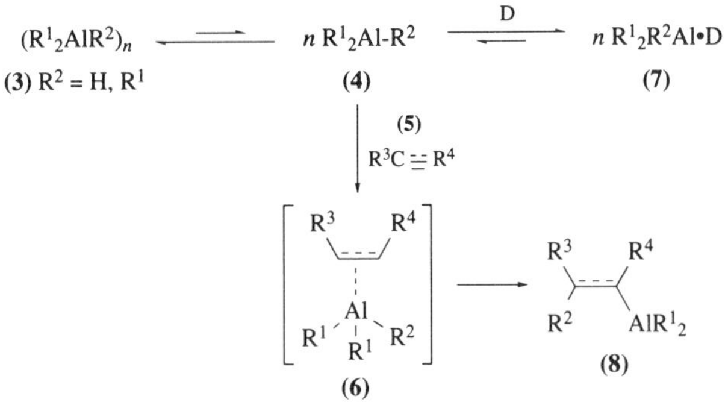

Scheme 3

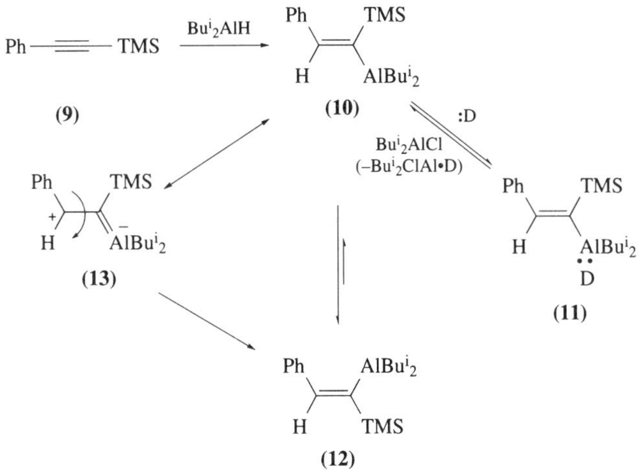

Scheme 4

6.1.1.4 Unexpected effects of transition metal salts

The dramatic increase in interest in organoaluminum compounds since the 1950s is not solely due to the discovery of the carbalumination and hydroalumination reactions. What has riveted the attention of the chemical industry worldwide are two serendipitously discovered effects of transition metal salts on such organoaluminum reactions. (The details of these discoveries and their historical relevance to the rise of "Ziegler chemistry" have already been related elsewhere.[31–3]) One effect has been termed the "nickel effect," while the other can be termed the "early transition metal effect"; both originally concerned the reaction of triethylaluminum (**1**) with ethene. In the nickel effect, the carbalumination of ethene, involving 1–12 insertions of ethene units into the Et–C bonds of (**1**) (Equation (11)), is halted by nickel salts after a single ethene insertion and the prevailing reaction now becomes a catalytic dimerization of ethene into 1-butene (Scheme 5).

Scheme 5

In the early transition metal effect the presence of salts of titanium, zirconium, hafnium, vanadium, or chromium does not decrease the rate and number of ethene insertions in the reaction with (**1**) but, instead, radically increases the number of ethene insertions to 10^3–10^5. These result in alkane chains or linear polyethene units having molecular masses over 10^6 (Equation (15)).

These two transition metal effects are seminal discoveries directly responsible for important unsaturated hydrocarbon oligomerization and polymerization reactions that have revolutionized the petrochemical industry and have evolved a large number of selective C– and C–H bond formation reactions for precise organic synthesis. In addition to these two effects, transition metal salts have now been shown to exert a considerable number of other unusual effects on organoaluminum compounds.

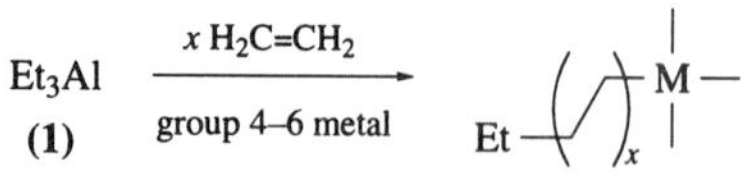

$$\text{Et}_3\text{Al} \xrightarrow[\text{group 4–6 metal}]{x\,\text{H}_2\text{C=CH}_2} \text{Et}-\left(\!\!\!\diagdown\!\!\!\right)_x\!\!-\!\text{M}- \qquad (15)$$
(1)

Active M center now known to be a group 4–6 metal

These influences were analyzed in some detail and their implications for organic synthesis and mechanistic insights were treated in the main review of aluminum in *COMC-I*.[34]

6.1.1.5 Impact on organometallics of other metals

Implicit in the foregoing effects of transition metals on organoaluminum chemistry is the ability of $R^1_n AlE_{3-n}$ (E = R^2, H, X, OR^2, NR^2_2) to interact with another metal salt, either transition or main-group, to generate an organometallic intermediate having a reactivity or mode of reaction different from the starting aluminum reagent. This synergistic effect exerted by a combination of $R_n AlE_{3-n}$ and another metal salt or organometallic explains why the applications of organoaluminum compounds in selective organic synthesis, in both homogeneous and heterogeneous industrial catalysis and in materials science, have burgeoned so amazingly since 1960.

In the light of the importance of such reactions between $R_n AlE_{3-n}$ and another metal compound, it is useful to classify the principal types of interactions that are known to occur. First, the aluminum reagent can transfer either an alkyl or a hydride group to the other metal (Equations (16),[35] (17),[36] and (18)).[37]

$$\text{GaCl}_3 + 3\,\text{Et}_3\text{Al} \xrightarrow[\Delta]{3\,\text{KCl}} \text{Et}_3\text{Ga} + 3\text{K}\,[\text{AlEt}_2\text{Cl}_2] \qquad (16)$$

$$\text{Cp}_2\text{TiCl}_2 + \text{Me}_3\text{Al} \xrightarrow{\text{CH}_2\text{Cl}_2} \text{Cp}_2\text{TiMeCl} + \text{Me}_2\text{AlCl} \qquad (17)$$

$$\text{PhSnCl}_3 + 3\,\text{Bu}^i_2\text{AlH} \longrightarrow \text{PhSnH}_3 + 3\,\text{Bu}^i_2\text{AlCl} \qquad (18)$$

Second, an "ate" complex can be formed with more saline organometallics, which then exhibits a reactivity of the resulting C–M bond intermediate between that of the C–M bonds in the components (Equation (19)).[38]

$$\text{PhLi} + \text{Ph}_3\text{Al} \longrightarrow \text{Li}[\text{AlPh}_4] \qquad (19)$$

Third, the strong Lewis acidity of $R_n AlCl_{3-n}$ can lead to the generation of ion pairs or an onium complex (Equation (20)).[39]

$$\text{Cp}_2\text{Ti}\!\!\begin{array}{c}\text{Me}\\ \diagup\\ \diagdown\\ \text{Cl}\end{array} + \text{MeAlCl}_2 \longrightarrow \text{Cp}_2\overset{+}{\text{Ti}}\!-\!\text{Me} + \text{AlMeCl}_3^- \qquad (20)$$

Fourth, the aluminum component can bring about a reduction of the metal center, via a labile intermediate undergoing homolysis (Scheme 6) (J. J. Eisch and S. I. Pombrik, unpublished work).

$$\text{TiCl}_4 + 2\,\text{Me}_3\text{Al} \xrightarrow{-\text{Me}_2\text{AlCl}} [\text{Me}_2\text{TiCl}_2] \xrightarrow{-\text{Me}\bullet} \text{TiCl}_2\!\cdot\!\text{Me}_2\text{AlCl}$$

Scheme 6

Fifth, the aluminum alkyl can effect a hydrogen–aluminum exchange at certain positively polarized bonds, as in the generation of the Tebbe reagent (Scheme 7).[40]

$$\text{Cp}_2\text{TiCl}_2 \xrightarrow{\text{Me}_3\text{Al}} \text{Cp}_2\overset{+}{\text{Ti}}\!\!\begin{array}{c}\text{Me}\\ \diagup\\ \diagdown\\ \text{AlMe}_2\text{Cl}_2^-\end{array} \xrightarrow[-\text{MeH}]{\text{Me}_3\text{Al}} \text{Cp}_2\text{Ti}\!\!\begin{array}{c}\diagup\,\text{AlMe}_2\\ \diagdown\\ \text{Cl}\end{array}$$

Scheme 7

These reaction types serve to show how such reactions greatly enrich the scope of organoaluminum chemistry.

6.1.1.6 Focus of the present treatment relative to recent reviews

In *COMC-I* the reactions of organoaluminum compounds were discussed in the general chapter on aluminum[34] and in a chapter devoted to compounds of aluminum in organic synthesis.[41] In both of those chapters and in Chapter 10, Volume 1, such reactions have been largely classified by the organic or inorganic substrate interacting with the aluminum reagent. Since the present chapter is intended mainly for the synthetic chemist, the focus here is placed on the availability of the desired aluminum reagents, the methods by which various C–C, C–H, C–O, C–X, and C–M bonds can be made by these reagents and, as an aid in controlling the selectivity of such reagents, the mechanistic principles of organoaluminum reactions.

In order to derive maximum benefit from the present treatment, the reader should also consult *COMC-I*[34,41] and Chapter 10, Volume 1 as well as the following recent reviews on more specialized aspects of aluminum chemistry: (i) the hydroalumination of C=C and C≡C bonds;[42] (ii) the chemistry of unsaturated organoaluminum compounds;[43] and (iii) the synthesis, structure, and reactions of aluminoxanes.[44]

Finally, since this review wishes to emphasize the best known methods for making covalent bonds with aluminum reagents, it is not limited to publications that have appeared since 1980. Many of the most versatile aluminum reagents were developed in the period 1955–1980.

6.1.2 Scope and Limitations of Organoaluminum Reactions

6.1.2.1 Addition reactions of aluminum–carbon bonds

The carbalumination of a wide variety of unsaturated functional groups, such as C=C, C≡C, C=O (and of CO_2 itself), C=S, C=N, C≡N, S=O, epoxide linkages, and O_2 and S_8, can generally be effected below 100 °C to give useful yields of the expected adducts (Equations (21),[45] (22),[46] and (23)[47]). As is evident from known experimental results, not all three Al–C bonds in $R^1R^2R^3Al$ are equally reactive toward such substrates: (i) C=C, C≡C, C=O, C=S, and C≡N bonds tend to insert readily into one Al–C bond, but the initial adduct, $R_2Al–E=C–R$, undergoes further carbalumination much more slowly; (ii) CO_2, SO_2, and SO_3 can react readily with two or even all three Al–C bonds; and (iii) such carbalumination can occur with high regioselectivity and/or high stereoselectivity. Note that only 1-phenyloctane is produced in Equation (22) and only the *trans*-ring-opened adduct in Equation (24).[48] A further pertinent example is the *syn*-carbalumination of ethyne itself, already cited in Equation (13).

$$R_3Al \xrightarrow[\text{ii, } H_2O]{\text{i, } O_2} 3\ ROH \qquad (21)$$

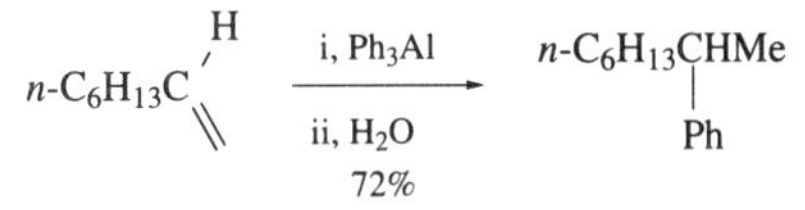

$$n\text{-}C_6H_{13}CH\text{=}CH_2 \xrightarrow[\text{ii, } H_2O]{\text{i, } Ph_3Al} \underset{72\%}{n\text{-}C_6H_{13}\underset{|}{\overset{}{C}}HMe\ Ph} \qquad (22)$$

$$(n\text{-}C_8H_{17})_3Al \xrightarrow[100\%]{SO_2} \left(C_8H_{17}\overset{\overset{O}{\|}}{-}S\overset{}{-}O\right)_3 Al \qquad (23)$$

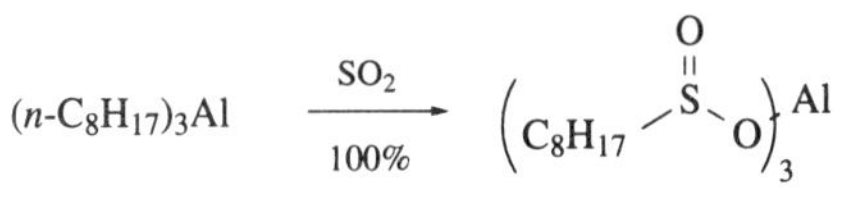

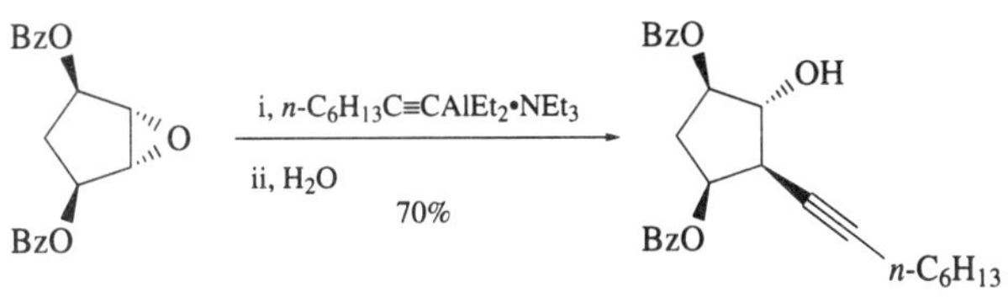

$$(24)$$

Partly offsetting these preparative advantages of carbalumination are the considerable limitations exhibited by R_3Al reagents having C–H bonds β to the aluminum center, such as Et_3Al and especially Bu^i_3Al. In situations where the desired carbalumination occurs too slowly and thus requires heating, the β-alkyl branching becomes significant in fostering alkene elimination. The resulting dialkylaluminum hydride can then cause hydroalumination to become a serious competing reaction with carbalumination. As shown in Scheme 8, such hydroalumination can often dominate.[49] Due to this facile elimination (especially promoted by nickel salts), Bu^i_3Al can often be used as a convenient hydroaluminating agent for the reduction of alkenes, alkynes, and carbonyl derivatives. Conversely, because of the absence of β-C–H bonds or the relative slowness of such eliminations, aluminum organyls containing methyl, *t*-butyl, neopentyl, trimethylsilylmethyl, allyl, 1-alkenyl, 1-alkynyl, aryl, or benzyl are suitable reagents for carbalumination reactions. Although such reagents are accordingly not subject to β-Al–H bond eliminations, their carbalumination activity is sometimes too low and hence must be enhanced by suitable catalysts. The titanium-promoted carbalumination of silylalkynes is a case in point (Equation (25)).[50]

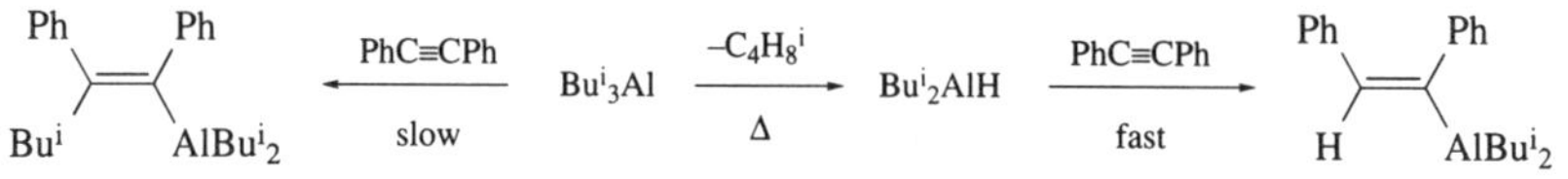

Scheme 8

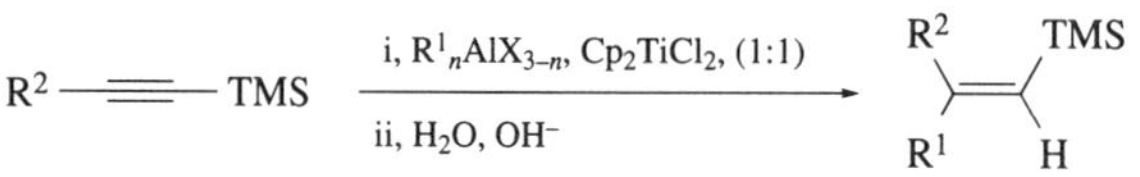

$$R^2 \longequal TMS \xrightarrow[\text{ii, } H_2O, OH^-]{\text{i, } R^1{}_nAlX_{3-n}, Cp_2TiCl_2, (1:1)} \underset{R^1 \quad H}{\overset{R^2 \quad TMS}{\diagup\!\!\!\!=\!\!\!\!\diagdown}} \quad (25)$$

The use of a silylalkyne in Equation (25) illustrates how another side reaction can be avoided in attempted carbaluminations. Despite the smooth carbalumination of ethyne (Equation (13)), monosubstituted alkynes undergo metallation at the alkynic C–H bond at rates greater than that of carbalumination (Equation (26)). Replacing the alkynic proton by the R_3Si group obviates this difficulty for carbalumination.

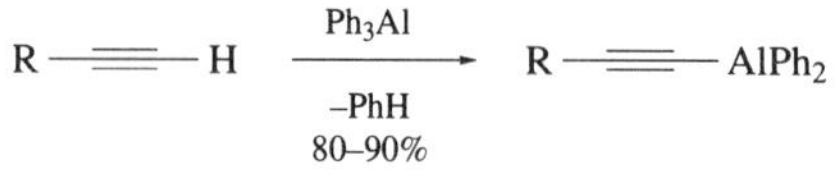

$$R \longequal H \xrightarrow[\substack{-PhH \\ 80–90\%}]{Ph_3Al} R \longequal AlPh_2 \quad (26)$$

6.1.2.2 Addition reactions of aluminum–hydrogen bonds

The hydroalumination reaction is applicable to the same wide variety of unsaturated substrates and epoxides that undergo carbalumination. The source of the Al–H bond can be unstable AlH_3 or the more stable and convenient R_2AlH reagents available commercially (R = Et and Bu^i). At elevated temperatures Bu^i_3Al behaves as a source of AlH_3, through the stepwise loss of isobutene as depicted in Scheme 8. In general, the addition of one or more Al–H bonds in actual (AlH_3) or potential (Bu^i_3Al) aluminum hydride to unsaturated substrates is faster and more complete than the corresponding Al–C bond additions. These kinetic and equilibrium aspects, as well as some facets of regio- and stereoselectivity, are illustrated in Equations (27)[20] and (28)[22] and Schemes 9 and 10.[51]

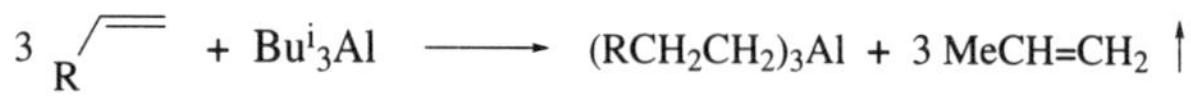

$$3\ \overset{}{\underset{R}{\diagup\!\!=}} + Bu^i_3Al \longrightarrow (RCH_2CH_2)_3Al + 3\ MeCH=CH_2 \uparrow \quad (27)$$

$$R\longequal H + Bu^i_2AlH \longrightarrow \underset{H \quad AlBu^i_2}{\overset{R \quad H}{\diagup\!\!\!\!=\!\!\!\!\diagdown}} \quad (28)$$

(14)

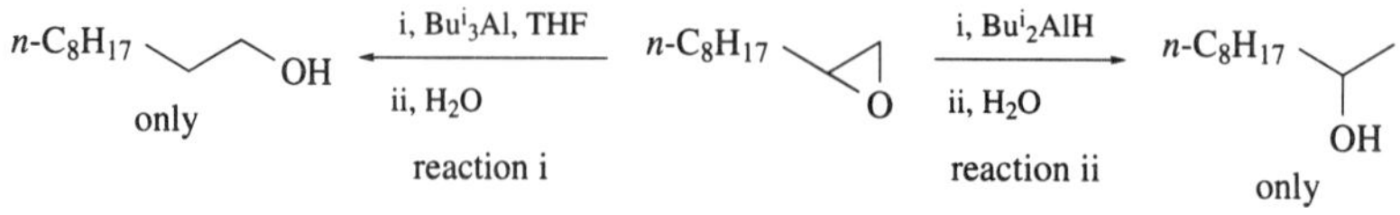

Scheme 9

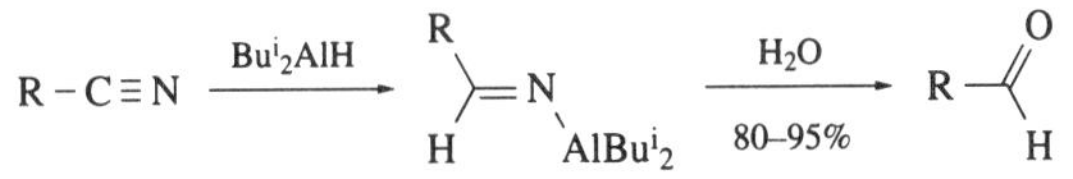

$$R-C\equiv N \xrightarrow{Bu^i_2AlH} \begin{matrix} R \\ \diagdown \\ H \end{matrix}\!\!=\!\!N\!\!\begin{matrix} \\ \diagdown \\ AlBu^i_2 \end{matrix} \xrightarrow[80-95\%]{H_2O} R-\!\!\begin{matrix} O \\ \diagup\!\!\diagdown \\ H \end{matrix}$$

(15)

Scheme 10

The attainment of exclusive hydroalumination with R_2AlH, with no trace of carbalumination, illustrates the great reactivity of the Al–H bond. With terminal alkenes or alkynes the 1-alumino regioisomer constitutes >95% of the product. Furthermore, as shown in Equations (14) and (28), *syn*-hydroalumination is the kinetically controlled process. Finally, metallation at the alkynic C–H bond is not a serious side reaction in hydroalumination, as it is in carbalumination (Equation (26)), especially if R_2AlH is pure and free of any donors, such as amines.[52]

The hydroalumination of alkenes and alkynes has been critically reviewed recently, especially with regard to the regiochemistry and stereochemistry exhibited by substituted unsaturated substrates and the useful applications of such hydroaluminated adducts in organic synthesis.[42] One aspect of the profound effects substituents can exert on the stereochemical course of hydroalumination has already been summarized (Scheme 4) and the reader should consult the previous review[42] for a fuller survey.

The advantages of conducting hydroaluminations of polar unsaturated groups with R_2AlH, rather than with LAH, are: (i) R_2AlH is freely soluble and highly reactive in hydrocarbon media, so that ether solvents can be avoided and low reaction temperatures can be chosen; and (ii) only one Al–H bond of uniform reactivity is present, rather than the four Al–H bonds of LAH, which have varying degrees of activity. The ability to control the reactivity and availability of the Al–H bond in R_2AlH permits one to minimize further hydroalumination of unsaturated adducts, such as (**14**) and (**15**) (Equation (28) and Scheme 10). If desired, these adducts could be purposely hydroaluminated by a second equivalent of R_2AlH and thus converted into $RCH_2CH(AlBu^i_2)_2$ and $RCH_2N(AlBu^i_2)_2$, respectively. Subsequent hydrolysis would then give practical yields of REt and RCH_2NH_2.

6.1.2.3 Metathesis reactions of Al–C and Al–H bonds with nonmetallic compounds

The Al–C and Al–H bonds of organoaluminum compounds enter into a large number of metathesis reactions with organic substrates containing polar covalent bonds (E^1–E^2, Equation (29)), such as those containing H–O, H–N, H–S, and even certain H–C groups, as well as R–O, R–X, N–X, X–CN, X–SO$_2$R, and X–COR functionalities. The ease of reaction seems to increase with either increasing electronegativity of E^2 (E^1–E^2: Cl–R, Cl–COR, Cl–SO$_2$R) or an increasing number of such linkages (E^1–E^2: R–O–R vs. (RO)$_2$CR$_2$; Cl–Me vs. CCl$_4$). CCl$_4$ can react explosively with Me$_3$Al but CH$_2$Cl$_2$ reacts only slowly. Similarly, ordinary ethers are not cleaved by Bu^i_2AlH, but ketals are smoothly cleaved at 70–80 °C[53] (Equation (30)). Due to these profound effects of substituents, the metathesis reactions of R^1_2Al–R^2 are strongly limited in their utility by feasibility and by safety factors. Nevertheless, where successful, such transformations provide ready access to diversely functionalized hydrocarbons (Equations (31)[54] and (32)[55] and Schemes 11[56] and 12). In the last reaction the mode of cleavage seems to ally the R_2Al group with the more electronegative fragment of X–CN.[57,58]

$$R^1_2Al\text{-}R^2 + E^1\text{-}E^2 \longrightarrow R^1_2Al\text{-}E^2 + R^2\text{-}E^1 \qquad (29)$$

R^2 = alkyl, H

$$R^1_2C(OR^2)_2 + Bu^i_2AlH \longrightarrow R^1_2C\!\!\begin{matrix}H \\ \diagdown \\ OR^2\end{matrix} + Bu^i_2AlOR^2 \qquad (30)$$

$$\underset{R^1}{\overset{O}{\|}}\!\!-\!\!Cl + R^2AlCl_2 \longrightarrow \underset{R^1}{\overset{O}{\|}}\!\!-\!\!R^2 + AlCl_3 \qquad (31)$$

$$\left(Ph\!\!\diagdown\!\!\diagup\!\!CH_2\right)_3Al + 3\,I_2 \xrightarrow[62\%]{Et_2O} 3\; Ph\!\!\diagdown\!\!\diagup\!\!I \qquad (32)$$

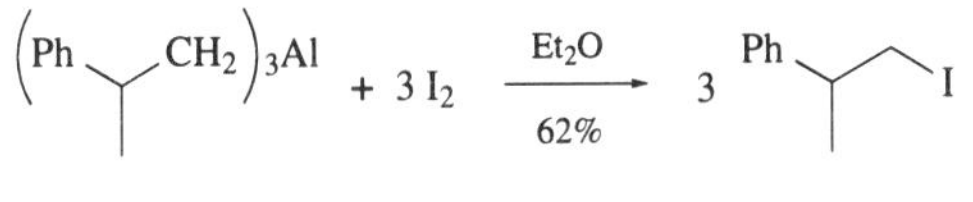

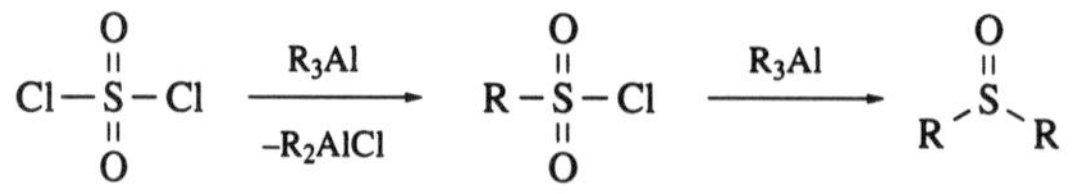

Scheme 11

$$Cl-S(=O)_2-Cl \xrightarrow[-R_2AlCl]{R_3Al} R-S(=O)_2-Cl \xrightarrow{R_3Al} R-S(=O)-R$$

Scheme 12

$$RBr + R_2AlCN \xleftarrow{BrCN} R_3Al \xrightarrow{ClCN} RCN + R_2AlCl$$

However, the most common and, in some ways, most useful metathesis of R_3Al is with a protic substrate, ROH, R_2NH, N≡C–H, and certain hydrocarbons. These protodealuminations are important routes to organoalumium alkoxides, amides, cyanides and acetylides which are useful as reagents in selective synthesis (Equations (33),[59] (34),[60] (35),[61] and (36)[52]). An impressive application of such a metathesis is that of Woodward and co-workers in the synthesis of cephalosporin C, through which the β-lactam ring was formed (Scheme 13).[62]

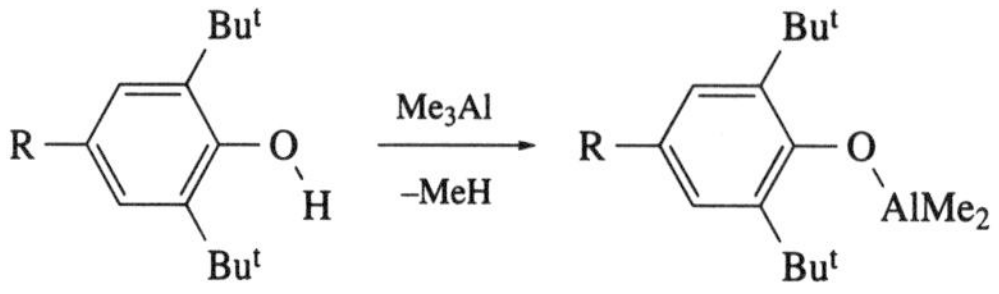

(33) 2,6-di-tert-butylphenol (R-substituted) + Me_3Al → aryl-O-$AlMe_2$ (−MeH)

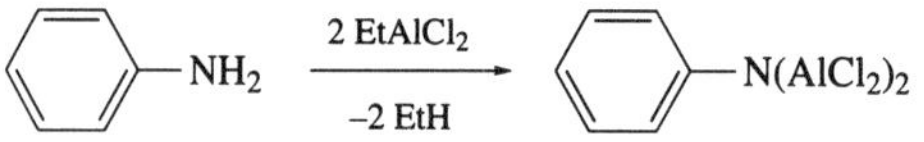

(34) $PhNH_2 + 2 EtAlCl_2 \rightarrow PhN(AlCl_2)_2$ (−2 EtH)

(35) $Et_3Al \xrightarrow[-EtH]{HCN} Et_2Al-C\equiv N$

(36) $Bu-C\equiv C-H \xrightarrow[-H_2]{Bu^i_2AlH\cdot NEt_3} Bu-C\equiv C-AlBu^i_2\cdot NEt_3$

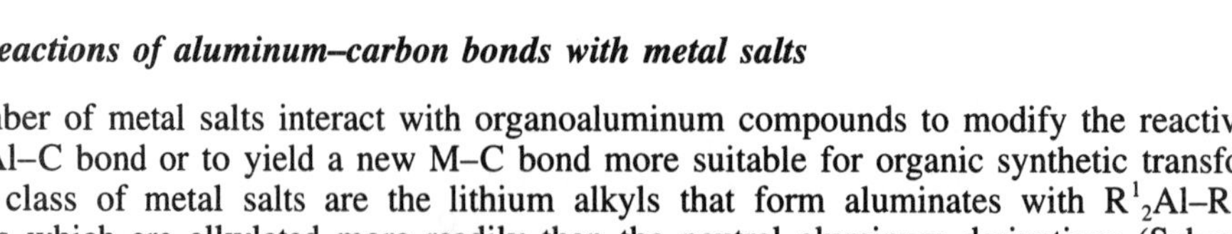

Scheme 13

β-lactam synthesis via Bu^i_3Al (−Bu^iH).

6.1.2.4 Reactions of aluminum–carbon bonds with metal salts

A number of metal salts interact with organoaluminum compounds to modify the reactivity of the original Al–C bond or to yield a new M–C bond more suitable for organic synthetic transformations. The first class of metal salts are the lithium alkyls that form aluminates with $R^1_2Al-R^2$, anionic complexes which are alkylated more readily than the neutral aluminum derivatives (Scheme 14). A further effect of aluminate formation with MeLi is to convert the *syn*-hydroaluminating agent, R_2AlH, into the *anti*-hydroaluminating agent, $LiAlR_2MeH$ (Scheme 15).

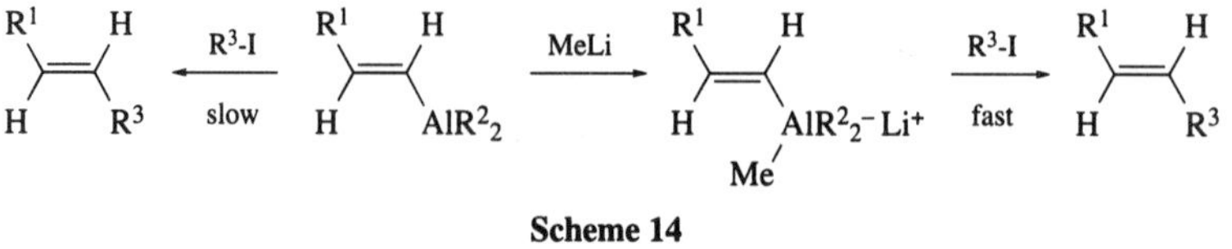

Scheme 14

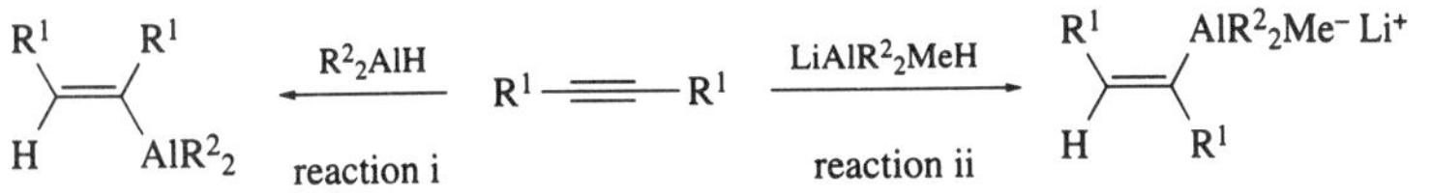

Scheme 15 shows:

$R^1R^1C=C(H)AlR^2_2$ ← R^2_2AlH (reaction i) — $R^1-\equiv-R^1$ — $LiAlR^2_2MeH$ (reaction ii) → $R^1(H)C=C(R^1)AlR^2_2Me^-Li^+$

Scheme 15

The transformation of vinylic aluminum compounds (**16**), readily obtained by the hydroalumination of alkynes, into other vinylmetallics is potentially of great value in synthesis. Interaction of (**16**) with alkylmetal chlorides of less active metals (B, Si, Sn, and transition metals) generally leads to a selective metal–metal exchange at the vinylic–aluminum bond (Scheme 16). The newly formed vinylmetallic can then feasibly undergo reactions for which the original vinylaluminum system would have been unsuitable: for example, referring to the reactions represented in Scheme 17, (i) vinylborane (**17**) can be smoothly oxidized with Me₃NO to the carbonyl derivative; (ii) vinylsilane (**18**) can be epoxidized to the extremely versatile epoxyalkylsilane; (iii) vinyltin (**19**) can be stereoselectively transformed to the more reactive lithium derivative; and (iv) vinyltitanium (**20**) can serve either as a stoichiometric carbometallating agent for alkynes or alkanes (Equation (25)), or (cf. Sections 6.1.1.4 and 6.1.2.5) as an initiator in Ziegler–Natta polymerizations.

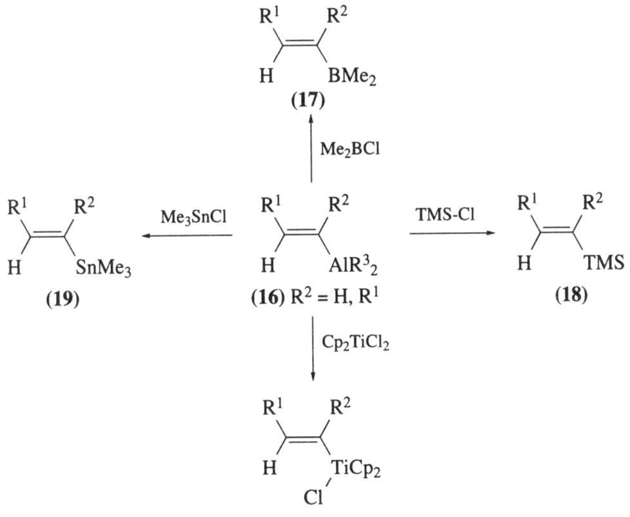

Scheme 16

Central compound (**16**) $R^2 = H, R^1$: vinyl-AlR³₂
- Me₃SnCl → (**19**) vinyl-SnMe₃
- Me₂BCl → (**17**) vinyl-BMe₂
- TMS-Cl → (**18**) vinyl-TMS
- Cp₂TiCl₂ → (**20**) vinyl-TiCp₂Cl

As already noted in Section 6.1.1.4, transition metals and their salts can profoundly alter the reactivity of C–Al bonds. Although some of these synergistic effects find their molecular basis in the type of ion pairs generated in Scheme 17, reaction iv, another origin of such effects are redox reactions that produce both subvalent transition metals and organic radicals (Scheme 18). The latter radicals need not always be completely "free" but seem to form metal complexes. As a result, transition metals are known to exert a significant promoting action on many organoaluminum reactions but the mechanistic nature of some of these mediated processes is not completely clear.

6.1.2.5 Reactions catalyzed by organoaluminum Lewis acids

Before 1950 aluminum compounds were prominent in organic chemistry, not for any potential reactions at Al–C or Al–H bonds, but rather for the Lewis acidity of the aluminum center. Aluminum chloride, not at all an organoaluminum derivative, has long maintained a towering position in organic synthesis. What organoaluminum compounds have added to the general field of Friedel–Crafts catalysis is the capability of fine-tuning Lewis acidity employed in a given reaction by a combination of polar and steric factors. Two illustrations may suffice: (i) use of alkylaluminum halides or alkoxides, RAlX₂, R₂AlX, RAl(OR)₂, and R₂AlOR, enables one to vary the Lewis acidity with compounds freely soluble in hydrocarbon media (contrast with AlX₃); (ii) use of stoichiometric amounts of aluminum tris(2,6-

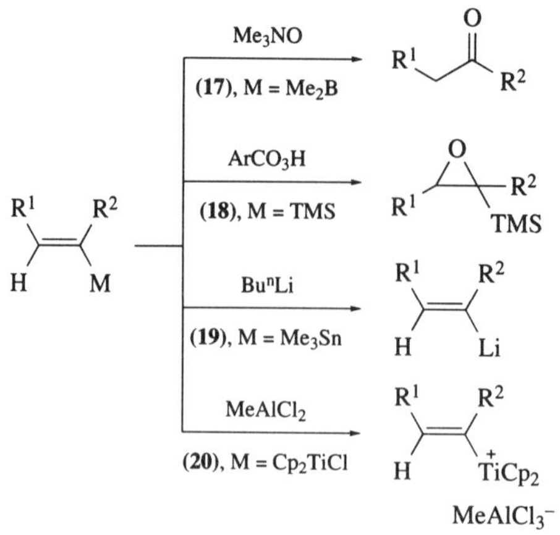

diphenylphenoxide) (atph) in reactions of group 2 metallics almost completely blocks formation of the 1,2-adduct and yields more than 96% of the 1,4-adduct (Scheme 19).[63] The most likely cause is the complexation and steric encapsulation of the carbonyl group by the atph. These comments should provide a foretaste of the great future awaiting organic-substituted Lewis acids in organic synthesis. As will be evident from Sections 6.3.1.4 and 6.3.3.2, the Ziegler–Natta polymerization of alkenes has already been shown to be superbly cocatalyzed by the Lewis acid obtained by the partial hydrolysis of Me_3Al, the polymeric "methylaluminoxane," $(MeAlO)_x$, MAO. Here also the highly effective Lewis acidic action is assumed to stem from an optimal interplay of polar and steric factors.[64]

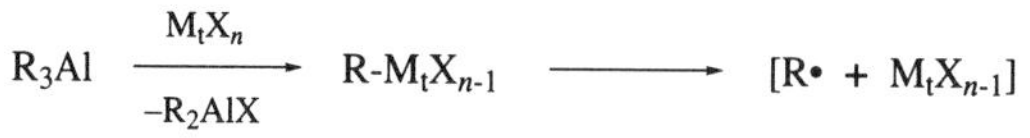

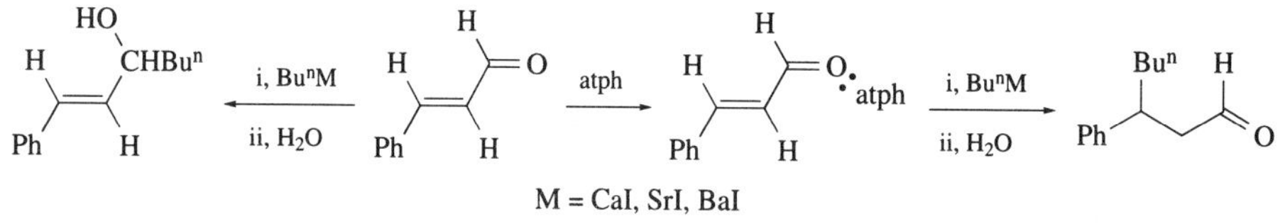

6.1.2.6 Thermal transformations of aluminum–carbon bonds

The various thermal interconversions of Al–C bonds have already been outlined in the earlier review[34] and in Chapter 10, Volume 1. Of interest here are only those thermal reactions of preparative importance. The hydroaluminations of C=C and C≡C linkages, for instance, occur in a kinetically controlled *syn* manner. In order to prevent isomerization to a *syn–anti* mixture, then this initial adduct must be complexed with a Lewis base ((**22**):D = R_2O or R_3N; cf. Scheme 4). Under such experimental conditions, high *syn* stereoselectivity is attainable (Scheme 20).[23,65] In the absence of a donor, equilibration of the *syn*- and *anti*-adduct sets in (**21**) and (**23**), and whether preparatively useful proportions of (**23**) can be obtained, depends on the steric demands of the R groups. For alkynes with R^1, R^2 = Ph, the proportion of (**23**) at equilibrium is >90%; for R^1 = Ph, R^2 = TMS, it is >95%; but for $R^1 = R^n$ and R^2 = TMS, it is ~60%. With alkenes or cycloalkenes, such equilibration leads to comparable amounts of (**21**) and (**23**) and hence makes stereoselective hydroalumination without a donor impractical in synthesis.

Due to the isomerizations of Pr^i_3Al and Bu^t_3Al to Pr^n_3Al and Bu^i_3Al, which occur above 100 °C, carbaluminations by these reagents occurring without rearrangement must be achievable below such temperatures.

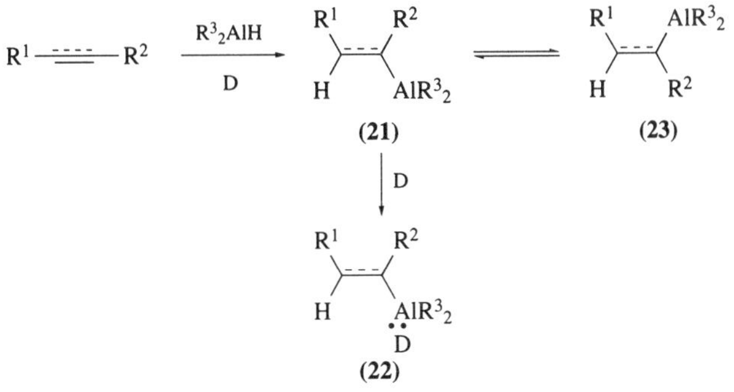

Scheme 20

$$R^1 \equiv\!\!\!= R^2 \xrightarrow[D]{R^3_2AlH} \underset{(21)}{\overset{R^1R^2}{\underset{HAlR^3_2}{\diagup\!\!\!=\!\!\!\diagdown}}} \rightleftharpoons \underset{(23)}{\overset{R^1AlR^3_2}{\underset{HR^2}{\diagup\!\!\!=\!\!\!\diagdown}}}$$

$$\downarrow D$$

$$\underset{(22)}{\overset{R^1R^2}{\underset{H\underset{D}{AlR^3_2}}{\diagup\!\!\!=\!\!\!\diagdown}}}$$

Scheme 20

6.2 REACTIVITY, STEREOCHEMISTRY, AND MECHANISMS OF REACTION FOR ALUMINUM REAGENTS

6.2.1 Gradations in Reactivity of Aluminum–Hydrogen and Aluminum–Carbon Bonds

6.2.1.1 General considerations

Detailed mechanistic discussions of organoaluminum reactions have been presented in *COMC-I*[34] and in Chapter 10, Volume 1. Only the principal mechanistic findings drawn from such studies are summarized here that have a direct bearing on the application of organoaluminum reagents in organic synthesis. Also included here are useful experimental observations for which no satisfactory mechanistic explanation has yet found wide acceptance, such as the "Ziegler nickel effect."

6.2.1.2 Polar factors

For aluminum reagents bearing Al–H or Al–C bonds, where carbon may be in different hybridization states, the empirical reactivity gradations have been observed to be Al–H ≫ Al–C and Al–C≡C > Al–C=C ≫ Al–C≡.

Consideration of Allred–Rochow electronegativities (H = 2.20; C = 2.50; Al = 1.47) shows that such covalent bonds to aluminum are highly polarized and thus likely to be reactive. Further, that the actual electronegativity of carbon increases with its f-orbital character would rationalize the greater Al–C polarity and hence reactivity of 1-alkynyl over 1-alkenyl over 1-alkynyl derivatives.

Since the greater reactivity of Al–H over Al–C bonds is not readily understandable in terms of ground-state bond polarity, the search for an explanation must extend to the likely transition states for hydroalumination (**24**) and carbalumination (**25**) of alkynes. From mechanistic studies on alkynes, the rate-determining steps for both reactions are known to involve electrophilic attack of monomeric, tricoordinate R_2Al–H or $R^1{}_2Al$–R^2 on the alkyne π-cloud.[26,27,66] From the negligible deuterium-isotope effect for hydroaluminum[67] and the small, but negative Hammett ρ-value for carbalumination,[26,27] bridging but no great Al–H or Al–C stretching is thought to occur in transition states (**24**) and (**25**). With these models in mind, then one could explain the greater Al–H bond reactivity over that of Al–C in terms of the superior bridging capability of hydridic H between two electron-deficient atoms (C_β and Al (**24**)), over that of carbon (R^2 in (**25**)). Similarly, the more electron-rich is R^2 in (**25**), the better the bridging between C_β and aluminum. Clearly, an alkynic group would bridge better than vinylic, and that group in turn better than alkyl. In all cases, in (**24**) and in (**25**), such bridging should lower the energy of the transition state.

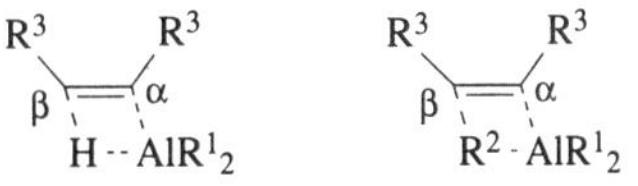

$$\underset{(24)}{\overset{R^3R^3}{\underset{H\cdots AlR^1{}_2}{\underset{\beta\alpha}{\diagup\!\!\!=\!\!\!\diagdown}}}} \qquad \underset{(25)}{\overset{R^3R^3}{\underset{R^2\cdot AlR^1{}_2}{\underset{\beta\alpha}{\diagup\!\!\!=\!\!\!\diagdown}}}}$$

One type of Al–C bond, formally involving an sp^3-hybridized carbon but possessing unusual reactivity in carbalumination, is an allylic system, such as in allyl-, benzyl-, or 1-acenaphthenylaluminum (26) derivatives. However, when such compounds effect carbalumination, the resulting kinetically controlled products are those expected via a six-membered transition state involving allylic rearrangement (Scheme 21). The synchronous Al–C bond-breaking and C–C bond-making in (27) leads to a lowering of activation energy and hence fostering of such allylic–aluminum bond reactivity.

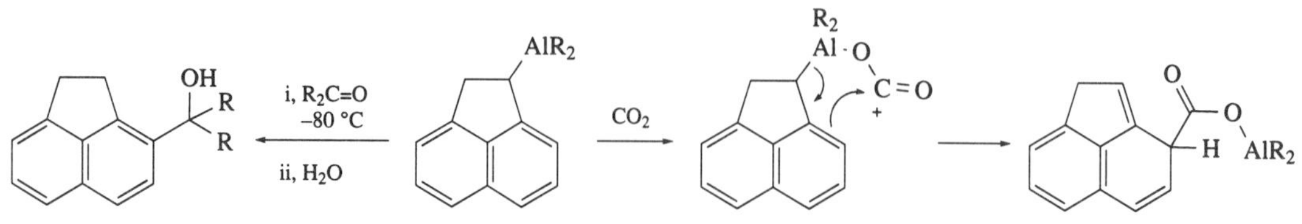

Scheme 21

6.2.1.3 Steric factors

In addition to polarity and bridging leading to a lowering of transition state (24) for hydroalumination, certainly the smaller size of the bridging hydride over that of the bridging R^2 in (25) provides a steric reason for the greater ease of Al–H additions over Al–C additions. Consequently, the smaller steric demands of R–C≡C over R–CH=CH and of the latter over RCH_2CH_2 in approaching and bonding to C_β in (25) would account for the reactivity gradations given in Section 6.2.1.2. Thus, the observed reactivities of Al–H and Al–C bonds can be equally well rationalized by either polar or steric factors operative in transition states (24) and (25).

Since aluminum–alkyl bonds are generally the least reactive in carbalumination, the β-elimination of Al–H from R_3Al may occur faster and the resulting R_2AlH may instead effect hydroalumination (cf. Scheme 8). In light of the importance of steric factors in determining the ease of Al–C bond additions, one might expect that branching at the α-carbon, as in Pr^i_3Al or in Bu^t_3Al, would lower the reactivity in carbalumination. However, studies with the carbalumination of ethene by the diethyl etherates of R_3Al at 110 °C give just the opposite outcome: one insertion of ethene occurs more extensively with Bu^t_3Al (>90%) than with Pr^i_3Al (66%), which occurs more extensively than with Et_3Al (14%) (Equation (37)).[68]

$$R_3Al \cdot OEt_2 \xrightarrow[110\,°C]{H_2C=CH_2} (RCH_2CH_2)_3Al \cdot OEt_2 \qquad (37)$$

The maximum reactivity of Bu^t_3Al in this insertion suggests that polar, rather than steric, factors are more important in stabilizing transition state (25). In this view the appended electron-releasing methyl groups in Bu^t, Me_2CH, or $MeCH_2$ successively provide the bridging R^2 in (25) with smaller electron densities which stabilizes (25) to a lesser degree.

6.2.1.4 Role of ligands around aluminum

Hydroaluminations and carbaluminations with $(R_2AlH)_3$ and $(R^1_2Al–R^2)_2$ proceed most readily in hydrocarbon or halocarbon media where the monomeric, weakly, or noncoordinated R_2AlH or $R^1_2Al–R^2$ monomers are the active intermediates (*caution*: $CHCl_3$ or CCl_4 are to be avoided as reaction solvents; cf. Section 6.1.2.3). Such reactions still proceed at lower (but useful) rates in the presence of donors, such as ethers and amines (cf. Scheme 3 and Equation (37)), where a higher reaction temperature is necessary to cause displacement of the donor :D by the organic substrate, C≐E, in order to form the transition state, such as (24) and (25).

When the donor D is a carbanionic ligand, as in the aluminates concerned in Schemes 14 and 15 ($Li^+[AlMeR_2CH=CHR]^-$ and $Li^+[AlMeR_2^2CR^1=CHR^1]^-$), evidence suggests that the resulting reagent is no longer electrophilic but now becomes a nucleophile toward organic substrates.

Finally, with transition metal salts or complexes there are indications that R_2AlH or $R^1_2Al–R^2$ consummate their reactions neither by electrophilic nor by nucleophilic initiation but by radicals generated via intermediate Al–M bonds (Scheme 22; cf. Scheme 18): hydroalumination of PhC≡CPh by $(cod)_2Ni$ and Bu^t_3Al has been shown to be photopromoted (J. J. Eisch and M. Singh, unpublished work). This and other experiments connected with the Ziegler nickel effect offer evidence that low-valent transition metals M_t can serve as transitory ligands about aluminum.[49]

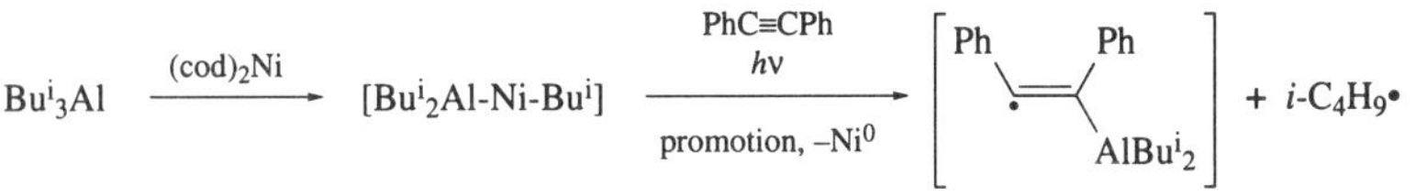

$$Bu^i_3Al \xrightarrow{(cod)_2Ni} [Bu^i_2Al\text{-}Ni\text{-}Bu^i] \xrightarrow[\text{promotion, }-Ni^0]{PhC\equiv CPh, \ h\nu} \left[\begin{array}{c} Ph \quad Ph \\ \diagup = \diagdown \\ \bullet \quad AlBu^i_2 \end{array} \right] + i\text{-}C_4H_9\bullet$$

Scheme 22

6.2.2 Stereochemistry

6.2.2.1 Locoselectivity

Generally, with dienes, diynes, and enynes, both carbalumination and hydroalumination can be selectively achieved at the terminal C=C and C≡C linkages over any internal unsaturated site. With terminal C≡C–H systems it may be necessary to silylate the terminal carbon to prevent the side reaction of alumination (cf. Equation (25)). Also, carbalumination may require promotion by an early transition metal salt (Equations (38),[69] (39),[70] and (40)[36]).

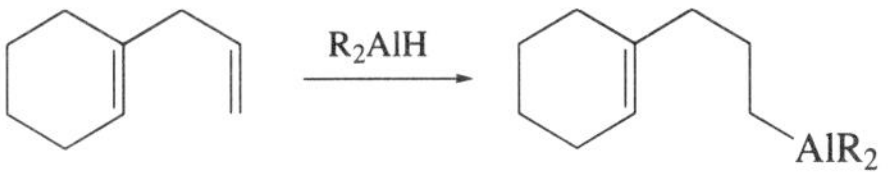

(38)

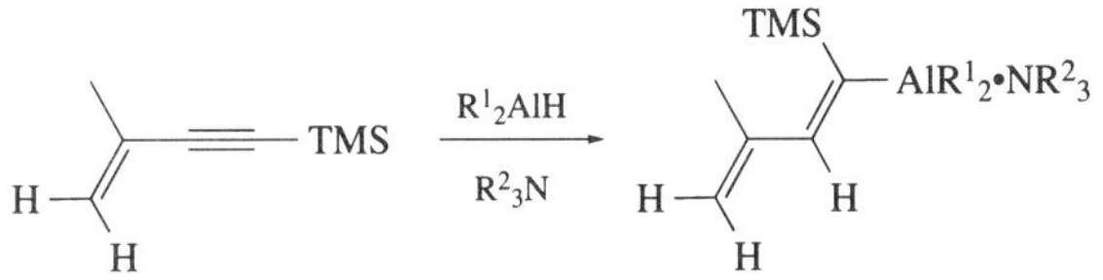

(39)

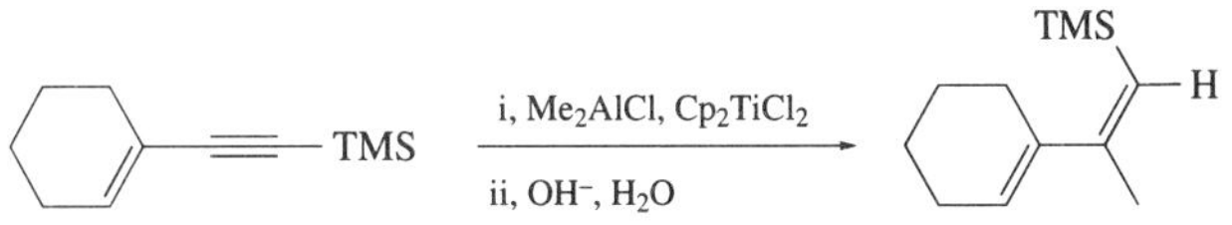

(40)

Such additions to α,β-unsaturated carbonyl substrates can give varying proportions of 1,2- and 1,4-adducts. Consider Equation (9), where Ph_3Al yields only the 1,4-adduct. However, Et_3Al or its etherate produce the 1,2-adduct as the main product (Equation (41)).[71] As already mentioned (cf. Scheme 19), organoaluminum Lewis acids show promise in suppressing 1,2-addition. Likewise, treatment of α,β-unsaturated aldehydes, ketones, or esters with R_2AlH gives the corresponding 1,2-adduct in high yield. By working with one equivalent of R_2AlH at low temperatures, such esters can provide the aldehyde in acceptable yield (Equation (42)).[72]

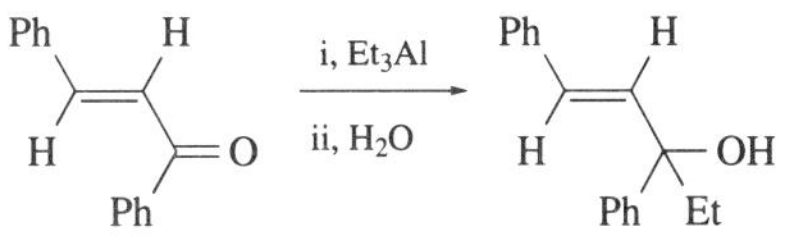

(41)

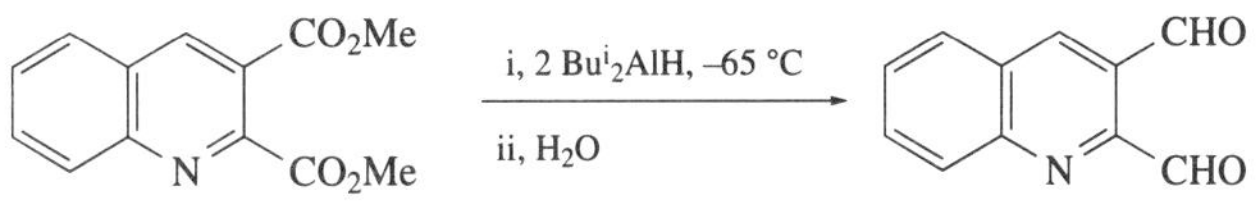

(42)

6.2.2.2 Regioselectivity

The carbalumination or hydroalumination of terminal C=C or C≡C linkages generally proceeds to produce 95% or more of the 1,2-adduct, that is, the 1-alumino derivative. This regioselectivity is undoubtedly determined by both polar and steric factors operative in such an electrophilic attack by R_3Al or R_2AlH. That a steric explanation alone is inadequate is seen in the significant amount of 2,1-adducts formed with styrene[73] and with silylethylenes[74] (Scheme 23).

However, that the steric effect can dominate in certain situations is shown by the regioselectivity exhibited by *t*-butyl(phenyl)acetylene in such additions. Carbalumination with Ph_3Al yields only (**28**), while hydroalumination produces only (**29**) (Scheme 24).[75]

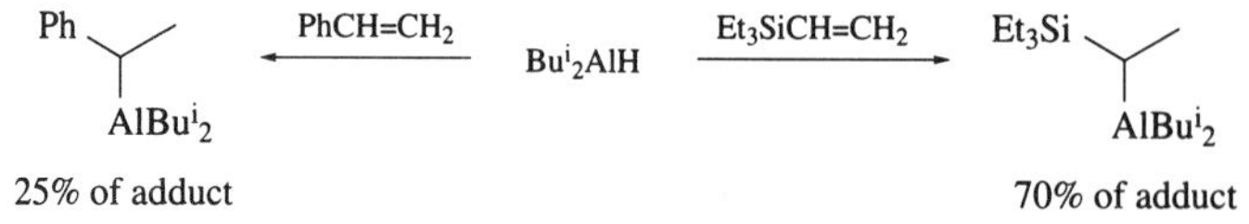

Scheme 23

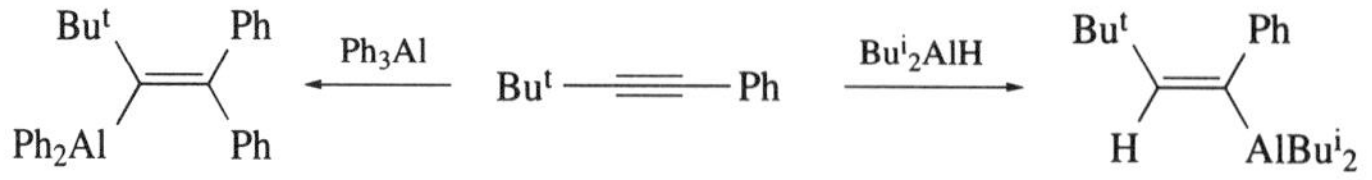

Scheme 24

Not only terminal alkynes, but R–C≡C–E derivatives when E = R$_3$Si, R$_3$Ge, RS, or R$_2$P, yield exclusively or preponderantly the 1,2-adduct with R$_2$AlH (Scheme 25, reaction i),[23] while when E = OR or NR$_2$ hydroalumination gives only the 2,1-adduct (Scheme 25, reaction ii).[24] Such reversal in regioselectivity may prove useful in synthesizing 1- or 2-substituted derivatives of terminal alkynes.

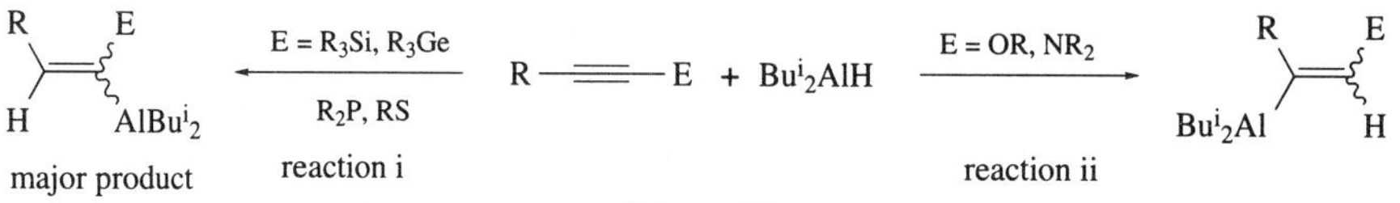

Scheme 25

6.2.2.3 Stereoselectivity

The possibility of achieving either selective *syn*- or *anti*-hydroalumination of silylalkynes has already been mentioned (Scheme 4), as has the kinetic *syn*- or thermodynamic *anti*-hydroalumination of ordinary alkynes (Scheme 20). The principles determining such stereoselectivity are also applicable to heterosubstituted alkynes and may prove useful in selective synthesis, as shown in Equations (43),[23] (44),[24] and (45).[24]

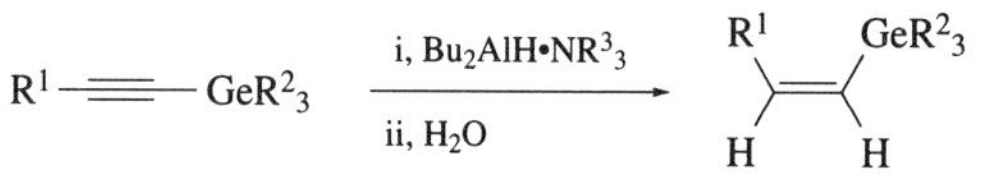

$$R^1 \equiv\!\!\!= GeR^2{}_3 \xrightarrow[\text{ii, H}_2\text{O}]{\text{i, Bu}_2\text{AlH}\cdot\text{NR}^3{}_3} \quad \begin{array}{c} R^1 \quad GeR^2{}_3 \\ \diagup\!\!\!\!=\!\!\!\!\diagdown \\ H \quad\quad H \end{array} \tag{43}$$

$$R^1 \equiv\!\!\!= NR^2{}_2 \xrightarrow[\text{ii, H}_2\text{O}]{\text{i, Bu}_2\text{AlH}\cdot\text{NR}^3{}_3} \quad \begin{array}{c} R^1 \quad H \\ \diagup\!\!\!\!=\!\!\!\!\diagdown \\ H \quad NR^2{}_2 \end{array} \tag{44}$$

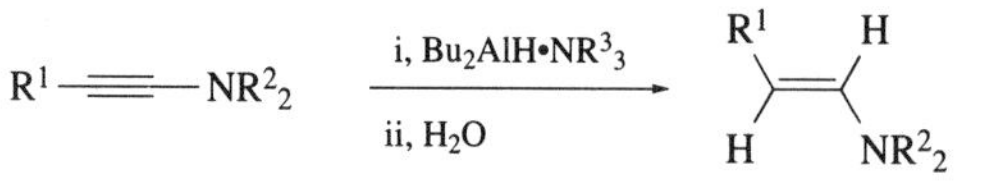

$$R^1 \equiv\!\!\!= OR^2 \xrightarrow[\text{ii, H}_2\text{O}]{\text{i, Bu}_2\text{AlH}\cdot\text{NR}^3{}_3} \quad \begin{array}{c} R^1 \quad OR^2 \\ \diagup\!\!\!\!=\!\!\!\!\diagdown \\ H \quad\quad H \end{array} \tag{45}$$

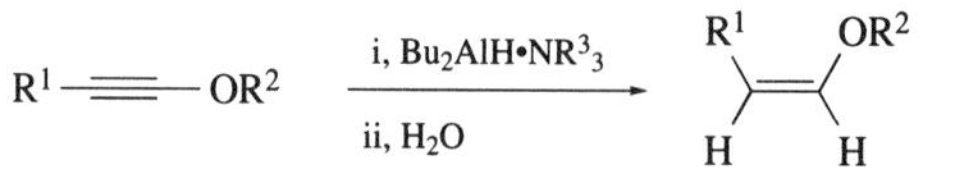

6.2.3 Mechanisms of Organoaluminum Reactions

6.2.3.1 Electrophilic attack by aluminum

Accumulated experimental observations of preparative organoaluminum chemistry and detailed mechanistic studies support the view that carbalumination or hydroalumination of unsaturated substrates (Sections 6.1.2.1 and 6.1.2.2) involve the crucial electrophilic attack of R$_3$Al or R$_2$AlH monomers on the unshared electron pairs or π-electron cloud of the substrates. The foregoing loco-, regio-, and stereoselectivities can be rationalized within such a mechanistic framework, as can the effects of Lewis acid catalysts, Lewis base retardants and solvent polarity. The composite evidence for this mechanistic viewpoint has been scrutinized previously.[34]

6.2.3.2 Nucleophilic attack by an aluminate complex

The chemical behavior of tetracoordinate aluminate complexes, such as LiAlPh$_4$ (**30**) and LiAlBu$^i_2$MeH (**31**), stands in sharp contrast with that of their tricoordinate counterparts, Ph$_3$Al (**32**) and Bu$^i_2$AlH (**33**). Although (**32**) and (**33**) readily add in a *syn* manner to alkynes (cf. the alternative reactions in Scheme 24), (**30**) does not add to alkynes (without nickel catalysis) and (**31**) adds only in an *anti* manner (cf. Scheme 15, reaction ii). These and other observations clearly require a mechanistic explanation other than electrophilic attack by aluminum. We suggest that these electron-rich aluminate reagents behave as nucleophiles toward the electrophilic substrates, such as X$_2$, CO$_2$ and R$_2$C=O, with which they do readily respond (Equation (46)).

$$\text{Li}^+ \text{Al}(\text{C}\equiv\text{CPh})_4^- \xrightarrow[\text{ii, H}_2\text{O}]{\text{i, PhCHO}} \text{Ph}-\equiv-\overset{\text{Ph}}{\underset{\text{H}}{\text{C}}}-\text{OH} \qquad (46)$$

6.2.3.3 Homolytic pathways

With paramagnetic substrates (O$_2$, NO$_x$, alkali metals), with reactants easily giving rise to radicals (S$_8$, R$_2$O$_2$, quinones, or transition metal salts), or under photochemical or electrochemical conditions, abundant evidence is available that supports homolytic bond ruptures and the involvement of free-radical intermediates. What is of interest in the present discussion are those reactions which proceed in good yield to give one principal product. Air oxidation is one such process that is most satisfactory for preparing alcohols from alkenes (Scheme 26),[45] but it is not feasible for converting aluminum aryls into phenols (Equation (47)).[76] Unfortunately, such homolytic processes are fraught with difficulties, such as producing many products and low yields.

$$3\ \text{RCH=CH}_2 \xrightarrow[-3\ \text{C}_4\text{H}_8^i]{\text{Bu}^i_3\text{Al}} (\text{RCH}_2\text{CH}_2)_3\text{Al} \xrightarrow[\text{ii, H}_2\text{O}]{\text{i, O}_2} 3\ \text{RCH}_2\text{CH}_2\text{OH}$$
$$>90\%$$

Scheme 26

$$\text{Ph}_3\text{Al} \xrightarrow[\text{ii, H}_2\text{O}]{\text{i, O}_2} \text{PhOH} + \text{Ph}_2 \qquad (47)$$
$$\sim 10\%$$

6.2.3.4 Intermediates

Besides the formation of electrophilic or nucleophilic aluminum intermediates or organic free radicals, efficient preparative routes have been discovered for generating reactive aluminum reagents capable of effecting valuable synthetic transformations. Samples of such reagents and their principal applications in synthesis are offered in Equations (48),[40] (49),[77] (50),[78] (51),[39] and (52).[79]

$$\text{Cp}_2\text{TiCl}_2 \xrightarrow[-\text{MeH},\ -\text{Me}_2\text{AlCl}]{\text{Me}_3\text{Al}} \text{Cp}_2\text{Ti}\begin{matrix}\text{CH}_2\text{AlMe}_2 \\ \diagdown \\ \text{Cl}\end{matrix} \quad \text{for } \text{R}_2\text{C=O} \longrightarrow \text{R}_2\text{C=CH}_2 \qquad (48)$$
Tebbe reagent

$$\text{CH}_2\text{Cl}_2 \xrightarrow[\text{ii, Me}_3\text{Al}]{\text{i, Al}} \text{CH}_2(\text{AlMeCl})_2 \quad \text{for } \text{R}_2\text{C=O} \longrightarrow \text{R}_2\text{C=CH}_2 \qquad (49)$$

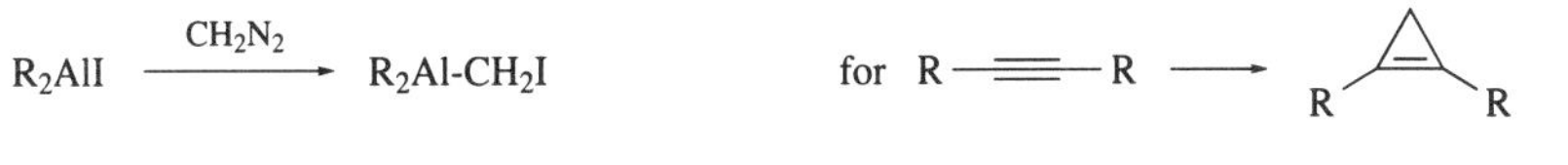

$$\text{R}_2\text{AlI} \xrightarrow{\text{CH}_2\text{N}_2} \text{R}_2\text{Al-CH}_2\text{I} \quad \text{for } \text{R}-\equiv-\text{R} \longrightarrow \underset{\text{R}\ \ \ \ \text{R}}{\triangle} \qquad (50)$$

$$\text{Cp}_2\text{TiCl}_2 \xrightarrow{\text{MeAlCl}_2} \text{Cp}_2\overset{+}{\text{Ti}}\text{Me}\ \overset{-}{\text{AlCl}}_4 \quad \text{for } \text{R}-\equiv-\text{TMS} \longrightarrow \underset{\text{H}}{\overset{\text{R}\quad\text{TMS}}{\diagup\!=\!\diagdown}} \qquad (51)$$

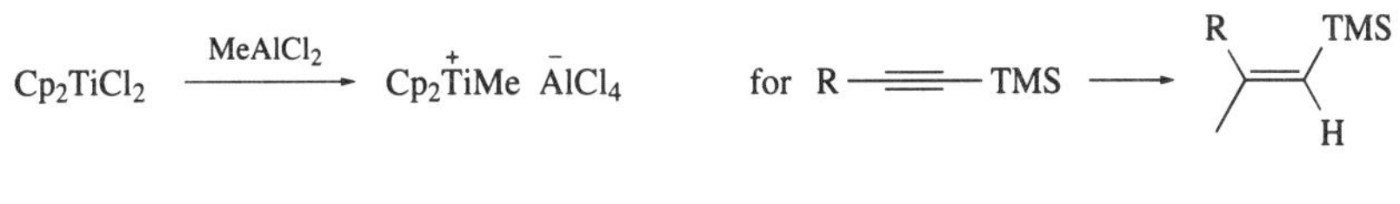

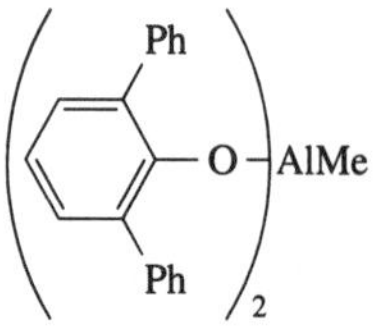

$$\left(\begin{array}{c} Ph \\ \\ Ph \end{array} \middle\rangle\!\!\!\!\!\!\!\!\!\!\!\!\!\!\!- O \right)_2 \!\!AlMe \xrightarrow{H_2C=O} H_2C=O\cdot maph \quad \text{for RM} \longrightarrow R\text{-}CH_2OH \qquad (52)$$

maph

6.3 AVAILABILITY OF ALUMINUM REAGENTS

6.3.1 Commercially Available Compounds

Several chemical companies worldwide produce organoaluminum compounds by variants of Ziegler's original reactions and market them in quantities desired for research as well as for large-scale applications in the petrochemical and polymer industries. Some of the principal producers are Akzo (formerly Texas Alkyls), Albemarle (formerly Ethyl), and Witco (formerly Schering). Such suppliers provide these compounds in well-defined grades of purity in cylinders protected by a nitrogen atmosphere. Technical brochures are also provided that furnish details for the safe, convenient, and nonhazardous transfer and manipulation of such air- and moisture-sensitive compounds, which can spontaneously inflame or explosively hydrolyze if improperly handled. Any synthetic chemist considering the use of organoaluminum reagents in synthesis should consult both the aforementioned technical brochures and available monographs[80] before commencing experimental work.

6.3.1.1 Aluminum alkyls

The most widely used alkyls available are Me_3Al, Et_3Al, Bu^i_3Al, and even-carbon tri-*n*-alkylaluminums, where $R = C_4$, C_6, C_8, and C_{10}. The commercial Me_3Al should be checked for chloride content and the higher R_3Al for hydride and alkoxide impurities. Specific analytical procedures for these impurities have been worked out.[80] A fractional distillation under vacuum and for higher R_3Al, in the presence of the corresponding alkene (Equation (53)), should reduce any R_2AlH impurity to a low level (being trimeric, $(R_2AlH)_3$ will boil at a higher temperature than R_3Al, which are monomeric or at most dimeric).

$$(RCH_2CH_2)_2AlH + RCH=CH_2 \xrightarrow{\Delta} (RCH_2CH_2)_3Al \qquad (53)$$

If the α-alkene of even- or odd-carbon number with b.p. >100 °C is available, then the Ziegler alkene displacement reaction with Bu^i_3Al can be used to produce the tri-*n*-alkylaluminum of acceptable purity (Equation (54)).

$$3\ C_9H_{19}CH=CH_2^n \xrightarrow[-C_4H_8^i]{Bu^i_3Al} (C_9H_{19}CH_2CH_2^n)_3Al \qquad (54)$$

6.3.1.2 Alkylaluminum hydrides

Commercially available Et_2AlH and Bu^i_2AlH fulfill almost all needs for a preparatively useful hydroaluminating agent. For demanding stoichiometric work such reagents should be redistilled and reanalyzed for hydride content. They can be used directly in hydrocarbon solution, combined with 1.1 equiv. of THF or *N*-methylpyrrolidine to moderate their reactivity or control their *syn* mode of addition (Section 6.1.1.3), or complexed with 1.1 equiv. of methyllithium to yield the *anti*-hydroaluminating agent, $LiAlR_2MeH$ (Section 6.1.2.4).

6.3.1.3 Alkylaluminum halides

The reagents, Me_2AlX, $MeAlX_2$, Et_2AlX, $EtAlX_2$, Bu^i_2AlCl, and Bu^iAlCl_2, where X = Cl, Br, and I, represent a broad array of commercially available hydrocarbon-soluble Lewis acids useful as promoters or catalysts for many valuable synthetic transformations and as starting materials for preparing other aluminum reagents, such as R_2Al-D, $ArN(AlCl_2)_2$, and $R^1_2Al-SR^2$. Fractional distillation or recrystallization, monitored by halide analysis, can lead to highly pure Lewis acids.

Where R_nAlX_{3-n} cannot be purchased, disproportionation reactions with the appropriate ratio of R_3Al and AlX_3 can be used to prepare the necessary reagent, as illustrated in Equation (55).

$$Bu^n_3Al + 2\,AlCl_3 \xrightarrow{\Delta} 3\,Bu^nAlCl_2 \qquad (55)$$

6.3.1.4 Alkylaluminum alkoxides and aluminoxanes

Aluminum alkoxides, such as $Al(OEt)_3$, $Al(OPr^i)_3$, $Al(OBu^s)_3$, $Al(OBu^t)_3$, Me_2AlOMe, and Et_2AlOEt are commercially available and through disproportionation reactions of R_3Al with $Al(OR)_3$, other alkylaluminum alkoxides of the type $R_nAl(OR)_{3-n}$ can be readily generated.

Due to its enormous importance in polymerization, a polymeric composition, $(AlMeO)_x$, known as methylaluminoxane (MAO), is readily available. This catalyst is a metastable mixture of MAO chains and rings resulting from the controlled, partial hydrolysis of Me_3Al under patented conditions and in its final form can contain up to 30% of residual Me_3Al. Although the molecular units responsible for its superb Lewis acidity are uncertain, MAO is the cocatalyst of choice in most Ziegler–Natta catalyst systems.

6.3.2 Accessible, Specialized Reagents

6.3.2.1 Aluminum aryls

Solvent-free, symmetrical aluminum aryls are best prepared from the corresponding ArLi or ArMgX via the mercury compound (Scheme 27). Naturally, any feasible substituent G must be tolerant of the reducing conditions involved. In some cases, the aluminum aryl (**36**) can be produced directly from (**34**), via (**35**), if the latter etherate dissociates upon heating without Al–C bond rupture.

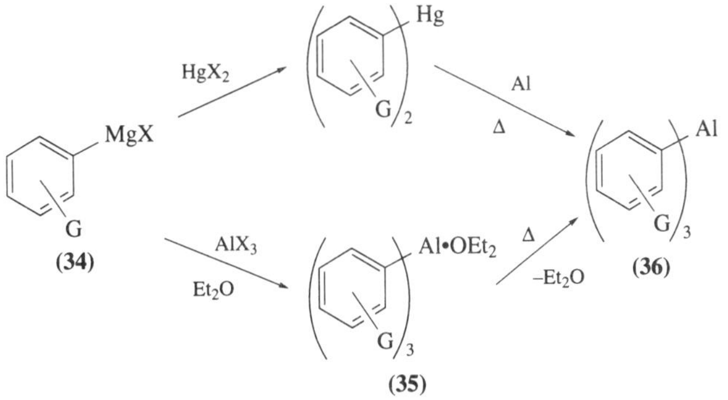

Scheme 27

Unsymmetrical aluminum aryls, $ArAlR_2$ or Ar_2AlR, can be prepared by a number of feasible ways (Equations (56)–(58)). Such alkyl(aryl)aluminum reagents are of interest because often selective carbalumination occurs only at the aluminum–aryl bond.

$$ArLi + R_2AlCl \longrightarrow R_2Al\text{-}Ar + LiCl \qquad (56)$$

$$R\equiv\!\!\!=\!\!\!=\!H + Ar_3Al \longrightarrow R\equiv\!\!\!=\!\!\!=\!AlAr_2 + ArH \qquad (57)$$

$$Ar_3Al + 2\,Me_3Al \longrightarrow 3\,Me_2Al\text{-}Ar \qquad (58)$$

6.3.2.2 Unsymmetrical vinylaluminums

Just as alkyl(aryl)aluminum compounds are of interest because of their selective aryl–aluminum bond reactivity, so are dialkyl(vinyl)aluminums. In many Al–C bond cleavages the vinyl–aluminum bond is much more reactive than the alkyl–aluminum bond (Section 6.2.1). Fortunately, such compounds are readily prepared in high yield by the hydroalumination of alkynes by R_2AlH. If due consideration is given to the regio- and stereoselectivity of the process (Section 6.2.2), the desired vinylaluminum isomer can often be obtained in preparatively feasible yields.

6.3.2.3 Unsymmetrical aluminum acetylides or cyanides

Such acetylides can be obtained by the alumination of terminal alkynes with aluminum aryls (Equation (57)) or with $R_2AlH \cdot NEt_3$ (Equation (36)), and by simple metathesis (Equation (59)). Again, in (**37**) the alkynic–aluminum bond can be much more readily cleaved (e.g., by CO_2, I_2, or RCHO) than the alkyl–aluminum bonds.

Analogously to Equation (57), the useful reagent diethylaluminum cyanide is accessible by protodealumination of triethylaluminum with HCN (Equation (60)).

$$R^1{\equiv\!\equiv}\text{—Li} + R^2{}_2AlCl \xrightarrow{\text{alkane}} R^1{\equiv\!\equiv}\text{—}AlR^2{}_2 + LiCl \quad (59)$$
$$(\mathbf{37})$$

$$Et_3Al + HC{\equiv}N \longrightarrow Et_3Al\text{-}C{\equiv}N + C_2H_6 \quad (60)$$

6.3.2.4 Aluminum alkoxides, aryloxides, oxides, and related sulfur, selenium, and nitrogen derivatives

This diverse class of compounds is unified by their common, principal method of preparation, namely protodealumination with a stoichiometric amount of the appropriate Brønsted acid. Illustrative reactions, given in Equations (61)–(63), can proceed in quantitative yield.

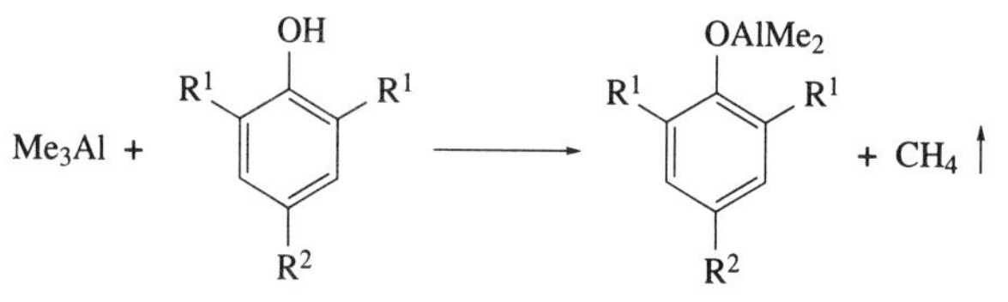

$$Me_3Al + \underset{\substack{R^1 \\ R^2}}{\text{Ar-OH}} \longrightarrow \underset{\substack{R^1 \\ R^2}}{\text{Ar-OAlMe}_2} + CH_4 \uparrow \quad (61)$$

$$Et_3Al + PhSH \longrightarrow Et_2Al\text{-}SPh + C_2H_6 \uparrow \quad (62)$$

$$2\ EtAlCl_2 + PhNH_2 \longrightarrow PhN(AlCl_2)_2 + 2\ C_2H_6 \uparrow \quad (63)$$

6.3.3 Specialized Organoaluminum Reagents for Carbon–Carbon Bond Formation

6.3.3.1 Stoichiometric reagents

A number of specifically designed reagents for selective C–C bond formation have already been mentioned in this review, but it seems appropriate to cite references to them here again: (i) Yamamoto's aluminum tris(2,6-diphenylphenoxide) as an aid to achieving 1,4-addition to α,β-unsaturated carbonyl derivatives with RM (Scheme 19);[63] (ii) Tebbe's reagent, $Cp_2Ti(Cl)CH_2AlMe_2$, for methylenating carbonyl derivatives (Equation (48));[40] (iii) the Eisch–Piotrowski reagent $CH_2(AlMeCl)_2$ for methylenating carbonyl derivatives (Equation (49));[77] (iv) the Hoberg reagent, $R_2Al\text{-}CH_2I$, for cyclopropanating alkenes and alkynes (Equation (50));[78] (v) the Natta–Breslow reagent, $Cp_2TiCl_2R_nAlX_{3-n}$, for alkylating alkynes and strained alkenes (Equation (51));[39] and Yamamoto's complex of $H_2C{=}O$ with methylaluminum bis(2,6-diphenylphenoxide) for the smooth insertion of $H_2C{=}O$ into C–M bonds (Equation (52)).[79] In addition, other reagents have been developed since 1980; their scope of reaction is discussed in Chapter 10, Volume 1.

6.3.3.2 Catalysts

A wide variety of inorganic and organic aluminum derivatives have risen to prominence as Lewis acid catalysts in polymerization reactions and in specific organic transformations of great value in selective synthesis. Mention is made in Chapter 10, Volume 1 of the preeminent position of MAO in the stereoselective polymerization of α-alkenes when combined with metallocenes of titanium or zirconium (cf. Section 10.6) and the interesting skeletal rearrangements promoted by R_3Al or $MeAl(OAr)_2$ (Section 10.5.5). The reader is referred to these sections for further details.

6.4 CARBON–HYDROGEN BONDS MADE BY ORGANOALUMINUM HYDRIDES

The preceding sections have dealt with the scope, limitations, selectivity, and mechanisms of organoaluminum reactions; this and succeeding sections will serve to illustrate those reactions of particular utility in organic synthesis. Reference 34 and Chapter 10, Volume 1 discuss the applications of organoaluminum chemistry to organic synthesis, but in terms of the bonds which are broken (e.g., in the carbalumination of a ketone, the Al–C and C=O linkages). In this chapter, such reactions will now be discussed in terms of the bonds which are made, such as C–H, C–O, and C–C. This approach seems most appropriate, for this is the viewpoint of the chemist seeking guidance in organic synthesis.

6.4.1 Alkanes from Alkenes

Although alkenes undergo ready catalytic hydrogenation, there often may be a real gain in selective reduction and the avoidance of high-pressure conditions by employing R_2AlH instead for such transformations. Where these hydroaluminations are slow or incomplete, as with internal or highly substituted C=C linkages, salts of titanium, zirconium, nickel or other transition metals can be used to promote the hydroalumination. Noteworthy, however, is the tendency of such salts to promote migration of the C=C bond and changes in both the stereoselectivity and the regioselectivity with which R_2AlH adds to an unsymmetrically substituted linkage, for example, cis-$R^1HC=CHR^2$ (**38**). If merely an overall addition of dihydrogen is desired, the formation of such isomeric aluminum intermediates is of no practical consequence (Scheme 28).

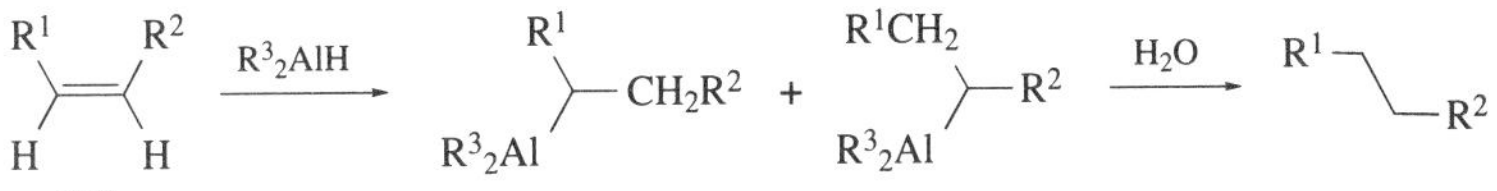

Scheme 28

However, complications such as those depicted in Scheme 28 do interfere with another valuable application of hydroalumination, the locoselective introduction of deuterium (and potentially tritium) into an alkane chain. For this purpose it is essential that the C=C bond should not migrate and that the *uncatalyzed* hydroalumination should occur with high regio- and stereoselectivity. These criteria are met with α-alkenes but not with internal alkenes: under these conditions the addition of R_2AlH or R_2AlD and workup with H_2O and D_2O can give >95% of specific mono- or dideuteriated alkanes[28,81] (Scheme 29 and Equation (64)).

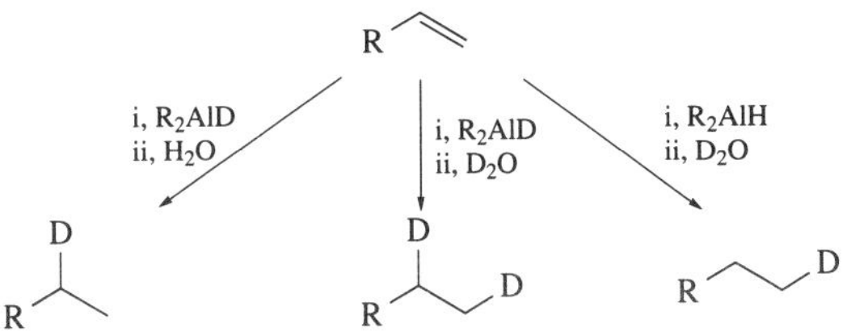

Scheme 29

Some alkenic linkages bearing heteroatoms, such as vinylic ethers, halides, and sulfides (**39**), undergo hydroalumination with partial or total loss of the heteroatomic substituent, most likely by a β-*cis*-elimination from intermediate (**40**) (Scheme 30).

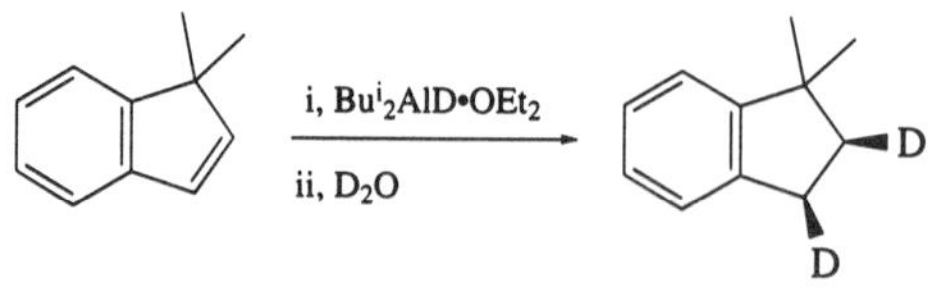

$$\text{(indene-like)} \xrightarrow[\text{ii, D}_2\text{O}]{\text{i, Bu}^i_2\text{AlD·OEt}_2} \text{(deuterated indane)} \qquad (64)$$

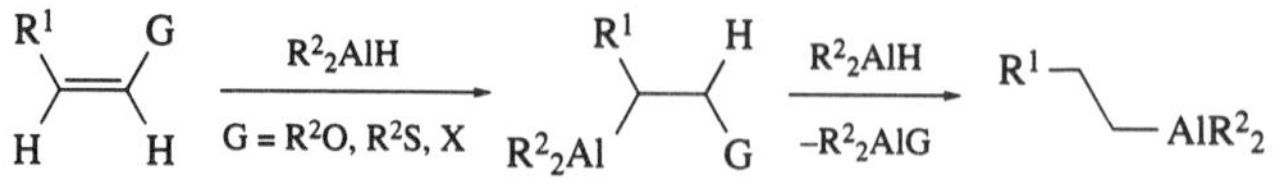

Scheme 30

6.4.2 Alkenes from Alkynes

Numerous instances of the hydroalumination of alkynes to yield alkenic derivatives have already been cited in previous discussions of mechanism and stereochemistry (e.g., Schemes 3, 4, 15, 20, and 24, and Equations (25), (28), and (39)). The important stereoselectivity principle is that *syn* addition is kinetically controlled and that *anti* addition (where attainable) is thermodynamically controlled (Scheme 31). Many of these hydroaluminations are also highly regioselective, so sequential reactions with R_2AlD and H_2O or with R_2AlH and D_2O, can yield selectively deuterated alkenes (Scheme 32).

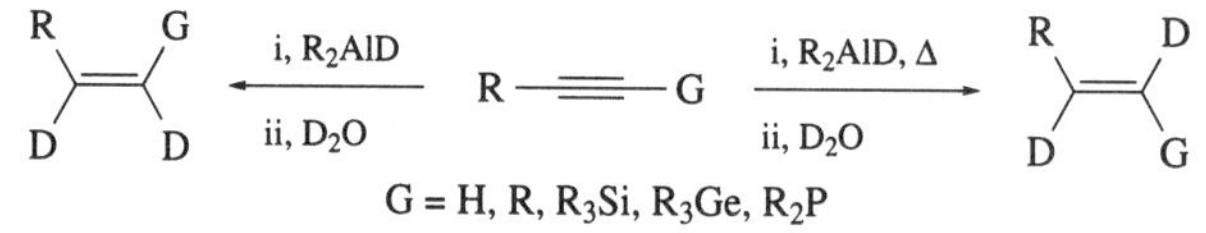

Scheme 31

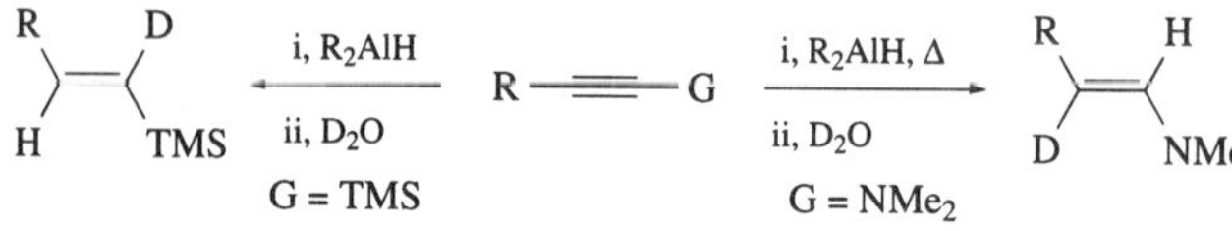

G = H, R, R_3Si, R_3Ge, R_2P

Scheme 32

The controlled reduction of alkynes to alkenes, often with the directed formation of the *cis* or the *trans* isomer in high selectivity, offers a number of clear advantages over alternatives like the Lindlar catalytic hydrogenation or the Birch reduction. The former method requires the tricky preparation of suitably poisoned palladium; the latter involves a strongly basic medium with a reducing agent that is nondiscriminating toward many functional groups. Furthermore, as is evident from Schemes 31 and 32, hydroalumination simultaneously provides routes to specifically labeled deuterated alkenes.

As with alkenes (Scheme 30), alkynic ethers, sulfides, and halides undergo hydroalumination to split out R_2AlG and yield first the alkyne and then its adduct (Scheme 33). Such hydroaluminations of 1-alkynyl(trimethyl)silanes can provide routes to either the *cis*-1-alkenyl- or the *trans*-1-alkenyl(trimethyl)silanes in high yield. Moreover, the 1-alkynyl(dialkyl)amines can serve as precursors to the corresponding aldehyde (Equation (65)).

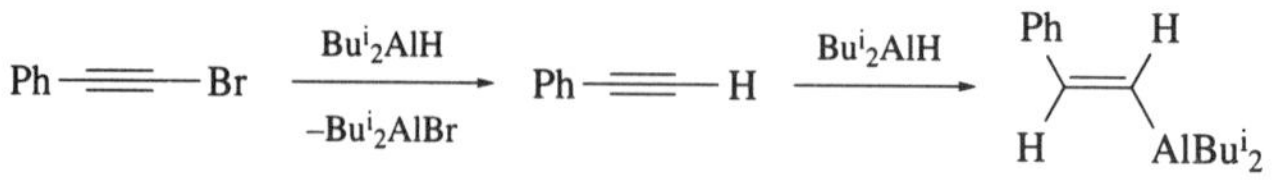

$$\text{Ph}-\!\!\!\equiv\!\!\!-\text{Br} \xrightarrow[-\text{Bu}^i_2\text{AlBr}]{\text{Bu}^i_2\text{AlH}} \text{Ph}-\!\!\!\equiv\!\!\!-\text{H} \xrightarrow{\text{Bu}^i_2\text{AlH}} \text{(Ph,H)C=C(H,AlBu}^i_2)$$

Scheme 33

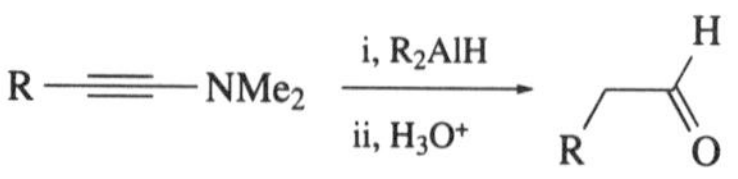

$$R-\!\!\!\equiv\!\!\!-\text{NMe}_2 \xrightarrow[\text{ii, H}_3\text{O}^+]{\text{i, R}_2\text{AlH}} R\text{CH}_2\text{CHO} \qquad (65)$$

6.4.3 Alcohols or Aldehydes from Carbonyl Compounds or Epoxides

The reducing action of R_2AlH or of R_3Al compounds, which are capable of the facile eliminations of alkene, upon carbonyl or epoxide linkages has already been cited (cf. Equations (10) and (42) and

Scheme 9). Compared with the alkali metal hydridoaluminates or borohydrides that generally require ether solvates, the R_2AlH and R_3Al reagents can be employed either in hydrocarbon media, in which they are more reactive, or in ether media, in which they are more selective. By variations in reagent (R_2AlH or R_3Al), solvent, and temperature, therefore, a number of important selective reductions can be achieved in high yield (Scheme 9 (reactions i and ii) and Equation (66)).[82] From this we see that the following reductions of esters can demonstrate the preparative advantages of these reagents: reductions of esters can be halted at the aldehyde stage (Equation (67));[83] α,β-unsaturated aldehydes can be made to undergo principally 1,2-reductions (Equation (68));[18] and terminal epoxides can be cleaved to yield preponderantly either the 1-alkanol (Scheme 9 (reaction i) and Equation (66)) or the 2-alkanol (Scheme 9 (reaction ii)).[82] By forcing conditions hydroalumination can be effected to a maximum extent at both C=C and C=O functions (Equation (69)).[84]

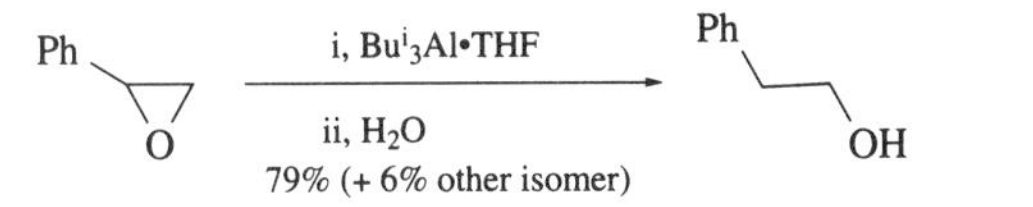

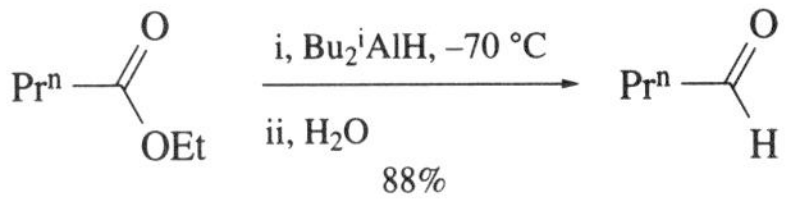

Certain internal epoxides, such as 1,2-epoxycyclododecane, undergo rearrangement (rather than reduction) with Bu^i_2AlH, and 3-hydroxycyclododecene results in high yield. Realizing that Bu^i_2AlH was causing a base-catalyzed reaction, Yamamoto and co-workers have employed diethylaluminum 2,2,6,6-tetramethylpiperidide (datmp) to effect this and other such rearrangements in high yield ((**41**), Scheme 34).[85]

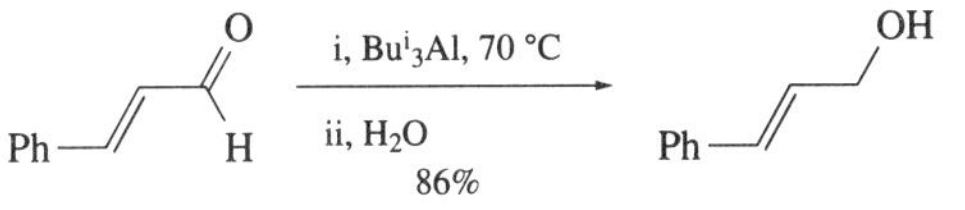

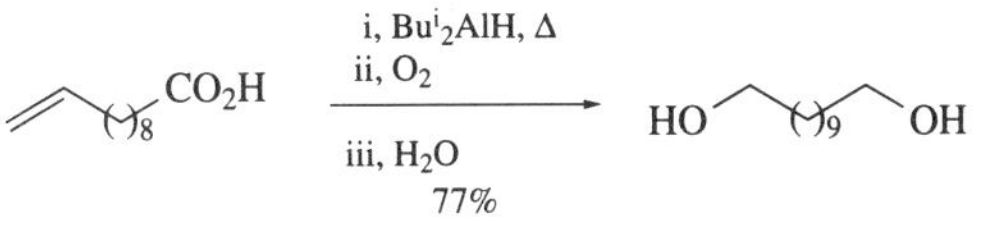

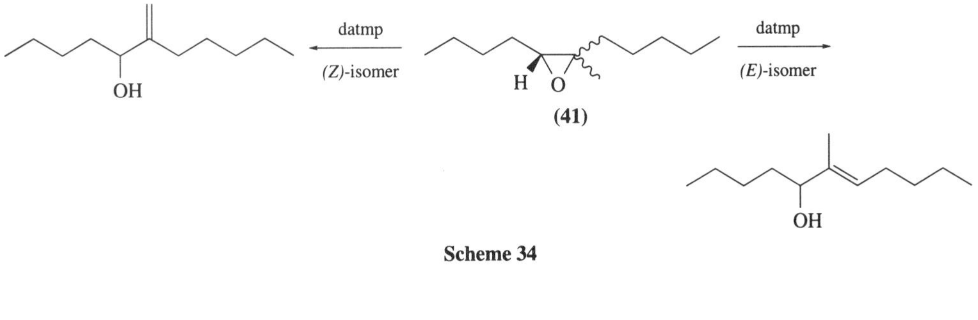

Scheme 34

6.4.4 Methylene Derivatives from Alcohols or Carbonyls

Complete deoxygenation of a carbon center to a methylene derivative ((**42**) → (**44**)) has traditionally been achieved by amalgamated zinc and HCl (Clemmensen) or by base-promoted deazotization of a hydrazone (Wolff–Kishner), conditions involving strong aqueous acid or base, respectively (Scheme 35). Conversions of primary CH_2OH groups to Me ((**43**) → (**44**)) are often feasible by reduction of the sulfonates, RCH_2OSO_2Ar, by LAH. Recently, the direct reduction of ketones or secondary alcohols by organoaluminum Lewis acids has proved to give good to excellent yields of the corresponding methylene compound, regardless of whether R^1 and R^2 are alkyl or aryl groups. Where there is no possibility of aldol condensation, as with diaryl ketones, a one-step reduction of the ketone with $(Bu^iClAl)_2O$, $Bu^i_2AlH–AlBr_3$, or $Bu^i_2AlH–Bu^iAlCl_2$ is most feasible (Equation (70)). With alkyl aryl or dialkyl ketones, better yields of (**44**) are obtained by first reducing (**42**) to the aluminum alkoxide of (**43**) with Bu^i_2AlH before adding the strong Lewis acid. With secondary alcohols, crucial intermediate (**45**) can be generated alternatively by alcoholyzing Bu^i_3Al with (**43**). This two-step procedure minimizes the undesired aldol condensation of the ketone (Scheme 36). Although this organoaluminum route to methylene derivatives does involve strongly reducing conditions, which may be intolerant of certain functional groups (NO_2, $C\equiv N$, SO_2R, etc.), it can permit reaction in the absence of water or other solvolyzing conditions. For this reason, it may prove advantageous in certain preparative situations and even superior to the Clemmensen or Wolff–Kishner methods.[86]

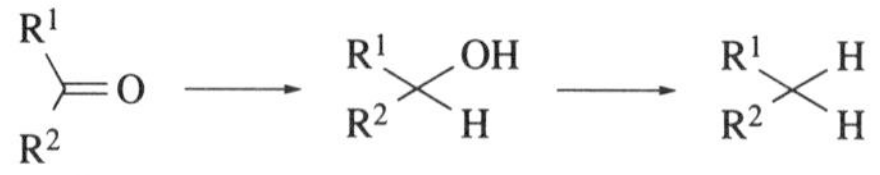

$$\underset{(42)}{\overset{R^1}{\underset{R^2}{>}}=O} \longrightarrow \underset{(43)}{\overset{R^1}{\underset{R^2}{>}}\overset{OH}{\underset{H}{<}}} \longrightarrow \underset{(44)}{\overset{R^1}{\underset{R^2}{>}}\overset{H}{\underset{H}{<}}}$$

Scheme 35

$$Ph_2C=O \xrightarrow[95\%]{Bu^i_2AlH-AlBr_3} Ph_2CH_2 \qquad (70)$$

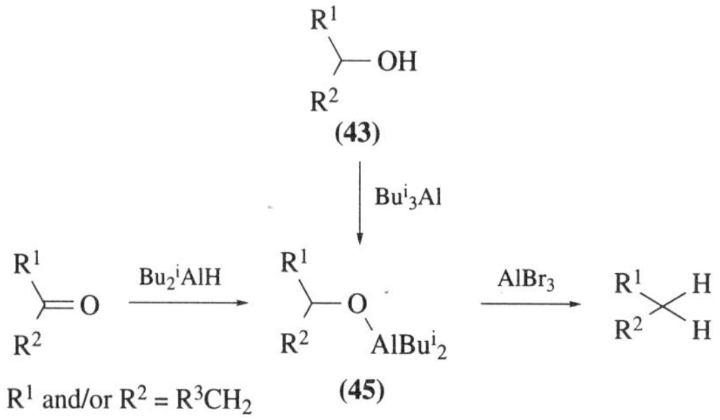

$$\overset{R^1}{\underset{R^2}{>}}-OH$$
(**43**)
$\downarrow Bu^i_3Al$

$$\underset{R^1 \text{ and/or } R^2 = R^3CH_2}{\overset{R^1}{\underset{R^2}{>}}=O} \xrightarrow{Bu^i_2AlH} \underset{(45)}{\overset{R^1}{\underset{R^2}{>}}\overset{O}{\underset{AlBu^i_2}{<}}} \xrightarrow{AlBr_3} \overset{R^1}{\underset{R^2}{>}}\overset{H}{\underset{H}{<}}$$

Scheme 36

6.4.5 Ethers from Acetals, Ketals, or Orthoformates

Analogously to the reduction of carbonyl derivatives to alcohols, the ether-like adducts of carbonyl groups, $C(OR)_2$ and $C(OR)_3$, can often be reduced to simple ethers by organoaluminum reagents. As discussed in Chapter 10, Volume 1 (Sections 10.5.5 and 10.7.3.3), such reductive cleavages are often accompanied by alkylations and skeletal rearrangements because of the carbenium-ion-like intermediates that are generated. However, many reductions occur cleanly and in useful yields (Scheme 37 and Equation (71)).[87]

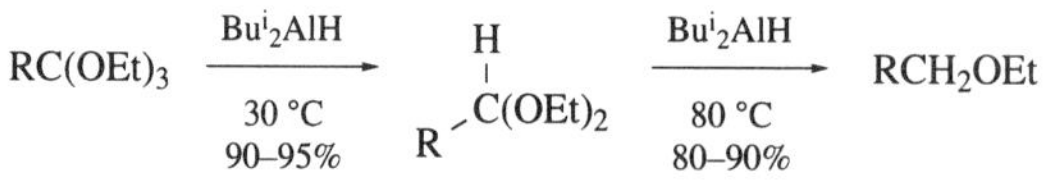

$$RC(OEt)_3 \xrightarrow[\substack{30\,°C \\ 90-95\%}]{Bu^i_2AlH} R\overset{H}{\underset{}{-}}C(OEt)_2 \xrightarrow[\substack{80\,°C \\ 80-90\%}]{Bu^i_2AlH} RCH_2OEt$$

Scheme 37

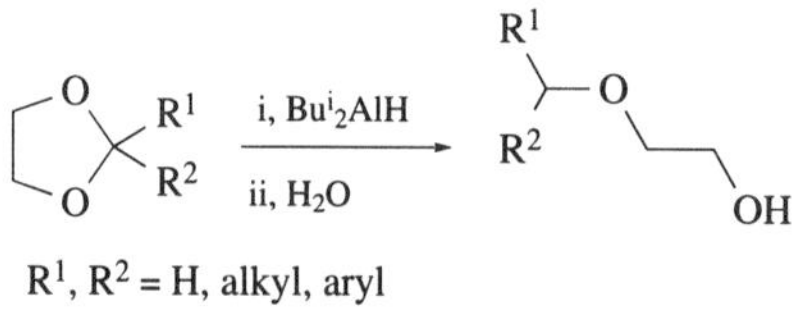

$$\underset{R^1, R^2 = H, \text{alkyl, aryl}}{\overset{O}{\underset{O}{\bigcirc}}\overset{R^1}{\underset{R^2}{<}}} \xrightarrow[\text{ii, } H_2O]{\text{i, } Bu^i_2AlH} \overset{R^1}{\underset{R^2}{>}}\overset{O\frown}{\underset{}{\phantom{<}}}\diagdown_{OH} \qquad (71)$$

6.4.6 Amines or Imines from Nitriles or Enamines

The hydroalumination of nitriles has already been mentioned as a stepwise process for producing either the monoadduct (**46**) or the diadduct (**47**) in high yield (Section 6.1.2.2). Mild hydrolysis or acidic hydrolysis of (**46**) affords the imine (**48**) or aldehyde (**49**), respectively. Simple hydrolysis of (**47**) provides the primary amines (**50**) (Scheme 38).

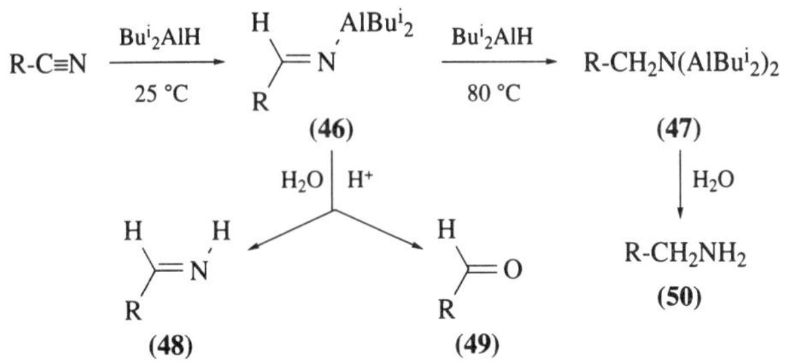

Scheme 38

$$R-C\equiv N \xrightarrow[25\,°C]{Bu^i_2AlH} \underset{(46)}{\overset{H}{\underset{R}{>}}=N\overset{AlBu^i_2}{}} \xrightarrow[80\,°C]{Bu^i_2AlH} \underset{(47)}{R-CH_2N(AlBu^i_2)_2}$$

With H_2O/H^+, (46) gives imine (48) $\underset{R}{\overset{H}{>}}=N\overset{H}{}$ or aldehyde (49) $\underset{R}{\overset{H}{>}}=O$.

(47) with H_2O gives $R-CH_2NH_2$ (50).

Although the behavior of α,β-unsaturated azomethine derivatives and enamines has been explored only in a preliminary manner, hydroaluminations can occur most readily (Equations (72) and (73)) and with high regioselectivity (Equation (74)). The latter reduction of enamines and the former hydroalumination of imines (Equations (72) and (73)) permit a novel alternative for the reductive amination of carbonyl compounds, namely the two-step sequence of imine or enamine formation followed by reduction with Bu^i_2AlH (Scheme 39) (J. J. Eisch and Z. R. Liu, unpublished results).

$$PhCH=NPh \xrightarrow[ii,\,H_2O]{i,\,Bu^iAlCl_2,\,\Delta} PhCH_2NHPh \tag{72}$$

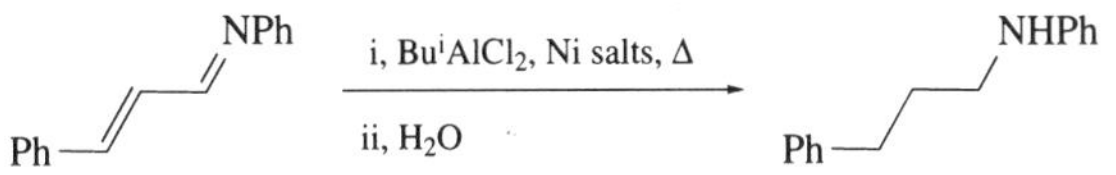

$$Ph\text{—CH=CH—CH=NPh} \xrightarrow[ii,\,H_2O]{i,\,Bu^iAlCl_2,\,Ni\text{ salts},\,\Delta} Ph\text{—CH}_2\text{CH}_2\text{CH(NHPh)} \tag{73}$$

$$Ph_2N\text{—CH=CHPh} \xrightarrow{Bu^i_2AlH} Ph_2N\text{—CH(AlBu}^i_2\text{)—CH}_2\text{Ph} \tag{74}$$

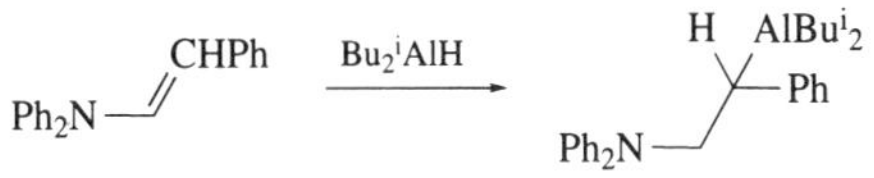

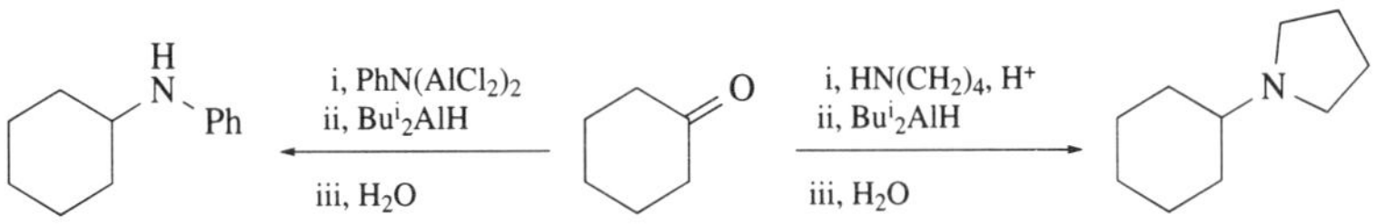

Scheme 39

Cyclohexanone $\xrightarrow[iii,\,H_2O]{i,\,PhN(AlCl_2)_2;\,ii,\,Bu^i_2AlH}$ N-phenylcyclohexylamine

Cyclohexanone $\xrightarrow[iii,\,H_2O]{i,\,HN(CH_2)_4,\,H^+;\,ii,\,Bu^i_2AlH}$ 1-cyclohexylpyrrolidine

6.4.7 Hydrocarbons from Halides

The uncatalyzed cleavage of C–X bonds by R_2AlH is an unpredictable reaction that can occur in a slow or in an explosive manner (cf. Section 6.1.2.3). The preparative utility of making C–H bonds in this manner has naturally therefore been little studied. In a more appealing and safer variant, the transition-metal-promoted metathesis shown in Equations (75) and (76) promises to become useful for the selective removal of halogen bonded to vinylic or aryl centers. Such reactions appear to proceed by the oxidative addition of the C–X bond to a subvalent transition metal intermediate (Ni^0 or Ti^{II}) and the hydride of such intermediates undergoes reductive elimination to give the final products (J. J. Eisch, S.-G. Rhee and X. Shi, unpublished studies).

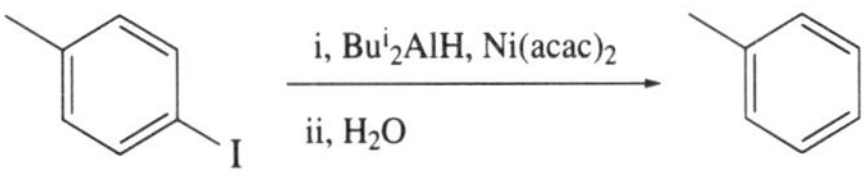

$$\text{4-iodotoluene} \xrightarrow[ii,\,H_2O]{i,\,Bu^i_2AlH,\,Ni(acac)_2} \text{toluene} \tag{75}$$

$$\underset{H}{\overset{Ph}{>}}=\underset{Ph}{\overset{Br}{<}} \xrightarrow[ii,\,H_2O]{i,\,Bu^i_2AlH,\,TiCl_4} \underset{H}{\overset{Ph}{>}}=\underset{Ph}{\overset{H}{<}} \tag{76}$$

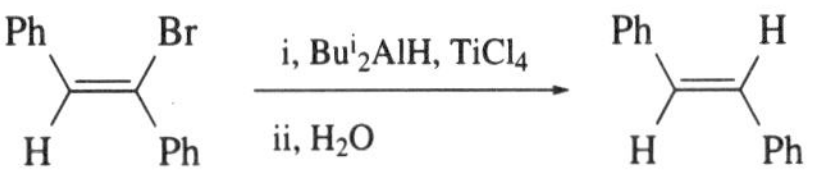

6.5 CARBON–HALOGEN BONDS

The halodealumination reaction is the principal and most valuable method for making C–Cl, C–Br, and C–I bonds with organoaluminum reagents. Its chief application is in converting terminal alkynes regio- and stereoselectively into the (E)-1-halo-1-alkene, when used in conjunction with hydroalumination (Scheme 40, reaction ii).[88] It should be noted that direct polar hydrohalogenation of the alkyne (practical only with HCl and HBr) gives the 2-halo-1-alkene (Scheme 40, reaction i). As previously indicated (Section 6.2.1.1), the heightened reactivity of vinylic–aluminum over alkyl–aluminum bonds permits the selective halodealumination in Scheme 40 (reaction ii) to be achieved with one equivalent of X_2.

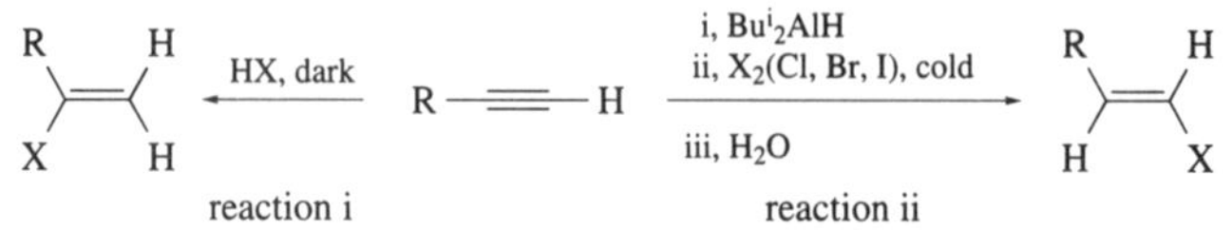

Scheme 40

Another valuable use of this approach is for the preparation of the individual pure (Z)- and (E)-halovinylic isomers from symmetrically disubstituted alkynes such as 4-octyne. Simple *syn*-hydroalumination with Bu^i_2AlH or *anti*-hydroalumination with $LiAlBu^i_2MeH$ (Scheme 15), followed by halodealumination by X_2 at temperatures below 0 °C, smoothly yields the (Z)- or (E)-haloalkene, respectively.[88]

6.6 CARBON–OXYGEN BONDS MADE BY OXIDATION OF ALUMINUM ORGANYLS

6.6.1 Alcohols from Aluminum Alkyls

On an industrial scale the oxidation of aluminum alkyls with dry air to aluminum alkoxides can be made to proceed smoothly and in high yield. This oxidation applied to mixtures of aluminum alkyls resulting from the insertion of ethene into triethylaluminum (Ziegler growth process) is the basis of the Ziegler higher linear alcohol synthesis (Section 6.2.3.3, Scheme 26). In the laboratory such air oxidations can readily be carried out on unsymmetrical aluminum alkyls resulting from the hydroalumination of alkenes with Et_2AlH or Bu^i_2AlH. In order to isolate the desired alcohol in a pure state, the by-product ethanol or isobutanol should be more volatile or more readily extractable (e.g., by water) than the main product (Equations (77) and (78)). Unfortunately, neither aryl– (Equation (47)) nor vinylic–aluminum systems undergo air oxidation in a feasible manner. Even with aluminum alkyls, air oxidation does generate some dangerous aluminum peroxide intermediates, especially during the latter stages of oxidation. Such peroxides can lead to shock-sensitive precipitates and thus may constitute considerable hazards for laboratory-scale operations in glass. Short of sudden decomposition, fragmentation of these peroxides can yield radicals capable of chain "backbiting" and producing 1,4- and 1,5-diol impurities in the final product (Scheme 41).

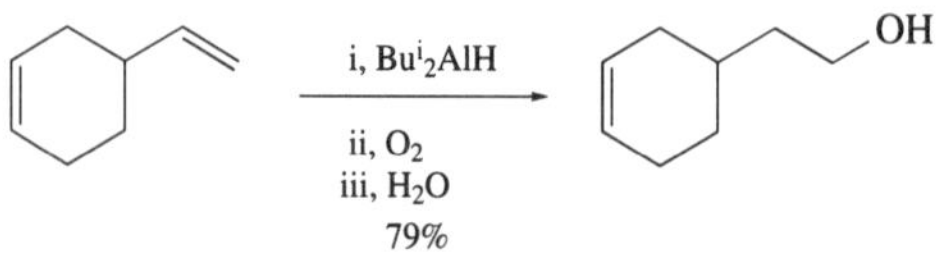

(77)

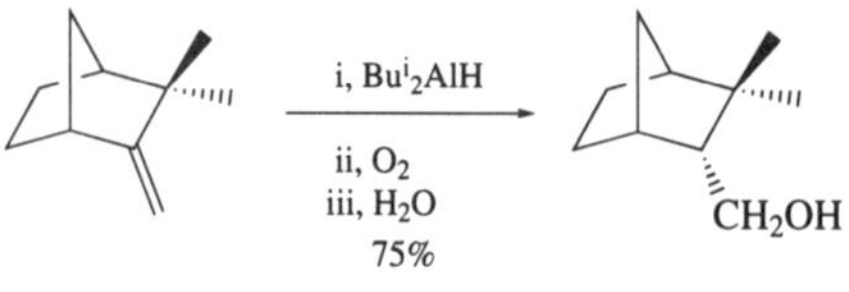

(78)

To avoid these hazards and impurities and to provide an oxidant suitable for alkyl- and arylaluminum derivatives, Kabalka and Newton have explored the utility of trimethylamine oxide,[89] which Köster and Morita employed with such great success for converting C–B bonds in alkyl-, benzyl-, and arylboranes (but not 1-alkynylboranes)[90] into C–O–B linkages. Oxidation by Me_3NO converts aluminum–alkyl and aluminum–phenyl bonds almost quantitatively into the corresponding alcohol or phenol after 4 h at

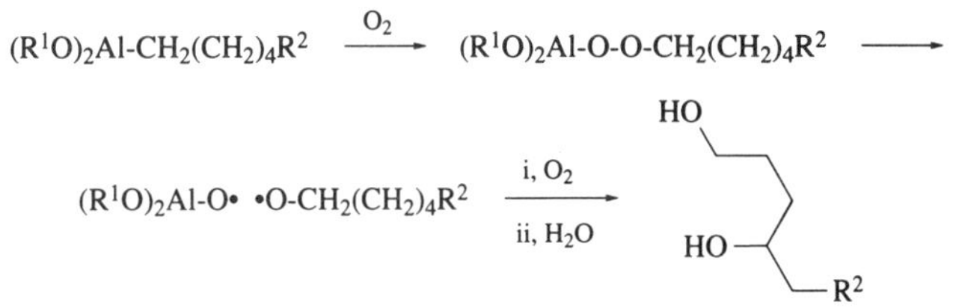

Scheme 41

138 °C and subsequent hydrolysis (Equation (79)). Such oxidations are not suitable for transforming vinylic– or alkynic–aluminum bonds, however. Oxidation of diisobutyl(2-phenylethenyl)aluminum yields essentially 100% of isobutanol but no phenylacetaldehyde. The formation of styrene and 1,3-diphenyl-1-butene suggests that intermediate styryl radicals were formed instead.

$$[Me(CH_2)_8CH_2]_3Al \xrightarrow[\text{ii, } H_3O^+]{\text{i, } Me_3NO, 138\ °C} Me(CH_2)_8CH_2OH \qquad 94\% \qquad (79)$$

6.6.2 Ethers from Aluminum Vinyls

The inability to oxidize vinyl–aluminum bonds smoothly is a severe limitation, for it would be highly desirable to be able to hydroaluminate a terminal alkyne or a symmetrically disubstituted alkyne and directly oxidize such adducts to a vinyloxy derivative, which can be hydrolyzed to the carbonyl derivative under mild conditions (Scheme 42). There are some preliminary indications that *t*-butyl perbenzoate may be a suitable reagent for such vinyl–aluminum bond oxidations. As shown in Chapter 10, Volume 1 (Section 10.5.2, Equation (120)), this reagent smoothly cleaves one phenyl–aluminum bond to provide *t*-butyl phenyl ether and a similar oxidation of the *anti*-hydroalumination product of trimethyl(phenyl)silane (51) yields about 50% of the *cis*- and *trans*-*t*-butyl ethers (52) (J. J. Eisch and D. A. Komar, unpublished work). Such aryl and vinyl *t*-butyl ethers are readily cleaved by acid hydrolysis (e.g., to (53), Scheme 43). It is clear that such a sequential hydroalumination–oxidation with Bu^{t}O$_3$CPh offers considerable promise for the practical realization of the transformations depicted in Scheme 42.

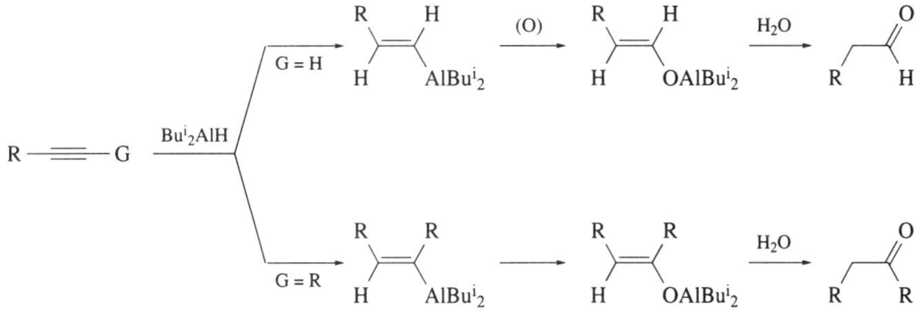

Scheme 42

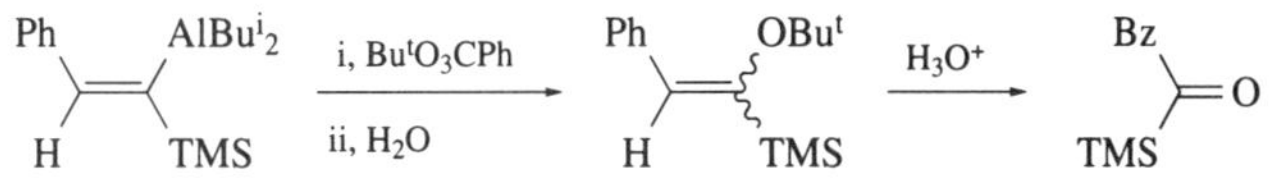

Scheme 43

6.7 CARBON–SULFUR AND CARBON–HETEROATOM BONDS

There are many substitution reactions by which C–S bonds in rich variety, as well as C–Se, C–Te, C–P, C–As and C–Sb bonds, can be made with highly selective organoaluminum reagents. As has been mentioned (Section 6.1.2.1), both SO$_2$ and SO$_3$ effect the sulfination or sulfonation of a wide range of triorganylaluminums with the involvement of all three C–Al bonds and with yields of 60–90%. The aluminum tris(organosulfinate) arising directly from the reaction with SO_2 can in turn be converted to the sulfonyl chloride with Cl_2[91] or to a sulfone with an active chloride such as benzyl chloride (J. J. Eisch and K. C. Fichter, unpublished work) (Scheme 44).

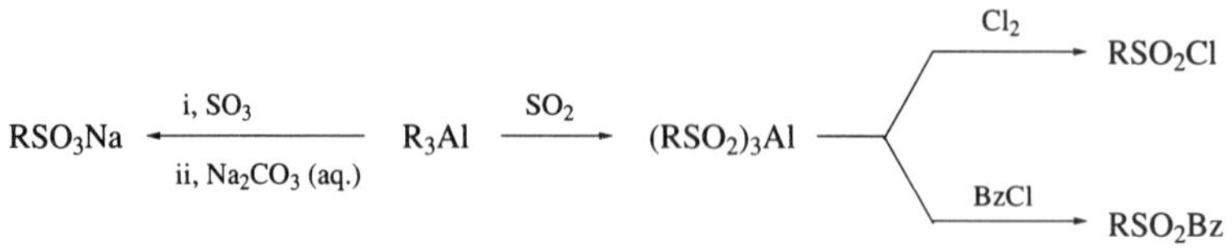

$$RSO_3Na \xleftarrow[\text{ii, } Na_2CO_3 \text{ (aq.)}]{\text{i, } SO_3} R_3Al \xrightarrow{SO_2} (RSO_2)_3Al \xrightarrow[BzCl]{Cl_2} \begin{array}{c} RSO_2Cl \\ RSO_2Bz \end{array}$$

Scheme 44

Arylsulfonyl chlorides react with either aluminum alkyls or alkylaluminum halides, ultimately to give alkyl aryl sulfoxides in 60–70% yield via a two-step process. For maximal use of alkyl groups and for reasons of economy, it is best to use cheap Et_nAlCl_{3-n} to effect the first step (Equation (80)) and then the desired $RAlCl_2$ (rather than R_3Al) for the next step (Equation (81)), where only one C–Al bond would be reactive.[92] The applicability of these reactions of SO_2, SO_3, and $ArSO_2Cl$ to selective cleavage of vinylaluminum derivatives seems not to have been studied.

$$PhSO_2Cl + EtAlCl_2 \longrightarrow PhSO_2AlCl_2 + EtCl \quad (80)$$

$$PhSO_2AlCl_2 + RAlCl_2 \longrightarrow PhSOR + O(AlCl_2)_2 \quad (81)$$

Mercaptans can be prepared in yields over 80% by the 1:1 interaction of R_3Al and sulfur at 60 °C, but only one C–Al bond is cleanly cleaved (Equation (82)).[93] A higher ratio of sulfur leads to the formation of R_2S, R_2S_2, and R_2S_x. Unexpectedly, carbon disulfide also gives mercaptans with R_3Al, rather than dithiocarbonic acids, which would form if CS_2 had reacted like CO_2.[94]

$$R_3Al \xrightarrow[\text{ii, } H_2O]{\text{i, S, 60 °C}} RSH + 2 RH \quad (82)$$

Finally, both alkylsulfonyl chlorides and symmetrical dialkyl sulfoxides can be obtained in yields of 55–75% by the interaction of R_3Al with SO_2Cl_2,[56] or $SOCl_2$,[95] respectively (Equations (83) and (84)). Essential for the successful preparation of RSO_2Cl (Equation (83)) is that R_3Al be introduced slowly to SO_2Cl_2; reverse addition yields principally RCl instead.

$$R_3Al + 3 SO_2Cl_2 \longrightarrow 3 RSO_2Cl + AlCl_3 \quad (83)$$

$$3 (n\text{-}C_8H_{17})_3Al + 3 SOCl_2 \longrightarrow 3\ R\overset{\overset{O}{\|}}{S}R + AlCl_3 \quad (84)$$

The dialkylaluminum salts of R–SH, R–SeH, and R–TeH derivatives have proved to be valuable reagents for preparing heteroatom-substituted esters by metathesis or 1,4-adducts by conjugate additions to α,β-unsaturated ketones. In all these conversions the oxophilic aluminum center undoubtedly provides the driving force (Scheme 45[96] and Equations (85)[97] and (86)[98]). Such salts also show considerable potential for the introduction of RS groups by clean displacement of phosphate groups (Equation (87)).[99]

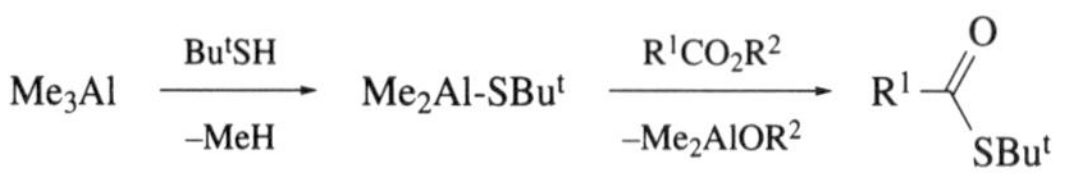

$$Me_3Al \xrightarrow[-MeH]{Bu^tSH} Me_2Al\text{-}SBu^t \xrightarrow[-Me_2AlOR^2]{R^1CO_2R^2} R^1\overset{O}{\underset{SBu^t}{\diagup\!\!\!\diagdown}}$$

Scheme 45

$$Me_2Al\text{-}SeMe + R^1CO_2R^2 \longrightarrow R^1\overset{O}{\underset{SeMe}{\diagup\!\!\!\diagdown}} + Me_2AlOR^2 \quad (85)$$

Symmetrically substituted trialkyl derivatives of phosphorus, arsenic, and antimony can be readily synthesized from the interaction of EF_3 or ECl_3 and the appropriate aluminum alkyl (Equation (88)).[100] These Lewis bases are valuable ligands used in the synthesis of the various transition metal catalysts and reagents that are finding increasing utility in organic synthesis (e.g., Scheme 46).[101]

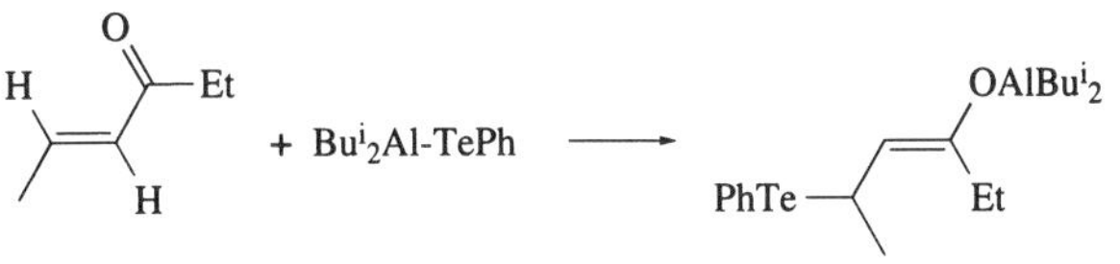

$$\text{(crotonaldehyde)} + Bu^i_2Al\text{-}TePh \longrightarrow \text{PhTe-CH(Me)-C(OAlBu}^i_2)=\text{CHEt} \tag{86}$$

$$\text{geranyl-O-P(O)(OEt)}_2 \xrightarrow[-Me_2AlOPO(OEt)_2]{Me_2Al\text{-}SBu^t} \text{geranyl-SBu}^t \quad 97\% \tag{87}$$

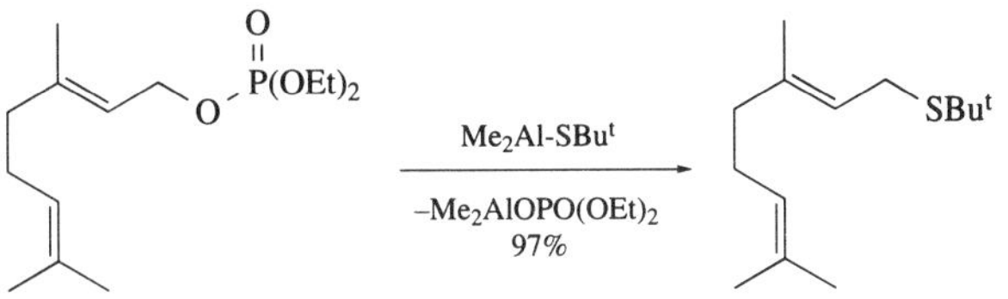

$$AsCl_3 + Et_3Al \longrightarrow Et_3As + AlCl_3 \tag{88}$$

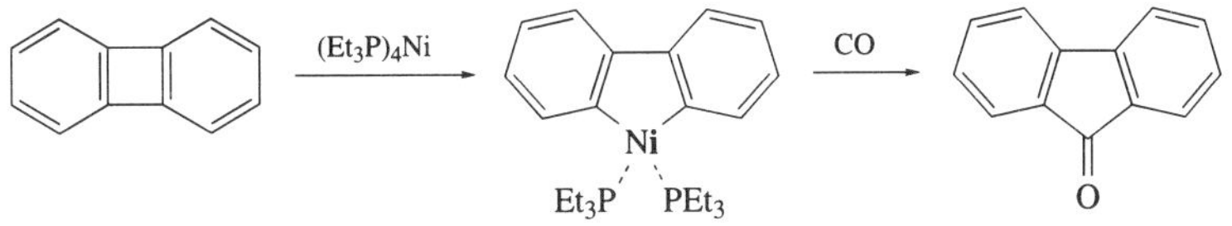

Scheme 46

6.8 CARBON–NITROGEN BONDS

A number of routes to carbon–nitrogen bonds involving organoaluminum reagents are discussed in Chapter 10, Volume 1 and also in this chapter (e.g., Scheme 13). For the convenience of the chemist seeking guidance in synthesis, however, a summary of these methods is presented here.

The two principal aluminum-based approaches to amines involve the regioselective and, where discernible, *trans*-stereoselective additions of $R^1_2Al\text{-}NR^2_2$ to epoxides (Equation (89)) (J. J. Eisch, Z. R. Liu and X. Shi, unpublished work) and the conjugate addition of $Et_2Al\text{-}N_3$ to α,β-unsaturated ketones (Equation (90)),[102] respectively. The azide produced in the latter reaction can be easily hydrogenolyzed to the primary amine.

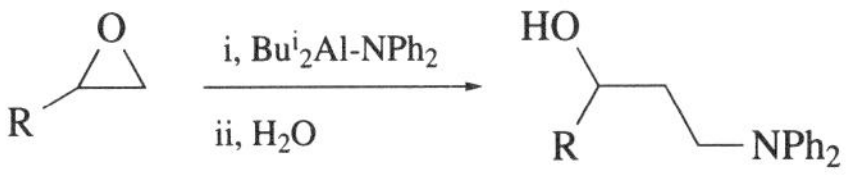

$$\text{R-epoxide} \xrightarrow[ii, H_2O]{i, Bu^i_2Al\text{-}NPh_2} \text{R-CH(OH)-CH}_2\text{-NPh}_2 \tag{89}$$

$$\text{CH}_2=\text{CH-C(O)R} \xrightarrow[ii, H_2O]{i, Et_2Al\text{-}N_3} N_3\text{-CH}_2\text{-CH}_2\text{-C(O)R} \tag{90}$$

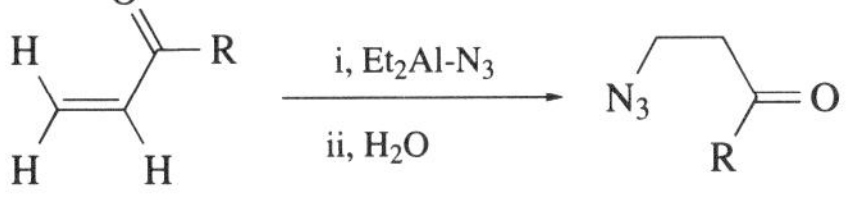

Conversion of a ketonic carbonyl group into an *N*-arylimine can be achieved by $ArN(AlCl_2)_2$ (Equation (91));[60] with α,β-unsaturated ketones such iminations occur locoselectively with the avoidance of any 1,4-addition of the amine salt (cf. Equation (34)).[60] Esters, in turn, can be smoothly converted into primary, secondary, or tertiary amides by the appropriate aluminum salt of ammonia or the amine (Equation (92)).[103,104]

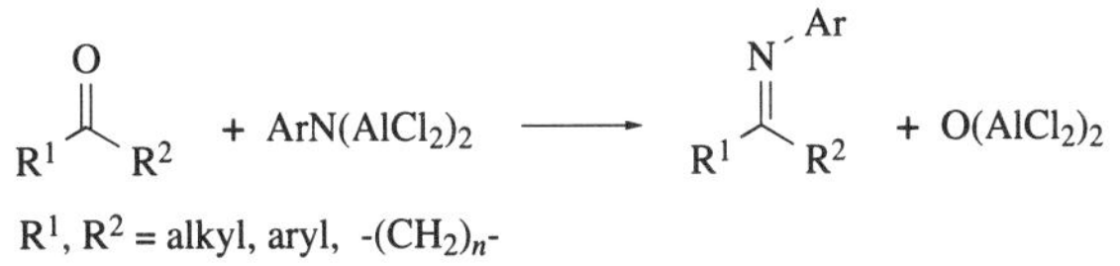

$$R^1\text{C(O)}R^2 + ArN(AlCl_2)_2 \longrightarrow R^1\text{C(=NAr)}R^2 + O(AlCl_2)_2 \tag{91}$$

R^1, R^2 = alkyl, aryl, -(CH$_2$)$_n$-

$$R^1\text{C(O)}OR^2 + Me_2Al\text{-}NR^3_2 \longrightarrow R^1\text{C(O)}NR^3_2 + Me_2AlOR^2 \tag{92}$$

R^3 = H, alkyl

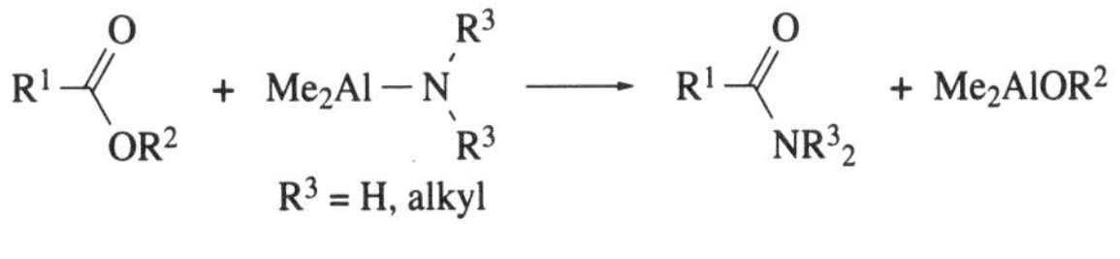

Finally, nitriles result from the reaction of chloroamine with R_3Al (Scheme 47, reaction i)[58] and these nitriles can readily form amidines from alkylaluminum amides formed *in situ* (Scheme 47, reactions i and ii).[104]

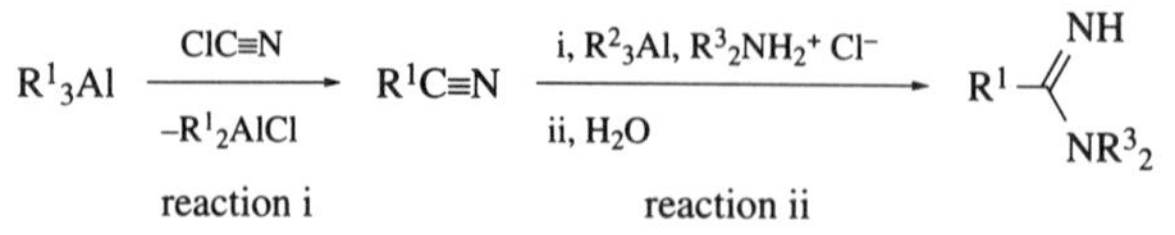

$$R^1_3Al \xrightarrow[-R^1_2AlCl]{ClC \equiv N} R^1C \equiv N \xrightarrow[ii, H_2O]{i, R^2_3Al, R^3_2NH_2^+ Cl^-} R^1 \underset{NR^3_2}{\overset{NH}{{\Large\diagup\kern-1em\diagdown}}}$$

reaction i   reaction ii

Scheme 47

6.9 CARBON–METAL BONDS

The preparative advantages of converting carbon–aluminum bonds into other carbon–metal bonds have already been examined in Section 6.1.2.4. If a given cleavage reaction of a C–Al bond is known not to proceed cleanly or rapidly, it may prove useful to generate a new C–M bond. This approach is especially appealing with the readily accessible vinylaluminums (obtainable by hydroalumination; see Section 6.1.2.2), which react selectively at the vinyl–aluminum bond with halides such as R_2BCl, R_3SiCl, or R_3SnCl to yield the vinylmetallic derivative. The vinylboranes thereby produced undergo many reactions not cleanly attainable with the vinylalane, such as oxidation with Me_3NO (Scheme 48; cf. the discussion in Section 6.6.1).

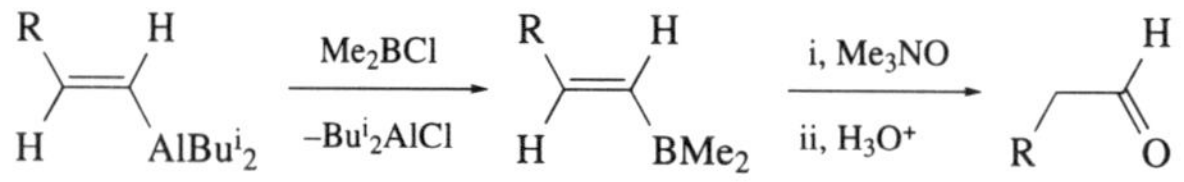

Scheme 48

6.10 CARBON–CARBON BONDS

In bringing out the mode of reaction and the stereochemistry of C–Al bond cleavages in the foregoing sections, dozens of equations involving C–C bond formation have already been cited. What remains to be done here is to organize these C–C bond-forming reactions into the types of greatest value to organic synthesis. We have already discussed how the reactivity and selectivity of a given C–Al bond in a triorganylaluminum depend upon the hybridization state and steric factors of the carbon attached to aluminum (Section 6.2.1.1) and how the substrate in the organoaluminum reaction can control whether one, two, or three of the available C–Al bonds are cleaved (Section 6.1.2). With these limitations in mind, then, such C–C bond-forming reactions can be conveniently subdivided into C–Al bond additions or carbaluminations, and carbon–aluminum center exchanges or carbodealuminations, respectively.

6.10.1 Carbaluminations

Carbon–aluminum bond additions to unsaturated substrates have been observed with CO_2, $H_2C=O$, $RHC=O$, $R_2C=O$, $R-C \equiv N$, $H_2C=CHR$, $R-C \equiv C-R$, $RCH=CHC(O)R$, and $RH\overline{C\overline{OC}}HR$ and thus offer feasible routes to the functionalized C–C bonds depicted in Scheme 49 (after appropriate hydrolytic workup). These are the principal applications, but by no means all the possible ones, of the carbalumination reaction for forming C–C bonds. The 1,2- or 1,4-addition of Et_2Al-CN to α,β-unsaturated carbonyl derivatives has also found significant utility in synthesis (Section 6.3.2.3). Where a specific example of the general equation is cited elsewhere in this chapter, the equation is cited for easy reference. For each individual conversion the reader should consult discussions in this chapter and Chapter 10, Volume 1 and in *COMC-I*[34] for further details on the scope, limitations, and selectivity of the process. It should be noted that where these reactions can be achieved intramolecularly the result will be carbocyclization, often a desired outcome in synthesis (cf. Equation (93)).[105] In a similar vein, the use of iodomethylaluminum reagents permits the direct conversion of alkenes or alkynes into the corresponding cyclopropane or cyclopropene derivatives through a sequence of carbalumination and diethylaluminum iodide elimination (Scheme 50).[78,106]

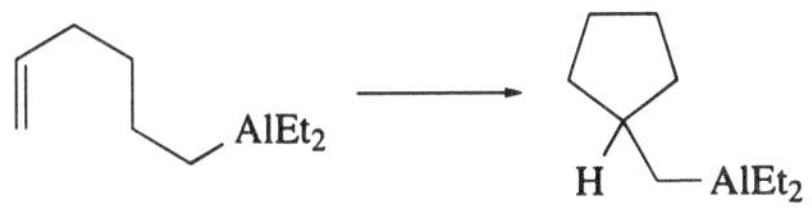

(93)

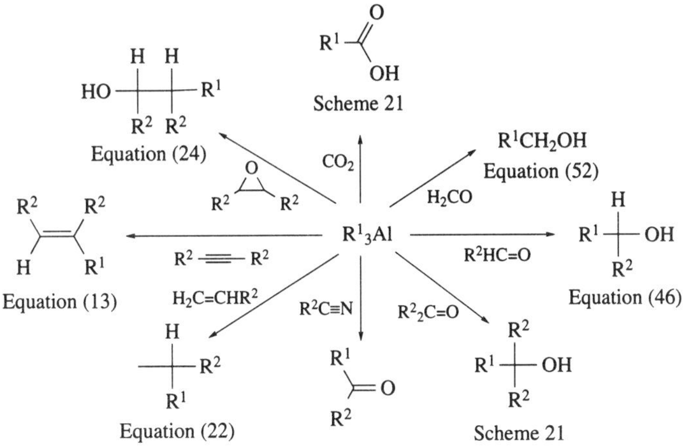

Scheme 49

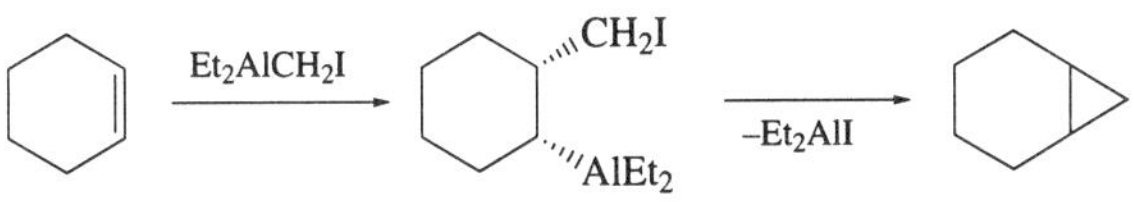

Scheme 50

6.10.2 Carbodealuminations

The exchange of aluminum for carbon can be effected by the use of organic halides, reactive ethers, or cyanogen. Although the halogen–alkyl exchange reaction (Equation (94)) might seem to be a straightforward way of making carbon–carbon bonds, it is often dominated by side reactions, such as hydride reduction of R^3X to R^3H, dehydrohalogenation of R^3X, and competitive coupling to yield R^1–R^3 as well. A reminder of the violent reaction between some polyhalides and aluminum alkyls is in order here.[107] Generally, only when R^1 and R^2 have different electronegativities (i.e., R^1 = alkyl and R^2 = 1- or 2-alkenyl) and R^3 is allylic, propargylic or simple n-alkyl, do such reactions give useful yields of R^2–R^3. Furthermore, formation of the aluminate of $R^1_2AlR^2$ is often necessary to promote reaction. Some examples are given in Equations (95)[108] and (96)[109] and Scheme 51.[110,111]

$$R^1_2AlR^2 + R^3X \longrightarrow R^2\text{-}R^3 + R^1_2AlX \qquad (94)$$

$$(Bu^nC{\equiv}C)_3Al + Bu^tCl \longrightarrow Bu^nC{\equiv}C\text{-}Bu^t + (Bu^nC{\equiv}C)_2AlCl \qquad (95)$$

$$Me_3Al + Ph_3CCl \longrightarrow Ph_3CMe + Me_2AlCl \qquad (96)$$

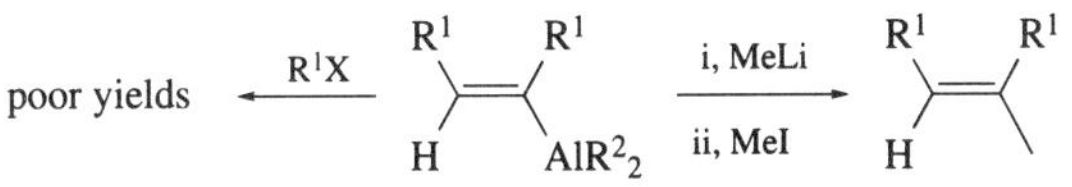

Scheme 51

That ordinary ethers are cleaved by organoaluminum halides is not surprising, since aluminum halides themselves cleave ethers. Cleavage of ethers by aluminum alkyls, on the other hand, usually requires higher temperatures where it appears that R_2AlH is the active agent (Scheme 52).[112] In a similar way, orthoesters, acetals, and ketals are reduced to ethers by Bu^i_2AlH (cf. Equations (30) and (71) and Scheme 37).

Carbodealumination, however, is observed with more reactive aluminum reagents and labile ethers, such as orthoesters. Allyl- or propargylaluminum bromides convert ethyl orthoformate into the corresponding diethyl acetals (Equation (97)).[113]

$$Bu^t_3Al \cdot OEt_2 \xrightarrow[-C_4H_8]{\Delta} Bu^t_2AlH \cdot OEt_2 \xrightarrow{-EtH} Bu^t_2AlOEt$$

Scheme 52

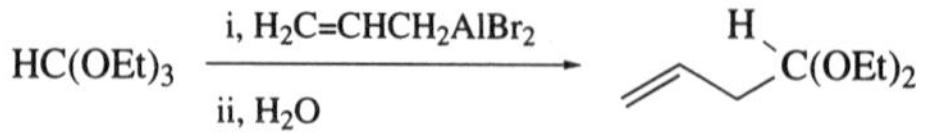

$$HC(OEt)_3 \xrightarrow[\text{ii, } H_2O]{\text{i, } H_2C=CHCH_2AlBr_2} \text{CH}_2=\text{CHCH}_2\text{CH(OEt)}_2 \quad (97)$$

Cyanogen, $(CN)_2$, reacts very selectively with the vinyl–aluminum bond in a 1-alkenyl-(dialkyl)aluminum compound to give excellent yields of the 1-alkenyl nitrile. The stereochemistry of the original C–Al bond is retained in the nitrile product (Equation (98)).[114]

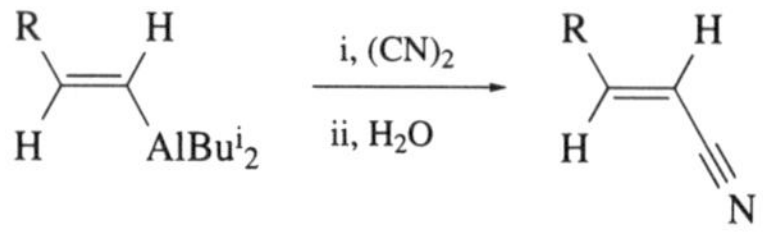

$$\underset{H}{\overset{R}{>}}=\underset{AlBu^i_2}{\overset{H}{<}} \xrightarrow[\text{ii, } H_2O]{\text{i, } (CN)_2} \underset{H}{\overset{R}{>}}=\underset{CN}{\overset{H}{<}} \quad (98)$$

Carbodealuminations with geminal dialuminoalkanes or with Tebbe's reagent ($Cp_2Ti(Cl)CH_2AlMe_2$; Equation (48)) lead directly to a C=C bond. This reaction is a useful alternative for the Wittig methylenation reaction with $R_3P=CH_2$ reagents (Equation (99)) and can provide access to internal alkenes as well (Scheme 53).[77]

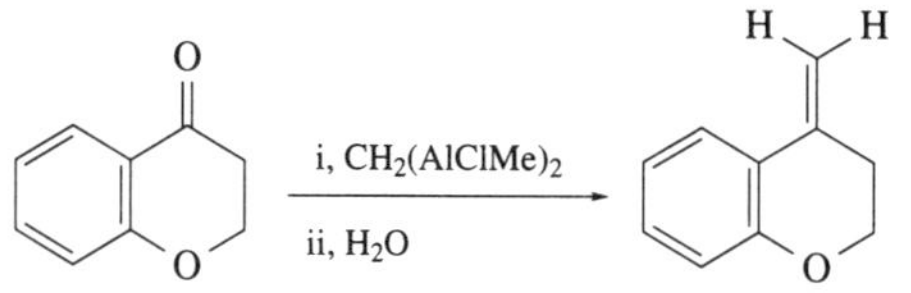

$$\text{(chroman-4-one)} \xrightarrow[\text{ii, } H_2O]{\text{i, } CH_2(AlClMe)_2} \text{(4-methylenechroman)} \quad (99)$$

$$Bu^n-\!\!\equiv\!\!-H \xrightarrow{2\ Cl_2AlH \cdot THF} Bu^nCH_2CH(AlCl_2)_2 \xrightarrow{Ph_2C=O} Ph_2C=CHCH_2Bu^n$$

Scheme 53

Finally, the homo- and cross-coupling of vinylaluminum compounds can be promoted by various transition metal salts of copper, nickel, or palladium (Equations (100) and (101)).[115,116]

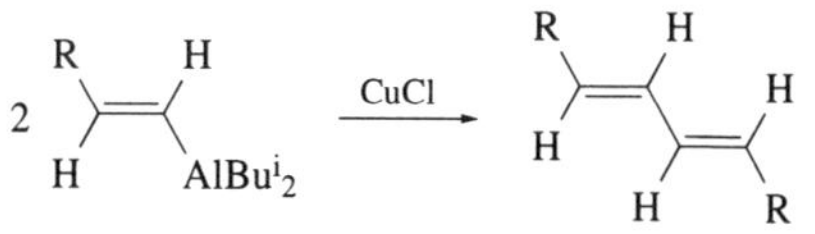

$$2\ \underset{H}{\overset{R}{>}}=\underset{AlBu^i_2}{\overset{H}{<}} \xrightarrow{CuCl} \underset{H}{\overset{R}{>}}=\underset{H}{\overset{H}{<}}-\underset{H}{\overset{H}{>}}=\underset{R}{\overset{H}{<}} \quad (100)$$

$$\underset{H}{\overset{R^1}{>}}=\underset{AlBu^i_2}{\overset{H}{<}} + \underset{H}{\overset{R^2}{>}}=\underset{Br}{\overset{H}{<}} \xrightarrow{Pd} \underset{H}{\overset{R^1}{>}}=\underset{H}{\overset{H}{<}}-\underset{H}{\overset{H}{>}}=\underset{R^2}{\overset{H}{<}} \quad (101)$$

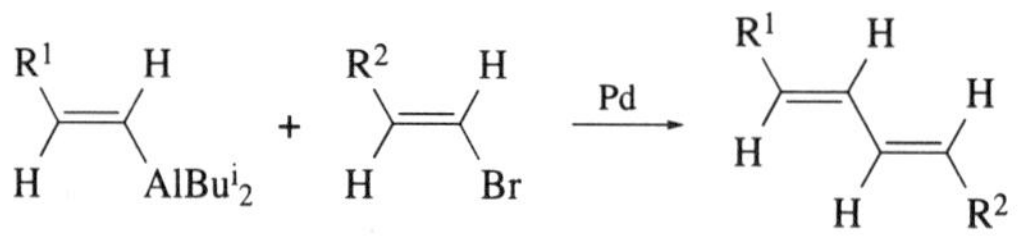

6.11 TRANSITION METAL-PROMOTED REACTIONS OF ALUMINUM REAGENTS

The multifarious ways by which transition metal compounds can interact with and change the nature of organoaluminum reagents are explored in detail in Chapter 10, Volume 1. The principal significance of such interactions in that discussion lies in the Ziegler–Natta process for the polymerization of α-alkenes and conjugated dienes and in the Ziegler nickel effect, which forms the basis for the Wilke alkene dimerization process and the Keim alkene oligomerization or SHOP technology.

The pertinence of transition metal-promoted organoaluminum reactions in this chapter is an appreciation of how such interactions can foster the selective formation of C–H and C–C bonds in organic synthesis. Transition metal salts, chiefly of titanium, zirconium, vanadium, cobalt, nickel, and palladium, have been found to catalyze and/or stoichiometrically promote both the hydrometallation and carbometallation of unsaturated organic substrates. In Section 6.10.2, such catalysis or promotion was likewise shown to be applicable to carbodealumination as well (Equations (100) and (101)).

For the transition metal-catalyzed process, the overall addition of an H–Al or C–Al bond must occur by way of a transition metal intermediate, for example (**54**), that is steadily regenerated or "turned over" from the catalyst (Scheme 54). In the nickel-catalyzed hydroalumination of alkenes or alkynes, intermediate (**54**) is most probably a nickel(0) complex but it is not yet certain whether (**55**) or (**56**) is the crucial catalytic carrier.

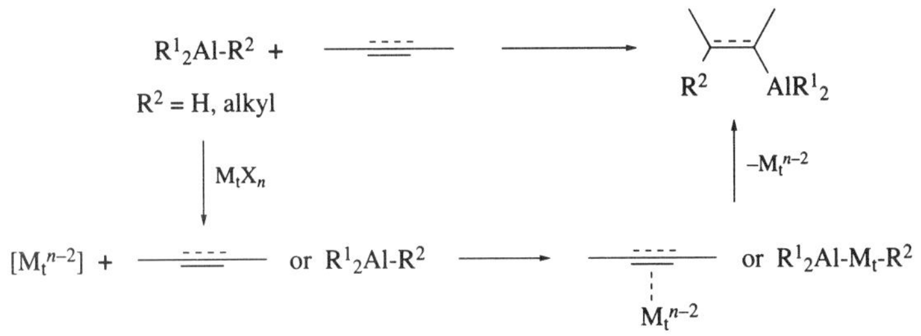

Scheme 54

For stoichiometrically promoted reactions, the aluminum compound is completely converted into the operative transition metal reagent. The formation of the Tebbe reagent (Equation (48)) is illustrative.[40] Furthermore, the titanocene dichloride-promoted addition of $RAlCl_2$ to alkynes has been shown to proceed via the formation of the ion pair, alkyltitanocenium tetrachloroaluminate (**57**) (Equation (102)).[39]

$$RAlCl_2 + Cp_2TiCl_2 \xrightarrow{CH_2Cl_2} [Cp_2Ti\text{-}R]^+ [AlCl_4]^- \qquad (102)$$
$$(\mathbf{57})$$

It is evident from the foregoing examples how greatly transition metal promotion extends the scope of organoaluminum reagents for organic synthesis. The relative slowness with which H–Al and C–Al bonds undergo insertions of the unsaturated substrates R–C≡E, and the tendency of only one C–Al bond of R_3Al to undergo such insertions, place serious limitations on the utility of pure aluminum reagents for synthesis. For this reason, the transition metal-promoted processes already discovered have proved most valuable in raising organoaluminum chemistry to such high prominence in both industrial processes and in modern organic synthesis. Further discoveries of transition metal catalysis or promotion can only enhance the versatility of aluminum reagents for selective organic transformations.

6.12 REFERENCES

1. E. Krause and A. von Grosse, 'Die Chemie der Metallorganischen Verbindungen', Gebrüder Borntraeger, Berlin, 1937, pp. 61–8.
2. V. Grignard, *Compt. Rend.*, 1900, **130**, 1322.
3. V. Grignard, *Ann. Chim. Phys.*, 1901, **24**, 454.
4. K. Ziegler and K. Bähr, *Chem. Ber.*, 1928, **61**, 253.
5. R. G. Jones and H. Gilman, in 'Organic Reactions', ed. R. Adams, Wiley, New York, 1951, vol. 6, pp. 339–66.
6. H. Gilman and J. W. Morton, Jr., in 'Organic Reactions', ed. R. Adams, Wiley, New York, 1954, vol. 8, pp. 258–304.
7. G. Wittig, U. Pockels and H. Dröge, *Chem. Ber.*, 1938, **71**, 1903.
8. K. Ziegler *et al.*, *Angew. Chem.*, 1952, **64**, 323.
9. C. R. Noller, *J. Am. Chem. Soc.*, 1929, **51**, 594.
10. H. E. Simmons, T. L. Cairns, S. A. Vladuchick and C. M. Hoiness, in 'Organic Reactions', ed. R. Adams, Wiley, New York, 1973, vol. 20, p.1.
11. J. Furukawa and N. Kawabata, *Adv. Organomet. Chem.*, 1974, **12**, 83.
12. J. J. Eisch and A. Piotrowski, *Tetrahedron Lett.*, 1983, **24**, 2043.
13. H. Gilman and R. H. Kirby, *J. Am. Chem. Soc.*, 1941, **63**, 2046.
14. H. Meerwein, G. Hinz, H. Majert and H. Sönke, *J. Prakt. Chem.*, 1937, **147**, 226.
15. K. Ziegler, H.-G. Gellert, H. Martin, K. Nagel and J. Schneider, *Liebigs Ann. Chem.*, 1954, **589**, 91.
16. K. Ziegler *et al.*, *Liebigs Ann. Chem.*, 1960, **629**, 121.
17. H. Lehmkuhl and K. Ziegler, in 'Methoden der Organischen Chemie', ed. E. Müller, Thieme, Stuttgart, 1970, vol. XIII/4: Metallorganische Verbindungen, pp. 1–314.
18. K. Ziegler, K. Schneider and J. Schneider, *Liebigs Ann. Chem.*, 1959, **623**, 9.
19. K. Ziegler *et al.*, *Liebigs Ann. Chem.*, 1960, **629**, 121.
20. K. Ziegler, H. Martin and F. Krupp, *Liebigs Ann. Chem.*, 1960, **629**, 14.
21. G. Wilke and H. Müller, *Liebigs Ann. Chem.*, 1960, **629**, 222.
22. G. Wilke and H. Müller, *Chem. Ber.*, 1956, **89**, 444.
23. J. J. Eisch and M. W. Foxton, *J. Org. Chem.*, 1971, **36**, 3520.
24. J. J. Eisch, H. Gopal and S.-G. Rhee, *J. Org. Chem.*, 1975, **40**, 2064.

25. K. Ziegler, *Brennst. Chem.*, 1952, **33**, 193.
26. J. J. Eisch and C. K. Hordis, *J. Am. Chem. Soc.*, 1971, **93**, 2974.
27. J. J. Eisch and C. K. Hordis, *J. Am. Chem. Soc.*, 1971, **93**, 4496.
28. J. J. Eisch and K. C. Fichter, *J. Organomet. Chem.*, 1983, **250**, 63.
29. J. J. Eisch and M. W. Foxton, *J. Organomet. Chem.*, 1968, **11**, P7.
30. J. J. Eisch and S. G. Rhee, *J. Am. Chem. Soc.*, 1975, **97**, 4673.
31. K. Ziegler, *Brennst. Chem.*, 1954, **35**, 321.
32. G. Wilke, *Angew. Chem.*, 1975, **87**, 805.
33. J. J. Eisch, *J. Chem. Educ.*, 1983, **60**, 1009.
34. J. J. Eisch, in 'COMC-I', vol. 1, p. 555.
35. J. J. Eisch, *J. Am. Chem. Soc.*, 1962, **84**, 3605.
36. J. J. Eisch, R. J. Manfre and D. J. Komar, *J. Organomet. Chem.*, 1978, **159**, C13.
37. W. P. Neumann and H. Niermann, *Liebigs Ann. Chem.*, 1962, **653**, 164.
38. G. Wittig and G. Keicher, *Naturwissenschaften*, 1947, **34**, 216.
39. J. J. Eisch, A. M. Piotrowski, S. K. Brownstein, E. J. Gabe and F. L. Lee, *J. Am. Chem. Soc.*, 1985, **107**, 7219.
40. F. N. Tebbe, G. W. Parshall and G. S. Reddy, *J. Am. Chem. Soc.*, 1978, **100**, 3611.
41. J. R. Zietz, Jr., G. C. Robinson and K. L. Lindsay, in 'COMC-I', vol. 7, p. 365.
42. J. J. Eisch, in 'Comprehensive Organic Synthesis', eds. B. M. Trost and I. Fleming, Pergamon, Oxford, 1991, vol. 8, p. 733.
43. G. Zweifel and J. A. Miller, in 'Organic Reactions', ed. W. G. Dauben, Wiley, New York, 1984, vol. 32, Chap. 2, p. 377.
44. S. Pasynkiewicz, *Polyhedron*, 1990, **9**, 429.
45. K. Ziegler, F. Krupp and K. Zosel, *Angew. Chem.*, 1955, **67**, 425.
46. J. J. Eisch and S. J. Y. Liu, *J. Organomet. Chem.*, 1970, **21**, 285.
47. H. Lehmkuhl and K. Ziegler, in 'Methoden der Organischen Chemie', ed. E. Müller, Thieme, Stuttgart, 1970, vol. XIII/4: Metallorganische Verbindungen, p. 246.
48. J. Fried et al., *J. Chem. Soc., Chem. Commun.*, 1968, 634.
49. J. J. Eisch, S. R. Sexsmith and K. C. Fichter, *J. Organomet. Chem.*, 1990, **382**, 273.
50. J. J. Eisch and R. J. Manfre, in 'Fundamental Research in Homogeneous Catalysis', Plenum, New York, 1979, vol. 3, p. 397.
51. L. A. Zakharkin and I. M. Khorlina, *Dokl. Akad. Nauk SSSR*, 1957, **116**, 422 (*Chem. Abstr.*, 1958, **52**, 8040).
52. P. Binger, *Angew. Chem.*, 1963, **75**, 918.
53. A. A. Volkov, S. S. Zlotskii, E. Kh. Kravetz, L. V. Spirikhin and D. L. Rakhmankulov, *Khim. Geterotsikl. Soedin.*, 1985, **8**, 1036.
54. H. Reinheckel, K. Haage and D. Jahnke, *Organomet. Chem. Rev. A*, 1969, **4**, 47.
55. K. Weyer, Doctoral Dissertation, Technische Hochschule, Aachen, Germany, 1956.
56. P. W. K. Flanagan, *Ger. Pat.* 1 124 033 (1962) (*Chem. Abstr.*, 1962, **57**, 11 023.)
57. G. D. Brindell and D. W. Marshall, *US Pat.* 3 304 317 (1967) (*Chem. Abstr.*, 1967, **66**, 115 788).
58. O. Scherer and K. Uhl, *Ger. Pat.* 1 179 194 (1964) (*Chem. Abstr.*, 1965, **62**, 453).
59. M. A. Petrie, M. M. Olmstead and P. P. Power, *J. Am. Chem. Soc.*, 1991, **113**, 8704.
60. J. J. Eisch and R. Sanchez, *J. Org. Chem.*, 1986, **51**, 1848.
61. W. Nagata and M. Yoshioka, *Tetrahedron Lett.*, 1966, 1913.
62. R. B. Woodward et al., *J. Am. Chem. Soc.*, 1966, **88**, 852.
63. K. Maruoka, H. Imoto, S. Saito and H. Yamamoto, *J. Am. Chem. Soc.*, 1994, **116**, 4131.
64. J. J. Eisch, S. I. Pombrik, S. Gürtzgen, R. Rieger and W. Uzick, 'Proceedings of the International Symposium on Catalyst Design for Tailor-Made Polymers', Hokuriku, Japan, March 10–12, 1994.
65. J. J. Eisch and K. C. Fichter, *J. Am. Chem. Soc.*, 1974, **96**, 6815.
66. J. J. Eisch and S.-G. Rhee, *Liebigs Ann. Chem.*, 1975, 565.
67. J. J. Eisch and S.-G. Rhee, *J. Am. Chem. Soc.*, 1974, **96**, 6815.
68. H. Lehmkuhl, *Liebigs Ann. Chem.*, 1968, **719**, 40.
69. J. J. Eisch and G. R. Husk, *J. Org. Chem.*, 1966, **31**, 3419.
70. J. J. Eisch and G. Gupta, *J. Organomet. Chem.*, 1979, **168**, 139.
71. Y. Baba, *Bull. Chem. Soc. Jpn.*, 1968, **41**, 928.
72. G. Queguiner, G. Joly and P. Pastour, *C. R. Hebd. Seances Acad. Sci.*, 1966, **263**, 307.
73. G. Natta, P. Pino, G. Mazzanti, P. Longi and F. Bernardini, *J. Am. Chem. Soc.*, 1959, **81**, 2561.
74. J. J. Eisch and G. R. Husk, *J. Org. Chem.*, 1964, **29**, 254.
75. J. J. Eisch and R. Amtmann, *J. Org. Chem.*, 1972, **37**, 3410.
76. R. Köster and G. Bruno, *Liebigs Ann. Chem.*, 1960, **629**, 89.
77. A. M. Piotrowski, D. B. Malpass, M. P. Boleslawski and J. J. Eisch, *J. Org. Chem.*, 1988, **53**, 2829.
78. H. Hoberg, *Liebigs Ann. Chem.*, 1962, **656**, 1.
79. K. Maruoka, A. B. Concepcion, N. Hirayama and H. Yamamoto, *J. Am. Chem. Soc.*, 1990, **112**, 7422.
80. J. J. Eisch, 'Organometallic Syntheses', Academic Press, New York, 1981, vol. 2, p. 194.
81. G. Wilke and H. Müller, *Liebigs Ann. Chem.*, 1958, **618**, 267.
82. J. J. Eisch, Z. R. Liu and M. Singh, *J. Org. Chem.*, 1992, **57**, 1618.
83. L. I. Zakharkin and I. M. Khorlina, *Tetrahedron Lett.*, 1962, 619.
84. K. Ziegler, F. Krupp and K. Zosel, *Liebigs Ann. Chem.*, 1960, **629**, 241.
85. A. Yasuda, S. Tanaka, K. Oshima, H. Yamamoto and H. Nozaki, *J. Am. Chem. Soc.*, 1974, **90**, 6513.
86. J. J. Eisch, Z. R. Liu and M. P. Boleslawski, *J. Org. Chem.*, 1992, **57**, 2143.
87. L. I. Zakharkin and I. M. Khorlina, *Izv. Akad. Nauk SSSR, Otd. Khim. Nauk*, 1959, 2255 (*Chem. Abstr.*, 1960, **54**, 10 837h).
88. G. Zweifel and R. B. Steele, *J. Am. Chem. Soc.*, 1967, **89**, 2754.
89. G. W. Kabalka and R. J. Newton, Jr., *J. Organomet. Chem.*, 1978, **156**, 65.
90. R. Köster and Y. Morita, *Liebigs Ann. Chem.*, 1967, **704**, 70.
91. K. Ziegler, F. Krupp, K. Weyer and W. Larbig, *Liebigs Ann. Chem.*, 1960, **629**, 251.
92. H. Reinheckel and D. Jahnke, *Angew. Chem., Int. Ed. Engl.*, 1966, **5**, 903.
93. L. I. Zakharkin and V. V. Gavrilenko, *Izv. Akad. Nauk SSSR, Otd. Khim. Nauk*, 1960, 1391 (*Chem. Abstr.*, 1961, **55**, 361).

94. H. Reinheckel and D. Jahnke, *Chem. Ber.*, 1966, **99**, 23.
95. P. W. K. Flanagan, *Br. Pat.* 940 381 (*Chem. Abstr.*, 1961, **55**, 380).
96. R. P. Hatch and S. M. Weinreb, *J. Org. Chem.*, 1977, **42**, 3960.
97. A. P. Kozikowski and A. Ames, *Tetrahedron*, 1985, **41**, 4821.
98. K. Sasaki, Y. Aso, T. Otsubo and F. Ogura, *Chem. Lett.*, 1989, **4**, 607.
99. H. Yamamoto and H. Nozaki, *Angew. Chem., Int. Ed. Engl.*, 1978, **17**, 169.
100. L. I. Zakharkin and O. Y. Okhlobystin, *Izv. Akad. Nauk SSSR, Otd. Khim. Nauk*, 1959, 1942 (*Chem. Abstr.*, 1960, **54**, 12 696).
101. J. J. Eisch, A. M. Piotrowski, K. I. Han, C. Krüger and Y. H. Tsay, *Organometallics*, 1985, **4**, 224.
102. B. Y. Chung, Y. S. Park, I. S. Cho and B. C. Hyun, *Bull. Korean Chem. Soc.*, 1988, **9**, 269.
103. J. I. Levin, E. Turos and S. M. Weinreb, *Synth. Commun.*, 1982, **12**, 989.
104. J. L. Wood, N. A. Khatri and S. M. Weinreb, *Tetrahedron Lett.*, 1979, **51**, 4907.
105. R. Rienäcker and G. Gofthel, *Angew. Chem.*, 1967, **79**, 862.
106. H. Hoberg, *Liebigs Ann. Chem.*, 1966, **695**, 1.
107. W. H. Thomas, *Ind. Eng. Chem. Prod. Res. Dev.*, 1982, **21**, 120.
108. E. I. Negishi and S. Baba, *J. Am. Chem. Soc.*, 1975, **97**, 7385.
109. R. F. Galiullina, V. N. Pankratova, L. P. Stepovik and A. D. Chernova, *J. Gen. Chem. USSR (Engl. Transl.)*, 1976, 100.
110. J. J. Eisch and G. A. Damaservitz, *J. Org. Chem.*, 1976, **41**, 2214.
111. K. Uchida, K. Utimoto and H. Nozaki, *J. Org. Chem.*, 1976, **41**, 2215.
112. H. Lehmkuhl, *Liebigs Ann. Chem.*, 1968, **719**, 40.
113. L. Groizeleau-Miginiac, *Ann. Chim. (Paris)*, 1961, **6**, 1071 (*Chem. Abstr.*, 1962, **57**, 7289).
114. G. Zweifel, J. T. Snow and C. C. Whitney, *J. Am. Chem. Soc.*, 1968, **90**, 7139.
115. G. Zweifel and R. L. Miller, *J. Am. Chem. Soc.*, 1970, **92**, 6678.
116. S. Baba and E. Negishi, *J. Am. Chem. Soc.*, 1976, **98**, 6729.

7
Silicon

ERNEST W. COLVIN
University of Glasgow, UK

7.1 INTRODUCTION	314
7.1.1 Literature	314
7.2 STRUCTURE AND REACTIVITY	314
7.3 VINYLSILANES	314
7.3.1 Preparation	314
7.3.2 Reactions	315
7.4 ALLYLSILANES	317
7.4.1 Preparation	317
7.4.2 Reactions	317
7.4.2.1 General	317
7.4.2.2 Oxidation	320
7.4.2.3 Hosomi–Sakurai and related reactions	321
7.4.2.4 Bifunctional annulating reagents	323
7.5 ALKYNYL- AND PROPARGYLSILANES	324
7.6 ARYL- AND HETEROARYLSILANES	326
7.7 α,β-EPOXYSILANES	326
7.8 PETERSON ALKENATION	327
7.9 KETOSILANES	328
7.9.1 Acylsilanes	328
7.9.2 α-Silylcarbonyl Compounds (β-Ketosilanes)	329
7.10 ALKYL SILYL ETHERS	329
7.11 SILICON TETHERING	330
7.12 SILYL ENOL ETHERS	333
7.12.1 Preparation	333
7.12.2 Lewis Acid Induced Reactions	333
7.12.3 Conjugate Addition	335
7.12.4 Cycloaddition Reactions	335
7.12.5 Sigmatropic Rearrangement Reactions	336
7.12.6 α-Oxidation and Related Reactions	336
7.12.7 Heterosubstituted Silyl Enol Ethers	337
7.12.8 Reactivity and Miscellaneous Reactions	337
7.13 TRIALKYLSILYL-X AND RELATED REAGENTS	338
7.13.1 Trialkylsilyl Halides	338
7.13.2 Trialkylsilyl Cyanides	339
7.13.3 Trialkylsilyl Trifluoromethanesulfonates (Triflates)	340
7.13.4 Trialkylsilyl Peroxides and Related Species	340
7.13.5 Trialkylsilyl Azides	341
7.13.6 TMS Diazomethane	341
7.13.7 Trialkylsilyl Thiols and Selenols and Related Species	341

7.13.8 Trialkylsilyl Stannanes	342
7.13.9 Miscellaneous	342
7.14 AMINOSILANES AND RELATED COMPOUNDS	342
7.15 SILANES AS REDUCING AGENTS	343
7.15.1 Hydrosilylation	343
7.15.1.1 Intermolecular processes	343
7.15.1.2 Intramolecular processes	343
7.15.1.3 Radical-based reducing agents	344
7.15.1.4 Carbon monoxide incorporation	344
7.15.1.5 Functional transformation	344
7.15.2 Ionic Hydrogenation	344
7.16 ORGANIC SYNTHESES	345
7.17 REFERENCES	346

7.1 INTRODUCTION

The purpose of this chapter is to review the advances made in organosilicon chemistry since the earlier review in *COMC-I*.[1] Throughout, emphasis is placed on reviews published on specific areas.

7.1.1 Literature

Several monographs[2-5] have been published, as have conference proceedings[6-9] and two edited multiauthor works[10,11] devoted entirely or significantly to organosilicon chemistry. Also there have been two[12,13] 'symposia-in-print', a review[14] and a number of annual surveys.[15]

7.2 STRUCTURE AND REACTIVITY

The stereochemistry of nucleophilic substitution at tetracoordinate silicon has been reviewed,[16] as has the reactivity of penta- and hexacoordinate silicon species as reaction intermediates.[17]

The steric influence[18] of the TMS group in organic reactions has been surveyed. The interaction of silicon with positively charged carbon has been reviewed,[19] with discussion of the α-effect (Si–C$^+$), the β-effect (Si–C–C$^+$) and the γ-effect (Si–C–C–C$^+$).

Chlorosilanes have a diminished but significant β-effect when compared with trialkylsilyl species and also a lower leaving group ability.[20]

7.3 VINYLSILANES

7.3.1 Preparation

The preparation and reactions of silyl-substituted conjugated dienes have been reviewed.[21]

An investigation of rhodium-catalysed hydrosilylation of hex-1-yne with triethylsilane has revealed a remarkable dependence on solvent in controlling the stereochemistry of the product vinylsilane;[22] use of [Rh(cod)Cl]$_2$ in EtOH or DMF preferentially produces the (Z)-vinylsilane, whereas use of [Rh(cod)Cl]$_2$–Ph$_3$P in nitrile solvents favours production of the (E)-isomer (Scheme 1).

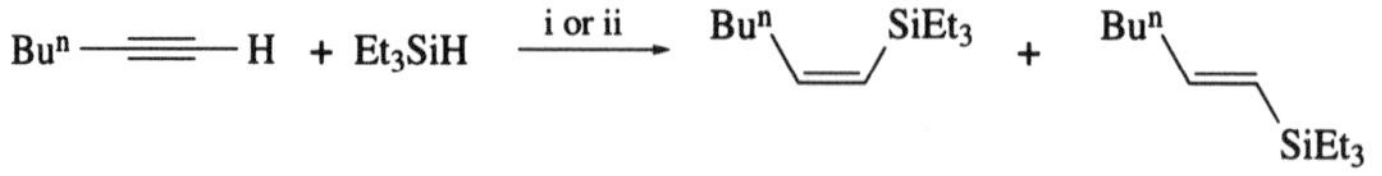

Bu^n—≡—H + Et$_3$SiH →(i or ii) Bu^n\=/SiEt$_3$ + Bu^n\=/ SiEt$_3$

i, [Rh(cod)Cl]$_2$, DMF 97:1
ii, [Rh(cod)Cl]$_2$, PPh$_3$, MeCN 2:97

Scheme 1

Full details have been provided on the palladium-catalysed Heck-type coupling of aryl halides with vinylsilanes;[23] a full equivalent of silver nitrate is present to enhance the reaction rate and also to depress desilylation. If the vinylsilane carries electronegative substituents on silicon, silver(I) is no longer necessary.[24] The stereospecific arylation of vinylsilanes with arylpalladium acetates, in which the product vinylsilanes are formed with inversion of original double bond geometry, has also been fully described.[25] The Suzuki reaction, the palladium(0)-catalysed cross-coupling reaction between vinylborons and aryl halides in the presence of hydroxide ion,[26] has been employed in a stereodefined route to β,β-disubstituted vinylsilanes.[27] Vinyl halides undergo a fluoride ion-induced, palladium(0)-catalysed cross-coupling reaction with disilanes.[28]

α-Hydroxy-γ-oxotrimethylsilanes undergo stereoselective dehydration (Scheme 2) to (Z)- and (E)-γ-oxovinylsilanes, potential precursors for the preparation of stereodefined tetrasubstituted alkenes using the above methodology (ppts, pyridinium p-toluenesulfonate).[29]

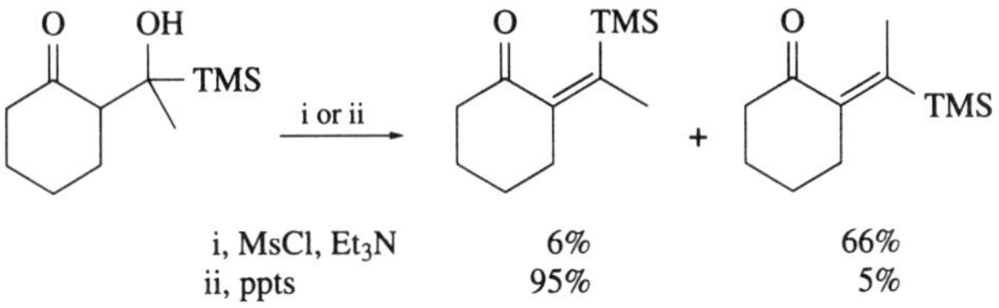

i, MsCl, Et₃N   6%   66%
ii, ppts   95%   5%

Scheme 2

TMS alkynes undergo a palladium-catalysed regio- and stereospecific hydroesterification, providing a convenient route to (E)-β-ethoxycarbonylvinylsilanes.[30]

Vinylsilane phosphates react with organocuprates to produce vinylsilanes;[31] the precursors are available by a variety of routes, and in acyclic cases, where yields are best, the original double bond stereochemistry is retained. Certain functionalized terminal alkynes undergo silyl-cupration followed by cyclization to provide functionalized vinylsilanes;[32] in many cases, however, cyclization is not a particularly favoured pathway.

As with the lithium salt of bis(trimethylsilyl)methane, the corresponding methoxy derivative converts carbonyl compounds into vinylsilanes via the Peterson reaction; additionally, it can be used successfully with enolizable substrates,[33] and adds in a 1,2-manner to cyclohexenone. The process is not stereoselective.

In a general process, which proceeds best with ketones, carbonyl compounds are converted nonstereoselectively into vinylsilanes using the titanium species (1).[34] By contrast, esters are converted into alkoxyvinylsilanes with Z selectivity using the complex reagent system shown (Scheme 3).[35]

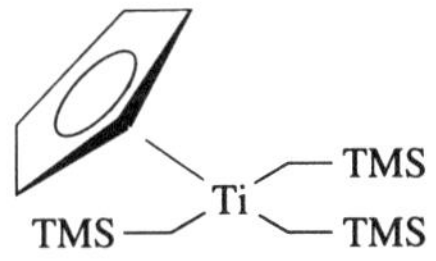

(1)

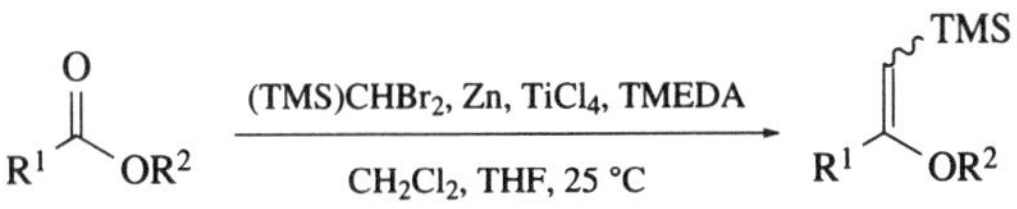

Scheme 3

β-(Silyl)allylic ethers undergo an S_N2'-type reaction with alkyllithiums, providing a stereoselective route to (E)-vinylsilanes.[36] Another stereoselective entry to such silanes (Scheme 4) involves nucleophilic attack on TMS-substituted allyliron cations.[37]

7.3.2 Reactions

Electrophilic substitution reactions of vinylsilanes have been the subject of an excellent and thorough review.[38] The normal *ipso*-substitution controlled by the β-effect is not observed with vinylsilanes substituted at the 1-position; in such cases attack occurs at the terminal position to form a tertiary α-silyl rather than a primary β-silyl cation.[39]

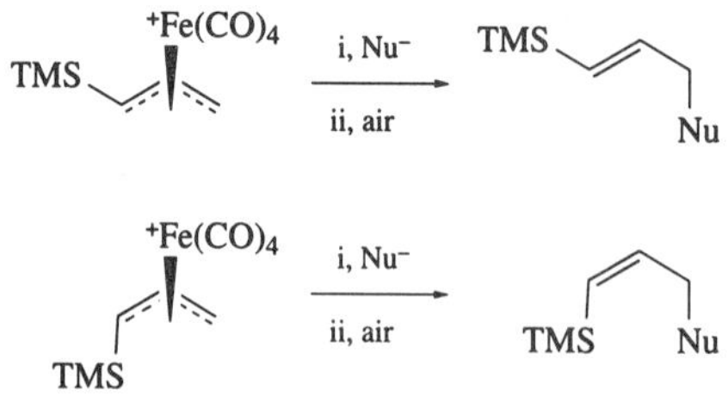

316 *Silicon*

Scheme 4

Further investigation of the reactions of medium-ring vinylsilanes (and epoxysilanes) has revealed the central feature of acid- and acyl-desilylation to be a facile and stereospecific 1,5-desilylative ring closure.[40]

The use of pyridinium bromide perbromide[41] (inversion) and xenon difluoride[42] in the presence of alkali metal halides (variable) for halogenodesilylation has been described.

Manipulation of the geometrically pure dibromovinylsilane (2) has provided a simple methodology for the construction of geometrically pure alkenes.[43]

Electrophilic substitution of α,ω-bis(TMS) conjugated (E)-polyenes has led to a stereocontrolled route to functionalized (E)-polyenes.[44]

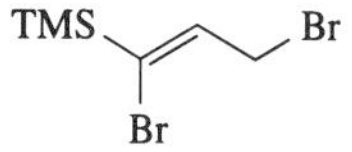

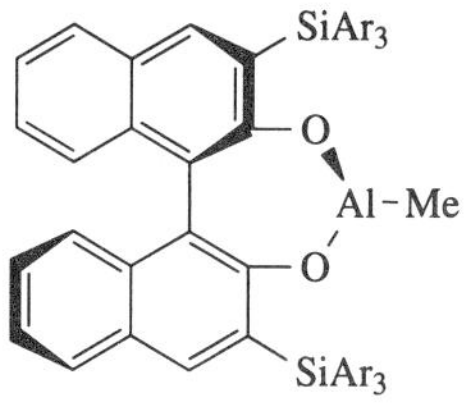

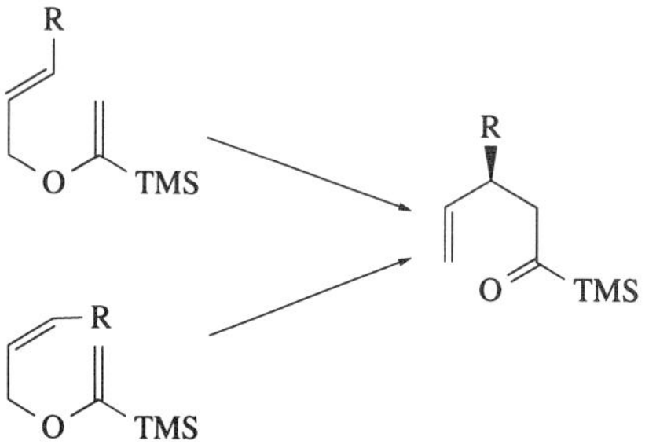

Scheme 5

Both (E)- and (Z)-allylic α-(trimethylsilyl)vinyl ethers undergo an asymmetric Claisen rearrangement in the presence of the binaphthol-derived chiral aluminum reagent (3) to produce optically active acylsilanes with the same absolute configuration (Scheme 5); chair- and boatlike transition states, respectively, have been proposed to account for this interesting observation.[45]

The Lewis acid-promoted carbonyl-ene reactions of vinylsilanes have been reviewed,[46] as have [2,3]-Wittig sigmatropic rearrangements.[47] An application of the latter process can be seen in a stereoselective route[48] to (Z)-disubstituted alkenes (Scheme 6), employed in an approach to leukotrienes.

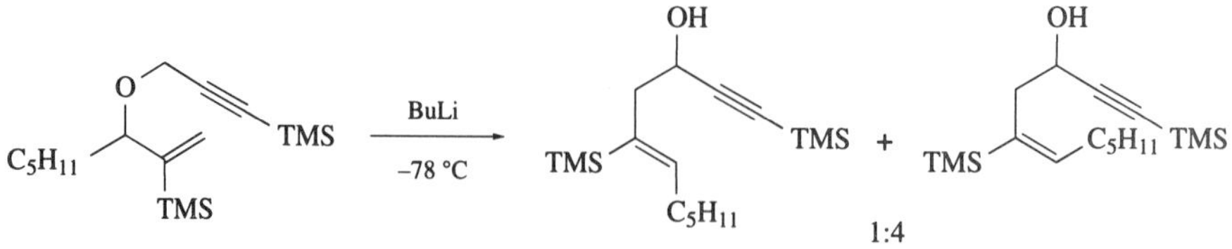

Scheme 6

Sugar-derived vinylsilane-aldehydes undergo stereospecific cyclization under the influence of Lewis acids to produce conduritols and other cyclitols; use of either chelating or nonchelating Lewis acids makes the cyclization stereodivergent.[49] Vinylsilanes have also been used in the stereoselective production[50] of *anti*-1,2-diols.

A synthesis of the α,β-disilyl enone (**4**) has been described,[51] as have some of its reactions as a Michael acceptor; incorporation of the dimethyl(phenyl)silyl group permits oxidative C–Si bond cleavage. α-TMS enones have been employed as radical traps in prostaglandin synthesis,[52] and also in a stereocontrolled synthesis of tetrasubstituted alkenes.[53] γ-Amino-α-triisopropylsilyl enones undergo smooth desilylation[54] when treated with oxalic acid in MeOH.

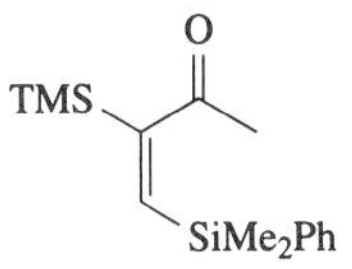

(**4**)

Certain β-silylacrylic esters undergo a stereoselective samarium(II)-induced coupling; further transformations, including oxidative C–Si bond cleavage, have provided a synthesis of (±)-2-deoxyribonolactone.[55]

The phenylselenovinylsilane (**5**) reacts with enones under Lewis acid catalysis to form cyclopropanes;[56] this formal [2 + 1]-cycloaddition, rather than the expected [2 + 2]-cycloaddition, is accompanied by a 1,2-silyl shift (Scheme 7).

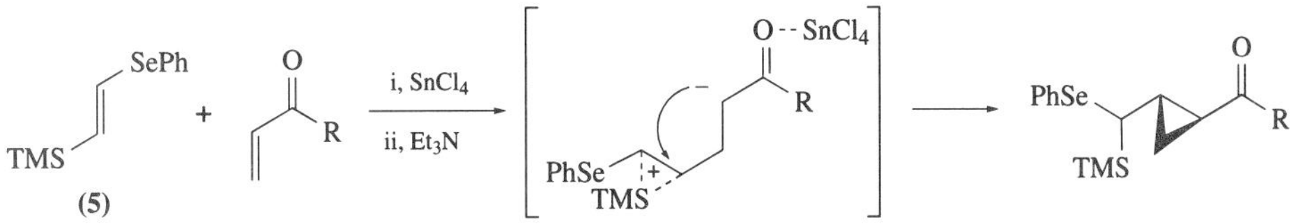

Scheme 7

Vinylsilanes undergo hydroesterification to yield either α-silyl or β-silyl esters in high yield and with high regioselectivity, which is controlled by the choice of catalyst.[57]

7.4 ALLYLSILANES

7.4.1 Preparation

Methods for the synthesis of allylsilanes have been comprehensively reviewed.[58,59] A selection of some more recent syntheses,[60–5] with emphasis on regio- and stereocontrol, is shown in Scheme 8.

7.4.2 Reactions

7.4.2.1 General

Thorough and excellent accounts of electrophilic substitution reactions of allylsilanes have been published.[38,66]

Allylsilanes undergo regio- and stereocontrolled electrophilic attack, with loss of the silyl group from the intermediate cation (**6**) leading to the product of substitution with overall shift of the double bond (Scheme 9). In cases where steric effects are not dominant, such $S_{E'}$ reactions have been shown to proceed with extremely high, if not complete, *anti*-stereoselectivity.[67]

A series of quantitative studies have shown that, in their reactions with carbonium ions, the reactivity order allylstannanes > silyl enol ethers > allylsilanes prevails.[68] Additionally, reactivity varies with substitution at silicon,[69] with, for example, allylphenyldimethylsilane being considerably less reactive than allyltrimethylsilane.

In addition to providing a most useful overview,[70] Fleming has provided full details of the work of his group on the synthesis of allylsilanes from both secondary allylic alcohols[64] and β-silyl enolates.[65] He has also given a full account of his studies on the regiochemistry and stereochemistry of their protonation, epoxidation, acylation[71] and hydroboration,[72] and their reaction with osmium tetroxide.[73]

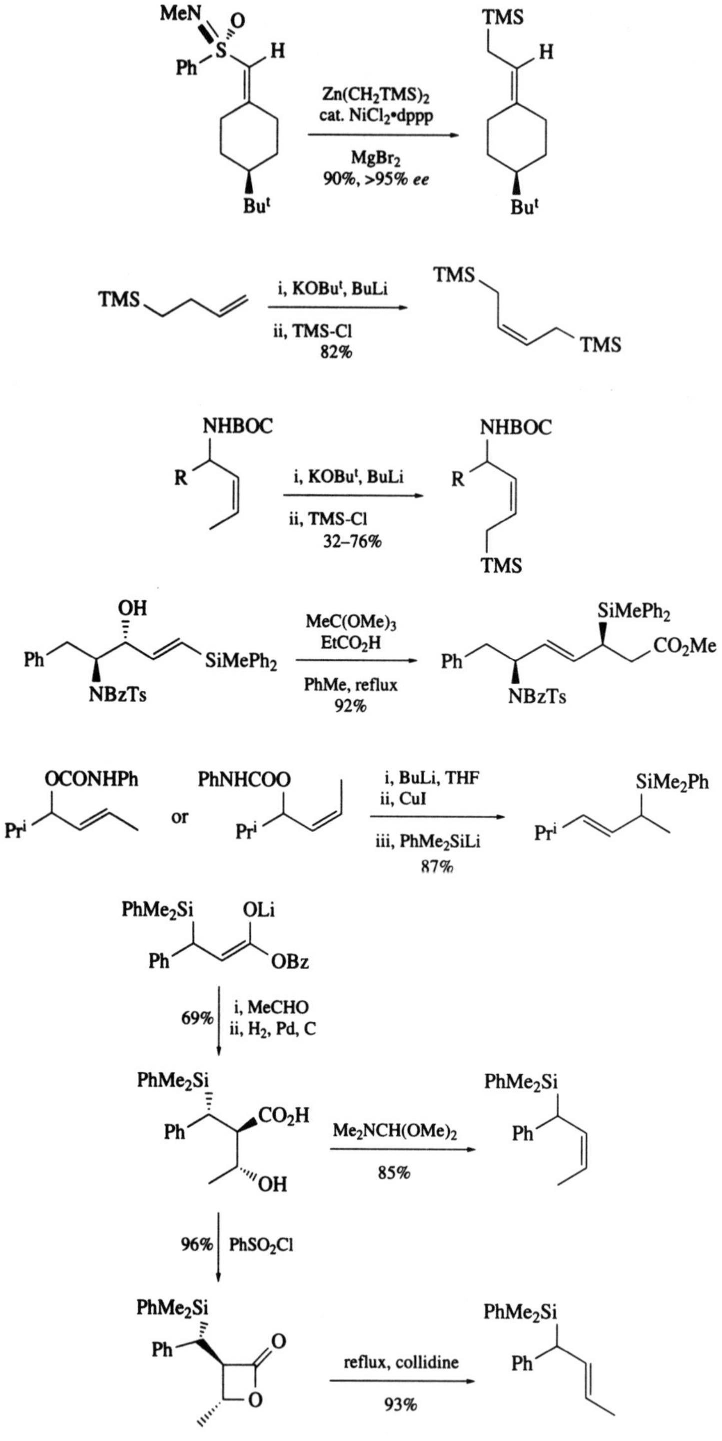

Scheme 8

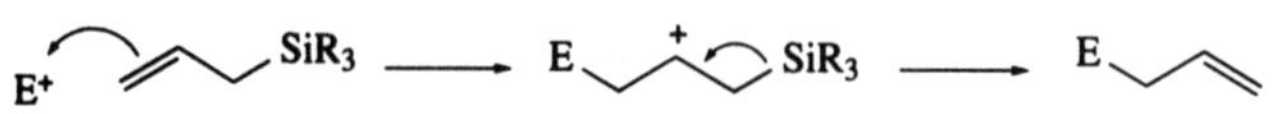

Scheme 9

The *N*-chlorosulfonyl β-lactams formed as intermediates in the regioselective reaction between chlorosulfonyl isocyanate and allylsilanes can be intercepted by aqueous sodium sulfite, allowing the preparation of a range of monocyclic *N*-protio β-lactams[74] (Scheme 10). (Allenylmethyl)silanes react in an analogous manner to yield 3-alkylidene β-lactams;[75] a chirally-enriched substrate transferred its axial chirality to the carbon-centred chirality of the product β-lactam with approximately 60% efficiency.

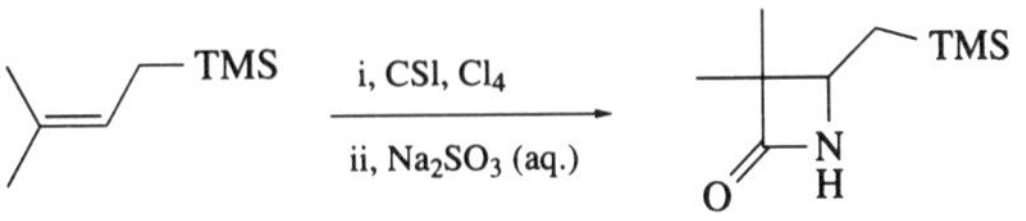

Scheme 10

N-Acyliminium ions[76] react efficiently with allylsilanes to give usefully functionalized products, as exemplified by a route[77] to racemic α-substituted α-amino acids (Scheme 11), and an enantiodivergent synthesis[78] of a key intermediate used in the preparation of the Geissman–Waiss lactone (Scheme 12). N-Acyliminium coupling has also been utilized in a route to enantiopure 4,5-disubstituted pyrrolidin-2-ones.[79]

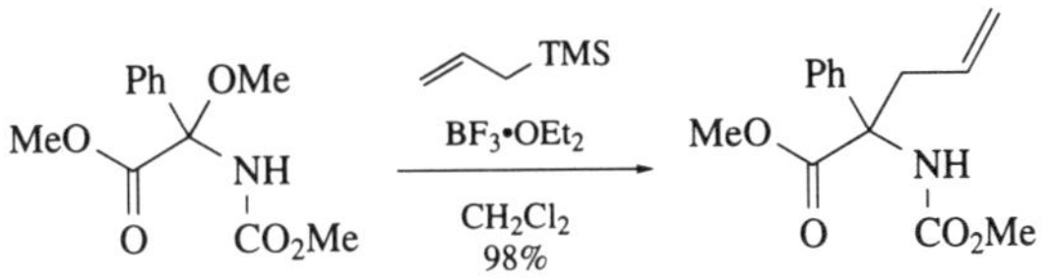

Scheme 11

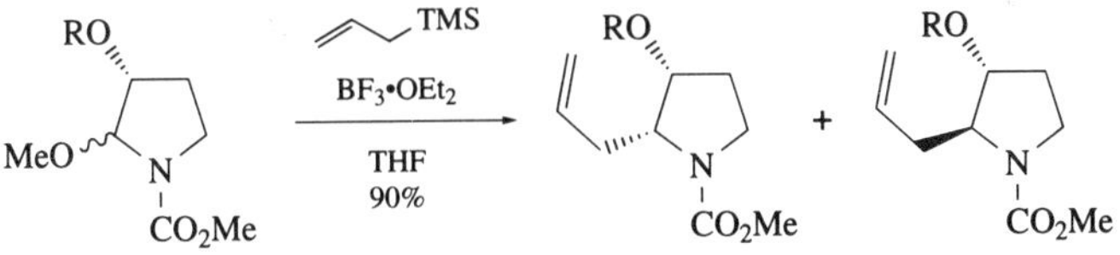

Scheme 12

Photochemically generated iminium ions from tertiary amines can be trapped by allyltrimethylsilane, providing a route for the direct α-allylation of such amines.[80]

Oxocarbenium ions are also good electrophiles towards allylsilanes. In the carbon nucleophile variant of the Ferrier rearrangement using crotylsilanes, Danishefsky and co-workers have shown[81] that while both geometric isomers attack in an axial sense, the geometry of the silane does influence the stereogenic centre of the butenyl group (Scheme 13). 4'-C-Branched nucleosides have also been accessed[82] via this rearrangement.

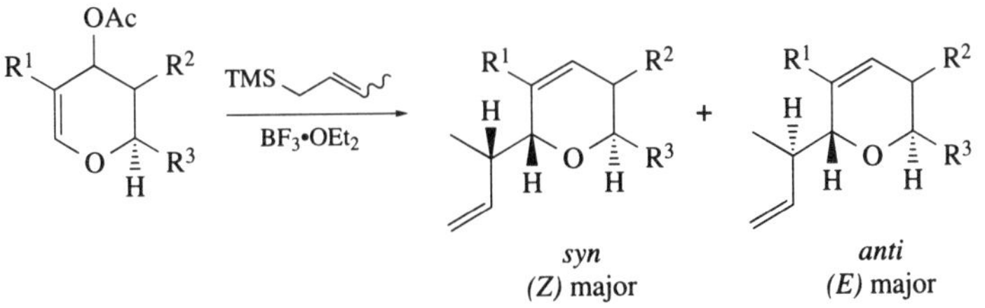

Scheme 13

2-Azido C-glycosyl sugars can be prepared by the Lewis acid mediated reaction[83] of allylsilanes with the nitrate glycosides of 2-azido sugars.

The Lewis acid mediated reaction of allylsilanes with γ-lactols has provided a highly diastereoselective route[84] to disubstituted tetrahydrofuran derivatives.

Denmark and co-workers have shown[85] that, in the silicon-directed Nazarov cyclizations shown in Scheme 14, the direction of conrotatory ring closure is controlled by silicon such that carbon–carbon bond formation occurs in a highly selective *anti* $S_{E'}$ manner.

Allylsilanes undergo a smooth cross-coupling reaction[86] with allylstannanes, particularly when the stannane carries a phenylthio group at the γ-position (Scheme 15).

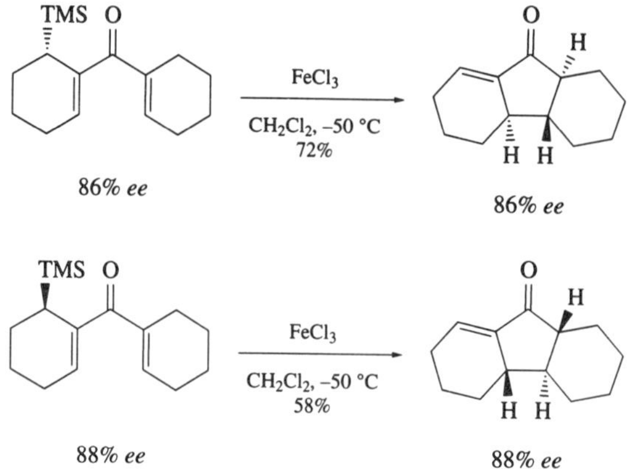

Scheme 14

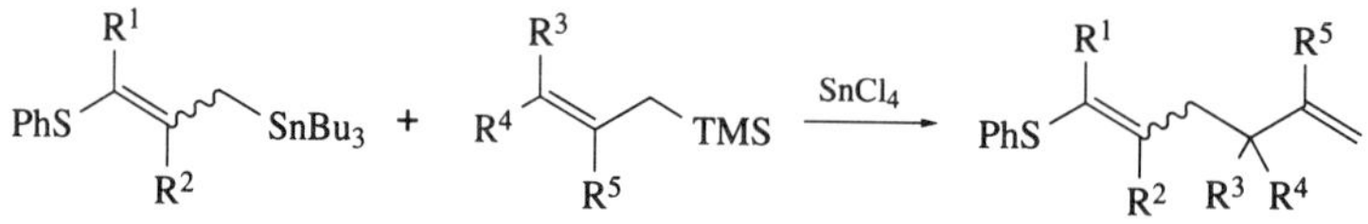

Scheme 15

α-Silylallyl anions in which the silyl moiety contains a chelating group react with carbonyl compounds at the γ-position; in certain solvents, good stereoselectivity can be attained (Scheme 16).[87]

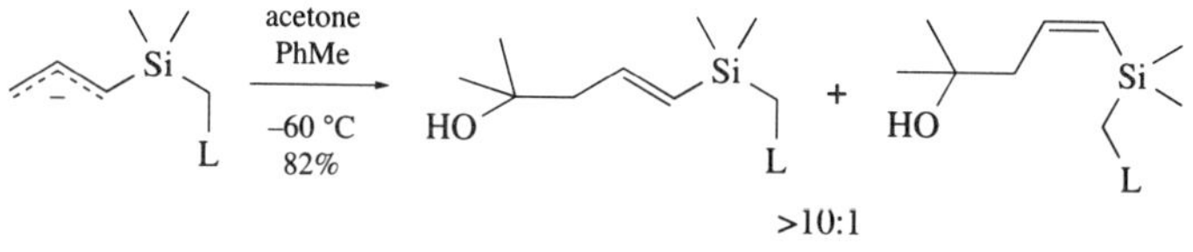

Scheme 16

Allylsilanes undergo a fluoride ion induced thiophilic addition to thioketones,[88] as exemplified in Scheme 17.

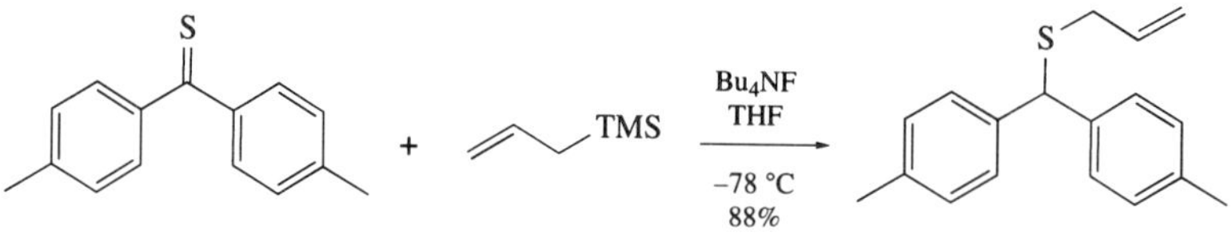

Scheme 17

7.4.2.2 Oxidation

Other workers,[89,90] in addition to Fleming,[73] have reported on the diastereoselectivity of osmylation of allylsilanes. Sharpless asymmetric dihydroxylation[91] (Sharpless AD) of allylsilanes yields optically active diols;[92,93] treatment of these with KH (Peterson alkenation, see Section 7.8) induces *syn* elimination to give allylic alcohols, often in excellent yield and enantiopurity (Scheme 18). The starting allylsilanes were prepared by a new two-step method[93] (Scheme 19). Double asymmetric induction can be used to increase diastereoselectivity[94] in Sharpless AD (Scheme 20).

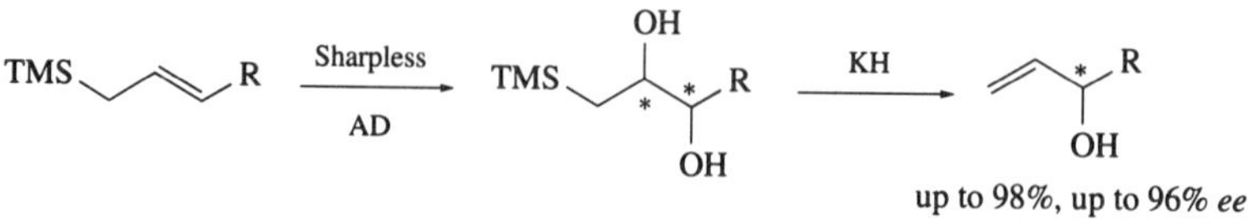

Scheme 18

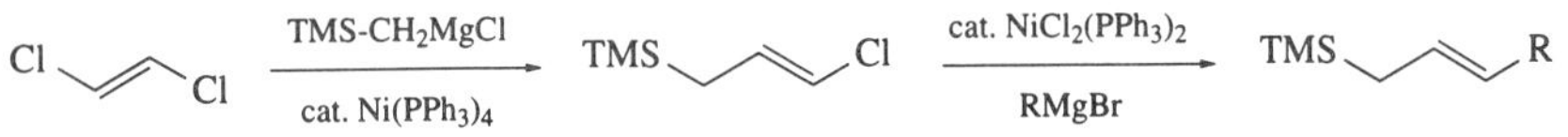

Scheme 19

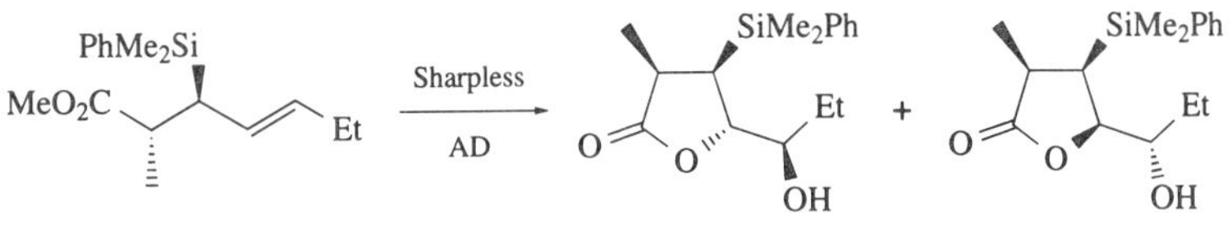

Scheme 20

(Z)-Allylsilanes (**7**) undergo epoxidation with a high degree of *erythro* selectivity (Scheme 21); subsequent treatment with either HF or fluoride ion then provides a good route to *syn*-1,3-diols (e.g., (**8**)).[95]

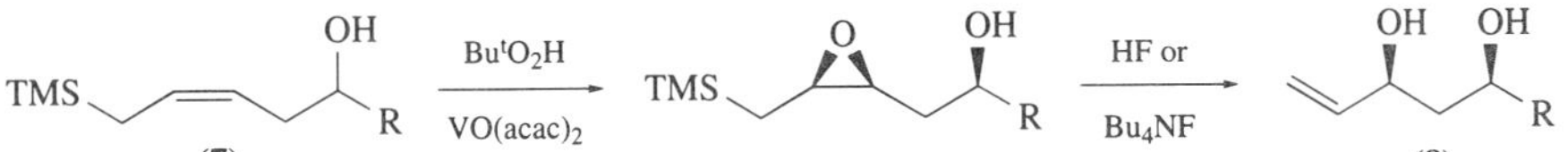

Scheme 21

Hydroboration of allylsilanes has provided a diastereocontrolled synthesis of 1,3,5-triols;[96] interestingly, treatment of the intermediate silyl diol under benzylation conditions resulted in Si–Ph cleavage to regioselectively produce the primary benzyl ether (Scheme 22).

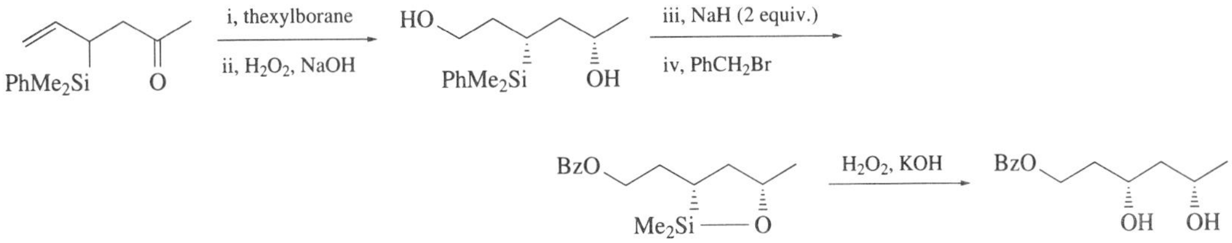

Scheme 22

7.4.2.3 Hosomi–Sakurai and related reactions

The Lewis acid mediated addition of allylsilanes to carbonyl and α,β-unsaturated carbonyl compounds and ketals, both inter- and intramolecularly, is commonly referred to as the Sakurai (or Hosomi–Sakurai) reaction. Its stereochemistry has been reviewed.[97] Sakurai[98] and Hosomi[99] have separately provided detailed accounts of their major contributions to allylsilane chemistry. Schinzer[100] and Majetich[101] have published overviews of their independent work on the synthetic utility of such intramolecular additions in natural product synthesis.

Reaction between allylsilanes, carbonyl compounds and trimethylsilyl ethers obviates the need for prior ketalization; this process can also be performed intramolecularly.[102]

Diastereoselective reactions of chiral crotylsilanes have been extensively investigated, with Panek[103] and his group having made major contributions. For example, racemic vinylsilane (**9**) can be resolved either by lipase promoted transesterification,[104] as shown, or by classical diastereoisomeric mandelate methods.[105] Ireland–Claisen rearrangement of the derived ester provides the optically active and diastereoisomerically pure (*E*)-crotylsilane (**10**), which reacts[106] with achiral α-alkoxy and β-alkoxy acetals with high diastereoselectivity (Scheme 23).

The diastereoselectivity in the reaction of crotylsilanes with acetals has been found[107] to be dependent on both the geometry of the silane and on the nature of the acetal; with acetals derived from aliphatic aldehydes, both (*Z*)- and (*E*)-crotylsilanes give preferentially *erythro/syn* products, whereas those derived from aromatic aldehydes give *erythro/syn* from (*E*)- but *threo/anti* products from (*Z*)-silanes.

The first example of catalytic asymmetric allylation of aldehydes with allylsilanes has been reported, employing a chiral boron catalyst derived from tartaric acid (Scheme 24). The reaction is highly *erythro*-selective, with up to 96% *ee* being attained.[108] The catalytic activity of the superacid system, $TfOH_2^+B(OTf)_4^-$, in the allylation of carbonyl compounds with allylsilanes has been investigated.[109]

Scheme 23

Scheme 24

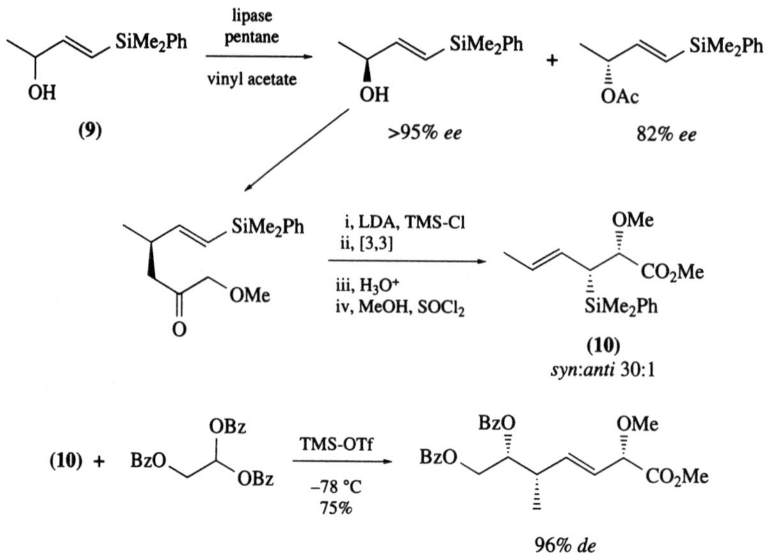

Pentacoordinate allylsilicates have been implicated in the reactions of allyltrifluoro-, allyltrialkoxy- and allyltrichlorosilanes[110] with aldehydes. High yields and high regio- and diastereoselectivities are observed (Scheme 25). With allyltrichlorosilanes,[111] no catalyst is required if a donor solvent such as DMF is employed.

Scheme 25

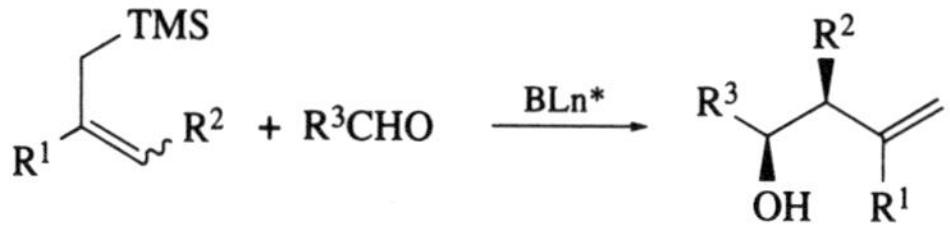

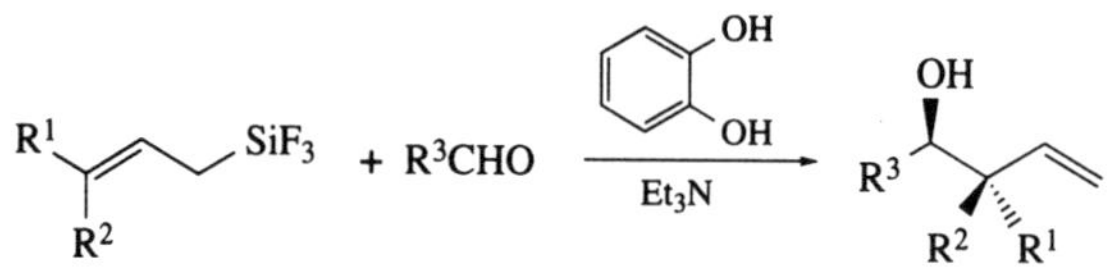

The conjugate addition of allylsilanes to cyclohexenones[112] and cyclopentenones[113] carrying chiral auxiliaries at the 2-position has been investigated. High diastereoselectivities were observed in both cases.

The Lewis acid promoted 1,4-addition of allyltrimethylsilane to chiral α,β-unsaturated *N*-acylamides also proceeds with good diastereoselectivity (Scheme 26).[114] A related ene reaction of such amides with allyltrimethylsilane has been investigated.[115] An ene reaction between an allylsilane and butyn-2-one has provided a new route[116] to (±)-γ-ionone; with the substrate shown (Scheme 27), no 'normal', that is, Sakurai product was formed.

Scheme 26

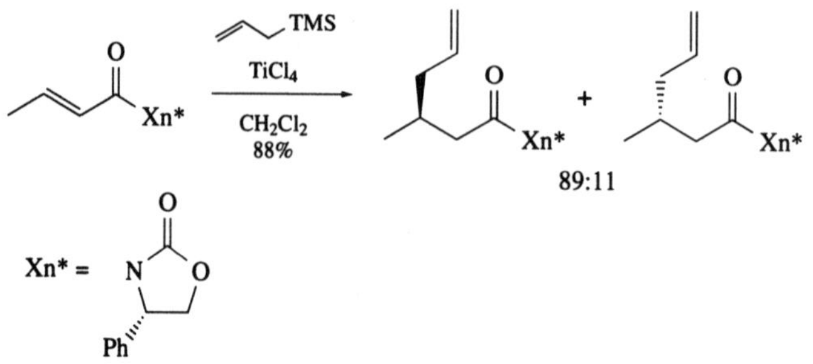

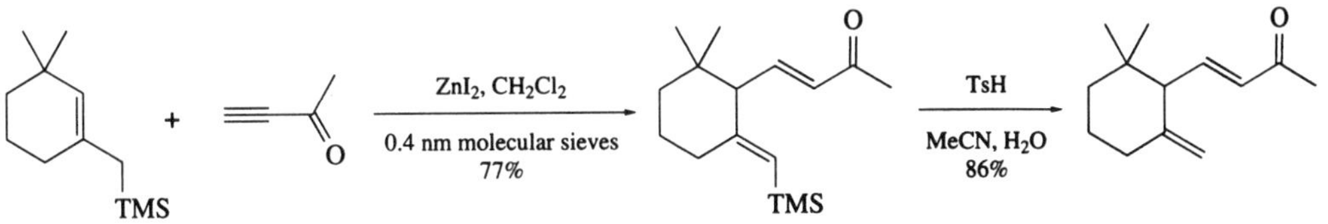

Scheme 27

A tandem intramolecular Sakurai-carbonyl ene process has been employed[117] in a synthesis of steroid derivatives. On the one hand, the allylsilane/enal (11) gave a single cyclized product (Scheme 28); this high selectivity is remarkable, since a mixture of all four geometric isomers was employed. On the other hand, and with bulky silyl substituents, allylsilanes react intermolecularly with cyclic enones to give [3 + 2] cycloadducts (Scheme 29) via a 1,2-silyl shift, with only traces of conjugate addition product being observed.[118]

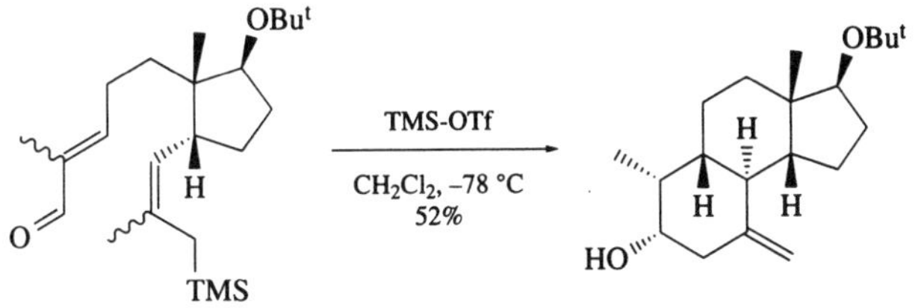

Scheme 28

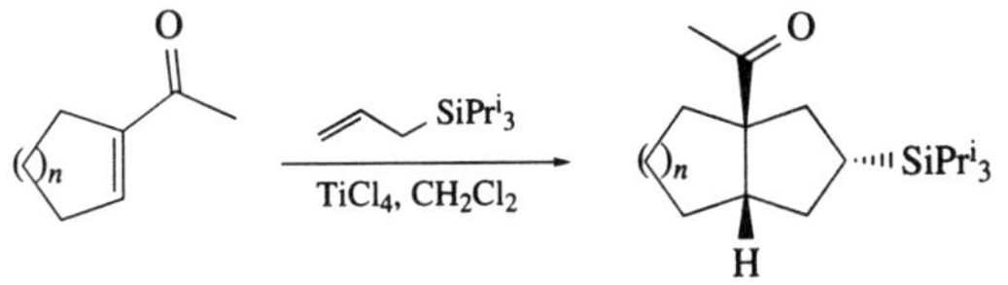

Scheme 29

7.4.2.4 Bifunctional annulating reagents

Trost has reviewed[119] his extensive studies on 1,3-bifunctional conjunctive reagents, which under Pd⁰ catalysis behave as trimethylenemethane equivalents. Two examples, demonstrating such a use of 2-[(trimethylsilyl)methyl]allyl acetate (12) and illustrating a [3 + 2] cycloaddition[120] and the first reported [6 + 3] cycloaddition,[121] respectively, are shown in Scheme 30. Lee, by contrast, concentrated on developing bifunctional acetal-allylsilanes (13) for the synthesis of fused and spirocyclic ring systems,[122] while Molander and Shubert utilized[123] 3-iodo-2-[(trimethylsilyl)methyl]propene (14) for the generation of various carbocyclic systems, as illustrated in Scheme 31.

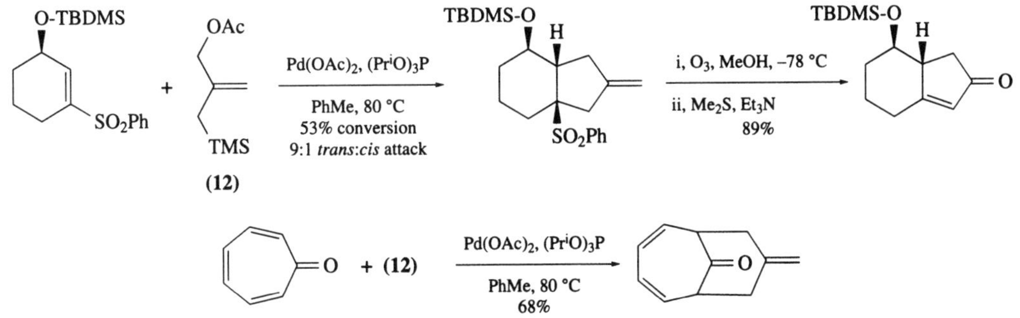

Scheme 30

Other bifunctional allylsilanes which have been used as annulating agents include (15),[124] (16),[125] (17),[126] (18)[127] and (19).[128]

(13)

Scheme 31

(15)

(16)

(17)

(18)

(19)

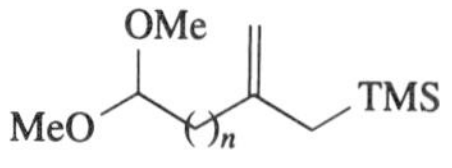

7.5 ALKYNYL- AND PROPARGYLSILANES

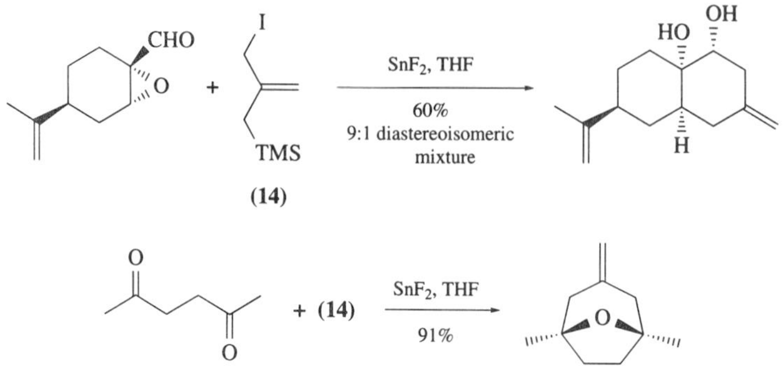

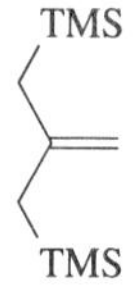

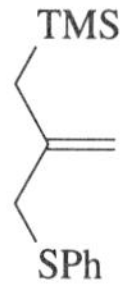

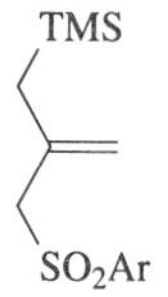

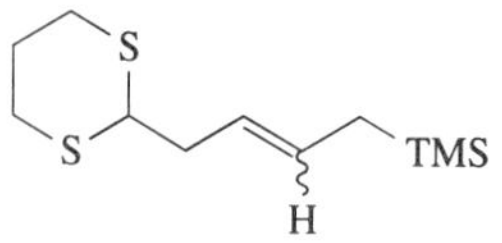

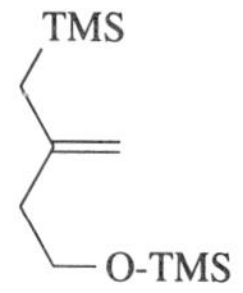

A variety of metal-mediated routes to alkynylsilanes have been described,[129] including stereoselective preparations of (*E*)- and (*Z*)-1,6-bis(TMS)hex-3-ene-1,5-diyne. Several routes to (*Z*)-enynes have been reported.[130] The synthesis and cycloaddition reactions of 1-nitro-2-(trialkylsilyl)alkynes have been described.[131]

Vinyl chloride (**20**) has been used in a synthesis of optically active enediynes[132] related to dynemicin A. TMS alkynes have been employed as nucleophiles towards acyliminium cations in a synthetic approach[133] to this enediyne antitumour antibiotic. They have also been utilized in a synthesis of the epoxydiyne core of the neocarzinostatin chromophore; this synthesis involves a remarkable selective desilylation (Scheme 32).[134]

Scheme 32

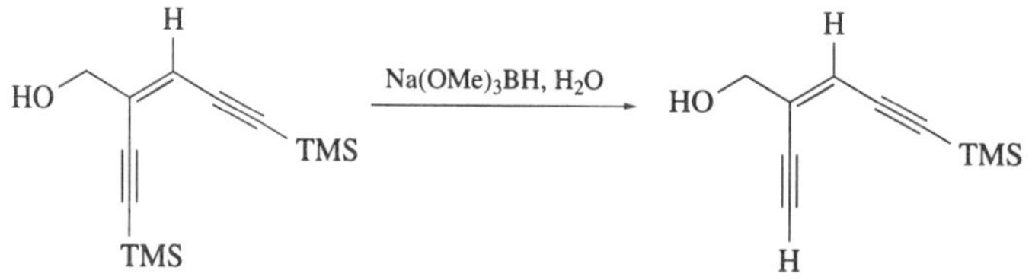

Alkynylsilanes have been employed in radical approaches to the pyrrolizidine skeleton[135] and to γ-butyrolactones,[136] and also as nucleophiles in the *C*-glycosidation[137] of D-glucals.

The silylated tetraalkyne (**21**) has been prepared;[138] desilylation gave the free tetraalkyne, which decomposed in a few minutes at ambient temperature.

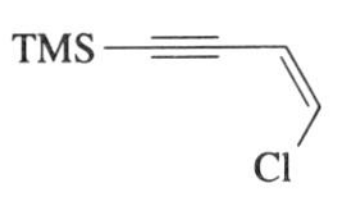

(20) 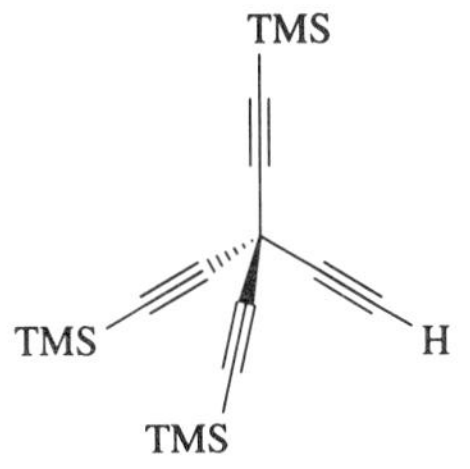   (21)

Two groups[139,140] have independently reported the synthesis and properties of the first isolable six-membered ring containing a carbon–carbon triple bond, the tetrasilacyclohexyne (22); when the silyl substituents are isopropyl groups, the species is a crystalline solid, allowing its x-ray structural analysis.

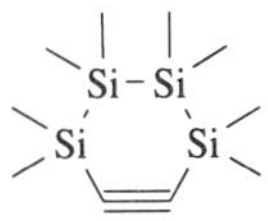

(22)

There have been several reports on the utility[141] of alkynylsilanes in the Pauson–Khand cyclization–carbonylation reaction.

The current state-of-the-art of cobalt-mediated [2 + 2 + 2] cycloadditions of alkynylsilanes in steroid synthesis has been reviewed.[142]

A nickel-catalysed reaction between monosilyl-dialkynes and carbon dioxide has been reported[143] to lead to bicyclic α-pyrones.

A remarkable effect of silyl substitution has been observed[144] in the Wittig rearrangement of 3-aryl-2-propenyl propargyl ethers; whereas the unsubstituted alkyne rearranges in the (1,2)/(1,4) mode, silyl substitution directs the course of the rearrangement in a (2,3) mode, with the product propargyl alcohols being formed with a high degree of *erythro* selectivity (Scheme 33).

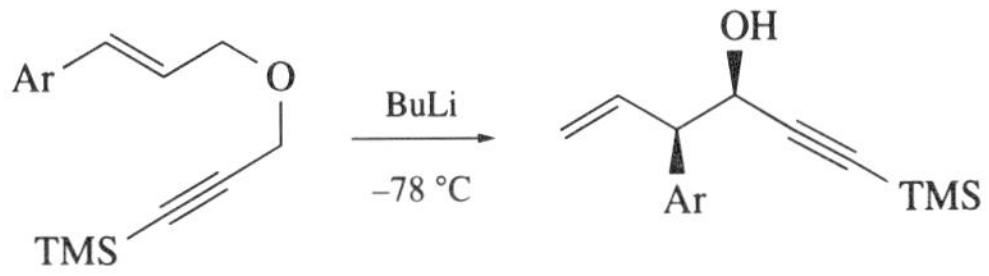

Scheme 33

TMS alkynylboranes undergo regiospecific Diels–Alder cycloaddition reactions with electron-rich dienes;[145] the products can be converted oxidatively into arylsilanes.

An efficient three-step transformation of alkynylsilanes into substituted acetic acids has been employed[146] in syntheses of (−)-isocitric acid lactone and (−)-homoisocitric acid.

Silylated propargyl/allenyl cations undergo a [3 + 2]-cycloaddition reaction with alkenes to produce cyclopentenes (Scheme 34).[147]

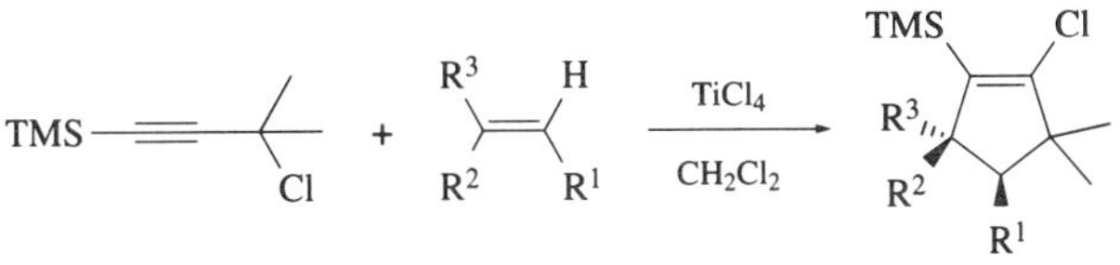

Scheme 34

Propargylsilanes act as allene precursors in the syntheses of the alkaloids nirurine and norsecurinine,[148] and also as terminator functions[149] in polyene cyclization using chiral acetal methodology.

Propargylsilanes with a chiral auxiliary on silicon can be alkylated with high diastereoselectivity; oxidative cleavage of the carbon–silicon bond yields propargyl alcohols in excellent enantiomeric excess.[150]

Propargylsilanes undergo an Amberlyst 15 catalysed intramolecular cyclization with enones to produce bicyclic allenyl ketones in good yield.[151]

7.6 ARYL- AND HETEROARYLSILANES

The preparation and electrophile-induced *ipso*-desilylation reactions of arylsilanes have been reviewed,[152] as has the utility of 2-(TMS)thiazole as a formyl anion equivalent.[153]

2-Silylated 3-furoic acids and 3-thiophenecarboxylic acids can be readily prepared from 3-(silyloxycarbonyl)furans and -thiophenes in a process[154] involving anionic 1,4 O → C migration.

Hexakis(TMS)benzene (23) has been synthesized;[155] x-ray analysis of the yellow crystals show it to be the most deformed benzene prepared so far. In solution, it exists as an equilibrium mixture of chair and boat forms. Photolysis converts it into the Dewar benzene (24), whereas thermolysis results in ring opening with formation of the bis(allene) (25) (Scheme 35).

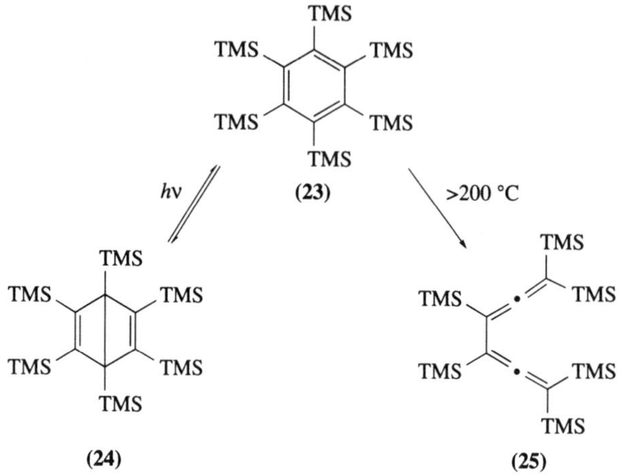

Scheme 35

A variety of unsaturated organosilanes, including arylsilanes, undergo a palladium-catalysed cross-coupling reaction[28] with organic halides or triflates in the presence of a fluoride ion promoter; the reaction is normally highly chemoselective and, if appropriate, stereospecific.

Stereoselective nucleophilic addition to homochiral tricarbonyl(η^6-*o*-triisopropylsilylbenzaldehyde)chromium(0) has led to an asymmetric synthesis[156] of α-substituted benzyl alcohols.

7.7 α,β-EPOXYSILANES

Epoxides directly substituted by a single unsaturated moiety (vinyl, alkynyl, aryl, alkoxycarbonyl or cyano) undergo lithiation of the geminal C–H bond. The resulting species have sufficient kinetic stability to be converted into α,β-epoxysilanes in good yields.[157] The preparation and applications of α,β-epoxysilanes have been described in detail.[158]

α,β-Epoxytriphenylsilanes undergo an unusual β-opening[159] when reacted with anions of hindered sulfones; further transformation leads to stable cyclopropene derivatives (Scheme 36). The more usual selective α-opening[160] can be seen in the silica gel catalysed reaction of α,β-epoxytrimethylsilanes with benzenethiol; the products lead to α-hydroxyaldehydes after sila-Pummerer rearrangement and hydrolysis.

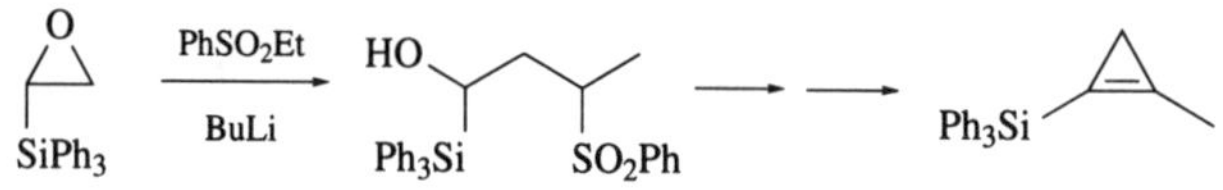

Scheme 36

Sharpless kinetic resolution of γ-TMS allylic alcohols can be performed with very high efficiency; in the example shown (Scheme 37), the epoxyalcohol and the remaining allylic alcohol were both formed in greater than 99% *ee*, indicating that the k_{rel} value is greater than 1056. This process, and further synthetic applications of the chiral epoxysilanes, has been reviewed.[161] Such kinetic resolution proved crucial to the successful synthesis,[162] in protected form, of (2*S*,9*S*)-2-amino-8-oxo-9,10-epoxydecanoic acid, the active component of the physiologically active family of cyclic tetrapeptides which includes chlamydocin. A further application can be seen in an iterative route[163] to optically active polyols.

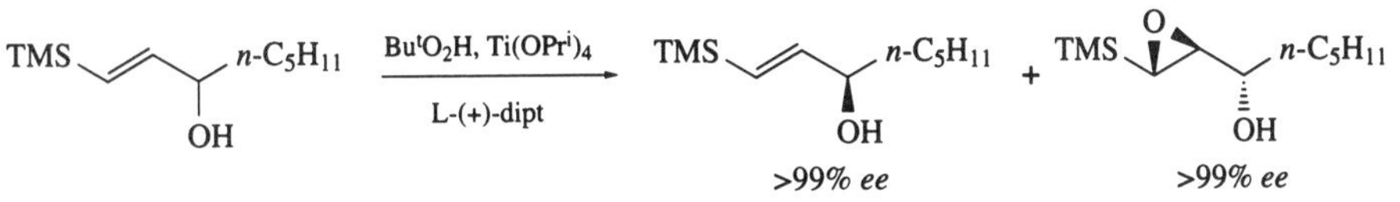

Alternatively, photooxygenation of vinylsilanes followed by Ti(OPr^i)$_4$-catalysed oxygen transfer leads directly[164] to such epoxyalcohols with good regio- and diastereoselectivity.

Chiral β-vinyl-α,β-epoxysilanes undergo a Pd^0-catalysed rearrangement to α-silyl-β,γ-unsaturated aldehydes[165] in high yield and with high stereoselectivity (Scheme 38).

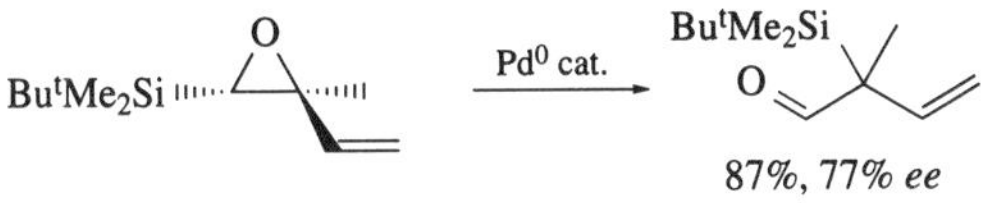

Scheme 38

7.8 PETERSON ALKENATION

The Peterson alkenation reaction has been comprehensively reviewed.[166] Based on analogy with the Wittig reaction and the *syn*-stereochemistry of elimination under basic conditions, a pentacoordinate 1,2-oxasiletanide has been considered as reaction intermediate or transition state; until recently, however, proof of such a species was lacking. Deprotonation of the β-hydroxysilane (**26**) with Bu^{n}Li gave the oxasiletanide (**27**), characterized by ^{19}F and ^{29}Si NMR spectroscopy; mild heating gave alkene (**29**), indicating that (**27**) is a true intermediate in the Peterson reaction (Scheme 39). Further, deprotonation of (**26**) using KH in the presence of 18-crown-6 gave the crystalline potassium salt (**28**), whose structure was confirmed by x-ray crystallography.[167]

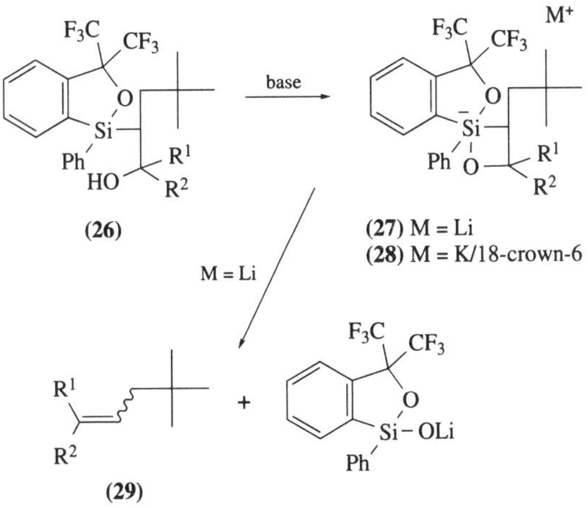

Scheme 39

Base-catalysed eliminations of vinylogous β-hydroxysilanes are also stereospecifically *syn*, producing (*E,E*) or (*Z,E*) dienes;[168] when a (*Z*) double bond is formed, it is placed selectively adjacent to the carbon which originally carried the hydroxyl group (Scheme 40).

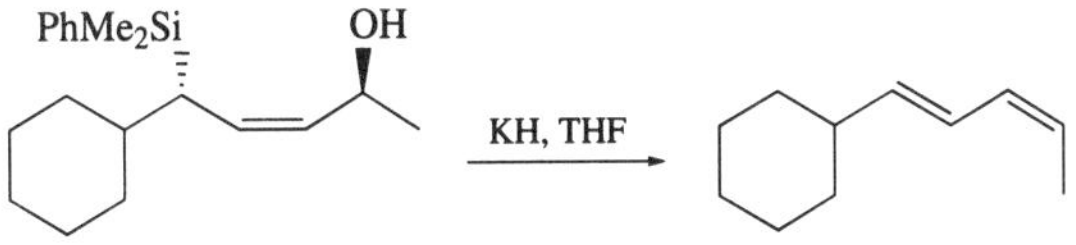

Scheme 40

The use of Nafion-H has been recommended[169] for the acid-catalysed elimination of β-hydroxysilanes, which proceeds with *anti*-stereochemistry; yields are high, the elimination proceeds at room temperature, and workup is simple filtration of the catalyst.

Alkenes can be protected from catalytic reduction by epoxidation followed by epoxide opening with TMS lithium to give β-hydroxysilanes;[170] when desired, the alkene can be regenerated as either the (E)- or (Z)-isomer by appropriate choice of conditions.

Condensation between a γ-(TMS)allylborane and a variety of N-heterocyclic aldehydes afforded, after base-induced elimination, N-heterocyclic buta-1,3-dienes, often with excellent stereoselectivity.[171]

Prolonged treatment of α-methylene-β-hydroxysilanes with NaH in octane at reflux results in elimination to form allenes in moderate yield.[172]

The isourea (30) has been recommended[173] for the direct conversion of benzyloxycarbonyl-protected amino acids into the protected 2-(TMS)ethyl esters.

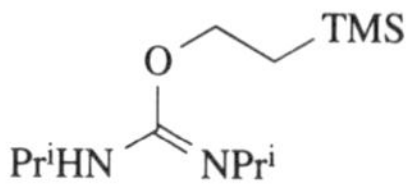

(30)

7.9 KETOSILANES

7.9.1 Acylsilanes

Two reviews on the synthesis, spectral properties and synthetic potential of acylsilanes have been published.[174] Prior to 1988, no formylsilane had been isolated. Formyltrimethylsilane itself had been generated by Swern oxidation of trimethylsilylmethanol and then reacted *in situ* with Wittig reagents[175] and organometallics.[176] In 1988, formyltris(TMS)silane was isolated[177] as a stable compound. More recently, dithiane methodology has been employed in the syntheses of formyltriisopropylsilane[178] (which ignites on exposure to air) and formyl-*t*-butyldimethylsilane,[179] the latter being isolated as its 2,4-dinitrophenylhydrazone (DNP) derivative. Formyltriphenylsilane[180] has been generated in solution from an organozirconium precursor.

α-(TMS)vinyl alcohol[181] has been generated by flash photolysis, and the kinetics of its ketonization to acetyltrimethylsilane determined.

Methoxy(phenylthio)(TMS)methyllithium adds conjugatively to α,β-unsaturated carbonyl compounds, acting as a masked formyl, carboxyl or acetyl(TMS) anion equivalent, depending on the cleavage conditions employed. Its utility can be seen in a synthesis (Scheme 41) of a prostaglandin intermediate.[182]

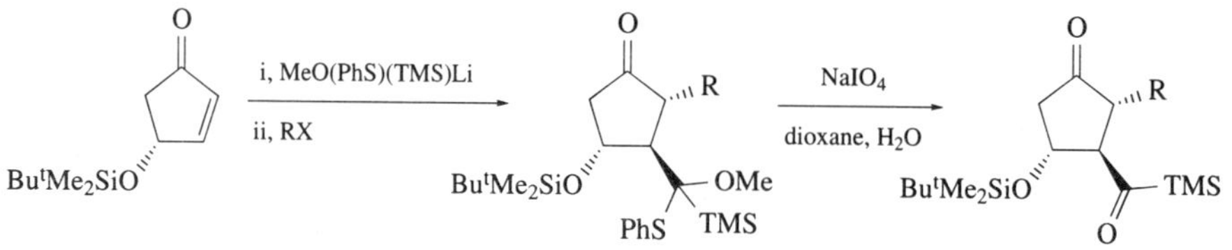

Scheme 41

A wide range of acylsilanes can be prepared in high yield by anodic oxidation[183] of 2-alkyl-2-trialkylsilyl-1,3-dithianes.

(E)-3-Chloroalkenoyltrimethylsilanes[184] are produced when 3,3-dichloropropenoyltrimethylsilane is treated with organocuprate reagents.

The synthesis of trifluoroacetyltriphenylsilane[185] has been reported; reaction with organolithium reagents, followed by Brook rearrangement and fluoride ion elimination, provide a range of potentially useful 2,2-difluoro silyl enol ethers (Scheme 42). Under carefully controlled conditions to avoid Brook rearrangement, perfluoroorgano Grignard reagents react with acyl- and aroylsilanes to give fluorinated silyl carbinols.[186]

Scheme 42

Organocuprates add conjugatively to ethynyl triphenylsilyl ketone; employment of vinyl cuprates, followed by desilylation, provides a new method[187] for the generation of stereodefined polyenals.

Acylsilanes undergo highly regioselective allylation and propargylation reactions; this methodology[188] has been applied to syntheses of PGE$_3$ and F$_{3\alpha}$ methyl ester.

The carbon–silicon bond of acylsilanes can be cleaved by electrochemical anodic oxidation, with concomitant introduction of oxygen and nitrogen nucleophiles at the carbonyl carbon.[189]

7.9.2 α-Silylcarbonyl Compounds (β-Ketosilanes)

Metallation of β-bromo TMS enol ethers, followed by 1,3-migration of silicon from oxygen to carbon, has resulted in the first isolation[190] of α-TMS aldehydes (Scheme 43).

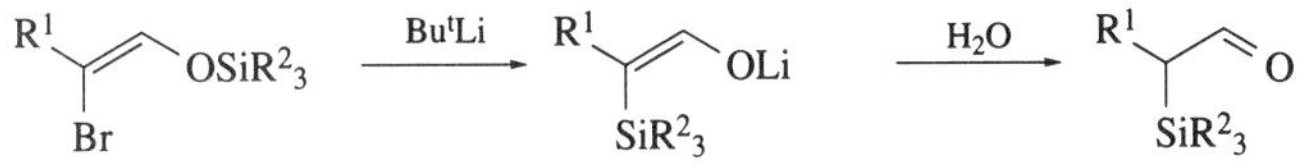

Scheme 43

Addition of organocerium reagents to silylketenes, with subsequent quenching with aqueous ammonium chloride or alkyl halides, provides a high yielding one-pot synthesis[191] of α-silyl ketones. The catalysed addition of hydroxyl compounds to silylketenes has been reinvestigated; using ZnCl$_2$ for alcohols and ZnI$_2$ for phenols, a wide range of esters of trialkylsilylacetic acids[192] were obtained in high yield.

α-Nitrocycloalkanones react with (TMS)methylmagnesium bromide to produce ω-nitro-β-ketosilanes;[193] the reaction fails with larger than six-membered rings.

(S)/(R)-(−)-1-amino-2-(methoxymethyl)pyrrolidine (SAMP/RAMP) hydrazone methodology has been employed in the synthesis of α-(isopropoxydimethyl)silyl ketones in high enantiopurity; diastereoselective hydride reduction followed by carbon–silicon bond oxidative cleavage completes a new route[194] to stereodefined 1,2-diols.

The α-silyl ketone derived silyl enol ether (**31**) behaves as the synthetic equivalent of acetone α,α'-dianion, undergoing Lewis acid mediated double alkylation[195] with electrophiles such as acetals and carbonyl compounds (Scheme 44).

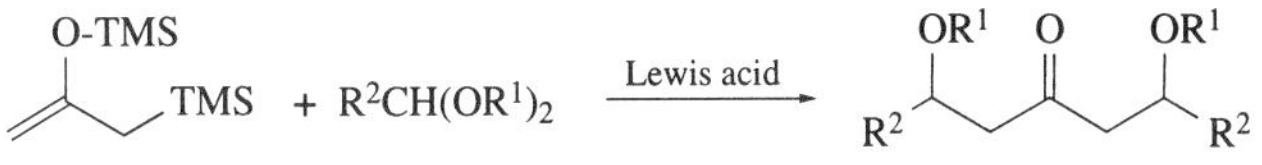

Scheme 44

Deprotonation of (Z)-α-TMS-α,β-unsaturated esters, followed by protonation of the intermediate dienolates with methanol, provides a stereoselective route[196] to desilylated (E)-β,γ-unsaturated esters.

Two different routes[197,198] to both enantiomers of 5-TMS-cyclohex-2-enone have been reported. These chiral enones undergo a highly diastereoselective 1,4-addition with organometallic reagents; elimination of the TMS group, using CuCl$_2$ in warm DMF, then generates 5-substituted cyclohex-2-enones in high enantiomeric excess.

7.10 ALKYL SILYL ETHERS

A crystallographic and *ab initio* investigation into the nature of the silicon–oxygen bond in alkyl silyl ethers, to account for widening of the bond angle and decreased basicity, has revealed that neither steric effects or $n \rightarrow 3d$ delocalization need to be invoked; consideration of the detailed nature of the HOMOs suffices.[199]

The use of alkyl silyl ethers as protective groups for alcohols has been detailed,[200] and methods for their oxidative deprotection have been reviewed.[201]

The transition metal catalysed direct reaction of hydridosilanes with alcohols has been further investigated[202] as a good alternative to the use of the more expensive chlorosilanes.

Diphenylmethylsilyl ethers show promise[203] as protective groups; they are readily prepared from primary, secondary and tertiary alcohols, and their stability lies between TMS and TBDMS ethers. They are stable to flash chromatography on silica gel, even as the related silyl enol ethers, from which enolates can be generated by normal methods.

Silyl ethers are cleaved by tetrafluorosilane gas in dichloromethane or acetonitrile at a rate which depends on the steric bulk of the silyl group, providing a promising method[204] for selective desilylation; the relative rates of cleavage are in the order $Et_3Si > TBDMS \gg [Me_2CHC(Me_2)]Me_2Si$ or Bu^tPh_2Si.

Sodium azide in DMF effects the selective cleavage[205] of $MePh_2Si$ ethers in the presence of TBDMS and Bu^tPh_2Si ethers. Aqueous HF in acetonitrile in the presence of a catalytic amount of fluorosilicic acid selectively cleaves TBDMS ethers in the presence of Pr^i_3Si ethers.[206] Other recent TBDMS ether cleavage systems include catalytic transfer hydrogenation,[207] ammonium fluoride in methanol,[208] potassium fluoride on basic alumina and ultrasound (selective for phenolic ethers),[209] partially deactivated neutral alumina,[210] and solvent-free alumina with microwave irradiation.[211] Whereas TBDMS ethers are unreactive towards diisobutylaluminum hydride at −78 °C, they do undergo quite rapid cleavage at room temperature.[212]

2-Trimethylsilyloxy-1,3-dioxolanes (silyl orthoesters) have been suggested[213] as protective groups for unbranched carboxylic acids; the moiety is stable towards organolithium reagents.

Further progress[214] has been reported on the organoaluminum-mediated rearrangement of epoxy silyl ethers; with chiral substrates, optically active β-silyloxyaldehydes are produced.

Stereocontrol in the Prins cyclization of unsaturated silyl ethers to tetrahydrofurans,[215] in the iodolactonization of silyloxy alkenoic acids to δ-valerolactones[216] and in the ene reaction[217] of 2-(alkylthio)allyl silyl ethers has been demonstrated.

7.11 SILICON TETHERING

The 'temporary silicon connection', pioneered by Stork, is based on the concept of temporarily bringing together two reaction partners by means of an eventually removable silicon atom. Such a strategy has proven a valuable synthetic 'trick' in intramolecular reactions of various types. For example, silylmethyl radicals,[218] generated from halomethylsilyl ethers of allylic alcohols, have been used in an indirect method of achieving acyclic stereocontrol. Depending on the substrate, either 6-endo- or 5-exo-trig cyclization can predominate. In the 5-exo-trig process shown in Scheme 45, transition state geometry demands a *cis* fusion of the newly formed five-membered ring. The resulting radical, being cup-shaped, permits access to stannane only from the convex face. This results, after oxidative cleavage of the C–Si bond,[219] in an excellent method[220] for the overall *anti* addition of a hydrogen and a hydroxymethyl group to the original alkene.

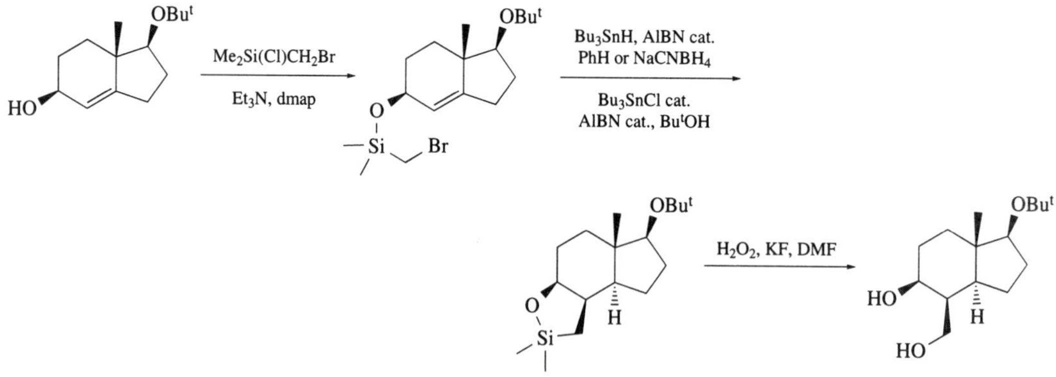

Scheme 45

Related silyl ethers of homoallylic alcohols undergo[221] a highly regio- and stereoselective 6-*exo* cyclization as shown (Scheme 46), but only if the allyl terminus carries one or two methyl groups; unsubstituted cases undergo a regioselective 7-*endo* closure. In certain steroidal systems, the 6-*endo* mode becomes preferred, ultimately producing (Scheme 47) 1,4-diols in a regio- and stereocontrolled manner.[222]

The Stork group has also described[223] a radical-mediated method for the synthesis of *C*-glycosides, exemplified in Scheme 48 and leading stereospecifically to α-glucosides; β-*C*-glucosides can be obtained by silicon tethering to the 3β-hydroxyl of glucose, even though the 3-silyloxy group must become axial. Related methodology has also provided stereocontrolled syntheses of disaccharides.[224]

The key step in a highly convergent synthesis[225] of tunicaminyluracil (**32**) involves a 7-*endo-trig* radical closure of the silyloxy radical generated from the selenoacetal (**33**).

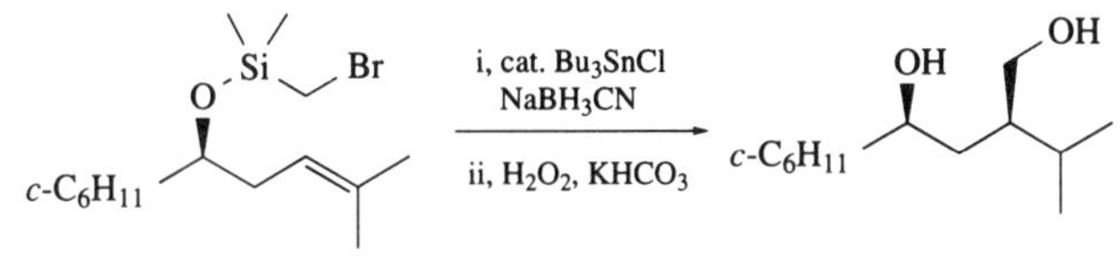

Scheme 46

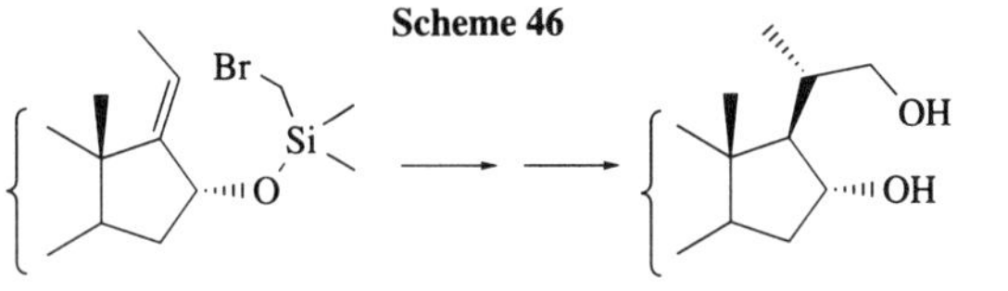

Scheme 47

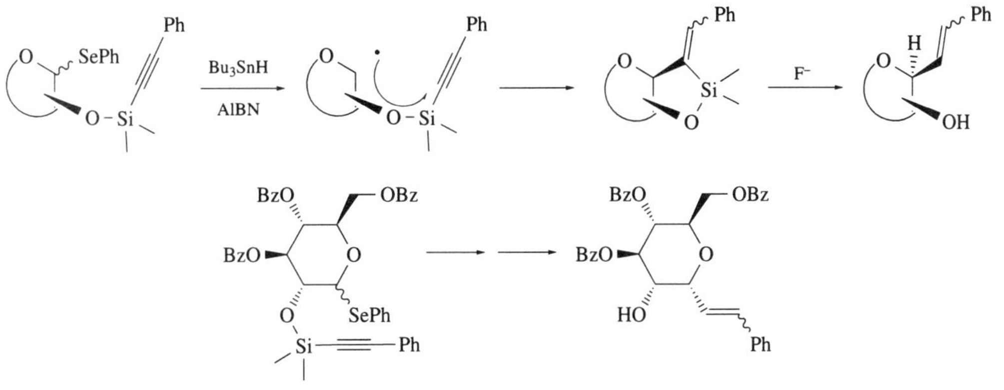

Scheme 48

Such temporary tethering has also been applied to control the regio- and stereochemistry of intramolecular [4 + 2]-cycloadditions.[226] For example, cyclization[227] of the silyl acetal (**34**) gave the product (**35**) as a single stereoisomer, whose stereochemistry was established by subsequent cleavage to (**36**) and x-ray analysis. In contrast, *inter*molecular reaction between (**37**) and (**38**) gave all four regio- and stereoisomers.

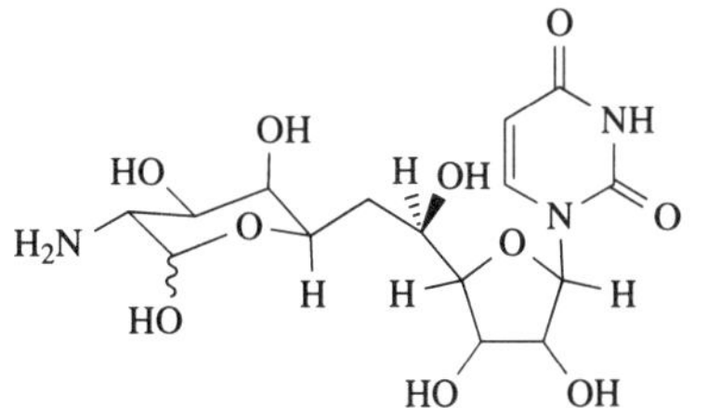

(**32**)

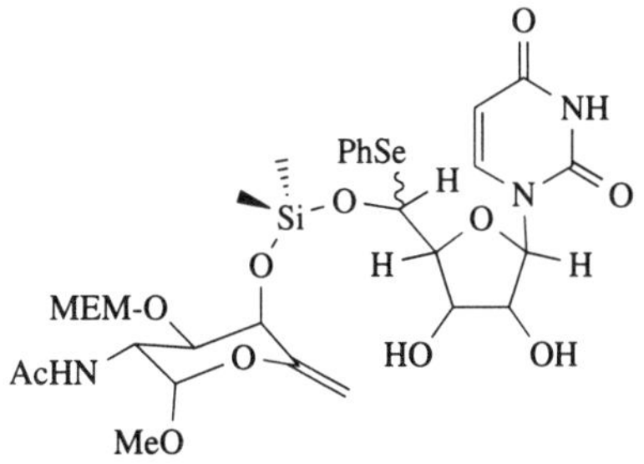

(**33**)

Vinylsilanes can also be induced to undergo 4 + 2 cycloaddition,[228] as exemplified by the formation of (**39**) and (**40**) (Scheme 49). The stereoselectivity of cyclization is influenced by the R groups on silicon, with Bu^t being better than Me or Ph; however, while oxidative cleavage of the last two was facile, the Bu^t substituted products could not be cleaved.

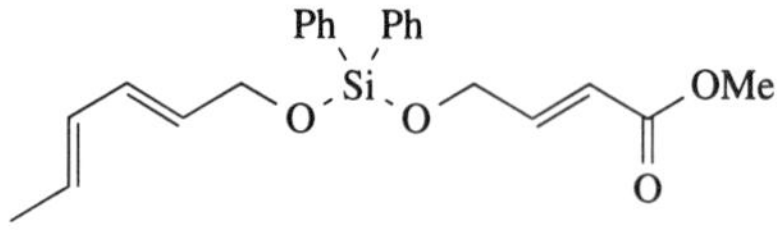

(34)

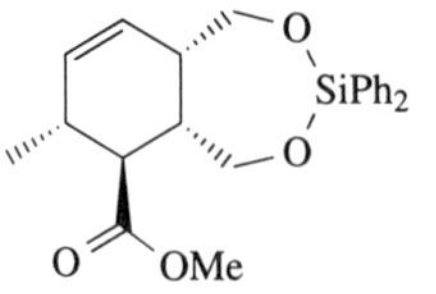

(35) 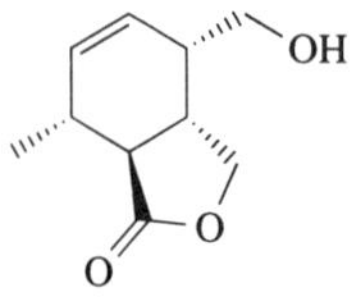   (36)

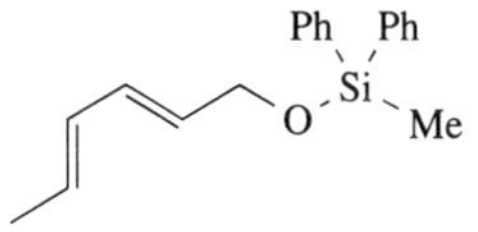

(37) 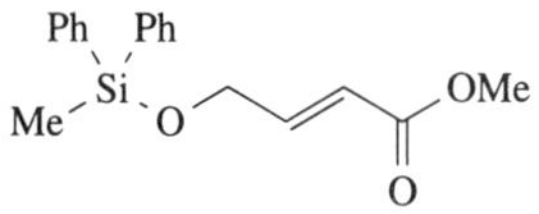   (38)

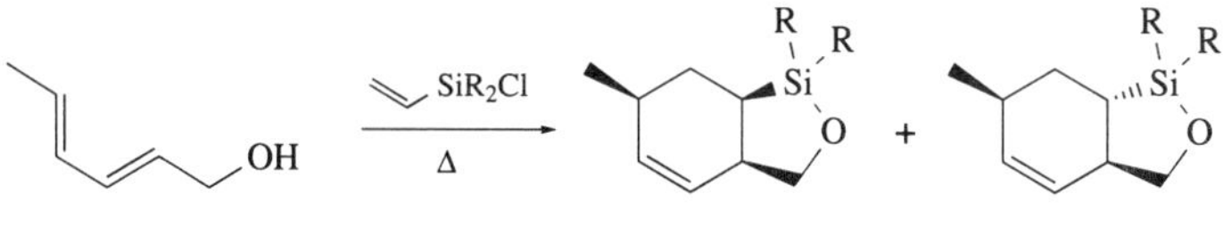

Scheme 49

Stork's group has also investigated[229] such cycloadditions, as shown in Scheme 50. Interestingly, the siloxane is produced as a 3:2 mixture of isomers, but (**41**) is obtained as a single isomer, presumably reflecting the intermediacy of carbanions or radicals in the fluoride-induced hydrodesilylation.

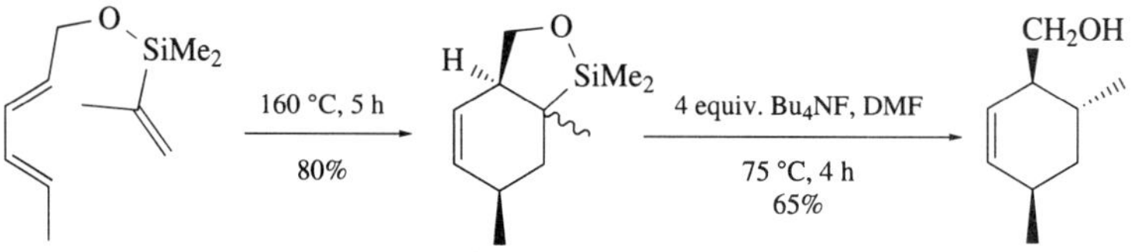

Scheme 50

Tethered vinylsilanes have been shown to undergo anodic alkene coupling reactions, resulting in the generation of quaternary carbon atoms (Scheme 51).[230]

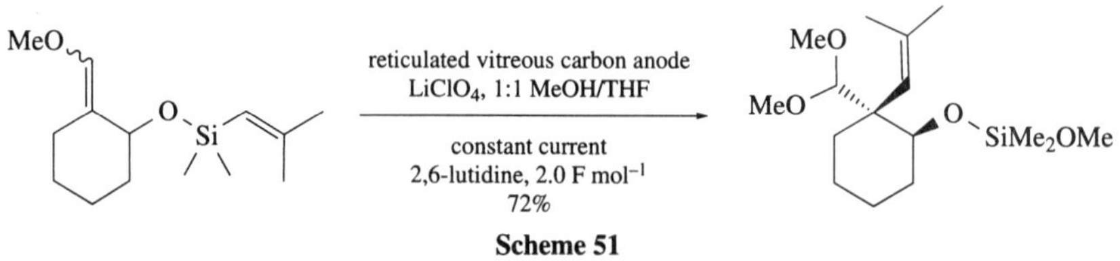

Scheme 51

Another interesting application[231] of temporary tethering can be seen in a stereocontrolled synthesis of a 1β-methylcarbapenem intermediate (Scheme 52); intermolecular variants proved unsatisfactory.

Finally,[232] silyl ethers have been employed in intramolecular hydrosilylation of alkenes[233] and ketones[234] (Section 7.15).

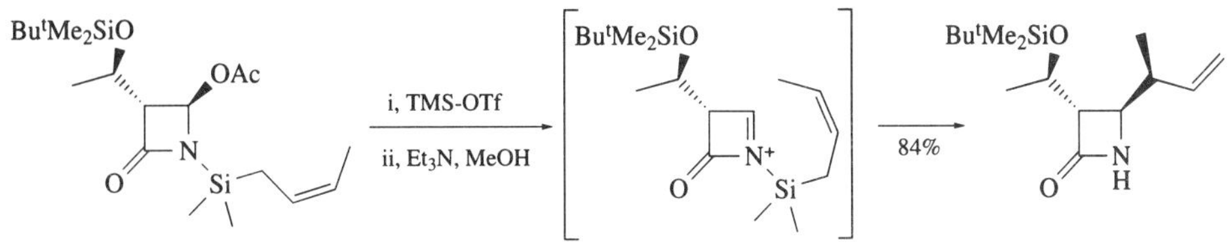

Scheme 52

7.12 SILYL ENOL ETHERS

7.12.1 Preparation

Full details have been published[235] on the preparation of silyl enol and dienol ethers from enolizable aldehydes and ketones using the reagent combination of trichloromethylsilane/sodium iodide/triethylamine in acetonitrile; the method is mild and quick, and shows (Z)-selectivity and a preference for formation of the kinetic isomer.

1,4-Hydrosilylation of α,β-unsaturated carbonyl compounds provides a reliable route to regiochemically defined silyl enol ethers. Choice of catalyst is important; $(Ph_3P)_4RhH$ has been found to be highly effective.[236]

Further studies on the generation of chirally-enriched silyl enol ethers using homochiral lithium amide bases have been communicated.[237]

The generation of regio- and stereoisomerically pure silyl enol ethers from acylsilanes has been reported in detail;[238] diastereoselective organometallic addition to the carbonyl group is followed by Brook rearrangement and fragmentation.

The homochiral 2-(alkylthio)allyl silyl ether (42) undergoes a highly enantioselective ene reaction[239] with a range of aldehydes under the influence of Me_2AlCl (Scheme 53); the product silyl enol ethers are obtained with high *ee*.

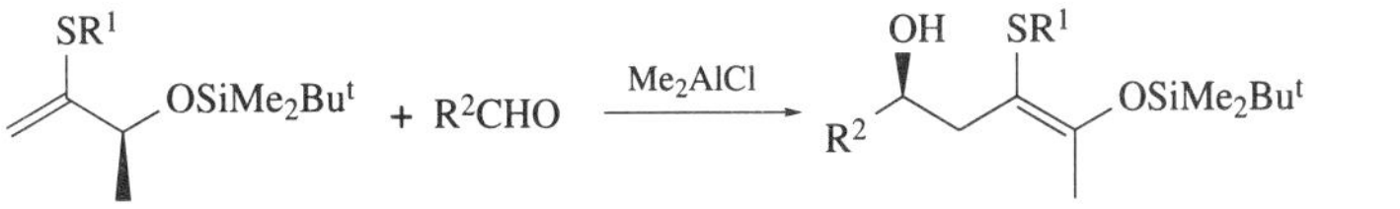

Scheme 53

Several stereocontrolled routes to silyl dienol ethers have been described.[240]

A reliable method for the assignment of geometry to trimethylsilyl enol ethers is reportedly provided[241] by ^1H NMR NOE (nuclear Overhauser enhancement) difference spectroscopy.

7.12.2 Lewis Acid Induced Reactions

The Lewis acid (usually titanium tetrachloride) mediated aldol reaction of silyl enol ethers and ketene acetals with aldehydes (the Mukaiyama reaction) has been the subject of much investigation, particularly in terms of chiral induction. Strategies applied include reaction between chiral ketene acetals derived from *N*-methylephedrine[242] and camphor derivatives[243] and achiral aldehydes, achiral silyl enol ethers and chiral acetals[244] and orthoesters,[245] and achiral silyl ketene acetals and α-chiral aldehydes such as glyceraldehyde acetonide[246] or α-benzyloxyaldehydes;[247] in this last case, it was found that lithium perchlorate suspended in dichloromethane functioned as a mild and efficient catalyst.

A β-chiral aldehyde has been employed in the synthesis of a bryostatin fragment.[248]

The (S)-prolinol-derived *O*-silyl ketene *N,O*-acetal (43) reacts with aldehydes such as benzaldehyde without apparent catalysis to provide a route (Scheme 54) to enantiomerically pure *anti* aldols; the minor *syn* diastereoisomer is easily removed by crystallization.[249]

Enoxysilacyclobutanes also react with aldehydes without catalysis; this has been ascribed[250] to 'strain release Lewis acidity'. The (E)-isomer reacts more rapidly to produce the aldol products with high *syn* diastereoselectivity (Scheme 55).

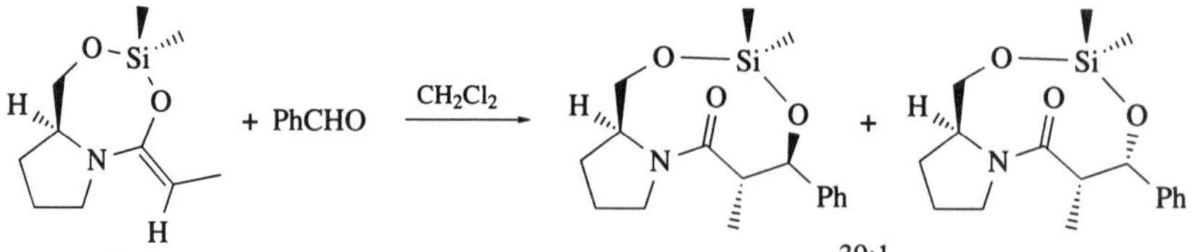

Scheme 54

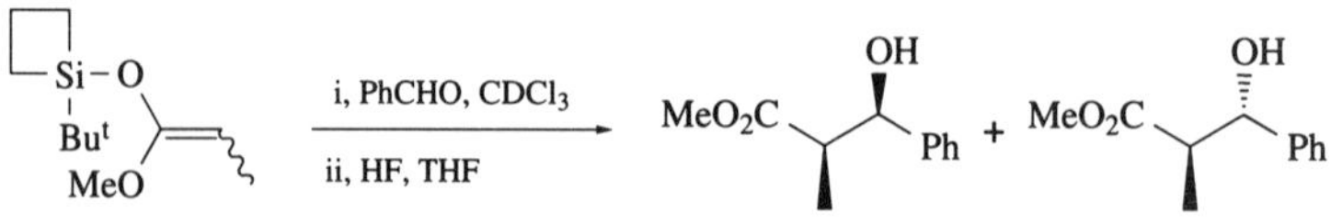

Scheme 55

Kobayashi *et al.* have reported[251] that achiral silyl ketene acetals of *S*-alkyl ethanethioates and achiral aldehydes undergo a highly enantioselective aldol reaction in the presence of a chiral promoter system consisting of chiral diamine-coordinated tin(II) triflate and tributyltin fluoride (or dibutyltin diacetate).

Corey's tryptophan-derived oxazaborolidine (**44**) also efficiently catalyses enantioselective aldol reactions between achiral silyl enol and dienol ethers and achiral aldehydes, with up to 93% *ee* being realized.[252]

Yamamoto's group has developed a chiral acyloxyborane catalyst,[253] with levels of enantioselectivity of up to 96% *ee* being attained. The catalyst (**45**) is tartrate-derived, allowing access to either enantiomer of the product. Interestingly, regardless of the stereochemistry (*E* or *Z*) of the starting silyl enol ethers, the aldol products were obtained with high *syn*-selectivity.

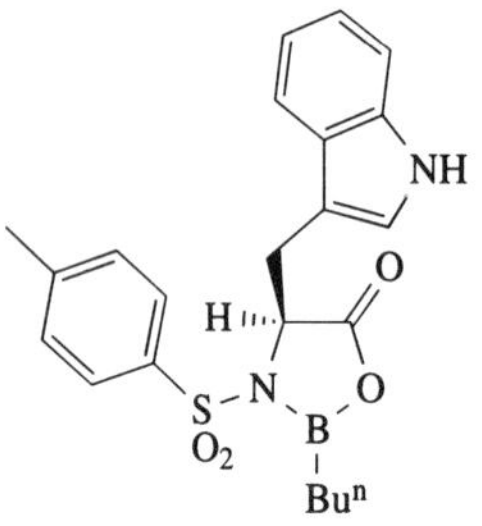

Reaction between an achiral silyl ketene acetal and a chiral imine in the presence of a chiral binaphthol-derived boron reagent has provided a route to a key carbapenem intermediate.[254] A lanthanide(III)-catalysed aldol between the (*R*)-3-hydroxybutanoate derived silyl ketene acetal (**46**) and benzyloxyacetaldehyde has been found to proceed[255] with a high level of rather unusual 2,3-*anti* diastereoselection, together with a high 1,2-*anti* facial selection (Scheme 56); the aldol adduct was also elaborated to a related carbapenem intermediate.

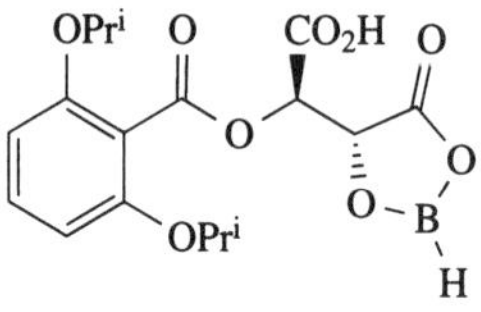

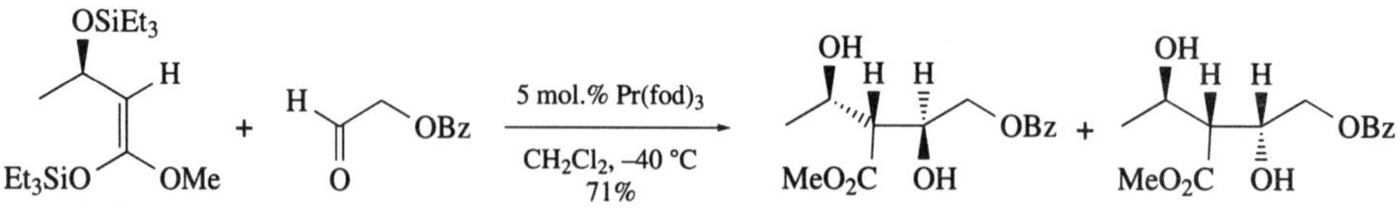

Scheme 56

Trimethylsilylpropynal reacts with *O*-silyl ketene *O*,*S*-acetals with high *anti*-selectivity, again regardless of ketene geometry; the corresponding cobalt-complexed propynal gives the *syn*-products exclusively.[256]

As part of a prostanoid synthesis,[257] the observation that a common silyl enol ether gave a completely reversed stereochemical outcome when reacted with oct-2-ynal as opposed to either (*E*)- or (*Z*)-oct-2-enal was used to good advantage.

Cationic iron carbonyl complexes have been shown[258] to catalyse the nonstereoselective condensation of silyl ketene acetals with benzaldehyde.

Generation of seven-, eight- and, with more difficulty, nine-membered rings can be achieved by intramolecular Mukaiyama aldol reaction of silyl enol or dienol ethers with acetals.[259]

Dibutyltin bis(triflate) appears to be a more chemoselective catalyst[260] than trimethylsilyl triflate for the reaction of silyl enol ethers with aldehydes and ketals.

Ruthenium-, titanium- and zirconium-based organometallic complexes have been reported to be efficient catalysts in the Mukaiyama reaction.[261]

Fluoride ion can also be used to effect this aldol condensation. Using a chiral quaternary ammonium fluoride derived from cinchonine, reasonable levels of diastereo- and enantioselectivity can be realized in the reaction of achiral silyl enol ethers with benzaldehyde.[262]

The generation of β-lactams by Lewis acid mediated reaction of, *inter alia*, silyl ketene acetals with imines has been reviewed.[263] Microwave irradiation in the presence of KF/18-crown-6 also achieves this transformation.[264]

7.12.3 Conjugate Addition

Silyl enol ethers undergo conjugate addition[265] to methyl vinyl ketone in a solvent-free reaction on alumina impregnated with anhydrous $ZnCl_2$. Silyl ketene acetals undergo an anion-catalysed sequence of inter- and intramolecular conjugate additions with dienoates;[266] in the example shown (Scheme 57), the product was obtained as a single diastereoisomer.

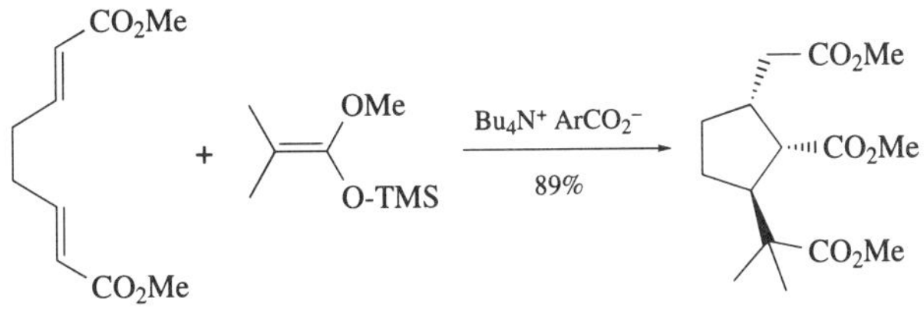

Scheme 57

Enone ketene acetals such as (**47**) undergo stereoselective cyclization on silica gel chromatography[267] to yield bicyclo[2.2.2]octanones (Scheme 58).

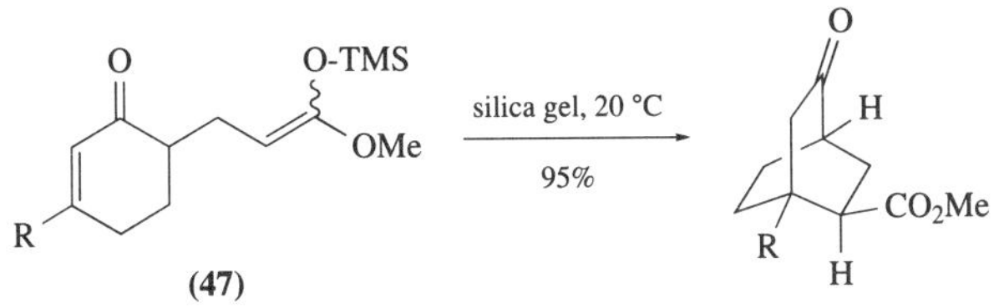

Scheme 58

7.12.4 Cycloaddition Reactions

The utility of 1-alkoxy-1-silyloxycyclopropanes, obtained *inter alia* by cyclopropanation of silyl ketene acetals, as ester homoenolate equivalents has been reviewed.[268] 1-Substituted-1-silyloxycyclopropanes undergo a Pd^0-catalysed reaction with aryl triflates in an atmosphere of carbon monoxide to produce 1,4-ketoesters and 1,4-diketones in moderate to good yields (Scheme 59).[269]

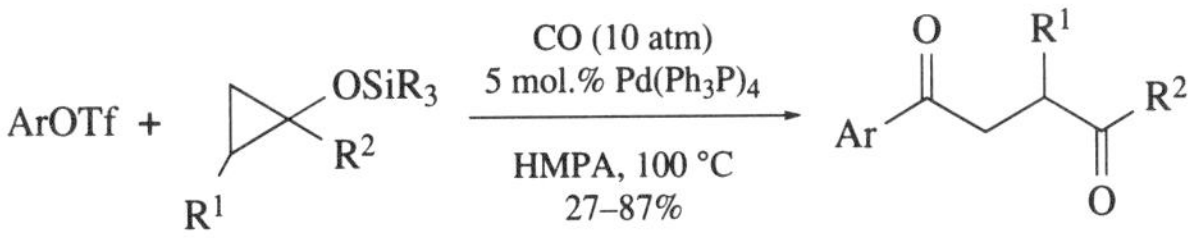

Scheme 59

Further applications of the oxidative cleavage[270] of silyloxycyclopropanes to ω-unsaturated carboxylic acids have been reported.[271] Such silyloxycyclopropanes undergo a smooth room temperature isomerization (Scheme 60) to allyl silyl ethers; this rearrangement, catalysed by Zeise's dimer, is quite general,[272] unlike the earlier ZnI_2-promoted isomerization which was restricted to strained systems.

Silyloxyalkynes undergo a [2 + 2]-cycloaddition with aryl ketenes to produce silyloxycyclobutenes; electrocyclic cleavage and ring closure then completes[273] an aromatic annulation strategy.

Silyl enol ethers take part in a tin(IV)-catalysed [3 + 2]-cycloaddition[274] with 2-alkoxycyclopropyl esters and ketones to produce functionalized cyclopentanes. Their stepwise catalysed [3 + 2]-

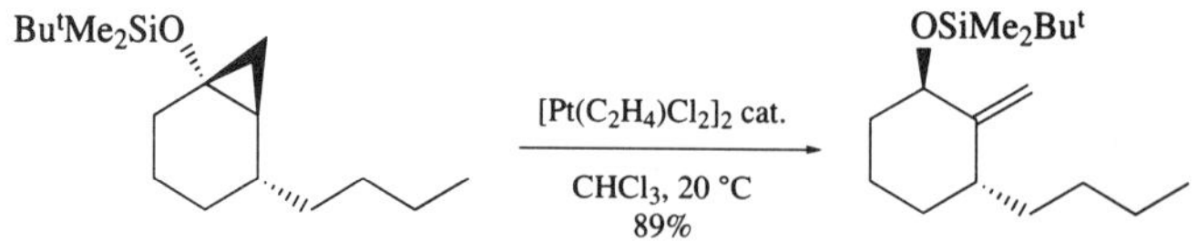

cycloaddition with terminal alkynes (Scheme 61) leads to cyclopentenones.[275] They also participate in a trimethylsilyl triflate-promoted [3 + 2]-cycloaddition with nitrones, providing ready access[276] to 5-silyloxyisoxazolidines.

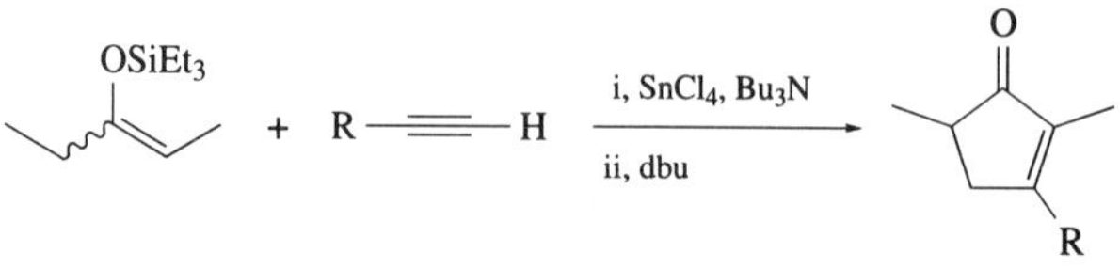

Scheme 61

More applications[277] of Danishefsky's diene (**48**) and related silyloxydienes in Diels–Alder and hetero-Diels–Alder cycloadditions, with particular application to natural product synthesis, have been described, as have more examples[278] of the use of chiral Lewis acid catalysts.

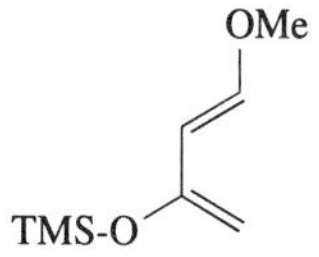

(**48**)

The preparation of 1,1-difluoro-2-triphenylsilyloxybuta-1,3-diene and a range of its [4 + 2]- and [2 + 2]-cycloaddition reactions have been reported.[279]

The bis(TMS) enol ether of methyl acetoacetate reacts in a Lewis acid promoted [3 + 4]-annulation process[280] with 1,4-dicarbonyl compounds, with the resulting regio- and stereoselective formation (Scheme 62) of 2-carbomethoxy-8-oxabicyclo[3.2.1]octan-3-ones.

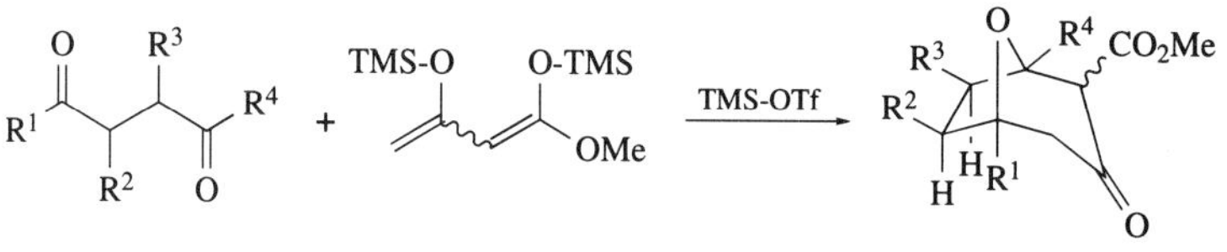

Scheme 62

7.12.5 Sigmatropic Rearrangement Reactions

The silyl ketal acetal variant[281] of the Claisen rearrangement of allyl alcohol ester enolates allows a predictable transfer of stereochemistry from starting material to product under very mild conditions. Ireland *et al.* have published full details[282] on the factors controlling the stereoselectivity in silyl ketene acetal formation, and on the preference[283] for a chair- or boatlike transition state in the actual [3,3]-sigmatropic shift.

A related rearrangement of alkynyl alcohol esters[284] leads to allenyl acetic acids; thermal decarboxylation then provides a convenient route to a range of buta-1,3-dienes.

7.12.6 α-Oxidation and Related Reactions

The oxidation of silyl enol ethers to α-hydroxycarbonyl compounds involves the intermediacy of silyloxyoxiranes. Several reports[285–9] of the detection and spectroscopic characterization of these normally fleeting intermediates have appeared.[290]

The possibility of conferring chirality to the product α-hydroxycarbonyl compounds has been investigated using carbon–chiral silicon substituents,[291] chiral 2-sulphonyloxaziridines,[287] and chiral manganese(salen) catalysts;[292] only the last method gave good stereoselectivity.

Certain substituted ketones, as their triisopropylsilyl enol ethers, undergo a remarkable double hydroxylation,[293] to produce α,α'-dihydroxyketones (Scheme 63); such methodology has been applied to a synthesis of (±)-cortisone.

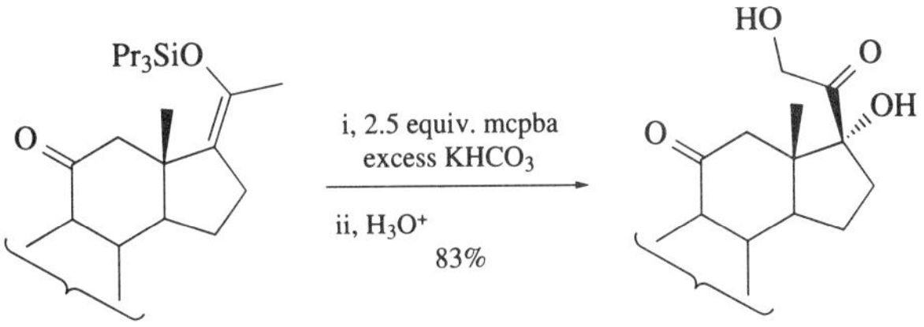

Scheme 63

Suitably substituted silyloxycyclohexenes can be transformed regio- and stereoselectively[294] into the corresponding α'-acetoxy silyl enol ethers (Scheme 64).

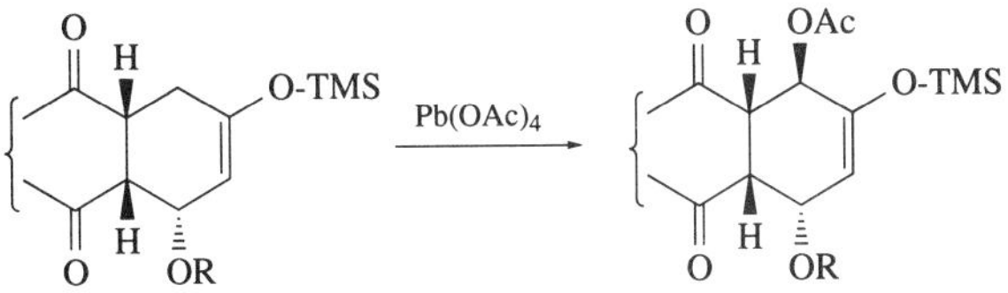

Scheme 64

Oxidative methods have been described for the α-methoxylation[295] and α-sulphonyloxylation[296] of ketones.

The asymmetric α-amination of chiral ketene silyl acetals has been reported; the chirality is ephedrine-[297] or camphor[298]-derived, and the nitrogen source is di(t-butyl) azodicarboxylate.[299]

Silyl enol ethers react with diphenyliodonium fluoride in a regiospecific route[300] to α-phenyl ketones. A simple method for the preparation of α-iodoketones has been described;[301] enol acetates can be similarly employed. When treated with the reagent combination of iodosobenzene and trimethysilyl azide, triisopropylsilyl enol ethers undergo a remarkable transformation (Scheme 65) into β-azido trisopropylsilyl enol ethers;[302] the azido group can be displaced readily by a range of nucleophiles to give products of formal conjugate addition, without involving enones.[303]

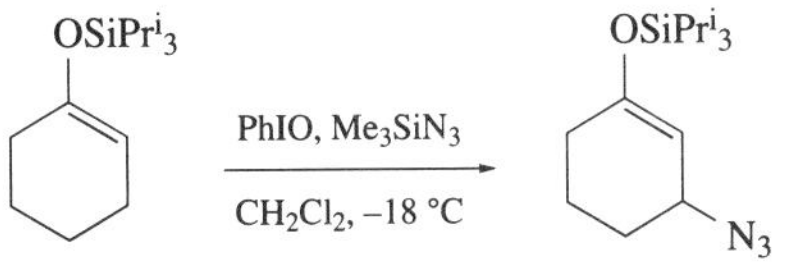

Scheme 65

Mild methods have been described[304] for the preparation of α-phenylthio- and α-phenylselenoketones via silyl enol ethers.

7.12.7 Heterosubstituted Silyl Enol Ethers

The bis(TMS) enol ether of N-phenylsuccinimide reacts with aromatic aldehydes to give, after elimination of trimethylsilanol, (EE)/(ZZ)-2,3-bis(arylmethylene)-N-phenylsuccinimides (fulgimides).[305]

There have been several further applications[306-9] of 2-silyloxyfurans as butenolide precursors and as potential anthracycline synthons.[310]

7.12.8 Reactivity and Miscellaneous Reactions

The nucleophilicity of silyl enol ethers has been determined[68] to lie between that of structurally analogous allylsilanes and -stannanes. Full details have been published on the generation of simple enols[311] and dienols[312] from the corresponding silyl enol ethers; the kinetics of ketonization were determined for both classes of enol.

Both 1,3-dithiolan-2-ylium tetrafluoroborate[313] and 2-ethoxy-1,3-dithiolane[314] are good alkylating agents, the latter reagent requiring zinc chloride catalysis for silyl enol ethers to produce half-protected 1,3-dicarbonyl compounds; the former reagent can be prepared more efficiently.

Silyl enol ethers react with acyliminium ions to afford satisfactory yields of β-ketocarbamates,[315] in some cases with good *de*s.

Acetate-derived TMS and triethylsilyl ketene acetals undergo a mild, organoaluminium-catalysed 1,3-migration[316] of silicon from oxygen to carbon to produce trialkylsilylacetic acid esters (Scheme 66); the same rearrangement with TMS enol ethers proved unsuccessful.

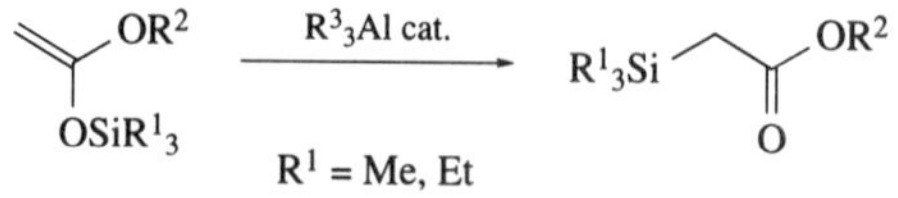

Scheme 66

2-Chloroalkanoic acids can be prepared with high *ee* by the diastereoselective halogenation[317] of sugar-derived silyl ketene acetals with *N*-chlorosuccinimide.

Silyl enol ethers react with phenylthio- or methylthiotrimethylsilane in the presence of boron trifluoride etherate to yield vinyl sulfides[318] in good yields.

In the presence of silver triflate, silyl enol ethers undergo a decarboxylative coupling reaction with benzylic chloroformates to produce benzylated ketones (Scheme 67), often in good yields.[319]

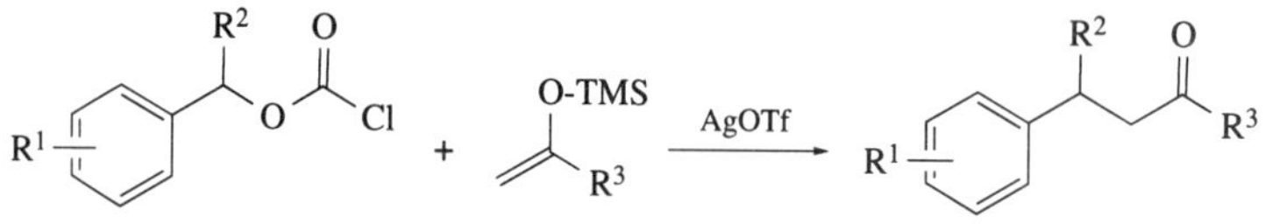

Scheme 67

A Lewis acid mediated reaction between trimethylsilyl enol ethers of (tricarbonyl)iron complexed dienones and aldehydes has provided a route to polyols of known configuration.[320]

Reports on the oxidative cyclization of unsaturated silyl enol ethers have been published;[321,322] δ,ε- and ε,ζ-unsaturated silyl enol ethers of aryl ketones provide tricyclic ketones stereoselectively. 3,4-Disubstituted 4-ketoaldehydes can be prepared in good yields by the oxidative addition[323] of 1,2-disubstituted silyl enol ethers to ethyl vinyl ether using ceric ammonium nitrate. Trimethylsilyl dienol ethers are readily oxidized by ceric ammonium nitrate to α-carbonyl allyl radicals, which in turn add to silyl enol ethers with very high γ-regioselectivity ultimately to produce 6-oxo-α,β-unsaturated carbonyl compounds.[324] Vanadium(V) species have been employed[325] in the oxidative homo- or cross-coupling of silyl enol ethers and ketene acetals to produce 1,4-dicarbonyl compounds.

Certain allylic alcohols undergo a remarkable carbon–carbon bond formation (Scheme 68) in lithium perchlorate–diethyl ether when reacted with two equivalents of the TBDMS ketene acetal of methyl acetate; it would appear that for success, the allylic alcohol should be secondary and possess γ,γ-disubstitution.[326]

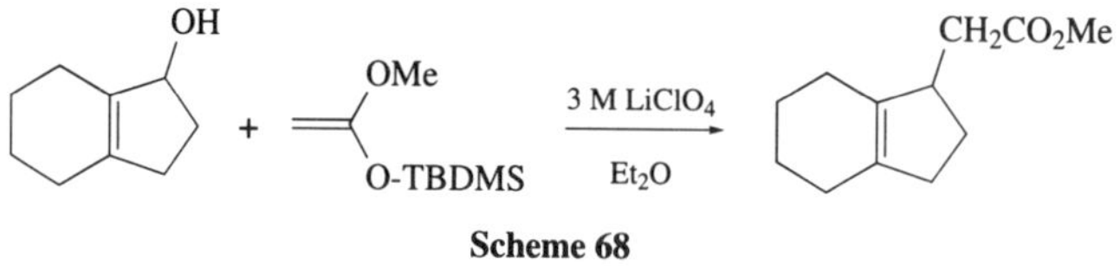

Scheme 68

7.13 TRIALKYLSILYL-X AND RELATED REAGENTS

7.13.1 Trialkylsilyl Halides

The role of TMS-Cl in accelerating organocuprate conjugate additions has been discussed,[327] and its utility in promoting diastereoselective 1,2-addition of organocuprates to α,β-unsaturated γ-alkoxyaldehydes has been reported.[328]

Motherwell's group has reported[329] further on applications of the TMS-Cl/zinc reagent. This system effects a symmetrical dicarbonyl coupling reaction of aryl and α,β-unsaturated carbonyl compounds, probably via trapping of an intermediate zinc carbenoid by a second molecule of carbonyl compound and subsequent deoxygenation of the resulting epoxide. Certain epoxides can be deoxygenated.[330] The dicarbonyl coupling reaction can be improved by substituting 1,2-bis(chlorodimethylsilyl)ethane[331] for TMS-Cl, and the intermediate organozinc carbenoids[332] can be trapped with alkenes to give cyclopropanes (Scheme 69).

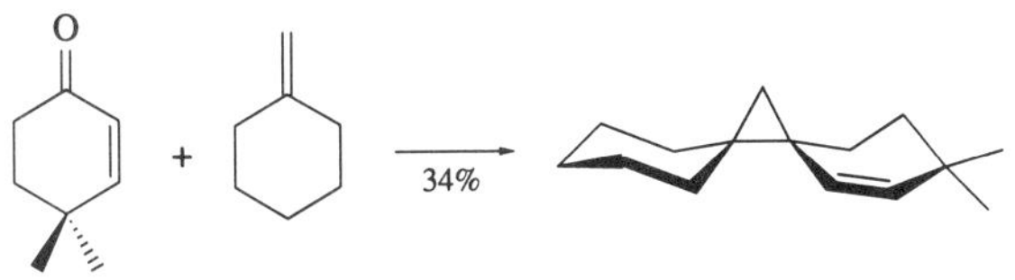

Scheme 69

The combination of TMS-Cl/NaI/water in acetonitrile generates HI *in situ*. Reaction with alkynes[333] gives vinyl iodides with *cis*-selectivity and Markovnikov regiochemistry. The same reagent system with the addition of metallic tin effects the α-selective coupling[334] of allylic alcohols with aldehydes to produce linear homoallylic alcohols (Scheme 70), probably via an allylic tin iodide intermediate. A new route[335] to vinyl iodides from ketones has been described, based on the reaction of vinyl phosphates with TMS-I.

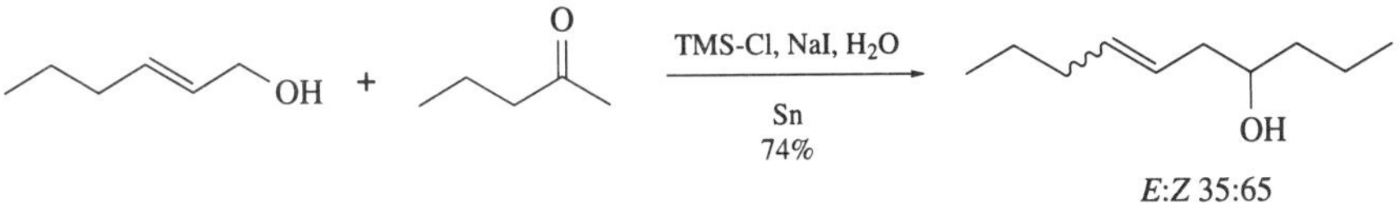

Scheme 70

The preparation and utility of diiodosilane, H_2SiI_2, has been reported.[336] In many respects, its reactions are similar to TMS iodide;[337] however, in contrast, it converts acid derivatives into acyl iodides, and selectively reduces acetals and ketals to iodides in the presence of unprotected carbonyl groups.

An improved method[338] for the *N*-formylation of secondary amines using TMS-Cl and imidazole in DMF has been described.

The reagent combination of TMS-Cl/NaNO$_2$ in CCl$_4$ converts anilines into aryl chlorides in high yield;[339] TMS-Br and TMS-I react similarly.

In the presence of a catalytic amount of selenium dioxide, TMS-Cl has been reported[340] to serve as an efficient chlorinating agent for a wide variety of alcohols; selenium oxychloride is believed to be the active species.

Allylic acetates are converted into allylic bromides using TMS-Br with ZnI$_2$ catalysis; the regiochemistry is determined by the relative stability of the product alkenes.[341]

The combination of TMS-Cl, anisole and a catalytic amount of tin(II) chloride cleaves[342] *p*-methoxybenzyl ethers selectively in the presence of benzyl ethers.

The system of TMS-Cl/benzyltriethylammonium borohydride transforms cyclic and acyclic enol ethers into diols and alcohols, respectively.[343]

Acetals and ketals can be cleaved reductively to ethers or hydroxyethers using TMS-Cl/zinc borohydride.[344]

7.13.2 Trialkylsilyl Cyanides

The chemistry of TMS-CN has been reviewed.[345]

Enantioselective TMS cyanohydrin formation[346] has been achieved with aldehydes using a variety of chirally-modified Lewis acid catalysts. A recyclable tungsten Lewis acid catalyst gives cyanohydrin ethers with comparable diastereoselectivity to that observed with ZnI$_2$; no aqueous workup is required.[347] Lewis bases such as amines and phosphines[348] also catalyse the reaction of TMS-CN with aldehydes.

4-Alkoxy substituted diisopropyl and dibenzyl acetals react with TMS-CN in the presence of TMS-OTf to give cyanohydrin ethers with high 1,4-relative asymmetric induction (Scheme 71); neighbouring group participation has been proposed[349] to account for this.

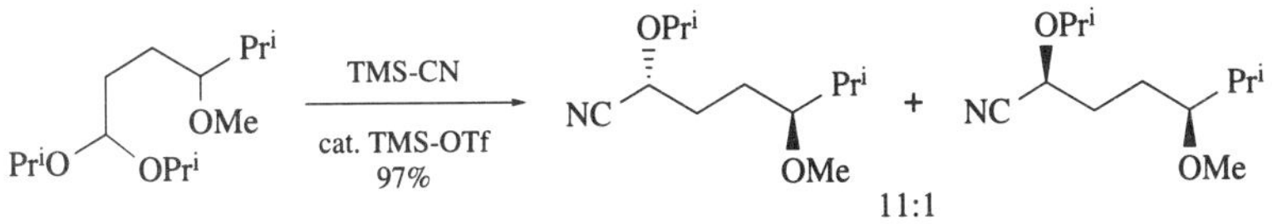

Scheme 71

α,β-Unsaturated ketones react regioselectively with TMS-CN to give 1,4- and 1,2-adducts in the presence of strongly acidic montmorillonite and strongly basic CaO or MgO, respectively.[350]

α-Aminonitriles[351] can be prepared in good yield by reaction of primary amines with aldehydes in the presence of TMS-CN.

There have been several reports on the Lewis acid mediated opening of epoxides with TMS-CN to give β-trimethylsilyloxy nitriles.[352]

Reaction between dicyanodimethylsilane and β-hydroxyketones proceeds with high diastereoselectivity to give, after hydrolysis, *syn*-β-hydroxycyanohydrins. The *syn* configuration has been ascribed[353] to intermolecular addition of cyanide via a chairlike six-membered transition state.

Full details have been published[354] on the palladium- and nickel-catalysed addition of TMS-CN to terminal alkynes to give cyano-substituted (Z)-vinylsilanes; the addition is regioselective, with the cyano group appearing on the internal carbon.

α-Cyano selenides and α-cyano selenoacetals can be prepared from selenoacetals and seleno *ortho*-esters, respectively, by Lewis acid mediated reaction with TMS-CN.[355]

7.13.3 Trialkylsilyl Trifluoromethanesulfonates (Triflates)

The uses of trialkylsilyl perfluoroalkane sulfonates have been reviewed.[356]

TMS-OTf in the presence of diisopropylethylamine converts ketals into enol ethers.[357] TMS fluorosulfonate, generated *in situ* by reaction of fluorosulfonic acid with either tetramethylsilane or allyltrimethylsilane, has been suggested[358] as an alternative to TMS triflate; its reactivity is essentially equivalent and it is cheaper to prepare.

Trialkylsilyl triflates in the presence of tertiary amines function as highly effective catalysts[359] for intramolecular Michael reactions.

Suitable 1,4-*endo*-peroxides undergo a diastereospecific Diels–Alder reaction[360] with dienes to produce 1,2-dioxanes (Scheme 72). The use of TBDMS-OTf as catalyst in Diels–Alder and hetero-Diels–Alder reactions to control *exo/endo* selectivity has been reported.[361]

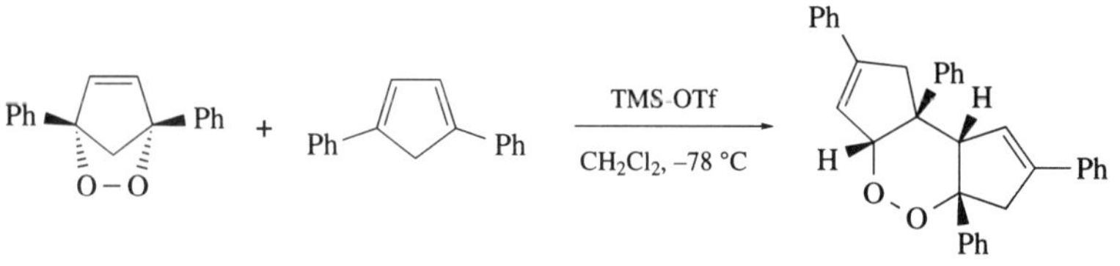

Scheme 72

The combination of TMS-OTf, dimethyl sulfide and 1,2-bis(trimethylsilyloxy)ethane provides a method[362] for the selective dioxolanation of ketones in the presence of aldehydes, via intermediate protection of the aldehyde as its silyloxysulfonium salt. 1,3-Dioxolan-4-ones can be readily prepared from carbonyl compounds by a TMS-OTf- or TMS-I-catalysed reaction with trimethylsilyl (trimethylsilyloxy)acetate.[363]

Ring opening of tetrahydrofuran with TBDMS–Mn(CO)$_5$, generated *in situ* from TBDMS triflate and NaMn(CO)$_5$, proved a key step[364] in the synthesis of spiroketal lactones, precursors of certain insect pheromones.

Dimethoxyamine reacts with electron-rich alkenes in the presence of TMS-OTf to produce N-methoxyaziridines in good yield; cyclic ketones react in a Beckmann-type process to yield N-methoxylactams.[365]

An unusual TBDMS triflate-promoted intramolecular hydride shift[366] has been observed in the reactions of a potential taxane intermediate.

The aluminoxy acetal intermediates in ester reduction using dibal-H can be trapped[367] with either TMS triflate or TMS imidazole to give monosilyl acetals.

7.13.4 Trialkylsilyl Peroxides and Related Species

An improved method[368] for the preparation of bis(TMS) peroxide, using the urea-hydrogen peroxide complex and N,N'-bis(TMS)urea, has been presented. Use of the same complex led to the synthesis[369] of the first simple cyclic silyl peroxide (**49**) (Scheme 73).

Bis(TMS) peroxide/triflic acid has been found to be a highly effective hydroxylating reagent[370] for aromatics, providing the corresponding phenols in high yields.

Triethylsilyl hydrotrioxide, Et₃SiOOOH, prepared by ozonolysis of triethylsilane, is a new chemical source[371] of singlet oxygen, $^1\Delta_g$ O₂. It also reacts directly with alkenes to give 1,2-dioxetanes and oxidatively cleaved carbonyl products.[372]

Selective inter- and intramolecular oxidations with α-silyloxyalkyl hydroperoxides have been reviewed.[373]

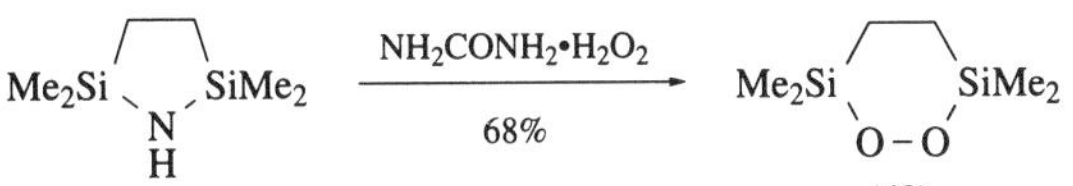

$$\underset{\underset{H}{N}}{Me_2Si\diagdown \diagup SiMe_2} \xrightarrow[68\%]{NH_2CONH_2 \cdot H_2O_2} \underset{O-O}{Me_2Si\diagdown \diagup SiMe_2} \quad (49)$$

Scheme 73

There has been some controversy over whether TMS perchlorate is covalent[374] or in equilibrium between covalent and ionic forms.[375]

7.13.5 Trialkylsilyl Azides

The reagent system TMS azide/triflic acid allows simple and near quantitative amination of aromatics,[376] and the system TMS azide/N-bromosuccinimide with Nafion-H catalysis performs an efficient azidobromination[377] of alkenes to give β-bromoalkyl azides.

The ring opening of functionalized epoxides with TMS azide in the presence of a catalytic amount of Ti(OPr^i)₄ or Al(OPr^i)₃ has been reported[378] in detail. Using chiral titanium catalysts, cyclohexene oxide gives trans-2-azidocyclohexanol in up to 63% ee.[379]

The combination of TMS azide and CrO₃ efficiently transforms aldehydes into acyl azides.[380]

The tin(II) chloride or zinc chloride catalysed addition of TMS azide to aldehydes and ketones has been found to produce gem-diazides, whereas catalysis using sodium azide and 18-crown-5 gave α-silyloxyazides exclusively.[381]

7.13.6 TMS Diazomethane

The chemistry of TMS diazomethane has been reviewed.[382]

The reaction of lithio(TMS)diazomethane with carbonyl compounds, which generates alkylidenecarbenes, has been subject to further investigation. Aldehydes and aryl alkyl ketones give alkynes,[383] whereas aliphatic ketones give cyclopentene derivatives.[384] The intermediate diazoalkenes from aliphatic ketones can be trapped with diisopropylamine to produce aldehyde enamines, providing a method[385] for the one carbon homologation of ketones to aldehydes.

Lithio(TMS)diazomethane reacts with ketenimines to produce 4-TMS-1,2,3-triazoles or 4-amino-3-TMS-pyrazoles,[386] depending on the ketenimine substitution pattern.

7.13.7 Trialkylsilyl Thiols and Selenols and Related Species

The preparation and reactions of organosilathianes and other mixed silicon–sulfur compounds have been reviewed.[387]

Triphenylsilanethiol, a white crystalline solid, has been recommended[388] for the ring opening of epoxides to β-hydroxymercaptans or β-dihydroxysulfides, as have sodium TMS thiolate and hexamethyldisilathiane for the bis-O-demethylation of aryl methyl ethers.[389]

Nitriles react with sodium TMS thiolate in 1,3-dimethyl-2-imidazolidinone (DMEU) to yield primary thioamides, often in good yield.[390]

Full details[391] have been published on the use of hexamethyldisilathiane in the generation of thioaldehydes and thioketones. The corresponding selenium reagent has been employed[392] in the generation of selenoaldehydes.

The TMS triflate catalysed reaction of (phenylthio)trimethylsilane with aldehydes to give hemithioacetals has been employed in a stereocontrolled route[393] to polyols.

Bis(trimethylsilylmethyl) sulfide reacts with carbonyl compounds in the presence of fluoride ion to produce β-substituted β-hydroxyethyl methyl thioethers.[394]

TMS isoselenocyanate, TMS-NCSe, prepared in situ from TMS-Cl and excess KSeCN, shows remarkable chemoselectivity[395] in its transformation of carbonyl compounds into O-TMS cyanohydrins.

Aliphatic aldehydes react smoothly, but aromatic and α,β-unsaturated aldehydes react very slowly; ketones are unreactive. In marked contrast,[396] TMS isothiocyanate reacts with aldehydes under Lewis acid catalysis to produce dimeric α,α'-isothiocyanato ethers.

The utility of 2-(TMS)ethanethiol for the preparation of, *inter alia*, carboxylic acid thiol esters and α-mercapto ketones has been reported.[397]

Further details[398] have been given on the preparation and synthetic utility of 1-(phenylsulfonyl)-2-(TMS)ethane. An application of its utility in alkene generation can be seen in an asymmetric total synthesis[399] of the methylenecyclopropane-containing α-amino acid hypoglycin A.

7.13.8 Trialkylsilyl Stannanes

The scope and limitations of the Pd⁰-catalysed addition of (trialkylsilyl)trialkylstannanes to alkynes has been investigated in detail.[400] The addition proceeds best with terminal alkynes, regio- and stereospecifically producing synthetically useful (Z)-1-silyl-2-stannylalkenes.

(Trialkylsilyl)trialkylstannanes react with higher-order cuprates to afford trialkylstannyl or trialkylsilyl mixed cuprates, depending on the nature of the alkyl groups; these cuprates participate in substitution, conjugate addition and metalcupration processes.[401]

Fluoride ion cleavage of (TMS)tributylstannane provides a new method[402] for the generation of the tributyltin anion.

The insertion of isonitriles into the Si–Sn bond of organosilylstannanes has been investigated.[403]

7.13.9 Miscellaneous

Trifluoromethyltrimethylsilane is a convenient trifluoromethylating agent[404] for carbonyl compounds, producing after hydrolysis trifluoromethylcarbinols.

The synthetic utility of silyl esters of phosphorus has been reviewed.[405] Diethyl TMS phosphite reacts with α,β-unsaturated imines to give 1,2-addition products, regardless of the electronic demand on the imine bond.[406] Modest diastereoselectivity in carbon–phosphorus bond formation is observed[407] in the addition of ethyl TMS phenylphosphite to *N*-benzylimines. Ephedrine-derived chiral *O*-TMS phosphites react with benzaldehyde to afford phosphonate esters with good diastereoselectivity.[408] *In situ* generated bis(TMS) phosphonite, (TMS-O)₂PH, adds conjugatively to α,β-unsaturated ketones to afford mono- and disubstituted phosphinic acids.[409]

The palladium-catalysed deprotection of allylic carbamates can be achieved under mild and selective conditions[410] by employing silylated amines such as TMS dimethylamine or TMS morpholine in the presence of TMS trifluoroacetate; competitive formation of allylamine is greatly suppressed.

The SiCl₄/ethanol induced self-condensation of ketones to yield tri- or hexasubstituted benzenes has been investigated.[411]

Iodotrichlorosilane, readily prepared *in situ* from SiCl₄ and NaI, has been recommended[412] for the regeneration of carbonyl compounds from acetals and ketals.

An efficient one-pot synthesis[413] of Mosher's acid, involving the addition of TMS trichloroacetate to α,α,α-trifluoroacetophenone and subsequent hydrolysis, has been described.

A text on organosilicon-based nucleoside synthesis has been published.[414]

7.14 AMINOSILANES AND RELATED COMPOUNDS

Anilines are afforded protection as 'benzostabase' derivatives by dehydrogenative silylation (Scheme 74) with 1,2-bis(dimethylsilyl)benzene.[415] The protective group is considerably more stable than the corresponding 'stabase' derivative towards acid hydrolysis and silica gel chromatography. Similar protection of primary aliphatic amines[416] gives a useful, but lesser, improvement in stability.

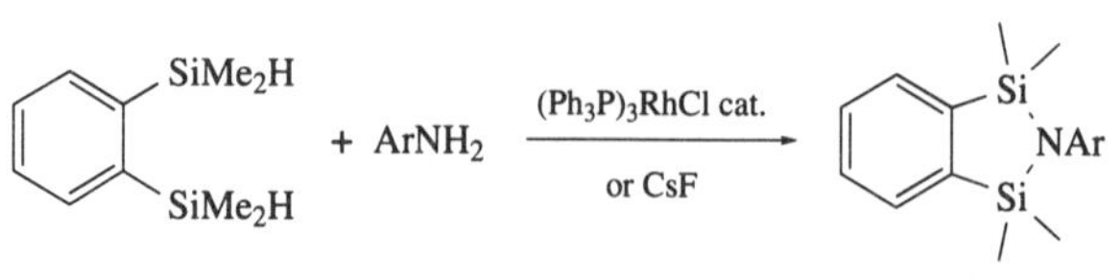

Scheme 74

Allylic chlorides undergo nucleophilic substitution with the mixed reagent lithium N,N-bis(TMS)amide/AgI to give silyl-protected primary allylic amines in good to excellent yields.[417]

Full details[418] have been published on the preparation and reactions of N,N-bis(silyl)enamines. Full details have also been given on N-(triisopropylsilyl)pyrrole and its reactions, mainly with electrophiles.[419] The bulky silyl substituent effectively blocks attack at the 2-position, leading to 3-substituted pyrroles.

Selective nucleophilic addition of alkyllithium reagents to N-(TMS)lactams gives cyclic ketimines in good to excellent yields.[420]

The preparation and some reactions of N-silyl imines have been reviewed,[421] as have their condensation reactions with ester enolates and equivalents to produce β-lactams.[263,422]

The N-(TMS) imines of O-protected α-hydroxyaldehydes react with organolithium and Grignard reagents to yield *syn* β-amino alcohols[423] with good diastereoselectivity. The *anti* diastereoisomers[424] are produced when the reagent is $RCuMgXI \cdot BF_3$.

N-(TMS)benzaldehyde imine has been alkylated with chirally modified organometallic reagents to give optically active primary amines in up to 62% *ee*. A remarkable solvent effect was observed,[425] with diethyl ether favouring the S enantiomer and hexane the R.

N-(TMS) imines have been employed in routes to 3,4-dihydro-1,3,5-2H-thiadiazines[426] and N-unsubstituted imidazoles.[427]

7.15 SILANES AS REDUCING AGENTS

7.15.1 Hydrosilylation

A major treatise has been published on hydrosilylation, covering the literature up to 1990.[428] Dehydrogenative silylation, a process in which alkenes are converted into vinylsilanes, has been reviewed.[429]

7.15.1.1 Intermolecular processes

Highly substituted alkenes undergo an $AlCl_3$-catalysed hydrosilylation in an unusual *trans* manner; this stereochemistry and regiochemistry has been ascribed[430] to operation of the β-effect.

As cited earlier,[22] the rhodium-catalysed hydrosilylation of hex-1-yne exhibits a remarkable dependence on solvent in controlling the stereochemistry of the vinylsilane products. The potential of relatively inexpensive organoneodymium complexes as catalysts has been explored; only mono- and diorganosilanes reacted satisfactorily.[431]

Hydrosilylation using triethoxysilane in a THF–water system gives silicon-free reduction products; internal alkynes are cleanly reduced to (Z)-alkenes, with many other functional groups being unaffected.[432]

Palladium-catalysed hydrosilylation of terminal alkenes using trichlorosilane and a chiral binaphthol-derived monodentate phosphine ligand proceeds with unusual regiochemistry, selectively producing the internal silane;[433] oxidative cleavage[219] then yields secondary alcohols in up to 97% *ee*. Chiral bisoxazolines and bisthiazolines show some promise[434] as ligands in metal-catalysed asymmetric hydrosilylation.

7.15.1.2 Intramolecular processes

Intramolecular hydrosilylation of silyl ethers derived from allylic and homoallylic alcohols, with subsequent oxidative cleavage[219] of the resulting C–Si bond, provides a regiocontrolled route[435] to 1,2- and 1,3-diols. This reaction can exhibit high levels of diastereoselectivity.[436] With allylic alcohols, this process can be achieved enantioselectively[437] using rhodium(I) chiral diphosphine complexes.

Hydroxyketones also take part in Lewis acid catalysed intramolecular hydrosilylation. With β-hydroxyketones, it is highly *anti*-selective, leading to *syn*-1,3-diols; this can also be performed enantioselectively.[234]

Intramolecular bis(silylation) of disilanyl ethers derived from homoallylic alcohols results in regioselective *exo*-ring closure (Scheme 75) to produce 1,2-oxasilolanes. Oxidative C–Si bond cleavage completes a stereoselective route to 1,2,4-triols;[438] this methodology has been extended to the stereoselective preparation of 1,2,4,5,7- and 1,2,4,6,7-pentaols. Interestingly, in one case iodine monochloride was used to cleave the Si–Ph bond to prevent phenyl migration. Other stereocontrolled routes to polyols have been reported;[439] although they do not involve hydrosilylation, they do employ oxidative C–Si cleavage.

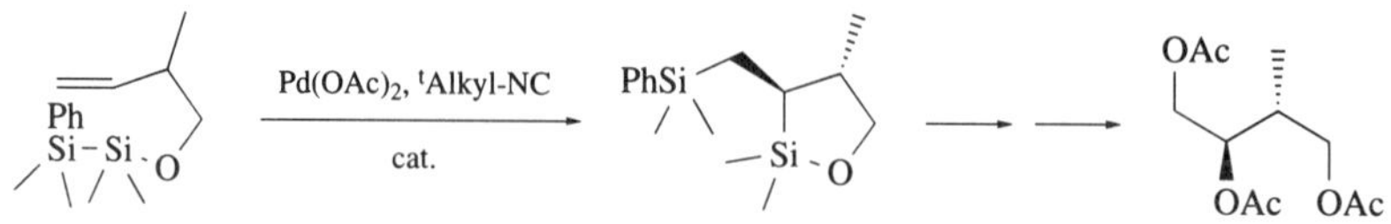

Scheme 75

7.15.1.3 Radical-based reducing agents

The applications of tris(TMS)silane have been reviewed;[440] this reagent is as effective a reducing agent as tributyltin hydride, and is ecologically more acceptable. Its use as a radical mediator has allowed the carbonylation of alkyl halides to aldehydes[441] under relatively low CO pressures; in the additional presence of electron-deficient alkenes, unsymmetrical ketones are produced in a formal double alkylation of CO.

Triphenylsilane,[442] tris(alkylthio)silanes[443] and triethylsilane in the presence of thiols[444] have shown promise as alternative radical-based reducing agents. Phenylsilane and diphenylsilane have been recommended[445] as alternatives to tributyltin hydride for the radical deoxygenation of primary and secondary alcohol thionocarbonates and xanthates, on both ecological and practical grounds.

7.15.1.4 Carbon monoxide incorporation

The combination of a silane and CO in the presence of a transition metal catalyst results in the incorporation of both the silyl group and CO into many substrates, including alkynes,[446] cyclic orthoesters,[447] oxiranes[448] and enamines.[449]

7.15.1.5 Functional transformation

A cautionary note[450] has been given on the use of the triethoxysilane–titanium tetraisopropoxide reagent system for the reduction of esters to primary alcohols; pyrophoric SiH_4 can be evolved.

The palladium-catalysed reaction of triethylsilane with thioesters efficiently produces aldehydes.[451]

7.15.2 Ionic Hydrogenation

A detailed kinetic study[452] of hydride transfer from hydridosilanes to carbenium ions using kinetic isotope effects has indicated a polar, rather than a single-electron transfer mechanism[453] to be operative.

Addition of ammonium fluoride to the reducing system of triethylsilane-trifluoroacetic acid makes product separation easier,[454] due to the formation of volatile fluorotriethylsilane rather than the normal triethylsilanol and hexaethyldisiloxane.

Intramolecular ionic hydrogenation has been employed to control benzylic stereochemistry; in the example shown,[455] conventional reduction gave the epimeric methyl stereochemistry (Scheme 76).

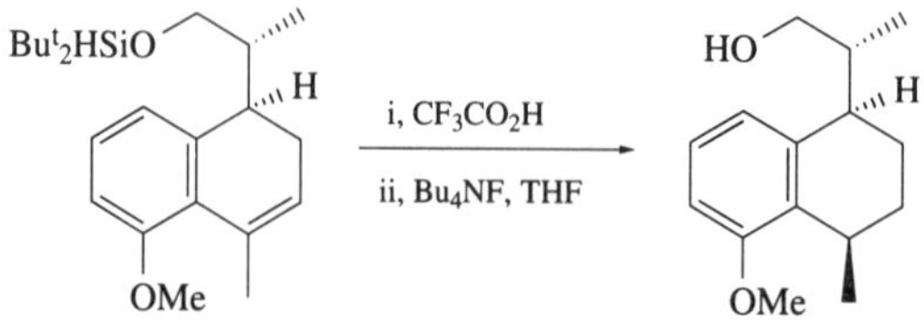

Scheme 76

Triethylsilane-trifluoroacetic acid has been recommended[456] for the selective cleavage of *t*-butyl esters and *t*-butoxycarbonyl groups in the presence of other common acid-sensitive peptide protective groups such as benzyloxycarbonyl.

Reductive coupling of carbonyl compounds with themselves or with alkyl silyl ethers using triethylsilane in the presence of either TMS triflate or iodide provides a mild route[457] to symmetrical and unsymmetrical ethers, respectively.

Ionic hydrogenation has been applied to the deoxygenation of sugar hemiacetals, providing a direct synthesis of *C*-glycosides[458] and sugar allylic acetates.[459]

7.16 ORGANIC SYNTHESES

Table 1 shows a compilation of those preparations and transformations involving silicon published in *Organic Syntheses* since the previous compilation.[460]

Table 1 Preparations and transformations involving silicon.

Detail	Reference
Trimethylsilyldiazomethane, a stable and safe substitute for diazomethane	*Org. Synth.*, 1990, **68**, 1
Acetyltrimethylsilane, the simplest acylsilane	*Org. Synth.*, 1990, **68**, 25
(Z)-4-(Trimethylsilyl)-3-buten-1-ol	*Org. Synth.*, 1990, **68**, 182
Regioselective synthesis of tetrahydropyridines using the above reagent: 1-(4-methoxyphenyl)-1,2,5,6-tetrahydropyridine	*Org. Synth.*, 1990, **68**, 188
The conversion of esters to allylsilanes: trimethyl(2-methylene-4-phenyl-3-butenyl)silane	*Org. Synth.*, 1990, **69**, 89
Nucleophilic hydroxymethylation of carbonyl compounds: 1-(hydroxymethyl)cyclohexanol	*Org. Synth.*, 1990, **69**, 96
2-Substituted pyrroles from *N-t*-butoxycarbonyl-2-bromopyrrole: *N-t*-butoxycarbonyl-2-TMS-pyrrole	*Org. Synth.*, 1992, **70**, 151
Tris(TMS)silane	*Org. Synth.*, 1992, **70**, 164
Preparation of (*E,Z*)-1-methoxy-2-methyl-3-(trimethylsilyloxy)-1,3-pentadiene	*Org. Synth.*, 1992, **70**, 231
Nickel-catalysed silylolefination of allylic dithioacetals: (*E,E*)-trimethyl-(4-phenyl-1,3-butadienyl)silane	*Org. Synth.*, 1992, **70**, 240
3-(S)-[(*t*-Butyldiphenylsilyl)oxy]-2-butanone, a useful chiral precursor	*Org. Synth.*, 1993, **71**, 56
Stereocontrolled preparation of 3-acyltetrahydrofurans using the above reagent	*Org. Synth.*, 1993, **71**, 63
Benzoannelation of ketones, silyl enol ether chemistry	*Org. Synth.*, 1993, **71**, 158
Methoxycarbonylmethylation of aldehydes via silyloxycyclopropanes: methyl 3,3-dimethyl-4-oxobutanoate	*Org. Synth.*, 1993, **71**, 189
Titanium-mediated addition of silyl dienol ethers to electrophilic glycine: a short synthesis of 4-ketopipecolic acid hydrochloride	*Org. Synth.*, 1993, **71**, 200

Asymmetric synthesis using chiral organosilicon compounds, where the chirality is either silicon- or carbon-centred, has been reviewed.[461]

The synthesis, physical properties and transition metal complexes of silacyclopentadienes (siloles) have been detailed.[462]

The composition and chemistry of a variety of mixed higher-order silyl cuprates has been investigated.[463] Intriguingly, it has been reported[464] that, under certain conditions, treatment of hexamethyldisilane with methyllithium results in carbon–silicon, and not silicon–silicon bond cleavage, generating TMS-Me$_2$SiLi. The preparation of the (aminosilyl)lithium reagents (**50**), (**51**) and (**52**) has been reported;[465] these are the first stable functional silyl anions to be generated. An example of the use of (**50**) is shown in Scheme 77.

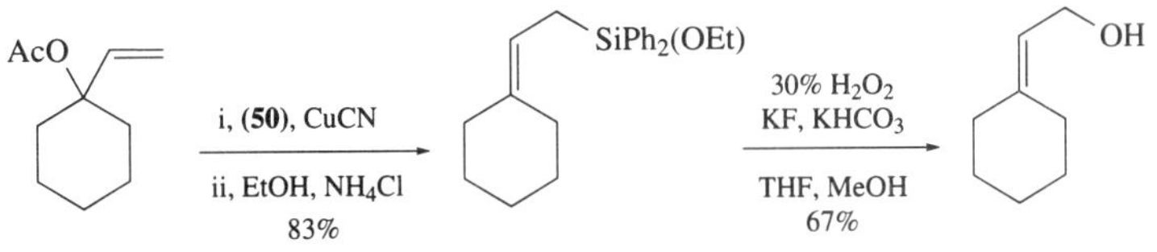

Scheme 77

(Et$_2$N)Ph$_2$SiLi   (Et$_2$N)$_2$PhSiLi   (Et$_2$N)MePhSiLi

(**50**)   (**51**)   (**52**)

7.17 REFERENCES

1. P. D. Magnus, T. Sarkar and S. Djuric, in 'COMC-I', vol. 7, p. 515.
2. W. P. Weber, 'Silicon Reagents for Organic Synthesis', Springer, Berlin, 1983.
3. S. Pawlenko, 'Organosilicon Chemistry', Walter de Gruyter, Berlin, 1986.
4. E. W. Colvin, 'Silicon Reagents in Organic Synthesis', Academic Press. London, 1988.
5. S. E. Thomas, 'Organic Synthesis. The Roles of Boron and Silicon', Oxford University Press, Oxford, 1991.
6. H. Sakurai (ed.), 'Organosilicon and Bioorganosilicon Chemistry—Structure, Bonding, Reactivity and Synthetic Application', Ellis Horwood, Chichester, 1985.
7. E. R. Corey, J. Y. Corey and P. P. Gaspar (eds.), 'Silicon Chemistry', Ellis Horwood, Chichester, 1988.
8. D. Schinzer (ed.), 'Selectivities in Lewis Acid Promoted Reactions', Kluwer, Dordrecht, 1989.
9. A. R. Bassindale and P. P. Gaspar (eds.), 'Frontiers of Organosilicon Chemistry', The Royal Society of Chemistry, Cambridge, 1991.
10. S. Patai and Z. Rappoport (eds.), 'The Chemistry of Organic Silicon Compounds', Wiley, Chichester, 1989.
11. B. M. Trost and I. Fleming (eds), 'Comprehensive Organic Synthesis', Pergamon, Oxford, 1991; see in particular vols. 1, 2, 3, 6, 7 and 8.
12. H. J. Reich (ed.), *Tetrahedron*, 1983, **39**, 839.
13. I. Fleming (ed.), *Tetrahedron*, 1988, **44**, 3761.
14. E. W. Colvin, in 'The Chemistry of the Metal–Carbon Bond', eds. S. Patai and F. R. Hartley, Wiley, Chichester, 1987, vol. 4, p. 539.
15. G. L. Larson, *J. Organomet. Chem.*, 1984, **274**, 29; *ibid.*, 1986, **313**, 141; *ibid.*, 1987, **337**, 195; *ibid.*, 1989, **360**, 39; *ibid.*, 1991, **416**, 1; *ibid.*, 1992, **422**, 1; G. L. Larson, *J. Organomet. Chem. Library*, 1984, **14**, 267; *ibid.*, 1985, **17**, 1; J. Y. Corey, *ibid.*, 1984, **14**, 1, 133; *ibid.*, 1985, **17**, 163; F. K. Cartledge, *ibid.*, 1984, **14**, 409.
16. R. R. Holmes, *Chem. Rev.*, 1990, **90**, 17.
17. C. Chuit, R. J. P. Corriu, C. Reye and J. C. Young, *Chem. Rev.*, 1993, **93**, 1371; see also Ref. 98.
18. J. R. Hwu and N. Wang, *Chem. Rev.*, 1989, **89**, 1599.
19. J. B. Lambert, *Tetrahedron*, 1990, **46**, 2677; J. B. Lambert and E. C. Chelius, *J. Am. Chem. Soc.*, 1990, **112**, 8120 and references therein.
20. M. A. Brook, C. Henry, R. Jueschke and P. Modi, *Synlett*, 1993, 97.
21. T.-Y. Luh and K.-T. Wong, *Synthesis*, 1993, 349.
22. R. Takeuchi and N. Tanouchi, *J. Chem. Soc., Perkins Trans. 1*, 1993, 1319.
23. K. Karabelas and A. Hallberg, *J. Org. Chem.*, 1986, **51**, 5286.
24. H. Yamashita, B. L. Roan and M. Tanaka, *Chem. Lett.*, 1990, 2175.
25. K. Ikenaga, K. Kikukawa and T. Matsuda, *J. Org. Chem.*, 1987, **52**, 1276.
26. N. Miyaura, T. Ishiyama, H. Sasaki, M. Ishikawa, M. Satoh and A. Suzuki, *J. Am. Chem. Soc.*, 1989, **111**, 314.
27. J. A. Soderquist and J. C. Colberg, *Synlett*, 1989, 25.
28. Y. Hatanaka and T. Hiyama, *Synlett*, 1991, 845.
29. K. Nakatani, T. Izawa, Y. Odagaki and S. Isoe, *J. Chem. Soc., Chem. Commun.*, 1993, 1365.
30. R. Takeuchi and M. Sugiura, *J. Chem. Soc., Perkin Trans. 1*, 1993, 1031.
31. F. L. Koerwitz, G. B. Hammond and D. F. Wiemer, *J. Org. Chem.*, 1989, **54**, 738, 743.
32. I. Fleming and E. M. de Marigorta, *Tetrahedron Lett.*, 1993, **34**, 1201 and references therein.
33. T. F. Bates and R. D. Thomas, *J. Org. Chem.*, 1989, **54**, 1784.
34. N. A. Petasis and I. Akritopoulou, *Synlett*, 1992, 665.
35. K. Takai, M. Tezuka, Y. Kataoka and K. Utimoto, *Synlett*, 1989, 27.
36. N. Kishi, H. Imma, K. Mikami and T. Nakai, *Synlett*, 1992, 189.
37. C. Gajda and J. R. Green, *Synlett*, 1992, 973.
38. I. Fleming, J. Dunoguès and R. Smithers, *Org. React. (NY)* 1989, **37**, 57.
39. T. A. Blumenkopf and L. E. Overman, *Chem. Rev.*, 1986, **86**, 857; T. A. Blumenkopf *et al.*, *J. Am. Chem. Soc.*, 1990, **112**, 4386; M. T. Reetz and P. Hois, *J. Chem. Soc., Chem. Commun.*, 1989, 1081.
40. A. P. Wells, B. H. Riches and W. Kitching, *J. Chem. Soc., Chem. Commun.*, 1992, 1575.
41. P. G. Baraldi, R. Bazzanini, S. Manfredini, D. Simoni and M. J. Robins, *Tetrahedron Lett.*, 1993, **34**, 3177.
42. M. J. Robins and S. Manfredini, *Tetrahedron Lett.*, 1990, **31**, 5633.
43. R. Angell, P. J. Parsons and A. Naylor, *Synlett*, 1993, 189.
44. F. Babudri, V. Fiandanese, F. Naso and A. Punzi, *Synlett*, 1992, 221 and references therein; V. Fiandanese and L. Mazzone, *Tetrahedron Lett.*, 1992, **33**, 7067.
45. K. Maruoka, H. Banno and H. Yamamoto, *J. Am. Chem. Soc.*, 1990, **112**, 7791; K. Maruoka, H. Banno and H. Yamamoto, *Tetrahedron Asymmetry*, 1991, **2**, 647; K. Maruoka and H. Yamamoto, *Synlett*, 1991, 793.
46. K. Mikami and M. Shimizu, *Chem. Rev.*, 1992, **92**, 1021.
47. K. Mikami and T. Nakai, *Synthesis*, 1991, 594.
48. J. E. Crawley, A. D. Kaye, G. Pattenden and S. M. Roberts, *J. Chem. Soc., Perkin Trans. 1*, 1993, 2001.
49. M. C. McIntosh and S. M. Weinreb, *J. Org. Chem.*, 1991, **56**, 5010.
50. A. G. M. Barrett and J. W. Malecha, *J. Org. Chem.*, 1991, **56**, 5243.
51. I. Fleming, T. W. Newton, V. Sabin and F. Zammattio, *Tetrahedron*, 1992, **48**, 7793.
52. G. Stork, P. M. Sher and H.-L. Chen, *J. Am. Chem. Soc.*, 1986, **108**, 6384.
53. J. Germanas and K. P. C. Vollhardt, *Synlett*, 1990, 505.
54. D. L. Comins and H. Hong, *J. Am. Chem. Soc.*, 1991, **113**, 6672; D. L. Comins and D. H. La Munyon, *J. Org. Chem.*, 1992, **57**, 5807.
55. I. Fleming and S. K. Ghosh, *J. Chem. Soc., Chem. Commun.*, 1992, 1775.
56. S. Yamazaki, S. Katoh and S. Yamabe, *J. Org. Chem.*, 1992, **57**, 4.
57. R. Takeuchi, N. Ishii and N. Sato, *J. Chem. Soc., Chem. Commun.*, 1991, 1247.
58. T. K. Sarkar, *Synthesis*, 1990, 969 and 1101.
59. I. Fleming, in 'Comprehensive Organic Synthesis', eds. B. M. Trost and I. Fleming, Pergamon, Oxford, 1991, vol. 2, p. 563.

60. H.-J. Gais and G. Bülow, *Tetrahedron Lett.*, 1992, **33**, 461, 465.
61. A. Degl'Innocenti, A. Mordini, L. Pagliai and A. Ricci, *Synlett*, 1991, 155.
62. M. Franciotti, A. Mordini and M. Taddei, *Synlett*, 1992, 137.
63. M. Heneghan and G. Procter, *Synlett*, 1992, 489.
64. I. Fleming, D. Higgins, N. J. Lawrence and A. P. Thomas, *J. Chem. Soc., Perkin Trans. 1*, 1992, 3331.
65. I. Fleming, S. Gil, A. K. Sarkar and T. Schmidlin, *J. Chem. Soc., Perkin Trans. 1*, 1992, 3351.
66. Y. Yamamoto and N. Asao, *Chem. Rev.*, 1993, **93**, 2207.
67. M. J. C. Buckle, I. Fleming and S. Gil, *Tetrahedron Lett.*, 1992, **33**, 4479 and references therein.
68. M. Patz and H. Mayr, *Tetrahedron Lett.*, 1993, **34**, 3393.
69. G. Hagen and H. Mayr, *J. Am. Chem. Soc.*, 1991, **113**, 4954.
70. I. Fleming, *J. Chem. Soc., Perkin Trans. 1*, 1992, 3363.
71. I. Fleming and J. J. Lewis, *J. Chem. Soc., Perkin Trans. 1*, 1992, 3267; I. Fleming and D. Higgins, *J. Chem. Soc., Perkin Trans. 1*, 1992, 3327.
72. I. Fleming and N. J. Lawrence, *J. Chem. Soc., Perkin Trans. 1*, 1992, 3309.
73. I. Fleming, N. J. Lawrence, A. K. Sarkar and A. P. Thomas, *J. Chem. Soc., Perkin Trans. 1*, 1992, 3303.
74. E. W. Colvin and M. Monteith, *J. Chem. Soc., Chem. Commun.*, 1990, 1230; see also C. Nativi, E. Perrotta, A. Ricci and M. Taddei, *Tetrahedron Lett.*, 1991, **32**, 2265.
75. E. W. Colvin, W. A. König, M. A. Loreto, J. Y. Rowden and I. Tommasini, *Bioorg. Medicinal Chem. Lett.*, 1993, **3**, 2405.
76. H. Hiemstra and W. N. Speckamp, in 'Comprehensive Organic Synthesis', eds. B. M. Trost and I. Fleming, Pergamon, Oxford, 1991, vol. 2, p. 1047.
77. E. C. Roos et al., *J. Org. Chem.*, 1993, **58**, 3259 and references therein.
78. M. Thaning and L.-G. Wistrand, *J. Org. Chem.*, 1990, **55**, 1406.
79. W.-J. Koot, H. Hiemstra and W. N. Speckamp, *Tetrahedron Lett.*, 1992, **33**, 7969.
80. G. Pandey, K. S. Rani and G. Lakshmaiah, *Tetrahedron Lett.*, 1992, **33**, 5107.
81. S. J. Danishefsky, H. G. Selnick, R. E. Zelle and M. P. DeNinno, *J. Am. Chem. Soc.*, 1988, **110**, 4368 and references therein.
82. K. Haraguchi, H. Tanaka, Y. Itoh, S. Saito and T. Miyasaka, *Tetrahedron Lett.*, 1992, **33**, 2841.
83. C. R. Bertozzi and M. D. Bednarski, *Tetrahedron Lett.*, 1992, **33**, 3109.
84. A. Schmitt and H.-U. Reissig, *Synlett*, 1990, 40.
85. S. E. Denmark, M. A. Wallace and C. B. Walker, Jr., *J. Org. Chem.*, 1990, **55**, 5543 and references therein; K.-T. Kang, S. S. Kim, J. C. Lee and J. S. U, *Tetrahedron Lett.*, 1992, **33**, 3495; D. El Abed, C. Santelli-Rouvier and M. Santelli, *Tetrahedron*, 1990, **46**, 5993.
86. T. Takeda, Y. Takagi, H. Takano and T. Fujiwara, *Tetrahedron Lett.*, 1992, **33**, 5381.
87. Y. Yamamoto, in 'Comprehensive Organic Synthesis', eds. B. M. Trost and I. Fleming, Pergamon, Oxford, 1991, vol. 2, p. 55; T. H. Chan and D. Labrecque, *Tetrahedron Lett.*, 1992, **33**, 7997.
88. A. Capperucci et al., *Synlett*, 1992, 880.
89. E. Vedejs and C. K. McClure, *J. Am. Chem. Soc.*, 1986, **108**, 1094.
90. J. S. Panek and P. F. Cirillo, *J. Am. Chem. Soc.*, 1990, **112**, 4873; J. S. Panek and J. Zhang, *J. Org. Chem.*, 1993, **58**, 294.
91. K. B. Sharpless et al., *J. Org. Chem.*, 1992, **57**, 2768 and references therein.
92. J. A. Soderquist, A. M. Rane and C. J. López, *Tetrahedron Lett.*, 1993, **34**, 1893.
93. S. Okamoto, K. Tani, F. Sato, K. B. Sharpless and D. Zargarian, *Tetrahedron Lett.*, 1993, **34**, 2509.
94. R. A. Ward and G. Procter, *Tetrahedron Lett.*, 1992, **33**, 3363.
95. P. Mohr, *Tetrahedron Lett.*, 1992, **33**, 2455; for other diastereoselective epoxidation studies see P. J. Murphy, A. T. Russell and G. Procter, *Tetrahedron Lett.*, 1990, **31**, 1055; P. J. Murphy and G. Procter, *Tetrahedron Lett.*, 1990, **31**, 1059.
96. T. Harada, S. Imanaka, Y. Ohyama, Y. Matsuda and A. Oku, *Tetrahedron Lett.*, 1992, **33**, 5807.
97. Y. Yamamoto and N. Sasaki, in 'Stereochemistry of Organometallic and Inorganic Compounds', ed. I. Bernal, Elsevier, Amsterdam, 1989, vol. 3, p. 363; J. S. Panek, in 'Comprehensive Organic Synthesis', eds. B. M. Trost and I. Fleming, Pergamon, Oxford, 1991, vol. 1, p. 579; see also Ref. 66.
98. H. Sakurai, *Synlett*, 1989, 1; H. Sakurai, *Pure Appl. Chem.*, 1985, **57**, 1759.
99. A. Hosomi, *Acc. Chem. Res.*, 1988, **21**, 200.
100. D. Schinzer, *Synthesis*, 1988, 263.
101. G. Majetich, in 'Organic Synthesis, Theory and Applications', ed. T. Hudlicky, Jai Press, Greenwich, CT, 1989, p. 173.
102. I. E. Markó and D. J. Bayston, *Tetrahedron Lett.*, 1993, **34**, 6595 and references therein.
103. J. S. Panek and M. A. Sparks, *J. Org. Chem.*, 1990, **55**, 5564; P. F. Cirillo and J. S. Panek, *J. Org. Chem.*, 1990, **55**, 6071; J. S. Panek, R. Beresis, F. Xu and M. Yang, *J. Org. Chem.*, 1991, **56**, 7341; J. S. Panek and M. Yang, *J. Am. Chem. Soc.*, 1991, **113**, 6594; J. S. Panek and M. Yang, *J. Am. Chem. Soc.*, 1991, **113**, 9868; J. S. Panek and R. Beresis, *J. Org. Chem.*, 1993, **58**, 809; J. S. Panek and P. F. Cirillo, *J. Org. Chem.*, 1993, **58**, 999; J. S. Panek, M. Yang and J. S. Solomon, *J. Org. Chem.*, 1993, **58**, 1003.
104. M. A. Sparks and J. S. Panek, *Tetrahedron Lett.*, 1991, **32**, 4085.
105. J. S. Panek and M. A. Sparks, *Tetrahedron Asymmetry*, 1990, **1**, 801.
106. M. A. Sparks and J. S. Panek, *J. Org. Chem.*, 1991, **56**, 3431; J. S. Panek and M. Yang, *J. Org. Chem.*, 1991, **56**, 5755.
107. A. Hosomi, M. Ando and H. Sakurai, *Chem. Lett.*, 1986, 365; see also ref. 81.
108. K. Furuta, M. Mouri and H. Yamamoto, *Synlett*, 1991, 561.
109. A. P. Davis and M. Jaspars, *Angew. Chem., Int. Ed. Engl.*, 1992, **31**, 470; *J. Chem. Soc., Perkin Trans. 1*, 1992, 2111.
110. H. Sakurai, *Synlett*, 1989, 1; K. Sato, M. Kira and H. Sakurai, *J. Am. Chem. Soc.*, 1989, **111**, 6429; M. Kira, K. Sato and H. Sakurai, *J. Am. Chem. Soc.*, 1990, **112**, 257.
111. S. Kobayashi and K. Nishio, *Tetrahedron Lett.*, 1993, **34**, 3453.
112. A. G. Schultz and H. Lee, *Tetrahedron Lett.*, 1992, **33**, 4397.
113. L.-R. Pan and T. Tokoroyama, *Tetrahedron Lett.*, 1992, **33**, 1469.
114. M.-J. Wu, C.-C. Wu and P.-C. Lee, *Tetrahedron Lett.*, 1992, **33**, 2547.
115. B. B. Snider and Q. Zhang, *J. Org. Chem.*, 1991, **56**, 4908.
116. G. Audran, H. Monti, G. Léandri and J.-P. Monti, *Tetrahedron Lett.*, 1993, **34**, 3417.
117. L. F. Tietze and M. Rischer, *Angew. Chem., Int. Ed. Engl.*, 1992, **31**, 1221.

118. H.-J. Knölker, N. Foitzik, H. Goesmann and R. Graf, *Angew. Chem., Int. Ed. Engl.*, 1993, **32**, 1081.
119. B. M. Trost, *Angew. Chem., Int. Ed. Engl.*, 1986, **25**, 1; B. M. Trost and M. C. Matelich, *Synthesis*, 1992, 151 and references therein.
120. B. M. Trost, P. Seoane, S. Mignani and M. Acemoglu, *J. Am. Chem. Soc.*, 1989, **111**, 7487.
121. B. M. Trost and P. R. Seoane, *J. Am. Chem. Soc.*, 1987, **109**, 615.
122. T. V. Lee and C. Cregg, *Synlett*, 1990, 317 and references therein.
123. G. A. Molander and D. C. Shubert, *J. Am. Chem. Soc.*, 1987, **109**, 576, 6877.
124. B. Guyot, J. Pornet and L. Miginiac, *Tetrahedron*, 1991, **47**, 3981.
125. M. Harmata and B. F. Herron, *Synthesis*, 1993, 202.
126. D. E. Ward and B. F. Kaller, *Tetrahedron Lett.*, 1993, **34**, 407.
127. S. Hatakeyama, M. Kawamura, E. Shimanuki, K. Saijo and S. Takano, *Synlett*, 1992, 114.
128. I. E. Markó and A. Mekhalfia, *Tetrahedron Lett.*, 1992, **33**, 1799.
129. J. K. Stille and J. H. Simpson, *J. Am. Chem. Soc.*, 1987, **109**, 2138; J. A. Walker, S. P. Bitler and F. Wudl, *J. Org. Chem.*, 1984, **49**, 4733; K. P. C. Vollhardt and L. S. Winn, *Tetrahedron Lett.*, 1985, **26**, 709.
130. A. B. Holmes and G. R. Pooley, *Tetrahedron*, 1992, **48**, 7775; M. H. P. J. Aerssens, R. van der Heiden, M. Heus and L. Brandsma, *Synth. Commun.*, 1990, **20**, 3421.
131. J. C. Bottaro, R. J. Schmitt, C. D. Bedford, R. Gilardi and C. George, *J. Org. Chem.*, 1990, **55**, 1916.
132. K. C. Nicolaou, Y. P. Hong, W.-M. Dai, Z.-J. Zeng and W. Wrasidlo, *J. Chem. Soc., Chem. Commun.*, 1992, 1542.
133. T. Nishikawa, M. Isobe and T. Goto, *Synlett*, 1991, 99.
134. A. G. Myers, P. M. Harrington and E. Y. Kuo, *J. Am. Chem. Soc.*, 1991, **113**, 694.
135. Y.-M. Tsai, B.-W. Ke, C.-T. Yang and C.-H. Lin, *Tetrahedron Lett.*, 1992, **33**, 7895.
136. J. P. Marino, E. Laborde and R. S. Paley, *J. Am. Chem. Soc.*, 1988, **110**, 966; M. D. Bachi and E. Bosch, *Tetrahedron Lett.*, 1988, **29**, 2581.
137. T. Tsukiyama and M. Isobe, *Tetrahedron Lett.*, 1992, **33**, 7911.
138. K. S. Feldman, C. M. Kraebel and M. Parvez, *J. Am. Chem. Soc.*, 1993, **115**, 3846; Y. Rubin, C. B. Knobler and F. Diederich, *Angew. Chem., Int. Ed. Engl.*, 1991, **30**, 698; L. T. Scott, M. J. Cooney and D. Johnels, *J. Am. Chem. Soc.*, 1990, **112**, 4054.
139. Y. Pang, A. Schneider, T. J. Barton, M. S. Gordon and M. T. Carroll, *J. Am. Chem. Soc.*, 1992, **114**, 4920; Y. Pang, S. A. Petrich, V. G. Young, Jr., M. S. Gordon and T. J. Barton, *J. Am. Chem. Soc.*, 1993, **115**, 2534; J. Lin, Y. Pang, V. G. Young, Jr. and T. J. Barton, *J. Am. Chem. Soc.*, 1993, **115**, 3794.
140. W. Ando, F. Hojo, S. Sekigawa, N. Nakayama and T. Shimizu, *Organometallics*, 1992, **11**, 1009.
141. G. Agnel and E. Negishi, *J. Am. Chem. Soc.*, 1991, **113**, 7424; M. E. Krafft, C. A. Juliano, I. L. Scott, C. Wright and M. D. McEachin, *J. Am. Chem. Soc.*, 1991, **113**, 1693; E. P. Johnson and K. P. C. Vollhardt, *J. Am. Chem. Soc.*, 1991, **113**, 381; P. Magnus, L. M. Principe and M. J. Slater, *J. Org. Chem.*, 1987, **52**, 1483.
142. K. P. C. Vollhardt, *Pure Appl. Chem.*, 1985, **57**, 1819; E. Negishi, M. Ay and T. Sugihara, *Tetrahedron*, 1993, **49**, 5471; R. Grigg, R. Scott and P. Stevenson, *J. Chem. Soc., Perkin Trans. 1*, 1988, 1357.
143. T. Tsuda, S. Morikawa, N. Hasegawa and T. Saegusa, *J. Org. Chem.*, 1990, **55**, 2978.
144. K. Hayakawa, A. Hayashida and K. Kanematsu, *J. Chem. Soc., Chem. Commun.*, 1988, 1108.
145. D. A. Singleton and S.-W. Leung, *J. Org. Chem.*, 1992, **57**, 4796.
146. C. Schmitz, A.-C. Rouanet-Dreyfuss, M. Tueni and J.-F. Biellmann, *Tetrahedron Lett.*, 1992, **33**, 4911.
147. H. Mayr, E. Bäuml, G. Cibura and R. Koschinsky, *J. Org. Chem.*, 1992, **57**, 768.
148. P. Magnus, J. Rodríguez-López, K. Mulholland and I. Matthews, *J. Am. Chem. Soc.*, 1992, **114**, 382.
149. D. Guay, W. S. Johnson and U. Schubert, *J. Org. Chem.*, 1989, **54**, 4731.
150. R. C. Hartley, S. Lamothe and T. H. Chan, *Tetrahedron Lett.*, 1993, **34**, 1449.
151. D. Schinzer, J. Kabbara and K. Ringe, *Tetrahedron Lett.*, 1992, **33**, 8017.
152. B. Bennetau and J. Dunoguès, *Synlett*, 1993, 171; V. Snieckus, *Chem. Rev.*, 1990, **90**, 879, *Pure Appl. Chem.*, 1990, **62**, 671.
153. A. Dondoni, *Pure Appl. Chem.*, 1990, **62**, 643; A. Dondoni and L. Colombo, in 'Advances in the Use of Synthons in Organic Chemistry', ed. A. Dondoni, Jai Press, Greenwich, CT, 1993, vol. 1, p. 1; A. Dondoni, J. Orduna and P. Merino, *Synthesis*, 1992, 201.
154. G. Beese and B. A. Keay, *Synlett*, 1991, 33.
155. H. Sakurai, K. Ebata, C. Kabuto and A. Sekiguchi, *J. Am. Chem. Soc.*, 1990, **112**, 1799.
156. S. G. Davies and C. L. Goodfellow, *Synlett*, 1989, 59.
157. J. J. Eisch and J. E. Galle, *J. Org. Chem.*, 1990, **55**, 4835.
158. P. F. Hudrlik and A. M. Hudrlik, in 'Advances in Silicon Chemistry', Jai Press, Greenwich, CT, 1993, vol. 2, p. 1.
159. P. Jankowski and J. Wicha, *J. Chem. Soc., Chem. Commun.*, 1992, 802.
160. P. Raubo and J. Wicha, *Synlett*, 1993, 25.
161. F. Sato and Y. Kobayashi, *Synlett* 1992, 849.
162. J. E. Baldwin, R. M. Adlington, C. R. A. Godfrey and V. K. Patel, *J. Chem. Soc., Chem. Commun.*, 1991, 1277.
163. J. Yoshida, T. Maekawa, Y. Morita and S. Isoe, *J. Org. Chem.*, 1992, **57**, 1321; B. Achmatowicz, P. Raubo and J. Wicha, *J. Org. Chem.*, 1992, **57**, 6593.
164. W. Adam and M. Richter, *Tetrahedron Lett.*, 1992, **33**, 3461.
165. F. Gilloir and M. Malacria, *Tetrahedron Lett.*, 1992, **33**, 3859.
166. D. J. Ager, *Org. React. (NY)*, 1990, **38**, 1; A. G. M. Barrett, J. M. Hill, E. M. Wallace and J. A. Flygare, *Synlett*, 1991, 764.
167. T. Kawashima, N. Iwama and R. Okazaki, *J. Am. Chem. Soc.*, 1992, **114**, 7598.
168. I. Fleming, I. T. Morgan and A. K. Sarkar, *J. Chem. Soc., Chem. Commun.*, 1990, 1575.
169. G. A. Olah, V. Prakash Reddy and G. K. Surya Prakash, *Synthesis*, 1991, 29.
170. J. E. Oliver, M. Schwarz, J. A. Klun, W. R. Lusby and R. M. Waters, *Tetrahedron Lett.*, 1993, **34**, 1593.
171. P. D. Sattsangi and K. K. Wang, *Tetrahedron Lett.*, 1992, **33**, 5025; see also K. K. Wang, C. Liu, Y. G. Gu, F. N. Burnett and P. D. Sattsangi, *J. Org. Chem.*, 1991, **56**, 1914.
172. P. F. Hudrlik, A. M. Kassim, E. L. O. Agwaramgbo, K. A. Doonquah, R. R. Roberts and A. M. Hudrlik, *Organometallics*, 1993, **12**, 2367.

173. E. K. Dolence and M. J. Miller, *J. Org. Chem.*, 1991, **56**, 492.
174. A. Ricci and A. Degl'Innocenti, *Synthesis*, 1989, 647; P. C. B. Page, S. S. Klair and S. Rosenthal, *Chem. Soc. Rev.*, 1990, **19**, 147.
175. R. E. Ireland and D. W. Norbeck, *J. Org. Chem.*, 1985, **50**, 2198.
176. R. J. Linderman and Y. Suhr, *J. Org. Chem.*, 1988, **53**, 1569.
177. F. H. Elsner, H.-G. Woo and T. D. Tilley, *J. Am. Chem. Soc.*, 1988, **110**, 313.
178. J. A. Soderquist and E. I. Miranda, *J. Am. Chem. Soc.*, 1992, **114**, 10078.
179. R. B. Silverman, X. Lu and G. M. Banik, *J. Org. Chem.*, 1992, **57**, 6617.
180. H.-G. Woo, W. P. Freeman and T. D. Tilley, *Organometallics*, 1992, **11**, 2198.
181. A. J. Kresge and J. B. Tobin, *J. Am. Chem. Soc.*, 1990, **112**, 2805.
182. J. Otera, Y. Niibo and H. Nozaki, *Tetrahedron Lett.*, 1992, **33**, 3655.
183. K. Suda, J. Watanabe and T. Takanami, *Tetrahedron Lett.*, 1992, **33**, 1355.
184. R. F. Cunico and C. Zhang, *Tetrahedron Lett.*, 1992, **33**, 6751.
185. F. Jin, B. Jiang and Y. Xu, *Tetrahedron Lett.*, 1992, **33**, 1221.
186. B. Dondy, P. Doussot and C. Portella, *Synthesis*, 1992, 995.
187. A. Degl'Innocenti, E. Stucchi, A. Capperucci, A. Mordini, G. Reginato and A. Ricci, *Synlett*, 1992, 329; for other conjugate additions to alkynyl silyl ketones, see, for example, A. Degl'Innocenti, A. Capperucci, G. Reginato, A. Mordini and A. Ricci, *Tetrahedron Lett.*, 1992, **33**, 1507; A. Degl'Innocenti, P. Ulivi, A. Capperucci, G. Reginato, A. Mordini and A. Ricci, *Synlett*, 1992, 883.
188. A. Yanagisawa, S. Habaue and H. Yamamoto, *J. Org. Chem.*, 1989, **54**, 5198.
189. J. Yoshida, M. Itoh, S. Matsunaga and S. Isoe, *J. Org. Chem.*, 1992, **57**, 4877 and references therein.
190. L. Duhamel, J. Gralak and A. Bouyanzer, *J. Chem. Soc., Chem. Commun.*, 1993, 1763.
191. Y. Kita, S. Matsuda, S. Kitagaki, Y. Tsuzuki and S. Akai, *Synlett*, 1991, 401; for a new convenient route to silylketenes from 1-*t*-butoxy-2-silylethynes, see E. Valentí, M. A. Pericàs and F. Serratosa, *J. Org. Chem.*, 1990, **55**, 395.
192. Y. Kita, J. Sekihachi, Y. Hayashi, Y.-Z. Da, M. Yamamoto and S. Akai, *J. Org. Chem.*, 1990, **55**, 1108.
193. R. Ballini, G. Bartoli, R. Giovannini, E. Marcantoni and M. Petrini, *Tetrahedron Lett.*, 1993, **34**, 3301.
194. D. Enders and S. Nakai, *Helv. Chim. Acta*, 1990, **73**, 1833; for a related oxidative cleavage route to terminal diols, see O. Andrey, Y. Landais and D. Planchenault, *Tetrahedron Lett.*, 1993, **34**, 2927.
195. A. Hosomi, H. Hayashida and Y. Tominaga, *J. Org. Chem.*, 1989, **54**, 3254.
196. M. R. Najafi, M.-L. Wang and G. Zweifel, *J. Org. Chem.*, 1991, **56**, 2468.
197. M. Asaoka, K. Shima and H. Takei, *J. Chem. Soc., Chem. Commun.*, 1988, 430; M. Asaoka, K. Nishimura and H. Takei, *Bull. Chem. Soc. Jpn.*, 1990, **63**, 407.
198. S. Takano, Y. Higashi, T. Kamikubo, M. Moriya and K. Ogasawara, *J. Chem. Soc., Chem. Commun.*, 1993, 788.
199. S. Shambayati, J. F. Blake, S. G. Wierschke, W. L. Jorgensen and S. L. Schreiber, *J. Am. Chem. Soc.*, 1990, **112**, 697.
200. T. W. Greene and P. G. M. Wuts, 'Protective Groups in Organic Synthesis', 2nd. edn., Wiley, Chichester, 1991; M. Lalonde and T. H. Chan, *Synthesis*, 1985, 817.
201. J. Muzart, *Synthesis*, 1993, 11.
202. D. H. R. Barton and J. T. Kelly, *Tetrahedron Lett.*, 1992, **33**, 5041; M. P. Doyle, K. G. High, V. Bagheri, R. J. Pieters, P. J. Lewis and M. M. Pearson, *J. Org. Chem.*, 1990, **55**, 6082.
203. S. E. Denmark, R. P. Hammer, E. J. Weber and K. L. Habermas, *J. Org. Chem.*, 1987, **52**, 165.
204. E. J. Corey and K. Y. Yi, *Tetrahedron Lett.*, 1992, **33**, 2289.
205. S. J. Monger, D. M. Parry and S. M. Roberts, *J. Chem. Soc., Chem. Commun.*, 1989, 381.
206. A. S. Pilcher, D. K. Hill, S. J. Shimshock, R. E. Waltermire and P. DeShong, *J. Org. Chem.*, 1992, **57**, 2492.
207. J. F. Cormier, *Tetrahedron Lett.*, 1991, **32**, 187; J. F. Cormier, M. B. Isaac and L.-F. Chen, *Tetrahedron Lett.*, 1993, **34**, 243.
208. W. Zhang and M. J. Robins, *Tetrahedron Lett.*, 1992, **33**, 1177.
209. E. A. Schmittling and J. S. Sawyer, *Tetrahedron Lett.*, 1991, **32**, 7207.
210. J. Feixas, A. Capdevila, F. Camps and A. Guerrero, *J. Chem. Soc., Chem. Commun.*, 1992, 1451.
211. R. S. Varma, J. B. Lamture and M. Varma, *Tetrahedron Lett.*, 1993, **34**, 3029.
212. E. J. Corey and G. B. Jones, *J. Org. Chem.*, 1992, **57**, 1028.
213. D. Waldmüller, M. Braun and A. Steigel, *Synlett*, 1991, 160.
214. K. Maruoka, T. Ooi, S. Nagahara and H. Yamamoto, *Tetrahedron*, 1991, **47**, 6983; S. Nagahara, K. Maruoka and H. Yamamoto, *Tetrahedron Lett.*, 1992, **33**, 527.
215. K. Mikami and M. Shimizu, *Tetrahedron Lett.*, 1992, **33**, 6315.
216. S. B. Bedford, G. Fenton, D. W. Knight and D. Shaw, *Tetrahedron Lett.*, 1992, **33**, 6505.
217. T. Nakamura, K. Tanino and I. Kuwajima, *Tetrahedron Lett.*, 1993, **34**, 477.
218. J. W. Wilt, *Tetrahedron*, 1985, **41**, 3979; D. P. Curran, *Synthesis*, 1988, 417 and 489.
219. For reviews, see E. W. Colvin, in 'Comprehensive Organic Synthesis', eds. B. M. Trost and I. Fleming, Pergamon, Oxford, 1991, vol. 7, chap. 4.3 p. 641; K. Tamao, *J. Synth. Org. Chem., Jpn.*, 1988, **46**, 861.
220. G. Stork and M. Kahn, *J. Am. Chem. Soc.*, 1985, **107**, 500; G. Stork and M. J. Sofia, *J. Am. Chem. Soc.*, 1986, **108**, 6826.
221. M. Koreeda and L. G. Hamann, *J. Am. Chem. Soc.*, 1990, **112**, 8175; H. Nishiyama, T. Kitajima, M. Matsumoto and K. Itoh, *J. Org. Chem.*, 1984, **49**, 2298.
222. M. Koreeda and I. A. George, *J. Am. Chem. Soc.*, 1986, **108**, 8098; see, however, A. Kurek-Tyrlik, J. Wicha and G. Snatzke, *Tetrahedron Lett.*, 1988, **29**, 4001.
223. G. Stork, H. S. Suh and G. Kim, *J. Am. Chem. Soc.*, 1991, **113**, 7054; D. Craig and V. R. N. Munasinghe, *J. Chem. Soc., Chem. Commun.*, 1993, 901.
224. G. Stork and G. Kim, *J. Am. Chem. Soc.*, 1992, **114**, 1087; M. Bols, *J. Chem. Soc., Chem. Commun.*, 1992, 913; *ibid.*, 1993, 791; Y. C. Xin, J.-M. Mallet and P. Sinaÿ, *J. Chem. Soc., Chem. Commun.*, 1993, 864.
225. A. G. Myers, D. Y. Gin and K. L. Widdowson, *J. Am. Chem. Soc.*, 1991, **113**, 9661; for non-tethered examples of cyclizations of radicals generated from α-silyloxy phenylselenides, see T. L. Fevig, R. L. Elliott and D. P. Curran, *J. Am. Chem. Soc.*, 1988, **110**, 5064; G. E. Keck and A. M. Tafesh, *Synlett*, 1990, 257.
226. K. Tamao, K. Kobayashi and Y. Ito, *J. Am. Chem. Soc.*, 1989, **111**, 6478; K. J. Shea, K. S. Zandi, A. J. Staab and R. Carr, *Tetrahedron Lett.*, 1990, **31**, 5885; J. W. Gillard *et al.*, *Tetrahedron Lett.*, 1991, **32**, 1145; K. J. Shea, A. J. Staab and K. S. Zandi, *Tetrahedron Lett.*, 1991, **32**, 2715.

227. D. Craig and J. C. Reader, *Tetrahedron Lett.*, 1990, **31**, 6585; *ibid.*, 1992, **33**, 4073; *ibid.*, 1992, **33**, 6165.
228. S. McN. Sieburth and L. Fensterbank, *J. Org. Chem.*, 1992, **57**, 5279.
229. G. Stork, T. Y. Chan and G. A. Breault, *J. Am. Chem. Soc.*, 1992, **114**, 7578.
230. K. D. Moeller, C. M. Hudson and L. V. Tinao-Woodridge, *J. Org. Chem.*, 1993, **58**, 3478.
231. S. Uyeo and H. Itani, *Tetrahedron Lett.*, 1991, **32**, 2143; J. Kang, J. Kim and K.-J. Lee, *Synlett*, 1991, 885.
232. S. A. Fleming and S. C. Ward, *Tetrahedron Lett.*, 1992, **33**, 1013; M. Journet, E. Magnol, W. Smadja and M. Malacria, *Synlett*, 1991, 58; M. Journet and M. Malacria, *Tetrahedron Lett.*, 1992, **33**, 1893; D. P. Curran, K. V. Somayajula and H. Yu, *ibid.*, 1992, **33**, 2295; J. Lejeune and J. Y. Lallemand, *ibid.*, 1992, **33**, 2977; M. Koreeda and D. C. Visger, *ibid.*, 1992, **33**, 6603; K. Matsumoto, K. Miura, K. Oshima and K. Utimoto, *ibid.*, 1992, **33**, 7031; S. Kim and J. S. Koh, *ibid.*, 1992, **33**, 7391; M. Suzuki, H. Koyano and R. Noyori, *J. Org. Chem.*, 1987, **52**, 5583; J. C. López, A. M. Gómez and B. Fraser-Reid, *J. Chem. Soc., Chem. Commun.*, 1993, 762.
233. K. Tamao, T. Tohma, N. Inui, O. Nakayama and Y. Ito, *Tetrahedron Lett*, 1990, **31**, 7333; M. R. Hale and A. H. Hoveyda, *J. Org. Chem.*, 1992, **57**, 1643.
234. S. Anwar, G. Bradley and A. P. Davis, *J. Chem. Soc., Perkin Trans. 1*, 1991, 1383; M. J. Burk and J. E. Feaster, *Tetrahedron Lett.*, 1992, **33**, 2099.
235. P. Cazeau, F. Duboudin, F. Moulines, O. Babot and J. Dunoguès, *Tetrahedron*, 1987, **43**, 2075, 2089.
236. T. H. Chan and G. Z. Zheng, *Tetrahedron Lett.*, 1993, **34**, 3095.
237. K. Bambridge, N. S. Simpkins and B. P. Clark, *Tetrahedron Lett.*, 1992, **33**, 8141 and references therein.
238. H. J. Reich, R. C. Holtan and C. Bolm, *J. Am. Chem. Soc.*, 1990, **112**, 5609.
239. K. Tanino, H. Shoda, T. Nakamura and I. Kuwajima, *Tetrahedron Lett.*, 1992, **33**, 1337.
240. K. Krägeloh, G. Simchen and K. Schweiker, *Liebigs Ann. Chem.*, 1985, 2352; C. S. K. Wan, A. C. Weedon and D. F. Wong, *J. Org. Chem.*, 1986, **51**, 3335; Y. Tominaga, C. Kamio and A. Hosomi, *Chem. Lett.*, 1989, 1761; M. Sodeoka, H. Yamada and M. Shibasaki, *J. Am. Chem. Soc.*, 1990, **112**, 4906.
241. T. H. Keller, E. G. Neeland and L. Weiler, *J. Org. Chem.*, 1987, **52**, 1870.
242. C. Gennari, L. Colombo, G. Bertolini and G. Schimperna, *ibid.*, 1987, **52**, 2754.
243. G. Helmchen, U. Leikauf and I. Taufer-Knöpfel, *Angew. Chem., Int. Ed. Engl.*, 1985, **24**, 874.
244. I. R. Silverman, C. Edington, J. D. Elliott and W. S. Johnson, *J. Org. Chem.*, 1987, **52**, 180.
245. T. Basile, L. Longobardo, E. Tagliavini, C. Trombini and A. Umani-Ronchi, *J. Chem. Soc., Chem. Commun.*, 1990, 759.
246. Y. Kita, *et al.*, *J. Org. Chem.*, 1988, **53**, 554; G. Casiraghi, G. Rassu, P. Spanu and L. Pinna, *ibid.*, 1992, **57**, 3760.
247. M. T. Reetz and D. N. A. Fox, *Tetrahedron Lett.*, 1993, **34**, 1119.
248. R. Roy and A. W. Rey, *Synlett*, 1990, 448.
249. A. G. Myers and K. L. Widdowson, *J. Am. Chem. Soc.*, 1990, **112**, 9672.
250. S. E. Denmark, B. D. Griedel and D. M. Coe, *J. Org. Chem.*, 1993, **58**, 988.
251. S. Kobayashi, H. Uchiro, Y. Fujishita, I. Shiina and T. Mukaiyama, *J. Am. Chem. Soc.*, 1991, **113**, 4247.
252. E. J. Corey, C. L. Cywin and T. D. Roper, *Tetrahedron Lett.*, 1992, **33**, 6907.
253. K. Furuta, T. Maruyama and H. Yamamoto, *J. Am. Chem. Soc.*, 1991, **113**, 1041; *Synlett*, 1991, 439.
254. K. Hattori and H. Yamamoto, *Synlett*, 1993, 239; for other chiral boron catalysts, see S. Kiyooka, Y. Kaneko and K. Kume, *Tetrahedron Lett.*, 1992, **33**, 4927; E. R. Parmee, Y. Hong, O. Tempkin and S. Masamune, *ibid.*, 1992, **33**, 1729.
255. J.-H. Gu, M. Terada, K. Mikami and T. Nakai, *Tetrahedron Lett.*, 1992, **33**, 1465; for the lanthanide(III)-catalysed reaction of silyl enol ethers with water-soluble aldehydes, see S. Kobayashi and I. Hachiya, *ibid.*, 1992, **33**, 1625; for Michael reactions, see S. Kobayashi, I. Hachiya, T. Takahori, M. Araki and H. Ishitani, *ibid.*, 1992, **33**, 6815.
256. C. Mukai, O. Kataoka and M. Hanaoka, *Tetrahedron Lett.*, 1991, **32**, 7553; R. Tester, V. Varghese, A. M. Montana, M. Khan and K. M. Nicholas, *J. Org. Chem.*, 1990, **55**, 186.
257. K. Chow and S. Danishefsky, *J. Org. Chem.*, 1989, **54**, 6016.
258. T. Bach, D. N. A. Fox and M. T. Reetz, *J. Chem. Soc., Chem. Commun.*, 1992, 1634.
259. Y. Kataoka, Y. Nakamura, K. Morihira, H. Arai, Y. Horiguchi and I. Kuwajima, *Tetrahedron Lett.*, 1992, **33**, 6979.
260. T. Sato, J. Otera and H. Nozaki, *J. Am. Chem. Soc.*, 1990, **112**, 901.
261. W. Odenkirk, J. Whelan and B. Bosnich, *Tetrahedron Lett.*, 1992, **33**, 5729; T. K. Hollis, N. P. Robinson and B. Bosnich, *ibid.*, 1992, **33**, 6423.
262. A. Ando, T. Miura, T. Tatematsu and T. Shioiri, *ibid.*, 1993, **34**, 1507.
263. D. J. Hart and D.-C. Ha, *Chem. Rev.*, 1989, **89**, 1447.
264. F. Texier-Boullet, R. Latouche and J. Hamelin, *Tetrahedron Lett.*, 1993, **34**, 2123.
265. B. C. Ranu, M. Saha and S. Bhar, *Tetrahedron Lett.*, 1993, **34**, 1989.
266. P. G. Klimko and D. A. Singleton, *J. Org. Chem.*, 1992, **57**, 1733.
267. D. Schinzer and M. Kalesse, *Synlett*, 1989, 34.
268. I. Kuwajima, *Pure Appl. Chem.*, 1988, **60**, 115.
269. S. Aoki and E. Nakamura, *Synlett*, 1990, 741.
270. G. M. Rubottom, E. C. Beedle, C.-W. Kim and R. C. Mott, *J. Am. Chem. Soc.*, 1985, **107**, 4230 and references therein.
271. See, for example, B. S. Davidson, K. A. Plavcan and J. Meinwald, *J. Org. Chem.*, 1990, **55**, 3912.
272. K. Ikura, I. Ryu, N. Kambe and N. Sonoda, *J. Am. Chem. Soc.*, 1992, **114**, 1520.
273. R. L. Danheiser, D. S. Casebier and J. L. Loebach, *Tetrahedron Lett.*, 1992, **33**, 1149.
274. M. Komatsu, I. Suehiro, Y. Horiguchi and I. Kuwajima, *Synlett*, 1991, 771.
275. M. Yamaguchi, M. Sehata, A. Hayashi and M. Hirama, *J. Chem. Soc., Chem. Commun.*, 1993, 1708; for a different strategy leading to cyclopentenes, see T. Hudlicky, N. E. Heard and A. Fleming, *J. Org. Chem.*, 1990, **55**, 2570.
276. D. D. Dhavale and C. Trombini, *J. Chem. Soc., Chem. Commun.*, 1992, 1268.
277. S. J. Danishefsky, *Aldrichimica Acta*, 1986, **19**, 59; S. J. Danishefsky and C. Vogel, *J. Org. Chem.*, 1986, **51**, 3915; S. J. Danishefsky, D. M. Armistead, F. E. Wincott, H. G. Selnick and R. Hungate, *J. Am. Chem. Soc.*, 1987, **109**, 8117; W. Pfrengle and H. Kunz, *J. Org. Chem.*, 1989, **54**, 4261; P. E. Vorndam, *ibid.*, 1990, **55**, 3693; D. W. Cameron, I. T. Crosby and G. I. Feutrill, *Tetrahedron Lett.*, 1992, **33**, 2855; D. W. Cameron and P. J. de Bruyn, *ibid.*, 1992, **33**, 5593; H. Nakamura, B. Ye and A. Murai, *ibid.*, 1992, **33**, 8113; T. H. Chan and C. V. C. Prasad, *J. Org. Chem.*, 1987, **52**, 110; C. V. C. Prasad and T. H. Chan, *ibid.*, **52**, 120; J. Gordon and R. Tabacchi, *ibid.*, 1992, **57**, 4728.
278. K. Maruoka, T. Itoh, T. Shirasaka and H. Yamamoto, *J. Am. Chem. Soc.*, 1988, **110**, 310; K. Hattori and H. Yamamoto, *J. Org. Chem.*, 1992, **57**, 3264.

279. F. Jin, Y. Xu and W. Huang, *J. Chem. Soc., Chem. Commun.*, 1993, 814.
280. G. A. Molander and K. O. Cameron, *J. Org. Chem.*, 1991, **56**, 2617.
281. R. E. Ireland, *Aldrichimica Acta*, 1988, **21**, 59; S. Pereira and M. Srebnik, *ibid.*, 1993, **26**, 17.
282. R. E. Ireland, P. Wipf and J. D. Armstrong, III, *J. Org. Chem.*, 1991, **56**, 650.
283. R. E. Ireland, P. Wipf and J.-N. Xiang, *J. Org. Chem.*, 1991, **56**, 3572.
284. J. E. Baldwin, P. A. R. Bennett and A. K. Forrest, *J. Chem. Soc., Chem. Commun.*, 1987, 250.
285. J. H. Dodd, J. E. Starrett, Jr. and S. M. Weinreb, *J. Am. Chem. Soc.*, 1994, **106**, 1811.
286. L. A. Paquette, H.-S. Lin and J. C. Gallucci, *Tetrahedron Lett.*, 1987, **28**, 1363.
287. F. A. Davis and A. C. Sheppard, *J. Org. Chem.*, 1987, **52**, 954.
288. H. K. Chenault and S. J. Danishefsky, *J. Org. Chem.*, 1989, **54**, 4249.
289. B. Pujol, R. Sabatier, P.-A. Driguez and A. Doutheau, *Tetrahedron Lett.*, 1992, **33**, 1447.
290. T. K. Jones and S. E. Denmark, *J. Org. Chem.*, 1985, **50**, 4037; P. T. W. Cheng and S. McLean, *Tetrahedron Lett.*, 1988, **29**, 3511.
291. R. D. Walkup and N. U. Obeyesekere, *J. Org. Chem.*, 1988, **53**, 920; P. T. Kaye and R. A. Learmonth, *Synth. Commun.*, 1990, **20**, 1333.
292. D. R. Reddy and E. R. Thornton, *J. Chem. Soc., Chem. Commun.*, 1992, 172.
293. Y. Horiguchi, E. Nakamura and I. Kuwajima, *J. Am. Chem. Soc.*, 1989, **111**, 6257; *Tetrahedron Lett.*, 1989, **30**, 3323.
294. M. M. L. Crilley, D. S. Larsen, R. J. Stoodley and Fernando Tomé, *Tetrahedron Lett.*, 1993, **34**, 3305.
295. R. M. Moriarty, O. Prakash, M. P. Duncan, R. K. Vaid and H. A. Musallam, *J. Org. Chem.*, 1987, **52**, 150.
296. R. V. Hoffman and H.-O. Kim, *J. Org. Chem.*, 1988, **53**, 3855; R. M. Moriarty, R. Penmasta, A. K. Awasthi, W. R. Epa and I. Prakash, *ibid.*, 1989, **54**, 1101; see also J. A. Vega, A. Molina, R. Alajarín, J. J. Vaquero, J. L. García Navío and J. Alvarez-Builla, *Tetrahedron Lett.*, 1992, **33**, 3677.
297. C. Gennari, L. Colombo and G. Bertolini, *J. Am. Chem. Soc.*, 1986, **108**, 6394.
298. W. Oppolzer and R. Moretti, *Helv. Chim. Acta*, 1986, **69**, 1923.
299. See also D. A. Evans, T. C. Britton, R. L. Dorow and J. F. Dellaria, *J. Am. Chem. Soc.*, 1986, **108**, 6395; L. A. Trimble and J. C. Vederas, *J. Am. Chem. Soc.*, 1986, **108**, 6397.
300. K. Chen and G. F. Koser, *J. Org. Chem.*, 1991, **56**, 5764.
301. A. Della Cort, *J. Org. Chem.*, 1991, **56**, 6708.
302. P. Magnus and J. Lacour, *J. Am. Chem. Soc.*, 1992, **114**, 3993 and references therein; P. Magnus, J. Lacour and P. A. Evans, *Janssen Chimica Acta*, 1993, **11**, 3; P. Magnus and L. Barth, *Tetrahedron Lett.*, 1992, **33**, 2777; P. Magnus, A. Evans and J. Lacour, *ibid.*, 1992, **33**, 2933.
303. S. Kim, J. H. Park, Y. G. Kim and J. M. Lee, *J. Chem. Soc., Chem. Commun.*, 1993, 1188.
304. P. Magnus and P. Rigollier, *Tetrahedron Lett.*, 1992, **33**, 6111.
305. R. Zetter and G. Simchen, *Synthesis*, 1992, 922; see also F. Mezger, G. Simchen and P. Fischer, *ibid.*, 1991, 375.
306. G. Jas, *Synthesis.*, 1991, 965.
307. S. F. Martin and J. W. Corbett, *Synthesis.*, 1992, 55.
308. G. Casiraghi, L. Colombo, G. Rassu and P. Spanu, *J. Org. Chem.*, 1990, **55**, 2565.
309. C. W. Jefford, P.-Z. Huang, J.-C. Rossier, A. W. Sledeski and J. Boukouvalas, *Synlett*, 1990, 745 and references therein.
310. J. L. Bloomer and M. E. Lankin, *Tetrahedron Lett.*, 1992, **33**, 2769.
311. Y. Chiang, A. J. Kresge and P. A. Walsh, *J. Am. Chem. Soc.*, 1986, **108**, 6314.
312. B. Capon and B. Guo, *J. Am. Chem. Soc.*, 1988, **110**, 5144.
313. V. P. Reddy, D. R. Bellew and G. K. S. Prakash, *Synthesis*, 1992, 1209.
314. K. Hatanaka, S. Tanimoto, T. Sugimoto and M. Okano, *Tetrahedron Lett.*, 1981, **22**, 3243.
315. D. S. Brown *et al.*, *Synlett*, 1990, 619.
316. K. Maruoka, H. Banno and H. Yamamoto, *Synlett*, 1991, 253.
317. L. Duhamel, P. Angibaud, J. R. Desmurs and J. Y. Valnot, *Synlett.*, 1991, 807.
318. A. Degl'Innocenti, P. Ulivi, A. Capperucci, A. Mordini, G. Reginato and A. Ricci, *Synlett.*, 1992, 499.
319. K. Takeda, A. Ayabe, H. Kawashima and Y. Harigaya, *Tetrahedron Lett.*, 1992, **33**, 951.
320. M. Franck-Neumann, P.-J. Colson, P. Geoffroy and K. M. Taba, *Tetrahedron Lett.*, 1992, **33**, 1903.
321. B. B. Snider and T. Kwon, *J. Org. Chem.*, 1990, **55**, 4786.
322. A. Heidbreder and J. Mattay, *Tetrahedron Lett.*, 1992, **33**, 1973.
323. E. Baciocchi, A. Casu and R. Ruzziconi, *Synlett*, 1990, 679.
324. A. B. Paolobelli, D. Latini and R. Ruzziconi, *Tetrahedron Lett.*, 1993, **34**, 721.
325. T. Fujii, T. Hirao and Y. Ohshiro, *Tetrahedron Lett.*, 1992, **33**, 5823.
326. P. A. Grieco, J. L. Collins and K. J. Henry, Jr., *Tetrahedron Lett.*, 1992, **33**, 4735.
327. E. Nakamura, *Synlett*, 1991, 539.
328. M. Arai, T. Nemoto, Y. Ohashi and E. Nakamura, *Synlett*, 1992, 309.
329. A. K. Banerjee, M. C. S. Carrasco, C. S. V. Frydrych-Houge and W. B. Motherwell, *J. Chem. Soc., Chem. Commun.*, 1986, 1803.
330. C. A. M. Alfonso, W. B. Motherwell and L. R. Roberts, *Tetrahedron Lett.*, 1992, **33**, 3367.
331. C. A. M. Alfonso, W. B. Motherwell, D. M. O'Shea and L. R. Roberts, *Tetrahedron Lett.*, 1992, **33**, 3899.
332. W. B. Motherwell and L. R. Roberts, *J. Chem. Soc., Chem. Commun.*, 1992, 1582.
333. N. Kamiya, Y. Chikami and Y. Ishii, *Synlett*, 1990, 675; T. Kanai, Y. Kanagawa and Y. Ishii, *J. Org. Chem.*, 1990, **55**, 3274; for an application, see F.-T. Luo, S.-L. Fwu and W.-S. Huang, *Tetrahedron Lett.*, 1992, **33**, 6839.
334. Y. Kanagawa, Y. Nishiyama and Y. Ishii, *J. Org. Chem.*, 1992, **57**, 6988.
335. K. Lee and D. F. Wiemer, *Tetrahedron Lett.*, 1993, **34**, 2433.
336. E. Keinan, M. Sahai and R. Shvily, *Synthesis*, 1991, 641 and references therein.
337. G. A. Olah, G. K. Surya Prakash and R. Krishnamurthi, in 'Advances in Silicon Chemistry', ed. G. L. Larson, Jai Press, Greenwich, CT, 1991, vol. 1, p. 1.
338. M. B. Berry, J. Blagg, D. Craig and M. C. Willis, *Synlett*, 1992, 659.
339. J. G. Lee and H. T. Cha, *Tetrahedron Lett.*, 1992, **33**, 3167 and references therein.
340. J. G. Lee and K. K. Kang, *J. Org. Chem.*, 1988, **53**, 3634.
341. H. H. Seltzman, M. A. Moody and M. K. Begum, *Tetrahedron Lett.*, 1992, **33**, 3443.

342. T. Akiyama, H. Shima and S. Ozaki, *Synlett*, 1992, 415.
343. S. Baskaran, N. Chidambaram, N. Narasimhan and S. Chandrasekaran, *Tetrahedron Lett.*, 1992, **33**, 6371.
344. H. Kotsuki, Y. Ushio, N. Yoshimura and M. Ochi, *J. Org. Chem.*, 1987, **52**, 2594.
345. J. K. Rasmussen, S. M. Heilman and L. Krepski, in 'Advances in Silicon Chemistry', ed. G. L. Larson, Jai Press, Greenwich, CT, 1991, vol. 1, p. 65.
346. S. Kobayashi, Y. Tsuchiya and T. Mukaiyama, *Chem. Lett.*, 1991, 541; M. Hayashi, T. Matsuda and N. Oguni, *J. Chem. Soc., Perkin Trans. 1*, 1992, 3135; H. Ohno, H. Nitta, K. Tanaka, A. Mori and S. Inoue, *J. Org. Chem.*, 1992, **57**, 6778; M. Hayashi, Y. Miyamoto, T. Inoue and N. Oguni, *ibid.*, 1993, **58**, 1515 and references therein.
347. J. W. Faller and L.-L. Gunderson, *Tetrahedron Lett.*, 1993, **34**, 2275.
348. S. Kobayashi, Y. Tsuchiya and T. Mukaiyama, *Chem. Lett.*, 1991, 537.
349. G. A. Molander and J. P. Haar, Jr., *J. Am. Chem. Soc.*, 1991, **113**, 3608.
350. K. Higuchi, M. Onaka and Y. Izumi, *J. Chem. Soc., Chem. Commun.*, 1991, 1035.
351. J.-P. Leblanc and H. W. Gibson, *Tetrahedron Lett.*, 1992, **33**, 6295.
352. M. Hayashi, M. Tamura and N. Oguni, *Synlett*, 1992, 663 and references therein.
353. M. S. Batra and E. Brunet, *Tetrahedron Lett.*, 1993, **34**, 711.
354. N. Chatani, T. Takeyasu, N. Horiuchi and T. Hanafusa, *J. Org. Chem.*, 1988, **53**, 3539.
355. M. Yoshimatsu, T. Yoshiuchi, H. Shimizu, M. Hori and T. Kataoka, *Synlett*, 1993, 121; for a synthesis of monoselenoacetals, see M. Sakakibara, K. Katsumata, Y. Watanabe, T. Toru and Y. Ueno, *Synlett*, 1992, 965.
356. G. Simchen, in 'Advances in Silicon Chemistry', ed. G. L. Larson, Jai Press, Greenwich, CT, 1991, vol. 1, p. 189.
357. P. G. Gassman and S. J. Burns, *J. Org. Chem.*, 1988, **53**, 5574.
358. B. H. Lipshutz, J. Burgess-Henry and G. P. Roth, *Tetrahedron Lett.*, 1993, **34**, 995.
359. M. Ihara, S. Suzuki, N. Taniguchi and K. Fukumoto, *J. Chem. Soc., Chem. Commun.*, 1992, 755 and references therein.
360. C. W. Jefford, S. Jin, P. Kamalaprija, U. Burger and G. Bernardinelli, *Tetrahedron Lett.*, 1992, **33**, 7129.
361. H. Lamy-Schelkens, D. Giomi and L. Ghosez, *Tetrahedron Lett.*, 1989, **30**, 5887; H. Lamy-Schelkens and L. Ghosez, *ibid.*, 1989, **30**, 5891; D. Nogue, R. Paugam and L. Wartski, *ibid.*, 1992, **33**, 1265.
362. S. Kim, Y. G. Kim, and D. Kim, *Tetrahedron Lett.*, 1992, **33**, 2565.
363. T. R. Hoye, B. H. Peterson and J. D. Miller, *J. Org. Chem.*, 1987, **52**, 1351; W. H. Pearson and M.-C. Cheng, *ibid.*, 1987, **52**, 1353.
364. P. DeShong and P. J. Rybczynski, *J. Org. Chem.*, 1991, **56**, 3207.
365. E. Vedejs and H. Sano, *Tetrahedron Lett.*, 1992, **33**, 3261.
366. L. A. Paquette, M. Zhao and D. Friedrich, *Tetrahedron Lett.*, 1992, **33**, 7311.
367. R. Polt, M. A. Peterson and L. DeYoung, *J. Org. Chem.*, 1992, **57**, 5469; K. Shibata, N. Tokitoh and R. Okazaki, *Tetrahedron Lett.*, 1993, **34**, 1495.
368. W. P. Jackson, *Synlett*, 1990, 536.
369. W. Adam and R. Albert, *Tetrahedron Lett.*, 1992, **33**, 8015.
370. G. A. Olah and T. D. Ernst, *J. Org. Chem.*, 1989, **54**, 1204.
371. E. J. Corey, M. M. Mehrotra and A. U. Khan, *J. Am. Chem. Soc.*, 1986, **108**, 2472.
372. G. H. Posner, K. S. Webb, W. M. Nelson, T. Kishimoto and H. H. Seliger, *J. Org. Chem.*, 1989, **54**, 3252.
373. R. Nagata and I. Saito, *Synlett*, 1990, 291.
374. G. A. Olah, L. Heiliger, X.-Y. Li and G. K. Surya Prakash, *J. Am. Chem. Soc.*, 1990, **112**, 5991.
375. J. B. Lambert and W. Schilf, *J. Am. Chem. Soc.*, 1988, **110**, 6364.
376. G. A. Olah and T. D. Ernst, *J. Org. Chem.*, 1989, **54**, 1203.
377. G. A. Olah, Q. Wang, X.-Y. Li and G. K. Surya Prakash, *Synlett*, 1990, 487.
378. K. I. Sutowardoyo, M. Emziane, P. Lhoste and D. Sinou, *Tetrahedron*, 1991, **47**, 1435.
379. M. Hayashi, K. Kohmura and N. Oguni, *Synlett*, 1991, 774.
380. J. G. Lee and K. H. Kwak, *Tetrahedron Lett.*, 1992, **33**, 3165.
381. K. Nishiyama and T. Yamaguchi, *Synthesis*, 1988, 106 and references therein.
382. T. Shioiri and T. Aoyama, 'Advances in the Use of Synthons in Organic Chemistry', ed. A. Dondoni, Jai Press, Greenwich, CT, 1993, vol. 1, p. 51; R. Anderson and S. B. Anderson, in 'Advances in Silicon Chemistry', ed. G. L. Larson, Jai Press, Greenwich, CT, 1991, vol. 1, p. 303.
383. K. Miwa, T. Aoyama and T. Shioiri, *Synlett*, 1994, 107.
384. S. Ohira, K. Okai and T. Moritani, *J. Chem. Soc., Chem. Commun.*, 1992, 721.
385. K. Miwa, T. Aoyama and T. Shioiri, *Synlett*, 1994, 109.
386. T. Aoyama, S. Katsuta and T. Shioiri, *Heterocycles*, 1989, **28**, 133; T. Aoyama, T. Nakano, K. Marumo, Y. Uno and T. Shioiri, *Synthesis*, 1991, 1163.
387. M. D. Mizhiritskii and V. O. Reikhsfel'd, *Russ. Chem. Rev. (Engl. Transl.)*, 1988, **57**, 447; E. Block and M. Aslam, *Tetrahedron*, 1988, **44**, 281.
388. J. Brittain and Y. Gareau, *Tetrahedron Lett.*, 1993, **34**, 3363.
389. J. R. Hwu and S.-C. Tsay, *J. Org. Chem.*, 1990, **55**, 5987.
390. P.-Y. Lin, W.-S. Ku and M.-J. Shiao, *Synthesis*, 1992, 1219.
391. A. Capperucci, A. Degl'Innocenti, A. Ricci, A. Mordini and G. Reginato, *J. Org. Chem.*, 1991, **56**, 7323; A. Degl'Innocenti, A. Capperucci, A. Mordini, G. Reginato, A. Ricci and F. Cerreta, *Tetrahedron Lett.*, 1993, **34**, 873.
392. M. Segi, *et al.*, *J. Am. Chem. Soc.*, 1988, **110**, 1976; for another silicon-based route to these reactive species, see G. A. Krafft and P. T. Meinke, *ibid.*, 1986, **108**, 1314.
393. S. D. Rychnovsky, *J. Org. Chem.*, 1989, **54**, 4982.
394. A. Hosomi, K. Ogata, M. Ohkuma and M. Hojo, *Synlett*, 1991, 557.
395. K. Sukata, *J. Org. Chem.*, 1989, **54**, 2015.
396. K. Nishiyama and M. Oba, *Bull. Chem. Soc. Jpn.*, 1987, **60**, 2289.
397. M. B. Anderson, M. G. Ranasinghe, J. T. Palmer and P. L. Fuchs, *J. Org. Chem.*, 1988, **53**, 3125.
398. C.-N. Hsiao and H. Shechter, *J. Org. Chem.*, 1988, **53**, 2688.
399. J. E. Baldwin, R. M. Adlington, D. Bebbington and A. T. Russell, *J. Chem. Soc., Chem. Commun.*, 1992, 1249; see also M. M. Kabat and J. Wicha, *Tetrahedron Lett.*, 1991, **32**, 531; M. Lai, E. Oh, Y. Shih and H. Liu, *J. Org. Chem.*, 1992, **57**, 2471.

400. B. L. Chenard and C. M. Van Zyl, *J. Org. Chem.*, 1986, **51**, 3561; T. N. Mitchell, R. Wickenkamp, A. Amamria, R. Dicke and U. Schneider, *ibid.*, 1987, **52**, 4868.
401. B. H. Lipshutz, D. C. Reuter and E. L. Ellsworth, *J. Org. Chem.*, 1989, **54**, 4975.
402. M. Mori, N. Isono, N. Kaneta and M. Shibasaki, *J. Org. Chem.*, 1993, **58**, 2972.
403. Y. Ito, *Pure Appl. Chem.*, 1990, **62**, 583 and references therein.
404. P. Ramaiah and G. K. Surya Prakash, *Synlett*, 1991, 643 and references therein; for a review on nucleophilic trifluoromethylation, see G. K. Surya Prakash, in 'Organometallics in Synthetic Fluorine Chemistry', eds. G. A. Olah, R. D. Chambers and G. K. Surya Prakash, Wiley, New York, 1992, p. 227.
405. L. Wozniak and J. Chojnowski, *Tetrahedron*, 1989, **45**, 2465.
406. K. Afarinkia, J. I. G. Cadogan and C. W. Rees, *Tetrahedron*, 1990, **46**, 7175; *Synlett*, 1992, 123.
407. K. Afarinkia, J. I. G. Cadogan and C. W. Rees, *Tetrahedron*, 1992, 124.
408. V. Sum, A. J. Davies and T. P. Kee, *J. Chem. Soc., Chem. Commun.*, 1992, 1771.
409. E. A. Boyd, A. C. Regan and K. James, *Tetrahedron Lett.*, 1992, **33**, 813 and references therein.
410. A. Merzouk, F. Guibé and A. Loffet, *Tetrahedron Lett.*, 1992, **33**, 477.
411. S. S. Elmorsy, A. Pelter, K. Smith, M. B. Hursthouse and D. Ando, *Tetrahedron Lett.*, 1992, **33**, 821.
412. S. S. Elmorsy, M. V. Bhatt and A. Pelter, *Tetrahedron Lett.*, 1992, **33**, 1657.
413. Y. Goldberg and H. Alper, *J. Org. Chem.*, 1992, **57**, 3731.
414. E. Lukevics and A. Zablocka, 'Nucleoside Synthesis: Organosilicon Methods', Ellis Horwood, Chichester, 1991.
415. R. P. Bonar-Law, A. P. Davis and B. J. Dorgan, *Tetrahedron Lett.*, 1990, **31**, 6721.
416. R. P. Bonar-Law, A. P. Davis, B. J. Dorgan, M. T. Reetz and A. Wehrsig, *Tetrahedron Lett.*, 1990, **31**, 6725.
417. T. Murai, M. Yamamoto and S. Kato, *J. Chem. Soc., Chem. Commun.*, 1990, 789; for the use of related amines in the synthesis of 2-aza-1,3-dienes, see A. Degl'Innocenti, A. Mordini, D. Pinzani, G. Reginato and A. Ricci, *Synlett*, 1991, 712; J. Barluenga, M. Tomás, A. Ballesteros and J.-S. Kong, *Synthesis*, 1992, 106.
418. R. J. P. Corriu, J. J. E. Moreau and M. Pataud-Sat, *J. Org. Chem.*, 1990, **55**, 2878.
419. B. L. Bray, P. H. Mathies, R. Naef, D. R. Solas, T. T. Tidwell, D. R. Artis and J. M. Muchowski, *J. Org. Chem.*, 1990, **55**, 6317.
420. D. H. Hua, S. W. Miao, S. N. Bharathi, T. Katsuhira and A. A. Bravo, *J. Org. Chem.*, 1990, **55**, 3682.
421. G. Cainelli, M. Panunzio, P. Andreoli, G. Martelli, G. Spunta, D. Giacomini and E. Bandini, *Pure Appl. Chem.*, 1990, **62**, 605.
422. I. Ojima, I. Habus, M. Zhao, G. I. Georg and L. R. Jayasinghe, *J. Org. Chem.*, 1991, **56**, 1681; S. Busato, G. Cainelli, M. Panunzio, E. Bandini, G. Martelli and G. Spunta, *Synlett*, 1991, 243.
423. G. Cainelli, D. Giacomini, E. Mezzina, M. Panunzio and P. Zarantonello, *Tetrahedron Lett.*, 1991, **33**, 2967.
424. G. Cainelli, D. Giacomini, M. Panunzio and P. Zarantonello, *Tetrahedron Lett.*, 1992, **33**, 7783.
425. S. Itsuno, H. Yanaka, C. Hachisuka and K. Ito, *J. Chem. Soc., Perkin Trans. 1*, 1991, 1341; correction *ibid.*, 1991, 2290.
426. J. Barluenga, M. Tomás, A. Ballesteros and L. A. López, *Synlett*, 1991, 93.
427. N.-Y. Shih, *Tetrahedron Lett.*, 1993, **34**, 595.
428. B. Marciniec (ed.), 'Comprehensive Handbook on Hydrosilylation', Pergamon Press, Oxford, 1992.
429. J. Y. Corey, in 'Advances in Silicon Chemistry', ed. G. L. Larson, Jai Press, Greenwich, CT, 1991, vol. 1, p. 327.
430. K. Yamamoto and M. Takemae, *Synlett*, 1990, 259.
431. T. Sakakura, H.-J. Lautenschlager and M. Tanaka, *J. Chem. Soc., Chem. Commun.*, 1991, 40.
432. J. M. Tour, S. L. Pendalwar, C. M. Kafka and J. P. Cooper, *J. Org. Chem.*, 1992, **57**, 4786.
433. Y. Oozumi and T. Hayashi, *J. Am. Chem. Soc.*, 1991, **113**, 9887; see also Y. Oozumi, S.-Y. Lee and T. Hayashi, *Tetrahedron Lett.*, 1992, **33**, 7185; Y. Oozumi and T. Hayashi, *ibid.*, 1993, **34**, 2335.
434. G. Helmchen, A. Krotz, K.-T. Ganz and D. Hansen, *Synlett*, 1991, 257.
435. K. Tamao, Y. Nakagawa and Y. Ito, *Organometallics*, 1993, **12**, 2297 and references therein.
436. N. R. Curtis, A. B. Holmes and M. G. Looney, *Tetrahedron Lett.*, 1992, **33**, 671; N. R. Curtis and A. B. Holmes, *ibid.*, 1992, **33**, 675; for examples involving protiodesilylation, see S. E. Denmark and D. C. Forbes, *ibid.*, 1992, **33**, 5037; M. R. Hale and A. H. Hoveyda, *J. Org. Chem.*, 1992, **57**, 1643.
437. S. H. Bergens, P. Noheda, J. Whelan and B. Bosnich, *J. Am. Chem. Soc.*, 1992, **114**, 2121, 2128.
438. M. Murakami, M. Suginome, K. Fujimoto, H. Nakamura, P. G. Andersson and Y. Ito, *J. Am. Chem. Soc.*, 1993, **115**, 6487.
439. T. Harada, S. Imanaka, Y. Ohyama, Y. Matsuda and A. Oku, *Tetrahedron Lett.*, 1992, **33**, 5807; K. Matsumoto, K. Miura, K. Oshima and K. Utimoto, *ibid.*, 1992, **33**, 7031.
440. C. Chatgilialoglu, *Acc. Chem. Res.*, 1992, **25**, 188.
441. I. Ryu, M. Hasegawa, A. Kurihara, A. Ogawa, S. Tsunoi and N. Sonoda, *Synlett*, 1993, 143.
442. M. Lesage, J. A. Martinho Simões and D. Griller, *J. Org. Chem.*, 1990, **55**, 5413.
443. C. Chatgilialoglu, A. Guerrini and G. Seconi, *Synlett*, 1990, 219.
444. R. P. Allen, B. P. Roberts and C. R. Willis, *J. Chem. Soc., Chem. Commun.*, 1989, 1387.
445. D. H. R. Barton, D. O. Jang and J. Cs. Jaszberenyi, *Synlett*, 1991, 435; *Tetrahedron Lett.*, 1992, **33**, 6629.
446. I. Ojima *et al.*, *Tetrahedron*, 1993, **49**, 5431; I. Matsuda, A. Ogiso and S. Sato, *J. Am. Chem. Soc.*, 1990, **112**, 6120.
447. N. Chatani, Y. Kajikawa, H. Nishimura and S. Murai, *Organometallics*, 1991, **10**, 21.
448. T. Murai *et al.*, E. Yasui, S. Kato, Y. Hatayama, S. Suzuki, Y. Yamasaki, N. Sonoda, H. Kurosawa, Y. Kawasaki and S. Murai, *J. Am. Chem. Soc.*, 1989, **111**, 7938.
449. S. Ikeda, N. Chatani, Y. Kajikawa, K. Ohe and S. Murai, *J. Org. Chem.*, 1992, **57**, 2.
450. S. C. Berk and S. L. Buchwald, *J. Org. Chem.*, 1993, **58**, 3221; S. C. Berk and S. L. Buchwald, *ibid.*, 1992, **57**, 3751.
451. T. Fukuyama, S.-C. Lin and L. Li, *J. Am. Chem. Soc.*, 1990, **112**, 7050; for an example of use, see D. A. Evans and H. P. Ng, *Tetrahedron Lett.*, 1993, **34**, 2229.
452. H. Mayr, N. Basso and G. Hagen, *J. Am. Chem. Soc.*, 1992, **114**, 3060.
453. J. Chojnowski, W. Fortuniak and W. Stańczyk, *J. Am. Chem. Soc.*, 1987, **109**, 7776.
454. G. A. Olah, Q. Wang and G. K. Surya Prakash, *Synlett*, 1992, 647.
455. S. W. McCombie, B. Cox, S.-I. Lin, A. K. Granguly and A. T. McPhail, *Tetrahedron Lett.*, 1991, **32**, 2083.
456. A. Mehta, R. Jaouhari, T. J. Benson and K. T. Douglas, *Tetrahedron Lett.*, 1992, **33**, 5441.
457. M. B. Sassaman, K. D. Kotian, G. K. Surya Prakash and G. A. Olah, *J. Org. Chem.*, 1987, **52**, 4314.
458. G. A. Kraus and M. T. Molina, *J. Org. Chem.*, 1988, **53**, 752.

459. N. Greenspoon and E. Keinan, *J. Org. Chem.*, 1988, **53**, 3723.
460. E. W. Colvin, 'Silicon Reagents in Organic Synthesis', Academic Press, London, 1988, chap. 18.
461. T. H. Chan and D. Wang, *Chem. Rev.*, 1992, **92**, 995.
462. J. Dubac, A. Laporterie and G. Manuel, *Chem. Rev.*, 1990, **90**, 215; E. Colomer, R. J. P. Corriu and M. Lheureux, *ibid.*, 1990, **90**, 265.
463. R. D. Singer and A. C. Oehlschlager, *J. Org. Chem.*, 1991, **56**, 3510 and references therein.
464. K. Krohn and K. Khanbabaee, *Angew. Chem., Int. Ed. Engl.*, 1994, **33**, 99.
465. K. Tamao, A. Kawachi and Y. Ito, *J. Am. Chem. Soc.*, 1992, **114**, 3989.

8
Tin

TADASHI SATO
Waseda University, Tokyo, Japan

8.1 INTRODUCTION	356
8.1.1 Literature	356
8.1.2 Overview	356
8.1.2.1 General scope of organotin chemistry	356
8.1.2.2 Classification of tin reagents	356
8.1.2.3 Reaction types of the carbon–tin bond	356
8.1.2.4 Handling	356
8.2 ORGANOTIN HYDRIDES AND ORGANODITINS	356
8.2.1 Reduction of Carbon–Heteroatom Bonds	356
8.2.1.1 Halogen compounds	356
8.2.1.2 Alcohols	356
8.2.1.3 Divalent heteroatom compounds	356
8.2.1.4 Nitrogen compounds	356
8.2.2 Reduction of Carbonyl and Carboxyl Groups	356
8.2.3 Addition to Multiple Bonds	356
8.2.4 Carbon–Carbon Bond Formation	356
8.2.5 Other Reactions	356
8.3 ANIONIC TIN REAGENTS	356
8.3.1 Stannyl Anions (Sn^-)	356
8.3.1.1 Preparation	356
8.3.1.2 Reactions	356
8.3.2 α-Stannyl Carbanions ($Sn–C^-$)	356
8.3.2.1 Preparation	356
8.3.2.2 Reactions	356
8.4 CATIONIC TIN REAGENTS	356
8.4.1 Reagents with Tin–Heteroatom Bonds (Sn–X)	356
8.4.1.1 Tin–oxygen bonds	356
8.4.1.2 Tin–halogen bonds	356
8.4.1.3 Tin–nitrogen bonds	356
8.4.1.4 Other heteroatoms	356
8.4.2 Carbocationic Tin Reagents ($Sn–(C)_n–X$)	356
8.5 ALKENYL- AND ARYLSTANNANES (Sn–C=C)	356
8.5.1 Preparation	356
8.5.2 Reactions	356
8.5.2.1 Transmetallation	356
8.5.2.2 Palladium-catalyzed cross-coupling reactions	356
8.5.2.3 Miscellaneous reactions	356
8.6 ALLYLSTANNANES (Sn–C–C=C)	356
8.6.1 Preparation	356
8.6.2 Reactions	356
8.6.2.1 Transmetallation	356
8.6.2.2 Coupling with halides	356
8.6.2.3 Reactions with electrophiles	356
8.6.2.4 Miscellaneous reactions	356

8.7 MISCELLANEOUS UNSATURATED STANNANES	356
8.7.1 Allenylstannanes	356
8.7.2 Alkynylstannanes	356
8.8 FUNCTIONALIZED STANNANES (Sn–(C)$_n$–F)	356
8.8.1 α-Stannyl Alcohols (n = 1, F = OH)	356
8.8.1.1 Preparation	356
8.8.1.2 Reactions	356
8.8.2 β-Stannyl Alcohols (n = 2, F = OH)	356
8.8.3 γ-Stannyl Alcohols (n = 3, F = OH)	356
8.8.3.1 Preparation	356
8.8.3.2 Reactions	356
8.8.4 Acylstannanes (n = 0, F = CO)	356
8.8.5 α-Stannyl Ketones (n = 1, F = CO)	356
8.8.6 β-Stannyl Ketones (n = 2, F = CO)	356
8.8.6.1 Preparation	356
8.8.6.2 Reactions	356
8.8.7 γ-Stannyl Ketones (n = 3, F = CO)	356
8.8.8 Other Alkylstannanes Functionalized at the α-Position	356
8.8.9 Alkylstannanes Functionalized at Remote Positions	356
8.9 PHOTOREACTIONS OF ORGANOTIN COMPOUNDS	356
8.10 SUMMARY	356
8.11 REFERENCES	356

8.1 INTRODUCTION

Since the introduction of boron and phosphorus to organic synthesis, as hydroboration and the Wittig reaction, the chemistry of many other elements has been pursued to develop novel selective reactions and reagents. Organotin chemistry has been playing an increasingly important role for this purpose. Since the last publication on organotin compounds in *COMC-I*,[1] applications to organic synthesis have expanded enormously. Among many types of organotin compounds, tin hydride, α-heteroatom-substituted tin compounds, and vinyl- and allylstannanes are the most widely known and, consequently, most extensively documented. In 1992 alone, approximately 230 papers appeared on "organotin in organic synthesis," excluding those dealing with tin reagents only in a particular step in multistep synthetic studies. More than 85% of these papers deal with one of the above four types of stannyl compounds. In this chapter, emphasis is on the less popular fields, as well as on the well-documented areas, to reveal new aspects of organotin chemistry. In the earlier sections (Sections 8.2–8.4), the chemistry of smaller molecules which are used as "tin reagents" is discussed and in later sections (Sections 8.5–8.9) the behavior of many types of organotin compounds is discussed.

8.1.1 Literature

Numerous specialized books and reviews are available; those published prior to 1984 have been listed[2] and are not duplicated here. A wide survey of organotin compounds from basic concepts to industrial applications, including synthetic applications, has been made.[3] Original papers in many fields in organotin chemistry have been collected in *Tetrahedron Symposia-in-Print*, No. 36.[4] In *Comprehensive Organic Synthesis*, the chemistry of allylstannanes has been compared with that of allyl compounds of related elements.[5] In this series, various types of other tin chemistries have been discussed in many of the related sections. Review articles dealing with special fields will be referred to in the appropriate sections. Due to the limited space, only the most recent references are cited here and the earlier papers cited in these references, particularly those from the same laboratories, are omitted.

8.1.2 Overview

8.1.2.1 General scope of organotin chemistry

The most common form of tin in organotin compounds is tetravalent with sp^3 hybridization. Divalent tin compounds, known as stannylenes, have been studied in only a few cases. When the tin atom bears electronegative substituents, its Lewis acidity increases and coordination with electron-rich sites leads to the hypervalent form of sp^3d (trigonal bipyramid) or sp^3d^2 (octahedral). The covalent radius of tin is large (0.214 nm) and bonds to the tin atom are long (Sn–C, 0.22 nm; Sn–H, 0.17 nm) and mostly covalent with facile polarizability. In view of the relatively low carbon–tin bond dissociation energy (~210 kJ mol^{-1}), it is no surprise that organotin compounds exhibit reactivities in both homolytic and heterolytic senses. Stabilization of an electron-deficient center at the β-position, known as the "β-effect" in silicon chemistry,[6] is also observed in tin chemistry[7] and defines many of the reaction patterns of organotin compounds.

While being considered hard, tin is softer than silicon. Hence, like silicon, it has a tendency to associate with hard bases, such as fluoride, but it is also apt to associate with much softer bases such as sulfides. Actually, tin is known as a thiophile.

8.1.2.2 Classification of tin reagents

There are several types of organotin compounds used as synthetic reagents. They can be classified into two categories, homolytic and heterolytic, depending on their reaction modes. Typical reagents of homolytic nature are trialkyltin hydrides, which are now the most frequently utilized tin reagents and they have been used as radical initiators, reducing reagents, or, to a lesser extent, as delivery reagents of tin atoms to organic substrates.

The heterolytic reagents can be either cationic or anionic. Anionic tin compounds are widely used to stannylate electrophilic centers or multiple bonds. The counter-cation most commonly employed is lithium, but sodium, potassium, and magnesium are also widely used. Copper(II) complexes also behave anionically. The most popular cationic reagents are tin halides, tin alkoxides, and tin triflate, which are used as Lewis acids or counter-species of enolates. Cationic tin reagents can also be used to introduce the stannyl group to electron-rich substrates. However, this type of stannylation has limited use only, because the electron-rich substrates are usually none other than the species which it is intended to prepare from the tin compounds.

8.1.2.3 Reaction types of the carbon–tin bond

Due to the electropositive character of tin compared with that of carbon and also because of the weak tin–carbon bond, the tin-bearing carbon reacts as a carbanion or radical. However, except in special cases,[8] the normal tin–carbon bond is so stable that it is necessary to activate the bond in some way to initiate reaction. One of the activation methods is to introduce an "activating group," such as a vinyl or allyl group or a heteroatom-bearing carbon, directly to the tin atom. In this way, the tin atom can be either transmetallated with more reactive metals or brought into reaction directly. Usually, the tin is replaced with lithium to enhance nucleophilicity or activated by palladium catalysis to facilitate cross-coupling with halides or triflates. Such cross-coupling reactions are among the best methods available for allyl or vinyl group transfer to organic substrates.

Another common activation method is to use other "activating groups," such as a carbonyl, hydroxy group, or some cationic center, located at various positions from the tin atom in the same molecule. Due to the very low reactivity of the carbon–tin bond, a wide variety of these functionalities can be introduced into organotin compounds, which are in many cases isolable as stable intermediates. The organotin compounds thus prepared in the first stage can be reacted with electrophiles in various ways by manipulating the second stage of the reaction through control of functionalities and reaction conditions. In contrast to reactions proceeding via tin–metal exchange, which are inherently those of the corresponding organometals, many of the reactions induced by the latter activation method have characteristics typical of organotin compounds.[9]

8.1.2.4 Handling

Most organotin compounds are stable in air, insensitive to moisture, and can be stored for long periods; ordinary distillation, chromatographic separation, and recrystallization procedures are tolerated. They are soluble in ordinary organic solvents, insoluble in water and can be treated without any special techniques other than those required for ordinary organic reactions. Generally, organotin compounds which are formed as by-products after a given reaction are tetraalkylstannanes or trialkyltin halides. These by-products can be separated from the reaction products by ordinary techniques, in most cases by distillation or column chromatography. If problems are encountered in separating tin by-products from desired products due to similar polarity, the separation can be done simply by shaking the reaction mixture with an aqueous solution of potassium fluoride and filtering off the insoluble tin fluoride thus formed[10] or by treating the reaction mixture with I_2/dbu (1,5-diazabicyclo[5.4.0]undec-5-ene) and removing the solid material by filtration.[11]

Most organotin compounds are toxic and the toxicity varies over a wide range depending on the types and nature of the substituents. Maximum toxicity occurs with methyl and ethyl derivatives, but decreases as the alkyl groups on tin become larger, such as butyl or phenyl. In the series R_nSnY_{4-n}, the toxicity decreases as the number of alkyl groups decreases and inorganic tin compounds are mostly nontoxic. Despite their higher toxicity, trimethylstannyl compounds are sometimes preferred as reagents to the tributyl compounds, because of the simplicity of their NMR spectra, in which the methyl signal appears as a singlet at $\delta \cong 0$, with small satellite signals due to ^{117}Sn (7.54% abundance, $J = 51$ Hz) and ^{119}Sn (8.62% abundance, $J = 53$ Hz). In some cases the reaction pattern differs between trimethyl- and tributylstannyl compounds.[12,13] Due to their toxicity, organotin compounds should be handled with adequate safety measures: use of protective gloves, good hoods, and caution in waste disposal.

8.2 ORGANOTIN HYDRIDES AND ORGANODITINS

Organotin hydrides, particularly tri-*n*-butyltin hydride, are the most frequently used organotin compounds in the laboratory. Some of the tin hydrides can be prepared *in situ* by the reaction of tin chloride and borohydride.[14] The chemistry of tri-*n*-butyltin hydride has been summarized in a review.[15] Reactions involving radical species are the most common pattern, since tin hydrides generate tin radicals easily by means of 2,2'-azobisisobutyronitrile (AIBN) or other radical sources, by UV, or simply by heat. Ultrasonic[16] and electrooxidative[17] initiation are also successful. Triethylborane, which is unlikely to behave as a radical source, greatly enhances the reactivity of triphenyltin hydride.[18] An involvement of molecular oxygen has been suggested for this initiation.[19]

Tin radicals either abstract heteroatoms from the C–X bond or add to carbon–carbon or carbon–heteroatom multiple bonds to generate carbon radicals. In some cases, the intermediate radicals undergo secondary reactions involving nearby functional groups which induces multistep C–C skeleton rearrangements.[20] In view of the recent advances in free-radical chemistry in organic synthesis,[21,22] the use of tin hydride in radical initiation is very important, while reaction pathways involving transition metal catalysis are also widely employed.[23]

Polymer-supported tin hydrides and ditins have been developed which control the succeeding reaction types and facilitate the separation of the reaction products.[24] Hexaalkylditins are cleaved to tin radicals under thermal or photochemical conditions[25] or in palladium-catalyzed reactions.[26,27]

8.2.1 Reduction of Carbon–Heteroatom Bonds

The most common reaction of carbon radicals generated from C–X bonds in the presence of tin hydrides is hydrogen abstraction from the tin hydride, which results in net reduction of the C–X bond to a C–H bond. This process can be carried out with several types of C–X compounds.

8.2.1.1 Halogen compounds

Free-radical halogen abstraction from a C–X bond presents a broad variety of synthetic possibilities. Almost all chlorine, bromine, or iodine (but not fluorine) atoms at saturated or alkenic carbons and bromine or iodine (but not chlorine) at aromatic carbons can be replaced by hydrogen by means of tin hydride. Reactivity is greatest with iodides and decreases in the order of bromides and then chlorides.

The initially formed radicals may undergo secondary reactions with neighboring groups prior to hydrogen abstraction. When acetoxyl or phosphoryl groups are present on the vicinal carbon, the initially formed radicals induce a 1,2-radical migration as shown in Scheme 1. This is an effective method for the synthesis of 2-deoxy sugars.[28] With α-chloro epoxides, triphenyltin hydride affords α-chloro carbonyl compounds, while the reagent gives exclusively allyl alcohols from β-chloro epoxides.[29] A ring expansion has been observed with iodo epoxides.[30] *o*-Bromobenzyl bromide derivatives can serve as "self-oxidizing protecting groups" since alcohols protected with this reagent are oxidized to carbonyl compounds under reductive conditions when treated with tributyltin hydride, as outlined in Scheme 2.[31]

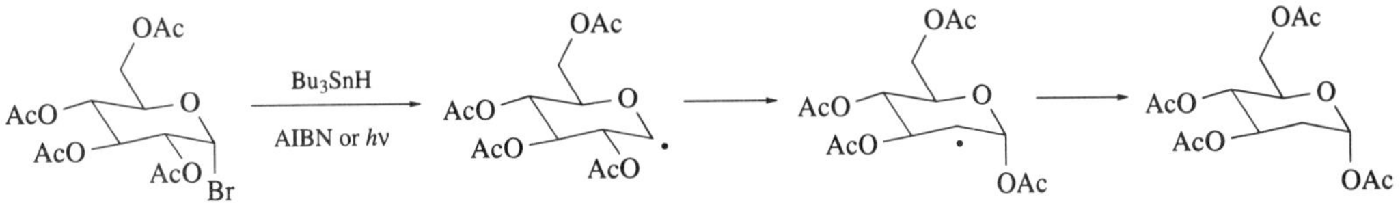

Scheme 1

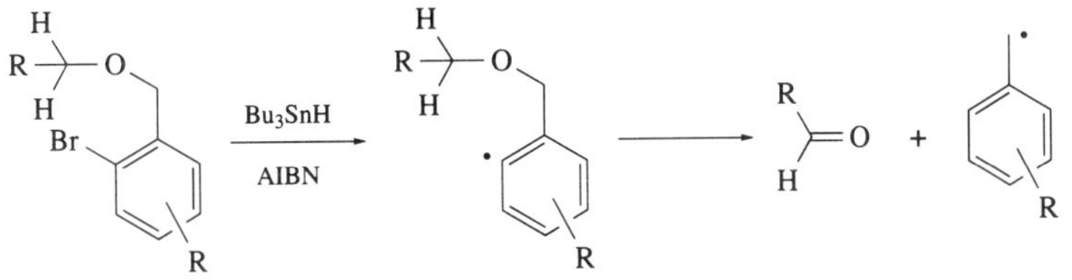

Scheme 2

The stereochemistry of halogen–hydrogen exchange was examined with diastereomeric 2-halo sugars in connection with the development of a synthetic method for 2-deoxy sugars.[32] The reaction proceeds via a common radical intermediate and the final stereochemistry is controlled by neighboring steric factors. Reaction with α-halo esters proceeds by inversion and this has been interpreted in terms of conformational analysis of the intermediate radical.[33]

The reduction of alkenyl iodides, which proceeds with a high degree of retention of configuration, can be achieved with tributyltin hydride under palladium catalysis. However, the triethylborane-induced reduction is not stereoselective.[34]

8.2.1.2 Alcohols

The selective and mild replacement of a hydroxy group by hydrogen is an important process, particularly in carbohydrate chemistry. Although a direct removal of the hydroxy group is not possible, alcohol reduction can be achieved via derivatization and conversion into halides is a convenient choice. An alternative is conversion of the alcohol into esters or thioesters, particularly to xanthates, which is known as the Barton–McCombie reaction.[35] Although the mechanism of R–O cleavage from xanthates is controversial, it is believed that the tin radical adds to the sulfur atom of the C=S bond to afford a thioketyl radical, which undergoes the desired R–O bond cleavage (Scheme 3).[19] Primary alcohols can be easily deoxygenated by derivatization as polyhalophenyl thiocarbonate esters.[36] In contrast to the general reaction pattern of hydrogen abstraction, xanthates derived from allyl alcohols undergo stannyl group abstraction to afford allylstannanes via [3,3]-sigmatropic rearrangement.[37] Thermal reaction of trimethyltin hydride with allyl alcohols, on the other hand, results in hydride addition to afford 3-(trimethylstannyl)propanol.[38]

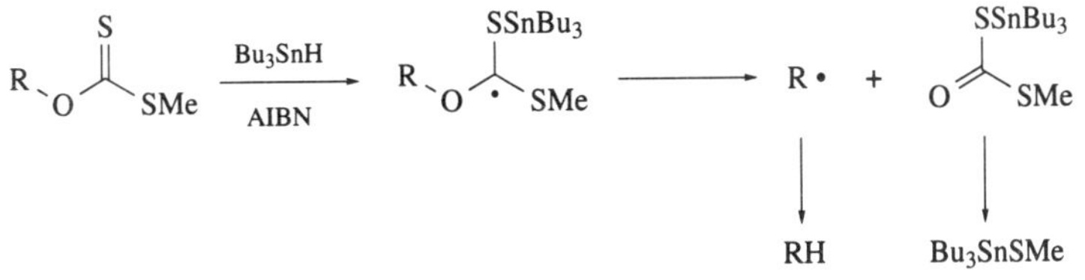

Scheme 3

In contrast to ketyl radical formation in the case of thioesters, the reaction of selenocarbonates proceeds by Se–C bond cleavage. The resulting alkoxycarbonyl radicals can give either formates, alcohols, or alkanes (Scheme 4).[39]

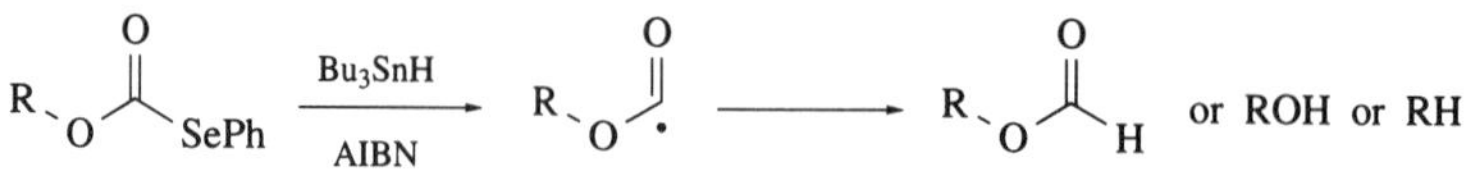

$$R-O-C(=O)-SePh \xrightarrow[\text{AIBN}]{Bu_3SnH} R-O-C(=O)\cdot \longrightarrow R-O-C(=O)-H \text{ or ROH or RH}$$

Scheme 4

8.2.1.3 Divalent heteroatom compounds

The bonds R^1–X–R^2, where X = S, Se, Te, or other heteroatom, are readily split by stannyl radicals. The reactivity increases in the order phenyl (not split) ≪ methyl < primary < secondary < tertiary < allyl ≅ benzyl,[15] and therefore phenyl is generally used as the R^2 group so that bond fission occurs only at the R^1–X bond.[40] The C–Hg bond in β-mercuriocycloalkanones is homolytically cleaved by tin hydride to afford either direct reduction products or one-carbon-expanded cycloalkanones, depending on the radical stability of the intermediate.[41]

8.2.1.4 Nitrogen compounds

The nitro group can be removed homolytically by tin hydride. Since the nitro group can activate α-carbon anionically to effect C–C bond formation with carbon electrophiles, the tandem reactions of C–C bond formation followed by removal of the nitro group can serve as a versatile synthetic method.[42]

The amino group can be reductively eliminated if first converted to an isonitrile. Aliphatic, alicyclic, and steroidal isonitriles give the corresponding hydrocarbons in 80–90% yields.[15]

8.2.2 Reduction of Carbonyl and Carboxyl Groups

Reduction of carbonyl groups to alcohols can be accomplished either through ionic or radical pathways. Ionic reactions can be induced by activating the carbonyl group with Lewis acids or by activating the tin hydride by coordination of Lewis bases to the tin atom to facilitate the liberation of hydride. A remarkable contrast in stereoselectivity has been observed in the reduction of α-alkoxy ketones (Equations (1) and (2)).[43] The use of a Lewis base induced a *syn* reduction (Equation (1)), while the use of a Lewis acidic tin hydride resulted in *anti* reduction (Equation (2)). These results have been rationalized by assuming that the Lewis base effectively traps the stannyl cation, thus inducing Cram-type reduction, while the chlorostannyl group coordinates with two oxygen atoms to induce the anti-Cram reduction by chelation.

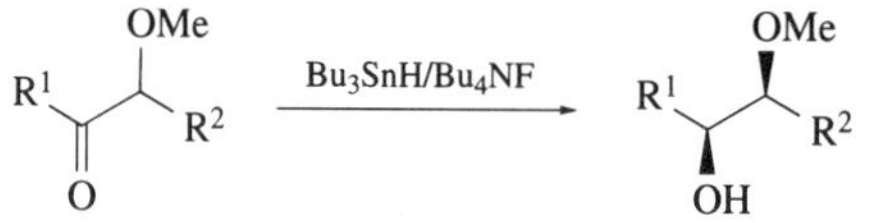

$$R^1-C(=O)-CH(OMe)-R^2 \xrightarrow{Bu_3SnH/Bu_4NF} R^1-CH(OH)-CH(OMe)-R^2 \quad (1)$$

$$R^1-C(=O)-CH(OMe)-R^2 \xrightarrow{Bu_2SnCl-H} R^1-CH(OH)-CH(OMe)-R^2 \quad (2)$$

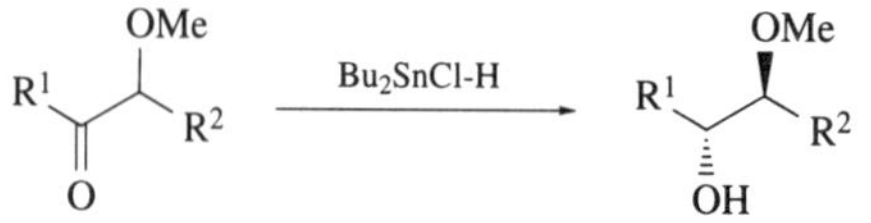

The radical reactions proceed by addition of the tin radical to the carbonyl oxygen, as shown in Scheme 5.[44] The stannyloxycarbon radicals thus formed abstract hydrogen directly to give alcohols or they may interact with a neighboring group prior to hydrogen abstraction. As an example of the latter case, an epoxide can participate in the reaction to give β-hydroxy ketones.[45]

α,β-Unsaturated aldehydes react with tin hydride in various ways: the conjugate reduction to saturated aldehydes occurs with palladium catalysis, one-carbon loss to give saturated ketones occurs with palladium/oxygen, while an aldehyde function is reduced to an allyl alcohol with $ZnCl_2$ catalysis.[46] Smooth conjugate reduction of α,β-enones is achieved by Et_3B-induced reaction of triphenyltin hydride[18] or by CuI/LiCl-assisted reaction of tributyltin hydride.[12] A number of other functional groups

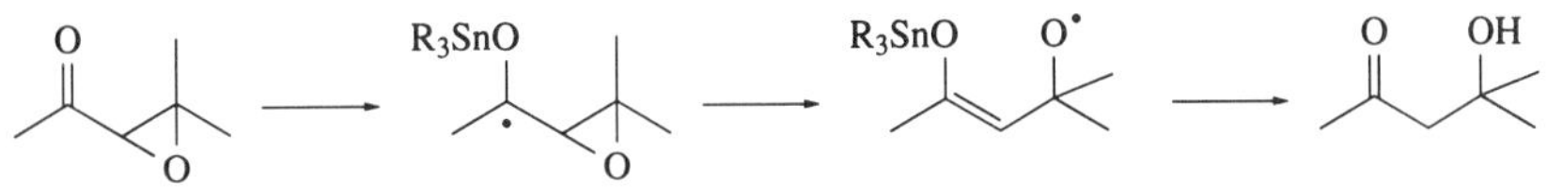

Scheme 5

including sulfoxide, ester, and nitrile are unaffected. The AIBN-induced reduction of 3-oxo-1,4-diene steroid derivatives results in B-ring fission, aromatization of the A-ring, and formation of 9,10-secosteroids.[47]

The reduction of carboxylic acid derivatives to aldehydes by tin hydride is carried out using acid chlorides[48] or thio- or selenoesters[49] under palladium catalysis or by AIBN initiation.

8.2.3 Addition to Multiple Bonds

Tin hydride adds easily to various types of multiple bonds, but the addition to alkynes or allenes is the most important, because it is a convenient method of vinylstannane preparation. The AIBN-induced radical method is frequently employed, while palladium catalysis is also used.[8,50] The regio- and stereochemistry are greatly influenced by the nature of the substrates and the reaction conditions.[8,51]

8.2.4 Carbon–Carbon Bond Formation

Hydrogen abstraction by primarily formed radicals can be competitive with other reaction types such as addition to an unsaturated group to form a C–C bond, which is known as the Giese reaction.[52,53] Reaction with allyl sulfides or allylstannanes results in a radical allylation.[54] Although intermolecular C–C couplings are certainly useful,[55] the intramolecular reactions have been used more frequently as powerful and versatile tools for ring construction. Much success has been achieved in stereochemical control in acyclic carbon–carbon bond formation[56] and in cyclization.[57] The unsaturated moiety is usually C=C,[58] but C≡C,[59] C=O,[60] imine,[61] or enamine[62] double bonds and aromatic rings[63] or acylsilanes[64] can also serve as radical acceptors. The majority of cyclizations involve five-membered ring formation from δ,ε-unsaturated radicals via a 5-*exo-trig* cyclization,[65] although the *exo:endo* ratio is sometimes controlled by the steric[66] or electronic environment.[11] The 5-*exo-trig* cyclization seems to be an extremely facile reaction, since the initially formed thiocarbonyl radical from xanthates of homoallylic alcohols cyclizes to thiolactones prior to the otherwise ensuing C–O bond cleavage.[67] In cases where the intermediate radical is stabilized, *endo* cyclization is exclusive.[68,69] The cyclized radicals can abstract hydrogen or a halogen atom to afford the final products.[70,71] The halogen abstraction reaction is often referred to as "atom-transfer cyclization."[11] Under aerobic conditions, the cyclized radicals are oxygenated to give alcohols.[72]

Since halogen atoms are abstracted most readily compared with other heteroatoms, γ-lactams having a sulfur substituent are obtained by the cyclization of the alkenic α-chloro-α-phenylthiocarbamates.[73] The reaction of selenocarbonates which have a triple bond to give the corresponding α-alkylidene-γ-butyrolactones is characterized by high *exo* cyclization and a high ratio of cyclization to reduction.[39]

When there is more than one double bond in a molecule, a variety of products, such as spiro and fused ring systems, can be prepared through tandem radical cyclizations.[74–7] An effective cyclization is possible by the Et₃B-induced process, since the reaction can be carried out at lower concentrations compared with other conventional radical-initiated reactions. This procedure has been applied to the stereoselective synthesis of α-methylene-γ-butyrolactones (**1**) (Scheme 6).[78]

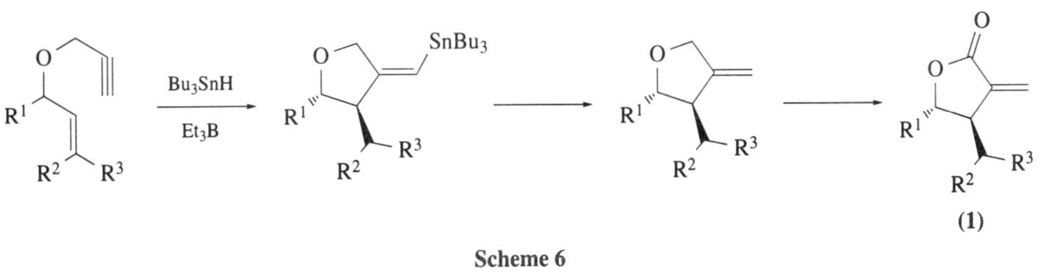

(1)

Scheme 6

Although the tin radical initiated cyclization proceeds satisfactorily with enynes, that of dienes or trienes has been far less successful, mainly because the reaction results in bistannylation without cyclization. However, a successful cyclization from various activated or unactivated dienes and trienes (**2**) has been reported.[79] A development of oxidative cleavage of the unactivated C–Sn bond in (**3**) to give an acetal definitely enhances the synthetic applicability of the method (Scheme 7; can, cerium(IV) ammonium nitrate).

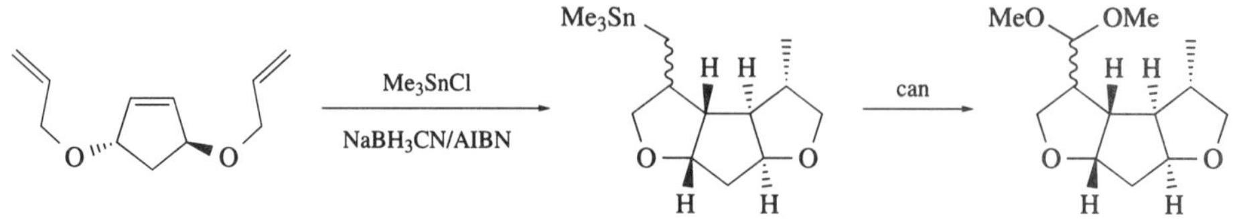

Scheme 7

ω-Iodoalkyl propiolate esters (**5**) of $n = 10$–12 cyclized to 14–16-membered *trans*-α,β-unsaturated macrocyclic lactones (**6**), while the reaction afforded only reduction products (**4**) when $n = 6$–9 (Scheme 8).[80] Due to slow hydrogen abstraction from the tin hydride, the primarily formed *cis*-radical intermediate inverts to give the thermodynamically more favored *trans* geometry. This macrocyclization technique has been extended to even a 22-membered ring.[81]

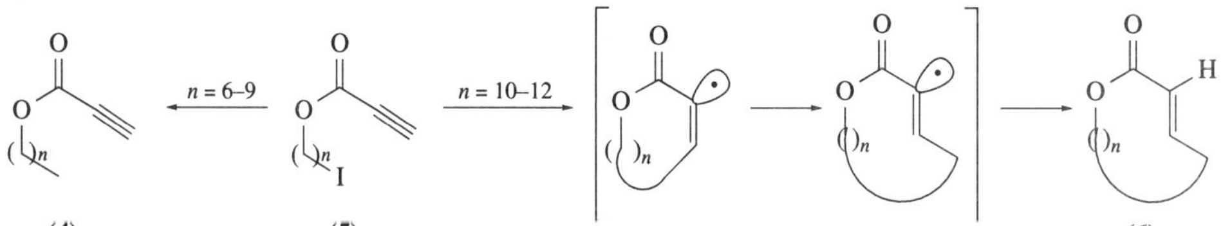

Scheme 8

Some radical intermediates undergo a couple of radical transfers by consecutive intramolecular hydrogen abstractions before affording final products. This has been developed into a useful way of constructing novel cyclic systems.[14,25,82-4]

The reaction of unsaturated acid chlorides with tin hydride is reagent dependent. Radical initiation by AIBN results in cyclization, while smooth reduction to aldehydes is realized by the palladium-catalyzed reaction.[48]

When the AIBN/Bu$_3$SnH-induced reaction was applied to β-allenic oxime ethers, intramolecular radical attack occurred to produce stannylated aminocyclopentenes. Acid treatment easily eliminated the stannyl group to produce aminocyclopentenes.[85]

8.2.5 Other Reactions

Heteroaromatic compounds having a suitable leaving group undergo *ipso*-substitution by the stannyl group in the AIBN-induced reaction with tributyltin hydride.[86] The *ipso*-substitution also proceeds cleanly with phenylthio-substituted furanones. Surprisingly, neither the conventional reduction nor addition–elimination products were identified.[87]

Carbene species undergo insertion into the Sn–H σ-bond. Aliphatic Fischer carbene complexes (**7**) having an α-stereogenic center afford the insertion products with considerable diastereoselectivity (Equation (3)). Notably, the stereochemistry was reversed compared to that obtained by Cram-type addition of Bu$_3$SnLi to the structurally comparable aldehydes.[88]

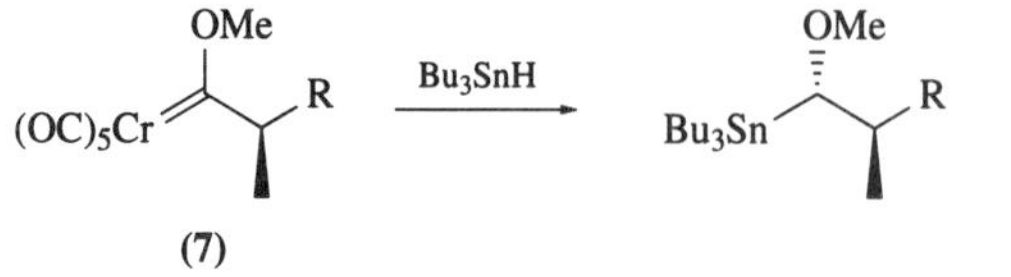

(3)

Aminyl radicals can be produced by abstraction of a thio group from sulfenamides. Although the aminyl radical thus produced does not cyclize with ordinary alkenes, it cyclizes with aryl-activated alkenes to afford pyrrolidines.[89] A stannyl radical initiated decomposition of azides results in cyclization with ensuing ring expansion.[90] Aryl radicals generated from aryl iodides by iodine atom abstraction can react with an azo group intramolecularly to afford N-aminocarbazoles.[91]

The allyloxycarbonyl group, a useful amine-protecting group, can be cleanly eliminated by palladium-catalyzed reaction with tributyltin hydride. Other conventional deprotections sometimes afford allylamines as by-products.[92]

8.3 ANIONIC TIN REAGENTS

8.3.1 Stannyl Anions (Sn⁻)

8.3.1.1 Preparation

The most typical anionic stannyl reagents are tributylstannyllithium and trimethylstannyllithium. In a few cases, triphenylstannyl derivatives are used. Tetrahydrofuran solutions of the stannyl anions can be prepared from the corresponding hexaalkylditins in yields over 95% by treatment with butyllithium or methyllithium.[93] Tributylstannyllithium can be prepared more conveniently by deprotonation of the tin hydride with lithium diisopropylamide.[94] Chlorine–lithium exchange of trialkyltin chlorides with lithium metal is also a convenient method.[95]

Magnesium derivatives are best accessible from Bu_3SnH by treatment with sterically hindered secondary alkylmagnesium halides.[96] Stannylcuprates having several "dummy" ligands, such as CN and RC≡C,[97] CN, and Me,[98,99] have been prepared. A convenient method for the preparation of higher order stannylcuprates from trimethyltin hydride or hexamethylditin has been described.[12]

The naked tributylstannyl anion can be prepared from (trimethylsilyl)tributylstannane by treatment with naked cyanide[100] or with fluoride anion.[101] These species are particularly interesting in two respects: first, the preparative method is very typical of tin chemistry and is not realizable with carbanions; second, the species generated have an entirely different chemistry to that of stannyl anions having metal counter-ions.

8.3.1.2 Reactions

(i) Substitution

Stannyl anions undergo nucleophilic substitution reactions with compounds containing leaving groups at the sp^3 carbon. The reaction mechanism has been extensively studied[102,103] and it is generally accepted that the reaction with acyclic[104] and cyclic[95] halides proceeds by an initial electron transfer, while the reaction with tosylates proceeds via an S_N2 mechanism with complete inversion of configuration. Reactions with oxirane or tetrahydrofuran rings proceed with ring opening to produce β- or δ-stannyl alcohols.[105]

Reaction between a stannyl anion and haloarenes in liquid ammonia can lead either to substitution or halogen–metal exchange, depending on the ligands on the tin atom and the nature of the halogen(s) of the haloarenes.[106]

(ii) Addition to multiple bonds

Since Piers found that trimethylstannylcuprates added to alkynic bonds, this method has been used as a useful preparative method for alkenylstannanes. In contrast to tin hydride, in which the partner of the tin atom is always hydrogen, reagents having tin–heteroatom bonds are potentially double functional, if both the heteroatom and tin atom are manipulated individually in succeeding steps. However, this has proved to be a rather difficult task with electrophiles other than the proton, because the stannylcupration step is reversible and, although the equilibrium is well to the product side, the original stannylcuprate reagents are more reactive towards most electrophiles than the vinylcuprate intermediates. The problem has been overcome by using $ZnCl_2$, which promotes the formation of vinylcopper–zinc intermediates,[107] or by using higher order (cyano)(methyl)(tributylstannyl)cuprates under specific conditions.[98,108]

The regio-, stereochemistry and characterization of the intermediates have been discussed.[109] The reaction is effectively controlled by neighboring polar substituents: terminal vinylstannanes are obtained from propargyl alcohols[110] or propargyl acetals,[111] while inner vinylstannanes are obtained from 1-alkynes.[98] The regioselectivity of the stannylcupration of propargylic sulfides[112] and amines[113] has also been discussed. The geometry of the vinylstannanes from 2-alkynoates is dependent upon the reaction conditions: (Z)-isomers were obtained at 0 °C under aprotic conditions, while (E)-isomers were obtained at −78 °C in the presence of alcohols.[97]

Various types of stannylcuprates add to allenes[114] or alkynes.[115] Further *in situ* manipulation of the cuprates offers an easy entry to polyfunctionalized compounds such as exomethylenic 1,3-diketones.[115] Stannylmetals other than stannylcuprates also undergo addition to triple bonds. The regio- and stereochemistry are controlled by the nature of the counter-metals.[116]

(iii) Reactions with carbonyl compounds

Trialkylstannyllithium adds to aldehydes to produce 1-hydroxyalkylstannanes.[94] Although the free hydroxy compounds are unstable, the alkoxy derivatives are isolable by distillation or chromatographic separation and can be stored for several months.

The stannyl anion reagent undergoes 1,2-addition to α,β-unsaturated aldehydes to give α-alkoxyallylstannanes.[117] On the other hand, the addition proceeds in a 1,4-fashion with α,β-enones in THF, which is a simple and efficient entry to functionalized tetraorganostannanes. The apparent 1,4-addition is believed to proceed through a primary 1,2-addition, followed by 1,3-migration of the stannyl group.[118]

Many types of stannylcuprates undergo 1,4-addition to α,β-enones. The delivery of the stannyl group proceeds effectively even with cuprates which contain an alkyl group because the Sn–Cu bond is weaker than the C–Cu bond.[119] The nature of the counter-cation controls the regio- and stereoselectivity. In contrast to the exclusive 1,4-addition by stannyllithium, both 1,2- and 1,4-adducts have been observed with trialkylstannylmagnesium chloride and the reaction of bicyclic enones (**9**) affords *cis* adduct (**8**) with stannylcuprate, while it affords the *trans* adduct (**10**) with stannyllithium (Scheme 9).[120] Some (E)- and (Z)-γ-silyloxyallylic stannanes can be prepared separately from enals by using stannylcuprates and stannyllithium, respectively.[121]

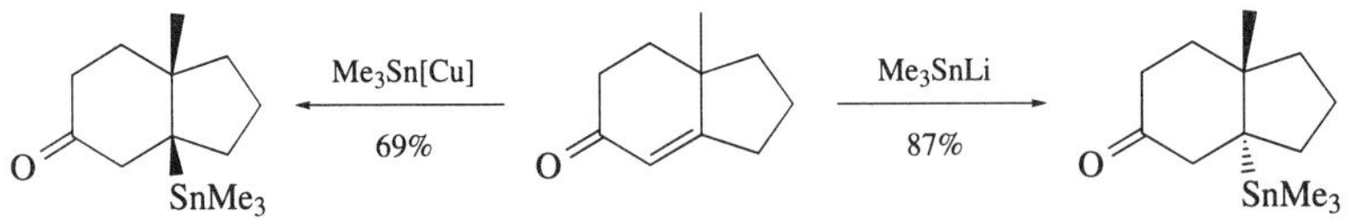

Scheme 9

The stereochemistry of stannyl addition to 2-cyclohexenone derivatives is characterized by the rigid axial approach of the reagent.[122] Subsequent treatment of the enolates with electrophiles results in high *anti* selectivity in the acyclic as well as in cyclic systems.[123] The electrophiles can be a proton, alkyl halides, or aldehydes.[124] Conjugate addition to vinyl sulfones, followed by trapping with aldehydes, gives γ-hydroxyvinylstannanes.[125]

The Michael addition of R^1_3SnLi to 2-cyclohexenone SAMP ((S)-1-amino-2-methoxymethyl-1-pyrrolidine) and RAMP ((R)-1-amino-2-methoxymethyl-1-pyrrolidine) hydrazones (**11**), followed by trapping of the resulting azaenolates with alkyl halides and oxidative removal of the chiral auxiliary, affords *trans*-tandem adducts (**12**) with high diastereomeric excess (Scheme 10). Remarkably, the quenching of the azaenolate with protons ($R^2 = H$) affords the product with only modest diastereomeric excess (*de* = 42–44%).[126]

The stannyl anion prepared from fluoride-induced desilylation of a silylstannane abstracts halogens from aryl or vinyl halides much faster than it adds to the carbonyl group of (**13**), and thus an interesting annulation has been accomplished as shown in Equation (4).[101]

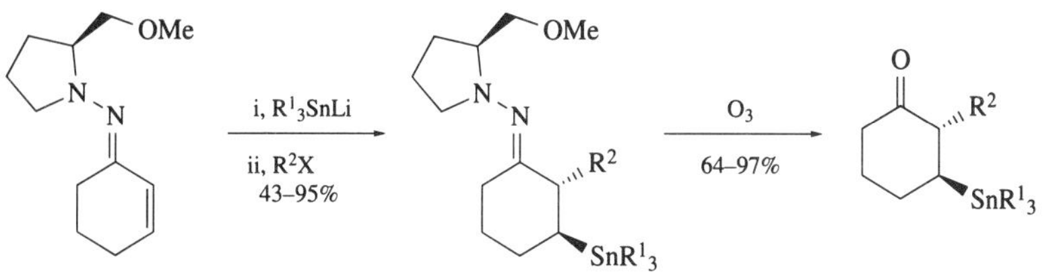

(11) *de* = 87– >96% **(12)** *de* > 98%
ee = 85– >96%

Scheme 10

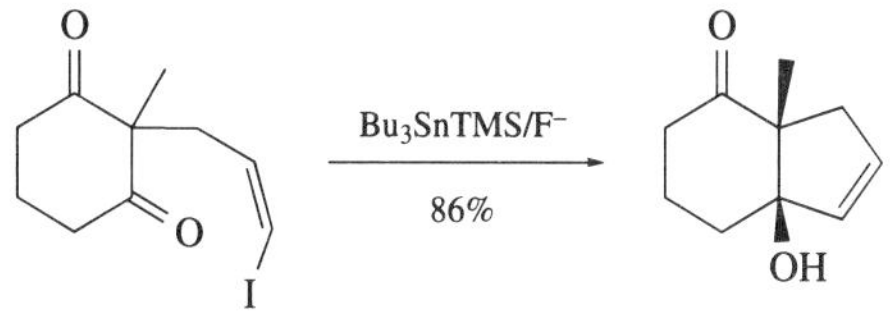

(13) (4)

8.3.2 α-Stannyl Carbanions (Sn–C$^-$)

8.3.2.1 Preparation

Trialkylstannylmethyllithiums are another kind of typical anionic reagent and a review article including other organoelement groups is available.[127] The reagents can be prepared by iodine–lithium exchange of trialkylstannylmethyl iodides, obtainable by reaction of trialkylstannyl chloride with the Simmons–Smith reagent. Trimethylstannylmethyllithium (**15**) can be prepared more conveniently and in larger scale by tin–lithium exchange of (**14**), which can be obtained from diiodomethane, tin(II) bromide, and methylmagnesium iodide, as shown in Scheme 11. The corresponding alkyl derivatives (**17**) can be obtained in the same way, but their application for synthetic purposes is not satisfactory because of the low yields obtained of the bis(stannyl) compounds (**16**). The corresponding zinc and copper reagents can be prepared from (**18**) through iodine–metal exchange, as shown in Scheme 12.[128] The cuprate (**19**) undergoes a conjugate addition to α,β-enones to afford γ-stannyl ketones.[129]

$$2\,SnBr_2 + R\text{–}CHI_2 \longrightarrow R\text{–}CH(SnBr_2I)_2 \xrightarrow{MeMgI} R\text{–}CH(SnMe_3)_2 \xrightarrow{BuLi} Me_3SnCH(R)Li$$

R = H (**14**) (**15**)
R = alkyl (**16**) (**17**)

Scheme 11

$$R_3Sn\text{–}CH(R)I \xrightarrow{Zn/Cu} R_3Sn\text{–}CH(R)[Zn] \xrightarrow{CuCN} R_3Sn\text{–}CH(R)[Cu]$$

(18) **(19)**

Scheme 12

8.3.2.2 Reactions

(i) Reactions with carbonyl compounds

Kauffmann *et al.*[130] observed that trialkyl- or triarylstannylmethyllithiums or their α-thio-substituted derivatives reacted with carbonyl compounds to afford β-stannyl alcohols, which gave alkenes upon acid treatment or heating. Further studies revealed that the reaction is sometimes superior to the Wittig reaction, particularly with enolizable ketones.[131] Unlike the Wittig or Peterson reaction, the tin-based reaction is subject to the effects of neighboring leaving groups. The β-stannyl alcohols (**21**) produced from epoxide (**20**) react with a second molecule of the reagent at the tin center to afford the diol (**22**) (Scheme 13). A comparison of the reactivity patterns between the stannyllithium and silyllithium reagents is noteworthy. Although the first-stage reaction proceeds similarly, the second molecule of the silyl reagent attacks the oxirane ring of (**23**) to produce the allylic alcohol (**24**).

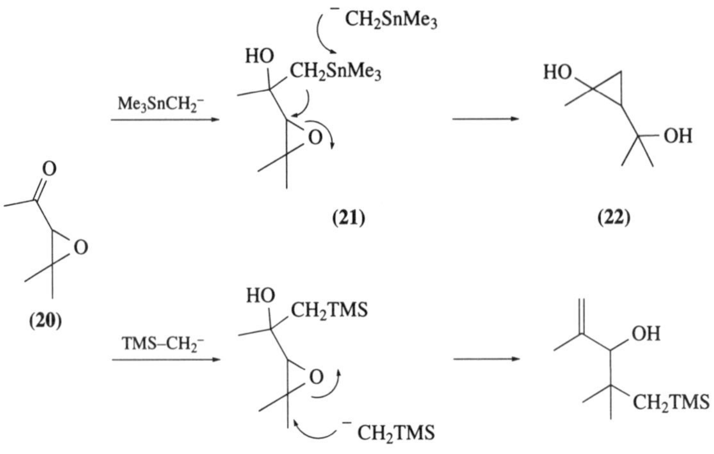

Scheme 13

(ii) Reactions with esters

The reaction of conventional alkyllithiums with esters of simple alcohols usually produces tertiary alcohols because the initially formed ketones are more reactive toward the reagents than the starting esters. However, the reaction of the anion (**25**) leads to a special situation, since the presence of the tin group in the intermediate product (**26**) induces nucleophilic attack by a second molecule of the reagent at the tin atom, rather than at the carbonyl carbon, thus effecting heterolytic bond cleavage of the tin–carbon bond. The net result is the formation of enolates from esters with two equivalents of the reagent.[132] Quenching of the enolates with electrophiles affords the ketone (**27**) (Scheme 14). The electrophile can be a proton, an aldehyde, carbon dioxide, or methyl formate. No racemization was observed with suitably protected chiral α-hydroxy and α-amino esters. Obviously, the anion (**25**) serves as a conjunctive reagent, connecting an acyl group and the E moiety by a methylene group. This reaction contrasts with that of an α-silyl carbanion, which either terminates at the stage of α-silyl ketone formation or proceeds further to produce allylsilanes through reagent attack at the carbonyl carbon. The reaction with methyl formate as electrophile affords a β-keto aldehyde and this process has been successfully applied to D-mycaroside synthesis.

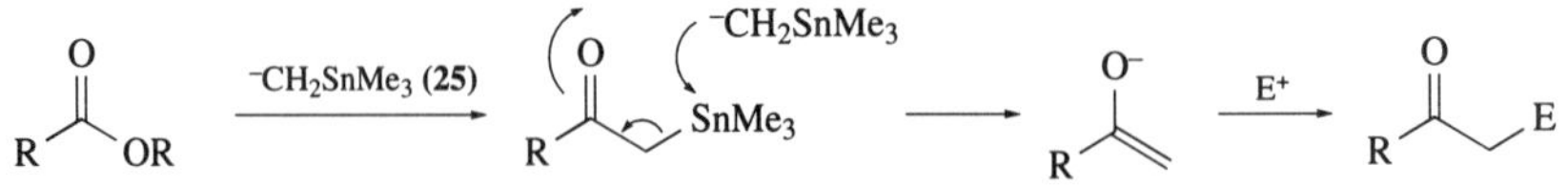

Scheme 14

8.4 CATIONIC TIN REAGENTS

8.4.1 Reagents with Tin–Heteroatom Bonds (Sn–X)

One of the characteristic features of organotin compounds is that tin atoms attached to heteroatoms easily coordinate Lewis bases to give hypervalent tin compounds in which the nucleophilicity of the heteroatoms bound to tin is greatly enhanced.

8.4.1.1 Tin–oxygen bonds

(i) Organotin alkoxides and oxides

Enhancement of the nucleophilicity of the alkoxide group by coordination of the tin atom with Lewis bases leads to a facile exchange of the alkoxy groups in esters and acetals.[133] The concept has also been successfully applied to glycosylation of carbohydrates.[134–6] Trifluoroethyl esters are particularly suitable for transesterification, which can be used for the conversion of ω-hydroxy esters to macrolactones.[137] 1,3-Disubstituted tetrabutyldistannoxanes have been developed as effective catalysts for esterification, acetalization, and deprotection of silyl ethers under mild conditions.[138]

Tin alkoxide induced reactions often display high chemoselectivity when applied to polyfunctional derivatives, due to the ability of the tin atom to coordinate with neighboring groups and thus to direct the reaction intramolecularly.[139] The coordination is subject to a very significant solvent effect. When an O-stannylated glycal (**28**) was treated with NIS (N-iodosuccinimide) in benzene, allylic oxidation was the major pathway (Equation (5)), while use of the same reagent in acetonitrile resulted in iodocyclization (Equation (6)).[140] The halostannyl group in γ-hydroxystannanes can be replaced by a hydroxy group with retention of configuration by use of hydrogen peroxide. The chelating effect of the halogen-containing tin atom with the internal hydroxy group controls the stereochemistry.[141,142]

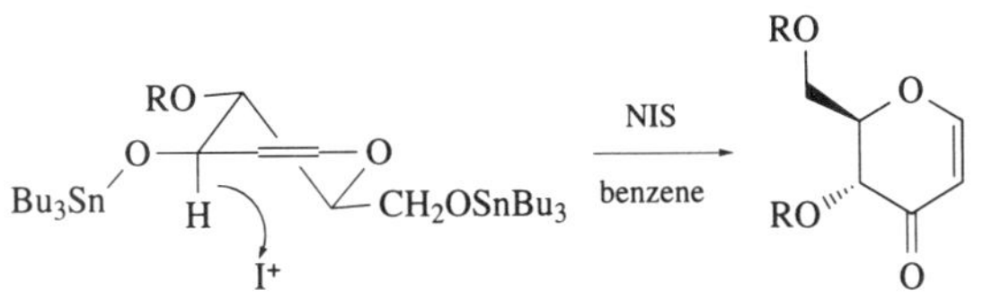

$$\text{(28)} \xrightarrow{\text{NIS, benzene}} \text{product} \quad (5)$$

$$\text{(28)} \xrightarrow{\text{NIS, acetonitrile}} \text{product} \quad (6)$$

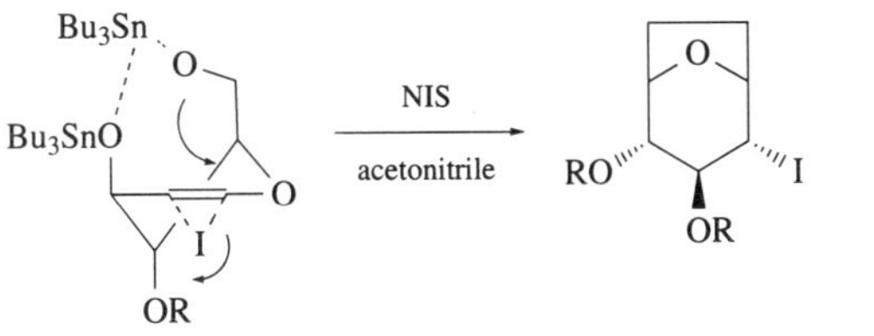

A remarkable "tin effect" has been observed, in which trimethylstannyl acetate greatly enhances the reactivity of palladium-catalyzed trimethylenemethane cycloaddition to carbonyl groups (Scheme 15). The effect has been rationalized by assumption of an enhancement of the nucleophilicity of the alcohol group for cyclization by formation of a tin alkoxide.[13]

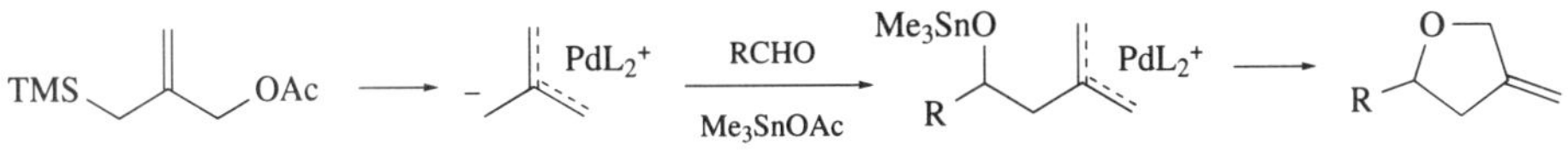

Scheme 15

The enhanced nucleophilicity of tin alkoxides is also shown in the nucleophilic substitution of a sulfoxide group by tin benzyloxide[143] and in tributyltin alkoxide addition to ketene[144] or isocyanates[145] to afford α-stannyl acetates or N-stannyl carbamates.

Tin oxides, particularly dibutyltin oxide, have evoked much interest due to their remarkable ability to protect specific hydroxy groups of polyhydroxy compounds.[146] The most important application is in the regioselective protection of saccharides. The position of protection differs among saccharides and protecting groups and the results have been discussed based on an equilibrium of the tin intermediate and the kinetic behavior of the protecting reagent.[147] This concept has been used for the regio- and stereospecific ring opening of carbohydrates to afford polyhydroxy-α,β-enals.[148]

Dibutyltin oxide has also been used for the preparation of macrocyclic oligolactones from glycosides and dicarboxylic acid dichlorides.[149] Dibutyltin oxide undergoes sulfur–oxygen exchange with thiolactones.[150] Dibutyltin bis(triflate) catalyzes the Michael addition of silyl enol ethers to α,β-enones under mild conditions.[151]

(ii) Enolates

Since the first introduction of tin(II) enolates for the *syn*-selective aldol reaction by Mukaiyama in 1982, many applications have been reported. A series of tin(II) enolate reactions which led to the catalytic asymmetric aldol reactions has been overviewed.[152] One of the recent and successful applications has been reported as a substrate-controlled aldol reaction, as shown in Scheme 16. The high selectivity is rationalized by selective formation of the (Z)-enolate and the rigidity of the tin(II) chelate (**29**).[153] Such a high selectivity cannot be achieved with the corresponding boron and titanium enolates.

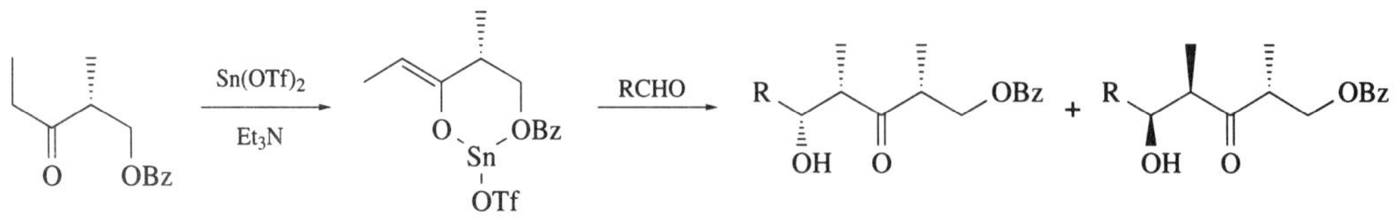

Scheme 16

Organotin(IV) enolates are generally prepared by transmetallation of the lithium enolate, by hydrostannation of α,β-enones, or by transesterification of enol acetates. Convenient methods for the preparation of organotin(IV) enolates have been developed using stannylcarbamates[145] or $Bu_3SnSnBu_3/Bu_2SnI_2$/HMPA[154] as stannylating reagents. The reactions of α-stannyl ketones with α-bromo ketones have been extensively studied and the results show that the addition of hexamethylphosphoramide (HMPA) favors direct coupling to 1,4-diketones. In contrast, β-keto oxirane formation occurs without HMPA, but under otherwise similar conditions.[155]

8.4.1.2 Tin–halogen bonds

Organotin halides of the type $R_{4-n}SnX_n$ have been shown to be useful catalysts for dehydration processes, as well as being tin delivery reagents for anionic species. Particularly useful is $BuSnCl_3$, which promotes bimolecular dehydration of allylic alcohols, cyclization of 1,n-diols to cyclic ethers, cyclization of 1,4-diketones to furans, dehydration of cyclic diols, and acetalization of aldehydes and diols.[156] As with the tin alkoxides, the nucleophilicity of the chloride is enhanced by use of coordinating Lewis bases. The complexes which result can be used as catalysts for regioselective ring opening of α,β-epoxy ketones by acyl chlorides.[157]

A hypervalent tin fluoride, obtained from triphenyltin fluoride and hydrated tetrabutylammonium fluoride, can be used as an equivalent to anhydrous fluoride.[158]

Tin(II) chloride induces the aldol-type condensation of enol esters with arenecarbaldehydes under palladium catalysis.[159] It can also reductively hydrolyze a nitro group to a keto group.[160] Diazo compounds activated by neighboring functional groups can be reacted with aldehydes to afford β-functionalized ketones by catalytic reaction with tin(II) chloride.[161] Tin(II) bromide combined with acetyl chloride converts benzyl ethers into acetates under mild conditions.[162]

8.4.1.3 Tin–nitrogen bonds

N-Stannylcarbamates are prepared by the reaction of tributyltin methoxide with ethyl isocyanate and have proved to be effective reagents for the preparation of organotin(IV) enolates.[145]

Although the chemistry of divalent tin compounds, known as stannylenes, has not been investigated as thoroughly as carbenes, a nitrogen compound was first synthesized in 1974 as a stable orange-yellow solid. Recently, ligand transfer was reported by the reaction of the tin(II) amide (**30**) with primary aldehydes to afford *trans*-enamines (**31**) (Scheme 17).[163] The reaction is believed to proceed through carbonyl insertion into the Sn–N bond, followed by β-elimination. Tin(II) amides have been used as convenient reagents for the conversion of esters into amides.[164]

Organotin azides are well known as 1,3-dipolar reagents that undergo [3 + 2]-cycloaddition with alkynes to give triazoles. Recently, reaction with isothiocyanates was found to proceed at the C=N functionality, in contrast to the reaction at the C=S moiety as observed with hydrogen azide.[165]

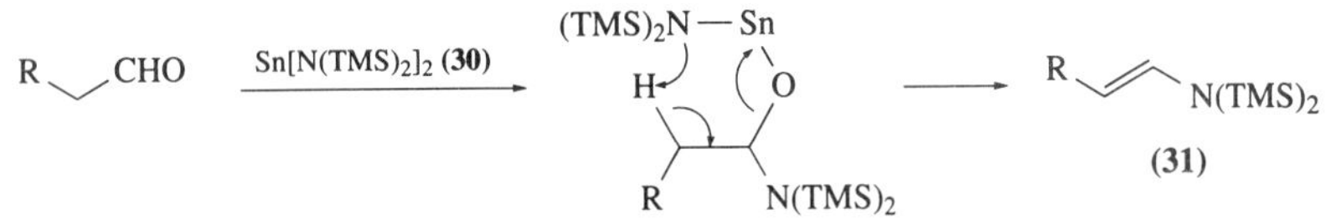

Scheme 17

8.4.1.4 Other heteroatoms

Allylthiostannanes undergo nucleophilic substitution of aryl iodides under palladium catalysis to afford aryl allyl sulfides.[166] Fluoride ions smoothly destannylate organotin chalcogenides to liberate highly nucleophilic chalcogenide ions. Thus, the tin compounds are equivalents to oxide-transfer (O^{2-}) or selenide-transfer (Se^{2-}) reagents.[167] Fluoride activation has been applied to a selenoaldehyde synthesis.

Although silylstannanes are barely regarded as anionic tin reagents, they behave as stannyl anions when activated by naked cyanide catalysis[100] or by $(Et_2N)_3S^+ \cdot SiMe_3F_2^-$ (tasf) or CsF in DMF.[101] Silylstannanes add to alkynes regio- and stereoselectively under palladium catalysis, while they undergo 1,4-addition to dienes with platinum catalysis.[168] Selective destannylation of the adducts by HI or desilylation by fluoride affords (E)-vinylsilanes or vinylstannanes, respectively.[169]

8.4.2 Carbocationic Tin Reagents ($Sn-(C)_n-X$)

The most popular carbocationic tin reagent in this category is Me_3SnCH_2I. It is used to prepare α-alkoxyalkylstannanes from alcohols as a synthon of α-alkoxy carbanions.[170] 3-Stannylpropanal dimethylacetal, another reagent in this category, undergoes the Mukaiyama aldol reaction with silyl enol ethers to afford ε-stannyl ketones, which cyclize *in situ* to cyclopentanol derivatives. The reaction is a formal, one-pot [3 + 2]-annulation.[171] γ-Chloroallylstannanes (**32**) react with aldehydes in the presence of Lewis acids to afford aldehydes (**33**) and 1,2-migration of the R group has been postulated (Scheme 18). The reaction is a formal insertion of allylic carbon into the C–C σ-bond of the aldehyde.[172]

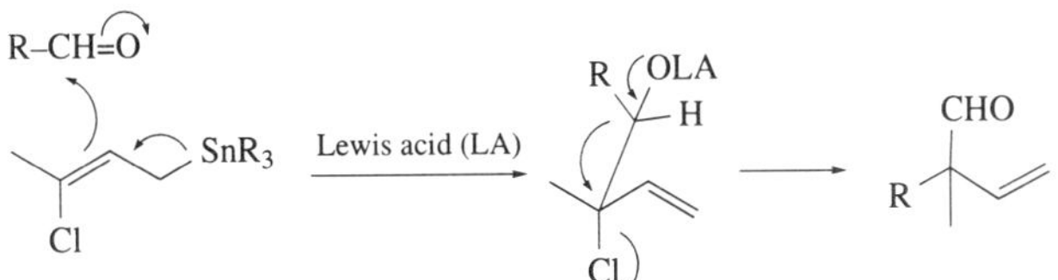

Scheme 18

8.5 ALKENYL- AND ARYLSTANNANES (Sn–C=C)

8.5.1 Preparation

Alkenylstannanes are usually prepared by the hydrostannation or stannylmetallation (R_3SnMX_n: M = Al, B, Cu, Mg, Si, and Zn) of alkynes[173] or by treatment of alkenylorganometallics with Bu_3SnOTf or R_3SnCl.[174,175] (Z)-Alkenylstannanes are prepared by the selective *cis*-addition of $Cp_2Zr(H)Cl$ to alkynylstannanes, followed by proton quenching.[176] The addition is regioselective to afford 1,1-dimetallo compounds. They are viewed as stereodefined 1,1-vinyl dianions, into which two distinct electrophiles can be introduced sequentially by using the difference in reactivities between vinylstannanes and vinylzirconates. Enol triflates or aryl/alkenyl halides couple with hexamethylditin under palladium catalysis to afford alkenylstannanes.[177] The reaction of vinyl triflates with high-order stannylcuprates offers a convenient preparative method for vinylstannanes.[178] (E)-Alkenylstannanes can be prepared from aldehydes by one-carbon homologation, using $Bu_3SnCHBr_2$ through $CrCl_2$-mediated alkenation[179] or by dehydroiodination of iodoalkylstannanes.[173]

The radical-initiated hydrostannylation of propargylic alcohols (**34**) affords (Z)-β-stannylated allyl alcohols (**35**), while titanium-catalyzed hydromagnesiation followed by treatment with tributyltin chloride affords (Z)-γ-stannylated allyl alcohols (36) (Scheme 19).[51] The radical-initiated addition of tin hydride to propargylamine affords (E)-γ-(trialkylstannyl)allylamine exclusively.[180] The preparation of the corresponding (Z)-isomer has been attained from the corresponding (Z)-vinyllithium with the cationic tin reagent, Bu$_3$SnCl.

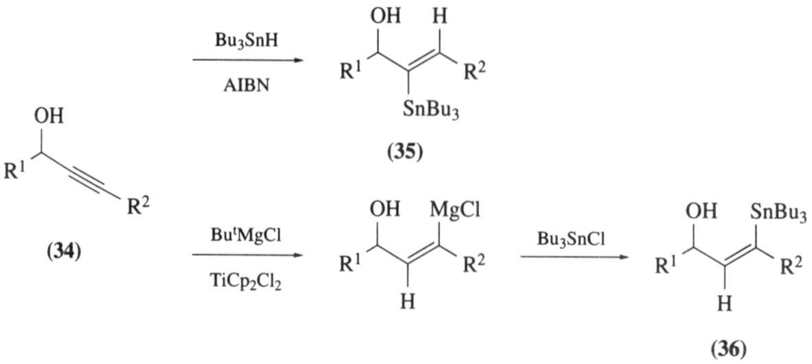

Scheme 19

The reaction of α-stannyl-α-silyl acetates is characteristic. Deprotonation by lithium diisopropylamide (LDA) and subsequent reaction with aldehydes results in deoxysilylation to afford α-alkoxycarbonyl-substituted vinylstannanes as (E)/(Z) mixtures. Remarkably, under these conditions desilylation precedes destannylation.[181] The nucleophilic nature of the tributyltin radical induces the displacement of sulfonyl groups from electron-deficient heteroaryl tosylates[86] or vinyl sulfones to produce aryl- or vinylstannanes.[182] Nucleophilic attack of Bu$_3$SnLi on chiral epoxysilanes affords enantiomerically pure alkenylstannanes.[183]

8.5.2 Reactions

8.5.2.1 Transmetallation

(i) Alkali metals

The tin atom of alkenylstannanes[184] or arylstannanes[185] can be replaced easily by lithium by treatment with alkyllithiums. The overall process is an equilibrium in which the driving force is the relative difference in base strengths of the organolithium species (Equation (7)). With the correct choice of solvents, tin compounds of allyl, benzyl, vinyl, α-heteroatom-substituted alkyl, or even cyclopropyl undergo complete tin–lithium exchange by alkyl- or aryllithiums. This method is far superior to other conventional methods for the preparation of organolithium compounds, such as lithium–halogen exchange, in avoiding contamination by lithium halides. The formation of by-product tetraalkylstannanes is not a problem, since these hydrocarbon-like species are virtually unreactive and in general are easily separated at the end of the process. Since the tin–lithium exchange in vinyl- or allylstannanes proceeds so fast, it can be accomplished successfully even in the presence of electrophilic centers within the molecule, such as a carbonyl group[186] or chlorine.[184] When a silyloxy[187] or silylamino group[180] is present in the molecule, silyl migration from these heteroatoms to carbon proceeds upon transmetallation of tin to lithium in vinylstannanes. The latter reaction has been used as a synthetic method for (Z)-allylamines and nitrogen heterocycles.

$$R^1{}_3SnR^2 + R^3Li \rightleftharpoons R^2Li + R^1{}_3SnR^3 \qquad (7)$$

Tin–lithium exchange of *o*-stannylbenzyl alcohol derivatives can be applied to the preparation of highly strained benzocyclopropanes.[188]

(ii) Other metals

Tin–boron exchange proceeds cleanly on treatment of alkenylstannanes with 9-BBN bromide. This reaction constitutes a simple method for the synthesis and manipulation of vinylboranes that avoids the problems otherwise encountered frequently.[189]

Alkenyl- or allylstannanes can be transmetallated to cuprates by treatment with higher order alkylcuprates. The resulting cuprates undergo smooth conjugate addition to α,β-enones[12] or nucleophilic substitution.[190]

Tin–lead exchange of vinylstannanes proceeds with lead tetraacetate to afford vinyllead derivatives, which react with carbon nucleophiles, such as β-dicarbonyl compounds, to give C-vinylated products.[191]

8.5.2.2 Palladium-catalyzed cross-coupling reactions

Stille[192] developed the palladium-catalyzed cross-coupling of organotin compounds (Equation (8)) and published a detailed review on this subject. Results obtained since then in this rapidly developing field have been summarized.[193] Alkenyl-, alkynyl-, and arylstannanes are by far the most important tin compounds, with decreasing reactivity in this order. Heteroaromatic stannanes have also been utilized.[194] Allylstannanes and alkylstannanes[8,195] also undergo the cross-coupling, albeit less frequently. The presence of oxygen[196-8] or nitrogen atoms[199] in the α-position does not interfere with the cross-coupling. The substrates R^3 are typically aryl, heteroaryl,[200] alkenyl, or alkynyl,[201] but ring-strained allyl[202] can also be used. Alkyl groups other than methyl cannot be used, because β-elimination occurs rapidly. Acyl chlorides,[203] α-halo ethers or sulfides,[204] and chloroformates or carbamates[205] can be used as the R^3X moiety.

$$R^1{}_3Sn-R^2 + R^3-X \xrightarrow{[Pd]} R^2-R^3 + R^1{}_3Sn-X \qquad (8)$$

The X group is usually a halide or triflate,[206,207] but hypervalent iodine,[208] arenesulfonate,[209] and epoxide[210] are also effective. The reaction of aryl triflates proceeds quite readily even with highly hindered o,o'-disubstituted aryl triflates.[211] Enol triflates of β-keto esters also react smoothly.[212] When the reactions of enol triflates with vinylic or alkynic stannanes are allowed to proceed in the presence of CO, high yields of divinyl ketones are obtained by carbonylative coupling, although the reaction conditions must be carefully controlled.[206] The carbonylation of vinylstannanes has also been realized with aryl iodides.[213] The reaction is generally stereospecific and the geometry of the vinyl halides is retained.[214]

The palladium complexes which result from the intramolecular Heck reaction of alkynic aryl triflates or iodides couple in situ with vinylstannanes to afford (Z)-indanylidenes with high stereoselectivity.[215] The cross-coupling of aryl- or allylstannanes with the palladium intermediates generated from a radical cyclization has been realized.[204]

The coupling reaction can be accomplished by employing various combinations of Pd/solvent/additive. Both Pd^0 and Pd^{2+} are effective as catalysts, but use of copper(I) iodide and triphenylarsine or silver(I) oxide[216] as cocatalyst and use of highly polar solvents and soft ligands are effective.[217,218] The presence of a tertiary amino group in the vicinity accelerates the reaction 100-fold through coordination to the tin atom.[219]

Cyclization can be accomplished by carrying out the cross-coupling intramolecularly[220] and can be applied to macrolide synthesis.[221] Tandem palladium-catalyzed addition of tin hydride to alkynes and cyclization of the resulting vinylstannanes has been used for the preparation of benzocyclobutanone derivatives (Scheme 20).[222]

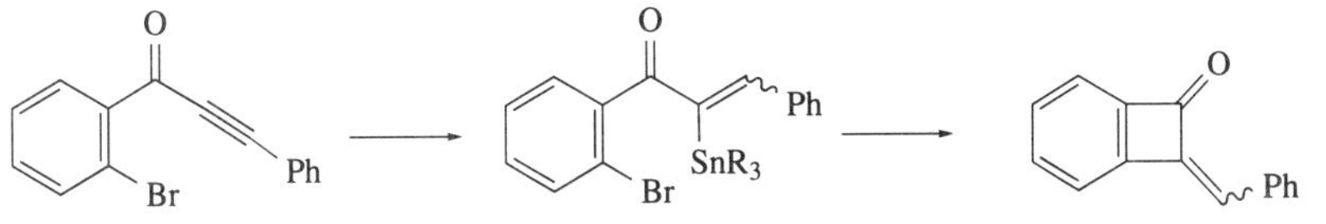

Scheme 20

The reaction of the bifunctional organometallic (**37**) with acid chlorides is noteworthy (Scheme 21).[223] When the reaction was induced by palladium catalysis, (**37**) reacted as a vinylstannane to afford the enone (**38**). However, when the reaction was induced by Lewis acid catalysis, (**37**) reacted as an allylsilane to produce initially the cation (**39**). This intermediate carbocation underwent a 1,2-silyl migration spontaneously, probably due to the double stabilization of the cationic center in (**40**) by the silyl and tin β-effect, and gave vinylsilane (**41**) as the final product. The preferential elimination of the stannyl group is a reflection of the more labile nature of the C–Sn bond as compared with the C–Si bond.

The reaction of (E)-1,2-bis(tri-n-butylstannyl)ethene with acyl chlorides yields 1,2-diacylethenes initially, but upon prolonged reaction the double bond is reduced to produce 1,4-diketones.[203] The reduction has been assumed to be induced by a palladium hydride formed from a butylpalladium intermediate through β-elimination.

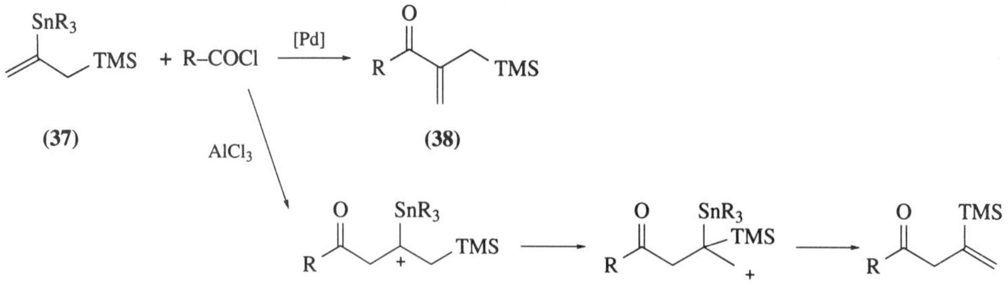

A formal Michael addition of a vinyl anion has been accomplished by reaction of an enal and tributylvinylstannane under catalysis by Ni(cod)$_2$ and trialkylsilyl chloride (Scheme 22). A mechanism involving vinyl insertion into the π-allylnickel intermediate (**42**) has been proposed.[224]

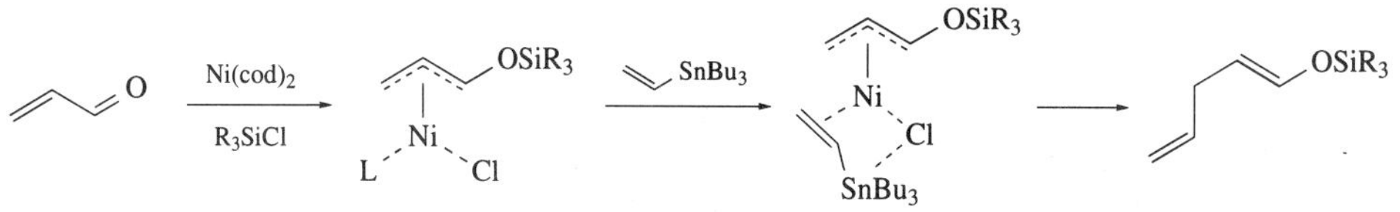

Scheme 22

8.5.2.3 Miscellaneous reactions

The stannyl group in vinylstannanes or arylstannanes can be replaced by fluorine,[177,182,225,226] chlorine,[227] bromine,[228] or iodine.[8]

The trialkylstannyl group in arylstannanes is a good leaving group in electrophilic substitution by isocyanates, diazonium salts, or sulfonyl chlorides.[229] Normally, *ipso* substitution takes place and the reaction has been applied to the preparation of regiochemically defined diaryl sulfones, arenesulfonamides, and arenesulfonic acids.

Alkenylstannanes undergo a Simmons–Smith-type reaction to afford cyclopropylstannanes. The reaction is stereocontrolled by the presence of hydroxy groups.[230] Alkenylstannanes substituted by electron-withdrawing groups act as Michael acceptors[125] or as dienophiles.[231] Hydrogenation of alkenylstannanes can be carried out with cationic rhodium complex catalysis. The accompanying reductive C–Sn bond cleavage can be minimized by correct choice of conditions. The stereochemistry of the hydrogenation can be controlled by a neighboring hydroxy group to afford γ-stannyl alcohols with high diastereoselectivity.[232]

8.6 ALLYLSTANNANES (Sn–C–C=C)

8.6.1 Preparation

Simple allylstannanes have been prepared by: (i) the reactions of conventional allylorganometallic compounds with organotin halides or oxides; (ii) reactions of stannyl anions with allylic halides or their equivalents; (iii) 1,4-hydrostannation of 1,3-dienes; and (iv) Wurtz-type cross-coupling between organotin halides and allylic halides with zinc.[233] Palladium/samarium-induced cross-coupling of tributyltin chloride with allyl acetate belongs to this category.[234] As with alkenylstannane preparation, nucleophilic radical attack by tin hydride on allyl sulfones affords allylstannanes.[235]

The direct coupling of allyl groups and aldehydes to give homoallylic alcohols via allylstannanes has been carried out with allyl alcohol[236,237] or with allyl chloride by treatment with SnCl$_2$ under catalysis by palladium or iodide,[238] respectively. Metallic tin can be used for the stannylation of allyl alcohols[239] or allyl halides[240] in aqueous solution and the reagents thus generated react *in situ* with carbonyl compounds.

α-Alkoxyallylstannanes can be prepared by 1,2-addition of a stannyllithium to α,β-enals; the addition products undergo a 1,3-stannyl rearrangement to afford γ-alkoxyallylstannanes upon treatment with BF_3 etherate.[241] Chiral products are available by BINAL-H (lithium(1,1'-binaphthalene-2,2'-diolato)(ethanolato)hydridoaluminate) reduction of the acylstannanes obtained by *in situ* oxidation of racemic α-hydroxyallylstannanes.[117]

8.6.2 Reactions

8.6.2.1 Transmetallation

Similarly to vinylstannanes, allylstannanes can be transmetallated with lithium even in the presence of a carbonyl group.[186] A smooth transmetallation of tin to copper was realized using higher order alkylcuprates. Although the chemistry of allylic cuprates has not been documented in as much detail as alkyl- or vinylcuprates, they do undergo typical coupling reactions with several electrophiles.[12]

8.6.2.2 Coupling with halides

Carbon–carbon coupling between organic halides and allylstannanes can be effected either by radical[235,242] or palladium-catalyzed pathways.[243] A comparison between these two activation methods has been made with α-chloro ketones.[244] An important feature is the compatibility of this allylation method with the presence of other functionalities such as alcohol, acetal, ether, ketone, and some others. A free-radical carbonylation of alkyl halides was extended to three- or four-component coupling involving allylstannanes and electron-deficient alkenes.[245] Acyl chlorides also couple with allylstannanes under dibutyltin dichloride catalysis.[246]

8.6.2.3 Reactions with electrophiles

Allylstannanes have been extensively used as allyl anion equivalents and a number of review articles are available.[5,247,248] The π-nucleophilicity of the alkenic bonds in the allylic groups towards cationic centers has been investigated and the relative activating effect of trialkylstannyl groups against hydrogen is 3×10^9, which is much larger than that of trialkylsilyl groups (5×10^5).[249] The most widely documented reactions of allylstannanes are S_E2' reactions with aldehydes. The reaction is induced by heat,[250] high pressure,[251] Lewis acids,[252] or cationic transition metal complexes.[253] The last method could open a way to an enantioselective preparation of homoallylic alcohols, albeit that asymmetric induction is still low (17% *ee*) at the moment. Spectroscopic investigation of the reaction intermediates has been reported.[254] The BF_3-induced reaction generally favors *syn* adducts regardless of the double-bond geometry of the starting allylstannanes and an open-chain transition state has been suggested.[255] On the other hand, the pressure-induced reaction proceeds through a cyclic six-membered transition state, affording *anti* products from the (*E*)-allylstannanes and *syn* products from the (*Z*)-isomers.[256] When $SnCl_4$, $BuSnCl_3$, and Bu_2SnCl_2 were used as Lewis acids, an initial S_E2'-type transmetallation proceeded to yield a reactive allylmetal halide (**43**), which then reacted with aldehydes, again in an S_E2' manner (Scheme 23). The regio- and stereoselectivity of this apparent α-substitution are, however, subject to the experimental conditions.[257–9] No such transmetallation was observed with silicon Lewis acids.[252]

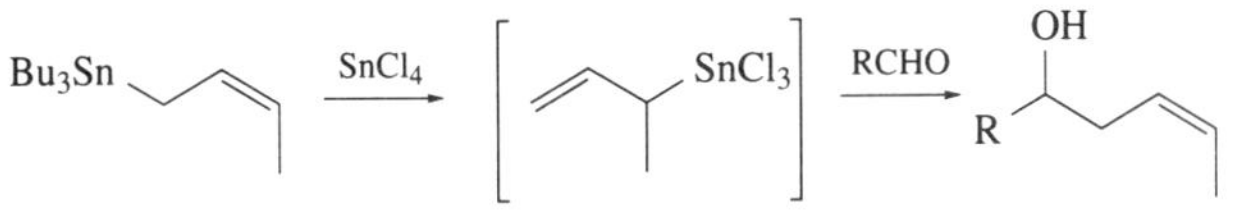

Scheme 23

Nucleophilic allylation by allylstannanes proceeds stereospecifically with cyclic acetals,[260] α-alkoxyaldehydes,[261] or 2-methoxyoxazolidines[262] containing stereogenic centers. Using the enhanced ability of a halogen-bearing tin atom to coordinate with Lewis bases, as discussed in Section 8.4.1.2, 1,5-remote asymmetric induction has been attained by neighboring ether substituents.[263] The reactions of α-[264] and γ-alkoxy-substituted[117] allylstannanes with aldehydes proceeds similarly through S_E2'

reaction to afford stereochemically defined vinyl ethers or 1,2-diol derivatives, respectively, under thermal or Lewis acid conditions. The stereochemistry has been investigated using enantiomerically enriched substrates.[265]

In contrast to the Lewis acid induced activation of C=O groups, activation of the C=N group can be effected by acid chlorides and this method has been elegantly applied to a one-pot, three-component bicycloannulation to construct a yohimban nucleus (Scheme 24).[266]

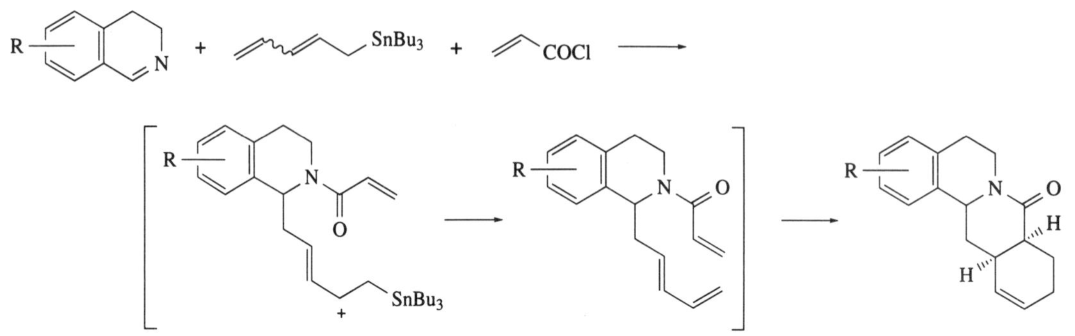

Scheme 24

The Lewis acid induced reaction of 2,4-pentadienyltrimethylstannane with aldehydes is characteristic in that the regiochemistry is dependent upon the (E)/(Z) geometry of the diene moiety.[267] Two-stage reaction of a double allylstannane is possible with (44), which can be used as a conjunctive reagent for imines and aldehydes, using the difference in reactivities of the two-stage reaction.[268] Allylstannanes undergo conjugate addition to α,β-enones under TMS-OTf (trimethylsilyltriflate) treatment.[269]

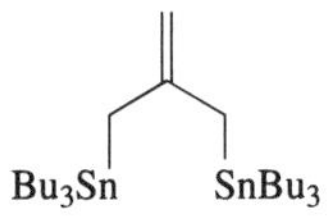

(44)

8.6.2.4 Miscellaneous reactions

Allylation of the imidazole ring with allylstannanes has been carried out by enhancing the electrophilicity of the ring by quaternization of the nitrogen atom.[270] Treatment of allylstannanes with mcpba affords allyl alcohols with 1,3-migration.[37] This is in sharp contrast to the mcpba oxidation of allylsilanes, which produces epoxysilanes. The reaction, when combined with S_H2-type stannylation of allyl alcohols, constitutes a 1,3-transposition of allyl alcohols. Allylstannanes, which are considered as nucleophiles, react with other nucleophiles in the presence of tin(IV) chloride, which probably functions as an oxidant.[271]

8.7 MISCELLANEOUS UNSATURATED STANNANES

8.7.1 Allenylstannanes

The synthesis of enantio-enriched allenylstannanes can be effected from the (R)- or (S)-propargylic alcohol (45) by S_N2' reaction (Scheme 25). The allenylstannane (46) reacts readily with aldehydes under Lewis acid catalysis to afford optically active alcohol (47) with good *syn* selectivity.[272]

Hydrozirconation of allene (48) generates the vinylstannane (49), which reacts with aldehydes and ketones in a highly regio- and stereoselective manner, and subsequent treatment with $BF_3 \cdot OEt_2$ effects β-elimination to give terminal 1,3-dienes with high (E)-selectivity (Scheme 26).[273]

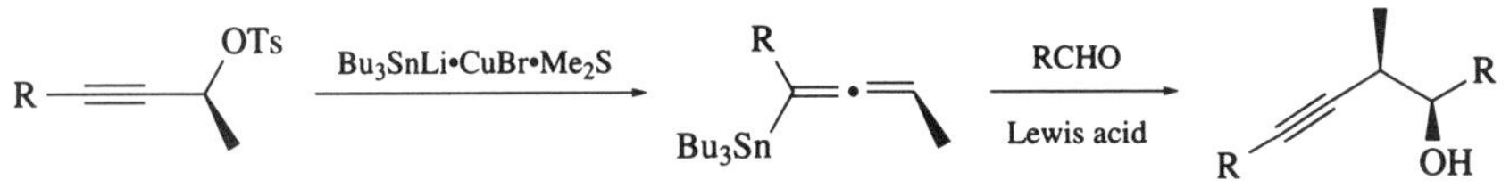

Scheme 25

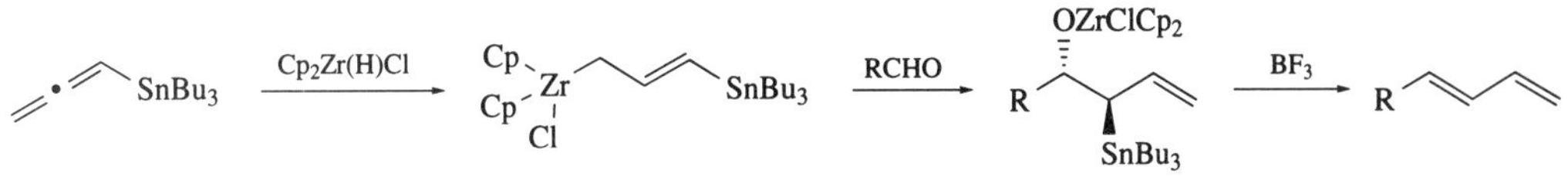

Scheme 26

8.7.2 Alkynylstannanes

Alkynylstannanes undergo palladium-catalyzed cross-coupling with many substrates in the same way as alkenylstannanes.[274] The cross-coupling of alkynylstannanes with the vinylpalladium intermediate (**50**), formed *in situ* from alkynes and trimethylsilyl iodide in the presence of a palladium catalyst, gives stereochemically defined enynes (**51**) (Scheme 27).[275]

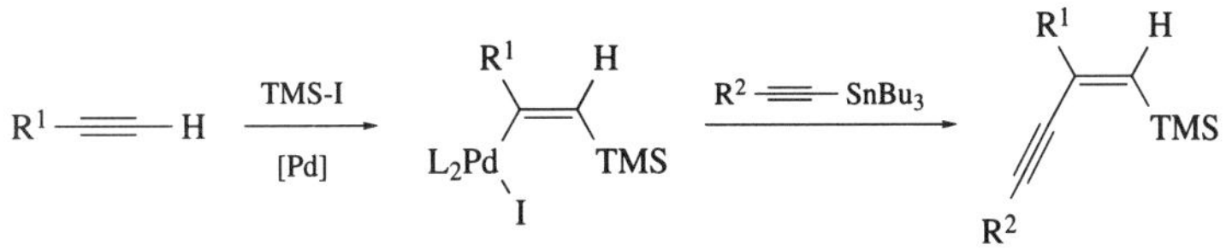

Scheme 27

Zirconocene hydride adds to alkynylstannanes and the products afford (Z)-alkenylstannanes upon proton quenching.[276] When the adducts of dialkylboranes to alkynylstannanes were treated with aryllithiums, tin–lithium exchange proceeded to give boron-stabilized alkenyl carbanions, which underwent boron–Wittig reaction with aldehydes to give allenes.[277] Alkynyltrichlorostannanes are prepared directly from a 1-alkyne, tin(IV) chloride, and a base and are efficient alkynylation reagents for aldehydes and α,β-enones.[278]

1,2-Distannylacetylene undergoes Diels–Alder reaction with oxazoles to give bis(stannylfurans), which can be converted into disubstituted furans.[279] Alkynylstannanes are subject to electrophilic substitution by highly electrophilic iodonium reagents.[280]

8.8 FUNCTIONALIZED STANNANES (Sn–(C)$_n$–F)

8.8.1 α-Stannyl Alcohols (*n* = 1, F = OH)

8.8.1.1 Preparation

Still[94] found that trialkylstannyllithiums add to aldehydes to produce 1-hydroxyalkylstannanes. Although the free hydroxy compounds are unstable, the alkoxy derivatives are isolable by distillation or chromatographic separation and can be stored for several months. The α-alkoxyalkylstannanes can be prepared from acetals by acetal cleavage with acetyl chloride to α-chloro ethers, followed by displacement of the chlorides by the stannyl group with Bu$_3$SnLi.[281]

8.8.1.2 Reactions

An alkoxy group at the α-position enhances the reactivity of the neighboring C–Sn bond and rapid tin–lithium exchange can be effected upon addition of butyllithium to yield the corresponding α-alkoxyorganolithium compounds. When reacted with an electrophile, such as cyclohexanone, the lithium compounds give the expected 1,2-diols. Significantly, no 1-butylcyclohexanol was detected in the crude product, indicating that the equilibrium in the tin–lithium exchange is far on the side of the 1-alkoxyalkyllithium.[94] It was found that both the tin–lithium exchange and trapping of the anion with electrophiles proceed with retention of configuration of the tin-bearing carbon.[282] Addition of the stannyl anion to 1-alkoxyaldehydes proceeds under chelation control to produce *syn* adducts as major products, particularly in the presence of zinc or copper salts.[283] Treatment of the resulting 1-stannyl-1,2-diol derivatives with butyllithium induced selective 1,2-*anti*-elimination of stannyl and alkoxy groups to produce vinyl ethers with (*E*)-stereochemistry.

1-Stannylalkyl allyl ethers (**52**) undergo stereocontrolled 2,3-Wittig rearrangement to give homoallylic alcohols (**53**) upon transmetallation (Scheme 28). It was shown that the tin–lithium exchange proceeded with retention of configuration, but the Wittig rearrangement proceeded with inversion of the carbanion carbon.[284,285] The reaction has been used for the synthesis of (*S*)-α-damascone.[286]

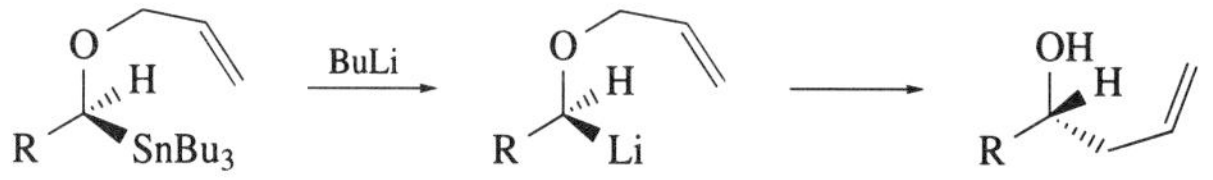

Scheme 28

α-Silyloxystannanes (**54**), upon transmetallation with butyllithium, undergo silyl migration from oxygen to carbon (reverse Brooke rearrangement) (Scheme 29).[287] The reaction is tolerant to several functional groups.

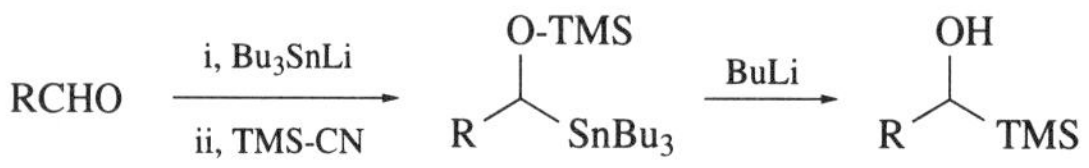

Scheme 29

α-Alkoxyorganocopper(I) compounds are prepared from the corresponding stannanes by successive transmetallations via lithium compounds.[288] The cuprates can transfer chiral ligands to triple bonds with retention of configuration to give products that are otherwise accessible only by multistep sequences. MOM-protected α-hydroxyorganocuprates behave as carbonyl ylide equivalents and undergo annulation with cyclic enones (MOM = methoxymethyl). When the hydroxy group is converted into a better leaving group by tosylation, nucleophilic substitution by alkylcopper occurs with the carbon–tin bond left intact.[289]

8.8.2 β-Stannyl Alcohols (*n* = 2, F = OH)

β-Stannyl alcohols can be prepared by the reaction of stannylmethyllithiums with carbonyl compounds[171] or of stannyllithiums with oxiranes,[105] and give alkenes upon acid treatment or by heating.[290] The deoxystannylation of alcohol (**55**) affords allenes.[291] Chiral allenes (**56**) have been prepared by this methodology (Scheme 30).

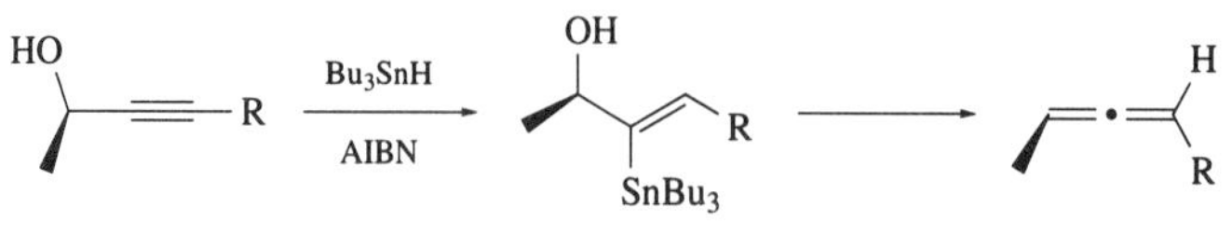

Scheme 30

8.8.3 γ-Stannyl Alcohols ($n = 3$, F = OH)

8.8.3.1 Preparation

Although γ-stannyl alcohols can be prepared in several ways, such as tin hydride addition to allyl alcohols,[38] the preparation from β-stannyl ketones by addition of nucleophiles R^3 (including hydride) to the carbonyl group, as shown in Scheme 31, represents a wide range of application[292] because *anti*-regulated introduction of various groups R^1 is possible with a large choice for R^1 and R^3. Enantioselective catalytic addition of functionalized dialkylzincs to β-stannylaldehydes affords chiral γ-stannyl alcohols.[293]

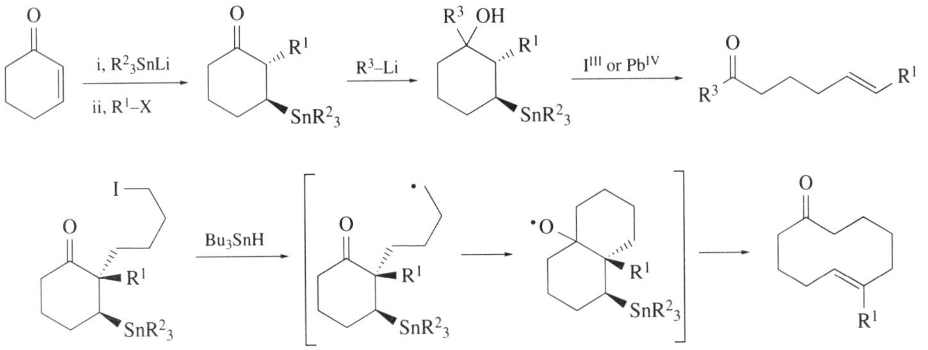

Scheme 31

8.8.3.2 Reactions

In 1970, Davis reported that γ-stannyl tertiary alcohols produced cyclopropanes upon treatment with thionyl chloride or phosphorus trichloride. In subsequent studies it was established that the reaction proceeded with inversion of configuration at both carbon centers.[294] The cyclopropanation is also observed by Lewis acid treatment of γ,δ-epoxystannanes.[120]

In 1984, Ochiai[292] and Isoe[295] independently found that γ-stannyl alcohols undergo C–C bond cleavage upon oxidation with iodonium(III) or with lead tetraacetate to produce alkenes (Scheme 31). Notably, the geometry of the double bond is unambiguously determined by the relative configuration of the stannyl group and R^1. Evidently the reaction proceeds under rigid stereoelectronic control of the antiperiplanar relationship with respect to the Sn–C bond and the C–C bond cleaved. Although the exact mechanism of the oxidative bond cleavage is unknown, a radical pathway is feasible because the process is induced by radical initiation, as shown in Scheme 31.[296] The reaction has a variety of applications for synthetic purposes.[297,298]

Treatment of 3-hydroxyalkylstannanes with BuLi provides γ-lithiated lithium alkoxides. Although the reactivity of the γ-position is low because of chelation, it can be enhanced by derivatization of the alcohols as methoxyethoxymethyl (MEM) ethers. The enantiomeric MEM-protected γ-lithio alcohols react with various electrophiles to afford polyfunctionalized alcohols in optically active form.[293]

8.8.4 Acylstannanes ($n = 0$, F = CO)

In contrast to the intensive studies reported on acylsilanes, only little is known about acylstannanes. Although there are several preparative methods, each has its own limitations. Palladium-catalyzed coupling between acyl chlorides and hexaalkylditins has been reported recently.[26]

Chiral cyclic α-stannylacetals undergo stereoselective ring cleavage upon treatment with organocopper reagents.[299] α-Tributylstannylthioacetals, prepared from the lithium salts of thioacetals with tributyltin chloride, react with silyl enol ethers to afford thioaldols, which are in turn converted stereoselectively into (Z)-β-tributylstannyl-α,β-enones upon thiol elimination by KH treatment.[300] The carbon–tin bond in α-tributylstannylthioacetals is oxidatively cleaved by cerium(IV) ammonium nitrate (can) to afford thionium cations, which react with silyl enol ethers to give thioacetals of β-ketoaldehydes.[301]

As acylstannane derivatives, α-silyloxyvinylstannanes[302] and C-alkyl-C-stannylimines[303] have been prepared and transmetallated with lithium and used as acyl anion precursors.

8.8.5 α-Stannyl Ketones (n = 1, F = CO)

α-Stannyl ketones are in equilibrium with tin enolates (Section 8.4.1.1(ii)). Although highly stable, ethyl tributylstannylacetate (57) reacts as a carbon nucleophile to cleave chiral oxazolidines under Lewis acid catalysis to afford β-amino esters with remarkable stereocontrol (Equation (9)).[304]

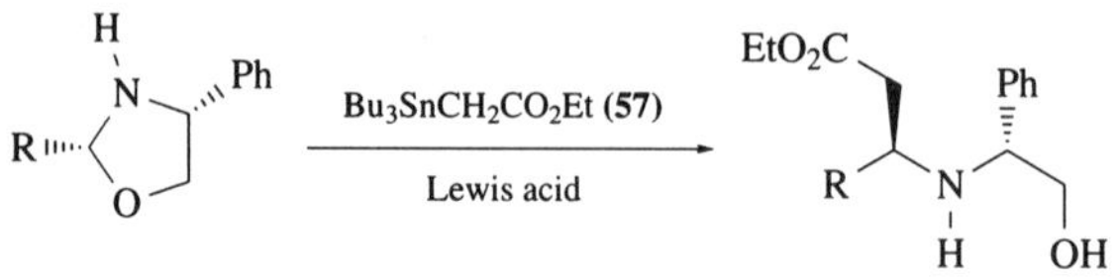

(9)

8.8.6 β-Stannyl Ketones (n = 2, F = CO)

8.8.6.1 Preparation

β-Stannyl ketones are prepared most conveniently by conjugate addition of stannyl anions to α,β-enones as discussed in Section 8.3.1.2(iii) The TMS-Cl-induced addition of $SnCl_2$ to α,β-enones affords β-trichlorostannyl ketones, which can be chemoselectively alkylated with Grignard reagents at the tin atom.[305]

8.8.6.2 Reactions

The Lewis acid induced reaction of β-stannyl ketones (59) proceeds via cyclopropanol intermediates (60) (Scheme 32, path I) and affords saturated ketones of type A (61) or type B (62), according to the position of the bond cleavage of the cyclopropanol ring of intermediate (60).[124] Generally, the type B reaction becomes predominant when the β-carbon is quaternized and TMS-OTf is used as the Lewis acid. Only when an alkyl group of sufficient migratory aptitude occupies a position antiperiplanar to the stannyl group does 1,2-alkyl migration compete with cyclopropanation to afford an allylic alcohol (58) (path II).[306] Since the type B reaction (Scheme 32) involves a carbon skeleton rearrangement, usually with high stereoselectivity and yields, it can be an effective synthetic reaction[124] and has been applied to the synthesis of (+)-β-cuparenone.[307]

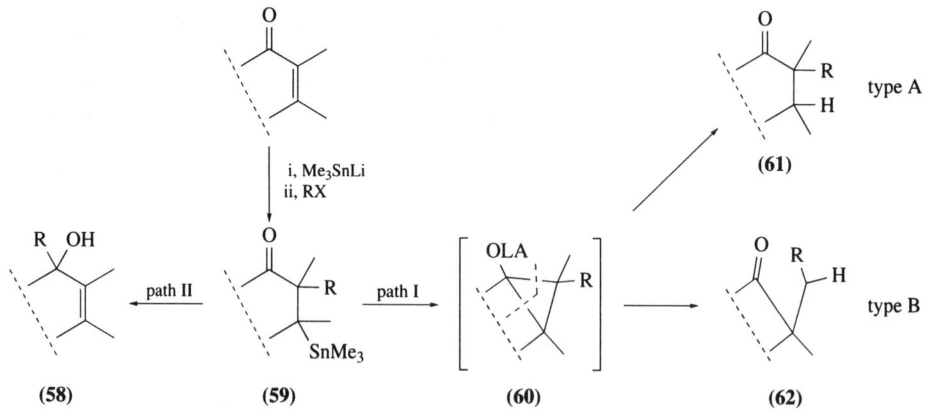

Scheme 32

In contrast to acidic activation of the carbonyl group, direct anionic activation of the tin–carbon bond can be realized only when the carbonyl group is suitably deactivated to nucleophilic attack. Deactivation of the carbonyl group is possible by conversion to an amide,[308] to a lithium[309] or silyl[310] enolate, or to an enamine.[311]

Conjugate addition of tributylstannyllithium to 2-phenylseleno-2-cycloalkenones, followed by trapping of the resulting enolate with allylic halides and subsequent destannylselenylation, gives 2-substituted 2-cycloalkenones in high yields in a one-pot procedure. The destannylselenylation can be performed by F$^-$, Lewis acids, or silica gel as well as by thermal or photochemical treatment.[312]

When silyl enol ethers of β-stannyl ketones are treated with mild oxidants, such as mcpba, periodic acid, anhydrous iron(III) chloride, or manganese dioxide, the parent α,β-enones are obtained. Clearly, this process represents a method for protection of the enone function and has been applied in a synthesis of (±)-periplanone-B.[313] In cases where the bulky trialkylstannyl group exerts an influence on the equilibrium among conformers of the silyl enol ether in a macrocyclic system, the net result of the addition–elimination reaction is *cis–trans* isomerization of the double bond in the α,β-enone.[314] With a stronger oxidant such as CrO$_3$/pyridine, the carbon–tin bond is oxidatively cleaved to produce a ketone. This process constitutes a dialkylative enone transposition, as shown in Scheme 33.[93]

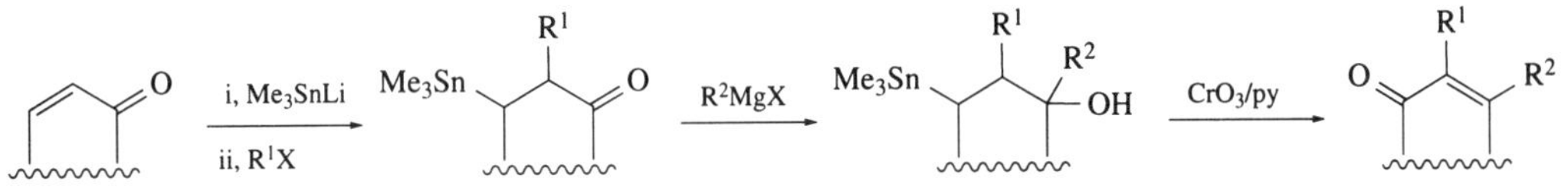

Scheme 33

8.8.7 γ-Stannyl Ketones (*n* = 3, F = CO)

γ-Stannyl ketones are prepared by conjugate addition of α-stannylalkylcuprates to α,β-enones, as discussed in Section 8.3.2.1. TMS-SPh/TiCl$_4$ treatment of γ-stannyl ketones proceeds via a possible cyclobutanation, while EtAlCl$_2$ treatment induces a C–C bond cleavage, as shown in Equation (10).[315] Notably, such a captodative-type bond cleavage occurs only with assistance of opening of the strained ring (three- or four-membered ring) in silyl compounds,[316] while the reaction with stannyl compounds proceeds in the six-membered system.

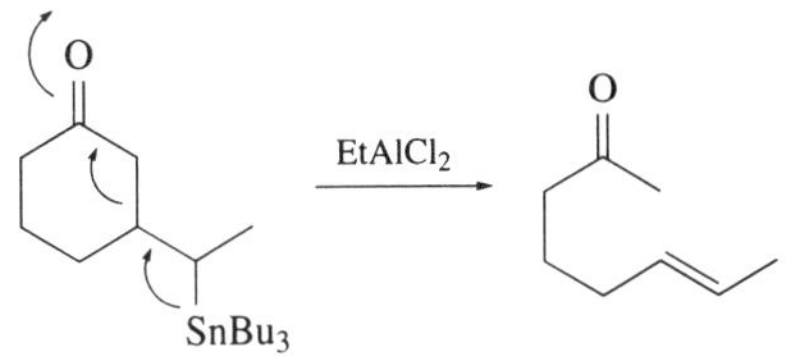

(10)

8.8.8 Other Alkylstannanes Functionalized at the α-Position

Chiral α-aminoalkylstannanes can be prepared by (*S*)-BINAL-H reduction of acylstannanes, followed by Mitsunobu reaction with phthalide.[317] The α-aminoalkylstannanes undergo facile transmetallation with lithium and the enantiomerically enriched α-aminoorganolithiums are configurationally stable at −95 °C and can be trapped with various electrophiles.[318] When the lithio compounds are treated with one or one half equivalent of CuCN, they afford lower or higher order cyanocuprates, respectively, which undergo effective conjugate addition to enones.[319]

2-Azaallylstannanes (**64**), easily prepared from azides (**63**) and aldehydes[320] or from α-azidostannanes by aza–Wittig reaction,[321] can be transmetallated to 2-azaallyl anions, which can serve as 4π systems to afford pyrrolidine derivatives through [4 + 2]-cycloaddition with alkenes (Scheme 34). α-Diazostannanes react smoothly with triphenylmethyl chloride to afford a nitrilimine as air-stable yellow crystals, which undergoes 1,3-dipolar addition with alkenes to give pyrazoline derivatives.[322]

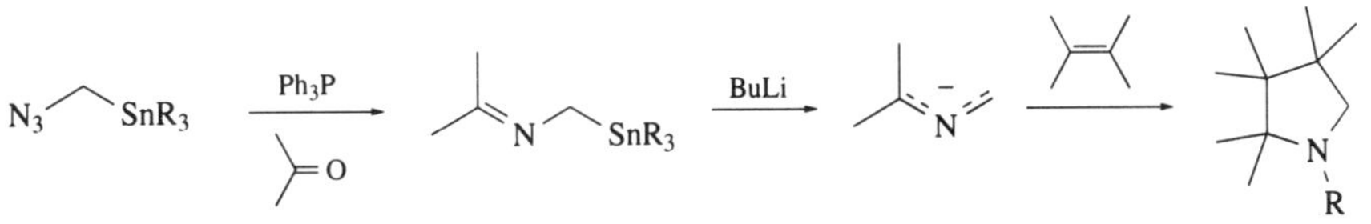

(63)   (64)

Scheme 34

Electron transfer from α-heteroatom-substituted (O, N, and S) stannanes is facile and chemical[323] and electrochemical[324,325] oxidation produces a cationic intermediate, which couples easily with heteroatom or carbon nucleophiles.

Treatment of diethyl alkylphosphonates (**65**) with two equivalents of lithium diisopropylamide (LDA) and Bu$_3$SnCl affords the α-lithiostannane (**66**), which reacts *in situ* with aldehydes to produce (*E*)- or (*Z*)-alkenylphosphonates (**67**) (Scheme 35).[326] The geometry of the product is dependent upon the nature of R^1, R^2, and R^3.

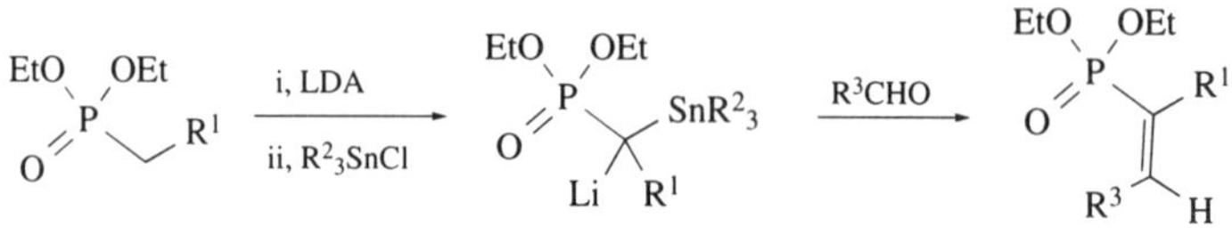

Scheme 35

8.8.9 Alkylstannanes Functionalized at Remote Positions

Macdonald *et al.*[327] observed that the Lewis acid induced reactions of stannyl ketones having the carbonyl group and tin atom separated by more than three carbons proceed by either cyclization or β-hydride shift, depending upon the types of the substrates and the reaction conditions. Thionium cation containing stannanes also undergo hydride shift or cyclization and the balance between the two is sensitive to the class of the tin-bearing carbon and the number of carbons between the tin atom and the cationic center.[328] With primary alkylstannanes (**68**), cyclization predominates when $n = 0$ or 2, while hydride shift predominates when $n = 3$ or 4 (Scheme 36). With secondary alkylstannanes (**69**), however, only hydride shift occurs and there exists a strict selectivity between H$_a$- and H$_b$-shifts, depending on the alkyl chain length: when $n = 1$ or 2, only H$_a$-shift proceeds, while only H$_b$-shift proceeds when $n = 3$ or 4. The high selectivity was explained by assuming a cyclic transition state with substantial rigidity, in which the trimethylstannyl group and the migrating hydrogen atom are antiperiplanar. This has been verified by a 1,4-chirality transfer in the 1,5-hydride shift of a chiral stannane.[329] The CuI-catalyzed 1,4-addition of the Grignard (**70**) to α,β-enones affords ketone (**71**), which undergoes 1,5-hydride shift stereospecifically to produce the alcohol (**72**) on treatment with TiCl$_4$ (Scheme 37). Remarkably, the introduction of two stereogenic centers at the 1- and 3-positions is controlled by the stereochemistry of the 5-position.[307]

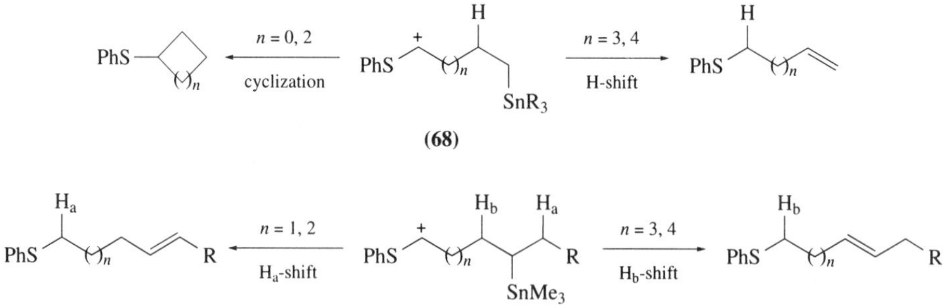

Scheme 36

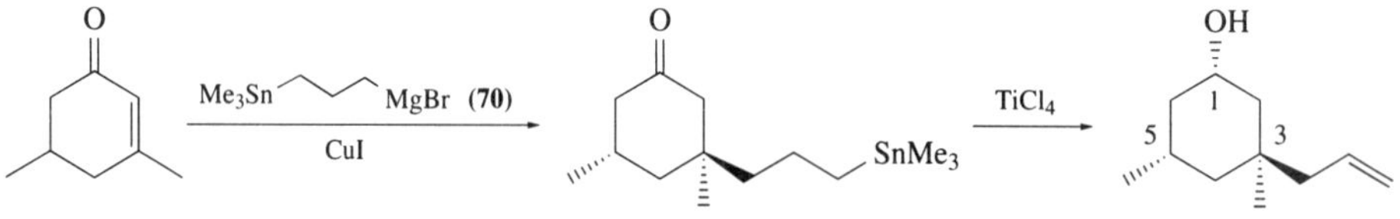

Scheme 37

Oximes of β-stannyl ketones undergo two types of carbon–carbon bond cleavage. Treatment with pyridine/SOCl$_2$ induces a Beckmann fragmentation to give unsaturated nitriles,[330] while treatment with lead tetraacetate affords nitrile oxides through the same type of bond cleavage as that of γ-hydroxystannanes.[331] Tin-directed regioselective bond cleavage is also observed in the Baeyer–Villiger fragmentation of β-stannyl ketones.[330]

β-Sulfonylstannanes can be prepared by Michael addition of Bu_3SnLi to vinyl sulfones[332] or by the reaction of R_3SnCH_2I with carbanions stabilized by a sulfone or nitrile group.[333] Stereospecific desulfonylstannylation with silica gel or fluoride ion affords alkenes. The stereochemistry in the cyclization and hydride shift of several types of stannanes having electrophilic centers has been investigated and it was found that the selectivity was dependent upon the ring size.[334]

8.9 PHOTOREACTIONS OF ORGANOTIN COMPOUNDS

Although the most convenient method for tin radical formation is to use tin hydride as mentioned in Section 8.2, a problem is sometimes encountered in that the intermediate radicals may be quenched by hydrogen abstraction prior to undergoing the desired slow processes such as cyclization or intermolecular additions, due to the high radical-scavenging power of the tin hydride. This drawback can be avoided by using a small amount of hexaalkylditins[335] or polymer-supported ditin[336] under UV irradiation. The tin radical thus generated serves as an initiator for the radical chain reaction. Even though light induced, the reactions cannot be strictly classified as real photochemical reactions of organotin compounds, since the light energy is used only to create a tin radical. A genuine type of photoreaction has been observed in the reactions of allylstannanes with dicyanobenzenes, electron-deficient alkenes, and several other electron-deficient substrates.[337] An extensive literature survey is given by Mizuno et al.,[337] and the reaction is believed to proceed through a single-electron-transfer mechanism.

As discussed in Section 8.6.2.3, the Lewis acid induced thermal reactions of allylstannanes with aldehydes proceed by an S_E2'-type mechanism and involve γ-attack. On the other hand, the reaction of allylstannanes with aldehydes proceeds by an S_E2-type mechanism under UV irradiation to afford products of α-attack.[338] The reaction has also been postulated to involve a single-electron transfer from allylstannanes to the triplet state of aldehydes.

Photochemical activation of β-stannyl ketones has been realized by carbonyl excitation. The reactivity pattern has been compared with those of the Lewis acid induced reactions.[339]

8.10 SUMMARY

The reaction patterns of organotin compounds can be classified as summarized in Table 1. The most typical reaction is transmetallation with lithium or palladium. Transmetallations with other metals have also been used frequently, but they are usually carried out secondarily via the organolithium compounds. Transmetallation with lithium is used primarily to enhance nucleophilicity, while that with palladium is used to induce reductive elimination to give cross-coupling products. Thus, reactions involving transmetallation are actually those of the corresponding organometals, rather than those of the organotin compounds themselves.

The direct reaction of carbon–tin bonds is another important reaction type of organotin compounds. Although intermolecular reactions are mostly confined to those of allylstannanes, intramolecular reactions can be realized with various types of compounds by activation of respective cationic functional groups present in the same molecule in various ways. The characteristic feature of this reaction is that the reaction types can be varied by changing the nature of the cationic centers, the reaction conditions, and the activation methods.

Among the many tin reagents described in this chapter, the tin hydrides are the most widely used. Although still less popular, anionic and cationic reagents are also promising as synthetic tools, because various types of organostannanes can be prepared by using these reagents. The organostannanes thus prepared are ideal substrates for performing the intramolecular activation described in the bottom row of Table 1. In view of the anionic or radical nature of the tin-bearing carbon, the cationic reagent (**73**) can be viewed as an anion cation (**74**) or a corresponding radical cation equivalent, while the anionic reagent (**75**) can be viewed as a double anion (**76**) or a radical anion equivalent (**77**). The characteristic features of these reagents can be represented more clearly by depicting some of the reactions described in the main text with these synthons as shown in Table 2. These types of reactions are typical of organotin compounds and their chemistry is a developing field embracing great potential for novel synthetic methodology.

Table 1 Reaction patterns of organotin compounds.

Reaction type	Reaction scheme	Typical substrate
I Transmetallations		
Lithium	R–Sn ⟶ R–Li $\xrightarrow{E}$ R–E	R = allyl, vinyl, α-heteroatom; E = carbonyl, halide
Palladium	R^1–Sn ⟶ R^1–PdL$_n$ ⟶ R^1–R^2 ; R^2–X ⟶ R^2–PdL$_n$	R^1 = vinyl; R^2 = vinyl, aryl; X = halides, triflate
II Direct reactions		
Homolytic C–C coupling	R^1–Sn $\xrightarrow{rad\bullet}$ $R^1\bullet$ $\xrightarrow{R^2-X}$ R^1–R^2	R^1 = allyl; R^2 = alkyl; X = halide
C–C fission	SnC–(C)$_n$–OH $\xrightarrow{Ox}$ SnC–(C)$_n$–O• ⟶ Bond fission	γ-Stannyl alcohol
Anionic intermolecular	R–Sn $\xrightarrow{E}$ R–E	R = allyl; E = carbonyl
intramolecular	SnC–(C)$_n$–E ⟶ Various types	E = various electrophiles

Table 2 Reaction patterns of heterolytic organotin reagents.

Reagent	Synthon	Reaction in text	Synthon representation of reaction
R_3SnLi (**75**) (n = 0)	2– (**76**) (n = 0)	Scheme 32 type B	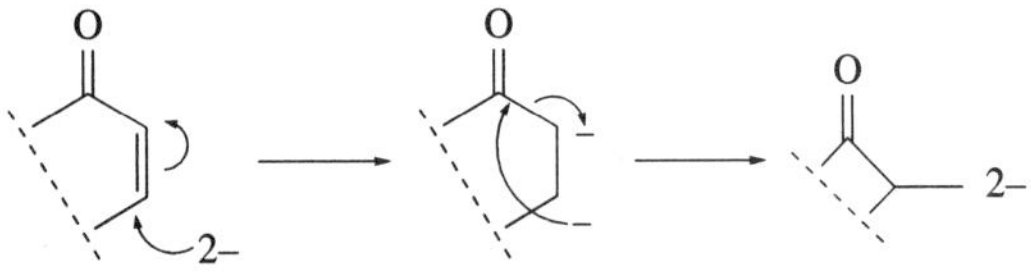 enone + 2– ⟶ enolate ⟶ cyclobutanone 2–
R_3SnCH$_2$Li (**75**) (n = 1)	CH$_2^{2-}$ (**76**) (n = 1)	Scheme 14	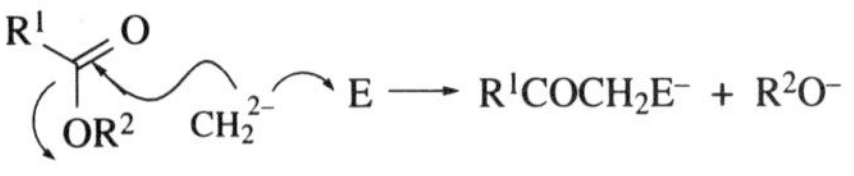 R^1C(O)(OR^2) + CH$_2^{2-}$ $\xrightarrow{E}$ R^1COCH$_2$E$^-$ + R^2O$^-$
R_3SnLi + Ox (**75**) (n = 0)	–• (**77**) (n = 0)	Scheme 31	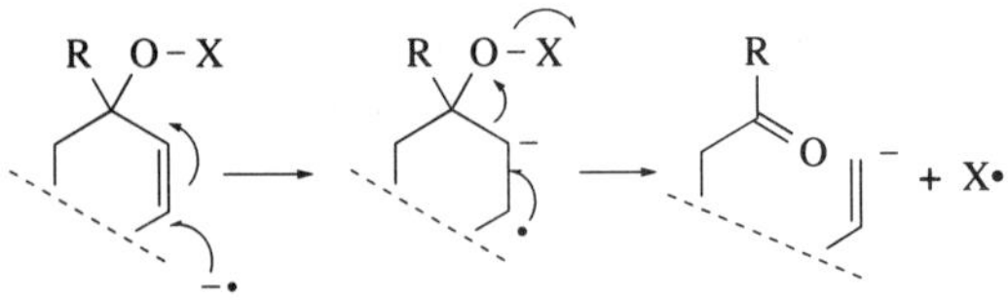 R–O–X cyclohexene ⟶ radical ⟶ R–C(O)–CH=CH– + X•
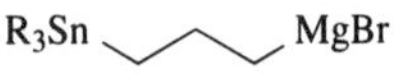 R_3Sn~~~MgBr (**75**) (n = 3)	H isopropyl – (**76**) (n = 3)	Scheme 37	cyclohexenone + allyl$^-$ ⟶ 3-allylcyclohexanol
Cl–C(Me)=CH–CH$_2$–SnR$_3$ (**73**) (n = 3)	+~~~– (**74**) (n = 3)	Scheme 18	 R–CHO + allyl ⟶ R–C(OH)(–)–CH=CH– CHO

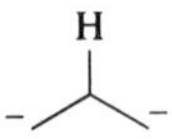

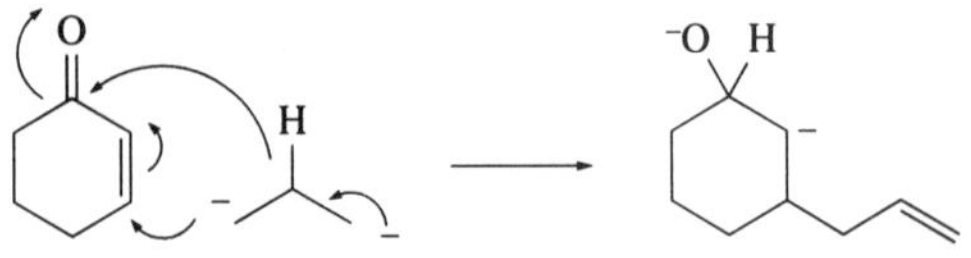

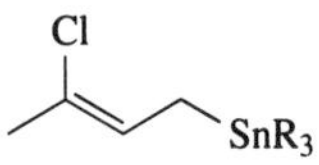

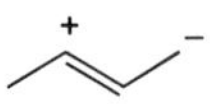

8.11 REFERENCES

1. A. G. Davies and P. J. Smith, in 'COMC-I', vol. 2, p. 519.
2. M. Pereyre, J.-P. Quintard and A. Rahm (eds.), 'Tin in Organic Synthesis', Butterworth, London, 1986.
3. I. Omae, 'Organotin Chemistry', *J. Organomet. Chem.*, Library 21, Elsevier, Amsterdam, 1989.
4. Y. Yamamoto, *Tetrahedron*, 1989, **45**, 909.
5. I. Fleming, in 'Comprehensive Organic Synthesis', ed. C. H. Heathcock, Pergamon, Oxford, 1991, vol. 2, p. 563.
6. P. D. Magnus, T. Sarkar and S. Djuric, in 'COMC-I', vol. 7, p. 515.
7. A. Alvanipour, C. Eaborn and D. R. M. Walton, *J. Organomet. Chem.*, 1980, **201**, 233.
8. F. Bellina, A. Carpita, D. Ciucci, M. De Santis and R. Rossi, *Tetrahedron*, 1993, **49**, 4677.
9. T. Sato, *Synthesis*, 1990, 259.
10. J. E. Leibner and J. Jacobus, *J. Org. Chem.*, 1979, **44**, 449.
11. D. P. Curran and C.-T. Chang, *J. Org. Chem.*, 1989, **54**, 3140.
12. B. H. Lipshutz, *Synlett*, 1990, 119.
13. B. M. Trost, S. A. King and T. Schmidt, *J. Am. Chem. Soc.*, 1989, **111**, 5902.
14. P. Dowd and S.-C. Choi, *Tetrahedron Lett.*, 1991, **32**, 565.
15. W. P. Neumann, *Synthesis*, 1987, 665.
16. E. Nakamura, D. Machii and T. Inubushi, *J. Am. Chem. Soc.*, 1989, **111**, 6849.
17. H. Tanaka, H. Suga, H. Ogawa, A. K. M. Abdul Hai and S. Torii, *Tetrahedron Lett.*, 1992, **33**, 6495.
18. K. Nozaki, K. Oshima and K. Utimoto, *Bull. Chem. Soc. Jpn.*, 1991, **64**, 2585.
19. D. H. R. Barton, D. O. Jang and J. Cs. Jaszberenyi, *Tetrahedron Lett.*, 1990, **31**, 3991.
20. M. Peeran, J. W. Wilt, R. Subramanian and D. S. Crumrine, *J. Org. Chem.*, 1993, **58**, 202.
21. C. P. Jasperse, D. P. Curran and T. L. Fevig, *Chem. Rev.*, 1991, **91**, 1237.
22. A. L. J. Beckwith, *Chem. Soc. Rev.*, 1993, **22**, 143.
23. T. N. Mitchell, *J. Organomet. Chem.*, 1986, **304**, 1.
24. M. Harendza, K. Leβmann and W. P. Neumann, *Synlett*, 1993, 283.
25. V. H. Rawal and S. Iwasa, *Tetrahedron Lett.*, 1992, **33**, 4687.
26. T. N. Mitchell and K. Kwetkat, *J. Organomet. Chem.*, 1992, **439**, 127.
27. Y. Tsuji and T. Kakehi, *J. Chem. Soc., Chem. Commun.*, 1992, 1000.
28. A. Koch, C. Lamberth, F. Wetterich and B. Giese, *J. Org. Chem.*, 1993, **58**, 1083.
29. K. W. Krosley, G. J. Gleicher and G. E. Clapp, *J. Org. Chem.*, 1992, **57**, 840.
30. P. Galatsis, S. D. Millan and T. Faber, *J. Org. Chem.*, 1993, **58**, 1215.
31. D. P. Curran and H. Yu, *Synthesis*, 1992, 123.
32. D. Horton, W. Priebe and M. L. Sznaidman, *J. Org. Chem.*, 1993, **58**, 1821.
33. K. Durkin, D. Liotta, J. Rancourt, J. F. Lavallée, L. Boisvert and Y. Guindon, *J. Am. Chem. Soc.*, 1992, **114**, 4912.
34. M. Taniguchi, Y. Takeyama, K. Fugami, K. Oshima and K. Utimoto, *Bull. Chem. Soc. Jpn.*, 1991, **64**, 2593.
35. D. Crich and L. Quintero, *Chem. Rev.*, 1989, **89**, 1413.
36. D. H. R. Barton, P. Blundell, J. Dorchak, D. O. Jang and J. Cs. Jaszberenyi, *Tetrahedron*, 1991, **47**, 8969.
37. Y. Ueno, H. Sano and M. Okawara, *Synthesis*, 1980, 1011.
38. Y. Ueno, M. Ohta and M. Okawara, *Tetrahedron Lett.*, 1982, **23**, 2577.
39. M. D. Bachi and E. Bosch, *J. Org. Chem.*, 1992, **57**, 4696.
40. V. H. Rawal, S. P. Singh, C. Dufour and C. Michoud, *J. Org. Chem.*, 1991, **56**, 5245.
41. S. Kim and K. H. Uh, *Tetrahedron Lett.*, 1992, **33**, 4325.
42. Y.-J. Chen and W.-Y. Lin, *Tetrahedron Lett.*, 1992, **33**, 1749.
43. I. Shibata, T. Yoshida, T. Kawakami, A. Baba and H. Matsuda, *J. Org. Chem.*, 1992, **57**, 4049.
44. Y. Zelechonok and R. B. Silverman, *J. Org. Chem.*, 1992, **57**, 5785.
45. E. Hasegawa, K. Ishiyama, T. Kato, T. Horaguchi and T. Shimizu, *J. Org. Chem.*, 1992, **57**, 5352.
46. A. J. Laurent and S. Lesniak, *Tetrahedron Lett.*, 1992, **33**, 3311.
47. H. Künzer, G. Sauer and R. Wiechert, *Tetrahedron Lett.*, 1991, **32**, 7247.
48. P. Four and F. Guibe, *J. Org. Chem.*, 1981, **46**, 4439.
49. H. Kuniyasu, A. Ogawa and N. Sonoda, *Tetrahedron Lett.*, 1993, **34**, 2491.
50. H. Miyake and K. Yamamura, *Chem. Lett.*, 1992, 507.
51. M. Lautens and A. H. Huboux, *Tetrahedron Lett.*, 1990, **31**, 3105.
52. D. P. Curran, *Synthesis*, 1988, 417; 489.
53. F. Foubelo, F. Lloret and M. Yus, *Tetrahedron*, 1992, **48**, 9531.
54. D. P. Curran and B. Yoo, *Tetrahedron Lett.*, 1992, **33**, 6931.
55. C. C. Huval and D. A. Singleton, *Tetrahedron Lett.*, 1993, **34**, 3041.
56. N. A. Porter, B. Giese and D. P. Curran, *Acc. Chem. Res.*, 1991, **24**, 296.
57. T. V. RajanBabu, *Acc. Chem. Res.*, 1991, **24**, 139.
58. T. Morikawa, Y. Kodama, J. Uchida, M. Takano, Y. Washio and T. Taguchi, *Tetrahedron*, 1992, **48**, 8915.
59. G. V. M. Sharma and S. R. Vepachedu, *Carbohydr. Res.*, 1992, **226**, 185.
60. D. L. J. Clive and M. H. D. Postema, *J. Chem. Soc., Chem. Commun.*, 1993, 429.
61. M. J. Tomaszewski and J. Warkentin, *Tetrahedron Lett.*, 1992, **33**, 2123.
62. S. A. Glover and J. Warkentin, *J. Org. Chem.*, 1993, **58**, 2115.
63. E. Lee, C. Lee, J. S. Tae, H. S. Whang and K. S. Li, *Tetrahedron Lett.*, 1993, **34**, 2343.
64. D. P. Curran, W.-T. Jiaang, M. Palovich and Y.-M. Tsai, *Synlett*, 1993, 403.
65. M. Koreeda and D. C. Visger, *Tetrahedron Lett.*, 1992, **33**, 6603.
66. P. M. Esch, H. Hiemstra, R. F. de Boer and W. N. Speckamp, *Tetrahedron*, 1992, **48**, 4659.
67. M. D. Bachi, E. Bosch, D. Denenmark and D. Girsh, *J. Org. Chem.*, 1992, **57**, 6803.
68. A. Padwa, D. N. Kline, S. S. Murphree and P. E. Yeske, *J. Org. Chem.*, 1992, **57**, 298.
69. E. Lee, C. H. Yoon and T. H. Lee, *J. Am. Chem. Soc.*, 1992, **114**, 10 981.
70. D. P. Curran and J. Tamine, *J. Org. Chem.*, 1991, **56**, 2746.
71. D. Batty and D. Crich, *Tetrahedron Lett.*, 1992, **33**, 875.

72. E. Nakamura, T. Inubushi, S. Aoki and D. Machii, *J. Am. Chem. Soc.*, 1991, **113**, 8980.
73. T. Sato, K. Tsujimoto, K.-i. Matsubayashi, H. Ishibashi and M. Ikeda, *Chem. Pharm. Bull.*, 1992, **40**, 2308.
74. E. J. Enholm and J. A. Burroff, *Tetrahedron Lett.*, 1992, **33**, 1835.
75. D. L. Boger and R. J. Mathvink, *J. Org. Chem.*, 1990, **55**, 5442.
76. P. J. Parsons, M. Stefinovic, P. Willis and F. Meyer, *Synlett*, 1992, 864.
77. W. Zhang and P. Dowd, *Tetrahedron Lett.*, 1992, **33**, 3285.
78. K. Nozaki, K. Oshima and K. Utimoto, *Tetrahedron*, 1989, **45**, 923.
79. S. Hanessian and R. Léger, *J. Am. Chem. Soc.*, 1992, **114**, 3115.
80. J. E. Baldwin, R. M. Adlington and S. H. Ramcharitar, *Tetrahedron*, 1992, **48**, 3413.
81. K. J. Shea, R. O'Dell and D. Y. Sasaki, *Tetrahedron Lett.*, 1992, **33**, 4699.
82. C. D. S. Brown, N. S. Simpkins and K. Clinch, *Tetrahedron Lett.*, 1993, **34**, 131.
83. W. B. Motherwell, A. M. K. Pennell and F. Ujjainwalla, *J. Chem. Soc., Chem. Commun.*, 1992, 1067.
84. M. Journet and M. Malacria, *Tetrahedron Lett.*, 1992, **33**, 1893.
85. J. Hatem, C. Henriet-Bernard, J. Grimaldi and R. Maurin, *Tetrahedron Lett.*, 1992, **33**, 1057.
86. S. Caddick and S. Joshi, *Synlett*, 1992, 805.
87. G. J. Hollingworth and J. B. Sweeney, *Tetrahedron Lett.*, 1992, **33**, 7049.
88. E. Nakamura, K. Tanaka and S. Aoki, *J. Am. Chem. Soc.*, 1992, **114**, 9715.
89. W. R. Bowman, D. N. Clark and R. J. Marmon, *Tetrahedron Lett.*, 1992, **33**, 4993.
90. S. Kim, G. H. Joe and J. Y. Do, *J. Am. Chem. Soc.*, 1993, **115**, 3328.
91. R. Leardini *et al.*, *J. Org. Chem.*, 1993, **58**, 2419.
92. E. C. Roos, P. Bernabé, H. Hiemstra and W. N. Speckamp, *Tetrahedron Lett.*, 1991, **32**, 6633.
93. W. C. Still, *J. Am. Chem. Soc.*, 1977, **99**, 4836.
94. W. C. Still, *J. Am. Chem. Soc.*, 1978, **100**, 1481.
95. W. Kitching, H. Olszowy and J. Waugh, *J. Org. Chem.*, 1978, **43**, 898.
96. H.-J. Albert and W. P. Neumann, *Synthesis*, 1980, 942.
97. E. Piers, T. Wong and K. A. Ellis, *Can. J. Chem.*, 1992, **70**, 2058.
98. A. Barbero, P. Cuadrado, I. Fleming, A. M. González and F. J. Pulido, *J. Chem. Soc., Chem. Commun.*, 1992, 351.
99. B. H. Lipshutz, S. Sharma and D. C. Reuter, *Tetrahedron Lett.*, 1990, **31**, 7253.
100. B. L. Chenard, E. D. Laganis, F. Davidson and T. V. RajanBabu, *J. Org. Chem.*, 1985, **50**, 3666.
101. M. Mori, N. Isono, N. Kaneta and M. Shibasaki, *J. Org. Chem.*, 1993, **58**, 2972.
102. E. C. Ashby, *Acc. Chem. Res.*, 1988, **21**, 414.
103. M. S. Alnajjar and H. G. Kuivila, *J. Am. Chem. Soc.*, 1985, **107**, 416.
104. J. S. Filippo, Jr. and J. Silbermann, *J. Am. Chem. Soc.*, 1982, **104**, 2831.
105. A. Ricci and M. Taddei, *J. Organomet. Chem.*, 1986, **306**, 23.
106. C. C. Yammal, J. C. Podestá and R. A. Rossi, *J. Org. Chem.*, 1992, **57**, 5720.
107. P. A. Magriotis, M. E. Scott and K. D. Kim, *Tetrahedron Lett.*, 1991, **32**, 6085.
108. J. P. Martino, M. V. M. Emonds, P. J. Stengel, A. R. M. Oliveira, F. Simonelli and J. T. B. Ferreira, *Tetrahedron Lett.*, 1992, **33**, 49.
109. R. D. Singer, M. W. Hutzinger and A. C. Oehlschlager, *J. Org. Chem.*, 1991, **56**, 4933.
110. K. I. Booker-Milburn, G. D. Heffernan and P. J. Parsons, *J. Chem. Soc., Chem. Commun.*, 1992, 350.
111. I. Marek, A. Alexakis and J.-F. Normant, *Tetrahedron Lett.*, 1991, **32**, 6337.
112. A. Casarini *et al.*, *Synlett*, 1992, 981.
113. L. Capella, A. Degl'Innocenti, A. Mordini, G. Reginato, A. Ricci and G. Seconi, *Synthesis*, 1991, 1201.
114. A. Barbero, P. Cuadrado, I. Fleming, A. M. González and F. J. Pulido, *J. Chem. Soc., Perkin Trans. 1*, 1992, 327.
115. A. Degl'Innocenti, E. Stucchi, A. Capperucci, A. Mordini, G. Reginato and A. Ricci, *Synlett*, 1992, 332.
116. I. Beaudet, J.-L. Parrain and J.-P. Quintard, *Tetrahedron Lett.*, 1991, **32**, 6333.
117. J. A. Marshall and G. S. Welmaker, *Synlett*, 1992, 537.
118. W. C. Still and A. Mitra, *Tetrahedron Lett.*, 1978, 2659.
119. S. Sharma and A. C. Oehlschlager, *J. Org. Chem.*, 1991, **56**, 770.
120. L. Plamondon and J. D. Wuest, *J. Org. Chem.*, 1991, **56**, 2066.
121. J. A. Marshall and G. S. Welmaker, *J. Org. Chem.*, 1992, **57**, 7158.
122. M. J. Chapdelaine and M. Hulce, in 'Organic Reactions', ed. L. A. Paquette, Wiley, New York, 1990, vol. 38, p. 225.
123. G. J. McGarvey and J. M. Williams, *J. Am. Chem. Soc.*, 1985, **107**, 1435.
124. T. Sato, M. Watanabe, T. Watanabe, Y. Onoda and E. Murayama, *J. Org. Chem.*, 1988, **53**, 1894.
125. M. Ochiai, T. Ukita and E. Fujita, *Tetrahedron Lett.*, 1983, **24**, 4025.
126. D. Enders, K.-J. Heider and G. Raabe, *Angew. Chem., Int. Ed. Engl.*, 1993, **32**, 598.
127. T. Kauffmann, *Angew. Chem., Int. Ed. Engl.*, 1982, **21**, 410.
128. T. Sato, A. Kawase and T. Hirose, *Synlett*, 1992, 891.
129. T. Sato, K. Tachibana, A. Kawase and T. Hirose, *Bull. Chem. Soc. Jpn.*, 1993, **66**, 3825.
130. T. Kauffmann, R. König, R. Kriegesmann and M. Wensing, *Tetrahedron Lett.*, 1984, **25**, 641.
131. T. Sato *et al.*, *Tetrahedron*, 1991, **47**, 3281.
132. T. Sato and S. Ariura, *Angew. Chem., Int. Ed. Engl.*, 1993, **32**, 105.
133. S. Ohuchi, H. Ayukawa and T. Hata, *Chem. Lett.*, 1992, 1501.
134. S. J. Danishefsky *et al.*, *J. Am. Chem. Soc.*, 1992, **114**, 8331.
135. K. Koide, M. Ohno and S. Kobayashi, *Tetrahedron Lett.*, 1991, **32**, 7065.
136. T. Mukaiyama and K. Matsubara, *Chem. Lett.*, 1992, 1755.
137. J. D. White, N. J. Green and F. F. Fleming, *Tetrahedron Lett.*, 1993, **34**, 3515.
138. J. Otera, N. Dan-oh and H. Nozaki, *Tetrahedron*, 1993, **49**, 3065.
139. V. Farina and S. Huang, *Tetrahedron Lett.*, 1992, **33**, 3979.
140. S. Czernecki, C. Leteux and A. Veyrières, *Tetrahedron Lett.*, 1992, **33**, 221.
141. M. Ochiai, S. Iwaki, T. Ukita, Y. Matsuura, M. Shiro and Y. Nagao, *J. Am. Chem. Soc.*, 1988, **110**, 4606.
142. J. W. Herndon and C. Wu, *Tetrahedron Lett.*, 1989, **30**, 6461.
143. Y. Kita, N. Shibata, N. Yoshida and T. Tohjo, *Chem. Pharm. Bull.*, 1992, **40**, 1044.

144. S. Akai, Y. Tsuzuki, S. Matsuda, S. Kitagaki and Y. Kita, *J. Chem. Soc., Perkin Trans. 1*, 1992, 2813.
145. I. Shibata, H. Yamasaki, A. Baba and H. Matsuda, *J. Org. Chem.*, 1992, **57**, 6909.
146. N. Nagashima and M. Ohno, *Chem. Pharm. Bull.*, 1991, **39**, 1972.
147. Y. Tsuda, M. Nishimura, T. Kobayashi, Y. Sato and K. Kanemitsu, *Chem. Pharm. Bull.*, 1991, **39**, 2883.
148. S. Köpper and A. Brandenburg, *Liebigs Ann. Chem.*, 1992, 933.
149. M. W. Bredenkamp, H. M. Flowers and C. W. Holzapfel, *Chem. Ber.*, 1992, **125**, 1159.
150. Y. Tsuda, Y. Sato, K. Kakimoto and K. Kanemitsu, *Chem. Pharm. Bull.*, 1992, **40**, 1033.
151. T. Sato, Y. Wakahara, J. Otera and H. Nozaki, *Tetrahedron*, 1991, **47**, 9773.
152. S. Kobayashi, H. Uchiro, I. Shiina and T. Mukaiyama, *Tetrahedron*, 1993, **49**, 1761.
153. I. Paterson and R. D. Tillyer, *Tetrahedron Lett.*, 1992, **33**, 4233.
154. I. Shibata, T. Yamaguchi, A. Baba and H. Matsuda, *Chem. Lett.*, 1993, 97.
155. M. Yasuda, T. Oh-hata, I. Shibata, A. Baba and H. Matsuda, *J. Chem. Soc., Perkin Trans. 1*, 1993, 859.
156. G. Tagliavini, *J. Organomet. Chem.*, 1992, **437**, 15.
157. I. Shibata, N. Yoshimura, A. Baba and H. Matsuda, *Tetrahedron Lett.*, 1992, **33**, 7149.
158. M. Gingras, *Tetrahedron Lett.*, 1991, **32**, 7381.
159. Y. Masuyama, T. Sakai and Y. Kurusu, *Tetrahedron Lett.*, 1993, **34**, 653.
160. N. B. Das, J. C. Sarma, R. P. Sharma and M. Bordoloi, *Tetrahedron Lett.*, 1993, **34**, 869.
161. C. R. Holmquist and E. J. Roskamp, *Tetrahedron Lett.*, 1992, **33**, 1131.
162. T. Oriyama, M. Kimura, M. Oda and G. Koga, *Synlett*, 1993, 437.
163. C. Burnell-Curty and E. J. Roskamp, *J. Org. Chem.*, 1992, **57**, 5063.
164. W.-B. Wang and E. J. Roskamp, *J. Org. Chem.*, 1992, **57**, 6101.
165. R. J. Deeth, K. C. Molloy, M. F. Mahon and S. Whittaker, *J. Organomet. Chem.*, 1992, **430**, 25.
166. M. J. Dickens, J. P. Gilday, T. J. Mowlem and D. A. Widdowson, *Tetrahedron*, 1991, **47**, 8621.
167. M. Segi, M. Kato and T. Nakajima, *Tetrahedron Lett.*, 1991, **32**, 7427.
168. Y. Tsuji and Y. Obora, *J. Am. Chem. Soc.*, 1991, **113**, 9368.
169. M. Mori, N. Watanabe, N. Kaneta and M. Shibasaki, *Chem. Lett.*, 1991, 1615.
170. J. A. Marshall and W. J. DuBay, *J. Am. Chem. Soc.*, 1992, **114**, 1450.
171. T. V. Lee, K. A. Richardson, K. L. Ellis and N. Visani, *Tetrahedron*, 1989, **45**, 1167.
172. J. Fujiwara and T. Sato, *J. Organomet. Chem.*, 1994, **473**, 63.
173. J. M. Chong and S. B. Park, *J. Org. Chem.*, 1993, **58**, 523.
174. B. L. Groh, *Tetrahedron Lett.*, 1991, **32**, 7647.
175. A. J. Bennett, J. M. Percy and M. H. Rock, *Synlett*, 1992, 483.
176. B. H. Lipshutz, R. Keil and J. C. Barton, *Tetrahedron Lett.*, 1992, **33**, 5861.
177. H. F. Hodson, D. J. Madge and D. A. Widdowson, *Synlett*, 1992, 831.
178. S. R. Gilbertson, C. A. Challener, M. E. Bos and W. D. Wulff, *Tetrahedron Lett.*, 1988, **29**, 4795.
179. D. M. Hodgson, *Tetrahedron Lett.*, 1992, **33**, 5603.
180. R. J. P. Corriu, B. Geng and J. J. E. Moreau, *J. Org. Chem.*, 1993, **58**, 1443.
181. A. J. Zapata, C. R. Fortoul and C. A. Acuña, *J. Organomet. Chem.*, 1993, **448**, 69.
182. D. P. Matthews, S. C. Miller, E. T. Jarvi, J. S. Sabol and J. R. McCarthy, *Tetrahedron Lett.*, 1993, **34**, 3057.
183. F. Sato and Y. Kobayashi, *Synlett*, 1992, 849.
184. E. Piers, B. W. A. Yeung and F. F. Fleming, *Can. J. Chem.*, 1993, **71**, 280.
185. Y. Yang and H. N. C. Wong, *J. Chem. Soc., Chem. Commun.*, 1992, 1723.
186. A. Barbero, P. Cuadrado, A. M. González, F. J. Pulido, R. Rubio and I. Fleming, *Tetrahedron Lett.*, 1992, **33**, 5841.
187. M. Lautens, P. H. M. Delanghe, J. B. Goh and C. H. Zhang, *J. Org. Chem.*, 1992, **57**, 3270.
188. A. T. McNichols and P. J. Stang, *Synlett*, 1992, 971.
189. D. A. Singleton, J. P. Martinez and G. M. Ndip, *J. Org. Chem.*, 1992, **57**, 5768.
190. B. Santiago and J. A. Soderquist, *J. Org. Chem.*, 1992, **57**, 5844.
191. C. J. Parkinson, J. T. Pinhey and M. J. Stoermer, *J. Chem. Soc., Perkin Trans. 1*, 1992, 1911.
192. J. K. Stille, *Angew. Chem., Int. Ed. Engl.*, 1986, **25**, 508.
193. T. N. Mitchell, *Synthesis*, 1992, 803.
194. E. V. Dehmlow and A. Sleegers, *Liebigs Ann. Chem.*, 1992, 953.
195. N. Tamayo, A. M. Echavarren and M. C. Paredes, *J. Org. Chem.*, 1991, **56**, 6488.
196. D. A. Elsley, D. Macleod, J. A. Miller, P. Quayle and G. M. Davies, *Tetrahedron Lett.*, 1992, **33**, 409.
197. E. Dubois and J.-M. Beau, *Carbohydr. Res.*, 1992, **228**, 103.
198. H.-C. Zhang, M. Brakta and G. D. Daves, Jr., *Tetrahedron Lett.*, 1993, **34**, 1571.
199. G. Palmisano and M. Santagostino, *Tetrahedron*, 1993, **49**, 2533.
200. Y. Yang and A. R. Martin, *Synth. Commun.*, 1992, **22**, 1757.
201. I. Beaudet, J.-L. Parrain and J.-P. Quintard, *Tetrahedron Lett.*, 1992, **33**, 3647.
202. D. J. Krysan, A. Gurski and L. S. Liebeskind, *J. Am. Chem. Soc.*, 1992, **114**, 1412.
203. M. Pérez, A. M. Castaño and A. M. Echavarren, *J. Org. Chem.*, 1992, **57**, 5047.
204. R. K. Bhatt, D.-S. Shin, J. R. Falck and C. Mioskowski, *Tetrahedron Lett.*, 1992, **33**, 4885.
205. B. Jousseaume, H. A. Kwon, J.-B. Verlhac, F. Denat and J. Dubac, *Synlett*, 1993, 117.
206. W. J. Scott and J. E. McMurry, *Acc. Chem. Res.*, 1988, **21**, 47.
207. I. N. Houpis, L. DiMichele and A. Molina, *Synlett*, 1993, 365.
208. R. M. Moriarty and W. R. Epa, *Tetrahedron Lett.*, 1992, **33**, 4095.
209. D. Badone, R. Cecchi and U. Guzzi, *J. Org. Chem.*, 1992, **57**, 6321.
210. D. R. Tueting, A. M. Echavarren and J. K. Stille, *Tetrahedron*, 1989, **45**, 979.
211. J. M. Sáa, G. Martorell and A. García-Raso, *J. Org. Chem.*, 1992, **57**, 678.
212. I. N. Houpis, *Tetrahedron Lett.*, 1991, **32**, 6675.
213. T. Sakamoto, A. Yasuhara, Y. Kondo and H. Yamanaka, *Chem. Pharm. Bull.*, 1992, **40**, 1137.
214. R. S. Paley, A. de Dios and R. Fernández de la Pradilla, *Tetrahedron Lett.*, 1993, **34**, 2429.
215. F.-T. Luo and R.-T. Wang, *Tetrahedron Lett.*, 1991, **32**, 7703.
216. J. Malm, P. Björk, S. Gronowitz and A.-B. Hörnfeldt, *Tetrahedron Lett.*, 1992, **33**, 2199.

217. C. R. Johnson, J. P. Adams, M. P. Braun and C. B. W. Senanayake, *Tetrahedron Lett.*, 1992, **33**, 919.
218. B. H. Lipshutz and M. Alami, *Tetrahedron Lett.*, 1993, **34**, 1433.
219. J. M. Brown, M. Pearson, J. T. B. H. Jastrzebski and G. v. Koten, *J. Chem. Soc., Chem. Commun.*, 1992, 1440.
220. E. Piers, R. W. Friesen and S. J. Rettig, *Can. J. Chem.*, 1992, **70**, 1385.
221. J. E. Baldwin, R. M. Adlington and S. H. Ramcharitar, *Synlett*, 1992, 875.
222. J.-C. Bradley, T. Durst and A. J. Williams, *J. Org. Chem.*, 1992, **57**, 6575.
223. K.-T. Kang, J. C. Lee and J. S. U, *Tetrahedron Lett.*, 1992, **33**, 4953.
224. B. A. Grisso, J. R. Johnson and P. B. Mackenzie, *J. Am. Chem. Soc.*, 1992, **114**, 5160.
225. M. A. Tius and J. K. Kawakami, *Synth. Commun.*, 1992, **22**, 1461.
226. M. A. Tius and J. K. Kawakami, *Synlett*, 1993, 207.
227. T. Takeda, F. Kanamori, H. Matsusita and T. Fujiwara, *Tetrahedron Lett.*, 1991, **32**, 6563.
228. M. Wada, H. Wakamatsu, A. Hiraiwa and T. Erabi, *Bull. Chem. Soc. Jpn.*, 1992, **65**, 1389.
229. W. P. Neumann and C. Wicenec, *Chem. Ber.*, 1993, **126**, 763.
230. M. Lautens and P. H. M. Delanghe, *J. Org. Chem.*, 1992, **57**, 798.
231. C. R. Johnson and J. F. Kadow, *J. Org. Chem.*, 1987, **52**, 1493.
232. M. Lautens, C.-H. Zhang and C. M. Crudden, *Angew. Chem., Int. Ed. Engl.*, 1992, **31**, 232.
233. T. Carofiglio, D. Marton and G. Tagliavini, *Organometallics*, 1992, **11**, 2961.
234. T. Tabuchi, J. Inanaga and M. Yamaguchi, *Tetrahedron Lett.*, 1987, **28**, 215.
235. J. E. Baldwin, R. M. Adlington, D. J. Birch, J. A. Crawford and J. B. Sweeney, *J. Chem. Soc., Chem. Commun.*, 1986, 1339.
236. J. P. Takahara, Y. Masuyama and Y. Kurusu, *J. Am. Chem. Soc.*, 1992, **114**, 2577.
237. Y. Masuyama, A. Hayakawa and Y. Kurusu, *J. Chem. Soc., Chem. Commun.*, 1992, 1102.
238. T. Imai and S. Nishida, *Synthesis*, 1993, 395.
239. Y. Kanagawa, Y. Nishiyama and Y. Ishii, *J. Org. Chem.*, 1992, **52**, 6988.
240. J.-Y. Zhou, G.-D. Lu and S.-H. Wu, *Synth. Commun.*, 1992, **22**, 481.
241. J. A. Marshall and G. P. Luke, *Synlett*, 1992, 1007.
242. B. Giese, T. Linker and R. Muhn, *Tetrahedron*, 1989, **45**, 935.
243. J.-P. Quintard, G. Dumartin and B. Elissondo, *Tetrahedron*, 1989, **45**, 1017.
244. M. Kosugi, H. Arai, A. Yoshino and T. Migita, *Chem. Lett.*, 1978, 795.
245. I. Ryu, H. Yamazaki, A. Ogawa, N. Kambe and N. Sonoda, *J. Am. Chem. Soc.*, 1993, **115**, 1187.
246. K. Yano, A. Baba and H. Matsuda, *Chem. Lett.*, 1991, 1181.
247. C. Hull, S. V. Mortlock and E. J. Thomas, *Tetrahedron*, 1989, **45**, 1007.
248. J. A. Marshall, *Chemtracts: Org. Chem.*, 1992, **5**, 66.
249. G. Hagen and H. Mayr, *J. Am. Chem. Soc.*, 1991, **113**, 4954.
250. A. H. McNeill and E. J. Thomas, *Tetrahedron Lett.*, 1990, **31**, 6239.
251. N. S. Isaacs, R. L. Marshall and D. J. Young, *Tetrahedron Lett.*, 1992, **33**, 3023.
252. R. L. Marshall and D. J. Young, *Tetrahedron Lett.*, 1992, **33**, 1365.
253. J. M. Nuss and R. A. Rennels, *Chem. Lett.*, 1993, 197.
254. S. E. Denmark, T. Wilson and T. M. Willson, *J. Am. Chem. Soc.*, 1988, **110**, 984.
255. Y. Yamamoto, *Acc. Chem. Res.*, 1987, **20**, 243.
256. Y. Yamamoto, *Chemtracts: Org. Chem.*, 1991, **4**, 255.
257. H. Miyake and K. Yamamura, *Chem. Lett.*, 1992, 1369.
258. K. Yano, A. Baba and H. Matsuda, *Bull. Chem. Soc. Jpn.*, 1992, **65**, 66.
259. R. L. Marshall and D. J. Young, *Tetrahedron Lett.*, 1992, **33**, 2369.
260. T. Sammakia and R. S. Smith, *J. Am. Chem. Soc.*, 1992, **114**, 10 998.
261. A. B. Charette, C. Mellon, L. Rouillard and E. Malenfant, *Synlett*, 1993, 81.
262. A. Pasquarello, G. Poli and C. Scolastico, *Synlett*, 1992, 93.
263. J. S. Carey and E. J. Thomas, *Synlett*, 1992, 585.
264. B. W. Gung, D. T. Smith and M. A. Wolf, *Tetrahedron*, 1992, **48**, 5455.
265. O. Zschage, J.-R. Schwark, T. Krämer and D. Hoppe, *Tetrahedron*, 1992, **48**, 8377.
266. R. Yamaguchi *et al.*, *J. Org. Chem.*, 1993, **58**, 1136.
267. Y. Nishigaichi, M. Fujimoto and A. Takuwa, *J. Chem. Soc., Perkin Trans. 1*, 1992, 2581.
268. G. E. Keck and A. Palani, *Tetrahedron Lett.*, 1993, **34**, 3223.
269. S. Kim and J. M. Lee, *Synth. Commun.*, 1991, **21**, 25.
270. T. Itoh, H. Hasegawa, K. Nagata, M. Okada and A. Ohsawa, *Tetrahedron Lett.*, 1992, **33**, 5399.
271. T. Takeda, Y. Takagi, H. Takano and T. Fujiwara, *Tetrahedron Lett.*, 1992, **33**, 5381.
272. J. A. Marshall and X. Wang, *J. Org. Chem.*, 1992, **57**, 1242.
273. H. Maeta, T. Hasegawa and K. Suzuki, *Synlett*, 1993, 341.
274. T. Sakamoto, A. Yasuhara, Y. Kondo and H. Yamanaka, *Synlett*, 1992, 502.
275. N. Chatani, N. Amishiro and S. Murai, *J. Am. Chem. Soc.*, 1991, **113**, 7778.
276. B. H. Lipshutz, R. Keil and J. C. Barton, *Tetrahedron Lett.*, 1992, **33**, 5861.
277. A. Pelter, K. Smith and K. D. Jones, *J. Chem. Soc., Perkin Trans. 1*, 1992, 747.
278. M. Yamaguchi, A. Hayashi and M. Hirama, *Chem. Lett.*, 1992, 2479.
279. Y. Yang and H. N. C. Wong, *J. Chem. Soc., Chem. Commun.*, 1992, 656.
280. R. R. Tykwinski and P. J. Stang, *Tetrahedron*, 1993, **49**, 3043.
281. R. Hoffmann and R. Brückner, *Chem. Ber.*, 1992, **125**, 2731.
282. W. C. Still and C. Sreekumar, *J. Am. Chem. Soc.*, 1980, **102**, 1201.
283. G. J. McGarvey, M. Kimura and A. Kucerovy, *Tetrahedron Lett.*, 1985, **26**, 1419.
284. R. Hoffmann and R. Brückner, *Angew. Chem., Int. Ed. Engl.*, 1992, **31**, 647.
285. X. Tong and J. Kallmerten, *Synlett*, 1992, 845.
286. K. Mori, M. Amaike and M. Itou, *Tetrahedron*, 1993, **49**, 1871.
287. R. J. Linderman and K. Chen, *Tetrahedron Lett.*, 1992, **33**, 6767.
288. R. S. Coleman and E. B. Grant, *Chemtracts: Org. Chem.*, 1992, **5**, 230.

289. J. Ye, D.-S. Shin, R. K. Bhatt, P. A. Swain and J. R. Falck, *Synlett*, 1993, 205.
290. B. Jousseaume, N. Noiret, M. Pereyre and J. M. Francès, *J. Chem. Soc., Chem. Commun.*, 1992, 739.
291. T. Konoike and Y. Araki, *Tetrahedron Lett.*, 1992, **33**, 5093.
292. M. Ochiai, T. Ukita, Y. Nagao and E. Fujita, *J. Chem. Soc., Chem. Commun.*, 1985, 637.
293. W. Brieden, R. Ostwald and P. Knochel, *Angew. Chem., Int. Ed. Engl.*, 1993, **32**, 582.
294. I. Fleming and C. J. Urch, *J. Organomet. Chem.*, 1985, **285**, 173.
295. K. Nakatani and S. Isoe, *Tetrahedron Lett.*, 1984, **25**, 5335.
296. J. E. Baldwin, R. M. Adlington and R. Singh, *Tetrahedron*, 1992, **48**, 3385.
297. G. H. Posner, K. S. Webb, E. Asirvatham, S.-s. Jew and A. Degl'Innocenti, *J. Am. Chem. Soc.*, 1988, **110**, 4754.
298. M. G. O'Shea and W. Kitching, *Tetrahedron*, 1989, **45**, 1177.
299. J.-L. Parrain, J.-C. Cintrat and J.-P. Quintard, *J. Organomet. Chem.*, 1992, **437**, C19.
300. T. Takeda, S.-i. Sugi, A. Nakayama, Y. Suzuki and T. Fujiwara, *Chem. Lett.*, 1992, 819.
301. K. Narasaka, T. Okauchi and N. Arai, *Chem. Lett.*, 1992, 1229.
302. J.-B. Verlhac, H. Kwon and M. Pereyre, *J. Organomet. Chem.*, 1992, **437**, C13.
303. B. Jousseaume, M. Pereyre, N. Petit, J.-B. Verlhac and A. Ricci, *J. Organomet. Chem.*, 1993, **443**, C1.
304. M. K. Mokhallalati, M.-J. Wu and L. N. Pridgen, *Tetrahedron Lett.*, 1993, **34**, 47.
305. H. Nakahira, I. Ryu, A. Ogawa, N. Kambe and N. Sonoda, *Organometallics*, 1990, **9**, 277.
306. J. Fujiwara, T. Yamamoto and T. Sato, *Chem. Lett.*, 1992, 1775.
307. T. Sato, M. Hayashi and T. Hayata, *Tetrahedron*, 1992, **48**, 4099.
308. R. Goswami and D. E. Corcoran, *J. Am. Chem. Soc.*, 1983, **105**, 7182.
309. H. Nakahira, I. Ryu, M. Ikebe, N. Kambe and N. Sonoda, *Angew. Chem., Int. Ed. Engl.*, 1991, **30**, 177.
310. B. L. Chenard, *Tetrahedron Lett.*, 1986, **25**, 2805.
311. H. Ahlbrecht and P. Weber, *Synthesis*, 1992, 1018.
312. S. Kusuda, Y. Watanabe, Y. Ueno and T. Toru, *J. Org. Chem.*, 1992, **57**, 3145.
313. W. C. Still, *J. Am. Chem. Soc.*, 1979, **101**, 2493.
314. W. C. Still, *J. Am. Chem. Soc.*, 1977, **99**, 4186.
315. T. Sato, K. Tachibana, A. Kawase and T. Hirose, *Chem. Lett.*, 1993, 937.
316. T. Fujiwara, A. Suda and T. Takeda, *Chem. Lett.*, 1991, 1619.
317. A. F. Burchat, J. M. Chong and S. B. Park, *Tetrahedron Lett.*, 1993, **34**, 51.
318. W. H. Pearson and A. C. Lindbeck, *J. Am. Chem. Soc.*, 1991, **113**, 8546.
319. R. K. Dieter and C. W. Alexander, *Synlett*, 1993, 407.
320. W. H. Pearson and M. J. Postich, *J. Org. Chem.*, 1992, **57**, 6354.
321. W. H. Pearson, D. P. Szura and M. J. Postich, *J. Am. Chem. Soc.*, 1992, **114**, 1329.
322. R. Réau, G. Veneziani and G. Bertrand, *J. Am. Chem. Soc.*, 1992, **114**, 6059.
323. K. Narasaka, Y. Kohno and S. Shimada, *Chem. Lett.*, 1993, 125.
324. J.-i. Yoshida, M. Itoh and S. Isoe, *J. Chem. Soc., Chem. Commun.*, 1993, 547.
325. J.-i. Yoshida, Y. Ishichi and S. Isoe, *J. Am. Chem. Soc.*, 1992, **114**, 7594.
326. N. Mimouni, E. About-Jaudet, N. Collignon and Ph. Savignac, *Synth. Commun.*, 1991, **21**, 2341.
327. T. L. Macdonald, C. M. Delahunty, K. Macdonald and D. E. O'Dell, *Tetrahedron Lett.*, 1989, **30**, 1473.
328. E. Murayama, M. Uematsu, H. Nishio and T. Sato, *Tetrahedron Lett.*, 1984, **25**, 313.
329. T. Sato, M. Haramura and N. Taka, *Tetrahedron Lett.*, 1989, **30**, 4983.
330. R. P. Bakale, M. A. Scialdone and C. R. Johnson, *J. Am. Chem. Soc.*, 1990, **112**, 6729.
331. H. Nishiyama, H. Arai, T. Ohki and K. Itoh, *J. Am. Chem. Soc.*, 1985, **107**, 5310.
332. M. Ochiai, T. Ukita and E. Fujita, *J. Chem. Soc., Chem. Commun.*, 1983, 619.
333. B. A. Pearlman, S. R. Putt and J. A. Fleming, *J. Org. Chem.*, 1985, **50**, 3625.
334. I. Fleming and M. Rowley, *Tetrahedron*, 1986, **42**, 3181.
335. B. B. Snider and B. O. Buckman, *J. Org. Chem.*, 1992, **57**, 4883.
336. M. Harendza, J. Junggebauer, K. Leßmann, W. P. Neumann and H. Tews, *Synlett*, 1993, 286.
337. K. Nakanishi, K. Mizuno and Y. Otsuji, *Bull. Chem. Soc. Jpn.*, 1993, **66**, 2371.
338. A. Takuwa, J. Shiigi and Y. Nishigaichi, *Tetrahedron Lett.*, 1993, **34**, 3457.
339. T. Sato and K. Takezoe, *Tetrahedron Lett.*, 1991, **32**, 4003.

9
Mercury

RICHARD C. LAROCK
Iowa State University, Ames, IA, USA

9.1 INTRODUCTION	390
9.2 PREPARATION OF ORGANOMERCURIALS	390
9.2.1 *Introduction*	390
9.2.2 *Mercuration of Organic Halides*	390
9.2.3 *Transmetallation*	391
9.2.4 *Mercuration of Carbon Monoxide*	392
9.2.5 *C–H Substitution in Activated Hydrocarbons*	392
9.2.6 *Mercuration of Cyclopropanes*	393
9.2.7 *Mercuration of Alkenes*	393
9.2.8 *Mercuration of Alkynes*	394
9.2.9 *Aromatic Mercuration*	394
9.2.10 *Substitution in Nitrogen-containing Compounds*	394
9.2.11 *Elimination Reactions*	395
9.3 HYDROGEN SUBSTITUTION	395
9.3.1 *Protonolysis*	395
9.3.2 *Metal Hydride Reduction*	396
9.4 HALOGENATION	396
9.5 HETEROATOM SUBSTITUTION	398
9.5.1 *Oxygen Compounds*	398
9.5.2 *Sulfur, Selenium, and Tellurium Compounds*	400
9.5.3 *Nitrogen and Phosphorus Compounds*	401
9.6 DIMERIZATION	401
9.7 ALKYLATION	402
9.8 ALKENE AND ALKYNE ADDITION AND SUBSTITUTION REACTIONS	403
9.9 CARBONYLATION	407
9.10 ACYLATION	408
9.11 DIVALENT CARBON TRANSFER REACTIONS	409
9.12 SOLVOMERCURATION/DEMERCURATION REACTIONS	411
9.12.1 *Hydroxymercuration*	411
9.12.2 *Alkoxymercuration*	414
9.12.3 *Peroxymercuration*	417
9.12.4 *Acyloxymercuration*	418
9.12.5 *Aminomercuration*	421
9.12.6 *Amidomercuration*	424
9.12.7 *Azidomercuration*	426
9.12.8 *Nitromercuration*	426
9.12.9 *Carbomercuration*	427
9.12.10 *Halomercuration*	428
9.12.11 *Miscellaneous Mercuration Reactions*	429
9.13 CONCLUSION	430

9.1 INTRODUCTION

Organomercurials are among the oldest organometallics known and the most useful in organic synthesis. The first organomercury compound, MeHgI, was reported in 1852 only 3 years after discovery of the first organometallic. Due to their toxicity and low reactivity, little subsequent chemistry of these compounds was reported until the turn of the century when the direct mercuration of aromatic compounds and the solvomercuration of alkenes were discovered. These very important reactions spurred considerable interest in the chemistry of organomercurials and today some of the most important applications of organometallic compounds in organic synthesis involve these two processes. Organomercurials have also found widespread application in the synthesis of other organometallics.

Numerous reviews on organomercury compounds have been published, dating all the way back to Whitmore's book *Organic Compounds of Mercury* published in 1921.[1] The books entitled *The Organic Compounds of Mercury* by Makarova and Nesmeyanov,[2] *Metallorganische Verbindungen–Hg* in the series Houben-Weyl: Methoden der Organischen Chemie,[3] and *The Chemistry of Mercury* by Bloodworth,[4] and the earlier reviews in *COMC-I*[5,6] have focused on the preparation of organomercurials, their structures, and some of their reactions. Earlier reviews by Makarova in *Organometallic Reactions*[7,8] and by the present author[9] surveyed the reactions of these organometallics. The present author's books entitled *Organomercury Compounds in Organic Synthesis*[10] and *Solvomercuration/Demercuration Reactions in Organic Synthesis*[11] provide the most recent and comprehensive survey of the many important applications of organomercurials in organic synthesis. Annual updates on the preparation, structure and reactivity of organomercurials can also be found in the Specialist Periodical Reports on Organometallic Chemistry published by the Royal Society of Chemistry.

This review will cover briefly the preparation of organomercurials, including some of the newer methodology which has appeared since the earlier reviews in *COMC-I*,[5,6] and the more important applications of these compounds in organic synthesis with particular emphasis on chemistry which has appeared since 1985 and 1986 when the author's two books were published.[10,11]

9.2 PREPARATION OF ORGANOMERCURIALS

9.2.1 Introduction

Organomercury compounds of a wide variety of structural types have been prepared by a multitude of methods. This methodology has been extensively covered earlier in Houben-Weyl,[3] *COMC-I*,[5,6] and *Organomercury Compounds in Organic Synthesis*,[10] so the primary emphasis here will be on newer developments and summarizing methods of particular utility in organic synthesis.

9.2.2 Mercuration of Organic Halides

Relatively few organic halides react directly with metallic mercury to afford the corresponding organomercury halides. Among the simple alkyl halides only methyl iodide produces good yields (Equation (1)).[12] These reactions appear to be free radical in nature since they are usually promoted by light or photolysis. Polyhalogenated alkanes appear to be more reactive towards mercury and can afford organomercurials useful in divalent carbon transfer reactions (see Section 9.11) (Scheme 1).[12,13] Unsaturated organic halides are significantly more reactive towards mercury and light. Allylic[14] and benzylic halides[15] react well (Equations (2) and (3)). Propargylic bromides and iodides afford either propargylic or allenic mercurials depending on the substitution pattern present in the organic halide (Scheme 2).[16]

$$\text{MeI} + \text{Hg} \xrightarrow[81\%]{h\nu} \text{MeHgI} \quad (1)$$

Dialkyl- and diarylmercurials have often been prepared by the reaction of the corresponding organic halide with sodium–mercury amalgam (Equation (4)),[17] while the electrolysis of alkyl, benzylic, and aryl halides affords another useful route to diorganomercurials (Equation (5)).[18]

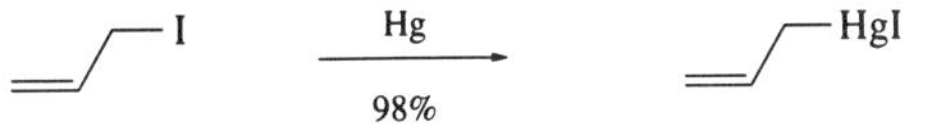

$$\text{CH}_2=\text{CH-CH}_2\text{I} \xrightarrow[98\%]{\text{Hg}} \text{CH}_2=\text{CH-CH}_2\text{HgI} \quad (2)$$

$$\text{PhCH}_2\text{I} \xrightarrow[95\%]{\text{Hg}} \text{PhCH}_2\text{HgI} \quad (3)$$

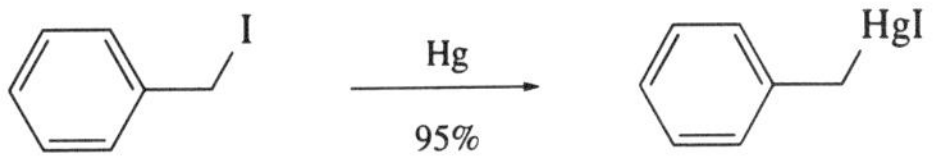

$$\text{H}_2\text{CI}_2 \xrightarrow{\text{Hg}} \text{IHgCH}_2\text{I} \xrightarrow{\text{Hg}} \text{CH}_2(\text{HgI})_2$$

Scheme 1

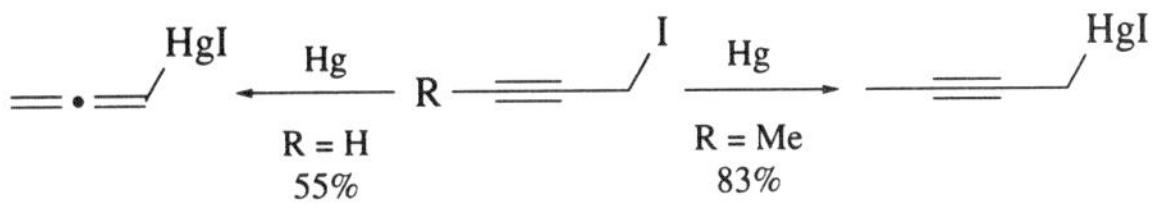

$$\text{CH}_2=\text{C}=\text{CH-HgI} \xleftarrow[\substack{R=H\\55\%}]{\text{Hg}} R-\text{C}\equiv\text{C-CH}_2\text{I} \xrightarrow[\substack{R=Me\\83\%}]{\text{Hg}} R-\text{C}\equiv\text{C-CH}_2\text{HgI}$$

Scheme 2

$$2\,\text{RX} + \text{Na(Hg)} \longrightarrow \text{R}_2\text{Hg} \quad (4)$$

$$\text{RX} \xrightarrow{\text{Hg cathode}} \text{R}_2\text{Hg} \quad (5)$$

9.2.3 Transmetallation

The transmetallation of other organometallics by mercury salts provides one of the most generally useful routes to either mono- or diorganomercury compounds (Scheme 3). The most useful process involves the readily available organometallics of magnesium and lithium (Scheme 3, M = Li, MgX), which react rapidly with mercuric halides to afford either the mono- or disubstituted mercurials. While alkyl, aryl, and vinylic mercurials have been thus prepared, the reactivity of these organometallics limits the types of functionality which can be accommodated. This approach is particularly useful for the preparation of organomercurials from readily prepared Grignard reagents, or from organolithium reagents prepared by the direct lithiation of organic substrates, such as cyclopentadienes,[19] indenes, carboranes, and hydrocarbons bearing certain functionality (Equation (6)).

$$\text{HgX}_2 \xrightarrow{\text{RM}} \text{RHgX} \xrightarrow{\text{RM}} \text{R}_2\text{Hg}$$

Scheme 3

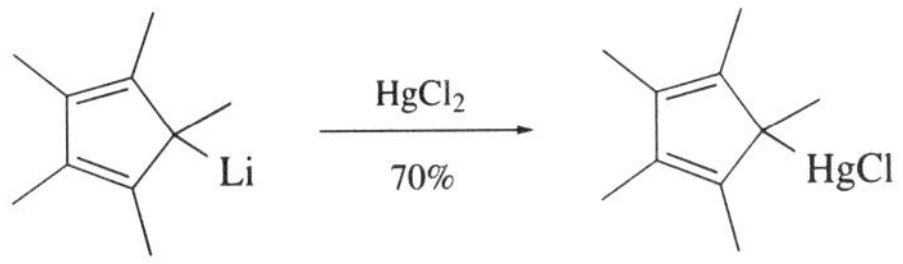

$$\text{(Me}_4\text{C}_5\text{)Li} \xrightarrow[70\%]{\text{HgCl}_2} \text{(Me}_4\text{C}_5\text{)HgCl} \quad (6)$$

While generally less reactive than the above organometallics, organoboranes have proven very useful in the synthesis of organomercurials because they are readily available and can accommodate considerably more functionality.[20] Thus, aryl- and vinylic boronic acids react with a variety of mercury salts to afford symmetrical or unsymmetrical mono- or diorganomercurials (Scheme 4).[21]

$$\text{HgX}_2 \xrightarrow{\text{R}^1\text{B(OH)}_2} \text{R}^1\text{HgX} \xrightarrow{\text{R}^2\text{B(OH)}_2} \text{R}^1\text{HgR}^2$$

Scheme 4

Organoboranes derived from alkenes or alkynes by hydroboration are especially useful for the preparation of organomercurials (Scheme 5–7).[20,22,23]

No other organometallics have achieved the importance of the organometallics of lithium, magnesium, and boron, but arylsilanes, vinylic stannanes, alkylplumbanes, allylic, benzylic and propargylic zinc reagents, and π-allylpalladium halides have seen limited use in the preparation of the corresponding organomercurials via transmetallation.

$$RCH=CH_2 \xrightarrow{BH_3} (RCH_2CH_2)_3B \xrightarrow{HgX_2} RCH_2CH_2HgX \text{ or } (RCH_2CH_2)_2Hg$$

Scheme 5

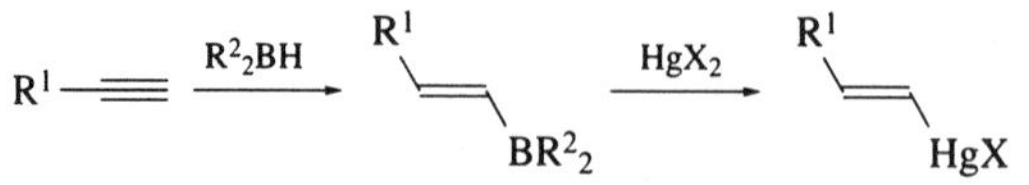

Scheme 6

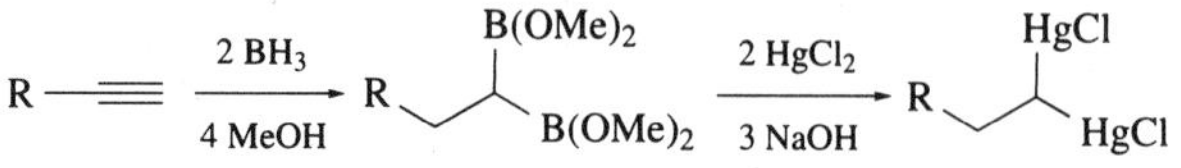

Scheme 7

9.2.4 Mercuration of Carbon Monoxide

Ester- and amide-containing organomercurials can be readily prepared by reacting mercuric acetate with carbon monoxide in the presence of an alcohol or amine (Scheme 8).[24]

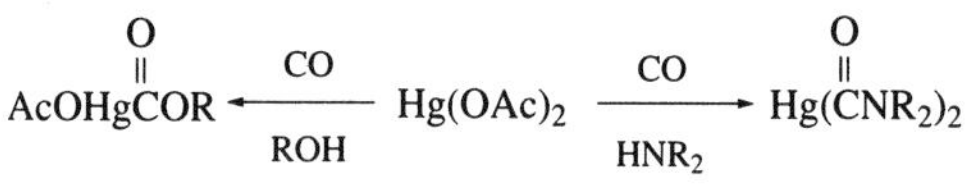

$$AcOHgCOR \xleftarrow[ROH]{CO} Hg(OAc)_2 \xrightarrow[HNR_2]{CO} Hg(CNR_2)_2$$

Scheme 8

9.2.5 C–H Substitution in Activated Hydrocarbons

Many organic compounds contain hydrogens sufficiently acidic that they undergo direct substitution by mercuric salts. Thus, a variety of mono- and polymercurated cyclopentadienes have been prepared in this fashion (Equation (7)). Organometallics such as ruthenocene[25] and porphyrins[26] have been mercurated similarly. Aldehydes, ketones, and carboxylic acids have been directly mercurated by electrophilic mercury salts to produce a variety of structurally complex mono- and polymercurated substrates. The reagent $Hg[C(NO_2)_3]_2$ appears most useful for the direct monomercuration of ketones (Equation (8)).[27] However, simple monomercurated products are probably still best prepared by mercuration of the corresponding enol silanes (Equation (9)).

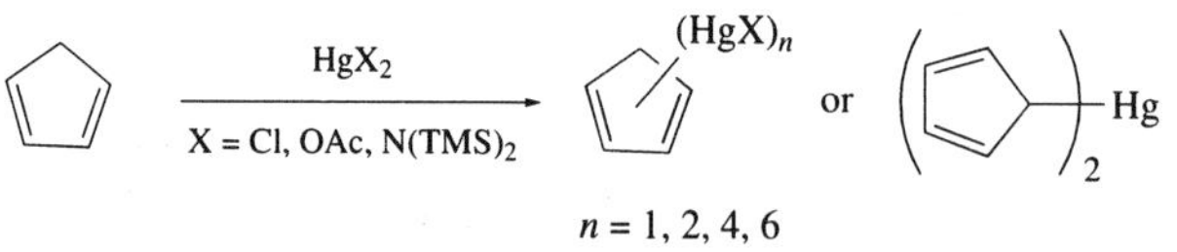

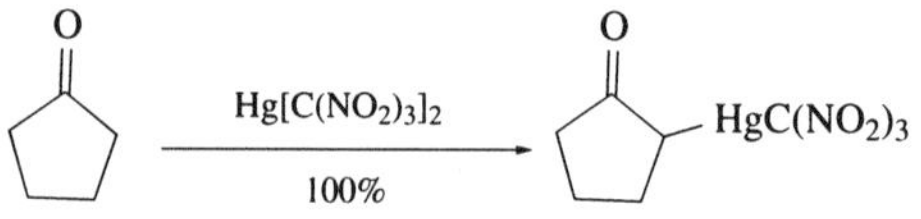

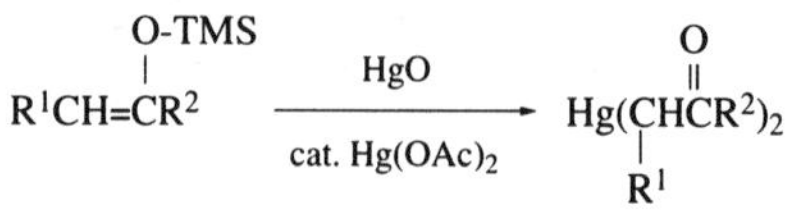

Many electron-withdrawing groups are sufficiently activating that the α-C–H bonds can be mercurated directly by mercury(II) salts. Thus, there are many examples of the direct mercuration of nitriles, nitroalkanes, diketones, diacids, diesters, diamides, ketoesters, ketoamides, nitroesters, disulfones, diphosphonates, phosphonate esters, cyanophosphonates, cyanoesters, cyanoamides, diphosphines, and diazoesters.

The very useful divalent carbon transfer reagents (see Section 9.11) are commonly prepared by the reaction of polyhaloalkanes with organomercuric halides and a base (Equation (10)).

$$RHgCl + HCXYZ \xrightarrow{KOBu^t} RHgCXYZ \quad (10)$$

9.2.6 Mercuration of Cyclopropanes

Cyclopropanes undergo facile ring opening in the presence of mercury(II) salts and a nucleophilic solvent to form a wide variety of γ-substituted alkylmercurials (Equation (11)).[10,28] When nitriles are employed as the solvent, amide-containing mercurials are produced (Equation (12)). The mechanism of these cyclopropane-opening reactions has been extensively studied.[28,29] Recent interest in this reaction has focused on the effect of oxygen-substitution on the ring and intramolecular entrapment of nucleophiles (Equations (13)–(15)).[30]

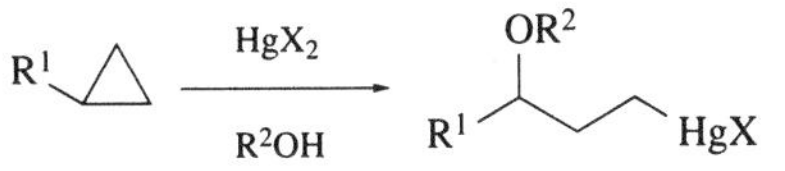

$$R^1\text{-cyclopropane} \xrightarrow[R^2OH]{HgX_2} R^1\text{-CH(OR}^2\text{)-CH}_2\text{-CH}_2\text{-HgX} \quad (11)$$

X = Cl, OAc, O$_2$CCF$_3$, NO$_3$, SO$_4$, ClO$_4$, C(NO$_2$)$_3$
R^2 = H, alkyl, OH, OR^1, Ac

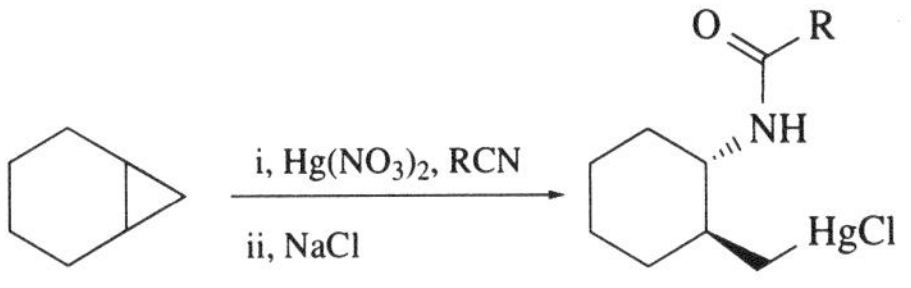

$$\text{norcarane} \xrightarrow[\text{ii, NaCl}]{\text{i, Hg(NO}_3)_2, \text{RCN}} \text{trans-2-(RCONH)cyclohexyl-CH}_2\text{HgCl} \quad (12)$$

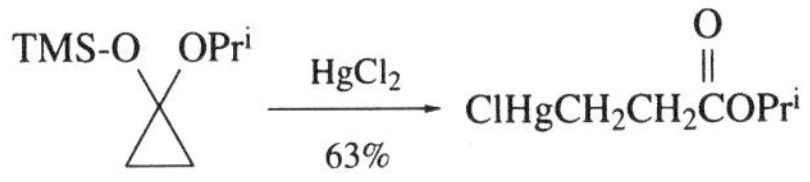

$$\text{TMS-O, OPr}^i\text{-cyclopropane} \xrightarrow{HgCl_2} ClHgCH_2CH_2\overset{O}{\underset{\|}{C}}OPr^i \quad (13)$$
63%

$$\text{(hydroxyalkyl-oxycyclopropane)} \xrightarrow[100\%]{Hg(OAc)_2} \text{spiroketal-HgOAc} \quad (14)$$

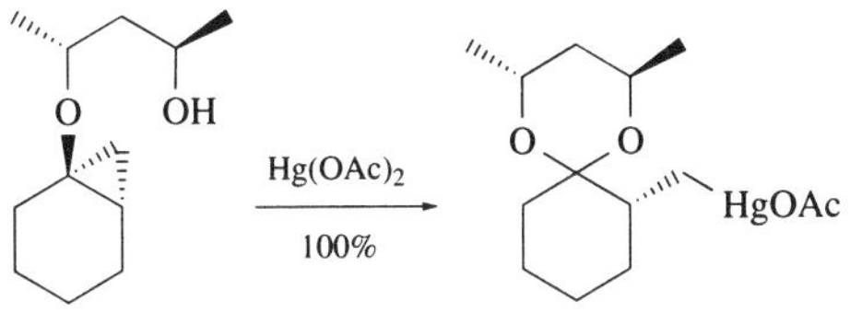

$$\text{cyclopropyl-C(OOH)(R)CH}_2 \xrightarrow[\text{ii, KBr}]{\text{i, Hg(OAc)}_2} BrHg\text{-CH}_2\text{CH}_2\text{-(O-O cyclic)-R} \quad (15)$$

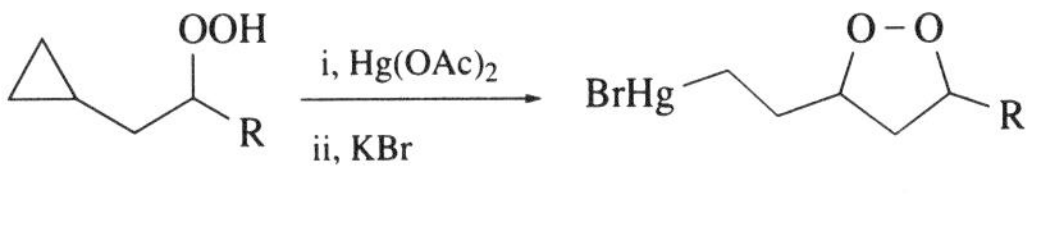

9.2.7 Mercuration of Alkenes

The reaction of alkenes with electrophilic mercury(II) salts in a nucleophilic solvent usually produces high yields of the corresponding β-substituted alkylmercurial (Equation (16)).[11] This provides the single most important application of organomercurials in organic synthesis and will be discussed in detail in Section 9.12.

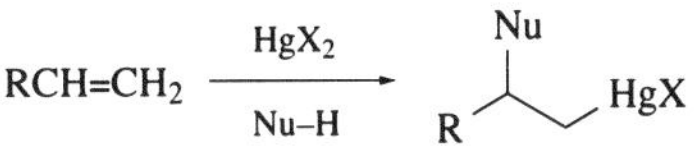

$$RCH=CH_2 \xrightarrow[Nu-H]{HgX_2} R\text{-CH(Nu)-CH}_2\text{HgX} \quad (16)$$

Some aryl- or heteroatom-substituted alkenes, steroids, and hindered polycyclic alkenes react with mercuric carboxylates to afford vinylic mercurials instead (Equations (17) and (18)).[31]

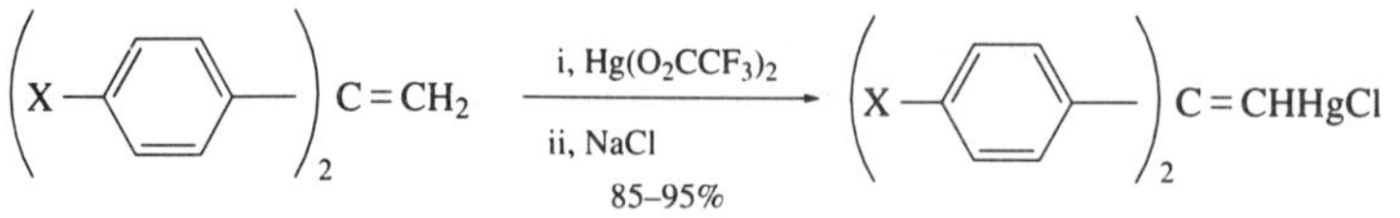

$$\left(X-\underset{2}{\underset{|}{\bigcirc}}\right)C=CH_2 \xrightarrow[\text{ii, NaCl}]{\text{i, Hg(O}_2\text{CCF}_3)_2} \left(X-\underset{2}{\underset{|}{\bigcirc}}\right)C=CHHgCl \quad (17)$$

$$\text{85-95\%}$$

X = H, Me, Cl, OMe

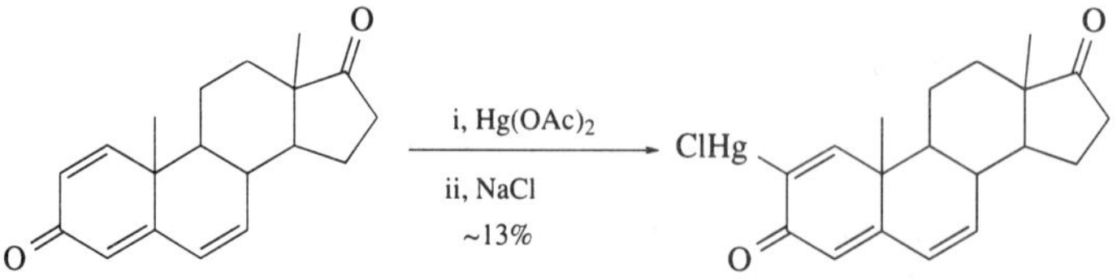

$$\text{(steroid)} \xrightarrow[\text{ii, NaCl}]{\text{i, Hg(OAc)}_2} \text{ClHg-(steroid)} \quad (18)$$

~13%

9.2.8 Mercuration of Alkynes

The reaction of terminal alkynes with mercuric salts usually affords dialkynylmercurials, while internal alkynes typically undergo *anti* addition to yield vinylmercurials (Equations (19) and (20)). The latter reaction has been observed with a variety of simple alkyl and aryl alkynes, as well as many alkynes substituted with halogen, ether, alcohol, amine, and carbonyl groups. Occasionally *syn* adducts are formed.

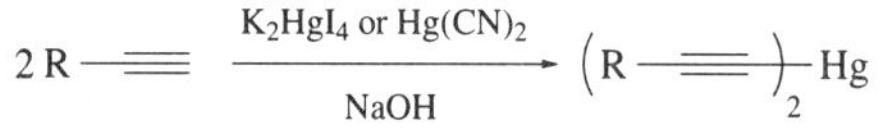

$$2\,R\!-\!\!\!\equiv\!\!\!-\!\!\! \xrightarrow[\text{NaOH}]{\text{K}_2\text{HgI}_4 \text{ or Hg(CN)}_2} \left(R\!-\!\!\!\equiv\!\!\!-\!\!\!\right)_2\!Hg \quad (19)$$

$$R\!-\!\!\!\equiv\!\!\!-\!R \xrightarrow{\text{HgX}_2} \underset{R\quad HgX}{\overset{X\quad R}{\diagdown\!=\!\diagup}} \quad (20)$$

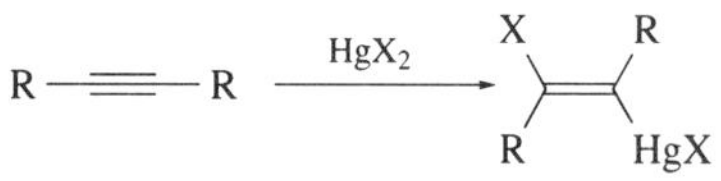

X = F, Cl, O$_2$CR, O$_3$SR, O$_2$SR, SCN, Bu^t(*syn*)

9.2.9 Aromatic Mercuration

The direct electrophilic substitution of arenes by mercuric salts is one of the most important reactions in organomercury chemistry (Equation (21)).[10] It has been effected using a variety of electrophilic mercury salts, including mercuric chloride, acetate, trifluoroacetate, and nitrate, and a large number of arenes, including simple arenes, alkylbenzenes, aryl halides, phenols, anilines, aryl ethers and amides, nitroarenes, polynuclear aromatics, numerous heterocycles, and even metallocenes. The resulting arylmercurials have found many applications in organic synthesis.

$$\text{ArH} + \text{HgX}_2 \longrightarrow \text{ArHgX} + \text{HX} \quad (21)$$

X = Cl, OAc, O$_2$CCF$_3$, NO$_3$

9.2.10 Substitution in Nitrogen-containing Compounds

Arylmercurials may be prepared by the reaction of aryldiazonium salts with mercuric chloride, followed by reduction with copper salts (Scheme 9) or by the reaction of arylhydrazines with HgO. Hydrazones afford a variety of products.

$$(\text{ArN}_2)\text{Cl} \xrightarrow{\text{HgCl}_2} (\text{ArN}_2)\text{HgCl}_3 \xrightarrow[(\text{NH}_3)]{\text{Cu or CuCl}} \text{ArHgCl or Ar}_2\text{Hg}$$

Scheme 9

Diazoalkanes react with mercury salts by either hydrogen or nitrogen substitution (Schemes 10 and 11).

$$(RC)_2Hg \xleftarrow{} \underset{RCH}{\overset{N_2}{\|}} \xrightarrow{} \underset{RCHHgX}{\overset{X}{|}}$$

Scheme 10

$$XCH_2HgX \xleftarrow{HgX_2} CH_2N_2 \xrightarrow{RHgX} RHgCH_2X$$

Scheme 11

9.2.11 Elimination Reactions

Elimination reactions afford a variety of organomercurials.[10] The thermal decarboxylation of mercuric carboxylates has provided a valuable route to mercurated ketones, arylmercurials, and trihalomethylmercurials (Equations (22)–(24)).[32] The latter compounds are useful for dihalocarbon transfer reactions (see Section 9.11).

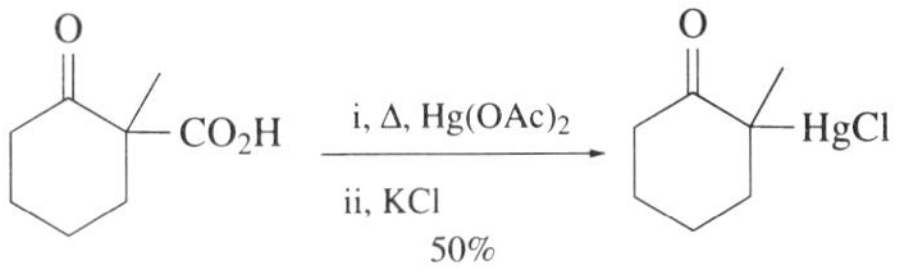

$$\text{cyclohexanone-2-CO}_2H \xrightarrow[\text{ii, KCl}]{i, \Delta, Hg(OAc)_2} \text{cyclohexanone-2-HgCl} \quad 50\% \tag{22}$$

$$RHgO_2CAr \xrightarrow{\Delta} RHgAr \tag{23}$$

$$PhHgCl + NaO_2CCCl_3 \xrightarrow[65\%]{\Delta} PhHgCCl_3 \tag{24}$$

Alkylmercuric carboxylates can be prepared in high yields by electrolytic, photochemical, or free radical decarboxylation (Equation (25)).

$$Hg(O_2CR)_2 \xrightarrow[\text{or peroxides}]{\text{electrolysis or } h\nu} RHgO_2CR \tag{25}$$

Sulfinate[33] and sulfonate[32] salts of mercury undergo thermal elimination of SO_2 and SO_3, respectively, to afford organomercurials (Equation (26) and Scheme 12).

$$RSO_2H(Na) + HgX_2 \xrightarrow{\Delta} RHgX \text{ or } R_2Hg \tag{26}$$

$$2 \text{ ArSO}_3H \xrightarrow{Hg(OAc)_2} (ArSO_3)_2Hg \xrightarrow{\Delta} Ar_2Hg$$

Scheme 12

9.3 HYDROGEN SUBSTITUTION

9.3.1 Protonolysis

The substitution of mercury in organomercurials by hydrogen is a synthetically very valuable process, particularly when combined with solvomercuration. While the protonolysis or deuterolysis of organomercurials has been extensively studied mechanistically,[34] this reaction has found only limited synthetic use and that has been primarily for the preparation of isotopically labeled compounds. Thus, heterocycles, such as furan and thiophene[35] (Equation (27)), and vinylmercurials derived from the solvomercuration of alkynes (Equation (28))[36] undergo clean deuterolysis by DCl.

The products of the solvomercuration of alkenes generally react with protic acids by reversion to the alkene. When the mercury moiety is next to a carbonyl group, however, reduction can be effected by H_2S, thiols or Na_2CS_3 in protic solvents.[37]

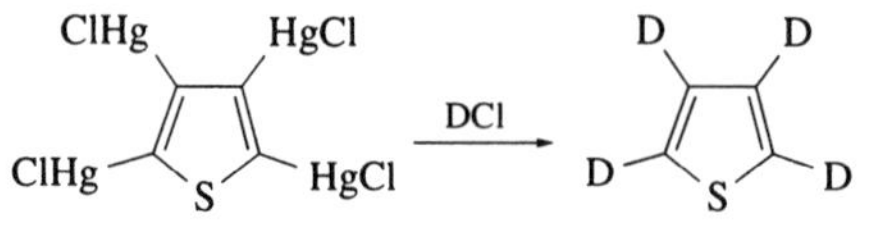

$$\text{(tetramercurated thiophene)} \xrightarrow{\text{DCl}} \text{(2,3,4,5-tetradeuterothiophene)} \quad (27)$$

$$\underset{(E) \text{ or } (Z)}{\text{ClCH=CHHgCl}} \xrightarrow[> 96\%]{\text{DCl}} \text{ClCH=CHD} \quad (28)$$

9.3.2 Metal Hydride Reduction

The very important substitution of mercury by hydrogen in solvomercuration products is most commonly carried out by reduction with metal hydrides of boron, aluminum, or tin. Sodium borohydride is the most important metal hydride used for this purpose. Under alkaline conditions it instantaneously reduces organomercuric salts to the corresponding hydrogen substitution product[38] by generation of an organomercury hydride and subsequent decomposition by a free radical chain process (Scheme 13).[39] This process is not stereospecific and reversion to alkene or skeletal rearrangement is sometimes observed. Some of these difficulties have been overcome by using phase transfer conditions or varying the solvent.[40]

$$\text{RHgX} \xrightarrow{\text{NaBH}_4} \text{RHgH} \longrightarrow \text{R} \bullet \longrightarrow \text{RH}$$

Scheme 13

Organomercurials have also been reduced to the corresponding hydrocarbons by Et_2AlD, $LiAlH_4$, $NaH_2Al(OCH_2CH_2OCH_3)_2$, LiH, Bu^n_3SnH, $DCuP(Bu^n)_3$, sodium naphthalene, N-benzyl-1,4-dihydronicotinamide, and hydrazine, but these reactions are of little synthetic utility. For the stereospecific reduction of organomercurials, sodium amalgam is the reagent of choice.[41]

9.4 HALOGENATION

The substitution of mercury by halogen is a facile reaction widely used in synthesis and for establishing the number and locations of the mercury moieties in an organomercurial.[10] It can be quite useful for the preparation of radiolabeled compounds.[42] Halogenation is most commonly effected using the halogens themselves. The mechanism of this process has been closely examined.[34] In the presence of pyridine or a polar solvent, such as methanol, halogenation proceeds by electrophilic substitution with retention of stereochemistry. In a nonpolar solvent, such as carbon tetrachloride, or upon photolysis, halogenation can proceed by a free radical chain process with loss of stereochemistry. Virtually all kinds of organomercurials undergo facile halogenation by bromine or iodine. Very little work has been carried out with chlorine or fluorine, due presumably to the difficulties in handling these reagents.

The halogenation of simple alkylmercurials is of more mechanistic interest than synthetic utility. However, the halogenation of organomercurials prepared by solvomercuration provides a highly convenient method for the addition of a nucleophile and a halogen across a cyclopropane,[43] or a carbon–carbon double bond[44] (Schemes 14 and 15).

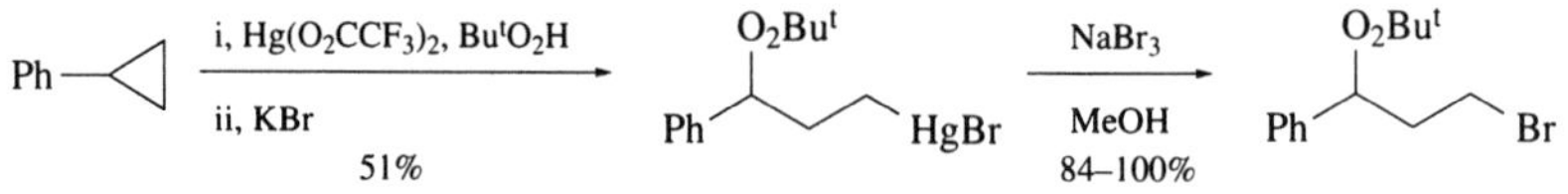

$$\text{Ph-}\triangleleft \xrightarrow[\text{ii, KBr}]{\text{i, Hg(O}_2\text{CCF}_3)_2, \text{Bu}^t\text{O}_2\text{H}} \underset{51\%}{\text{Ph}\underset{\text{O}_2\text{Bu}^t}{\diagdown}\text{HgBr}} \xrightarrow[\underset{84-100\%}{\text{MeOH}}]{\text{NaBr}_3} \text{Ph}\underset{\text{O}_2\text{Bu}^t}{\diagdown}\text{Br}$$

Scheme 14

The halogenation of vinylmercurials provides a useful route to vinylic halides. The solvomercuration and subsequent halogenation of alkynes produces functionally substituted vinylic halides (Scheme 16). The stereochemistry of the bromination of simple vinylmercurials has been found to be highly dependent on the solvent employed (Scheme 17).[45] Allenic and propargylic mercurials undergo halogenation by pyridinium hydrobromide perbromide and iodine in pyridine with rearrangement to the corresponding propargylic and allenic halides, respectively (Equations (29) and (30)).[46]

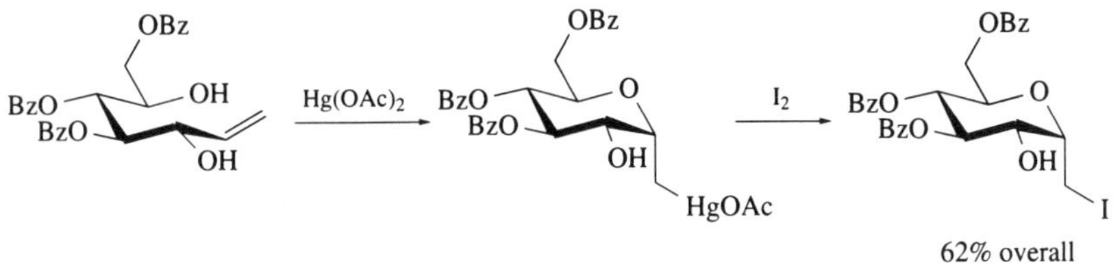

Scheme 15

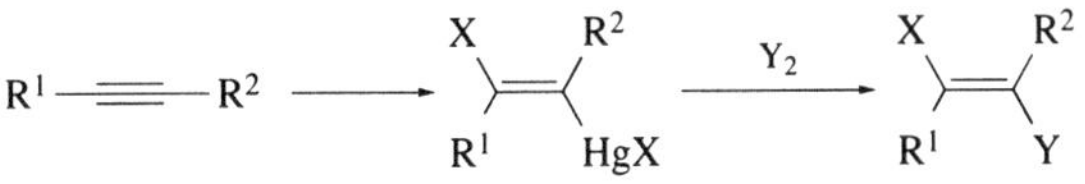

Scheme 16

X = Cl, OAc, SCN; Y = Br, I

Scheme 17

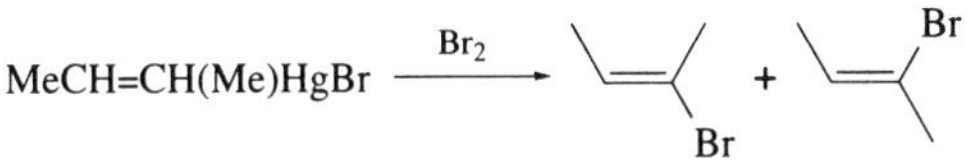

Stereochemistry	Solvent	Product ratio	
(E)	pyridine	99.6	: 0.4
(E)	CS$_2$	12	: 88
(Z)	pyridine	8.1	: 91.9
(Z)	CS$_2$	93	: 7

MeCH=C=CHHgI →

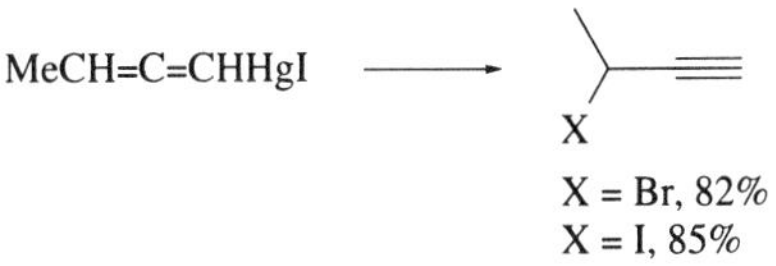

X = Br, 82%
X = I, 85%   (29)

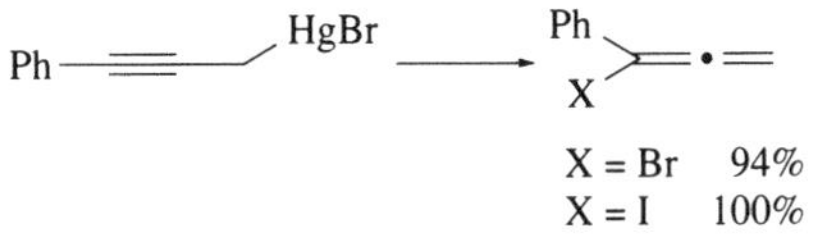

Ph—≡—HgBr →

X = Br 94%
X = I   100%   (30)

The halogenation of arylmercurials has proven to be a valuable route to aryl bromides and iodides, as well as a useful method for determining the number of mercury atoms and their location within a molecule. Arene mercuration and subsequent halogenation gains practical utility when direct halogenation of the arene is difficult or leads to mixtures of isomers (Scheme 18).[47] One can also take advantage of some of the unique methods for preparing arylmercurials (Scheme 19).[47] This methodology has proven useful for the preparation of halides from simple arenes, a large number of different functionally-substituted arenes, polycyclic aromatic hydrocarbons, heterocycles, and metallocenes (Scheme 20).[48] It can also be employed to prepare polyhaloarenes (Equation (31)).[10,49]

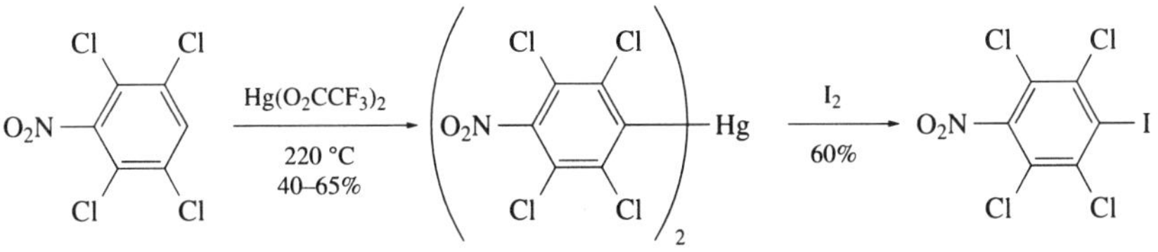

Scheme 18

Although the vast majority of halogenation reactions have been effected using bromine or iodine, miscellaneous other reagents, including ICl, IBr, NCS, NBS, NIS, CuCl$_2$, and CuBr$_2$, have also been employed. Acetyl hypofluorite has been used to prepare aryl fluorides[50] and β-fluoroethers (Scheme 21).[51]

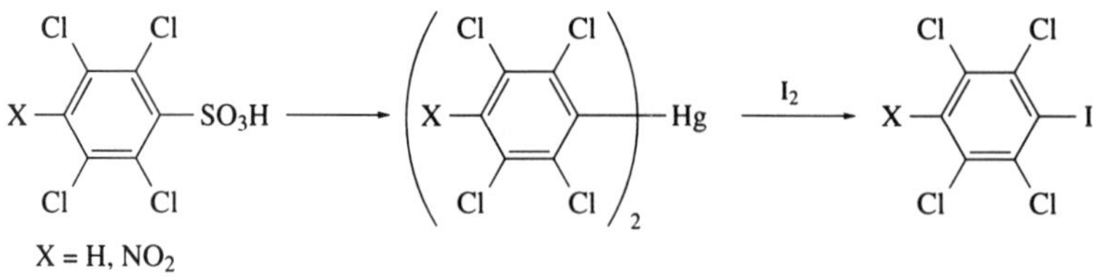

Scheme 19

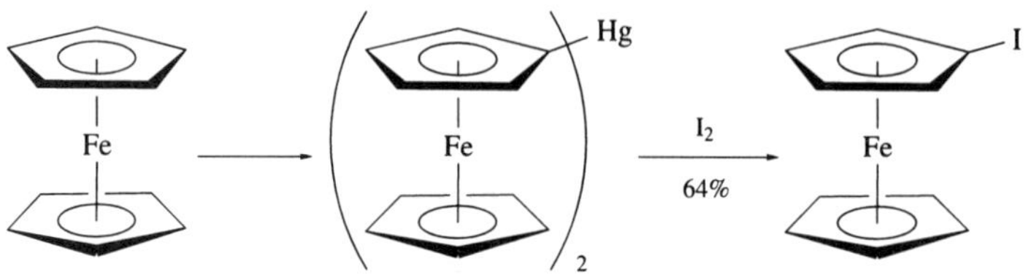

Scheme 20

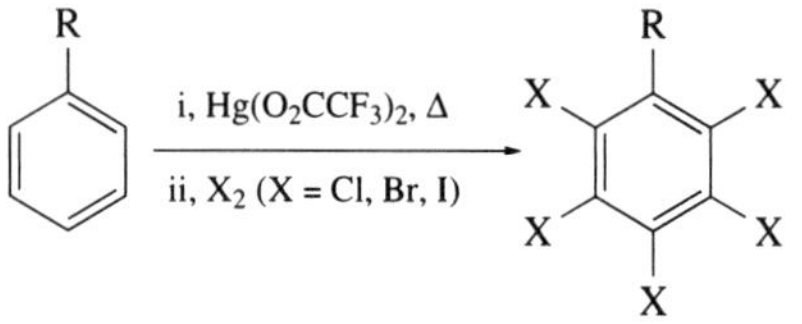

(31)

R = H, NH$_2$, NHAc, MeO, Me, CF$_3$, F, Cl,
Br, NO$_2$, CO$_2$H, CO$_2$Na, CONH$_2$

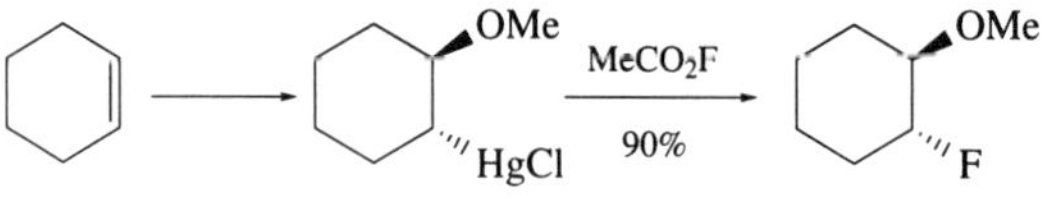

Scheme 21

9.5 HETEROATOM SUBSTITUTION

9.5.1 Oxygen Compounds

Organomercurials can be useful as intermediates in the synthesis of heteroatom-containing compounds, particularly those containing oxygen. While organomercurials are generally quite stable to oxygen, the reduction of alkylmercurials with NaBH$_4$ in the presence of oxygen provides a useful synthesis of alcohols. This process is quite valuable when combined with solvomercuration, and has been employed in the synthesis of hydroxytetrahydrofurans, carbohydrates, prostaglandins (Equation (32)), aphidicolin, and nucleosides (Equation (33)).[11,52]

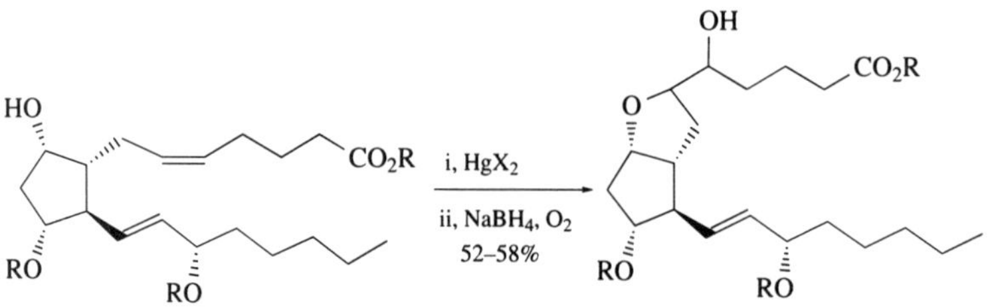

(32)

Alkylmercurials are very reactive towards ozone. Alkylmercurials bearing 1° alkyl groups afford carboxylic acids, but significant degradation to shorter chain carboxylic acids is observed. The reaction of 2° alkylmercurials is more synthetically useful since it affords reasonable yields of ketones (Scheme 22).[53] Ketones can also be obtained by oxidizing 2° alkylmercurials with peracetic acid.[54]

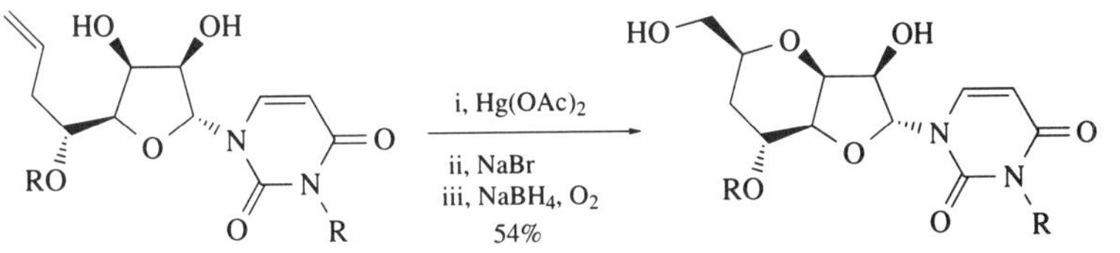

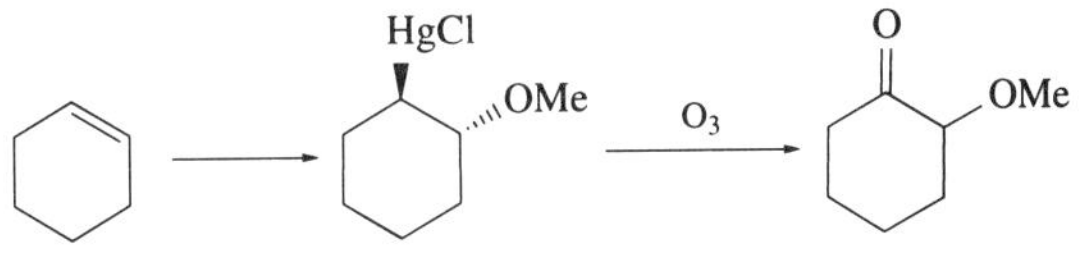

Scheme 22

Certain organomercuric salts undergo demercuration to produce useful oxygen-containing products. Thus, divinylmercury reacts with carboxylic acids and phenols to produce vinylmercuric salts which thermally decompose to vinyl esters and aryl vinyl ethers (Scheme 23).[55] Vinylic esters can also be prepared by reacting vinylmercuric halides with mercuric acetate and catalytic amounts of Pd(OAc)$_2$ (Equation (34)).[56]

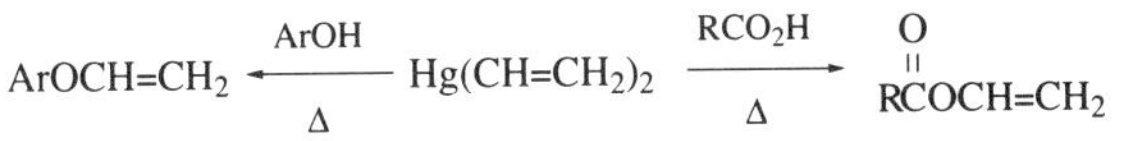

Scheme 23

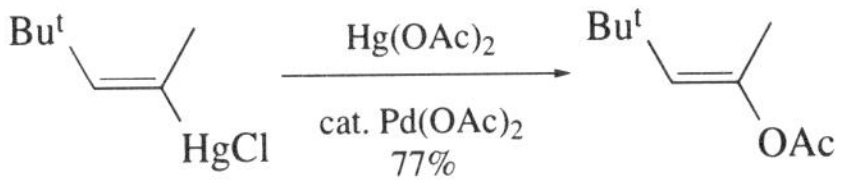

The solvolysis of organomercuric salts has been extensively studied mechanistically,[34] but only the thermal decomposition of allylmercuric acetates formed by heating alkenes and mercuric acetate appears to be synthetically useful (Equation (35)).[57] This reaction, known as the Treibs reaction, has also been extended to the synthesis of α-acetoxy ketones, although the yields are quite variable (Equation (36)).[58]

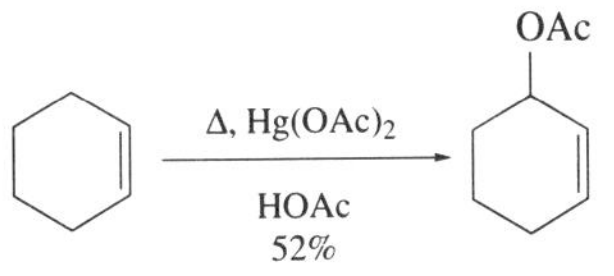

The hydroxy- or alkoxymercuration of alkenes, followed by treatment with Li$_2$PdCl$_4$, affords ketones and acetals (Scheme 24).[59]

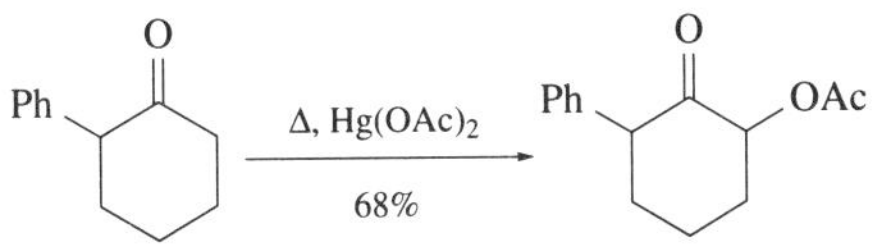

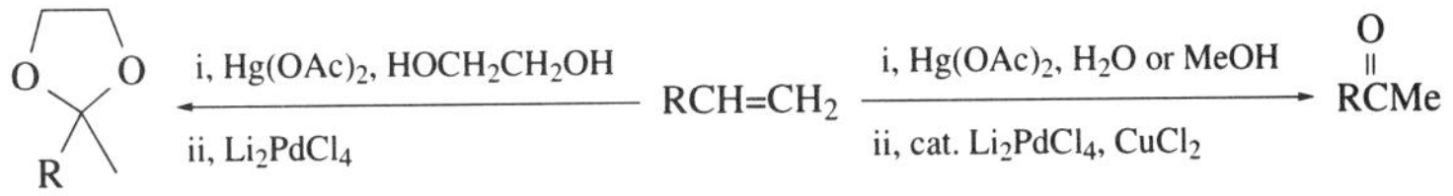

Scheme 24

Synthesis via organoboranes has provided new routes to phenols and esters. Phenols can be prepared from arylmercuric salts by treatment with BH$_3$·THF, followed by oxidation of the resulting arylborane with alkaline hydrogen peroxide (Scheme 25).[60] Transmetallation of organoboranes derived from terminal alkenes with mercuric carboxylates, followed by *in situ* iodination, affords high yields of esters, rather than the anticipated iodides (Equation (37)).[61]

$$\text{ArHgCl} \xrightarrow{\text{BH}_3} \text{ArB} \begin{array}{c} \\ \end{array} \xrightarrow[\text{NaOH}]{\text{H}_2\text{O}_2} \text{ArOH}$$

Scheme 25

$$R^1CH{=}CH_2 \xrightarrow[\substack{\text{ii, Hg(O}_2\text{CR}^2)_2 \\ \text{iii, I}_2 \\ 73\text{-}93\%}]{\text{i, BH}_3} R^1CH_2CH_2O_2CR^2 \quad (37)$$

9.5.2 Sulfur, Selenium, and Tellurium Compounds

A wide variety of sulfur, selenium, and tellurium compounds have been prepared from organomercurials. Diarylsulfides, -selenides, and -tellurides have been prepared by heating above 200 °C the corresponding diarylmercurials with sulfur, selenium, or tellurium (Equation (38)).[10] At lower temperatures dialkyl and diaryl disulfides are formed in good yields from RHgCl and sulfur (Equation (39)).[62]

$$Ar_2Hg + X \xrightarrow{\Delta} Ar_2X \quad (38)$$
$$X = S, Se, Te$$

$$RHgCl + S \xrightarrow{140\text{-}180\,°C} R-S-S-R \quad (39)$$

Alkyl and vinylic sulfides, selenides, and tellurides have been prepared by the photostimulated reaction of alkyl- and vinylic mercuric chlorides and disulfides, diselenides, and ditellurides (Equation (40)).[63] Vinylic sulfides have also been prepared by the photolysis of vinylmercuric halides and mercaptides[63d] or by heating divinylmercury with thiols.[55] Diarylmercurials and vinylmercuric chlorides also react with arylselenenyl halides to afford selenides.[64]

$$R^1HgCl + (R^2X)_2 \xrightarrow{h\nu} R^1XR^2 \quad (40)$$

R^1 = alkyl, vinylic; R^2 = alkyl, aryl; X = S, Se, Te

Sulfones can be prepared by the reaction of diarylmercurials and arylsulfonyl halides plus AlBr$_3$,[65] by heating divinylmercury and sulfinic acids,[55a,66] or by photolysis of vinylmercuric chlorides in the presence of RSO$_2$Cl[63b] or sulfinate salts (Equation (41)).[67]

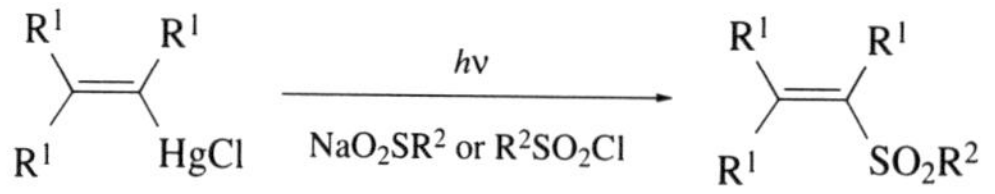

$$\underset{R^1}{\overset{R^1}{>}}{=}\underset{HgCl}{\overset{R^1}{<}} \xrightarrow[\text{NaO}_2\text{SR}^2 \text{ or } R^2\text{SO}_2\text{Cl}]{h\nu} \underset{R^1}{\overset{R^1}{>}}{=}\underset{SO_2R^2}{\overset{R^1}{<}} \quad (41)$$

Copper salts have been used to convert arylmercuric chlorides to aryl thiocyanates and sulfones (Equation (42)).[68]

$$ArHgCl + CuX_2 \longrightarrow ArX \quad (42)$$
$$X = SCN, PhSO_2$$

Alkenyl thiocyanates have also been prepared by the addition of mercuric thiocyanate to internal alkynes, followed by protonolysis (Equation (43)).[69]

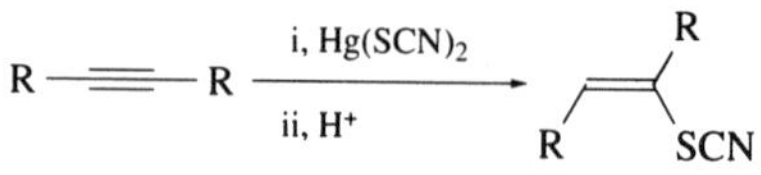

$$R{\equiv\!\equiv}R \xrightarrow[\text{ii, H}^+]{\text{i, Hg(SCN)}_2} \underset{R}{\overset{R}{>}}{=}\underset{SCN}{\overset{R}{<}} \quad (43)$$

α-Mercurated carbonyl compounds react with thionyl chloride and sulfonyl chlorides by attack on oxygen (Scheme 26).[70]

$$(H_2C=CHO)_2SO \xleftarrow[45\%]{SOCl_2} Hg(CH_2CHO)_2 \xrightarrow[42-75\%]{RSO_2Cl} \underset{O}{\overset{O}{\underset{\|}{\overset{\|}{R}}}}SOCH=CH_2$$

Scheme 26

9.5.3 Nitrogen and Phosphorus Compounds

Organomercurials have found only limited use in the synthesis of nitrogen and phosphorus compounds. Nitroso compounds have been prepared by reacting perfluoroalkylmercurials, dialkynylmercurials, and arylmercurials with nitrosyl reagents (Equation (44)).[71] Aryl nitroso compounds are apparently first formed upon reacting arylmercurials with nitric acid, but the corresponding nitroarenes are usually isolated.[10]

$$R_2Hg \text{ or } RHgX \xrightarrow[Y = Cl, Br, BF_4]{NOY} RNO \qquad (44)$$

Organomercurials were once fairly useful for the synthesis of phosphorus compounds, but have now been largely supplanted by other organometallics. Today, organomercurials are probably only useful for the synthesis of alkenylphosphorus compounds from vinylmercurials (Equations (45) and (46)).[63d,72] Carbonyl-containing mercurials react with many phosphorus reagents to produce enol derivatives (Equations (47)–(50)).[73]

$$R^1CH=CHHgCl + KOPR^2(OR^3) \xrightarrow{h\nu} R^1CH=CHPR^2(OR^3) \qquad (45)$$

$$R^2 = Ph \text{ or } OR^3$$

$$(H_2C=CH)_2Hg + HOPR^1(OR^2) \xrightarrow{\Delta} H_2C=CHPR^1(OR^2) \qquad (46)$$

$$R^1 = Me \text{ or } OR^2$$

$$Hg(CH_2CHO)_2 + PCl_n(OR)_{3-n} \longrightarrow (H_2C=CHO)_nP(OR)_{3-n} \qquad (47)$$

$$n = 1-3$$

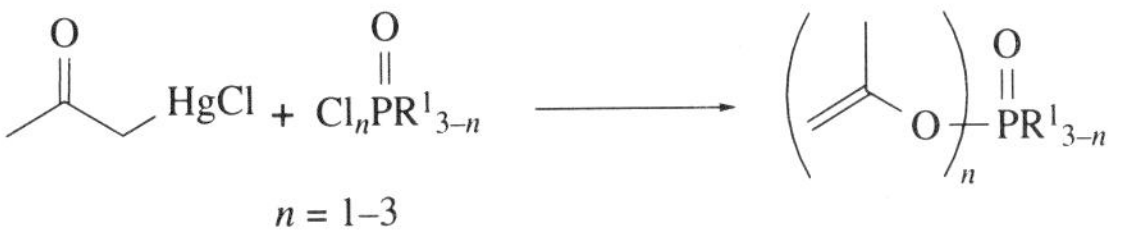

$$\text{(acetonyl)}HgCl + Cl_nPR^1_{3-n} \longrightarrow \left(\overset{}{\underset{}{}}\right)_n \text{OPR}^1_{3-n} \qquad (48)$$

$$n = 1-3$$
$$R^1 = Et, OR^2$$

$$R^1CCH_2HgCl + P(OR^2)_3 \xrightarrow{60-72\%} (H_2C=C)OP(OR^2)_2 \qquad (49)$$

$$(R^1CCH_2)_2Hg + HP(OR^2)_2 \longrightarrow H_2C=COP(OR^2)_2 \qquad (50)$$

9.6 DIMERIZATION

The dimerization of organomercurials is limited in scope, but can be used to generate bibenzyls, biaryls, and 1,3-dienes and polyenes.[10] Alkylmercurials are not easily dimerized, but the thermolysis or photolysis of dibenzylmercury affords good yields of bibenzyl. Many transition metals also catalyze this process, particularly palladium. 1,3-Dienes are formed by the palladium- or rhodium-promoted dimerization of vinylmercurials. Vinylic mercuric chlorides and divinylmercurials are catalytically dimerized to 1,3-dienes in the presence of Pd(PPh$_3$)$_4$, but some loss of stereochemistry is observed.

Dilithium tetrachloropalladate in HMPA stereospecifically dimerizes vinylic mercuric chlorides to 1,3-dienes, tetraenes, diene-diynes, and 1,4-diacetoxy-1,3-dienes, but the reaction is stoichiometric in palladium (Equation (51)).[74] In a nonpolar solvent, this reaction provides "head-to-tail" dienes and by adding $CuCl_2$ it can be made catalytic in palladium (Equation (52)).[75] The symmetrical coupling can be effected catalytically if one simply uses 0.5% $[ClRh(CO)_2]_2$.[76]

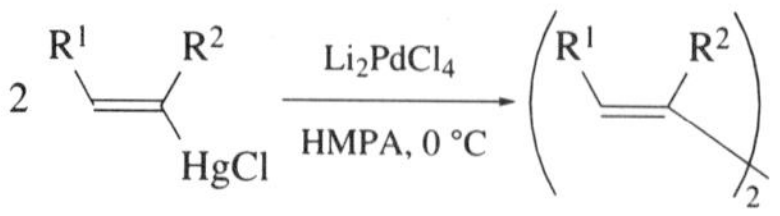

$$2 \underset{HgCl}{\overset{R^1 \quad R^2}{\diagup\!\!\!\diagdown}} \xrightarrow[\text{HMPA, 0 °C}]{Li_2PdCl_4} \left(\underset{}{\overset{R^1 \quad R^2}{\diagup\!\!\!\diagdown}} \right)_2 \qquad (51)$$

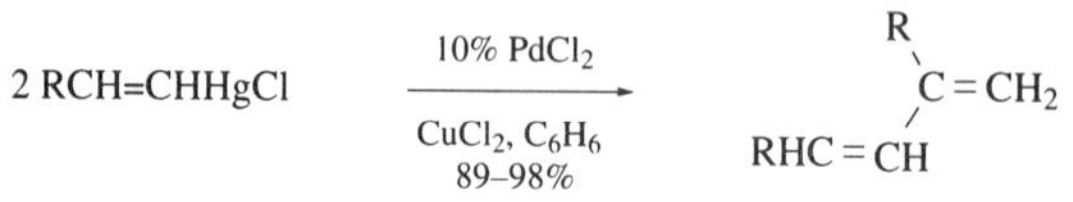

$$2\, RCH=CHHgCl \xrightarrow[\substack{CuCl_2,\, C_6H_6 \\ 89-98\%}]{10\%\ PdCl_2} \underset{RHC=CH}{\overset{R}{\diagdown}}C=CH_2 \qquad (52)$$

Arylmercurials can also be dimerized to biaryls by using Li_2PdCl_4, catalytic amounts of Li_2PdCl_4 plus Cu or $CuCl_2$, or probably best by using 0.5% $[ClRh(CO)_2]_2$.[76]

9.7 ALKYLATION

The direct alkylation of organomercurials is not a very general process and is quite limited synthetically. Only benzylic halides seem to react with organomercurials and the yields of cross-coupling are usually quite low. Friedel–Crafts type alkylations have been effected on diphenylmercury, but the yields here are also low. However, alkylmercuric chlorides will alkylate benzene in the presence of $AlCl_3$ (Equation (53)).[77]

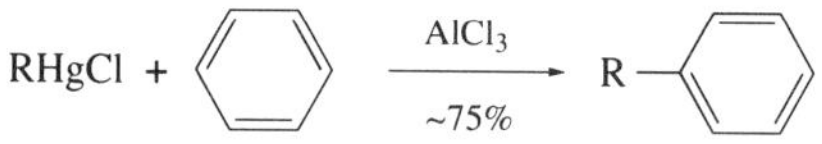

$$RHgCl + C_6H_6 \xrightarrow[\sim 75\%]{AlCl_3} R\text{–}C_6H_5 \qquad (53)$$

Under photochemical conditions, 1° and 2° alkyl and benzylic mercurials, as well as α-mercurated ketones, react with nitronates to produce nitro products (Equation (54)).[78] Malonate anions react similarly.

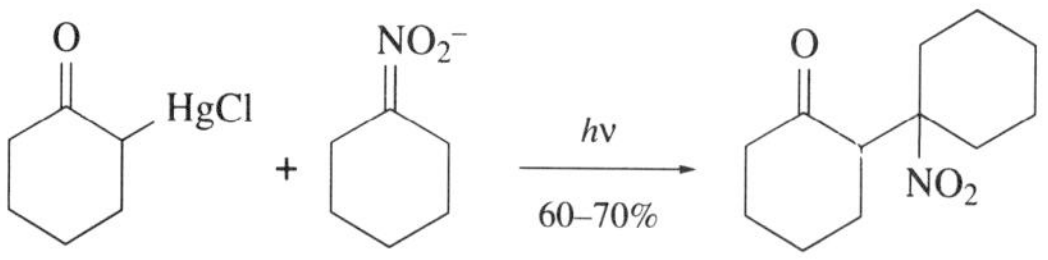

$$\underset{}{\text{cyclohexanone-HgCl}} + \underset{}{\text{cyclohexanone-NO}_2^-} \xrightarrow[60-70\%]{h\nu} \underset{}{\text{product}} \qquad (54)$$

2° and 3° Alkylmercuric chlorides react with vinylic iodides, sulfones, stannanes, and mercurials upon photolysis to afford substituted alkenes (Equation (55)).[79] Analogous coupling is observed with allylic and propargylic substrates (Equations (56) and (57)).[80]

$$R^1CH=CHX + R^2HgCl \xrightarrow{h\nu} R^1CH=CHR^2 \qquad (55)$$

$$X = I,\ SO_2Ph,\ HgCl,\ Bu^n_3Sn$$

$$RHgX + CH_2=CHCH_2Y \xrightarrow{h\nu} RCH_2CH=CH_2 \qquad (56)$$

$$Y = Cl,\ Br,\ I,\ SPh,\ SO_2Ph,\ HgCl,\ Bu^n_3Sn$$

$$RHgX + \underset{}{\equiv\!\!-\!\!\diagup^{Cl}} \xrightarrow{h\nu} \underset{}{\overset{R}{\diagdown}\!=\!\!\bullet\!\!=} \qquad (57)$$

These free radical chain processes have been employed in the alkylation of pyridine and N,N,N',N'-tetramethyl-p-phenylenediamine (Equation (58)).[81]

Other organometallics can also be used to alkylate organomercurials. The complex $MeRhI(PPh_3)_2$ will methylate vinylic, alkynyl, and arylmercurials, and $ClRh(PPh_3)_3$ will catalyze the cross-coupling of arylmercuric chlorides and vinyl bromide (Equation (59)).[82] Palladium catalyzes the cross-coupling of

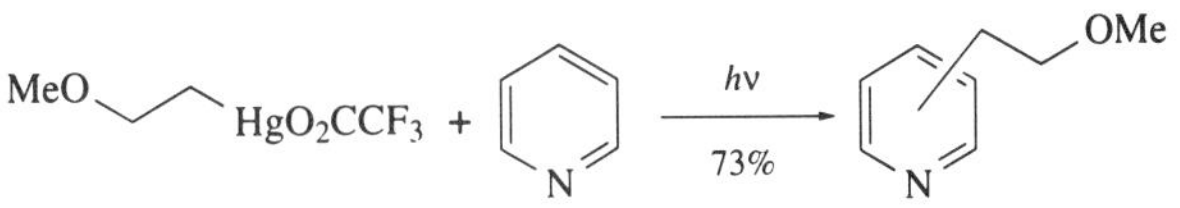

$$\text{MeO} \diagup \diagdown \text{HgO}_2\text{CCF}_3 + \text{pyridine} \xrightarrow[73\%]{h\nu} \text{pyridine-CH}_2\text{CH}_2\text{OMe} \quad (58)$$

diarylmercurials and aryl halides (Equation (60)),[83] while organocopper reagents react with alkyl, aryl, and vinylic mercuric chlorides to give good yields of cross-coupled products (Equation (61)).[84]

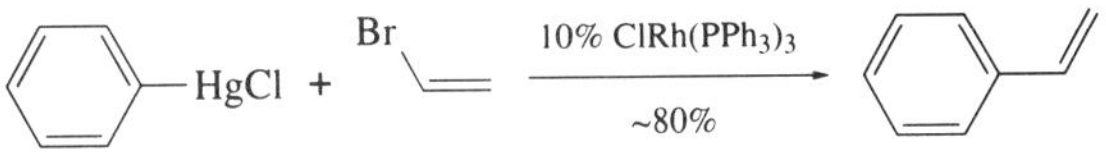

$$\text{PhHgCl} + \text{BrCH=CH}_2 \xrightarrow[\sim 80\%]{10\% \text{ ClRh(PPh}_3)_3} \text{PhCH=CH}_2 \quad (59)$$

$$\text{Ar}^1{}_2\text{Hg} + \text{Ar}^2\text{I} \xrightarrow[\text{NaI}]{\text{cat. } p\text{-NO}_2\text{C}_6\text{H}_4\text{PdI(PPh}_3)_2} \text{Ar}^1-\text{Ar}^2 \quad (60)$$

$$\text{R}^1\text{HgCl} + \text{R}^2{}_2\text{CuLi} \longrightarrow \text{R}^1-\text{R}^2 \quad (61)$$

Aryl-, vinylic, and alkynyllead compounds prepared by transmetallation of the corresponding diorganomercury compounds by $Pb(OAc)_4$ can alkylate β-dicarbonyl anions and nitronates (Scheme 27).[85]

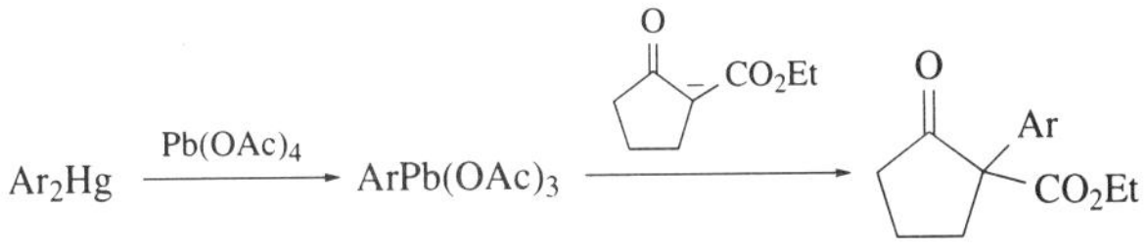

$$\text{Ar}_2\text{Hg} \xrightarrow{\text{Pb(OAc)}_4} \text{ArPb(OAc)}_3 \xrightarrow{\text{cyclopentanone-CO}_2\text{Et}} \text{Ar-substituted cyclopentanone-CO}_2\text{Et}$$

Scheme 27

Finally, solvomercuration products can be alkylated by conversion to the corresponding alkali metal reagents and subsequent electrophilic attack (Equation (62)).[86]

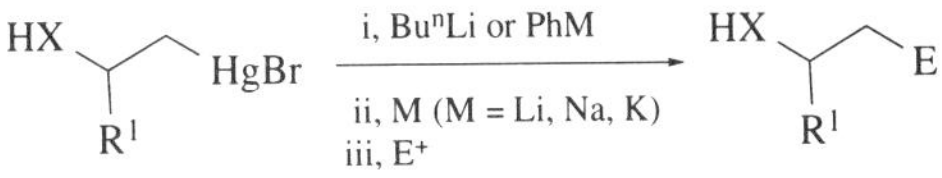

$$\underset{R^1}{HX\diagdown\diagup HgBr} \xrightarrow[\text{iii, } E^+]{\text{i, Bu}^n\text{Li or PhM} \atop \text{ii, M (M = Li, Na, K)}} \underset{R^1}{HX\diagdown\diagup E} \quad (62)$$

$$X = O, NR^2; E = D_2O, O_2, CO_2, R^3{}_2CO, \text{TMS-Cl}$$

9.8 ALKENE AND ALKYNE ADDITION AND SUBSTITUTION REACTIONS

Organomercurials will add to electron-deficient alkenes[87] and alkynes[87d,88] either thermally or photochemically to generate new organomercurials which can subsequently be reduced or halogenated (Equations (63) and (64)). Recently, the borohydride-induced alkylation of alkenes by alkylmercurials has become a process of major synthetic importance (Equation (65)).[10,89] The reaction can be run using a wide variety of solvents, alkylmercurials, and alkenes. Sodium borohydride is most commonly used as the reducing agent to produce the free radical, but $NaHB(OMe)_3$ has also occasionally been employed. A wide variety of 1°, 2°, and 3° alkylmercurials can be utilized, including those prepared by the solvomercuration of alkenes and cyclopropanes (Scheme 28). Since this process involves free radicals, best results are usually obtained using alkenes bearing electron-withdrawing groups. The advantage of this approach to free radical conjugate addition over the tin hydride reduction of organic halides is the fact that the reaction can be run at room temperature in the absence of light for very short reaction times. A number of significant applications of this process have appeared, but one simple example of intramolecular solvomercuration and free radical entrapment will suffice to illustrate the power and versatility of this methodology (Equation (66)).[90] Recently, silyl hydrides in DMSO have been employed in these conjugate addition reactions (Equation (67)).[91] This approach has the advantage that the alkene need not be used in excess as commonly done when using boron hydrides.

$$R^1{}_2Hg + R^2{}_2C=C(CN)_2 \longrightarrow R^1CR^2{}_2C(CN)_2HgR^1 \quad (63)$$

Nicely complementing these alkene addition reactions is the Heck reaction of organomercurials and alkenes, which produces alkene substitution products (Equation (68)).[10,92] This reaction can be useful for

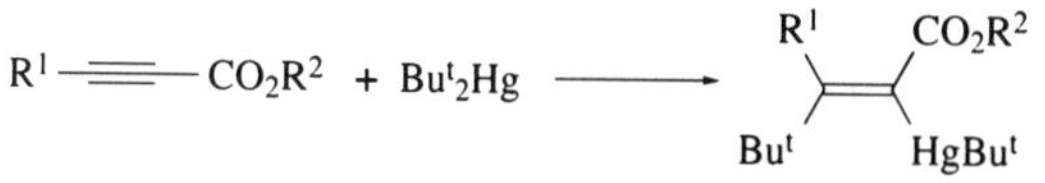

$$R^1-\!\!\!\equiv\!\!\!-CO_2R^2 + Bu^t_2Hg \longrightarrow \underset{Bu^t\quad HgBu^t}{\overset{R^1\quad CO_2R^2}{\diagup\!\!\!=\!\!\!\diagdown}} \quad (64)$$

$$RHgX + H_2C=CHY \xrightarrow[\text{NaHB(OMe)}_3]{\text{NaBH}_4 \text{ or}} RCH_2CH_2Y \quad (65)$$

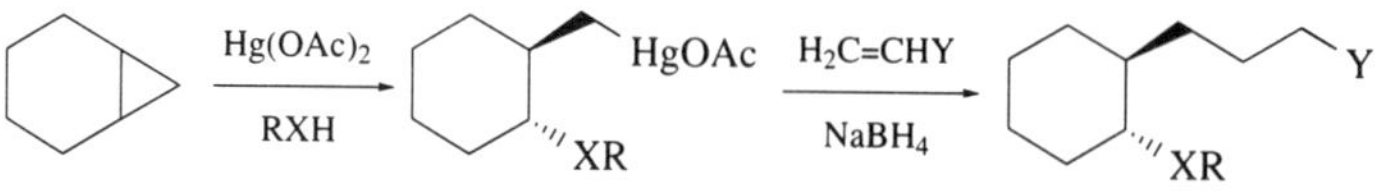

XR = OH, OMe, OAc, NHAc

Scheme 28

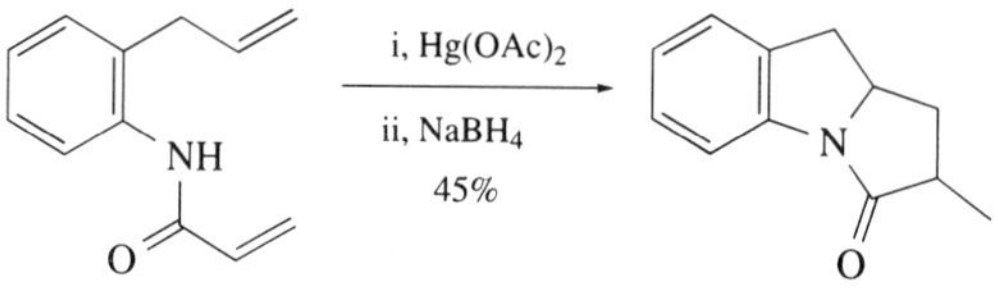

(66)

$$4\ Bu^nHgCl + H_2C=CHCO_2Et \xrightarrow[\text{DMSO}]{\text{Et}_3SiH} Bu^nCH_2CH_2CO_2Et \quad (67)$$
$$93\%$$

heterocyclic and arylmercurials or alkylmercurials which do not possess β-hydrogens, such as R = Me, CH$_2$Ar, and CO$_2$Me. Best results are obtained using less substituted alkenes which do not contain allylic hydrogens. The process proceeds by transmetallation to form an organopalladium intermediate which adds *syn* across the carbon–carbon double bond of the alkene. The resulting alkylpalladium intermediate then undergoes a *syn* β-hydride palladium elimination to produce the substituted alkene and palladium(0). This process can be made catalytic in palladium if reoxidants such as CuCl$_2$ are added, but the yields are usually low. This overall process is now more commonly effected using aryl or vinylic halides, and catalytic amounts of palladium.[92]

$$RHgCl + H_2C=CHX \xrightarrow{Li_2PdCl_4 \text{ or } Pd(OAc)_2} RCH=CHX \quad (68)$$

X = CHO, COR, CO$_2$R, CN, Ar, etc.

Mercurated esters bearing acyclic alkenes react with Li$_2$PdCl$_4$ to afford butenolides (Equation (69)).[93]

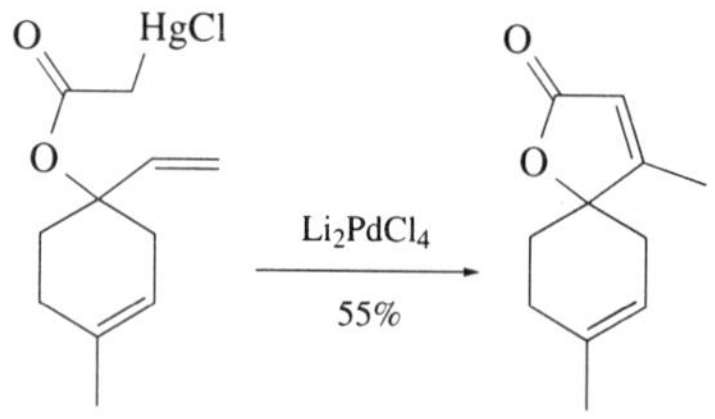

(69)

Andirolactone

When cyclic alkenes are employed in the Heck reaction, the allylic product is produced, but this product is oftentimes contaminated with double bond regioisomers (Equations (70) and (71)).[94] With styrenes, oxyarylation can occur. This process has been employed in the synthesis of aryl alcohols,[95] pterocarpans,[96] and lactones[97] (Equations (72)–(74)). Certain functional groups can have a profound effect on the nature of the product of the organomercurial Heck reaction. In the presence of strong HCl, conjugate addition to enones is observed (Equation (75)).[98] Conjugate addition using *ortho*-mercurated phenols affords 2-chromanols and 2-chromenes, while mercurated anilines generate intermediates useful in quinoline synthesis (Equations (76) and (77)).[99]

Quinones react with arylmercuric chlorides and LiPdCl$_3$ to give substituted quinones (Equation (78)).[100]

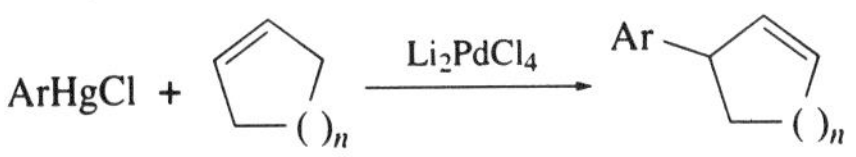

The arylation of allylic alcohols affords saturated carbonyl products by palladium hydride elimination to an enol (Equation (79)).[101] This reaction has been used in an efficient approach to prostaglandins (Equation (80)).[102] With good allylic leaving groups, arylation and vinylation afford elimination products. Thus, allylic halides produce simple allylic arenes or 1,4-dienes,[103] and unsaturated epoxides, oxetanes, and β-lactams produce unsaturated alcohols and amides[104] (Equations (81) and (82)).

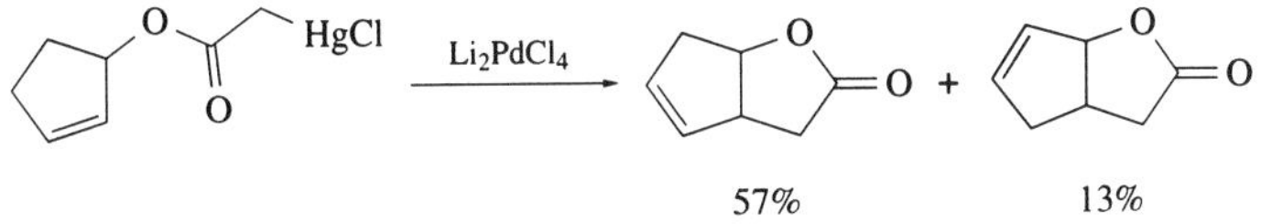

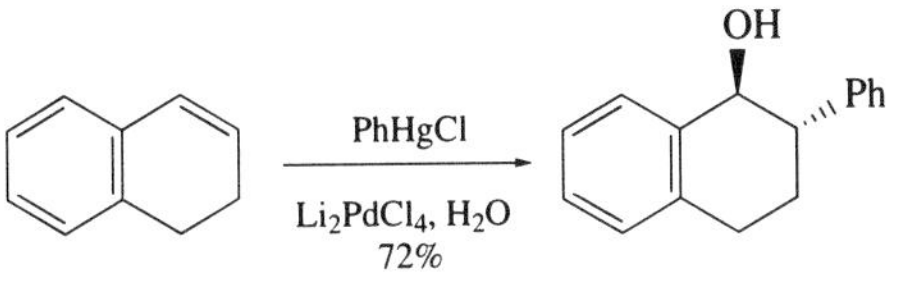

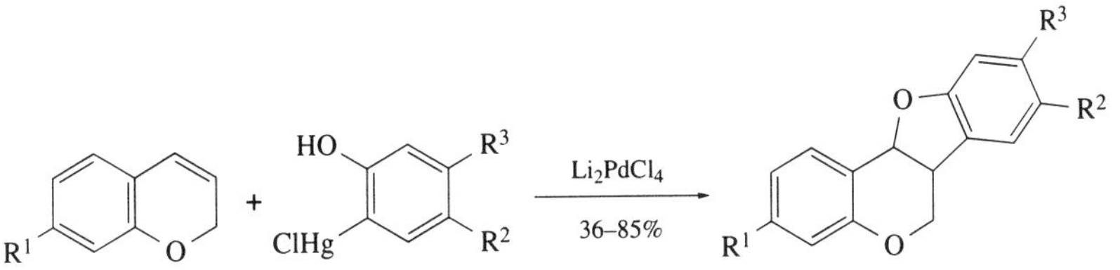

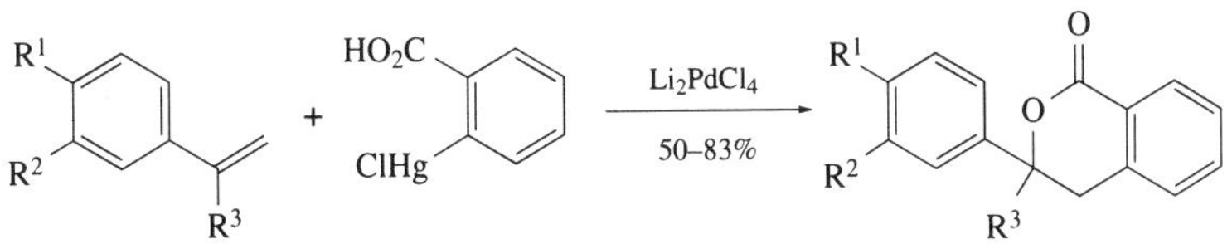

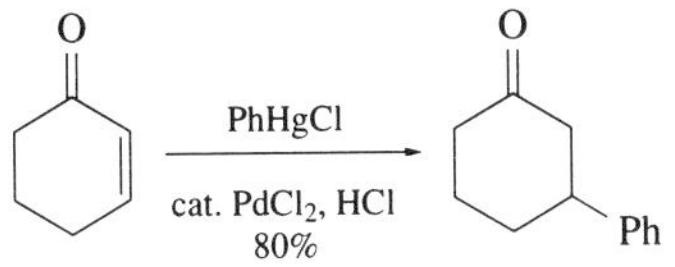

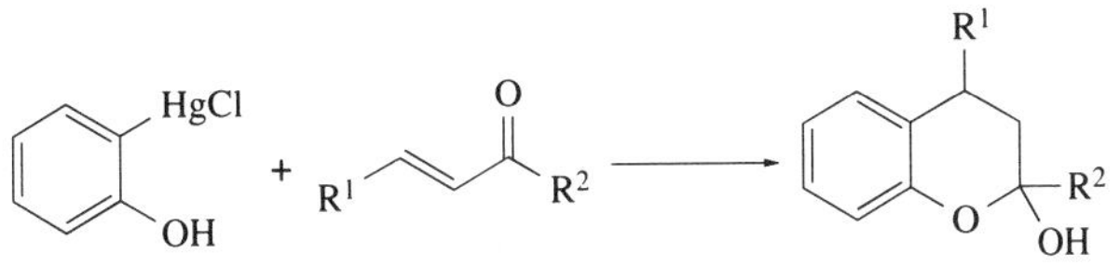

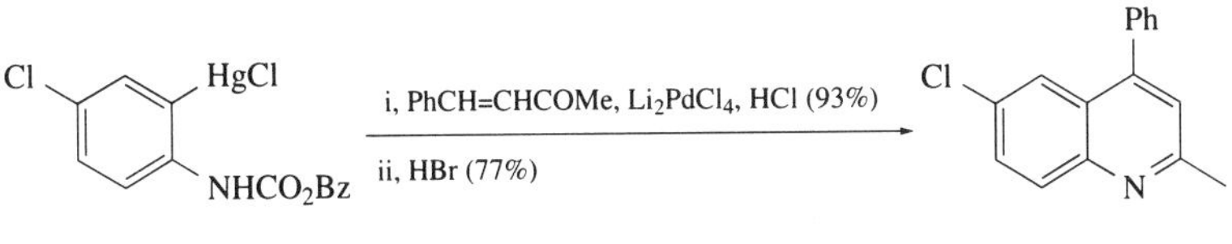

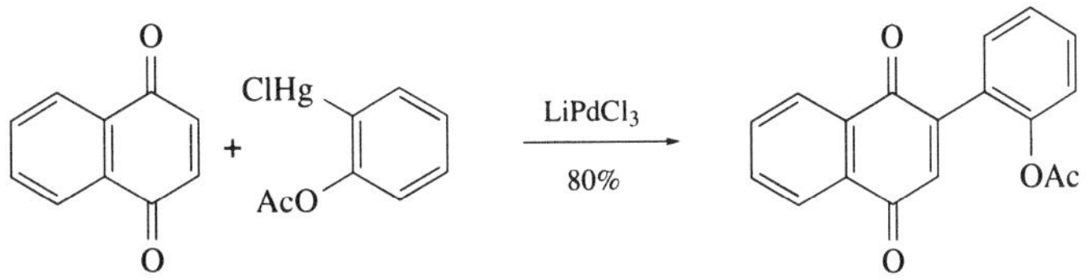

When the Heck arylation is run in the presence of excess cupric chloride, low to good yields of alkyl chlorides are produced (Equation (83)).[105]

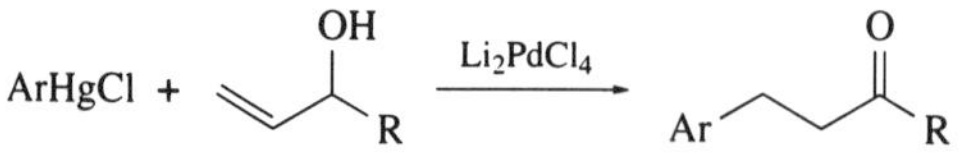

$$ArHgCl + \underset{R}{\overset{OH}{\overset{|}{CH_2=CH}}} \xrightarrow{Li_2PdCl_4} Ar\underset{}{CH_2CH_2}\overset{O}{\overset{\|}{C}}R \quad (79)$$

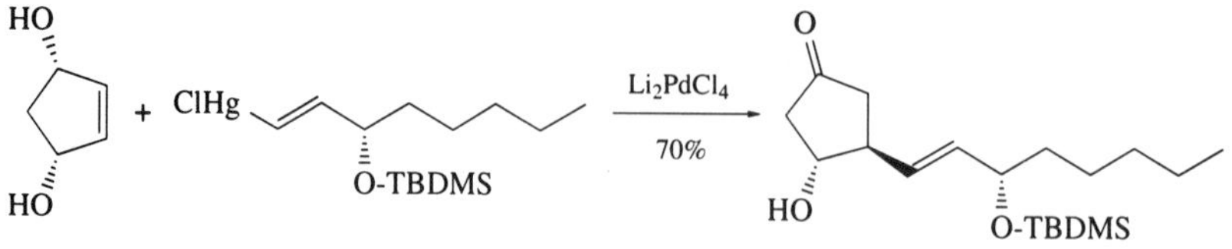

(80)

$$R^1HgCl + H_2C=CHCHR^2 \xrightarrow{Li_2PdCl_4} R^1CH_2CH=CHR^2 \quad (81)$$

$$\overset{Cl}{\underset{}{|}}$$
(position above CHCHR²)

$$RHgCl + \text{(vinyl epoxide X)} \xrightarrow[H_2O]{Li_2PdCl_4} RCH_2CH=CHCH_2XH \quad (82)$$

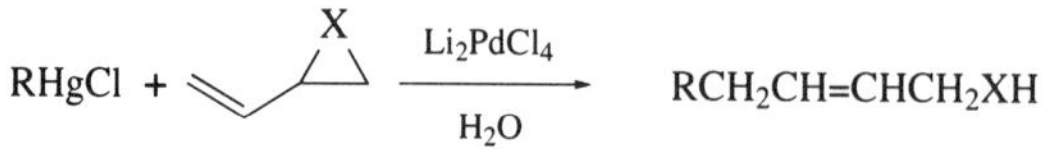

R = aryl, vinylic; X = O, CH$_2$O, CONH

$$PhHgCl + H_2C=CHR \xrightarrow[CuCl_2]{cat.\ Li_2PdCl_4} PhCH_2\underset{}{\overset{Cl}{\overset{|}{C}}}HR \quad (83)$$

R = H (76%), Me (50%), CHO (63%), Ac (~80%)

The addition of aryl- or vinylic palladium compounds to bicyclic alkenes affords stable addition compounds which have been further functionalized in a variety of ways,[106] and even carried on to prostaglandin analogues[107] (Schemes 29 and 30).

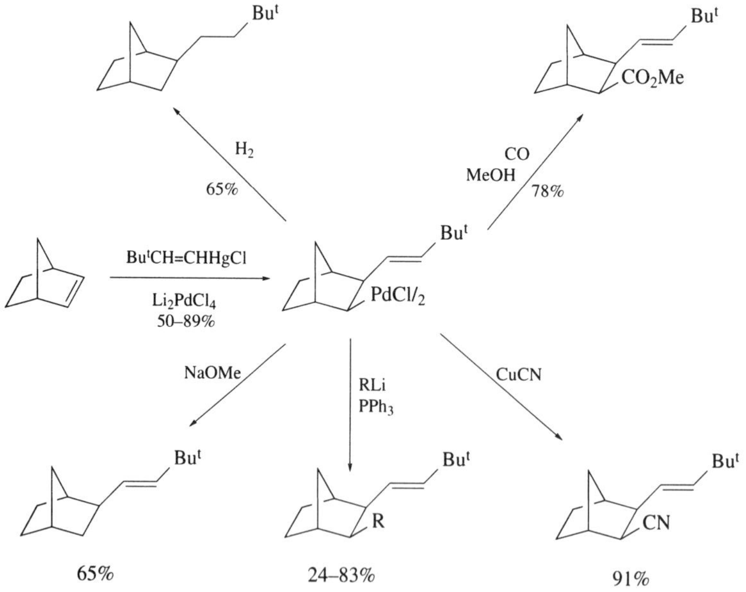

Scheme 29

The addition of vinylic palladium compounds to simple alkenes affords stable π-allylpalladium compounds by a palladium hydride rearrangement (Equation (84)).[108] When alkenoic acids are employed and a base is then added, unsaturated lactones are produced in good yield (Equation (85)).[109]

The reaction of cyclic alkenes and vinylmercurials produces either π-allylpalladium compounds or 1,4-dienes depending on how the reaction is run (Scheme 31).[110] π-Allylpalladium compounds are also formed in the reaction of organomercurials and unsaturated cyclopropanes and cyclobutanes (Equation (86)),[111] and in the addition of arylpalladium compounds to conjugated and nonconjugated dienes

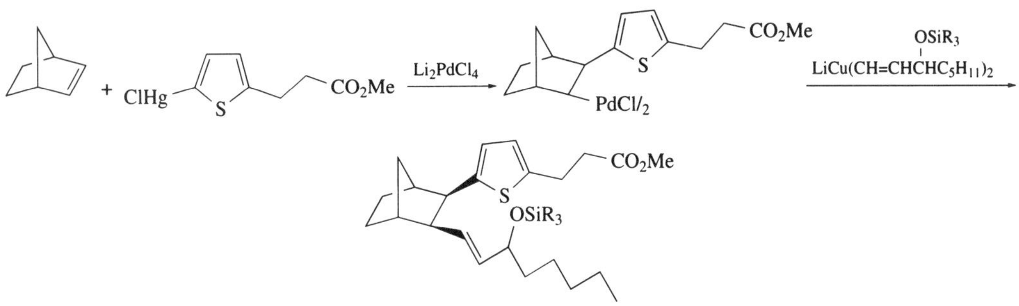

Scheme 30

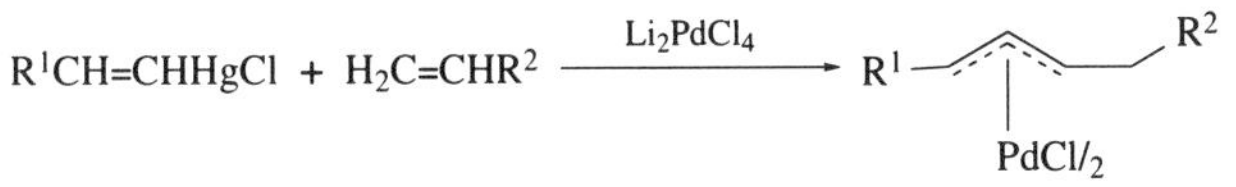

$$R^1CH=CHHgCl + H_2C=CHR^2 \xrightarrow{Li_2PdCl_4} R^1 \diagup\!\!\!\diagdown R^2 \quad PdCl/2 \tag{84}$$

$$RCH=CHHgCl + H_2C=CH(CH_2)_nCO_2H \xrightarrow[ii, K_2CO_3]{i, Li_2PdCl_4} R\text{—CH=CH—}\underset{n=1,2}{\text{(butenolide)}} \tag{85}$$

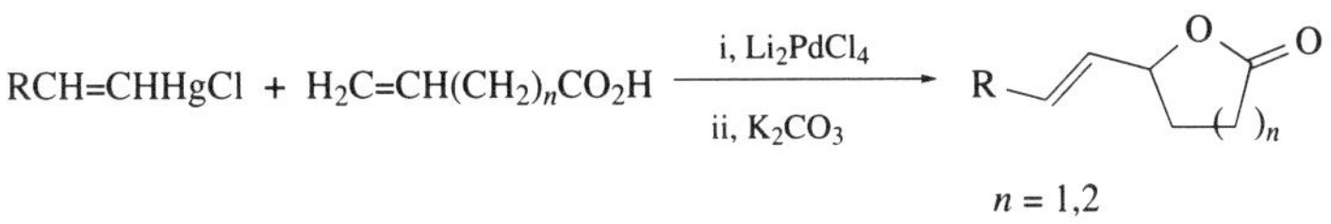

(Equation (87)).[112] By combining these approaches to π-allylpalladium compounds with intramolecular displacement of palladium, a useful annulation process ensues (Scheme 32).[113]

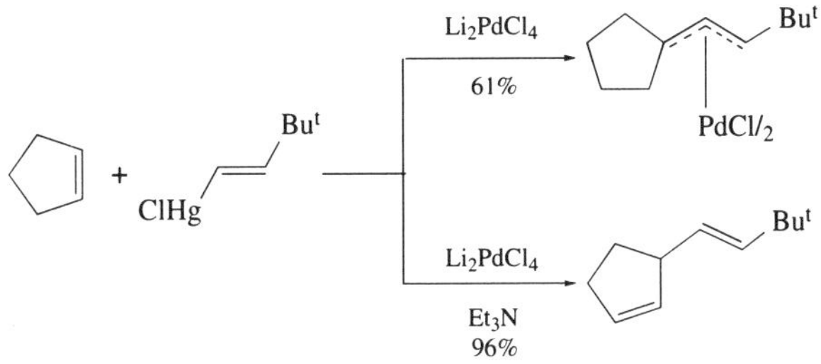

Scheme 31

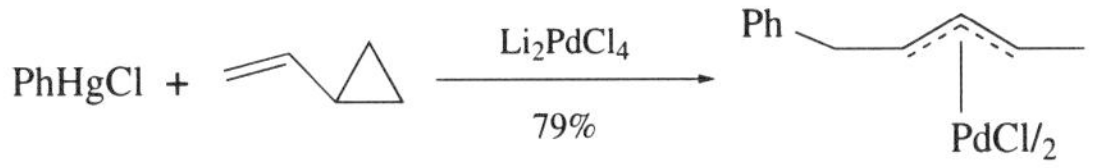

$$PhHgCl + \text{(vinylcyclopropane)} \xrightarrow[79\%]{Li_2PdCl_4} Ph\diagup\!\!\!\diagdown \quad PdCl/2 \tag{86}$$

$$ArHgCl + H_2C=CH(CH_2)_nCH=CH_2 \xrightarrow{Li_2PdCl_4} Ar(CH_2)_{n+1}\diagup\!\!\!\diagdown \quad PdCl/2 \tag{87}$$

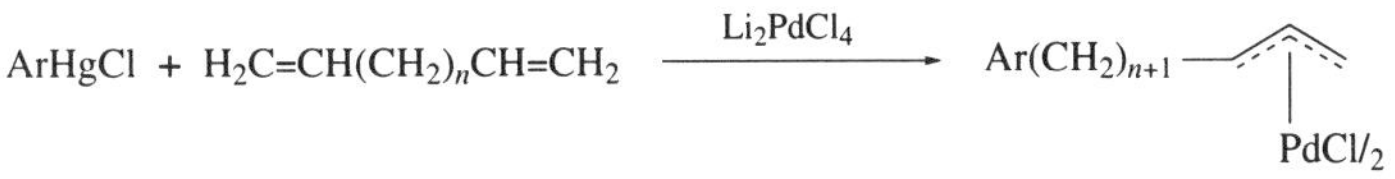

9.9 CARBONYLATION

The carbonylation of organomercurials provides a useful synthesis of a variety of carbonyl products.[114] Organomercurials do not react readily directly with carbon monoxide, but in the presence of transition metals, particularly palladium and rhodium, facile carbonylation takes place. Alkyl- and arylmercurials undergo carbonylation to esters in the presence of either a rhodium or palladium catalyst or stoichiometric amounts of Li_2PdCl_4 (Equation (88)).[115] The palladium-catalyzed carbonylation of vinylic mercurials provides a convenient synthesis of α,β-unsaturated carboxylic acids and esters (Equation (89)).[116] This provides a useful approach to butenolides from alkynols (Equation (90)).[116,117] Ketones can also be prepared via carbonylation of organomercurials. Alkyl- and arylmercurials undergo carbonylation to symmetrical ketones in the presence of $Ni(CO)_4$[118] or $Co_2(CO)_8$,[119] while catalytic $[ClRh(CO)_2]_2$[120] is particularly useful for the synthesis of diaryl and divinyl ketones (Equation (91)).

Unsymmetrical ketones can also be prepared using transition metal reagents. Aryl iodides and arylmercurials will cross-couple to unsymmetrical ketones in the presence of $Ni(CO)_4$ (Equation (92)).[118,121] Alkylmercurials and aryl iodides also undergo palladium-catalyzed carbonylation to unsymmetrical ketones (Equation (93)).[122]

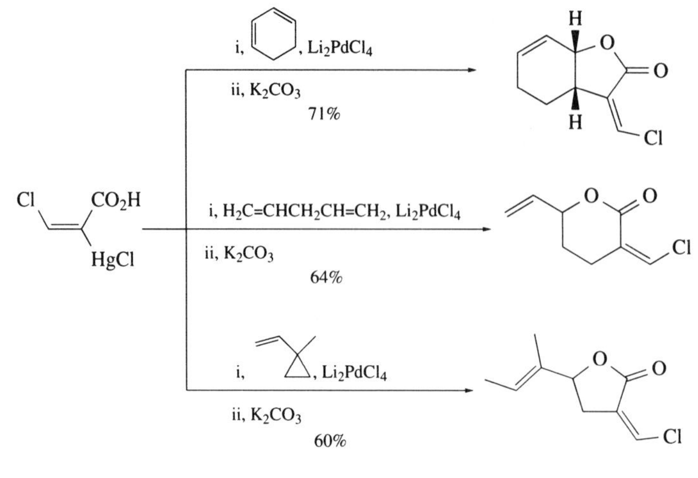

Scheme 32

$$R^1HgX \xrightarrow[\substack{Li_2PdCl_4 \text{ or} \\ \text{cat. } PdCl_2(PPh_3)_2 \text{ or} \\ \text{cat. } RhCl(PPh_3)_3}]{CO,\ R^2OH} R^1CO_2R^2 \quad (88)$$

$$R^1CH=CHHgCl \xrightarrow[\text{cat. } Li_2PdCl_4,\ CuCl_2]{CO,\ R^2OH} R^1CH=CHCOR^2 \quad (89)$$

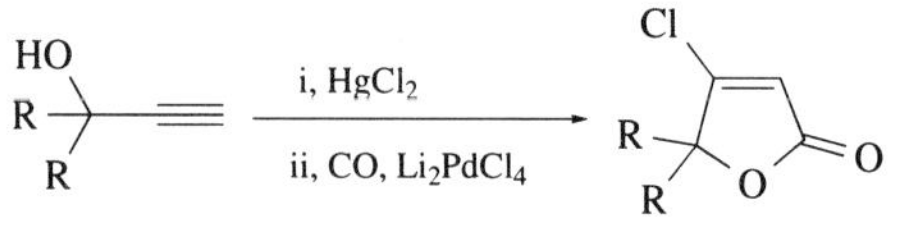

(90)

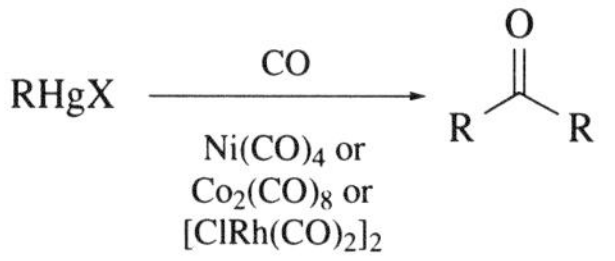

$$RHgX \xrightarrow[\substack{Ni(CO)_4 \text{ or} \\ Co_2(CO)_8 \text{ or} \\ [ClRh(CO)_2]_2}]{CO} R\text{-CO-}R \quad (91)$$

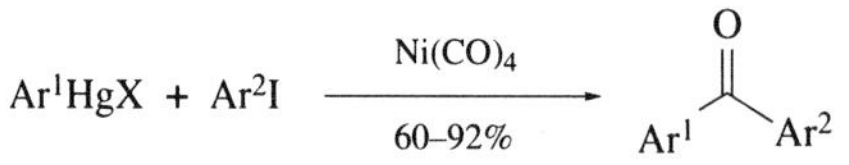

$$Ar^1HgX + Ar^2I \xrightarrow[60-92\%]{Ni(CO)_4} Ar^1\text{-CO-}Ar^2 \quad (92)$$

$$MeHgI + I\text{-C}_6H_4\text{-}NO_2 \xrightarrow[\substack{\text{cat. } PdCl_2(MeCN)_2,\ Bu^n_4NI \\ 92\%}]{CO} Me\text{-CO-C}_6H_4\text{-}NO_2 \quad (93)$$

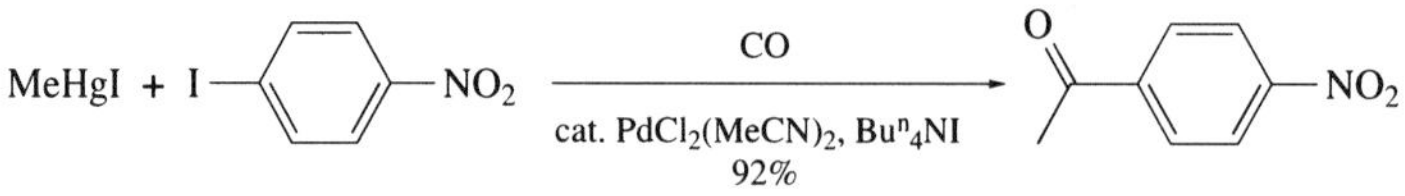

Arylmercurials react with carbon monoxide and alkenes in the presence of $PdCl_2(PPh_3)_2$ to afford enones (Equation (94)).[123]

$$ArHgCl + CO + H_2C=CHR \xrightarrow[CuCl_2,\ Bu^n_3N]{PdCl_2(PPh_3)_2} ArCOCH=CHR \quad (94)$$

9.10 ACYLATION

The acylation of organomercurials is usually effected in the presence of either a Lewis acid or a palladium catalyst to provide a wide variety of ketones. While dialkynylmercurials react directly with acyl halides,[124] aluminum chloride or bromide is required for the acylation of alkyl,[125] vinylic,[126] allylic,[14] allenic,[127] and propargylic[127] mercurials (Equation (95)). Allylic, allenic, and propargylic mercurials react by S_E2' substitution at the remote end of the unsaturated system to provide allylic, propargylic, and allenic ketones, respectively.

$$R^1HgCl + \underset{R^2}{\underset{\|}{\overset{O}{C}}}Cl \xrightarrow{AlCl_3} \underset{R^1}{\underset{\|}{\overset{O}{C}}}R^2 \qquad (95)$$

Unsymmetrical ketones have also been prepared by the palladium-catalyzed acylation of dialkyl- and arylmercurials (Equation (96)).[122,128]

$$R^1{}_2Hg + \underset{Cl}{\underset{\|}{\overset{O}{C}}}R^2 \xrightarrow{cat.\ Pd} \underset{R^1}{\underset{\|}{\overset{O}{C}}}R^2 \qquad (96)$$

α-Mercurated carbonyl compounds undergo facile acylation on oxygen to produce simple enol esters, enol carbonates, divinyl carbonates, enol chlorocarbonates, and ketene acetals (Equation (97)).[10]

$$\underset{R^1}{\underset{R^1}{\overset{XHg}{\diagdown}}}\overset{O}{\underset{\|}{C}}R^2 + \underset{Cl}{\underset{\|}{\overset{O}{C}}}R^3 \longrightarrow \underset{R^1}{\overset{R^1}{\diagdown}}C=C\underset{R^2}{\overset{O_2CR^3}{\diagup}} \qquad (97)$$

R^2 = H, alkyl, OR^1; R^3 = Cl, alkyl, OR^1

9.11 DIVALENT CARBON TRANSFER REACTIONS

Organomercurials have proven to be useful as divalent carbon transfer reagents, particularly for the conversion of alkenes to cyclopropanes.[3,10,129] The most useful of such reagents have generally been prepared by the reaction of organomercuric halides with polyhaloalkanes and a base, such as potassium *t*-butoxide (Equation (10)). Most important of these reagents are the phenyl trihalomethylmercury reagents (RHgCXYZ, where R = Ph and X, Y, Z = halogen), but a large number of other related compounds, where R = benzyl, cyclohexyl and other aryl groups, and CXYZ contains H, alkyl, CF_3, CO_2R, and SiR_3 have also been prepared. These reagents can also be prepared using preformed organomercuric alkoxides or amides and the appropriate hydrocarbon, or by reacting mercuric salts with appropriate organometallics of lithium, zinc, and chromium (Equations (98) and (99)). Some useful reagents of this type can also be prepared by (i) the direct reaction of mercury with the appropriate organic halide (see Scheme 1), (ii) the decarboxylation of sodium carboxylates in the presence of an organomercuric halide (see Equation (24)), or (iii) chemical modification of RHgCXYZ itself.

$$R^1HgX + HCXYZ \longrightarrow R^1HgCXYZ \qquad (98)$$

$$X = OR^2, NR^2{}_2$$

$$RHgX + MCXYZ \longrightarrow RHgCXYZ \qquad (99)$$

$$M = Li, Zn, Cr$$

The most synthetically important reaction of these divalent carbon transfer reagents is their reaction with alkenes to form cyclopropanes (Equation (100)). This reaction has been widely used to prepare simple cyclopropanes (Y, Z = H) and virtually all possible combinations of halocyclopropanes (Y = H and halogen, Z = halogen). This methodology is attractive because the reaction proceeds under relatively mild (25–130 °C), neutral conditions and gives good yields for a wide variety of alkenes. The process is particularly advantageous for normally unreactive alkenes, such as ethylene, sterically hindered and highly substituted alkenes, silyl alkenes, styrenes and functionally-substituted alkenes, such as haloalkenes, vinylic ethers and esters, enol silanes, α,β-unsaturated esters and nitriles, and alkenols. Many of these types of alkenes are either unreactive, or decompose or polymerize under the basic conditions usually used to convert alkenes to halocyclopropanes using halocarbons and base. Allenes and conjugated and nonconjugated dienes and polyenes generally undergo clean mono-, di-, or polycyclopropanation.

This cyclopropanation process is observed to proceed with complete retention of the alkene stereochemistry. When unsymmetrical CYZ groups are transferred, *syn/anti* mixtures are usually observed. Addition tends to occur from the less hindered face of the alkene.

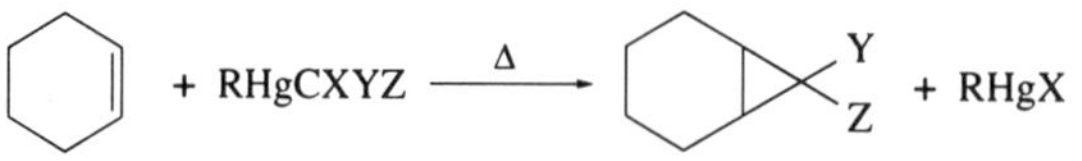

$$\text{cyclohexene} + \text{RHgCXYZ} \xrightarrow{\Delta} \text{bicyclic product with } Y, Z + \text{RHgX} \quad (100)$$

The reactivity of various organomercurials has been examined. Cyclohexylmercurials are more reactive than their phenylmercurial counterparts, although the latter are more easily prepared and more often used. The elimination of X with mercury is observed to follow the order X = I > Br > Cl > F. Some organomercurials are activated by reaction with Ph_2Hg or NaI.

These reactions appear to proceed by three different mechanisms. Most reactions appear to involve the formation of a carbene, but direct transfer of the divalent carbon species to the alkene has also been observed, as has the formation of a halomethyl anion which decomposes to a carbene which then undergoes cyclopropanation.

Besides the cyclopropanation of alkenes, organomercurials have also proven useful for the formation of cyclopropenones via divalent carbon transfer to alkynes (Scheme 33).[130]

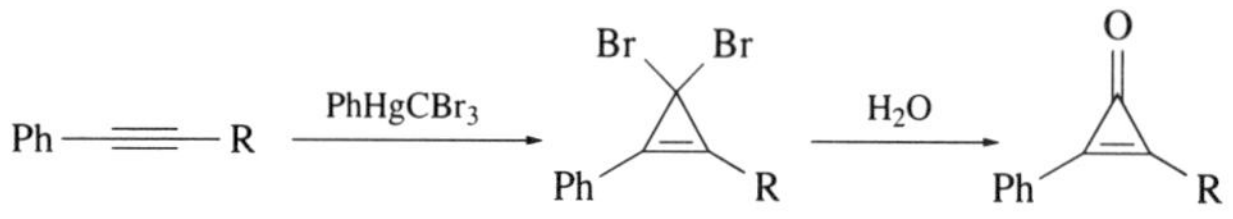

$$Ph-\!\!\!\equiv\!\!\!-R \xrightarrow{PhHgCBr_3} \text{Ph, R cyclopropane with Br, Br} \xrightarrow{H_2O} \text{Ph, R cyclopropenone}$$

Scheme 33

The cyclopropanes formed by organomercurial divalent carbon transfer processes are also useful for a variety of ring expansion processes. In some cases the organomercurial reaction with cycloalkenes leads directly to rearranged ring-expanded products (Equation (101)).[131]

This process has also been used to ring-expand indenes,[132] a variety of heterocycles,[133] and naphthyl ethers[134] (Equations (102) and (103)). Ketones can be ring-expanded through cyclopropanation of the corresponding enol ester (Equation (104)).[135]

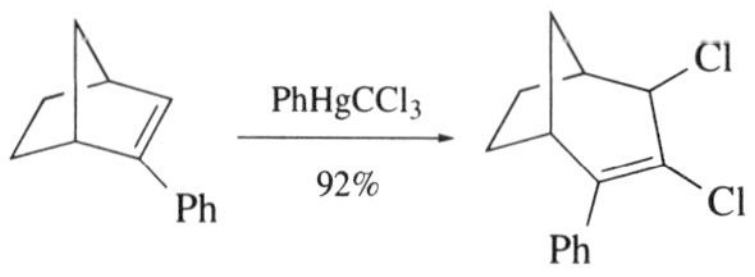

$$\text{norbornene-Ph} \xrightarrow[92\%]{PhHgCCl_3} \text{ring-expanded product with Cl, Cl, Ph} \quad (101)$$

$$\text{indene-}(CH_2)_n\text{-X} \xrightarrow{PhHgCY_3} \text{ring-expanded product with Y, }(CH_2)_n, X \quad (102)$$

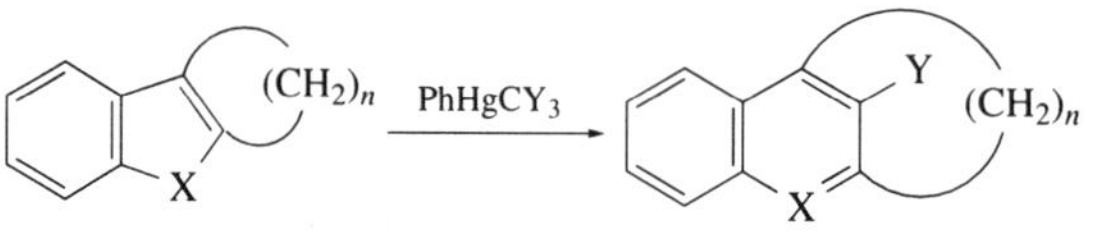

X = CH_2, NH, NAc; Y = Cl, Br; n = 6, 8, 10

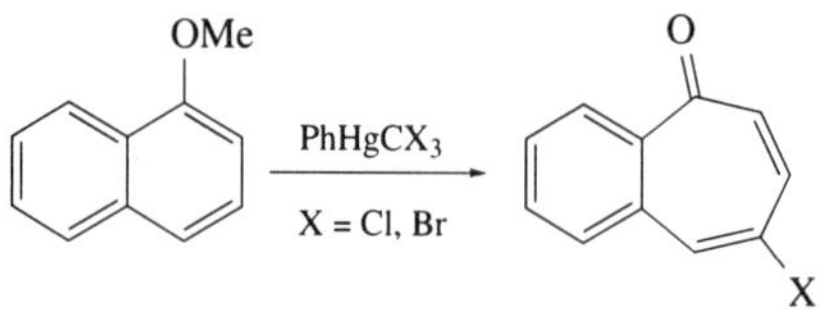

$$\text{1-methoxynaphthalene} \xrightarrow[X = Cl, Br]{PhHgCX_3} \text{benzocycloheptenone with X} \quad (103)$$

$$\text{AcO-cholesterol derivative} \xrightarrow[\text{ii, KOH}]{\text{i, }PhHgCCl_2Br} \text{ring-expanded ketone with Cl} \quad (104)$$

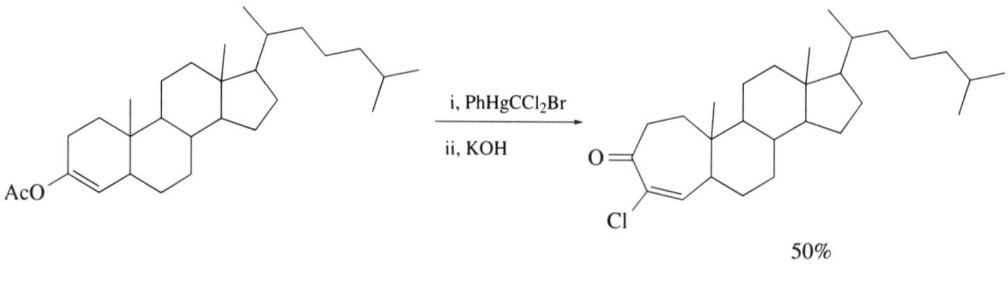

50%

The phenyltrihalomethylmercurials can also be useful for insertion reactions. Thus, carboxylic acids are readily converted to their dichloromethyl esters by these reagents, even when a carbon–carbon double bond is present in the acid (Equation (105)).[136] Insertion into a C–H bond of simple alkanes, arylalkanes, ethers, and alkylsilanes can be synthetically useful (Equation (106)).[137] The Si–H bond is even more reactive than the C–H bond and many examples of this insertion process are known.[129]

$$RCO_2H \xrightarrow{PhHgCCl_2Br} RCO_2CCl_2H \qquad (105)$$

$$R-H \xrightarrow{PhHgCCl_2Br} RCCl_2H \qquad (106)$$

9.12 SOLVOMERCURATION/DEMERCURATION REACTIONS

9.12.1 Hydroxymercuration

The reaction of an alkene or alkyne with a mercuric salt and a nucleophile, which often is the solvent, is known as solvomercuration. This process when combined with *in situ* demercuration provides a very valuable synthetic method for the Markovnikov addition of a hydrogen and a nucleophile to a carbon–carbon multiple bond. This sequence is one of the more important applications of organometallics in all of organic synthesis.

The addition of an hydroxyl group and mercury across a carbon–carbon double bond, otherwise known as oxymercuration, or preferably hydroxymercuration, was first reported in 1892 by Kucherov (Equation (107)). Although numerous examples of this reaction were subsequently reported, it was not until 1967 when Brown[38] closely examined the reaction conditions and reported a convenient procedure for hydroxymercuration and *in situ* alkaline $NaBH_4$ reduction that this process became widely used by synthetic organic chemists. This reaction is best conducted by adding the alkene to an equal amount of mercuric acetate suspended in 1:1 H_2O/THF. After stirring the reaction for 10–15 min at room temperature, an alkaline $NaBH_4$ solution is added which instantaneously reduces off the mercury moiety in a free radical chain process (see Section 9.3.2). This valuable process has been briefly reviewed several times, but most completely and recently in 1986.[11]

$$RCH=CH_2 \xrightarrow[H_2O-THF]{Hg(OAc)_2} \underset{\underset{|}{OH}}{RCHCH_2HgOAc} \qquad (107)$$

Hydroxymercuration/demercuration is a very versatile process. Although a wide variety of mercury salts, including mercuric trifluoroacetate, mercuric nitrate, mercuric sulfate, mercuric perchlorate, and mercuric methanesulfonate, have been employed in this process, none affords yields as high as mercuric acetate with a range of alkenes.[138]

The regio- and stereochemistry of the hydroxymercuration of a wide range of acyclic, monocyclic, bicyclic, and polycyclic alkenes has been studied. Alkenes of virtually all substitution patterns undergo hydroxymercuration in a highly Markovnikov manner. With similar substitution, the regiochemistry is controlled by steric effects (Equation (108)).[38] The stereochemistry of addition is usually *trans* diaxial, except with certain strained cyclic alkenes, such as norbornene, where *cis*, *exo* adducts are observed (Equations (109) and (110)).[139] Attempts to prepare optically active alcohols by using chiral mercuric carboxylates have afforded generally less than 30% asymmetric induction.

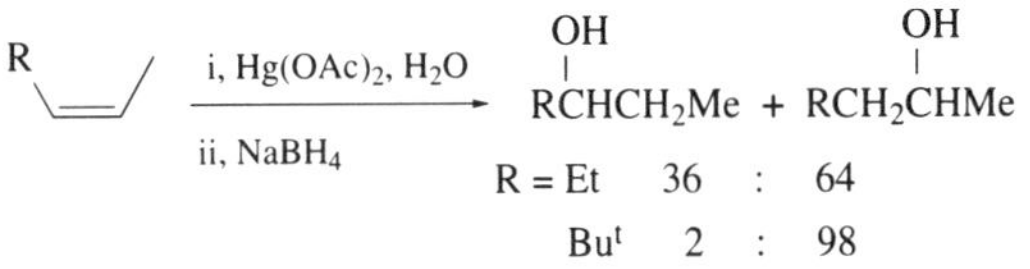

$$R\diagdown\!\!=\diagup \xrightarrow[\text{ii, NaBH}_4]{\text{i, Hg(OAc)}_2,\ H_2O} \underset{|}{\overset{OH}{RCHCH_2Me}} + \underset{|}{\overset{OH}{RCH_2CHMe}} \qquad (108)$$

$$\begin{array}{ccc} R = Et & 36 & : & 64 \\ Bu^t & 2 & : & 98 \end{array}$$

$$\text{cyclohexene} \xrightarrow[\substack{H_2O \\ 100\%}]{Hg(OAc)_2} \text{trans-2-(acetoxymercurio)cyclohexanol} \qquad (109)$$

$$\text{norbornene} \xrightarrow[\substack{H_2O,\ THF \\ 96\%}]{Hg(OAc)_2} \text{exo-cis product} \qquad (110)$$

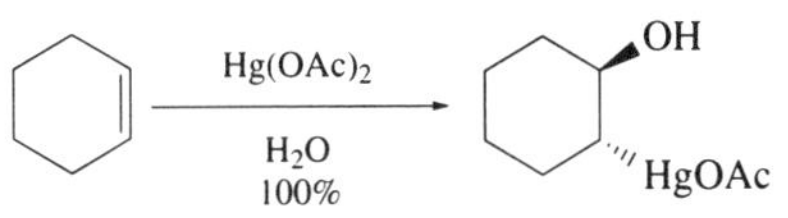

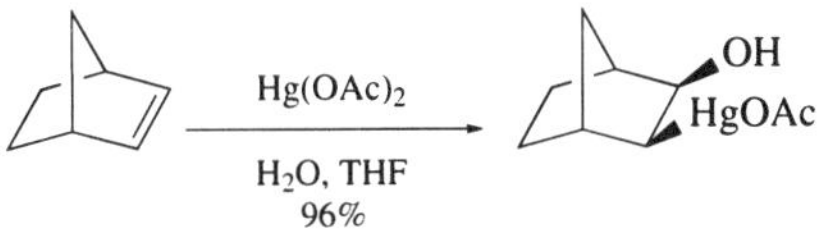

The relative rates of hydroxymercuration follow the general trend $R_2C=CH_2 > RCH=CH_2 > cis$-$RCH=CHR > trans$-$RCH=CHR$, and the process is observed to be first order in mercuric salt and alkene.[140] The kinetics have been used to argue for and against bridged mercurinium and unsymmetrical mercury-stabilized carbocations.[141] Unlike bromination, the hydroxymercuration of alkenes exhibits significant steric effects similar to those observed in hydroboration.[142] The rate-limiting step appears to be solvent attack on the mercury-coordinated alkene.

The presence of functional groups seldom interferes with hydroxymercuration, but can have a substantial effect on the regiochemistry[143] and diastereoselectivity[144] of addition (Equations (111) and (112)). Vinylic ethers and esters, or enol silanes undergo hydroxymercuration to afford α-mercurated carbonyl compounds (Equation (113)). This process has found utility in carbohydrate chemistry where rearrangement and carbon–carbon bond formation are observed (Equation (114)),[145] although carbon skeleton rearrangements are normally very rare.

$$\text{MeCH=CHCH}_2\text{OH} \longrightarrow \underset{94}{\text{MeCHCH}_2\text{CH}_2\text{OH}}\overset{\text{OH}}{|} + \underset{6}{\text{MeCH}_2\text{CHCH}_2\text{OH}}\overset{\text{OH}}{|} \quad (111)$$

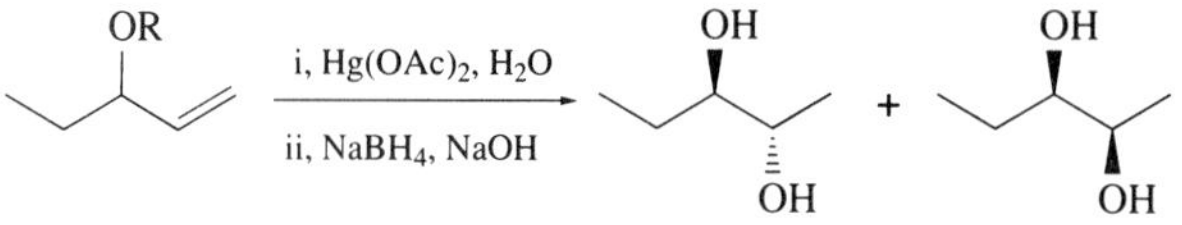

R		
H	76 :	24
MeCO	39 :	61
PhCO	23 :	77

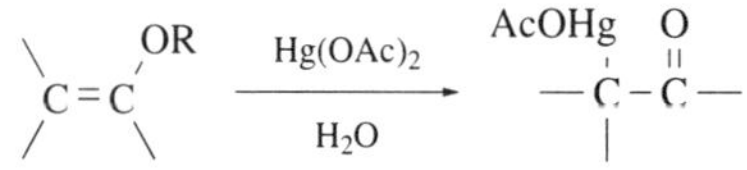

(112)

(113)

R = alkyl, silyl, acyl

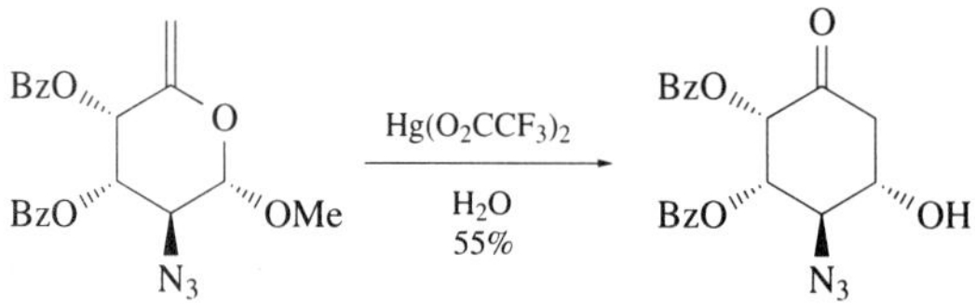

(114)

The β-hydroxyalkylmercurials formed by hydroxymercuration can be easily isolated, halogenated, transmetallated, reduced *in situ* to the corresponding alcohols, or reduced by borohydrides in the presence of alkenes to effect conjugate addition. Halogenation is quite facile (see Section 9.4). Reduction is most commonly effected by alkaline $NaBH_4$, but $Na(Hg)/H_2O$ has proven useful for stereospecific reductions (see Section 9.3.2). Conjugate addition is perhaps best effected in the presence of surfactants.[146] The transmetallation of hydroxymercurials by alkali metals provides a convenient method for electrophilic substitution of the mercury moiety (Equation (62)), or one can transmetallate by palladium to produce carbonyl products (Scheme 24).

The hydroxy- and alkoxymercuration of alkenes using HgO/HBF_4 produces vicinal diols and ethers (Equation (115)).[147]

$$\text{PhCH=CH}_2 \xrightarrow[\text{HBF}_4, \text{ROH}]{\text{HgO}} \text{PhCHCH}_2\text{OR}\overset{\text{OR}}{|} \quad (115)$$

Dienes and polyenes also undergo facile hydroxymercuration. 1,2-Dienes directly afford either ketones or allylic alcohols depending on their structures (Equations (116) and (117)). Conjugated dienes usually produce the 1,3-diols expected from applying what is known about the relative rates of alkene hydroxymercuration and the directing effects of a neighboring alcohol group (Scheme 34). The hydroxymercuration of nonconjugated dienes can afford alkenols when the isolated double bonds are remote and exhibit significantly different reactivities (Equation (118)). This selectivity is enhanced by using $Hg(O_2CCF_3)_2$,[148] or $Hg(BF_4)_2$,[149] or adding sodium lauryl sulfate,[150] or a combination of $Hg(O_2CCF_2CF_2CF_3)_2$ and sodium lauryl sulfate.[151] When the initially formed alkenol can cyclize to a five- or six-membered ring ether by intramolecular alkoxymercuration, these products are often observed in high yield (Scheme 35).

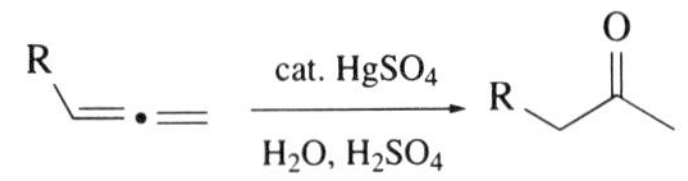

$$R\text{—}{\cdot}\text{=} \xrightarrow[H_2O, H_2SO_4]{\text{cat. } HgSO_4} R\text{—}C(=O)\text{—}CH_3 \quad (116)$$

$$\text{(cyclononadiene)} \xrightarrow[H_2SO_4, THF \\ 74\%]{HgSO_4, H_2O} \text{(cyclononenol)} \quad (117)$$

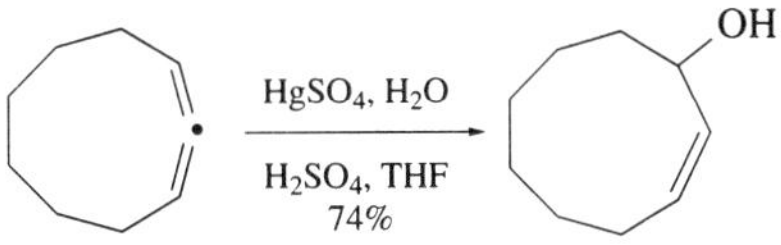

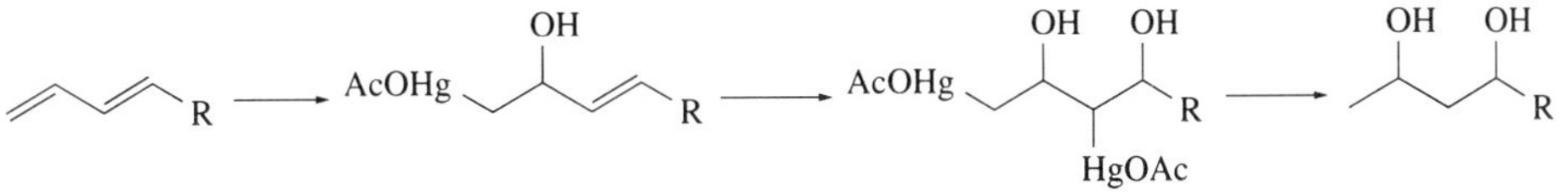

Scheme 34

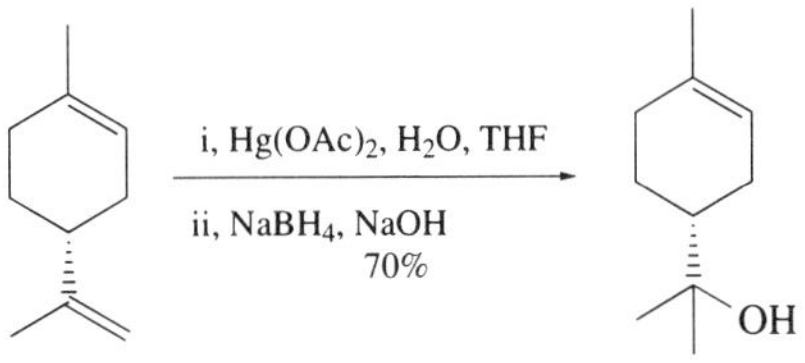

$$\text{(limonene)} \xrightarrow[\text{ii, NaBH}_4, \text{NaOH}]{\text{i, Hg(OAc)}_2, H_2O, THF} \text{(α-terpineol)} \quad (118)$$
$$70\%$$

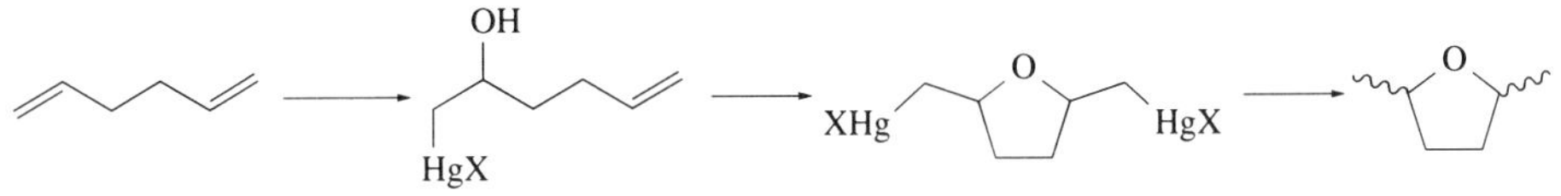

Scheme 35

The hydroxymercuration of alkynes is a very valuable method for the hydration of alkynes. Although many mercury salts have been employed in this process, the reaction is most commonly run using catalytic amounts of $HgSO_4$ in aqueous H_2SO_4.[152] The hydration of terminal alkynes generally affords exclusively methyl ketones (Equation (119)), while internal alkynes usually afford mixtures of regioisomers. However, certain alkynones undergo hydration regioselectively (Equation (120)).[153] A wide variety of functionality is accommodated by this process. When the mercury-catalyzed hydration is carried out in the presence of hydrogen peroxide and Na_2WO_4, Na_2MoO_4 or MoO_5(HMPA), α-ketoaldehydes or α-diketones are formed (Equation (121)).[154]

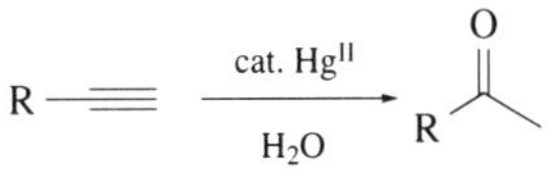

$$R\text{—}{\equiv}\text{—} \xrightarrow[H_2O]{\text{cat. Hg}^{II}} R\text{—}C(=O)\text{—}CH_3 \quad (119)$$

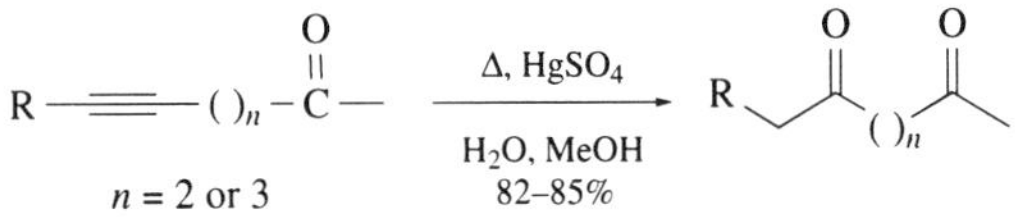

$$R\text{—}{\equiv}\text{—}()_n\text{—}C(=O)\text{—} \xrightarrow[H_2O, MeOH \\ 82\text{–}85\%]{\Delta, HgSO_4} R\text{—}C(=O)\text{—}CH_2\text{—}()_n\text{—}C(=O)\text{—} \quad (120)$$
$$n = 2 \text{ or } 3$$

$$R\text{—}{\equiv}\text{—} \xrightarrow[\text{cat. Hg(OAc)}_2 \\ MoO_5(HMPA)]{H_2O} R\text{—}C(=O)\text{—}CHO \quad (121)$$

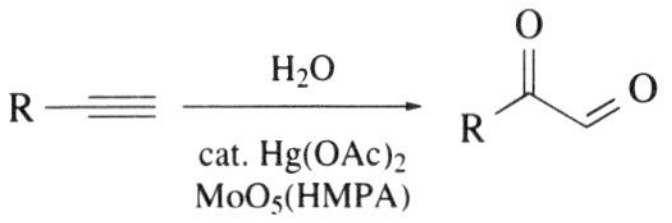

9.12.2 Alkoxymercuration[11]

When an alcohol is employed as the solvent in the solvomercuration process, β-alkoxyalkylmercurials are formed which again are easily reduced by alkaline $NaBH_4$ to the corresponding ethers (Scheme 36). This process has been extensively studied and has proven very versatile. It is most commonly effected using mercuric acetate, but more-sterically hindered alcohols require the use of the more electrophilic $Hg(O_2CCF_3)_2$,[155] and it is sometimes desirable to run the reaction at 0 °C. The alcohol is commonly used as the solvent, but limited amounts of alcohol can be employed if sodium dodecyl sulfate in water is added.[156] The process can also be run using only equivalent amounts of alkene and alcohol in the presence of $Hg(ClO_4)_2$ and collidine in acetonitrile as solvent (Equation (122)).[157]

$$R^1CH=CH_2 \xrightarrow[R^2OH]{Hg(OAc)_2} R^1\overset{OR^2}{\underset{|}{C}}HCH_2HgOAc \xrightarrow[NaOH]{NaBH_4} R^1\overset{OR^2}{\underset{|}{C}}HMe$$

Scheme 36

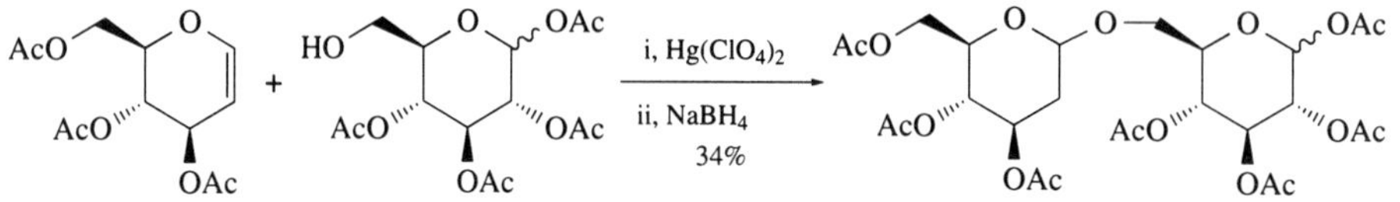

$$(122)$$

Alkoxymercuration is a very general process accommodating a wide range of alcohols and acyclic, cyclic, bicyclic, and polycyclic alkenes, and one which also tolerates considerable functionality. The stereo- and regioselectivity are analogous to those of hydroxymercuration and consistent with electrophilic addition (Equations (123) and (124)).[158,159]

$$MeCH=CHCO_2Et \xrightarrow[ii, Br_2]{i, Hg(OAc)_2, MeOH} Me\overset{MeO}{\underset{|}{C}}H\overset{Br}{\underset{|}{C}}HCO_2Et \quad (123)$$

90%

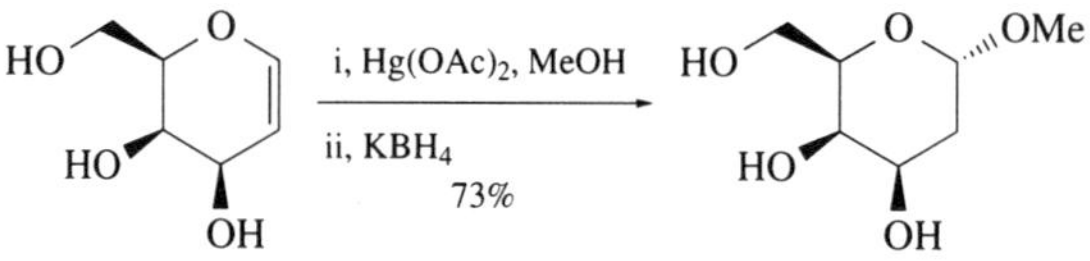

$$(124)$$

73%

Alkoxymercurials undergo a number of useful reactions. The most important is clearly reduction, which is commonly carried out using alkaline $NaBH_4$. Reduction has also been effected using KBH_4, $NaHB(OMe)_3$, Ca–MeOH, and Na(Hg). α-Mercurated esters are reduced by $NaBH_4$ with 70–86% inversion of stereochemistry, while 1,3-propanedithiol gave 80–95% retention (Equation (125)).[160] Sodium borohydride reduction in an O_2 atmosphere affords alcohols, while reduction in the presence of electron-deficient alkenes produces conjugate addition products.

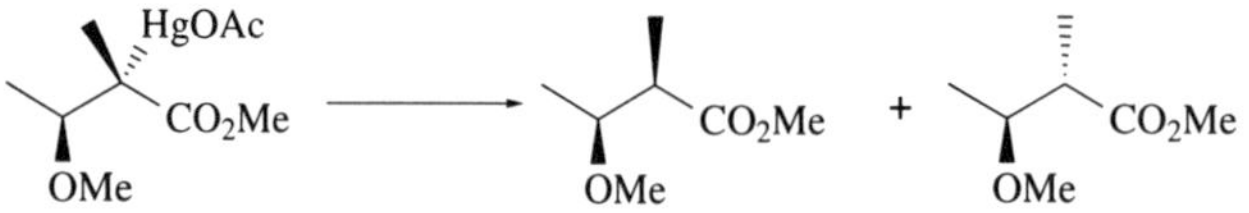

$$(125)$$

$NaBH_4$	20	: 80
$HS(CH_2)_3SH$	95	: 5

Alkoxymercuration of vinylic ethers and subsequent reduction using $NaBH_4$ or Na_2CS_3 provides a useful synthesis of mixed acetals (Equation (126)).[161] Bromination of these alkoxymercurials produces bromoacetals (Equation (127)).[162]

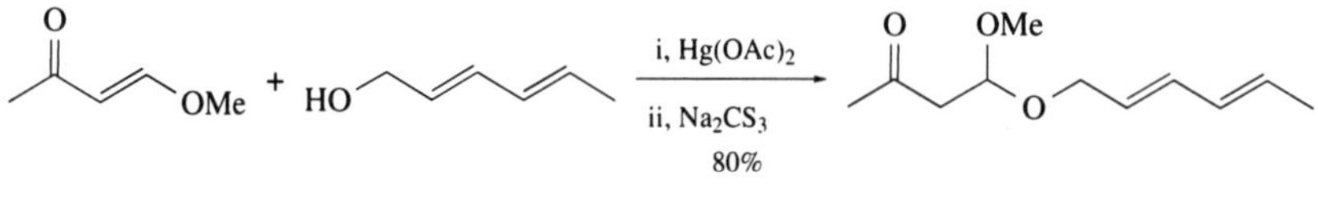

$$(126)$$

80%

$$H_2C=CHOR \xrightarrow[\text{ii, Br}_2]{\text{i, Hg(OAc)}_2, \text{ROH}} BrCH_2CH(OR)_2 \quad (127)$$

Numerous alkoxymercurials have been reported to undergo bromination or iodination. This has provided α-halo-β-alkoxy esters useful for the synthesis of amino acids (see Equation (123)). Fluorination by AcOF provides fluoro ethers (see Scheme 21).[51]

The intramolecular alkoxymercuration of alkenols and unsaturated phenols provides a general synthesis of five- and six-membered ring ethers (Equations (128) and (129)).[163,164] Many different mercuric salts have been utilized for this process and the nature of the anion can substantially effect the regio- and stereochemistry of the cyclization (Scheme 37).[165] The stereochemistry and substitution pattern of the C=C double bond in the alkenol also effects the regiochemistry of ring closure. With cyclic systems, steric factors and ring strain also play an important role. Occasionally the more strained of these cyclic alkoxymercurials revert to starting alkenols when quenched with aqueous NaCl or KBr. Treatment with CuCl allows these chloromercurials to be isolated (Equation (130)).[166]

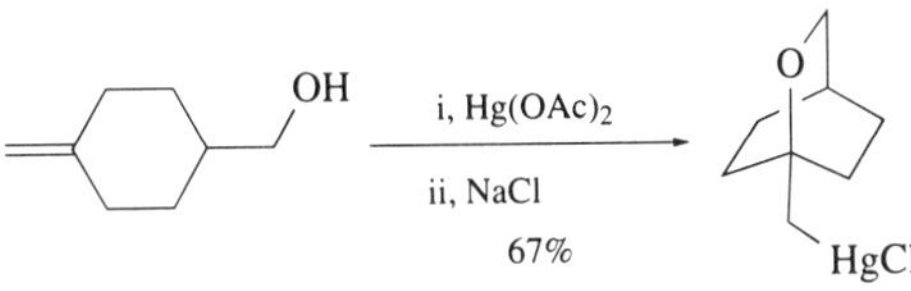

(128)

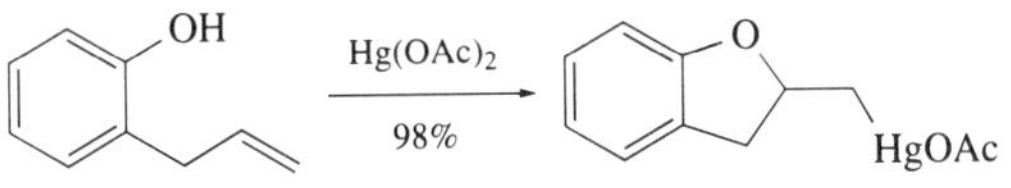

(129)

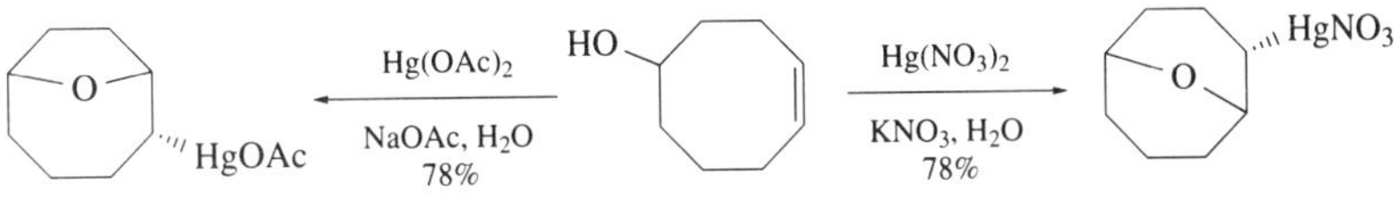

Scheme 37

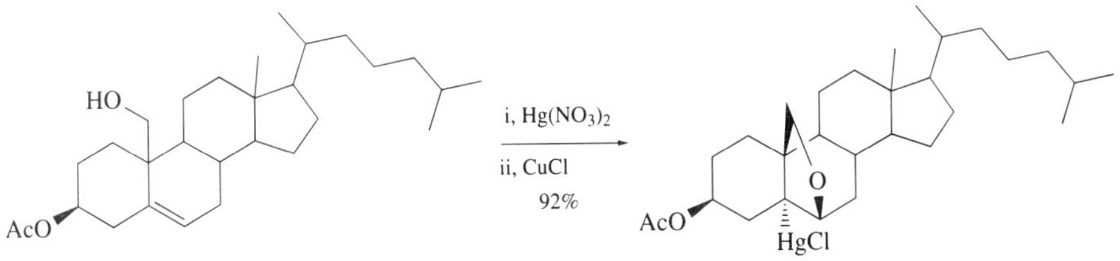

(130)

Intramolecular alkoxymercuration has been employed in the synthesis of cyclic acetals[167] and 1,2,4-trioxanes[168] (Equations (131) and (132)). Exogonic acid has recently been synthesized using an interesting variation of this process (Scheme 38).[169]

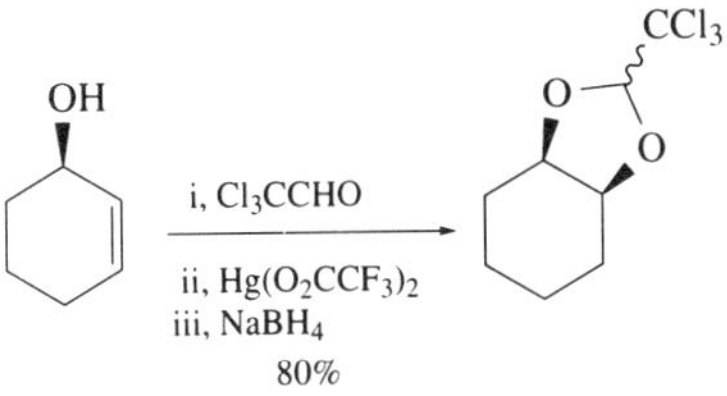

(131)

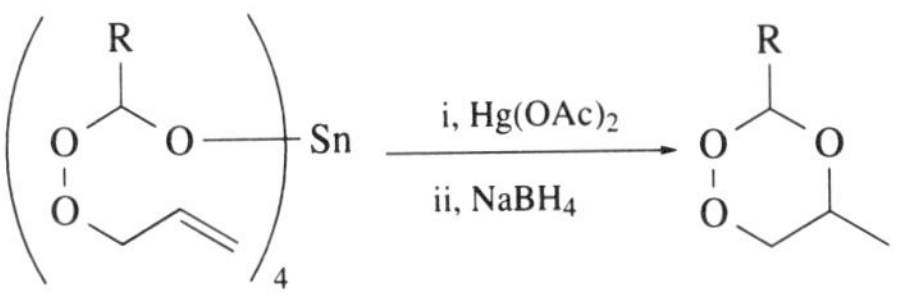

(132)

The cyclic alkoxymercurials derived from alkenols undergo all of the same reactions as simple acyclic alkoxymercurials. They are prone to ring open sometimes during reduction, but this can generally be avoided using NaBH$_4$, aqueous NaOH, CH$_2$Cl$_2$, and tetra-alkylammonium chlorides.[40,170] Sodium borohydride-induced oxidation and conjugate addition are feasible as well as halogenation.

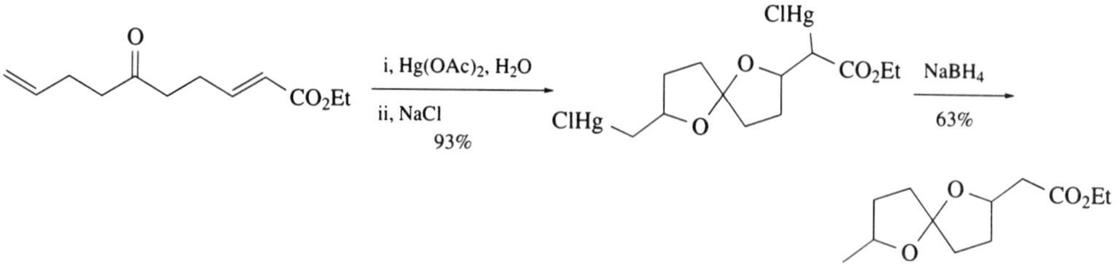

This intramolecular alkoxymercuration chemistry has been employed in the synthesis of carbohydrates (Equations (133) and (134)).[44,170b,171]

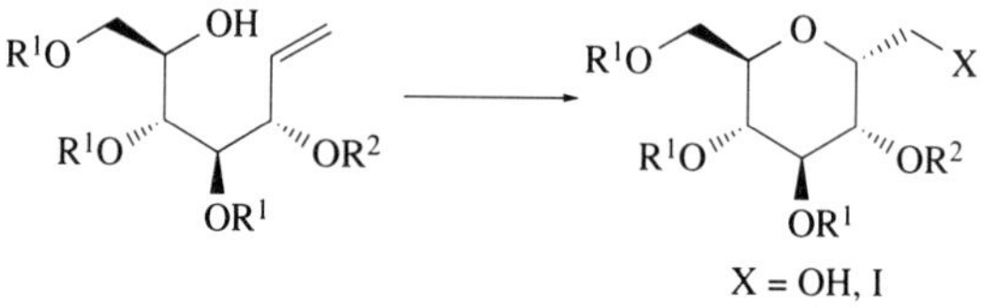

Dienes and polyenes undergo facile alkoxymercuration. 1,2-Dienes generally afford mercurated acetals[172] or vinylic mercurials[173] depending on their substitution pattern (Equations (135) and (136)). An intramolecular variant of this process has been used to prepare nonactate esters (Equation (137)).[174] 1,3-Dienes produce either mono[175]- or dimercurated[176] products (Scheme 39). The former undergo conjugate addition processes.[177] Conjugated dienes react with HgO·HBF$_4$ in an alcohol solvent to form 1,4-dialkoxy-2-alkenes by solvolysis (Equation (138)).[178] The alkoxymercuration of nonconjugated dienes is predictable from what is known about simple isolated alkenes. Mercuric tetrafluoroborate provides useful selectivity at −20 °C (Equation (139)).[149]

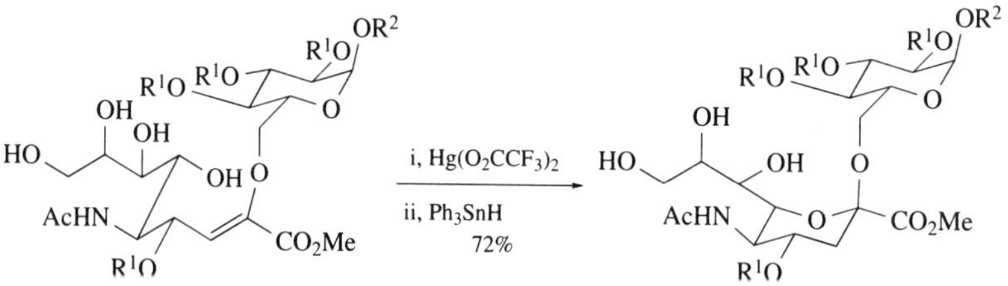

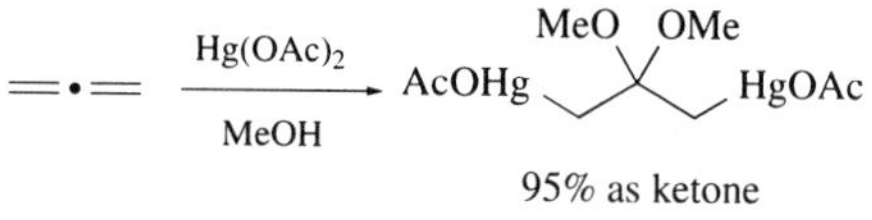

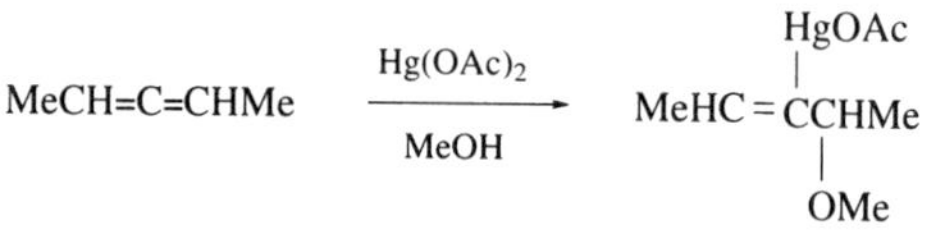

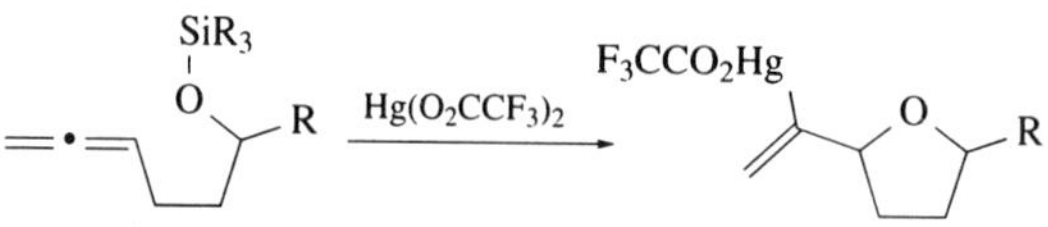

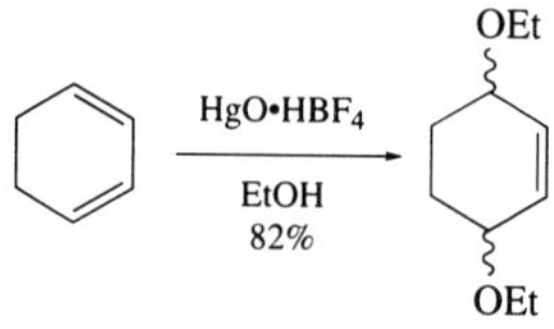

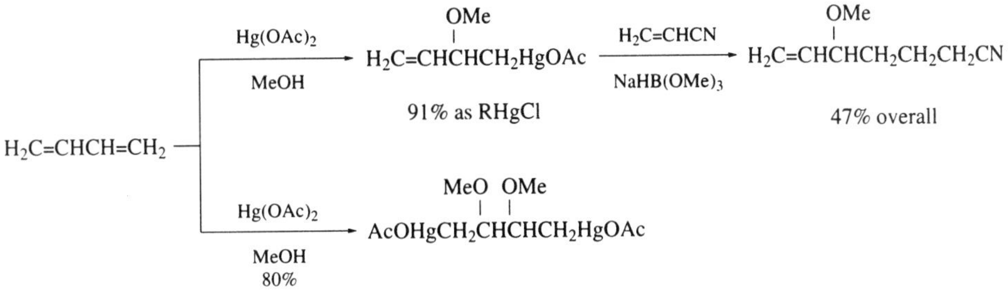

Scheme 39

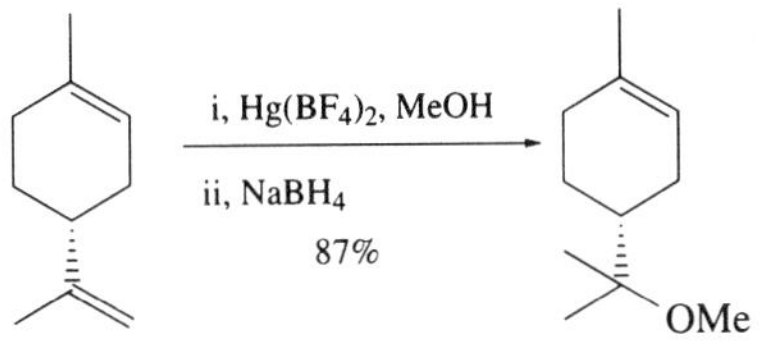

(139)

There are few examples of the isolation of organomercurials from the alkoxymercuration of alkynes (Scheme 40).[179] This reaction usually affords either vinylic ethers or acetals via protonolysis (Equations (140) and (141)). This process has provided an interesting approach to mercurated benzofurans (Equation (142)).[115b]

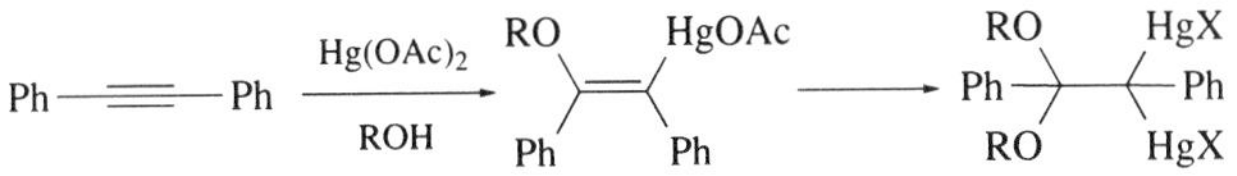

Scheme 40

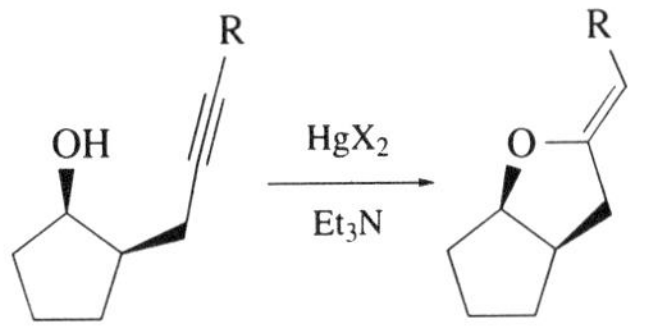

(140)

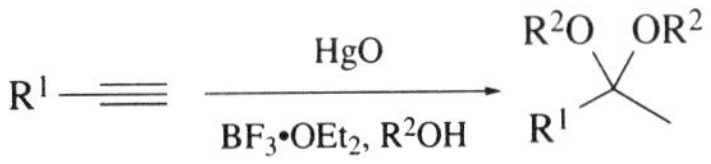

(141)

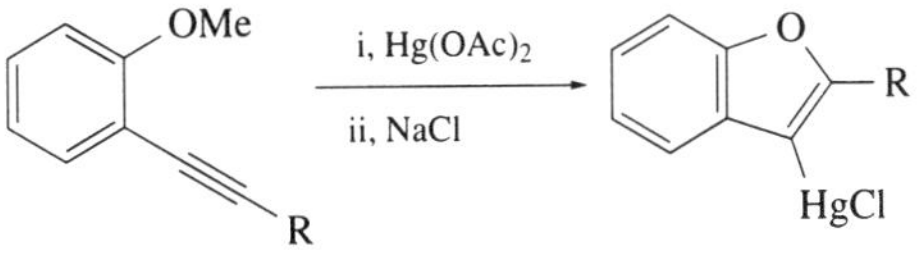

(142)

9.12.3 Peroxymercuration[11]

When H_2O_2 or alkyl hydroperoxides are employed as the nucleophile in solvomercuration, new alkylated peroxides are formed.[180] Hydrogen peroxide usually affords mixtures of hydroperoxide and dialkylperoxide products, but alkyl hydroperoxides generally produce good yields of the anticipated dialkylperoxides (Equation (143)). Mercuric nitrate or trifluoroacetate are the most commonly used mercuric salts for this process, although Hg(OAc)$_2$ with or without HClO$_4$ has occasionally been used. Methylene chloride is usually the solvent. This process exhibits all the characteristics of hydroxy- and alkoxymercuration.

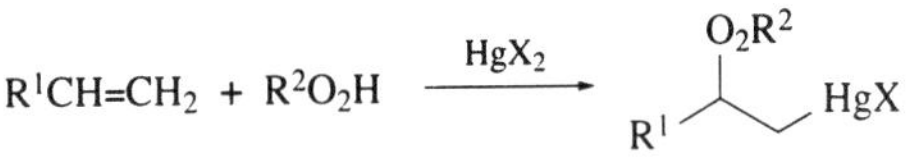

(143)

Several unsaturated hydroperoxides have been subjected to intramolecular peroxymercuration (Equation (144)).[181]

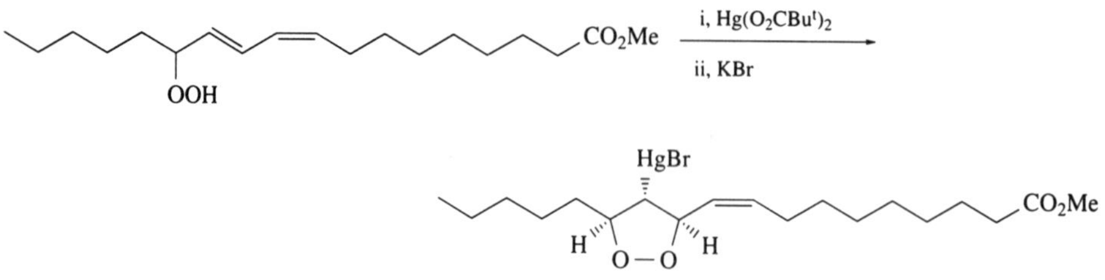

$$\text{(144)}$$

Nonconjugated dienes can be mono- or dimercurated (Equations (145) and (146)).[182,183]

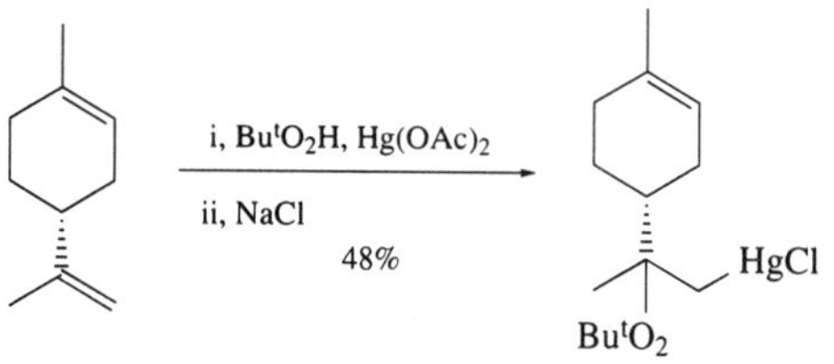

$$\text{(145)}$$

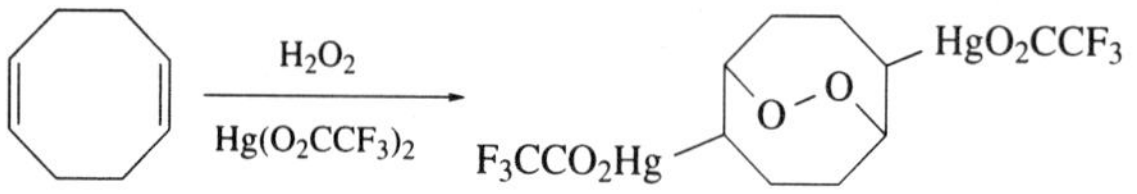

$$\text{(146)}$$

The mercurated peroxides prepared by peroxymercuration undergo many of the reactions discussed in previous sections. While reduction with alkaline $NaBH_4$ often produces significant amounts of epoxide side-products, Bu^n_3SnH gives much improved yields.[184] The successful chlorination, bromination, and iodination of these mercurials have all been reported.[185] Transmetallation with palladium salts affords ketone products (Equation (147)).[186]

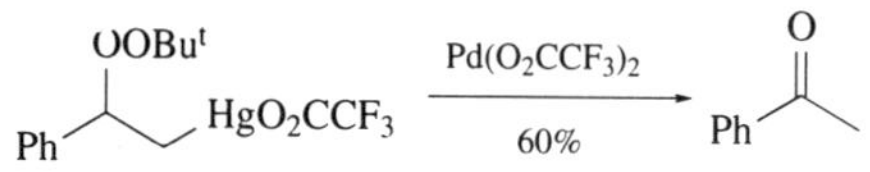

$$\text{(147)}$$

9.12.4 Acyloxymercuration

The reaction of alkenes and alkynes with a mercuric carboxylate salt produces β-acyloxymercurials (Equation (148)).[11] This reaction exhibits the same regio- and stereoselectivity as the previously mentioned solvomercuration processes. The vast majority of acyloxymercurations have been effected using $Hg(OAc)_2$ in HOAc. Equilibria have been observed when using $Hg(O_2CCF_3)_2$ in inert solvents.[187] Relatively few functional groups interfere with the acyloxymercuration process, although acid-sensitive groups may rearrange. Vinylic halides can be converted to ketones using mercuric carboxylates.[188] The regioselectivity of this reaction depends on the solvent employed (Equation (149)).[188b]

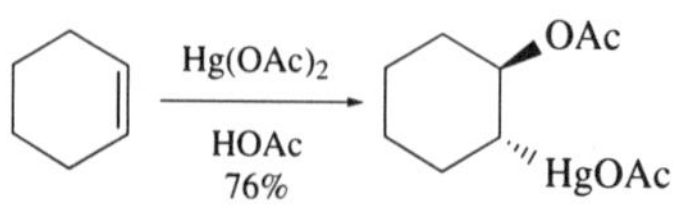

$$\text{(148)}$$

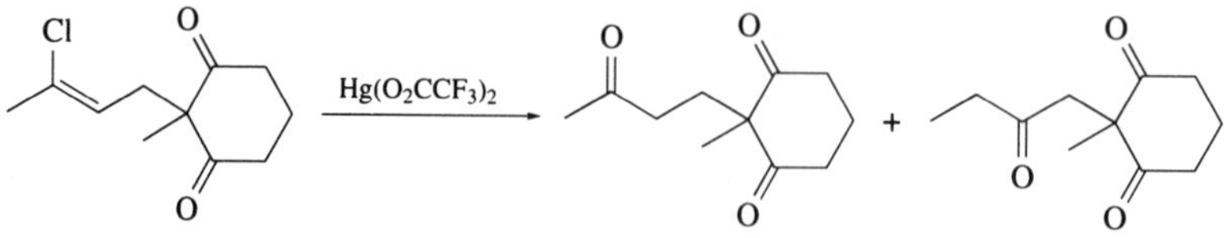

$$\text{(149)}$$

$MeNO_2$	~100%	:	0%
MeOH	0%	:	87%

Mercuric salts have been used as catalysts to effect ester interchange in vinylic esters (Equation (150)).

$$H_2C=CHOCR^1 + HOCR^2 \xrightarrow{cat.\ Hg^{II}} H_2C=CHOCR^2 + HOCR^1 \qquad (150)$$

Mercuric trifluoroacetate has been used to effect the ring expansion of 1-alkenyl-1-cycloalkanols via their silyl ethers (Scheme 41).[189]

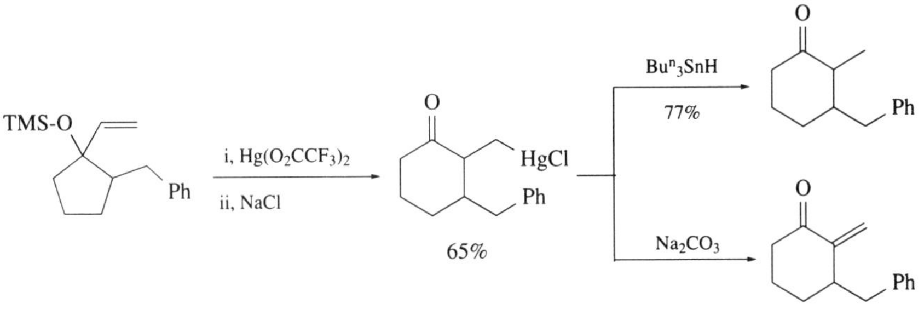

Scheme 41

Acyloxymercurials undergo several useful reactions. Alkaline NaBH$_4$ reduction has been effected, but saturated alcohols and starting alkene have been observed as side products. Halogenation has been reported, although relatively few examples exist. Acyloxymercurials also react with arenes and β-diketones with carbon–carbon bond formation (Scheme 42).[190]

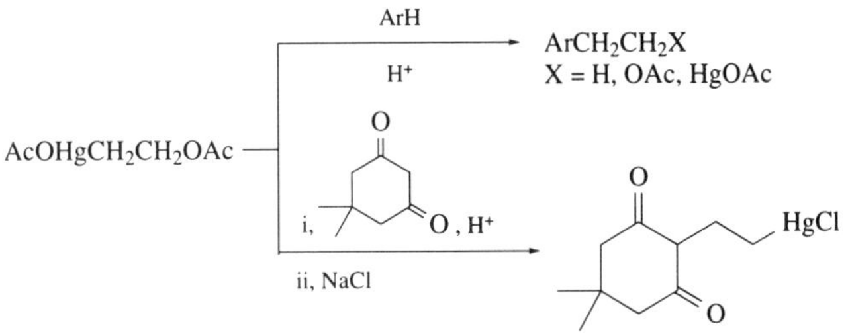

Scheme 42

Alkenoic acids and esters undergo facile cyclization to mercurated lactones in the presence of HgCl$_2$ or Hg(OAc)$_2$ (Equation (151)).[191] Good 1,2-stereoselectivity has been observed (Equation (152)).[192]

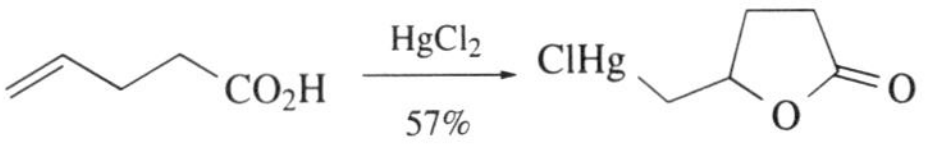

A number of dienes and polyenes have been subjected to acyloxymercuration. Simple allenes have been reported to generate vinylic mercurials (Equation (153)).[193] Alternatively, allenic ketones produce oxidized 3(2H)-furanones (Equation (154)).[194] Little work has been reported on the acyloxymercuration of 1,3-dienes, but non-conjugated dienes can often be selectively converted to either mono- or dimercuration products by proper choice of the reaction conditions (Scheme 43).[195] Alkaline NaBH$_4$ reduction of the monomercurial provides a useful route to the corresponding alkenol which cannot be prepared in good yield by hydroxymercuration of the corresponding diene due to tetrahydrofuran formation. By using Hg(BF$_4$)$_2$ at −20 °C, the selective monomercuration of nonconjugated dienes can also be achieved.[149]

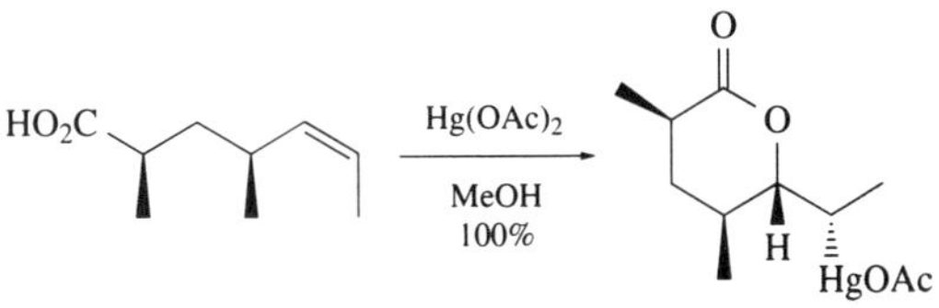

$$\text{MeCH=C=CHMe} \xrightarrow{\text{Hg(OAc)}_2} \text{MeHC=}\overset{\overset{\text{HgOAc}}{|}}{\text{C}}\text{CH(OAc)Me} \quad (153)$$

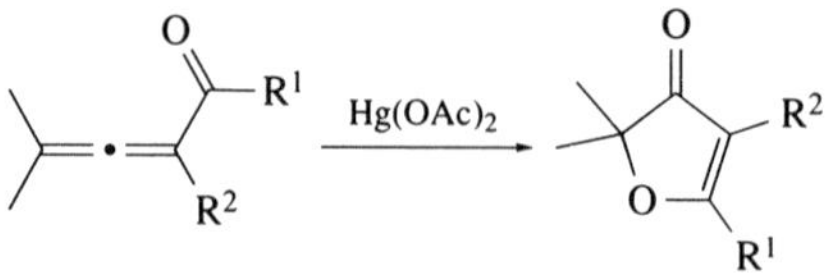

(154)

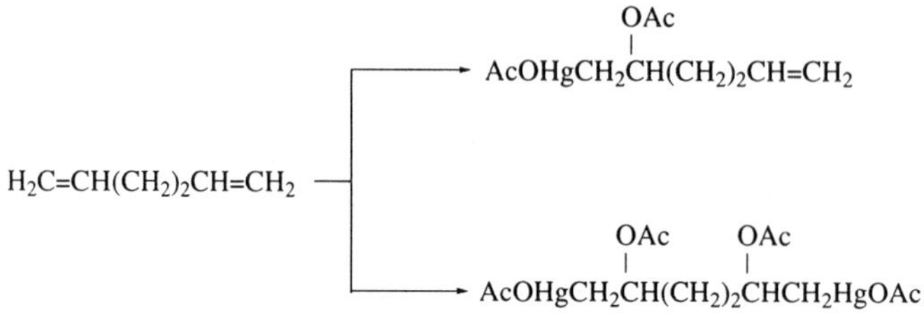

Scheme 43

The reaction of alkynes and mercuric carboxylates produces a variety of products depending on the alkyne and the reaction conditions. Terminal alkynes react with $Hg(OAc)_2$ in inert solvents to form dialkynylmercurials of complex polymercurated products (see Section 9.2.8). In the presence of carboxylic acids, mercuric salts catalyze vinylic ester formation (Equation (155)). Internal alkynes react with $Hg(OAc)_2$ to produce vinylmercurials[196] which have been halogenated[197] or oxidized to enol acetates by $Pd(OAc)_2$[56] (Scheme 44). *Anti*-addition is usually observed. Internal alkynes also react with iodine and mercury(II) salts of chloride, acetate, methanesulfonate, *p*-toluenesulfinate, and thiocyanate to afford (*E*)-β-functionalized vinylic iodides (Equation (156)).[198]

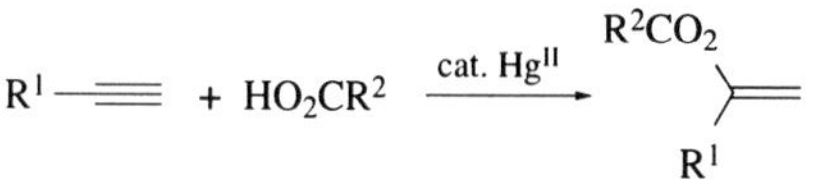

$$R^1 {=\!\!=} + HO_2CR^2 \xrightarrow{\text{cat. Hg}^{II}} \underset{R^1}{\overset{R^2CO_2}{\diagup\!\!\!\!\diagdown}} \quad (155)$$

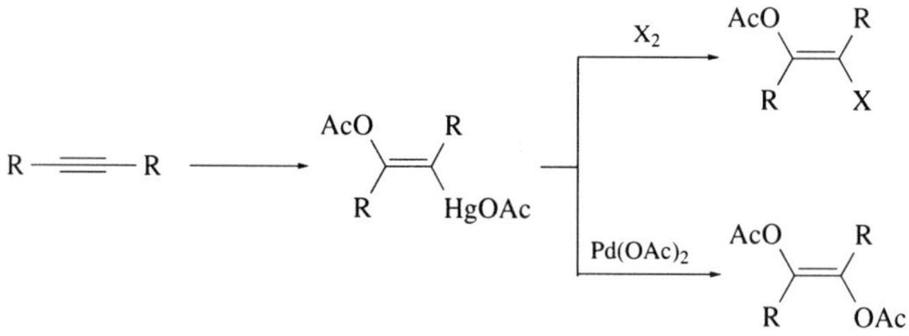

Scheme 44

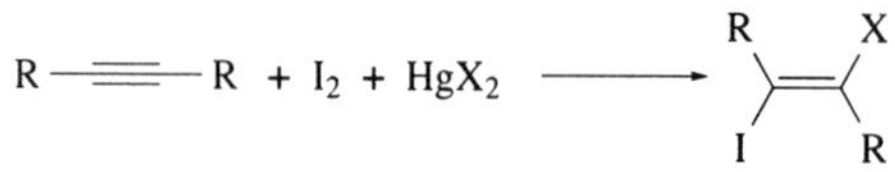

$$R{=\!\!=\!\!=}R + I_2 + HgX_2 \longrightarrow \underset{I\quad R}{\overset{R\quad X}{\diagup\!\!\!\!\diagdown}} \quad (156)$$

X = Cl, OAc, MeSO₃, *p*-MeC₆H₄SO₃, PhCH₂S, *p*-MeC₆H₄SO₂, NCS

4- and 5-Alkynoic acids have been cyclized to unsaturated lactones using only catalytic amounts of HgO, $Hg(OAc)_2$, or $Hg(O_2CCF_3)_2$ (Equation (157)).[199] In the presence of primary alcohols, the process affords alkoxyfuranones (Equation (158)).[200]

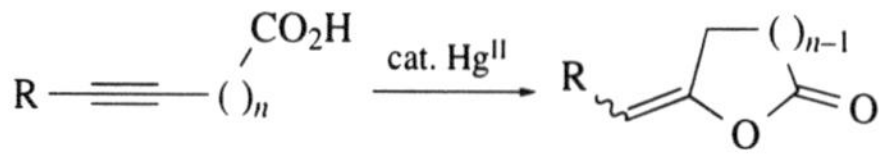

(157)

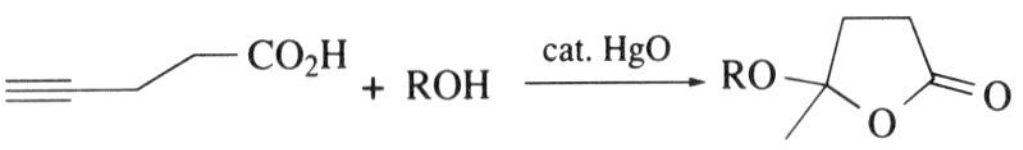

$$\overset{\equiv}{}\!\!\!\!\diagup\!\!\!\!\text{CO}_2\text{H} + \text{ROH} \xrightarrow{\text{cat. HgO}} \text{RO}\diagup\!\!\!\!\diagdown\!\!\!\text{O}\!\!=\!\!\text{O} \qquad (158)$$

9.12.5 Aminomercuration

Solvomercuration in the presence of an amine provides a very valuable method for the preparation of substituted amines and a wide variety of nitrogen heterocycles.[11] Although the relative reactivities $Hg(ClO_4)_2 \sim Hg(NO_3)_2 \gg Hg(OAc)_2 > HgCl_2$ toward aminomercuration have been reported,[201] almost all work on this reaction has involved the latter two salts (Equation (159)). Aminomercuration has not been as extensively studied as solvomercuration using oxygen nucleophiles. The nature of the amines that can be employed appears to be limited to 2° aliphatic amines and 1° or 2° anilines. A range of alkenes successfully react with amines and $HgCl_2$ or $Hg(OAc)_2$. Mechanistically, the exact structure of the mercurating agent is not clear.[202] The process has been found to be reversible or irreversible depending on the nature of the mercury(II) salt employed.[203]

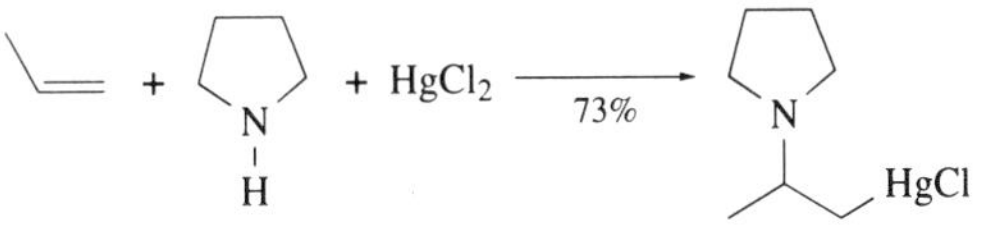

$$\diagup\!\!\!=\!\!\!\diagdown + \text{pyrrolidine-NH} + HgCl_2 \xrightarrow{73\%} \text{pyrrolidine-N-CH(CH}_3\text{)HgCl} \qquad (159)$$

Aminomercurials are most commonly reduced to amines by $NaBH_4$, but this process can be accompanied by alkene formation and nitrogen migration. The extent of alkene formation is dependent on the structure of the mercurial and the reaction conditions (Scheme 45).[204] The amount of alkene is limited by adding NaOH to the reduction[201] or by using a phase transfer catalyst.[40] Rearrangement during reduction is another major problem. The extent of rearrangement depends on the structure of the alkene and the amine,[201] and the reaction conditions employed (Equation (160)).[205] Reduction is best carried out using alkaline $NaBH_4$ under phase transfer conditions.[40] Demercuration has also been effected using $LiAlH_4$, which gives more rearrangement than $NaBH_4$, and Na(Hg) in water, which reduces stereospecifically, but affords low yields and substantial amounts of starting alkene. Demercuration has also been reported using magnesium, lithium, sodium, or calcium in $PhNH_2$, MeOH, and/or NH_3, but alkaline $NaBH_4$ under phase transfer conditions is superior.

In the presence of $NaBH_4$ and an alkene, aminomercurials can effect conjugate addition (Equation (161)).[206]

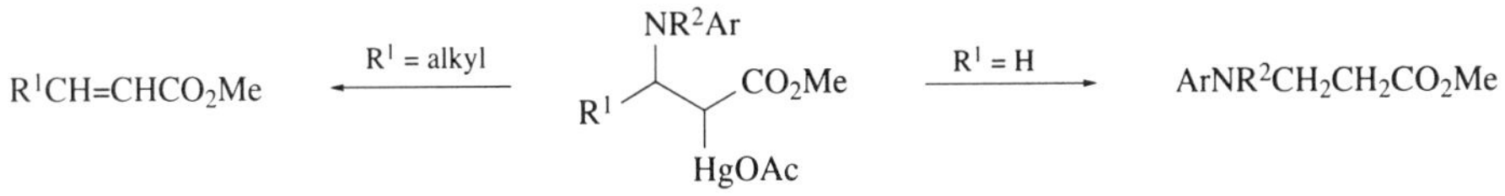

$$R^1CH=CHCO_2Me \xleftarrow{R^1 = \text{alkyl}} \underset{\text{HgOAc}}{\overset{NR^2Ar}{R^1\diagup\!\!\diagdown CO_2Me}} \xrightarrow{R^1 = H} ArNR^2CH_2CH_2CO_2Me$$

Scheme 45

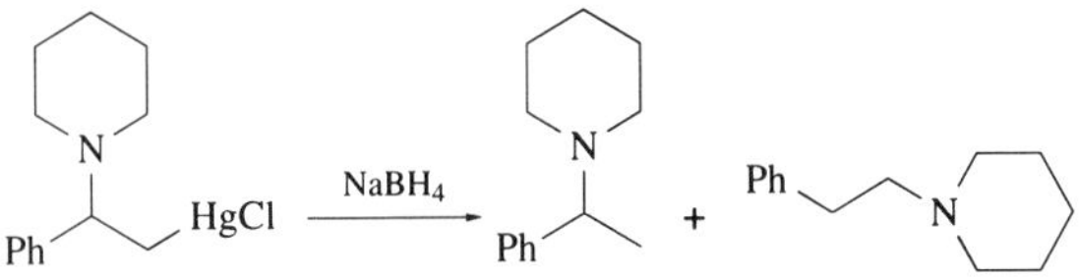

$$\underset{Ph}{\text{piperidinyl-CH-CH}_2\text{HgCl}} \xrightarrow{NaBH_4} \underset{Ph}{\text{piperidinyl-CH(CH}_3)} + \text{Ph-CH}_2\text{CH}_2\text{-piperidinyl} \qquad (160)$$

H_2O–THF	30%	30%
NaOH	65%	0%

$$\underset{\text{RCHCH}_2\text{HgX}}{\overset{NHAr}{|}} + H_2C=CHY \xrightarrow{NaBH_4} \underset{\text{RCHCH}_2\text{CH}_2\text{CH}_2Y}{\overset{NHAr}{|}} \qquad (161)$$

X = OAc, Cl; Y = CO_2Et, CN

Aminomercurials can also be reduced to alkali metal alkyls which undergo a variety of useful substitution processes (see Equation (62)).[207]

The reaction of alkenes and anilines with $HgO \cdot HBF_4$ yields vicinal diamines in high yield (Equation (162)).[208]

$$R^1CH=CH_2 \xrightarrow[ArNHR^2]{HgO \cdot HBF_4} R^1\overset{\overset{NR^2Ar}{|}}{C}HCH_2NR^2Ar \qquad (162)$$

Allylic alcohols react with aromatic amines and a catalytic amount of HgO·HBF$_4$ to produce allylic anilines (Equation (163)).[209]

$$ArNH_2 + H_2C=CHCH_2OH \xrightarrow{cat.\ HgO \cdot HBF_4} ArNHCH_2CH=CH_2 \qquad (163)$$

The mercuration of alkenylamines provides a valuable synthesis of five- and six-membered ring nitrogen heterocycles (Equations (164) and (165)).[210,211] While the mercuration step appears highly regio- and stereoselective, alkene formation and/or rearrangement during demercuration can be a problem (Scheme 46). Reduction is best effected using alkaline NaBH$_4$ under phase transfer conditions.[170a]

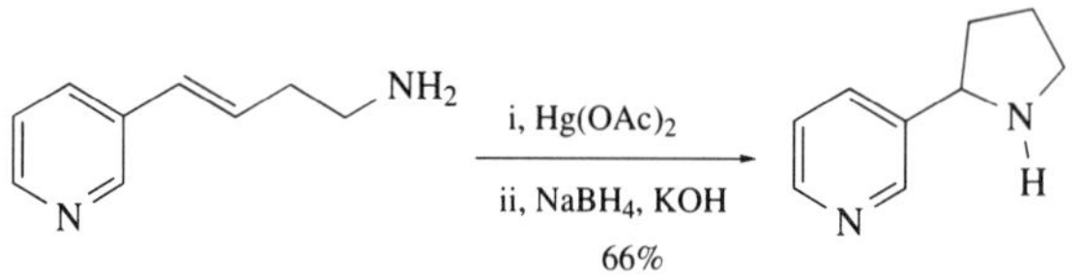

(164)

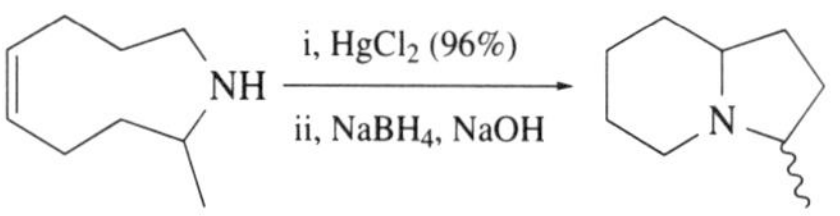

(165)

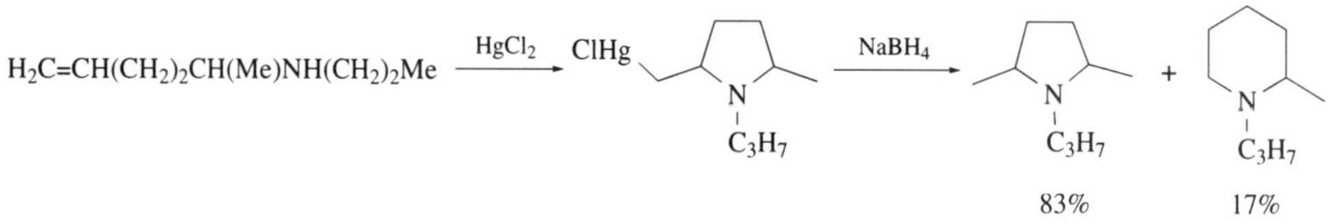

Scheme 46

Nitrogen heterocycles derived from alkenylamines by aminomercuration can undergo conjugate addition to electron-deficient alkenes in the presence of borohydride reagents (Equation (166)).[212]

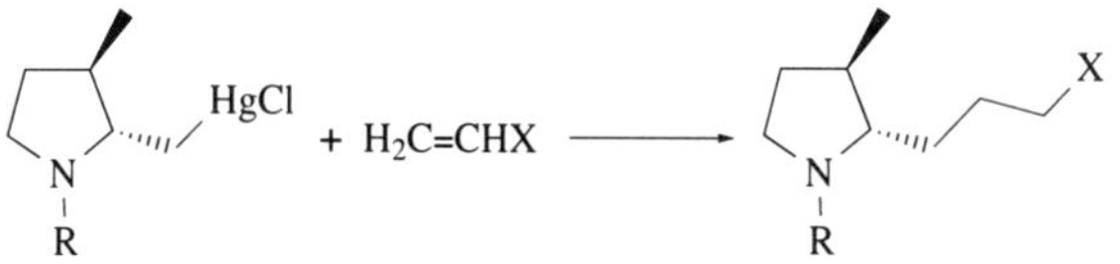

(166)

Sodium borohydride-induced oxidation has been employed in the conversion of an alkenylamine to an aza analogue of D-galacturonic acid (Equation (167)).[213]

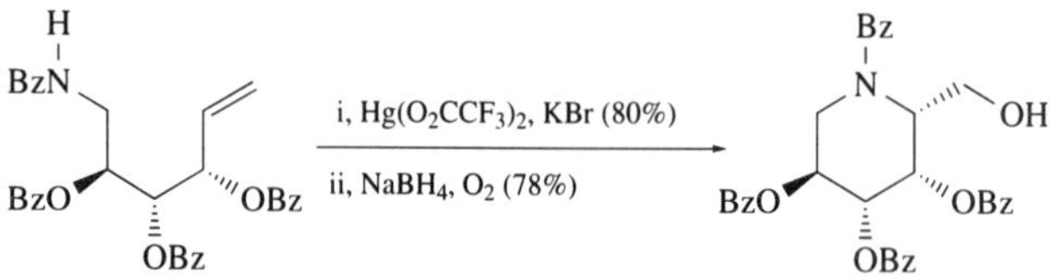

(167)

The mercury-promoted cyclization of cyclohexenylanilines in the presence of an alcohol leads to tetrahydrocarbazoles (Equation (168)).[214]

Dienes undergo aminomercuration to afford a variety of products. Allenes generally add the mercury moiety to the central carbon of the allene and the amine to a terminal carbon, but these products can rearrange during NaBH$_4$ reduction. Allenic amines can be cyclized, but AgNO$_3$ effects this same process catalytically (Equation (169)).[215] 1,3-Dienes can be dimercurated to produce vicinal diamines[216] or cyclized to unsaturated heterocycles[217] depending on the diene and reagents employed (Equations (170)

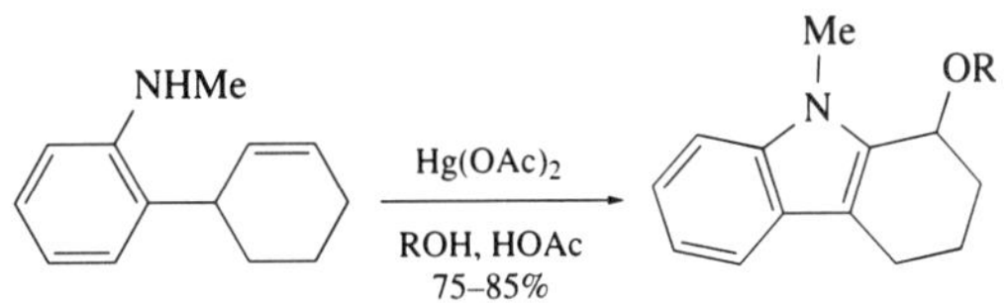

$$\text{(168)}$$

and (171)). The aminomercuration of nonconjugated dienes provides a useful route to pyrrolidines[218] (Scheme 47) and six-membered ring nitrogen heterocycles[219] (Equation (172)).

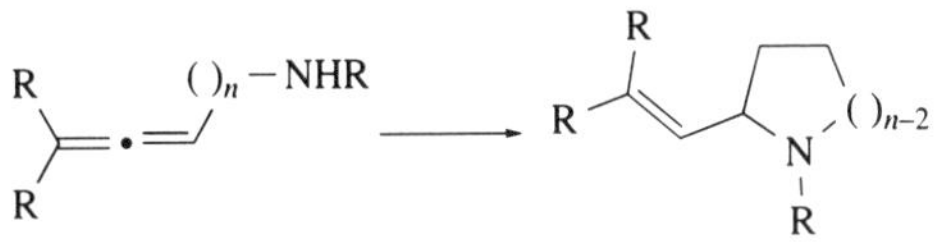

$$\text{(169)}$$

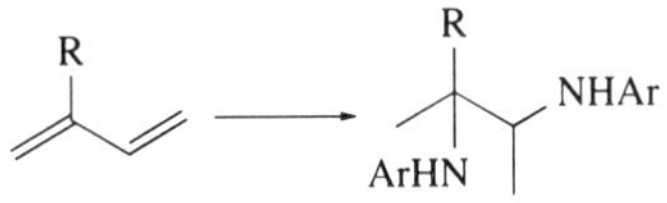

$$\text{(170)}$$

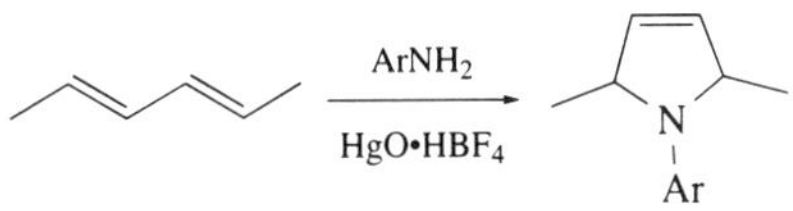

$$\text{(171)}$$

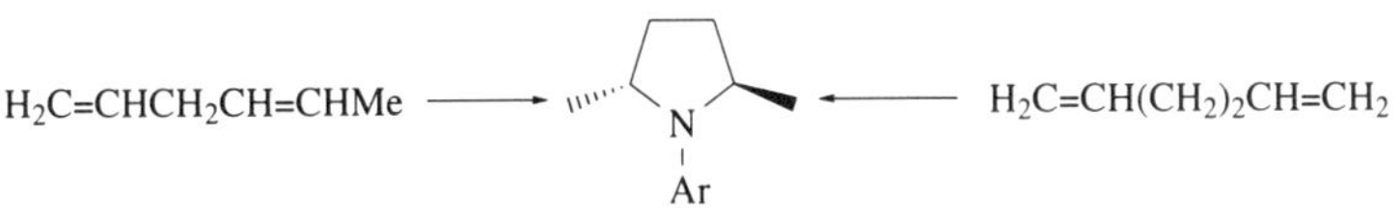

Scheme 47

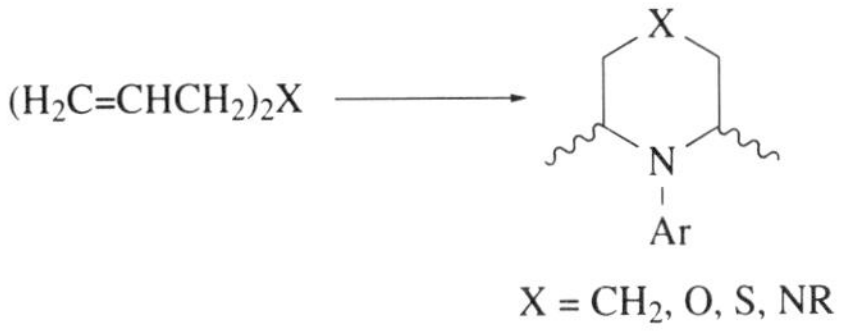

$$\text{(172)}$$

X = CH$_2$, O, S, NR

The aminomercuration of 1,5-cyclooctadiene affords different bicyclic amines depending on the reaction conditions (Scheme 48).[203] Selective mono- and diamination of limonene has also been reported (Scheme 49).[220]

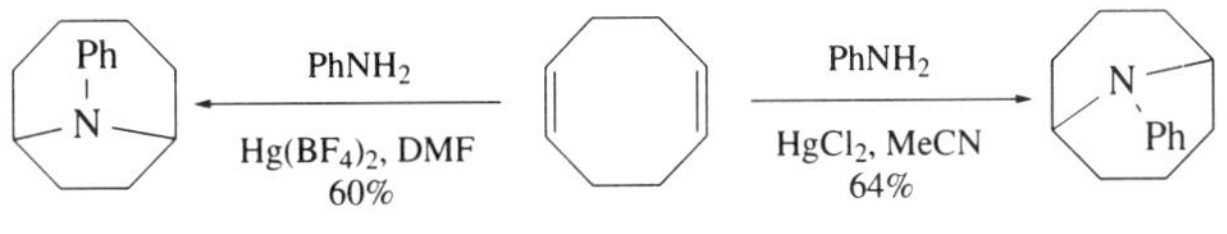

Scheme 48

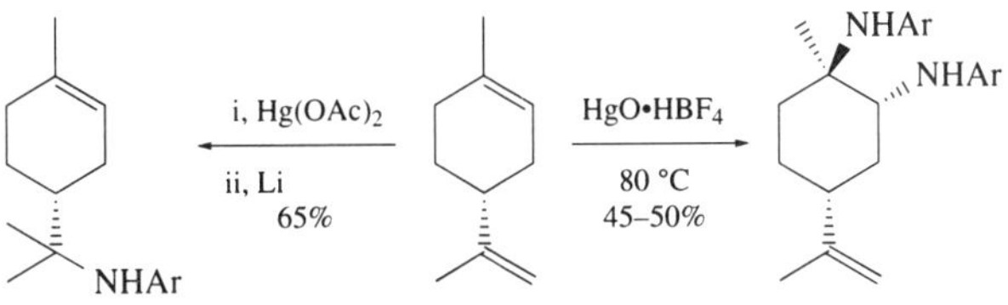

Scheme 49

Alkynes react with 1° and 2° amines in the presence of mercury salts to produce imines[221] and enamines,[222] respectively (Scheme 50). By the proper choice of amine and reaction conditions, enynes can be converted to either 1-aza-1,3-dienes or 2-amino-1,3-dienes (Scheme 51).[223]

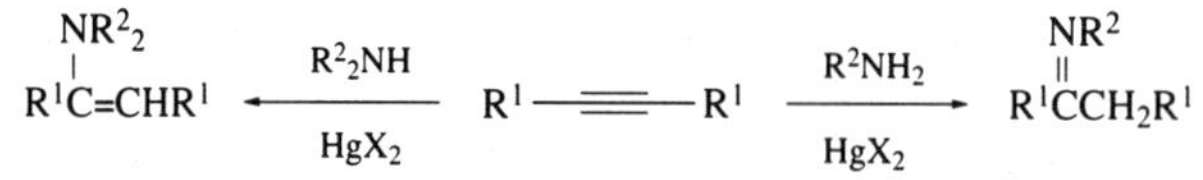

Scheme 50

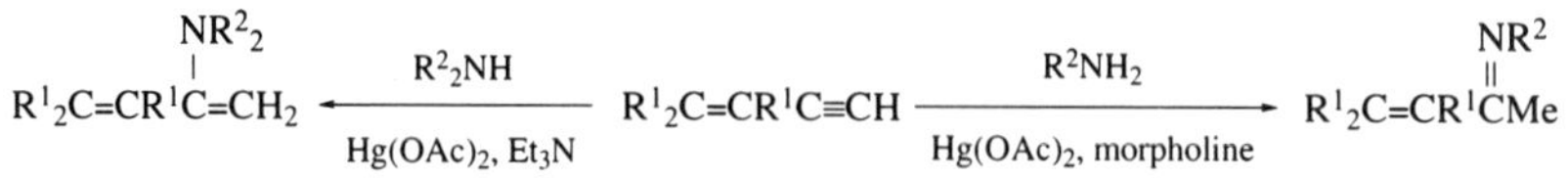

Scheme 51

9.12.6 Amidomercuration

Alkenes react with $Hg(NO_3)_2$ and either nitriles[224] or amides[225] to generate β-amidoalkylmercurials (Equation (173)). A convenient procedure for aminomercuration using nitriles was reported in 1969 by Brown.[224] While most examples have employed acetonitrile, other simple nitriles work well. This process affords Markovnikov *anti*-addition products, but alkenes of the type $R_2C=CH_2$ or $R_2C=CHR$ often produce allylic amides or other rearranged products. These rearrangements have been exploited for efficient syntheses of the alkaloids (+)-makomakine and hobartine (Scheme 52).[226] Amidomercuration using amides exhibits similar limitations in the type of alkenes which can be employed, but a wider range of amides have been utilized, including 1° but not 2° amides, sulfonamides, urea, and urethanes.[225] This process has been used to substitute first on one nitrogen of urea and then on the other (Scheme 53).[225a]

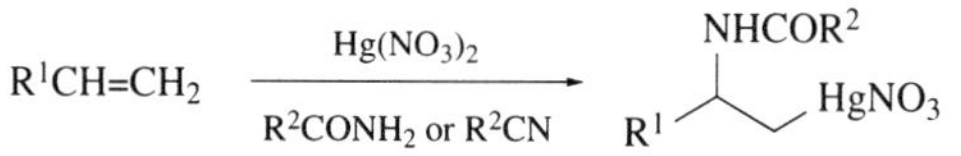

$$R^1CH=CH_2 \xrightarrow[R^2CONH_2 \text{ or } R^2CN]{Hg(NO_3)_2} \begin{array}{c} NHCOR^2 \\ R^1 \end{array} HgNO_3 \qquad (173)$$

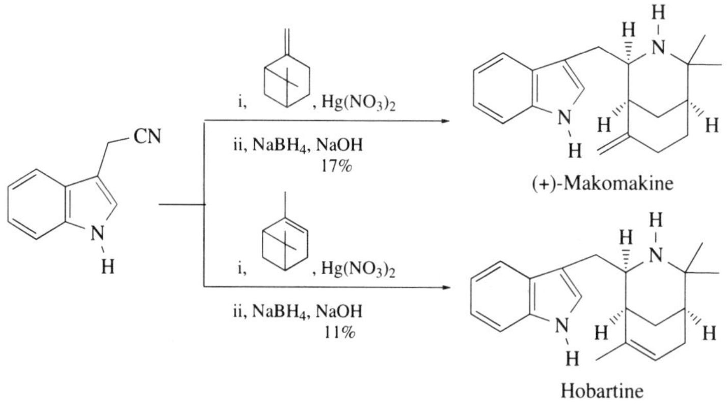

Scheme 52

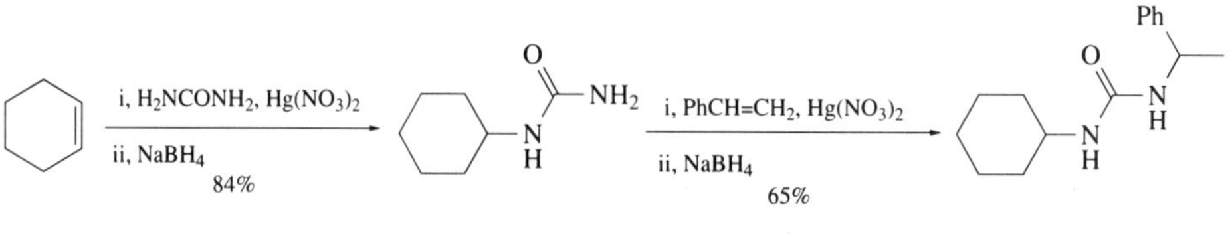

Scheme 53

Amidomercurials are generally reduced to the corresponding amides using alkaline $NaBH_4$,[224] but Na(Hg) works well.[227] Reduction with $LiAlH_4$ produces the corresponding amines[227] and halogenation affords β-haloamides.[227] Conjugate addition to alkenes has been effected using $NaHB(OMe)_3$.[212]

Intramolecular amidomercuration has been reported using alkenamides, but there appears to be only one example in which an unsaturated nitrile has been employed. Mercuric acetate has been utilized in most such reactions. The intramolecular amidomercuration of unsaturated amidals provides a useful route for the stereocontrolled preparation of protected aminoalcohols,[228] pyrrolidines, and piperidines[229] (Equations (174) and (175)). Intramolecular ureidomercuration has been more extensively studied. The stereoselectivity of this process depends on whether the reaction is run under equilibrating ($Hg(O_2CCF_3)_2$ in $MeNO_2$) or nonequilibrating ($Hg(OAc)_2$ in THF) conditions (Scheme 54).[230] This methodology has been employed in the stereoselective synthesis of precursors to amino acids[231] and amino alcohols[231c,232] (Equations (176) and (177)).

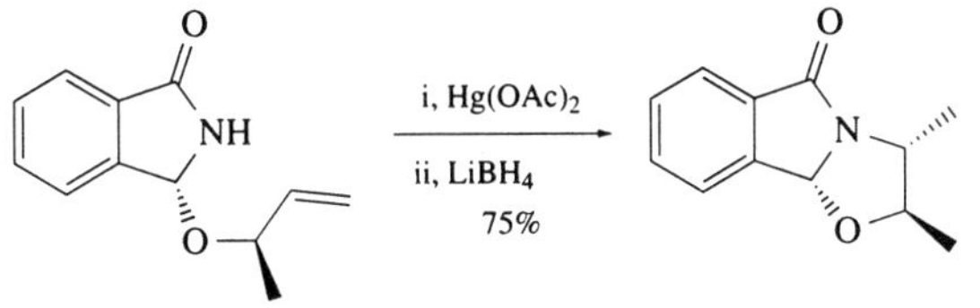

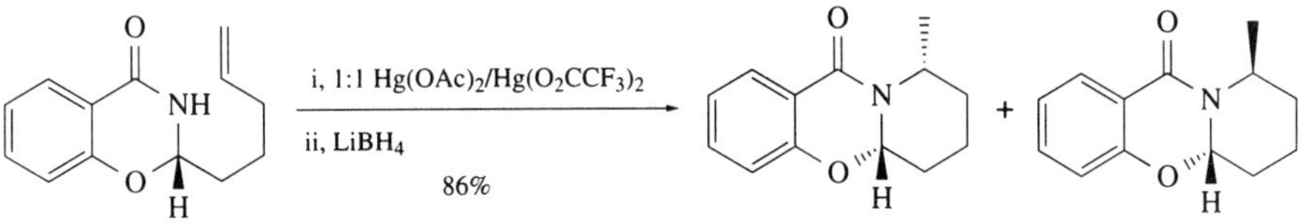

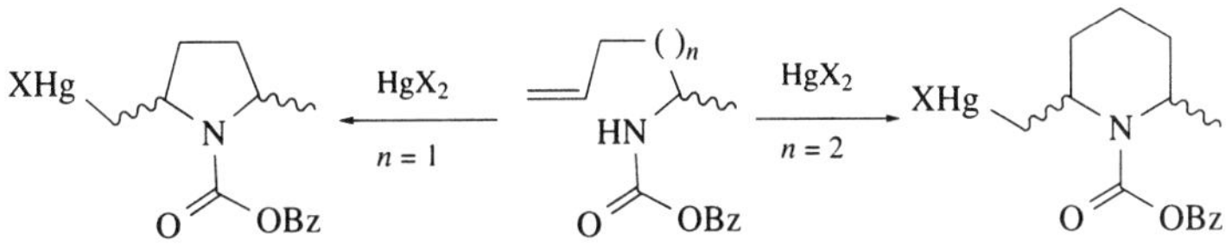

trans/cis	X	trans/cis
98/2	OAc	60/40
< 50/> 50	O_2CCF_3	< 2/> 98

Scheme 54

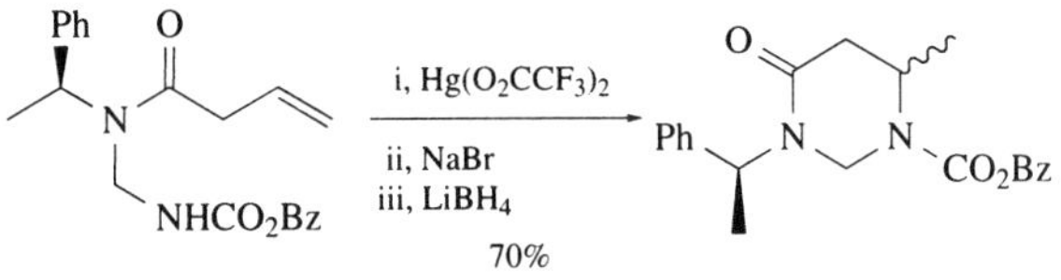

The mercurials derived from intramolecular ureidomercuration undergo facile halogenation (synthesis of pseudoconhydrine),[233] $NaBH_4$-induced oxidation[231b,c] and borohydride-induced conjugate addition (Equation (178)).[234] Simple cyclic amidomercurials are more prone to ring-open during conjugate reduction, but can be trapped by an internal alkene (see Equation (66)).

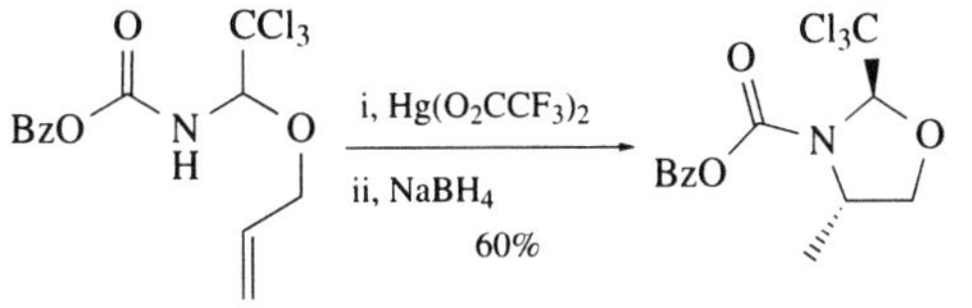

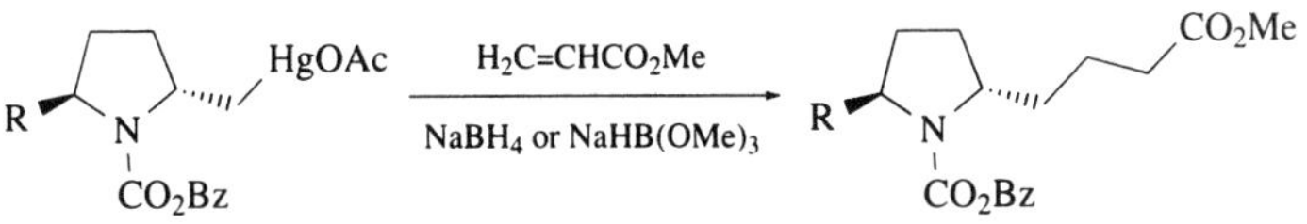

Nonconjugated dienes undergo amidomercuration to form mono- or disubstituted products depending on the diene and the procedure (Equations (179) and (180), Scheme 55).[149,225]

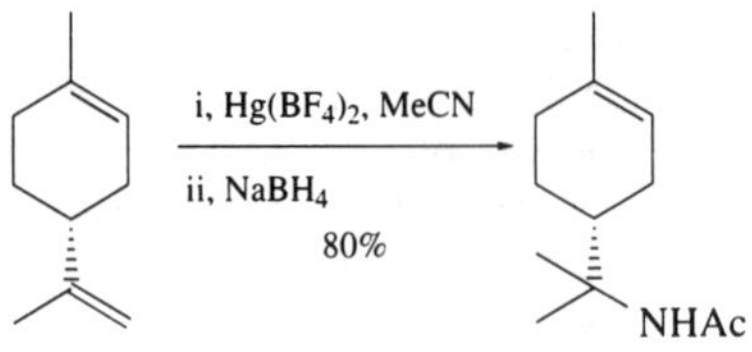
i, Hg(BF4)2, MeCN
ii, NaBH4
80%
NHAc

(179)

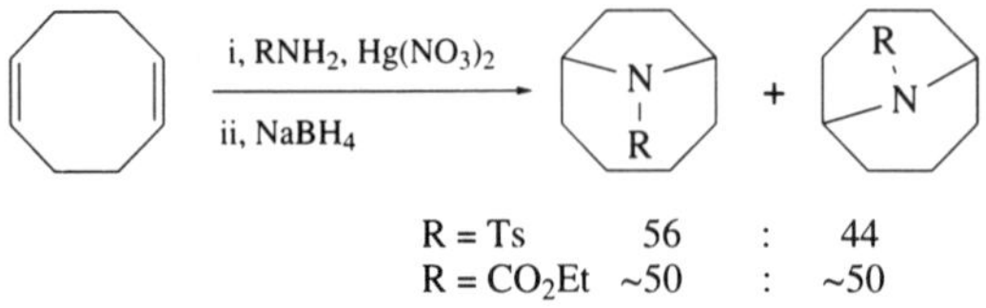
i, RNH2, Hg(NO3)2
ii, NaBH4
N
R
+
R
N
R = Ts 56 : 44
R = CO2Et ~50 : ~50

(180)

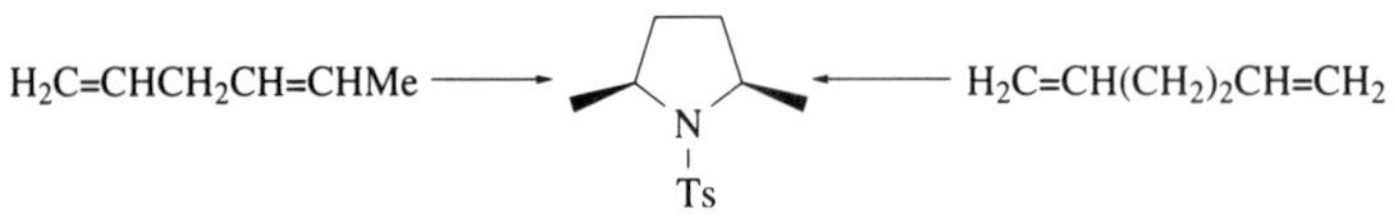
H2C=CHCH2CH=CHMe
N
Ts
H2C=CH(CH2)2CH=CH2

Scheme 55

9.12.7 Azidomercuration

The reaction of simple alkenes, NaN_3 and $Hg(OAc)_2$, $Hg(O_2CCF_3)_2$, or $Hg(NO_3)_2$ affords β-azidoalkylmercurials (Equation (181)).[11,235] The reaction appears limited to terminal alkenes and strained cyclic alkenes. Nonconjugated dienes have been reported to undergo selective monomercuration (Equation (182)).

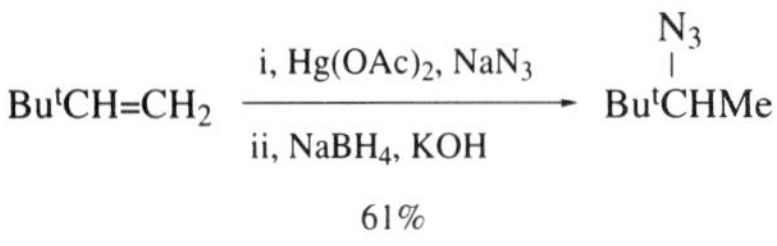
Bu^tCH=CH2
i, Hg(OAc)2, NaN3
ii, NaBH4, KOH
61%
N3
Bu^tCHMe

(181)

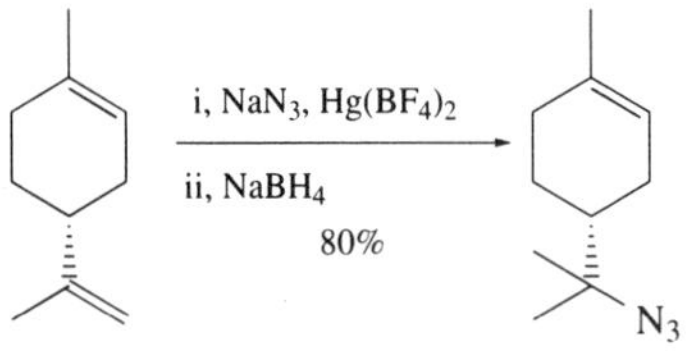
i, NaN3, Hg(BF4)2
ii, NaBH4
80%
N3

(182)

The azidomercurials can be iodinated,[236] demercurated using alkaline $NaBH_4$ or Na(Hg),[235] or reduced fully to the corresponding primary amine by refluxing with $NaBH_4$ in tetraglyme.[149]

9.12.8 Nitromercuration

The nitromercuration of alkenes can be effected using $NaNO_2$ plus $HgCl_2$ or $Hg(ClO_4)_2$ in water (Equation (183)).[11,237] A variety of alkenes react to form β-nitromercurials, but tri- and tetrasubstituted alkenes, plus those bearing electron-withdrawing groups are unreactive. For substituted cyclohexenes the reaction is highly regioselective and the resulting mercurials undergo facile elimination to nitroalkenes upon treatment with base (Equation (184)).[237c] No procedure exists for the reduction of β-nitroalkylmercurials to the corresponding nitroalkanes.

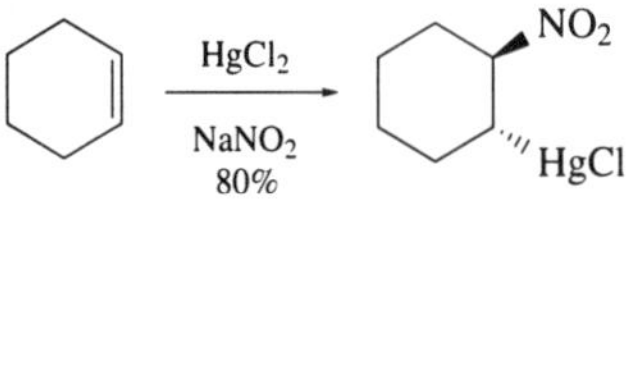
HgCl2
NaNO2
80%
NO2
HgCl

(183)

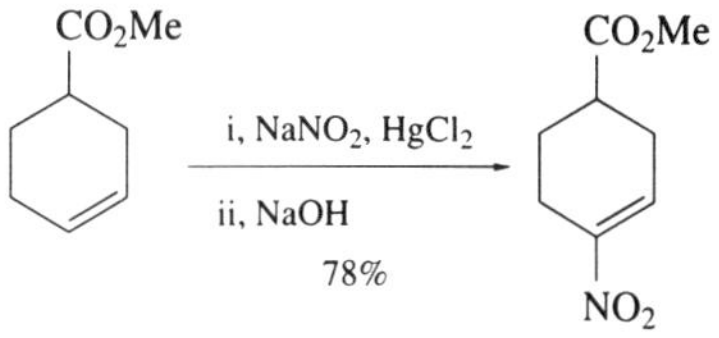

9.12.9 Carbomercuration[11]

The addition of a carbon group and mercury across a C–C double bond is a fairly limited process. The organomercurial $Hg[C(NO_2)_3]_2$ adds to a variety of mono- and disubstituted alkenes (Equation (185)),[238] while $(Bu^t)_2Hg$ reacts only with alkenes bearing two strong electron-withdrawing cyano, ester, or ketone groups on one end of the double bond (see Equation (63)).[88a] There are a few other reactions of alkenes which lead directly to carbon–carbon and carbon–mercury bond formation, but most of these do not appear to be very general. Arenes react with alkenes and mercuric salts in the presence of a strong acid to produce β-arylalkylmercurials (Equation (186)).[239] β-Dicarbonyl compounds react with alkenes, $Hg(OAc)_2$ and acid to yield mercurated diketones (Scheme 56).[240] Subsequent treatment with base affords cyclopropanes.

$$MeCH=CH_2 \xrightarrow[97\%]{Hg[C(NO_2)_3]_2} MeCHCH_2HgC(NO_2)_3 \quad | \quad C(NO_2)_3 \tag{185}$$

$$ArH + RCH=CH_2 + HgX_2 \longrightarrow ArCHCH_2HgX \quad | \quad R \tag{186}$$

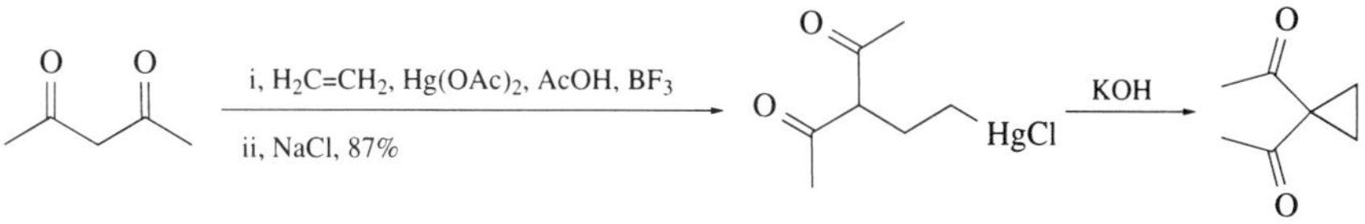

Scheme 56

Many of the solvomercuration reactions of dienes and polyenes afford cyclization products in which a new carbon–carbon bond is formed. A few of the more general are the cyclization of vinylallenes (Equation (187)),[241] the oxy-Cope rearrangement of 1,5-hexadien-3-ols (Equation (188)),[242] and the cyclization of cyclic dienes to bicyclic products (Equation (189)).[243] A number of farnesyl derivatives have been cyclized to polycyclic products using $Hg(O_3SCF_3)_2 \cdot PhNMe_2$ (Equation (190)).[244]

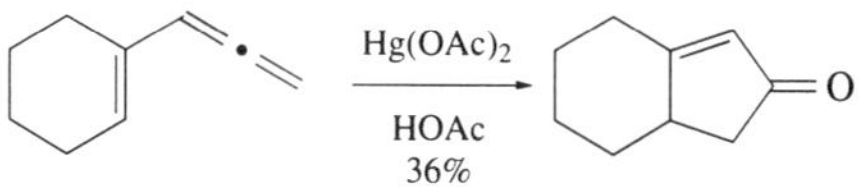

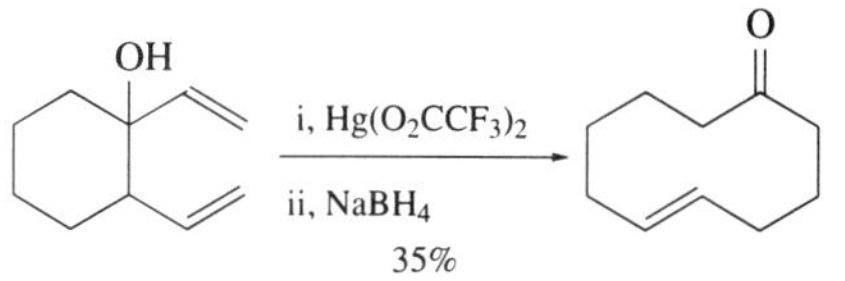

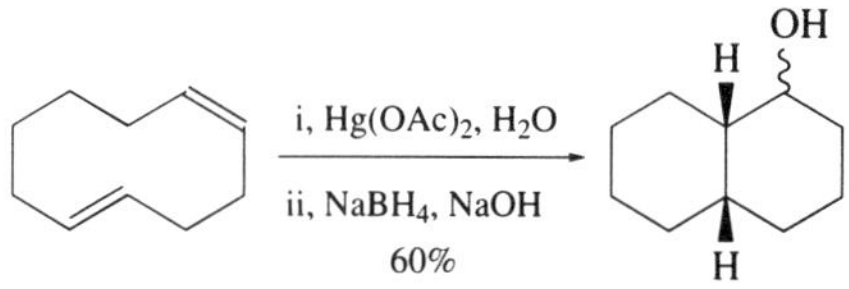

There are a few general reactions of alkynes and mercury reagents which lead to carbomercuration. Electron-deficient alkynes add $(Bu^t)_2Hg$ to produce vinylmercurials (see Equation (64)).[88a] Aryl-substituted alkynes undergo cyclization to vinylmercurials (Equation (191)).[115b] Alkynes containing neighboring enol silanes undergo similar carbomercuration processes (Equation (192)).[245]

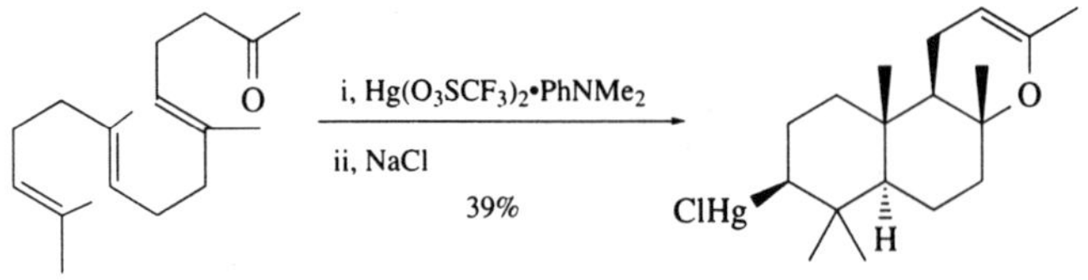

(190)

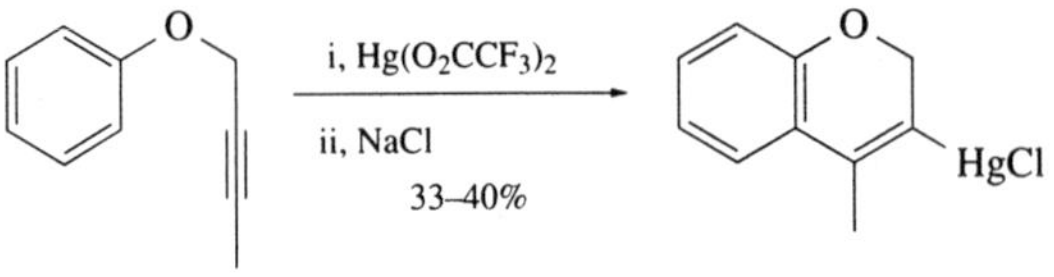

(191)

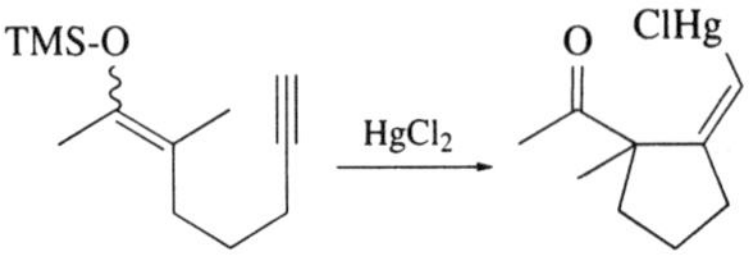

(192)

9.12.10 Halomercuration

There are numerous examples of the addition of HgF_2 across the C–C double bond of polyhaloalkenes in the presence of HF, KF, AsF_3, or $HgCl_2$ (Equation (193)),[11] and virtually no examples of the chloro-, bromo-, or iodomercuration of alkenes. However, HgX_2 (X = F, Cl, Br) reacts with alkenes and Br_2 or I_2 to produce the corresponding *trans*-dihalide (Equation (194)).[246]

$$CF_3CF{=}CF_2 \xrightarrow[\text{HF, } 60\text{–}80\%]{HgF_2} [(CF_3)_2CF]_2Hg \qquad (193)$$

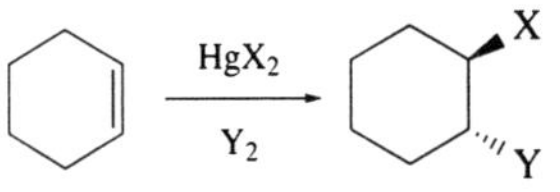

(194)

X = F, Cl, Br
Y = Br, I

There are many examples of the addition of mercuric halides to terminal and internal alkynes to produce β-halovinylmercurials. While there is only one example of the fluoromercuration of an alkyne (Equation (195)),[247] mercuric chloride adds readily to acetylene itself, as well as to vinyl acetylene, propargylic alcohols, alkynones, alkynoic acids and esters, alkynyl ethers, and propargylic amines.[11] The *trans*-adducts are formed in almost all cases. Few subsequent reactions of these β-chlorovinylmercurials have been reported. Protonolysis and halogenation are known, and the propargylic alcohol products undergo facile carbonylation to butenolides (Scheme 57; see also Equation (90)).

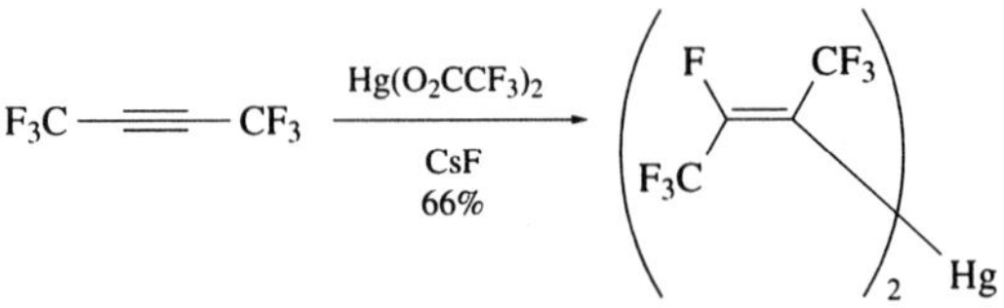

(195)

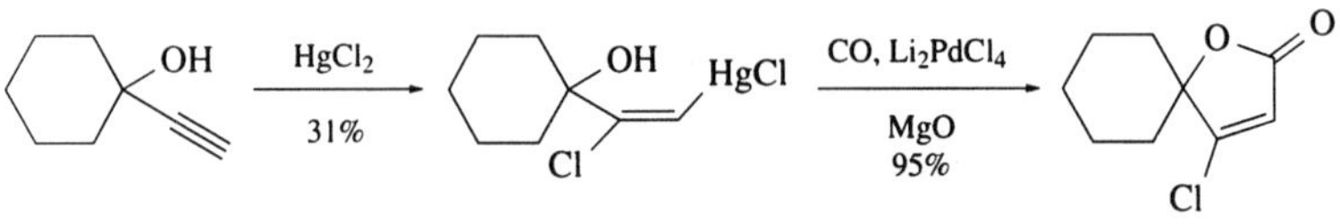

Scheme 57

9.12.11 Miscellaneous Mercuration Reactions

There are a few other solvomercuration reactions which are known, but few appear to be very general. Mercuric nitrate reacts with alkenes to generate β-nitratoalkylmercurials, which react with Br_2 to produce the corresponding bromides (Equation (196)).[248] Alkenes also react directly with $Hg(NO_3)_2$, $Hg(O_3SMe)_2$, $Hg(O_3SC_6H_5Me-p)_2$, and $Hg(O_2SC_6H_5Me-p)_2$ plus Br_2 or I_2 to produce β-substituted organic halides (Equation (197)).[246] Similar reactions have been reported for I_2 and internal alkynes, where benzyl sulfide and thiocyanate have also been used as nucleophiles (see Equation (156)).[198]

$$\text{MeCH=CH}_2 \xrightarrow[\text{ii, Br}_2]{\text{i, Hg(NO}_3)_2 \cdot 1/2\, \text{H}_2\text{O}} \text{MeCHCH}_2\text{Br} \quad\quad \overset{\text{ONO}_2}{|} \tag{196}$$

$$\text{cyclohexene} + \text{HgX}_2 + \text{Y}_2 \longrightarrow \text{trans-cyclohexane-X,Y} \tag{197}$$

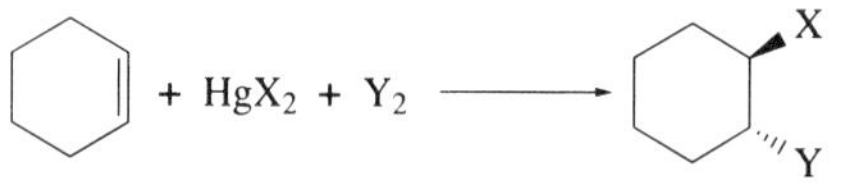

X = NO_3, $MeSO_3$, $p\text{-}MeC_6H_4SO_3$, $p\text{-}MeC_6H_4SO_2$
Y = Br, I

Mercuric acetate is reported to cyclize alkenyl oximes to nitrones (Equation (198)).[249]

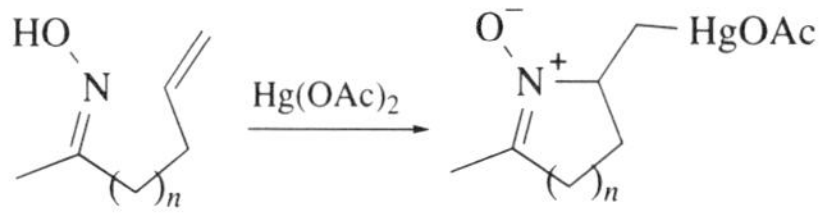

$$\text{(alkenyl oxime)} \xrightarrow{\text{Hg(OAc)}_2} \text{(nitrone-HgOAc)} \tag{198}$$

n = 1,2

Conjugated dienes react with $HgCl_2$ and NaO_2SPh to yield sulfonyl-containing mercurials (Equation (199)).[250]

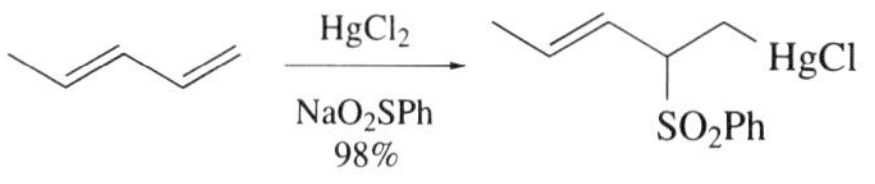

$$\text{(diene)} \xrightarrow[\text{NaO}_2\text{SPh}]{\text{HgCl}_2} \text{(sulfonyl mercurial)} \quad 98\% \tag{199}$$

Alkynes undergo addition of mercury and SCN[88a,251] or phosphorus[252] (Equation (200)).

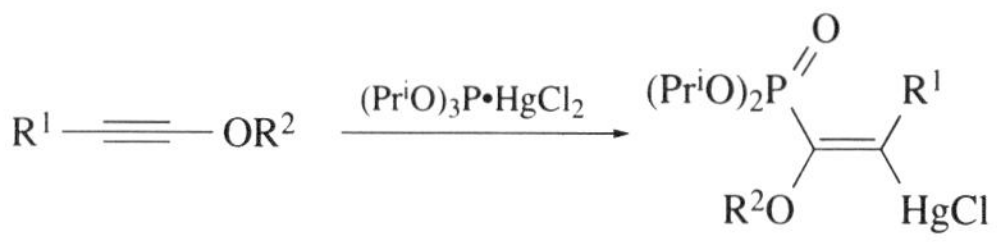

$$R^1\!\!\equiv\!\!\text{—}OR^2 \xrightarrow{(Pr^iO)_3P \cdot HgCl_2} (Pr^iO)_2P(=O)\text{—C(R}^1\text{)=C(OR}^2\text{)(HgCl)} \tag{200}$$

Mercurated benzothiophenes have been produced by intramolecular thiomercuration (Equation (201)).[115b]

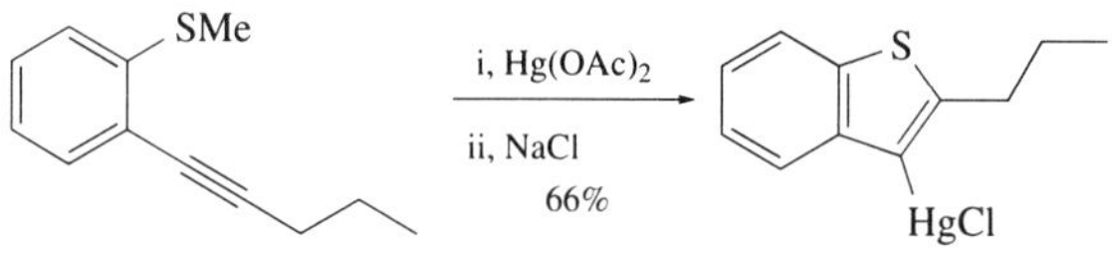

$$\text{(o-SMe aryl alkyne)} \xrightarrow[\text{ii, NaCl}]{\text{i, Hg(OAc)}_2} \text{(mercurated benzothiophene)} \quad 66\% \tag{201}$$

Finally, mercurated alkynes react with hydrazines to afford hydrazones (Scheme 58).[253]

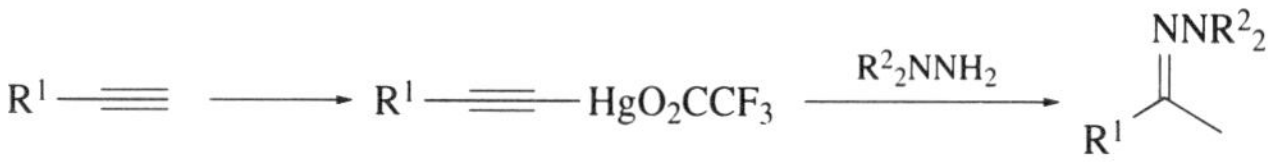

$$R^1\!\!\equiv\!\! \longrightarrow R^1\!\!\equiv\!\!\text{—HgO}_2\text{CCF}_3 \xrightarrow{R^2_2\text{NNH}_2} R^1\text{C(=NNR}^2_2\text{)CH}_3$$

Scheme 58

It is a safe bet that further, new applications of the solvomercuration process will continue to be reported in the future and the better known processes will continue to be applied to increasingly complex molecules and natural products.

9.13 CONCLUSION

Organomercurials are among the most useful organometallics for organic synthesis. They are readily available by a range of methodology, including the direct mercuration of unsaturated organic compounds. Numerous synthetic applications of these organometallics have been reported, including reduction, halogenation, heteroatom displacements, alkylation, dimerization, alkene and alkyne addition, and substitution processes, carbonylation, and acylation. Many of these reactions are promoted by transition metals and proceed under very mild reaction conditions. Organomercurials have proven useful as divalent carbon transfer agents, particularly for the conversion of alkenes to cyclopropanes. Finally, the solvomercuration process has provided a very valuable method for the Markovnikov functionalization of alkenes and alkynes, which is one of the most useful of all applications of organometallics in organic synthesis.

9.14 REFERENCES

1. F. C. Whitmore, 'Organic Compounds of Mercury', Chemical Catalog Co., New York, 1921.
2. L. G. Makarova and A. N. Nesmeyanov, (eds.), 'The Organic Compounds of Mercury, Methods of Elemento-Organic Chemistry', North-Holland, Amsterdam, 1967, vol. 4.
3. H. Staub, K. P. Zeller and H. Leditsche, 'Metallorganische Verbindungen–Hg, Houben-Weyl: Methoden der Organischen Chemie', 4th edn., G. Thieme, Stuttgart, 1974, vol. 13/2b.
4. A. J. Bloodworth, in 'The Chemistry of Mercury', ed. C. A. McAuliffe, Macmillan, Toronto, 1977.
5. J. L. Wardell, in 'COMC-I', vol. 2, p. 863.
6. W. Carruthers, in 'COMC-I', vol. 7, p. 671.
7. L. G. Makarova, *Organomet. React.*, 1970, **1**, 119.
8. L. G. Makarova, *Organomet. React.*, 1971, **2**, 335.
9. R. C. Larock, *Tetrahedron*, 1982, **38**, 1713.
10. R. C. Larock, 'Organomercury Compounds in Organic Synthesis', Springer, New York, 1985.
11. R. C. Larock, 'Solvomercuration/Demercuration Reactions in Organic Synthesis', Springer, New York, 1986.
12. A. A. Baldoni and J. J. Miyashiro, *US Pat.* 2 914 451 (1959) (*Chem. Abstr.*, 1960, **54**, 5467e).
13. E. P. Blanchard, Jr., D. C. Blomstrom and H. E. Simmons, *J. Organomet. Chem.*, 1965, **3**, 97.
14. R. C. Larock and Y.-d. Lu, *J. Org. Chem.*, 1993, **58**, 2846.
15. J. L. Maynard, *J. Am. Chem. Soc.*, 1932, **54**, 2108.
16. R. C. Larock and M.-S. Chow, *Tetrahedron Lett.*, 1984, **25**, 2727.
17. H. F. Lewis and E. Chamberlin, *J. Am. Chem. Soc.*, 1929, **51**, 291.
18. G. A. Tedoradze, *J. Organomet. Chem.*, 1975, **88**, 1.
19. A. Razavi, M. D. Rausch and H. G. Alt, *J. Organomet. Chem.*, 1987, **329**, 281.
20. R. C. Larock, *Intra-Sci. Chem. Rep.*, 1973, **7**(1), 95.
21. R. S. Varma, S. A. Kunda and G. W. Kabalka, *J. Organomet. Chem.*, 1984, **276**, 311.
22. R. C. Larock and K. Narayanan, *J. Org. Chem.*, 1984, **49**, 3411.
23. H. C. Brown, R. C. Larock, S. K. Gupta, S. Rajagopalan and N. G. Bhat, *J. Org. Chem.*, 1989, **54**, 6079.
24. U. Schöllkopf and F. Gerhart, *Angew. Chem.*, 1966, **78**, 675 (*Angew. Chem., Int. Ed. Engl.*, 1966, **5**, 664).
25. C. H. Winter, Y.-H. Han and M. J. Heeg, *Organometallics*, 1992, **11**, 3169.
26. K. M. Smith, K. C. Langry and O. M. Minnetian, *J. Org. Chem.*, 1984, **49**, 4602.
27. S. S. Novikov, T. I. Godovikova and V. A. Tartakovskii, *Izv. Akad. Nauk SSSR, Otdel. Khim. Nauk*, 1960, 669 (*Bull. Acad. Sci. USSR, Div. Chem. Sci.*, 1960, 632).
28. C. H. DePuy, *Topics Curr. Chem.*, 1973, **40**, 74.
29. J. B. Lambert, E. C. Chelius, R. H. Bible, Jr. and E. Hajdu, *J. Am. Chem. Soc.*, 1991, **113**, 1331.
30. (a) E. Nakamura, J.-i. Shimada and I. Kuwajima, *Organometallics*, 1985, **4**, 641; (b) T. Sugimura, K. Koguro and A. Tai, *Tetrahedron Lett.*, 1993, **34**, 509; (c) D. B. Collum, F. Mohamadi and J. S. Hallock, *J. Am. Chem. Soc.*, 1983, **105**, 6882; (d) A. J. Bloodworth and D. Korkodilos, *Tetrahedron Lett.*, 1991, **32**, 6953; (e) D. B. Collum, W. C. Still and F. Mohamadi, *J. Am. Chem. Soc.*, 1986, **108**, 2094.
31. (a) V. I. Sokolov, V. V. Bashilov and O. A. Reutov, *J. Organomet. Chem.*, 1978, **162**, 271; (b) R. G. Smith, H. E. Ensley and H. E. Smith, *J. Org. Chem.*, 1972, **37**, 4430.
32. G. B. Deacon, *Organomet. Chem. Rev. A*, 1970, **5**, 355.
33. W. Peters, *Ber. Dtsch. Chem. Ges.*, 1905, **38**, 2567.
34. F. R. Jensen, 'Electrophilic Substitution of Organomercurials', McGraw-Hill, New York, 1968.
35. W. Steinkopf and M. Boëtius, *Liebigs Ann. Chem.*, 1941, **546**, 208.
36. I. P. Beletskaya, V. I. Karpov, V. A. Moskalenko and O. A. Reutov, *Dokl. Akad. Nauk SSSR*, 1965, **162**, 86 (*Proc. Acad. Sci. USSR, Chem. Sec.*, 1965, **162**, 414).
37. F. H. Gouzoules and R. A. Whitney, *J. Org. Chem.*, 1986, **51**, 2024.
38. (a) H. C. Brown and P. J. Geoghegan, Jr., *J. Org. Chem.*, 1970, **35**, 1844; (b) H. C. Brown and P. Geoghegan, Jr., *J. Am. Chem. Soc.*, 1967, **89**, 1522.
39. D. J. Pasto and J. A. Gontarz, *J. Am. Chem. Soc.*, 1969, **91**, 719.
40. G. Etemad-Moghadam, M. C. Benhamou, V. Spéziale, A. Lattes and A. Bielawska, *Nouv. J. Chim.*, 1980, **4**, 727.
41. (a) F. R. Jensen, J. J. Miller, S. J. Cristol and R. S. Beckley, *J. Org. Chem.*, 1972, **37**, 4341; (b) W. Kitching, A. R. Atkins, G. Wickham and V. Alberts, *ibid.*, 1981, **46**, 563.
42. G. W. Kabalka and R. S. Varma, *Tetrahedron*, 1989, **45**, 6601.
43. A. J. Bloodworth, K. H. Chan and C. J. Cooksey, *J. Org. Chem.*, 1986, **51**, 2110.
44. O. R. Martin, F. Xie, R. Kakarla and R. Benhamza, *Synlett*, 1993, 165.

45. C. P. Casey, G. M. Whitesides and J. Kurth, *J. Org. Chem.*, 1973, **38**, 3406.
46. R. C. Larock and M.-S. Chow, *Organometallics*, 1986, **5**, 603.
47. G. B. Deacon, J. M. Miller and B. S. F. Taylor, *Aust. J. Chem.*, 1975, **28**, 1499.
48. P. V. Roling and M. D. Rausch, *J. Org. Chem.*, 1974, **39**, 1420.
49. E. B. Merkushev, *Synthesis*, 1988, 923.
50. G. W. M. Visser, B. W. v. Halteren, J. D. M. Herscheid, G. A. Brinkman and A. Hoekstra, *J. Chem. Soc., Chem. Commun.*, 1984, 655.
51. D. Hebel and S. Rozen, *J. Org. Chem.*, 1987, **52**, 2588.
52. (a) S. Hanessian, J. Kloss and T. Sugawara, *J. Am. Chem. Soc.*, 1986, **108**, 2758; (b) J. C. Sih, R. A. Johnson, E. G. Nidy and D. R. Graber, *Prostaglandins*, 1978, **15**, 409.
53. (a) P. E. Pike, P. G. Marsh, R. E. Erickson and W. L. Waters, *Tetrahedron Lett.*, 1970, 2679; (b) W. L. Waters, P. E. Pike and J. G. Rivera, *Adv. Chem. Ser.*, 1972, **112**, 78.
54. J. H. Robson and G. F. Wright, *Can. J. Chem.*, 1960, **38**, 1.
55. (a) D. J. Foster and E. Tobler, *J. Am. Chem. Soc.*, 1961, **83**, 851. (b) D. J. Foster and E. Tobler, *J. Org. Chem.*, 1962, **27**, 834.
56. R. C. Larock, K. Oertle and K. M. Beatty, *J. Am. Chem. Soc.*, 1980, **102**, 1966.
57. (a) H. Arzoumanian and J. Metzger, *Synthesis*, 1971, 527; (b) D. J. Rawlinson and G. Sosnovsky, *ibid.*, 1973, 567.
58. (a) W. Treibs, *Naturwiss.*, 1948, **35**, 125; (b) W. Treibs and M. Weissenfels, *Chem. Ber.*, 1960, **93**, 1374.
59. (a) G. T. Rodeheaver and D. F. Hunt, *J. Chem. Soc., Chem. Commun.*, 1971, 818; (b) D. F. Hunt and G. T. Rodeheaver, *Tetrahedron Lett.*, 1972, 3595.
60. S. W. Breuer, M. J. Leatham and F. G. Thorpe, *J. Chem. Soc., Chem. Commun.*, 1971, 1475.
61. R. C. Larock, *J. Org. Chem.*, 1974, **39**, 834.
62. G. M. La Roy and E. C. Kooyman, *J. Organomet. Chem.*, 1967, **7**, 357.
63. (a) G. A. Russell, P. Ngoviwatchai, H. I. Tashtoush, A. Pla-Dalmau and R. K. Khanna, *J. Am. Chem. Soc.*, 1988, **110**, 3530; (b) G. A. Russell, P. Ngoviwatchai, H. Tashtoush and J. Hershberger, *Organometallics*, 1987, **6**, 1414; (c) M. D. Erion and J. E. McMurry, *Tetrahedron Lett.*, 1985, **26**, 559; (d) G. A. Russell and J. Hershberger, *J. Am. Chem. Soc.*, 1980, **102**, 7603.
64. (a) T. W. Campbell and J. D. McCullough, *J. Am. Chem. Soc.*, 1945, **67**, 1965; (b) S. Raucher, M. R. Hansen and M. A. Colter, *J. Org. Chem.*, 1978, **43**, 4885.
65. I. P. Beletskaya, V. B. Vol'eva and O. A. Reutov, *Dokl. Akad. Nauk SSSR*, 1972, **204**, 93 (*Proc. Acad. Sci. USSR, Chem. Sec.*, 1972, **204**, 383).
66. E. Tobler and D. J. Foster, *Z. Naturforsch.*, 1962, **17B**, 135.
67. J. Hershberger and G. A. Russell, *Synthesis*, 1980, 475.
68. V. A. Nefedov, *Zh. Obshch. Khim.*, 1968, **38**, 2184 (*J. Gen. Chem. USSR*, 1968, **38**, 2115).
69. M. Giffard and J. Cousseau, *J. Organomet. Chem.*, 1980, **201**, C1.
70. A. N. Nesmeyanov, I. F. Lutsenko, R. M. Khomutov and V. A. Dubovitskii, *Zh. Obshch. Khim.*, 1959, **29**, 2817 (*J. Gen. Chem. USSR*, 1959, **29**, 2776).
71. (a) P. Tarrant and D. E. O'Connor, *J. Org. Chem.*, 1964, **29**, 2012; (b) E. Robson, J. M. Tedder and D. J. Woodcock, *J. Chem. Soc. C*, 1968, 1324; (c) A. Baeyer, *Ber. Dtsch. Chem. Ges.*, 1874, **7**, 1638. (d) L. I. Smith and F. L. Taylor, *J. Am. Chem. Soc.*, 1935, **57**, 2460. (e) R. J. Cross and N. H. Tennent, *J. Organomet. Chem.*, 1973, **61**, 33. (f) L. M. Stock and T. L. Wright, *J. Org. Chem.*, 1977, **42**, 2875.
72. Yu. G. Gololobov, T. F. Dmitrieva and L. Z. Soborovskii, *Probl. Org. Sinteza, Akad. Nauk SSSR, Otdel. Obshch. i Tekhn. Khim.*, 1965, **314**, (*Chem. Abstr.*, 1966, **64**, 6683h).
73. (a) A. N. Nesmeyanov, I. F. Lutsenko, Z. S. Kraits and A. P. Bokovoi, *Dokl. Akad. Nauk SSSR*, 1959, **124**, 1251 (*Proc. Acad. Sci. USSR, Chem. Sec.*, 1959, **124**, 155); (b) I. F. Lutsenko and Z. S. Kraits, *Zh. Obshch. Khim.*, 1962, **32**, 1663 (*J. Gen. Chem. USSR*, 1962, **32**, 1645); (c) I. F. Lutsenko and Z. S. Kraits, *Dokl. Akad. Nauk SSSR*, 1960, **135**, 860 (*Proc. Acad. Sci. USSR, Chem. Sec.*, 1960, **135**, 1371); (d) P. S. Magee, *Tetrahedron Lett.*, 1965, 3995; (e) E. M. Gaydou, A. Guillemonat and M. T. Bendayan, *Bull. Soc. Chim. France*, 1975, 805; (f) Z. S. Novikova, N. P. Sadovnikov, S. N. Zdorova and I. F. Lutsenko, *Zh. Obshch. Khim.*, 1974, **44**, 2233 (*J. Gen. Chem. USSR*, 1974, **44**, 2189).
74. R. C. Larock, *J. Org. Chem.*, 1976, **41**, 2241.
75. R. C. Larock and B. Riefling, *J. Org. Chem.*, 1978, **43**, 1468.
76. R. C. Larock and J. C. Bernhardt, *J. Org. Chem.*, 1977, **42**, 1680.
77. (a) J. H. Robson and G. F. Wright, *Can. J. Chem.*, 1960, **38**, 21. (b) J. Barluenga and A. M. Mastral, *An. Quim.*, 1977, **73**, 1032 (*Chem. Abstr.*, 1978, **88**, 105 502k).
78. (a) G. A. Russell, J. Hershberger and K. Owens, *J. Am. Chem. Soc.*, 1979, **101**, 1312; (b) G. A. Russell, S. V. Kulkarni and R. K. Khanna, *J. Org. Chem.*, 1990, **55**, 1080.
79. (a) G. A. Russell, P. Ngoviwatchai and H. I. Tashtoush, *Organometallics*, 1988, **7**, 696; (b) G. A. Russell, H. Tashtoush and P. Ngoviwatchai, *J. Am. Chem. Soc.*, 1984, **106**, 4622; (c) G. A. Russell and P. Ngoviwatchai, *Tetrahedron Lett.*, 1987, **28**, 6113.
80. G. A. Russell, P. Ngoviwatchai and Y. W. Wu, *J. Am. Chem. Soc.*, 1989, **111**, 4921.
81. G. A. Russell, D. Guo and R. K. Khanna, *J. Org. Chem.*, 1985, **50**, 3423.
82. R. C. Larock and S. S. Hershberger, *Tetrahedron Lett.*, 1981, **22**, 2443.
83. N. A. Bumagin, P. G. More and I. P. Beletskaya, *J. Organomet. Chem.*, 1989, **364**, 231.
84. R. C. Larock and D. R. Leach, *Organometallics*, 1982, **1**, 74.
85. (a) R. P. Kozyrod and J. T. Pinhey, *Tetrahedron Lett.*, 1982, **23**, 5365. (b) M. G. Moloney and J. T. Pinhey, *J. Chem. Soc., Perkin Trans. 1*, 1988, 2847. (c) M. G. Moloney, J. T. Pinhey and E. G. Roche, *J. Chem. Soc., Perkin Trans. 1*, 1989, 333. (d) M. G. Moloney and J. T. Pinhey, *J. Chem. Soc., Chem. Commun.*, 1984, 965.
86. J. Barluenga, F. J. Fañanás, J. Villamaña and M. Yus, *J. Chem. Soc., Perkin Trans. 1*, 1984, 2685.
87. (a) V. A. Nikanorov, V. I. Rozenberg, G. V. Gavrilova, Yu. G. Bundel and O. A. Reutov, *Izv. Akad. Nauk SSSR, Ser. Khim.*, 1975, 1675 (*Bull. Acad. Sci. USSR, Div. Chem. Sci.*, 1975, 1568); (b) H. C. Gardner and J. K. Kochi, *J. Am. Chem. Soc.*, 1976, **98**, 2460; (c) O. A. Reutov, V. I. Rozenberg, G. A. Gavrilova and V. A. Nikanorov, *J. Organomet. Chem.*, 1979, **177**, 101; (d) G. A. Russell, W. Jiang, S. S. Hu and R. K. Khanna, *J. Org. Chem.*, 1986, **51**, 5498; (e) G. A. Russell, R. K. Khanna and D. Guo, *J. Chem. Soc., Chem. Commun.*, 1986, 632.

88. (a) U. Blaukat and W. P. Neumann, *J. Organomet. Chem.*, 1973, **49**, 323; (b) W. P. Neumann and U. Blaukat, *Angew. Chem.*, 1969, **81**, 625 (*Angew. Chem., Int. Ed. Engl.*, 1969, **8**, 611).
89. B. Giese, *Angew. Chem., Int. Ed. Engl.*, 1985, **24**, 553.
90. S. Danishefsky and E. Taniyama, *Tetrahedron Lett.*, 1983, **24**, 15.
91. G. A. Russell and B. Z. Shi, *Synlett*, 1993, 701.
92. R. F. Heck, *Organic Reactions*, 1982, **27**, 345.
93. R. C. Larock, D. E. Stinn and M.-Y. Kuo, *Tetrahedron Lett.*, 1990, **31**, 17.
94. R. C. Larock and D. E. Stinn, *Tetrahedron Lett.*, 1989, **30**, 2767.
95. (a) H. Horino and N. Inoue, *Bull. Chem. Soc. Jpn.*, 1971, **44**, 3210; (b) H. Horino, M. Arai and N. Inoue, *ibid.*, 1974, **47**, 1683.
96. (a) H. Horino and N. Inoue, *J. Chem. Soc., Chem. Commun.*, 1976, 500; (b) M. Ishiguro, T. Tatsuoka and N. Nakatsuka, *Tetrahedron Lett.*, 1982, **23**, 3859.
97. H. Horino and N. Inoue, *Heterocycles*, 1978, **11**, 281.
98. S. Cacchi, F. La-Torre and D. Misiti, *Tetrahedron Lett.*, 1979, **20**, 4591.
99. (a) S. Cacchi and D. Misiti, *J. Org. Chem.*, 1982, **47**, 2995; (b) S. Cacchi and G. Palmieri, *Tetrahedron*, 1983, **39**, 3373.
100. P. K. Singh, B. K. Rohtagi and R. N. Khanna, *Synth. Commun.*, 1992, **22**, 987.
101. R. F. Heck, *J. Am. Chem. Soc.*, 1968, **90**, 5526.
102. R. C. Larock, F. Kondo, K. Narayanan, L. K. Sydnes and M.-F. H. Hsu, *Tetrahedron Lett.*, 1989, **30**, 5737.
103. (a) R. F. Heck, *J. Am. Chem. Soc.*, 1968, **90**, 5531; (b) R. C. Larock, J. C. Bernhardt and R. J. Driggs, *J. Organomet. Chem.*, 1978, **156**, 45.
104. (a) R. C. Larock and S. J. Ilkka, *Tetrahedron Lett.*, 1986, **27**, 2211; (b) R. C. Larock and S. K. Stolz-Dunn, *ibid.*, 1988, **29**, 5069; (c) R. C. Larock and S. Ding, *J. Org. Chem.*, 1993, **58**, 2081.
105. R. F. Heck, *J. Am. Chem. Soc.*, 1968, **90**, 5538.
106. R. C. Larock, S. S. Hershberger, K. Takagi and M. A. Mitchell, *J. Org. Chem.*, 1986, **51**, 2450.
107. (a) R. C. Larock, D. R. Leach and S. M. Bjorge, *J. Org. Chem.*, 1986, **51**, 5221. (b) R. C. Larock, K. Narayanan, R. K. Carlson and J. A. Ward, *ibid.*, 1987, **52**, 1364. (c) R. C. Larock, M. H. Hsu and K. Narayanan, *Tetrahedron*, 1987, **43**, 2891.
108. R. C. Larock and M. A. Mitchell, *J. Am. Chem. Soc.*, 1976, **98**, 6718.
109. R. C. Larock, D. J. Leuck and L. W. Harrison, *Tetrahedron Lett.*, 1987, **28**, 4977.
110. R. C. Larock and K. Takagi, *J. Org. Chem.*, 1988, **53**, 4329.
111. R. C. Larock and S. Varaprath, *J. Org. Chem.*, 1984, **49**, 3432.
112. (a) R. F. Heck, *J. Am. Chem. Soc.*, 1968, **90**, 5542; (b) A. Kasahara and T. Izumi, *Bull. Chem. Soc. Jpn.*, 1972, **45**, 1256; (c) R. C. Larock and K. Takagi, *J. Org. Chem.*, 1984, **49**, 2701.
113. (a) R. C. Larock, L. W. Harrison and M. H. Hsu, *J. Org. Chem.*, 1984, **49**, 3662; (b) R. C. Larock and H. Song, *Synth. Commun.*, 1989, **19**, 1463.
114. C. Narayana and M. Periasamy, *Synthesis*, 1985, 253.
115. (a) W. C. Baird, Jr., R. L. Hartgerink and J. H. Surridge, *J. Org. Chem.*, 1985, **50**, 4601; (b) R. C. Larock and L. W. Harrison, *J. Am. Chem. Soc.*, 1984, **106**, 4218.
116. R. C. Larock, *J. Org. Chem.*, 1975, **40**, 3237.
117. R. C. Larock, B. Riefling and C. A. Fellows, *J. Org. Chem.*, 1978, **43**, 131.
118. A. Kasahara, T. Izumi and S. Ohnishi, *Bull. Chem. Soc. Jpn.*, 1972, **45**, 951.
119. D. Seyferth, J. S. Merola and C. S. Eschbach, *J. Am. Chem. Soc.*, 1978, **100**, 4124.
120. R. C. Larock and S. S. Hershberger, *J. Org. Chem.*, 1980, **45**, 3840.
121. I. Rhee, M. Ryang, T. Watanabe, H. Omura, S. Murai and N. Sonada, *Synthesis*, 1977, 776.
122. N. A. Bumagin, P. G. More and I. P. Beletskaya, *J. Organomet. Chem.*, 1989, **365**, 379.
123. D. S. Ryu, K. H. Song and J. T. Lee, *Bull. Korean Chem. Soc.*, 1992, **13**, 354 (*Chem. Abstr.*, 1992, **117**, 212 093s).
124. B. P. Gusev, E. A. El'perina and V. F. Kucherov, *Izv. Akad. Nauk SSSR, Ser. Khim.*, 1980, 603 (*Bull. Acad. Sci. USSR, Div. Chem. Sci.*, 1980, 418).
125. A. L. Kurts, I. P. Beletskaya, I. A. Savchenko and O. A. Reutov, *J. Organomet. Chem.*, 1969, **17**, P21.
126. R. C. Larock and J. C. Bernhardt, *J. Org. Chem.*, 1978, **43**, 710.
127. R. C. Larock, M.-S. Chow and S. J. Smith, *J. Org. Chem.*, 1986, **51**, 2623.
128. (a) K. Takagi, T. Okamoto, Y. Sakakibara, A. Ohno, S. Oka and N. Hayama, *Chem. Lett.*, 1975, 951; (b) N. A. Bumagin, A. B. Ponomaryov and I. P. Beletskaya, *J. Organomet. Chem.*, 1985, **291**, 129.
129. (a) D. Seyferth, *Pure Appl. Chem.*, 1970, **23**, 391; (b) D. Seyferth, *Acc. Chem. Res.*, 1972, **5**, 65; (c) O. M. Nefedov, A. I. D'yachenko and A. K. Prokof'ev, *Russ. Chem. Rev.*, 1977, **46**, 941.
130. (a) E. V. Dehmlow, *J. Organomet. Chem.*, 1966, **6**, 296; (b) D. Seyferth and R. Damrauer, *J. Org. Chem.*, 1966, **31**, 1660.
131. C. W. Jefford, D. T. Hill, J. Goré and B. Waegell, *Helv. Chim. Acta*, 1972, **55**, 790.
132. (a) W. E. Parham and J. K. Rinehart, *J. Am. Chem. Soc.*, 1967, **89**, 5668; (b) W. E. Parham, D. R. Johnson, C. T. Hughes, M. K. Meilahn and J. K. Rinehart, *J. Org. Chem.*, 1970, **35**, 1048.
133. W. E. Parham, R. W. Davenport and J. B. Biasotti, *Tetrahedron Lett.*, 1969, 557.
134. (a) S. D. Saraf, *Synthesis*, 1971, 264; (b) M. V. Moncur and J. B. Grutzner, *J. Chem. Soc., Chem. Commun.*, 1972, 667.
135. G. Stork, M. Nussim and B. August, *Tetrahedron Suppl. 8 Part 1*, 1966, 105.
136. (a) D. Seyferth *et al.*, *J. Am. Chem. Soc.*, 1965, **87**, 4259; (b) D. Seyferth and J. Y.-P. Mui, *ibid.*, 1966, **88**, 4672.
137. (a) D. Seyferth and Y.-M. Cheng, *Synthesis*, 1974, 114; (b) D. Seyferth, V. A. Mai and M. E. Gordon, *J. Org. Chem.*, 1970, **35**, 1993.
138. H. C. Brown, P. J. Geoghegan, Jr. and J. T. Kurek, *J. Org. Chem.*, 1981, **46**, 3810.
139. (a) H. C. Brown and W. J. Hammar, *J. Am. Chem. Soc.*, 1967, **89**, 1524. (b) H. C. Brown, J. H. Kawakami and S. Ikegami, *ibid.*, 1967, **89**, 1525.
140. (a) J. Halpern and H. B. Tinker, *J. Am. Chem. Soc.*, 1967, **89**, 6427; (b) P. Abley, J. E. Byrd and J. Halpern, *ibid.*, 1973, **95**, 2591; (c) H. C. Brown and P. J. Geoghegan, Jr., *J. Org. Chem.*, 1972, **37**, 1937.
141. (a) H. C. Brown and J. H. Kawakami, *ibid.*, 1973, **95**, 8665; (b) H. C. Brown and K.-T. Liu, *J. Am. Chem. Soc.*, 1970, **92**, 3502.
142. (a) D. J. Nelson, P. J. Cooper and R. Soundararajan, *J. Am. Chem. Soc.*, 1989, **111**, 1414; (b) H. B. Vardhan and R. D. Bach, *J. Org. Chem.*, 1992, **57**, 4948.

143. H. C. Brown, P. J. Geoghegan, Jr., J. T. Kurek and G. J. Lynch, *Organomet. Chem. Syn.*, 1970, **1**, 7.
144. B. Giese and D. Bartmann, *Tetrahedron Lett.*, 1985, **26**, 1197.
145. (a) R. Blattner, R. J. Ferrier and S. R. Haines, *J. Chem. Soc., Perkin Trans. 1*, 1985, 2413. (b) F. Chretien and Y. Chapleur, *J. Chem. Soc., Chem. Commun.*, 1984, 1268. (c) R. J. Ferrier, *J. Chem. Soc., Perkin Trans. 1*, 1979, 1455; (d) R. J. Ferrier and P. Prasit, *Carbohydr. Res.*, 1980, **82**, 263.
146. J. Barluenga, J. López-Prado, P. J. Campos and G. Asensio, *Tetrahedron*, 1983, **39**, 2863.
147. J. Barluenga, L. Alonso-Cires, P. J. Campos and G. Asensio, *Tetrahedron*, 1984, **40**, 2563.
148. H. C. Brown, P. J. Geoghegan, Jr., G. J. Lynch and J. T. Kurek, *J. Org. Chem.*, 1972, **37**, 1941.
149. M. C. S. de Mattos, W. B. Kover, F. Aznar and J. Barluenga, *Tetrahedron Lett.*, 1992, **33**, 4863.
150. (a) C. M. Link, D. K. Jansen and C. N. Sukenik, *J. Am. Chem. Soc.*, 1980, **102**, 7798; (b) J. K. Sutter and C. N. Sukenik, *J. Org. Chem.*, 1982, **47**, 4174.
151. J. Einhorn, C. Einhorn and J. L. Luche, *J. Org. Chem.*, 1989, **54**, 4479.
152. R. R. Vogt and J. A. Nieuwland, *J. Am. Chem. Soc.*, 1921, **43**, 2071.
153. (a) G. Stork and R. Borch, *J. Am. Chem. Soc.*, 1964, **86**, 935. (b) S. Padmanabhan and K. M. Nicholas, *Synth. Commun.*, 1980, **10**, 503.
154. F. P. Ballistreri, S. Failla and G. A. Tomaselli, *J. Org. Chem.*, 1988, **53**, 830.
155. (a) H. C. Brown, J. T. Kurek, M.-H. Rei and K. L. Thompson, *J. Org. Chem.*, 1984, **49**, 2551; (b) H. C. Brown, J. T. Kurek, M.-H. Rei and K. L. Thompson, *ibid.*, 1985, **50**, 1171; (c) H. C. Brown and M.-H. Rei, *J. Am. Chem. Soc.*, 1969, **91**, 5646.
156. M. Livneh, J. K. Sutter and C. N. Sukenik, *J. Org. Chem.*, 1987, **52**, 5039.
157. S. Honda, K. Kakehi, H. Takai and K. Takiura, *Carbohydr. Res.*, 1973, **29**, 477.
158. M. L. Wood, R. J. Madden and H. E. Carter, *J. Biol. Chem.*, 1937, **117**, 1.
159. J. Bain and M. M. Harding, *J. Chem. Soc.*, 1965, 4025.
160. F. H. Gouzoules and R. A. Whitney, *Tetrahedron Lett.*, 1985, **26**, 3441.
161. R. K. Boeckman, Jr. and C. J. Flann, *Tetrahedron Lett.*, 1983, **24**, 4923.
162. A. N. Nesmeyanov, I. F. Lutsenko and R. M. Khomutov, *Izv. Akad. Nauk SSSR, Otdel. Khim. Nauk*, 1957, 942 (*Bull. Acad. Sci. USSR, Div. Chem. Sci. USSR*, 1957, 971).
163. N. L. Weinberg and G. F. Wright, *Can. J. Chem.*, 1965, **43**, 24.
164. A. Lethbridge, R. O. C. Norman and C. B. Thomas, *J. Chem. Soc., Perkin Trans. 1*, 1975, 2465.
165. F. G. Bordwell and M. L. Douglass, *J. Am. Chem. Soc.*, 1966, **88**, 993.
166. P. Kočovský, *Organometallics*, 1993, **12**, 1969.
167. L. E. Overman and C. B. Campbell, *J. Org. Chem.*, 1974, **39**, 1474.
168. J. Cai and A. G. Davies, *J. Chem. Soc., Perkin Trans. 1*, 1992, 3383.
169. T. Nishiyama, J. F. Woodhall, E. N. Lawson and W. Kitching, *J. Org. Chem.*, 1989, **54**, 2183.
170. (a) M. C. Benhamou, G. Etemad-Moghadam, V. Spéziale and A. Lattes, *Synthesis*, 1979, 891; (b) J.-R. Pougny, M. A. M. Nassr and P. Sinaÿ, *J. Chem. Soc., Chem. Commun.*, 1981, 375.
171. F. Paquet and P. Sinaÿ, *Tetrahedron Lett.*, 1984, **25**, 3071.
172. W. L. Waters and E. F. Kiefer, *J. Am. Chem. Soc.*, 1967, **89**, 6261.
173. D. J. Pasto and K. D. Sugi, *J. Org. Chem.*, 1991, **56**, 4157.
174. R. D. Walkup and G. Park, *J. Am. Chem. Soc.*, 1990, **112**, 1597.
175. (a) K. H. McNeely and G. F. Wright, *J. Am. Chem. Soc.*, 1955, **77**, 2553; (b) A. J. Bloodworth, M. G. Hutchings and A. J. Sotowicz, *J. Chem. Soc., Chem. Commun.*, 1976, 578.
176. J. R. Johnson, W. H. Jobling and G. W. Bodamer, *J. Am. Chem. Soc.*, 1941, **63**, 131.
177. B. Giese, K. Heuck and U. Lüning, *Tetrahedron Lett.*, 1981, **22**, 2155.
178. J. Barluenga, J. Pérez-Prieto and G. Asensio, *J. Chem. Soc., Perkin Trans. 1*, 1984, 629.
179. (a) M. Bassetti and G. Bocelli, *J. Chem. Soc., Chem. Commun.*, 1990, 257. (b) M. Bassetti and M. P. Trovato, *Organometallics*, 1990, **9**, 2292.
180. (a) V. I. Sokolov and O. A. Reutov, *J. Org. Chem. USSR*, 1969, **5**, 168; (b) D. H. Ballard, A. J. Bloodworth and R. J. Bunce, *J. Chem. Soc., Chem. Commun.*, 1969, 815.
181. N. A. Porter, P. J. Zuraw and J. A. Sullivan, *Tetrahedron Lett.*, 1984, **25**, 807.
182. E. Schmitz, A. Rieche and O. Brede, *J. Prakt. Chem.*, 1970, **312**, 30.
183. W. Adam, A. J. Bloodworth, H. J. Eggelte and M. E. Loveitt, *Angew. Chem.*, 1978, **90**, 216 (*Angew. Chem., Int. Ed. Engl.*, 1978, **17**, 209).
184. A. J. Bloodworth and J. L. Courtneidge, *J. Chem. Soc., Chem. Commun.*, 1981, 1117.
185. A. J. Bloodworth and I. M. Griffin, *J. Chem. Soc., Perkin Trans. 1*, 1974, 688.
186. H. Mimoun, R. Charpentier, A. Mitschler, J. Fischer and R. Weiss, *J. Am. Chem. Soc.*, 1980, **102**, 1047.
187. H. C. Brown, M.-H. Rei and K.-T. Liu, *J. Am. Chem. Soc.*, 1970, **92**, 1760.
188. (a) M. Julia and C. Blasioli, *Bull. Soc. Chim. France*, 1976, 1941. (b) H. Yoshioka, K. Takasaki, M. Kobayashi and T. Matsumoto, *Tetrahedron Lett.*, 1979, **20**, 3489. (c) S. F. Martin and T.-S. Chou, *ibid.*, 1978, 1943. (d) G. Fráter, *ibid.*, 1981, **22**, 425.
189. S. Kim and K. H. Uh, *Tetrahedron Lett.*, 1992, **33**, 4325.
190. (a) K. Ichikawa, O. Itoh and T. Kawamura, *Bull. Chem. Soc. Jpn.*, 1968, **41**, 1240; (b) B. K. Nefedov, N. S. Sergeeva and Ya. T. Eidus, *Izv. Akad. Nauk SSSR, Ser. Khim.*, 1972, 2497 (*Bull. Acad. Sci. USSR, Div. Chem. Sci.*, 1972, 2429).
191. O. A. El Seoud, A. T. do Amaral, M. Moura Campos and L. do Amaral, *J. Org. Chem.*, 1974, **39**, 1915.
192. P. A. Bartlett, D. P. Richardson and J. Myerson, *Tetrahedron*, 1984, **40**, 2317.
193. W. S. Linn, W. L. Waters and M. C. Caserio, *J. Am. Chem. Soc.*, 1970, **92**, 4018.
194. S. Wolff and W. C. Agosta, *Tetrahedron Lett.*, 1985, **26**, 703.
195. V. Gómez Aranda, J. Barluenga, M. Yus and G. Asensio, *Synthesis*, 1974, 806.
196. M. Bassetti and B. Floris, *J. Org. Chem.*, 1986, **51**, 4140.
197. (a) G. Drefahl, G. Hueblein and A. Wintzer, *Angew. Chem.*, 1958, **70**, 166; (b) G. Drefahl and S. Schaaf, *Chem. Ber.*, 1957, **90**, 148; (c) S. Uemura, H. Miyoshi and M. Okano, *J. Chem. Soc., Perkin Trans. 1*, 1980, 1098.
198. J. Barluenga, J. M. Martinez-Gallo, C. Nájera and M. Yus, *J. Chem. Soc., Perkin Trans. 1*, 1987, 1017.

199. (a) G. A. Krafft and J. A. Katzenellenbogen, *J. Am. Chem. Soc.*, 1981, **103**, 5459; (b) R. A. Amos and J. A. Katzenellenbogen, *J. Org. Chem.*, 1978, **43**, 560. (c) M. Yamamoto, *J. Chem. Soc., Perkin Trans. 1*, 1981, 582. (d) M. Yamamoto, *J. Chem. Soc., Chem. Commun.*, 1978, 649.
200. M. Yamamoto, M. Yoshitake and K. Yamada, *J. Chem. Soc., Chem. Commun.*, 1983, 991.
201. R. C. Griffith, R. J. Gentile, T. A. Davidson and F. L. Scott, *J. Org. Chem.*, 1979, **44**, 3580.
202. J. Barluenga, A. M. Bayón, J. Pérez-Prieto and G. Asensio, *Tetrahedron*, 1984, **40**, 5053.
203. J. Barluenga, J. Pérez-Prieto, A. M. Bayón and G. Asensio, *Tetrahedron*, 1984, **40**, 1199.
204. J. Barluenga, J. Villamaña and M. Yus, *Synthesis*, 1981, 375.
205. J. J. Perie and A. Lattes, *Bull. Soc. Chim. Fr.*, 1971, 1378.
206. J. Barluenga, P. J. Campos and J. López-Prado, *Synthesis*, 1985, 1125.
207. (a) J. Barluenga, F. J. Fañanás and M. Yus, *J. Org. Chem.*, 1981, **46**, 1281. (b) J. Barluenga, F. J. Fañanás and M. Yus, *ibid.*, 1979, **44**, 4798.
208. J. Barluenga, L. Alonso-Cires and G. Asensio, *Synthesis*, 1979, 962.
209. J. Barluenga, J. Pérez-Prieto and G. Asensio, *Tetrahedron*, 1990, **46**, 2453.
210. W. C. Frank, Y. C. Kim and R. F. Heck, *J. Org. Chem.*, 1978, **43**, 2947.
211. S. R. Wilson and R. A. Sawicki, *J. Org. Chem.*, 1979, **44**, 330.
212. A. P. Kozikowski and J. Scripko, *Tetrahedron Lett.*, 1983, **24**, 2051.
213. M. K. Tong, E. M. Blumenthal and B. Ganem, *Tetrahedron Lett.*, 1990, **31**, 1683.
214. K. C. Majumdar, R. N. De and S. Saha, *Tetrahedron Lett.*, 1990, **31**, 1207.
215. S. Arseniyadis and J. Gore, *Tetrahedron Lett.*, 1983, **24**, 3997.
216. V. Gómez Aranda, J. Barluenga Mur, M. Yus Astiz and F. Aznar, *Rev. Acad. Cienc. Exactas, Fis.-Quim. Natur. Zaragoza*, 1974, **29**, 231 (*Chem. Abstr.*, 1976, **85**, 20 716w).
217. J. Barluenga, J. Pérez-Prieto and G. Asensio, *J. Chem. Soc., Chem. Commun.*, 1982, 1181.
218. J. Barluenga, C. Nájera and M. Yus, *J. Heterocyclic Chem.*, 1981, **18**, 1297.
219. (a) J. Barluenga, C. Nájera and M. Yus, *Synthesis*, 1979, 896; (b) J. Barluenga, C. Nájera and M. Yus, *ibid.*, 1978, 911; (c) J. Barluenga, C. Nájera and M. Yus, *J. Heterocyclic Chem.*, 1980, **17**, 917.
220. J. Barluenga, F. Aznar, M. C. S. de Mattos, W. B. Kover, S. Garcia-Granda and E. Pérez-Carreño, *J. Org. Chem.*, 1991, **56**, 2930.
221. (a) J. Barluenga, F. Aznar, R. Liz and R. Rodes, *J. Chem. Soc., Perkin Trans. 1*, 1980, 2732; (b) H. E. Stavely, *J. Am. Chem. Soc.*, 1940, **62**, 489; (c) J. Barluenga and F. Aznar, *Synthesis*, 1975, 704.
222. J. Barluenga, F. Aznar, R. Liz and R. Rodes, *J. Chem. Soc., Perkin Trans. 1*, 1983, 1087.
223. J. Barluenga, F. Aznar, C. Valdés and M.-P. Cabal, *J. Org. Chem.*, 1991, **56**, 6166.
224. H. C. Brown and J. T. Kurek, *J. Am. Chem. Soc.*, 1969, **91**, 5647.
225. (a) J. Barluenga, C. Jiménez, C. Nájera and M. Yus, *J. Chem. Soc., Perkin Trans. 1*, 1983, 591; (b) J. Barluenga, C. Jiménez, C. Nájera and M. Yus, *J. Chem. Soc., Chem. Commun.*, 1981, 1178.
226. R. V. Stevens and P. M. Kenney, *J. Chem. Soc., Chem. Commun.*, 1983, 384.
227. J. Beger and D. Vogel, *J. Prakt. Chem.*, 1969, **311**, 737.
228. J. M. Takacs, M. A. Helle and L. Yang, *Tetrahedron Lett.*, 1989, **30**, 1777.
229. (a) J. M. Takacs, M. A. Helle and F. Takusagawa, *Tetrahedron Lett.*, 1989, **30**, 7321; (b) J. M. Takacs, M. A. Helle, B. J. Sanyal, T. A. Eberspacher, *ibid.*, 1990, **31**, 6765.
230. K. E. Harding and T. H. Marman, *J. Org. Chem.*, 1984, **49**, 2838.
231. (a) R. Amoroso, G. Cardillo and C. Tomasini, *Heterocycles*, 1992, **34**, 349; (b) R. Amoroso, G. Cardillo, C. Tomasini and P. Tortoreto, *J. Org. Chem.*, 1992, **57**, 1082; (c) K. E. Harding, T. H. Marman and D. Nam, *Tetrahedron*, 1988, **44**, 5605.
232. K. E. Harding, D. R. Hollingsworth and J. Reibenspies, *Tetrahedron Lett.*, 1989, **30**, 4775.
233. K. E. Harding and S. R. Burks, *J. Org. Chem.*, 1984, **49**, 40.
234. (a) S. Danishefsky, E. Taniyama and R. R. Webb II, *Tetrahedron Lett.*, 1983, **24**, 11; (b) W. Carruthers, M. J. Williams and M. T. Cox, *J. Chem. Soc., Chem. Commun.*, 1984, 1235.
235. (a) J. E. Galle and A. Hassner, *J. Am. Chem. Soc.*, 1972, **94**, 3930; (b) C. H. Heathcock, *Angew. Chem.*, 1969, **81**, 148.
236. G. Mehta and P. N. Pandey, *J. Org. Chem.*, 1975, **40**, 3631.
237. (a) G. B. Bachman and M. L. Whitehouse, *J. Org. Chem.*, 1967, **32**, 2303; (b) S. Shinoda and Y. Saito, *J. Organomet. Chem.*, 1975, **90**, 1; (c) E. J. Corey and H. Estreicher, *J. Am. Chem. Soc.*, 1978, **100**, 6294; (d) M. Matsuo and Y. Saito, *J. Organomet. Chem.*, 1971, **27**, C41.
238. (a) S. S. Novikov, T. I. Godovikova and V. A. Tartakovskii, *Dokl. Akad. Nauk SSSR*, 1959, **124**, 834 (*Proc. Acad. Sci. USSR, Chem. Sec.*, 1959, **124**, 59); (b) V. A. Tartakovskii, S. S. Novikov and T. I. Godovikova, *Izv. Akad. Nauk SSSR, Otdel. Khim. Nauk*, 1961, 1042 (*Bull. Acad. Sci. USSR, Div. Chem. Sci.*, 1961, 963).
239. (a) K. Ichikawa, S. Fukushima, H. Ouchi and M. Tsuchida, *J. Am. Chem. Soc.*, 1958, **80**, 6005; (b) M. Julia, E. Colomer Gasquez and R. Labia, *Bull. Soc. Chim. France*, 1972, 4145; (c) M. Julia and R. Labia, *Compt. Rend. C*, 1969, **268**, 104.
240. (a) K. Ichikawa, O. Itoh, T. Kawamura, M. Fujiwara and T. Ueno, *J. Org. Chem.*, 1966, **31**, 447; (b) K. Ichikawa, H. Ouchi and S. Fukushima, *ibid.*, 1959, **24**, 1129.
241. R. Baudouy, F. Delbecq and J. Gore, *Tetrahedron*, 1980, **36**, 189.
242. N. Bluthe, M. Malacria and J. Gore, *Tetrahedron Lett.*, 1982, **23**, 4263.
243. J. G. Traynham, G. R. Franzen, G. A. Knesel and D. J. Northington, Jr., *J. Org. Chem.*, 1967, **32**, 3285.
244. (a) M. Nishizawa, H. Takenaka, H. Nishide and Y. Hayashi, *Tetrahedron Lett.*, 1983, **24**, 2581; (b) A. S. Gopalan, R. Prieto, B. Mueller and D. Peters, *ibid.*, 1992, **33**, 1679; (c) M. Nishizawa, H. Takenaka and Y. Hayashi, *J. Am. Chem. Soc.*, 1985, **107**, 522; (d) M. Nishizawa, H. Takenaka and Y. Hayashi, *J. Org. Chem.*, 1986, **51**, 806.
245. (a) L. Drouin, M.-A. Boaventura and J.-M. Conia, *J. Am. Chem. Soc.*, 1985, **107**, 1726; (b) C. J. Forsyth and J. Clardy, *ibid.*, 1990, **112**, 3497.
246. J. Barluenga, J. M. Martinez-Gallo, C. Nájera and M. Yus, *J. Chem. Soc., Chem. Commun.*, 1985, 1422.
247. B. I. Martynov, S. R. Sterlin and B. L. Dyatkin, *Izv. Akad. Nauk SSSR, Ser. Khim.*, 1974, 1642 (*Bull. Acad. Sci. USSR, Div. Chem. Sci.*, 1974, 1564).
248. A. J. Bloodworth and P. N. Cooper, *J. Chem. Soc., Chem. Commun.*, 1986, 709.
249. R. Grigg, M. Hadjisoteriou, P. Kennewell, J. Markandu and M. Thornton-Pett, *J. Chem. Soc., Chem. Commun.*, 1992, 1388.
250. O. S. Andell and J.-E. Bäckvall, *Tetrahedron Lett.*, 1985, **26**, 4555.

251. M. Giffard, J. Cousseau, L. Gouin and M.-R. Crahe, *Tetrahedron*, 1986, **42**, 2243.
252. M. A. Kazankova, T. Ya. Satina and I. F. Lutsenko, *Zh. Obshch. Khim.*, 1979, **49**, 2414 (*J. Gen. Chem. USSR*, 1979, **49**, 2131).
253. J. Barluenga, F. Aznar, R. Liz and M. Bayod, *J. Chem. Soc., Chem. Commun.*, 1988, 121.

10
Thallium

ISTVAN E. MARKO AND CHIU W. LEUNG
Université Catholique de Louvain, Louvain-la-Neuve, Belgium

10.1 INTRODUCTION	437
10.2 THALLIUM(III) COMPOUNDS	437
10.2.1 Triorganothallium Compounds	437
10.2.1.1 Preparation and properties	437
10.2.1.2 Reactions	438
10.2.2 Diorganothallium Compounds	444
10.2.3 Monoorganothallium Compounds	446
10.2.3.1 Preparation and properties	446
10.2.3.2 Reactions	447
10.3 THALLIUM(I) COMPOUNDS	452
10.3.1 Cyclopentadienylthallium	452
10.4 MISCELLANEOUS	455
10.4.1 Conclusions and Perspectives	458
10.5 REFERENCES	458

10.1 INTRODUCTION

There has been an extensive growth in the development and use of organothallium compounds in organic synthesis since the mid-1980s. This chapter summarizes the chemistry of thallium(I) and thallium(III) species from 1982 to 1994 and their utility as organic and inorganic reagents. Excellent reviews have been published and the reader is referred to them for extensive coverage of the previous literature.[1]

10.2 THALLIUM(III) COMPOUNDS

The organic derivatives of thallium(III) compounds are classified according to the number of organic groups bound to the thallium metal via their carbon atoms.

10.2.1 Triorganothallium Compounds

10.2.1.1 Preparation and properties

Triorganothallium compounds R_3Tl (R = methyl, ethyl, phenyl, etc.) are usually highly reactive species, sensitive to air, water, light and mild acids. Triorganothallium reagents are monomeric, weak Lewis acids, which are highly soluble in organic solvents such as benzene, ether and THF. Their reactivity can be illustrated by the weakness of the thallium–carbon bond. For example, the mean value for the thallium–carbon bond dissociation energy in trimethylthallium is 114.7 kJ mol^{-1} compared with 295 kJ mol^{-1} for monomeric trimethylaluminum.[2]

Triorganothallium compounds (**2**) are readily prepared from the corresponding diorganothallium halides (**1**), which are rather inert organometallic reagents, by the addition of either an organolithium derivative in ether or a Grignard reagent in THF (Equation (1)).[3] Unlike triorganothallium reagents, diorganothallium halides (**1**) are unaffected by water, air, light and mild acids. They are also insoluble in most organic solvents and in water.

$$\underset{(1)}{\underset{R^1}{\overset{R^1}{\diagdown}}Tl-X} \xrightarrow[R^2MgX, THF, 20\,°C]{R^2Li, ether, 20\,°C\text{ or}} \underset{(2)}{\underset{R^1}{\overset{R^1}{\diagdown}}Tl-R^2} \qquad (1)$$

$R^1, R^2 =$ Me, Et, Ph, -C≡Ph, etc.
X = Cl, Br, I

Homo and mixed tetraorganothallium 'ate' complexes (**5**) have also been generated *in situ*, either by the addition of 1 equivalent of RLi to a triorganothallium derivative or 2 equivalents of RLi to a diorganothallium halide in ether at –40 °C (Scheme 1).[4,5]

$$\underset{(3)}{Me_2TlCl} \xrightarrow[\underset{20\,°C}{Et_2O}]{MeLi} \underset{(4)}{Me_3Tl} \xrightarrow[\underset{-40\,°C}{Et_2O}]{RLi} \underset{(5)}{Me_3Tl \cdot RLi}$$

R = Me, Bu, -C≡CPh, -C≡CC$_5$H$_{11}$

Scheme 1

Tris(trifluoromethyl)thallium Tl(CF$_3$)$_3$ has been prepared by condensing thallium vapour with trifluoromethyl radicals on a cryogenic surface.[6] In the absence of a donor ligand or if self-association was prohibited, Tl(CF$_3$)$_3$ was found to be the most stable derivative and the weakest Lewis acid among the group 13 metal series. The stability of the CF$_3$ complexes decreases in the order Tl > In > Ga > Al. The reactions of Cd(CF$_3$)$_2$ complexes[7] with GaCl$_3$, InCl$_3$ and TlX$_3$ (X = Cl, OCOMe, OCOCF$_3$) in aprotic basic solvents have also yielded the compounds Ga(CF$_3$)$_3$·DMF, Ga(CF$_3$)$_2$Cl·DMF, [Cd(MeCN)$_2$][Ga(CF$_3$)$_4$]$_2$, In(CF$_3$)$_3$·2MeCN, In(CF$_3$)$_2$Cl·DMF and Tl(CF$_3$)$_3$·2DMF.

10.2.1.2 Reactions

The formation of carbon–carbon bonds constitutes one of the most fundamental processes in organic chemistry and the use of organometallic reagents to perform this transformation has evolved tremendously. For example, the conversion of an acid chloride into the corresponding ketone, a useful functional group transformation, can be brought about using a plethora of organometallic compounds. However, most of the reported procedures suffer from major drawbacks. These are typically:
  (i) the overaddition of the organometallic reagent, leading to the corresponding tertiary alcohol which is often difficult to separate from the desired ketone;
  (ii) the decomposition or racemization of the starting material under the vigorous conditions required for some reagents, for example organocadmiums;
  (iii) the instability of the organometallic reagent, for example, organocuprates;
  (iv) the use of very low temperatures and excess reagents; and
  (v) the difficulties in preparing branched organometallic reagents and therefore in forming branched ketones. Numerous branched triorganothallium compounds are known.[1] They are much more stable than their copper, cadmium and zinc counterparts towards β-hydride elimination.

A simple and efficient method has been developed whereby triorganothallium compounds react smoothly with acid chlorides at room temperature to afford alkyl, aryl and acetylenic ketones in high yields.[8] This protocol is probably one of the most versatile routes to ketones. Addition of an organolithium compound to an ethereal suspension of a diorganothallium halide (**6**) generates the soluble and highly reactive triorganothallium derivative (**7**). On addition of an acid chloride, rapid precipitation of the highly insoluble diorganothallium chloride (**6**) takes place with concomitant formation of the ketone (**8**). Filtration of the thallium(III) salt gives virtually pure ketones in excellent yields (Scheme 2). The reaction, which applies to both aliphatic and aromatic substrates (Table 1), is also highly chemoselective. The triorganothallium reagents react selectively with acid chlorides in the presence of other functional groups such as alkenes, esters and even ketones. Under these mild reaction conditions, no tertiary alcohols are detected even in the presence of an excess of the reagent.

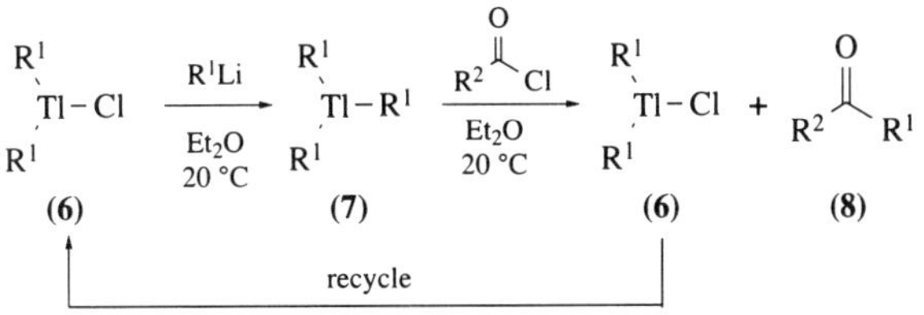

$$R^1_2Tl\text{-}Cl \; (\mathbf{6}) \xrightarrow{R^1Li, \; Et_2O, \; 20\,°C} R^1_2Tl\text{-}R^1 \; (\mathbf{7}) \xrightarrow{R^2COCl, \; Et_2O, \; 20\,°C} R^1_2Tl\text{-}Cl \; (\mathbf{6}) + R^2COR^1 \; (\mathbf{8})$$

recycle

R^1 = Me, Et, Ph; R^2 = alkyl, aryl

Scheme 2

Table 1 Synthesis of ketones from acid chlorides and triorganothallium reagents.

Substrate	Reagent	Product	Yield (%)
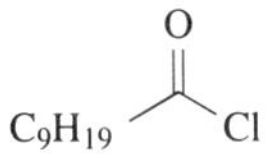 $C_9H_{19}COCl$	Me_3Tl	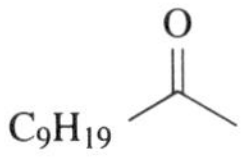 $C_9H_{19}COCH_3$	85
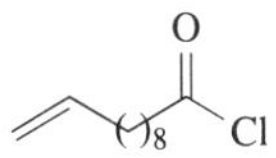 $CH_2=CH(CH_2)_8COCl$	Me_3Tl	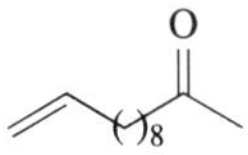 $CH_2=CH(CH_2)_8COCH_3$	73
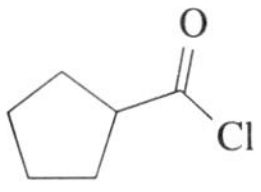 cyclopentyl-COCl	Me_3Tl	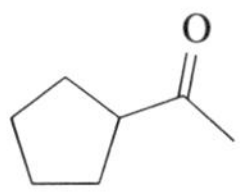 cyclopentyl-COCH_3	76
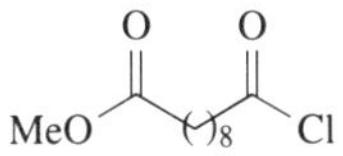 $MeOOC(CH_2)_8COCl$	Me_3Tl	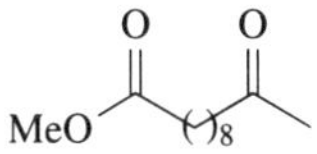 $MeOOC(CH_2)_8COCH_3$	92
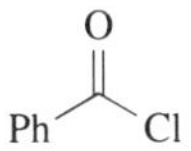 $PhCOCl$	Me_3Tl	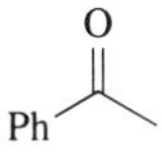 $PhCOCH_3$	78
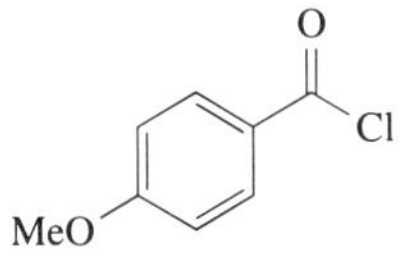 4-MeO-C_6H_4-COCl	Me_3Tl	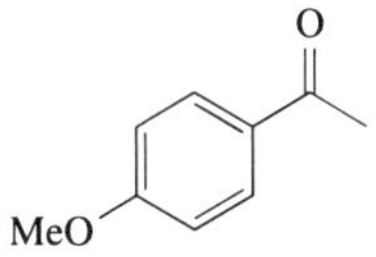 4-MeO-C_6H_4-COCH_3	82
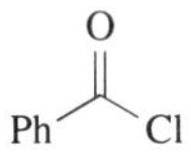 $PhCOCl$	Et_3Tl	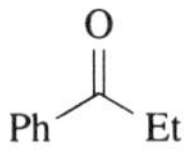 $PhCOEt$	88
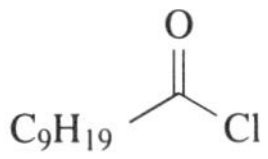 $C_9H_{19}COCl$	Et_3Tl	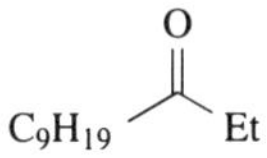 $C_9H_{19}COEt$	91
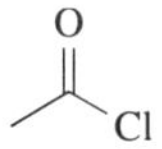 CH_3COCl	Ph_3Tl	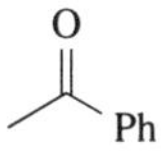 CH_3COPh	87
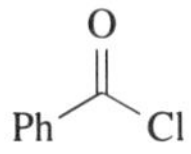 $PhCOCl$	Ph_3Tl	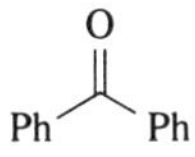 $PhCOPh$	85

The by-product of this ketone synthesis is the insoluble diorganothallium(III) chloride (**6**) which is employed as the starting material for the preparation of the triorganothallium reagent (**7**). Simple filtration followed by washing and drying of the salt leads to efficient recovery (>95%) of the diorganothallium halide which can be engaged in another reaction. Since only minute quantities of the original organometallic are lost, the whole sequence can be, at least formally, considered as a catalytic process.

Further investigation into the synthetic utility of triorganothallium reagents has led to the preparation of mixed triorganothallium derivatives. For example, dimethyl(phenylethynyl)thallium(III) (**9**), previously made from Me_3Tl and $PhC\equiv CH$, can be easily prepared from dimethylthallium chloride and 1-lithio-2-phenylacetylene. Reaction with decanoyl chloride (**10**, $R = C_9H_{19}$) and acetyl chloride (**10**, R = Me) affords in high yields the acetylenic ketones (**11**) and (**12**), respectively (Equation (2)).[8] Again, the reaction is performed at room temperature and rapid formation of the corresponding propargylic ketones is observed. Neither 2-undecanone nor acetone, resulting from a competitive methyl transfer, could be detected by ^1H NMR or capillary GC. The chemoselectivity shown by the thallium reagent contrasts with that displayed by copper complexes, for which the alkyne unit acts as a 'dummy' ligand.

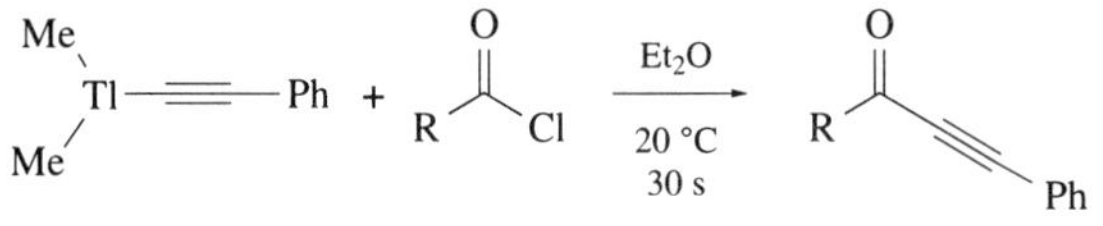

$$\underset{(\mathbf{9})}{\underset{Me}{\overset{Me}{>}}Tl\text{---}\!\!\equiv\!\!\text{---}Ph} + \underset{(\mathbf{10})}{R\overset{O}{\underset{\|}{C}}Cl} \xrightarrow[\substack{20\,°C \\ 30\,s}]{Et_2O} \underset{\substack{(\mathbf{11})\,R = C_9H_{19},\,79\% \\ (\mathbf{12})\,R = Me,\,73\%}}{R\overset{O}{\underset{\|}{C}}\!\!-\!\!C\!\!\equiv\!\!C\!\!-\!\!Ph} \qquad (2)$$

This versatile methodology for the generation of carbon–carbon bonds has also been applied to the alkylation of activated tertiary and secondary halides.[9] Hence, treatment of trityl chloride (**13**) with an equimolar amount of trimethylthallium (**4**) leads to rapid formation of 1,1,1-triphenylethane (**14**) (Equation (3)). Such a rapid formation of a quarternary carbon centre by the direct reaction of a tertiary halide with an organometallic reagent is unusual, highlighting some unique features of organothallium reagents. Further examples of the reaction of triphenyl- and trimethylthallium with activated halides are collected in Table 2.

$$\underset{(\mathbf{4})}{Me_3Tl} + \underset{(\mathbf{13})}{Ph_3CCl} \xrightarrow[\substack{5\,min \\ 78\%}]{\substack{Et_2O \\ 20\,°C}} \underset{(\mathbf{14})}{Ph_3CMe} + \underset{(\mathbf{3})}{Me_2TlCl} \qquad (3)$$

Table 2 Reaction of simple triorganothallium derivatives with activated halides.

Substrate	Reagent	Product	Yield (%)
Ph_3CCl	Ph_3Tl	CPh_4	80
Ar_2PhCCl	Me_3Tl	Ar_2PhCMe	85
Ar_2PhCCl	Ph_3Tl	Ar_2CPh_2	79
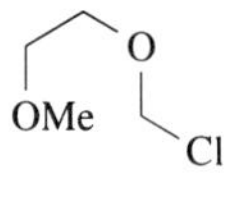 2-chloro-2-methoxytetrahydropyran	Ph_3Tl	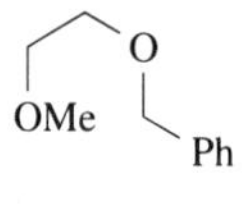 2-methoxy-2-phenyltetrahydropyran	80
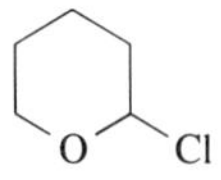 2-chlorotetrahydropyran	Ph_3Tl	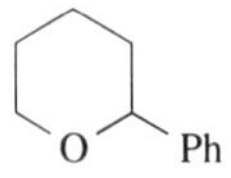 2-phenyltetrahydropyran	82
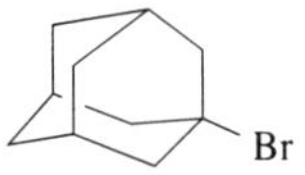 1-adamantyl bromide	Me_3Tl	No reaction	

Ar = p-$MeOC_6H_4$

It is interesting to note that the alkylations are only successful if the substrates employed are activated halides. These halides are prone to undergo facile dissociation of the carbon–halogen bond, indicating that in the mechanism of this reaction significant cationic character might be building up on the carbon centre of the starting halide. So far, simple tertiary halides, such as 1-adamantyl bromide, have proved unreactive. This could be due to the relatively weak Lewis acid character of the triorganothallium derivatives, compared with their aluminum counterparts. They are unable to coordinate strongly enough to the halogen atom to heterolytically cleave the carbon–bromine bond of 1-adamantyl bromide.

Similarly, reaction of the mixed dimethylalkynyl thallium derivatives (**15**) and (**16**) with activated tertiary and secondary halides leads to the formation of the corresponding substituted alkynes (Table 3).[9] High chemoselectivity is observed and compounds resulting from the transfer of the methyl ligand could not be detected. Again, the by-product of this reaction is the diorganothallium halide which can be reused after simple filtration and washing. However, in contrast with the 'pseudocatalytic' process pertaining to the ketone preparations, a truly catalytic procedure could be developed for the alkylation of activated secondary and tertiary halides. Thus, when a catalytic quantity of a diorganothallium halide is employed, along with molar equivalents of an activated halide, and an alkyl- or alkynyllithium derivative is added dropwise to the reaction mixture, smooth ligand transfer takes place and the alkylated product can be obtained in good to excellent yields (Table 4).[10] The catalytic cycle, featuring the preparation of 1,1,1-triphenylethane, is shown in Scheme 3.

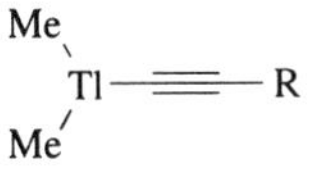

Me\\Tl—≡—R / Me

(**15**) R = Ph
(**16**) R = C_5H_{11}

Table 3 Reaction of mixed triorganothallium derivatives with activated halides.

Substrate	Reagent	Product	Yield (%)
Ph_3CCl	 Me\\Tl—≡—Ph / Me	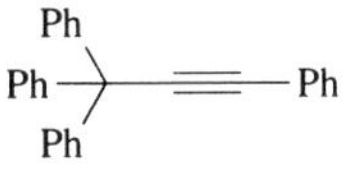 Ph\\Ph—≡—Ph / Ph	78
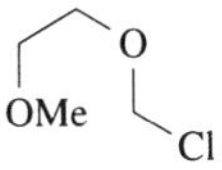 2-(2-chloroethoxy)... OMe, Cl	 Me\\Tl—≡—Ph / Me	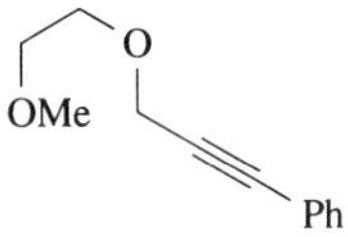 OMe-O-CH₂CH₂-C≡C-Ph	81
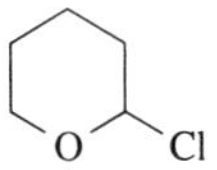 2-chlorotetrahydropyran	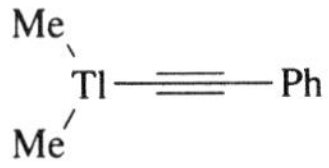 Me\\Tl—≡—Ph / Me	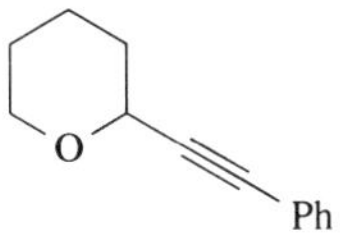 tetrahydropyran-C≡C-Ph	91
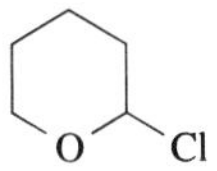 2-chlorotetrahydropyran	 Me\\Tl—≡—C_5H_{11} / Me	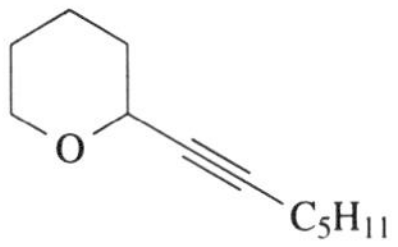 tetrahydropyran-C≡C-C_5H_{11}	66

MeLi (1 equiv.) → Me₂TlCl (0.1 equiv.) → Ph₃CMe 78%

LiCl ← Me₃Tl (**4**) ← Ph₃CCl (1 equiv.)

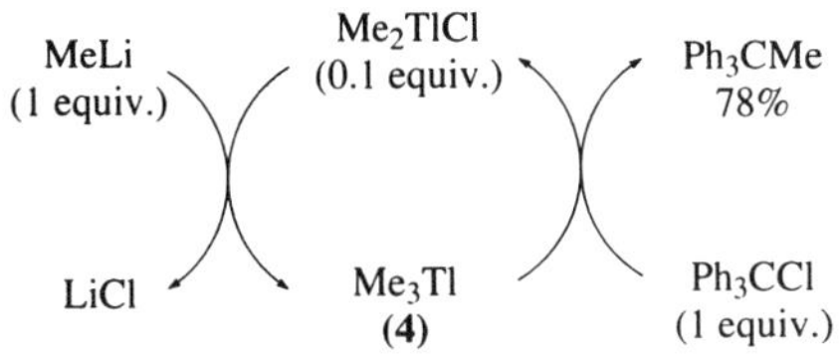

Scheme 3

Besides acyl and activated alkyl halides, triorganothallium compounds also react with enones, leading generally to the polymerization of α,β-unsaturated carbonyl derivatives. Thus, attempted alkylation of cyclohexenone using trimethylthallium (**4**) results in rapid polymerization of the enone at room temperature. The starting enone is recovered unchanged when the reaction is carried out at lower temperature. However, addition of 1 equiv. of methyllithium to trimethylthallium at −40 °C generates the tetraorganothallium ate complex, Me₃Tl·MeLi *in situ*.[4] When cyclohexenone is added to this reagent, at room temperature, a 1:2 mixture of 1,2- and 1,4-addition products, (**18**) and (**19**), respectively, is

Table 4 Catalytic carbon–carbon bond formation using triorganothallium reagents.

Substrate	Catalyst	Reagent	Product	Yield (%)
Ph_3CCl	Ph_2TlBr	PhLi	CPh_4	74
Ar_2PhCCl	Me_2TlCl	MeLi	Ar_2PhCMe	75
Ph_3CCl	Me_2TlCl	Ph—≡—Li	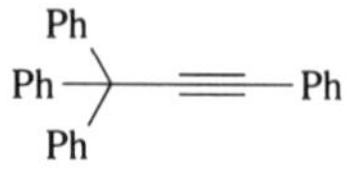 Ph(Ph)(Ph)C—≡—Ph	82
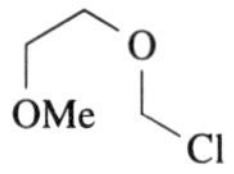 MeO-CH₂CH₂-O-CH(Cl)-	Ph_2TlBr	PhLi	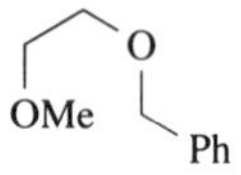 MeO-CH₂CH₂-O-CH(Ph)-	78
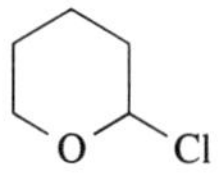 2-chlorotetrahydropyran	Ph_2TlBr	PhLi	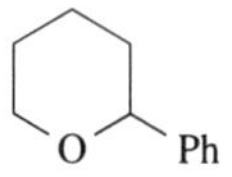 2-phenyltetrahydropyran	63
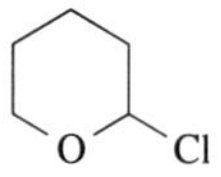 2-chlorotetrahydropyran	Me_2TlCl	Ph—≡—Li	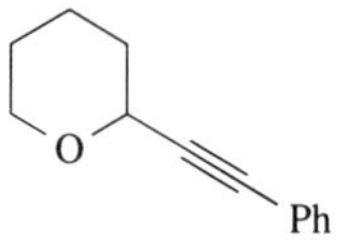 2-(phenylethynyl)tetrahydropyran	96
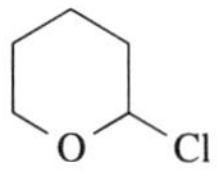 2-chlorotetrahydropyran	Me_2TlCl	C_5H_{11}—≡—Li	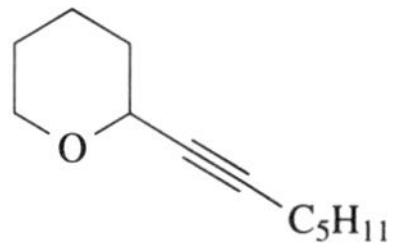 2-(C_5H_{11}-ethynyl)tetrahydropyran	57

Ar = p-MeOC$_6$H$_4$

produced. When the reaction is performed at –40 °C, the saturated ketone (**19**), resulting from the conjugate addition of the methyl group, is formed almost solely (Scheme 4).

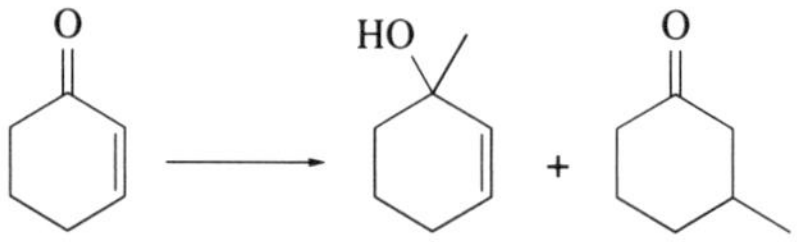

(**17**) → (**18**) + (**19**)

Me_3Tl, Et_2O, 20 °C	Polymerization
Me_3Tl, Et_2O, –40 °C	No reaction
Me_3Tl·MeLi, Et_2O, 0 °C	1:2 (75%)
Me_3Tl·MeLi, Et_2O, –40 °C	<5:>95 (70%)

Scheme 4

Some other examples are collected in Table 5. While simple cyclic enones lead to 1,4-addition products (entries 1 and 5), acyclic enones are converted into the corresponding allylic alcohols (entry 2). The regioselectivity observed in the addition of the thallium ate complex is thus totally opposite to that found with aluminum reagents. Further reactions of the ate complex with some representative enones are shown in Table 5.

The continuous investigation of the synthetic potential of the triorganothallium ate complexes has led to a unique transformation in which the ate complex is able to react preferentially in a 1,2-fashion with enones in the presence of the more reactive ketones.[5] Thus, when equimolar amounts of an enone (**20**) and a ketone (**21**) are treated with 1 equiv. of Me_3Tl·MeLi in ether, at –50 °C (inverse addition of the ate complex to the enone–ketone mixture), a rapid reaction takes place, giving rise to the preferential formation of the allylic alcohol (**22**) over the saturated tertiary alcohol (**23**) (Equation (4)). Selected examples of this chemoselective process are shown in Table 6. One of the most striking features is that the selectivity of the addition increases as the substrate becomes more conjugated and as the intrinsic reactivity of the carbonyl function towards nucleophiles decreases. When a donor group such as a p-methoxy function is present on the aromatic system (cyclohexyl methyl ketone vs. p-methoxy

Table 5 Reaction of lithium tetramethylthallate with selected enones.

Entry	Substrate	Product	Yield (%)
1	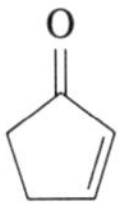 cyclopent-2-enone	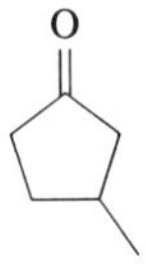 3-methylcyclopentanone	73
2	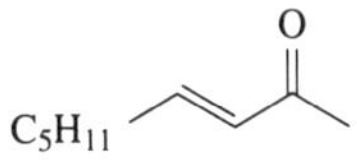 C_5H_{11}-CH=CH-C(O)-CH₃	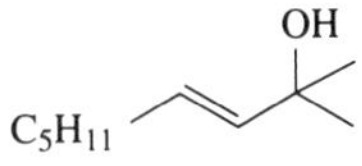 C_5H_{11}-CH=CH-C(OH)(CH₃)₂	81
3	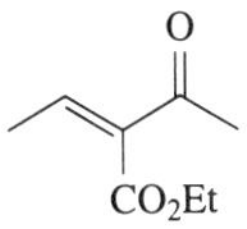 ethyl 2-acetyl-2-butenoate (CO₂Et)	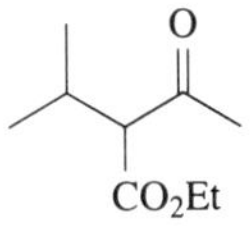 ethyl 2-acetyl-3-methylbutanoate (CO₂Et)	85
4	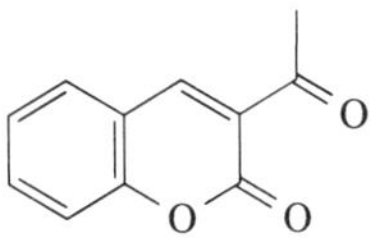 3-acetylcoumarin	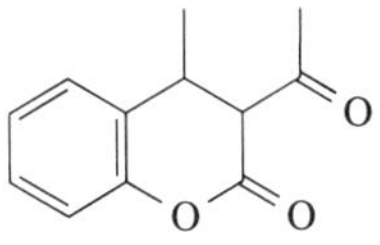 4-methyl-3-acetylchroman-2-one	70
5	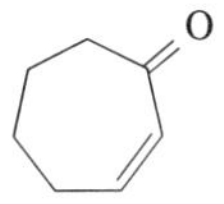 cyclohept-2-enone	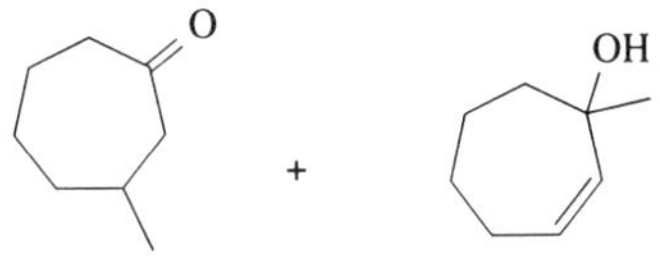 3-methylcycloheptanone + 1-methylcyclohept-2-enol (1 : 4)	80

acetophenone), the two possible alcohols are obtained in essentially 1:1 ratio (entry 6b). In sharp contrast, when the aromatic ring is substituted by an electron-withdrawing group such as a nitrile, the selectivity becomes exquisite and favours the conjugated ketone (entry 6c).

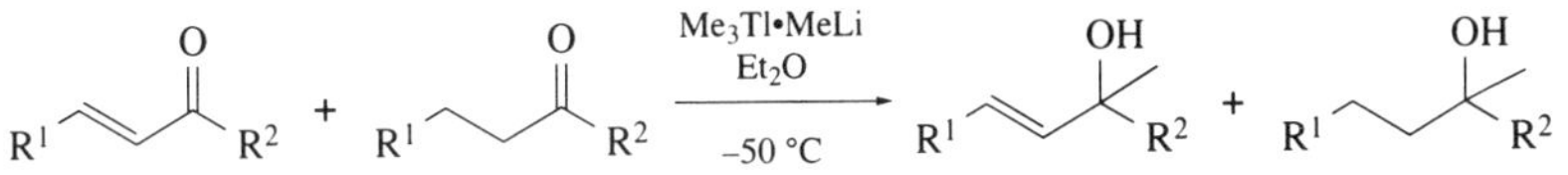

$$R^1\text{-CH=CH-C(O)-}R^2 \;+\; R^1\text{-CH}_2\text{-CH}_2\text{-C(O)-}R^2 \xrightarrow[-50\,°C]{Me_3Tl\cdot MeLi,\; Et_2O} R^1\text{-CH=CH-C(OH)(Me)-}R^2 \;+\; R^1\text{-CH}_2\text{-CH}_2\text{-C(OH)(Me)-}R^2 \quad (4)$$

(20)   (21)   (22)   (23)

R^1, R^2 = alkyl, aryl

It appears from the above observations that the chemoselectivity of the reaction increases as the reduction potential of the enone substrate decreases, that is, as the ability of the enone towards accepting a single electron (hence forming a radical anion) increases. It is believed that the triorganothallium ate complex (**24**) is involved in the transfer of a single electron to the enone (**20**), forming the radical anion (**26**) and the unusual 'thallium(IV)' species (**25**) (Scheme 5). Although thallium(IV) formation is energetically unfavoured, the species might be better represented by a triorganothallium(III) derivative closely associated with a methyl radical. A rapid recombination of these two radicals then takes place, leading to Me₃Tl (**4**) and the lithium alkoxide (**27**) which is hydrolysed to (**22**) during workup (Scheme 5). This is a little precedented transformation and its synthetic utility should be further established.

It is interesting to note that the judicious positioning of the electron-withdrawing group can have profound influences on the outcome of the thallium(III) addition reaction. Thus, whereas enone (**28**), in a competition experiment with ketone (**29**), affords the allylic alcohol (**30**) with only ~5:1 selectivity, the activated enone (**32**) gives a far better selectivity (>35:1). Remarkably, by placing the ester function at the α-position, the course of the reaction is altered and a smooth Michael reaction takes place ((**34**) → (**35**), Scheme 6).

Table 6 Chemoselective addition of Me₃Tl·MeLi to enones/ketones.

Entry	Enone	Ketone	Unsat. alcohol: Sat. alcohol	Yield (%)
1	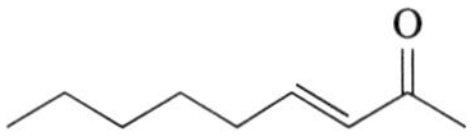 CH₃(CH₂)₃CH=CHCOCH₃	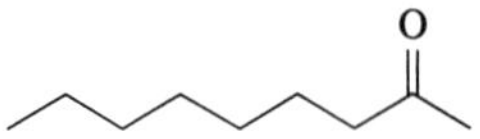 CH₃(CH₂)₅COCH₃	5:1	79
2	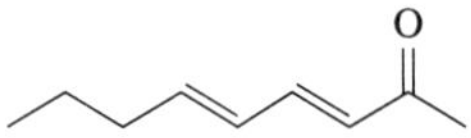 dienone	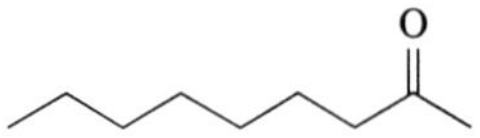 CH₃(CH₂)₅COCH₃	10:1	89
3	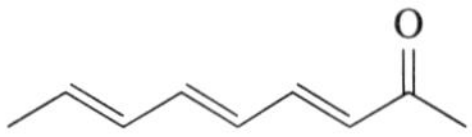 dienone	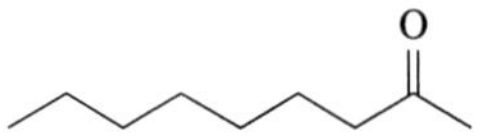 CH₃(CH₂)₆COCH₃	20:1	79
4	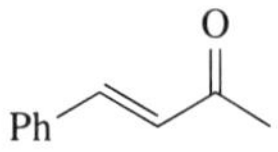 PhCH=CHCOCH₃	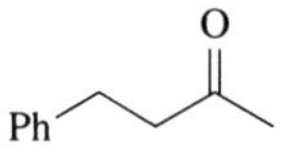 PhCH₂CH₂COCH₃	40:1	84
5	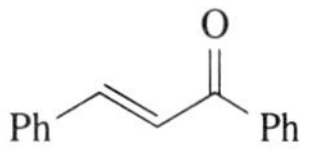 PhCH=CHCOPh	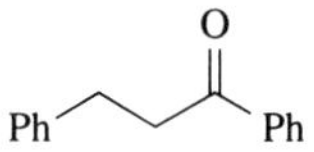 PhCH₂CH₂COPh	>75:1	98
6	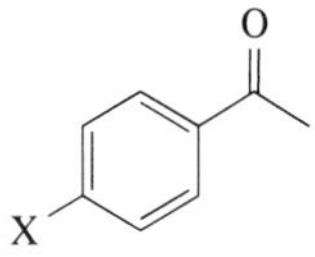 4-X-C₆H₄-COCH₃	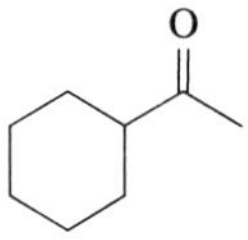 cyclohexyl-COCH₃		
a	X = H		15:1	84
b	X = OMe		1:1	86
c	X = CN		>75:1	95
7	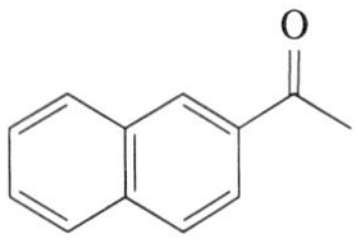 2-acetylnaphthalene	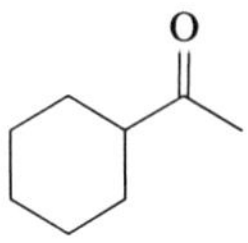 cyclohexyl-COCH₃	45:1	95

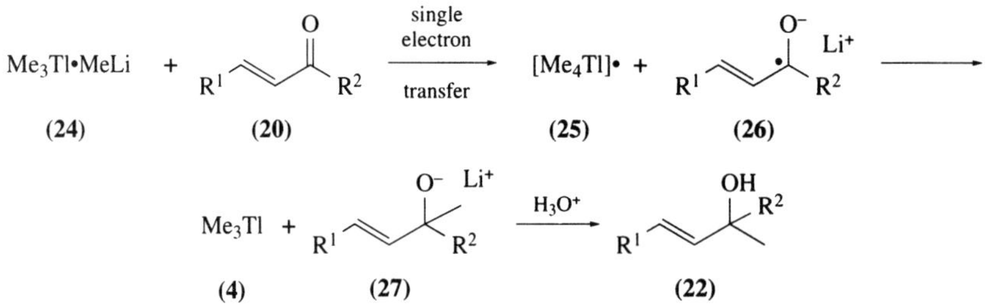

Scheme 5

$$\text{Me}_3\text{Tl·MeLi} + \underset{R^1}{\overset{O}{\text{CH=CHCOR}^2}} \xrightarrow{\text{single electron transfer}} [\text{Me}_4\text{Tl}]^\bullet + \underset{R^1}{\overset{O^- \text{Li}^+}{\text{CH=CH–C}^\bullet R^2}}$$

(24) (20) (25) (26)

$$\text{Me}_3\text{Tl} + \underset{R^1}{\overset{O^- \text{Li}^+}{\text{CH=CH–CR}^2}} \xrightarrow{\text{H}_3\text{O}^+} \underset{R^1}{\overset{OH}{\text{CH=CH–CR}^2}}$$

(4) (27) (22)

10.2.2 Diorganothallium Compounds

Although considerable interest has been shown in the preparation and characterization of diorganothallium compounds, their application in synthesis is somewhat limited. Complexes such as (**36**), containing an N₆-macrocyclic ligand, have been synthesized by a Schiff base condensation of 2,6-diacetylpyridine with ethylenediamine using the thallium salt as a template.[11]

Dimethylthallium(III) complexes of diphenyldithiophosphinate have been prepared[12] by the reaction of TlMe₂X and [R₄N][S₂Ph₂] (X = OH or NO₃, R = H or Et) in water/EtOH or CHCl₃. In (**37**), the metal is coordinated to two methyl groups, to two sulfur atoms belonging to a single bidentate dithiophosphinate ligand and, more weakly, to two sulfur atoms belonging to two neighbouring molecules. In contrast, no detectable intermolecular interactions are observed in (**38**).

The preparation of potentially electroactive thallium polymers (**40**), derived from the acidic copolymer (**39**), has been reported (Equation (5)).[13]

Scheme 6

(36)
R = Me, Et, Ph

(37)

(38)

(39) **(40)** (5)

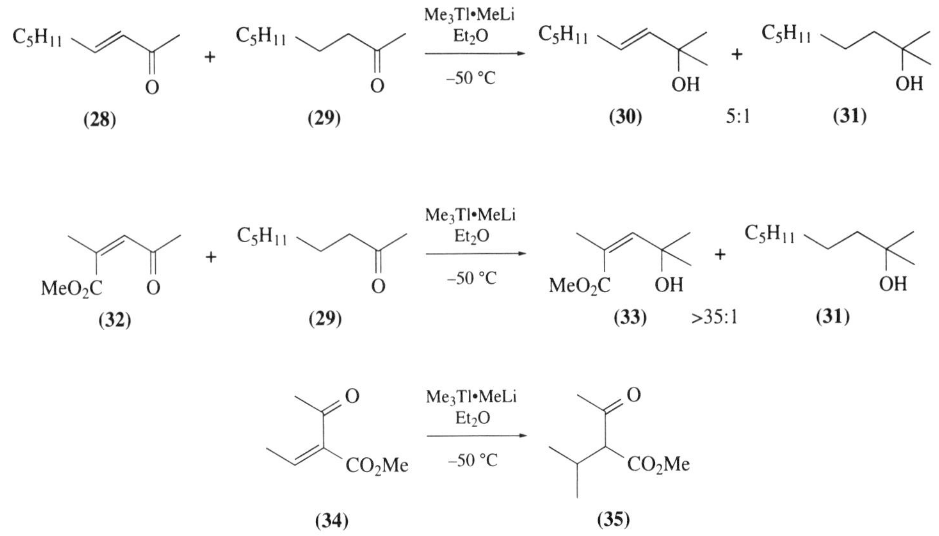

Arylsilver(I) complexes, AgR (R = mesityl, $C_6F_3H_2$, C_6H_5) react with thallium(III) chloride to yield arylthallium(III) complexes (**41**) of the type $[TlR_2][TlCl_3R]$, $TlClR_2$ or TlR_3.[14]

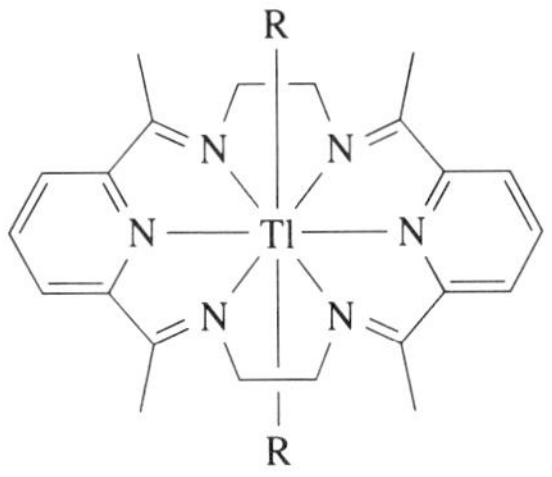

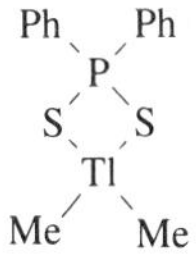

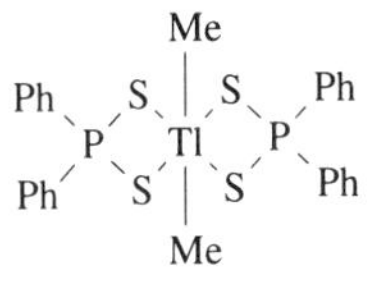

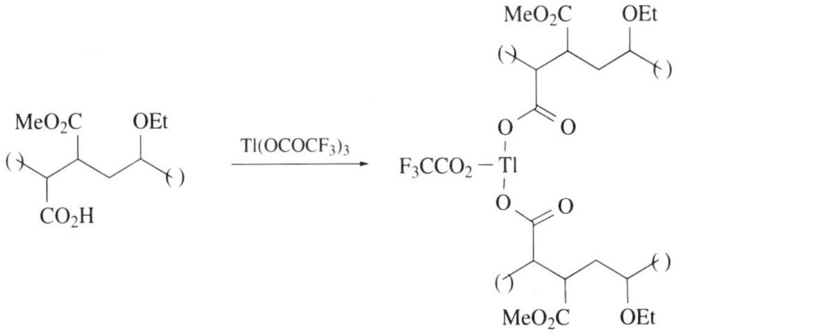

The synthesis of α-metallo ketones of the type $RCOCH_2Tl^{III}$tolyl($OCOCF_3$) (**42**), where R = phenyl, substituted phenyl, thienyl, pyridinyl and *t*-butyl, has been reported.[15] These compounds can serve as anion equivalents in the $TiCl_4$-mediated aldol condensation. Metal–metal exchange with iodosylbenzene or lead tetraacetate leads to α-nucleophilic substitution derived from the formal α-keto carbonium ion equivalent. Nitrogen compounds of thallium such as (**43**), possessing intra- and intermolecular donor–acceptor bonds, have also been synthesized.[16]

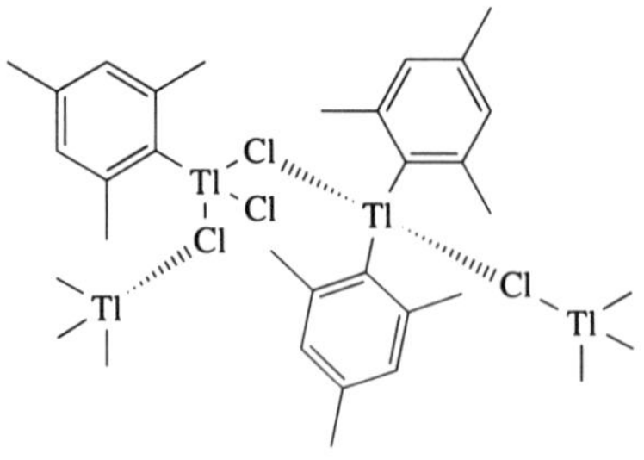

(41)

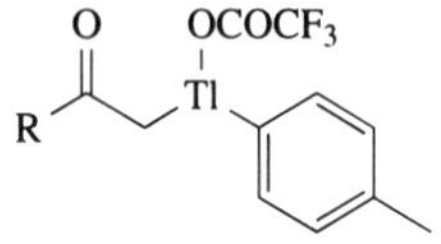

(42) 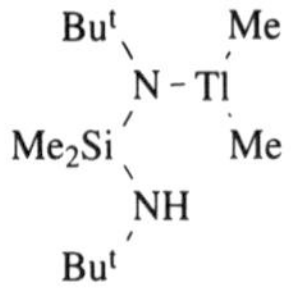 **(43)**

Derivatives of 2-mercaptobenzoxazole of the type MR$_n$L (M = Hg (**44**) or Tl (**45**), R = Me or Ph and n = 1 (Hg) or 2 (Tl)) have been prepared.[17] Compounds of the type R$_n$MH$_2$Tb (R = Me, Ph; n = 1 (M = Hg), 2 (M = Tl), H$_3$Tb = 2-thiobarbituric acid) have also been described.[18] In the organomercury derivatives, the metal is found to be bound to the thiolate sulfur atom of the ligand in both the solid state and in DMSO. The organothallium compounds (**46**), however, while having the metal bound to the sulfur atom and possibly to one of the nitrogen atoms of the pyrimidine ring in the solid state, appears to dissociate and form a conducting solution containing H$_2$Tb$^-$ and R$_2$Tl$^+$ ions in DMSO.

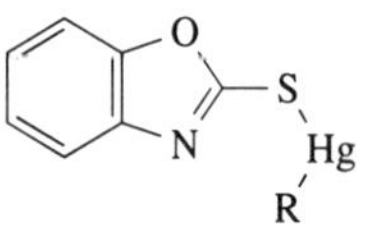

(44) 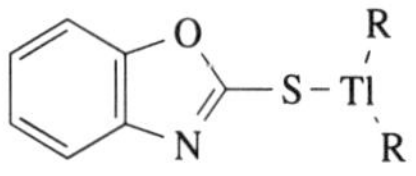 **(45)** 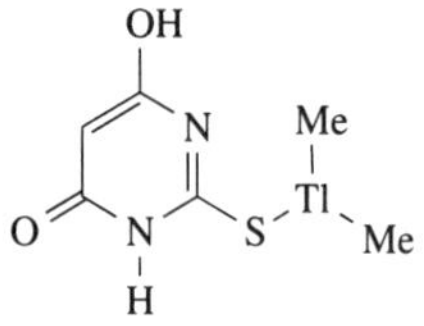 **(46)**

Other examples of heterocyclic compounds containing diorganothallium moieties include the triazole compound (**47**) and pyrazole derivative (**48**).[19,20]

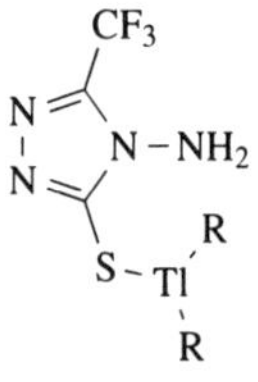

R = Me, Ph
(47) **(48)**

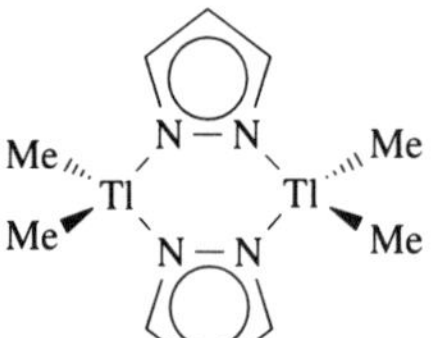

10.2.3 Monoorganothallium Compounds

10.2.3.1 Preparation and properties

The utilization of thallium chemistry, especially in the field of heterocyclic synthesis, has become increasingly important, and organothallium compounds have proven to be valuable intermediates in organic synthesis. Electrophilic thallation of arenes are highly regiospecific and many novel methods have been developed whereby the thallated intermediate is substituted by a variety of functional groups, offering convenient routes to a plethora of heterocyclic ring systems.

In contrast with mercuration, thallation of trithiadiazepine (**49**) with thallium(III) trifluoroacetate gives only the monometallated product (**50**),[21] which can be readily converted into the corresponding cyano-, iodo- and methoxycarbonyltrithiadiazepines (Equation (6)).

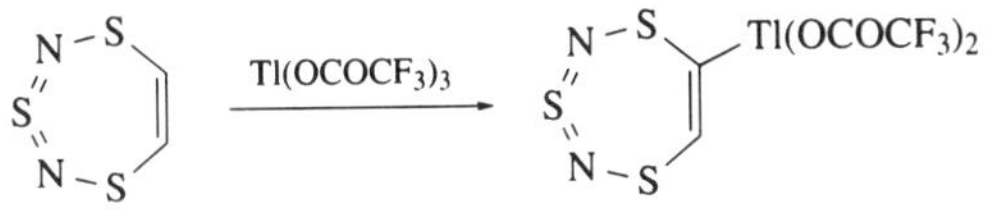

Monoorganothallium(III) species of type (**53**) are prepared by reaction of acetophenone with Tl$_2$O$_3$ in acidic aqueous solution (Equation (7)).[22]

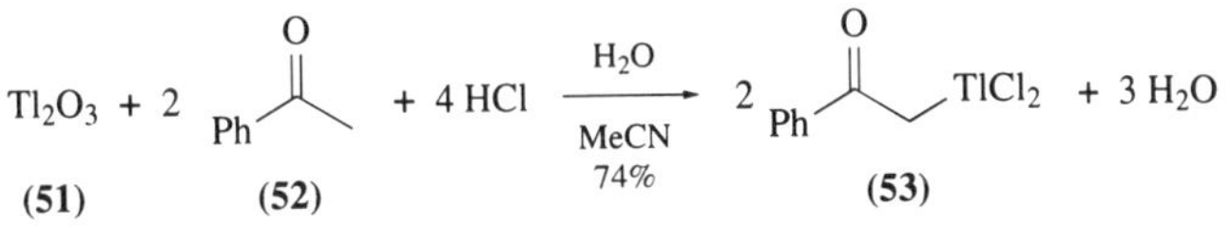

$$Tl_2O_3 + 2 \; PhC(O)CH_3 + 4 \; HCl \xrightarrow[MeCN]{H_2O} 2 \; PhC(O)CH_2TlCl_2 + 3 \; H_2O \quad 74\% \quad (7)$$

(**51**)　(**52**)　　　　(**53**)

Thallium adducts derived from an allene would generally be stable as heterolysis of the carbon–thallium bond would result in a relatively unstable vinyl cation. The oxythallation reaction of (−)-1,2-cyclononadiene (**54**) with Tl(OAc)$_3$ in glacial acetic acid containing a catalytic amount of boron trifluoride etherate has afforded the thallium adduct (**55**). On reduction of (**55**) with sodium borohydride, (Z)-(S)-(+)-3-acetoxycyclononene (**56**) is obtained (Scheme 7). The methoxythallation of (**54**) with Tl(OAc)$_3$ thus proceeded by an antarafacial addition with greater than 70% stereoselectivity.[23] The methoxythallation of the allene (**57**) has been found to occur at the more substituted terminal double bond, giving rise to the adduct (**58**).[24] The preparation of (C$_6$F$_5$)$_3$GeTlEt$_2$ and [(C$_6$F$_5$)$_3$Ge]$_2$TlEt by reaction of tris(pentafluorophenyl)germane with triethylthallium has also been reported.[25]

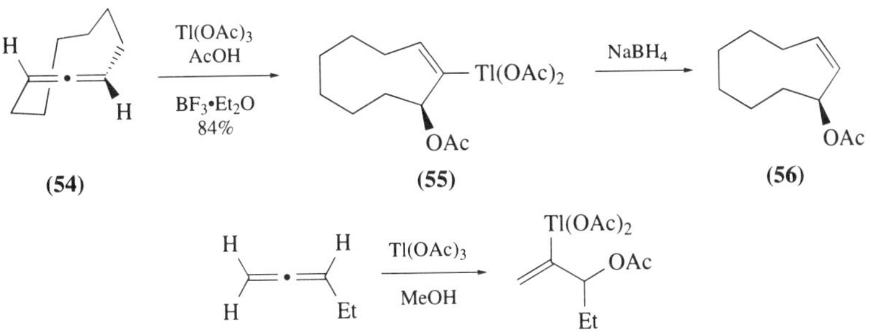

Scheme 7

10.2.3.2 Reactions

Benzoxazoles can be readily formed by photolysis of *ortho*-thallated anilides (**61**) in cyclohexane suspension.[26] Initially, (**59**) is converted to (**60**) with thallium(III) nitrate and trimethyl orthoformate in methanol. The amide (**60**) can then be thallated with Tl(OCOCF$_3$)$_3$ in a mixture of ether and TFA. The arylthallium bis(trifluoroacetate) (**61**) precipitates directly from the reaction mixture in 38% overall yield. Photolysis of (**61**) affords benzoxaprofen methyl ester (**62**) in 90% yield (Scheme 8).

Benzoic acid and substituted benzoic acids are readily thallated by thallium(III) trifluoroacetate. Subsequently, reaction with palladium chloride and simple alkenes, allylic halides, vinyl halides or vinyl esters gives isocoumarins.[27] Similarly, the thallation–palladium-promoted alkenation of *p*-tolylacetic acid, *N*-methylbenzamide, benzamide and acetanilide provides a convenient entry into a variety of oxygen and nitrogen heterocycles (Scheme 9).[28]

Aromatic nitriles can be readily formed by the reaction of arylthallium bis(trifluoroacetates) with CuCN in acetonitrile (Equation (8)).[29]

The synthesis of aromatic carbonyl compounds can also be effected via thallation–carbonylation of arenes (Scheme 10).[30] Simple arenes, substituted benzylic and β-phenylethyl alcohols, benzoic acid, phenylacetic acid, benzamide, acetanilide, phenylurea and benzophenone have all been thallated under a variety of reaction conditions. Subsequent reaction with 10% PdCl$_2$, 2 equiv. of LiCl and MgO in either MeOH or THF under 1 atm of CO gives rise to the corresponding aromatic esters, substituted phthalides and 3,4-dihydroisocoumarins, phthalic and homophthalic anhydride, phthalimide, and the *ortho*-substituted methyl esters of acetanilide, phenylurea and benzophenone, respectively.

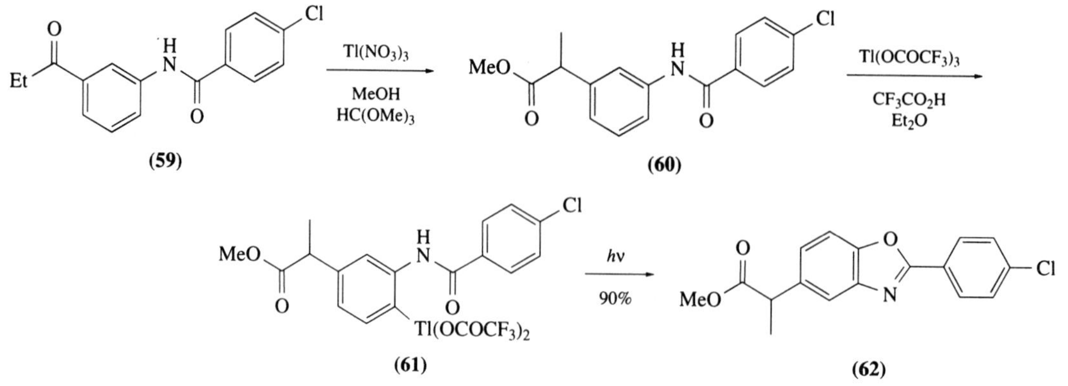

Scheme 8

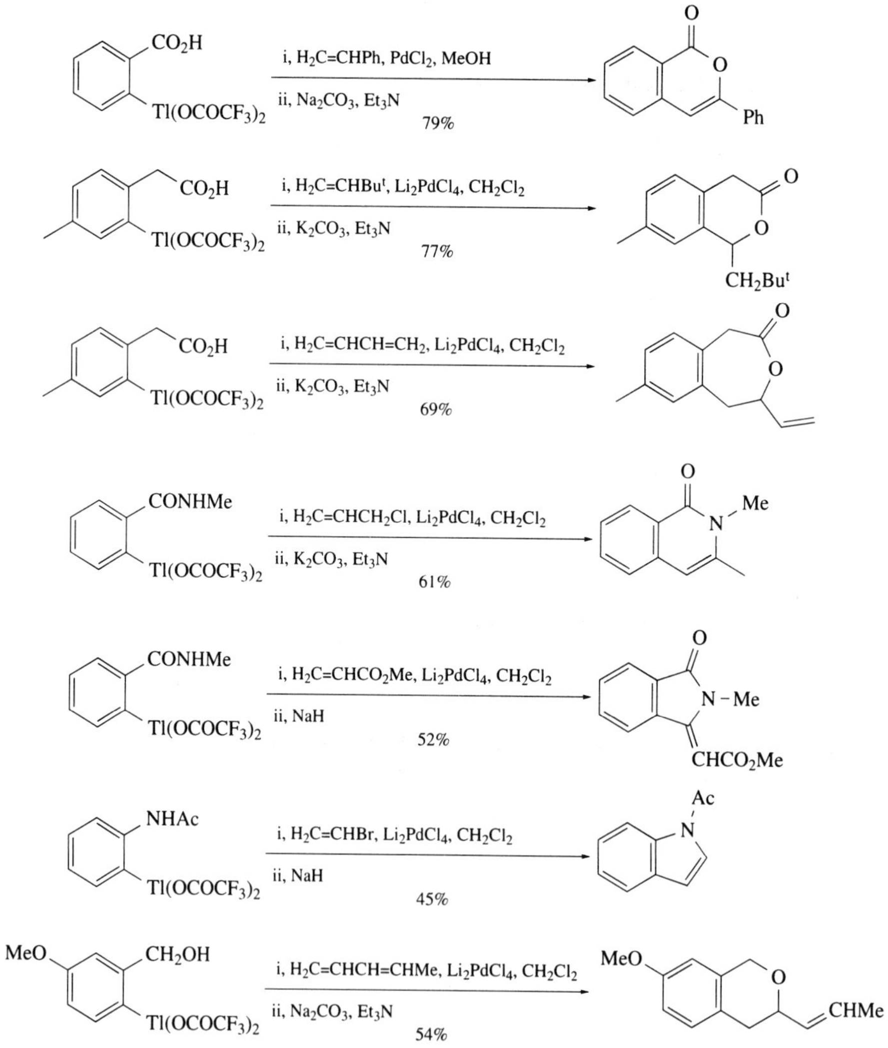

Scheme 9

$$ArTl(OCOCF_3)_2 + CuCN \xrightarrow{MeCN} ArCN \qquad (8)$$
$$(63) \qquad\qquad (64)$$
Ar = Ph, 72%; 4-MeC$_6$H$_4$, 64%; 4-MeOC$_6$H$_4$, 66%

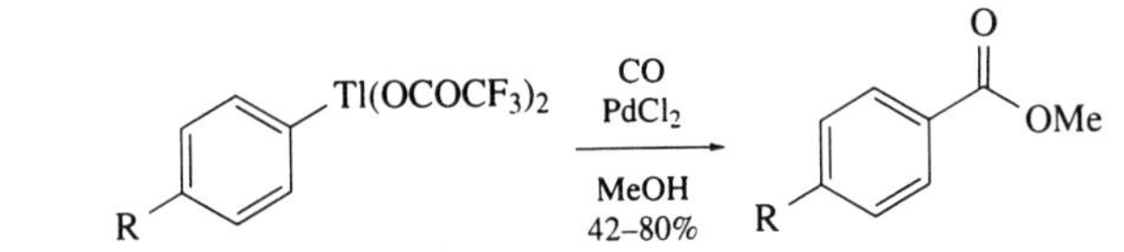

Thallium

Scheme 10

(65) → (66), R = H, F, OMe, Bu^t, 42–80%

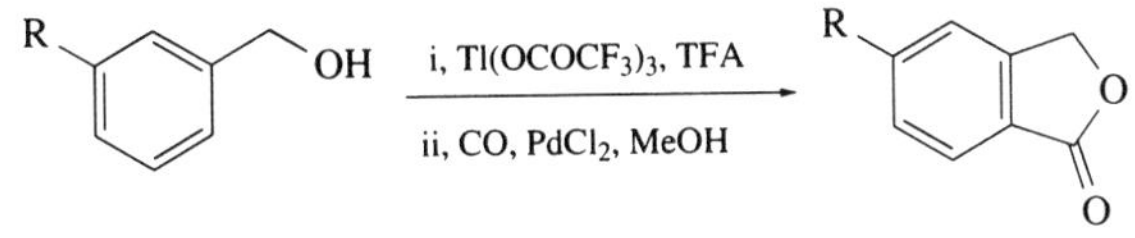

(67) → (68), R = H, OMe, OH, Cl

Symmetrical biaryls are synthesized via coupling of arylthallium bis(trifluoroacetates) with catalytic amounts of lithium tetrachloropalladate (Scheme 11).[31]

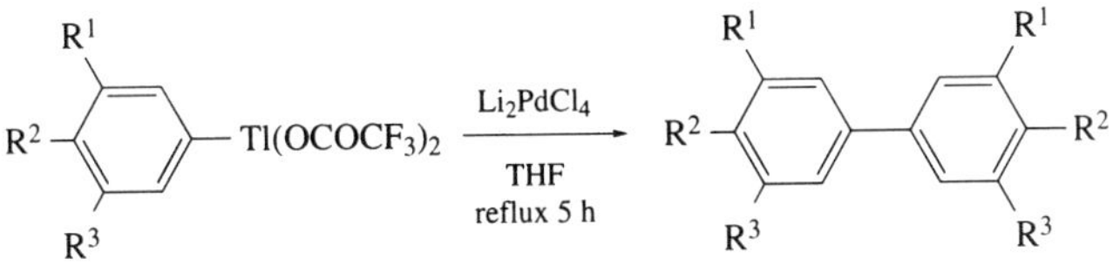

R^1	R^2	R^3	Yield (%)
H	H	H	88
H	Me	H	97
H	Bu^n	H	88
H	Bu^t	H	72
H	OMe	H	89
H	(CH$_2$)$_3$CO$_2$H	H	90
Me	Me	H	81
Pr^i	H	Pr^i	66
H	Cl	H	20

Scheme 11

Electron-rich arenes, which normally undergo oxidative coupling to biaryls with Tl(OCOCF$_3$)$_3$ in TFA, can be smoothly thallated in high yields with Tl(OCOCF$_3$)$_3$ in a 1:1 mixture of TFA and ether (Scheme 12).[32]

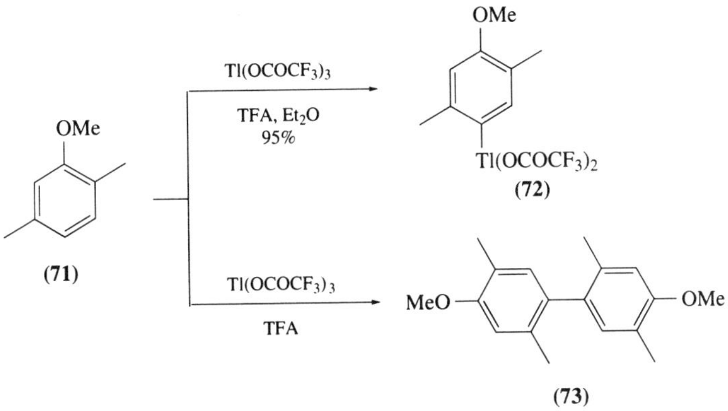

Scheme 12

Methyl vinyl ketone can be β-arylated by a Li_2PdCl_4-catalysed reaction of various thallated aromatic compounds (Equation (9)).[33] The conditions employed in these couplings are very mild.

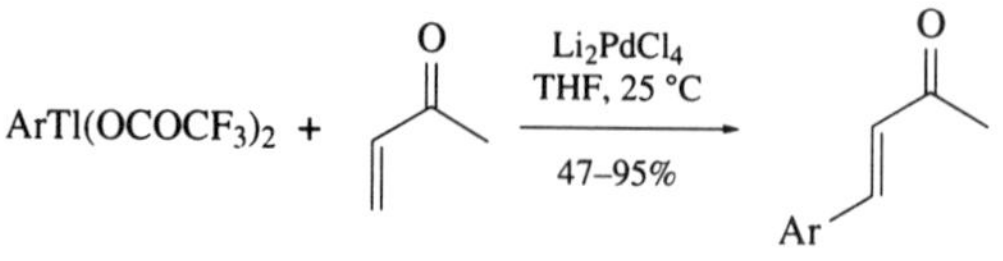

$$ArTl(OCOCF_3)_2 + \text{(74)} \xrightarrow[47-95\%]{Li_2PdCl_4, THF, 25°C} \text{(75)} \qquad (9)$$

(63)　(74)　(75)

Ar = Ph, p-tolyl, o/p-anisyl, o/p-chlorophenyl

Monoorganothallium(III) acetates, $RTl(OAc)_2$ (76) (R = $PhCH(OMe)CH_2$, $p\text{-}MeC_6H_4$, (E)- and (Z)-PhCH=CH), react with the anion of 2-nitropropane (77) to give the coupling products RMe_2CNO_2 (78) in moderate to good yields.[34,35] The formation of the carbon–carbon bond has been suggested to proceed through radical intermediates which are generated by electron transfer activation of the thallium–carbon bond (Equation (10)).

$$RTl(OAc)_2 + Me_2CNO_2^- \longrightarrow RMe_2CNO_2 \qquad (10)$$
(76)　(77)　(78)

R = PhCH(OMe)CH₂–, p-MeC₆H₄–, PhCH=CH– (E), PhCH=CH– (Z)

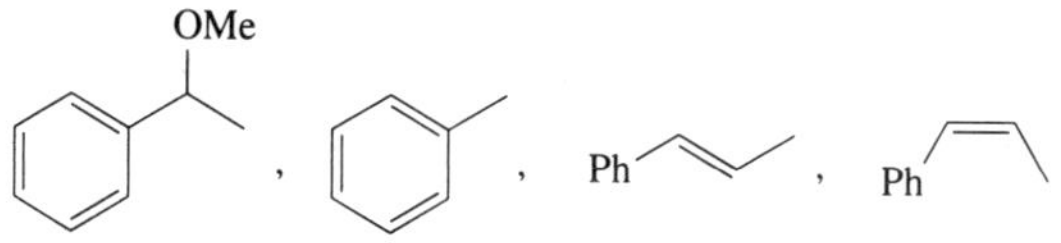

On treatment of $ArTl(OCOCF_3)_2$ with 2.5 mol. equiv. of diborane in THF, transmetallation occurs to give organoboron intermediates which, on hydrolysis or oxidation, afford arylboronic acids or phenols, respectively (Scheme 13).[36]

$$ArTl(OCOCF_3)_2 + B_2H_6 \longrightarrow [Ar-B\langle] \xrightarrow{H_2O_2} ArOH$$
(63)　　　　　　　　　(79)　　　　(80)

$$\downarrow H_2O$$

$$ArB(OH)_2$$
(81)

Ar	Yield (%)
p-ClC₆H₄	67
p-EtC₆H₄	59
2,5-Me₂C₆H₃	75
p-MeC₆H₄	50

Scheme 13

A direct alkylation of aromatic compounds using allylsilane (82), allylgermane or allylstannane has been developed.[37] The actual transmetallation reagent is proposed to be thallium(III) trifluoroacetate, which is produced together with diarylthallium trifluoroacetate by the disproportionation of arylthallium bis(trifluoroacetate) (Equation (11)).

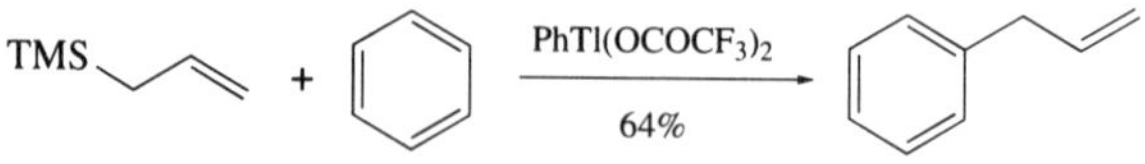

$$TMS\text{-allyl} + C_6H_6 \xrightarrow[64\%]{PhTl(OCOCF_3)_2} PhCH_2CH=CH_2 \qquad (11)$$

(82)　　　　　　　　　　　　　　　(83)

The polyfluoroarene $m\text{-}H_2C_6F_4$ (84) and other derivatives have been thallated by thallium(III) trifluoroacetate in TFA in the presence of antimony pentafluoride.[38] The dithallated product (85) was identified by conversion into the corresponding polyfluoroiodoarene (86) by reaction with aqueous sodium iodide (Scheme 14).

Thallation of benzanilide with $Tl(OCOCF_3)_3$ in a mixture of TFA and ether gives the ortho-thallated derivative (87) which yields 2-benzamidotolane (88) upon reaction with copper(I) phenylacetylide in acetonitrile. Treatment of (88) with palladium(II) chloride results in ring closure to give 1-benzoyl-2-phenylindole (89). On hydrolysis, 2-phenylindole (90) is obtained in quantitative yield (Scheme 15).[39]

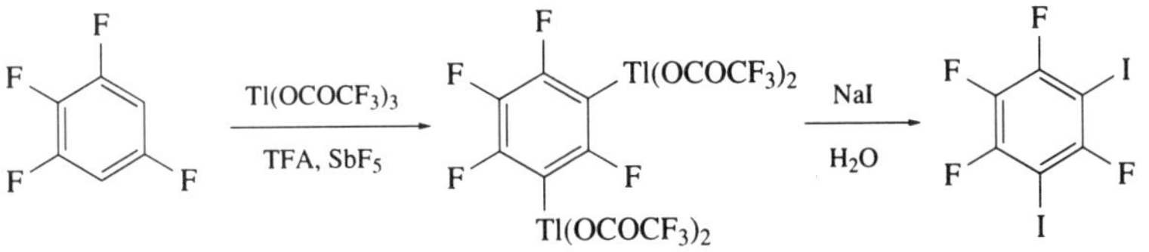

Scheme 14

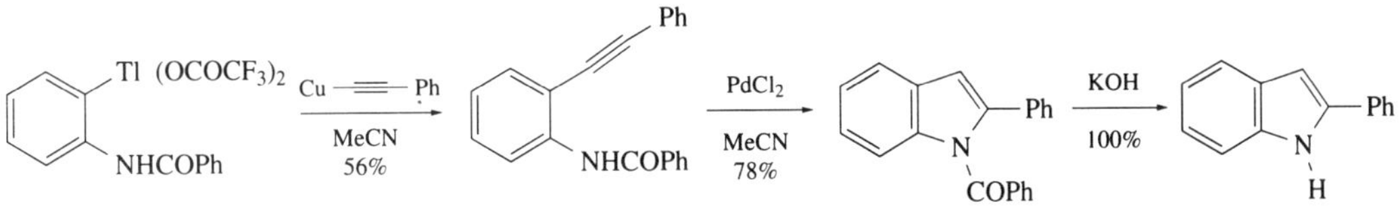

Scheme 15

4-Substituted-3-formylindoles (**93**) can be prepared in high yields by means of a thallation–palladation method.[40] Thus, thallation of 3-formylindole (**91**) has been found to occur at the 4-position, giving (3-formylindol-4-yl)thallium bis(trifluoroacetate) (**92**). Subsequent treatment of (**92**) with methyl acrylate in DMF in the presence of palladium acetate produces methyl 3-(3-formylindol-4-yl)acrylate (**93**). Similar methods of introducing functionality into the 4-position of the indole systems have been well documented (Scheme 16).[41–7] Conversely, thallation of 1-acetyl-2,3-dihydroindole (**94**) with Tl(OCOCF$_3$)$_3$ in TFAA (trifluoroacetic anhydride) has been found to proceed at the 7-position.[48] The regioselectivity of this metallation step is rationalized by invoking coordination of the metal centre by the acetyl carbonyl oxygen. Reaction of (**95**) with CuBr$_2$ affords the 7-bromodihydroindole (**96**) (Scheme 17).

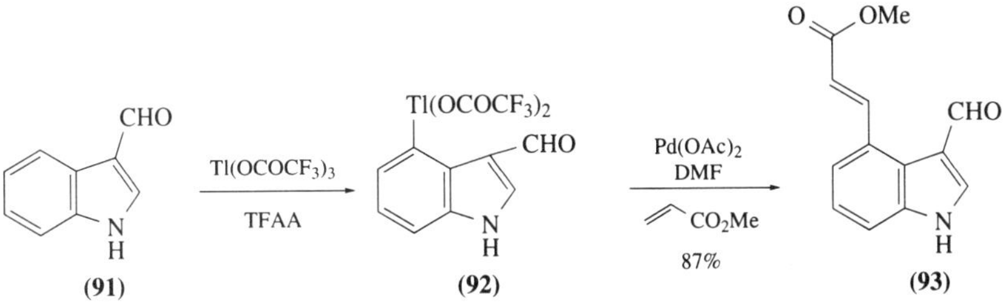

Scheme 16

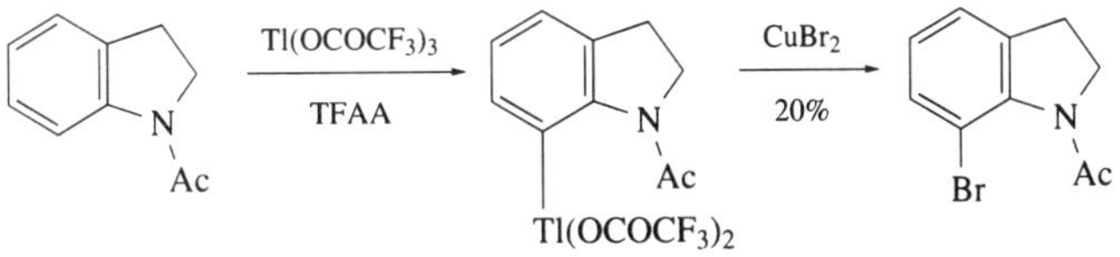

Scheme 17

Control of the regioselectivity in the thallation of indole derivatives has enabled various functionalities to be introduced at the 4- and 7-positions of these heterocyclic compounds. For example, 4-nitroindole-3-carboxaldehyde (**97**) and 1-acetyl-2,3-dihydro-7-nitroindole (**98**) are readily prepared, in a one-pot synthesis, from indole-3-carboxaldehyde (**91**) and 1-acetyl-2,3-dihydroindole (**94**), respectively (Scheme 18).[49]

Thallation of the bis(indole) dione (**99**) at positions C-4 and C-11 has afforded the dithallated compound (**100**). This can be halogenated, alkylated and hydroxylated, leading to an easy preparation of various derivatives of the indole alkaloid bipolaramide (**101**) and (**102**) (Scheme 19).[50]

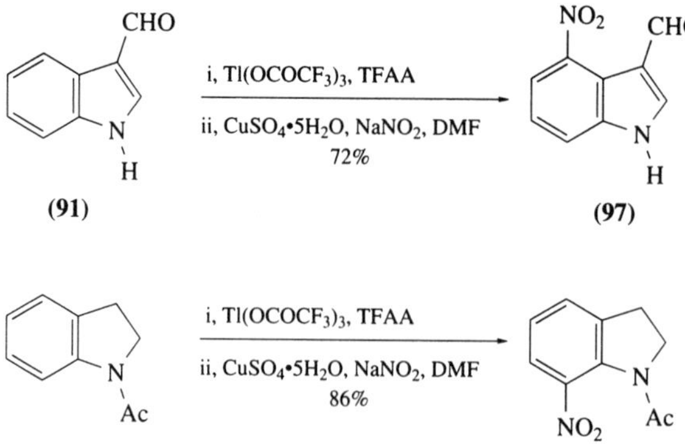

452   Thallium

Scheme 18

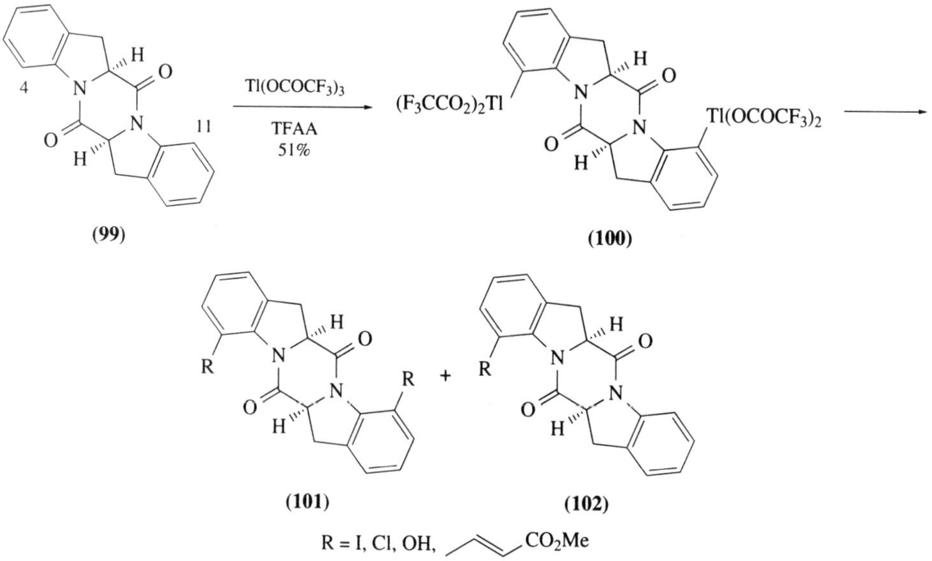

Scheme 19

R = I, Cl, OH, ⌐⌐CO₂Me

10.3 THALLIUM(I) COMPOUNDS

10.3.1 Cyclopentadienylthallium

Although monoalkyl- and monoarylthallium(I) derivatives have been postulated as possible reaction intermediates, they have not yet been isolated due to their inherent instability. Cyclopentadienylthallium and its analogues are the only examples of organic derivatives of thallium(I) which can be isolated.

Formyl- (**104**), acetyl- (**105**), carbomethoxy- (**106**) and carboethoxycyclopentadienylthallium (**107**) have been prepared by reacting the appropriately functionalized cyclopentadienyl sodium salts (**103**) with thallium ethoxide in ethanol. Products are readily isolated as air stable solids (Equation (12)).[51] Reactions of (**104**)–(**107**) with $Mn(CO)_5Br$, $Re(CO)_5Br$, or $[Rh(CO)_2Cl]_2$ provide convenient routes to the respective substituted cyclopentadienyl derivatives of manganese (**108**), rhenium (**109**) or rhodium (**110**).

Similarly, benzyl- and phenylcyclopentadiene have been shown to react with thallium ethoxide in ethanol to provide the corresponding benzyl- and phenylcyclopentadienylthallium compounds (**111**) and (**112**), respectively.[52] Such cyclopentadienylthallium complexes have proved to be very useful reservoirs of the important Cp moiety. They are readily prepared in high yields and can be conveniently stored and handled in air. However, they react readily with most metal halides as well as with many organic substrates.

The use of cyclopentadienylthallium reagents with metal halides, metal carbonyl halides and cyclopentadienyl metal halides to yield new cyclopentadienyl metal complexes is well established. Representative examples of this reaction with several metallic reagents are collected in Table 7.

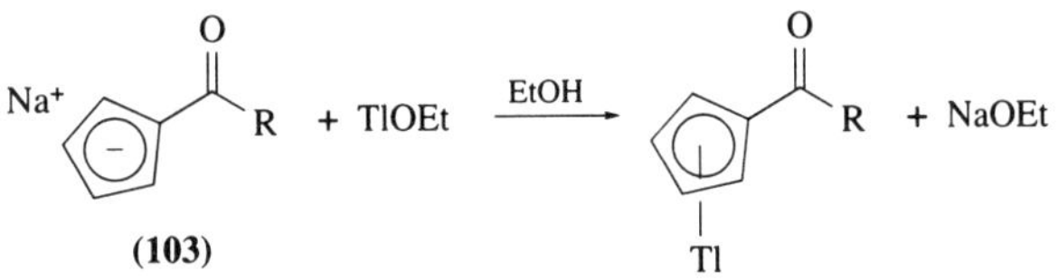

Thallium

(12)

(103)

(104) R = H, ~100%
(105) R = Me, 79%
(106) R = OMe, 93%
(107) R = OEt, ~100%

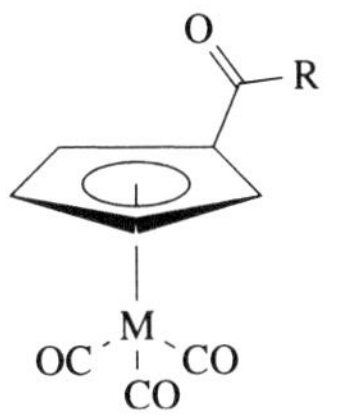

(108) M = Mn
(109) M = Re

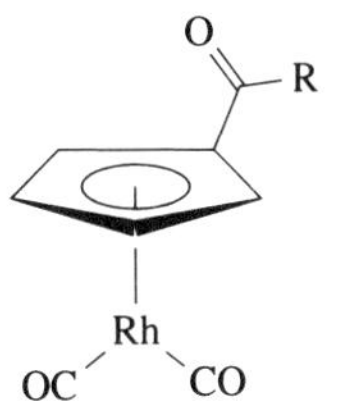

(110)

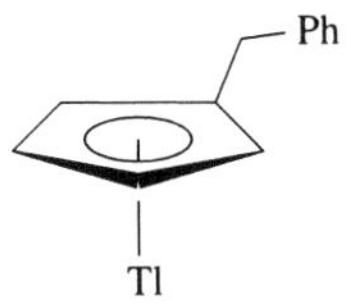

(111)

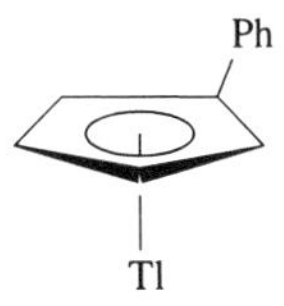

(112)

Table 7 Preparation of new cyclopentadienyl metal complexes by transmetallation from thallium(I) salts.

RCpTl + X_mML ⟶ RCpML

R	Reagent	Product		
Ph	$BrMn(CO)_5$	$PhCpMn(CO)_3$		
PhCO	$BrRe(CO)_5$	$PhCOCpRe(CO)_3$		
PhCO	$FeCl_2$	$(PhCOCp)_2Fe$		
PhCO	$TiCl_4$	$(PhCOCp)_2TiCl_2$		
Ph	$[Cl_2Ru(CO)_3]_2$	$[PhCpRu(CO)_2]_2$	$PhCpRu(CO)_2Cl$	
PhCO	$[ClRh(CO)_2]_2$	$PhCOCpRh(CO)_2$	$[PhCOCpRh(CO)]_3$	$(PhCOCp)_2Rh_2(CO)_3$
PhCO	'$ICo(CO)_4$'	$PhCOCpCo(CO)_2$		

The reaction of [(diphenylphosphino)cyclopentadienyl]thallium (**113**) with cyclopentadienyl-trichlorotitanium (**114**) affords the dichlorotitanium complex (**115**). Subsequent coordination of (**115**) with $(\eta\text{-}C_5H_5)(CO)_2(THF)Mn$ (**116**) yields the heterobimetallic titanium–manganese complex (**117**) (Scheme 20).[53] Reaction of (**115**) with group 9 metal carbonyl halides has also been shown to afford new homobimetallic complexes, for example, $[M(\eta^5C_5H_4)PPh_2(CO)]_2$ where M = Co, Rh and Ir.[54]

Chloro-, bromo-, and iodocyclopentadiene (**119**) can be generated in diethyl ether solution from cyclopentadienylthallium (**118**) and N-halosuccinimides or iodine. Reaction of (**119**) with thallium ethoxide affords the respective (halocyclopentadienyl)thallium complexes (**120**)–(**122**) in 72–96% yield (Equation (13)).[55] These new reagents are useful intermediates in the formation of many other cyclopentadienyl metal derivatives containing halogen substituents (Structures (**123**)–(**127**)).

Interaction of TlOEt with $C_5Ph_4H_2$, C_5Ph_5H, $C_5Ph_4(Bu^tPh)H$ or C_5Bz_5H gives the new cyclopentadienylthallium(I) compounds C_5Ph_4HTl, C_5Ph_5Tl, $C_5Ph_4(Bu^tPh)Tl$ and C_5Bz_5Tl.[56] Reaction of (dimethylphenylsilyl)- and (benzyldimethylsilyl)tetramethylcyclopentadiene with TlOEt yields $[C_5Me_4(SiMe_2Ph)]Tl$ (**128**) and $[C_5Me_4(SiMe_2Bz)]Tl$ (**129**), respectively.[57] Treatment of halide-free solutions of dihydrofulvalene with TlOEt has produced $(\eta^5{:}\eta^5\text{-fulvalene})$dithallium (**130**) in 85–100% yield.[58,59] Subsequent reactions with carbonyl halides of cobalt, rhodium, iridium, manganese and rhenium have produced the corresponding homobimetallic fulvalene metal carbonyls.

The alkylation of cyclopentadienylthallium(I) was reported by Corey and co-workers in their prostaglandin work.[60] It is generally accepted that the thallium(I) counterion is superior to lithium or

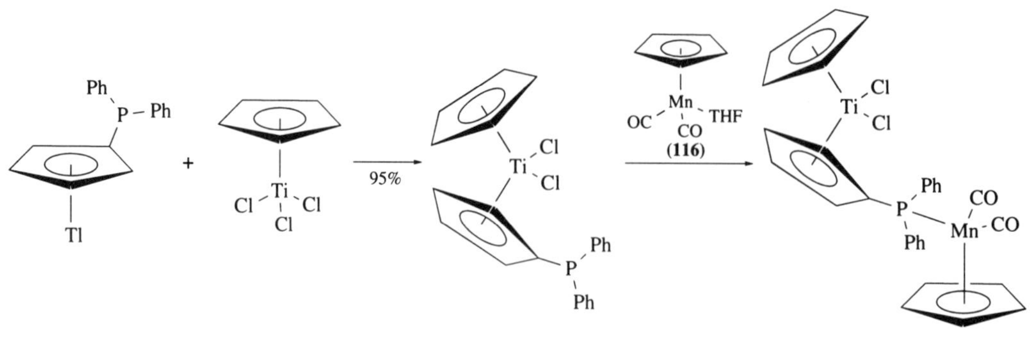

Scheme 20

(113) (114) (115) (116) (117)

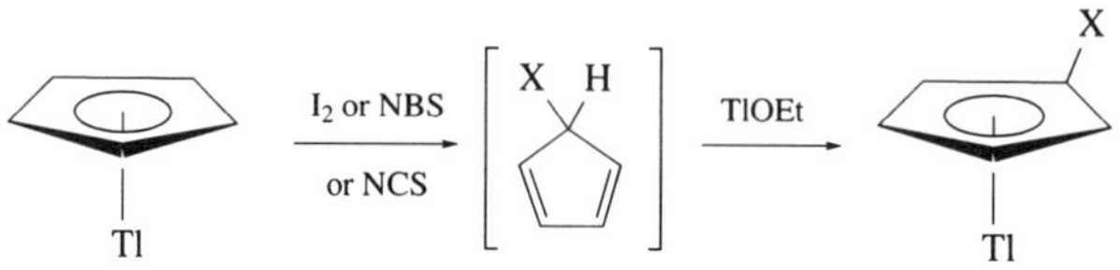

$$\text{(118)} \xrightarrow[\text{or NCS}]{\text{I}_2 \text{ or NBS}} \text{(119)} \xrightarrow{\text{TlOEt}} \text{(120) X = Cl, 72%} \quad (13)$$
(121) X = Br, 96%
(122) X = I, 89%

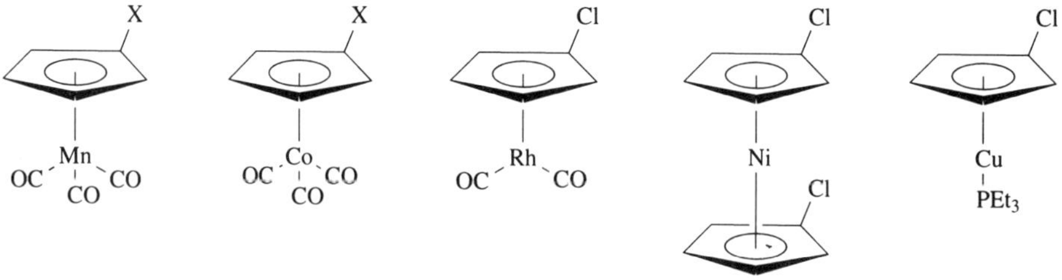

(119) X = Cl, Br, I

(123) X = Cl, Br   (124)   (125)   (126)   (127)

(128) R = Ph
(129) R = CH$_2$Ph

(130)

sodium since it minimizes the isomerization of the initially formed product to other cyclopentadiene isomers. Thus, alkylation of (**118**) with benzyl chloromethyl ether afforded (**131**) exclusively.[61] Cyclopentadienylthallium (**118**) in CH$_2$Cl$_2$ or ether reacts readily with various acyl iodides to give the corresponding acylcyclopentadienes (**132**) (Scheme 21).[62]

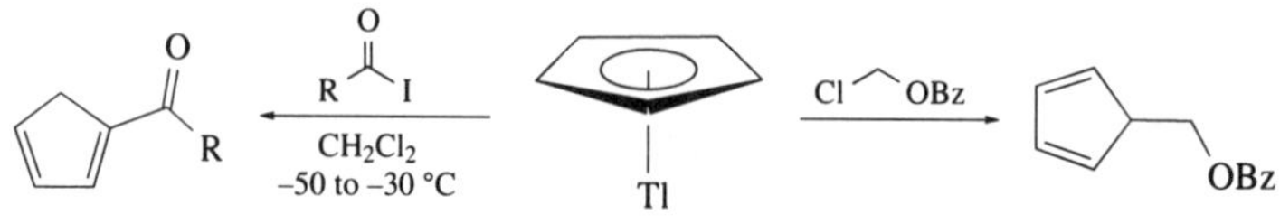

(132) ← (118) → (131)

Scheme 21

The Diels–Alder cycloaddition of dimethylacetylenedicarboxylate to 5-halocyclopentadienes (**119**), generated *in situ* from cyclopentadienylthallium (**118**) suspended in the acetylenic diester, has been reported.[63] The *anti*-7-bromo-norbornadiene (**134**) (X = Br) is formed stereoselectively (Equation (14)).

However, when the chlorine substituted cyclopentadiene derivative is employed, the formation of the *anti*-adduct (**134**) (X = Cl) is accompanied by a small amount of the *syn*-adduct (**135**) (X = Cl).

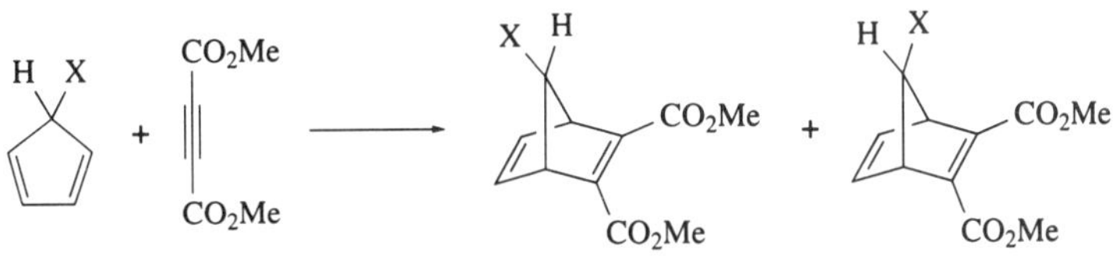

$$
\begin{array}{c}
(119) + (133) \longrightarrow (134) + (135) \\
X = Cl, Br \qquad X = Cl \; 36\% \qquad 24\% \\
\qquad\qquad\qquad X = Br \; 35\%
\end{array} \tag{14}
$$

5-Fluorocyclopentadiene (**137**) has been prepared by the reaction of cyclopentadienylthallium (**118**) with the F$^+$ source (**136**).[64] Treatment of (**137**) with dienophiles leads to adducts possessing exclusively the *syn*-orientation (Equation (15)).

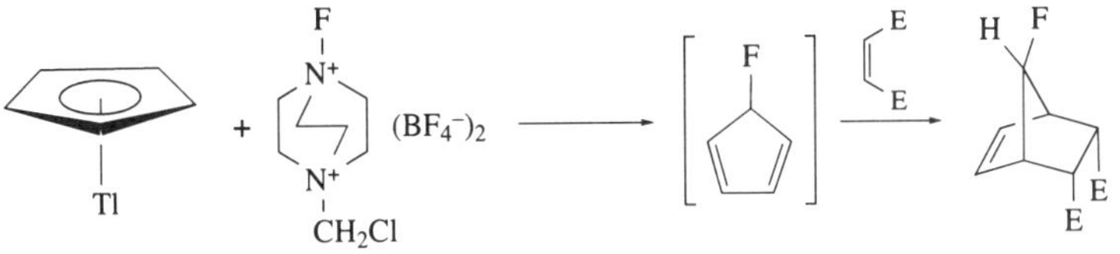

$$(118) + (136)\;(BF_4^-)_2 \longrightarrow (137) \longrightarrow (138) \tag{15}$$

E = CO$_2$Me

Finally, condensation of cyclopentadienylthallium (**118**) with triphenylcyclopropenyl perchlorate (**139**), followed by [1,5]-H-shift of (**140**), affords the adduct (**141**) (Equation (16)).[65]

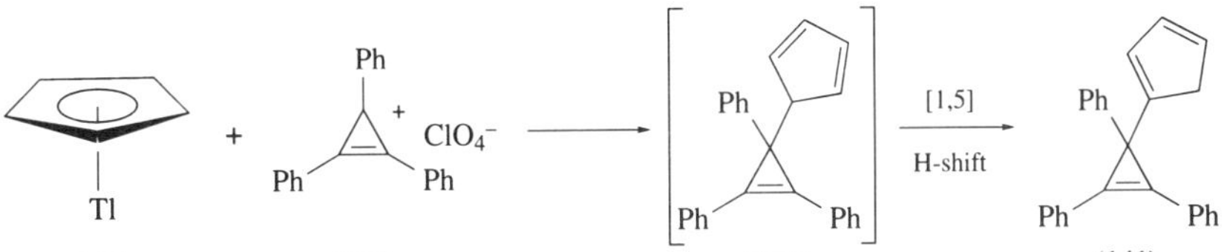

$$(118) + (139) \longrightarrow (140) \xrightarrow{[1,5]\;H\text{-shift}} (141) \tag{16}$$

10.4 MISCELLANEOUS

Thallium(I) and thallium(III) salts are widely used in many organic transformations. Although the thallated intermediates are usually not isolated, mainly due to their toxic nature, they have nonetheless found useful applications in the synthesis of many organic and organometallic compounds.

The reaction of the thallium salt of pentane-2,4-dione (**142**) with dimethyldichlorosilane yields the pyrylium salt (**143**) (Equation (17)).[66]

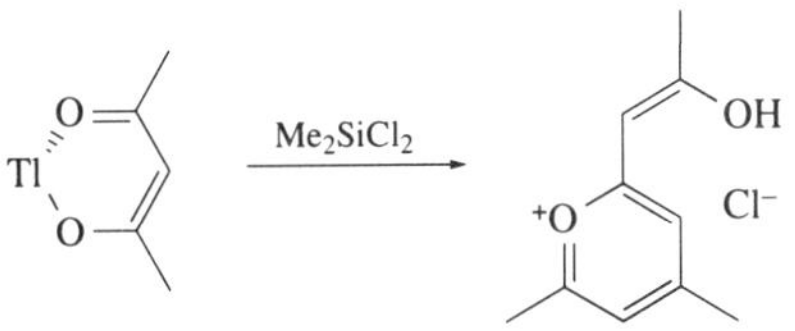

$$(142) \xrightarrow{Me_2SiCl_2} (143) \tag{17}$$

Oxidative transformation of 5-aminouracil (**144**) into imidazolone (**146**) is effected by the use of thallium(III) nitrate trihydrate in methanol.[67] Coumarin (**147**) has been found to react with thallium(III) nitrate in dry methanol to give the substituted nitrocoumarin adduct (**148**) (Scheme 22).[68]

Thallium 2,4,6-tris(trifluoromethyl)phenoxide[69] (**149**) (thallium(I) compound with coordination number two at the thallium atom) can be prepared by the reaction of C$_6$H$_2$(CF$_3$)$_3$OH with TlOEt.

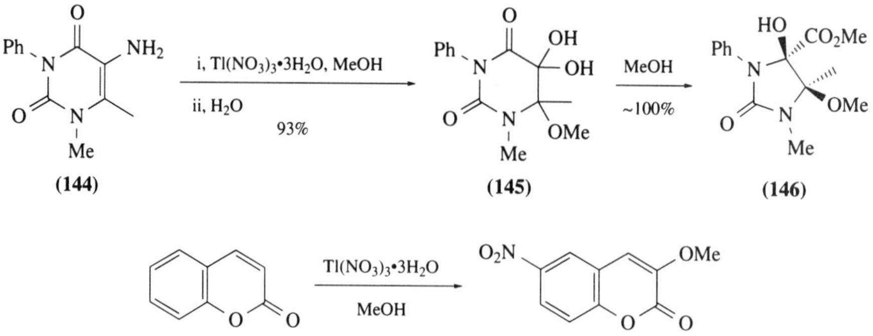

(144) → (145) → (146)

(147) → (148)

Scheme 22

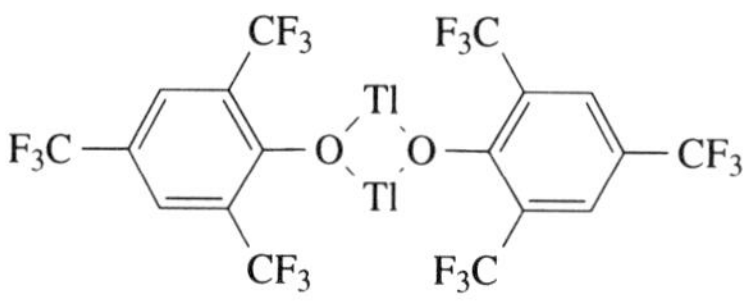

(149)

A vast amount of work has dealt with the reactions and characterizations of complexes obtained by reacting TlX_3 (X = Cl, Br, I) with substituted imidazoles and pyridine N-oxides (Structures (150)–(152)).[70-7]

(150)   (151)   (152)

L = substituted imadazoles, pyridine N-oxide

Oxidation of aromatics substituted either by electron-donating or moderately electron-withdrawing groups using $Tl(OCOCF_3)_3$, in the presence of catalytic amounts of $Pd(OAc)_2$, affords biaryls in good yields, 4,4'-biaryls being the major products.[78] Competition experiments and kinetic studies using arenes and arylthallium derivatives as starting materials, combined with quenching experiments, have demonstrated the first step of this reaction to be a fast thallation of the arene to form the arylthallium intermediate $ArTl(OCOCF_3)_2$. This organothallium derivative transmetallates to $Pd(OAc)_2$ in the rate-determining step. Subsequent, fast decomposition of the diarylpalladium species eventually gives the final reaction products (Equation (18)).

$$2\,ArH + Tl(OCOCF_3)_3 \xrightarrow{Pd^{II}} ArAr \qquad (18)$$

ArH = PhOMe, PhMe, PhEt, PhH, PhF, PhBr, PhCl

A stereospecific synthesis of 9-demethylretinoids (155) and (157), possessing either all *trans*- or 11-*cis*-geometries, has been described. This methodology utilizes the thallium-accelerated palladium-catalysed[79] cross-coupling reaction of an (*E*)-1-alkenylboronic acid (153) with either the (*E*)- or (*Z*)-alkenyl iodides (154) and (156), respectively (Scheme 23).

Similarly, thallium(I) salts have been found to greatly enchance the rates of a series of cyclization–carbonylation processes.[80] The reaction of (158), in methanol under 1 atm of CO and in the presence of TlOAc, catalysed by $PdCl_2(PPh_3)_2$, affords (159) in 86% yield. In the absence of TlOAc, the reaction is much slower (~15% conversion after 17 h) and leads to the formation of a ~10:1 mixture of (159) and (160) (Equation (19)).

When terminally unsubstituted alkenols such as 4-penten-1-ol (161) are treated with $Tl(OAc)_3$ in benzene, acetoxymethyltetrahydrofurans (162) are obtained. However, when the reaction is carried out in acetic acid, the products are the regioisomeric five- and six-membered cyclic ethers (162) and (163), respectively (Scheme 24).[81]

Flavanones (164) have been found to undergo facile dehydrogenation to afford flavones (165) when treated with $Tl(OAc)_3$ in AcOH or acetonitrile. However, on treatment with thallium(III) toluene-*p*-

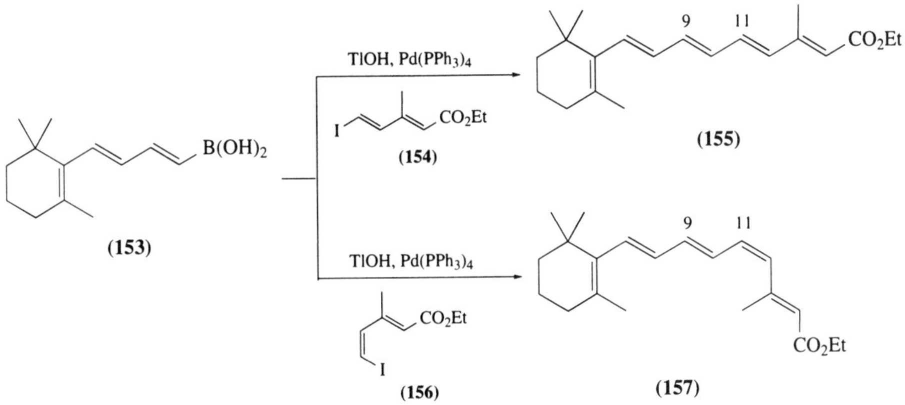

Scheme 23

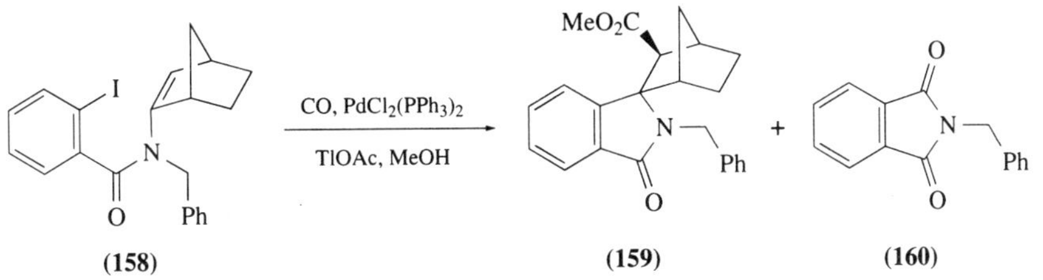

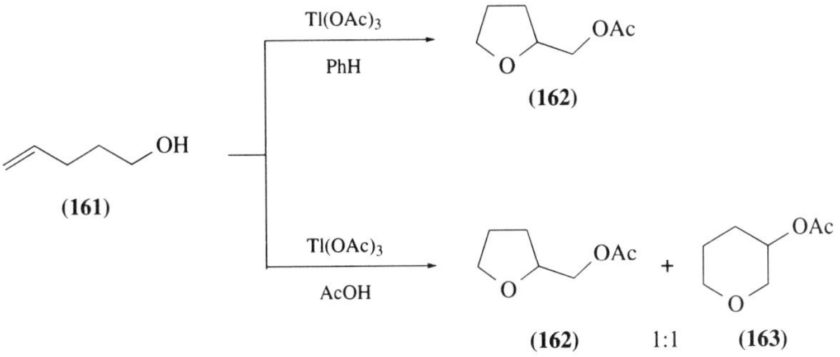

Scheme 24

sulfonate or $Tl(NO_3)_3$ in propionitrile or acetonitrile, respectively, they undergo oxidative 2,3-aryl migration to give isoflavones (**166**) (Scheme 25).[82]

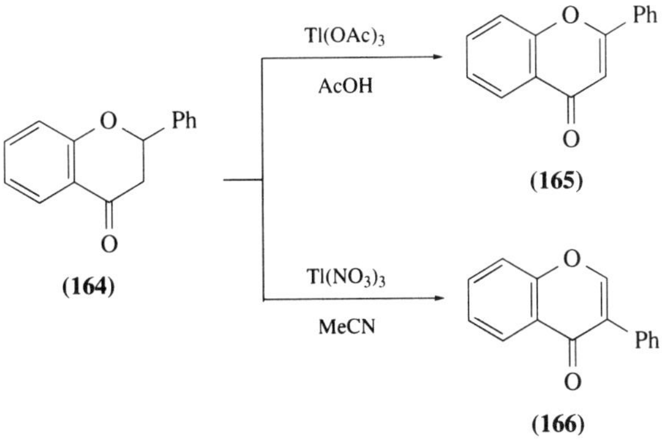

Scheme 25

10.4.1 Conclusions and Perspectives

The unique nature of organothallium reagents has allowed the successful realization of a range of transformations either difficult or impossible to perform otherwise. The synthetic utility of thallium compounds, initially revealed by Taylor and McKillop in their seminal contribution to the area, is being gradually recognized and exploited and the next few years will see increased interest in these organometallic reagents and in their use in organic synthesis.

10.5 REFERENCES

1. H. Kurosawa in 'COMC-I', vol. 1, p. 725 and references cited therein.
2. A. G. Lee, 'The Chemistry of Thallium', Elsevier, Barking, 1971.
3. O. Y. Okhlobystin, K. A. Bilevitch and L. I. Zakharkin, *J. Organomet. Chem.*, 1964, **2**, 281.
4. I. E. Markó and F. Rebière, *Tetrahedron Lett.*, 1992, **33**, 1763.
5. I. E. Markó and C. W. Leung, *J. Am. Chem. Soc.*, 1994, **116**, 371.
6. T. R. Bierschenk, T. J. Juhlke, W. I. Bailey and R. J. Lagow, *J. Organomet. Chem.*, 1984, **277**, 1.
7. D. Naumann, W. Strauss and W. Tyrra, *J. Organomet. Chem.*, 1991, **407**, 1.
8. I. E. Markó and J. M. Southern, *J. Org. Chem.*, 1990, **55**, 3368.
9. I. E. Markó, J. M. Southern and M. L. Kantam, *Synlett.*, 1991, 235.
10. I. E. Markó and M. L. Kantam, *Tetrahedron Lett.*, 1991, **32**, 2255.
11. Y. Kawasaki and N. Okuda, *Chem. Lett.*, 1982, 1161.
12. J. S. Casas, A. Sánchez, J. Sordo, E. M. Vazquez-Lopez, E. E. Castellano and J. Zukerman-Schpector, *Polyhedron*, 1992, **11**, 2889.
13. B. S. R. Reddy, R. Arshady and M. H. George, *Polymer*, 1984, **25**, 115.
14. A. Laguna, E. J. Fernández, A. Mendia, M. E. Ruiz-Romero and P. G. Jones, *J. Organomet. Chem.*, 1989, **365**, 201.
15. R. M. Moriarty, R. Penmasta, I. Prakash and A. K. Awasthi, *J. Org. Chem.*, 1988, **53**, 1022.
16. M. Veith, H. Lange, A. Belo and O. Recktenwald, *Chem. Ber.*, 1985, **118**, 1600.
17. M. V. Castano et al., *J. Organomet. Chem.*, 1991, **417**, 327.
18. M. S. G. Tasende, M. I. S. Gimeno, A. Sánchez, J. S. Casas and J. Sordo, *J. Organomet. Chem.*, 1990, **390**, 293.
19. Y. P. Mascarenhas, I. Vencato, M. C. Carrascal, J. M. Varela, J. S. Casas and J. Sordo, *J. Organomet. Chem.*, 1988, **344**, 137.
20. P. K. Byers, A. J. Canty, K. Mills and L. Titcombe, *J. Organomet. Chem.*, 1985, **295**, 401.
21. (a) J. L. Morris, C. W. Rees and D. J. Rigg, *J. Chem. Soc., Chem. Commun.*, 1985, 396; (b) C. W. Rees and J. R. J. Surtees, *J. Chem. Soc., Perkin Trans. 1*, 1991, 2945.
22. B. Györi, A. Sánchez and J. Glaser, *J. Organomet. Chem.*, 1989, **361**, 1.
23. R. D. Bach, J. W. Holubka and C. L. Willis, *J. Am. Chem. Soc.*, 1982, **104**, 3980.
24. R. K. Sharma, E. D. Martinez and H. Zolfaghari, *J. Chem. Res. (S)*, 1990, 237.
25. M. N. Bochkarev, T. A. Basalgina, G. S. Kalinina and G. A. Razuvaev, *J. Organomet. Chem.*, 1983, **243**, 405.
26. E. C. Taylor, A. H. Katz, S. I. Alvarado and A. McKillop, *J. Org. Chem.*, 1986, **51**, 1607.
27. R. C. Larock, S. Varaprath, H. H. Lau and C. A. Fellows, *J. Am. Chem. Soc.*, 1984, **106**, 5274.
28. R. C. Larock, C.-L. Liu, H. H. Lau and S. Varaprath, *Tetrahedron Lett.*, 1984, **25**, 4459.
29. E. C. Taylor, A. H. Katz and A. McKillop, *Tetrahedron Lett.*, 1984, **25**, 5473.
30. R. C. Larock and C. A. Fellows, *J. Am. Chem. Soc.*, 1982, **104**, 1900.
31. R. A. Kjonaas and D. C. Shubert, *J. Org. Chem.*, 1983, **48**, 1924.
32. E. C. Taylor, A. H. Katz, S. I. Alvarado and A. McKillop, *J. Organomet. Chem.*, 1985, **285**, C9.
33. R. A. Kjonaas, *J. Org. Chem.*, 1986, **51**, 3708.
34. H. Kurosawa, M. Sato and H. Okada, *Tetrahedron Lett.*, 1982, **23**, 2965.
35. H. Kurosawa, H. Okada, M. Sato and T. Hattori, *J. Organomet. Chem.*, 1983, **250**, 83.
36. G. M. Pickles, T. Spencer, F. G. Thorpe, A. D. Ayala and J. C. Podestá, *J. Chem. Soc., Perkin Trans. 1*, 1982, 2949.
37. M. Ochiai, E. Fujita, M. Arimoto and H. Yamaguchi, *Chem. Pharm. Bull.*, 1982, **30**, 3994.
38. G. B. Deacon and R. N. M. Smith, *Aust. J. Chem.*, 1982, **35**, 1587.
39. E. C. Taylor, A. H. Katz, H. Salgado-Zamora and A. McKillop, *Tetrahedron Lett.*, 1985, **26**, 5963.
40. M. Somei, T. Hasegawa and C. Kaneko, *Heterocycles*, 1983, **20**, 1983.
41. M. M. Goodman, G. Kirsch and F. F. Knapp, Jr., *J. Med. Chem.*, 1984, **27**, 390.
42. M. Somei, H. Amari and Y. Makita, *Chem. Pharm. Bull.*, 1986, **34**, 3971.
43. M. Somei, F. Yamada and K. Naka, *Chem. Pharm. Bull.*, 1987, **35**, 1322.
44. M. Somei, E. Iwasa and F. Yamada, *Heterocycles*, 1986, **24**, 3065.
45. F. Yamada and M. Somei, *Heterocycles*, 1987, **26**, 1173.
46. M. Somei, T. Ohta, J. Shinoda and Y. Somada, *Heterocycles*, 1989, **29**, 653.
47. A. Arduini, A. Pochini, A. Rizzi, A. R. Sicuri and R. Ungaro, *Tetrahedron Lett.*, 1990, **31**, 4653.
48. M. Somei, Y. Saida, T. Funamoto and T. Ohta, *Chem. Pharm. Bull.*, 1987, **35**, 3146.
49. M. Somei, F. Yamada, H. Hamada and T. Kawasaki, *Heterocycles*, 1989, **29**, 643.
50. M. Somei and T. Kawasaki, *Chem. Pharm. Bull.*, 1989, **37**, 3426.
51. S. S. Jones, M. D. Rausch and T. E. Bitterwolf, *J. Organomet. Chem.*, 1990, **396**, 279.
52. P. Singh, M. D. Rausch and T. E. Bitterwolf, *J. Organomet. Chem.*, 1988, **352**, 273.
53. M. D. Rausch, B. H. Edwards, R. D. Rogers and J. L. Atwood, *J. Am. Chem. Soc.*, 1983, **105**, 3882.
54. M. D. Rausch, W. C. Spink, J. L. Atwood, A. J. Baskar and S. G. Bott, *Organometallics*, 1989, **8**, 2627.
55. B. G. Conway and M. D. Rausch, *Organometallics*, 1985, **4**, 688.
56. H. Schumann, C. Janiak and H. Khani, *J. Organomet. Chem.*, 1987, **330**, 347.
57. H. Schumann, H. Kucht, A. Dietrich and L. Esser, *Chem. Ber.*, 1990, **123**, 1811.
58. W. C. Spink and M. D. Rausch, *J. Organomet. Chem.*, 1986, **308**, C1.

59. M. D. Rausch, W. C. Spink, B. G. Conway, R. D. Rogers and J. L. Atwood, *J. Organomet. Chem.*, 1990, **383**, 227.
60. E. J. Corey, U. Koelliker and J. Neuffer, *J. Am. Chem. Soc.*, 1971, **93**, 1489.
61. G. V. B. Madhavan and J. C. Martin, *J. Org. Chem.*, 1986, **51**, 1287.
62. G. Grundke and H. M. R. Hoffmann, *J. Org. Chem.*, 1981, **46**, 5428.
63. M. Franck-Neumann and M. Sedrati, *Tetrahedron Lett.*, 1983, **24**, 1391.
64. M. A. McClinton and V. Sik, *J. Chem. Soc., Perkin Trans. 1*, 1992, 1891.
65. A. Padwa, Y. S. Kulkarni and L. W. Terry, *J. Org. Chem.*, 1990, **55**, 2478.
66. N. Serpone, T. Ignacz, B. G. Sayer and M. J. McGlinchey, *Inorg. Chim. Acta*, 1984, **89**, 139.
67. I. Matsuura *et al.*, *J. Chem. Soc., Chem. Commun.*, 1992, 1474.
68. A. Banerji and G. Nandi, *Heterocycles*, 1987, **26**, 1221.
69. H. W. Roesky, M. Scholz, M. Noltemeyer and F. T. Edelmann, *Inorg. Chem.*, 1989, **28**, 3829.
70. M. R. Bermejo, J. Irisarri and M. Gayoso, *Synth. React. Inorg. Metal-Org. Chem.*, 1985, **15**, 197.
71. M. R. Bermejo, M. I. Fernández, J. G. Taboada and M. Gayoso, *Synth. React. Inorg. Metal-Org. Chem.*, 1986, **16**, 327.
72. M. R. Bermejo, M. T. Lage, M. I. Fernández and A. Castiñeiras, *Synth. React. Inorg. Metal-Org. Chem.*, 1987, **17**, 79.
73. M. R. Bermejo, A. G. Deibe, A. Rodríguez and A. Castiñeiras, *Synth. React. Inorg. Metal-Org. Chem.*, 1987, **17**, 693.
74. M. R. Bermejo, E. Solleiro, A. Rodríguez and A. Castiñeiras, *Polyhedron*, 1987, **6**, 315.
75. M. R. Bermejo, A. Fernández, M. Gayoso, A. Castiñeiras, W. Hiller and J. Strähle, *Polyhedron*, 1988, **7**, 2561.
76. M. R. Bermejo, M. B. Fernández, M. I. Fernández and M. E. Gómez, *Synth. React. Inorg. Metal-Org. Chem.*, 1991, **21**, 915.
77. M. R. Bermejo, M. I. Fernández, B. Fernández and M. E. Gómez, *Synth. React. Inorg. Metal-Org. Chem.*, 1992, **22**, 759.
78. A. K. Yatsimirsky, S. A. Deiko and A. D. Ryabov, *Tetrahedron.*, 1983, **39**, 2381.
79. A. R. de Lera, A. Torrado, B. Iglesias and S. López, *Tetrahedron Lett.*, 1992, **33**, 6205.
80. R. Grigg, P. Kennewell and A. J. Teasdale, *Tetrahedron Lett.*, 1992, **33**, 7789.
81. M. L. Mihailović, R. Vukićević, S. Konstantinović, S. Milosavljević and G. Schroth, *Liebigs Ann. Chem.*, 1992, 305.
82. M. S. Khanna, O. V. Singh, C. P. Garg and R. P. Kapoor, *J. Chem. Soc., Perkin Trans. 1*, 1992, 2565.

11
Lead

JOHN T. PINHEY
University of Sydney, NSW, Australia

11.1 INTRODUCTION	461
11.2 GROUP TRANSFER FROM R_4Pb TO ALDEHYDES	462
11.2.1 Alkylation, Vinylation and Alkynylation	462
11.2.2 Stereoselectivity in Alkylation	462
11.2.3 Stereoselectivity in α-Methoxyorganolead–Aldehyde Condensations	463
11.3 PALLADIUM-CATALYSED COUPLING OF ACID CHLORIDES AND TETRAALKYLLEADS	464
11.4 RADICAL-MEDIATED TRANSFER OF AN ALLYL GROUP FROM ALLYLTRIPHENYLLEAD	464
11.5 CHEMISTRY OF ORGANOLEAD TRICARBOXYLATES	465
11.5.1 General Properties	465
11.5.2 Synthesis by Direct Plumbation	466
11.5.3 Synthesis by Metal–Metal Exchange	467
11.5.4 Carbon–Carbon Bond Forming Reactions in the Absence of Acid Catalysis	468
11.5.4.1 Reactions with phenols	469
11.5.4.2 Reactions with β-diketones, β-keto esters and their vinylogues	471
11.5.4.3 Reactions of α-hydroxymethylene ketones	474
11.5.4.4 Reactions with derivatives of malonic acid	475
11.5.4.5 Reactions with nitroalkanes	477
11.5.4.6 Reactions of ketone enolates and enamines	477
11.5.5 Replacement of Lead in Aryllead Triacetates by Iodide and Azide	479
11.5.6 Mechanism of the Arylation Reactions of Sections 11.5.4 and 11.5.5	480
11.5.7 Copper-catalysed Reactions of Aryllead Triacetates	480
11.5.8 Acid-catalysed Reactions of Aryllead Tricarboxylates	481
11.6 REFERENCES	484

11.1 INTRODUCTION

This review is confined to those organolead compounds which contain a C–Pb σ-bond. It is, therefore, restricted to lead(IV) compounds since organolead(II) compounds, with a few exceptions,[1a,2,3] are highly unstable and have only been postulated as transient intermediates. The reactions of lead tetraacetate, which is one of a small number of stable inorganic lead(IV) compounds and which is widely used in organic synthesis, are also beyond the scope of this chapter; its only reactions covered here are those in which it is employed in the synthesis of organolead(IV) reagents.

If compounds containing Pb–Pb bonds and lead bonded to another metal are excluded, the vast majority of organolead compounds belong to one of the groups, R_4Pb, R_3PbX, R_2PbX_2 or $RPbX_3$, where R is a σ-bonded organic residue and X is an electronegative species. A very large number of compounds fall within this series, and their synthesis and properties have been the subject of a major monograph[1] and extensive review articles.[4,5]

Despite the long history of the study of organolead compounds, and the commercial importance of one of them, namely tetraethyllead, it is only since about 1980 that they have been developed as reagents for use in organic synthesis. These reagents, which belong to either the tetraorganolead (R_4Pb) or the monoorganolead ($RPbX_3$) groups, are reviewed in this chapter.

11.2 GROUP TRANSFER FROM R_4Pb TO ALDEHYDES

11.2.1 Alkylation, Vinylation and Alkynylation

Tetraalkylleads react readily with aldehydes in the presence of $TiCl_4$ to produce alcohols in good yield (Equation (1)).[6] Other Lewis acids, such as BF_3, bring about the alkyl transfer but are less effective than $TiCl_4$. The nature of the aldehyde has little effect on the yield of the secondary alcohol; benzaldehyde, octanal and cyclohexanecarbaldehyde give similar high yields with both $Et_4Pb/TiCl_4$ and $Bu_4Pb/TiCl_4$. Only one of the four alkyl groups is transferred, and by using mixed tetraalkylleads and $TiCl_4$ it was shown that the transfer order of groups in the reaction of aldehydes is $Me > Et > Pr^i \gg Bu^n$. Reactions of Et_4Pb were very rapid and practically quantitative, while the transfer of butyl and *s*-alkyl groups was sluggish, giving significantly lower yields. The reaction is quite chemoselective; aldehydes underwent alkylation in the presence of ketone s, and no reaction occurred when cyclohexanone and other ketones were treated with $Bu_4Pb/TiCl_4$ for a prolonged time at an elevated temperature.

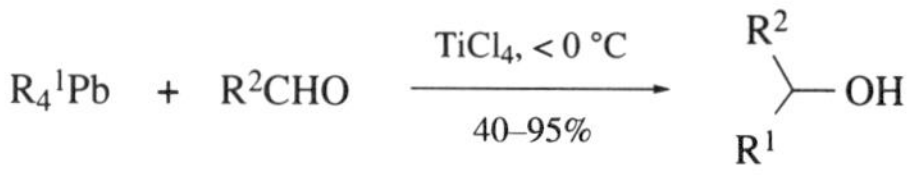

R4¹Pb + R²CHO
TiCl4, < 0 °C
40–95%
R²
OH
R¹

(1)

The mechanism of the reaction is unknown. Although transmetallation can occur in the $R_4Pb/TiCl_4$ system, as shown in Equation (2), it has been established that $RTiCl_3$ is not a reactive intermediate. When R_4Sn compounds were substituted for R_4Pb, no alkyl group transfer occurred, and the higher reactivity of R_4Pb has been attributed to the weak C–Pb bond. The transfer of a vinyl group or an alkynyl group occurs in preference to an alkyl group, as indicated in Equations (3) and (4); however, the reaction appears to be less efficient than alkylation and the conditions required modification. For the vinylation (Equation (3)), it was necessary to protect the aldehyde as the ketal, while replacement of $TiCl_4$ by $BF_3{\cdot}OEt_2$ was required for alkynylation (Equation (4)).

$$R_4Pb + TiCl_4 \longrightarrow RTiCl_3 + R_3PbCl \qquad (2)$$

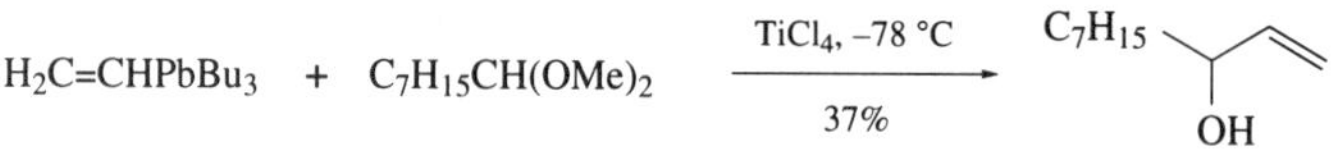

H2C=CHPbBu3 + C7H15CH(OMe)2
TiCl4, –78 °C
37%
C7H15
OH

(3)

Bu3Pb
Bu
+ PhCHO
BF3•OEt2, –78 to –30 °C
64%
Ph
HO
Bu

(4)

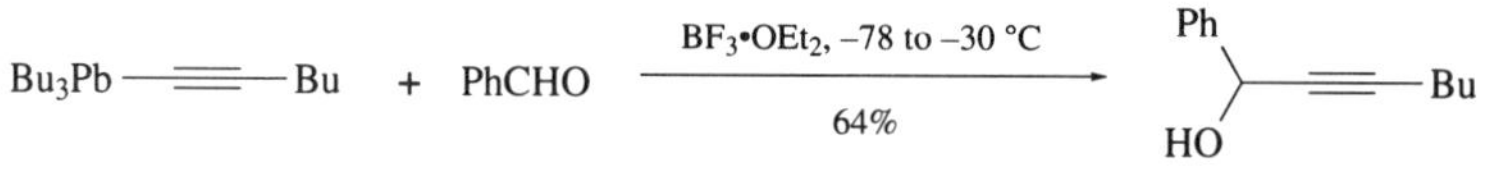

11.2.2 Stereoselectivity in Alkylation

The reactions of aldehydes with the $R_4Pb/TiCl_4$ system can result in very high asymmetric induction; for example, reaction of α-phenylpropionaldehyde with $Et_4Pb/TiCl_4$ gave high Cram selectivity (Equation (5)), while 2-benzyloxypropanal reacted under the same conditions to yield a 98:2 ratio of chelation to nonchelation product (Equation (6)).[6] Significant selectivity was also achieved in a similar reaction with 3-benzyloxybutanal (Equation (7)); here the chelation to nonchelation product ratio was 78:22 for Et_4Pb and 91:9 for Bu_4Pb.

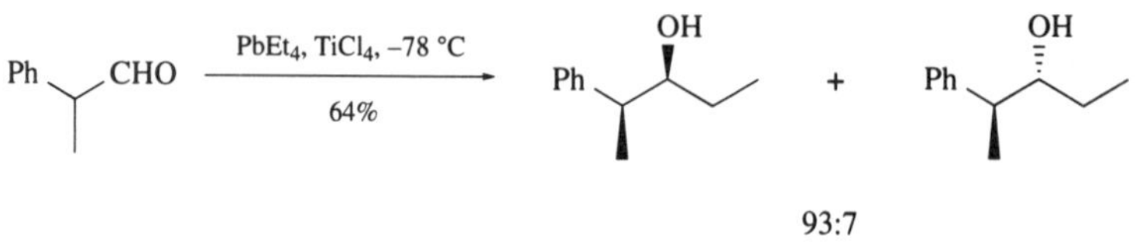

Ph
CHO
PbEt4, TiCl4, –78 °C
64%
OH
Ph
+
OH
Ph
93:7

(5)

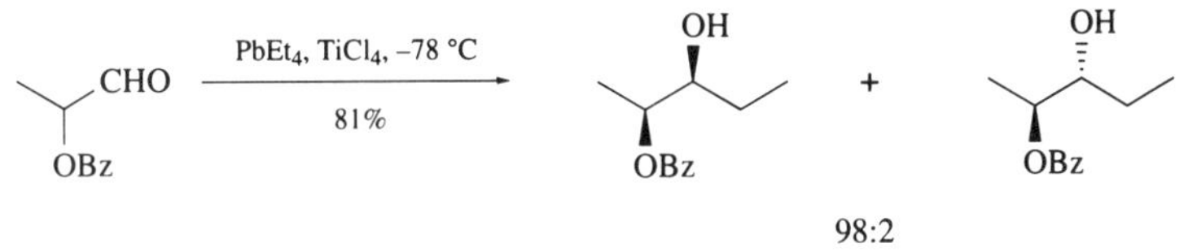

(6) [reaction scheme: isopropyl α-OBz aldehyde + PbEt₄, TiCl₄, −78 °C, 81% → two diastereomeric OH/OBz products, 98:2]

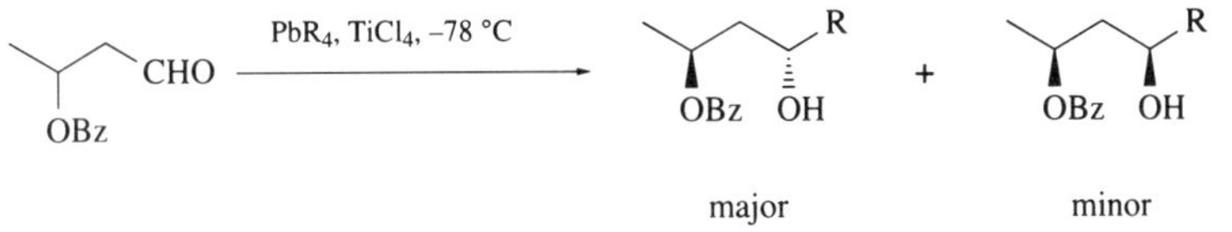

(7) [reaction scheme: α-OBz aldehyde + PbR₄, TiCl₄, −78 °C → major and minor syn/anti OBz,OH products]

11.2.3 Stereoselectivity in α-Methoxyorganolead–Aldehyde Condensations

Tributyl(α-methoxyalkyl)lead compounds (**1**) may be prepared from the corresponding stannanes by transmetallation with BuLi and subsequent trapping with Bu₃PbBr (Equation (8)).[7] Although the lead compounds (**1**) are somewhat unstable, transfer of the α-methoxyalkyl group from lead to an aldehyde can be achieved in good yield in the presence of a Lewis acid.[7,8] Very high diastereoselectivity is attained in the reaction with TiCl₄ as the Lewis acid. The ratio of *syn* product (**2**) to *anti* product (**3**) was 97:3 in the reaction of (**1a**) with octanal and 99.5:0.5 when the lead compound (**1b**) was reacted with cyclohexanecarbaldehyde (Scheme 1).[7] Interestingly, the diastereoselectivity of the reaction was reversed when BF₃·OEt₂ was used instead of TiCl₄; however, the selectivity was lower than in the TiCl₄-mediated reactions and yields were generally not as high. A reaction of the lead compound (**1a**) with octanal produced an 18:82 ratio of *syn* product (**2**) to *anti* product (**3**), while lower selectivity was achieved with the lead compound (**1b**) (Scheme 1). These results have been rationalized in terms of chelation control (TiCl₄) and nonchelation control (BF₃) of the reaction.[7]

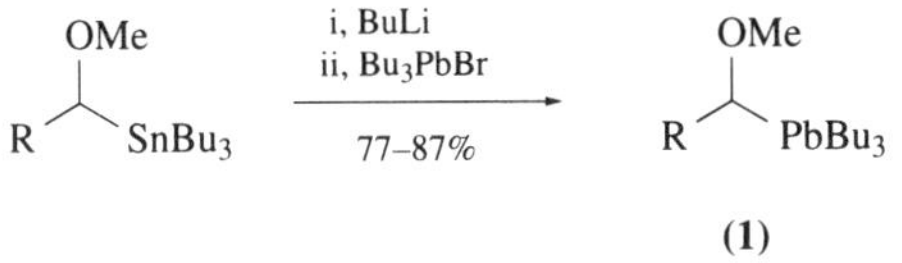

$$\text{R}\overset{\text{OMe}}{\underset{}{\diagup}}\text{SnBu}_3 \xrightarrow[77-87\%]{\text{i, BuLi; ii, Bu}_3\text{PbBr}} \text{R}\overset{\text{OMe}}{\underset{}{\diagup}}\text{PbBu}_3 \quad (1) \tag{8}$$

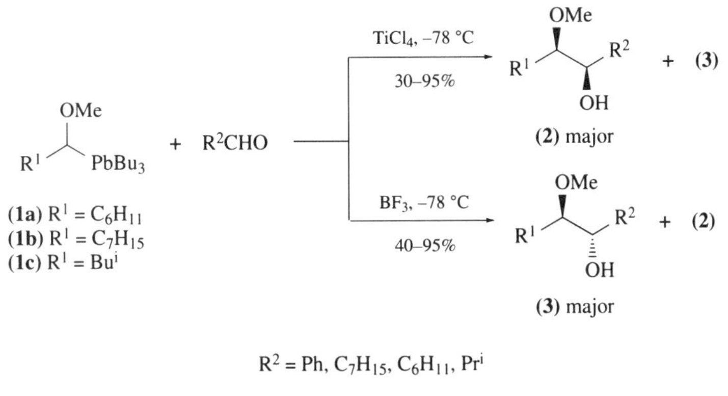

Scheme 1

(**1a**) R¹ = C₆H₁₁
(**1b**) R¹ = C₇H₁₅
(**1c**) R¹ = Buⁱ

R² = Ph, C₇H₁₅, C₆H₁₁, Prⁱ

TiCl₄, −78 °C, 30–95% → (**2**) major + (**3**)
BF₃, −78 °C, 40–95% → (**3**) major + (**2**)

The TiCl₄-mediated reaction of α-alkoxylead compounds (**1**) with aldehydes has been extended to include an aldehyde with a chiral centre at the α-position.[8] Here again, very high diastereofacial stereoselectivity could be achieved. The reaction of (±)-2-phenylpropanal with the α-alkoxylead compound (**1c**) in the presence of TiCl₄ resulted in very high 1,2-*syn* and 2,3-*syn* selectivity (Equation (9)); however, when 2-methylbutanal was employed the diastereofacial selection was low, presumably due to the small difference, steric and electronic, between the substituents.

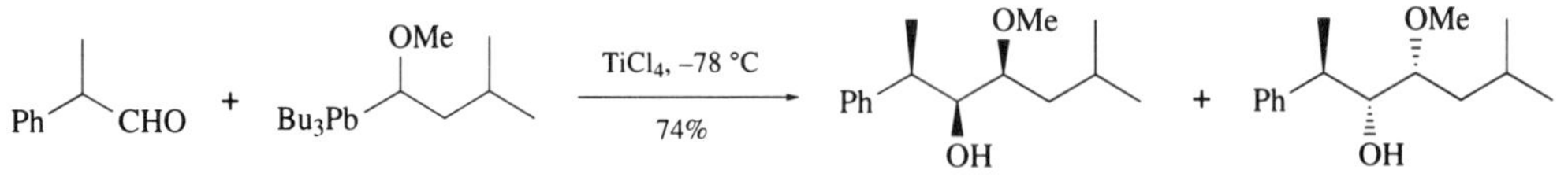

$$\text{Ph-CHO} + \text{Bu}_3\text{Pb}\overset{\text{OMe}}{\underset{}{\diagup}} \xrightarrow[74\%]{\text{TiCl}_4, -78\ °C} \text{Ph}\overset{\text{OMe}}{\underset{\text{OH}}{\diagup\diagup}} + \text{Ph}\overset{\text{OMe}}{\underset{\text{OH}}{\diagup\diagup}} \quad (9)$$

95:5

11.3 PALLADIUM-CATALYSED COUPLING OF ACID CHLORIDES AND TETRAALKYLLEADS

Ketones are formed in high yield when a tetraalkyllead and an acid chloride are heated in the presence of a palladium(0) catalyst (1 mol.%) (Equation (10)).[9] The reaction proceeds rapidly and would appear to be quite general. Two of the alkyl groups of R$_4$Pb are transferred and a high yield of ketone results from the use of 0.6 equiv. of lead compound. This is in contrast to the analogous reaction of R$_4$Sn, which is sluggish and in which only one of the four alkyl groups is utilized.

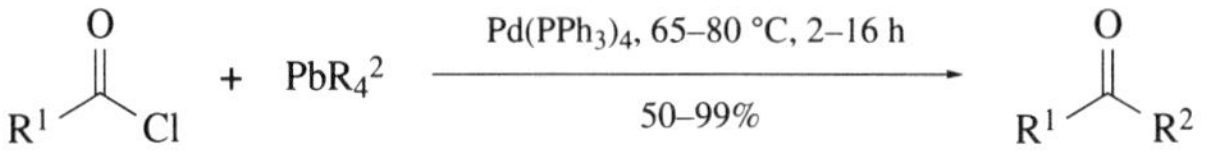

$$R^1\text{COCl} + \text{PbR}_4^2 \xrightarrow[50-99\%]{\text{Pd(PPh}_3)_4,\ 65-80\ °C,\ 2-16\ h} R^1\text{COR}^2 \quad (10)$$

R^1 = Ph, C_7H_{15}, PhC=CH; R^2 = Et, Bu

11.4 RADICAL-MEDIATED TRANSFER OF AN ALLYL GROUP FROM ALLYLTRIPHENYLLEAD

Allyltriphenyllead shows very high reactivity towards carbon radicals α to a carbonyl group. For example, UV irradiation of a mixture of α-(phenylseleno)-γ-butyrolactone (**4**) and allyltriphenyllead produced α-allyl-γ-butyrolactone (**5**). In the presence of a catalytic amount of Ph$_3$Pb–PbPh$_3$ the reaction is very fast, leading to a practically quantitative yield of the lactone (**5**) (Equation (11)).[10] High yields of analogous products were also obtained with the α-phenylseleno derivatives of a range of ketones.

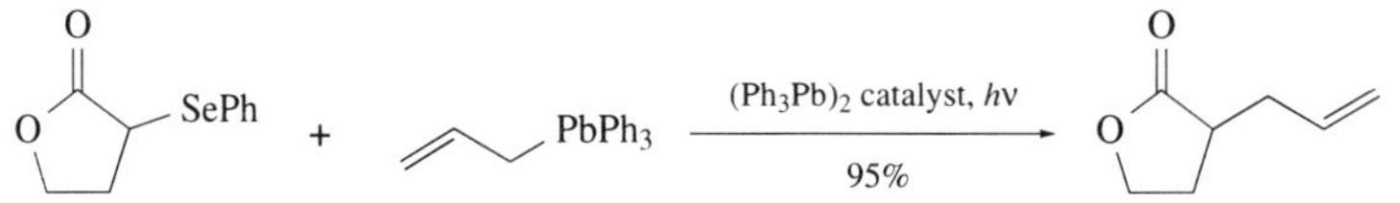

$$\text{(butyrolactone-SePh)} + \diagup\diagdown\text{PbPh}_3 \xrightarrow[95\%]{(\text{Ph}_3\text{Pb})_2\ \text{catalyst},\ h\nu} \text{(allyl-butyrolactone)} \quad (11)$$

(**4**)   (**5**)

The high reactivity of allyltriphenyllead to radical addition has been exploited in a potentially useful one-pot dialkylation of α,β-unsaturated carbonyl compounds.[10] The reaction, which is illustrated in Equation (12), involves the UV irradiation of a mixture of an α,β-unsaturated ketone, allyltriphenyllead and an alkyl bromide, with a catalytic amount of (Ph$_3$Pb)$_2$. Alkyl iodides may also be used, but the yields are lower. The reactions of cyclopentenone and cyclohexenone were examined with a variety of alkyl bromides which contained primary, secondary and tertiary alkyl groups; all gave a mixture of *cis* and *trans* isomers, in which the major product was the *trans* compound. The reaction is fast, is generally high yielding and has been extended to the α,β-unsaturated compounds acrylonitrile and methyl crotonate. The proposed mechanism for the reaction is shown in Scheme 2.

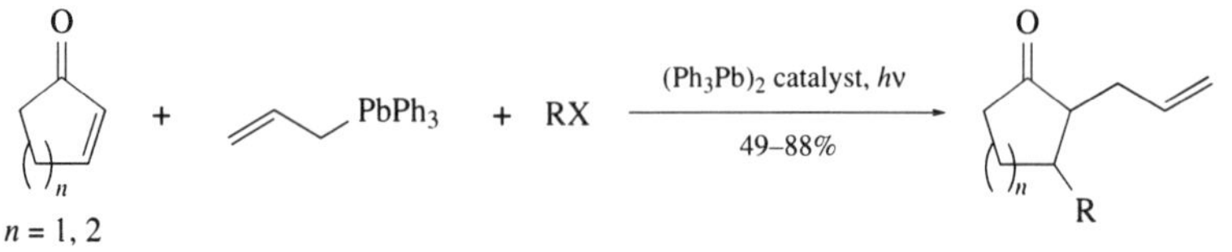

$$\text{cycloenone} + \diagup\diagdown\text{PbPh}_3 + RX \xrightarrow[49-88\%]{(\text{Ph}_3\text{Pb})_2\ \text{catalyst},\ h\nu} \text{product} \quad (12)$$

n = 1, 2

R = C_6H_{11}, Bu, Pr^i, Bu^t (**6**) *trans* >> *cis*

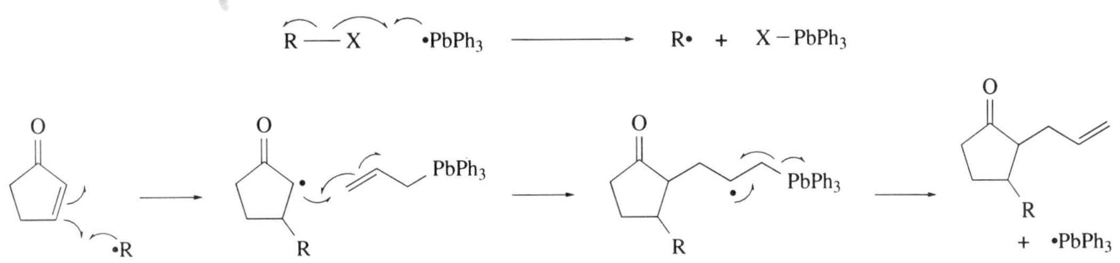

Scheme 2

11.5 CHEMISTRY OF ORGANOLEAD TRICARBOXYLATES

11.5.1 General Properties

For many years β-acetoxyalkyllead triacetates (**7**) have been proposed as reactive intermediates in the reactions of alkenes with lead tetraacetate (Scheme 3).[11,12] In general, these intermediates are highly unstable because of the facile heterolysis of the weak C–Pb bond with concomitant change in the oxidation state of lead from +4 to +2 (Equation (13)). This cleavage of the C–Pb bond can lead to formation of an incipient carbocation, or it may be accompanied by nucleophilic attack by acetate or solvent, or by an intramolecular displacement, as in the formation of an acetoxonium ion.

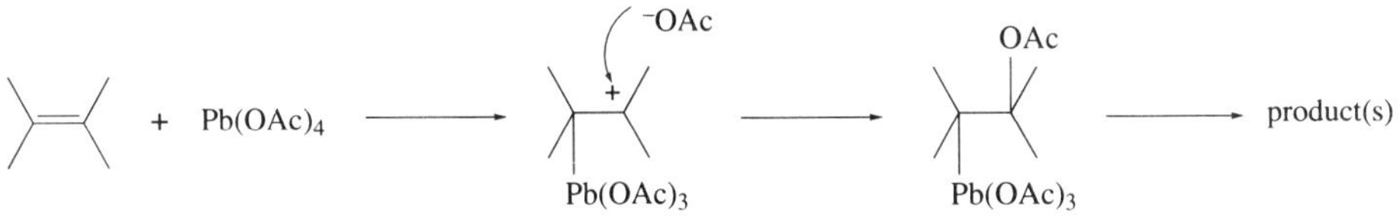

Scheme 3

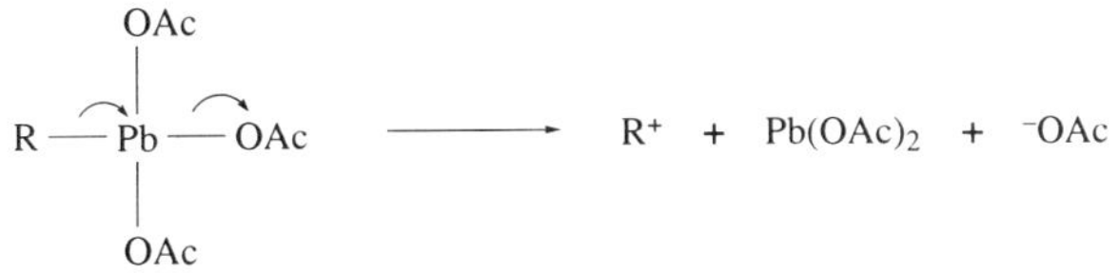

Aryllead tricarboxylates are the only compounds of this class which are relatively stable, and a large number of such compounds have been prepared (see Sections 11.5.2 and 11.5.3).[13,14] Their stability is thought to be due either to the high energy of the aryl cation, making the dissociation shown in Equation (13) much less favourable,[14] or to the unfavourable nature of a nucleophile-assisted decomposition. A single-crystal x-ray analysis of *p*-methoxyphenyllead triacetate showed that the lead atom was heptacoordinate, being bonded unsymmetrically to both oxygens of the three acetate units, and with approximate pentagonal-bipyramidal geometry.[15]

Vinyllead tricarboxylates are intermediate in stability between the alkyllead and aryllead compounds. In general, they cannot be isolated; however, they can be generated in solution by metal–metal exchange (see Section 11.5.3) and a considerable number of vinyllead triacetates have been characterized by ^{1}H NMR spectroscopy.[14–16] The stability order, aryl- > vinyl- > alkyllead tricarboxylates, has been attributed to the ease of the decomposition depicted in Equation (13), which is dependent on the corresponding carbocation stabilities.[14]

Alk-1-ynyllead tricarboxylates are similar in stability to vinyllead tricarboxylates, and as yet none has proved to be sufficiently stable for isolation and characterization. Evidence for their existence comes from reactions of solutions containing them immediately after their formation in a metal–metal exchange reaction[14,17] (see Section 11.5.3). The instability of these compounds is unrelated to the decomposition depicted in Equation (13). Instead, they undergo a rapid disproportionation, which is outlined for an alkynyllead triacetate in the equilibria of Scheme 4 and results in the formation of a tetraalkynyllead and Pb(OAc)$_4$.

$$2\ RC{\equiv}CPb(OAc)_3 \rightleftharpoons (RC{\equiv}C)_2Pb(OAc)_2 + Pb(OAc)_4$$
$$RC{\equiv}CPb(OAc)_3 + (RC{\equiv}C)_2Pb(OAc)_2 \rightleftharpoons (RC{\equiv}C)_3Pb(OAc) + Pb(OAc)_4$$
$$RC{\equiv}CPb(OAc)_3 + (RC{\equiv}C)_3PbOAc \rightleftharpoons (RC{\equiv}C)_4Pb + Pb(OAc)_4$$

Scheme 4

11.5.2 Synthesis by Direct Plumbation

Electrophilic plumbation is the most direct route to aryllead tricarboxylates; however, it is limited to a small range of aromatic compounds and cannot be applied to the synthesis of vinyllead or alkynyllead tricarboxylates.[13,14] Unlike the analogous aromatic mercuration[18] and thallation[19] reactions, direct plumbation is restricted to substrates which are more electron rich than the halobenzenes; for example, nitrobenzene does not undergo the reaction.

The first reported synthesis of an aryllead compound by this method was in 1960; the compound, 2,4-dimethoxyphenyllead triacetate (**8**), was obtained in 20% yield by reacting 1,3-dimethoxybenzene with $Pb(OAc)_4$ in benzene.[20] Later, it was shown that the lead compound (**8**) could be obtained in considerably higher yield by the use of acetic acid as the solvent (Equation (14)).[21] At the same time, it was found that mercury(II) acetate was a useful catalyst for aromatic plumbation, producing higher yields of aryllead triacetates in some cases;[21] for example, 2,4,6-trimethoxyphenyllead triacetate was obtained in 80% yield with $Hg(OAc)_2$ catalysis. The scope of direct plumbation was further extended by the use of haloacetic acids in place of acetic acid.[13,22] This results from ligand exchange and stronger acid catalysis, which increase the electrophilicity of the lead(IV) reagent. Although this allows the plumbation of fluorobenzene, chlorobenzene and bromobenzene when trifluoroacetic[23] or trichloroacetic[22] acids are employed (Equation (15)), aryllead tricarboxylates of less-reactive aromatics are not accessible by this route, and metal–metal exchange (see Section 11.5.3) must be used for their synthesis. As in the case of the halobenzenes (Equation (15)), other monosubstituted benzenes yield predominantly the *para*-substituted product, presumably due to the bulk of the electrophile.

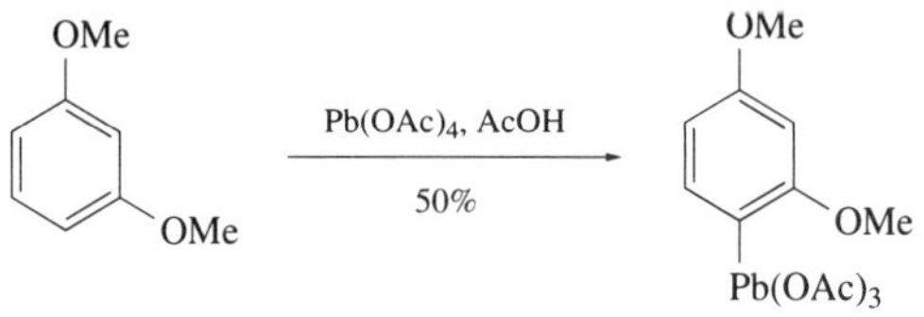

(14)

(8)

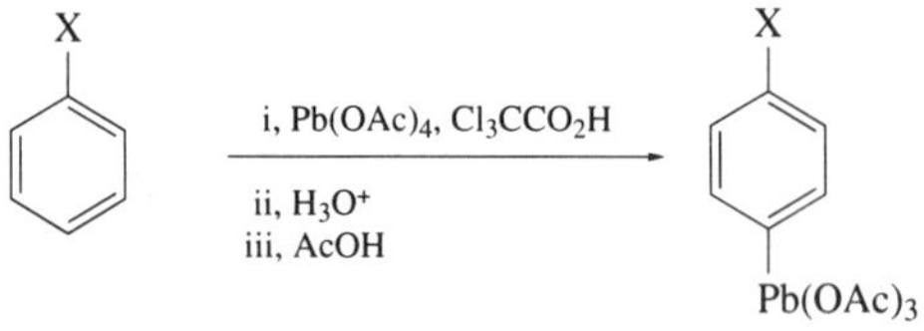

(15)

X = F, 52%; Cl, 34%; Br, 19%

For aromatics intermediate in reactivity between 1,3-dimethoxybenzene (Equation (14)) and the halobenzenes (Equation (15)), it is necessary to use one of the three chlorinated acetic acids to optimize yields in the plumbation;[13,22] *p*-methoxyphenyllead triacetate[24] was best prepared with dichloroacetic acid, while the plumbation of toluene,[13] biphenyl[13] and isobutylbenzene[25] gave highest yields with trichloracetic acid. It should be noted that aryllead tricarboxylates yield an insoluble oligomeric plumboxane (**9**) when reacted with water (Equation (16)), the reaction being faster with the more electron withdrawing haloacetates. Therefore, the initial product after workup of the plumbation reaction is generally an oligomer of type (**9**), which may be converted into the aryllead triacetate by treatment with excess acetic acid (Equation (17)).

$$2\ ArPb(OCOR)_3 + H_2O \longrightarrow [ArPb(OCOR)_2]_2O + 2\ RCO_2H \quad (16)$$

(9)

$$[ArPb(OCOR)_2]_2O \ + \ 6\,HOAc \ \longrightarrow \ 2\,ArPb(OAc)_3 \ + \ 4\,HOCOR \ + \ H_2O \quad (17)$$

11.5.3 Synthesis by Metal–Metal Exchange

Electrophilic substitutions of this type provide the most general methods of synthesis of organolead tricarboxylates, and may involve Hg–Pb, Tl–Pb, Si–Pb, Sn–Pb and B–Pb exchange. The methods allow access to a great variety of aryllead and heteroaryllead tricarboxylates, and provide the only routes to o- and m-substituted aryllead tricarboxylates.[1b,14] Metal–metal exchange is also the only route to vinyllead and alk-1-ynyllead tricarboxylates.[14]

The first synthesis of an aryllead tricarboxylate was achieved in 1952 by Russian workers,[26] who obtained phenyllead triacetate by reacting diphenyllead diacetate with mercury acetate (Equation (18)). A more versatile synthesis of phenyllead triacetate, also involving Hg–Pb exchange, was reported a few years later;[27] in this synthesis, shown in Equation (19), diphenylmercury was reacted with Pb(OAc)$_4$. The latter method has been applied to a wide range of diarylmercury compounds;[1b] however, although the reaction proceeds cleanly, isolated yields of aryllead triacetates are only moderate due to difficulties in separating them from the arylmercury acetate. This is generally achieved by making use of the difference in the solubilities of the corresponding chlorides, the arylmercury chlorides being rather insoluble. Due to this separation problem, and since the Hg–Pb exchange is rapid, the reaction of Equation (19) is probably best employed for the generation of a solution of an aryllead triacetate which is then used for the *in situ* arylation of a nucleophile.[28] The reaction has been shown to be applicable to a wide range of diarylmercury compounds; however, the failure of bis(m-nitrophenyl)mercury to react indicates that the presence of strongly electron withdrawing groups may inhibit the Hg–Pb exchange.[28] The procedure has been extended to heteroaromatics, and solutions of both 2-thienyllead triacetate and tribenzoate have been generated from reactions of bis(2-thienyl)mercury with Pb(OAc)$_4$ and Pb(OCOPh)$_4$.[29]

$$Ph_2Pb(OAc)_2 \ + \ Hg(OAc)_2 \ \xrightarrow[\text{RT}]{\text{CHCl}_3} \ PhPb(OAc)_3 \ + \ PhHgOAc \quad (18)$$

$$Ph_2Hg \ + \ Pb(OAc)_4 \ \xrightarrow[\text{RT}]{\text{CHCl}_3} \ PhPb(OAc)_3 \ + \ PhHgOAc \quad (19)$$

Aryl(trimethyl)silanes undergo Si–Pb exchange with Pb(OAc)$_4$ in the presence of CF$_3$CO$_2$H[23,30,31] and BF$_3$[32] (see Section 11.5.8); however, since aryllead triacetates undergo C–Pb heterolytic cleavage on treatment with CF$_3$CO$_2$H and BF$_3$, the reaction is of greatest use when the products resulting from the replacement of the Pb(OCOR)$_3$ group are required.

The most useful syntheses of aryllead tricarboxylates are those involving Sn–Pb[33,34] and B–Pb[35] exchange. Tin–lead exchange is generally conducted by stirring an aryl(tributyl)stannane with Pb(OAc)$_4$ with mercury(II) trifluoroacetate (or acetate) catalysis, as shown in Equation (20).[33,34] Isolation of the aryllead compound is simply achieved by precipitation with light petroleum, and, as indicated, high yields are obtained for aromatics containing electron withdrawing or electron releasing groups. Although the required metal–metal exchange reaction is faster for aryl(trimethyl)stannanes, they are less satisfactory precursors due to a competing Pb(OAc)$_4$-induced Me–Sn cleavage.[14,33] Reactions analogous to those of Equation (20) have been employed to produce the heterocyclic derivatives 2- and 3-thienyllead triacetate, and 2- and 3-furyllead triacetate.[29]

Arylboronic acids and their esters undergo a rapid B–Pb exchange with Pb(OAc)$_4$,[35] and, as indicated in Equation (21), the reaction is performed with mercury(II) catalysis and under the conditions developed for the stannanes. It should be noted that unless a mercury(II) catalyst is included, a considerable amount of the corresponding diaryllead diacetate is also formed. Yields are high for both electron-rich and electron-deficient aromatics, and, because of the speed of the reaction and the ease of removal of the boron-containing by-product, it lends itself to the *in situ* generation of aryllead triacetates.[35]

The greater efficiency of the B–Pb exchange is well illustrated in Equations (22) and (23); p-phenylenebis(tributylstannane) undergoes a single Sn–Pb exchange quite cleanly to give p-tributylstannylphenyllead triacetate in high yield (Equation (22)). Even with a large excess of Pb(OAc)$_4$ there was no further reaction, whereas p-phenylenediboronic acid was converted in good yield into p-bis(triacetoxyplumbyl)benzene (Equation (23)).

$$\text{ArSnBu}_3 + \text{Pb(OAc)}_4 \xrightarrow[40–60\%]{\text{CHCl}_3,\ \text{Hg(OCOR)}_2\ \text{catalyst}} \text{ArPb(OAc)}_3 + \text{Bu}_3\text{SnOAc} \quad (20)$$

Ar	Yield ArPb(OAc)$_3$ (%)	Ref.
Ph	69	34
o-MeOC$_6$H$_4$	92	34
m-MeOC$_6$H$_4$	81	35
p-FC$_6$H$_4$	76	34
p-F$_3$CC$_6$H$_4$	75	34
3,4-(MeO)$_2$C$_6$H$_3$	84	34
3,4-(OCH$_2$O)C$_6$H$_3$	99	35
6-methoxy-2-naphthyl	87	34

$$\text{ArB(OR}^1)_2 + \text{Pb(OAc)}_4 \xrightarrow[40–60\ °C]{\text{CHCl}_3,\ \text{Hg(OCOR}^2)_2\ \text{catalyst}} \text{ArPb(OAc)}_3 + \text{AcOB(OR}^1)_2 \quad (21)$$

R^1 = H or alkyl

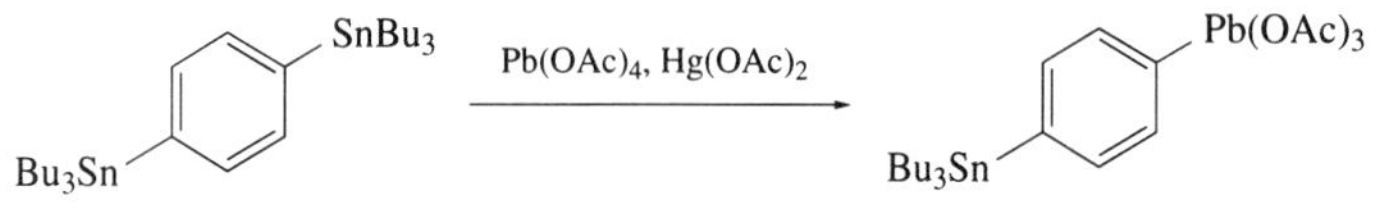

$$\underset{\text{Bu}_3\text{Sn}}{\text{SnBu}_3\text{-C}_6\text{H}_4\text{-}} \xrightarrow{\text{Pb(OAc)}_4,\ \text{Hg(OAc)}_2} \underset{\text{Bu}_3\text{Sn}}{\text{Pb(OAc)}_3\text{-C}_6\text{H}_4\text{-}} \quad (22)$$

$$\underset{(\text{HO})_2\text{B}}{\text{B(OH)}_2\text{-C}_6\text{H}_4\text{-}} \xrightarrow{\text{Pb(OAc)}_4,\ \text{Hg(OAc)}_2} \underset{(\text{AcO})_3\text{Pb}}{\text{Pb(OAc)}_3\text{-C}_6\text{H}_4\text{-}} \quad (23)$$

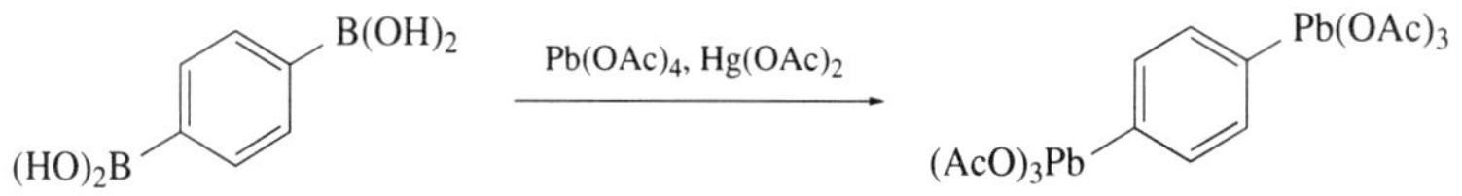

The only general routes to vinyllead[14–16,36,37] and alk-1-ynyllead[14,17] tricarboxylates are by Hg–Pb and Sn–Pb exchange (Equations (24)–(27)), and since both reactions are very fast, it has been possible to employ these unstable organolead species in organic synthesis. Unlike the Sn–Pb exchange with arylstannanes, the trimethylstannyl group is preferred for these reactions since Me–Sn cleavage is slow compared with the reactions indicated in Equations (25) and (27), and the other product of the reactions, Me$_3$SnOAc, is readily removed. A further advantage is the greater speed of the metal–metal exchange for the trimethylstannyl derivatives compared with the tributyltin compounds.

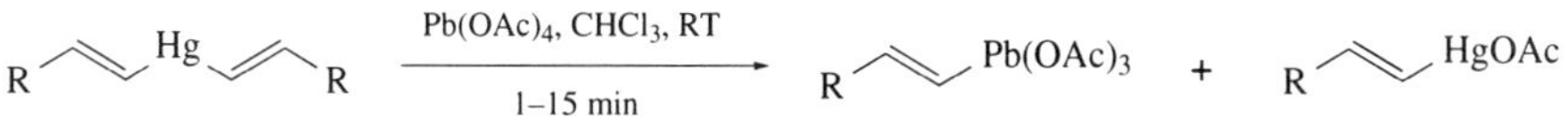

$$R\text{-CH=CH-Hg-CH=CH-}R \xrightarrow[1–15\ \text{min}]{\text{Pb(OAc)}_4,\ \text{CHCl}_3,\ \text{RT}} R\text{-CH=CH-Pb(OAc)}_3 + R\text{-CH=CH-HgOAc} \quad (24)$$

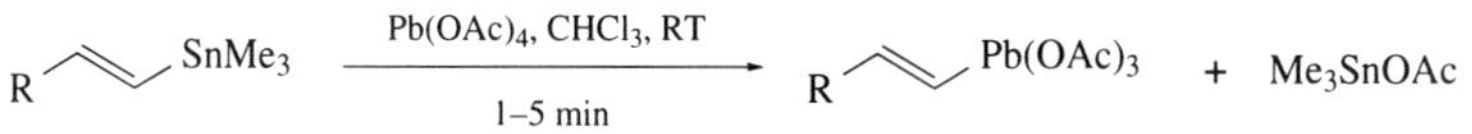

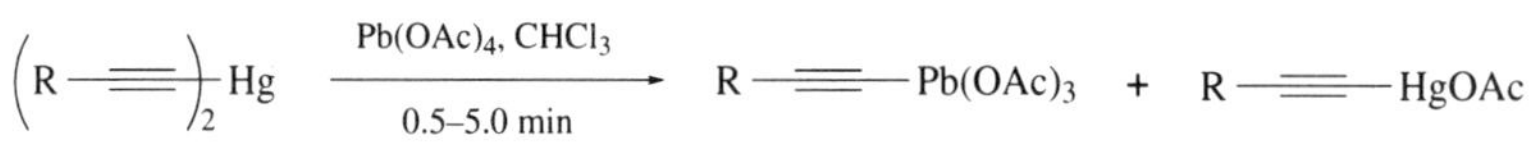

$$R\text{-CH=CH-SnMe}_3 \xrightarrow[1–5\ \text{min}]{\text{Pb(OAc)}_4,\ \text{CHCl}_3,\ \text{RT}} R\text{-CH=CH-Pb(OAc)}_3 + \text{Me}_3\text{SnOAc} \quad (25)$$

$$(R\text{-C}\equiv\text{C-})_2\text{Hg} \xrightarrow[0.5–5.0\ \text{min}]{\text{Pb(OAc)}_4,\ \text{CHCl}_3} R\text{-C}\equiv\text{C-Pb(OAc)}_3 + R\text{-C}\equiv\text{C-HgOAc} \quad (26)$$

$$R\text{-C}\equiv\text{C-SnMe}_3 \xrightarrow[0.5–2.0\ \text{min}]{\text{Pb(OAc)}_4,\ \text{CHCl}_3} R\text{-C}\equiv\text{C-Pb(OAc)}_3 + \text{Me}_3\text{SnOAc} \quad (27)$$

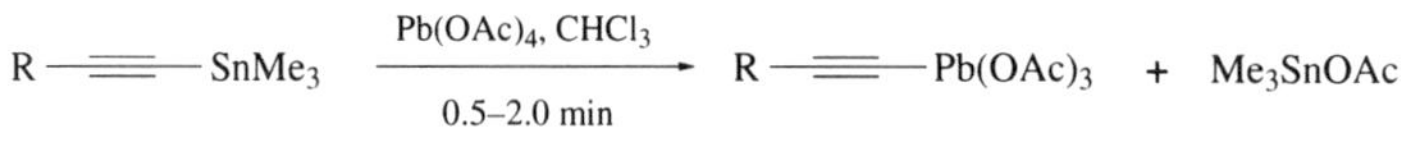

11.5.4 Carbon–Carbon Bond Forming Reactions in the Absence of Acid Catalysis

Aryllead, vinyllead, and alk-1-ynyllead tricarboxylates behave as aryl, vinyl and alkynyl cation equivalents, with their most synthetically interesting reactions being those with 'soft' carbon nucleophiles, which result in C-arylation, C-vinylation and C-alkynylation, respectively.[14] The substrates which undergo these reactions include phenols, β-dicarbonyl compounds, α-cyano esters, malononitriles, nitroalkanes and enamine derivatives of ketones. In general the reactions proceed cleanly in chloroform at 40–60 °C, and for a majority of the substrates, especially in the case of arylation, the presence of a tertiary base which can coordinate to lead(IV), such as py, results in considerably higher yields and the suppression of acetoxylation. These conditions, which were developed during a study of the arylation of phenols and β-dicarbonyls, are discussed more fully in the sections devoted to those substrates (Sections 11.5.4.1–11.5.4.4).

11.5.4.1 Reactions with phenols

The first study of C-arylation by aryllead triacetates under nonacidic conditions was carried out on phenols.[38] It was found that mesitol (**10**) reacted with p-methoxyphenyllead triacetate (**11**) in chloroform containing py (3 equiv.) at 40 °C to yield the arylated dienones (**12**) and (**13**) in almost quantitative yield, with the ratio of (**12**):(**13**) being about 4:1 (Equation (28)). This is analogous to the Wessely acetoxylation reaction which substituted phenols undergo with Pb(OAc)$_4$. If py is excluded from the reaction of Equation (28), dienones (**12**) and (**13**) are produced in considerably lower yield, and they are accompanied by the Wessely products (**14**) and (**15**). The reaction is very sensitive to the type of substitution and position of substitution in the phenolic ring. As a synthetic method, it is only useful for the production of dienones of type (**12**), and with methylated phenols, high yields were obtained only when both *ortho* positions were substituted. For example, 2,6-dimethylphenol reacted with the aryllead compound (**11**) to produce the dienone (**16**) in about 80% yield, whereas 2,4-dimethylphenol gave only 18% of dienone (**17**) when treated with the same reagent. As in the case of the Wessely acetoxylation, methoxy groups increase the rate of arylation; for example, catechol monomethyl ether gave the dienone (**18**) in 48% yield when treated with lead compound (**11**), whereas o-cresol gave less than 10% of the analogous 2,4-cyclohexadienone. Although methoxy groups are activating for the reaction, the dienone (**19**) resulting from *o*-arylation was the only product obtained on reacting 4-methoxy-2,6-dimethylphenol with the aryllead compound (**11**).[39]

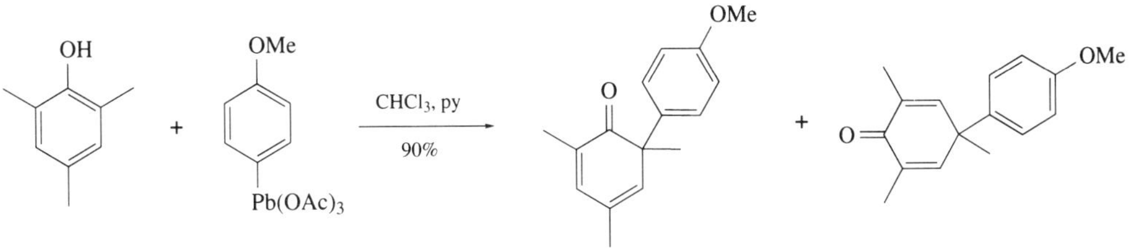

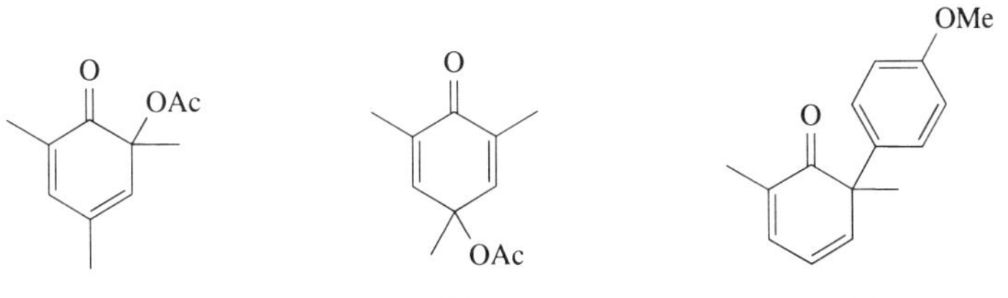

The finding that 3,5-di-*t*-butylphenol reacts with 2,4,6-trimethoxyphenyllead triacetate to give the very sterically crowded 2,6-diarylated phenol (**20**) in excellent yield (Equation (29)), shows that *ortho* substituents are not necessary for C-arylation to occur in certain cases.[40] This, together with the fact that 3,4,5-trimethylphenol gave the o-arylated product (**21**) in modest yield when treated with the aryllead reagent (**11**) (Equation (30)),[38] leads to speculation that perhaps in certain cases steric compression in an intermediate may accelerate the coupling reaction.

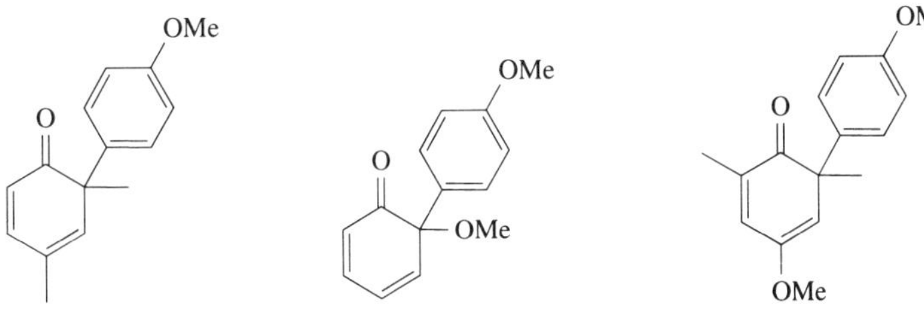

The reaction has also been applied to vinyllead and alkynyllead triacetates,[41] which were generated *in situ* from the appropriate trimethylstannyl derivatives as indicated earlier (Section 11.5.3). The *ortho* vinylations shown in Equation (31) all proceeded in synthetically useful yields, and unlike some of the arylations, none of the isomeric 2,5-cyclohexadienones were produced. In a similar manner,

(29)

(20) 87%

(30)

(21)

phenylethynyllead triacetate (**22**) reacted with mesitol (**10**), to give only the 2,4-cyclohexadienone (**23**) in high yield (Equation (32)). A feature of the dienone (**23**), which may limit the usefulness of this alkynylation in synthesis, was its rapid dimerization in a [2 + 4] cycloaddition to yield a single diastereoisomer (**24**).

(31)

(31)

R^1	R^2	*Yield* (%)
H	Me	82
OMe	Me	83
OMe	H	41
H	Br	67
OMe	Br	74

(32)

(22) **(23)**

(24)

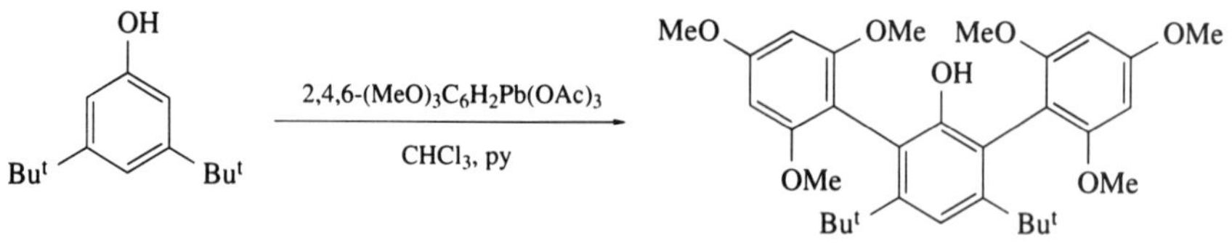

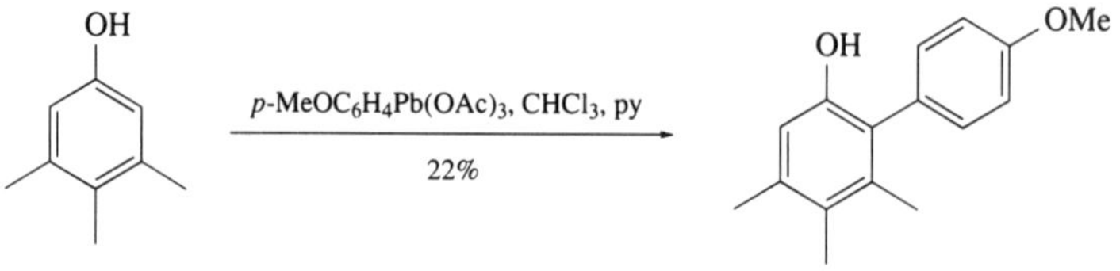

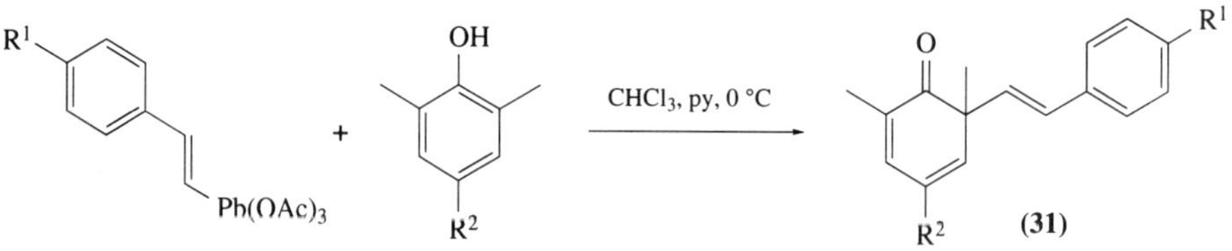

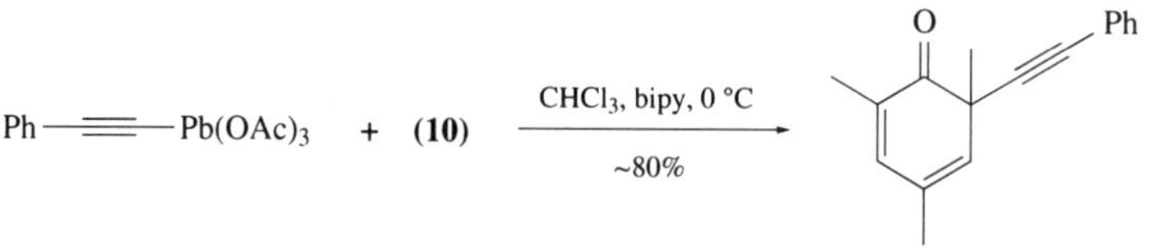

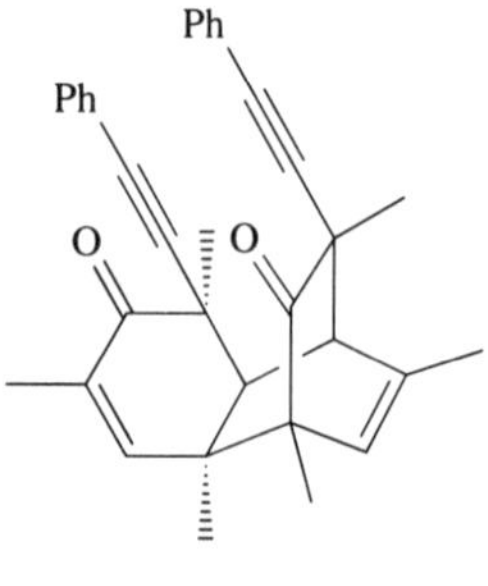

11.5.4.2 Reactions with β-diketones, β-keto esters and their vinylogues

β-Dicarbonyl compounds with pK_a values similar to or less than that of phenol were found to undergo C-substitution under conditions similar to those developed for phenols (Section 11.5.4.1), that is, a ratio of substrate to organollead compound to py of 1:1:3 in a concentrated chloroform solution at 40–60 °C.[14] A major variation, which has been found to be useful, is the replacement of py by bipy, phen or 4-dimethylaminopyridine (dmap). These bases, which probably complex more strongly to lead(IV), usually result in faster reactions and can lead to higher yields (see also Section 11.5.4.4.).[14] As in the case of phenols, O-arylation, O-vinylation or O-alkynylation were not competing reactions. In general, β-dicarbonyls with one α-hydrogen undergo C-substitution in high yield, whereas those compounds with two replaceable hydrogens behave unpredictably. Although dimedone (**25**) underwent diarylation in high yield with p-methoxyphenyllead triacetate and p-tolyllead triacetate (Equation (33)),[42] acetylacetone (**26**)[42] and ethyl acetoacetate (**27**)[43] reacted with 2 equiv. of p-methoxyphenyllead triacetate to give both mono- and diarylated products in poor yield (Equation (34)). Although the two reactions depicted in Equation (34) produced a greater amount of the monoarylated dicarbonyl, in the case of dimedone (**25**) (Equation (33)) the second arylation was faster than the first, and even with 1 equiv. of ArPb(OAc)$_3$ none of the monoaryl derivative could be detected.[42]

$$\text{(25)} + 2\text{ArPb(OAc)}_3 \xrightarrow[75-82\%]{\text{CHCl}_3,\text{ py, 40 °C}} \text{diarylated product} \quad (33)$$

Ar = p-MeC$_6$H$_4$, p-MeOC$_6$H$_4$

$$\text{MeCOCH}_2\text{COR} + 2\text{ArPb(OAc)}_3 \xrightarrow[40\text{ °C}]{\text{CHCl}_3,\text{ py}} \text{monoaryl} + \text{diaryl} \quad (34)$$

(**26**) R = Me
(**27**) R = OEt

| | R = Me | 19% | 2.5% |
| | R = OEt | 31% | 15% |

Ar = p-MeOC$_6$H$_4$

As mentioned earlier, β-diketones and β-keto esters with one α-hydrogen generally react smoothly with organolead tricarboxylates, and some examples of arylations, heteroarylations, vinylations and alkynylations, as indicated in the general Equation (35), are listed in Table 1. It should be noted that in most cases it is unnecessary to employ an enolate salt of the dicarbonyl compound; however, with the unstable vinyllead and alkynyllead reagents better yields may be obtained in some cases by use of the salt in a solvent such as DMSO.

$$R^1\text{COCHR}^2\text{COR}^3 + R^4\text{Pb(OAc)}_3 \longrightarrow R^1\text{COCR}^2R^4\text{COR}^3 + \text{Pb(OAc)}_2 + \text{AcOH} \quad (35)$$

(**28**)

Vinylogues of β-diketones and β-keto esters have also been employed as substrates, and, like phenols (Section 11.5.4.1), provide useful information regarding the preferred regiochemistry of their reactions. From the examples studied, there would appear to be a marked preference for the generation of a quaternary carbon centre rather than a tertiary centre, and since the quaternary centres developed in such reactions are highly functionalized, there is considerable potential for the use of organolead tricarboxylates in synthesis. Examples of this regiochemical preference are shown in Equations (36) and (37). In Equation (36), 6β-acetylcholest-4-en-3-one (**29**) reacted with p-methoxyphenyllead triacetate to give a mixture of the 6α and 6β arylated derivatives; no substitution at C-4 was detected.[42] Hagemann's ester (**30**) behaved similarly to the diketone (**29**), reacting with p-methoxyphenyllead triacetate under similar conditions to give only the arylated keto ester (**31**) in high yield (Equation (37)).[44] A further example is to be found in the vinylation of a closely related substrate by (E)-styryllead triacetate (Equation (38)).[37]

Table 1 Products from reactions of β-diketones and β-keto esters with organolead triacetates (RPb(OAc)$_3$), as indicated in Equation (35).

β-Dicarbonyl	R	Product (**28**) (%)	Ref.
2-Methyl-1,3-cyclohexanedione	p-MeOC$_6$H$_4$	95	43
	p-MeC$_6$H$_4$	82	43
	3-Thienyl	82	30
	3-Furyl	50	30
2-Methyl-1,3-cyclohexanedione sodium salt	2-Furyl	42	30
2-Acetyl-3,4-dihydronaphthalen-1(2H)-one	p-MeOC$_6$H$_4$	53	43
	(E)-Styryl	70	38
	PhC≡C	87	18
	1-Octynyl	75	18
	HC≡C	91	18
Ethyl 2-oxocyclopentanecarboxylate	p-MeOC$_6$H$_4$	91	44
	p-FC$_6$H$_4$	92	44
	3-Thienyl	79	30
	2-Furyl	70	30
	(E)-Styryl	78	38
	1-Cyclohexenyl[a]	67	38
	PhC≡C	73	18
	1-Octynyl	77	18
	TMS-C≡C	78	18
Methyl 2-oxocyclopentanecarboxylate	(E)-1-Octenyl	51	38
Ethyl 2-oxocyclohexanecarboxylate	p-MeOC$_6$H$_4$	87	44
Ethyl 3-oxo-2-methylbutanoate	p-MeOC$_6$H$_4$	86	44
	PhC≡C	14	18
Ethyl 3-oxo-2-methylbutanoate sodium salt	PhC≡C	58	18
Methyl 2-benzoylpropionate	p-MeO$_6$H$_4$	77	44
α-Acetyl-γ-butyrolactone	p-MeOC$_6$H$_4$	90	44
	PhC≡C	60	18
Methyl 3-oxo-5α-cholestane-2-carboxylate	p-FC$_6$H$_4$	95(α-aryl)	44

[a] Organolead tribenzoate was used.

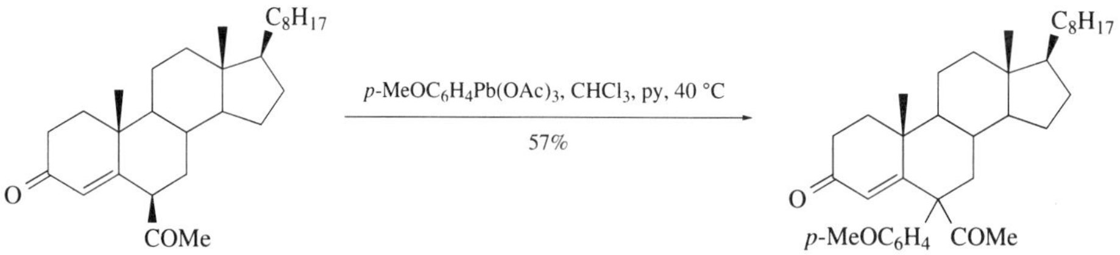

(29) → (36) p-MeOC$_6$H$_4$Pb(OAc)$_3$, CHCl$_3$, py, 40 °C, 57%; α-Ar:β-Ar = 7:4

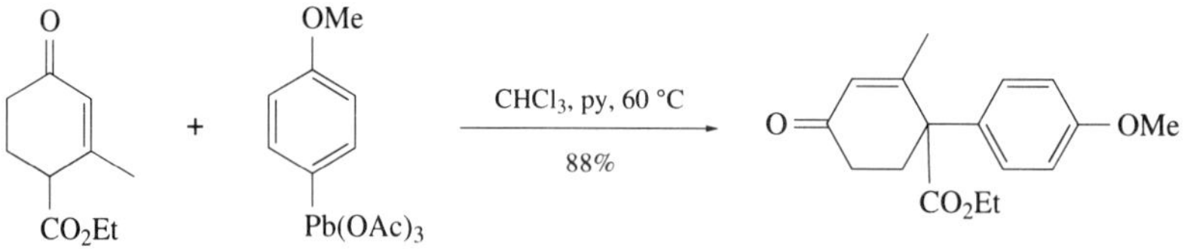

(30) + Pb(OAc)$_3$-C$_6$H$_4$-OMe → (31) CHCl$_3$, py, 60 °C, 88% (37)

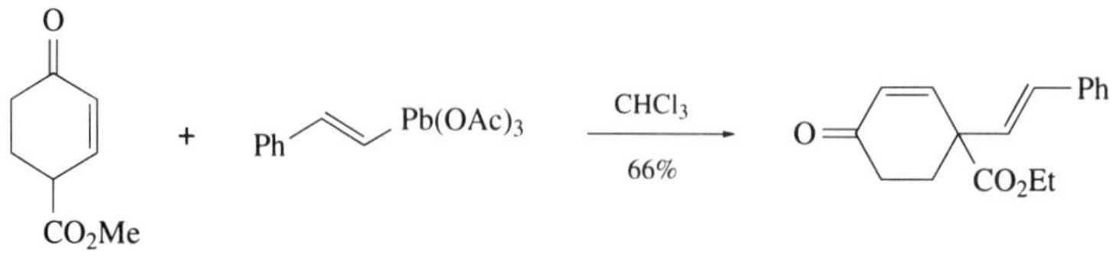

cyclohexenone-CO$_2$Me + PhCH=CH-Pb(OAc)$_3$ → CHCl$_3$, 66% (38)

The reactions of organolead triacetates with keto esters have featured in a number of target syntheses. For example, the key intermediate (**32**) in a synthesis of (+)-isocarbacyclin was produced in this way,[45] while a bisarylation (Equation (39)) has been employed in a novel approach to the synthesis of the lignan, (±)-sesamin.[46] The three arylations depicted in Equation (40) have been employed in short high-yielding syntheses of the alkaloids (±)-lycoramine,[47] (±)-O-methyljoubertiamine[48] and (±)-mesembrine.[48] During this work,[48] and that concerned with the isocarbacyclin synthesis,[45] it was found that for some reactions of keto esters with the organolead reagents, methyl and benzyl esters give significantly higher yields than the corresponding ethyl esters. It has been suggested that this difference in reactivity is probably due to a steric interaction in a lead(IV) intermediate.[14]

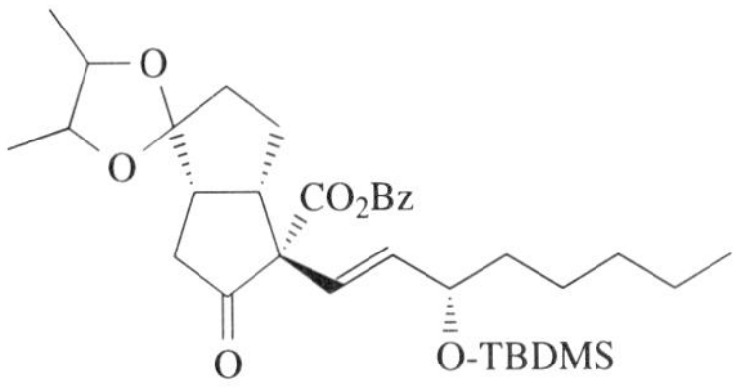

(32)

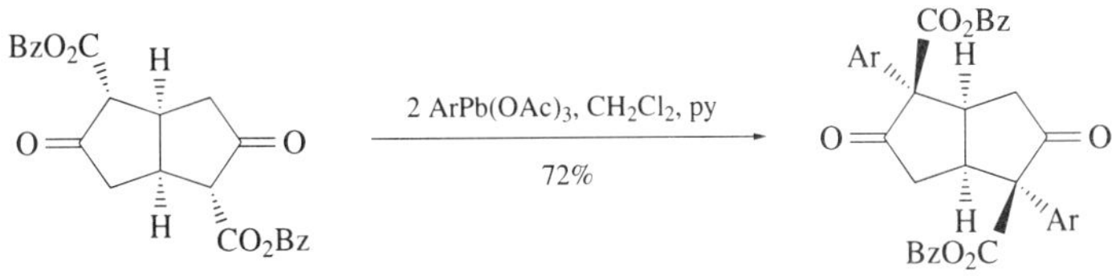

$$\text{(39)}$$

Ar = 3,4-(OCH$_2$O)C$_6$H$_3$

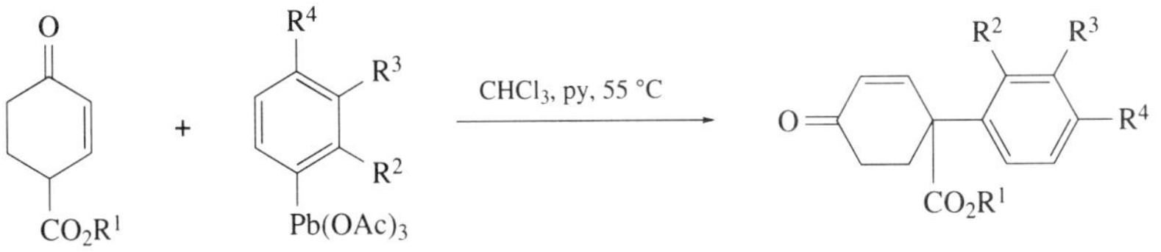

$$\text{(40)}$$

R^1 = Et, R^2 = R^3 = OMe, R^4 = H, 96%
R^1 = Me, R^2 = R^3 = H, R^4 = OMe, 90%
R^1 = Me, R^2 = H, R^3 = R^4 = OMe, 90%

4-Hydroxycoumarins, which are special examples of enolized β-keto esters, undergo C-arylation with aryllead triacetates under the usual conditions; however, unlike simple keto esters, a second arylation does not occur and the reaction provides a convenient route to 3-aryl-4-hydroxycoumarins,[34] members of the isoflavanoid class of natural products. This reaction has been put to good use in the synthesis of a large number of such compounds, including isorobustin (33),[34] which is shown in Equation (41). This work has been extended to include 3-hydroxycoumarins, compounds which could in theory also undergo diarylation; however, monoarylation was achieved in high yield (Equation (42)), providing a useful route to 4-aryl-3-hydroxycoumarins.[49]

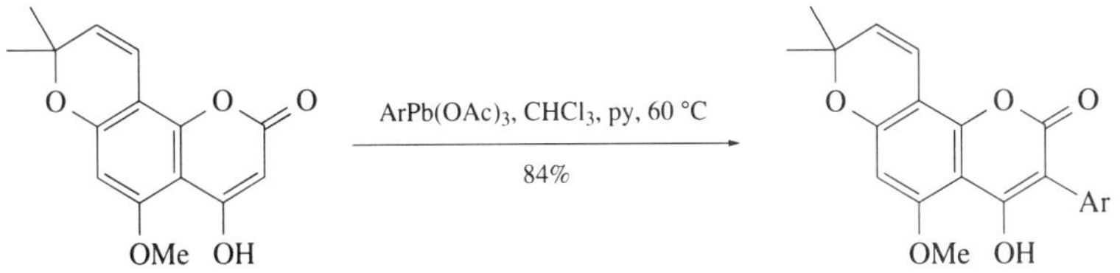

$$\text{(41)}$$

Ar = 3,4-(OCH$_2$O)C$_6$H$_3$ (33)

Two heterocyclic β-keto esters have also been employed in target syntheses. The first involved the synthesis of a 2-arylbenzofuran, a norneolignan present in a medicinal Mexican shrub, for which the key arylation step is shown in Equation (43).[50] The second was concerned with the development of a new route to 2-aryl-(2H)-indoles possessing anti-inflammatory and oestrogen antagonist activity.[51] The arylation (Equation (44)), which is analogous to that in Equation (43), proceeded in high yield either with or without protection of the nitrogen.

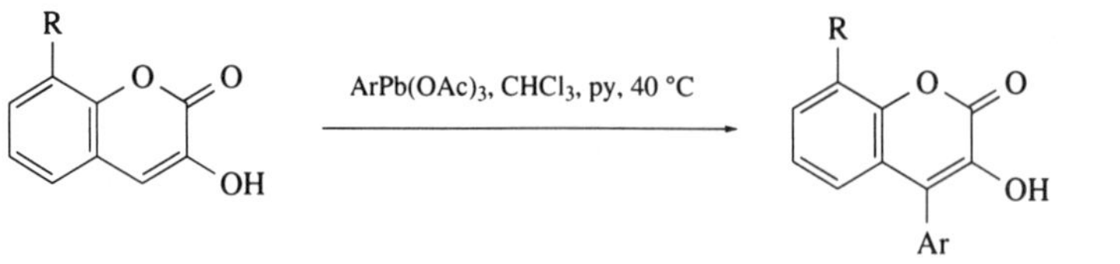

$$\text{(42)}$$

R = H, Ar = *m*-MeOC$_6$H$_4$, 72%
R = H, Ar = 2,4-(MeO)$_2$C$_6$H$_3$, 92%
R = OMe, Ar = 2,5-(MeO)$_2$C$_6$H$_3$, 69%
R = OMe, Ar = Ph, 59%

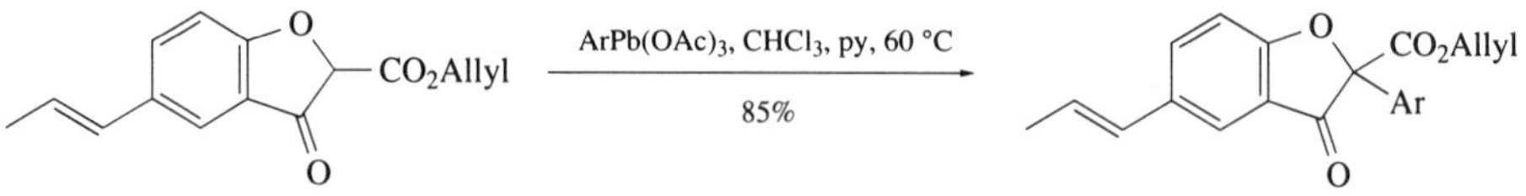

$$\text{(43)}$$

Ar = 2,4,6-(MeO)$_3$C$_6$H$_2$

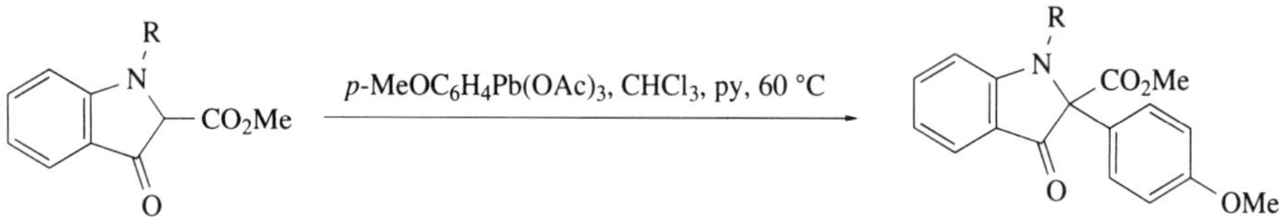

$$\text{(44)}$$

R = H, 95%
R = COMe, 84%

The last example is of a β-diketone arylation which provides an interesting route to a limited number of α-aryl-α-methylene ketones and esters (Equation (45)). The three steps involved, arylation, β-dicarbonyl cleavage and β-elimination, occur under the conditions of the reaction.[52]

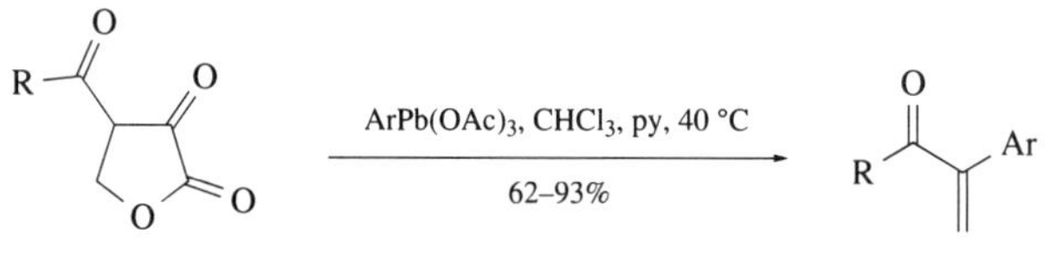

$$\text{(45)}$$

R = Ph, EtO, Me$_2$C=CH

11.5.4.3 Reactions of α-hydroxymethylene ketones

Aryllead triacetates are the only organolead triacetates which have been reacted with α-hydroxymethylene ketones. The initial *C*-arylation occurs as in the case of β-diketones; however, under the usual chloroform/py conditions the formyl group is generally lost, and this can be followed by a second arylation to yield an α,α-diaryl ketone. A change in the reaction conditions was found to overcome this problem, and, as shown in Equation (46), good yields of α-aryl ketones may be obtained.[53] The reaction has been used in an interesting approach to the synthesis of hexahydrochrysenes with potential oestrogenic activity.[54]

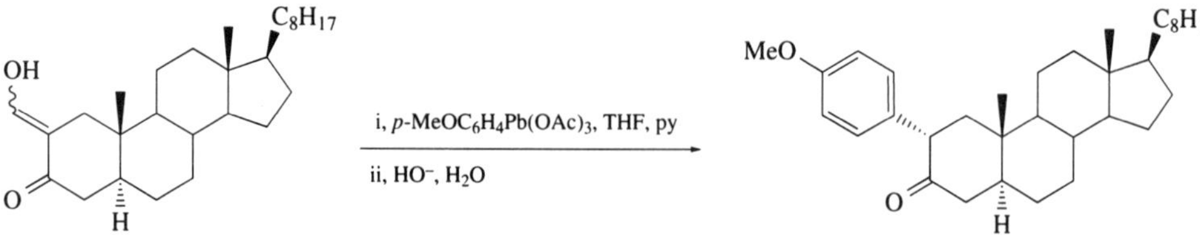

$$\text{(46)}$$

80%

11.5.4.4 Reactions with derivatives of malonic acid

Diethyl malonate failed to react with aryllead triacetates, while even the reaction of diethyl methylmalonate with *p*-methoxyphenyllead triacetate gave only 25% of diethyl α-(*p*-methoxyphenyl)-α-methylmalonate under the usual chloroform/py conditions. For these compounds, and some less acidic substrates, it has been found that satisfactory yields can be obtained by employing enolate salts in the reaction, as shown in Equation (47).[25] As indicated earlier, the reactivity of the CH acid is controlled largely by the pK_a, and therefore not surprisingly derivatives of Meldrum's acid were found to be among the most reactive β-dicarbonyl substrates; they undergo rapid arylation with aryllead triacetates in high yield (Equation (48)).[25] The Meldrum's acid derivatives provided compounds suitable for a study of the effect of the size of the α-substituent on the arylation reaction. In reactions carried out under comparable conditions there was a decrease in reactivity with increasing size of the group methyl > ethyl > isopropyl (Equations (48) and (49)). The isopropyl derivative (**34**) gave a low yield of the arylated compound (**35**), and, even with an extended reaction time, the yield of product (**35**) was only moderate. During an examination of ways of improving the yield of the derivative (**35**), it was found that the rate of the reaction, and yield of the product, could be improved by replacing py by bipy or phen, with the effect of base being py < bipy < phen (Equation (49)). Since the arylation reaction does not occur in the presence of a stronger base such as triethylamine, it appears that the reaction is promoted by coordination of the base to lead, which would explain the above order of reactivity.[14,25] There is a precedent for this proposal in the modification of the oxidation potential of Pb(OAc)$_4$ by coordination with the same three heteroaromatic bases.[14]

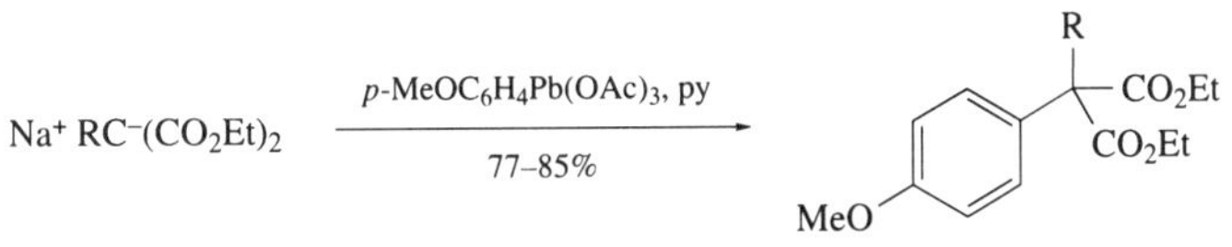

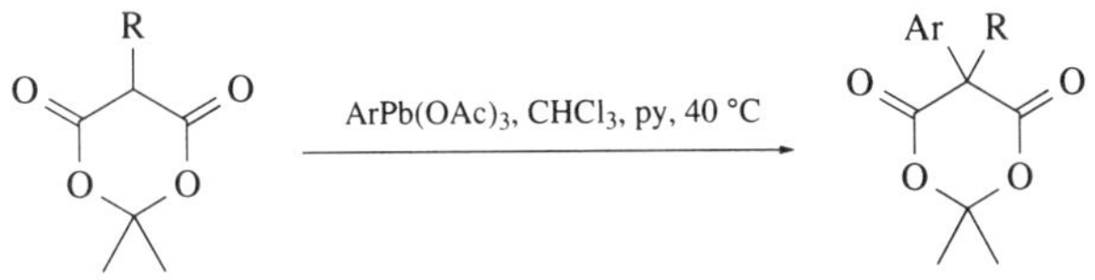

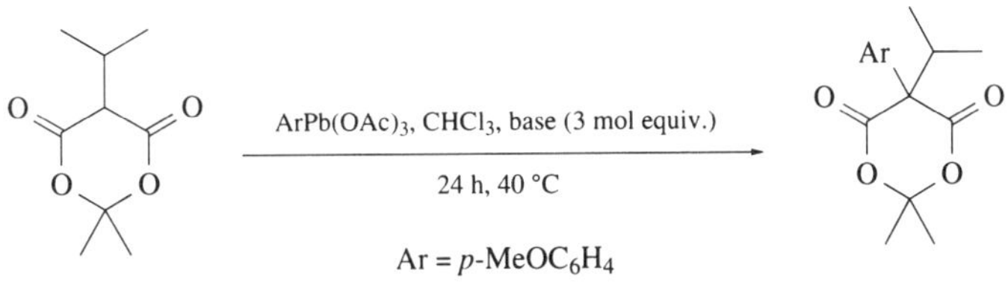

Base	Yield (%)
py	29
bipy	51
phen	87

Barbituric acid (**36**) is another malonic acid derivative with enhanced acidity, and its 5-substituted derivatives behave similarly to the Meldrum's acid analogues with the organolead reagents (Equation (50)).[25,29,36] The unpredictable behaviour of compounds with two α-hydrogens is well illustrated by the cyclic malonic acid derivatives. Meldrum's acid was very slow to react, and, even after a long reaction time with *p*-methoxyphenyllead triacetate, less than 10% of the diarylated product was obtained; however, barbituric acid (**36**) behaved like dimedone, giving only products of diarylation (**37**) in good yield with a number of aryllead triacetates (Equation (51)).[25] This provides a simple route to the 5,5-

diaryl derivatives (**37**), which are produced in poor yield by the usual substituted malonic ester–urea condensation.

$$\text{(50)}$$

R = Ph, 91%
R = p-MeOC$_6$H$_4$, 81%   R = 2-thienyl, 51%
R = 2-furyl, 76%   R = 3-thienyl, 51%
R = 3-furyl, 62%   R = (E)-styryl, 66%

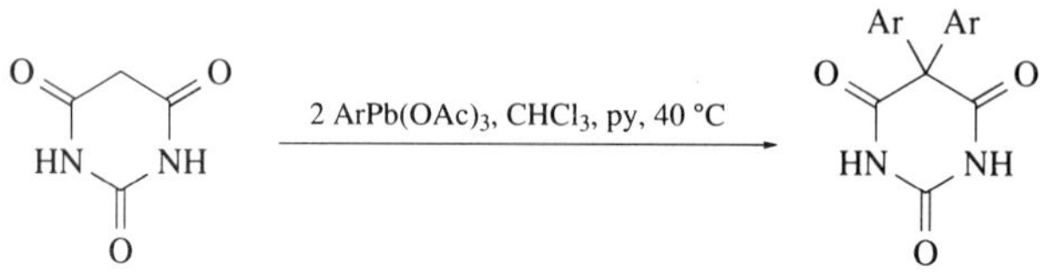

$$\text{(51)}$$

(**36**) → (**37**)

Ar = Ph, 65%
Ar = p-MeC$_6$H$_4$, 61%
Ar = p-MeOC$_6$H$_4$, 55%

In keeping with the lower pK_a and higher kinetic acidity of ethyl cyanoacetate compared with those of diethyl malonate,[55] α-substituted derivatives of ethyl cyanoacetate showed a higher reactivity to aryllead triacetates than the corresponding derivatives of diethyl malonate. Their reactivity was lower than that of the β-keto esters; however, by the use of the stronger complexing bases dmap and phen, in place of py, good yields of α-aryl derivatives could be obtained (Equation (52)).[55] A rate enhancement and improvement in yield could also be achieved by replacing the chloroform/base system by DMSO, which is also known to form strong complexes with organolead compounds.[1c] Substituted malononitriles were also investigated, but not in the same detail as the cyano esters; they behaved similarly but reacted more slowly under comparable conditions, and the α-arylmalononitriles (**38**)–(**40**) have been prepared in moderate yields.

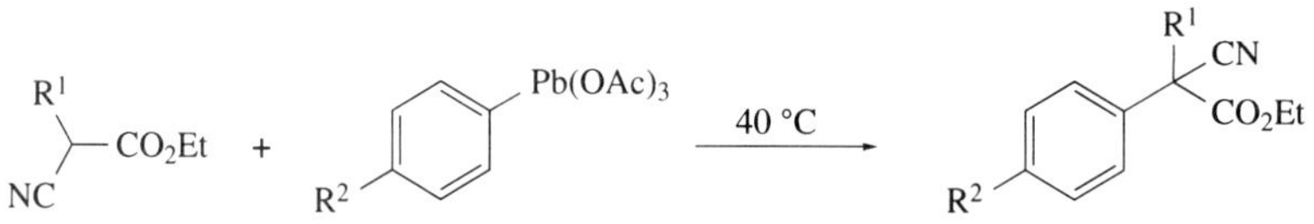

$$\text{(52)}$$

R^1	R^2	Solvent	Yield (%)
Me	OMe	CHCl$_3$, dmap	72
Me	OMe	DMSO	70
Ph	F	DMSO	74
EtCHMe	Me	CHCl$_3$, phen	54

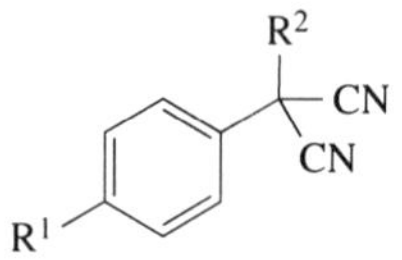

(**38**) R^1 = OMe, R^2 = Et
(**39**) R^1 = F, R^2 = Et
(**40**) R^1 = OMe, R^2 = Pr^i

A general high-yielding route to N-benzoyl-α-arylglycines has been developed from the study of the arylation of malonic acid derivatives.[56] Diethyl acetamidomalonate, a convenient substrate for the synthesis of α-substituted glycines, was unreactive towards aryllead triacetates and even its sodium salt failed to behave like diethyl sodiomethylmalonate (Equation (47)). However, the more acidic 5-oxazolone derivative (**41**), which is readily prepared from diethyl benzamidomalonate, reacted rapidly with a wide range of aryllead triacetates under the usual chloroform/py conditions to produce the arylated oxazolones (**42**) in excellent yield. These compounds are moisture sensitive and are best converted directly without purification into the *N*-benzoyl-α-arylglycine (**43**) (Scheme 5).[56]

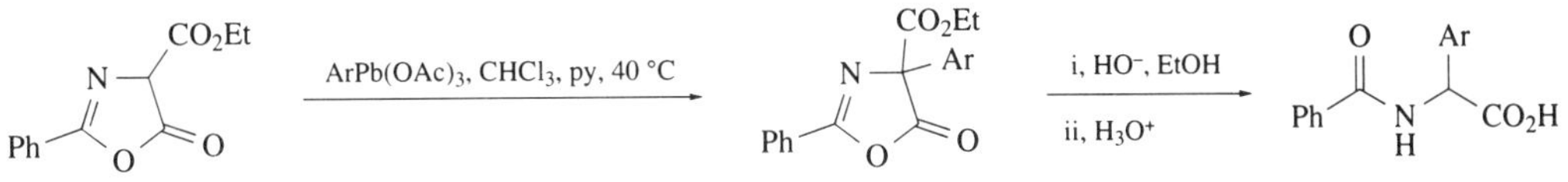

Ar	Yield (**43**) (%)
Ph	87
o-FC$_6$H$_4$	93
p-FC$_6$H$_4$	88
p-F$_3$CC$_6$H$_4$	75
o-MeOC$_6$H$_4$	83
p-MeOC$_6$H$_4$	87
2,4-(MeO)$_2$C$_6$H$_3$	90
3,4-(MeO)$_2$C$_6$H$_3$	89

Scheme 5

11.5.4.5 Reactions with nitroalkanes

Nitroalkanes, which have pK_a values similar to those of the reactive β-dicarbonyls (Sections 11.5.4.2–11.5.4.4), also react with organolead triacetates, either as the free nitro compound and/or as the nitronate salt, to provide access to a wide range of α-alkynyl,[17] α-aryl[14,57] and α-vinyl[37] nitroalkanes. The reaction of nitroalkanes is very slow under the conditions developed for the β-dicarbonyls. This is presumably due to the lower kinetic acidity of these compounds and, therefore, the reactions are best conducted in DMSO, a solvent in which deprotonation is accelerated. Nevertheless, the reaction is only applicable in the case of the stable reagents, the aryllead triacetates, for which a number of examples are given in Equation (53). Yields were unaffected by the addition of py or other complexing base, which is thought to be due to the ability of DMSO to complex strongly to lead(IV) as previously mentioned (Section 11.5.4.4). Conditions similar to those of Equation (53) were also employed for the reaction of nitronate salts,[17,37,57] and since the reactions were rapid at room temperature, vinylation, alkynylation and arylation could be effected with these substrates (Equation (54)). In contrast to β-dicarbonyls bearing two α-hydrogens, nitroethane and ethyl nitroacetate could be reacted with 1 equiv. of phenyllead triacetate to give good yields of monoarylated products (Equation (55)).

R^1, R^2	Ar	Yield (%)
(CH$_2$)$_4$	Ph	70
(CH$_2$)$_5$	*p*-MeC$_6$H$_4$	71
(CH$_2$)$_5$	*p*-MeOC$_6$H$_4$	65
Me, Me	*p*-MeC$_6$H$_4$	75

(53)

11.5.4.6 Reactions of ketone enolates and enamines

Except in special cases, simple ketones do not react with organolead reagents under the conditions developed for the dicarbonyl compounds. Compounds which were found to react slowly with *p*-methoxyphenyllead triacetate in chloroform/py are 2-phenylcyclohexanone and benzyl methyl

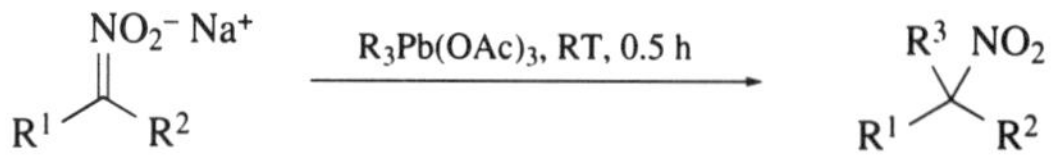

$$R^1\underset{R^2}{\overset{NO_2^- Na^+}{\diagup\!\!\!\diagdown}} \xrightarrow{R_3Pb(OAc)_3, RT, 0.5\ h} R^1\underset{R^2}{\overset{R^3\ NO_2}{\diagup\!\!\!\diagdown}} \quad (54)$$

R^1, R^2	R^3	Yield (%)
Me, Me	p-MeC$_6$H$_4$	71
(CH$_2$)$_5$	(E)-styryl	71
Me, Me	PhC≡C	72
(CH$_2$)$_5$	TMS-C≡C	60

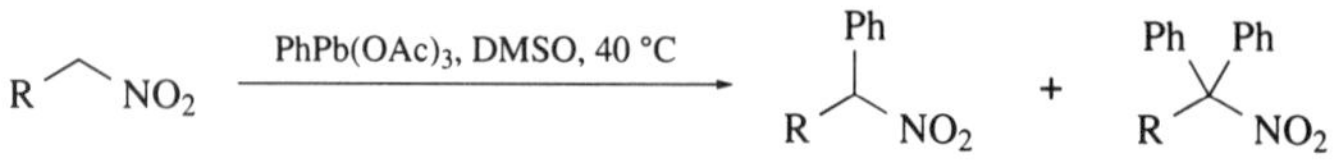

$$R\frown NO_2 \xrightarrow{PhPb(OAc)_3, DMSO, 40\ °C} R\underset{NO_2}{\overset{Ph}{\diagup\!\!\!\diagdown}} + R\underset{NO_2}{\overset{Ph\ Ph}{\diagup\!\!\!\diagdown}} \quad (55)$$

R = Me, 58–65%, 0–5%
R = CO$_2$Me, 70%

ketone;[14,59] that is, compounds in which the acidity of the α-hydrogens is enhanced. After 24 h reactions, yields of 2-(p-methoxyphenyl)-2-phenylcyclohexanone and 1-(p-methoxyphenyl)-1-phenyl-2-propanone were 46% and 20%, respectively, indicating that the method will be of little practical value. The enolate salts of simple ketones show greater reactivity, but a brief study of their reactions with aryllead triacetates indicated that they will also be of little use as nucleophiles for the organolead reagents. Here again, a strong preference was shown for attack at a tertiary centre, while there was no evidence for attack at oxygen.[59] For example, the potassium enolate of cyclohexanone was completely unreactive towards p-methoxyphenyllead triacetate, whereas the potassium enolate of 2,6-dimethylcyclohexanone gave the 2-arylated derivative in good yield with the same reagent (Equation (56)). Similarly, the potassium enolate of acetone failed to react with p-methoxyphenyllead triacetate, while the enolate of isobutyrophenone produced a 41% yield of the α-aryl derivative when treated with the same reagent under the conditions of Equation (56).

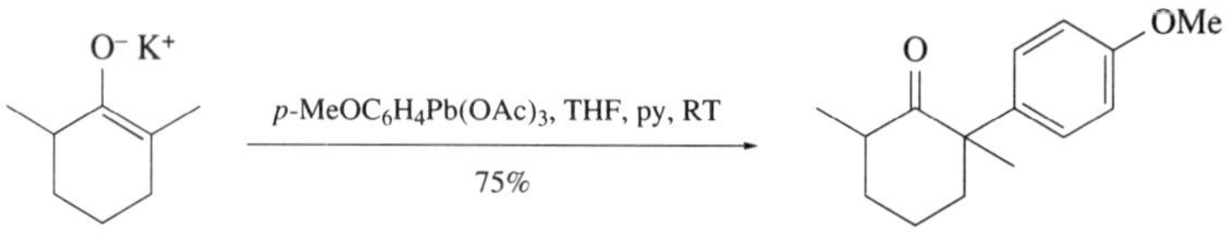

$$\text{[2,6-dimethylcyclohexenolate K}^+\text{]} \xrightarrow[75\%]{p\text{-MeOC}_6H_4Pb(OAc)_3, THF, py, RT} \text{[2-aryl-2,6-dimethylcyclohexanone]} \quad (56)$$

N-Substituted-3-oxo-2,3-dihydroindoles are a special group of ketones which undergo α-arylation with aryllead triacetates under chloroform/py conditions. Monoarylation was achieved in good yield either with or without a substituent at C-2 (Equation (57)).[51]

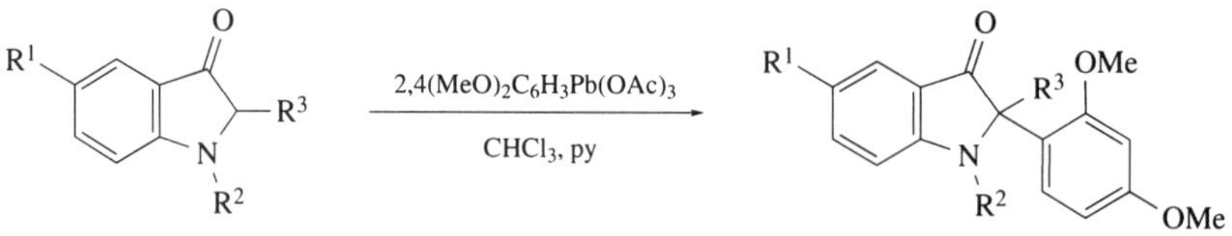

$$\text{[indolinone with R}^1, R^2, R^3\text{]} \xrightarrow{2,4(MeO)_2C_6H_3Pb(OAc)_3, CHCl_3, py} \text{[2-aryl indolinone]} \quad (57)$$

R^1	R^2	R^3	Yield (%)
H	COMe	H	70
H	COMe	Me	72
H	SO$_2$Ph	H	60
OMe	COMe	H	81

The reactions of enamines have been examined in some detail, but only with the aryllead triacetates.[58] The enamines of aldehydes failed to react, whereas ketone derivatives underwent a very exothermic arylation. Moderate to good yields of α-aryl ketones were obtained from reactions of p-methoxyphenyllead triacetate with the morpholine enamines of cyclopentanone, cyclohexanone, 3-pentanone, 4-heptanone and 5-nonanone (Equation (58)); however, yields fell markedly with an increase in the ring size and with substitution in the vicinity of the double bond. Unlike other substrates, α-acetoxylation was a major competing reaction, especially in those cases where arylation was inhibited, as shown in Equation (59).

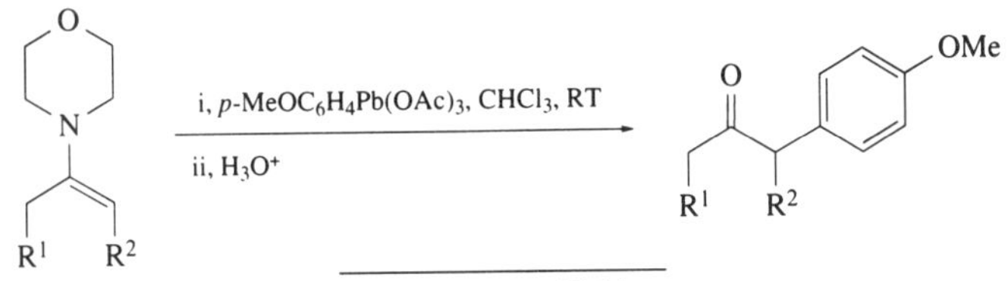

(58)

R^1, R^2	Yield (%)
Me, Me	75
Et, Et	59
Pr, Pr	50
(CH$_2$)$_2$	82
(CH$_2$)$_3$	51

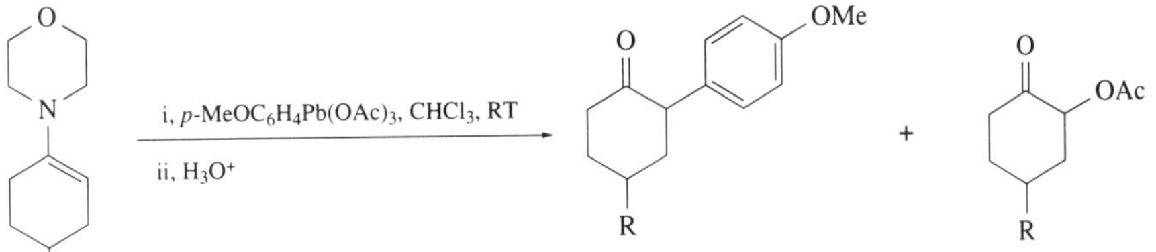

(59)

R = Me   36%   54%
R = Bu^t 10%   65%

11.5.5 Replacement of Lead in Aryllead Triacetates by Iodide and Azide

In the absence of copper or acid catalysis (Sections 11.5.7 and 11.5.8), the only noncarbon nucleophiles which have been found to undergo arylation with aryllead triacetates are iodide and azide. Both reactions are quite general and proceed in very high yield at room temperature. The iodides are produced simply by stirring the aryllead triacetate with aqueous potassium iodide,[1d] while the formation of aryl azides is best achieved by reacting the aryllead compound with sodium azide in DMSO (Scheme 6).[59] The *in situ* generation of aryllead triacetates by metal–metal exchange has been put to use in a convenient one-pot conversion of the readily available arylboronic acids into aryl azides (Equation (60)).[59]

ArI ←(KI, H$_2$O, RT, >90%)— ArPb(OAc)$_3$ —(NaN$_3$, DMSO, RT, >80%)→ ArN$_3$

Scheme 6

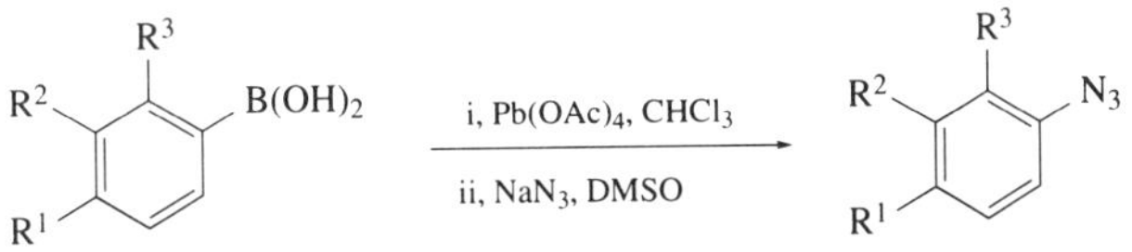

(60)

R^1 = R^2 = R^3 = H, 59%
R^1 = OMe, R^2 = R^3 = H, 71%
R^1 = Me, R^2 = R^3 = H, 69%
R^1 = F, R^2 = R^3 = H, 73%
R^1 = R^2 = OMe, R^3 = H, 82%

There is a superficial similarity between aryllead tricarboxylates and arenediazonium salts in their reactions with iodide and azide ions; however, this similarity does not extend to the mechanisms of the reactions (see Section 11.5.6). It has been established that aryl radicals are not involved in either of these reactions of the aryllead compounds,[60] while evidence is at present pointing to a ligand coupling process.[61]

11.5.6 Mechanism of the Arylation Reactions of Sections 11.5.4 and 11.5.5

The fact that aryllead triacetates react readily with iodide, azide and the salts of nitroalkanes, but fail to react with other noncarbon nucleophiles in the absence of a copper catalyst, points to the possibility that the mechanism of arylation may involve a single-electron transfer, leading to a radical process, such as an $S_{RN}1$ reaction. This has now been discounted in two separate investigations which led to the conclusion that none of the reactions outlined in Sections 11.5.4 and 11.5.5 involve aryl radicals.[39,60]

In the first of these studies, it was shown that radical trapping agents had no effect on the arylation of a β-keto ester,[39] while in the later work an internal trapping, which had been successfully used as a probe for radicals in arenediazonium salt reactions, was employed.[60] This approach involved the use of *o*-allyloxyphenyllead triacetate (**44**), which was readily prepared from the corresponding arylboronic acid. All of the reactions of the lead compound (**44**) resulted in the nucleophilic displacement indicated in Equation (61). The nucleophiles included the sodium salt of nitropropane, ethyl 2-oxocyclopentane-carboxylate, mesitol (**10**), iodide and azide. None of the 3-substituted dihydrobenzofurans expected from a radical cyclization in the *exo* mode could be detected in any of these reactions.[60]

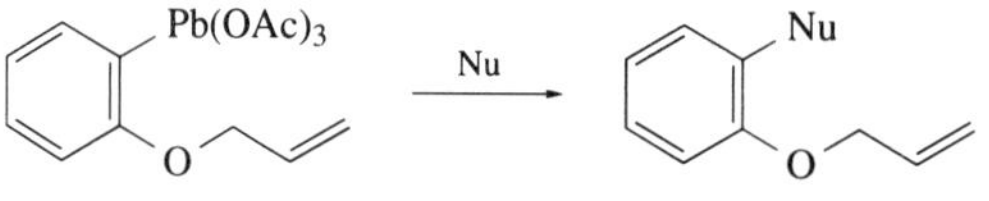

$$\text{(44)} \xrightarrow{\text{Nu}} \text{product} \tag{61}$$

A ligand coupling mechanism has been proposed for these reactions,[14,34,39,40] but, despite attempts to demonstrate such a pathway, no lead(IV) intermediates have been detected in any of the reactions of Sections 11.5.4 and 11.5.5. There is, however, some indirect evidence which supports the ligand coupling hypothesis. This derives from a study of the acid-catalysed reactions of the trimethylsilyl enol ethers (**45**) of some alkyl phenyl ketones with *p*-methoxyphenyllead triacetate, which is depicted in Scheme 7.[61] The (aryl)phenacyllead diacetates (**46**), which are obtained in high yield in these reactions, are relatively unstable, and in most cases undergo thermal decomposition at about 60 °C to give the product of ligand coupling (**47**) in modest yield. The ease with which the thermal collapse of the lead compounds (**46**) occurred depended on the degree of substitution at the carbon attached to lead.[61] For example, in the case of the acetophenone derivative (**46a**) none of the expected α-aryl ketone (**47a**) was produced even after 3 d in refluxing chloroform, whereas the propiophenone- and butyrophenone-derived compounds (**46b**) and (**46c**) produced modest amounts of the ketones (**47b**) and (**47c**), respectively, after similar treatment. The compound (**46d**), derived from isobutyrophenone, was even more thermally labile. It yielded the ketone (**47d**) and was completely decomposed after 14 h. No symmetrical products, such as biaryls or 1,4-diketones, were produced in these reactions and radical processes were again excluded by use of the *o*-allyloxyphenyllead compound (**44**) in place of *p*-methoxyphenyllead triacetate in a reaction sequence of the type shown in Scheme 7. It has, therefore, been proposed that the ketones arise by ligand coupling.

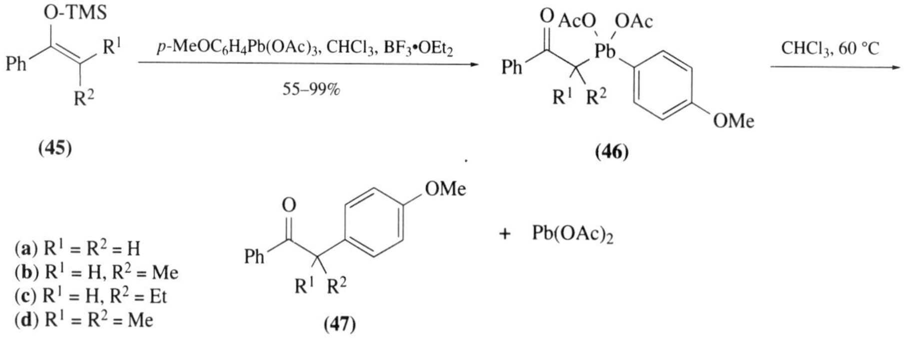

(a) $R^1 = R^2 = H$
(b) $R^1 = H, R^2 = Me$
(c) $R^1 = H, R^2 = Et$
(d) $R^1 = R^2 = Me$

Scheme 7

11.5.7 Copper-catalysed Reactions of Aryllead Triacetates

The finding that 'soft' carbon nucleophiles such as the enolates of β-dicarbonyls and phenolates undergo preferential *C*-arylation with aryllead triacetates resulted in some unsuccessful attempts to employ these reagents in the arylation of stabilized nitrogen anions. For example, both potassium phthalimide and trifluoroacetanilide failed to react with p-methoxyphenyllead triacetate under a variety of conditions.[59] However, with copper(II) acetate catalysis the N-arylation of amines can be achieved with these reagents, although the reactive intermediates are possibly arylcopper(III) species.[62] The reaction has been used to effect the N-arylation of aniline and a variety of aniline derivatives in excellent yield, as indicated by the examples shown in Equation (62). The arylation is dependent on the basicity of nitrogen, as illustrated by the failure of the reaction in the case of p-nitroaniline and ethyl p-aminobenzoate and also the considerably lower yields obtained with aliphatic amines such as piperidine, tetrahydroisoquinoline and benzylamine. Simple aliphatic amines such as butylamine and t-butylamine gave particularly poor yields.

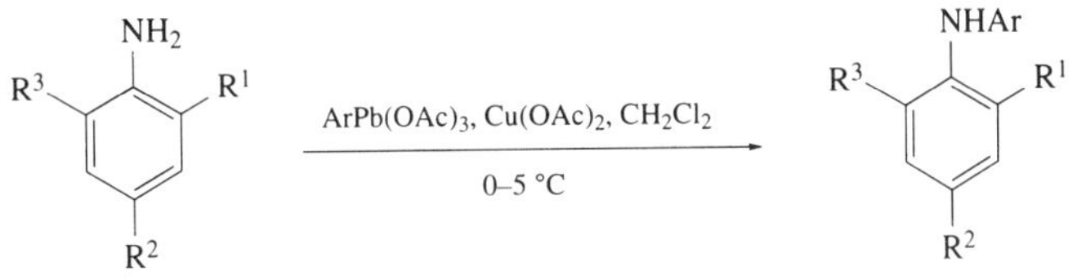

$$\text{Ar-NH}_2 \xrightarrow{\text{ArPb(OAc)}_3, \text{Cu(OAc)}_2, \text{CH}_2\text{Cl}_2}_{0-5\,°C} \text{Ar-NHAr} \quad (62)$$

R^1	R^2	R^3	Ar	Yield (%)
H	H	H	p-MeOC$_6$H$_4$	72
H	H	H	2,4-(MeO)$_2$C$_6$H$_3$	78
H	OMe	H	o-MeC$_6$H$_4$	85
H	OMe	H	2,4-(MeO)$_2$C$_6$H$_3$	25
Me	Me	Me	p-MeOC$_6$H$_4$	76
Me	Me	Me	2,4-(MeO)$_2$C$_6$H$_3$	74

Copper has also been shown to catalyse the reductive displacement of lead(II) by Cl$^-$, Br$^-$ and CN$^-$ in aryllead triacetates (Equation (63)); however, although the yields are moderate, the reactions are only of mechanistic interest.[14]

$$\text{ArPb(OAc)}_3 \xrightarrow{\text{CuX, DMF}} \text{ArX} \quad (63)$$
$$X = \text{Cl, Br, CN}$$

11.5.8 Acid-catalysed Reactions of Aryllead Tricarboxylates

The Pb(OCOR)$_3$ group is an exceptionally good leaving group due to the thermodynamically favourable change in oxidation state of lead(IV) to lead(II) which occurs in this process. The heterolytic C–Pb bond cleavage in such a reaction has been examined in considerable detail for vinyllead triacetates, where the generation of vinyl cations under nonsolvolytic conditions, as in Equation (64), has been firmly established.[15,16] As indicated earlier (Section 11.5.1), the greater stability of alkyl cations is the reason for the ease of this C–Pb bond cleavage in alkyllead triacetates and for their especially high instability.

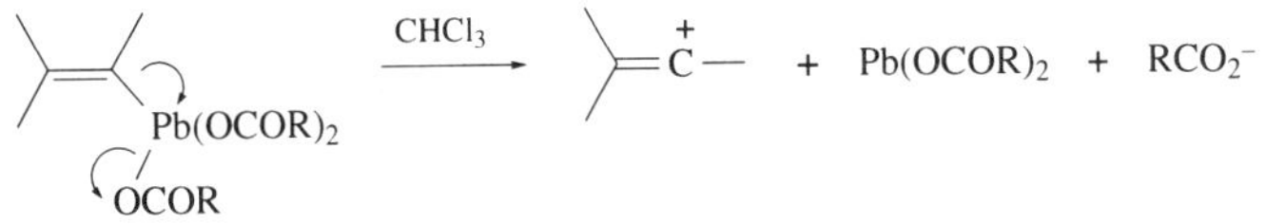

$$\xrightarrow{\text{CHCl}_3} \quad =\overset{+}{C}- \;+\; \text{Pb(OCOR)}_2 \;+\; \text{RCO}_2^- \quad (64)$$

The high energy of aryl cations has been given as the main reason for the relative stability of aryllead tricarboxylates (Section 11.5.1); however, it has been shown that in the presence of a strong acid, heterolytic C–Pb bond cleavage can be induced, and three synthetically useful reactions involving aryl cations (or incipient aryl cations) generated in this way have been developed.[14]

The first reaction, which provides a new route to phenols, involves the treatment of an aryllead tricarboxylate with a large excess of trifluoroacetic acid (Equation (65)).[31] The reaction presumably involves ligand exchange to produce an aryllead tristrifluoroacetate, which on O-protonation undergoes C–Pb heterolytic cleavage of the type shown in Equation (64). The aryl trifluoroacetate would then arise as the aryl cation (or incipient cation) reacted with the solvent.[63] Variations of the reaction involve plumbation in trifluoroacetic acid[64] and Hg–Pb or Si–Pb exchange with Pb(OAc)$_4$ in trifluoroacetic acid,[23,31] conditions under which the initially formed aryllead tricarboxylate yields the aryl trifluoroacetate. The reaction of aryl(trimethyl)silanes with $Pb(OCOCF_3)_4$ in trifluoroacetic acid is a very useful route to phenols, and has been employed as a final step in an interesting oestrone synthesis (Equation (66)).[30] The reaction shown in Equation (65) proceeds for compounds with electron donating and electron withdrawing groups in the aromatic ring; however, protodeplumbation can be a competing reaction for compounds of the former type. For example, 2,4-dimethoxyphenyllead triacetate gives approximately 30% of 1,3-dimethoxybenzene when dissolved in trifluoroacetic acid.

$$ArPb(OAc)_3 \xrightarrow[75-95\,°C]{CF_3CO_2H,\ RT} ArOCOCF_3 \qquad (65)$$

$$Ar = p\text{-}MeC_6H_4,\ p\text{-}MeOC_6H_4,\ m\text{-}MeOC_6H_4,\ o\text{-}MeOC_6H_4$$

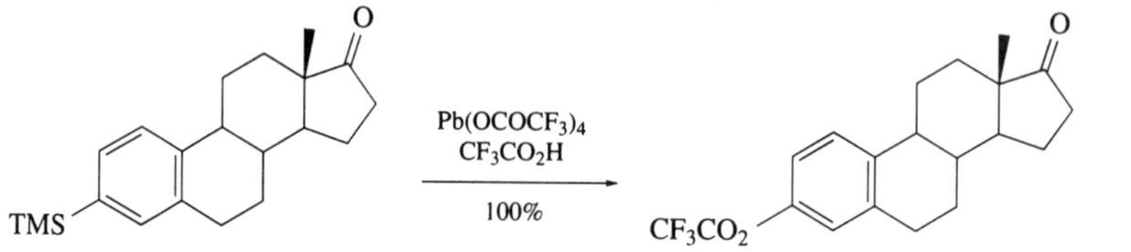

During a study of the mechanism of the reaction depicted in Equation (65) in which arenes were successfully employed to trap intermediate aryl cations, the second reaction, a useful route to unsymmetrical biaryls, was discovered.[63] When arenes such as benzene and toluene were employed as aryl cation traps, small amounts (<10%) of biaryls were produced together with the aryl trifluoroacetate, as anticipated for such a trapping experiment. However, when p-xylene or more highly methylated benzenes were employed, unexpectedly high yields of biaryls were obtained, even in reactions where relatively small amounts (1.5–2 mol equiv.) of the arene were used, as in Equation (67).[63] Significantly, the rate of biaryl formation exceeded the rate of collapse of the aryllead compound to the aryl trifluoroacetate (Equation (65)). A number of unsymmetrical biaryls were produced in synthetically useful yields from similar reactions of aryllead triacetates containing either electron withdrawing or electron donating groups with p-xylene, mesitylene, durene, hemimellitene and pentamethylbenzene. In seeking an explanation for the high yields of biaryls produced in these reactions, it was found that the rates of the reactions increased with the increasing π-donor properties of the arene, and a mechanism in which formation of a π-complex of the type (48) is the rate-determining step has been proposed. Therefore, for p-xylene and more highly methylated benzenes, it would appear that free aryl cations are not involved.

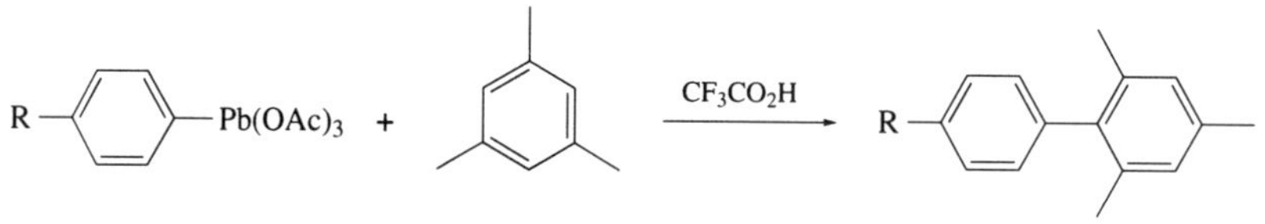

R = OMe   84%
R = NO$_2$ 72%

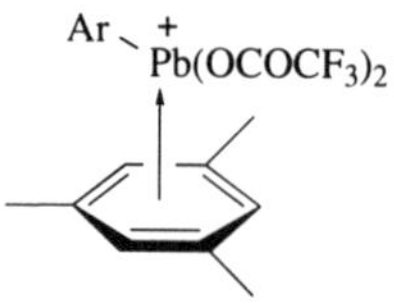

(48)

A third reaction of aryllead triacetates which proceeds under acidic conditions has industrial potential for the production of fluoroaromatics. They are very simply produced by stirring an aryllead triacetate with excess $BF_3 \cdot OEt_2$ at room temperature (Equation (68)).[32] For the compounds examined (Table 2) yields were moderate to good except for the o-methoxy- and o-fluoro-substituted phenyllead triacetates. Best yields were obtained with electron-rich aromatics, but even in the case of the p-trifluoromethylphenyllead compound a moderate yield of aryl fluoride was obtained. The reaction requires the rigorous exclusion of moisture, since protodeplumbation competes with fluorodeplumbation, as indicated in Table 2. A useful variation on this route to aryl fluorides involves the *in situ* generation of the aryllead compound by Si–Pb or B–Pb exchange, or by direct plumbation, in a Pb(OAc)$_4$/BF$_3$·OEt$_2$ mixture.[32] These three reactions are illustrated by examples in Equations (69)–(71), and further results are included in Table 3. It can be seen from Table 3 that the direct fluorination of arenes will be limited to a small number of compounds because of the formation of inseparable mixtures in many cases. Also, it would appear that there is greater scope for the B–Pb exchange route as it is faster than Si–Pb exchange and results in better yields of aryl fluorides if the aromatic ring contains electron withdrawing groups. As in the conversion of the aryllead compounds into aryl trifluoroacetates (Equation (65)), the addition of benzene to a mixture of an aryllead triacetate and BF$_3$·OEt$_2$ produced some of the biaryl, indicating that fluorodeplumbation also proceeds by aryl cation (or incipient cation) formation. When benzene was replaced by mesitylene as the trapping agent, a similar low yield of a biaryl was produced. Therefore, it would appear that the unusually high yields of biaryls obtained when trifluoroacetic acid was the catalyst (Equation (67)) require specific acid conditions.

$$\text{ArPb(OAc)}_3 \xrightarrow{\text{BF}_3\cdot\text{OEt}_3,\ \text{RT}} \text{ArF} + \text{Pb(OAc)}_2 + \text{AcOBF}_2 \qquad (68)$$

Table 2 Reactions of aryllead triacetates (ArPb(OAc)$_3$), with BF$_3$·OEt$_2$, as in Equation (68).

Ar	ArF (%)	ArH (%)
Ph	62	6
p-MeC$_6$H$_4$	82	<1
p-PhC$_6$H$_4$	74	2
p-MeOC$_6$H$_4$	66	<5
p-FC$_6$H$_4$	49	7
o-MeOC$_6$H$_4$	14	7
o-FC$_6$H$_4$	0	<1
p-CF$_3$C$_6$H$_4$	68	11
α-Naphthyl	78	11
β-Naphthyl	78	0

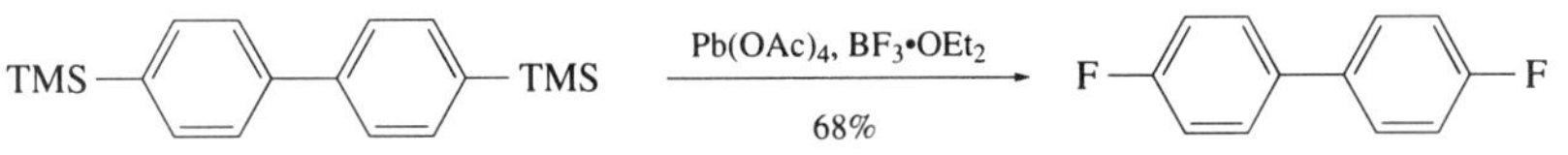

$$\text{TMS}-\!\!\bigcirc\!\!-\!\!\bigcirc\!\!-\text{TMS} \xrightarrow[68\%]{\text{Pb(OAc)}_4,\ \text{BF}_3\cdot\text{OEt}_2} \text{F}-\!\!\bigcirc\!\!-\!\!\bigcirc\!\!-\text{F} \qquad (69)$$

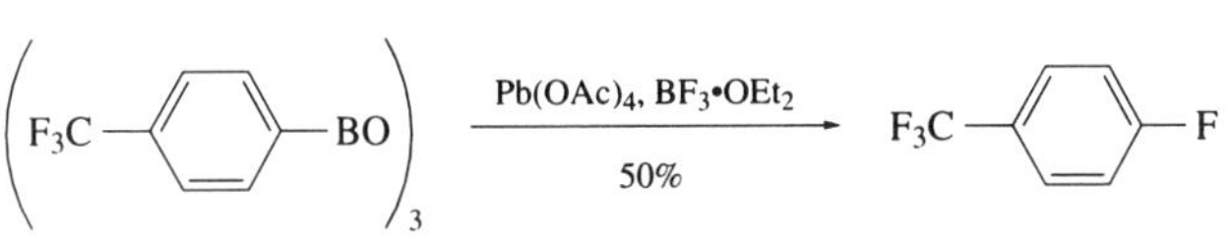

$$\left(\text{F}_3\text{C}-\!\!\bigcirc\!\!-\text{BO}\right)_3 \xrightarrow[50\%]{\text{Pb(OAc)}_4,\ \text{BF}_3\cdot\text{OEt}_2} \text{F}_3\text{C}-\!\!\bigcirc\!\!-\text{F} \qquad (70)$$

$$\text{(3,5-dimethylbenzene)} \xrightarrow[73\%]{\text{Pb(OAc)}_4,\ \text{Hg(OAc)}_2\ \text{catalyst},\ \text{BF}_3\cdot\text{OEt}_2} \text{(2,4-dimethyl-1-fluorobenzene)} \qquad (71)$$

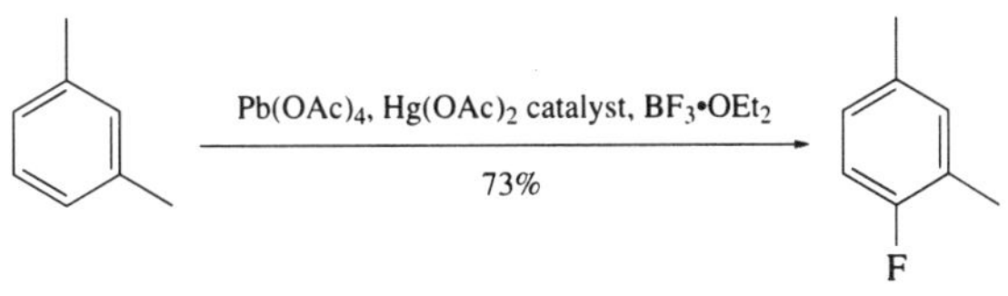

Table 3 Reactions of aryltrimethylsilanes, triarylboroxines and arenes with Pb(OAc) in $BF_3 \cdot OEt_2$ (Equations (69)–(71)).

Substrate	ArF (%)	ArH (%)
Ph-TMS	83	6
p-FC$_6$H$_4$TMS	43	0
p-CF$_3$C$_6$H$_4$TMS	<5	<5
α-C$_{10}$H$_7$TMS	73	4
β-C$_{10}$H$_7$TMS	64	0
(PhBO)$_3$	68	1
(p-MeOC$_6$H$_4$BO)$_3$	47	0
(p-CF$_3$C$_6$H$_4$BO)$_3$	50	1
PhMe	77 (o:m:p = 19:4:17)	7
Mesitylene	76	0

11.6 REFERENCES

1. H. Shapiro and F. W. Frey, 'The Organic Compounds of Lead', Wiley, New York, 1968: (a) p. 185; (b) p. 293; (c) p. 298; (d) p. 297.
2. P. J. Davidson and M. F. Lappert, *J. Chem. Soc., Chem. Commun.*, 1973, 317.
3. K. Shibata, N. Tokitoh and R. Okazaki, *Tetrahedron Lett.*, 1993, **34**, 1495.
4. L. C. Willemsens 'Organolead Chemistry', International Lead and Zinc Research Organization, New York, 1964.
5. L. C. Willemsens and G. J. M. van der Kerk, 'Investigations in the Field of Organolead Chemistry', International Lead and Zinc Research Organization, New York, 1965.
6. Y. Yamamoto and J. Yamada, *J. Am. Chem. Soc.*, 1987, **109**, 4395; Y. Yamamoto, J. Yamada and T. Asano, *Tetrahedron*, 1992, **48**, 5587.
7. J. Yamada, H. Abe and Y. Yamamoto, *J. Am. Chem. Soc.*, 1990, **112**, 6118.
8. T. Furuta and Y. Yamamoto, *J. Chem. Soc., Chem. Commun.*, 1992, 863.
9. J. Yamada and Y. Yamamoto, *J. Chem. Soc., Chem. Commun.*, 1987, 1302.
10. T. Toru *et al.*, *Tetrahedron Lett.*, 1992, **33**, 4037.
11. R. Criegee, in 'Oxidation in Organic Chemistry', Part A, ed. K. B. Wiberg, Academic Press, New York, 1965, p. 277.
12. G. M. Rubottom, in 'Oxidation in Organic Chemistry', Part D, ed. W. S. Trahanovsky, Academic Press, New York, 1982, p. 1.
13. H. C. Bell, J. R. Kalman, J. T. Pinhey and S. Sternhell, *Aust. J. Chem.*, 1979, **32**, 1521.
14. J. T. Pinhey, *Aust. J. Chem.*, 1991, **44**, 1353.
15. M. G. Moloney, J. T. Pinhey and M. J. Stoermer, *J. Chem. Soc., Perkin Trans. 1*, 1990, 2645.
16. J. T. Pinhey and M. J. Stoermer, *J. Chem. Soc., Perkin Trans. 1*, 1991, 2455.
17. M. G. Moloney, J. T. Pinhey and E. G. Roche, *J. Chem. Soc., Perkin Trans. 1*, 1989, 333.
18. R. C. Larock, 'Organomercury Compounds in Organic Synthesis', Springer, Berlin, 1985, p. 60.
19. E. C. Taylor and A. McKillop, *Acc. Chem. Res.*, 1970, **3**, 338.
20. F. R. Preuss and I. Janshen, *Arch. Pharm. (Weinheim, Ger.)*, 1960, **293**, 933 (*Chem. Abstr.*, 1961, **55**, 5396).
21. L. C. Willemsens, D. de Vos, J. Spierenburg and J. Wolters, *J. Organomet. Chem.*, 1972, **39**, C61.
22. D. de Vos *et al.*, *Recl. Trav. Chim. Pays-Bas*, 1975, **94**, 97.
23. J. R. Kalman, J. T. Pinhey and S. Sternhell, *Tetrahedron Lett.*, 1972, 5369.
24. R. P. Kozyrod and J. T. Pinhey, *Org. Synth.*, 1984, **62**, 24.
25. R. P. Kopinski, J. T. Pinhey and B. A. Rowe, *Aust. J. Chem.*, 1984, **37**, 1245.
26. E. M. Panov and K. A. Kocheshkov, *Dokl. Akad. Nauk SSSR*, 1952, **85**, 1037 (*Chem. Abstr.*, 1953, **47**, 6365).
27. R. Criegee, P. Dimroth and R. Schempf, *Chem. Ber.*, 1957, **90**, 1337.
28. R. P. Kozyrod and J. T. Pinhey, *Aust. J. Chem.*, 1985, **38**, 1155.
29. J. T. Pinhey and E. G. Roche, *J. Chem. Soc., Perkin Trans. 1*, 1988, 2415.
30. R. L. Funk and K. P. C. Vollhardt, *J. Am. Chem. Soc.*, 1979, **101**, 215.
31. H. C. Bell, J. R. Kalman, J. T. Pinhey and S. Sternhell, *Tetrahedron Lett.*, 1974, 853.
32. G. V. De Meio and J. T. Pinhey, *J. Chem. Soc., Chem. Commun.*, 1990, 1065; G. V. De Meio, J. Morgan and J. T. Pinhey, *Tetrahedron*, 1993, **49**, 8129.
33. R. P. Kozyrod and J. T. Pinhey, *Tetrahedron Lett.*, 1983, **24**, 1301; R. P. Kozyrod, J. Morgan and J. T. Pinhey, *Aust. J. Chem.*, 1985, **38**, 1147.
34. D. H. R. Barton, D. M. X. Donnelly, J.-P. Finet and P. J. Guiry, *J. Chem. Soc., Perkin Trans. 1*, 1992, 1365.
35. J. Morgan and J. T. Pinhey, *J. Chem. Soc., Perkin Trans. 1*, 1990, 715.
36. M. G. Moloney and J. T. Pinhey, *J. Chem. Soc., Perkin Trans. 1*, 1988, 2847.
37. C. J. Parkinson, J. T. Pinhey and M. J. Stoermer, *J. Chem. Soc., Perkin Trans. 1*, 1992 1911.
38. H. C. Bell, J. T. Pinhey and S. Sternhell, *Aust. J. Chem.*, 1979, **32**, 1551.
39. D. H. R. Barton, J.-P. Finet, C. Giannotti and F. Halley, *J. Chem. Soc., Perkin Trans. 1*, 1987, 241.
40. D. H. R. Barton, D. M. X. Donnelly, P. J. Guiry and J. H. Reibenspies, *J. Chem. Soc., Chem. Commun.*, 1990, 1110.
41. T. W. Hambley, R. J. Holmes, C. J. Parkinson and J. T. Pinhey, *J. Chem. Soc., Perkin Trans. 1*, 1992, 1917.
42. J. T. Pinhey and B. A. Rowe, *Aust. J. Chem.*, 1979, **32**, 1561.
43. J. T. Pinhey and B. A. Rowe, *Aust. J. Chem.*, 1980, **33**, 113.

44. D. J. Ackland and J. T. Pinhey, *J. Chem. Soc., Perkin Trans. 1*, 1987, 2689.
45. S.-I. Hashimoto, T. Shinoda and S. Ikegami, *J. Chem. Soc., Chem. Commun.*, 1988, 1137.
46. K. Orito, K. Yorita and H. Suginome, *Tetrahedron Lett.*, 1991, **32**, 5999.
47. D. J. Ackland and J. T. Pinhey, *J. Chem. Soc., Perkin Trans. 1*, 1987, 2695.
48. C. J. Parkinson and J. T. Pinhey, *J. Chem. Soc., Perkin Trans. 1*, 1991, 1053.
49. D. M. X. Donnelly, J.-P. Finet, P. J. Guiry and R. M. Hutchinson, *J. Chem. Soc., Perkin Trans. 1*, 1990, 2851.
50. D. M. X. Donnelly, J.-P. Finet and J. M. Kielty, *Tetrahedron Lett.*, 1991, **32**, 3835.
51. J.-Y. Merour, L. Chichereau and J.-P. Finet, *Tetrahedron Lett.*, 1992, **33**, 3867.
52. R. P. Kopinski and J. T. Pinhey, *Aust. J. Chem.*, 1983, **36**, 311.
53. J. T. Pinhey and B. A. Rowe, *Aust. J. Chem.*, 1983, **36**, 789.
54. D. J. Collins, J. D. Cullen, G. D. Fallon and B. M. Gatehouse, *Aust. J. Chem.*, 1984, **37**, 2279.
55. R. P. Kozyrod, J. Morgan and J. T. Pinhey, *Aust. J. Chem.*, 1991, **44**, 369.
56. M. J. Koen, J. Morgan and J. T. Pinhey, *J. Chem. Soc., Perkins Trans. 1*, 1993, 2383.
57. R. P. Kozyrod and J. T. Pinhey, *Aust. J. Chem.*, 1985, **38**, 713.
58. G. L. May and J. T. Pinhey, *Aust. J. Chem.*, 1982, **35**, 1859.
59. M.-L. Huber and J. T. Pinhey, *J. Chem. Soc., Perkin Trans. 1*, 1990, 721.
60. J. Morgan and J. T. Pinhey, *J. Chem. Soc., Perkin Trans. 1*, 1993, 1673.
61. J. Morgan, I. Buys, T. W. Hambley and J. T. Pinhey, *J. Chem. Soc., Perkin Trans. 1*, 1993, 1677.
62. D. H. R. Barton, D. M. X. Donnelly, J.-P. Finet and P. J. Guiry, *J. Chem. Soc., Perkin Trans. 1*, 1991, 2095.
63. H. C. Bell, J. R. Kalman, G. L. May, J. T. Pinhey and S. Sternhell, *Aust. J. Chem.*, 1979, **32**, 1531.
64. J. R. Campbell, J. R. Kalman, J. T. Pinhey and S. Sternhell, *Tetrahedron Lett.*, 1972, 1763.

12
Antimony and Bismuth

YAO-ZENG HUANG and ZHANG-LIN ZHOU
Shanghai Institute of Organic Chemistry, People's Republic of China

12.1 INTRODUCTION	488
12.1.1 Literature	488
12.1.2 General Considerations	488
12.2 ORGANOANTIMONY COMPOUNDS IN ORGANIC SYNTHESIS	488
12.2.1 Alkenation Reactions Mediated by Organoantimony Reagents	488
12.2.1.1 Trialkylstibine-mediated alkenation of aldehydes and ketones	489
12.2.1.2 Alkenation reaction through antimony ylides	489
12.2.1.3 Alkenation through addition–elimination reactions	491
12.2.1.4 Miscellaneous alkenations	491
12.2.2 Formation of Alcohols via Organoantimony Compounds	492
12.2.2.1 Trialkylstibine-mediated formation of alcohols	492
12.2.2.2 Synthesis of alcohols via pentaorganylstiboranes	493
12.2.2.3 Antimony- and $SbCl_3$-metal-mediated synthesis of alcohols	495
12.2.2.4 Diphenylstibine-mediated reduction of carbonyl compounds	496
12.2.2.5 Miscellaneous formation of alcohols via organoantimony compounds	496
12.2.3 Synthesis of Ketones via Organoantimony Compounds	496
12.2.3.1 Formation of ketones via pentaorganylstiboranes	496
12.2.3.2 Pentavalent organoantimony compounds as oxidizing agents	497
12.2.3.3 Tributylstibine-assisted synthesis of alkyl aryl sulfones	497
12.2.4 Organoantimony Compound-mediated Acetalization Reactions	498
12.2.4.1 Trialkoxystibine-promoted acetalization of aldehydes	498
12.2.4.2 $SbCl_3$-metal-mediated acetalization	498
12.2.5 Cyclopropanation via Organoantimony Compounds	498
12.2.6 Organoantimony(V) Halides as Catalysts in Organic Synthesis	499
12.2.6.1 Synthesis of silylated enol ethers	499
12.2.6.2 Cycloaddition of substituted aziridines or oxiranes with heterocumulenes	500
12.2.7 Antimony Salts as Catalysts in Organic Synthesis	500
12.2.7.1 $SbCl_5$-catalyzed reactions	500
12.2.7.2 $SbCl_3$-catalyzed reactions	501
12.2.8 Miscellaneous Reactions	501
12.3 ORGANOBISMUTH COMPOUNDS IN ORGANIC SYNTHESIS	502
12.3.1 Organobismuth(III) Compounds	502
12.3.1.1 Barbier-type reactions via organobismuth(III) compounds	502
12.3.1.2 Synthesis of homoalkylated amines	502
12.3.1.3 Aldol and Michael reactions catalyzed by $BiCl_3$	502
12.3.1.4 Miscellaneous reactions	503
12.3.2 Organobismuth(V) Compounds	503
12.3.2.1 Oxidation reactions with organobismuth(V) compounds	504
12.3.2.2 Use of bismuthonium ylides in organic synthesis	506
12.3.2.3 Arylation reactions with pentavalent organobismuth compounds	506
12.3.2.4 Miscellaneous reactions	511
12.4 REFERENCES	511

12.1 INTRODUCTION

12.1.1 Literature

Organoantimony and organobismuth compounds have been known for well over a century, but there were few studies on their synthetic applications prior to the 1980s. A number of recent developments, however, have demonstrated that these compounds can have considerable utility in organic synthesis. The preparation, reactions and properties of organoantimony and organobismuth compounds were thoroughly reviewed in *COMC-I*.[1] Since then, a number of review articles on organoantimony and/or organobismuth chemistry have appeared. Reviews devoted solely to the chemistry of organobismuth compounds are by Freedman and Doak,[2] and Barton and co-workers.[3] More recent developments in the applications of organobismuth compounds have been partly reviewed by Barton and Finet.[4,5] A review mainly covering our own work and concerned solely with the synthetic applications of organoantimony compounds has been published.[6] Freedman and Doak have also reviewed organoantimony and organobismuth chemistry on an annual basis from 1982 to 1990, and review literature has appeared during this period in the *Journal of Organometallic Chemistry* and the *Journal of Organometallic Chemistry, Library*.[7-24] Wardell, in *Organometallic Chemistry*, has also summarized organoantimony and organobismuth compounds annually from 1982 to 1990.[25-33] The uses of organoantimony and organobismuth compounds in organic synthesis up to 1987 have been reviewed.[34]

12.1.2 General Considerations

Research on the synthetic applications of organoantimony compounds has been carried out sporadically. Lloyd and co-workers prepared a number of stable stibonium ylides with strong α-electron-withdrawing substituents in the alkylidene moiety from diazo precursors and found that all of them were inert, even toward very reactive substrates such as 2,4-dinitrobenzaldehyde.[35] In contrast to triphenylstibine, which forms stibonium salts with halo compounds only with difficulty, trialkylstibines react readily with halo compounds, especially with α-halocarboxylic derivatives, at room temperature to form the substituted stibonium halides. A change in the substituent on antimony from phenyl to alkyl results in an acute change in the ease of formation of the stibonium salts and in the reactivities thereof. Thus, it has been shown that organoantimony compounds can be used for alkenation, the oxidation and reduction of various functional groups, formation of alcohols and ketones, cyclopropanation reactions, acetalization, the acylation of amines, and a number of other synthetically interesting transformations.

Trivalent organobismuth compounds have been used only to a very limited extent in organic synthesis. They can mediate a number of reactions, such as Barbier-type reactions and certain other reactions. Of particular significance are the synthetic applications of organobismuth(V) compounds. Pentavalent organobismuth compounds have found considerable use in organic synthesis in recent years. Organobismuth(V) reagents have served as oxidizing and arylating agents. They have proved to be particularly valuable where the substrates are sensitive natural products.

It seems likely that organoantimony and organobismuth compounds may soon be more widely recognized as important reagents in synthetic organic chemistry. This review focuses on the synthetic applications of trivalent and pentavalent organoantimony and organobismuth compounds, with strong emphasis on the synthetically useful transformations which have appeared in the literature since 1982.

12.2 ORGANOANTIMONY COMPOUNDS IN ORGANIC SYNTHESIS

12.2.1 Alkenation Reactions Mediated by Organoantimony Reagents

Formation of carbon–carbon double bonds is a very important reaction in synthetic organic chemistry, especially for the synthesis of natural products, a topic which continues to attract considerable attention as milder, more selective methods are developed. Alkenation of carbonyl compounds is one of the most important methods of forming carbon–carbon double bonds. A very common approach to the preparation of an alkene from an aldehyde or ketone is via the Wittig reaction, which involves three steps: preparation of a phosphonium salt, base treatment to produce an ylide, and reaction of the latter with the carbonyl compound. Recent developments, however, have shown that organoantimony compounds can be used successfully for the conversion of aldehydes and ketones to alkenes.

12.2.1.1 Trialkylstibine-mediated alkenation of aldehydes and ketones

In studies on the uses of organoantimony compounds in organic synthesis, it was reported that trialkylstibines are very effective reagents for the formation of carbon–carbon double bonds between α-halocarboxylic derivatives, including esters, amides and nitriles, and carbonyl compounds. α,β-Unsaturated esters, amides and nitriles were obtained.[36–8] In the reactions of ethyl bromoacetate, or N,N-diethylacetamide with aldehydes, the double bonds of all the products formed were *trans*, whereas for substituted α-halocarboxylic acid esters and haloacetonitriles, mixtures of (E) and (Z) isomers were obtained (Equations (1)–(4)).

$$Me(CH_2)_8CHO + BrCH_2CO_2Et \xrightarrow[88\%]{Bu_3Sb,\ 100\ °C} Me(CH_2)_8CH=CHCO_2Et \quad (E) \tag{1}$$

$$Me_2CHCH_2CHO + BrCH_2CONEt_2 \xrightarrow[64\%]{Bu_3Sb,\ 80\ °C} Me_2CHCH_2CH=CHCONEt_2 \quad (E) \tag{2}$$

$$Me(CH_2)_7CHO + ClCH_2CN \xrightarrow[97\%]{Bu_3Sb,\ 120\ °C} Me(CH_2)_7CH=CHCN \quad (E):(Z)=60:40 \tag{3}$$

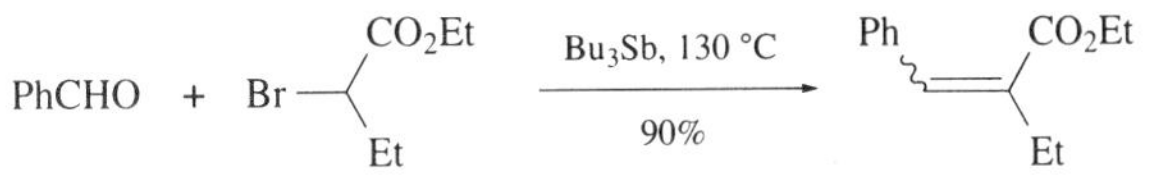

$$PhCHO + Br\!\!\underset{Et}{\overset{CO_2Et}{\diagup}} \xrightarrow[90\%]{Bu_3Sb,\ 130\ °C} \underset{Et}{\overset{Ph\quad CO_2Et}{C=C}} \tag{4}$$

α-Enones are important intermediates in organic synthesis. Generally, they can be synthesized by aldol condensation, but unsymmetrical mixtures are produced. A new method for the synthesis of α-enones by the reaction of aldehydes and bromoacetone with tributylstibine has been reported. The (E)-alkene was formed exclusively (Equation (5)).[39]

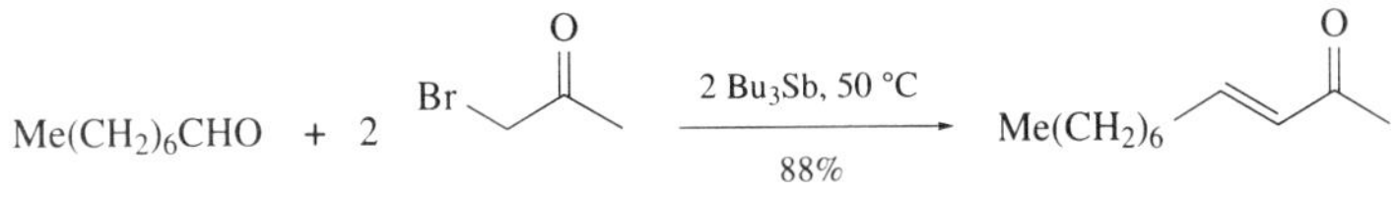

$$Me(CH_2)_6CHO + 2\ Br\!\!\overset{O}{\diagup}\!\!\diagdown \xrightarrow[88\%]{2\ Bu_3Sb,\ 50\ °C} Me(CH_2)_6\overset{O}{\diagdown}\!\!\diagup \tag{5}$$

12.2.1.2 Alkenation reaction through antimony ylides

Lloyd *et al.* prepared a number of phenyl-ligand stable stibonium ylides with strong α-electron-withdrawing groups in the alkylidene moiety from diazo precursors, but found that all of them were inert, even towards 2,4-dinitrobenzaldehyde.[35] The isolation of an antimony ylide in the solid state and its conversion to an alkene was first accomplished by Lloyd and co-workers in 1967. The ylide was obtained by heating a mixture of triphenylstibine and diazotetraphenylcyclopentadiene under nitrogen at 140 °C. It does not react with benzaldehyde when refluxed in chloroform for 4 h, but it does react with 4-nitrobenzaldehyde to give the expected fulvene in high yield.[40]

Tributylstibine reacts readily with bromoacetic acid derivatives at room temperature to form the substituted stibonium bromides ($Bu_3Sb^+CH_2E\ Br^-$; E = CO_2Me, CO_2Et, CN, $CONEt_2$, $CO(NC_5H_{10})$). When treated with potassium *t*-butoxide, these stibonium salts react with a variety of carbonyl compounds to yield α,β-unsaturated acrylic acid derivatives in moderate to high yields via the stibonium ylides, together with dihydroxystiborane, formed by hydration of the tributylstibine oxide (Scheme 1).[41] The alkenation products from the aldehydes all had the (E) configuration (Equations (6) and (7)). These ylides are air sensitive and could not be isolated under atmospheric conditions. However, evidence for an ylide intermediate was obtained from the tandem reaction summarized in Scheme 2.

$$Bu_3Sb^+CH_2CO_2Me\ Br^- \xrightarrow[90\%]{Bu^tOK/THF,\ PhCHO} PhCH=CHCO_2Me \tag{6}$$

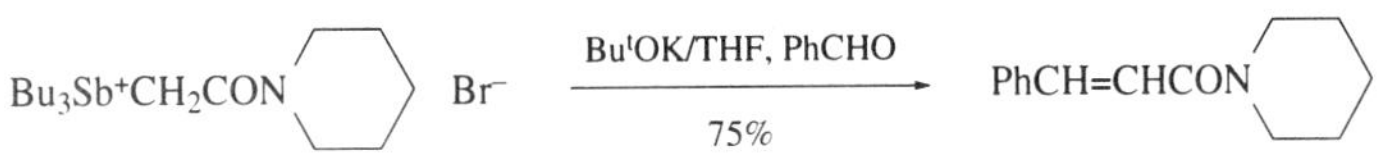

$$Bu_3Sb^+CH_2CON\!\!\diagup\!\!\diagdown\ Br^- \xrightarrow[75\%]{Bu^tOK/THF,\ PhCHO} PhCH=CHCON\!\!\diagup\!\!\diagdown \tag{7}$$

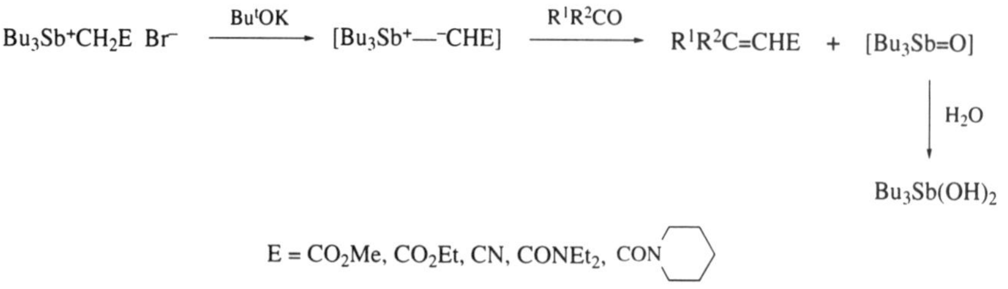

$Bu_3Sb^+CH_2E\ Br^- \xrightarrow{Bu^tOK} [Bu_3Sb^+\text{—}^-CHE] \xrightarrow{R^1R^2CO} R^1R^2C=CHE + [Bu_3Sb=O]$

$\downarrow H_2O$

$Bu_3Sb(OH)_2$

$E = CO_2Me, CO_2Et, CN, CONEt_2, CON\bigcirc$

Scheme 1

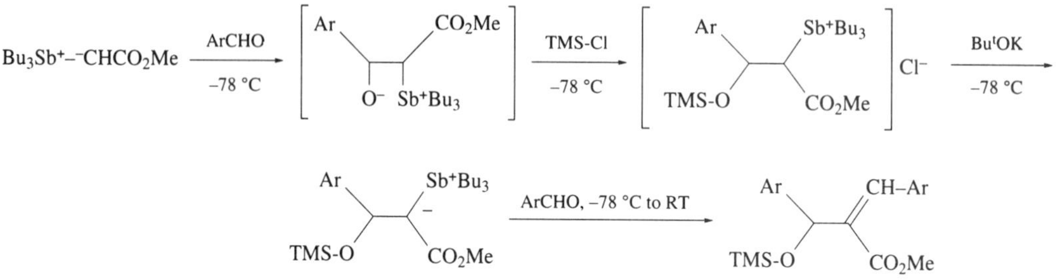

$Bu_3Sb^+\text{—}^-CHCO_2Me \xrightarrow[-78\ °C]{ArCHO} \left[\begin{array}{c} Ar \quad CO_2Me \\ \diagdown\ /\ \\ /\ \diagdown\ \\ O^-\ Sb^+Bu_3 \end{array}\right] \xrightarrow[-78\ °C]{TMS-Cl} \left[\begin{array}{c} Ar\ \ Sb^+Bu_3 \\ \diagdown/ \\ /\diagdown \\ TMS-O\ CO_2Me \end{array}\right] Cl^- \xrightarrow[-78\ °C]{Bu^tOK}$

$\begin{array}{c} Ar\ \ Sb^+Bu_3 \\ \diagdown/ \\ /\diagdown\text{—} \\ TMS-O\ CO_2Me \end{array} \xrightarrow{ArCHO,\ -78\ °C\ to\ RT} \begin{array}{c} Ar\ \ CH-Ar \\ \diagdown\ \|\ \\ /\ \diagdown \\ TMS-O\ CO_2Me \end{array}$

Scheme 2

One of the most common synthetic routes to heteroatom ylides is from diazo compounds.[42] By use of this method, several stable phenyl-ligand stibonium ylides with strong electron-withdrawing substituents, such as COR, SO_2R, and so on, in the alkylidene moiety, which were inert towards carbonyl compounds, have been synthesized. However, diazomalonic esters, including the diazo derivatives from Meldrum's acid, ethyl diazoacetate, and so on, did not give stibonium ylides with triphenylstibine.[35] A one-pot reaction of tributylstibine, a diazo compound (including dimethyl diazomalonate, ethyl diazoacetate and diazoacetylacetone), a carbonyl compound and a catalytic amount of Cu^II which resulted in alkenation in high yields via stibonium ylides has been reported.[43] In the absence of either tributylstibine or the CuI catalyst, alkenation did not occur at all. Since the alkyl ligand is an electron-donating group, while the phenyl ligand is an electron-withdrawing group, the lone pairs of electrons of Bu_3Sb are more accessible than those of Ph_3Sb. Therefore, Bu_3Sb can react with the carbene derived from the diazo compound more easily than Ph_3Sb to generate the stibonium ylide. Moreover, the alkyl-ligand stibonium ylide is reactive toward carbonyl compounds. Various carbonyl compounds react easily with such ylides to afford alkenation products (Equations (8) and (9)). This one-pot reaction can also be used with ethyl diazoacetate and carbonyl compounds at 40–50 °C to give ethyl β-substituted acrylates in high yields (Equation (10)).[43]

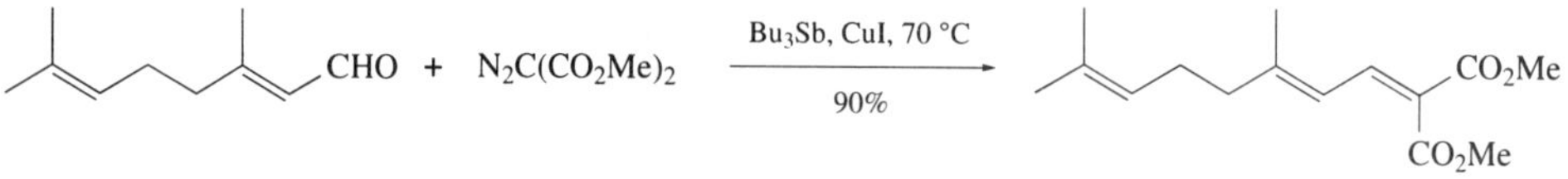

$\text{(geranial)CHO} + N_2C(CO_2Me)_2 \xrightarrow[90\%]{Bu_3Sb,\ CuI,\ 70\ °C} \text{(geranyl)}=C(CO_2Me)_2 \quad (8)$

$PhCHO + N_2C(COMe)_2 \xrightarrow[92\%]{Bu_3Sb,\ CuI,\ 80\ °C} PhCH=C(COMe)_2 \quad (9)$

$Me(CH_2)_4CHO + N_2CHCO_2Et \xrightarrow[96\%]{Bu_3Sb,\ CuI,\ 40\ °C} Me(CH_2)_4CH=CHCO_2Et \quad (10)$

Tributylstibine can mediate the alkenation of carbonyl compounds with bromomalonic esters and with dibromomalonic esters, which also probably proceeds via stibonium ylides (Equations (11) and (12)).[44] Heating was needed to initiate the reaction when the alkenation was carried out with triphenylstibine as mediator. For example, triphenylstibine-mediated alkenation of benzaldehyde with dimethyl bromomalonate should be performed at 70 °C (Equation (13)).[44]

$Me_2CHCH_2CHO + 2BrCH(CO_2Me)_2 \xrightarrow[97\%]{Bu_3Sb,\ CuI,\ 50\ °C} Me_2CHCH_2CH=C(CO_2Me)_2 \quad (11)$

$$Ph_2CO + 2BrCH(CO_2Me)_2 \xrightarrow[91\%]{2Bu_3Sb,\ 50\ °C} Ph_2C=C(CO_2Me)_2 \quad (12)$$

$$PhCHO + 2BrCH(CO_2Me)_2 \xrightarrow[80\%]{2Ph_3Sb,\ 70\ °C} PhCH=C(CO_2Me)_2 \quad (13)$$

12.2.1.3 Alkenation through addition–elimination reactions

It has been reported that certain α-lithio derivatives of alkyldiphenylstibines can be employed for the conversion of carbonyl compounds to alkenes (Scheme 3 and Equation (14)).[45,46]

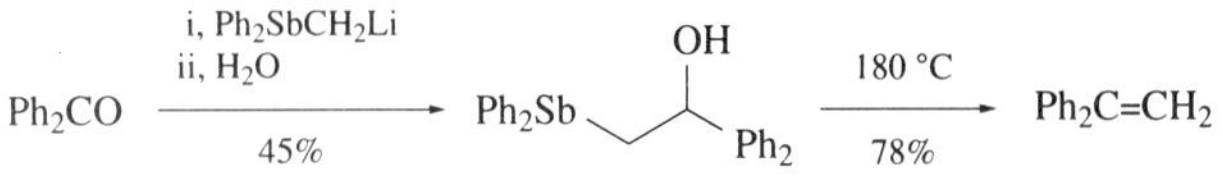

$$Ph_2CO \xrightarrow[45\%]{\text{i, } Ph_2SbCH_2Li;\ \text{ii, } H_2O} Ph_2Sb\text{-}CH_2\text{-}CH(OH)Ph_2 \xrightarrow[78\%]{180\ °C} Ph_2C=CH_2$$

Scheme 3

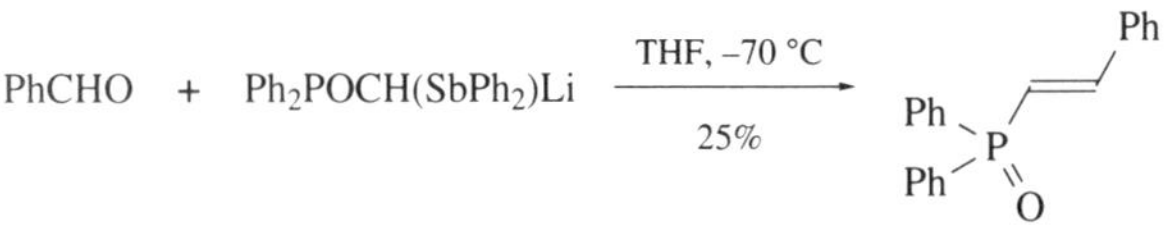

$$PhCHO + Ph_2POCH(SbPh_2)Li \xrightarrow[25\%]{THF,\ -70\ °C} Ph_2P(O)\text{-}CH=CH\text{-}Ph \quad (14)$$

In the presence of diphenylantimonymagnesium, various aldehydes react with ω-bromoacetophenone to form α,β-unsaturated ketones in good yields (Equation (15)).[47]

$$4\text{-}MeC_6H_4CHO + BrCH_2COPh \xrightarrow[86\%]{(Ph_2Sb)_2Mg} 4\text{-}MeC_6H_4CH=CHCOPh \quad (15)$$

12.2.1.4 Miscellaneous alkenations

Reaction of triphenylstibine with certain α-dicarbonyl compounds has been found to lead to partial deoxygenation and results in the formation of alkenes which are widely used as dyestuffs.[48] Thus, when triphenylstibine reacts with isatin or *N*-methylisatin in dry toluene for 25 h, indirubin or dimethylindirubin precipitates. Similar treatment of naphtho[2,1-*b*]furan-1,2-dione with triphenylstibine gave the alkene in 20% yield (Equations (16) and (17)).

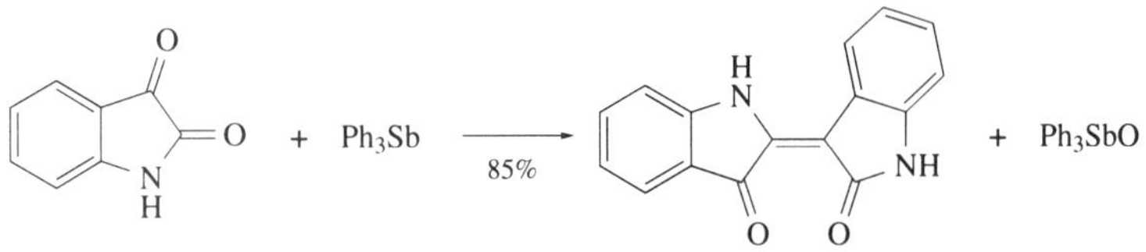

$$\text{isatin} + Ph_3Sb \xrightarrow{85\%} \text{indirubin} + Ph_3SbO \quad (16)$$

$$\text{naphtho[2,1-}b\text{]furan-1,2-dione} + Ph_3Sb \xrightarrow{20\%} \text{alkene} + Ph_3SbO \quad (17)$$

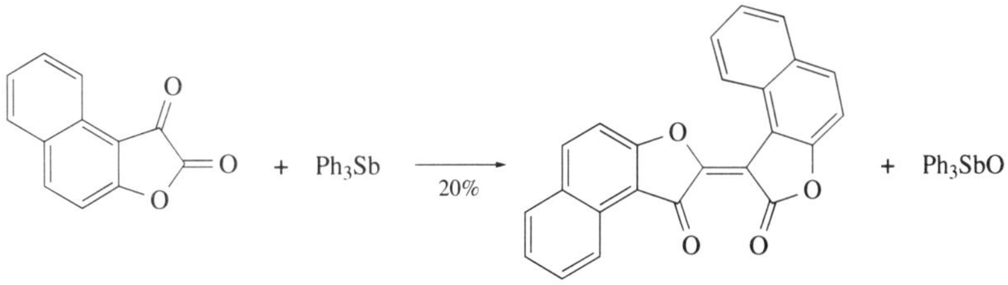

In the presence of tributyl- or triphenylstibine, certain vicinal dibromides undergo 1,2-elimination to give the corresponding alkene in high yield (Equation (18)).[49]

It was discovered that palladium(II) acetate can promote the transfer of a phenyl group from triphenylstibine to an unsubstituted alkenic carbon atom to give disubstituted alkenes (Equation (19)).[50] Similarly, when equimolar amounts of triphenylstibine and palladium(II) acetate are allowed to react with a 10-fold excess of oct-1-ene, a 57% yield of phenylated octenes is obtained. When ethyl acrylate is used instead of oct-1-ene, the product is ethyl cinnamate.[51]

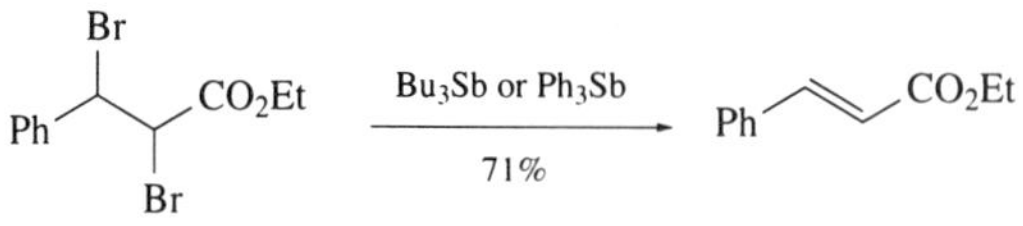

$$Ph\underset{Br}{\overset{Br}{\underset{|}{C}}}H-\underset{}{\overset{}{C}}H(CO_2Et) \xrightarrow[71\%]{Bu_3Sb \text{ or } Ph_3Sb} Ph-CH=CH-CO_2Et \quad (18)$$

$$Ph_3Sb + PhCH=CH_2 \xrightarrow[67\%]{Pd(OAc)_2} PhCH=CHPh \quad (19)$$

12.2.2 Formation of Alcohols via Organoantimony Compounds

Alcohols are an important class of compounds in organic synthesis since they readily transform to other functional groups. Generally, they can be easily prepared by the simple reduction of carbonyl compounds with an appropriate reducing agent, by nucleophilic addition of a suitable reagent to carbonyl compounds, and also by a number of other methods. Recent developments have shown that alcohols can also be synthesized via organoantimony compounds.

12.2.2.1 Trialkylstibine-mediated formation of alcohols

Allylation of carbonyl compounds to homoallylic alcohols is an important synthetic operation. Many metals have been used to promote this Barbier-type reaction with allyl halides,[52] and it has been reported that trialkylstibines can be used.[53]

Triethylstibine and tributylstibine appear to be equally effective as promoters. The reactivity of allyl iodide is almost the same as the bromide, while that of allyl chloride is somewhat lower. A variety of aldehydes undergo the reaction readily (Equation (20)) and the condensation is chemoselective. Ketones such as cyclohexanone and acetophenone did not react at all under similar conditions.

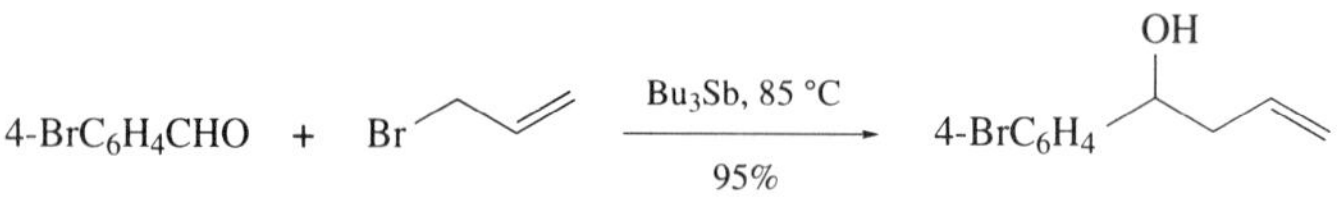

$$4\text{-}BrC_6H_4CHO + Br\text{-}CH_2CH=CH_2 \xrightarrow[95\%]{Bu_3Sb,\ 85\ °C} 4\text{-}BrC_6H_4\text{-}CH(OH)\text{-}CH_2\text{-}CH=CH_2 \quad (20)$$

The reaction of substituted allyl bromides with aldehydes, mediated by tributylstibine, results in predominant formation of the γ-adduct, as in the case of other crotylmetallic reagents.[54] Some α-adduct is also formed as a by-product. It is noteworthy that reaction of the allylantimony-derived reagents with aldehydes provides homoallylic alcohols with high *threo* selectivity in the (*E*)-4-methyl-2-pentenylantimony case and with preferential *erythro* selectivity in the crotylantimony case. The result with (*E*)-2-hexenylantimony is not regular (Equation (21)).

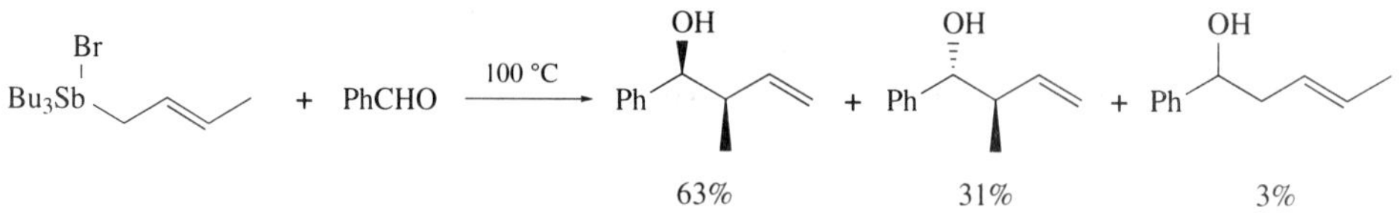

$$Bu_3Sb\text{-}CH(Br)\text{-}CH=CH\text{-}CH_3 + PhCHO \xrightarrow{100\ °C} Ph\text{-}CH(OH)\text{-}CH(CH_3)\text{-}CH=CH_2 \text{ (63\%)} + Ph\text{-}CH(OH)\text{-}CH(CH_3)\text{-}CH=CH_2 \text{ (31\%)} + Ph\text{-}CH(OH)\text{-}CH_2\text{-}CH=CH\text{-}CH_3 \text{ (3\%)} \quad (21)$$

It has been reported that trichloroacetonitrile reacts with aldehydes in the presence of tributylstibine to give α,α-dichloro-β-hydroxynitriles in excellent yields.[55] A variety of aldehydes react readily to give the expected alcohols (Equation (22)), but ketones and imines do not react under similar conditions.

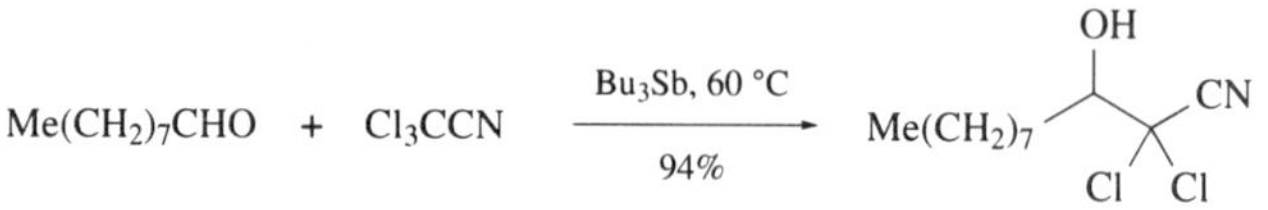

$$Me(CH_2)_7CHO + Cl_3CCN \xrightarrow[94\%]{Bu_3Sb,\ 60\ °C} Me(CH_2)_7\text{-}CH(OH)\text{-}CCl_2\text{-}CN \quad (22)$$

A novel method for synthesis of β-hydroxyketones from aldehydes and α-bromoketones has been reported (Equation (23)).[56] This reaction is mediated by trialkylstibines and catalyzed by I_2 (2–4 mol.%).

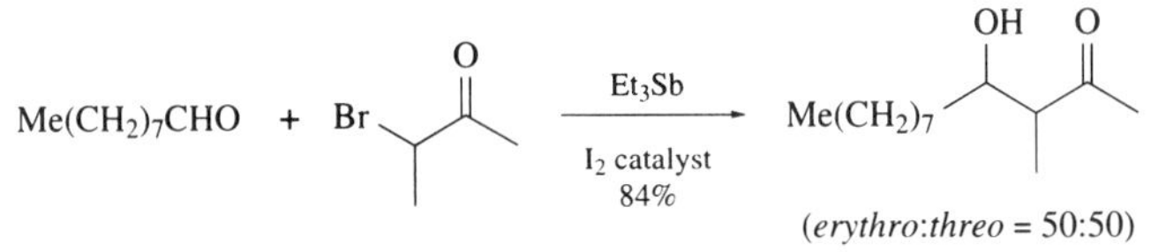

$$Me(CH_2)_7CHO \ + \ Br\underset{}{\overset{O}{\diagup\!\!\!\diagdown}} \xrightarrow[\text{84\%}]{\text{Et}_3\text{Sb}, \ \text{I}_2 \text{ catalyst}} Me(CH_2)_7\underset{}{\overset{OH \ \ O}{\diagup\!\!\!\diagdown}} \quad (23)$$

(*erythro:threo* = 50:50)

12.2.2.2 Synthesis of alcohols via pentaorganylstiboranes

Although pentaalkylstiboranes have long been known, their application in organic synthesis has not been widely exploited. It has been found that quaternary stibonium salts $[Bu^n_3SbCH_2E]^+ \ X^-$ (E = Ph, CH=CH$_2$, CH=CHCO$_2$Et, CH=CH-TMS, C≡CMe, C≡C-TMS, CO$_2$Et, CN; X = Br, BPh$_4$) on treatment with phenyl- or alkyllithiums give pentaalkylstiboranes (Bu$^n_3$Sb(R)CH$_2$E) which react with carbonyl compounds to give various alcohols in good to excellent yields. For example, benzyltrialkylstibonium bromide, after being treated with phenyllithium or an alkyllithium, reacts with carbonyl compounds to give homobenzylic alcohols in high yields (Equations (24) and (25)).[57] The reaction is chemoselective for aldehydes; cyclic ketones, benzophenone and acetophenone do not react under the same conditions.

$$PhCHO \ + \ Et_3Sb^+CH_2Ph \ Br^- \xrightarrow[\text{92\%}]{\text{BuLi/THF, } -78 \ °C \ \text{to RT}} Ph\underset{}{\overset{OH}{\diagup\!\!\!\diagdown}}Ph \quad (24)$$

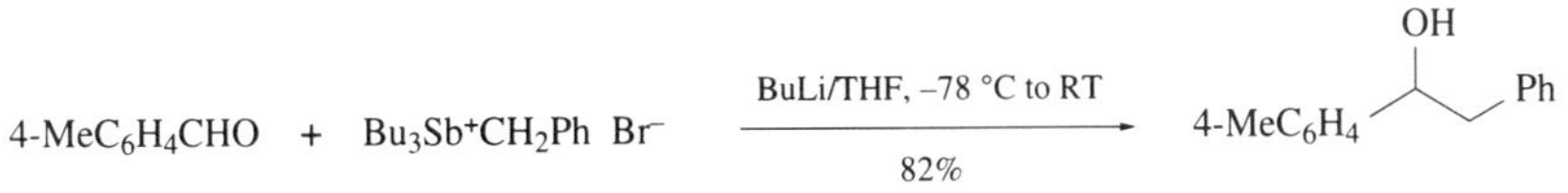

$$4\text{-MeC}_6H_4CHO \ + \ Bu_3Sb^+CH_2Ph \ Br^- \xrightarrow[\text{82\%}]{\text{BuLi/THF, } -78 \ °C \ \text{to RT}} 4\text{-MeC}_6H_4\underset{}{\overset{OH}{\diagup\!\!\!\diagdown}}Ph \quad (25)$$

An alkyllithium is not only a strong base but also a strong nucleophile. The antimony atom of the stibonium salt is an electrophile with a large atomic radius, so the alkyllithium can attack the antimony atom preferentially and displace the anion X$^-$, instead of abstracting a proton and forming an antimony ylide, as occurs in the case of the phosphonium or arsonium analogues. A pentaorganylstiborane may thus be formed as shown in Scheme 4. This species may become polarized in the presence of Li$^+$ and nucleophilic addition to aldehydes gives homobenzylic alcohols.

$$Bu_3Sb^+CH_2Ph \ Br^- \xrightarrow[-\text{LiBr}]{\text{BuLi}} [Bu_4SbCH_2Ph] \ \rightleftharpoons \ \left[\begin{array}{c} \overset{\delta+}{Bu_4Sb}\text{---}\overset{\delta-}{CH_2}\diagdown_{Li^+}^{Ph} \end{array} \right] Br^-$$

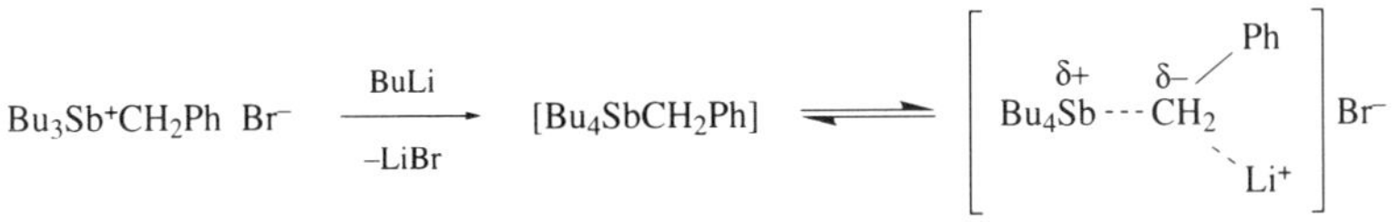

Scheme 4

Treatment of allyltributylstibonium bromide and allyltributylstibonium iodide with phenyllithium or an alkyllithium followed by reaction with aldehydes gives homoallylic alcohols in good to excellent yields (Equation (26)).[58] The reaction is again chemoselective for aldehydes. However, in the presence of AlCl$_3$, allyl tetrabutylstiborane or crotyltetrabutylstiborane react with ketones to give tertiary allylic alcohols in good yields (Equation (27)).[59] Of particular interest is the PhLi mediated synthesis of ethyl 5-aryl-5-hydroxypent-2-enoates in high yield from the corresponding stibonium salt and aldehydes (Equation (28)).[58]

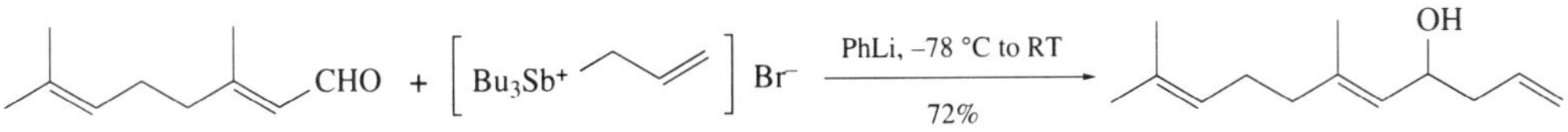

$$\text{(geranyl)CHO} \ + \ [Bu_3Sb^+\diagdown\!\!\!\diagup]Br^- \xrightarrow[\text{72\%}]{\text{PhLi, } -78 \ °C \ \text{to RT}} \text{product with OH} \quad (26)$$

$$Bu_4Sb\diagdown\!\!\!\diagup \ + \ Ph\underset{}{\overset{O}{\diagup\!\!\!\diagdown}} \xrightarrow[\text{85\%}]{\text{AlCl}_3} \underset{}{\overset{Ph \ \ OH}{\diagdown\!\!\!\diagup}} \quad (27)$$

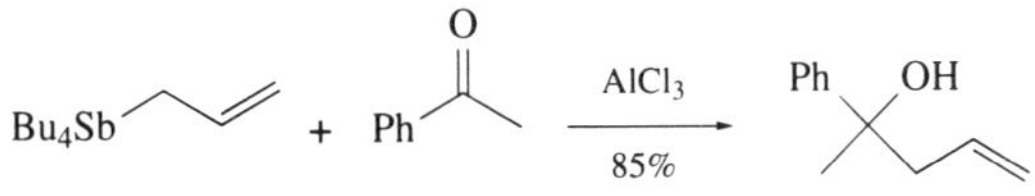

The regioselectivity in the reactions of silylated allylic organoantimony compounds has been investigated.[60] It was found that the reaction of trimethylsilylallylic organoantimony compounds with aldehydes gave α-adducts exclusively when either BuLi or BuMgBr was used, and the products were

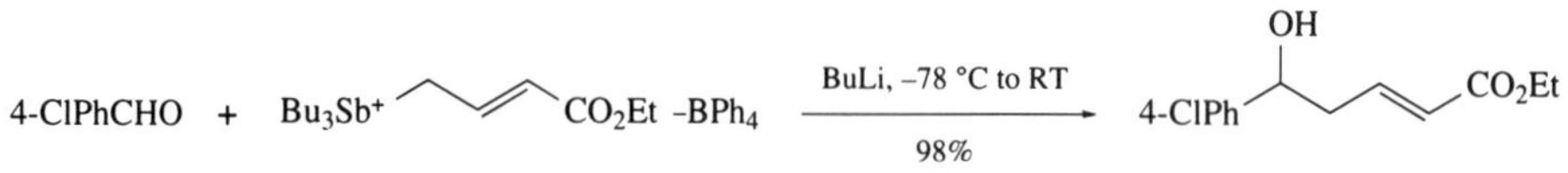

$$4\text{-ClPhCHO} + \text{Bu}_3\text{Sb}^+\diagdown\!\diagup\!\diagdown\text{CO}_2\text{Et} \; -\text{BPh}_4 \xrightarrow[98\%]{\text{BuLi, }-78\,°\text{C to RT}} 4\text{-ClPh}\underset{\text{OH}}{\diagdown}\!\diagup\!\diagdown\text{CO}_2\text{Et} \quad (28)$$

the (*E*) isomers, irrespective of whether the organoantimony reagent was the (*E*) or (*Z*) isomer. The reaction is chemoselective for aldehydes (Equation (29)).

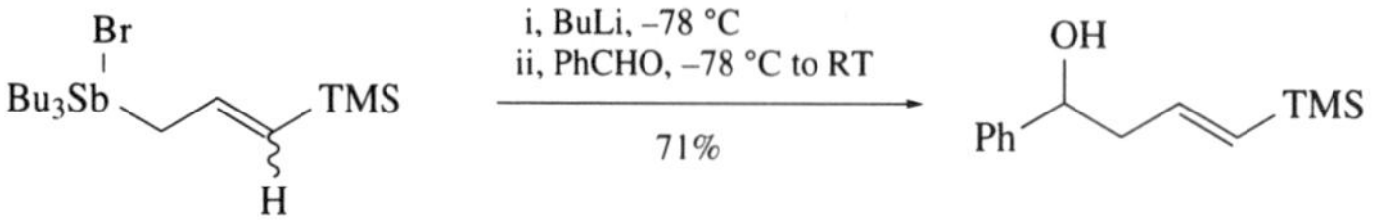

$$\underset{\text{H}}{\overset{\text{Br}}{\text{Bu}_3\text{Sb}\diagdown\!\diagup\!\diagdown\text{TMS}}} \xrightarrow[71\%]{\text{i, BuLi, }-78\,°\text{C}\;\;\text{ii, PhCHO, }-78\,°\text{C to RT}} \text{Ph}\underset{\text{OH}}{\diagdown}\!\diagup\!\diagdown\text{TMS} \quad (29)$$

Reaction of tributylstibine with propargyl bromide gives the allenyltributylstibonium bromide and not the propargyltributylstibonium bromide.[61] The structure of the allenyl stibonium salt was fully characterized by ^{1}H NMR and IR spectroscopy. The allenic salt is remarkably stable and does not react with aldehydes even on heating to 120 °C. However, it can be readily converted into the allenyl tetrabutylstiborane by reaction with butylmagnesium bromide in THF at low temperature. The pentaorganylstiborane thus produced reacts smoothly with various aldehydes to give homopropargylic alcohols exclusively in excellent yields (Scheme 5). By contrast, the reaction of tributylstibine with 1-bromo-2-butyne and with trimethylsilylpropargyl bromide gives the corresponding alkynic stibonium salts, and not the allenic stibonium salts. Treatment of the stibonium salts with BuLi or BuMgBr gives the corresponding pentaorganylstiboranes, which react with various aldehydes to give allenic alcohols and alkynic alcohols, respectively (Schemes 6 and 7).[62]

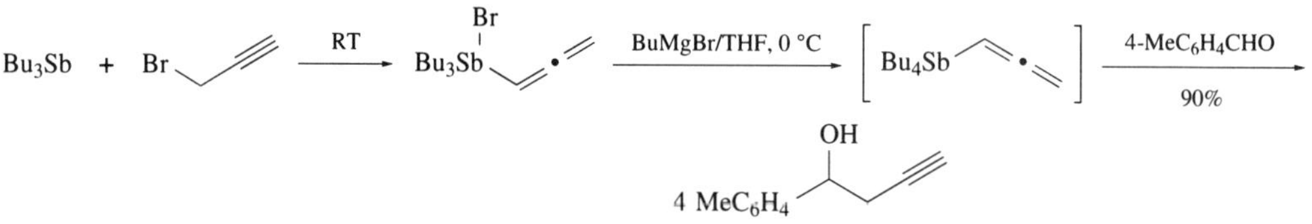

Scheme 5

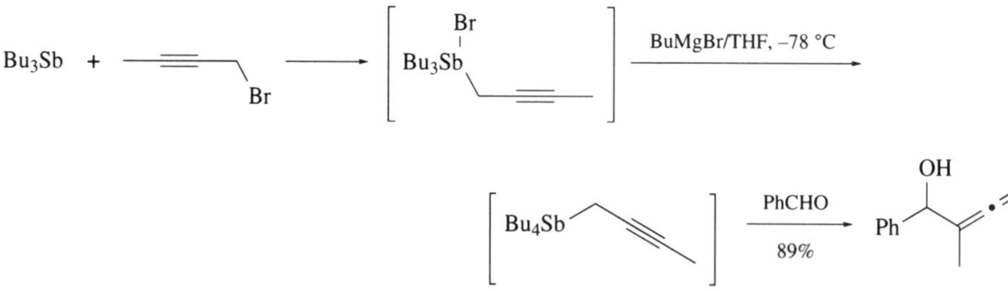

Scheme 6

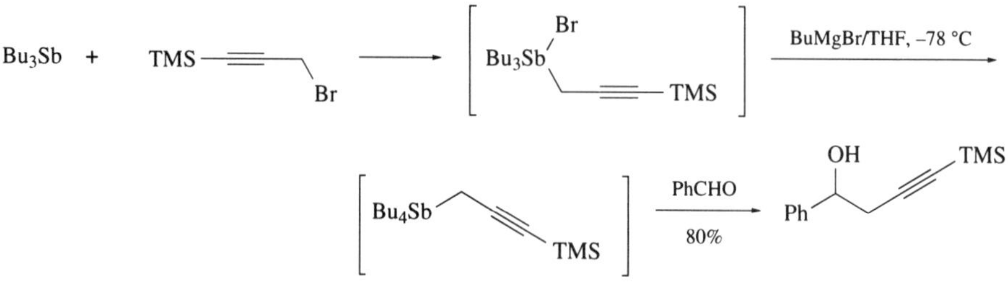

Scheme 7

[(Methoxycarbonyl)methyl]tributylstibonium bromide, after treatment with BuLi in THF at −78 °C, reacts with benzaldehyde to give a mixture of methyl β-phenyl-β-hydroxypropionate (65%) and methyl β-phenylacrylate (35%) in good yield. Under similar conditions, the corresponding stibonium tetraphenylborate gave methyl β-phenyl-β-hydroxypropionate exclusively (Equation (30)).[58] This method can also be applied to the synthesis of β-aryl-β-hydroxypropionitriles, starting from (cyanomethyl)tributylstibonium bromide. In addition to the expected β-hydroxypropionitrile, α,β-unsaturated nitriles are also produced as minor products (Equation (31)).[58]

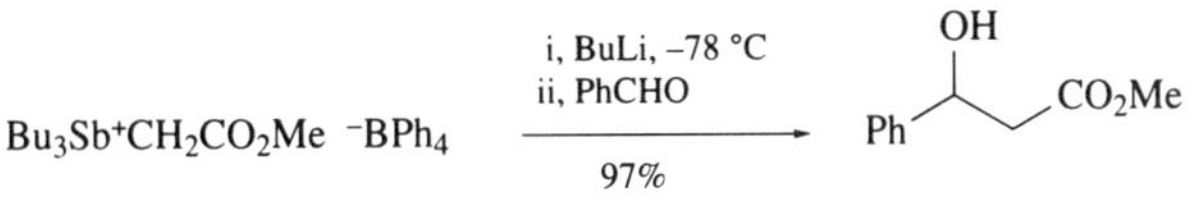

$$Bu_3Sb^+CH_2CO_2Me \ ^-BPh_4 \xrightarrow[97\%]{\text{i, BuLi, }-78\,°C; \text{ ii, PhCHO}} Ph\underset{OH}{\overset{}{\diagup}}CO_2Me \quad (30)$$

$$Bu_3Sb^+CH_2CN \ Br^- \xrightarrow[]{\text{i, BuLi; ii, PhCHO}} Ph\underset{OH}{\overset{}{\diagup}}CN \ (95\%) \ + \ Ph\diagup\!\!\!\diagdown CN \ (3\%) \quad (31)$$

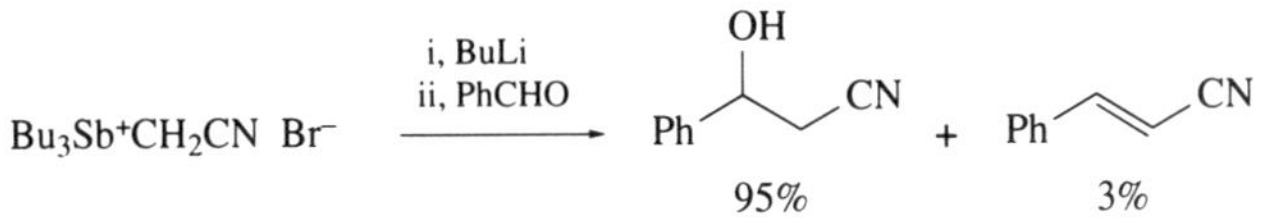

12.2.2.3 Antimony- and SbCl$_3$-metal-mediated synthesis of alcohols

The allylation of aldehydes to homoallylic alcohols is of synthetic importance and a number of metals have been used to promote Barbier-type allylation of aldehydes with allyl halides.

Metallic antimony has been found to induce allylation of aldehydes by allylic halides or phosphates (Equation (32)).[63] α,β-Unsaturated aldehydes afford 1,2-addition products selectively. The reaction is highly regiospecific, giving products coupled only at the γ-position of the allylic system, and is chemoselective.

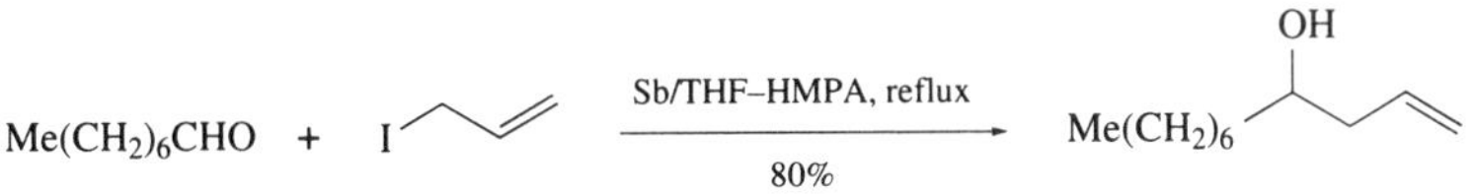

$$Me(CH_2)_6CHO \ + \ I\diagup\!\!\!\diagdown \xrightarrow[80\%]{\text{Sb/THF–HMPA, reflux}} Me(CH_2)_6\underset{OH}{\overset{}{\diagup}}\!\!\!\diagdown \quad (32)$$

The active zero-valent antimony generated *in situ* from SbCl$_3$–Fe or SbCl$_3$–Al can induce allylation of aldehydes with allylic iodides at room temperature to give high yields of the corresponding homoallylic alcohols with high regio- and chemoselectivity.[64] The less reactive allyl bromide reacts with the same aldehydes to give homoallylic alcohols at 60 °C. However, addition of sodium iodide allowed the reaction temperature to be reduced to ambient. Metallic aluminum, when it replaced metallic iron, was also effective in a DMF–H$_2$O (3:1) medium (Equations (33) and (34)). Similarly, metallic antimony can induce the reaction of cinnamyl bromide with aldehydes, giving products coupled at the γ-position of the allylic system (Equation (35)).[63]

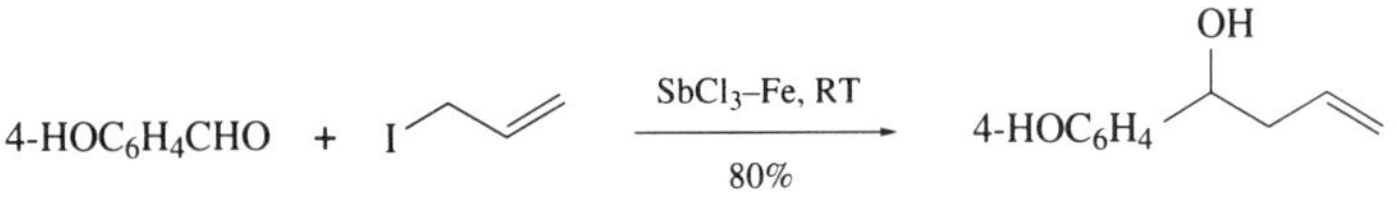

$$4\text{-HOC}_6H_4CHO \ + \ I\diagup\!\!\!\diagdown \xrightarrow[80\%]{\text{SbCl}_3\text{–Fe, RT}} 4\text{-HOC}_6H_4\underset{OH}{\overset{}{\diagup}}\!\!\!\diagdown \quad (33)$$

$$Ph\diagup\!\!\!\diagdown CHO \ + \ Br\diagup\!\!\!\diagdown \xrightarrow[95\%]{\text{NaI/SbCl}_3\text{–Fe, DMF–H}_2\text{O}} Ph\diagup\!\!\!\diagdown\underset{OH}{\overset{}{\diagup}}\!\!\!\diagdown \quad (34)$$

$$PhCHO \ + \ Br\diagup\!\!\!\diagdown Ph \xrightarrow[65\%]{\text{Sb}} Ph\underset{Ph}{\overset{OH}{\diagup}}\!\!\!\diagdown \quad (35)$$

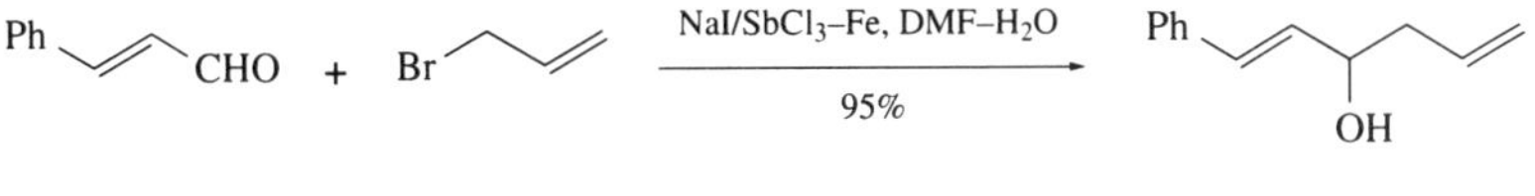

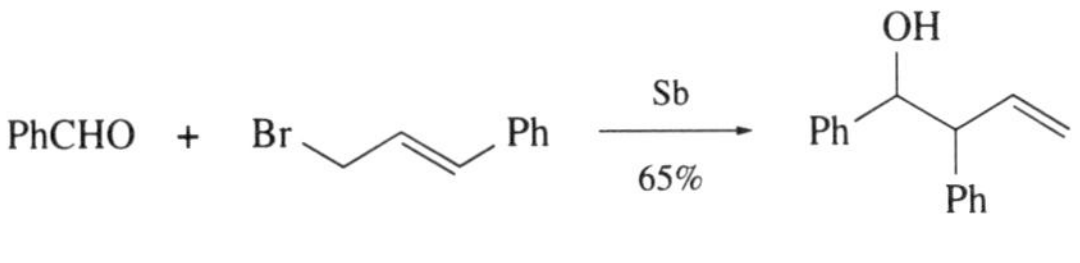

SbCl$_3$–Al/DMF–H$_2$O and SbCl$_3$–Zn/DMF–H$_2$O have been found to be efficient reduction systems for the conversion of a variety of aldehydes to alcohols.[65,66] When the reaction was carried out in DMF–D$_2$O instead of DMF–H$_2$O, the corresponding deuterium-labeled alcohol was obtained (Equations (36) and (37)).

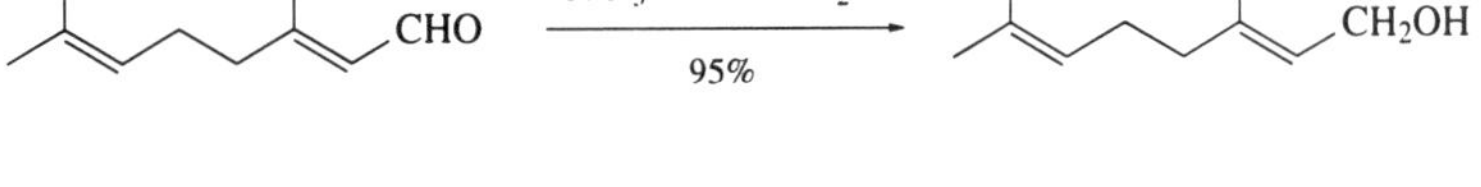

$$\text{geranial-CHO} \xrightarrow[95\%]{\text{SbCl}_3\text{–Zn/DMF–H}_2\text{O}} \text{geranyl-CH}_2\text{OH} \quad (36)$$

$$4\text{-ClC}_6\text{H}_4\text{CHO} \xrightarrow[\substack{\text{or SbCl}_3\text{-Zn/DMF-D}_2\text{O} \\ 98\%}]{\text{SbCl}_3\text{-Al/DMF-D}_2\text{O}} 4\text{-ClC}_6\text{H}_4\text{CHDOH} \qquad (37)$$

12.2.2.4 Diphenylstibine-mediated reduction of carbonyl compounds

Many metal hydrides and organometallic hydrides can reduce carbonyl compounds, and each hydride has its own characteristic features. In the presence of Lewis acids, diphenylstibine reacts with carbonyl compounds under mild conditions to give alcohols in excellent yields (Equations (38) and (39)).[67]

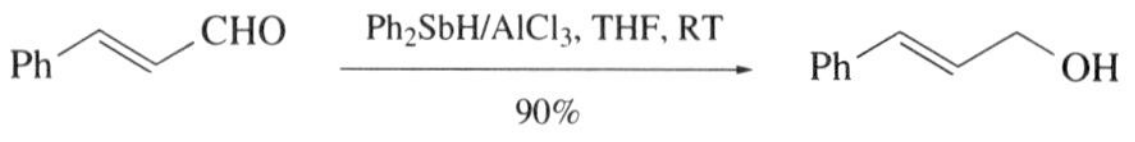

$$\text{Ph-CH=CH-CHO} \xrightarrow[90\%]{\text{Ph}_2\text{SbH/AlCl}_3, \text{THF, RT}} \text{Ph-CH=CH-CH}_2\text{OH} \qquad (38)$$

$$\text{Ph-CH(OH)-C(=O)-Ph} \xrightarrow[89\%]{\text{Ph}_2\text{SbH/AlCl}_3, \text{THF, RT}} \text{Ph-CH(OH)-CH(OH)-Ph} \qquad (39)$$

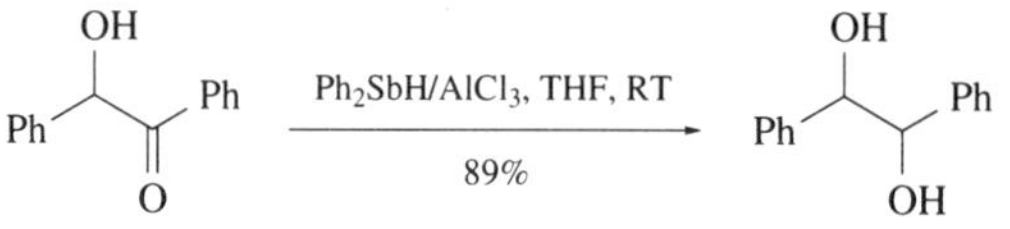

12.2.2.5 Miscellaneous formation of alcohols via organoantimony compounds

Esters of carboxylic acids and thiohydroxamic acid react with trisphenylthioantimony in the presence of oxygen and water to give high yields of nor-alcohols.[68] The reaction appears to proceed by a radical chain mechanism in which a carbon radical attacks the stibine and forms an oxygen-sensitive organoantimony compound. Aerial oxidation and subsequent hydrolysis of the intermediate organometallic species produces the nor-alcohols. This method can be applied to a variety of substrates (Equation (40)).

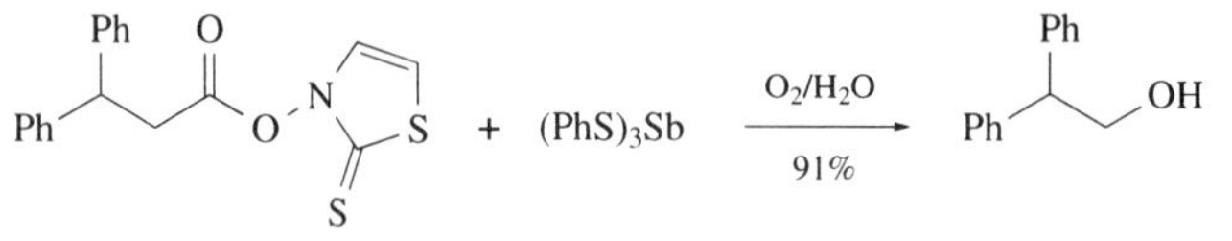

$$\text{PhCH(Ph)CH}_2\text{C(O)O-N(thiazole-S)} + (\text{PhS})_3\text{Sb} \xrightarrow[91\%]{\text{O}_2/\text{H}_2\text{O}} \text{PhCH(Ph)CH}_2\text{OH} \qquad (40)$$

The reaction of (dimethylamino)dimethylstibine with cyclic carbonates or thiolocarbonates has been shown to take place by addition of the Sb–N bond across the oxygen–carbonyl of carbonates to give insertion products, which afford carbonates containing the hydroxy group after hydrolysis.[69] (Dimethylamino)dimethylstibine reacts with lactones and epoxides to give 2-substituted alkoxy metal derivatives, hydrolysis of which gives amide alcohols or amine alcohols (Scheme 8).[70]

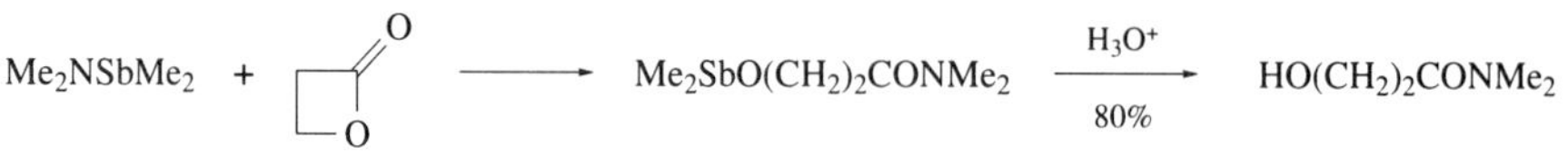

$$\text{Me}_2\text{NSbMe}_2 + \text{β-propiolactone} \longrightarrow \text{Me}_2\text{SbO(CH}_2)_2\text{CONMe}_2 \xrightarrow[80\%]{\text{H}_3\text{O}^+} \text{HO(CH}_2)_2\text{CONMe}_2$$

Scheme 8

12.2.3 Synthesis of Ketones via Organoantimony Compounds

12.2.3.1 Formation of ketones via pentaorganylstiboranes

Pentaorganylstiboranes are normally unreactive toward ketones but are very reactive toward aldehydes. The reaction of pentaorganylstiboranes with acyl chlorides should therefore produce the corresponding ketones.[59] This is so, and in the absence of any additional catalyst, benzyltetrabutylstiborane reacts with acyl chlorides to give the corresponding benzyl ketones in good yields (Equation (41)). Only the benzyl ketone was obtained from this reaction; no butyl ketone, resulting from coupling of a butyl group with the acyl chloride, was detected. Similarly, crotyltetrabutylstiborane reacts with acyl chlorides to give the corresponding α-methyl allyl ketones in good yields, rather than the crotyl ketones (Equation (42)). This result is different from that seen with crotyltins. Other pentaorganylstiboranes also react with acyl chlorides to give good yields of the corresponding ketones (Equations (43) and (44)).[59]

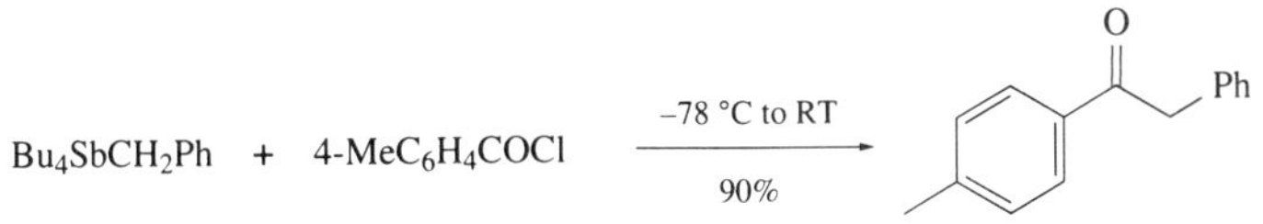

$$Bu_4SbCH_2Ph + 4\text{-}MeC_6H_4COCl \xrightarrow[90\%]{-78\ °C\ to\ RT} 4\text{-}MeC_6H_4C(O)CH_2Ph \qquad (41)$$

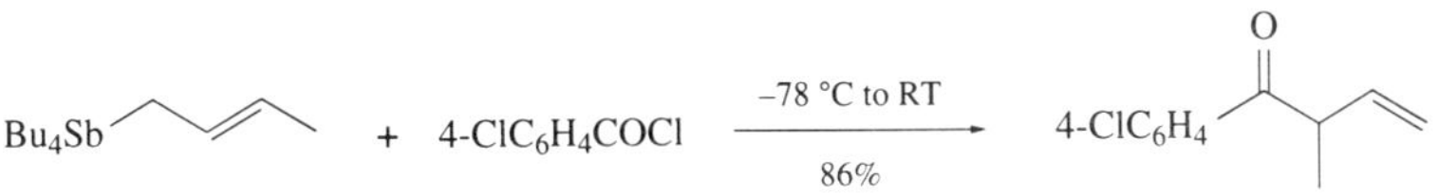

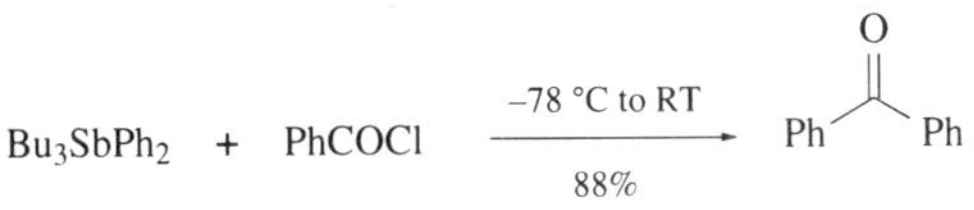

$$Bu_4Sb\text{-}CH_2CH=CHMe + 4\text{-}ClC_6H_4COCl \xrightarrow[86\%]{-78\ °C\ to\ RT} 4\text{-}ClC_6H_4C(O)CH(Me)CH=CH_2 \qquad (42)$$

$$Bu_3SbPh_2 + PhCOCl \xrightarrow[88\%]{-78\ °C\ to\ RT} PhC(O)Ph \qquad (43)$$

$$Bu_3SbMe_2 + PhCOCl \xrightarrow[79\%]{-78\ °C\ to\ RT} PhC(O)Me \qquad (44)$$

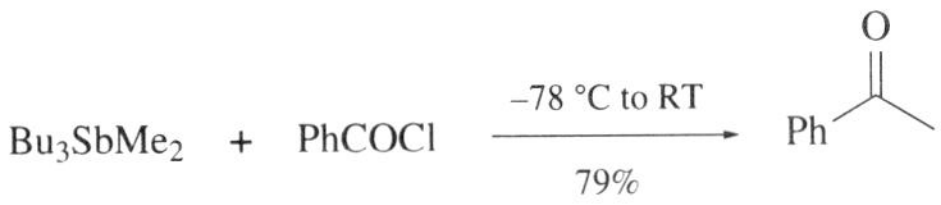

In the presence of a catalytic amount of palladium, dipropynyltrimethylantimony couples with acid chlorides to give only propynyl ketones (Equation (45)).[70]

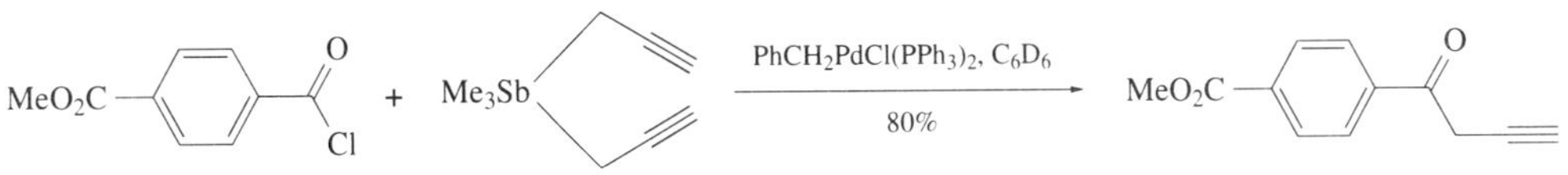

$$MeO_2C\text{-}C_6H_4\text{-}COCl + Me_3Sb(CH_2C{\equiv}CH)_2 \xrightarrow[80\%]{PhCH_2PdCl(PPh_3)_2,\ C_6D_6} MeO_2C\text{-}C_6H_4\text{-}C(O)CH_2C{\equiv}CH \qquad (45)$$

12.2.3.2 Pentavalent organoantimony compounds as oxidizing agents

It has been found that, in the presence of two equivalents of base, α-hydroxyketones are smoothly oxidized to the corresponding α-dicarbonyl compounds by triphenylantimony dibromide (Equation (46)).[71] Triphenylantimony diacetate will also oxidize benzoin to benzil in high yield under mild conditions. However, organoantimony(V) compounds such as Ph_3SbBr_2 or Ph_2SbBr_3 do not oxidize benzyl alcohol, even in the presence of a variety of bases. In the presence of diethylamine, benzyl alcohols are readily transformed to benzyloxydiphenylstibines, which are oxidized to benzylaldehydes by addition of bromine (Equation (47)).[72]

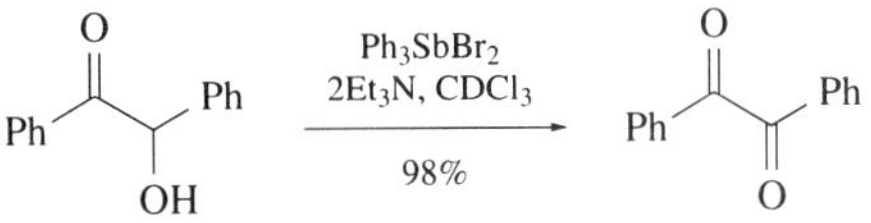

$$PhC(O)CH(OH)Ph \xrightarrow[98\%]{Ph_3SbBr_2,\ 2Et_3N,\ CDCl_3} PhC(O)C(O)Ph \qquad (46)$$

$$3\text{-}MeC_6H_4CH_2OH \xrightarrow[86\%]{i,\ Ph_2SbBr/Et_2NH;\ ii,\ Br_2} 3\text{-}MeC_6H_4CHO \qquad (47)$$

12.2.3.3 Tributylstibine-assisted synthesis of alkyl aryl sulfones

In the presence of tributylstibine, toluene-4-sulfonyl chloride reacts with various alkyl halides to give alkyl aryl sulfones in good yields (Equations (48) and (49)).[73] In the case of alkyl iodides, smooth reactions are observed, while bromides are less reactive and chlorides are not effective at all. No O-alkylation products are obtained under the conditions employed.

$$4\text{-MeC}_6\text{H}_4\text{SO}_2\text{Cl} \quad + \quad \text{MeI} \quad \xrightarrow[90\%]{\text{Bu}_3\text{Sb, RT}} \quad 4\text{-MeC}_6\text{H}_4\text{SO}_2\text{Me} \quad (48)$$

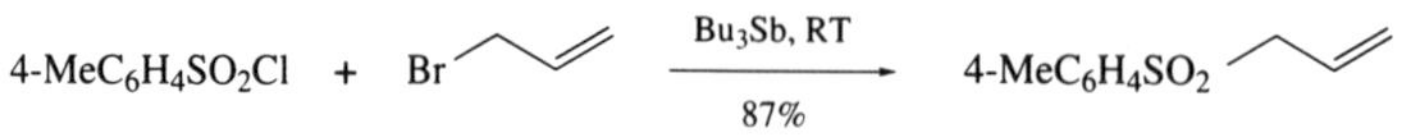

$$4\text{-MeC}_6\text{H}_4\text{SO}_2\text{Cl} \quad + \quad \text{Br}\diagup\!\!\!\diagdown \quad \xrightarrow[87\%]{\text{Bu}_3\text{Sb, RT}} \quad 4\text{-MeC}_6\text{H}_4\text{SO}_2\diagup\!\!\!\diagdown \quad (49)$$

12.2.4 Organoantimony Compound-mediated Acetalization Reactions

Although acetalization has been performed by many methods,[74] the use of organometallic compounds has seldom been reported. Organoantimony compounds can be used to effect this transformation.

12.2.4.1 Trialkoxystibine-promoted acetalization of aldehydes

Acetalization is normally accomplished with alcohols or orthoformate esters in an acidic medium. In the presence of allyl bromide, triethoxystibine reacts with aldehydes to afford the corresponding diethyl acetals. This method can be applied to acid-sensitive or protic-solvent-sensitive aldehydes, and is chemoselective for aldehydes and gives high yields in a variety of cases (Equation (50)).[75]

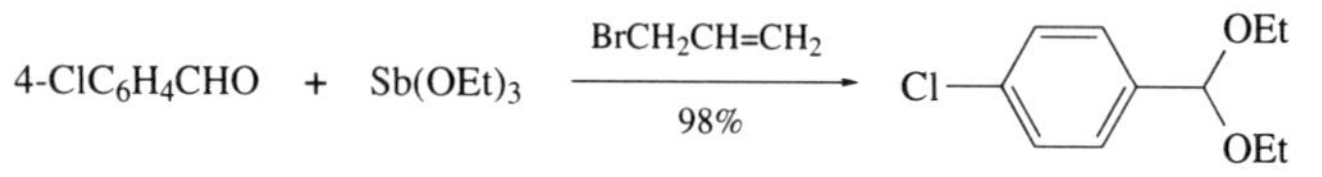

$$4\text{-ClC}_6\text{H}_4\text{CHO} \quad + \quad \text{Sb(OEt)}_3 \quad \xrightarrow[98\%]{\text{BrCH}_2\text{CH=CH}_2} \quad \text{Cl-C}_6\text{H}_4\text{-CH(OEt)}_2 \quad (50)$$

12.2.4.2 SbCl₃-metal-mediated acetalization

SbCl$_3$–Al or SbCl$_3$–Fe can promote the acetalization of carbonyl compounds with methyl or ethyl alcohol, but isopropyl alcohol does not react. The acetals are obtained in almost quantitative yield using a catalytic amount of SbCl$_3$ (Equations (51) and (52)).[66]

$$\text{hexan-2-one} \quad + \quad \text{MeOH} \quad \xrightarrow[98\%]{\text{SbCl}_3 \text{ catalyst–Al}} \quad \text{2,2-dimethoxyhexane} \quad (51)$$

$$\text{citronellal-type CHO} \quad + \quad \text{MeOH} \quad \xrightarrow[95\%]{\text{SbCl}_3 \text{ catalyst–Al}} \quad \text{dimethyl acetal} \quad (52)$$

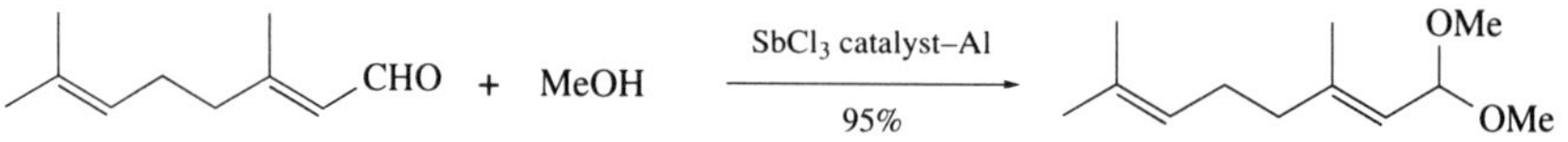

12.2.5 Cyclopropanation via Organoantimony Compounds

Cyclopropane derivatives activated by two electron-withdrawing substituents at a geminal position, the so-called electrophilic cyclopropanes, are useful intermediates in organic synthesis. Their inter- and intramolecular ring-opening reactions by various nucleophiles have been studied intensively and applied to the synthesis of several natural products.[76]

The reaction of dibromomalonic ester with tributylstibine is exothermic and gives alkene products in good yield when carried out in the presence of carbonyl compounds. In cases where the substrate was an electron-deficient terminal alkene, cyclopropane derivatives could be obtained.[77,78]

Terminal alkenes, such as methyl vinyl ketone, acrolein, acrylonitrile and acrylic esters, and cyclic α,β-unsaturated ketones, such as 2-cyclopentenone and 2-cyclohexenone, also react with dibromomalonate to give bicyclic compounds (Equations (53) and (54)). The proposed reaction mechanism is that an ion pair is the reactive intermediate.[78] In order to prove that the reaction pathway is not through a carbene intermediate, the following reaction was performed. Dimethyl dibromomalonate was reacted with 1-dicyclopentadienone promoted by tributylstibine to give tetracyclo[4.4.0^{3,5}.1^{7,10}]-4,4-bis(methoxycarbonyl)undec-8-en-2-one exclusively (Equation (55)). No product of addition at the isolated double bond was obtained.

Dibromocyanoacetic ester also reacts with electron-deficient alkenes to give cyclopropane derivatives, although the yields are slightly lower (Equation (56)). Both isomeric cyclopropanes were obtained from acrylonitrile (Equation (57)).[78]

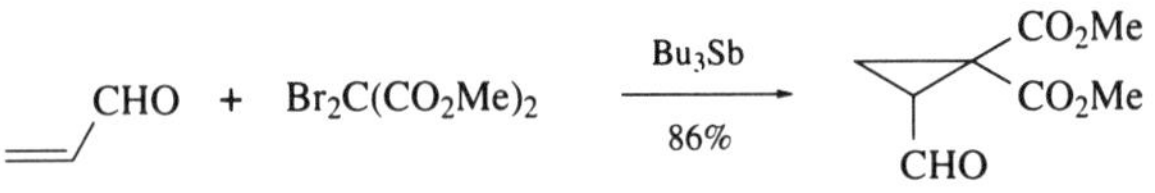

$$\text{CHO} + \text{Br}_2\text{C}(\text{CO}_2\text{Me})_2 \xrightarrow[86\%]{\text{Bu}_3\text{Sb}} \text{cyclopropane-CHO with C(CO}_2\text{Me})_2 \quad (53)$$

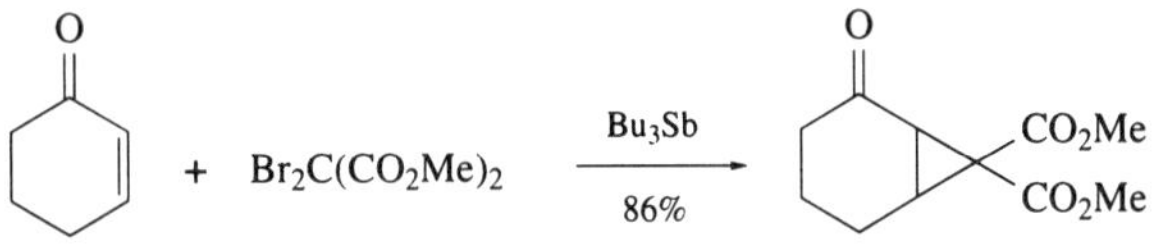

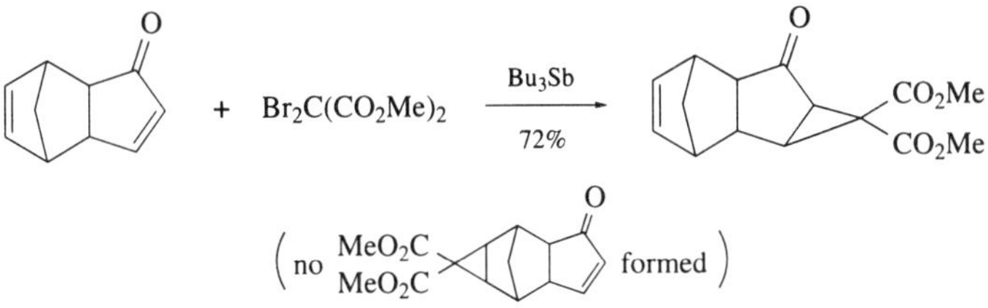

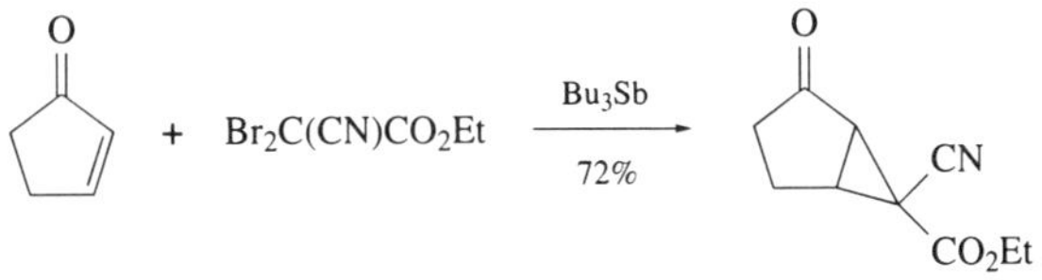

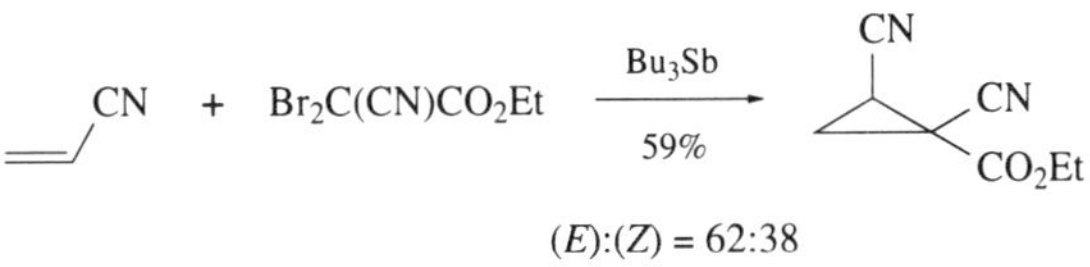

$$\text{cyclohexenone} + \text{Br}_2\text{C}(\text{CO}_2\text{Me})_2 \xrightarrow[86\%]{\text{Bu}_3\text{Sb}} \text{bicyclic product} \quad (54)$$

$$\text{tricyclic enone} + \text{Br}_2\text{C}(\text{CO}_2\text{Me})_2 \xrightarrow[72\%]{\text{Bu}_3\text{Sb}} \text{tetracyclic product} \quad (55)$$

(no MeO$_2$C / MeO$_2$C isomer formed)

$$\text{cyclopentenone} + \text{Br}_2\text{C}(\text{CN})\text{CO}_2\text{Et} \xrightarrow[72\%]{\text{Bu}_3\text{Sb}} \text{bicyclic CN/CO}_2\text{Et product} \quad (56)$$

$$\text{CH}_2=\text{CHCN} + \text{Br}_2\text{C}(\text{CN})\text{CO}_2\text{Et} \xrightarrow[59\%]{\text{Bu}_3\text{Sb}} \text{cyclopropane with CN, CN, CO}_2\text{Et} \quad (57)$$

$(E):(Z) = 62:38$

Reaction of ethyl dibromobenzeneacetate with electron-deficient alkenes gave cyclopropanes in low to moderate yield, and both isomers were obtained (Equation (58)).[69] The lower yield in this case has been shown to be the result of side reactions which lead to the formation of two reduced products as well as a coupled compound.

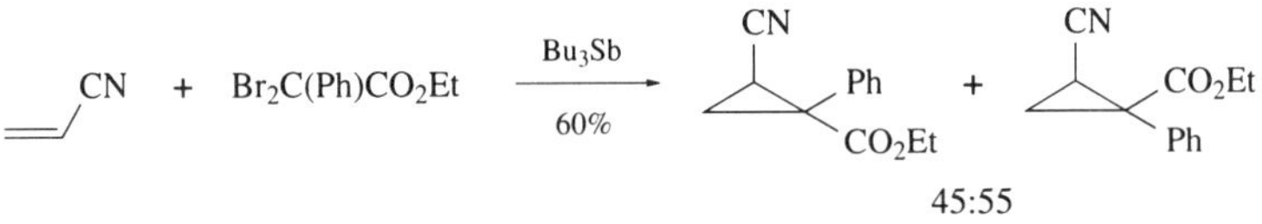

$$\text{CH}_2=\text{CHCN} + \text{Br}_2\text{C}(\text{Ph})\text{CO}_2\text{Et} \xrightarrow[60\%]{\text{Bu}_3\text{Sb}} \text{CN/Ph/CO}_2\text{Et cyclopropane} + \text{CN/CO}_2\text{Et/Ph cyclopropane} \quad (58)$$

45:55

12.2.6 Organoantimony(V) Halides as Catalysts in Organic Synthesis

12.2.6.1 Synthesis of silylated enol ethers

Tetraphenylantimony bromide has been found to be an excellent catalyst for the synthesis of trimethylsilyl enol ethers from ketones, trimethylsilyl bromide and aziridines (Equation (59)).[79] In the absence of the antimony catalyst, the yield of the enol ether was only 63%. The reactions are quite regioselective and the more highly substituted enol ether is favored (Equation (60)).

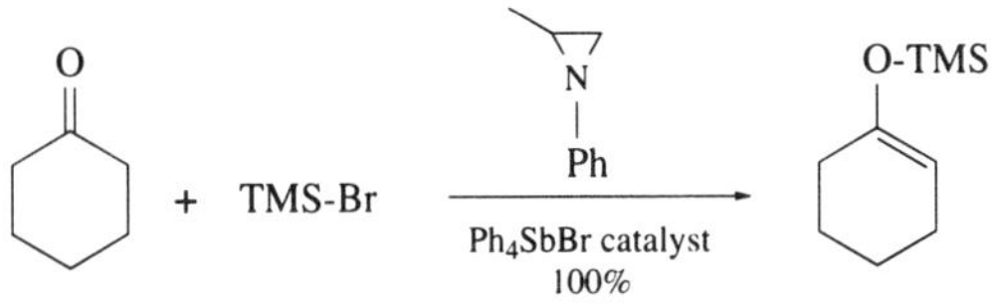

$$\text{cyclohexanone} + \text{TMS-Br} \xrightarrow[\substack{\text{Ph}_4\text{SbBr catalyst} \\ 100\%}]{\text{N-Ph aziridine}} \text{cyclohexenyl-O-TMS} \quad (59)$$

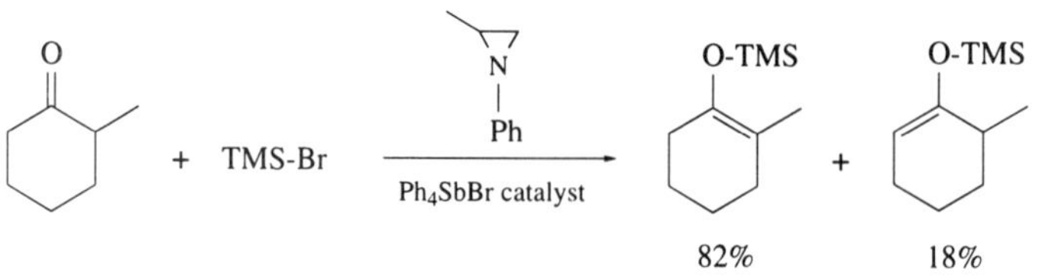

12.2.6.2 Cycloaddition of substituted aziridines or oxiranes with heterocumulenes

In the presence of a catalytic amount of tetraphenylantimony iodide, the condensation of monosubstituted oxiranes with heterocumulenes such as isocyanates or carbodiimides leads to the predominant formation of 3,4-disubstituted oxazolidine-2-ones or oxazolidine-2-imines, respectively (Equation (61)).[80,81] Similarly, in the presence of catalytic amounts of organoantimony(V) halides such as Ph_4SbI Ph_4SbBr, Ph_3SbBr_2 and Ph_3SbCl_2, the cycloaddition of aziridines with heterocumulenes selectively gives ring-expanded cycloadducts by cleavage of the aziridine rings (Equation (62)).[82]

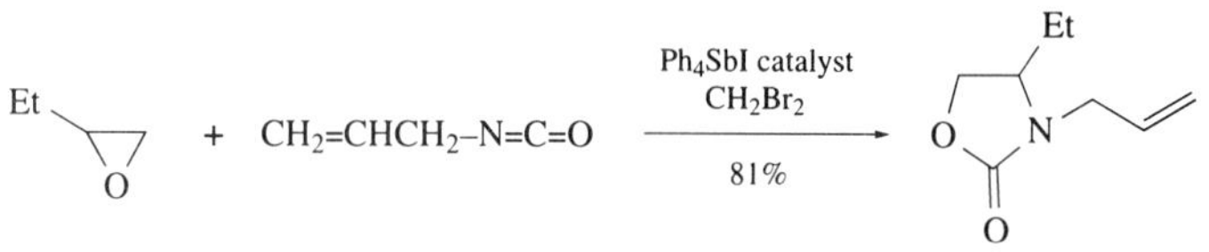

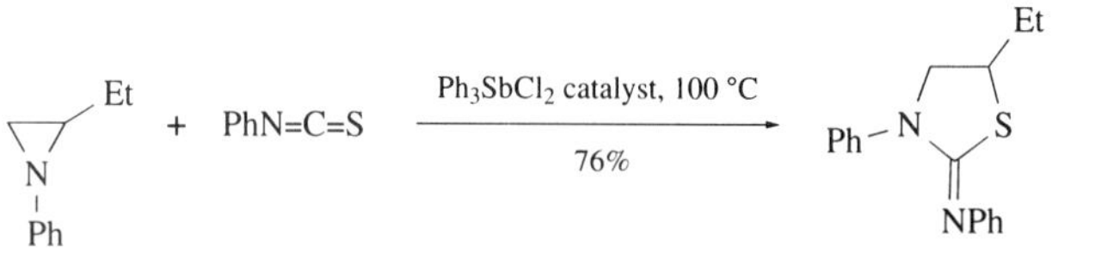

12.2.7 Antimony Salts as Catalysts in Organic Synthesis

12.2.7.1 $SbCl_5$-catalyzed reactions

In the presence of a catalytic amount of antimony(V) chloride or the antimony(V) salt generated from antimony(V) chloride and silver hexafluoroantimonate, the pinacol rearrangement of a variety of 1,2-diols or their trimethylsilyl ethers proceeds smoothly to give the corresponding ketones in good yields (Equations (63) and (64)).[83]

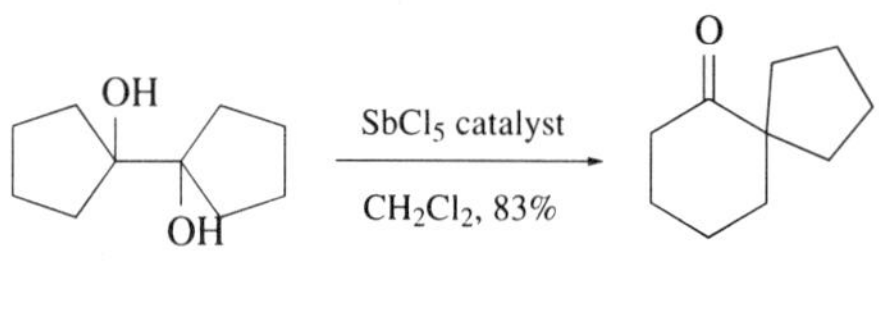

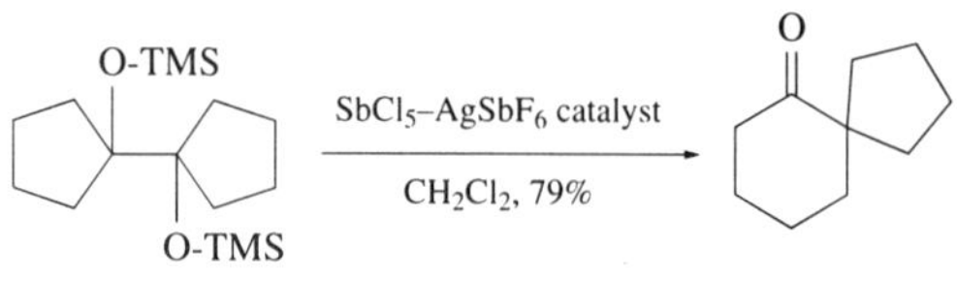

α,β-Unsaturated thioesters react with silyl enol ethers in the presence of antimony(V) chloride and tin(III) triflate to give the corresponding Michael adducts stereoselectively in high yields (Equation (65)).[84]

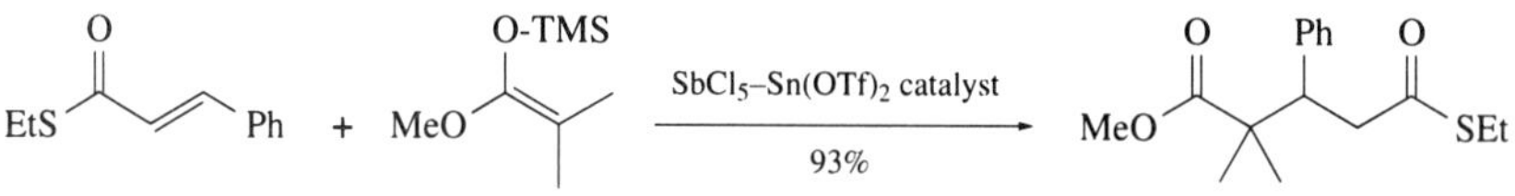

In the presence of a catalytic amount of an antimony(V) salt, the Beckmann rearrangement of several ketoxime trimethylsilyl ethers proceeds smoothly to give good yields of the corresponding amides or lactams (Equation (66)).[85]

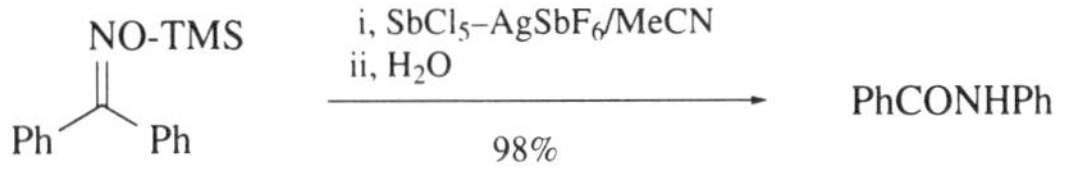

$$\text{Ph}_2\text{C=N-O-TMS} \xrightarrow[\text{98\%}]{\substack{\text{i, SbCl}_5\text{-AgSbF}_6/\text{MeCN} \\ \text{ii, H}_2\text{O}}} \text{PhCONHPh} \quad (66)$$

Similarly, in the presence of a catalytic amount of trityl hexafluoroantimonate, sequential reactions, rearrangement and reductive condensation of epoxides proceed smoothly to give the corresponding ethers in fairly good yields (Equation (67)).[86]

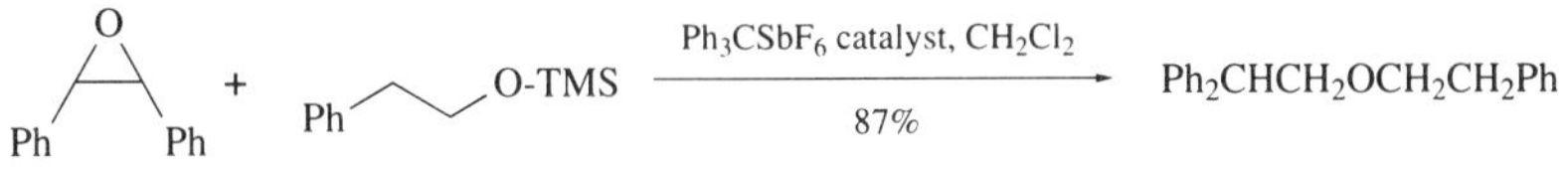

$$\text{Ph-epoxide-Ph} + \text{PhCH}_2\text{CH}_2\text{-O-TMS} \xrightarrow[\text{87\%}]{\text{Ph}_3\text{CSbF}_6 \text{ catalyst, CH}_2\text{Cl}_2} \text{Ph}_2\text{CHCH}_2\text{OCH}_2\text{CH}_2\text{Ph} \quad (67)$$

12.2.7.2 SbCl$_3$-catalyzed reactions

Lewis acid–base reaction of SbCl$_3$ with DAST ((diethylamino)sulfur trifluoride) can be used to catalyze the conversion of sulfoxides to α-fluorothioethers (Equation (68)).[87] The reagent combination LAH–SbCl$_3$ has been found to be more efficient for the conjugate reduction of 2-butene-1,4-diones than the combination of LAH with other metal halides (Equation (69)).[88]

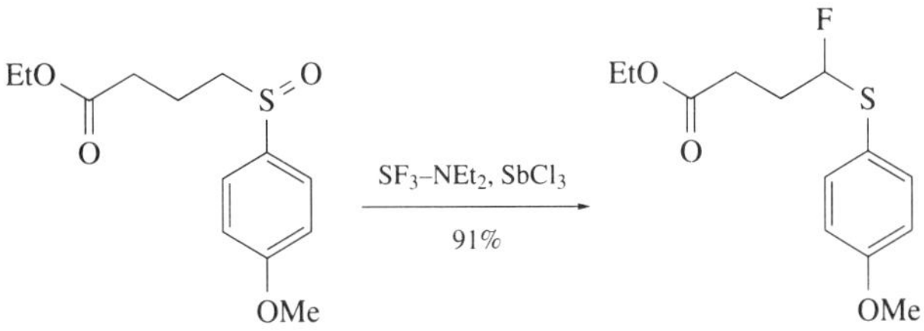

$$\text{(sulfoxide)} \xrightarrow[\text{91\%}]{\text{SF}_3\text{-NEt}_2, \text{ SbCl}_3} \text{(α-fluorothioether)} \quad (68)$$

$$\text{(enedione)} \xrightarrow[\text{94\%}]{\text{LAH–SbCl}_3} \text{(diketone)} \quad (69)$$

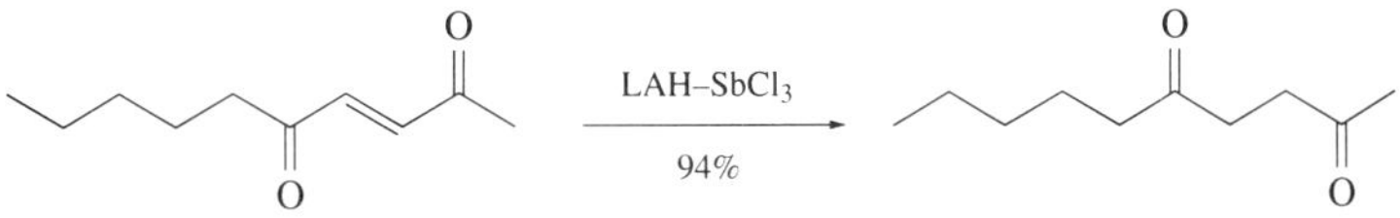

12.2.8 Miscellaneous Reactions

Triphenylantimony dicarboxylates react with a number of primary amines to give excellent yields of amides (Equation (70)).[89] These amides can also be prepared by using a catalytic amount of triphenylantimony oxide or a triphenylantimonyl dicarboxylate.

$$\text{Ph}_3\text{Sb(OAc)}_2 + \text{PhCH}_2\text{NH}_2 \xrightarrow[\text{97\%}]{50\,°\text{C}} \text{MeCONHCH}_2\text{Ph} \quad (70)$$

Triphenylstibine has been used both for dechlorination and for the formation of sulfur–sulfur linkages. Thus, the preparation of a 12-membered heterocycle with a transannular sulfur–sulfur bond has been accomplished by employing the stibine as a reducing agent (Equation (71)).[90,91] The interaction of the stibine and a dichlorodithiatriazine has been found to yield an interesting dimeric product.[92]

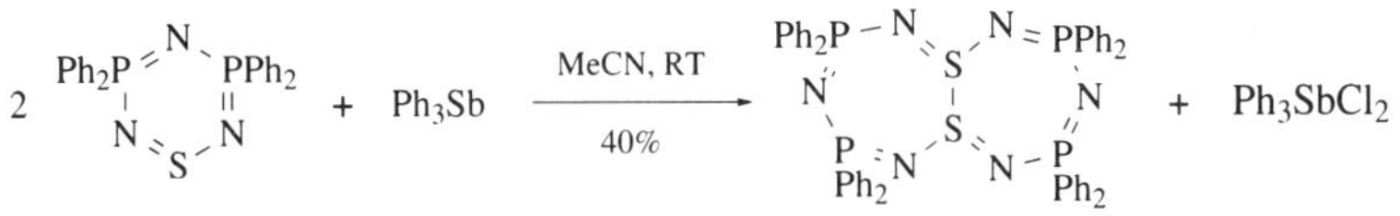

$$2\ \text{Ph}_2\text{P(=N)(N=S-N)PPh}_2 + \text{Ph}_3\text{Sb} \xrightarrow[\text{40\%}]{\text{MeCN, RT}} \text{(dimeric product)} + \text{Ph}_3\text{SbCl}_2 \quad (71)$$

The use of tributyl- or triphenylstibine for the replacement of halogen by hydrogen in aryl and benzylic bromides has been reported.[49]

Triphenylantimony has been used to mediate the Lewis acidity of TiCl$_4$.[93]

12.3 ORGANOBISMUTH COMPOUNDS IN ORGANIC SYNTHESIS

12.3.1 Organobismuth(III) Compounds

Organobismuth(III) compounds have found relatively little use in organic synthesis, because of their chemical and physiological properties. Trialkylbismuthines are readily oxidized. The triarylbismuth compounds, by contrast, are stable crystalline solids which can be handled with ease; however, they are poor nucleophiles or donors. They do not react with aldehydes or ketones or with halides. Even so, organobismuth(III) compounds can mediate several useful reactions.

12.3.1.1 Barbier-type reactions via organobismuth(III) compounds

In the presence of metallic bismuth, allyl halides react with aldehydes under mild conditions to give good yields of the corresponding homoallylic alcohols (Equation (72)).[94] This reaction is chemoselective. Similarly, in the presence of $BiCl_3$–metallic zinc or $BiCl_3$–metallic iron, allylic halides react with aldehydes under mild conditions to give the corresponding homoallylic alcohols in high yields with high chemoselectivity.[95,96] In this $BiCl_3$-mediated allylation, the desired homoallylic alcohol is not obtained at all in THF when metallic aluminum is used. However, a Barbier-type allylation of aldehydes with allylic halides can be easily effected in THF–H_2O using $BiCl_3$–metallic aluminum, even though organometallic compounds usually have to be prepared and handled in anhydrous solvents owing to rapid protonolysis.[96,97] Both aromatic and aliphatic aldehydes react smoothly to afford the desired products in good yields. When an α,β-unsaturated aldehyde is used, the 1,2-addition product is obtained selectively (Equation (73)).

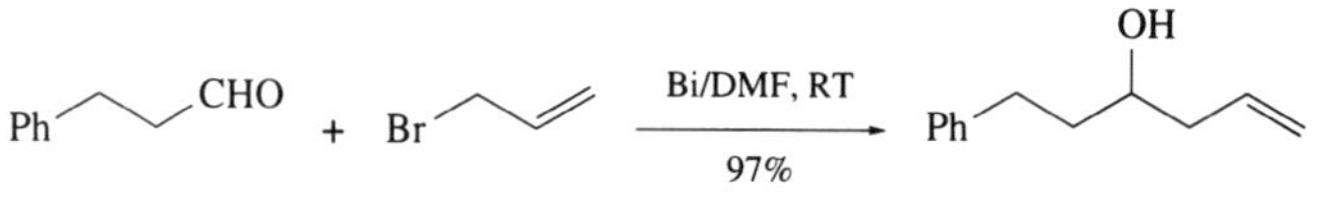

$$Ph\text{-}CH_2CH_2\text{-}CHO + Br\text{-}CH_2CH=CH_2 \xrightarrow[97\%]{\text{Bi/DMF, RT}} Ph\text{-}CH_2CH_2\text{-}CH(OH)\text{-}CH_2CH=CH_2 \quad (72)$$

$$CH_3CH=CH\text{-}CHO + Br\text{-}CH_2CH=CH_2 \xrightarrow[82\%]{\text{BiCl}_3\text{-Al} \atop \text{THF-H}_2\text{O}} CH_3CH=CH\text{-}CH(OH)\text{-}CH_2CH=CH_2 \quad (73)$$

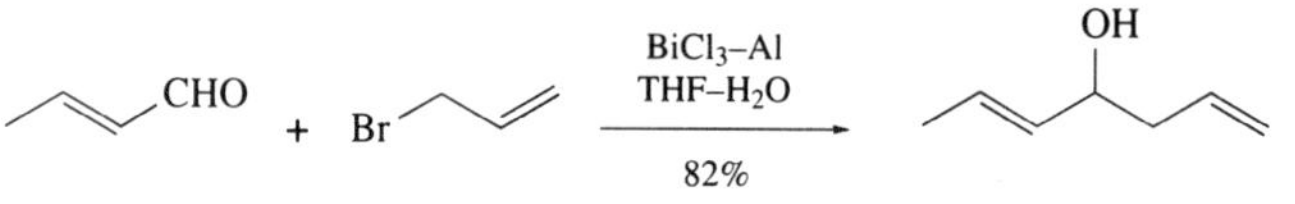

In the presence of tributylbismuthine, both ketones and aldehydes react with allyl bromide to form homoallylic alcohols together with their allylic ethers. Various aldehydes react smoothly with allyl bromide to give the desired products (Equations (74) and (75)).[98]

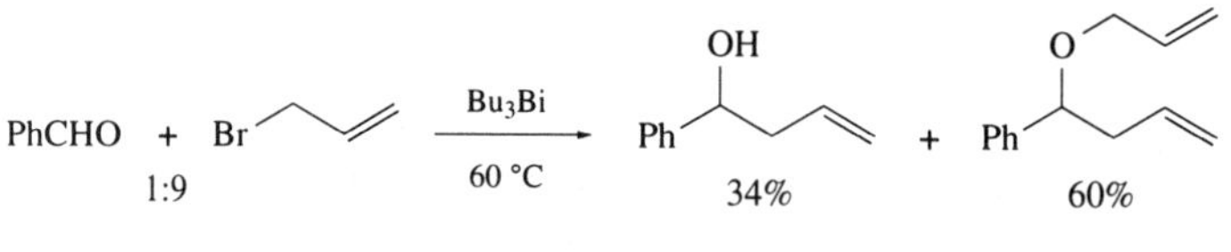

$$PhCHO + Br\text{-}CH_2CH=CH_2 \xrightarrow[60\ ^\circ\text{C}]{\text{Bu}_3\text{Bi}} Ph\text{-}CH(OH)\text{-}CH_2CH=CH_2 + Ph\text{-}CH(OCH_2CH=CH_2)\text{-}CH_2CH=CH_2 \quad (74)$$
$$\text{1:9} \qquad\qquad 34\% \qquad\qquad 60\%$$

$$\text{cyclohexanone} + Br\text{-}CH_2CH=CH_2 \xrightarrow[60\ ^\circ\text{C}]{\text{Bu}_3\text{Bi}} \text{1-allylcyclohexanol} + \text{1-allyloxy-1-allylcyclohexane} \quad (75)$$
$$\text{1:9} \qquad\qquad 58\% \qquad\qquad 21\%$$

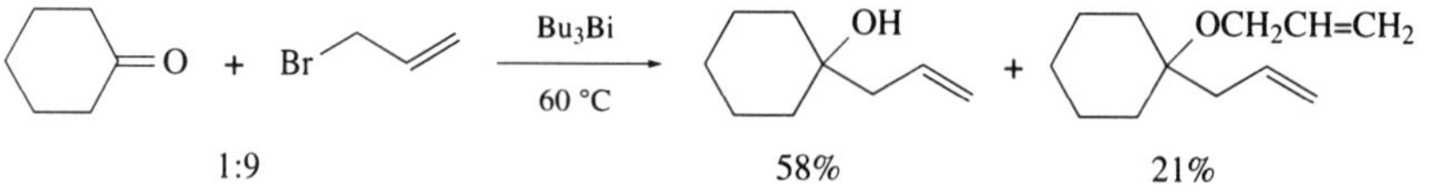

12.3.1.2 Synthesis of homoalkylated amines

In the presence of $BiCl_3$–metallic aluminum, alkyl and allyl halides react with N-(alkylamino)-benzotriazoles at room temperature in DMF–H_2O to give the corresponding homoalkylated amines in high yields (Equation (76)).[99,100]

12.3.1.3 Aldol and Michael reactions catalyzed by $BiCl_3$

A catalytic amount of $BiCl_3$ can promote aldol reactions between silyl enol ethers and aldehydes to give the corresponding aldols in good yields (Equation (77)).[101] Under the same conditions, silyl enol

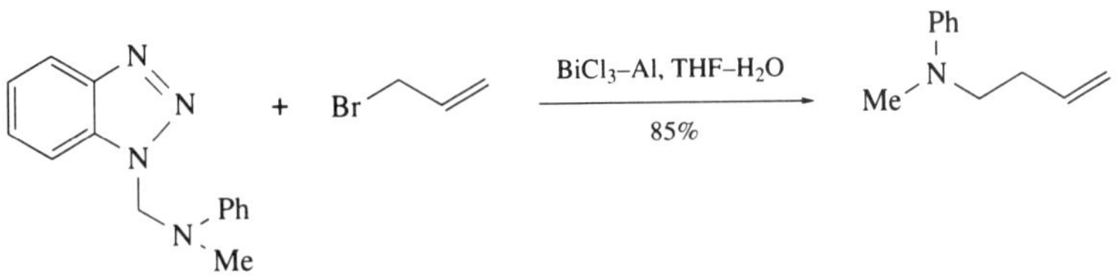

$$\text{(benzotriazole-CH}_2\text{-N(Me)Ph)} + \text{Br-CH}_2\text{CH=CH}_2 \xrightarrow[85\%]{\text{BiCl}_3\text{-Al, THF-H}_2\text{O}} \text{Me-N(Ph)-CH}_2\text{CH}_2\text{CH=CH}_2 \quad (76)$$

ethers have also been found to react with α,β-unsaturated ketones at room temperature to afford the corresponding 1,5-dicarbonyl compounds, the Michael adducts, in good yields (Equation (78)).[102]

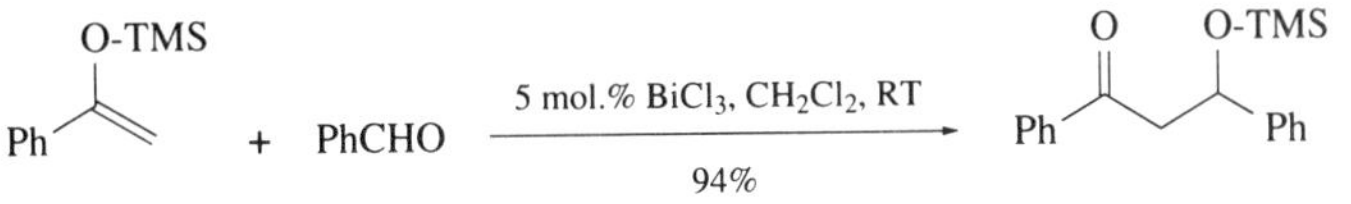

$$\text{Ph-C(O-TMS)=CH}_2 + \text{PhCHO} \xrightarrow[94\%]{5\text{ mol.\% BiCl}_3, \text{CH}_2\text{Cl}_2, \text{RT}} \text{Ph-C(O)-CH}_2\text{-CH(O-TMS)-Ph} \quad (77)$$

$$\text{Ph-C(O-TMS)=CH}_2 + \text{Ph-CH=CH-C(O)Me} \xrightarrow[80\%]{5\text{ mol.\% BiCl}_3, \text{CH}_2\text{Cl}_2, \text{RT}} \text{Ph-C(O)-CH}_2\text{-CH(Ph)-CH}_2\text{-C(O)Me} \quad (78)$$

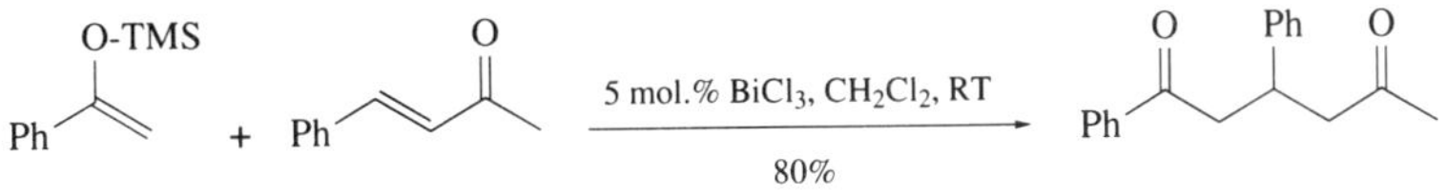

12.3.1.4 Miscellaneous reactions

Low yields of ketones have been obtained by reaction of triarylbismuthines with acyl halides.[103] When the reaction was performed in the presence of a palladium catalyst, excellent yields of phenyl ketones were obtained (Equation (79)).[104]

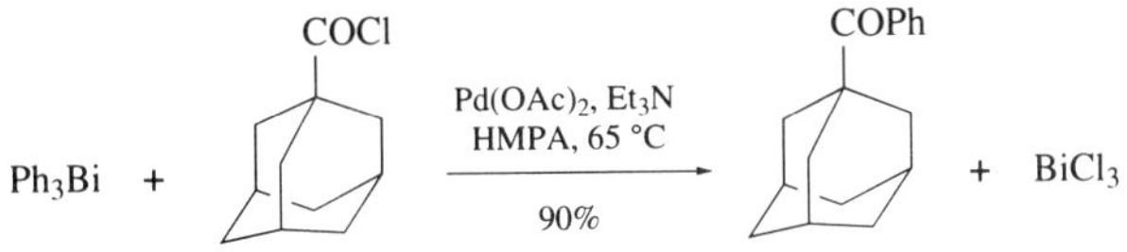

$$\text{Ph}_3\text{Bi} + \text{adamantyl-COCl} \xrightarrow[90\%]{\text{Pd(OAc)}_2, \text{Et}_3\text{N}, \text{HMPA, 65 °C}} \text{adamantyl-COPh} + \text{BiCl}_3 \quad (79)$$

The compound $(\text{Ph}_2\text{Bi})_2\text{CH}_2$, after treatment with phenyllithium, reacts with carbonyl compounds to give the bismuth-containing alcohols, which can be further transformed into the terminal alkenes (Scheme 9).[105,106]

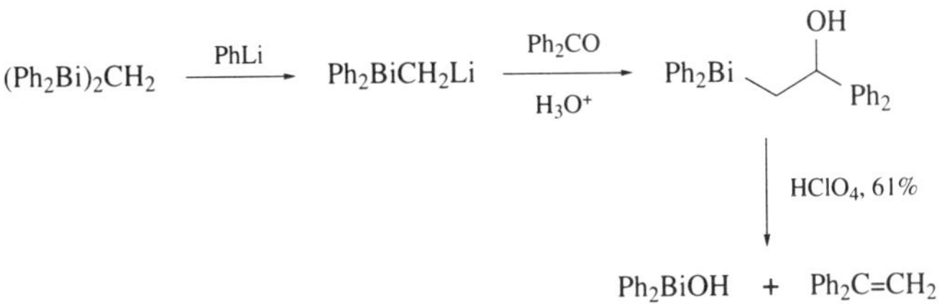

$$(\text{Ph}_2\text{Bi})_2\text{CH}_2 \xrightarrow{\text{PhLi}} \text{Ph}_2\text{BiCH}_2\text{Li} \xrightarrow[\text{H}_3\text{O}^+]{\text{Ph}_2\text{CO}} \text{Ph}_2\text{Bi-CH}_2\text{-C(OH)Ph}_2 \xrightarrow{\text{HClO}_4, 61\%} \text{Ph}_2\text{BiOH} + \text{Ph}_2\text{C=CH}_2$$

Scheme 9

12.3.2 Organobismuth(V) Compounds

Organobismuth(V) compounds have found considerable use in organic synthesis. These compounds can serve as oxidizing agents or arylation agents, depending upon the nature of the organobismuth compound that is used and the reaction conditions. They can be used for the conversion of primary or secondary alcohols to the corresponding aldehydes or ketones, for the oxidation of thiols to disulfides, and for the oxidative cleavage of vicinal glycols. On the other hand, they can be used for *O*-arylation, *N*-arylation and *C*-arylation of a variety of substrates.

12.3.2.1 Oxidation reactions with organobismuth(V) compounds

(i) Oxidation of alcohols

Organobismuth(V) compounds are valuable reagents for the oxidation of alcohols to aldehydes or ketones. The first organobismuth(V) reagent used for this purpose was μ-oxo-bis(chlorotriphenylbismuth), $(Ph_3BiCl)_2O$.[107] In the presence of $(Ph_3BiCl)_2O$, both primary and secondary alcohols are readily oxidized to carbonyl compounds in high yields. In particular, the reagent is especially effective for the oxidation of allylic alcohols (Equation (80)). Many natural product alcohols are readily oxidized by $(Ph_3BiCl)_2O$, such as geraniol, vitamin A alcohol, cholest-1-en-3-ol, cholest-4-en-3β-ol, 3β-cholestanol, tigogenin, testosterone, α-amyrin and cholestan-3β,6β-diol.[107] The oxidation of methyl hederagenin to methyl hederagonate represents a significant improvement over the published yield.[108]

$$\text{(CH}_3\text{)}_2\text{C=CHCH}_2\text{OH} \xrightarrow[90\%]{(Ph_3BiCl)_2O, \ 60\ °C} \text{(CH}_3\text{)}_2\text{C=CHCHO} \qquad (80)$$

More recently, triphenylbismuth carbonate (Ph_3BiCO_3) was shown to oxidize alcohols efficiently under neutral conditions[108,109] and the reaction is selective. Thus, cholest-4-en-3β-ol in the presence of thiophenol is oxidized to cholest-4-en-3-one without oxidation of the thiol. Similarly, 8-methylselenotetradecan-7-ol gives the corresponding 8-methylselenotetradecan-7-one without oxidation of the selenium.

In addition to Ph_3BiCO_3, arylbismuth reagents of the type Ar_3BiX_2 are mild and efficient oxidizing agents for a wide range of primary, secondary, allylic and benzylic alcohols. Oxidations with reagents of the Ar_3BiX_2 type are best performed under basic conditions. Under neutral or acidic conditions, or in the presence of a weaker base such as Et_3N, the yields of oxidation products are usually smaller, and O-arylation to give ethers is a competing reaction.[108]

In addition to compounds of the type Ar_3BiX_2, arylbismuth reagents of the type Ar_4BiX and Ar_5Bi have also been used to oxidize alcohols to the corresponding carbonyl compounds under basic conditions (Equation (81)).[110,111]

$$\text{cholesterol} \longrightarrow \text{cholest-5-en-3-one} \qquad (81)$$

Ph$_4$BiOTs–BTMG, 92%
Ph$_5$Bi, 72%

Organobismuth(V) compounds are excellent reagents for the oxidation of primary and secondary alcohols to aldehydes and ketones because they give high yields under mild reaction conditions and they are especially effective for sensitive natural product alcohols.

(ii) Oxidative cleavage of vicinal glycols

The oxidative cleavage of vicinal glycols to yield two molecules of aldehyde or ketone is a synthetically useful transformation. Organobismuth(V) compounds can be used for this purpose, and 1,2-glycols are cleaved by triphenylbismuth carbonate and μ-oxo-bis(chlorotriphenylbismuth). Thus, *meso*-hydrobenzoin is converted to benzaldehyde in 80% yield by $(Ph_3BiCl)_2O$,[107] and *cis*-cyclohexane-1,2-diol is cleaved to the dicarbonyl compound in 100% yield when Ph_3BiCO_3 is used as oxidizing agent.[108,109]

The oxidative cleavage of 1,2-glycols can be carried out catalytically, with N-bromosuccinimide as oxidant, in the presence of potassium carbonate and a trace of water in acetonitrile (Equations (82) and (83)).[112,113]

A very mild oxidizing ability has been observed with methyl diarylbismuthinates, as only activated 1,2-diaryl-1,2-glycols were affected, while nonactivated glycols such as 1,2-cyclohexanediol and 1-phenyl-1,2-ethanediol were recovered unchanged (Equation (84)).[114] Stannylene derivations of vicinal glycols are readily cleaved by $Ph_3Bi(OAc)_2$ to the corresponding aldehydes.[115]

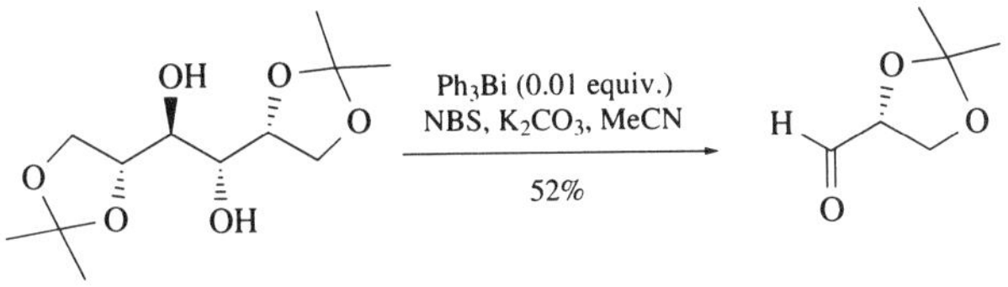

[Equation (82): diol with acetonide groups → aldehyde, Ph₃Bi (0.01 equiv.), NBS, K₂CO₃, MeCN, 52%]

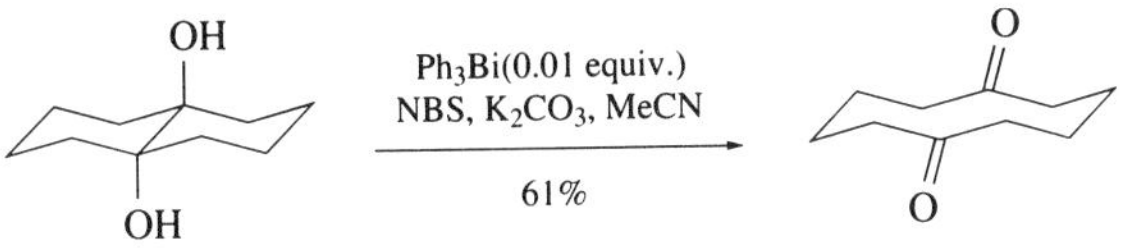

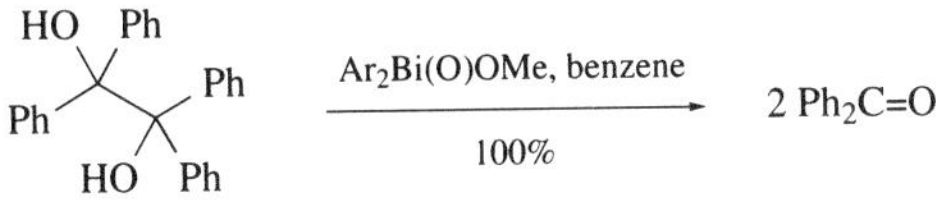

[Equation (83): cyclohexane-1,2-diol → 1,2-cyclohexanedione, Ph₃Bi (0.01 equiv.), NBS, K₂CO₃, MeCN, 61%]

[Equation (84): Ph₂C(OH)–C(OH)Ph₂ → 2 Ph₂C=O, Ar₂Bi(O)OMe, benzene, 100%]

(iii) Oxidation of substrates other than alcohols

Thiols are oxidized to the corresponding disulfides by Ph_3BiCO_3,[108,109] but when Ph_5Bi is used as an oxidant, mixed sulfides are obtained.[111] Treatment of thiophenylate anion with Ph_3BiCl_2 gave a nearly quantitative yield of the disulfide (Equation (85)).[116]

$$2\ PhSH\ +\ NaH\ \xrightarrow[99\%]{Ph_3BiCl_2}\ Ph_2S_2 \qquad (85)$$

Although the thiono group in xanthates, di-*t*-butyl thioketone and dialkylaminothionocarbonates are not oxidized by Ph_3BiCO_3, oxidation of the thiono group in 1,2:5,6-di-*O*-isopropylidene-3-(*N*-4-nitrophenylthionocarbamato)-α-D-glucofurane gave the corresponding disulfide in 81% yield (Equation (86)).[109,117]

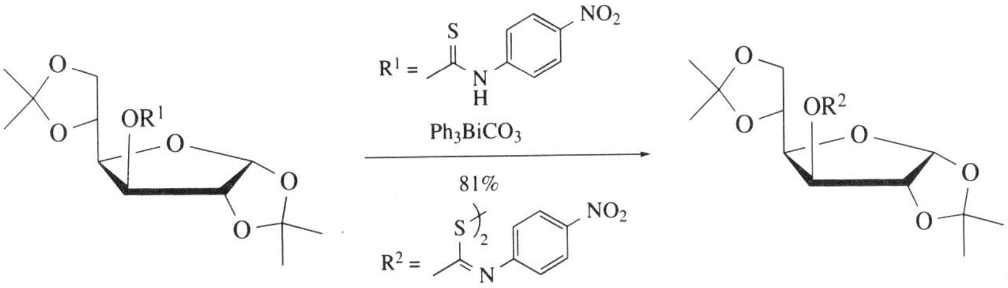

[Equation (86): glucofuranose derivative with OR¹ (R¹ = C(=S)NH-C₆H₄-NO₂) → OR² (R² = [C(=N-C₆H₄-NO₂)S]₂/2), Ph₃BiCO₃, 81%]

A number of phenolic compounds have been oxidized by triphenylbismuth carbonate and triphenylbismuth dichloride under basic conditions. 2,6-Dimethylphenol is oxidized by triphenylbismuth carbonate under neutral or basic conditions to the 3,3',5,5'-tetramethyldiphenoquinone.[118] 2,6-Di-*t*-butylphenol reacts with triphenylbismuth dichloride or carbonate in the presence of a base which favors electron transfer, such as *N*-*t*-butyl-*N'*,*N'*,*N''*,*N''*-tetramethylguanidine (BTMG), to give the corresponding diphenoquinone (Equation (87)).[119,120]

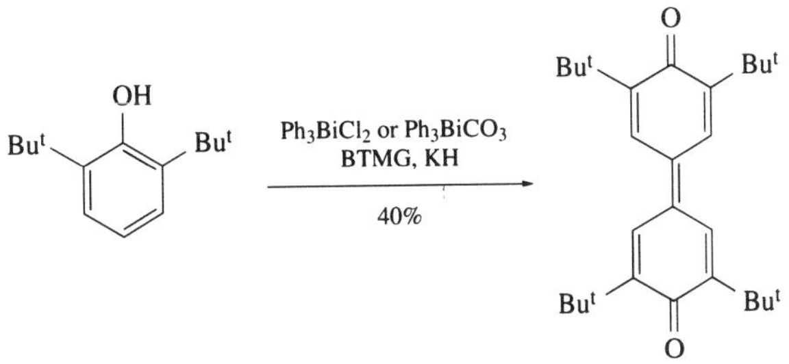

[Equation (87): 2,6-di-*t*-butylphenol → 3,3',5,5'-tetra-*t*-butyldiphenoquinone, Ph₃BiCl₂ or Ph₃BiCO₃, BTMG, KH, 40%]

12.3.2.2 Use of bismuthonium ylides in organic synthesis

Several stabilized bismuthonium ylides have been described,[42] but little attention has been paid to their synthetic applications. Suzuki *et al.* reported that unstabilized bismuthonium ylides derived from straight chain 1,3-dicarbonyl compounds reacted with sulfenes to give 1,3-oxathiole-3,3-dioxide derivatives in moderate yield (Scheme 10).[121]

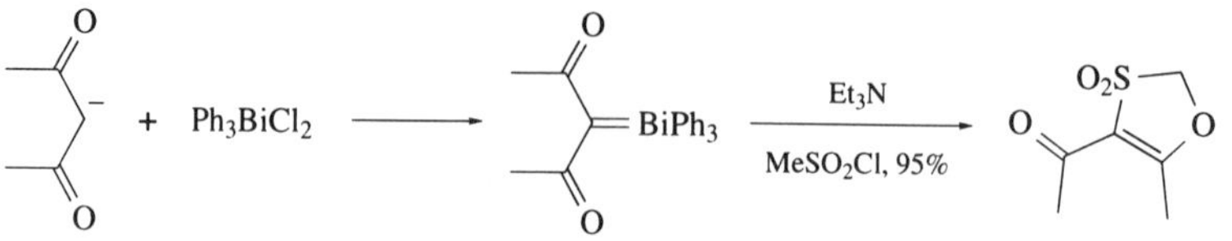
O
O
+ Ph3BiCl2
O
O
BiPh3
Et3N
MeSO2Cl, 95%
O2S
O
O

Scheme 10

N-Tosyltriarylbismuthimines are easily prepared by the reaction of triarylbismuthines with chloramine-T. These bismuthimines react with aromatic aldehydes, acid chlorides and isocyanates to form *N*-arylidenesylamides, *N*-aroyltosylamides and *N*-aryl-*N'*-tosylureas, respectively (Equation (88)).[122]

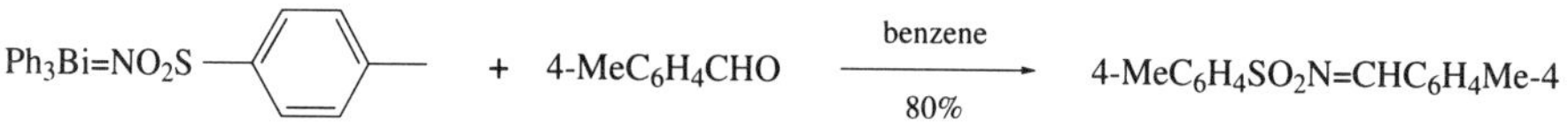
Ph3Bi=NO2S
+ 4-MeC6H4CHO
benzene
80%
4-MeC6H4SO2N=CHC6H4Me-4

(88)

A stabilized bismuthonium ylide has been found to react with phenyl isothiocyanate to give the 1,3-oxathiole and an amide in 55% and 10% yield, respectively (Equation (89)).[123] Stabilized bismuthonium ylides react with aldehydes to afford three types of products: tetraacylcyclopropanes, dihydrofurans and α,β-unsaturated carbonyl compounds, depending on the ylides, the aldehydes, and the reaction conditions employed.[118]

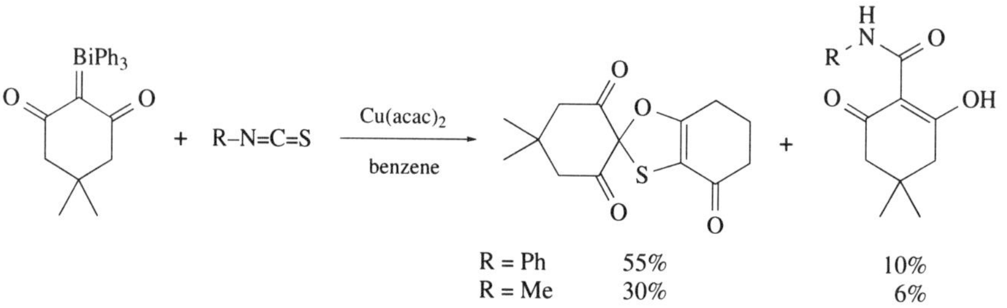
BiPh3
O
O
+ R–N=C=S
Cu(acac)2
benzene
O
O
S
O
O
+
H
N
R
O
O
OH
R = Ph 55% 10%
R = Me 30% 6%

(89)

Triphenylbismuthonio-4,4-dimethyl-2,6-dioxocyclohexan-1-ide reacts with 1-alkynes in the presence of small amounts of copper(I) chloride to form bicyclic furan derivatives, probably via a carbenoid intermediate (Equation (90)).[124]

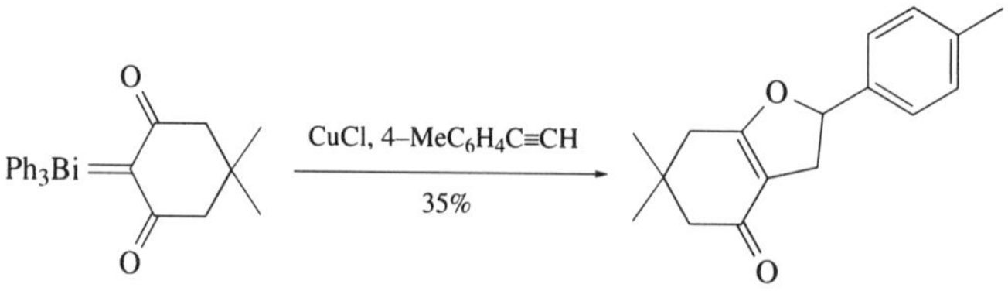
O
Ph3Bi
O
CuCl, 4–MeC6H4C≡CH
35%
O
O

(90)

12.3.2.3 Arylation reactions with pentavalent organobismuth compounds

Arylation by pentavalent organobismuth compounds has been known for many years. However, it is only recently that the versatility and importance of this process as a mild and selective method for the arylation of a wide variety of substrates has been recognized. The application of bismuth(V) reagents to the arylation of a wide range of nucleophiles to give (even very hindered) compounds in good yields has been reviewed.[3,5] The use of copper catalysis enables a range of oxygen, nitrogen, and sulfur functions, including hindered systems, to be arylated under very mild conditions.

(i) C-arylation with bismuth(V) reagents

The first arylation came while studying the use of bismuth(V) reagents for the oxidation of alcohols under very mild conditions. The alkaloid quinine gave the phenylated quininone in high yield.[117]

The C-arylation of phenolic compounds can be performed by triphenylbismuth carbonate and also by a variety of triaryl and tetraphenylbismuthonium derivatives in the presence of a base and in various solvents such as methylene chloride, benzene, toluene, and THF.[111,117,125,126] Similar C-phenylation is also realized with pentaphenylbismuth under neutral conditions (Equations (91) and (92) and Scheme 11).[111]

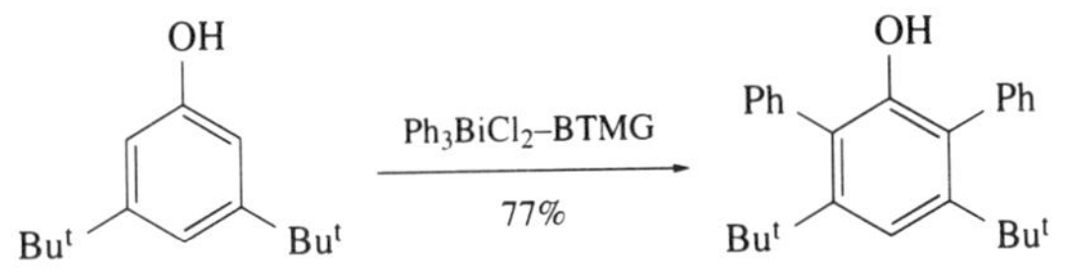

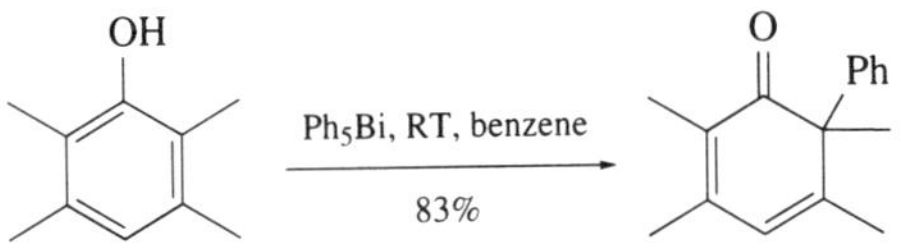

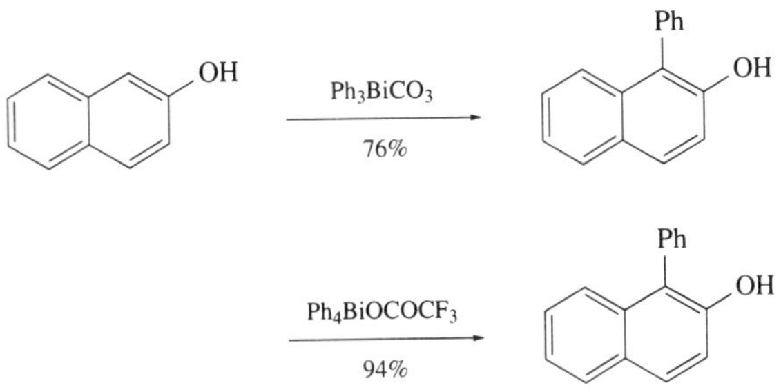

Scheme 11

As a variety of substituted pentavalent triarylbismuth derivatives can be prepared readily, arylation can be carried out to introduce aromatic groups which have either electron-donating or electron-withdrawing substituents (Equations (93) and (94)).[127] The nature of the substituents on the phenol appear to govern the regioselectivity of the arylation reactions: phenols bearing electron-donating substituents are essentially *ortho* C-phenylated, while those bearing electron-withdrawing substituents are mostly, or selectively, O-phenylated.[119]

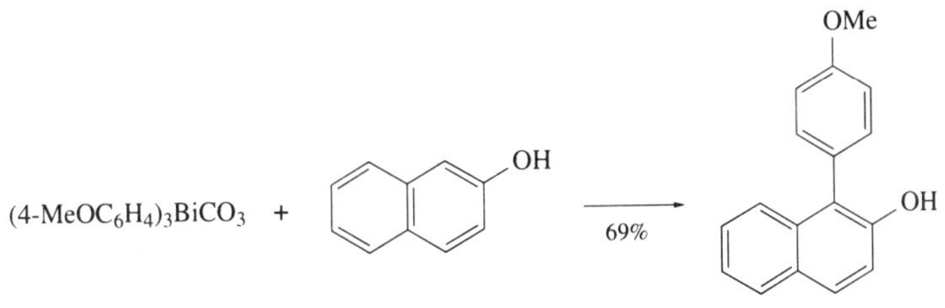

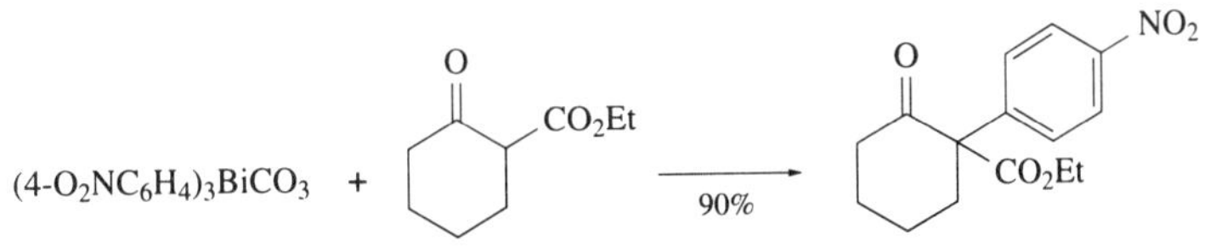

α-C-Arylation of 1,3-dicarbonyl compounds such as 1,3-diketones, β-keto esters, or malonic esters is easily performed with a variety of pentavalent organobismuth reagents under neutral or basic conditions (Equations (95) and (96)).[117,125,126] This type of reaction has also been used in a new approach for the synthesis of isoflavanones and 3-aryl-4-hydroxycoumarins.[128] Phenylation of activated chroman-4-ones to isoflavanones has been achieved in moderate to good yields. Selective monophenylation was performed in low-temperature reactions with pentaphenylbismuth. Arylation of 4-hydroxycoumarins by various bismuth(V) reagents gives rise to functionally substituted 3-aryl-4-hydroxycoumarins in high yields (Equations (97) and (98)).[128]

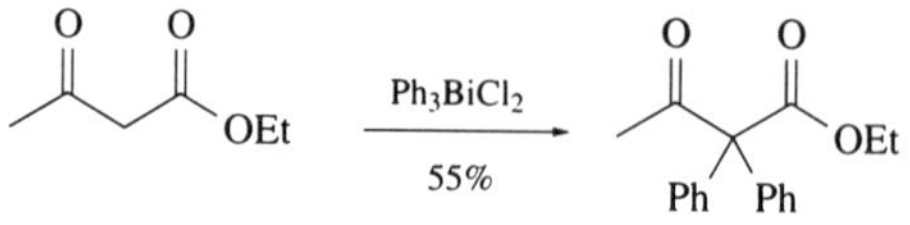

$$\text{MeCOCH}_2\text{CO}_2\text{Et} \xrightarrow[55\%]{\text{Ph}_3\text{BiCl}_2} \text{MeCOCPh}_2\text{CO}_2\text{Et} \quad (95)$$

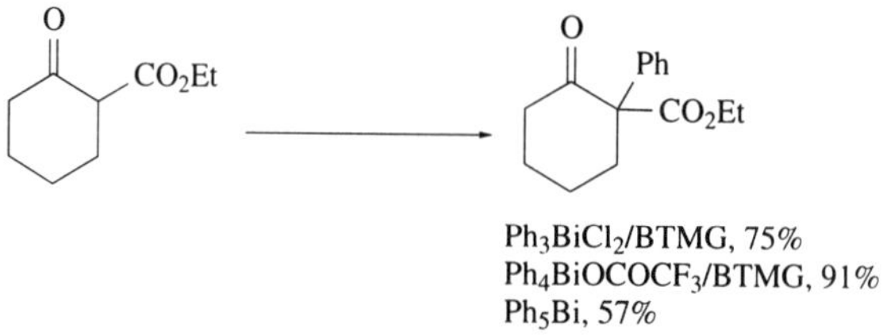

(96) 2-(ethoxycarbonyl)cyclohexanone → 2-phenyl-2-(ethoxycarbonyl)cyclohexanone
Ph$_3$BiCl$_2$/BTMG, 75%
Ph$_4$BiOCOCF$_3$/BTMG, 91%
Ph$_5$Bi, 57%

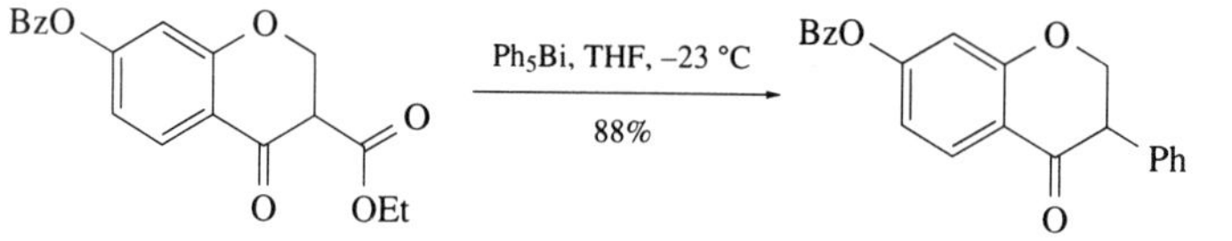

(97) 7-BzO-3-(ethoxycarbonyl)chroman-4-one $\xrightarrow[88\%]{\text{Ph}_5\text{Bi, THF, }-23\,°\text{C}}$ 7-BzO-3-phenylchroman-4-one

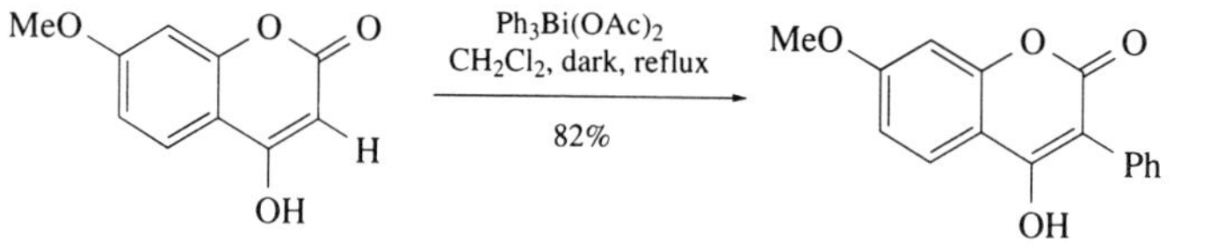

(98) 7-MeO-4-hydroxycoumarin $\xrightarrow[82\%]{\text{Ph}_3\text{Bi(OAc)}_2,\ \text{CH}_2\text{Cl}_2,\ \text{dark, reflux}}$ 7-MeO-3-phenyl-4-hydroxycoumarin

Nonenolized substrates are not arylated by pentavalent organobismuth compounds under neutral conditions. However, the derived enolates react easily with bismuth(V) reagents to give the corresponding α-arylated products.[116,117,126,129,130] Generally, only polyarylated derivatives are obtained.

Other stabilized anions, such as the anions of nitroalkanes, α-methyl-α-nitrocarboxylic acid derivatives, esters, triphenylmethane, indole, and skatole, and so on, can be successfully α-arylated (Equations (99) and (100)).[131,132]

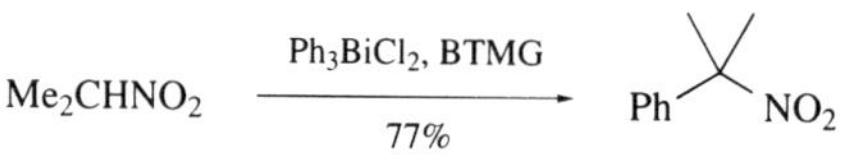

$$\text{Me}_2\text{CHNO}_2 \xrightarrow[77\%]{\text{Ph}_3\text{BiCl}_2,\ \text{BTMG}} \text{Me}_2\text{C(Ph)NO}_2 \quad (99)$$

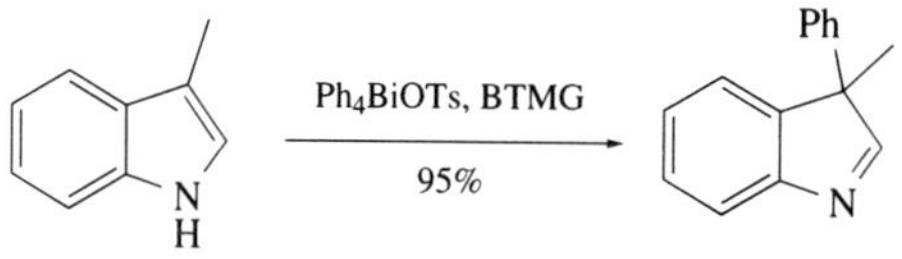

(100) 3-methylindole $\xrightarrow[95\%]{\text{Ph}_4\text{BiOTs, BTMG}}$ 3-methyl-3-phenyl-3H-indole

Selective monophenylation of an active methylene compound has been reported (Equation (101)).[133]

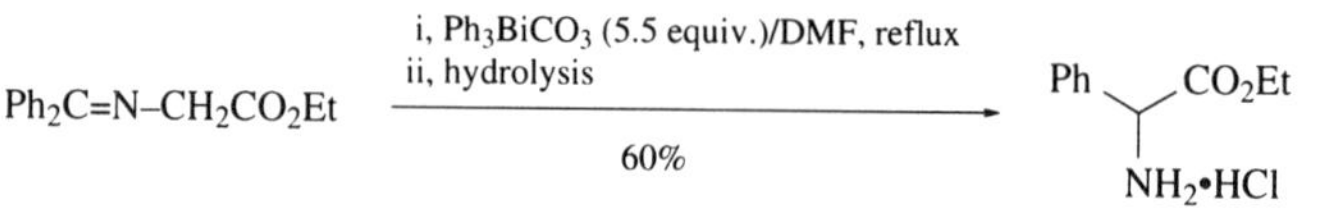

$$\text{Ph}_2\text{C=N-CH}_2\text{CO}_2\text{Et} \xrightarrow[60\%]{\substack{\text{i, Ph}_3\text{BiCO}_3\ (5.5\ \text{equiv.})/\text{DMF, reflux} \\ \text{ii, hydrolysis}}} \text{PhCH(NH}_2\cdot\text{HCl)CO}_2\text{Et} \quad (101)$$

(no diphenylation observed)

(ii) O-arylation

The reaction of 4-nitrophenol with pentaphenylbismuth gives 4-nitrodiphenyl ether exclusively.[111] This observation developed into a study of the reaction of electron-rich phenols with pentavalent organobismuth reagents bearing electron-withdrawing groups under neutral conditions.[134] Tetraphenylbismuth trifluoroacetate is useful for selective formation of the aryl–oxygen bond (Equation (102)).

Good yields are obtained with Ph$_4$BiOCOCF$_3$ in benzene under reflux, but acid catalysis significantly increases the yields.[125,126] Under copper catalysis, a variety of phenols are O-phenylated by triphenylbismuth diacetate under neutral conditions (Equation (103)).[135] Similary, arylation of enolic systems is also subject to copper powder catalysis and exclusive O-phenylation is observed (Equation (104)).

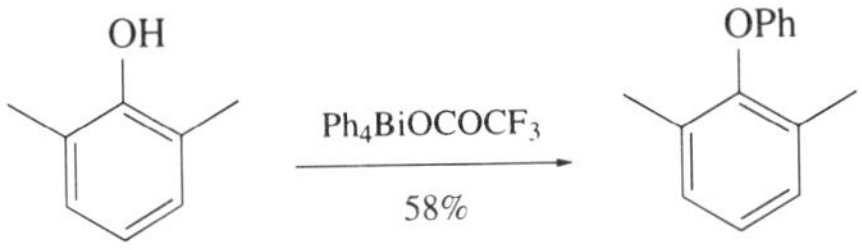

$$\text{2,6-dimethylphenol} \xrightarrow[58\%]{\text{Ph}_4\text{BiOCOCF}_3} \text{2,6-dimethylphenyl phenyl ether} \quad (102)$$

$$\text{3,5-dimethoxyphenol} \xrightarrow[90\%]{\text{Ph}_3\text{Bi(OAc)}_2,\ \text{Cu}} \text{3,5-dimethoxyphenyl phenyl ether} \quad (103)$$

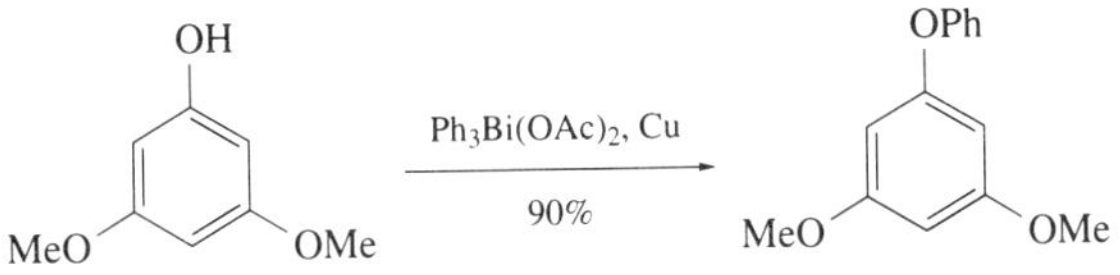

$$\text{dimedone} \xrightarrow[88\%]{\text{Ph}_3\text{Bi(OCOCF}_3)_2 \atop \text{CH}_2\text{Cl}_2,\ \text{Cu (0.1 equiv.)}} \text{O-phenyl enol ether} \quad (104)$$

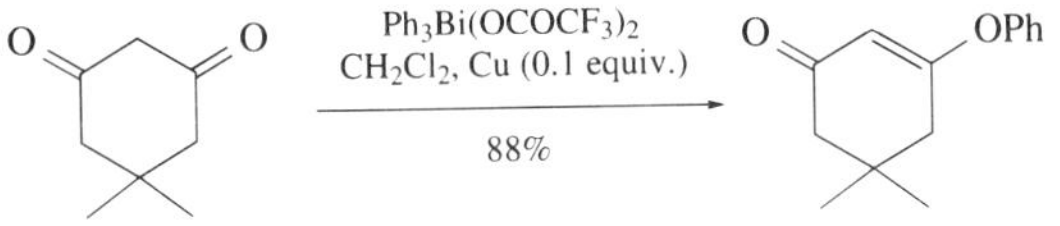

Reaction of the anions of electron-deficient phenols with triphenylbismuth dichloride results in O-phenylation.[119] The reaction is regioselective with 4-substituted phenols, but a mixture of O- and C-arylated products is formed in the case of the 3-substituted phenols (Equation (105)).[119]

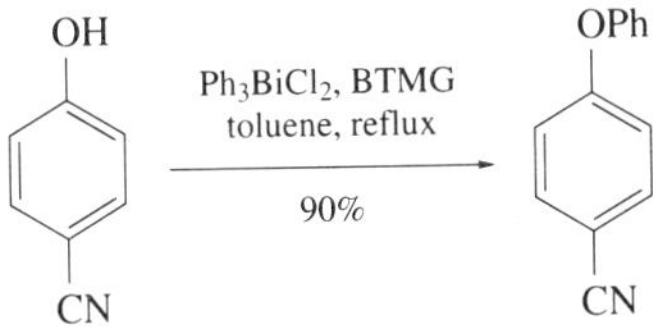

$$\text{4-cyanophenol} \xrightarrow[90\%]{\text{Ph}_3\text{BiCl}_2,\ \text{BTMG} \atop \text{toluene, reflux}} \text{4-cyanophenyl phenyl ether} \quad (105)$$

Reactions of primary and secondary alcohols with tetraphenylbismuthonium trifluoroacetate or related tetraphenylbismuthonium reagents under basic conditions result in oxidation to the carbonyl derivatives. However, under neutral or acidic conditions, the O-phenyl ethers of primary and secondary alcohols can be obtained in moderate to good yields.[110] In contrast to simple alcohols, glycols are monophenylated by triphenylbismuth diacetate because of the presence of the neighboring group.[115] Thus, a variety of glycols can be selectively mono-O-phenylated in good to excellent yields (Equation (106)).[115,136,137] With compounds where either a secondary or a tertiary hydroxy group can be phenylated, the product is usually that in which the secondary hydroxy group is phenylated (Equation (107)). Furthermore, an axial preference is found in conformationally rigid glycols (Equation (108)). Neighboring groups other than hydroxy can promote phenylation. For example, 2-phenoxyethanol and 2-methoxyethanol are easily O-phenylated (Equation (109)).[138]

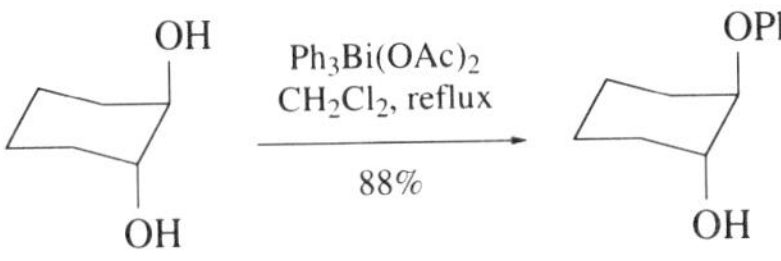

$$\text{trans-1,2-cyclohexanediol} \xrightarrow[88\%]{\text{Ph}_3\text{Bi(OAc)}_2 \atop \text{CH}_2\text{Cl}_2,\ \text{reflux}} \text{mono-O-phenyl ether} \quad (106)$$

$$\text{1-phenyl-1,2-cyclohexanediol} \xrightarrow[74\%]{\text{Ph}_3\text{Bi(OAc)}_2 \atop \text{CH}_2\text{Cl}_2,\ \text{reflux}} \text{2-phenoxy-1-phenylcyclohexanol} \quad (107)$$

$$\text{4-t-butyl-1,2-cyclohexanediol} \xrightarrow{\text{Ph}_3\text{Bi(OAc)}_2 \atop \text{CH}_2\text{Cl}_2,\ \text{reflux}} \text{axial-OPh product (73\%)} + \text{equatorial-OPh product (2\%)} \quad (108)$$

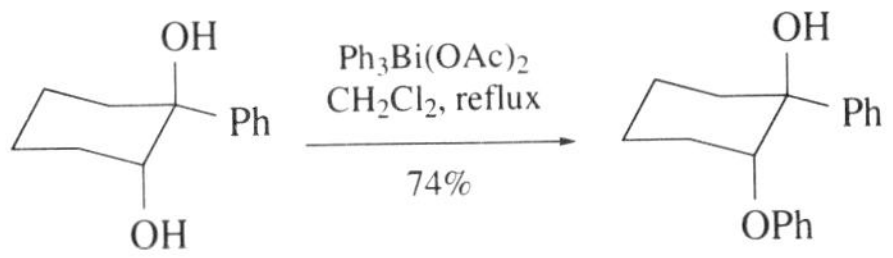

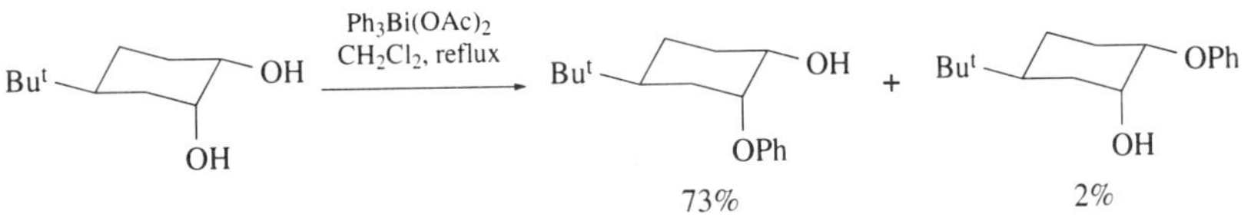

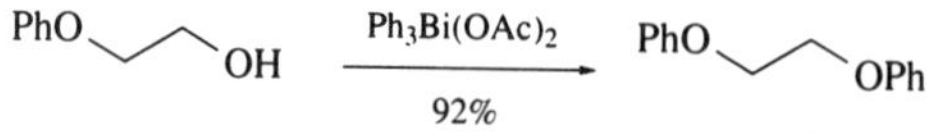

$$\text{PhO}\diagdown\diagdown\text{OH} \xrightarrow[92\%]{\text{Ph}_3\text{Bi}(\text{OAc})_2} \text{PhO}\diagdown\diagdown\text{OPh} \tag{109}$$

Enantioselective monophenylation has been observed in the presence of chiral ligands and optical inductions of up to 30% *ee* have been attained.[139] The Cu(OAc)$_2$-catalyzed monophenylation of *meso*-diols with Ph$_3$Bi(OAc)$_2$ was rendered enantioselective by use of optically active pyridinyloxazoline ligands as cocatalysts, and enantioselectivities in the middle range up to 50.4% *ee* were obtained (Equation (110)).[140]

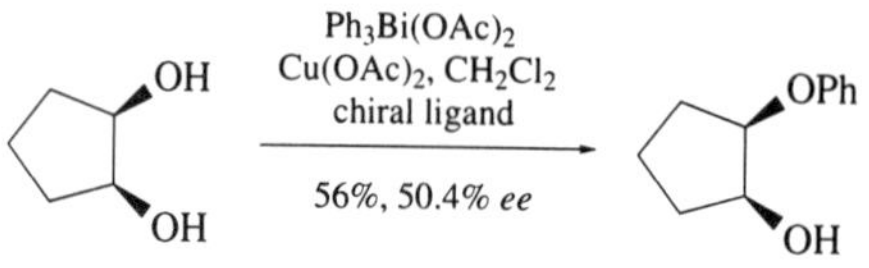

$$\text{cis-cyclopentane-1,2-diol} \xrightarrow[56\%, 50.4\% \ ee]{\substack{\text{Ph}_3\text{Bi}(\text{OAc})_2 \\ \text{Cu}(\text{OAc})_2, \text{CH}_2\text{Cl}_2 \\ \text{chiral ligand}}} \text{trans-2-phenoxycyclopentanol} \tag{110}$$

(iii) N-arylation

The reaction of amino alcohols with triphenylbismuth diacetate leads preferentially to *N*-phenylation[138] and this type of reaction is strongly catalyzed by copper and its salts. Simple primary aliphatic and aromatic amines and secondary aliphatic amines are readily phenylated by Ph$_3$Bi(OAc)$_2$ at room temperature in the presence of Cu(OAc)$_2$, the yield varying from 60% to 85%. No reaction occurs in the absence of the copper catalyst. When the reaction is performed in methylene chloride at room temperature with metallic copper as catalyst, the monophenylamine derivatives are obtained in high yields. A variety of aliphatic, alicyclic, heterocyclic, and aromatic amines as well as hydrazines have been *N*-arylated (Equation (111)).[141]

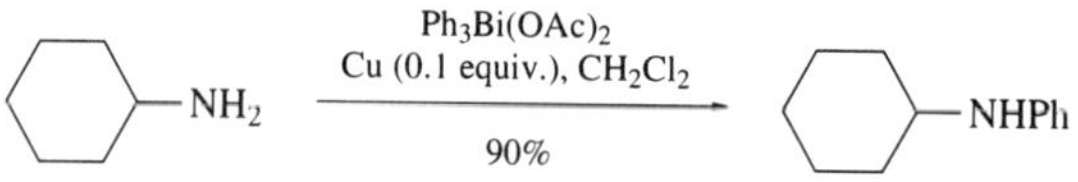

$$\text{Cy-NH}_2 \xrightarrow[90\%]{\substack{\text{Ph}_3\text{Bi}(\text{OAc})_2 \\ \text{Cu (0.1 equiv.), CH}_2\text{Cl}_2}} \text{Cy-NHPh} \tag{111}$$

Selective mono-*N*-phenylation of α-amino acid derivatives has been realized in the presence of a catalytic amount of copper metal or copper diacetate. No reaction takes place with the free α-amino acids, but esters are readily mono-*N*-phenylated (Equation (112)).[142]

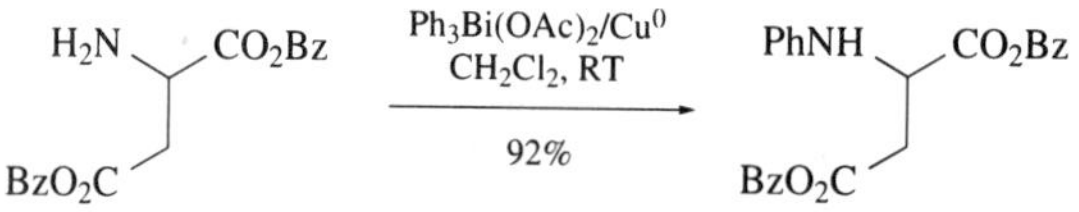

$$\underset{\text{BzO}_2\text{C}}{\text{H}_2\text{N}}\diagup\text{CO}_2\text{Bz} \xrightarrow[92\%]{\substack{\text{Ph}_3\text{Bi}(\text{OAc})_2/\text{Cu}^0 \\ \text{CH}_2\text{Cl}_2, \text{RT}}} \underset{\text{BzO}_2\text{C}}{\text{PhNH}}\diagup\text{CO}_2\text{Bz} \tag{112}$$

Triphenylbismuth bis(trifluoroacetate) is an efficient reagent for the *N*- or *C*-3-phenylation of indole derivatives under copper catalysis. *C*-3-Phenyl derivatives are obtained with *C*-3-unsubstituted indoles and *N*-phenyl derivatives with *C*-3-substituted indoles (Equation (113)).[143]

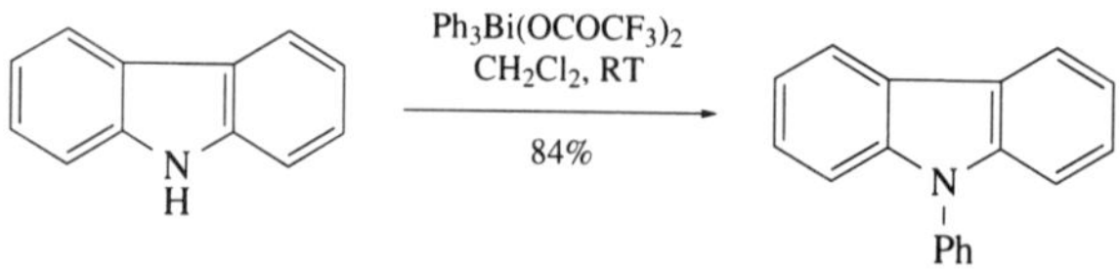

$$\text{carbazole (NH)} \xrightarrow[84\%]{\substack{\text{Ph}_3\text{Bi}(\text{OCOCF}_3)_2 \\ \text{CH}_2\text{Cl}_2, \text{RT}}} \text{carbazole (N-Ph)} \tag{113}$$

Trivalent bismuth compounds in the presence of cupric acetate can be used for the alkylation of amines (Equation (114)).[144] Bismuth(V) intermediates generated *in situ* may be involved.

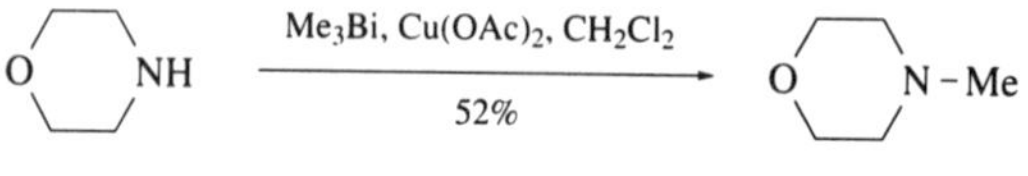

$$\text{morpholine (NH)} \xrightarrow[52\%]{\text{Me}_3\text{Bi}, \text{Cu}(\text{OAc})_2, \text{CH}_2\text{Cl}_2} \text{morpholine (N-Me)} \tag{114}$$

12.3.2.4 Miscellaneous reactions

Organobismuth(V) reagents of the type $Ar_2Bi(O)OMe$ can be used for the acylation of amides and thioamides[145] and yields vary from 76% to 89%. In addition to simple amides, 2-piperidone has been acylated to N-acetyl-2-piperidone in 84% yield. Two substituted ureas and thioureas have also been acylated.[145]

12.4 REFERENCES

1. J. L. Wardell, in 'COMC-I', vol. 2, p. 681 and references therein.
2. L. D. Freedman and G. O. Doak, *Chem. Rev.*, 1982, **82**, 15.
3. D. H. R. Barton and J.-P. Finet, *Pure Appl. Chem.*, 1987, **59**, 937.
4. R. A. Abramovitch, D. H. R. Barton and J.-P. Finet, *Tetrahedron*, 1988, **44**, 3039.
5. J.-P. Finet, *Chem. Rev.*, 1989, **89**, 1487.
6. Y.-Z. Huang, *Acc. Chem. Res.*, 1992, **25**, 182.
7. L. D. Freedman and G. O. Doak, *J. Organomet. Chem.*, 1984, **261**, 31.
8. G. O. Doak and L. D. Freedman, *J. Organomet. Chem.*, 1984, **261**, 59.
9. L. D. Freedman and G. O. Doak, *J. Organomet. Chem., Library* 1985, **17**, 311.
10. G. O. Doak and L. D. Freedman, *J. Organomet. Chem., Library* 1985, **17**, 353.
11. L. D. Freedman and G. O. Doak, *J. Organomet. Chem.*, 1986, **298**, 37.
12. G. O. Doak and L. D. Freedman, *J. Organomet. Chem.*, 1986, **298**, 67.
13. L. D. Freedman and G. O. Doak, *J. Organomet. Chem.*, 1987, **324**, 1.
14. G. O. Doak and L. D. Freedman, *J. Organomet. Chem.*, 1987, **324**, 39.
15. L. D. Freedman and G. O. Doak, *J. Organomet. Chem.*, 1988, **351**, 25.
16. G. O. Doak and L. D. Freedman, *J. Organomet. Chem.*, 1988, **351**, 62.
17. L. D. Freedman and G. O. Doak, *J. Organomet. Chem.*, 1989, **360**, 263.
18. G. O. Doak and L. D. Freedman, *J. Organomet. Chem.*, 1989, **360**, 297.
19. L. D. Freedman and G. O. Doak, *J. Organomet. Chem.*, 1990, **380**, 1.
20. G. O. Doak and L. D. Freedman, *J. Organomet. Chem.*, 1990, **380**, 35.
21. L. D. Freedman and G. O. Doak, *J. Organomet. Chem.*, 1991, **404**, 49.
22. G. O. Doak and L. D. Freedman, *J. Organomet. Chem.*, 1991, **404**, 87.
23. L. D. Freedman and G. O. Doak, *J. Organomet. Chem.*, 1992, **442**, 1.
24. G. O. Doak and L. D. Freedman, *J. Organomet. Chem.*, 1992, **442**, 61.
25. J. L. Wardell, *Organomet. Chem.*, 1982, **10**, 129.
26. J. L. Wardell, *Organomet. Chem.*, 1983, **11**, 157.
27. J. L. Wardell, *Organomet. Chem.*, 1984, **12**, 127.
28. J. L. Wardell, *Organomet. Chem.*, 1985, **13**, 133.
29. J. L. Wardell, *Organomet. Chem.*, 1986, **14**, 141.
30. J. L. Wardell, *Organomet. Chem.*, 1987, **15**, 138.
31. J. L. Wardell, *Organomet. Chem.*, 1988, **16**, 142.
32. J. L. Wardell, *Organomet. Chem.*, 1989, **17**, 130.
33. J. L. Wardell, *Organomet. Chem.*, 1990, **18**, 124.
34. L. D. Freedman and G. O. Doak, in 'The Chemistry of the Metal–Carbon Bond', ed. F. R. Hartley, Wiley, New York, 1989, vol. 5, p. 397.
35. C. Glidewell, D. Lloyd and S. Metcalfe, *Tetrahedron*, 1986, **42**, 3887.
36. Y.-Z. Huang, C. Chen and Y. C. Shen, *Tetrahedron Lett.*, 1986, **27**, 2903.
37. Y.-Z. Huang, C. Chen and Y. C. Shen, *J. Organomet. Chem.*, 1989, **366**, 87.
38. Y.-Z. Huang, Y. C. Shen and C. Chen, *Synth. Commun.*, 1989, **19**, 83.
39. Y.-Z. Huang, C. Chen and Y. C. Shen, *Synth. Commun.*, 1989, **19**, 501.
40. D. Lloyd and M. I. C. Singer, *Chem. Ind. (London)*, 1967, 787.
41. Y. Liao, Y.-Z. Huang, L. J. Zhang and C. Chen, *J. Chem. Res. (S)*, 1990, 388.
42. C. Glidewell, D. Lloyd and S. Metcalfe, *Synthesis*, 1988, 319.
43. Y. Liao and Y.-Z. Huang, *Tetrahedron Lett.*, 1990, **31**, 5897.
44. C. Chen, Y.-Z. Huang, Y. C. Shen and Y. Liao, *Heteroatom Chem.*, 1990, **1**, 49.
45. T. Kauffmann, *Angew. Chem., Int. Ed. Engl.*, 1982, **21**, 410.
46. T. Kauffmann, *Top. Curr. Chem.*, 1980, **92**, 109.
47. L. J. Zhang and Y.-Z. Huang, *J. Organomet. Chem.*, 1993, **454**, 101.
48. L. S. Boulos and A. A. El-Kateb, *Chem. Ind. (London)*, 1983, 864.
49. K. Akiba, A. Shimizu, H. Ohnari and K. Ohkata, *Tetrahedron Lett.*, 1985, **26**, 3211.
50. R. Asano, I. Moritani, Y. Fujiwara and S. Teranishi, *Bull. Chem. Soc. Jpn.*, 1973, **46**, 2910.
51. T. Kawamura, K. Kikukawa, M. Takagi and T. Matsuda, *Bull. Chem. Soc. Jpn.*, 1977, **50**, 2021.
52. C. Petrier and J. L. Luche, *J. Org. Chem.*, 1985, **50**, 910.
53. C. Chen, Y. C. Shen and Y.-Z. Huang, *Tetrahedron Lett.*, 1988, **29**, 1395.
54. Y.-Z. Huang, L. J. Zhang, C. Chen and G. Z. Guo, *J. Organomet. Chem.*, 1991, **412**, 47.
55. Y.-Z. Huang, C. Chen and Y. C. Shen, *Tetrahedron Lett.*, 1988, **29**, 5275.
56. Y.-Z. Huang, C. Chen and Y. C. Shen, *J. Chem. Soc., Perkin Trans. 1*, 1988, 2855.
57. Y.-Z. Huang, Y. Liao and C. Chen, *J. Chem. Soc., Chem. Commun.*, 1990, 85.
58. Y.-Z. Huang, and L. Liao, *J. Org. Chem.*, 1991, **56**, 1381.
59. L. J. Zhang, Y.-Z. Huang, H. X. Jiang, L. J. Duan-Mu and Y. Liao, *J. Org. Chem.*, 1992, **57**, 774.
60. L. J. Zhang, X. S. Mo, J. L. Huang and Y.-Z. Huang, *Tetrahedron Lett.*, 1993, **34**, 1621.

61. L. J. Zhang, Y.-Z. Huang and Z.-H. Huang, *Tetrahedron Lett.*, 1991, **32**, 6579.
62. L. J. Zhang, X. S. Mo and Y.-Z. Huang, *J. Organomet. Chem.*, 1994, **471**, 77.
63. Y. Butsugan, H. Ito and S. Araki, *Tetrahedron Lett.*, 1987, **28**, 3707.
64. W. B. Wang, L. L. Shi, R. H. Xu and Y.-Z. Huang, *J. Chem. Soc., Perkin Trans. 1*, 1990, 424.
65. W. B. Wang, L. L. Shi and Y.-Z. Huang, *Tetrahedron Lett.*, 1990, **31**, 1185.
66. W. B. Wang, L. L. Shi and Y.-Z. Huang, *Tetrahedron*, 1990, **46**, 3315.
67. Y.-Z. Huang, Y. C. Shen and C. Chen, *Tetrahedron Lett.*, 1986, **27**, 2903.
68. D. H. R. Barton, D. Bridon and S. Z. Zard, *J. Chem. Soc., Chem. Commun.*, 1985, 1066.
69. J. Koketsu, S. Kokjma and Y. Ishii, *J. Organomet. Chem.*, 1972, **35**, 69.
70. Y. Yamamoto, T. Okinaka, M. Nakatani and K. Akiba, *Nippon Kagaku Kaishi*, 1987, 1286 (*Chem. Abstr.*, 1988, **109**, 22 614z).
71. K. Akiba, H. Ohnari and K. Ohkata, *Chem. Lett.*, 1985, 1577.
72. Y.-Z. Huang, Y. C. Shen, L. J. Zhang and S. Zhang, *Synthesis*, 1985, 57.
73. C. Chen, F.-H. Zhu and Y.-Z. Huang, *J. Chem. Res.*, 1989, 381.
74. F. A. J. Meskens, *Synthesis*, 1981, 501.
75. Y. Liao, Y.-Z. Huang and F.-H. Zhu, *J. Chem. Soc., Chem. Commun.*, 1990, 493.
76. S. Danishefsky, *Acc. Chem. Res.*, 1979, **12**, 66.
77. C. Chen, Y.-Z. Huang and Y. C. Shen, *Tetrahedron Lett.*, 1988, **29**, 1033.
78. C. Chen, Y. Liao and Y.-Z. Huang, *Tetrahedron*, 1989, **45**, 3011.
79. M. Fujiwara, A. Baba and H. Matsuda, *Chem. Lett.*, 1989, 1247.
80. M. Fujiwara, M. Imada, A. Baba and H. Matsuda, *J. Org. Chem.*, 1988, **53**, 5974.
81. M. Fujiwara, A. Baba and H. Matsuda, *Bull. Chem. Soc. Jpn.*, 1990, **63**, 1069.
82. R. Nomura, T. Nakano, Y. Nishio, S. Ogawa, A. Ninagawa and H. Matsuda, *Chem. Ber.*, 1989, **122**, 2407.
83. T. Harada and T. Mukaiyama, *Chem. Lett.*, 1992, 81.
84. S. Kobayashi, M. Tamura and T. Mukaiyama, *Chem. Lett.*, 1988, 91.
85. T. Mukaiyama and T. Harada, *Chem. Lett.*, 1991, 1653.
86. T. Harada and T. Mukaiyama, *Chem. Lett.*, 1992, 1901.
87. S. F. Wnuk and M. J. Robins, *J. Org. Chem.*, 1990, **55**, 4757.
88. S. Sayama and Y. Inamura, *Bull. Chem. Soc. Jpn.*, 1991, **64**, 306.
89. R. Nomura, T. Wada, Y. Yamada and H. Matsuda, *Chem. Lett.*, 1986, 1901.
90. T. Chivers, M. N. S. Rao and J. F. Richardson, *J. Chem. Soc., Chem. Commun.*, 1983, 186.
91. N. Burford, T. Chivers, M. N. S. Rao and J. F. Richardson, *Inorg. Chem.*, 1984, **23**, 1946.
92. R. T. Boeré *et al.*, *J. Am. Chem. Soc.*, 1985, **107**, 7710.
93. I. Suzuki and Y. Yamamoto, *J. Org. Chem.*, 1993, **58**, 4783.
94. M. Wada and K. Akiba, *Tetrahedron Lett.*, 1985, **26**, 4211.
95. M. Wada, II. Ohki and K. Akiba, *Tetrahedron Lett.*, 1986, **27**, 4771.
96. M. Wada, H. Ohki and K. Akiba, *Bull. Chem. Soc. Jpn.*, 1990, **63**, 1738.
97. M. Wada, H. Ohki and K. Akiba, *J. Chem. Soc., Chem. Commun.*, 1987, 708.
98. Y.-Z. Huang and Y. Liao, *Heteroatom Chem.*, 1991, **2**, 297.
99. A. R. Katritzky, N. Shobana and P. A. Harris, *Tetrahedron Lett.*, 1991, **32**, 4247.
100. A. R. Katritzky, N. Shobana and P. A. Harris, *Organometallics*, 1992, **11**, 1381.
101. H. Ohki, M. Wada and K. Akiba, *Tetrahedron Lett.*, 1988, **29**, 4719.
102. M. Wada, E. Takeichi and T. Matsumoto, *Bull. Chem. Soc. Jpn.*, 1991, **64**, 990.
103. F. Challenger and L. R. Ridgway, *J. Chem. Soc.*, 1922, **121**, 104.
104. D. H. R. Barton, N. Ozbalik and M. Ramesh, *Tetrahedron*, 1988, **44**, 5661.
105. F. Steinseifer and T. Kauffmann, *Angew Chem., Int. Ed. Engl.*, 1980, **19**, 723.
106. T. Kauffmann, F. Steinseifer and N. Klas, *Chem. Ber.*, 1985, **118**, 1039.
107. D. H. R. Barton, J. P. Kitchin and W. B. Motherwell. *J. Chem. Soc., Chem. Commun.*, 1978, 1099.
108. D. H. R. Barton, D. J. Lester, W. B. Motherwell and M. T. B. Papoula, *J. Chem. Soc., Chem. Commun.*, 1979, 705.
109. D. H. R. Barton, J. P. Kitchin, D. J. Lester, W. B. Motherwell and M. T. B. Papoula, *Tetrahedron*, 1981, **37**, W 73.
110. D. H. R. Barton, J.-P. Finet, W. B. Motherwell and C. Pichon, *J. Chem. Soc., Perkin Trans. 1*, 1987, 251.
111. D. H. R. Barton, J.-C. Blazejewski, B. Charpiot, D. J. Lester, W. B. Motherwell and M. T. B. Papoula, *J. Chem. Soc., Chem. Commun.*, 1980, 827.
112. D. H. R. Barton, W. B. Motherwell and A. Stobie, *J. Chem. Soc., Chem. Commun.*, 1981, 1232.
113. D. H. R. Barton, J.-P. Finet, W. B. Motherwell and C. Pichon, *Tetrahedron*, 1986, **42**, 5627.
114. T. Ogawa, T. Murafuji, K. Iwata and H. Suzuki, *Chem. Lett.*, 1988, 2021.
115. S. David and A. Thiéffry, *Tetrahedron Lett.*, 1981, **22**, 2885.
116. D. H. R. Barton *et al.*, *J. Chem. Soc., Perkin Trans. 1* 1985, 2667.
117. D. H. R. Barton, D. J. Lester, W. B. Motherwell and M. T. B. Papoula, *J. Chem. Soc., Chem. Commun.*, 1980, 246.
118. T. Ogawa, T. Murafuji and H. Suzuki, *Chem. Lett.*, 1988, 849.
119. D. H. R. Barton, N. Yadav-Bhatnagar, J.-P. Finet, J. Khamsi, W. B. Motherwell and S. P. Stanforth, *Tetrahedron*, 1987, **43**, 323.
120. D. H. R. Barton, J.-P. Finet, C. Giannotti and F. Halley, *Tetrahedron*, 1988, **44**, 4483.
121. T. Ogawa, T. Murafuji and H. Suzuki, *J. Chem. Soc., Chem. Commun.*, 1989, 1749.
122. H. Suzuki, C. Nakaya, Y. Matano and T. Ogawa, *Chem. Lett.*, 1991, 105.
123. H. Suzuki, T. Murafuji and T. Ogawa, *Chem. Lett.*, 1988, 847.
124. T. Ogawa, T. Murafuji, K. Iwata and H. Suzuki, *Chem. Lett.*, 1989, 325.
125. D. H. R. Barton *et al.*, *J. Chem. Soc., Perkin Trans. 1*, 1985, 2657.
126. D. H. R. Barton, B. Charpiot and W. B. Motherwell, *Tetrahedron Lett.*, 1982, **23**, 3365.
127. D. H. R. Barton, N. Y. Bhatnagar, J.-P. Finet and W. B. Motherwell, *Tetrahedron*, 1986, **42**, 3111.
128. D. H. R. Barton, D. M. X. Donnelly, J.-P. Finet and P. H. Stenson, *Tetrahedron* 1988, **44**, 6387.
129. D. H. R. Barton, M. T. B. Papoula, J. Guilhem, W. B. Motherwell, C. Pascard and E. T. H. Dau, *J. Chem. Soc., Chem. Commun.*, 1982, 732.

130. D. H. R. Barton et al., *J. Am. Chem. Soc.*, 1985, **107**, 3607.
131. J. J. Lalonde, D. E. Bergbreiter and C.-H. Wong, *J. Org. Chem.*, 1988, **53**, 2323.
132. D. H. R. Barton, J.-P. Finet, C. Giannotti and F. Halley, *J. Chem. Soc., Perkin Trans. 1*, 1987, 241.
133. M. J. O'Donnell, W. D. Bennett, W. N. Jacobsen, Y. Ma and J. C. Huffman, *Tetrahedron Lett.*, 1989, **30**, 3909.
134. D. H. R. Barton, J.-C. Blazejewski, B. Charpiot and W. B. Motherwell *J. Chem. Soc., Chem. Commun.*, 1981, 503.
135. D. H. R. Barton, J.-P. Finet, J. Khamsi and C. Pichon, *Tetrahedron Lett.*, 1986, **27**, 3619.
136. S. David and A. Thiéffry, *Tetrahedron Lett.*, 1981, **22**, 5063.
137. S. David and A. Thiéffry, *J. Org. Chem.*, 1983, **48**, 441.
138. D. H. R. Barton, J.-P. Finet and C. Pichon, *J. Chem. Soc., Chem. Commun.*, 1986, 65.
139. H. Brunner, U. Obermann and P. Wimmer, *J. Organomet. Chem.*, 1986, **316**, C1.
140. H. Brunner, U. Obermann and P. Wimmer, *Organometallics*, 1989, **8**, 821.
141. D. H. R. Barton, J.-P. Finet and J. Khamsi, *Tetrahedron Lett.*, 1986, **27**, 3615.
142. D. H. R. Barton, J.-P. Finet and J. Khamsi, *Tetrahedron Lett.*, 1989, **30**, 937.
143. D. H. R. Barton, J.-P. Finet and J. Khamsi, *Tetrahedron Lett.*, 1988, **29**, 1115.
144. D. H. R. Barton, N. Ozbalik and M. Ramesh, *Tetrahedron Lett.*, 1988, **29**, 857.
145. T. Ogawa, K. Miyazaki and H. Suzuki, *Chem. Lett.*, 1990, 1651.

13
Selenium

ALAIN KRIEF
Facultés Universitaires Notre-Dame de la Paix, Namur, Belgium

13.1 INTRODUCTION	516
13.2 SELENIUM: PHYSICAL PROPERTIES AND TYPICAL STRUCTURES OF SELENIUM-CONTAINING MOLECULES	516
13.2.1 The Radionuclides of Selenium and their Uses	517
13.2.2 Location of Selenium in the Periodic Table	518
13.2.3 Electronegativity, Carbon–Selenium Bond Values and Typical Reactivity of Organoselenium Compounds	518
13.2.4 Nature and Structure of some Selenium Compounds and Reagents	519
13.2.4.1 Nature and structure of some inorganic selenium compounds and reagents	519
13.2.4.2 Nature and structure of some organoselenium compounds and reagents	519
13.2.5 Nature and Structure of some α-Selenoalkyl Intermediates	520
13.2.5.1 α-Selenoalkylmetals	521
13.2.5.2 α-Seleno carbenium ions	521
13.2.5.3 α-Selenoalkyl radicals	521
13.2.6 Reviews on Selenium Chemistry	522
13.3 ORGANOSELENIUM CHEMISTRY	522
13.3.1 Selenium in Organic Synthesis: Strategy and Practice	522
13.3.1.1 Introduction of selenium into organic compounds: general considerations	522
13.3.1.2 Removal of the seleno moiety from organic compounds: general considerations	522
13.3.2 Useful Inorganic and Organic Reagents for Organoselenium-monitored Reactions	527
13.3.2.1 Useful inorganic reagents for organoselenium-monitored reactions	527
13.3.3 Useful Organic Reagents for Organoselenium-monitored Reactions	528
13.3.3.1 Strategy	528
13.3.3.2 Synthesis of organoselenium reagents	529
13.3.4 Functional Group Manipulations Involving Inorganic Selenium and Organoselenium Reagents	531
13.3.4.1 Reactions involving hydrogen selenide, selenols and selenolates	531
13.3.4.2 Reactions involving metallic selenium	532
13.3.4.3 Reactions involving selenoxides and related derivatives	533
13.3.4.4 Reactions involving diselenides and related reagents	533
13.3.4.5 Reactions involving areneselenenyl halides	534
13.3.4.6 Reactions involving benzeneseleninic acid and anhydride and related reagents	534
13.3.4.7 Reactions involving benzeneseleninyl halides	535
13.3.4.8 Reactions involving perseleninic acids	535
13.3.4.9 Reactions involving selenium dioxide	536
13.3.4.10 Reactions of imido selenium derivatives	537
13.3.4.11 Reactions of selenium tetrafluoride	537
13.3.5 Synthesis of Organoselenium Compounds	538
13.3.5.1 Synthesis of selenides and functionalized selenides from nucleophilic and electrophilic organoselenium reagents	538
13.3.5.2 Syntheses of selenides involving diselenides as radical initiators or as radical traps	543
13.3.5.3 Syntheses of selenides and functionalized selenides involving concomitant formation of a new C–C bond	543
13.3.5.4 Syntheses of carbonyl compounds bearing a selenium atom attached to the sp^2 carbon of the carbonyl group	545
13.3.6 Reactions of Selenides and Functionalized Selenides: Synthesis of Selenium-free Compounds	547
13.3.6.1 Reactions involving reduction of the C–Se bond	547
13.3.6.2 Reactivity of products resulting from the alkylation, halogenation or complexation of selenides and functionalized selenides	552

13.3.6.3 Reactions involving oxidation of selenides and functionalized selenides	553
13.4 CONCLUSION	557
13.5 REFERENCES	558

13.1 INTRODUCTION

The versatility and the overall synthetic utility of various reagents containing selenium[1-3] have been exemplified in their increasing uses in synthesis, especially for the synthesis of complex natural products.[4] Several thousand papers dealing with the use of selenium-based reagents and reactions have appeared since about 1980. Why has this exotic element, a member of the well-known family of chalcogens, a close neighbour to sulfur and bromine, enjoyed such interest even though the 'known toxicity of organoselenium compounds called for caution and their malodorous reputation made them doubly unattractive to any chemist who wanted a social life outside the laboratory'?[4] This is due to the mixed metallic–nonmetallic nature of the element, its unique reactivity towards oxidation and reduction and the capacity of selenium-based reagents to perform highly chemoselective transformations.

Selenium plays an important role in living systems[5-11] and is involved in serious problems at both ends of the supply: deficiency (<0.05–0.1 mg kg^{-1})[12-14] and toxicity (>5–15 mg kg^{-1})[12,15] have been recognized. Selenocysteine,[5] the 21st essential amino acid,[10,16] has been discovered in the active site of glutathione peroxidase, an enzyme which catalyses the destruction of peroxides once formed, and type I iodothyronine deiodinase, an enzyme responsible for the synthesis of thyroxin, the active principle which regulates metabolism in mammals.[17] Several reports stress the essential nature of selenium as a micronutrient[13] and as a protecting agent against radiations,[18,19] certain degenerative diseases[13,14,20] and oxidative stress,[21] as well as a potential chemotherapeutic agent.[22]

The chemistry of heterocycles[23-5] containing selenium has been reviewed elsewhere and will not be discussed here. This is also the case with tetraselenofulvalenes and related heterocycles, which have been successfully used for the production of organic conductors[26-9] and selenium-containing polymers.[30,31]

Selenium-based reagents were not popular in organic synthesis until 1973. Selenium dioxide (SeO_2), although used extensively[32-45] by organic chemists since the discovery in 1932 of its unique aptitude to oxidize methine and methylene groups linked to various functional groups, did not enjoy a good reputation owing to the bad smell associated with the organoselenium by-products and the need for the use of mercury as scavenger (Equation (1), (i)). It nevertheless remained almost the unique selenium reagent to be used until 1973. At that time five important events occurred: (i) the intimate mechanism of allylic oxidation of alkenes[46,47] using this reagent was discovered; (ii) the catalytic version, which avoids the production of malodorous by-products and involved t-butyl hydroperoxide as the active oxidant,[48] was disclosed (Equation (1), (ii); compare with (i)); (iii) the value of organoselenium chemistry for alkene syntheses was recognized by Jones et al.[49] and Barton and co-workers[50] (Scheme 1 (a) and (b)); (iv) the cleavage of the C–Se bond in bis-, tris- and tetrakis(phenylseleno)methane was observed by Seebach and Peleties[51,52] (Scheme 2); and (v) a valuable monograph fully covering the chemistry and biology of organic selenium compounds was published.[53]

$$Bu\diagup\diagdown \xrightarrow[\text{i or ii}]{0.5 \text{ equiv. } SeO_2} Bu\diagup\diagdown_{OH} + Bu\diagup\diagdown^{OH} \quad (1)$$

i, Bu^{t}OH, reflux   16%   2%
ii, 2 equiv. Bu^{t}O$_2$H, CH$_2$Cl$_2$, 20 °C   53%   11%

13.2 SELENIUM: PHYSICAL PROPERTIES AND TYPICAL STRUCTURES OF SELENIUM-CONTAINING MOLECULES

For a survey of the literature on the physical properties and typical structures of selenium-containing molecules see Refs. 11 and 54–7.

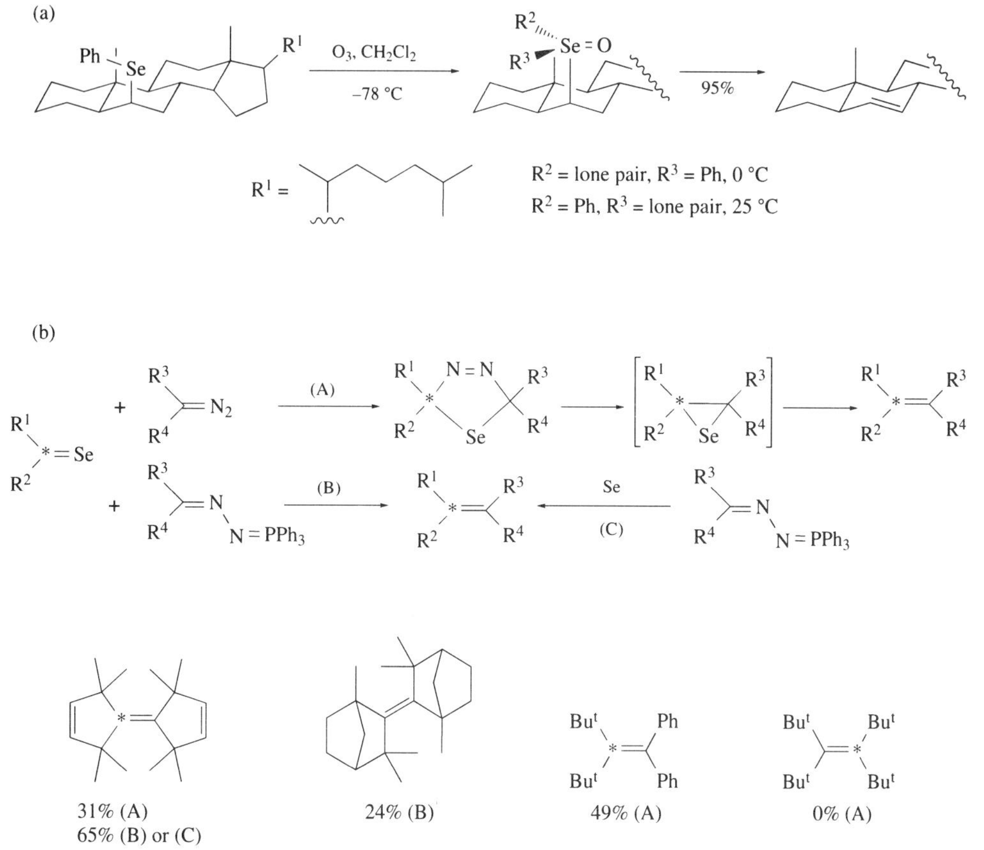

Scheme 1

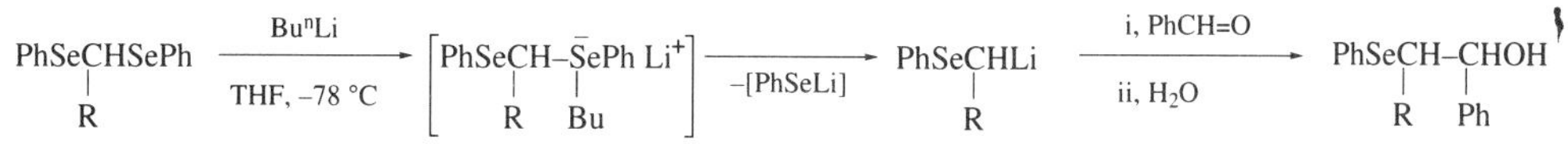

Scheme 2

13.2.1 The Radionuclides of Selenium and their Uses

Grey, metallic selenium is the only crystalline[3,57,58] and the ordinary commercial form of the element. It is widely used for the direct synthesis of various organoselenium derivatives since it is, for example, reduced to metal selenides or diselenides on reaction with metal[59] or metal hydrides, and to organoselenolates or selenols on reaction with organometallics. Red, amorphous selenium, originally known as α-selenium, is precipitated when H_2Se is oxidized by air, when selenous acid is reduced, for example, with sulfur dioxide or when metal selenocyanates are reacted with acids.

Selenium possesses an atomic weight of 76.86 owing to the presence of six stable isotopes[11,58] (mass (ratio, %) 74 (0.87), 76 (9.02), 77 (7.58), 78 (23.52), 80 (49.82) and 82 (9.19)) which are responsible for the complex patterns characteristic of the element observed in the mass spectra of organoselenium compounds.[60,61] Among these, the ^{77}Se isotope has a nuclear spin of $\frac{1}{2}$ and a reasonably good receptivity (2.98 times that of ^{13}C), and has been extensively used[62,63] in ^{77}Se NMR for the structure elucidation of organoselenium compounds, the determination of the reaction pathways in organoselenium chemistry and the elucidation of the structure and conformation of peptides into which seleno analogues of sulfur-containing amino acids have been deliberately introduced. Several unstable radionuclides are also known (mass 70, 72, 73, 75, 79, 81, 83, 85, 86 and 87).[64] Among these, ^{75}Se is commonly available (gamma ray emission, half-life 122 d) and is used extensively as a tracer in biochemical studies and as a radiopharmaceutical for diagnostic purposes.[10,65,66] [^{75}Se]Selenomethionine,[10] for example, has been used for pancreatic imaging.[10]

Analytical methods,[67] including detection and structure determination,[68] thermochemistry[69] and other physicochemical investigations of selenium compounds have been reviewed. They include UV–visible light spectroscopy,[70] infrared spectroscopy,[71] chiroptical properties,[72,73] proton magnetic resonance,[62,74] EPR spectroscopy on organoselenium radicals[75] and x-ray diffraction.[60]

13.2.2 Location of Selenium in the Periodic Table

Selenium belongs to the well-known family of *chalcogens* and lies in the fourth row of the periodic table. Its inner shells ($1s^2$, $2s^2$, $2p^6$, $3s^2$, $3p^6$, $3d^{10}$) are completely filled, and its outer shell ($4s^2$, $4p^4$) configuration approaches that of the next noble gas (krypton).[3,55] The electrons present in the $4p$ orbitals of selenides (lone pairs) and metal selenides confer upon them very good nucleophilic properties (sodium phenylselenolate is a soft and particularly nucleophilic species (PhSeNa > PhSNa > INa > MeONa; PhSeNa: ×7 PhSNa, ×10^3 INa, ×10^4 MeONa;[76] estimated nucleophilicities toward MeI, n_{MeI}: Ph$_2$Bi$^-$ 14.3, Co(dmgH)(2py)$^-$ 13.3, Ph$_3$Sn$^-$ 11.5, CpFe(CO)$_2^-$ 11.3, PhSe$^-$ 10.2, CN$^-$ 7.4, I$^-$ 7.3, OH$^-$ 6.5, N$_3^-$ 6.3, NH$_3$ 5.5, Mn(CO)$_5^-$ 5.4, MoCp(CO)$_3^-$ 5.3, Co(CO)$_4^-$ 3.5, MeOH 0.00)),[77] whereas the $4d$ unfilled orbitals, which are close in energy to the $4p$ orbitals, can be easily populated,[11,78,79] thus allowing the synthesis of the stable bis(4,4-dimethyl-2,2'-biphenylene)selenanes,[79] the formation of organoselenium ate complexes[80] (Scheme 3) and the stabilization of α-seleno carbanions (see below).[78,81-7]

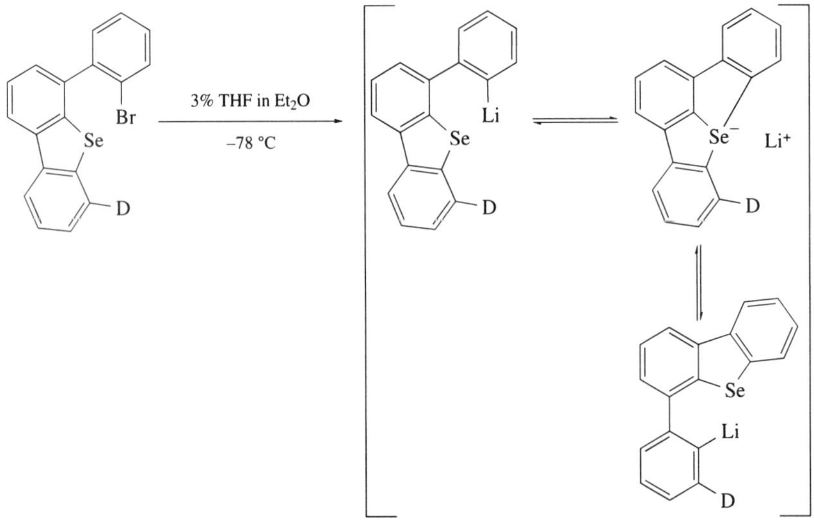

Scheme 3

Selenium chemistry is predominantly nonmetallic, but the position of this element in the period confers on it increasing metallic properties.[58] Consequently, one of its most apparent characteristics is its 'schizophrenic personality', and it behaves as a nonmetallic nonmetal and as a nonmetallic metal.[53,88] Thus selenium, unlike sulfur, possesses one metallic among the different allotropic forms, whereas metallic forms are more common with tellurium,[58] and butyllithiums show a reasonably high propensity to react on the selenium atom of selenides to produce novel alkyllithiums and butyl selenides through selenium–metal exchange (Scheme 2).[78,81-6,89]

13.2.3 Electronegativity, Carbon–Selenium Bond Values and Typical Reactivity of Organoselenium Compounds

The electronegativity of selenium is close to that of carbon but lower than that of bromine (Se 2.48, C 2.50, S 2.44, Br 2.74).[90] The carbon–selenium single bond (234 kJ mol^{-1}) is weaker than the carbon–sulfur bond (272 kJ mol^{-1}) or the carbon–bromine bond (285 kJ mol^{-1}) (covalent radii: C 0.077, O 0.074, S 0.104, Se 0.117, Te 0.137 nm).[91] Hydrogen selenide and selenols[92] are far more acidic than their thio and oxygen analogues (pK_a: H$_2$X[93,94] 3.74 (Se), 7 (S) and PhXH[11] 5.9 (Se), 6.25 (S)). The strength of the carbon–selenium double bond, which involves a $2p$–$4p$ overlap, is much lower than that of the carbon–oxygen double bond but related to that of the carbon–sulfur double bond (covalent radii: C 0.067, O 0.062, S 0.094, Se 0.107, Te 0.127 nm).[91] Comparison of the ^{17}O and ^{77}Se chemical shifts in NMR and the absorption spectra of carbonyl and seleno carbonyl compounds suggests that there is a true 'double-bond character' in the latter functional group.[91,95,96] Selenoaldehydes and -ketones are unstable derivatives unless sterically shielded and although other seleno carbonyl compounds are reasonably stable, they are all characterized by a high tendency to undergo replacement of the selenium by an oxygen, nitrogen or sulfur atom and to be converted into compounds in which the more stable carbon–selenium single bond is produced (Scheme 4).[91]

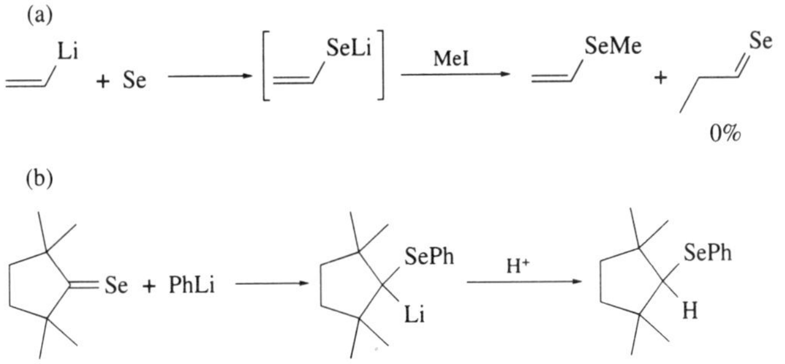

Scheme 4

The polarizability of the carbon–selenium double bond is greater than that of the carbon–oxygen double bond. Seleno carbonyl compounds are therefore weaker bases than their oxygen or sulfur analogues but show increased nucleophilicity at the selenium atom and increased electrophilicity at carbon. Hence the tendency of the former reagents to react with nucleophiles at the heteroatom rather than at the carbon centre can be explained by the polarity of the seleno carbonyl group, which is much lower than that of the carbonyl group, and by the stabilization by the seleno group of the incipient anion or radical intermediates (Scheme 4 (b)).[91]

13.2.4 Nature and Structure of some Selenium Compounds and Reagents

Selenium compounds can accommodate numerous structural variations since the selenium atom can possess an oxidation level from −2 to +6. The most representative are listed below.

13.2.4.1 Nature and structure of some inorganic selenium compounds and reagents

Commonly encountered inorganic selenium compounds and reagents[37,54] are hydrogeno (**1a**) (H_2Se),[32,97,98] hydrogeno metal (**1b**) (HSeM)[32,97,98] and dimetal (**1c**) (M_2Se)[32,97] selenides, hydrogeno (**2a**) (H_2Se_2) and metal (**2b**) (M_2Se_2) diselenides,[32,97] hydrogeno (**3a**) (HSeCN) and metal (**3b**) (MSeCN) selenocyanates,[32,97,99] metal selenosulfates (**4**) (SO_3SeM_2),[32,100] selenium dihalides (X_2Se, X = Cl, Br) (**5**),[32,97] selenium halides (X_2Se_2, X = Cl, Br) (**6**),[37,97] selenium tetrahalides (**7**) (X_4Se, X = F, Cl, Br),[32,101] selenium hexafluoride (**8**) (F_6Se), selenium oxodihalides (X_2SeO, X = F, Cl, Br) (**9**),[32] selenium dioxide (**10**) (SeO_2),[32,33,36,37,39–45,102,103] imidoselenium compounds (**11**) (Se(=NTs)$_2$), selenium trioxide (**12**) (SeO_3),[32] selenous (**13a**) (H_2SeO_3) and selenic (**14a**) acids, their salts ((**13b**) and (**14b**)) ($H_{2-n}M_nSeO_4$) and their esters (**13b**) and (**14b**),[104,105] peroxoselenous (**15**) (H_2SeO_4) and peroxoselenic (**16**) (H_2SeO_5) acids, carbon oxoselenide (**17a**) (O=C=Se),[106,107] carbon diselenide (**17b**) (Se=C=Se)[32,97,106] and phosphorus pentaselenide (P_2Se_5).[32]

13.2.4.2 Nature and structure of some organoselenium compounds and reagents

Commonly encountered organoselenium compounds and reagents[54] are selenols (**19a**) and metal selenolates (**19b**),[37,59,98,108–11] selenides (**20**),[56,112–16] diselenides (**21**),[37,56,108,109,117–21] selenocyanates (**22**),[99,108,109,122,123] selenenyl halides (**23**),[109,121,124] selenenic acids (**24**) (unstable) and related derivatives,[37,108–10,120,124–6] selenenylamines (**25a**),[109] selenenylamides (**25b**) and selenosulfonates (**26**),[108,127,128] selenocarboxylic acids (**27a**) and their salts (**27b**), selenol esters (acyl selenides)[96,129] (**28**), diacyl selenides (**29**) and diselenides (**30**), selenocarbonates (alkoxycarbonyl selenides) (**31**),[91,95] selenonium salts (**32**),[37,130-2] selenide dihalides (**33**),[37,112,113,133] selenoxides (**34**),[37,39,114,115,120,134-6] selenimines (**35**),[137,138] selenones (**36**),[37,114,115,135,139] seleninic acids (**37**),[37,108-10,120,140] seleninic anhydrides (**38**),[109,110,141] perseleninic acids (**39**),[142] selenonic acids (**40**),[37,108-10,143] selones (**41**),[37,91,95,96,144,145] selenocarboxylic acids, their salts and esters (**42**),[91,95,96,146-8] selenoamides (**43**),[91,95,96,129,148,149] selenocarbonates (**44a**)[148] and selenoureas (**44b**),[32,91,95,106,149] seleno ketenes (**45**),[91,95,96,148] isoselenocyanates (**46**),[91,95,96,149,150] diselenoic acid esters (**47**),[151] selenium tetraalkoxides (**48**),[105,152] selenious acid esters (**49**),[105,152] selenic acid esters (**50**)[105] and organoselenophosphorus compounds.[153]

R¹SeH (**19a**)   R¹SeM (**19b**)   R¹SeR² (**20**)   R¹SeSeR² (**21**)   R¹SeCN (**22**)

R¹SeX, X = Cl, Br, I (**23**)   R¹SeOH (**24**)   R¹SeNR²R³ (**25a**)   R¹SeN(C(O)R²)(C(O)R³) (**25b**)

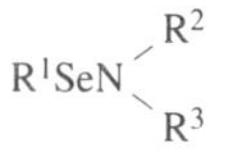

R¹SeSO₂R² (**26**)   R¹C(O)SeH (**27a**)   R¹C(O)SeM (**27b**)   R¹C(O)SeR² (**28**)

R¹C(O)SeC(O)R² (**29**)   R¹C(O)Se–SeC(O)R² (**30**)   R¹C(O)SeOR² (**31**)   R¹–Se⁺–R² R³ X⁻ (**32**)

R¹–Se⁺(X)–R² X⁻, X = halogen (**33**)   R¹–Se(=O)–R² (**34**)   R¹–Se(=NR³)–R² (**35**)   R¹–Se(=O)₂–R² (**36**)

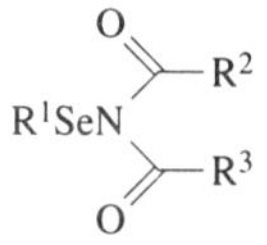

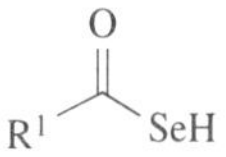

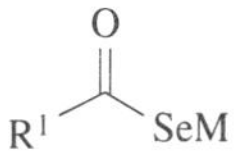

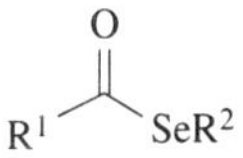

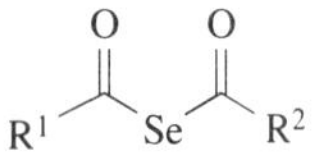

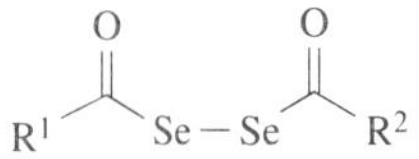

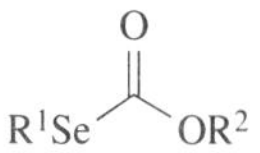

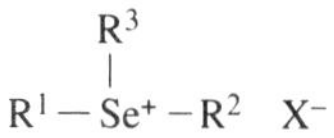

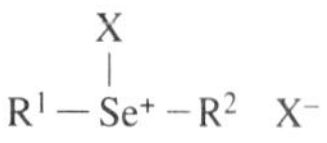

R¹–Se(=O)–OH (**37**)   R¹–Se(=O)–O–Se(=O)–R² (**38**)   R¹–Se(=O)–O–OH (**39**)   R¹–Se(=O)₂–OH (**40**)

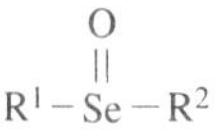

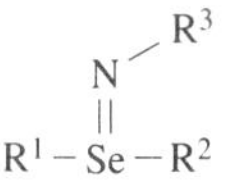

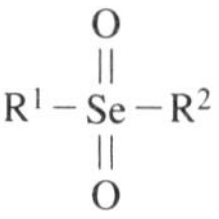

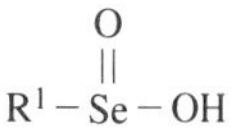

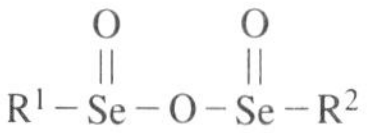

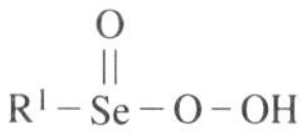

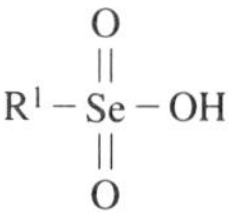

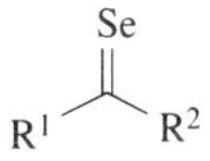

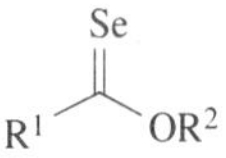

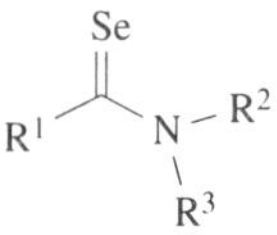

R¹R²C=Se (**41**)   R¹(OR²)C=Se (**42**)   R¹C(=Se)NR²R³ (**43**) 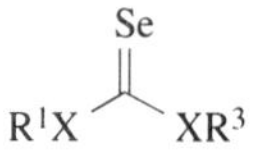   (**44a**) X = O: R¹X–C(=Se)–XR³; (**44b**) X = NR²

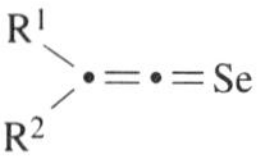

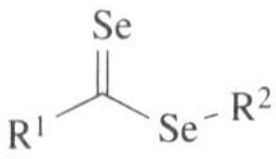

R¹R²C=•=Se (**45**)   R¹N=C=Se (**46**)   R¹(Se⁻R²)C=Se (**47**)   Se(OR)₄ (**48**)   (RO)₂SeO (**49**)   (RO)₂SeO₂ (**50**)

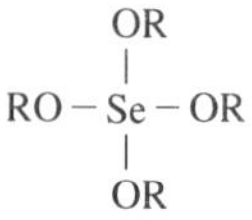

13.2.5 Nature and Structure of some α-Selenoalkyl Intermediates

α-Seleno carbanions, α-seleno carbenium ions and α-seleno carbon-centred radicals have been postulated as intermediates in organoselenium chemistry. The ease with which they are formed or react, as compared, for example, with the related sulfur derivatives, has been used to explain their relative reactivity. Information on their synthesis, stability and physical data, and in some cases comparison with related derivatives bearing a sulfur or an oxygen moiety, is summarized below. Their synthetic applications, especially of α-selenoalkylmetals, are discussed in Section 13.3.5.3(i).

13.2.5.1 α-Selenoalkylmetals

α-Selenoalkylmetals have been prepared[1,78,81-7,89] (i) from selenoacetals and alkyllithiums by selenium–metal exchange, (ii) by addition of alkyllithiums to vinyl selenides, (iii) by metallation, with suitable bases such as LDA (lithium diisopropylamide), KDA (potassium diisopropylamide), LiDBA (lithium diisobutylamine) or LiTMP (lithium 2,2,6,6-tetramethyl piperidide), of selenides (a) bearing an electron-withdrawing group in the α-position such as α-seleno carbonyl compounds, nitriles and nitroalkanes, or (b) another seleno moiety such as seleno orthoesters and selenoacetals derived from aromatic aldehydes and (iv) by halogen–metal exchange of α-bromoalkyl selenides. The stabilization of the α-seleno carbanions has been ascribed to delocalization to the $4d$ unfilled orbitals of the selenium atom or to the high polarization of the element allowing the delocalization of the charge through the atom on the closely located bonds. In the former event the carbanionic centre in α-seleno carbanions would have been more stabilized than in α-thio carbanions owing to the better $2p(C)-3d(S)$ than $2p(C)-4d(Se)$ overlap, whereas the reverse is expected in the latter case owing to the expected higher polarization of the selenium than the sulfur atom. Experimentally, α-thioalky lithiums proved to be slightly more stable than their seleno analogues (60:40 ratio under thermodynamically controlled conditions).[81] The ^{77}Se NMR spectra of some α-selenoalkylmetals have been reported and data supporting delocalization of the carbanion through the selenium atom have been obtained.[81]

13.2.5.2 α-Seleno carbenium ions

α-Seleno carbenium ions have been postulated[154-6] as intermediates in *inter alia* (i) the synthesis of selenoacetals from carbonyl compounds[78,154] and in the acid-catalysed hydrolysis of selenoacetals, vinyl selenides and ketene selenoacetals, (ii) the reactions of α-haloalkyl selenides, selenoacetals and orthoesters with silyl enol ethers, furan and pyrrole and (iii) a series of reactions involving a seleno-Pummerer rearrangement.[155] They are believed to be stabilized by the lone pairs lying in the $4p$ orbitals of the selenium atom.[155] Detailed insight into the nature of 'α-seleno carbeniums' was gained through ^1H, ^{13}C and ^{77}Se NMR and x-ray studies of some representative compounds such as bis(methylseleno)benzylcarbenium hexachloroantimonate.[157] It appears that these ions exhibit very similar thermodynamic stabilities to the sulfur analogues, which is less pronounced than that of the oxygen analogues, and that a relatively high sp^2 character can be attributed to the selenium atom in these species. Kinetic studies of the acidic hydrolysis of various methyl vinyl selenides have shown, however, that the mechanism differs from that of the sulfur and oxygen analogues in that the protonation step is a partially reversible process. The kinetic behaviour cannot be rationalized on the basis of the intrinsic stability of the intermediate α-seleno carbenium ions, but rather by the existence of a high kinetic barrier towards nucleophilic attack due to the high polarization of the electron clouds surrounding the selenium atom.

13.2.5.3 α-Selenoalkyl radicals

The field of α-selenoalkyl radicals is much less well documented than that of the related anions and cations.[155] They have been postulated as intermediates in the reduction of selenoacetals with lithium in ethylamine or with tin hydrides[158] and in the addition of carbon-centred radicals to α-seleno-α,β-unsaturated nitriles and esters.[155] Hindered α-selenoalkyl radicals have also been easily generated from various radicals (Me, Bu^t, CF$_3$, Bu^{t}O, Me$_3$Sn) and di-*t*-butyl selenoketone, and some of their spectroscopic data have been reported.[155] All attempts at the direct observation of nonhindered α-selenoalkyl radicals formed by hydrogen abstraction from dimethyl selenide, diethyl selenide and methyl phenyl selenide in solution proved to be unsuccessful, but trapping experiments of the methylselenomethyl radical with di-*t*-butyl thioketone proved that it had indeed been produced.[155]

13.2.6 Reviews on Selenium Chemistry

Various aspects of organoselenium chemistry have been covered since the 1940s as reviews in books[4,23,36,40-2,48,53,58,82,84,85,88,99,159-72] and as review articles.[1,26,35,37,44,47,78,89,132,154,173-95]

Selenium and Tellurium Abstracts have been edited by the Selenium and Tellurium Association. They have been published since 1955 (vols. 1–7 cover 1955–66) and since 1967 have been published monthly by the Chemical Abstracts Services of the American Chemical Society.

13.3 ORGANOSELENIUM CHEMISTRY

13.3.1 Selenium in Organic Synthesis: Strategy and Practice

The success of selenium reagents is related to their capacity to perform highly chemoselective transformations[4,48,88,160,162,163] such as reduction (Scheme 5 (a) and (b) (HMPA, hexamethylphosphoramide))[88,111,120,158,196] or oxidation (Equation (1); Scheme 5 (c)–(h)),[34,39,88,102,103,120,195,197,198] nucleophilic displacements (Scheme 5 (i))[39,53,76,190,194,198] or electrophilic reactions (Scheme 6 (d) and (e)).[39,128,132,164,194,198,199] It is also related to the ease with which the selenium atom can be introduced[32] into organic molecules and then expelled after proper activation of the seleno moiety.[1,4,39,48,76,78,81-4,89,132,174,192,199]

13.3.1.1 Introduction of selenium into organic compounds: general considerations

Introduction of selenium into organic molecules has been usually achieved using (i) the element (Scheme 6 (a)), (ii) inorganic nucleophilic or electrophilic reagents (Scheme 6 (b) and (c)), (iii) nucleophilic or electrophilic organoselenium reagents (Scheme 6 (d)–(f)) and (iv) building blocks bearing a seleno moiety directly attached to a carbanion, a carbocation or a radical (Schemes 7 and 8 (AIBN, 2,2'-azobisisobutyronitrile).[89] In methods (i)–(iii) a new carbon–selenium bond is generated, whereas in approach (iv) a new carbon–carbon bond is produced concomitantly with introduction of the seleno moiety.[89] In practice, the most widely used methods are those involving organoselenium reagents (method (iii)) and α-selenoalkylmetals (method (iv)).

13.3.1.2 Removal of the seleno moiety from organic compounds: general considerations

C–Se bond cleavage in selenides and functionalized selenides has been successfully accomplished under various conditions. Those involving a nucleophilic substitution, a β-elimination reaction, a reduction via an organometallic or a radical intermediate have been the most widely used.

(i) Removal of the seleno moiety involving a nucleophilic substitution or a β-elimination reaction

This general approach has been achieved by transformation of the seleno moiety into a better leaving group by reaction with (i) Lewis acids,[84,191,192,200,201] alkyl halides,[1,84,202-4] carbenes[84,191,192] or halogens (Schemes 7 and 9),[1,85,113] taking advantage of the high nucleophilicity of the selenium atom, or (ii) oxidants, leading to selenoxides[39,114,115,120,135] or selenones,[135,192] and using the great propensity of the selenium atom to be oxidized (Scheme 1 (a); Scheme 6; Scheme 8 (a)).

The selenenyl group in selenones is very easily substituted, as is the seleno group in selenonium salts[132,192] (Scheme 7 (d) and (e); Scheme 9 (a)). On reaction with bases, however, selenonium salts give alkenes via an elimination reaction[205] (Scheme 9 (b)). Alkenes are also produced from selenoxides (Scheme 1 (a)).[1,4,39,48,49,76,78,128,135,160,173,190,206] This reaction is probably the most frequently used in selenium chemistry owing to the very mild and neutral conditions (20 °C) involved (Scheme 1 (a); Scheme 6 (a), (d), (e); Scheme 8 (b)).

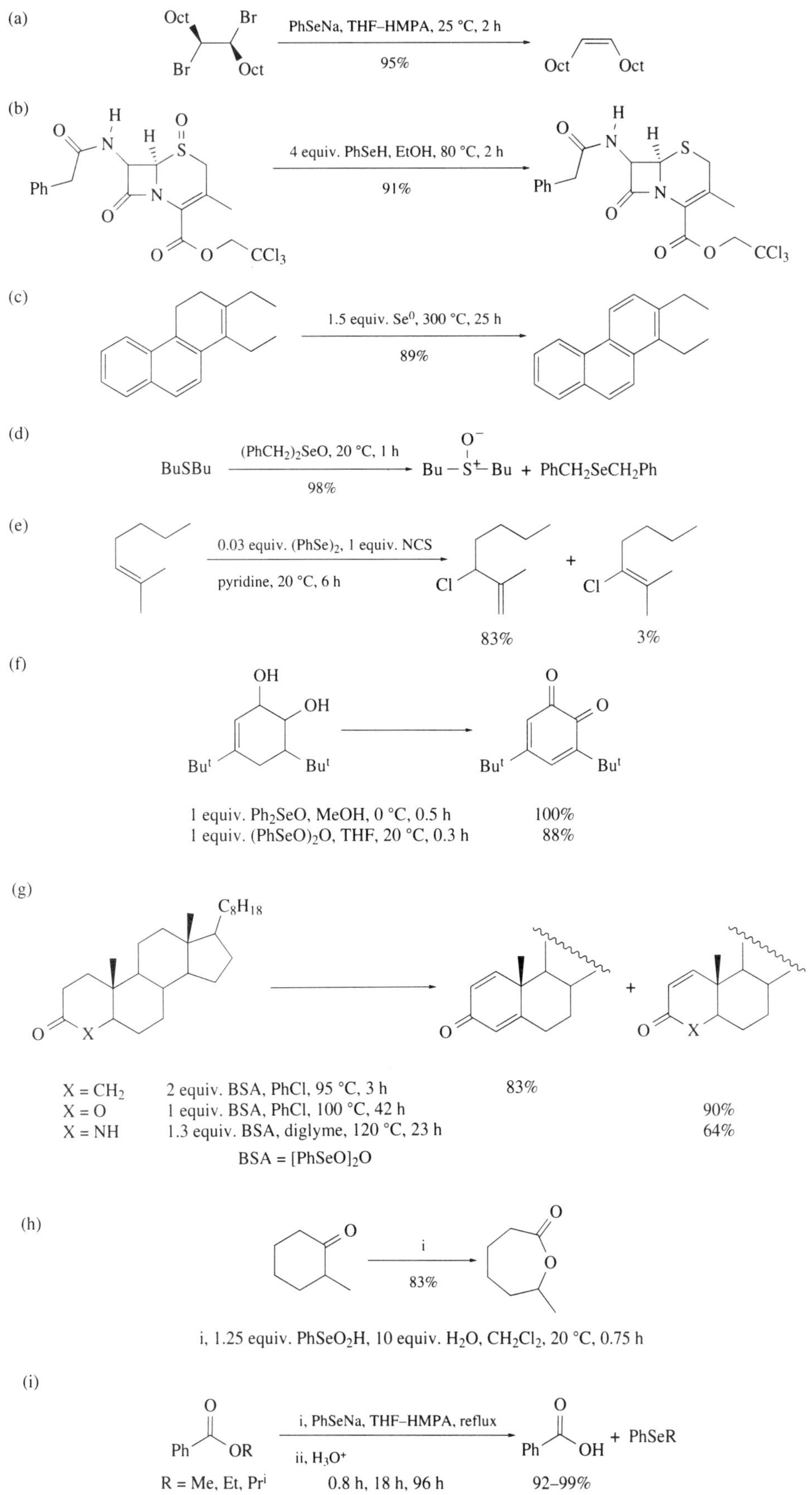

Selenium

(a)

Oct—CHBr—CHBr—Oct $\xrightarrow{\text{PhSeNa, THF–HMPA, 25 °C, 2 h}}$ Oct–CH=CH–Oct (cis)

95%

(b)

$\xrightarrow{\text{4 equiv. PhSeH, EtOH, 80 °C, 2 h}}$

91%

(Reduction of sulfoxide to sulfide in cephem derivative)

(c)

$\xrightarrow{\text{1.5 equiv. Se}^0\text{, 300 °C, 25 h}}$

89%

(d)

BuSBu $\xrightarrow{\text{(PhCH}_2\text{)}_2\text{SeO, 20 °C, 1 h}}$ Bu–S$^+$(O$^-$)–Bu + PhCH$_2$SeCH$_2$Ph

98%

(e)

$\xrightarrow{\text{0.03 equiv. (PhSe)}_2\text{, 1 equiv. NCS}}_{\text{pyridine, 20 °C, 6 h}}$

83% + 3%

(f)

1 equiv. Ph$_2$SeO, MeOH, 0 °C, 0.5 h — 100%
1 equiv. (PhSeO)$_2$O, THF, 20 °C, 0.3 h — 88%

(g)

X	Conditions	Yield
X = CH$_2$	2 equiv. BSA, PhCl, 95 °C, 3 h	83%
X = O	1 equiv. BSA, PhCl, 100 °C, 42 h	90%
X = NH	1.3 equiv. BSA, diglyme, 120 °C, 23 h	64%

BSA = [PhSeO]$_2$O

(h)

$\xrightarrow{\text{i}}$ 83%

i, 1.25 equiv. PhSeO$_2$H, 10 equiv. H$_2$O, CH$_2$Cl$_2$, 20 °C, 0.75 h

(i)

PhCO–OR $\xrightarrow[\text{ii, H}_3\text{O}^+]{\text{i, PhSeNa, THF–HMPA, reflux}}$ PhCO–OH + PhSeR

R = Me, Et, Pr^i 0.8 h, 18 h, 96 h   92–99%

Scheme 5

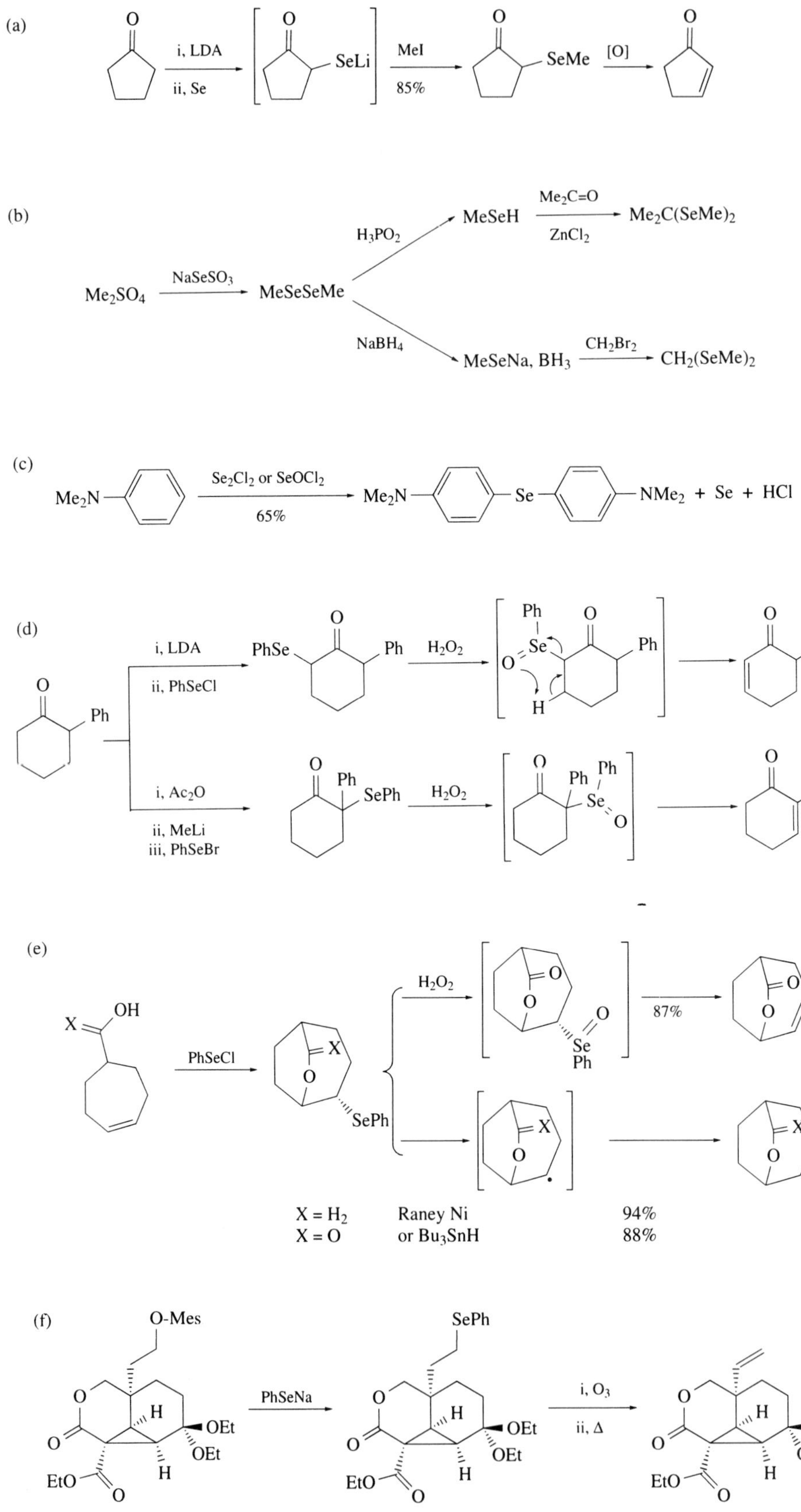
(a)
i, LDA
ii, Se
SeLi
MeI
85%
SeMe
[O]
(b)
Me2SO4
NaSeSO3
MeSeSeMe
H3PO2
MeSeH
Me2C=O
ZnCl2
Me2C(SeMe)2
NaBH4
MeSeNa, BH3
CH2Br2
CH2(SeMe)2
(c)
Me2N
Se2Cl2 or SeOCl2
65%
Se
NMe2 + Se + HCl
(d)
i, LDA
ii, PhSeCl
PhSe
Ph
H2O2
i, Ac2O
ii, MeLi
iii, PhSeBr
SePh
(e)
OH
PhSeCl
SePh
87%
X = H2
X = O
Raney Ni
or Bu3SnH
94%
88%
(f)
O-Mes
PhSeNa
OEt
EtO
i, O3
ii, Δ

Scheme 6

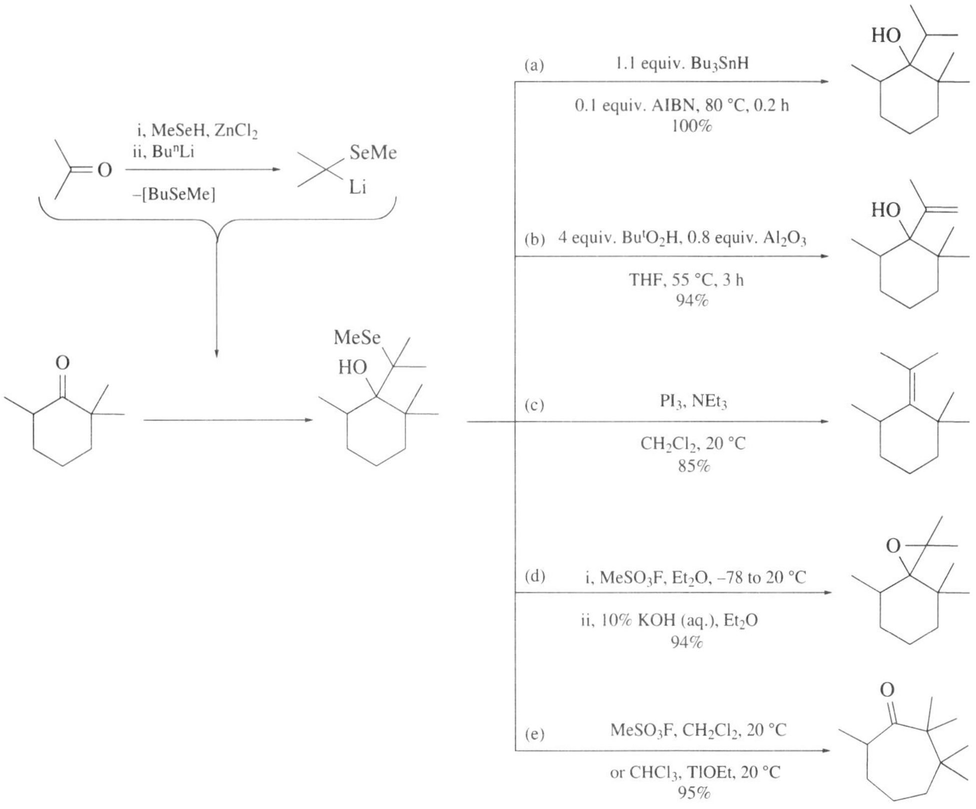

Scheme 7

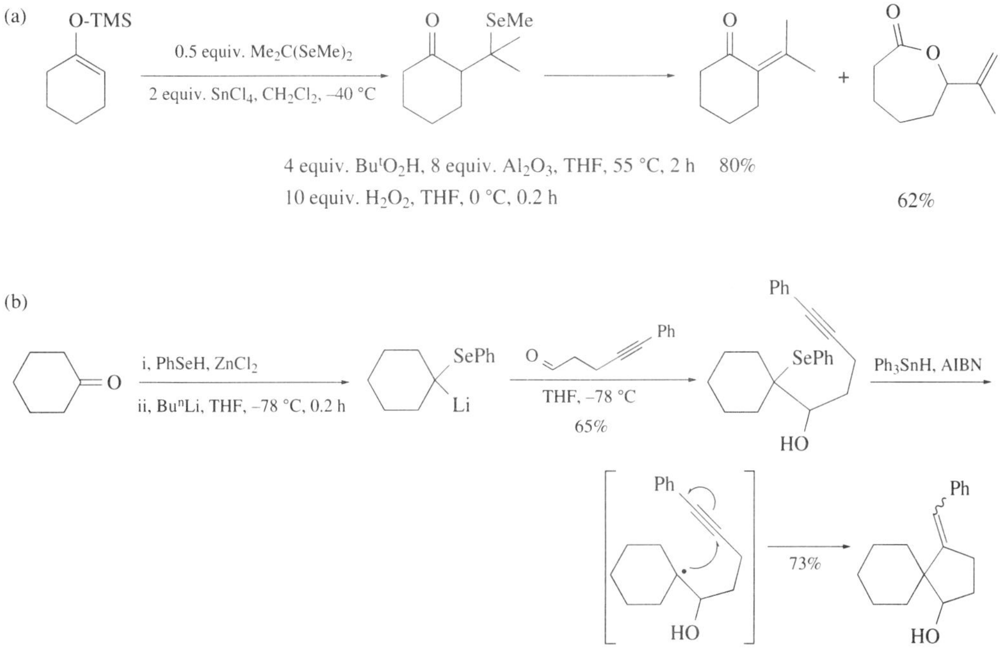

Scheme 8

(ii) Removal of the seleno moiety involving organolithium compounds

These reactions have usually been performed with butyllithiums and lead to the formation of a novel organometallic, but only proceed when the organometallic compound produced possesses a carbanionic

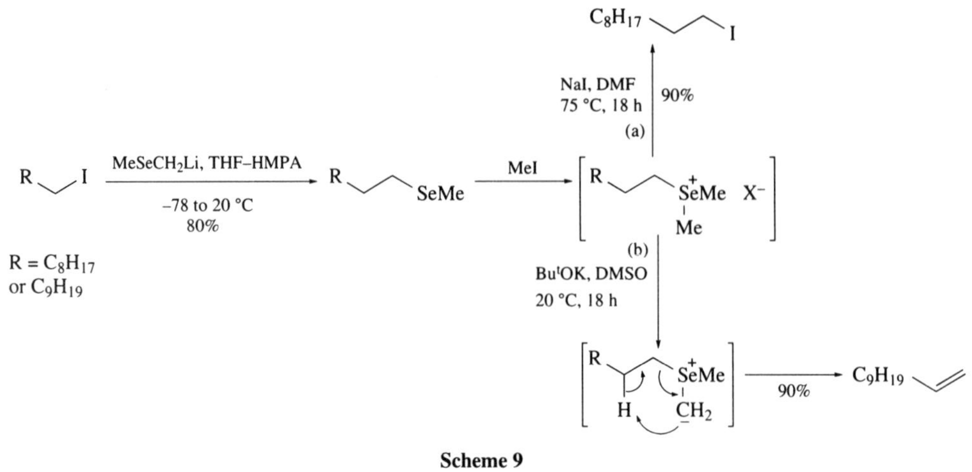

Scheme 9

centre that is well stabilized in comparison with those of butyllithiums (Schemes 2 and 7; Scheme 8 (a) and (b)). Thus, butyllithiums react even at −78 °C with methyl phenyl selenide[78,81,83] to produce phenyllithium (and methyl butyl selenide) via selenium–phenyl bond cleavage, whereas allyl and benzyl selenides give allyl-[207,208] and benzyllithiums[209,210] (and a butyl selenide) via selenium–allyl and selenium–benzyl bond cleavage. The reaction has been successfully extended to the synthesis of α-selenoalkyllithiums[1,48,78,81-3,85,174] from selenoacetals, of α-thioalkyllithiums[82,85,211,212] from α-selenoalkyl sulfides, and of α-silylalkyl lithiums[82] from α-selenoalkylsilanes. These organometallics, especially those which bear two alkyl groups at the carbanionic centre, often cannot be prepared efficiently by classical methods, and have been used in further unusual reactions. This is effectively the case with α-selenoalkyllithiums, which show a very high propensity to react at the carbonyl group of highly hindered or enolizable ketones (such as 2,2,6,6-tetramethylcyclohexanone,[213] permethylcyclopentanone and deoxybenzoin),[78,81,84] and with α-selenoalkylbenzyllithiums, which readily add to nonfunctionalized C=C double bonds (Scheme 10).[209,214-16]

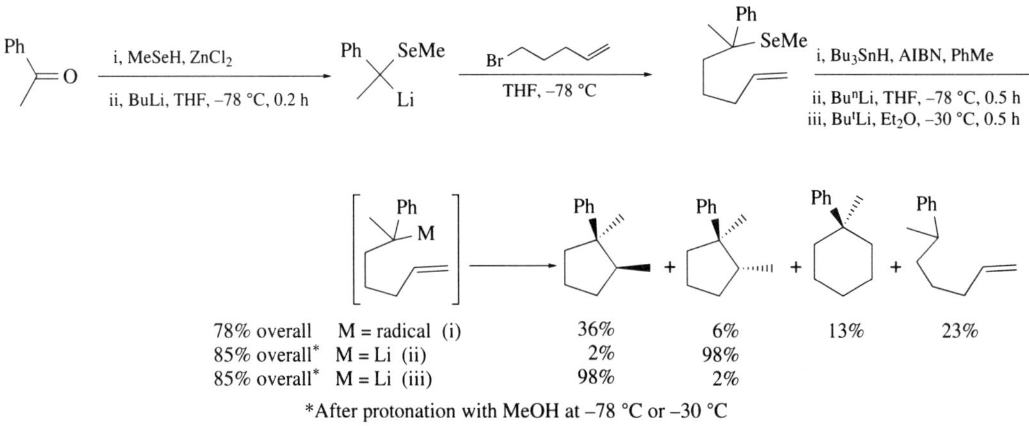

Scheme 10

(iii) Removal of the seleno moiety involving the intermediate formation of a radical

Selenides are cleaved on reaction with metals or tin hydrides (see below).[158] These reagents are expected to produce radical intermediates. The reactivity towards S_H2 attack by tributyltin radicals of various groups X in different molecular environments (RCH_2X, RCO_2CH_2X, EtO_2CCH_2X and RCH_2OCH_2X) was found by competition experiments to be Br > PhSe > Cl > 4-NCC_6H_4S > PhS, and therefore aryl selenides are highly valuable precursors of radicals.[217]

As general trends, the C–Se(phenyl) or the C–Se(vinyl) bond is more readily cleaved than the Se–phenyl or the Se–vinyl bond. One of the most selective systems is probably that involving lithium,[218] lithium di-t-butylbiphenyl (LiDBB) or lithium naphthalenide in THF, which does not cleave dialkyl selenides, although it cleanly reduces alkyl phenyl selenides, benzyl phenyl selenides and benzyl methyl selenides to alkyl- or benzyllithiums. Lithium in ethylamine is much less selective since it will even cleave the C–Se bond of dialkyl selenides. The reaction with tin hydrides has proved to be particularly valuable, especially for the construction of cyclic compounds by trapping of the radical intermediate on a suitably positioned C=C double bond (Scheme 8 (b); Scheme 10). In order to achieve valuable 'selenium-monitored' synthetic transformations, the means of introduction of the seleno moiety into the organic molecule and its subsequent removal must be different. The transformation described in Scheme 6 (f) is a good example of such a strategy since the introduction of the seleno moiety is achieved via a substitution reaction which takes advantage of the high nucleophilicity of sodium phenylselenolate, whereas its removal is performed by selenoxide formation and phenylselenenic acid elimination, which gives the terminal alkene. This general transformation must therefore be of greater synthetic value than those involving the direct elimination of water from an alcohol, its sulfonate or an alkyl halide precursor. These latter methods usually require strongly acidic or basic conditions and often lead to substitution reactions instead of elimination.[219,220] The value of the selenium-based approach is now well established and the overall transformation is widely used, especially for the synthesis of complex molecules.[4,181]

13.3.2 Useful Inorganic and Organic Reagents for Organoselenium-monitored Reactions

13.3.2.1 Useful inorganic reagents for organoselenium-monitored reactions

Although some useful inorganic selenium reagents such as selenium dioxide, hydrogen selenide, potassium selenocyanate and aluminum selenide are commercially available, certain others are not, but they are easy to prepare and it is reasonable to briefly describe their synthesis, which usually involves the use of grey elemental selenium.[221] Many reactions suffer from the concomitant formation of polyselenides, which must be avoided to prevent problems during reagent purification.[222]

(i) Synthesis of metal selenides (M_2Se_2) and dimetal selenides (M_2Se)

Metal selenides[32,97,223] and dimetal selenides[32,97,112] have been prepared from selenium and stoichiometric amounts of the following.
  (a) *Metal*, in particular elemental lithium, sodium or potassium, in either (i) liquid ammonia, (ii) THF in the presence of an electron carrier such as naphthalene under sonication[224] or diphenylethyne[225] or (iii) DMF. Similar reductions have been successfully carried out with zinc powder in 5 mol L^{-1} aqueous KOH (K_2Se_2),[226] with magnesium turnings in methanol (MeOMgSeSeMgOMe) or with aluminum when initiated with ignited magnesium strip (Al_2Se_3).[53]
  (b) *Metal hydrides*, such as lithium and sodium borohydride.[97] These are probably the best and most efficient reagents for the synthesis of both metal selenides and dimetal selenides. The synthesis of the former can be achieved in ethanol or water using the proper amounts of reagents. In such solvents sodium hydrogen selenide (NaSeH), and not disodium selenide, is produced by increasing the amount of reducing agent.[97] The synthesis of the latter and of metal selenides can be performed if the reaction is carried out instead in aprotic solvents in the presence of an alcohol (1 mol equiv. of $LiBH_4$ or $NaBH_4$, 3 mol equiv. of MeOH for M_2Se_2 or 2 mol equiv. of Li or $NaBH_4$, 6 mol equiv. of MeOH for M_2Se, THF or DMF, 20 °C).[227] The role of the hydroxylic coreagent or solvent is not clear, but its presence is fundamental for the success of the reactions. It probably transforms the boron by-products and undoubtedly alters the original reactivity of the expected selenium reagent. Lithium triethylborohydride[228,229] (2 mol equiv., THF, 20 °C) reacts with selenium shot and produces Li_2Se_2.
  (c) *Hydrazine and sodium or potassium hydroxide*, which allows the synthesis, in a wide range of solvents, of M_2Se_2 (0.25–1.1 equiv., NaOH, water to light petroleum including DMF, 20 °C).[97,230] It does not, however, lead to Na_2Se, but this can be produced by *in situ* reduction of the preformed Na_2Se_2 with sodium dithionite ($Na_2S_2O_4$).
  (d) *Sodium formaldehyde sulfoxylate* ($HOCH_2SO_2Na$) or rongalite, the Tchugaiev–Chlopin reagent. This has been successfully used for the synthesis of sodium selenide (1 mol equiv., NaOH, H_2O, reflux) or disodium selenide (2.4 mol equiv., 5.5 mol equiv. of NaOH, H_2O, 20 °C) depending on the amount of reducing agent used.[230]
  (e) *Potassium sulfite* (K_2SO_3) in aqueous or ethanolic solution, which readily dissolves selenium at reflux to yield potassium selenosulfate ($KSeSO_3K$),[32,100] a particularly valuable nucleophilic species. This, on heating with alkyl halides and alkyl sulfonates, gives dialkyl diselenides directly.

This has proved to be, in our hands, the most convenient reagent for the synthesis of dimethyl diselenide (after further alkylation with dimethyl sulfate).[104]

(f) *2,2,4,4,6,6-Hexaalkyl-1,3,5-triselena-2,4,6-tristannacyclohexanes*, used for the synthesis of 1,3,5-triselenanes, have been synthesized from dialkyldichlorostannanes, sodium borohydride and selenium (1.1:2:1);[231] distannylselenides have been prepared from trialkylstannanes and elemental selenium.[183] Radiolabelled reagents have been also prepared from ^{75}SeO$_2$ or from ^{75}Se.[65]

Metal selenides have been used for the synthesis of diselenides,[108,109,119] which are valuable precursors of selenols and selenolates. Dimetal selenides are efficient reducing agents[98] and good precursors of symmetrical dialkyl[113,119,232] and diaryl[113,119,232] selenides.

(ii) Synthesis of sodium and potassium selenocyanates

Sodium and potassium selenocyanates,[32,97] including radiolabelled derivatives,[65] are readily synthesized from elemental selenium and sodium or potassium cyanide in alcohols or in dimethylacetamide. These are valuable reagents which give alkyl and aryl selenocyanates[37,76,99,108,109,123] on reaction with alkyl halides and aryldiazonium salts, respectively.

(iii) Synthesis of selenourea

Selenourea[32,106,149] is efficiently synthesized from isothiouronium salts and sodium hydrogen selenide, usually produced from sodium hydroxide and hydrogen selenide, itself generated by reaction of aluminum selenide with acids.[91,149] Selenourea has proved to be a valuable nucleophilic reagent for the preparation of isoselenouronium salts, whose alkaline hydrolysis gives the corresponding selenols.[149] Selenourea has been used for the synthesis of selenium-containing carbohydrates,[233,234] purine bases[189] and some heterocycles containing both nitrogen and selenium atoms.[149]

(iv) Synthesis of selenocyanogen

Selenocyanogen[122] has been prepared by oxidation of potassium selenocyanate, for example, with bromine (KSeCN, 0.5 equiv. Br$_2$, MeOH, −70 °C to −50 °C, 2.5 mol L^{-1} MeOH).

(v) Synthesis of selenium halides

Selenium halides, including selenium chloride and bromide, are produced on reaction of stoichiometric quantities of the halogen and selenium suspended in CS$_2$.[58] Selenium tetrahalides[58] can be prepared at 0 °C from stoichiometric quantities of the halogen and selenium.[58,32,101] Selenium tetrafluoride especially has proved to be a valuable reagent for the transformation of carbonyl compounds into difluoroalkanes, alcohols into alkyl fluorides and carboxylic acids and anhydrides into acyl fluorides.[101]

(vi) Synthesis of amino and imido selenium reagents

A reagent expected to be dibenzotriazolylselenium has been synthesized from diallyl selenide and *N*-chlorobenzotriazole,[235] and an imido selenium reagent of unknown structure has been readily prepared from chloramine-T and selenium and shown to be particularly useful for the allylic amination of alkenes (0.63–0.83 equiv. TsNClNa, 0.63–0.83 equiv. Se, CH$_2$Cl$_2$, 20 °C).[236]

13.3.3 Useful Organic Reagents for Organoselenium-monitored Reactions

13.3.3.1 *Strategy*

Most of the organoselenium reagents used in synthesis bear at least a phenyl, a methyl or a benzyl group directly attached to the selenium atom. In some very limited cases *o*-nitrophenyl and *p*-methoxyphenyl derivatives have also been used.[4,48,78,82,160,162,163,174] The most salient and interesting features of the phenylseleno series are without doubt the low volatility and the comparatively less pungent smell of the reagents and by-products as compared, for example, with those of the methylseleno series. The latter series, although often involving at one stage bad-smelling or volatile compounds, offers the advantage in some instances that, once the overall transformation has been completed, separation of the seleno by-products is easy by simple vacuum removal from the crude mixture (Me_2Se, MeSeBu, b.p. < 120 °C).[203] Benzylseleno reagents have been used much less. They nevertheless not only possess similar reactivity to the methylseleno derivatives but also offer the advantage that selective cleavage of the 'selenium–benzyl bond' is possible (Na, NH_3, producing toluene and a selenolate).[5,66] Further oxidation by oxygen, acid hydrolysis or alkylation of the selenolate allows the one-pot synthesis of diselenides, selenols or selenides, respectively.[5,66] Benzyl selenolates can therefore be used as masked metal selenides, dimetal selenides or metal selenolates. The 'selenium–benzyl' bond in benzyl alkyl selenides can also be cleaved with bromine to selectively produce the alkylselenenyl bromide, which can then be transformed into the diselenide on further reaction with hydrazine.[237]

As general trends, the selenium atoms in methylseleno or benzylseleno derivatives are more nucleophilic than those in the arylseleno counterparts and therefore all the transformations which involve such reactions as the key step will be favoured. Furthermore, α-methylseleno organolithiums show a higher propensity than their phenylseleno analogues to react at the carbonyl group of ketones, especially the more enolizable or hindered ketones.[78,84,175,191,192] Among the selenides, alkyl aryl selenides, especially the *o*-nitrophenyl derivatives,[238] are those which are the most prone to produce alkenes via selenoxide elimination reaction.

13.3.3.2 Synthesis of organoselenium reagents

Organoselenium reagents[1,4,48,53,82,88,160-3,190] have been prepared from metallic selenium, inorganic reagents containing selenium and organoselenium derivatives. These transformations have been efficiently used for the synthesis of phenylseleno, *o*-nitrophenylseleno, methylseleno and benzylseleno reagents. We shall present them as models. Several reagents of the phenylseleno and methylseleno series are now commercially available. Related diselenides, which are stable compounds and are commercially available, have proved to be key intermediates from which the whole series of reagents can be produced. Polymer-supported[239] and electrogenerated[240-2] organoselenium reagents have also been reported.

(i) Synthesis of organoselenium reagents from elemental selenium

As expected, the most efficient route to organoselenium reagents remains the direct synthesis from the element itself and an organic residue. This approach has been effective with various organometallics, including alkyl, aryl, vinyl, acetylenic and phosphonate carbanions, organomagnesium, -sodium, -aluminum,[243,244] -mercury[79] and -lithium[245,246] derivatives and metal enolates,[53,76,110,162] which produce directly the corresponding selenolates in reasonably good yields. This is the method usually used to access the phenylseleno series of reagents and compounds (see above). Otherwise, phosphine selenides (R_3PSe) have been prepared by simple heating of the related phosphine with elemental selenium.[153]

(ii) Synthesis of organoselenium reagents

Organoselenium reagents have been prepared (see leading references to review articles in Section 13.2.6).

(a) *From inorganic reagents containing selenium.*[4,53,160-3] Thus, for example, the following organoselenium compounds have been synthesized as follows:
   (i) organoselenolates from alkyl or aryl halides and sodium hydrogen selenide;
   (ii) organoselenocyanates from alkyl, benzyl[14] and aryl halides or aryldiazonium salts and sodium or potassium selenocyanate; the latter reaction has been successfully used for the synthesis of optically active 2,2'-diselenocyanato-1,1'-binaphthyl from the corresponding diamine;[247] and
   (iii) diselenides from alkyl or aryl halides or aryldiazonium salts and metal selenolates or from selenenyl halides and hydrazine.[237]

(b) *From other organoselenium reagents.* Examples of these reagents are as follows:
(i) selenols by acid hydrolysis of metal selenolates or selenocyanates (H_3O^+) or by reduction of diorgano diselenides or organoselenocyanates (H_3PO_2,[248] Zn–H^+);
(ii) metal organoselenolates[53,76] from:
   (a) selenols and bases (NR_3, M_2CO_3,[103] MOH, MOR, MNR_2, MH, RM, etc.) by metal–metal interchange from silyl or stannyl selenides (F^-[249] or RLi[250]);
   (b) organo diselenides by reduction with metals (Na–NH_3,[5,66] Na–THF or K–THF[251]), in the presence of hexamethylphosphoramide (HMPA)[76] or under sonication,[252] electrolysis in methanol containing tetrabutylammonium tosylate, using platinum electrodes,[253] or metal hydrides such as NaH or KH (but not LiH)[254] or $NaBH_4$,[255,256] with either NaOH under phase-transfer conditions (these conditions allow the concomitant formation of 25% seleninic acid by disproportionation[257]) KOH–NH_2NH_2,[258] KOH–PPh_3,[259] $HOCH_2SO_2Na$[237,260] or $(Ph_3Sn)_2Te$–CsF[261];
   (c) organoselenocyanates by reduction ($NaBH_4$, Na_2S);[123]
   (d) selenol esters or selenopseudoureido derivatives (from an alkyl halide and a selenourea) with nucleophiles such as metal alkoxides or hydroxides, respectively.
   Metal organoselenolates are potent nucleophiles (see Section 13.2.2) in protic solvents but their nucleophilicity is enhanced in apolar aprotic solvents such as DMF or DMSO.[76] Although sodium borohydride reduction of diselenides is the most convenient route to organoselenolates (because the reducing medium required for the selenolate synthesis avoids its further oxidation to diselenide, a side reaction often observed on reaction of selenols with bases), the presence of boron derivatives can greatly influence the reactivity of the selenolate, especially towards lactones[76,111,173,190,262,263] and cyclopropane derivatives bearing electron-withdrawing groups.[76,173] In such cases NaH can advantageously replace $NaBH_4$;[254]
(iii) diselenides from organoselenocyanates under acidic (H_3O^+–O_2) or basic (KOH or MeONa–MeOH) conditions, or by oxidation of selenols or selenolates (O_2, H_2O_2,[264] $Na_2B_4O_7$[265]);
(iv) organoselenocyanates from selenenyl halides and silyl cyanides;[266]
(v) selenenyl halides by reaction of alkyl benzyl selenides,[237] diselenides or organoselenocyanates with halogens (RSeSeR and Cl_2, SO_2Cl_2, Br_2 or XeF_2[132] or *N*-phenylselenophthalimide (*N*-PSP) and an HF–amine complex,[267] or from another selenenyl halide and a halide ion (PhSeBr and AgF under sonication)).[268–70] Interaction between diphenyl diselenide and iodine (PhSePh, I_2, MeCN) does not lead to benzeneselenenyl iodide[271] but produces a complex which reacts like benzeneselenenyl halides;
(vi) selenenoamines and -amides from selenenyl halides and amines or metalloimides;[272]
(vii) seleninic acids by oxidation of diselenides;
(viii) seleninic anhydrides by ozonolysis of diselenides or by dehydration of seleninic acids;
(ix) organoselenosulfonates by reduction of seleninic acids by sulfonylhydrazides or sulfinic acids.

(iii) Strategy and practice in the synthesis of the most used organoselenium reagents

(a) Synthesis of organoselenium reagents belonging to the phenylseleno series. Entry to the phenylseleno series is best achieved from metal benzeneselenolates, which are readily available from phenylmetals (MgBr, Li) and grey selenium (Scheme 11).[273,274] Benzeneselenolates can be then transformed into phenylselenol[273] or diphenyl diselenide[274] on acidic hydrolysis or oxidation with bromine, respectively. Useful reagents in this series and their syntheses are given in Scheme 11.

(b) Synthesis of organoselenium reagents belonging to the o-*nitrophenylseleno series.* o-Nitrophenylselenocyanate is the starting material for the synthesis of all the members of the series. It is usually synthesized from the corresponding diazo compound and sodium or potassium selenocyanate.[275]

(c) Synthesis of organoselenium reagents belonging to the methylseleno series. Dimethyl diselenide is the starting material for the synthesis of all the members of the series (Scheme 12). It is usually synthesized from sodium selenosulfate, sodium hydroxide and dimethyl sulfate in water and is obtained in good yield by azeotropic distillation. Lithium methylselenolate, which is readily available from methyllithium and elemental selenium, is also a valuable starting material for the synthesis of all the members of the series,[276,277] including the ^{77}Se derivative.[246] The valuable dimethylaluminum methylselenolate ($MeSeAlMe_2$) has been synthesized[243,278] from trimethylaluminum and elemental selenium.

(d) Synthesis of organoselenium reagents belonging to the benzylseleno series. Entry to the benzylseleno series[32] involves the synthesis of dibenzyl diselenide by alkylation of metal selenides with benzyl halides[237,279] and further transformation to the corresponding selenolate by reaction with KH[254]

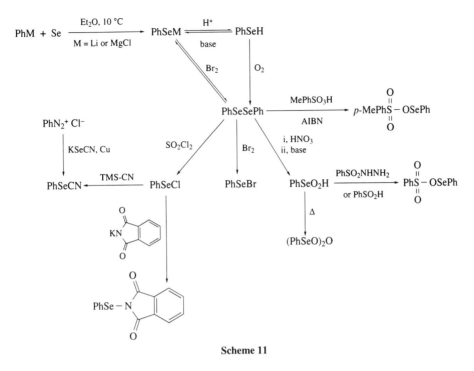

Scheme 11

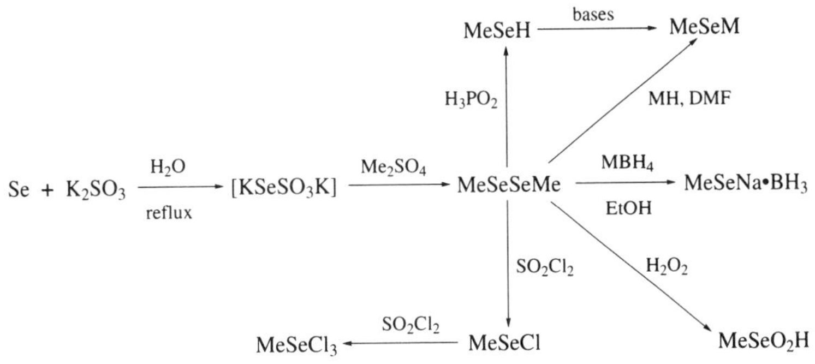

Scheme 12

(but not NaH or LiH) or LAH. Bromomagnesium benzylselenolate has been synthesized from benzylmagnesium bromide and elemental selenium.[280]

13.3.4 Functional Group Manipulations Involving Inorganic Selenium and Organoselenium Reagents

13.3.4.1 Reactions involving hydrogen selenide, selenols and selenolates

Hydrogen selenide, selenols and their salts are particularly good nucleophiles and reducing agents.[88]

(i) Hydrogen selenide, selenols and selenolates as dealkylating agents

Selenols and selenolates have been successfully used for the dealkylation[39,76,98,111,174,196] of:
(a) quaternary ammonium salts to amines;
(b) the whole range of amines to less *N*-alkyl-substituted amines (primary to tertiary). Although the method is limited to significantly basic amines which possess an alkyl group susceptible to S_N2 reaction, it offers interesting advantages over existing methods;
(c) alkyl aryl ethers, thioethers and selenoethers to phenols, aryl thiols and selenols, respectively;

(d) esters to the corresponding carboxylic acids (Scheme 5 (i)). This reaction has proved to be very efficient when carried out on methyl esters with sodium phenylselenolate under suitable conditions (DMF, HMPA or crown ethers)[254,263] or with phenylselenotrimethylsilane–KF–crown ether combinations.[249] The reaction is chemoselective under the former conditions since methyl ethers and amides do not react. The method used for the synthesis of the selenolate is crucial for the success of its further reaction with esters. Typically, the presence of BH_3 must be avoided and therefore generation of the selenolate by reduction of diphenyl diselenide with sodium borohydride must be avoided.[263] The reaction has been extended to five-membered lactones (see below); and

(e) urethanes to amines via C–O cleavage and subsequent decarboxylation.[281]

(ii) Hydrogen selenide, selenols and selenolates and related derivatives as reducing agents

Hydrogen selenide, selenols and their salts, trialkylsilyl selenides, metalloselenocyanates and phosphine selenides[39,98,111] have been successfully used for the reduction of:

(a) α-halo- (especially the iodo derivatives) and α-selenoketones to ketones and benzyl halides bearing electron-withdrawing groups on the aromatic ring to the corresponding arylalkanes;

(b) alkyl halides bearing a heteroatom in the β-position and *vic*-ditosylates, which are transformed into alkenes. This is the case with β-dihaloalkanes, β-selenobromoalkanes and β-hydroxybromoalkanes and their esters (Na_2Se, NaHSe, RSeNa (Me, Ph) or KSeCN). These reactions are best with bromo derivatives and proceed via an *anti* elimination of the two heteroatomic moieties from *vic*-dibromoalkanes and via a *syn* elimination from the dichloro analogues and from β-hydroxybromoalkanes (Scheme 5 (a)). Anti addition of chlorine or BrOH to alkenes followed by treatment of the resulting adducts with PhSeNa or KSeCN allows their stereospecific isomerization;

(c) epoxides (K or NH_4, SeCN; $Ph_3PSe–CF_3CO_2H$; $Bu_3PSe–CF_3CO_2H$; 3-methyl-2-selenoxobenzothiazole; $PhC(=Se)NH_2–CF_3CO_2H$)[282] and thiiranes (3-methyl-2-selenoxobenzothiazole) to alkenes, usually by stereospecific *syn* elimination;

(d) amine oxides (on reaction with $B(SePh)_3$), sulfoxides (on reaction with PhSeH, 4-PhC_6H_4SeH,[107] $B(SePh)_3$, $B(SeMe)_3$, $(BuB)_2Se_3$, $(EtO)_2P(O)SeH$, PhSe(TMS) or $Se(TMS)_2$), selenoxides and telluroxides (PhSe(TMS) or $Se(TMS)_2$) to amines, sulfides (Scheme 5 (b)), selenides and tellurides, respectively;

(e) disulfides to thiols with NaHSe (from $NaBH_4$, Se);

(f) benzyl phenyl selenide to toluene (PhSeH, $h\nu$);

(g) imines[39] and oximes derived from aromatic aldehydes to benzylamines or to N-benzylhydroxylamines (PhSeH, thermal or photochemical), nitro-, nitroso-, azo-, azoxy-, hydrazo- and hydroxylaminoaromatic compounds to aromatic amines (H_2Se or PhSeH) and arenediazonium tetrafluoroborates to arylhydrazines;

(h) (*E*)-cinnamaldehyde, chalcone and (*E*)-stilbene to the related derivatives resulting from the reduction of the C=C double bond (PhSeH, $h\nu$); and

(i) radicals; hydrogen selenide and phenylselenol are efficient radical traps. Thus hydrogen selenide quantitatively traps the 1,4-biradical intermediates involved in the photocycloaddition between 2-cyclopentenone and allene, and phenylselenol is able to trap cyclopropylcarbinyl radicals as well as the 4-butenyl radicals resulting from their ring opening.[283] This is very fast when the cyclopropyl ring is phenyl substituted (at 25 °C: rate constant 10^{11} s^{-1}, a few picoseconds, much faster than the diffusion-controlled process).

13.3.4.2 Reactions involving metallic selenium

Metallic selenium[111,284] at high temperature (>200 °C) induces the (*Z*)–(*E*) isomerization of alkenes[284] such as stilbenes and unsaturated fatty acids and their esters and is also used for the oxidation of cyclic and polycyclic hydrocarbons and heterocyclics to aromatic compounds (Scheme 5 (c)).[285] Compounds with quaternary centres lose a substituent; an ethyl is lost in preference to a methyl group, and an angular alkyl group rather than ring cleavage.[284]

Metallic selenium is reduced under mild conditions to hydrogen selenide by carbon monoxide and water in the presence of a base.[1,39,111,173,284] This extremely convenient route to such a toxic and malodorous compound is particularly useful for the synthesis of ureas, seleno carboxamides and seleno carbamate salts (precursors of various carbonic acid derivatives) from amines and of carbonohydrazides, semicarbazides, carbazates and carbonates from hydrazines.

13.3.4.3 Reactions involving selenoxides and related derivatives

(i) Reactions involving selenoxides

Selenoxides[39,136,197] are mild oxidants which have been used for the transformation of:
(a) tertiary amines to amine oxides, trivalent phosphorus derivatives to their oxides with inversion of configuration and sulfides, including hindered ones, to sulfoxides (Scheme 5 (d));
(b) anilines to azo derivatives, acyl hydrazines to symmetrical diacylhydrazines and thiols to disulfides;
(c) *o*- and *p*-hydroquinones to the corresponding quinones (Scheme 5 (f)). In some cases the quinone is intramolecularly trapped by a nucleophile present in a favourable position in the molecule;
(d) thiocarbonyl compounds such as thioamides, thioureas and thiouracil to carbonyl compounds, and thio and seleno phosphorus compounds to their oxides; and
(e) osmium(VI) to osmium(VII) derivatives. The reaction requires a basic medium and has been used for the oxidation of alkenes to diols. The catalytic version uses the aptitude of singlet oxygen to oxidize the selenide produced in the process back to selenoxide.

Most of the reactions have been performed with dimethyl or diphenyl selenoxide. The latter was not efficient for the oxidation of strychnine, but this has been efficiently achieved with 2-(phthalimidoethyl) methyl selenoxide. Bis(*p*-methoxyphenyl) selenoxide is a milder oxidant which behaves similarly except toward amines, which are not oxidized to their oxides, and thiocarbonyl compounds, which are oxidatively dimerized rather than transformed into their oxides.

(ii) Reactions involving reagents related to selenoxides

Dimethyl selenide-*N*-chlorosuccinimide in the presence of 1,5-diazabicyclo[5.4.0]undec-5-ene (dbu) efficiently oxidizes alcohols to carbonyl compounds. This reagent, closely related to the Corey–Kim reagent (Me$_2$S–NCS–NEt$_3$), offers over the related sulfur reagent the advantage of cleanly oxidizing allyl alcohols to α,β-unsaturated carbonyl compounds. It is also able to oxidize β-hydroxyalkyl selenides to α-selenoketones, leaving the readily oxidizable seleno moiety untouched. Interestingly, the presence of dimethyl selenide is not crucial for the success of the last reaction.

Diphenylselenium bis(trifluoroacetate), prepared from diphenylselenium dibromide and silver trifluoroacetate or more conveniently from diphenyl selenoxide and trifluoroacetic anhydride,[103] chemoselectively oxidizes[102] tetrahydrocarbolines to dihydrocarbolines or to carbolines depending on the number of equivalents used. This reagent, even if used in excess, is also able to smoothly oxidize tetrahydroisoquinolines, unsubstituted on the nitrogen, to dihydroquinolines except for cases where there is an ester group at the 2-position, when the corresponding isoquinoline is obtained. Tetrahydroisoquinolines bearing an alkyl group on the nitrogen are oxidized to the iminium salts; trapping with potassium cyanide gives the α-aminonitriles. Diphenylselenium bis(trifluoroacetate) and diphenylselenium hydroxyacetate promote the formation of alkoxy radicals from the corresponding alcohols on photolysis in the presence of iodine.[286] These undergo intramolecular hydrogen abstraction to afford cyclic ethers in good yields (Equation (2)).[286]

$$\text{EtO}_2\text{C}\text{—[cyclohexyl-OH]} \xrightarrow[\substack{1\text{ equiv. I}_2,\ 80\ °\text{C, 5 h} \\ 95\%}]{1.7\text{ equiv. Ph}_2\text{Se(OAc)OH}} \text{EtO}_2\text{C}\text{—[bicyclic ether]} \qquad (2)$$

13.3.4.4 Reactions involving diselenides and related reagents

Diorganic diselenides, especially 2,2'-dipyridyldiselenide, with triphenylphosphite in pyridine, have proved to be efficient coupling reagents in nucleotide syntheses since they permit phosphate ester synthesis via internucleotide bond formation.[121] Use of the selenium reagent avoids the concomitant formation of symmetrical pyrophosphates which is often observed with the related 2,2'-dipyridyl disulfide. The combination of diorganic diselenide and N-chlorosuccinimide promotes the allylic oxidation of alkenes to allyl chlorides via a process which does not involve free radicals (Scheme 5 (e))[121] and combination of a diselenide with t-butyl hydroperoxide efficiently oxidizes alcohols to carbonyl compounds.[103] The best results have been obtained with bis(2,4,6-trimethylphenyl) diselenide at reflux in benzene. Primary alcohols are more reactive than secondary alcohols and allylic alcohols are more reactive than the saturated analogues. The reagent is mild and does not oxidize carbon–carbon double bonds or seleno moieties if present. The reagent combination of diorganic diselenide with N_2O_3, PhIO or 3-$IOC_6H_4CO_2H$, which is believed to produce benzeneseleninic anhydride (BSA), promotes the dehydrogenation of ketones to enones and enediones (see Section 13.3.4.6(ii)), whereas with hydrogen peroxide it allows the synthesis of perseleninic acids, powerful oxidants whose reactivity is discussed in Section 13.3.4.8

13.3.4.5 Reactions involving areneselenenyl halides

Areneselenenyl halides are used for the ring expansion of 1,3-dithiolans and 1,3-dithianes to dihydro-1,4-dithiins and dithiepins, respectively, the synthesis of α-chloro- or α-bromoenones from enones, and the halogenation at the *ortho* position of phenols.[121] Thus, oestrogens such as estrone, estradiol and their analogues are transformed into the 2-chloro or 2-bromo derivatives on reaction with phenylselenenyl chloride or bromide.[287,288] The reaction failed with the selenenyl iodide.[287,288] Areneselenenyl halides have also been used for the synthesis of tetrahalo *meso*-tetraphenylporphyrins from the related porphyrins.[289]

13.3.4.6 Reactions involving benzeneseleninic acid and anhydride and related reagents

Benzeneseleninic acid and BSA are efficient and mild oxidants which have been used for the transformation of various functional groups.[39,141,197,290]

(i) Oxidation of phenols to hydroxydienones or to quinophenylselenoimines and of hydroquinones to quinones

BSA is a particularly mild and efficient oxidant for phenols.[107,177,188] Either alone or in the presence of a base, it oxidizes *o,o*-dialkylphenols to *o*-hydroxydienones (often isolated as their dimers) and their nonsubstituted analogues to *o*-quinones, even where the *para* position is unsubstituted (Scheme 5 (f)).[1,291,292] Both reactions involve the intermediate formation of seleninic esters which rearrange by a [2,3] sigmatropic shift. When the reaction is carried out in the presence of lithium hexamethyldisilazide,[293] *o*- or *p*-quinophenylselenoimines[293-6] are produced, which are valuable precursors of aminophenols (PhSH or Zn–Ac_2O,[293] SmI_2[295]). This method has proved useful for the selective amination of steroids.[290] BSA smoothly transforms hydroquinones into *o*- or *p*-quinones.

(ii) Dehydrogenation of ketones, lactones and lactams to their α,β-unsaturated analogues

BSA and its corresponding acid act as potent dehydrogenating reagents.[103,188] Reaction with carbonyl compounds gives the α,β-unsaturated analogues.[103] The oxidation, which probably involves the intermediate formation of an α-selenoxy carbonyl compound, has been successfully applied to the dehydrogenation of steroidal and triterpenoid ketones, lactones and lactams (Scheme 5 (g)). In the case of ketones, the degree of unsaturation introduced depends on the number of equivalents of oxidant used (Scheme 5 (g)) and overoxidized products resulting from ring A contraction have been noticed when an excess of reagent is used for an extended period of time. Remote double bonds are not affected by BSA under the usual conditions. This method offers a great improvement over existing procedures such as with SeO_2 (see Section 13.3.4.9(iii)) or ddq (dichloro dicyano quinone) and has been improved by the use of catalytic amounts of BSA in the presence of iodoxybenzene or *m*-iodoxybenzoic acid in excess or generated *in situ* from diphenyl diselenide.

Reaction of BSA with certain enolizable ketones whose carbonyl groups are adjacent to a nonenolizable α'-carbon gives α-hydroxy ketones. This reaction proceeds best when carried out in the presence of NaH and usually delivers the thermodynamically more stable product. Oxidation of alcohols with BSA gives the corresponding carbonyl compounds. The reaction occurs more rapidly with benzyl than with saturated or allyl alcohols and allows, in the presence of an excess of reagent, an efficient access to enones or enediones, especially those derived from ring A ketones.

(iii) Oxidation of thiols, amines, hydrazines and hydroxylamines

BSA oxidizes thiols to unstable thiolseleninates, which in turn decompose to a mixture of disulfides and diselenides or selenosulfides, depending on the conditions used. Primary amines which do not readily form enamines are oxidized by BSA to ketones, whereas secondary amines give α-cyanoamines (viewed as protected imines) if the oxidation is carried out in the presence of NaCN or TMS-CN. Indolines give the corresponding indoles or 3-phenylselenenylindoles. The synthesis of indoles is best carried out in the presence of the parent indole, which is able to trap the intermediate Se^{II} species.

N,N'-Disubstituted hydrazines are oxidized by BSA to diimides, N-acylhydrazines to selenol esters and N-substituted hydroxylamines to the nitroso compounds.[188]

(iv) Oxidation of thioketones, hydrazones, oximes and thio- and selenoacetals and related derivatives to their oxo analogues

BSA can be used for the transformation of thiocarbonyl compounds, including thioketones, and of carbonyl derivatives such as semicarbazones, oximes (except O-methyloximes), hydrazones, thioacetals and selenoacetals into the corresponding carbonyl compounds. Side reactions include the formation of diketones in the former case and vinyl selenides in the latter. Xanthates, thioesters and thionocarbonates, and thioamides and some of their seleno and telluro analogues are also efficiently transformed into the oxo analogues.[188]

(v) Oxidation of alkylated aromatic and heteroaromatic compounds

BSA efficiently oxidizes the benzylic carbon of a large variety of aromatic and heteroaromatic compounds, leading to the corresponding ketones, aldehydes or acids if an excess (>0.33 equiv.) of reagent is used.[188,297]

13.3.4.7 Reactions involving benzeneseleninyl halides

Benzeneseleninyl chloride, prepared by ozonolysis of the corresponding benzeneselenenyl chloride, oxidizes aldoximes to nitriles and carbonyl compounds to α,β-unsaturated carbonyl compounds.[298] The latter transformation, which proceeds via an *in situ* selenoxide elimination reaction, offers an advantage over the stepwise process (which involves the oxidation of a functionalized selenide) when competing oxidation can occur elsewhere in the molecule.

13.3.4.8 Reactions involving perseleninic acids

Organoseleninic acids (PhSeO$_2$H, 2,4-(NO$_2$)$_2$C$_6$H$_3$SeO$_2$H) or diselenides (oxidized *in situ* to perseleninic acids)[142,197] catalyse the hydrogen peroxide oxidation of alkenes to epoxides, ketones to esters or lactones, sulfides to sulfoxides or sulfones and selenides to selenoxides or selenones,[135] depending on the amount of oxidant used, and phosphines to phosphine oxides. The order of reactivity of alkenes parallels their nucleophilicity but the reaction is less chemoselective than that involving peracids for allyl alcohols. In the case of ketones the reaction behaves as the well-known Baeyer–Villiger reaction[47] since the most substituted carbon migrates with retention of configuration (Scheme 5 (h); Scheme 8 (a)). Finally, the oxidation of sulfides to sulfones occurs chemoselectively in the presence of a hydroxy, a carbonyl or a C=C double bond. The simplicity of the procedure and the low price of hydrogen peroxide compared with peracids are the attractive aspects of this reaction.

Polystyrene-bound seleninic acid also promotes similar reactions.[299] In the presence of t-BuO$_2$H it chemoselectively oxidizes benzyl alcohols to aromatic aldehydes and ketones, leaving primary alkanols untouched, and hydroquinones and catechols to the corresponding quinones. It also efficiently oxidizes, in the presence of hydrogen peroxide, ketones to esters and lactones and trisubstituted alkenes to epoxides. Terminal and α,β-disubstituted alkenes, however, give diols rather than epoxides.

13.3.4.9 Reactions involving selenium dioxide

Selenium dioxide has been widely used in synthesis to oxidize methyl, methylene and methine groups directly linked to vinyl, aryl, acetylenic and carbonyl functional groups.[33,35–7,40–2,44,48,102,103,197] The reactions are usually carried out with 0.5–1 equiv. of the reagent in t-butanol, dioxane or acetic acid at reflux, or more conveniently in the case of alkene compounds with stoichiometric or catalytic amounts of selenium dioxide and t-butyl hydroperoxide in methylene dichloride (Equation (1)).[300,301] Particularly valuable results have been obtained if the reaction is performed with selenium dioxide supported on silica gel[302] or under sonication.[303] This reaction occurs in two discrete steps, namely an ene reaction followed by a [2,3]-sigmatropic shift and hydrolysis of the resulting Se^{II} ester.[1,88,301]

(i) Reactions of selenium dioxide with alkenes: synthesis of allyl alcohols

This reaction is applicable to a wide range of alkenes and gives allyl alcohols and/or α,β-unsaturated carbonyl compounds without scrambling of the C=C double bond. Overoxidation is usually observed when the reactions are carried out with 1 equiv. of the reagent for a long time (10 h), but further reduction or oxidation of the resulting mixture, for example, with NaBH$_4$ or MnO$_2$, allows the synthesis of only one of the compounds. The reactivity of alkenes follows their nucleophilicity. Terminal alkenes, the least reactive, produce allyl alcohols bearing a terminal C=C bond and the hydroxy group on the most substituted carbon of the side chain with alicyclic alkenes and on the least substituted carbon with alkylidenecycloalkanes. A methylene is oxidized in preference to a methyl or a methine group in α,β-disubstituted alkenes, and in the case of trisubstituted alkenes not only does oxidation occur on the most substituted side of the C=C double bond but also a methyl group is oxidized preferentially over a methylene or a methine group. Interestingly, the reaction proceeds in the latter case with high stereocontrol, producing almost exclusively the (E)-stereoisomer. This reaction also takes place with functionalized alkenes such as α,β-unsaturated carbonyl compounds, allyl alcohols (transformed into aldehydes)[304] and allyl ethers (transformed into aldehydes), and has been used successfully as a key step in the synthesis of several natural products.[4,33] It has been reported that selenium dioxide is depolymerized by trimethylsilyl polyphosphate in CCl$_4$ and that it aromatizes cyclohexene and cyclohexadienes under mild conditions.[305]

(ii) Reactions of selenium dioxide with alkylated aromatic and heteroaromatic compounds: synthesis of aryl and heteroaryl carbonyl compounds

Selenium dioxide in dioxane promotes the selective oxidation of methyl groups attached to an aromatic or heteroaromatic ring to aldehydes. The acid can be produced if the reaction is carried out in pyridine with an excess of reagent.

(iii) Reactions of selenium dioxide with carbonyl compounds: synthesis of α-dicarbonyl compounds and α,β-unsaturated carbonyl compounds

Ketones susceptible to oxidation to enones by dehydrogenation react in this way with SeO$_2$, especially if t-butanol is used as the solvent (Scheme 13 (a)). In some cases α-diketones are also produced. These can be further oxidized to the corresponding enediones if excess reagent is used. Otherwise, α-diketones are obtained if, for structural reasons, the enones cannot be produced. Excess reagent must again be avoided since overoxidation may give α-diones as a result of loss of carbon monoxide from the intermediate triketones. The formation of acetals has been reported in a few cases when the reaction is carried out in methanol or in ethylene glycol.

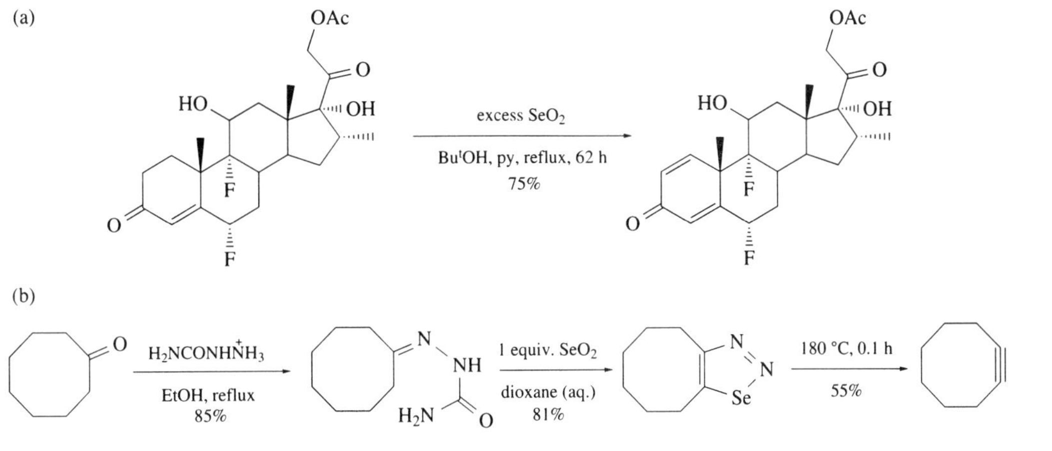

Scheme 13

(iv) Miscellaneous reactions of selenium dioxide

Selenium dioxide promotes the aromatization[285] of cyclic and heterocyclic derivatives, the selective oxidation[304] of allyl and benzyl alcohols in the presence of aliphatic derivatives, the oxidation of hydrazones, imines and O-alkyloximes to α-carbonylated or α,β-unsaturated derivatives, the dehydration of aldoximes to nitriles and the transformation of semicarbazones into 1,2,3-selenadiazoles, which in turn have proved to be valuable precursors of acetylenic compounds, including particularly strained (Scheme 13 (b))[1,33,306-8] and functionalized alkynes, which have been used for the synthesis of anti-Bredt alkenes.[308] It has also been used, in the presence of triethylamine, for the oxidation of nitroalkanes to N-hydroxycarboxamides and for the conversion of thio and seleno carbonyl derivatives into carbonyl compounds and of phosphoroselenolates and phosphites into phosphates.

(v) Reactions of selenium dioxide–hydrogen peroxide

Selenium dioxide in the presence of hydrogen peroxide generates perselenious acid, which reacts in a different way to selenium dioxide with alkenes, aldehydes and ketones. It leads to diols in modest yields in the first case, to carboxylic acids or esters from aldehydes depending on the solvent used (tertiary or primary alcohols) and to carboxylic esters or acids resulting from a Baeyer–Villiger reaction or a rearrangement.

13.3.4.10 Reactions of imido selenium derivatives

The reagents obtained from $SeCl_4$ and t-butylamine or p-toluenesulfonamide in the presence of a base or from selenium and chloramine-T, and believed to be RN=Se=NR, react with alkenes and alkynes at the allylic or propargylic site in a manner closely related to SeO_2, to produce allyl- or propargylamines. The same reagent is believed to cycloadd to dienes and this process leads, after sigmatropic shift of the seleno moiety of the intermediate, to α,β-*cis*-di-N-tosylamines by addition of two N-tosylamino groups across the less substituted C=C double bond of dienes.[1,101,173,197,236]

13.3.4.11 Reactions of selenium tetrafluoride

Selenium tetrafluoride,[1,101] neat or in methylene dichloride, reacts at the carbonyl group of aldehydes, ketones and of N,N-dimethylbenzamide to give the corresponding geminal difluorides in high yields. It also allows, especially when complexed to pyridine, the synthesis of alkyl, allyl and acyl fluorides from the corresponding alcohols and carboxylic acids or anhydrides, respectively, under conditions milder than those required for the sulfur analogue.[309] Cyclopropylcarbinols and 2-methylpropanol produce rearranged fluorocyclobutanes and t-butyl fluoride, respectively (20 °C, CH_2Cl_2, 60% each).

13.3.5 Synthesis of Organoselenium Compounds

13.3.5.1 Synthesis of selenides and functionalized selenides from nucleophilic and electrophilic organoselenium reagents

(i) Synthesis of selenides and functionalized selenides from nucleophilic organoselenium reagents

(a) Synthesis of selenides from alcohols. Alcohols and lactols have been successfully transformed into methyl or aryl selenides on reaction with the corresponding selenols in acidic media. Although various protic or Lewis acids have been used, the best results have been obtained using zinc chloride in 1,2-dichloroethane in the case of alcohols[310–12] and boron trifluoride etherate or *p*-toluenesulfonic acid with lactols.[313] The reaction proceeds only with those derivatives which possess a carbon framework able to stabilize a carbenium ion at the substituted centre, such as secondary, tertiary, benzyl and allyl alcohols, and the method can be used for the synthesis of benzylic alcohols[312] bearing bulky groups at the benzylic carbon.

Alternatively, arylselenocyanates[39,76,173] or *N*-phenylselenophthalimide (*N*-PSP)[39,76,314] and tributylphosphine may be used. These reactions are complementary to that described above, and are efficient with primary, secondary, neopentyl[282] and benzyl[315] alcohols, but have not yet been described with tertiary derivatives. It failed at least once[316] but fortunately the desired phenyl selenide could be synthesized from the mesylate.[316] The phenylselenocyanate method is successful with allyl alcohols,[317] whereas the corresponding reaction with *N*-PSP has proved inadequate and leads to an allylphthalimido derivative.[317] The reaction with arylselenocyanates is stereospecific with secondary alcohols,[318] including allyl derivatives,[317] and proceeds by inversion of configuration at the substituted carbon atom. This allows the selective transformation of primary alcohols in the presence of secondary or tertiary alkyl derivatives. Although both methods are highly chemoselective, leaving untouched C=C double bonds, acetals, epoxides and ketones, the arylselenocyanate–tributylphosphine procedure gives α-arylselenonitriles,[319] γ-arylseleno-α,β-unsaturated nitriles[319] and selenol esters[107] from aldehydes, enones and carboxylic acids, respectively. Phenylselenophthalimide–tributylphosphine is also able to transform carboxylic acids into selenol esters,[107] and phenylselenophthalimide alone is able to cyclize suitably positioned unsaturated alcohols and acids to β-seleno ethers and lactones.[76,199,320,321]

(b) Synthesis of selenides from selenolates and alkyl halides, alkyl sulfonates and related substrates. The synthesis of selenides[56,112–16] has been efficiently achieved from alkyl halides or sulfonates and phenyl-, benzyl-, 2-nitrophenyl- and methylselenolates,[1,37,76,112,120] including samarium arylselenolates.[322,323] Sodium alkynylselenolates and alkenylselenolates have also been used from time to time. Methylseleno sugars,[233,234] methylseleno nucleosides,[189,233] methylselenoamino acids,[5,66] α-methylseleno ketones, esters, lactones and β-dicarbonyl compounds have all been synthesized by alkylation of more complex selenolates with methyl iodide (Scheme 6 (a)).

Reaction of selenolates with *s*-alkyl halides occurs with complete inversion of configuration at the substituted carbon atom irrespective of the conditions used[77,220,318,324,325] and no scrambling is observed with primary or secondary allyl halides bearing terminal,[326] α,β-disubstituted (with control of the stereochemistry)[253] or trisubstituted double bonds.[253] Isomerization nevertheless occurs with allyl halides bearing a fully substituted allylic centre but a terminal C=C double bond.[327,328] Phenyl and methyl allyl selenides rearrange[326,329] photochemically (fluorescent bulb), thermally and in acidic media (even on SiO₂ for allyl methyl selenides).

These reactions allow the synthesis of a wide range of selenides and functionalized selenides from the corresponding halides.[1] These include alkyl aryl selenides, alkyl pyridyl selenides,[330] propargylic selenides, α-selenenyl esters,[331] acids, lactones, nitriles and ketones, α-hetero-substituted alkyl selenides such as α-selenoalkyl ethers, thioethers, silanes, amines and phosphonates and α-haloalkyl selenides, α,α-diselenoalkanes[84] from α,α-dihaloalkanes and seleno orthoesters from trihalomethanes.[258]

Finally, 1-phenylselenoadamantane has been synthesized from the corresponding trifluoroacetate and phenylselenol or diphenyl diselenide[332] and phenylselenododecahedrane has been generated from bromododecahedrane, phenyl trimethylsilyl selenide and zinc iodide (Scheme 14).[333]

(c) Synthesis of selenides from selenols or selenolates and vinyl and aryl halides and aryldiazonium salts. Selenolates,[76] especially the lithium salts, also react with nonactivated vinyl halides, including β-bromostyrenes and 1,2-dichloroethene, in dipolar aprotic solvents (DMF, dimethylacetamide (dma), HMPA) to give the corresponding vinyl selenides with retention of configuration.[334] The same type of reaction occurs with nonactivated aromatic or heteroaromatic halides to afford the corresponding aryl or heteroaryl selenides,[335] but requires HMPA and the presence of copper iodide. Substitution by sodium phenylselenolate of iodobenzene (55% yield), 1-chloronaphthalene (70% yield), 9-bromophenanthrene

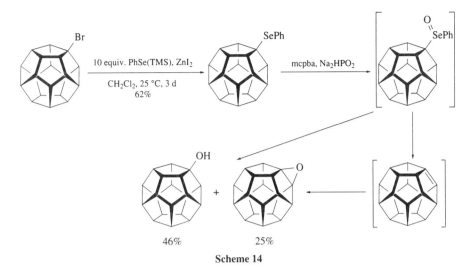

Scheme 14

(72% yield) and 4-chlorobiphenyl (37% yield) occurs via a photostimulated $S_{RN}1$ mechanism in liquid ammonia.[336-8]

Only a few reports have dealt with the reactions of selenols or selenolates with diazo compounds.[37,109,159] This type of reaction has been used, *inter alia*, for the synthesis of diaryl selenides, including diphenyl selenide, benzyl phenyl selenide and 6α-phenylselenenyl penicillanate (in the presence of $BF_3 \cdot OEt_2$).

(d) Synthesis of functionalized selenides from selenols and selenolates.

(i) Synthesis of β-heterosubstituted selenides have been achieved from epoxides, thiiranes and aziridines:[1,56,76,112-16,132,190,194,339] β-hydroxyalkyl selenides, which are available from β-bromohydrins and selenolates,[340] are more efficiently synthesized by ring opening of epoxides with selenols,[339] in some cases in the presence of catalytic amounts of alumina[341] or sodium[342] (RSeM; M = Na, from $NaBH_4$–PhSeSePh), ammonium[253] or magnesium[343] selenolates, as well as with $B(SeR)_3$,[344] $MeSeAlMe_2$,[243,244] or with TMS-SeR in the presence of zinc diiodide[345] or catalytic amounts of butyllithium.[345] The reaction has been carried out efficiently with a large variety of epoxides, including trisubstituted and α,β-unsaturated epoxides, oxaspiropentanes[346,347] and epichlorohydrins.[348] The reaction of selenols and metal selenolates usually occurs at the least hindered side of the molecule and by inversion of the configuration there[47,256] unless stereoelectronic factors reverse the tendency.[256] It is highly chemoselective and, for example, a terminal epoxide reacts much faster than a trisubstituted epoxide.[344]

N-Tosylaziridines[349,350] have also been transformed into the corresponding β-selenoalkylamines on reaction with phenylselenolates. Asymmetric ring opening of prochiral epoxides with chiral selenols has been reported.[132,194]

(ii) Synthesis of ω-selenocarboxylic acids and related derivatives from lactones and cyclopropyl esters:[76,173] selenolates, including alkali metal selenolates and $PhSe(TMS)–ZnI_2$,[351] are powerful nucleophiles able to promote the ring opening of lactones,[262,263,352,353] oxazolines (PhSe(TMS)),[354] oxazolinium ions[355] and cyclopropane derivatives bearing one (keto or cyano) or, better, two electron-withdrawing groups (two esters or an ester and a keto group),[356,357] leading to formation of the corresponding ω-functionalized selenides (Scheme 15). The reactions usually proceed better in DMF, THF–HMPA or THF–crown ethers rather than in ethanol and have been performed with potassium, sodium or lithium selenolates. They are much less efficient when carried out in the presence of BR^2_3 (R^2 = H or OMe, resulting from reaction of R^1SeSeR^1 with $NaHBR^2_3$).[262,263,352] The best results have been obtained on five- or six-membered cyclic lactones, but substitution at C-5 disfavours the reaction, especially for six-membered lactones. The reaction occurs chemoselectively on the lactone ring even if it is fused to a cyclopropane ring[358] or possesses an α-*exo*-methylene group,[251] and in any case epimerization α to the carboxyl group is observed.

(iii) Synthesis of β-seleno carbonyl compounds from α,β-unsaturated carbonyl compounds:[1,76,359] selenols in acidic media or in the presence of catalytic amounts of amines and selenolates as well as neutral samarium phenylselenolate[323] add to the C=C double bond of α,β-unsaturated amides,[360] acids, anhydrides, esters, lactones, aldehydes and ketones to give β-seleno carbonyl compounds. The reaction proceeds enantioselectively in the presence of a chiral amine at room

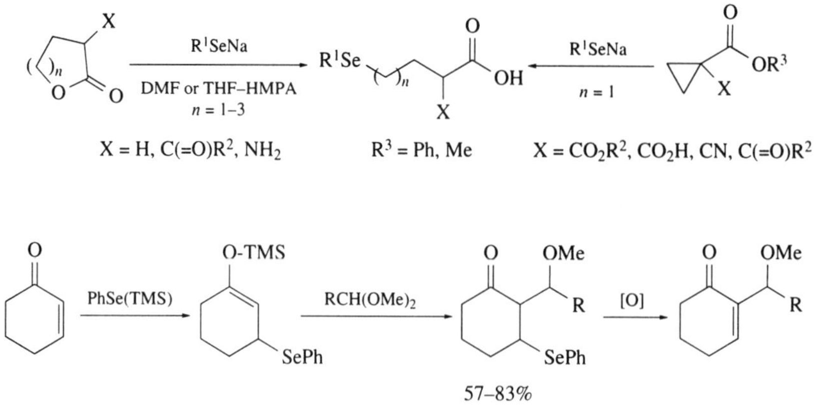

Scheme 15

temperature.[361] Addition of chiral selenols to enones has also been reported.[132,194] Other conditions involve the use of selenolates in alcohols or dimethylaluminum methylselenolate.[243] The reaction proved to be chemoselective and highly stereoselective.[362] It allows trapping of the enolate intermediates by the counterion present, leading to isolable allyl selenides (from SiR_3 enolates),[363–5] or by aldehydes, leading to α-hydroxyalkyl-β-selenoketones (from BR_2,[366] $AlMe_2$[367] or titanium[367] enolates) (Scheme 15). Phenylselenocyanate also adds to enones in the presence of tributylphosphine to give γ-phenylseleno-α,β-unsaturated nitriles.[319]

(iv) *Synthesis of selenoacetals and related derivatives:*[1,78,81,84,89,113,173,368] selenols add reversibly to aldehydes (neat or in nonpolar solvents) and the reaction leads almost instantaneously to α-hydroxyalkyl selenides, whereas the equilibrium is completely shifted towards the starting materials if ketones are involved. The reaction takes another course when carried out in the presence of protic or Lewis acids. With aldehydes and ketones,[368] including aromatic[368] and α,β-unsaturated ones,[369] the corresponding selenoacetals are obtained (Scheme 6 (b)). The best results have been obtained from phenyl- or methylselenol using zinc chloride (0.5 equiv.) at 20 °C (usual cases) or titanium tetrachloride (0.35 equiv.) at −78 °C (aromatic or hindered ketones).[368,370] The use of tris(methylseleno)borane allows the selenoacetalization of aldehydes without the requirement for an acid catalyst and permits the selenoacetalization of citronellal without concomitant cyclization.[368] Selenoacetals have also been prepared from acetals and selenols in the presence of $BF_3 \cdot OEt_2$ or with selenoboranes in the presence of trifluoroacetic acid. The reaction proved to be particularly useful for the synthesis of 1,1-bis(methylseleno)cyclopropane from 1-ethoxy-1-silyloxycyclopropane.[82]

Mixed *O,Se*-acetals can be prepared from *O,O*-acetals and related derivatives by use of 1 equiv. of selenol in the presence of $BF_3 \cdot OEt_2$,[313,371–4] or zinc chloride,[372] from trimethylsilyl phenyl selenide and TMS-OTf (Tf, triflyl (trifluoromethanesulfonyl)),[375] from tributyltin phenyl selenide and $BF_3 \cdot OEt_2$,[376] or from tris(phenylseleno)borane if the acid catalyst is omitted.[377] This reaction has been successfully used for the synthesis of α-seleno sugars and glycosides.[372,373,378] The synthesis of related α-silyloxyalkyl selenides has been achieved from aldehydes, alcohols and silyl chlorides.[379,380] Otherwise, *O,Se*-acetals and related derivatives have been prepared by oxidation of phenylseleno acetates and phenylselenoacetonitrile via Pummerer-type rearrangements.[381]

(ii) Synthesis of functionalized selenides from electrophilic organoselenium reagents

Diselenides[132,331] or, better, phenylselenenyl halides (PhSeX, X = F,[132,382] Cl,[1,39] Br,[1,39] I[271,383]),[194] and related phenylselenenyl pseudohalides such as PhSeOC(=O)Me, PhSeOC(=O)CF_3,[1] *N*-PSP,[132,314] selenamides ($R^1SeNR^2_2$),[194] unstable selenenic acids (RSeOH) and anhydrides ((RSe)$_2$O)),[125,126,384–6] putative PhSeOSePh (from PhSeSePh and (PhSeO)$_2$O or *t*-butyl hydroperoxide),[387] PhSeOSnBu_3 (from PhSeSePh,Br_2 and (Bu_3Sn)$_2$O),[387] PhSeOSO_3^- (from PhSeSePh and $S_2O_8^{2-}$),[388,389] benzeneseleninic anhydride ((PhSeO)$_2$O),[331] PhSeCN–$CuCl_2$[39,123] and some aryl- and methylseleno analogues have proved to be valuable and easily available reagents able to transfer the electrophilic selenium atom to organometallics, alkenes, carbonyl compounds and their enolates[331] and phenols.[287,288] Among these reagents, phenylselenenyl chloride and bromide[1,164,287] and *N*-PSP[314] have been by far the most widely used.[1,48,128,162,164,173,190,199,290,331,359]

(a) Synthesis of selenides from electrophilic organoselenium reagents and organometallics. Selenides and functionalized selenides have been synthesized[39] from electrophilic organoselenium reagents (RSeSeR or RSeX, X = halogen) and organometallics including Grignard reagents,[390] mercury diaryls,[390] alkyllithiums,[245] α-alkoxy-,[391,392] α-silyl-,[393] and α-selenoalkyllithiums,[394] 1-zircono-1,3-dienes[395] (*N*-PSP) and phosphoranes,[396] α-lithiovinyl sulfides[397] and α-lithioalkyl sulfoxides.[398,399] The reaction has been successfully extended to enolates derived from aldehydes, esters, lactones, amides and nitriles,[1,331] and leads to the corresponding α-seleno carbonyl compounds (see Section 13.3.5.1(ii)(b)).

(b) Synthesis of α-seleno carbonyl compounds from electrophilic organoselenium reagents and carbonyl compounds. α-Seleno carbonyl compounds[48,128,164,190,331,339] are valuable derivatives easily prepared from the corresponding carbonyl compounds and transformed under mild conditions into (*E*)-α,β-unsaturated carbonyl compounds after oxidation via the corresponding selenoxide (Scheme 6 (a); Scheme 26).[1,331,400]

Selenenylation of aldehydes and ketones has been successfully achieved directly by reaction with PhSeCl in AcOEt or in diethyl ether (but not with PhSeBr owing to competing bromination).[1,331] Reactions under these conditions proved highly chemoselective and the selective selenenylation of aldehydes and ketones in the presence of alcohols, C=C double bonds, epoxides, esters, ketals, lactones, silyl ethers or ketals is possible.[128,331] 3-Keto steroids and 17-keto steroids usually give 2α-phenylseleno[401] and 16α-phenylseleno[47] derivatives, respectively, and enones are generally selenenylated at the α'-position.[291,402] α-Phenylselenoaldehydes are also efficiently synthesized from aldehydes and *N,N*-diethylbenzeneselenamide (PhSeNEt$_2$).[331,339] Related reactions[331] have been carried out successfully on β-diketones, β-keto esters,[403] β-keto lactams, malonates[404,405] and malononitriles, esters and lactones including β-lactones,[406] nitriles and lactams.[407] In the last case the *N*-unprotected derivatives give the unstable *N*-SePh compounds even if the reaction is carried out with 2 equiv. of base. *N*-Benzyl-protected lactams give bis(phenylselenenyl) derivatives owing to selenenylation at both the carbon α to the carbonyl group and at the benzylic carbon. *N*-Butoxycarbonyl protection avoids the problem.[407]

Asymmetric selenenylation of aldehydes and ketones using chiral selenamides has been reported.[132,194]

(c) Synthesis of β-hetero-substituted selenides from electrophilic organoselenium reagents and alkenes. Selenenyl halides add to alkenes in the absence of other nucleophiles and give β-haloalkyl selenides, probably via a seleniironium salt intermediate (Scheme 16).[132,173,190,194,198] The reaction proceeds regioselectively with terminal alkenes and leads to β-haloalkyl selenides bearing at the terminus either the halogen atom under kinetically controlled conditions (anti-Markovnikov adduct; for example, PhSeCl, CCl$_4$ or CH$_2$Cl$_2$, −78 to −10 °C; or PhSeBr, THF, −78 °C) or the seleno moiety under thermodynamically controlled conditions (Markovnikov adduct; for example, PhSeCl, MeCN, 25 °C, 24 h; or PhSeBr, MeCN, 25 °C) (Scheme 16). The Markovnikov adduct (SeR on the least substituted carbon) is obtained from trisubstituted alkenes irrespective of the conditions used. The addition is *exo*-selective with norbornene and *endo*-selective with nobornadiene, takes place chemoselectively on the least substituted C=C double bond of differently substituted polyalkenes,[408] but leads to a mixture of 1,2- and 1,4-adducts with 1,3-dienes.[409] The reaction is highly *trans*-stereospecific and compatible with the intermediate formation of an episelenurane (Scheme 16).

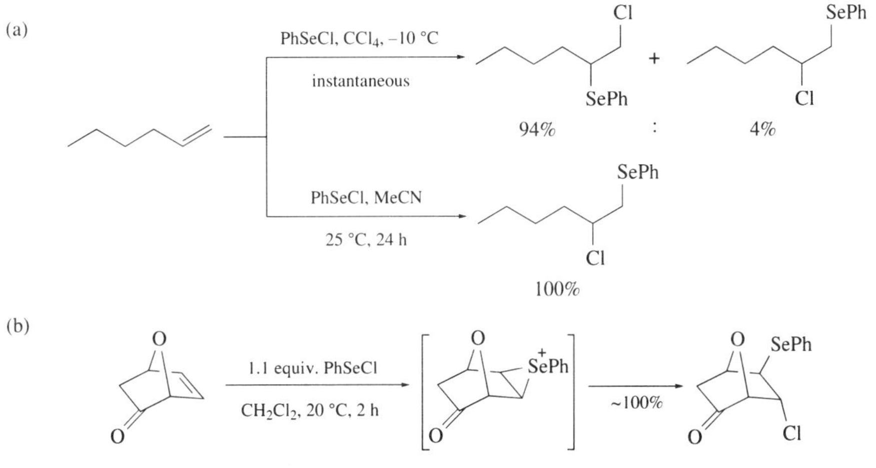

Scheme 16

Selenenyl halides have also been reacted with α-heterosubstituted alkenes such as 1,1-difluoroethene[410] and 1-chloro-1-cyclohexene, with enol ethers, enol acetates and enamines,[1,132,331] with vinylsilanes[411] and with electrophilic alkenes bearing an electron-withdrawing group such as α,β-unsaturated aldehydes,[412] ketones,[412] esters,[1,412] amides,[412] nitriles,[1,331,412] sulfones[412] and nitro compounds.[412] Selenenyl halides in chloroform add to acrolein, methyl acrylate and acrylamide to give mixtures of regioisomers in which that bearing the seleno moiety α to the carbonyl group usually predominates (93:7 to 55:45). The reaction takes another course if carried out in the presence of an amine (NEt$_3$ or pyridine), when α-seleno-α,β-unsaturated aldehydes, ketones, esters, amides and nitriles are formed selectively (Scheme 17).[413,414] However, if the reactions are performed with an excess of selenenyl halide–pyridine (3 equiv., 20–37 °C, 12–96 h), α-halo-α,β-unsaturated aldehydes or ketones are formed in good yields (Scheme 17).[415,416]

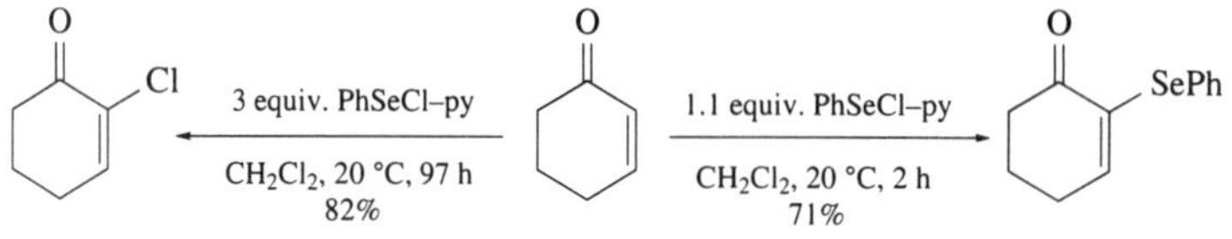

Scheme 17

α-Phenylselenoketones have also been synthesized from trimethylsilyl enol ethers and the reagent generated from diphenyl diselenide and chloramine-T (PhSe(=NTs)NPhTs),[417] and by hydrolysis of the adduct between CF$_3$CO$_2$/SePh and alkynes.[418]

Optically active α-phenylseleno carbonyl compounds have been obtained with poor *ee* (13–26%) from 4-alkylcyclohexanones and a chiral selenamide derived from α-methylbenzylamine,[419] and with reasonably high diastereoselectivity from homochiral silyl enol ethers derived from oxazolidinones and phenylselenenyl chloride.[420]

If the above-mentioned reactions are carried out in the presence of a nucleophile (including in some cases the solvent), or if the reaction is instead performed with the electrophilic reagents listed below, the seleno moiety is introduced together with another heteroatomic moiety.[132] This is the case, for example, with RSeO(C=O)Me,[421] RSeO(C=O)CF$_3$ (RSeX–AgOC(=O)CF$_3$),[1,331] RSeOR (RSeX–ROH),[422] RSeOH (RSeX–H$_2$O, RSeO$_2$H–RSeSeR, RSeO$_2$H–PPh$_3$, RSeO$_2$H–H$_3$PO$_2$), RSeX–MeCN–H$_2$O, RSeCN, R^{1}SeNR$^2_2$, RSeNH(C=O)Me, RSeX–MeCN–CF$_3$SO$_3$H, RSeNO$_2$ (RSeX–AgNO$_2$),[194] RSeN$_3$ (RSeX–NaN$_3$ or PhI(OAc)$_2$–NaN$_3$–PhSeSePh),[388] RSeSCN, RSe(=NTs)N(Ts)SePh and RSeSO$_2$Ar (RSeX–AgBF$_4$–purine base).[423] PhSeBr reacts with dienes in the presence of silver nitrate and produces, on treatment with base, the extremely reactive 2-nitro-1,3-dienes.[424] Asymmetric methoxyselenenylation of alkenes by optically active binaphthylselenenyl bromides in methanol proceeds with good yield but relatively poor diastereoselectivity.[132,194,425-7]

Finally, diphenyl and dimethyl diselenide add, in the presence of tin tetrachloride, to alkenes and alkynes with high stereocontrol (*anti* mode),[428] and selenosulfonates react[128] with terminal and disubstituted alkenes in the presence of boron trifluoride etherate to give β-seleno sulfones with Markovnikov orientation. The reaction also occurs with terminal, disubstituted alkenes and allenes under free-radical conditions when carried out thermally, at reflux in benzene, or photochemically.[127] It is tolerant to a wide range of functional groups, occurs with anti-Markovnikov orientation, its rate is enhanced by AIBN, it does not occur with tetrasubstituted alkenes and it is *trans*-stereospecific with cyclic alkenes. Ring opening takes place with vinylcyclopropanes and addition occurs on alkynes, leading to β-seleno vinyl sulfones.[429] Reaction of a number of 1,6-dienes with selenosulfonates under free-radical conditions[127] results in selenosulfonylation with concomitant C–C bond formation, delivering 1-selenomethyl-2-methylsulfonylcyclopentane derivatives.[430,431] PhSeSC(=O)R[432] also allows, under free-radical conditions, the photoinduced selenoesterification of alkenes. Among the various reactions described the alkoxyselenation of alkenes[184] has proved to be particularly useful, especially for the synthesis of natural products.[198] It occurs when the reaction is carried out in alcohols, and has been used efficiently for the synthesis of allyl vinyl ethers, valuable starting materials in the Claisen rearrangement (Scheme 18).

Asymmetric methoxyselenylation reaction between bis(1,1'-binaphthalene-2-yl) diselenide, bromine and α,β-disubstituted alkenes has been achieved with optical yields of up to 49% (styrene).[132,426]

(d) Cyclization reactions involving electrophilic organoselenium reagents and alkenes. When a hydroxy group of an alcohol or an acid is suitably localized in a molecule it can promote selenoetherification[287,321,388,434] to give oxygen and sulfur heterocycles[435] or selenolactonization[436] (Scheme 6 (e)).[39,128,173,181,199,320,433] Although phenylselenenyl chloride has been used extensively for this

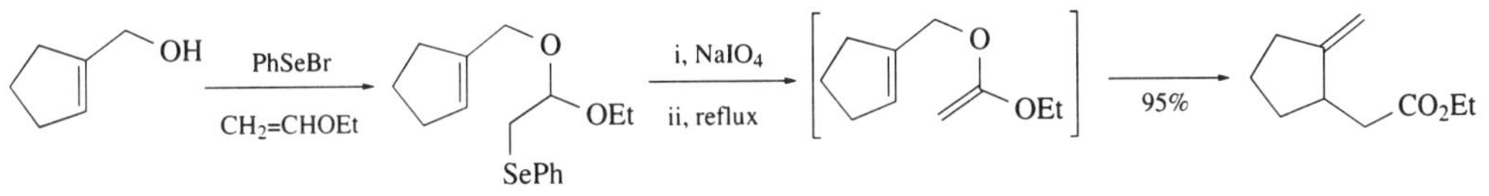

Scheme 18

purpose, *N*-phenylselenophthalimide (*N*-PSP) has proved to be a valuable alternative.[314,437,438] Other seleno-monitored cyclizations which involve the participation of NR_2[383] and SR groups, β-dicarbonyl moieties[39] α-amino esters,[439] acylsilanes[440] or another C=C double bond[314] have been reported. The lithium enolate of 4-chloro-4,5-dimethylhex-5-en-3-one also reacts with PhSeBr and produces, at −78 °C, 2,3,5-trimethyl-5-phenylselenocyclopent-2-enone, which can be converted into the α'-*exo*-methylene-α,β-unsaturated cyclopentanone on oxidation.[441]

Transformations involving PhSeCl- or *N*-PSP-monitored cyclization of ε-hydroxyalkenes have been used successfully for the synthesis of *C*-aryl glycosides,[437] 2,6-disubstituted dihydropyrans[437] and tetrahydrofurans[438] whose stereochemistry can be controlled by the conditions used. Several γ- and ε-alkenylstannanes smoothly cyclize in the presence of *N*-PSP to phenylselenomethylenecycloalkanes.[442] Limitations of the method are described.[442]

13.3.5.2 Syntheses of selenides involving diselenides as radical initiators or as radical traps

The high reactivity of PhS• (10–50 × PhSe•) toward alkenes and the high capturing ability of PhSeSePh (160 × PhSSPh) allow[443] the highly selective photostimulated thioselenation of alkenes by the disulfide–diselenide system. Under related conditions diphenyl diselenide adds to the C≡C triple bond of alkynes and enynes.[444] Allyltributylstannanes undergo S_H2' substitution on photostimulated reaction with diphenyl diselenide and allylic selenides.[445]

Diphenyl diselenide is a particularly valuable radical trap.[446-8] For example, it can efficiently trap the radical produced on reaction of bicyclo[4.1.0]heptan-1-ol derivatives bearing a butenyl group at C-5 with Mn(pic)₃ (manganese(III) tris(2-pyridine carboxylate)) (Equation (3))[446] and, in the presence of $(NH_4)_2S_2O_8$,[449] converts allyl-substituted pyrrolidones into azabicyclo[3.3.0]octanes.[75,127,359]

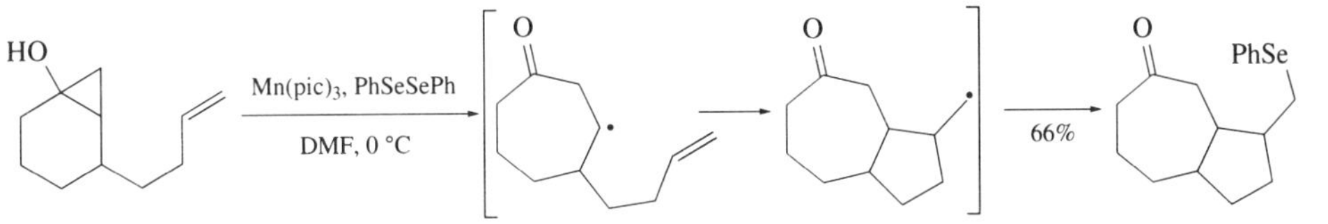

13.3.5.3 Syntheses of selenides and functionalized selenides involving concomitant formation of a new C–C bond

(i) Syntheses involving α-selenoalkylmetals

(a) Synthesis of α-selenoalkyllithiums via Se–Li exchange. The synthesis of α-selenoalkylmetals bearing hydrogen or alkyl groups at the carbanionic centre is best achieved by selenium–metal exchange from selenoacetals and butyllithiums.[1,47,48,78,81-6,89,113,154,174] The reaction has been efficiently carried out on phenylseleno and methylseleno derivatives,[450] usually at −78 °C with Bu^{n}Li in THF, and provides a large variety of α-selenoalkyllithiums, including products where the carbanionic centre is nonsubstituted, mono- or dialkyl-substituted including cyclopropyl and cyclobutyl compounds,[82] or products with an aryl, a vinyl or a heteroaromatic moiety[451] (Schemes 2 and 7). Bu^{t}Li is more reactive than Bu^{s}Li or Bu^{n}Li, which is the least reactive reagent of the series. The reaction is more efficient if carried out in THF rather than in diethyl ether and only those α-selenoacetals derived from aromatic carbonyl compounds are cleaved in pentane. Phenylselenoacetals react faster than the methylseleno analogues, probably because the carbanionic centre on phenylselenoalkyllithiums is more efficiently stabilized.[211] Surprisingly, *n*-butyllithium metallates and does not cleave the C–Se bond of α-sulfenylalkyl selenides.[399]

Under the conditions used for their synthesis (THF, −78 °C), these organometallics are stable for long periods but are destroyed at higher temperatures (around −40 °C) by metallation of the solvent. The corresponding cuprates have been synthesized by addition at −78 °C of 0.5 equiv. of CuI–SMe$_2$ complex. They are stable at this temperature but give alkenes by oxidative coupling at −40 °C.

Primary α-selenoalkyllithiums epimerize at temperatures down to −105 °C in THF[452] but are configurationally stable for at least 6 h at −125 °C in 2-methyl-THF.[106] Their addition to chiral aldehydes nevertheless occurs faster than enantiomer equilibration[106] and similar behaviour was observed at −78 °C for α-selenoalkyllithiums having a chiral centre in the β-position.[452] Selenoacetals derived from 4-t-butylcyclohexanone[394] and phenylselenol or methylselenol and that derived from cis-3,5-diphenylcyclohexanone and phenylselenol[212] react with butyllithium selectively on the axial group. The related mixed (SePh, SeMe) acetals bearing the phenylseleno moiety in the axial position react mainly on the equatorial methylseleno moiety but epimerization leads to the equatorially oriented phenylseleno group.[394,453] Selenoacetals derived from cis-3,5-diphenylcyclohexanone[212] or from 4-t-butylcyclohexanone[454] and 1,3-propanediselenol, in contrast, react with t-butyllithium at the equatorial seleno group. This has been accounted for by the favourable conformation, which is expected to better delocalize the anion on the carbon framework through the selenium atom.[212,454]

α-Selenoalkyllithiums react with electrophiles to give selenides or functionalized selenides. Typical electrophiles are chlorotrimethylsilane, alkyl halides (MeI, primary alkyl only),[82,85,89] chloromethyl isopropyl ether,[450] epoxides (terminal),[85] oxetanes,[85,89] tetrahydrofurans (requires the use of BF$_3$·OEt$_2$),[438] aldehydes or ketones (Scheme 7),[82,84,191,192,339] acid chlorides,[89] carboxylic anhydrides,[89] dimethylformamide, nitriles,[455] chlorocarbonates,[89] carbonic anhydride[89] and α,β-unsaturated lactones.[456] In several instances HMPA favours the alkylation of α-selenoalkyllithiums. Allylation is not very efficient,[457] but allylation of the corresponding cuprate proceeds very efficiently with conservation of the regio- and stereochemistry of the allylic moiety. The reaction of α-selenoalkyllithiums with ketones (the best results being obtained from the methylseleno series) is exceptional[84] since it allows the high-yield synthesis of β-hydroxyalkyl selenides, even from particularly hindered[84,89,213,339] or enolizable[84] ketones, and with high stereochemical control in the case of rigid cyclohexanones (equatorial attack predominates).[81] 1,2-Addition usually takes place with enones and enals.[78,339]

α-Selenobenzyllithiums[458] are prone to add to alkenes (Scheme 10). Although these organometallics are stable around −78 °C, addition reaction takes place around 0 °C and produces cyclopropane derivatives.[216]

Selenoacetals, including functionalized derivatives, have in turn been prepared from selenols and carbonyl compounds,[113,368] or by reaction of 1,1-bis(seleno)alkyllithiums with chlorotrimethylsilane, alkyl halides, epoxides and aldehydes or ketones. 1,2-Addition usually takes place with enones and enals if the reaction is performed in diethyl ether[78,87] and 1,4-addition is favoured if the reaction is carried out in the presence of HMPA.[78,87] 1,1-Bis(seleno)alkyllithiums are easily synthesized from seleno orthoesters and n-butyllithium.

(b) Synthesis of α-selenoalkyllithiums by hydrogen–lithium exchange. Other functionalized α-selenoalkylmetals have been prepared by metallation with LiDBA, LDA, LiTMP or KDA in THF solution of the corresponding carbon acids and have been successfully reacted with a large array of electrophiles.[78,81–4,162,174] They include aryl alkyl selenides whose alkyl group is substituted at C-1 by (i) an aryl,[458] vinyl,[458] ethynyl, methoxy, seleno, phosphonato, keto, ester, lactone, lactam, acid, nitrile, sulfenyl, silyl or nitro group or a chlorine atom, including tetrametallotetraselenofulvalene,[459] and (ii) two seleno moieties or an aryl and a seleno moiety as well as alkylselenonium salts,[192] alkyl selenoxides and alkyl selenones.[84,192,460,461] Alkylation of α-selenoketones has been studied and shown to occur directly at the carbon of the enolate.[462]

(ii) Syntheses involving α-selenoalkylcarbenium ions and related reagents

Selenoacetals react in the presence of Lewis acids (usually SnCl$_4$) with silyl enol ethers,[463-6] allylsilanes[453,467] or allylstannanes,[453] aromatic hydrocarbons[468] and trimethylsilyl cyanide[469] and produce via α-seleno carbenium ions (Section 13.2.5.2)[155] β-selenoketones, homoallyl selenides, arylated α-seleno esters and α-cyanoalkyl selenides, respectively. Similarly, seleno orthoesters produce α-cyanoselenoacetals on reaction with trimethylsilyl cyanide.[469] If selenoacetals are reacted with SnCl$_4$ and triethylamine, vinyl selenides or products derived from an aldol-type coupling are formed, depending on the rate of addition of the reagents.[470] Vinyl selenides are also formed, probably through the same intermediates, from selenoacetals and methyl iodide[471] or P$_2$I$_4$,[472] or from α-haloalkyl selenides and DMF.[473] Selenoacetals derived from α,β-unsaturated carbonyl compounds react in a Friedel–Crafts reaction with N-methylpyrrole[474] and with pyrrole.[463,475] The latter reaction allows the one-pot synthesis of meso-tetravinylporphyrins[463] through α-selenoallyl[475] or α-selenopropargylic cationic species,[476] and those derived from δ,ε-unsaturated carbonyl compounds react with tin tetrachloride to give 3-chlorocyclohexyl selenides, often with high stereocontrol.[477] Finally, under the same conditions seleno orthoesters react with P_2I_4[472] or with silyl enol ethers and $SnCl_4$[478] to generate ketene selenoacetals or monoselenoacetals derived from β-ketoaldehydes, respectively. Few reactions so far have taken advantage of the reactivity of bromomethyl benzyl selenide, a valuable precursor of selenium-containing α-amino acids.[237]

13.3.5.4 Syntheses of carbonyl compounds bearing a selenium atom attached to the sp^2 carbon of the carbonyl group

(i) Syntheses of selenol esters (acyl selenides)

Selenol esters[107] ($R^1C(=O)SeR^2$) have been prepared from the corresponding acids either (a) directly by reaction of various selenenyl-transfer reagents, such as phenylselenocyanate (PhSeCN),[479] benzeneselenenyl chloride,[480,481] diphenyl diselenide[480,482] or N-phenylselenophthalimide (N-PSP) and trialkylphosphines,[314,482] or by reaction with selenols either in acidic media (the cyclic version allows the synthesis of seleno γ-butyrolactones) or with 1,1-carbonyldiimidazole or 1,2,4-triazole,[483,484] or (b) via their ammonium salts and alkyl- or arylselenophosphonium salts (from alkylation of trialkyl- or triphenylphosphine selenides (R_3PSe, Section 13.3.3.2(i)), (c) their chlorides and selenols,[392,482,485,486] or alkali, thallium[327] or samarium diiodide,[322,323] or trimethylsilyl selenolates[487] or their dialkyl phosphates and sodium phenylselenolate[482] or (d) their esters or lactones and dimethylaluminum methylselenolate ($MeSeAlMe_2$, Section 13.3.3.2(i)).[243,244,488] The reaction does not always occur with five-membered lactones and discriminates between axial and equatorial esters, which afford selectively the equatorial selenol esters;[489] (e) their hydrazides and benzeneseleninic acid ($PhSeO_2H$);[386,490] (f) their enol carboxylates and alkali selenolates;[491] (g) their tellurol esters (acyl tellurides) and diphenyl diselenide on photolysis at low temperature (benzene, 8 °C, 1.5 h) or thermally in the dark (benzene, reflux, 2 h).[492] Although tellurol esters bearing a C=C double bond cyclize under photolysis, the radical intermediates are trapped when the reaction is carried out with diphenyl diselenide and lead to the corresponding selenol esters. Selenol esters have also been prepared from metal selenocarboxylates and alkyl halides.[493] The synthesis of β-seleno-α,β-unsaturated selenol esters has been achieved from ethyne and diselenides in the presence of a palladium catalyst.[494]

Selenol esters are useful as acyl equivalents. They react with alkenylcopper reagents in the presence of HMPA to give α-enones[488] and have been used for macrocycle formation.[480] They are also valuable acyl radical precursors which lead to formates or alkanes, or can be trapped with a C=C double bond to produce ketones with concomitant C–C bond formation (reactivity, see Section 13.3.6.1).

(ii) Syntheses of selenol carbonates

Selenol carbonates are valuable building blocks available from the corresponding alcohols via the corresponding chloroformates by sequential reaction with phosgene and selenols (Scheme 19).[495–7] The reaction proceeds with primary, secondary and tertiary alcohols (but is difficult in the last case) and with β,γ-unsaturated alcohols.

(iii) Syntheses of seleno carbonyl compounds

Seleno carbonyl compounds are being widely used for the synthesis of both organoselenium derivatives and of selenium-free compounds.[91,95,145,498] Among the former, selones (selenoketones) and selenoaldehydes are of particular interest. They are unstable, except for the specific cases of particularly hindered and nonenolizable derivatives. They often trimerize or react in situ with one of the reagents used for their preparation or they can be trapped by transition metals such as chromium or tungsten derivatives[376] or by a 1,3-diene introduced in the reaction medium. Transition metal-coordinated selenoaldehydes and selenoketones also undergo [4 + 2]-cycloadditions with dienes.[499]

Reaction of aldehydes or their acetals with either H_2Se[145] or 2,2,4,4,6,6-hexamethyl-1,3,5-triselena-2,4,6-tristannacyclohexane and Lewis acids[231] gives 2,4,6-trisubstituted 1,3,5-triselenanes and it is still unclear if selenoaldehydes are transient intermediates in the process. Those which possess a C=Se group which is either particularly hindered[500] (2,4,6-tri-t-butylphenyl) or is stabilized by electron-donating

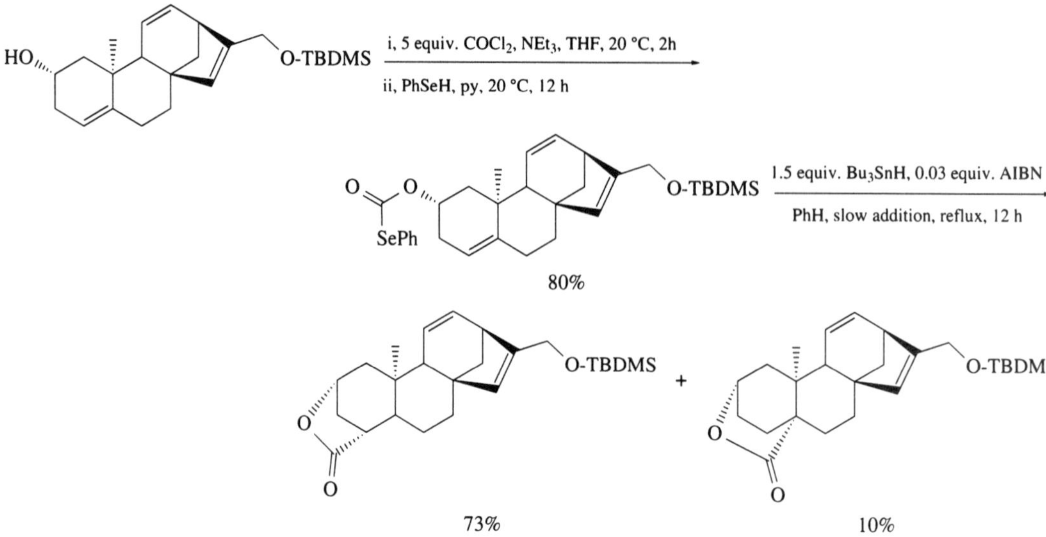

Scheme 19

groups[501] can be isolated if prepared under suitable conditions. Thus, 2,4,6-tri-*t*-butylselenobenzaldehyde has been synthesized[500] as the first stable selenoaldehyde (44% yield) by β-elimination of the silicon and cyano moieties from TMS-CH(R)SeCN (R = 2,4,6-tri-*t*-butylphenyl) using tetrabutylammonium fluoride but cannot be generated from *O*-alkyl selenoformates (ROCH(=Se)) and 2,4,6-tri-*t*-butylphenyllithium. Pyrrolo[2,1-*b*]thiazole-7-carboselenaldehydes are stable crystalline derivatives which have been prepared from the corresponding aldehydes and phosphorus pentaselenide or by selenoformylation of pyrrolo[2,1-*b*]thiazoles.[501]

Unstable selenoaldehydes have been prepared from TMS-CH(R)SeCN and tetrabutylammonium fluoride,[498,500,502] RO_2CCH_2Se-phthalimido on reaction with triethylamine as base,[503] bis-(dimethylaluminum) selenide and either acetals,[148] including 1-alkoxy epoxides, tetrahydrofurans and tetrahydropyrans, or α,β-unsaturated aldehydes,[504] which undergo[504] regioselective self-condensation to cyclic diselenide derivatives if not trapped, Se^{2-} species (produced from fluoride ion or butyllithium catalysed reaction of bis(trimethylsilyl) or bis(tributylstannyl) selenides ($(TMS)_2Se$ or $(Bu_3Sn)_2Se$) and aldehydes bearing an electron-withdrawing group[505] or α,α-dichloroacetates,[506] elemental selenium and phosphoranes (Ph_3P=CHR, R = CN, CO_2Me, COMe, Ph; toluene, reflux, 110 °C, 3 h)[507] (the selenoaldehyde intermediate usually couples with the ylide to produce alkenes)[507] and allyl vinyl selenides via a seleno-Claisen rearrangement.[508]

Selenoaldehydes have been trapped *in situ* with 1,3-dienes such as 2,3-dimethylbutadiene, cyclopentadiene and even anthracene to give the corresponding Diels–Alder adducts (2*H*-selenapyrans) in good to very good yields.[148,498,502,503,506,508–10] Their α,β-unsaturated derivatives and enones undergo[504] intermolecular hetero-Diels–Alder reaction with cyclopentadiene and [2 + 4]-cycloaddition with norbornadiene.

Selones are also unstable derivatives. However, some compounds which are very sterically hindered and are nonenolizable have been isolated and characterized and are valuable building blocks for the synthesis of hindered C=C double bonds bearing particularly bulky substituents (see below).

Selones have been efficiently prepared (Scheme 20) from bis(dimethylaluminum) selenide and α,β-unsaturated enones[504] which undergo[504] regioselective self-condensation to cyclic diselenide derivatives if not trapped, hindered ketones and bis(tricyclohexyltin) selenide and boron trichloride,[511] hydrazones derived from hindered ketones and selenium bromide (Se_2Br_2, Section 13.2.4.1),[511] phosphoranylidene hydrazones derived from hindered ketones and elemental selenium,[511] diazoalkanes and elemental selenium,[511] phosphonium ylides and elemental selenium[512] and allyl vinyl selenides via a seleno-Claisen rearrangement.[508] The resulting selones have been transformed into the corresponding carbonyl compounds or trapped with cyclopentadiene.[508]

Selenobenzophenones[507] are also fairly unstable and, although they can be isolated, are very reactive.[507] Unusual cycloadditions occur with acetylenedicarboxylates and norbornadiene, which produce benzoselenapyrans or seven-membered cyclic selenides.[507]

Particularly crowded alkenes have been prepared (Scheme 1 (b))[50,95,511,513–15] from selones and diazo compounds via the intermediate formation of 1,3,4-selenadiazolinephosphoranylidene hydrazone, or more directly from phosphoranylidene hydrazone and elemental selenium. They are also available from

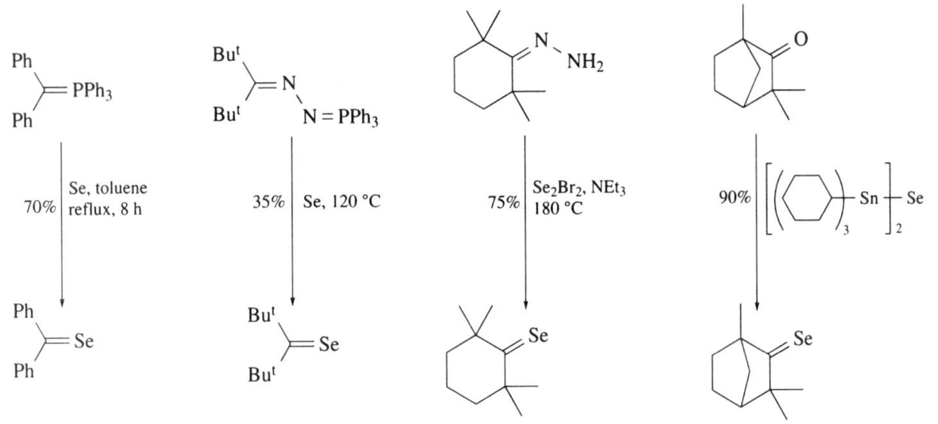

Scheme 20

Wittig reagents (Ph$_3$P=CHR, R = CN, CO$_2$Me, COMe, Ph) and elemental selenium, or from benzylselenocyanates and triethylamine.[516] None of these methods allowed the synthesis of tetra-*t*-butylethene.[514,515] Selones participate efficiently in Diels–Alder[516] and 1,3-dipolar cycloaddition reactions.[516]

13.3.6 Reactions of Selenides and Functionalized Selenides: Synthesis of Selenium-free Compounds

13.3.6.1 *Reactions involving reduction of the C–Se bond*

(i) Reduction of the C–Se bond involving radicals

(a) Reduction of the C–Se bond leading to C–H bond formation. Selenides and functionalized selenides are easily reduced with a large array of reagents (Scheme 6 (e); Scheme 7).[1,4,78,89,111,113,127,158,196,199,517,518] Raney nickel,[1,78,89,282,519] in one case under pressure,[282] nickel boride, lithium in ethylamine,[1,78,89] organolithium reagents,[1] tributyl- or triphenyltin hydrides thermally in the presence of AIBN, photochemically,[392,435,520-2] or in the presence of triethylborane,[391,523-5] tris(trimethylsilyl)silane[374,453,526,527] and cobalamin, including cyclodextrin-B$_{12}$,[528] all reduce selenides and selenoacetals to alkanes. Selective reduction of selenoacetals to selenides is also possible with tin hydrides. In the case of selenoacetals derived from 4-*t*-butylcyclohexanone the reaction is not stereoselective. Axial and equatorial attack both take place and provide a 60–70:40–30 *cis–trans* mixture of stereoisomers.[453] The C–Se bond of phenyl selenides such as adamantyl phenyl selenide is also cleaved homolytically on irradiation.[332] Tin hydrides and nickel boride allow the highly chemoselective reduction of selenides in the presence of chlorides, nitriles, esters, sulfides, sulfones, ketones and alkenes, but iodides and sulfinate esters undergo concomitant or preferential reduction with nickel boride.[529] Certain other reducing agents such as lithium triethylborohydride in the presence of a Pd^0 catalyst[530] and P$_2$I$_4$[531] are more specific and have been used to reduce the C–Se bond of allyl selenides and α-selenoketones, respectively. Selenides,[127] including acyl selenides,[482] and α-selenoketones[532] react with allylstannane or -plumbane under photostimulation[532] and are reduced with concomitant allylation.

Tri-*n*-butyltin hydride has been, without doubt, the most widely used reducing agent. Several types of radicals including alkyl,[360] β-hydroxyalkyl and β-alkoxyalkyl (ROCC•),[4,199] β-thioalkyl (RSCC•),[199] α-carboethoxy (EtOC(=O)C•),[217] furan-3-yl,[217] α-alkoxyalkyl (ROC•),[217] acyloxyalkyl (RC(=O)OC•),[217] acyl (RC•(=O))[481] and alkoxycarbonyl (ROC•(=O)) have thus been efficiently produced and trapped with hydrogen. Under the same conditions β-chloro and β-sulfonyl selenides afford alkenes by a β-elimination reaction.[127]

Selenol esters (acyl selenides), on reaction with tri-*n*-butyltin hydride and AIBN or Pd^0 catalysts[158,494] under thermal or under photochemical conditions, are reduced to aldehydes, alkanes or functionalized alkanes as a result of reductive decarbonylation.[392,482,486] Reduction of β-seleno-α,β-unsaturated selenol esters has been achieved chemoselectively in the presence of a palladium catalyst, and gives β-seleno-α,β-unsaturated aldehydes.[494] This reduction is also applicable to selenocarbonates derived from primary, secondary, tertiary and β,γ-unsaturated alcohols and affords[482,486,495] formates, alcohols or/and hydrocarbons by decarbonylation or decarboxylation of the intermediate radical, respectively. The nature of the products depends on the experimental conditions used.[486,495] High temperature is required for decarbonylation (144 °C) of the alkoxycarbonyl radical intermediate, otherwise the formyl derivative is formed. Esters and α,β-unsaturated ketones are unaffected.[486]

(b) Reduction of the C–Se bond leading to inter- and intramolecular addition to alkenes, alkynes and aldehydes. Radicals derived from alkyl,[353,360,524,536-8] including tertiary alkyl,[539] β-hydroxyalkyl, β-alkoxyalkyl,[4] and β-silyloxyalkyl,[540] α-silyloxyalkyl,[379] and α-alkoxyalkyl selenides,[378,391] α-aminoalkyl selenides,[158,541] acyl[481,482,484,542,543] and alkoxycarbonyl[482,496] selenides and 2-phenylseleno-1,3-dicarbonyl compounds such as 2-(phenylseleno)malonates,[404,405,544] and 2-phenylseleno-1,3-keto esters,[545] methyl(phenylseleno)malononitrile,[546] 2-phenylseleno-1,3-keto sulfones,[545] and 2-phenylseleno-1,3-malonoamides,[404] α-selenoketones,[532,545,547,548] α-seleno esters,[545,547] α-seleno lactones,[532,547] α-selenonitriles,[545,549] α-seleno sulfones[549,550] and selenoacetals[524] have been efficiently produced and trapped *in situ* in an inter-[158,533] or intramolecular[158,534,535] mode by alkenes, functionalized alkenes such as enol ethers[546] and enamines,[546,550] alkynes,[353,391,496,524,539,548,549] enones,[484,538] and aldehydes.[39,127,517,523,536] In most of the cases five- rather than six-membered derivatives are formed via 5-*exo* rather than 6-*endo* ring closure. The large number of methods for the synthesis of selenides has stimulated the use of these selenium-based synthetic method for the synthesis of complex molecules.[536]

Selenides bearing one or two electron-withdrawing groups on the α-carbon react on photolysis[544,545] or in the presence of a catalytic amount of AIBN[546] to give the corresponding radicals possessing electrophilic reactivity. These react with 1-alkenes to give the adducts in reasonably good yield (50–78%, 16 h–8 d) (Scheme 21 (a)).[545] Selenomalonates react in the same way not only with 1-octene but also with isopropenyl acetate, cyclohexene, norbornene allyl ether and ethyne to produce, if at all, a mixture of stereoisomers (1.5–1.8:1).[544] The reaction is easier than that of monoactivated compounds[405,544] but styrene derivatives are inert since the intermediate alkene would generate a relatively nonreactive benzylic radical.[544] Methyl(phenylseleno)malononitrile, when heated at 60 °C in the presence of a catalytic amount of AIBN, adds[546] not only to 1-hexene, cyclopentene, *trans*-methylstyrene and styrene (which form adducts with the related iodomalononitriles), but also to *O*-alkyl-, *O*-acyl-, *S*-phenyl-, nitrogen heterocyclic and *N*-acylalkenes (which do not form adducts with the related iodomalononitriles) (Scheme 21 (b)). Although the reaction time can be very long (3–8 d), the yields and regioselectivities are very high and follow the trends for reactions of electrophilic radicals (Scheme 21 (b)).[546] Stereoselectivity also follows the expected trend: *anti* addition is observed almost exclusively with cyclopentene (*de* 100%) and phenylthio enol ethers (*de* 80%), whereas *syn* addition is found with *N*-arylenamines (*de* 80%). These stereoselectivities have been rationalized in terms of Felkin–Anh and A-strain models, respectively.[546] Finally, bis(phenylseleno)malonates couple on photolysis but are unreactive towards alkenes, probably owing to their capto-dative nature.[544]

These reactions have been used for the synthesis of carbocyclic compounds, including spiro derivatives (Scheme 8 (b)), polycyclic structures via two consecutive radical cyclizations (Scheme 22 (a)),[524] heterocyclic compounds[391,404,541] (Scheme 22 (b)) including spirolactones[539] and azaspirocyclic ketones as found in the alkaloids cephalotaxine and histrionicotoxin,[360] alkylidene- and benzylidenecyclopentanes (Scheme 8 (b); Scheme 22; Scheme 23 (c)), precursors of cyclopentanones via ozonolysis,[539,549] *C*-disaccharides, using a temporary ketal connection,[378] and *C*-branched-β-D-nucleosides, using stereocontrolled intramolecular free-radical cyclization–opening reactions based on temporary silicon connection (Scheme 24).[379,540]

Primary alkyl-, vinyl- and aryl-substituted acyl radicals, generated from the corresponding selenol ester (acyl selenide) and Bu$_3$SnH, show[482] nucleophilic character and add intermolecularly to alkenes bearing electron-withdrawing or radical-stabilizing groups at rates that exceed those of competitive decarbonylation or reduction. Intramolecular addition reactions of acyl radicals generated as above proved insensitive to the nature of the alkene acceptor since electron-deficient (C=CHCO$_2$R), nonactivated (C=CH$_2$) and electron-rich (C=CHOR) π-systems serve as suitable substrates (Scheme 23).[482] In the absence of directing functionality the acyl radical–alkene addition reactions follow a well-defined and useful level of regioselectivity: 5-*exo*-trig > 6-*endo*-trig > 6-*exo*-trig > 7-*endo*-trig > 7-*exo*-trig > 8-*endo*-trig. In some cases (Scheme 23 (a)) decarbonylation becomes a competitive pathway, illustrating that the intramolecular 6-*exo*-trig cyclization of acyl radicals with the acrylate acceptor proceeds at a rate nearly identical to that of decarbonylation, leading to a benzylic radical (rate: $2.5 \times 10^7 \text{ s}^{-1}$).[482]

Acyl radicals have been used in a tandem 7-*endo*–5-*exo* cyclization resulting in the formation of bicyclo[5.3.0]decanones as a mixture of all four possible stereoisomers (Scheme 23 (c))[542] and in a radical cyclization–fragmentation sequence involving the intermediate formation of a radical centred on

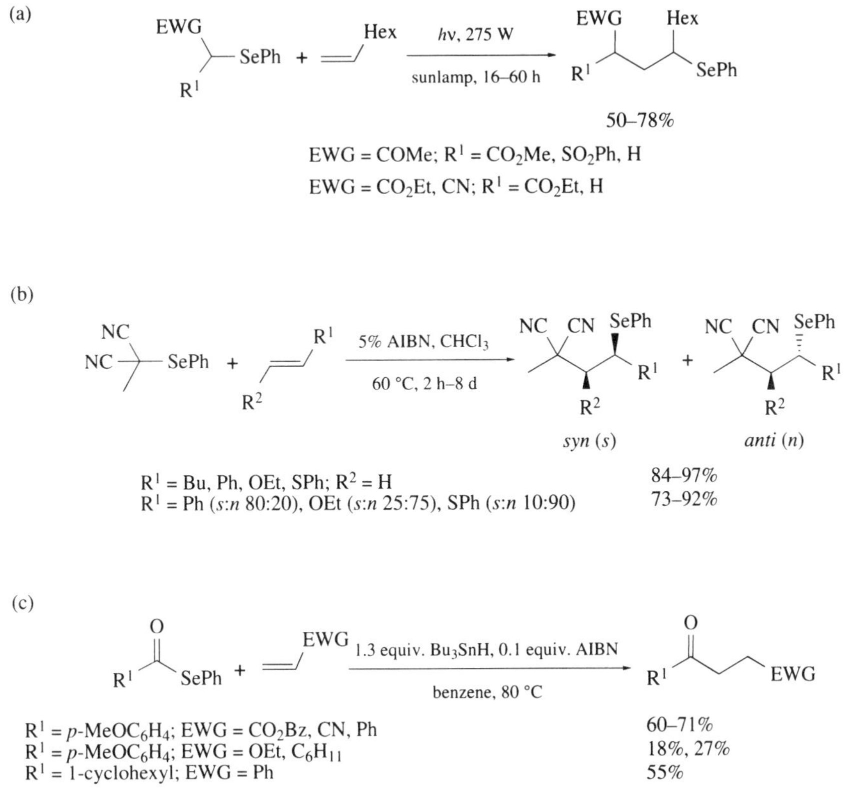

Scheme 21

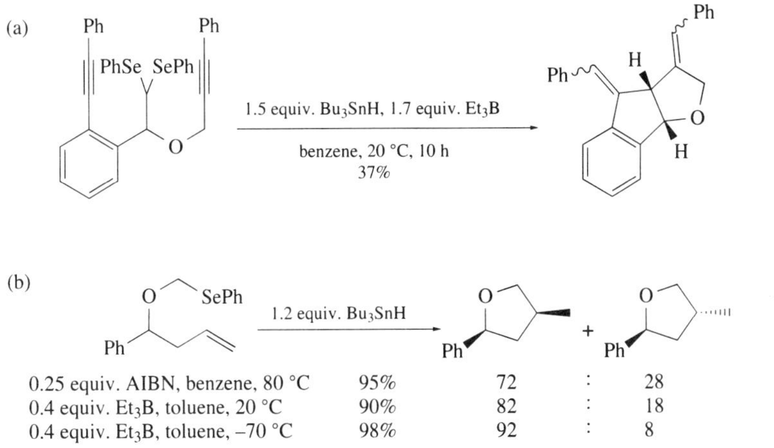

(a) EWG-CH(R¹)-SePh + CH₂=CH-Hex, hv, 275 W, sunlamp, 16–60 h → EWG-CH(R¹)-CH₂-CH(Hex)-SePh, 50–78%

EWG = COMe; R¹ = CO₂Me, SO₂Ph, H
EWG = CO₂Et, CN; R¹ = CO₂Et, H

(b) (NC)₂C(Me)-SePh + R¹CH=CHR² → syn + anti products, 5% AIBN, CHCl₃, 60 °C, 2 h–8 d

R¹ = Bu, Ph, OEt, SPh; R² = H — 84–97%
R¹ = Ph (s:n 80:20), OEt (s:n 25:75), SPh (s:n 10:90) — 73–92%

(c) R¹C(O)-SePh + CH₂=CH-EWG, 1.3 equiv. Bu₃SnH, 0.1 equiv. AIBN, benzene, 80 °C → R¹C(O)-CH₂-CH₂-EWG

R¹ = p-MeOC₆H₄; EWG = CO₂Bz, CN, Ph — 60–71%
R¹ = p-MeOC₆H₄; EWG = OEt, C₆H₁₁ — 18%, 27%
R¹ = 1-cyclohexyl; EWG = Ph — 55%

Scheme 22

(a) 1.5 equiv. Bu₃SnH, 1.7 equiv. Et₃B, benzene, 20 °C, 10 h, 37%

(b) 1.2 equiv. Bu₃SnH

Conditions	Yield	ratio
0.25 equiv. AIBN, benzene, 80 °C	95%	72 : 28
0.4 equiv. Et₃B, toluene, 20 °C	90%	82 : 18
0.4 equiv. Et₃B, toluene, −70 °C	98%	92 : 8

a methylenecyclopropane.[481] Selenol carbonates derived from primary, secondary or even tertiary alcohols[497] add intramolecularly to alkenes (Scheme 19)[497,551] and alkynes[496,497] in the presence of Bu₃SnH and AIBN to give very high-yield syntheses of lactones, including α-alkylidene-γ-butyrolactones.[496]

(ii) Reduction of the C–Se bond involving carbanions

The C–Se bond cleavage of selenides and functionalized selenides can be carried out with alkyllithiums. Although not general, this reaction has proved to be particularly useful for the synthesis of selenium-free organolithium compounds whose carbanionic centre is fairly well stabilized. For example, it allows the synthesis, in almost quantitative yields, of aryllithiums[78,89] and also of allyl,[87,207,552]

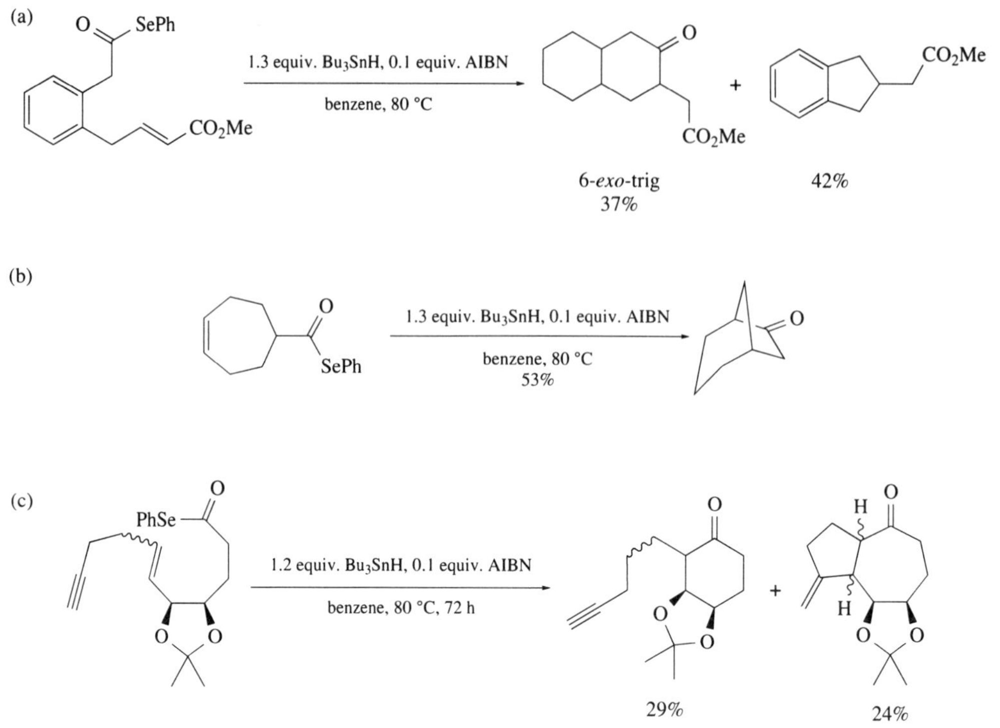

Scheme 23

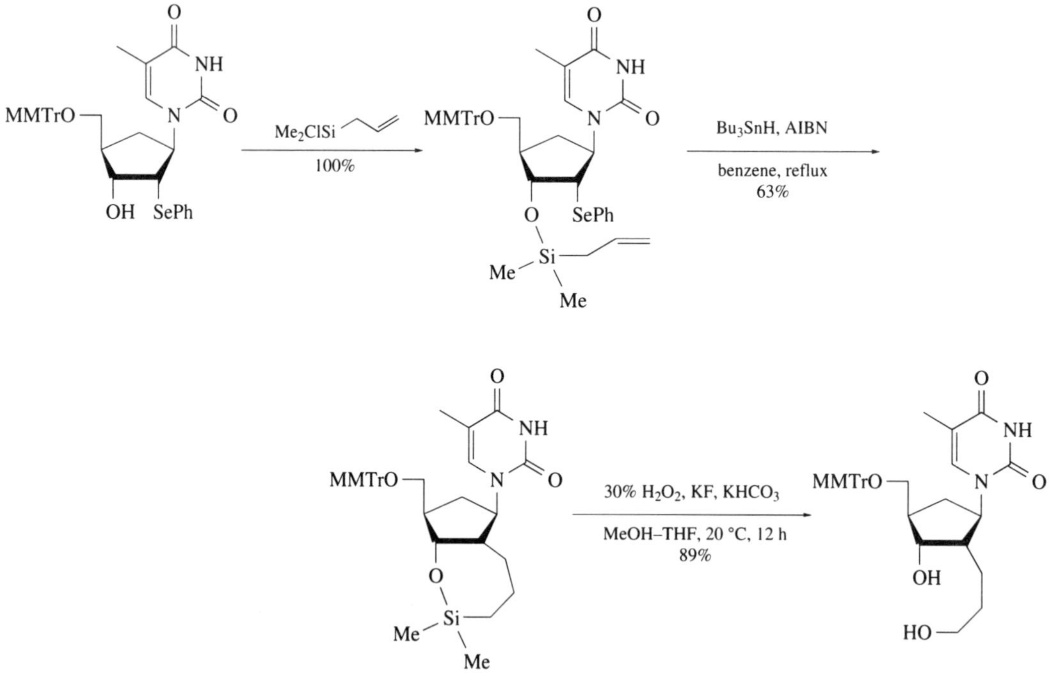

Scheme 24

MMTr = 4-monomethoxytrityl

benzyl[209,210,214–16,312,370,553–9] and α-thioalkyllithiums[211,371] whose carbanionic centre can be even dialkyl substituted, of α-silylalkyllithiums whose carbanionic centre is monoalkyl substituted or part of a cyclopropane ring and of α-thiovinyllithiums from the corresponding selenides. The resulting organometallics have been reacted with a large variety of electrophiles including water, carbonyl compounds and epoxides. The reaction is not suitable, however, for the preparation of alkyllithiums, α-alkoxyalkyllithiums or α-silylalkyllithiums whose carbanionic centre is dialkyl substituted.[81]

Benzyllithiums, even those which bear a fully alkyl-substituted carbanionic centre, can be efficiently alkylated even with secondary alkyl halides. This reaction is part of the transformation which allows the geminal dialkylation of the carbonyl group of aromatic aldehydes and ketones.[312,370,555] Furthermore, γ-tosyloxy benzyl selenides, available from α-selenobenzyllithiums and epoxides, also react with butyllithium to give arylcyclopropanes with high stereocontrol.[556-9] The reaction proceeds with inversion of configuration at both ends and allows the synthesis of enantiomerically pure compounds from optically active epoxides.[559] Allyl- and benzyllithiums possess the unusual propensity to add across unactivated C=C double bonds.[209,214,215] Benzyllithiums add intermolecularly to ethene and vinylsilanes, sulfides and selenides[214] and intramolecularly to suitably positioned C=C double bonds to give five- and six-membered rings with very high stereocontrol (Scheme 10).[85,209,215,555] A *cis* relationship between the phenyl and the lithiomethyl groups results if the reaction is carried out in pentane or diethyl ether, and *trans* if THF is used.[209,215] This reaction has been extended to the synthesis of arylcyclopentane derivatives bearing two quaternary centres linked to one another and has been used for the synthesis of cuparene.[215,555] Under the same conditions, the related α-selenobenzyllithiums produce in one step arylcyclopropanes from ethene, styrene and 1,3-butadiene, and ω-alkenyl-α-selenobenzyllithiums give bicyclo[3.1.0]cyclohexanes and bicyclo[4.1.0]cycloheptanes in very good yields.[216]

The cleavage of the C–Se bond in alkyl aryl selenides can also be efficiently achieved with elemental lithium in ethylamine[457] or in THF in the absence or the presence of arenyllithiums.[218] The former reaction takes another course with β-thioalkyl[522] and β-acetoxyalkyl[435] selenides which instead lead to an alkene via an elimination reaction, whereas lithium in THF allows the synthesis of alkyllithiums from alkyl phenyl selenides[218] by selective cleavage of the alkyl–selenium bond. This reaction has been extended to the synthesis of α-alkoxyalkyllithiums[372,560,561] from α-alkoxyalkyl selenides and of β-lithio silyl enol ethers[562] from β-phenylseleno silyl enol ethers and arenyllithiums.

α-Selenoketones are deselenylated on reaction with soft nucleophiles such as lithium phenylselenolate, phenylthiol in the presence of triethylamine, phenylthiolate in the presence of 18-crown-6, dimedone in the presence of triethylamine[525] or LDA in the presence of HMPA.[462,563]

(iii) Reduction of β-hydroxyalkyl selenides to alkenes

β-Hydroxyalkyl selenides[1,78,81-4,174,175,198,339] are valuable precursors of alkenes (Scheme 7 (c)).[89] This reduction reaction, whose mechanism is tentatively presented in Equation (4), is observed[564] when the hydroxy group is transformed into a better leaving group and occurs by formal *anti* elimination of the hydroxy and selenenyl moieties. It takes advantage of the difference in reactivity between the hard oxygen and the soft selenium atom, which allows the selective activation of the hydroxy group any time a hard electrophile is used. The high propensity of the selenium atom to favour the substitution of the oxygen-containing moiety by participation allows the intermediate formation of a seleniiranium ion, and the attack of an external nucleophile on the charged selenium atom of the seleniiranium ions finally leads to the alkene.

$$\underset{R^1Se}{\overset{R^2}{\underset{R^4}{\longrightarrow}}}\overset{OH}{\underset{X}{\longrightarrow}} \xrightarrow[\text{or }(CF_3CO)_2O]{\underset{\text{or thiocarbonyldiimidazole}}{\text{MsCl–NEt}_3 \text{ or SOCl}_2\text{–NEt}_3}} \left[\underset{R^1}{\overset{R^2}{\underset{Se^+}{\overset{X}{\underset{R^4}{\triangle}}}}}{R^3} \right] \xrightarrow{-[R^1SeSeR^1]} \underset{R^3}{\overset{R^2}{\longrightarrow}}\overset{X}{\underset{R^4}{\longrightarrow}} \quad (4)$$

R^1 = Ph, $R^5C_6H_4$, Me
R^2, R^3, R^4 = H, alkyls, R^2, R^3 = cycloalkyl; R^3, R^4 = cycloalkyl
X = H, alkyl, EWG, EDG, SiR^1_3

This reaction has been performed on various β-hydroxyalkyl selenides[78,81-4,89,175,198,339] using different reagents able to react selectively at the hydroxy group, such as perchloric or *p*-toluenesulfonic acid in pentane or diethyl ether, mesyl chloride, thionyl chloride, trifluoroacetic anhydride, phosphorus oxychloride, diphosphorus tetraiodide or phosphorus triiodide in the presence of an amine, or trimethylsilyl chloride and sodium iodide in acetonitrile, or carbonyl- or thiocarbonyldiimidazole.

All these conditions work equally well and allow the synthesis of almost all different types of alkenes such as terminal,[565] α,α- and α,β-disubstituted and tri- and tetrasubstituted, including cycloalkenes and alkylidenecyclopropanes and -cyclobutanes.[82] In general, the best combination involves phosphorus triiodide or diphosphorus tetraiodide and a β-hydroxyalkyl methyl selenide. These are the only conditions which allow the synthesis of highly hindered alkenes,[78,81,84,89] allenes[566] and alkylidenecyclopropanes.[82,89] These conditions take advantage of the high affinity of trivalent phosphorus for oxygen, the ease with which a methylseleno moiety (compared with phenylseleno) is able to participate in order to expel the activated oxygen-containing group and the softness of the iodide counterion, which favours reaction at the selenium rather than at the carbon atom of the intermediate seleniiranium ion (see Equation (4)). The method has been applied successfully[78,81,82,84,175] to the synthesis of various functionalized alkenes such as dienes, including α,ω-dienes and 1,3-dienes,[78,567] vinyl ethers, vinylsilanes, vinyl selenides and 1,3-dienyl selenides, ketene selenoacetals, allyl sulfides,[175] allylsilanes, allyl selenides[82,175] and α,β-unsaturated esters and lactones. β-Hydroxy-α-trimethylsilylalkyl selenides give vinylsilanes exclusively.[568]

13.3.6.2 Reactivity of products resulting from the alkylation, halogenation or complexation of selenides and functionalized selenides

(i) Reactions involving selenonium salts and functionalized selenonium salts

(a) Synthesis of selenonium salts. Selenonium salts are easily available from selenides and alkyl halides,[59,78,81,84,89,130,131,191,192] dihalocarbenes[191] or halogens.[78,112,113,115] Functionalized derivatives have been synthesized by reaction of aryl- or alkylselenonium trihalides and alkenes and by reaction of selenonium ylides with electrophiles. β-Alkoxyalkylselenonium salts are unstable and cyclize to epoxides.[81,89,192] Selenides are efficiently complexed by metal salts derived from soft metals.[191,570] In all these cases the seleno group is transformed into a better leaving group.[81]

The structure of the 1:1 adduct between dimethyl selenide and halogens depends on the nature of the latter.[571] Chlorine and bromine form trigonal bipyramidal structures, whereas a molecular complex is formed with iodine. When mixed with sodium borohydride, the bromo adduct selectively reduces N,N-dialkylated amides to amines but not monoalkylated amides.[571]

Alkylation of methylseleno derivatives is easy and occurs with methyl iodide, dimethyl sulfate and methyl fluorosulfonate. The synthesis of their phenylseleno analogues requires the combined use of alkyl halides and silver tetrafluoroborates.[78,81,192] The β-, γ-,[185] δ- and ε-hydroxyalkyl[438] selenides give the corresponding selenonium salts.[81,192] However, β-hydroxyalkyl phenyl selenides fully substituted at the carbon bearing the heteroatoms and methylseleno compounds derived from very hindered ketones show a high tendency to rearrange to carbonyl compounds via a pinacol-type rearrangement if the alkylation is carried out in methylene dichloride.[191,192]

(b) Reactivity of selenonium salts resulting from alkylation of selenides. Trialkyl- and alkylmethyl-phenylselenonium iodides decompose on heating to alkyl iodides[59,130] and this reaction has been used for the homologation of alkyl halides.[78,82] Alkylation of ε-hydroxyalkyl selenides with methyl fluorosulfonate leads to the stable ε-hydroxyalkylselenonium salts, whereas use of methyl iodide smoothly generates the corresponding iodide if the substituted site is benzylic.[438]

Selenonium salts with at least one alkyl group are powerful alkylating agents[1,82,132,572] and have been used for asymmetric alkylation of prochiral ketones,[132] while diphenylselenonium salts are effective phase-transfer catalysts.[573] Alkyldimethyl-, alkylmethylphenyl- and dimethylphenylselenonium salts react with bases to give the corresponding ylides which react at low temperature with nonenolizable carbonyl compounds to give epoxides,[1,47,48,78,81,84,86,130,154,192] but decompose at room temperature to alkenes, probably via an intramolecular process (Scheme 9). This reaction is particularly useful for the synthesis of alkylidene- and allylidenecyclopropanes.[78,82] Ylides derived from allylic derivatives rearrange to homoallyl selenides.[574]

β-Hydroxyalkylselenonium salts react with bases to give epoxides (Schemes 7 (d) and 25). The reaction is stereospecific, is best with methylseleno derivatives when carried out with potassium hydroxide in diethyl ether and allows a synthesis of a whole range of epoxides, including those derived from very hindered and α,β-unsaturated salts.[192] This process has been used for the synthesis of epoxides from carbonyl compounds and competes favourably with the related sulfur ylide reaction.[192]

The γ-, δ- and ε-hydroxyalkylselenonium salts also react with bases (Scheme 25). The γ-compounds selectively give oxetanes[78,81,85,89,185] when the seleno moiety is attached to a methylene group, and homoallylic alcohols otherwise.[78,81,85,86,89,185] The formation of tetrahydrofurans from δ-hydroxyalkyl-selenonium salts is much easier and takes place on heating without added bases,[89] whereas tetrahydropyrans are produced in only modest yields from ε-hydroxyalkylselenonium fluoroborates.[438]

β-Hydroxyalkyl selenides also react with dihalocarbenes very efficiently and produce selectively in both the methyl- and phenylseleno series either epoxides[192] if the carbon bearing the seleno moiety is attached to a methylene or a methine group, or ketones[191] resulting from a pinacol-type rearrangement

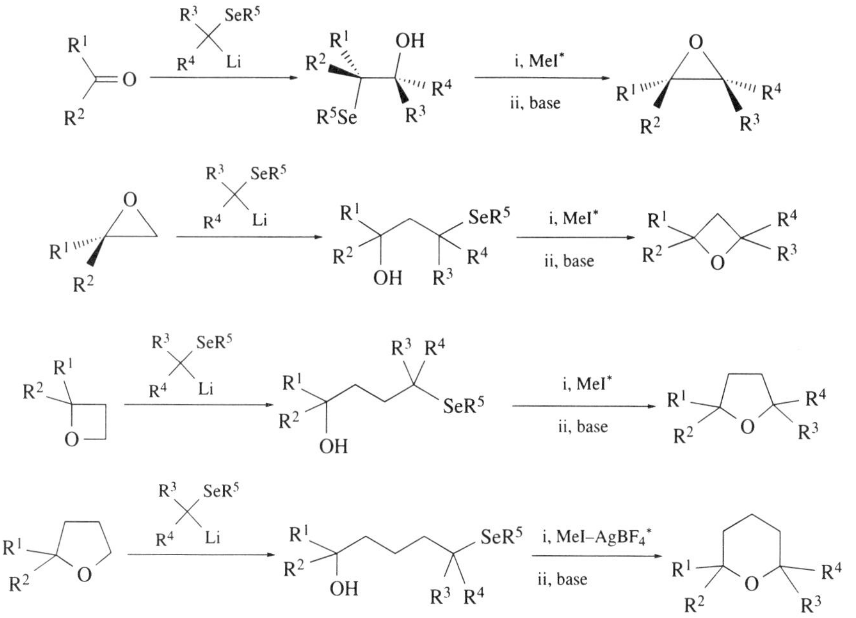

*See text for other conditions

Scheme 25

if the carbon is fully substituted (Scheme 7). The reactions are particularly smooth when carried out in chloroform with thallium(I) ethoxide and also occur under phase-transfer conditions.[191,192]

(c) Reactivity of selenonium salts resulting from halogenation of selenides. Selenonium salts formed from alkyl methyl and alkyl phenyl selenides and bromine[112] are smoothly and stereoselectively transformed into alkyl bromides[89] with inversion of configuration.[318] This reaction has been successfully extended to functionalized alkyl phenyl selenides such as 3-hydroxynonyl phenyl selenide[575] and has been used in the transformation of β-hydroxyalkyl selenides[576] and of β-chloroalkyl selenides[577] into β-halohydrins and vinyl chlorides, respectively.

(d) Reactivity of selenonium salts resulting from selenides and metal salts. Selenides give stable complexes on reaction with soft metal salts.[578] However, tertiary and benzylic selenides bearing a β- or ε-hydroxy group lead, in very high yield, to ketones (AgBF$_4$) or tetrahydrofurans (Hg(CF$_3$CO$_2$)$_2$) via pinacolic rearrangement[191] or cyclization,[438] respectively.

Seleno glycosides are versatile intermediates for glycosidation using silver triflate with potassium or silver carbonate (Equation (5)). The reaction can be carried out in the presence of thio glycosides, which remain untouched.[373] Liberation of the anomeric hydroxy group from seleno glycosides has been achieved by reaction of silver triflate followed by quenching with water.[373] Related glycosidations have been achieved[569] from seleno glycosides and either *N*-iodosuccinimide and catalytic amounts of triflic acid or *sym*-collidine perchlorate.

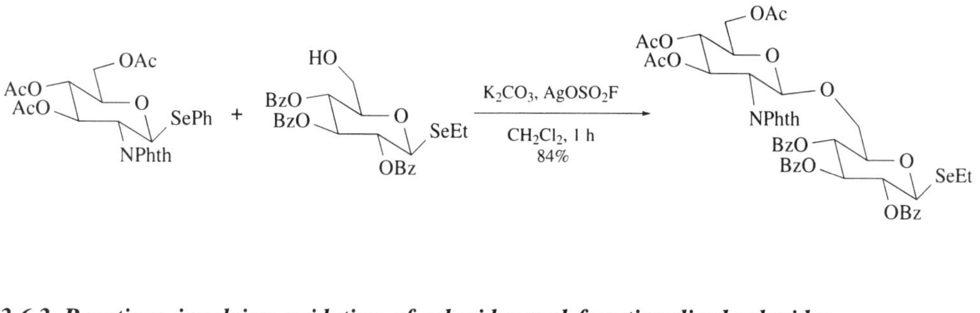

13.3.6.3 Reactions involving oxidation of selenides and functionalized selenides

Selenides can be oxidized to selenoxides under mild conditions and in the presence of various functional groups with a large variety of oxidants.[1,39,47,48,78,81,89,115,135,173,174,195,198,238,579]

Overoxidation to selenones is not easy[115,135,192,460,580] owing to the inherent difficulty of the reaction, the ease with which selenoxides undergo either a *syn* elimination reaction to alkenes if a β-hydrogen is present or [2,3]-sigmatropic rearrangement to give allyl alcohols from allyl selenides[39,135,581,582] and the high reactivity of selenones in substitution reactions.[135,192,583]

(i) Oxidation of selenides

(a) Oxidation of selenides to selenoxides. Aqueous hydrogen peroxide, ozone,[400] mcpba and oxaziridines[429,466,579,584-7] are among the most widely used oxidants.[1,89,134,173] Sodium metaperiodate,[588] potassium permanganate, singlet oxygen, *t*-butyl hydroperoxide and *t*-butyl hydroperoxide–alumina[589] have also been successfully used for the synthesis of diaryl, alkyl phenyl, alkyl 2-nitrophenyl and alkyl methyl selenoxides.[134,408] Iodobenzene dichloride seems to be very specific for diaryl selenides, which are almost instantaneously oxidized to their oxides.[588] Otherwise, selenoxides have been synthesized in a two-step sequence which involves sequential reaction with halogens and silver in water or methanol.[134] Optically active selenoxides have been successfully prepared from chiral oxaziridines,[429,466,579,585-7,590] using the Sharpless method[429,591] or by using enzyme-catalysed oxidation of aryl selenides.[592]

(b) Oxidation of selenides to selenones. Selenones are thermally more stable than selenoxides but are much more difficult to prepare.[115,135,139,195,593] Only a few of the reagents used for the synthesis of selenoxides are also able to produce selenones.[135] This is especially the case, if used in proper amounts, with peracids in chlorinated solvents, potassium permanganate in water and hydrogen peroxide in the presence of catalytic amounts of seleninic acid (perseleninic acid). Use of peracids allows the synthesis of dialkyl and alkyl aryl selenones bearing primary alkyl groups only. Potassium permanganate is complementary and offers the advantage of easy recovery of the product. It is efficient for the synthesis of dialkyl and alkyl aryl selenones including those bearing secondary alkyl groups, but is unable to promote the synthesis of long-chain primary alkyl aryl selenones.[461,580] Cyclopropyl selenones, whose corresponding selenoxides are not prone to elimination, have been efficiently produced from the corresponding selenides using one of the above-mentioned methods[460,461] and especially with perseleninic acid.[460] These reactions are sensitive to steric bulk and tetramethylcyclopropyl phenyl selenone cannot be prepared in this way.[461] Vinyl selenones have been prepared by mcpba oxidation of the corresponding selenides.[594]

(ii) Reactivity of selenoxides and selenones

(a) Typical reactivity of selenoxides. Selenoxides are reactive species and extremely valuable synthetic intermediates.[1,39,47,48,78,81,134,135,173,174,195,198,238,331,579] They are usually unstable and decompose to alkenes, via a β-elimination reaction, by loss of selenenic acid or rearrange rapidly (below 50 °C) (see Section 13.3.6.3(ii)(b)). Isolable selenoxides are those in which these reactions, and especially the β-elimination reaction, are impossible or difficult to perform. This is the case with selenoxides which either do not bear hydrogens on the β-carbon,[134] such as dimethyl, diaryl, dibenzyl, aryl benzyl and aryl methyl selenoxides, or whose conformation is unsuitable for *syn* elimination reactions, such as terminal selenoxides bearing on the β-carbon a hydroxy, alkoxy or amino group[256,595,596] or which would result in formation of strained C=C double bonds (e.g., cyclopropyl selenoxides),[205,597] an alkyne or an allene (from vinyl selenoxides).[114,187,397,471,598]

Although selenoxides which do not bear two identical substituents are chiral, optically active selenoxides[587,592,599,600] are very difficult to obtain owing to poor configurational stability,[599] the reasons for which are not yet clearly understood.[599]

Selenoxides are highly polar and basic. They possess a high propensity to react with both protic acids and water. The adducts from protic acids have been transformed in high yield into the corresponding alkyl halides (chlorides from HCl, bromides from HBr),[601] while the reaction with water results in equilibration of optically active selenoxides.[599] Reaction with carboxylic anhydrides gives highly hygroscopic oxygenated seleniuranes.[602]

Selenoxides can be subjected to Pummerer-type rearrangement (Scheme 26),[104,381,603] especially those which do not bear a hydrogen on the β-carbon and therefore cannot produce alkenes. The reaction is usually carried out with carboxylic acids[604] or anhydrides[287,605-7] as electrophiles, thus producing α-alkoxyalkyl or α-acetoxyalkyl selenides, or carbonyl compounds (Scheme 26 (b)) or their acetals. It is particularly facile with benzyl[608] and α-silyl selenoxides,[609-11] which are so unstable that they rearrange, even below room temperature, via a sila-Pummerer rearrangement.[609-11]

Selenoxides are easily reduced to selenides by reaction with phosphorus pentasulfide[612] or phosphorus tri- or tetraiodide[613] and their ability to perform oxidation reactions has already been described (Section 13.3.4.3(i)). Methyl selenoxides are labile and slowly release the oxygen even at 0 °C.[471,589]

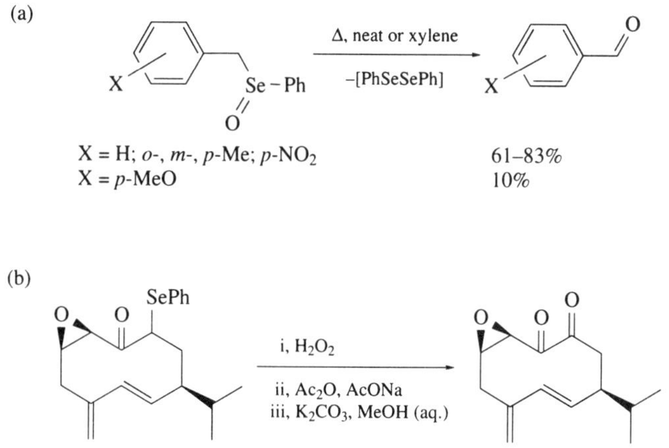

(a)

X—C₆H₄—CH₂—Se(=O)—Ph →(Δ, neat or xylene; –[PhSeSePh])→ X—C₆H₄—CHO

X = H; o-, m-, p-Me; p-NO₂   61–83%
X = p-MeO   10%

(b)

i, H₂O₂
ii, Ac₂O, AcONa
iii, K₂CO₃, MeOH (aq.)

Scheme 26

Selenoxides are easily metallated (Section 13.3.5.3(i)(b)),[1,48,78,81,83–6,154,174,192] and in some special cases easily removed by substitution[346,601,614] or elimination[615] reactions. Thus, alkyl selenoxides have been transformed into alkyl chlorides or bromides on reaction with HCl or HBr,[601] and β-alkoxy selenoxides derived from α-lithioalkyl selenoxides and strained ketones such as cyclobutanones rearrange to cyclopentanones by expulsion of the selenoxy group.[614]

(b) The selenoxide elimination reaction. The selenoxide elimination reaction[1,39,47,48,78,81,173,174,198,238,331] is probably the most efficient method for the production of C=C double bonds. It is related to the well-known ester, amine oxide (Cope elimination)[616] and sulfoxide[617] elimination reactions and, like those, occurs stereoselectively in a *syn* mode but offers the advantage that it takes place at lower temperatures (usually <40 °C, instead of 150–200 °C as for the others). This is the result of the higher polarization and length of the Se–O bond and of the longer C–Se bond, compared with the bonds in the related processes.[618]

Selenoxides are readily available by mild oxidation of selenides, which are available by selenation of organic molecules with nucleophilic (Section 13.3.5.1(i)) or electrophilic (Section 13.3.5.1(ii)) selenium reagents, or by reaction of α-seleno carbanions (Section 13.2.5.1) or carbenium ions (Section 13.3.4.3(ii)) with electrophilic or nucleophilic reagents, respectively (Scheme 1 (a); Scheme 6 (d); Scheme 7).

The selenoxide elimination reaction was first observed by Jones et al.[49] (Scheme 1 (a)) and later by Roy and co-workers[619,620] and its synthetic potential was recognized and exploited by Sharpless and co-workers,[47,256,401,421,621] Reich and co-workers[48,174,400,622] and Clive[1,173,623] in particular.

The structure. Most selenoxide elimination reactions have been carried out on phenylseleno derivatives and work efficiently. The elimination of methylseleno derivatives is more difficult since the selenoxy group loses oxygen to give back the selenides.[589] In such cases use of *t*-butyl hydroperoxide–alumina restores the efficiency of the reaction.[589] In the most difficult cases, which involve *inter alia* primary alkyl[316,624] and vinyl selenoxides, it was found that *o*-nitrophenylselenoxy,[238] *m*-chlorophenylselenoxy,[238] *m*-[598] or *p*-(trifluoromethyl)phenylselenoxy[238] and pyridylselenoxy[625] groups are eliminated much faster than phenylselenoxy groups.[238,621,626] Among these, the *o*-nitrophenylselenoxy group has proved to be the most widely used for the production of terminal alkenes required in the synthesis of natural products.

The conditions. The reaction occurs on the selenoxide first prepared (usually on reaction with O₃ or mcpba) or in the pot where oxidation of the selenide is carried out (H₂O₂ in aqueous THF or NaIO₄, in aqueous alcohols, mcpba, in CH₂Cl₂, for example). In both cases it leads to formation of an alkene and a selenenic acid. The latter is unstable and can undergo a variety of reactions. Thus, it can add to the alkene formed, especially if it is strained or highly nucleophilic such as in an enol ether,[627] to produce a β-hydroxyalkyl selenide.[595,627,628] It can also be dehydrated to the anhydride[126] or undergo disproportionation to the seleninic acid and diselenide,[53,398,595] or it can be further oxidized to the seleninic acid[398] or to the perseleninic acid if an excess of oxidant (H₂O₂) is used.[595] The latter is a powerful oxidant (Section 13.3.4.8) which can epoxidize alkenes (Scheme 14), including that formed in the selenoxide fragmentation reaction, and carry out Baeyer–Villiger reaction on ketones (Scheme 5 (h)) (see Section 13.3.4.8). These reactions are particularly important when strained alkenes are formed. In order to avoid such side reactions, the selenoxide decomposition is best achieved with an excess of oxidant in order to produce seleninic acids,[1,47,48,78,331] better in the presence of an amine which neutralizes seleninic acids,[286,407] or with *t*-butyl hydroperoxide–alumina.[464,589] Finally, the formation of terminal alkenes from primary alkyl phenyl selenoxides is greatly favoured if the reaction is performed in the presence of magnesium sulfate, which is able to dehydrate the selenoxide hydrate.[621]

Uemura and co-workers succeeded in the synthesis of a chiral allenic sulfone (42% *ee*)[429] and of an axially substituted cyclohexylidene derivative (83% *ee*)[466] by enantioselective elimination of chiral selenoxides prepared from the corresponding selenides and a chiral oxaziridine.

Stereochemical and regiochemical implications.[1] The reaction occurs in a *syn* fashion[49,621,629] and this stereochemical requirement often directs the regiochemistry of the selenoxide elimination since only those β-hydrogens which allow *syn* elimination of seleninic acid can be removed (Scheme 27 (b))[49,629] (with one exception).[220] If hydrogens are available on both the β- and β'-carbons, a regioisomeric mixture of alkenes is produced in which often the least substituted prevails.[325,621] Furthermore, although a stereoisomeric mixture of alkenes is formed, it mainly contains the (*E*)-isomer.[621]

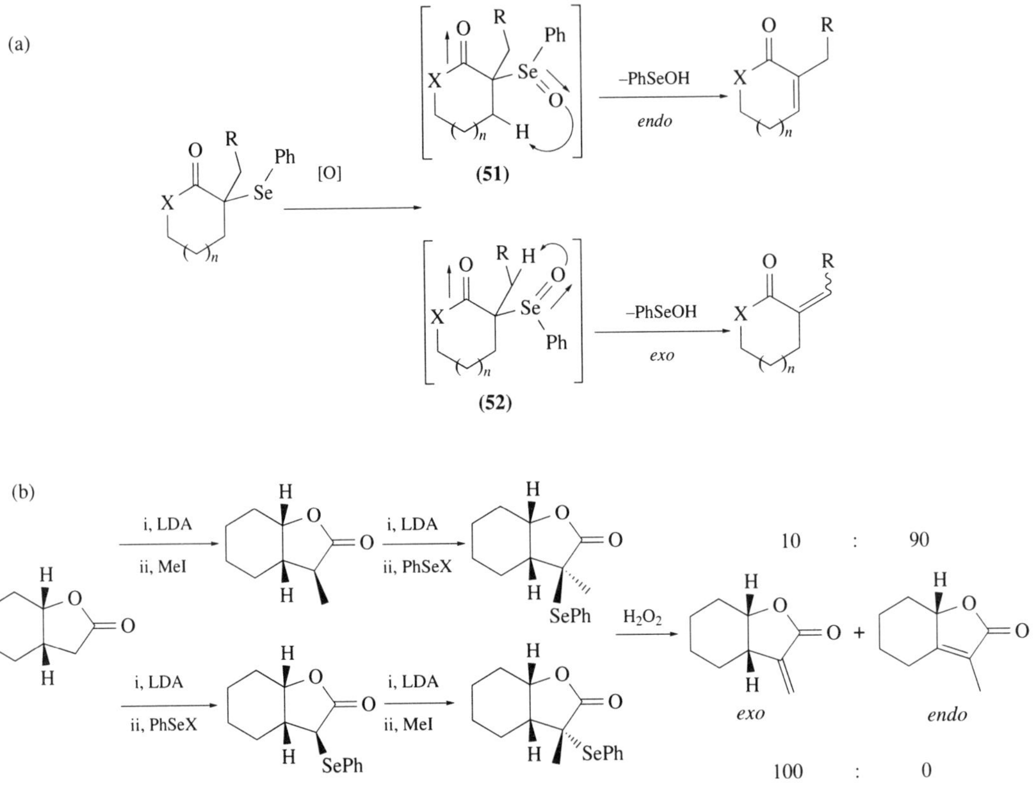

Scheme 27

The reaction is often site specific. For example,[48] an allylic or a propargylic hydrogen is more easily removed than, in order, a benzylic, a methyl, a methylene or a methine hydrogen,[621,622] and a methylene or a methine hydrogen in a five- or six-membered cycle[622,630] is more easily removed than a hydrogen from an exocyclic methyl or methylene group. Thus, 1-seleno-1-alkylcycloalkanones and lactones give endocyclic rather than exocyclic alkenes on oxidation (Scheme 27). This has been attributed (as for the thio analogues) to the relative stabilities of the two conformations involved and to the more stable and more populated conformation (**51**) over (**52**) owing to the more closely aligned C=O and Se=O bonds, which leads to smaller dipole–dipole repulsion (Scheme 27 (a)).[198,631] Regioselective synthesis of *exo*-methylene lactones has nevertheless been successfully achieved by disfavouring the endocyclic alkene formation by taking advantage of the unfavourable arrangement for *syn*-selenoxide elimination toward this site (Scheme 27).

A related proposal can explain the selective removal of the hydrogen away from the heteroatomic moiety, leading to the allyl-substituted compound, rather than towards it, which would have instead generated the vinyl-substituted derivative. This is particularly the case with β-hydroxyalkyl selenoxides, their acetates or ethers and allyl amides, which selectively lead to allyl alcohols, acetates, ethers[84,78,174,421,632-4] and amides.[635] However, other atoms or groups of atoms (F, Cl, Br, N₃[198,206,270] or SO₂Ar, NO₂) which acidify the α-hydrogen lead to mixtures of the two regioisomers or selectively to the vinyl derivatives, respectively.[198,206] (E)-α,β-Unsaturated carbonyl compounds[331] and (E)-allyl alcohols are usually obtained but mixtures of both stereoisomers of α,β-unsaturated nitriles ((E):(Z) = 50:50)[319] and of primary allyl alcohols ((E):(Z) = 75:25)[636] are formed. β-Alkoxyalkyl aryl selenoxides derived from cyclobutanones rearrange to cyclopentanones[614] and do not eliminate seleninic acid, whereas the corresponding cyclobutanols instead give allyl alcohols.[614] Increasing the strain as in 2-selenoalkylcyclopropanols favours the ring expansion to cyclobutanones at the expense of elimination leading to vinylcyclopropanols.[346]

Scope and limitations. The reaction can be applied to the synthesis of a wide range of alkenes[1,39,48,78,81,163,174,198,206,238,331] including (for the most recent references) terminal,[213] di- and tri-substituted alkenes, functionalized alkenes such as allyl alcohols, ethers[78,174,286] and amines,[324] enone ketals (aqueous H_2O_2),[400,422] α,β-unsaturated aldehydes,[331] ketones,[331,441,637] including α'-*exo*-alkylidene-α,β-unsaturated cyclopentanones,[441] esters,[331] nitriles,[319,331] acylsilanes ($NaIO_4$, MeOH),[638] lactones,[331] including α-methylene lactones[406,629,639] and α-methylene β-lactones,[406] lactams,[331,407] alkylidene α-keto esters[403] and extremely reactive 3-acetyl-2(5H)-furanones,[640] α-allenic sulfones,[429] vinyl ethers,[313] vinyl halides (Br, Cl, F),[641,642] vinyl phosphonates,[643] phosphonium salts (mcpba), phosphonates[644,645] and vinylnitroalkanes.[439]

[2,3]-Sigmatropic rearrangements of allyl and propargyl selenoxides. [2,3]-Sigmatropic rearrangement of allyl and propargyl selenoxides[1,83,135,174,194,198,256,317,329,574,579,581,582,646] to allyl and allenyl alcohols is very easy and takes place more efficiently than that of the sulfur analogues.[621] Not only is the activation barrier much lower for selenium compounds but also the cleavage of the Se–C bond in the starting material and of the Se–O bond in the resulting seleninate is easier. The presence of a *cis*-substitutent at C-3 on the C=C double bond favours a high *endo–exo* selectivity and an additional *trans*-substituent is not detrimental.[599] [2,3]-Sigmatropic rearrangement of allyl selenoxides does not occur when the resulting alkene is conjugated with a carbonyl group, and selenoxide elimination leading to a conjugated diene takes place instead.[647] Oxidative rearrangement of allylic selenides in the presence of various amine nucleophiles provides synthetic access to a variety of allylamine derivatives.[648] Related butadienyl, propargylic and allenyl selenoxides and allylselenimides and allyl selenonium ylides all show a high propensity to rearrange.[47]

Asymmetric oxidation of achiral allyl selenides leads to optically active alcohols after [2,3]-sigmatropic rearrangement of the intermediate allyl selenoxides.[132,194,586,590,591] Thus, Sharpless oxidation of cinnamyl selenides affords[591] chiral 1-aryl-2-propen-1-ols (up to 92% *ee*, especially if the *o*-nitrophenylseleno group and the diisopropyl tartrate ligand are used) via asymmetric [2,3]-sigmatropic rearrangement. Asymmetric induction in the oxidation of [2.2]paracyclophane-substituted allyl selenides and chirality transfer have also been observed[599] during the sigmatropic rearrangement which allowed the enantioselective synthesis of (S)-linalool (up to 83% *ee*).

(c) Reactivity of selenones. The selenonyl group is a particularly good leaving group[81,83,84,114,135,192,195,460,461,583,594] and acidifies α-hydrogens, which can be easily removed with $LiN(TMS)_2$,[83] KDA[460] or Bu^{t}OK.[460,461,583] Metallation even occurs in the presence of carbonyl compounds[461,583] or α,β-unsaturated esters[460] and provides, in reasonably good yields, epoxides including oxaspiropentanes or cyclopropyl esters, respectively.

n-Decyl phenyl selenone[580] reacts with a large variety of nucleophiles and allows the high-yield synthesis[583] of *n*-decyl iodide (NaI), bromide ($MgBr_2$ or $EtMgBr$), chloride (BuMgCl), azide (NaN_3), alcohol (H_2O or KOH), methyl ether (MeONa), phenyl sulfide (PhSNa) and cyanide (KCN) on reaction with the proper nucleophile (Scheme 28).[583] A few of these transformations have been performed directly on alkyl and β-methoxyalkyl phenyl selenides and involve the transient formation of a selenone.[649] In the transformations shown in Scheme 28 phenylselenomethyllithium plays alternatively the role of $LiCH_2X$ (X = halogen), $LiCH_2N_3$, $LiCH_2OH$, $LiCH_2OMe$, $LiCH_2SPh$ and $LiCH_2CN$. Otherwise, reaction of β-hydroxyalkyl selenides derived from aromatic ketones with mcpba affords ketones resulting from the rearrangement of the intermediate β-hydroxyalkyl selenones.[195,650,651] Extension of this reaction to 1-phenyl-2-phenylselenoethanol affords methyl phenyl acetate in very good yield (86%), resulting from oxidation of the phenyl acetaldehyde intermediate by the excess of mcpba used (3–5 mol equiv., MeOH, reflux, 3 h; Scheme 28).[195,652]

13.4 CONCLUSION

Representative syntheses and typical reactivities of some of the most valuable organoselenium compounds have been described. Several of these reactions are already established and routinely used in organic synthesis. They take advantage of the easy synthesis, often from the element, of

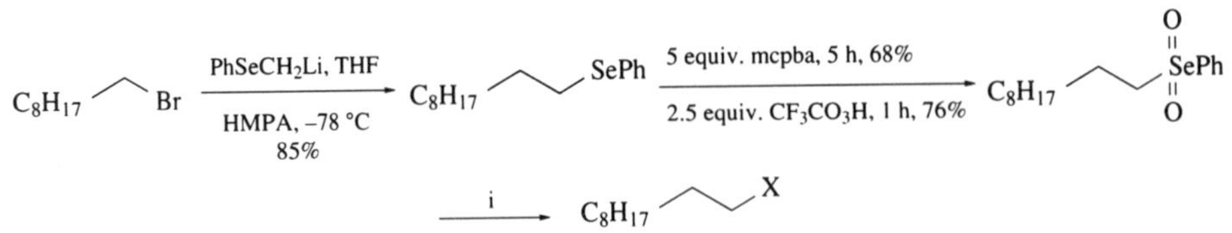

$$C_8H_{17}\text{—Br} \xrightarrow[\text{HMPA, }-78\,^\circ\text{C}]{\text{PhSeCH}_2\text{Li, THF}} C_8H_{17}\text{—}\diagup\text{—SePh} \xrightarrow[\text{2.5 equiv. CF}_3\text{CO}_3\text{H, 1 h, 76\%}]{\text{5 equiv. mcpba, 5 h, 68\%}} C_8H_{17}\text{—}\diagup\text{—Se(O)}_2\text{Ph}$$

$$\xrightarrow{\text{i}} C_8H_{17}\text{—}\diagup\text{—X}$$

i, X = I (2 equiv. NaI, 20 °C, 1 h, 97%); X = Br (1.5 equiv. MgBr$_2$ or EtMgBr, 20 °C, 1 h, 77%); X = Cl (1.5 equiv. BuMgCl, 20 °C, 1 h, 51%); X = N$_3$ (5 equiv. NaN$_3$, 20 °C, 0.7 h, 93%); X = OH (H$_2$O, 80 °C, 1.25 h, 95% or KOH, 80 °C, 22 h, 76%); X = OMe (1.1 equiv. MeONa, 20 °C, 3 h, 73%); X = PhS (1.1 equiv. PhSNa, 20 °C, 3 h, 94%); X = CN (2 equiv. KCN, 75 °C, 2 h, 66%)

Scheme 28

organoselenium reagents and compounds and the diversity and high-yielding reactions they accommodate. It has not been possible to describe the whole area in detail, but only to give a reasonable overview of the chemistry. Reference should be made to review articles for more specialized topics such as the syntheses and reactivity of vinylic[107,114,187,566,653] and acetylenic[114,187] seleno compounds, including α-seleno-α,β-unsaturated carbonyl compounds[654] and β-hydroxyalkyl selenides,[81,82,84,339] the physical organic chemistry of organoselenium derivatives[71,92,161,655,656] and the use of organoselenium chemistry for the synthesis of complex molecules and natural products,[4,657] including selenium-containing amino acids and peptides,[5,66] carbohydrates[233,234] and nucleosides.[189,233,352]

13.5 REFERENCES

1. D. L. J. Clive, *Tetrahedron*, 1978, **34**, 1049.
2. K. C. Nicolaou and N. A. Petasis, in 'Selenium in Natural Products Synthesis', CIS, Philadelphia, PA, 1984.
3. R. G. Crystal, in 'Organic Selenium Compounds: Their Chemistry and Biology', eds. D. L. Klayman and W. H. H. Günther, Wiley, Chichester, 1973, p. 13.
4. K. C. Nicolaou and N. A. Petasis, in 'Selenium in Natural Products Synthesis', CIS, Philadelphia, PA, 1984.
5. G. Zdansky, R. Walter and J. Roy, in 'Organic Selenium Compounds: Their Chemistry and Biology', eds. D. L. Klayman and W. H. H. Günther, Wiley, Chichester, 1973, p. 579.
6. J. R. Shapiro, in 'Organic Selenium Compounds: Their Chemistry and Biology', eds. D. L. Klayman and W. H. H. Günther, Wiley, Chichester, 1973, p. 693.
7. A. Shrift, in 'Organic Selenium Compounds: Their Chemistry and Biology', eds. D. L. Klayman and W. H. H. Günther, Wiley, Chichester, 1973, p. 763.
8. M. L. Scott, J. L. Martin, J. R. Shapiro, D. L. Klayman and A. Shrift, in 'Organic Selenium Compounds: Their Chemistry and Biology', eds. D. L. Klayman and W. H. H. Günther, Wiley, Chichester, 1973, p. 629.
9. K. Soda, H. Tanaka and N. Esaki, in 'The Chemistry of Organic Selenium and Tellurium Compounds', eds. S. Patai and Z. Rappoport, Wiley, Chichester, 1987, vol. 2, p. 349.
10. T. Masukawa, in 'The Chemistry of Organic Selenium and Tellurium Compounds', eds. S. Patai and Z. Rappoport, Wiley, Chichester, 1987, vol. 2, p. 377.
11. A. Krief and L. Hevesi, 'Organoselenium Chemistry I—Functional Group Transformations', Springer, Heidelberg, 1988, p. 1.
12. E. T. Thompson-Eagle and W. T. J. Frankenberger, *Adv. Soil Sci.*, 1992, **17**, 261.
13. A. Krief, *Janssen Chim. Acta*, 1993, **1**, 1; 10.
14. B. S. Reddy, A. Rivenson, N. Kulkarni, P. Upadhyaya and K. El-Bayoumy, *Cancer Res.*, 1992, **52**, 5635.
15. M. Mihailovic, G. Matic, P. Lindberg and B. Zigic, *Biol. Trace Elem. Res.*, 1992, **33**, 63.
16. A. Böck *et al.*, *Mol. Microbiol.*, 1991, 515.
17. M. J. Berry and P. R. Larsen, *Am. J. Clin. Nutr.*, 1993, 249.
18. J. F. Weiss, V. Srinivasan, K. S. Kumar and M. R. Landauer, *Adv. Space Res.*, 1992, **12**, 223.
19. K. E. Burke, G. F. J. Combs, E. G. Gross, K. C. Bhuyan and H. Abu-Libdeh, *Nutr. Cancer*, 1992, **17**, 123.
20. K. El-Bayoumy, Y.-H. Chae, P. Upadhyaya, C. Meschter, L. A. Cohen and B. S. Reddy, *Cancer Res.*, 1992, **52**, 2402.
21. P. G. Geiger, F. Lin and A. W. Girotti, *Free Rad. Biol. Med.*, 1993, **14**, 251.
22. D. L. Klayman, in 'Organic Selenium Compounds: Their Chemistry and Biology', eds. D. L. Klayman and W. H. H. Günther, Wiley, Chichester, 1973, p. 727.
23. A. R. Katritzky and C. W. Rees (eds.), 'Comprehensive Heterocyclic Chemistry', Pergamon, Oxford, 1984.
24. L. Mortillaro *et al.*, in 'Organic Selenium Compounds: Their Chemistry and Biology', eds. D. L. Klayman and W. H. H. Günther, Wiley, Chichester, 1973, p. 379.
25. M. Renson, in 'The Chemistry of Organic Selenium and Tellurium Compounds', eds. S. Patai and Z. Rappoport, Wiley, Chichester, 1986, vol. 1, p. 399.
26. A. Krief, *Tetrahedron*, 1986, **42**, 1209.
27. F. Wudl, in 'Organoselenium Chemistry', ed. D. Liotta, Wiley, Chichester, 1987, p. 395.
28. D. Cowan and A. Kini, in 'The Chemistry of Organic Selenium and Tellurium Compounds', eds. S. Patai and Z. Rappoport, Wiley, Chichester, 1987, vol. 2, p. 463.
29. M. R. Bryce, *Aldrichim. Acta*, 1985, **18**, 73.
30. L. Mortillaro and M. Russo, in 'Organic Selenium Compounds: Their Chemistry and Biology', eds. D. L. Klayman and W. H. H. Günther, Wiley, Chichester, 1973, p. 815.

31. Y. Okamoto, in 'The Chemistry of Organic Selenium and Tellurium Compounds', eds. S. Patai and Z. Rappoport, Wiley, Chichester, 1986, vol. 1, p. 331.
32. W. H. H. Günther, in 'Organic Selenium Compounds: Their Chemistry and Biology', eds. D. L. Klayman and W. H. H. Günther, Wiley, Chichester, 1973, p. 29.
33. A. Krief and L. Hevesi, 'Organoselenium Chemistry I—Functional Group Transformations', Springer, Heidelberg, 1988, p. 115.
34. H. J. Reich, in 'Oxidation in Organic Chemistry', eds. W. S. Trahanovsky and H. H. Wasserman, Academic Press, New York, NY, 1978, vol. 5, p. 111.
35. G. R. Waitkins and C. W. Clark, *Chem. Rev.*, 1945, **36**, 235.
36. N. Rabjohn, *Org. React.*, 1949, **5**, 331.
37. T. W. Campbell, H. G. Walker and G. M. Coppinger, *Chem. Rev.*, 1952, **50**, 279.
38. S.-I. Hirayama, *Chem. Rev. Jpn.*, 1939, **5**, 134.
39. T. G. Back, in 'The Chemistry of Organic Selenium and Tellurium Compounds', eds. S. Patai and Z. Rappoport, Wiley, Chichester, 1987, vol. 2, p. 91.
40. E. N. Trachtenberg, in 'Oxidation', vol. 1: 'Techniques and Applications in Organic Synthesis Series', vol. 3, ed. R. A. Augustine, Dekker, New York, NY, 1969, p. 119.
41. R. A. Jerussi, in 'Selective Organic Transformations', ed. B. S. Thyagarajan, Wiley, Chichester, 1970, p. 301.
42. N. Rabjohn, *Org. React.*, 1976, **24**, 261.
43. H. O. House, 'Modern Synthetic Reactions', Benjamin, Elmsford, NY, 1972.
44. Y. Mayor, *Chim. Ind. (Paris)*, 1940, **43**, 188.
45. G. Stein, *Angew. Chem.*, 1941, **54**, 146.
46. K. B. Sharpless and R. F. Lauer, *J. Am. Chem. Soc.*, 1972, **94**, 7154.
47. K. B. Sharpless, K. M. Gordon, R. F. Lauer, D. W. Patrick, S. P. Singer and M. W. Young, *Chem. Scr.*, 1975, **8**, 9.
48. H. J. Reich, in 'Oxidation in Organic Chemistry: Organoselenium Oxidations', eds. W. S. Trahanovsky and H. H. Wasserman, Academic Press, New York, NY, 1978.
49. D. N. Jones, D. Mundy and R. D. Whitehouse, *J. Chem. Soc., Chem. Commun.*, 1970, 86.
50. T. G. Back, D. H. R. Barton, M. R. Britten-Kelly and F. S. Guziec, Jr., *J. Chem. Soc., Perkin Trans. 1*, 1976, 2079.
51. D. Seebach and N. Peleties, *Angew. Chem., Int. Ed. Engl.*, 1969, **8**, 450.
52. D. Seebach and N. Peleties, *Chem. Ber.*, 1972, **105**, 511.
53. D. L. Klayman and W. H. H. Günther (eds.), 'Organic Selenium Compounds: Their Chemistry and Biology', Wiley, Chichester, 1973.
54. K. A. Jensen and A. Kjaer, in 'The Chemistry of Organic Selenium and Tellurium Compounds', eds. S. Patai and Z. Rappoport, Wiley, Chichester, 1986, vol. 1, p. 1.
55. R. A. Poirier and I. G. Csizmadia, in 'The Chemistry of Organic Selenium and Tellurium Compounds', eds. S. Patai and Z. Rappoport, Wiley, Chichester, 1986, vol. 1, p. 21.
56. I. Hargittai and B. Rozsondai, in 'The Chemistry of Organic Selenium and Tellurium Compounds', eds. S. Patai and Z. Rappoport, Wiley, Chichester, 1986, vol. 1, p. 63.
57. W. H. H. Günther, in 'Organic Selenium Compounds: Their Chemistry and Biology', eds. D. L. Klayman and W. H. H. Günther, Wiley, Chichester, 1973, p. 1.
58. K. W. Bagnall, in 'Comprehensive Inorganic Chemistry', eds. J. C. Bailar, H. J. Emeléus, R. Nyholm and A. F. Trotman-Dickenson, Pergamon, Oxford, 1973, vol. 2, p. 935.
59. H. Rheinboldt, in 'Methoden der Organische Chemie (Houben-Weyl)', ed. E. Müller, Thieme, Stuttgart, 1967, vol. 9, p. 1034.
60. L.-B. Agenäs, in 'Organic Selenium Compounds: Their Chemistry and Biology', eds. D. L. Klayman and W. H. H. Günther, Wiley, Chichester, 1973, p. 963.
61. G. D. Sturgeon and M. L. Gross, in 'The Chemistry of Organic Selenium and Tellurium Compounds', eds. S. Patai and Z. Rappoport, Wiley, Chichester, 1986, vol. 1, p. 243.
62. M. A. Lardon, in 'Organic Selenium Compounds: Their Chemistry and Biology', eds. D. L. Klayman and W. H. H. Günther, Wiley, Chichester, 1973, p. 933.
63. N. P. Luthra and J. D. Odom, in 'The Chemistry of Organic Selenium and Tellurium Compounds', eds. S. Patai and Z. Rappoport, Wiley, Chichester, 1986, vol. 1, p. 189.
64. R. Badiello, in 'The Chemistry of Organic Selenium and Tellurium Compounds', eds. S. Patai and Z. Rappoport, Wiley, Chichester, 1986, vol. 1, p. 287.
65. K. Fujimori and S. Oae, in 'The Chemistry of Organic Selenium and Tellurium Compounds', eds. S. Patai and Z. Rappoport, Wiley, Chichester, 1986, vol. 1, p. 369.
66. T. C. Stadtman, *Annu. Rev. Biochem.*, 1980, **49**, 93.
67. J. F. Alicino and J. A. Kowald, in 'Organic Selenium Compounds: Their Chemistry and Biology', eds. D. L. Klayman and W. H. H. Günther, Wiley, Chichester, 1973, p. 1049.
68. K. J. Irgolic and D. Chakraborti, in 'The Chemistry of Organic Selenium and Tellurium Compounds', eds. S. Patai and Z. Rappoport, Wiley, Chichester, 1986, vol. 1, p. 161.
69. L. Batt, in 'The Chemistry of Organic Selenium and Tellurium Compounds', eds. S. Patai and Z. Rappoport, Wiley, Chichester, 1986, vol. 1, p. 157.
70. J. E. Kuder, in 'Organic Selenium Compounds: Their Chemistry and Biology', eds. D. L. Klayman and W. H. H. Günther, Wiley, Chichester, 1973, p. 865.
71. K. A. Jensen *et al.*, in 'Organic Selenium Compounds: Their Chemistry and Biology', eds. D. L. Klayman and W. H. H. Günther, Wiley, Chichester, 1973, p. 835.
72. G. Snatzke and M. Kajtàr, in 'Organic Selenium Compounds: Their Chemistry and Biology', eds. D. L. Klayman and W. H. H. Günther, Wiley, Chichester, 1973, p. 885.
73. G. Snatzke, in 'The Chemistry of Organic Selenium and Tellurium Compounds', eds. S. Patai and Z. Rappoport, Wiley, Chichester, 1986, vol. 1, p. 667.
74. U. Svanholm, in 'Organic Selenium Compounds: Their Chemistry and Biology', eds. D. L. Klayman and W. H. H. Günther, Wiley, Chichester, 1973, p. 903.

75. H. J. Shine, in 'Organic Selenium Compounds: Their Chemistry and Biology', eds. D. L. Klayman and W. H. H. Günther, Wiley, Chichester, 1973, p. 941.
76. R. Monahan, D. Brown, L. Waykole and D. Liotta, in 'Organoselenium Chemistry', ed. D. Liotta, Wiley, Chichester, 1987, p. 207.
77. P. L. Bock and G. M. Whitesides, *J. Am. Chem. Soc.*, 1974, **96**, 2826.
78. A. Krief, *Tetrahedron*, 1980, **36**, 2531.
79. D. Hellwinkel and G. Fahrbach, *Liebigs Ann. Chem.*, 1968, **715**, 68.
80. H. J. Reich, B. O. Gudmundsson and R. R. Dykstra, *J. Am. Chem. Soc.*, 1992, **114**, 7937.
81. A. Krief, in 'The Chemistry of Organic Selenium and Tellurium Compounds', eds. S. Patai and Z. Rappoport, Wiley, Chichester, 1987, vol. 2, p. 675.
82. A. Krief, in 'Topics in Current Chemistry, vol. 135. Small Ring Compounds in Organic Synthesis II', ed. A. De Meijere, Springer, Heidelberg, 1987, p. 1.
83. H. J. Reich, in 'Organoselenium Chemistry', ed. D. Liotta, Wiley, Chichester, 1987, p. 243.
84. A. Krief, in 'Comprehensive Organic Synthesis', eds. B. M. Trost and I. Fleming, Pergamon, Oxford, 1991, vol. 1, p. 629.
85. A. Krief, in 'Comprehensive Organic Synthesis', eds. B. M. Trost and I. Fleming, Pergamon, Oxford, 1991, vol. 3, p. 85.
86. K. C. Nicolaou and N. A. Petasis, 'Selenium in Natural Products Synthesis', CIS, Philadelphia, PA, 1984, p. 176.
87. C. Paulmier, in 'Selenium Reagents and Intermediates in Organic Synthesis', ed. J. E. Baldwin, Pergamon, Oxford, 1986, vol. 5, p. 257.
88. A. Krief and L. Hevesi, 'Organoselenium Chemistry I—Functional Group Transformations', Springer, Heidelberg, 1988.
89. A. Krief, in 'Proceedings of the 3rd International Symposium on Selenium and Tellurium Compounds', eds. D. Cagniant and G. Kirch, Metz, 1979, p. 13.
90. A. L. Allred and E. G. Rochow, *J. Inorg. Nucl. Chem.*, 1958, **5**, 264.
91. F. S. Guziec, in 'The Chemistry of Organic Selenium and Tellurium Compounds', eds. S. Patai and Z. Rappoport, Wiley, Chichester, 1987, vol. 2, p. 215.
92. T. B. Rauchfuss, in 'The Chemistry of Organic Selenium and Tellurium Compounds', eds. S. Patai and Z. Rappoport, Wiley, Chichester, 1987, vol. 2, p. 339.
93. A. J. Bard (ed.), 'Encyclopedia of Electrochemistry of the Elements', Dekker, New York, NY, 1975, vol. 4, p. 280.
94. A. J. Bard (ed.) 'Encyclopedia of Electrochemistry of the Elements', Dekker, New York, NY, 1975, vol. 4, p. 363.
95. J. S. Guziec, in 'Organoselenium Chemistry', ed. D. Liotta, Wiley, Chichester, 1987, p. 277.
96. C. Paulmier, in 'Selenium Reagents and Intermediates in Organic Synthesis', ed. J. E. Baldwin, Pergamon, Oxford, 1986, vol. 5, p. 58.
97. C. Paulmier, in 'Selenium Reagents and Intermediates in Organic Synthesis', ed. J. E. Baldwin, Pergamon, Oxford, 1986, vol. 5, p. 4.
98. A. Krief and L. Hevesi, 'Organoselenium Chemistry I—Functional Group Transformations', Springer, Heidelberg, 1988, p. 12.
99. E. Bulka, in 'The Chemistry of Cyanates and their Thio Derivatives, Part I', ed. S. Patai, Wiley, Chichester, 1977, p. 887.
100. F. Förster, F. Lange, O. Drossbach and W. Seidel, *Z. Anorg. Allg. Chem.*, 1923, **128**, 312.
101. A. Krief and L. Hevesi, 'Organoselenium Chemistry I—Functional Group Transformations', Springer, Heidelberg, 1988, p. 187.
102. K. C. Nicolaou and N. A. Petasis, 'Selenium in Natural Products Synthesis', CIS, Philadelphia, PA, 1984, p. 5.
103. K. C. Nicolaou and N. A. Petasis, 'Selenium in Natural Products Synthesis', CIS, Philadelphia, PA, 1984, p. 45.
104. M. L. Bird and F. Challenger, *J. Chem. Soc.*, 1942, 570.
105. R. Paetzold and M. Reichenbächer, in 'Organic Selenium Compounds: Their Chemistry and Biology', eds. D. L. Klayman and W. H. H. Günther, Wiley, Chichester, 1973, p. 305.
106. R. W. Hoffmann, M. Julius and K. Oltmann, *Tetrahedron Lett.*, 1990, **31**, 7419.
107. K. C. Nicolaou and N. A. Petasis, 'Selenium in Natural Products Synthesis', CIS, Philadelphia, PA, 1984, p. 248.
108. C. Paulmier, in 'Selenium Reagents and Intermediates in Organic Synthesis', ed. J. E. Baldwin, Pergamon, Oxford, 1986, vol. 5, p. 25.
109. D. L. Klayman, in 'Organic Selenium Compounds: Their Chemistry and Biology', eds. D. L. Klayman and W. H. H. Günther, Wiley, Chichester, 1973, p. 67.
110. N. Sonoda and A. Ogawa, in 'The Chemistry of Organic Selenium and Tellurium Compounds', eds. S. Patai and Z. Rappoport, Wiley, Chichester, 1986, vol. 1, p. 619.
111. C. Paulmier, in 'Selenium Reagents and Intermediates in Organic Synthesis', ed. J. E. Baldwin, Pergamon, Oxford, 1986, vol. 5, p. 387.
112. H. Rheinboldt, in 'Methoden der Organische Chemie (Houben-Weyl)', ed. E. Müller, Thieme, Stuttgart, 1967, vol. 9, p. 972.
113. C. Paulmier, in 'Selenium Reagents and Intermediates in Organic Synthesis', ed. J. E. Baldwin, Pergamon, Oxford, 1986, vol. 5, p. 84.
114. C. Paulmier, in 'Selenium Reagents and Intermediates in Organic Synthesis', ed. J. E. Baldwin, Pergamon, Oxford, 1986, vol. 5, p. 285.
115. L.-B. Agenäs, in 'Organic Selenium Compounds: Their Chemistry and Biology', eds. D. L. Klayman and W. H. H. Günther, Wiley, Chichester, 1973, p. 173.
116. L. Mortillaro and M. Russo, in 'Organic Selenium Compounds: Their Chemistry and Biology', eds. D. L. Klayman and W. H. H. Günther, Wiley, Chichester, 1973, p. 379.
117. H. Rheinboldt, in 'Methoden der Organische Chemie (Houben-Weyl)', ed. E. Müller, Thieme, Stuttgart, 1967, vol. 9, p. 1086.
118. G. Kirsch and L. Christiaens, in 'The Chemistry of Organic Selenium and Tellurium Compounds', eds. S. Patai and Z. Rappoport, Wiley, Chichester, 1987, vol. 2, p. 421.
119. W. R. McWhinnie, in 'The Chemistry of Organic Selenium and Tellurium Compounds', eds. S. Patai and Z. Rappoport, Wiley, Chichester, 1987, vol. 2, p. 495.
120. H. J. Reich, in 'Oxidation in Organic Chemistry', eds. W. S. Trahanovsky and H. H. Wasserman, Academic Press, New York, NY, 1978, vol. 5, p. 5.

121. A. Krief and L. Hevesi, 'Organoselenium Chemistry I—Functional Group Transformations', Springer, Heidelberg, 1988, p. 72.
122. H. Rheinboldt, in 'Methoden der Organische Chemie (Houben-Weyl)', ed. E. Müller, Thieme, Stuttgart, 1967, vol. 9, p. 930.
123. A. Toshimitsu and S. Uemura, in 'The Chemistry of Organic Selenium and Tellurium Compounds', eds. S. Patai and Z. Rappoport, Wiley, Chichester, 1987, vol. 2, p. 541.
124. H. Rheinboldt, in 'Methoden der Organische Chemie (Houben-Weyl)', ed. E. Müller, Thieme, Stuttgart, 1967, vol. 9, p. 1161.
125. D. Labar and A. Krief, *Bull. Soc. Chim. Belg.*, 1984, **93**, 1005.
126. H. J. Reich, W. W. Willis, Jr. and S. Wollowitz, *Tetrahedron Lett.*, 1982, **23**, 3319.
127. T. G. Back, in 'Organoselenium Chemistry', ed. D. Liotta, Wiley, Chichester, 1987, p. 325.
128. T. G. Back, in 'Organoselenium Chemistry', ed. D. Liotta, Wiley, Chichester, 1987, p. 1.
129. H. Rheinboldt, in 'Methoden der Organische Chemie (Houben-Weyl)', ed. E. Müller, Thieme, Stuttgart, 1967, vol. 9, p. 1204.
130. C. Paulmier, in 'Selenium Reagents and Intermediates in Organic Synthesis', ed. J. E. Baldwin, Pergamon, Oxford, 1986, vol. 5, p. 162.
131. R. J. Shine, in 'Organic Selenium Compounds: Their Chemistry and Biology', eds. D. L. Klayman and W. H. H. Günther, Wiley, Chichester, 1973, p. 223.
132. S. Tomoda, Y. Usuki, K. I. Fujita and M. Iwaoka, *Rev. Heteroatom Chem.*, 1991, **4**, 249.
133. J. Bergman, L. Engman and J. Sidén, in 'The Chemistry of Organic Selenium and Tellurium Compounds', eds. S. Patai and Z. Rappoport, Wiley, Chichester, 1986, vol. 1, p. 517.
134. H. Rheinboldt, in 'Methoden der Organische Chemie (Houben-Weyl)', ed. E. Müller, Thieme, Stuttgart, 1967, vol. 9, p. 1020.
135. C. Paulmier, in 'Selenium Reagents and Intermediates in Organic Synthesis', ed. J. E. Baldwin, Pergamon, Oxford, 1986, vol. 5, p. 124.
136. A. Krief and L. Hevesi, 'Organoselenium Chemistry I—Functional Group Transformations', Springer, Heidelberg, 1988, p. 60.
137. K. B. Sharpless, T. Hori, L. K. Truesdale and C. O. Dietrich, *J. Am. Chem. Soc.*, 1976, **98**, 269.
138. S. Tamagaki, S. Oae and K. Sakaki, *Tetrahedron Lett.*, 1975, 649.
139. H. Rheinboldt, in 'Methoden der Organische Chemie (Houben-Weyl)', ed. E. Müller, Thieme, Stuttgart, 1967, vol. 9, p. 1030.
140. H. Rheinboldt, in 'Methoden der Organische Chemie (Houben-Weyl)', ed. E. Müller, Thieme, Stuttgart, 1967, vol. 9, p. 1129.
141. A. Krief and L. Hevesi, 'Organoselenium Chemistry I—Functional Group Transformations', Springer, Heidelberg, 1988, p. 76.
142. A. Krief and L. Hevesi, 'Organoselenium Chemistry I—Functional Group Transformations', Springer, Heidelberg, 1988, p. 106.
143. H. Rheinboldt, in 'Methoden der Organische Chemie (Houben-Weyl)', ed. E. Müller, Thieme, Stuttgart, 1967, vol. 9, p. 1124.
144. H. Rheinboldt, in 'Methoden der Organische Chemie (Houben-Weyl)', ed. E. Müller, Thieme, Stuttgart, 1967, vol. 9, p. 1195.
145. R. B. Silverman, in 'Organic Selenium Compounds: Their Chemistry and Biology', eds. D. L. Klayman and W. H. H. Günther, Wiley, Chichester, 1973, p. 245.
146. K. A. Jensen, in 'Organic Selenium Compounds: Their Chemistry and Biology', eds. D. L. Klayman and W. H. H. Günther, Wiley, Chichester, 1973, p. 263.
147. S. Kato, T. Murai and M. Ishida, *Org. Prep. Proceed. Int.*, 1986, **18**, 369.
148. M. Segi, T. Takahashi, H. Ichinose, G. M. Li and T. Nakajima, *Tetrahedron Lett.*, 1992, **33**, 7865.
149. R. J. Shine, in 'Organic Selenium Compounds: Their Chemistry and Biology', eds. D. L. Klayman and W. H. H. Günther, Wiley, Chichester, 1973, p. 273.
150. Z. J. Witczak, *Tetrahedron*, 1985, **41**, 4781.
151. T. Murai, T. Mizutani, T. Kanda and S. Kato, *J. Am. Chem. Soc.*, 1993, **115**, 5823.
152. H. Rheinboldt, in 'Methoden der Organische Chemie (Houben-Weyl)', ed. E. Müller, Thieme, Stuttgart, 1967, vol. 9, p. 1118.
153. J. Michalski and A. Markowska, in 'Organic Selenium Compounds: Their Chemistry and Biology', eds. D. L. Klayman and W. H. H. Günther, Wiley, Chichester, 1973, p. 325.
154. A. Krief and L. Hevesi, *Janssen Chim. Acta*, 1984, **2**, 3.
155. L. Hevesi, in 'The Chemistry of Organic Selenium and Tellurium Compounds', eds. S. Patai and Z. Rappoport, Wiley, Chichester, 1986, vol. 1, p. 307.
156. L. Hevesi, *Bull. Soc. Chim. Fr.*, 1990, **127**, 697.
157. L. Hevesi *et al.*, *J. Am. Chem. Soc.*, 1984, **106**, 3784.
158. R. Muthyala, *Tetrahedron*, 1987, **43**, 3541.
159. H. Rheinboldt, in 'Methoden der Organische Chemie (Houben-Weyl)', ed. E. Müller, Thieme, Stuttgart, 1967, vol. 9, p. 917.
160. S. Patai and Z. Rappoport, in 'The Chemistry of Organic Selenium and Tellurium Compounds', eds. S. Patai and Z. Rappoport, Wiley, Chichester, 1987, vol. 2.
161. S. Patai and Z. Rappoport, in 'The Chemistry of Organic Selenium and Tellurium Compounds', eds. S. Patai and Z. Rappoport, Wiley, Chichester, 1986, vol. 1.
162. C. Paulmier, in 'Selenium Reagents and Intermediates in Organic Synthesis', ed. J. E. Baldwin, Pergamon, Oxford, 1986, vol. 5.
163. D. Liotta, in 'Organoselenium Chemistry' ed. D. Liotta, Wiley, Chichester, 1987.
164. G. H. Schmid and D. G. Garratt, in 'The Chemistry of Double-bonded Functional Groups, Part 2, Suppl. A', ed. S. Patai, Wiley, Chichester, 1977, p. 855.

165. P. D. Magnus, in 'Comprehensive Organic Chemistry', eds. D. H. R. Barton and D. Ollis, Pergamon, Oxford, 1979, vol. 3, p. 491.
166. R. W. Zingaro and W. C. Cooper, 'Selenium', Van Nostrand Reinhold, Princeton, NJ, 1974, p. 200.
167. D. H. Reid, in 'Specialist Periodical Reports: Organic Compounds', Royal Society of Chemistry, London, 1970, vol. 1.
168. D. H. Reid, in 'Specialist Periodical Reports: Organic Compounds', Royal Society of Chemistry, London, 1973, vol. 2.
169. D. H. Reid, in 'Specialist Periodical Reports: Organic Compounds', Royal Society of Chemistry, London, 1975, vol. 3.
170. D. R. Hogg, in 'Specialist Periodical Reports: Organic Compounds', Royal Society of Chemistry, London, 1977, vol. 4.
171. D. R. Hogg, in 'Specialist Periodical Reports: Organic Compounds', Royal Society of Chemistry, London, 1979, vol. 5.
172. D. H. Hogg, in 'Specialist Periodical Reports: Organic Compounds', Royal Society of Chemistry, London, 1981, vol. 6.
173. D. L. J. Clive, *Aldrichim. Acta*, 1978, **11**, 43.
174. H. J. Reich, *Acc. Chem. Res.*, 1979, **12**, 22.
175. A. Krief *et al.*, *Bull. Soc. Chim. Fr.*, 1980, II-519.
176. J. Gosselck, *Angew. Chem.*, 1963, **75**, 831.
177. D. H. R. Barton and S. V. Ley, *Further Perspect. Org. Chem.*, 1978, 53.
178. J. V. Comasseto, J. T. B. Ferreira and M. Marcuzzo do Canto, *Quim. Nova*, 1979, 58.
179. K. B. Sharpless and T. R. Verhoeven, *Aldrichim. Acta*, 1979, **12**, 63.
180. S. V. Ley, *Annu. Rep. Chem. Soc.*, 1980, **77**, 233.
181. K. C. Nicolaou, *Tetrahedron*, 1981, **37**, 4097.
182. S. I. Pennanen, *Kem.-Kemi*, 1981, 275.
183. H. Schumann and M. Schmidt, *Angew. Chem., Int. Ed. Engl.*, 1965, **4**, 1007.
184. A. Toshimitsu and S. Uemura, *Yuki Gosei Kagaku Kyokai Shi*, 1981, **32**, 1210.
185. K. Hermann, *Nachr. Chem. Tech. Lab.*, 1978, **26**, 209.
186. S. I. Pennanen, *Kem.-Kemi*, 1981, 501.
187. J. V. Comasseto, *J. Organomet. Chem.*, 1983, **253**, 131.
188. R. Okazaki and K. Kang, *Yuki Gosei Kagaku Kyokai Shi*, 1980, **38**, 1223.
189. Z. J. Witczak, *Nucleosides Nucleotides*, 1983, **2**, 295.
190. D. Liotta, *Acc. Chem. Res.*, 1984, **17**, 28.
191. A. Krief, J. L. Laboureur, W. Dumont and D. Labar, *Bull. Soc. Chim. Fr.*, 1990, **127**, 681.
192. A. Krief *et al.*, *Heterocycles*, 1989, **28**, 1203.
193. D. H. R. Barton and S. Z. Zard, *Pure Appl. Chem.*, 1986, **58**, 675.
194. K. Ösapay, J. Delhalle, K. M. Nsunda, E. Rolli, R. Houriet and L. Hevesi, *J. Am. Chem. Soc.*, 1989, **111**, 5028.
195. S. Uemura, *Rev. Heteroatom Chem.*, 1990, **3**, 105.
196. K. C. Nicolaou and N. A. Petasis, 'Selenium in Natural Products Synthesis', CIS, Philadelphia, PA, 1984, p. 167.
197. C. Paulmier, in 'Selenium Reagents and Intermediates in Organic Synthesis', ed. J. E. Baldwin, Pergamon, Oxford, 1986, vol. 5, p. 353.
198. K. C. Nicolaou and N. A. Petasis, 'Selenium in Natural Products Synthesis', CIS, Philadelphia, PA, 1984, p. 66.
199. K. C. Nicolaou, N. A. Petasis and D. A. Claremon, in 'Organoselenium Chemistry', ed. D. Liotta, Wiley, Chichester, 1987, p. 127.
200. J. L. Laboureur, W. Dumont and A. Krief, *Tetrahedron Lett.*, 1984, **25**, 4569.
201. D. Labar, J. L. Laboureur and A. Krief, *Tetrahedron Lett.*, 1982, **23**, 983.
202. W. Dumont and A. Krief, *Angew. Chem., Int. Ed. Engl.*, 1975, **14**, 350.
203. D. Van Ende, W. Dumont and A. Krief, *Angew. Chem., Int. Ed. Engl.*, 1975, **14**, 700.
204. D. Labar and A. Krief, *J. Chem. Soc., Chem. Commun.*, 1982, 564.
205. S. Halazy and A. Krief, *Tetrahedron Lett.*, 1979, 4233.
206. H. J. Reich, in 'Oxidation in Organic Chemistry', eds. W. S. Trahanovsky and H. H. Wasserman, Academic Press, New York, NY, 1978, vol. 5, p. 15.
207. M. Clarembeau and A. Krief, *Tetrahedron Lett.*, 1984, **25**, 3629.
208. K. Nishitani, Y. Mimaki, K. Sato and K. Yamakawa, *Chem. Pharm. Bull.*, 1992, **40**, 288.
209. A. Krief and P. Barbeaux, *J. Chem. Soc., Chem. Commun.*, 1987, 1214.
210. M. Clarembeau and A. Krief, *Tetrahedron Lett.*, 1985, **26**, 1093.
211. A. Krief, W. Dumont, M. Clarembeau and E. Badaoui, *Tetrahedron*, 1989, **45**, 2023.
212. H. J. Reich and M. D. Bowe, *J. Am. Chem. Soc.*, 1990, **112**, 8994.
213. A. Krief and Z. Milkova, *Janssen Chim. Acta*, 1993, **11**, 1; 8.
214. A. Krief, P. Barbeaux and E. Guittet, *Synlett*, 1990, 509.
215. A. Krief and P. Barbeaux, *Synlett*, 1990, 511.
216. A. Krief and P. Barbeaux, *Tetrahedron Lett.*, 1991, **32**, 417.
217. A. L. J. Beckwith and P. E. Pigou, *Aust. J. Chem.*, 1986, **39**, 77.
218. A. Krief and A.-M. Laval, *Janssen Chimica Acta*, 1993, **11**, 26.
219. S. J. Foster, C. W. Rees and D. J. Williams, *J. Chem. Soc., Perkin Trans. 1*, 1985, 711.
220. E. Carceller, M. L. Garcia, A. Moyano and F. Serratosa, *J. Chem. Soc., Chem. Commun.*, 1984, 825.
221. W. H. H. Günther, in 'Organic Selenium Compounds: Their Chemistry and Biology', eds. D. L. Klayman and W. H. H. Günther, Wiley, Chichester, 1973, p. 29.
222. H. Rheinboldt, in 'Methoden der Organische Chemie (Houben-Weyl)', ed. E. Müller, Thieme, Stuttgart, 1967, vol. 9, p. 1109.
223. H. Rheinboldt, in 'Methoden der Organische Chemie (Houben-Weyl)', ed. E. Müller, Thieme, Stuttgart, 1967, vol. 9, p. 951.
224. D. P. Thompson and P. Boudjouk, *J. Org. Chem.*, 1988, **53**, 2109.
225. L. Syper and J. Mlochowski, *Tetrahedron*, 1988, **44**, 6119.
226. L. Ping and Z. Xunjun, *Synth. Commun.*, 1993, **23**, 1721.
227. D. Harpp and M. Gingras, *J. Am. Chem. Soc.*, 1988, **110**, 7737.
228. J. A. Gladysz, J. L. Hornby and J. E. Garbe, *J. Org. Chem.*, 1978, **43**, 1204.
229. M. R. Detty and M. D. Seidler, *J. Org. Chem.*, 1982, **47**, 1354.
230. O. M. Jakiwczyk, E. M. Kristoff and D. J. McPhee, *Synth. Commun.*, 1993, **23**, 195.

231. K. Shimada, S. Okuse and Y. Takikawa, *Bull. Chem. Soc. Jpn.*, 1992, **65**, 2848.
232. L.-B. Agenäs, in 'Organic Selenium Compounds: Their Chemistry and Biology', eds. D. L. Klayman and W. H. H. Günther, Wiley, Chichester, 1973, p. 173.
233. Z. Witczak, in 'The Chemistry of Organic Selenium and Tellurium Compounds', eds. S. Patai and Z. Rappoport, Wiley, Chichester, 1987, vol. 2, p. 765.
234. Z. J. Witczak and R. L. Whistler, *Heterocycles*, 1982, **19**, 1719.
235. M. D. Ryan and D. N. Harpp, *Tetrahedron Lett.*, 1992, **33**, 2129.
236. A. Krief and L. Hevesi, 'Organoselenium Chemistry I—Functional Group Transformations', Springer, Heidelberg, 1988, p. 181.
237. H. J. Reich, C. P. Jasperse and J. M. Renga, *J. Org. Chem.*, 1986, **51**, 2981.
238. K. B. Sharpless and M. W. Young, *J. Org. Chem.*, 1975, **40**, 947.
239. R. Michels, M. Kato and W. Heitz, *Makromol Chem.*, 1976, **177**, 2311.
240. S. Torii, K. Uneyama, M. Ono and T. Bannou, *J. Am. Chem. Soc.*, 1981, **103**, 4606.
241. S. Torii, K. Uneyama and K. Handa, *Tetrahedron Lett.*, 1980, **21**, 1863.
242. K. Uneyama, M. Ono and S. Torii, *Phosphorus Sulfur*, 1983, **16**, 35.
243. A. P. Kozikowski and A. Ames, *J. Org. Chem.*, 1978, **43**, 2735.
244. A. P. Kozikowski and A. Ames, *Tetrahedron*, 1985, **41**, 4821.
245. S. O. De Silva, J. N. Reed, R. J. Billedeau, X. Wang, D. J. Norris and V. Snieckus, *Tetrahedron*, 1992, **48**, 4863.
246. J. Römer, P. Mäding and F. Rösch, *Appl. Radiat. Isot.*, 1992, **43**, 495.
247. S. Tomoda, M. Iwaoka, K. Yakushi, A. Kawamoto and J. Tanaka, *J. Phys. Org. Chem.*, 1988, **1**, 179.
248. W. H. H. Günther, *J. Org. Chem.*, 1966, **31**, 1202.
249. M. R. Detty, *Tetrahedron Lett.*, 1978, 5087.
250. R. A. Zingaro, in 'The Chemistry of Organic Selenium and Tellurium Compounds', eds. S. Patai and Z. Rappoport, Wiley, Chichester, 1986, vol. 1, p. 343.
251. T. R. Hoye and A. J. Caruso, *Tetrahedron*, 1978, **47**, 4611.
252. S. V. Ley, I. A. O'Neil and C. M. R. Low, *Tetrahedron*, 1986, **42**, 5363.
253. S. Torii, T. Inokuchi, G. Asanuma, N. Sayo and H. Tanaka, *Chem. Lett.*, 1980, 867.
254. A. Krief, M. Trabelsi and W. Dumont, *Synthesis*, 1992, 933.
255. B. Sjöberg and S. Herdevall, *Acta Chem. Scand.*, 1958, **12**, 1347.
256. K. B. Sharpless and R. F. Lauer, *J. Am. Chem. Soc.*, 1973, **95**, 2697.
257. J. V. Comasseto, J. T. B. Ferreira, C. A. Brandt and N. Petragnani, *J. Chem. Res.*, 1982, 212.
258. L. Syper and J. Mlochowski, *Synthesis*, 1984, 439.
259. M. Sakakibara, K. Katsumata, Y. Watanabe, T. Toru and Y. Ueno, *Synthesis*, 1992, 377.
260. H. J. Reich, F. Chow and S. K. Shah, *J. Am. Chem. Soc.*, 1979, **101**, 6638.
261. C. J. Li and D. N. Harpp, *Sulfur Lett.*, 1992, **154**, 155.
262. M. L. Pedersen and D. B. Berkowitz, *Tetrahedron Lett.*, 1992, **33**, 7315.
263. A. Krief and M. Trabelsi, *Synth. Commun.*, 1989, **19**, 1203.
264. A. Krief, A. F. De Mahieu, W. Dumont and M. Trabelsi, *Synthesis*, 1988, 131.
265. A. McKillop, D. Koyuncu, A. Krief, W. Dumont, P. Renier and M. Trabelsi, *Tetrahedron Lett.*, 1990, **31**, 5007.
266. S. Tomoda, Y. Takeuchi and Y. Nomura, *Chem. Lett.*, 1981, 1069.
267. C. Saluzzo, G. Alvernhe and D. Anker, *Tetrahedron Lett.*, 1990, **31**, 663.
268. Y. Usuki, M. Iwaoka and S. Tomoda, *J. Chem. Soc., Chem. Commun.*, 1992, 1148.
269. S. Tomoda and Y. Usuki, *Chem. Lett.*, 1989, 1235.
270. J. R. McCarthy, D. P. Matthews and C. L. Barney, *Tetrahedron Lett.*, 1990, **31**, 973.
271. A. Toshimitsu, S. Uemura and M. Okano, *J. Chem. Soc., Chem. Commun.*, 1982, 87.
272. T. Hosay and L. Christiaens, *Tetrahedron Lett.*, 1990, **31**, 873.
273. D. G. Foster, *Org. Synth.*, 1955, **3**, 771.
274. H. J. Reich, M. L. Cohen and P. S. Clark, *Org. Synth.*, 1979, **59**, 141.
275. H. Bauer, *Chem. Ber.*, 1913, **46**, 92.
276. M. Tiecco, L. Testaferri, M. Tingoli, D. Chianelli and M. Montanucci, *J. Org. Chem.*, 1983, **48**, 4289.
277. M. Tiecco, L. Testaferri, M. Tingoli, D. Chianelli and M. Montanucci, *Synth. Commun.*, 1983, **13**, 617.
278. A. R. Barron, *Chem. Soc. Rev.*, 1993, **22**, 93.
279. S. S. Chu, W. H. H. Goenther and H. G. Mautner, *Biochem. Prep.*, 1963, **10**, 153.
280. E. P. Painter, *J. Am. Chem. Soc.*, 1947, **69**, 232.
281. D. Liotta, U. Sunay, H. Santiesteban and W. Markiewicz, *J. Org. Chem.*, 1981, **46**, 2605.
282. P. Gaskin, J. Macmillan, I. K. Makinson and C. L. Willis, *J. Chem. Soc., Perkin Trans. 1*, 1992, 1359.
283. M. Newcomb, C. C. Johnson, M. B. Manek and T. R. Varick, *J. Am. Chem. Soc.*, 1992, **114**, 10 915.
284. A. Krief and L. Hevesi, 'Organoselenium Chemistry I—Functional Group Transformations', Springer, Heidelberg, 1988, p. 46.
285. P. P. Fu and R. G. Harvey, *Chem. Rev.*, 1978, **78**, 317.
286. R. L. Dorta, C. G. Francisco, R. Freire and E. Suarez, *Tetrahedron Lett.*, 1988, **29**, 5429.
287. H. Ali and J. E. Van Lier, *J. Chem. Soc., Perkin Trans. 1*, 1991, 269.
288. H. Ali and J. E. Van Lier, *J. Chem. Soc., Perkin Trans. 1*, 1991, 2485.
289. H. Ali and J. E. Van Lier, *Tetrahedron Lett.*, 1991, **32**, 5015.
290. S. V. Ley, in 'Organoselenium Chemistry', ed. D. Liotta, Wiley, Chichester, 1987, p. 163.
291. J. H. Zaidi and A. J. Waring, *J. Chem. Soc., Chem. Commun.*, 1980, 618.
292. Y. Miyahara, T. Inazu and T. Yoshino, *Tetrahedron Lett.*, 1982, **23**, 2189.
293. D. H. R. Barton and S. I. Parekh, *J. Am. Chem. Soc.*, 1993, **115**, 948.
294. D. H. R. Barton, A. G. Brewster, S. V. Ley and M. N. Rosenfeld, *J. Chem. Soc., Chem. Commun.*, 1977, 147.
295. D. Díez-Martin et al., *Tetrahedron*, 1992, **48**, 7899.
296. D. H. R. Barton, M. B. Hall, Z. Y. Lin, S. I. Parekh and J. Reibenspies, *J. Am. Chem. Soc.*, 1993, **115**, 5056.
297. D. H. R. Barton, R. A. H. Hui and S. V. Ley, *J. Chem. Soc., Perkin Trans. 1*, 1982, 2179.

298. A. Krief and L. Hevesi, 'Organoselenium Chemistry I—Functional Group Transformations', Springer, Heidelberg, 1988, p. 104.
299. R. T. Taylor and L. A. Flood, *J. Org. Chem.*, 1983, **48**, 5160.
300. M. A. Warpehoski, B. Chabaud and K. B. Sharpless, *J. Org. Chem.*, 1982, **47**, 2897.
301. G. Pourcelot and J. Cense, *Bull. Soc. Chim. Fr.*, 1976, **9**, 1578.
302. B. R. Chhabra, K. Hayano, T. Ohtsuka, H. Shirahama and T. Matsumoto, *Chem. Lett.*, 1981, 1703.
303. K. V. Bhaskar et al., *Tetrahedron Lett.*, 1992, **32**, 6203.
304. P. S. Kalsi, B. R. Chhabra, J. Singh and R. Vig, *Synlett*, 1992, 425.
305. J. G. Lee and K. C. Kim, *Tetrahedron Lett.*, 1992, **33**, 6363.
306. B. A. Keay and R. Rodrigo, *J. Am. Chem. Soc.*, 1982, **104**, 4725.
307. W. W. Sander and O. L. Chapman, *J. Org. Chem.*, 1985, **50**, 543.
308. H. Detert, C. Anthony-Mayer and H. Meier, *Angew. Chem., Int. Ed. Engl.*, 1992, **31**, 791.
309. W. A. Sheppard and C. M. Sharts, 'Organic Fluorine Chemistry', Benjamin, New York, NY, 1969.
310. M. Clarembeau and A. Krief, *Tetrahedron Lett.*, 1984, **25**, 3625.
311. M. Renard and L. Hevesi, *Tetrahedron Lett.*, 1986, **27**, 1.
312. A. Krief, M. Clarembeau and Ph. Barbeau, *J. Chem. Soc., Chem. Commun.*, 1986, 457.
313. D. J. Goldsmith, D. C. Liotta, M. Volmer, W. Hoekstra and L. Waykole, *Tetrahedron*, 1985, **41**, 4873.
314. K. C. Nicolaou, N. A. Petasis and D. A. Claremon, *Tetrahedron*, 1985, **41**, 4835.
315. P. Grieco, S. Gilman and M. Nishizawa, *J. Org. Chem.*, 1976, **41**, 1485.
316. M. Ihara, T. Suzuki, M. Katogi, N. Taniguchi and K. Fukumoto, *J. Chem. Soc., Perkin Trans. 1*, 1992, 865.
317. T. G. Back and D. J. McPhee, *J. Org. Chem.*, 1984, **49**, 3842.
318. M. Sevrin and A. Krief, *J. Chem. Soc., Chem. Commun.*, 1980, 656.
319. P. A. Grieco and Y. Yokoyama, *J. Am. Chem. Soc.*, 1977, **99**, 5210.
320. K. C. Nicolaou and N. A. Petasis, 'Selenium in Natural Products Synthesis', CIS, Philadelphia, PA, 1984, p. 208.
321. S. Konstantinovic, Z. Bugarcic, S. Milosavljevic, G. Schroth and M. L. Mihailovic, *Liebigs Ann. Chem.*, 1992, 261.
322. Y. Zhang, Y. Yu and R. Lin, *Synth. Commun.*, 1993, **23**, 189.
323. S. Fukuzawa, Y. Niimoto, T. Fujinami and S. Sakai, *Heteroatom Chem.*, 1990, **16**, 491.
324. H. Rüger and M. H. Benn, *Can. J. Chem.*, 1982, **60**, 2918.
325. S. H. Kang and S. A. Monti, *J. Org. Chem.*, 1984, **49**, 3830.
326. K. B. Sharpless and R. F. Lauer, *J. Org. Chem.*, 1972, **37**, 3973.
327. M. R. Detty and G. P. Wood, *J. Org. Chem.*, 1980, **45**, 80.
328. E. G. Kataev, G. A. Chmutova and E. G. Yarkova, *Zh. Org. Khim.*, 1967, **3**, 2188.
329. T. Di Giamberardino, S. Halazy, W. Dumont and A. Krief, *Tetrahedron Lett.*, 1983, **24**, 3413.
330. A. Toshimitsu, H. Owada, S. Uemura and M. Okano, *Tetrahedron Lett.*, 1980, **21**, 5037.
331. H. J. Reich, *Org. React.*, 1993, **44**, 4.
332. P. J. Kropp et al., *J. Am. Chem. Soc.*, 1991, **113**, 7300.
333. L. A. Paquette, D. R. Lagerwall, J. L. King, S. Niwayama and R. Skerlj, *Tetrahedron Lett.*, 1991, **32**, 5259.
334. M. Tiecco, L. Testaferri, M. Tingoli, D. Chianelli and M. Montanucci, *Tetrahedron Lett.*, 1984, **25**, 4975.
335. H. Suzuki, H. Abe and A. Osuka, *Chem. Lett.*, 1981, 151.
336. A. B. Pierini and R. A. Rossi, *J. Org. Chem.*, 1979, **44**, 4667.
337. A. B. Pierini and R. A. Rossi, *J. Organomet. Chem.*, 1978, **144**, C12.
338. A. B. Peñéñory, A. B. Pierini and R. A. Rossi, *J. Org. Chem.*, 1984, **49**, 3834.
339. C. Paulmier, in 'Selenium Reagents and Intermediates in Organic Synthesis', ed. J. E. Baldwin, Pergamon, Oxford, 1986, vol. 5, p. 319.
340. F. C. McIntire and E. P. Painter, *J. Am. Chem. Soc.*, 1947, **69**, 1834.
341. G. H. Posner and D. Z. Rogers, *J. Am. Chem. Soc.*, 1977, **99**, 8208.
342. L. A. Khazemova and V. M. Al'bitskaya, *Zh. Org. Khim.*, 1970, **6**, 935.
343. I. M. Akhmedov, F. G. Gasanov, S. B. Kurbanov and M. M. Guseinov, *Zh. Org. Khim.*, 1978, **14**, 881.
344. A. Cravador and A. Krief, *Tetrahedron Lett.*, 1981, **22**, 2491.
345. N. Miyoshi, K. Kondo, S. Murai and N. Sonoda, *Chem. Lett.*, 1979, 909.
346. B. M. Trost and P. H. Scudder, *J. Am. Chem. Soc.*, 1977, **99**, 7601.
347. B. M. Trost, Y. Nishimura, K. Yamamoto and S. S. McElvain, *J. Am. Chem. Soc.*, 1979, **101**, 1328.
348. S. E. Kurbanov, I. M. Akhmedov, F. G. Gasanov and D. T. Radzhabov, *Dokl. Akad. Nauk Az. SSR*, 1977, **33**, 23.
349. D. H. R. Barton, M. R. Britten-Kelly and D. Ferreira, *J. Chem. Soc., Perkin Trans. 1*, 1978, 1090.
350. D. H. R. Barton, M. R. Britten-Kelly and D. Ferreira, *J. Chem. Soc., Perkin Trans. 1*, 1978, 1682.
351. N. Miyoshi, H. Ishii, S. Murai and N. Sonoda, *Chem. Lett.*, 1979, 873.
352. T. Koch and O. Buchardt, *Synthesis*, 1993, 1065.
353. D. L. J. Clive, H. W. Manning and T. L. B. Boivin, *J. Chem. Soc., Chem. Commun.*, 1990, 972.
354. S. Saito, H. Tamai, Y. Usui, M. Inaba and T. Moriwake, *Chem. Lett.*, 1984, 1243.
355. M. Inaba, T. Moriwake and S. Saito, *Tetrahedron Lett.*, 1985, **26**, 3235.
356. A. Krief and M. Trabelsi, *Tetrahedron Lett.*, 1987, **28**, 4225.
357. H. Iio, M. Isobe, T. Kawai and T. Goto, *Tetrahedron*, 1979, **35**, 941.
358. A. B. Smith, III and R. M. Scarborough, Jr., *Tetrahedron Lett.*, 1978, 1649.
359. C. Paulmier, in 'Selenium Reagents and Intermediates in Organic Synthesis', ed. J. E. Baldwin, Pergamon, Oxford, 1986, vol. 5, p. 182.
360. D. S. Middleton and N. S. Simpkins, *Tetrahedron Lett.*, 1989, **30**, 3865.
361. H. Pluim and H. Wynberg, *Tetrahedron Lett.*, 1979, 1251.
362. M. Miyashita and A. Yoshikoshi, *Synthesis*, 1980, **8**, 664.
363. D. Liotta, P. B. Paty, J. Johnston and G. Zima, *Tetrahedron Lett.*, 1978, 5091.
364. M. R. Detty, *Tetrahedron Lett.*, 1979, **43**, 4189.
365. J. Lucchetti and A. Krief, *C. R. Hebd. Seances Acad. Sci., Ser. C*, 1979, **289**, 287.
366. W. R. Leonard and T. Livinghouse, *J. Org. Chem.*, 1985, **50**, 730.
367. W. R. Leonard and T. Livinghouse, *Tetrahedron Lett.*, 1985, **26**, 6431.

368. M. Clarembeau et al., Tetrahedron, 1985, **41**, 4793.
369. R. Dieden and L. Hevesi, Synthesis, 1988, 616.
370. M. Clarembeau and A. Krief, Tetrahedron Lett., 1986, **27**, 1723.
371. D. Seebach, N. Meyer and A. K. Beck, Liebigs Ann. Chem., 1977, 846.
372. A. Krief, M. Hobe, E. Badaoui, J. Bousbaa, W. Dumont and A. Nazih, Synlett, 1993, 707.
373. S. Mehta and B. M. Pinto, J. Org. Chem., 1993, **58**, 3269.
374. A. Bouali, G. Descotes, D. F. Ewing, A. Grouiller and J. Lefkidou, J. Carbohydr. Chem., 1992, **11**, 159.
375. M. Sakakibara, K. Katsumata, Y. Watanabe, T. Toru and Y. Ueno, Synlett, 1992, 965.
376. Y. Nishiyama, S. Aoyama and S. Hamanaka, Phosphorus Sulfur Silicon, 1992, **67**, 267.
377. D. L. J. Clive and S. M. Menchen, J. Org. Chem., 1979, **44**, 1883.
378. B. Vauzeilles, D. Cravo, J. M. Mallet and P. Sinay, Synlett, 1993, 522.
379. A. G. Myers, D. Y. Gin and K. L. Widdowson, J. Am. Chem. Soc., 1991, **113**, 9661.
380. W. Dumont and A. Krief, Angew. Chem., Int. Ed. Engl., 1977, **16**, 540.
381. G. Galambos and V. Simonidesz, Tetrahedron Lett., 1982, **23**, 4371.
382. S. A. Lermontov, S. I. Zavorin, A. N. Pushin, A. N. Chekhlov, N. S. Zefirov and P. J. Stang, Tetrahedron Lett., 1993, **34**, 703.
383. A. Toshimitsu, K. Terao and S. Uemura, J. Org. Chem., 1986, **51**, 1724.
384. D. Labar, A. Krief and L. Hevesi, Tetrahedron Lett., 1978, 3967.
385. R. A. Gancarz and J. L. Kice, Tetrahedron Lett., 1981, **22**, 1661.
386. T. G. Back, S. Collins and R. G. Kerr, J. Org. Chem., 1981, **46**, 1564.
387. M. Shimizu and I. Kuwajima, Bull. Chem. Soc. Jpn., 1981, **54**, 3100.
388. M. Tiecco, M. Tingoli and L. Testaferri, Pure Appl. Chem., 1993, **65**, 715.
389. M. Tiecco, L. Testaferri, M. Tingoli, L. Bagnoli and C. Santi, J. Chem. Soc., Chem. Commun., 1993, 637.
390. T. W. Campbell and J. D. McCullough, J. Am. Chem. Soc., 1945, **67**, 1965.
391. V. H. Rawal, S. P. Singh, C. Dufour and C. Michoud, J. Org. Chem., 1991, **56**, 5245.
392. D. Friedrich and L. A. Paquette, J. Chem. Soc., Perkin Trans. 1, 1991, 1621.
393. H. Imanieh, P. Quayle, M. Voaden and S. D. A. Street, Tetrahedron Lett., 1992, **33**, 543.
394. A. Krief, G. Evrard, E. Badaoui, V. De Beys and R. Dieden, Tetrahedron Lett., 1989, **30**, 5635.
395. M. D. Fryzuk, G. S. Bates and C. Stone, J. Org. Chem., 1991, **56**, 7201.
396. N. Petragnani, R. Rodrigues and J. V. Comasseto, J. Organomet. Chem., 1976, **114**, 281.
397. B. Harirchian and P. Magnus, J. Chem. Soc., Chem. Commun., 1977, 522.
398. H. J. Reich, J. M. Renga and I. L. Reich, J. Am. Chem. Soc., 1975, **97**, 5434.
399. Y. Arai, S. Kuwayama, Y. Takeuchi and T. Koizumi, Tetrahedron Lett., 1985, **26**, 6205.
400. H. J. Reich, J. M. Renga and I. L. Reich, J. Org. Chem., 1974, **39**, 2133.
401. K. B. Sharpless, R. F. Lauer and A. Y. Teranishi, J. Am. Chem. Soc., 1973, **95**, 6137.
402. M. Miyano, J. N. Smith and C. R. Dorn, Tetrahedron, 1982, **38**, 3447.
403. M. Sasaki et al., Bull. Chem. Soc. Jpn., 1988, **61**, 3587.
404. J. H. Byers, T. G. Gleason and K. S. Knight, J. Chem. Soc., Chem. Commun., 1991, 354.
405. J. H. Byers and G. C. Lane, Tetrahedron Lett., 1990, **31**, 5697.
406. R. L. Danheiser, Y. M. Choi, M. Menichincheri and E. J. Stoner, J. Org. Chem., 1993, **58**, 322.
407. K.-C. Woo and K. Jones, Tetrahedron Lett., 1991, **32**, 6949.
408. J. N. Denis, J. Vicens and A. Krief, Tetrahedron Lett., 1979, 2697.
409. R. S. Brown, S. C. Eyley and P. J. Parsons, J. Chem. Soc., Chem. Commun., 1984, 438.
410. A. E. Feiring, J. Org. Chem., 1980, **45**, 1958.
411. T. Hayama, S. Tomoda, Y. Takeuchi and Y. Nomura, J. Org. Chem., 1984, **49**, 3235.
412. S. Piettre, Z. Janousek, R. Merenyi and H. G. Viehe, Tetrahedron, 1985, **41**, 2527.
413. Z. Janousek, S. Piettre, F. Gorissen-Hervens and H. G. Viehe, J. Organomet. Chem., 1983, **250**, 197.
414. G. Zima and D. Liotta, Synth. Commun., 1979, **9**, 697.
415. S. V. Ley and A. J. Whittle, Tetrahedron Lett., 1981, **22**, 3301.
416. L. Engman and K. W. Tornroos, J. Organomet. Chem., 1990, **391**, 165.
417. P. Magnus and P. Rigollier, Tetrahedron Lett., 1992, **33**, 6111.
418. H. J. Reich, J. Org. Chem., 1974, **39**, 428.
419. K. Hiroi and S. Sato, Synthesis, 1985, 635.
420. A. B. Holmes, A. Nadin, P. J. O'Hanlon and N. D. Pearson, Tetrahedron: Asymmetry, 1992, **3**, 1289.
421. K. B. Sharpless and R. F. Lauer, J. Org. Chem., 1974, **39**, 429.
422. E. A. Mash, J. B. Arterburn and J. A. Fryling, Tetrahedron Lett., 1989, **30**, 7145.
423. D. Wolf-kugel and S. Halazy, Tetrahedron Lett., 1991, **32**, 6341.
424. J.-E. Bäckvall, U. Karlsson and R. Chinchilla, Tetrahedron Lett., 1991, **32**, 5607.
425. S. Tomoda, K. Fujita and M. Iwaoka, Phosphorus Sulfur Silicon, 1992, **67**, 247.
426. S. Tomoda and M. Iwaoka, Chem. Lett., 1988, 1895.
427. S. Tomoda, K. Fujita and M. Iwaoka, J. Chem. Soc., Chem. Commun., 1990, 129.
428. B. Hermans, N. Colard and L. Hevesi, Tetrahedron Lett., 1992, **33**, 4629.
429. N. Komatsu, T. Murakami, Y. Nishibayashi, T. Sugita and S. Uemura, J. Org. Chem., 1993, **58**, 3697.
430. J. E. Brumwell, N. S. Simpkins and N. K. Terrett, Tetrahedron Lett., 1993, **34**, 1219.
431. C.-P. Chuang, Synth. Commun., 1992, **22**, 3151.
432. T. Toru, T. Seko, E. Maekawa and M. Ueno, J. Chem. Soc., Perkin Trans. 1, 1989, 1927.
433. C. Paulmier, in 'Selenium Reagents and Intermediates in Organic Synthesis', ed. J. E. Baldwin, Pergamon, Oxford, 1986, vol. 4, p. 228.
434. S. H. Kang and S. B. Lee, Tetrahedron Lett., 1993, **34**, 1955.
435. K. C. Nicolaou, R. L. Magolda, W. J. Sipio, W. E. Barnette, Z. Lysenko and M. M. Joullie, J. Am. Chem. Soc., 1980, **102**, 3784.
436. R. S. Huber and G. B. Jones, J. Org. Chem., 1992, **57**, 5778.
437. D. J. Hart, V. Leroy, G. H. Merriman and D. G. J. Young, J. Org. Chem., 1992, **57**, 5670.

438. A. Krief, J. Bousbaa and M. Hobe, *Synlett*, 1992, 320.
439. S. Hanessian, D. Desilets and Y. L. Bennani, *J. Org. Chem.*, 1990, **55**, 3098.
440. Y.-M. Tsai, H.-C. Nieh and C. D. Cherng, *J. Org. Chem.*, 1992, **57**, 7010.
441. J. Mathew, *J. Chem. Soc., Perkin Trans. 1*, 1991, 2039.
442. J. W. Herndon and J. J. Harp, *Tetrahedron Lett.*, 1992, **33**, 6243.
443. A. Ogawa, H. Tanaka, H. Yokoyama, R. Obayashi, K. Yokoyama and N. Sonoda, *J. Org. Chem.*, 1992, **57**, 111.
444. A. Ogawa, H. Yokoyama, K. Yokoyama, T. Masawaki, N. Kambe and N. Sonoda, *J. Org. Chem.*, 1991, **56**, 5721.
445. G. A. Russell and L. L. Herold, *J. Org. Chem.*, 1985, **50**, 1037.
446. N. Iwasawa, M. Funahashi, S. Hayakawa and K. Narasaka, *Chem. Lett.*, 1993, 545.
447. J. L. Esker and M. Newcomb, *J. Org. Chem.*, 1993, **58**, 4933.
448. J. L. Esker and M. Newcomb, *Tetrahedron Lett.*, 1992, **33**, 5913.
449. L. E. Burgess and A. I. Meyers, *J. Am. Chem. Soc.*, 1991, **113**, 9858.
450. M. J. Calverley, S. Strugnell and G. Jones, *Tetrahedron*, 1993, **49**, 739.
451. I. Kuwajima, S. Hoshino, T. Tanaka and M. Shimizu, *Tetrahedron Lett.*, 1980, **21**, 3209.
452. R. W. Hoffmann and M. Bewersdorf, *Liebigs Ann. Chem.*, 1992, 643.
453. A. Krief, E. Badaoui, W. Dumont, L. Hevesi, B. Hermans and R. Dieden, *Tetrahedron Lett.*, 1991, **32**, 3231.
454. A. Krief, E. Badaoui and W. Dumont, *Tetrahedron Lett.*, 1993, **34**, 8517.
455. S. Raucher and G. A. Koolpe, *J. Org. Chem.*, 1978, **43**, 4252.
456. T. Willson, P. Kocienski, A. Faller and S. Campbell, *J. Chem. Soc., Chem. Commun.*, 1987, 106.
457. M. Sevrin, D. Van Ende and A. Krief, *Tetrahedron Lett.*, 1976, 2643.
458. J.-F. Biellman and J. B. Ducep, *Org. React.*, 1982, **27**, 1.
459. G. Cooke, M. R. Bryce, M. C. Petty, D. J. Ando and M. B. Hursthouse, *Synthesis*, 1993, 465.
460. A. Krief, W. Dumont and A. F. De Mahieu, *Tetrahedron Lett.*, 1988, **29**, 3269.
461. A. Krief, W. Dumont and J. L. Laboureur, *Tetrahedron Lett.*, 1988, **29**, 3265.
462. M. Solomon, W. Hoekstra, G. Zima and D. Liotta, *J. Org. Chem.*, 1988, **53**, 5058.
463. H. Ishibashi and M. Ikeda, *Yuki Gosei Kagaku Kyokai Shi*, 1989, **47**, 330.
464. K. M. Nsunda and L. Hevesi, *Tetrahedron Lett.*, 1984, **25**, 4441.
465. L. Hevesi and A. Lavoix, *Tetrahedron Lett.*, 1989, **30**, 4433.
466. N. Komatsu, S. Matsunaga, T. Sugita and S. Uemura, *J. Am. Chem. Soc.*, 1993, **115**, 5847.
467. B. Hermans and L. Hevesi, *Tetrahedron Lett.*, 1990, **31**, 4363.
468. C. C. Silveira, E. J. Lenardão, J.-V. Commasseto and M. J. Dabdoub, *Tetrahedron Lett.*, 1991, **32**, 5741.
469. M. Yoshimatsu, T. Yoshiuchi, H. Shimizu, M. Hori and T. Kataoka, *Synlett*, 1993, 121.
470. K. M. Nsunda and L. Hevesi, *J. Chem. Soc., Chem. Commun.*, 1987, 1518.
471. M. Sevrin, W. Dumont and A. Krief, *Tetrahedron Lett.*, 1977, 3835.
472. J. N. Denis and A. Krief, *Tetrahedron Lett.*, 1982, **23**, 3407.
473. W. Dumont, M. Sevrin and A. Krief, *Tetrahedron Lett.*, 1978, 183.
474. M. Renard and L. Hevesi, *J. Chem. Soc., Chem. Commun.*, 1986, 688.
475. L. Hevesi, M. Renard and G. Proess, *J. Chem. Soc., Chem. Commun.*, 1986, 1725.
476. G. Proess, D. Pankert and L. Hevesi, *Tetrahedron Lett.*, 1992, **33**, 269.
477. T. Kataoka, M. Yoshimatsu, H. Shimizu and M. Hori, *Tetrahedron Lett.*, 1991, **32**, 105.
478. L. Hevesi and K. M. Nsunda, *Tetrahedron Lett.*, 1985, **26**, 6513.
479. P. A. Grieco, Y. Yokoyama and E. Williams, *J. Org. Chem.*, 1978, **43**, 1283.
480. S. Masamune, Y. Hayase, W. Schilling, W. K. Chan and G. S. Bates, *J. Am. Chem. Soc.*, 1977, **99**, 6756.
481. D. Batty and D. Crich, *J. Chem. Soc., Perkin Trans. 1*, 1992, 3205.
482. D. L. Boger and R. J. Mathvink, *J. Org. Chem.*, 1992, **57**, 1429.
483. H.-J. Gais, *Angew. Chem., Int. Ed. Engl.*, 1977, **16**, 244.
484. M. P. Astley and G. Pattenden, *Synthesis*, 1992, 101.
485. M. Renson and C. Draguet, *Bull. Soc. Chim. Belg.*, 1962, **71**, 260.
486. J. Pfenninger, C. Heuberger and W. Graf, *Helv. Chim. Acta*, 1980, **63**, 2328.
487. N. Ya. Derkach and N. P. Tishchenko, *Zh. Org. Khim.*, 1977, **13**, 100.
488. A. F. Sviridov, M. S. Ermolenko, D. V. Yashunsky and N. K. Kochetkov, *Tetrahedron Lett.*, 1983, **24**, 4359.
489. A. F. Sviridov, M. S. Ermolenko, D. V. Yashunsky and N. K. Kochetkov, *Tetrahedron Lett.*, 1983, **24**, 4355.
490. T. G. Back and S. Collins, *Tetrahedron Lett.*, 1979, 2661.
491. H. J. Gais and T. Lied, *Angew. Chem., Int. Ed. Engl.*, 1978, **17**, 267.
492. C. Chen, D. Crich and A. Papadatos, *J. Am. Chem. Soc.*, 1992, **114**, 8313.
493. H. Ishihara and Y. Hirabayashi, *Chem. Lett.*, 1978, **9**, 1007.
494. H. Kuniyasu, A. Ogawa, K. Higaki and N. Sonoda, *Organometallics*, 1992, **11**, 3937.
495. W. Hartwig, *Tetrahedron*, 1983, **39**, 2609.
496. M. D. Bachi and E. Bosch, *Tetrahedron Lett.*, 1986, **27**, 641.
497. M. D. Bachi and E. Bosch, *J. Org. Chem.*, 1992, **57**, 4696.
498. M. R. Bryce, J. Blecher and B. Fält-Hansen, *Adv. Heterocycl. Chem.*, 1992, **55**, 1.
499. H. Fischer, U. Gerbing, J. Riede and R. Benn, *Angew. Chem., Int. Ed. Engl.*, 1986, **25**, 78.
500. R. Okazaki, N. Kumon and N. Inamoto, *J. Am. Chem. Soc.*, 1989, **111**, 5949.
501. D. H. Reid, R. G. Webster and S. McKenzie, *J. Chem. Soc., Perkin Trans. 1*, 1979, 2334.
502. G. A. Krafft and P. T. Meinke, *J. Am. Chem. Soc.*, 1986, **108**, 1314.
503. G. W. Kirby and A. N. Trethewey, *J. Chem. Soc., Chem. Commun.*, 1986, 1152.
504. G. M. Li, M. Segi and T. Nakajima, *Tetrahedron Lett.*, 1992, **33**, 3515.
505. M. Segi *et al.*, *J. Am. Chem. Soc.*, 1988, **110**, 1976.
506. M. Segi, M. Kato and T. Nakajima, *Tetrahedron Lett.*, 1991, **32**, 7427.
507. K. Okuma, K. Kojima, I. Kaneko and H. Ohta, *Tetrahedron Lett.*, 1992, **33**, 1333.
508. Y. Vallée and M. Worrell, *J. Chem. Soc., Chem. Commun.*, 1992, 1680.
509. P. T. Meinke and G. A. Krafft, *Tetrahedron Lett.*, 1987, **28**, 5121.
510. M. M. Abelman, *Tetrahedron Lett.*, 1991, **32**, 7389.

511. F. S. Guziec, L. J. SanFilippo, C. J. Murphy, C. A. Moustakis and E. R. Cullen, *Tetrahedron*, 1985, **41**, 4843.
512. K. Okuma, J. Sakata, Y. Tachibana, T. Honda and H. Ohta, *Tetrahedron Lett.*, 1987, **28**, 6649.
513. E. R. Cullen and F. S. Guziec, Jr., *J. Org. Chem.*, 1986, **51**, 1212.
514. E. R. Cullen, F. S. Guziec, M. I. Hollander and C. J. Murphy, *Tetrahedron Lett.*, 1981, **22**, 4563.
515. A. Krebs, W. Rüger and W.-U. Nickel, *Tetrahedron Lett.*, 1981, **22**, 4937.
516. P. T. Meinke, G. A. Krafft and J. T. Spencer, *Tetrahedron Lett.*, 1987, **28**, 3887.
517. L. Castle and M. J. Perkins, in 'The Chemistry of Organic Selenium and Tellurium Compounds', eds. S. Patai and Z. Rappoport, Wiley, Chichester, 1987, vol. 2, p. 657.
518. B. Giese, in 'Radicals in Organic Synthesis: Formation of Carbon–Carbon Bonds', ed. J. E. Baldwin, Pergamon, Oxford, 1986, vol. 4, p. 1.
519. S. V. Ley, P. J. Murray and B. D. Palmer, *Tetrahedron*, 1985, **41**, 4765.
520. D. L. J. Clive, G. Chittattu and C. K. Wong, *J. Chem. Soc., Chem. Commun.*, 1978, 41.
521. D. L. J. Clive *et al.*, *J. Am. Chem. Soc.*, 1980, **102**, 4438.
522. K. C. Nicolaou, S. P. Seitz, W. J. Sipio and J. F. Blount, *J. Am. Chem. Soc.*, 1979, **101**, 3884.
523. D. L. J. Clive and M. H. D. Postema, *J. Chem. Soc., Chem. Commun.*, 1993, 429.
524. D. L. J. Clive and D. C. Cole, *J. Chem. Soc., Perkin Trans. 1*, 1991, 3263.
525. H. Arai and M. Kasai, *J. Org. Chem.*, 1993, **58**, 4151.
526. M. Ballestri, C. Chatgilialoglu, K. B. Clark, D. Griller, B. Giese and B. Kopping, *J. Org. Chem.*, 1991, **56**, 678.
527. C. Chatgilialoglu, A. Guerrini and M. Lucarini, *J. Org. Chem.*, 1992, **57**, 3405.
528. R. Breslow, P. J. Duggan and J. P. Light, *J. Am. Chem. Soc.*, 1992, **114**, 3982.
529. T. G. Back, V. I. Birss, M. Edwards and M. V. Krishna, *J. Org. Chem.*, 1988, **53**, 3815.
530. R. O. Hutchins and K. Learn, *J. Org. Chem.*, 1982, **47**, 4380.
531. J. N. Denis and A. Krief, *J. Chem. Soc., Chem. Commun.*, 1983, 229.
532. T. Toru *et al.*, *Tetrahedron Lett.*, 1992, **33**, 4037.
533. B. Giese, in 'Radicals in Organic Synthesis: Formation of Carbon–Carbon Bonds', ed. J. E. Baldwin, Pergamon, Oxford, 1986, vol. 4, p. 36.
534. B. Giese, in 'Radicals in Organic Synthesis: Formation of Carbon–Carbon Bonds', ed. J. E. Baldwin, Pergamon, Oxford, 1986, vol. 4, p. 141.
535. C. Thebtaranonth and Y. Thebtaranonth, *Tetrahedron*, 1990, **46**, 1385.
536. D. P. Curran, *Synthesis*, 1988, 489.
537. A. Y. Mohammed and D. L. J. Clive, *J. Chem. Soc., Chem. Commun.*, 1986, 588.
538. Z. Xi, P. Agback, A. Sandstrom and J. Chattopadhyaya, *Tetrahedron*, 1991, **47**, 9675.
539. L. Set, D. R. Cheshire and D. L. J. Clive, *J. Chem. Soc., Chem. Commun.*, 1985, 1205.
540. X. Zhen, P. Agback, J. Plavec, A. Sandström and J. Chattopadhyaya, *Tetrahedron*, 1992, **48**, 349.
541. M. D. Bachi and C. Hoornaert, *Tetrahedron Lett.*, 1981, **22**, 2693.
542. D. Batty and D. Crich, *J. Chem. Soc., Perkin Trans. 1*, 1992, 3193.
543. D. Batty and D. Crich, *Tetrahedron Lett.*, 1992, **33**, 875.
544. J. H. Byers and G. C. Lane, *J. Org. Chem.*, 1993, **58**, 3355.
545. J. H. Byers and B. C. Harper, *Tetrahedron Lett.*, 1992, **33**, 6953.
546. D. P. Curran and G. Thoma, *J. Am. Chem. Soc.*, 1992, **114**, 4436.
547. Y. Watanabe, T. Yoneda, Y. Ueno and T. Toru, *Tetrahedron Lett.*, 1990, **31**, 6669.
548. D. L. J. Clive *et al.*, *J. Chem. Soc., Chem. Commun.*, 1992, 1489.
549. D. L. J. Clive, T. L. B. Boivin and A. G. Angoh, *J. Org. Chem.*, 1987, **52**, 4943.
550. P. Renaud, P. Björup, P. A. Carrupt, K. Schenk and S. Schubert, *Synlett*, 1992, 211.
551. A. K. Singh, R. K. Bakshi and E. J. Corey, *J. Am. Chem. Soc.*, 1987, **109**, 6187.
552. A. Krief, D. Derouane and W. Dumont, *Synlett*, 1992, 907.
553. M. Clarembeau and A. Krief, *Tetrahedron Lett.*, 1986, **27**, 4917.
554. M. Clarembeau and A. Krief, *Tetrahedron Lett.*, 1986, **27**, 1719.
555. Atta-ur-Rahman (ed.), 'Studies in Natural Products Chemistry', Elsevier, Amsterdam, 1991, vol. 8.
556. A. Krief and M. Hobe, *Synlett*, 1992, 317.
557. A. Krief, M. Hobe, W. Dumont, E. Badaoui, E. Guittet and G. Evrard, *Tetrahedron Lett.*, 1992, **33**, 3381.
558. A. Krief and M. Hobe, *Tetrahedron Lett.*, 1992, **33**, 6527.
559. A. Krief and M. Hobe, *Tetrahedron Lett.*, 1992, **33**, 6529.
560. R. Hoffmann, T. Ruckert and R. Brückner, *Tetrahedron Lett.*, 1993, **34**, 297.
561. R. Hoffmann and R. Brückner, *Chem. Ber.*, 1992, **125**, 1957.
562. I. Kuwajima and R. Takeda, *Tetrahedron Lett.*, 1981, **22**, 2381.
563. D. Liotta, M. Saindane, C. Barnum and G. Zima, *Tetrahedron*, 1985, **41**, 4881.
564. J. Rémion and A. Krief, *Tetrahedron Lett.*, 1976, 3743.
565. W. M. Fan and J. B. White, *Tetrahedron Lett.*, 1993, **34**, 957.
566. J. N. Denis and A. Krief, *Tetrahedron Lett.*, 1982, **23**, 3411.
567. R. V. Bonnert and P. R. Jenkins, *J. Chem. Soc., Chem. Commun.*, 1987, 1540.
568. W. Dumont, D. Van Ende and A. Krief, *Tetrahedron Lett.*, 1979, 485.
569. H. M. Zuurmond, P. A. M. Vanderklein, P. H. Vandermeer, G. A. Vandermarel and J. H. Vanboom, *Recl. Trav. Chim. Pays-Bas*, 1992, **111**, 365.
570. H. J. Gysling, in 'The Chemistry of Organic Selenium and Tellurium Compounds', eds. S. Patai and Z. Rappoport, Wiley, Chichester, 1986, vol. 1, p. 679.
571. S. Akabori, Y. Takanohashi, S. Aoki and S. Sato, *J. Chem. Soc., Perkin Trans. 1*, 1991, 3121.
572. W. Dumont, P. Bayet and A. Krief, *Angew. Chem., Int. Ed. Engl.*, 1974, **13**, 274.
573. S. Kondo, A. Shibata, H. Kunisada and Y. Yuki, *Bull. Chem. Soc. Jpn.*, 1992, **65**, 2555.
574. S. Halazy and A. Krief, *Tetrahedron Lett.*, 1981, **22**, 2135.
575. M. Sevrin and A. Krief, *Tetrahedron Lett.*, 1978, 187.
576. M. Sevrin, W. Dumont, L. Hevesi and A. Krief, *Tetrahedron Lett.*, 1976, 2647.
577. L. Engman, *Tetrahedron Lett.*, 1987, **28**, 1463.

578. K. A. Jensen and C. K. Jørgensen, in 'Organic Selenium Compounds: Their Chemistry and Biology', eds. D. L. Klayman and W. H. H. Günther, Wiley, Chichester, 1973, p. 1017.
579. F. A. Davis, R. T. Reddy, W. Han and R. E. Reddy, *Pure Appl. Chem.*, 1993, **65**, 633.
580. A. Krief, W. Dumont, J. N. Denis, G. Evrard and B. Norberg, *J. Chem. Soc., Chem. Commun.*, 1985, 569.
581. H. J. Reich, in 'Organoselenium Chemistry', ed. D. Liotta, Wiley, Chichester, 1987, p. 365.
582. H. J. Reich, in 'Oxidation in Organic Chemistry', eds. W. S. Trahanovsky and H. H. Wasserman, Academic Press, New York, 1978, vol. 5, p. 102.
583. A. Krief, W. Dumont and J. N. Denis, *J. Chem. Soc., Chem. Commun.*, 1985, 571.
584. S. A. Davis, O. D. Stringer and J. M. Billmers, *Tetrahedron Lett.*, 1983, **24**, 1213.
585. F. A. Davis, J. M. Billmers and O. D. Stringer, *Tetrahedron Lett.*, 1983, **24**, 3191.
586. F. A. Davis, O. D. Stringer and J. P. J. McCauley, *Tetrahedron*, 1985, **41**, 4747.
587. F. A. Davis, R. T. Reddy and M. C. Weismiller, *J. Am. Chem. Soc.*, 1989, **111**, 5964.
588. M. Cinquini, S. Colonna and R. Giovini, *Chem. Ind. (London)*, 1969, 1737.
589. D. Labar, L. Hevesi, W. Dumont and A. Krief, *Tetrahedron Lett.*, 1978, 1141.
590. F. A. Davis and R. T. Reddy, *J. Org. Chem.*, 1992, **57**, 2599.
591. N. Komatsu, Y. Nishibayashi and S. Uemura, *Tetrahedron Lett.*, 1993, **34**, 2339.
592. J. A. Latham, B. P. Branchaud, Y.-C. J. Chen and C. Walsh, *J. Chem. Soc., Chem. Commun.*, 1986, 528.
593. L.-B. Agenäs, in 'Organic Selenium Compounds: Their Chemistry and Biology', eds. D. L. Klayman and W. H. H. Günther, Wiley, Chichester, 1973, p. 209.
594. M. Shimizu and I. Kuwajima, *J. Org. Chem.*, 1980, **45**, 4063.
595. T. Hori and K. B. Sharpless, *J. Org. Chem.*, 1978, **43**, 1689.
596. T. Kametani, H. Kurobe and H. Nemoto, *J. Chem. Soc., Perkin Trans. 1*, 1981, 756.
597. Y. Masuyama, Y. Ueno and M. Okawara, *Chem. Lett.*, 1977, 835.
598. H. J. Reich and W. W. J. Willis, *J. Am. Chem. Soc.*, 1980, **102**, 5967.
599. H. J. Reich and K. E. Yelm, *J. Org. Chem.*, 1991, **56**, 5672.
600. T. Shimizu and M. Kobayashi, *Chem. Lett.*, 1986, 161.
601. L. Hevesi, M. Sevrin and A. Krief, *Tetrahedron Lett.*, 1976, 2651.
602. H. J. Reich, *J. Am. Chem. Soc.*, 1973, **95**, 964.
603. O. K. Edwards, W. R. Gaythwaite, J. Kenyon and H. Phillips, *J. Chem. Soc.*, 1928, 2293.
604. N. Miyoshi, S. Murai and N. Sonoda, *Tetrahedron Lett.*, 1977, 851.
605. N. Ikota and B. Ganem, *J. Org. Chem.*, 1978, **43**, 1607.
606. T. Fukuyama, B. D. Robins and R. A. Sachleben, *Tetrahedron Lett.*, 1981, **22**, 4155.
607. K. Uneyama, Y. Tokunaga and K. Maeda, *Tetrahedron Lett.*, 1993, **34**, 1311.
608. I. D. Entwistle, R. Johnstone and J. H. Varley, *J. Chem. Soc., Chem. Commun.*, 1976, 61.
609. K. Sachdev and H. S. Sachdev, *Tetrahedron Lett.*, 1976, 4223.
610. H. J. Reich and S. K. Shah, *J. Org. Chem.*, 1977, **42**, 1773.
611. D. Van Ende, W. Dumont and A. Krief, *J. Orgmet. Chem.*, 1978, **149**, C10.
612. I. W. Still, S. K. Hasan and K. Turnbull, *Can. J. Chem.*, 1978, **56**, 1423.
613. J. N. Denis and A. Krief, *J. Chem. Soc., Chem. Commun.*, 1980, 544.
614. R. C. Gadwood, I. M. Mallick and A. J. DeWinter, *J. Org. Chem.*, 1987, **52**, 774.
615. T. G. Back and M. V. Krishna, *J. Org. Chem.*, 1987, **52**, 4265.
616. D. J. Cram and J. E. McCarty, *J. Am. Chem. Soc.*, 1954, **76**, 5740.
617. C. A. Kingsbury and D. J. Cram, *J. Am. Chem. Soc.*, 1960, **82**, 1810.
618. L. D. Kwart, A. G. Horgan and H. Kwart, *J. Am. Chem. Soc.*, 1981, **103**, 1232.
619. R. Walter and J. Roy, *J. Org. Chem.*, 1971, **36**, 2561.
620. R. Walter, I. L. Schwartz and J. Roy, *Ann. N.Y. Acad. Sci.*, 1972, **192**, 175.
621. K. B. Sharpless, M. W. Young and R. F. Lauer, *Tetrahedron Lett.*, 1973, 1979.
622. H. J. Reich, I. L. Reich and J. M. Renga, *J. Am. Chem. Soc.*, 1973, **95**, 5813.
623. D. L. J. Clive, *J. Chem. Soc., Chem. Commun.*, 1973, 695.
624. R. D. Clark and C. H. Heathcock, *J. Org. Chem.*, 1976, **41**, 1396.
625. A. Toshimitsu, H. Owada, S. Uemura and M. Okano, *Tetrahedron Lett.*, 1982, **23**, 2105.
626. P. A. Grieco, J. A. Noguez and Y. Masaki, *Tetrahedron Lett.*, 1975, 4213.
627. T. Takahashi, H. Nagashima and J. Tsuji, *Tetrahedron Lett.*, 1978, 799.
628. H. J. Reich, S. Wollowitz, J. E. Trend, F. Chow and D. F. Wendelborn, *J. Org. Chem.*, 1978, **43**, 1697.
629. P. A. Grieco, *Synthesis*, 1975, 67.
630. P. A. Grieco, C. S. Pogonowski and S. Burke, *J. Org. Chem.*, 1975, **40**, 542.
631. B. M. Trost, T. N. Salzmann and K. Hiroi, *J. Am. Chem. Soc.*, 1976, **98**, 4887.
632. A. P. Kozikowski, K. L. Sorgi and R. J. Schmiesing, *J. Chem. Soc., Chem. Commun.*, 1980, 477.
633. H. J. Reich, S. K. Shah and F. Chow, *J. Am. Chem. Soc.*, 1979, **101**, 6648.
634. S. David, A. Lubineau and J. M. Vatèle, *J. Chem. Soc., Chem. Commun.*, 1975, 701.
635. A. Toshimitsu, T. Aoai, H. Owada, S. Uemura and M. Okano, *J. Org. Chem.*, 1981, **46**, 4727.
636. D. Labar, W. Dumont, L. Hevesi and A. Krief, *Tetrahedron Lett.*, 1978, 1145.
637. B. Fernández, J. A. M. Pérez, J. R. Granja, L. Castedo and A. Mourino, *J. Org. Chem.*, 1992, **57**, 3173.
638. J. Yoshida, S. Matsunaga, Y. Ishichi, T. Maekawa and S. Isoe, *J. Org. Chem.*, 1991, **56**, 1307.
639. G. Ladouceur and L. A. Paquette, *Synthesis*, 1992, 185.
640. D. A. Horne, B. Fugmann, K. Yakushijin and G. Buchi, *Tetrahedron Lett.*, 1993, **58**, 62.
641. D. Fattori, E. Guchteneere and P. Vogel, *Tetrahedron Lett.*, 1989, **30**, 7415.
642. D. Fattori and P. Vogel, *Tetrahedron Lett.*, 1993, **34**, 1017.
643. M. Mikolajczyk, S. Grzejszczak and K. Korbacz, *Tetrahedron Lett.*, 1981, **22**, 3097.
644. G. Saleh, T. Minami, Y. Ohshiro and T. Agawa, *Chem. Ber.*, 1979, 355.
645. W. A. Kleschick and C. H. Heathcock, *J. Org. Chem.*, 1978, **43**, 1256.
646. D. L. J. Clive, G. Chittattu, N. J. Curtis and S. M. Menchen, *J. Chem. Soc., Chem. Commun.*, 1978, 770.
647. C. A. Wilson and T. A. Bryson, *J. Org. Chem.*, 1975, **40**, 800.

648. R. G. Shea et al., *J. Org. Chem.*, 1986, **51**, 5243.
649. S. Uemura, S. Fukuzawa and A. Toshimitsu, *J. Chem. Soc., Chem. Commun.*, 1983, 1501.
650. S. Uemura, K. Ohe and N. Sugita, *J. Chem. Soc., Chem. Commun.*, 1988, 111.
651. S. Uemura, K. Ohe and N. Sugita, *J. Chem. Soc., Perkin Trans. 1*, 1990, 1697.
652. S. Uemura and K. Ohe, *J. Chem. Soc., Perkin Trans. 1*, 1990, 907.
653. S. Yamazaki, H. Fujitsuka, S. Yamabe and H. Tamura, *J. Org. Chem.*, 1992, **57**, 5610.
654. S. A. Woski and M. Koreeda, *J. Org. Chem.*, 1992, **57**, 5736.
655. C. Cauletti and G. Distefano, in 'The Chemistry of Organic Selenium and Tellurium Compounds', eds. S. Patai and Z. Rappoport, Wiley, Chichester, 1987, vol. 2, p. 1.
656. F. J. Berry, in 'The Chemistry of Organic Selenium and Tellurium Compounds', eds. S. Patai and Z. Rappoport, Wiley, Chichester, 1987, vol. 2, p. 51.
657. C. Paulmier, in 'Selenium Reagents and Intermediates in Organic Synthesis', ed. J. E. Baldwin, Pergamon, Oxford, 1986, vol. 5, p. 402.

14
Tellurium

NICOLA PETRAGNANI
Universidade de São Paulo, Brazil

14.1 INTRODUCTION	571
14.2 INORGANIC TELLURIUM REAGENTS EMPLOYED IN ORGANIC SYNTHESIS	571
14.3 THE PREPARATION OF THE PRINCIPAL CLASSES OF ORGANOTELLURIUM COMPOUNDS	572
14.4 TELLURIUM REAGENTS IN ORGANIC SYNTHESIS	576
14.4.1 Reductions	576
14.4.2 Tellurium-mediated Formation of Anionic Species and Their Reactions with Electrophiles	581
14.4.3 Deprotection of Organic Functionalities	582
14.4.4 Oxidations	583
14.4.5 Organotellurium-based Ring-closure Reactions	585
14.4.6 Conversion of Organotellurium Compounds into Tellurium-free Organic Compounds	587
14.4.7 Alkene Synthesis	591
14.4.8 Transmetallation Reactions of Organotellurium Compounds	592
14.4.9 Free Radical Chemistry	594
14.4.10 Other Important Reactions Involving Tellurium Reagents	596
14.5 REFERENCES	597

14.1 INTRODUCTION

Applications of both inorganic and organometallic tellurium reagents to organic synthesis follow on, to a significant extent, from the enormous interest shown since the early 1970s, particularly in the development of organoselenium chemistry, and remarkable progress has been made in the early 1990s. Most organic chemists are becoming familiar with tellurium, which is now no longer considered as a perverse and exotic element but as the basis of a variety of versatile and useful reagents for organic synthesis. This rapidly developing role for tellurium is well illustrated by the observation that in the mid-1950s Rheinboldt reviewed the organic chemistry of tellurium in comparison with selenium in the Houben–Weyl series,[1] and required only around 50 pages for tellurium. By contrast, in 1990 and in the same series, Irgolic dedicated more than 1000 pages to tellurium.[2] In the early 1990s many papers and a number of review articles and books have been devoted to the organic chemistry of tellurium.[3–18]

In this chapter, it is clearly not possible to cover in detail all of the aspects of organotellurium chemistry. The intention, rather, is to give a brief outline of methods for the preparation of the main classes of organotellurium compounds and an overview of the recent applications of tellurium compounds, both organic and inorganic, to synthetic methodology.

14.2 INORGANIC TELLURIUM REAGENTS EMPLOYED IN ORGANIC SYNTHESIS

A variety of tellurium reagents are employed in organic synthesis, the main ones being elemental tellurium, alkali tellurides and ditellurides, sodium hydrogen telluride, hydrogen telluride, tellurium tetrachloride and tellurium dioxide, all of which are discussed below.

Elemental tellurium, a black solid, is commercially available as ingot, pieces, granules and powder (60–200 mesh). It is the obvious starting material for the preparation of inorganic tellurium reagents and for several organotellurium compounds. For these purposes it is generally used as a finely ground powder.

Alkali tellurides and ditellurides (the most familiar are Na_2Te and Na_2Te_2) are of great importance for the synthesis of diorganotellurides and diorganoditellurides and are also used as reagents for the manipulation of a number of functional groups. They are prepared *in situ* by the methods described in Section 14.3 and allowed to react with the organic substrates without isolation.

Sodium hydrogen telluride, a valuable reagent for the reduction of several types of functional groups (see Section 14.4), is prepared *in situ* by the reduction of elemental tellurium with $NaBH_4$. The original procedure[19] uses ethanol as solvent, followed by addition of a defined amount of acetic acid after consumption of the tellurium. Hydrogen telluride is prepared *in situ* by the hydrolysis of Al_2Te_3.[20-2] Tellurium tetrachloride, which is prepared from the elements,[23] is a crystalline, pale yellow solid, m.p. 224 °C. It is highly hygroscopic and is readily hydrolysed to tellurium oxychloride. It dissolves in concentrated hydrochloric acid to give the chlorotellurates $HTeCl_5$ and H_2TeCl_6. The Raman data for tellurium tetrachloride suggest the ionic structure $[TeCl_3]^+ Cl^-$ in the solid and liquid states.[24] Tellurium dioxide is prepared by the oxidation of elemental tellurium with nitric acid.[25]

14.3 THE PREPARATION OF THE PRINCIPAL CLASSES OF ORGANOTELLURIUM COMPOUNDS

Diorganotellurides and diorganoditellurides are prepared from alkali tellurides and ditellurides and organic halides (Equations (1) and (2)). Due to the high nucleophilicity of the telluride anions, even reactions with aryl halides work well. The most widely employed telluride reagent is sodium telluride, generated *in situ* from the elements in liquid ammonia[26,27] or DMF,[28] or by treatment of elemental tellurium with reducing agents such as Rongalite $(HOCH_2SO_2Na)$.[29-36] More recently, hydride sources such as $NaBH_4$[37-9] or thiourea dioxide $(HN=C(NH_2)S(O)OH)$ have been used.[40-2]

$$M_2Te + 2\,RX \longrightarrow RTeR \qquad (1)$$
$$R = alkyl, aryl; M = Na, K(Me_4N)$$

$$M_2Te_2 + 2\,RX \longrightarrow RTeTeR \qquad (2)$$
$$R = alkyl, aryl; M = Na, K(Me_4N)$$

Aryltellurium trichlorides and diaryltellurium dichlorides can be prepared from $TeCl_4$. The electrophilic $TeCl_4$ reacts with arenes bearing electron-releasing groups such as RO, PhO, HO, PhS to give aryltellurium trichlorides[43-7] or dichlorides[43,48,49] depending on the experimental conditions (Scheme 1). Nonactivated arenes react only under more severe conditions or in the presence of a Lewis acid. In this case, however, dichlorides are produced.[48-50] When the arene is not sufficiently reactive for direct condensation with $TeCl_4$, arylmercury chlorides in 1:1 or 1:2 ratio are employed to prepare both trichlorides[46,51,52] or dichlorides[53] (Scheme 2).

$$TeCl_4 + ArH \xrightarrow{CHCl_3 \text{ or } CCl_4,\,reflux} ArTeCl_3$$

$$TeCl_4 + ArH \xrightarrow{\Delta} [ArTeCl_3] \xrightarrow[excess]{ArH} Ar_2TeCl_2$$

Scheme 1

A widely used method for the preparation of diaryl ditellurides and tellurides involves the reduction of the corresponding aryltellurium trichlorides and dichlorides (Equations (3) and (4)). Reducing agents used include $Na_2S \cdot 9H_2O$,[54,46] $KHSO_3$,[45] $(H_2N)_2 \cdot H_2O$[48] and TUDO (thiourea dioxide).[55] Conversely, ditellurides[46,56-8] and tellurides[59,60] are converted to trihalides and dihalides by treatment with halogens or halogenating reagents (Equations (5) and (6)).

Arenetellurolates and arenetellurols can be converted into other tellurium compounds. Elemental tellurium inserts easily into the C-sp^2–Mg bonds of aryl-[61-3] and vinylmagnesium[64,65] halides (but not the

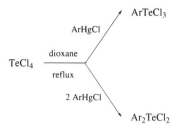

Scheme 2

$$ArTeCl_3 \xrightarrow{\text{reducing agent}} ArTeTeAr \quad (3)$$

reducing agents: $Na_2S \cdot 9 H_2O$, $KHSO_3$, $(H_2N)_2 \cdot H_2O$, TUDO

$$Ar_2TeCl_2 \xrightarrow{\text{reducing agent}} Ar_2Te \quad (4)$$

reducing agents: $Na_2S \cdot 9 H_2O$, $(H_2N)_2 \cdot H_2O$, TUDO

$$ArTeTeAr \xrightarrow{X_2} 2\, ArTeX_3 \quad (5)$$

$X_2 = SO_2Cl_2$, Br_2, I_2

$$ArTeAr \xrightarrow{X_2} Ar_2TeX_2 \quad (6)$$

$X_2 = SO_2Cl_2$, Br_2, I_2

alkyl reagents[61]) and into the C–Li bond of organolithiums to give organotellurolates (Equation (7)).[66–70] Arenetellurolates are very important synthetic intermediates. They are converted into the parent tellurols by protonation or by methanolysis of the corresponding trimethylsilyl derivatives (Equation (8) and Scheme 3). Important transformations of arenetellurolates are their oxidation to diaryl ditellurides[61–3,68–70] and alkylation to give aryl alkyl tellurides[66,67] (Equations (9) and (10)).

$$RM + Te \xrightarrow{\text{ether, THF}} RTeM \quad (7)$$

M = MgX; R = aryl, vinyl: M = Li; R = aryl, alkyl

$$ArTeM \xrightarrow{H^+} ArTeH \quad (8)$$

M = Li

$$ArTeM \xrightarrow{TMS\text{-}Cl} ArTe\text{-}TMS \xrightarrow{MeOH} ArTeH$$

Scheme 3

$$ArTeM \xrightarrow[\text{ii, }O_2\text{ (air)}]{\text{i, }H^+} ArTeTeAr \quad (9)$$

M = MgX, Li

$$ArTeM \xrightarrow[\text{ether, THF}]{RX} ArTeR \quad (10)$$

M = Li

Diaryl ditellurides can be reductively converted into aryl tellurolates by treatment with sodium in liquid ammonia[71,72] and aprotic solvents,[28,73,74] or with lithium in THF.[75] A more practical method, however, is treatment with reducing agents such as $NaBH_4$.[76–9] A very useful and inexpensive reagent system is thiourea dioxide in aqueous sodium hydroxide.[80,81] Trapping of the tellurolates thus formed with appropriate alkylating or arylating agents (alkyl and aryl halides, oxiranes[66,75–83]) constitutes the most important route to unsymmetrical tellurides (Scheme 4).

$$\text{ArTeTeAr} \xrightarrow[\text{reducing agents}]{\text{alkali metals or}} \text{ArTe}^- \xrightarrow{E} \text{ArTeE}$$

E = alkyl and aryl halides, oxiranes

Scheme 4

Arenetellurenyl halides are used as starting materials for unsymmetrical tellurides. They are the products of controlled halogenolysis of the parent ditellurides,[57,84-8] and are rather unfamiliar compounds in comparison with the well-known selenium analogues. They are, however, important intermediates, as they are easily converted into different types of unsymmetrical telluride by *in situ* treatment with appropriate Grignard reagents (Scheme 5).[56,84-6,89,90]

$$\text{ArTeTeAr} \xrightarrow{X_2, \text{benzene/THF}} [\text{ArTeX}] \xrightarrow{\text{RMgX}} \text{ArTeR}$$

R = alkyl, aryl, vinyl, ethynyl

Scheme 5

Vinylic and alkynic tellurides are important synthetic intermediates and are prepared by the methods summarized in Schemes 6–9 and Equation (11). (i) Addition of alkali tellurides to alkynes. The telluride reagent is generated by treatment of elemental tellurium with nonreducing bases with heating,[91-3] or by the familiar $NaBH_4$ method[94] (Scheme 6). (ii) Addition of organotellurols to terminal alkynes (Scheme 7).[95-8] (iii) Reaction of tellurolate[64] and telluride[95] anions with vinyl bromides (Scheme 8). (iv) Alkylation of ethynyltellurolate anions (Scheme 9).[99-103] (v) Reaction of alkynyllithiums with arenetellurenyl halides (Equation (11))[103] (vinyl- and ethynylmagnesium halides react similarly).[88]

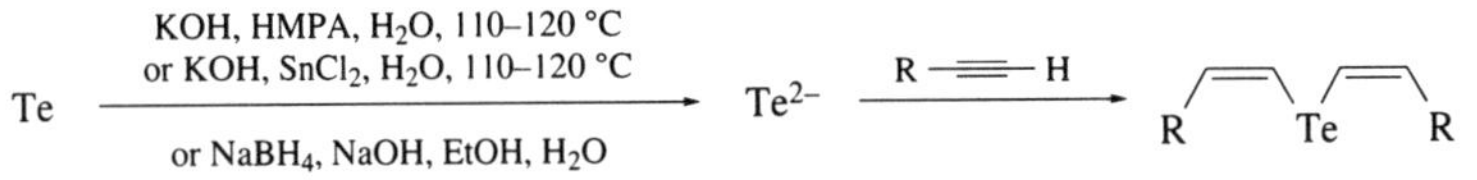

$$\text{Te} \xrightarrow[\text{or } NaBH_4, NaOH, EtOH, H_2O]{\substack{\text{KOH, HMPA, } H_2O, 110-120\,°C \\ \text{or KOH, } SnCl_2, H_2O, 110-120\,°C}} \text{Te}^{2-} \xrightarrow{R \equiv H} R\text{—CH=CH—Te—CH=CH—}R$$

Scheme 6

$$R^1\text{TeTe}R^1 \xrightarrow{NaBH_4, EtOH} [R^1\text{TeH}] \xrightarrow{R^2 \equiv H} \underset{H\quad H}{R^1Te\text{—C=C—}R^2}$$

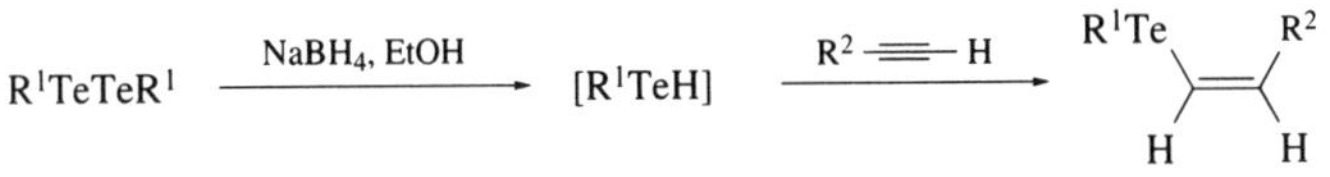

Scheme 7

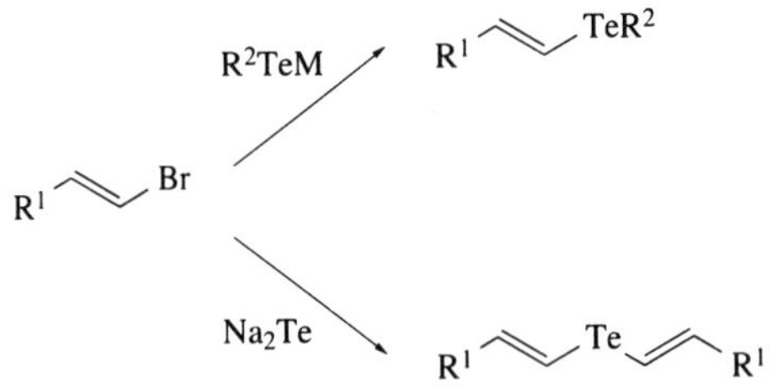

M = Li, Na

Scheme 8

$$R^1 \equiv H \xrightarrow[Bu^nLi, THF]{Na, NH_3 \text{ or}} \xrightarrow{Te} R^1 \equiv \text{TeNa(Li)} \xrightarrow{R^2X} R^1 \equiv \text{Te}R^2$$

Scheme 9

$$\text{Ph} \equiv \text{Li} + \text{ArTeX} \longrightarrow \text{Ph} \equiv \text{TeAr} \qquad (11)$$

The introduction of tellurium into the α-position of ketones can be achieved by the reaction of $TeCl_4$ with aliphatic or cycloaliphatic ketones, which gives α-trichlorotelluro- or α-dichlorotelluroketones, depending on the structure of the substrate (Equation (12)).[104,105] Silyl enol ethers have been used similarly (Equation (13)),[106] and organotellurium trichlorides also give α-aryldichlorotelluroketones.[107-9]

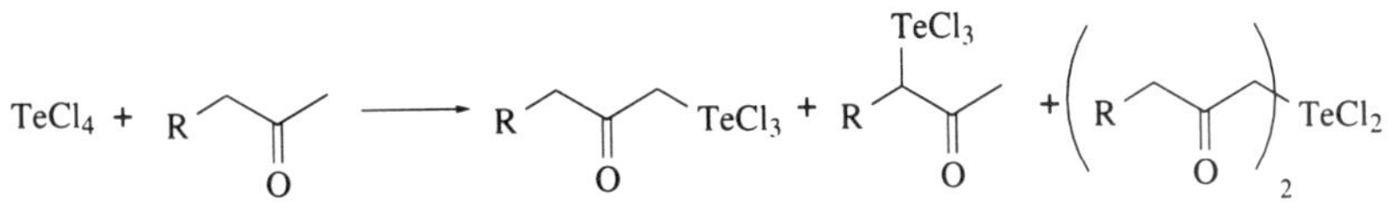

$$TeCl_4 + R\text{-CO-CH}_3 \longrightarrow R\text{-CO-CH}_2\text{-}TeCl_3 + R\text{-CO-CH}(TeCl_3)\text{-} + (R\text{-CO-CH}_2\text{-})_2 TeCl_2 \quad (12)$$

$$\text{(O-TMS alkene)} + TeCl_4 \xrightarrow{\text{ether}} (\text{ketone})_2 TeCl_2 \quad (13)$$

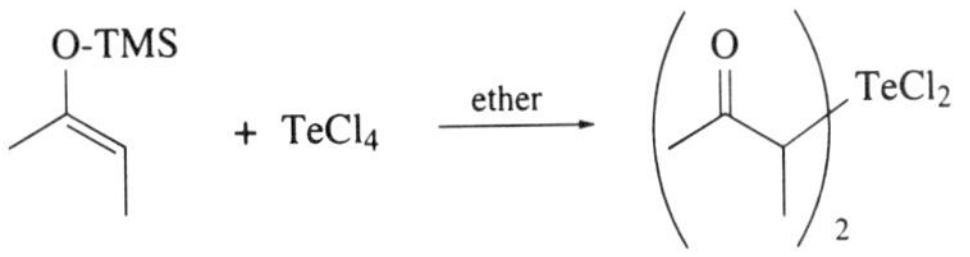

TeCl$_4$ adds to alkenes such as butenes, 1-decene, cycloalkenes and 3-substituted-1-propenes to give chloroalkyltellurium trichlorides as mixtures of *cis* and *trans* adducts (Equation (14)).[110–14] Depending on the TeCl$_4$/alkene ratio and the solvent polarity, bis(dichloroalkyl)tellurium dichlorides can be formed.[110,112,115,116] A related reaction is the addition of the TeO$_2$/HCl/ROH[117] or TeCl$_4$(TeBr$_4$)/ROH/CCl$_4$[118] systems to alkenes or cycloalkenes, which produces alkoxyalkyltellurium trichlorides (Scheme 10 and Equation (15)). The reaction is regiospecific (the tellurium moiety adds to the less substituted carbon) and *anti*-stereospecific (*trans* adducts are formed with cycloalkenes). The trichlorides can be reduced *in situ* to the corresponding ditellurides.

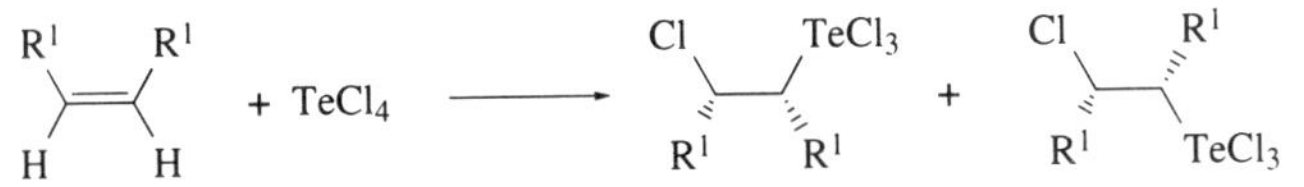

$$R^1R^1C=CH_2 + TeCl_4 \longrightarrow \begin{array}{c} Cl \quad TeCl_3 \\ R^1 \quad R^1 \end{array} + \begin{array}{c} Cl \quad R^1 \\ R^1 \quad TeCl_3 \end{array} \quad (14)$$

$$R^1\text{-CH=CH}_2 \xrightarrow{TeO_2, HCl, R^2OH} \left[\begin{array}{c} OR^2 \\ R^1\text{-CH-CH}_2\text{-}TeCl_3 \end{array} \right] \xrightarrow{\text{reduction}} R^1\text{-CH}(OR^2)\text{-CH}_2\text{-}TeTe\text{-}CH_2\text{-CH}(OR^2)\text{-}R^1$$

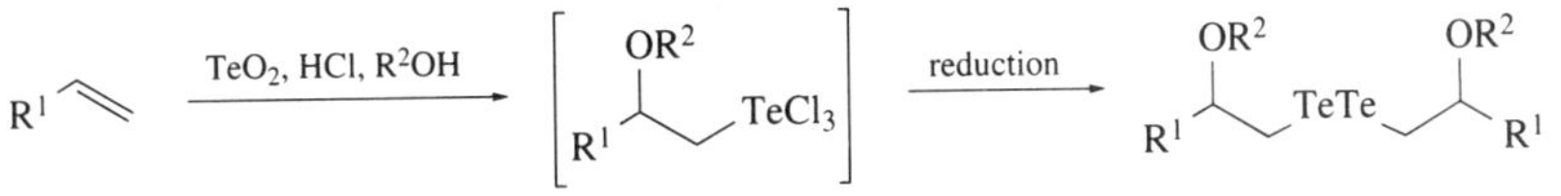

Scheme 10

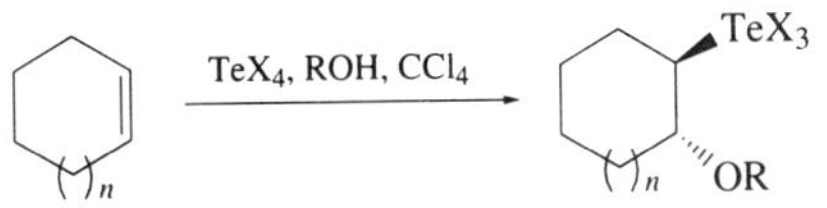

$$\text{cycloalkene} \xrightarrow{TeX_4, ROH, CCl_4} \text{cycloalkane-}TeX_3, OR \quad (15)$$

X = Cl, Br; n = 1, 2

Aryltellurium trichlorides add to alkenes in an *anti*-stereospecific manner,[113,114,119,120] in contrast to *syn* and *anti* addition of TeCl$_4$, to give chloroalkylaryltellurium dichlorides (Equation (16)). In the presence of alcohols, *anti* addition gives alkoxytellurination[81,121,122] products, the tellurium adding exclusively to the terminal carbon. The reaction is usually carried out with aryltellurium tribromides generated *in situ* from the parent ditellurides and bromine (Equation (17)).

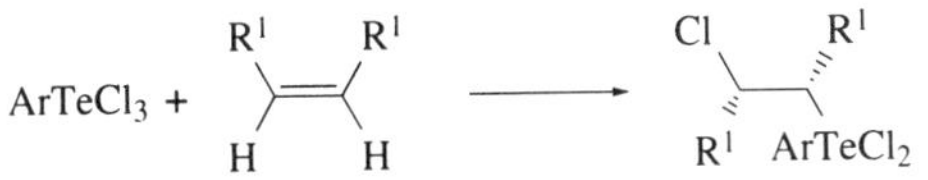

$$ArTeCl_3 + R^1R^1C=CH_2 \longrightarrow \begin{array}{c} Cl \quad R^1 \\ R^1 \quad ArTeCl_2 \end{array} \quad (16)$$

$$R^1R^1C=CH_2 \xrightarrow{PhTeBr_3, R^2OH} \begin{array}{c} R^2O \quad R^1 \\ R^1 \quad PhTeBr_2 \end{array} \quad (17)$$

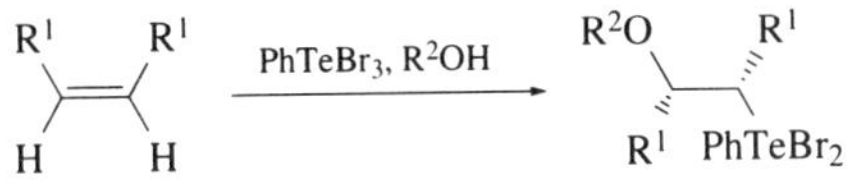

TeCl$_4$ adds to terminal and disubstituted[90,123] alkynes (Equation (18)) and aryltellurium trichlorides to terminal alkynes[124] (Equation (19)), giving (Z)-adducts.

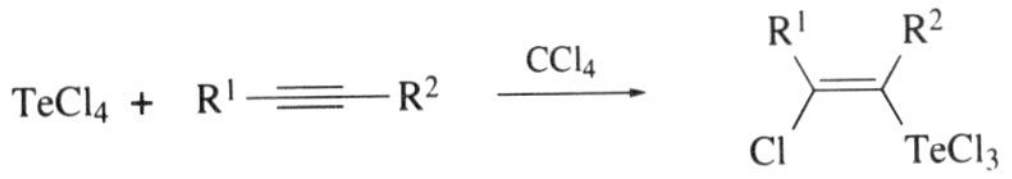

$$TeCl_4 + R^1\text{-C≡C-}R^2 \xrightarrow{CCl_4} \begin{array}{c} R^1 \quad R^2 \\ Cl \quad TeCl_3 \end{array} \quad (18)$$

$$ArTeCl_3 + R\text{-C≡C-H} \xrightarrow{\text{benzene, reflux}} \begin{array}{c} R \quad H \\ Cl \quad ArTeCl_2 \end{array} \quad (19)$$

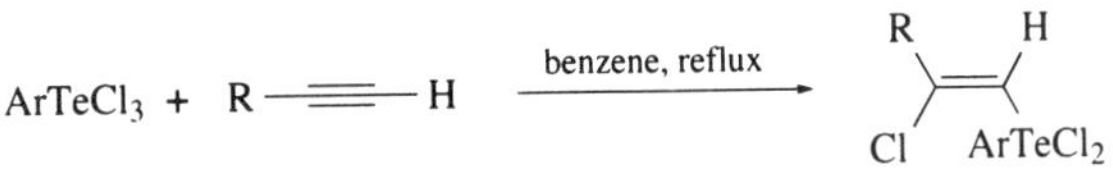

Aryltellurinic anhydrides are prepared by alkaline hydrolysis of the corresponding trichlorides and are obtained as stable, colourless, high-melting solids (Scheme 11).[125-7] Mixed anhydrides are generated *in situ* by reaction of tellurinic anhydrides with carboxylic acids (Equation (20)).[128] Tellurinic anhydrides have attracted attention recently as synthetic reagents (see Sections 14.4.4 and 14.4.5).

$$ArTeX_3 \xrightarrow[\text{ii, H}^+]{\text{i, H}_2\text{O, OH}^-} [ArTe(O)OH] \longrightarrow (ArTeO)_2O$$

Scheme 11

$$(ArTeO)_2O + 2\,RCO_2H \longrightarrow 2\,ArTe(O)OCOR + H_2O \qquad (20)$$

Diaryl telluroxides, Ar_2TeO, recognized recently as useful and selective oxidizing reagents for several functionalities (see Section 14.4.4), can be prepared by two main methods, namely by hydrolysis of the parent dihalides Ar_2TeX_2,[129,130] or by oxidation of diaryl tellurides.[62,131]

14.4 TELLURIUM REAGENTS IN ORGANIC SYNTHESIS

14.4.1 Reductions

Inorganic reagents, such as H_2Te, $NaHTe$ and Na_2Te, and organic derivatives, such as aryltellurols and diorganotellurides, are widely employed for the reduction of different functional groups and for the reductive cleavage of carbon–heteroatom bonds. In many cases, the use of tellurium reagents is clearly advantageous compared with conventional methods because of the mildness of the conditions employed, the high yields obtained and the selectivity possible, coupled with the possibility of generating the reagent *in situ* and the ease of recovery of tellurium materials.

Carbonyl compounds are reduced to the corresponding alcohols by H_2Te[20,132] and PhTeH,[133,134] and the reagents are generated *in situ* (Equation (21) and Scheme 12). Aliphatic ketones are hardly reduced at all under the conditions of method C, but in the presence of zinc iodide alkyl ethers are formed in high yields. Aldehydes and aromatic ketones react similarly.[135] The system $Bu^i_2Te/TiCl_4$, which generates a tellurium(III) species (diorganotellurides are known to reduce metal salts[136]), has also been employed for the reduction of carbonyl compounds, (Equation (22) and Scheme 13).[137a,b,7]

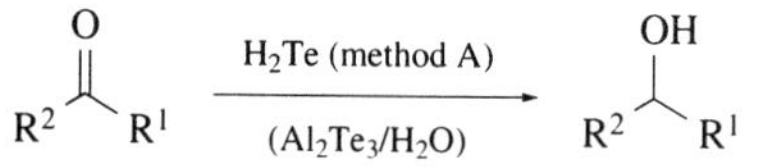

$$\underset{R^2}{\overset{O}{\|}}\!\!\!-\!\!R^1 \xrightarrow[(Al_2Te_3/H_2O)]{H_2Te\;(\text{method A})} \underset{R^2}{\overset{OH}{|}}\!\!\!-\!\!R^1 \qquad (21)$$

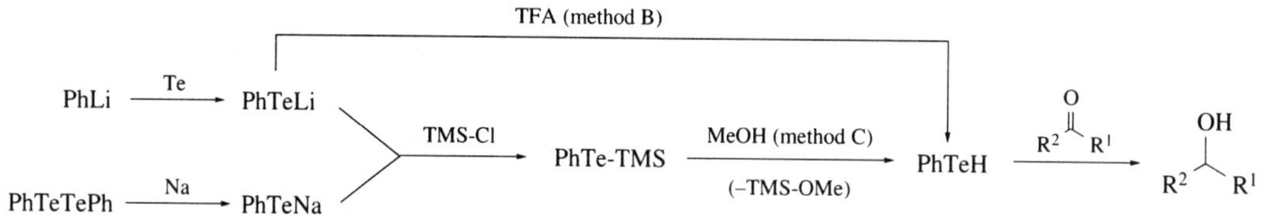

Scheme 12

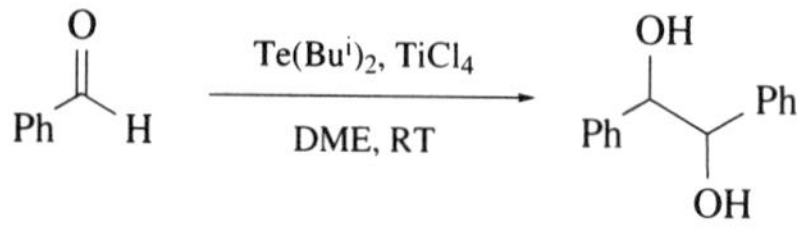

$$\underset{Ph}{\overset{O}{\|}}\!\!\!-\!\!H \xrightarrow[\text{DME, RT}]{Te(Bu^i)_2,\;TiCl_4} \underset{Ph}{\overset{OH}{|}}\!\!\!\underset{OH}{\overset{|}{-}}\!\!\!Ph \qquad (22)$$

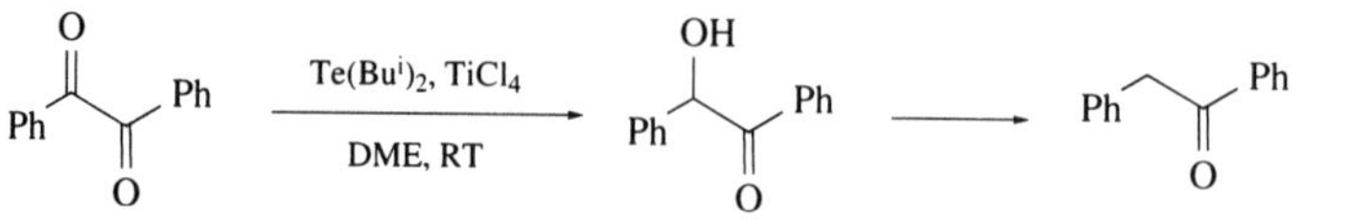

Scheme 13

Regioselective reduction of the carbon–carbon double bond of α,β-unsaturated carbonyl compounds can be carried out with either H_2Te,[20] PhTeH (prepared by protonolysis of PhTeLi[133]), or NaHTe

(Equation (23)).[133,138–40] If PhTeH, generated by the methanolysis of PhTe-TMS, is used, both the carbon–carbon double bond and the CO group are reduced.[134] The synthetically important isopropyl derivatives of Meldrum's acid are accessible by a 'one pot' procedure using this method (Equation (24)).[141] NaHTe and PhTeH are useful reagents for the selective reduction of many carbon–carbon double and triple bonds conjugated to aromatic systems (Equations (25) and (26)),[133,134,138] although stilbene and β-methylstyrene are almost inert to these reagents.

$$R^2\text{-CH=CH-C(O)}R^1 \xrightarrow{\text{H}_2\text{Te or PhTeH or NaHTe}} R^2\text{-CH}_2\text{-CH}_2\text{-C(O)}R^1 \quad (23)$$

$$R^2\text{-C(O)}R^1 + \text{Meldrum's acid} \xrightarrow{\text{NaHTe, AcOH, piperidine}} \text{product} \quad (24)$$

$$\text{Ar-CH=CH-R} \xrightarrow{\text{NaHTe or PhTeH}} \text{Ar-CH}_2\text{-CH}_2\text{-R} \quad (25)$$

$$\text{Ph-C}\equiv\text{C-H} \xrightarrow{\text{PhTeH}} \text{Ph-CH=CH}_2 \quad (26)$$

Imines are reduced to secondary amines by H_2Te[21,132], NaHTe[133,142] and PhTeH[133] (Equation (27)). A useful extension of these reductions is the one-step conversion of carbonyl compounds into secondary amines, without isolation of the intermediate imine, by reductive alkylation of amines with carbonyl compounds (Equation (28)).[21,143] Enamines are reduced in high yield by H_2Te (Equation (29)).[21]

$$R^2\text{-CH=N-}R^2 \xrightarrow{\text{H}_2\text{Te or NaHTe or PhTeH}} R^2\text{-CH(H)-NH-}R^1 \quad (27)$$

$$R^2\text{-C(O)}R^1 + R^3NH_2 \text{ (or } R^3_2NH) \xrightarrow{\text{H}_2\text{Te or NaHTe}} R^2R^1\text{CH-NHR}^3 \quad (28)$$

$$\text{pyrrolidine-cyclohexenyl} \xrightarrow{\text{H}_2\text{Te, 85\%}} \text{pyrrolidine-cyclohexyl} \quad (29)$$

Aromatic thioketones are reduced by Na_2Te to the corresponding hydrocarbons.[145] The success of the reaction is dependent on the use of aprotic conditions.

Various types of nitroarene reduction product can be obtained by the use of tellurium reagents. The product is dependent upon the nature of the reagents and the conditions employed. Scheme 14 summarizes the results obtained with H_2Te, Na_2Te and NaHTe. H_2Te[22,132] and Na_2Te in protic solvents[144] easily reduce nitroarenes to anilines, whereas Na_2Te in aprotic solvents shows milder reducing properties and the reduction proceeds to the azo compound stage.[145] NaHTe reduces unhindered nitroarenes to azoxy derivatives, whereas sterically hindered compounds are reduced to anilines.[133,146] Under alkaline conditions, however, NaHTe reduces nitrobenzene to N-phenylhydroxylamine.[147] N-Arylhydroxylamines are also obtained in a more practical procedure employing $NaBH_4$ and catalytic tellurium.[147] Nitroalkanes and dinitroalkanes are reduced by NaHTe to diazenes (dimers of nitrosoalkenes) and alkenes.[146]

Phenyltellurol generated by methods B and C (see Scheme 12) or by reduction of diphenyl ditelluride with $NaBH_4$ reduces nitrobenzenes to anilines (Scheme 15). The complex $Na^+[PhTeB(OEt)_3]^-$ was proposed as the effective reagent (a milder reducing agent than the free PhTeH) when the PhTeH was generated from $NaBH_4$/benzene/ethanol/H_2O.[148] This same method can also be carried out catalytically in ditelluride.[149] Treatment of diphenyl ditelluride with $NaBH_4$ in the presence of NaOH produces the tellurolate PhTeNa, a reducing agent even milder than the complex borate. It reduces nitroarenes in either stoichiometric or catalytical processes to azoxy or to azo compounds, depending on whether the reaction is performed at room temperature or under reflux.[148] The yields are medium to high (Scheme 15). An additional method for the reduction of nitrobenzene to aniline uses the previously described $Bu^i_2Te/TiCl_4$ system.[137a,b,38]

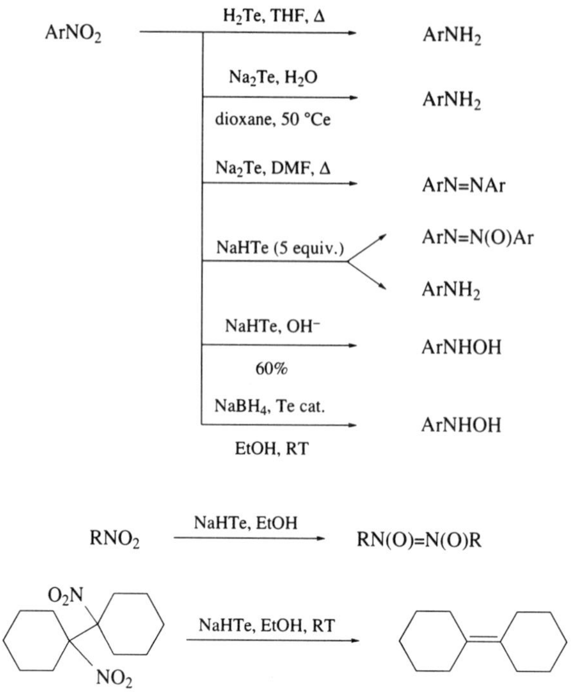

Scheme 14

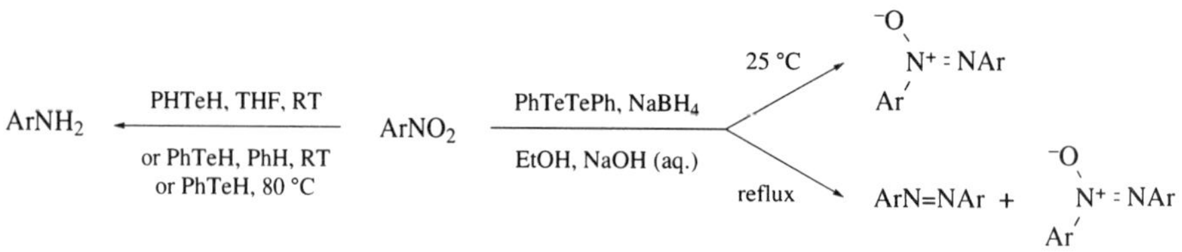

ArNO$_2$ →(H$_2$Te, THF, Δ)→ ArNH$_2$

ArNO$_2$ →(Na$_2$Te, H$_2$O, dioxane, 50 °Ce)→ ArNH$_2$

ArNO$_2$ →(Na$_2$Te, DMF, Δ)→ ArN=NAr

ArNO$_2$ →(NaHTe (5 equiv.))→ ArN=N(O)Ar / ArNH$_2$

ArNO$_2$ →(NaHTe, OH$^-$, 60%)→ ArNHOH

ArNO$_2$ →(NaBH$_4$, Te cat., EtOH, RT)→ ArNHOH

RNO$_2$ →(NaHTe, EtOH)→ RN(O)=N(O)R

[Dinitrodicyclohexyl compound] →(NaHTe, EtOH, RT)→ [dicyclohexylidene]

Scheme 14

Scheme 15

ArNH$_2$ ←(PHTeH, THF, RT or PhTeH, PhH, RT or PhTeH, 80 °C)— ArNO$_2$ —(PhTeTePh, NaBH$_4$, EtOH, NaOH (aq.))→

- 25 °C: Ar–N$^+$(O$^-$)=NAr
- reflux: ArN=NAr + Ar–N$^+$(O$^-$)=NAr

Scheme 15

Tellurium reagents can also be used for the reduction of other nitrogen compounds. Hydrogen telluride reduces arylhydroxylamines to anilines, and nitroso, azo and azoxy compounds to the corresponding hydrazo compounds.[20,132] Further reducing agents of azo to hydrazo compounds are aryl tellurols and NaHTe.[133] NaHTe (generated from NaBH$_4$ and catalytic tellurium) reduces amine N-oxides to the parent amines[150] and azides to primary amines.[151]

Deoxygenation of oxiranes to alkenes[152] is achieved with alkali metal O,O-diethylphosphorotellurolates (Equation (30)).[153,154] The reagent can be generated in situ from (EtO)$_2$PONa and catalytic tellurium. Terminal oxiranes are the most reactive, and (Z)-isomers more reactive than (E)-isomers. The reaction is stereospecific, (Z)-oxiranes giving (Z)-alkenes.

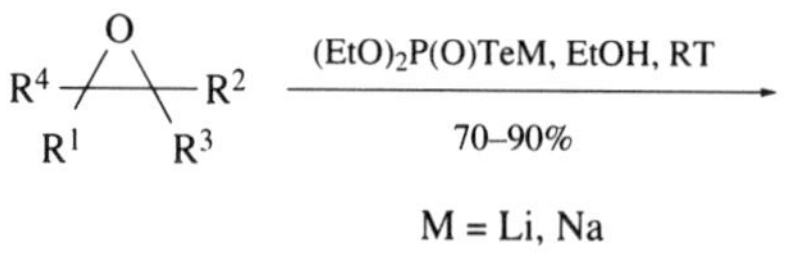

$$\text{R}^4\text{R}^1\text{C}-\text{O}-\text{CR}^2\text{R}^3 \xrightarrow[70-90\%]{(\text{EtO})_2\text{P(O)TeM, EtOH, RT}} \text{R}^4\text{R}^1\text{C}=\text{CR}^2\text{R}^3 \quad (30)$$

M = Li, Na

Treatment of oxiranes with NaHTe gives telluroalcohols, which can be easily converted into the corresponding alcohols and ketones by sequential detelluration with nickel boride and oxidation (path (A)) or to alkenes via the corresponding tosylates (path (B), Scheme 16).[155] α,β-Epoxyketones are simply reduced to β-hydroxyketones by treatment with NaHTe (Scheme 17).[156] Epoxides bearing a leaving group in a suitable position, such as chloromethyl epoxides, react with Na$_2$Te (prepared by the Rongalite method) to give allylic alcohols via an unstable epitelluride (Scheme 18).[157] Bischloromethylcarbinols are converted directly to allylic alcohols (Equation (31)).[157]

A combination of the method described above for the preparation of allylic alcohols with the Sharpless kinetic resolution (SKR) of s-allylic alcohols (which like all resolutions is limited to a theoretical yield of 50% of one enantiomer) provides a procedure for the conversion of the original

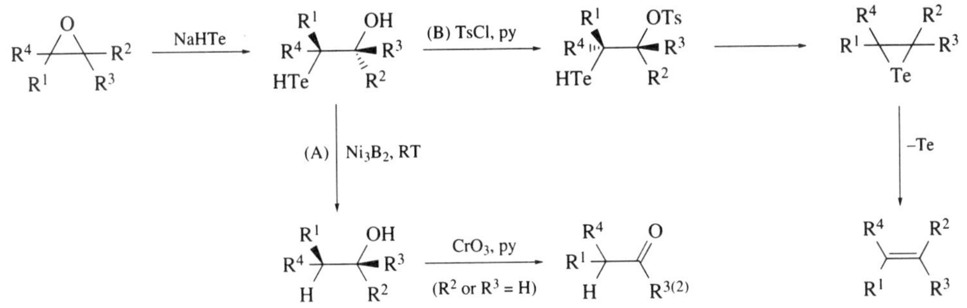

Scheme 16

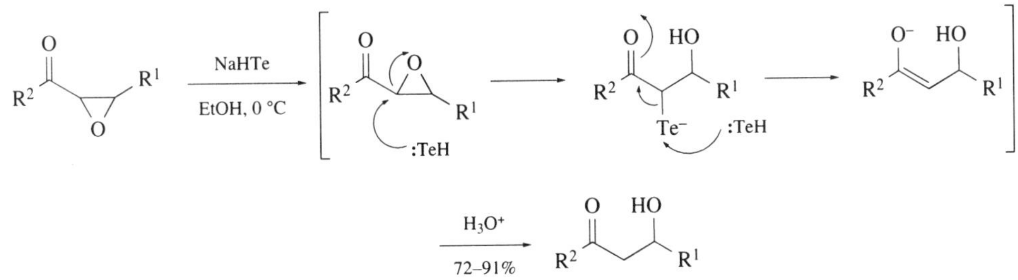

Scheme 17

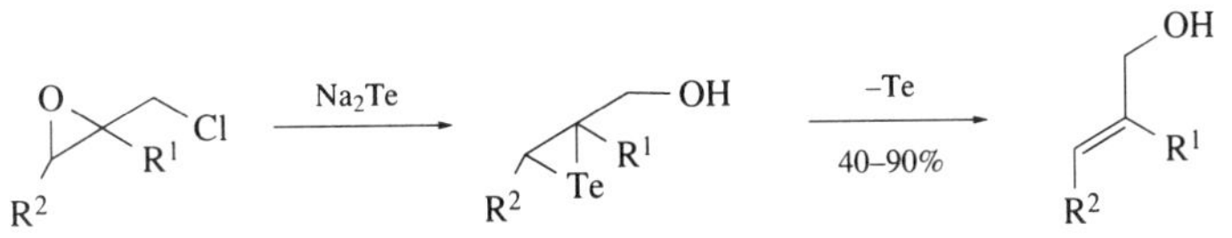

Scheme 18

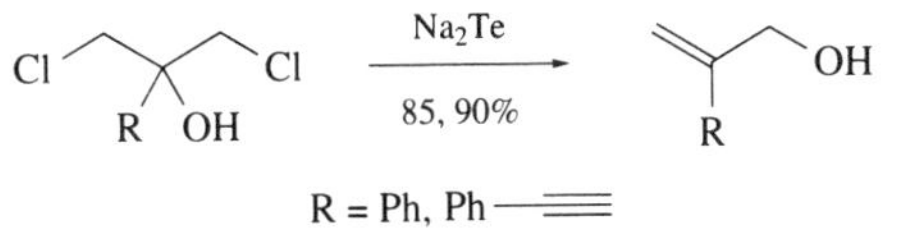

$$R = Ph, Ph-\!\!\equiv\!\!\!-\quad\quad\quad\quad\quad\quad\quad\quad (31)$$

racemic mixture to a single enantiomer with a 100% theoretical yield.[158] The method is based on treatment of the mesylate of the glycidol resulting from the SKR epoxidation with Na$_2$Te, which gives the allylic alcohol with the same configuration as the unreacted enantiomer in the SKR (Scheme 19). As part of a recent systematic investigation of the process, a wide range of substrates, such as 1,1-, 1,2- and 1,3-(*cis* and *trans*) disubstituted allylic alcohols have been submitted to a similar conversion. One of the most useful results is the obtention of sterically crowded 1,3-*cis*-disubstituted *s*-allylic alcohols from the *trans*-alcohols, which overcomes the known limitation of *cis*-*s*-allylic alcohols to undergo the SKR (Scheme 20).[159,160] The transposition of the carbon–carbon double bond and the alcohol functionality is highly stereospecific, *erythro*- and *threo*-glycidol sulfonates giving, respectively, *cis*- and *trans*-allylic alcohols (Equation (32)).

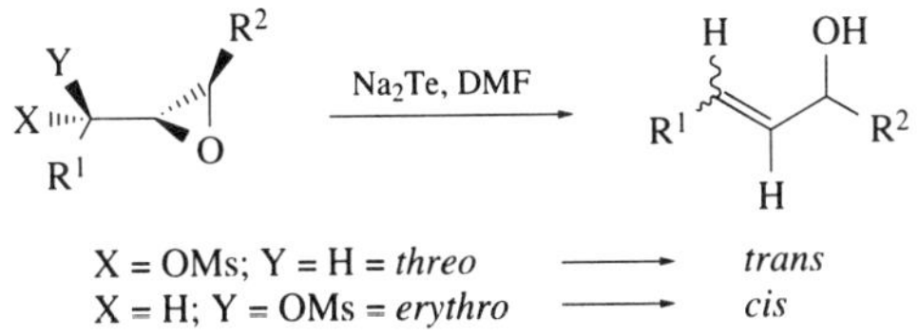

$$X = OMs; Y = H = threo \longrightarrow trans$$
$$X = H; Y = OMs = erythro \longrightarrow cis \quad\quad (32)$$

The debromination of *vic*-dibromides to alkenes can be carried out with several different types of tellurium reagent. In accordance with a typical E2 type of elimination, alkenes with (*Z*) and (*E*) geometry are formed, respectively, from *threo*- and *erythro*-dibromides. These methods can be especially advantageous over the known conventional procedures,[161] because of the mildness of the experimental conditions, the good to quantitative yields, the lack of side reactions and the inertness of

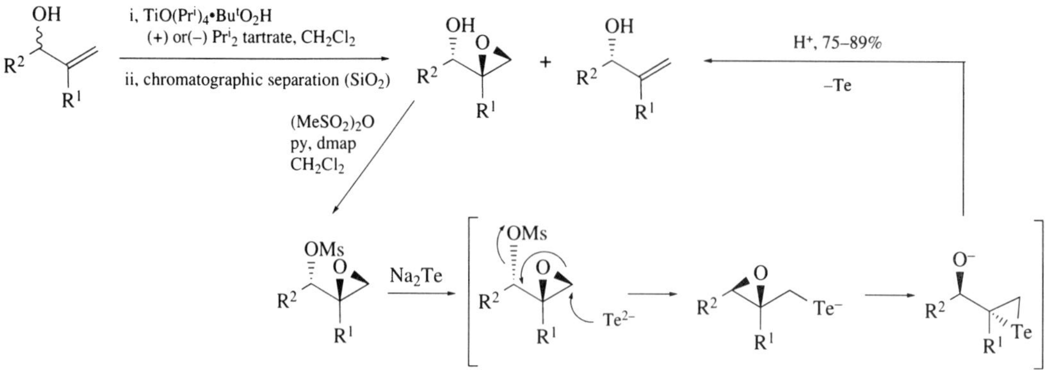

Scheme 19

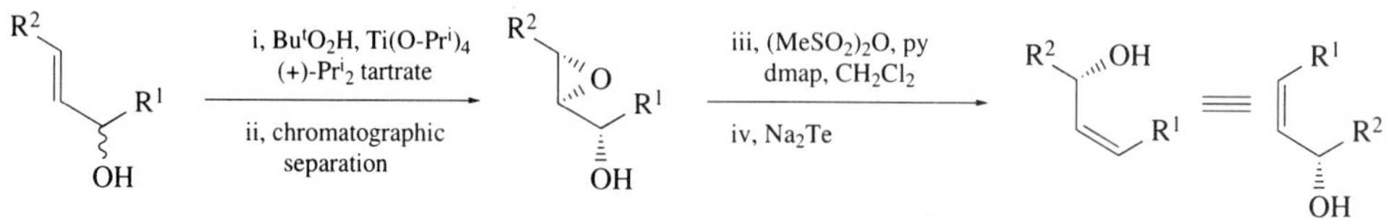

Scheme 20

certain functional groups to the tellurium reagents (Scheme 21).[162-9] Important features of methods A[162] and D[138,165] are the recovery of Ar_2TeBr_2 and elemental tellurium, respectively, and these can be recycled. Method C[164] is attractive because of the easy, one-step preparation of the starting ditelluride. The reaction can be rationalized as proceeding via a tellurenyl halide intermediate. The tellurolate debromination by method G[168] has been applied to the preparation of conjugated dienes from 1,2,3,4-tetrabromoalkanes (and cycloalkanes), 1,4-dibromo-2-alkenes and allylic dibromides (Scheme 22).[170] An interesting related reaction is the debromination of α,α'-dibromoxylene to o-quinodimethane.[171]

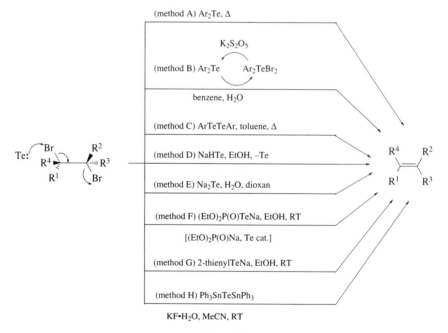

Scheme 21

Tellurium reagents are also involved in the reductive fission of carbon–heteroatom bonds. These reactions are very important transformations in organic synthesis.[172] α-Haloketones are reduced to the parent ketones by NaHTe,[173] $(EtO)_2P(O)TeNa$,[174] alkali metal 2-thienyltellurolates[175] and bistriphenylstannyl telluride/KF.[176] 2-Thienyltellurolates can also remove several different types of substituent such as acetoxy, mesyloxy, phenylthio and 2-thienyltelluro groups (Equation (33)). The same reagents can be used for the reductive dehalogenation of bromoacetanilide and α-haloacids such as α-bromophenyl-, α-bromo-1-naphthyl- and α-chlorodiphenylacetic acid.[175]

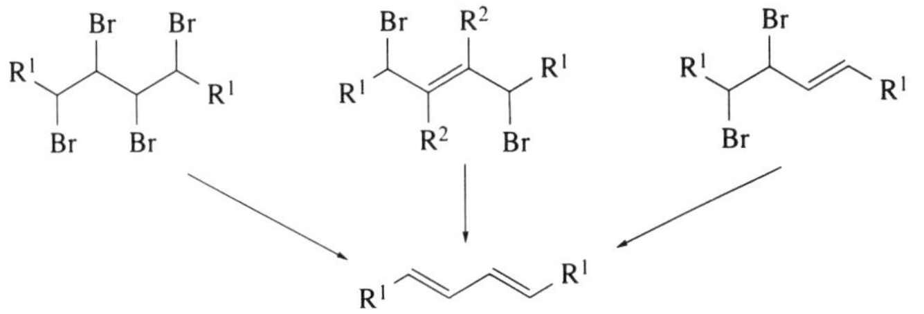

Tellurium

Scheme 22

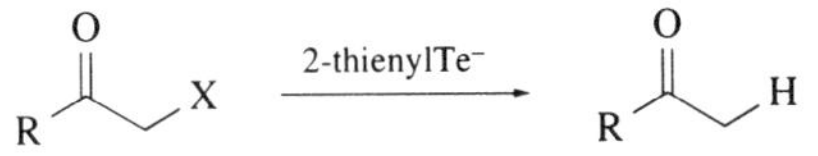

$$R\text{-COCH}_2\text{X} \xrightarrow{\text{2-thienylTe}^-} R\text{-COCH}_2\text{H} \quad (33)$$

R = Ph; X = OAc, OSO$_2$Me, SPh: R = PhNH; X = 2-thienylTe

The nitro group is reductively removed from tertiary nitroalkanes by NaHTe.[177] The success of this reaction is dependent on the presence of an electron-attracting group at the α- or β-position to the nitro group. α-Nitrocumene derivatives undergo similar reductive fission. NaHTe also dealkylates quaternary ammonium salts to give tertiary amines in high yields.[155]

β-Ketosulfones are readily desulfonylated by NaHTe in aprotic solvent[140] (they are inert in ethanol), or under phase-transfer catalysis (Equation (34)).[178] This reductive desulfonylation combined with the reduction of α,β-unsaturated carbonyl compounds constitutes a direct 'one pot' synthesis of unsymmetrical ketones starting from the easily accessible alkylidene β-ketosulfones (Scheme 23).[140] α-Alkylidene-β-cyanosulfones are desulfonylated by NaHTe in protic solvent (Equation (35)).[179] The same vinylic cyanides can be obtained by a Knovenagel-type reaction between aldehydes and β-cyanosulfones, promoted by Na$_2$Te acting as base. The NaHTe formed during the reaction is responsible for the desulfonylation.[179] Treatment of 1-nitro-1-phenylsulfonylcyclohexane with NaHTe/EtOH results in quantitative preferential displacement of the sulfonyl group.[177]

Diaryl thioketals and bissulfenylated α-dicarbonyl compounds are converted into sulfides by Na$_2$Te in aprotic solvents.[180]

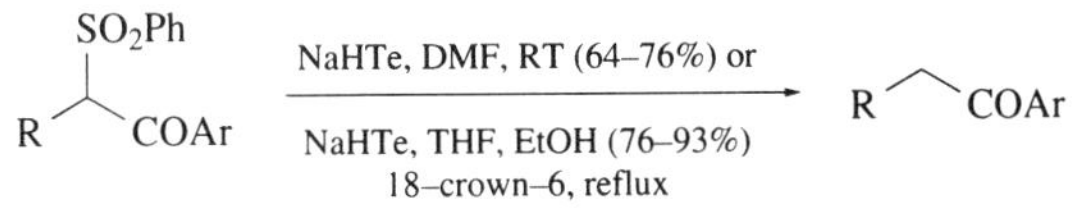

$$\text{R-CH(SO}_2\text{Ph)-COAr} \xrightarrow[\text{18-crown-6, reflux}]{\substack{\text{NaHTe, DMF, RT (64–76\%) or} \\ \text{NaHTe, THF, EtOH (76–93\%)}}} \text{R-CH}_2\text{-COAr} \quad (34)$$

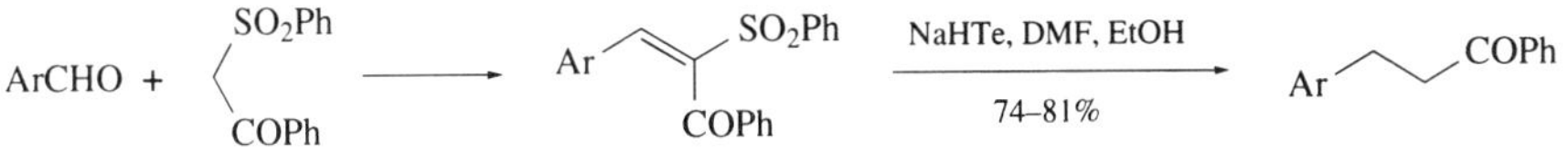

Scheme 23

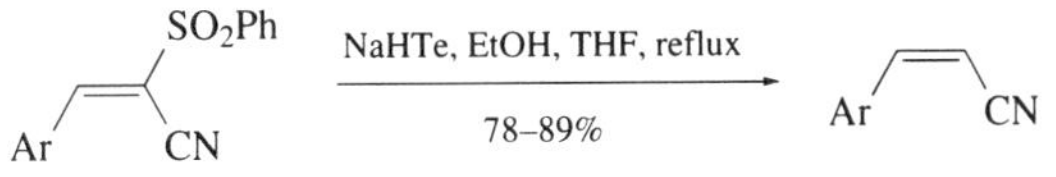

$$\text{ArCHO} + \text{CH(SO}_2\text{Ph)(COPh)} \longrightarrow \text{Ar-C(SO}_2\text{Ph)=CH-COPh} \xrightarrow[74-81\%]{\text{NaHTe, DMF, EtOH}} \text{Ar-CH}_2\text{CH}_2\text{-COPh}$$

$$\text{Ar-C(SO}_2\text{Ph)=CH-CN} \xrightarrow[78-89\%]{\text{NaHTe, EtOH, THF, reflux}} \text{Ar-CH=CH-CN} \quad (35)$$

14.4.2 Tellurium-mediated Formation of Anionic Species and Their Reactions with Electrophiles

On treatment with Na$_2$Te in aprotic solvents, haloacetic esters and haloacetonitriles react with aldehydes to give α,β-unsaturated compounds (Equation (36)).[181] The reaction involves the intermediacy of an enolate formed by attack of the telluride ion at the halogen atom.

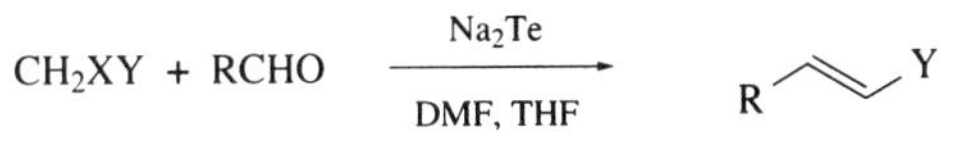

$$\text{CH}_2\text{XY} + \text{RCHO} \xrightarrow[\text{DMF, THF}]{\text{Na}_2\text{Te}} \text{R-CH=CH-Y} \quad (36)$$

X = Cl or Br; Y = CO$_2$R or CN

Alkylidene and cyclopropane malonates can be prepared by reaction of aldehydes and α,β-unsaturated carbonyl compounds with dibromomalonates promoted by Bu$_2$Te. As shown in Scheme

(24), these reactions involve the attack of a bromomalonate anion at either (a) the carbonyl group or (b) the β-carbon, followed by either an elimination or a cyclopropanation step.[182]

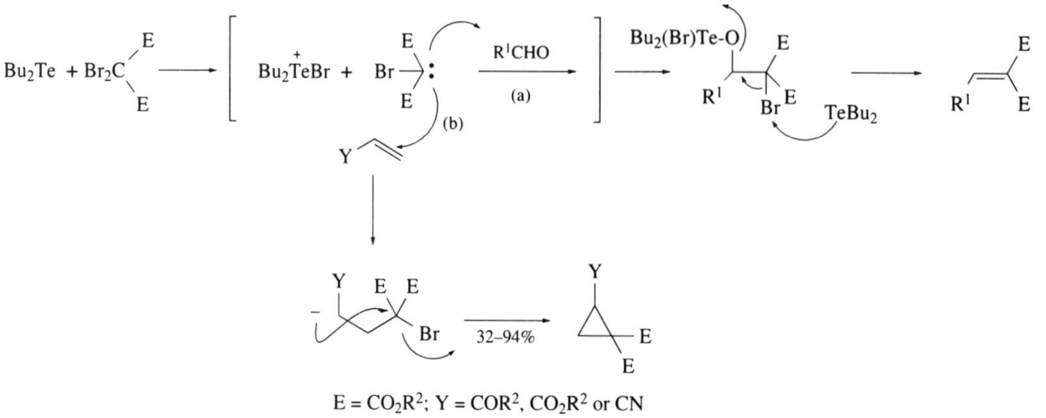

Scheme 24

$E = CO_2R^2$; $Y = COR^2$, CO_2R^2 or CN

Treatment of diphenyl disulfide and arylsulfonyl chlorides with Na_2Te gives thiolate and sulfinate anions. These react *in situ* with α-halocarbonyl compounds and alkyl halides to give, respectively, α-thiocarbonyl compounds[183] and sulfones[184] (Scheme 25). In the case of arylsulfonyl chlorides, if diethylphosphorotellurolate is employed instead of Na_2Te, the subsequent alkylation can be carried out successfully even with halides carrying an α-electron-withdrawing group.[185]

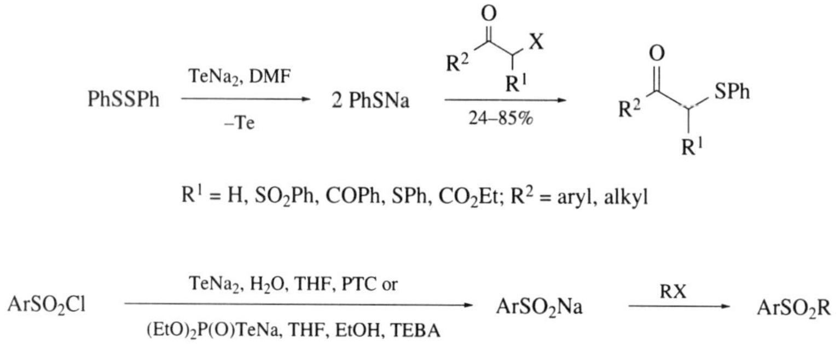

$R^1 = H$, SO_2Ph, $COPh$, SPh, CO_2Et; R^2 = aryl, alkyl

Scheme 25

Aldehyde methylenation can be achieved in the presence of dibutyl telluride. Thus, iodomethyltriphenylphosphonium iodide reacts with aldehydes in a typical Wittig alkenation via generation of methylenephosphorane (Scheme 26).[186]

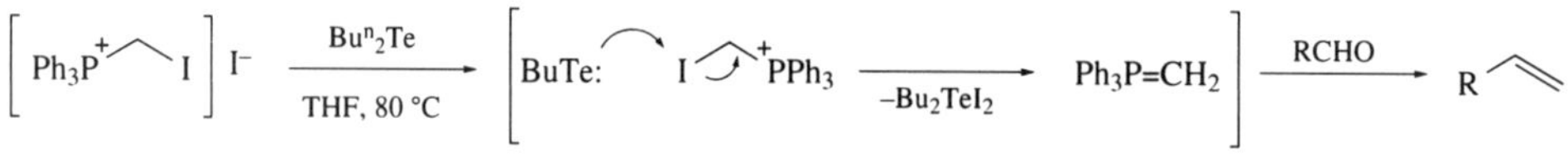

Scheme 26

14.4.3 Deprotection of Organic Functionalities

Some tellurium anionic reagents are very useful for the deblocking of protected functional groups, a fundamental operation in synthetic methodology.[187] The reactions proceed under mild experimental conditions and in nonhydrolytic media.

NaHTe, Na_2Te and Na_2Te_2 can be used to effect the high-yield regeneration of carboxylic acids both from common alkyl and benzyl esters[188] and from the less familiar phenacyl,[189] allyl[190] and 2-haloethyl carboxylates (Equation (37)).[191,192] In the last case, only catalytic amounts of NaHTe are employed, the reagent being continuously regenerated from the tellurium—which separates during the reaction—and excess $NaBH_4$.[193] Phenols are regenerated from phenyl allyl ethers,[190] aryl carboxylates[194] and carbonates[195] by NaHTe, and from aryl haloacetates by Na_2Te.[191] The last method involves the elimination of ketene from an intermediate enolate. Aryl alkyl ethers are formed by quenching the reaction with alkyl halides.

$$R^2COR^1 \xrightarrow{i, ii} R^2CO_2H \qquad (37)$$

i, NaHTe, Na_2Te or Na_2Te_2 in hot DMF (R^1 = alkyl, benzyl); NaHTe, DMF, RT (R^1 = $PhCOCH_2$);
NaHTe, EtOH, heat (R^1 = allyl); Na_2Te, DMF, RT or NaHTe, EtOH, RT or NaHTe, cat. $NaBH_4$, EtOH (R^1 = $(CH_2)_2Cl(Br)$);
ii, H_3O^+

The regeneration of amines from trichloro-*t*-butyl carbamates is readily achieved with sodium 2-thienyltellurolate. The process is catalytic in tellurium, the reagent being generated from a catalytic amount of the parent ditelluride and $NaBH_4$ (Equation (38)).[196]

$$R_2N-C(=O)-O-C(CH_3)_2-CCl_2Cl \xrightarrow[ii, H_2O]{i, ArTeNa, THF, \Delta} R_2NH + CO_2 + Cl_2C=CMe_2 \qquad (38)$$

14.4.4 Oxidations

Bis(*p*-methoxyphenyl)telluroxide (An_2TeO), an easily accessible (see Section 14.3), stable crystalline compound, has found wide use as a mild and selective oxidizing reagent for several types of functional groups and has the advantage that the reagent can be easily regenerated from the parent telluride produced during the oxidation process.[130,197] Typical oxidation reactions are those of tertiary phosphines to the phosphine oxides, phenylisocyanate to diphenylurea, thiols to disulphides, catechols and hydroquinones to quinones (simple phenols are unreactive), acylhydrazines to acylhydrazides, *N*-phenylhydroxylamine to nitrosobenzene, benzophenone hydrazone to diphenyldiazomethane and glycols to aldehydes (monohydric alcohols are unaffected).[130] Noteworthy are the oxidations of thio- and selenocarbonyl compounds such as thionoesters, dithioesters, selenoesters, xanthates, thionocarbonates, trithiocarbonates, thioketones, thioamides and thioureas to the corresponding oxoanalogues.[130] In the case of thiocarbonyl oxidation, a procedure catalytic in telluroxide has been established, where the reagent is continuously generated *in situ* from the parent telluride via bromination using an excess of 1,2-dibromotetrachloroethane as the halogen source.[198] A useful modification of the An_2TeO reagent is its immobilization on a polymeric resin prepared from *p*-methoxyphenyltellurocyanate and poly-*p*-lithiostyrene.[199]

Sodium tellurite selectively oxidizes thiols to disulfides under phase-transfer conditions at room temperature. The ease of oxidation decreases progressively in the order PhSH ~ $PhCH_2SH > RCH_2SH > R_2CHSH > R_3CSH$, short-chain thiols being more reactive than long-chain thiols.[200] The reagent is inert towards many functional groups, for example, amino, hydroxy, azo, hydrazo, phenols, sulfides, disulfoxides and aldehydes, and to alkenic and alkynic carbon–carbon bonds. Unsymmetrical disulfides can be prepared by this method starting from a mixture of two different thiols.

Arenetellurinic anhydrides (Section 14.3) are oxidizing reagents which are similar in many respects to diaryl telluroxides.[127,201,202] The relative oxidizing ability follows the order: $(2-naphthTeO)_2O > p-An_2TeO > (p-AnTeO)_2O > (p-PhOC_6H_4TeO)_2O$. In contrast to telluroxides, tellurinic anhydrides oxidize benzylic alcohols to carbonyl compounds (bis(*p*-methoxyphenyl)tellurone exhibits similar behaviour[203]) and *N,N'*-diphenylthiourea to *N,N'*-diphenylcarbodiimide. Mixed anhydrides have been recognized as valuable reagents and, in certain cases, appear to be superior to the parent benzenetellurinic anhydrides.[204]

Epoxidation of alkenes can be catalysed by polystyrene-bound tellurinic acid. Anchored aryltellurinic acid, prepared by the hydrolysis of polymeric aryltellurium trichlorides, catalyses the epoxidation of alkenes (in contrast to the free acid, which does not exhibit similar activity).[205] The epoxidation is stereospecific, as demonstrated by the retention of the configuration at the alkenic bond.

The diacetoxylation of alkenes is readily achieved by using the following systems: benzenetellurinic anhydride/HOAc (catalysed by H_2SO_4) (method A),[206] TeO_2/LiBr/HOAc (method B)[207] and $TeCl_4$/LiOAc/HOAc 120 °C (method C)[208] (Equation (39)). In general these reactions exhibit a predominant *syn* stereochemistry. The results have been rationalized for method A as involving the addition of a tellurenyl acetate (formed via reduction of the starting reagent by the ditelluride produced as the reaction proceeds, Scheme 27) to give a telluronium intermediate, followed by two sequential *anti*-acetolyses (the latter susceptible to acid catalysis). The diacetoxylation using TeO_2 or $TeCl_4$ (methods B and C) proceeds via an intermediate *trans* adduct and an S_N2-type acetolysis. Exceptions to the *syn* stereochemical course are found with *trans* alkenes, which follow concomitant *syn* and *anti* additions (in method B), and with cyclopentene, which gives mainly the *trans* adduct (in method C). If the reaction using $TeCl_4$/LiOAc/HOAc is performed at 80 °C in the presence of BF_3·ether, and the mixture treated with $Na_2S_2O_3$, acetoxyalkyl ditellurides are obtained. These can be converted to the corresponding tribromide by addition of bromine, and the stereochemistry of the addition is *anti*, as expected by an attack of the acetate ion at an intermediate telluronium ion (Scheme 28).[208] By submitting the acetoxytellurium tribromide to treatment with HOAc at 120 °C, diacetoxylated products are obtained.

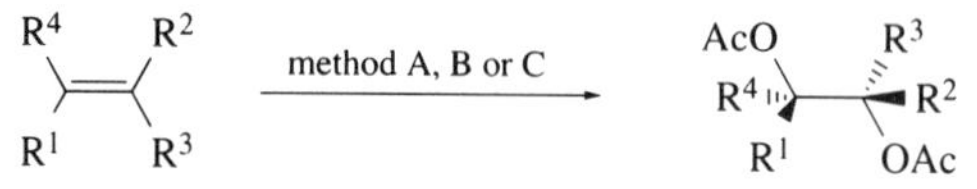

$$\text{(39)}$$

method A, $Ph(TeO)_2$, HOAc, reflux; method B, TeO_2, LiBr, HOAc, reflux; method C, $TeCl_4$, LiOAc, HOAc, 120 °C

Scheme 27

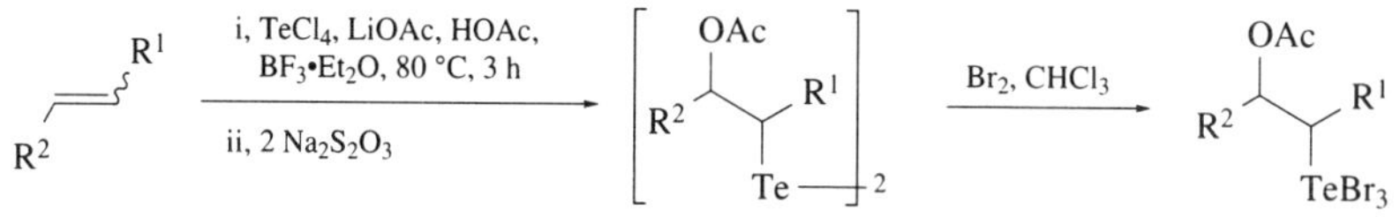

Scheme 28

Conjugated dienes give mixtures of 1,2- and *cis*- and *trans*-1,4-diacetoxylated adducts on treatment with the TeO_2/LiBr/HOAc system,[209] via acetolysis of an intermediate homoallylic and allylic telluroacetate or tellurohalide (Equation (40)). The yield and the isomeric ratio depend on the LiBr/TeO_2 ratio, the 1,4-isomer being favoured by a ratio of 5. The *trans*-isomer is the major product in the 1,4-addition to 1,3-butadiene. The reaction can also be carried out catalytically in TeO_2 if a reoxidant like H_2O_2 or Bu^tO_2H is used. Higher and cyclic dienes give unsatisfactory results.

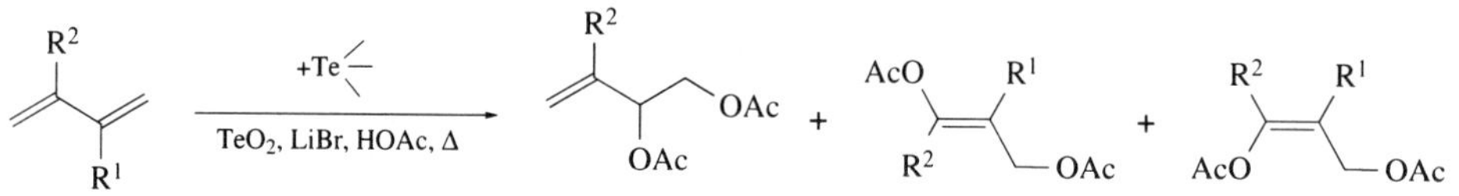

$$\text{(40)}$$

Aliphatic and cycloaliphatic alkenes are converted into methoxytellurenylated products[210] on treatment with equivalent amounts of diphenyl ditelluride and Bu^tO_2H in MeOH in the presence of H_2SO_4. The reaction is regioselective, the tellurium moiety adding exclusively to the terminal carbon of terminal alkenes, and *syn*-stereospecific (*trans*-adduct from cyclohexene). Under identical conditions, aromatic alkenes such as styrene and *p*-methylstyrene give dimethoxylated adducts as the major products. The same results are achieved using a catalytic amount of ditelluride.[210,211]

The *anti*-addition of a carbamate and a tellurinyl moiety to the carbon–carbon double bond of alkenes results from treatment of the alkene with the mixed anhydrides phenyltellurinyl acetate and trifluoroacetate in the presence of excess ethyl carbamate and BF_3·Et_2O.[128,202,212,213] The amidotellurinylation of alkenes can also be carried out without ethyl carbamate if acetonitrile, acting as solvent and nucleophile, is used.[214,215] With terminal alkenes, the electrophilic tellurium residue adds to the terminal carbon. As a result of purification difficulties the tellurinyl products are normally reduced with hydrazine and isolated as tellurides (Scheme 29). Amidoalkenes can be obtained by reaction of the tellurinyl carbamate with base, which results in elimination.[214] If these two amidotellurinylation procedures are carried out at higher temperatures, the carbonyl oxygen or the imidoyl hydroxy group reacts at the rear side of the carbon–tellurium bond to give tellurium-free products, 2-oxazolidinones[128,213] and 2-oxazolines,[214,215] respectively (Scheme 30).

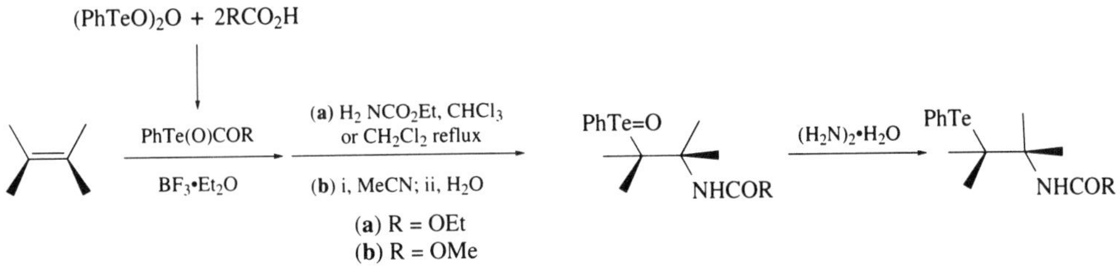

Scheme 29

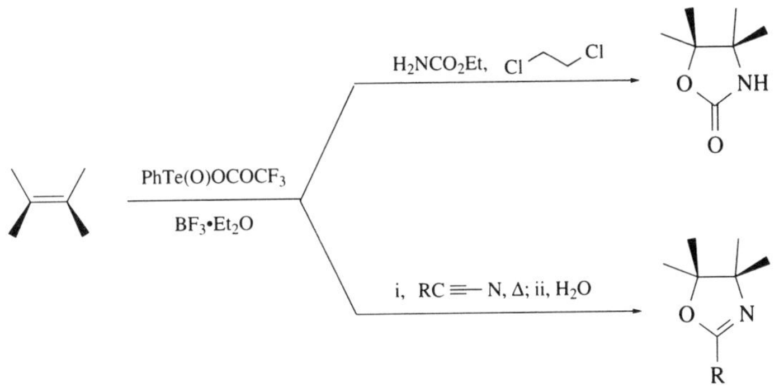

Scheme 30

14.4.5 Organotellurium-based Ring-closure Reactions

The addition of an electrophilic tellurium reagent to alkenes bearing an effective nucleophilic group at a suitable position, followed by intramolecular capture of the resulting telluronium intermediate, results in ring formation. Similar ring-closure reactions are well known with electrophilic selenium reagents.[216] The combination of these ring-closure procedures, named tellurocyclofunctionalizations, with subsequent removal of tellurium by well-established methods, constitutes a useful approach to ring synthesis, often adaptable to natural-product synthesis (Scheme 31).

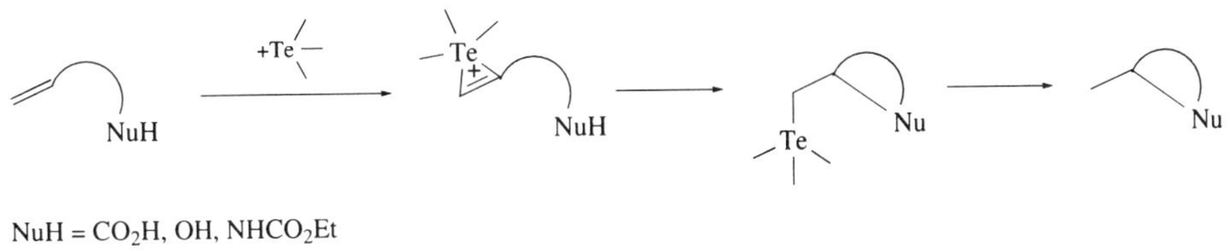

NuH = CO$_2$H, OH, NHCO$_2$Et

Scheme 31

Tellurolactonization of unsaturated carboxylic acids can be achieved with aryltellurium trichlorides and benzenetellurenyl nitrobenzenesulfonate. Thus, aryltellurium trichlorides react with γ,δ-unsaturated carboxylic acids in refluxing chloroform to give aryldichlorotellurium butyrolactones (Scheme 32).[119,217,218] Benzenetellurenyl nitrobenzenesulfonate, generated *in situ* from diphenyl ditelluride and *p*-nitrobenzenesulfonyl peroxide (NBSP), reacts with 4-pentenoic acid to give the corresponding tellurobutyrolactone (Equation (41)).[219] The reductive removal of the tellurium moiety from aryldichlorotellurolactones is readily effected by treatment with tributyltin hydride in refluxing deoxygenated toluene.[220] The dichlorotellurolactone can also be reduced to the parent tellurolactone by treatment with NaBH$_4$,[218] but treatment with excess NaBH$_4$ results in fission of the tellurium–carbon bond and recovery of the starting γ,δ-unsaturated carboxylic acid.[218]

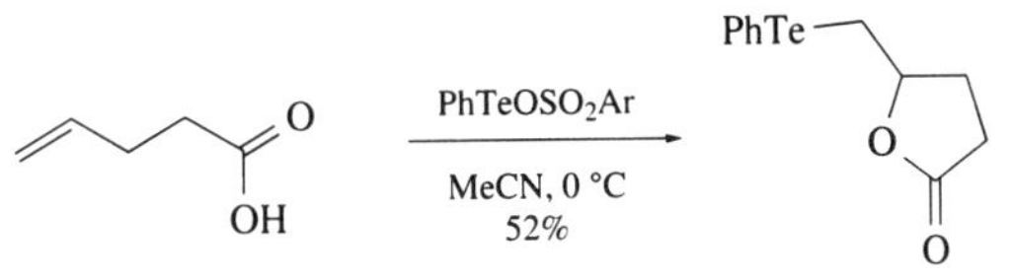

(41)

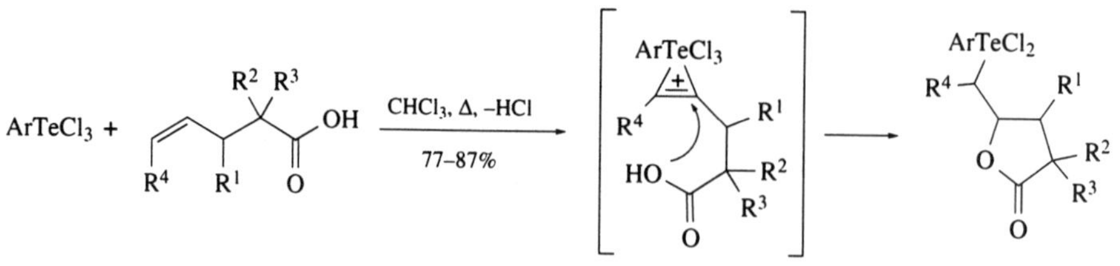

Scheme 32

Cyclotelluroetherification of unsaturated alcohols and allylphenols can be performed with aryltellurium trichlorides, benzenetellurenyl nitrobenzenesulfonate (which also effectively promotes the tellurocyclization of unsaturated acids), aryltellurinyl acetates and with TeO$_2$/HOAc/LiCl or TeO$_2$/HCl. Typical cyclotelluroetherifications with aryltellurium trichlorides are illustrated in Scheme 33.[221] The dichlorotelluroethers are readily reduced to the parent telluroethers by treatment with thiourea dioxide.[220] Aryltellurinyl acetates, a source of electrophilic tellurium, are good reagents for the tellurocyclization of hydroxy alkenes, and the reaction is carried out in HOAc or in chloroform in the presence of BF$_3$·Et$_2$O.[202,222,223] A disadvantage, however, is that the products are hygroscopic and intractable, and *in situ* reduction with hydrazine hydrate is required to obtain the stable tellurides. Allylphenols undergo analogous cyclizations.

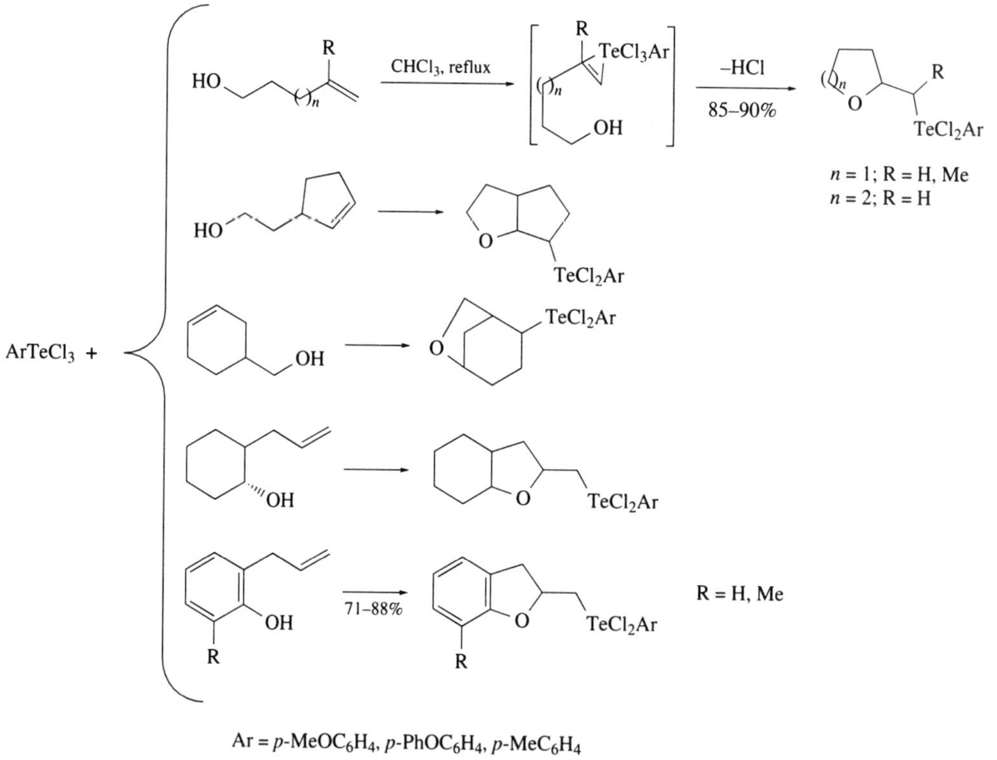

Scheme 33

The TeO$_2$/HOAc/LiCl system reacts with two equivalents γ- or δ-hydroxy alkenes to give five or six-membered cyclic derivatives (Equation (42)).[224] The reaction can be carried out with an equimolar ratio of reagents, using TeO$_2$ in alcoholic aqueous HCl. The trichloride formed is normally reduced without isolation to the corresponding ditelluride.[57] Allylphenols undergo analogous cyclizations.

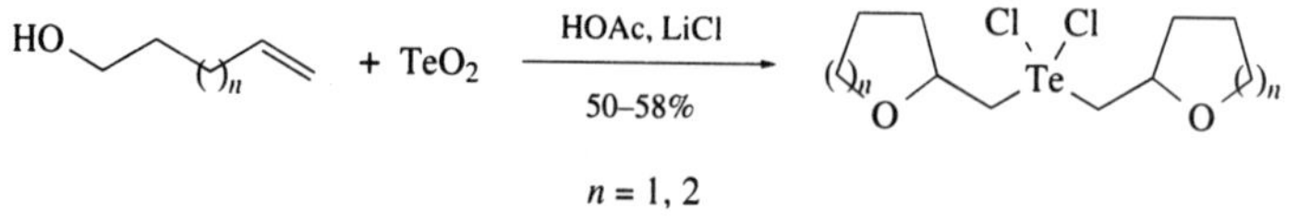

The products obtained from these telluroetherification reactions can be subjected to reductive detelluration with Bu_3SnH,[220,223] or converted into alkene derivatives by elimination of either PhTeH, by treatment with chloramine-T, or PhTeOH (after oxidation) by alkaline treatment.[223] The tellurium residue can also be replaced by bromine, through sequential treatment with bromine and sodium bromide.[223]

Pyrrolidine and piperidine derivatives are obtained by treatment of suitable alkenic carbamates with phenyltellurinyl acetate in the presence of $BF_3 \cdot Et_2O$ (Equations (43) and (44)).[212]

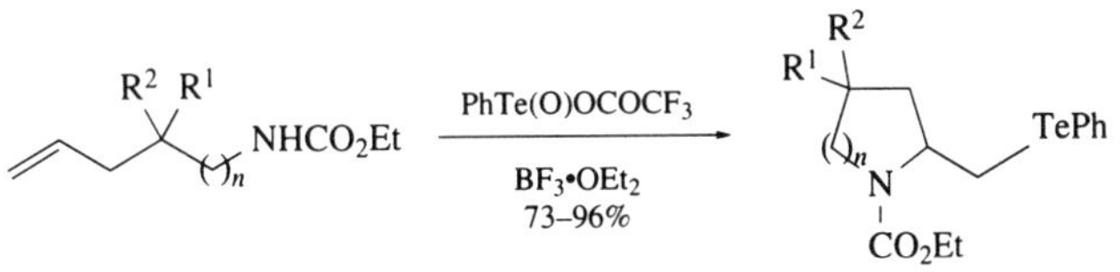

$$\text{alkenyl-NHCO}_2\text{Et} \xrightarrow[\substack{BF_3 \cdot OEt_2 \\ 73-96\%}]{PhTe(O)OCOCF_3} \text{pyrrolidine-TePh} \quad (43)$$

$$\text{o-allyl-aniline-NHCO}_2\text{Et} \xrightarrow[\substack{BF_3 \cdot OEt_2 \\ 87\%}]{PhTe(O)OCOCF_3} \text{indoline-TePh} \quad (44)$$

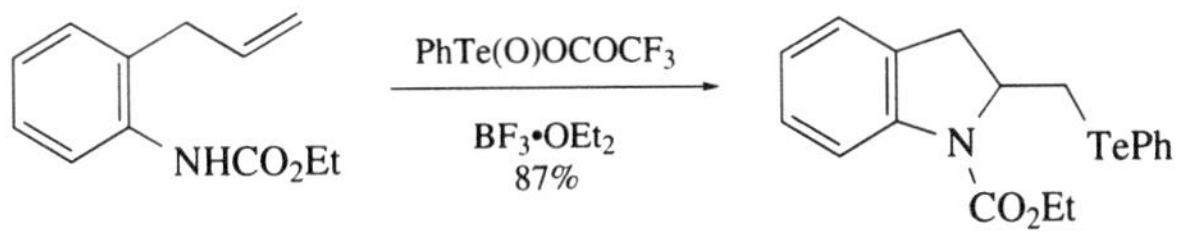

14.4.6 Conversion of Organotellurium Compounds into Tellurium-free Organic Compounds

Detelluration of organotellurium compounds can be carried out with the formation of new C–C bonds (carbodetelluration) or by replacement of the tellurium moiety by other functionalities such as halogens, methoxy groups or hydrogen. Carbodetelluration of diaryltellurium dichlorides[48] and aryltellurium trichlorides[120] with Raney-nickel in an appropriate solvent constitutes a useful route to biaryls (Scheme 34). Aryl alkyl and dialkyl tellurides also undergo carbodetelluration in high yield on treatment with one equivalent of Pd(O) in acetonitrile.[225] Diaryl ditellurides react similarly, but require an additional equivalent of the Pd(O) catalyst. Styryl tellurides and distyryl telluride are detellurated with lithium chloropalladate (2 equiv.), giving stereoisomeric mixtures of the coupled 1,3-butadiene derivatives.[226] Use of palladium acetate as catalyst with distyryl telluride gives styryl acetate.[226]

A number of carbocyclic systems have been obtained by pyrolytic extrusion of tellurium (Equations (45) and (46)).[227,228]

$$ArTe(Cl_2)Ar \xrightarrow[\substack{\Delta, 8\,h \\ 61-91\%}]{Ra-Ni,\ diglyme} Ar-Ar \xleftarrow[\substack{\Delta, 2\,h \\ 98\%}]{Ra-Ni,\ diglyme} ArTeCl_3$$

Scheme 34

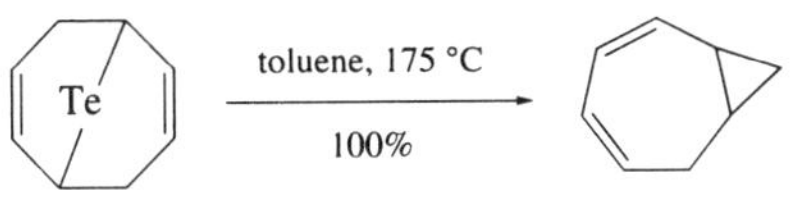

$$\text{Te-bridged bicycle} \xrightarrow[\substack{100\%}]{toluene,\ 175\ ^\circ C} \text{bicyclic alkene} \quad (45)$$

$$\text{R}^1,\text{R}^2\text{-tellurophene} \xrightarrow[\substack{0.5\ torr}]{500\ ^\circ C} \text{R}^1,\text{R}^2\text{-cyclobutene} \quad (46)$$

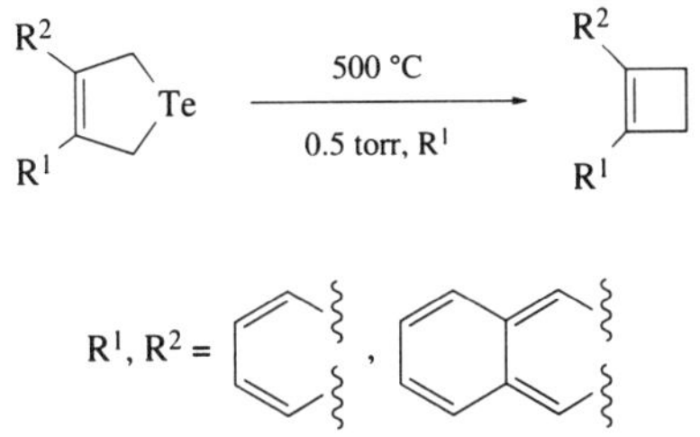

$R^1, R^2 =$ cyclohexadienyl, naphthalenediyl

Alkenes are arylated in poor to moderate yield by treatment with aryltellurium trichlorides or diaryltellurium dichlorides in the presence of $PdCl_2/NaOAc$ (Equation (47)).[229] The reaction can be carried out catalytically in palladium(II) if a suitable oxidant such as Bu^tO_2H or $CuCl_2$ is added to the system. The products show (E) configuration with the exception of acrylonitrile and methacrylonitrile derivatives. In the absence of the alkene, the corresponding biaryls are produced in moderate yields (in analogy with the Raney-Ni method).

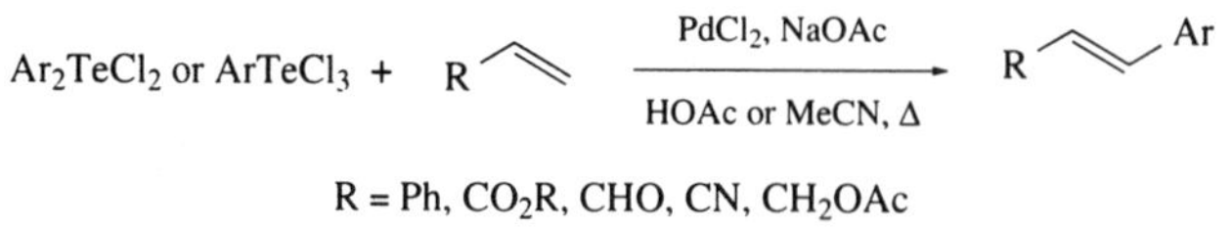

$$Ar_2TeCl_2 \text{ or } ArTeCl_3 + R\diagup\!\!\!\diagdown \xrightarrow[\text{HOAc or MeCN, }\Delta]{PdCl_2, NaOAc} R\diagup\!\!=\!\!\diagdown Ar \qquad (47)$$

R = Ph, CO_2R, CHO, CN, CH_2OAc

Under nickel(II)– or cobalt(II)–phosphine complex catalysis, some organic tellurides react with Grignard reagents to give mixtures of cross-coupled and homo-coupled products in good to moderate yields (Equations (48) and (49)).[97] Under the same conditions, tellurophene gives 1,4-disubstituted (Z)(Z)-1,3-butadienes.[230] Selenophene and thiophene react similarly, but yields are lower.

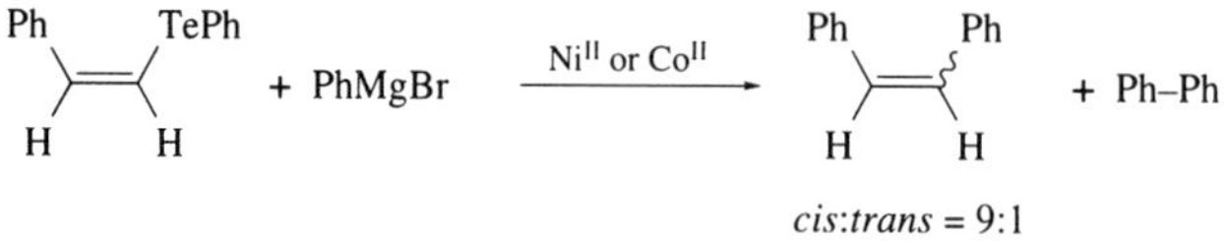

$$\underset{H\qquad H}{\overset{Ph\qquad TePh}{>\!\!=\!\!<}} + PhMgBr \xrightarrow{Ni^{II} \text{ or } Co^{II}} \underset{H\qquad H}{\overset{Ph\qquad Ph}{>\!\!=\!\!<}} + Ph\text{–}Ph \qquad (48)$$

cis:trans = 9:1

$$Ar_2Te + RMgX \xrightarrow{Ni^{II} \text{ or } Co^{II}} Ar\text{–}R + Ar\text{–}Ar + R\text{–}R \qquad (49)$$

The detellurative carbonylation of organotellurium compounds gives carboxylic acids. Benzoic acids are obtained from the reaction of aryltellurium trichlorides and diaryltellurium dichlorides with $Ni(CO)_4$ in DMF.[231] Alkyl aryl tellurides react with CO under palladium(II) catalysis in methanol/triethylamine to give methyl carboxylates.[74,95,96,232] Butenolides are formed from hydroxyvinyl phenyl tellurides.

Organotellurium compounds can be converted into tellurium-free compounds by replacement of the tellurium moiety by other functionalities such as halogens, methoxy groups and hydrogen. For example, aryltellurium trichlorides and diaryltellurium dichlorides bearing a *p*-electron-releasing group, such as methoxy, undergo substitution of the tellurium moiety for iodine or bromine, furnishing the corresponding haloarenes. The substitution by iodine is greatly enhanced by the addition of fluoride ion. In the case of bromine substitution, *o,p*-dibromoderivatives are formed as the main products.[233] An electrophilic mechanism is operative in these reactions, as indicated by dependence on the *p*-electron-donating group, as well as by the rate enhancement by fluoride ions, which form $ArTeCl_3F_2^{2-}$ or $Ar_2TeCl_2F_2^{2-}$, facilitating electrophilic attack of halogens.

Treatment of (Z)-2-chlorovinyltellurium trichlorides, readily prepared by the addition of $TeCl_4$ to phenylalkynes (see Section 14.3), with iodine or NBS–$AlCl_3$ gives the corresponding (Z)-iodo- or (Z)-bromoalkenes.[123] The chlorotelluration/iododetelluration of propargyl alcohol proceeds analogously. This two-step procedure offers, therefore, a useful method for the *syn* iodo- and bromochlorination of alkynes (Scheme 35).

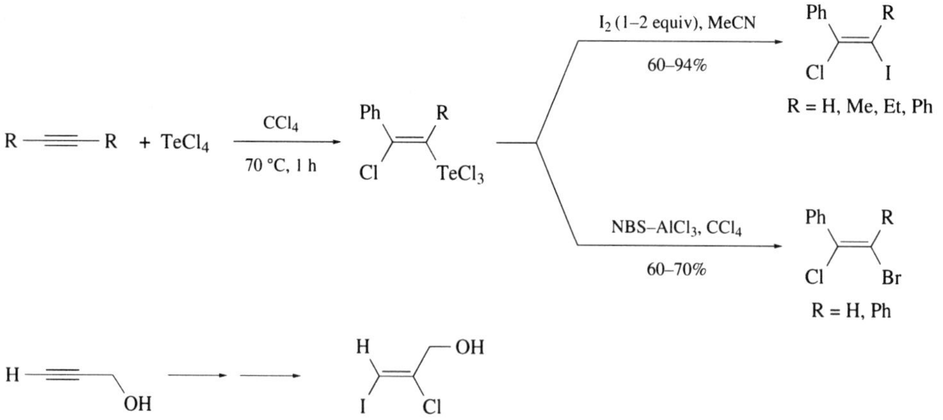

Scheme 35

Organotellurium tri- and dihalides undergo α-elimination and *ipso*-substitution of the tellurium moiety by halogen, as shown in Equations (50)–(52). The oxidative procedure, method A, is based on treatment with oxidants, preferably Bu^tO_2H, in solvents such as dioxane, acetic acid and acetonitrile.[114,234] The photolytic procedure, method B, is based on irradiation in benzene with a high pressure mercury lamp at room temperature, in the presence of atmospheric oxygen.[114,235] The thermal procedure, method C, is where the substrate is either heated at 230–250 °C in a Kugelrohr distillation apparatus (C^1)[114] or heated in DMF at 70–110 °C in the presence of an additional alkali halide (C^2).[236] The following remarks are pertinent. The order of trichloride reactivity assessed for method A is alkyl > aryl > alkenyl and for method B is aryl > alkenyl > alkyl. Retention of configuration is observed in all the reactions involving cyclohexyl and alkenyl derivatives. These oxidative α-elimination reactions have been rationalized as proceeding through a 1,2-tellurium halogen shift. Considering that alkyl phenyl tellurides are easily prepared from alkyl halides and phenyltelluromethyllithium, method C constitutes a homologation of alkyl halides (Scheme 36). 1,1-Dibromoalkanes can be prepared similarly using lithium diphenyltelluromethane (Scheme 37).[236] An alternative iodination method involves treatment of the telluride with methyl iodide and NaI in DMF (Scheme 38).[236]

$$RTeCl_3 \xrightarrow[\text{(see text)}]{\text{method A or B}} RCl \quad (50)$$

R = alkyl, cycloalkyl, vinyl, aryl

$$Ar_2TeCl_2 \xrightarrow[\text{(see text)}]{\text{method B}} ArCl \quad (51)$$

$$RTeX_2Ar \xrightarrow[\text{(see text)}]{\text{method A, C}^1\text{, C}^2} RX \quad (52)$$

R = alkyl, cycloalkyl; X = Cl, Br

$$RX \xrightarrow{PhTeCH_2Li} PhTeCH_2R \xrightarrow{X_2} RCH_2X$$

Scheme 36

$$RX \xrightarrow{(PhTe)_2CHLi} (PhTe)_2CHR \xrightarrow{X_2} RCHBr_2$$

Scheme 37

$$PhTeR + MeI \longrightarrow \begin{bmatrix} Ph \\ {}^+Te-Me \\ R' \end{bmatrix} I^- \xrightarrow[\substack{55\,°C \\ 75–90\%}]{NaI, DMF} RI + PhTeMe$$

Scheme 38

On treatment with equimolar amounts of chlorine in dichloromethane, α-aryldichlorotelluroketones undergo halogenolysis to the corresponding α-chloroketones and aryltellurium trichlorides.[109] Alkynic tellurides, when treated with three equivalents of bromine or iodine, undergo both halogenolysis of the C–Te bond and addition of halogen to the alkynic C–C bond, yielding trihaloalkenes and organotellurium trihalides.[237]

Treatment of alkylphenyltelluroxides with oxidants in alcohols results in replacement of the phenyltellurium moiety by an alkoxy group and formation of the corresponding alkyl ethers. This detellurative alkoxylation can be carried out either directly on the parent tellurides by treatment with excess (3–5 equiv.) mcpba (method A),[238,239] by oxidation of the telluroxides with the same reagent (2 equiv.) (method B),[239] or with trifluoroperacetic acid (method C).[239] Oxidation of tellurones (method D) gives the same results (Scheme 39).[239] Competitive telluroxide syn-elimination has seldom been observed.

When a phenyl group is linked at a vicinal position to tellurium, the replacement of tellurium by a methoxy group is accompanied by phenyl migration (Scheme 40).[238,239] 2-Methoxycycloalkyl and 2-hydroxycycloalkylphenyl tellurides (see Sections 14.3 and 14.4.4), when submitted to the above conditions, are converted into the dimethylacetals of the ring-contracted aldehydes (Scheme 41). Noteworthy is the difference in stability between the cyclic six-membered methoxytelluroxide, which is stable and isolable and undergoes ring contraction on treatment with mcpba, and the seven-membered analogue, which is unstable and undergoes telluroxide elimination. A synthesis of α-arylpropanoic acids, which are important antiinflammatory pharmaceuticals, employs this methodology as a key step (Scheme 42).[240]

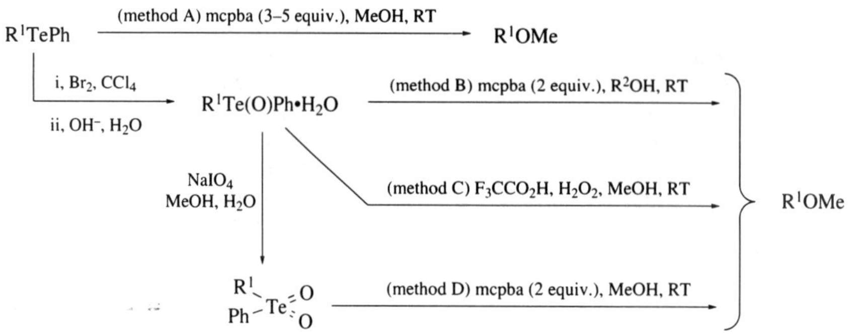

Scheme 39

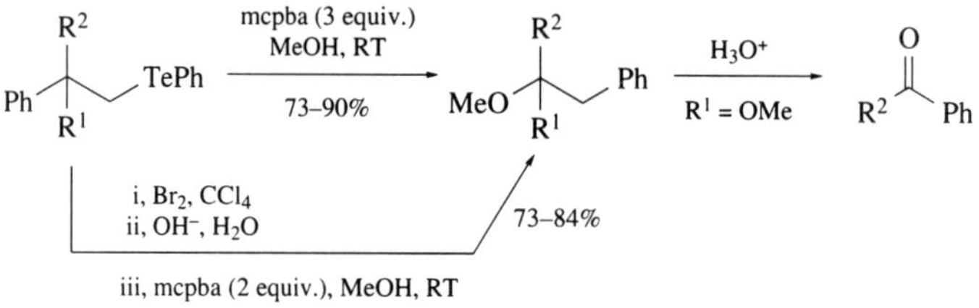

Scheme 40

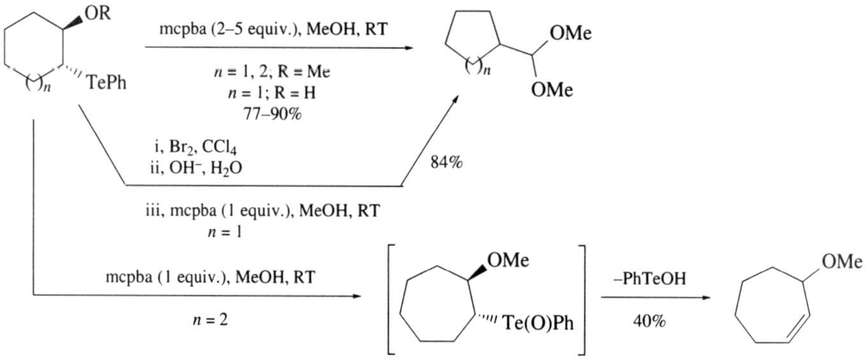

Scheme 41

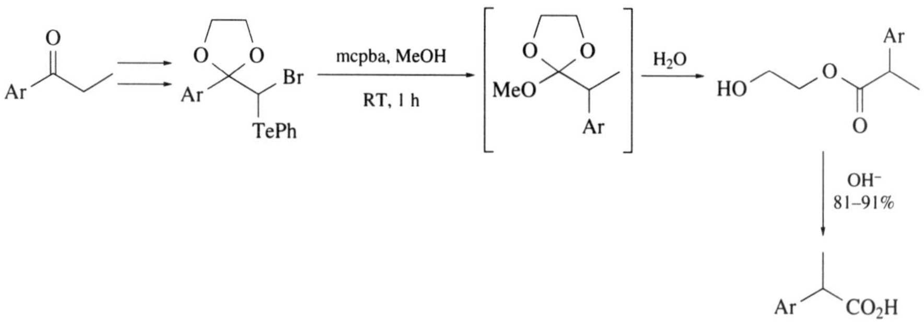

Scheme 42

Phenyl alkyl tellurides are reduced by triphenyltin hydride under mild conditions to the corresponding alkanes in high yields.[77] The corresponding tellurium dichlorides undergo similar reduction; an excess of the reagent is needed, but the reaction is faster. The analogous reductions of selenides require more severe conditions.

14.4.7 Alkene Synthesis

The synthesis of alkenes can be achieved by telluroxide elimination, the reaction of alkyl phenyl tellurides with chloramine-T, from telluronium ylides and by tellurium-catalysed decomposition of α-lithiated benzylic sulfones.

Although not so widely used as the selenoxide *syn*-elimination reaction, the telluroxide elimination (Equation (53)) has recently become more widely accepted for alkene synthesis.[79,241-3] The starting telluroxides are prepared by well-established methods, for example, the hydrolysis of the parent dibromides,[79] or oxidation of the telluride with mcpba.[243] Using the former method, the telluroxides are obtained as hydrates, which normally give higher yields in the elimination. The phenyltellurinic acid produced as the reaction proceeds can be detected, at least partially, as diphenyl ditelluride.

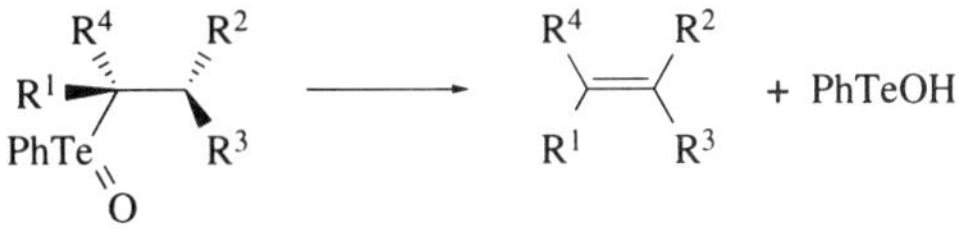

$$R^1 \underset{\underset{O}{\overset{|}{PhTe}}}{\overset{R^4}{\diagup}} \underset{R^3}{\overset{R^2}{\diagdown}} \longrightarrow \underset{R^1}{\overset{R^4}{\diagdown}}=\underset{R^3}{\overset{R^2}{\diagup}} + PhTeOH \qquad (53)$$

The telluroxide elimination reaction, combined with the ease of preparation of alkyl phenyl tellurides and β-hydroxyalkyl phenyl tellurides and with the alkoxytelluration of alkenes, constitutes a straightforward route for the dehydrobromination of alkyl bromides, for the conversion of epoxides into allylic alcohols and for the preparation of allylic and vinylic ethers. There are several important points. (i) *s*-Alkyl phenyl and cycloalkyl phenyl telluroxides (with the exception of cyclohexyl derivatives) undergo the elimination reaction at room temperature, whereas *n*-alkyl phenyl and cyclohexyl phenyl telluroxides are stable compounds and undergo elimination only on prolonged heating in solvent (toluene or THF) or on pyrolysis at 200–40 °C. (ii) In comparison with selenoxide elimination, the telluroxide elimination is much slower and shows greater preference for the less substituted carbon atom. Moreover, the geometry of the alkene formed is markedly dependent upon the amount of oxidant employed. For example, cyclododecyl phenyl telluride gives exclusively the *trans*-alkene with one equivalent of oxidant and a *cis–trans* mixture with excess oxidant.[244]

Treatment of alkyl phenyl tellurides with chloramine-T in refluxing THF gives alkenes in high yields via intermediate tellurosulfimino derivatives (Scheme 43).[245] Vinylsilanes, can be prepared from the readily accessible 1-phenyltelluro-1-trimethylsilylalkanes.[246]

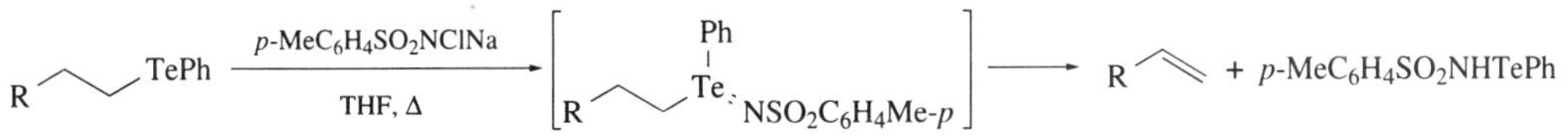

$$R\text{\textbackslash}TePh \xrightarrow[\text{THF, }\Delta]{p\text{-MeC}_6\text{H}_4\text{SO}_2\text{NClNa}} \left[R\text{\textbackslash}\underset{\text{NSO}_2\text{C}_6\text{H}_4\text{Me-}p}{\overset{\overset{Ph}{|}}{Te}} \right] \longrightarrow R\text{\textbackslash} + p\text{-MeC}_6\text{H}_4\text{SO}_2\text{NHTePh}$$

Scheme 43

Dibutyltelluronium methylides stabilized by electron-attracting groups, such as carbethoxy,[247] phenacyl,[248] cyano[248] and carbamoyl,[249] and dibutyltelluronium benzylide[250] (easily prepared by the reaction of dibutyl telluride with the appropriate substituted methyl halide, followed by base treatment), undergo typical Wittig alkenations with a variety of aldehydes and ketones to give the expected alkenes in satisfactory yields (Scheme 44). It must be pointed out that, by contrast, stabilized sulfonium ylides are inert toward carbonyl compounds. Noteworthy features of these tellurium reactions are the high predominance of (*E*)-stereochemistry in the case of aldehydes and the good results obtained even with highly enolizable ketones (cyclopentanone), α,β-epoxyketones (isophorone oxide) and α,β-unsaturated compounds such as benzalacetophenone and cinnamaldehyde. In the case of carbamoyl derivatives (Y = CONHBu^i), phase-transfer catalysis conditions can be employed by simply treating aromatic aldehydes with the telluronium salt (itself behaving as a phase-transfer catalyst in wet THF) in the presence of K_2CO_3 at 50–60 °C.[251] Two simplified high-yielding procedures have been described, exploiting the low energy and high polarity of the Te–C bond.[248,252] In the first, the reaction is performed with the telluronium salts under neutral conditions, instead of the ylides. In the second, the preparation of the telluronium salts and the alkenation step are combined in a 'one-pot' operation. A further procedure is also effective, which uses dibutyl telluride in catalytic amount; addition of triphenyl phosphite reduces the telluroxide formed.[253]

Semi- and nonstabilized telluronium ylides such as di-*i*-butyltelluronium allylide,[254] diphenyltelluronium methylide[255] (the first nonstabilized telluronium ylide) and di-*i*-butyltelluronium trimethylsilylpropynylide[256] react with aldehydes and ketones to give epoxides (in analogy to nonstabilized sulfonium and selenonium ylides) (Equations (54)–(56) TMP = 2,2,6,6-tetramethylpiperidine).

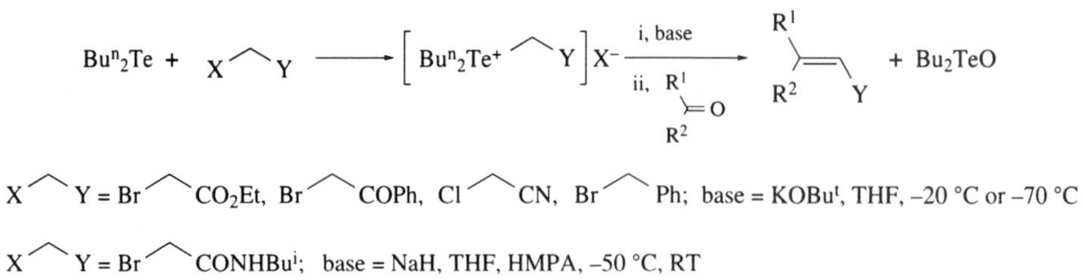

$Bu^n_2Te + X\frown Y \longrightarrow [Bu^n_2Te^+\frown Y]X^- \xrightarrow[\text{ii, } R^1R^2C=O]{\text{i, base}} \begin{array}{c} R^1 \\ \diagdown \\ R^2 \end{array}=\begin{array}{c} \\ \diagup \\ Y \end{array} + Bu_2TeO$

$X\frown Y = Br\frown CO_2Et,\ Br\frown COPh,\ Cl\frown CN,\ Br\frown Ph;\ \text{base} = KOBu^t, THF, -20\ °C\ \text{or}\ -70\ °C$

$X\frown Y = Br\frown CONHBu^i;\ \text{base} = NaH, THF, HMPA, -50\ °C, RT$

Scheme 44

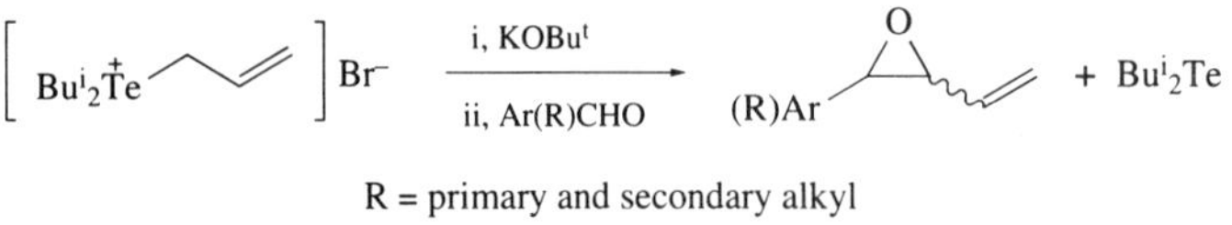

$[Bu^i_2\overset{+}{T}e\frown\hspace{-0.3em}\diagup]Br^- \xrightarrow[\text{ii, Ar(R)CHO}]{\text{i, KOBu}^t} (R)Ar\overset{O}{\triangle}\hspace{-0.3em}\diagup + Bu^i_2Te \quad (54)$

R = primary and secondary alkyl

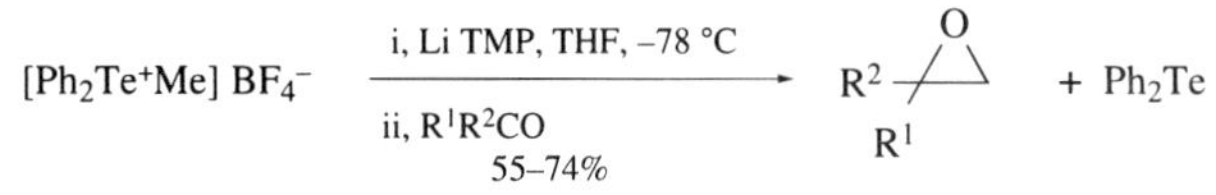

$[Ph_2Te^+Me]\ BF_4^- \xrightarrow[\substack{\text{ii, R}^1R^2CO \\ 55-74\%}]{\text{i, Li TMP, THF, -78 °C}} R^2\underset{R^1}{\overset{O}{\triangle}} + Ph_2Te \quad (55)$

$[Bu^i_2\overset{+}{T}e\diagup\hspace{-0.3em}=\hspace{-0.3em}\text{—TMS}]Br^- \xrightarrow[\substack{\text{ii, R}^1R^2CO \\ 76-96\%}]{\text{i, Li TMP, THF, -78 °C}} R^2\underset{R^1}{\overset{O}{\triangle}}\hspace{-0.3em}\equiv\hspace{-0.3em}\text{TMS} + Bu^i_2Te \quad (56)$

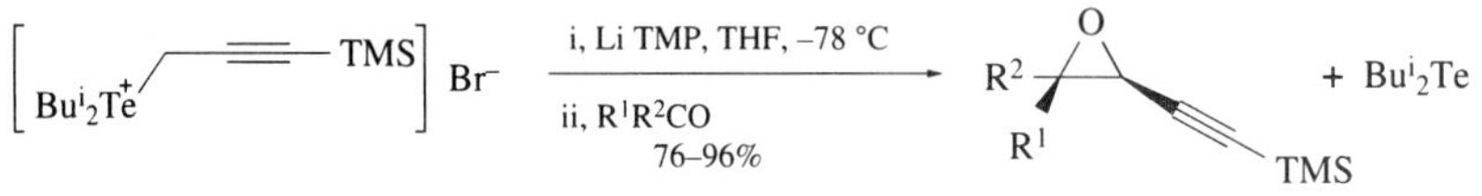

3-Trimethylsilyl diisobutylteluronium prop-2-enylide reacts with α,β-unsaturated esters to give trimethylsilylvinylcyclopropane derivatives (Equation (57)).[257]

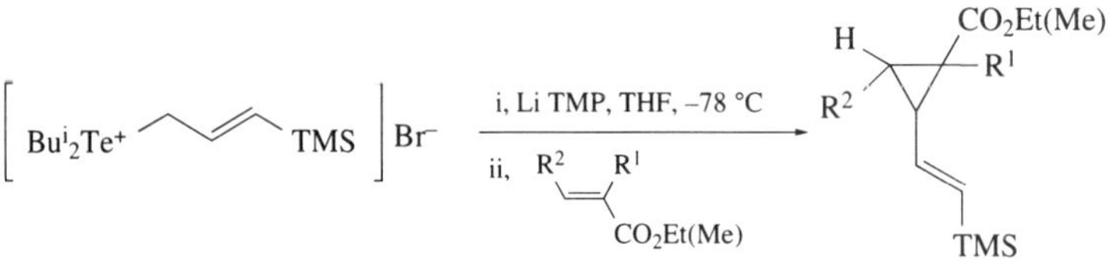

$[Bu^i_2Te^+\diagup\hspace{-0.3em}=\hspace{-0.3em}\text{TMS}]Br^- \xrightarrow[\text{ii, } R^2\diagup\hspace{-0.3em}=\hspace{-0.3em}\underset{CO_2Et(Me)}{R^1}]{\text{i, Li TMP, THF, -78 °C}} R^2\overset{H}{\underset{\text{TMS}}{\triangle}}\overset{CO_2Et(Me)}{\underset{\diagup=}{R^1}} \quad (57)$

Many different types of telluronium salt react with carbonyl compounds on treatment with organolithiums to give alcohols instead of alkenes.[250,256,258,259] This reaction proceeds via the intermediacy of a tetraorganotellurium, which generates a carbanionic species suitable for addition to the carbonyl group (Equation (58)). Phenyl, benzyl, cyanomethyl and trimethylsilylpropargyl groups are transferred in preference to other groups. The β-hydroxynitriles obtained from the cyanomethyltelluronium salts are of great synthetic importance.[260] Similarly, the importance of propargylic anions in synthetic chemistry enhances the importance of the propargyl-substituted telluronium salt.[261]

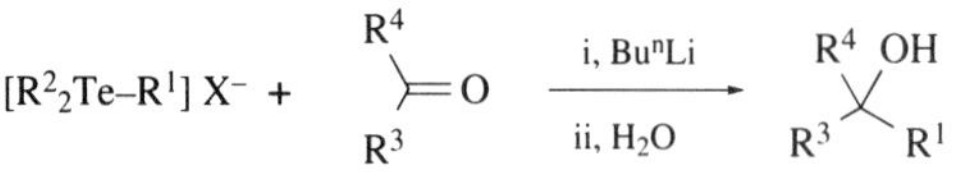

$[R^2_2Te-R^1]\ X^- + \underset{R^3}{\overset{R^4}{\diagdown}}=O \xrightarrow[\text{ii, H}_2O]{\text{i, Bu}^nLi} \underset{R^3}{\overset{R^4\ OH}{\diagdown\diagup}}\overset{}{\underset{R^1}{\diagdown}} \quad (58)$

Benzylic sulfones are converted into diarylethylenes on treatment with butyllithium in the presence of elemental tellurium; an epitelluride is the intermediate (Scheme 45).[262] The *cis–trans* mixture which is obtained is converted into the pure *trans*-isomer by treatment with $TeCl_4$.

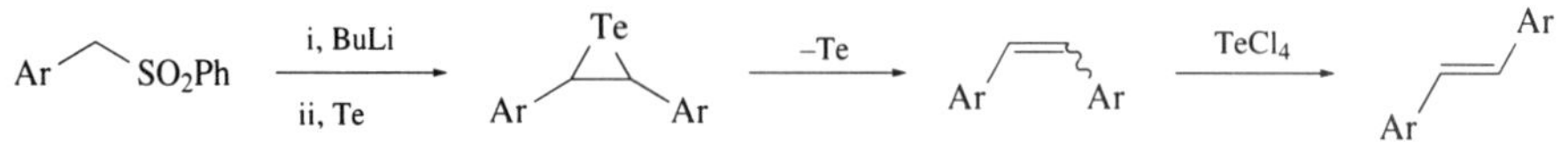

$Ar\frown SO_2Ph \xrightarrow[\text{ii, Te}]{\text{i, BuLi}} \underset{Ar\ \ \ Ar}{\overset{Te}{\triangle}} \xrightarrow{-Te} \underset{Ar\ \ \ Ar}{\diagup=\diagdown} \xrightarrow{TeCl_4} \underset{Ar}{\diagup}=\overset{Ar}{\diagdown}$

Scheme 45

14.4.8 Transmetallation Reactions of Organotellurium Compounds

Treatment of diorganotellurides with alkyllithiums in THF at −78 °C results in lithium–tellurium exchange with displacement of one of the groups originally attached to the tellurium as an organolithium

(Scheme 46, method A). In the case of symmetrical diorganotellurides, the displacement of both the groups is achieved by use of two equivalents of alkyllithium (method B), Moreover, since organotellurides are prepared from lithium tellurolates and organic halides, a 'one-pot' procedure can be employed, which overcomes the problem of isolation of the telluride (method C). Using these procedures, alkyl-, vinyl-, aryl- and ethynyllithiums, and even allyl- and benzyllithiums, which are difficult to access by conventional methods because of the interference of Wurtz-type coupling, can be easily generated and trapped *in situ* with electrophiles (aldehydes, ketones, alkyl halides, Me$_2$SO$_4$, H$^+$).[99,263,264] In this general context, it should be noted that vinylic tellurides prepared by *anti*-addition of tellurols to alkynes have the (Z)-configuration and therefore generate (Z)-vinyllithiums. These results are in sharp contrast to those from the lithium–tin exchange of vinylstannanes (characterized by (E)-configuration), which give (E)-vinyllithiums.[265] Similar sodium–, potassium–, calcium– and magnesium–tellurium exchanges have recently been described.[266]

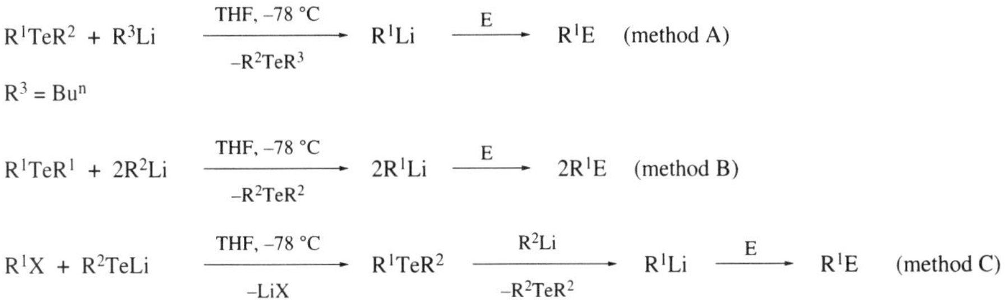

Scheme 46

Acyl- and aroyllithiums can be generated from telluroesters (easily prepared from tellurolate anions and acyl chlorides[267]) by a similar lithium–tellurium exchange reaction (Scheme 47).[268] The corresponding selenoesters and acylstannanes are not suitable for this type of reaction.

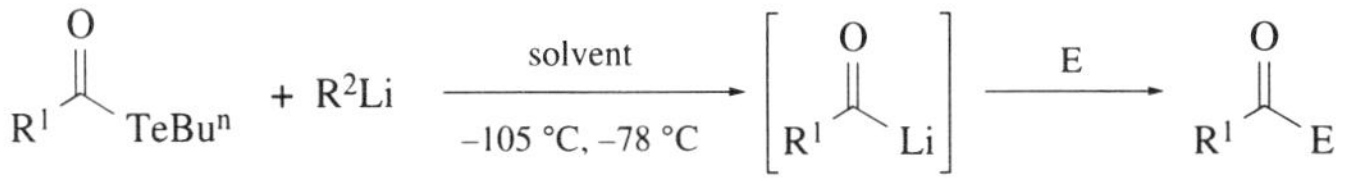

Scheme 47

A further development of these exchange reactions is the generation of heteroatom-substituted methyllithiums (Scheme 48).[269] Tin and selenium[270] derivatives undergo lithium–tellurium exchange in preference to lithium–tin and lithium–selenium exchange. Scheme 49 shows how (BuTe)$_2$CH$_2$ can be submitted to two successive lithium/tellurium exchange-trapping sequences.

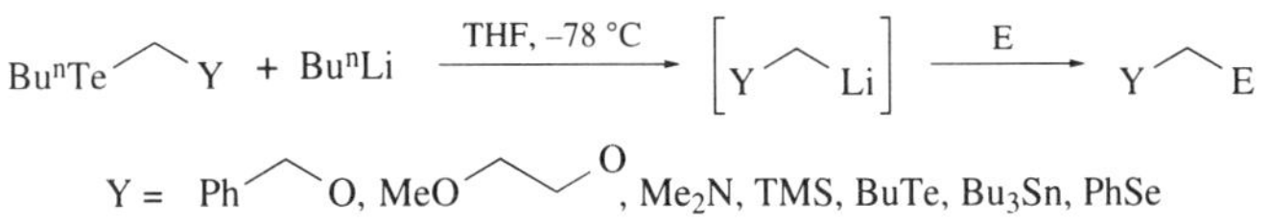

Scheme 48

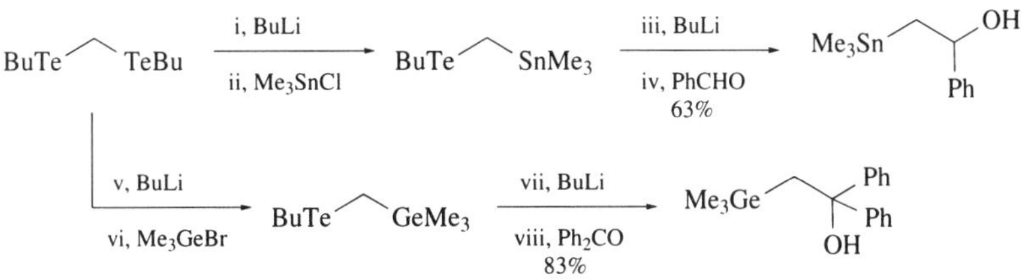

Scheme 49

Copper–tellurium exchange with vinyl tellurides is achieved by using higher-order cuprates, and the resulting vinylcuprates readily undergo conjugate additions to α,β-unsaturated carbonyl compounds (Scheme 50).[271] These results show that vinyl groups are transferred with high preference over other groups. If, however, a phenyl group is linked to tellurium, phenyl transfer becomes important, and phenyl vinyl tellurides are therefore not suitable for these reactions. An especially efficient organocuprate reagent is the thienyl derivative $Bu^n(2-Th)CuLi_2$, since the presence of the nontransferable 2-thienyl group avoids the transfer of the butyl group to the enone, which occurs when $Bu_2Cu(CN)Li_2$ is used.[272] Otherwise, bisvinylic tellurides are very useful reagents for transmetallation reactions, since only one equivalent of the telluride is required to generate two equivalents of the mixed vinylcuprate (Scheme 51).[272] The reaction is sensitive to steric factors, as shown by the low yields of 1,4-adducts obtained with β,β-disubstituted enones like mesityl oxide. The reagent $Me_2Cu(CN)Li_2$ has also been used successfully.

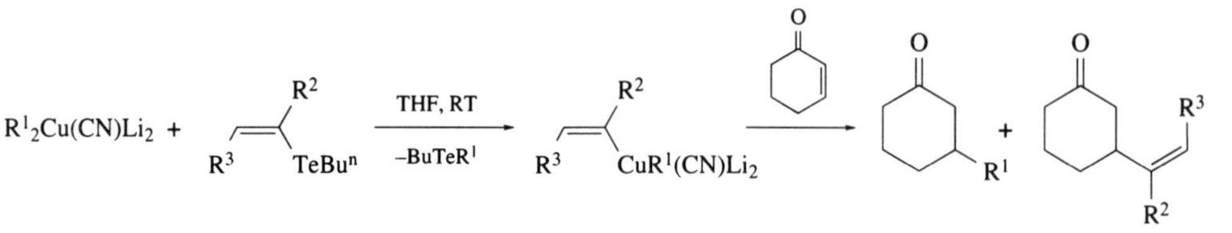

	(a)	(b)
R^1 = Me, Bu^n; R^2 = H; R^3 = Ph	trace	88–90%
R^1 = Me; R^2 = Ph; R^3 = H	23%	75%
R^1 = Bu^s; R^2 = Ph; R^3 = H	trace	89%

Scheme 50

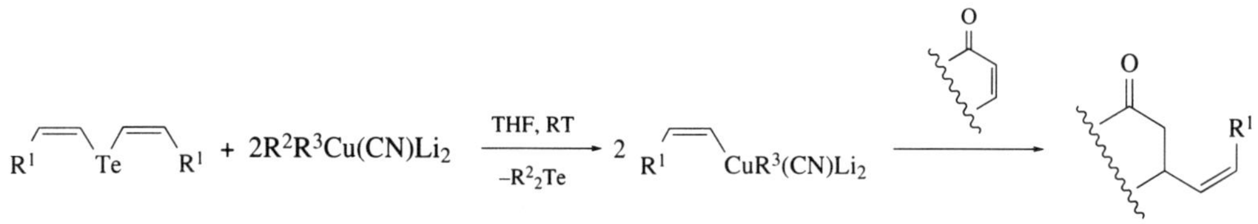

Scheme 51

Vinyl 2-thienyl tellurides are valuable reagents for the transfer of the vinyl group. Indeed, the presence of the nontransferable 2-thienyl group makes these reagents more attractive than the corresponding phenyl derivatives, which are susceptible to phenyl transfer (Equation (59)).[272] Tellurodienes and telluroenynes are also suitable for similar transmetallation reactions.[272]

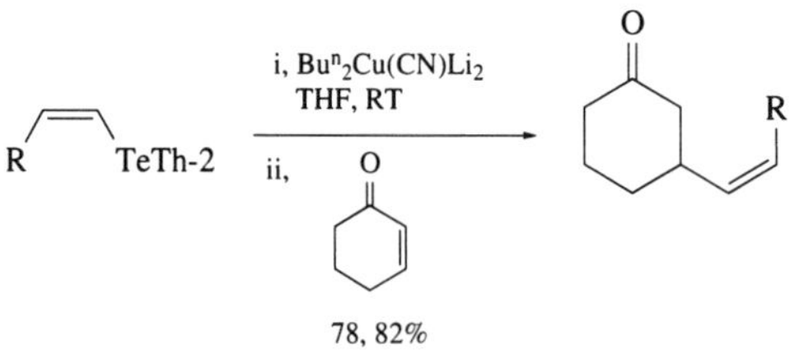

78, 82%

R = H, Ph   (59)

14.4.9 Free Radical Chemistry

Organic tellurides have recently become important in free radical chemistry. Allylic halides are converted into coupled 1,5-dienes by a radical mechanism on treatment with lithium telluride prepared *in situ* from elemental tellurium and lithium triethylborohydride (Scheme 52).[273]

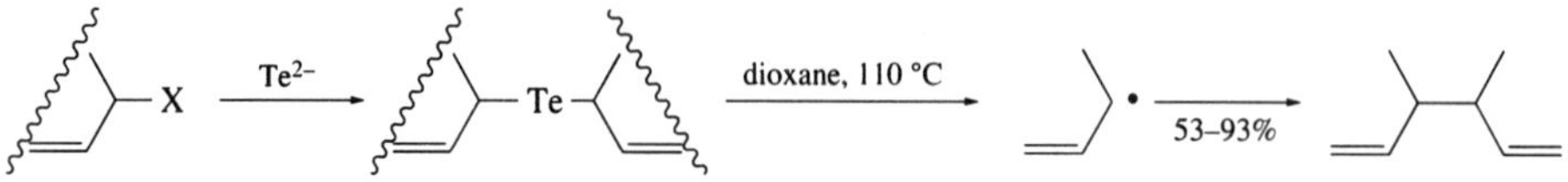

Scheme 52

If *N*-acetoxy-2-thiopyridone, a suitable source of methyl radicals, is irradiated with a simple tungsten lamp in the presence of an alkyl anisyl telluride, a radical exchange takes place, giving methyl anisyl telluride and a new alkyl radical. In the presence of an electrophilic alkene, a tandem addition occurs with the participation of the thiocarbonyl function of the starting reagent. A methyl radical is regenerated and a new cycle begins (Scheme 53).[274,275] This type of radical chemistry is useful for the synthesis or manipulation of complex natural products, such as additions of carbohydrates to alkenes[275] and the synthesis of six-membered carbocycles[276] and cyclonucleosides.[277]

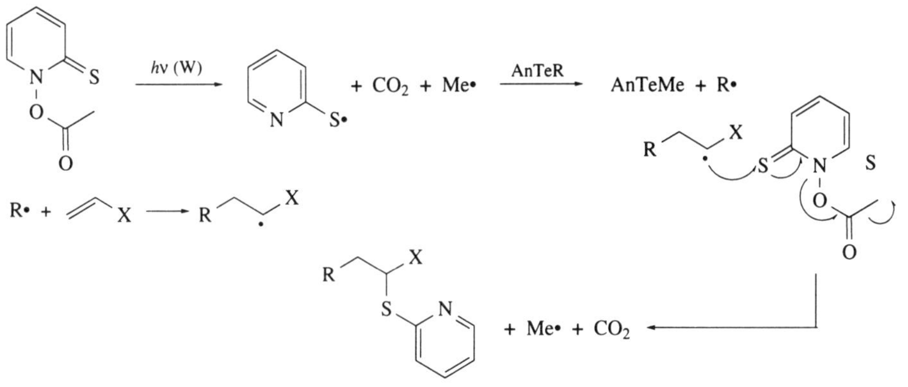

Scheme 53

Tetraalkyltelluriums (R_4Te, $R \neq Me$) react with arylalkynes ($ArC \equiv CH$) to give cis-adducts $ArCH=CHR$ as the major products, together with a telluride (R_2Te) and an alkene $RCH=CH_2$, the last originating from an R group of the R_4Te.[278] $Pr^i_2TeBu_2$ transfers exclusively the isopropyl group, whereas $Bu^n_2TeDec_2$ gives mixtures, and Ph_4Te is unreactive. The reaction has been rationalized as a radical addition of R_4Te ($\equiv R_3Te\bullet + R\bullet$) to the alkyne, followed by a β-hydrogen transfer from an R group on tellurium to the vinyl carbon.

Acyl aryl tellurides have been used recently as clean photolytic and thermal sources of acyl radicals (Scheme 54).[279,280] The results are best rationalized in terms of photochemically or thermally initiated homolytic cleavage of the acyl–Te bond, followed by the inter- or intramolecular attack of the acyl radical thus formed.

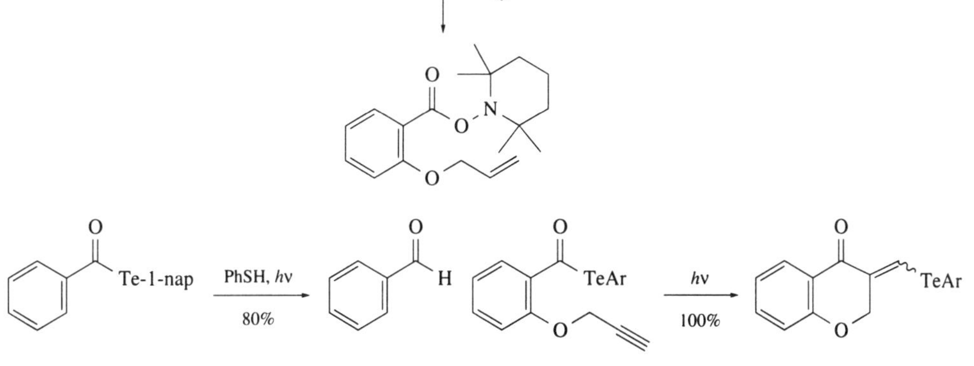

Scheme 54

14.4.10 Other Important Reactions Involving Tellurium Reagents

α-Hydroxyalkylation of α,β-unsaturated carbonyl compounds, dehydrotelluration of β-trichlorotelluroketones, conversion of allylsilanes into allylamines via phenyltellurinylation, and alkene inversion by *syn*-chlorotelluration/*anti*-dechlorotelluration are all important reactions involving tellurium reagents. Also significant is the use of TeCl$_4$ as a Lewis acid catalyst.

Addition of diisobutylaluminiumphenyltellurolate to α,β-unsaturated carbonyl compounds followed by aldol reaction with aldehydes and telluroxide elimination results in α-hydroxyalkylation of the enone (Scheme 55).[281]

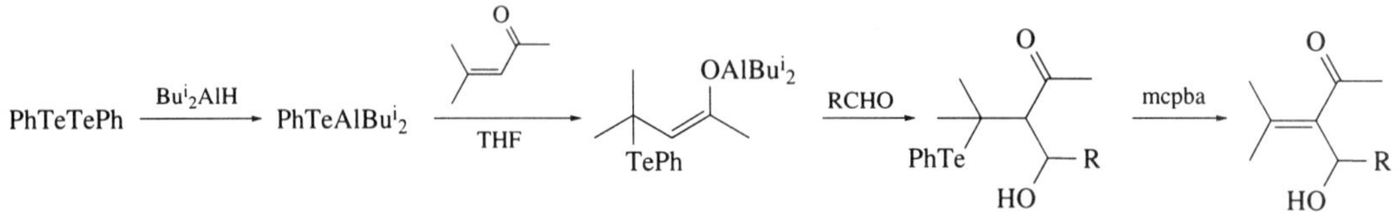

Scheme 55

Both β-trichlorotelluroketones and the corresponding dichlorides, which are easily prepared as shown in Scheme 56, undergo dehydrotelluration on treatment with DMSO or bases such as TMEDA, triethylamine or pyridine.[282] The overall procedure constitutes a useful method for the α-methylenation of ketones under mild conditions.

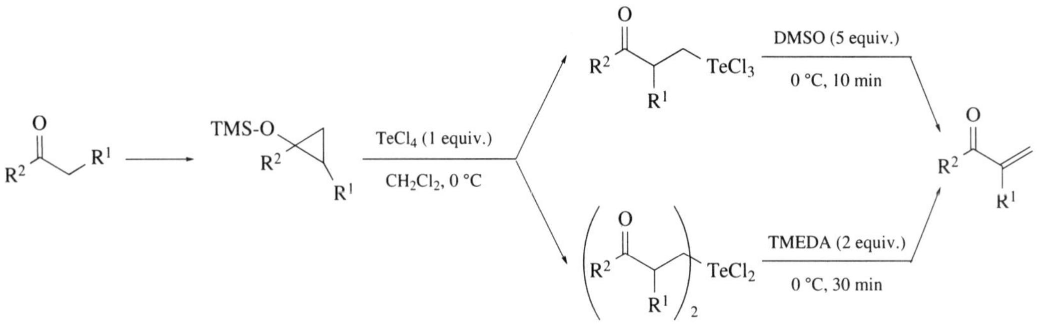

Scheme 56

Allylamines are obtained by sequential treatment of allylsilanes[283] with PhTe(O)OCOCF$_3$ in the presence of BF$_3$·Et$_2$O and an alkylamine. A phenyl allyltelluroxide is formed as intermediate (Scheme 57).[284]

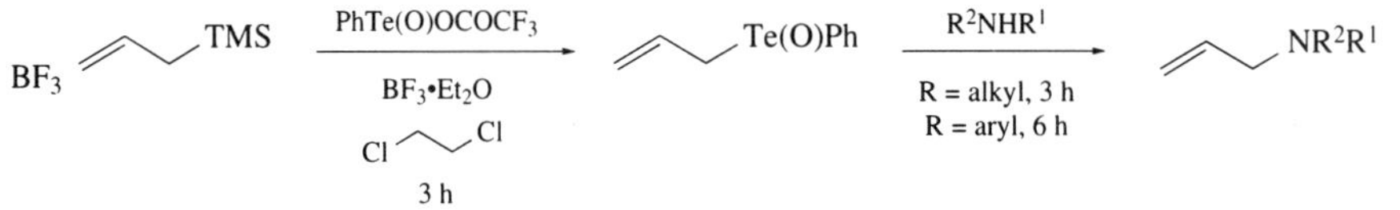

Scheme 57

The geometry of carbon–carbon double bonds can be inverted by a 'one-pot' *syn*-chlorotelluration/*anti*-dechlorotelluration sequence.[285] The proposed mechanism involves an intermediate epitelluride (Scheme 58).

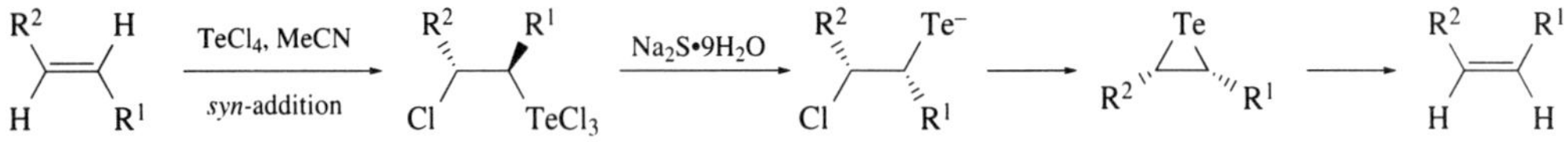

Scheme 58

TeCl$_4$ functions as a Lewis acid catalyst for reactions such as thioacetalization and thioketalization of aldehydes and ketones,[286] aromatic alkylation[287] and cationic oligo- and polymerization of phenyl-substituted ethenes and benzyl chlorides.[288]

14.5 REFERENCES

1. H. Rheinboldt, in 'Houben-Weyl Methoden der Organischen Chemie', 4th edn., ed. E. Müller, Thieme, Stuttgart, 1955, vol. 4.
2. K. Y. Irgolic, in 'Houben-Weyl Methoden der Organischen Chemie', 4th edn., ed. D. Klamann, Thieme, Stuttgart, 1990, vol. E12b.
3. W. C. Cooper (ed.) 'Tellurium', Van Nostrand Rheinhold, New York, 1971.
4. K. Y. Irgolic, 'The Organic Chemistry of Tellurium', Gordon and Breach, New York, 1974.
5. N. Petragnani and M. Moura Campos, *Organomet. Chem. Rev.*, 1967, **2**, 61.
6. K. Y. Irgolic and R. Zingaro, in 'Organometallic Reaction', eds. E. I. Becker and M. Tsutsui, Wiley, New York, 1971.
7. K. J. Irgolic, *J. Organomet. Chem.*, 1975, **103**, 91; 1977, **130**, 411; 1978, **158**, 235; 1980, **189**, 65; 1980, **203**, 368.
8. S. Uemura; *Kagaku (Kyoto)*, 1981, **36**, 381.
9. N. Petragnani and J. V. Comasseto, in 'Proceedings of the 4th International Conference of the Organic Chemistry of Selenium and Tellurium', eds. F. Y. Berry and W. R. McWhinnie, University of Aston, Birmingham 1983, pp. 98–241.
10. S. Uemura, *J. Synth. Org. Chem. Jpn.*, 1983, **41**, 804.
11. L. Engman, *Acc. Chem. Res.*, 1985, **18**, 274.
12. N. Petragnani and J. V. Comasseto, *Synthesis*, 1986, 1.
13. H. Suzuki, *J. Synth. Org. Chem. Jpn.*, 1987, **45**, 603.
14. I. D. Sadekov, B. B. Rivkin and V. I. Minkin, *Russ. Chem. Rev.*, 1987, **56**, 343.
15. S. Patai and Z. Rappoport (eds.), 'The Chemistry of Organic Selenium and Tellurium Compounds', Wiley, New York, 1986, vol. 1; 1987, vol. II.
16. L. Engman, *Phosphorus Sulfur*, 1988, **38**, 105.
17. N. Petragnani and J. V. Comasseto, *Synthesis*, 1991, 793.
18. N. Petragnani and J. V. Comasseto, *Synthesis*, 1991, 897.
19. D. H. R. Barton and S. W. McCombie, *J. Chem. Soc., Perkin Trans 1*, 1975, 1574.
20. N. Kambe, K. Kondo, S. Morita, S. Murai and N. Sonoda, *Angew Chem., Int. Ed. Engl.*, 1980 **19**, 1008.
21. N. Kambe, T. Inagaki, N. Miyoshi, A. Ogawa and N. Sonoda, *Chem. Lett.*, 1987, 1275.
22. N. Kambe, K. Kondo and N. Sonoda, *Angew Chem.*, 1980 **19**, 1009.
23. J. F. Suttle and C. R. F. Smith, 'Inorganic Synthesis', McGraw Hill, New York, 1950, vol. 3, p. 140.
24. H. Gerding and H. Houtgraaff, *Recl. Trav. Chim. Pays-Bas* 1954 **73**, 737.
25. H. Marshall, 'Inorganic Synthesis' McGraw Hill, New York, 1950, vol. 3, p. 143.
26. L. Brandsma and H. E. Wijers, *Recl. Trav. Chim. Pays. Bas.* 1963 **82**, 68.
27. M. T. Chen and J. W. George, *J. Organomet. Chem.*, 1968 **12**, 401.
28. L. Engman, *Organometallics*, 1986, **5**, 427.
29. L. Tschugaeff and W. Chlopin, *Ber. Dtsch. Chem. Ges.*, 1914 **47**, 1269.
30. M. P. Balfe and K. N. Nandi, *J. Chem. Soc.*, 1941, 70.
31. H. K. Spencer and M. P. Cava, *J. Org. Chem.*, 1977, **42**, 2937.
32. J. D. McCullough, *Inorg. Chem.*, 1965, **4**, 862.
33. F. G. Holliman and F. G. Mann, *J. Chem. Soc.*, **37**, 1945.
34. F. F. Knapp, Jr, *J. Labelled Comp. Radiopharm.*, 1980, **17**, 81.
35. H. Suginome, S. Yamada and J. B. Wang, *J. Org. Chem.*, 1990, **55**, 2170.
36. I. Davies, W. R. McWhinnie, N. S. Dance and C. H. Jones, *Inorg. Chim. Acta*, 1978, **29L**, 217.
37. K. Ramasamy and P. Shanmugam, *Z. Naturforsch., Teil B*, 1977, **32**, 605.
38. M. M. Goodman and F. F. Knapp, *Organometallics*, 1983, **2**, 1106.
39. S. W. Lee, M. S. Thesis, Texas A & M University 1986 (see ref. 2, p. 259).
40. B. G. Gribov, V. L. Bregadze, L. M. Golubinskaya, L. G. Tonoyan and B. I. Kozyrkin, *Otkrytiya Izobret. Prom. Obraztsy, Tovarnya Znaki*, 1977, **54**(1), 108 (*Chem. Abstr.*, 1977, **87**, 5394; USSR, **541**, 851, Cl C0711/001).
41. B. I. Kozyrkin, B. A. Salamtin, L. L. Ivanov, I. A. Kuzovlev, B. G. Gribov and V. A. Federov, *Poluch. Anal. Veshchestv Osoboi Chist, [Dokl. Vses. Konf.]*, 5th, 1976, p. 142 (*Chem. Abstr.*, 1979, **91**, 140 278).
42. J. T. B. Ferreira, A. R. M. de Oliveira and J. V. Comasseto, *Synth. Commun.*, 1989, **19**, 239.
43. G. T. Morgan and R. E. Kellett, *J. Chem. Soc.*, 1926, 1080.
44. H. D. Drew, *J. Chem. Soc.*, 1926, 223.
45. L. Reichel and E. Kirschbaum, *Liebigs Ann. Chem.*, 1936 **523**, 211.
46. N. Petragnani, *Tetrahedron*, 1960, **11**, 15.
47. I. D. Sadekov, L. M. Sayapina, A. Y. Bushkov and V. I. Minkin, *J. Gen. Chem. USSR (Engl. Transl.)*, 1971, **41**, 2747.
48. J. Bergman, *Tetrahedron*, 1972, **28**, 3323.
49. K. Y. Irgolic in 'Houben-Weyl Methoden der Organischen Chemie', 4th edn., ed. D. Klamann, Thieme, Stuttgart, 1990, Vol. E12b, p. 529, Table 16.
50. W. H. H. Günther, J. Nepywoda and J. Y. C. Chu, *J. Organomet. Chem.*, 1974, **74**, 79.
51. W. V. Farrar, *Research*, 1951, **4**, 177.
52. I. G. M. Campbell and E. E. Turner, *J. Chem. Soc.*, 1938, 37.
53. N. Petragnani, J. V. Comasseto and N. H. Varella, *J. Organomet. Chem.*, 1976, **120**, 375.
54. L. Reichel and E. Kirschbaum, *Ber. Dtsch. Chem. Ges.*, 1943, **76**, 1105.
55. E. S. Lang and J. V. Comasseto, *Synth. Commun.*, 1988, **18**, 301.
56. G. Vicentini, E. Giesbrecht and L. R. M. Pitombo, *Chem. Ber.*, 1959, **92**, 40.
57. L. Engman, *J. Org. Chem.*, 1989, **54**, 2964.
58. K. Y. Irgolic, in 'Houben-Weyl Methoden der Organischen Chemie', 4th edn., ed. D. Klamann, Thieme, Stuttgart, 1990, vol. E12b, p. 315.
59. H. Rheinboldt and G. Vicentini, *Chem. Ber.*, 1956, **89**, 624.
60. K. Y. Irgolic, in 'Houben-Weyl Methoden der Organischen Chemie', 4th edn., ed. D. Klamann, Thieme, Stuttgart, NY 1990, vol. E12b, p. 553.
61. W. S. Haller and K. J. Irgolic, *J. Organomet. Chem.*, 1972, **38**, 97.
62. M. Akiba, M. V. Lakshmikantham, K. Jen and M. P. Cava, *J. Org. Chem.*, 1984, **49**, 4819.

63. M. R. Detty, B. J. Murray, D. L. Smith and N. Zumbulyadis, *J. Am. Chem. Soc.*, 1983, **105**, 875.
64. M. J. Dabdoub, V. B. Dabdoub, J. V. Comasseto and N. Petragnani, *J. Organomet. Chem.*, 1986, **308**, 211.
65. M. J. Dabdoub and J. V. Comasseto, *J. Organomet. Chem.*, 1988, **344**, 167.
66. E. G. Hope, T. Kemmitt and W. Levason, *Organometallics*, 1988, **7**, 78.
67. D. Seebach and A. K. Beck, *Chem. Ber.*, 1975, **108**, 314.
68. L. Engman and M. P. Cava, *Synth. Commun.*, 1982, **12**, 163.
69. L. Engman and M. P. Cava, *Organometallics*, 1982, **1**, 470.
70. L. Engman and J. Persson, *J. Organomet. Chem.*, 1990, **388**, 71.
71. A. B. Pierini and R. A. Rossi, *J. Organomet. Chem.*, 1979, **168**, 163.
72. A. B. Pierini and R. A. Rossi, *J. Org. Chem.*, 1979, **44**, 4667.
73. K. Lederer, *Ber. Dtsch. Chem. Ges.*, 1915, **48**, 1345.
74. K. Ohe, H. Takahashi, S. Uemura and N. Sugita, *J. Organomet. Chem.*, 1987, **326**, 35.
75. K. J. Irgolic, P. J. Busse, R. A. Grigsby and M. R. Smith, *J. Organomet. Chem.*, 1975, **88**, 175.
76. J. L. Piette and M. Renson, *Bull. Soc. Chim. Belg.*, 1970, **79**, 383.
77. D. L. J. Clive *et al.*, *J. Am. Chem. Soc.*, 1980, **102**, 4438.
78. F. F. Knapp, K. R. Ambrose and A. P. Callahan, *J. Med. Chem.*, 1981, 24, 794.
79. S. Uemura and S. Fukuzawa, *J. Am. Chem. Soc.*, 1983, **105**, 2748.
80. J. V. Comasseto, E. S. Lang, J. T. B. Ferreira, F. Simonelli and V. R. Correia, *J. Organomet. Chem.*, 1987, **334**, 329.
81. J. T. Ferreira, A. R. M. de Oliveira and J. V. Comasseto, *Tetrahedron Lett.*, 1992, **33**, 915.
82. T. Kemmitt and W. Levason, *Organometallics*, 1989, **8**, 1303.
83. K. Y. Irgolic, in 'Houben-Weyl Methoden in Organischen Chemie', 4th edn., ed. D. Klamann, Thieme, Stuttgart, NY, 1990, vol. E12b, p. 389.
84. J. L. Piette, R. Lysy and M. Renson, *Bull. Soc. Chim. Fr.*, 1972, 3559.
85. P. Wiriyachitra, S. J. Falcone and M. P. Cava, *J. Org. Chem.*, 1979, **44**, 3957.
86. L. Engman, *J. Org. Chem.*, 1989, **54**, 2964.
87. N. Petragnani, L. Torres and K. J. Wynne, *J. Organomet. Chem.*, 1975, **92**, 175.
88. M. Moura Campos and N. Petragnani, *Tetrahedron*, 1962, **18**, 527.
89. P. Schulz and G. Klar, *Z. Naturforsch. Teil B*, 1975, **30**, 40, 43.
90. J. L. Piette and M. Renson, *Bull. Soc. Chim. Belg.*, 1971, **80**, 669.
91. B. A. Trofimov, S. V. Amosova, N. K. Gusarova and G. K. Musorin, *Tetrahedron*, 1982, **38**, 713.
92. N. K. Gusarova, A. A. Tatarinova and L. M. Sinegovskaya, *Sulfur Reports*, 1991, **11**, 1.
93. B. A. Trofimov, S. V. Amosova, N. K. Gusarova, V. A. Potapov and A. A. Tatarinova, *Sulfur Lett.*, 1983, **1**, 151.
94. S. M. Barros, M. J. Dabdoub, V. R. Dabdonb and J. V. Comasseto, *Organometallics*, 1989, **8**, 1661.
95. K. Ohe, H. Takahashi, S. Uemura and N. Sugita, *Nippon. Kagaku Kaishi*, 1987, 1469.
96. K. Ohe, H. Takahashi, S. Uemura and N. Sugita, *J. Org. Chem.*, 1987, **52**, 4859.
97. S. Uemura, S. I. Fukuzawa and S. R. Patil, *J. Organomet. Chem.*, 1989, **243**, 9.
98. V. A. Potapov *et al.*, *J. Org. Chem. USSR*, 1983, **23**, 596.
99. L. Brandsma, H. E. Wijers and J. F. Arens, *Recl. Trav. Chim. Pays-Bas*, 1962, **81**, 583.
100. S. A. Radchenko and K. S. Mingaleva, *J. Org. Chem. USSR (Engl. Transl.)*, 1977, **13**, 2303.
101. Y. A. Boiko, B. S. Kupin and A. A. Petrov, *J. Org. Chem. USSR (Engl. Transl.)*, 1969, **5**, 1516.
102. S. L. Bender, M. R. Detty and N. F. Haley, *Tetrahedron Lett.*, 1982, **23**, 1531.
103. M. G. Dabdoub and J. V. Comasseto, *Organometallics*, 1988, **7**, 84.
104. G. T. Morgan and H. D. Drew, *J. Chem. Soc.*, 1925, **127**, 2307.
105. H. Rheinboldt, in 'Houben-Weyl Methoden der Organischen Chemie', 4th edn., ed. E. Müller, Thieme, Stuttgart, 1955, p. 1061.
106. I. D. Sadekov, A. A. Maksimenko and B. B. Rivkin, *J. Org. Chem. USSR (Engl. Transl.)*, 1978, **14**, 810.
107. N. Petragnani, *Tetrahedron*, 1961, **12**, 219.
108. H. A. Stefani, J. V. Comasseto and N. Petragnani, *Synth. Commun.*, 1987, **17**, 443.
109. H. A. Stefani, A. Chieffi and J. V. Comasseto, *Organometallics*, 1991, **10**, 1178.
110. H. Funk and H. Weiss, *J. Prakt. Chem.*, 1954, **1**(4), 33.
111. M. Moura Campos and N. Petragnani, *Tetrahedron Lett.*, 1959, 11.
112. M. Ogawa and R. Ishioka, *Bull. Chem. Soc. Jpn.*, 1970, **43**, 496.
113. J. E. Bäckwall, J. Bergman, L. Engman, *J. Org. Chem.*, 1983, **48**, 3918.
114. S. Uemura, S. Fukuzawa, *J. Organomet. Chem.*, 1984, **268**, 223.
115. H. J. Arpe, H. Kuckertz, *Angew Chem. Int. Ed. Engl.* 1971, **10**, 73.
116. M. Ogawa, *Bull. Chem. Soc. Jpn.*, 1968 **41**, 3031.
117. L. Engman, *Organometallics*, 1989, **8**, 1997.
118. M. E. S. Ali, M. A. Malik and S. Smith, *Inorg. Chim. Acta*, 1989, **162**, 157.
119. M. Moura Campos and N. Petragnani, *Tetrahedron*, 1962, **18**, 527.
120. J. Bergman and L. Engman, *Tetrahedron.*, 1980, **36**, 1275.
121. S. Uemura, S. Fukuzawa, A. Toshimitsu and M. Okano, *Tetrahedron Lett.*, 1982, **23**, 1177.
122. S. Uemura, S. Fukuzawa and A. Toshimitsu, *J. Organomet. Chem.*, 1983, **250**, 203.
123. S. Uemura, H. Miyoshi and M. Okano, *Chem. Lett.*, 1979, 1357.
124. J. V. Comasseto, H. A. Stefani and A. Chieffi, *Organometallics*, 1991, **10**, 845.
125. N. Petragnani and G. Vicentini, *Universidade de São Paulo, F.F.C.L. Bol. Quim.*, 1959, **5**, 75 (Chem. Abst., 1963 **58**, 11256).
126. P. Thavornyutikarn, and W. R. Whinnie, *J. Organomet. Chem.*, 1973, **50**, 135 and references therein.
127. D. H. Barton, J. P. Finet and M. Thomas, *Tetrahedron*, 1986, **42**, 2319.
128. N. X. Hu, Y. Aso, T. Otsubo and F. Ogura, *J. Org. Chem.*, 1989, **54**, 4398.
129. K. Lederer, *Ber. Dtsch. Chem. Ges.*, 1916, **49**, 345, 2532, 2663; 1920, **53B**, 1674.
130. S. V. Ley, C. A. Merholz and D. H. R. Barton, *Tetrahedron*, 1981, **37**, W 213.
131. M. R. Detty, *J. Org. Chem.*, 1980, **45**, 274.
132. N. Kambe, T. Inagaki, N. Miyoshi, A. Ogawa and N. Sonoda, *Nippon Kagaku Kaishi*, 1987, 1152.

133. M. Akiba and M. P. Cava, *Synth. Commun.*, 1984, **14**, 1119.
134. Y. Aso et al., *Nippon Kagaku Kashi*, 1987, 1490.
135. K. Nagakawa, M. Osuka, K. Sasaki, Y. Aso, T. Otsubo and F. Ogura, *Chem. Lett.*, 1987, 1331.
136. M. Moura Campos, E. L. Suranyi, H. de Andrade, Jr. and N. Petragnani, *Tetrahedron*, 1964, **20**, 2797.
137. (a) H. Suzuki, H. Manabe, R. Enokiya and Y. Hanazaki, *Chem. Lett.*, 1986, 1339; (b) H. Suzuki and Y. Hanazaki, *Chem. Lett.*, 1986, 549.
138. K. Ramasamy, S. K. Kalyanasundaram and P. Shanmugam, *Synthesis*, 1978, 545.
139. M. Yamashita, Y. Kato, R. Suemitsu, *Chem. Lett.*, 1980, 847.
140. X. Huang, H. Zhang, *Synth. Commun.* 1989, **19**, 97.
141. X. Huang, L. Xie, *Synth. Commun.*, 1986, **16**, 1701.
142. D. H. R. Barton, A. Fekih and X. Lusinchi, *Tetrahedron Lett.*, 1985, **26**, 3693.
143. M. Yamashita, M. Kadokura and R. Suemitsu, *Bull. Chem. Soc. Jpn.*, 1984, **57**, 3359.
144. H. Suzuki, H. Manabe, T. Kawaguchi and M. Inouye, *Bull. Chem. Soc. Jpn.*, 1987, **60**, 771.
145. H. Suzuki, H. Manabe and M. Inouye, *Chem. Lett.*, 1985, 1671.
146. A. Osuka, H. Shimizu and H. Suzuki, *Chem. Lett.*, 1983, 1373.
147. S. Uchida, K. Yanada, H. Yamaguchi and H. Meguri, *Chem. Lett.*, 1986, 1069.
148. K. Ohe, S. Uemura, N. Sugita, H. Masuda and T. Taga, *J. Org. Chem.*, 1989, **54**, 4169.
149. N. Ohira, Y. Aso, T. Otsubo and F. Ogura, *Chem. Lett.*, 1984, 853.
150. D. H. Barton, A. Fekih and X. Lusinchi, *Tetrahedron Lett.*, 1985, **26**, 4603.
151. H. Suzuki and K. Takaoka, *Chem. Lett.*, 1984, 1733.
152. I. T. Harrison and S. Harrison (eds.) 'Compenduim of Organic Synthetic Methods', Wiley, London, 1971, vol. 1 sect. 204 p. 501; 1974, vol. 2, sect. 204, p. 200.
153. D. L. J. Clive and S. M. Menchen, *J. Chem. Soc. Chem. Commun.*, 1977, 658.
154. D. L. J. Clive and S. M. Menchen, *J. Org. Chem.*, 1980, **45**, 2347.
155. D. H. R. Barton, A. Fekih and X. Lusinchi, *Tetrahedron Lett.*, 1985, **26**, 6197.
156. A. Osuka, K. Takaoka and H. Suzuki, *Chem. Lett.*, 1984, 271.
157. G. Polson and D. C. Dittmer, *Tetrahedron Lett.*, 1986, **27**, 5579.
158. R. P. Discordia and D. C. Dittmer, *J. Org. Chem.*, 1990, **55**, 1414.
159. R. P. Discordia, C. K. Murphy and D. C. Dittmer, *Tetrahedron Lett.*, 1990, **31**, 5603.
160. D. C. Dittmer et al., *J. Org. Chem.*, 1993, **58**, 718.
161. (a) 'Houben-Weyl Methoden der Organischen Chemie', 4th edn., Thieme, Stuttgart, 1972, vol. 5/1b 180; (b) D. W. Young, 'Protective Groups in Organic Chemistry', Plenum, London, 1973, p. 309; (c) I. M. Mathai, K. Schug and S. I. Miller, *J. Org. Chem.*, 1970, **35**, 1733; (d) J. E. Gordon and V. S. K. Chang, *J. Org. Chem.*, 1973, **38**, 3062 and references therein.
162. M. Moura Campos and N. Petragnani, *Tetrahedron Lett.*, 1960, 5.
163. H. Suzuki, A. Kondo and A. Osuka, *Bull. Chem. Soc. Jpn.*, 1985, **58**, 1335.
164. N. Petragnani and M. Moura Campos, *Chem. Ber.*, 1961, **94**, 1759.
165. K. Ramasamy, S. K. Kalyanasundaram and P. Shanmugam, 1978, *Synthesis*, 311.
166. S. Suzuki and M. Inouye, *Chem. Lett.*, 1985, 225.
167. X. Huang and Y. Q. Hou, *Synth. Commun.*, 1988, **18**, 2201.
168. L. Engman, *Tetrahedron Lett.*, 1982, **23**, 3601.
169. C. J. Li and D. N. Harpp, *Tetrahedron Lett.*, 1990, **31**, 6219.
170. L. Engman, S. E. Byström, *J. Org. Chem.*, 1985, **50**, 3170.
171. N. Kambe, T. Tsukamoto, N. Miyoshi, S. Murai and N. Sonoda, *Bull. Chem. Soc. Jpn.*, 1986, **59**, 3013.
172. A. Osuka and H. Suzuki, *Chem. Lett.* 1983, 173–6.
173. A. Osuka and H. Suzuki, *Chem. Lett.*, 1983, 119.
174. D. L. J. Clive and P. L. Beaulieu, *J. Org. Chem.*, 1982, **47**, 1124.
175. L. Engman and M. P. Cava, *J. Org. Chem.*, 1982, **47**, 3946.
176. C. J. Li and D. N. Harpp, *Tetrahedron Lett.*, 1991, **32**, 1545.
177. H. Suzuki, K. Takaoka and A. Osuka, *Bull. Chem. Soc. Jpn.*, 1985, **58**, 1067.
178. X. Huang and J. H. Pi, *Synth. Commun.*, 1990, **20**, 2297.
179. X. Huang, J. H. Pi and Z. Z. Huang, *Phosphorus, Sulfur and Silicon*, 1992, **67**, 177.
180. S. Padmanabhan, T. Ogawa and H. Suzuki, *J. Chem. Res. (S)*, 1989, 266.
181. (a) H. Suzuki and M. Inouye, *Chem. Lett.*, 1986, 403; (b) H. Suzuki, H. Manabe and M. Inouye, *Nippon Kagaku Kaishi*, 1987, 1485.
182. T. Matsuki, N. X. Hu, Y. Aso, T. Otsubo and F. Ogura, *Bull. Chem. Soc. Jpn.*, 1989, **62**, 2105.
183. S. Padmanabhan, T. Ogawa and H. Suzuki, *Bull. Chem. Soc. Jpn.*, 1989, **62**, 1358.
184. H. Suzuki, Y. Nishioka, S. Padmanabhan and T. Ogawa, *Chem. Lett.*, 1988, 727.
185. X. Huang and J. H. Pi, *Synth. Commun.*, 1990, **20**, 2291.
186. S. W. Li, Y. Z. Huang and L. L. Shi, *Chem. Ber.*, 1990, **123**, 1441.
187. T. W. Greene, 'Protective Groups in Organic Synthesis', Wiley, New York, 1981.
188. J. Chen and X. J. Zhou, *Synthesis*, 1987, 586.
189. Z. Huang, L. Xie and X. Huang, *Synth. Commun.*, 1988, **18**, 1167.
190. N. Shobana and P. Shanmugam, *Indian J. Chem.*, 1986, **25B**, 658.
191. H. Suzuki, S. Padmanabhan and T. Ogawa, *Chem. Lett.*, 1989, 1017.
192. J. Chen and X. J. Zhou, *Synth. Commun.*, 1987, **17**, 161.
193. Z. Huang, X. J. Zhou, *Synthesis*, 1990, 633.
194. N. Shobana and P. Shanmugam, *Indian J. Chem.*, 1985, **24B**, 690.
195. N. Shobana, M. Amirthavalli, V. Deepa and P. Shanmugam, *Indian J. Chem.*, 1988, **27B**, 965.
196. M. V. Lakshmikantham, Y. A. Jackson, R. J. Jones, G. J. O'Malley, K. Ravichandran and M. P. Cava, *Tetrahedron Lett.*, 1986, **27**, 4687.
197. D. H. R. Barton, S. V. Ley and C. A. Meerholz, *J. Chem. Soc., Chem. Commun.*, 1979, 755.
198. S. V. Ley, C. A. Meerholz and D. H. R. Barton, *Tetrahedron Lett.*, 1980, **21**, 1785.
199. N. X. Hu, Y. Aso and T. Otsubo and F. Ogura, *Bull. Chem. Soc. Jpn.*, 1986, **59**, 879.

200. H. Suzuki, S. Kawato and A. Nasu, *Bull. Chem. Soc. Jpn.*, 1992, **65**, 626.
201. N. X. Hu, Y. Aso, T. Otsubo and F. Ogura, *Tetrahedron Lett.*, 1986, **27**, 6099.
202. N. X. Hu, Y. Aso, T. Otsubo and F. Ogura, *Phosphorus Sulfur*, 1988, **38**, 177.
203. L. Engman and M. P. Cava, *J. Chem. Soc. Chem. Commun.*, 1982, 164.
204. T. Fukumoto, T. Matsuki, N. X. Hu, Y. Aso, T. Otsubo and F. Ogura, *Chem. Lett.*, 1990, 2269.
205. W. F. Brill, *J. Org. Chem.*, 1986, **51**, 1149.
206. N. Kambe, T. Tsukamoto, N. Miyoshi, S. Murai and N. Sonoda, *Chem. Lett.*, 1987, 269.
207. S. Uemura, K. Ohe, S. I. Fukuzawa, S. R. Patil and N. Sugita, *J. Organomet. Chem.*, 1986, **316**, 67.
208. S. I. Fukuzawa, K. J. Irgolic and D. H. O'Brien, *Organometallics*, 1990, **9**, 3073.
209. S. Uemura, S. Fukuzawa, S. R. Patil and M. Okano, *J. Chem. Soc., Perkin Trans. 1*, 1985, 499.
210. N. Kambe, T. Fujioka, A. Ogawa, N. Miyoshi and N. Sonoda, *Phosphorus Sulfur*, 1988, **38**, 167.
211. N. Kambe, T. Fujioka, A. Ogawa, N. Miyoshi and N. Sonoda, *Chem. Lett.*, 1987, 2077.
212. N. X. Hu, Y. Aso, T. Otsubo and F. Ogura, *Chem. Lett.*, 1987, 1327.
213. N. X. Hu, Y. Aso, T. Otsubo and F. Ogura, *J. Chem. Soc. Chem. Commun.*, 1987, 1447.
214. N. X. Hu, Y. Aso, T. Otsubo and F. Ogura, *Tetrahedron Lett.*, 1988, **29**, 1049.
215. N. X. Hu, Y. Aso, T. Otsubo and F. Ogura, *J. Chem. Soc., Perkin Trans. 1*, 1989, 1775.
216. (a) C. Paulmier, 'Selenium Reagents and Intermediates in Organic Synthesis', Chap. VIII, Pergamon, Oxford 1986; (b) K. C. Nicolaou, N. A. Petasis and D. A. Claremon, in 'Organoselenium Chemistry', ed. D. Liotta, Wiley, New York, 1987, ch. 2.
217. M. Moura Campos and N. Petragnani, *Chem. Ber.*, 1960, **93**, 317.
218. J. V. Comasseto and N. Petragnani, *Synth. Commun.*, 1983, **13**, 889.
219. M. Yoshida, T. Suzuki and N. Kamigata, *J. Org. Chem.*, 1992, **57**, 383.
220. J. V. Comasseto, H. M. C. Ferraz, C. A. Brandt and K. K. Gaeta, *Tetrahedron Lett.*, 1989, **30**, 1209.
221. J. V. Comasseto, H. M. C. Ferraz and N. Petragnani, *Tetrahedron Lett.*, 1987, **28**, 5611.
222. N. X. Hu, Y. Aso, T. Otsubo and F. Ogura, *Tetrahedron Lett.*, 1987, **28**, 1281.
223. N. X. Hu and Y. Aso, T. Otsubo and F. Ogura, *J. Org. Chem.*, 1989, **54**, 4391.
224. J. Bergman and L. Engman, *J. Am. Chem. Soc.*, 1981, **103**, 5196.
225. D. H. R. Barton, N. Ozbalik and M. Ramesh, *Tetrahedron Lett.*, 1988, **29**, 3533.
226. S. Uemura, H. Takahashi and K. Ohe, *J. Organomet. Chem.*, 1992, **423**, C9.
227. E. Cuthbertson and D. D. MacNicol, *J. Chem. Soc., Chem. Commun.*, 1974, 498.
228. E. Cuthbertson and D. D. MacNicol, *Tetrahedron Lett.*, 1975, 1893.
229. S. Uemura, M. Wakasugi and M. Okano, *J. Organomet. Chem.*, 1980, **194**, 277.
230. E. Wenkert, M. H. Leftin and E. L. Michelotti, *J. Chem. Soc., Chem. Commun.*, 1984, 617.
231. J. Bergman and L. Engman, *J. Organomet. Chem.*, 1979, **175**, 233.
232. S. Uemura, K. Ohe, J. R. Kim, K. Kudo and N. Sugita, *J. Chem. Soc., Chem. Commun.*, 1985, 271.
233. S. Uemura, S. I. Fukuzawa, M. Wakasugi and M. Okano, *J. Organomet. Chem.*, 1981, **214**, 319.
234. S. Uemura and S. I. Fukuzawa, *J. Chem. Soc., Chem. Commun.*, 1980, 1033.
235. S. Uemura and S. I. Fukuzawa, *Chem. Lett.*, 1980, 943.
236. K. Chikamatsu, T. Otsubo, F. Ogura and H. Yamaguchi, *Chem. Lett.*, 1982, 1081.
237. M. J. Dabdoub, J. V. Comasseto, S. M. Barros and F. Moussa, *Synth. Commun.*, 1990, **20**, 2181.
238. S. Uemura and S. I. Fukuzawa, *Tetrahedron Lett.*, 1983, **24**, 4347.
239. S. Uemura and S. I. Fukuzawa, *J. Chem. Soc., Perkin Trans. 1*, 1985, 471.
240. S. Uemura and S. I. Fukuzawa, *J. Chem. Soc., Perkin Trans 1*, 1986, 1983.
241. K. B. Sharpless, K. M. Gordon, R. F. Lauer, D. W. Patrick, S. P. Singer and M. W. Young, *Chem. Scr.*, 1975, **8A**, 9.
242. H. Lee and M. P. Cava, *J. Chem. Soc., Chem. Commun.*, 1981, 277.
243. S. Uemura, K. Ohe and S. I. Fukuzawa, *Tetrahedron Lett.*, 1985, **26**, 895.
244. S. Uemura, Y. Hirai, K. Ohe and N. Sugita, *J. Chem. Soc., Chem. Commun.*, 1985, 1037.
245. T. Otsubo, F. Ogura and H. Yamaguchi, *Chem. Lett.*, 1981, 447.
246. F. Ogura, T. Otsubo and N. Ohira, *Synthesis*, 1983, 1006.
247. A. Osuka, Y. Mori, H. Shimizu and H. Suzuki, *Tetrahedron Lett.*, 1983, **24**, 2599.
248. X. Huang, L. Xie and H. Wu, *J. Org. Chem.*, 1988, **53**, 4862.
249. A. Osuka, Y. Hanazaki and H. Suzuki, *Nippon Kagaku Kaishi*, 1987, 1505.
250. S. W. Li, Z. L. Zhou, Y. Z. Huang and L. L. Shi, *J. Chem. Soc., Perkin Trans 1*, 1991, 1099.
251. Z. Z. Huang, L. W. Wen and X. Huang, *Synth. Commun.*, 1990, **20**, 2579.
252. X. Huang, L. Xie and H. Wu, *Tetrahedron Lett.*, 1987, **28**, 801.
253. Y. Z. Huang, L. L. Shi, S. W. Li and X. Q. Wen, *J. Chem. Soc., Perkin Trans 1*, 1989, 2397.
254. A. Osuka and H. Suzuki, *Tetrahedron Lett.*, 1983, **24**, 5109.
255. L. L. Shi, Z. L. Zhou and Y. Z. Huang, *Tetrahedron Lett.*, 1990, **31**, 4173.
256. Z. L. Zhou, Y. Z. Huang, L. L. Shi and J. Hu, *J. Org. Chem.*, 1992, **57**, 6598.
257. Y. Z. Huang, Y. Tang, Z. L. Zhou, J. L. Huang, *J. Chem. Soc., Chem. Commun.*, 1993, 7.
258. L. L. Shi, Z. L. Zhou and Y. Z. Huang, *J. Chem. Soc., Perkin Trans 1*, 1990, 2847.
259. Z. L. Zhou, L. L. Shi and Y. Z. Huang, *J. Chem. Soc., Perkin Trans 1*, 1991, 1931.
260. 'Houben-Weyl Methoden der Organischen Chemie', 4th edn., Thieme, Stuttgart, 1952, vol. 8, p. 427; 661.
261. (a) S. Patai, (ed.) 'The Chemistry of the Carbon–Carbon Triple Bond', Wiley, New York, 1978; (b) L. Brandsma, H. D. Verkruijsse, 'Synthesis of Acetylenes, Allenes and Cumulenes', Elsevier, Amsterdam 1981.
262. L. Engman, *J. Org. Chem.*, 1984, **49**, 3559.
263. T. Hiiro, N. Kambe, A. Ogawa, N. Miyoshi, S. Murai and N. Sonoda, *Angew. Chem., Int. Ed. Engl.*, 1987, **26**, 1187.
264. S. M. Barros, J. V. Comasseto and J. Berriel, *Tetrahedron Lett.*, 1989, **30**, 7353.
265. M. Pereyre, J. P. Quintard, A. Rahm, 'Tin in Organic Synthesis', Butterworths, London, 1987, p. 23; 153.
266. T. Kanda, T. Sugino, N. Kambe and N. Sonoda, *Phosphorus, Sulfur and Silicon*, 1992, **67**, 103.
267. (a) J. Bergman and L. Engman, *Z. Naturforsch., Teil B*, 1980, **35**, 217; (b) J. L. Piette and M. Renson, *Bull. Soc. Chim. Belg.*, 1970, **79**, 383; (c) J. L. Piette, D. Debergh, M. Baiwir and G. Llabses, *Spectrochim. Acta*, 1980, **36A**, 769; (d) T. Kanda, S. Nakaiida, T. Murai and S. Kato, *Tetrahedron Lett.*, 1989, **30**, 1829.

268. T. Hiiro et al., *J. Am. Chem. Soc.*, 1990, **112**, 455.
269. T. Hiiro et al., *Organometallics*, 1990, **9**, 1355.
270. C. A. Brandt, J. V. Comasseto, W. Nakamura and N. Petragnani, *J. Chem. Res. (S)*, 1983, 156.
271. J. V. Comasseto and J. N. Berriel, *Synth. Commun.*, 1990, **20**, 1681.
272. F. C. Tucci, A. Chieffi and J. V. Comasseto, *Tetrahedron Lett.*, 1992, **33**, 5721.
273. D. L. J. Clive, P. C. Anderson, N. Moss and A. Singh, *J. Org. Chem.*, 1982, **47**, 1641.
274. D. H. R. Barton, N. Ozbalik and J. C. Sarma, *Tetrahedron Lett.*, 1988, **29**, 6581.
275. D. H. R. Barton and M. Ramesh, *J. Am. Chem. Soc.*, **112**, 1990, 891.
276. D. H. R. Barton, P. I. Dalko and S. D. Géro, *Tetrahedron Lett.*, 1991, **32**, 4713.
277. D. H. R. Barton, S. D. Géro, B. Q. Sire, M. Samadi and C. Vincent, *Tetrahedron*, 1991, **47**, 9383.
278. L. B. Han, N. Kambe, A. Ogawa, I. Ryu and N. Sonoda, *Organometallics*, 1993, **12**, 473.
279. C. Chen, D. Crich and A. Papadatos, *J. Am. Chem. Soc.*, 1992, **114**, 8313.
280. C. Chen and D. Crich, *Tetrahedron Lett.*, 1993, **34**, 1545.
281. K. Sasaki, Y. Aso, T. Otsubo and F. Ogura, *Chem. Lett.*, 1989, 607.
282. H. Nakahira, I. Ryu, L. Han, N. Kambe and N. Sonoda, *Tetrahedron Lett.*, 1991, **32**, 229.
283. E. W. Colvin, in 'Silicon Reagents in Organic Synthesis', Academic Press, London, 1988, p. 25.
284. N. X. Hu, Y. Aso, T. Otsubo and F. Ogura, *Tetrahedron Lett.*, 1988, **29**, 4949.
285. J. E. Bäckwall and L. Engman, *Tetrahedron Lett.*, 1981, **22**, 1919.
286. H. Tani, K. Masumoto and T. Inamasu, *Tetrahedron Lett.*, 1991, **32**, 2039.
287. T. Yamauchi, K. Hattori, S. Mizutaki, K. Tamaki and S. Uemura, *Bull. Chem. Soc. Jpn.*, 1986, **59**, 3617.
288. M. Albeck and T. Tamari, *J. Organomet. Chem.*, 1982, **238**, 357.

Author Index

This Author Index comprises an alphabetical listing of the names of the authors cited in the text and the references listed at the end of each chapter in this volume.

Each entry consists of the author's name, followed by a list of numbers, for example

Templeton, J. L., 366, 385[233] (350, 366), 387[370] (363)

For each name, the page numbers for the citation in the reference list are given, followed by the reference number in superscript and the page number(s) in parentheses of where that reference is cited in the text. Where a name is referred to in text only, the page number of the citation appears with no superscript number. References cited both in the text and in the tables are included.

Although much effort has gone into eliminating inaccuracies resulting from the use of different combinations of initials by the same author, the use by some journals of only one initial, and different spellings of the same name as a result of the transliteration processes, the accuracy of some entries may have been affected by these factors.

Abdul Hai, A. K. M., 383[17] (358)
Abdullah, A. H., 141, 156[106] (141), 156[108] (141), 156[113] (142)
Abe, H., 89[505] (55), 484[7] (463), 564[335] (538)
Abe, S., 271[529] (233), 271[531] (234)
Abecassis, J., 87[382] (39)
Abel, S., 273[680] (244)
Abelman, M. M., 566[510] (546)
Abenhaïm, D., 187[106] (169)
Abley, P., 432[140b] (412)
Aboujaoude, E. E., 86[311] (26)
About-Jaudet, E., 387[326] (380)
Abramovitch, A., 270[466] (229)
Abramovitch, R. A., 70, 91[617] (70), 511[4] (488)
Abu-Libdeh, H., 558[19] (516)
Acemoglu, M., 348[120] (323)
Achi, S., 274[703] (246)
Achmatowicz, B., 348[163] (326)
AchyuthaRao, S., 186[40] (162, 172, 174), 186[66] (165, 179), 186[75] (166), 188[137c] (174, 177), 188[137d] (174, 177), 189[163] (177), 189[184a] (179), 189[185] (180), 190[201c] (182), 190[202] (182)
Ackland, D. J., 485[44] (471), 485[47] (473)
Acuña, C. A., 385[181] (370)
Adam, M. A., 272[572] (237)
Adam, W., 348[164] (327), 352[369] (340), 433[183] (418)
Adams, J. P., 386[217] (371)
Adams, R., 309[5] (278), 309[6] (278), 309[10] (279)
Adlington, R. M., 348[162] (326), 352[399] (342), 384[80] (362), 386[221] (371), 386[235] (372, 373), 387[296] (377)
Aeberli, P., 128[199] (122)
Aerssens, M. H. P. J., 348[130] (324)
Afarinkia, K., 353[406] (342), 353[407] (342)
Agami, C., 187[114a] (171), 187[114b] (171)

Agawa, T., 568[644] (557)
Agback, P., 567[538] (548), 567[540] (548)
Agenäs, L.-B., 559[60] (517, 518), 560[115] (519, 520, 522, 538, 539, 552–4), 563[232] (528), 568[593] (554)
Ager, D. A., 86[309] (26)
Ager, D. J., 82[28] (2, 26), 86[291] (25), 86[292] (25, 28), 86[298] (26), 91[663] 264[48] (195), 348[166] (327)
Aggarwal, V. K., 85[217] (15)
Agnel, G., 348[141] (325)
Agosta, W. C., 433[194] (419)
Agwaramgbo, E. L. O., 348[172] (328)
Ahlbrecht, H., 83[88] (5), 85[218] (15), 85[221] (15), 115, 127[144] (115), 127[145] (115), 127[146] (115), 127[184] (120), 387[311] (378)
Ahmed, M. K., 157[176] (152)
Ahrens, H., 86[272] (23)
Ahuja, J. R., 90[604] (67)
Aidhen, I. S., 90[603] (67), 90[604] (67)
Aigbirhio, F., 186[57b] (165)
Aizpurua, J. M., 156[169] (151)
Akabori, S., 567[571] (552)
Akahane, A., 85[241] (19)
Akai, S., 349[191] (329), 349[192] (329), 385[144] (367)
Akermark, B., 187[111] (170)
Akers, J. A., 264[47] (194)
Akhmedov, I. M., 564[343] (539), 564[348] (539)
Akiba, K., 511[49] (491, 501), 512[70] (496, 497), 512[71] (497), 512[94] (502), 512[95] (502), 512[96] (502), 512[97] (502), 512[101] (502)
Akiba, K.-Y., 156[110] (141)
Akiba, M., 597[62] (572, 573, 576), 599[133] (576–8)
Akiyama, M., 156[149] (149)
Akiyama, T., 352[342] (339)
Akiyama, Y., 190[209] (184)
Akiyoshi, K., 189[183a] (179), 189[183b]

(179)
Akkerman, O. S., 185[21f] (160)
Akritopoulou, I., 346[34] (315)
Akutagawa, K., 85[243] (19), 85[256] (22), 92[691] (72)
Alajarín, R., 351[296] (337)
Alami, M., 386[218] (371)
Alami, N. E., 186[38d] (162, 169)
Al-Arnaout, A., 156[105] (141)
Al-awar, R. S., 156[112] (142), 156[120] (144)
Albeck, M., 601[288] (596)
Albert, H.-J., 384[96] (363)
Albert, R., 352[369] (340)
Alberts, V., 430[41b] (396)
Al'bitskaya, V. M., 564[342] (539)
Albizati, K. F., 91[629] (71), 91[636] (71)
Alex, R. F., 267[200] (204)
Alexakis, A., 155[86] (137), 188[149d] (175, 182), 188[149e] (175, 182), 384[111] (364)
Alexander, C. W., 387[319] (379)
Alfonso, C. A. M., 351[330] (338), 351[331] (338)
Ali, H., 563[287] (534, 540, 542, 554), 563[288] (534, 540), 563[289] (534)
Ali, M. E. S., 598[118] (575)
Alicino, J. F., 559[67] (518)
Al-Jabar, N. A. A., 83[67] (4)
Allcock, H. R., 90[577] (65)
Allen, D. J., 84[168] (8)
Allen, R. P., 353[444] (344)
Allred, A. L., 560[90] (518)
Almena, J., 85[204] (14), 85[205] (14)
Alnajjar, M. S., 384[103] (363)
Alonso-Cires, L., 433[147] (412), 434[208] (421)
Alper, H., 353[413] (342)
Alt, H. G., 430[19] (391)
Altman, J., 266[168] (203)
Alvanipour, A., 383[7] (357)
Alvarado, S. I., 458[26] (447), 458[32] (449)
Alvarez-Builla, J., 351[296] (337)
Alvernhe, G., 563[267] (530)
Amaike, M., 386[286] (376)

Amamria, A., 353[400] (342)
Amano, T., 126[111] (109)
Amaratunga, W., 156[164] (151)
Amari, H., 458[42] (451)
Ambrose, K. R., 598[78] (573)
Ames, A., 311[97] (304), 563[243] (529, 530, 539, 540, 545), 563[244] (529, 539, 545)
Ames, M. M., 267[249] (208)
Amice, P., 189[174c] (178)
Amirthavalli, M., 599[195] (582)
Amishiro, N., 386[275] (375)
Amoroso, R., 434[231a] (425), 434[231b] (425)
Amos, M. F., 89[494] (53)
Amos, R. A., 434[199b] (420)
Amosova, S. V., 598[91] (574), 598[93] (574)
Amstrong, J. D., III, 83[113] (5)
Amt, H., 267[232] (206)
Amtmann, R., 310[75] (291)
Ananthanarayan, T. P., 88[462] (49)
Andell, O. S., 434[250] (429)
Andersen, M. W., 272[570] (237)
Andersen, N. H., 87[387] (40)
Anderson, B. A., 265[95] (199)
Anderson, C. L., 268[333] (215)
Anderson, M. B., 352[397] (342)
Anderson, P. C., 273[679] (244), 601[273] (594)
Anderson, R., 86[290] (25), 352[382] (341)
Anderson, S. B., 352[382] (341)
Andersson, P. G., 353[438] (343)
Ando, A., 350[262] (335)
Ando, D., 353[411] (342)
Ando, D. J., 566[459] (544)
Ando, M., 347[107] (321)
Ando, W., 84[141] (6), 348[140] (325)
Andreini, B. P., 187[88b] (167), 187[88c] (167)
Andreoli, P., 83[117] (5), 353[421] (343)
Andrés, C., 187[114d] (171)
Andrews, G. C., 30, 83[82] (4, 29, 30, 36)
Andrey, O., 349[194] (329)
Andringa, H., 88[446] (48, 53, 59), 128[206] (123, 124)
Angeles, E., 82[44] (3)
Angell, R., 346[43] (316)
Angibaud, P., 351[317] (338)
Angoh, A. G., 567[549] (548)
Anh, N. T., 146
Anker, D., 563[267] (530)
Anklekar, T. V., 85[246] (20)
Annoura, H., 148
Annunziata, R., 87[367] (36), 87[368] (36)
Ansorge, A., 270[415] (223)
Antel, J., 155[41] (132)
Anthonsen, T., 156[137] (147)
Anthony-Mayer, C., 564[308] (537)
Anwar, S., 350[234] (332, 343)
Aoai, T., 568[635] (556)
Aoki, S., 185[20a] (160, 165), 350[269] (335), 384[72] (361), 384[88] (362), 567[571] (552)
Aoyama, S., 565[376] (540, 545)
Aoyama, T., 352[382] (341), 352[383] (341), 352[385] (341), 352[386] (341)
Apostolopoulos, C. D., 91[632] (71)
Apparu, M., 127[187] (120)
Applequist, D. E., 84[161] (8)

Arai, H., 350[259] (335), 386[244] (373), 387[331] (381), 567[525] (547, 551)
Arai, I., 189[177b] (178), 274[760] (252)
Arai, M., 351[328] (338), 432[95b] (404)
Arai, N., 387[301] (377)
Arai, Y., 565[399] (541, 543)
Araki, M., 350[255] (334)
Araki, S., 90[584] (65), 90[587] (65), 512[63] (495)
Araki, Y., 387[291] (376)
Arase, A., 265[116] (200), 269[378] (218), 269[407] (221), 269[408] (221), 269[410] (221), 269[411] (221)
Arduini, A., 458[47] (451)
Arens, J. F., 598[99] (574, 593)
Argade, A. B., 267[210] (205)
Arguelles, R., 270[423] (224)
Arimoto, M., 458[37] (450)
Ariura, S., 384[132] (366)
Armer, R., 84[132] (6, 33)
Armistead, D. M., 350[277] (336)
Armstrong, J. D., III, 351[282] (336)
Armstrong, R. W., 271[488] (232)
Armstrong-Chong, R. J., 89[532] (60)
Arne, K., 268[272] (209, 210), 268[273] (209, 210, 240)
Arnold, M., 90[567] (64), 90[568] (64)
Arnold, R. T., 185[21d] (160)
Arnstedt, M., 186[51b] (164)
Arora, S. K., 268[301] (212)
Arpe, H. J., 598[115] (575)
Arseniyadis, S., 434[215] (422)
Arshady, R., 458[13] (444)
Arterburn, J. B., 565[422] (542, 557)
Arthur, M.-P., 266[161] (203), 266[162] (203)
Artis, D. R., 353[419] (343)
Arvanaghi, M., 156[163] (151)
Arzoumanian, H., 265[63] (196), 431[57a] (399)
Asami, M., 156[135] (147), 156[143] (148), 188[130c] (171)
Asano, R., 511[50] (491)
Asano, T., 484[6] (462)
Asanuma, G., 563[253] (530, 538, 539)
Asao, N., 83[127] (5), 127[141] (114), 271[551] (235, 236, 238, 251, 252), 347[66] (317)
Asaoka, M., 349[197] (329)
Asberom, T., 127[182] (119)
Asensio, G., 433[146] (412), 433[147] (412), 433[178] (416), 433[195] (419), 434[202] (421), 434[203] (421, 423), 434[208] (421), 434[209] (422), 434[217] (422)
Ashby, E. C., 84[169] (8, 11), 384[102] (363)
Ashe, A. J., III, 266[196] (204)
Asirvatham, E., 387[297] (377)
Aslam, M., 352[387] (341)
Aso, Y., 82[43] (3), 311[98] (304), 598[128] (576, 584), 599[134] (576, 577), 599[135] (576, 584), 599[149] (577), 599[182] (582), 599[199] (583), 600[201] (583), 600[202] (583, 584, 586), 600[204] (583), 600[212] (584, 587), 600[213] (584), 600[214] (584), 600[215] (584), 600[222] (586), 600[223] (586, 587), 601[281] (596), 601[284] (596)
Asprey, L. B., 186[56b] (164)
Asselin, A., 89[480] (51)
Astley, M. P., 566[484] (545, 548)

Atkins, A. R., 430[41b] (396)
Atwood, J. L., 458[53] (453), 458[54] (453), 459[59] (453)
Aubé, J., 90[566] (64)
Aubert, T., 88[408] (42, 53)
Audran, G., 347[116] (322)
Auger, J., 189[156] (176)
August, B., 432[135] (410)
Augustine, R. A., 559[40] (516, 519, 522, 536)
Aurrekoetxea, N., 156[169] (151)
Auvray, P., 187[98e] (169)
Avasthi, K., 270[472] (229)
Avery, M. A., 187[98a] (169)
Awad, M. M. A., 84[198] (12)
Awasthi, A. K., 351[296] (337), 458[15] (445)
Awra, S., 268[331] (215)
Ay, M., 348[142] (325)
Ayabe, A., 351[319] (338)
Ayala, A. D., 458[36] (450)
Ayers, T. A., 84[191] (11), 84[192] (12)
Ayukawa, H., 384[133] (367)
Azhar Saliente, T., 91[625] (71)
Azimioara, M. D., 276[855] (263)
Aznar, F., 433[149] (413, 416, 419, 425, 426), 434[216] (422), 434[220] (423), 434[221a] (423), 434[221c] (423), 434[222] (423), 434[223] (423), 435[253] (429)

Baba, A., 383[43] (360), 385[145] (367, 368), 385[154] (368), 385[155] (368), 385[157] (368), 386[246] (373), 386[258] (373), 512[79] (499), 512[80] (500), 512[81] (500)
Baba, S., 311[108] (307), 311[116] (308)
Baba, T., 270[472] (229)
Baba, Y., 310[71] (291)
Babot, O., 350[235] (333)
Baboulène, M., 264[53] (195), 264[54] (195), 264[55] (195)
Babudri, F., 155[58] (134), 155[60] (134), 155[65] (134), 346[44] (316)
Baceiredo, A., 266[161] (203), 266[162] (203)
Bach, R. D., 126[113] (109), 432[142b] (412), 458[23] (447)
Bach, R. O., 82[13] (2)
Bach, T., 350[258] (334)
Bachi, M. D., 348[136] (324), 383[39] (360, 361), 383[67] (361), 566[496] (545, 548, 549), 566[497] (545, 549), 567[541] (548)
Bachman, G. B., 434[237a] (426)
Bachovchin, W. W., 264[21] (192)
Baciocchi, E., 351[323] (338)
Back, T. G., 86[332] (28), 559[39] (516, 519, 520, 522, 531-4, 538, 540-3, 548, 553-5, 557), 559[50] (516, 546), 561[127] (519, 542, 543, 547, 548), 561[128] (519, 522, 540-2), 564[317] (538, 557), 565[386] (540, 545), 566[490] (545), 567[529] (547), 568[615] (555)
Backes, J., 82[20] (2)
Backlund, S. J., 272[588] (238)
Bäckvall, J.-E., 434[250] (429), 565[424] (542), 598[113] (575), 601[285] (596)
Baclawski, L. M., 82[41] (3)
Badaoui, E., 89[512] (56), 562[211] (526, 543, 550), 565[372] (540, 551),

565[394] (541, 544), 566[453] (544, 547), 566[454] (544), 567[557] (550, 551)
Badiello, R., 559[64] (517)
Badone, D., 385[209] (371)
Baeyer, A., 431[71c] (401)
Bagheri, V., 158[8d] (159, 166), 349[202] (329)
Bagnall, K. W., 559[58] (517, 518, 522, 528)
Bagnoli, L., 565[389] (540)
Bahl, J. J., 126[88] (103, 107)
Bähr, K., 309[4] (278)
Bailar, J. C., Jr., 91[633] (71), 559[58] (517, 518, 522, 528)
Bailey, T. R., 85[225] (16, 17), 85[248] (20)
Bailey, W. F., 10, 82[23] (2), 84[165] (8), 84[174] (10), 84[183] (10), 84[184] (12), 84[185] (10, 11, 14), 84[186] (10), 84[187] (10), 84[189] (11), 84[190] (11), 84[193] (12), 84[195] (12), 154[20] (130)
Bailey, W. I., 458[6] (438)
Bailie, I. C., 125[47] (95)
Bain, J., 433[159] (414)
Baird, W. C., Jr., 432[115a] (407)
Baiwir, M., 600[267c] (593)
Bakale, R. P., 387[330] (380, 381)
Baker, D. R., 92[676]
Baker, K. V., 154[3] (130)
Baker, R. J., 156[147] (148)
Baker, R. T., 265[93] (199), 265[96] (199), 265[97] (199), 265[99] (199), 265[100] (199)
Baker, R. W., 89[493] (53), 155[53] (133), 155[54] (133), 155[55] (133)
Bakshi, R. K., 267[215] (205), 267[216] (205, 206), 267[223] (206), 272[623] (240), 275[817] (259), 275[821] (259), 567[551] (549)
Baldoni, A. A., 430[12] (390)
Baldridge, K. K., 88[448] (48)
Baldwin, J. E., 348[162] (326), 351[284] (336), 352[399] (342), 384[80] (362), 386[221] (371), 386[235] (372, 373), 387[296] (377), 560[87] (518, 521, 544, 549), 560[96] (519, 520), 560[97] (519, 527, 528), 560[108] (519, 520, 528), 560[111] (519, 522, 530–2, 547), 560[113] (519, 520, 522, 528, 538–40, 543, 544, 547, 552), 560[114] (519, 520, 522, 538, 539, 554, 557, 558), 561[130] (520, 552), 561[135] (520, 522, 535, 553, 554, 557), 561[162] (522, 529, 540, 544), 562[197] (522, 533–7), 564[339] (539, 541, 544, 551, 558), 564[359] (539, 540, 543), 565[433] (542), 567[518] (547), 567[533] (548), 567[534] (548), 569[657] (558)
Baldwin, R. M., 266[165] (203)
Bales, S. E., 154[6] (130)
Balfe, M. P., 597[30] (572)
Ball, D., 274[749] (251)
Ballard, D. H., 433[180b] (417)
Ballesteros, A., 353[417] (343), 353[426] (343)
Ballestri, M., 567[526] (547)
Ballini, R., 155[89] (139), 349[193] (329)
Ballistreri, F. P., 433[154] (413)
Balmer, M., 91[661]

Bambridge, K., 350[237] (333)
Bamford, C. H., 264[31] (192)
Bandini, E., 353[421] (343), 353[422] (343)
Banerjee, A. K., 351[329] (338)
Banerji, A., 459[68] (455)
Banfi, L., 274[742] (251, 253)
Banik, G. M., 349[179] (328)
Bannai, K., 186[68] (165)
Banno, H., 346[45] (316), 351[316] (338)
Bannou, T., 563[240] (529)
Baquet, C., 83[95] (5)
Barak, A. V., 126[76] (100, 103)
Baraldi, P. G., 346[41] (316)
Barbeaux, P., 562[209] (526, 550, 551), 562[214] (526, 550, 551), 562[215] (526, 550, 551), 562[216] (526, 544, 550, 551), 564[312] (538, 550, 551)
Barbero, A., 384[98] (363, 364), 384[114] (364), 385[186] (370, 373)
Barbey, S., 92[721] (81)
Barcock, R. A., 91[654]
Bard, A. J., 560[93] (518), 560[94] (518)
Bardin, F., 92[720] (81)
Baret, P., 273[692] (245)
Barluenga, J., 84[155] (7), 353[417] (343), 353[426] (343), 431[77b] (402), 431[86] (403), 433[146] (412), 433[147] (412), 433[149] (413, 416, 419, 425, 426), 433[178] (416), 433[195] (419), 433[198] (420, 429), 434[202] (421, 423), 434[203] (421, 423), 434[204] (421), 434[206] (421), 434[207a] (421), 434[207b] (421), 434[208] (421), 434[209] (422), 434[216] (422), 434[217] (422), 434[218] (423), 434[219a] (423), 434[219b] (423), 434[219c] (423), 434[220] (423), 434[221a] (423), 434[221c] (423), 434[222] (423), 434[223] (423), 434[225a] (424, 425), 434[225b] (424, 425), 434[246] (428, 429), 435[253] (429)
Barnes, J. L., 187[103b] (169)
Barnette, W. E., 565[435] (542, 547, 551)
Barney, C. L., 563[270] (530, 556)
Barnier, J. P., 189[174c] (178)
Barnum, C., 567[563] (551)
Barrett, A. G. M., 274[752] (252), 274[757] (252), 275[775] (255), 346[50] (316), 348[166] (327)
Barrett, S. D., 186[67] (165)
Barrish, J. C., 264[49] (195)
Barron, A. R., 563[278] (530)
Barros, S. M., 598[94] (574), 600[237] (589), 600[264] (593)
Barry, C. E., III, 126[119] (109)
Bartels, M., 266[191] (204)
Barth, L., 351[302] (337)
Barth, R. F., 264[16] (192)
Bartle, K. D., 128[200] (122)
Bartlett, P. A., 433[192] (419)
Bartmann, D., 433[144] (412)
Bartmann, E., 84[196] (12)
Bartoli, G., 86[318] (27), 154[26] (130), 154[27] (130), 349[193] (329)
Bartolotti, L. J., 85[222] (16), 85[247] (20)
Barton, D., 83[109] (5), 84[163] (8)
Barton, D. H. R., 126[81] (101), 267[258] (208), 273[655] (243),
349[202] (329), 353[445] (344), 383[19] (358, 359), 383[36] (359), 484[34] (467, 473, 480), 484[39] (469, 480), 484[40] (469, 480), 485[62] (481), 488, 511[3] (488, 506), 511[4] (488), 512[68] (496), 512[104] (503), 512[107] (504), 512[108] (504, 505), 512[109] (504, 505), 512[110] (504, 509), 512[111] (504, 505, 507, 508), 512[112] (504), 512[113] (504), 512[116] (505, 508), 512[117] (505, 507, 508), 512[120] (505, 507, 509), 512[125] (507, 509), 512[126] (507–9), 512[127] (507), 512[128] (507, 508), 512[129] (508), 513[130] (508), 513[132] (508), 513[134] (508), 513[135] (509), 513[138] (509, 510), 513[141] (510), 513[142] (510), 513[143] (510), 513[144] (510), 516, 559[50] (516, 546), 562[165] (522), 562[177] (522, 534), 562[193] (522), 563[293] (534), 563[294] (534), 563[296] (534), 563[297] (535), 564[349] (539), 564[350] (539), 597[19] (572), 598[127] (576, 583), 598[130] (576, 583), 599[142] (577), 599[150] (578), 599[155] (578, 581), 599[197] (583), 599[198] (583), 600[225] (587), 601[274] (595), 601[275] (595), 601[276] (595), 601[277] (595)
Barton, J. C., 385[176] (369), 386[276] (375)
Barton, T. J., 348[139] (325)
Bartroli, J., 275[782] (256), 275[783] (256)
Basalgina, T. A., 458[25] (447)
Basavaiah, D., 265[103] (199), 265[120] (200), 265[127] (200, 221, 242), 266[128] (200), 266[129] (200), 267[209] (205), 270[440] (225), 270[461] (229), 270[470] (229)
Bashilov, V. V., 430[31a] (393)
Basile, T., 272[604] (240, 241), 350[245] (333)
Baskar, A. J., 458[54] (453)
Baskaran, S., 265[70] (196), 265[71] (196), 352[343] (339)
Bassetti, M., 433[179a] (417), 433[179b] (417), 433[196] (420)
Bassindale, A. R., 86[296] (25), 86[308] (26), 346[9] (314)
Basso, N., 353[452] (344)
Bates, G. S., 266[199] (204), 565[395] (541), 566[480] (545)
Bates, R. B., 126[88] (103, 107), 126[93] (103), 126[113] (109), 126[114] (109), 126[115] (109), 126[116] (109), 126[117] (109), 126[120] (109), 126[121] (109)
Bates, T. F., 84[176] (10), 346[33] (315)
Batra, M. S., 352[353] (340)
Bats, J. W., 90[568] (64)
Batt, L., 559[69] (518)
Batty, D., 383[71] (361), 566[481] (545, 547–9), 567[542] (548), 567[543] (548)
Baudouy, R., 434[241] (427)
Bauer, H., 563[275] (530)
Bauer, T., 156[138] (147)
Bauer, W., 82[11] (2), 87[398] (42), 125[43] (95, 112), 125[44] (95, 112),

125⁴⁵ (95)
Baumgärtel, O., 84¹⁴⁰ (6)
Bäuml, E., 348¹⁴⁷ (325)
Bayet, P., 567⁵⁷² (552)
Bayod, M., 435²⁵³ (429)
Bayón, A. M., 434²⁰² (421), 434²⁰³ (421, 423)
Bayston, D. J., 347¹⁰² (321)
Bazzanini, R., 346⁴¹ (316)
Beak, P., 16, 17, 56, 83⁹⁹ (5), 84¹³⁸ (6), 84¹⁶⁸ (8), 85²¹⁶ (15, 16), 85²²³ (16, 17), 85²²⁴ (16), 85²²⁹ (17), 85²³² (17), 85²³⁴ (18), 87³⁹⁴ (42), 87⁴⁰¹ (42), 87⁴⁰³ (42), 87⁴⁰⁴ (42), 88⁴²³ (42), 89⁵⁰⁹ (56), 89⁵¹⁰ (56), 89⁵¹¹ (56), 91⁶¹⁵ (70), 127¹⁴³ (114), 127¹⁴⁸ (115)
Beatty, K. M., 431⁵⁶ (399, 420)
Beau, J.-M., 85²⁶³ (22), 271⁴⁸⁸ (232), 385¹⁹⁷ (371)
Beaudet, I., 384¹¹⁶ (364), 385²⁰¹ (371)
Beaulieu, F., 89⁴⁷⁴ (51), 89⁴⁷⁵ (51)
Beaulieu, P. L., 599¹⁷⁴ (580)
Bebb, R. L., 127¹⁹⁰ (121)
Bebbington, D., 352³⁹⁹ (342)
Beck, A. K., 188¹²⁸ᵃ (171, 172), 188¹²⁸ᶜ (171, 172), 565³⁷¹ (540, 550), 598⁶⁷ (573)
Becker, E. I., 597⁶ (571)
Beckley, R. S., 430⁴¹ᵃ (396)
Beckwith, A. L. J., 383²² (358), 562²¹⁷ (526, 547)
Bedford, C. D., 348¹³¹ (324)
Bedford, S. B., 349²¹⁶ (330)
Bednarski, M. D., 347⁸³ (319)
Beedle, E. C., 274⁷²⁸ (250), 350²⁷⁰ (335)
Beese, G., 91⁶⁷², 348¹⁵⁴ (326)
Beetz, I., 271⁵²² (232)
Beger, J., 434²²⁷ (424)
Begley, M. J., 84¹³² (6, 33)
Begum, M. K., 351³⁴¹ (339)
Behrendt, L., 188¹²⁸ᵇ (171, 172)
Belaud, C., 186³⁸ᵃ (162, 169), 186³⁸ᵇ (162, 169), 186³⁸ᶜ (162, 169), 186³⁸ᵈ (162, 169), 187¹¹⁶ᶜ (171), 272⁵⁶¹ (236)
Belen'kii, L. I., 86³²⁸ (28)
Beletskaya, I. P., 271⁵⁰⁷ (232), 271⁵⁰⁸ (232), 430³⁶ (395), 431⁶⁵ (400), 431⁸³ (403), 432¹²² (407, 409), 432¹²⁵ (408), 432¹²⁸ᵇ (409)
Bell, A. S., 187⁸⁵ (167)
Bell, H. C., 484¹³ (465, 466), 484³¹ (467, 481), 484³⁸ (469), 485⁶³ (481, 482)
Bell, S. H., 89⁵³² (60)
Bellamy, F., 92⁶⁹² (72)
Bellassoued, M., 188¹⁴⁸ᶜ (175, 182), 189¹⁵⁸ (176)
Bellew, D. R., 351³¹³ (337)
Bellina, F., 383⁸ (357, 361, 371, 372)
Belo, A., 458¹⁶ (445)
Belyk, K., 186⁴¹ (162, 172)
Ben Rayana, E., 127¹⁸⁶ (120)
Benalil, A., 269³⁶⁷ (218), 269³⁶⁸ (218)
Bendayan, M. T., 431⁷³ᵉ (401)
Bender, S. L., 598¹⁰² (574)
Benderly, A., 272⁶⁰⁵ (240)

Benefice, S., 185³²ᵇ (161)
Benefice-Malouet, S., 185³²ᶜ (161)
Benezra, C., 187⁹⁸ᶜ (169)
Bengelsdorf, I. S., 268²⁹⁹ (212)
Bengtsson, S., 89⁴⁸⁵ (51)
Benhamou, M. C., 430⁴⁰ (396, 416, 421), 433¹⁷⁰ᵃ (416, 422)
Benhamza, R., 430⁴⁴ (396, 416)
Benjelloun, N. R., 156¹²² (144)
Benkeser, R. A., 94, 108, 125⁹ (94), 125¹⁹ (94), 125²¹ (94, 108), 125³¹ (94)
Benmaarouf-Khallaayoun, Z., 264⁵⁴ (195)
Benn, M. H., 564³²⁴ (538, 557)
Benn, R., 566⁴⁹⁹ (545)
Bennani, Y. L., 566⁴³⁹ (543, 557)
Bennetau, B., 88⁴¹⁰ (42, 44, 72), 348¹⁵² (326)
Bennett, A. J., 385¹⁷⁵ (369)
Bennett, P. A. R., 351²⁸⁴ (336)
Bennett, S. G., 154¹⁸ (130)
Bennett, W. D., 513¹³³ (508)
Benson, T. J., 353⁴⁵⁶ (344)
Beppu, K., 87³⁸⁴ (39)
Berenguer, R., 275⁸²⁴ (259)
Beresis, R., 347¹⁰³ (321)
Bergbreiter, D. E., 513¹³¹ (508)
Bergens, S. H., 353⁴³⁷ (343)
Bergman, J., 92⁶⁹³ (73), 155⁹⁸ (140), 561¹³³ (520), 597⁴⁸ (572, 587), 598¹¹³ (575), 598¹²⁰ (575, 587), 600²²⁴ (586), 600²³¹ (588), 600²⁶⁷ᵃ (593)
Bergmann, H.-J., 86³⁰² (26)
Berk, S. C., 185⁹ (159, 160, 165, 168, 174), 185¹³ᵃ (160, 162), 188¹³⁸ (174), 353⁴⁵⁰ (344)
Berkowitz, D. B., 563²⁶² (530, 539)
Berkulin, W., 125⁶⁷
Bermejo, M. R., 459⁷⁰ (456), 459⁷¹ (456), 459⁷² (456), 459⁷³ (456), 459⁷⁴ (456), 459⁷⁵ (456), 459⁷⁶ (456), 459⁷⁷ (456)
Bernabé, P., 384⁹² (363)
Bernal, I., 347⁹⁷ (321)
Bernard, H., 273⁶⁹³ (245)
Bernardi, A., 273⁶⁴⁶ (241, 258), 274⁷⁶⁷ (253), 275⁷⁹⁸ (258), 275⁷⁹⁹ (258), 275⁸⁰⁰ (258)
Bernardi, F., 241, 273⁶⁴⁷ (241)
Bernardinelli, G., 352³⁶⁰ (340)
Bernardini, F., 310⁷³ (291)
Bernardon, C., 154¹⁴ (130)
Bernardou, F., 188¹⁵⁵ᵇ (176)
Berndt, A., 264² (192)
Berriel, J. N., 600²⁶⁴ (593), 601²⁷¹ (593)
Berridge, M. S., 268³²⁵ (215)
Berry, F. J., 569⁶⁵⁶ (558)
Berry, F. Y., 597⁹ (571)
Berry, M. B., 351³³⁸ (339)
Berry, M. J., 558¹⁷ (516)
Berthiaume, S., 273⁶⁷⁹ (244)
Bertolini, G., 350²⁴² (333), 351²⁹⁷ (337)
Bertozzi, C. R., 347⁸³ (319)
Bertozzi, S., 155¹⁰³ (141)
Bertrand, G., 266¹⁶¹ (203), 266¹⁶² (203), 387³²² (379)
Bertrand, M. T., 189¹⁵⁷ᵇ (176)

Beruben, D., 188¹⁴⁹ᶜ (175, 182)
Beswick, P. J., 92⁶⁹⁸ (74)
Bettiol, J.-L., 188¹¹⁹ᵇ (171)
Betz, R., 127¹⁸¹ (119)
Bewersdorf, M., 84¹⁷³ (9), 87³⁹⁰ (40), 566⁴⁵² (544)
Bhanu Prasad, A. S., 273⁶⁷⁰ (244)
Bhar, S., 350²⁶⁵ (335)
Bharathi, S. N., 87³⁶³ (33), 353⁴²⁰ (343)
Bhaskar, K. V., 564³⁰³ (536)
Bhaskar Kanth, J. V., 273⁶⁶⁹ (244), 273⁶⁷⁰ (244), 275⁸¹⁴ (258)
Bhat, K. S., 125⁶⁸ (103), 126⁸⁴ (101), 267²²⁶ (206), 274⁷¹⁵ (247, 252), 274⁷⁴⁵ (251), 274⁷⁴⁶ (251), 274⁷⁵³ (252)
Bhat, N. G., 265¹⁰² (199), 266¹²⁹ (200), 266¹⁵⁶ (202), 267²⁰⁹ (205), 267²¹⁹ (205), 267²⁵⁴ (208), 267²⁵⁵ (208), 269³⁸⁵ (219), 270⁴⁶¹ (229), 270⁴⁶² (229), 270⁴⁷⁰ (229), 270⁴⁷¹ (229), 430²³ (391)
Bhatnagar, N. Y., 512¹²⁷ (507)
Bhatt, M. V., 353⁴¹² (342)
Bhatt, R. K., 385²⁰⁴ (371), 387²⁸⁹ (376)
Bhupathy, M., 84¹⁹⁸ (12), 85²¹⁴ (15, 21), 187¹⁰⁴ (169)
Bhushan, V., 275⁸⁰⁴ (258, 259)
Bhuyan, K. C., 558¹⁹ (516)
Bianchi, G., 268²⁹¹ (212)
Biasotti, J. B., 432¹³³ (410)
Bible, R. H., Jr., 430²⁹ (393)
Bickelhaupt, F., 154²² (130), 185²¹ᶠ (160), 186⁵⁹ (165), 187⁹⁷ᵃ (169), 189¹⁶¹ᶜ (176), 189¹⁶¹ᵇ (176), 190²⁰¹ᵃ (182), 190²⁰¹ᵇ (182), 274⁷⁵⁸ (252)
Biehl, E. R., 89⁵²⁶ (57)
Bielawska, A., 430⁴⁰ (396, 416, 421)
Biellmann, J.-F., 348¹⁴⁶ (325), 566⁴⁵⁸ (544)
Bieri, J. H., 91⁶⁶¹
Bierschenk, T. R., 186³⁷ᵇ (162), 458⁶ (438)
Bigelow, S. S., 267²⁵² (208), 271⁵⁵⁴ (235), 271⁵⁵⁵ (235)
Bijpost, E. A., 274⁷⁵⁸ (252)
Bilevitch, K. A., 458³ (438)
Billedeau, R. J., 89⁵³³ (60), 563²⁴⁵ (529, 541)
Billington, D. C., 187⁷⁹ᵃ (166, 167)
Billmers, J. M., 568⁵⁸⁴ (554), 568⁵⁸⁵ (554)
Binder, J., 86³¹⁰ (26)
Binger, P., 270⁴⁵¹ (228), 270⁴⁵³ (228), 310⁵² (285, 286)
Bingu, P., 268³³¹ (215)
Binnewirtz, R.-J., 266¹⁴⁹ (202)
Biondi, S., 272⁶⁰⁴ (240, 241)
Bir, G., 276⁸⁵⁸ (263)
Birch, D. J., 386²³⁵ (372, 373)
Bird, C. W., 91⁶²⁰ (71), 91⁶²⁶ (71)
Bird, M. L., 560¹⁰⁴ (519, 528, 554)
Birkinshaw, S., 270⁴²⁷ (224, 229)
Birss, V. I., 567⁵²⁹ (547)
Bisaha, J., 90⁵⁴³ (63)
Bischofberger, N., 155⁹¹ (139)
Bischoff, L., 269³⁵⁷ (217)
Bishop, J. J., 90⁵⁷⁶ (65)
Bitler, S. P., 348¹²⁹ (324)

Bitterwolf, T. E., 458[51] (452), 458[52] (452)
Bjorge, S. M., 432[107a] (406)
Björk, P., 385[216] (371)
Björup, P., 567[550] (548)
Black, M., 89[489] (52, 59)
Blackborow, J. R., 266[145] (201, 202), 266[147] (202)
Blagg, J., 85[257] (22), 88[438] (46), 90[555] (63), 275[785] (256), 351[338] (339)
Blagoev, B., 154[13] (130)
Blake, J. F., 349[199] (329)
Blanchard, C., 266[160] (203)
Blanchard, E. P., Jr., 430[13] (390)
Blancou, H., 185[15b] (160), 185[32a] (161), 185[32b] (161), 185[32c] (161), 185[32d] (161)
Blasioli, C., 433[188a] (418)
Blattner, R., 433[145a] (412)
Blaukat, U., 432[88a] (403, 427, 429), 432[88b] (403)
Blaum, J., 268[307] (213)
Blazejewski, J.-C., 512[111] (504, 505, 507, 508), 513[134] (508)
Blecher, J., 566[498] (545, 546)
Bloch, R., 86[305] (26), 87[382] (39)
Block, E., 89[491] (53), 128[208] (124), 352[387] (341)
Blom, H. P., 265[96] (199), 265[99] (199)
Blomberg, C., 189[164] (178)
Blomstrom, D. C., 430[13] (390)
Bloodworth, A. J., 390, 430[4] (390), 430[30d] (393), 430[43] (396), 433[175b] (416), 433[180b] (417), 433[183] (418), 433[184] (418), 433[185] (418), 434[248] (429)
Bloom, J. D., 88[416] (42)
Bloomer, J. L., 351[310] (337)
Blount, J. F., 567[522] (547, 551)
Blue, C. D., 268[313] (213)
Blümel, J., 85[219] (15)
Blumenkopf, T. A., 346[39] (315)
Blumenthal, E. M., 434[213] (422)
Blumenthal, M., 126[100] (107)
Blundell, P., 383[36] (359)
Bluthe, N., 434[242] (427)
Boardman, L. D., 270[459] (228)
Boaventura, M.-A., 434[245a] (427)
Bobbitt, K. L., 273[686] (244)
Bobosik, V., 271[504] (232)
Bocelli, G., 433[179a] (417)
Boche, G., 82[15] (2), 85[252] (21), 86[281] (24), 125[34] (94)
Bochkarev, M. N., 458[25] (447)
Böck, A., 558[16] (516)
Bock, P. L., 560[77] (518, 538)
Bodamer, G. W., 433[176] (416)
Boeckman, R. K., Jr., 127[182] (119), 433[161] (414)
Boehnke, H., 266[168] (203)
Boeré, R. T., 512[92] (501)
Boersma, J., 187[115a] (171)
Boes, M., 85[236] (18)
Boese, R., 276[841] (262)
Boëtius, M., 430[35] (395)
Bogdanović, B., 154[2] (130)
Boger, D. L., 384[75] (361), 566[482] (545, 547, 548)
Boiko, Y. A., 598[101] (574)
Boireau, G., 187[106] (169)
Boisvert, L., 383[33] (359)
Boivin, T. L. B., 564[353] (539, 548), 567[549] (548)

Bokovoi, A. P., 431[73a] (401)
Boldorini, G. P., 268[334] (215)
Boldrini, G. P., 185[23b] (160), 265[89] (198), 272[604] (240, 241), 272[610] (240), 272[611] (240)
Boleslawski, M. P., 310[77] (293, 296, 308), 310[86] (300)
Bolm, C., 86[287] (25), 188[141d] (174), 188[141e] (174), 188[141f] (174), 275[836] (260), 350[238] (333)
Bols, M., 349[224] (330)
Bonar-Law, R. P., 353[415] (342), 353[416] (342)
Bonato, M., 264[41] (194, 247)
Bonfiglio, J. N., 89[496] (53), 270[477] (230)
Bönnemann, H., 185[21e] (160)
Bonnert, R. V., 86[293] (25), 567[567] (552)
Bons, P., 154[2] (130)
Booker-Milburn, K. I., 384[110] (364)
Borch, R. F., 83[112] (5), 433[153a] (413)
Borchardt, R. T., 89[498] (54)
Borden, J. H., 126[76] (100, 103)
Bordoloi, M., 385[160] (368)
Bordwell, F. G., 127[159] (116), 433[165] (415)
Borer, B. C., 84[172] (8)
Börner, A., 275[810] (258, 261)
Bortolotti, M., 265[89] (198), 272[611] (240)
Bos, M. E., 385[178] (369)
Bosch, E., 348[136] (324), 383[39] (360, 361), 383[67] (361), 566[496] (545, 548, 549), 566[497] (545, 549)
Bosco, M., 86[318] (27)
Bosnich, B., 350[261] (335), 353[437] (343)
Bosold, F., 85[252] (21)
Bossert, H., 126[79] (101)
Bosshardt, H., 126[74] (100, 104), 126[90] (103, 105)
Bott, S. G., 458[54] (453)
Bottaro, J. C., 348[131] (324)
Bouali, A., 565[374] (540, 547)
Boudin, A., 154[7] (130)
Boudjouk, P., 90[597] (66), 185[28a] (161), 185[28b] (161), 185[28c] (161), 562[224] (527)
Boudreaux, G. J., 87[373] (36)
Boukouvalas, J., 351[309] (337)
Boulos, L. S., 511[48] (491)
Bousbaa, J., 565[372] (540, 551), 566[438] (543, 544, 552, 553)
Bouyanzer, A., 349[190] (329)
Bowe, M. D., 562[212] (526, 544)
Bowen, P., 91[639] (71)
Bowman, W. R., 384[89] (363)
Boyd, E. A., 353[409] (342)
Brachman, A. E., 125[32] (94)
Bradley, G., 350[234] (332, 343)
Bradley, J.-C., 386[222] (371)
Bradsher, C. K., 91[611] (67)
Braga, D., 185[23a] (160)
Brakta, M., 385[198] (371)
Branchaud, B. P., 568[592] (554)
Brand, S., 155[41] (132)
Brandão, M. A. F., 91[671]
Brandenburg, A., 385[148] (367)
Brandsma, L., 3, 46, 81[7] (2, 3), 81[8] (2, 3), 82[34] (2, 3, 42), 83[85] (4), 88[429] (46), 88[446] (48, 53, 59), 91[622] (71), 93, 107, 112,

124[3] (93, 98, 116, 119), 126[101] (107), 126[102] (107), 126[123] (110), 126[124] (110), 126[125] (110), 126[126] (110), 126[127] (111, 112), 126[131] (112), 127[136] (112), 127[137] (112), 127[138] (112), 127[155] (116), 127[158] (116), 127[175] (119), 127[178] (119), 128[195] (121), 128[196] (119), 128[206] (123, 124), 154[23] (130), 348[130] (324), 597[26] (572), 598[99] (574, 593), 600[261b] (592)
Brandt, C. A., 563[257] (530), 600[220] (585-7), 601[270] (593)
Brauer, D. J., 270[415] (223)
Braun, J., 268[285] (211)
Braun, L. L., 157[176] (152)
Braun, M., 82[17] (2), 82[19] (2), 272[598] (240, 241), 349[213] (330)
Braun, M. P., 386[217] (371)
Bravo, A., 353[420] (343)
Bray, B. L., 353[419] (343)
Braye, E., 91[637] (71)
Breault, G. A., 350[229] (332)
Brede, O., 433[182] (418)
Bredenkamp, M. W., 385[149] (367)
Bregadze, V. L., 264[5] (192), 597[40] (572)
Breslow, R., 567[528] (547)
Breuer, S. W., 431[60] (399)
Brewster, A. G., 563[294] (534)
Brich, Z., 125[46] (95)
Bridges, A. J., 88[409] (42, 54), 89[500] (54), 89[501] (54)
Bridon, D., 512[68] (496)
Bridson, J. N., 269[413] (223)
Brieden, W., 188[132a] (172), 188[132b] (172), 387[293] (377)
Brijoux, W., 185[21e] (160)
Brill, W. F., 600[205] (583)
Brindell, G. D., 310[57] (285)
Bringmann, G., 276[871] (264)
Brinkman, H. A., 431[50] (397)
Brisset, H., 274[702] (246)
Brittain, J. M., 91[625] (71), 352[388] (341)
Britten-Kelly, M. R., 559[50] (516, 546), 564[349] (539), 564[350] (539)
Britton, T. C., 351[299] (337)
Brix, B., 125[44] (95, 112)
Broaddus, C. D., 125[20] (94, 108, 109), 125[48] (95), 126[110] (109)
Brocard, J., 83[83] (4)
Broka, C. A., 14, 85[206] (14, 22), 87[385] (39)
Bronk, B. S., 188[145] (175)
Brook, M. A., 346[50] (314)
Brooks, D. W., 272[620] (240)
Brossi, A., 156[173] (152)
Brown, B. B., 156[152] (150), 271[489] (232), 271[490] (232)
Brown, C. D. S., 384[82] (362)
Brown, D., 560[76] (518, 522, 528-31, 538, 539)
Brown, D. S., 351[315] (338)
Brown, E., 155[40] (132)
Brown, H. C., 107, 125[68] (103), 126[84] (101), 126[103] (107), 188[131] (171), 193, 202, 240, 264[1] (192, 193, 196, 214), 264[27] (192, 198, 199, 201, 202, 208, 214–16, 220, 222, 223, 225,

227–30, 235, 238, 241, 244, 247), 264³⁷ᵃ (193), 264³⁷ᵇ (193), 264⁴⁰ (193, 198, 247), 264⁴³ (194), 264⁴⁴ (194), 264⁴⁵ (194), 265⁵⁹ (195), 265⁶⁰ (195), 265⁶⁴ (196, 198), 265⁶⁵ (196), 265⁶⁶ (196), 265⁷³ (196), 265⁷⁴ (196), 265⁷⁶ (196, 205, 247), 265⁷⁸ (196), 265⁷⁹ (196, 206), 265⁸⁰ (196), 265⁸¹ (196, 197), 265⁸² (196), 265⁸³ (196), 265⁸⁴ (198, 200), 265⁸⁵ (198, 200), 265⁸⁸ (198), 265¹⁰¹ (199), 265¹⁰² (199), 265¹⁰³ (199), 265¹⁰⁴ (200), 265¹⁰⁶ (200, 248), 265¹¹¹ (200), 265¹²⁰ (200), 265¹²¹ (200), 265¹²⁷ (200, 221, 242), 266¹²⁸ (200), 266¹²⁹ (200), 266¹³⁰ (200, 208), 266¹⁵² (202, 203), 266¹⁵³ (202, 203), 266¹⁵⁵ (202), 266¹⁵⁶ (202), 266¹⁷¹ (203), 266¹⁷² (203), 266¹⁷³ (203), 266¹⁷⁴ (203), 266¹⁷⁵ (203), 266¹⁷⁷ (203), 267²⁰⁷ (205), 267²⁰⁸ (205), 267²⁰⁹ (205), 267²¹⁰ (205), 267²¹² (205), 267²¹³ (205), 267²¹⁴ (205, 206), 267²¹⁵ (205), 267²¹⁶ (205, 206), 267²¹⁷ (205, 206), 267²¹⁹ (205), 267²²⁰ (205), 267²²¹ (205), 267²²² (206), 267²²³ (206), 267²²⁴ (206), 267²²⁵ (206, 218), 267²²⁶ (206), 267²²⁸ (206), 267²²⁹ (206), 267²⁴⁰ (207), 267²⁴² (207), 267²⁴³ (207), 267²⁴⁴ (207), 267²⁵⁰ (208), 267²⁵⁴ (208), 267²⁵⁵ (208), 267²⁶⁴ (208), 268³¹⁰ (213), 268³¹² (213), 268³¹⁷ (214), 268³²¹ (214, 215), 268³²² (215), 268³²⁷ (215), 268³²⁸ (215), 269³⁴⁴ (216), 269³⁴⁵ (216), 269³⁴⁶ (216), 269³⁴⁷ (216), 269³⁵¹ (217), 269³⁵⁶ (217), 269³⁵⁹ (217), 269³⁶⁰ (217), 269³⁶¹ (217), 269³⁶⁴ (217), 269³⁶⁶ (218), 269³⁷⁵ (218), 269³⁷⁷ (218), 269³⁷⁹ (219), 269³⁸¹ (219), 269³⁸³ (219), 269³⁸⁵ (219), 269³⁸⁶ (219), 269³⁹⁵ (220), 269⁴⁰² (221), 269⁴⁰⁴ (221), 269⁴⁰⁶ (221), 269⁴¹³ (223), 269⁴¹⁴ (223, 240), 270⁴²⁰ (224), 270⁴³⁵ (225), 270⁴³⁶ (225), 270⁴³⁸ (225), 270⁴³⁹ (225), 270⁴⁴⁰ (225), 270⁴⁴² (226), 270⁴⁴⁵ (227), 270⁴⁴⁷ (227), 270⁴⁴⁸ (227), 270⁴⁶¹ (229), 270⁴⁶² (229), 270⁴⁷⁰ (229), 270⁴⁷¹ (229), 272⁵⁶⁷ (236), 272⁵⁸⁷ (238), 272⁶⁰⁹ (240), 272⁶¹³ (240), 272⁶²³ (240), 272⁶²⁴ (240), 272⁶²⁵ (240), 273⁶³² (240), 273⁶⁴⁸ (242), 273⁶⁷⁶ (244), 274⁷⁰⁶ (247), 274⁷⁰⁸ (247, 258, 260), 274⁷⁰⁹ (247), 274⁷¹⁰ (247), 274⁷¹² (247), 274⁷¹³ (247), 274⁷¹⁵ (247, 252), 274⁷⁴⁵ (251), 274⁷⁴⁶ (251), 274⁷⁴⁷ (251), 274⁷⁵¹ (251), 274⁷⁵³ (252), 274⁷⁵⁹ (252), 275⁸⁰² (258, 260), 275⁸³⁵ (260), 275⁸³⁷ (260), 276⁸⁷⁰ (264), 411, 424, 430²³ (391), 430³⁸ᵃ (396, 411), 430³⁸ᵇ (396, 411), 432¹³⁸ (411), 432¹³⁹ᵃ (411), 432¹³⁹ᵇ (411), 432¹⁴⁰ᶜ (412), 432¹⁴¹ᵃ (412), 432¹⁴¹ᵇ (412), 433¹⁴³ (412), 433¹⁴⁸ (413), 433¹⁵⁵ᵃ (414), 433¹⁵⁵ᵇ (414), 433¹⁵⁵ᶜ (414), 433¹⁸⁷ (418), 434²²⁴ (424)
Brown, J. D., 88⁴³¹ (46), 88⁴³² (46)
Brown, J. M., 89⁵⁰⁶ (55), 91⁶²³ (71), 154³ (130), 386²¹⁹ (371)
Brown, M. L., 125¹⁷ (94), 125¹⁹ (94)
Brown, P. A., 86²⁹³ (25)
Brown, R. A., 87⁴⁰¹ (42), 87⁴⁰³ (42)
Brown, R. S., 565⁴⁰⁹ (541)
Brownstein, S., 125⁷⁰ (97)
Brownstein, S. K., 310³⁹ (282, 293, 296, 309)
Brubaker, C. H., Jr., 90⁵⁸⁹ (65)
Brückner, R., 86²⁸² (24), 386²⁸¹ (375), 386²⁸⁴ (376), 567⁵⁶⁰ (551), 567⁵⁶¹ (551)
Brugger, R. M., 264¹⁶ (192)
Brumwell, J. E., 565⁴³⁰ (542)
Brunel, J. M., 275⁸⁰⁸ (258, 261), 275⁸⁰⁹ (258, 261), 275⁸²⁷ (259)
Brunet, E., 352³⁵³ (340)
Brunner, H., 513¹³⁹ (510), 513¹⁴⁰ (510)
Bruno, G., 267²⁰⁶ (204), 310⁷⁶ (293)
Bryce, M. R., 558²⁹ (516), 566⁴⁵⁹ (544), 566⁴⁹⁸ (545, 546)
Bryce-Smith, D., 125²² (94, 112), 125²⁴ (94, 112)
Brynolf, A., 155⁹⁸ (140)
Bryson, T. A., 264⁴⁷ (194), 568⁶⁴⁷ (557)
Bube, T., 274⁷³⁷ (251, 253)
Bubnov, Y. N., 264²⁶ (192, 196, 212–14, 235, 238), 269³⁷² (218), 271⁵⁴⁹ (235, 236, 238, 239), 271⁵⁵⁰ (235, 236, 238), 271⁵⁵³ (235), 272⁵⁷³ (237, 251), 272⁵⁹¹ (239), 274⁷⁶¹ (252)
Buchardt, O., 564³⁵² (539, 558)
Buchi, G., 568⁶⁴⁰ (557)
Buchwald, S. L., 353⁴⁵⁰ (344)
Buckle, M. J. C., 347⁶⁷ (317)
Buckman, B. O., 387³³⁵ (381)
Budhram, R. S., 86³⁰¹ (26)
Budries, N., 186⁷⁶ (166)
Bugarcic, Z., 564³²¹ (538, 542)
Bulka, E., 560⁹⁹ (519, 522, 528)
Bulman Page, P. C., 86²⁸⁴ (24)
Bülow, G., 347⁶⁰ (317)
Bumagin, N. A., 271⁵⁰⁷ (232), 271⁵⁰⁸ (232), 431⁸³ (403), 432¹²² (407, 409), 432¹²⁸ᵇ (409)
Bunce, R. J., 433¹⁸⁰ᵇ (417)
Bundel, Yu. G., 431⁸⁷ᵃ (403)
Bunn, B. J., 84¹³³ (6)
Buono, G., 275⁸⁰⁸ (258, 261), 275⁸⁰⁹ (258, 261), 275⁸²⁷ (259)
Buonora, P. T., 85²⁴⁰ (19)
Burchat, A. F., 387³¹⁷ (379)
Bures, E. J., 92⁷⁰⁵
Burford, C., 86³¹² (26)
Burford, N., 512⁹¹ (501)
Bürger, H., 270⁴¹⁵ (223)
Burger, U., 352³⁶⁰ (340)
Burgess, K., 264⁵⁰ (195), 265⁵⁶ (195), 265⁹² (199, 248), 265⁹³ (199), 274⁷²¹ (249), 275⁷⁷³ (255), 275⁷⁷⁴ (255)
Burgess, L. E., 566⁴⁴⁹ (543)
Burgess-Henry, J., 352³⁵⁸ (340)
Burk, M. J., 350²³⁴ (332, 343)
Burke, K. E., 558¹⁹ (516)
Burke, S., 568⁶³⁰ (556)
Burkhardt, E. R., 190²⁰⁵ (183)
Burkholder, W. G., 126⁷⁸ (100)
Burks, S. R., 434²³³ (425)
Burnell-Curty, C., 385¹⁶³ (368)
Burnett, F. N., 265¹⁰⁸ (200, 236, 237), 348¹⁷¹ (328)
Burns, B., 275⁸¹² (258)
Burns, E. G., 267²⁰³ (204)
Burns, M. R., 189¹⁸⁴ᵇ (179)
Burns, S. A., 88⁴³⁶ (46)
Burns, S. J., 352³⁵⁷ (340)
Burns, T. P., 154⁵ (130), 154⁶ (130), 155²¹ᵇ (160)
Burroff, J. A., 384⁷⁴ (361)
Burton, D. J., 185³¹ (161), 185³³ᵃ (161), 185³³ᵇ (161), 186³⁴ᵃ (161, 163), 186³⁴ᵇ (161, 163), 186³⁴ᶜ (161, 163), 189¹⁹⁰ (181)
Burton, R., 90⁵⁷³ (64)
Busato, S., 353⁴²² (343)
Bushkov, A. Y., 597⁴⁷ (572)
Buss, D., 265⁷⁵ (196), 271⁵⁴⁴ (235), 271⁵⁴⁵ (235)
Busse, P. J., 598⁷⁵ (573)
Butler, W. M., 188¹³⁸ (174)
Butlerow, A., 278
Butsugan, Y., 90⁵⁸⁴ (65), 90⁵⁸⁷ (65), 512⁶³ (495)
Buttery, C. D., 92⁷⁰¹ (74)
Buys, I., 485⁶¹ (479, 480)
Byers, J. H., 565⁴⁰⁴ (541, 548), 565⁴⁰⁵ (541, 548), 567⁵⁴⁴ (548), 567⁵⁴⁵ (548)
Byers, P. K., 458²⁰ (446)
Bykov, V. V., 271⁵⁰⁷ (232), 271⁵⁰⁸ (232)
Byrd, J. E., 432¹⁴⁰ᵇ (412)
Byström, S. E., 599¹⁷⁰ (580)
Bywater, S., 125⁷⁰ (97)

Cabal, M.-P., 434²²³ (423)
Cabiddu, S., 83⁶¹ (4), 128²⁰⁹ (124)
Cabrera, A., 82⁴⁴ (3)
Cacchi, S., 432⁹⁸ (404), 432⁹⁹ᵃ (404), 432⁹⁹ᵇ (404)
Caddick, S., 384⁸⁶ (362, 370)
Cadiot, P., 268³⁰⁷ (213)
Cadogan, J. I. G., 353⁴⁰⁶ (342), 353⁴⁰⁷ (342)
Cagniant, D., 560⁸⁹ (518, 521, 522, 540, 543, 544, 547, 549, 551–4)
Cahiez, G., 186⁴⁹ᵇ (163)
Cai, J., 433¹⁶⁸ (415)
Cain, P. A., 88⁴¹⁰ (42, 44, 72)
Caine, D., 83⁸⁰ (4), 84¹⁵³ (7)
Cainelli, G., 155⁸² (137), 267²⁶² (208), 271⁵⁴² (235), 353⁴²¹ (343), 353⁴²² (343), 353⁴²³ (343), 353⁴²⁴ (343)
Cairns, T. L., 185³ (159, 178), 309¹⁰ (279)
Calabrese, J. C., 265⁹³ (199), 265⁹⁷ (199), 265⁹⁹ (199), 267²⁰³ (204), 271⁵¹⁰ (232)
Calderon, S. N., 89⁴⁹⁹ (54)

Callahan, A. P., 598[78] (573)
Callen, G. R., 267[251] (208)
Calverley, M. J., 87[346] (30), 566[450] (543, 544)
Calzada, J. G., 269[413] (223)
Cameau, A., 126[77] (100)
Cameron, D. W., 350[277] (336)
Cameron, K. O., 351[280] (336)
Camou, F. A., 126[115] (109)
Campbell, A. L., 88[433] (46)
Campbell, C. B., 433[167] (415)
Campbell, I. G. M., 597[52] (572)
Campbell, J. B., Jr., 187[83] (166), 265[121] (200)
Campbell, J. D., 264[35] (192), 274[734] (251)
Campbell, J. R., 485[64] (481)
Campbell, S., 87[347] (30), 566[456] (544)
Campbell, T. W., 431[64a] (400), 559[37] (516, 519, 520, 522, 528, 536, 538, 539), 565[390] (541)
Campos, P. J., 433[146] (412), 433[147] (412), 434[206] (421)
Campostrini, R., 186[37a] (162)
Camps, F., 349[210] (330)
Canali, C., 127[153] (116)
Canella, K. A., 88[443] (48), 128[207]
Canet, J. L., 273[689] (245)
Canning, L. R., 91[623] (71)
Canty, A. J., 458[20] (446)
Capdevila, A., 349[210] (330)
Capella, L., 384[113] (364)
Capelli, A. M., 273[646] (241, 258)
Capon, B., 351[312] (337)
Capozzi, G., 127[163] (117)
Capperucci, A., 347[88] (320), 349[187] (329), 351[318] (338), 352[391] (341), 384[115] (364)
Carboni, B., 266[133] (201), 268[289] (211, 212), 268[296] (212), 268[297] (212, 216), 268[303] (213), 269[362] (217), 269[363] (217), 269[367] (218), 269[368] (218), 269[369] (218), 269[370] (218), 269[371] (218), 269[374] (218)
Carceller, E., 562[220] (527, 538, 556)
Card, R. J., 90[544] (63)
Cardani, S., 273[640] (241, 264)
Cardillo, G., 434[231a] (425), 434[231b] (425)
Carey, J. S., 386[263] (373)
Carlson, B. A., 270[435] (225), 270[436] (225)
Carlson, J. A., 271[525] (233)
Carlson, R. K., 432[107b] (406)
Carofiglio, T., 189[173] (178), 386[233] (372)
Carpenter, A. J., 91[653] (73), 91[660], 91[666], 91[669], 187[103a] (169)
Carpita, A., 187[88a] (167), 187[88b] (167), 187[88c] (167), 383[8] (357, 361, 371, 372)
Carr, R., 349[226] (331)
Carrascal, M. C., 458[19] (446)
Carrasco, M. C. S., 351[329] (338)
Carretero, J. C., 87[378] (36)
Carrié, R., 268[296] (212), 269[362] (217), 269[363] (217), 269[370] (218)
Carroll, M. T., 348[139] (325)
Carrupt, P. A., 567[550] (548)
Carruthers, W., 185[1] (159, 178, 183), 430[6] (390), 434[234b] (425)

Carstens, A., 85[264] (22, 24)
Carter, D. S., 83[108] (5)
Carter, H. E., 433[158] (414)
Cartledge, F. K., 346[15] (314)
Caruso, A. J., 563[251] (530, 539)
Casalnuovo, A. L., 271[510] (232)
Casarini, A., 384[112] (364)
Casas, J. S., 458[12] (444), 458[18] (446), 458[19] (446)
Casebier, D. S., 350[273] (335)
Caserio, M. C., 433[193] (419)
Casey, C. P., 431[45] (396)
Casey, M., 82[50] (3)
Casiraghi, G., 350[246] (333), 351[308] (337)
Cassidy, J., 264[50] (195)
Castaldi, G., 84[139] (6)
Castaño, A. M., 385[203] (371)
Castano, M. V., 458[17] (446)
Castedo, L., 91[613] (67), 568[637] (557)
Casteel, D. A., 88[469] (50, 68)
Castellano, E. E., 458[12] (444)
Castiñeiras, A., 459[72] (456), 459[73] (456), 459[74] (456), 459[75] (456)
Castle, L., 567[517] (547, 548)
Castro, B., 127[151] (116)
Casu, A., 351[323] (338)
Casucci, D., 84[137] (6)
Caubère, P., 83[60] (4), 94–6, 125[15] (94–6), 125[42] (95), 125[52] (96)
Cauletti, C., 569[655] (558)
Caulton, K. G., 267[204] (204)
Cava, M. P., 597[31] (572), 597[62] (572, 573, 576), 598[68] (573), 598[69] (573), 598[85] (574), 599[133] (576–8), 599[175] (580), 599[196] (583), 600[203] (583), 600[242] (591)
Cavazza, M., 156[133] (146)
Cazeau, P., 350[235] (333)
Cecchi, R., 385[209] (371)
Cederbaum, F. E., 84[194] (12, 67)
Cegla, M. T., 155[93] (139, 140)
Cense, J., 564[301] (536)
Cerreta, F., 352[391] (341)
Cerveau, G., 154[7] (130)
Cha, H. T., 351[339] (339)
Cha, J. S., 265[126] (200), 273[676] (244), 273[677] (244)
Chabaud, B., 564[300] (536)
Chadha, R. K., 267[200] (204)
Chadwick, D. J., 91[621] (71), 91[648] (72), 91[653] (73), 91[654], 91[660], 91[666], 91[669], 92[679] (73), 92[690] (72), 127[173] (119)
Chadwick, L., 91[619] (70)
Chae, Y.-H., 558[20] (516)
Chakraborti, D., 559[68] (518)
Chalk, A. J., 126[110] (109)
Challener, C. A., 385[178] (369)
Challenger, F., 512[103] (503), 560[104] (519, 528, 554)
Chamberlin, E., 430[17] (390)
Chambers, R. D., 70, 91[618] (70), 91[619] (70), 353[404] (342)
Chan, C. M., 126[107] (109)
Chan, K. H., 430[43] (396)
Chan, P. C.-M., 86[277] (24)
Chan, S., 190[207b] (183)
Chan, T. H., 27, 82[29] (2, 27), 86[299] (26), 86[300] (26), 86[320] (27), 86[321] (27), 86[322] (27, 28), 86[324] (28), 127[165] (117), 127[168] (118),

347[87] (320), 348[150] (325), 349[200] (329), 350[236] (333), 350[277] (336), 354[461] (345)
Chan, T. Y., 350[229] (332)
Chan, W. K., 566[480] (545)
Chandrakumar, N. S., 276[852] (263)
Chandrasekaran, S., 265[70] (196), 265[71] (196), 352[343] (339)
Chang, C.-T., 383[11] (358, 361)
Chang, E., 86[299] (26)
Chang, L. S., 90[602] (67)
Chang, L.-J., 87[381] (39)
Chang, V. S. K., 599[161d] (579)
Chang, Z.-Y., 155[75] (136)
Channon, J. A., 154[25] (130)
Chanon, M., 154[11] (130)
Chapdelaine, M. J., 384[122] (364)
Chapleo, C. B., 90[557] (63)
Chapleur, Y., 83[68] (4), 433[145b] (412)
Chaplin, D. A., 275[774] (255)
Chapman, O. L., 564[307] (537)
Charette, A. B., 127[182] (119), 189[177a] (178), 386[261] (373)
Charleson, D. A., 156[153] (150)
Charpentier, R., 433[186] (418)
Charpiot, B., 512[111] (504, 505, 507, 508), 512[126] (507–9), 513[134] (508)
Charreau, P., 87[370] (36)
Chatani, N., 352[354] (340), 353[447] (344), 353[449] (344), 386[275] (375)
Chatgilialoglu, C., 353[440] (344), 353[443] (344), 567[526] (547), 567[527] (547)
Chatterjee, S., 185[8d] (159, 166)
Chattopadhyay, S., 89[472] (50)
Chattopadhyaya, J., 567[538] (548), 567[540] (548)
Chau, T.-Y., 187[118b] (171)
Chaussard, J., 186[36e] (162, 174)
Chavant, P.-Y., 266[159] (203), 269[365] (217)
Cheeseman, G. W. H., 91[620] (71), 91[626] (71)
Chekhlov, A. N., 565[382] (540)
Chelius, E. C., 346[19] (314), 430[29] (393)
Chemburkar, S., 85[245] (20)
Chen, C., 511[36] (489), 511[37] (489), 511[38] (489), 511[39] (489), 511[41] (489), 511[44] (490), 511[53] (492), 511[54] (492), 511[55] (492), 511[56] (492), 511[57] (493), 512[67] (496), 512[73] (497), 512[77] (498), 512[78] (498), 566[492] (545), 601[279] (595), 601[280] (595)
Chen, C.-W., 87[404] (42)
Chen, G., 89[499] (54)
Chen, H. G., 185[13b] (160, 162), 186[67] (165), 186[77a] (166, 183), 187[108b] (169, 170), 188[143a] (175)
Chen, H.-L., 346[52] (317)
Chen, J., 599[188] (582), 599[192] (582)
Chen, J. S., 155[79] (136)
Chen, K., 351[300] (337), 386[287] (376)
Chen, K. M., 273[684] (244), 273[685] (244)
Chen, L. J., 91[667]
Chen, L.-F., 349[207] (330)
Chen, M. T., 597[27] (572)
Chen, Y., 84[188] (11), 155[101a] (140)

Chen, Y.-C. J., 568[592] (554)
Chen, Y.-J., 383[42] (360)
Chenard, B. L., 353[400] (342), 384[100] (363, 369), 387[310] (378)
Chenault, H. K., 351[288] (336)
Cheng, J. W., 154[21] (130)
Cheng, M.-C., 352[363] (340)
Cheng, P. T. W., 351[290] (336)
Cheng, T.-C., 267[245] (208)
Cheng, W., 266[167] (203, 204)
Cheng, Y.-M., 432[137a] (411)
Cherng, C. D., 566[440] (543)
Chernova, A. D., 311[109] (307)
Cheshire, D. R., 567[539] (548)
Chhabra, B. R., 564[302] (536), 564[304] (536, 537)
Chia, W.-L., 156[107] (141), 187[118a] (171), 187[118b] (171)
Chianelli, D., 563[276] (530), 563[277] (530), 564[334] (538)
Chiang, Y., 351[311] (337)
Chiba, Y., 271[499] (232)
Chichereau, L., 485[51] (473, 478)
Chida, Y., 155[76] (136)
Chidambaram, N., 265[70] (196), 352[343] (339)
Chieffi, A., 598[109] (574, 589), 598[124] (575), 601[272] (594)
Chikamatsu, K., 600[236] (589)
Chikami, Y., 351[333] (339)
Chikashita, H., 87[386] (40), 88[414] (42)
Chinchilla, R., 565[424] (542)
Chiou, B. L., 270[416] (223)
Chiquete, L. M., 267[233] (206)
Chittattu, G., 567[520] (547), 568[646] (557)
Chiu, K. W., 270[443] (227)
Chivers, T., 512[90] (501), 512[91] (501)
Chlopin, W., 597[29] (572)
Chmutova, G. A., 564[328] (538)
Cho, B. T., 275[822] (259), 275[833] (260)
Cho, C. S., 271[527] (233)
Cho, I.-S., 89[486] (51), 311[102] (305)
Choi, J. H., 127[135] (112, 118)
Choi, S.-C., 383[14] (358, 362)
Choi, Y. M., 565[406] (541, 557)
Chojnowski, J., 353[405] (342), 353[453] (344)
Chong, J. M., 86[276] (24), 86[277] (24), 385[173] (369), 387[317] (379)
Chou, T.-S., 87[381] (39), 186[74a] (166), 186[77a] (166, 183), 186[77b] (166, 183), 186[77c] (166, 183), 188[137d] (174, 177), 433[188c] (418)
Chounan, Y., 186[65c] (165)
Chow, F., 563[260] (530), 568[628] (555), 568[633] (556)
Chow, H. F., 273[635] (240)
Chow, J., 268[301] (212)
Chow, K., 350[257] (334)
Chow, M.-S., 430[16] (390), 431[46] (396), 432[127] (408)
Choy, W., 272[597] (240, 241, 256), 275[781] (256)
Chrétien, F., 83[68] (4), 433[145b] (412)
Christiaens, L., 560[118] (519), 563[272] (530)
Christmann, K. F., 127[152] (116)
Chu, J., 85[234] (18)
Chu, J. Y. C., 597[50] (572)
Chu, K.-H., 269[394] (220, 228), 270[450] (228)
Chu, S. S., 563[279] (530)

Chuang, C.-N., 186[71c] (166)
Chuang, C.-P., 565[431] (542)
Chuang, L.-W., 186[71d] (166)
Chuit, C., 154[7] (130), 346[17] (314)
Chun, Y. S., 275[822] (259), 275[833] (260)
Chung, B. Y., 311[102] (305)
Ciattini, P. G., 271[526] (233)
Cibura, G., 348[147] (325)
Cinquini, M., 87[367] (36), 87[368] (36), 568[588] (554)
Cintrat, J.-C., 387[299] (377)
Cirillo, P. F., 347[90] (320), 347[103] (321)
Ciucci, D., 383[8] (357, 361, 371, 372)
Claff, C. E., 125[29] (94), 125[30] (94, 112)
Clapp, G. E., 383[29] (359)
Clardy, J., 88[414] (42), 434[245b] (427)
Clarembeau, M., 562[207] (526, 549), 562[210] (526, 550), 562[211] (526, 543, 550), 564[310] (538), 564[312] (538, 550, 551), 565[368] (540, 544), 565[370] (540, 550, 551), 567[553] (550), 567[554] (550)
Claremon, D. A., 562[199] (522, 538, 540, 542, 547), 564[314] (538, 540, 543, 545), 600[216b] (585)
Clark, B. P., 350[237] (333)
Clark, C. W., 559[35] (516, 522, 536)
Clark, D. N., 384[89] (363)
Clark, J. E., 82[47] (3)
Clark, J. S., 275[793] (258)
Clark, K. B., 567[526] (547)
Clark, M. C., 87[355] (32)
Clark, P. S., 563[274] (530)
Clark, R. D., 82[36] (2, 4, 56), 88[426] (43), 89[517] (56), 89[518] (56), 89[527] (59), 89[528] (59), 89[536] (61), 568[624] (555)
Clarke, A. J., 127[177] (119)
Clarke, M. T., 84[176] (10)
Clément, J. C., 273[693] (245)
Clinch, K., 384[82] (362)
Clinet, J.-C., 89[520] (56)
Clive, D. L. J., 273[654] (243), 383[60] (361), 555, 558[1] (516, 521, 522, 526, 529, 532, 534, 536–43, 547, 551–7), 562[173] (522, 530, 532, 537–42, 553–5), 564[353] (539, 548), 565[377] (540), 567[520] (547), 567[521] (547), 567[523] (547, 548), 567[524] (547, 548), 567[537] (548), 567[539] (548), 567[548] (548), 567[549] (548), 568[623] (555), 568[646] (557), 598[77] (573, 590), 599[153] (578), 599[154] (578), 599[174] (580), 601[273] (594)
Clough, J. M., 269[376] (218)
Coates, R. M., 136, 155[75] (136)
Coe, D. M., 350[250] (333)
Coffey, D. S., 83[62] (4)
Cogoli, A., 268[291] (212)
Cohen, L. A., 558[20] (516)
Cohen, M. L., 563[274] (530)
Cohen, T., 84[134] (6), 84[135] (6, 21), 84[147] (7, 12), 84[198] (12), 84[199] (12), 84[200] (13), 85[202] (14), 85[214] (15, 21), 187[104] (169)
Coindard, G., 268[285] (211)
Colard, N., 565[428] (542)
Colberg, J. C., 265[113] (200), 271[530] (233), 346[27] (315)
Colclough, M. E., 270[455] (228), 271[545] (235)
Cole, D. C., 567[524] (547, 548)
Cole, T. E., 265[74] (196), 265[79] (196, 206), 266[152] (202, 203), 266[153] (202, 203), 266[155] (202), 266[158] (202), 266[172] (203), 267[201] (204), 267[202] (204), 267[205] (204), 267[211] (205), 267[214] (205, 206), 267[220] (205), 267[221] (205), 267[223] (206), 267[224] (206), 269[356] (217), 270[462] (229)
Coleman, R. S., 187[103a] (169), 386[288] (376)
Coll, G., 88[444] (48)
Collignon, N., 86[311] (26), 387[326] (380)
Collins, D. J., 485[54] (474)
Collins, F. W., 125[30] (94, 112)
Collins, J. L., 351[326] (338)
Collins, S., 565[386] (540, 545), 566[490] (545)
Collum, D. B., 4, 83[63] (4), 83[70] (4), 83[105] (5), 83[106] (5), 430[30c] (393), 430[30e] (393)
Colombo, L., 273[633] (240), 273[640] (241, 264), 273[643] (241), 348[153] (326), 350[242] (333), 351[297] (337), 351[308] (337)
Colomer, E., 354[462] (345)
Colomer Gasquez, E., 434[239b] (427)
Colonna, S., 568[588] (554)
Colson, P.-J., 351[320] (338)
Colter, M. A., 431[64h] (400)
Colvin, E. W., 83[118] (5, 25), 346[4] (314), 346[14] (314), 347[74] (318), 347[75] (319), 349[219] (330, 343), 354[460] (345), 601[283] (596)
Comasseto, J.-V., 86[307] (26), 562[178] (522), 562[187] (522, 554, 558), 563[257] (530), 565[396] (541), 566[468] (544), 597[9] (571), 597[12] (571), 597[17] (571), 597[18] (571), 597[42] (572), 597[53] (572), 597[55] (572), 598[64] (572, 574), 598[65] (572), 598[80] (573), 598[81] (573, 575), 598[94] (574), 598[103] (574), 598[108] (574), 598[109] (574, 589), 598[124] (575), 600[218] (585), 600[220] (585–7), 600[221] (586), 600[237] (589), 600[264] (593), 601[270] (593), 601[271] (593), 601[272] (594)
Combs, G. F. J., 558[19] (516)
Comins, D. L., 46, 66, 83[125] (5, 46, 66, 72), 88[431] (46), 88[432] (46), 91[655], 141, 142, 144, 151, 156[106] (141), 156[108] (141), 156[111] (142), 156[112] (142), 156[113] (142), 156[114] (142), 156[115] (142), 156[116] (143), 156[117] (143), 156[118] (143), 156[119] (144), 156[120] (144), 156[121] (144), 156[122] (144), 156[123a] (144), 156[123b] (144), 156[171] (151), 187[117] (171), 188[119a] (171), 346[54] (317)
Commeyras, A., 185[15b] (160), 185[32a] (161), 185[32b] (161), 185[32c] (161), 185[32d] (161)
Conan, A., 186[36c] (162, 174)

Concepcion, A. B., 310[79] (293, 296)
Conia, J.-M., 178, 187[93b] (168), 189[174a] (178), 189[174b] (178), 189[174c] (178), 189[174d] (178), 434[245a] (427)
Contreras, R., 267[230] (206), 267[233] (206), 275[820] (259)
Conway, B. G., 458[55] (453), 459[59] (453)
Cook, K. L., 82[29] (2, 27)
Cooke, F., 86[312] (26)
Cooke, G., 566[459] (544)
Cooke, G. E., 86[309] (26)
Cooke, M. P., Jr., 84[177] (10), 84[180] (10), 84[181] (10), 84[182] (10), 268[279] (210), 268[306] (213)
Cooksey, C. J., 430[43] (396)
Cooney, M. J., 157[176] (152), 348[138] (324)
Cooper, J. P., 353[432] (343)
Cooper, K., 155[85] (137)
Cooper, P. J., 432[142a] (412)
Cooper, P. N., 434[248] (429)
Cooper, W. C., 562[166] (522), 597[3] (571)
Coote, S. J., 90[560] (63)
Copper, T., 91[615] (70)
Coppinger, G. M., 559[37] (516, 519, 520, 522, 528, 536, 538, 539)
Corbett, J. W., 351[307] (337)
Corcoran, D. E., 387[308] (378)
Corey, E. J., 83[123] (5, 42), 83[124] (5), 85[258] (22), 185[26d] (161, 178), 188[136] (172), 193, 240, 257, 259, 263, 266[197] (204, 235, 238, 251), 266[198] (204, 252), 270[458] (228), 273[631] (240, 257), 273[638] (240, 257), 275[789] (257), 275[790] (257), 275[817] (259), 275[821] (259), 275[828] (259), 276[853] (263), 276[854] (263), 276[855] (263), 276[862] (263), 334, 349[204] (330), 349[212] (330), 350[252] (334), 352[371] (341), 434[237c] (426), 453, 459[60] (453), 567[551] (549)
Corey, E. R., 86[320] (27), 346[7] (314)
Corey, J. Y., 86[320] (27), 90[602] (67), 346[7] (314), 346[15] (314), 353[429] (343)
Cormier, J. F., 349[207] (330)
Correia, V. R., 598[80] (573)
Corriu, R. J. P., 83[116] (5), 88[436] (46), 127[166] (117), 154[7] (130), 346[17] (314), 353[418] (343), 354[462] (345), 385[180] (370)
Cortez, C., 88[462] (49)
Costa, A., 88[444] (48), 88[449] (48), 88[452] (48)
Côté, B., 189[177a] (178)
Cottens, S., 125[60] (97)
Couladouros, E. A., 91[628] (71)
Courgeon, T., 269[363] (217)
Courtemanche, G., 188[153a] (175)
Courtneidge, J. L., 433[184] (418)
Courtois, G., 155[63] (134), 156[105] (141), 188[147a] (175), 188[153b] (175), 188[155a] (176), 189[156] (176), 189[157b] (176)
Cousseau, J., 431[69] (400), 435[251] (429)
Coutts, R. T., 91[617] (70)
Couture, A., 91[614] (68)

Couty, F., 187[114a] (171), 187[114b] (171)
Cowan, D., 558[28] (516)
Coward, J. K., 189[184b] (179)
Cowling, M. P., 155[85] (137)
Cox, B., 353[455] (344)
Cox, M. T., 434[234b] (425)
Cox, P. J., 84[132] (6, 33), 85[233] (17)
Cozzi, F., 87[367] (36), 87[368] (36)
Cragg, G. M. L., 269[399] (220)
Crahe, M.-R., 435[251] (429)
Craig, D., 349[223] (330), 350[227] (331), 351[338] (339)
Cram, D. J., 88[451] (48), 146, 568[616] (555), 568[617] (555)
Crandall, J. K., 84[191] (11), 84[192] (12), 127[187] (120)
Cravador, A., 564[344] (539)
Cravo, D., 565[378] (540, 548)
Crawford, J. A., 386[235] (372, 373)
Crawford, T. C., 270[473] (229)
Crawley, J. E., 346[48] (316)
Creger, P. L., 57, 89[522] (57)
Cregg, C., 154[25] (130), 348[122] (323)
Cremer, D., 272[603] (240, 241)
Crenshaw, L., 89[526] (57)
Crich, D., 383[35] (359), 383[71] (361), 566[481] (545, 547–9), 566[492] (545), 567[542] (548), 567[543] (548), 601[279] (595), 601[280] (595)
Criegee, R., 484[11] (465), 484[27] (467)
Crilley, M. M. L., 351[294] (337)
Crimmin, M. J., 84[133] (6)
Crimmins, M. T., 189[162a] (177), 189[162b] (177), 189[162c] (177)
Crimmins, T. F., 125[31] (94), 126[107] (109), 126[108] (109)
Cristol, S. J., 430[41a] (396)
Crosby, I. T., 350[277] (336)
Cross, R. J., 431[71e] (401)
Croteau, A. A., 154[20] (130)
Crowe, D. F., 83[110] (5)
Crowther, G. P., 88[454] (48)
Crudden, C. M., 386[232] (372)
Crumrine, D. S., 383[20] (358)
Crystal, R. G., 558[3] (516–18)
Csáky, A. G., 91[649] (72)
Csizmadia, I. G., 559[55] (518)
Csuk, R., 185[24a] (160)
Cuadrado, P., 384[98] (363, 364), 384[114] (364), 385[186] (370, 373)
Cuevas, J.-C., 86[304] (26)
Cullen, E. R., 567[511] (546), 567[513] (546), 567[514] (546, 547)
Cullen, J. D., 485[54] (474)
Cunico, R. F., 349[184] (328)
Cunkle, G. T., 84[137] (6)
Cuomo, J., 92[676]
Curran, D. P., 349[218] (330), 349[225] (330), 350[232] (332), 383[11] (358, 361), 383[21] (358), 383[31] (359), 383[52] (361), 383[54] (361), 383[56] (361), 383[64] (361), 383[70] (361), 567[536] (548), 567[546] (548)
Curtin, D. Y., 128[198] (122)
Curtis, N. J., 568[646] (557)
Curtis, N. R., 353[436] (343)
Cuthbertson, E., 600[227] (587), 600[228] (587)
Cuvigny, T., 83[90] (5)
Cywin, C. L., 276[862] (263), 350[252] (334)
Czarny, M., 90[543] (63)

Czernecki, S., 384[140] (367)

Da, Y.-Z., 349[192] (329)
Dabdonb, V. R., 598[94] (574)
Dabdoub, M. G., 598[103] (574)
Dabdoub, M. J., 566[468] (544), 598[64] (572, 574), 598[65] (572), 598[94] (574), 600[237] (589)
Dabdoub, V. B., 598[64] (572, 574)
Dahan, R., 125[60] (97)
Dai, L. X., 269[358] (217), 274[720] (249)
Dai, W.-M., 348[132] (324)
Dal Bello, G., 267[262] (208), 271[542] (235)
Dalko, P. I., 601[276] (595)
Dalpozzo, R., 86[318] (27)
Damaservitz, G. A., 311[110] (307)
Damm, W., 189[171a] (178)
Damrauer, R., 432[130b] (410)
Dance, N. S., 597[36] (572)
Danda, H., 275[794] (258)
Dang, H. S., 154[16] (130), 273[662] (243)
Danheiser, R. L., 86[285] (24), 188[145] (175), 268[339] (215), 350[273] (335), 565[406] (541, 557)
Daniels, R. G., 84[137] (6)
Danishefsky, S. J., 36, 87[379] (36), 89[473] (50), 319, 347[81] (319), 350[257] (334), 350[277] (336), 351[288] (336), 384[134] (367), 432[90] (403), 434[234a] (425), 512[76] (498)
Dankwardt, J. W., 89[472] (50), 89[515] (56)
Dan-oh, N., 384[138] (367)
Daran, J.-C., 187[114a] (171)
Das, B. C., 156[124] (144)
Das, N. B., 385[160] (368)
Das, V. G. K., 264[44] (194)
Date, M., 83[121] (5), 88[407] (42, 50, 51)
Date, T., 270[476] (229)
Dau, E. T. H., 512[129] (508)
Dauben, W. G., 87[351] (30), 310[43] (283)
Dauelsberg, C., 275[826] (259)
Davenport, R. W., 432[133] (410)
Davenport, T. W., 187[83] (166)
Daves, G. D., Jr., 385[198] (371)
David, S., 512[115] (504, 509), 513[136] (509), 513[137] (509), 568[634] (556)
David, V., 125[39] (94, 100)
Davidson, A., 90[576] (65)
Davidson, B. S., 350[271] (335)
Davidson, F., 384[100] (363, 369)
Davidson, J., 265[86] (198–200)
Davidson, P. J., 484[2] (461)
Davidson, T. A., 434[201] (421)
Davies, A. G., 268[337] (215), 272[616] (240), 272[617] (240), 383[1] (356), 433[168] (415)
Davies, A. J., 353[408] (342)
Davies, G. M., 385[196] (371)
Davies, I., 597[36] (572)
Davies, S. G., 85[257] (22), 89[537] (63), 90[542] (63), 90[549] (63), 90[550] (63), 90[555] (63), 90[558] (63), 90[560] (63), 348[156] (326)
Davis, A. P., 347[109] (321), 350[234] (332, 343), 353[415] (342), 353[416] (342)
Davis, D. D., 377

Davis, F. A., 351²⁸⁷ (336), 568⁵⁷⁹ (553, 554, 557), 568⁵⁸⁵ (554), 568⁵⁸⁶ (554, 557), 568⁵⁸⁷ (554), 568⁵⁹⁰ (554, 557)
Davis, S. A., 568⁵⁸⁴ (554)
De, R. N., 434²¹⁴ (422)
de Andrade., H., Jr, 599¹³⁶ (576)
De Beys, V., 565³⁹⁴ (541, 544)
de Boer, R. F., 383⁶⁶ (361)
de Bruyn, P. J., 350²⁷⁷ (336)
de Dios, A., 385²¹⁴ (371)
De Haan, R. A., 91⁶³⁶ (71)
de Jong, R. L. P., 126¹²³ (110), 126¹²⁴ (110), 126¹²⁵ (110), 127¹³⁸ (112), 127¹⁷⁸ (119)
de Kanter, F. J. J., 189¹⁶¹ᶜ (176), 274⁷⁵⁸ (252)
de Lera, A. R., 271⁴⁹² (232), 459⁷⁹ (456)
De Lima, C., 87³⁷¹ (36)
De Lucchi, O., 87³⁴¹ (29)
De Lue, N. R., 269³⁷⁷ (218), 269³⁸¹ (219), 269³⁸³ (219), 269⁴⁰⁶ (221)
De Mahieu, A. F., 563²⁶⁴ (530), 566⁴⁶⁰ (544, 553, 554, 557)
de Marigorta, E. M., 346³² (315)
de Mattos, M. C. S., 433¹⁴⁹ (413, 416, 419, 425, 426), 434²²⁰ (423)
de Meijere, A., 83¹¹¹ (5), 560⁸² (518, 521, 522, 526, 529, 540, 543, 544, 551, 552, 558)
De Meio, G. V., 484³² (467, 482, 483)
de Oliveira, A. B., 91⁶⁷¹
de Oliveira, A. R. M., 597⁴² (572), 598⁸¹ (573, 575)
De Santis, M., 383⁸ (357, 361, 371, 372)
De Silva, S. O., 563²⁴⁵ (529, 541)
De Sousa, P. T., Jr., 91⁶⁵⁸
de Souza Barbosa, J. C., 185²⁶ᶜ (161, 178), 189¹⁶⁸ᵃ (178)
de Vos, D., 484²¹ (466), 484²² (466)
Deacon, G. B., 430³² (395), 431⁴⁷ (397), 458³⁸ (450)
Dean, F. M., 91⁶²⁶ (71), 91⁶⁴⁶ (72), 91⁶⁴⁷ (72)
Debergh, D., 600²⁶⁷ᶜ (593)
Deberly, A., 187¹⁰⁶ (169)
DeCamp, A. E., 187¹⁰⁷ᶜ (169)
Deepa, V., 599¹⁹⁵ (582)
Deeth, R. J., 385¹⁶⁵ (368)
Degl'Innocenti, A., 86²⁸³ (24), 127¹⁴⁷ (115), 127¹⁷⁰ (118), 127¹⁸⁸ (120), 347⁶¹ (317), 349¹⁷⁴ (328), 349¹⁸⁷ (329), 351³¹⁸ (338), 352³⁹¹ (341), 353⁴¹⁷ (343), 384¹¹³ (364), 384¹¹⁵ (364), 387²⁹⁷ (377)
Dehghani, A., 156¹¹⁷ (143)
Dehmlow, E. V., 385¹⁹⁴ (371), 432¹³⁰ᵃ (410)
Deibe, A. G., 459⁷³ (456)
Deiko, S. A., 459⁷⁸ (456)
Del Cima, F., 156¹³³ (146)
del Rosario, J. D., 156¹⁵⁰ (150)
Del Valle, L., 265¹¹³ (200)
Delahunty, C. M., 387³²⁷ (380)
Delanghe, P. H. M., 385¹⁸⁷ (370), 386²³⁰ (372)
Delbecq, F., 434²⁴¹ (427)
Delhalle, J., 562¹⁹⁴ (522, 539–42, 557)
Dell, C. P., 92⁶⁸⁵ (72)

Della, E. W., 84¹⁷⁰ (8), 84¹⁷¹ (8)
Della Cort, A., 351³⁰¹ (337)
Dellaria, J. F., 351²⁹⁹ (337)
Deloux, L., 275⁸⁰¹ (258, 259, 262)
Demailly, G., 87³⁵⁹ (33)
Dembélé, Y. A., 187¹¹⁶ᶜ (171)
Demole, C., 126⁸¹ (101)
Demole, E., 126⁸¹ (101)
Denat, F., 385²⁰⁵ (371)
Denenmark, D., 383⁶⁷ (361)
Deng, M. Z., 272⁶⁰⁸ (240)
Deniau, E., 91⁶¹⁴ (68)
DeNinno, M. P., 347⁸¹ (319)
Denis, J. N., 189¹⁷⁴ᵃ (178), 565⁴⁰⁸ (541, 554), 566⁴⁷² (544, 545), 567⁵³¹ (547), 567⁵⁶⁶ (551, 558), 568⁵⁸⁰ (553, 554, 557), 568⁵⁸³ (554, 557), 568⁶¹³ (554)
Denmark, S. E., 154²⁸ (130), 155⁷² (136), 188¹⁴⁴ (175), 189¹⁷⁹ᵃ (178), 189¹⁷⁹ᵇ (178), 189¹⁷⁹ᶜ (178), 275⁷⁹⁷ (258), 319, 347⁸⁵ (319), 349²⁰³ (329), 350²⁵⁰ (333), 351²⁹⁰ (336), 353⁴³⁶ (343), 386²⁵⁴ (373)
Denne, I., 266¹³³ (201)
Denniston, A. D., 87³⁸⁷ (40)
DePuy, C. H., 430²⁸ (393)
Derkach, N. Ya., 566⁴⁸⁷ (545)
Derouane, D., 85²⁰⁸ (15), 567⁵⁵² (549)
des Abbayes, H., 273⁶⁹³ (245)
Desai, M. C., 92⁶⁷⁸, 264⁴⁰ (193, 198, 247), 267²¹² (205)
Desai, R. C., 127¹⁸⁰ (119)
Deschênes, D., 274⁷⁵⁰ (251)
Descotes, G., 565³⁷⁴ (540, 547)
DeShong, P., 349²⁰⁶ (330), 352³⁶⁴ (340)
Desilets, D., 566⁴³⁹ (543, 557)
Desio, P. J., 190²⁰⁴ (183)
Desmurs, J. R., 351³¹⁷ (338)
Desponds, O., 125¹² (94, 95, 99, 100, 104, 105, 108), 126⁹⁶ (105, 106)
Detert, H., 564³⁰⁸ (537)
Detty, M. R., 189¹⁷⁶ (178), 562²²⁹ (527), 563²⁴⁹ (530, 532), 564³²⁷ (538, 545), 564³⁶⁴ (540), 598⁶³ (572, 573), 598¹⁰² (574), 598¹³¹ (576)
Devasagayaraj, A., 270⁴⁴¹ (226)
DeWinter, A. J., 568⁶¹⁴ (555, 557)
Deyá, P. M., 88⁴⁴⁵ (48), 88⁴⁴⁹ (48)
DeYoung, L., 352³⁶⁷ (340)
Dhar, R. K., 272⁶²³ (240), 272⁶²⁴ (240), 272⁶²⁵ (240), 273⁶³² (240)
Dhavale, D. D., 350²⁷⁶ (336)
Dhawan, B., 88⁴⁷⁰ (50)
Dhillon, R. S., 264³⁸ (193, 231, 232), 265⁶⁷ (196), 270⁴¹⁹ (223)
Di Giamberardino, T., 564³²⁹ (538, 557)
Dickason, W. C., 270⁴⁴² (226)
Dicke, R., 353⁴⁰⁰ (342)
Dickens, M. J., 92⁶⁹⁹ (74), 385¹⁶⁶ (369)
Dickens, P. J., 90⁵⁴⁶ (63), 90⁵⁵² (63)
Dickerman, S., 83⁹⁴ (5)
Dickman, D. A., 85²³⁶ (18), 85²⁴⁸ (20)
Dicko, A., 264⁵⁵ (195)
Didiuk, M. T., 154³⁰ (131), 154³³

(131), 154³⁴ (131)
Dieden, R., 565³⁶⁹ (540), 565³⁹⁴ (541, 544), 566⁴⁵³ (544, 547)
Diederich, F., 348¹³⁸ (324)
Dieter, R. K., 387³¹⁹ (379)
Dietrich, A., 458⁵⁷ (453)
Dietrich, C. O., 561¹³⁷ (520)
Dietrich, H., 125⁷⁰ (97)
Díez-Martin, D., 563²⁹⁵ (534)
Dikic, B., 87³⁶⁵ (34)
Dillon, K. B., 266¹⁶² (203)
DiMichele, L., 385²⁰⁷ (371)
Dimitrov, V., 266¹⁷⁶ (203)
Dimroth, P., 484²⁷ (467)
Ding, S., 432¹⁰⁴ᶜ (405)
Diorazio, L. J., 269³⁷⁶ (218)
Discordia, R. P., 599¹⁵⁸ (579), 599¹⁵⁹ (579)
Distefano, G., 569⁶⁵⁵ (558)
Ditrich, K., 272⁶⁰² (240), 272⁶⁰³ (240, 241), 273⁶⁴¹ (241), 274⁷³⁷ (251, 253), 274⁷⁶⁶ (253)
Dittmer, D. C., 599¹⁵⁷ (578), 599¹⁵⁸ (579), 599¹⁵⁹ (579), 599¹⁶⁰ (579)
Djerassi, C., 101, 125⁶⁴ (101), 156¹³² (146)
Djuric, S., 346¹ (314), 383⁶ (357)
Dmitrieva, T. F., 431⁷² (401)
Do, J. Y., 384⁹⁰ (363)
do Amaral, A. T., 433¹⁹¹ (419)
do Amaral, L., 433¹⁹¹ (419)
Doadt, E. G., 91⁶⁶⁵
Doak, G. O., 488, 511² (488), 511⁷ (488), 511⁸ (488), 511⁹ (488), 511¹⁰ (488), 511¹¹ (488), 511¹² (488), 511¹³ (488), 511¹⁴ (488), 511¹⁵ (488), 511¹⁶ (488), 511¹⁷ (488), 511¹⁸ (488), 511¹⁹ (488), 511²⁰ (488), 511²¹ (488), 511²² (488), 511²³ (488), 511²⁴ (488), 511³⁴ (488)
Dodd, J. H., 351²⁸⁵ (336)
Doherty, A. M., 92⁶⁸⁸ (72)
Dojo, H., 266¹⁴⁴ (201, 214), 270⁴⁵⁷ (228)
Doktycz, S. J., 185²⁷ᵃ (161)
Dolbier, W. R., Jr., 84¹⁸⁸ (11)
Dolence, E. K., 349¹⁷³ (328)
Dollinger, H., 85²¹⁸ (15), 127¹⁸⁴ (120)
Domaille, P. J., 268³¹⁵ (213)
Domínguez, C., 91⁶⁴⁹ (72)
Domínguez, D., 91⁶¹³ (67)
Donahue, K. M., 88⁴³⁵ (46)
Dondoni, A., 348¹⁵³ (326), 352³⁸² (341)
Dondy, B., 349¹⁸⁶ (328)
Donnelly, D. M. X., 484³⁴ (467, 473, 480), 484⁴⁰ (469, 480), 485⁴⁹ (473), 485⁵⁰ (473), 485⁶² (481), 512¹²⁸ (507, 508)
Donohoe, T. J., 90⁵⁴² (63)
Doolittle, R. E., 126⁹⁸ (106)
Doonquah, K. A., 348¹⁷² (328)
Dorchak, J., 383³⁶ (359)
Dorgan, B. J., 353⁴¹⁵ (342), 353⁴¹⁶ (342)
Döring, I., 188¹⁴⁶ᵇ (175), 188¹⁴⁷ᵇ (175)
Dorn, C. R., 565⁴⁰² (541)
Dörner, W., 186⁴² (162, 175)
Dorow, R. L., 351²⁹⁹ (337)

Dorta, R. L., 563²⁸⁶ (533, 556, 557)
Douglas, K. T., 353⁴⁵⁶ (344)
Douglass, M. L., 433¹⁶⁵ (415)
Doussot, P., 349¹⁸⁶ (328)
Doutheau, A., 351²⁸⁹ (336)
Dowd, P., 383¹⁴ (358, 362), 384⁷⁷ (361)
Doyle, M. P., 349²⁰² (329)
Draguet, C., 566⁴⁸⁵ (545)
Drake, R. A., 270⁴⁶⁵ (229)
Drakesmith, F. G., 91⁶¹⁸ (70)
Drefahl, G., 433¹⁹⁷ᵃ (420), 433¹⁹⁷ᵇ (420)
Dresely, S., 265¹¹⁵ (200, 215), 274⁷⁶⁵ (253)
Drew, H. D., 597⁴⁴ (572), 598¹⁰⁴ (574)
Drewlies, R., 155⁸⁴ (137)
Driggs, R. J., 432¹⁰³ᵇ (405)
Driguez, P.-A., 351²⁸⁹ (336)
Dröge, H., 84¹⁵⁹ (7), 309⁷ (278)
Dromzee, Y., 186⁴⁸ᶜ (163)
Drossbach, O., 560¹⁰⁰ (519, 527)
Drouin, J., 187⁹³ᶜ (168), 434²⁴⁵ᵃ (427)
Drozda, S. E., 271⁴⁸⁹ (232)
Druliner, J. D., 268³¹⁵ (213)
Drysdale, N. E., 83¹⁰³ (5)
D'Silva, C., 266¹⁷⁰ (203, 245)
Du, C.-J. F., 155⁴⁶ (133), 155⁴⁷ (133)
Du, H., 89⁵¹⁰ (56)
Duan-Mu, L. J., 511⁵⁹ (493, 496, 497)
Dubac, J., 354⁴⁶² (345), 385²⁰⁵ (371)
DuBay, W. J., 385¹⁷⁰ (369)
Dubois, E., 385¹⁹⁷ (371)
Duboudin, F., 350²³⁵ (333)
Dubovitskii, V. A., 431⁷⁰ (400)
Ducep, J. B., 566⁴⁵⁸ (544)
Dufour, C., 383⁴⁰ (360), 565³⁹¹ (541, 547, 548)
Duggan, P. J., 567⁵²⁸ (547)
Duhamel, L., 82⁴⁶ (3), 349¹⁹⁰ (329), 351³¹⁷ (338)
Dumartin, G., 386²⁴³ (373)
Dumont, W., 85²⁰⁸ (15), 86²⁹⁴ (25), 89⁵¹² (56), 562¹⁹¹ (522, 529, 544, 552, 553), 562²⁰⁰ (522), 562²⁰² (522), 562²⁰³ (522, 529), 562²¹¹ (526, 543, 550), 563²⁵⁴ (530, 532), 563²⁶⁴ (530), 563²⁶⁵ (530), 564³²⁹ (538, 557), 565³⁷² (540, 551), 565³⁸⁰ (540), 566⁴⁵³ (544, 547), 566⁴⁵⁴ (544), 566⁴⁶⁰ (544, 553, 554, 557), 566⁴⁶¹ (544, 554, 557), 566⁴⁷¹ (544, 554), 566⁴⁷³ (544), 567⁵⁵² (549), 567⁵⁵⁷ (550, 551), 567⁵⁶⁸ (552), 567⁵⁷² (552), 567⁵⁷⁶ (553), 568⁵⁸⁰ (553, 554, 557), 568⁵⁸³ (554, 557), 568⁵⁸⁹ (554–6), 568⁶¹¹ (554), 568⁶³⁶ (557)
Duñach, E., 89⁵²⁰ (56)
Duncan, M. P., 351²⁹⁵ (337)
Dunn, M. J., 186⁶⁹ᵃ (166), 186⁷² (166)
Dunoguès, J., 346³⁸ (315, 317), 348¹⁵² (326), 350²³⁵ (333)
Duppa, D. F., 186⁵⁴ (164)
Dupuy, C., 185²⁶ᵇ (161, 178), 185²⁶ᶜ (161, 178), 189¹⁶⁸ᵇ (178)
Durkin, K., 383³³ (359)
Dürner, G., 90⁵⁶⁷ (64)
Durst, T., 386²²² (371)

Duthaler, R. O., 185¹⁰ᵇ (160, 162, 171, 172)
D'yachenko, A. I., 432¹²⁹ᶜ (409, 411)
Dyatkin, B. L., 434²⁴⁷ (428)
Dykstra, R. R., 560⁸⁰ (518)
Dziadulewicz, E., 87³⁴⁹ (30)

Eaborn, C., 126¹³⁰ (111), 186⁵⁷ᵃ (165), 186⁵⁷ᵇ (165), 383⁷ (357)
East, M. B., 86³⁰⁹ (26), 264⁴⁸ (195)
Eastham, J. F., 82⁴¹ (3)
Eaton, P. E., 83¹⁰⁰ (5), 84¹³⁷ (6), 84¹³⁹ (6), 88⁴²² (42)
Ebata, K., 348¹⁵⁵ (326)
Eberspacher, T. A., 434²²⁹ᵇ (425)
Ebihara, K., 188¹²⁵ (171), 188¹³³ (172), 188¹⁴¹ᵇ (174)
Echavarren, A. M., 385¹⁹⁵ (371), 385²⁰³ (371), 385²¹⁰ (371)
Eckrich, T. M., 85²⁵⁸ (22)
Edelmann, F. T., 459⁶⁹ (455)
Edington, C., 350²⁴⁴ (333)
Edmunds, J. J., 274⁷⁵² (252)
Edwards, B. H., 458⁵³ (453)
Edwards, J. P., 154²⁸ (130), 189¹⁷⁹ᵃ (178), 189¹⁷⁹ᵇ (178), 189¹⁷⁹ᶜ (178)
Edwards, M., 567⁵²⁹ (547)
Edwards, M. P., 92⁶⁸⁸ (72)
Edwards, O. K., 568⁶⁰³ (554)
Edwards, P. D., 85²²⁵ (16, 17)
Edwin, J., 266¹⁵⁰ (202)
Eggelte, H. J., 433¹⁸³ (418)
Eguchi, M., 187¹¹⁴ᶜ (171)
Eichler, J., 127¹⁴⁴ (115)
Eidus, Ya. T., 433¹⁹⁰ᵇ (419)
Eigendorf, U., 266¹⁸⁸ (203), 266¹⁸⁹ (203)
Einholz, W., 266¹⁹⁰ (204)
Einhorn, C., 89⁵¹⁶ (56), 185²⁶ᵃ (161, 178), 185²⁷ᵇ (161), 189¹⁶⁶ᶜ (178), 433¹⁵¹ (413)
Einhorn, J., 66, 82⁵¹ (3–5, 7, 48, 71), 89⁵¹⁶ (56), 90⁵⁹⁶ (66), 185²⁶ᵃ (161, 178), 185²⁷ᵇ (161), 189¹⁶⁶ᵇ (178), 433¹⁵¹ (413)
Eisch, J. J., 189¹⁹⁶ᵃ (182), 280, 282, 290, 301, 303–5, 309¹² (279), 309²³ (280, 288, 292), 309²⁴ (280, 292), 310²⁶ (280, 289), 310²⁷ (280, 289), 310²⁸ (280, 297), 310²⁹ (280), 310³⁰ (280), 310³³ (281), 310³⁴ (282, 283, 288, 289, 292), 310³⁵ (282), 310³⁶ (282, 291), 310³⁹ (282, 293, 296, 309), 310⁴² (283, 285), 310⁴⁶ (283), 310⁴⁹ (284, 290), 310⁵⁰ (284), 310⁶⁰ (286, 305), 310⁶⁴ (288), 310⁶⁵ (288), 310⁶⁶ (289), 310⁶⁷ (289), 310⁶⁹ (291), 310⁷⁰ (291), 310⁷⁴ (291), 310⁷⁵ (291), 310⁷⁷ (293, 296, 308), 310⁸⁰ (294), 310⁸² (299), 310⁸⁶ (300), 311¹⁰¹ (304), 311¹¹⁰ (307), 348¹⁵⁷ (326)
Eisenberg, C., 186⁴¹ (162, 172), 188¹³²ᵇ (172)
Eisenberg, R. L., 89⁴⁸² (51)
Eisenbraun, E. J., 86³⁰¹ (26)
Eisenmann, J. L., 125¹⁸ (94)
El Abed, D., 347⁸⁵ (319)
El Alami, N., 186³⁸ᵃ (162, 169), 186³⁸ᵇ (162, 169), 186³⁸ᶜ (162, 169)

El Louzi, A., 268³⁰⁴ (213)
El Seoud, O. A., 433¹⁹¹ (419)
El-Bayoumy, K., 558¹⁴ (516, 529), 558²⁰ (516)
Elbein, A. D., 275⁷⁷⁴ (255)
Elgendy, S., 271⁵⁴⁶ (235)
Eliel, E. L., 150, 156¹⁵⁶ (150)
Elissondo, B., 85²⁵⁹ (22), 386²⁴³ (373)
El-Jazouli, M., 155⁵⁶ (134, 140)
El-Kateb, A. A., 511⁴⁸ (491)
Elliott, J. D., 350²⁴⁴ (333)
Elliott, R. L., 349²²⁵ (330)
Ellis, K. A., 384⁹⁷ (363, 364)
Ellis, K. L., 385¹⁷¹ (369, 376)
Ellis, R. J., 86²⁹⁶ (25), 86³⁰⁸ (26)
Ellsworth, E. L., 353⁴⁰¹ (342)
Elman, B., 155⁹⁸ (140)
Elmorsy, S. S., 353⁴¹¹ (342), 353⁴¹² (342)
El'perina, E. A., 432¹²⁴ (408)
Elsley, D. A., 385¹⁹⁶ (371)
Elsner, F. H., 349¹⁷⁷ (328)
El-Telbany, F., 273⁶⁶⁸ (244)
Emeléus, H. J., 559⁵⁸ (517, 518, 522, 528)
Emonds, M. V. M., 384¹⁰⁸ (364)
Emziane, M., 352³⁷⁸ (341)
Enders, D., 155⁷³ (136), 349¹⁹⁴ (329), 384¹²⁶ (364)
Endesfelder, A., 272⁵⁸² (237)
Engel, P., 84¹³⁷ (6)
Engelhardt, G., 186⁵¹ᵇ (164)
Enggist, P., 126⁸¹ (101)
Englert, G., 274⁷¹⁷ (248)
Englert, U., 266¹⁸⁶ (203), 266¹⁸⁹ (203)
Engman, L., 88⁴¹⁹ (42), 561¹³³ (520), 565⁴¹⁶ (542), 567⁵⁷⁷ (553), 597¹¹ (571), 597¹⁶ (571), 597²⁸ (572, 573), 597⁵⁷ (572, 574, 586), 598⁶⁸ (573), 598⁶⁹ (573), 598⁷⁰ (573), 598⁸⁶ (574), 598¹¹³ (575), 598¹¹⁷ (575), 598¹²⁰ (575, 587), 599¹⁶⁸ (580), 599¹⁷⁰ (580), 599¹⁷⁵ (580), 600²⁰³ (583), 600²²⁴ (586), 600²³¹ (588), 600²⁶² (592), 600²⁶⁷ᵃ (593), 601²⁸⁵ (596)
Enholm, E. J., 384⁷⁴ (361)
Ennis, D. S., 91⁶⁴⁸ (72), 91⁶⁶²
Enokiya, R., 599¹³⁷ᵃ (576, 577)
Ensley, H. E., 430³¹ᵇ (393)
Entwistle, I. D., 568⁶⁰⁸ (554)
Epa, W. R., 351²⁹⁶ (337), 385²⁰⁸ (371)
Epifani, E., 155⁶⁶ (134), 155⁶⁹ (135)
Epsztajn, J., 82³⁸ (2, 4, 42, 70, 75, 76, 79, 81)
Erabi, T., 386²²⁸ (372)
Erdelmeier, I., 87³⁹³ (41)
Erdik, E., 185¹² (160), 187⁷⁹ᵉ (166, 167), 274⁷²⁶ (249)
Erdmann, P., 189¹⁷¹ᵇ (178)
Erickson, G. W., 84¹⁴⁵ (6)
Erickson, R. E., 431⁵³ᵃ (398)
Erion, M. D., 431⁶³ᶜ (400)
Ermert, P., 155⁸⁸ (137)
Ermolenko, M. S., 566⁴⁸⁸ (545), 566⁴⁸⁹ (545)
Ernst, T. D., 352³⁷⁰ (340), 352³⁷⁶ (341)

Esaki, N., 558⁹ (516)
Esch, P. M., 383⁶⁶ (361)
Eschbach, C. S., 432¹¹⁹ (407)
Esker, J. L., 566⁴⁴⁷ (543), 566⁴⁴⁸ (543)
Es-Sayed, M., 83¹¹¹ (5)
Esser, L., 458⁵⁷ (453)
Estel, L., 92⁷¹² (79)
Estreicher, H., 434²³⁷ᶜ (426)
Etemad-Moghadam, G., 430⁴⁰ (396, 416, 421), 433¹⁷⁰ᵃ (416, 422)
Etinger, M. Y., 274⁷⁶¹ (252)
Etzrodt, H., 125³⁴ (94)
Eugster, C. H., 91⁶⁶¹
Evans, A., 351³⁰² (337)
Evans, D. A., 30, 83⁸² (4, 29, 30, 36), 188¹²⁰ᵃ (171, 172), 240, 241, 265⁹⁴ (199), 265⁹⁵ (199), 268²⁸⁶ (211), 268²⁸⁷ (211), 269³⁷³ (218), 270⁴⁷³ (229), 272⁵⁹⁵ (240, 241), 272⁶²¹ (240), 272⁶²² (240), 275⁷⁸² (256), 275⁷⁸³ (256), 275⁷⁹³ (258), 351²⁹⁹ (337), 353⁴⁵¹ (344)
Evans, I. A., 88⁴⁴⁷ (48)
Evans, P. A., 351³⁰² (337)
Everhardus, R. H., 127¹⁵⁸ (116)
Evrard, G., 89⁵¹² (56), 565³⁹⁴ (541, 544), 567⁵⁵⁷ (550, 551), 568⁵⁸⁰ (553, 554, 557)
Ewald, M., 188¹⁴¹ᵈ (174), 275⁸³⁶ (260)
Ewing, D. F., 565³⁷⁴ (540, 547)
Eyley, S. C., 565⁴⁰⁹ (541)

Faber, T., 383³⁰ (359)
Fabicon, R. M., 189¹⁹¹ᵇ (182), 189¹⁹¹ᶜ (182)
Fadel, A., 273⁶⁸⁹ (245)
Fagan, P. J., 267²⁰³ (204)
Fahrbach, G., 560⁷⁹ (518, 529)
Faigl, F., 89⁵⁰⁸ (55), 125¹³ (94, 95), 126¹¹² (109), 126¹²² (109), 127¹⁷⁶ (119)
Failla, S., 433¹⁵⁴ (413)
Fairchild, R. G., 264¹⁶ (192)
Falck, J. R., 87³⁵⁷ (33), 385²⁰⁴ (371), 387²⁸⁹ (376)
Falcone, S. J., 598⁸⁵ (574)
Faller, A., 87³⁴⁷ (30), 566⁴⁵⁶ (544)
Faller, J. W., 352³⁴⁷ (339)
Fallon, G. D., 485⁵⁴ (474)
Fallwell, F., Jr., 126¹³³ (112)
Falorni, M., 156¹⁵¹ (150)
Fält-Hansen, B., 566⁴⁹⁸ (545, 546)
Fan, W. M., 567⁵⁶⁵ (551)
Fan, W.-Q., 85²⁵⁶ (22), 88⁴¹⁷ (42), 89⁴⁸⁹ (52, 59)
Fañanás, F. J., 431⁸⁶ (403), 434²⁰⁷ᵃ (421), 434²⁰⁷ᵇ (421)
Farfan, N., 267²³³ (206)
Farina, V., 384¹³⁹ (367)
Farkhani, D., 187¹⁰⁵ (169)
Farrar, W. V., 597⁵¹ (572)
Fattori, D., 568⁶⁴¹ (557), 568⁶⁴² (557)
Fattuoni, C., 83⁶¹ (4), 128²⁰⁹ (124)
Faure, B., 275⁸⁰⁸ (258, 261), 275⁸⁰⁹ (258, 261)
Favre, E., 272⁵⁸⁵ (238), 272⁵⁹⁰ (238)
Feaster, J. E., 350²³⁴ (332, 343)
Federov, V. A., 597⁴¹ (572)
Fehr, C., 151, 156¹⁶² (151)
Fehrentz, J.-A., 127¹⁵¹ (116)

Feiring, A. E., 565⁴¹⁰ (542)
Feit, B.-A., 83⁹⁴ (5)
Feixas, J., 349²¹⁰ (330)
Fekih, A., 599¹⁴² (577), 599¹⁵⁰ (578), 599¹⁵⁵ (578, 581)
Felder, M., 188¹⁴¹ᶠ (174), 275⁸³⁶ (260)
Feldkamp, J., 189¹⁵⁷ᵃ (176)
Feldman, K. S., 348¹³⁸ (324)
Felix, C., 155⁶⁷ (134)
Felix, D., 188¹²⁸ᵇ (171, 172)
Felkin, H., 146
Fellows, C. A., 432¹¹⁷ (407), 458²⁷ (447), 458³⁰ (447)
Felman, S. W., 84¹²⁹ (6)
Fensterbank, L., 350²²⁸ (331)
Fenton, G., 349²¹⁶ (330)
Fenyes, J. G., 92⁶⁷⁶
Fenzl, W., 272⁵⁹³ (240)
Feringa, B. L., 91⁶²² (71), 91⁶²⁷ (71), 127¹⁷⁵ (119), 155³⁷ (132), 156¹⁶⁰ (151), 157¹⁷⁷ (152), 188¹⁴¹ᵍ (174), 189¹⁹⁴ᵃ (182), 189¹⁹⁴ᵇ (182), 189¹⁹⁴ᶜ (182)
Ferles, M., 264⁵² (195)
Fernández, A., 459⁷⁵ (456)
Fernández, B., 459⁷⁷ (456), 568⁶³⁷ (557)
Fernández, E. J., 458¹⁴ (445)
Fernández, M. B., 459⁷⁶ (456)
Fernández, M. I., 459⁷¹ (456), 459⁷² (456), 459⁷⁶ (456), 459⁷⁷ (456)
Fernández de la Pradilla, R., 385²¹⁴ (371)
Ferraz, H. M. C., 600²²⁰ (585–7), 600²²¹ (586)
Ferreira, D., 564³⁴⁹ (539), 564³⁵⁰ (539)
Ferreira, J. T. B., 384¹⁰⁸ (364), 562¹⁷⁸ (522), 563²⁵⁷ (530), 597⁴² (572), 598⁸⁰ (573), 598⁸¹ (573, 575)
Ferrier, R. J., 433¹⁴⁵ᵃ (412), 433¹⁴⁵ᶜ (412), 433¹⁴⁵ᵈ (412)
Feutrill, G. I., 350²⁷⁷ (336)
Fevig, T. L., 349²²⁵ (330), 383²¹ (358)
Fewkes, E. J., 186⁴⁷ᵃ (163)
Fiandanese, V., 346⁴⁴ (316)
Fiaschi, R., 88⁴⁴² (47)
Fichter, K. C., 304, 310²⁸ (280, 297), 310⁴⁹ (284, 290), 310⁶⁵ (288)
Fields, L. B., 275⁸¹¹ (258, 261)
Fieser, L. F., 82⁵³ (3)
Fieser, M., 82⁵³ (3)
Figuly, G. D., 128²⁰⁸ (124)
Filippakis, S. E., 91⁶²⁸ (71)
Filippo, J. S., Jr., 384¹⁰⁴ (363)
Finet, J.-P., 484³⁴ (467, 473, 480), 484³⁹ (469, 480), 485⁴⁹ (473), 485⁵⁰ (473), 485⁵¹ (473, 478), 485⁶² (481), 488, 511³ (488, 506), 511⁴ (488), 511⁵ (488, 506), 512¹¹⁰ (504, 509), 512¹¹³ (504), 512¹¹⁹ (505, 507, 509), 512¹²⁰ (505), 512¹²⁷ (507), 512¹²⁸ (507, 508), 513¹³² (508), 513¹³⁵ (509), 513¹³⁸ (509, 510), 513¹⁴¹ (510), 513¹⁴² (510), 513¹⁴³ (510), 598¹²⁷ (576, 583)
Fink, D. M., 86²⁸⁵ (24)
Finnegan, R. A., 125²⁵ (94, 114), 126¹²⁸ (111)

Firnau, G., 92⁶⁹² (72)
Firor, J. W., 187⁸³ (166)
Fischer, H., 566⁴⁹⁹ (545)
Fischer, J., 433¹⁸⁶ (418)
Fischer, M.-R., 86³¹⁹ (27)
Fischer, P., 351³⁰⁵ (337)
Fish, R. H., 268³⁰⁰ (212)
Fisher, G. B., 265⁶¹ (195), 265⁶² (195), 273⁶⁶⁵ (244), 273⁶⁶⁷ (244)
Fisher, L. E., 88⁴²⁶ (43), 89⁵²⁸ (59)
Fisher, R. P., 270⁴⁶⁷ (229)
Fishman, A., 83⁹⁴ (5)
Fitjer, L., 189¹⁷⁴ᵉ (178)
Fitzgerald, J. J., 83¹⁰³ (5)
Flanagan, P. W. K., 310⁵⁶ (285, 304), 311⁹⁵ (304)
Flann, C. J., 433¹⁶¹ (414)
Flautt, T. J., 125⁴⁸ (95)
Fleischhauer, I., 87³⁵¹ (30)
Fleming, A., 350²⁷⁵ (336)
Fleming, F. F., 384¹³⁷ (367), 385¹⁸⁴ (370)
Fleming, I., 82¹² (2, 4), 82¹⁸ (2), 82²¹ (2), 82²⁴ (2, 15), 82²⁵ (2, 15), 82²⁷ (2, 25), 83⁸⁰ (4), 86²⁸⁰ (24), 86³²⁵ (28), 86³²⁶ (28), 86³²⁷ (28), 87³⁷⁵ (36), 89⁵³⁹ (63), 187⁷⁹ᵃ (166, 167), 187⁹⁴ (169, 171, 172, 183), 188¹⁵⁰ (175, 182), 264³⁹ (193, 201, 213, 247), 264⁴⁶ (194), 267²⁶¹ (208, 234, 235), 268³²⁰ (214–16, 218), 271⁵⁴⁸ (235, 236, 251, 252), 272⁵⁸⁴ (238), 272⁶⁰⁰ (240, 241, 256), 273⁶⁶³ (244), 310⁴² (283, 285), 317, 320, 346¹¹ (314), 346¹³ (314), 346³² (315), 346³⁸ (315, 317), 346⁵¹ (317), 346⁵⁵ (317), 346⁵⁹ (317), 347⁶⁴ (317), 347⁶⁵ (317), 347⁶⁷ (317), 347⁷⁰ (317), 347⁷¹ (317), 347⁷² (317), 347⁷³ (317, 320), 347⁷⁶ (319), 347⁸⁷ (320), 347⁹⁷ (321), 348¹⁶⁸ (327), 349²¹⁹ (330, 343), 383⁵ (356, 373), 384⁹⁸ (363, 364), 384¹¹⁴ (364), 385¹⁸⁶ (370, 373), 387²⁹⁴ (377), 387³³⁴ (381), 560⁸⁴ (518, 521, 522, 526, 529, 538, 540, 543, 544, 551, 552, 555–8), 560⁸⁵ (518, 521, 522, 526, 543, 544, 551, 552, 555)
Fleming, J. A., 87³³⁷ (29), 387³³³ (381)
Fleming, S. A., 350²³² (332)
Fletcher, H., III, 88⁴⁵⁵ (49)
Flippin, L. A., 83¹⁰⁸ (5), 88⁴²⁷ (44), 89⁵²⁸ (59), 89⁵³⁵ (60)
Flood, L. A., 156¹²⁵ (144), 564²⁹⁹ (536)
Flórez, J., 84¹⁵⁵ (7)
Florio, S., 134, 135, 155⁵⁸ (134), 155⁵⁹ (134), 155⁶⁰ (134), 155⁶⁵ (134), 155⁶⁶ (134), 155⁶⁹ (135)
Floris, B., 433¹⁹⁶ (420)
Floris, C., 83⁶¹ (4), 128²⁰⁹ (124)
Flowers, H. M., 385¹⁴⁹ (367)
Flygare, J. A., 348¹⁶⁶ (327)
Foitzik, N., 348¹¹⁸ (323)
Folest, J.-C., 186³⁶ᵉ (162, 174)
Foley, M. A., 188¹¹⁹ᵃ (171)
Follet, M., 264⁴¹ (194, 247)
Fontani, P., 268²⁹⁶ (212), 268²⁹⁷ (212,

Forbes, D. C., 353[436] (343)
Ford, W. T., 126[93] (103)
Forero-Kelly, Y., 83[109] (5)
Forrest, A. K., 351[284] (336)
Förster, F., 560[100] (519, 527)
Forsyth, C. J., 434[245b] (427)
Forth, M. A., 88[430] (46, 60)
Fortoul, C. R., 385[181] (370)
Fortuniak, W., 353[453] (344)
Fossatelli, M., 127[136] (112), 128[195] (121), 128[196] (121)
Foster, D. G., 563[273] (530)
Foster, D. J., 125[9] (94), 431[55a] (399, 400), 431[55b] (399, 400), 431[66] (400)
Foster, S. J., 562[219] (527)
Foubelo, F., 85[204] (14), 85[205] (14), 383[53] (361)
Fouquet, G., 126[79] (101)
Four, P., 383[48] (361, 362)
Fourquez, J. M., 271[517] (232)
Fowler, F. W., 91[624] (71, 72)
Fox, D. E., 91[612] (67)
Fox, D. N. A., 350[247] (333), 350[258] (334)
Foxton, M. W., 309[23] (280, 288, 292), 310[29] (280)
Fraenkel, G., 82[13] (2)
Francès, J. M., 387[290] (376)
Franceschini, M. P., 268[325] (215)
Franciotti, M., 127[150] (116), 347[62] (317)
Francisco, C. G., 563[286] (533, 556, 557)
Franck-Neumann, M., 351[320] (338), 459[63] (454)
François, J.-P., 155[96] (140)
Frangin, Y., 188[148b] (175, 182), 188[148c] (175, 182), 189[158] (176)
Frank, W. C., 434[210] (422)
Frankenberger, W. T. J., 558[12] (516)
Frankland, E., 186[54] (164), 278
Franzen, G. R., 434[243] (427)
Franzini, L., 125[13] (94, 95)
Fraser, R. R., 83[72] (4), 83[79] (4, 5), 83[98] (5)
Fraser-Reid, B., 350[232] (332)
Fráter, G., 433[188d] (418)
Fréchet, J. M. J., 156[164] (151), 188[122b] (171, 172), 188[122c] (171, 172)
Freedman, L. D., 488, 511[2] (488), 511[7] (488), 511[8] (488), 511[9] (488), 511[10] (488), 511[11] (488), 511[12] (488), 511[13] (488), 511[14] (488), 511[15] (488), 511[16] (488), 511[17] (488), 511[18] (488), 511[19] (488), 511[20] (488), 511[21] (488), 511[22] (488), 511[23] (488), 511[24] (488), 511[34] (488)
Freeman, P. K., 7, 84[146] (7)
Freeman, W. P., 349[180] (328)
Frei, B., 86[286] (24), 87[388] (40)
Freijee, F. J. M., 185[21f] (160)
Freire, R., 563[286] (533, 556, 557)
Fresneda, P. M., 269[369] (218)
Freund, A., 278
Frey, F. W., 484[1a] (461)
Fried, J., 310[48] (283)
Friedmann, R. C., 91[612] (67)
Friedrich, D., 352[366] (340), 565[392] (541, 545, 547)

Friesen, R. W., 386[220] (371)
Fritz, S., 187[105] (169)
Fröch, S., 272[602] (240), 272[603] (240, 241)
Fröhlich, H.-O., 186[45b] (163)
Fronza, G., 187[101a] (169)
Fry, J. L., 84[145] (6)
Frydrych-Houge, C. S. V., 351[329] (338)
Fryling, J. A., 565[422] (542, 557)
Fryzuk, M. D., 266[199] (204), 267[200] (204), 565[395] (541)
Fu, G. C., 265[94] (199), 265[95] (199)
Fu, H.-W., 186[71d] (166)
Fu, J.-M., 88[420] (42)
Fu, P. P., 563[285] (532, 537)
Fuchs, P. L., 90[607] (67), 352[397] (342)
Fugami, K., 187[92] (168), 273[651] (243), 383[34] (359)
Fuganti, C., 187[101a] (169), 187[101b] (169)
Fugmann, B., 568[640] (557)
Fuhrer, W., 89[477] (51)
Fuhrmann, G., 127[191] (121)
Fujihara, H., 155[50] (133), 155[51] (133), 155[52] (133)
Fujii, T., 351[325] (338)
Fujimori, K., 559[65] (517, 528)
Fujimoto, K., 353[438] (343)
Fujimoto, M., 386[267] (374)
Fujimura, O., 190[199c] (182)
Fujinami, T., 564[323] (538, 539, 545)
Fujioka, H., 156[144] (148)
Fujioka, T., 600[210] (584), 600[211] (584)
Fujisawa, T., 156[142] (148), 187[101d] (169), 187[101e] (169)
Fujishita, Y., 350[251] (334)
Fujita, A., 87[384] (39)
Fujita, E., 384[125] (364, 372), 387[292] (377), 387[332] (381), 458[37] (450)
Fujita, K. I., 125[57] (101), 125[72] (98, 101), 561[132] (520, 522, 539–42, 552, 557), 565[425] (542), 565[427] (542)
Fujitsuka, H., 569[653] (558)
Fujiwara, J., 276[851] (263), 385[172] (369), 387[306] (378)
Fujiwara, K., 85[220] (15)
Fujiwara, M., 434[240a] (427), 512[79] (499), 512[80] (500), 512[81] (500)
Fujiwara, T., 347[86] (319), 386[227] (372), 386[271] (374), 387[300] (377), 387[316] (379)
Fujiwara, Y., 511[50] (491)
Fukuhara, K., 90[600] (67)
Fukui, H., 126[78] (100)
Fukui, M., 156[142] (148)
Fukumoto, K., 83[114] (5), 84[157] (7), 352[359] (340), 564[316] (538, 555)
Fukumoto, T., 600[204] (583)
Fukushima, S., 434[239a] (427), 434[240b] (427)
Fukuyama, T., 353[451] (344), 568[606] (554)
Fukuzawa, S. I., 564[323] (538, 539, 545), 569[649] (557), 598[79] (573, 591), 598[97] (574, 588), 598[114] (575, 588, 589), 598[121] (575), 598[122] (575), 600[207] (583), 600[208] (583, 584), 600[209] (584), 600[233] (588), 600[234] (588), 600[235] (588), 600[238] (589),
600[239] (589), 600[240] (589), 600[243] (591)
Full, R., 266[150] (202)
Fuller, D. J., 83[70] (4), 83[106] (5)
Fuller, J. C., 273[665] (244), 273[666] (244), 273[667] (244)
Fuller, L. S., 91[654]
Funahashi, M., 566[446] (543)
Funamoto, T., 458[48] (451)
Funk, H., 598[110] (575)
Funk, R. L., 155[73] (136), 484[30] (467, 482)
Furlano, D. C., 89[499] (54)
Furlong, M. T., 186[66] (165, 179), 188[137a] (174, 177)
Fürrer, J., 125[59]
Fürstner, A., 154[1] (130), 185[2b] (159, 170), 185[24a] (160), 185[24b] (160), 185[25] (160)
Furukawa, J., 162, 186[39] (162, 178), 309[11] (279)
Furukawa, N., 89[492] (53), 155[49] (133), 155[50] (133), 155[51] (133), 155[52] (133)
Furukawa, S., 83[121] (5), 88[407] (42, 50, 51), 89[533] (60)
Furuta, K., 276[845] (262), 276[846] (262), 276[847] (262), 276[860] (263), 276[865] (264), 347[108] (321), 350[253] (334)
Furuta, T., 484[8] (463)
Futagawa, T., 189[177g] (178)
Fwu, S.-L., 351[333] (339)

Gabe, E. J., 310[39] (282, 293, 296, 309)
Gacek, M., 155[70] (135)
Gadru, K., 84[169] (8, 11)
Gadwood, R. C., 85[255] (21), 87[345] (29), 568[614] (555, 557)
Gaeta, K. K., 600[220] (585–7)
Gage, J. L., 186[67] (165)
Gagnier, R. P., 84[183] (10)
Gai, Y.-Z., 266[165] (203)
Gais, H.-J., 87[383] (39), 87[393] (41), 347[60] (317), 566[483] (545), 566[491] (545)
Gajda, C., 346[37] (315)
Galambos, G., 565[381] (540, 554)
Galatsis, P., 383[30] (359)
Galiullina, R. F., 311[109] (307)
Gallagher, D. J., 84[138] (6), 91[615] (70)
Gallagher, T., 87[349] (30)
Galle, J. E., 348[157] (326), 434[235a] (426)
Gallucci, J. C., 351[286] (336)
Gálvez, C., 92[677]
Gammill, R. B., 92[681] (72)
Gancarz, R. A., 565[385] (540)
Ganem, B., 434[213] (422), 568[605] (554)
Ganesan, K., 272[624] (240), 272[625] (240), 273[632] (240)
Gangloff, A. R., 187[111] (170)
Ganguly, A. K., 92[683]
Gangwar, S., 126[121] (109)
Gant, T. G., 88[411] (42, 43), 155[43] (133)
Ganter, C., 266[188] (203)
Ganz, K.-T., 353[434] (343)
Gao, Q., 276[848] (262)
Garad, M. V., 86[315] (26), 268[277] (210, 211)
Garanti, L., 126[97] (106)

Garbe, J. E., 562[228] (527)
Garcia, E., 127[185] (120)
Garcia, J., 126[84] (101), 275[824] (259)
Garcia, M. L., 562[220] (527, 538, 556)
García Navío, J. L., 351[296] (337)
Garcia-Granda, S., 434[220] (423)
García-Raso, A., 385[211] (371)
Gardner, H. C., 431[87b] (403)
Gardner, M., 274[767] (253)
Gareau, Y., 352[388] (341)
Garg, C. P., 269[344] (216), 459[82] (457)
Garratt, D. G., 561[164] (522, 540, 541)
Garrett, C. G., 84[138] (6)
Garrigues, B., 267[231] (206)
Garst, J. F., 154[8] (130)
Garst, M. E., 91[640] (71), 91[650] (72), 270[477] (230)
Gasanov, F. G., 564[343] (539), 564[348] (539)
Gasanz, Y., 92[677]
Gasche, J., 92[706] (78)
Gaskin, P., 563[282] (532, 538, 547)
Gaspar, P. P., 86[320] (27), 346[7] (314), 346[9] (314)
Gassman, P. G., 84[179] (10), 352[357] (340)
Gatehouse, B. M., 485[54] (474)
Gaudemar, M., 83[86] (4, 5), 175, 185[11] (160, 162, 169), 187[95a] (169), 187[99b] (169), 188[148a] (175, 182), 188[148b] (175, 182), 188[148c] (175, 182), 189[158] (176), 272[585] (238), 272[590] (238)
Gaudemar-Bardone, F., 83[86] (4, 5), 154[13] (130)
Gaudino, J. J., 157[183] (153)
Gavaskar, K. V., 84[185] (10, 11, 14), 84[190] (11)
Gavrilenko, V. V., 310[93] (304)
Gavrilova, G. A., 431[87c] (403)
Gavrilova, G. V., 431[87a] (403)
Gawley, R. E., 20, 21, 82[24] (2, 15), 82[25] (2, 15), 85[222] (16), 85[245] (20), 85[246] (20), 85[247] (20), 85[250] (20, 21), 85[266] (22, 47), 156[154] (150)
Gaydou, E. M., 431[73e] (401)
Gayoso, M., 459[70] (456), 459[71] (456), 459[75] (456)
Gaythwaite, W. R., 568[603] (554)
Gayton-Garcia, R., 83[62] (4)
Gee, S. K., 92[676]
Geiger, P. G., 558[21] (516)
Geissler, H., 87[391] (41)
Gellert, H.-G., 309[15] (279)
Gelli, G., 128[209] (124)
Gemal, A. L., 84[158] (7)
Geneste, H., 125[13] (94, 95)
Genêt, J.-P., 269[357] (217), 274[703] (246), 274[704] (246), 274[705] (246)
Geng, B., 385[180] (370)
Génisson, Y., 144, 156[124] (144)
Gennari, C., 241, 257, 273[629] (240, 257), 273[633] (240), 273[640] (241, 264), 273[643] (241), 273[646] (241, 258), 274[767] (253), 275[800] (258), 350[242] (333), 351[297] (337)
Gentile, R. J., 434[201] (421)
Gentry, C. R., 126[98] (106)
Geoffroy, P., 351[320] (338)

Geoghegan, P. J., Jr., 430[38a] (396, 411), 430[38b] (396, 411) 432[138] (411), 432[140c] (412), 433[143] (412), 433[148] (413)
Georg, G. I., 353[422] (343)
George, A. V. E., 127[178] (119)
George, C., 348[131] (324)
George, C. F., 189[191d] (182)
George, I. A., 349[222] (330)
George, J. W., 597[27] (572)
George, M. H., 458[13] (444)
Georgiadis, M. P., 91[628] (71), 91[632] (71), 91[633] (71)
Gerbing, U., 566[499] (545)
Gerding, H., 597[24] (572)
Gerhart, F., 155[96] (140), 430[24] (392)
Germanas, J., 346[53] (317)
Géro, S. D., 601[276] (595), 601[277] (595)
Gewert, J. A., 87[391] (41)
Ghanimi, A., 274[701] (246)
Ghannam, A., 86[278] (24)
Gharpure, M., 92[692] (72)
Ghavshou, M., 90[551] (63)
Ghosez, L., 87[378] (36), 352[361] (340)
Ghosh, S. K., 346[55] (317)
Ghosh, T., 84[143] (6), 155[48] (133)
Giacomelli, G., 156[151] (150)
Giacomini, D., 155[82] (137), 353[421] (343), 353[423] (343), 353[424] (343)
Giannone, E., 88[442] (47)
Giannotti, C., 484[39] (469, 480), 512[120] (505), 513[132] (508)
Gibson, H. W., 352[351] (340)
Gielen, M., 264[44] (194)
Giesbrecht, E., 597[56] (572, 574)
Giese, B., 189[171a] (178), 189[171b] (178), 273[680] (244), 383[28] (359), 383[56] (361), 386[242] (373), 432[89] (403), 433[144] (412), 433[177] (416), 567[518] (547), 567[526] (547), 567[533] (548), 567[534] (548)
Giffard, M., 431[69] (400), 435[251] (429)
Gil, S., 347[65] (317), 347[67] (317)
Gilardi, R., 348[131] (324)
Gilbert, J. C., 83[107] (5)
Gilbertson, S. R., 385[178] (369)
Gilchrist, J. H., 83[70] (4), 83[105] (5), 83[106] (5)
Gilchrist, T. L., 91[662]
Gilday, J. P., 90[545] (63), 90[546] (63), 90[553] (63), 90[554] (63), 385[166] (369)
Gill, M., 155[95] (139)
Gillard, J. W., 349[226] (331)
Gillet, J.-P., 186[34d] (161, 163), 186[48a] (163)
Gilloir, F., 348[165] (327)
Gilman, H., 82[48] (3), 82[49] (3, 8), 82[56] (3), 84[160] (7), 84[162] (8), 121, 125[47] (95), 127[190] (121), 127[193] (121, 122), 278, 309[5] (278), 309[6] (278), 309[13] (279)
Gilman, S., 564[315] (538)
Gimeno, M. I. S., 458[18] (446)
Gin, D. Y., 349[225] (330), 565[379] (540, 548)
Giner, J. L., 125[64] (101)
Gingras, M., 385[158] (368), 562[227] (527)
Gingrich, H. L., 84[143] (6)

Giomi, D., 352[361] (340)
Giovannini, R., 349[193] (329)
Giovini, R., 568[588] (554)
Girard, C., 189[174a] (178), 189[174b] (178), 189[174c] (178)
Girard, Y., 273[679] (244)
Girotti, A. W., 558[21] (516)
Girsh, D., 383[67] (361)
Gladysz, J. A., 562[228] (527)
Glänzer, B. I., 185[24a] (160)
Glaser, I., 458[22] (447)
Gleason, T. G., 565[404] (541, 548)
Gleave, D. M., 83[77] (4)
Gleicher, G. J., 383[29] (359)
Glidewell, C., 511[35] (488–90), 511[42] (490, 506)
Glover, S. A., 383[62] (361)
Gluzinski, P., 91[630] (71)
Godard, A., 92[712] (79), 92[713] (79), 271[514] (232), 271[515] (232), 271[516] (232), 271[517] (232), 271[518] (232), 271[519] (232)
Goddard, R., 273[696] (245)
Godfrey, C. R. A., 348[162] (326)
Godovikova, T. I., 430[27] (392), 434[238a] (427), 434[238b] (427)
Goehring, R. R., 156[123b] (144)
Goenther, W. H. H., 563[279] (530)
Goesmann, H., 348[118] (323)
Gofthel, G., 311[105] (306)
Goh, J. B., 385[187] (370)
Goicoechea-Pappas, M., 85[246] (20)
Goindord, G., 268[307] (213)
Gokel, G. W., 90[580] (65), 90[581] (65), 90[582] (65)
Gold, J. M., 88[441] (47)
Goldberg, Y., 353[413] (342)
Goldsmith, D. J., 91[639] (71), 564[313] (538, 540, 557)
Golebiowski, A., 91[630] (71)
Gollinger, W., 266[190] (204)
Golob, A. M., 268[287] (211)
Gololobov, Yu. G., 431[72] (401)
Golubinskaya, L. M., 597[40] (572)
Gombatz, K., 88[430] (46, 60)
Gómez, A. M., 350[232] (332)
Gómez, M. E., 459[76] (456), 459[77] (456)
Gómez Aranda, V., 433[195] (419), 434[216] (422)
Gong, L., 89[486] (51)
Gontarz, J. A., 430[39] (396)
González, A. M., 384[98] (363, 364), 384[114] (364), 385[186] (370, 373)
Gonzalez, M. A., 85[241] (19), 85[242] (19), 275[824] (259)
Goodfellow, C. L., 85[257] (22), 90[558] (63), 90[560] (63), 348[156] (326)
Goodman, J. M., 240, 273[627] (240), 273[630] (240, 258), 273[646] (241, 258), 274[767] (253), 275[791] (257), 275[796] (258), 275[800] (258)
Goodman, M. M., 269[353] (217), 458[41] (451), 597[38] (572, 577)
Goodwin, H. P., 266[162] (203)
Gopal, H., 309[24] (280, 292)
Gopalan, A. S., 87[388] (40), 434[244b] (427)
Goralski, C. T., 265[59] (195), 265[61] (195), 265[62] (195), 273[665] (244), 273[666] (244), 273[667] (244)

Gordon, B., III, 126[88] (103, 107), 126[100] (107), 126[118] (109)
Gordon, J., 350[277] (336)
Gordon, J. E., 599[161d] (579)
Gordon, K. M., 559[47] (516, 522, 535, 539, 541, 543, 552–5, 557), 600[241] (591)
Gordon, M. E., 432[137b] (411)
Gordon, M. S., 348[139] (325)
Goré, J., 432[131] (410), 434[215] (422), 434[241] (427), 434[242] (427)
Gorissen-Hervens, F., 565[413] (542)
Gornowicz, G. A., 86[289] (25)
Gorrell, I. B., 185[5] (159)
Gorys, V., 273[681] (244)
Gosselck, J., 562[176] (522)
Gosselink, D. W., 126[93] (103)
Gosser, L. W., 268[315] (213)
Goswami, R., 387[308] (378)
Goto, T., 87[380] (36), 189[192] (182), 348[133] (324), 564[357] (539)
Gottlieb, L., 85[237] (18)
Goubeau, J., 268[308] (213)
Goudgaon, N. M., 268[340] (216), 269[352] (217), 269[355] (217)
Gouin, L., 435[251] (429)
Gould, K. J., 270[454] (228)
Gould, K. L., 268[325] (215)
Gould, S. J., 89[479] (51), 89[482] (51)
Gourdel, Y., 274[701] (246), 274[702] (246)
Gouzoules, F. H., 430[37] (395), 433[160] (414)
Goyalski, C. T., 265[60] (195)
Graber, D. R., 431[52b] (398)
Graf, R., 348[118] (323)
Graf, W., 566[486] (545, 547, 548)
Gräfing, R., 127[158] (116)
Graham, J., 90[609] (67)
Graham, S. L., 91[656]
Gralak, J., 349[190] (329)
Grandclaudon, P., 91[614] (68)
Granguly, A. K., 353[455] (344)
Granja, J. R., 568[637] (557)
Grant, E. B., 386[288] (376)
Grassberger, M. A., 155[97] (140)
Grasselli, P., 187[101a] (169), 187[101b] (169)
Gratkowski, C., 83[111] (5)
Gray, J. L., 189[162c] (177)
Gream, G. E., 83[96] (5)
Greck, C., 87[359] (33), 269[357] (217)
Green, D. L. C., 87[336] (29)
Green, D. P., 84[167] (8)
Green, J. F., 269[353] (217), 269[388] (219)
Green, J. R., 90[562] (63, 64), 346[37] (315)
Green, N. J., 384[137] (367)
Greenberg, A., 264[3] (192, 208)
Greene, A. E., 187[89b] (167, 176), 188[140] (174)
Greene, T. W., 349[200] (329), 599[187] (582)
Greenspoon, N., 354[459] (345)
Greenwood, C. S., 92[698] (74)
Grey, R. A., 185[8e] (159, 166)
Gribble, G. W., 92[702] (74)
Gribov, B. G., 597[40] (572), 597[41] (572)
Gridnev, I. D., 265[119] (200), 271[553] (235)
Grieco, P. A., 92[687] (72), 351[326]

(338), 564[315] (538), 564[319] (538, 540, 557), 566[479] (545), 568[626] (555), 568[629] (556, 557), 568[630] (556)
Griedel, B. D., 350[250] (333)
Griffen, E. J., 92[704] (74)
Griffin, I. M., 433[185] (418)
Griffith, R. C., 434[201] (421)
Grigg, R., 348[142] (325), 434[249] (429), 459[80] (456)
Grignard, V., 278, 309[2] (278), 309[3] (278)
Grigsby, R. A., 598[75] (573)
Griller, D., 353[442] (344), 567[526] (547)
Grimaldi, J., 384[85] (362)
Grimes, R. N., 264[7] (192)
Grimm, E. L., 82[32] (2, 38)
Grisso, B. A., 386[224] (372)
Grocock, D. E., 128[201] (122)
Groh, B. L., 385[174] (369)
Groizeleau-Miginiac, L., 311[113] (307)
Grondin, J., 185[15b] (160)
Gronowitz, S., 70, 91[616] (70, 72), 91[644] (71), 91[645] (72), 232, 271[503] (232), 271[504] (232), 271[505] (232), 271[511] (232), 385[216] (371)
Gross, A. W., 83[123] (5, 42), 83[124] (5)
Gross, E. G., 558[19] (516)
Gross, J. L., 88[415] (42, 79)
Gross, M. L., 559[61] (517)
Gross, U. M., 266[191] (204)
Grouiller, A., 565[374] (540, 547)
Group, E. F., Jr., 90[574] (64)
Grover, P. T., 274[755] (252)
Gruber, R. J., 89[495] (53)
Gruenanger, P., 268[291] (212)
Grumelec, C. Le, 155[40] (132)
Grundke, G., 459[62] (454)
Grutzner, J. B., 432[134b] (410)
Grzejszczak, S., 568[643] (557)
Gschwend, H. W., 51, 82[33] (2, 42, 48, 72), 89[477] (51), 127[142] (114, 121)
Gu, J.-H., 350[255] (334)
Gu, Y. G., 265[108] (200, 236, 237), 265[109] (200, 237), 348[171] (328)
Guay, D., 348[149] (325)
Guazzaroni, M. E., 189[165] (178)
Guchteneere, E., 568[641] (557)
Gudmundsson, B. O., 560[80] (518)
Guennouni, N., 268[303] (213), 268[305] (213)
Guerra, M. A., 186[37b] (162)
Guerrero, A., 349[210] (330)
Guerriero, A., 156[133] (146)
Guerrini, A., 353[443] (344), 567[527] (547)
Gugel, H., 185[3] (159, 178)
Guggisberg, Y., 126[122] (109)
Guibé, F., 353[410] (342), 383[48] (361, 362)
Guijarro, D., 82[52] (3, 15), 85[215] (15)
Guiles, J., 85[241] (19)
Guilhem, J., 512[129] (508)
Guillemonat, A., 431[73e] (401)
Guillena, G., 82[52] (3, 15)
Guillier, F., 92[707] (78, 79)
Guindon, Y., 273[679] (244), 273[681] (244), 383[33] (359)
Guinosso, C., 89[480] (51)
Guiry, P. J., 484[34] (467, 473, 480),

484[40] (469, 480), 485[49] (473), 485[62] (481)
Guittet, E., 89[512] (56), 562[214] (526, 550, 551), 567[557] (550, 551)
Gunderson, K. G., 273[684] (244)
Gunderson, L.-L., 352[347] (339)
Gung, B. W., 386[264] (373)
Günther, W. H. H., 558[3] (516–18), 558[5] (516, 529, 530, 538, 558), 558[6] (516), 558[7] (516), 558[8] (516), 558[22] (516), 558[24] (516), 558[30] (516), 559[32] (516, 519, 520, 522, 527, 528, 530), 559[53] (516, 518, 522, 527, 529, 530, 555), 559[57] (517), 559[60] (517, 518), 559[62] (517, 518), 559[67] (518), 559[70] (518), 559[71] (518, 558), 559[72] (518), 559[74] (518), 560[75] (518, 543), 560[105] (519, 520), 560[109] (519, 520, 528, 539), 560[115] (519, 520, 522, 538, 539, 552–4), 560[116] (519, 538, 539, 552), 561[131] (520, 552), 561[145] (520, 545), 561[146] (520), 561[149] (520, 528), 561[153] (520, 529), 562[221] (527), 563[232] (528), 563[248] (530), 568[578] (553), 568[593] (554), 597[50] (572)
Guo, B., 351[312] (337)
Guo, D., 431[81] (402), 431[87e] (403)
Guo, G., 274[720] (249)
Guo, G. Z., 511[54] (492)
Gupta, A. K., 267[215] (205)
Gupta, G., 310[70] (291)
Gupta, S. K., 267[207] (205), 267[208] (205), 430[23] (391)
Gupta, V., 265[70] (196)
Gurski, A., 385[202] (371)
Gurskii, M. E., 269[372] (218), 271[553] (235), 272[591] (239)
Gürtzgen, S., 310[64] (288)
Gurumurthy, R., 84[169] (8, 11)
Gusarova, N. K., 598[91] (574), 598[92] (574), 598[93] (574)
Guseinov, M. M., 564[343] (539)
Gusev, B. P., 432[124] (408)
Gutheil, W. G., 264[21] (192)
Guyot, B., 272[578] (237), 348[124] (323)
Guziec, F. S., Jr., 559[50] (516, 546), 560[91] (518–20, 528, 545), 567[511] (546), 567[513] (546), 567[514] (546, 547)
Guziec, J. S., 560[95] (519, 520, 545, 546)
Guzzi, U., 385[209] (371)
Györi, B., 458[22] (447)
Gysling, H. J., 567[570] (552)

Ha, D.-C., 350[263] (335, 343)
Haack, J. L., 89[495] (53)
Haage, K., 310[54] (285)
Haar, J. P., 352[349] (339)
Habaue, S., 156[127] (145, 188[154] (175), 349[188] (329)
Habeeb, J. J., 186[36a] (162, 174)
Habermas, K. L., 349[203] (329)
Habtemariam, A., 186[57b] (165)
Habus, I., 353[422] (343)
Hachisuka, C., 353[425] (343)
Hachiya, I., 350[255] (334)
Hackett, S., 86[306] (26)
Hadjisoteriou, M., 434[249] (429)
Hafner, A., 185[10b] (160, 162, 171, 172)

Hagelee, L. A., 270[452] (228)
Hagen, G., 347[69] (317), 353[452] (344), 386[249] (373)
Hagen, T., 270[415] (223)
Hagiwara, T., 84[141] (6), 90[590] (65), 187[84] (167)
Hahn, G. R., 187[100] (169), 266[131] (201)
Haines, S. R., 433[145a] (412)
Hajdu, E., 430[29] (393)
Hajicek, J., 269[357] (217)
Halazy, S., 87[344] (29), 562[205] (522, 554), 564[329] (538, 557), 565[423] (542), 567[574] (552, 557)
Hale, M. R., 350[233] (332), 353[436] (343)
Haley, N. F., 598[102] (574)
Hall, E. S., 264[17] (192)
Hall, I. H., 264[17] (192), 264[18] (192)
Hall, M. B., 563[296] (534)
Hall, P. L., 83[105] (5), 83[106] (5)
Hallas, G., 128[200] (122), 128[201] (122)
Hallberg, A., 88[419] (42), 89[488] (52), 91[673], 346[23] (315)
Haller, J., 85[269] (23)
Haller, W. S., 597[61] (572, 573)
Halley, F., 484[39] (469, 480), 512[120] (505), 513[132] (508)
Hallock, J. S., 430[30c] (393)
Hallowell, A. T., 127[134] (112)
Halpern, J., 432[140a] (412), 432[140b] (412)
Halteren, B. W. v., 431[50] (397)
Halterman, R. L., 126[84] (101), 269[405] (221)
Halut-Desportes, S., 274[705] (246)
Haly, B. D., 266[158] (202)
Hamada, H., 458[49] (451)
Hamada, S., 269[396] (220)
Hamana, H., 273[634] (240)
Hamanaka, S., 565[376] (540, 545)
Hamann, L. G., 349[221] (330)
Hamaoka, T., 269[385] (219)
Hambley, T. W., 484[41] (469), 485[61] (479, 480)
Hamdouchi, C., 87[362] (33)
Hamelin, J., 350[264] (335)
Hammar, W. J., 432[139a] (411)
Hammer, R. P., 349[203] (329)
Hammond, G. B., 83[119] (5), 346[31] (315)
Hamon, D. P. G., 155[87] (137)
Han, B.-H., 90[597] (66), 185[28a] (161), 185[28b] (161), 185[28c] (161)
Han, K. I., 311[101] (304)
Han, L. B., 601[278] (595), 601[282] (596)
Han, W., 568[579] (553, 554, 557)
Han, Y.-H., 430[25] (392)
Hanack, M., 81[2] (2), 82[19] (2), 82[20] (2), 82[34] (2, 3, 42), 83[66] (4)
Hanafusa, T., 352[354] (340)
Hanagan, M. A., 156[147] (148)
Hanaoka, M., 350[256] (334)
Hanazaki, Y., 599[137a] (576, 577), 599[137b], 600[249] (591)
Handa, K., 563[241] (529)
Handel, H., 273[693] (245)
Hanekamp, J. C., 88[429] (46), 127[137] (112)
Hanessian, S., 127[180] (119), 384[79] (362), 431[52a] (398), 566[439] (543, 557)
Hannon, F. J., 188[136] (172)

Hansen, D., 353[434] (343)
Hansen, M. M., 275[794] (258)
Hansen, M. R., 431[64b] (400)
Hansen, S. W., 186[34b] (161, 163)
Hanson, J. R., 82[21] (2)
Hänssle, P., 127[157] (116)
Hao, N. K., 271[494] (232)
Hara, D., 189[187a] (180, 182)
Hara, S., 187[80] (166), 266[144] (201, 214), 266[146] (201), 266[148] (202), 266[151] (202), 269[382] (219), 270[419] (223), 270[457] (228), 271[499] (232), 271[500] (232), 271[501] (232), 271[502] (232), 272[565] (236), 272[612] (240), 272[615] (240)
Harada, K., 155[46] (133)
Harada, T., 189[186] (180, 182), 189[187a] (180, 182), 189[187b] (180, 182), 189[188a] (180, 182), 189[188b] (180, 182), 264[51] (195), 273[687] (245), 273[688] (245), 276[859] (263), 347[96] (321), 353[439] (344), 512[83] (500), 512[85] (501), 512[86] (501)
Haraguchi, K., 347[82] (319)
Haramura, M., 387[329] (380)
Harbach, J., 85[221] (15), 85[265] (22)
Harcourt, D. A., 91[640] (71), 91[650] (72)
Harden, D. B., 155[93] (139, 140)
Harding, K. E., 434[230] (425), 434[231c] (425), 434[232] (425), 434[233] (425)
Harding, M. M., 433[159] (414)
Hardtmann, G. E., 273[684] (244), 273[685] (244)
Harendza, M., 383[24] (358), 387[336] (381)
Harigaya, Y., 351[319] (338)
Harirchian, B., 565[397] (541, 554)
Harley, F. R., 186[58] (165, 168–70, 176)
Harmat, N., 156[131] (146)
Harmata, M., 348[125] (323)
Harms, K., 155[41] (132), 273[697] (245)
Harnisch, J., 84[140] (6)
Haroutounian, S. A., 91[632] (71), 91[633] (71)
Harp, J. J., 566[442] (543)
Harper, B. C., 567[545] (548)
Harpp, D. N., 562[227] (527), 563[235] (528), 563[261] (530), 599[169] (580), 599[176] (580)
Harrington, P. J., 89[538] (63)
Harrington, P. M., 348[134] (324)
Harris, D. J., 272[572] (237)
Harris, P. A., 187[116b] (171), 512[99] (502), 512[100] (502)
Harrison, A. T., 83[70] (4), 83[106] (5)
Harrison, C. R., 270[449] (227), 270[454] (228), 270[456] (228)
Harrison, I. T., 599[152] (578)
Harrison, J., 273[665] (244), 273[667] (244)
Harrison, L. W., 432[109] (406), 432[113a] (407), 432[115b] (407, 417, 427, 429)
Harrison, S., 599[152] (578)
Hart, D. J., 155[100] (140), 350[263] (335, 343), 565[437] (543)
Hart, G. C., 85[246] (20), 85[247] (20)

Hart, H., 155[46] (133), 155[47] (133), 155[48] (133)
Hartgerink, R. L., 432[115a] (407)
Hartley, F. R., 82[9] (2, 42), 264[30] (192, 193, 201, 220, 222, 225, 230, 235, 238, 247), 346[14] (314), 511[34] (488)
Hartley, R. C., 86[322] (27, 28), 348[150] (325)
Hartmann, J., 84[136] (6), 125[38] (94, 98), 125[39] (94, 100), 125[40] (94, 98, 99, 117), 125[49] (95, 108, 112, 114), 125[50] (95, 112–14), 125[55] (97), 125[56] (103, 116), 125[58], 127[154] (116)
Hartog, F. A., 189[164] (178)
Hartung, T., 276[871] (264)
Hartwig, J. F., 265[98] (199)
Hartwig, W., 566[495] (545, 548)
Hartzell, S. C., 92[676]
Harvey, R. G., 88[462] (49), 563[285] (532, 537)
Hasan, I., 91[624] (71, 72)
Hasan, S. K., 568[612] (554)
Hase, T., 82[45] (3)
Hasegawa, E., 383[45] (360)
Hasegawa, H., 386[270] (374)
Hasegawa, M., 353[441] (344)
Hasegawa, N., 348[143] (325)
Hasegawa, T., 386[273] (374), 458[40] (451)
Hasegawa, Y., 84[131] (6)
Haseltine, J., 89[530] (59)
Hasha, D. L., 265[60] (195)
Hashimoto, C., 89[529] (59)
Hashimoto, H., 189[180] (179)
Hashimoto, N., 90[587] (65)
Hashimoto, S.-I., 485[45] (472, 473)
Hassan, D., 86[305] (26)
Hassner, A., 265[122] (200, 215), 434[235a] (426)
Hata, E.-I., 156[104] (141)
Hata, T., 384[133] (367)
Hatajima, T., 154[24] (130)
Hatakeyama, S., 348[127] (323)
Hatanaka, K., 351[314] (337)
Hatanaka, T., 187[116a] (171)
Hatanaka, Y., 346[28] (315, 326)
Hatayama, Y., 353[448] (344)
Hatch, R. P., 311[96] (304)
Hatem, J., 384[85] (362)
Hattori, K., 155[68] (134), 189[186] (180, 182), 189[187a] (180, 182), 276[849] (262), 276[850] (262), 276[861] (263), 350[254] (334), 350[278] (336), 601[287] (596)
Hattori, T., 458[35] (450)
Haubold, W., 204, 266[190] (204)
Hauck, T., 85[221] (15)
Haugwitz, R. D., 88[450] (48)
Häusel, R., 90[592] (65)
Hauser, A., 187[95b] (169)
Havenith, R. W. A., 187[115a] (171)
Hawkins, J. M., 87[404] (42), 276[857] (263)
Hawthorne, M. F., 264[15] (192)
Hay, D. R., 85[224] (16)
Hayakawa, A., 386[237] (372)
Hayakawa, K., 348[144] (325)
Hayakawa, S., 566[446] (543)
Hayama, N., 432[128a] (409)
Hayama, T., 565[411] (542)
Hayano, K., 564[302] (536)

Hayasaka, T., 188^{125} (171), 188^{133} (172), 188^{141a} (174), 188^{141b} (174), 188^{141c} (174)
Hayase, Y., 566^{480} (545)
Hayashi, A., 350^{275} (336), 386^{278} (375)
Hayashi, M., 188^{129} (171), 274^{698} (246), 352^{346} (339), 352^{352} (340), 352^{379} (341), 387^{307} (378, 380)
Hayashi, T., 87^{353} (31), 89^{505} (55), 90^{578} (65), 90^{590} (65), 187^{84} (167), 265^{90} (198), 266^{136} (201), 274^{719} (248, 249), 274^{722} (249), 274^{723} (249), 274^{724} (249), 353^{433} (343)
Hayashi, Y., 90^{564} (63, 64), 90^{570} (64), 188^{141h} (174), 349^{192} (329), 434^{244a} (427), 434^{244c} (427), 434^{244d} (427)
Hayashida, A., 348^{144} (325)
Hayashida, H., 349^{195} (329)
Hayata, T., 387^{307} (378, 380)
Hazato, A., 186^{68} (165)
Heard, N. E., 350^{275} (336)
Heath, R. R., 126^{98} (106)
Heathcock, C. H., 82^{18} (2), 83^{81} (4), 88^{412} (42), 127^{140} (114), 258, 272^{596} (240, 241, 256), 275^{779} (256, 258), 275^{794} (258), 383^{5} (356, 373), 434^{235b} (426), 568^{624} (555), 568^{645} (557)
Heaton, C. A., 91^{619} (70)
Heaton, S. B., 90^{565} (63)
Hebel, D., 431^{51} (397, 415)
Hechenbleikner, I., 125^{19} (94), 126^{133} (112)
Heck, R. F., 270^{480} (231), 432^{92} (403, 404), 432^{101} (405), 432^{103a} (405), 432^{105} (405), 432^{112a} (407), 434^{210} (422)
Hedgecock, H. C., Jr., 268^{332} (215)
Heeg, M. J., 91^{636} (71), 430^{25} (392)
Heffernan, G. D., 384^{110} (364)
Heffron, P. J., 269^{348} (216)
Hegedus, L. S., 89^{540} (63)
Heidbreder, A., 351^{322} (338)
Heider, K.-J., 384^{126} (364)
Heiliger, L., 352^{374} (341)
Heilman, S. M., 352^{345} (339)
Heinz, K. J., 154^{18} (130)
Heitmann, P., 273^{637} (240)
Heitz, W., 563^{239} (529)
Helle, M. A., 434^{228} (425), 434^{229a} (425), 434^{229b} (425)
Helling, J. F., 90^{573} (64)
Hellwinkel, D., 88^{471} (50, 51, 68), 560^{79} (518, 529)
Helmchen, G., 262, 276^{843} (262), 276^{844} (262), 350^{243} (333), 353^{434} (343)
Helquist, P., 187^{111} (170)
Hemperly, S. B., 189^{177e} (178), 189^{177f} (178)
Henderson, D. A., 269^{354} (217)
Henderson, H. E., 126^{75} (100)
Henderson, I., 275^{773} (255), 275^{774} (255)
Hendrickson, J. B., 87^{372} (36), 87^{373} (36)
Hendrix, J. P., 128^{198} (122)
Heneghan, M., 347^{63} (317)
Henke, B. R., 275^{797} (258)

Henriet-Bernard, C., 384^{85} (362)
Henry, C., 346^{20} (314)
Henry, K. J., 351^{326} (338)
Hense, A., 272^{559} (235, 237)
Hense, T., 85^{269} (23)
Heppert, J. A., 90^{566} (64)
Hepworth, J. D., 128^{200} (122), 128^{201} (122)
Herberich, G. E., 266^{185} (203), 266^{186} (203), 266^{187} (203), 266^{188} (203), 266^{189} (203)
Herdevall, S., 563^{255} (530)
Herdtle, J., 266^{190} (204)
Heřmánek, S., 264^{24} (192)
Hermann, K., 562^{185} (522, 552)
Hermans, B., 565^{428} (542), 566^{453} (544, 547), 566^{467} (544)
Hermes, M., 273^{696} (245)
Hernández, A., 82^{44} (3)
Herndon, J. W., 384^{142} (367), 566^{442} (543)
Herold, L. L., 566^{445} (543)
Herold, T., 274^{738} (251), 274^{739} (251)
Herrick, J. J., 156^{116} (143)
Herron, B. F., 348^{125} (323)
Herscheid, J. D. M., 431^{50} (397)
Hershberger, J., 431^{63b} (400), 431^{63d} (400, 401), 431^{67} (400), 431^{78a} (402)
Hershberger, S. S., 431^{82} (402), 432^{106} (406), 432^{120} (407)
Hertkorn, N., 85^{219} (15), 126^{99} (107)
Hess, B. A., 126^{116} (109)
Hesse, A., 83^{91} (5)
Hessner, B., 266^{187} (203)
Heuberger, C., 566^{486} (545, 547, 548)
Heuck, K., 433^{177} (416)
Heus, M., 348^{130} (324)
Hevesi, L., 558^{11} (516–18), 559^{33} (516, 519, 536, 537), 560^{88} (518, 522, 529, 531, 536), 560^{98} (519, 528, 531, 532), 560^{101} (519, 528, 537), 561^{121} (519, 534), 561^{136} (520, 533), 561^{141} (520, 534), 561^{142} (520, 535), 561^{154} (521, 522, 543, 552, 555), 561^{155} (521, 544), 561^{156} (521), 561^{157} (521), 562^{194} (522, 539–42, 557), 563^{236} (528, 537), 563^{284} (532), 564^{298} (535), 564^{311} (538), 565^{369} (540), 565^{384} (540), 565^{428} (542), 566^{453} (544, 547), 566^{464} (544, 556), 566^{465} (544), 566^{467} (544), 566^{470} (544), 566^{474} (544), 566^{475} (544), 566^{476} (544), 566^{478} (545), 567^{576} (553), 568^{589} (554–6), 568^{601} (554, 555), 568^{636} (557)
Hewawasam, P., 89^{487} (51)
Hibino, H., 157^{184} (153)
Hibino, S., 265^{69} (196)
Hiemstra, H., 347^{76} (319), 347^{79} (319), 383^{66} (361), 384^{92} (363)
Higaki, K., 566^{494} (545, 547, 548)
Higashi, Y., 349^{198} (329)
Higashino, T., 155^{94} (139)
Higashiyama, K., 156^{146} (148), 157^{180} (153), 157^{181} (153), 157^{182} (153)
Higgins, D., 347^{64} (317), 347^{71} (317)
High, K. G., 349^{202} (329)
Highsmith, T. K., 85^{240} (19)
Higuchi, K., 352^{350} (339)

Hiiro, T., 85^{209} (15), 600^{263} (593), 601^{268} (593), 601^{269} (593)
Hildebrandt, B., 272^{570} (237)
Hill, D. K., 349^{206} (330)
Hill, D. T., 432^{131} (410)
Hill, J. M., 348^{166} (327)
Hiller, W., 186^{45b} (163), 459^{75} (456)
Hillis, L. R., 89^{479} (51)
Hino, T., 275^{834} (260)
Hintze, F., 85^{267} (23), 85^{268} (23)
Hinz, G., 309^{14} (279)
Hirabayashi, Y., 566^{493} (545)
Hirai, Y., 600^{244} (591)
Hiraiwa, A., 386^{228} (372)
Hirama, M., 350^{275} (336), 386^{278} (375)
Hirano, Y., 156^{132} (146)
Hirao, A., 275^{815} (259), 275^{816} (259)
Hirao, T., 351^{325} (338)
Hirashima, T., 265^{72} (196)
Hirayama, N., 272^{619} (240), 310^{79} (293, 296)
Hirayama, S.-I., 559^{38} (516)
Hiroi, K., 565^{419} (542), 568^{631} (556)
Hirose, T., 384^{128} (365), 384^{129} (365), 387^{315} (379)
Hirose, Y., 187^{110} (170)
Hitchcock, P., 187^{116c} (171)
Hiyama, T., 346^{28} (315, 326)
Ho, C. D., 272^{586} (238)
Ho, J., 90^{562} (63, 64)
Hobe, M., 89^{512} (56), 565^{372} (540, 551), 566^{438} (543, 544, 552, 553), 567^{556} (550, 551), 567^{557} (550, 551), 567^{558} (550, 551), 567^{559} (550, 551)
Hoberg, H., 310^{78} (293, 296, 306), 311^{106} (306)
Hodgeman, D. K. C., 83^{96} (5)
Hodgson, D. M., 385^{179} (369)
Hodgson, S. T., 92^{690} (72)
Hodson, H. F., 385^{177} (369, 372)
Hoechstetter, C., 185^{13b} (160, 162)
Hoekstra, A., 431^{50} (397)
Hoekstra, W., 564^{313} (538, 540, 557), 566^{462} (544, 551)
Hoffer, R. K., 189^{193a} (182)
Hoffman, R. V., 351^{296} (337)
Hoffmann, H. M. R., 83^{128} (5), 459^{62} (454)
Hoffmann, P., 90^{580} (65), 90^{581} (65)
Hoffmann, R. W., 84^{173} (9), 85^{265} (22), 86^{282} (24), 87^{390} (40), 237, 241, 265^{115} (200, 215), 266^{133} (201), 267^{253} (208), 271^{552} (235), 271^{556} (235), 272^{557} (235), 272^{559} (235, 237), 272^{560} (236), 272^{562} (236), 272^{568} (237), 272^{569} (237), 272^{570} (237), 272^{571} (237), 272^{582} (237), 272^{599} (240, 241, 256, 258), 272^{602} (240), 272^{603} (240, 241), 273^{641} (241), 274^{737} (251, 253), 274^{738} (251), 274^{739} (251), 274^{762} (252), 274^{763} (253), 274^{764} (253), 274^{765} (253), 274^{766} (253), 275^{770} (255), 275^{771} (255), 386^{281} (375), 386^{284} (376), 560^{106} (519, 520, 528, 544), 566^{452} (544, 547), 567^{560} (551), 567^{561} (551)
Hofmann, G., 88^{471} (50, 51, 68)

Hofmann, H. P., 90[573] (64)
Högberg, T., 89[485] (51)
Hogg, D. H., 562[171] (522), 562[172] (522)
Hogg, D. R., 562[170] (522)
Hoiness, C. M., 185[3] (159, 178), 309[10] (279)
Hois, P., 346[39] (315)
Hojo, F., 348[140] (325)
Hojo, M., 352[394] (341)
Holden, M. E. T., 125[16] (94, 95), 125[17] (94)
Hollander, J., 90[567] (64), 90[568] (64)
Hollander, M. I., 567[514] (546, 547)
Holliman, F. G., 597[33] (572)
Hollingsworth, D. R., 434[232] (425)
Hollingworth, G. J., 384[87] (362)
Hollis, T. K., 350[261] (335)
Holm, T., 154[10] (130), 154[12] (130), 154[15] (130)
Holman, N. J., 90[555] (63)
Holmes, A. B., 348[130] (324), 353[436] (343), 565[420] (542)
Holmes, J. M., 91[640] (71), 91[650] (72)
Holmes, R. J., 484[41] (469)
Holmes, R. R., 346[16] (314)
Holmquist, C. R., 385[161] (368)
Holtan, R. C., 86[287] (25), 87[356] (32), 350[238] (333)
Holtzmann, R. T., 264[36] (192)
Holubka, J. W., 458[23] (447)
Holubová, N., 264[52] (195)
Holzapfel, C. W., 385[149] (367)
Hombini, C., 268[334] (215)
Hommes, H. H., 126[101] (107), 126[125] (110), 127[138] (112)
Honda, S., 433[157] (414)
Honda, T., 91[631] (71), 92[682] (72), 275[838] (260), 567[512] (546)
Hong, H., 156[119] (144), 156[121] (144), 346[54] (317)
Hong, Y. P., 348[132] (324), 350[254] (334)
Hoogeboom, T. J., 126[110] (109)
Hoong, L. K., 274[741] (251, 253), 274[768] (253)
Hoornaert, C., 567[541] (548)
Hooz, J., 125[19] (94), 125[21] (94, 108), 269[412] (222, 240), 269[413] (223), 272[605] (240)
Hope, E. G., 598[66] (573)
Hopkins, P. B., 92[708] (78)
Hoppe, D., 85[264] (22, 24), 85[267] (23), 85[268] (23), 85[269] (23), 85[270] (23), 85[271] (23), 86[272] (23), 86[273] (23), 86[274] (23), 127[143] (114), 386[265] (374)
Horaguchi, T., 383[45] (360)
Horan, N. R., 154[30] (131)
Hordis, C. K., 310[26] (280, 289), 310[27] (280, 289)
Horgan, A. G., 568[618] (555)
Hori, H., 188[134b] (172)
Hori, M., 352[355] (340), 566[469] (544), 566[477] (545)
Hori, T., 88[407] (42, 50, 51), 561[137] (520), 568[595] (554, 555)
Horiguchi, Y., 350[259] (335), 350[274] (335), 351[293] (337)
Horino, H., 432[95a] (404), 432[95b] (404), 432[96a] (404), 432[97] (404)
Horita, K., 274[752] (252)
Horiuchi, N., 352[354] (340)

Hornby, J. L., 562[228] (527)
Horne, D. A., 568[640] (557)
Hörnfeldt, A.-B., 91[641] (71), 91[642] (71), 91[644] (71), 271[511] (232), 385[216] (371)
Horton, D., 383[32] (359)
Horvath, A., 155[97] (140)
Horvath, R. F., 86[321] (27)
Hosay, T., 563[272] (530)
Hoshi, M., 265[116] (200), 269[378] (218), 269[407] (221), 269[408] (221), 269[410] (221), 269[411] (221)
Hoshino, S., 566[451] (543)
Hosmane, N. S., 264[10] (192)
Hosoi, A., 156[168] (151)
Hosomi, A., 321, 347[99] (321), 347[107] (321), 349[195] (329), 350[240] (333), 352[394] (341)
Hostalek, M., 266[187] (203)
Hotta, Y., 190[199a] (182)
Hou, Y. Q., 599[167] (580)
Houk, K. N., 241, 266[138] (201), 266[140] (201), 273[644] (241), 273[645] (241, 258), 275[795] (258), 275[807] (258, 261)
Houlihan, W. J., 88[428] (44), 128[199] (122)
Houpis, I. N., 84[177] (10), 385[207] (371), 385[212] (371)
Houri, A. F., 154[29] (131), 154[30] (131), 154[31] (131)
Houriet, R., 562[194] (522, 539–42, 557)
House, H. O., 82[55] (3, 6), 83[84] (4), 559[43] (516, 519)
Houtgraaff, H., 597[24] (572)
Hovestreydt, E. R., 155[73] (136)
Hoveyda, A. H., 154[29] (131), 154[30] (131), 154[31] (131), 154[32] (131), 154[33] (131), 154[34] (131), 265[94] (199), 350[233] (332), 353[436] (343)
Hoye, T. R., 352[363] (340), 563[251] (530, 539)
Hsiao, C.-N., 87[339] (29), 352[398] (342)
Hsieh, B. R., 89[495] (53)
Hsu, H. C., 82[13] (2), 269[380] (219)
Hsu, M.-F. H., 432[102] (405), 432[107c] (406), 432[113a] (407)
Hu, J., 600[256] (591, 592)
Hu, N. X., 598[128] (576, 584), 599[182] (582), 599[199] (583), 600[201] (583), 600[202] (583, 584, 586), 600[212] (584, 587), 600[213] (584), 600[214] (584), 600[215] (584), 600[222] (586), 600[223] (586, 587), 601[284] (596)
Hu, S. S., 431[87d] (403)
Hu, X., 155[62] (134)
Hua, D. H., 82[30] (2, 33, 36), 87[363] (33), 155[79] (136), 353[420] (343)
Huang, H. C., 275[790] (257)
Huang, J. L., 511[60] (493), 600[257] (592)
Huang, P.-Z., 351[309] (337)
Huang, S., 384[139] (367)
Huang, W.-S., 351[279] (336), 351[333] (339)
Huang, X., 599[140] (577, 581), 599[141] (577), 599[167] (580), 599[178] (581), 599[179] (581), 599[185] (582), 599[189] (582), 600[248] (591), 600[251] (591), 600[252] (591)

Huang, Y.-Z., 272[608] (240), , 511[6] (488), 511[36] (489), 511[37] (489), 511[38] (489), 511[39] (489), 511[41] (489), 511[43] (490), 511[44] (490), 511[47] (491), 511[53] (492), 511[54] (492), 511[55] (492), 511[56] (492), 511[57] (493), 511[58] (493–5), 511[59] (493, 496, 497), 511[60] (493), 512[61] (494), 512[62] (494), 512[64] (495), 512[65] (495), 512[66] (495, 498), 512[67] (496), 512[72] (497), 512[73] (497), 512[75] (498), 512[77] (498), 512[78] (498), 512[98] (502), 599[186] (582), 600[250] (591, 592), 600[253] (591), 600[255] (591, 592), 600[256] (591, 592), 600[257] (592), 600[258] (592), 600[259] (592)
Huang, Z. Z., 599[179] (581), 599[189] (582), 599[193] (582), 600[251] (591)
Huang, Z.-H., 512[61] (494)
Hubbard, J. L., 270[438] (225)
Hübel, M., 155[84] (137)
Huber, I. M. P., 85[228] (17)
Huber, M.-L., 485[59] (477–9, 481)
Huber, R. S., 565[436] (542)
Huber, S., 265[98] (199)
Huboux, A. H., 383[51] (361, 370)
Hudlicky, T., 347[101] (321), 350[275] (336)
Hudnall, P. M., 185[21a] (160)
Hudrlik, A. M., 348[158] (326), 348[172] (328)
Hudrlik, P. F., 348[158] (326), 348[172] (328)
Hudson, C. M., 350[230] (332)
Hueblein, G., 433[197a] (420)
Huet, F., 155[40] (132)
Huffman, J. C., 513[133] (508)
Hughes, C. T., 432[132b] (410)
Hughes, L., 268[329] (215)
Hughes, N., 154[3] (130)
Hughes, R. J., 270[421] (224)
Hughes, R. L., 264[36] (192)
Hui, R. A. H., 563[297] (535)
Hulce, M., 384[122] (364)
Hull, C., 386[247] (373)
Hulme, A. N., 275[792] (258)
Hulst, R., 91[622] (71), 127[175] (119)
Hundnall, P. M., 154[6] (130)
Hungate, R., 350[277] (336)
Hunt, D. F., 431[59a] (399), 431[59b] (399)
Hunziker, D., 188[128c] (171, 172)
Hurst, D., 267[211] (205)
Hurst, G. D., 266[154] (202), 274[731] (250), 274[733] (251)
Hursthouse, M. B., 353[411] (342), 566[459] (544)
Husk, G. R., 310[69] (291), 310[74] (291)
Husson, H.-P., 89[529] (59)
Hutchings, M. G., 126[89] (103), 126[92] (103), 433[175b] (416)
Hutchings, R. H., 85[239] (19)
Hutchins, R. O., 244, 273[668] (244), 567[530] (547)
Hutchins, R. R., 84[134] (6)
Hutchinson, L. L., 7, 84[146] (7)
Hutchinson, R. M., 485[49] (473)
Huth, A., 271[522] (232)
Huttner, G., 187[101c] (169)
Hutzinger, M. W., 384[109] (364)

Huval, C. C., 383[55] (361)
Huynh, V., 83[116] (5), 88[436] (46)
Hwu, J. R., 87[342] (29), 346[18] (314), 352[389] (341)
Hyde, B. R., 92[681] (72)
Hylarides, M. D., 269[380] (219)
Hyuga, S., 187[80] (166), 271[499] (232), 271[501] (232), 271[502] (232)
Hyun, B. C., 311[102] (305)

Ibuka, T., 186[65a] (165), 186[65b] (165), 186[65c] (165)
Ichikawa, J., 269[396] (220)
Ichikawa, K., 433[190a] (419), 434[239a] (427), 434[240a] (427), 434[240b] (427)
Ichikawa, Y., 87[380] (36)
Ichinose, H., 561[148] (520, 546)
Ichinose, Y., 273[651] (243)
Idacavage, M. J., 264[29] (192)
Igarashi, Y., 84[141] (6)
Iglesias, B., 271[492] (232), 459[79] (456)
Ignacz, T., 459[66] (455)
Ignatenko, A. V., 271[553] (235), 272[573] (237, 251)
Iguchi, S., 155[79] (136)
Ihara, M., 83[114] (5), 84[157] (7), 352[359] (340), 564[316] (538, 555)
Iihama, T., 88[406] (42, 53), 88[408] (42, 53)
Iio, H., 564[357] (539)
Ikebe, M., 85[207] (14), 387[309] (378)
Ikeda, K., 88[415] (42, 79)
Ikeda, M., 384[73] (361), 566[463] (544)
Ikeda, N., 274[760] (252)
Ikeda, S., 157[178] (153), 353[449] (344)
Ikegami, S., 87[389] (40), 432[139b] (411), 485[45] (472, 473)
Ikenaga, K., 346[25] (315)
Ikeuchi, M., 90[599] (67)
Ikota, N., 568[605] (554)
Ikura, K., 350[272] (335)
Ilkka, S. J., 432[104a] (405)
Imada, M., 512[80] (500)
Imai, M., 91[631] (71)
Imai, T., 189[178] (178), 266[130] (200, 208), 267[242] (207), 267[254] (208), 268[298] (212), 270[420] (224), 275[806] (258, 261), 386[238] (372)
Imamoto, T., 130, 154[24] (130), 274[699] (246)
Imanaka, S., 273[687] (245), 273[688] (245), 347[96] (321), 353[439] (344)
Imanieh, H., 565[393] (541)
Imma, H., 346[36] (315)
Imoto, H., 310[63] (288, 296)
Imperiali, B., 275[781] (256)
Imwinkelried, R., 275[789] (257)
Inaba, M., 564[354] (539), 564[355] (539)
Inagaki, T., 597[21] (572, 577), 598[132] (576-8)
Inamasu, T., 601[286] (596)
Inamoto, N., 566[500] (545, 546)
Inamura, Y., 512[88] (501)
Inanaga, J., 386[234] (372)
Inazu, T., 563[292] (534)
Inbasekaran, M., 87[369] (36)
Ingrosso, G., 155[58] (134), 155[59] (134), 155[66] (134), 155[69] (135)
Inguscio, G., 155[65] (134)
Inokawa, H., 266[184] (203)
Inokuchi, T., 563[253] (530, 538, 539)

Inomata, K., 272[592] (240), 272[614] (240)
Inomata, N., 275[813] (258)
Inoue, H., 157[180] (153), 157[181] (153)
Inoue, N., 432[95a] (404), 432[95b] (404), 432[96a] (404), 432[97] (404)
Inoue, S., 155[57] (134), 188[130c] (171), 352[346] (339)
Inoue, T., 352[346] (339)
Inouye, M., 599[144] (577), 599[145] (577), 599[166] (580), 599[181a] (581), 599[181b] (581)
Inouye, Y., 189[175] (178)
Inubushi, T., 383[16] (358), 384[72] (361)
Inui, N., 350[233] (332)
Iqbal, J., 83[116] (5)
Ireland, R. E., 83[113] (5), 336, 349[175] (328), 351[281] (336), 351[282] (336), 351[283] (336)
Irgolic, K. J., 559[68] (518), 597[7] (571, 576), 597[61] (572, 573), 598[75] (573), 600[208] (583, 584)
Irgolic, K. Y., 597[2] (571), 597[4] (571), 597[6] (571), 597[49] (572), 597[58] (572), 597[60] (572), 598[83] (573)
Irisarri, J., 459[70] (456)
Isaac, M. B., 349[207] (330)
Isaacs, N. S., 386[251] (373)
Isaacson, E. I., 157[176] (152)
Isaka, M., 275[791] (257)
Iseki, Y., 156[110] (141)
Ishibashi, H., 384[73] (361), 566[463] (544)
Ishichi, Y., 387[325] (380), 568[638] (557)
Ishida, M., 561[147] (520)
Ishifune, M., 189[167a] (178)
Ishiguro, M., 432[96b] (404)
Ishihara, H., 566[493] (545)
Ishihara, K., 156[155] (150)
Ishii, H., 564[351] (539)
Ishii, N., 346[57] (317)
Ishii, Y., 351[333] (339), 351[334] (339), 386[239] (372), 512[69] (496, 499)
Ishikawa, H., 157[178] (153)
Ishikawa, M., 90[605] (67), 271[528] (233, 234), 271[532] (234), 346[26] (315)
Ishikawa, N., 185[29a] (161), 185[29b] (161), 185[29c] (161), 185[32e] (161)
Ishikawa, T., 273[690] (245)
Ishikura, M., 270[476] (229)
Ishioka, R., 598[112] (575)
Ishitani, H., 350[255] (334)
Ishiyama, K., 383[45] (360)
Ishiyama, T., 271[495] (232), 271[528] (233, 234), 271[531] (234), 271[532] (234), 271[536] (234), 271[537] (234), 271[538] (234), 346[26] (315)
Isobe, M., 87[380] (36), 189[192] (182), 348[133] (324), 348[137] (324), 564[357] (539)
Isoe, S., 85[261] (22), 346[29] (315), 348[163] (326), 349[189] (329), 377, 387[295] (377), 387[324] (380), 387[325] (380), 568[638] (557)
Isono, N., 353[402] (342), 384[101] (363, 364, 369)
Itani, H., 350[231] (332)
Ito, H., 512[63] (495)
Ito, K., 188[122c] (171, 172), 275[816] (259), 275[832] (260), 353[425] (343)
Itô, S., 85[220] (15), 87[350] (30)
Ito, W., 155[71] (136, 137), 272[580] (237)
Ito, Y., 89[505] (55), 89[531] (59), 90[593] (65), 127[169] (118), 155[80] (137), 274[719] (248, 249), 349[226] (331), 350[233] (332), 353[403] (342), 353[435] (343), 353[438] (343), 354[465] (345)
Itoh, K., 87[386] (40), 349[221] (330), 387[331] (381)
Itoh, M., 269[359] (217), 269[364] (217), 270[429] (224), 270[446] (227), 349[189] (329), 387[324] (380)
Itoh, O., 433[190a] (419), 434[240a] (427)
Itoh, T., 156[130] (146), 187[101d] (169), 350[278] (336), 386[270] (374)
Itoh, Y., 347[82] (319)
Itotani, K., 271[527] (233)
Itou, M., 386[286] (376)
Itsuno, S., 188[122b] (171, 172), 188[122c] (171, 172), 259, 260, 275[815] (259), 275[816] (259), 275[830] (260), 275[831] (260), 275[832] (260), 353[425] (343)
Ivanov, L. L., 597[41] (572)
Iwaki, S., 384[141] (367)
Iwama, N., 348[167] (327)
Iwanaga, K., 276[845] (262)
Iwao, M., 74, 88[406] (42, 53), 89[483] (51), 89[497] (54), 92[700] (74), 92[703] (74), 92[714] (79)
Iwaoka, M., 561[132] (520, 522, 539–42, 552, 557), 563[247] (529), 563[268] (530), 565[425] (542), 565[426] (542), 565[427] (542)
Iwasa, E., 458[44] (451)
Iwasa, S., 383[25] (358, 362)
Iwasawa, N., 155[39] (132), 273[694] (245), 566[446] (543)
Iwata, K., 512[114] (504), 512[124] (506)
Iyer, R. R., 267[255] (208)
Iyer, V. S., 156[165] (151)
Izawa, T., 346[29] (315)
Izumi, T., 432[112b] (407), 432[118] (407)
Izumi, Y., 352[350] (339)

Jablonowski, J. A., 274[743] (251)
Jackson, R. F. W., 186[69a] (166), 186[69b] (166), 186[72] (166), 187[91a] (168), 187[91b] (168), 187[91c] (168), 187[91d] (168), 187[91e] (168)
Jackson, W. P., 274[718] (248), 352[368] (340)
Jackson, Y. A., 599[196] (583)
Jacobsen, E. N., 275[811] (258, 261)
Jacobsen, W. N., 513[133] (508)
Jacobus, J., 268[311] (213), 383[10] (358)
Jacoby, A. L., 84[160] (7)
Jadhav, P. K., 264[40] (193, 198, 247), 267[212] (205), 267[226] (206), 274[713] (247), 274[715] (247, 252), 274[745] (251), 274[753] (252)
Jaeger, R., 155[84] (137)
Jahangir, 89[517] (56), 89[518] (56), 89[536] (61)
Jahnke, D., 310[54] (285), 310[92] (304), 311[94] (304)
Jakiwczyk, O. M., 562[230] (527)
Jakobi, U., 87[391] (41)
James, K., 186[69b] (166), 187[91a] (168), 353[409] (342)

Janakiram Rao, C., 185^{15c} (160), 185^{15d} (160)
Jang, D. O., 273^{655} (243), 353^{445} (344), 383^{19} (358, 359), 383^{36} (359)
Janiak, C., 458^{56} (453)
Jankowski, P., 86^{288} (25), 348^{159} (326)
Janousek, Z., 565^{412} (542), 565^{413} (542)
Jansen, D. K., 433^{150a} (413)
Jansen, J. F. G. A., 155^{37} (132), 156^{160} (151), 157^{177} (152), 188^{141g} (174), 189^{194a} (182), 189^{194b} (182), 189^{194c} (182)
Janshen, I., 484^{20} (466)
Jaouhari, R., 353^{456} (344)
Jardine, A. M., 88^{468} (50, 51)
Jarman, M., 89^{481} (51), 128^{205}
Jarvi, E. T., 385^{182} (370, 372)
Jas, G., 351^{306} (337)
Jaspars, M., 347^{109} (321)
Jasperse, C. P., 383^{21} (358), 563^{237} (529, 530, 545)
Jastrzebski, J. T. B. H., 187^{115b} (171), 386^{219} (371)
Jaszberenyi, J. C., 273^{655} (243), 353^{445} (344), 383^{19} (358, 359), 383^{36} (359)
Jayaraman, S., 267^{243} (207)
Jayasinghe, L. R., 353^{422} (343)
Jazouli, M., 268^{292} (212)
Jeannin, Y., 186^{48c} (163)
Jefferies, I., 83^{102} (5)
Jefford, C. W., 351^{309} (337), 352^{360} (340), 432^{131} (410)
Jeganathan, S., 88^{459} (49)
Jego, J.-M., 269^{370} (218), 269^{371} (218), 269^{374} (218)
Jen, K., 597^{62} (572, 573, 576)
Jenkins, P. R., 86^{293} (25), 155^{85} (137), 567^{567} (552)
Jennings, C. A., 87^{400} (42)
Jenny, T., 125^{66}
Jensen, F. R., 430^{34} (395, 396, 399), 430^{41a} (396)
Jensen, K. A., 559^{54} (519), 559^{71} (518, 558), 561^{146} (520), 568^{578} (553)
Jeong, I.-H., 84^{198} (12)
Jeong, N., 185^{13a} (160, 162), 189^{182b} (179)
Jeoung, M. K., 273^{677} (244)
Jerussi, R. A., 559^{41} (516, 519, 522, 536)
Jesthi, P. K., 267^{265} (208), 268^{282} (210), 274^{732} (251)
Jew, S.-s., 387^{297} (377)
Jiaang, W.-T., 383^{64} (361)
Jiang, B., 186^{35} (162), 349^{185} (328)
Jiang, H. X., 511^{59} (493, 496, 497)
Jiang, J., 155^{101a} (140), 155^{101b} (140)
Jiang, W., 431^{87d} (403)
Jibril, I., 187^{101c} (169)
Jiménez, C., 434^{225a} (424, 425), 434^{225b} (424, 425)
Jin, F., 349^{185} (328), 351^{279} (336)
Jin, S., 352^{360} (340)
Jin, Z., 265^{75} (196)
Jinno, K., 90^{599} (67)
Jobling, W. H., 433^{176} (416)
Joe, G. H., 384^{90} (363)
Johnels, D., 348^{138} (324)

Johnson, B. F., 87^{385} (39)
Johnson, C. C., 563^{283} (532)
Johnson, C. R., 41, 87^{392} (41), 386^{217} (371), 386^{231} (372), 387^{330} (380, 381)
Johnson, D. R., 432^{132b} (410)
Johnson, E. P., 348^{141} (325)
Johnson, F., 89^{503} (54), 89^{504} (54)
Johnson, J. R., Jr., 128^{198} (122), 386^{224} (372), 433^{176} (416)
Johnson, R. A., 431^{52b} (398)
Johnson, W. S., 87^{388} (40), 348^{149} (325), 350^{244} (333)
Johnston, B. H., 127^{182} (119)
Johnston, E. R., 264^{19} (192, 208)
Johnston, J., 564^{363} (540)
Johnstone, R., 568^{608} (554)
Joly, G., 310^{72} (291)
Jona, H., 88^{463} (49)
Jones, A. B., 87^{379} (36)
Jones, C. H., 597^{36} (572)
Jones, D. K., 275^{819} (259)
Jones, D. N., 516, 555, 559^{49} (516, 522, 555, 556)
Jones, G. B., 90^{565} (63), 349^{212} (330), 565^{436} (542), 566^{450} (543, 544)
Jones, J. H., 271^{509} (232)
Jones, K. D., 267^{260} (208), 267^{268} (208, 234), 271^{547} (235), 386^{277} (375), 565^{407} (541, 556, 557)
Jones, M., Jr., 84^{143} (6), 84^{166} (8)
Jones, P. G., 458^{14} (445)
Jones, P. R., 190^{204} (183)
Jones, R. A., 91^{625} (71)
Jones, R. G., 82^{49} (3, 8), 84^{162} (8), 92^{701} (74), 309^{5} (278)
Jones, R. J., 599^{196} (583)
Jones, S. S., 458^{51} (452)
Jones, T. K., 128^{201} (122), 351^{290} (336)
Jørgensen, C. K., 568^{578} (553)
Jorgensen, W. L., 349^{199} (329)
Joseph, S. P., 156^{120} (144), 156^{123b} (144)
Joseph-Nathan, P., 82^{44} (3), 267^{233} (206)
Joshi, N. N., 188^{131} (171), 266^{174} (203), 267^{213} (205), 267^{217} (205, 206), 274^{710} (247), 276^{870} (264)
Joshi, S., 384^{86} (362, 370)
Joullie, M. M., 565^{435} (542, 547, 551)
Journet, M., 350^{232} (332), 384^{84} (362)
Jousseaume, B., 385^{205} (371), 387^{290} (376), 387^{303} (378)
Joussen, T., 185^{21e} (160)
Joy, F., 266^{142} (201)
Juarez-Brambila, J. J., 265^{62} (195)
Jubert, C., 185^{14} (160, 162), 186^{77c} (166, 183), 186^{78} (166, 175)
Jueschke, R., 346^{20} (314)
Juge, S., 274^{703} (246), 274^{704} (246), 274^{705} (246)
Juhlke, T. J., 458^{6} (438)
Julia, M., 83^{102} (5), 87^{370} (36), 87^{371} (36), 87^{376} (36), 433^{188a} (418), 434^{239b} (427), 434^{239c} (427)
Juliano, C. A., 348^{141} (325)
Julius, M., 560^{106} (519, 520, 528, 544)
Jung, D. K., 189^{162c} (177)
Jung, Y. W., 272^{579} (237), 272^{583} (237)
Junggebauer, J., 387^{336} (381)

Jurczak, J., 91^{630} (71), 156^{138} (147)
Kaba, T., 188^{123a} (171), 188^{123b} (171)
Kabalka, G. W., 264^{28} (192), 265^{117} (200), 266^{164} (203), 266^{165} (203), 268^{311} (213), 268^{321} (214, 215), 268^{323} (215), 268^{324} (215), 268^{326} (215), 268^{332} (215), 268^{340} (216), 268^{341} (216), 269^{342} (216), 269^{349} (217), 269^{350} (217), 269^{352} (217), 269^{353} (217), 269^{354} (217), 269^{355} (217), 269^{380} (219), 269^{384} (219, 226), 269^{387} (219), 269^{388} (219), 272^{574} (237), 272^{575} (237), 273^{649} (242), 273^{673} (244), 273^{674} (244), 273^{675} (244), 302, 310^{89} (302), 430^{21} (391), 430^{42} (396)
Kabat, M. M., 352^{399} (342)
Kabbara, J., 348^{151} (325)
Kabe, Y., 84^{141} (6)
Kabuto, C., 83^{114} (5), 348^{155} (326)
Kaczynski, J. A., 126^{93} (103)
Kadokura, M., 599^{143} (577)
Kadow, J. F., 386^{231} (372)
Kafka, C. M., 353^{432} (343)
Kagan, H. B., 65, 90^{572} (64), 90^{594} (65), 90^{595} (65), 275^{810} (258, 261), 275^{839} (262)
Kahn, M., 349^{220} (330)
Kahn, S. D., 275^{796} (258)
Kaiho, T., 274^{718} (248)
Kaino, M., 156^{155} (150)
Kaiser, E. M., 83^{71} (4)
Kaji, A., 86^{329} (28, 53)
Kajikawa, Y., 353^{447} (344), 353^{449} (344)
Kajtàr, M., 559^{72} (518)
Kakarla, R., 430^{44} (396, 416)
Kakehi, A., 156^{145} (148)
Kakehi, K., 433^{157} (414)
Kakehi, T., 383^{27} (358)
Kakigi, H., 88^{464} (49)
Kakikawa, T., 275^{834} (260)
Kakimoto, K., 385^{150} (367)
Kalesse, M., 350^{267} (335)
Kalinina, G. S., 458^{25} (447)
Kalinowski, H.-O., 85^{221} (15)
Kaller, B. F., 348^{126} (323)
Kallmerten, J., 386^{285} (376)
Kalman, J. R., 484^{13} (465, 466), 484^{23} (466, 467, 481), 484^{31} (467, 481), 485^{63} (481, 482), 485^{64} (481)
Kalsi, P. S., 564^{304} (536, 537)
Kalyanasundaram, S. K., 599^{138} (577, 580), 599^{165} (580)
Kam, C. K., 84^{143} (6)
Kamabuchi, A., 265^{124} (200), 265^{125} (200)
Kamalaprija, P., 352^{360} (340)
Kambe, N., 85^{207} (14), 85^{209} (15), 350^{272} (335), 386^{245} (373), 387^{305} (378), 387^{309} (378), 566^{444} (543), 597^{20} (572, 576, 578), 597^{21} (572, 577), 597^{22} (572, 577), 598^{132} (576–8), 599^{171} (580), 600^{206} (583), 600^{210} (584), 600^{211} (584), 600^{263} (593), 600^{266} (593), 601^{278} (595), 601^{282} (596)

Kamenka, J.-M., 264[41] (194, 247)
Kametani, T., 84[157] (7), 568[596] (554)
Kamigata, N., 600[219] (585)
Kamikawa, T., 88[458] (49), 88[461] (49)
Kamikubo, T., 349[198] (329)
Kamio, C., 350[240] (333)
Kamiya, N., 351[333] (339)
Kamiya, S., 90[600] (67)
Kamiya, Y., 154[24] (130)
Kampf, J. W., 85[249] (20, 21), 266[196] (204)
Kanagawa, Y., 351[333] (339), 351[334] (339), 386[239] (372)
Kanai, T., 351[333] (339)
Kanamori, F., 386[227] (372)
Kanatani, R., 86[297] (25)
Kanda, T., 561[151] (520), 600[266] (593), 600[267d] (593)
Kandil, A. A., 274[729] (250), 274[735] (251)
Kane, V. V., 126[115] (109), 126[121] (109)
Kanehira, K., 90[590] (65)
Kaneko, C., 458[40] (451)
Kaneko, I., 566[507] (546)
Kaneko, T., 188[121b] (171, 172), 188[129] (171)
Kaneko, Y., 276[863] (263), 276[864] (263), 350[254] (334)
Kanemasa, S., 86[313] (26)
Kanematsu, A., 276[847] (262)
Kanemitsu, K., 348[144] (325), 385[147] (367), 385[150] (367)
Kaneta, N., 353[402] (342), 384[101] (363, 364, 369), 385[169] (369)
Kang, J., 350[231] (332)
Kang, K., 562[188] (522, 534, 535)
Kang, K. K., 351[340] (339)
Kang, K.-T., 347[85] (319), 386[223] (371)
Kang, S. H., 564[325] (538, 556), 565[434] (542)
Kano, S., 265[69] (196)
Kantam, M. L., 458[9] (440, 441), 458[10] (441)
Kapoor, R. P., 459[82] (457)
Kappert, M., 187[101c] (169)
Karabelas, K., 346[23] (315)
Karlsson, U., 565[424] (542)
Karpov, V. I., 430[36] (395)
Kasahara, A., 432[112b] (407), 432[118] (407)
Kasai, M., 567[525] (547, 551)
Kashimoto, M., 90[598] (67), 90[599] (67)
Kashimura, S., 189[167a] (178)
Kassim, A. M., 348[172] (328)
Kataev, E. G., 564[328] (538)
Katagiri, T., 156[174] (152)
Kataoka, O., 350[256] (334)
Kataoka, T., 352[355] (340), 566[469] (544), 566[477] (545)
Kataoka, Y., 190[199c] (182), 190[199d] (182), 190[199e] (182), 346[35] (315), 350[259] (335)
Katcher, M. L., 90[576] (65)
Kathawala, F. G., 273[683] (244)
Kato, M., 385[167] (369), 563[239] (529), 566[506] (546)
Kato, S., 83[115] (5), 353[417] (343), 353[448] (344), 561[147] (520), 561[151] (520), 600[267d] (593)
Kato, T., 266[148] (202), 269[382] (219), 383[45] (360)

Kato, Y., 186[68] (165), 599[139] (577)
Katogi, M., 84[157] (7), 564[316] (538, 555)
Katoh, S., 346[56] (317)
Katritzky, A. R., 52, 82[37] (2, 42, 71, 73, 75, 76, 79, 81), 85[243] (19), 85[256] (22), 86[314] (26, 27), 88[417] (42), 88[456] (49), 89[489] (52, 59), 91[646] (72), 91[647] (72), 92[691] (72), 187[116b] (171), 270[480] (231), 512[99] (502), 512[100] (502), 558[23] (516, 522)
Katsoulos, G., 83[73] (4, 54), 88[453] (48, 54), 125[13] (94, 95), 127[194] (121, 122), 128[202] (122, 123), 128[203] (123)
Katsuhira, T., 189[186] (180, 182), 189[188a] (180, 182), 189[188b] (180, 182), 353[420] (343)
Katsuki, M., 88[463] (49)
Katsumata, K., 352[355] (340), 563[259] (530), 565[375] (540)
Katsuro, Y., 187[84] (167)
Katsuta, S., 352[386] (341)
Katz, A. H., 458[26] (447), 458[29] (447), 458[32] (449), 458[39] (450)
Katz, H. E., 266[169] (203)
Katz, J. J., 265[78] (196)
Katzenellenbogen, J. A., 434[199a] (420), 434[199b] (420)
Kauffmann, T., 365, 384[127] (365), 384[130] (365), 511[45] (491), 511[46] (491), 512[105] (503), 512[106] (503)
Kaufman, C. R., 88[437] (46)
Kaufmann, D., 266[191] (204), 266[192] (204), 276[841] (262), 276[858] (263)
Kaur, N. P., 85[251] (21)
Kausch, C. M., 266[196] (204)
Kawabata, N., 186[39] (162, 178), 309[11] (279)
Kawachi, A., 354[465] (345)
Kawada, M., 87[338] (29), 87[352] (30)
Kawaguchi, A. T., 187[107c] (169)
Kawaguchi, T., 599[144] (577)
Kawahara, M., 188[134b] (172)
Kawai, K., 188[129] (171)
Kawai, T., 155[49] (133), 564[357] (539)
Kawakami, J. H., 432[139b] (411), 432[141a] (412)
Kawakami, J. K., 386[225] (372), 386[226] (372)
Kawakami, T., 383[43] (360)
Kawakita, T., 188[127a] (171, 172), 188[127b] (171, 172), 188[127c] (171, 172)
Kawamoto, A., 563[247] (529)
Kawamura, M., 348[127] (323)
Kawamura, T., 433[190a] (419), 434[240a] (427), 511[51] (491)
Kawanishi, K., 88[407] (42, 50, 51)
Kawanisi, M., 156[104] (141)
Kawasaki, H., 84[131] (6)
Kawasaki, M., 190[209] (184), 458[49] (451), 458[50] (451)
Kawasaki, Y., 353[448] (344), 458[11] (444)
Kawase, A., 384[128] (365), 384[129] (365), 387[315] (379)
Kawase, M., 89[498] (54)
Kawase, Y., 156[161] (151)
Kawashima, E., 273[661] (243)

Kawashima, H., 351[319] (338)
Kawashima, T., 348[167] (327)
Kawate, T., 275[834] (260)
Kawato, S., 600[200] (583)
Kaye, A. D., 346[48] (316)
Kaye, P. T., 351[291] (336)
Kayser, F., 91[610] (67)
Kazankova, M. A., 435[252] (429)
Kazubski, A., 264[42] (194, 258, 261)
Ke, B.-W., 348[135] (324)
Keay, B. A., 91[672] (92[675], 348[154] (326), 564[306] (537)
Keck, G. E., 349[225] (330), 386[268] (374)
Kee, T. P., 353[408] (342)
Keicher, G., 310[38] (282)
Keil, R., 385[176] (369), 386[276] (375)
Keinan, E., 351[336] (339), 354[459] (345)
Keino, K., 91[631] (71)
Keller, T. H., 350[241] (333)
Kellett, R. E., 597[43] (572)
Kelly, M. J., 87[356] (32), 88[438] (46), 349[202] (329)
Kelly, T. A., 83[107] (5)
Kelly, T. R., 89[532] (60), 276[852] (263)
Kemmitt, T., 598[66] (573), 598[82] (573)
Kemper, B., 271[556] (235)
Kempf, D. J., 127[149] (115)
Kennedy, R. M., 275[806] (258, 261), 275[807] (258, 261)
Kennedy, V. O., 92[674]
Kennewell, P., 434[249] (429), 459[80] (456)
Kenney, P. M., 434[226] (424)
Kenyon, J., 568[603] (554)
Kerdesky, F. A. J., 275[781] (256)
Kerr, M. A., 88[415] (42, 79)
Kerr, R. G., 565[386] (540, 545)
Kerrick, S. T., 85[234] (18), 89[509] (56)
Kessar, S. V., 85[251] (21)
Kettner, C. A., 264[19] (192, 208)
Key, M. S., 186[37c] (162)
Khaldi, M., 83[68] (4)
Khamsi, J., 512[119] (505, 507, 509), 513[135] (509), 513[141] (510), 513[142] (510), 513[143] (510)
Khan, A. U., 352[371] (341)
Khan, M., 350[256] (334)
Khanapure, S. P., 89[526] (57)
Khanbabaee, K., 354[464] (345)
Khani, H., 458[56] (453)
Khanna, I. K., 88[433] (46)
Khanna, M. S., 459[82] (457)
Khanna, R. K., 431[63a] (400), 431[78b] (402), 431[81] (402), 431[87d] (403), 431[87e] (403)
Khanna, R. N., 432[100] (404)
Khanna, V. V., 269[346] (216), 269[347] (216)
Khanolkar, A. D., 84[187] (10), 84[190] (11)
Khatri, N. A., 311[104] (305, 306)
Khazemova, L. A., 564[342] (539)
Khise, U. R., 272[587] (238)
Khishan Reddy, C., 273[672] (244)
Khomutov, R. M., 431[70] (400), 433[162] (414)
Khorlina, I. M., 310[51] (284), 310[83] (299), 310[87] (300)
Kice, J. L., 565[385] (540)
Kiefel, M. J., 155[95] (139)
Kiefer, E. F., 433[172] (416)
Kielty, J. M., 485[50] (473)

Kiesewetter, R., 91⁶⁴³ (71)
Kiesgen de Richter, R., 264⁴¹ (194, 247)
Kihara, M., 90⁵⁹⁸ (67), 90⁵⁹⁹ (67)
Kikukawa, K., 346²⁵ (315), 511⁵¹ (491)
Kilaas, L., 156¹³⁷ (147)
Kiljunen, H., 82⁴⁵ (3)
Killpack, M. O., 91⁶⁵⁵, 156¹²³ᵃ (144)
Kim, B. M., 126⁸⁴ (101), 272⁶⁰⁰ (240, 241, 256), 273⁶³⁶ (240, 257), 274⁷¹⁴ (247)
Kim, C.-W., 350²⁷⁰ (335)
Kim, D., 352³⁶² (340)
Kim, G., 349²²³ (330), 349²²⁴ (330)
Kim, H.-O., 351²⁹⁶ (337)
Kim, J., 350²³¹ (332)
Kim, J. M., 265¹²⁶ (200), 273⁶⁷⁷ (244)
Kim, J. R., 600²³² (588)
Kim, K. C., 564³⁰⁵ (536)
Kim, K. D., 384¹⁰⁷ (364)
Kim, K.-W., 267²²⁴ (206), 269³⁵¹ (217), 269³⁵⁶ (217)
Kim, S., 188¹³⁷ᵉ (174, 177), 350²³² (332), 351³⁰³ (337), 352³⁶² (340), 383⁴¹ (360), 384⁹⁰ (363), 386²⁶⁹ (374), 433¹⁸⁹ (419)
Kim, S. S., 240, 266¹⁹⁷ (204, 235, 238, 251), 273⁶³¹ (240, 257), 347⁸⁵ (319)
Kim, Y. C., 434²¹⁰ (422)
Kim, Y. G., 351³⁰³ (337), 352³⁶² (340)
Kimura, M., 385¹⁶² (368), 386²⁸³ (376)
Kimura, R., 156¹³⁵ (147)
Kinberger, K., 265⁷⁷ (196, 198), 266¹⁵⁰ (202)
Kinder, D. H., 267²⁴⁹ (208)
King, J. L., 85²¹¹ (15), 564³³³ (538)
King, S. A., 383¹³ (358, 367)
Kingsbury, C. A., 568⁶¹⁷ (555)
Kini, A., 558²⁸ (516)
Kira, M., 347¹¹⁰ (322)
Kirby, G. W., 566⁵⁰³ (546)
Kirby, R. H., 309¹³ (279)
Kirch, G., 560⁸⁹ (518, 521, 522, 540, 543, 544, 547, 549, 551–4)
Kirk, K. L., 89⁴⁹⁹ (54)
Kirkpatrick, D., 270⁴⁴⁹ (227)
Kirsch, G., 458⁴¹ (451), 560¹¹⁸ (519)
Kirschbaum, E., 597⁴⁵ (572), 597⁵⁴ (572)
Kirschning, A., 86³¹⁹ (27)
Kishi, N., 346³⁶ (315)
Kishi, Y., 39, 87³⁸⁵ (39), 232, 271⁴⁸⁸ (232)
Kishimoto, T., 352³⁷² (341)
Kishimura, K., 270⁴¹⁹ (223)
Kita, Y., 349¹⁹¹ (329), 349¹⁹² (329), 350²⁴⁶ (333), 384¹⁴³ (367), 385¹⁴⁴ (367)
Kitagaki, S., 349¹⁹¹ (329), 385¹⁴⁴ (367)
Kitajima, T., 349²²¹ (330)
Kitamura, M., 156¹⁵⁸ (151), 188¹²⁰ᵇ (171, 172), 188¹²²ᵃ (171, 172), 188¹²⁹ (171)
Kitayama, H., 156¹⁴⁵ (148)
Kitazume, T., 185²⁹ᵃ (161), 185²⁹ᵇ (161)
Kitchin, J. P., 512¹⁰⁷ (504), 512¹⁰⁹ (504, 505)
Kitching, W., 346⁴⁰ (316), 384⁹⁵ (363), 387²⁹⁸ (377), 430⁴¹ᵇ (396), 433¹⁶⁹ (415)
Kiyooka, S., 276⁸⁶³ (263), 276⁸⁶⁴ (263), 350²⁵⁴ (334)
Kjaer, A., 559⁵⁴ (519)
Kjonaas, R. A., 155³⁸ (132), 189¹⁹³ᵃ (182), 189¹⁹³ᵇ (182), 458³¹ (449), 458³³ (450)
Klaar, M., 89⁵²³ (57)
Klabunde, K. J., 186³⁷ᵃ (162), 186³⁷ᶜ (162), 186³⁷ᵈ (162)
Klair, S. S., 86²⁸⁴ (24), 349¹⁷⁴ (328)
Klamann, D., 597² (571), 597⁴⁹ (572), 597⁵⁸ (572), 597⁶⁰ (572), 598⁸³ (573)
Klar, G., 598⁸⁹ (574)
Klas, N., 512¹⁰⁶ (503)
Klatt, M., 155⁷³ (136)
Klayman, D. L., 558³ (516–18), 558⁵ (516, 529, 530, 538, 558), 558⁶ (516), 558⁷ (516), 558⁸ (516), 558²² (516), 558²⁴ (516), 558³⁰ (516), 559³² (516, 519, 520, 522, 527, 528, 530), 559⁵³ (516, 518, 522, 527, 529, 530, 555), 559⁵⁷ (517), 559⁶⁰ (517, 518), 559⁶² (517, 518), 559⁶⁷ (518), 559⁷⁰ (518), 559⁷¹ (518, 558), 559⁷² (518), 559⁷⁴ (518), 560⁷⁵ (518, 543), 560¹⁰⁵ (519, 520), 560¹⁰⁹ (519, 520, 528, 539), 560¹¹⁵ (519, 520, 522, 538, 539, 552–4), 560¹¹⁶ (519, 538, 539), 561¹³¹ (520, 552), 561¹⁴⁵ (520, 545), 561¹⁴⁶ (520), 561¹⁴⁹ (520, 528), 561¹⁵³ (520, 529), 562²²¹ (527), 563²³² (528), 568⁵⁷⁸ (553), 568⁵⁹³ (554)
Kleijn, H., 187¹¹⁵ᵇ (171)
Klein, K.-D., 84¹⁴⁴ (6)
Klement, I., 185⁷ (159, 165)
Kleschick, W. A., 568⁶⁴⁵ (557)
Kliegel, W., 267²³² (206)
Klimko, P. G., 350²⁶⁶ (335)
Kline, D. N., 383⁶⁸ (361)
Klingenberger, H., 266¹⁴⁹ (202)
Klix, R. C., 126¹¹³ (109)
Klopman, G., 268³⁰⁹ (213)
Kloss, J., 431⁵²ᵃ (398)
Klumpp, G. W., 187⁹⁷ᵃ (169), 189¹⁶¹ᵇ (176), 189¹⁶¹ᶜ (176), 274⁷⁵⁸ (252)
Klun, J. A., 348¹⁷⁰ (328)
Klunder, A. J. H., 82³¹ (2, 33), 87³⁴³ (29)
Klusacek, H., 90⁵⁸⁰ (65), 90⁵⁸¹ (65)
Klusener, P. A. A., 88⁴²⁹ (46), 126¹⁰¹ (107), 126¹⁰² (107), 126¹²⁵ (110), 127¹³⁷ (112)
Knapp, F. F., Jr., 266¹⁶⁴ (203), 458⁴¹ (451), 597³⁴ (572), 597³⁸ (572, 577), 598⁷⁸ (573)
Knesel, G. A., 434²⁴³ (427)
Knight, D. W., 91⁶⁵¹, 91⁶⁵⁹, 92⁶⁸⁵ (72), 92⁷⁰¹ (74), 187⁷⁹ᶜ (166, 167), 349²¹⁶ (330)
Knight, K. S., 565⁴⁰⁴ (541, 548)
Knobler, C. B., 348¹³⁸ (324)
Knochel, P., 82²⁶ (2, 22), 185⁴ (159, 160), 185⁷ (159, 165), 185⁹ (159, 160, 165, 168, 174), 185¹³ᵃ (160, 162), 185¹³ᵇ (160, 162), 185¹⁴ (160, 162), 185¹⁵ᵃ (160), 185¹⁵ᶜ (160), 185¹⁵ᵈ (160), 185²⁵ (160), 185²⁶ᵉ (161, 178), 186⁴⁰ (162, 172, 174), 186⁴¹ (162, 172), 186⁴² (162, 175), 186⁴³ (162, 175), 186⁴⁹ᵃ (163), 186⁴⁹ᵇ (163), 186⁵³ (164), 186⁵⁵ (164), 186⁶⁰ (165), 186⁶⁴ (165), 186⁶⁶ (165, 179), 186⁶⁷ (165), 186⁷³ (166, 167), 186⁷⁴ᵃ (166), 186⁷⁴ᵇ (166), 186⁷⁵ (166), 186⁷⁶ (166), 186⁷⁷ᵃ (166, 183), 186⁷⁷ᵇ (166, 183), 186⁷⁷ᶜ (166, 183), 186⁷⁸ (166, 175), 187⁸⁹ᵃ (167, 176), 187⁹³ᵃ (168), 187⁹⁴ (169, 171, 172, 183), 187⁹⁸ᵈ (169), 187⁹⁸ᵉ (169), 187¹⁰⁸ᵃ (169, 170), 187¹⁰⁸ᵇ (169, 170), 188¹³²ᵃ (172), 188¹³²ᵇ (172), 188¹³⁷ᵃ (174, 177), 188¹³⁷ᵇ (174, 177), 188¹³⁷ᶜ (174, 177), 188¹³⁷ᵈ (174, 177), 188¹³⁸ (174), 188¹⁴³ᵃ (175), 188¹⁴³ᵇ (175), 188¹⁴⁹ᵃ (175, 182), 188¹⁴⁹ᵇ (175, 182), 188¹⁵⁰ (175, 182), 188¹⁵¹ (175, 182), 188¹⁵² (175, 182), 189¹⁶³ (177), 189¹⁸²ᵃ (179), 189¹⁸²ᵇ (179), 189¹⁸⁴ᵃ (179), 189¹⁸⁵ (180), 189¹⁸⁹ (180, 181), 189¹⁹⁷ᵃ (182), 189¹⁹⁷ᵇ (182), 189¹⁹⁸ (182), 190²⁰² (182), 211, 265⁸⁶ (198–200), 268²⁸⁰ (210, 211), 269³⁹² (220), 271⁵⁴¹ (235), 387²⁹³ (377)
Knoess, H. P., 188¹³⁷ᵃ (174, 177)
Knölker, H.-J., 348¹¹⁸ (323)
Kobayashi, H., 269³⁹⁶ (220)
Kobayashi, K., 89⁵³¹ (59), 349²²⁶ (331)
Kobayashi, M., 185⁸ᵃ (159, 166), 185⁸ᵇ (159, 166), 187⁸¹ (166), 433¹⁸⁸ᵇ (418), 568⁶⁰⁰ (554)
Kobayashi, S., 90⁵⁹⁸ (67), 188¹²⁷ᵇ (171, 172), 188¹²⁷ᶜ (171, 172), 276⁸⁵⁹ (263), 334, 347¹¹¹ (322), 350²⁵¹ (334), 350²⁵⁵ (334), 352³⁴⁶ (339), 352³⁴⁸ (339), 384¹³⁵ (367), 385¹⁵² (368), 512⁸⁴ (500)
Kobayashi, T., 385¹⁴⁷ (367)
Kobayashi, Y., 87³⁵⁰ (30), 90⁵⁹⁸ (67), 90⁵⁹⁹ (67), 92⁶⁸² (72), 155⁸⁰ (137), 348¹⁶¹ (326), 385¹⁸³ (370)
Köbrich, G., 186⁴⁷ᵇ (163)
Kobuke, Y., 274⁶⁹⁸ (246)
Koch, A., 383²⁸ (359)
Koch, K. R., 269³⁹⁹ (220)
Koch, P., 87³⁷⁴ (36)
Koch, T., 564³⁵² (539, 558)
Kocheshkov, K. A., 484²⁶ (467)
Kochetkov, N. K., 566⁴⁸⁸ (545), 566⁴⁸⁹ (545)
Kochi, J. K., 431⁸⁷ᵇ (403)
Kocienski, P., 87³⁴⁷ (30), 566⁴⁵⁶ (544)
Kociénski, P., 186⁴⁶ (163), 270⁴²⁷ (224, 229)
Kočovský, P., 433¹⁶⁶ (415)
Kodama, M., 87³⁵⁰ (30)

Kodama, Y., 383[58] (361)
Koelliker, U., 459[60] (453)
Koen, M. J., 485[56] (476, 477)
Koerwitz, F. L., 346[31] (315)
Kofron, W. G., 82[41] (3)
Koga, G., 385[162] (368)
Koga, K., 84[131] (6), 155[78] (136), 156[159] (151), 273[690] (245)
Koga, M., 186[68] (165)
Kogure, T., 156[156] (150)
Koguro, K., 430[30b] (393)
Koh, J. S., 350[232] (332)
Köhler, F. H., 85[219] (15), 126[99] (107)
Köhler, J., 186[51b] (164)
Kohmura, K., 352[379] (341)
Kohno, Y., 387[323] (380)
Koide, K., 384[135] (367)
Koizumi, T., 565[399] (541, 543)
Kojima, E., 187[101d] (169)
Kojima, K., 566[507] (546)
Koketsu, J., 512[69] (496, 499)
Kokjma, S., 512[69] (496, 499)
Kolb, M., 155[96] (140)
Kölle, P., 264[22] (192)
Komar, D. A., 303
Komar, D. J., 310[36] (282, 291)
Komatsu, M., 350[274] (335)
Komatsu, N., 565[429] (542, 554, 556, 557), 566[466] (544, 554, 556), 568[591] (554, 557)
Komatsu, T., 155[71] (136, 137), 155[83] (137, 139), 187[112a] (171), 272[576] (237), 272[577] (237)
Komura, M., 276[864] (263)
Konakahara, T., 86[316] (27), 86[317] (27)
Kondo, A., 599[163] (580)
Kondo, F., 432[102] (405)
Kondo, K., 564[345] (539), 597[20] (572, 576, 578), 597[22] (572, 577)
Kondo, S., 83[115] (5), 189[192] (182), 567[573] (552)
Kondo, Y., 90[562] (63, 64), 385[213] (371), 386[274] (375)
Kong, J.-S., 353[417] (343)
König, R., 384[130] (365)
König, W. A., 347[75] (319)
Konishi, H., 266[196] (204)
Kono, H., 269[412] (222, 240)
Konoike, T., 387[291] (376)
Konstantinović, S., 154[2] (130), 459[81] (456), 564[321] (538, 542)
Koolpe, G. A., 566[455] (544)
Koot, W.-J., 347[79] (319)
Kooyman, E. C., 431[62] (400)
Kopanski, L., 89[523] (57)
Kopinski, R. P., 484[25] (466, 475), 485[52] (474)
Kopka, I. E., 83[104] (5)
Kopola, N., 92[680]
Köpper, S., 385[148] (367)
Kopping, B., 567[526] (547)
Korbacz, K., 568[643] (557)
Korda, A., 187[114c] (171)
Koreeda, M., 349[221] (330), 349[222] (330), 350[232] (332), 383[65] (361), 569[654] (558)
Korkodilos, D., 430[30d] (393)
Kortus, K., 275[810] (258, 261)
Kosan, B., 186[45b] (163)
Koschinsky, R., 348[147] (325)
Koser, G. F., 351[300] (337)
Koshino, H., 266[146] (201)

Koshino, J., 270[428] (224)
Köster, G., 272[570] (237), 274[766] (253)
Köster, R., 192, 264[25] (192, 206), 267[206] (204), 268[330] (215), 268[331] (215), 270[451] (228), 270[452] (228), 270[453] (228), 272[593] (240), 302, 310[76] (293), 310[90] (302)
Kosugi, M., 386[244] (373)
Kotani, Y., 189[188a] (180, 182)
Kotecha, N. R., 92[689] (72)
Kotian, K. D., 353[457] (345)
Kotsuki, H., 352[344] (339)
Koumaglo, K., 127[165] (117)
Kouwenhoven, C. G., 83[93] (5)
Köver, A., 83[128] (5)
Kover, W. B., 433[149] (413, 416, 419, 425, 426), 434[220] (423)
Kow, R., 208, 267[269] (208, 235), 267[270] (208)
Kowald, J. A., 559[67] (518)
Koyano, H., 350[232] (332)
Koyano, M., 188[134a] (172)
Koyuncu, D., 563[265] (530)
Kozikowski, A. P., 311[97] (304), 434[212] (422, 424), 563[243] (529, 530, 539, 540, 545), 563[244] (529, 539, 545), 568[632] (556)
Kozyrkin, B. I., 597[40] (572), 597[41] (572)
Kozyrod, R. P., 431[85a] (403), 484[24] (466), 484[28] (467), 484[33] (467), 485[55] (476), 485[57] (477)
Kraebel, C. M., 348[138] (324)
Krafft, G. A., 352[392] (341), 434[199a] (420), 566[502] (546), 566[509] (546), 567[516] (547)
Krafft, M. E., 348[141] (325)
Krägeloh, K., 350[240] (333)
Kraits, Z. S., 431[73a] (401), 431[73b] (401), 431[73c] (401)
Krajewski, J. W., 91[630] (71)
Kramař, J., 84[136] (6), 125[50] (95, 112–14), 126[127] (111, 112)
Kramer, G. W., 270[445] (227)
Krämer, T., 85[264] (22, 24), 127[143] (114), 386[265] (374)
Krass, N., 83[111] (5)
Kraus, G. A., 88[460] (49), 353[458] (345)
Krause, E., 264[33] (192), 309[1] (278)
Krause, L. J., 190[208] (184)
Kravetz, E. Kh., 310[53] (285)
Krebs, A., 567[515] (546, 547)
Krepski, L., 352[345] (339)
Kresge, A. J., 349[181] (328), 351[311] (337)
Krief, A., 85[208] (15), 86[294] (25), 86[326] (28), 86[327] (28), 86[333] (28), 87[344] (29), 87[375] (36), 89[512] (56), 558[11] (516–18), 558[13] (516), 558[26] (516, 522), 559[33] (516, 519, 536, 537), 560[78] (518, 521, 522, 526, 529, 540, 543, 544, 547, 549, 551–7), 560[81] (518, 521, 522, 526, 540, 543, 544, 550–5, 557, 558), 560[82] (518, 521, 522, 526, 529, 540, 543, 544, 551, 552, 558), 560[84] (518, 521, 522, 526, 529, 538, 540, 543, 544, 551, 552, 555–8), 560[85] (518, 521, 522, 526, 543, 544, 551, 552, 555), 560[88] (518, 522, 529, 531, 536), 560[89] (518, 521, 522, 540, 543, 544, 547, 549, 551–4), 560[98] (519, 528, 531, 532), 560[101] (519, 528, 537), 561[121] (519, 534), 561[125] (519, 540), 561[136] (520, 533), 561[141] (520, 534), 561[142] (520, 535), 561[154] (521, 522, 543, 552, 555), 562[175] (522, 529, 551, 552), 562[191] (522, 529, 544, 552, 553), 562[192] (522, 529, 544, 552–5, 557), 562[200] (522), 562[201] (522), 562[202] (522), 562[203] (522, 529), 562[204] (522), 562[205] (522, 554), 562[207] (526, 549), 562[209] (526, 550, 551), 562[210] (526, 550), 562[211] (526, 543, 550), 562[213] (526, 544, 557), 562[214] (526, 550, 551), 562[215] (526, 550, 551), 562[216] (526, 544, 550, 551), 562[218] (526, 551), 563[236] (528, 537), 563[254] (530, 532), 563[263] (530, 532, 539, 551), 563[264] (530), 563[265] (530), 563[284] (532), 564[298] (535), 564[310] (538), 564[312] (538, 550, 551), 564[318] (538, 553), 564[329] (538, 557), 564[344] (539), 564[356] (539, 550, 551), 564[365] (540), 565[370] (540, 550, 551), 565[372] (540, 551), 565[380] (540), 565[384] (540), 565[394] (541, 544), 565[408] (541, 554), 566[438] (543, 544, 552, 553), 566[453] (544, 547), 566[454] (544), 566[457] (544, 551), 566[460] (544, 553, 554, 557), 566[461] (544, 554, 557), 566[471] (544, 554), 566[472] (544, 545), 566[473] (544), 567[531] (547), 567[552] (549), 567[553] (550), 567[554] (550), 567[556] (550, 551), 567[557] (550, 551), 567[558] (550, 551), 567[559] (550, 551), 567[564] (551), 567[566] (551, 558), 567[568] (552), 567[572] (552), 567[574] (552, 557), 567[575] (553), 567[576] (553), 568[580] (553, 554, 557), 568[583] (554, 557), 568[589] (554–6), 568[601] (554, 555), 568[611] (554), 568[613] (554), 568[636] (557)
Kriegesmann, R., 384[130] (365)
Krishna, M. V., 567[529] (547), 568[615] (555)
Krishnamurthi, R., 351[337] (339)
Kristen, M. O., 266[196] (204)
Kristoff, E. M., 562[230] (527)
Krizan, T. D., 83[97] (5, 42, 53)
Krohn, K., 354[464] (345)
Kroker, J., 266[196] (204)
Kronenthal, D. R., 156[148] (149)
Kropf, H., 272[601] (240)
Kropp, P. J., 564[332] (538, 547)
Krosley, K. W., 383[29] (359)
Krotz, A., 353[434] (343)
Krüger, C., 311[101] (304)
Krüger, M., 84[173] (9)
Krupp, F., 309[20] (279, 284), 310[45] (283, 293), 310[84] (299), 310[91] (304)
Kruse, C. G., 83[78] (4, 5)
Krysan, D. J., 385[202] (371)
Ku, W.-S., 352[390] (341)
Kubelka, V., 264[52] (195)

Kubo, I., 88[458] (49), 88[461] (49)
Kucerovy, A., 85[254] (21, 22), 386[283] (376)
Kucherov, V. F., 411, 432[124] (408)
Kucht, H., 458[57] (453)
Kuckertz, H., 598[115] (575)
Kuder, J. E., 559[70] (518)
Kudo, K., 600[232] (588)
Kuivila, H. G., 384[103] (363)
Kulenovic, S. T., 185[21d] (160)
Kulik, W., 126[125] (110)
Kulkarni, N., 558[14] (516, 529)
Kulkarni, S. U., 265[80] (196), 265[83] (196), 265[84] (198, 200), 265[120] (200), 266[128] (200), 269[345] (216), 269[347] (216), 270[470] (229)
Kulkarni, S. V., 431[78b] (402)
Kulkarni, Y. S., 459[65] (455)
Kulp, S.S., 155[99] (140)
Kumada, M., 86[297] (25), 90[578] (65), 90[590] (65), 187[84] (167), 187[86] (167)
Kumar, K. S., 558[18] (516)
Kume, K., 276[863] (263), 350[254] (334)
Kumon, N., 566[500] (545, 546)
Kumpf, R. J., 126[100] (107)
Kunda, S. A., 268[323] (215), 430[21] (391)
Kündig, E. P., 90[547] (63)
Kunisada, H., 567[573] (552)
Kunisch, F., 273[637] (240)
Kuniyasu, H., 383[49] (361), 566[494] (545, 547, 548)
Kunng, F.-A., 91[635] (71)
Kunz, H., 350[277] (336)
Künzer, H., 383[47] (361)
Kuo, E. Y., 348[134] (324)
Kuo, M.-Y., 432[93] (404)
Kupin, B. S., 598[101] (574)
Kuraishi, T., 92[714] (79)
Kurbanov, S. B., 564[343] (539)
Kurbanov, S. E., 564[348] (539)
Kurek, J. T., 432[138] (411), 433[143] (412), 433[148] (413), 433[155a] (414), 433[155b] (414), 434[224] (424)
Kurek-Tyrlik, A., 349[222] (330)
Kurihara, A., 353[441] (344)
Kurobe, H., 568[596] (554)
Kuroda, A., 273[690] (245)
Kurosaki, Y., 86[317] (27)
Kurosawa, H., 353[448] (344), 458[1] (437, 438), 458[34] (450), 458[35] (450)
Kurosky, J. M., 274[733] (251)
Kurozumi, M., 186[68] (165)
Kurth, J., 431[45] (396)
Kurts, A. L., 432[125] (408)
Kurusu, Y., 385[159] (368), 386[236] (372), 386[237] (372)
Kusabayashi, S., 154[19] (130)
Kusuda, S., 387[312] (379)
Kutazume, T., 185[29b] (161)
Kuwajima, I., 89[513] (56), 91[638] (71), 185[8f] (159, 166), 185[20a] (160, 165), 185[20b] (160, 165), 185[20c] (160, 165), 185[20d] (160, 165), 349[217] (330), 350[239] (333), 350[259] (335), 350[268] (335), 350[274] (335), 351[293] (337), 430[30a] (393), 565[387] (540), 566[451] (543), 567[562] (551), 568[594] (554, 557)
Kuwano, R., 90[593] (65)
Kuwayama, S., 565[399] (541, 543)
Kuzovlev, I. A., 597[41] (572)
Kwak, K. H., 352[380] (341)
Kwart, H., 568[618] (555)
Kwart, L. D., 568[618] (555)
Kwetkat, K., 383[26] (358, 377)
Kwon, H. A., 385[205] (371), 387[302] (378)
Kwon, O. O., 265[126] (200), 273[677] (244)
Kwon, T., 351[321] (338)

La Munyon, D. H., 346[54] (317)
La Roy, G. M., 431[62] (400)
Labar, D., 561[125] (519, 540), 562[191] (522, 529, 544, 552, 553), 562[201] (522), 562[204] (522), 565[384] (540), 568[589] (554–6), 568[636] (557)
Labaudinière, L., 188[149c] (175, 182)
Labia, R., 434[239b] (427), 434[239c] (427)
Laborde, E., 348[136] (324)
Laboureur, J. L., 562[191] (522, 529, 544, 552, 553), 562[200] (522), 562[201] (522), 566[461] (544, 554, 557)
Labrecque, D., 347[87] (320)
Lacour, J., 351[302] (337)
Ladouceur, G., 568[639] (557)
Laffitte, J. A., 274[704] (246)
Laganis, E. D., 384[100] (363, 369)
Lage, M. T., 459[72] (456)
Lagerwall, D. R., 85[211] (15), 564[333] (538)
Lagow, R. J., 186[37b] (162), 458[6] (438)
Laguna, A., 458[14] (445)
Lai, M., 352[399] (342)
Laine, R. M., 186[76] (166)
Lakshmaiah, G., 347[80] (319)
Lakshmikantham, M. V., 597[62] (572, 573, 576), 599[196] (583)
Lallemand, J. Y., 350[232] (332)
Lally, D. A., 155[95] (139)
Lalonde, J. J., 513[131] (508)
Lalonde, M., 349[200] (329)
Lamas, C., 91[613] (67)
Lambert, C., 83[66] (4)
Lambert, H., 276[844] (262)
Lambert, J. B., 346[19] (314), 352[375] (341), 430[29] (393)
Lamberth, C., 383[28] (359)
Lämmerzahl, F., 88[471] (50, 51, 68)
Lamothe, S., 82[29] (2, 27), 86[322] (27, 28), 348[150] (325)
Lamture, J. B., 349[211] (330)
Lamy-Schelkens, H., 352[361] (340)
Lan, X., 88[456] (49)
Lancelin, J.-M., 85[253] (21)
Landais, Y., 84[130] (6), 349[194] (329)
Landauer, M. R., 558[18] (516)
Landmann, B., 267[253] (208), 272[568] (237), 274[763] (253), 274[764] (253)
Lane, C. A., 269[350] (217)
Lane, C. F., 265[117] (200), 269[379] (219), 269[381] (219)
Lane, G. C., 565[405] (541, 548), 567[544] (548)
Lang, E. S., 597[55] (572), 598[80] (573)

Lang, H., 88[456] (49)
Lange, F., 560[100] (519, 527)
Lange, H., 186[44] (162), 458[16] (445)
Langer, F., 186[53] (164), 269[392] (220)
Langer, W., 189[195] (182)
Langguth, E., 186[45a] (163)
Langham, W., 82[48] (3), 84[160] (7)
Langry, K. C., 430[26] (392)
Langston, J. A., 89[536] (61)
Lankin, M. E., 351[310] (337)
Lanpher, E. J., 125[19] (94), 125[26] (94, 113)
Lansard, J.-P., 188[140] (174)
Laporterie, A., 354[462] (345)
Lappert, M. F., 266[141] (201), 266[142] (201), 484[2] (461)
Larbig, W., 310[91] (304)
Lardicci, L., 156[151] (150)
Lardon, M. A., 559[62] (517, 518)
Larock, R. C., 430[9] (390), 430[10] (390, 393–7, 400, 401, 403, 409), 430[11] (390, 393, 398, 411, 414, 417, 418, 421, 426–8), 430[14] (390, 408), 430[16] (390), 430[20] (391), 430[22] (391), 430[23] (391), 431[46] (396), 431[56] (399, 420), 431[61] (399), 431[74] (402), 431[75] (402), 431[76] (402), 431[82] (402), 431[84] (403), 432[93] (404), 432[94] (404), 432[102] (405), 432[103b] (404), 432[104a] (405), 432[104b] (405), 432[104c] (405), 432[106] (406), 432[107a] (406), 432[107b] (406), 432[107c] (406), 432[108] (406), 432[109] (406), 432[110] (406), 432[111] (406), 432[112c] (407), 432[113a] (407), 432[113b] (407), 432[115b] (407, 417, 427, 429), 432[116] (407), 432[117] (407), 432[120] (407), 432[126] (408), 432[127] (408), 458[27] (447), 458[28] (447), 458[30] (447), 484[18] (466)
Larsen, D. S., 351[294] (337)
Larsen, P. R., 558[17] (516)
Larson, G. L., 83[122] (5), 270[423] (224), 346[15] (314), 351[337] (339), 352[345] (339), 352[356] (340), 352[382] (341), 353[429] (343)
Lashley, L. K., 154[18] (130)
Laskovics, M., 155[96] (140)
Latham, J. A., 568[592] (554)
Latini, D., 351[324] (338)
La-Torre, F., 432[98] (404)
Latouche, R., 350[264] (335)
Lattes, A., 264[53] (195), 264[54] (195), 430[40] (396, 416, 421), 433[170a] (416, 422), 434[205] (421)
Lau, H. H., 458[27] (447), 458[28] (447)
Lau, J. C.-Y., 86[308] (26)
Lau, P. W. K., 86[300] (26)
Lauer, R. F., 559[46] (516), 559[47] (516, 522, 535, 539, 541, 543, 552-5, 557), 563[256] (530, 539, 554, 555, 557), 564[326] (538), 565[401] (541, 555), 565[421] (542, 555, 556), 568[621] (555-7), 600[241] (591)
Laughton, C. A., 90[555] (63)
Laurent, A. J., 155[67] (134), 383[46] (360)
Lautens, M., 383[51] (361, 370), 385[187] (370), 386[230] (372), 386[232]

(372)
Lautenschlager, H.-J., 353[431] (343)
Laval, A.-M., 562[218] (526, 551)
Lavallée, J. F., 383[33] (359)
Lavin, K. D., 90[577] (65)
Lavoix, A., 566[465] (544)
Lavrinovich, L. I., 272[573] (237, 251)
Lawitz, K., 271[503] (232), 271[504] (232), 271[505] (232)
Lawless, E. W., 264[36] (192)
Lawrence, N. J., 264[46] (194), 347[64] (317), 347[72] (317), 347[73] (317, 320)
Lawson, E. N., 433[169] (415)
Lazrak, T., 187[105] (169)
Lazzaroni, R., 155[103] (141)
Le Corre, M., 274[701] (246), 274[702] (246)
Le Gall, T., 187[108c] (169, 170)
Leßmann, K., 383[24] (358), 387[336] (381)
Leach, D. R., 431[84] (403), 432[107a] (406)
Léandri, G., 347[116] (322)
Leardini, R., 384[91] (363)
Learmonth, R. A., 351[291] (336)
Learn, K., 273[668] (244), 567[530] (547)
Leatham, M. J., 431[60] (399)
Leblanc, J.-P., 352[351] (340)
LeBorgne, J.-F., 155[96] (140)
Lederer, K., 598[73] (573), 598[129] (576)
Leditsche, H., 430[3] (390, 409)
Lee, A., 88[409] (42, 54), 89[500] (54), 89[501] (54)
Lee, A. G., 458[2] (438)
Lee, B., 83[99] (5)
Lee, C., 383[63] (361)
Lee, C.-H., 88[422] (42)
Lee, D.-H., 266[198] (204, 252), 273[638] (240, 257)
Lee, E., 383[63] (361), 383[69] (361)
Lee, F. L., 310[39] (282, 293, 296, 309)
Lee, G. C. M., 91[640] (71), 91[650] (72), 92[686] (72)
Lee, H., 347[112] (322), 600[242] (591)
Lee, H. D., 265[84] (198, 200)
Lee, J., 275[828] (259)
Lee, J. C., 347[85] (319), 386[223] (371)
Lee, J. G., 351[339] (339), 351[340] (339), 352[380] (341), 564[305] (536)
Lee, J. M., 188[137e] (174, 177), 351[303] (337), 386[269] (374)
Lee, J. T., 432[123] (408)
Lee, J. Y., 89[473] (50)
Lee, K., 351[335] (339)
Lee, K.-J., 350[231] (332)
Lee, L. T. C., 272[606] (240)
Lee, P.-C., 347[114] (322)
Lee, S., 89[511] (56)
Lee, S. B., 565[434] (542)
Lee, S. W., 597[39] (572)
Lee, S.-Y., 353[433] (343)
Lee, T. H., 383[69] (361)
Lee, T. V., 90[557] (63), 154[25] (130), 323, 348[122] (323), 385[171] (369, 376)
Lee, W. J., 85[206] (14, 22)
Lee, W. K., 85[223] (16, 17)
Lee, Y. S., 266[137] (201)
Lefkidou, J., 565[374] (540, 547)
Lefrançois, J.-M., 188[149g] (175, 182), 188[149h] (175, 182), 188[149i] (175, 182)

Leftin, M. H., 600[230] (588)
Léger, R., 384[79] (362)
Lehmann, R., 125[12] (94, 95, 99, 100, 104, 105, 108), 125[51] (95, 96, 120), 125[66], 126[127] (111, 112)
Lehmkuhl, H., 154[35] (131), 188[146a] (175), 188[146b] (175), 188[146c] (175), 188[147b] (175), 309[17] (279), 310[47] (283), 310[68] (290), 311[112] (307)
Leibner, J. E., 383[10] (358)
Leigh, A. J., 90[557] (63)
Leikauf, U., 350[243] (333)
Leising, F., 266[160] (203)
Leites, L. A., 264[9] (192)
Lejeune, J., 350[232] (332)
Lemieux, R. P., 84[138] (6), 273[681] (244)
Lemus, R. H., 88[415] (42, 79)
Lenardão, E. J., 566[468] (544)
Lennick, K., 185[7] (159, 165)
Lensen, N., 155[86] (137)
Lentz, R., 186[42] (162, 175)
Leonard, J., 82[50] (3)
Leonard, W. R., 564[366] (540), 564[367] (540)
Leong, W., 266[137] (201)
Lermontov, S. A., 565[382] (540)
Leroy, V., 565[437] (543)
Lesage, M., 353[442] (344)
Lesimple, P., 85[263] (22)
Lesniak, S., 155[67] (134), 383[46] (360)
Lester, D. J., 512[108] (504, 505), 512[109] (504, 505), 512[111] (504, 505, 507, 508), 512[117] (505, 507, 508)
Leteux, C., 384[140] (367)
Lethbridge, A., 433[164] (415)
LeTourneau, M. E., 87[369] (36)
Letourneux, Y., 186[38c] (162, 169)
Letsinger, R. L., 125[17] (94), 125[27] (94)
Leuck, D. J., 432[109] (406)
Leung, C. W., 458[5] (438, 442)
Leung, S.-W., 348[145] (325)
Leung, T., 272[588] (238)
Levac, S., 82[32] (2, 38)
Levason, W., 598[66] (573), 598[82] (573)
Levin, J. I., 311[103] (305)
Levine, R., 127[174] (119)
Levy, A. B., 91[624] (71, 72), 269[360] (217), 269[361] (217), 269[364] (217), 269[366] (218), 269[413] (223), 270[425] (224), 270[447] (227), 270[474] (229), 270[475] (229)
Lew, G., 269[409] (221), 270[466] (229)
Lewis, H. F., 430[17] (390)
Lewis, J. J., 347[71] (317)
Lewis, P. J., 349[202] (329)
Ley, S. V., 84[156] (7), 92[688] (72), 92[689] (72), 562[177] (522, 534), 562[180] (522), 563[252] (530), 563[290] (534, 540), 563[294] (534), 563[291] (535), 565[415] (542), 567[519] (547), 598[130] (576, 583), 599[197] (583), 599[198] (583)
Lhermitte, F., 269[365] (217)
Lheureux, M., 354[462] (345)
Lhoste, P., 352[378] (341)
Li, B., 265[57] (195)
Li, C. J., 563[261] (530), 599[169] (580),

599[176] (580)
Li, G. M., 561[148] (520, 546), 566[504] (546)
Li, J. D., 83[62] (4)
Li, K. S., 383[63] (361)
Li, L., 353[451] (344)
Li, L.-H., 127[168] (118)
Li, N. S., 272[608] (240)
Li, P. T.-J., 185[21b] (160)
Li, S. W., 599[186] (582), 600[250] (591, 592), 600[253] (591)
Li, X.-Y., 352[374] (341), 352[377] (341)
Li, Y., 266[140] (201), 273[644] (241), 273[645] (241, 258), 275[795] (258)
Liang, C. D., 88[439] (47)
Liang, S., 272[566] (236)
Liang, Y., 269[355] (217)
Liao, L., 511[58] (493-5)
Liao, Y., 274[751] (251), 511[41] (489), 511[43] (490), 511[44] (490), 511[57] (493), 511[59] (493, 496, 497), 512[75] (498), 512[78] (498), 512[98] (502)
Licandro, E., 85[201] (13)
Lichtenberg, D. W., 90[576] (65)
Liebeskind, L. S., 90[559] (63), 90[560] (63), 90[561] (63), 385[202] (371)
Liebman, J. F., 264[3] (192, 208)
Lied, T., 566[491] (545)
Liétjé, S., 86[311] (26)
Light, J. P., 567[528] (547)
Lím, D., 94, 125[35] (94)
Lim, M. S. L., 264[19] (192, 208)
Lin, C.-H., 348[135] (324)
Lin, C.-R., 84[153] (7)
Lin, F., 558[21] (516)
Lin, H.-C., 186[71c] (166)
Lin, H.-S., 351[286] (336)
Lin, J., 348[139] (325)
Lin, K.-C., 265[107] (200)
Lin, L.-C. C., 91[624] (71, 72)
Lin, L.-G., 187[118c] (171)
Lin, M.-T., 84[135] (6, 21)
Lin, P., 86[303] (26)
Lin, P.-Y., 352[390] (341)
Lin, R., 564[322] (538, 545)
Lin, S.-C., 353[451] (344)
Lin, S.-I., 353[455] (344)
Lin, S.-K., 186[71a] (166)
Lin, W.-Y., 383[42] (360)
Lin, X., 274[748] (251)
Lin, Z. Y., 563[296] (534)
Linard, F., 92[712] (79)
Lindbeck, A. C., 85[249] (20, 21), 387[318] (379)
Lindberg, P., 558[15] (516)
Linde, R. G., II, 87[379] (36)
Linderman, R. J., 349[176] (328), 386[287] (376)
Lindermann, R. J., 86[278] (24)
Lindsay, C. M., 83[92] (5), 128[208] (124)
Lindsay, K. L., 310[41] (283)
Link, C. M., 433[150a] (413)
Link, J. O., 185[26d] (161, 178), 275[821] (259)
Linker, T., 273[680] (244), 386[242] (373)
Linn, W. S., 433[193] (419)
Liotta, D. C., 86[331] (28), 91[639] (71), 275[819] (259), 383[33] (359), 558[27] (516), 560[76] (518, 522, 528-31, 538, 539), 560[83] (518, 521, 522, 526, 543, 544, 551, 555,

557), 560⁹⁵ (519, 520, 545, 546), 561¹²⁷ (519, 542, 543, 547, 548), 561¹²⁸ (519, 522, 540–2), 561¹⁶³ (522, 529, 557), 562¹⁹⁰ (522, 529, 530, 539–41), 562¹⁹⁹ (522, 538, 540, 542, 547), 563²⁸¹ (532), 563²⁹⁰ (534, 540), 564³¹³ (538, 540, 557), 564³⁶³ (540), 565⁴¹⁴ (542), 566⁴⁶² (544, 551), 567⁵⁶³ (551), 568⁵⁸¹ (554, 557), 600²¹⁶ᵇ (585)
Liotta, R., 265¹¹¹ (200)
Lippard, S. J., 188¹⁴⁵ (175)
Lipshutz, B. H., 91⁶¹⁰ (67), 127¹⁸⁵ (120), 352³⁵⁸ (340), 353⁴⁰¹ (342), 383¹² (358, 360, 363, 371, 373), 384⁹⁹ (363), 385¹⁷⁶ (369), 386²¹⁸ (371), 386²⁷⁶ (375)
Lister, M. A., 273⁶²⁶ (240), 273⁶²⁸ (240), 275⁷⁸⁶ (256)
Liston, T. V., 125²¹ (94, 108)
Little, E. L., 125²³ (94, 112)
Littlefield, B. A., 89⁵⁰¹ (54)
Liu, C., 89⁵¹¹ (56), 91⁶¹⁵ (70), 265¹⁰⁸ (200, 236, 237), 272⁵⁸⁹ (238), 348¹⁷¹ (328)
Liu, C.-L., 458²⁸ (447)
Liu, E. K. S., 186⁵⁶ᵃ (164), 186⁵⁶ᵇ (164)
Liu, H., 352³⁹⁹ (342)
Liu, K.-H., 187¹¹⁸ᶜ (171)
Liu, K.-T., 432¹⁴¹ᵇ (412), 433¹⁸⁷ (418)
Liu, S. J. Y., 310⁴⁶ (283)
Liu, Y. C., 154¹⁶ (130)
Liu, Z. R., 301, 305, 310⁸² (299), 310⁸⁶ (300)
Liu, Z.-P., 91⁶¹⁰ (67)
Livinghouse, T., 86³⁰⁶ (26), 564³⁶⁶ (540), 564³⁶⁷ (540)
Livneh, M., 433¹⁵⁶ (414)
Liz, R., 434²²¹ᵃ (423), 434²²² (423), 435²⁵³ (429)
Llabses, G., 600²⁶⁷ᶜ (593)
Lloret, F., 383⁵³ (361)
Lloyd, D., 488, 489, 511³⁵ (488–90), 511⁴⁰ (489), 511⁴² (490, 506)
Lochmann, L., 4, 55, 82⁵⁸ (4), 88⁴⁴⁶ (48, 53, 59), 89⁵⁰⁷ (55), 125³⁵ (94), 125³⁶ (94, 109), 125³⁷ (94, 112), 125⁴⁵ (95), 126¹⁰⁹ (109), 127¹³⁶ (112), 128²⁰⁶ (123, 124)
Lodi, L., 268³³⁴ (215)
Loebach, J. L., 350²⁷³ (335)
Loffet, A., 353⁴¹⁰ (342)
Lofthagen, M., 88⁴⁴⁸ (48)
Loftus, J. E., 126¹¹⁸ (109)
Logan, T. J., 125⁴⁸ (95)
Loh, T. P., 276⁸⁵³ (263), 276⁸⁵⁵ (263)
Lohmann, J.-J., 85²²⁷ (17, 19)
Lohray, B. B., 275⁸⁰⁴ (258, 259)
Lohrenz, J. C. W., 85²⁵² (21)
Lombardo, L., 182, 190²⁰⁰ (182)
Longi, P., 310⁷³ (291)
Longobardo, L., 350²⁴⁵ (333)
Looney, A., 185⁵ (159)
Looney, M. G., 353⁴³⁶ (343)
Loop, C. K., 128²⁰⁸ (124)
Loosli, H. R., 125⁴⁶ (95)
Lopes, C. C., 88⁴¹² (42)
Lopes, R. S. C., 88⁴¹² (42)
López, C. J., 347⁹² (320)
López, J. C., 350²³² (332)

López, L. A., 353⁴²⁶ (343)
López, M. C., 156¹⁶⁹ (151)
López, S., 271⁴⁹² (232), 459⁷⁹ (456)
López-Prado, J., 433¹⁴⁶ (412), 434²⁰⁶ (421)
Loren, S., 276⁸⁵⁷ (263)
Loreto, M. A., 347⁷⁵ (319)
Lou, B., 274⁷²⁰ (249)
Loveitt, M. E., 433¹⁸³ (418)
Low, C. M. R., 84¹⁵⁶ (7), 563²⁵² (530)
Low, J. Y. F., 186³⁷ᶜ (162)
Lu, G.-D., 386²⁴⁰ (372)
Lu, H., 186³⁴ᵃ (161, 163)
Lu, L. D. L., 274⁷¹⁸ (248)
Lu, X., 349¹⁷⁹ (328)
Lu, Y.-d., 430¹⁴ (390, 408)
Lubineau, A., 568⁶³⁴ (556)
Lucarini, M., 567⁵²⁷ (547)
Lucchesini, F., 91⁶⁵⁷
Lucchetti, J., 564³⁶⁵ (540)
Luche, J.-L., 48, 66, 82⁵¹ (3–5, 7, 48, 71), 84¹⁵⁸ (7), 89⁵¹⁶ (56), 90⁵⁹⁶ (66), 174, 185²⁶ᵃ (161, 178), 185²⁶ᵇ (161, 178), 185²⁶ᶜ (161, 178), 185²⁷ᵇ (161), 189¹⁶⁶ᵃ (178), 189¹⁶⁶ᵇ (178), 189¹⁶⁶ᶜ (178), 189¹⁶⁸ᵃ (178), 189¹⁶⁸ᵇ (178), 188¹⁴⁰ (174), 189¹⁷² (178), 433¹⁵¹ (413), 511⁵² (492)
Luh, T.-Y., 346²¹ (314)
Lukas, K. L., 87³⁸³ (39)
Luke, G. P., 386²⁴¹ (373)
Lukevics, E., 353⁴¹⁴ (342)
Lüning, U., 433¹⁷⁷ (416)
Luo, F.-T., 154²¹ (130), 185⁸ᵈ (159, 166), 351³³³ (339), 385²¹⁵ (371)
Luo, W., 272⁵⁶⁶ (236)
Lusby, W. R., 348¹⁷⁰ (328)
Lusch, M. J., 82⁵⁵ (3, 6)
Lusinchi, X., 599¹⁴² (577), 599¹⁵⁰ (578), 599¹⁵⁵ (578, 581)
Luteyn, J. M., 83⁷⁸ (4, 5)
Luthra, N. P., 559⁶³ (517)
Lütjens, H., 186⁴¹ (162, 172)
Lutomski, K., 87⁴⁰² (42)
Lutsenko, I. F., 431⁷⁰ (400), 431⁷³ᵃ (401), 431⁷³ᵇ (401), 431⁷³ᶜ (401), 431⁷³ᶠ (401), 433¹⁶² (414), 435²⁵² (429)
Lutz, G., 275⁸⁴⁰ (262)
Lutz, G. P., 89⁵⁰⁹ (56)
Luu, B., 87³⁷⁴ (36)
Lyga, J. W., 88⁴¹⁸ (42)
Lygo, J. L., 82⁵⁰ (3)
Lyle, P. A., 271⁵⁰⁹ (232)
Lynch, G. J., 433¹⁴³ (412), 433¹⁴⁸ (413)
Lynch, J., 87³⁷⁷ (36)
Lysenko, Z., 565⁴³⁵ (542, 547, 551)
Lysy, R., 598⁸⁴ (574)

Ma, M. C., 126⁷⁸ (100)
Ma, Y., 513¹³³ (508)
Maas, G., 268²⁹⁷ (212, 216)
McAuliffe, C. A., 430⁴ (390)
McCallum, J. S., 91⁶³⁵ (71)
McCarthy, J. R., 87³⁶⁹ (36), 385¹⁸² (370, 372), 563²⁷⁰ (530, 556)
McCarthy, P. A., 275⁷⁷⁷ (256, 258)
McCarty, J. E., 568⁶¹⁶ (555)
McCauley, J. P. J., 568⁵⁸⁶ (554, 557)

McClinton, M. A., 459⁶⁴ (455)
McClure, C. K., 273⁶²⁶ (240), 273⁶²⁸ (240), 275⁷⁸⁷ (257), 275⁷⁸⁸ (257), 347⁸⁹ (320)
McCollum, G. W., 269³⁴⁹ (217), 269³⁵⁰ (217)
McCombie, S. W., 82⁴⁷ (3), 92⁶⁸³, 353⁴⁵⁵ (344), 597¹⁹ (572)
McCrae, D. A., 87³⁸⁷ (40)
McCullough, J. D., 431⁶⁴ᵃ (400), 565³⁹⁰ (541), 597³² (572)
Macdonald, J. E., 89⁴⁸⁴ (51)
McDonald, J. H., III, 146, 156¹³⁴ (146)
Macdonald, T. L., 85²⁵⁴ (21, 22), 380, 387³²⁷ (380)
McDowell, D. C., 270⁴³² (224)
McEachin, M. D., 348¹⁴¹ (325)
McElvain, S. S., 564³⁴⁷ (539)
McGarrity, J. F., 125⁴⁶ (95)
McGarvey, G. J., 85²⁵⁴ (21, 22), 187¹⁰² (169), 384¹²³ (364), 386²⁸³ (376)
McGlinchey, M. J., 459⁶⁶ (455)
McHenry, B. M., 84¹³⁴ (6)
Machii, D., 383¹⁶ (358), 384⁷² (361)
Maciejewski, L., 83⁸³ (4)
McIntire, F. C., 564³⁴⁰ (539)
McIntosh, M. C., 346⁴⁹ (316)
Mackenzie, P. B., 386²²⁴ (372)
McKenzie, S., 566⁵⁰¹ (545, 546)
McKie, J. A., 84¹⁷⁸ (10)
McKillop, A., 458²⁶ (447), 458²⁹ (447), 458³² (449), 458³⁹ (450), 484¹⁹ (466), 563²⁶⁵ (530)
McKinley, S. V., 126⁷³ (98)
McKnight, M. V., 92⁶⁷⁹ (73)
McLane, R., 188¹⁴⁷ᵇ (175)
McLean, S., 351²⁹⁰ (336)
MacLeod, A. M., 90⁵⁶⁹ (64)
Macleod, D., 385¹⁹⁶ (371)
Macmillan, J., 563²⁸² (532, 538, 547)
McMurry, J. E., 385²⁰⁶ (371), 431⁶³ᶜ (400)
McNamara, S., 127¹⁷⁷ (119)
McNeely, K. H., 433¹⁷⁵ᵃ (416)
McNees, R. S., 125²⁵ (94, 114), 126¹²⁸ (111)
McNeill, A. H., 386²⁵⁰ (373)
McNichols, A. T., 385¹⁸⁸ (370)
MacNicol, D. D., 600²²⁷ (587), 600²²⁸ (587)
McPhail, A. T., 353⁴⁵⁵ (344)
McPhee, D. J., 562²³⁰ (527), 564³¹⁷ (538, 557)
McQuillin, F. J., 89⁵⁴¹ (63)
McWhinnie, W. R., 560¹¹⁹ (519, 528), 597⁹ (571), 597³⁶ (572)
Madden, R. J., 433¹⁵⁸ (414)
Maddocks, P. J., 270⁴⁷⁸ (230)
Madge, D. J., 385¹⁷⁷ (369, 372)
Madhavan, G. V. B., 459⁶¹ (454)
Mäding, P., 563²⁴⁶ (529, 530)
Maduakor, E. C., 88⁴⁰⁹ (42, 54), 89⁵⁰⁰ (54)
Maeda, K., 568⁶⁰⁷ (554)
Maekawa, E., 565⁴³² (542)
Maekawa, T., 348¹⁶³ (326), 568⁶³⁸ (557)
Maercker, A., 84¹⁴⁴ (6), 125⁶⁷
Maeta, H., 386²⁷³ (374)
Maffei, M., 275⁸²⁷ (259)
Magat, E. E., 125¹⁷ (94), 125²⁷ (94)

Magee, P. S., 431[73d] (401)
Magnol, E., 350[232] (332)
Magnus, P., 86[312] (26), 348[141] (325), 348[148] (325), 351[302] (337), 351[304] (337), 565[397] (541, 554), 565[417] (542)
Magnus, P. D., 346[1] (314), 383[6] (357), 562[165] (522)
Magolda, R. L., 565[435] (542, 547, 551)
Magriotis, P. A., 384[107] (364)
Mahalanabis, K. K., 88[406] (42, 53)
Mahdi, W., 125[70] (97)
Mahidol, C., 87[358] (33)
Mahindroo, V. K., 266[173] (203), 267[213] (205)
Mahon, M. F., 275[812] (258), 385[165] (368)
Mahrwald, R., 156[131] (146)
Mai, V. A., 432[137b] (411)
Maier, W. F., 83[87] (4)
Maiorana, S., 85[201] (13)
Majert, H., 309[14] (279)
Majetich, G., 321, 347[101] (321)
Majewski, M., 83[77] (4)
Majid, T. N., 185[15a] (160), 186[49a] (163), 186[49b] (163), 186[74b] (166), 188[137b] (174, 177)
Majumdar, D., 267[237] (207, 209, 224, 240), 267[238] (207, 236), 267[246] (208), 268[274] (209), 272[564] (236)
Majumdar, K. C., 434[214] (422)
Makarova, L. G., 390, 430[2] (390), 430[7] (390), 430[8] (390)
Makinson, I. K., 563[282] (532, 538, 547)
Makita, Y., 458[42] (451)
Maksimenko, A. A., 598[106] (574)
Malacria, M., 348[165] (327), 350[232] (332), 384[84] (362), 434[242] (427)
Malecha, J. W., 275[775] (255), 346[50] (316)
Malenfant, E., 386[261] (373)
Mali, R. S., 87[395] (42), 88[434] (46), 127[192] (121)
Malik, M. A., 598[118] (575)
Mallet, J.-M., 349[224] (330), 565[378] (540, 548)
Mallick, I. M., 568[614] (555, 557)
Malm, J., 385[216] (371)
Malpass, D. B., 310[77] (293, 296, 308)
Manabe, H., 599[137a] (576, 577), 599[144] (577), 599[145] (577), 599[181b] (581)
Mancheño, B., 82[52] (3, 15)
Mancilla, T., 267[230] (206)
Mancini, F., 155[77] (136), 265[89] (198), 272[610] (240)
Mandai, T., 87[338] (29), 87[352] (30)
Mandard, X., 86[305] (26)
Mandel, G. S., 268[287] (211)
Mandel, N. S., 268[287] (211)
Manek, M. B., 563[283] (532)
Manfre, R. J., 310[36] (282, 291), 310[50] (284)
Manfredini, S., 346[41] (316), 346[42] (316)
Mangeney, P., 155[86] (137)
Mann, A., 127[150] (116)
Mann, F. G., 597[33] (572)
Mann, T. A., 82[47] (3)
Männig, D., 199, 265[91] (199)

Manning, H. W., 564[353] (539, 548)
Mansour, T. S., 83[72] (4), 83[79] (4, 5)
Mansuri, M. M., 275[776] (256, 258)
Mantegani, S., 92[689] (72)
Mantlo, N. B., 156[113] (142), 156[114] (142)
Manuel, G., 354[462] (345)
Mar, E. K., 86[276] (24)
Marazano, C., 156[124] (144)
Marcantoni, E., 154[26] (130), 154[27] (130), 155[89] (139), 349[193] (329)
Marchesini, A., 126[97] (106)
Marcin, L. R., 188[144] (175)
Marciniec, B., 353[428] (343)
Marcoux, J.-F., 189[177a] (178)
Marcuzzo do Canto, M., 562[178] (522)
Marczak, S., 86[288] (25)
Marder, T. B., 265[93] (199), 265[96] (199), 265[97] (199), 265[99] (199), 265[100] (199)
Marek, I., 188[149c] (175, 182), 188[149g] (175, 182), 188[149h] (175, 182), 188[149i] (175, 182), 188[149j] (175, 182), 188[153a] (175), 384[111] (364)
Margaretha, P., 91[643] (71)
Margot, C., 125[53] (96, 120), 125[64] (101), 125[65] (120), 126[80] (101), 127[183] (120), 127[186] (120)
Marinelli, E. R., 91[624] (71, 72)
Marino, J. P., 348[136] (324)
Markandu, J., 434[249] (429)
Markiewicz, W., 563[281] (532)
Markó, I. E., 347[102] (321), 348[128] (323), 458[4] (438, 441), 458[5] (438, 442), 458[8] (438, 440), 458[9] (440, 441), 458[10] (441)
Markowska, M., 561[153] (520, 529)
Marks, R. C., 264[28] (192)
Marman, T. H., 434[230] (425), 434[231c] (425)
Marmon, R. J., 384[89] (363)
Marquarding, D., 90[580] (65), 90[581] (65), 90[582] (65)
Marron, T. G., 274[756] (252, 256)
Marsais, F., 82[38] (2, 4, 42, 70, 75, 76, 79, 81), 92[712] (79), 92[713] (79), 271[514] (232), 271[515] (232), 271[516] (232), 271[517] (232), 271[518] (232), 271[519] (232)
Marsella, J. A., 267[204] (204)
Marsh, P. G., 431[53a] (398)
Marshall, D. W., 310[57] (285)
Marshall, H., 597[25] (572)
Marshall, J. A., 86[280] (24), 91[664] (72), 276[866] (264), 384[117] (364, 373), 384[121] (364), 385[170] (369), 386[241] (373), 386[248] (373), 386[272] (374)
Marshall, R. L., 386[251] (373), 386[252] (373), 386[259] (373)
Marsili, A., 88[442] (47)
Martelli, G., 353[421] (343), 353[422] (343)
Martens, D., 275[805] (258, 259, 262), 275[826] (259)
Martin, A. R., 232, 270[483] (231, 232), 271[521] (232), 385[200] (371)
Martin, G. E., 126[89] (103)
Martin, H., 309[15] (279), 309[20] (279, 284)
Martin, J. C., 83[97] (5, 42, 53), 128[208]

(124), 459[61] (454)
Martin, J. L., 558[8] (516)
Martin, M., 83[83] (4), 127[180] (119)
Martin, O. R., 430[44] (396, 416)
Martin, P., 92[721] (81)
Martin, R. M., 83[100] (5)
Martin, S. F., 351[307] (337), 433[188c] (418)
Martinet, P., 186[48d] (163)
Martinez, E. D., 458[24] (447)
Martinez, G. R., 92[687] (72)
Martinez, J. P., 265[112] (200), 266[193] (204, 211), 266[194] (204), 385[189] (370)
Martinez-Fresneda, P., 265[123] (200, 205, 211), 268[288] (211)
Martinez-Gallo, J. M., 433[198] (420, 429), 434[246] (428, 429)
Martinho Simões, J. A., 353[442] (344)
Martino, J. P., 384[108] (364)
Marton, D., 189[173] (178), 386[233] (372)
Martorell, G., 385[211] (371)
Martynov, B. I., 434[247] (428)
Marumo, K., 352[386] (341)
Marumoto, S., 89[513] (56)
Maruoka, K., 155[68] (134), 156[130] (146), 272[619] (240), 273[691] (245), 276[851] (263), 310[63] (288, 296), 310[79] (293, 296), 346[45] (316), 349[214] (330), 350[278] (336), 351[316] (338)
Maruyama, K., 155[71] (136, 137), 155[83] (137, 139), 187[112a] (171), 269[406] (221), 272[576] (237), 272[577] (237), 272[580] (237)
Maruyama, T., 188[122c] (171, 172), 276[848] (262), 276[860] (263), 350[253] (334)
Masaki, Y., 568[626] (555)
Masamune, S., 126[84] (101), 240, 257, 272[597] (240, 241, 256), 272[600] (240, 241, 256), 272[620] (240), 273[636] (240, 257), 273[639] (241), 274[714] (247), 274[718] (248), 274[744] (251), 275[777] (256, 258), 275[781] (256), 275[806] (258, 261), 275[807] (258, 261), 350[254] (334), 566[480] (545)
Masawaki, T., 566[444] (543)
Mascarenhas, Y. P., 458[19] (446)
Mash, E. A., 189[177c] (178), 189[177d] (178), 189[177e] (178), 189[177f] (178), 565[422] (542, 557)
Masnyk, M., 86[288] (25)
Masrawe, D., 83[94] (5)
Masse, J., 127[166] (117)
Massengale, J. T., 125[19] (94)
Massey, A. G., 83[67] (4)
Massey, S. M., 88[425] (43)
Masson, A., 188[153b] (175)
Masson, S., 140, 155[56] (134, 140), 155[102] (140)
Massy-Westropp, R. A., 155[87] (137)
Masters, N. F., 92[696] (74)
Mastral, A. M., 431[77b] (402)
Masuda, C., 155[76] (136)
Masuda, H., 599[148] (577)
Masuda, Y., 188[149d] (175, 182), 188[149e] (175, 182), 265[116] (200), 269[378] (218), 269[407] (221), 269[408] (221), 269[410]

(221), 269⁴¹¹ (221)
Masukawa, T., 558¹⁰ (516, 517)
Masumi, F., 155⁷⁸ (136)
Masumoto, K., 601²⁸⁶ (596)
Masuyama, Y., 385¹⁵⁹ (368), 386²³⁶ (372), 386²³⁷ (372), 568⁵⁹⁷ (554)
Matano, Y., 512¹²² (506)
Matelich, M. C., 348¹¹⁹ (323)
Mathai, I. M., 599¹⁶¹ᶜ (579)
Mathew, J., 566⁴⁴¹ (543, 557)
Mathews, N., 92⁶⁹⁶ (74)
Mathies, P. H., 353⁴¹⁹ (343)
Mathre, D. J., 275⁸¹⁹ (259), 275⁸²⁹ (259)
Mathvink, R. J., 384⁷⁵ (361), 566⁴⁸² (545, 547, 548)
Matic, G., 558¹⁵ (516)
Matsubara, K., 384¹³⁶ (367)
Matsubara, Y., 265⁷² (196)
Matsubayashi, K.-i., 384⁷³ (361)
Matsuda, H., 383⁴³ (360), 385¹⁴⁵ (367, 368), 385¹⁵⁴ (368), 385¹⁵⁵ (368), 385¹⁵⁷ (368), 386²⁴⁶ (373), 386²⁵⁸ (373), 512⁷⁹ (499), 512⁸⁰ (500), 512⁸¹ (500), 512⁸² (500), 512⁸⁹ (501)
Matsuda, I., 353⁴⁴⁶ (344)
Matsuda, S., 349¹⁹¹ (329), 385¹⁴⁴ (367)
Matsuda, T., 346²⁵ (315), 352³⁴⁶ (339), 511⁵¹ (491)
Matsuda, Y., 188¹²¹ᵇ (171, 172), 188¹²⁹ (171), 264⁵¹ (195), 273⁶⁸⁷ (245), 273⁶⁸⁸ (245), 347⁹⁶ (321), 353⁴³⁹ (344)
Matsui, T., 275⁸³⁴ (260)
Matsuki, T., 156¹⁰⁴ (141), 599¹⁸² (582), 600²⁰⁴ (583)
Matsumoto, K., 90⁶⁰⁵ (67), 273⁶⁶⁰ (243), 350²³² (332), 353⁴³⁹ (344)
Matsumoto, M., 349²²¹ (330)
Matsumoto, T., 88⁴⁶³ (49), 88⁴⁶⁴ (49), 155⁸¹ (137), 156¹⁴¹ (147), 433¹⁸⁸ᵇ (418), 512¹⁰² (503), 564³⁰² (536)
Matsumoto, Y., 189¹⁸⁰ (179), 265⁹⁰ (198), 266¹³⁶ (201), 274⁷¹⁹ (248, 249), 274⁷²² (249), 274⁷²³ (249), 274⁷²⁴ (249)
Matsumura, F., 126⁷⁸ (100)
Matsunaga, S., 349¹⁸⁹ (329), 566⁴⁶⁶ (544, 554, 556), 568⁶³⁸ (557)
Matsunaga, S. I., 85²⁶¹ (22), 273⁶⁵¹ (243)
Matsuo, H., 276⁸⁶⁴ (263)
Matsuo, M., 434²³⁷ᵈ (426)
Matsuo, S., 189¹⁶⁷ᶜ (178)
Matsushita, H., 187⁸¹ (166)
Matsusita, H., 386²²⁷ (372)
Matsuura, I., 459⁶⁷ (455)
Matsuura, Y., 384¹⁴¹ (367)
Mattay, J., 351³²² (338)
Mattes, H., 187⁹⁸ᶜ (169)
Matteson, D. S., 86²⁷⁵ (24), 193, 212, 264³⁰ (192, 193, 201, 220, 222, 225, 230, 235, 238, 247), 264³² (192), 264³⁵ (192), 265⁸⁷ (198, 200), 266¹⁵⁴ (202), 266¹⁵⁷ (202), 267²²⁷ (206, 249, 251), 267²³⁵ (207, 208, 212, 249), 267²³⁶ (207, 208, 249), 267²³⁷ (207, 209, 224, 240), 267²³⁸ (207, 236), 267²³⁹ (207), 267²⁴¹ (207), 267²⁴⁵ (208), 267²⁴⁶ (208), 267²⁴⁸ (208), 267²⁵⁶ (208), 267²⁵⁷ (208, 210), 267²⁶⁵ (208), 267²⁶⁶ (208), 267²⁶⁷ (208), 268²⁷¹ (209), 268²⁷² (209, 210), 268²⁷³ (209, 210, 240), 268²⁷⁴ (209), 268²⁸¹ (210), 268²⁸² (210), 268²⁸³ (210), 268²⁸⁴ (211), 268²⁹⁴ (211), 268²⁹⁵ (212, 213), 268³⁰² (212), 272⁵⁶⁴ (236), 274⁷²⁵ (249), 274⁷²⁶ (249), 274⁷²⁷ (249), 274⁷²⁸ (250), 274⁷²⁹ (250), 274⁷³⁰ (250), 274⁷³¹ (250), 274⁷³² (251), 274⁷³³ (251), 274⁷³⁴ (251), 274⁷³⁵ (251), 274⁷³⁶ (251)
Matthews, D. P., 385¹⁸² (370, 372), 563²⁷⁰ (530, 556)
Matthews, I., 348¹⁴⁸ (325)
Matthews, W. S., 127¹⁵⁹ (116)
Mattson, R. J., 92⁷¹⁵ (79)
Matz, J. R., 84¹³⁴ (6)
Maurer, B., 187⁹⁵ᵇ (169)
Maurin, R., 384⁸⁵ (362)
Mautner, H. G., 563²⁷⁹ (530)
Mavrov, M. V., 271⁴⁹⁴ (232)
May, G. L., 485⁵⁸ (478), 485⁶³ (481, 482)
Mayer, H., 274⁷¹⁶ (248), 274⁷¹⁷ (248)
Maynard, J. L., 430¹⁵ (390)
Mayor, Y., 559⁴⁴ (516, 519, 522, 536)
Mayr, H., 347⁶⁸ (317, 337), 347⁶⁹ (317), 348¹⁴⁷ (325), 353⁴⁵² (344), 386²⁴⁹ (373)
Mayrhofer, R., 86³⁰² (26)
Mazzanti, G., 310⁷³ (291)
Mazzone, L., 346⁴⁴ (316)
Mead, K., 387³²⁷ (380)
Meanwell, N. A., 89⁴⁸⁷ (51)
Mears, R. J., 267²⁴⁷ (208)
Meerholz, C. A., 599¹⁹⁷ (583), 599¹⁹⁸ (583)
Meerwein, H., 309¹⁴ (279)
Meguri, H., 599¹⁴⁷ (577)
Mehrotra, M. M., 352³⁷¹ (341)
Mehta, A., 353⁴⁵⁶ (344)
Mehta, G., 434²³⁶ (426)
Mehta, S., 565³⁷³ (540, 553)
Meier, A., 155⁴⁴ (133)
Meier, G. A., 88⁴¹⁸ (42)
Meier, H., 564³⁰⁸ (537)
Meilahn, M. K., 432¹³²ᵇ (410)
Meinke, P. T., 352³⁹² (341), 566⁵⁰² (546), 566⁵⁰⁹ (546), 567⁵¹⁶ (547)
Meinwald, J., 350²⁷¹ (335)
Mekhalfia, A., 348¹²⁸ (323)
Melis, S., 83⁶¹ (4), 128²⁰⁹ (124)
Mellon, C., 386²⁶¹ (373)
Melnyk, P., 92⁷⁰⁶ (78)
Meltz, C. N., 155⁹⁰ (139)
Menchen, S. M., 565³⁷⁷ (540), 568⁶⁴⁶ (557), 599¹⁵³ (578), 599¹⁵⁴ (578)
Mendia, A., 458¹⁴ (445)
Meng, X. J., 272⁵⁷⁴ (237), 272⁵⁷⁵ (237)
Menge, W., 91⁶²⁷ (71)
Menichincheri, M., 565⁴⁰⁶ (541, 557)
Mentzafos, D., 91⁶²⁸ (71)

Mera, A. E., 126¹⁰⁰ (107)
Merdes, R., 274⁷⁰⁵ (246)
Merenyi, R., 565⁴¹² (542)
Merholz, C. A., 598¹³⁰ (576, 583)
Merifield, E., 275⁸¹² (258)
Merino, P., 348¹⁵³ (326)
Merkle, H. R., 186⁴⁷ᵇ (163)
Merkushev, E. B., 431⁴⁹ (397)
Merola, J. S., 432¹¹⁹ (407)
Merour, J.-Y., 485⁵¹ (473, 478)
Merrill, R. E., 90⁵⁷⁶ (65)
Merriman, G. H., 565⁴³⁷ (543)
Merzouk, A., 353⁴¹⁰ (342)
Meschter, C., 558²⁰ (516)
Meskens, F. A. J., 512⁷⁴ (498)
Messer, J. R., 264³⁴ (192)
Metcalfe, S., 511³⁵ (488–90), 511⁴² (490, 506)
Meth-Cohn, O., 127¹⁷⁷ (119), 270⁴⁸⁰ (231)
Metternich, R., 272⁵⁵⁷ (235)
Metz, J. T., 266¹³⁸ (201)
Metzger, J., 431⁵⁷ᵃ (399)
Meyer, C., 188¹⁵³ᵃ (175)
Meyer, F., 384⁷⁶ (361)
Meyer, J., 155⁸⁸ (137)
Meyer, N., 565³⁷¹ (540, 550)
Meyers, A. I., 7, 16, 18, 83¹¹¹ (5), 84¹⁴⁸ (7), 85²²⁵ (16, 17), 85²³⁰ (17), 85²³⁵ (18), 85²³⁶ (18), 85²³⁷ (18), 85²³⁸ (18), 85²³⁹ (19), 85²⁴⁰ (19), 85²⁴¹ (19), 85²⁴² (19), 85²⁴⁸ (20), 87⁴⁰² (42), 88⁴¹¹ (42, 43), 88⁴²³ (42), 88⁴²⁴ (43), 127¹⁴³ (114), 133, 155⁴³ (133), 155⁴⁴ (133), 156¹⁴⁷ (148), 256, 275⁷⁸⁴ (256, 257), 566⁴⁴⁹ (543)
Mezger, F., 351³⁰⁵ (337)
Mezzina, E., 155⁸² (137), 353⁴²³ (343)
Miah, M. A. J., 88⁴⁰⁵ (42, 59), 92⁷⁰⁹ (78)
Miao, S. W., 155⁷⁹ (136), 353⁴²⁰ (343)
Michalak, R. S., 88⁴⁵⁰ (48)
Michalski, J., 561¹⁵³ (520, 529)
Micha-Screttas, M., 83⁷⁴ (4)
Michel, S. T., 85²⁵⁵ (21)
Michel, T., 86³¹⁹ (27)
Michelotti, E. L., 600²³⁰ (588)
Michels, R., 563²³⁹ (529)
Michnick, T. J., 267²⁴¹ (207), 267²⁴⁸ (208)
Michoud, C., 383⁴⁰ (360), 565³⁹¹ (541, 547, 548)
Middleton, D. S., 564³⁶⁰ (539, 547, 548)
Midland, M. M., 264⁴² (194, 258, 261), 268³²¹ (214, 215), 268³²² (215), 269³⁵⁹ (217), 269³⁶⁰ (217), 269³⁶¹ (217), 269³⁶⁴ (217), 269⁴⁰⁵ (221), 269⁴¹³ (223), 270⁴³¹ (224), 270⁴³² (224), 270⁴³³ (224), 270⁴⁴⁷ (227), 270⁴⁴⁸ (227), 273⁶⁴⁸ (242)
Miginiac, L., 156¹⁰⁵ (141), 186⁵⁸ (165, 168–70, 176), 187¹¹²ᵇ (171), 188¹⁴⁷ᵃ (175), 188¹⁵³ᵇ (175), 188¹⁵⁵ᵃ (176), 188¹⁵⁵ᵇ (176), 189¹⁵⁶ (176), 189¹⁵⁷ᵇ (176), 272⁵⁷⁸ (237), 348¹²⁴

(323)
Miginiac, P., 155[63] (134), 187[95c] (169)
Miginiac-Groizeleau, L., 187[95c] (169)
Migita, T., 386[244] (373)
Mignani, G., 266[160] (203)
Mignani, S., 348[120] (323)
Mihailović, M. L., 459[81] (456), 558[15] (516), 564[321] (538, 542)
Mikami, K., 86[279] (24), 346[36] (315), 346[46] (316), 346[47] (316), 349[215] (330), 350[255] (334)
Mikhaĭlov, B. M., 264[26] (192, 196, 212–14, 235, 238)
Mikhail, I., 266[192] (204)
Mikolaiski, W., 84[173] (9)
Mikolajczyk, M., 568[643] (557)
Miles, W. H., 150, 156[150] (150)
Miligan, M. L., 90[566] (64)
Milkova, Z., 562[213] (526, 544, 557)
Millan, S. D., 383[30] (359)
Millar, J. G., 126[76] (100, 103)
Miller, J. A., 88[467] (50), 185[8d] (159, 166), 188[155c] (176), 310[43] (283), 385[196] (371)
Miller, J. D., 352[363] (340)
Miller, J. J., 430[41a] (396)
Miller, J. M., 431[47] (397)
Miller, M. C., III, 264[17] (192)
Miller, M. J., 349[173] (328)
Miller, R. L., 311[115] (308)
Miller, S. C., 385[182] (370, 372)
Miller, S. I., 599[161c] (579)
Mills, K., 458[20] (446)
Mills, N. S., 126[89] (103)
Mills, R. J., 89[521] (56)
Mills, S., 87[404] (42)
Milosavljević, S., 459[81] (456), 564[321] (538, 542)
Mimaki, Y., 562[208] (526)
Mimoun, H., 433[186] (418)
Mimouni, N., 387[326] (380)
Min, S. J., 265[126] (200), 273[677] (244)
Minami, T., 568[644] (557)
Minato, A., 187[86] (167)
Mineta, H., 189[178] (178), 268[298] (212)
Mingaleva, K. S., 598[100] (574)
Minkin, V. I., 597[14] (571), 597[47] (572)
Minnetian, O. M., 430[26] (392)
Mioskowski, C., 385[204] (371)
Miotti, U., 87[341] (29)
Miranda, E. I., 349[178] (328)
Miranda, R., 82[44] (3)
Mishra, P. K., 126[115] (109)
Misiti, D., 432[98] (404), 432[99a] (404)
Mison, P., 155[67] (134)
Mitchell, E. R., 126[98] (106)
Mitchell, M. A., 432[106] (406), 432[108] (406)
Mitchell, M. B., 88[430] (46, 60), 271[506] (232)
Mitchell, R. H., 156[165] (151)
Mitchell, T. N., 353[400] (342), 383[23] (358), 383[26] (358, 377), 385[193] (371)
Mitra, D., 384[118] (364)
Mitschler, A., 433[186] (418)
Miura, I., 88[461] (49)
Miura, K., 273[653] (243), 273[658] (243), 273[659] (243), 273[660] (243),

350[232] (332), 353[439] (344)
Miura, T., 350[262] (335)
Miwa, K., 352[383] (341), 352[385] (341)
Miwa, Y., 276[845] (262), 276[846] (262)
Miyahara, Y., 563[292] (534)
Miyake, H., 383[50] (361), 386[257] (373)
Miyake, R., 90[564] (63, 64), 188[141h] (174)
Miyamoto, Y., 352[346] (339)
Miyano, M., 565[402] (541)
Miyano, S., 156[149] (149), 189[180] (179)
Miyasaka, T., 347[82] (319)
Miyashiro, J. J., 430[12] (390)
Miyashita, M., 564[362] (540)
Miyata, M., 276[861] (263)
Miyata, N., 90[600] (67)
Miyaura, N., 265[119] (200), 265[124] (200), 265[125] (200), 270[429] (224), 270[446] (227), 270[479] (231), 270[484] (231, 232), 270[485] (231, 232), 270[486] (232), 271[487] (232), 271[493] (232), 271[495] (232), 271[496] (232), 271[497] (232), 271[498] (232), 271[500] (232), 271[513] (232), 271[523] (232), 271[524] (232, 234), 271[528] (233, 234), 271[529] (233), 271[531] (234), 271[532] (234), 271[536] (234), 271[537] (234), 271[538] (234), 272[558] (235), 272[563] (234), 346[26] (315)
Miyazaki, K., 513[145] (511)
Miyazawa, T., 187[116a] (171)
Miyoshi, H., 433[197c] (420), 598[123] (575, 588)
Miyoshi, N., 85[209] (15), 564[345] (539), 564[351] (539), 568[604] (554), 597[21] (572, 577), 598[132] (576–8), 599[171] (580), 600[206] (583), 600[210] (584), 600[211] (584), 600[263] (593)
Mizhiritskii, M. D., 352[387] (341)
Mizuno, K., 381, 387[337] (381)
Mizutaki, S., 601[287] (596)
Mizutani, T., 561[151] (520)
Mladenova, M., 154[13] (130)
Mlochowski, J., 562[225] (527), 563[258] (530, 538)
Mo, X. S., 511[60] (493), 512[62] (494)
Mo, Y. K., 268[309] (213)
Mobbs, B. E., 90[555] (63)
Moberg, W. K., 92[676] (676)
Mobilio, S., 87[354] (31)
Mobraaten, E. C., 84[143] (6)
Modena, G., 87[341] (29)
Modi, P., 346[20] (314)
Modro, T. A., 88[468] (50, 51)
Moeller, K. D., 350[230] (332)
Mohamadi, F., 430[30c] (393), 430[30e] (393)
Mohammed, A. Y., 567[537] (548)
Mohr, P., 347[95] (321)
Moiseenkov, A. M., 126[80] (101)
Moiseev, S. K., 90[575] (64)
Mojovic, L., 92[716] (79), 92[718] (79)
Mok, P. L. H., 276[867] (264), 276[868] (264)
Mokhallalati, M. K., 387[304] (378)
Mokrosz, J. L., 155[93] (139, 140)
Mokrosz, M. J., 155[93] (139, 140)
Molander, G. A., 84[178] (10), 187[96] (169), 188[155d] (176), 268[312]

(213), 269[395] (220), 272[613] (240), 273[686] (244), 323, 348[123] (323), 351[280] (336), 352[349] (339)
Mole, S. J., 86[309] (26)
Molina, A., 351[296] (337), 385[207] (371)
Molina, M. T., 353[458] (345)
Molloy, K. C., 275[812] (258), 385[165] (368)
Moloney, M. G., 431[85b] (403), 431[85c] (403), 431[85d] (403), 484[15] (465, 468, 481), 484[17] (1, 465, 468, 477), 484[36] (468, 475)
Monahan, R., 560[76] (518, 522, 528–31, 538, 539)
Moncur, M. V., 432[134b] (410)
Moneta, W., 273[692] (245)
Monger, S. J., 349[205] (330)
Montana, A. M., 350[256] (334)
Montanucci, M., 563[276] (530), 563[277] (530), 564[334] (538)
Monteith, M., 347[74] (318)
Monti, H., 347[116] (322)
Monti, J.-P., 347[116] (322)
Monti, S. A., 564[325] (538, 556)
Montserrat, J. M., 84[155] (7)
Montury, M., 264[55] (195)
Moody, C. M., 156[172] (152)
Moody, M. A., 351[341] (339)
Moody, R. J., 267[265] (208), 268[271] (209), 268[281] (210)
Moon, R., 83[62] (4)
Moore, F. W., 82[48] (3)
Moore, H. W., 91[634] (71)
Mordini, A., 83[59] (4), 84[136] (6), 94, 125[14] (94, 95), 125[50] (95, 112–14), 127[139] (114), 127[147] (115), 127[150] (116), 127[153] (116), 127[163] (117), 127[167] (117), 127[170] (118), 127[171] (118), 127[172] (118), 127[186] (120), 127[188] (120), 127[189] (120), 347[61] (317), 347[62] (317), 349[187] (329), 351[318] (338), 352[391] (341), 353[417] (343), 384[113] (364), 384[115] (364)
More, P. G., 431[83] (403), 432[122] (407, 409)
Moreau, J. J. E., 83[116] (5), 88[436] (46), 353[418] (343), 385[180] (370)
Moreau, J.-L., 187[99a] (169), 187[99b] (169), 187[99c] (169)
Moreau, P., 185[32d] (161)
Morera, E., 271[526] (233)
Moret, E., 125[12] (94, 95, 99, 100, 104, 105, 108), 125[59] (120), 125[65] (120), 126[83] (101), 126[91] (103, 105), 126[96] (105, 106)
Moretti, R., 351[298] (337)
Morey, J., 88[444] (48), 88[449] (48)
Morgan, G. T., 597[43] (572), 598[104] (574)
Morgan, I. T., 348[168] (327)
Morgan, J., 484[32] (467, 482, 483), 484[33] (467), 484[35] (467), 485[55] (476), 485[56] (476, 477), 485[60] (479, 480), 485[61] (479, 480)
Morgan, L. A., 156[118] (143)
Mori, A., 189[177b] (178), 352[346] (339)
Mori, K., 126[82] (101), 386[286] (376)
Mori, M., 353[402] (342), 384[101] (363, 364, 369), 385[169] (369)

Mori, S., 272[620] (240)
Mori, Y., 600[247] (591)
Moriarty, K. J., 271[490] (232)
Moriarty, R. M., 351[295] (337), 351[296] (337), 385[208] (371), 458[15] (445)
Morihira, K., 350[259] (335)
Morikawa, S., 348[143] (325)
Morikawa, T., 383[58] (361)
Morikawe, T., 273[690] (245)
Morimoto, M., 87[348] (30)
Morin-Allory, L., 85[253] (21)
Morita, S., 597[20] (572, 576, 578)
Morita, Y., 187[99d] (169), 268[330] (215), 302, 310[90] (302), 348[163] (326)
Moritani, I., 511[50] (491)
Moritani, T., 352[384] (341)
Moriwake, T., 564[354] (539), 564[355] (539)
Moriya, M., 349[198] (329)
Moriya, T., 265[125] (200)
Moriyama, T., 87[338] (29), 87[352] (30)
Morken, J. P., 154[29] (131), 154[31] (131), 154[32] (131), 154[33] (131), 154[34] (131)
Morken, P. A., 186[34a] (161, 163), 186[34c] (161, 163)
Morris, J. L., 458[21a] (446)
Morrison, J. A., 190[208] (184), 264[13] (192), 270[459] (228)
Morrison, J. D., 272[596] (240, 241, 256)
Morrison, R. C., 82[54] (3, 10)
Morrone, R., 90[583] (65), 90[585] (65)
Morse, P., 126[89] (103)
Mortier, J., 88[410] (42, 44, 72)
Mortillaro, L., 558[24] (516), 558[30] (516), 560[116] (519, 538, 539)
Mortlock, S. V., 386[247] (373)
Morton, A. A., 94, 95, 124[8] (94), 125[16] (94, 95), 125[17] (94), 125[18] (94), 125[19] (94), 125[23] (94, 112), 125[26] (94, 113), 125[27] (94), 125[28] (94, 95), 125[29] (94), 125[30] (94, 112), 125[32] (94), 126[133] (112), 127[134] (112)
Morton, H. E., 273[679] (244)
Morton, J. W., Jr., 82[56] (3), 309[6] (278)
Moskalenko, V. A., 430[36] (395)
Moss, G. P., 274[717] (248)
Moss, N., 601[273] (594)
Motherwell, W. B., 338, 351[329] (338), 351[330] (338), 351[331] (338), 351[332] (338), 384[83] (362), 512[107] (504), 512[108] (504, 505), 512[109] (504, 505), 512[110] (504, 509), 512[111] (504, 505, 507, 508), 512[112] (504), 512[113] (504), 512[117] (505, 507, 508), 512[119] (505, 507, 509), 512[126] (507–9), 512[127] (507), 512[129] (508), 513[134] (508)
Mothies, V., 155[102] (140)
Mott, R. C., 350[270] (335)
Motyka, L. A., 92[684] (72)
Moulines, F., 350[235] (333)
Moura Campos, M., 433[191] (419), 597[5] (571), 598[88] (574), 598[111] (575), 598[119] (575, 585), 599[136] (576), 599[162] (580), 599[164] (580), 600[217] (585)

Mouri, M., 276[848] (262), 276[865] (264), 347[108] (321)
Mouriño, A., 189[172] (178), 568[637] (557)
Moussa, F., 600[237] (589)
Moustakis, C. A., 567[511] (546)
Mowlem, T. J., 92[698] (74), 92[699] (74), 385[166] (369)
Moyano, A., 562[220] (527, 538, 556)
Moyroud, J., 88[410] (42, 44, 72)
Mstislavsky, V. I., 271[553] (235)
Muchowski, J. M., 51, 83[108] (5), 88[426] (43), 89[478] (51), 89[486] (51), 89[527] (59), 89[528] (59), 89[535] (60), 353[419] (343)
Mudryk, B., 84[147] (7, 12), 84[198] (12), 84[199] (12), 84[200] (13), 85[202] (14)
Mueller, B., 434[244b] (427)
Mueller-Westerhoff, U. T., 156[166] (151)
Muhn, R., 386[242] (373)
Mui, J. Y.-P., 432[136b] (410)
Mukai, C., 350[256] (334)
Mukaiyama, T., 132, 155[39] (132), 156[143] (148), 157[178] (153), 193, 263, 267[263] (208), 270[417] (223), 272[592] (240), 272[594] (240), 272[614] (240), 272[618] (240), 276[859] (263), 350[251] (334), 352[346] (339), 352[348] (339), 368, 369, 384[136] (367), 385[152] (368), 512[83] (500), 512[84] (500), 512[85] (501), 512[86] (501)
Mukhopadhyay, T., 83[69] (4)
Mulholland, K., 348[148] (325)
Müller, B., 186[45b] (163)
Müller, E., 124[1] (93), 185[6] (159, 162–5, 168, 169, 173), 190[203] (183), 309[17] (279), 310[47] (283), 559[59] (517, 519, 552), 560[112] (519, 520, 527, 538, 539, 552, 553), 560[117] (519), 561[122] (519, 528), 561[124] (519), 561[129] (519, 520), 561[134] (520, 554), 561[139] (520, 554), 561[140] (520), 561[143] (520), 561[144] (520), 561[152] (520), 561[159] (522, 539), 562[222] (527), 562[223] (527), 597[1] (571), 598[105] (574)
Müller, G., 127[152] (116)
Müller, G. E., 186[45a] (163)
Müller, H., 309[21] (280), 309[22] (280, 284), 310[81] (297)
Müller, J., 188[141f] (174)
Mulliez, M., 267[231] (206)
Mulzer, J., 187[101c] (169)
Munasinghe, V. R. N., 349[223] (330)
Mundy, P., 559[49] (516, 522, 555, 556)
Murafuji, T., 512[114] (504), 512[118] (505, 506), 512[121] (506), 512[123] (506), 512[124] (506)
Murahashi, S.-I., 87[366] (34), 270[444] (227)
Murai, A., 350[277] (336)
Murai, S., 85[209] (15), 353[447] (344), 353[448] (344), 353[449] (344), 386[275] (375), 432[121] (407), 564[345] (539), 564[351] (539), 568[604] (554), 597[20] (572, 576, 578), 599[171] (580), 600[206] (583), 600[263] (593)

Murai, T., 83[115] (5), 353[417] (343), 353[448] (344), 561[147] (520), 561[151] (520), 600[267d] (593)
Murakami, M., 267[263] (208), 272[618] (240), 276[859] (263), 353[438] (343)
Murakami, T., 565[429] (542, 554, 556, 557)
Muraki, M., 272[592] (240)
Muraoka, K., 154[19] (130)
Murayama, E., 384[124] (364, 378), 387[328] (380)
Murphree, S. S., 383[68] (361)
Murphy, B. L., 82[47] (3)
Murphy, C. J., 567[511] (546), 567[514] (546, 547)
Murphy, C. K., 599[159] (579)
Murphy, P. J., 347[95] (321)
Murray, B. J., 598[63] (572, 573)
Murray, K. J., 268[310] (213)
Murray, P. J., 567[519] (547)
Murray, R. W., 268[318] (214)
Murray, W., 83[109] (5)
Murtiashaw, C. W., 91[612] (67)
Musallam, H. A., 351[295] (337)
Musgrave, W. K. R., 91[618] (70), 91[619] (70)
Musick, T. J., 91[615] (70)
Musorin, G. K., 598[91] (574)
Muthukrishnan, R., 125[56] (103, 116), 127[156] (116)
Muthyala, R., 561[158] (521, 522, 526, 547, 548)
Muzart, J., 349[201] (329)
Myers, A. G., 348[134] (324), 349[225] (330), 350[249] (333), 565[379] (540, 548)
Myers, D. R., 84[166] (8), 88[450] (48)
Myerson, J., 433[192] (419)
Myoung, Y. C., 156[115] (142)

Nadin, A., 565[420] (542)
Naef, R., 353[419] (343)
Nagahama, H., 86[313] (26)
Nagahara, S., 349[214] (330)
Nagai, M., 157[183] (153)
Nagakawa, K., 599[135] (576)
Nagao, Y., 90[599] (67), 384[141] (367), 387[292] (377)
Nagarajan, S. C., 85[255] (21)
Nagasawa, N., 189[192] (182)
Nagase, H., 90[605] (67), 275[838] (260)
Nagase, S., 154[19] (130)
Nagashima, H., 568[627] (555)
Nagashima, N., 385[146] (367)
Nagata, K., 386[270] (374)
Nagata, R., 352[373] (341)
Nagata, W., 310[61] (286)
Nagel, K., 309[15] (279)
Naito, M., 266[136] (201), 274[723] (249), 274[724] (249)
Najafi, M. R., 349[196] (329)
Nájera, C., 84[197] (12), 433[198] (420, 429), 434[218] (423), 434[219a] (423), 434[219b] (423), 434[219c] (423), 434[225a] (424, 425), 434[225b] (424, 425), 434[246] (428, 429)
Naka, K., 458[43] (451)
Nakadaira, Y., 266[195] (204)
Nakagami, Y., 157[179] (153)
Nakagawa, M., 275[834] (260)
Nakagawa, Y., 353[435] (343)

Nakahama, S., 188[122c] (171, 172), 275[815] (259), 275[816] (259)
Nakahara, Y., 87[384] (39)
Nakahata, K., 157[182] (153)
Nakahira, H., 85[207] (14), 387[305] (378), 387[309] (378), 601[282] (596)
Nakai, S., 349[194] (329)
Nakai, T., 86[279] (24), 346[36] (315), 346[47] (316), 350[255] (334)
Nakaiida, S., 600[267d] (593)
Nakajima, M., 156[159] (151)
Nakajima, T., 385[167] (369), 561[148] (520, 546), 566[504] (546), 566[506] (546)
Nakajo, E., 127[169] (118)
Nakamura, A., 126[94] (103), 126[95] (103), 186[34a] (161, 163)
Nakamura, E., 185[8f] (159, 166), 185[20a] (160, 165), 185[20b] (160, 165), 185[20c] (160, 165), 185[20d] (160, 165), 186[63] (165, 166), 350[269] (335), 351[293] (337), 351[327] (338), 351[328] (338), 383[16] (358), 384[72] (361), 384[88] (362), 430[30a] (393)
Nakamura, H., 350[277] (336), 353[438] (343)
Nakamura, K., 154[24] (130)
Nakamura, T., 185[16a] (160), 185[17] (160), 185[19] (160, 167), 187[90] (168), 187[107b] (169), 349[217] (330), 350[239] (333)
Nakamura, W., 601[270] (593)
Nakamura, Y., 350[259] (335)
Nakanishi, K., 387[337] (381)
Nakano, M., 276[864] (263)
Nakano, T., 352[386] (341), 512[82] (500)
Nakatani, K., 155[80] (137), 346[29] (315), 387[295] (377)
Nakatani, M., 512[70] (496, 497)
Nakatani, Y., 87[374] (36)
Nakatsuka, N., 432[96b] (404)
Nakaya, C., 512[122] (506)
Nakayama, A., 387[300] (377)
Nakayama, K., 90[564] (63, 64), 188[141h] (174)
Nakayama, N., 348[140] (325)
Nakayama, O., 350[233] (332)
Nakayama, Y., 87[352] (30)
Nakazono, Y., 156[104] (141)
Nam, D., 434[231c] (425)
Nandi, G., 459[68] (455)
Nandi, K. N., 597[30] (572)
Nantermet, P. G., 189[162a] (177)
Napolitano, E., 88[442] (47)
Narasaka, K., 267[218] (205, 211), 273[682] (244), 273[694] (245), 273[695] (245), 387[301] (377), 387[323] (380), 566[446] (543)
Narasimhan, N. S., 87[395] (42), 90[603] (67), 127[192] (121), 265[66] (196), 352[343] (339)
Narayama Rao, M. L., 270[441] (226)
Narayana, C., 273[673] (244), 273[674] (244), 273[675] (244), 432[114] (407)
Narayanan, K., 430[22] (391), 432[102] (405), 432[107b] (406), 432[107c] (406)
Narayanan, V. L., 88[450] (48)
Narula, C. K., 264[14] (192)
Nasipuri, D., 154[17] (130)

Näsman, J. H., 92[680]
Naso, F., 346[44] (316)
Nassr, M. A. M., 433[170b] (416)
Nasu, A., 600[200] (583)
Nativi, C., 126[129] (111), 347[74] (318)
Natta, G., 310[73] (291)
Naumann, D., 186[44] (162), 458[7] (438)
Naylor, A., 346[43] (316)
Nayyar, K., 265[67] (196)
Nazer, B., 273[676] (244)
Nazih, A., 565[372] (540, 551)
Ncube, S., 270[418] (223), 270[421] (224), 270[422] (224)
Ndip, G. M., 266[194] (204), 385[189] (370)
Ndzi, B., 83[101] (5, 81)
Nechvatal, G., 92[694] (74), 92[695] (74), 92[696] (74), 92[698] (74)
Nédélec, J.-Y., 186[36e] (162, 174), 186[36f] (162, 174)
Neeland, E. G., 350[241] (333)
Nefedov, B. K., 433[190b] (419)
Nefedov, O. M., 432[129c] (409, 411)
Nefedov, V. A., 431[68] (400)
Negishi, E. I., 84[175] (10), 84[194] (12, 67), 166, 185[8a] (159, 166), 185[8b] (159, 166), 185[8c] (159, 166), 185[8d] (159, 166), 187[81] (166), 187[82] (166), 188[155c] (176), 189[159] (176), 189[160a] (176), 189[160b] (176), 189[183a] (179), 189[183b] (179), 264[1] (192, 193, 196, 214), 264[29] (192), 265[65] (196), 265[73] (196), 265[78] (196), 265[82] (196), 269[397] (220), 269[398] (220), 269[409] (221), 270[416] (223), 270[421] (224), 270[439] (225), 270[443] (227), 270[459] (228), 270[466] (229), 311[108] (307), 311[116] (308), 348[141] (325), 348[142] (325)
Negri, J. T., 90[545] (63), 90[546] (63)
Nehl, H., 188[146b] (175), 188[146c] (175), 188[147b] (175)
Nelson, D. J., 91[664] (72), 268[313] (213), 432[142a] (412)
Nelson, J. V., 272[595] (240, 241), 272[621] (240), 272[622] (240)
Nelson, K. A., 189[177c] (178), 189[177e] (178)
Nelson, W. M., 352[372] (341)
Nemoto, H., 568[596] (554)
Nemoto, T., 351[328] (338)
Nepywoda, J., 597[50] (572)
Nesmeyanov, A. N., 90[575] (64), 390, 430[2] (390), 431[70] (400), 431[73a] (401), 433[162] (414)
Neuffer, J., 459[60] (453)
Neumann, W. P., 310[37] (282), 383[15] (358, 360), 383[24] (358), 384[96] (363), 386[229] (372), 387[336] (381), 432[88a] (403, 427, 429), 432[88h] (403)
Nevalainen, V., 275[818] (259)
Newcomb, M., 126[93] (103), 563[283] (532), 566[447] (543), 566[448] (543)
Newitt, D. M., 264[31] (192)
Newton, R. J., Jr., 268[311] (213), 302, 310[89] (302)
Newton, T. W., 346[51] (317)

Ng, H. P., 275[793] (258), 353[451] (344)
Ng, S. W., 264[44] (194)
Ngochindo, R., 92[679] (73)
Ngoviwatchai, P., 431[63a] (400), 431[63b] (400), 431[79a] (402), 431[79b] (402), 431[79c] (402), 431[80] (402)
Nguyen, P., 265[97] (199)
Nguyen, T., 270[459] (228), 274[749] (251)
Nicaise, O., 154[28] (130)
Nicholas, K. M., 350[256] (334), 433[153b] (413)
Nickel, W.-U., 567[515] (546, 547)
Nicolaou, K. C., 88[415] (42, 79), 271[491] (232), 348[132] (324), 558[2] (516), 558[4] (516, 522, 527, 529, 536, 547, 548, 558), 560[86] (518, 521, 543, 552, 555), 560[102] (519, 522, 533, 536), 560[103] (519, 522, 530, 533, 534, 536), 560[107] (519, 532, 534, 538, 545, 558), 562[181] (522, 527, 542), 562[196] (522, 531, 547), 562[198] (522, 541, 542, 551, 553–7), 562[199] (522, 538, 540, 542, 547), 564[314] (538, 540, 543, 545), 564[320] (538, 542), 565[435] (542, 547, 551), 567[522] (547, 551), 600[216b] (585)
Nicolosi, G., 90[583] (65), 90[585] (65)
Nidy, E. G., 431[52] (398)
Nieh, H.-C., 566[440] (543)
Niel, G., 272[562] (236)
Niemeyer, C. M., 273[696] (245), 273[697] (245)
Niermann, H., 310[37] (282)
Nieuwland, J. A., 433[152] (413)
Niibo, Y., 349[182] (328)
Niimoto, Y., 564[323] (538, 539, 545)
Nikam, S. S., 272[586] (238)
Nikanorov, V. A., 431[87a] (403), 431[87c] (403)
Ninagawa, A., 512[82] (500)
Ninan, A., 90[609] (67)
Nishibayashi, Y., 565[429] (542, 554, 556, 557), 568[591] (554, 557)
Nishida, S., 189[178] (178), 268[298] (212), 386[238] (372)
Nishide, H., 434[244a] (427)
Nishigaichi, Y., 386[267] (374), 387[338] (381)
Nishiguchi, I., 189[169] (178), 265[72] (196)
Nishihara, T., 187[107a] (169)
Nishii, S., 155[71] (136, 137), 272[581] (237)
Nishikawa, T., 348[133] (324)
Nishimura, H., 353[447] (344)
Nishimura, J., 349[197] (329)
Nishimura, M., 385[147] (367)
Nishimura, Y., 564[347] (539)
Nishio, H., 387[328] (380)
Nishio, K., 347[111] (322)
Nishio, Y., 512[82] (500)
Nishioka, Y., 599[184] (582)
Nishitani, K., 562[208] (526)
Nishiyama, H., 90[605] (67), 90[606] (67), 349[221] (330), 387[331] (381)
Nishiyama, K., 352[381] (341), 352[396] (342)
Nishiyama, T., 433[169] (415)
Nishiyama, Y., 351[334] (339), 386[239]

(372), 565³⁷⁶ (540, 545)
Nishizawa, M., 434²⁴⁴ᵃ (427), 434²⁴⁴ᶜ (427), 434²⁴⁴ᵈ (427), 564³¹⁵ (538)
Nitsche, R., 264³³ (192)
Nitta, H., 352³⁴⁶ (339)
Niwa, S., 185¹⁰ᵃ (160, 162, 171, 172)
Niwayama, S., 85²¹¹ (15), 564³³³ (538)
Nixon, G. A., 264³⁵ (192)
Nobis, J. F., 125⁹ (94)
Noe, M. C., 276⁸⁵⁵ (263)
Nogrady, T., 127¹⁶⁴ (117)
Nogue, D., 352³⁶¹ (340)
Noguez, J. A., 568⁶²⁶ (555)
Noheda, P., 353⁴³⁷ (343)
Noiret, N., 268²⁸⁹ (211, 212), 387²⁹⁰ (376)
Nojima, M., 154¹⁹ (130)
Noller, C. R., 309⁹ (279)
Noltemeyer, M., 459⁶⁹ (455)
Nomoto, Y., 272⁵⁶³ (236)
Nomura, G. S., 82⁵⁵ (3, 6)
Nomura, N., 157¹⁸⁴ (153)
Nomura, R., 512⁸² (500), 512⁸⁹ (501)
Nomura, Y., 563²⁶⁶ (530), 565⁴¹¹ (542)
Nonoshita, K., 156¹³⁰ (146)
Norbeck, D. W., 349¹⁷⁵ (328)
Norberg, B., 568⁵⁸⁰ (553, 554, 557)
Norbury, A., 273⁶⁷⁸ (244)
Norcross, R. D., 273⁶²⁸ (240)
Nordlander, J. E., 125⁷¹ (97), 126⁷³ (98)
Norman, R. O. C., 433¹⁶⁴ (415)
Normant, H., 83⁹⁰ (5)
Normant, J.-F., 83⁹⁵ (5), 185²⁶ᵉ (161, 178), 186³⁴ᵈ (161, 163), 186⁴⁸ᵃ (163), 186⁴⁸ᵇ (163), 186⁴⁸ᶜ (163), 186⁴⁸ᵈ (163), 186⁶⁰ (165), 187⁸⁷ (167), 187⁹³ᵃ (168), 187⁹⁸ᵈ (169), 187⁹⁸ᵉ (169), 188¹⁴⁹ᵃ (175, 182) 188¹⁴⁹ᶜ (175, 182), 188¹⁴⁹ᵈ (175, 182), 188¹⁴⁹ᵉ (175, 182), 188¹⁴⁹ᶠ (175, 182), 188¹⁴⁹ᵍ (175, 182), 188¹⁴⁹ʰ (175, 182), 188¹⁴⁹ⁱ (175, 182), 188¹⁴⁹ʲ (175, 182), 188¹⁵³ᵃ (175, 182), 384¹¹¹ (364)
Norris, D. J., 563²⁴⁵ (529, 541)
Northington, D. J., Jr., 434²⁴³ (427)
Nöth, H., 193, 199, 264²² (192), 264²³ (192), 265⁹¹ (199)
Nott, A. P., 91⁶⁵¹, 91⁶⁵⁹
Novikov, S. S., 430²⁷ (392), 434²³⁸ᵃ (427), 434²³⁸ᵇ (427)
Novikova, Z. S., 431⁷³ᶠ (401)
Nowak, M. A., 83¹⁰⁴ (5)
Noyori, R., 90⁵⁷⁹ (65), 151, 156¹⁵⁸ (151), 187⁹⁹ᵈ (169), 188¹²⁰ᵇ (171, 172), 188¹²²ᵃ (171, 172), 188¹²⁹ (171), 350²³² (332)
Nozaki, H., 127¹⁵⁷ (116), 189¹⁹¹ᵃ (182), 190¹⁹⁹ᵃ (182), 270⁴³⁰ (224), 270⁴⁶⁸ (229), 270⁴⁶⁹ (229), 310⁸⁵ (299), 311⁹⁹ (304), 311¹¹¹ (307), 349¹⁸² (328), 350²⁶⁰ (335), 384¹³⁸ (367), 385¹⁵¹ (367)
Nozaki, K., 273⁶⁵⁰ (242), 273⁶⁵² (243), 273⁶⁵⁶ (243), 273⁶⁵⁷ (243), 273⁶⁵⁸ (243), 383¹⁸ (358,

360), 384⁷⁸ (361)
Nozoe, S., 156¹⁶⁸ (151)
Nsunda, K. M., 562¹⁹⁴ (522, 539–42, 557), 566⁴⁶⁴ (544, 556), 566⁴⁷⁰ (544), 566⁴⁷⁸ (545)
Nuss, J. M., 386²⁵³ (373)
Nussim, M., 432¹³⁵ (410)
Nützel, K., 185⁶ (159, 162–5, 168, 169, 173), 190²⁰³ (183)
Nyholm, R., 559⁵⁸ (517, 518, 522, 528)
Nysted, L. N., 189¹⁹⁶ᵇ (182), 189¹⁹⁶ᶜ (182)
Nyzam, V., 272⁵⁶¹ (236)

Oae, S., 155⁴⁹ (133), 559⁶⁵ (517, 528), 561¹³⁸ (520)
Oba, M., 352³⁹⁶ (342)
Obata, S., 92⁷⁰⁰ (74)
Obayashi, M., 270⁴³⁰ (224)
Obayashi, R., 566⁴⁴³ (543)
Obermann, U., 513¹³⁹ (510), 513¹⁴⁰ (510)
Obeyesekere, N. U., 351²⁹¹ (336)
Obora, Y., 385¹⁶⁸ (369)
Obrecht, J.-P., 155⁸⁸ (137)
O'Brien, D. F., 84¹⁶¹ (8)
O'Brien, D. H., 600²⁰⁸ (583, 584)
Ochi, M., 352³⁴⁴ (339)
Ochiai, H., 185¹⁶ᵃ (160), 185¹⁶ᵇ (160), 185¹⁷ (160), 185¹⁹ (160, 167), 185⁶¹ᵃ (165, 166, 174), 187⁹⁰ (168), 187¹⁰⁷ᵃ (169), 187¹⁰⁷ᵇ (169)
Ochiai, M., 377, 384¹²⁵ (364, 372), 384¹⁴¹ (367), 387²⁹² (377), 387³³² (381), 458³⁷ (450)
O'Connor, B., 189¹⁸³ᵇ (179)
O'Connor, D. E., 431⁷¹ᵃ (401)
O'Connor, S., 156¹¹¹ (142), 156¹²³ᵇ (144), 187¹¹⁷ (171)
Oda, M., 266¹⁸³ (203), 266¹⁸⁴ (203), 385¹⁶² (368)
Oda, Y., 189¹⁶⁷ᶜ (178)
Odagaki, Y., 346²⁹ (315)
O'Dell, D. E., 387³²⁷ (380)
O'Dell, R., 384⁸¹ (362)
Odenkirk, W., 350²⁶¹ (335)
Odom, J. D., 559⁶³ (517)
O'Donnell, M. J., 513¹³³ (508)
Oehlschlager, A. C., 126⁷⁶ (100, 103), 354⁴⁶³ (345), 384¹⁰⁹ (364), 384¹¹⁹ (364)
Oertle, K., 431⁵⁶ (399, 420)
Oeveren, A. v., 91⁶²⁷ (71)
Ofori-Okai, G., 89⁴⁹¹ (53)
Ogasawara, K., 87³⁴⁸ (30), 349¹⁹⁸ (329)
Ogata, K., 352³⁹⁴ (341)
Ogawa, A., 85²⁰⁹ (15), 353⁴⁴¹ (344), 383⁴⁹ (361), 386²⁴⁵ (373), 387³⁰⁵ (378), 560¹¹⁰ (519, 520, 529), 566⁴⁴³ (543), 566⁴⁴⁴ (543), 566⁴⁹⁴ (545, 547, 548), 597²¹ (572, 577), 598¹³² (576–8), 600²¹⁰ (584), 600²¹¹ (584), 600²⁶³ (593), 601²⁷⁸ (595)
Ogawa, H., 383¹⁷ (358)
Ogawa, K., 188¹²³ᵃ (171), 188¹²³ᵇ (171)
Ogawa, M., 598¹¹² (575), 598¹¹⁶ (575)

Ogawa, S., 89⁴⁹² (53), 512⁸² (500)
Ogawa, T., 87³⁸⁴ (39), 512¹¹⁴ (504), 512¹¹⁸ (505, 506), 512¹²¹ (506), 512¹²² (506), 512¹²³ (506), 512¹²⁴ (506), 513¹⁴⁵ (511), 599¹⁸⁰ (581), 599¹⁸³ (582), 599¹⁸⁴ (582), 599¹⁹¹ (582, 583)
Ogay, P., 84¹³⁰ (6)
Ogilvie, W., 273⁶⁸¹ (244)
Ogiso, A., 353⁴⁴⁶ (344)
Ogle, C. A., 125⁴⁶ (95), 126¹¹³ (109), 126¹¹⁶ (109)
Ogoshi, H., 274⁶⁹⁸ (246)
Oguni, N., 171, 188¹²¹ᵃ (171, 172), 188¹²¹ᵇ (171, 172), 188¹²⁹ (171), 276⁸⁶⁹ (264), 352³⁴⁶ (339), 352³⁵² (340), 352³⁷⁹ (341)
Ogura, F., 82⁴³ (3), 311⁹⁸ (304), 598¹²⁸ (576, 584), 599¹³⁵ (576), 599¹⁴⁹ (577), 599¹⁸² (582), 599¹⁹⁹ (583), 600²⁰¹ (583), 600²⁰² (583, 584, 586), 600²⁰⁴ (583), 600²¹² (584, 587), 600²¹³ (584), 600²¹⁴ (584), 600²¹⁵ (584), 600²²² (586), 600²²³ (586, 587), 600²³⁶ (589), 600²⁴⁵ (591), 600²⁴⁶ (591), 601²⁸¹ (596), 601²⁸⁴ (596)
Ogura, K., 86³²⁵ (28)
Oh, E., 352³⁹⁹ (342)
O'Hanlon, P. J., 565⁴²⁰ (542)
Ohannesian, L., 156¹⁶³ (151)
Ohashi, N., 89⁵³² (60)
Ohashi, Y., 351³²⁸ (338)
Ohe, K., 88⁴¹⁵ (42, 79), 353¹⁴⁹ (344), 569⁶⁵⁰ (557), 569⁶⁵¹ (557), 569⁶⁵² (557), 598⁷⁴ (573, 588), 598⁹⁵ (574, 588), 598⁹⁶ (574, 588), 599¹⁴⁸ (577), 600²⁰⁷ (583), 600²²⁶ (587), 600²³² (588), 600²⁴³ (591), 600²⁴⁴ (591)
Oh-e, T., 271⁴⁹³ (232), 271⁵²³ (232), 271⁵²⁴ (232, 234), 271⁵³⁸ (234)
Oh-hata, T., 385¹⁵⁵ (368)
Ohira, N., 599¹⁴⁹ (577), 600²⁴⁶ (591)
Ohira, S., 352³⁸⁴ (341)
Ohkata, K., 511⁴⁹ (491, 501), 512⁷¹ (497)
Ohki, H., 512⁹⁵ (502), 512⁹⁶ (502), 512⁹⁷ (502), 512¹⁰¹ (502)
Ohki, T., 387³³¹ (381)
Ohkuma, M., 352³⁹⁴ (341)
Öhler, E., 187⁹⁸ᵇ (169)
Ohlmeyer, M. J., 264⁵⁰ (195), 265⁵⁶ (195), 265⁹² (199, 248), 274⁷²¹ (249)
Ohnari, H., 511⁴⁹ (491, 501), 512⁷¹ (497)
Ohnishi, S., 432¹¹⁸ (407)
Ohno, A., 432¹²⁸ᵃ (409)
Ohno, H., 352³⁴⁶ (339)
Ohno, K., 90⁶⁰⁵ (67)
Ohno, M., 188¹²⁷ᵃ (171, 172), 188¹²⁷ᵇ (171, 172), 188¹²⁷ᶜ (171, 172), 384¹³⁵ (367), 385¹⁴⁶ (367)
Ohnuma, Y., 126⁹⁴ (103), 126⁹⁵ (103)
Ohrbom, W. H., 185²⁸ᶜ (161)
Ohsawa, A., 386²⁷⁰ (374)
Ohshiro, Y., 351³²⁵ (338), 568⁶⁴⁴ (557)
Ohta, H., 566⁵⁰⁷ (546), 567⁵¹² (546)

Ohta, M., 383[38] (359, 377)
Ohta, T., 156[168] (151), 458[46] (451), 458[48] (451)
Ohtsuka, T., 564[302] (536)
Ohuchi, S., 384[133] (367)
Ohyama, Y., 273[688] (245), 347[96] (321), 353[439] (344)
Oishi, T., 87[353] (31)
Ojima, I., 187[114c] (171), 353[422] (343), 353[446] (344)
Oka, S., 432[128a] (409)
Okabe, H., 87[389] (40)
Okada, H., 458[34] (450), 458[35] (450)
Okada, K., 266[183] (203), 266[184] (203)
Okada, M., 386[270] (374)
Okada, S., 188[122a] (171, 172), 188[129] (171)
Okai, K., 352[384] (341)
Okamoto, M., 188[142] (174)
Okamoto, S., 156[140] (147), 188[139a] (174), 188[139b] (174), 347[93] (320)
Okamoto, T., 432[128a] (409)
Okamoto, Y., 559[31] (516)
Okamura, A., 156[149] (149)
Okamura, K., 270[476] (229)
Okano, K., 86[285] (24)
Okano, M., 351[314] (337), 433[197c] (420), 563[271] (530, 540), 564[330] (538), 568[625] (555), 568[635] (556), 598[121] (575), 598[123] (575, 588), 600[209] (584), 600[229] (587), 600[233] (588)
Okauchi, T., 387[301] (377)
Okawara, M., 383[37] (359, 374), 383[38] (359, 377), 568[597] (554)
Okazaki, R., 348[167] (327), 352[367] (340), 484[3] (461), 562[188] (522, 534, 535), 566[500] (545, 546)
Okazoe, T., 190[199b] (182), 190[199d] (182)
Okhlobystin, O. Y., 311[100] (304), 458[3] (438)
Ōki, M., 267[234] (206)
Okinaka, T., 512[70] (496, 497)
Okoroafor, M. O., 90[589] (65)
Oku, A., 189[186] (180, 182), 189[187a] (180, 182), 189[188a] (180, 182), 189[188b] (180, 182), 264[51] (195), 273[687] (245), 273[688] (245), 347[96] (321), 353[439] (344)
Okuda, N., 458[11] (444)
Okudo, M., 188[142] (174)
Okuma, K., 566[507] (546), 567[512] (546)
Okumura, K., 87[350] (30)
Okuse, S., 563[231] (528, 545)
Olah, G. A., 156[163] (151), 268[309] (213), 348[169] (327), 351[337] (339), 352[370] (340), 352[374] (341), 352[376] (341), 352[377] (341), 353[404] (342), 353[454] (344), 353[457] (345)
Olbrysch, O., 188[146a] (175)
Oliveira, A. R. M., 384[108] (364)
Oliver, J. E., 348[170] (328)
Ollis, D., 562[165] (522)
Ollis, W. D., 84[163] (8), 267[258] (208)
Olmstead, M. M., 310[59] (286)
Olofson, R. A., 83[103] (5)
Olson, R. E., 87[356] (32)
Olszowy, H., 384[95] (363)
Oltmann, K., 560[106] (519, 520, 528, 544)
Omae, I., 383[3] (356)
O'Malley, G. J., 599[196] (583)
Omi, T., 188[121a] (171, 172), 276[869] (264)
Omura, H., 432[121] (407)
Omura, S., 275[777] (256, 258)
Onak, T., 268[319] (214)
Onaka, M., 352[350] (339)
O'Neil, I. A., 563[252] (530)
Ongstad, L., 155[70] (135)
Ono, M., 563[240] (529), 563[242] (529)
Ono, N., 188[139a] (174)
Onoda, Y., 384[124] (364, 378)
Oohara, T., 87[364] (33)
Ooi, T., 349[214] (330)
Ookawa, A., 188[123a] (171), 188[123b] (171), 188[130a] (171), 188[130b] (171)
Oozumi, Y., 353[433] (343)
Opel, A., 85[252] (21)
Oppolzer, W., 88[438] (46), 131, 155[36a] (132), 155[36b] (132), 186[52a] (164, 172), 188[126a] (171), 188[126b] (171), 188[135] (172), 256, 269[390] (220), 269[391] (220), 275[785] (256), 351[298] (337)
Orchin, M., 154[4] (130)
Orduna, J., 348[153] (326)
O'Reilly, N. J., 84[179] (10)
Organ, H. M., 92[688] (72)
Orito, K., 485[46] (473)
Oriyama, T., 267[263] (208), 272[618] (240), 385[162] (368)
Ortar, G., 271[526] (233)
Osada, K., 273[695] (245)
Ösapay, K., 562[194] (522, 539–42, 557)
O'Shea, D. M., 351[331] (338)
O'Shea, M. G., 387[298] (377)
Oshima, K., 127[157] (116), 187[95d] (169), 189[191a] (182), 190[199a] (182), 190[199b] (182), 273[650] (242), 273[651] (243), 273[652] (243), 273[653] (243), 273[656] (243), 273[657] (243), 273[658] (243), 273[659] (243), 273[660] (243), 310[85] (299), 350[232] (332), 353[439] (344), 383[18] (358, 360), 383[34] (359), 384[78] (361)
Oshino, H., 185[20a] (160, 165), 185[20b] (160, 165)
Oshio, A., 156[161] (151)
Osman, A., 186[36a] (162, 174)
Osoda, K., 273[694] (245)
Ostwald, R., 186[41] (162, 172), 188[132a] (172), 387[293] (377)
Osuka, A., 564[335] (538), 599[146] (577), 599[156] (578), 599[163] (580), 599[172] (580), 599[173] (580), 599[177] (581), 600[247] (591), 600[249] (591), 600[254] (591)
Osuka, M., 599[135] (576)
Otera, J., 87[338] (29), 87[352] (30), 349[182] (328), 350[260] (335), 384[138] (367), 385[151] (367)
Otsubo, T., 82[43] (3), 311[98] (304), 598[128] (576, 584), 599[135] (576), 599[149] (577), 599[182] (582), 599[199] (583), 600[201] (583), 600[202] (583, 584, 586), 600[204] (583), 600[212] (584, 587), 600[213] (584), 600[214] (584), 600[215] (584), 600[222] (586), 600[223] (586, 587), 600[236] (589), 600[245] (591), 600[246] (591), 601[281] (596), 601[284] (596)
Otsuji, Y., 387[337] (381)
Otto, C., 275[840] (262)
Otto, H.-H., 86[302] (26)
Ouchi, H., 434[239a] (427), 434[240b] (427)
Oudenes, J., 272[605] (240)
Ourisson, G., 87[374] (36)
Out, G. J. J., 187[97a] (169)
Ovaska, T. V., 84[184] (12), 84[193] (12)
Overly, K. R., 187[102] (169)
Overman, L. E., 346[39] (315), 433[167] (415)
Owada, H., 564[330] (538), 568[625] (555), 568[635] (556)
Owczarczyk, Z., 187[82] (166)
Owens, K., 431[78a] (402)
Ozaki, S., 352[342] (339)
Ozbalik, N., 512[104] (503), 513[144] (510), 600[225] (587), 601[274] (595)

Paddon-Row, M. N., 266[138] (201), 266[140] (201), 273[644] (241), 273[645] (241, 258), 275[795] (258)
Padmanabhan, S., 273[673] (244), 273[674] (244), 433[153b] (413), 599[180] (581), 599[183] (582), 599[184] (582), 599[191] (582, 583)
Padwa, A., 383[68] (361), 459[65] (455)
Paetow, M., 86[272] (23)
Paetzold, P., 266[149] (202)
Paetzold, R., 560[105] (519, 520)
Page, P. C. B., 349[174] (328)
Paget, W. E., 126[89] (103), 126[92] (103)
Pagliai, L., 127[170] (118), 347[61] (317)
Pagni, R. M., 269[388] (219)
Pagnoni, U. M., 126[97] (106)
Pai, F. C., 273[682] (244)
Paine, R. T., 264[14] (192)
Painter, E. P., 563[280] (531), 564[340] (539)
Pajerski, A. D., 189[191c] (182)
Palani, A., 386[268] (374)
Palaniswamy, V. A., 86[301] (26)
Palazon, J. M., 271[491] (232)
Paleo, M. R., 91[613] (67)
Paley, R. S., 348[136] (324), 385[214] (371)
Palio, G., 127[167] (117)
Palmer, B. D., 567[519] (547)
Palmer, J. T., 352[397] (342)
Palmer, L., 126[133] (112)
Palmer, M. A. J., 274[741] (251, 253)
Palmieri, G., 86[318] (27), 432[99b] (404)
Palmisano, G., 385[199] (371)
Palomo, C., 156[155] (151)
Palovich, M., 383[64] (361)
Palumbo, P. S., 87[372] (36), 87[373] (36)
Pan, L.-R., 347[113] (322)
Pan, Y. T., 275[774] (255)
Panda, C. S., 126[86] (102)
Pandey, G., 347[80] (319)
Pandey, P. N., 434[236] (426)
Pandiarajan, P. K., 272[623] (240)
Panek, J. S., 82[27] (2, 25), 265[58] (195), 321, 347[90] (320), 347[97] (321), 347[103] (321), 347[104] (321), 347[105] (321), 347[106] (321)

Pang, Y., 348[139] (325)
Pankert, D., 566[476] (544)
Pankratova, V. N., 311[109] (307)
Panov, E. M., 484[26] (467)
Panunzio, M., 155[82] (137), 353[421] (343), 353[422] (343), 353[423] (343), 353[424] (343)
Panyachotipun, C., 87[358] (33)
Paolobelli, A. B., 351[324] (338)
Papadatos, A., 566[492] (545), 601[279] (595)
Papagni, A., 85[201] (13)
Papoula, M. T. B., 512[108] (504, 505), 512[109] (504, 505), 512[111] (504, 505, 507, 508), 512[117] (505, 507, 508), 512[129] (508)
Paquet, F., 433[171] (416)
Paquette, L. A., 82[22] (2, 3), 85[211] (15), 189[176] (178), 351[286] (336), 352[366] (340), 384[122] (364), 564[333] (538), 565[392] (541, 545, 547), 568[639] (557)
Pardasani, R., 86[315] (26)
Pardigon, O., 275[808] (258, 261), 275[809] (258, 261)
Paredes, M. C., 385[195] (371)
Parekh, S. I., 563[293] (534), 563[296] (534)
Parham, W. E., 91[611] (67), 432[132a] (410), 432[132b] (410), 432[133] (410)
Park, B., 84[169] (8, 11)
Park, G., 433[174] (416)
Park, J. C., 274[741] (251, 253), 274[754] (252), 275[772] (255)
Park, J. H., 351[303] (337)
Park, P.-u., 87[385] (39)
Park, S. B., 385[173] (369), 387[317] (379)
Park, Y. S., 311[102] (305)
Parker, D. G., 89[541] (63)
Parkin, G., 185[5] (159)
Parkinson, C. J., 274[752] (252), 385[191] (371), 484[37] (1, 468, 471, 477), 484[41] (469), 485[48] (473)
Parmee, E. R., 350[254] (334)
Parrain, J.-L., 384[116] (364), 385[201] (371), 387[299] (377)
Parrino, V. A., 88[428] (44)
Parry, D. E., 267[268] (208, 234)
Parry, D. M., 349[205] (330)
Parshall, G. W., 310[40] (282, 293, 296, 309)
Parsons, J. L., 88[450] (48)
Parsons, P. J., 346[43] (316), 384[76] (361), 384[110] (364), 565[409] (541)
Parvez, M., 189[191b] (182), 348[138] (324)
Pascard, C., 512[129] (508)
Pasquarello, A., 386[262] (373)
Pasto, D. J., 268[301] (212), 272[607] (240), 430[39] (396), 433[173] (416)
Pastor, S. D., 90[588] (65), 90[591] (65)
Pastour, P., 310[72] (291)
Pasynkiewicz, S., 310[44] (283)
Patai, S., 86[329] (28, 53), 86[330] (28), 86[332] (28), 86[333] (28), 186[58] (165, 168–70, 176), 187[99a] (169), 346[10] (314), 346[14] (314), 558[9] (516), 558[10] (516, 517), 558[25] (516), 558[28] (516), 559[31] (516),

559[39] (516, 519, 520, 522, 531–4, 538, 540–3, 548, 553–5, 557), 559[54] (519), 559[55] (518), 559[56] (519, 538, 539), 559[61] (517), 559[63] (517), 559[64] (517), 559[65] (517, 528), 559[68] (518), 559[69] (518), 559[73] (518), 560[81] (518, 521, 522, 526, 540, 543, 544, 550–5, 557, 558), 560[91] (518–20, 528, 545), 560[92] (518, 558), 560[99] (519, 522, 528), 560[110] (519, 520, 529), 560[118] (519), 560[119] (519, 528), 561[123] (519, 528, 530, 540), 561[133] (520), 561[155] (521, 544), 561[160] (522, 529), 561[161] (522, 529, 558), 561[164] (522, 540, 541), 563[233] (528, 538, 558), 563[250] (530), 567[517] (547, 548), 567[570] (552), 569[655] (558), 569[656] (558), 597[15] (571)
Pataud-Sat, M., 353[418] (343)
Patel, V. K., 348[162] (326)
Paterson, I., 240, 273[626] (240), 273[627] (240), 273[628] (240), 273[630] (240, 258), 273[646] (241, 258), 274[767] (253), 275[776] (256, 258), 275[780] (256, 258), 275[786] (256), 275[787] (257), 275[788] (257), 275[791] (257), 275[792] (258), 275[796] (258), 275[800] (258), 385[153] (368)
Patil, G. S., 84[169] (8, 11)
Patil, P., 86[304] (26)
Patil, S. D., 88[434] (46)
Patil, S. L., 88[434] (46)
Patil, S. R., 598[97] (574, 588), 600[207] (583), 600[209] (584)
Patil, V. D., 269[345] (216), 269[347] (216)
Patricia, J. J., 84[165] (8), 84[183] (10)
Patrick, D. W., 559[47] (516, 522, 535, 539, 541, 543, 552–5, 557), 600[241] (591)
Pattenden, G., 187[79a] (166, 167), 346[48] (316), 566[484] (545, 548)
Patti, A., 90[583] (65), 90[585] (65)
Paty, P. B., 564[363] (540)
Patz, M., 347[68] (317, 337)
Paugam, R., 352[361] (340)
Paulmier, C., 87[334] (28), 560[87] (518, 521, 544, 549), 560[96] (519, 520), 560[97] (519, 527, 528), 560[108] (519, 520, 528), 560[111] (519, 522, 530–2, 547), 560[113] (519, 520, 522, 528, 538–40, 543, 544, 547, 552), 560[114] (519, 520, 522, 538, 539, 554, 557, 558), 561[130] (520, 552), 561[135] (520, 522, 535, 553, 554, 557), 561[162] (522, 529, 540, 544), 562[197] (522, 533–7), 564[339] (539, 541, 544, 551, 558), 564[359] (539, 540, 543), 565[433] (542), 569[657] (558), 600[216a] (585)
Paulson, K. L., 156[167] (151)
Pauluth, D., 83[128] (5)
Pawelke, G., 270[415] (223)
Pawlenko, S., 346[3] (314)
Peach, J. M., 90[558] (63)
Peacock, K., 266[157] (202)
Pearlman, B. A., 87[337] (29), 387[333] (381)

Pearson, A. J., 190[207a] (183)
Pearson, M. M., 349[202] (329), 386[219] (371)
Pearson, N. D., 565[420] (542)
Pearson, N. R., 265[105] (200), 270[434] (224)
Pearson, W. H., 85[249] (20, 21), 265[107] (200), 352[363] (340), 387[318] (379), 387[320] (379), 387[321] (379)
Pecchi, S., 127[153] (116), 127[163] (117), 127[188] (120)
Pedaja, P., 89[488] (52), 91[673]
Pedersen, M. L., 563[262] (530, 539)
Peeran, M., 383[20] (358)
Peet, N. P., 87[369] (36)
Peleties, N., 87[335] (28), 516, 559[51] (516), 559[52] (516)
Pellechia, P. J., 272[567] (236)
Pellon, P., 86[324] (28), 274[700] (246), 274[701] (246), 274[702] (246)
Pelter, A., 86[315] (26), 202, 264[27] (192, 198, 199, 201, 202, 208, 214–16, 220, 222, 223, 225, 227–30, 235, 238, 241, 244, 247), 264[39] (193, 201, 213, 247), 265[75] (196), 265[88] (198), 267[258] (208), 267[259] (208), 267[260] (208), 267[261] (208, 234, 235), 267[268] (208, 234), 268[276] (210), 268[277] (210, 211), 268[278] (210), 268[320] (214–16, 218), 268[329] (215), 268[338] (215), 270[418] (223), 270[421] (224), 270[422] (224), 270[437] (225), 270[449] (227), 270[454] (228), 270[455] (228), 270[456] (228), 270[465] (229), 270[478] (230), 271[539] (234), 271[540] (234), 271[543] (235), 271[544] (235), 271[545] (235), 271[546] (235), 271[547] (235), 272[601] (240), 273[678] (244), 353[411] (342), 353[412] (342), 386[277] (375)
Pendalwar, S. L., 353[432] (343)
Peñéñory, A. B., 564[338] (539)
Penmasta, R., 351[296] (337), 458[15] (445)
Pennanen, S. I., 562[182] (522), 562[186] (522)
Pennell, A. M. K., 384[83] (362)
Penning, T. D., 87[392] (41)
Pensar, G., 92[680]
Percy, J. M., 385[175] (369)
Pereira, S., 351[281] (336)
Pereyre, M., 85[259] (22), 383[2] (356), 387[290] (376), 387[302] (378), 387[303] (378), 600[265] (593)
Pérez, J. A. M., 568[637] (557)
Pérez, M., 385[203] (371)
Pérez-Carreño, E., 434[220] (423)
Pérez-Prieto, J., 433[178] (416), 434[202] (421), 434[203] (421, 423), 434[209] (422), 434[217] (422)
Peri, S. P., 88[469] (50, 68)
Periasamy, M., 190[201c] (182), 265[68] (196), 265[118] (200), 270[441] (226), 273[669] (244), 273[670] (244), 273[671] (244), 273[672] (244), 275[814] (258), 432[114] (407)
Pericàs, M. A., 349[191] (329)
Périchon, J., 186[36c] (162, 174),

186[36d] (162, 174), 186[36e] (162, 174), 186[36f] (162, 174)
Perie, J. J., 434[205] (421)
Perkins, M. J., 567[517] (547, 548)
Perret, C., 90[547] (63)
Perrier, H., 88[406] (42, 53), 88[408] (42, 53)
Perron, F., 91[629] (71)
Perrotta, E., 347[74] (318)
Persad, A., 84[132] (6, 33)
Pershin, D. G., 269[372] (218)
Persson, J., 598[70] (573)
Perumal, P. T., 270[420] (224), 274[745] (251)
Petasis, N. A., 346[34] (315), 558[2] (516), 558[4] (516, 522, 527, 529, 536, 547, 548, 558), 560[86] (518, 521, 543, 552, 555), 560[102] (519, 522, 533, 536), 560[103] (519, 522, 530, 533, 534, 536), 560[107] (519, 532, 534, 538, 545, 558), 562[196] (522, 531, 547), 562[198] (522, 541, 542, 551, 553–7), 562[199] (522, 538, 540, 542, 547), 564[314] (538, 540, 543, 545), 564[320] (538, 542), 600[216b] (585)
Peters, D., 434[244b] (427)
Peters, W., 430[33] (395)
Petersen, J. S., 272[597] (240, 241, 256), 274[714] (247), 275[807] (258, 261)
Peterson, B. H., 352[363] (340)
Peterson, M. A., 352[367] (340)
Peterson, M. L., 274[727] (249), 274[730] (250)
Peterson, P. E., 274[711] (247)
Petit, N., 387[303] (378)
Petragnani, N., 563[257] (530), 565[396] (541), 597[5] (571), 597[9] (571), 597[12] (571), 597[17] (571), 597[18] (571), 597[46] (572), 597[53] (572), 598[64] (572, 574), 598[87] (574), 598[88] (574), 598[107] (574), 598[108] (574), 598[111] (575), 598[119] (575, 585), 598[125] (576), 599[136] (576), 599[162] (580), 599[164] (580), 600[217] (585), 600[218] (585), 600[221] (586), 601[270] (593)
Petránek, J., 89[507] (55), 126[109] (109)
Petrich, S. A., 348[139] (325)
Petrie, M. A., 310[59] (286)
Pétrier, C., 84[158] (7), 185[26b] (161, 178), 185[26c] (161, 178), 188[140] (174), 189[166a] (178), 189[166b] (178), 189[168a] (178), 189[168b] (178), 511[52] (492)
Petrini, M., 86[318] (27), 154[26] (130), 154[27] (130), 155[89] (139), 349[193] (329)
Petrov, A. A., 598[101] (574)
Petter, W., 188[128c] (171, 172)
Petty, M. C., 566[459] (544)
Pfammatter, E., 85[244] (20)
Pfeiffer, T., 155[41] (132)
Pfenninger, J., 566[486] (545, 547, 548)
Pfrengle, W., 350[277] (336)
Phadke, A. S., 267[244] (207)
Philipp, F., 84[140] (6)
Phillips, H., 568[603] (554)
Phillips, N. H., 84[167] (8)
Phillips, W. V., 82[55] (3, 6), 83[84] (4)
Pi, J. H., 599[178] (581), 599[179] (581), 599[185] (582)

Pi, R., 125[44] (95, 112)
Piattelli, M., 90[583] (65), 90[585] (65)
Piazza, M. G., 155[77] (136)
Pichon, C., 512[110] (504, 509), 512[113] (504), 513[135] (509), 513[138] (509, 510)
Pickering, D., 85[260] (22)
Pickles, G. M., 458[36] (450)
Pierce, A. M., 126[76] (100, 103)
Pierce, H. D., 126[76] (100, 103)
Pierini, A. B., 564[336] (539), 564[337] (539), 564[338] (539), 598[71] (573), 598[72] (573)
Pierre, J. L., 273[692] (245)
Piers, E., 363, 384[97] (363, 364), 385[184] (370), 386[220] (371)
Pieters, R. J., 349[202] (329)
Pietra, F., 156[133] (146)
Piette, J. L., 598[76] (573), 598[84] (574), 598[90] (574, 575), 600[267b] (593), 600[267c] (593)
Piettre, S., 565[412] (542), 565[413] (542)
Pigou, P. E., 562[217] (526, 547)
Pike, P. E., 431[53a] (398), 431[53b] (398)
Pikul, S., 156[138] (147), 275[789] (257)
Pilcher, A. S., 349[206] (330)
Pimm, A., 186[46] (163)
Pindur, U., 275[840] (262)
Ping, L., 562[226] (527)
Pinhey, J. T., 385[191] (371), 431[85a] (403), 431[85b] (403), 431[85c] (403), 431[85d] (403), 484[13] (465, 466), 484[14] (465–8, 471, 473, 475, 477, 480, 481), 484[15] (465, 468, 481), 484[16] (465, 468, 481), 484[17] (1, 465, 468, 477), 484[23] (466, 467, 481), 484[24] (466), 484[25] (466, 475), 484[28] (467), 484[29] (1, 467, 475), 484[31] (467, 481), 484[32] (467, 482, 483), 484[33] (467), 484[35] (467), 484[36] (468, 475), 484[37] (1, 468, 471, 477), 484[38] (469), 484[41] (469), 484[42] (1, 471), 484[43] (1, 471), 485[44] (473), 485[47] (471), 485[48] (473), 485[52] (474), 485[53] (474), 485[55] (476), 485[56] (476, 477), 485[57] (477), 485[58] (478), 485[59] (477–9, 481), 485[60] (479, 480), 485[61] (479, 480), 485[63] (481, 482), 485[64] (481)
Pini, D., 155[103] (141)
Pinna, L., 350[246] (333)
Pino, P., 310[73] (291)
Pinto, B. M., 565[373] (540, 553)
Pinzani, D., 127[147] (115), 127[188] (120), 353[417] (343)
Piotrowski, A. M., 189[196a] (182), 309[12] (279), 310[39] (282, 293, 296, 309), 310[77] (293, 296, 308), 311[101] (304)
Piotrowski, D. W., 155[72] (136)
Pirie, D. K., 155[92] (139)
Pitchford, A., 271[544] (235)
Pitombo, L. R. M., 597[56] (572, 574)
Pitteloud, R., 155[36a] (132)
Pla-Dalmau, A., 431[63a] (400)
Plamondon, L., 384[120] (364, 377)
Planchenault, D., 349[194] (329)
Plaquevent, J.-C., 82[46] (3)
Plattner, D. A., 188[128c] (171, 172)
Plavcan, K. A., 350[271] (335)
Plavec, J., 567[540] (548)

Plé, N., 83[101] (5, 81), 92[713] (79), 92[716] (79), 92[718] (79), 92[720] (81), 92[721] (81), 271[514] (232), 271[516] (232)
Plešek, J., 264[8] (192)
Pluim, H., 564[361] (540)
Plumet, J., 91[649] (72)
Plzak, K., 85[260] (22)
Pochini, A., 458[47] (451)
Pockels, U., 84[159] (7), 309[7] (278)
Pocock, G. R., 89[493] (53), 155[53] (133), 155[54] (133)
Podestá, J. C., 384[106] (363), 458[36] (450)
Pogonowski, C. S., 568[630] (556)
Poindexter, G. S., 89[484] (51), 89[514] (56), 154[6] (130)
Poirier, R. A., 559[55] (518)
Poli, G., 273[633] (240), 386[262] (373)
Polissiou, M. G., 91[628] (71)
Polívka, Z., 264[52] (195)
Polniaszek, R. P., 88[437] (46)
Polson, G., 599[157] (578)
Polston, N. L., 268[316] (213), 268[336] (215)
Polt, R., 352[367] (340)
Pombrik, S. I., 282, 310[64] (288)
Ponomaryov, A. B., 432[128b] (409)
Pooley, G. R., 348[130] (324)
Poon, Y.-F., 265[107] (200)
Porco, J. A., Jr., 88[414] (42)
Pornet, J., 187[112b] (171), 272[578] (237), 348[124] (323)
Portella, C., 349[186] (328)
Porter, J. R., 154[25] (130)
Porter, N. A., 383[56] (361), 433[181] (418)
Posner, G. H., 86[330] (28), 88[443] (48), 128[207], 352[372] (341), 387[297] (377), 564[341] (539)
Pospíšil, J., 125[35] (94)
Poss, M. A., 86[303] (26)
Postema, M. H. D., 273[654] (243), 383[60] (361), 567[523] (547, 548)
Postich, M. J., 387[320] (379), 387[321] (379)
Potapov, V. A., 598[93] (574), 598[98] (574)
Potapova, T. V., 272[591] (239)
Pougny, J.-R., 433[170b] (416)
Pourcelot, G., 564[301] (536)
Poursoulis, M., 187[114b] (171)
Power, P. P., 310[59] (286)
Prager, R. H., 83[96] (5), 270[435] (225)
Prajapati, D., 190[206] (183)
Prakash, G. K. S., 351[313] (337)
Prakash, I., 351[296] (337), 458[15] (445)
Prakash, O., 351[295] (337)
Prakash Reddy, V., 348[169] (327)
Prapansiri, V., 87[358] (33)
Prasad, A. S. B., 265[118] (200)
Prasad, C. V. C., 350[277] (336)
Prasad, K., 273[684] (244), 273[685] (244)
Prasit, P., 433[145d] (412)
Prats, M., 92[677]
Preston, S. B., 270[433] (224)
Preuss, F. R., 484[20] (466)
Preuss, R., 127[181] (119)
Prévost, C., 187[95c] (169)
Prewo, R., 91[661]
Price, D. A., 90[569] (64)
Pridgen, L. N., 387[304] (378)
Priebe, W., 383[32] (359)

Prieto, J. A., 83[122] (5)
Prieto, R., 434[244b] (427)
Prince, B., 187[114a] (171)
Principe, L. M., 348[141] (325)
Pritchard, G. J., 128[208] (124)
Procter, G., 82[50] (3), 347[63] (317), 347[94] (320), 347[95] (321)
Proess, G., 566[475] (544), 566[476] (544)
Prokai, B., 266[141] (201), 266[142] (201)
Prokof'ev, A. K., 432[129c] (409, 411)
Puchot, C., 187[114a] (171)
Pujol, B., 351[289] (336)
Pulido, F. J., 384[98] (363, 364), 384[114] (364), 385[186] (370, 373)
Punzalan, E. R., 84[174] (10)
Punzi, A., 346[44] (316)
Purdy, A. P., 189[191d] (182)
Purmort, J. I., 126[73] (98)
Pushin, A. N., 565[382] (540)
Putt, S. R., 87[337] (29), 387[333] (381)
Pyne, S. G., 87[365] (34)
Pyun, C., 267[213] (205), 267[217] (205, 206)

Quallich, G. J., 91[612] (67), 275[823] (259), 275[825] (259)
Quayle, P., 385[196] (371), 565[393] (541)
Quéguiner, G., 79, 82[38] (2, 4, 42, 70, 75, 76, 79, 81), 83[101] (5, 81), 91[668], 92[712] (79), 92[713] (79), 92[716] (79), 92[718] (79), 92[720] (81), 92[721] (81), 271[514] (232), 271[515] (232), 271[516] (232), 271[517] (232), 271[518] (232), 271[519] (232), 310[72] (291)
Quesnelle, C., 88[408] (42, 53)
Quintanilla, R., 267[201] (204), 267[202] (204), 267[211] (205)
Quintard, J.-P., 85[259] (22), 383[2] (356), 384[116] (364), 385[201] (371), 386[243] (373), 387[299] (377), 600[265] (593)
Quintero, L., 383[35] (359)
Quinton, P., 187[108c] (169, 170)
Quirion, J.-C., 188[149d] (175, 182), 188[149e] (175, 182), 188[149f] (175, 182)

Raabe, G., 155[73] (136), 384[126] (364)
Rabjohn, N., 559[36] (516, 519, 522, 536), 559[42] (516, 519, 522, 536)
Racherla, U. S., 265[101] (199), 265[104] (200), 266[175] (203), 266[177] (203), 267[222] (206), 269[346] (216), 269[347] (216), 270[440] (225), 272[567] (236), 272[587] (238), 274[746] (251), 274[747] (251), 274[751] (251)
Raczko, J., 91[630] (71)
Radchenko, S. A., 598[100] (574)
Radinov, R. N., 186[52a] (164, 172), 188[126a] (171), 188[126b] (171), 188[135] (172), 269[390] (220), 269[391] (220)
Radzhabov, D. T., 564[348] (539)
Raharinirina, A., 267[231] (206)
Rahm, A., 383[2] (356), 600[265] (593)
Rahman, Atta-ur-, 274[707] (247), 567[555] (550, 551)
Raimondi, L., 87[367] (36), 87[368] (36)
Rajagopal, D., 186[55] (164), 186[77c] (166, 183)
Rajagopalan, S., 266[134] (201), 268[314] (213), 430[23] (391)
RajanBabu, T. V., 383[57] (361), 384[100] (363, 369)
Rakhmankulov, D. L., 310[53] (285)
Ramachandran, P. V., 265[79] (196, 206), 274[708] (247, 258, 260), 275[802] (258, 260), 275[835] (260), 275[837] (260)
Ramaiah, P., 353[404] (342)
Ramanathan, V., 127[174] (119)
Ramasamy, K., 597[37] (572), 599[138] (577, 580), 599[165] (580)
Ramcharitar, S. H., 384[80] (362), 386[221] (371)
Ramesh, M., 512[104] (503), 513[144] (510), 600[225] (587), 601[275] (595)
Ramin Najafi, M., 266[134] (201), 268[335] (215)
Ramón, D. J., 84[149] (7, 15), 84[150] (7), 84[151] (7), 84[152] (7), 84[154] (7), 85[203] (14)
Rampersaud, A., 86[309] (26)
Ramphal, J. Y., 271[491] (232)
Ranasinghe, M. G., 352[397] (342)
Rancourt, J., 383[33] (359)
Rand, C. L., 187[81] (166), 189[159] (176)
Randad, R. S., 107, 126[103] (107), 274[746] (251), 274[759] (252)
Rane, A. M., 347[92] (320)
Rangaishenvi, M. V., 264[44] (194), 265[60] (195), 265[106] (200, 248), 267[222] (206), 267[240] (207), 267[243] (207), 267[244] (207), 275[835] (260)
Rani, K. S., 347[80] (319)
Ranu, B. C., 350[265] (335)
Rao, C. G., 269[345] (216)
Rao, J. M., 268[329] (215), 270[437] (225)
Rao, M. N. S., 512[90] (501), 512[91] (501)
Rappoport, Z., 86[329] (28, 53), 86[330] (28), 346[10] (314), 558[9] (516), 558[10] (516, 517), 558[25] (516), 558[28] (516), 559[31] (516), 559[39] (516, 519, 520, 522, 531–4, 538, 540-3, 548, 553–5, 557), 559[54] (519), 559[55] (518), 559[56] (519, 538, 539), 559[61] (517), 559[63] (517), 559[64] (517), 559[65] (517, 528), 559[68] (518), 559[69] (518), 559[73] (518), 560[81] (518, 521, 522, 526, 540, 543, 544, 550–5, 557, 558), 560[91] (518–20, 528, 545), 560[92] (518, 558), 560[110] (519, 520, 529), 560[118] (519), 560[119] (519, 528), 561[123] (519, 528, 530, 540), 561[133] (520), 561[155] (521, 544), 561[160] (522, 529), 561[161] (522, 529, 558), 563[233] (528, 538, 558), 563[250] (530), 567[517] (547, 548), 567[570] (552), 569[655] (558), 569[656] (558), 597[15] (571)
Rasmussen, J. K., 352[345] (339)
Rasset-Deloge, C., 265[123] (200, 205, 211), 268[290] (211, 212), 268[303] (213)
Rassu, G., 350[246] (333), 351[308] (337)
Rather, E. M., 126[108] (109)
Rathke, J., 269[403] (221)
Rathke, M. W., 83[104] (5, 83[120] (5),
185[2a] (159, 170), 208, 267[269] (208, 235), 267[270] (208), 269[383] (219), 272[609] (240)
Rathman, T. L., 82[54] (3, 10)
Ratovelomanana, V., 186[36d] (162, 174), 186[36f] (162, 174)
Ratz, A. M., 274[743] (251)
Raubo, P., 348[160] (326), 348[163] (326)
Raucher, S., 431[64b] (400), 566[455] (544)
Rauchfuss, T. B., 560[92] (518, 558)
Rauchschwalbe, G., 125[12] (94, 95, 99, 100, 104, 105, 108), 125[41] (94, 103), 125[62] (101)
Rausch, M. D., 430[19] (391), 431[48] (397), 458[51] (452), 458[52] (452), 458[53] (453), 458[54] (453), 458[55] (453), 458[58] (453), 459[59] (453)
Ravi Kumar, K. S., 265[71] (196)
Ravichandran, K., 599[196] (583)
Ravidà, N., 126[129] (111)
Ravindran, N., 265[80] (196), 269[385] (219)
Rawal, V. H., 383[25] (358, 362), 383[40] (360), 565[391] (541, 547, 548)
Rawlinson, D. J., 431[57b] (399)
Rawson, D. J., 7, 84[148] (7), 155[44] (133)
Ray, R., 267[227] (206, 249, 251)
Razavi, A., 430[19] (391)
Razino, P., 155[87] (137)
Razuvaev, G. A., 458[25] (447)
Rea, S. O., 155[55] (133)
Read, J. M., Jr., 268[315] (213)
Reader, J. C., 350[227] (331)
Real, S. D., 156[148] (149)
Réau, R., 387[322] (379)
Rebière, F., 90[572] (64), 90[595] (65), 458[4] (438, 441)
Recktenwald, O., 458[16] (445)
Reddy, B. S. R., 458[13] (444), 558[14] (516, 529), 558[20] (516)
Reddy, D. R., 351[292] (336)
Reddy, G. S., 310[40] (282, 293, 296, 309)
Reddy, N. K., 273[675] (244)
Reddy, R. E., 568[579] (553, 554, 557)
Reddy, R. T., 568[579] (553, 554, 557), 568[587] (554), 568[590] (554, 557)
Reddy, V. P., 351[313] (337)
Redman, L. M., 125[26] (94, 113)
Redmore, D., 88[470] (50)
Reed, J. N., 563[245] (529, 541)
Reed, L. A., 156[153] (150)
Reed, M. W., 91[634] (71)
Reed, T. J., 268[323] (215)
Rees, C. W., 270[480] (231), 353[406] (342), 353[407] (342), 458[21a] (446), 458[21b] (446), 558[23] (516, 522), 562[219] (527)
Reetz, M. T., 83[87] (4), 137, 146, 155[84] (137), 156[129] (146), 156[131] (146), 187[109a] (170), 273[637] (240), 273[696] (245), 273[697] (245), 274[740] (251), 275[778] (256, 258), 346[39] (315), 350[247] (333), 350[258] (334), 353[416] (342)
Reeve, W., 90[574] (64)
Reformatsky, S., 278
Regan, A. C., 89[525] (57), 353[409] (342)
Reginato, G., 127[147] (115), 127[153]

(116), 127[188] (120), 349[187] (329), 351[318] (338), 352[391] (341), 353[417] (343), 384[113] (364), 384[115] (364)
Regitz, M., 185[3] (159, 178)
Rei, M.-H., 433[155a] (414), 433[155b] (414), 433[155c] (414), 433[187] (418)
Reibenspies, J. H., 434[232] (425), 484[40] (469, 480), 563[296] (534)
Reich, H. J., 84[167] (8), 86[287] (25), 86[331] (28), 87[355] (32), 87[356] (32), 91[652], 156[170] (151), 346[12] (314), 350[238] (333), 555, 559[34] (516, 522), 559[48] (516, 522, 526, 529, 536, 540, 541, 543, 552–7), 560[80] (518), 560[83] (518, 521, 522, 526, 543, 544, 551, 557), 560[120] (519, 520, 522, 538), 561[126] (519, 540, 555), 562[174] (522, 526, 529, 531, 543, 544, 551, 553–7), 562[206] (522, 556, 557), 562[212] (526, 544), 563[237] (529, 530, 545), 563[260] (530), 563[274] (530), 564[331] (538, 540–2, 554, 555, 557), 565[398] (541, 555), 565[400] (541, 554, 555, 557), 565[418] (542), 568[581] (554, 557), 568[582] (554, 557), 568[598] (554, 555), 568[599] (554, 557), 568[602] (554), 568[610] (554), 568[622] (555, 556), 568[628] (555), 568[633] (556)
Reich, I. L., 565[398] (541, 555), 565[400] (541, 554, 555, 557), 568[622] (555, 556)
Reichel, L., 597[45] (572), 597[54] (572)
Reichenbächer, M., 560[105] (519, 520)
Reid, D. H., 562[167] (522), 562[168] (522), 562[169] (522), 566[501] (545, 546)
Reikhsfel'd, V. O., 352[387] (341)
Rein, K., 82[24] (2, 15), 82[25] (2, 15), 85[245] (20), 85[246] (20)
Reinecke, M. G., 91[667]
Reinheckel, H., 310[54] (285), 310[92] (304), 311[94] (304)
Reinhoudt, D. N., 83[93] (5)
Reininger, K., 187[98b] (169)
Reisdorf, D., 83[90] (5)
Reissig, H.-U., 347[84] (319)
Reitstøen, B., 156[137] (147)
Reitz, D. B., 85[216] (15, 16), 88[425] (43)
Rémion, J., 567[564] (551)
Renard, M., 564[311] (538), 566[474] (544), 566[475] (544)
Renaud, P., 567[550] (548)
Renaut, P., 87[377] (36)
Renga, J. M., 563[237] (529, 530, 545), 565[398] (541, 555), 565[400] (541, 554, 555, 557), 568[622] (555, 556)
Renier, P., 563[265] (530)
Rennels, R. A., 386[253] (373)
Renson, M., 558[25] (516), 566[485] (545), 598[76] (573), 598[84] (574), 598[90] (574, 575), 600[267b] (593)
Repić, O., 185[30] (161), 273[684] (244), 273[685] (244)
Repke, D. B., 89[527] (59), 89[528] (59)
Resek, J. E., 127[148] (115)
Retherford, C., 186[66] (165, 179), 186[74a] (166), 188[143a] (175),

188[143b] (175)
Retta, N., 186[57a] (165)
Rettig, S. J., 267[232] (206), 386[220] (371)
Reuman, M., 88[424] (43)
Reuter, D. C., 353[401] (342), 384[99] (363)
Reutov, O. A., 430[31a] (393), 430[36] (395), 431[65] (400), 431[87a] (403), 431[87c] (403), 432[125] (408), 433[180a] (417)
Reutrakul, V., 87[358] (33)
Rewcastle, G. W., 82[37] (2, 42, 71, 73, 75, 76, 79, 81)
Rey, A. W., 350[248] (333)
Reye, C., 154[7] (130), 346[17] (314)
Reza, K., 90[609] (67)
Rhee, I., 432[121] (407)
Rhee, S.-G., 309[24] (280, 292), 310[30] (280), 310[66] (289), 310[67] (289)
Rheinboldt, H., 559[59] (517, 519, 552), 560[112] (519, 520, 527, 538, 539, 552, 553), 560[117] (519), 561[122] (519, 528), 561[124] (519, 520), 561[129] (519, 520), 561[134] (520, 554), 561[139] (520, 554), 561[140] (520), 561[143] (520), 561[144] (520), 561[152] (520), 561[159] (522, 539), 562[222] (527), 562[223] (527), 571, 597[1] (571), 597[59] (572), 598[105] (574)
Riant, O., 90[594] (65), 90[595] (65), 275[839] (262)
Ribéreau, P., 91[668]
Ricard, L., 90[595] (65)
Ricart, G., 83[83] (4)
Ricci, A., 86[283] (24), 126[129] (111), 127[147] (115), 127[153] (116), 127[167] (117), 127[170] (118), 127[188] (120), 347[61] (317), 347[74] (318), 349[174] (328), 349[187] (329), 351[318] (338), 352[391] (341), 353[417] (343), 384[113] (363, 376), 384[113] (364), 384[115] (364), 387[303] (378)
Richards, C. J., 276[844] (262)
Richardson, D. P., 433[192] (419)
Richardson, G. M., 127[134] (112)
Richardson, J. F., 512[90] (501), 512[91] (501)
Richardson, K. A., 385[171] (369, 376)
Richardson, S., 86[303] (26)
Richey, H. G., Jr., 189[191b] (182), 189[191c] (182)
Richter, M., 348[164] (327)
Ridgway, L. R., 512[103] (503)
Riding, G. H., 90[577] (65)
Rieche, A., 433[182] (418)
Riede, J., 566[499] (545)
Riefling, B., 431[75] (402), 432[117] (407)
Rieger, D. L., 275[793] (258)
Rieger, R., 310[64] (288)
Rieke, R. D., 154[5] (130), 154[6] (130), 185[21a] (160), 185[21b] (160), 185[21c] (160), 185[22] (160), 186[62] (165, 167, 168), 186[70] (166), 190[205] (183)
Rieker, W. F., 85[225] (16, 17)
Rienäcker, R., 311[105] (306)
Rigg, D. J., 458[21a] (446)
Rigollier, P., 351[304] (337), 565[417] (542)

Rikers, R., 91[622] (71), 127[175] (119)
Rinehart, J. K., 432[132a] (410), 432[132b] (410)
Ringe, K., 348[151] (325)
Ripamonti, A., 185[23a] (160)
Risbood, P. A., 88[450] (48)
Rischer, M., 347[117] (323)
Rise, F., 155[70] (135)
Ritter, K., 275[770] (255)
Rittmeyer, P., 87[399] (42)
Rivenson, A., 558[14] (516, 529)
Rivera, I., 270[424] (224), 271[530] (233), 271[533] (234), 271[535] (234)
Rivera, J. G., 431[53b] (398)
Rivera, S. L., 156[150] (150)
Rivers, G. T., 88[441] (47)
Rivkin, B. B., 597[14] (571), 598[106] (574)
Rizzacasa, M. A., 155[45] (133)
Rizzi, A., 458[47] (451)
Rizzolio, M., 127[183] (120)
Roan, B. L., 346[24] (315)
Robb, M. A., 273[647] (241)
Roberts, B. P., 268[337] (215), 272[617] (240), 273[662] (243), 276[867] (264), 276[868] (264), 353[444] (344)
Roberts, D. A., 187[85] (167)
Roberts, J. D., 125[71] (97), 126[73] (98), 128[198] (122)
Roberts, J. L., 272[605] (240)
Roberts, L. R., 351[330] (338), 351[331] (338), 351[332] (338)
Roberts, R. R., 348[172] (328)
Roberts, S. M., 346[48] (316), 349[205] (330)
Robins, B. D., 568[606] (554)
Robins, M. J., 346[41] (316), 346[42] (316), 349[208] (330), 512[87] (501)
Robinson, G. C., 310[41] (283)
Robinson, N. P., 350[261] (335)
Robinson, P. D., 87[363] (33)
Robinson, R. P., 88[435] (46)
Robson, E., 431[71b] (401)
Robson, J. H., 431[54] (398), 431[77a] (402)
Rocca, P., 92[705] (78), 271[515] (232), 271[517] (232), 271[518] (232), 271[519] (232)
Roche, E. G., 431[85c] (403), 484[17] (1, 465, 468, 477), 484[29] (1, 467, 475)
Rochow, E. G., 560[90] (518)
Rock, M. H., 385[175] (369)
Rocks, R. R., 267[227] (206, 249, 251)
Rodeheaver, G. T., 431[59a] (399), 431[59b] (399)
Roden, F. S., 154[25] (130)
Rodes, R., 434[221a] (423), 434[222] (423)
Rodewald, S., 267[201] (204), 267[205] (204)
Rodin, J. O., 126[104] (107)
Rodrigo, R., 564[306] (537)
Rodrigues, R., 565[396] (541)
Rodríguez, A., 92[677], 459[73] (456), 459[74] (456)
Rodriguez, H. R., 82[33] (2, 42, 48, 72), 127[142] (114, 121)
Rodriguez, I., 275[785] (256)
Rodríguez-López, J., 348[148] (325)
Roe, D. G., 92[704] (74)
Roelofs, W. L., 126[77] (100)
Roesky, H. W., 459[69] (455)

Rogers, D. Z., 564[341] (539)
Rogers, R. D., 458[53] (453), 459[59] (453)
Rogić, M. M., 269[383] (219), 272[609] (240)
Rohtagi, B. K., 432[100] (404)
Roling, P. V., 431[48] (397)
Rolli, E., 562[194] (522, 539–42, 557)
Romanelli, A., 155[99] (140)
Römer, J., 563[246] (529, 530)
Romesberg, F. E., 83[70] (4)
Ronald, R. C., 88[457] (49)
Rondan, N. G., 266[138] (201), 266[140] (201)
Ronman, P., 89[490] (53)
Ronzini, L., 155[60] (134)
Roos, E. C., 347[77] (319), 384[92] (363)
Roper, T. D., 276[855] (263), 276[862] (263), 350[252] (334)
Rosario, O., 270[423] (224)
Rösch, F., 563[246] (529, 530)
Rosenfeld, M. N., 563[294] (534)
Rosenthal, S., 86[284] (24), 349[174] (328)
Roskamp, E. J., 385[161] (368), 385[163] (368), 385[164] (368)
Rosser, R. M., 271[540] (234)
Rossi, K., 84[189] (11)
Rossi, R., 187[88a] (167), 187[88b] (167), 187[88c] (167), 383[8] (357, 361, 371, 372)
Rossi, R. A., 384[106] (363), 564[336] (539), 564[337] (539), 564[338] (539), 598[71] (573), 598[72] (573)
Rossier, J.-C., 351[309] (337)
Rossiter, B. E., 155[42] (132, 151)
Roth, G. P., 352[358] (340)
Roth, M., 189[171a] (178)
Rothen, L., 126[105] (108)
Rouanet-Dreyfuss, A.-C., 348[146] (325)
Rouillard, L., 386[261] (373)
Roush, W. R., 126[84] (101), 271[489] (232), 271[490] (232), 271[548] (235, 236, 251, 252), 272[572] (237), 274[741] (251, 253), 274[742] (251, 253), 274[743] (251), 274[748] (251), 274[754] (252), 274[755] (252), 274[756] (252, 256), 274[768] (253), 274[769] (253), 275[772] (255)
Roussakis, C., 186[38c] (162, 169)
Rousseau, G., 187[93b] (168), 187[93c] (168), 189[181] (179)
Rousset, C. J., 84[175] (10)
Rovera, J.-C., 92[713] (79), 271[516] (232), 271[517] (232)
Rowden, J. Y., 347[75] (319)
Rowe, B. A., 484[25] (466, 475), 484[42] (1, 471), 484[43] (1, 471), 485[53] (474)
Rowlands, M., 271[546] (235)
Rowley, M., 387[334] (381)
Roy, G., 86[312] (26)
Roy, J., 555, 558[5] (516, 529, 530, 538, 558), 568[619] (555), 568[620] (555)
Roy, P., 350[248] (333)
Rozema, M. J., 186[40] (162, 172, 174), 186[41] (162, 172), 186[55] (164), 186[66] (165, 179), 186[77a] (166, 183), 188[132b] (172), 188[137a] (174, 177), 189[182a] (179), 189[182b] (179), 189[189] (180, 181)

Rozen, S., 431[51] (397, 415)
Rozenberg, V. I., 431[87a] (403), 431[87c] (403)
Rozsondai, B., 559[56] (519, 538, 539)
Rubin, Y., 348[138] (324)
Rubino, M. R., 85[255] (21)
Rubio, R., 385[186] (370, 373)
Rubottom, G. M., 350[270] (335), 484[12] (465)
Ruch, E., 90[581] (65)
Rücker, C., 85[213] (15, 25)
Rücker, G., 189[157a] (176)
Ruckert, T., 567[560] (551)
Ruddock, K. S., 187[85] (167)
Rudolph, B., 90[547] (63)
Rüger, H., 564[324] (538, 557)
Rüger, W., 567[515] (546, 547)
Rühl, T., 85[265] (22)
Ruiz-Romero, M. E., 458[14] (445)
Rupani, P., 268[338] (215)
Ruppert, J. F., 187[98a] (169)
Rusinko, A., 126[89] (103)
Russell, A. T., 347[95] (321), 352[399] (342)
Russell, G. A., 431[63a] (400), 431[63b] (400), 431[63d] (400, 401), 431[67] (400), 431[78a] (402), 431[78b] (402), 431[79a] (402), 431[79b] (402), 431[79c] (402), 431[80] (402), 431[81] (402), 431[87d] (403), 431[87e] (403), 432[91] (403), 566[445] (543)
Russo, M., 558[30] (516), 560[116] (519, 538, 539)
Rüttimann, A., 274[716] (248), 274[717] (248)
Ruzziconi, R., 351[323] (338), 351[324] (338)
Ryabov, A. D., 459[78] (456)
Ryan, M. D., 563[235] (528)
Ryang, M., 432[121] (407)
Rybczynski, P. J., 352[364] (340)
Rychnovsky, S. D., 85[212] (15, 25), 85[260] (22), 352[393] (341)
Ryu, D. S., 432[123] (408)
Ryu, I., 85[207] (14), 350[272] (335), 353[441] (344), 386[245] (373), 387[305] (378), 387[309] (378), 601[278] (595), 601[282] (596)

Saá, J. M., 88[444] (48), 88[445] (48), 88[449] (48), 88[452] (48), 385[211] (371)
Saavedra, J. E., 16, 85[226] (16)
Sabatier, R., 351[289] (336)
Sabin, V., 346[51] (317)
Sabol, J. S., 385[182] (370, 372)
Saccomano, N. A., 88[435] (46)
Sachdev, H. S., 568[609] (554)
Sachdev, K., 568[609] (554)
Sachleben, R. A., 568[606] (554)
Sadekov, I. D., 597[14] (571), 597[47] (572), 598[106] (574)
Sadhu, K. M., 86[275] (24), 267[239] (207), 267[256] (208), 274[727] (249), 274[733] (251), 274[736] (251)
Sadovnikov, N. P., 431[73f] (401)
Saegusa, T., 89[531] (59), 348[143] (325)
Saffrich, J., 276[843] (262), 276[844] (262)
Saha, A., 154[17] (130)
Saha, M., 91[617] (70), 350[265] (335)

Saha, S., 434[214] (422)
Sahai, M., 351[336] (339)
Sahara, M., 89[533] (60)
Said, F. F., 186[36b] (162, 174)
Saida, Y., 458[48] (451)
Saijo, K., 348[127] (323)
Sain, B., 190[206] (183)
Saindane, M., 91[639] (71), 567[563] (551)
Sainsbury, M., 90[609] (67)
Saito, I., 352[373] (341)
Saito, K., 189[167c] (178)
Saito, S., 273[690] (245), 310[63] (288, 296), 347[82] (319), 564[354] (539), 564[355] (539)
Saito, Y., 434[237b] (426), 434[237d] (426)
Sakaguchi, M., 187[107b] (169)
Sakai, N., 270[430] (224)
Sakai, S., 564[323] (538, 539, 545)
Sakai, T., 385[159] (368)
Sakaki, K., 561[138] (520)
Sakakibara, M., 352[355] (340), 563[259] (530), 565[375] (540)
Sakakibara, Y., 432[128a] (409)
Sakakura, T., 353[431] (343)
Sakamoto, M., 190[209] (184)
Sakamoto, T., 385[213] (371), 386[274] (375)
Sakata, J., 567[512] (546)
Sakata, S., 187[110] (170)
Sakuraba, H., 275[813] (258)
Sakurai, H., 185[18] (160), 266[195] (204), 321, 346[6] (314), 347[98] (321), 347[107] (321), 347[110] (322), 348[155] (326)
Sakurai, M., 156[130] (146), 276[851] (263)
Sakurai, Y., 188[122c] (171, 172), 275[816] (259), 275[832] (260)
Salamtin, B. A., 597[41] (572)
Salaun, J., 273[689] (245)
Saleh, G., 568[644] (557)
Salgado-Zamora, H., 458[39] (450)
Salmón, M., 82[44] (3)
Salunkhe, A. M., 267[210] (205), 267[225] (206, 218), 269[375] (218), 269[414] (223, 240)
Saluzzo, C., 563[267] (530)
Salvador, J. M., 156[119] (144)
Salvadori, P., 155[103] (141)
Salzmann, T. N., 568[631] (556)
Samadi, M., 601[277] (595)
Samate, D., 127[166] (117)
Sammakia, T., 386[260] (373)
Sampson, P., 83[119] (5)
Samuel, O., 90[572] (64), 90[594] (65)
Sánchez, A., 458[12] (444), 458[18] (446), 458[22] (447)
Sanchez, R., 310[60] (286, 305)
Sanda, F., 185[16b] (160)
Sander, T., 237, 272[559] (235, 237), 272[571] (237)
Sander, W. W., 564[307] (537)
Sandhu, J. S., 190[206] (183)
Sandosham, J., 92[719] (81)
Sandoval, S., 270[423] (224)
Sandström, A., 567[538] (548), 567[540] (548)
SanFilippo, L. J., 567[511] (546)
Sanner, M. A., 85[231] (17)
Sano, H., 352[365] (340), 383[37] (359, 374)
Santa, L. E., 187[108a] (169, 170)

Santagostino, M., 385[199] (371)
Santelli, M., 156[126] (145), 347[85] (319)
Santelli-Rouvier, C., 156[126] (145), 347[85] (319)
Santi, C., 565[389] (540)
Santiago, B., 265[114] (200), 271[534] (234), 271[535] (234), 385[190] (371)
Santiesteban, H., 563[281] (532)
Sanyal, B. J., 434[229b] (425)
Saraf, S. D., 432[134a] (410)
Sarandeses, L. A., 189[172] (178)
Sargent, M. V., 89[493] (53), 91[626] (71), 155[45] (133), 155[53] (133), 155[54] (133), 155[55] (133)
Sarkar, A. K., 86[275] (24), 274[736] (251), 347[65] (317), 347[73] (317, 320), 348[168] (327)
Sarkar, T. K., 346[1] (314), 346[58] (317), 383[6] (357)
Sarma, J. C., 385[160] (368), 601[274] (595)
Sarpeshkar, A. M., 88[454] (48)
Sarshar, S., 276[854] (263)
Sartor, D., 276[843] (262), 276[844] (262)
Sasaki, D. Y., 384[81] (362)
Sasaki, H., 271[528] (233, 234), 346[26] (315)
Sasaki, K., 311[98] (304), 599[135] (576), 601[281] (596)
Sasaki, M., 189[169] (178), 565[403] (541, 557)
Sasaki, N., 270[446] (227), 347[97] (321)
Sasakura, K., 273[634] (240)
Sassaman, M. B., 353[457] (345)
Sastry, K. A. R., 266[164] (203), 269[349] (217), 269[350] (217), 269[380] (219), 269[387] (219)
Sastry, U., 266[164] (203)
Satina, T. Ya., 435[252] (429)
Sato, F., 156[139] (147), 156[140] (147), 156[141] (147), 188[139a] (174), 188[139b] (174), 347[93] (320), 348[161] (326), 385[183] (370)
Satô, H., 187[113] (171)
Sato, K., 155[57] (134), 347[110] (322), 562[208] (526)
Sato, M., 272[565] (236), 458[34] (450), 458[35] (450)
Sato, N., 346[57] (317)
Sato, R., 266[195] (204)
Sato, S., 353[446] (344), 565[419] (542), 567[571] (552)
Sato, T., 187[101d] (169), 273[636] (240, 257), 274[714] (247), 275[806] (258, 261), 350[260] (335), 383[9] (357), 384[73] (361), 384[124] (364, 378), 384[128] (365), 384[129] (365), 384[131] (365), 384[132] (366), 385[151] (367), 385[172] (369), 387[306] (378), 387[307] (378, 380), 387[315] (379), 387[328] (380), 387[329] (380), 387[339] (381)
Sato, Y., 385[147] (367), 385[150] (367)
Satoh, M., 271[487] (232), 271[496] (232), 271[528] (233, 234), 272[563] (236), 346[26] (315)
Satoh, N., 271[495] (232)
Satoh, S., 87[348] (30)
Satoh, T., 87[364] (33)
Satoh, Y., 266[146] (201), 271[500] (232), 272[612] (240)

Sattsangi, P. D., 265[108] (200, 236, 237), 265[110] (200), 348[171] (328)
Satyanarayana, N., 265[68] (196)
Sauer, G., 383[47] (361)
Saugier, R. K., 87[351] (30)
Sauve, D. M., 125[9] (94)
Sauvêtre, R., 186[34d] (161, 163), 186[48a] (163), 186[48b] (163), 186[48c] (163), 186[48d] (163), 187[87] (167)
Savard, S., 83[98] (5)
Savchenko, I. A., 432[125] (408)
Savignac, Ph., 86[311] (26), 387[326] (380)
Savoca, A. C., 268[339] (215)
Savoia, D., 185[23a] (160), 185[23b] (160)
Sawada, H., 189[160a] (176), 270[459] (228)
Sawada, S., 189[175] (178)
Sawamura, M., 90[593] (65)
Sawicki, R. A., 434[211] (422)
Sawyer, J. S., 85[254] (21, 22), 349[209] (330)
Saxena, A. K., 264[10] (192)
Sayama, S., 512[88] (501)
Sayapina, L. M., 597[47] (572)
Sayer, B. G., 459[66] (455)
Sayer, T. S. B., 83[84] (4)
Sayo, N., 563[253] (530, 538, 539)
Sazonova, V. A., 90[575] (64)
Scarborough, R. M., Jr., 564[358] (539)
Schaad, L. J., 126[116] (109)
Schaaf, S., 433[197b] (420)
Schade, C., 125[43] (95, 112), 125[44] (95, 112), 126[131] (112)
Schaeffer, R., 269[403] (221)
Schäfer, J., 189[171b] (178)
Schaub, B., 126[80] (101)
Schaumann, E., 86[319] (27), 272[601] (240)
Schaumberg, G. D., 268[302] (212)
Scheeren, H. W., 276[856] (263)
Scheffold, R., 125[11] (94, 95, 97, 100)
Schelkun, R. M., 186[74a] (166)
Scheller, M. E., 86[286] (24)
Schempf, R., 484[27] (467)
Schenk, K., 567[550] (548)
Schenkelaars, E. M. C., 155[55] (133)
Scherer, O., 310[58] (285, 306)
Schilf, W., 352[375] (341)
Schilling, W., 566[480] (545)
Schimperna, G., 350[242] (333)
Schinzer, D., 321, 346[8] (314), 347[100] (321), 348[151] (325), 350[267] (335)
Schipor, I., 188[137d] (174, 177), 188[143a] (175)
Schlapbach, A., 272[560] (236), 275[771] (255)
Schlessinger, R. H., 86[303] (26)
Schleyer, P. v. R., 125[43] (95, 112), 125[44] (95, 112), 126[131] (112), 127[137] (112), 127[155] (116)
Schlingloff, G., 275[836] (260)
Schlögl, K., 90[586] (65)
Schlosser, M., 4, 55, 81[3] (2), 82[14] (2), 82[16] (2), 82[57] (4), 83[64] (4), 83[65] (4), 83[73] (4, 54), 84[136] (6), 87[399] (42), 88[440] (47), 88[453] (48, 54), 88[459] (49), 89[502] (54), 89[508] (55), 93–6, 124[2] (93, 95), 124[4] (93), 124[7] (94, 119),
125[10] (94, 100, 105), 125[11] (94, 95, 97, 100), 125[12] (94, 95, 99, 100, 104, 105, 108), 125[13] (94, 95), 125[33] (94, 95, 108, 109, 112), 125[38] (94, 98), 125[39] (94, 100), 125[40] (94, 98, 99, 117), 125[41] (94, 103), 125[49] (95, 108, 112, 114), 125[50] (95, 112–14), 125[51] (95, 96, 120), 125[53] (96, 120), 125[54] (96, 113), 125[55] (97), 125[56] (103, 116), 125[57] (101), 125[58] 125[59], 125[60] (97), 125[61], 125[62] (101), 125[63] (101), 125[65] (120), 125[66], 125[69] (97), 125[72] (98, 101), 126[74] (100, 104), 126[79] (101), 126[80] (101), 126[83] (101), 126[90] (103, 105), 126[91] (103, 105), 126[96] (105, 106), 126[105] (108), 126[106] (108), 126[112] (109), 126[122] (109), 126[127] (111, 112), 127[135] (112, 118), 127[139] (114), 127[152] (116), 127[154] (116), 127[156] (116), 127[161] (117), 127[162] (117), 127[172] (118), 127[176] (119), 127[179] (119), 127[183] (120), 127[186] (120), 127[194] (121, 122), 128[197] (121, 122), 128[202] (122, 123), 128[203] (123), 128[204] (123)
Schlüter, A.-D., 84[142] (6)
Schmalz, H.-G., 90[567] (64), 90[568] (64)
Schmelzer, E. R., 268[315] (213)
Schmid, G. H., 561[164] (522, 540, 541)
Schmid, W., 156[136] (147)
Schmidlin, T., 347[65] (317)
Schmidt, B., 186[50b] (163, 171, 172), 186[50c] (163, 171, 172)
Schmidt, M., 562[183] (522, 528)
Schmidt, R. R., 127[181] (119)
Schmidt, T., 383[13] (358, 367)
Schmidt, U., 187[98b] (169)
Schmiesing, R. J., 568[632] (556)
Schmitt, A., 347[84] (319)
Schmitt, R. J., 348[131] (324)
Schmittling, E. A., 349[209] (330)
Schmitz, C., 348[146] (325)
Schmitz, E., 433[182] (418)
Schmolka, S., 88[462] (49)
Schneebeli, J., 155[88] (137)
Schneider, A., 348[139] (325)
Schneider, G., 83[88] (5)
Schneider, J., 309[15] (279), 309[18] (279, 299)
Schneider, K., 309[18] (279, 299)
Schneider, P., 125[54] (96, 113), 125[65] (120), 127[179] (119), 155[36b] (132)
Schneider, U., 353[400] (342)
Schöllkopf, U., 93, 124[1] (93), 127[157] (116), 127[160] (117), 430[24] (392)
Scholz, M., 459[69] (455)
Scholz, T. H., 91[656]
Schorigin, P., 112, 126[132] (112)
Schow, S. R., 88[416] (42)
Schreiber, S. L., 88[414] (42), 187[94] (169, 171, 172, 183), 349[199] (329)
Schroth, G., 459[81] (456), 564[321] (538, 542)
Schrott, U., 274[738] (251)

Schubert, S., 567[550] (548)
Schubert, U., 348[149] (325)
Schug, K., 599[161c] (579)
Schulte, K. E., 189[157a] (176)
Schultz, A. G., 92[684] (72), 144, 156[125] (144), 347[112] (322)
Schulz, G., 155[97] (140)
Schulz, P., 598[89] (574)
Schumacher, D. P., 82[47] (3)
Schumann, H., 458[56] (453), 458[57] (453), 562[183] (522, 528)
Schumann, I., 271[522] (232)
Schumann, R. C., 273[628] (240), 275[788] (257)
Schümann, U., 125[70] (97)
Schuster, G. B., 190[207b] (183)
Schwark, J.-R., 127[143] (114), 386[265] (374)
Schwartz, C. E., 88[409] (42, 54), 89[500] (54), 89[501] (54)
Schwartz, I. L., 568[620] (555)
Schwartz, M. A., 155[62] (134)
Schwartz, S. J., 270[425] (224)
Schwarz, M., 348[170] (328)
Schweiker, K., 350[240] (333)
Schwerdtfeger, J., 85[271] (23)
Schwering, J. E., 187[103b] (169)
Schwickardi, M., 154[2] (130)
Schwinden, M. D., 88[460] (49)
Scialdone, M. A., 387[330] (380, 381)
Scolastico, C., 273[640] (241, 264), 273[643] (241), 386[262] (373)
Scott, F. L., 434[201] (421)
Scott, I. L., 348[141] (325)
Scott, L. T., 348[138] (324)
Scott, M. E., 384[107] (364)
Scott, M. L., 558[8] (516)
Scott, R., 348[142] (325)
Scott, W. J., 385[206] (371)
Scott, W. L., 268[286] (211)
Scouten, C. G., 265[111] (200)
Screttas, C. G., 83[74] (4), 83[75] (4, 7), 157[175] (152)
Scripko, J., 434[212] (422, 424)
Scudder, P. H., 564[346] (539, 555, 557)
Sebban, M., 185[15b] (160)
Seconi, G., 126[129] (111), 353[443] (344), 384[113] (364)
Sedova, N. N., 90[575] (64)
Sedrati, M., 459[63] (454)
Seebach, D., 16, 83[69] (4), 83[76] (4), 83[89] (5), 84[197] (12), 85[227] (17, 19), 85[228] (17), 85[244] (20), 87[335] (28), 145, 151, 156[128] (145), 156[157] (151), 186[50a] (163, 171, 172), 186[50b] (163, 171, 172), 186[50c] (163, 171, 172), 187[109b] (170), 188[128b] (171, 172), 188[128c] (171, 172), 189[195] (182), 273[635] (240), 516, 559[51] (516), 559[52] (516), 565[371] (540, 550), 598[67] (573)
Seefeld, M. A., 274[757] (252)
Seerden, J. P. G., 276[856] (263)
Seetz, J. W. F. L., 185[21f] (160)
Segi, M., 352[392] (341), 385[167] (369), 561[148] (520, 546), 566[504] (546), 566[505] (546), 566[506] (546)
Sehata, M., 350[275] (336)
Seibel, W. L., 270[458] (228)
Seidel, W., 560[100] (519, 527)
Seidler, M. D., 562[229] (527)
Seitz, D. E., 86[295] (25)

Seitz, S. P., 567[522] (547, 551)
Sekigawa, S., 348[140] (325)
Sekiguchi, A., 348[155] (326)
Sekihachi, J., 349[192] (329)
Sekiya, A., 185[32e] (161)
Sekiya, K., 185[20a] (160, 165), 185[20d] (160, 165), 186[63] (165, 166)
Seko, N., 89[531] (59)
Seko, T., 565[432] (542)
Seliger, H. H., 352[372] (341)
Selim, M. R., 86[293] (25)
Selnick, H. G., 347[81] (319), 350[277] (336)
Seltzman, H. H., 351[341] (339)
Semmelhack, M. F., 89[539] (63), 90[543] (63), 90[548] (63), 186[47a] (163), 188[150] (175, 182)
Senanayake, C. B. W., 386[217] (371)
Sengupta, S., 86[314] (26, 27), 88[465] (49)
Seoane, P. R., 348[120] (323), 348[121] (323)
Sepulveda Arques, J., 91[625] (71)
Serci, A., 83[61] (4), 128[209] (124)
Serebkyakov, E. P., 271[494] (232)
Sergeeva, N. S., 433[190b] (419)
Serizawa, H., 271[500] (232), 272[612] (240)
Serpone, N., 459[66] (455)
Serratosa, F., 349[191] (329), 562[220] (527, 538, 556)
Servi, S., 187[101b] (169)
Set, L., 567[539] (548)
Sevrin, M., 564[318] (538, 553), 566[457] (544, 551), 566[471] (544, 554), 566[473] (544), 567[575] (553), 567[576] (553), 568[601] (554, 555)
Sexsmith, S. R., 310[49] (284, 290)
Sexton, A., 154[3] (130)
Seyden-Penne, J., 273[664] (244)
Seyferth, D., 90[573] (64), 432[119] (407), 432[129a] (409, 411), 432[129b] (409, 411), 432[130b] (410), 432[136a] (410), 432[136b] (410), 432[137a] (411), 432[137b] (411)
Sgarra, R., 155[59] (134), 155[69] (135)
Shah, S. K., 563[260] (530), 568[610] (554), 568[633] (556)
Shambayati, S., 349[199] (329)
Shankar, B. B., 92[683]
Shanmugam, P., 597[37] (572), 599[138] (577, 580), 599[165] (580), 599[190] (582), 599[194] (582), 599[195] (582)
Shao, Y., 185[26d] (161, 178)
Shapiro, H., 484[1a] (461)
Shapiro, J. R., 558[6] (516), 558[8] (516)
Shapiro, M. J., 273[684] (244), 273[685] (244)
Sharma, G. V. M., 383[59] (361)
Sharma, R., 156[167] (151)
Sharma, R. K., 458[24] (447)
Sharma, R. P., 385[160] (368)
Sharma, S., 384[99] (363), 384[119] (364)
Sharp, J. T., 90[601] (67)
Sharp, M. J., 88[420] (42), 266[166] (203), 266[167] (203, 204)
Sharpless, K. B., 347[91] (320), 347[93] (320), 555, 559[46] (516), 559[47] (516, 522, 535, 539, 541, 543, 552–5, 557), 561[137] (520), 562[179] (522), 563[238] (529), 563[256] (530, 539,

554, 555, 557), 564[300] (536), 564[326] (538), 565[401] (541, 555), 565[421] (542, 555, 556), 568[595] (554, 555), 568[621] (555–7), 600[241] (591)
Sharts, C. M., 564[309] (537)
Shaw, D., 349[216] (330)
Shawe, T. T., 85[230] (17)
Shea, K. J., 349[226] (331), 384[81] (362)
Shea, R. G., 569[648] (557)
Shechter, H., 87[339] (29), 352[398] (342)
Sheldrick, G. M., 155[41] (132)
Shen, T., 85[206] (14, 22)
Shen, Y. C., 511[36] (489), 511[37] (489), 511[38] (489), 511[39] (489), 511[44] (490), 511[53] (492), 511[55] (492), 511[56] (492), 512[67] (496), 512[72] (497), 512[77] (498)
Sheppard, A. C., 351[287] (336)
Sheppard, W. A., 564[309] (537)
Sher, P. M., 346[52] (317)
Sherbine, J. P., 84[134] (6)
Sherman, D., 274[749] (251)
Shertzer, H. G., 90[609] (67)
Sheu, B.-A., 186[71d] (166)
Shi, B. Z., 432[91] (403)
Shi, L. L., 512[64] (495), 512[65] (495), 512[66] (495, 498), 599[186] (582), 600[250] (591, 592), 600[253] (591), 600[255] (591), 600[256] (591, 592), 600[258] (592), 600[259] (592)
Shi, X., 305
Shiao, M.-J., 156[107] (141), 187[118a] (171), 187[118b] (171), 187[118c] (171), 352[390] (341)
Shibasaki, M., 90[563] (63), 350[240] (333), 353[402] (342), 384[101] (363, 364), 385[169] (369)
Shibata, A., 567[573] (552)
Shibata, I., 383[43] (360), 385[145] (367, 368), 385[154] (368), 385[155] (368), 385[157] (368)
Shibata, K., 352[367] (340), 484[3] (461)
Shibata, N., 384[143] (367)
Shibata, S., 275[817] (259)
Shibutani, T., 155[50] (133), 155[51] (133), 155[52] (133)
Shida, N., 83[126] (5)
Shieh, W. C., 271[525] (233)
Shih, N.-Y., 353[427] (343)
Shih, T. L., 275[782] (256)
Shih, Y., 352[399] (342)
Shiigi, J., 387[338] (381)
Shiina, I., 350[251] (334), 385[152] (368)
Shima, H., 352[342] (339)
Shima, K., 349[197] (329)
Shimada, J.-i., 430[30a] (393)
Shimada, K., 563[231] (528, 545)
Shimada, S., 273[694] (245), 273[695] (245), 387[323] (380)
Shimanuki, E., 348[127] (323)
Shimizu, A., 511[49] (491, 501)
Shimizu, H., 269[382] (219), 272[615] (240), 352[355] (340), 566[469] (544), 566[477] (545), 599[146] (577), 600[247] (591)
Shimizu, K., 275[832] (260)
Shimizu, M., 346[46] (316), 349[215] (330), 565[387] (540), 566[451] (543), 568[594] (554, 557)
Shimizu, S., 276[846] (262)
Shimizu, T., 348[140] (325), 383[45] (360), 568[600] (554)

Shimoji, K., 127[157] (116)
Shimshock, S. J., 349[206] (330)
Shin, D.-S., 385[204] (371), 387[289] (376)
Shine, H. J., 560[75] (518, 543)
Shine, R. J., 561[131] (520, 552), 561[149] (520, 528)
Shing, T.-L., 187[118b] (171)
Shinkai, I., 187[107c] (169), 275[819] (259)
Shinoda, J., 458[46] (451)
Shinoda, S., 434[237b] (426)
Shinoda, T., 485[45] (472, 473)
Shioiri, T., 350[262] (335), 352[382] (341), 352[383] (341), 352[385] (341), 352[386] (341)
Shiono, M., 270[417] (223)
Shirahama, H., 564[302] (536)
Shirasaka, T., 350[278] (336)
Shirley, D. A., 128[198] (122)
Shiro, M., 90[564] (63, 64), 384[141] (367)
Shobana, N., 512[99] (502), 512[100] (502), 599[190] (582), 599[194] (582), 599[195] (582)
Shoda, H., 350[239] (333)
Shono, T., 189[167a] (178), 189[169] (178), 189[170] (178)
Shook, C. A., 84[200] (13)
Short, R. P., 274[744] (251)
Shoup, T. M., 266[131] (201), 266[132] (201), 266[135] (201), 268[340] (216), 268[341] (216), 269[342] (216)
Shrift, A., 558[7] (516), 558[8] (516)
Shubert, D. C., 187[96] (169), 323, 348[123] (323), 458[31] (449)
Shundo, R., 265[72] (196)
Shvily, R., 351[336] (339)
Siahaan, T. J., 126[114] (109), 126[117] (109), 126[120] (109)
Sibi, M. P., 88[466] (49), 89[472] (50), 89[515] (56), 156[167] (151), 265[57] (195)
Sibille, S., 186[36c] (162, 174), 186[36d] (162, 174), 186[36e] (162, 174), 186[36f] (162, 174)
Sicuri, A. R., 458[47] (451)
Sidduri, A., 186[76] (166), 189[182a] (179), 189[197b] (182), 271[541] (235)
Sidén, J., 561[133] (520)
Siebert, W., 264[12] (192), 265[77] (196, 198), 266[150] (202)
Sieburth, S. McN., 350[228] (331)
Siegel, J. S., 88[448] (48)
Siegeland, A., 83[62] (4)
Sielecki, T. H., 85[238] (18)
Sieloff, R. F., 82[55] (3, 6)
Signer, M., 88[438] (46)
Sih, J. C., 431[52b] (398)
Sik, V., 459[64] (455)
Sikorski, J. A., 265[84] (198, 200), 265[85] (198, 200), 270[462] (229), 273[676] (244)
Silbermann, J., 384[104] (363)
Silveira, A., Jr., 270[416] (223)
Silveira, C. C., 566[468] (544)
Silverman, I. R., 350[244] (333)
Silverman, R. B., 349[179] (328), 383[44] (360), 561[145] (520, 545)
Silverstein, R. M., 126[104] (107)
Simchen, G., 155[64] (134), 350[240] (333), 351[305] (337), 352[356] (340)
Simig, G., 88[440] (47), 128[204] (123)
Simmons, H. E., 178, 185[3] (159, 178), 309[10] (279), 430[13] (390)
Simon, H., 127[146] (115)
Simoneau, B., 273[681] (244)
Simonelli, F., 384[108] (364), 598[80] (573)
Simoni, D., 346[41] (316)
Simonidesz, V., 565[381] (540, 554)
Simonsen, J. L., 126[81] (101)
Simpkins, N. S., 84[132] (6, 33), 84[133] (6), 85[233] (17), 90[569] (64), 350[237] (333), 384[82] (362), 564[360] (539, 547, 548), 565[430] (542)
Simpson, J. H., 348[129] (324)
Sinaÿ, P., 85[253] (21), 85[263] (22), 349[224] (330), 433[170b] (416), 433[171] (416), 565[378] (540, 548)
Sinegovskaya, L. M., 598[92] (574)
Singaram, B., 265[59] (195), 265[60] (195), 265[61] (195), 265[62] (195), 265[76] (196, 205, 247), 267[213] (205), 267[214] (205, 206), 267[216] (205, 206), 267[217] (205, 206), 267[220] (205), 267[221] (205), 267[223] (206), 267[225] (206, 218), 268[276] (210), 268[277] (210, 211), 268[278] (210), 269[351] (217), 269[356] (217), 270[420] (224), 271[543] (235), 272[623] (240), 272[624] (240), 272[625] (240), 273[665] (244), 273[666] (244), 273[667] (244), 274[706] (247), 274[749] (251)
Singaram, S., 265[88] (198)
Singer, M. I. C., 511[40] (489)
Singer, R. D., 185[4] (159, 160), 185[25] (160), 211, 268[280] (210, 211), 354[463] (345), 384[109] (364)
Singer, S. P., 559[47] (516, 522, 535, 539, 541, 543, 552–5, 557), 600[241] (591)
Singh, A., 601[273] (594)
Singh, A. K., 567[551] (549)
Singh, J., 265[67] (196), 564[304] (536, 537)
Singh, K. N., 85[251] (21)
Singh, M., 290, 310[82] (299)
Singh, O. V., 459[82] (457)
Singh, P., 85[251] (21), 458[52] (452)
Singh, P. K., 432[100] (404)
Singh, R., 387[296] (377)
Singh, S. M., 265[104] (200), 267[240] (207), 267[250] (208)
Singh, S. P., 383[40] (360), 565[391] (541, 547, 548)
Singh, V. K., 275[803] (258–60)
Singleton, D. A., 265[112] (200), 266[193] (204, 211), 266[194] (204), 348[145] (325), 350[266] (335), 383[55] (361), 385[189] (370)
Sinhababu, A. K., 89[498] (54)
Sinisterra-Gago, J. V., 89[516] (56), 185[27b] (161)
Sinou, D., 352[378] (341)
Sipio, W. J., 565[435] (542, 547, 551), 567[522] (547, 551)
Sire, B. Q., 601[277] (595)
Siriwardane, U., 89[526] (57)
Sisko, J., 155[61] (134)
Sita, L. R., 272[597] (240, 241, 256)

Sjöberg, B., 563[255] (530)
Skarnulis, A. J., 154[3] (130)
Skelton, B. W., 155[55] (133)
Skerlj, R., 564[333] (538)
Skinner, C. E. D., 90[601] (67)
Slater, M. J., 348[141] (325)
Slawin, A. M. Z., 90[552] (63), 92[699] (74)
Slayden, S. W., 270[463] (229), 270[464] (229)
Sledeski, A. W., 351[309] (337)
Sleegers, A., 385[194] (371)
Sloan, C. P., 92[715] (79)
Slocum, D. W., 83[62] (4), 87[400] (42), 90[571] (64)
Slocum, M. G., 83[62] (4)
Slougui, N., 189[181] (179)
Smadja, W., 350[232] (332)
Smart, J. C., 90[576] (65)
Smith, A. B., III, 88[416] (42), 89[530] (59), 564[358] (539)
Smith, B. M., 267[211] (205)
Smith, C. R. F., 597[23] (572)
Smith, D. L., 598[63] (572, 573)
Smith, D. T., 386[264] (373)
Smith, E. M., 91[617] (70)
Smith, G. A., 85[246] (20)
Smith, H. E., 430[31b] (393)
Smith, I. C., 264[36] (192)
Smith, J. D., 186[57a] (165), 186[57b] (165)
Smith, J. N., 565[402] (541)
Smith, K., 126[89] (103), 126[92] (103), 128[208] (124), 264[27] (192, 198, 199, 201, 202, 208, 214–16, 220, 222, 223, 225, 227–30, 235, 238, 241, 244, 247), 264[39] (193, 201, 213, 247), 265[75] (196), 267[258] (208), 267[260] (208), 267[261] (208, 234, 235), 267[268] (208, 234), 268[320] (214–16, 218), 270[418] (223), 270[421] (224), 270[422] (224), 270[438] (225), 270[478] (230), 271[546] (235), 271[547] (235), 273[678] (244), 353[411] (342), 386[277] (375)
Smith, K. M., 430[26] (392)
Smith, L. I., 431[71d] (401)
Smith, M. R., 598[75] (573)
Smith, P. J., 383[1] (356)
Smith, R. G., 430[31b] (393)
Smith, R. N. M., 458[38] (450)
Smith, R. S., 386[260] (373)
Smith, S., 598[118] (575)
Smith, S. A. C., 88[430] (46, 60)
Smith, S. G., 85[224] (16)
Smith, S. J., 432[127] (408)
Smithers, R., 346[38] (315, 317)
Snatzke, G., 349[222] (330), 559[72] (518), 559[73] (518)
Snider, B. B., 347[115] (322), 351[321] (338), 387[335] (381)
Snieckus, V., 81[4] (2), 82[11] (2), 82[17] (2), 82[30] (2, 33, 36), 82[35] (2, 4, 42, 43, 46, 48, 49), 82[38] (2, 4, 42, 70, 75, 76, 79, 81), 83[59] (4), 86[304] (26), 87[394] (42), 87[396] (42, 43), 87[397] (42), 88[406] (42, 53), 88[408] (42, 53), 88[413] (42), 88[420] (42), 88[421] (42), 88[465] (49), 88[466] (49), 89[472] (50), 89[474] (51), 89[476] (51), 89[515] (56), 89[519] (56), 89[521] (56),

89[533] (60), 89[534] (60), 91[665], 91[670], 91[671], 92[692] (72), 92[704] (74), 92[709] (78), 92[710] (78), 92[711] (78), 125[14] (94, 95), 127[192] (121), 204, 266[166] (203), 266[167] (203, 204), 348[152] (326), 563[245] (529, 541)
Snow, J. T., 311[114] (308)
Snyder, C., 268[328] (215)
Snyder, L., 88[430] (46, 60)
Soai, K., 156[161] (151), 185[10a] (160, 162, 171, 172), 187[110] (170), 187[116a] (171), 188[123a] (171), 188[123b] (171), 188[124] (171), 188[125] (171), 188[130a] (171), 188[130b] (171), 188[133] (172), 188[134a] (172), 188[134b] (172), 188[141a] (174), 188[141b] (174), 188[141c] (174), 188[142] (174)
Soborovskii, L. Z., 431[72] (401)
Soda, K., 558[9] (516)
Soddy, T. S., 127[193] (121, 122)
Sodeoka, M., 90[563] (63), 350[240] (333)
Soderquist, J. A., 264[45] (194), 265[64] (196, 198), 265[113] (200), 265[114] (200), 265[122] (200, 215), 268[333] (215), 268[335] (215), 270[424] (224), 271[530] (233), 271[533] (234), 271[534] (234), 271[535] (234), 346[27] (315), 347[92] (320), 349[178] (328), 385[190] (371)
Sofia, M. J., 349[220] (330)
Sokolov, V. I., 430[31a] (393), 433[180a] (417)
Solas, D. R., 353[419] (343)
Soll, R. M., 89[480] (51)
Solladié, G., 33, 82[31] (2, 33), 87[359] (33), 87[360] (33), 87[361] (33), 87[362] (33)
Solladié-Cavallo, A., 90[559] (63), 187[105] (169)
Solleiro, E., 459[74] (456)
Solomon, J. S., 347[103] (321)
Solomon, M. S., 92[708] (78), 566[462] (544, 551)
Solooki, D., 92[674]
Solow, M., 274[749] (251)
Soloway, A. H., 264[16] (192), 264[34] (192), 264[35] (192)
Somada, Y., 458[46] (451)
Somayaji, V., 265[66] (196), 265[102] (199), 267[219] (205), 269[385] (219), 269[387] (219)
Somayajula, K. V., 350[232] (332)
Somei, M., 458[40] (451), 458[42] (451), 458[43] (451), 458[44] (451), 458[45] (451), 458[46] (451), 458[48] (451), 458[49] (451), 458[50] (451)
Sommerfeld, P., 85[270] (23)
Sonada, N., 432[121] (407)
Song, H., 432[113b] (407)
Song, K. H., 432[123] (408)
Song, Z., 85[224] (16)
Sönke, H., 309[14] (279)
Sono, S., 269[359] (217), 269[364] (217)
Sonoda, A., 270[444] (227)
Sonoda, N., 85[207] (14), 85[209] (15), 350[272] (335), 353[441] (344), 353[448] (344), 383[49] (361), 386[245] (373), 387[305] (378), 387[309] (378), 560[110] (519, 520, 529), 564[345] (539), 564[351] (539), 566[443] (543), 566[444]

(543), 566[494] (545, 547, 548), 568[604] (554), 597[20] (572, 576, 578), 597[21] (572, 577), 597[22] (572, 577), 598[132] (576–8), 599[171] (580), 600[206] (583), 600[210] (584), 600[211] (584), 600[263] (593), 600[266] (593), 601[278] (595), 601[282] (596)
Sonoda, T., 269[396] (220)
Sonogashira, K., 187[79d] (166, 167)
Sood, A., 264[17] (192)
Sooriyakumaran, R., 90[597] (66)
Sordo, J., 458[12] (444), 458[18] (446), 458[19] (446)
Sörensen, H., 187[89b] (167, 176)
Sorgi, K. L., 568[632] (556)
Sosnovsky, G., 431[57b] (399)
Sotowicz, A. J., 433[175b] (416)
Souchet, M., 89[527] (59), 89[528] (59)
Soufiaoui, M., 268[304] (213)
Soundararajan, R., 265[87] (198, 200), 274[729] (250), 432[142a] (412)
Southern, J. M., 458[8] (438, 440), 458[9] (440, 441)
Spanevello, R. A., 271[491] (232)
Spanu, P., 350[246] (333), 351[308] (337)
Sparks, M. A., 347[103] (321), 347[104] (321), 347[105] (321), 347[106] (321)
Spavold, Z., 84[133] (6)
Spawn, T. D., 186[34b] (161, 163)
Speckamp, W. N., 347[76] (319), 347[79] (319), 383[66] (361), 384[92] (363)
Spencer, H. K., 597[31] (572)
Spencer, J. T., 567[516] (547)
Spencer, T., 458[36] (450)
Spéziale, V., 264[53] (195), 264[54] (195), 430[40] (396, 416, 421), 433[170a] (416, 422)
Spielvogel, B. F., 264[17] (192)
Spierenburg, J., 484[21] (466)
Spink, W. C., 458[54] (453), 458[58] (453), 459[59] (453)
Spirikhin, L. V., 310[53] (285)
Sponholtz, W. R., III, 269[388] (219)
Spreafico, F., 187[101b] (169)
Springer, J. P., 156[125] (144)
Springer, R., 189[171b] (178)
Spunta, G., 353[421] (343), 353[422] (343)
Srebnik, M., 186[52b] (164, 172), 188[131] (171), 265[74] (196), 265[79] (196, 206), 266[155] (202), 266[156] (202), 266[171] (203), 266[172] (203), 267[223] (206), 267[224] (206), 269[351] (217), 269[389] (220), 275[801] (258, 259, 262), 276[870] (264), 351[281] (336)
Sreekumar, C., 22, 85[262] (22, 24), 386[282] (376)
Srinivasan, C. V., 92[687] (72)
Srinivasan, V., 558[18] (516)
Srivastava, P. C., 266[164] (203)
Staab, A. J., 349[226] (331)
Stacino, J.-P., 87[376] (36)
Stadtman, T. C., 559[66] (517, 529, 530, 538, 558)
Stadtmüller, H., 186[42] (162, 175), 186[43] (162, 175)
Stähle, M., 84[136] (6), 125[50] (95, 112–14), 125[58] (120), 125[61] (120), 125[65] (120), 125[69] (97), 126[90] (103, 105), 126[127] (111, 112), 127[154]

(116)
Stańczyk, W., 353[453] (344)
Stanforth, S. P., 512[119] (505, 507, 509)
Stang, P. J., 385[188] (370), 386[280] (375), 565[382] (540)
Stangeland, E. L., 273[666] (244)
Starrett, J. E., Jr., 351[285] (336)
Stasi, F., 155[69] (135)
Staub, H., 430[3] (390, 409)
Staunton, J., 89[525] (57)
Stavely, H. E., 434[221b] (423)
Steele, B. R., 83[75] (4, 7), 157[175] (152)
Steele, R. B., 310[88] (302)
Stefanelli, S., 87[367] (36), 87[368] (36)
Stefani, H. A., 598[108] (574), 598[109] (574, 589), 598[124] (575)
Stefinovic, M., 384[76] (361)
Steglich, W., 89[523] (57)
Steigel, A., 266[168] (203), 349[213] (330)
Stein, G., 559[45] (516, 519)
Steinkopf, W., 430[35] (395)
Steinman, D. H., 87[377] (36)
Steinseifer, F., 512[105] (503), 512[106] (503)
Stengel, P. J., 384[108] (364)
Stenson, P. H., 512[128] (507, 508)
Stepanian, M., 274[711] (247)
Stephan, M., 274[703] (246), 274[704] (246), 274[705] (246)
Stephenson, G. R., 89[541] (63), 186[72] (166)
Stepovik, L. P., 311[109] (307)
Stercho, Y. P., 273[668] (244)
Sterlin, S. R., 434[747] (428)
Sternhell, S., 484[13] (465, 466), 484[23] (466, 467, 481), 484[31] (467, 481), 484[38] (469), 485[63] (481, 482), 485[64] (481)
Stevens, R. V., 434[226] (424)
Stevens, W. R., 268[315] (213)
Stevenson, P., 348[142] (325)
Stewart, P., 268[338] (215)
Štíbr, B., 264[6] (192, 211)
Stichter, H., 187[97a] (169)
Still, I. W. J., 568[612] (554)
Still, W. C., 22, 85[262] (22, 24), 87[354] (31), 146, 156[134] (146), 264[49] (195), 375, 384[93] (363, 379), 384[94] (363, 364, 375, 376), 384[108] (364), 386[282] (376), 387[313] (379), 387[314] (379), 430[30e] (393)
Stille, J. K., 348[129] (324), 371, 385[192] (371), 385[210] (371)
Stinn, D. E., 432[93] (404), 432[94] (404)
Stirling, C., 86[329] (28, 53), 86[330] (28)
Stirling, C. J. M., 91[645] (72)
Stobie, A., 512[112] (504)
Stock, L. M., 431[71f] (401)
Stoermer, M. J., 385[191] (371), 484[15] (465, 468, 481), 484[16] (465, 468, 481), 484[37] (1, 468, 471, 477)
Stoll, A. T., 185[8d] (159, 166)
Stoller, A., 92[692] (72)
Stolz-Dunn, S. K., 432[104b] (405)
Stone, C., 266[199] (204), 267[200] (204), 565[395] (541)
Stoner, E. J., 565[406] (541, 557)
Stoodley, R. J., 351[294] (337)
Stork, G., 85[212] (15), 87[340] (29), 330, 332, 346[52] (317), 349[220] (330),

349²²³ (330), 349²²⁴ (330), 350²²⁹ (332), 432¹³⁵ (410), 433¹⁵³ᵃ (413)
Storr, R. C., 91⁶⁵⁴
Stout, T. J., 88⁴¹⁴ (42)
Stoyanovich, F. M., 86³²⁸ (28)
Strähle, J., 459⁷⁵ (456)
Straub, J. A., 274⁷⁴⁸ (251), 274⁷⁶⁹ (253)
Strauss, W., 458⁷ (438)
Street, S. D. A., 186⁴⁶ (163), 565³⁹³ (541)
Strekowski, L., 155⁹³ (139, 140)
Stringer, O. D., 568⁵⁸⁴ (554), 568⁵⁸⁵ (554), 568⁵⁸⁶ (554, 557)
Strong, W. O., 125²³ (94, 112)
Stroud, E. D., 156¹¹⁶ (143)
Strugnell, S., 566⁴⁵⁰ (543, 544)
Strunk, S., 82⁵⁷ (4), 126¹⁰⁶ (108), 127¹⁶¹ (117)
Struzka, V., 85²⁴¹ (19)
Stucchi, E., 349¹⁸⁷ (329), 384¹¹⁵ (364)
Stucki, C., 155⁸⁸ (137)
Stüdemann, T., 186⁴² (162, 175)
Sturgeon, G. D., 559⁶¹ (517)
Stürmer, R., 84¹⁷³ (9), 266¹⁶³ (203, 253), 274⁷³⁷ (251, 253), 274⁷⁶⁶ (253), 275⁷⁷⁰ (255)
Su, B. M., 82¹³ (2)
Suarez, E., 563²⁸⁶ (533, 556, 557)
Suarez, J., 83¹²² (5)
Subba Rao, B. C., 264³⁷ᵃ (193), 264³⁷ᵇ (193), 268³²⁸ (215)
Subrahmanyam, C., 269³⁸⁵ (219), 270⁴⁴⁹ (227)
Subramanian, R., 89⁵⁰³ (54), 89⁵⁰⁴ (54), 383²⁰ (358)
Suda, A., 387³¹⁶ (379)
Suda, K., 349¹⁸³ (328)
Sudheendranath, C. S., 127¹⁴⁵ (115)
Suehiro, I., 350²⁷⁴ (335)
Suemitsu, R., 599¹³⁹ (577), 599¹⁴³ (577)
Suffert, J., 82⁴² (3), 187¹⁰⁵ (169)
Suga, H., 383¹⁷ (358)
Suga, S., 188¹²²ᵃ (171, 172), 188¹²⁹ (171)
Sugasawa, T., 273⁶³⁴ (240)
Sugawa, T., 266¹⁸³ (203)
Sugawara, T., 270⁴²⁸ (224), 431⁵²ᵃ (398)
Sugi, K. D., 433¹⁷³ (416)
Sugi, S.-i., 387³⁰⁰ (377)
Sugihara, T., 348¹⁴² (325)
Sugimoto, J., 273⁶⁵⁹ (243)
Sugimoto, T., 351³¹⁴ (337)
Sugimura, T., 189¹⁷⁷ᵍ (178), 430³⁰ᵇ (393)
Sugino, K., 87³⁵² (30)
Sugino, T., 600²⁶⁶ (593)
Suginome, H., 270⁴⁸⁴ (231, 232), 270⁴⁸⁶ (232), 485⁴⁶ (473), 597³⁵ (572)
Suginome, M., 353⁴³⁸ (343)
Sugita, N., 569⁶⁵⁰ (557), 569⁶⁵¹ (557), 598⁷⁴ (573, 588), 598⁹⁵ (574, 588), 598⁹⁶ (574, 588), 599¹⁴⁸ (577), 600²⁰⁷ (583), 600²³² (588), 600²⁴⁴ (591)
Sugita, T., 565⁴²⁹ (542, 554, 556, 557), 566⁴⁶⁶ (544, 554, 556)
Sugiura, M., 346³⁰ (315)

Suh, H. S., 349²²³ (330)
Suhr, Y., 349¹⁷⁶ (328)
Sukata, K., 352³⁹⁵ (341)
Sukenik, C. N., 433¹⁵⁰ᵃ (413), 433¹⁵⁰ᵇ (413), 433¹⁵⁶ (414)
Sullivan, J. A., 433¹⁸¹ (418)
Sum, V., 353⁴⁰⁸ (342)
Sumida, Y., 274⁶⁹⁸ (246)
Sun, J., 87³⁶⁶ (34)
Sun, M.-L., 186⁷¹ᵃ (166)
Sunay, U., 563²⁸¹ (532)
Sundberg, P.-O., 91⁶⁴² (71)
Sundberg, R. J., 88⁴⁵⁴ (48), 188¹¹⁹ᵇ (171)
Suñer, G. A., 88⁴⁴⁵ (48), 88⁴⁴⁹ (48)
Suranyi, E. L., 599¹³⁶ (576)
Surridge, J. H., 432¹¹⁵ᵃ (407)
Surtees, J. R. J., 458²¹ᵇ (446)
Surya Prakash, G. K., 348¹⁶⁹ (327), 351³³⁷ (339), 352³⁷⁴ (341), 352³⁷⁷ (341), 353⁴⁰⁴ (342), 353⁴⁵⁴ (344), 353⁴⁵⁷ (345)
Suseela, Y., 265¹¹⁸ (200), 273⁶⁷¹ (244)
Suslick, K. S., 185²⁷ᵃ (161)
Suszko, P. R., 90⁵⁷⁷ (65)
Sutowardoyo, K. I., 352³⁷⁸ (341)
Sutter, J. K., 433¹⁵⁰ᵇ (413), 433¹⁵⁶ (414)
Suttle, J. F., 597²³ (572)
Sutton, K. H., 85²⁵⁷ (22)
Suvannachut, K., 126¹¹⁵ (109), 126¹¹⁷ (109), 126¹²⁰ (109), 126¹²¹ (109)
Suzuki, A., 187⁸⁰ (166), 193, 231, 232, 264³⁸ (193, 231, 232), 265¹¹⁹ (200), 265¹²⁴ (200), 265¹²⁵ (200), 266¹⁴³ (201, 231), 266¹⁴⁴ (201, 214), 266¹⁴⁶ (201), 266¹⁴⁸ (202), 266¹⁵¹ (202), 269³⁵⁹ (217), 269³⁶⁴ (217), 269³⁸² (219), 269⁴⁰⁰ (220), 269⁴⁰¹ (220, 227, 231, 232), 270⁴¹⁹ (223), 270⁴²⁶ (224), 270⁴²⁹ (224), 270⁴⁴⁶ (227), 270⁴⁵⁷ (228), 270⁴⁶⁰ (229), 270⁴⁷² (229), 270⁴⁷⁹ (231), 270⁴⁸¹ (231, 232), 270⁴⁸² (231), 270⁴⁸⁴ (231, 232), 270⁴⁸⁵ (231, 232), 270⁴⁸⁶ (232), 271⁴⁸⁷ (232), 271⁴⁹³ (232), 271⁴⁹⁶ (232), 271⁴⁹⁷ (232), 271⁴⁹⁸ (232), 271⁴⁹⁹ (232), 271⁵⁰⁰ (232), 271⁵⁰¹ (232), 271⁵⁰² (232), 271⁵¹³ (232), 271⁵²³ (232), 271⁵²⁴ (232, 234), 271⁵²⁸ (233, 234), 271⁵²⁹ (233), 271⁵³¹ (234), 271⁵³² (234), 271⁵³⁶ (234), 271⁵³⁷ (234), 271⁵³⁸ (234), 272⁵⁵⁸ (235), 272⁵⁶³ (236), 272⁵⁶⁵ (236), 272⁶¹² (240), 272⁶¹⁵ (240), 346²⁶ (315)
Suzuki, H., 506, 512¹¹⁴ (504), 512¹¹⁸ (505, 506), 512¹²¹ (506), 512¹²² (506), 512¹²³ (506), 512¹²⁴ (506), 513¹⁴⁵ (511), 564³³⁵ (538), 597¹³ (571), 599¹³⁷ᵃ (576, 577), 599¹³⁷ᵇ, 599¹⁴⁴ (577), 599¹⁴⁵ (577), 599¹⁴⁶ (577), 599¹⁵¹ (578), 599¹⁵⁶ (578), 599¹⁶³ (580), 599¹⁷² (580), 599¹⁷³ (580),

599¹⁷⁷ (581), 599¹⁸⁰ (581), 599¹⁸¹ᵃ (581), 599¹⁸¹ᵇ (581), 599¹⁸³ (582), 599¹⁸⁴ (582), 599¹⁹¹ (582, 583), 600²⁰⁰ (583), 600²⁴⁷ (591), 600²⁴⁹ (591), 600²⁵⁴ (591)
Suzuki, I., 512⁹³ (501)
Suzuki, K., 88⁴⁶³ (49), 88⁴⁶⁴ (49), 187⁸⁶ (167), 386²⁷³ (374)
Suzuki, M., 83¹¹⁴ (5), 187⁹⁹ᵈ (169), 350²³² (332)
Suzuki, O., 155⁵⁷ (134)
Suzuki, S., 352³⁵⁹ (340), 353⁴⁴⁸ (344), 599¹⁶⁶ (580)
Suzuki, T., 155⁷⁶ (136), 564³¹⁶ (538, 555), 600²¹⁹ (585)
Suzuki, Y., 155⁷⁴ (136), 155⁷⁶ (136), 387³⁰⁰ (377)
Suzzi-Valli, G., 273⁶⁴⁷ (241)
Svanholm, U., 559⁷⁴ (518)
Sviridov, A. F., 566⁴⁸⁸ (545), 566⁴⁸⁹ (545)
Swain, P. A., 387²⁸⁹ (376)
Swanson, D. R., 84¹⁷⁵ (10)
Sweeney, J. B., 384⁸⁷ (362), 386²³⁵ (372, 373)
Swingle, N. M., 155⁴² (132, 151)
Syage, E. T., 91⁶⁴⁰ (71)
Sydnes, L. K., 432¹⁰² (405)
Syfrig, M. A., 85²²⁷ (17, 19), 156¹²⁸ (145)
Syper, L., 562²²⁵ (527), 563²⁵⁸ (530, 538)
Szczepanski, S. W., 86²⁸⁵ (24)
Szeimies, G., 84¹⁴⁰ (6)
Sznaidman, M. L., 383³² (359)
Szura, D. P., 387³²¹ (379)

Taba, K. M., 351³²⁰ (338)
Tabacchi, R., 350²⁷⁷ (336)
Taber, T. R., 272⁵⁹⁵ (240, 241), 272⁶²² (240)
Taboada, J. G., 459⁷¹ (456)
Tabuchi, T., 386²³⁴ (372)
Tachibana, K., 384¹²⁹ (365), 387³¹⁵ (379)
Tachibana, Y., 567⁵¹² (546)
Taddei, M., 126¹²⁹ (111), 127¹⁵⁰ (116), 127¹⁶⁷ (117), 347⁶² (317), 347⁷⁴ (318), 384¹⁰⁵ (363, 376)
Tae, J. S., 383⁶³ (361)
Tafesh, A. M., 349²²⁵ (330)
Taga, T., 599¹⁴⁸ (577)
Tagliavini, E., 185²³ᵇ (160), 265⁸⁹ (198), 268³³⁴ (215), 272⁶⁰⁴ (240, 241), 272⁶¹⁰ (240), 272⁶¹¹ (240), 273⁶⁴⁷ (241), 350²⁴⁵ (333)
Tagliavini, G., 189¹⁷³ (178), 385¹⁵⁶ (368), 386²³³ (372)
Taguchi, T., 383⁵⁸ (361)
Tahbaz, P., 82⁴⁷ (3)
Tai, A., 189¹⁷⁷ᵍ (178), 430³⁰ᵇ (393)
Taka, N., 387³²⁹ (380)
Takabe, K., 156¹⁴⁵ (148), 156¹⁷⁴ (152), 157¹⁷⁹ (153)
Takacs, J. M., 434²²⁸ (425), 434²²⁹ᵃ (425), 434²²⁹ᵇ (425)
Takagi, K., 189¹⁸³ᵇ (179), 432¹⁰⁶ (406), 432¹¹⁰ (406), 432¹¹²ᶜ (407), 432¹²⁸ᵃ (409)
Takagi, M., 511⁵¹ (491)
Takagi, Y., 86³¹⁶ (27), 347⁸⁶ (319),

386[271] (374)
Takagishi, S., 83[73] (4, 54), 88[453] (48, 54), 89[502] (54), 127[135] (112, 118), 127[194] (121, 122), 128[197] (121, 122), 128[202] (122, 123), 128[203] (123)
Takahara, J. P., 386[236] (372)
Takahashi, H., 127[157] (116), 136, 155[74] (136), 155[76] (136), 156[146] (148), 157[180] (153), 157[181] (153), 157[182] (153), 188[127b] (171, 172), 188[127c] (171, 172), 598[74] (573, 588), 598[95] (574, 588), 598[96] (574, 588), 600[226] (587)
Takahashi, K., 156[173] (152)
Takahashi, T., 561[148] (520, 546), 568[627] (555)
Takahori, T., 350[255] (334)
Takai, H., 433[157] (414)
Takai, K., 182, 190[199a] (182), 190[199b] (182), 190[199c] (182), 190[199d] (182), 190[199e] (182), 346[35] (315)
Takanabe, H., 187[92] (168)
Takanami, T., 349[183] (328)
Takano, H., 347[86] (319), 386[271] (374)
Takano, M., 383[58] (361)
Takano, S., 87[348] (30), 348[127] (323), 349[198] (329)
Takanohashi, Y., 567[571] (552)
Takaoka, K., 599[151] (578), 599[156] (578), 599[177] (581)
Takaoka, S., 276[847] (262)
Takasaki, K., 433[188b] (418)
Takasu, M., 262, 276[842] (262)
Takeda, K., 351[319] (338)
Takeda, R., 567[562] (551)
Takeda, T., 347[86] (319), 386[227] (372), 386[271] (374), 387[300] (377), 387[316] (379)
Takeda, Y., 156[141] (147)
Takehara, H., 92[700] (74)
Takei, H., 349[197] (329)
Takeichi, E., 512[102] (503)
Takemae, M., 353[430] (343)
Takemura, I., 187[101e] (169)
Takenaka, H., 434[244a] (427), 434[244c] (427), 434[244d] (427)
Takeuchi, R., 346[22] (314, 343), 346[30] (315), 346[57] (317)
Takeuchi, Y., 563[266] (530), 565[399] (541, 543), 565[411] (542)
Takeyama, Y., 383[34] (359)
Takeyasu, T., 352[354] (340)
Takezoe, K., 387[339] (381)
Takikawa, Y., 563[231] (528, 545)
Takinami, S., 266[144] (201, 214)
Takiura, K., 433[157] (414)
Takiyama, N., 154[24] (130)
Takusagawa, F., 90[566] (64), 434[229a] (425)
Takuwa, A., 386[267] (374), 387[338] (381)
Talbert, J., 185[9] (159, 160, 165, 168, 174)
Talbot, M. L., 268[284] (211)
Talley, J. J., 88[447] (48)
Tamagaki, S., 561[138] (520)
Tamai, H., 564[354] (539)
Tamai, Y., 156[149] (149)
Tamaki, K., 601[287] (596)
Tamao, K., 86[297] (25), 89[505] (55),
127[169] (118), 187[79b] (166, 167), 187[86] (167), 349[219] (330, 343), 349[226] (331), 350[233] (332), 353[435] (343), 354[465] (345)
Tamari, T., 601[288] (596)
Tamaru, Y., 185[16a] (160), 185[16b] (160), 185[17] (160), 185[18] (160), 185[19] (160, 167), 186[61a] (165, 166, 174), 186[61b] (165, 166, 174), 187[90] (168), 187[92] (168), 187[107a] (169), 187[107b] (169)
Tamayo, N., 385[195] (371)
Tamine, J., 383[70] (361)
Tamura, H., 569[653] (558)
Tamura, M., 352[352] (340), 512[84] (500)
Tamura, T., 275[806] (258, 261)
Tan, L. C., 269[388] (219)
Tanabe, M., 83[110] (5)
Tanaka, H., 347[82] (319), 383[17] (358), 558[9] (516), 563[253] (530, 538, 539), 566[443] (543)
Tanaka, J., 86[313] (26), 563[247] (529)
Tanaka, K., 86[329] (28, 53), 352[346] (339), 384[88] (362)
Tanaka, M., 186[65c] (165), 346[24] (315), 353[431] (343)
Tanaka, S., 187[92] (168), 310[85] (299)
Tanaka, T., 186[68] (165), 566[451] (543)
Tanaka, Y., 265[69] (196), 275[813] (258)
Tang, Y., 276[866] (264), 600[257] (592)
Tani, H., 126[95] (103), 601[286] (596)
Tani, K., 347[93] (320)
Tanigawa, H., 186[61b] (165, 166, 174)
Taniguchi, M., 187[95d] (169), 273[658] (243), 383[34] (359)
Taniguchi, N., 352[359] (340), 564[316] (538, 555)
Tanimoto, S., 351[314] (337)
Tanino, K., 349[217] (330), 350[239] (333)
Taniyama, E., 432[90] (403), 434[234a] (425)
Tanji, K.-I., 155[94] (139)
Tanouchi, N., 346[22] (314, 343)
Tapparelli, C., 264[20] (192, 208)
Tarnchompoo, B., 89[524] (57)
Tarrant, P., 431[71a] (401)
Tartakovskii, V. A., 430[27] (392), 434[238a] (427), 434[238b] (427)
Taschenberg, E. F., 126[77] (100)
Tasende, M. S. G., 458[18] (446)
Tashtoush, H. I., 431[63a] (400), 431[63b] (400), 431[79a] (402), 431[79b] (402)
Tatarinova, A. A., 598[92] (574), 598[93] (574)
Tatematsu, T., 350[262] (335)
Tatsuoka, T., 432[96b] (404)
Tau, S.-I., 186[71b] (166), 186[71d] (166)
Taufer-Knöpfel, I., 350[243] (333)
Tayano, T., 266[146] (201)
Taylor, B. S. F., 431[47] (397)
Taylor, D. K., 84[170] (8), 84[171] (8)
Taylor, E. C., 91[616] (70, 72), 458[26] (447), 458[29] (447), 458[32] (449), 458[39] (450), 484[19] (466)
Taylor, F. L., 431[71d] (401)
Taylor, P. G., 86[296] (25), 86[308] (26)
Taylor, R. J. K., 84[172] (8), 91[658]
Taylor, R. T., 564[299] (536)
Taylor, S. K., 154[18] (130)
Taylor, S. R., 126[121] (109)
Teasdale, A. J., 459[80] (456)
Tebbe, F. N., 310[40] (282, 293, 296, 309)
Tebben, P., 85[267] (23)
Tedder, J. M., 431[71b] (401)
Tedoradze, G. A., 430[18] (390)
Tellier, F., 186[48b] (163), 186[48c] (163), 187[87] (167)
Tempkin, O., 350[254] (334)
ten Hoeve, W., 83[78] (4, 5)
Tennent, N. H., 431[71e] (401)
Teodorović, A. V., 275[835] (260), 275[837] (260)
Terada, M., 350[255] (334)
Teranishi, A. Y., 565[401] (541, 555)
Teranishi, S., 511[50] (491)
Terao, K., 565[383] (540, 543)
Terashima, M., 270[476] (229)
Terashima, S., 155[80] (137)
Terrett, N. K., 565[430] (542)
Terry, L. W., 459[65] (455)
Terzis, A., 91[628] (71)
Tessier, C. A., 92[674]
Testaferri, L., 563[276] (530), 563[277] (530), 564[334] (538), 565[388] (540, 542), 565[389] (540)
Tester, R., 350[256] (334)
Tette, J. P., 126[77] (100)
Teulade, M. P., 86[311] (26)
Teunissen, H. T., 186[59] (165)
Tews, H., 387[336] (381)
Tewson, T. J., 268[325] (215)
Texier-Boullet, F., 350[264] (335)
Tezuka, M., 190[199e] (182), 346[35] (315)
Thal, C., 92[706] (78)
Thaning, M., 347[78] (319)
Thavornyutikarn, P., 598[126] (576)
Thayumanavan, S., 89[511] (56)
Thebtaranonth, C., 89[524] (57), 567[535] (548)
Thebtaranonth, Y., 89[524] (57), 567[535] (548)
Thiecke, J. R. G., 83[78] (4, 5)
Thiéffry, A., 512[115] (504, 509), 513[136] (509), 513[137] (509)
Thiele, K.-H., 186[45a] (163), 186[51a] (164), 186[51b] (164), 266[176] (203)
Thieme, G., 430[3] (390, 409)
Thies, J. E., 271[509] (232)
Thoma, G., 567[546] (548)
Thomas, A. P., 347[64] (317), 347[73] (317, 320)
Thomas, C. B., 433[164] (415)
Thomas, E. J., 386[247] (373), 386[250] (373), 386[263] (373)
Thomas, M., 598[127] (576, 583)
Thomas, P. J., 88[460] (49)
Thomas, R. C., 270[473] (229)
Thomas, R. D., 84[176] (10), 346[33] (315)
Thomas, S. E., 346[5] (314)
Thomas, W. H., 311[107] (307)
Thomas-Miller, M. E., 90[566] (64)
Thompson, A. S., 88[416] (42)
Thompson, C. M., 87[336] (29)
Thompson, D. P., 185[28c] (161), 562[224] (527)
Thompson, J., 83[62] (4)
Thompson, K. L., 433[155a] (414), 433[155b] (414)
Thompson, W. J., 187[103b] (169), 271[509] (232)

Thompson-Eagle, E. T., 558[12] (516)
Thomson, P. A., 267[251] (208)
Thornton, E. R., 351[292] (336)
Thornton, T. J., 89[481] (51), 128[205]
Thornton-Pett, M., 434[249] (429)
Thorpe, F. G., 431[60] (399), 458[36] (450)
Thuillier, A., 155[56] (134, 140), 155[102] (140)
Thyagarajan, B. S., 559[41] (516, 519, 522, 536)
Tidwell, T. T., 353[419] (343)
Tiecco, M., 563[276] (530), 563[277] (530), 564[334] (538), 565[388] (540, 542), 565[389] (540)
Tietze, L. F., 87[391] (41), 155[41] (132), 347[117] (323)
Tilley, T. D., 349[177] (328), 349[180] (328)
Tillyer, R. D., 385[153] (368)
Tinao-Woodridge, L. V., 350[230] (332)
Tingoli, M., 563[276] (530), 563[277] (530), 564[334] (538), 565[388] (540, 542), 565[389] (540)
Tinker, H. B., 432[140a] (412)
Tip, L., 126[102] (107)
Tishchenko, N. P., 566[487] (545)
Titcombe, L., 458[20] (446)
Titouani, S. L., 268[304] (213)
Tius, M. A., 386[225] (372), 386[226] (372)
Tlahuext, H., 275[820] (259)
Tobin, J. B., 349[181] (328)
Tobler, E., 431[55a] (399, 400), 431[55b] (399, 400), 431[66] (400)
Todeschini, R., 273[643] (241)
Togni, A., 90[588] (65), 90[591] (65), 90[592] (65)
Tohjo, T., 384[143] (367)
Tohma, T., 350[233] (332)
Toi, H., 270[444] (227)
Tokitoh, N., 352[367] (340), 484[3] (461)
Tokoroyama, T., 347[113] (322)
Tokunaga, Y., 568[607] (554)
Tomás, A., 353[417] (343), 353[426] (343)
Tomaselli, G. A., 433[154] (413)
Tomasini, C., 434[231a] (425), 434[231b] (425)
Tomaszewski, M. J., 383[61] (361)
Tomé, F., 351[294] (337)
Tominaga, Y., 349[195] (329), 350[240] (333)
Tomioka, K., 155[78] (136), 156[159] (151)
Tomlinson, D. W., 264[35] (192)
Tommasini, I., 347[75] (319)
Tomoda, S., 561[132] (520, 522, 539–42, 552, 557), 563[247] (529), 563[266] (530), 563[268] (530), 563[269] (530), 565[411] (542), 565[425] (542), 565[426] (542), 565[427] (542)
Tong, M. K., 434[213] (422)
Tong, W. H., 125[31] (94)
Tong, X., 386[285] (376)
Tonoyan, L. G., 597[40] (572)
Torii, S., 383[17] (358), 563[240] (529), 563[241] (529), 563[242] (529), 563[253] (530, 538, 539)
Torisawa, Y., 87[389] (40)
Tornroos, K. W., 565[416] (542)
Torok, D. S., 189[177d] (178)

Torrado, A., 271[492] (232), 459[79] (456)
Torregrosa, J.-L., 264[53] (195)
Torres, L., 598[87] (574)
Tortoreto, P., 434[231b] (425)
Toru, T., 352[355] (340), 387[312] (379), 484[10] (464), 563[259] (530), 565[375] (540), 565[432] (542), 567[532] (547, 548), 567[547] (548)
Toshimitsu, A., 561[123] (519, 528, 530, 540), 562[184] (522, 542), 563[271] (530, 540), 564[330] (538), 565[383] (540, 543), 568[625] (555), 568[635] (556), 569[649] (557), 598[121] (575), 598[122] (575)
Toth, J. E., 90[607] (67)
Totter, F., 87[399] (42)
Touet, J., 155[40] (132)
Toullec, J., 154[13] (130)
Tour, J. M., 189[160a] (176), 353[432] (343)
Towle, M. J., 89[501] (54)
Toya, T., 275[838] (260)
Toyoda, T., 274[718] (248)
Toyota, S., 267[234] (206)
Trabelsi, M., 563[254] (530, 532), 563[263] (530, 532, 539), 563[264] (530), 563[265] (530), 564[356] (539)
Trachtenberg, E. N., 559[40] (516, 519, 522, 536)
Trahanovsky, W. S., 90[544] (63), 484[12] (465), 559[34] (516, 522), 559[48] (516, 522, 526, 529, 536, 540, 541, 543, 552–7), 560[120] (519, 520, 522, 538), 562[206] (522, 556, 557), 568[582] (554, 557)
Trancher, J.-P., 155[86] (137)
Trave, R., 126[97] (106)
Traxler, M. D., 241, 273[642] (241)
Traylor, T. G., 269[343] (216)
Traynham, J. G., 434[243] (427)
Trefonas, L. M., 156[147] (148)
Treibs, W., 431[58a] (399), 431[58b] (399)
Trekoval, J., 82[58] (4), 125[36] (94, 109)
Trend, J. E., 568[628] (555)
Trethewey, A. N., 566[503] (546)
Trevillyan, A. E., 125[19] (94), 125[21] (94, 108)
Trimble, L. A., 82[32] (2, 38), 351[299] (337)
Tripathy, P. B., 86[275] (24), 274[736] (251)
Trofimenko, S., 264[11] (192)
Trofimov, B. A., 598[91] (574), 598[93] (574)
Trohay, D., 83[101] (5, 81)
Trombini, C., 155[77] (136), 185[23a] (160), 185[23b] (160), 265[89] (198), 272[604] (240, 241), 272[610] (240), 272[611] (240), 273[647] (241), 350[245] (333), 350[276] (336)
Trost, B. M., 81[5] (2), 82[12] (2, 4), 82[18] (2), 82[21] (2), 82[24] (2, 15), 82[25] (2, 15), 82[27] (2, 25), 83[80] (4), 86[280] (24), 86[325] (28), 86[326] (28), 86[327] (28), 87[375] (36), 87[377] (36), 88[441] (47), 89[539] (63), 187[79a] (166, 167), 187[94] (169, 171, 172, 183), 188[150] (175, 182), 264[39] (193, 201, 213, 247), 267[261] (208, 234,

235), 268[320] (214–16, 218), 271[548] (235, 236, 251, 252), 272[584] (238), 272[600] (240, 241, 256), 273[663] (244), 310[42] (283, 285), 323, 346[11] (314), 346[59] (317), 347[76] (319), 347[87] (320), 347[97] (321), 348[119] (323), 348[120] (323), 348[121] (323), 349[219] (330, 343), 383[13] (358, 367), 560[84] (518, 521, 522, 526, 529, 538, 540, 543, 544, 551, 552, 555–8), 560[85] (518, 521, 522, 526, 543, 544, 551, 552, 555), 564[346] (539, 555, 557), 564[347] (539), 568[631] (556)
Trotman-Dickenson, A. F., 559[58] (517, 518, 522, 528)
Trotter, J., 267[232] (206)
Troupel, M., 186[36e] (162, 174)
Trovato, M. P., 433[179b] (417)
Truce, W. E., 89[494] (53)
Truchet, F., 266[133] (201)
Truesdale, L. K., 268[286] (211), 561[137] (520)
Tsai, D. J. S., 267[227] (206, 249, 251), 267[267] (208), 274[732] (251)
Tsai, Y.-M., 86[285] (24), 348[135] (324), 383[64] (361), 566[440] (543)
Tsanaktsidis, J., 84[170] (8)
Tsay, S.-C., 352[389] (341)
Tsay, Y. H., 311[101] (304)
Tschugaeff, L., 597[29] (572)
Tse, A., 87[404] (42)
Tso, H.-H., 87[381] (39)
Tsou, C.-J., 186[71c] (166)
Tsubaki, K., 185[16a] (160), 186[61a] (165, 166, 174)
Tsubuki, M., 91[631] (71), 92[682] (72)
Tsubuki, T., 156[146] (148)
Tsuchida, M., 434[239a] (427)
Tsuchiya, Y., 352[346] (339), 352[348] (339)
Tsuda, T., 87[366] (34), 348[143] (325)
Tsuda, Y., 385[147] (367), 385[150] (367)
Tsuge, O., 86[313] (26)
Tsuji, J., 568[627] (555)
Tsuji, Y., 383[27] (358), 385[168] (369)
Tsujimoto, A., 87[363] (33)
Tsujimoto, K., 87[338] (29), 384[73] (361)
Tsujiyama, H., 156[140] (147), 188[139a] (174)
Tsukamoto, M., 88[459] (49)
Tsukamoto, T., 599[171] (580), 600[206] (583)
Tsukazaki, M., 83[121] (5), 92[710] (78), 92[711] (78)
Tsukiyama, T., 348[137] (324)
Tsunoda, T., 85[220] (15), 87[350] (30)
Tsunoi, S., 353[441] (344)
Tsutsui, M., 597[6] (571)
Tsutsumi, Y., 89[505] (55)
Tsuzuki, Y., 349[191] (329), 385[144] (367)
Tucci, F. C., 601[272] (594)
Tuck, D. G., 186[36a] (162, 174), 186[36b] (162, 174)
Tucker, C. E., 185[7] (159, 165), 186[42] (162, 175), 186[43] (162, 175), 186[49a] (163), 186[49b] (163), 186[55] (164), 186[66] (165, 179), 186[73] (166, 167), 186[75] (166), 188[137c] (174, 177), 188[152] (175, 182), 189[198] (182), 265[86]

Tucker, T. J., 187[103b] (169)
Tückmantel, W., 189[191a] (182)
Tueni, M., 348[146] (325)
Tueting, D. R., 385[210] (371)
Tufariello, J. J., 272[606] (240)
Tuladhar, S. M., 266[170] (203, 245)
Tumlinson, J. H., 126[98] (106)
Turck, A., 83[101] (5, 81), 92[716] (79), 92[717] (79), 92[718] (79), 92[720] (81), 92[721] (81), 271[514] (232)
Turco, A. M., 155[58] (134)
Turnbull, K., 568[612] (554)
Turner, E. E., 125[24] (94, 112), 597[52] (572)
Turos, E., 311[103] (305)
Twiss, E., 89[493] (53), 155[54] (133)
Tykwinski, R. R., 386[280] (375)
Tyrra, W., 458[7] (438)

U, J. S., 347[85] (319), 386[223] (371)
Uccello-Barretta, G., 156[151] (150)
Uchida, J., 383[58] (361)
Uchida, K., 270[468] (229), 270[469] (229), 311[111] (307)
Uchida, S., 599[147] (577)
Uchimura, J., 264[51] (195)
Uchiro, H., 350[251] (334), 385[152] (368)
Ueda, Y., 87[364] (33)
Uematsu, M., 387[328] (380)
Uemura, M., 90[561] (63), 90[564] (63, 64), 90[570] (64), 90[584] (65), 188[141h] (174)
Uemura, S., 271[527] (233), 433[197c] (420), 561[123] (519, 528, 530, 540), 562[184] (522, 542), 562[195] (522, 553, 554, 557), 563[271] (530, 540), 564[330] (538), 565[383] (540, 543), 565[429] (542, 554, 556, 557), 566[466] (544, 554, 556), 568[591] (554, 557), 568[625] (555), 568[635] (556), 569[649] (557), 569[650] (557), 569[651] (557), 569[652] (557), 597[8] (571), 597[10] (571), 598[74] (573, 588), 598[79] (573, 591), 598[95] (574, 588), 598[96] (574, 588), 598[97] (574, 588), 598[114] (575, 588, 589), 598[121] (575), 598[122] (575), 598[123] (575, 588), 599[148] (577), 600[207] (583), 600[209] (584), 600[226] (587), 600[229] (587), 600[232] (588), 600[233] (588), 600[234] (588), 600[235] (588), 600[238] (589), 600[239] (589), 600[240] (589), 600[243] (591), 600[244] (591), 601[287] (596)
Uenishi, J., 271[488] (232)
Ueno, M., 565[432] (542)
Ueno, T., 434[240a] (427)
Ueno, Y., 352[355] (340), 383[37] (359, 374), 383[38] (359, 377), 387[312] (379), 563[259] (530), 565[375] (540), 567[547] (548), 568[597] (554)
Ugajin, S., 188[141a] (174), 188[141c] (174)
Ugi, I. K., 65, 90[580] (65), 90[581] (65), 90[582] (65)
Uh, K. H., 383[41] (360), 433[189] (419)
Uhl, K., 310[58] (285, 306)
Uhm, S. J., 185[21a] (160), 185[21b] (160), 185[21c] (160)
Ujjainwalla, F., 384[83] (362)
Ukaji, Y., 156[142] (148), 187[101e] (169)
Ukita, T., 384[125] (364, 372), 384[141] (367), 387[292] (377), 387[332] (381)
Ulivi, P., 349[187] (329), 351[318] (338)
Ulmschneider, D., 268[308] (213)
Umani-Ronchi, A., 185[23a] (160), 185[23b] (160), 265[89] (198), 268[334] (215), 272[604] (240, 241), 272[610] (240), 272[611] (240), 273[647] (241), 350[245] (333)
Undheim, K., 92[719] (81), 155[70] (135)
Uneyama, K., 563[240] (529), 563[241] (529), 563[242] (529), 568[607] (554)
Ungaro, R., 458[47] (451)
Uno, H., 86[323] (27)
Uno, Y., 352[386] (341)
Uozumi, Y., 274[724] (249)
Upadhyaya, P., 558[14] (516, 529), 558[20] (516)
Urabe, H., 91[638] (71), 156[139] (147)
Urch, C. J., 387[294] (377)
Urdaneta, N. A., 271[494] (232)
Ushio, Y., 352[344] (339)
Ustynyuk, Y. A., 271[553] (235)
Usui, Y., 564[354] (539)
Usuki, Y., 561[132] (520, 522, 539–42, 552, 557), 563[268] (530), 563[269] (530)
Utimoto, K., 187[95d] (169), 190[199b] (182), 190[199c] (182), 190[199d] (182), 190[199e] (182), 270[430] (224), 270[468] (229), 270[469] (229), 273[650] (242), 273[651] (243), 273[652] (243), 273[653] (243), 273[656] (243), 273[657] (243), 273[658] (243), 273[659] (243), 273[660] (243), 311[111] (307), 346[35] (315), 350[232] (332), 353[439] (344), 383[18] (358, 360), 383[34] (359), 384[78] (361)
Uyehara, T., 83[126] (5), 83[127] (5)
Uyeo, S., 350[231] (332)
Uzick, W., 310[64] (288)

Vaid, R. K., 351[295] (337)
Vaissermann, J., 187[114b] (171)
Valdés, C., 434[223] (423)
Valente, L. F., 185[8a] (159, 166)
Valentí, E., 349[191] (329)
Vallée, Y., 566[508] (546)
Valnot, J. Y., 351[317] (338)
van der Baan, J. L., 187[97a] (169), 189[161b] (176), 189[161c] (176), 274[758] (252)
van der Donk, W. A., 265[93] (199), 274[721] (249)
Van der Heide, T. A. J., 274[758] (252)
van der Heiden, R., 348[130] (324)
van der Kerk, G. J. M., 484[5] (461)
van der Louw, J., 187[97a] (169), 187[97b] (169), 189[161a] (176), 189[161b] (176), 189[161c] (176), 189[161d] (176)
van der Steen, F. H., 187[115b] (171)
Van Deusen, S., 189[177e] (178)
Van Eikema Hommes, N. J. R., 266[139] (201)
Van Ende, D., 562[203] (522, 529), 566[457] (544, 551), 567[568] (552), 568[611] (554)
Van Horn, D. E., 189[159] (176), 272[620] (240), 273[639] (241)
van Koten, G., 187[115a] (171), 187[115b] (171), 386[219] (371)
Van Lier, J. E., 563[287] (534, 540, 542, 554), 563[288] (534, 540), 563[289] (534)
Van Nieuwenhze, M. S., 274[769] (253)
Van Zyl, C. M., 353[400] (342)
Vanboom, J. H., 567[569] (553)
Vanderklein, P. A. M., 567[569] (553)
Vandermarel, G. A., 567[569] (553)
Vandermeer, P. H., 567[569] (553)
Vanier, N. R., 127[159] (116)
Vaquero, J. J., 351[296] (337)
Vara Prasad, J. V. N., 267[215] (205), 267[228] (206), 267[229] (206), 270[470] (229), 274[712] (247)
Varaprath, S., 432[111] (406), 458[27] (447), 458[28] (447)
Vardhan, H. B., 432[142b] (412)
Varela, J. M., 458[19] (446)
Varella, N. H., 597[53] (572)
Varghese, V., 350[256] (334)
Varick, T. R., 563[283] (532)
Varley, J. H., 568[608] (554)
Varma, M., 349[211] (330)
Varma, R. S., 266[165] (203), 269[354] (217), 269[384] (219, 226), 349[211] (330), 430[21] (391), 430[42] (396)
Vasey, S. K., 126[117] (109)
Vasumathi, N., 266[174] (203)
Vatèle, J. M., 568[634] (556)
Vather, S. M., 88[468] (50, 51)
Vaughan-Williams, G. F., 271[540] (234)
Vaultier, M., 265[123] (200, 205, 211), 266[133] (201), 266[159] (203), 266[160] (203), 268[288] (211), 268[289] (211, 212), 268[290] (211, 212), 268[296] (212), 268[297] (212, 216), 268[303] (213), 268[304] (213), 269[362] (217), 269[363] (217), 269[365] (217), 269[367] (218), 269[368] (218), 269[369] (218), 269[370] (218), 269[371] (218), 269[374] (218)
Vaupel, A., 186[43] (162, 175)
Vauzeilles, B., 565[378] (540, 548)
Vawter, E. J., 155[38] (132), 189[193b] (182)
Vazquez-Lopez, E. M., 458[12] (444)
Vedejs, E., 87[343] (29), 347[89] (320), 352[365] (340)
Vederas, J. C., 351[299] (337)
Veenstra, S. J., 274[714] (247)
Vega, J. A., 351[296] (337)
Veith, M., 458[16] (445)
Vencato, I., 458[19] (446)
Venegas, P., 186[49b] (163)
Venemalm, L., 92[693] (73)
Veneziani, G., 387[322] (379)
Venuti, M. C., 89[478] (51)
Vepachedu, S. R., 383[59] (361)
Verbeek, J., 127[178] (119)
Verbit, L., 269[348] (216)
Verhoeven, T. R., 562[179] (522)
Verkruijsse, H. D., 81[7] (2, 3), 83[85] (4), 88[446] (48, 53, 59), 93, 124[3] (93, 98, 116, 119), 126[101] (107),

126[125] (110), 126[127] (111, 112), 126[131] (112), 127[138] (112), 127[155] (116), 128[195] (121), 128[206] (123, 124), 154[23] (130), 600[261b] (592)
Verlhac, J.-B., 385[205] (371), 387[302] (378), 387[303] (378)
Vernhet, C., 83[116] (5)
Vernon-Clark, R., 88[448] (48)
Verpeaux, J.-N., 83[102] (5), 87[370] (36), 87[371] (36)
Veyrières, A., 384[140] (367)
Vicens, J., 565[408] (541, 554)
Vicentini, G., 597[56] (572, 574), 597[59] (572), 598[125] (576)
Viehe, H. G., 565[412] (542), 565[413] (542)
Vig, R., 564[304] (536, 537)
Vilarrasa, J., 275[824] (259)
Villalobos, A., 87[379] (36)
Villamaña, J., 431[86] (403), 434[204] (421)
Villieras, J., 83[95] (5), 272[561] (236)
Villiéras, J., 186[38a] (162, 169), 186[38b] (162, 169), 186[38c] (162, 169), 186[38d] (162, 169), 187[116c] (171)
Vincent, C., 601[277] (595)
Visani, N., 385[171] (369, 376)
Visger, D. C., 350[232] (332), 383[65] (361)
Visnick, M., 89[530] (59)
Visser, G. W. M., 431[50] (397)
Visser, M. S., 154[34] (131)
Vladuchick, S. A., 185[3] (159, 178), 309[10] (279)
Voaden, M., 565[393] (541)
Vogel, C., 350[277] (336)
Vogel, D., 434[227] (424)
Vogel, E., 272[621] (240), 272[622] (240)
Vogel, P., 568[641] (557), 568[642] (557)
Vogt, R. R., 433[152] (413)
Vogt, S., 185[30] (161)
Vohra, R., 85[251] (21)
Volante, R. P., 187[107c] (169)
Vol'eva, V. B., 431[65] (400)
Volkmann, R. A., 150, 155[90] (139), 155[92] (139), 156[152] (150), 156[153] (150)
Volkov, A. A., 310[53] (285)
Vollhardt, J., 87[383] (39)
Vollhardt, K. P. C., 89[520] (56), 346[53] (317), 348[129] (324), 348[141] (325), 348[142] (325), 484[30] (467, 482)
Volmer, M., 564[313] (538, 540, 557)
von dem Bussche-Hünnefeld, J. L., 186[50a] (163, 171, 172)
von Grosse, A., 309[1] (278)
von Ragué Schleyer, P., 82[11] (2) 83[66] (4), 87[398] (42), 88[429] (46), 266[139] (201)
Vorndam, P. E., 350[277] (336)
Vottero, P., 185[15b] (160)
Vukićević, R., 459[81] (456)
Vulpetti, A., 274[767] (253), 275[800] (258)
Vuorinen, E., 155[98] (140)

Waas, J. R., 186[53] (164), 189[197b] (182), 269[392] (220), 271[541] (235)
Wachter, M., 83[109] (5)
Wada, M., 156[110] (141), 386[228] (372),
512[94] (502), 512[95] (502), 512[96] (502), 512[97] (502), 512[101] (502), 512[102] (503)
Wada, T., 512[89] (501)
Waddell, S. T., 85[210] (15)
Wadgaonkar, P. P., 268[341] (216), 269[342] (216)
Waegell, B., 432[131] (410)
Waitkins, G. R., 559[35] (516, 522, 536)
Wakahara, Y., 385[151] (367)
Wakamori, H., 386[228] (372)
Wakasugi, M., 600[229] (587), 600[233] (588)
Wakefield, B. J., 2, 81[1] (2, 42), 81[6] (2, 3, 8), 84[163] (8), 93, 124[6] (93, 94)
Walborsky, H. M., 89[490] (53), 154[9] (130)
Waldbillig, J. O., 268[283] (211)
Walde, A., 84[136] (6), 125[50] (95, 112–14), 126[90] (103, 105)
Waldmann, H., 189[167b] (178)
Waldmüller, D., 349[213] (330)
Walker, C. B., Jr., 347[85] (319)
Walker, H. G., 559[37] (516, 519, 520, 522, 528, 536, 538, 539)
Walker, J. A., 270[473] (229), 348[129] (324)
Walkup, R. D., 351[291] (336), 433[174] (416)
Wallace, D. J., 275[792] (258)
Wallace, E. M., 348[166] (327)
Wallace, M. A., 347[85] (319)
Wallace, R. H., 268[293] (212)
Wallbank, P. J., 271[506] (232)
Wallbaum, S., 275[805] (258, 259, 262)
Waller, A., 90[558] (63)
Wallin, A. P., 89[509] (56)
Wally, H., 90[586] (65)
Walsh, C., 568[592] (554)
Walsh, P. A., 351[311] (337)
Walter, R., 558[5] (516, 529, 530, 538, 558), 568[619] (555), 568[620] (555)
Waltermire, R. E., 349[206] (330)
Walther, E., 275[785] (256)
Walton, D. R. M., 383[7] (357)
Walts, A. E., 272[572] (237), 274[768] (253)
Wan, C. S. K., 350[240] (333)
Wang, D., 127[168] (118), 354[461] (345)
Wang, J. B., 597[35] (572)
Wang, K. K., 265[108] (200, 236, 237), 265[109] (200, 237), 265[110] (200), 268[317] (214), 269[394] (220, 228), 270[450] (228), 270[462] (229), 272[586] (238), 272[589] (238), 348[171] (328)
Wang, M.-L., 349[196] (329)
Wang, N., 346[18] (314)
Wang, Q., 352[377] (341), 353[454] (344)
Wang, R.-T., 385[215] (371)
Wang, S., 266[186] (203)
Wang, W., 89[476] (51)
Wang, W.-B., 385[164] (368), 512[64] (495), 512[65] (495), 512[66] (495, 498)
Wang, X., 88[413] (42), 89[519] (56), 89[534] (60), 266[140] (201), 386[272] (374), 563[245] (529, 541)
Wang, Y. M., 188[128c] (171, 172)
Wang, Z., 269[352] (217), 269[353] (217), 272[574] (237), 272[575] (237),
274[750] (251)
Ward, D. E., 348[126] (323)
Ward, D. L., 90[589] (65)
Ward, J., 275[810] (258, 261)
Ward, J. A., 432[107b] (406)
Ward, R. A., 347[94] (320)
Ward, S. C., 350[232] (332)
Wardell, J. L., 82[9] (2, 42), 84[164] (8), 93, 124[5] (93), 430[5] (390), 488, 511[1] (488), 511[25] (488), 511[26] (488), 511[27] (488), 511[28] (488), 511[29] (488), 511[30] (488), 511[31] (488), 511[32] (488), 511[33] (488)
Ware, J. C., 269[343] (216)
Waring, A. J., 563[291] (534, 541)
Warkentin, J., 383[61] (361), 383[62] (361)
Warmus, J. S., 85[241] (19)
Warpehoski, M. A., 564[300] (536)
Warren, L., 271[539] (234)
Wartski, L., 352[361] (340)
Washio, Y., 383[58] (361)
Wasmuth, D., 83[89] (5)
Wasserman, H. H., 559[34] (516, 522), 559[48] (516, 522, 526, 529, 536, 540, 541, 543, 552–7), 560[120] (519, 520, 522, 538), 562[206] (522, 556, 557), 568[582] (554, 557)
Watanabe, J., 349[183] (328)
Watanabe, M., 83[121] (5), 88[407] (42, 50, 51), 89[533] (60), 90[584] (65), 90[587] (65), 91[670], 92[700] (74), 188[124] (171), 188[134a] (172), 384[124] (364, 378)
Watanabe, N., 385[169] (369)
Watanabe, T., 271[513] (232), 272[558] (235), 384[124] (364, 378), 432[121] (407)
Watanabe, Y., 352[355] (340), 387[312] (379), 563[259] (530), 565[375] (540), 567[547] (548)
Waters, R. M., 348[170] (328)
Waters, W. L., 431[53a] (398), 431[53b] (398), 433[172] (416), 433[193] (419)
Watson, C. L., 267[205] (204)
Watson, J. V., 266[194] (204)
Watson, S. C., 82[41] (3)
Watt, A. P., 90[569] (64)
Watthey, J. W. H., 92[678]
Waugh, J., 384[95] (363)
Waykole, L., 91[639] (71), 560[76] (518, 522, 528–31, 538, 539), 564[313] (538, 540, 557)
Webb, K. S., 352[372] (341), 387[297] (377)
Webb, R. R., II, 434[234a] (425)
Webb, T. R., 141, 156[109] (141)
Webb, V. J., 86[309] (26)
Weber, A. E., 269[373] (218)
Weber, B., 156[157] (151)
Weber, E. J., 349[203] (329)
Weber, P., 387[311] (378)
Weber, T., 155[72] (136)
Weber, W. P., 346[2] (314)
Webster, R. G., 566[501] (545, 546)
Weedon, A. C., 350[240] (333)
Weedon, B. C. L., 274[717] (248)
Weeks, P. D., 155[92] (139)
Weglarz, M. A., 156[111] (142)
Wehmeyer, R. M., 186[62] (165, 167,

Wehrsig, A., 353⁴¹⁶ (342)
Wei, Y., 189¹⁶⁰ᵃ (176)
Weidmann, B., 187¹⁰⁹ᵇ (170)
Weiler, L., 350²⁴¹ (333)
Weinberg, N. L., 433¹⁶³ (415)
Weinreb, S. M., 155⁶¹ (134), 311⁹⁶ (304), 311¹⁰³ (305), 311¹⁰⁴ (305, 306), 346⁴⁹ (316), 351²⁸⁵ (336)
Weisberger, A. R., 91⁶¹⁶ (70, 72)
Weismiller, M. C., 568⁵⁸⁷ (554)
Weiss, E., 125⁷⁰ (97)
Weiss, H., 598¹¹⁰ (575)
Weiss, J. F., 558¹⁸ (516)
Weiss, R., 433¹⁸⁶ (418)
Weissenfels, M., 431⁵⁸ᵇ (399)
Weissensteiner, W., 90⁵⁸⁶ (65)
Welch, W. M., 155⁹² (139)
Wells, A. P., 346⁴⁰ (316)
Welmaker, G. S., 384¹¹⁷ (364, 373), 384¹²¹ (364)
Welte, R., 266¹⁴⁹ (202)
Wen, L. W., 600²⁵¹ (591)
Wen, X. Q., 600²⁵³ (591)
Wendelborn, D. F., 568⁶²⁸ (555)
Wender, P. A., 90⁶⁰⁸ (67)
Wenkert, E., 600²³⁰ (588)
Wensing, M., 384¹³⁰ (365)
West, P., 126⁷³ (98)
West, R., 86²⁸⁹ (25)
Westcott, S. A., 265⁹³ (199), 265⁹⁶ (199), 265⁹⁷ (199), 265⁹⁹ (199), 265¹⁰⁰ (199)
Westeppe, U., 154² (130)
Westerlund, C., 89⁴⁸⁸ (52)
Westerman, P. W., 268³⁰⁹ (213)
Wetterich, F., 383²⁸ (359)
Weyer, K., 310⁵⁵ (285), 310⁹¹ (304)
Whang, H. S., 383⁶³ (361)
Whelan, J., 350²⁶¹ (335), 353⁴³⁷ (343)
Whinnie, W. R., 598¹²⁶ (576)
Whipple, W. L., 91⁶⁵². 156¹⁷⁰ (151)
Whistler, R. L., 563²³⁴ (528, 538, 558)
White, A. H., 155⁵⁵ (133)
White, A. W., 90⁶⁰⁸ (67)
White, J. B., 567⁵⁶⁵ (551)
White, J. D., 187⁹⁸ᵃ (169), 384¹³⁷ (367)
White, J. J., 126¹¹⁵ (109)
Whitehouse, M. L., 434²³⁷ᵃ (426)
Whitehouse, R. D., 559⁴⁹ (516, 522, 555, 556)
Whiteley, C. G., 266¹⁷⁸ (203), 266¹⁷⁹ (203), 266¹⁸⁰ (203), 266¹⁸¹ (203)
Whitesell, J. K., 84¹²⁹ (6)
Whitesides, G. M., 125⁷¹ (97), 156¹³⁶ (147), 431⁴⁵ (396), 560⁷⁷ (518, 538)
Whiting, A., 267²⁴⁷ (208), 276⁸⁵² (263)
Whitman, B., 264³⁴ (192)
Whitmore, F. C., 390, 430¹ (390)
Whitney, C. C., 268³³⁶ (215), 311¹¹⁴ (308)
Whitney, R. A., 430³⁷ (395), 433¹⁶⁰ (414)
Whittaker, S., 385¹⁶⁵ (368)
Whittle, A. J., 565⁴¹⁵ (542)
Whittle, R. R., 90⁵⁷⁷ (65)
Wiberg, K. B., 85²¹⁰ (15), 484¹¹ (465)
Wicenec, C., 386²²⁹ (372)
Wicha, J., 86²⁸⁸ (25), 348¹⁵⁹ (326), 348¹⁶⁰ (326), 348¹⁶³ (326), 349²²² (330), 352³⁹⁹ (342)
Wickenkamp, R., 353⁴⁰⁰ (342)
Wickham, G., 430⁴¹ᵇ (396)
Widdowson, D. A., 83⁹² (5), 90⁵⁴⁵ (63), 90⁵⁴⁶ (63), 90⁵⁵¹ (63), 90⁵⁵² (63), 90⁵⁵³ (63), 90⁵⁵⁴ (63), 92⁶⁹⁴ (74), 92⁶⁹⁵ (74), 92⁶⁹⁶ (74), 92⁶⁹⁷ (74), 92⁶⁹⁸ (74), 92⁶⁹⁹ (74), 269³⁷⁶ (218), 385¹⁶⁶ (369), 385¹⁷⁷ (369, 372)
Widdowson, K. L., 349²²⁵ (330), 350²⁴⁹ (333), 565³⁷⁹ (540, 548)
Widener, R. K., 84¹⁸⁰ (10), 268²⁷⁹ (210)
Widhalm, M., 90⁵⁸⁶ (65)
Wiechert, R., 383⁴⁷ (361)
Wieland, G., 155⁶⁴ (134)
Wiemer, D. F., 83¹¹⁹ (5), 346³¹ (315), 351³³⁵ (339)
Wiemers, D. M., 185³³ᵃ (161), 185³³ᵇ (161), 189¹⁹⁰ (181)
Wierda, D. A., 188¹³⁶ (172)
Wierschke, S. G., 349¹⁹⁹ (329)
Wijers, H. E., 597²⁶ (572), 598⁹⁹ (574, 593)
Wilcox, C. S., 157¹⁸³ (153)
Wilke, G., 280, 309²¹ (280), 309²² (280, 284), 310³² (281), 310⁸¹ (297)
Wilkinson, G., 199
Willbe, C., 91⁶²¹ (71), 127¹⁷³ (119)
Willemsens, L. C., 484¹ (461), 484⁵ (461), 484²¹ (466)
Willey, P. R., 84¹³⁴ (6)
Williams, A. J., 386²²² (371)
Williams, D. J., 90⁵⁵² (63), 92⁶⁹⁴ (74), 92⁶⁹⁹ (74), 562²¹⁹ (527)
Williams, E., 566⁴⁷⁹ (545)
Williams, G. J., 268³¹⁸ (214)
Williams, J. M., 187¹⁰² (169), 384¹²³ (364)
Williams, L., 268²⁷⁶ (210)
Williams, M. J., 434²³⁴ᵇ (425)
Williams, R. E., 264³ (192, 208), 264⁴ (192)
Williams, R. M., 269⁴⁰⁹ (221)
Williams, R. W., 270⁴⁶⁶ (229)
Williams, S. F., 272⁶⁰⁰ (240, 241, 256)
Williard, P. G., 82¹² (2, 4)
Willis, C. L., 458²³ (447), 563²⁸² (532, 538, 547)
Willis, C. R., 353⁴⁴⁴ (344)
Willis, M. C., 351³³⁸ (339)
Willis, P., 384⁷⁶ (361)
Willis, W. W., Jr., 87³⁵⁵ (32), 561¹²⁶ (519, 540, 555), 568⁵⁹⁸ (554, 555)
Wills, M., 88⁴³⁸ (46), 275⁸¹² (258)
Willson, T. M., 87³⁴⁷ (30), 386²⁵⁴ (373), 566⁴⁵⁶ (544)
Wilson, C. A., 568⁶⁴⁷ (557)
Wilson, J. W., 86³¹⁵ (26), 210, 267²⁶⁶ (208), 268²⁷⁵ (210), 268²⁷⁶ (210), 268²⁷⁷ (210, 211), 268²⁷⁸ (210), 271⁵³⁹ (234), 271⁵⁴³ (235)
Wilson, S. R., 189¹⁶⁵ (178), 189¹⁷⁹ᵃ (178), 189¹⁷⁹ᶜ (178), 434²¹¹ (422)
Wilson, T., 386²⁵⁴ (373)
Wilt, J. W., 349²¹⁸ (330), 383²⁰ (358)
Wimmer, P., 513¹³⁹ (510), 513¹⁴⁰ (510)
Wincott, F. E., 350²⁷⁷ (336)
Winkle, M. R., 88⁴⁵⁷ (49)
Winn, L. S., 348¹²⁹ (324)
Winter, C. H., 430²⁵ (392)
Wintzer, A., 433¹⁹⁷ᵃ (420)
Winzenberg, K. N., 88⁴¹⁶ (42)
Wipf, P., 83¹¹³ (5), 351²⁸² (336), 351²⁸³ (336)
Wipke, W. T., 265⁶² (195)
Wiriyachitra, P., 598⁸⁵ (574)
Wise, L. D., 271⁵²⁰ (232)
Wishart, N., 186⁶⁹ᵇ (166), 187⁹¹ᵈ (168), 187⁹¹ᵉ (168)
Wissing, E., 187¹¹⁵ᵃ (171)
Wistrand, L.-G., 347⁷⁸ (319)
Witczak, Z. J., 561¹⁵⁰ (520), 562¹⁸⁹ (522, 528, 538, 558), 563²³³ (528, 538, 558), 563²³⁴ (528, 538, 558)
Wittig, G., 83⁹¹ (5), 84¹⁵⁹ (7), 121, 127¹⁶⁰ (117), 127¹⁹¹ (121), 278, 309⁷ (278), 310³⁸ (282)
Wnuk, S. F., 512⁸⁷ (501)
Wojtkowski, P. W., 272⁶⁰⁶ (240), 272⁶⁰⁷ (240)
Wolf, M. A., 386²⁶⁴ (373)
Wolff, J. J., 272⁵⁶⁹ (237)
Wolff, S., 433¹⁹⁴ (419)
Wolf-kugel, D., 565⁴²³ (542)
Wölfle, I., 190²⁰⁷ᵇ (183)
Wollmann, T. A., 273⁶³⁶ (240, 257), 275⁸⁰⁶ (258, 261)
Wollowitz, S., 561¹²⁶ (519, 540, 555), 568⁶²⁸ (555)
Wolters, J., 484²¹ (466)
Wong, C. K., 567⁵²⁰ (547)
Wong, C.-H., 513¹³¹ (508)
Wong, D. F., 350²⁴⁰ (333)
Wong, H. N. C., 385¹⁸⁵ (370), 386²⁷⁹ (375)
Wong, K.-T., 346²¹ (314)
Wong, T., 384⁹⁷ (363, 364)
Woo, H.-G., 349¹⁷⁷ (328), 349¹⁸⁰ (328)
Woo, K.-C., 565⁴⁰⁷ (541, 556, 557)
Wood, A., 186⁶⁹ᵇ (166), 187⁹¹ᵃ (168), 187⁹¹ᵇ (168), 187⁹¹ᶜ (168)
Wood, D. L., 126¹⁰⁴ (107)
Wood, G. P., 564³²⁷ (538, 545)
Wood, J. L., 311¹⁰⁴ (305, 306)
Wood, M. L., 433¹⁵⁸ (414)
Woodall, T. M., 275⁸²³ (259), 275⁸²⁵ (259)
Woodcock, D. J., 431⁷¹ᵇ (401)
Woodhall, J. F., 433¹⁶⁹ (415)
Woods, W. G., 268²⁹⁹ (212)
Woodward, R. B., 286, 310⁶² (286)
Woodward, S., 89⁵⁰⁶ (55)
Worrell, M., 566⁵⁰⁸ (546)
Worsfold, D. J., 125⁷⁰ (97)
Woski, S. A., 569⁶⁵⁴ (558)
Wozniak, L., 353⁴⁰⁵ (342)
Wrackmeyer, B., 264²³ (192), 267²³⁰ (206)
Wrasidlo, W., 348¹³² (324)
Wright, C., 348¹⁴¹ (325)
Wright, G. F., 431⁵⁴ (398), 431⁷⁷ᵃ (402), 433¹⁶³ (415), 433¹⁷⁵ᵃ

(416)
Wright, M. E., 90[556] (63)
Wright, T. L., 431[71f] (401)
Wu, C., 384[142] (367)
Wu, C.-C., 347[114] (322)
Wu, G., 84[194] (12, 67), 189[183b] (179)
Wu, H., 600[248] (591), 600[252] (591)
Wu, H. Y., 156[148] (149)
Wu, M.-J., 347[114] (322), 387[304] (378)
Wu, S., 85[234] (18)
Wu, S.-H., 386[240] (372)
Wu, Y.-D., 266[138] (201), 266[140] (201), 275[807] (258, 261)
Wu, Y. W., 431[80] (402)
Wudl, F., 348[129] (324), 558[27] (516)
Wuest, J. D., 384[120] (364, 377)
Wulff, G., 266[168] (203)
Wulff, W. D., 91[635] (71), 385[178] (369)
Wustrow, D. J., 271[520] (232)
Wuts, P. G. M., 267[251] (208), 267[252] (208), 271[554] (235), 271[555] (235), 272[579] (237), 272[583] (237), 349[200] (329)
Wynberg, H., 83[78] (4, 5), 564[361] (540)
Wynne, K. J., 598[87] (574)
Wythes, M. J., 186[69b] (166), 187[91a] (168), 187[91b] (168), 187[91c] (168), 187[91d] (168), 187[91e] (168)

Xi, Z., 567[538] (548)
Xia, J., 157[176] (152)
Xiang, J.-N., 351[283] (336)
Xiang, Y. B., 275[789] (257)
Xiao, C., 188[149b] (175, 182), 188[151] (175, 182)
Xie, F., 430[44] (396, 416)
Xie, L., 599[141] (577), 599[189] (582), 600[248] (591), 600[252] (591)
Xin, Y. C., 349[224] (330)
Xiong, Y., 88[422] (42)
Xu, F., 265[58] (195), 347[103] (321)
Xu, R. H., 512[64] (495)
Xu, Y., 186[35] (162), 349[185] (328), 351[279] (336)
Xu, Z., 154[29] (131), 154[30] (131), 154[31] (131)
Xunjun, Z., 562[226] (527)

Yadav-Bhatnagar, N., 512[119] (505, 507, 509)
Yager, K. M., 126[117] (109)
Yakushi, K., 563[247] (529)
Yakushijin, K., 568[640] (557)
Yamabe, S., 346[56] (317), 569[653] (558)
Yamada, F., 458[43] (451), 458[44] (451), 458[45] (451), 458[49] (451)
Yamada, H., 275[834] (260), 350[240] (333)
Yamada, J., 187[113] (171), 484[6] (462), 484[7] (463), 484[9] (464)
Yamada, K., 270[479] (231), 270[484] (231, 232), 434[200] (420)
Yamada, S., 597[35] (572)
Yamada, Y., 512[89] (501)
Yamaguchi, H., 458[37] (450), 599[147] (577), 600[236] (589), 600[245] (591)
Yamaguchi, M., 267[263] (208), 272[618] (240), 350[275] (336), 386[234] (372), 386[278] (375)
Yamaguchi, R., 141, 156[104] (141),
386[266] (374)
Yamaguchi, T., 352[381] (341), 385[154] (368)
Yamakawa, K., 87[364] (33), 562[208] (526)
Yamamoto, H., 127[157] (116), 155[68] (134), 156[127] (145), 156[130] (146), 156[155] (150), 157[184] (153), 188[154] (175), 189[177b] (178), 262, 272[619] (240), 273[691] (245), 274[760] (252), 276[842] (262), 276[845] (262), 276[846] (262), 276[847] (262), 276[848] (262), 276[849] (262), 276[850] (262), 276[851] (263), 276[860] (263), 276[861] (263), 276[865] (264), 296, 299, 310[63] (288, 296), 310[79] (293, 296), 310[85] (299), 311[99] (304), 334, 346[45] (316), 347[108] (321), 349[188] (329), 349[214] (330), 350[253] (334), 350[254] (334), 350[278] (336), 351[316] (338)
Yamamoto, I., 267[218] (205, 211)
Yamamoto, K., 156[142] (148), 353[430] (343), 564[347] (539)
Yamamoto, M., 83[115] (5), 349[192] (329), 353[417] (343), 434[199c] (420), 434[199d] (420), 434[200] (420)
Yamamoto, S., 270[417] (223)
Yamamoto, T., 186[61b] (165, 166, 174), 387[306] (378)
Yamamoto, Y., 83[81] (4), 83[126] (5), 83[127] (5), 127[140] (114), 127[141] (114), 155[71] (136, 137), 155[83] (137, 139), 186[65c] (165), 187[112a] (171), 187[113] (171), 256, 269[402] (221), 269[404] (221), 269[406] (221), 270[444] (227), 271[551] (235, 236, 238, 251, 252), 272[565] (236), 272[576] (237), 272[577] (237), 272[580] (237), 272[581] (237), 272[584] (238), 275[784] (256, 257), 347[66] (317), 347[87] (320), 347[97] (321), 383[4] (356), 386[255] (373), 386[256] (373), 484[6] (462), 484[7] (463), 484[8] (463), 484[9] (464), 512[70] (496, 497), 512[93] (501)
Yamamoto, Y.-I., 85[220] (15)
Yamamura, K., 383[50] (361), 386[257] (373)
Yamamuro, A., 275[806] (258, 261)
Yamanaka, H., 385[213] (371), 386[274] (375)
Yamasaki, H., 385[145] (367, 368)
Yamasaki, Y., 353[448] (344)
Yamashina, N., 187[80] (166), 271[499] (232), 271[501] (232)
Yamashita, H., 82[43] (3), 346[24] (315)
Yamashita, M., 599[139] (577), 599[143] (577)
Yamashita, T., 155[78] (136)
Yamauchi, M., 126[94] (103), 126[95] (103)
Yamauchi, T., 601[287] (596)
Yamazaki, H., 386[245] (373)
Yamazaki, N., 275[815] (259)
Yamazaki, S., 346[56] (317), 569[653] (558)
Yammal, C. C., 384[106] (363)
Yanada, K., 599[147] (577)

Yanagi, T., 270[485] (231, 232), 271[493] (232)
Yanagisawa, A., 156[127] (145), 157[184] (153), 188[154] (175), 349[188] (329)
Yanaka, H., 353[425] (343)
Yanaura, S., 155[76] (136)
Yang, C.-T., 348[135] (324)
Yang, L., 434[228] (425)
Yang, M., 347[103] (321), 347[106] (321)
Yang, R. Y., 269[358] (217)
Yang, T.-K., 155[100] (140)
Yang, Y., 232, 270[483] (231, 232), 271[511] (232), 271[512] (232), 271[521] (232), 385[185] (370), 385[200] (371), 386[279] (375)
Yang, Y.-L., 87[357] (33)
Yang, Z.-Y., 185[31] (161), 189[190] (181)
Yano, K., 386[246] (373), 386[258] (373)
Yano, T., 271[497] (232)
Yao, H., 89[505] (55)
Yaouang, J. J., 273[693] (245)
Yardley, J. P., 88[455] (49)
Yarkova, E. G., 564[328] (538)
Yashunsky, D. V., 566[488] (545), 566[489] (545)
Yasuda, A., 310[85] (299)
Yasuda, H., 126[94] (103), 126[95] (103)
Yasuda, M., 385[155] (368)
Yasue, K., 156[127] (145)
Yasuhara, A., 385[213] (371), 386[274] (375)
Yasui, E., 353[448] (344)
Yasui, K., 187[92] (168)
Yatagai, H., 269[393] (220)
Yatsimirsky, A. K., 459[78] (456)
Yau, C.-C., 83[84] (4)
Ye, B., 350[277] (336)
Ye, J., 387[289] (376)
Yeh, M.-C. P., 185[9] (159, 160, 165, 168, 174), 185[13a] (160, 162), 186[64] (165), 186[71a] (166), 186[71b] (166), 186[71c] (166), 186[71d] (166), 186[74b] (166), 186[77a] (166, 183), 187[89a] (167, 176), 187[108a] (169, 170), 187[108b] (169, 170), 188[138] (174), 188[143a] (175), 188[149b] (175, 182), 188[151] (175, 182), 189[182b] (179)
Yelm, K. E., 568[599] (554, 557)
Yeoh, T.-L., 154[25] (130)
Yeske, P. E., 383[68] (361)
Yeung, B. W. A., 385[184] (370)
Yi, K. Y., 349[204] (330)
Yoakim, C., 273[679] (244), 273[681] (244)
Yoda, H., 156[145] (148), 156[174] (152), 157[179] (153)
Yogo, T., 270[426] (224)
Yokoyama, H., 566[443] (543), 566[444] (543)
Yokoyama, K., 566[443] (543), 566[444] (543)
Yokoyama, S., 188[125] (171), 188[133] (172), 188[141a] (174), 188[141b] (174)
Yokoyama, Y., 564[319] (538, 540, 557), 566[479] (545)
Yonce, C. E., 126[98] (106)
Yoneda, T., 567[547] (548)
Yoo, B., 383[54] (361)
Yoon, C. H., 383[69] (361)

Yoon, J., 190[207a] (183)
Yorita, K., 485[46] (473)
Yoshida, J., 348[163] (326), 349[189] (329), 568[638] (557)
Yoshida, J.-i., 85[261] (22), 387[324] (380), 387[325] (380)
Yoshida, M., 600[219] (585)
Yoshida, N., 384[143] (367)
Yoshida, T., 189[159] (176), 269[409] (221), 270[416] (223), 270[421] (224), 270[466] (229), 383[43] (360)
Yoshida, Z., 185[16a] (160), 185[16b] (160), 185[17] (160), 185[19] (160, 167), 186[61a] (165, 166, 174), 186[61b] (165, 166, 174), 187[90] (168), 187[107a] (169), 187[107b] (169)
Yoshifuji, M., 85[227] (17, 19)
Yoshikawa, M., 189[177g] (178)
Yoshikoshi, A., 564[362] (540)
Yoshimatsu, M., 352[355] (340), 566[469] (544), 566[477] (545)
Yoshimura, N., 352[344] (339), 385[157] (368)
Yoshino, A., 386[244] (373)
Yoshino, T., 156[140] (147), 188[139a] (174), 188[139b] (174), 563[292] (534)
Yoshioka, H., 269[349] (217), 433[188b] (418)
Yoshioka, M., 188[127a] (171, 172), 188[127b] (171, 172), 188[127c] (171, 172), 310[61] (286)
Yoshitake, M., 434[200] (420)
Yoshiuchi, T., 352[355] (340), 566[469] (544)
Yosida, T., 270[443] (227)
Young, D. G. J., 565[437] (543)
Young, D. J., 386[251] (373), 386[252] (373), 386[259] (373)
Young, D. W., 156[172] (152), 599[161b] (579)
Young, J. C., 346[17] (314)
Young, J. H., 91[654]
Young, M. W., 559[47] (516, 522, 535, 539, 541, 543, 552-5, 557), 563[238] (529, 553-5, 557), 568[621] (555-7), 600[241] (591)
Young, V. G., Jr., 348[139] (325)
Youngs, W. J., 92[674]
Youssofi, A., 268[289] (211, 212), 269[374] (218)
Yu, C.-M., 266[197] (204, 235, 238, 251), 266[198] (204, 252)
Yu, H., 350[232] (332), 383[31] (359)
Yu, S., 272[566] (236)
Yu, Y., 564[322] (538, 545)
Yuasa, T., 87[386] (40)
Yuen, P.-W., 188[136] (172)
Yuki, Y., 567[573] (552)
Yum, E. K., 17, 85[232] (17)
Yun, H. H., 83[71] (4)
Yus, M., 82[52] (3, 15), 84[149] (7, 15), 84[150] (7), 84[151] (7), 84[152] (7), 84[154] (7), 84[197] (12), 85[203] (14), 85[204] (14), 85[205] (14), 85[215] (15), 383[53] (361), 431[86] (403), 433[195] (419), 433[198] (420, 429), 434[204] (421), 434[207a] (421), 434[207b] (421), 434[216] (422), 434[218] (423), 434[219a] (423), 434[219b] (423), 434[219c] (423), 434[225a] (424, 425), 434[225b] (424, 425), 434[246] (428, 429)

Zablocka, A., 353[414] (342)
Zahneisen, T., 83[102] (5)
Zaidi, J. H., 563[291] (534, 541)
Zaidlewicz, M., 101, 125[68] (103), 126[85] (101, 102), 126[86] (102), 126[87] (102), 264[1] (192, 193, 196, 214), 265[81] (196, 197), 266[182] (203), 274[712] (247), 274[746] (251)
Zair, T., 156[126] (145)
Zaitsev, A., 278
Zajdel, W. J., 85[216] (15, 16), 85[229] (17)
Zakharkin, L. A., 310[51] (284)
Zakharkin, L. I., 310[83] (299), 310[87] (300), 310[93] (304), 311[100] (304), 458[3] (438)
Zammattio, F., 346[51] (317)
Zandi, K. S., 349[226] (331)
Zanotti-Gerosa, A., 85[201] (13)
Zapata, A., 86[295] (25)
Zapata, A. J., 385[181] (370)
Zarantonello, P., 155[82] (137), 353[423] (343), 353[424] (343)
Zarcone, L. M. J., 84[195] (12)
Zard, S. Z., 512[68] (496), 562[193] (522)
Zargarian, D., 347[93] (320)
Zask, A., 90[548] (63)
Zavorin, S. I., 565[382] (540)
Zbiral, E., 86[310] (26)
Zdansky, G., 558[5] (516, 529, 530, 538, 558)
Zdorova, S. N., 431[73f] (401)
Zdunneck, P., 186[51a] (164)
Zefirov, N. S., 565[382] (540)
Zehnder, M., 189[171a] (178)
Zeitz, H.-G., 189[171b] (178)
Zelechonok, Y., 383[44] (360)
Zelle, R. E., 347[81] (319)
Zeller, K.-P., 185[3] (159, 178), 430[3] (390, 409)
Zeng, Q., 187[114c] (171)
Zeng, Z.-J., 348[132] (324)
Zetter, R., 351[305] (337)
Zhang, C., 349[184] (328)
Zhang, C.-H., 385[187] (370), 386[232] (372)
Zhang, H., 599[140] (577, 581)
Zhang, H.-C., 385[198] (371)
Zhang, J., 274[720] (249), 347[90] (320)
Zhang, L. J., 511[41] (489), 511[47] (491), 511[54] (492), 511[59] (493, 496, 497), 511[60] (493), 512[61] (494), 512[62] (494), 512[72] (497)
Zhang, P., 85[266] (22, 47), 156[154] (150)
Zhang, Q., 85[250] (20, 21), 347[115] (322)
Zhang, S., 512[72] (497)
Zhang, W., 349[208] (330), 384[77] (361)
Zhang, Y., 155[101a] (140), 155[101b] (140), 564[322] (538, 545)
Zhang, Z., 155[101b] (140)
Zhao, B.-P., 88[420] (42)
Zhao, K., 87[340] (29)
Zhao, M., 352[366] (340), 353[422] (343)
Zhen, X., 567[540] (548)
Zheng, G. Z., 350[236] (333)
Zheng, Q., 271[521] (232)
Zhong, G.-F., 125[13] (94, 95), 125[63] (101), 127[162] (117)

Zhou, J.-Y., 386[240] (372)
Zhou, M., 156[166] (151)
Zhou, W., 272[566] (236)
Zhou, X. J., 599[188] (582), 599[192] (582), 599[193] (582)
Zhou, Z. L., 600[250] (591, 592), 600[255] (591), 600[256] (591, 592), 600[257] (592), 600[258] (592), 600[259] (592)
Zhu, F.-H., 512[73] (497), 512[75] (498)
Zhu, L., 186[62] (165, 167, 168), 186[70] (166)
Ziani-Chérif, C., 87[362] (33)
Zieger, H. E., 126[73] (98)
Ziegler, K., 278-80, 294, 309[4] (278), 309[8] (278), 309[15] (279), 309[16] (279), 309[17] (279), 309[18] (279, 299), 309[19] (279), 309[20] (279, 284), 310[25] (280), 310[31] (281), 310[45] (283, 293), 310[47] (283), 310[84] (299), 310[91] (304)
Zierke, T., 274[740] (251)
Zietz, J. R., Jr., 310[41] (283)
Zigic, B., 558[15] (516)
Zima, G., 564[363] (540), 565[414] (542), 566[462] (544, 551), 567[563] (551)
Zimmerman, H. E., 241, 273[642] (241)
Zingaro, R., 597[6] (571)
Zingaro, R. A., 563[250] (530)
Zingaro, R. W., 562[166] (522)
Zirotti, C., 187[101b] (169)
Zlotskii, S. S., 310[53] (285)
Zolfaghari, H., 458[24] (447)
Zolopa, A. R., 269[405] (221)
Zong, K. K., 268[293] (212)
Zosel, K., 310[45] (283, 293), 310[84] (299)
Zschage, O., 86[273] (23), 127[143] (114), 386[265] (374)
Zschunke, A., 266[176] (203)
Zubiani, G., 267[262] (208), 271[542] (235)
Zubieta, J., 89[491] (53)
Zukerman-Schpector, J., 458[12] (444)
Zulauf, P., 85[252] (21)
Zumbulyadis, N., 598[63] (572, 573)
Zuraw, P. J., 433[181] (418)
Zutterman, F., 87[344] (29)
Zuurmond, H. M., 567[569] (553)
Zwane, I., 266[180] (203)
Zwanenburg, B., 82[31] (2, 33), 87[343] (29)
Zweifel, G., 187[100] (169), 264[43] (194), 265[63] (196), 265[105] (200), 266[131] (201), 266[132] (201), 266[134] (201), 266[135] (201), 266[137] (201), 267[264] (208), 268[314] (213), 268[316] (213), 268[327] (215), 268[328] (215), 268[336] (215), 270[434] (224), 270[467] (229), 272[588] (238), 310[43] (283), 310[88] (302), 311[114] (308), 311[115] (308), 349[196] (329)
Zykov, A. Y., 272[573] (237, 251)

Subject Index

JOHN NEWTON
David John (Services), Slough, UK

This Subject Index contains individual entries to the text pages of Volume 11. The index covers general types of organometallic compound, specific organometallic compounds, general and specific organic compounds where their synthesis or use involves organometallic compounds, types of reaction (insertion, oxidative addition, etc.), spectroscopic techniques (NMR, IR, etc.), and topics involving organometallic compounds.

Because authors may have approached similar topics from different viewpoints, index entries to those topics may not always appear under the same headings. Both synonyms and alternatives should therefore be considered to obtain all the entries on a particular topic. Commonly used synonyms include alkyne/acetylene, compound/complex, preparation/synthesis, etc. Entries where the oxidative state of a metal has been specified occur after all the entries for the unspecified oxidation state, and the same or similar compounds may occur under both types of heading. Thus $Cr(C_6H_6)_2$ occurs under Chromium, bis(η-benzene) and again under Chromium(0), bis(η-benzene). Similar ligands may also occur in different entries. Thus a carbene–metal complex may occur under Carbene complexes, Carbene ligands, or Carbenes, as well as under the specific metal. Individual organometallic compounds may also be listed in the Cumulative Formula Index in Volume 14.

Acetalization
    antimony trichloride, 498
    organoantimony compounds, 498
Acetals
    conversion to ethers
        organoaluminum hydrides, 300
    α,β-unsaturated
        reaction with allylic zinc reagents, 175
Acetylene, phenyl-
    double deprotonation
        lithiation, 46
    metallation, 112
Acetylene, silyl-
    hydroalumination
        stereoselectivity, 292
Acetylenes, thio-
    hydroboration
        nickel catalyst, 200
Acid chlorides
    coupling with tetraalkyllead
        palladium-catalyzed, 464
    coupling with zinc–copper reagents, 168
Acrylates
    β-substituted
        synthesis, 490
Acrylic acid
    α,β-unsaturated derivatives
        preparation, 489
Acyl anions
    equivalents
        lithiation, 39–41
Acylation
    organomercury compounds, 408, 409

Acyloxymercuration
    alkenes, 418–20
Addition-elimination reactions
    alkenation
        organoantimony compounds, 491
Alanine, 4-arylphenyl-
    synthesis, 233
Alcohols
    conversion to methylene compounds
        organoaluminum hydrides, 299
    oxidation
        organobismuth(V) compounds, 504
    polycyclic
        preparation, organoboranes, 226
    preparation
        organoantimony compounds, 492–6
        organoboron compounds, 214
        trialkylstibines in, 492
    reduction
        organotin compounds, 359, 360
    synthesis
        antimony in, 495
        organoboron compounds, 215, 216
        organomercury compounds, 398
        pentaorganylstiboranes, 493–5
    synthesis from aluminum alkyls, 302

    synthesis from carbonyl compounds
        organoaluminum hydrides, 298
Aldehydes
    addition of diethylzinc, 171
    alkenation
        trialkylstibine, 489
    allylation
        antimony, 495
        catalytic asymmetric, 321
    chiral
        addition of zinc organometallics, 169, 170
    diastereoselective addition to Grignard reagents, 145–50
    enantioselective addition to Grignard reagents, 150
    group transfer from tetraalkyl-lead, 462
    α-methoxyorganolead condensation
        stereoselectivity, 463
    synthesis
        organoboron compounds, 214
    synthesis from carbonyl compounds
        organoaluminum hydrides, 298
anti-Aldol products
    preparation, 175
Aldol reactions
    anti-selective, 257, 258
    asymmetric, 256–8
    boron mediated, 240–2

Aldol reactions
- Mukaiyama, 263
- *syn*-selective, 256, 257

Aldolization
- boron mediation
  - diastereoselectivity, 241, 242

Alkadienes
- conjugated
  - synthesis, 232

Alkanes
- heterocyclic
  - reductive cleavage, alkyllithium preparation, 12–15
- synthesis from alkenes
  - organoaluminum hydrides, 297

Alkanes, 1,1-diboryl-
- reactions, 208

Alkanes, nitro-
- reactions with organolead compounds, 477

Alkenamides
- amidomercuration
  - intramolecular, 424

Alkenation
- addition-elimination reactions
  - organoantimony compounds, 491
- antimony ylides, 489, 490
- carbonyl compounds
  - trialkylstibine, 489
- organoantimony reagents, 488

Alkenes
- alkynes from
  - organoboron compounds in, 213
- arylation
  - organotellurium compounds, 587
- carbomagnesiation
  - zirconium-catalyzed, 131
- carbomercuration, 427
- carbozincation, 175–7
- conversion to alkanes
  - organoaluminum hydrides, 297
- diacetoxylation
  - organotellurium compounds, 583
- epoxidation
  - organotellurium compounds, 583
- exocyclic
  - preparation, organoboron compounds, 228
- heterocyclic
  - asymmetric hydroboration, 248
- hydroboration, 200
- heterosubstituted
  - metallation, 114–24
- hydroalumination, 285
- mercuration, 393
- metallation, 97, 110–14
- nitrogen-containing
  - metallation, 115
- oxygen-containing
  - metallation, 116, 117
- reactions
  - with organomercury compounds, 403–7
- silicon-containing
  - metallation, 117, 118
- sulfur-containing
  - metallation, 116, 117
- synthesis
  - organotellurium compounds, 591, 592
- synthesis from alkynes
  - organoaluminum hydrides, 298

Alkenes, nitro-
- Michael acceptors, 174

Alkenols
- alkoxymercuration
  - intramolecular, 415

Alkenyl bromides
- reaction with allylic zinc reagents, 166

Alkoxymercuration
- alcohols, 414–17
- regioselectivity, 414
- stereoselectivity, 414

Alkyl halides
- lithiation, 7

Alkylaluminum alkoxides
- availability, 295

Alkylaluminum halides
- availability, 295

Alkylaluminum hydrides
- availability, 294

Alkylation
- tetraalkyllead to aldehydes, 462
- stereoselectivity, 462

Alkyne, silyl-
- carbalumination, 284

Alkynes
- alkoxymercuration, 417
- aminomercuration, 423
- carbomercuration, 427
- carbozincation, 175–7
- conversion to alkenes
  - organoaluminum hydrides, 298
- hydroalumination, 285
- hydroboration, 196
- hydroxymercuration, 413
- mercuration, 394
- metallation, 110
- reactions
  - with mercuric halides, 428
  - organomercury compounds, 429
  - with organomercury compounds, 403–7
  - with tin hydrides, 361

2-Alkynes
- metallation, 110

Alkynes, 1-bromo-
- reaction with zinc–copper organometallics, 167

Alkynes, 1-iodo-
- reaction with zinc–copper organometallics, 167

Alkynic alcohols
- synthesis, 494

Alkynic esters
- carbocupration, 179

Alkynylation
- organolead compounds in, 468
- tetraalkyllead to aldehydes, 462, 463

Allenes
- bromoboration, 202
- hydroboration, 200
- metallation, 110
- reaction with tin hydrides, 361

Allenic alcohols
- synthesis, 238

Alloyohimbane
- synthesis
  - lithiation, 19

Allyl alcohols
- synthesis
  - organoselenium compounds, 536

Allyl ethers
- metallation, 116

Allyl group
- transfer from allyltriphenyllead
  - radical-mediated transfer, 464

Allyl phenyl ether
- metallation, 116

Allyl selenoxides
- sigmatropic rearrangements, 557

Allyl sulfides
- metallation, 116

Allylamines
- hydroboration, 195
- nitrogen-containing, 115
- synthesis
  - organotellurium compounds, 596

Allylation
- aldehydes
  - antimony, 495

Allylboration, 236
- asymmetric, 251–6
- double asymmetric, 253
- intramolecular, 237, 238

Allylic alcohols
- secondary
  - hydroboration, 195
- tertiary
  - synthesis, 493

Aluminoxanes
- availability, 295

Aluminum acetylides
- availability, 296

Aluminum alkoxides
- availability, 296

Aluminum alkyls
- availability, 294, 295
- conversion to alcohols, 302

Aluminum aryloxides
- availability, 296

Aluminum aryls
- availability, 295

Aluminum cyanides
- availability, 296

Aluminum oxides
- availability, 296

Aluminum reagents
- availability, 294–7
- reactions
  - mechanisms, 289–93
  - stereochemistry, 289–93
- reactivity, 289–93
- transition metal-promoted reactions, 308, 309

Aluminum, vinyl-
- availability, 296
- conversion to ethers, 303

Amaryllidacea alkaloids
- synthesis

lithiation, 74
Amide alcohols
  synthesis, 496
Amides
  acylation
    organobismuth(V)
      compounds, 511
  addition to Grignard reagents,
    151
  amidomercuration, 424
  aryl
    lithiation, 42
  α,β-unsaturated
    alkenation, trialkylstibine,
      489
Amidomercuration, 424, 425
Amine alcohols
  synthesis, 496
Amines
  homoalkylated
    synthesis, 502
  lithiation, 15
  oxidation
    organoselenium
      compounds, 535
  primary
    synthesis, organoboron
      compounds, 216, 217
  secondary
    synthesis, organoboron
      compounds, 217, 218
  synthesis from nitriles or
    enamines
      organoaluminum hydrides,
        300
α-Amino acids
  synthesis, 249
  2,4,6-triisopropylbenzenesul-
    fonamide
      oxazaborolidines from, 262
Amino selenium reagents
  synthesis, 528
Aminomercuration, 421–3
Aminyl radicals
  production
    organotin compounds, 363
Amphimedine
  synthesis
    lithiation, 79
Aniline, difluoro-
  metallation, 123
Anilines
  lithiation, 51–3
Anisole
  lithiation, 48
Anisole, fluoro-
  metallation, 122
Anthemol
  synthesis, 105
Anthrenus flavipes
  pheromone
    synthesis, 100
Antimony
  metal
    alcohol synthesis, 495
Antimony dicarboxylates,
    triphenyl-
      reactions with amines, 501
Antimony pentachloride
  catalysts
    in organic synthesis, 500
Antimony salts
  catalysts

in organic synthesis, 500,
  501
Antimony trichloride
  acetalization, 498
  alcohol synthesis, 495
  catalysts
    in organic synthesis, 501
Antimony ylides
  alkenation, 489, 490
Aplysiatoxin
  synthesis
    lithiation, 39
Aplysiatoxin, debromo-
  synthesis
    lithiation, 39
Arenes
  electrophilic thallation, 446
  lithiation
    lithium–hydrogen exchange,
      42–65
  metallation, 112
Arenes, alkyl-
  metallation, 108, 109
Arenes, allyl-
  metallation, 107
Argentilactone
  synthesis
    lithiation, 36
Aristocularine alkaloids
  synthesis
    lithiation, 67
Aromatic compounds
  alkylated
    oxidation, organoselenium
      componds, 535
    reactions with selenium
      dioxide, 536
  fluoro
    production, 482
Aryl ethers
  lithiation, 48–51
Aryl fluorides
  synthesis
    organoboron compounds,
      218
Aryl halides
  lithiation, 54
Aryl thioethers
  lithiation, 53
Aryl thiols
  lithiation, 53
Arylation
  organobismuth(V) compounds,
    506–8
  organolead compounds 468
    mechanism, 480
Arylselenenyl halides
  reactions, 534
Asymmetric amplification
  organozinc reagents, 172
Asymmetric synthesis
  organoboron compounds,
    247–64
  organomagnesium compounds,
    130
1,2-Azaboracyclohexanes
  borane complexes
    reduction, ketones, 261
Azetidines
  reductive cleavage
    organolithium reagents, 14
Azidomercuration, 426
Aziridines

cycloaddition
  heterocumulenes, 500
reductive cleavage
  organolithium reagents, 14
synthesis
  organoboron compounds,
    218
Aziridines, *N*-methoxy-
  synthesis
    organosilicon compounds,
      340

Bacimethrin
  synthesis
    lithiation, 81
Barbier-type reactions
  organobismuth(III) compounds,
    502
  organozinc compounds, 178
  trialkylstibine, 492
Barbituric acid
  reactions with organolead
    compounds, 475
Barton–McCombie reaction
  organotin compounds, 359
9-BBN, *B*-alkyl-
  cross-coupling reaction, 233,
    234
9-BBN, vinyl-
  Diels–Alder cycloadditions,
    211
  synthesis, 204
9-BBN-H
  synthesis, 196
Benzazaphospholine
  synthesis
    lithiation, 68
Benzene, *p*-bis(triacetoxy-
    plumbyl)-
      preparation, 467
Benzene, 1,3-bis(trifluoromethyl)-
  metallation, 122
Benzene, *t*-butyl-
  metallation, 112
Benzene, 1,3-di(*t*-butyl)-
  metallation, 112
Benzene, 1,2-difluoro-
  reaction with Schlosser's base,
    121
Benzene, 1,4-difluoro-
  reaction with Schlosser's base,
    121
Benzene, (3,3-dimethyl-1-
    butenyl)-
      metallation, 112
Benzene, ethyl-
  metallation, 109
Benzene, fluoro-
  metallation, 121
Benzene, fluoro(trifluoromethyl)-
  metallation, 122
Benzene, hexakis(trimethylsilyl)-
  synthesis, 326
Benzene, (3-hydroxy-3-methyl-1-
    butynyl)-
      metallation, 112
Benzene, (3-hydroxy-3,4,4-
    trimethyl-1-pentynyl)-
      metallation, 113, 114
Benzene, (1-methylcyclopropyl)-
  metallation, 112
Benzene, nitro-
  reduction

Benzene, nitro-
  tellurium reagents, 577, 591
Benzene, trimethyl-
  trimetallation, 109
Benzeneseleninic acid
  reactions, 534
Benzeneseleninic anhydride
  reactions, 534
Benzeneseleninyl halides
  reactions, 535
Benzo-1,4-dioxin
  chromium tricarbonyl
    complexes
      lithiation, 63
Benzofuran, 2-aryl-
  synthesis, 473
Benzofurans
  synthesis
    lithiation, 54
Benzoic acids
  o-lithiation, 42
Benzonaphthyridinones
  synthesis
    lithiation, 78
Benzo[b]thiophene
  synthesis
    lithiation, 54
Benzothiophenes
  synthesis
    lithiation, 54
Benzyl alcohols
  lithiation, 22
Benzyl ketones
  synthesis, 496
Benzylic halides
  reaction with zinc–copper
    reagents, 166
Biaryl couplings
  Grignard reagents, 133
Biaryls
  synthesis
    organoboron compounds, 232
    organomercury compounds, 401
    organothallium compounds, 456
    unsymmetrical biaryls, 482
Bibenzyls
  synthesis
    organomercury compounds, 401
Bicyclo[2.2.1]hepta-2,5-diene
  metallation, 111
Bicyclo[3.2.0]hepta-2,6-diene
  metallation, 111
Bicyclo[4.1.0]heptane
  metallation, 114
Bicyclo[3.2.0]hept-6-ene
  metallation, 111
Bicyclo[3.2.1]octa-2,6-diene
  metallation, 107
Bicyclo[3.3.0]octanes
  synthesis
    organoboranes, 227
Bicyclo[2.2.2]oct-2-ene
  metallation, 112
Biphenyl, dimethyl-
  dimetallation, 109
Biphenyl, tetramethyl-
  tetrametallation, 109
Bisabolene
  synthesis
    organoboron compounds, 228
Bismuth trichloride
  aldol reactional Michael
    reaction, 502, 503
Bismuthimines, N-tosyltriaryl-
  preparation, 506
Bismuthonium ylides
  in organic synthesis, 506
Bombykol
  synthesis, 232
Boracyclanes, b-methoxy-
  ring enlargement, 207
Borane, 2-alkene-
  synthesis, 101
Borane, alkenyl-
  conjugate addition, 210
  oxidation, 214
Borane, alkyl-
  autoxidation, 242
  deprotonation, 210
Borane, alkyldibromo-
  synthesis, 205
Borane, B-alkyldimesityl-
  anions, 234
Borane, 1-alkynyldiisopropoxy-
  synthesis, 202
Borane, allenyl-
  chiral
    synthesis, 252
  rearrangement, 237, 238
Borane, allyl-
  addition to aldehydes, 236–8
  addition to ketones, 236–8
  reactions, 235–9
  reactivity, 238, 239
Borane, aryldimesityl-
  synthesis, 203
Borane, β-binapthyl-
  synthesis, 204
Borane, (3-chloro-1-iodo-1-
    propenyl)dialkyl-
  preparation, 221
Borane, cyclohexenyl-
  synthesis, 211
Borane, dialkenylchloro-
  synthesis, 197
Borane, dialkylallyl-
  preparation, 101
Borane, dialkylbromo-
  synthesis, 205
Borane, dialkylhalo-
  mixed, 196
Borane, dialkylvinyl-
  synthesis, 204
Borane, dibromoalkyl-
  synthesis, 199
Borane, diiodo-
  reactions, 244
Borane, diisopinocampheyl-
  hydroboration, 247
Borane, dilongifolyl-
  hydroboration, 247
Borane, (2-ethylapoisopinocam-
    pheyl)-
  hydroboration, 247
Borane, 1-halo-1-alkenyl-
  preparation, 221
Borane, [(methoxymethoxy)-
    allyl]diisopinocampheyl-
  allylboration, 255
Borane, monoalkylthexyl-
  synthesis, 196
Borane, monoisopinocampheyl-
  hydroboration, 247
Borane, thexyldialkyl-
  synthesis, 196
Borane, thexylmonochloro-
  preparation, 199
Borane, thexylmonohalogeno-
  reactions, 244
Borane, trialkyl-
  mixed
    synthesis, 196
Borane, triallyl-
  reactions with alkynes, 239
Borane, triethyl-
  free radical additions, 242
Borane, triiodo-
  reactions, 244
Borane, vinyl-
  deprotonation, 210
  dipolar cycloaddition, 212
Borane, vinyldichloro-
  Diels–Alder cycloadditions, 211
Boranes
  complexes
    hydroboration, 193–6
  disubstituted
    hydroboration, 198–201
  methanolysis, 213
  monosubstituted
    hydroboration, 196, 197
  oxidation
    chemoselectivity, 215
Borates, alkenyl-
  intramolecular transfer
    reactions, 227
  reactivity, 229
Borates, alkynyl-
  chemistry, 227–9
  intramolecular transfer
    reactions, 227
Borates, aryl-
  chemistry, 229
Borates, heteroaryl-
  chemistry, 229
Borepins
  synthesis, 204
Borinane, 3,5-dimethyl-
  synthesis, 196
Borinates
  cyclic
    synthesis, 205
Borinic esters
  synthesis
    from boronic esters, 203
    from Grignard reagents, 202
2-Bornene
  metallation, 111
Borohydrides
  reagents, 244
Borolane, B-allyl-2-(trimethyl-
    silyl)-
  reactions, 251
Borolane, 2,5-dimethyl-
  reduction
    ketones, 261
  synthesis, 247
Borolanyl mesylate, 2,5-dimethyl-
  reduction
    ketones, 261
1H-Borole, 2,5-dihydro-
  synthesis, 203

Boron enolates
  Aldol reactions, 240
Boron, fluorodimethoxy-
  in oroganic synthesis, 101
Boron, vinyl-
  cross-coupling reactions, 232
Boronates
  alkenyl
    synthesis, 205
  alkyl
    synthesis, 205
  cyclic
    synthesis, 205
Boronates, 2-alkene-
  synthesis, 101
Boronates, allyl-
  α-hetero-substituted, 252
Boronates, butyl-
  2,3-butanediol esters
    homologation, 251
Boronates, cyclopropyl-
  synthesis, 212
Boronates, pentenyl-
  enantiomerically pure, 253
Boronic acid, 1-alkenyl-
  cross-coupling reactions
    thallium-accelerated
      palladium-catalyzed, 456
Boronic acid, aryl-
  cross-coupling reactions, 232, 233
  synthesis, 203
Boronic esters
  synthesis
    from trialkoxyboranes, 202
Boronic esters, allyl-
  isomerization, 235
  synthesis, 207
Boronic esters, allylic
  synthesis, 208–11
Boronic esters, 1-chloroalkyl-
  homologated, 249
Boronic esters, 1-chlorocrotyl-
  synthesis, 253
Boronic esters, α-haloalkyl-
  preparation, 221
Boronic esters, α-halogeno-
  synthesis, 207
Boronic esters, α-iodoalkyl-
  reaction with zinc dust, 235
Boronic esters, α-organyl-
  substituted
    synthesis, 207
Boronic esters, α-phenylthio-
  carbanions, 210
Boronic esters, vinyl-
  Diels–Alder cycloadditions, 211
Boryl compounds
  dimesityl
    alkylation, 210
Brefeldin
  synthesis
    lithiation, 36
Bromine–zinc exchange
  zincates, 180
Bromoform
  reaction with diethylzinc and alkenes, 179
Brook rearrangement
  lithiation, 24
1-Butene, 3,3-dimethyl-
  metallation, 112

Butenolides
  asymmetric synthesis
    Grignard reagents, 148
γ-Butyrolactones
  preparation, 178

Cadmium
  highly reactive
    preparation, 183
Cadmium compounds
  in organic synthesis, 159–84
Caesium, allyl-
  rotational barrier, 98
Calamene, 7,8-dihydroxy-
  synthesis
    lithiation, 64
Calcimycin
  synthesis
    lithiation, 39
Calomel
  reaction with alkylzinc halides, 164
Camphene
  metallation, 112
Canadensolide
  synthesis
    lithiation, 72
Cannabinol, tetrahydro-
  synthesis
    lithiation, 52
Carbalumination
  carbon–carbon bond
    formation, 306
  discovery, 279, 280
  regioselectivity, 291
  scope and limitations, 283
Carbanions
  α-selenium stabilized
    generation, 28, 29
    lithiation, 28–41
    reactions, 29
  selenyl
    lithiation, 29, 30
  selenyl allylic
    lithiation, 30–2
  sp^3
    lithiation, 6–41
  α-nitrogen, lithiation, 15–21
  α-oxygen, configurational stability, 22–4
  α-oxygen, enantioinduction, 22–4
  α-oxygen, lithiation, 21–5
  α-silicon, lithiation, 25–8
  unstabilized, lithiation, 6–15
  α-sulfenyl
    lithiation, 29, 30
  α-sulfenyl allylic
    lithiation, 30–2
  α-sulfinyl
    lithiation, 32–4
  α-sulfinyl allylic
    lithiation, 35, 36
  α-sulfonyl
    lithiation, 36–8
  α-sulfonyl allylic
    lithiation, 39
  α-sulfur stabilized
    generation, 28, 29
    lithiation, 28–41
    reactions, 29

Carbazole, tetrahydro-
  synthesis
    lithiation, 67
Carbodealumination
  carbon–carbon bond
    formation, 307
Carbodiimides
  cycloaddition
    aziridines or oxiranes, 500
β-Carboline, dihydro-
  synthesis, 260
Carbomagnesiation
  alkenes
    zirconium-catalyzed, 131
Carbomercuriation
  alkenes, 427
Carbon monoxide
  acylation of organozinc
    halides, 168
  mercuriation, 392
Carbon-arylation
  organobismuth(V) compounds, 507, 508
Carbon–carbon bonds
  formation
    organoaluminum
      compounds, 306–8
Carbon–halogen bonds
  formation
    halodealumination, 302
Carbon–heteroatom bonds
  formation
    organoaluminum
      compounds, 303, 304
Carbon–hydrogen bonds
  formation by organoaluminum
    hydrides, 297–301
Carbon–metal bonds
  formation
    organoaluminum
      compounds, 306
Carbon–nitrogen bonds
  formation
    organoaluminum
      compounds, 305, 306
Carbon–oxygen bonds
  formation by oxidation of
    organoaluminum
      compounds, 302, 303
Carbon–sulfur bonds
  formation
    organoaluminum
      compounds, 303, 304
Carbon–tin bonds
  reaction types, 357
Carbonyl compounds
  conversion to alcohols or
    aldehydes
      organoaluminum hydrides, 298
  conversion to methylene
    compounds
      organoaluminum hydrides, 299
  reactions with selenium
    dioxide, 536
  reactions with stannyl anions, 364
  reactions with α-stannyl
    carbanions, 365
  reduction
    diphenylstibine, 496

Carbonyl compounds
    organotin compounds, 360, 361
    tellurium reagents, 576
  α-silyl-
    in organic synthesis, 329
  synthesis
    organoboron compounds, 216
  α,β-unsaturated
    reduction, tellurium reagents, 576
Carbonylation
  organoboranes, 225, 226
  organomercury compounds, 407, 408
o-Carborane
  lithiation, 6
Carboxylic acids
  addition to Grignard reagents, 151
  synthesis
    organoboranes, 223
    organomercury compounds, 398
Carboxylic compounds
  reduction
    organotin compounds, 360, 361
Carbozincation
  alkynes and alkenes, 175–7
Catechol monomethyl ether
  reaction with aryllead compounds, 469
Chemzyme, 259
Cholest-4-en-3-one, 6β-acetyl-
  reaction with p-methoxyphenyllead triacetate, 471
Cholesterol, 7,8-dehydro-
  metallation, 105
Chromium complexes
  arene tricarbonyl
    lithiated, 63, 64
Chuangxinmycin methyl ester
  synthesis
    lithiation, 74
Complex induced proximity effect
  lithiation
    arenes, 42
Confertin
  synthesis
    lithiation, 72
Coniferyl alcohol
  synthesis, 108
Coniine hydrochloride
  synthesis, 144
Coniine, N-methyl-
  synthesis, 144
Copper
  bimetallic reagents
    reactions, 182
  catalysis
    aryllead triacetate reactions, 480, 481
Copper methyltrialkylborates
  reactions with acrylates, 227
Cossus cossus
  5,13-tetradecadien-1-ol
    synthesis, 106
Coumarins, 3-aryl-4-hydroxy-
  synthesis, 507
Coumarins, 3-hydroxy-
  diarylation

    aryllead compounds, 473
Coumarins, 4-hydroxy-
  arylation
    aryllead compounds, 473
Coumaryl alcohol
  synthesis, 108
o-Cresol
  reaction with aryllead compounds, 469
Cresols
  dimetallation, 109
  lithiation, 59
Cross-coupling reactions
  palladium-catalyzed, 371, 372
Crown ethers
  boron-containing
    synthesis, 245
Cryptolestes pusillus
  pheromone
    synthesis, 100
Cryptopine, dihydro-
  chromium tricarbonyl complexes
    lithiation, 63
Cubanes
  lithiation, 6
Cumene
  metallation, 108
Cuprates, trimethylstannyl-
  reaction with acetylenic bonds, 363
Cyanoborate process, 230
Cyanogen, seleno-
  synthesis, 528
Cyclization–carbonylation
  thallium(I) salts, 456
Cycloaddition
  aziridines
    heterocumulenes, 500
Cycloalka-1,3-dienes
  hydroboration, 247
1,3,5-Cycloheptatriene
  metallation, 112
Cycloheptatriene, 2-trimethylsilyl-
  metallation, 118
2-Cyclohexenone, 3-iodo-
  reaction with zinc–copper reagents, 166
Cyclopentadiene, 5-fluoro-
  synthesis
    cyclopentadienylthallium, 455
meta-Cyclophanes
  synthesis, 109
Cyclopropanation
  organoantimony compounds, 498, 499
  zinc carbenoids, 178, 179
Cyclopropane, phenyl-
  metallation, 113
Cyclopropane, silyloxy-
  ring opening, 160
Cyclopropanes
  lithiation, 6
  mercuration, 393
  metallation, 113, 114
  synthesis
    carbomercuration, 427
    organomercury compounds, 409
Cyclopropenones
  synthesis

    organomercury compounds, 410
Cyclotelluroetherification, 586
Damascone
  synthesis, 102
Danishefsky's diene
  α-oxidation, 336
  reactions, 336
  sigmatropic rearrangement reactions, 336
Dehydrogenative silylation, 343
  anilines, 342
Demercuration
  solvomercuration, 411–30
ipso-Desilylation
  arylsilanes, 326
Desmosterol
  synthesis, 101
Diamines
  synthesis
    organomercury compounds, 422
Diazines
  DMG containing
    lithiation, 79–81
Diazo compounds
  heteroatom ylides from, 490
  synthesis
    organoboranes, 222
Diborane
  hydroboration, 193–6
Dibromides
  debromination
    tellurium reagents, 579, 592
  vicinal
    elimination reactions, alkenes from, 491
1,3-Dicarbonyl compounds
  carbon-arylation, 507
Diels–Alder reaction
  asymmetric aza-, 262
  hetero-, 262
Diene aldehydes
  allylboration, 255
Dienes
  alkoxymercuration, 416
  amidomercuration, 425
  aminomercuration, 422
  cyclization
    tin radical-initiated, 362
  diacetoxylation
    organotellurium compounds, 584
  dimetallation, 105
  hydroboration, 196
  hydroxymercuration, 412
  metallation, 103–5, 110–14
  silyl-
    preparation, 314
  solvomercuration, 427
1,3-Dienes
  synthesis
    organomercury compounds, 401
Diimines
  reaction with dialkylzincs, 171
β-Diketones
  reactions with organolead compounds, 471–4
Dimedone
  diarylation

organolead compounds, 471
Dimethyl selenide N-chlorosuc-
    cinimide
  reactions, 533
Diorganomercurials
  preparation of organozinc
    compounds, 164
Diorganothallium compounds
  in organic synthesis, 444–6
1,3-Dioxolanes, 2-trimethyl-
    silyloxy)-, 330
Diphenylselenium
    bis(trifluoroacetate)
  reactions, 533
Directed metallation groups
  heteroatom-based
    lithiation, 48–55
Directed *ortho*-metallation
  reactions, 72–5
  heteroaromatics
    lithiation, 70–81
Diselenides
  reactions, 533
Disparlure
  synthesis
    lithiation, 33
1,3-Dithia-2-borolane, 2-alkyl-
  transmetallation, 220
3-Dodecen-12-olide
  synthesis, 100
Dynemicin A
  synthesis
    lithiation, 79

Early transition metal effect, 281
β-Effect
  tin chemistry, 357
Eicosatetraenoic acid
  synthesis
    lithiation, 33
Eisch–Piotrowski reagent
  stoichiometric reagent, 296
Elaeokanine
  synthesis, 144
Elaeokanine B
  synthesis
    lithiation, 33
Ellipticine alkaloids
  synthesis
    lithiation, 74
Enamines
  conversion to amines or
    imines
  organoaluminum hydrides,
    300
Ene ethers
  metallation, 116
Ene reactions
  magnesium, 131
  silyl ethers, 330
Ene thioethers
  metallation, 116
Enol ethers
  silylated
    synthesis, 499
Enolates
  tin
    in organic synthesis, 368
α-Enones
  synthesis
    trialkylstibine, 489
Enoxysilacyclobutanes
  reactions with aldehydes, 333

Enynes
  cyclization
    tin radical-initiated, 362
1,3-Enynes
  monohydroboration, 201
Epoxides
  conversion to alcohols or
    aldehydes
  organoaluminum hydrides,
    298
  reductive cleavage
    organolithium reagents, 12
Erythronolide A, dihydro-
  synthesis, 255
Esters
  reactions with α-stannyl
    carbanions, 366
  α,β-unsaturated
    alkenation, trialkylstibine,
      489
Esters, α-halo-
  anions
    preparation, organoboranes,
      222
Ethane, 1,2-dibromo-
  for zinc activation, 160
Ethanol, 2-methoxy-
  O-phenylation, 509
Ethanol, 2-phenoxy-
  O-phenylation, 509
Ethers
  lithiation, 21
  silyl enol
    in organic synthesis, 333–8
    preparation, 333
  synthesis from acetals, ketals
    or orthoformates
  organoaluminum hydrides,
    300
  synthesis from aluminum
    vinyls, 303
Ethers, alkyl silyl, 329, 330
Ethyl propiolate
  carbozincation, 176
Ethylene, ethylsulfonyl-1-nitro-
  reaction with zinc–copper
    reagents, 166
Ethylenediamine, N-lithio-
  in organic synthesis, 5
Ethylenediamine, N-lithio-
    N,N',N'-trimethylene-
  in organic synthesis, 5
Ethyne
  carbalumination, 283
Exchange reactions
  lithium–tellurium, 15
  lithium–tin, 14

Fascaplysin
  synthesis
    organoboron compounds,
      232
Ferrocenes
  lithiated, 64, 65
FK-506
  synthesis
    lithiation, 36
Flavanones
  dehydrogenation
    thallium(I) salts, 456
Fluorenone
  synthesis
    lithiation, 42

Formamidines
  chiral
    synthesis, lithiation, 18
  lithiation, 17
Fredericamycin
  synthesis
    lithiation, 60
Fridamycin E
  synthesis
    lithiation, 49
Functionalities, 581
Furan, tetrahydro-
  reductive cleavage
    organolithium reagents, 14
Furans
  DMG-containing
    lithiation, 72
  metallation, 119
  non-DMG-containing
    lithiation, 71
Furochromone
  synthesis
    lithiation, 72
Furukawa reaction
  organozinc compounds, 178

Geraniol, 4,5-didehydro-
  synthesis, 105
Giese reaction
  organotin compounds, 361
Glycerol
  asymmetrically deuterated, 250
Glycols
  monophenylation
    organobismuth(V)
      compounds, 509
  vicinal
    oxidative cleavage, 504
Gnididione
  synthesis
    lithiation, 72
Grignard reactions
  1,4-addition, 132
Grignard reagents
  in organic synthesis, 129
Gymnopusin
  synthesis
    lithiation, 60

Hagemann's ester
  reaction with aryllead
    triacetates, 471
Halides
  conversion to hydrocarbons
    organoaluminum hydrides,
      301
Haloboration, 201, 202
Haloform
  preparation
    organoboranes, 223
Halomercuriation, 428
Heck reaction
  organomercury compounds,
    403
Heliannuol A
  synthesis
    lithiation, 38
Heteroaromatic compounds
  alkylated
    oxidation, organoselenium
      componds, 535
    reactions with selenium
      dioxide, 536

Heteroaromatic compounds
lithiation, 70–81
Heterocumulenes
  cycloaddition
    aziridines or oxiranes, 500
Heterocyclic compounds
  metallation, 119–21
Hexoses, 2-deoxy-
  synthesis, 253
HMPA
  lithiation
    in organic synthesis, 4
Hoberg reagent
  stoichiometric reagent, 296
Homoallylic alcohols
  synthesis, 235, 237, 251, 493, 502
Homobenzylic alcohols
  synthesis, 493
Homologation
  organoboron compounds, 249–51
Homopropargylic alcohols
  preparation, 181
  synthesis, 238, 494
Hosomi–Sakurai reactions, 321
Hydrangenol
  synthesis
    lithiation, 60
Hydrazines
  oxidation
    organoselenium compounds, 535
Hydrazones
  oxidation
    organoselenium compounds, 535
Hydroalkylation
  α,β-unsaturated carbonyl compounds, 596
Hydroalumination
  discovery, 279, 280
  regioselectivity, 291
  scope and limitations, 284
  stereoselectivity, 292
Hydroboration
  alkoxy-directed
    esters, 195
  allyl compounds, 195
  asymmetric synthesis, 247–9
  disubstituted boranes, 198–201
  enantioselective
    rhodium catalysts, 248
  functional alkenes, 196
  mechanism, 201
  monosubstituted boranes, 196, 197
  organoboron compounds
    synthesis, 193–201
  transition metal promoted, 199
Hydrocarbons
  metallation, 96–114
  organomercury compounds
    preparation, 392, 393
  synthesis from halides
    organoaluminum hydrides, 301
Hydroperoxides
  alkyl
    preparation, organoboron compounds, 214
Hydroquinones
  oxidation
    organoselenium compounds, 534
Hydrosilylation
  silanes, 343
Hydroxyalkylselenides
  reduction, 551
Hydroxylamines
  oxidation
    organoselenium compounds, 535
Hydroxymercuriation
  alkenes, 411–13
  regiochemistry, 411
  stereochemistry, 411

Ibuprofen
  synthesis, 109
  lithiation, 55
Imidazolidinone, stannyl-
  lithiation, 21
Imidazoline
  lithiation, 44
Imides
  cyclic
    addition to Grignard reagents, 152
Imido selenium reagents
  synthesis, 528
Imines
  reactions with Grignard reagents, 134, 135
  reduction
    organoboron compounds, 260
    tellurium reagents, 577
  synthesis from nitriles or enamines
    organoaluminum hydrides, 300
Imines, alkoxy-
  addition of allylic zinc reagents, 171
Imines, N-diphenylphosphinoyl-
  reaction with dialkylzincs, 171
Indirubin
  synthesis, 491
Indirubin, dimethyl-
  synthesis, 491
Indole, octahydro-
  synthesis
    organoboron compounds, 218
Indole, perhydro-
  synthesis
    organoboron compounds, 218
Indolecarbamates
  lithiation, 74
Indoles
  DMG containing
    lithiation, 73–5
  phenylation
    organobismuth(V) compounds, 510
  synthesis
    lithiation, 54
β-Ionol
  synthesis, 104
Ips paraconfusus
  pheromone
    synthesis, 108
Ipsdienol
  synthesis, 107

Ipsenol
  synthesis, 107
Ireland–Claisen rearrangement
  organosilicon compounds, 321
Iron complexes
  cationic pentadienyl
    reaction with zinc–copper reagents, 166
Iron, cyclohexadienyltricarbonyl tetrafluoroborate
  reaction with organocadmium reagents, 183
Isatins
  lithiation, 51
Isocarbacyclins
  preparation, 165
  synthesis, 472
  lithiation, 40
Isocyanates
  cycloaddition
    aziridines or oxiranes, 500
Isoflavanones
  synthesis, 507
Isoquinolines, tetrahydro-
  synthesis
    lithiation, 19
Isorobustin
  synthesis, 473
Isothiocyanates
  reactions with Grignard reagents, 139
Isoxazolines
  synthesis, 212

Joubertiamine, O-methyl-
  synthesis, 473

Kalloide
  synthesis
    lithiation, 72
Ketals
  conversion to ethers
    organoaluminum hydrides, 300
Ketene silyl acetals
  chiral
    asymmetric α-amination, 337
β-Keto esters
  reactions with organolead compounds, 471–4
Ketone enamines
  reactions with organolead compounds, 477, 478
Ketone enolates
  reactions with organolead compounds, 477, 478
Ketones
  alkenation
    trialkylstibine, 489
  dehydrogenation
    organoselenium compounds, 534
  diastereoselective addition to Grignard reagents, 145–50
  enantioselective addition to Grignard reagents, 150
  β-hydroxy-
    synthesis, 492
  α-iodo-
    synthesis, 337
  α-methoxylation

organosilicon compounds, 337
α-methyl allyl
  synthesis, 496
α-phenyl
  synthesis, 337
preparation
  organoboranes, 222
prochiral
  enantioselective reduction, organoboron compounds, 258
propynyl
  synthesis, 497
ring expansion
  organomercury compounds, 410
selective dioxanation, 340
α-sulfonyloxylation, 337
synthesis
  organoantimony compounds, 496, 497
  organoboron compounds, 214, 216
  organomercury compounds, 398
α,β-unsaturated
  hydroboration, 198
  synthesis, organoantimony compounds, 491
Ketones, amido-
  synthesis
    Grignard reagents, 152
Ketones, α-hydroxymethylene
  reactions with organolead compounds, 474
Khellin
  synthesis
    lithiation, 72
Lactams
  dehydrogenation
    organoselenium compounds, 534
Lactones
  dehydrogenation
    organoselenium compounds, 534
Larendamycin
  analogues
    synthesis, 232
Lasiodiplodin
  synthesis
    lithiation, 33
Lead compounds
  alk-1-ynyl tricarboxylates
    stability, 465
  aryl tricarboxylates
    chemistry, 465
  tributyl(α-methoxyalkyl)
    preparation, 463
Lead, alk-1-ynyl tricarboxylates
  preparation, 468
Lead, allyltriphenyl-
  allyl group transfer
    radical-mediated transfer, 464
Lead, aryl triacetates
  reactions
    copper catalysis, 480, 481
    replacement of lead
      by azides, 479
      by iodide, 479

Lead, aryl tricarboxylates
  reactions
    acid catalysis, 481–3
Lead, (2,4-dimethoxyphenyl) triacetate
  preparation, 466
Lead, furyl triacetate
  preparation, 467
Lead, (p-methoxyphenyl) triacetate
  preparation, 466
  structure, 465
Lead, phenyl triacetate
  preparation, 467
Lead, tetraalkyl-
  coupling with acid chlorides
    palladium-catalyzed, 464
  group transfer to aldehydes, 462, 463
Lead, tetraethyl-
  group transfer to aldehydes, 462
Lead, thienyl triacetate
  preparation, 467
Lead, p-tributylstannylphenyl triacetate
  preparation, 467
Lead, (2,4,6-trimethoxyphenyl) triacetate
  preparation, 466
Lead, vinyl tricarboxylates
  preparation, 468
  stability, 465
LIDAKOR
  applications, 96
Lithium amides
  homochiral
    in organic synthesis, 6
  in organic synthesis, 4
Lithium aminoborohydrides
  synthesis, 244
Lithium N-benzyltrimethyl-silylamide
  in organic synthesis, 5
Lithium bis(trimethylsilyl)amide
  in organic synthesis, 5
Lithium t-butyl(diphenylmethyl-silyl)amide
  in organic synthesis, 5
Lithium cyclohexylisopropylamide
  in organic synthesis, 5
Lithium dialkylborohydrides
  syntehsis, 205
Lithium diethylamide
  in organic synthesis, 5
Lithium diisopropylamide
  in organic synthesis, 4
Lithium 1-(dimethylamino)-naphthalenide
  in organic synthesis, 6
Lithium dimethylbenzylamide
  in organic synthesis, 5
Lithium di-(2,4,6-triisopropyl-phenyl)ethyl-
  reactions, 244
Lithium hexamethyldisilazane
  in organic synthesis, 5
Lithium monoorganylborohydrides
  syntehsis, 205
Lithium morpholide
  in organic synthesis, 5
Lithium naphthalide
  in organic synthesis, 4

Lithium t-octyl-t-butylamide
  in organic synthesis, 5
Lithium piperidide
  in organic synthesis, 5
Lithium pyrrolidide
  in organic synthesis, 5
Lithium reagents
  in organic synthesis, 3–6
Lithium 2,2,6,6-tetramethyl-piperidide
  in organic synthesis, 5
Lithium triorganozincates
  preparation, 182
Lithium zincates
  reaction with
    dibromocyclopropanes, 180
Lithium, alkenyl-
  structure, 97
Lithium, alkyl-
  unstabilized sp^3
    preparation using organolithium reagents, 7
Lithium, allyl-
  isomerization, 98
Lithium, aryl-
  anions
    synthesis, 42–81
  synthesis
    lithium–halogen exchange, 66–70
Lithium, benzyl-
  synthesis, 55–61
Lithium, butyl-
  in organic synthesis, 3
  superbase reagent, 94
Lithium, s-butyl-
  in organic synthesis, 3
Lithium, t-butyl-
  in organic synthesis, 3
Lithium, 2,4-dimethyl-2,4-pentadienyl-
  structure, 103
Lithium, heteroaryl-
  anions
    synthesis, 42–81
Lithium, methyl-
  in organic synthesis, 3
Lithium, 2-methyl-2,4-pentadienyl-
  structure, 103
Lithium, pentadienyl-
  structure, 103–5
Lithium, phenyl-
  in organic synthesis, 3
Lithium, phenylallyl-
  conformation, 108
Lycoramine
  synthesis, 473

Madelung indole synthesis
  lithiation, 59
Magnesium
  activated
    preparation, 130
Magnesium compounds
  dimetallic
    preparation, 175
Magnesium, allyl-
  isomerization, 98
Malonates, alkylidene-

Malonates, alkylidene-
  reaction with organozinc
    compounds, 175
Malonic acid
  reactions with organolead
    compounds, 475–7
Manoalide
  synthesis
    lithiation, 72
Mercuriation
  aromatic, 394
  organic halides, 390
Mesembrine
  synthesis, 473
Mesitol
  reaction with p-methoxy-
    phenyllead triacetate,
    469
  reaction with phenylethynyl-
    lead triacetate, 470
Mesitylene
  trimetallation, 109
Metathesis
  Al–C and Al–H bonds, 285
Methane, borylstannyl-
  reactions, 208
Methane, boryltrimethylsilyl-
  reactions, 208
Methane, dihalo-
  synthesis
    organoboranes, 223
Methane, diiodo-
  reaction with diethylzinc, 162
Methylene compounds
  synthesis from alcohols or
    carbonyls
    organoaluminum hydrides,
    299
Methylene homologation
  zinc carbenoids, 179–81
Methylseleno compounds
  synthesis, 530
Mevinolin
  synthesis
    lithiation, 33
Michael addition
  organozinc compounds, 173–5
Michael reaction
  aldol reations
    bismuth trichloride, 502,
    503
Minaprine
  synthesis
    lithiation, 79
Molybdenum complexes
  cationic pentadienyl
    reaction with zinc–copper
    reagents, 166
Monomorine
  synthesis, 153
Monoorganothallium compounds
  in organic synthesis, 446–51
  preparation, 446, 447
  properties, 446, 447
  reactions, 447–51
Muscalure
  synthesis, 101
Muscone
  synthesis, 172

Natta–Breslow reagent
  stoichiometric reagent, 296
Nazarov cyclization
  silicon-directed, 319

Nickel effect, 281
Nitriles
  aminomercuriation, 424
  conversion to amines or
    imines
    organoaluminum hydrides,
    300
  α,α-dichloro-β-hydroxy-
    preparation, 492
  preparation
    organoboranes, 222
  reactions with Grignard
    reagents, 139
  α,β-unsaturated
    alkenation, trialkylstibine,
    489
Nitro arenes
  reduction
    tellurium reagents, 577
Nitrogen compounds
  reduction
    organotin compounds, 360
  synthesis
    organomercury compounds,
    401
Nitrogen-arylation
  organobismuth(V) compounds,
  510
Nitromercuriation, 426
Nitrones
  reactions with Grignard
    reagents, 136
  synthesis
    halomercuration, 429
Nitrophenylseleno compounds
  synthesis, 530
Nitrosoamines
  lithiation, 16
Nor-alcohols
  synthesis, 496
Norbornadiene, 2-trimethylsilyl-
  metallation, 118
Norbornene
  metallation, 111
Norcarane
  metallation, 114
Nortricyclane
  metallation, 113
Nudaurelia cytherea cytherea
  sex attractant
    synthesis, 100

Ochratoxin B
  synthesis
    lithiation, 50
β-Ocimene
  metallation, 105
Okadaic acid
  synthesis
    lithiation, 36
Organoalkali compounds
  organoboron compounds from,
    202, 203
Organoaluminum reagents
  in organic synthesis, 278–309
Organoaluminum compounds
  oxidation
    carbon–oxygen bonds,
    302–3
  reactions with metal salts, 286
Organoaluminum hydrides
  carbon–hydrogen bond
    formation, 297–301

Organoantimony compounds
  in organic synthesis, 487–511
  pentavalent
    oxidizing agents, 497
Organoantimony(V) halides
  catalysts
    in organic synthesis, 499,
    500
Organobismuth compounds
  in organic synthesis, 487–511
Organobismuth(III) compounds
  in organic synthesis, 502, 503
Organobismuth(V) compounds
  arylation, 506–8
  in organic synthesis, 503–11
  oxidation, 504, 505
Organoboranes
  chiral
    α-pinene, 260
  cross-coupling reactions,
    231–4
    mechanism, 231
    palladium-catalyzed, 231
  organozinc compound prepara-
    tion, 164
  synthesis
    transmetallation, 202–4
Organoborates
  reactions with electrophiles,
    227–30
Organoboron compounds
  abstraction of α-hydrogen, 241
  autoxidation, 214, 215
  bimolecular homolytic
    substitution, 241
  carbanions
    in organic synthesis, 234,
    235
    ractions, 208–11
  chemistry, 192–264
  Diels–Alder cycloaddition, 211
  four coordinate boron
    carbon–carbon formation,
    220–42
  α-halo-
    rearrangements, 221–5
  halogenolysis, 218, 219
  oxidation, 214–16
  protonolysis, 213, 214
  radical reactions, 241–3
  safety, 192
  synthesis, 193–213
  synthetic applications, 213–46
  transmetallation, 220
  α,β-unsaturated
    addition reactions, 211–13
Organocerium compounds
  Grignard reagent reactivity,
    130
Organoditin compounds
  in organic synthesis, 358–63
Organolead compounds
  α-methoxy
    aldehyde condensation,
      stereoselectivity, 463
  in organic synthesis, 461–83
  tricarboxylates
    synthesis by metal–metal
      exchange, 467, 468
Organolead(IV) compounds
  tricarboxylates
    chemistry, 465–83
Organolithium compounds

in organic synthesis, 1–81
Organomagnesium compounds
  in organic synthesis, 129–53
  organoboron compounds from, 202, 203
Organomercury compounds
  alkylation, 402–7
  allenic
    preparation, 390
  dialkyl
    preparation, 390
  diaryl
    preparation, 390
  dimerization, 401, 402
  halogenation, 396, 397
  nitrogen-containing
    preparation, 394
  in organic synthesis, 389–430
  preparation, 390–5
    transmetallation, 391
  propargylic
    preparation, 390
  protonolysis, 395
    metal hydride reduction, 396
  sulfonyl, 429
Organopotassium compounds
  in organic synthesis, 93–124
Organoselenium chemistry, 515–58
Organoselenium complexes
  ate, 518
Organoselenium compounds
  in living systems, 516
  structures, 516–22
  synthesis, 538
Organoselenium monitored reactions
  inorganic reagents, 527
  organic reagents, 528–33
Organoselenium reagents
  electrophilic
    cyclization, 542
  synthesis, 529
Organoselenocyanates
  synthesis, 530
Organosilicon compounds
  in organic synthesis, 313–45
Organosodium compounds
  in organic synthesis, 93–124
Organotellurium compounds
  in deprotection of organic functionalities, 582, 583
  free radical chemistry, 594, 595
  in organic synthesis, 571–96
  preparation, 572–6
  tellurium-free compounds from, 587–90
Organothallium compounds
  in organic synthesis, 437–58
Organotin alkoxides
  in organic synthesis, 367
Organotin compounds
  halogen bonds
    in organic synthesis, 368
  halogens
    in organic synthesis, 358, 359
  handling, 358
  nitrogen bonds
    in organic synthesis, 368
  in organic synthesis, 355–81

photoreactions, 381
toxicity, 358
Organotin hydrides
  in organic synthesis, 358–63
Organotin oxides
  in organic synthesis, 367
Organozinc compounds
  addition reactions, 169–77
    to carbonyl derivatives, 169–72
  allylic
    addition to alkenes, 175
  cyclopentyl-
    preparation, 175
  preparation, 160–4
  transmetallation, 163, 164
  reactivity, 165–78
  substitution reactions, 165–8
Orthoformates
  conversion to ethers
    organoaluminum hydrides, 300
Oxasiletanides
  Peterson olefination, 327, 328
Oxazaborolidines
  ketones
    reduction, 259, 260
Oxazaphospholidines
  reduction, 261
N-Oxazolidinoyl activators
  lithiation, 20
Oxazolines
  lithiation, 42
N-Oxazolinoyl activators
  lithiation, 20
Oxetanes
  reductive cleavage
    organolithium reagents, 12
Oxidation
  organoaluminum compounds
    carbon–oxygen bonds, 302, 303
  organobismuth(V) compounds, 504, 505
  organotellurium compounds, 583, 584
Oximes
  oxidation
    organoselenium compounds, 535
Oxiranes
  cycloaddition
    heterocumulenes, 500
  deoxygenation
    tellurium reagents, 578, 591
  isomerization
    superbase, 120
Oxiranes, silyloxy-
  in oxidation of silyl enol ethers, 336
Oxygen-arylation
  organobismuth(V) compounds, 508–10
α-Oxylithio rearrangements, 24, 25
Oxythallation
  cyclononadiene, 447

Palladium
  catalysis
    acid chloride coupling with tetraalkyllead, 464

catalysts
  in benzylic zinc halide reactions, 167
  cross coupling reactions, 371, 372
Palytoxin
  synthesis, 232
Pancratistatin
  synthesis, 50
Paralobesia viteana
  pheromone
    synthesis, 100
Parham cyclicacylation
  lithiation, 67
Pauson–Khand cyclization–carbonylation
  alkynylsilanes, 325
Pentaorganylstiboranes
  alcohol synthesis, 493–5
  ketones from, 496, 497
Pent-2-enoic acid, 5-aryl-5-hydroxy-
  synthesis, 493
Peroxymercuriation, 417, 418
Perseleninic acid
  reactions, 535
Peterson olefination
  organosilicon compounds, 327, 328
Peterson reaction
  lithiation, 26–8
Phenol, 3,5-di-*t*-butyl-
  reaction with 2,4,6-trimethoxyphenyllead triacetate, 469
Phenol, 2,4-dimethyl-
  reaction with aryllead compounds, 469
Phenol, 2,6-dimethyl-
  reaction with aryllead compounds, 469
Phenol, dimethyl-
  dimetallation, 109
Phenol, 4-methoxy-2,6-dimethyl-
  reaction with aryllead compounds, 469
Phenol, 4-nitro-
  reaction with pentaphenylbismuth, 508
Phenol, thio-
  dimetallation, 124
Phenol, 3,4,5-trimethyl-
  reaction aryllead compounds, 469
Phenolic compounds
  carbon-arylation
    organobismuth(V) compounds, 507
  oxidation
    organobismuth(V) compounds, 505
Phenols
  C-arylation
    aryllead triacetates, 469, 470
  dimetallation, 124
  oxidation
    organoselenium compounds, 534
  synthesis
    aryllead tricarboxylates, 481

organoboron compounds, 214
Phenylation
  O-
    organobismuth(V) compounds, 509
  N-
    organobismuth(V) compounds, 510
Phenylseleno compounds
  synthesis, 530
Phosphorus compounds
  synthesis
    organomercury compounds, 401
N-Phosphorylamides
  lithiation, 19
Phthalides
  synthesis
    lithiation, 67
α-Pinene
  conversion to β-pinene, 103
  organoboranes, 260
Piperidines
  synthesis
    organoboron compounds, 218
Piperidines, α,α'-dialkyl-
  synthesis by lithiation, 17
2-Piperidone
  acylation
    organobismuth(V) compounds, 511
Plumbation
  electrophilic, 466
Polyenes
  alkoxymercuriation, 416
  hydroxymercuriation, 412
  metallation, 103
  solvomercuration, 427
  synthesis
    organomercury compounds, 401
Potassium t-butoxide
  superbase reagent, 94
Potassium selenocyanates, 528
Potassium-graphite
  preparation of highly activated zinc, 160
Potassium, alkenyl-
  structure, 97
Potassium, allyl-
  reactions, 97–103
  rotational barrier, 98
Potassium, butyl-
  structure, 95
Potassium, 2,4-dimethyl-2,4-pentadienyl-
  structure, 103
Potassium, heptatrienyl-
  conformation, 104
Potassium, 2-methyl-2,4-pentadienyl-
  conformation, 104
Potassium, pentadienyl-
  structure, 103
Potassium, phenylallyl-
  structure, 108
Potassium, prenyl-
  reaction with steroidal iodides, 101
Potassium, 2-tetrahydrofuryl-
  structure, 96

Propargyl selenoxides
  sigmatropic rearrangements, 557
Propargylic halides
  reaction with zinc–copper organometallic compounds, 166
Propargylic tosylates
  reaction with zinc–copper organometallic compounds, 166
[1.1.1]Propellane
  lithiation, 6
Propene, bromotrifluoro-
  zinc reagents, 161
Propene, 1,3-dichloro-
  reaction with organozinc compounds, 165
Propiolamide
  carbozincation, 177
Propionitriles, β-aryl-β-hydroxy-
  synthesis, 494
Prostaglandin
  synthesis
    organosilicon compounds, 317
Prostaglandin B_1 methyl ester
  synthesis, 232
2H-Pyran, 4-methyl-3,6-dihydro-
  metallation, 119
4H-Pyran
  metallation, 119
4H-Pyran, 4-methyl-
  metallation, 119
Pyridazine
  lithiation, 79
Pyridine, 1,4-dimethyldihydro-
  metallation, 119
Pyridine, 1,4-dimethyl-1,2,3,6-tetrahydro-
  metallation, 119
Pyridine, 1-methyl-1,4-dihydro-
  metallation, 119
Pyridine, 1,2,5,6-tetrahydro-
  synthesis, 144
Pyridines
  DMG containing
    lithiation, 76–9
  metallation, 119
  reactions with Grignard reagents, 141–4
Pyridinium salts
  reaction with zinc organometallics, 171
3-Pyridinols
  synthesis
    lithiation, 78
4-Pyridones, 2,3-dihydro-
  regiocontrolled synthesis
    Grignard reagents, 143
3,4-Pyridynes
  synthesis
    lithiation, 78
Pyrimidine
  lithiation, 81
Pyrrole, N-methyl-
  metallation, 119
Pyrrole, N-phenyl-
  metallation, 119
Pyrroles
  DMG-containing
    lithiation, 72, 73

Pyrrolidine, 2-(diphenylhydroxymethyl)-
  preparation, 259
Pyrrolidines
  lithiation, 17
  synthesis
    aminomercuration, 423
    organoboron compounds, 218
Pyrrolizidines
  synthesis
    organoboron compounds, 218

Quercus lactone
  synthesis, 101
o-Quinodimethane, α-hydroxy-
  Diels–Alder reaction, 245
Quinoline, decahydro-
  synthesis
    organoboron compounds, 218
Quinoline, 2-fluoro-
  lithiation, 78
Quinolines
  lithiation, 51
Quinophenylselenoimines
  synthesis from phenols, 534

Reagents, 576
Reduction
  enantioselective
    organoboron compounds, 258–61
  tellurium reagents, 576–81
Reformatsky reaction
  organozinc compounds, 159, 170
Reverse-Brook rearrangement
  lithiation, 24
Ribonolactone, 2-deoxy-
  synthesis
    organosilicon compounds, 317
Ring closure reactions
  organotellurium compounds, 585–7
Routiennocin
  synthesis
    lithiation, 72
Sanninoidea exitiosa
  sex attractant
    synthesis, 106
α-Santalene
  synthesis, 101
α-Santalol
  synthesis, 101

Sawada reaction
  organozinc compounds, 178
Schlosser's base
  structure, 96
Scolytus multistriatus
  ephromone
    synthesis, 101
Selenides
  alkylation, 552
  complexation, 552
  halogenation, 552
  heterosubstituted
    synthesis, 541
  oxidation, 553, 554
    to selenoxides, 554

synthesis, 527, 538
Selenium
    electronegativity, 518
    introduction into organic
        compounds, 522
    metallic
        reactions, 532
    in Periodic Table, 518
    physical properties, 516–22
    radionucleotides, 517, 518
    removal from organic
        compounds, 522
Selenium compounds
    structure, 519, 520
    synthesis
        organomercury compounds,
            400
Selenium dioxide
    hydrogen peroxide mixtures
        reactions, 537
    reactions, 536
Selenium halides
    synthesis, 528
Selenium tetrafluoride
    reactions, 537
Selenoacetals
    oxidation
        organoselenium
            compounds, 535
    synthesis, 540
Selenoalkyl compounds
    structure, 520–2
Selenoalkyl lithium
    synthesis, 543
Selenoalkyl metals
    structure, 521
Selenoalkyl radicals, 521
Selenocarbenium ions, 521
Selenocarbonyl compounds
    synthesis, 539, 541, 545
Selenocarboxylic acids
    synthesis, 539
Selenocysteine
    occurrence, 516
Selenoketones
    α-phenyl-
        synthesis, 337
Selenol esters
    synthesis, 530, 545
Selenolcarbonates
    synthesis, 545
Selenones
    reactivity, 557
    synthesis from selenides, 554
Selenonium salts
    reactivity, 552
    synthesis, 552
Selenopseudo ureido compounds
    synthesis, 530
Selenoxides
    elimination reactions, 555
    reactions, 533
    reactivity, 554
    synthesis from selenides, 554
Sesamin
    synthesis, 473
Sesbanine
    synthesis
        lithiation, 79
Sharpless dihydroxylation
    asymmetric
        allylsilanes, 320
Sharpless kinetic resolution, 326

Silacyclopentadienes, 345
Silanes
    reducing agents, 343–5
Silanes, acyl-
    in organic synthesis, 328, 329
Silanes, 2-alkenyltrimethyl-
    metallation, 117
Silanes, alkynyl-, 324, 325
Silanes, (allenylmethyl)-
    reactions, 318
Silanes, allyl-
    electrophilic substitution
        reactions, 317
    osmylation, 320
    oxidation, 320
    preparation, 317–23
    reactions, 317–23
    reactions with oxocarbenium
        ions, 319
    synthesis, 317
Silanes, allyltrimethyl-
    metallation, 117
Silanes, amino-
    in organic synthesis, 342, 343
Silanes, aryl-
    in organic synthesis, 326
Silanes, chlorotrimethyl-
    for zinc activation, 160
Silanes, crotyl
    chiral
        reactions, 321
Silanes, α,β-epoxy-
    in organic synthesis, 326, 327
Silanes, heteroaryl-
    in organic synthesis, 326
Silanes, keto-
    in organic synthesis, 328, 329
Silanes, β-keto-, 329
Silanes, phenyltrimethyl-
    metallation, 118
Silanes, propargyl-, 324, 325
    allene precursors, 325
Silanes, vinyl-
    carbonyl–ene reactions, 316
    electrophilic substitution
        reactions, 315
    oxidative cleavage, 317
    preparation, 314, 315
    reactions, 315–17
Silicon tethering
    in organic synthesis, 330–2
Silyl enol ethers
    conjugate addition, 335
    cycloaddition reactions, 335,
        336
    heterosubstituted
        synthesis, 337
    Lewis acid induced reactions,
        333
    nucleophilicty, 337
    oxidative cyclization, 338
Simmons–Smith reaction
    organozinc compounds, 159,
        178
Sinapyl alcohol
    synthesis, 108
Sodium t-butoxide
    superbase reagent, 94
Sodium selenocyanates
    synthesis, 528
Sodium, allyl-
    reactions, 97–103
    rotational barrier, 98

Sodium, butyl-
    structure, 95
Solenopsin-a
    synthesis, 144
Solvomercuration
    demercuration, 411–30
Spermine alkaloids
    synthesis
        organoboron compounds,
            245
Spiro[4.4]nona-1,3-diene
    metallation, 112
Spiropentane
    metallation, 114
Spiro[2.2]pentane
    metallation, 114
Stannanes
    functionalized
        in organic synthesis,
            375–81
Stannanes, acyl-
    in organic synthesis, 377, 378
Stannanes, alkenyl-
    in organic synthesis, 369–72
    preparation, 369
    reactions, 370–2
    transmetallation, 370, 371
Stannanes, alkyl-
    functionalized at remote
        positions, 380
Stannanes, alkynyl-
    cross-coupling
        palladium-catalyzed, 375
Stannanes, allenyl-
    synthesis, 374
Stannanes, allyl-
    coupling with halides, 373
    in organic synthesis, 372–4
    preparation, 372, 373
    reactions, 373, 374
    reactions with electrophiles,
        373, 374
    transmetallation, 373
Stannanes, α-aminoalkyl-
    preparation, 379
Stannanes, aryl-
    in organic synthesis, 369–72
    reactions, 370–2
    transmetallation, 370, 371
Stannanes, 2-azaallyl-
    preparation, 379
Stannanes, α-diazo-
    in organic synthesis, 379
Stannanes, trialkylchloro-
    reaction with organozinc
        compounds, 165
Stannanes, tributyl-
    in organic synthesis, 358
Stannanes, triphenyl-
    in organic synthesis, 358
Stannanes, vinyl-
    synthesis
        organoboron compounds,
            243
α-Stannyl alcohols
    in organic synthesis, 375–6
β-Stannyl alcohols
    preparation, 376
γ-Stannyl alcohols
    preparation, 377
Stannyl anions
    addition to multiple bonds,
        363–4

Stannyl anions
   in organic synthesis, 363–4
   reactions, 363–4
   substitution, 363
α-Stannyl carbanions
   in organic synthesis, 365–6
α-Stannyl ketones
   in organic synthesis, 378
β-Stannyl ketones
   preparation, 378
γ-Stannyl ketones
   in organic synthesis, 379
Stannyl lithium, tributyl-
   in organic synthesis, 363
Stannyl lithium, trimethyl-
   in organic synthesis, 363
Stibine, alkyldiphenyl-
   lithio derivatives
     alkenation, 491
Stibine, diphenyl-
   carbonyl compound reduction, 496
Stibine, trialkyl-
   alcohol preparation, 492
   alkenation
     carbonyl compounds, 489
Stibine, tributyl-
   alkyl aryl sulfone synthesis, 497
Stibine, triphenyl-
   dechlorination, 501
Stibonium bromide, allenyltributyl-
   synthesis, 494
Stibonium bromide, benzyltrialkyl-
   homobenzylic alcohol synthesis, 493
Streptonigrin
   synthesis
     lithiation, 79
     organoboron compounds, 232
Styrene
   hydroboration, 248
Styrene, nitro-
   reaction with zincates, 182
Sulfones
   alkyl aryl
     synthesis, 497
   β-keto
     desulfonylation, tellurium reagents, 581
   preparation
     organoboranes, 222
   synthesis
     organomercury compounds, 400
Sulfones, *t*-butyl
   lithiation, 53
Sulfones, diaryl
   lithiation, 53
Sulfoxides, *t*-butyl
   lithiation, 53
Sulfoximines
   lithiation, 41
Sulfur compounds
   synthesis
     organomercury compounds, 400
Superbase reagents
   organosodium and organopotassium compounds, 94

Superbases
   structure, 95, 96
   unimetal, 96
Tebbe's reagent
   stoichiometric reagent, 296
Tellimagradin
   synthesis
     lithiation, 67
Tellurides, vinyl 2-thienyl
   in organic synthesis, 594
Tellurium compounds
   synthesis
     organomercury compounds, 400
Tellurium reagents
   inorganic
     in organic synthesis, 571, 572
   in organic synthesis, 576–96
Tellurium, tetraalkyl-
   in organic synthesis, 595
Tellurolactonization
   unsaturated carboxylic acids, 585
Telluronium methylides, dibutyl
   stabilization, 591
α-Terpinene
   metallation, 105
Tetraazamacrocycles
   synthesis
     organoboron compounds, 245
Tetrasilacyclohexyne
   synthesis, 325
Tetrazole
   lithiation, 44
Thallation
   arenes, 456
Thallation–carbonylation
   arenes, 447
Thallation–palladation
   indole synthesis, 451
Thallium(I) compounds
   in organic synthesis, 452–5
Thallium(II) compounds
   in organic synthesis, 437–51
Thallium(III) compounds
   aryl
     preparation, 445
Thallium hydroxide
   cross-coupling reactions
     vinylboron compounds, 232
Thallium(III) nitrate
   in organic synthesis, 455
Thallium polymers
   electroactive
     preparation, 444
Thallium, cyclopentadienyl-
   in organic synthesis, 452–5
Thioacetals
   oxidation
     organoselenium compounds, 535
Thioamides
   acylation
     organobismuth(V) compounds, 511
Thiocyanates
   aryl
     synthesis, organomercury compounds, 400
Thioketones

oxidation
   organoselenium compounds, 535
α-phenyl-
   synthesis, 337
reduction
   tellurium reagents, 577
Thiols
   oxidation
     organobismuth(V) compounds, 505
     organoselenium compounds, 535
Thiophenes
   DMG containing
     lithiation, 72
   metallation, 119
   non-DMG containing
     lithiation, 71
Tin hydrides
   polymer supported
     in organic synthesis, 358
Tin reagents
   anionic
     in organic synthesis, 363–6
   carbocationic
     in organic synthesis, 369
   cationic
     in organic synthesis, 366–9
   classification, 357
Tin, hexaalkyldi-
   in organic synthesis, 358
Titanocene, dichloro-
   transmetallation
     with zincates, 182
TMEDA
   lithiation
     in organic synthesis, 4
Toluene
   lithiation, 55
Toluene, fluoro-
   metallation, 121
Toluidines
   lithiation, 59
Tosyl cyanide
   reaction with alkylzinc halides, 165
Transfer reactions
   divalent carbon
     organomercury compounds, 409–11
Transmetallation
   organotellurium compounds, 592–4
Trialkylsilyl azides
   chemistry, 341
Trialkylsilyl cyanides
   chemistry, 339, 340
Trialkylsilyl halides
   in organic synthesis, 338, 339
Trialkylsilyl peroxides
   chemistry, 340, 341
Trialkylsilyl selenols
   chemistry, 341, 342
Trialkylsilyl stannanes
   chemistry, 342
Trialkylsilyl thiols
   chemistry, 341, 342
Trialkylsilyl trifluoromethanesulfonates
   chemistry, 340
9-Tricosene
   synthesis, 101

Tricyclo[2.2.1.0^{2,6}]heptanes
  metallation, 113
Tricyclo[4.1.0.0^{2,7}]heptanes
  lithiation, 6
Triflates
  chemistry, 340
Trimethoprim
  synthesis
    lithiation, 81
Trimethylsilyl diazomethane
  chemistry, 341
Triorganothallium compounds
  in organic synthesis, 437–43
  preparation, 437, 438
  properties, 437, 438
  reactions, 438–43
Triorganozincates
  halogen–zinc exchange, 182
Triphenylene
  synthesis, 121
Tryptophols
  synthesis
    lithiation, 59

Ultrasonics
  zinc activation, 161
Urea, seleno-
  synthesis, 528
Ureidomercuration
  intramolecular, 425
Urethanes
  lithiation, 17

Vinylation
  organolead compounds in, 468
  tetraalkyllead to aldehydes, 462
[2,3]-Wittig rearrangements
  lithiation, 24

Wurtz-coupling
  zinc organometallic
    compounds, 162

Xylenes
  dimetallation, 109

Yamamoto's aluminum tris(2,6-diphenylphenoxide)
  stoichiometric reagent, 296
Yenhusomidine
  synthesis
    lithiation, 21
Yuehchukene
  synthesis
    lithiation, 73

Zinc
  bimetallic reagents
    preparation, 182
    reactions, 182
    reactivity, 182
Zinc carbenoids
  reactivity, 178–81
Zinc compounds
  dimetallic
    preparation, 175
  in organic synthesis, 159–84
Zinc electrodes
  sacrificial, 174
Zinc homoenolates
  preparation, 160
Zinc malonates
  reaction with alkynes, 176
Zinc reagents
  allenic
    preparation, 169
  allylic
    reactivity, 165
  propargylic
    preparation, 169
Zinc, bis(alkenyl)-
  synthesis
    organoboron compounds, 220
Zinc, α-(dialkoxyboryl)alkyl-
  preparation, 210
Zinc, dialkyl-
  reaction with alkynes, 176
Zinc, diallyl-
  addition to strained alkenes, 175
  preparation, 164
Zinc, dibenzyl-
  preparation, 164
Zinc, diethyl-
  reaction with diiodomethane, 162
Zinc, (iodomethyl)iodo-
  reaction with alkynylcopper, 180
Zincates
  preparation, 182
  reactions, 182